ARCTIC OCEAN FLOOR
252

EUROPE 136–163

NORWAY
SWEDEN FINLAND

**NORTHERN
EUROPE
142** EST.
LATV.
LITH.

DENMARK

NETH.
POLAND BELARUS

**EASTERN EUROPE
158**

**CENTRAL EUROPE
150**
GER.
BELG. CZECHIA
SLOVAKIA UKRAINE
SWITZ. AUST. HUNG.
MOLDOVA
FRANCE SLOV. CROATIA ROM.
BOSN. & HERZG. SERB.

**ITALY
AND
SWITZ.
152** MONTEN. ALBAN.
KOS. BULG.
N. MACED.
**GREECE
156**

**THE BALKANS
154**
GEORGIA
ASIA MINOR ARM. AZERB.
AND TRANSCAUCASIA
TURKEY **170**

RUSSIA
160

KAZAKHSTAN

**CENTRAL ASIA
178**
UZBEKISTAN

KYRGYZSTAN

TURKMENISTAN
TAJIKISTAN

MONGOLIA

**CHINA AND MONGOLIA
184**

CHINA

NORTH KOREA
**KOREAN
PENINSULA
186** SOUTH
KOREA

**JAPAN
188**

ASIA 164–193

ITALY

MALTA

TUNISIA

CYPRUS
LEB. SYRIA
**EASTERN
MEDITERRANEAN**
ISRAEL IRAQ **176**
JORDAN

**IRAQ AND IRAN
176** AFGHANISTAN
IRAN

KUWAIT

**SOUTHERN AFRICA
200** LIBYA

EGYPT

SAUDI
ARABIA
**NILE
VALLEY
202** BAHRAIN
QATAR
**ARABIAN
PENINSULA
174** U.A.E.
OMAN
SUDAN

ALGERIA

NIGER CHAD

ERITREA
YEMEN

**WEST-CENTRAL
AFRICA
206** DJIBOUTI

BENIN

NIGERIA

SOUTH
SUDAN **ETHIOPIA**

CENTRAL
AFRICAN
REPUBLIC **EAST
AFRICA
210** SOMALIA

CAMEROON

**AFGHANISTAN
AND PAKISTAN
180**
PAKISTAN

NEPAL BHUTAN

**SOUTH ASIA
182**

INDIA

BANGLADESH

MYANMAR LAOS
**INDOCHINA
190**

THAILAND VIETNAM

CAMBODIA

TAIWAN

**PACIFIC OCEAN FLOOR
246**

Northern
Mariana
Islands

PHILIPPINES

PALAU

FEDERATED STATES OF MICRONESIA

MARSHALL
ISLANDS

**OCEANIA
224–231** KIRIBATI

NAURU

EQ. GUINEA

SAO TOME
AND
PRINCIPE GABON CONGO

UGANDA KENYA

DEMOCRATIC
REPUBLIC
OF THE CONGO RWANDA
BURUNDI

TANZANIA

**CONGO BASIN
208**

ANGOLA

ZAMBIA

SRI LANKA

MALDIVES

AFRICA 194–215

SEYCHELLES

COMOROS

**INDIAN OCEAN FLOOR
250**

BRUNEI
MALAYSIA
**SOUTHEAST ASIA
192**
SINGAPORE

INDONESIA

TIMOR-LESTE

**PAPUA NEW
GUINEA
222**

SOLOMON
ISLANDS

TUVALU

MALAWI

ZIMBABWE

MOZAMBIQUE

MADAGASCAR

MAURITIUS

VANUATU

FIJI

NAMIBIA

BOTSWANA

**SOUTHERN AFRICA
212** ESWATINI

SOUTH
AFRICA LESOTHO

**AUSTRALIA
220**

SELECTED OTHER MAPS

NEW ZEALAND
223

AUSTRALIA & OCEANIA 216–231

OCEAN FLOOR AROUND ANTARCTICA
254

ANTARCTICA 232–239

NATIONAL GEOGRAPHIC

FAMILY REFERENCE ATLAS OF THE WORLD

FIFTH EDITION

FAMILY REFERENCE ATLAS OF THE WORLD

FIFTH EDITION

NATIONAL GEOGRAPHIC
WASHINGTON, D.C.

New Zealand's Mount Cook and its serene surroundings

CONTENTS

8 About This Atlas | 12 World Introduction | 16 Physical World Map | 18 Political World Map | 20 Political Poles

PAGES 22–79

CHANGING PLANET

22 TECTONICS: Structure of the Earth
Earth's layers; plate boundaries; earthquakes, volcanoes, and hot spots; Pangaea and Pangaea Proxima; tectonic processes

24 LANDFORMS: The Physical Landscape
Major landforms; tectonic uplifts; volcanic landscapes; rift and faulting; eroded landforms

26 LANDFORM PROCESSES: Sculpting Earth's Surface
Landforms created by water; landforms created by wind; landforms created by ice; features of Earth's surface

28 EARTH'S ROCKY EXTERIOR: Birth Story of Rocks
The rock cycle; distribution of surface rock; age of oceanic crust; types of rocks: igneous, sedimentary, and metamorphic; reading Earth's history from rocks

30 LAND COVER: Mosaic of Landscapes
Global vegetation; evergreen forests; deciduous forests; shrubland; savanna and grasslands

32 FRESHWATER: Earth's Freshwater
World's perennial and intermittent rivers; major watersheds; water availability

34 OCEANS: Our Blue Planet
Ocean floor features; sea surface temperature; currents; coral reefs

36 CLIMATE: Defining Climate
Climate zones; average temperatures; precipitation extremes

38 CLIMATE CHANGE: A Warming Planet
Temperature trends; melting permafrost; increases in carbon dioxide and methane; charting sea-level change

40 WEATHER: The World's Weather
Weather fronts; cloud types; cyclone tracks; lightning strikes

African elephants graze in Ngorongoro Conservation Area, Tanzania.

WILDLIFE & WILD PLACES

42 BIOSPHERE: Zones of Life on Earth
Size of the biosphere; land vegetation, ocean chlorophyll; chemosynthesis and photosynthesis; world biomass

44 BIODIVERSITY: Diversity of Life
Marine and terrestrial species richness; species diversity; hotspots

46 WILD PLACES: Safe Havens for Nature
Low-impact areas; protected areas; low-impact biomes

48 ENVIRONMENTAL STRESSES: Threats to Our Planet
Land degradation and desertification; air pollution; tree cover loss; global water risk

THE HUMAN JOURNEY

50 HUMAN IMPACT: Humans Change the Planet
Anthromes; human ocean impact; plastics in the oceans; plastic waste buildup; human impact over time

52 POPULATION: A World Full of People
Population density; gender disparity; most populous countries; world population growth—past, present, and future

54 URBANIZATION: Cities on the Rise
Urban centers; urban population growth; largest cities

56 MIGRATION & REFUGEES: Population on the Move
Migrant populations; climate migrants; refugees: origin of asylum seekers; internally displaced persons

58 LANGUAGES: The World Speaks
Language locations and level of endangerment; major language families; language speakers by the number; evolution of languages

60 RELIGION: Faith Beliefs & Practices
Predominant religions; world religion numbers; religious diversity; Buddhism; Hinduism; Judaism; Christianity; Islam; blended religions

62 HUMAN CONDITION: Quality of Life
Education by gender; adult literacy; happiness scores; press freedom

64 HEALTH: Seeking World Wellness
Life expectancy; top causes of deaths; adult obesity; communicable diseases; noncommunicable diseases; safe drinking water

66 AGRICULTURE: The World's Harvest
Cropland and pasture land; irrigated areas; crop allocation

68 FOOD: Feeding the World
Meat consumption; fisheries; calories per capita; food security; regional diets

70 BORDERS: Borders & Boundaries
Boundaries; United Nations member states; maritime divisions; maritime limits

72 ECONOMY: Wealth & Prosperity
GDP per capita; national debt; millionaires; wealthiest nations

74 TRADE: Buying & Selling
Trading nations; merchandise exports; regional trade agreements; imports; exports

76 ENERGY: Power Sources Worldwide
Energy consumption; energy use trends; alternative energy; renewable energy

78 GLOBAL CONNECTIVITY: Our Digital World
Access to the internet; government restrictions to the internet; mobile phone subscriptions; undersea telecommunications cables

PAGES 80–123

NORTH AMERICA

82 **INTRODUCTION**
Geography, history, culture, economy, facts & features

84 **NORTH AMERICA PHYSICAL AND POLITICAL**

86 **CANADA**
Canada, St. Pierre and Miquelon (Fr.)

88 **GREENLAND**

90 **PHYSICAL UNITED STATES**

92 **POLITICAL UNITED STATES**

94 **U.S. NATIONAL PARK SERVICE**

96 **NORTHEAST U.S.**
Connecticut, Maine, Massachusetts, New Hampshire, New Jersey,
New York, Pennsylvania, Rhode Island, Vermont

98 **SOUTH ATLANTIC U.S.**
D.C., Delaware, Florida, Georgia, Maryland, North Carolina, South
Carolina, Virginia, West Virginia

100 **GREAT LAKES U.S.**
Illinois, Indiana, Michigan, Ohio, Wisconsin

102 **MIDDLE SOUTH U.S.**
Alabama, Arkansas, Louisiana, Kentucky, Mississippi, Tennessee

104 **TEXAS AND OKLAHOMA U.S.**

106 **NORTHERN PLAINS U.S.**
Iowa, Kansas, Minnesota, Missouri, Nebraska, North Dakota, South Dakota

108 **ROCKY MOUNTAINS U.S.**
Arizona, Colorado, Idaho, Montana, New Mexico, Utah, Wyoming

110 **WEST COAST U.S.**
California, Nevada, Oregon, Washington

112 **ALASKA U.S.**

114 **HAWAI'I AND PUERTO RICO U.S.**

116 **MEXICO AND CENTRAL AMERICA**
Belize, Costa Rica, El Salvador, Guatemala, Honduras, Mexico,
Nicaragua, Panama

118 **GREATER ANTILLES AND THE BAHAMAS**
The Bahamas, Bermuda (U.K.), Cayman Islands (U.K.), Cuba,
Dominican Republic, Haiti, Jamaica, Turks and Caicos Is. (U.K.)

120 **LESSER ANTILLES**
Anguilla (U.K.), Antigua and Barbuda, Aruba (Neth.), Barbados,
British Virgin Islands, Curaçao (Neth.), Dominica, Grenada, Montserrat
(U.K.), St. Barthelemy (Fr.), St. Kitts and Nevis, St. Lucia, St. Martin (Fr.),
St. Vincent and the Grenadines, Sint Maarten (Neth.), Trinidad and
Tobago, U.S. Virgin Islands

122 **ISLANDS OF THE ANTILLES**
Abaco Islands, Anguilla (U.K.), Antigua, Aruba (Neth.), Barbados, Bimini
Islands, Bonaire (Neth.), Eleuthera, Exuma, Grand Bahama Island,
Grand Cayman, Curaçao (Neth.), Dominica, Grenada, Guadeloupe
(Fr.), Jamaica, Martinique (Fr.), Montserrat (U.K.) New Providence,
St. Croix, St. Kitts and Nevis, St. Lucia, St. Martin/St. Maarten
(Fr./Neth.), St. Thomas and St. John, St. Vincent, Tobago, Tortola,
Trinidad, Virgin Gorda

PAGES 124–135

SOUTH AMERICA

126 **INTRODUCTION**
Geography, history, culture, economy, facts & features

128 **SOUTH AMERICA PHYSICAL AND POLITICAL**

130 **NORTHERN SOUTH AMERICA**
Colombia, Ecuador, French Guiana (France), Guyana, Suriname, Venezuela

132 **CENTRAL SOUTH AMERICA**
Bolivia, Brazil, Paraguay, Peru

134 **SOUTHERN SOUTH AMERICA**
Argentina, Chile, Falkland Islands (Islas Malvinas) (U.K.), Uruguay

Glacier-fed Moraine Lake in Canada's Banff National Park

PAGES 136–163

EUROPE

138 **INTRODUCTION**
Geography, history, culture, economy, facts & features

140 **EUROPE PHYSICAL AND POLITICAL**

142 **NORTHERN EUROPE**
Denmark, Estonia, Faroe Islands (Den.), Finland, Iceland, Latvia,
Lithuania, Norway, Sweden

144 **BRITAIN AND IRELAND**
Ireland (Éire), United Kingdom

146 **IBERIAN PENINSULA**
Portugal, Spain

148 **BELGIUM, FRANCE, AND THE NETHERLANDS**

150 **CENTRAL EUROPE**
Austria, Czechia (Czech Republic), Germany, Hungary, Poland,
Slovakia

152 **ITALY AND SWITZERLAND**

154 **THE BALKANS**
Albania, Bosnia and Herzegovina, Bulgaria, Croatia, Kosovo,
North Macedonia, Montenegro, Romania, Serbia, Slovenia

156 **GREECE**

158 **EASTERN EUROPE**
Belarus, Moldova, Ukraine

160 **RUSSIA**

162 **SMALLEST COUNTRIES AND TERRITORIES**
Andorra, Cyprus, Gibraltar (U.K.), Liechtenstein, Luxembourg, Malta,
Monaco, San Marino, Vatican City

PAGES 164–193

ASIA

166 INTRODUCTION
Geography, history, culture, economy, facts & features

168 ASIA PHYSICAL AND POLITICAL

170 ASIA MINOR AND TRANSCAUCASIA
Abkhazia, Armenia, Azerbaijan, Georgia, *Nagorno-Karabakh*, *South Ossetia*, Turkey

172 EASTERN MEDITERRANEAN
Gaza Strip, Israel, Jordan, Lebanon, Syria, *West Bank*

174 ARABIAN PENINSULA
Bahrain, Kuwait, Oman, Qatar, Saudi Arabia, United Arab Emirates, Yemen

176 IRAN AND IRAQ

178 CENTRAL ASIA
Kazakhstan, Kyrgyzstan, Tajikistan, Turkmenistan, Uzbekistan

180 AFGHANISTAN AND PAKISTAN

182 SOUTH ASIA
Bangladesh, Bhutan, India, Maldives, Nepal, Sri Lanka

184 CHINA AND MONGOLIA
China, Mongolia, *Taiwan*

186 KOREAN PENINSULA
North Korea, South Korea

188 JAPAN

190 INDOCHINA
Cambodia, Laos, Myanmar (Burma), Thailand, Vietnam

192 SOUTHEAST ASIA
Brunei, Indonesia, Malaysia, Philippines, Singapore, Timor-Leste (East Timor)

PAGES 194–215

AFRICA

196 INTRODUCTION
Geography, history, culture, economy, facts & features

198 AFRICA PHYSICAL AND POLITICAL

200 NORTHERN AFRICA
Algeria, Libya, Morocco, Tunisia, *Western Sahara*

202 NILE VALLEY
Egypt, South Sudan, Sudan

204 WESTERN AFRICA
Benin, Burkina Faso, Côte d'Ivoire (Ivory Coast), The Gambia, Ghana, Guinea, Guinea-Bissau, Liberia, Mali, Mauritania, Senegal, Sierra Leone, Togo

206 WEST-CENTRAL AFRICA
Central African Republic, Cameroon, Chad, Equatorial Guinea, Niger, Nigeria

208 CONGO BASIN
Angola, Congo, Democratic Republic of the Congo, Gabon, Zambia

210 EAST AFRICA
Burundi, Djibouti, Eritrea, Ethiopia, Kenya, Rwanda, Somalia, *Somaliland*, Tanzania, Uganda

212 SOUTHERN AFRICA
Botswana, Eswatini (Swaziland), Lesotho, Madagascar, Malawi, Mozambique, Namibia, South Africa, Zimbabwe

214 ISLANDS OF AFRICA
Cabo Verde, Comoros, Mauritius, Mayotte (Fr.), Réunion (Fr.), Sao Tome and Principe, Seychelles, St. Helena, Ascension, Tristan da Cunha (U.K.)

PAGES 216–231

AUSTRALIA & OCEANIA

218 INTRODUCTION
Geography, history, culture, economy, facts & features

220 AUSTRALIA PHYSICAL AND POLITICAL

222 AUSTRALIA, PAPUA NEW GUINEA, NEW ZEALAND

224 OCEANIA

226 ISLANDS OF THE PACIFIC
Fifty-three inset maps of islands and island groups that correspond to numbered locations on the preceding overview map of Oceania

PAGES 232–239

ANTARCTICA

234 INTRODUCTION
Geography, exploration, Antarctic Treaty, climate, facts & features

236 DYNAMIC ANTARCTICA
Flow of the Antarctic ice sheet, surface elevation, ice sheet thickness, cross-section profile

238 ANTARCTICA

A futuristic municipal park in Singapore features gardens and an elevated skyway.

An octopus approaches the shore near Naples, Italy.

PAGES 240–255

OCEANS

242 INTRODUCTION
Geography, history, culture, economy, facts & features
244 LIMITS OF THE OCEANS AND SEAS
246 PACIFIC OCEAN FLOOR
248 ATLANTIC OCEAN FLOOR
250 INDIAN OCEAN FLOOR
252 ARCTIC OCEAN FLOOR
254 OCEAN FLOOR AROUND ANTARCTICA

PAGES 256–275

SPACE

258 INTRODUCTION
Measuring space, the race into space, space technology, exoplanets, facts & features
260 MOON: NEAR SIDE
262 MOON: FAR SIDE
264 INNER SOLAR SYSTEM
Sun, Mercury, Venus, Earth, Mars
266 MARS: WESTERN HEMISPHERE
268 MARS: EASTERN HEMISPHERE
270 OUTER SOLAR SYSTEM
Jupiter, Saturn, Uranus, Neptune
272 COSMIC JOURNEYS
274 MILKY WAY

PAGES 276–286

APPENDIX

276 ABBREVIATIONS AND METRIC CONVERSIONS
278 GEOGRAPHIC COMPARISONS
Earth, land, human geography, water
280 FOREIGN TERMS
282 TIME ZONES
284 GLOSSARY OF GEOGRAPHIC TERMS
287 KEY TO FLAGS AND FACTS

PAGES 287–399

INDEX & CREDITS

288 PLACE-NAME INDEX
393 MOON AND MARS INDEXES
397 CREDITS

ABOUT THIS ATLAS

W e hope this atlas will open many family conversations about our planet, its habitats and life-forms, and how its rich, diverse, and remarkable geography continues to shape the human story.

In this new edition of the *Family Reference Atlas*, National Geographic continues its commitment to make up-to-date knowledge of the world engaging for learners of all ages. Those who study geography, biodiversity, and climate today are the explorers, problem solvers, and planetary stewards of tomorrow.

Maps are a rich, useful, and—as far as possible—accurate means of depicting the world. Yet maps inevitably must simplify the world, and cartographic decisions may run at odds with individual perceptions. A neat border line on a map may actually cut through a contested war zone; the government-sanctioned name of an ethnically diverse region may not be the one local citizens prefer. Our cartographers rely on constant scrutiny, considerable discussion, and help from many experts to resolve these challenges. On political maps, local names are used, with the conventional name often added in parentheses. Place-name conventions are established by the U.S. Board on Geographic Names, a federal body dating back to 1890.

▶ PHYSICAL MAPS

Physical maps of the world, the continents, and the ocean floor reveal land-forms and vegetation in stunning detail. Detailed digital relief is rendered and combined with prevailing land cover based on global satellite data.

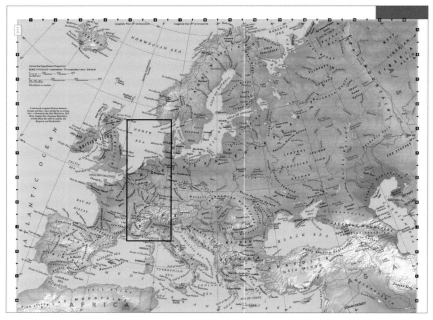

Physical Features:
Colors and shading illustrate variations in elevation, landforms, and vegetation. Patterns indicate specific landscape features, such as sand, glaciers, and swamps.

Water Features:
Blue lines indicate rivers; other water bodies are shown as areas of blue.

Boundaries and Political Divisions
are shown in red. Dotted lines indicate disputed or uncertain boundaries.

▶ POLITICAL MAPS

Political maps portray features such as international boundaries, the locations of cities, road networks, and other important elements of the world's human geography. Most index entries are keyed to the political maps, listing the page numbers and then the specific locations on the pages. (See page 287 for details on how to use the index.)

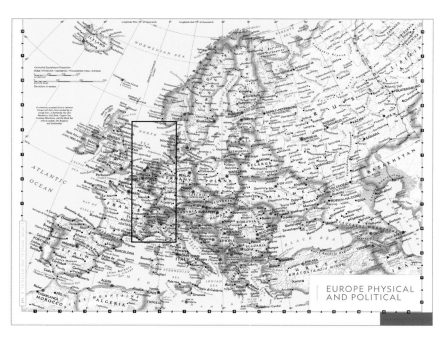

EUROPE PHYSICAL AND POLITICAL

Physical Features:
Gray relief shading depicts surface features such as mountains, hills, and valleys.

Water Features
are shown in blue. Solid lines and filled-in areas indicate perennial water features; dashed lines and patterns indicate intermittent features.

Boundaries and Political Divisions
are defined with both lines and colored bands; they vary according to whether a boundary is internal or international (for details, see map symbols key opposite).

Cities:
The regional political maps that form the bulk of this atlas depict four categories of cities or towns. The largest cities are shown in all capital letters (e.g., LONDON).

▶ THEMATIC MAPS

Thematic maps reveal the rich patchwork and infinite interrelationships of our changing planet. The thematic section at the beginning of the atlas focuses on physical and biological topics such as geology, landforms, land cover, and biodiversity. It also charts human patterns, with information on population, languages, religions, and the world economy. Throughout this section of the atlas, maps are coupled with satellite imagery, charts, diagrams, and photographs, for studying geographic patterns.

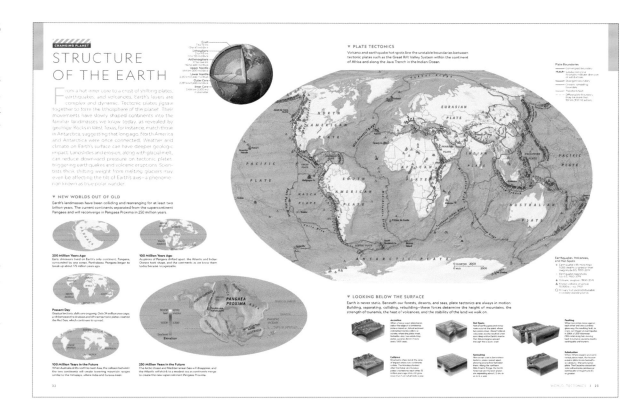

▶ REGIONAL MAPS

This atlas divides the continents into several subregions, each displayed on a two-page spread. Large-scale maps capture the political divisions and major surface features, whereas accompanying regional thematic maps lend insight into natural and human factors that give character to a region. Fact boxes, which include flag designs and information on area, populations, languages, and religions, appear alongside the maps as practical reference tools.

▼ MAP SYMBOLS

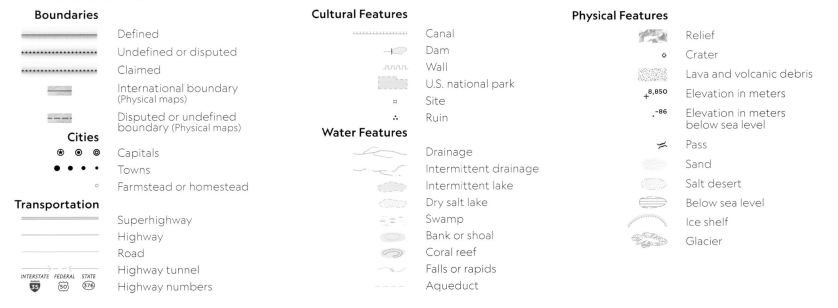

Boundaries

───────	Defined
•••••••••••••	Undefined or disputed
•••••••••••••	Claimed
	International boundary (Physical maps)
	Disputed or undefined boundary (Physical maps)

Cities

⊛ ◉ ◎	Capitals
• • • •	Towns
○	Farmstead or homestead

Transportation

═══════	Superhighway
───────	Highway
───────	Road
─ ─ ─ ─	Highway tunnel
INTERSTATE FEDERAL STATE 35 50 376	Highway numbers

Cultural Features

┈┈┈┈┈┈	Canal
⊸	Dam
⌐⌐⌐⌐	Wall
▒▒▒	U.S. national park
▫	Site
∴	Ruin

Water Features

～～～	Drainage
～ ～ ～	Intermittent drainage
⬭	Intermittent lake
⬭	Dry salt lake
⬭	Swamp
⬭	Bank or shoal
⬭	Coral reef
≈	Falls or rapids
─ ─ ─	Aqueduct

Physical Features

🮲	Relief
⊙	Crater
▒▒▒	Lava and volcanic debris
+8,850	Elevation in meters
•-86	Elevation in meters below sea level
⤫	Pass
░░	Sand
▒	Salt desert
⬭	Below sea level
⬭	Ice shelf
▒	Glacier

ABOUT THIS ATLAS

▶ **LOCATORS**

Each regional spread contains a locator map showing where the featured region lies within a continent. The region of interest is highlighted with a black line. The continent in which the featured region resides, appears in the continental section's color (in this case, olive green, for Africa).

▶ **POPULATION DENSITY MAPS**

Colors indicate relative population density, with the most crowded areas shown in the darkest orange color.

▶ **CLIMATE MAPS**

The climate maps are colored by the Köppen-Geiger system that classifies the world into five climate zones based on criteria like temperature and precipitation—features that promote different vegetation patterns.

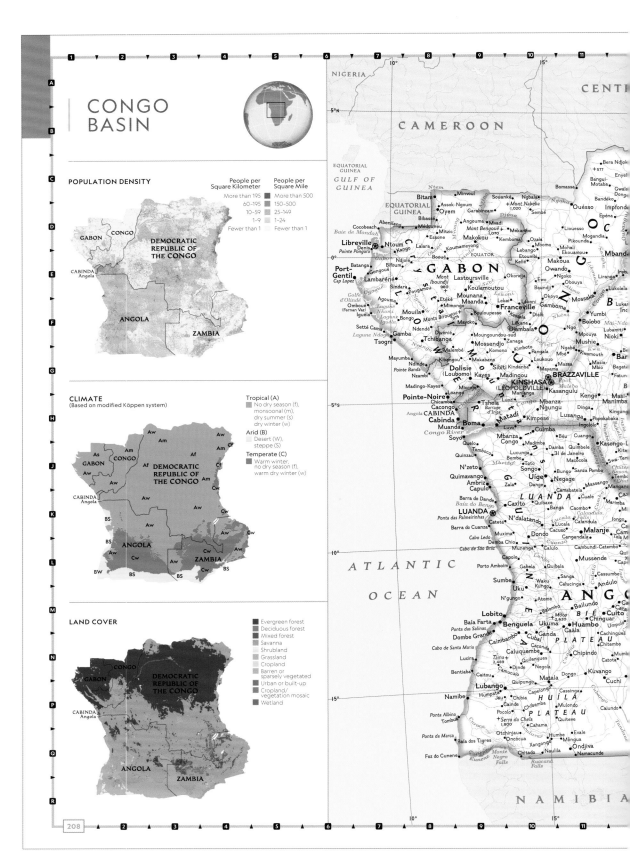

▲ **LAND COVER MAPS**

The colors on these maps indicate predominant land use and land-cover types—showing, for example, whether an area comprises mainly grassland or forest.

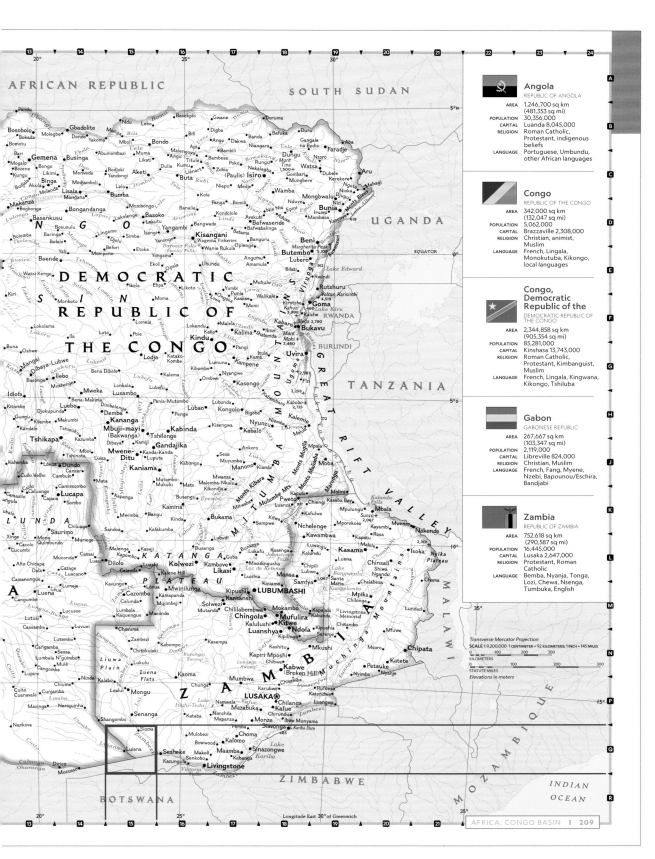

AFRICA: CONGO BASIN | 209

Flags and Facts Boxes

Angola
REPUBLIC OF ANGOLA
AREA 1,246,700 sq km (481,353 sq mi)
POPULATION 30,356,000
CAPITAL Luanda 8,045,000
RELIGION Roman Catholic, Protestant, indigenous beliefs
LANGUAGE Portuguese, Umbundu, other African languages

Congo
REPUBLIC OF THE CONGO
AREA 342,000 sq km (132,047 sq mi)
POPULATION 5,062,000
CAPITAL Brazzaville 2,308,000
RELIGION Christian, animist, Muslim
LANGUAGE French, Lingala, Monokutuba, Kikongo, local languages

Congo, Democratic Republic of the
DEMOCRATIC REPUBLIC OF THE CONGO
AREA 2,344,858 sq km (905,354 sq mi)
POPULATION 85,281,000
CAPITAL Kinshasa 13,743,000
RELIGION Roman Catholic, Protestant, Kimbanguist, Muslim
LANGUAGE French, Lingala, Kingwana, Kikongo, Tshiluba

Gabon
GABONESE REPUBLIC
AREA 267,667 sq km (103,347 sq mi)
POPULATION 2,119,000
CAPITAL Libreville 824,000
RELIGION Christian, Muslim
LANGUAGE French, Fang, Myene, Nzebi, Bapounou/Eschira, Bandjabi

Zambia
REPUBLIC OF ZAMBIA
AREA 752,618 sq km (290,587 sq mi)
POPULATION 16,445,000
CAPITAL Lusaka 2,647,000
RELIGION Protestant, Roman Catholic
LANGUAGE Bemba, Nyanja, Tonga, Lozi, Chewa, Nsenga, Tumbuka, English

Transverse Mercator Projection
SCALE 1:9,200,000 1 CENTIMETER = 92 KILOMETERS; 1 INCH = 145 MILES
KILOMETERS
STATUTE MILES
Elevations in meters

◄ FLAGS AND FACTS

This atlas recognizes 195 independent nations. All of these countries, along with dependencies and U.S. states, are pro-filed in the continental regional sections of the atlas. Accompanying each entry are highlights of geographic and demo-graphic data. These details provide a brief overview of each country, state, or territory; they are not intended to be comprehensive. A detailed description of the sources and policies used in com-piling the listings is included in the Key to Flags and Facts on page 399.

Angola
REPUBLIC OF ANGOLA
AREA 1,246,700 sq km (481,353 sq mi)
POPULATION 30,356,000
CAPITAL Luanda 8,045,000
RELIGION Roman Catholic, Protestant, indigenous beliefs
LANGUAGE Portuguese, Umbundu, other African languages

◄ MAP SCALES AND PROJECTIONS

Scale information indicates the distance on Earth represented by a given length on the map. Here, map scale is expressed in three ways: (1) as a representative fraction where scale is shown as a fraction or ratio as in 1:9,200,000; (2) as a verbal statement: 1 centimeter equals 92 kilo-meters or 1 inch equals 145 miles; and (3) as a bar scale, a linear graph symbol subdivided to show map lengths in kilometers and miles in the real world.

Map projections determine how land shapes are distorted when transferred from a sphere (the Earth) to a flat piece of paper. Many different projections are used in this atlas—each carefully chosen for a map's particular coverage area and purpose.

▲ INDEX AND GRID

Beginning on page 288 is a full index of place-names found in this atlas. The edge of each map is marked with letters (in rows) and numbers (in columns), to which the index entries are referenced. As an example, "Luiana, *Angola* **209** Q15" (see inset section, right) refers to the grid section on page 209 where row Q and column 15 meet. More examples and additional details about the index are included on page 288.

Luguruka, *Tanzania* **211** Y6
Luhans'k, *Ukr.* **159** Q10
Luhuo, *China* **184** K9
Lui, river, *Angola* **208** L12
Lui, river, *Zambia* **209** P15
Luiana, *Angola* **209** Q15
Luiana, river, *Angola* **209** Q15
Luilaka, river, *D.R.C.* **209** E13
Luilu, river, *D.R.C.* **209** K15
Luimneach *see* Limerick, *Ire.* **145** S4
Luing, island, *Scot., U.K.* **144** L8

THE
WORLD

WORLD PHYSICAL ● WORLD POLITICAL ● POLITICAL POLES ● TECTONICS
LANDFORMS ● LANDFORM PROCESSES ● EARTH'S ROCKY EXTERIOR ● LAND COVER
FRESHWATER ● OCEANS ● CLIMATE ● CLIMATE CHANGE ● WEATHER
BIOSPHERE ● BIODIVERSITY ● WILD PLACES ● ENVIRONMENTAL STRESSES
HUMAN IMPACT ● POPULATION ● URBANIZATION ● MIGRATION & REFUGEES
LANGUAGES ● RELIGION ● HUMAN CONDITION ● HEALTH ● AGRICULTURE
FOOD ● BORDERS ● ECONOMY ● TRADE ● ENERGY ● GLOBAL CONNECTIVITY

The view from Iceland's Seljalandsfoss waterfall at sunset

THE PLACE WE CALL HOME

A PLANET HANGING IN THE BALANCE

Some 150 million kilometers (93 million miles) from the sun, Earth whirls through space, its origins shrouded in deep time. Natural processes have given it shape and form; human activities define and harm it. We are still exploring this complex, dynamic home of ours, and as much as we've learned, there is still so much we have yet to discover.

GEOGRAPHY

THE SHAPE OF THE EARTH

Earth's hot and solid central core, rocky mantle, and outer crust formed not long after our solar system took shape, about 4.5 billion years ago. Our planetary axis tilts, giving us seasons. Whichever hemisphere is tilted closer to the sun experiences summer, while the other gets winter. The composition of the atmosphere and ocean created the perfect chemical cocktail required for life: The first organisms appeared about 3.8 billion years ago, the start of a long evolutionary process building a web of ecosystems supporting more and more life-forms in the sea and on land. *Homo sapiens* has only been around for about 300,000 years—the blink of an eye in geologic time.

By understanding how natural processes—along with geography, history, economics, and climate—shape the realities of life from time to time and place to place, we can better learn how to achieve a planet in balance. Accurate cartography is essential to picturing the planet, and geography means even more than that. Geographic literacy helps us explore extreme phenomena, how plants and animals are distributed, and how humans interact with the environment. How and why do people migrate? What cities are growing the fastest and what is their primary food source? How does trash discarded far inland end up in a swirling gyre in the North Pacific? These are important questions as we look to support an ever larger human population with Earth's rich—but finite—resources.

Drought and desertification threaten the livelihoods of millions of people.

CHANGING PLANET

LIVING IN A NEW GEOLOGIC AGE

Earth's geologic changes may seem to move at a crawl, but our world is constantly in flux. Rivers carve deep canyons, volcanic islands rise from the sea, and wind weathers rugged sea stacks on the ocean shore. The movement of plate tectonics over billions of years even shaped new continents. Recent changes have been so transformative, however, that scientists have defined a new geologic age: the Anthropocene. Beginning around 1950, this era is defined by increased emission of greenhouse gases into the atmosphere and other ways human activity is disrupting natural cycles.

Not even the extremes of our world have stayed constant. Mount Everest's glaciers are retreating, so scientists and cartographers are again mapping Earth's tallest mountain. These glaciers feed some of Asia's biggest rivers— the Ganges, Yellow, and Yangtze. Almost a fifth of the world's people depend on ice melt from the Himalaya, and as those glaciers retreat, water supplies dwindle. The Campaign for Nature, a group of conservation advocates including the National Geographic Society, is calling on policymakers to protect 30 percent of the planet by 2030. Only by reaching this milestone can the planet support its global population.

Hikers navigate the mountains of Nahuel Huapí National Park in Patagonia.

WILDLIFE & WILD PLACES

SAVING WHAT WE NEED MOST

Earth is losing much of its wilderness, but there are still spots that have escaped the human hand. We cherish them not just because they are pristine and majestic but also because they are central to our well-being: Forests reinvigorate the air, removing 2.6 billion tons of carbon dioxide from the atmosphere each year; coral reefs protect 200 million coastal dwellers from storm surge and flooding. These wild places support an intricate network of organisms and systems that maintain Earth's equilibrium and guarantee our quality of life. We need them to survive.

The rescue of natural places is one of the most encouraging narratives of our time. In southern Africa, for example, conservationists are exploring and restoring balance to a river basin vital to human and animal health. The Okavango is the continent's largest watershed and largest intact wetland wilderness. Straddling three countries, it is the main water source for a million people and home to some of Africa's most endangered animals, including the world's largest remaining elephant population. Traveling by canoe, scientists and explorers are transecting the massive basin— damaged and isolated during Angola's civil war, and now only partially protected—to map the region, record species new to science, and advocate among civic leaders for permanent protection.

Journalist Paul Salopek travels across Ethiopia's Afar desert on foot.

THE HUMAN JOURNEY

SINCE THE BEGINNING, AN URGE TO WANDER

The human footprint has undeniably shaped the planet ever since our *Homo sapiens* ancestors ventured out of Africa some 60,000 years ago. In 2013, storyteller Paul Salopek took the first steps in a 33,800-kilometer (21,000-mile), multiyear quest to retrace the path of human migration that populated the world. At a pace of roughly 24 kilometers (15 miles) a day, he chronicles along the way how people work, seek community, and dream of better lives for their children. As Salopek walks past monuments of ancient civilizations, along the rivers that power world trade, and through refugee camps and populous cities, many of the broad global trends that influence 21st-century life come into focus.

The spread of industrialization and large-scale agriculture across the globe are but two trends that have transformed economies and landscapes and lifted many people out of poverty. But behind the broad strokes of progress are enduring disparities in health, wealth, and security as well as deteriorating environments. In many places, the response is to strike out for new lands. We are experiencing the largest diaspora in human history, with nearly one billion people migrating within their own countries or across international borders. By documenting change and reflecting the diversity of cultures around the world, geography helps us understand and appreciate our planet and those who live on it.

A lioness at rest in Botswana's Okavango Delta

SNAPSHOT | *The World* **FACTS & FEATURES**

BIGGEST NATIONAL PARK Northeast Greenland National Park: 971,245 sq km (375,000 sq mi)

LONGEST INTERNATIONAL BORDER Canada-United States: 8,891 km (5,525 mi): almost one-third is the border with Alaska.

WETTEST PLACE Mawsynram, Meghalaya, India: average 1,187 cm (467 in) of rainfall

LONGEST RECORDED ANIMAL MIGRATION A gray whale swam 22,511 km (14,000 mi) round-trip from Russia to Mexico

BIGGEST ACTIVE VOLCANO Mauna Loa, Hawaii: erupting 33 times from 1843 to 1984

FASTEST ANIMAL Peregrine falcon: reaching speeds up to 321 km (200 mi) per hour

LARGEST HOT DESERT Sahara, Africa: 9,000,000 sq km (3,475,000 sq mi)

MOST SPOKEN LANGUAGE English: over 1.1 billion speakers, including non-native speakers

MOST TRADED CURRENCY U.S. dollar

THE LIMESTONE BEDROCK BENEATH THE **RAINFOREST IN BORNEO** HOLDS SOME OF THE MOST **EXTENSIVE CAVE SYSTEMS**—AND THE MOST **UNDISCOVERED TERRITORY**— IN THE WORLD.

Longitude East of Greenwich

15° 30° 45° 60° 75° 90° 105° 120° 135° 150° 165° 90° 180°

ARCTIC OCEAN

BARENTS SEA

NORWEGIAN SEA

George Land
Graham Bell Island
Komsomolets Island
October Revolution Island
Bol'shevik Island

Franz Josef Land
+606
Vize I.

North East Land
Spitsbergen +1,712
Edge Island
Svalbard

North Land
Cape Chelyuskin

Taymyr Peninsula
Lake Taymyr

LAPTEV SEA

New Siberian Islands

EAST SIBERIAN SEA

Wrangel I.

Chukchi Range

Bear Island
Kola Pen.
Novaya Zemlya
Yamal Pen.
Gyda Peninsula

North Siberian Lowland
+656

Kolyma

ARCTIC CIRCLE

A
R
C
T
I
C

Koryak Range

Central Range
+4,750

BERING SEA

NORWEGIAN SEA
Kebnekaise 2,097
North Cape
White Sea
Timan Ridge
Narodnaya +1,895

Ob'
West
Yenisey
Central Siberian Plateau

+1,701

Verkhoyansk Range
Cherskiy Range

Gora Mus Khaya +2,959
+1,830

Kolyma Range

Attu

Aleutian Is.

45°

NORTH

Scandinavia
Kjølen
Lake Onega
Lake Ladoga
Kola Pen.
SIBERIA
Siberian
Plain
Angara
Lake Baikal

Stanovoy Range
+2,412

Kamchatka Peninsula

SEA OF OKHOTSK

Sakhalin

Sikhote Alin Range

Kuril Islands

Hokkaido

PACIFIC

Galdhøpiggen 2,469
North Sea
Jutland
E
U
R
O
P
E
Northern European Plain
Volga
Ural Mountains
Irtysh
Ob
Belukha 4,506
Eastern Sayan Mts.
Yablonovyy Range
Amur

Greater Khingan Range

Manchurian Plain

SEA OF JAPAN (EAST SEA)

Honshu

OCEAN

30°

Mont Blanc 4,808
Rhine
Alps
Danube
Pripet Marshes
Central Russian Upland
The Steppes
Kazakh Uplands
Lake Balkhash
Dzungarian Basin +5,445
Mongolian Plateau
+3,957
Altay Mountains +4,506

Korea
Yellow Sea

Fuji 3,776
Kyushu
Shikoku

Nanpo Islands

Corsica
Sardinia
Apennines
Mt. Olympus +2,917
Balkan Peninsula
Carpathian Mts.
Crimea
Sea of Azov
Caspian Depression
Highest point in Europe
El'brus +5,642
Caspian Sea -28
Aral Sea
Kyzylkum
Syr Darya
Amu Darya
Tian Shan
Victory Peak 7,439
+2,584
Turpan Depression -154
Gobi

North China Plain
Qinghai Hu

Yellow
EAST CHINA SEA

TROPIC OF CANCER

Wake I.

15°

Sicily
Ionian Sea
Crete
Cyprus
ANATOLIA (ASIA MINOR)
Mt. Ararat +5,137
Caucasus Mts.
Zagros Mountains
Elburz Mts.
Ustyurt Plateau
Turan Lowland
Garagum
Hindu Kush
K2 +8,611
Kunlun Mountains
Muztag +6,973
Altun Shan
Taklimakan Desert
Plateau of Tibet
Qin Ling
Yangtze
G. of Tonkin
Hainan
Taiwan

Luzon Strait

PHILIPPINE SEA

Mariana Islands

MEDITERRANEAN SEA
Great Eastern Erg
Qattara Depression -133
Western Desert
Libyan Desert
Nile R. Delta
Sinai
Mt. Sinai +2,285
Syrian Desert
Mesopotamia
Tigris
Euphrates
Dead Sea -414
Lowest point in the world
Persian G.
Küh-e Taftan +4,042
Great Indian Desert
Indus
Ganges
Mount Everest 8,850
Highest point in the world
HIMALAYA
Brahmaputra
Mekong
INDOCHINA PENINSULA
SOUTH CHINA

Luzon
Mount Pulog 2,934
Guam

OCEAN

MICRONESIA

Ahaggar Mts.
+Mt. Tahat 2,908
Tibesti Mts.
Emi Koussi +3,445
SAHARA
+1,893
Nubian Desert
RED SEA
ARABIAN PENINSULA
Rub' al Khali
Lowest point in Africa
ARABIAN SEA
Deccan Plateau
INDIA
Western Ghats
BAY OF BENGAL
Eastern Ghats
Andaman Islands
Andaman Sea
Gulf of Thailand
Malay Pen.
CHINA SEA
Mount Pinatubo 1,486
Philippine Islands
Mindanao

Bikini Atoll
Enewetak Atoll
Taongi Atoll
Kwajalein Atoll
Marshall Islands

15°

Massif de l'Aïr
Marra Mts. 3,042
SAHEL
Lake Chad
+872
Blue Nile
White Nile
Ras Dejen 4,550
Lake Tana +4,185
Gulf of Aden
Socotra
Ethiopian Highlands
Somali Peninsula
Maldive Islands
Sri Lanka (Ceylon)
Nicobar Islands
Strait of Malacca
Sumatra
Kerinci 3,805
Kinabalu +4,095
Borneo
INDONESIA
Celebes
Buru
Moluccas

Chuuk (Ponape) Pohnpei
Caroline Islands

Admiralty Is.
Bismarck Archipelago
New Ireland +2,334
New Britain +4,509

EQUATOR

Nauru
Banaba (Ocean I.)
Beru
Gilbert Islands

0°

Gulf of Guinea
Bioko
São Tomé
Congo
Lower Guinea
Congo Basin
Lake Albert
Lake Victoria
Lake Turkana (L. Rudolf)
Mount Kenya +5,199
Kilimanjaro +5,895
Highest point in Africa
Mitumba Mts.
Lake Tanganyika
Zanzibar I.
Seychelles
Amirante Isles
Chagos Archipelago
Diego Garcia
Greater Sunda Islands
Java Sea
Flores
Java
Lesser Sunda Is.
Timor
Banda Sea
Arafura Sea
Timor Sea
New Guinea

Bougainville
Solomon Islands
Solomon Sea
Guadalcanal

MELANESIA

Nanumea
Tuvalu
Nukufetau

Katanga Plateau
Maromokotro +2,876
Lake Nyasa (Lake Malawi)
Comoro Is.
Mascarene Islands
Rodrigues
Mauritius
Réunion
Cocos Islands
Christmas I.

Cape York Pen.
Gulf of Carpentaria
Great Barrier Reef

CORAL SEA

New Caledonia +1,628

Vanuatu

Fiji Islands

15°

Brandberg 2,606
Namib Desert
Zambezi
Lake Kariba
+1,340
Kalahari Desert
+2,419
Madagascar
Mozambique Channel
INDIAN OCEAN
Mount Ord 947+
Kimberley Plateau
Great Sandy Desert
North West Cape
Mount Meharry +1,253
Western Plateau
MacDonnell Ranges
AUSTRALIA
Central Lowlands
Great Artesian Basin

Great Dividing Range

TROPIC OF CAPRICORN

Great Karoo
+2,202
Drakensberg
Cape of Good Hope
Cape Agulhas
Cape Inscription
Lowest point in Australia
Cape Naturaliste
Great Victoria Desert
Lake Eyre -15
Nullarbor Plain
Great Australian Bight
Murray
Darling
Mt. Kosciuszko +2,228
Highest point in Australia
Lord Howe I.

SOUTH

30°

Amsterdam
St. Paul
Bass Strait
Tasmania
TASMAN SEA
North Island (Te Ika-a-Māui)
NEW

PACIFIC
OCEAN

Prince Edward Islands
Crozet Islands
Kerguelen Islands
+1,850
(Mt. Cook) Aoraki +3,724
South Island (Te Waipounamu)
ZEALAND
+2,797

45°

Heard Island
Stewart Island (Rakiura)

Macquarie I.

Auckland Is.

Ross-Larsen Peninsula
Cosmonaut Sea
Cape Ann
Enderby Land
Prydz Bay
Wilkes Land
Balleny Is.

ANTARCTIC CIRCLE

60°

Queen Maud Land
TRANSANTARCTIC MOUNTAINS
Victoria Land
Mt. Erebus +3,794
-2,870
Lowest point in Antarctica
Ross Ice Shelf
Ross Sea

A
N
T
A
R
C
T
I
C
A

15° 30° 45° 60° 75° 90° 105° 120° 135° 150° 165° 180°

Winkel Tripel Projection

SCALE 1:81,657,000 1 CENTIMETER = 817 KILOMETERS; 1 INCH = 1,289 MILES AT THE EQUATOR

0 500 1,000 1,500 2,000 2,500
KILOMETERS

0 500 1,000 1,500 2,000 2,500
STATUTE MILES

Elevations in meters

POLITICAL
WORLD

Winkel Tripel Projection

SCALE 1:81,657,000 **1 CENTIMETER** = 817 KILOMETERS; **1 INCH** = 1,289 MILES AT THE EQUATOR

0   500   1,000   1,500   2,000   2,500
KILOMETERS

0   500   1,000   1,500   2,000   2,500
STATUTE MILES

Elevations in meters

NORTH POLE

SOUTH POLE

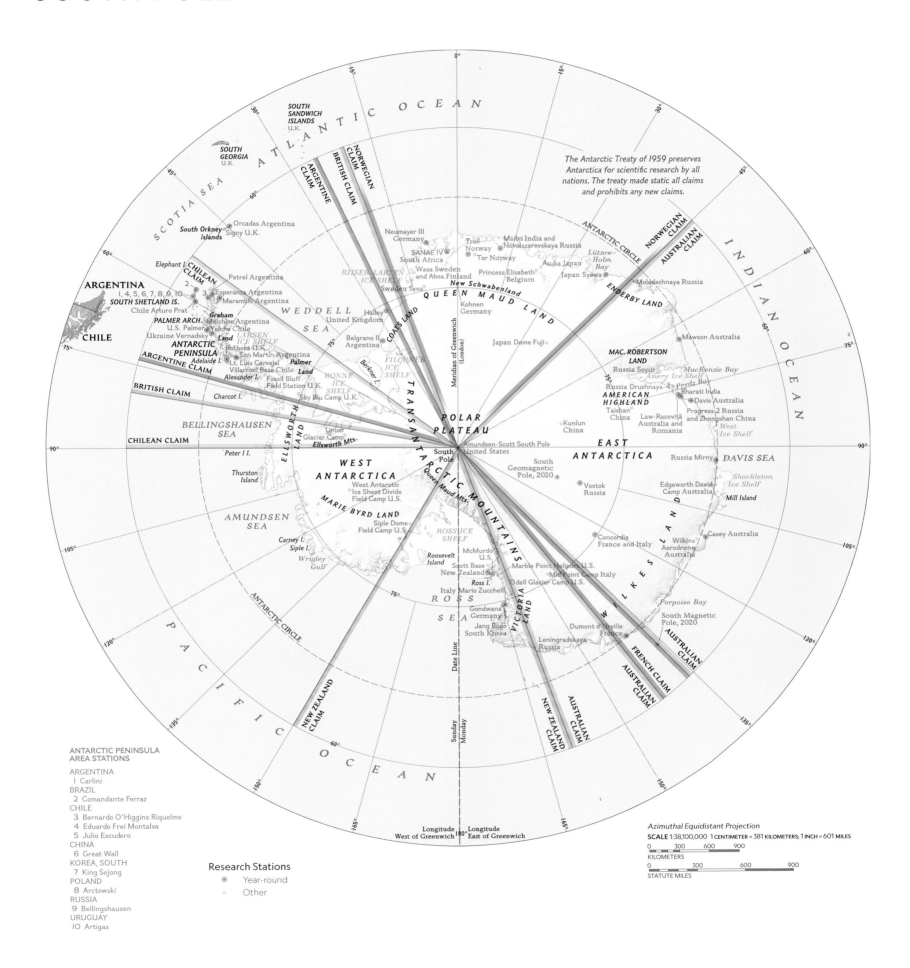

The Antarctic Treaty of 1959 preserves Antarctica for scientific research by all nations. The treaty made static all claims and prohibits any new claims.

ATLANTIC OCEAN

SCOTIA SEA

SOUTH SANDWICH ISLANDS
U.K.

SOUTH GEORGIA
U.K.

INDIAN OCEAN

NORWEGIAN CLAIM

ARGENTINE CLAIM

BRITISH CLAIM

AUSTRALIAN CLAIM

ANTARCTIC CIRCLE

Orcadas Argentina
Signy U.K.
South Orkney Islands

Neumayer III Germany
Troll Norway
Tor Norway
SANAE IV South Africa
Maitri India and Novolazarevskaya Russia
Asuka Japan
Japan Syowa
Lützow-Holm Bay

Elephant I.
CHILEAN CLAIM
Petrel Argentina
Wasa Sweden and Aboa Finland
Princess Elisabeth Belgium

ARGENTINA
1, 4, 5, 6, 7, 8, 9, 10
SOUTH SHETLAND IS.
Chile Arturo Prat
PALMER ARCH.
U.S. Palmer
Ukraine Vernadsky
ANTARCTIC PENINSULA
Esperanza Argentina
Marambio Argentina
Graham
Melchior Argentina
Yelcho Chile
Land
RIISER-LARSEN ICE SHELF
QUEEN MAUD LAND
New Schwabenland Germany
Kohnen Germany
Molodezhnaya Russia

NORWEGIAN CLAIM

ENDERBY LAND

Mawson Australia

CHILE

WEDDELL SEA
United Kingdom
Halley
COATS LAND
Belgrano II Argentina
Meridian of Greenwich (London)
Japan Dome Fuji

MAC. ROBERTSON LAND
Russia Soyuz
MacKenzie Bay
Amery Ice Shelf
Prydz Bay
Bharati India

ARGENTINE CLAIM
LARSEN ICE SHELF
San Martin Argentina
Rothera U.K.
Lt. Luis Carvajal
Villarroel Base Chile
Land
Palmer
BERKNER I.
RONNE ICE SHELF
FILCHNER ICE SHELF
AMERICAN HIGHLAND
Russia Druzhnaya
Taishan China
Davis Australia
Progress 2 Russia and Zhongshan China
West Ice Shelf

BRITISH CLAIM
Adelaide I.
Alexander I.
Charcot I.
Fossil Bluff Field Station U.K.
Sky Blu Camp U.K.
Kunlun China
Law-Racoviță Australia and Romania

CHILEAN CLAIM
BELLINGSHAUSEN SEA
ELLSWORTH LAND
Union Glacier Camp
Ellsworth Mts.
POLAR PLATEAU
EAST ANTARCTICA
DAVIS SEA
Russia Mirny

Peter I I.
WEST ANTARCTICA
West Antarctic Ice Sheet Divide Field Camp U.S.
South Pole
Amundsen-Scott South Pole United States
South Geomagnetic Pole, 2020
Vostok Russia
Edgeworth David Camp Australia
Shackleton Ice Shelf
Mill Island

Thurston Island
MARIE BYRD LAND
TRANSANTARCTIC MOUNTAINS
Queen Maud Mts.

AMUNDSEN SEA
Siple Dome Field Camp U.S.
ROSS ICE SHELF
Concordia France and Italy
Wilkins Aerodrome Australia
Casey Australia

Carney I.
Siple I.
Wrigley Gulf
Roosevelt Island
McMurdo U.S.
Scott Base New Zealand
Marble Point Helipots U.S.
Mid Point Camp Italy
Odell Glacier Camp U.S.
Porpoise Bay

NEW ZEALAND CLAIM
Ross I.
Italy Mario Zucchelli
ROSS SEA
Gondwana Germany
Jang Bogo South Korea
VICTORIA LAND
Leningradskaya Russia
Dumont d'Urville France
South Magnetic Pole, 2020
AUSTRALIAN CLAIM
FRENCH CLAIM
AUSTRALIAN CLAIM
NEW ZEALAND CLAIM

ANTARCTIC CIRCLE
Sunday Monday
Date Line
Longitude West of Greenwich | Longitude East of Greenwich

PACIFIC OCEAN

ANTARCTIC PENINSULA AREA STATIONS

ARGENTINA
 1 Carlini
BRAZIL
 2 Comandante Ferraz
CHILE
 3 Bernardo O'Higgins Riquelme
 4 Eduardo Frei Montalva
 5 Julio Escudero
CHINA
 6 Great Wall
KOREA, SOUTH
 7 King Sejong
POLAND
 8 Arctowski
RUSSIA
 9 Bellingshausen
URUGUAY
 10 Artigas

Research Stations
⊚ Year-round
○ Other

Azimuthal Equidistant Projection
SCALE 1:38,100,000 1 CENTIMETER = 381 KILOMETERS; 1 INCH = 601 MILES

0   300   600   900
KILOMETERS

0   300   600   900
STATUTE MILES

STRUCTURE OF THE EARTH

From a hot inner core to a crust of shifting plates, earthquakes, and volcanoes, Earth's layers are complex and dynamic. Tectonic plates jigsaw together to form the lithosphere of the planet. Their movements have slowly shaped continents into the familiar landmasses we know today, as revealed by geology: Rocks in West Texas, for instance, match those in Antarctica, suggesting that long ago, North America and Antarctica were once connected. Weather and climate on Earth's surface can have deeper geologic impact. Landslides and erosion, along with glacial melt, can reduce downward pressure on tectonic plates, triggering earthquakes and volcanic eruptions. Scientists think shifting weight from melting glaciers may even be affecting the tilt of Earth's axis—a phenomenon known as true polar wander.

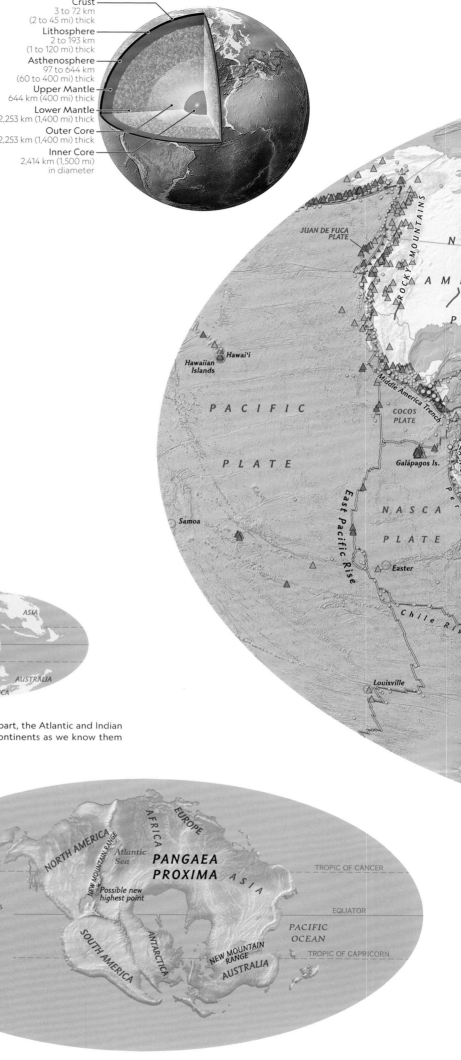

Crust
3 to 72 km
(2 to 45 mi) thick
Lithosphere
2 to 193 km
(1 to 120 mi) thick
Asthenosphere
97 to 644 km
(60 to 400 mi) thick
Upper Mantle
644 km (400 mi) thick
Lower Mantle
2,253 km (1,400 mi) thick
Outer Core
2,253 km (1,400 mi) thick
Inner Core
2,414 km (1,500 mi)
in diameter

▼ NEW WORLDS OUT OF OLD

Earth's landmasses have been colliding and rearranging for at least two billion years. The current continents separated from the supercontinent Pangaea and will reconverge in Pangaea Proxima in 250 million years.

200 Million Years Ago
Early dinosaurs lived on Earth's only continent, Pangaea, surrounded by one ocean, Panthalassa. Pangaea began to break up about 175 million years ago.

100 Million Years Ago
As pieces of Pangaea drifted apart, the Atlantic and Indian Oceans took shape, and the continents as we know them today became recognizable.

Present Day
Gradual tectonic shifts are ongoing. Only 34 million years ago, a rift between the Arabian and African tectonic plates created the Red Sea, which continues to spread.

100 Million Years in the Future
When Australia drifts north to meet Asia, the collision between the two continents will create towering mountain ranges similar to the Himalaya, where India and Eurasia meet.

PANGAEA PROXIMA

Atlantic Sea
Possible new highest point
NEW MOUNTAIN RANGE
TROPIC OF CANCER
EQUATOR
PACIFIC OCEAN
NEW MOUNTAIN RANGE
TROPIC OF CAPRICORN

Elevation	
30,000 feet	9,000 meters
15,000	4,500
Sea level	

250 Million Years in the Future
The Arctic Ocean and Mediterranean Sea will disappear, and the Atlantic will shrink to a modest sea as continents merge to create the new supercontinent Pangaea Proxima.

▼ PLATE TECTONICS

Volcano and earthquake hot spots line the unstable boundaries between tectonic plates such as the Great Rift Valley System within the continent of Africa and along the Java Trench in the Indian Ocean.

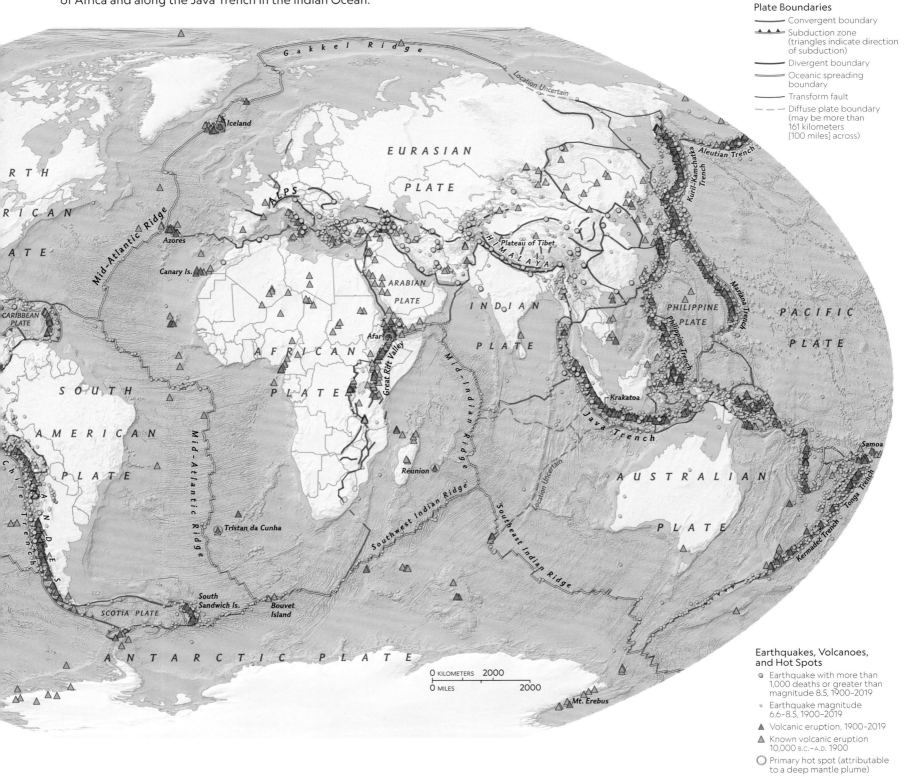

▼ LOOKING BELOW THE SURFACE

Earth is never static. Beneath our forests, deserts, and seas, plate tectonics are always in motion. Building, separating, colliding, rebuilding—these forces determine the height of mountains, the strength of tsunamis, the heat of volcanoes, and the stability of the land we walk on.

Accretion
When a heavy ocean plate bends below the edge of a continental plate or island arc, bits of sediment and melted mantle collect, or accrete, where the plates meet. Barbados, atop two subducting plates, accretes 25 mm (1 inch) every 1,000 years.

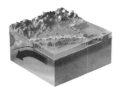

Collision
Mountains often rise at the zone of impact where two continents collide. The Himalaya formed after the Indian and Eurasian plates crashed into each other 55 million years ago; they still grow more than 1 cm (a half inch) a year.

Hot Spots
Not all earthquakes and volcanoes occur at the point where two plates meet. Hawai'i's island volcanoes are the result of a hot spot deep within Earth's mantle that drives magma upward through the oceanic crust.

Spreading
New ocean crust is born where tectonic plates spread apart, allowing lava to flow between them. Along the northern Mid-Atlantic Ridge, the North American and Eurasian plates are separating about 2.5 cm, or an inch, a year.

Faulting
When two plates move against each other and one suddenly gives way, the resulting fault, or crack, can trigger an earthquake. In 2004 a 1,500-kilometer (900-mile)-long fast-moving fault in Sumatra caused a deadly earthquake and tsunami.

Subduction
Where Where oceanic and continental plates meet, the heavier oceanic plate moves beneath—or subducts—the continental plate. The Cascadia subduction zone will someday produce an earthquake of magnitude 8.5 or greater.

THE PHYSICAL LANDSCAPE

Mountains, hills, plateaus, and plains are major landforms that make up the varied terrain of Earth. Buttes, valleys, and other minor landforms add further diversity to the landscape. Plate tectonics, weathering, and erosion create these landforms over long stretches of time, sometimes millions of years, pushing up the land to make hills or wearing it away to create valleys and canyons. Mountains often mark the meeting places of tectonic plates and are categorized as low mountains, at 91 to 400 meters (300 to 1,300 feet), or high mountains, taller than 400 meters (1,300 feet). Widely spaced mountains like those in the Basin and Range Province of the western United States are actually the tops of heavily eroded, faulted mountains; debris from erosion filled adjacent valleys, giving a spread-out appearance to these old summits.

Extensive, relatively flat lands that are higher than surrounding areas are plateaus. Formed by uplift, they include the terraced mesas of the Central Siberian Plateau and the Plateau of Tibet, popularly called the "roof of the world." Hills and low plateaus, like those of North America's Canadian Shield, are rounded natural elevations of land. The rolling, treeless country of flat plains cover much of Earth's surface, from the steppes of Eurasia to South America's Pampas.

Major Landforms
- High mountains
- Low mountains
- Plateaus
- Hills
- Flat plains
- No data

▲ THE FACE OF THE EARTH

The highs and lows, or terrain, of Earth's continents shape human life. Covering a third of the world's land area, flat plains are ideal for farming or raising livestock. High mountains, including the Rocky Mountains and the Himalaya, are sources of freshwater for millions.

▼ TECTONIC UPLIFTS

Some mountains are born when continental plates collided and folded up the land like a rumpled blanket. This is what created the highest landform on Earth: Mount Everest. Uplift also creates some types of plateaus.

New Zealand's lofty Southern Alps are shaped by tectonic uplift.

▼ VOLCANIC LANDSCAPES

Volcanic mountains form when molten rock erupts through Earth's crust and piles up on itself. The tiny island of Surtsey, near Iceland, was formed when a volcano poured hot lava into the Atlantic Ocean in the 1960s.

Iceland's Maelifell volcano

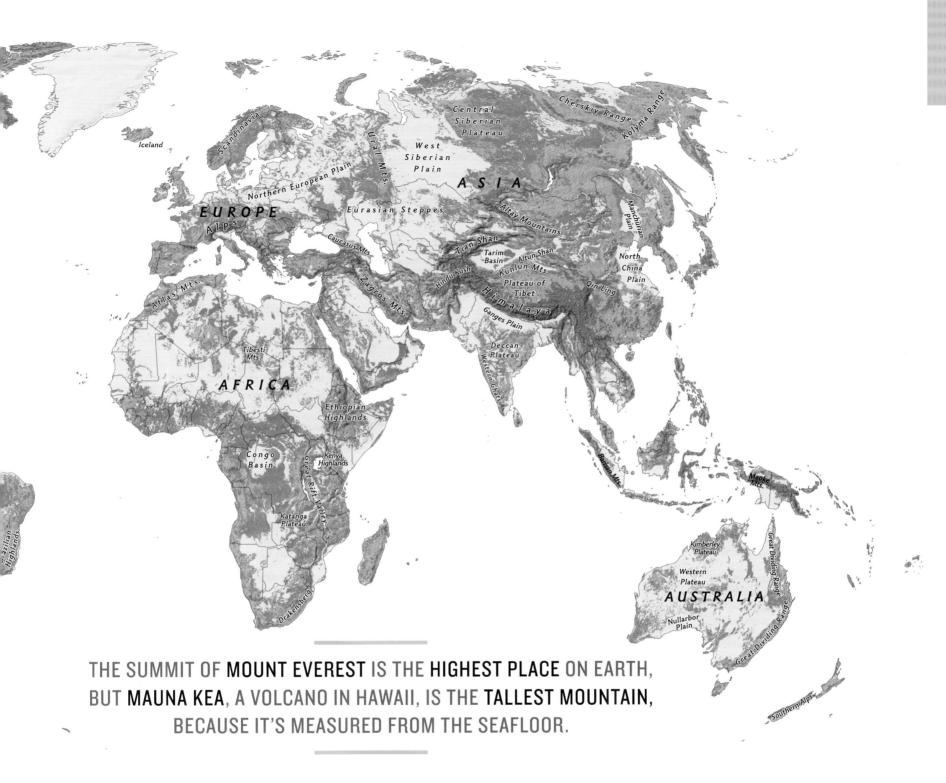

THE SUMMIT OF **MOUNT EVEREST** IS THE **HIGHEST PLACE** ON EARTH, BUT **MAUNA KEA**, A VOLCANO IN HAWAII, IS THE **TALLEST MOUNTAIN**, BECAUSE IT'S MEASURED FROM THE SEAFLOOR.

▼ **RIFT & FAULTING**

Rifting occurs when tectonic plates move apart. When plates move against each other, they create faults. Along the Sierra Nevada Fault they move up and down, creating mountains.

Kenya's Great Rift Valley is part of a large rift system.

▼ **ERODED LANDFORMS**

Weathering gradually carves rocks into distinctive formations as their surface is exposed to water, air, or chemicals. Erosion removes the weathered material and transports it by the forces of wind, water, or ice.

Eroded caves and cliffs shape the Algarve coast of Portugal.

SCULPTING EARTH'S SURFACE

Earth's features can change abruptly—a human-made dam, or a crater gouged by a falling meteor—but landscapes are usually formed gradually over millions of years. Wind, water, ice, and human activity create and continually change the terrain. Ocean currents transport sand and gravel from one part of a shore to another. Pounding waves undercut coastal cliffs, leaving behind sea stacks and sea arches. Water in the ground slowly dissolves limestone, a highly soluble rock. Over time, caves form and underground streams flow through the rock; sinkholes develop at the surface as underlying rock gives way. Many extraordinary landforms, such as the rock spires in the desert called buttes, were weathered and eroded by a combination of wind and water.

Among the legacies of Earth's most recent ice age are landforms shaped by glaciers. These large, slow-moving masses of ice can crush or topple anything in their paths; they can even stop rivers in their tracks, creating ice-dammed lakes. Ice sheets leave an even larger legacy simply because they cover more territory, creating giant lake basins like the ones now filled by the Great Lakes of North America.

▼ LANDFORMS CREATED BY WATER

Rivers flow downhill and around obstacles, but they also shape the surrounding land. Some rivers form broad loops, known as meanders, as faster currents erode their outer banks and slower currents deposit materials along inner banks. When a river breaks through the narrow neck of a meander, the abandoned curve becomes an oxbow lake.

The winding Snake River in Wyoming's Grand Teton National Park

River Meanders

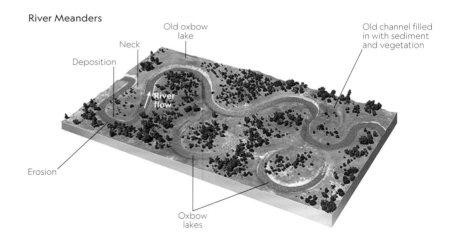

▼ LANDFORMS CREATED BY WIND

Wind can alter a landform by blowing dust and sand from dry soil, by sandblasting rock into new shapes, and by laying down sediments. Among desert landforms, sand dunes may be the most spectacular, and they come in several types. On beaches, dunes form when wind and waves deposit sediments along the shores of large bodies of water.

Barchan sand dunes in Death Valley National Park, United States

Star dunes in Sossusvlei, Namibia

Types of Dunes

Barchan dunes   Transverse dunes   Star dune   Seif dunes   Parabolic dunes

Labels on illustration: Mountain range, Mountain peak, Glacier, Dormant volcano, Iceberg, Archipelago, Basin, Mesa, Desert, Oasis, Island, Strait, Cape, Divide, Valley, Sound, Escarpment, Canal, Lake, Waterfall, Plateau, Plain, Hills, Point, Lagoon, Beach, Bay, Isthmus, Delta, River, Canyon, Peninsula, Cliff, Breakwater, Fork, Reef, Spit, Harbor, Ocean, Gulf, Tributary

▼ LANDFORMS CREATED BY ICE

Glaciers leave lasting imprints on the land. Meltwater deposits sand and gravel in long, narrow ridges called eskers. Ice embedded in the ground melts and forms depressions called kettles. Glaciers overrun rock debris to shape it into hills, or drumlins; they also pick up and carry huge amounts of rock and soil to create deposits called moraines.

▲ FEATURES OF EARTH'S SURFACE

In 41 natural and human-made features, this illustration shows typical landform locations and relationships. Note the way humans altered the physical geography such as with the breakwater, which protects the harbor from waves, and the canal, an artificial waterway that allows ships to navigate inland or take a shorter route between two bodies of water.

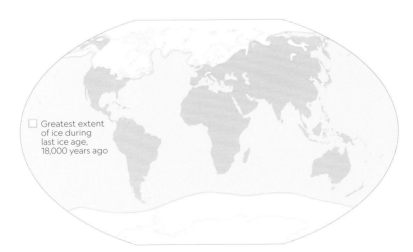

Greatest extent of ice during last ice age, 18,000 years ago

The Aletsch Glacier slowly glides downhill in the Swiss Alps.

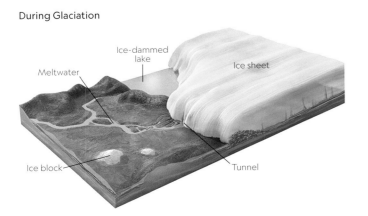

During Glaciation

Labels: Meltwater, Ice-dammed lake, Ice sheet, Ice block, Tunnel

After Glaciation

Labels: Esker formed by stream under ice sheet, Drumlin shaped by overriding glacier, Kettle lakes formed when ice blocks melt, Terminal moraines formed at margins of ice

BIRTH STORY OF ROCKS

Earth's outermost layer, averaging about 30 kilometers (19 miles) thick on land and five kilometers (3 miles) thick under the ocean, is called the crust. It is made of a wide variety of rocks, categorized into three main classes. Igneous and metamorphic rocks make up 95 percent of the crust's volume. Sedimentary rocks, although only a small percentage of the total, are the type most commonly found exposed on the surface.

As a result of plate tectonics, the crust is in constant slow motion. Rocks change positions over time, while heat and pressure change their compositions. Rocks contain materials that sustain civilization: Making steel requires the processing of iron, mainly from ancient sedimentary rocks. Rocks were carved into early tools to hunt and cook, and they became building blocks for our homes and for the world's grandest pyramids, cathedrals, and temples. Fossils and the geologic record hold valuable information to help scientists study plants and animals from long ago.

▼ THE ROCK CYCLE

Magma cools below or at Earth's surface, producing igneous rocks. Wind and water break down and transport those rocks to a new place. Over time, they become sedimentary rock. Pressure and heat from plate tectonics push rocks down, creating metamorphic rock or magma. The cycle begins anew.

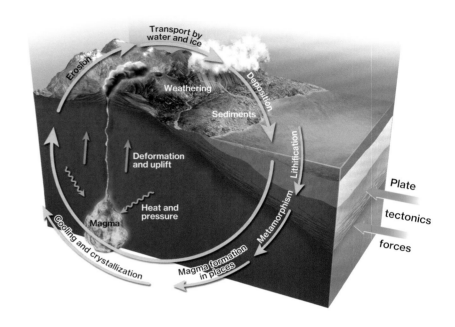

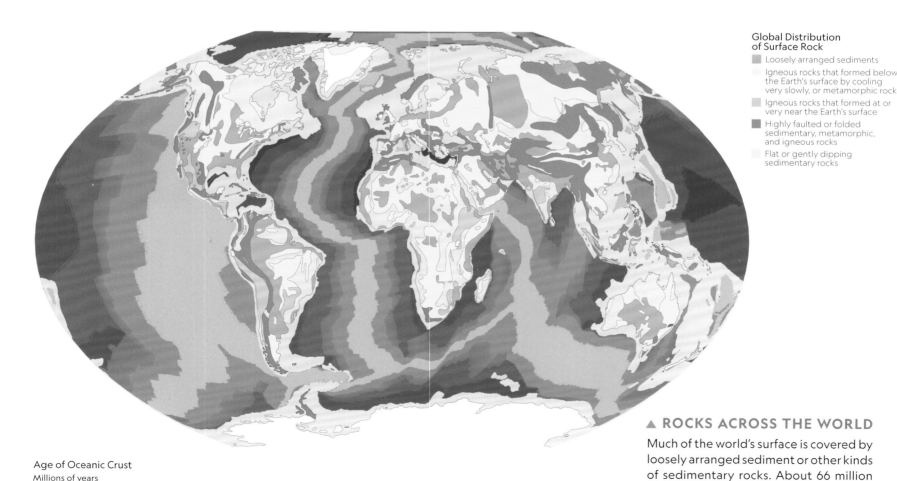

Global Distribution of Surface Rock

- Loosely arranged sediments
- Igneous rocks that formed below the Earth's surface by cooling very slowly, or metamorphic rock
- Igneous rocks that formed at or very near the Earth's surface
- Highly faulted or folded sedimentary, metamorphic, and igneous rocks
- Flat or gently dipping sedimentary rocks

Age of Oceanic Crust
Millions of years

0	23	66	100.5	145	200	255	280
Neogene	Paleogene	Upper Cretaceous	Lower Cretaceous	Jurassic	Triassic	Permian	

Geologic period

Age uncertain

▲ ROCKS ACROSS THE WORLD

Much of the world's surface is covered by loosely arranged sediment or other kinds of sedimentary rocks. About 66 million years ago, one of the largest volcanic eruptions in history flooded India's Deccan Plateau in lava, which is now igneous basalt.

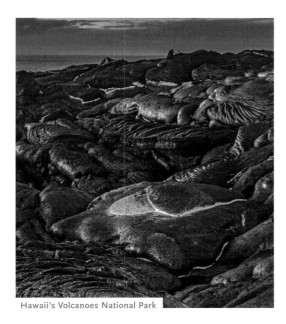

Hawaii's Volcanoes National Park

Layered stone in Red Rock Canyon, Nevada

A rocky shoreline in Maine

▲ IGNEOUS

Igneous rock forms when molten rock cools and solidifies. Cooling can happen quickly at Earth's surface, often as a lava flow, or more slowly underground in existing rock chambers. Examples: granite, basalt, obsidian.

▲ SEDIMENTARY

With their distinctive layering or bedding, sedimentary rocks form from preexisting rocks or from once living organisms. Wind or water erode or dissolve those rocks and deposit the pieces at Earth's surface, where it is compacted into new sedimentary rock. Examples: coal, chalk, sandstone.

▲ METAMORPHIC

Metamorphic rocks form when rocks of any origin (igneous, sedimentary, or metamorphic) are subjected to very high temperature and pressure. They also form as rocks react with fluids deep within the crust. Examples: marble, slate, quartzite.

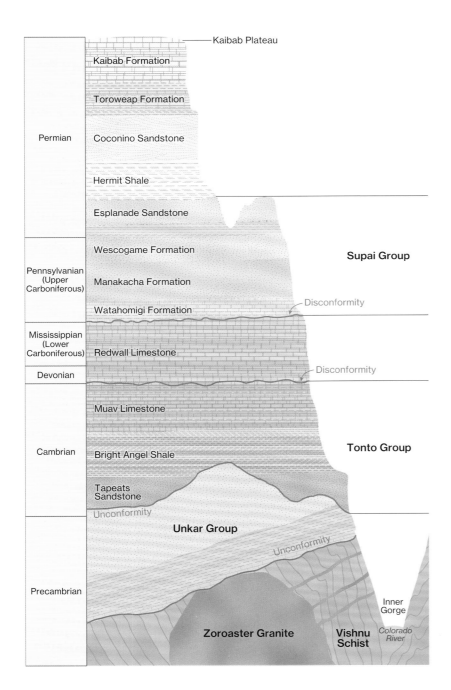

◀ READING EARTH'S HISTORY FROM ROCKS

Earth's complex history is written in layers of rock. This cross section of the walls of the Grand Canyon shows the oldest rocks, dating back 1.8 billion years, at the bottom of the sequence. The youngest rocks are at the top. Unconformities and disconformities represent major changes in rock layers, where older rocks were deformed or eroded before younger rock layers were laid down.

The Colorado River carved Arizona's Grand Canyon over millions of years.

MOSAIC OF LANDSCAPES

Rock is the continents' foundation, but most of the surface is blanketed by some type of land cover, from the immense grasslands of Central Asia to the forests of South America and ice sheets of Greenland. These different types of landscapes fall into about a dozen categories, shown on the map at right. They include wetlands, which skirt the edges of lakes, oceans, and the mouths of rivers; the pastures, fields, orchards, and vineyards of cropland that produce the world's food; and urban or developed areas. Barren lands such as exposed rock, salt flats, or beaches never have more than 10 percent vegetation.

Researchers track changes in land cover using satellite pictures of Earth. Information on spreading deserts, melting glaciers, and deforestation can help predict how Earth will change with climate warming and inform decisions on how to manage land and resources. For example, the pictures show that wetlands are shrinking faster than forests and that shrubland is spreading in the Siberian tundra due to warming temperatures.

NORTH AMERICA

SOUTH AMERICA

Land Cover Classification
- Evergreen forest
- Deciduous forest
- Mixed forest
- Savanna
- Shrubland
- Grassland
- Cropland
- Barren or sparsely vegetated
- Urban or built-up
- Snow and ice
- Cropland/vegetation mosaic
- Wetland

▲ GLOBAL VEGETATION

No single type of land cover dominates Earth's surface. Roughly one-fifth is covered in forest, while about one-third is either open grassland, shrubland, or savanna. Wetlands represent less than one percent.

▼ EVERGREEN FORESTS

These forests can be made up of conifers—trees with needles and cones—that grow in temperate regions of the Northern Hemisphere. Tropical rainforests include broadleaf evergreens.

The evergreen rainforest canopy of Malaysian Borneo

▼ DECIDUOUS FORESTS

Deciduous trees drop their leaves in autumn. Oaks, elms, and beeches are some of the deciduous trees in temperate forests like the United States' Eastern Deciduous Forest, which stretches from Florida to Maine.

Autumn foliage along the Eastern United States' Delaware River

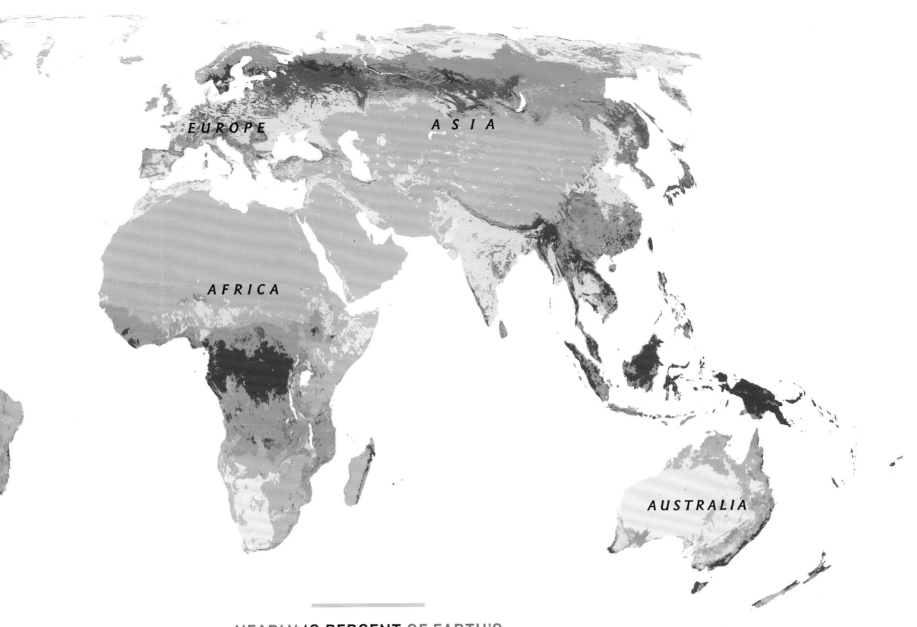

EUROPE

ASIA

AFRICA

AUSTRALIA

NEARLY **10 PERCENT** OF EARTH'S
4.6 BILLION ACRES OF **CROPLAND** IS IN **INDIA.**

▼ SHRUBLAND

These open, semiarid lands can occur in both hot and cold regions and are found in Australia, Chile, Mexico, northern Siberia, and southern Africa. Short woody plants called shrubs dominate; grasses and bushes can also thrive.

The shrubs of Western Australia's Kimberley region have adapted to low rainfall.

▼ SAVANNAS & GRASSLANDS

Savannas and grasslands are large plains, organized into different categories based on the relative density of grasses and woody plants or trees. Savannas make up large swaths of Africa, North and South America, and Siberia.

Elephants on the open plains of Tanzania's Serengeti National Park

EARTH'S FRESHWATER

Austria's Stausee Mooserboden Dam holds back a massive reservoir of freshwater.

Found in rivers, lakes, ponds, wetlands, springs, and belowground in aquifers, freshwater is essential to life. Humans can't live without it; many creatures call it home; agriculture depends on it. The amount of freshwater on the planet has remained steady over time, continually recycled through the atmosphere and returned to Earth with rainfall and snowmelt. But as the population grows, demand for freshwater has intensified: In Mexico City, taps are turned off periodically. Drinking water can be contaminated by runoff from fertilized fields, or chemicals from hydraulic fracking for oil and gas. Consuming less water, saving rainwater, reconsidering agricultural practices, and investing in seawater desalination are all ways to conserve and amplify our precious reserves.

Perennial River
Average Discharge 1961–1990,
Hectoliters per Second
(gallons per second)
—— More than 4,921 (130,000)
—— 284–4,921 (7,500–130,000)
—— 47–283 (1,250–7,499)
—— 9–46 (250–1,249)
—— Fewer than 9 (250)

Intermittent River
Average Discharge 1961–1990,
Hectoliters per Second
(gallons per second)
—— 284–4,921 (7,500–130,000)
—— 47–283 (1,250–7,499)
—— 9–46 (250–1,249)
—— Fewer than 9 (250)
🌀 Glaciated area or ice sheet

▲ STOPPING THE FLOW

Perennial rivers flow year-round, while intermittent rivers have channels that are periodically dry. When formerly perennially flowing rivers such as Central Asia's Syr Darya are dammed, they can become intermittent.

Watersheds
▮ The worlds's ten largest watersheds
▯ Other major watersheds

◀ MAJOR WATERSHEDS

A watershed describes an area drained by a river and its tributaries. Earth's largest watershed is the Amazon, which discharges 17 trillion liters (4.5 trillion gallons) into the Atlantic every second.

TWENTY PERCENT OF THE WORLD'S TOTAL SURFACE FRESHWATER IS CONTAINED IN RUSSIA'S LAKE BAIKAL.

▼ WATER AVAILABILITY

Only 2.5 percent of the planet's water is fresh, and most of that is frozen. Freshwater in the atmosphere has evaporated into vapor and clouds but will return to Earth as precipitation.

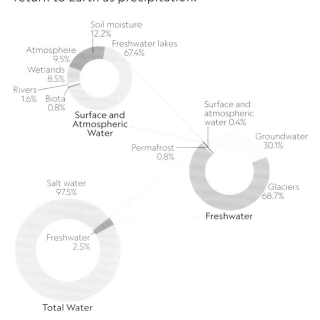

Soil moisture
12.2%

Atmosphere
9.5%

Freshwater lakes
67.4%

Wetlands
8.5%

Rivers
1.6%

Biota
0.8%

Surface and
Atmospheric
Water

Surface and
atmospheric
water 0.4%

Groundwater
30.1%

Permafrost
0.8%

Glaciers
68.7%

Salt water
97.5%

Freshwater

Freshwater
2.5%

Total Water

On a Zambia farm, irrigation infrastructure waters crops.

OUR BLUE PLANET

Earth is a watery planet. More than 70 percent of its surface is covered by interconnected bodies of salt water that together make up a continuous, global ocean. These are restless waters. Tidal movement—the regular ebb and flow of the ocean surface—results from gravitational forces exerted by the sun and the moon. Where marine and terrestrial realms meet, coral reefs—often called the rainforests of the sea—teem with life, while incessant waves shape cliffs and sea stacks. In the deep ocean lie vast, untouched plains; high mountains and ridges; and valleys plunging into trenches as deep as 11 kilometers (7 miles)—the deepest point on Earth.

The ocean is home to some of the oldest species on Earth. The horseshoe crab, for example, has outlived five mass extinctions in its 450-million-year history. But horseshoe crabs and other marine life are facing their greatest threat yet. Overharvesting has pushed some fish populations to the brink of collapse, and many coastal zones suffer from overdevelopment and pollution, killing corals. Although underwater exploration is progressing rapidly, maps of Mars are more comprehensive than our current maps of the ocean. So much more remains to be discovered.

A wave breaks over a coral reef on Rarotonga, in the Cook Islands.

Tropical fish and a sea star in a healthy coral reef in Thailand

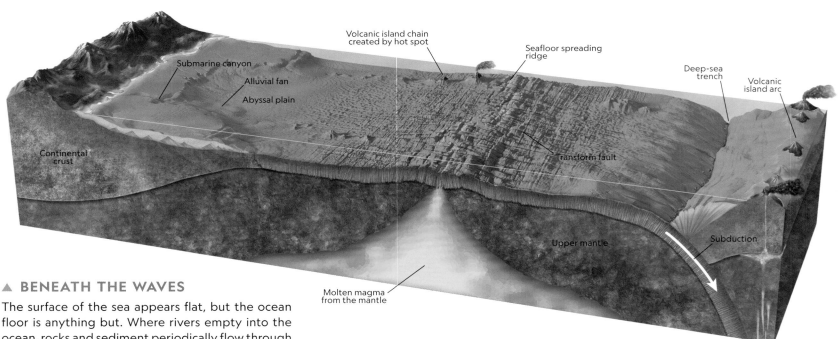

▲ BENEATH THE WAVES

The surface of the sea appears flat, but the ocean floor is anything but. Where rivers empty into the ocean, rocks and sediment periodically flow through submarine canyons to form alluvial fans. Hot magma pushes up from Earth's mantle and builds new seafloor through spreading; subducting tectonic plates form deep trenches.

▼ OCEAN DYNAMICS

Surface currents circle the major ocean basins, creating large systems that rotate warm water from equatorial regions to polar regions. Vertical currents, driven by wind, temperature, and salinity, take surface waters down to the depths and draw up cold water rich in nutrients.

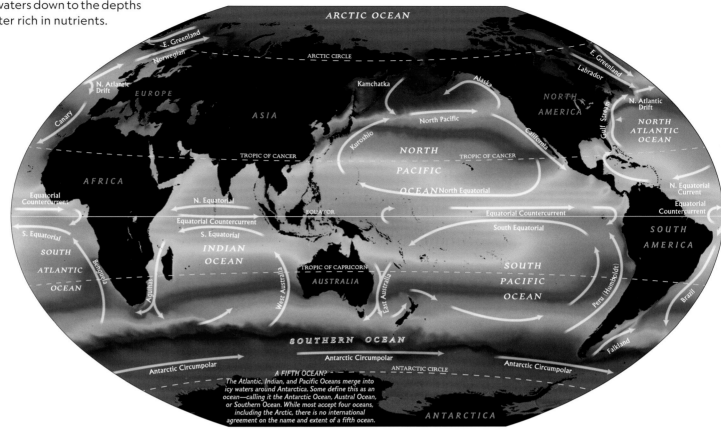

Sea Surface Temperature, July 2002–Aug. 2013 Average

High

Low

CORAL REEFS OCCUPY LESS THAN ONE PERCENT OF THE OCEAN FLOOR BUT ARE HOME TO ONE-QUARTER OF ALL MARINE SPECIES.

▼ CORAL REEFS

Coral reefs form over millennia as tiny invertebrates known as coral polyps bind themselves to submerged rock and grow slowly. The rich reefs of the Northwest Hawaiian Islands support more than 7,000 species of fish, sea turtles, birds, and mammals.

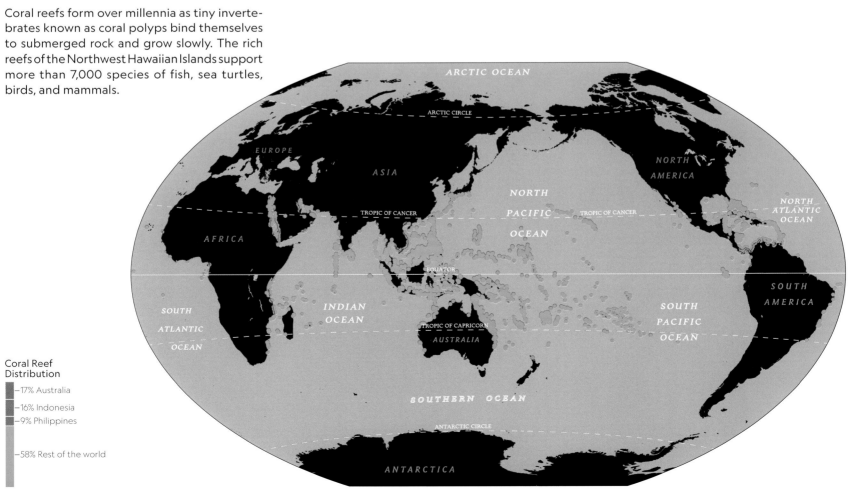

Coral Reef Distribution

–17% Australia

–16% Indonesia

–9% Philippines

–58% Rest of the world

DEFINING CLIMATE

Climate is not the same as weather. Climate describes a region's average weather conditions, measured over many years and influenced by geographic location, topography, nearness to oceans or mountains, and elevation. Weather, on the other hand, changes daily, and on a given day may differ dramatically from what's expected of the prevailing climate. Energy from the sun drives the global climate system.

The tropics absorb much of this incoming solar energy, while outgoing heat radiation tends to leave Earth at high latitudes. To achieve a balance across the globe, the atmosphere and the oceans both move huge amounts of heat from the tropics to polar regions. The tilt of the Earth's axis determines how much solar energy hits different latitudes. When it's winter in the Northern Hemisphere (in January), that part of the world is tilting away from the sun, and more solar energy is hitting South America, southern Africa, and Australia in the Southern Hemisphere. When it's summer in the Northern Hemisphere (in July), the Earth's tilt toward the sun is reversed, heating up North America and most of Eurasia. The surrounding atmosphere traps solar energy and makes the Earth warmer than it would be without an atmosphere; this is called the greenhouse effect.

Tropical (A)
- No dry season (Af)
- Monsoonal (Am)
- Dry summer (As)
- Dry winter (Aw)

Arid (B)
Desert (BW)
- Hot (BWh)
- Cold (BWk)

Steppe (BS)
- Hot (BSh)
- Cold (BSk)

Temperate (C)
Warm Winter, No Dry Season (Cf)
- Hot summer (Cfa)
- Warm summer (Cfb)
- Cool summer (Cfc)

▲ CLIMATE ZONES

The Köppen-Geiger system's five classes and 31 subcategories label different climates based on seasonal temperatures and precipitation. Latitude, proximity to the ocean, whether a mountain range blocks prevailing winds—all these factors determine how hot, cold, dry, or wet seasonal conditions are.

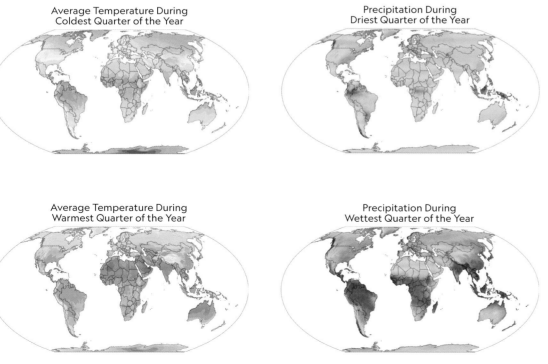

Average Temperature During Coldest Quarter of the Year

Precipitation During Driest Quarter of the Year

Average Temperature During Warmest Quarter of the Year

Precipitation During Wettest Quarter of the Year

-65°C (-85°F) 0°C (32°F) 40°C (104°F)

0 mm (0 in) 575 mm (23 in) More than 1,200 mm (47 in)

◄ CLIMATE EXTREMES

Tropical climates may experience a wet-dry cycle with long dry or rainy periods, while temperate regions have a wider range of temperatures over four seasons. Antarctica is an outlier—always cold, always dry—as is north-central Australia, which can be as hot as equatorial regions.

Warm Winter,
Dry Summer (Cs)
- Hot summer (Csa)
- Warm summer (Csb)
- Cool summer (Csc)

Warm Dry Winter (Cw)
- Hot summer (Cwa)
- Warm summer (Cwb)
- Cool summer (Cwc)

Cold (D)
Snow, No Dry Season (Df)
- Hot summer (Dfa)
- Warm summer (Dfb)
- Cool summer (Dfc)
- Extremely continental (Dfd)

Snow, Dry Summer (Ds)
- Hot summer (Dsa)
- Warm summer (Dsb)
- Cool summer (Dsc)
- Extremely continental (Dsd)

Snow, Dry Winter (Dw)
- Hot summer (Dwa)
- Warm summer (Dwb)
- Cool summer (Dwc)
- Extremely continental (Dwd)

Polar (E)
- Ice cap climate (EF)
- Tundra (ET)

Ocean Current
→ Warm
→ Cold

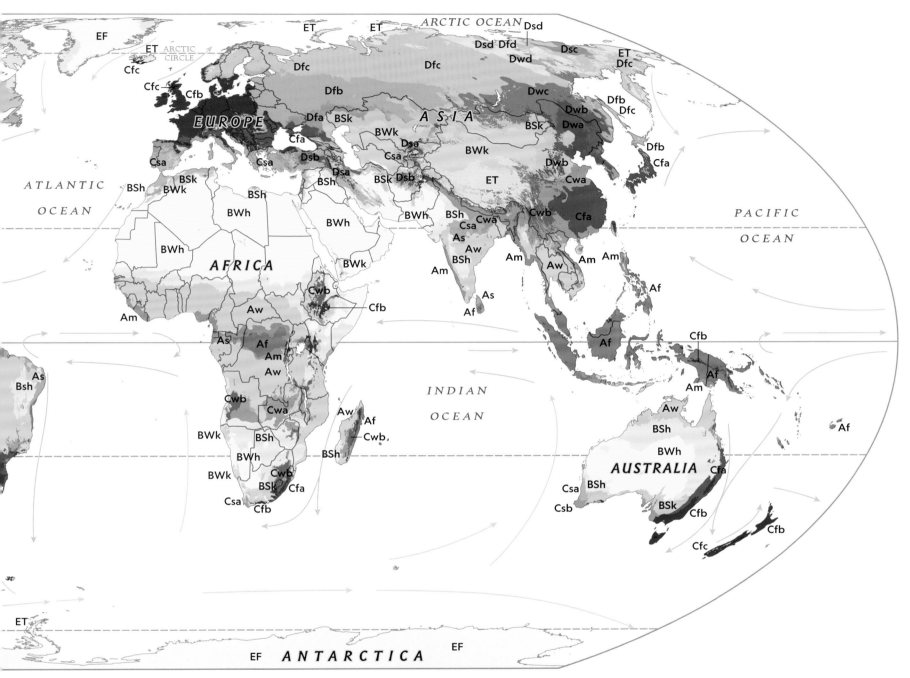

Brazil's Ilha Grande, where a tropical climate (Aw) means rainy summers and a drier winter.

Southwestern Iceland has a more temperate climate (Cfc) than the polar north.

A WARMING PLANET

Earth is getting warmer, and human activities are the cause. A global reliance on fossil fuels, in particular, has dramatically increased the emission of carbon dioxide into the atmosphere, which accentuates the greenhouse effect that traps heat. Across the planet, this warming trend changes precipitation patterns, melts glaciers, and intensifies storms. Threats to human health and livelihoods include water scarcity, a heightened risk of crop loss and disease, and rising sea levels swamping coastlines. Recent projections show large swaths of Bangkok, Ho Chi Minh City, Shanghai, and Mumbai underwater by 2050. On the current trajectory, heat waves in the Persian Gulf and elsewhere will become too intense for people to spend more than a few hours outdoors. To limit the effects, transformative changes must be made to keep global temperatures within 1.5 or 2 degrees Celsius (2.7 or 3.6 degrees Fahrenheit) above their preindustrial levels. This is an achievable goal, but to meet the target the world needs to zero-out fossil fuel emissions by 2050 and change to alternative sources of energy.

▼ MELTING PERMAFROST

When soil that had formerly been permanently frozen thaws, it releases long-buried carbon dioxide and methane. Permafrost could soon be as big a source of greenhouse gases as China.

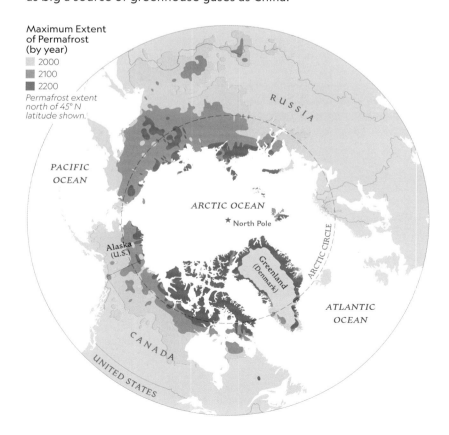

Maximum Extent of Permafrost (by year)
- 2000
- 2100
- 2200

Permafrost extent north of 45° N latitude shown.

▼ TEMPERATURE ON THE RISE

The top five hottest years ever recorded were 2015–19. Temperatures in the Arctic are increasing two to three times faster than anywhere else in the world. As polar ice melts, less solar radiation is reflected away from Earth, causing more melting and warming elsewhere.

Temperature Trends, January 1960–August 2016 (Change in degrees)

Fahrenheit	Celsius
-4.9°	-2.7°
0°	-0°
5°	-2°
10°	-4°
15.1°	-6°
	-8.4°

▼ UNPRECEDENTED INCREASES

Electric power plants, airplanes, livestock, landfills, and industry release carbon dioxide and even more potent methane. The concentration of these gases in the atmosphere has skyrocketed due to industrial development.

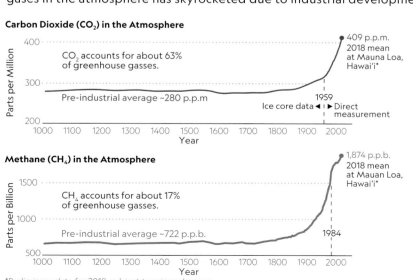

Carbon Dioxide (CO_2) in the Atmosphere

CO_2 accounts for about 63% of greenhouse gasses.

Pre-industrial average ~280 p.p.m.

409 p.p.m. 2018 mean at Mauna Loa, Hawai'i*

1959 Ice core data ◄ I ► Direct measurement

Methane (CH_4) in the Atmosphere

CH_4 accounts for about 17% of greenhouse gasses.

Pre-industrial average ~722 p.p.b.

1,874 p.p.b. 2018 mean at Mauan Loa, Hawai'i*

1984

Preliminary data for 2018, subject to minor changes.

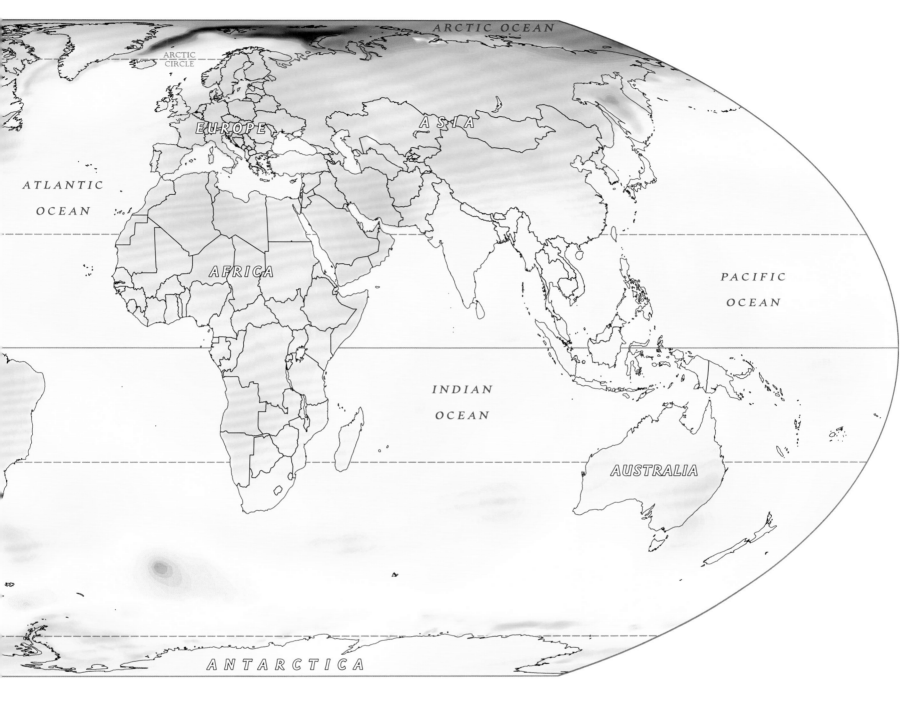

▼ CHARTING SEA-LEVEL CHANGE

Seas are rising because higher ocean temperatures cause water to expand, and because ice sheets and glaciers—particularly in Greenland and Antarctica—are now melting and creating an influx of water.

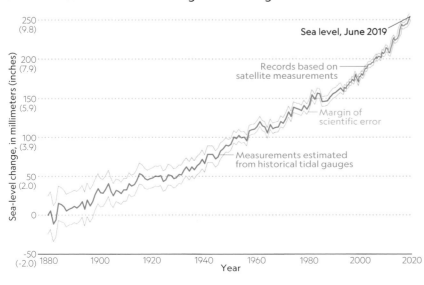

Sea level, June 2019

Records based on satellite measurements

Margin of scientific error

Measurements estimated from historical tidal gauges

Sea-level change, in millimeters (inches)

250 (9.8)
200 (7.9)
150 (5.9)
100 (3.9)
50 (2.0)
0
-50 (-2.0)

1880   1900   1920   1940   1960   1980   2000   2020

Year

The North Sea lashes the English town of Cromer during a storm surge.

THE WORLD'S WEATHER

Humidity, air temperature and pressure, wind speed and direction, cloud cover and type, and the amount and form of precipitation are all characteristics of the fleeting atmospheric conditions we call weather. The sun is ultimately responsible for the weather. Land and water surfaces absorb its rays differently, causing variations in the temperature and pressure of the air masses above. As an air mass warms, it lightens and rises higher into the atmosphere. As an air mass cools, it becomes heavier and sinks. The pressure differences between air masses generate winds, which tend to blow from high-pressure areas to areas of low pressure. Fast-moving upper-atmosphere winds known as jet streams help move weather systems around the world.

When a warm and cold air mass meet, weather conditions can change quickly. Intense storms may form along cold fronts if air is unstable; widespread clouds and rain, snow, or sleet may accompany warm or cold fronts. Many factors shape weather, including latitude, elevation, and proximity to water bodies. Even urban development, where pavement, buildings, and people create "heat islands," and the amount of snow cover, which chills an overlying air mass, can play important roles in influencing weather.

A supercell thunderstorm swirls above an Oklahoma town.

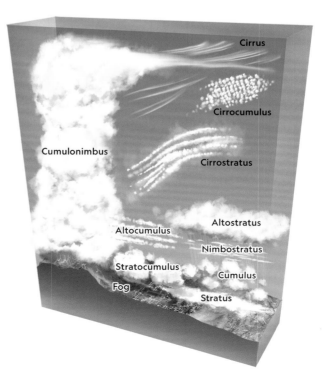

Lightning illuminates the sky over Perth, Australia.

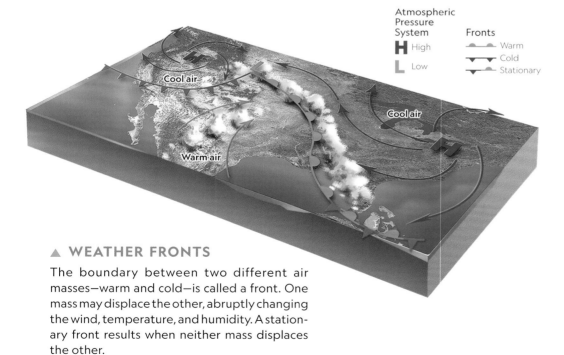

Atmospheric Pressure System

H High
L Low

Fronts
Warm
Cold
Stationary

▲ WEATHER FRONTS

The boundary between two different air masses—warm and cold—is called a front. One mass may displace the other, abruptly changing the wind, temperature, and humidity. A stationary front results when neither mass displaces the other.

▼ CLOUD TYPES

Clouds are collections of water droplets or ice crystals classified by shape and altitude: Low-level stratus clouds are flat or layered; cumulonimbus clouds can tower up to more than 10 kilometers (6 miles) tall.

ONE **LIGHTNING BOLT** CAN CONTAIN UP TO A **BILLION VOLTS** OF ELECTRICITY—A **CHARGE** THAT WOULD POWER MORE THAN **EIGHT MILLION LIGHT BULBS.**

▼ **CYCLONE TRACKS**

Cyclones, also known as hurricanes or typhoons, are large low-pressure systems that suck air in forcefully, resulting in circulating winds of 119 to over 249 kilometers (74 to over 155 miles) an hour. They usually form over warm ocean water in the tropics and subtropics.

Storm Tracks, 2000–2019
— Disturbance
— Subtropical
— Tropical in nature
— Extratropical
— Unknown

▼ **LIGHTNING STRIKES**

Lightning flashes are more likely to occur over land than water, and most often in the tropics, where more heat provides more energy to produce thunderstorms. The hot spot in Venezuela occurs where warm, moist air from the Caribbean Sea encounters cold air from the Andes Mountains.

Flashes
(per sq km per year)
■ More than 29
■ 2–29
■ 0.4–1.9
□ Less than 0.4
□ 0

ZONES OF LIFE ON EARTH

The biosphere is the relatively thin sliver of Earth's surface and atmosphere that is home to all living things. The planet is one global ecosystem made up of many smaller ecosystems where organisms, air, land, and water interact in dynamic relationships that drive Earth's self-sustaining natural processes. Billions of years ago, for instance, single-celled bacteria developed an ability to use sunlight to convert carbon dioxide and water into oxygen. Over three billion years ago, this process of photosynthesis gradually changed the biosphere as oxygen allowed plants and animals to evolve and thrive. The food chain is another vital cycle: Plants are consumed by herbivores who fall prey to carnivores. Dead organic matter from plants and animals is broken down by decomposers so nutrients can go back to the soil.

The organisms that make up the biosphere vary greatly in size and number. Small life-forms such as tiny bacteria reach very high numbers, whereas large ones like blue whales—the largest animals on Earth—are relatively rare. Humans make up less than one percent of the life on Earth by weight, but our outsized impact has disrupted ecosystems on a geologic scale. Over the last 10,000 years, human activity has diminished plant biomass by half and reduced the number of wild mammals by 85 percent.

Cloud forests of Cocos Island, Costa Rica

▶ THE SIZE OF THE BIOSPHERE

Since most of the planet is rock and magma, life only exists on the thin outer layer. The biosphere stretches roughly 20 kilometers (12 miles) in all, from the ocean floor to more than 10,000 meters (33,000 feet) above sea level. Most life occurs in a much slimmer zone.

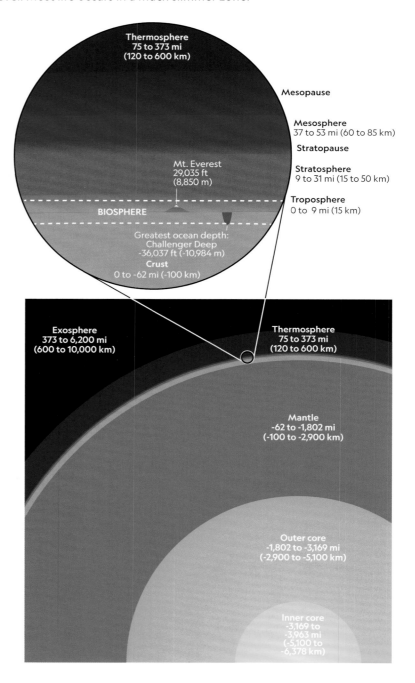

Thermosphere
75 to 373 mi
(120 to 600 km)

Mesopause

Mesosphere
37 to 53 mi (60 to 85 km)

Stratopause

Mt. Everest
29,035 ft
(8,850 m)

Stratosphere
9 to 31 mi (15 to 50 km)

BIOSPHERE

Troposphere
0 to 9 mi (15 km)

Greatest ocean depth:
Challenger Deep
-36,037 ft (-10,984 m)

Crust
0 to -62 mi (-100 km)

Exosphere
373 to 6,200 mi
(600 to 10,000 km)

Thermosphere
75 to 373 mi
(120 to 600 km)

Mantle
-62 to -1,802 mi
(-100 to -2,900 km)

Outer core
-1,802 to -3,169 mi
(-2,900 to -5,100 km)

Inner core
-3,169 to
-3,963 mi
(-5,100 to
-6,378 km)

Coral reefs in the Solomon Islands

▼ PLANT LIFE ON LAND & AT SEA

Satellite data paints a portrait of Earth's vegetation. In the ocean, dark blue to violet shows warmer areas with low nutrients and little plant life; greens and reds are cooler, nutrient-rich regions, including coastal areas. On land, forests, grasslands, and other areas where plant life is abundant are green; tan and white represent deserts or snow.

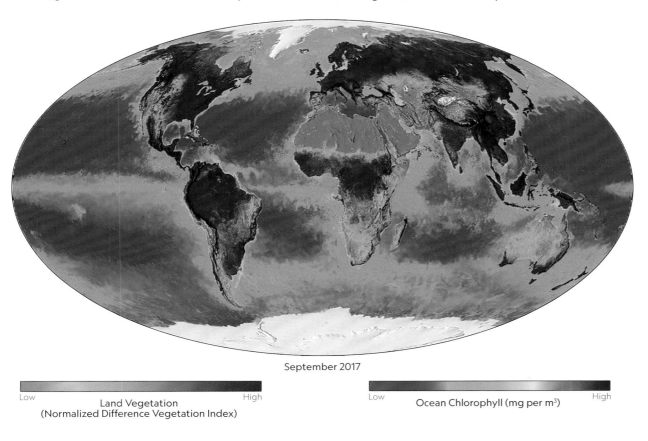

September 2017

Low ——— High
Land Vegetation
(Normalized Difference Vegetation Index)

Low ——— High
Ocean Chlorophyll (mg per m³)

BACTERIA MAY BE MICROSCOPIC, BUT THEIR COMBINED BIOMASS IS 1,166 TIMES GREATER THAN HUMANS.

▼ ENERGY IN THE OCEAN

There are two ways to produce food in the ocean. Near the surface, microbes and some plants use photosynthesis just like vegetation on land, absorbing sunlight to convert carbon dioxide and water into oxygen. At depths out of reach of sunlight, microbes run on chemical energy in a process called chemosynthesis, often processing compounds released from hydrothermal vents on the ocean floor.

Photosynthesis

Light energy

Plants use solar energy to make organic molecules

Plants are eaten by animals

Chemosynthesis

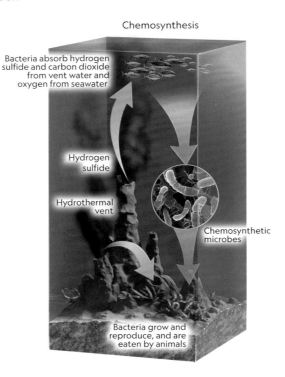

Bacteria absorb hydrogen sulfide and carbon dioxide from vent water and oxygen from seawater

Hydrogen sulfide

Hydrothermal vent

Chemosynthetic microbes

Bacteria grow and reproduce, and are eaten by animals

▶ WORLD BIOMASS

Combine all Earth's living organisms and the total amount of matter is its biomass. Plants dominate the biosphere, and live mainly on land. Animals— wild and domesticated, and including humans—make up only a tiny fraction of all living creatures, and most of those are insects, crustaceans, or fish.

World's Biomass

Animals range from microscopic to the blue whale, and include everyone reading this chart (see below).

- Humans
- Domesticated animals
- Wild mammals, birds, reptiles, and amphibians
- Fish
- Arthropods, includes insects, spiders, and crustaceans
- Annelids
- Molluscs
- Cnidarians, includes coral and jellyfish
- Nematodes

Plants

Fungi (includes mold and mushrooms)

Protists are miscellaneous one-celled organisms with a nucleus (eukaryotes)

Bacteria are one-celled organisms without a nucleus (prokaryotes)

Archaea are physically similar to bacteria but have distinct molecular characteristics

Viruses

DIVERSITY OF LIFE

Life on Earth is a finely tuned symphony of natural processes, and it is one that humans can't live without. Over three billion years, Earth's inhabitants—currently an estimated eight to 10 million species—have evolved to create complex ecosystems. Wetlands filter water and prevent erosion; plants provide us with oxygen to breathe. One of the most valuable gifts of nature is lifesaving medicine: A popular class of drugs used to treat high blood pressure, ACE inhibitors, were developed in the 1970s from the venom of the Brazilian pit viper. Every new extinction could mean the loss of a future medical discovery. Despite vast catalogs of flora and fauna, roughly 80 percent of Earth's species remain unknown to science—and we are losing them, fast. One million species may go extinct in the next few years; some biologists warn that entire branches of the evolutionary family tree, formed over millennia, could be lost.

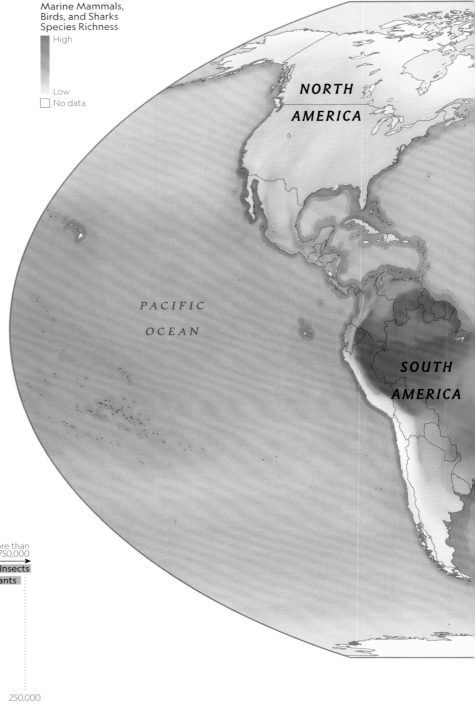

Marine Mammals, Birds, and Sharks Species Richness

High

Low

No data

NORTH AMERICA

SOUTH AMERICA

PACIFIC OCEAN

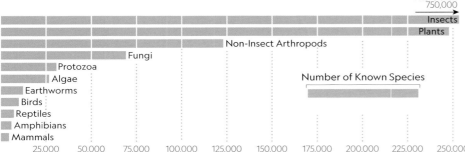

More than 750,000

Insects

Plants

Non-Insect Arthropods

Fungi

Protozoa

Algae

Earthworms

Birds

Reptiles

Amphibians

Mammals

Number of Known Species

25,000   50,000   75,000   100,000   125,000   150,000   175,000   200,000   225,000   250,000

▲ SPECIES DIVERSITY

Insects make up more than half of all the known or described species on Earth. Mammals, reptiles, and amphibians together make up less than 2 percent.

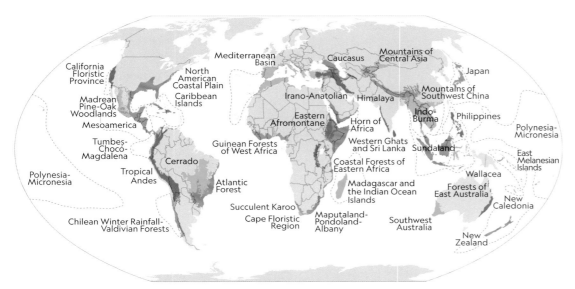

California Floristic Province
North American Coastal Plain
Madrean Pine-Oak Woodlands
Mesoamerica
Caribbean Islands
Mediterranean Basin
Caucasus
Mountains of Central Asia
Japan
Irano-Anatolian
Himalaya
Mountains of Southwest China
Indo-Burma
Philippines
Eastern Afromontane
Horn of Africa
Western Ghats and Sri Lanka
Sundaland
Polynesia-Micronesia
Guinean Forests of West Africa
Tumbes-Chocó-Magdalena
Cerrado
Tropical Andes
Atlantic Forest
Polynesia-Micronesia
Madagascar and the Indian Ocean Islands
Coastal Forests of Eastern Africa
East Melanesian Islands
Wallacea
Forests of East Australia
New Caledonia
Chilean Winter Rainfall-Valdivian Forests
Succulent Karoo
Cape Floristic Region
Maputaland-Pondoland-Albany
Southwest Australia
New Zealand

Biodiversity Hotspots

Hotspots

Hotspot outer limit

◀ BIODIVERSITY HOTSPOTS

These irreplaceable places are rich with life and are also at risk for destruction. The Western Ghats in India contain a third of the world's Asian elephants, and in the Tumbes-Chocó-Magdalena forests of South America live over 11,000 plant species, 900 bird species, and 320 reptile species.

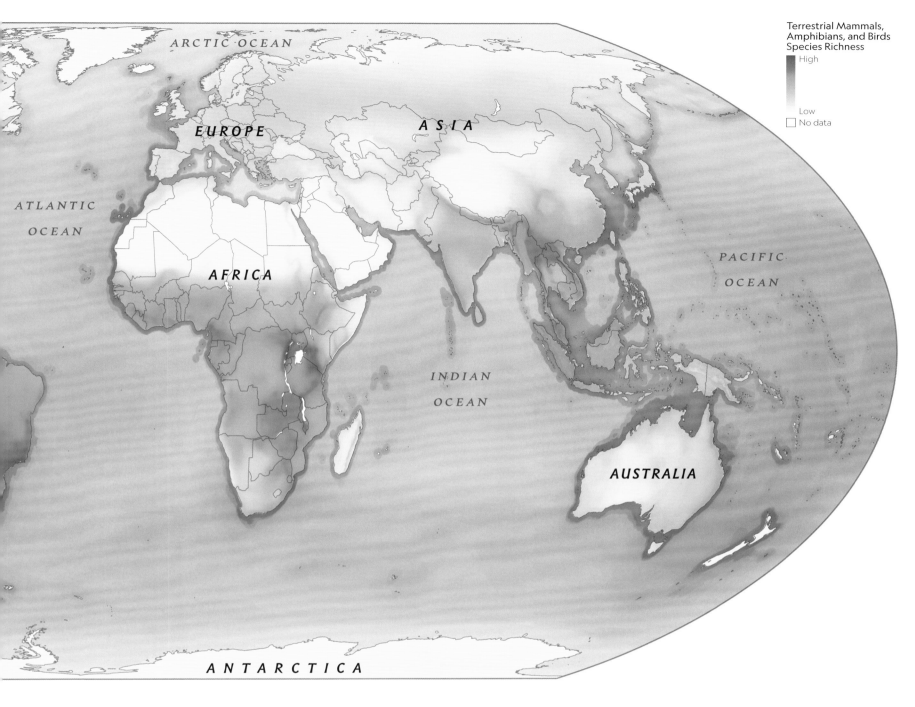

Terrestrial Mammals,
Amphibians, and Birds
Species Richness

High

Low
No data

ARCTIC OCEAN

EUROPE

ASIA

ATLANTIC
OCEAN

AFRICA

PACIFIC
OCEAN

INDIAN
OCEAN

AUSTRALIA

ANTARCTICA

▲ BIODIVERSITY ON LAND & SEA

This map shows the distribution of many animal species around the world. Some environments, like
coral reefs, are much more hospitable than others, resulting in the concentration of species. Land
mammals and birds occur in the greatest numbers in the wet forests of South America and central
Africa, East Africa, and Southeast Asia.

A green sea turtle cruises by butterflyfish and a pair of angelfish in a reef in Indonesia.

Macaws in flight in Manú National Park, a biosphere reserve in Peru

SAFE HAVENS FOR NATURE

Reintroduced African wild dogs in Mozambique's Gorongosa National Park

Humans have profoundly changed the natural world with activities such as overfishing the oceans and plowing rainforests into palm oil plantations. Yet there are still some places on Earth that have mostly avoided human impact. More than 70 percent of this true wilderness, on land and at sea, lies within the borders of five countries: Australia, Brazil, Canada, Russia, and the United States. Conservationists work to protect pristine landscapes and heal fragmented or degraded ones. Worldwide, UNESCO has identified nearly 700 "biosphere reserves," internationally recognized sites where conservation and development efforts work in tandem to protect and learn from natural areas without impeding economic progress.

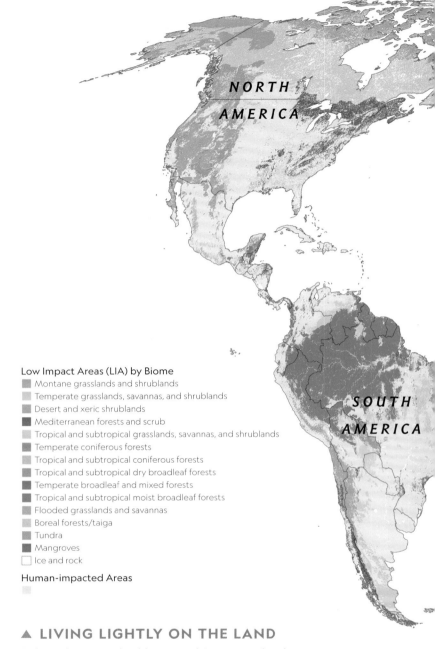

Low Impact Areas (LIA) by Biome

- Montane grasslands and shrublands
- Temperate grasslands, savannas, and shrublands
- Desert and xeric shrublands
- Mediterranean forests and scrub
- Tropical and subtropical grasslands, savannas, and shrublands
- Temperate coniferous forests
- Tropical and subtropical coniferous forests
- Tropical and subtropical dry broadleaf forests
- Temperate broadleaf and mixed forests
- Tropical and subtropical moist broadleaf forests
- Flooded grasslands and savannas
- Boreal forests/taiga
- Tundra
- Mangroves
- Ice and rock

Human-impacted Areas

▲ LIVING LIGHTLY ON THE LAND

Relatively untouched biomes—biomes are landscapes that share specific ecological characteristics—exist on every continent, but they are most apparent in deserts, tropical forests, and northern forests. The areas of greatest human impact include sprawling cities and extensive farmland.

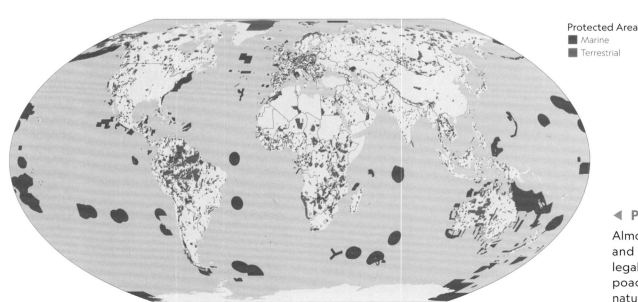

Protected Area
- Marine
- Terrestrial

◄ PROTECTED AREAS

Almost 15 percent of the world's land area and 7 percent of the global ocean are legally protected from development and poaching in the form of national parks, nature reserves, and marine monuments.

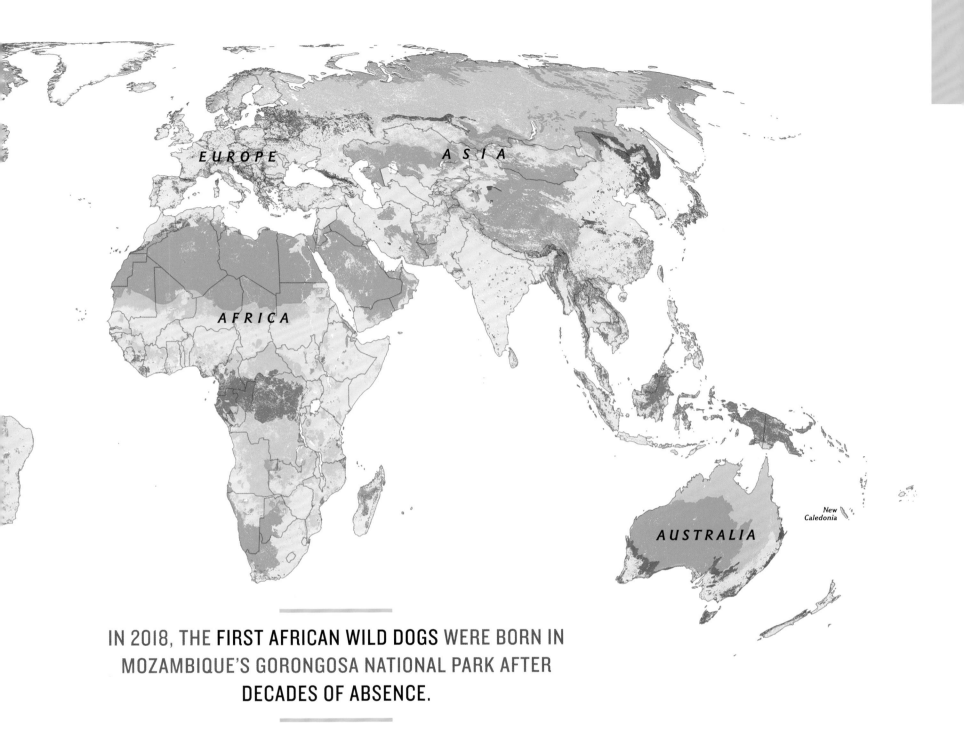

EUROPE

ASIA

AFRICA

AUSTRALIA

New
Caledonia

IN 2018, THE **FIRST AFRICAN WILD DOGS** WERE BORN IN
MOZAMBIQUE'S GORONGOSA NATIONAL PARK AFTER
DECADES OF ABSENCE.

▼ FADING LANDSCAPES

Low-impact biomes are landscapes that are relatively unaffected
by people. They represent just over half of Earth's land. But, like the
temperate grasslands of North America's Great Plains, many lack
protection, and they are shrinking because of human activities.

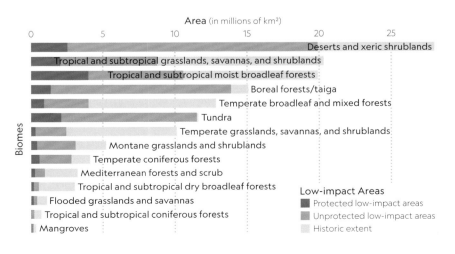

Area (in millions of km²)

0   5   10   15   20   25

Deserts and xeric shrublands
Tropical and subtropical grasslands, savannas, and shrublands
Tropical and subtropical moist broadleaf forests
Boreal forests/taiga
Temperate broadleaf and mixed forests
Tundra
Temperate grasslands, savannas, and shrublands
Montane grasslands and shrublands
Temperate coniferous forests
Mediterranean forests and scrub
Tropical and subtropical dry broadleaf forests
Flooded grasslands and savannas
Tropical and subtropical coniferous forests
Mangroves

Biomes

Low-impact Areas
▮ Protected low-impact areas
▮ Unprotected low-impact areas
▯ Historic extent

The San Rafael waterfall in Ecuador is threatened by hydroelectric dam construction.

THREATS TO OUR PLANET

Like all living things on Earth, people need natural systems to thrive. We rely on bees to pollinate our food crops, trees to grip soil and prevent erosion, and hungry bats to keep insects in check. Over-exploitation of natural resources can provoke a domino effect of hazardous consequences. For instance, high temperatures and desertification in northern India has intensified dust storms, illegal logging triggered a deadly landslide in the Philippines, and diverting Central Asia's Aral Sea to irrigate crops nearly drained one of the planet's largest bodies of freshwater.

Three-quarters of Earth's land surface is under pressure from human activity, and some of the most intense pressure is being felt in places with the highest diversity of plant and animal life. While some wealthier regions are showing a modest decrease in human impact, other parts of the world—the New Guinea mangroves, the Purus Várzea rainforest in the Amazon, the Baffin coastal tundra—have been more significantly changed. By some estimates, our current resource demands already exceed what Earth can provide. But with technology and sustainability, we can achieve a planet in balance that provides for humanity and for the millions of species with which we live.

Beijing's Tiananmen Square during a "red alert" for air pollution

The Mojave Desert of North America has been getting less and less rain.

A SEVERE DROUGHT IN SYRIA KILLED LIVESTOCK, INCREASED FOOD PRICES, AND DROVE PEOPLE FROM THEIR FARMS INTO JAM-PACKED CITIES. THESE FACTORS **HELPED SPARK A CIVIL WAR.**

◀ **LANDS GETTING DRIER**

If semiarid places (in yellow) are subject to long or frequent drought, they can turn to permanent desert (in brown), especially if overgrazing of livestock, clearing of trees, and depletion of groundwater have already degraded the land.

SYRIA—

Land Degradation and Desertification

Dryland systems

Land degradation in drylands

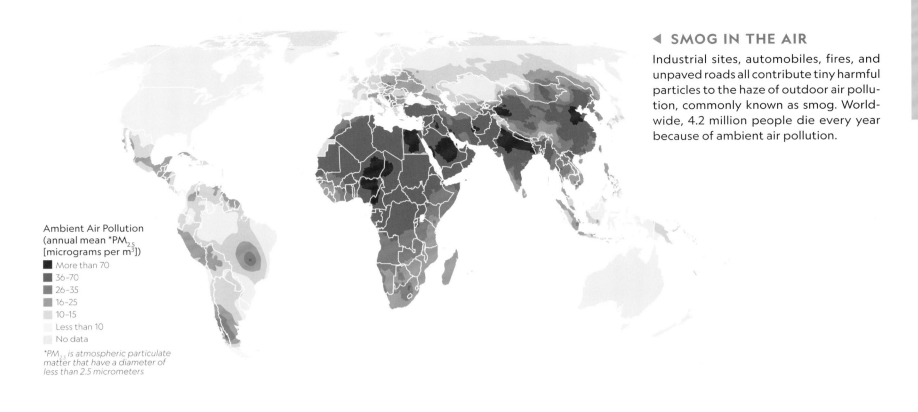

◀ SMOG IN THE AIR

Industrial sites, automobiles, fires, and unpaved roads all contribute tiny harmful particles to the haze of outdoor air pollution, commonly known as smog. Worldwide, 4.2 million people die every year because of ambient air pollution.

Ambient Air Pollution
(annual mean *PM$_{2.5}$ [micrograms per m^3])
- ■ More than 70
- ■ 36–70
- ■ 26–35
- ■ 16–25
- ■ 10–15
- ☐ Less than 10
- ☐ No data

*PM$_{2.5}$ is atmospheric particulate matter that have a diameter of less than 2.5 micrometers

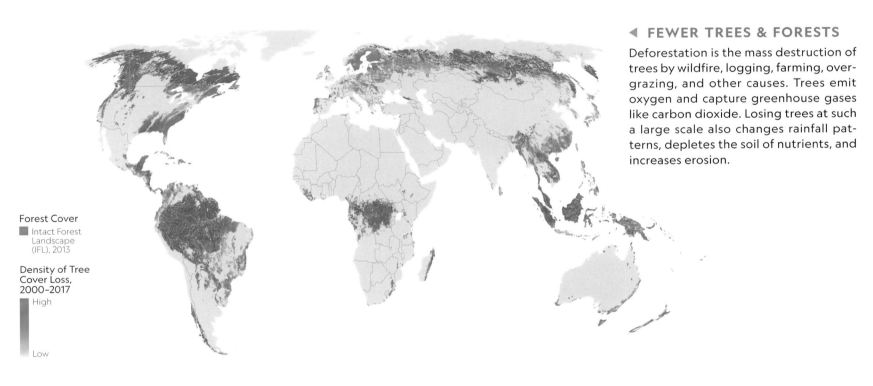

◀ FEWER TREES & FORESTS

Deforestation is the mass destruction of trees by wildfire, logging, farming, overgrazing, and other causes. Trees emit oxygen and capture greenhouse gases like carbon dioxide. Losing trees at such a large scale also changes rainfall patterns, depletes the soil of nutrients, and increases erosion.

Forest Cover
- ■ Intact Forest Landscape (IFL), 2013

Density of Tree Cover Loss, 2000–2017
- High
- Low

◀ WATER AT RISK

In certain regions, the availability of freshwater is an urgent problem. Sources of stress include watering crops, raising animals for meat, industry, energy production, growing populations, droughts, and climate change.

Global Water Risk
- ■ Very high
- ■ High
- ■ Moderate
- ☐ Low
- ☐ Very low

HUMANS CHANGE THE PLANET

Trash islands in the ocean, habitats destroyed, pollution in our air, and debris scattered across our lands: This is the Anthropocene, an unofficial geologic epoch defined by the broad and irreversible ways humans are transforming Earth. Satellite technology allows scientists to track the impact of human activities on the physical landscape. They can assess over time the extent of built environments, cropland and pastureland, population density, nighttime lights, roads, and railways, and they can map and monitor deforestation and oil drilling.

Yet global environmental problems are not impossible to solve. In the 1980s, when scientists discovered that the release of chlorofluorocarbons (CFCs) created a giant hole in the protective ozone layer into the atmosphere, the global community immediately took action. Dozens of countries quickly banned many of the products containing CFCs. The ozone layer has healed gradually ever since and is expected to fully recover by 2070. Since 2010, 190 countries have created plans to integrate the conservation of biodiversity into national priorities. Governments and conservationists have collaborated to bring endangered species back from the brink and regulate emissions of greenhouse gases.

▼ ALTERED LANDSCAPES

Human impact on the land since 1700 has been so significant as to create entirely new ecological areas. These new human-influenced biomes, known as anthromes, include the cropland, rangeland, and villages seen across much of Asia and Africa, as well as dense cities of North America, Europe, and Japan.

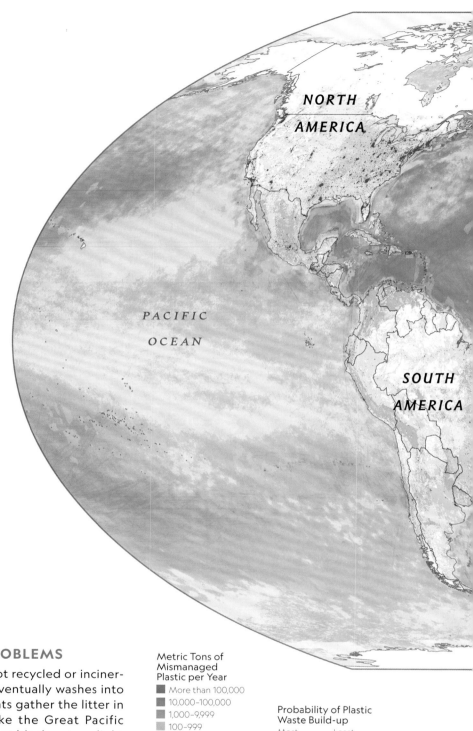

NORTH AMERICA

PACIFIC OCEAN

SOUTH AMERICA

▼ OCEANS FULL OF PLASTIC

Macroplastics are pieces of plastic debris found in the ocean (as opposed to tiny microplastic particles). Because plastic breaks down so slowly—over hundreds of years, if at all—scientists project an ever larger volume of litter if humans do not reduce emissions.

▼ PLASTIC PROBLEMS

When plastic is not recycled or incinerated on land, it eventually washes into the ocean. Currents gather the litter in swirling gyres like the Great Pacific Garbage Patch that block out sunlight and entangle animals.

Metric Tons of Mismanaged Plastic per Year
- More than 100,000
- 10,000–100,000
- 1,000–9,999
- 100–999
- 10–99
- Less than 10

Probability of Plastic Waste Build-up
Most likely — Least likely

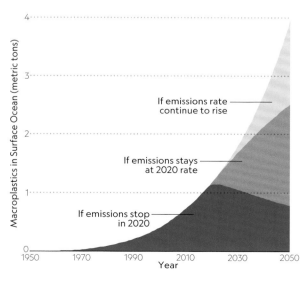

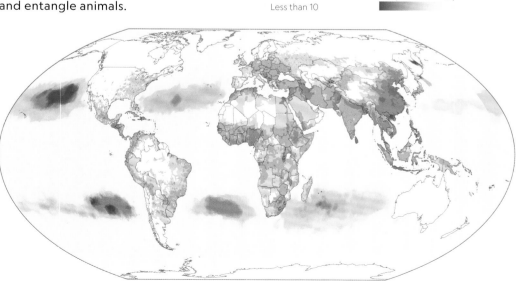

Anthromes

Map summarizes eight anthropogenic biome categories.

☐ **Wildlands**
Lands without human populations or substantial land use

■ **Dense settlements**
Densely populated cities and suburbs

■ **Villages**
Densely populated agricultural settlements

☐ **Croplands**
☐ **Remote croplands***
Lands used mainly for annual crops

☐ **Rangelands**
☐ **Remote rangelands***
Lands used mainly for livestock grazing

☐ **Seminatural landscapes**
Inhabited lands with minor use for permanent agriculture and settlements

☐ **No data**
**Remote rangelands and croplands lack significant human populations.*

Human Ocean Impact

High ▬▬▬▬▬ Low

☐ **No data**
Map shows human impact to marine ecosystems based on 19 stressors.

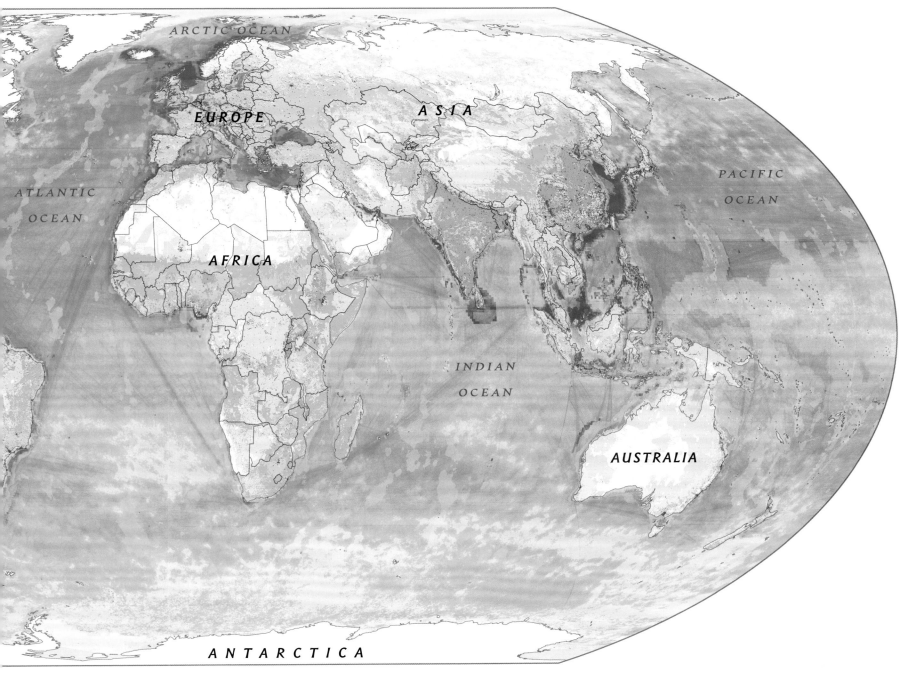

ARCTIC OCEAN

EUROPE

ASIA

ATLANTIC OCEAN

AFRICA

PACIFIC OCEAN

INDIAN OCEAN

AUSTRALIA

ANTARCTICA

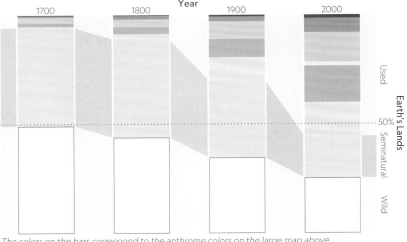

An acacia tree in Djibouti covered with plastic bags

▼ HUMAN IMPACT OVER TIME

Over three centuries of industrialization, Earth has gone from being mostly wild to mostly altered by human activities. From 1700 to 2000, humans have transformed more than half of ice-free land into rangeland, cropland, villages, and densely settled anthromes.

Year

1700   1800   1900   2000

Used

50%

Earth's Lands

Seminatural

Wild

The colors on the bars correspond to the anthrome colors on the large map above.

A WORLD FULL OF PEOPLE

Tokyo's Shibuya Crossing, the busiest intersection in the world

The world population has grown more than five-fold in 150 years (from 1.3 billion in 1870 to 7.7 billion in 2019). Today, 15,000 babies are born every hour, most of them in African and Asian developing nations. Meanwhile, Europe, Japan, and some other rich industrial areas show little population growth or may actually be shrinking: Higher incomes and educational levels correlate with lower birth rates. In some countries where half the population is under the age of 25, governments confront the overwhelming tasks of providing adequate education and jobs while encouraging family planning. By contrast, places with low birthrates have growing numbers of elderly people who need health care and pensions—but fewer workers to contribute to the tax base that funds such programs.

Population Density, 2018

(people per square kilometer)	(people per square mile)
More than 2,500	More than 6,475
500–2,500	1,300–6,475
100–499	260–1,299
25–99	65–259
1–24	3–64
Fewer than 1	Fewer than 3

○ Megacity (10 million or more inhabitants) based on UN urban agglomeration data

▲ WHERE PEOPLE LIVE

Integrating current census data with information about where people live and work, this map shows the density of population across the globe—from packed cities to empty deserts.

◀ GENDER IMBALANCE

Preferences for male children in China and India have created an extreme gender imbalance: Each nation has more than 30 million more men than women. Around the world, on average, women live about four years longer than men. The large gender imbalance on the Arabian Peninsula is the result of millions of male migrant workers.

Gender of the Population

- More than 53% female
- 51%–53% female
- Nearly equal
- 51%–53% male
- More than 53% male
- No data

BY 1800, THE WORLD'S POPULATION HAD PASSED ONE BILLION. TODAY, IT IS 7.7 BILLION.

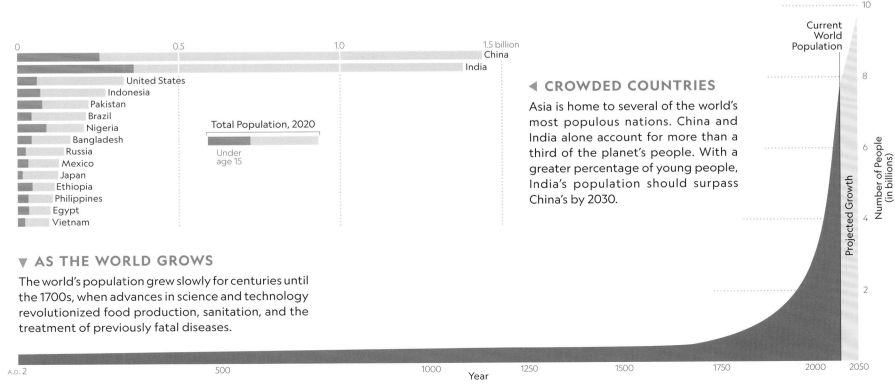

Total Population, 2020

Under age 15

◀ CROWDED COUNTRIES

Asia is home to several of the world's most populous nations. China and India alone account for more than a third of the planet's people. With a greater percentage of young people, India's population should surpass China's by 2030.

▼ AS THE WORLD GROWS

The world's population grew slowly for centuries until the 1700s, when advances in science and technology revolutionized food production, sanitation, and the treatment of previously fatal diseases.

CITIES ON THE RISE

The busy Chaoyang District of Beijing is the city's business center.

More people on Earth live in cities than not, and researchers estimate that by 2050, more than two-thirds of the world's population will be urban dwellers. It hasn't always been this way. As recently as 1950, 70 percent of the world's people lived in rural areas. As the world population grows ever larger, cities have been absorbing the increase. Sustainable growth for housing and sanitation is paramount to keep vulnerable populations safe from pollution, disease, and the threat of storms, floods, and fires. Yet bigger, denser cities are also more energy efficient than dispersed small communities and make less of an impact on surrounding natural habitats. Green buildings and public transit can dramatically reduce carbon emissions.

Urban Agglomerations, 2020

Largest City in Country	Other Cities Above 5 Million	
⬤ **City name**	⬤ **City name**	Urban agglomeration with more than 10 million people (megacity)
⬤ City name	⬤ City name	Urban agglomeration with more than 5 million people

Dots are scaled proportionally based on city population sizes

▲ URBAN CENTERS

Cities are typically on coasts or rivers, because of their importance as hubs of trade and transportation, both historically and today.

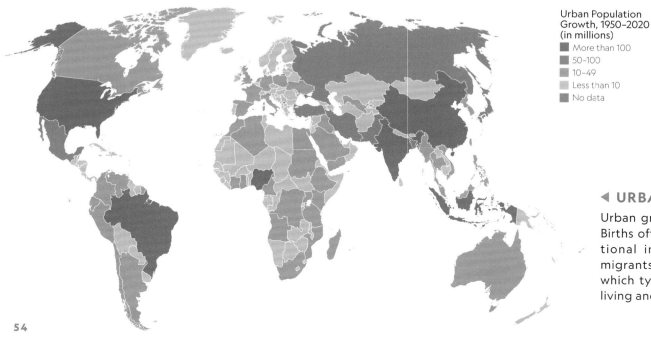

Urban Population Growth, 1950–2020 (in millions)

- ▣ More than 100
- ▣ 50–100
- ▣ 10–49
- ▣ Less than 10
- ▣ No data

◀ URBAN GROWTH RATES

Urban growth is driven by several factors. Births often outpace deaths, most international immigrants settle in cities, and migrants move from rural to urban areas, which typically have a higher standard of living and more economic opportunity.

St. Petersburg

Moscow

London
Paris

EUROPE

Istanbul

Madrid
Barcelona

Ankara

ASIA

Tehran

Baghdad

Lahore

Delhi

Chengdu

Shenyang

Harbin
Dalian
Qingdao
Seoul

Tianjin
Beijing

Jinan
Zhengzhou
Xi'an

Tokyo

Nagoya
Osaka

Kitakyushu-Fukuoka

Shanghai

Hangzhou
Suzhou
Nanjing

Wuhan

Dhaka

Alexandria

Cairo

Riyadh

Karachi

Ahmadabad

Surat

Mumbai

Pune

Chongqing
Guangzhou

Chittagong
Foshan

Dongguan

Shenzen
Hong Kong

AFRICA

Khartoum

Kolkata
Hyderabad

Yangon

Chennai

Bengaluru

Bangkok

Ho Chi
Minh City

Manila

Lagos

Abidjan

Kuala
Lumpur

Singapore

Kinshasa

Luanda

Dar es Salaam

Jakarta

Belo Horizonte
**Rio de
Janeiro**

Johannesburg

AUSTRALIA

▼ LARGEST CITIES

In 1950, only New York City and Tokyo were megacities. By 2030, growth patterns project there will be 43 megacities. The 10 fastest-growing cities are all in Africa.

Largest Urban Agglomerations
by Population (2020)

1.	Tokyo, Japan	37,400,000
2.	Delhi, India	30,300,000
3.	Shanghai, China	27,100,000
4.	São Paulo, Brazil	22,000,000
5.	Mexico City, Mexico	21,800,000
6.	Dhaka, Bangladesh	21,000,000
7.	Cairo, Egypt	20,900,000
8.	Beijing, China	20,500,000
9.	Mumbai, India	20,400,000
10.	Osaka, Japan	19,200,000
11.	New York, U.S.	18,800,000
12.	Karachi, Pakistan	16,100,000
13.	Chongqing, China	15,900,000
14.	Istanbul, Turkey	15,200,000
	Buenos Aires, Argentina	15,200,000
16.	Kolkata, India	14,900,000
17.	Lagos, Nigeria	14,400,000
18.	Kinshasa, Dem. Rep. Congo	14,300,000
19.	Manila, Philippines	13,900,000
20.	Tianjin, China	13,600,000

IN INDIA'S FINANCIAL HUB, **MUMBAI,** HALF OF THE **POPULATION** LIVES AND WORKS IN **SLUMS.**

Crowds at Churchgate Railway Station in Mumbai, India

POPULATION ON THE MOVE

Central American migrants and asylum seekers ride through Mexico atop a freight train.

When the earliest humans left Africa, they set in motion a migratory pattern that has persisted for millennia. People have always been on the move, seeking safety, opportunity, or freedom. Some migrants leave their birth country by choice. Others—refugees—flee under desperate circumstances to escape violence, war, or persecution. The United Nations estimates that the number of people forcibly displaced—either relocated within their country or as refugees abroad—is higher than at any time since World War II. Refugees often endure harrowing journeys to asylum—clinging to inflatable rafts or walking thousands of miles, children in tow, through hostile areas. Two-thirds of the world's 26 million refugees come from just five countries: Syria, Afghanistan, South Sudan, Myanmar, and Somalia. More than half of all refugees are children.

Migrant Population
(percent of total population)

- More than 25
- 15.1–25
- 5.5–15
- 2.5–5.4
- 1.–2.4
- Less than 1
- No data

▲ SEEKING A NEW HOME

In 2019, 272 million people worldwide were living outside the countries where they were born. This map shows the percentage of foreign-born residents of a country.

Displaced People due to Natural Disasters in 2018
(percent of people displaced worldwide due to natural disasters)

- More than 15
- 2.5–15
- 0.2–2.4
- 0.02–0.1
- Less than 0.02
- No data

(major natural disaster)

- ● Drought
- ● Earthquake
- ● Flood
- ● Storm
- ▪ Wildfire

◄ CLIMATE MIGRANTS

In 2018, natural disasters forced more than 17 million people to leave their homes. By 2050, the World Bank estimates there will be more than 140 million climate migrants as global warming intensifies floods, droughts, and wildfires.

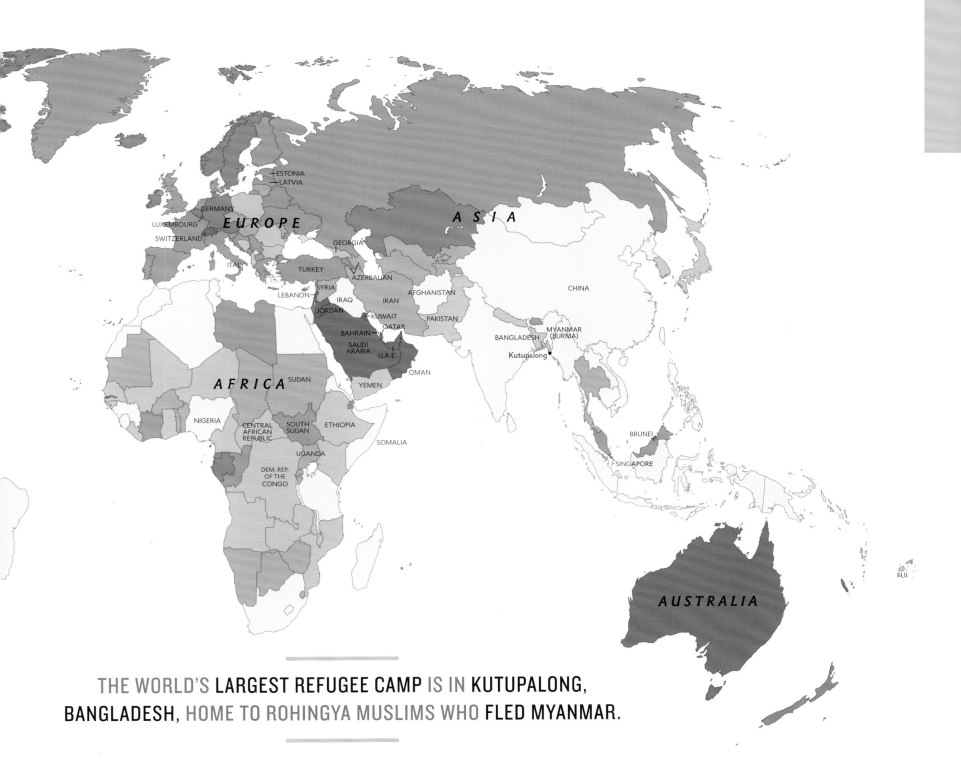

THE WORLD'S **LARGEST REFUGEE CAMP** IS IN **KUTUPALONG, BANGLADESH,** HOME TO ROHINGYA MUSLIMS WHO **FLED MYANMAR.**

▼ BY THE NUMBERS

The United Nations, governments, and other official groups regulate which people receive refugee status, protection, and aid. Not every asylum seeker will be recognized as a refugee, but all refugees were initially asylum seekers.

Countries with the Most Refugees, 2018

1.	Turkey	3,681,658
2.	Pakistan	1,404,008
3.	Uganda	1,165,636
4.	Sudan	1,078,275
5.	Germany	1,063,765
6.	Iran	979,435
7.	Lebanon	949,653
8.	Bangladesh	906,635
9.	Ethiopia	903,211
10.	Jordan	715,293

Countries with the Most Internally Displaced Persons (IDPs), 2018

1.	Colombia	7,816,472
2.	Syria	6,183,920
3.	Dem. Rep. of the Congo	4,516,865
4.	Somalia	2,648,000
5.	Ethiopia	2,615,800
6.	Nigeria	2,167,924
7.	Yemen	2,144,718
8.	Afghanistan	2,106,893
9.	South Sudan	1,878,153
10.	Sudan	1,864,195

Countries of Origin with the most Asylum Seekers, 2018

1.	Venezuela	464,209
2.	Afghanistan	310,094
3.	Iraq	256,687
4.	Syria	139,534
5.	Dem. Rep. of the Congo	133,401
6.	Ethiopia	133,240
7.	El Salvador	119,257
8.	China	94,339
9.	Mexico	89,771
10.	Iran	87,342

Countries with the most Asylum Seekers, 2018

1.	United States	718,970
2.	Germany	369,236
3.	Turkey	311,682
4.	Peru	230,790
5.	South Africa	184,188
6.	Brazil	152,600
7.	Italy	105,571
8.	France	89,035
9.	Canada	78,688
10.	Spain	78,685

A Syrian man tends the oasis he created in a refugee camp in Turkey.

THE WORLD SPEAKS

A language class in a Quechua high school in Peru

As the world grows more homogenized through globalization, linguistic diversity is increasingly at risk, yet few traditions reveal more about a society than its language. For example, in northern Finland, Russia, Sweden, and Norway, the Sami languages have several hundred words for snow. Cultural expression takes shape and persists through words, stories, and songs. Without spoken language, this heritage is lost. More than one in 10 people speak Mandarin Chinese as their first language; Spanish, English, and Hindi account for a billion more speakers. Although only 360 million people are native English speakers, up to a billion people use it to navigate the global economy, entertainment, technology, and science.

Language Location and Endangerment Level

× Extinct
● High
◦ Medium
▬ Language hotspot
(Regions with high linguistic density, severe endangerment, and lack of documentation)

▲ VANISHING LANGUAGES

Many of the world's 4,000 indigenous languages are fading, spoken only by older people and not taught to children; four in 10 are at risk of disappearing altogether

Major Language Families

◦ Afro-Asiatic	● Otomanguean
● Atlantic-Congo	● Pama-Nyungan
◦ Austroasiatic	● Sino-Tibetan
● Austronesian	◦ Tai-Kadai
● Indo-European	◦ Other*
● Nuclear Trans New Guinea	

*Contains over 400 language families

◄ LANGUAGE FAMILIES

Languages are grouped into families based on a common ancestral language. The Atlantic-Congo family is by far the largest, with more than 1,400 languages. Next largest is the Austronesian language family.

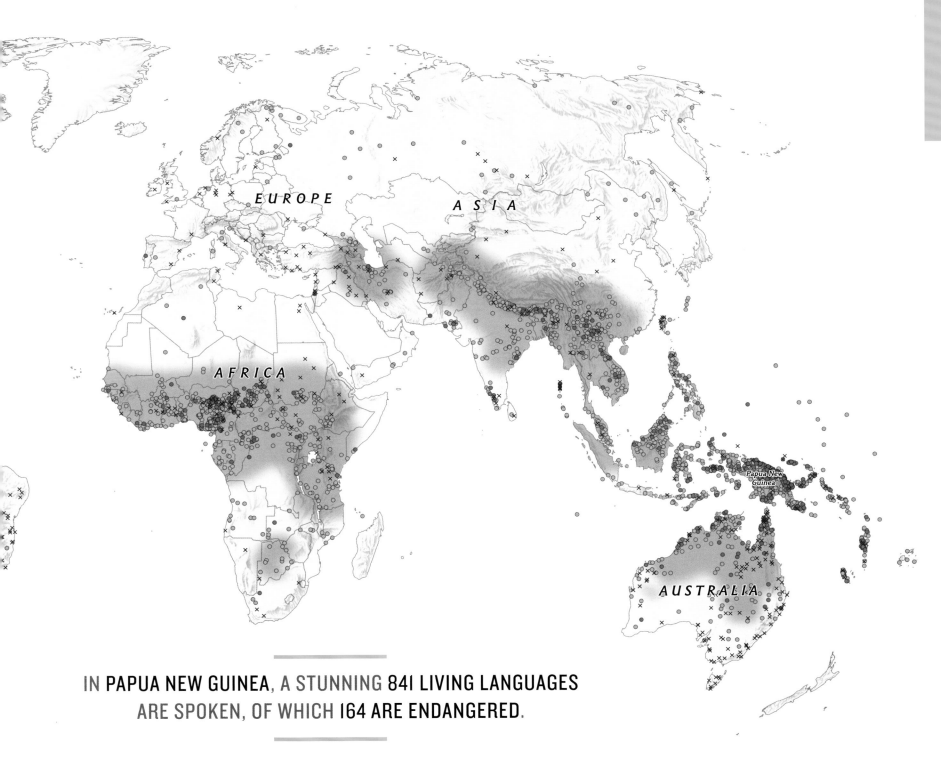

IN **PAPUA NEW GUINEA**, A STUNNING **841** LIVING LANGUAGES ARE SPOKEN, OF WHICH **164** ARE ENDANGERED.

▼ A MULTILINGUAL WORLD

Although the world has roughly 7,000 languages, 40 percent of its people count one of the top 10 most spoken first languages as their native tongue.

Largest Population of First Language Speakers	
1. Mandarin	955,000,000
2. Spanish	405,000,000
3. English	360,000,000
4. Hindi	310,000,000
5. Arabic	295,000,000
6. Portuguese	215,000,000
7. Bengali	205,000,000
8. Russian	155,000,000
9. Japanese	125,000,000
10. Punjabi	100,000,000
11. German	95,000,000
12. Javanese	82,000,000
13. Wu	80,000,000
14. Malay	77,000,000
15. Telugu	76,000,000
16. Vietnamese	76,000,000
17. Korean	76,000,000
18. French	75,000,000
19. Marathi	73,000,000
20. Tamil	70,000,000

▼ EVOLUTION OF LANGUAGES

Languages with shared characteristics can be traced back to a theoretical origin called a proto-language. Latin, for instance, derived from Proto-Indo-European and spread by the Roman Empire, diversified into the five Romance languages spoken by over a billion people today.

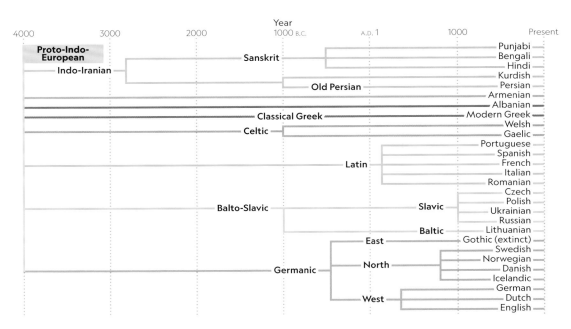

FAITH BELIEFS & PRACTICES

Beliefs in supreme beings and rituals that honor life and death have existed since the birth of human culture. Today, billions of people follow Hinduism, Buddhism, Judaism, Christianity, and Islam, all of which began in Asia or the Middle East. Common elements of these faiths include worship, moral guidelines, sacred sites, ritual clothing, dietary laws and fasting, festivals and holy days, and ceremonies for life's milestones. Participation varies regionally—for example, Christians in sub-Saharan Africa are much more likely to attend worship services than Christians in Europe. In the Asia-Pacific region, people surveyed in Muslim-majority countries, including Pakistan and Indonesia, are most likely to rate religion as very important to their daily lives, while people in Japan and China rate it as not very important. There are many belief systems outside the major religions, and roughly 15 percent of the world population does not claim any religious affiliation.

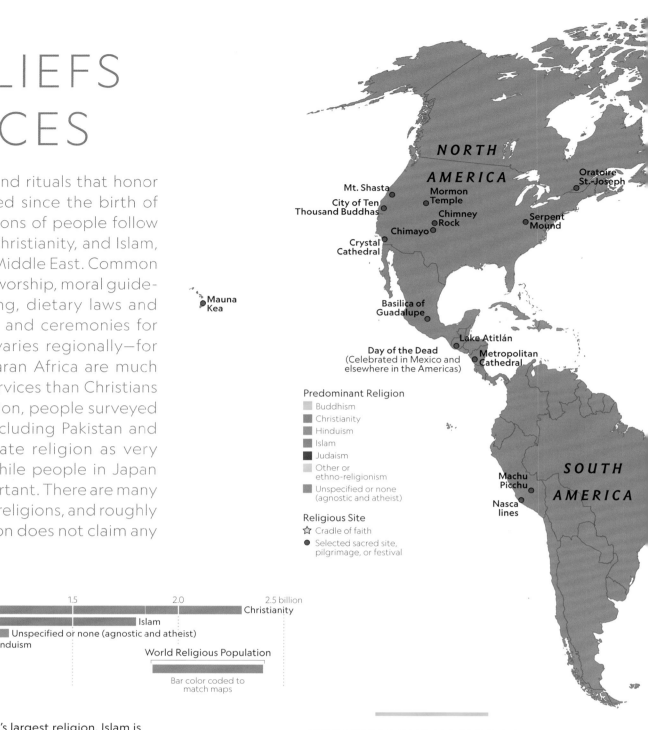

NORTH AMERICA

Oratoire St.-Joseph
Mt. Shasta
City of Ten Thousand Buddhas
Mormon Temple
Chimney Rock
Serpent Mound
Chimayo
Crystal Cathedral
Mauna Kea
Basilica of Guadalupe
Lake Atitlán
Day of the Dead (Celebrated in Mexico and elsewhere in the Americas)
Metropolitan Cathedral

SOUTH AMERICA

Machu Picchu
Nasca lines

Predominant Religion
- Buddhism
- Christianity
- Hinduism
- Islam
- Judaism
- Other or ethno-religionism
- Unspecified or none (agnostic and atheist)

Religious Site
- ☆ Cradle of faith
- ● Selected sacred site, pilgrimage, or festival

[Bar chart]
0   0.5   1.0   1.5   2.0   2.5 billion
Christianity
Islam
Unspecified or none (agnostic and atheist)
Hinduism
Buddhism
Other or ethno-religionism
Judaism

World Religious Population
Bar color coded to match maps

▲ NUMBERS OF THE FAITHFUL

Christianity, in its varying forms, is the world's largest religion. Islam is the second largest, and it is increasing at a faster rate because of population growth. Though more and more people don't identify with any religion, their overall share of the global population is decreasing due to the birth rate in religious areas.

HINDUISM BEGAN MORE THAN **4,000 YEARS AGO** IN THE INDUS RIVER VALLEY. **INDIA** IS NOW HOME TO **94 PERCENT** OF THE WORLD'S HINDUS.

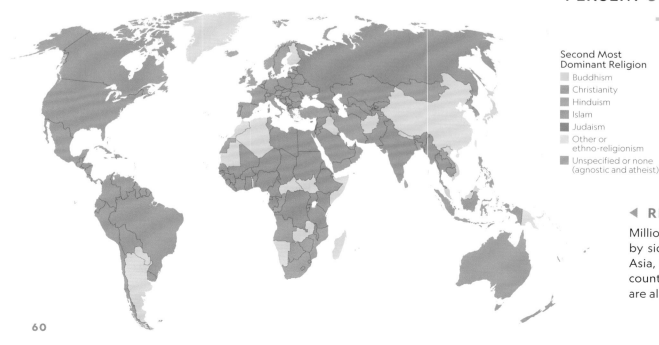

Second Most Dominant Religion
- Buddhism
- Christianity
- Hinduism
- Islam
- Judaism
- Other or ethno-religionism
- Unspecified or none (agnostic and atheist)

◄ RELIGIOUS DIVERSITY

Millions of Muslims and Christians live side by side in Africa, the Middle East, Central Asia, and Russia. In many Southeast Asian countries, Buddhism, Christianity, and Islam are all secondary faiths.

Sites Within the Holy Land

❶ Bethlehem
❷ Cave of Machpelah
❸ Church of the Holy Sepulchre
❹ Dome of the Rock
❺ Jordan River
❻ Sea of Galilee
❼ Temple Mount
❽ Western Wall

▲ **GLOBAL FAITHS**

Christianity spread throughout the ancient Roman Empire, and then globally as a result of European conquests and missionaries. Conquest and trade have spread Islam from the Middle East into Africa and Asia, while Buddhism and Hinduism are the majority faiths primarily in Asia, close to where they originated.

A Buddhist monk in Myanmar

■ **Buddhism**

Buddhism teaches liberation from suffering through morality, meditation, and wisdom. Buddhists revere the Three Treasures: the Buddha (the Awakened One), the Dharma (the teachings of the Buddha), and the Sangha (the community).

A Hindu Tamil bride in Sri Lanka

■ **Hinduism**

Hindus see God as one entity in many forms, including Brahma the creator, Vishnu the preserver, and Shiva the destroyer. They believe in reincarnation and hold that actions in this life affect circumstances of the next.

Jewish men gather in a synagogue.

■ **Judaism**

Judaism is a monotheistic religion that began about 4,000 years ago in present-day Israel. The Torah is Judaism's book of sacred law and literature, which Jews believe God gave to the prophet Moses on Mount Sinai.

A Roman Catholic nun in England

■ **Christianity**

The New Testament of the Christian Bible is based on Jesus Christ: his monotheistic teaching; persecution, Crucifixion and resurrection; and belief in eternal life. Christianity includes Catholicism, Eastern Orthodoxy, and Protestantism.

Muslims pray at a mosque in Istanbul.

■ **Islam**

Muslims believe the Quran records the spoken word of God (Allah) as revealed to the Prophet Muhammad. Strict followers pray five times a day, fast during the month of Ramadan, and make a pilgrimage to Mecca, Islam's holiest city.

Russian Orthodox at a church in Yekaterinburg

■ **Blended Religions**

Many cultures blend two or more religious traditions. For example, Haitian vodoo incorporates Christian, western African, and Amerindian traditions. Some South Koreans blend Buddhist, Confucian, and Catholic customs.

QUALITY OF LIFE

People are living better today than at any other time in history. Vaccines have nearly eradicated many fatal diseases. Men and women have narrowed the gap between them in education. The proportion of people living in extreme poverty continues to drop. Agricultural advances have boosted food supplies. Literacy has opened minds; in most countries, at least half of all adults are literate. Yet we have far to go before all people are living well. Income inequality is one disparity that is on the rise. The richest one percent of the population took home 22 percent of global income in 2016. One in seven people still lacks electricity, and more than one in three women has experienced either physical or sexual violence in her life.

Social scientists are continuing to explore the underpinnings of what makes people—or societies—happy, assessing the roles of technology, institutions, and social norms in influencing how people relate to one another. Countries with high literacy rates and a free press are home to happier people, while people are predictably unhappy in countries governed by oppressive political regimes or battered by sustained conflicts. The United Nations and its member countries have committed to achieving 17 Sustainable Development Goals by 2030. By addressing discrimination, violence, pollution, maternal mortality, fair elections, food waste, and many more of our world's deeply rooted problems, the gap between happy and unhappy nations could grow narrower still.

A young girl in Hong Kong

Men read the newspaper in Havana, Cuba, where the literacy rate is 99.8 percent.

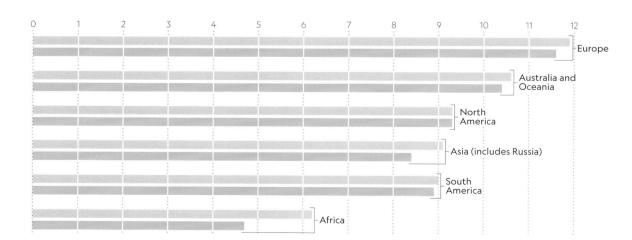

Europe
Australia and Oceania
North America
Asia (includes Russia)
South America
Africa

*Average years of education

Men
Women

*Continental averages calculated based on national averages.

◀ GENDER EDUCATION

Men attain greater levels of education than women on every continent except North America. The gap is closing everywhere except in Africa, where on average women receive fewer than five years of schooling.

WOMEN IN ESTONIA RECEIVE AN AVERAGE OF 14 YEARS OF EDUCATION—
NEARLY DOUBLE THE GLOBAL AVERAGE.

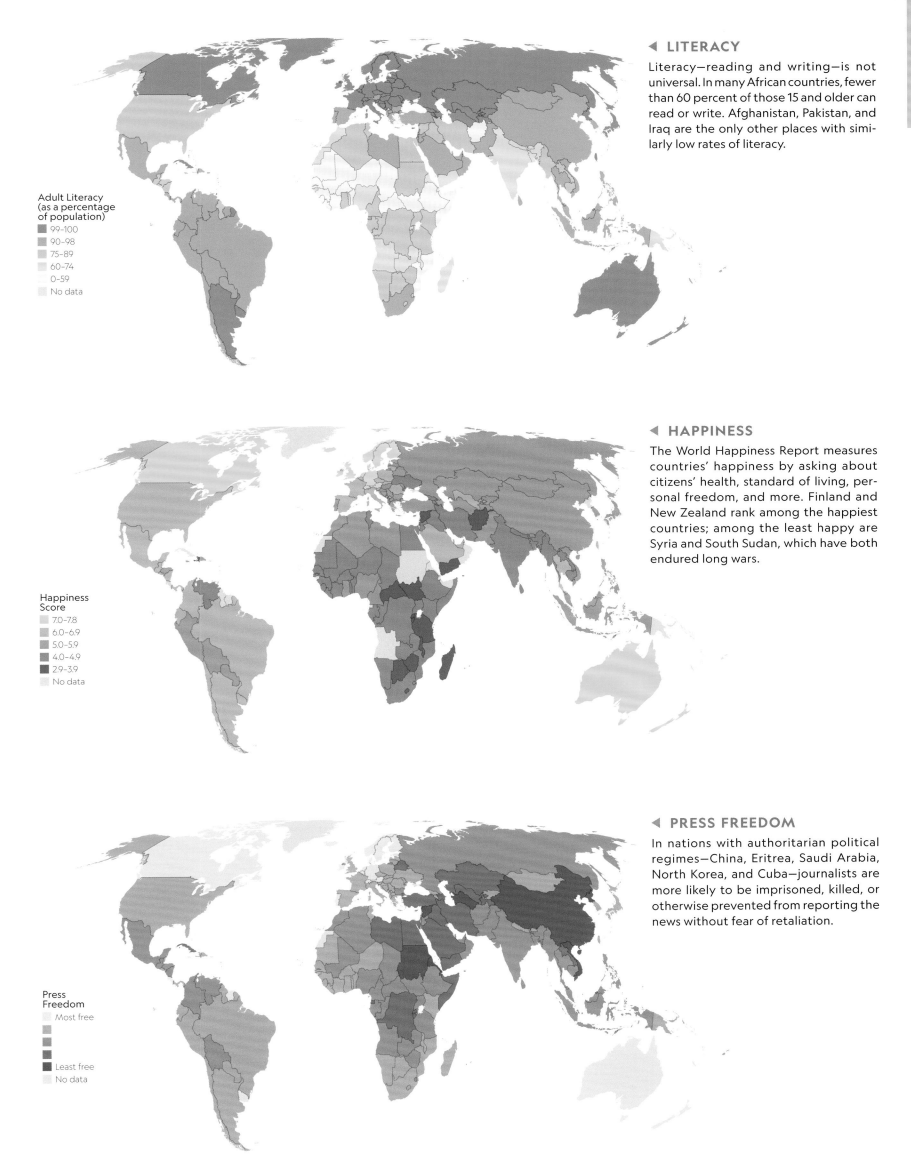

◀ LITERACY

Literacy—reading and writing—is not universal. In many African countries, fewer than 60 percent of those 15 and older can read or write. Afghanistan, Pakistan, and Iraq are the only other places with similarly low rates of literacy.

Adult Literacy
(as a percentage
of population)
- 99–100
- 90–98
- 75–89
- 60–74
- 0–59
- No data

◀ HAPPINESS

The World Happiness Report measures countries' happiness by asking about citizens' health, standard of living, personal freedom, and more. Finland and New Zealand rank among the happiest countries; among the least happy are Syria and South Sudan, which have both endured long wars.

Happiness
Score
- 7.0–7.8
- 6.0–6.9
- 5.0–5.9
- 4.0–4.9
- 2.9–3.9
- No data

◀ PRESS FREEDOM

In nations with authoritarian political regimes—China, Eritrea, Saudi Arabia, North Korea, and Cuba—journalists are more likely to be imprisoned, killed, or otherwise prevented from reporting the news without fear of retaliation.

Press
Freedom
- Most free
-
-
- Least free
- No data

SEEKING WORLD WELLNESS

I n the past 50 years, health conditions have improved dramatically. With better living conditions and access to immunization and other health services, global life expectancy has risen to 72 years; the death rate for children under five years old has fallen by more than half in the past 28 years; and many infectious and parasitic diseases have been eradicated, eliminated, or greatly reduced in impact. Yet depending on where a person lives in the world, the chances of living to adulthood—let alone old age—vary dramatically. In the Central African Republic, for example, the average life expectancy of men doesn't clear 50, while the average woman in Singapore can live well into her late 80s. Sex and gender shape health, too: Girls who marry at a young age suffer increased risks associated with early pregnancy, while men are exposed to greater risk of death from road accidents because of their higher rates of employment in the transport industry. Many risk factors and afflictions that shorten the lives of people in low- and middle-income countries remain—lack of basic care or vaccines; polluted air and contaminated water; maternal death—and pose some of the greatest public health challenges for policymakers.

Morning exercises in Shanghai, China

ALMOST **ONE IN FOUR DEATHS** IN THE **UNITED STATES** IS RELATED TO **HEART DISEASE,** THE **LEADING CAUSE OF DEATH** FOR BOTH MEN AND WOMEN.

▼ LIFE EXPECTANCY

Life expectancy from country to country can vary as much as 38 years, with the shortest expected life spans in low-income African nations and Papua New Guinea. Overall, women outlive men.

Female Life Expectancy, 2017

Highest		Lowest	
1. Singapore	87.6	1. Cen. Af. Rep.	55.0
2. Japan	87.2	2. Lesotho	59.4
Kuwait	87.2	3. Somalia	60.7
4. Iceland	85.9	4. Papua New Guinea	61.3
5. Spain	85.8	5. Sierra Leone	61.4
6. France	85.7	6. Chad	61.7
Switzerland	85.7	7. South Sudan	61.9
8. South Korea	85.5	8. Mozambique	62.0
9. Italy	85.3	9. Guinea	62.3
10. Cyprus	85.2	10. Guinea-Bissau	62.7

Male Life Expectancy, 2017

Highest		Lowest	
1. Switzerland	82.1	1. Cen. Af. Rep.	49.2
2. Singapore	81.9	2. Lesotho	50.3
3. Israel	81.3	3. Mozambique	54.9
4. Japan	81.1	4. Swaziland	55.0
5. Italy	80.8	5. Papua New Guinea	56.3
Sweden	80.8	6. Somalia	56.6
7. Kuwait	80.7	7. South Sudan	57.0
8. Andorra	80.6	8. Guinea-Bissau	57.4
9. Norway	80.5	9. Zimbabwe	58.2
10. Australia	80.2	10. Kiribati	58.6

▼ CAUSES OF DEATH

Women are more likely to die from Alzheimer's disease, in part because they live to more advanced ages when that disease occurs, while men are more likely to die from road injuries.

Top Global Causes of Death in Females, 2017

1. Coronary artery disease
2. Stroke
3. Alzheimer's disease and other dementias
4. Chronic obstructive pulmonary disease
5. Lower respiratory infections
6. Diarrheal diseases
7. Neonatal disorders
8. Diabetes
9. Breast cancer
10. Tracheal, bronchus, and lung cancer

Top Global Causes of Death in Males, 2017

1. Coronary artery disease
2. Stroke
3. Chronic obstructive pulmonary disease
4. Lower respiratory infections
5. Tracheal, bronchus, and lung cancer
6. Neonatal disorders
7. Road injuries
8. Cirrhosis and other chronic liver diseases
9. Alzheimer's disease and other dementias
10. Tuberculosis

▼ THE OBESITY PANDEMIC

Poor nutrition and lack of exercise have made obesity and diabetes a global problem. Hardest hit are Pacific island nations, where processed imported foods are replacing traditional diets.

Countries with the Highest Adult Obesity Rates, 2016

1. Nauru	61.0%
2. Palau	55.9
3. Marshall Islands	52.9
4. Tuvalu	51.6
5. Tonga	48.2
6. Samoa	47.3
7. Kiribati	46.0
8. Micronesia	45.8
9. Kuwait	37.9
10. United States	36.2

Countries with the Lowest Adult Obesity Rates, 2016

1. Vietnam	2.1%
2. Bangladesh	3.6
3. Timor-Leste	3.8
4. India	3.9
Cambodia	3.9
6. Nepal	4.1
7. Japan	4.3
8. Ethiopia	4.5
9. South Korea	4.7
10. Eritrea	5.0

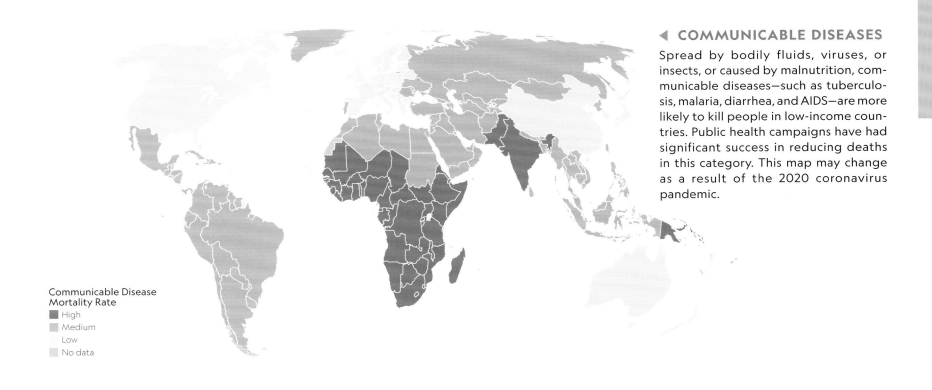

◀ COMMUNICABLE DISEASES

Spread by bodily fluids, viruses, or insects, or caused by malnutrition, communicable diseases—such as tuberculosis, malaria, diarrhea, and AIDS—are more likely to kill people in low-income countries. Public health campaigns have had significant success in reducing deaths in this category. This map may change as a result of the 2020 coronavirus pandemic.

**Communicable Disease
Mortality Rate**
- High
- Medium
- Low
- No data

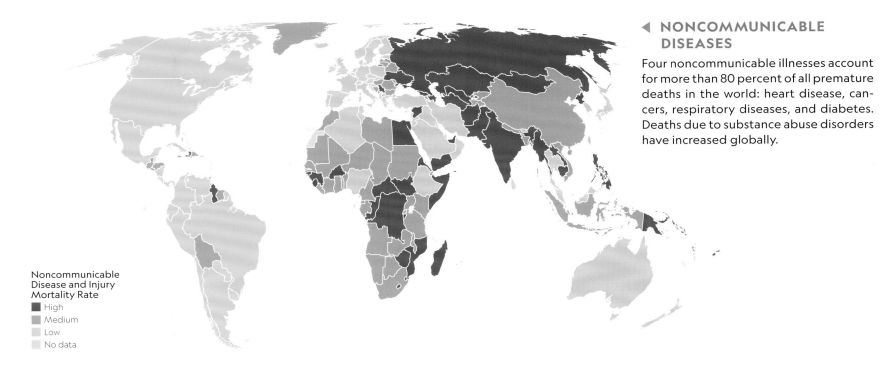

◀ NONCOMMUNICABLE DISEASES

Four noncommunicable illnesses account for more than 80 percent of all premature deaths in the world: heart disease, cancers, respiratory diseases, and diabetes. Deaths due to substance abuse disorders have increased globally.

**Noncommunicable
Disease and Injury
Mortality Rate**
- High
- Medium
- Low
- No data

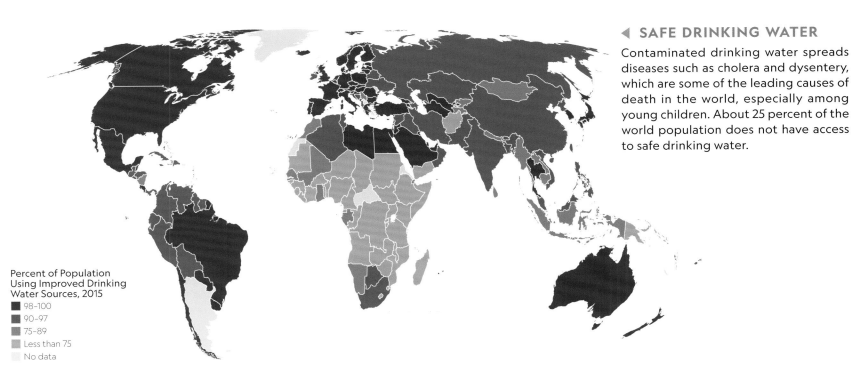

◀ SAFE DRINKING WATER

Contaminated drinking water spreads diseases such as cholera and dysentery, which are some of the leading causes of death in the world, especially among young children. About 25 percent of the world population does not have access to safe drinking water.

**Percent of Population
Using Improved Drinking
Water Sources, 2015**
- 98–100
- 90–97
- 75–89
- Less than 75
- No data

THE WORLD'S HARVEST

A field of celery in California's fertile Salinas Valley

Agriculture has developed in many different ways—from terraced rice paddies to slash-and-burn plots and cattle ranches. It takes up 50 percent of Earth's habitable land and is the single largest employer in the world. Many governments offer farming subsidies to drive down the cost of food and to use surplus crops for biofuels or exports. But agriculture is also a demanding industry: It uses 70 percent of available freshwater, results in a quarter of human-produced greenhouse gases, and is a major cause of deforestation. Scaling up food production on the land already in use requires innovation: increasing yields, making more effective use of resources, reducing waste, and changing our diets.

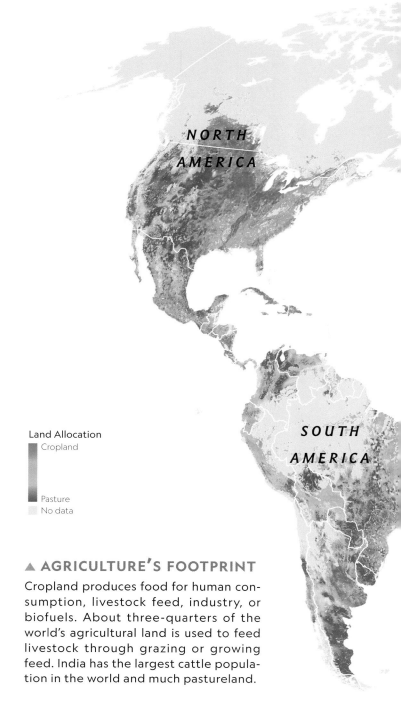

Land Allocation
Cropland
Pasture
No data

▲ AGRICULTURE'S FOOTPRINT

Cropland produces food for human consumption, livestock feed, industry, or biofuels. About three-quarters of the world's agricultural land is used to feed livestock through grazing or growing feed. India has the largest cattle population in the world and much pastureland.

Percent Irrigated Area
60-100   Less than 5
30-59   No data
5-29

◄ MAPPING IRRIGATION

Irrigation uses channels, pipes, and sprinklers to bring water from reservoirs to fields. In places where naturally occurring water is limited, agriculture can only succeed thanks to irrigation.

GREENHOUSES IN THE NETHERLANDS PRODUCE ENORMOUS YIELDS, MAKING THIS
SMALL NATION THE WORLD'S SECOND LARGEST EXPORTER OF FOOD BY VALUE AFTER THE U.S.

▶ HOW CROPS ARE USED

The United States, Europe, and much of China and
South America are major producers of crops for
biofuels, while Africa, India, and parts of Southeast
Asia grow more crops for human consumption.

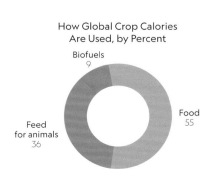

How Global Crop Calories
Are Used, by Percent

Biofuels
9

Food
55

Feed
for animals
36

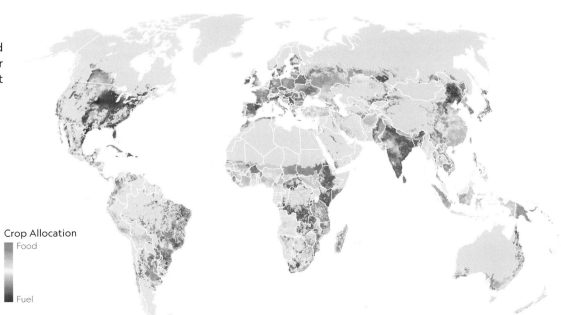

Crop Allocation

Food

Fuel

FEEDING THE WORLD

Food fuels bodies, shapes cultures, and drives local economies. Humans have a taste for sweeteners, whether they are called *jaggery* (India), *panela* (South America), or treacle (Great Britain). Broadly speaking, humans rely on plant sources for carbohydrates. Grains—the edible parts of cereal plants—provide 80 percent of the food energy, or calorie, supply. This means the major grains—corn, wheat, and rice—are the foods that fuel humanity. In many parts of the world, people are consuming more calories: Rising standards of living in developing countries have recently increased the worldwide appetite for meat, eggs, and dairy.

Worldwide, hunger is a serious problem. Food production will need to double by 2050 to feed a nourishing diet to the planet's projected nine billion people. One major issue to tackle is food waste: A third of what is produced is not consumed. In 2019, more than 820 million people, many of them in developing countries, were chronically undernourished. In striking contrast, patterns of consumption in developed countries reveal that changing diets, which incorporate more fat, sugar, and salt, are leading to both malnourishment and widespread obesity.

Cured meats for sale in Florence, Italy

▼ MEAT CONSUMPTION

Countries amid major economic development, like China, eat more meat now than 50 years ago. India's low rate of meat consumption reflects vegetarian preferences. Meat consumption in Brazil is increasing, though it also has a higher rate of vegetarians than the U.S.

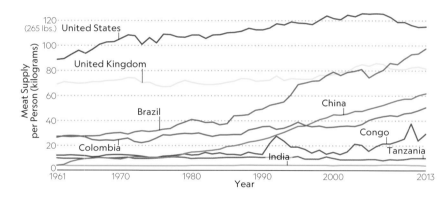

▼ SEA OF FOOD

Aquaculture, or the farming of fish, mollusks, and crustaceans, is an important food source—although wild-caught fish still account for slightly more tonnage. Fishing is becoming less and less sustainable, with a third of ocean fish stock overfished.

Global Fisheries, 2017
- ➤ Marine wild catch
- ➤ Marine aquaculture
- ➤ Inland wild catch
- ➤ Inland aquaculture

Each symbol denotes 500,000 metric tons of seafood produced by FAO Major Fishing Area.

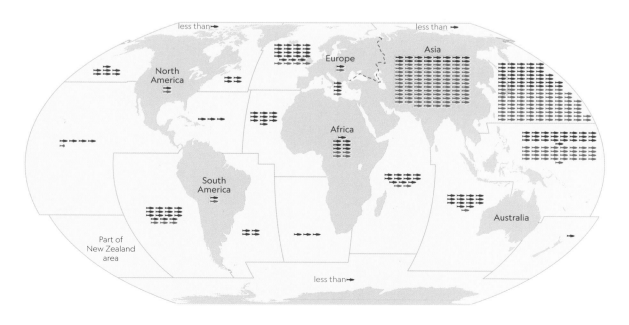

Top Aquaculture Countries, 2017 (in metric tons)*

1.	China	64,358,480
2.	Indonesia	15,896,100
3.	India	6,182,000
4.	Vietnam	3,831,241
5.	Bangladesh	2,333,352
6.	South Korea	2,306,280
7.	Philippines	2,237,787
8.	Egypt	1,451,841
9.	Norway	1,308,634
10.	Chile	1,219,747

*Includes inland capture.

Top Wild Catch Countries, 2017 (in metric tons)*

1.	China	15,373,197
2.	Indonesia	6,689,361
3.	India	5,427,678
4.	United States	5,036,233
5.	Russia	4,869,360
6.	Peru	4,157,414
7.	Vietnam	3,277,574
8.	Japan	3,204,347
9.	Norway	2,368,438
10.	Myanmar	2,150,400

*Only includes marine capture.

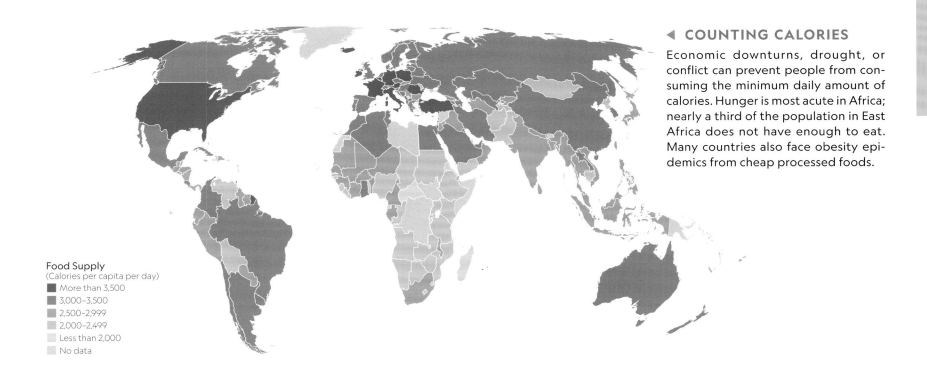

◄ COUNTING CALORIES

Economic downturns, drought, or conflict can prevent people from consuming the minimum daily amount of calories. Hunger is most acute in Africa; nearly a third of the population in East Africa does not have enough to eat. Many countries also face obesity epidemics from cheap processed foods.

Food Supply
(Calories per capita per day)
- More than 3,500
- 3,000–3,500
- 2,500–2,999
- 2,000–2,499
- Less than 2,000
- No data

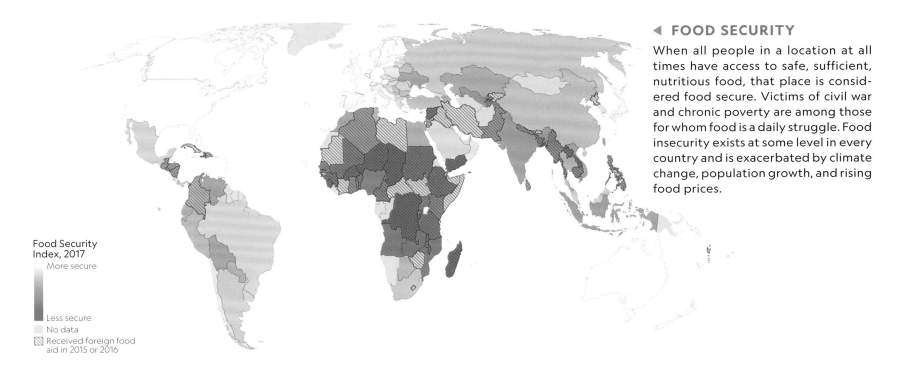

◄ FOOD SECURITY

When all people in a location at all times have access to safe, sufficient, nutritious food, that place is considered food secure. Victims of civil war and chronic poverty are among those for whom food is a daily struggle. Food insecurity exists at some level in every country and is exacerbated by climate change, population growth, and rising food prices.

Food Security Index, 2017
More secure
Less secure
No data
Received foreign food aid in 2015 or 2016

BIODIVERSITY IN AGRICULTURE IS VERY LOW: 75 PERCENT OF THE WORLD'S FOOD COMES FROM 5 ANIMAL SPECIES AND 12 PLANTS.

▼ ANALYZING REGIONAL DIETS

Diets vary around the world. In Asia and Africa, grains make up more than 50 percent of the typical diet. On all other continents, meat and dairy form a larger percentage of calorie consumption.

World Diets, 2017
☐ 1% of daily calorie consumption

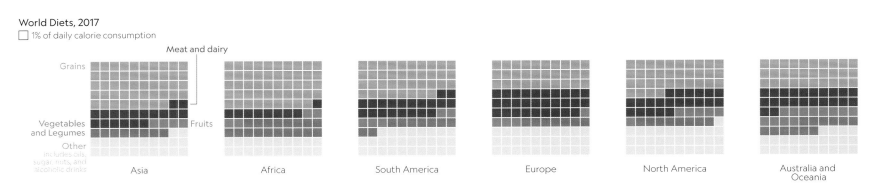

Meat and dairy
Grains
Vegetables and Legumes
Fruits
Other includes oils, sugar, nuts, and alcoholic drinks

Asia   Africa   South America   Europe   North America   Australia and Oceania

BORDERS & BOUNDARIES

Soldiers in the Demilitarized Zone separating North and South Korea

Geography can influence borders, but they are ultimately determined by politics. Native Americans in the United States still petition for the return of land promised in broken treaties. Within the last century, political borders were redrawn following the two world wars, the dissolution of the Soviet Union, and the retreat of colonial powers. Imperialists who drew boundaries in the Middle East during and after World War I, for instance, often used a ruler—designating straight lines that did not reflect cultural, ethnic, and political realities on the ground. The resulting discord continues to this day. And the rising number of migrants and refugees seeking asylum shows how contentious political borders can be in the face of crisis.

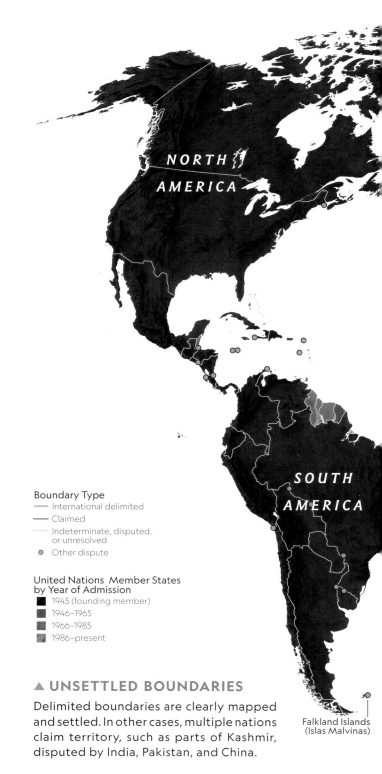

Boundary Type
— International delimited
— Claimed
— Indeterminate, disputed, or unresolved
● Other dispute

United Nations Member States by Year of Admission
■ 1945 (founding member)
■ 1946–1965
■ 1966–1985
■ 1986–present

▲ UNSETTLED BOUNDARIES

Delimited boundaries are clearly mapped and settled. In other cases, multiple nations claim territory, such as parts of Kashmir, disputed by India, Pakistan, and China.

Falkland Islands (Islas Malvinas)

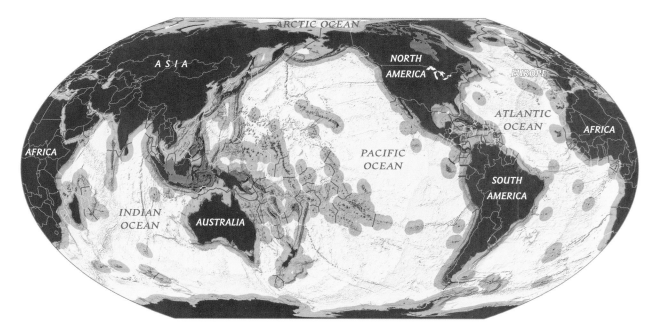

◄ WHO OWNS THE OCEAN?

Any nation can navigate and fish the high seas. Closer to shore, coastal countries have sole rights to the waters of their Exclusive Economic Zones (EEZs). When zones overlap, the dividing border is called a median line.

Maritime Divisions
— Maritime boundary
— Median line
— Joint Development Area boundary
■ Territorial waters
■ EEZ
High seas

EUROPE
ASIA
Transnistria
Kosovo
Crimea
Abkhazia
South Ossetia
Nagorno-Karabakh
Cyprus
Kashmir
Western Himalaya
Kuril Islands
Dokdo (Takeshima)
Senkaku Shoto (Daioyu Islands)
Western Sahara
AFRICA
Kafia Kingi
Abyei
Somaliland
Paracel Islands
South China Sea
Spratly Islands
AUSTRALIA

South Georgia and the South Sandwich Islands

OVERLAPPING CLAIMS OF RESOURCE-RICH AREAS HAVE ESCALATED THE MILITARIZATION OF THE SOUTH CHINA SEA.

▼ BORDERS IN THE SEA

Territorial waters extend a short distance offshore, but countries retain authority to use all resources in the Exclusive Economic Zone (EEZ) off their coastline.

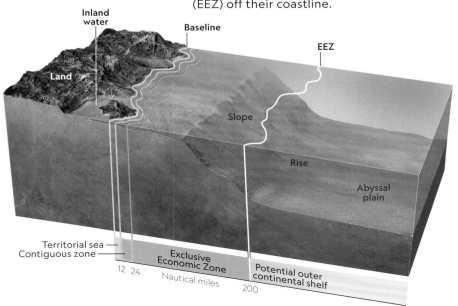

Inland water
Baseline
EEZ
Land
Slope
Rise
Abyssal plain
Territorial sea
Contiguous zone
12 24
Exclusive Economic Zone
Potential outer continental shelf
Nautical miles
200

Tunisian officials check boats for illegal migrants trying to reach Europe.

WEALTH & PROSPERITY

The trading floor of the New York Stock Exchange

Economic vitality falls on a spectrum between advanced economies, with high GDP and mature industrialization, and least developed countries, which often have poorer living conditions and may rely on export commodities and agriculture. Wealth stratification is also apparent within a country's population. The global middle class has grown significantly, but inequality has risen as well: The richest one percent controls nearly 50 percent of global wealth. The world's economy has become increasingly integrated due to improved communication and transportation technology and lower trade barriers. This means fortunes can change quickly: China's rapid economic growth added 158,000 millionaires in 2019, while the recession of 2008 caused joblessness and bankruptcies around the world.

GDP per capita, 2017
(in U.S. dollars)

64,000

32,000

16,000

8,000

4,000
2,000
1,000

No data

▲ COMPARING ECONOMIES

Gross domestic product (GDP) is the most used method of calculating a country's economic strength. It represents the total value of goods and services produced in a given year.

National Government Debt
(as a percent of annual GDP
2017 estimate)

More than 150   25–49
100–150   Less than 25
75–99   No data
50–74

◄ NATIONAL DEBT

Nearly every country has some level of debt. Nations seek out creditors—other countries, but also organizations like the World Bank—to pay for expenses such as infrastructure, health care, and defense.

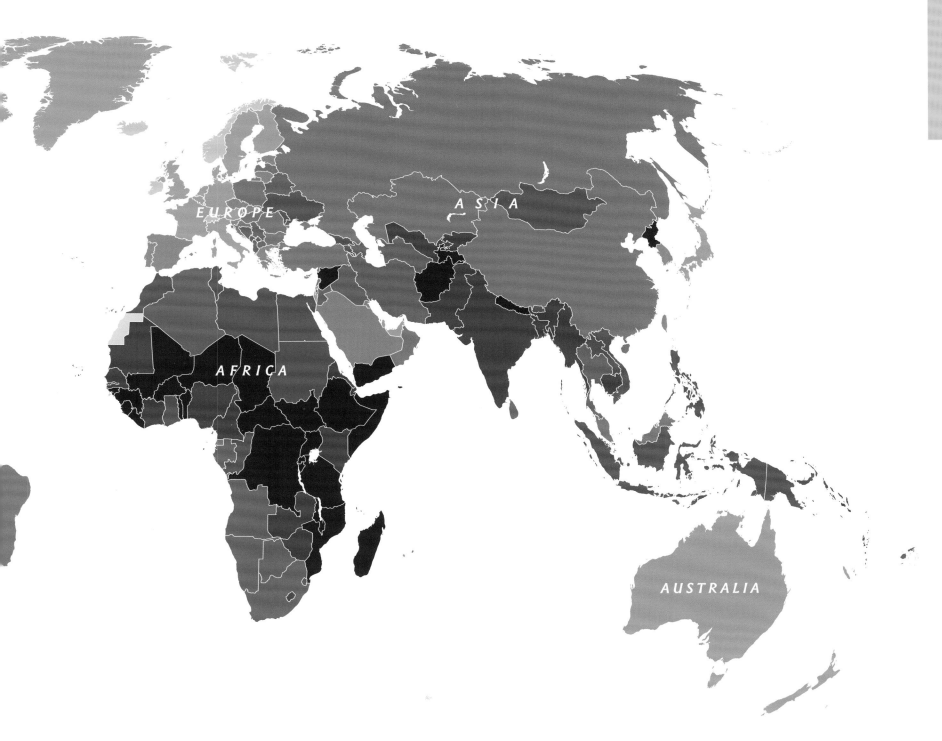

EUROPE

ASIA

AFRICA

AUSTRALIA

▶ MILLIONAIRES IN THE WORLD

About 47 million people worldwide have personal wealth exceeding a million dollars. Although the United States, western Europe, and several Asian countries are home to large numbers of millionaires, their middle class is also shrinking.

Canada

United Kingdom
Ireland

United States

Mexico

Chile Brazil

Portugal Spain

Norway Finland
Sweden
Russia
Denmark

Netherlands
Germany Poland
Belgium
Czechia
Austria
France
Switzerland Turkey Singapore
Italy Kuwait
Greece Pakistan
Israel
Egypt U.A.E.
Saudi Arabia
South Africa

China (mainland)
India

South Korea Japan

1 2 Hong Kong
3
4 5 Taiwan
Australia
New Zealand

1. Thailand
2. Vietnam
3. Malaysia
4. Indonesia
5. Philippines

World Millionaires, 2019 (in U.S. dollars)

Each hexagon represents 10,000 millionaires (individual net worth of $1,000,000 or more).

Gray hexagons indicate regional accumulations of countries with fewer than 5,000 millionaires.

Most Healthy National Economies
(gross domestic product per person, adjusted for PPP or cost of living, in U.S. dollars)

1. Qatar	130,475	11. San Marino	60,313
2. Luxembourg	106,705	12. Netherlands	56,383
3. Singapore	100,345	13. Saudi Arabia	55,944
4. Brunei	79,530	14. Iceland	55,917
5. Ireland	78,785	15. Sweden	52,984
6. Norway	74,356	16. Germany	52,559
7. United Arab Emirates	69,382	17. Australia	52,373
8. Kuwait	67,000	18. Austria	52,137
9. Switzerland	64,649	19. Denmark	52,121
10. United States	62,606	20. Bahrain	50,057

Nations with the Most Private Wealth
(average adult net worth, in U.S. dollars)

1. Switzerland	564,653	11. Netherlands	279,077
2. United States	432,365	12. France	276,121
3. Australia	386,058	13. Austria	274,919
4. Iceland	380,868	14. Ireland	272,310
5. Luxembourg	358,003	15. Norway	267,348
6. New Zealand	304,124	16. Sweden	265,260
7. Singapore	297,873	17. Belgium	246,135
8. Canada	294,255	18. Japan	238,104
9. Denmark	284,022	19. Italy	234,139
10. United Kingdom	280,049	20. Andorra	218,321

◀ WEALTHIEST NATIONS

The healthiest economies per capita and greatest concentrations of personal wealth per adult are found overwhelmingly in northern and western Europe, the United States, and a few other locales, including Singapore, Qatar, Australia, and New Zealand.

BUYING & SELLING

World trade has expanded at a dizzying pace since World War II. New international agreements encouraged free-trade policies that lowered or eliminated quotas and tariffs on goods exchanged across borders. The dollar value of world merchandise exports rose from $61 billion in 1950 to $19.7 trillion in 2018. Trade in manufactured goods now reigns; many emerging markets such as China, South Korea, and Mexico have become major exporters as well as consumers. But there are still many nations—primarily in Africa and the Middle East—that are dependent on a few primary commodities for their export earnings.

The rapid rise in commercial services, including information and communication technology and tourism services, has changed the face of world trade. World exports in this sector have increased by 46 percent since 2008. Globalization also facilitates the growth of more multinational enterprises, companies with operations and employees in more than one country. Some of the largest include Royal Dutch Shell oil company from the U.K., Toyota motor vehicles from Japan, and Apple computers from the U.S.

A cargo ship loaded with containers

MARITIME TRANSPORT DRIVES THE GLOBAL ECONOMY: MORE THAN **90 PERCENT** OF THE WORLD'S TRADE IS **CARRIED BY SEA.**

▼ MAIN TRADING NATIONS

Economists closely monitor the balance of imports versus exports in the world's main trading nations. The United States has a large imbalance, but the deficit's value makes up a small percentage of total GDP.

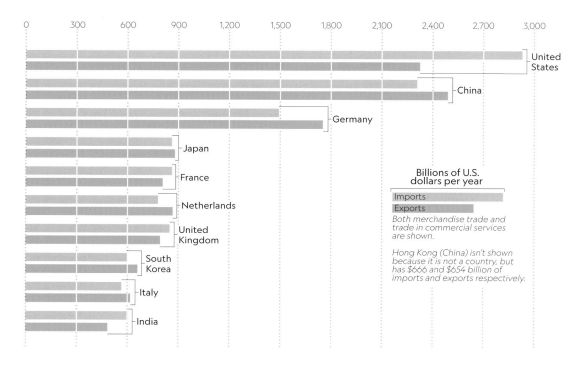

United States
China
Germany
Japan
France
Netherlands
United Kingdom
South Korea
Italy
India

Billions of U.S. dollars per year

Imports
Exports

Both merchandise trade and trade in commercial services are shown.

Hong Kong (China) isn't shown because it is not a country, but has $666 and $654 billion of imports and exports respectively.

▼ MERCHANDISE EXPORTS

Machinery and petroleum are particularly profitable commodities on the world market. Trade in information technology products, such as computers and integrated circuits, has tripled in the past two decades.

Ranking Global Merchandise Exports (in millions of U.S. dollars)

1. Machinery (not office/telecom or transport equipment)	2,224,223
2. Transport equipment	2,163,066
3. Fuels	1,960,413
4. Chemicals (not pharmaceuticals)	1,421,982
5. Food products (not fish)	1,321,952
6. Semi-manufactures (not chemicals, pharmaceutical, iron or steel)	1,116,799
7. Miscellaneous manufactures	832,850
8. Telecommunications equipment	722,041
9. Integrated circuits and electronic components	618,427
10. Pharmaceuticals	570,561
11. Electronic data processing and office equipment	525,307
12. Clothing	471,594
13. Iron and steel	414,424
14. Scientific and controlling instruments	402,930
15. Personal and household goods	376,106
16. Non-ferrous metals	346,061
17. Ores and other minerals	327,677
18. Textiles	300,226
19. Other raw materials	270,212
20. Fish	143,860

◄ TRADE BLOCS

Agreements among neighboring countries—or even across oceans, as with the 21 members of APEC—create larger markets where goods and labor cross borders more easily. Trade blocs are also political: intertwined economic interests can promote peaceful alliances, but outsourcing labor is controversial.

Major Regional Trade Agreements

APEC: Asia-Pacific Economic Cooperation

ASEAN: Association of Southeast Asian Nations

APEC and ASEAN

COMESA: Common Market for Eastern and Southern Africa

ECOWAS: Economic Community of West African States

EU: European Union

MERCOSUR: Southern Common Market

APEC and USMCA: United States-Mexico-Canada Agreement

SAFTA: South Asian Free Trade Area

◄ IMPORTS

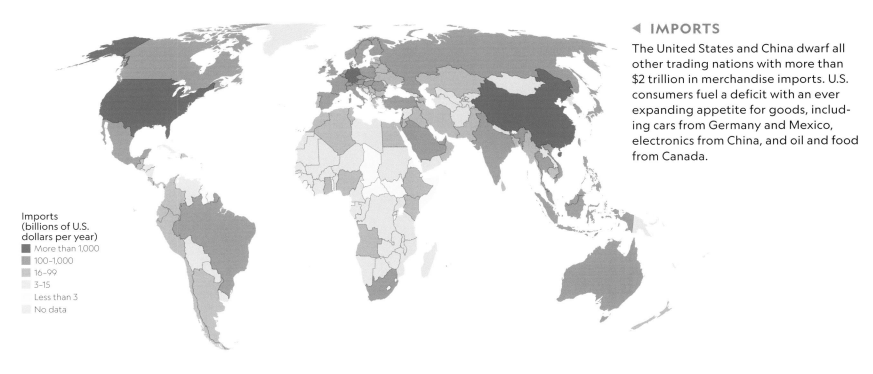

The United States and China dwarf all other trading nations with more than $2 trillion in merchandise imports. U.S. consumers fuel a deficit with an ever expanding appetite for goods, including cars from Germany and Mexico, electronics from China, and oil and food from Canada.

Imports (billions of U.S. dollars per year)

More than 1,000

100–1,000

16–99

3–15

Less than 3

No data

◄ EXPORTS

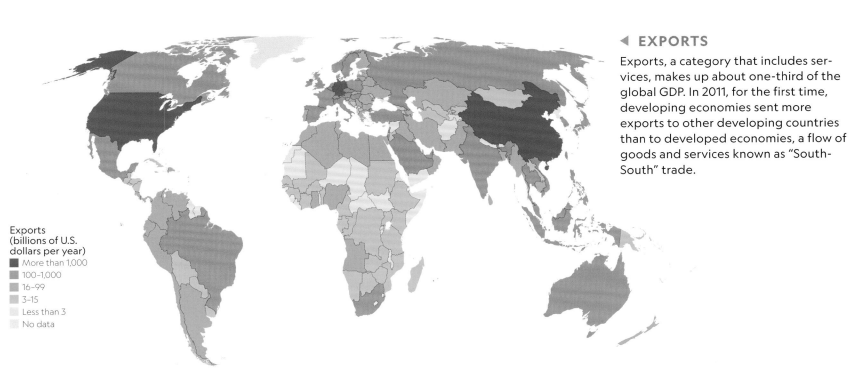

Exports, a category that includes services, makes up about one-third of the global GDP. In 2011, for the first time, developing economies sent more exports to other developing countries than to developed economies, a flow of goods and services known as "South-South" trade.

Exports (billions of U.S. dollars per year)

More than 1,000

100–1,000

16–99

3–15

Less than 3

No data

POWER SOURCES WORLDWIDE

Electricity is life-changing—especially for people in remote areas. Refrigeration gets more food to market, allows health clinics to store perishable medications, and provides light for children to study and learn. Yet around the world, nearly one billion people are still waiting for electricity to reach them. Historically, access to electricity required expanding a grid managed by a government or power company. But much of today's growth depends on microgrids: easily installed household- or community-size systems that generate power (usually solar) and store it.

Most countries, regardless of how firmly they are connected to the grid, face a pressing need to develop renewable energy sources such as wind or solar power. Fossil fuels such as coal, oil, and natural gas account for 80 percent of the world's energy, but these reserves are finite. Top producers of oil and natural gas are the U.S., Saudi Arabia, and Russia; China and India produce the most coal. Other countries heavily dependent on oil for GDP include Kuwait, Libya, Iraq, Angola, Oman, and Venezuela. Burning fossil fuels releases carbon dioxide and other greenhouse gases, which trap heat in Earth's atmosphere and accelerate global warming and climate change. If countries can transition to renewable energy paired with nimble infrastructure, electricity can continue to lift people out of poverty without endangering the environment.

▼ ENERGY CONSUMPTION

This distorted map enlarges countries that are the biggest energy consumers. Fossil fuels reign in most countries; some small nations, such as Tajikistan and Bhutan, which rely on hydropower, have had success with renewables.

Energy Consumption by Country, 2016
The area of each place shows total energy consumption, and the color shows the proportion of that energy that came from coal, petroleum, and natural gas sources.

■ 50 trillion BTUs of energy consumed (about 14.7 billion kilowatt-hours, or 52.8 million gigajoules)

Percent of energy consumed from fossil fuel sources
- More than 90
- 70–90
- 50–69
- 30–49
- Less than 30

Countries where 50% or more of energy consumption came from non-fossil fuel sources are labeled in green on the map.

▼ ENERGY USE TRENDS

Energy use has nearly doubled since 1980. Coal fed the recent economic revolutions in India and China, and natural gas is the most in-demand fuel today. Renewable energy has had persistent slow growth.

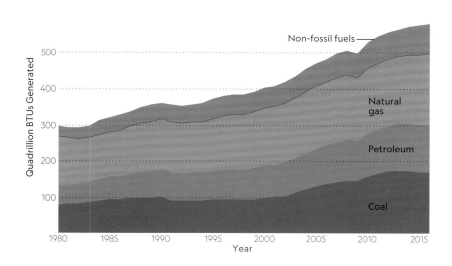

Turbines harness the power of wind.

FOSSIL FUEL CONSUMPTION IS DECLINING IN THE U.S., BUT IT IS STILL THREE TIMES THAT OF CHINA AND MORE THAN FOUR TIMES THE WORLD'S PER CAPITA DEMAND.

1 Burkina Faso
Mali
Mauritania
2 Cabo Verde
The Gambia
Guinea
Guinea-Bissau
Liberia
Sierra Leone
3 Central African
Republic
Chad
Niger
4 Djibouti
Eritrea
Somalia
South Sudan

5 Equatorial Guinea
Gabon
Sao Tome
and Principe
6 Burundi
Rwanda
7 Eswatini
8 Lesotho
9 Comoros
Maldives
Réunion
Seychelles

▼ POWER WITHOUT FOSSIL FUELS

Hydroelectricity, where flowing water turns the blades of a turbine, is the most common alternative to fossil fuels. Nuclear power is plateauing, but other renewable energy sources have grown significantly.

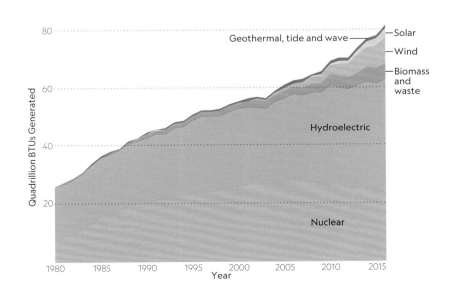

▼ RENEWABLE ENERGY PRODUCERS

Linked power grids allow countries to export electricity. Paraguay, home to one of the largest hydropower plants in the world, produces all of its own electricity and sells its large surplus.

Largest Generators of Renewable Electricity
(total 2016 renewable generation in billion BTUs, and percent of total electrical generation)

1. China	5254	(26.2%)
2. United States	2191	(15.7)
3. Brazil	1575	(81.2)
4. Canada	1466	(66.1)
5. India	764	(16.1)
6. Germany	665	(31.8)
7. Japan	658	(19.5)
8. Russia	636	(18.1)
9. Norway	497	(98.5)
10. Italy	375	(39.9)
11. Spain	352	(39.9)
12. France	340	(18.8)
13. Sweden	308	(59.1)
14. Turkey	305	(34.2)
15. United Kingdom	300	(27.7)
16. Venezuela	229	(61.3)
17. Vietnam	218	(40.3)
18. Paraguay	216	(100.0)
19. Colombia	172	(67.1)
20. Austria	168	(81.0)

Exporters of Renewable Electricity
Renewable electrical generation as a percent of domestic electricity consumption

1. Paraguay	579%
2. Bhutan	361
3. Laos	327
4. Albania	140
5. Mozambique	134
6. Tajikistan	127
7. Ethiopia	123
8. Dem. Rep. of the Congo	122
9. Norway	119
10. Uruguay	118
11. Kyrgyzstan	108
12. Costa Rica	108
13. Central African Republic	107
14. Malawi	106
15. Uganda	105
16. Congo	104
17. Iceland	103

OUR DIGITAL WORLD

Somali migrants gather on a beach in Djibouti to capture cell signals.

More than half the world's people use the internet, and most use their mobile phones to access it. As cell phones have become less expensive, internet access is increasing in Asia and Africa. In many developing countries, companies have made cell service available without ever having established landline telephone infrastructure. This has allowed many more people to use mobile money services to bank their savings, pay bills, and transfer payments and remittances. Over 70 percent of youth 15 to 24 years old are online, and younger people are more likely to have smartphones and use social media. Yet there remain communities that have lower access to the internet: Women and girls, indigenous populations, and older people are all more likely to be off-line.

NORTH AMERICA

SOUTH AMERICA

Access to the Internet (percent of the population)
- 80–100
- 60–79
- 40–59
- 20–39
- 1–19
- 0 or no data

▲ ONLINE & OFF-LINE

Improving access to the internet is one of the United Nations' Sustainable Development Goals for 2030. Only 28 percent of Africa's population has access to the internet.

Government Restrictions to the Internet
- Restricted internet access through blocking websites and internet blackouts
- Restricted internet access through monitoring email and social media and/or regulating published content
- A combination of both
- None
- No data

◀ RESTRICTED ACCESS

China has the strictest internet censorship and surveillance of any government in the world, while blackouts and blocked websites restrict access across much of Asia and Africa. The United States has laws allowing the government to regulate and access private electronic communications.

▼ MORE PHONES THAN PEOPLE

Mobile phone subscriptions increase every year. In some countries (top list), many people have more than one cell phone. In others (bottom list), few people have cell phones.

Countries with the Most Mobile-Cellular Subscriptions per 100 Inhabitants

1. United Arab Emirates	209
2. Antigua and Barbuda	193
3. Maldives	181
4. Costa Rica	179
5. Kuwait	176
Thailand	176
7. Seychelles	174
8. Montenegro	166
9. Indonesia	164
10. Turkmenistan	163

Countries with the Least Mobile-Cellular Subscriptions per 100 Inhabitants

1. North Korea	15
2. Eritrea	20
3. Federated States of Micronesia	21
4. South Sudan	26
Central African Republic	26
6. Marshall Islands	28
7. Madagascar	34
8. Belize	36
9. Ethiopia	37
10. Djibouti	40

Undersea Telecommunication Cables

▬▬	10
▬	5
—	2
—	1

Line darkness and width indicates number of cables roughly following that route.

▼ UNDERSEA CABLES

Approximately 1.2 million kilometers (750,000 miles) of submarine fiber-optic cables the diameter of a garden hose carry data around the world at a speed of up to 26 terabits a second.

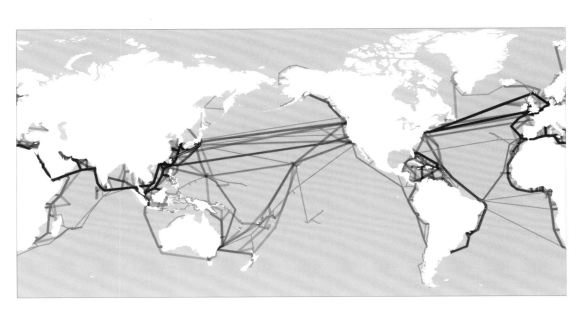

NORTH
AMERICA

Complex geologic processes created the striped Navajo sandstone in Zion National Park, Utah.

OLD STONES & NEW WORLD

FROM THE FRIGID ARCTIC CIRCLE TO THE STEAMY CARIBBEAN TROPICS

North America is cold and heat. Skyscrapers and ancient ruins. Parched deserts and fast-flowing rivers. Home to natives and newcomers, thinkers and makers, its stones are old and its human history young. Prosperity has flowed from its success in industry, innovation, and trade; poverty remains a visible failure. This is a continent of extremes.

GEOGRAPHY

LAND OF OPPOSITES

The world's third largest continent is home to a kaleidoscope of landscapes and climates. In Alaska, snowcapped Denali punches above the clouds. In Nicaragua, steamy rainforests teem with life. In California, a forbidding desert ranks among the hottest spots in the world. Canada's coastline—the longest of any country in the world—includes mighty glaciers hugging Baffin Island, while on shores farther south, mangrove swamps buffer Florida from tropical storms and whales breed in the lagoons of the Baja Peninsula.

It all started with a breakup. Roughly 200 million years ago, a landmass that included the future North America split off from the supercontinent Pangaea and seas filled the widening gap that would become the Atlantic Ocean. Clues to Earth's geologic history rest beneath Canada's tundra, where we find stones from four billion years ago—among the oldest rocks in the world. Relatively isolated from other landmasses, North America's mountains, plains, lakes, forests, and deserts gave rise to a unique array of plants and animals. The continent has the world's tallest trees (the sequoias and redwoods of California) and some of its largest land animals (grizzly bears, moose, and bison). North America has valuable natural resources such as timber, coal, oil, and natural gas, and agriculture benefits from rich soil and plentiful freshwater.

Palenque, an ancient Maya city-state near the Mexico-Guatemala border

HISTORY

A CHECKERED LEGACY OF COLONIALISM

Humans have inhabited North America's diverse landscapes for 14,000 years, perhaps much longer. The Olmec of Mexico built the continent's first civilization, around 1200 B.C., and developed an advanced calendar, writing system, and stone sculptures. Then came the Maya, with an elaborate religion and sprawling temple cities, and the militaristic Aztec. Around A.D. 1050 the Mississippian people built the sophisticated city of Cahokia atop giant earthen mounds in today's U.S. Midwest.

In the late 15th century, the arrival of Europeans ushered in a tumultuous era. New languages, new economic drivers (including a reliance on the labor of enslaved Africans), new governing structures: The colonial period was often violent, and, for some Europeans, unimaginably profitable. The rich mosaic of American cultures was worn away as European disease and firepower decimated native communities. In their place grew colonies ruled by the Netherlands, France, England, and Spain where indigenous languages, religion, art, and other forms of cultural expression were brutally suppressed. Centuries later, the aftereffects of colonialism persist. Many Central American and Caribbean countries have continued to wrestle with social unrest, despotic governments, and poverty, long after gaining independence from their colonial rulers. In the U.S., racial inequality remains complex and deeply rooted, a sobering reminder of slavery's dark legacy.

Alaska's Denali is the tallest mountain in North America.

CULTURE

A TAPESTRY OF MANY COLORS

Until the 16th century, North America's cultural giants were the great civilizations of Mexico and Central America, the Pueblo builders of the southwestern United States, and the highly organized cultivators of the Great Lakes region and the Mississippi Valley. European newcomers built towns—San Juan, New York, Toronto—based on Old World models. They brought new languages. Their religions soon superseded or blended with the ancient beliefs of indigenous societies. And they developed ideas—democracy, capitalism, religious choice, and free speech—that continue to shape the political, intellectual, and economic life of the continent and the world.

Today, although the continent's cultural fabric is laced tightly together, it retains distinctly individual colors. Mexico and Central America are shaped by Hispano-Indian culture and tend to have more in common with South America than with their northern neighbors. The Caribbean islands have nurtured cultures that blend European, African, and Amerindian traditions, from the Rastafarians of Jamaica to the Creoles of Martinique. And in the U.S. and Canada, European heritage now mixes with traditions introduced by waves of immigrants from Latin America, Africa, Asia, the Middle East, and the Pacific islands. In the U.S. alone, residents speak at least 350 different languages from all over the world.

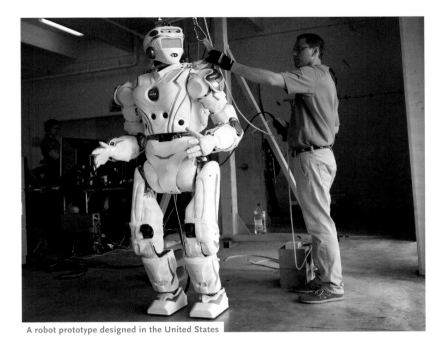
A robot prototype designed in the United States

ECONOMY

THINKERS & MAKERS

From Silicon Valley to the New York Stock Exchange, North America is a world leader in technology, finance, entertainment, medicine, defense, and agriculture. The continent produces more goods and services per person than any other region in the world. This relentless drive to produce, build, and succeed traces back to the earliest European immigrants. They saw a land of limitless opportunity and abundant natural resources: rich, expansive agricultural lands, huge coal deposits, vast petroleum reserves. By the end of the 19th century, North America's forests, minerals, and farmlands had stoked an industrial revolution that made the United States one of the world's richest nations and a superpower in geopolitics. The U.S. dollar is the standard reserve currency for banks around the world.

North America's most valuable commodity has always been innovation. Many devices that revolutionized modern life—telephones, electric lighting, cars, airplanes, computers, television, the internet, smartphones—were invented or first mass-produced in the United States. Free-trade agreements between Canada, Mexico, and the United States have fueled an export and import economy, while many Caribbean and Central American countries still rely on tourism and agriculture. Poverty, as well as war and violence, has driven millions of people northward in search of better prospects.

Worshippers gather in Cuba, where religion often blends Catholicism and African Santería.

SNAPSHOT
North America FACTS & FEATURES

HIGHEST ELEVATION Denali (Mount McKinley), United States: 6,190 m (20,310 ft)	**LARGEST COUNTRY BY AREA** Canada: 9,984,670 sq km (3,855,101 sq mi)
LOWEST ELEVATION Death Valley, United States: –86 m (–282 ft)	**SMALLEST COUNTRY BY AREA** Saint Kitts and Nevis: 261 sq km (101 sq mi)
LONGEST RIVER Mississippi-Missouri River: 5,971 km (3,710 mi)	**LONGEST RAIL LINE** Toronto to Vancouver: 4,466 km (2,775 mi)
BIGGEST URBAN AREA BY POPULATION Mexico City, Mexico: 21,581,000 (2018)	**WORLD'S BUSIEST AIRPORT** Atlanta, Georgia, U.S.: 107 million annual passengers
MOST POPULOUS COUNTRY United States: 329,256,000	**LARGEST LAND ANIMAL** American bison: weighs up to 1,000 kg (2,200 lbs)

LAKE SUPERIOR IS THE WORLD'S LARGEST FRESHWATER LAKE IN AREA. THE NORTHERNMOST OF THE FIVE GREAT LAKES, ITS SURFACE AREA MEASURES 82,100 SQUARE KILOMETERS (31,700 SQ MI).

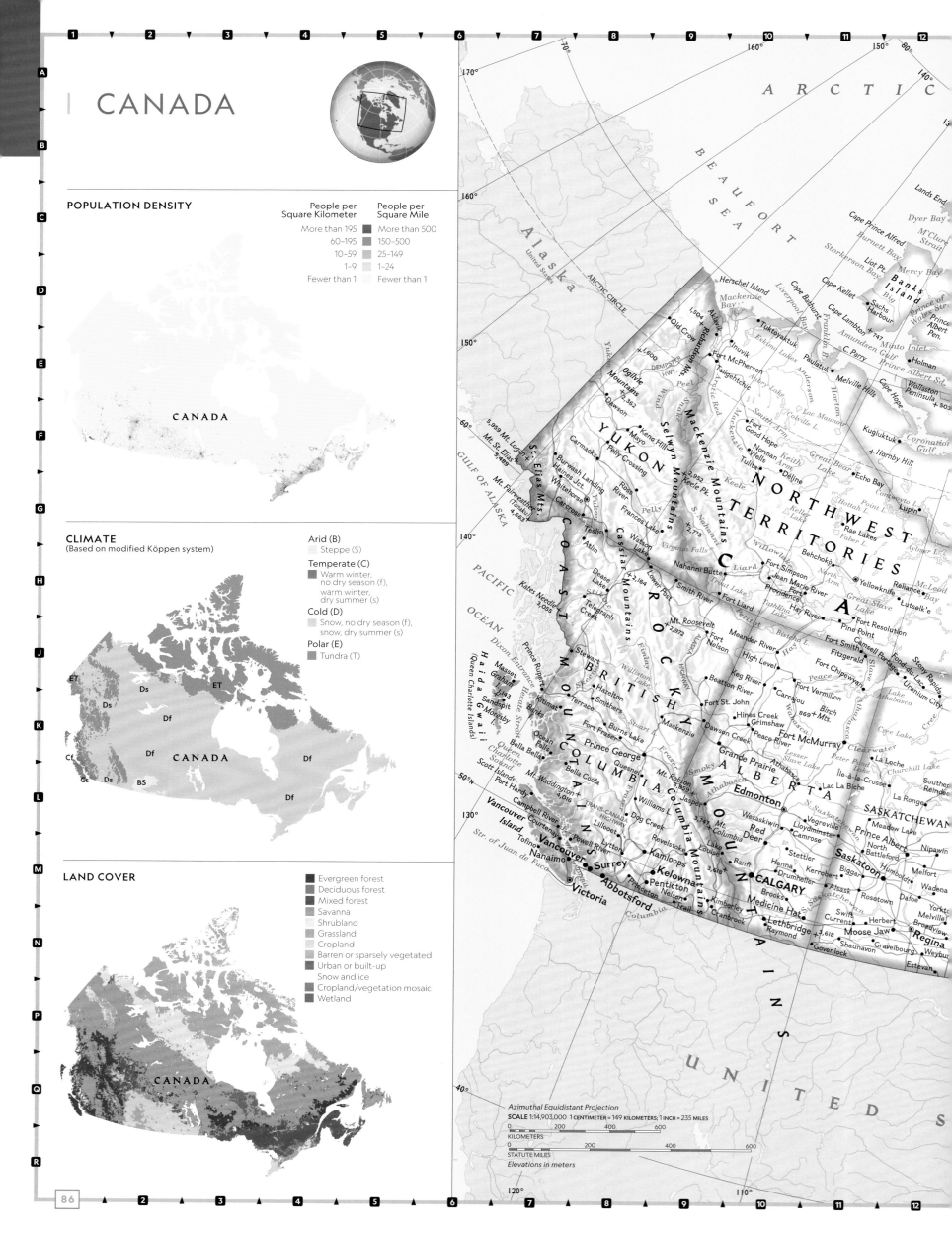

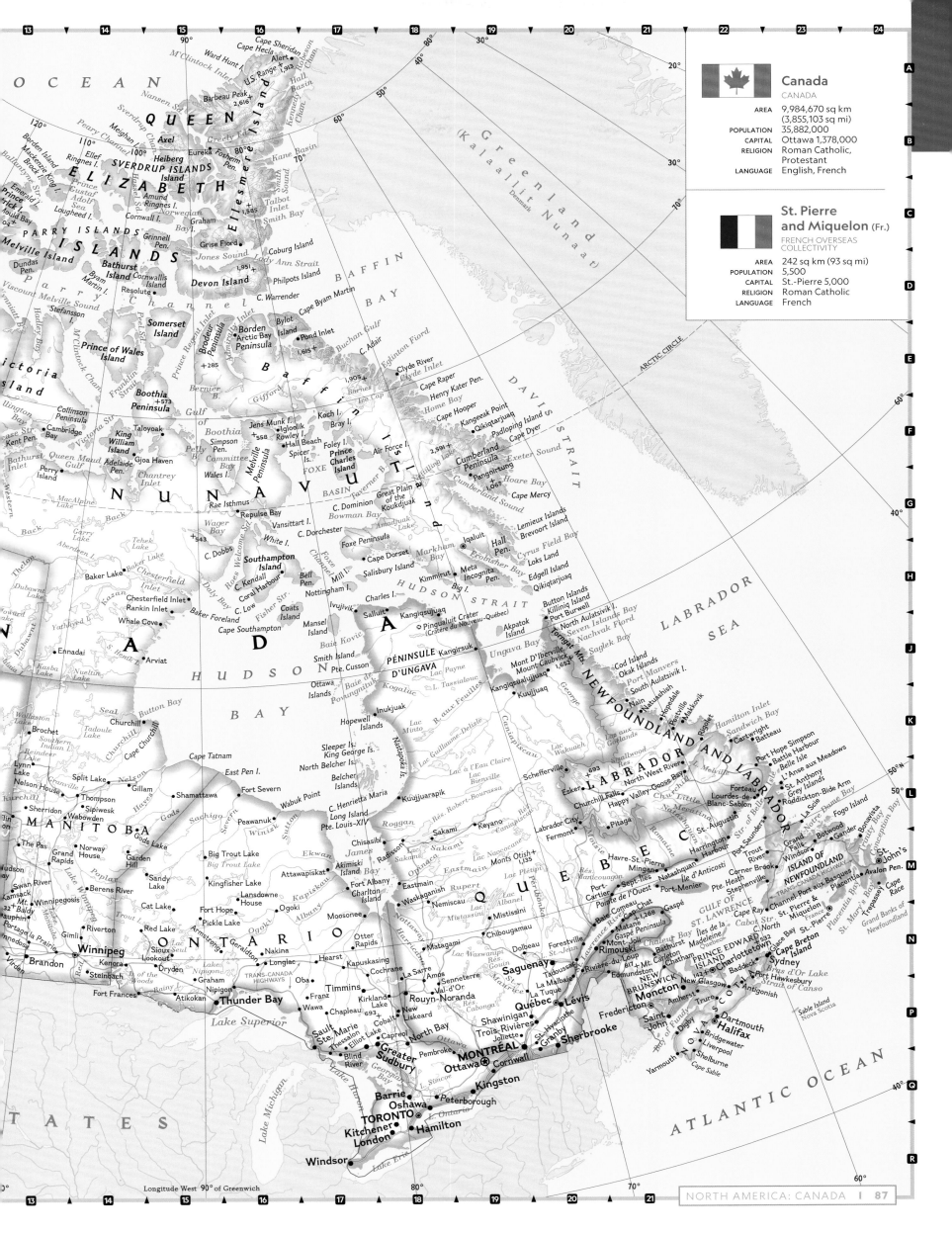

ICE SHEET

Greenland's immense, contiguous ice sheet is more than 3,000 meters (10,000 feet) thick in places and contains nearly three million cubic kilometers (720,000 cubic miles) of ice which accounts for one-tenth of the total freshwater on Earth. That's enough water to raise the global sea level by 73 meters (24 feet) if it were to melt completely.

THE CRAWL TOWARD INDEPENDENCE

Greenland has been a Danish possession since 1814 (after the breakup of the kingdom of Denmark-Norway) but has slowly taken steps toward increased autonomy. Greenland gained representation in the Danish Parliament in 1953 and was awarded home rule in 1979. In 2009, all matters except defense and foreign policy were transferred to Greenland after a self-government referendum was passed, although Greenland still relies heavily on subsidies from Denmark.

LARGEST ISLAND

Easily the world's largest island, Greenland has an area of 2,175,600 square kilometers (840,004 square miles). Gunnbjørn Field, at 3,694 meters (12,119 feet), is the island's highest point, although the massive ice sheet gently rises to a maximum elevation of about 3,234 meters (10,610 feet). With approximately 1,801,000 square kilometers (695,370 square miles) of ice, nearly 83 percent of Greenland is ice covered.

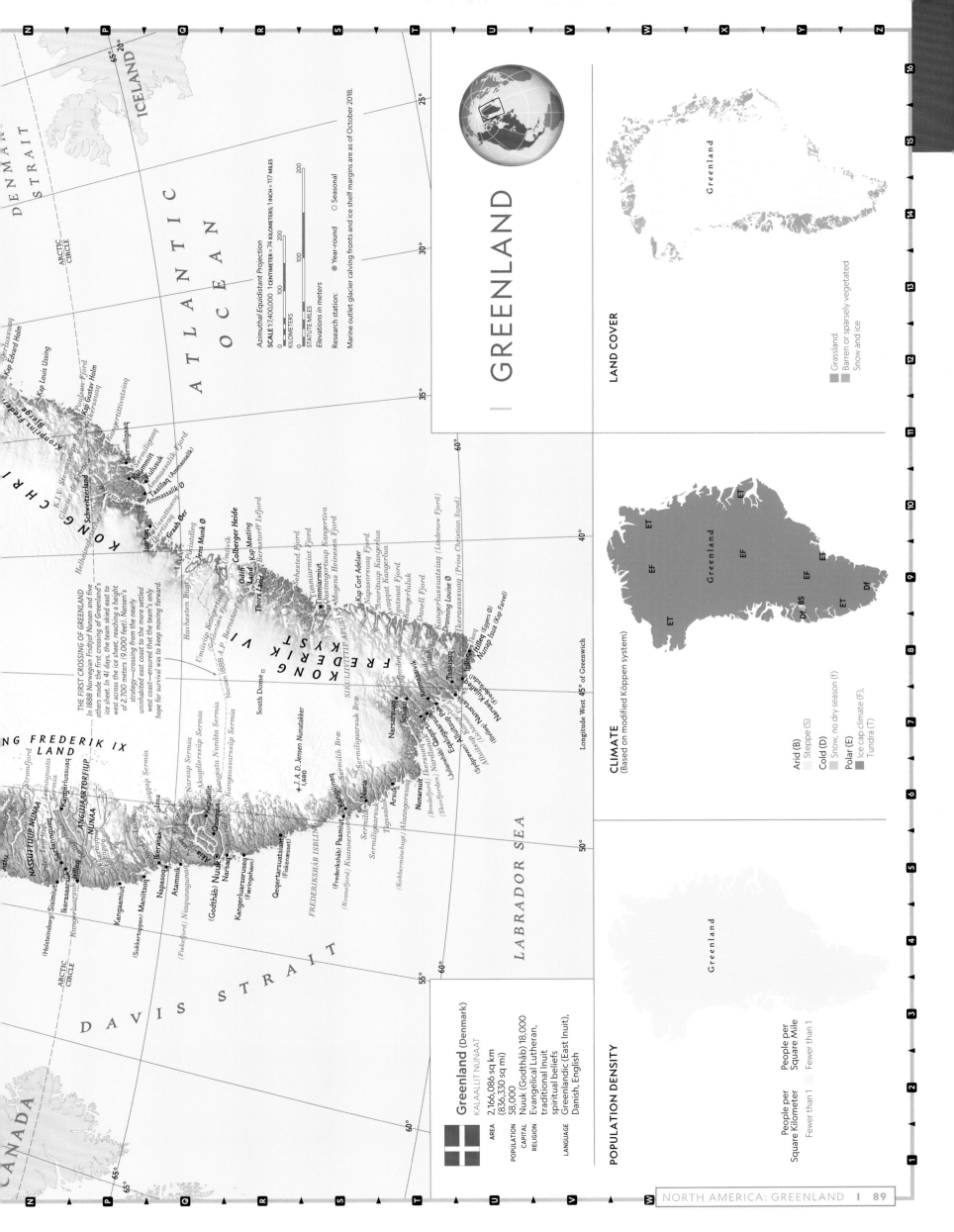

GREENLAND

Greenland (Denmark)
KALAALLIT NUNAAT

AREA	2,166,086 sq km (836,330 sq mi)
POPULATION	58,000
CAPITAL	Nuuk (Godthåb) 18,000
RELIGION	Evangelical Lutheran, traditional Inuit spiritual beliefs
LANGUAGE	Greenlandic (East Inuit), Danish, English

Azimuthal Equidistant Projection
SCALE 1:7,400,000 1 CENTIMETER = 74 KILOMETERS; 1 INCH = 117 MILES
Elevations in meters

Research station: ⊠ Year-round ○ Seasonal

Marine outlet glacier calving fronts and ice shelf margins are as of October 2018.

THE FIRST CROSSING OF GREENLAND In 1888 Norwegian Fridtjof Nansen and five others made the first crossing of Greenland's ice sheet. In 41 days, the team skied east to west across the ice sheet, reaching a height of 2,700 meters (9,000 feet). Nansen's strategy—crossing from the nearly uninhabited east coast to the more settled west coast—ensured that the team's only hope for survival was to keep moving forward.

POPULATION DENSITY

People per Square Kilometer	People per Square Mile
Fewer than 1	Fewer than 1

LAND COVER

- Grassland
- Barren or sparsely vegetated
- Snow and ice

CLIMATE
(Based on modified Köppen system)

Arid (B)
- Steppe (S)

Cold (D)
- Snow, no dry season (f)

Polar (E)
- Ice cap climate (F)
- Tundra (T)

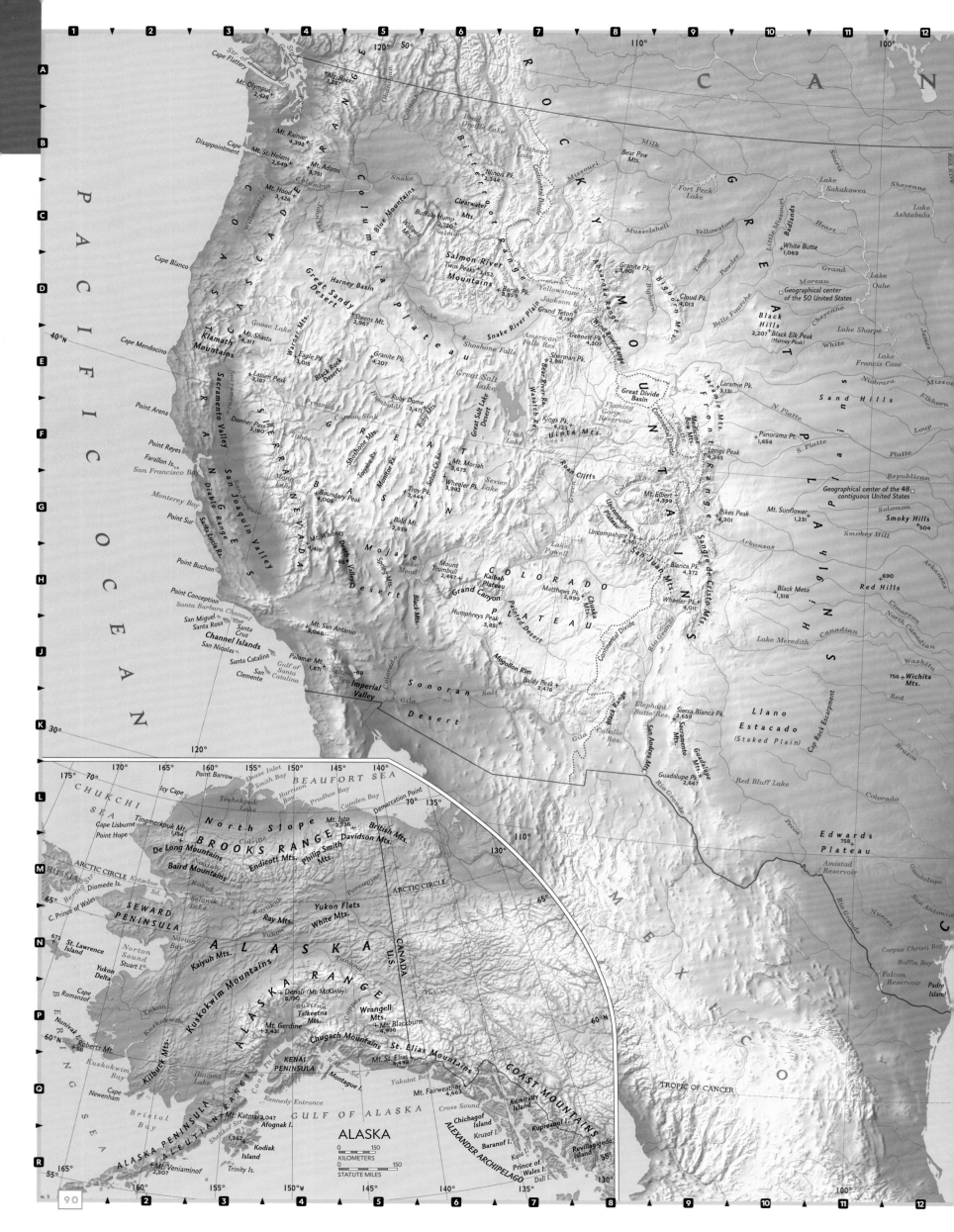

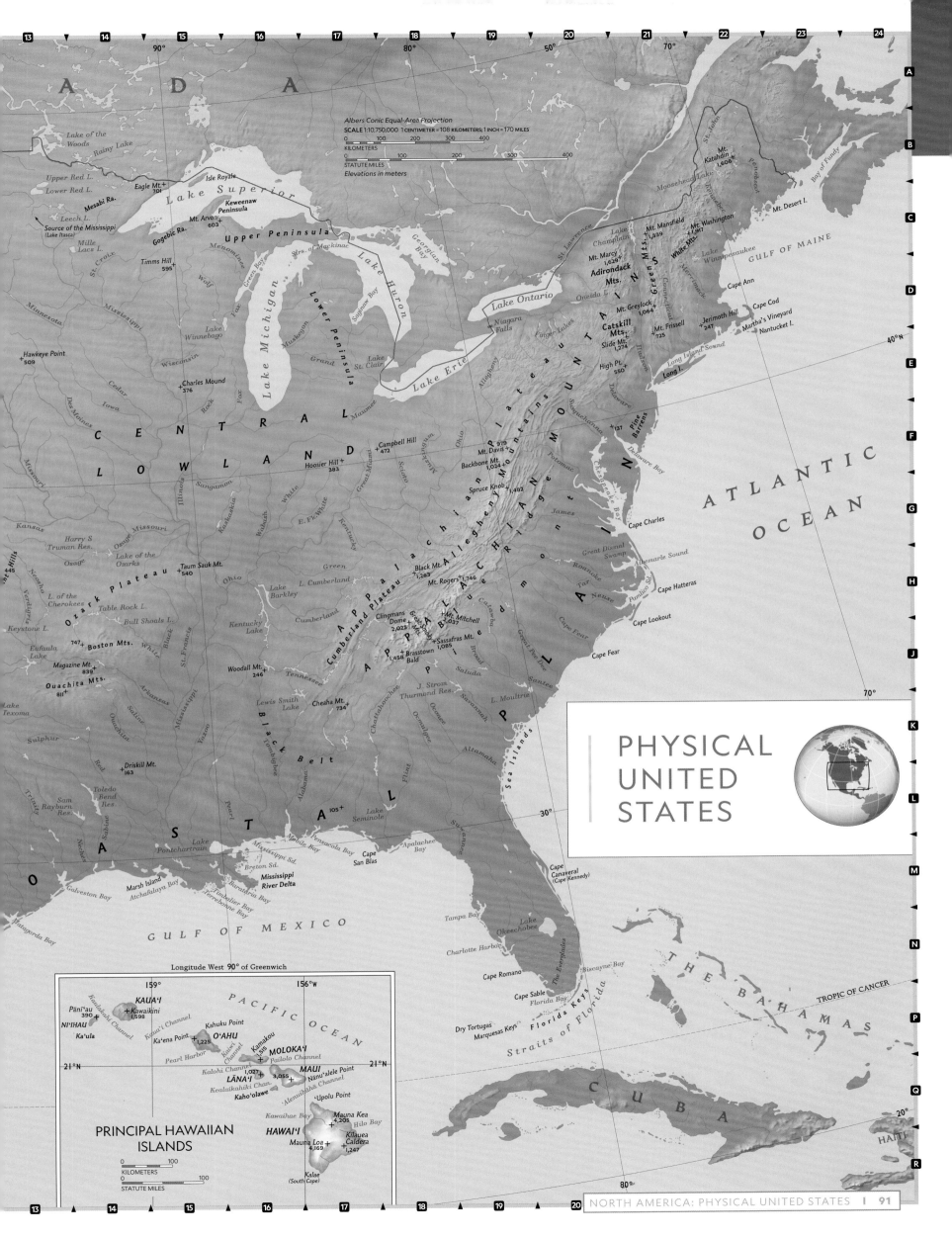

PHYSICAL UNITED STATES

PRINCIPAL HAWAIIAN ISLANDS

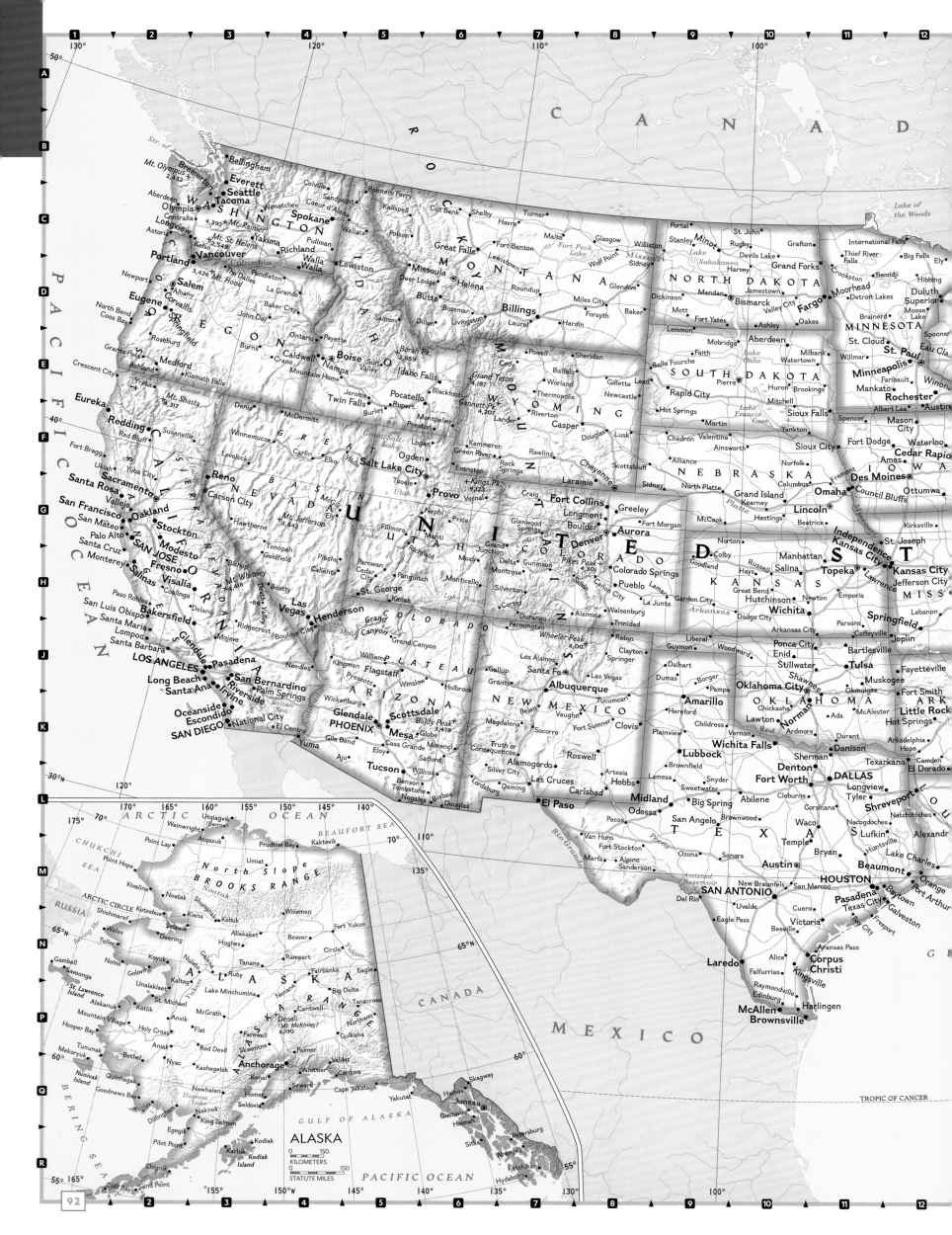

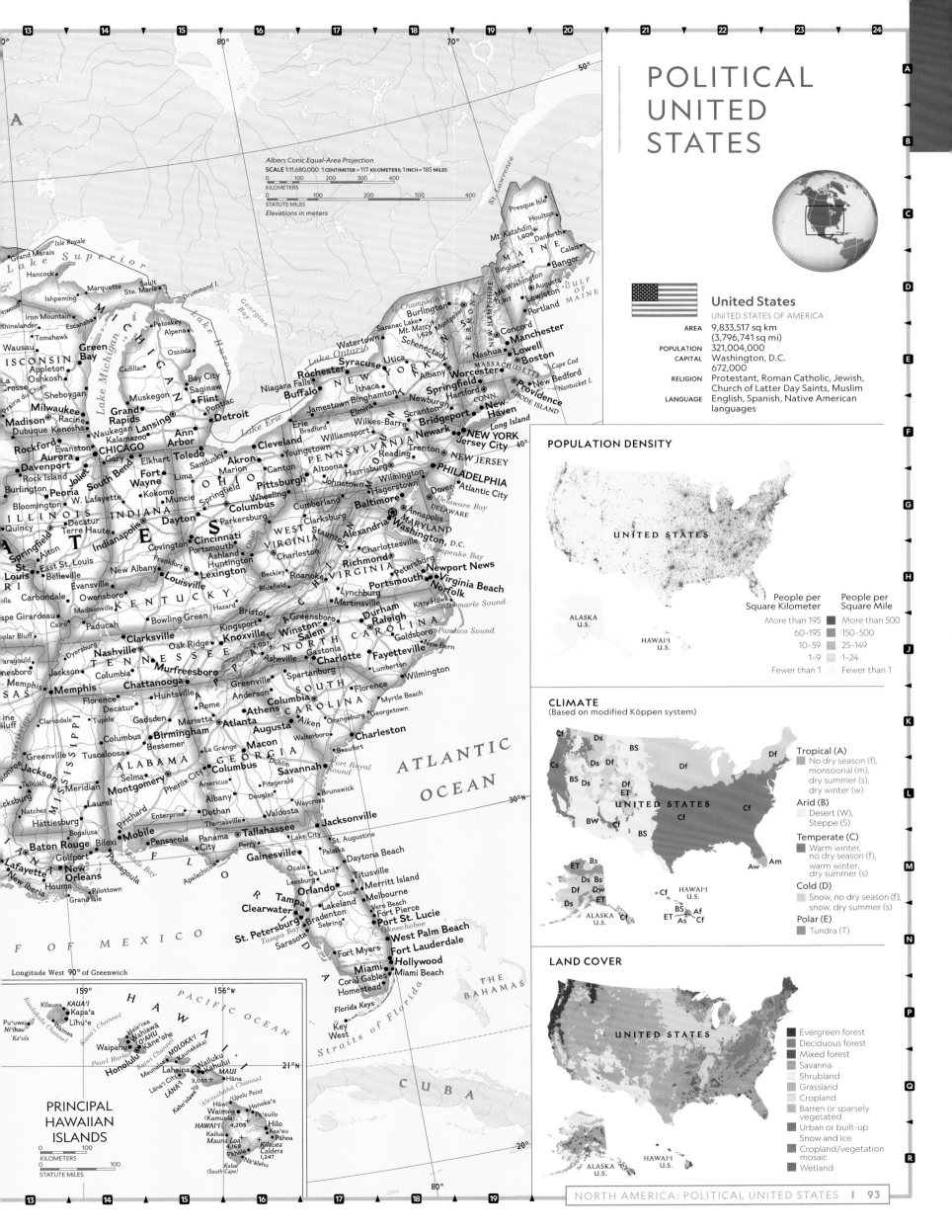

POLITICAL UNITED STATES

Albers Conic Equal-Area Projection

SCALE 1:11,680,000 1 CENTIMETER = 117 KILOMETERS; 1 INCH = 185 MILES

KILOMETERS
0   100   200   300   400

STATUTE MILES
0   100   200   300   400

Elevations in meters

United States
UNITED STATES OF AMERICA

AREA 9,833,517 sq km
(3,796,741 sq mi)
POPULATION 321,004,000
CAPITAL Washington, D.C.
672,000
RELIGION Protestant, Roman Catholic, Jewish,
Church of Latter Day Saints, Muslim
LANGUAGE English, Spanish, Native American
languages

POPULATION DENSITY

UNITED STATES

ALASKA
U.S.

HAWAI'I
U.S.

People per
Square Kilometer
More than 195
60–195
10–59
1–9
Fewer than 1

People per
Square Mile
More than 500
150–500
25–149
1–24
Fewer than 1

CLIMATE
(Based on modified Köppen system)

UNITED STATES

HAWAI'I
U.S.

ALASKA
U.S.

Tropical (A)
No dry season (f),
monsoonal (m),
dry summer (s),
dry winter (w)

Arid (B)
Desert (W),
Steppe (S)

Temperate (C)
Warm winter,
no dry season (f),
warm winter,
dry summer (s)

Cold (D)
Snow, no dry season (f),
snow, dry summer (s)

Polar (E)
Tundra (T)

LAND COVER

UNITED STATES

ALASKA
U.S.

HAWAI'I
U.S.

Evergreen forest
Deciduous forest
Mixed forest
Savanna
Shrubland
Grassland
Cropland
Barren or sparsely
vegetated
Urban or built-up
Snow and ice
Cropland/vegetation
mosaic
Wetland

PRINCIPAL HAWAIIAN ISLANDS

KILOMETERS
0   100

STATUTE MILES
0   100

PACIFIC TERRITORIES

American Memorial Park,
Northern Mariana Islands

War in the Pacific
National Historical Park,
Guam

National Park of
American Samoa

SAN FRANCISCO AREA PARKS

Fort Point National Historic Site

Presidio of San Francisco

Rosie the Riveter/
World War II Home Front
National Historical Park

San Francisco Maritime
National Historical Park

ALASKAN
PARKS

Albers Conic Equal-Area Projection
SCALE 1:10,824,000 1 CENTIMETER = 108 KILOMETERS; 1 INCH = 171 MILES

National Park Service property

Park boundary

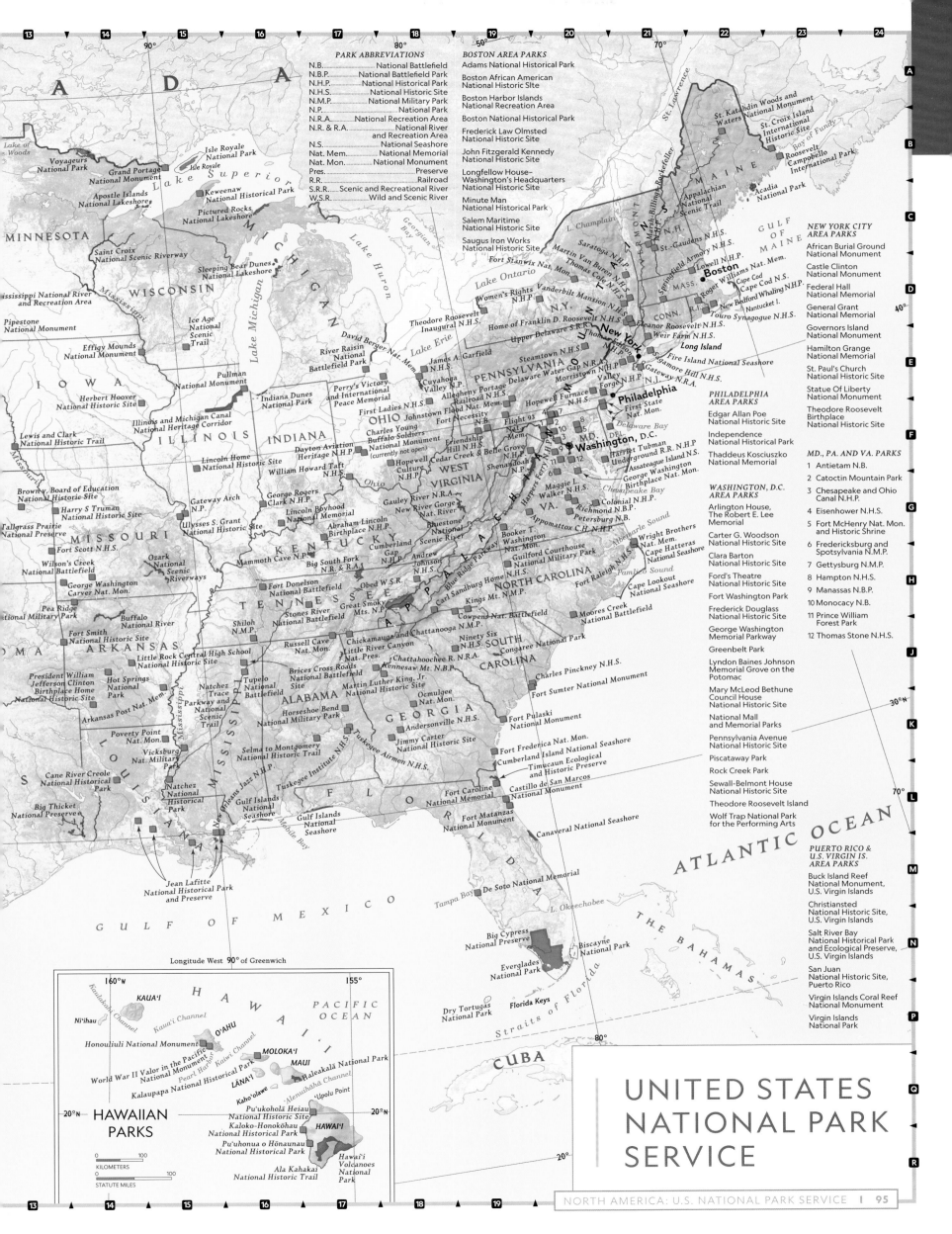

PARK ABBREVIATIONS

N.B. — National Battlefield
N.B.P. — National Battlefield Park
N.H.P. — National Historical Park
N.H.S. — National Historic Site
N.M.P. — National Military Park
N.P. — National Park
N.R.A. — National Recreation Area
N.R. & R.A. — National River and Recreation Area
N.S. — National Seashore
Nat. Mem. — National Memorial
Nat. Mon. — National Monument
Pres. — Preserve
R.R. — Railroad
S.R.R. — Scenic and Recreational River
W.S.R. — Wild and Scenic River

BOSTON AREA PARKS

Adams National Historical Park
Boston African American National Historic Site
Boston Harbor Islands National Recreation Area
Boston National Historical Park
Frederick Law Olmsted National Historic Site
John Fitzgerald Kennedy National Historic Site
Longfellow House–Washington's Headquarters National Historic Site
Minute Man National Historical Park
Salem Maritime National Historic Site
Saugus Iron Works National Historic Site

NEW YORK CITY AREA PARKS

African Burial Ground National Monument
Castle Clinton National Monument
Federal Hall National Memorial
General Grant National Memorial
Governors Island National Monument
Hamilton Grange National Memorial
St. Paul's Church National Historic Site
Statue Of Liberty National Monument
Theodore Roosevelt Birthplace National Historic Site

PHILADELPHIA AREA PARKS

Edgar Allan Poe National Historic Site
Independence National Historical Park
Thaddeus Kosciuszko National Memorial

WASHINGTON, D.C. AREA PARKS

Arlington House, The Robert E. Lee Memorial
Carter G. Woodson National Historic Site
Clara Barton National Historic Site
Ford's Theatre National Historic Site
Fort Washington Park
Frederick Douglass National Historic Site
George Washington Memorial Parkway
Greenbelt Park
Lyndon Baines Johnson Memorial Grove on the Potomac
Mary McLeod Bethune Council House National Historic Site
National Mall and Memorial Parks
Pennsylvania Avenue National Historic Site
Piscataway Park
Rock Creek Park
Sewall-Belmont House National Historic Site
Theodore Roosevelt Island
Wolf Trap National Park for the Performing Arts

MD., PA. AND VA. PARKS

1 Antietam N.B.
2 Catoctin Mountain Park
3 Chesapeake and Ohio Canal N.H.P.
4 Eisenhower N.H.S.
5 Fort McHenry Nat. Mon. and Historic Shrine
6 Fredericksburg and Spotsylvania N.M.P.
7 Gettysburg N.M.P.
8 Hampton N.H.S.
9 Manassas N.B.P.
10 Monocacy N.B.
11 Prince William Forest Park
12 Thomas Stone N.H.S.

PUERTO RICO & U.S. VIRGIN IS. AREA PARKS

Buck Island Reef National Monument, U.S. Virgin Islands
Christiansted National Historic Site, U.S. Virgin Islands
Salt River Bay National Historical Park and Ecological Preserve, U.S. Virgin Islands
San Juan National Historic Site, Puerto Rico
Virgin Islands Coral Reef National Monument
Virgin Islands National Park

HAWAIIAN PARKS

KAUA'I
Ni'ihau
O'AHU
Honouliuli National Monument
World War II Valor in the Pacific National Monument
MOLOKA'I
LĀNA'I
Kaho'olawe
MAUI
Haleakalā National Park
Kalaupapa National Historical Park
Pu'ukoholā Heiau National Historic Site
Kaloko-Honokōhau National Historical Park
Pu'uhonua o Hōnaunau National Historical Park
HAWAI'I
Hawai'i Volcanoes National Park
Ala Kahakai National Historic Trail
Upolu Point

0 — 100 KILOMETERS
0 — 100 STATUTE MILES

Longitude West 90° of Greenwich

UNITED STATES NATIONAL PARK SERVICE

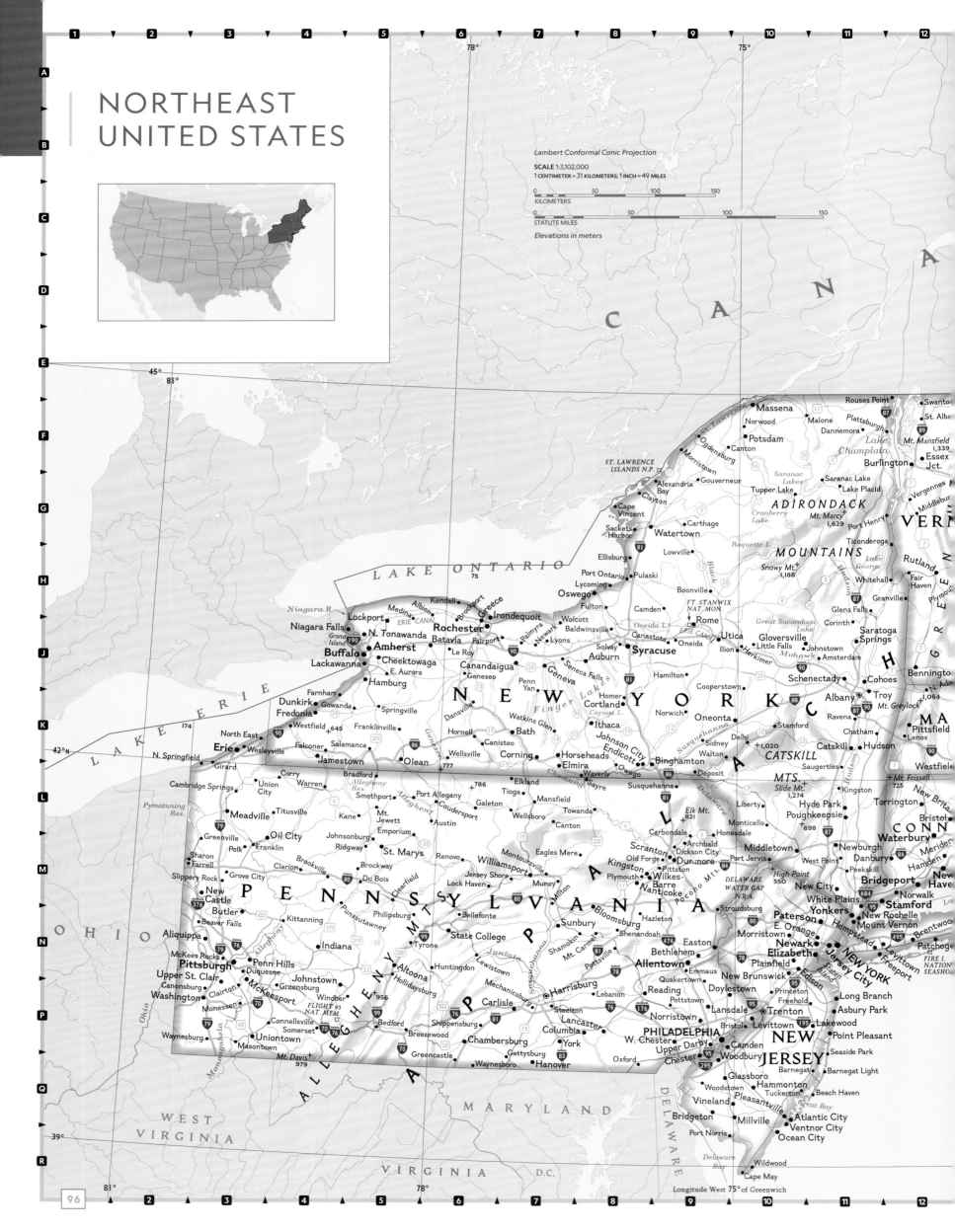

Lambert Conformal Conic Projection

SCALE 1:3,102,000
1 CENTIMETER = 31 KILOMETERS; 1 INCH = 49 MILES

KILOMETERS

STATUTE MILES

Elevations in meters

C A N A D A

LAKE ONTARIO

LAKE ERIE

ADIRONDACK MOUNTAINS

NEW YORK

PENNSYLVANIA

ALLEGHENY MTS.

CATSKILL MTS.

OHIO

NEW JERSEY

MARYLAND

WEST VIRGINIA

VIRGINIA

D.C.

VERMONT

CONN

MASS

Rouses Point
Swanton
St. Albans
Massena
Norwood
Malone
Plattsburgh
Ogdensburg
Canton
Dannemora
Mt. Mansfield 1,339
Potsdam
Essex Jct.
Morristown
Burlington
St. Lawrence Islands N.P.
Gouverneur
Saranac Lake
Alexandria Bay
Lake Placid
Clayton
Tupper Lake
Vergennes
Cape Vincent
Carthage
Mt. Marcy 1,629
Port Henry
Middlebury
Sackets Harbor
Watertown
Ticonderoga
Snowy Mt. 1,188
Lake George
Ellisburg
Lowville
Boonville
FT. STANWIX NAT. MON.
Rutland
Port Ontario
Pulaski
Whitehall
Fair Haven
Lycoming
Camden
Great Sacandaga Lake
Glens Falls
Oswego
Rome
Utica
Corinth
Fulton
Canastota
Gloversville
Saratoga Springs
Baldwinsville
Oneida
Little Falls
Johnstown
Solvay
Ilion
Herkimer
Amsterdam
Syracuse
Auburn
Seneca Falls
Hamilton
Schenectady
Geneva
Cohoes
Bennington
Canandaigua
Cooperstown
Troy
Penn Yan
Homer
Norwich
Albany
Mt. Greylock 1,064
Cortland
Oneonta
Ravena
Stamford
Pittsfield
Watkins Glen
Ithaca
Delhi 1,020
Catskill
Lenox
Dansville
Sidney
Chatham
Hudson
Hornell
Bath
Johnson City
Walton
Saugerties
Westfield
Canisteo
Endicott
Binghamton
Slide Mt. 1,274
Kingston 725
Wellsville
Corning
Horseheads
Owego
Deposit
Liberty
Elmira
Waverly
Monticello
Hyde Park
Torrington
Elkland
Sayre
Honesdale
Poughkeepsie 698
Bristol
786
Tioga
Susquehanna
Elk Mt. 821
Carbondale
Waterbury
Bradford
Galeton
Mansfield
Dickson City
Middletown
Newburgh
Danbury
Meriden
Port Allegany
Scranton
West Point
Hamden
Smethport
Towanda
Old Forge
Dunmore
Peekskill
New Haven
Canton
Pittston
Port Jervis
Coudersport
Wellsboro
High Point 550
Bridgeport
Mt. Jewett
Austin
Williamsport
Kingston
Plymouth
Wilkes-Barre
White Plains
New City
Stamford
Kane
Emporium
Montoursville
Muncy
Nanticoke
DELAWARE WATER GAP N.R.A.
Norwalk
St. Marys
Renovo
Lock Haven
Jersey Shore
Pocono Mts.
Paterson
Yonkers
New Rochelle
Ridgway
Brockway
Stroudsburg
E. Orange
Mount Vernon
Du Bois
Clearfield
Milton
Bloomsburg
Hazleton
Morristown
NEW YORK
Brockway
Philipsburg
Bellefonte
Sunbury
Easton
Newark
Hempstead
Freeport
Shamokin
Shenandoah
Elizabeth
Jersey City
Patchogue
State College
Mt. Carmel
Allentown
Plainfield
FIRE I. NATIONAL SEASHORE
Tyrone
Pottsville
Bethlehem
Emmaus
New Brunswick
Altoona
Huntingdon
Lewistown
Quakertown
Edison
Long Branch
Holidaysburg
Mechanicsburg
Harrisburg
Lebanon
Reading
Doylestown
Princeton
Freehold
FLIGHT 93 NAT. MEM.
956
Windber
Lewistown
Lansdale
Asbury Park
Somerset
Bedford
Carlisle
Pottstown
Trenton
Lakewood
Breezewood
Steelton
Lancaster
Norristown
Levittown
Point Pleasant
Shippensburg
Columbia
York
PHILADELPHIA
Camden
Mt. Davis 979
Chambersburg
Gettysburg
W. Chester
NEW JERSEY
Seaside Park
Greencastle
Hanover
Upper Darby
Woodbury
Barnegat
Waynesboro
Oxford
Chester
Glassboro
Barnegat Light
Woodstown
Hammonton
Beach Haven
Vineland
Pleasantville
Great Bay
Bridgeton
Millville
Atlantic City
Port Norris
Ventnor City
Ocean City
Delaware Bay
Wildwood
Cape May

Erie
N. Springfield
Girard
Wesleyville
Falconer
Salamanca
North East
Jamestown
Olean
Dunkirk
Fredonia
Westfield 645
Franklinville
Gowanda
Springville
Cambridge Springs
Union City
Warren
Titusville
Meadville
Kane
Oil City
Greenville
Johnsonburg
Franklin
Polk
Clarion
Sharon
Farrell
Grove City
Slippery Rock
Brookville
New Castle
Butler
Kittanning
Aliquippa
Indiana
McKees Rocks
Penn Hills
Pittsburgh
Upper St. Clair
Duquesne
Johnstown
Canonsburg
Clairton
Greensburg
Washington
McKeesport
Monessen
Connellsville
Waynesburg
Uniontown
Masontown

Niagara R.
Niagara Falls
Lockport
Medina
Albion
Kendall
Brockport
Greece
Irondequoit
Grand Island
N. Tonawanda
Batavia
Fairport
Palmyra
Newark
Wolcott
Amherst
Rochester
Le Roy
Lyons
Buffalo
Cheektowaga
Lackawanna
E. Aurora
Geneseo
Hamburg
Farnham
ERIE CANAL
174

Pymatuning Res.
Corry
Bradford
Allegheny Res.
Elkland
Sharon
Meadville

Ohio
Monongahela
Allegheny
Juniata
Susquehanna

Lake Champlain
Saranac Lakes
Cranberry Lake
Raquette L.
Black R.
Oneida L.
Mohawk
Hudson
Finger Lakes
Cayuga L.
Seneca L.
Genesee R.
Chemung R.
Delaware R.

45°
81°
42°N
39°
81°
78°
75°
Longitude West 75° of Greenwich

Massachusetts
BAY STATE

AREA	27,336 sq km (10,555 sq mi)
POPULATION	6,902,000
CAPITAL	Boston 669,000
LARGEST CITY	Boston
STATEHOOD	February 6, 1788; 6th state
STATE BIRD	Black-capped chickadee
STATE FLOWER	Mayflower

New Hampshire
GRANITE STATE

AREA	24,216 sq km (9,350 sq mi)
POPULATION	1,356,000
CAPITAL	Concord 43,000
LARGEST CITY	Manchester 111,000
STATEHOOD	June 21, 1788; 9th state
STATE BIRD	Purple finch
STATE FLOWER	Purple lilac

New Jersey
GARDEN STATE

AREA	22,588 sq km (8,271 sq mi)
POPULATION	8,909,000
CAPITAL	Trenton 85,000
LARGEST CITY	Newark 283,000
STATEHOOD	December 18, 1787; 3rd state
STATE BIRD	American goldfinch
STATE FLOWER	Violet

New York
EMPIRE STATE

AREA	141,299 sq km (54,556 sq mi)
POPULATION	19,542,000
CAPITAL	Albany 98,000
LARGEST CITY	New York 8,560,000
STATEHOOD	July 28, 1788; 11th state
STATE BIRD	Eastern bluebird
STATE FLOWER	Rose

Pennsylvania
KEYSTONE STATE

AREA	119,283 sq km (46,055 sq mi)
POPULATION	12,807,000
CAPITAL	Harrisburg 49,000
LARGEST CITY	Philadelphia 1,570,000
STATEHOOD	December 12, 1787; 2nd state
STATE BIRD	Ruffed grouse
STATE FLOWER	Mountain laurel

Rhode Island
OCEAN STATE

AREA	4,002 sq km (1,545 sq mi)
POPULATION	1,057,000
CAPITAL	Providence 180,000
LARGEST CITY	Providence
STATEHOOD	May 29, 1790; 13th state
STATE BIRD	Rhode Island red
STATE FLOWER	Violet

Connecticut
CONSTITUTION STATE

AREA	14,357 sq km (5,543 sq mi)
POPULATION	3,573,000
CAPITAL	Hartford 124,000
LARGEST CITY	Bridgeport 148,000
STATEHOOD	January 9, 1788; 5th state
STATE BIRD	American robin
STATE FLOWER	Mountain laurel

Maine
PINE TREE STATE

AREA	91,646 sq km (35,385 sq mi)
POPULATION	1,338,000
CAPITAL	Augusta 19,000
LARGEST CITY	Portland 67,000
STATEHOOD	March 15, 1820; 23rd state
STATE BIRD	Black-capped chickadee
STATE FLOWER	White pine cone and tassel

Vermont
GREEN MOUNTAIN STATE

AREA	24,901 sq km (9,614 sq mi)
POPULATION	626,000
CAPITAL	Montpelier 8,000
LARGEST CITY	Burlington 43,000
STATEHOOD	March 4, 1791; 14th state
STATE BIRD	Hermit thrush
STATE FLOWER	Red clover

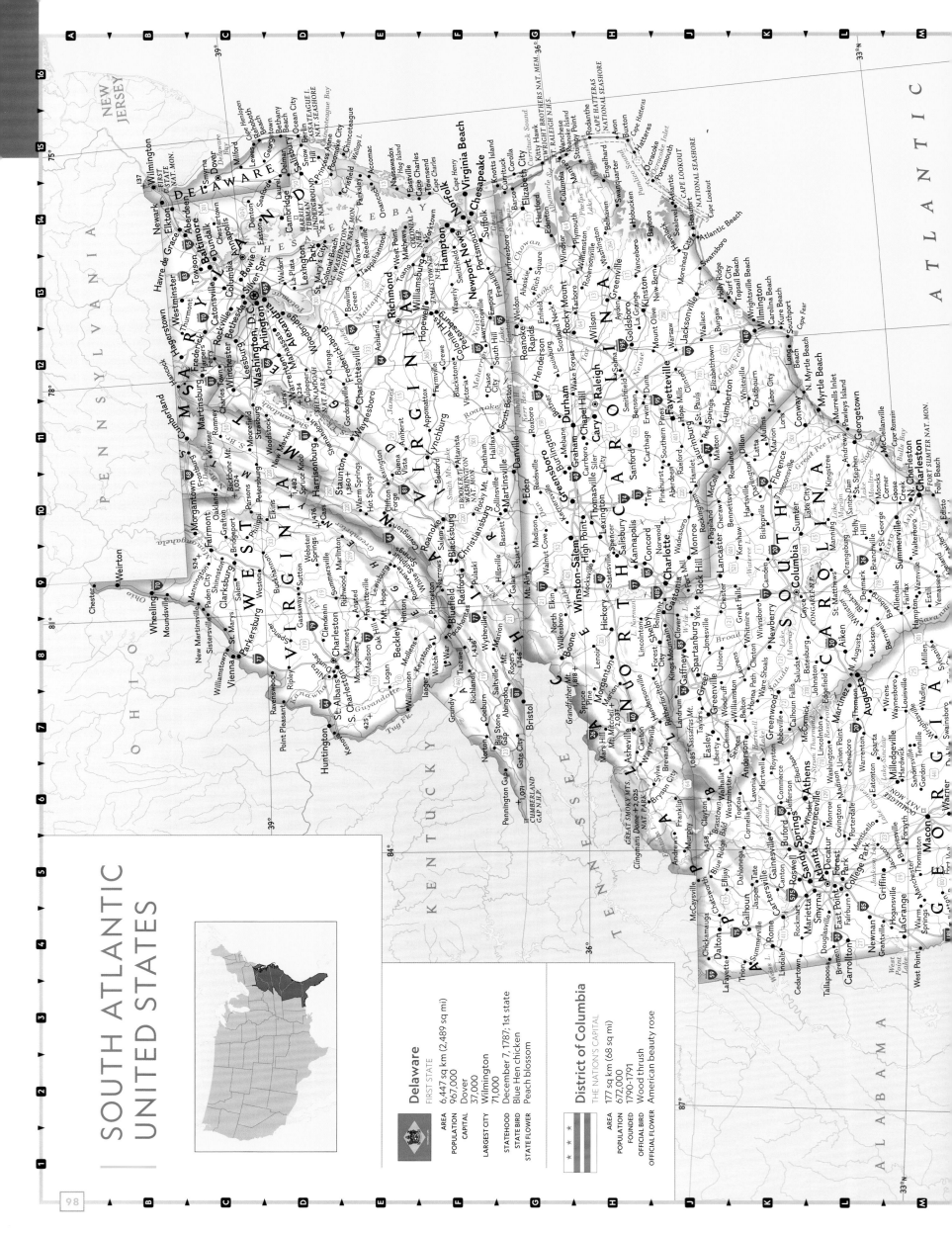

SOUTH ATLANTIC
UNITED STATES

Delaware
FIRST STATE

AREA	6,447 sq km (2,489 sq mi)
POPULATION	967,000
CAPITAL	Dover
	37,000
LARGEST CITY	Wilmington
	71,000
STATEHOOD	December 7, 1787, 1st state
STATE BIRD	Blue Hen chicken
STATE FLOWER	Peach blossom

District of Columbia
THE NATION'S CAPITAL

AREA	177 sq km (68 sq mi)
POPULATION	672,000
FOUNDED	1790–1791
OFFICIAL BIRD	Wood thrush
OFFICIAL FLOWER	American beauty rose

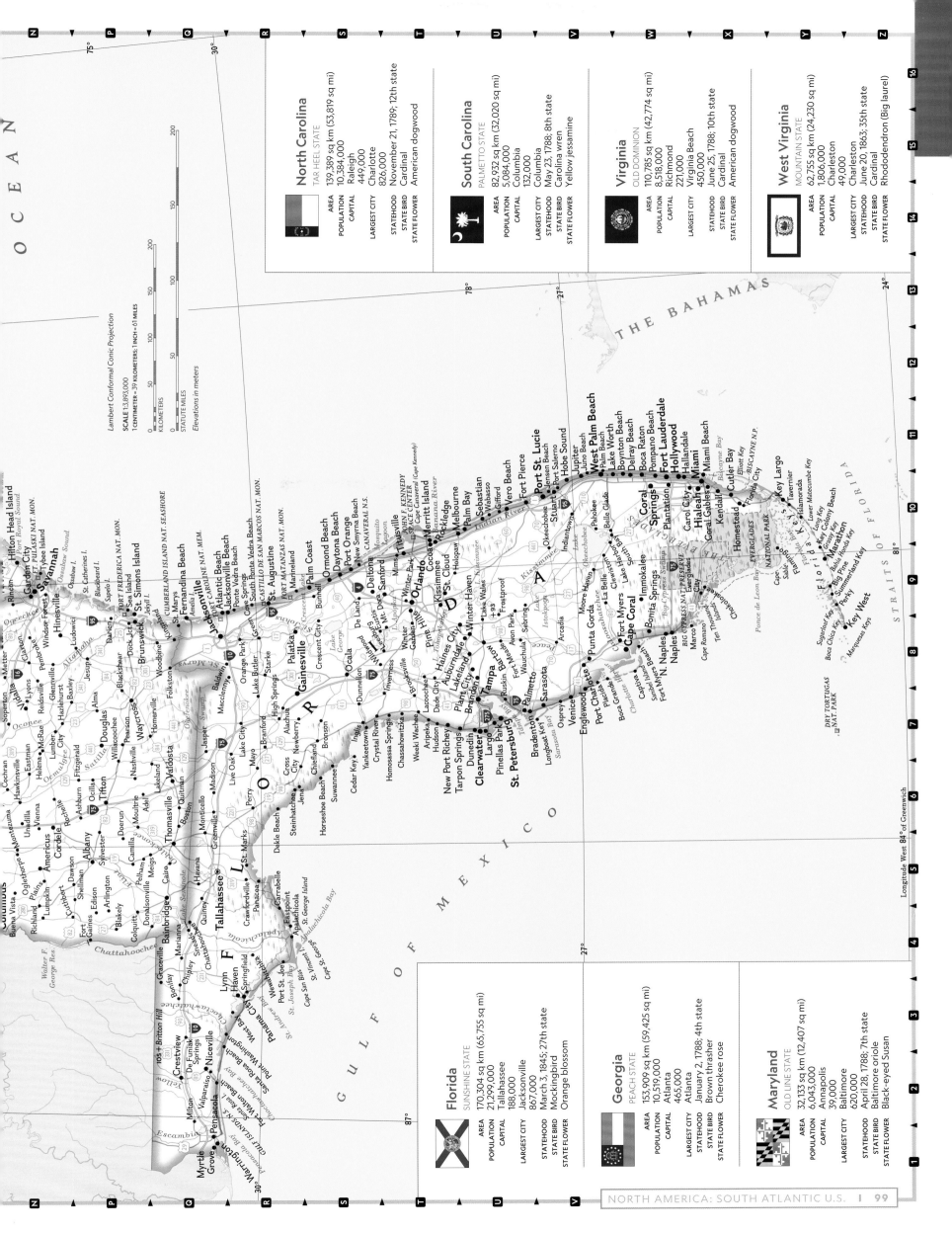

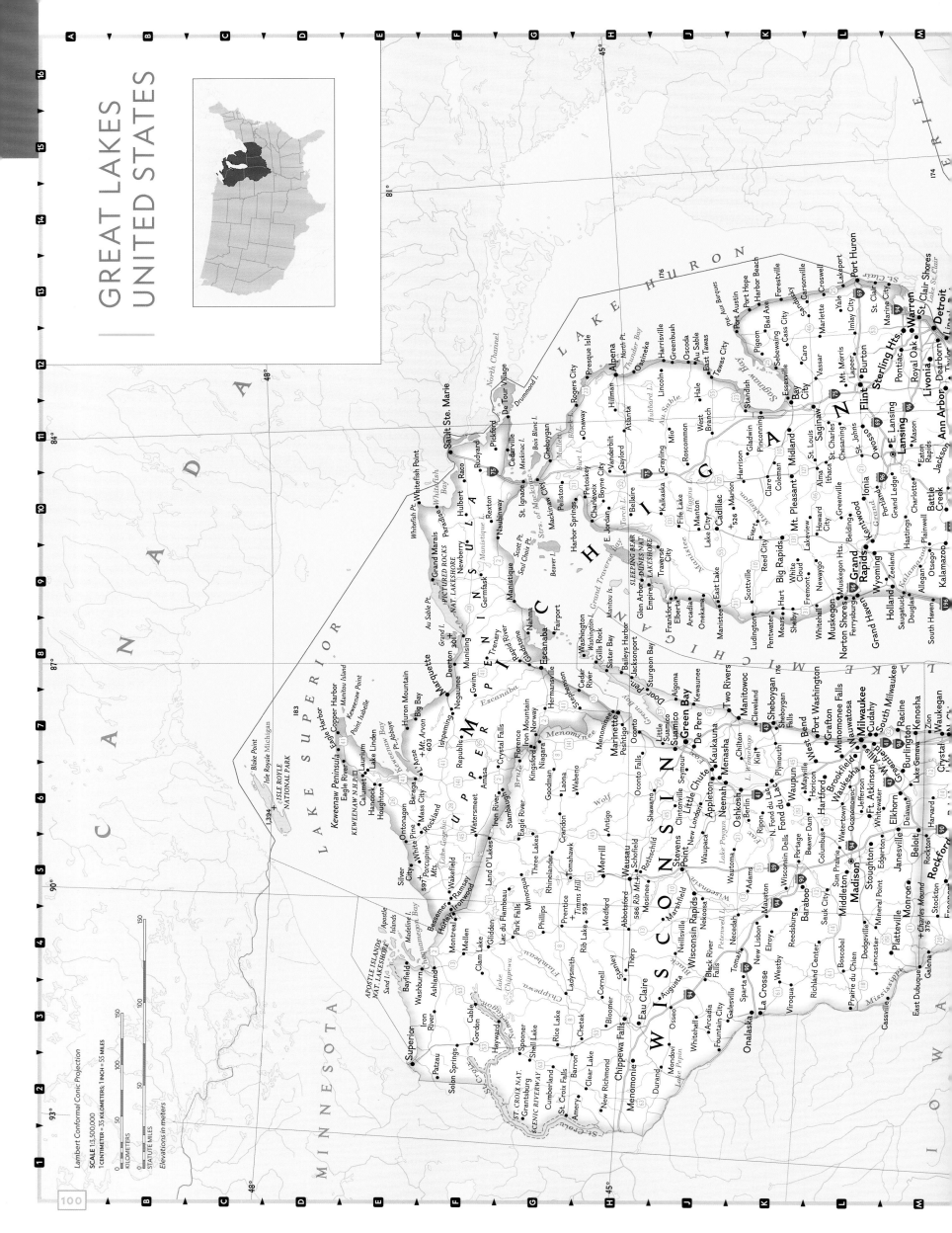

CANADA

MINNESOTA

LAKE SUPERIOR

ISLE ROYALE
NATIONAL PARK
1,394

Blake Point

183

APOSTLE ISLANDS
NAT. LAKESHORE
Sand I.
Madeline I.
Chequamegon Bay

Bayfield
Washburn
Ashland
Iron River
Superior
Patzau
Solon Springs

Cable
Gordon
Hayward
Clam Lake
Mellen

Spooner
Shell Lake
Rice Lake
Chetek

ST. CROIX NAT.
SCENIC RIVERWAY

Grantsburg
Cumberland
St. Croix Falls
Amery
New Richmond

Barron
Clear Lake

Bloomer
Chippewa Falls
Eau Claire
Menomonie
Durand

Mondovi
Osseo
Whitehall
Arcadia
Fountain City
Galesville
La Crosse

Onalaska

WISCONSIN

Lake Gogebic
Ontonagon
White Pine
Silver City
Porcupine Mts.
597
Wakefield
Bessemer
Hurley
Ironwood
Ramsay
Montreal

Mass City
Rockland

Baraga
L'Anse
Watersmeet
Land O'Lakes

Keweenaw Peninsula
Eagle River
Copper Harbor
Eagle Harbor
Keweenaw Point
Point Isabelle
Calumet
Laurium
Hancock
Houghton
Lake Linden
KEWEENAW N.H.P.

Huron Mountain
Big Bay
+Mt. Arvon
603

UPPER

Ishpeming
Negaunee
Marquette
Deerton
Gwinn

Republic
Amasa
Iron River
Stambaugh
Crystal Falls
Kingsford
Norway
Iron Mountain
Niagara

Goodman
Wabeno
Laona

MICHIGAN

PENINSULA

Au Sable Pt.
Grand I.
Munising
PICTURED ROCKS
NAT. LAKESHORE
301
Grand Marais
Trenary
Rapid River
Nahma
Trenary
Germfask

Whitefish Pt.
Whitefish Point
Whitefish Bay
Paradise
Newberry
Raco
Hulbert

Sault Ste. Marie
Pickford
Rudyard
De Tour Village
North Channel
Drummond I.

Escanaba
Gladstone
Fairport
Washington
Gills Rock
Sister Bay
Baileys Harbor

Menominee
Marinette
Peshtigo
Stephenson
Cedar River
Hermansville
Powers

Oconto Falls
Oconto
Little Suamico
Suamico
Seymour

Green Bay
De Pere
Kaukauna
Appleton
Neenah
Menasha

Two Rivers
Manitowoc
Cleveland
Kewaunee
Algoma

LAKE MICHIGAN

SCALE 1:3,500,000
1 CENTIMETER = 35 KILOMETERS; 1 INCH = 55 MILES
Lambert Conformal Conic Projection

KILOMETERS
STATUTE MILES
Elevations in meters

MICHIGAN

LAKE HURON

North Channel

St. Ignace
Mackinac City
Mackinac I.
Bois Blanc I.
Cheboygan
Cedarville

Harbor Springs
Petoskey
Boyne City
Charlevoix
Beaver I.
East Jordan
Bellaire
Torch L.

Presque Isle
Rogers City
Onaway
Atlanta
Hillman
Alpena
Ossineke
Harrisville
Greenbush
Oscoda
Au Sable
East Tawas
Tawas City

Gaylord
Vanderbilt
Grayling
Mio
Roscommon
West Branch
Standish

Higgins L.
Houghton L.
Hale
Lincoln

Kalkaska
Fife Lake
Manton
Cadillac
+526
Evart
Marion
Clare
Harrison
Gladwin
Coleman
Midland
Pinconning
Bay City
Saginaw

Traverse City
SLEEPING BEAR
DUNES NAT.
LAKESHORE
Glen Arbor
Empire
Frankfort
Elberta
Arcadia
Onekama
Manistee
Ludington
Pentwater
Mears
Shelby

Lake City
Reed City
Big Rapids
White Cloud
Newaygo
Fremont
Hart

Scottville
Whitehall
Muskegon Hts.
Muskegon
Norton Shores
Ferrysburg
Grand Haven
Holland
Saugatuck
Douglas
South Haven

Howard City
Greenville
Belding
Kentwood
Grand Rapids
Wyoming
Zeeland
Allegan
Otsego
Kalamazoo

St. Louis
Alma
Ithaca
St. Charles
Chesaning
Owosso
St. Johns
Ionia
Portland
Grand Ledge
Charlotte
Eaton Rapids
Battle Creek

Mt. Morris
Flint
Burton
Lansing
E. Lansing
Mason
Jackson

WISCONSIN

Florence
Crandon
Three Lakes
Eagle River
Rhinelander
Tomahawk
Minocqua
Lac du Flambeau
Park Falls
Phillips
Glidden

Prentice
+Timms Hill
595
Medford
Rib Lake
Ladysmith
Cornell
Stanley
Thorp
Abbotsford
+Rib Mt.
586
Wausau
Schofield
Rothschild
Mosinee
Merrill
Antigo
Shawano
Clintonville
Marshfield
Neillsville

Wisconsin Rapids
Stevens Point
Waupaca
New London
Clintonville

Black River Falls
Augusta

Sparta
Tomah
New Lisbon
Mauston
Necedah
Adams
Wautoma
Wisconsin Dells
Baraboo
Reedsburg
Richland Center
Viroqua
Westby
Cashton

Prairie du Chien
Boscobel
Lancaster
Platteville
+Charles Mound
376
Cassville
East Dubuque
Galena

Mineral Point
Dodgeville
Sauk City
Sun Prairie
Middleton
Madison
Stoughton
Edgerton
Monroe
Janesville
Beloit
Rockton
Rockford

L. Winnebago
Oshkosh
Ripon
Berlin
Fond du Lac
N. Fond du Lac
Waupun
Beaver Dam
Columbus
Portage

Plymouth
Sheboygan Falls
Sheboygan
Kiel
Chilton
Kaukauna
Little Chute

West Bend
Grafton
Port Washington
Cedarburg
Menomonee Falls
Mayville
Horicon
Hartford
Jackson

Waukesha
Brookfield
Wauwatosa
Milwaukee
Cudahy
South Milwaukee
Oconomowoc
Watertown
Jefferson
Ft. Atkinson
Whitewater
Elkhorn
Delavan
Lake Geneva
Burlington
Harvard

Racine
Kenosha
Waukegan
Zion

IOWA

ILLINOIS

LAKE ST. CLAIR

Port Huron
Marysville
St. Clair
Marine City
New Baltimore
Mt. Clemens
St. Clair Shores
Warren
Sterling Hts.
Royal Oak
Livonia
Dearborn
Detroit
Pontiac
Ann Arbor

Croswell
Lexington
Yale
Imlay City
Lapeer
Vassar
Caro
Cass City
Sebewaing
Bad Axe
Harbor Beach
Port Hope
Port Austin
Pte. Aux Barques
Pigeon
Sandusky
Marlette
Carsonville
Forestville
Caseville

Saginaw Bay
Essexville
Bay City

174
100

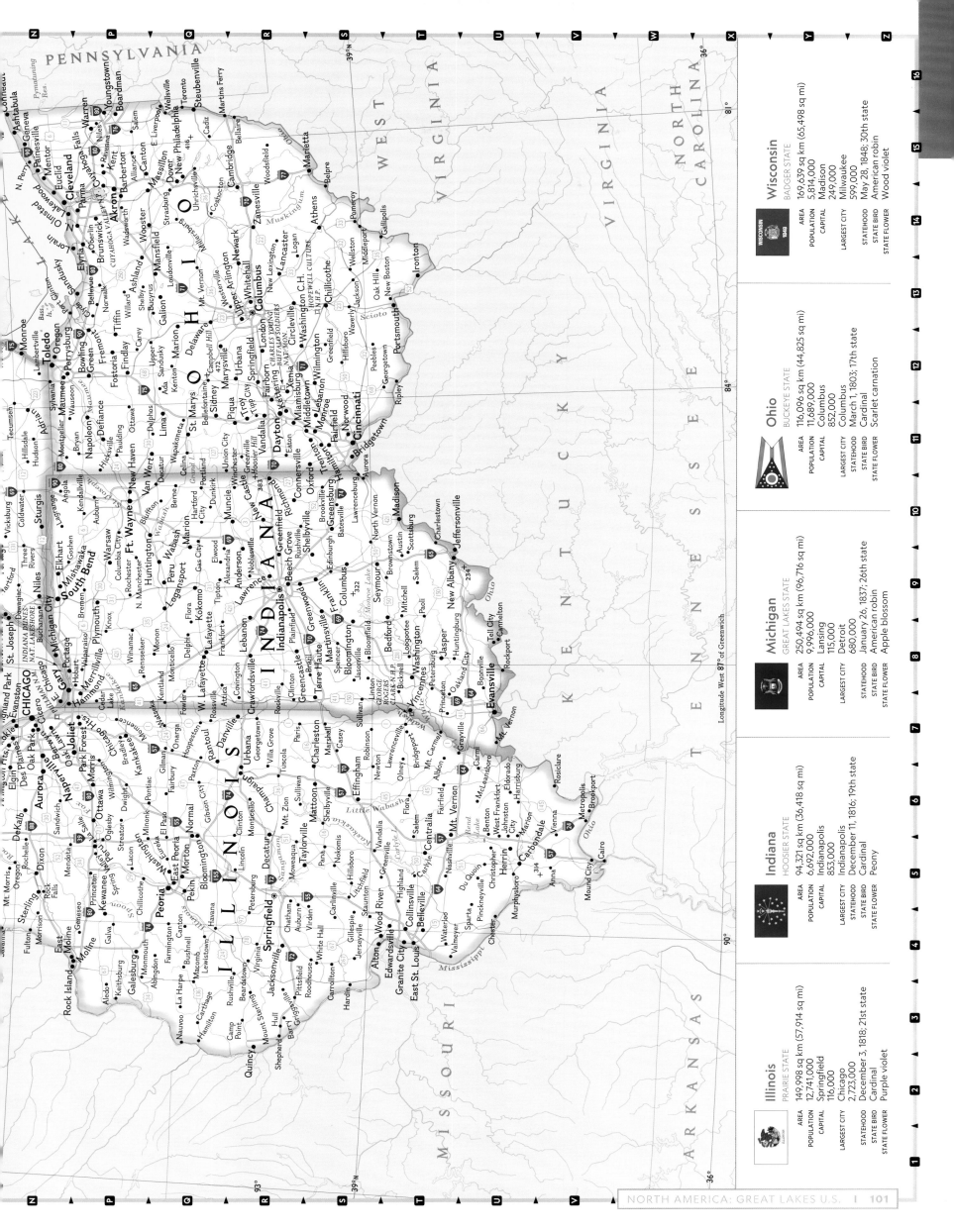

Illinois — PRAIRIE STATE
- AREA 149,998 sq km (57,914 sq mi)
- POPULATION 12,741,000
- CAPITAL Springfield 116,000
- LARGEST CITY Chicago 2,723,000
- STATEHOOD December 3, 1818; 21st state
- STATE BIRD Cardinal
- STATE FLOWER Purple violet

Indiana — HOOSIER STATE
- AREA 94,321 sq km (36,418 sq mi)
- POPULATION 6,692,000
- CAPITAL Indianapolis 853,000
- LARGEST CITY Indianapolis
- STATEHOOD December 11, 1816; 19th state
- STATE BIRD Cardinal
- STATE FLOWER Peony

Michigan — GREAT LAKES STATE
- AREA 250,494 sq km (96,716 sq mi)
- POPULATION 9,996,000
- CAPITAL Lansing 115,000
- LARGEST CITY Detroit 680,000
- STATEHOOD January 26, 1837; 26th state
- STATE BIRD American robin
- STATE FLOWER Apple blossom

Ohio — BUCKEYE STATE
- AREA 116,096 sq km (44,825 sq mi)
- POPULATION 11,689,000
- CAPITAL Columbus 852,000
- LARGEST CITY Columbus
- STATEHOOD March 1, 1803; 17th state
- STATE BIRD Cardinal
- STATE FLOWER Scarlet carnation

Wisconsin — BADGER STATE
- AREA 169,639 sq km (65,498 sq mi)
- POPULATION 5,814,000
- CAPITAL Madison 249,000
- LARGEST CITY Milwaukee 599,000
- STATEHOOD May 28, 1848; 30th state
- STATE BIRD American robin
- STATE FLOWER Wood violet

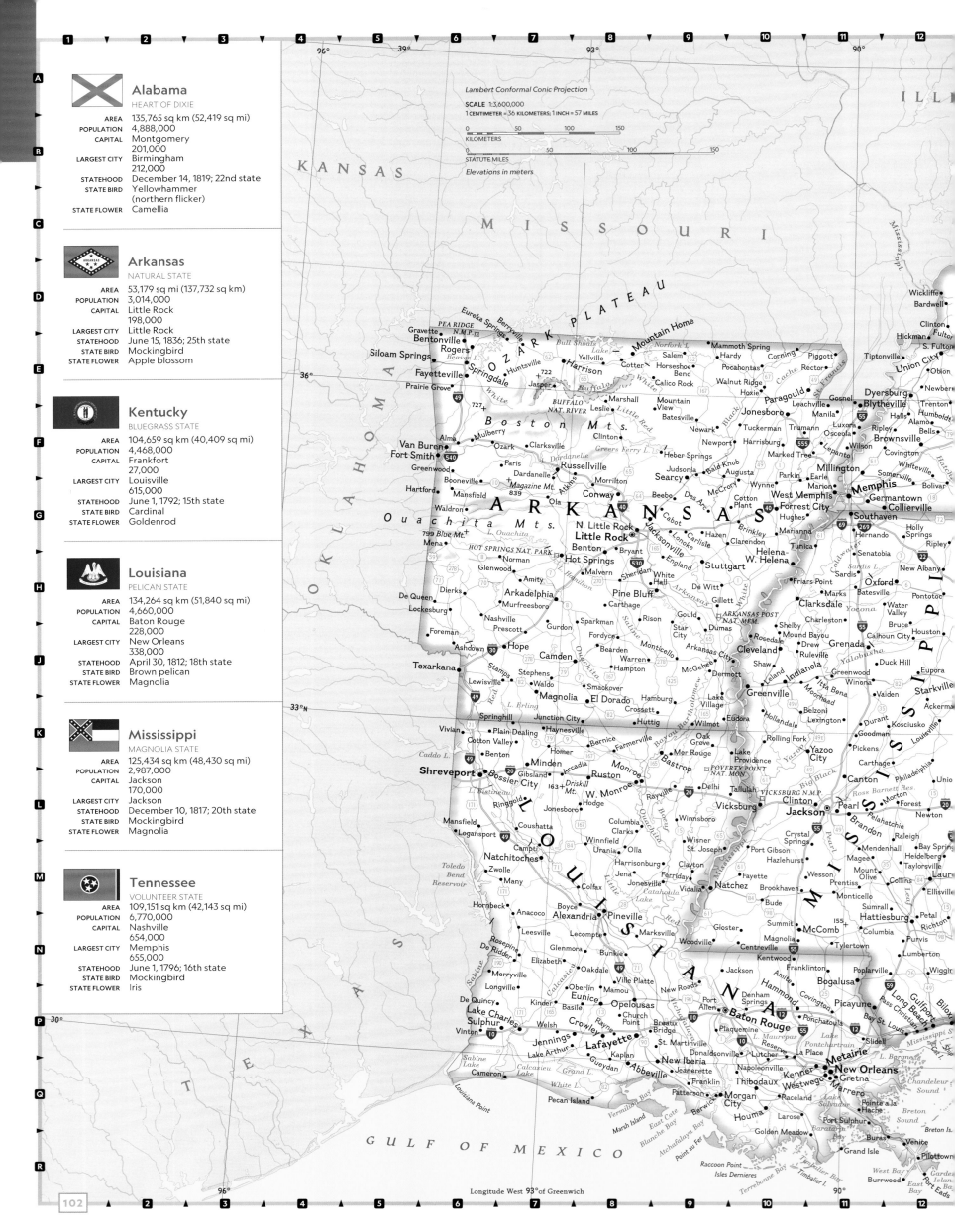

Alabama
HEART OF DIXIE

AREA	135,765 sq km (52,419 sq mi)
POPULATION	4,888,000
CAPITAL	Montgomery 201,000
LARGEST CITY	Birmingham 212,000
STATEHOOD	December 14, 1819; 22nd state
STATE BIRD	Yellowhammer (northern flicker)
STATE FLOWER	Camellia

Arkansas
NATURAL STATE

AREA	53,179 sq mi (137,732 sq km)
POPULATION	3,014,000
CAPITAL	Little Rock 198,000
LARGEST CITY	Little Rock
STATEHOOD	June 15, 1836; 25th state
STATE BIRD	Mockingbird
STATE FLOWER	Apple blossom

Kentucky
BLUEGRASS STATE

AREA	104,659 sq km (40,409 sq mi)
POPULATION	4,468,000
CAPITAL	Frankfort 27,000
LARGEST CITY	Louisville 615,000
STATEHOOD	June 1, 1792; 15th state
STATE BIRD	Cardinal
STATE FLOWER	Goldenrod

Louisiana
PELICAN STATE

AREA	134,264 sq km (51,840 sq mi)
POPULATION	4,660,000
CAPITAL	Baton Rouge 228,000
LARGEST CITY	New Orleans 338,000
STATEHOOD	April 30, 1812; 18th state
STATE BIRD	Brown pelican
STATE FLOWER	Magnolia

Mississippi
MAGNOLIA STATE

AREA	125,434 sq km (48,430 sq mi)
POPULATION	2,987,000
CAPITAL	Jackson 170,000
LARGEST CITY	Jackson
STATEHOOD	December 10, 1817; 20th state
STATE BIRD	Mockingbird
STATE FLOWER	Magnolia

Tennessee
VOLUNTEER STATE

AREA	109,151 sq km (42,143 sq mi)
POPULATION	6,770,000
CAPITAL	Nashville 654,000
LARGEST CITY	Memphis 655,000
STATEHOOD	June 1, 1796; 16th state
STATE BIRD	Mockingbird
STATE FLOWER	Iris

Lambert Conformal Conic Projection

SCALE 1:3,600,000
1 CENTIMETER = 36 KILOMETERS; 1 INCH = 57 MILES

Elevations in meters

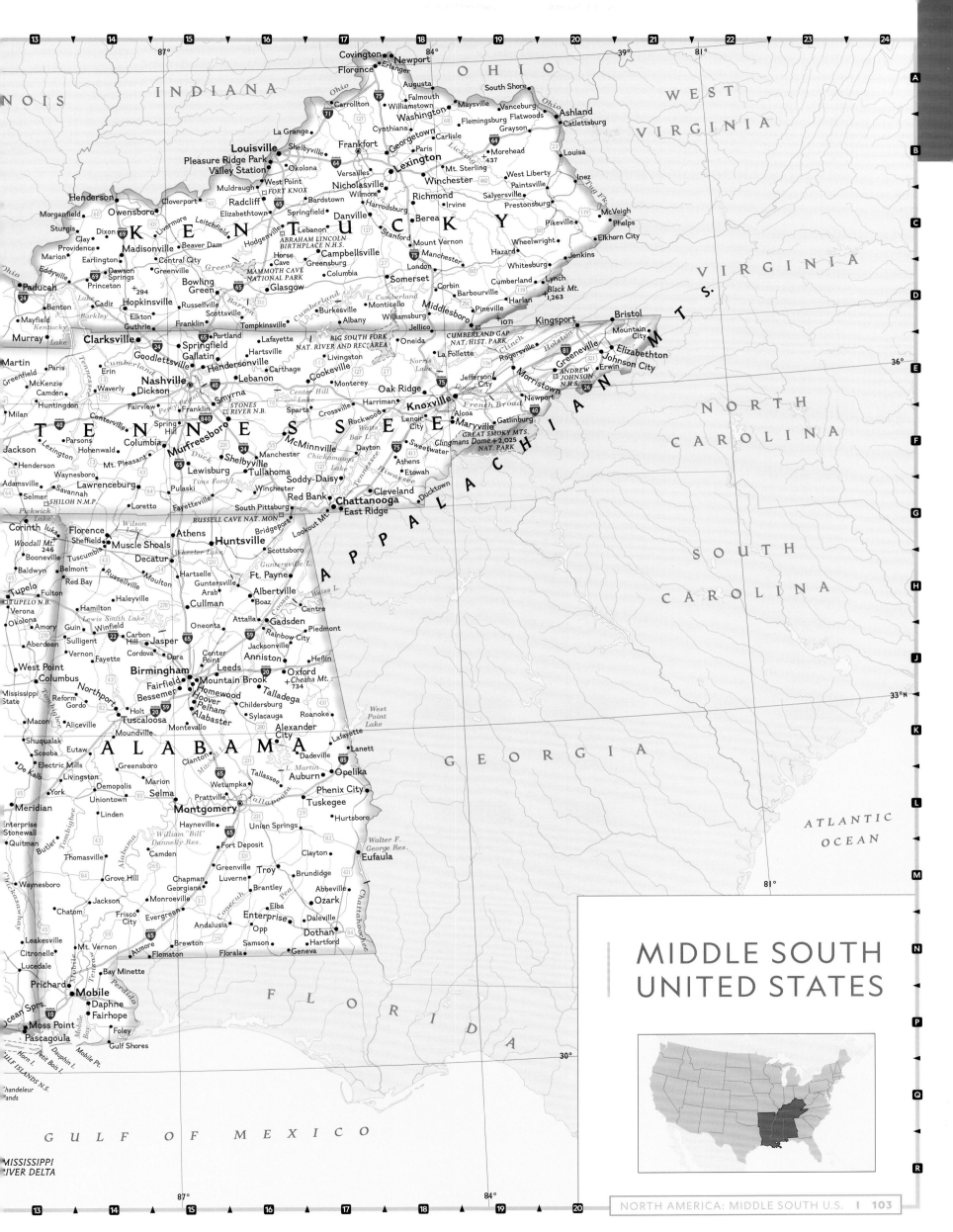

MIDDLE SOUTH UNITED STATES

TEXAS AND OKLAHOMA
UNITED STATES

COLORADO

NEW MEXICO

ARIZONA

108° 105° 102°

36°

33°

111°

30°N

111°

27°

108° 105° 102°

Lambert Conformal Conic Projection

SCALE 1:4,070,000
1 CENTIMETER = 41 KILOMETERS; 1 INCH = 64 MILES

0 50 100 150
KILOMETERS

0 50 100 150
STATUTE MILES

Elevations in meters

Oklahoma
SOONER STATE

AREA — 181,036 sq km
(69,898 sq mi)
POPULATION — 3,943,000
CAPITAL — Oklahoma City
629,000
LARGEST CITY — Oklahoma City
STATEHOOD — November 16, 1907;
46th state
STATE BIRD — Scissor-tailed flycatcher
STATE FLOWER — Mistletoe

Texas
LONE STAR STATE

AREA — 695,621 sq km
(268,581 sq mi)
POPULATION — 28,702,000
CAPITAL — Austin
917,000
LARGEST CITY — Houston
2,267,000
STATEHOOD — December 29, 1845;
28th state
STATE BIRD — Mockingbird
STATE FLOWER — Bluebonnet

MEXICO

Map labels (Texas panhandle and west Texas)

Black Mesa 1,516
Keyes
Boise City
Guymon
HIGH
Texline
Coldwater Cr.
Texhoma
Stratford
Gruver
Dalhart
Cactus
Sunray
1,356
Hartley
Dumas
Stinnett
Borger
Lake Meredith
Fritch
LAKE MEREDITH N.R.A.
ALIBATES FLINT QUARRIES NAT. MON.
Adrian
Vega
Panhandle
Amarillo
Claude
Canyon
Hereford
Happy
Friona
Dimmitt
Tulia
Silverton
Bovina
Hart
Kress
Farwell
Earth
PLAINS
Muleshoe
Olton
Plainview
Sudan
Amherst
Hale Center
Littlefield
Anton
Floydada
Morton
Shallowater
Abernathy
Levelland
Idalou
LLANO
Sundown
Wolfforth
Lubbock
ESTACADO
Ropesville
Slaton
Meadow
Post
(STAKED PLAIN)
Plains
Tahoka
Brownfield
Denver City
Sulphur Springs Draw
O'Donnell
Seagraves
Lamesa
Seminole
Andrews
Coahoma
Big Spring
Stanton
Canutillo 2,192
San Antonio Mt. 2,143
Guadalupe Mts.
GUADALUPE MTS. N.P.
Red Bluff L.
Goldsmith
El Paso
CHAMIZAL NAT. MEM.
Guadalupe Peak 2,667
Pecos
Kermit
Odessa
Midland
Socorro
Clint
1,015
Wink
Penwell
St. Lawrence
Fabens
Tornillo
Pecos
Wickett
Monahans
Fort Hancock
2,100 Sierra Blanca
Toyah
20
Pyote
T
Crane
Stiles
Sierra Blanca
Grandfalls
957
Rankin
Eagle Peak 2,285
Van Horn
Kent
Balmorhea
King Mt.
McCamey
Big Lake
1,725 Boracho Pk.
Toyahvale
Fort Stockton
10
Iraan
Mt. Livermore 2,555
Davis Mts.
FORT DAVIS N.H.S.
Valentine
Fort Davis
1,402
Sheffield
Alpine
Glass Mts.
Marfa
2,091
Marathon
Big Canyon
Chinati Peak 2,353
Cathedral Mt.
1,663
Sanderson
Ruidosa
Dryden
Shafter
Santiago Peak 1,988
RIO GRANDE WILD & SCENIC RIVER
Pecos
Presidio
Chisos Mts.
1,784
Amistad Res.
Lajitas
Emory Peak 2,385
BIG BEND NATIONAL PARK
Langtry
Rio Grande

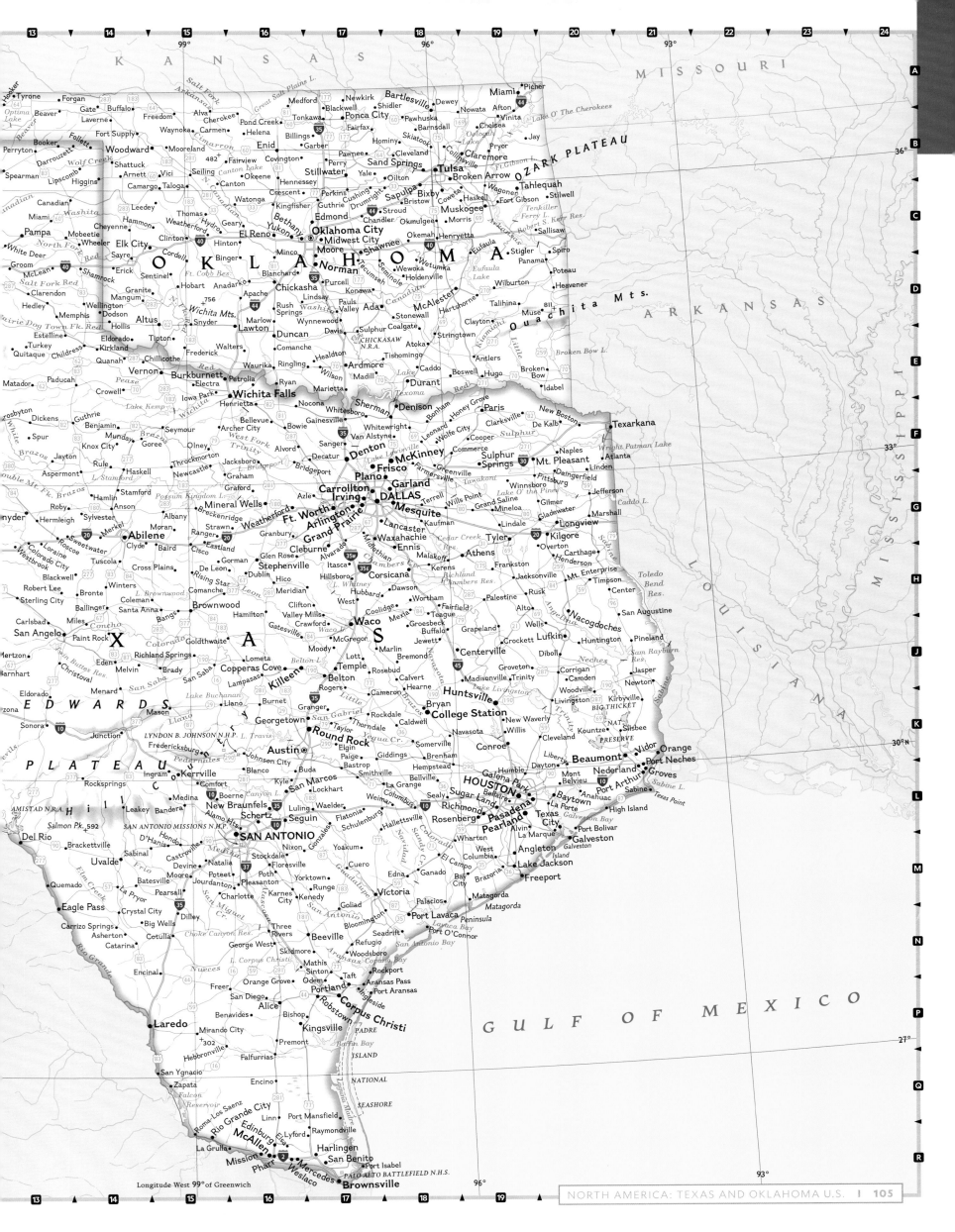

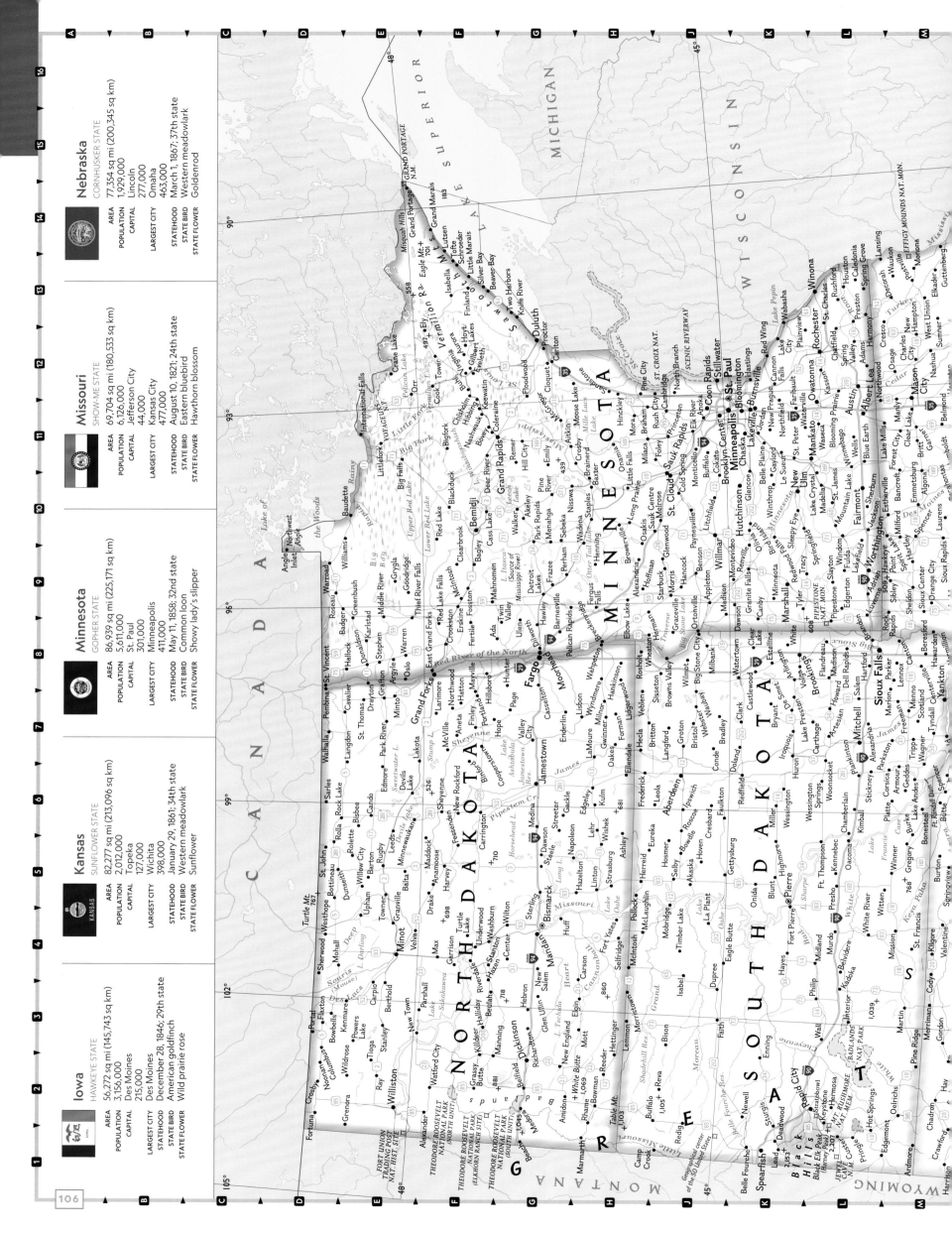

Iowa

HAWKEYE STATE

AREA	56,272 sq mi (145,743 sq km)
POPULATION	3,156,000
CAPITAL	Des Moines
	215,000
LARGEST CITY	Des Moines
STATEHOOD	December 28, 1846; 29th state
STATE BIRD	American goldfinch
STATE FLOWER	Wild prairie rose

Kansas

SUNFLOWER STATE

AREA	82,277 sq mi (213,096 sq km)
POPULATION	2,012,000
CAPITAL	Topeka
	127,000
LARGEST CITY	Wichita
	398,000
STATEHOOD	January 29, 1861; 34th state
STATE BIRD	Western meadowlark
STATE FLOWER	Sunflower

Minnesota

GOPHER STATE

AREA	86,939 sq mi (225,171 sq km)
POPULATION	5,611,000
CAPITAL	St. Paul
	301,000
LARGEST CITY	Minneapolis
	411,000
STATEHOOD	May 11, 1858; 32nd state
STATE BIRD	Common loon
STATE FLOWER	Showy lady's slipper

Missouri

SHOW-ME STATE

AREA	69,704 sq mi (180,533 sq km)
POPULATION	6,126,000
CAPITAL	Jefferson City
	44,000
LARGEST CITY	Kansas City
	477,000
STATEHOOD	August 10, 1821; 24th state
STATE BIRD	Eastern bluebird
STATE FLOWER	Hawthorn blossom

Nebraska

CORNHUSKER STATE

AREA	77,354 sq mi (200,345 sq km)
POPULATION	1,929,000
CAPITAL	Lincoln
	277,000
LARGEST CITY	Omaha
	463,000
STATEHOOD	March 1, 1867; 37th state
STATE BIRD	Western meadowlark
STATE FLOWER	Goldenrod

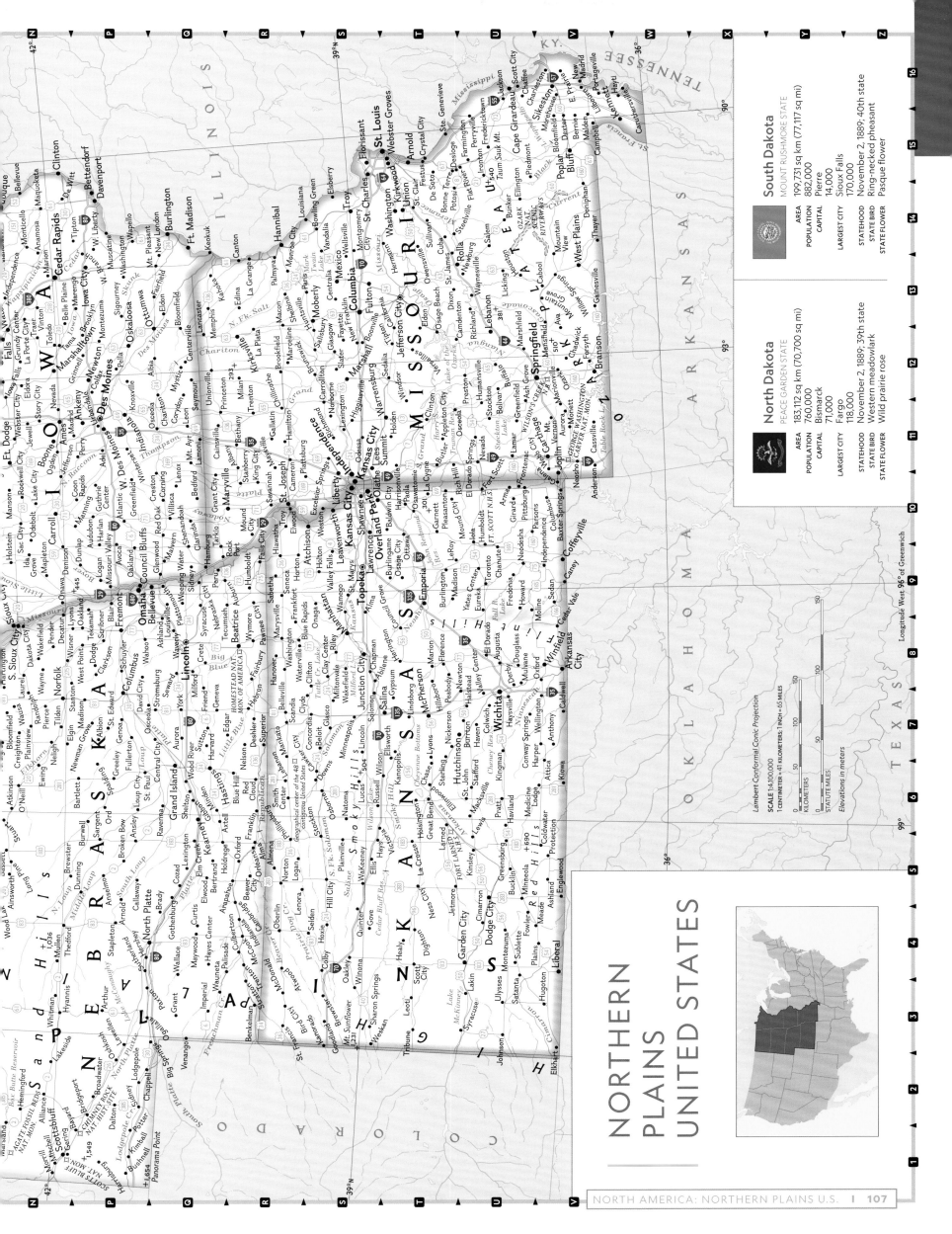

South Dakota
MOUNT RUSHMORE STATE

AREA	199,731 sq km (77,117 sq mi)
POPULATION	882,000
CAPITAL	Pierre
	14,000
LARGEST CITY	Sioux Falls
	170,000
STATEHOOD	November 2, 1889; 40th state
STATE BIRD	Ring-necked pheasant
STATE FLOWER	Pasque flower

North Dakota
PEACE GARDEN STATE

AREA	183,112 sq km (70,700 sq mi)
POPULATION	760,000
CAPITAL	Bismarck
	71,000
LARGEST CITY	Fargo
	118,000
STATEHOOD	November 2, 1889; 39th state
STATE BIRD	Western meadowlark
STATE FLOWER	Wild prairie rose

Lambert Conformal Conic Projection

SCALE 1:4,100,000
1 CENTIMETER = 41 KILOMETERS; 1 INCH = 65 MILES

KILOMETERS
STATUTE MILES

Elevations in meters

NORTHERN
PLAINS
UNITED STATES

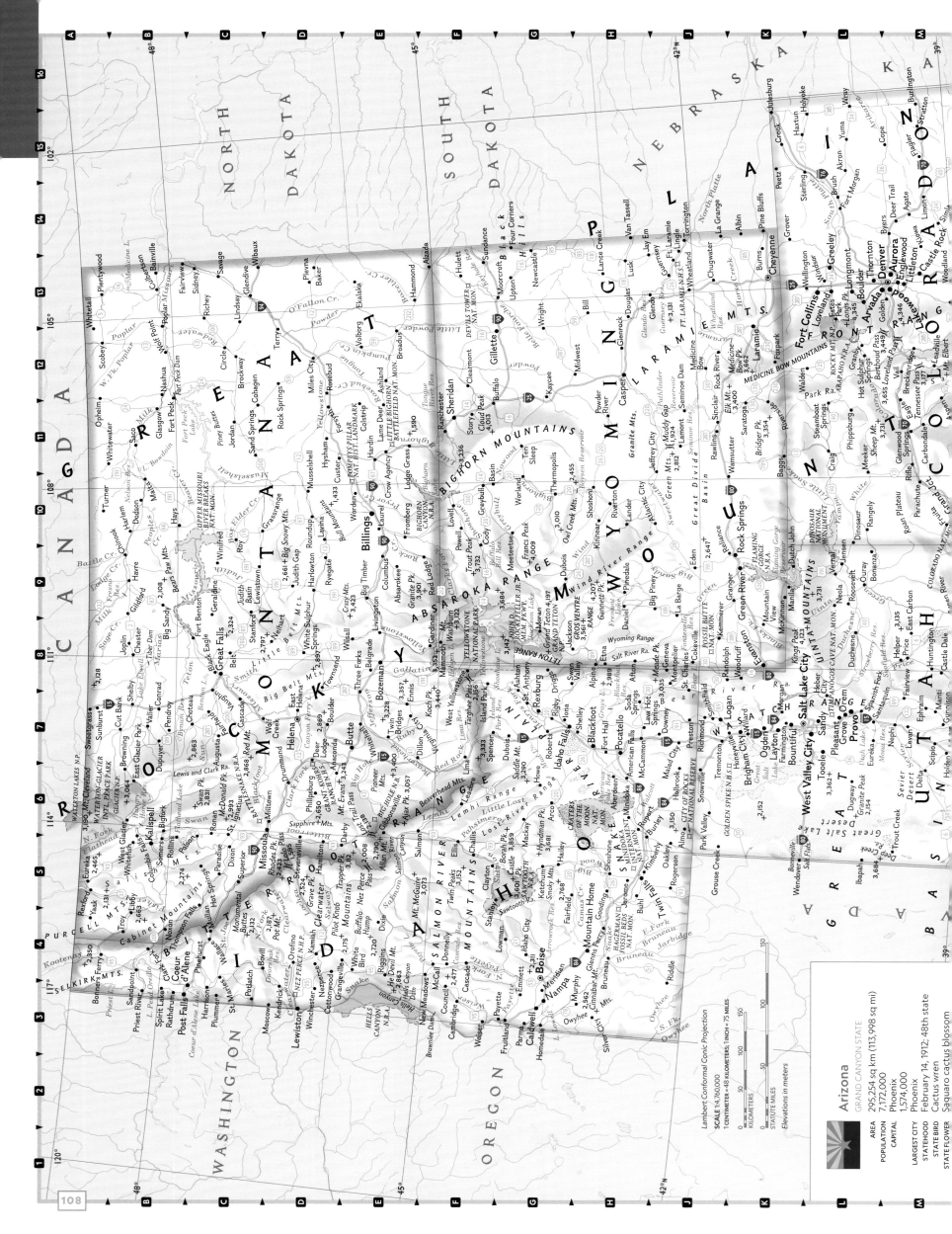

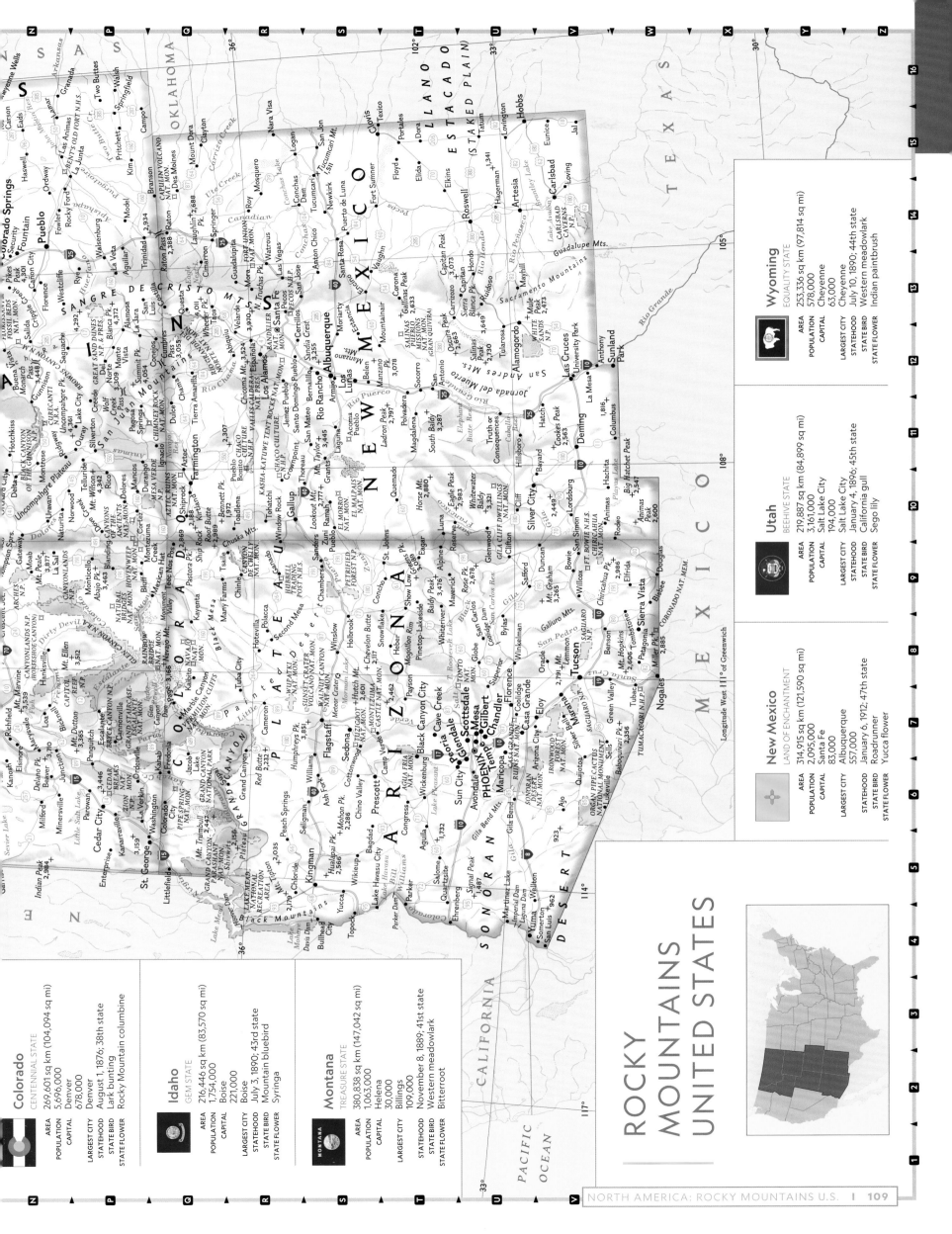

Colorado
CENTENNIAL STATE

AREA	269,601 sq km (104,094 sq mi)
POPULATION	5,696,000
CAPITAL	Denver
	678,000
LARGEST CITY	Denver
STATEHOOD	August 1, 1876; 38th state
STATE BIRD	Lark bunting
STATE FLOWER	Rocky Mountain columbine

Idaho
GEM STATE

AREA	216,446 sq km (83,570 sq mi)
POPULATION	1,754,000
CAPITAL	Boise
	221,000
LARGEST CITY	Boise
STATEHOOD	July 3, 1890; 43rd state
STATE BIRD	Mountain bluebird
STATE FLOWER	Syringa

Montana
TREASURE STATE

AREA	380,838 sq km (147,042 sq mi)
POPULATION	1,063,000
CAPITAL	Helena
	30,000
LARGEST CITY	Billings
	109,000
STATEHOOD	November 8, 1889; 41st state
STATE BIRD	Western meadowlark
STATE FLOWER	Bitterroot

New Mexico
LAND OF ENCHANTMENT

AREA	314,915 sq km (121,590 sq mi)
POPULATION	2,095,000
CAPITAL	Santa Fe
	83,000
LARGEST CITY	Albuquerque
	557,000
STATEHOOD	January 6, 1912; 47th state
STATE BIRD	Roadrunner
STATE FLOWER	Yucca flower

Utah
BEEHIVE STATE

AREA	219,887 sq km (84,899 sq mi)
POPULATION	3,161,000
CAPITAL	Salt Lake City
	194,000
LARGEST CITY	Salt Lake City
STATE BIRD	California gull
STATE FLOWER	Sego lily

Wyoming
EQUALITY STATE

AREA	253,336 sq km (97,814 sq mi)
POPULATION	578,000
CAPITAL	Cheyenne
	63,000
LARGEST CITY	Cheyenne
STATEHOOD	July 10, 1890; 44th state
STATE BIRD	Western meadowlark
STATE FLOWER	Indian paintbrush

ROCKY MOUNTAINS UNITED STATES

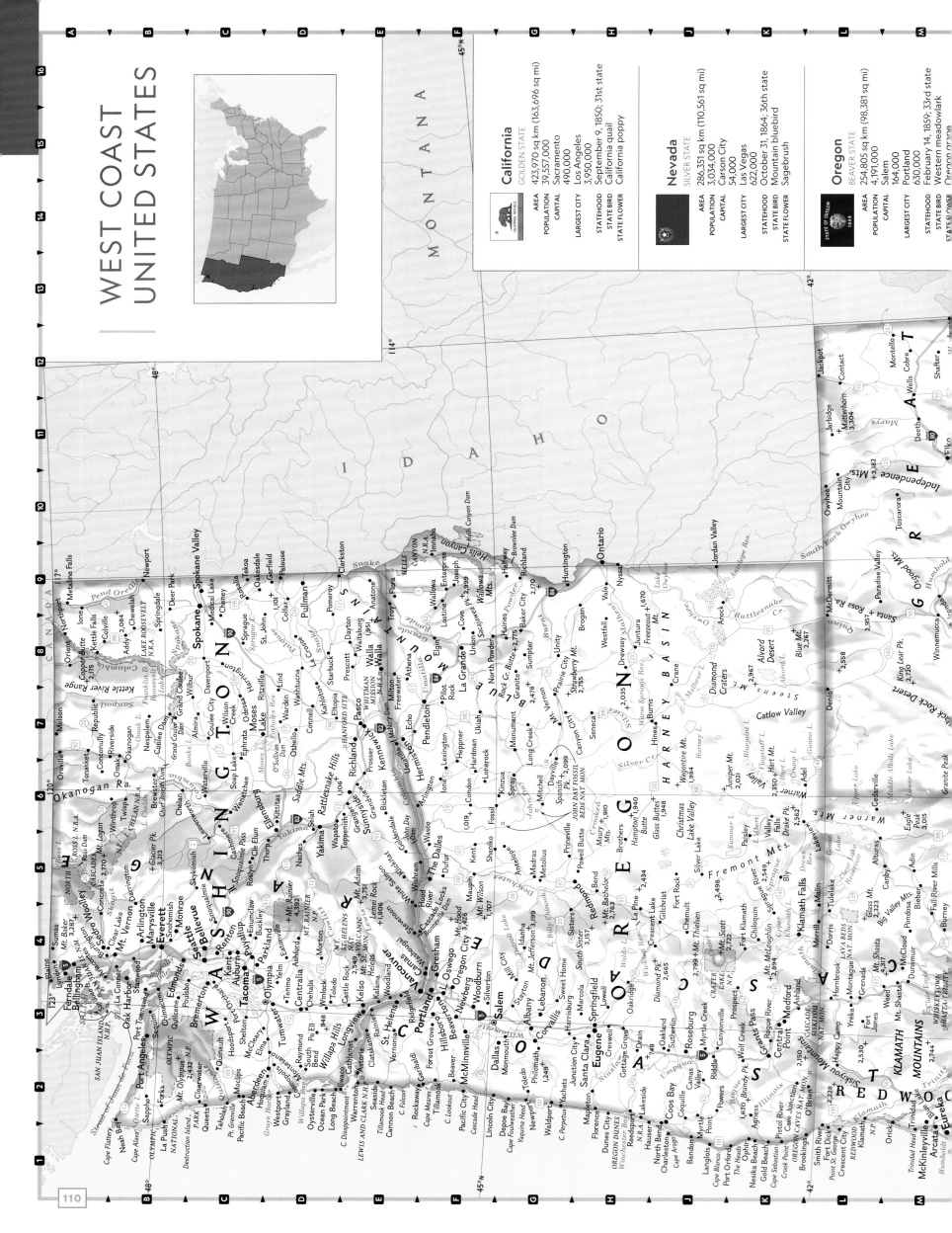

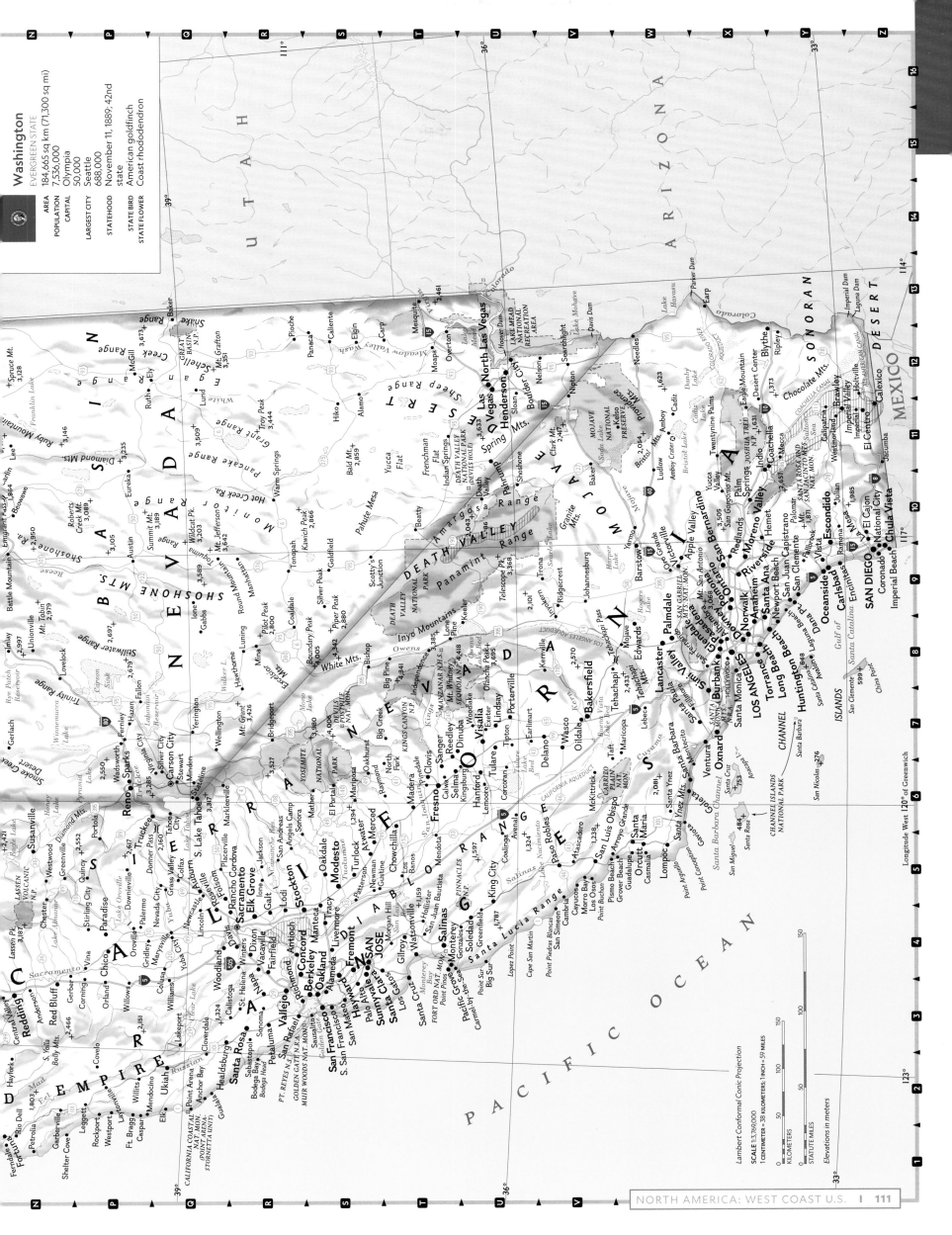

ALASKA
UNITED
STATES

171° 69° 174° 177° 180° 177° 174° 171° 72°

A

171°

B

C H U K C H I

S E A

Monday
Sunday

Cape Lisburne

Point
Hope

Kivalina

66°

ARCTIC CIRCLE

CAPE KRUSENSTERN
NAT. MO.

163° 166° 63° 169°

Date Line

60°N

R U S S I A

Shishmaref

BERING LAND
BRIDGE
NAT. PRESERV.

Diomede Is.

Wales

Seward
Peninsul

BERING STRAIT

Cape Prince of Wales

Teller

White Mountain

1,437

King I.

Nome

Golovin

166°

Gambell

Norton

57°

AREA COMPARISON OF ALASKA AND THE CONTIGUOUS U.S.

Savoonga

St. Lawrence Island

Kookooligit Mts.
673

Sound

C A N A D A

ME.

Stebbin

169°

WASH.

MONTANA

N. DAK.

MINN.

Yukon
Delta

Kotlik

VT.

N.H.

OREG.

IDAHO

WIS.

S. DAK.

MICH.

MASS.

Alakanuk

Emmonak

Sheldon Point

Yukon

H

NEVADA

WYO.

NEBR.

IOWA

N.Y.

PA.

R.I.
CONN.

Mountain Village

Pilot Station

UTAH

COLO.

ILL.

IND.

OHIO

N.J.

MD.

D.C.

DEL.

Scammon Bay

Russian Missio

San Francisco

CALIF.

KANS.

MO.

KY.

W.VA.

VA.

Hooper Bay

Chevak

Marshall

St. Matthew
Island

459

ARIZ.

N. MEX.

OKLA.

TENN.

N.C.

Nelson Island

Atmautluak

Akiachak

169°

ARK.

MISS.

ALA.

S.C.

GA.

Mekoryuk

Tununak

Baird Inlet

Bethel

Kwethluk

TEXAS

LA.

Jacksonville

Nunivak
Island

511

Toksook Bay

54°

MEXICO

FLA.

Kipnuk

Eek

171°

0 250 500 750
KILOMETERS

Kwigillingok

Kongiganak

K

Cape
Wrangell

Attu
Island

0 250 500 750
STATUTE MILES

Kuskokwim
Bay

Quinhagak

945

Semichi
Islands

Goodnews Bay

Near Islands

A

Agattu Str.

Shemya I.

L E

Cape Newenham

Agattu Island

U

St. Paul
Island

BERING

Hagemeister I.

Buldir I.

T

203

Pribilof

I

Rat Islands

Islands

M

51°

Kiska I.

A

St. George

308

Hawadax I.

Island

N

174°

1,221

Semisopochnoi
Island

1,806

I

S

L

Amchitka
Island

Delarof
Islands

Tanaga I.

A

N

Nelson Lagoon

Amchitka
Pass

Kanaga I.

Pavlof Volcano

ALAS

N

Adak I.

Great Sitkin I.

D

Unimak Island

Cold Bay

2,518

Sand Point

Andreanof Islands

1,740

1,533

Atka

Shishaldin Volcano

King Cove

Atka
Island

S

2,857

False Pass

Shumagin
Islands

Amlia I.

Islands of Four
Mountains

Umnak
Island

Unalaska
Island

2,036

Akutan

Sanak Islands

P

Seguam I.

Amukta I.

Nikolski

2,149

Unalaska a

Unimak
Pass

48°

P A C I F I C

Yunaska I.

Chuginadak I.

F o x

I s l

Q

177°

O C E A N

R

Alaska
LAST FRONTIER

AREA	1,717,854 sq km (663,267 sq mi)
POPULATION	737,000
CAPITAL	Juneau
	32,000
LARGEST CITY	Anchorage
	298,000
STATEHOOD	January 3, 1959; 49th state
STATE BIRD	Willow ptarmigan
STATE FLOWER	Forget-me-not

ARCTIC OCEAN

BEAUFORT SEA

(Barrow) Utqiaġvik
Point Barrow
Wainwright
Icy Cape
Point Lay
INUPIAT HERITAGE CENTER
Peard Bay
Dease Inlet
Smith Bay
Cape Halkett
Harrison Bay
Prudhoe Bay
Deadhorse
Kaktovik

Amatusuk
Kuskokwim Hills
NORTH SLOPE
Meade
Teshekpuk Lake
Colville

De Long Mountains
BROOKS RANGE
Lookout Ridge +714
Mt. Chamberlin + 2,712
Mt. Isto + 2,736

NOATAK NAT. PRESERVE
2,446
2,662+
Anaktuvuk Pass
Endicott Mts.
Philip Smith Mts.
Davidson Mts.

Noatak Baird Mountains
KOBUK VALLEY N.P.
Mt. Igikpak +2,523
GATES OF THE ARCTIC N.P. AND PRESERVE
Wiseman
Arctic Village

Kotzebue
Kiana
Ambler
Kobuk
Shungnak
Noorvik
Selawik
Deering
Kotzebue Sound
Selawik Lake
Kobuk
Evansville
Venetie
Fort Yukon
Chalkyitsik
Porcupine
Teedriinjik

ARCTIC CIRCLE
Hughes
Allakaket
Beaver
Stevens Village
Circle

St. Michael
Koyuk
Elim
Norton Bay
Shaktoolik
Unalakleet
Kaltag
Huslia
Rampart
Central
WHITE MTS. N.R.A.
YUKON-CHARLEY RIVERS NATIONAL PRESERVE
Eagle
Koyukuk
Nulato
Galena
Tanana
Manley Hot Springs
STEESE HWY.
Yukon
ALASKA

Grayling
Anvik
Shageluk
Holy Cross
Kaltag
Kaiyuh Mts.
Innoko
Ruby
College Fairbanks
Nenana
Anderson
Healy
Denali Park
Delta Junction
Mt. Hayes + 4,216
Tanacross
Tok
ALASKA HWY.
Tetlin
Northway Junction
TAYLOR HWY.

Kalskag
Anjak
Red Devil
Sleetmute
Nikolai
McGrath
+4,508
(Mt. McKinley) Denali 6,190+
DENALI NATIONAL PARK AND PRESERVE
Cantwell
DENALI HWY.
Mentasta Lake
RICHARDSON HWY.

Kilbuck Mts.
Kiokluk Mts. +1,248
Kuskokwim Mountains
Susitna
Talkeetna
Talkeetna Mts.
Gulkana
Gakona
Glennallen
Copper Center
WRANGELL-ST. ELIAS
Mt. Blackburn + 4,996
NATIONAL PARK
Mt. Bona + 5,005

Stony
Mulchatna
Wasilla
Palmer
Mt. Marcus Baker 4,016
AND PRESERVE

Tikchik Lakes
Mt. Torbert 3,479
Birchwood
Chugach Mountains
St. Elias

Wood River Lakes
Koliganek
New Stuyahok
Nondalton
Lake Clark
LAKE CLARK N.P. AND PRESERVE
Redoubt Volcano + 3,108
Tyonek
ANCHORAGE
Valdez
Whittier
Prince William Sound
Mt. Tom White + 3,417
Cordova
Bering Glacier
Mt. St. Elias 5,489
Mt. Foster + 2,172
KLONDIKE GOLD RUSH N.H.P.
Skagway
St. Elias Mountains

Iliamna Lake
Kenai
Soldotna
KENAI FJORDS N.P.
Seward
Montague Island
Hinchinbrook Island
Malaspina Glacier
Yakutat Bay
Yakutat
Mt. Fairweather (Tanaku) 4,663
Haines
GLACIER BAY N.P. AND PRESERVE
Devils Paw + 2,616
JUNEAU

Dillingham
Manokotak
Naknek
King Salmon
Egegik
Becharof Lake
KATMAI N.P. AND PRESERVE
Mount Katmai 2,047
Kenai Peninsula
Homer
Seldovia
Port Graham
Ninilchik

Bristol Bay
Pilot Point
Port Heiden
ANIAKCHAK N.M. AND PRESERVE
Sutwik I.
Chignik Lagoon
Mt. Veniaminof 2,507
Perryville
PENINSULA
Shelikof Strait
Shuyak I.
Afognak Island
Port Lions
Larsen Bay
Kodiak
Old Harbor
Akhiok
Kodiak Island
Trinity Islands
Chirikof Island

GULF OF ALASKA

Hoonah
Chichagof Island
Pelican
Chichagof Island
ADMIRALTY ISLAND N.M.
Angoon
Kates Needle + 3,055
Baranof Island
Sitka
SITKA N.H.P.
Kake
Kupreanof Island
Petersburg
Wrangell
Thorne Bay
Prince of Wales Island
Klawock
Craig
Ketchikan
Metlakatla
MISTY FIORDS NAT. MON.

ALEXANDER ARCHIPELAGO
COAST MTS.
Dixon Entrance

CANADA

PACIFIC OCEAN

Azimuthal Equidistant Projection

SCALE 1:7,650,000
1 CENTIMETER = 77 KILOMETERS; 1 INCH = 121 MILES

0	100	200	300

KILOMETERS

0	100	200	300

STATUTE MILES

Elevations in meters

ARCTIC CIRCLE

HAWAI'I AND PUERTO RICO UNITED STATES

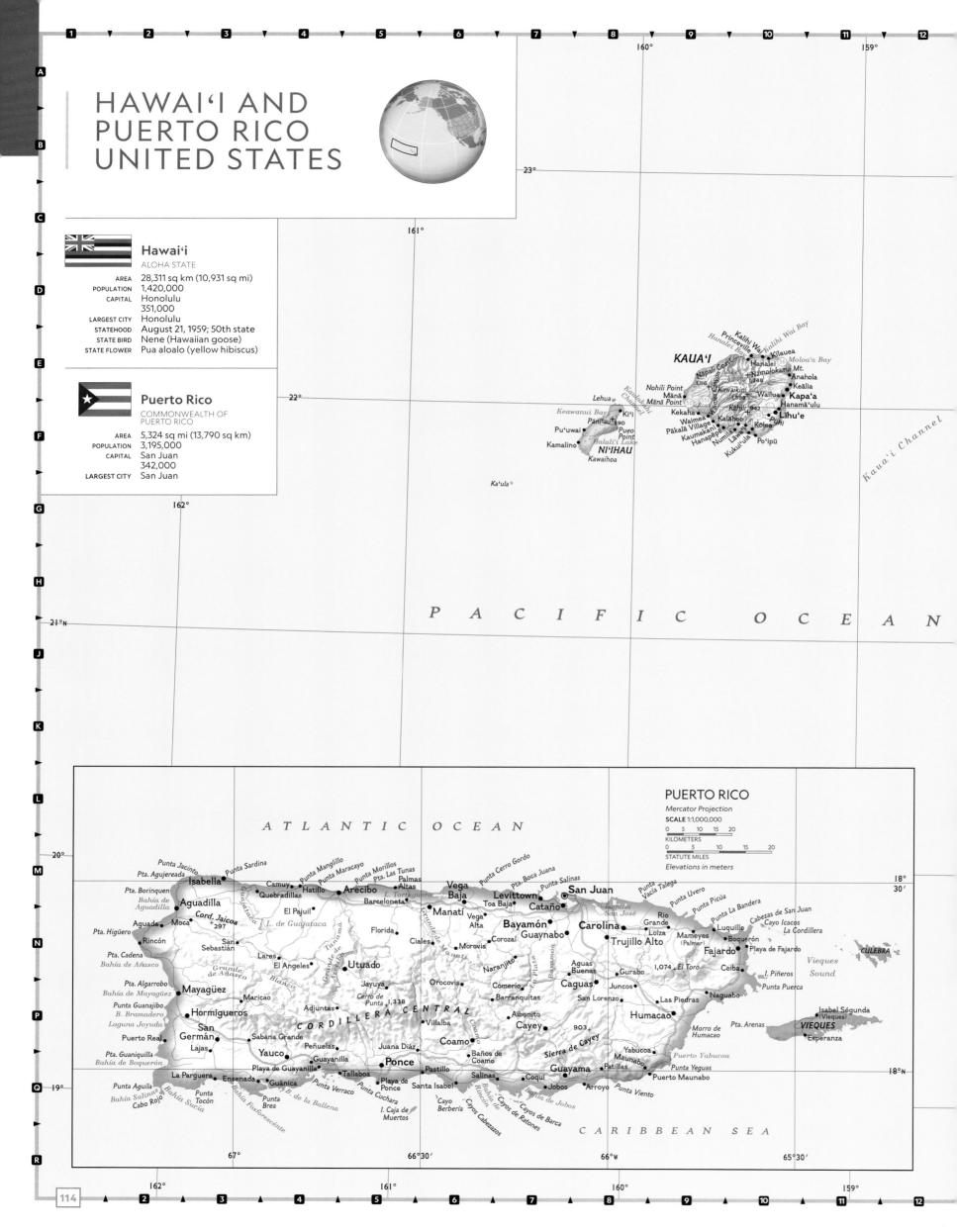

Hawai'i
ALOHA STATE

AREA	28,311 sq km (10,931 sq mi)
POPULATION	1,420,000
CAPITAL	Honolulu
	351,000
LARGEST CITY	Honolulu
STATEHOOD	August 21, 1959; 50th state
STATE BIRD	Nene (Hawaiian goose)
STATE FLOWER	Pua aloalo (yellow hibiscus)

Puerto Rico
COMMONWEALTH OF PUERTO RICO

AREA	5,324 sq mi (13,790 sq km)
POPULATION	3,195,000
CAPITAL	San Juan
	342,000
LARGEST CITY	San Juan

PACIFIC OCEAN

KAUA'I

Kalihi Wai · Kalihi Wai Bay
Princeville
Kilauea
Hanalei · Hanalei Bay
Nāpali Coast
1,116
Moloa'a Bay
Nāmolokama Mt.
1,348
Anahola
Keālia
Kawaikini
1,598
Wailua
Kapa'a
Nohili Point
Mānā
Mānā Point
Waimea Canyon
Hanamā'ulu
Kekaha
Waimea
Kalāheo
Līhu'e
Pākalā Village
Kaumakani
Numila
Lāwa'i
'Ele'ele
Hanapēpē
Kōloa
Po'ipū
Kūku'ula

Lehua
Kaulakahi Channel
Keawanui Bay
Pu'uwai
Ki'i
Pānī'au 390
Pueo Point
Halāli'i Lake
Kamalino
Kawaihoa
NI'IHAU

Ka'ula

PUERTO RICO
Mercator Projection
SCALE 1:1,000,000
0 5 10 15 20
KILOMETERS
0 5 10 15 20
STATUTE MILES
Elevations in meters

ATLANTIC OCEAN

Punta Jacinto
Punta Sardina
Punta Manglillo
Punta Morillos
Punta Cerro Gordo
Pta. Agujereada
Punta Maracayo
Pta. Las Tunas
Pta. Boca Juana
Isabella
Camuy
Palmas
Vega
Punta Salinas
Pta. Borinquen
Hatillo
Arecibo
Baja
Levittown
San Juan
Punta Uvero
Aguadilla
Quebradillas
Barceloneta
Toa Baja
Cataño
Laguna San José
Punta Vacía Talega
Bahía de Aguadilla
El Pajuil
L. Tortuguero
Manatí
Punta Picúa
Cord. Jaicoa 297
Vega Alta
Punta La Bandera
Aguada
Moca
Florida
Bayamón
Carolina
Río Grande
Cabezas de San Juan
Pta. Higüero
L. de Guajataca
Ciales
Morovis
Guaynabo
Loíza
Luquillo
Cayo Icacos
Rincón
San Sebastián
Corozal
Trujillo Alto
Mameyes (Palmer)
La Cordillera
Pta. Cadena
Lares
El Angeles
Utuado
Naranjito
Aguas Buenas
Gurabo
1,074 El Toro
Fajardo
Playa de Fajardo
Boquerón
Ceiba
Vieques Sound
Bahía de Añasco
Jayuya
Comerío
Caguas
Juncos
I. Piñeros
CULEBRA
Mayagüez
Maricao
Cerro de Punta 1,338
Orocovis
Barranquitas
San Lorenzo
Las Piedras
Naguabo
Punta Puerca
Punta Guanajibo
Adjuntas
Aibonito
Cayey
Humacao
Isabel Segunda
Vieques
B. Bramadero
Hormigueros
Villalba
903
Morro de Humacao
VIEQUES
Laguna Joyuda
CORDILLERA CENTRAL
Pta. Arenas
Esperanza
San Germán
Sabana Grande
Juana Díaz
Baños de Coamo
Yabucoa
Puerto Real
Coamo
Sierra de Cayey
Maunabo
Puerto Yabucoa
Pta. Guaniquilla
Lajas
Yauco
Peñuelas
Guayanilla
Ponce
Pastillo
Salinas
Guayama
Punta Yeguas
Bahía de Boquerón
Guánica
Playa de Guayanilla
Santa Isabel
Coquí
Jobos
Arroyo
Punta Viento
Punta Maunabo
Punta Aguila
La Parguera
Ensenada
Punta Verraco
Punta Cuchara
Baños de Coamo
Punta de Jobos
Bahía Salinas
Cabo Rojo
Punta Tocón
Bahía Sucia
Punta Brea
Playa de Ponce
Cayos de Barca
Cayos de Ratones
Bahía Fosforescente
I. Caja de Muertos
Cayo Berbería
Cayos Cabezazos

CARIBBEAN SEA

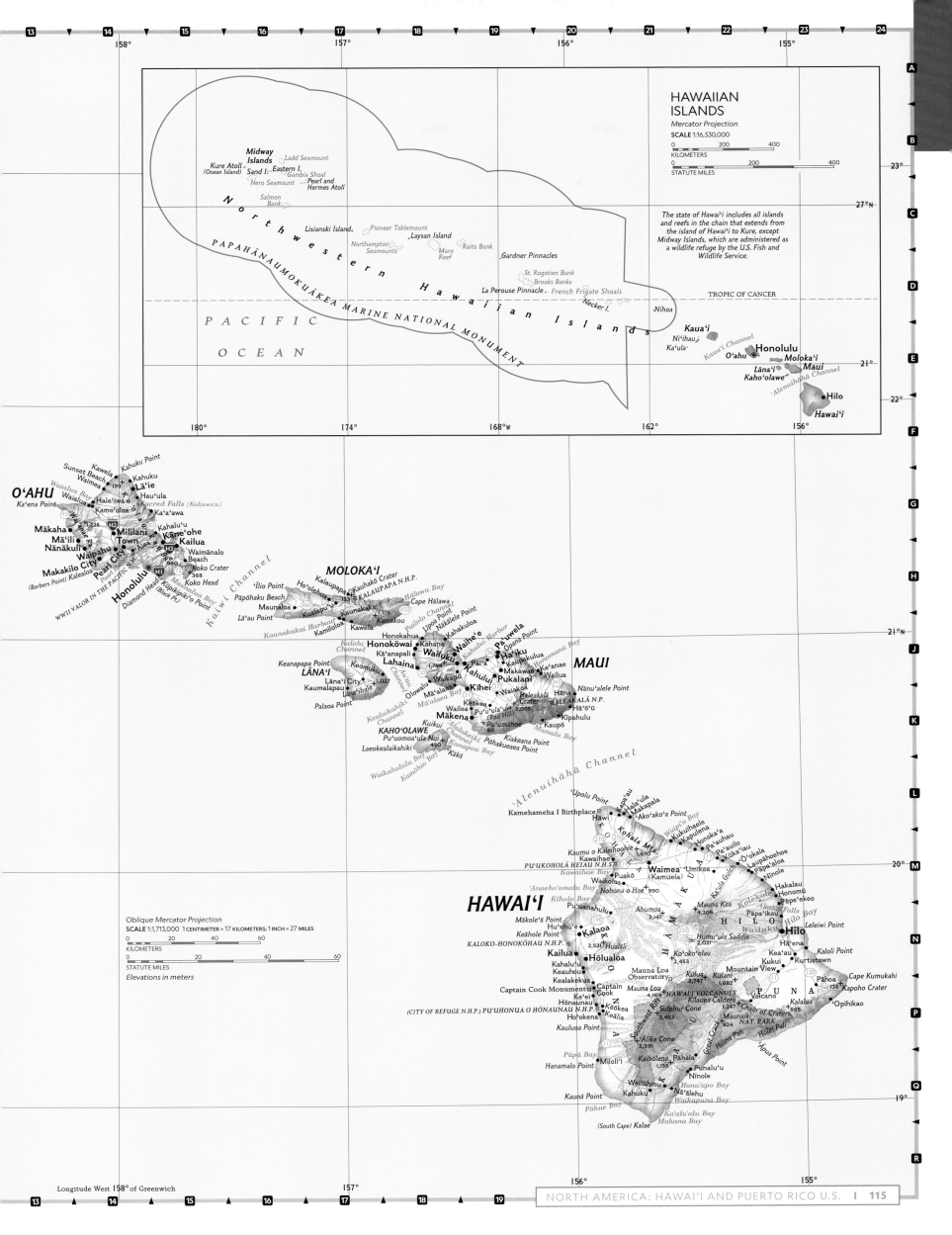

HAWAIIAN ISLANDS

Mercator Projection

SCALE 1:16,530,000

The state of Hawai'i includes all islands and reefs in the chain that extends from the island of Hawai'i to Kure, except Midway Islands, which are administered as a wildlife refuge by the U.S. Fish and Wildlife Service.

Oblique Mercator Projection

SCALE 1:1,713,000 1 CENTIMETER = 17 KILOMETERS; 1 INCH = 27 MILES

Elevations in meters

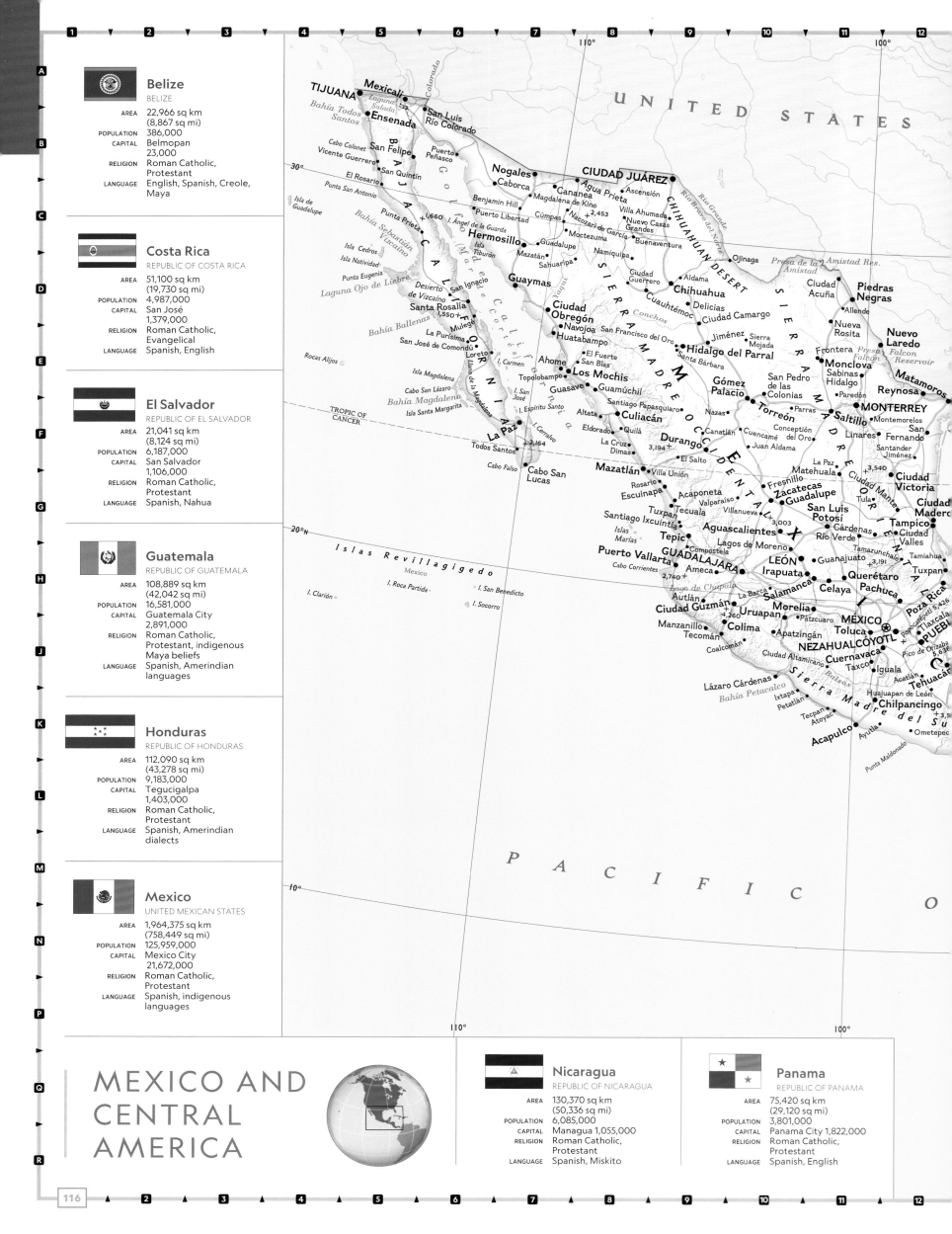

Belize
BELIZE

AREA	22,966 sq km (8,867 sq mi)
POPULATION	386,000
CAPITAL	Belmopan 23,000
RELIGION	Roman Catholic, Protestant
LANGUAGE	English, Spanish, Creole, Maya

Costa Rica
REPUBLIC OF COSTA RICA

AREA	51,100 sq km (19,730 sq mi)
POPULATION	4,987,000
CAPITAL	San José 1,379,000
RELIGION	Roman Catholic, Evangelical
LANGUAGE	Spanish, English

El Salvador
REPUBLIC OF EL SALVADOR

AREA	21,041 sq km (8,124 sq mi)
POPULATION	6,187,000
CAPITAL	San Salvador 1,106,000
RELIGION	Roman Catholic, Protestant
LANGUAGE	Spanish, Nahua

Guatemala
REPUBLIC OF GUATEMALA

AREA	108,889 sq km (42,042 sq mi)
POPULATION	16,581,000
CAPITAL	Guatemala City 2,891,000
RELIGION	Roman Catholic, Protestant, indigenous Maya beliefs
LANGUAGE	Spanish, Amerindian languages

Honduras
REPUBLIC OF HONDURAS

AREA	112,090 sq km (43,278 sq mi)
POPULATION	9,183,000
CAPITAL	Tegucigalpa 1,403,000
RELIGION	Roman Catholic, Protestant
LANGUAGE	Spanish, Amerindian dialects

Mexico
UNITED MEXICAN STATES

AREA	1,964,375 sq km (758,449 sq mi)
POPULATION	125,959,000
CAPITAL	Mexico City 21,672,000
RELIGION	Roman Catholic, Protestant
LANGUAGE	Spanish, indigenous languages

MEXICO AND CENTRAL AMERICA

Nicaragua
REPUBLIC OF NICARAGUA

AREA	130,370 sq km (50,336 sq mi)
POPULATION	6,085,000
CAPITAL	Managua 1,055,000
RELIGION	Roman Catholic, Protestant
LANGUAGE	Spanish, Miskito

Panama
REPUBLIC OF PANAMA

AREA	75,420 sq km (29,120 sq mi)
POPULATION	3,801,000
CAPITAL	Panama City 1,822,000
RELIGION	Roman Catholic, Protestant
LANGUAGE	Spanish, English

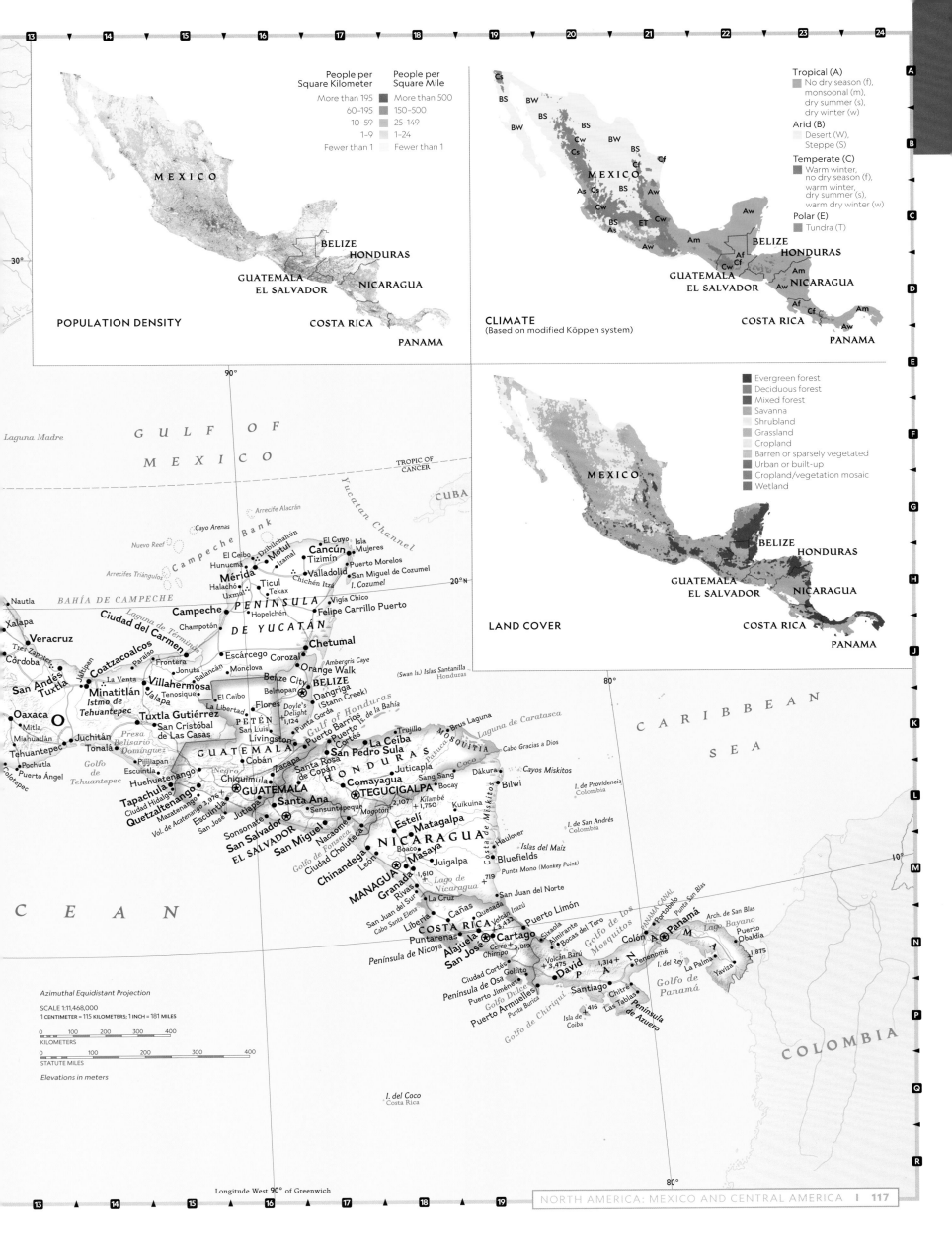

POPULATION DENSITY

People per Square Kilometer:
- More than 195
- 60–195
- 10–59
- 1–9
- Fewer than 1

People per Square Mile:
- More than 500
- 150–500
- 25–149
- 1–24
- Fewer than 1

MEXICO

BELIZE
HONDURAS
GUATEMALA
EL SALVADOR
NICARAGUA
COSTA RICA
PANAMA

CLIMATE
(Based on modified Köppen system)

Tropical (A)
- No dry season (f), monsoonal (m), dry summer (s), dry winter (w)

Arid (B)
- Desert (W), Steppe (S)

Temperate (C)
- Warm winter, no dry season (f), warm winter, dry summer (s), warm dry winter (w)

Polar (E)
- Tundra (T)

MEXICO

BELIZE
HONDURAS
GUATEMALA
EL SALVADOR
NICARAGUA
COSTA RICA
PANAMA

LAND COVER

- Evergreen forest
- Deciduous forest
- Mixed forest
- Savanna
- Shrubland
- Grassland
- Cropland
- Barren or sparsely vegetated
- Urban or built-up
- Cropland/vegetation mosaic
- Wetland

MEXICO

BELIZE
HONDURAS
GUATEMALA
EL SALVADOR
NICARAGUA
COSTA RICA
PANAMA

GULF OF MEXICO

Laguna Madre

TROPIC OF CANCER

CUBA

Yucatan Channel

Arrecife Alacrán

Cayo Arenas

Nuevo Reef

Campeche Bank

Arrecifes Triángulos

BAHÍA DE CAMPECHE

Nautla

Xalapa
Veracruz
Córdoba
Tres Zapotes
San Andrés Tuxtla
Oaxaca
Mitla
Miahuatlán
Tehuantepec
Pochutla
Puerto Ángel
Colotepec
Juchitán
Tonalá
Golfo de Tehuantepec

Coatzacoalcos
Minatitlán
Istmo de Tehuantepec
La Venta
Jáltipan
Paraíso
Frontera
Jonuta
Villahermosa
Tenosique
Balancán
Jalapa
Tuxtla Gutiérrez
San Cristóbal de Las Casas
Presa Belisario Domínguez
Negro
Tapachula
Ciudad Hidalgo
Quetzaltenango
Huehuetenango
Mazatenango
Vol. de Acatenango 3,976
Escuintla
San José
Jutiapa

Ciudad del Carmen
Laguna de Términos
Champotón
Campeche
Hopelchén
Escárcego
Corozal
Monclova
El Ceibo
Flores
PETÉN
Doyle's Delight (Stann Creek) 1,124
La Libertad
Tenosique

PENÍNSULA DE YUCATÁN

El Cuyo
Motul
Cancún
Isla Mujeres
Dzibilchaltún
Tizimín
Izamal
Valladolid
Mérida
Ticul
Tekax
Chichén Itzá
Uxmal
Halachó
Hunucmá
El Ceibo

I. Cozumel
Puerto Morelos
San Miguel de Cozumel
Vigía Chico
Felipe Carrillo Puerto
Chetumal

Ambergris Caye
Orange Walk
BELIZE
Belize City
Belmopan
Dangriga
(Stann Creek)
Punta Gorda

GUATEMALA
Cobán
Chiquimula
Zacapa
Santa Rosa de Copán
Lívingston
Puerto Barrios
Puerto Cortés
Gulf of Honduras
Islas de la Bahía
La Ceiba
Trujillo
San Pedro Sula
HONDURAS
Comayagua
TEGUCIGALPA
Juticalpa
Kilambé 1,750
2,107

Santa Ana
GUATEMALA
EL SALVADOR
SAN SALVADOR
Sensuntepeque
Mogotón
Sonsonate
Nacaome
San Miguel
Ciudad Choluteca
Golfo de Fonseca

Brus Laguna
MOSQUITIA
Patuca
Coco
Sang Sang
Bocay
Kuikuina
Cabo Gracias a Dios
Laguna de Caratasca
Dákura
Bilwi
Costa de Miskitos
Cayos Miskitos

CARIBBEAN SEA

I. de Providencia Colombia

Estelí
Matagalpa
Chinandega
León
NICARAGUA
MANAGUA
Granada
Masaya
Rivas
Boaco
Juigalpa
Lago de Nicaragua
719
Bluefields
Haulover
Islas del Maíz
Punta Mono (Monkey Point)
1,610

I. de San Andrés Colombia

Liberia
Cañas
Quesada
Puntarenas
COSTA RICA
Volcán Irazú 3,432
Alajuela
Cartago
SAN JOSÉ
Cerro 3,819
Chirripó 3,819
Ciudad Cortés
Golfito
Península de Nicoya
Cabo Santa Elena
La Cruz
San Juan del Sur
San Juan del Norte
Puerto Limón
Sixaola
Almirante
Bocas del Toro
Golfo de los Mosquitos
Volcán Barú 3,475
David 1,314
Penonomé
Colón
PANAMA
Portobelo
PANAMA CANAL
Punta San Blas
Arch. de San Blas
Lago Bayano
Puerto Obaldía
Yaviza 1,875
I. del Rey
La Palma
Golfo de Panamá
Santiago
Chitré
Las Tablas
Península de Azuero
416
Isla de Coiba
Golfo de Chiriquí
Punta Burica
Puerto Armuelles
Golfo Dulce
Puerto Jiménez
Península de Osa

OCEAN

I. del Coco
Costa Rica

Azimuthal Equidistant Projection

SCALE 1:11,468,000
1 CENTIMETER = 115 KILOMETERS; 1 INCH = 181 MILES

0 100 200 300 400
KILOMETERS

0 100 200 300 400
STATUTE MILES

Elevations in meters

Longitude West 90° of Greenwich

COLOMBIA

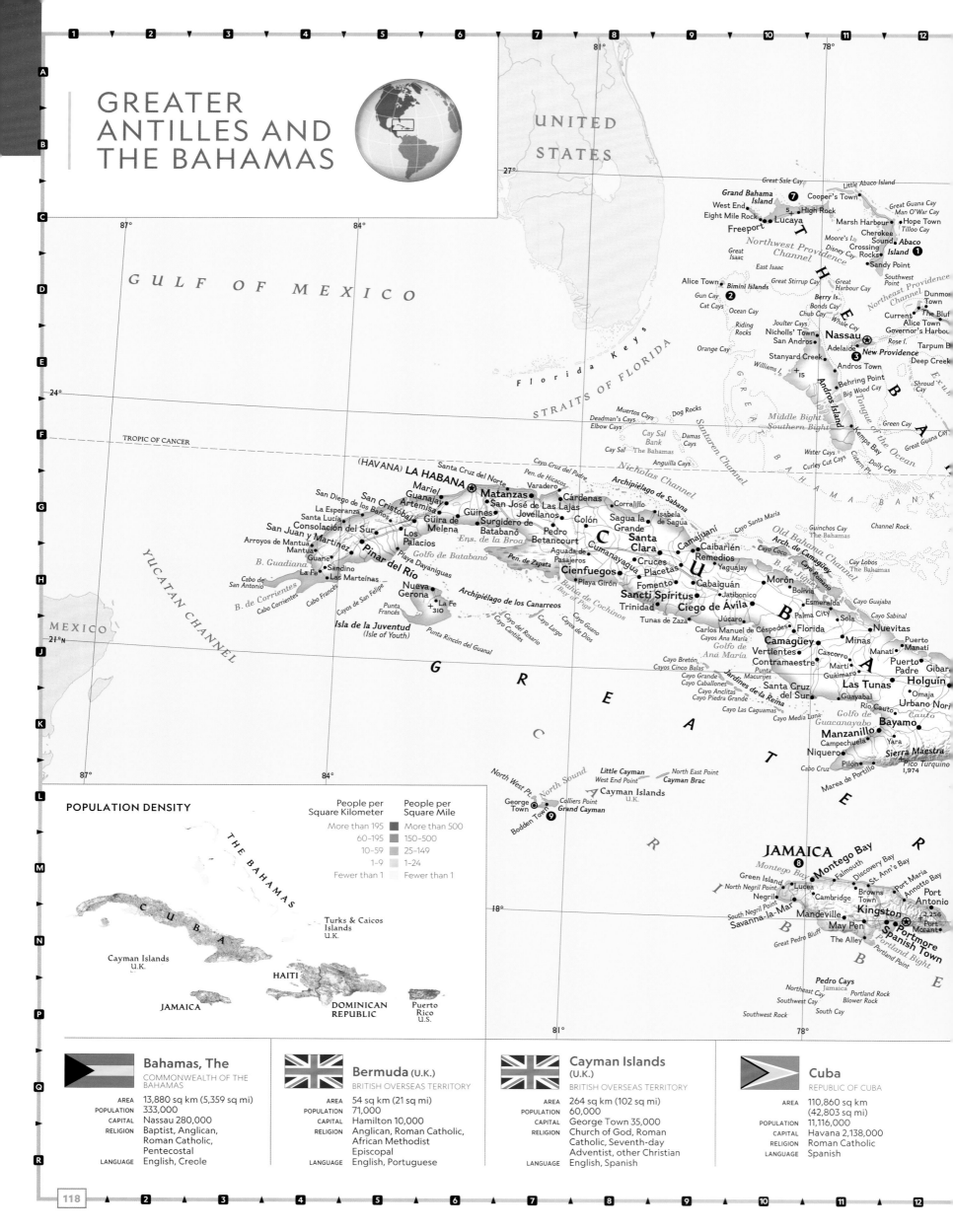

GREATER ANTILLES AND THE BAHAMAS

POPULATION DENSITY

People per Square Kilometer	People per Square Mile
More than 195	More than 500
60–195	150–500
10–59	25–149
1–9	1–24
Fewer than 1	Fewer than 1

Bahamas, The
COMMONWEALTH OF THE BAHAMAS
AREA 13,880 sq km (5,359 sq mi)
POPULATION 333,000
CAPITAL Nassau 280,000
RELIGION Baptist, Anglican, Roman Catholic, Pentecostal
LANGUAGE English, Creole

Bermuda (U.K.)
BRITISH OVERSEAS TERRITORY
AREA 54 sq km (21 sq mi)
POPULATION 71,000
CAPITAL Hamilton 10,000
RELIGION Anglican, Roman Catholic, African Methodist Episcopal
LANGUAGE English, Portuguese

Cayman Islands (U.K.)
BRITISH OVERSEAS TERRITORY
AREA 264 sq km (102 sq mi)
POPULATION 60,000
CAPITAL George Town 35,000
RELIGION Church of God, Roman Catholic, Seventh-day Adventist, other Christian
LANGUAGE English, Spanish

Cuba
REPUBLIC OF CUBA
AREA 110,860 sq km (42,803 sq mi)
POPULATION 11,116,000
CAPITAL Havana 2,138,000
RELIGION Roman Catholic
LANGUAGE Spanish

118

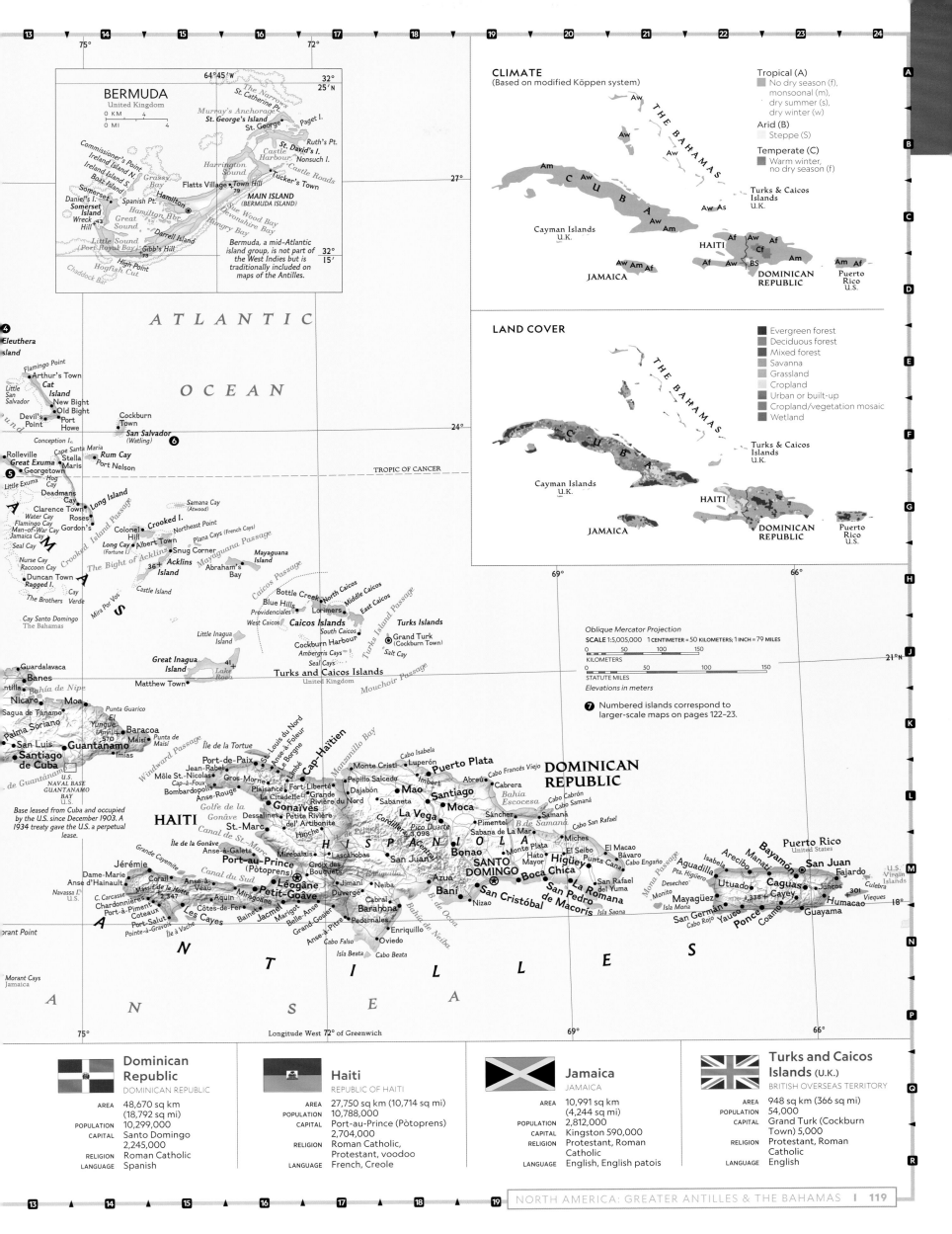

BERMUDA
United Kingdom

0 KM 4
0 MI 4

Bermuda, a mid-Atlantic island group, is not part of the West Indies but is traditionally included on maps of the Antilles.

MAIN ISLAND
(BERMUDA ISLAND)

ATLANTIC OCEAN

CLIMATE
(Based on modified Köppen system)

Tropical (A) — No dry season (f), monsoonal (m), dry summer (s), dry winter (w)

Arid (B) — Steppe (S)

Temperate (C) — Warm winter, no dry season (f)

THE BAHAMAS
CUBA
Cayman Islands U.K.
JAMAICA
HAITI
DOMINICAN REPUBLIC
Turks & Caicos Islands U.K.
Puerto Rico U.S.

LAND COVER

- Evergreen forest
- Deciduous forest
- Mixed forest
- Savanna
- Grassland
- Cropland
- Urban or built-up
- Cropland/vegetation mosaic
- Wetland

THE BAHAMAS
CUBA
Cayman Islands U.K.
JAMAICA
HAITI
DOMINICAN REPUBLIC
Puerto Rico U.S.
Turks & Caicos Islands U.K.

Oblique Mercator Projection
SCALE 1:5,005,000   1 CENTIMETER = 50 KILOMETERS; 1 INCH = 79 MILES
KILOMETERS
STATUTE MILES
Elevations in meters

7 Numbered islands correspond to larger-scale maps on pages 122–23.

TURKS AND CAICOS ISLANDS
United Kingdom

HAITI
HISPANIOLA
DOMINICAN REPUBLIC
SANTO DOMINGO
Port-au-Prince (Pòtoprens)
PUERTO RICO
United States
San Juan

Base leased from Cuba and occupied by the U.S. since December 1903. A 1934 treaty gave the U.S. a perpetual lease.

U.S. NAVAL BASE GUANTÁNAMO BAY U.S.

GREATER ANTILLES

CARIBBEAN SEA

Dominican Republic
DOMINICAN REPUBLIC

AREA: 48,670 sq km (18,792 sq mi)
POPULATION: 10,299,000
CAPITAL: Santo Domingo 2,245,000
RELIGION: Roman Catholic
LANGUAGE: Spanish

Haiti
REPUBLIC OF HAITI

AREA: 27,750 sq km (10,714 sq mi)
POPULATION: 10,788,000
CAPITAL: Port-au-Prince (Pòtoprens) 2,704,000
RELIGION: Roman Catholic, Protestant, voodoo
LANGUAGE: French, Creole

Jamaica
JAMAICA

AREA: 10,991 sq km (4,244 sq mi)
POPULATION: 2,812,000
CAPITAL: Kingston 590,000
RELIGION: Protestant, Roman Catholic
LANGUAGE: English, English patois

Turks and Caicos Islands (U.K.)
BRITISH OVERSEAS TERRITORY

AREA: 948 sq km (366 sq mi)
POPULATION: 54,000
CAPITAL: Grand Turk (Cockburn Town) 5,000
RELIGION: Protestant, Roman Catholic
LANGUAGE: English

LESSER ANTILLES

POPULATION DENSITY

British Virgin Islands U.K.
U.S. Virgin Islands
U.S.
Anguilla U.K.
St. Martin Fr. · St. Barthelemy Fr.
Neth. ST. MAARTEN
ST. KITTS & NEVIS
ANTIGUA AND BARBUDA
U.K. Montserrat
Fr. Guadeloupe
DOMINICA
Fr. Martinique
ST. LUCIA
ST. VINCENT AND THE GRENADINES
BARBADOS
GRENADA
TRINIDAD AND TOBAGO

ARUBA Neth.
CURAÇAO Neth.
Bonaire Neth.

People per Square Kilometer	People per Square Mile
More than 195	More than 500
60–195	150–500
10–59	25–149
1–9	1–24
Fewer than 1	Fewer than 1

CLIMATE
(Based on modified Köppen system)

Af British Virgin Islands
U.S. Virgin Islands
U.S.
Am U.K.
Neth. ST. MAARTEN
Anguilla U.K.
St. Martin Fr. · St. Barthelemy Fr.
ST. KITTS & NEVIS
Am
ANTIGUA AND BARBUDA
U.K. Montserrat
Fr. Guadeloupe
Am
Af DOMINICA
Fr. Martinique
ST. LUCIA
Am
ST. VINCENT AND THE GRENADINES
Am BARBADOS
GRENADA
Am TRINIDAD AND TOBAGO

ARUBA Neth.
BS
CURAÇAO Neth.
Aw Bonaire Neth.

Tropical (A)
No dry season (f), monsoonal (m), dry winter (w)

Arid (B)
Steppe (S)

LAND COVER

British Virgin Islands
U.S. Virgin Islands
U.S.
U.K.
Neth. ST. MAARTEN
Anguilla U.K.
St. Martin Fr. · St. Barthelemy Fr.
ST. KITTS & NEVIS
ANTIGUA AND BARBUDA
U.K. Montserrat
Fr. Guadeloupe
DOMINICA
Fr. Martinique
ST. LUCIA
ST. VINCENT AND THE GRENADINES
BARBADOS
GRENADA
TRINIDAD AND TOBAGO

ARUBA Neth.
CURAÇAO Neth.
Bonaire Neth.

- Evergreen forest
- Savanna
- Grassland
- Cropland
- Urban or built-up
- Cropland/vegetation mosaic
- Wetland

HAITI
DOMINICAN REPUBLIC
Puerto Rico U.S.

72° 69° 66°
18°

Oblique Mercator Projection
SCALE 1:5,196,000 1 CENTIMETER = 52 KILOMETERS; 1 INCH = 82 MILES
0 50 100 150
KILOMETERS
0 50 100 150
STATUTE MILES
Elevations in meters

22 Numbered islands correspond to larger-scale maps on pages 122–23.

15°
72°

C A R I B B E A N

L E S S E R A N

Oranjestad
29 ARUBA Neth.
188 Sint Nicolaas
372 CURAÇAO Neth.
Willemstad 30 Kralendijk 28
Bonaire Neth.
240

COLOMBIA
12°N

V E N E Z

9°
69° 66°

Anguilla (U.K.)
BRITISH OVERSEAS TERRITORY
AREA 91 sq km (35 sq mi)
POPULATION 17,000
CAPITAL The Valley 1,000
RELIGION Anglican, Methodist, other Protestant, Roman Catholic
LANGUAGE English

Aruba (Netherlands)
AUTONOMOUS COUNTRY OF THE NETHERLANDS
AREA 180 sq km (69 sq mi)
POPULATION 117,000
CAPITAL Oranjestad 30,000
RELIGION Roman Catholic, Protestant
LANGUAGE Papiamento, Spanish, English, Dutch

Antigua and Barbuda
ANTIGUA AND BARBUDA
AREA 443 sq km (171 sq mi)
POPULATION 86,000
CAPITAL St. John's 21,000
RELIGION Anglican, Methodist, other Protestant, Roman Catholic
LANGUAGE English, Antiguan creole

Barbados
BARBADOS
AREA 430 sq km (166 sq mi)
POPULATION 293,000
CAPITAL Bridgetown 89,000
RELIGION Protestant, Roman Catholic
LANGUAGE English, Bajan

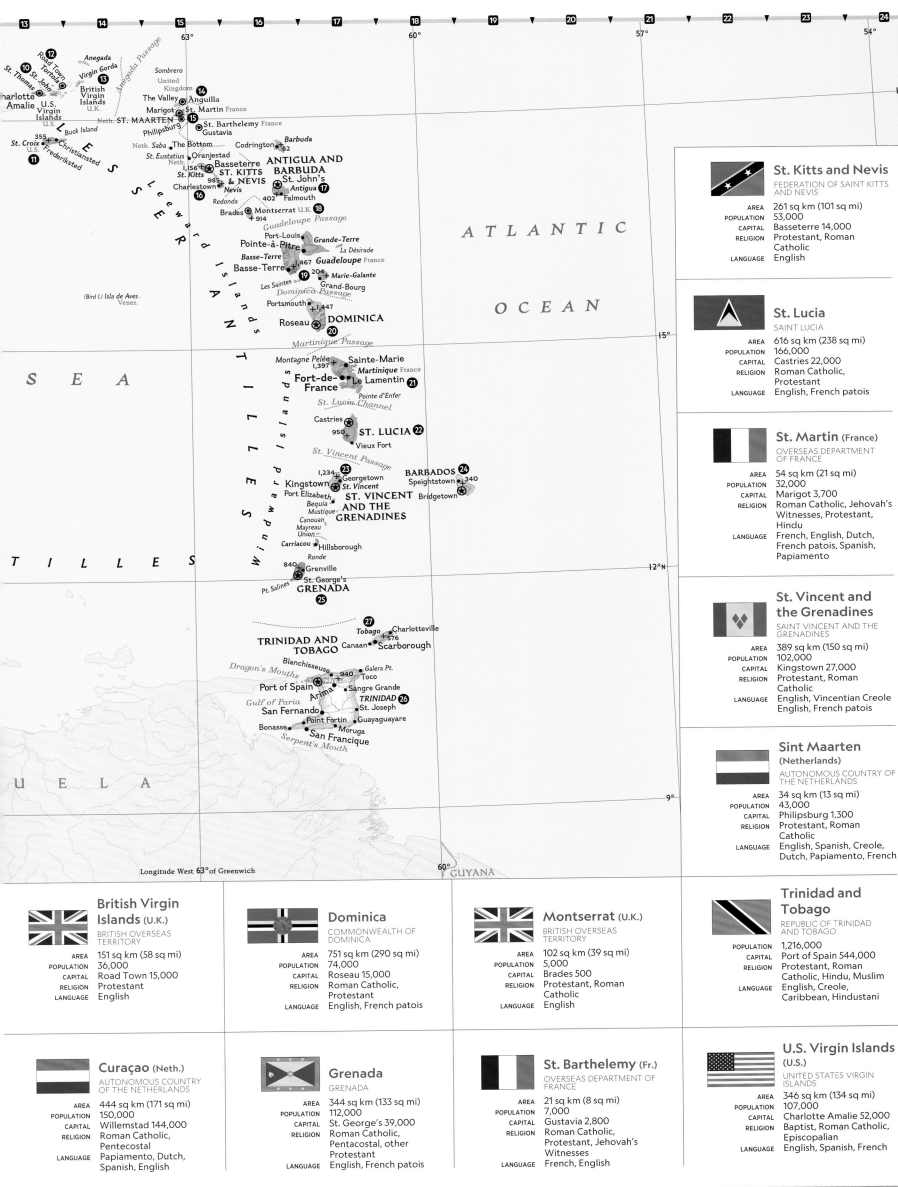

St. Kitts and Nevis
FEDERATION OF SAINT KITTS AND NEVIS

AREA	261 sq km (101 sq mi)
POPULATION	53,000
CAPITAL	Basseterre 14,000
RELIGION	Protestant, Roman Catholic
LANGUAGE	English

St. Lucia
SAINT LUCIA

AREA	616 sq km (238 sq mi)
POPULATION	166,000
CAPITAL	Castries 22,000
RELIGION	Roman Catholic, Protestant
LANGUAGE	English, French patois

St. Martin (France)
OVERSEAS DEPARTMENT OF FRANCE

AREA	54 sq km (21 sq mi)
POPULATION	32,000
CAPITAL	Marigot 3,700
RELIGION	Roman Catholic, Jehovah's Witnesses, Protestant, Hindu
LANGUAGE	French, English, Dutch, French patois, Spanish, Papiamento

St. Vincent and the Grenadines
SAINT VINCENT AND THE GRENADINES

AREA	389 sq km (150 sq mi)
POPULATION	102,000
CAPITAL	Kingstown 27,000
RELIGION	Protestant, Roman Catholic
LANGUAGE	English, Vincentian Creole English, French patois

Sint Maarten
(Netherlands)
AUTONOMOUS COUNTRY OF THE NETHERLANDS

AREA	34 sq km (13 sq mi)
POPULATION	43,000
CAPITAL	Philipsburg 1,300
RELIGION	Protestant, Roman Catholic
LANGUAGE	English, Spanish, Creole, Dutch, Papiamento, French

Trinidad and Tobago
REPUBLIC OF TRINIDAD AND TOBAGO

POPULATION	1,216,000
CAPITAL	Port of Spain 544,000
RELIGION	Protestant, Roman Catholic, Hindu, Muslim
LANGUAGE	English, Creole, Caribbean, Hindustani

British Virgin Islands (U.K.)
BRITISH OVERSEAS TERRITORY

AREA	151 sq km (58 sq mi)
POPULATION	36,000
CAPITAL	Road Town 15,000
RELIGION	Protestant
LANGUAGE	English

Dominica
COMMONWEALTH OF DOMINICA

AREA	751 sq km (290 sq mi)
POPULATION	74,000
CAPITAL	Roseau 15,000
RELIGION	Roman Catholic, Protestant
LANGUAGE	English, French patois

Montserrat (U.K.)
BRITISH OVERSEAS TERRITORY

AREA	102 sq km (39 sq mi)
POPULATION	5,000
CAPITAL	Brades 500
RELIGION	Protestant, Roman Catholic
LANGUAGE	English

Curaçao (Neth.)
AUTONOMOUS COUNTRY OF THE NETHERLANDS

AREA	444 sq km (171 sq mi)
POPULATION	150,000
CAPITAL	Willemstad 144,000
RELIGION	Roman Catholic, Pentecostal
LANGUAGE	Papiamento, Dutch, Spanish, English

Grenada
GRENADA

AREA	344 sq km (133 sq mi)
POPULATION	112,000
CAPITAL	St. George's 39,000
RELIGION	Roman Catholic, Pentecostal, other Protestant
LANGUAGE	English, French patois

St. Barthelemy (Fr.)
OVERSEAS DEPARTMENT OF FRANCE

AREA	21 sq km (8 sq mi)
POPULATION	7,000
CAPITAL	Gustavia 2,800
RELIGION	Roman Catholic, Protestant, Jehovah's Witnesses
LANGUAGE	French, English

U.S. Virgin Islands (U.S.)
UNITED STATES VIRGIN ISLANDS

AREA	346 sq km (134 sq mi)
POPULATION	107,000
CAPITAL	Charlotte Amalie 52,000
RELIGION	Baptist, Roman Catholic, Episcopalian
LANGUAGE	English, Spanish, French

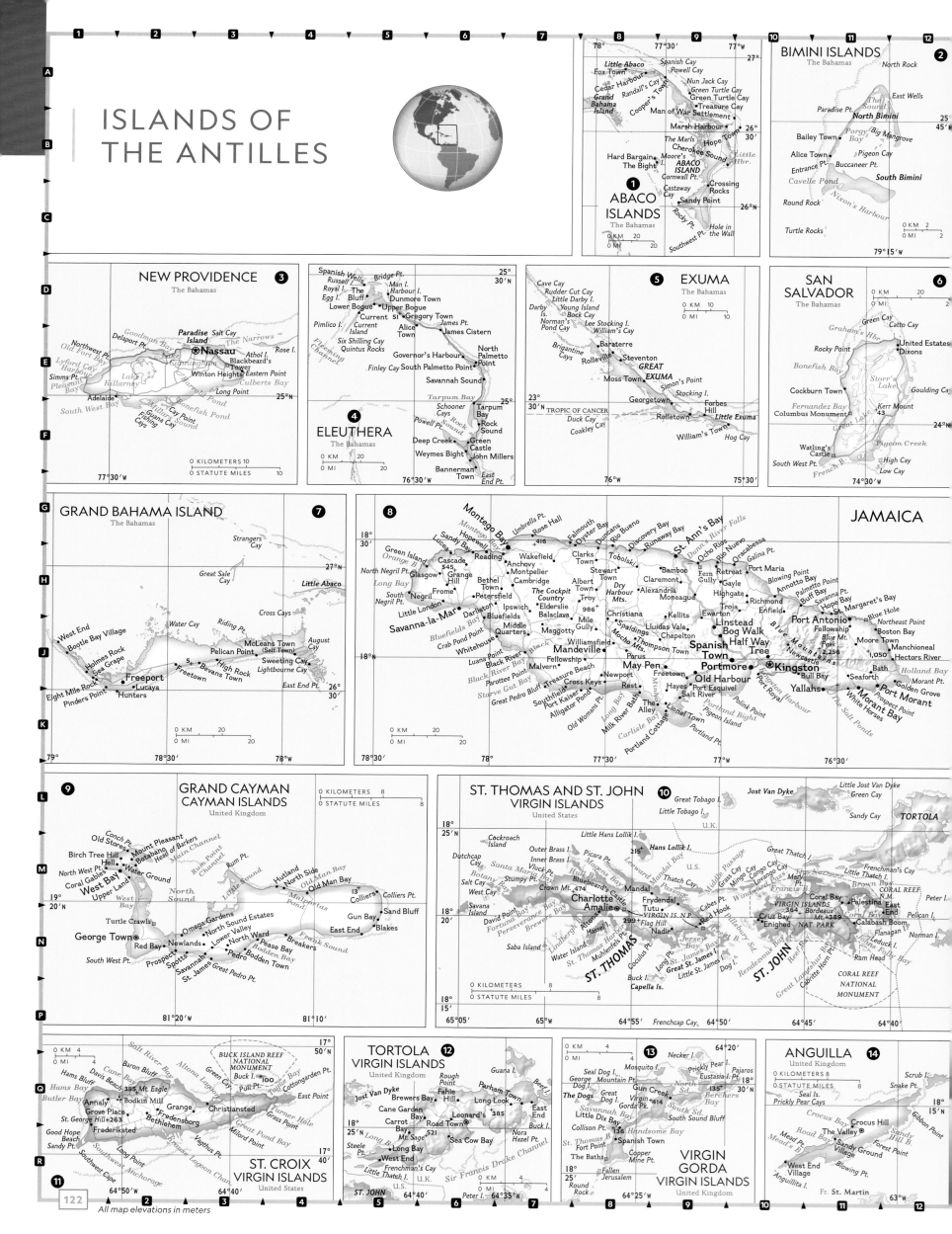

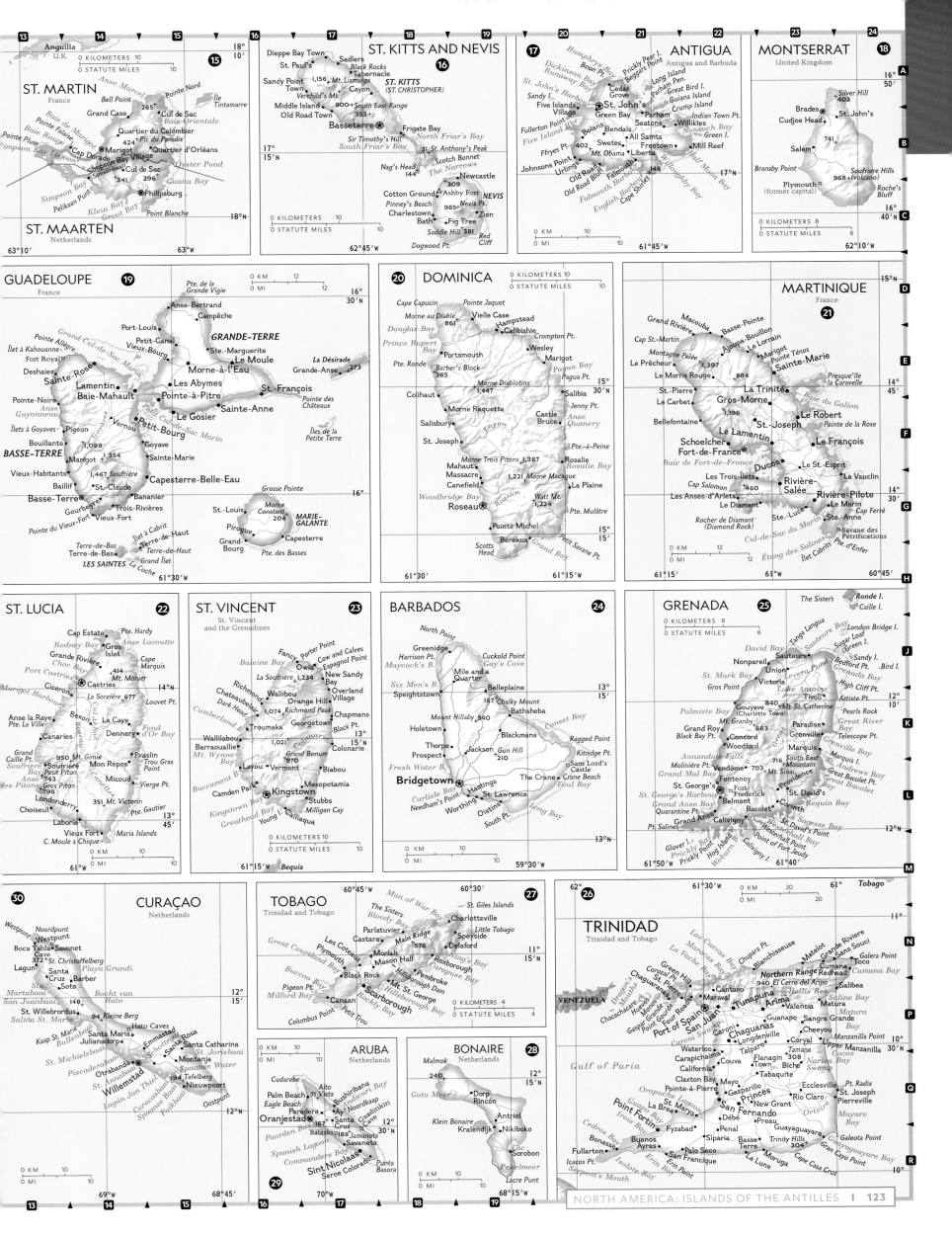

SOUTH AMERICA

SOUTH AMERICA PHYSICAL AND POLITICAL
NORTHERN SOUTH AMERICA
CENTRAL SOUTH AMERICA
SOUTHERN SOUTH AMERICA

The Andes Mountains tower over a guanaco in Chilean Patagonia.

TEEMING WITH LIFE & LEGENDS

RARE LANGUAGES, MIGHTY RIVERS, STORIED PAST

South American wilderness is an array of wonders: mountain, river, prairie, jungle, desert, glacier. Its megacities flourish, yet its storied history is never far—pre-Columbian monuments lie cloaked by jungle or mountaintop mist. South America today is engaged in a struggle: to confront its past and respond to the complexities of modern times.

GEOGRAPHY

WILD AT HEART

Travel inland from South America's crowded coasts, and the numbers of people gradually diminish. The sparsely populated or uninhabited land of this immense interior includes the open Pampas—or prairies—of Argentina, Brazil, and Uruguay; the forested Guiana Highlands of Venezuela, Guyana, Suriname, and French Guiana; and the great wetland of Brazil's Pantanal. If the Andes mountain range—the world's longest mountain chain—is South America's backbone, the Amazon River is its pulsing heart. Rainwater spilling off the Andes' jagged peaks tumbles thousands of feet to feed the mighty river, which carries more water than any other river on the planet. Its thousand or so tributaries sustain Earth's largest rainforest. The Amazon region harbors at least 10 percent of the world's plant and animal species and more than 300 indigenous groups of people. The forest is vital to the health of our planet, capturing carbon that can mitigate the effects of climate change—but it faces tremendous threats from agriculture and timber harvesting.

Offshore elements are distinctive, too: The Galápagos Islands in the Pacific are a biological wonderland where primeval giant tortoises and marine iguanas roam among waddling penguins. In the far south, the untamed Tierra del Fuego archipelago boasts lush kelp forest in the sea and massive glaciers on land.

Peru's Machu Picchu shows the Inca Empire at the peak of its power.

HISTORY

GREAT CITIES OF STONE

North and South America's earliest human inhabitants likely crossed a land bridge that once linked Siberia and modern-day Alaska. They migrated south, splitting into numerous different tribal groups. Among these were the Chavin, who thrived from roughly 900 to 200 B.C. in the Andes. In 1,500 years, they would be superseded by the sophisticated Inca empire, which stretched from present-day Colombia to Chile and Argentina, ruling up to 12 million subjects. Europeans arrived in the late 15th century, and by 1494 Spain and Portugal had divided the continent between them. Driven off their land, decimated by disease, and pressed into slavery, the population of native people declined. Today, there are more than 32 million people who claim indigenous heritage on the continent, and they make up more than 40 percent of the population in Peru and Bolivia.

Three distinct groups—the military, wealthy families, and the Roman Catholic Church—dominated South America's colonies. Sugarcane plantations and gold and silver mines were worked by millions of enslaved native people and Africans. In the early 19th century, the colonies revolted against their European overlords and formed independent democratic states. Military rule and Cold War politics stifled many of these nations in the mid-20th century. In the 1980s democracy flowered across the continent, but the 21st century has seen a resurgence of more extreme populist leaders.

A jaguar cub in the Pantanal region of Brazil

CULTURE

LATIN RHYTHMS, LITERARY GIANTS & A BEAUTIFUL GAME

Whether they are blond-haired, blue-eyed Germans in Paraguay, descendants of enslaved Africans on the Caribbean coast, or mixed-race mestizos in Brazil, most South Americans don't need a translator to talk to one another. The majority of the continent's 422 million people trace their ancestors back to Iberia; roughly half the population speaks Spanish, the other half Portuguese. In Suriname and Guyana, you can hear Dutch and English, as well as French in French Guiana. In the Amazon, Andes, and other remote areas, Amerindian dialects, among the most endangered languages in the world, are still spoken. Catholicism is the predominant religion.

A vibrant modern culture—blending Iberian, African, and Amerindian traditions—has proved to be influential far beyond the bounds of South America's borders. Argentina's beloved tango, born of the Buenos Aires barrios, is now a symbol of romance around the world. Brazil's coastal cities hatched upbeat Afro-Latino rhythms such as samba and bossa nova. And Peruvian panpipe music has become synonymous with the Andes. South America's rich literary map, meanwhile, includes the imaginative labyrinths of Jorge Luis Borges, the magical realism of Gabriel García Márquez, and the sensual poems of Pablo Neruda. A similar passion flows through soccer, the region's favorite game.

Dancing the tango in Buenos Aires, Argentina

Banana farming in Peru, where nearly all exported bananas are organic

ECONOMY

FINDING PROMISE IN THE AMAZON

South America's independence from European powers did not translate quickly into economic autonomy. Most countries remained dependent on commodity exports to Europe or the United States: bananas, soybeans, rubber, sugar, coffee, timber, emeralds, copper, oil, and beef. Oligarchs and wealthy families possessed the vast majority of the land and profits, and the resulting economic and social stratification sparked decades of struggle and civil war. Few South American economies made a full transition from resource extraction to modern business and industry. By the 1960s, Brazil and Venezuela were seeing strong growth, but other countries were mired in negative or neutral economic growth and were plagued by corruption, military rule, drug cartels, and mismanaged governments.

Economic prosperity has been impeded by world recessions and huge foreign debts. Many governments have amplified social welfare spending to break the paralyzing cycles of poverty and chart a path to development. The Amazon could be one of the keys to the region's economic future. Many researchers believe treatments for cancer and other ailments may lie within South America's rainforest. This largely untapped potential for medical, chemical, and nutritional products has heightened calls for sustainable management and commercial development in the region.

SNAPSHOT
South America
FACTS & FEATURES

LONGEST RIVER Amazon River: 6,679 km (4,150 mi)

LARGEST LAKE Lake Titicaca: 8,400 sq km (3,200 sq mi)

BIGGEST URBAN AREA BY POPULATION São Paulo, Brazil: 21,650,000 (2018)

MOST POPULOUS COUNTRY Brazil: 208,847,000 (2018)

LEAST POPULOUS COUNTRY Suriname: 598,000 (2018)

WORLD'S HIGHEST CAPITAL CITY La Paz, Bolivia: approximately 3,600 m (11,800 ft)

WORLD'S DRIEST PLACE Arica, Atacama Desert, Chile: rainfall barely measurable

HOTTEST PLACE Rivadavia, Argentina: record of 48.9° Celsius (120° Fahrenheit)

WORLD'S TALLEST WATERFALL Angel Falls, Venezuela: 979 m (3,212 ft)

WORLD'S MOST PRODUCTIVE SILVER MINES Potosí, Bolivia

THE **STORMY DRAKE PASSAGE** THAT SEPARATES SOUTH AMERICA FROM THE ANTARCTIC CONTINENT PRESENTS SOME OF THE **ROUGHEST SEAS** ON THE PLANET.

Azimuthal Equidistant Projection
SCALE 1:24,180,000 1 CENTIMETER = 242 KILOMETERS; 1 INCH = 382 MILES

0   200   400   600
KILOMETERS

0   200   400   600
STATUTE MILES

Elevations in meters

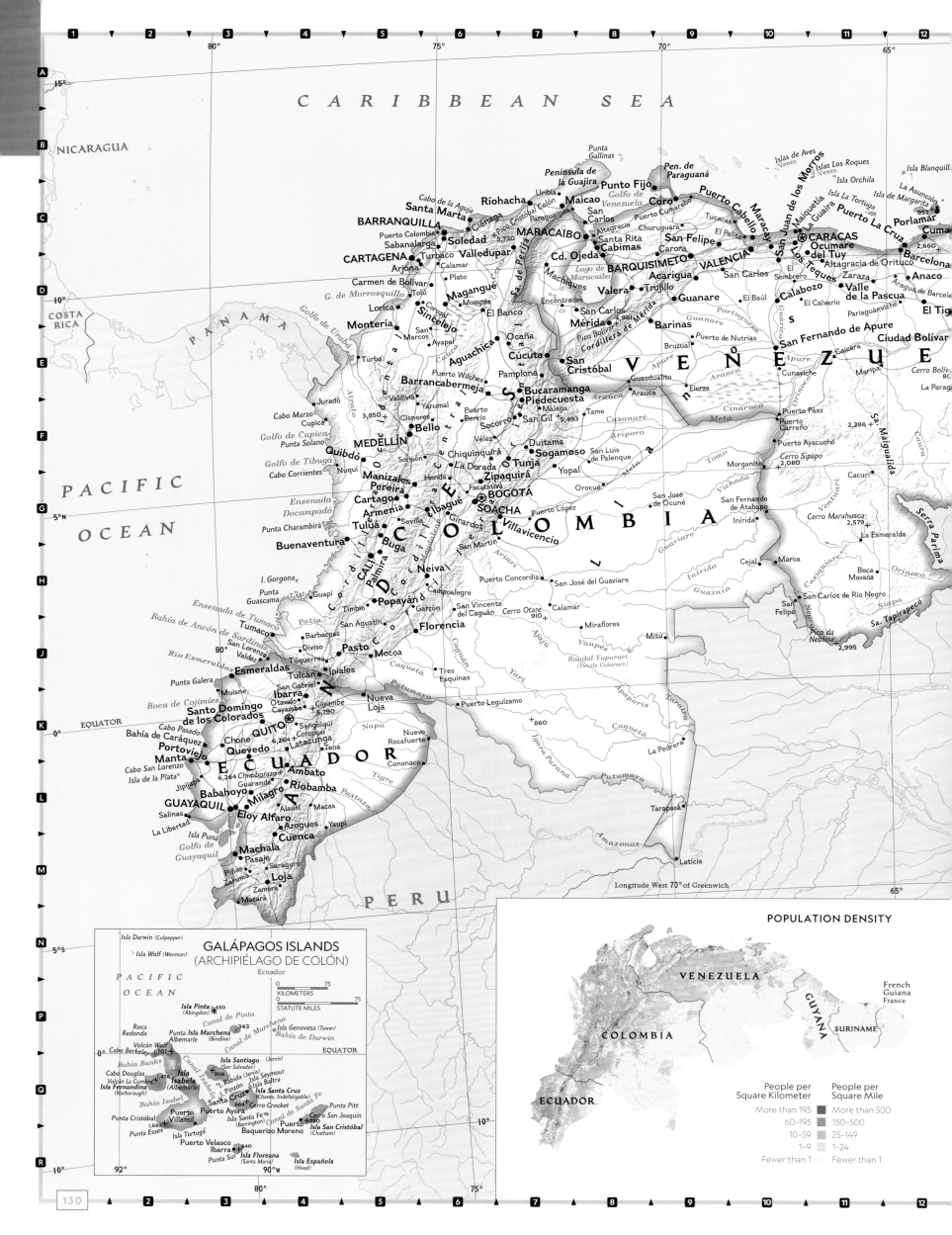

NORTHERN SOUTH AMERICA

60° 55° 15° 50°

ATLANTIC OCEAN

SAINT LUCIA
BARBADOS
SAINT VINCENT AND THE GRENADINES
GRENADA
TRINIDAD AND TOBAGO

Dragon's Mouths
Carúpano
Macuro Güiria
Gulf of Paria
Caripito
Serpent's Mouth
Maturín
Tucupita
Barrancas
Orinoco
Curiapo
Boca Grande
San José de Amacuro
Morawhanna
Shell Beach
CIUDAD GUAYANA
Mabaruma
Upata
El Pao
Guri Dam
Port Kaituma
Barima
Ciudad Piar
El Callao
Matthew's Ridge
Charity
Georgetown
Tumeremo
Suddie
Buxton
New Amsterdam
El Dorado
Cuyuni
Parika
Bartica
Linden
Mara
Corriverton
Nieuw Nickerie
Totness
Paramaribo
Angel Falls
Total drop 979 meters
Mazaruni
Issano
Ituni
Nieuw Amsterdam
Pointe Isère
La Gran Sabana
Tiboku Falls
Great Fall
Mandia
Berbice
Zanderij
Moengo
Mana
Iracoubo
Mt. Roraima 2,810
Orinduik
Santa Elena
Avanavero
Brokopondo
Brownsweg
Afobaka
St-Laurent du Maroni
Kourou
Sinnamary
Île du Diable (Devils I.)
Cayenne
Pakaraima Mts.
Wilhelmina Gebergte
Juliana Top 1,230
France
Bellevue de l'Inini
Matoury
Rémire
Roura
Régina
Lethem
Kanuku Mts.
861
Kayser Gebergte
851
Boundary claimed by Venezuela
Boundary claimed by Suriname
Mont Saint-Marcel 635
Acarai Mts.
Serra de Tumucumaque
1,177

GUYANA
SURINAME
French Guiana

EQUATOR 0°

BRAZIL

5°N
5°S

Azimuthal Equidistant Projection
SCALE 1:9,550,000 1 CENTIMETER = 96 KILOMETERS; 1 INCH = 151 MILES
0 100 200 300
KILOMETERS
0 100 200 300
STATUTE MILES
Elevations in meters

Colombia
REPUBLIC OF COLOMBIA
AREA 1,138,910 sq km (439,735 sq mi)
POPULATION 48,169,000
CAPITAL Bogotá 10,779,000
RELIGION Roman Catholic, Protestant
LANGUAGE Spanish

Ecuador
REPUBLIC OF ECUADOR
AREA 283,561 sq km (109,483 sq mi)
POPULATION 16,499,000
CAPITAL Quito 1,848,000
RELIGION Roman Catholic, Evangelical
LANGUAGE Spanish, Quechua, other Amerindian languages

Guyana
CO-OPERATIVE REPUBLIC OF GUYANA
AREA 214,969 sq km (83,000 sq mi)
POPULATION 741,000
CAPITAL Georgetown 110,000
RELIGION Protestant, Hindu, Roman Catholic, Muslim
LANGUAGE English, Guyanese Creole, Amerindian dialects, Caribbean Hindustani

Suriname
REPUBLIC OF SURINAME
AREA 163,820 sq km (63,251 sq mi)
POPULATION 598,000
CAPITAL Paramaribo 239,000
RELIGION Protestant, Hindu, Roman Catholic, Muslim
LANGUAGE Dutch, English, Sranang Tongo, Caribbean Hindustani, Javanese

Venezuela
BOLIVARIAN REPUBLIC OF VENEZUELA
AREA 912,050 sq km (352,144 sq mi)
POPULATION 31,689,000
CAPITAL Caracas 2,936,000
RELIGION Roman Catholic
LANGUAGE Spanish, numerous indigenous dialects

CLIMATE
(Based on modified Köppen system)

BW
BS
ET
Cf
Aw
Af
Am
As
Cw
VENEZUELA
GUYANA
SURINAME
French Guiana France
COLOMBIA
ECUADOR

Tropical (A)
No dry season (f), monsoonal (m), dry summer (s), dry winter (w)

Arid (B)
Desert (W), Steppe (S)

Temperate (C)
Warm winter, no dry season (f), warm dry winter (w)

Polar (E)
Tundra (T)

LAND COVER

VENEZUELA
COLOMBIA
ECUADOR
GUYANA
SURINAME
French Guiana France

Evergreen forest
Deciduous forest
Savanna
Shrubland
Grassland
Cropland
Barren or sparsely vegetated
Urban or built-up
Cropland/vegetation mosaic
Wetland

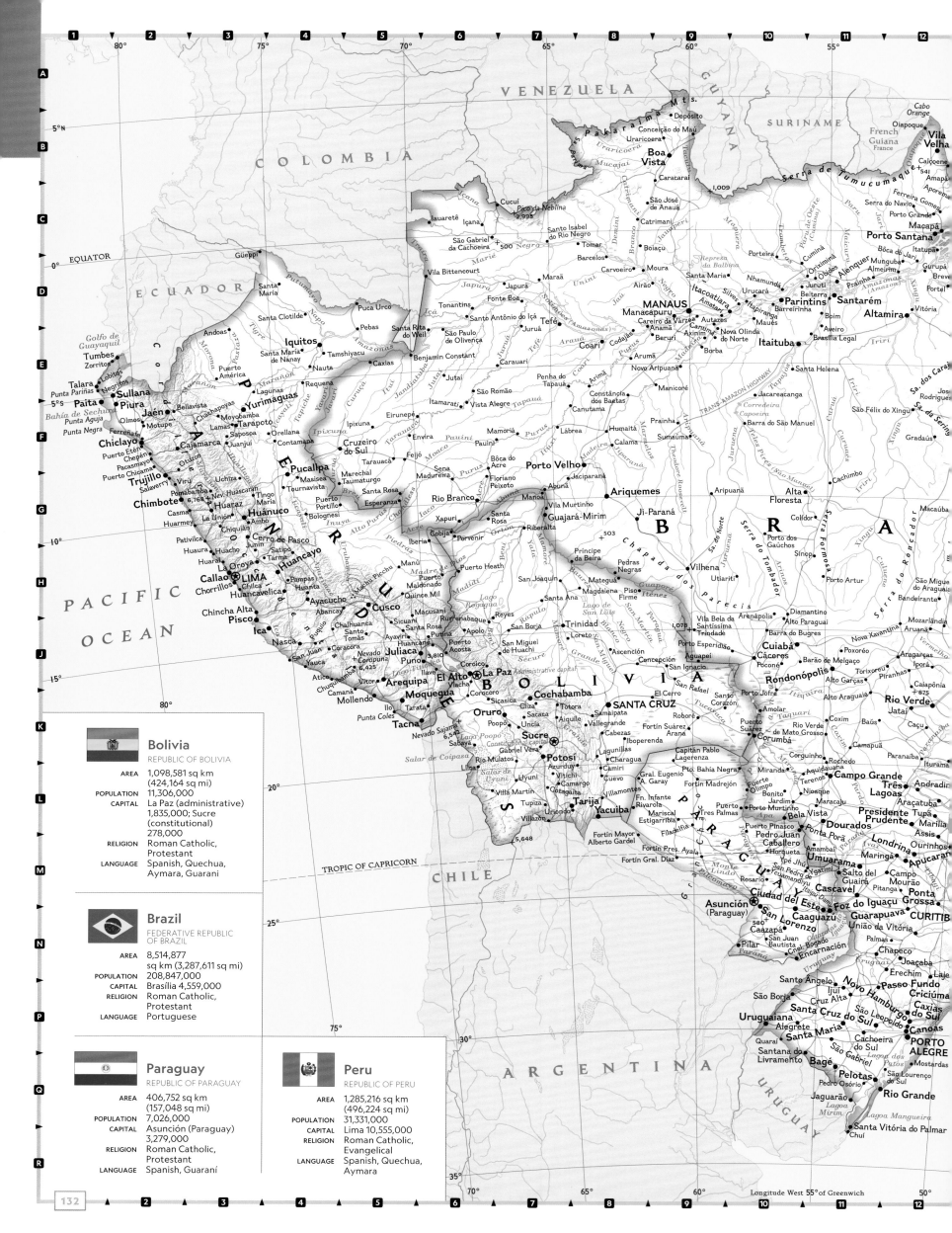

Bolivia
REPUBLIC OF BOLIVIA

AREA 1,098,581 sq km (424,164 sq mi)
POPULATION 11,306,000
CAPITAL La Paz (administrative) 1,835,000; Sucre (constitutional) 278,000
RELIGION Roman Catholic, Protestant
LANGUAGE Spanish, Quechua, Aymara, Guarani

Brazil
FEDERATIVE REPUBLIC OF BRAZIL

AREA 8,514,877 sq km (3,287,611 sq mi)
POPULATION 208,847,000
CAPITAL Brasília 4,559,000
RELIGION Roman Catholic, Protestant
LANGUAGE Portuguese

Paraguay
REPUBLIC OF PARAGUAY

AREA 406,752 sq km (157,048 sq mi)
POPULATION 7,026,000
CAPITAL Asunción (Paraguay) 3,279,000
RELIGION Roman Catholic, Protestant
LANGUAGE Spanish, Guaraní

Peru
REPUBLIC OF PERU

AREA 1,285,216 sq km (496,224 sq mi)
POPULATION 31,331,000
CAPITAL Lima 10,555,000
RELIGION Roman Catholic, Evangelical
LANGUAGE Spanish, Quechua, Aymara

CENTRAL SOUTH AMERICA

POPULATION DENSITY

People per Square Kilometer	People per Square Mile
More than 195	More than 500
60–195	150–500
10–59	25–149
1–9	1–24
Fewer than 1	Fewer than 1

CLIMATE
(Based on modified Köppen system)

Tropical (A)
No dry season (f), monsoonal (m), dry summer (s), dry winter (w)

Arid (B)
Desert (W), Steppe (S)

Temperate (C)
Warm winter, no dry season (f), warm dry winter (w)

Polar (E)
Tundra (T)

LAND COVER

- Evergreen forest
- Deciduous forest
- Mixed forest
- Savanna
- Shrubland
- Grassland
- Cropland
- Barren or sparsely vegetated
- Urban or built-up
- Snow and ice
- Cropland/vegetation mosaic
- Wetland

Azimuthal Equidistant Projection
SCALE 1:15,025,000 1 CENTIMETER = 150 KILOMETERS; 1 INCH = 237 MILES
KILOMETERS
STATUTE MILES
Elevations in meters

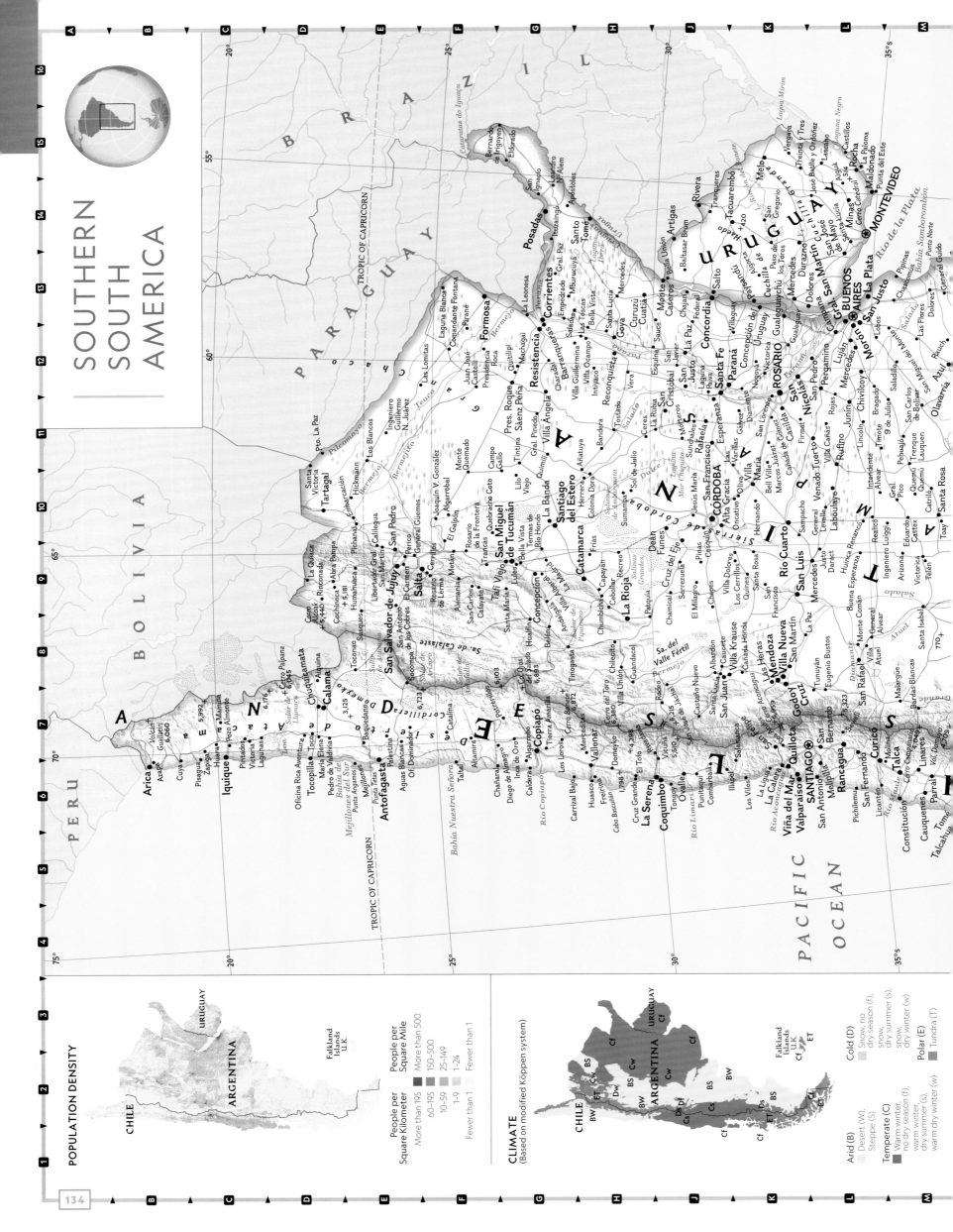

SOUTHERN SOUTH AMERICA

POPULATION DENSITY

People per Square Kilometer
- More than 195
- 60–195
- 10–59
- 1–9
- Fewer than 1

People per Square Mile
- More than 500
- 150–500
- 25–149
- 1–24
- Fewer than 1

CHILE
URUGUAY
ARGENTINA
Falkland Islands U.K.

CLIMATE
(Based on modified Köppen system)

Arid (B)
- Desert (W).
- Steppe (S)

Temperate (C)
- Warm winter, no dry season (f).
- warm winter, dry summer (s),
- warm dry winter (w)

Cold (D)
- Snow, no dry season (f),
- snow, dry summer (s),
- snow, dry winter (w)

Polar (E)
- Tundra (T)

CHILE
URUGUAY
ARGENTINA
Falkland Is. U.K.

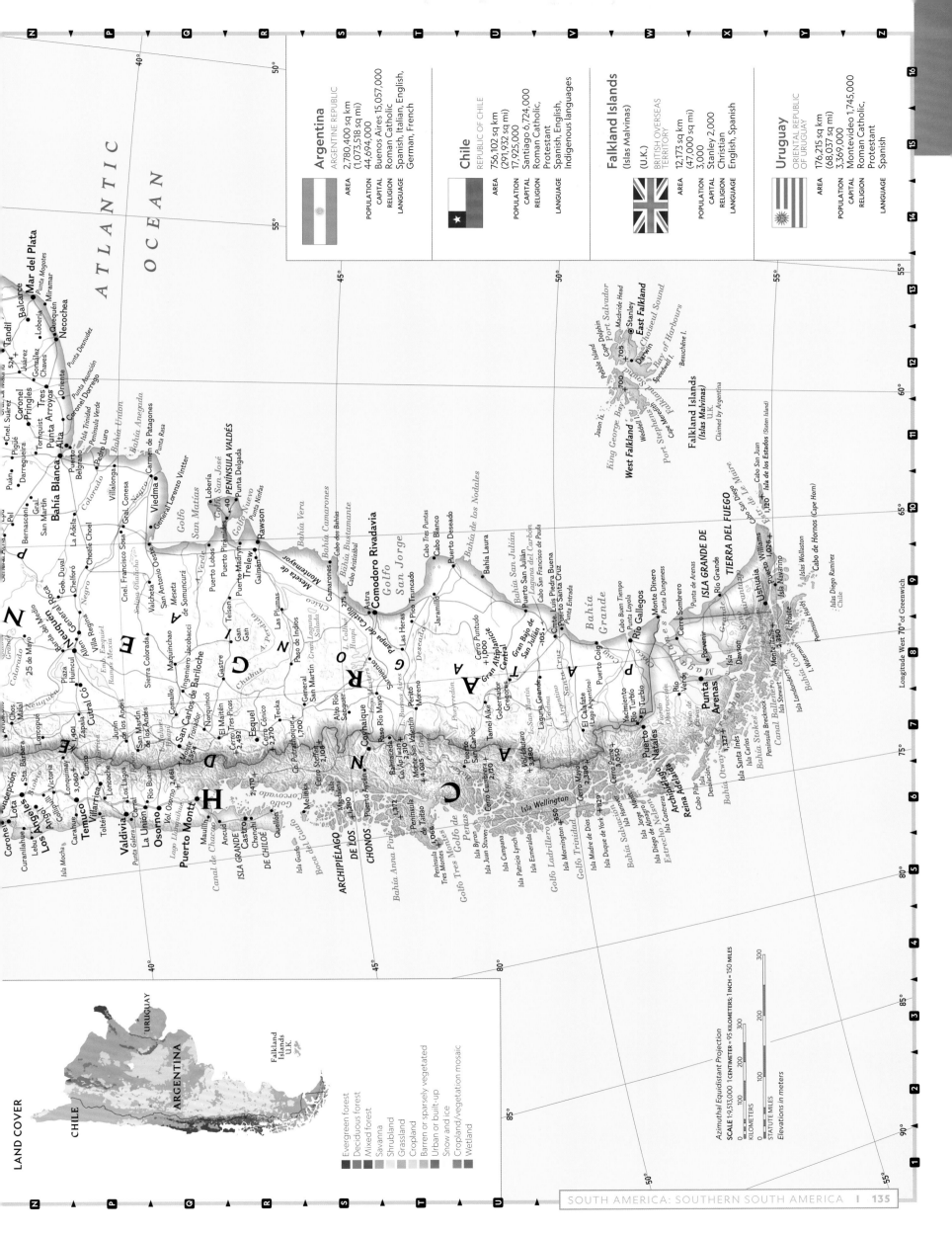

Argentina
ARGENTINE REPUBLIC

AREA 2,780,400 sq km (1,073,518 sq mi)
POPULATION 44,694,000
CAPITAL Buenos Aires 15,057,000
RELIGION Roman Catholic
LANGUAGE Spanish, Italian, English, German, French

Chile
REPUBLIC OF CHILE

AREA 756,102 sq km (291,932 sq mi)
POPULATION 17,925,000
CAPITAL Santiago 6,724,000
RELIGION Roman Catholic, Protestant
LANGUAGE Spanish, English, Indigenous languages

Falkland Islands
(Islas Malvinas)
(U.K.)
BRITISH OVERSEAS TERRITORY

AREA 12,173 sq km (47,000 sq mi)
POPULATION 3,000
CAPITAL Stanley 2,000
RELIGION Christian
LANGUAGE English, Spanish

Uruguay
ORIENTAL REPUBLIC OF URUGUAY

AREA 176,215 sq km (68,037 sq mi)
POPULATION 3,369,000
CAPITAL Montevideo 1,745,000
RELIGION Roman Catholic, Protestant
LANGUAGE Spanish

LAND COVER

- Evergreen forest
- Deciduous forest
- Mixed forest
- Savanna
- Shrubland
- Grassland
- Cropland
- Barren or sparsely vegetated
- Urban or built-up
- Snow and ice
- Cropland/vegetation mosaic
- Wetland

CHILE
URUGUAY
ARGENTINA
Falkland Islands
U.K.

Azimuthal Equidistant Projection
SCALE 1:9,515,000 1 CENTIMETER = 95 KILOMETERS 1 INCH = 150 MILES

KILOMETERS
0 100 200 300

STATUTE MILES
0 100 200 300

Elevations in meters

ATLANTIC OCEAN

EUROPE

EUROPE PHYSICAL AND POLITICAL ● NORTHERN EUROPE
BRITAIN AND IRELAND ● IBERIAN PENINSULA
BELGIUM, FRANCE, AND THE NETHERLANDS
CENTRAL EUROPE ● ITALY AND SWITZERLAND
THE BALKANS ● GREECE ● EASTERN EUROPE
RUSSIA ● SMALLEST COUNTRIES AND TERRITORIES

Santa Maria del Fiore, also known as the Duomo, in Florence, Italy

THE CROSSROADS OF POWER

A GLOBAL FORCE, FROM ANCIENT GREECE TO THE EUROPEAN UNION

Europe's densely populated lands, inhabited for more than 40,000 years, are home to 46 countries and more than 250 languages. Its mountains, rivers, plains, and seas have borne the brilliance of ancient civilizations and the horror of two modern world wars. Europe is bold and innovative, yet steeped in tradition—a continent of complexities.

GEOGRAPHY

DEFINED BY ICE & WATER

A cluster of peninsulas and islands frames the small but mighty continent of Europe. At its northernmost tip, icy tundra and boreal forests of Scandinavia give way to ancient, rugged highlands. Worn down by Ice Age glaciers, what remains of these mountains arc southwest through the British Isles. Several clusters of sharp peaks reach for the sky across southern Europe: the Pyrenees, the Alps, the Carpathians. The Alps are sometimes called the "water tower of Europe" because the glaciers feed major rivers, including the Danube, the Rhine, and the Rhône. Far to the east, the continent's longest river, the Volga, flows southeast across Russia to the Caspian Sea. Europe is a quieter place, seismically speaking, than most other continents, but rumblings in the Earth's crust trigger volcanic eruptions and earthquakes across the warm, hilly Mediterranean south.

Prominent physical features also help to define boundaries between countries and continents. The traditional land boundary with Asia (marked in yellow on pages 140–141) is a line following Russia's Ural Mountains from the Arctic Ocean south to the Ural River and on to the Caspian Sea, placing 23 percent of Russia in Europe. The continental border then continues west along the crest of the Caucasus Mountains between the Caspian and Black Seas, placing a small part of Turkey in Europe.

Italy's Roman Forum, now in ruins, was the heart of ancient Rome.

HISTORY

SMALL BUT MIGHTY

The story of Europe is a head-spinning forward march through time. Classical Greece made achievements in philosophy, mathematics, and science and bequeathed its legacy to the Roman Empire, which extended from Britain to Syria at the apex of its 500 years of power. The Byzantine Empire followed the Romans and in the eighth century, Muslims conquered the Iberian Peninsula and expanded as far as parts of Russia. Political power eventually shifted to Western Europe, where powerful monarchs of the Dutch Republic, England, France, Spain, and Portugal launched worldwide explorations to build mercantile empires. At home, ideas of the Age of Enlightenment promoted democracy and equality, illustrated most notably in the American and French Revolutions.

The 20th century brought seismic changes. Imperialism, militarism, defense alliances, and nationalism ignited two world wars; the Russian Revolution introduced communism. The Cold War, between the superpowers United States and Russia and their allies, swept across the continent. The conflict pitted capitalism and democracy against communism and state control. The Berlin Wall, built in 1961 to divide East and West Germany, came to symbolize these new, stark divisions until the collapse of the Soviet Union in 1991 and the fall of the wall reunited the continent.

In Norway's Geirangerfjord, snowcapped mountain peaks feed tumbling waterfalls.

CULTURE

FROM ANTIQUITY TO THE INTERNET

The artistic treasures of Europe span centuries: classical Greek sculpture, Gothic cathedrals, Renaissance art, Shakespearean plays, the music of Mozart, and Cubist paintings by Picasso. Despite Europeans' diverse tastes and traditions, they have many commonalities. Most of their languages have Germanic, Romance, or Slavic roots, and social structures nearly everywhere are based on economic class. Christianity binds Europeans together, with Orthodoxy dominating in the east, Protestantism in the north, and Roman Catholicism in the west and south. Epic rivalries and the Protestant Reformation generated centuries of war, but today the former European empires are allies. Peace prevails, though nationalism and long-simmering ethnic hostilities have at times flared into violence and xenophobia.

Europeans have a rich cultural heritage, whether it is expressed in Irish step dancing, Balkan hand-woven textiles, or thin-crust Neapolitan pizza. More recently, immigrants and refugees, many from Europe's former territories, have brought their own habits, religions, and languages to the continent. Along with the strong influence of American culture, these changes have political as well as social consequences. Today, people move freely throughout the continent, sharing the same pop culture, urbanized and often secular lifestyles, and heavy reliance on the internet.

Business workers in London's financial district

ECONOMY

EMPIRES FALL, FORTUNES SHIFT

Europe has fertile soil, ample natural resources, and access to large rivers and the sea. But for centuries, politics has done just as much as nature to shape Europe's economic destiny. Propelled by tremendous wealth stolen from their faraway colonies, by the end of the 19th century Europeans dominated world commerce, led the world in science and invention, and sparked the industrial revolution.

In the 20th century, weakened by two world wars fought largely on its land, Europe lost its colonial stance of dominance yet retained its economic influence. Western Europe, backed by the U.S., prospered with market economies, democracy, and free speech, while Eastern and Central European countries, their centrally controlled economies closely tied to the Soviet Union's, fell behind. As the Cold War ended in the late 20th century, the European Community matured from a small economic alliance among six countries into a powerful economic and political union called the European Union (EU). Today, its 28 member nations form a single economic market and conduct nearly two-thirds of their trade with other EU countries. With a population of more than half a billion, the EU is one of the largest economies in the world but is not without challenges: The United Kingdom, a member since 1973, voted in 2016 to leave the Union.

A market in Tarragona, Spain, in the Catalonia region

SNAPSHOT | *Europe* FACTS & FEATURES

LONGEST COASTLINE Norway: 58,133 km (36,122 mi)

HIGHEST ELEVATION El'brus, Russia: 5,642 m (18,510 ft)

LOWEST ELEVATION Caspian Sea: –28 m (–92 ft)

LONGEST RIVER Volga, Russia: 3,692 km (2,294 mi)

BIGGEST URBAN AREA BY POPULATION Moscow, Russia: 12.4 million (in 2018)

MOST POPULOUS COUNTRY ENTIRELY IN EUROPE Germany: 80,458,000 (2018)

WORLD'S LEAST POPULOUS COUNTRY Vatican City: 618 (2019)

WORLD'S HAPPIEST PLACE Finland: ranks #1 in World Happiness Report, 2019

OLDEST CITY Plovdiv, Bulgaria: inhabited for more than 6,000 years

OLDEST OPERATING RAIL STATION Leipzig, Germany: Bayerischer Bahnhof

CYPRUS LIES AT A **STRATEGIC JUNCTURE** OF THE MEDITERRANEAN. THE ISLAND HAS BEEN RULED BY **ASSYRIA, EGYPT, PERSIA, MACEDONIA, ROME, BYZANTIUM, VENICE, THE OTTOMAN TURKS, AND THE UNITED KINGDOM.**

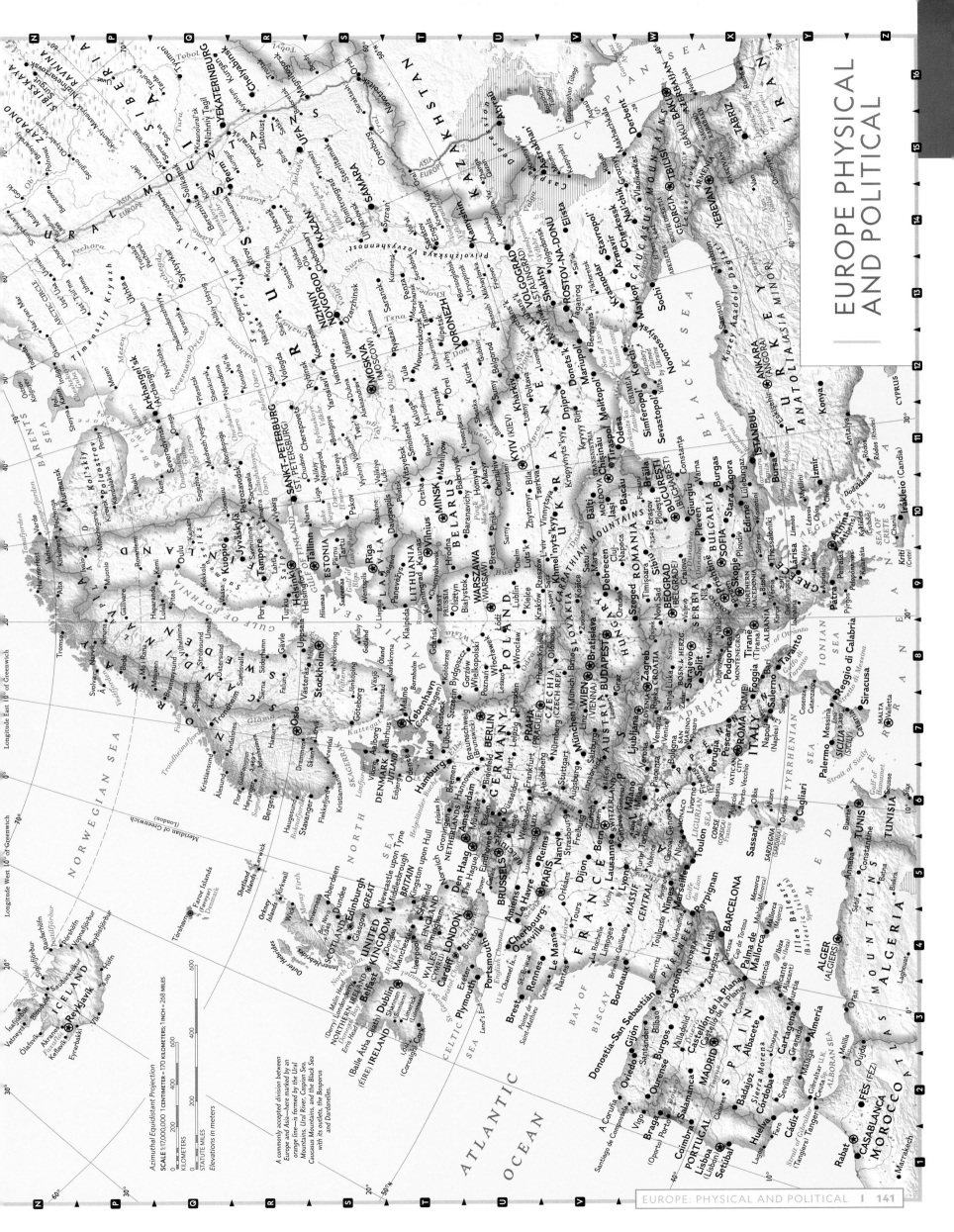

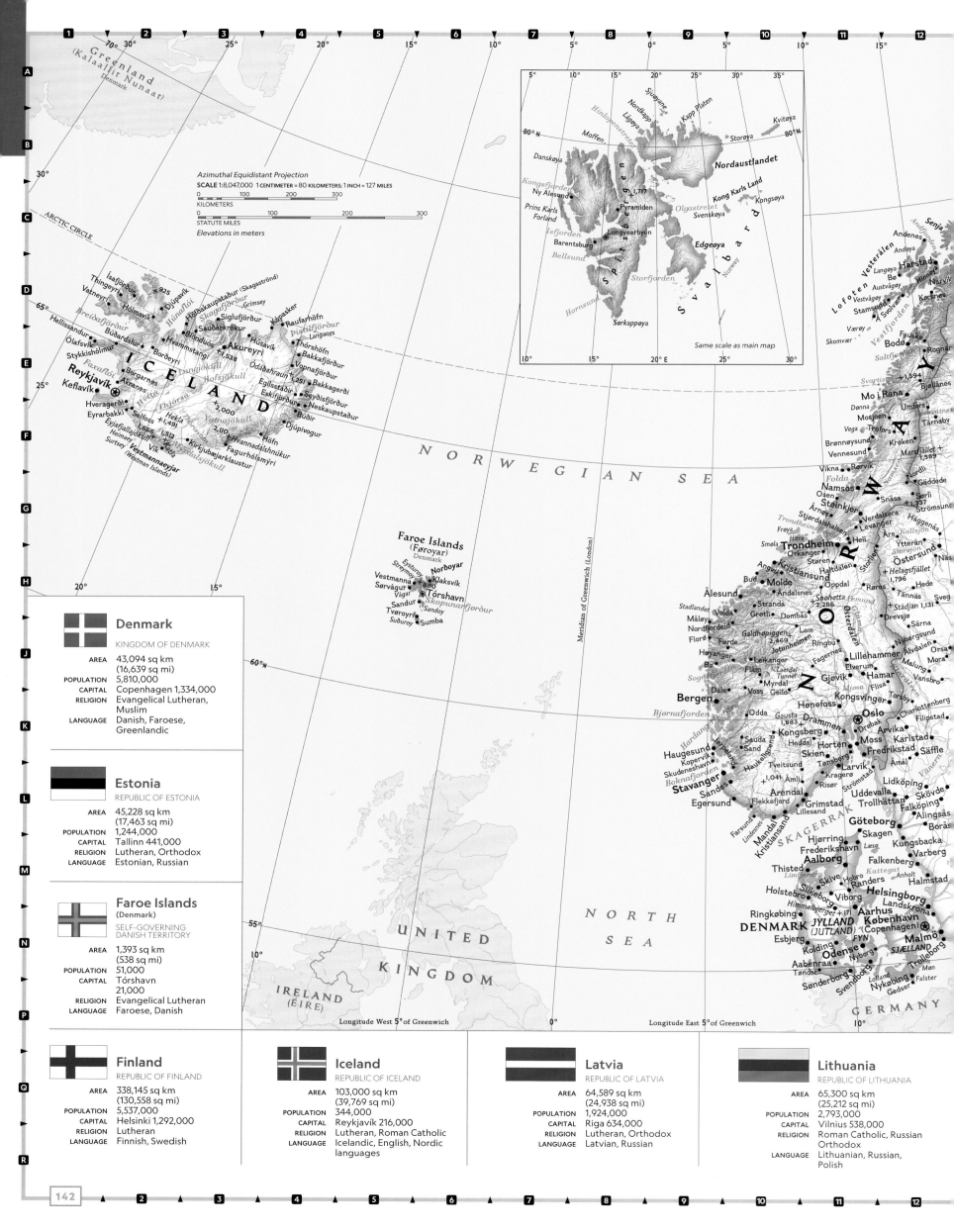

Azimuthal Equidistant Projection
SCALE 1:8,047,000 1 CENTIMETER = 80 KILOMETERS; 1 INCH = 127 MILES
KILOMETERS
STATUTE MILES
Elevations in meters

Denmark
KINGDOM OF DENMARK
AREA	43,094 sq km (16,639 sq mi)
POPULATION	5,810,000
CAPITAL	Copenhagen 1,334,000
RELIGION	Evangelical Lutheran, Muslim
LANGUAGE	Danish, Faroese, Greenlandic

Estonia
REPUBLIC OF ESTONIA
AREA	45,228 sq km (17,463 sq mi)
POPULATION	1,244,000
CAPITAL	Tallinn 441,000
RELIGION	Lutheran, Orthodox
LANGUAGE	Estonian, Russian

Faroe Islands
(Denmark)
SELF-GOVERNING DANISH TERRITORY
AREA	1,393 sq km (538 sq mi)
POPULATION	51,000
CAPITAL	Tórshavn 21,000
RELIGION	Evangelical Lutheran
LANGUAGE	Faroese, Danish

Finland
REPUBLIC OF FINLAND
AREA	338,145 sq km (130,558 sq mi)
POPULATION	5,537,000
CAPITAL	Helsinki 1,292,000
RELIGION	Lutheran
LANGUAGE	Finnish, Swedish

Iceland
REPUBLIC OF ICELAND
AREA	103,000 sq km (39,769 sq mi)
POPULATION	344,000
CAPITAL	Reykjavík 216,000
RELIGION	Lutheran, Roman Catholic
LANGUAGE	Icelandic, English, Nordic languages

Latvia
REPUBLIC OF LATVIA
AREA	64,589 sq km (24,938 sq mi)
POPULATION	1,924,000
CAPITAL	Riga 634,000
RELIGION	Lutheran, Orthodox
LANGUAGE	Latvian, Russian

Lithuania
REPUBLIC OF LITHUANIA
AREA	65,300 sq km (25,212 sq mi)
POPULATION	2,793,000
CAPITAL	Vilnius 538,000
RELIGION	Roman Catholic, Russian Orthodox
LANGUAGE	Lithuanian, Russian, Polish

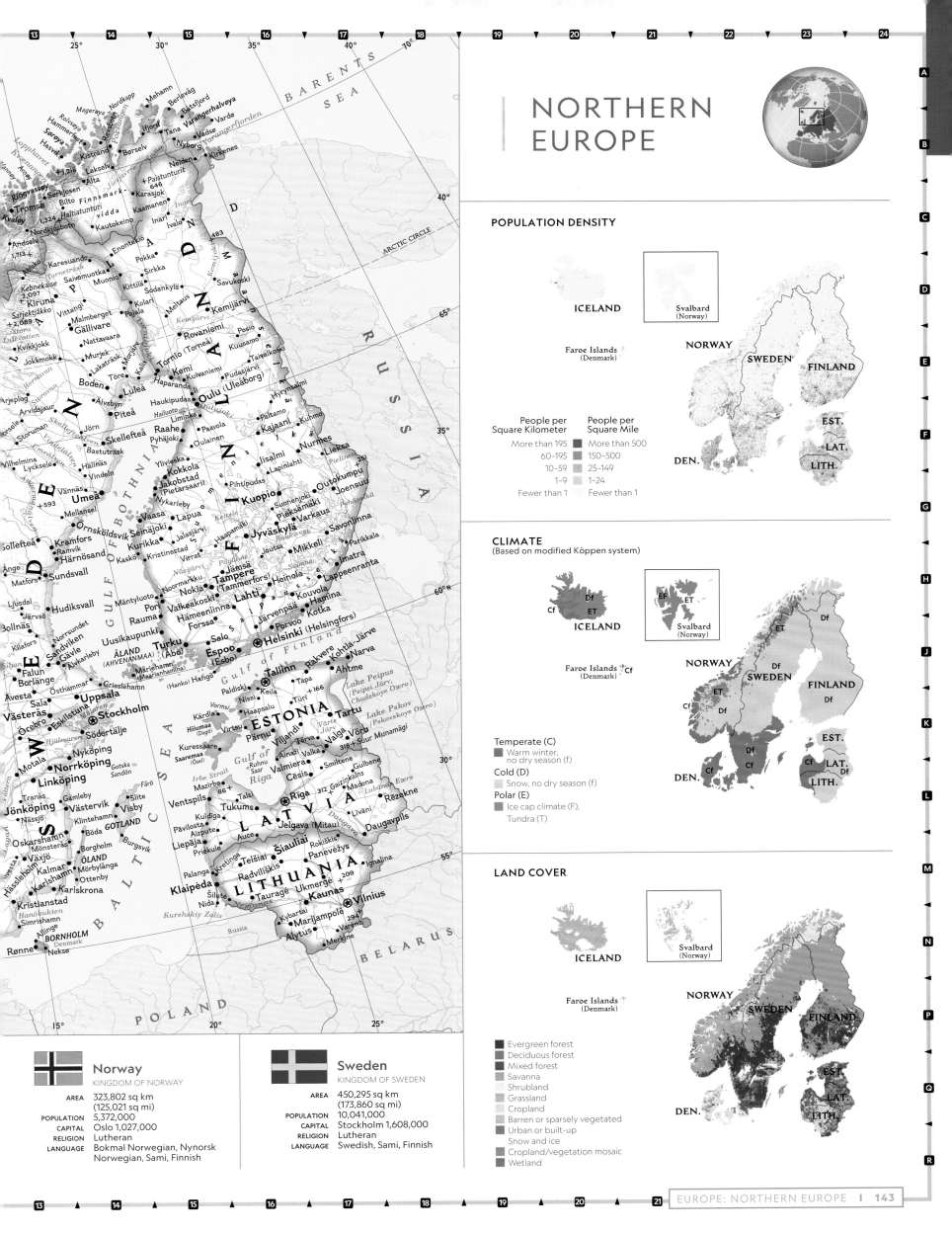

NORTHERN EUROPE

POPULATION DENSITY

People per Square Kilometer	People per Square Mile
More than 195	More than 500
60–195	150–500
10–59	25–149
1–9	1–24
Fewer than 1	Fewer than 1

ICELAND

Svalbard (Norway)

Faroe Islands (Denmark)

NORWAY · SWEDEN · FINLAND · EST. · LAT. · LITH. · DEN.

CLIMATE
(Based on modified Köppen system)

ICELAND · Df · ET · Cf

EF · ET · Svalbard (Norway)

Faroe Islands (Denmark) Cf

NORWAY · SWEDEN · FINLAND · DEN. · EST. · LAT. · LITH.

Temperate (C)
Warm winter, no dry season (f)

Cold (D)
Snow, no dry season (f)

Polar (E)
Ice cap climate (F), Tundra (T)

LAND COVER

ICELAND

Svalbard (Norway)

Faroe Islands (Denmark)

NORWAY · SWEDEN · FINLAND · DEN. · EST. · LAT. · LITH.

Evergreen forest
Deciduous forest
Mixed forest
Savanna
Shrubland
Grassland
Cropland
Barren or sparsely vegetated
Urban or built-up
Snow and ice
Cropland/vegetation mosaic
Wetland

Norway
KINGDOM OF NORWAY

AREA 323,802 sq km (125,021 sq mi)
POPULATION 5,372,000
CAPITAL Oslo 1,027,000
RELIGION Lutheran
LANGUAGE Bokmal Norwegian, Nynorsk Norwegian, Sami, Finnish

Sweden
KINGDOM OF SWEDEN

AREA 450,295 sq km (173,860 sq mi)
POPULATION 10,041,000
CAPITAL Stockholm 1,608,000
RELIGION Lutheran
LANGUAGE Swedish, Sami, Finnish

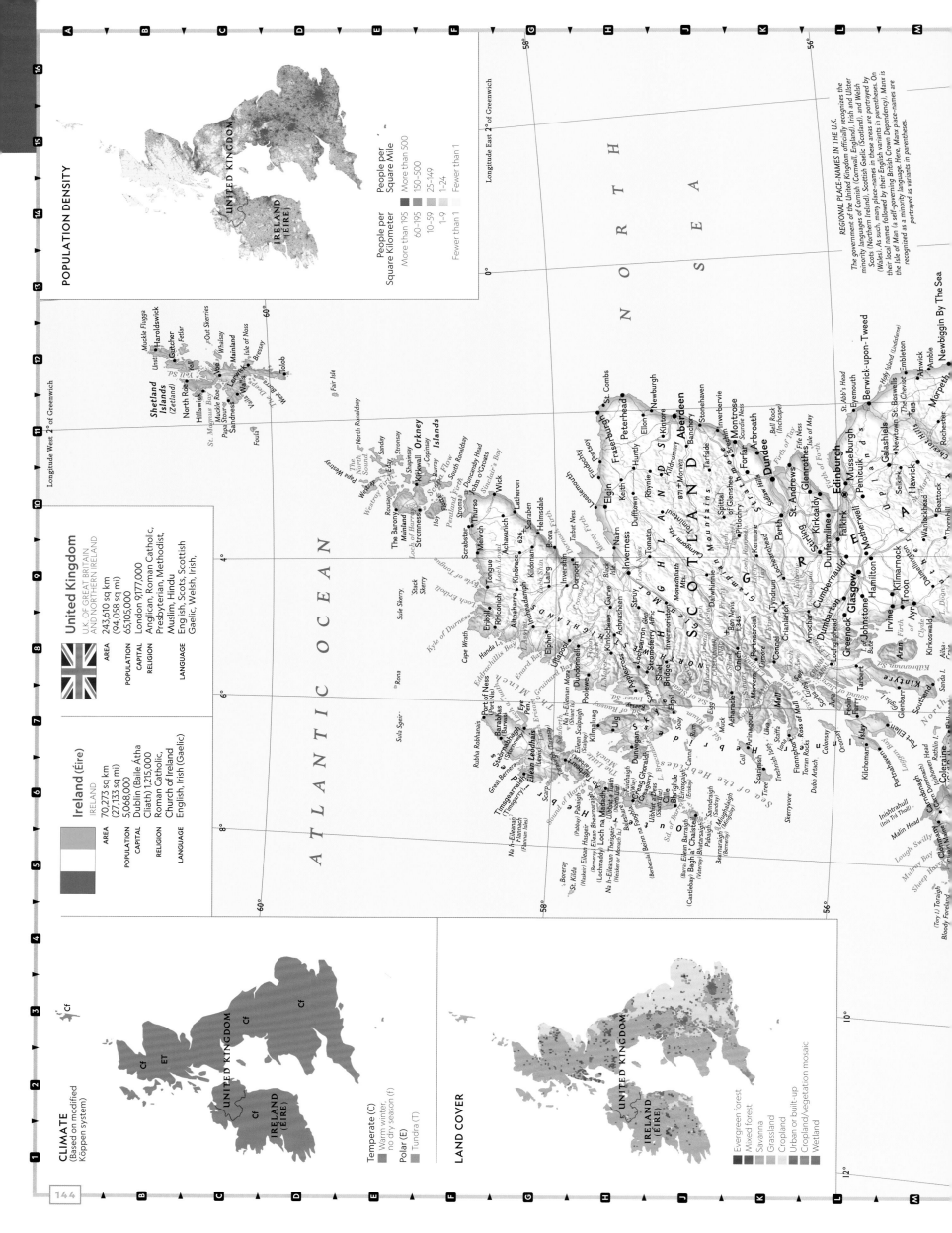

POPULATION DENSITY

People per Square Kilometer
- More than 195
- 60–195
- 10–59
- 1–9
- Fewer than 1

People per Square Mile
- More than 500
- 150–500
- 25–149
- 1–24
- Fewer than 1

UNITED KINGDOM
IRELAND (ÉIRE)

REGIONAL PLACE-NAMES IN THE U.K.
The government of the United Kingdom officially recognizes the minority languages of Cornish (Cornwall, England), Irish and Ulster Scots (Northern Ireland), Scottish Gaelic (Scotland), and Welsh (Wales). As such, many place-names in these areas are portrayed by their local names followed by their English variants in parentheses). Manx is the language of the Isle of Man (a self-governing British Crown Dependency). Here, Manx place-names are recognized as a minority language, portrayed as variants in parentheses.

Ireland (Éire)
IRELAND

AREA	70,273 sq km (27,133 sq mi)
POPULATION	5,068,000
CAPITAL	Dublin (Baile Átha Cliath) 1,215,000
RELIGION	Roman Catholic, Church of Ireland
LANGUAGE	English, Irish (Gaelic)

United Kingdom
U.K. OF GREAT BRITAIN AND NORTHERN IRELAND

AREA	243,610 sq km (94,058 sq mi)
POPULATION	65,105,000
CAPITAL	London 9,177,000
RELIGION	Anglican, Roman Catholic, Presbyterian, Methodist, Muslim, Hindu
LANGUAGE	English, Scots, Scottish Gaelic, Welsh, Irish,

CLIMATE
(Based on modified Köppen system)

Temperate (C)
- Cf Warm winter, no dry season (f)

Polar (E)
- ET Tundra (T)

UNITED KINGDOM
IRELAND (ÉIRE)

LAND COVER
- Evergreen forest
- Mixed forest
- Savanna
- Grassland
- Cropland
- Urban or built-up
- Cropland/vegetation mosaic
- Wetland

UNITED KINGDOM
IRELAND (ÉIRE)

ATLANTIC OCEAN

NORTH SEA

SCOTLAND

HIGHLANDS

Shetland Islands (Zetland)
Orkney Islands

144

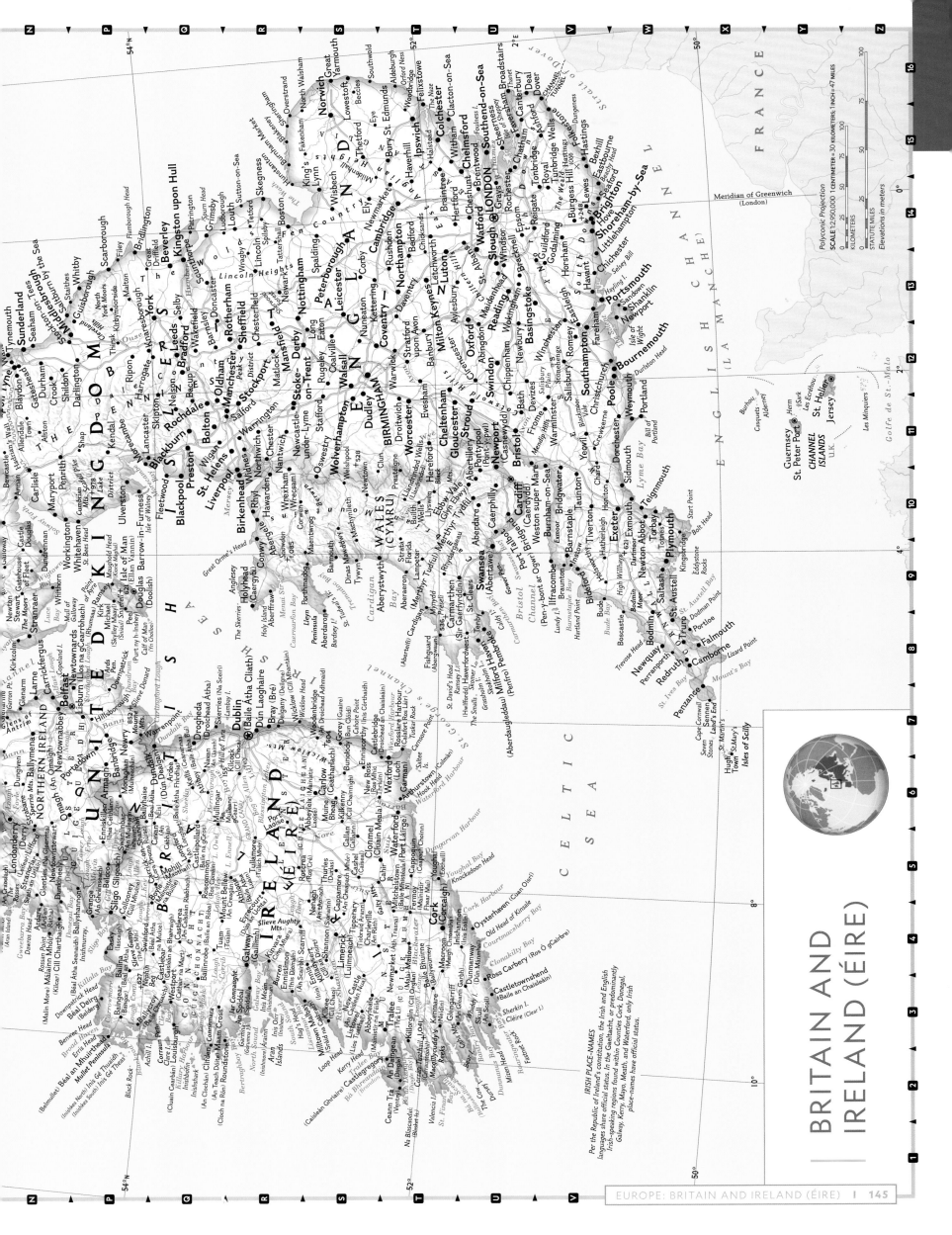

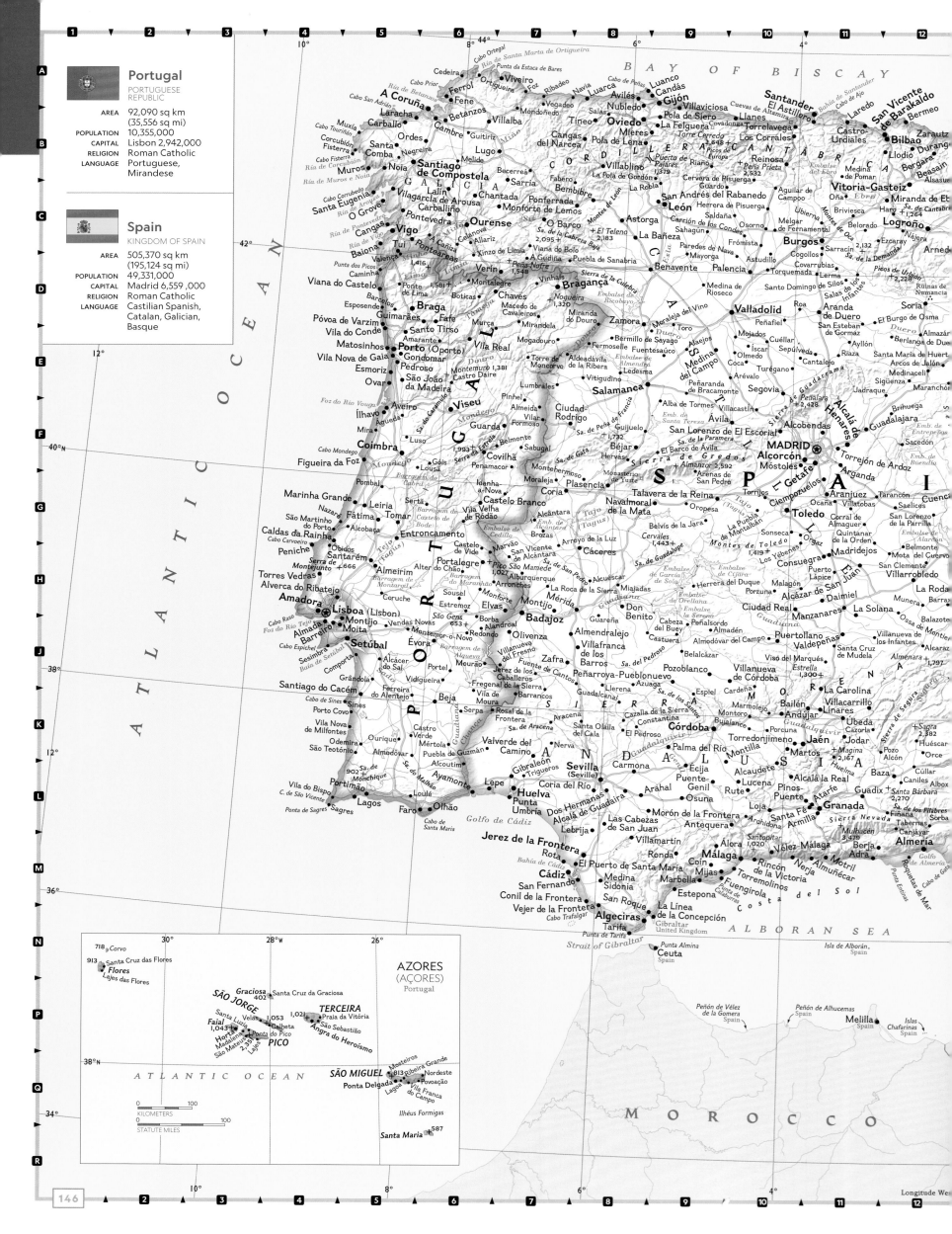

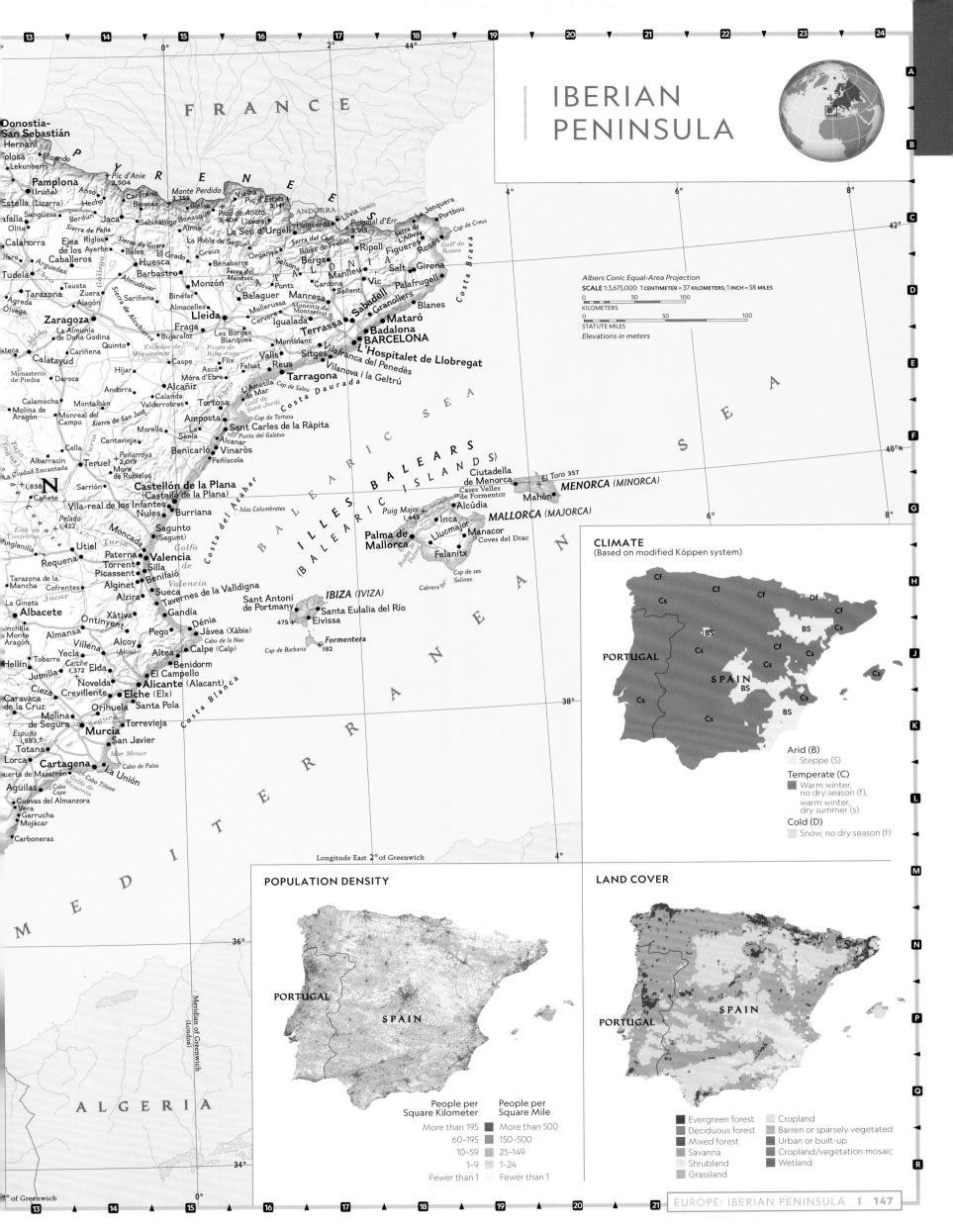

IBERIAN PENINSULA

FRANCE

SCALE 1:3,675,000 1 CENTIMETER = 37 KILOMETERS; 1 INCH = 58 MILES
Albers Conic Equal-Area Projection

KILOMETERS
0   50   100

STATUTE MILES
0   50   100

Elevations in meters

BALEARIC SEA

ILLES BALEARS
BALEARIC ISLANDS

MENORCA (MINORCA)
MALLORCA (MAJORCA)
IBIZA (IVIZA)

MEDITERRANEAN SEA

ALGERIA

CLIMATE
(Based on modified Köppen system)

PORTUGAL
SPAIN

Arid (B)
Steppe (S)

Temperate (C)
Warm winter, no dry season (f),
warm winter, dry summer (s)

Cold (D)
Snow, no dry season (f)

POPULATION DENSITY

PORTUGAL
SPAIN

People per Square Kilometer	People per Square Mile
More than 195	More than 500
60–195	150–500
10–59	25–149
1–9	1–24
Fewer than 1	Fewer than 1

LAND COVER

PORTUGAL
SPAIN

Evergreen forest
Deciduous forest
Mixed forest
Savanna
Shrubland
Grassland
Cropland
Barren or sparsely vegetated
Urban or built-up
Cropland/vegetation mosaic
Wetland

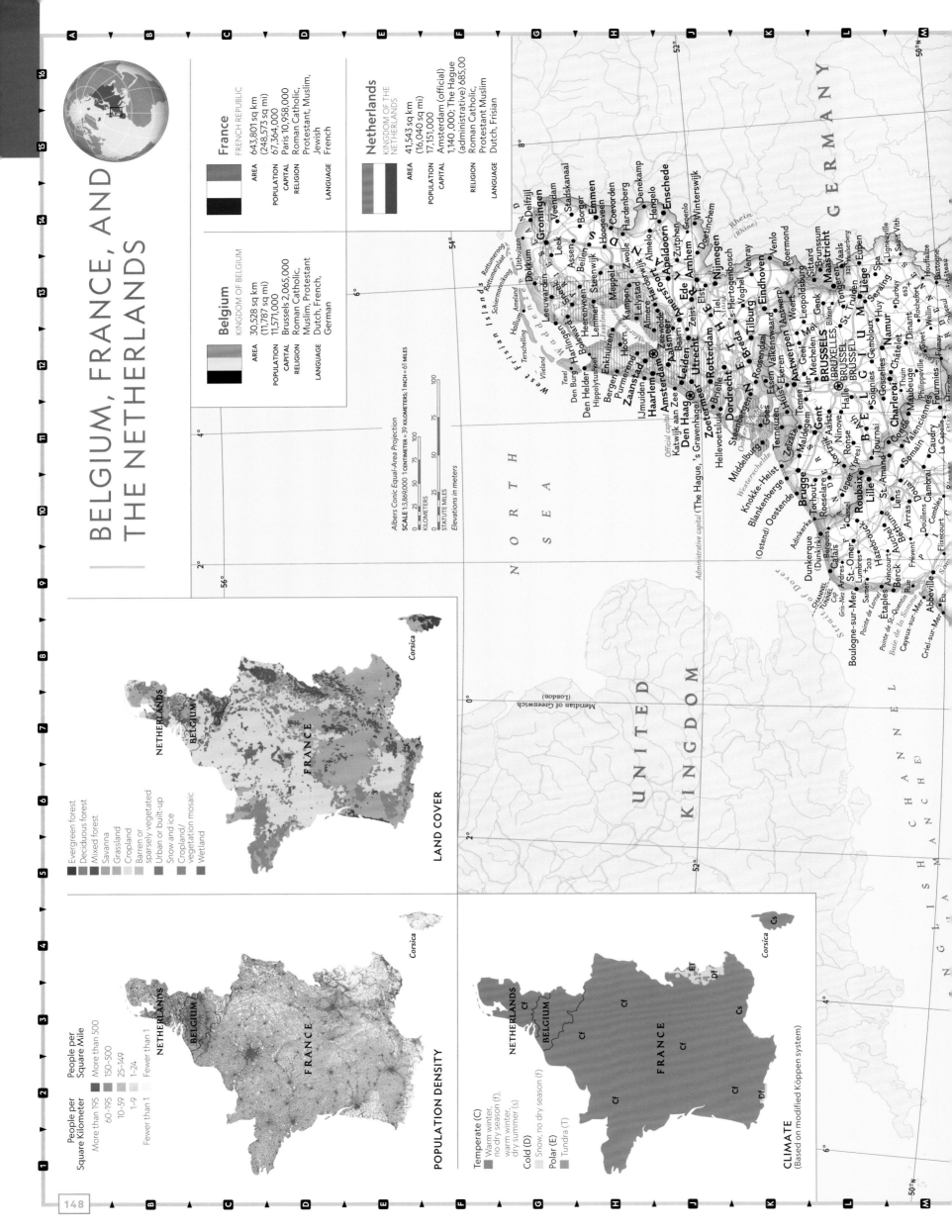

BELGIUM, FRANCE, AND THE NETHERLANDS

France
FRENCH REPUBLIC
AREA 643,801 sq km (248,573 sq mi)
POPULATION 67,364,000
CAPITAL Paris 10,958,000
RELIGION Roman Catholic, Protestant, Muslim, Jewish
LANGUAGE French

Netherlands
KINGDOM OF THE NETHERLANDS
AREA 41,543 sq km (16,040 sq mi)
POPULATION 17,151,000
CAPITAL Amsterdam (official) 1,140,000; The Hague (administrative) 685,00
RELIGION Roman Catholic, Protestant Muslim
LANGUAGE Dutch, Frisian

Belgium
KINGDOM OF BELGIUM
AREA 30,528 sq km (11,787 sq mi)
POPULATION 11,571,000
CAPITAL Brussels 2,065,000
RELIGION Roman Catholic, Muslim, Protestant
LANGUAGE Dutch, French, German

Albers Conic Equal-Area Projection
SCALE 1:3,869,000 1 CENTIMETER = 39 KILOMETERS; 1 INCH = 61 MILES
KILOMETERS
STATUTE MILES
Elevations in meters

LAND COVER

Evergreen forest
Deciduous forest
Mixed forest
Savanna
Grassland
Cropland
Barren or sparsely vegetated
Urban or built-up
Snow and ice
Cropland/vegetation mosaic
Wetland

POPULATION DENSITY

People per Square Kilometer
People per Square Mile
More than 195 — More than 500
60-195 — 150-500
10-59 — 25-149
1-9 — 1-24
Fewer than 1 — Fewer than 1

CLIMATE
(Based on modified Köppen system)

Temperate (C)
Warm winter, no dry season (f), warm winter, dry summer (s)
Cold (D)
Snow, no dry season (f)
Polar (E)
Tundra (T)

NORTH SEA

GERMANY

UNITED KINGDOM

BELGIUM

NETHERLANDS

FRANCE

Corsica

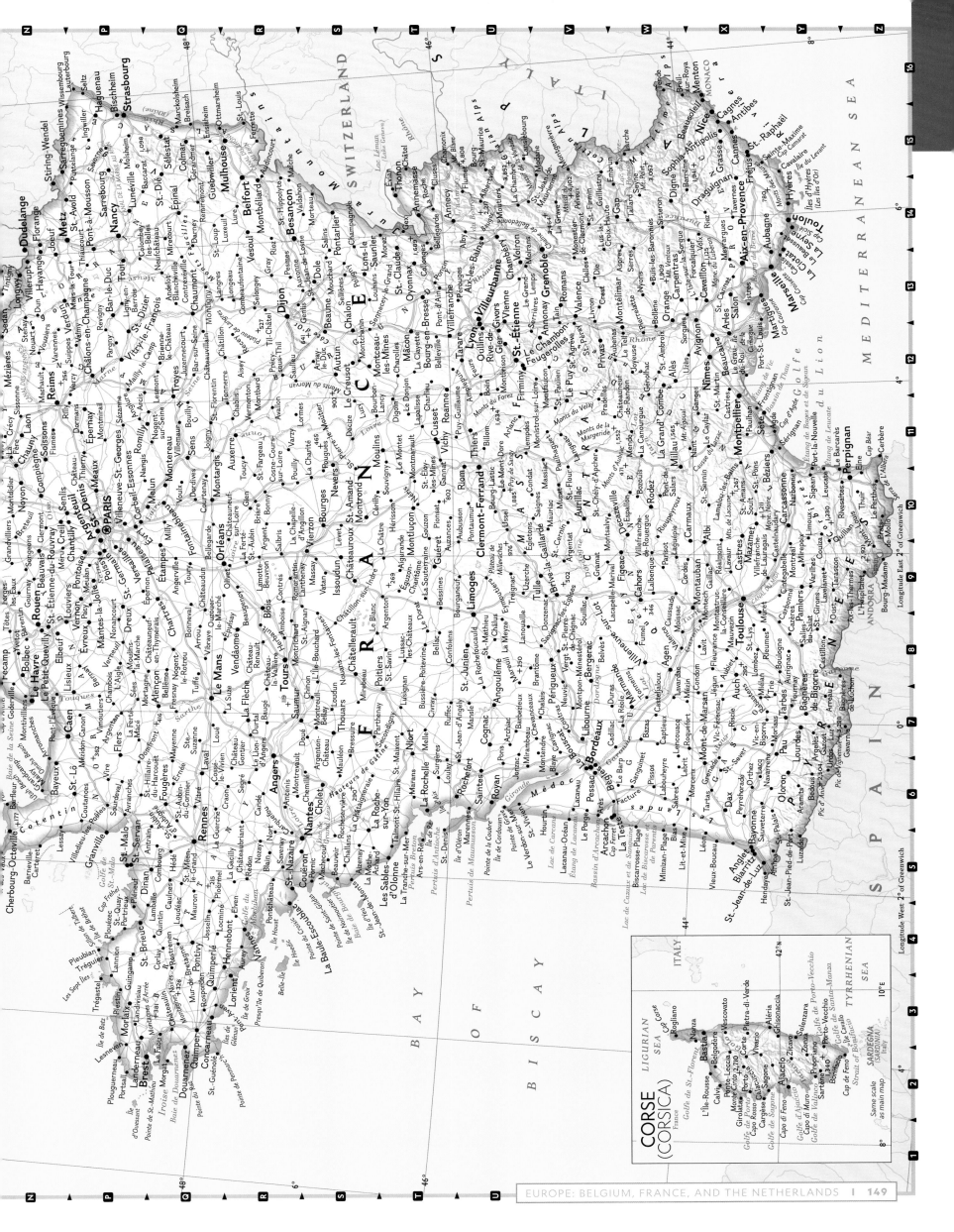

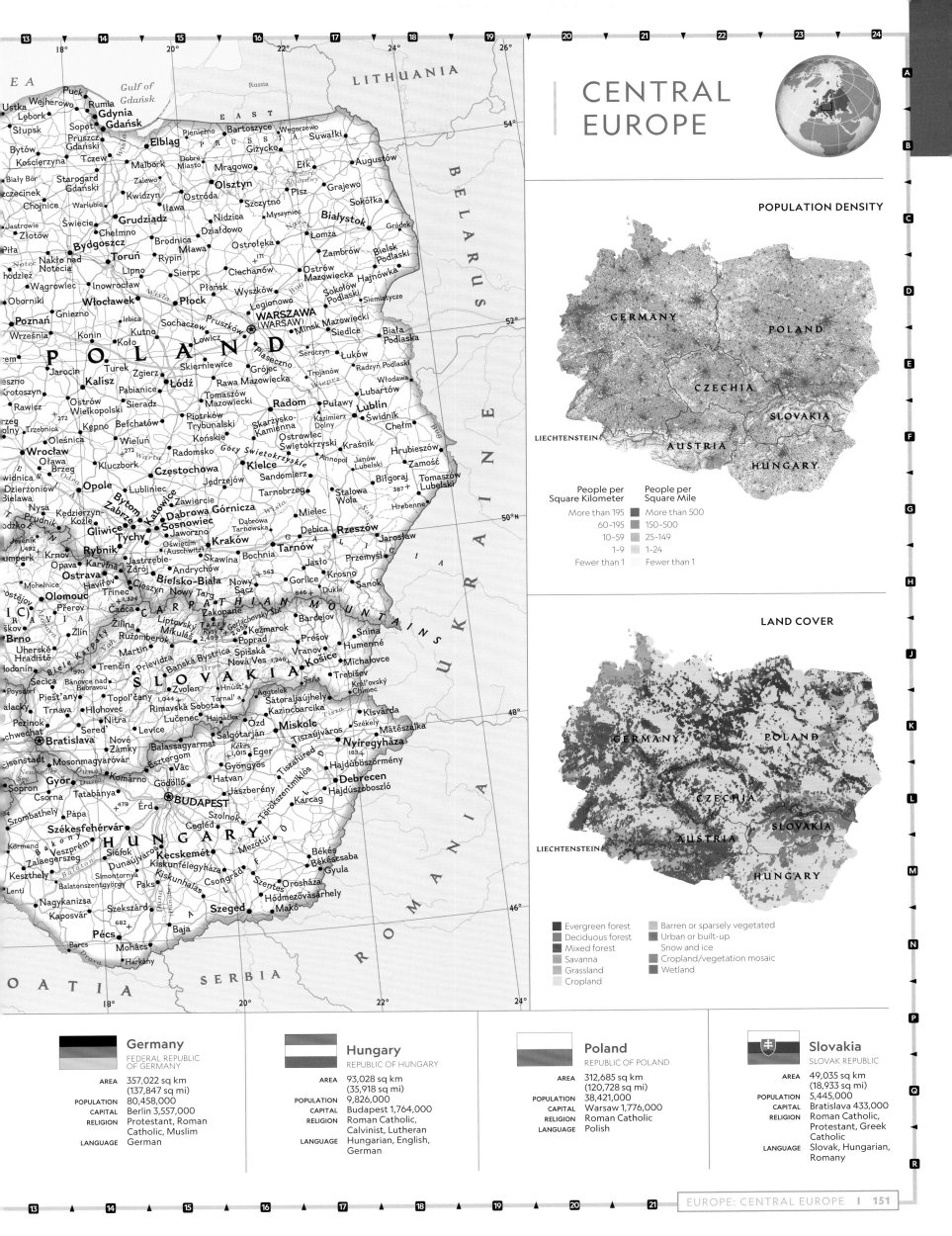

CENTRAL EUROPE

POPULATION DENSITY

People per Square Kilometer	People per Square Mile
More than 195	More than 500
60–195	150–500
10–59	25–149
1–9	1–24
Fewer than 1	Fewer than 1

LAND COVER

- Evergreen forest
- Deciduous forest
- Mixed forest
- Savanna
- Grassland
- Cropland
- Barren or sparsely vegetated
- Urban or built-up
- Snow and ice
- Cropland/vegetation mosaic
- Wetland

Germany
FEDERAL REPUBLIC OF GERMANY
- AREA 357,022 sq km (137,847 sq mi)
- POPULATION 80,458,000
- CAPITAL Berlin 3,557,000
- RELIGION Protestant, Roman Catholic, Muslim
- LANGUAGE German

Hungary
REPUBLIC OF HUNGARY
- AREA 93,028 sq km (35,918 sq mi)
- POPULATION 9,826,000
- CAPITAL Budapest 1,764,000
- RELIGION Roman Catholic, Calvinist, Lutheran
- LANGUAGE Hungarian, English, German

Poland
REPUBLIC OF POLAND
- AREA 312,685 sq km (120,728 sq mi)
- POPULATION 38,421,000
- CAPITAL Warsaw 1,776,000
- RELIGION Roman Catholic
- LANGUAGE Polish

Slovakia
SLOVAK REPUBLIC
- AREA 49,035 sq km (18,933 sq mi)
- POPULATION 5,445,000
- CAPITAL Bratislava 433,000
- RELIGION Roman Catholic, Protestant, Greek Catholic
- LANGUAGE Slovak, Hungarian, Romany

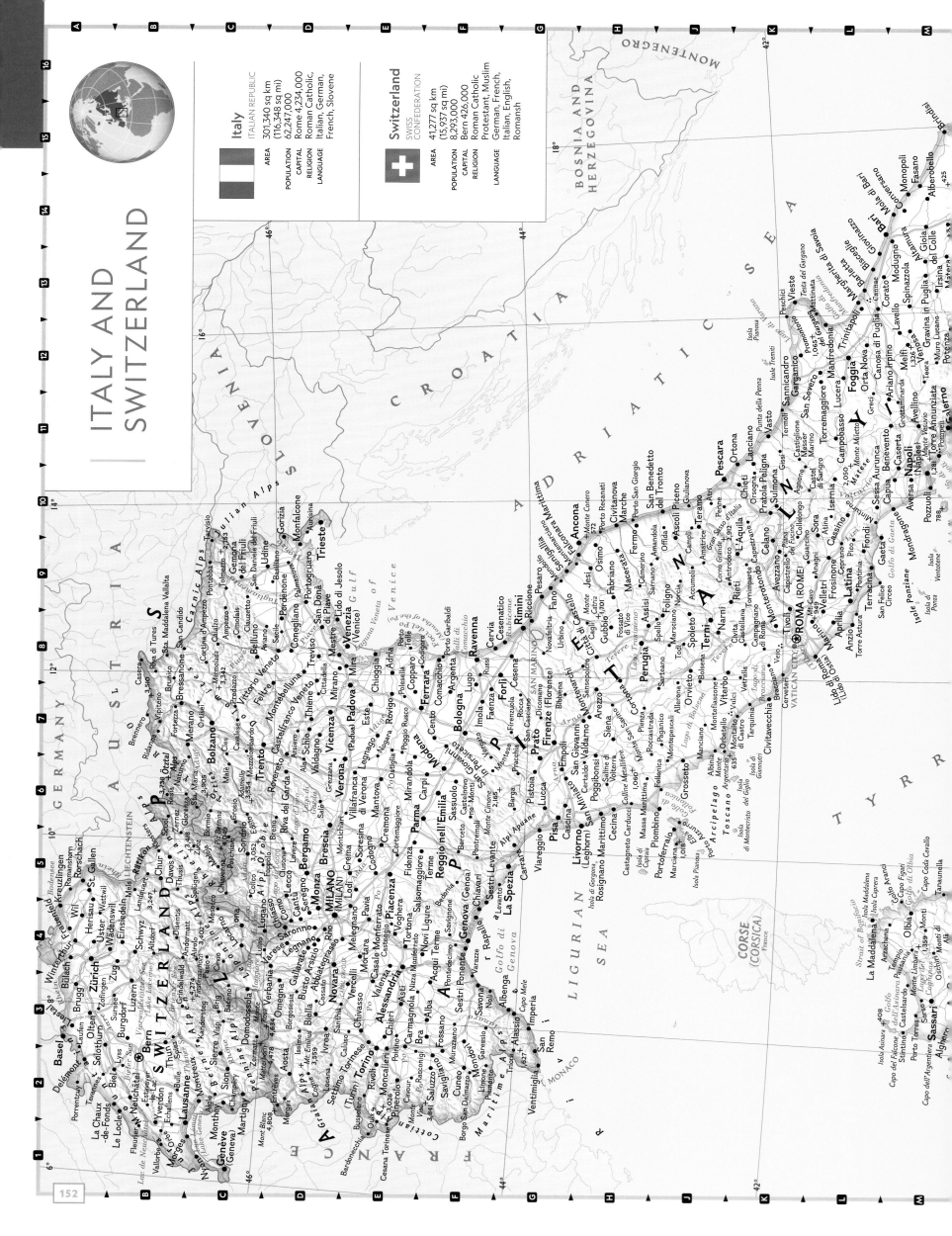

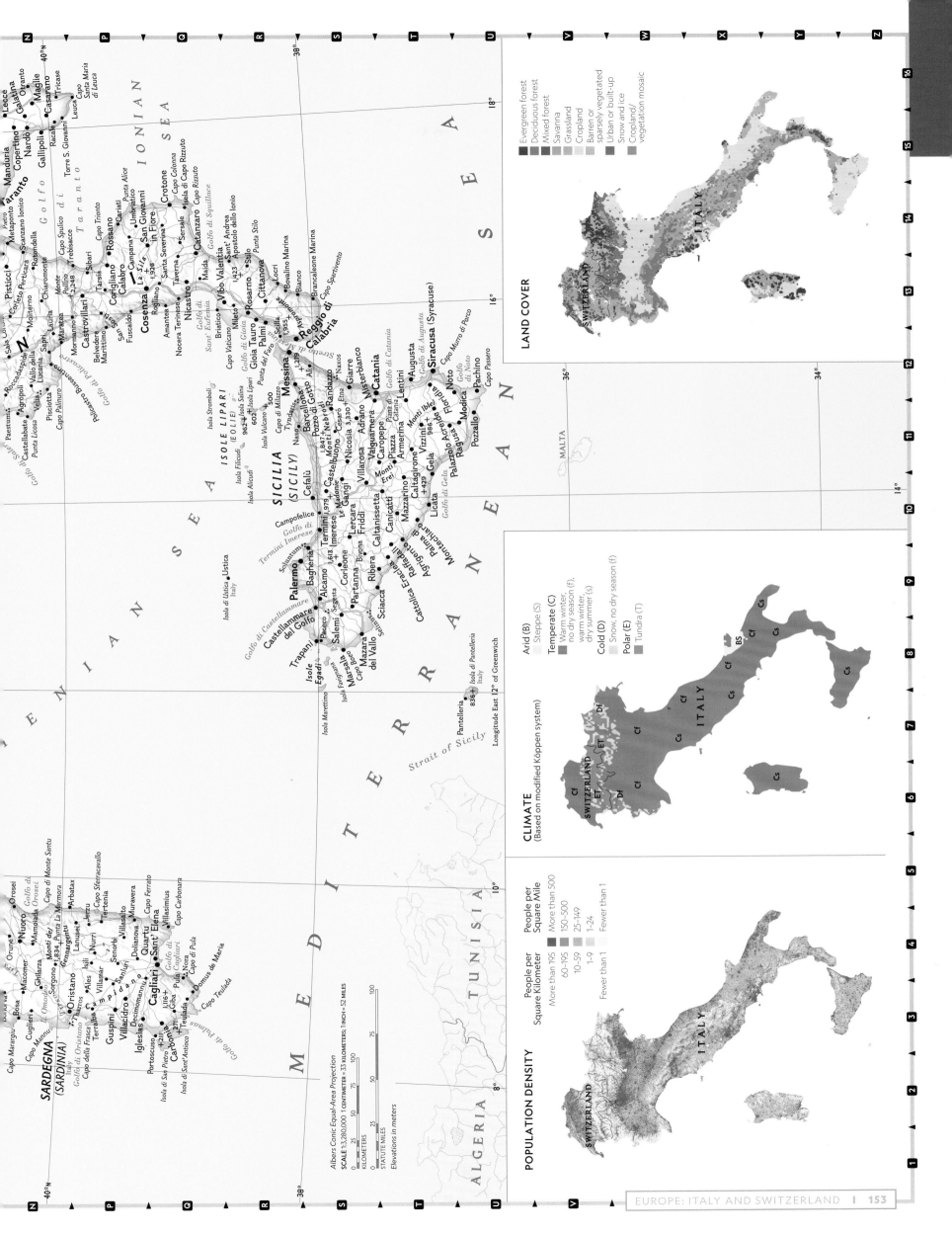

POPULATION DENSITY

People per Square Kilometer	People per Square Mile
More than 195	More than 500
60–195	150–500
10–59	25–149
1–9	1–24
Fewer than 1	Fewer than 1

CLIMATE
(Based on modified Köppen system)

Arid (B)
Steppe (S)

Temperate (C)
Warm winter, no dry season (f),
warm winter, dry summer (s)

Cold (D)
Snow, no dry season (f)

Polar (E)
Tundra (T)

LAND COVER

Evergreen forest
Deciduous forest
Mixed forest
Savanna
Grassland
Cropland
Barren or sparsely vegetated
Urban or built-up
Snow and ice
Cropland/ vegetation mosaic

Albers Conic Equal-Area Projection
SCALE 1:3,280,000 1 CENTIMETER = 33 KILOMETERS; 1 INCH = 52 MILES
KILOMETERS
STATUTE MILES
Elevations in meters

THE BALKANS
The Balkan states consist of Albania, Bosnia and Herzegovina, Bulgaria, Croatia, Greece, Kosovo, Montenegro, North Macedonia, Romania, Serbia, Slovenia, and the European part of Turkey.

Albania
REPUBLIC OF ALBANIA

AREA	28,748 sq km (11,100 sq mi)
POPULATION	3,057,000
CAPITAL	Tirana 485,000
RELIGION	Muslim, Roman Catholic, Orthodox
LANGUAGE	Albanian, Greek

Bosnia and Herzegovina
BOSNIA AND HERZEGOVINA

AREA	51,197 sq km (19,767 sq mi)
POPULATION	3,850,000
CAPITAL	Sarajevo 343,000
RELIGION	Muslim, Orthodox, Roman Catholic
LANGUAGE	Bosnian, Serbian, Croatian

Bulgaria
REPUBLIC OF BULGARIA

AREA	110,879 sq km (42,811 sq mi)
POPULATION	7,058,000
CAPITAL	Sofia 1,277,000
RELIGION	Eastern Orthodox, Muslim
LANGUAGE	Bulgarian, Turkish, Romany

Kosovo
REPUBLIC OF KOSOVO

AREA	10,887 sq km (4,203 sq mi)
POPULATION	1,908,000
CAPITAL	Pristina 207,000
RELIGION	Muslim, Roman Catholic, Serbian Orthodox
LANGUAGE	Albanian, Serbian, Bosnian

KOSOVO
In 2008 Kosovo declared its independence. Since then more than 100 UN member nations have recognized Kosovo, but Serbia still claims it as a province.

Longitude East 20° of Greenwich

Croatia
REPUBLIC OF CROATIA

AREA	56,594 sq km (21,851 sq mi)
POPULATION	4,270,000
CAPITAL	Zagreb 685,000
RELIGION	Roman Catholic, Orthodox
LANGUAGE	Croatian, Serbian

Montenegro
MONTENEGRO

AREA	13,812 sq km (5,333 sq mi)
POPULATION	614,000
CAPITAL	Podgorica 174,000
RELIGION	Orthodox, Muslim, Roman Catholic
LANGUAGE	Serbian, Montenegrin, Bosnian, Albanian

North Macedonia
REPUBLIC OF NORTH MACEDONIA

AREA	25,713 sq km (9,928 sq mi)
POPULATION	2,119,000
CAPITAL	Skopje 590,000
RELIGION	Macedonian Orthodox, Muslim
LANGUAGE	Macedonian, Albanian, Turkish, Romany, Aromanian, Serbian

Romania
ROMANIA

AREA	238,391 sq km (92,043 sq mi)
POPULATION	21,457,000
CAPITAL	Bucharest 1,812,000
RELIGION	Eastern Orthodox, Protestant, Roman Catholic
LANGUAGE	Romanian, Hungarian

THE BALKANS

POPULATION DENSITY

SLOVENIA
CROATIA
ROMANIA
BOSNIA AND HERZEGOVINA
SERBIA
BULGARIA
MONTENEGRO
KOSOVO
ALBANIA
NORTH MACEDONIA

People per Square Kilometer
More than 195
60–195
10–59
1–9
Fewer than 1

People per Square Mile
More than 500
150–500
25–149
1–24
Fewer than 1

CLIMATE
(Based on modified Köppen system)

ET
SLOVENIA
Cf
CROATIA
Cf
Df
Cf
ROMANIA
Df
BS
BOSNIA AND HERZEGOVINA
Cs
Df
SERBIA
Cf
Cf
MONTENEGRO
Df
Cs
Cf
KOSOVO
BULGARIA
Cf
Df
NORTH MACEDONIA
BS
ALBANIA
Cs
Cs

Arid (B)
Steppe (S)

Temperate (C)
Warm winter, no dry season (f), warm winter, dry summer (s)

Cold (D)
Snow, no dry season (f)

Polar (E)
Tundra (T)

LAND COVER

SLOVENIA
CROATIA
ROMANIA
BOSNIA AND HERZEGOVINA
SERBIA
MONTENEGRO
KOSOVO
BULGARIA
NORTH MACEDONIA
ALBANIA

Evergreen forest
Deciduous forest
Mixed forest
Savanna
Grassland
Cropland
Urban or built-up
Cropland/ vegetation mosaic
Wetland

Albers Conic Equal-Area Projection
SCALE 1:4,118,000 1 CENTIMETER = 41 KILOMETERS; 1 INCH = 65 MILES
0 25 50 75 100 125 150
KILOMETERS
0 25 50 75 100 125 150
STATUTE MILES
Elevations in meters

Serbia
REPUBLIC OF SERBIA

AREA 77,474 sq km (29,913 sq mi)
POPULATION 7,078,000
CAPITAL Belgrade 1,394,000
RELIGION Serbian Orthodox, Roman Catholic, Protestant
LANGUAGE Serbian, Hungarian, Bosniak, Romany

Slovenia
REPUBLIC OF SLOVENIA

AREA 20,273 sq km (7,827 sq mi)
POPULATION 2,102,000
CAPITAL Ljubljana 286,000
RELIGION Roman Catholic, Orthodox, Muslim
LANGUAGE Slovene, Serbo-Croatian, Italian, Hungarian

Greece
HELLENIC REPUBLIC

AREA	131,957 sq km (50,949 sq mi)
POPULATION	10,762,000
CAPITAL	Athens 3,154,000
RELIGION	Greek Orthodox, Muslim
LANGUAGE	Greek

POPULATION DENSITY

GREECE

People per Square Kilometer
- More than 195
- 60–195
- 10–59
- 1–9
- Fewer than 1

People per Square Mile
- More than 500
- 150–500
- 25–149
- 1–24
- Fewer than 1

GREECE

Longitude East 22° of Greenwich

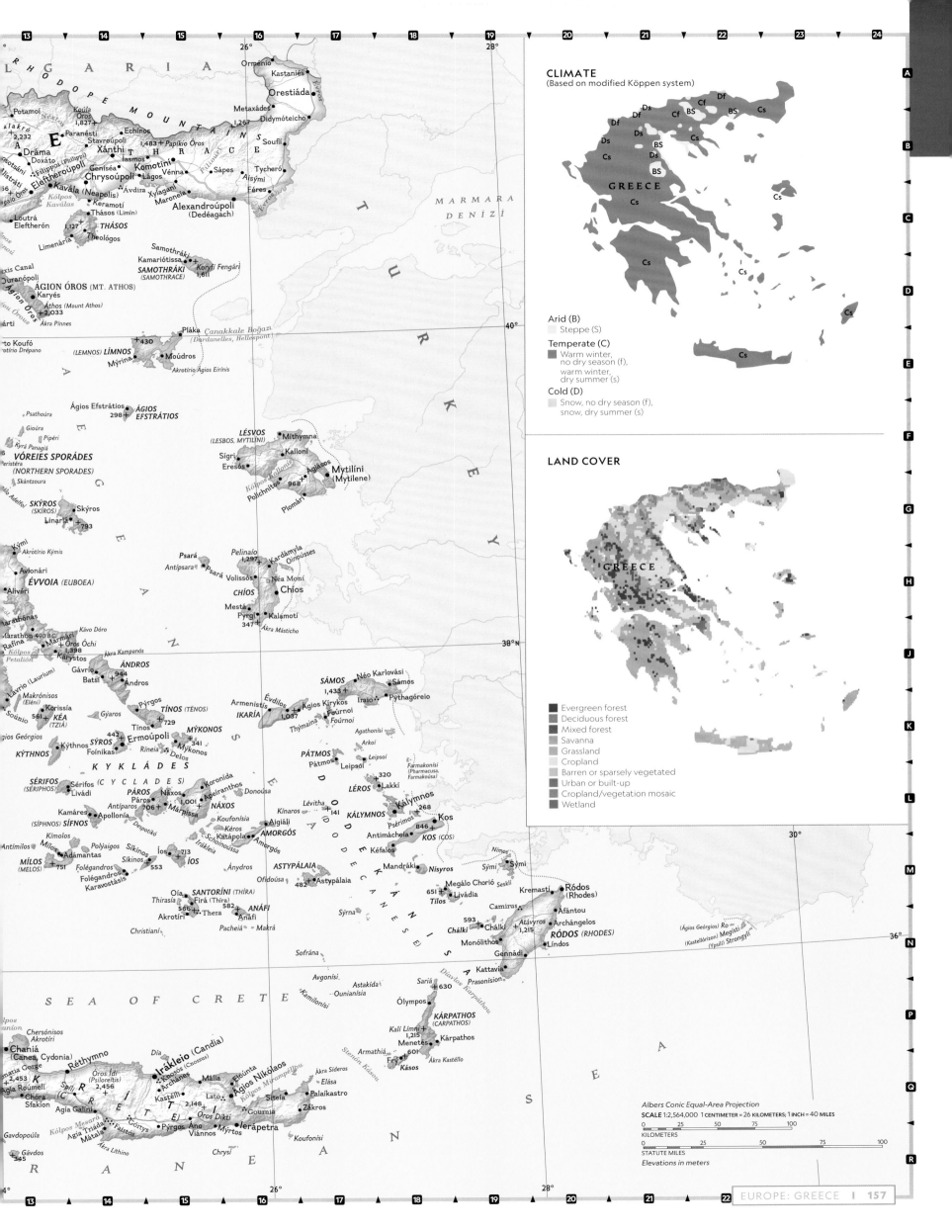

CLIMATE
(Based on modified Köppen system)

GREECE

Arid (B)
Steppe (S)

Temperate (C)
Warm winter,
no dry season (f),
warm winter,
dry summer (s)

Cold (D)
Snow, no dry season (f),
snow, dry summer (s)

LAND COVER

GREECE

- Evergreen forest
- Deciduous forest
- Mixed forest
- Savanna
- Grassland
- Cropland
- Barren or sparsely vegetated
- Urban or built-up
- Cropland/vegetation mosaic
- Wetland

Albers Conic Equal-Area Projection
SCALE 1:2,564,000 1 CENTIMETER = 26 KILOMETERS; 1 INCH = 40 MILES

KILOMETERS 0 25 50 75 100
STATUTE MILES 0 25 50 75 100
Elevations in meters

BULGARIA
RHODOPE MOUNTAINS
THRACE
TURKEY
MARMARA DENIZI

Orménio
Kastaniés
Orestiáda
Metaxádes
Didymóteicho
Soúfli
Tycheró
Aísymi
Féres
Alexandroúpoli (Dedéagach)
Maróneia
Xylaganí
Komotiní
Vénna
Sápes
Lágos
Ávdira
Chrysoúpoli
Geniséa
Iasmos
Komotiní
Xánthi
Papíkio Óros 1,483
Echínos
Paranésti
Stavroúpoli
Doxáto
Fílippoi (Philippi)
Eleftheroúpoli
Kavála (Neapolis)
Kólpos Kaválas
Keramotí
Thásos (Limín)
THÁSOS 1,127
Limenária
Theológos
Dráma
Koúla Óros 1,827
2,232

Samothráki
Kamariótissa
SAMOTHRÁKI (SAMOTHRACE)
Koryfí Fengári 1,611

AGION ÓROS (MT. ATHOS)
Karyés
Áthos (Mount Athos) 2,033
Ákra Pínnes

Pláka
Çanakkale Boğazı (Dardanelles, Hellespont)
430
(LEMNOS) LÍMNOS
Mýrina
Moúdros
Akrotírio Ágios Eirínis

Ágios Efstrátios 298
ÁGIOS EFSTRÁTIOS

LÉSVOS (LESBOS, MYTILÍNI)
Mithymna
Kalloní
Sígri
Eresós
Agiásos
Polichnítos 968
Plomári
Mytilíni (Mytilene)

VÓREIES SPORÁDES (NORTHERN SPORADES)
Kyrá Panagiá
Peristéra
Skántzoura

SKÝROS (SKÍROS)
Skýros
Linariá 793

Psará
Pelinaío 1,297
Kardámyla
Oinoússes
Antípsara
Psará
Volissós
Néa Moní
CHÍOS
Chíos
Mestá
Pyrgí 347
Kalamotí
Ákra Másticho

ÉVVOIA (EUBOEA)
Avlonári
Alivéri
Marathónas 490 B.C.
Rafína
Oros Óchi 1,398
Kárystos
Ákra Kampanós

ÁNDROS
Gávrio 944
Ándros
Batsí

TÍNOS (TÉNOS)
Pýrgos
Tínos 729

MÝKONOS
Mýkonos 341
Delos
Ríneia

SÝROS
Ermoúpoli 442
Foínikas

KÉA (TZIÁ)
Korissía 561
Kýthnos
KÝTHNOS

KYKLÁDES (CYCLADES)

SÉRIFOS (SÉRIPHOS)
Sérifos
Livádi

SÍFNOS (SÍPHNOS)
Kamáres
Apollonía
Antíparos

PÁROS
Páros
Náxos 1,001
NÁXOS
Márpissa 706

MÍLOS (MELOS)
Adámantas 751
Kímolos
Polýaigos
Síkinos

FolégandrosÍOS
Karavostásis
Folégandros
ÍOS 713
Síkinos 553
Ánydros

SANTORÍNI (THÍRA)
Oía
Thirasía
Firá (Thíra)
Akrotíri 566
Thera 582
ANÁFI
Anáfi
Pacheiá
Makrá
Christianí
Sofrána

SÁMOS
Néo Karlovási
Sámos 1,433
Armenistís
Évdilos
Ágios Kírykos
Iraío
Pythagóreio

IKARÍA 1,037
Thýmaina
Foúrnoi
Agathonísi
Arkoí

PÁTMOS
Pátmos
Leipsoí
Farmakonísi (Pharmacusa, Farmakoúsa)

LÉROS
Lakkí 320
Kínaros
Lévitha

KÁLYMNOS 268
Kálymnos
Psérimos
Kos 846
KOS (CÓS)
Antimácheia
Kéfalos

AMORGÓS
Aigiáli
Amorgós 141
Kéros
Koufonísia
Schoinoússa
Katápola
Iráklia
Donoúsa
Koronída
Apeíranthos

ASTYPÁLAIA 482
Ofidoúsa
Astypálaia
Sýrna

Nímos
Sými
Sými
Nísyros
Mandráki
Megálo Chorió
Seskli
Kremastí

DODEKÁNISA (DODECANESE)

Tílos 651
Livádia

Chálki 593
Chálki
Monólithos
Atávyros 1,215
RÓDOS (RHODES)
Ródos (Rhodes)
Afántou
Camirus
Archángelos
Líndos
Gennádi
Kattavía
Prasonísion
(Ágios Geórgios) Ro
(Kastellórizo) Megísti
(Ýpsili) Strongyli

Diávlos Karpáthou

Sofrána
Avgonísi
Astakída
Ounianísia
Kamílonísi
Olympos
Káli Límni 1,215
KÁRPATHOS (CARPATHOS)
Menetés
Kárpathos
Frý 601
Kásos
Sariá 630
Ákra Síderos
Ákra Kastéllo
Armathiá
Stenón Kásou

SEA OF CRETE
MEDITERRANEAN

Chersónisos Akrotíri
Chaniá (Canea, Cydonia)
Réthymno
Irákleio (Candia)
Knosós (Cnossus)
Archánes
Mália
Eloúnta
Ágios Nikólaos
Óros Ídi (Psiloreítis) 2,456
Óros Díkti 2,148
KRÍTI (CRETE)
Kastélli
Lató
Gournia
Kólpos Mirampéllou
Sitía
Palaíkastro
Zákros
Ierápetra
Myrtos
Áno Viánnos
Pýrgos
Górtys
Faistós
Mátala
Agía Triáda
Ákra Líthino
Kólpos Mesará
Agía Galíni
Spíli
Chóra Sfakíon
Agía Rouméli
Samariá Gorge
2,453
Koufonísi
Chrysí
Gávdos 345
Gavdopoúla

AIGAÍO PÉLAGOS (AEGEAN SEA)

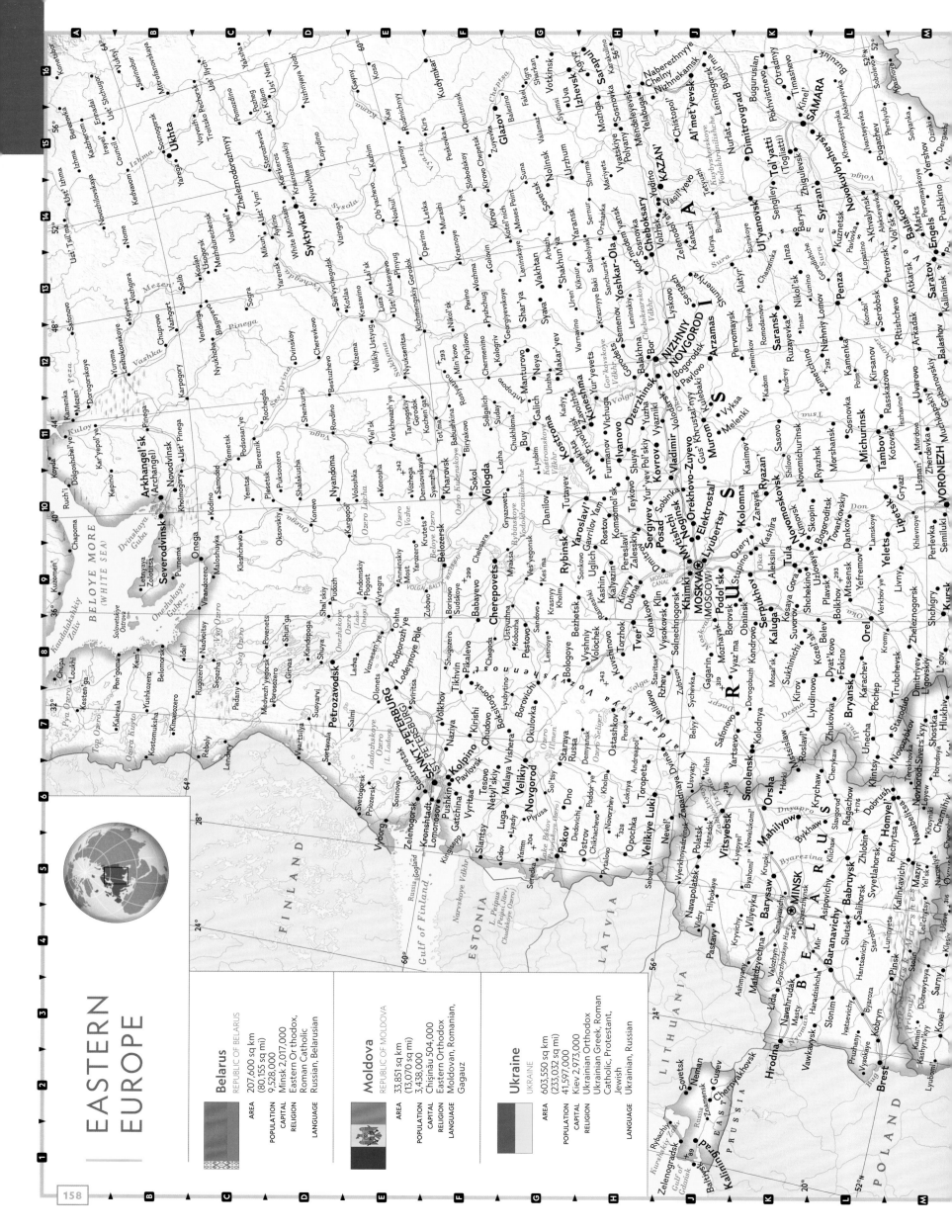

EASTERN EUROPE

Belarus
REPUBLIC OF BELARUS

AREA	207,600 sq km (80,155 sq mi)
POPULATION	9,528,000
CAPITAL	Minsk 2,017,000
RELIGION	Eastern Orthodox, Roman Catholic
LANGUAGE	Russian, Belarusian

Moldova
REPUBLIC OF MOLDOVA

AREA	33,851 sq km (13,070 sq mi)
POPULATION	3,438,000
CAPITAL	Chişinău 504,000
RELIGION	Eastern Orthodox, Roman Catholic
LANGUAGE	Moldovan, Romanian, Gagauz

Ukraine
UKRAINE

AREA	603,550 sq km (233,032 sq mi)
POPULATION	41,597,000
CAPITAL	Kiev 2,973,000
RELIGION	Ukrainian Orthodox, Ukrainian Greek, Roman Catholic, Protestant, Jewish
LANGUAGE	Ukrainian, Russian

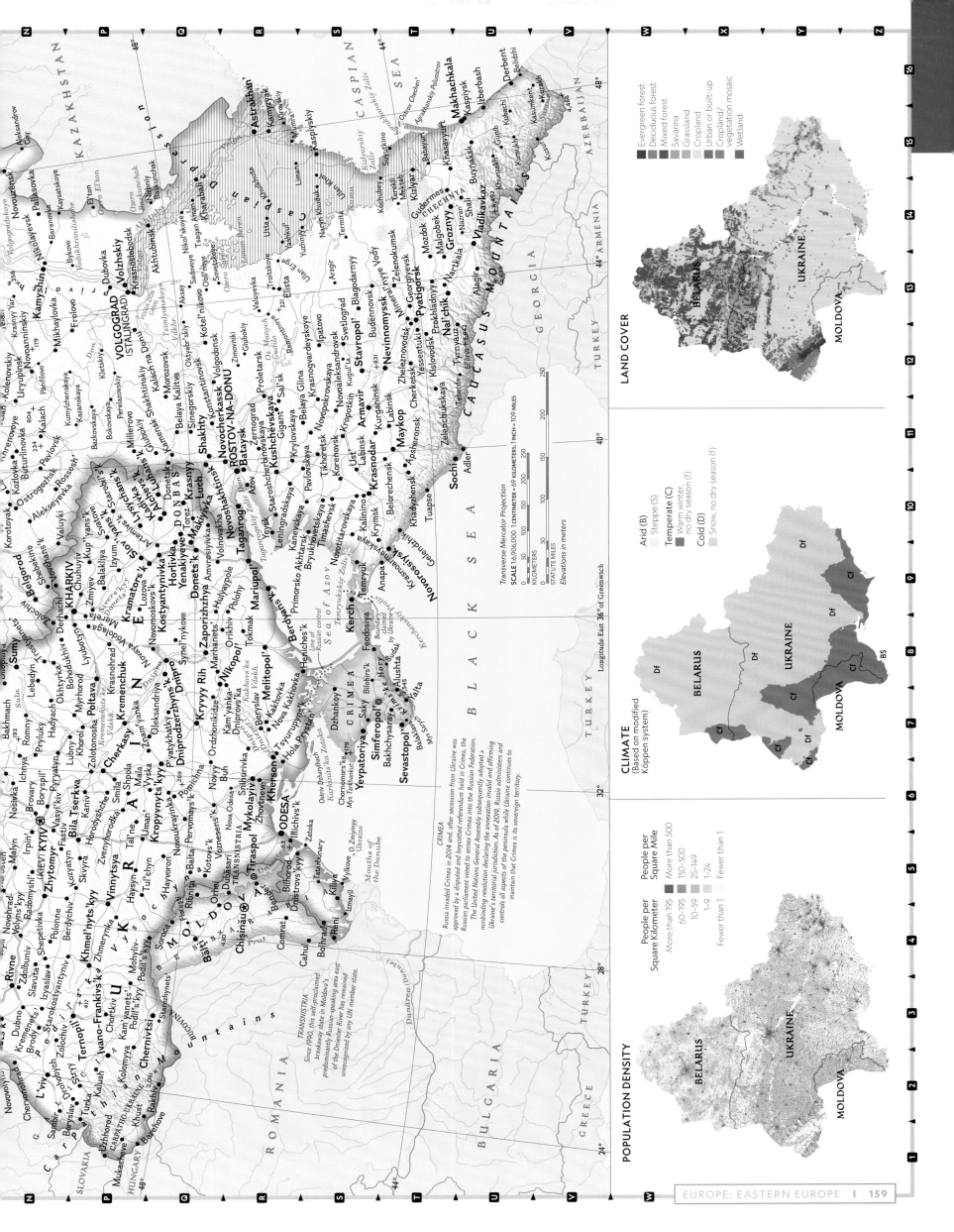

Transverse Mercator Projection
SCALE 1:6,906,000 1 CENTIMETER = 69 KILOMETERS; 1 INCH = 109 MILES

Elevations in meters

Longitude East 36° of Greenwich

CRIMEA
Russia invaded Crimea in 2014 and, after secession from Ukraine was approved by a disputed and boycotted referendum held in Crimea, the Russian parliament voted to annex Crimea into the Russian Federation. The United Nations General Assembly subsequently adopted a nonbinding resolution declaring the annexation invalid and affirming Ukraine's territorial jurisdiction. As of 2019, Russia administers and controls all aspects of the peninsula while Ukraine continues to maintain that Crimea is its sovereign territory.

TRANSNISTRIA
Since 1990, this self-proclaimed breakaway state in Moldova's predominantly Russian-speaking area east of the Dniester River has remained unrecognized by any UN member state.

POPULATION DENSITY

People per
Square Kilometer

People per
Square Mile

	More than 195	More than 500
	60–195	150–500
	10–59	25–149
	1–9	1–24
	Fewer than 1	Fewer than 1

CLIMATE
(Based on modified
Köppen system)

Arid (B)
  Steppe (S)

Temperate (C)
  Warm winter,
  no dry season (f)

Cold (D)
  Snow, no dry season (f)

LAND COVER

Evergreen forest
Deciduous forest
Mixed forest
Savanna
Grassland
Cropland
Urban or built-up
Cropland/
  vegetation mosaic
Wetland

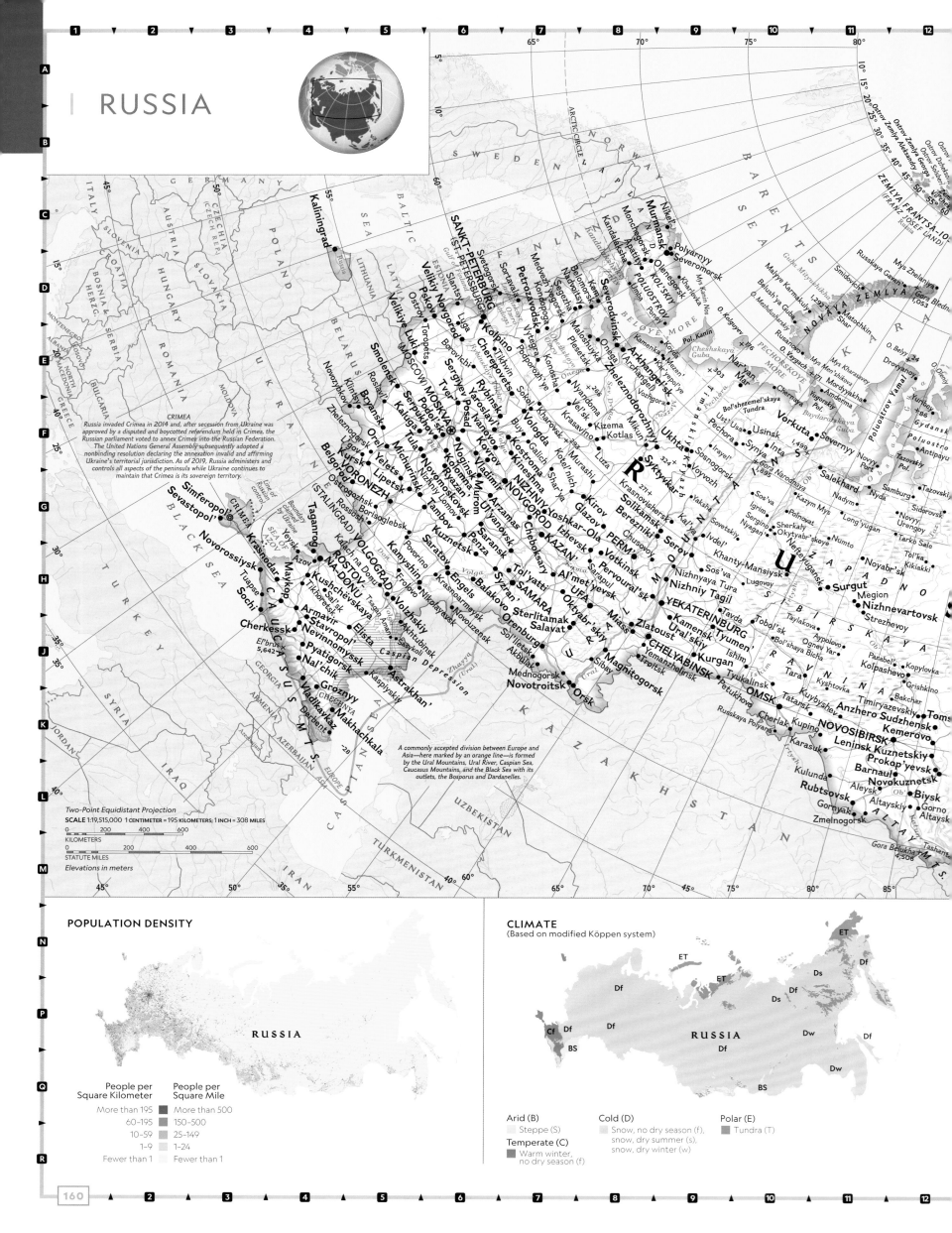

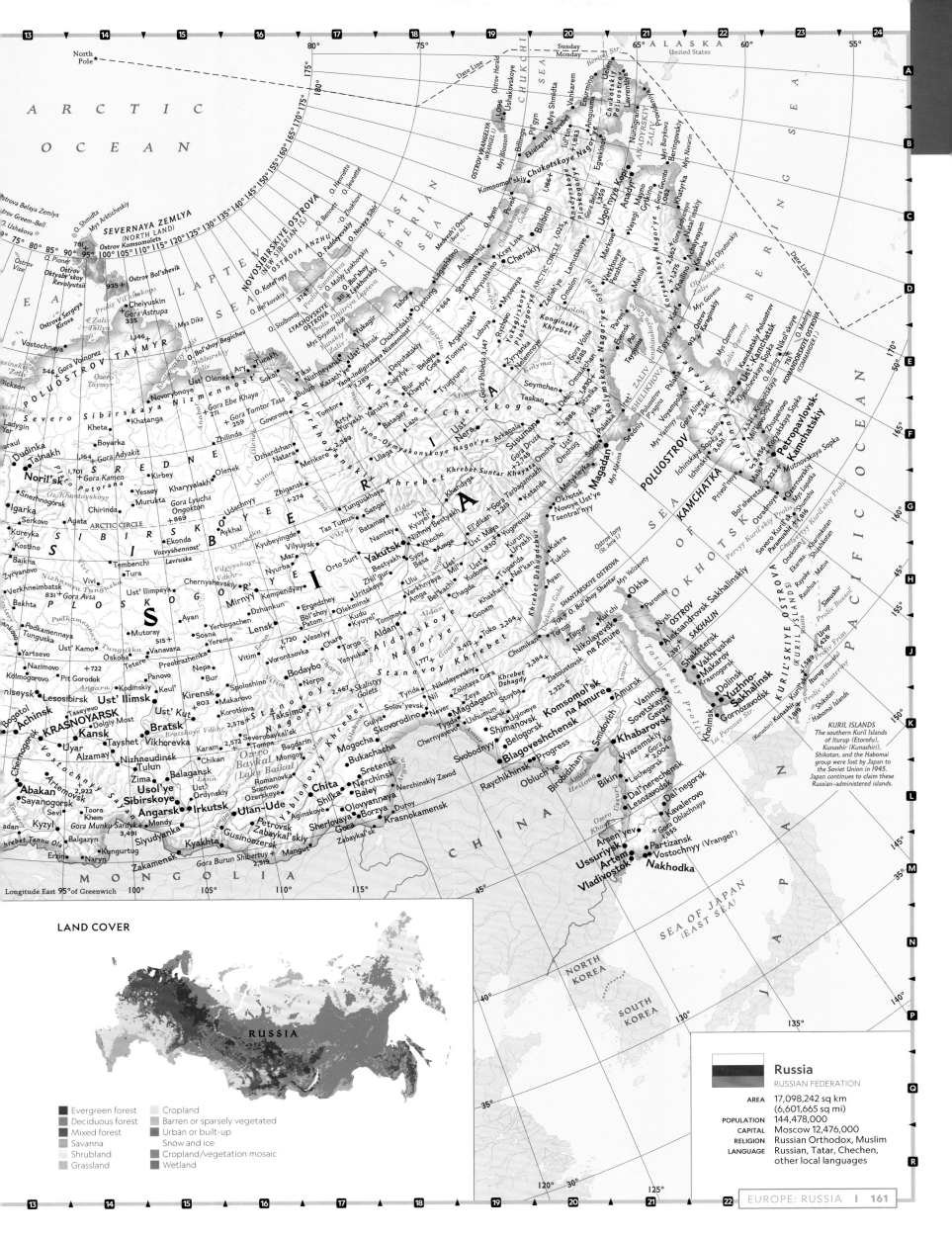

SMALLEST COUNTRIES AND TERRITORIES

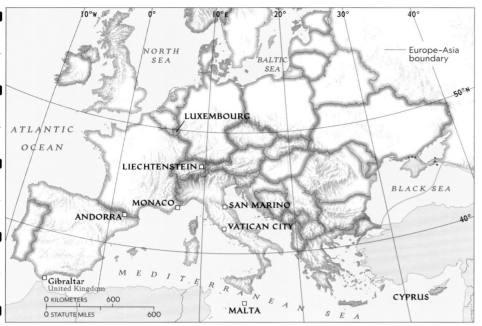

NORTH SEA
BALTIC SEA
ATLANTIC OCEAN
LUXEMBOURG
LIECHTENSTEIN
MONACO
SAN MARINO
ANDORRA
VATICAN CITY
BLACK SEA
Gibraltar
United Kingdom
MEDITERRANEAN SEA
MALTA
CYPRUS

Europe-Asia boundary

0 KILOMETERS 600
0 STATUTE MILES 600

Andorra
PRINCIPALITY OF ANDORRA

AREA	468 sq km (181 sq mi)
POPULATION	86,000
CAPITAL	Andorra la Vella 23,000
RELIGION	Roman Catholic
LANGUAGE	Catalan, French, Castilian, Portuguese

Luxembourg
GRAND DUCHY OF LUXEMBOURG

AREA	2,586 sq km (998 sq mi)
POPULATION	606,000
CAPITAL	Luxembourg 120,000
RELIGION	Roman Catholic
LANGUAGE	Luxembourgish, German, French, Portuguese

Liechtenstein
PRINCIPALITY OF LIECHTENSTEIN

AREA	160 sq km (62 sq mi)
POPULATION	39,000
CAPITAL	Vaduz 5,000
RELIGION	Roman Catholic, Protestant
LANGUAGE	German, Italian

Cyprus
REPUBLIC OF CYPRUS

AREA	9,251 sq km (3,572 sq mi)
POPULATION	1,237,000
CAPITAL	Nicosia 269,000
RELIGION	Greek Orthodox, Muslim
LANGUAGE	Greek, Turkish, English

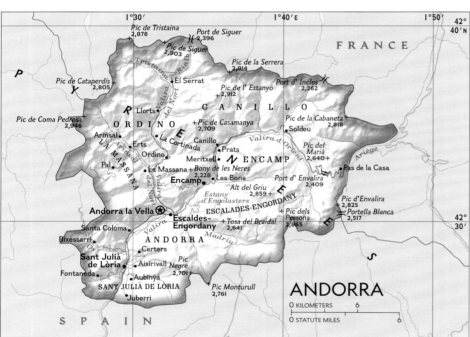

ANDORRA

FRANCE

Pic de Tristaina 2,878
Port de Siguer 2,396
Pic de Siguer 2,903
Pic de la Serrera 2,914
Pic de Cataperdís 2,805
El Serrat
Pic de l'Estanyó 2,912
Port d'Incles 2,262
CANILLO
Pic de Coma Pedrosa 2,946
Llorts
ORDINO
Pic de Casamanya 2,709
Pic de la Cabaneta 2,818
Arinsal
La Cortinada
Soldeu
Erts
Canillo
LA MASSANA
Meritxell
Ordino
Prats
Pal
ENCAMP
Pic del Maria 2,640
La Massana
Bony de les Neres
2,228
Encamp
Les Bons
Port d'Envalira 2,409
Pas de la Casa
Alt del Griu 2,859
Estany d'Engolasters
Pic d'Envalira 2,825
Andorra la Vella
ESCALDES-ENGORDANY
Escaldes-Engordany
Pic dels Pessons 3,065
Portella Blanca 2,517
Santa Coloma
Tosa del Braidal 2,641
Bixessarri
ANDORRA
Certers
Sant Julià de Lòria
Aixirivall
Pic Negre 2,701
Fontaneda
Aubinyà
SANT JULIA DE LÒRIA
Pic Monturull 2,761
Juberri

SPAIN

0 KILOMETERS 6
0 STATUTE MILES 6

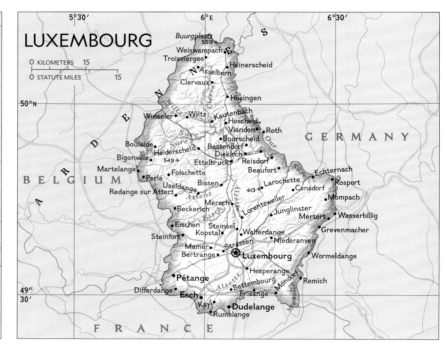

LUXEMBOURG

0 KILOMETERS 15
0 STATUTE MILES 15

Buurgplaatz 559
Weiswampach
Troisvierges
Asselborn
Heinerscheid
Clervaux
Hosingen
Winseler
Wiltz
Kautenbach
Hoscheid
Vianden
Roth
Boulaide
Bourscheid
Bigonville
Heiderscheid 549
Bastendorf
Diekirch
Martelange
Ettelbruck
Reisdorf
Perlé
Folschette
Beaufort
Larochette
Echternach
Rosport
Redange sur Attert
Useldange
413
Consdorf
Mompach
Beckerich
Mersch
Lorentzweiler
Junglinster
Mertert
Wasserbillig
Eischen
Steinsel
Walferdange
Grevenmacher
Kopstal
Niederanven
Steinfort
Mamer
Luxembourg
Wormeldange
Bertrange
Hesperange
Pétange
Bettembourg
Mondorf
Remich
Differdange
Frisange
Esch
Kayl
Dudelange
Rumelange

BELGIUM
GERMANY
FRANCE

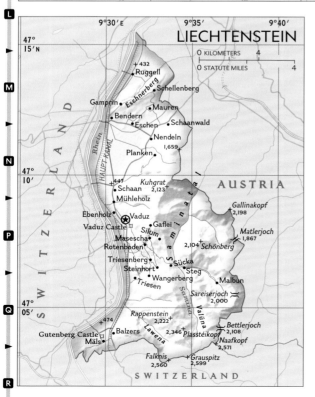

LIECHTENSTEIN

0 KILOMETERS 4
0 STATUTE MILES 4

Ruggell 432
Schellenberg
Eschnerberg
Gamprin
Mauren
Bendern
Schaanwald
Eschen
Nendeln
Planken 1,659
Schaan 447
Kuhgrat 2,123
Mühleholz
Ebenholz
Vaduz
Gallinakopf 2,198
Gaflei
Vaduz Castle
Sücka
Matlerjoch 1,867
Masescha
Rotenboden
Schönberg 2,104
Triesenberg
Steg
Steinfort
Wangerberg
Malbun
Triesen
Sareiserjoch 2,000
Rappenstein 2,222
474
Bettlerjoch 2,108
Gutenberg Castle
Balzers
2,346
Plassteikopf
Mäls
Naafkopf 2,571
Falknis
Grauspitz 2,599

SWITZERLAND
AUSTRIA
SWITZERLAND

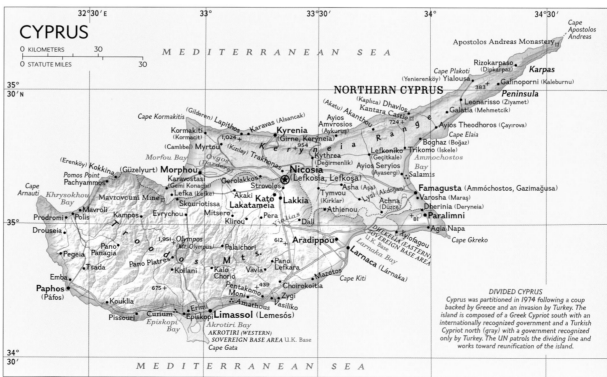

CYPRUS

0 KILOMETERS 30
0 STATUTE MILES 30

MEDITERRANEAN SEA

Apostolos Andreas Monastery
Cape Apostolos Andreas
Rizokarpaso (Dipkarpaz)
Cape Plakoti
Karpas
(Yenierenköy) Yialousa
Galinoporni (Kaleburnu)
NORTHERN CYPRUS
Peninsula
Cape Kormakitis (Gilderen) Lapithos
Leonarisso (Ziyamet)
383
Karavas (Alsancak)
(Akatu) Akanthou
Galátia (Mehmetcik)
Kormakiti (Kormacit)
Ayios Theodoros (Çayırova)
1,024
Kyrenia (Girne, Keryneia)
724
Ayios
Amvrosios (Aykurus)
Kantara Castle
Cape Elaia
(Camlibel) Myrtou (Kızılay)
954
Kythrea
Boghaz (Boğaz)
(Erenköy) Kokkina
Trikomo (İskele)
Ayios
Ammochostos Bay
(Güzelyurt) Morphou
Ovgos (Değirmenlik)
Lefkoniko (Geçitkale)
Seryios
Salamis
Morfou Bay
Akaki
Kythrea
Pomos Point
Karavostasi
Gerolakkos
Nicosia
Ayios Servios (Ayasergi)
Pachyammos
(Gemi Konagbi)
Strovolos
Lefkosia, Lefkoşa
Asha (Aşa)
Famagusta (Ammóchostos, Gazimağusa)
Cape Arnauti
Khrysokhou Bay
Lefka (Lefke)
Lakatameia
Tymvou (Kırklar)
Lysi
Varosha (Maraş)
Mavrovouni Mine
Skouriotissa
Lakkia
Achna (Düzce)
Dherinia (Deryneia)
Prodromi
Mavroli
Kampos
Evrychou
Mitsero
Pera
Dali
81
Paralimni
Polis
Drouseia
Klirou
Athienou
Agia Napa
1,951 Olympos
Palaichori
Pano (Mt. Olympus)
DHEKELIA SOVEREIGN (EASTERN) U.K. Base
Cape Gkreko
Pegeia
Tsada
Panagia
612
Aradippou
Pano Platres
Kollani
675
Kalo Chorio
Larnaca (Lárnaka)
Emba
439
Vavla
Choirokoitia
Mazotos
Larnaca Bay
Troodos Mts
Kalo Lefkara
Paphos (Páfos)
Pentakomo
Cape Kiti
Kouklia
Moni
Zygi
Amathous
Vasiliko
Pissouri
Erimi
Curium
Limassol (Lemesós)
Episkopi
Akrotiri Bay
AKROTIRI (WESTERN)
SOVEREIGN BASE AREA U.K. Base
Cape Gata
Episkopi Bay

DIVIDED CYPRUS
Cyprus was partitioned in 1974 following a coup backed by Greece and an invasion by Turkey. The island is composed of a Greek Cypriot south with an internationally recognized government and a Turkish Cypriot north (gray) with a government recognized only by Turkey. The UN patrols the dividing line and works toward reunification of the island.

MEDITERRANEAN SEA

All map elevations in meters

San Marino
REPUBLIC OF SAN MARINO

AREA	61 sq km (24 sq mi)
POPULATION	34,000
CAPITAL	San Marino 4,000
RELIGION	Roman Catholic
LANGUAGE	Italian

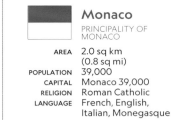

Monaco
PRINCIPALITY OF MONACO

AREA	2.0 sq km (0.8 sq mi)
POPULATION	39,000
CAPITAL	Monaco 39,000
RELIGION	Roman Catholic
LANGUAGE	French, English, Italian, Monegasque

Vatican City
THE HOLY SEE (STATE OF THE VATICAN CITY)

AREA	0.4 sq km (0.2 sq mi)
POPULATION	1,000
CAPITAL	Vatican City 1,000
RELIGION	Roman Catholic
LANGUAGE	Italian, Latin, French

Malta
REPUBLIC OF MALTA

AREA	316 sq km (122 sq mi)
POPULATION	449,000
CAPITAL	Valletta 213,000
RELIGION	Roman Catholic
LANGUAGE	Maltese, English

Gibraltar (U.K.)
BRITISH OVERSEAS TERRITORY

AREA	7 sq km (3 sq mi)
POPULATION	29,000
CAPITAL	Gibraltar 29,000
RELIGION	Roman Catholic, Anglican, Muslim, Jewish
LANGUAGE	English, Spanish, Italian, Portuguese

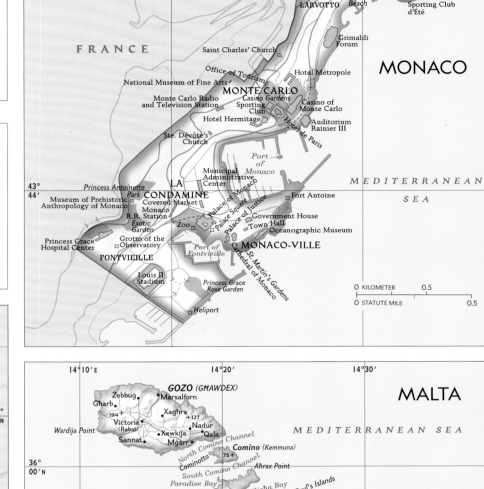

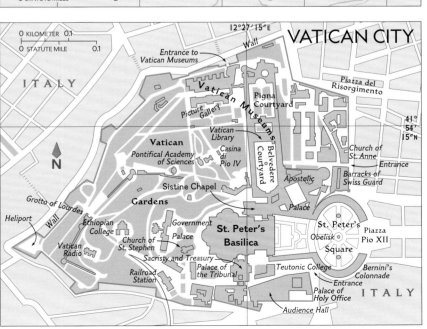

ASIA

Defensive watchtowers and guard stations dot the Great Wall of China.

HOME TO HALF THE WORLD

AN EPIC LAND WHERE CIVILIZATION BEGAN

The world's largest continent is a place of unrivaled magnitude. From Everest, world's highest mountain peak, to Shanghai, world's busiest port, the superlatives go on. Asia stretches across 10 time zones—deserts of Saudi Arabia in the west, Bering Sea ports of Siberia in the north—and contains six in ten people on Earth.

| GEOGRAPHY

VOLATILE & VARIED

Thirty percent of the Earth's land surface lies in Asia. Within that massive spread we find several huge countries and a dazzling collection of extremes. Asia is home to the planet's wettest spot (Meghalaya, India), largest cave chamber (in Gunung Mulu National Park, Malaysia), and largest unbroken woodlands (the taiga of Siberia). Yet much of Asia is too forbidding in landscape or climate for people to live. The deserts of the Arabian Peninsula, the thick jungles of Myanmar, and the high-altitude peaks of the Himalaya have all driven Asian people for millennia to settle instead along the region's great rivers—the Tigris, Euphrates, Yellow, Ganges, and Indus—and its seacoasts. Infrastructure has not kept pace with the booming population. Fifteen of the world's 20 most polluted rivers are in Asia. China's Yangtze River is the most polluted river on Earth.

Asia is also turbulent. A chain of volcanoes stretches along the Ring of Fire in the Pacific Ocean; significant earthquakes regularly rattle China, Japan, and Indonesia, occasionally triggering tsunamis. In 2004, a tsunami left almost 228,000 dead or missing in more than a dozen countries. Seasonal monsoons bring heavy, often torrential rains that replenish aquifers and wells to irrigate crops, power hydroelectric dams, and provide water to thirsty livestock.

Striped sandstone mountains of China's Zhangye Danxia National Geological Park

The Ottoman-era Ishak Pasha Palace in eastern Turkey

| HISTORY

ANCIENT ORIGINS, MODERN TURBULENCE

If Africa is the birthplace of human evolution, Asia is where human civilizations first blossomed. This long and sprawling story started some 10,000 years ago with early settlements in the Middle East's Fertile Crescent. The Sumerians produced the first wheel, the first system of writing, and the first cities. Around 130 B.C. an ancient trade route known as the Silk Road created a link between East Asia and the rest of the world. The trade route offered a way for China to spread some of its most noteworthy inventions and exports—gunpowder, paper, rice—and made Central Asia a crossroads of flourishing cultures for centuries.

Today Asia's 46 nations are home to 4.5 billion people. Full democracies are few; authoritarian governments or military regimes are more common, and stability has been elusive, particularly along disputed borders. The Korean Peninsula remains divided, and tensions simmer between nuclear-armed India and Pakistan. After the fall of the Ottoman Empire at the end of World War I, European powers drew political borders in the Middle East without regard to sectarian, tribal, and ethnic differences, which laid a volatile foundation for generations of conflict. More recent intervention by Western powers in supporting or opposing political regimes, along with civil wars and the spread of Islamic extremism, has further destabilized the region.

CULTURE

ON THE SHOULDERS OF GIANTS

The societies of Asia represent more ethnic and national groups than any other continent. India and China, which together account for almost two-thirds of the continent's population, loom large in Asian culture. Separated by mountains and jungles, both countries have for millennia cultivated art, literature, and philosophy of the highest order. India is the birthplace of both Hinduism and Buddhism, the leading religions in Sri Lanka, Nepal, Cambodia, and Thailand. China's 3,000-year-old civilization has spread its written language, architecture, and even chopsticks to all of East Asia, much of Southeast Asia, and parts of Central Asia. Islam is a third great cultural influence. From the seventh century on, Arabs spread their religion and culture, particularly Arabic writing, to the east as well as the west. Indonesia is now the world's largest Muslim country, with more than 220 million believers; Pakistan has more than 200 million and India has more than 180 million Muslims.

Japan and South Korea have made major contributions to modern pop culture, accounting for significant soft power on a global scale. The visual intensity and complex narratives of Japanese animation, known as anime, now represent a global multibillion-dollar industry. And the catchy tunes and videos of Korean pop music, or K-pop, top music charts around the world and attract tens of millions of fans on YouTube.

The skyscrapers of Shanghai, one of China's most populous cities

ECONOMY

THE RISE OF MEGACITIES

Services like information technology, health care, tourism, and retail sales play a major role in Asian economies. More jobs in services and in factories and fewer in farming have led hundreds of millions to flock to cities. The growth of megacities (population 10 million or above) from Jakarta to Shanghai represents a dramatic change over the past 50 years. As of 2018, 21 of the world's 33 megacities were in Asia; by 2030, the population of Delhi, India, is projected to be 39 million. Meanwhile, many farmers still practice traditional methods, whether they plow fields with water buffalo in Vietnam or handpick grapes in the vineyards of Lebanon. As many as 240 million Chinese form a "floating" population, moving wherever there is work.

Economic power in Asia is also shifting, with China, not technology heavyweight Japan, now reigning as the continent's largest economy—and by some measures, the largest in the world. In the 1970s, 90 percent of Chinese lived below the extreme poverty line; today, 99 percent are above that line. India has a large, growing middle class thanks to a liberalized economy. However, child health remains a critical priority in South Asia; 38 percent of children under five years of age experienced persistent malnutrition in 2018. Countries in the Persian Gulf region have grown rich thanks to the concentration of petroleum resources there.

Mumbai glows during celebrations for Diwali, the Hindu festival of lights.

SNAPSHOT

Asia FACTS & FEATURES

WORLD'S HIGHEST ELEVATION Mount Everest, China/Nepal: 8,850 m (29,035 ft)

WORLD'S LOWEST ELEVATION Dead Sea, Israel/Jordan: –427 m (–1,401 ft)

LONGEST RIVER Yangtze (Chang Jiang), China: 6,244 km (3,880 mi)

LARGEST LAKE ENTIRELY IN ASIA Lake Baikal, Russia: 31,500 sq km (12,200 sq mi)

WORLD'S BIGGEST URBAN AREA BY POPULATION Tokyo: 37,400,000 (2020)

WORLD'S MOST POPULOUS COUNTRY China: 1,384,689,000

SMALLEST COUNTRY BY AREA Maldives: 298 sq km (115 sq m)

WORLD'S LARGEST RELIGIOUS MONUMENT Angkor Wat, Cambodia: 401 acres

LONGEST BRIDGE Danyang-Kunshan Grand Bridge, China: 164 km (102 mi) long

WORLD'S BUSIEST CONTAINER PORT Shanghai, China: 40,233 TEU (2017)

THE WORLD'S LARGEST HYDROELECTRIC POWER STATION IS THREE GORGES DAM IN CHINA. 1.4 MILLION PEOPLE WERE RELOCATED TO MAKE WAY FOR ITS CONSTRUCTION.

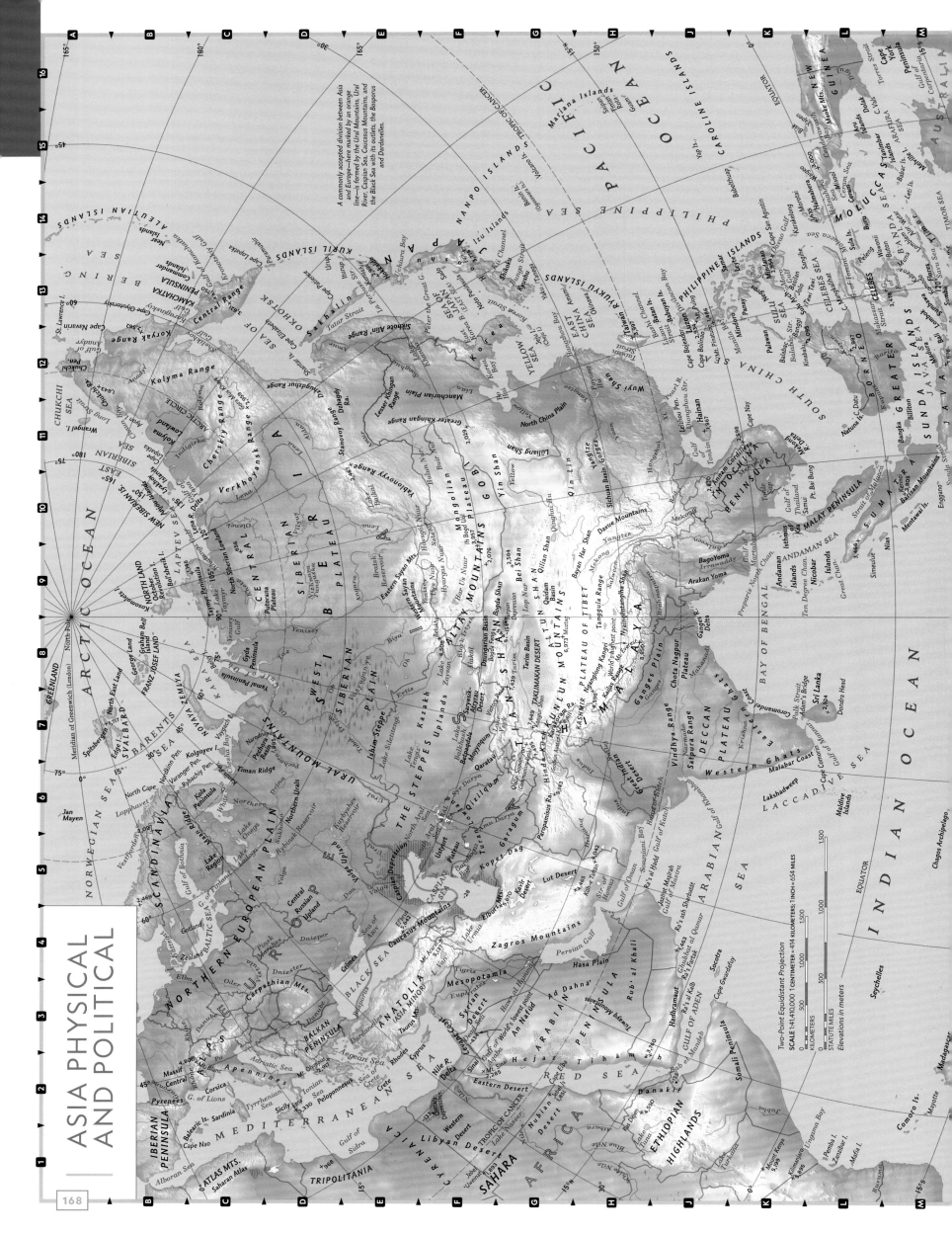

ASIA PHYSICAL AND POLITICAL

A commonly accepted division between Asia and Europe—here marked by an orange line—is formed by the Ural Mountains, Ural River, Caspian Sea, Caucasus Mountains, and the Black Sea with its outlets, the Bosporus and the Dardanelles.

Two-Point Equidistant Projection
SCALE 1:41,410,000 1 CENTIMETER = 414 KILOMETERS; 1 INCH = 654 MILES

KILOMETERS

STATUTE MILES

Elevations in meters

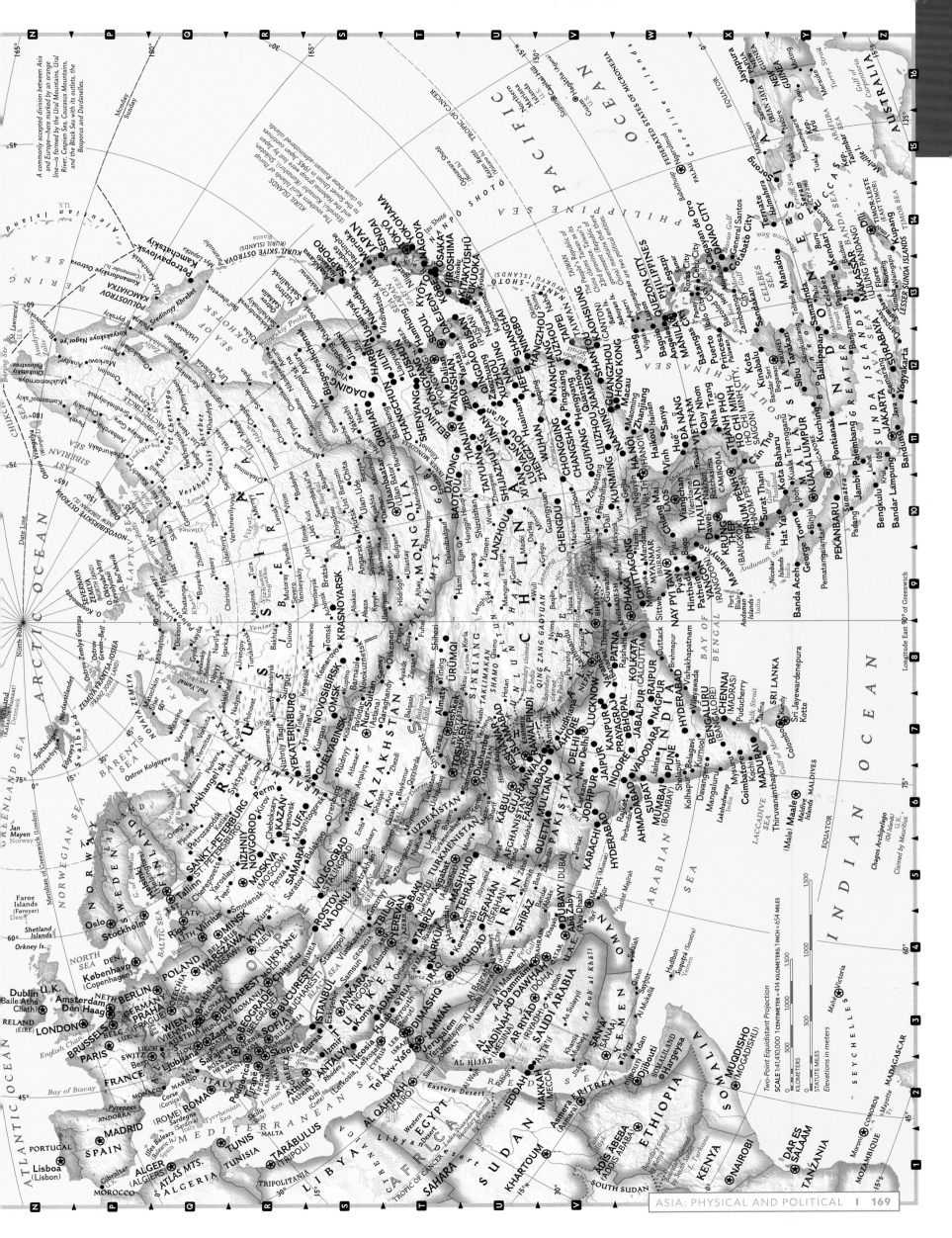

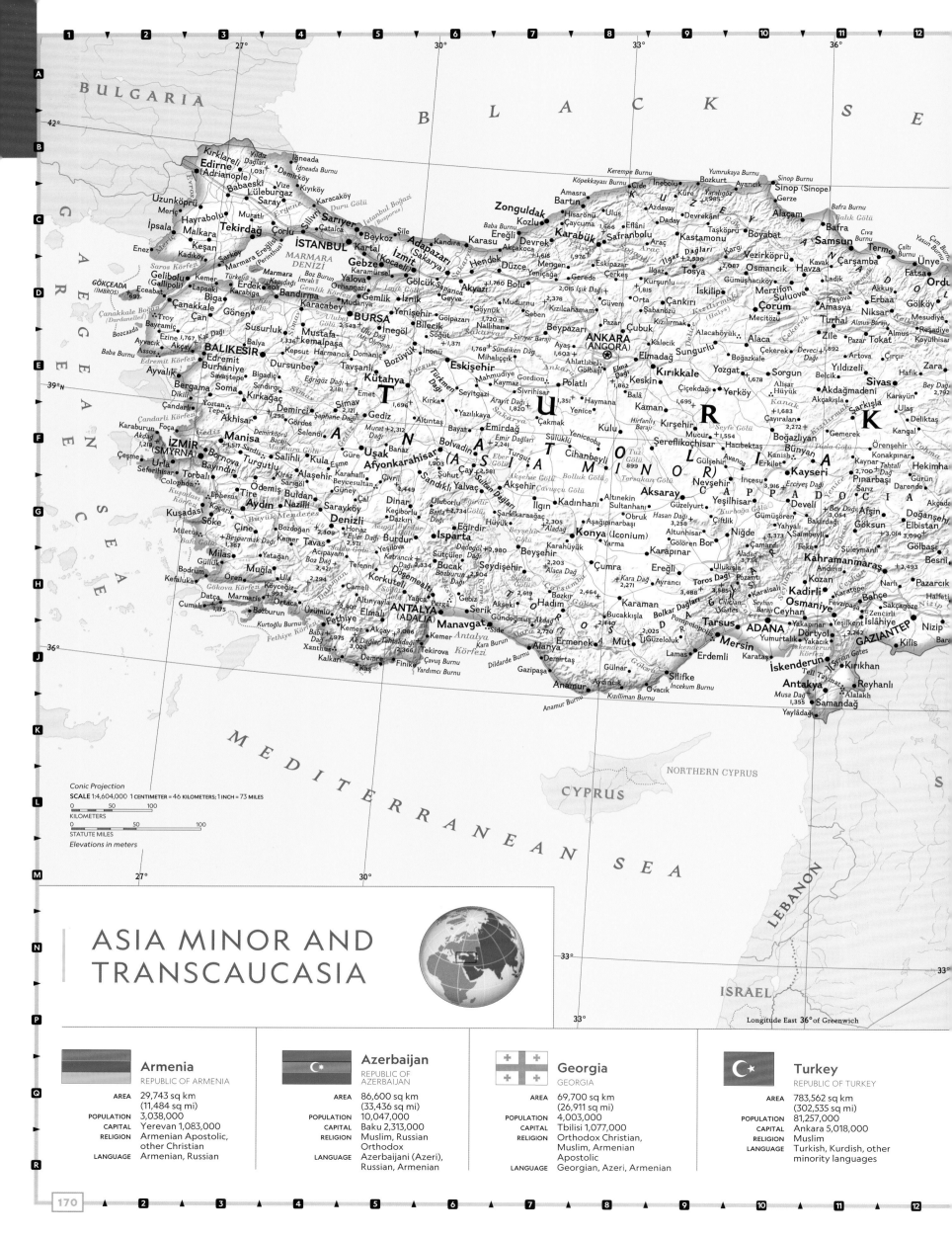

ASIA MINOR AND TRANSCAUCASIA

Armenia
REPUBLIC OF ARMENIA
AREA 29,743 sq km (11,484 sq mi)
POPULATION 3,038,000
CAPITAL Yerevan 1,083,000
RELIGION Armenian Apostolic, other Christian
LANGUAGE Armenian, Russian

Azerbaijan
REPUBLIC OF AZERBAIJAN
AREA 86,600 sq km (33,436 sq mi)
POPULATION 10,047,000
CAPITAL Baku 2,313,000
RELIGION Muslim, Russian Orthodox
LANGUAGE Azerbaijani (Azeri), Russian, Armenian

Georgia
GEORGIA
AREA 69,700 sq km (26,911 sq mi)
POPULATION 4,003,000
CAPITAL Tbilisi 1,077,000
RELIGION Orthodox Christian, Muslim, Armenian Apostolic
LANGUAGE Georgian, Azeri, Armenian

Turkey
REPUBLIC OF TURKEY
AREA 783,562 sq km (302,535 sq mi)
POPULATION 81,257,000
CAPITAL Ankara 5,018,000
RELIGION Muslim
LANGUAGE Turkish, Kurdish, other minority languages

Conic Projection
SCALE 1:4,604,000 1 CENTIMETER = 46 KILOMETERS; 1 INCH = 73 MILES
Elevations in meters

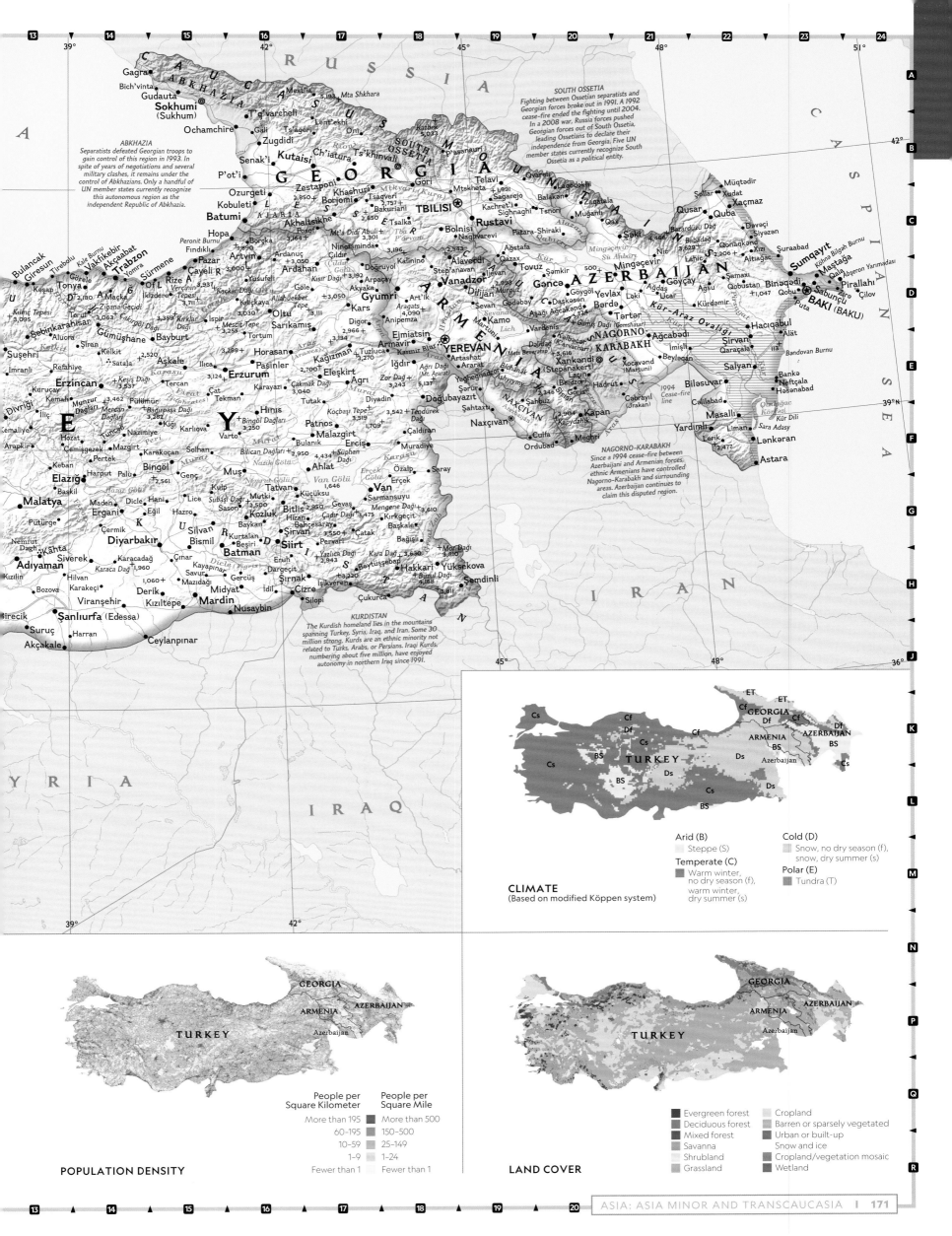

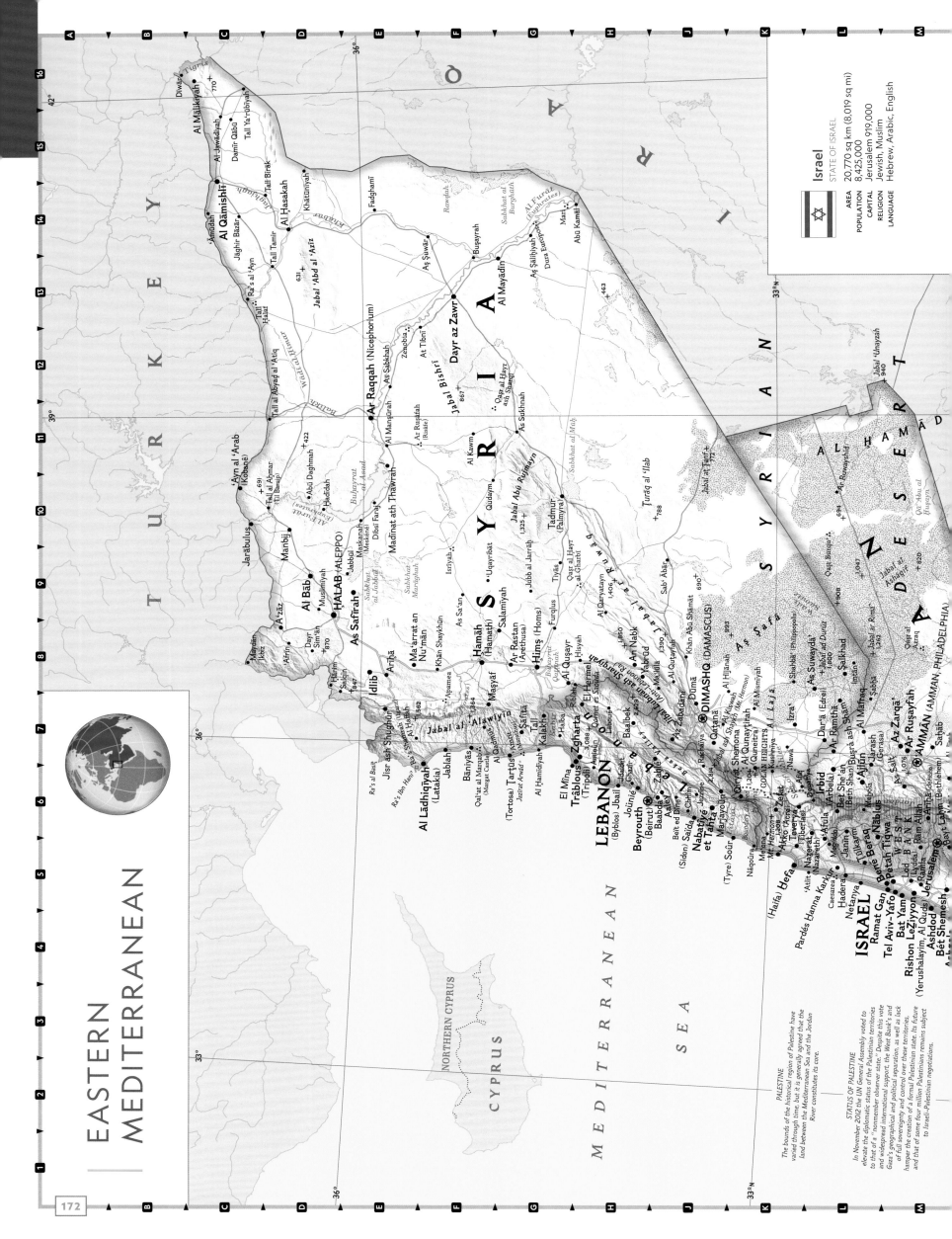

EASTERN MEDITERRANEAN

Israel
STATE OF ISRAEL

AREA 20,770 sq km (8,019 sq mi)
POPULATION 8,425,000
CAPITAL Jerusalem 919,000
RELIGION Jewish, Muslim
LANGUAGE Hebrew, Arabic, English

PALESTINE
The bounds of the historical region of Palestine have varied through time, but it is generally agreed that the land between the Mediterranean Sea and the Jordan River constitutes its core.

STATUS OF PALESTINE
In November 2012 the UN General Assembly voted to elevate the diplomatic status of the Palestinian territories to that of a "nonmember observer state." Despite this vote and widespread international support, the West Bank's and Gaza's geographical and political separation, as well as lack of full sovereignty and control over these territories, hamper the creation of a formal Palestinian state. Its future and that of some four million Palestinians remains subject to Israeli-Palestinian negotiations.

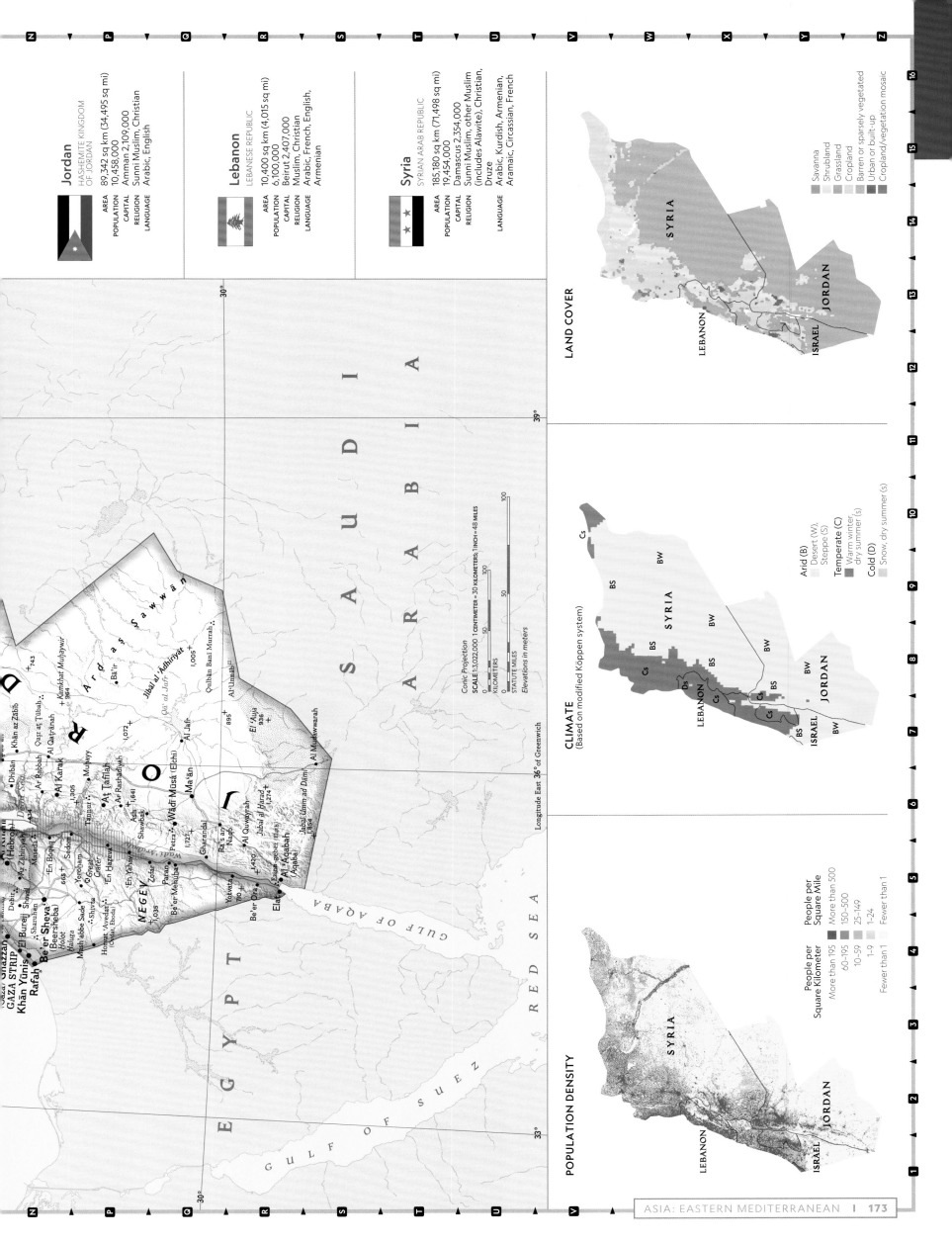

Jordan
HASHEMITE KINGDOM OF JORDAN

AREA	89,342 sq km (34,495 sq mi)
POPULATION	10,458,000
CAPITAL	Amman 2,109,000
RELIGION	Sunni Muslim, Christian
LANGUAGE	Arabic, English

Lebanon
LEBANESE REPUBLIC

AREA	10,400 sq km (4,015 sq mi)
POPULATION	6,100,000
CAPITAL	Beirut 2,407,000
RELIGION	Muslim, Christian
LANGUAGE	Arabic, French, English, Armenian

Syria
SYRIAN ARAB REPUBLIC

AREA	185,180 sq km (71,498 sq mi)
POPULATION	19,454,000
CAPITAL	Damascus 2,354,000
RELIGION	Sunni Muslim, other Muslim (includes Alawite), Christian, Druze
LANGUAGE	Arabic, Kurdish, Armenian, Aramaic, Circassian, French

Conic Projection
SCALE 1:3,022,000 1 CENTIMETER = 30 KILOMETERS; 1 INCH = 48 MILES

KILOMETERS
STATUTE MILES
Elevations in meters

Longitude East 36° of Greenwich

LAND COVER

Savanna
Shrubland
Grassland
Cropland
Barren or sparsely vegetated
Urban or built-up
Cropland/vegetation mosaic

SYRIA
LEBANON
ISRAEL
JORDAN

CLIMATE
(Based on modified Köppen system)

Arid (B)
Desert (W),
Steppe (S)

Temperate (C)
Warm winter,
dry summer (s)

Cold (D)
Snow, dry summer (s)

SYRIA
LEBANON
ISRAEL
JORDAN

POPULATION DENSITY

People per Square Kilometer	People per Square Mile
More than 195	More than 500
60–195	150–500
10–59	25–149
1–9	1–24
Fewer than 1	Fewer than 1

SYRIA
LEBANON
ISRAEL
JORDAN

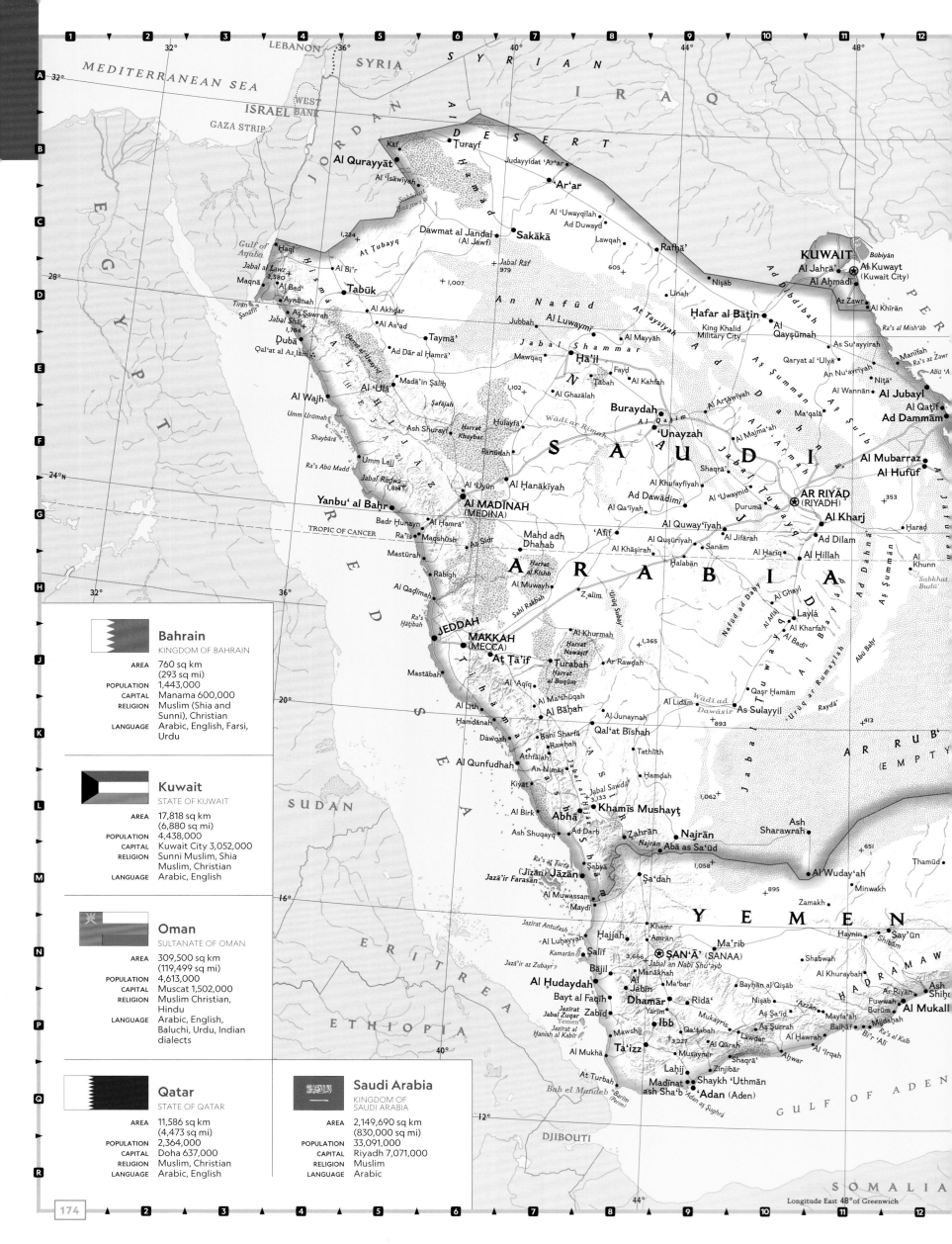

Bahrain
KINGDOM OF BAHRAIN
AREA 760 sq km (293 sq mi)
POPULATION 1,443,000
CAPITAL Manama 600,000
RELIGION Muslim (Shia and Sunni), Christian
LANGUAGE Arabic, English, Farsi, Urdu

Kuwait
STATE OF KUWAIT
AREA 17,818 sq km (6,880 sq mi)
POPULATION 4,438,000
CAPITAL Kuwait City 3,052,000
RELIGION Sunni Muslim, Shia Muslim, Christian
LANGUAGE Arabic, English

Oman
SULTANATE OF OMAN
AREA 309,500 sq km (119,499 sq mi)
POPULATION 4,613,000
CAPITAL Muscat 1,502,000
RELIGION Muslim Christian, Hindu
LANGUAGE Arabic, English, Baluchi, Urdu, Indian dialects

Qatar
STATE OF QATAR
AREA 11,586 sq km (4,473 sq mi)
POPULATION 2,364,000
CAPITAL Doha 637,000
RELIGION Muslim, Christian
LANGUAGE Arabic, English

Saudi Arabia
KINGDOM OF SAUDI ARABIA
AREA 2,149,690 sq km (830,000 sq mi)
POPULATION 33,091,000
CAPITAL Riyadh 7,071,000
RELIGION Muslim
LANGUAGE Arabic

Longitude East 48° of Greenwich

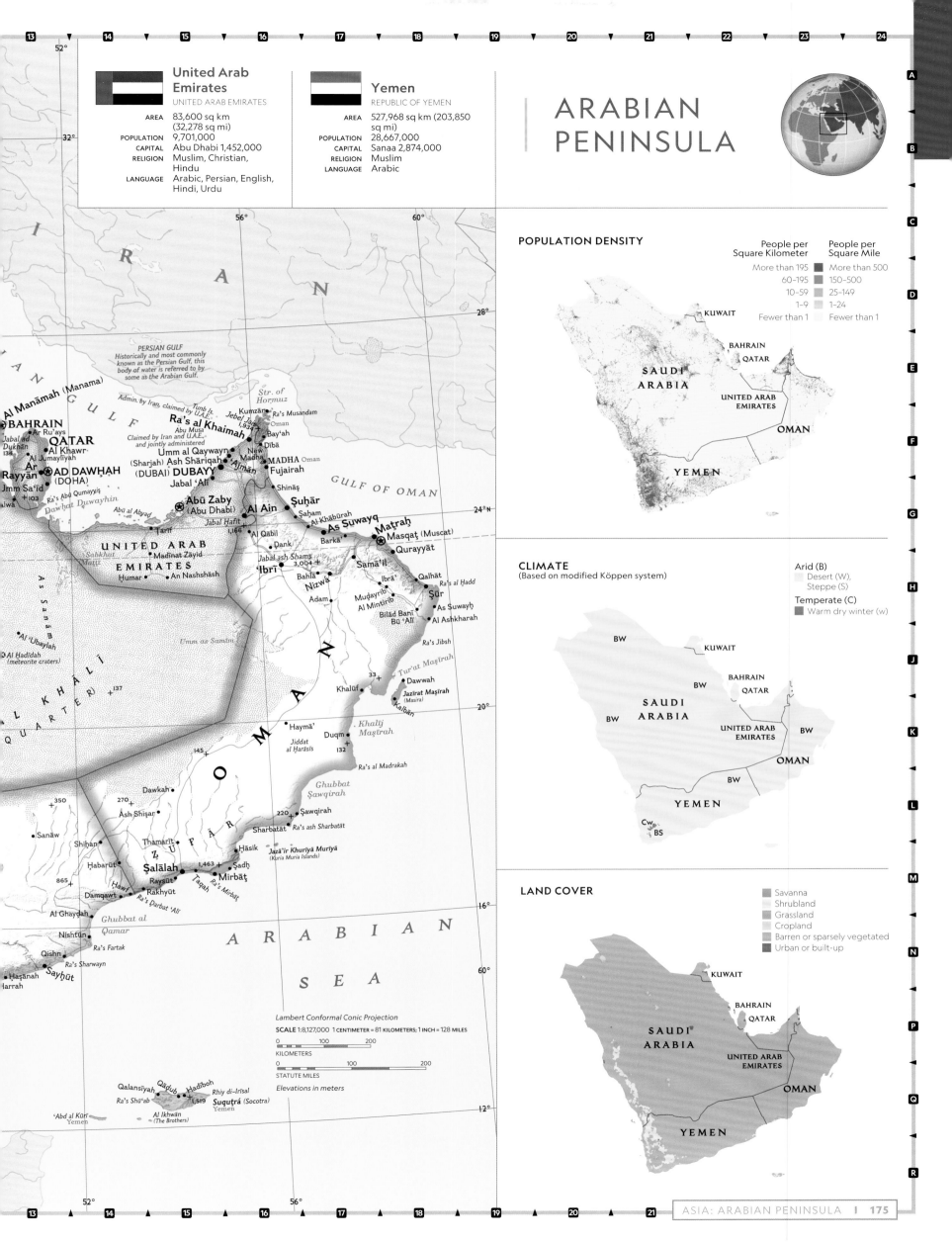

United Arab Emirates
UNITED ARAB EMIRATES

AREA	83,600 sq km (32,278 sq mi)
POPULATION	9,701,000
CAPITAL	Abu Dhabi 1,452,000
RELIGION	Muslim, Christian, Hindu
LANGUAGE	Arabic, Persian, English, Hindi, Urdu

Yemen
REPUBLIC OF YEMEN

AREA	527,968 sq km (203,850 sq mi)
POPULATION	28,667,000
CAPITAL	Sanaa 2,874,000
RELIGION	Muslim
LANGUAGE	Arabic

ARABIAN PENINSULA

POPULATION DENSITY

People per Square Kilometer	People per Square Mile
More than 195	More than 500
60–195	150–500
10–59	25–149
1–9	1–24
Fewer than 1	Fewer than 1

KUWAIT
BAHRAIN
QATAR
SAUDI ARABIA
UNITED ARAB EMIRATES
OMAN
YEMEN

CLIMATE
(Based on modified Köppen system)

Arid (B)
Desert (W),
Steppe (S)

Temperate (C)
Warm dry winter (w)

BW
KUWAIT
BW
BAHRAIN
QATAR
SAUDI ARABIA
UNITED ARAB EMIRATES
BW
BW
OMAN
BW
YEMEN
Cw
BS

LAND COVER

Savanna
Shrubland
Grassland
Cropland
Barren or sparsely vegetated
Urban or built-up

KUWAIT
BAHRAIN
QATAR
SAUDI ARABIA
UNITED ARAB EMIRATES
OMAN
YEMEN

Map

I R A N

PERSIAN GULF
Historically and most commonly known as the Persian Gulf, this body of water is referred to by some as the Arabian Gulf.

Str. of Hormuz
Admin. by Iran, claimed by U.A.E.
Tunb Is.
Jebel Iran 1,948
Kumzār
Ra's Musandam
Oman
Bay'ah
Al Manāmah (Manama)
Abu Musa
Ra's al Khaimah
Claimed by Iran and U.A.E. and jointly administered
Dībā
BAHRAIN
Ar Ru'ays
New Madha
MADHA Oman
QATAR
Al Khawr
Umm al Qaywayn
(Sharjah) Ash Shāriqah
Ajmān
Fujairah
Al Jumaylīyah
(DUBAI) DUBAYY
Ar Rayyān
AD DAWḤAH (DOHA)
Jabal 'Alī
Shināṣ
Jmm Sa'īd
Ra's Abu Qumayyiş
Abū al Abyaḑ
Abū Zaby (Abu Dhabi)
Al Ain
GULF OF OMAN
+103
Dawhat Duwayhin
Abū al Abyaḑ
Ṣuḥār
Iwā
Ţarīf
Jabal Ḩafīt 1,166
Saḩam
Al Khābūrah
As Suwayq
Maţraḩ
Masqaţ (Muscat)
Jabal ad Dukhān 134
Sabkhat Matti
Madīnat Zāyid
Al Qabil
Dank
Barkā
Qurayyāt
UNITED ARAB EMIRATES
Ḩumar
An Nashshāsh
Jabal ash Shams 3,004
Samā'il
Ibrā'
'Al 'Ubaylah
As Sanām
Ibrī
Bahlā
Nizwā
Muḑayrib
As Suwayḩ
Qalḩāt
Ra's al Ḩadd
Al Ḩadīdah (meteorite craters)
Adam
Al Mintirib
Bilād Banī Bū 'Alī
Al Ashkharah
Umm as Samīm
O M A N
Ra's Jibsh
AL KHĀLĪ (QUARTER)
+137
33 +
Dawwah
Tur'at Maşīrah
Khalūf
Jazīrat Maşīrah (Masira)
Kalbān
Haymā'
Khalīj Maşīrah
Duqm
132
Jiddat al Ḩarāsīs
Ra's al Madrakah
145 +
Ghubbat Şawqirah
Dawkah
220
Şawqirah
270
Ash Shişar
Sharbatāt
Ra's ash Sharbatāt
350
Thamarīt
Ḩāsik
Jazā'ir Khurīyā Murīyā (Kuria Muria Islands)
Sanāw
Shiḩan
Ẕ U F Ā R
1,463 +
Şadḩ
Mirbāţ
Ḩabarūt
Şalālah
Raysūt
Ṭaqah
Ra's Mirbāţ
865 +
Ḩawf
Ra's Darbat 'Alī
Damqawt
Rakhyūt
Al Ghaydah
Ghubbat al Qamar
A R A B I A N S E A
Nishṭūn
Ra's Fartak
Qishn
Ra's Sharwayn
Ḩaşānah
Sayḩūt
Ḩarrah

Qalansīyah
Qāḑub
Ḩadīboh
Rhiy di-Irīsal
Ra's Shū'ab
1,519
Suquṭrá (Socotra)
Yemen
'Abd al Kūrī
Yemen
Al Ikhwān (The Brothers)

Lambert Conformal Conic Projection

SCALE 1:8,127,000 1 CENTIMETER = 81 KILOMETERS; 1 INCH = 128 MILES

0   100   200
KILOMETERS

0   100   200
STATUTE MILES

Elevations in meters

IRAN AND IRAQ

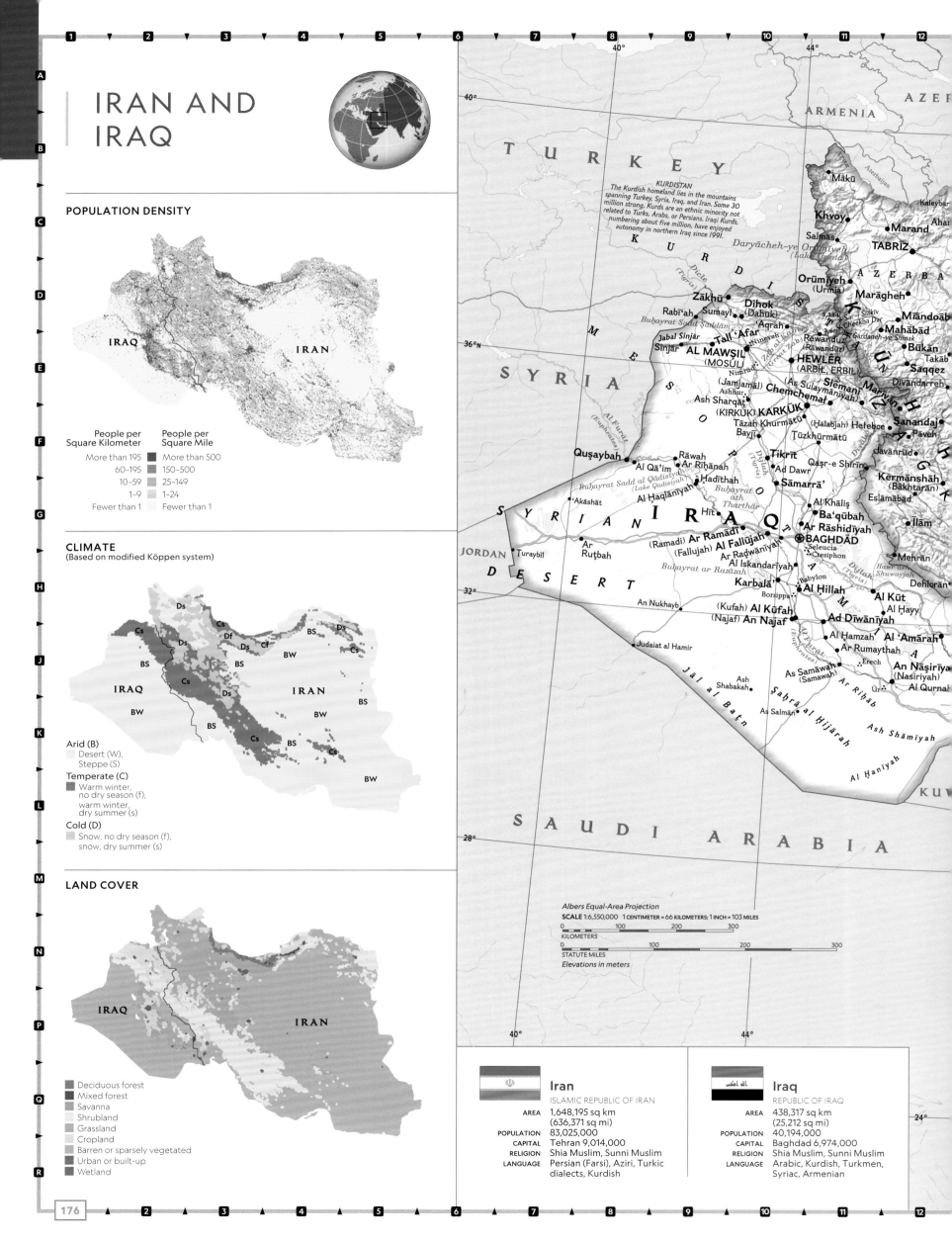

POPULATION DENSITY

IRAQ

IRAN

People per Square Kilometer	People per Square Mile
More than 195	More than 500
60–195	150–500
10–59	25–149
1–9	1–24
Fewer than 1	Fewer than 1

CLIMATE
(Based on modified Köppen system)

Ds · Cs · Df · Cf · BS · Ds · BW · Cs · BS · Cs · BS · BW · BS · Cs · BS · Cs · BW

IRAQ · IRAN

BW

Arid (B)
Desert (W), Steppe (S)

Temperate (C)
Warm winter, no dry season (f), warm winter, dry summer (s)

Cold (D)
Snow, no dry season (f), snow, dry summer (s)

LAND COVER

IRAQ

IRAN

- Deciduous forest
- Mixed forest
- Savanna
- Shrubland
- Grassland
- Cropland
- Barren or sparsely vegetated
- Urban or built-up
- Wetland

AZER

ARMENIA

TURKEY

KURDISTAN
The Kurdish homeland lies in the mountains spanning Turkey, Syria, Iraq, and Iran. Some 30 million strong, Kurds are an ethnic minority not related to Turks, Arabs, or Persians. Iraqi Kurds, numbering about five million, have enjoyed autonomy in northern Iraq since 1991.

Mākū

Kaleybar

Khvoy

Marand

Ahar

Salmās

Daryācheh-ye Orūmīyeh (Lake Urmia)

TABRĪZ

Orūmīyeh (Urmia)

Marāgheh

Zākhū

Dihok (Dahuk)

3,334

Sar-i Slakiv

Mīāndoāb

AZERBA

Rabī'ah

Sumayl

Aqrah

Cheekha Dar 3,611

Gardaneh-ye Shinak

Mahābād

Būkān

Jabal Sinjār

Tall 'Afar

Nineveh

Rewanduz (Rawanduz)

Takāb

KURDIS

Sinjār

AL MAWṢIL (MOSUL)

HEWLĒR (ARBĪL, ERBIL)

Divandarreh

(Jamjamāl)

Ashhur

Slēmani

Saqqez

SYRIA

Nimrud

Ash Sharqāṭ

(Halabjah)

Hefebce

Sanandaj

(KIRKUK) **KARKŪK**

Chemchemat

Pāveh

Tāzah Khurmātū

Bayjī

Tūzkhūrmātū

davānrūd

Qușaybah

Rāwah

Ar Rīḥānah

Tikrīt

Qaṣr-e Shīrīn

Kermānshāh (Bākhtarān)

Al Qā'im

Hadīthah

Ad Dawr

Sāmarrā'

Eslāmābād

Buḥayrat Sadd al Qādisīyah (Lake Qadisiya)

Buḥayrat ath Tharthār

Al Ḥaqlānīyah

Al Khāliṣ

Ba'qūbah

Īlām

'Akāshāt

Hīt

Ar Rāshidīyah

SYRIAN

(Ramadi) **Ar Ramādī**

Ar Radwānīyah

Seleucia

Mehrān

Ar Ruṭbah

(Fallujah) **Al Fallūjah**

⊛ **BAGHDAD**

Cresiphon

Hawr ash Shuwayjah

JORDAN

Turaybīl

Al Iskandarīyah

Buḥayrat ar Razāzah

Babylon

Karbalā'

Borsippa

Al Ḥillah

Al Kūt

Dehlorān

An Nukhayb

(Kufah) **Al Kūfah**

Al Hayy

(Najaf) **An Najaf**

Ad Dīwānīyah

Al Hamzah

Al 'Amārah

DESERT

Judaiat al Hamir

Ar Rumaythah

An Nāṣirīyah (Nasiriyah)

As Samāwah (Samawah)

Erech

Jāl al Baṭn

Ash Shabakah

Ṣaḥrā al Ḥijārah

As Salmān

Ūr

Ar Rīḥāb

Al Qurnah

Ash Shāmīyah

Ash Shāmīyah

Al Ḥanīyah

KU

SAUDI ARABIA

Albers Equal-Area Projection
SCALE 1:6,550,000 1 CENTIMETER = 66 KILOMETERS; 1 INCH = 103 MILES

0 — 100 — 200 — 300
KILOMETERS

0 — 100 — 200 — 300
STATUTE MILES

Elevations in meters

Iran
ISLAMIC REPUBLIC OF IRAN

AREA	1,648,195 sq km (636,371 sq mi)
POPULATION	83,025,000
CAPITAL	Tehran 9,014,000
RELIGION	Shia Muslim, Sunni Muslim
LANGUAGE	Persian (Farsi), Aziri, Turkic dialects, Kurdish

Iraq
REPUBLIC OF IRAQ

AREA	438,317 sq km (25,212 sq mi)
POPULATION	40,194,000
CAPITAL	Baghdad 6,974,000
RELIGION	Shia Muslim, Sunni Muslim
LANGUAGE	Arabic, Kurdish, Turkmen, Syriac, Armenian

A
B
C
D
E
F
G
H
J
K
L
M
N
P
Q
R

13 14 15 16 17 18 19 20 21 22 23 24

48° 52° 56° 60° 64°

40°

36°N

32°

28°

24°

UZBEKISTAN

AIJAN

CASPIAN
SEA

T U R K M E N I S T A N

Parsābād
Ja'farābād
Bīleh Savār
Germī

Kūh-e
Sabalān
4,811

Āstārā
Ardabīl
Hashtpar
Rezvānshahr
Bandar-e Anzalī
Rasht
Lāhījān

Kopet Dag

Marāveh
Tappeh
Ashkhāneh
Shīrvān
Bājgirān
Darreh
Gaz

Nīr
Sarāb
Mīāneh

Gonbad-e
Kāvūs
Minū Dasht
Bojnūrd
Fārūj
Qūchān

Chenārān
Sarakhs

Qā'emshahr
Behshahr
Gorgān
Āzād Shahr
Garmeh
Jājarm

Mayāmey
Soltānābād
Kūh-e Siāh Khvānī
3,314

MASHHAD

Nīk
Pey
Zanjān

Qazvīn

Abhar

Qīdar

Bījār

Behshahr

Sārī
Bābol
Nūr
Āmol

Shāhrūd

Sabzevār
Neyshābūr

Marzdārān

Farīmān

Tāybād

Kūhhā-ye Bakharz

Torbat-e
Jām

A F G H A N I S T A N

RESHTEH KŪHHĀ-YE ALBORZ (MTS.)

Tākestān

Ābyek

Āvej

Tajrīsh
5,670 Kūh-e Damāvand
KARAJ
Eslāmshahr TEHRĀN
Rey
Qarchak Palasht
Varāmīn

Semnān

Torūd

Torbat-e Heydarīyeh
Bardaskan
Kāshmar

Khvāf

Qorveh

Hamadān
(Ecbatana)
Sāveh

Tafresh

Sonqor

Tūyserkān

Malāyer

Arāk

QOM (QUM)

Daryā-ye
Namak

Dasht-e Kavīr
(Kavir Desert)

K H O R A S Ā N

Kavīr-e Namakī
Bejestān
Gonābād

Ferdows

Sarāyān

Qā'en

Sedeh

Āvāz

Sarbīsheh

Shūsf

Kūh-e Garīn
3,630
Nūrābād
Borūjerd
Khorramābād
Dorūd

Mahallāt
Khomeyn
Golpāyegān

Delījān
Ārān
Kāshān

Qom

Natanz

Ardestān

Jandaq

Khor

Tabas

Robāt-e
Khān

Kūhdasht

Alīgūdarz

Khvānsār

Dārān

Shāhīn Shahr
Khomeynīshahr
Najafābād ESFAHĀN (ISFAHAN)
Zarrīn Shahr

Na'īn

Kharānaq

Bīrjand

32°

Dezfūl
Shūsh Susa
Shūshtar
Sūsangerd

Kūh-e Zard
4,548
Masjed Soleymān
Īzeh

Shahr-e Kord

Borujen

Shahrezā
(Qomsheh)

Ardakān
Meybod
Ashkezār

Taft
Yazd
Mehrīz

Bāfq

Kūh Banen
Rāvar

Nehbandān

Daryācheh-ye
Hāmūn Ṣābert

SĪSTĀN

Zābol
Zehak
Lūtak

PAKISTAN

AHVĀZ
(AHWĀZ)

AL BAṢRAH (BASRA)
Abū al Khaṣīb
Khorramshahr
Ābādān

Az Zubayr

Rāmhormoz

Semīrom

Shīr Kūh
4,075

Kūh-e Dīnār
4,409

Ābādeh

Abarkūh

Eqlīd

Deh Bīd

Kermānshāhān

Anār

Zarand

Zābol

Rafsanjān

Shahr-e
Bābak

Bāghīn

Bardsīr

Harūz-e
Bālā

Kermān
Māhān

Noṣratābād

Hormak

Zāhedān

Mīrjāveh

Ladīz

Kūh-e
Taftān
4,042

Khāsh

Al Faw

Shatt al 'Arab

Bandar-e Māhshahr
Behbahān
Dow Gonbadān
Hendījān

Bandar-e
Deylam

Yāsūj

Nūrābād

Ardakān

Pasargadae

Persepolis

Daryācheh-ye
Bakhtegān

Shahr-e
Bābak

Sīrjān
(Sa'īdābād)

Bāft
Rābor

Kūh-e Hezār
4,465
Rāyen

Jīroft
(Sabzvārān)

'Anbarābād

Bam

Fahraj
Rīgān

Kūh-e Bazmān
3,489

Gazak

Sūrān

Sarāvān

Zāboli

Esfandak

Kūhak

BALUCHISTAN

Jazīreh-ye Khārk
(Khārg Island)

Bandar-e Būshehr

Bandar-e
Ganāveh

Borāzjān

Ahram

Khormūj

Farrāshband

Qīr

Marv Dasht

SHIRĀZ

Fasā
Dārāb

Jahrom

Estahbān

Neyrīz

Aliabad

Hājjīābād
Dowlatābād

Kahnūj

Kūh-e Fāreghān
3,240

Manūjān

Qal'eh-ye
Ganj

Hāmūn-e
Jaz Mūriān

Bampūr
Īrānshahr

Espakeh

Kūh-e
Nokhoch
2,093

Sarbāz

Hamun-i
Mashkel

Jalq

Fārsī Iran

Khalīj-e Nāy Band

Tāherī
Nāy Band

Galleh Dār
Lāmerd

Gāvbandī

Avaz

Gerāsh

Bastak

Khonj
Lār

Kūrān Dap

Sa'ādatābād
Tārom

Bandar-e
Khamīr

Bandar-e
Chārak
Lāvān
Hendorābī

Qeshm
Lārak

Bandar-e
Abbās

Ḥasan Langī
Mīnāb

Hormoz
Qeshm
Strait of
Hormuz

Sīrīk

Bandar-e
Lengeh

Kish Kish

Forūr

Sīrrī

Tunb Is.
Administered by Iran,
claimed by U.A.E.

Abu Musa
Claimed by Iran and U.A.E.
jointly administered

Oman

Fanūj

Bent

Nīk Shahr

Qaṣr-e
Qand

Pīshīn

Bāhū Kalāt
Polān

Sūrak

Ḥūmedān

Chāh Bahār

Gavāter

Khalīj-e Chāh Bahār

Gwatar Bay

BAHRAIN

P E R S I A N
G U L F

Bandar-e
Kangān

Jāsk

PERSIAN GULF
Historically and most commonly
known as the Persian Gulf, this
body of water is referred to by
some as the Arabian Gulf.

QATAR

G U L F   O F   O M A N

U N I T E D   A R A B
E M I R A T E S

O M A N

CENTRAL ASIA

R U S S
Ishim Steppe
Petropavlovsk   Būlaevo
Komsomolets
Uzynköl
Fedorovka   Borovskoy   Tayynsha
Vladimirovka   (Krasnoarmeysk)
Qostanay   Zätobyl   Sarygöl   Säumalköl   Kökshetaū
Rūdnyy   Novoishimskiy   Shchūchīnsk
Tobyl   Rūzaevka   +730
Līsakovsk   Qusmuryn   Makīnsk
Zhetiqara   Tobyl   Aqköl
Zhaqsy   Esil   Zhaltyr
Buzuluk   Nur-Sultan
Yuzhnyy   Zhayylma   (Astana)
312+   Qayghy   Derzhavīnsk   Ladyzhino
Shīli   Arqalyq
THE   Amangeldi   Tengīz
Kamenka   Oral   Aqsay   Shynqghyrlau   Köli
Alghabas   +263
Chapaev   Zhympity   Mortyq   Kurashassayskiy   Aqtöbe   Khromtaū   Araltobe
Mergenevo   Qarabutaq
Kalenyy   Bazartöbe   Qaratöbe   Algha   Saryqopa Köli   Qarabutaq   Qosköl
Taypaq   Shubarqudyq   Oktyabr'sk   Torghay   Arqalyq   KA
Külägino   Zhuryn   S   Qosköl
Zelenoe   Inderbor   Bayghanīn   Embi   T
Makhambet   Saghyz   Kozhasay   Birshoghyr   Yrghyz   E
Aqqystaū   −11   Saghyz   P   Zhayrang
Balyqshy   Atyraū   Maqat   Komsomol   +251   Sarysay   P   Sätbaev   UP
Dossor   Zharkamys   Togyz   1,133+   Qarsaqbay   E   Qarazhal
Ganyushkino   Qulsary   +408   Shalqar   S   Zhezqazghan
Qosshaghyl   −56   Dongyztaū   Sekseūil   Aral Mangy   Qarazhal
Qaraton   +215   Qaraqumy
Tengīz   Shaghan   North Aral Sea   BAIKONUR
Sarygamys   Tushchybas Shyghanaghy   COSMODROME
Former shoreline,   Qulandy   Aral   Russian-administered
2000   Aralsul'fat   Qamystybas   BETPA
Qazaly   Äyteke Bi   Bayqongyr   Zhosaly
Beyneu   Zhanatal   (Baikonur)
South   Qyzylorda
Aral   Tasböget   Shīeli   +2,176
USTYURT   Sea   ARAL   +164   Zhangatas
Komsomol'sk-na-Ustyurte   SEA   Zhangaqorghan   Qaratau
Zhaslyk   Once the world's fourth largest lake,   Türkistan
the Aral Sea today is one-tenth of its   Arys
PLATEAU   QORAQALPOGISTON   1960 extent. Soviet-era irrigation canals   Shymkent
divert river water—causing the sea to   Lenge
Münoq   shrink and changing the former
lake bed into a desert.
Qo'ng'irot   Chirchiq
Chimboy   Takhtaküpir   +61   Iskanda
Shumanay   UZBEKISTAN   (TASHKENT) TOSHKENT
Khūjayli   Nukus   Kulduduk   +764   Angren
Köneürgench   Takhiatosh   Mynbulaq   Guliston
Boldumsaz   Manghit   Uchquduq   Konibodom
Dashoguz   Beruniy   Zarafshon   Bekobod   Khūjand
Urganch   Türtkül   Sūlūktū
Khiwa   Gojaqk   Jizzax   Istaravshan
Gazojak   Lebap   (Ura-Tyube)
Darganata   Navoiy   Bulung'ur   Istani
Zarafshon   Gizhduvan   Payshanba   Panjakent   Samarqand
Buxoro   Qorakül   Kogon   Koson   Dushanbe
(Bukhara)   Qarshi

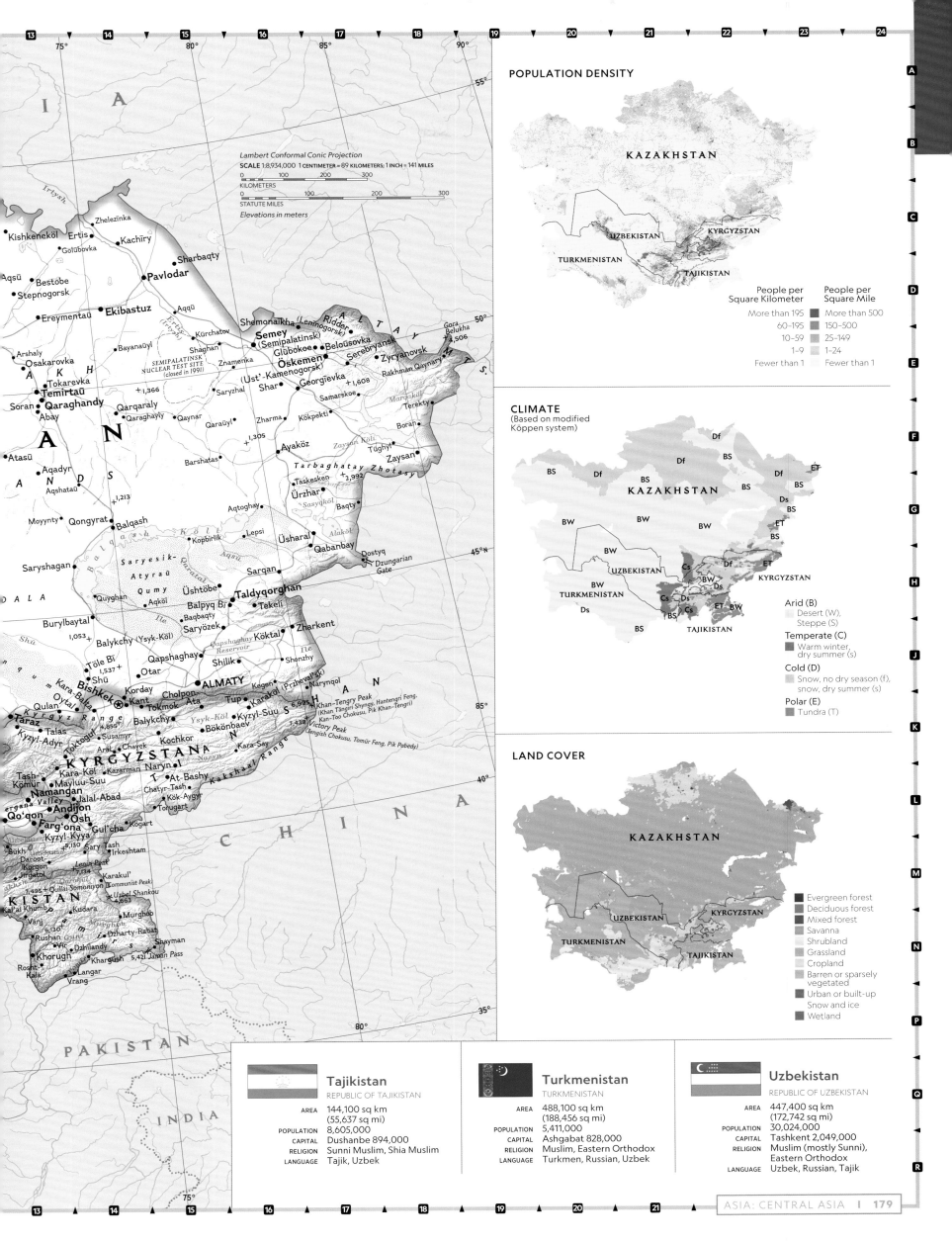

POPULATION DENSITY

KAZAKHSTAN

UZBEKISTAN

KYRGYZSTAN

TURKMENISTAN

TAJIKISTAN

People per Square Kilometer	People per Square Mile
More than 195	More than 500
60–195	150–500
10–59	25–149
1–9	1–24
Fewer than 1	Fewer than 1

CLIMATE
(Based on modified Köppen system)

KAZAKHSTAN

UZBEKISTAN

TURKMENISTAN

KYRGYZSTAN

TAJIKISTAN

Arid (B)
Desert (W), Steppe (S)

Temperate (C)
Warm winter, dry summer (s)

Cold (D)
Snow, no dry season (f), snow, dry summer (s)

Polar (E)
Tundra (T)

LAND COVER

KAZAKHSTAN

UZBEKISTAN

KYRGYZSTAN

TURKMENISTAN

TAJIKISTAN

- Evergreen forest
- Deciduous forest
- Mixed forest
- Savanna
- Shrubland
- Grassland
- Cropland
- Barren or sparsely vegetated
- Urban or built-up
- Snow and ice
- Wetland

Tajikistan
REPUBLIC OF TAJIKISTAN

AREA 144,100 sq km (55,637 sq mi)
POPULATION 8,605,000
CAPITAL Dushanbe 894,000
RELIGION Sunni Muslim, Shia Muslim
LANGUAGE Tajik, Uzbek

Turkmenistan
TURKMENISTAN

AREA 488,100 sq km (188,456 sq mi)
POPULATION 5,411,000
CAPITAL Ashgabat 828,000
RELIGION Muslim, Eastern Orthodox
LANGUAGE Turkmen, Russian, Uzbek

Uzbekistan
REPUBLIC OF UZBEKISTAN

AREA 447,400 sq km (172,742 sq mi)
POPULATION 30,024,000
CAPITAL Tashkent 2,049,000
RELIGION Muslim (mostly Sunni), Eastern Orthodox
LANGUAGE Uzbek, Russian, Tajik

Lambert Conformal Conic Projection
SCALE 1:8,934,000 1 CENTIMETER = 89 KILOMETERS; 1 INCH = 141 MILES
KILOMETERS
0   100   200   300
STATUTE MILES
0   100   200   300
Elevations in meters

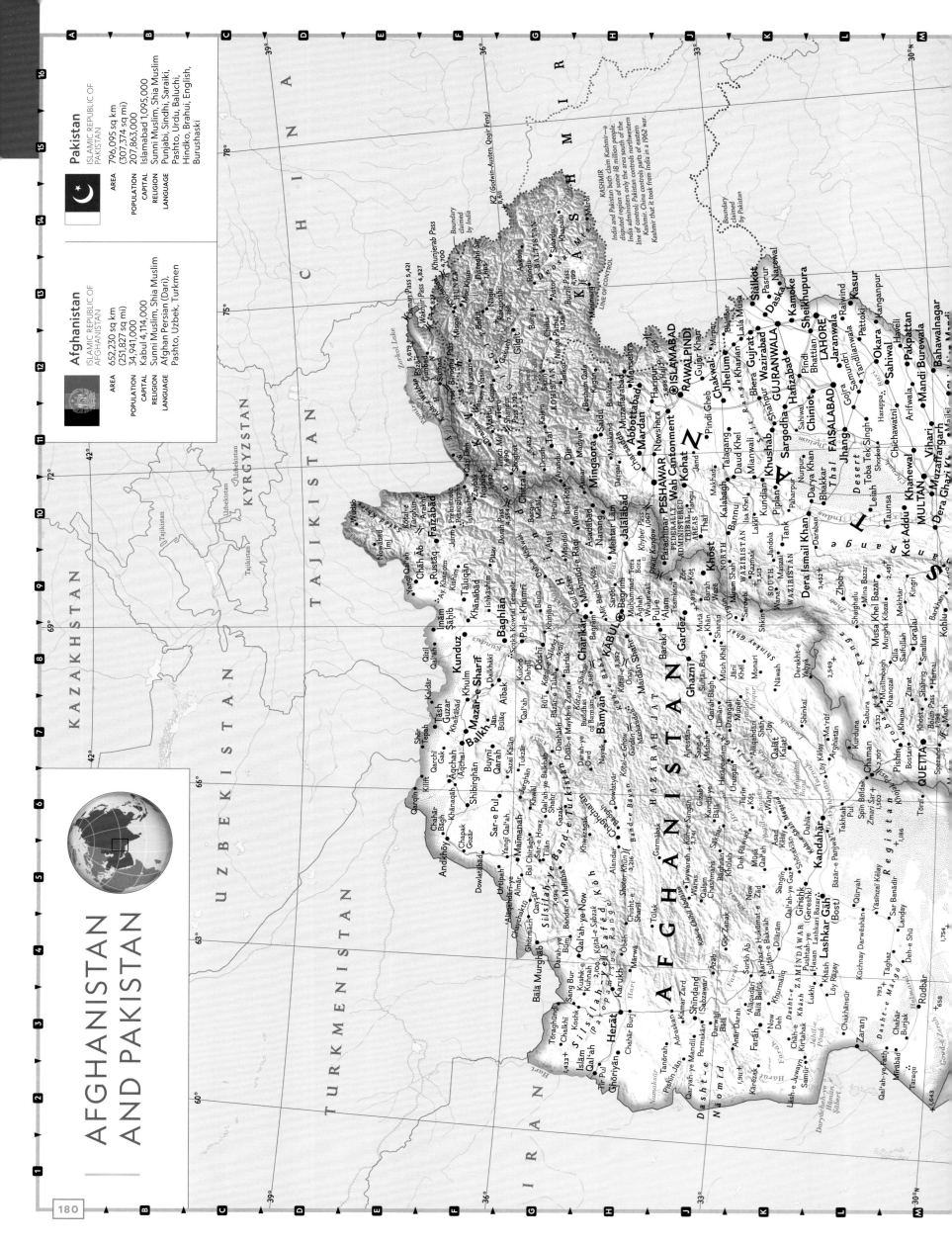

AFGHANISTAN AND PAKISTAN

Pakistan

ISLAMIC REPUBLIC OF PAKISTAN

AREA	796,095 sq km (307,374 sq mi)
POPULATION	207,863,000
CAPITAL	Islamabad 1,095,000
RELIGION	Sunni Muslim, Shia Muslim
LANGUAGE	Punjabi, Sindhi, Saraiki, Pashto, Urdu, Baluchi, Hindko, Brahui, English, Burushaski

Afghanistan

ISLAMIC REPUBLIC OF AFGHANISTAN

AREA	652,230 sq km (251,827 sq mi)
POPULATION	34,941,000
CAPITAL	Kabul 4,114,000
RELIGION	Sunni Muslim, Shia Muslim
LANGUAGE	Afghan Persian (Dari), Pashto, Uzbek, Turkmen

KASHMIR India and Pakistan both claim Kashmir—a disputed region of some 18 million people. India administers only the area south of the line of control; Pakistan controls parts of northwestern Kashmir. China controls parts of eastern Kashmir that it took from India in a 1962 war.

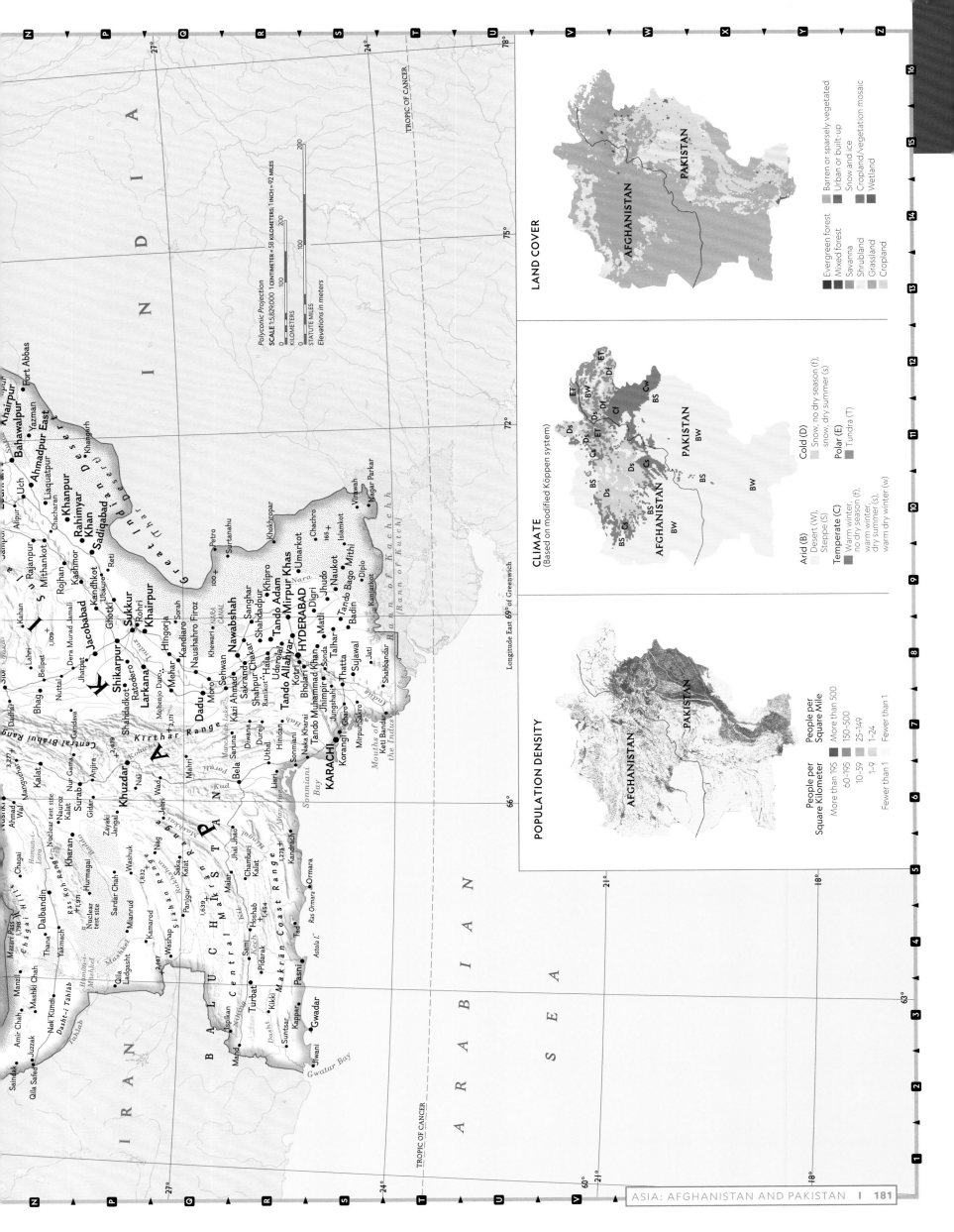

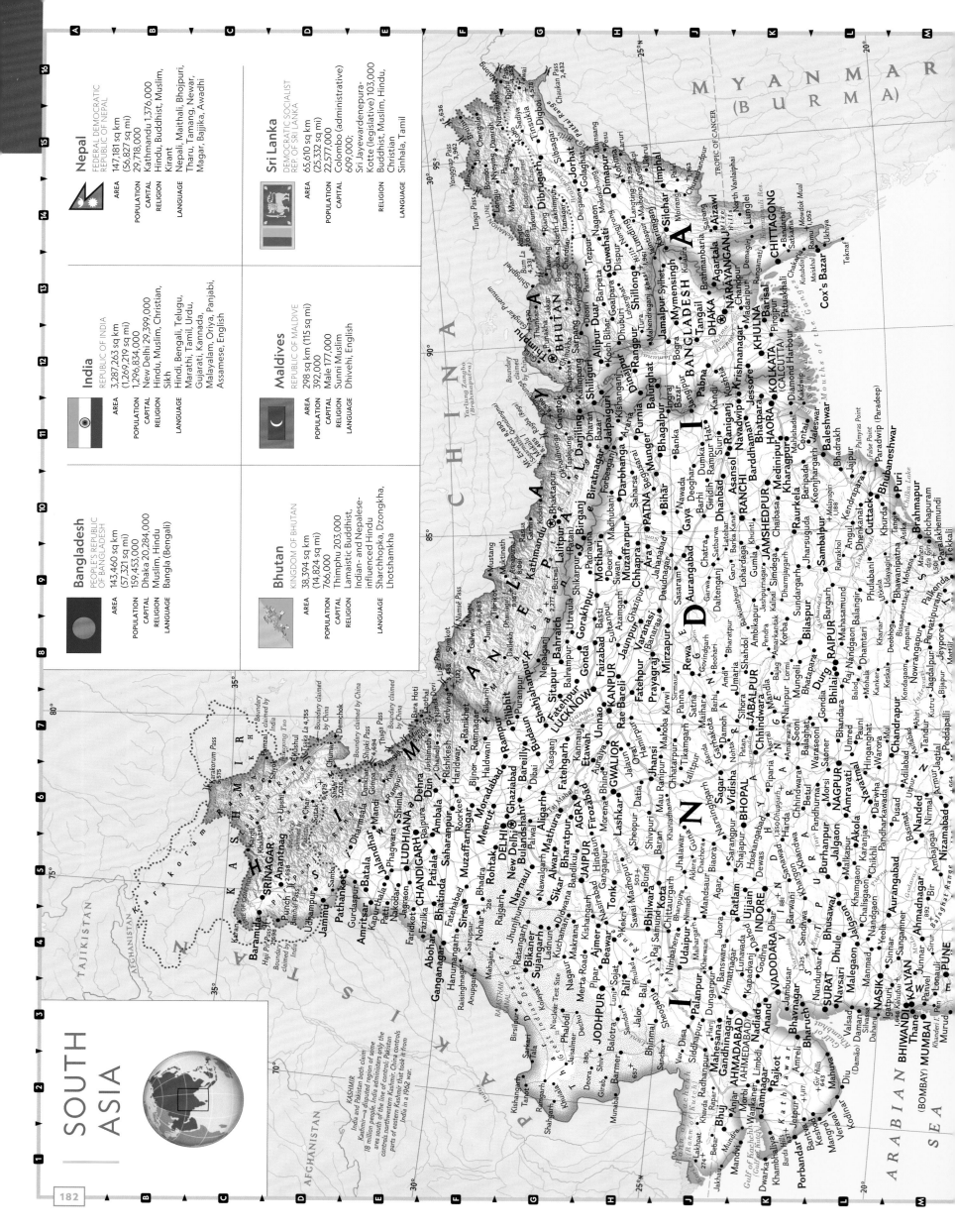

Nepal
FEDERAL DEMOCRATIC
REPUBLIC OF NEPAL

AREA | 147,181 sq km
(56,827 sq mi)
POPULATION | 29,718,000
CAPITAL | Kathmandu 1,376,000
RELIGION | Hindu, Buddhist, Muslim, Kirant
LANGUAGE | Nepali, Maithali, Bhojpuri, Tharu, Tamang, Newar, Magar, Bajjika, Awadhi

Sri Lanka
DEMOCRATIC SOCIALIST
REP OF SRI LANKA

AREA | 65,610 sq km
(25,332 sq mi)
POPULATION | 22,577,000
CAPITAL | Colombo (administrative)
609,000;
Sri Jayewardenepura-
Kotte (legislative) 103,000
RELIGION | Buddhist, Muslim, Hindu, Christian
LANGUAGE | Sinhala, Tamil

India
REPUBLIC OF INDIA

AREA | 3,287,263 sq km
(1,269,219 sq mi)
POPULATION | 1,296,834,000
CAPITAL | New Delhi 29,399,000
RELIGION | Hindu, Muslim, Christian, Sikh
LANGUAGE | Hindi, Bengali, Telugu, Marathi, Tamil, Urdu, Gujarati, Kannada, Malayalam, Oriya, Panjabi, Assamese, English

Maldives
REPUBLIC OF MALDIVE

AREA | 298 sq km (115 sq mi)
POPULATION | 392,000
CAPITAL | Male 177,000
RELIGION | Sunni Muslim
LANGUAGE | Dhivehi, English

Bangladesh
PEOPLE'S REPUBLIC
OF BANGLADESH

AREA | 143,460 sq km
(57,321 sq mi)
POPULATION | 159,453,000
CAPITAL | Dhaka 20,284,000
RELIGION | Muslim, Hindu
LANGUAGE | Bangla (Bengali)

Bhutan
KINGDOM OF BHUTAN

AREA | 38,394 sq km
(14,824 sq mi)
POPULATION | 766,000
CAPITAL | Thimphu 203,000
RELIGION | Lamaistic Buddhist, Indian- and Nepalese-influenced Hindu
LANGUAGE | Sharchhopka, Dzongkha, Lhotshamkha

KASHMIR
India and Pakistan both claim
Kashmir—a disputed region of some
18 million people, India administers only the
area south of the line of control; Pakistan
controls northwestern Kashmir. China
controls parts of eastern Kashmir that took it from
India in a 1962 war.

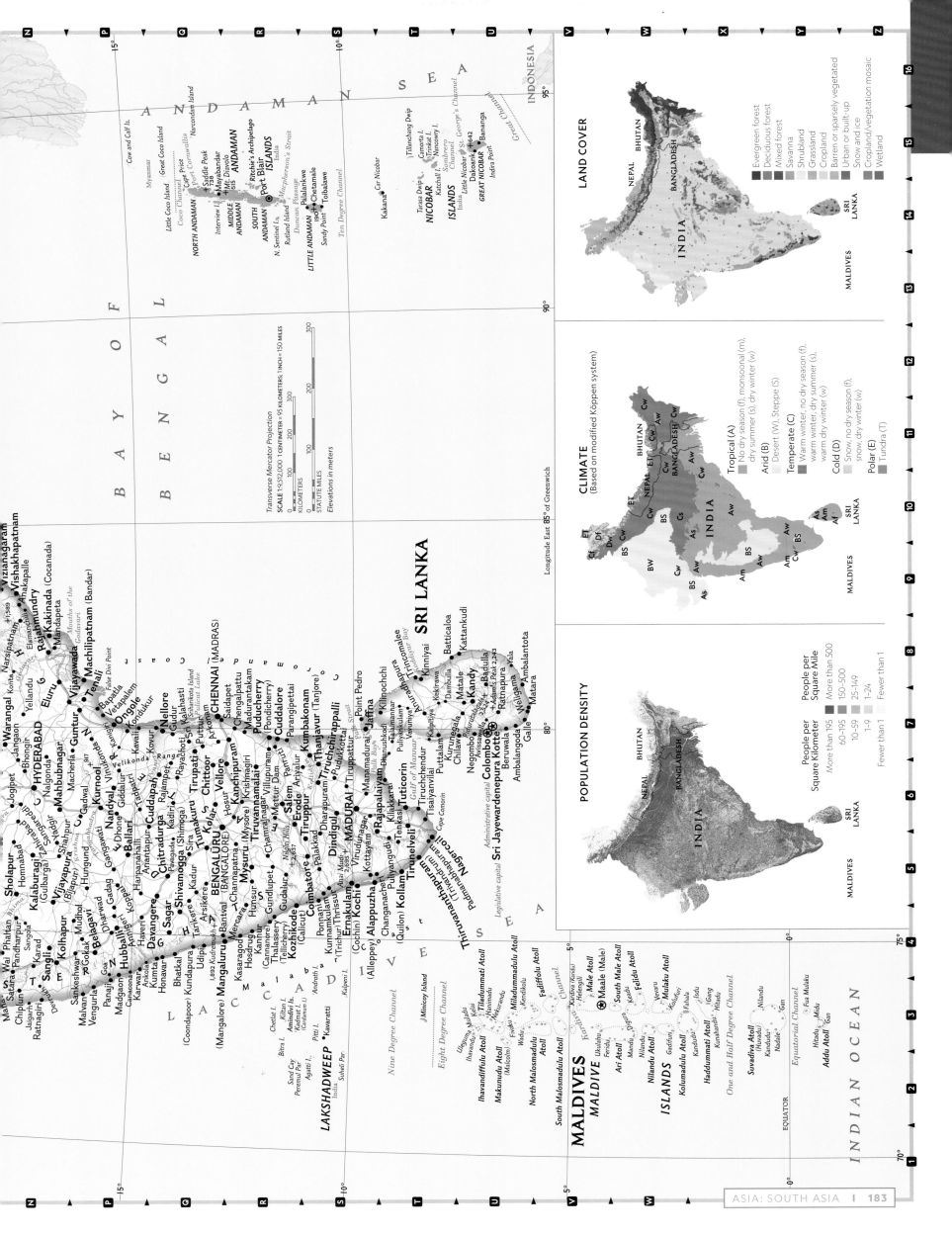

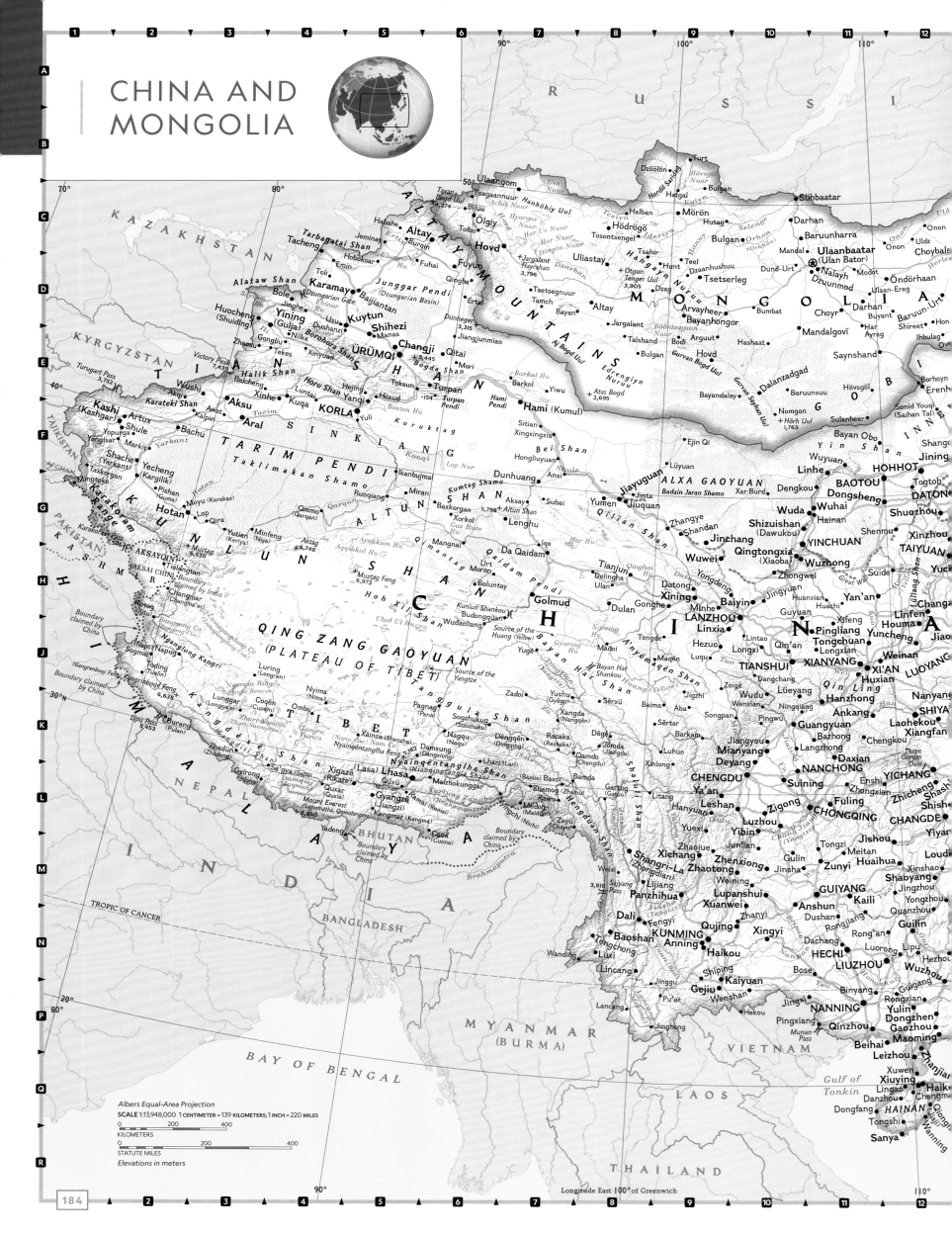

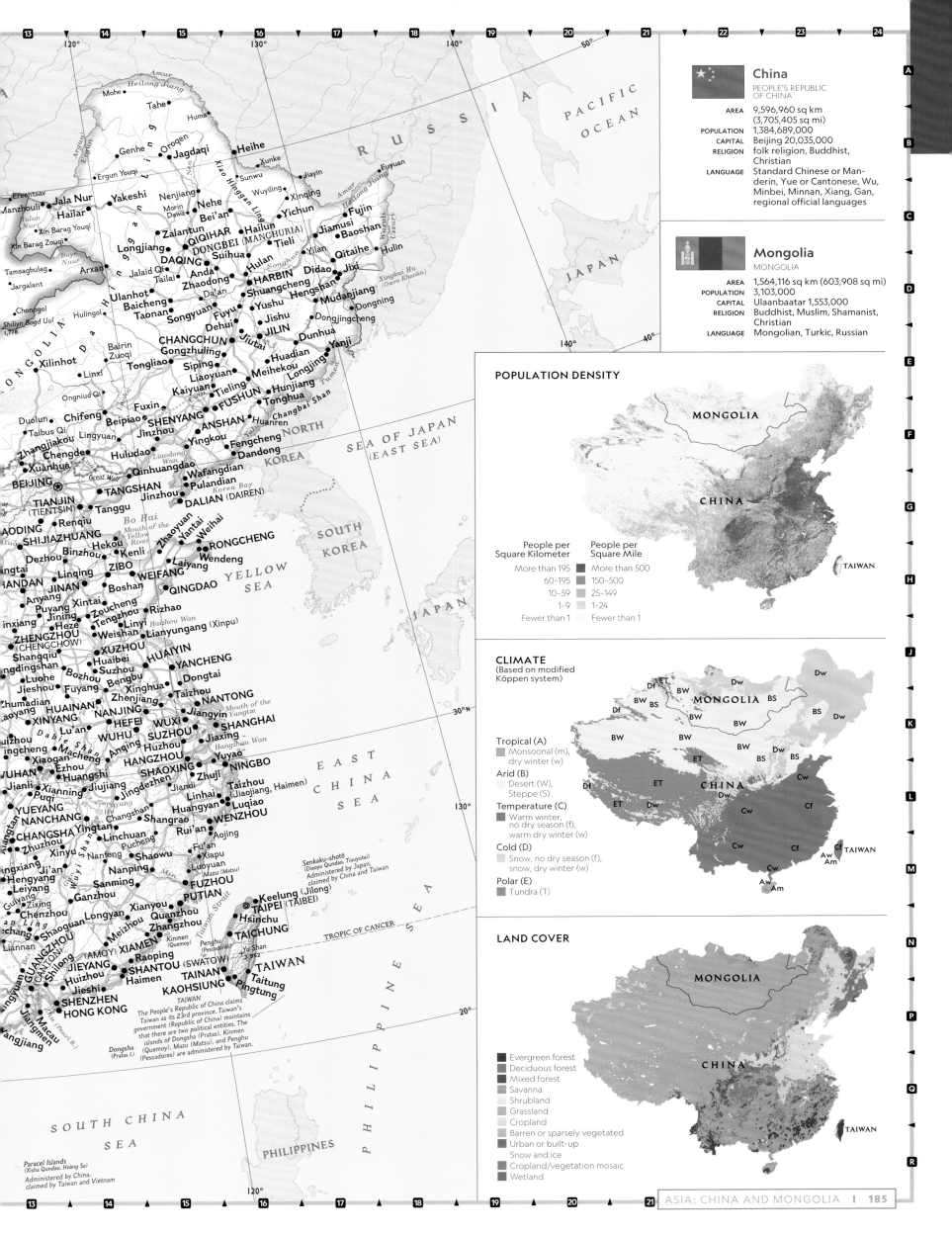

Map labels (China and Mongolia)

Mohe · Amur / Heilong Jiang · Tahe · Huma · Heihe · Xunke · Fuyuan

RUSSIA

PACIFIC OCEAN

Efeentsav · Genhe · Oroqen · Jagdaqi · Sunwu · Jiayin · Wuyiling · Xinqing

Argun / Ergun · Yakeshi · Nenjiang · Nehe · Bei'an · Yichun · Fujin · Jiamusi · Baoshan

Manzhouli · Jala Nur · Hailar · Morin Dawa · Zalantun · QIQIHAR · Hailun · (MANCHURIA) · Tieli · Yilan · Qitaihe · Hulin

Hulun Nur · Xin Barag Youqi · Longjiang · DONGBEI · Suihua · Hulan · Songhua · Didao · Jixi

Xin Barag Zuoqi · DAQING · Anda · Zhaodong · HARBIN · Shuangcheng · Hengshan · Mudanjiang · Dongning

Tamsagbulag · Jalaid Qi · Tailai · Da'an · Yushu · Jishu · Dongjingcheng · Xingkai Hu (Ozero Khanka)

Jargalant · Arxan · Ulanhot · Baicheng · Taonan · Songyuan · Fuyu · Dehui · JILIN · Dunhua

Chonogol · Hulingol · CHANGCHUN · Jiutai · Huadian · Yanji

Shiliyn Bogd Uul 1,778 · Bairin Zuoqi · Gongzhuling · Siping · Meihekou · Longjing · Yanji

MONGOLIA · Xilinhot · Tongliao · Liaoyuan · Hunjiang · Tonghua · Changbai Shan · Tumen

Linxi · Ongniud Qi · Kaiyuan · Tieling · FUSHUN · Huanren · Yalu

Duolun · Chifeng · Fuxin · Beipiao · SHENYANG · ANSHAN · Fengcheng · JAPAN

Taibus Qi · Lingyuan · Jinzhou · Yingkou · Dandong · NORTH KOREA

Zhangjiakou · Chengde · Huludao · Qinhuangdao · Wafangdian · SEA OF JAPAN (EAST SEA)

Xuanhua · Liaodong Wan · Jinzhou · Pulandian · Korea Bay

BEIJING · Great Wall · TANGSHAN · DALIAN (DAIREN) · Pulandian

TIANJIN (TIENTSIN) · Tanggu · Zhaoyuan · Yantai · Weihai

AODING · Renqiu · Bo Hai · Mouth of the Yellow River · Hekou · Kenli · RONGCHENG · Wendeng

SHIJIAZHUANG · Dezhou · Linqing · Boshan · Laiyang · SOUTH KOREA

Binzhou · ZIBO · WEIFANG · JAPAN

HANDAN · JINAN · Boshan · QINGDAO · YELLOW SEA

Anyang · Xintai · Zoucheng · Rizhao

Puyang · Jining · Heze · Linyi · Haizhou Wan

Xinxiang · ZHENGZHOU (CHENGCHOW) · Weishan · Lianyungang (Xinpu)

Shangqiu · XUZHOU · HUAIYIN

Dingshan · Huaibei · Suzhou · YANCHENG

Luohe · Bozhou · Bengbu · Dongtai

Jieshou · Fuyang · Xinghua · Taizhou

Zhoumadian · HUAINAN · Zhenjiang · NANTONG · Mouth of the Yangtze

XINYANG · XINGANG · HEFEI · Jiangyin

Lu'an · WUHU · WUXI · SHANGHAI

Macheng · SUZHOU · Jiaxing

Xiaogan · Anqing · HANGZHOU · Yuyao

WUHAN · Ezhou · Huangshi · SHAOXING · NINGBO · Hangzhou Wan

Jianli · Xianning · Jiujiang · Jingdezhen · Zhuji · EAST CHINA SEA

Puqi · Changshan · Shangrao · Taizhou (Jiaojiang, Haimen)

YUEYANG · NANCHANG · Yingtan · Linhai · Luqiao

CHANGSHA · Linchuan · WENZHOU · Rui'an · Aojing

Zhuzhou · Xinyu · Nanfeng · Pucheng · Fu'an · Xiapu

Xiangxiang · Ji'an · Shaowu · Luoyuan · Mazu (Matsu)

Hengyang · Nanping · Min · FUZHOU

Leiyang · Sanming · PUTIAN · Keelung (Jilong)

Chenzhou · Zixing · Ganzhou · Xianyou · Quanzhou · TAIPEI (TAIBEI)

Longyan · Meizhou · Zhangzhou · Hsinchu

Shaoguan · XIAMEN (AMOY) · Kinmen (Quemoy) · TAICHUNG

GUANGZHOU (CANTON) · JIEYANG · Raoping · TROPIC OF CANCER

Shilong · Huizhou · SHANTOU (SWATOW) · TAINAN · Yu Shan 3,952

Jieshi · Haimen · KAOHSIUNG · Taitung · Pingtung

SHENZHEN · HONG KONG · TAIWAN

Macau · Jiangmen · Yangjiang · (Pearl R.)

Dongsha (Pratas I.)

SOUTH CHINA SEA · PHILIPPINE SEA

PHILIPPINES

Paracel Islands (Xisha Qundao, Hoàng Sa) Administered by China, claimed by Taiwan and Vietnam

Senkaku-shotō (Diaoyu Qundao, Tiauyutai) Administered by Japan, claimed by China and Taiwan

Penghu (Pescadores)

TAIWAN
The People's Republic of China claims Taiwan as its 23rd province. Taiwan's government (Republic of China) maintains that there are two political entities. The islands of Dongsha (Pratas), Kinmen (Quemoy), Mazu (Matsu), and Penghu (Pescadores) are administered by Taiwan.

Dongsha (Pratas I.)

China
PEOPLE'S REPUBLIC OF CHINA

AREA	9,596,960 sq km (3,705,405 sq mi)
POPULATION	1,384,689,000
CAPITAL	Beijing 20,035,000
RELIGION	folk religion, Buddhist, Christian
LANGUAGE	Standard Chinese or Manderin, Yue or Cantonese, Wu, Minbei, Minnan, Xiang, Gan, regional official languages

Mongolia
MONGOLIA

AREA	1,564,116 sq km (603,908 sq mi)
POPULATION	3,103,000
CAPITAL	Ulaanbaatar 1,553,000
RELIGION	Buddhist, Muslim, Shamanist, Christian
LANGUAGE	Mongolian, Turkic, Russian

POPULATION DENSITY

People per Square Kilometer	People per Square Mile
More than 195	More than 500
60–195	150–500
10–59	25–149
1–9	1–24
Fewer than 1	Fewer than 1

MONGOLIA · CHINA · TAIWAN

CLIMATE
(Based on modified Köppen system)

Tropical (A)
Monsoonal (m), dry winter (w)

Arid (B)
Desert (W), Steppe (S)

Temperature (C)
Warm winter, no dry season (f), warm dry winter (w)

Cold (D)
Snow, no dry season (f), snow, dry winter (w)

Polar (E)
Tundra (T)

MONGOLIA · CHINA · TAIWAN

LAND COVER

- Evergreen forest
- Deciduous forest
- Mixed forest
- Savanna
- Shrubland
- Grassland
- Cropland
- Barren or sparsely vegetated
- Urban or built-up
- Snow and ice
- Cropland/vegetation mosaic
- Wetland

MONGOLIA · CHINA · TAIWAN

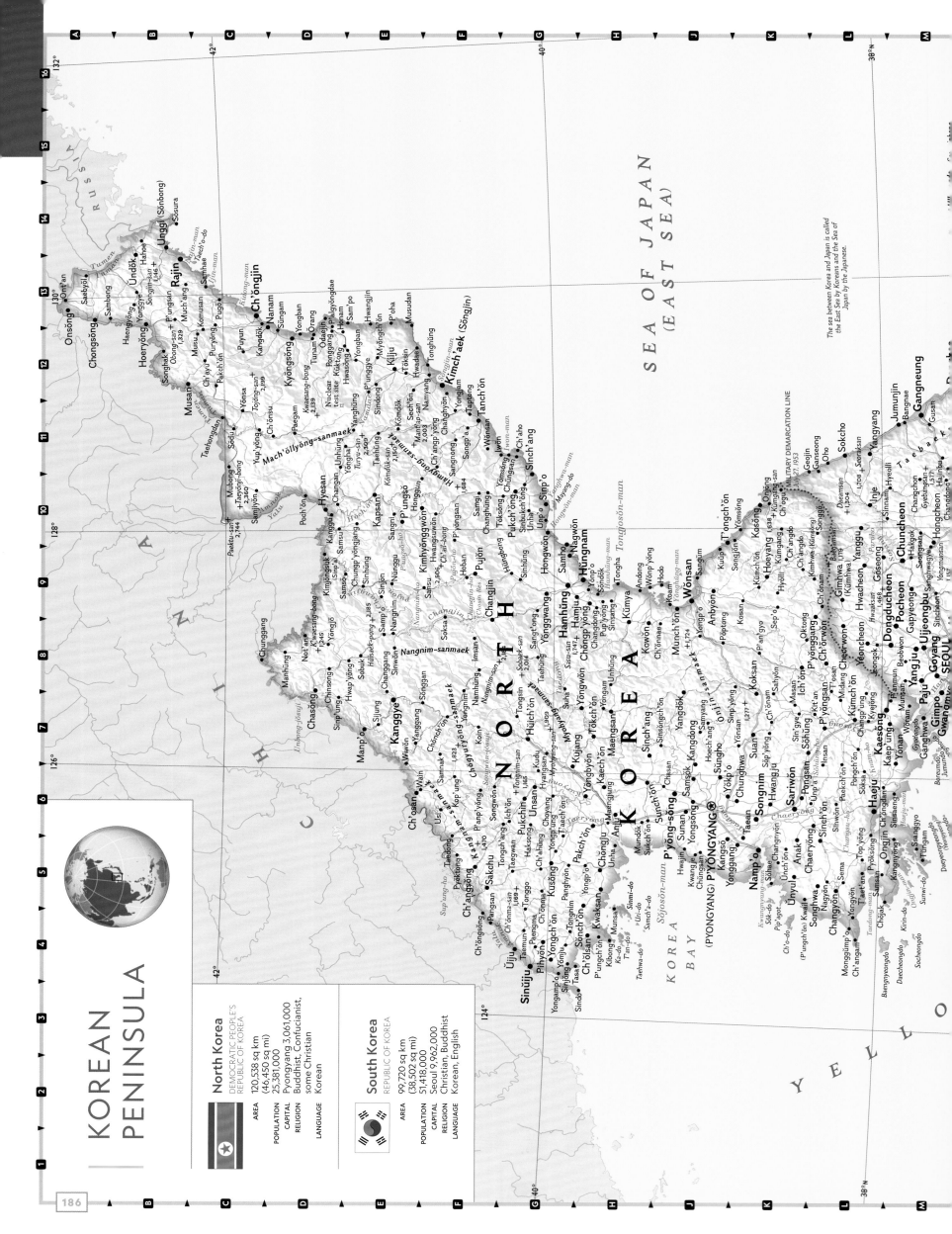

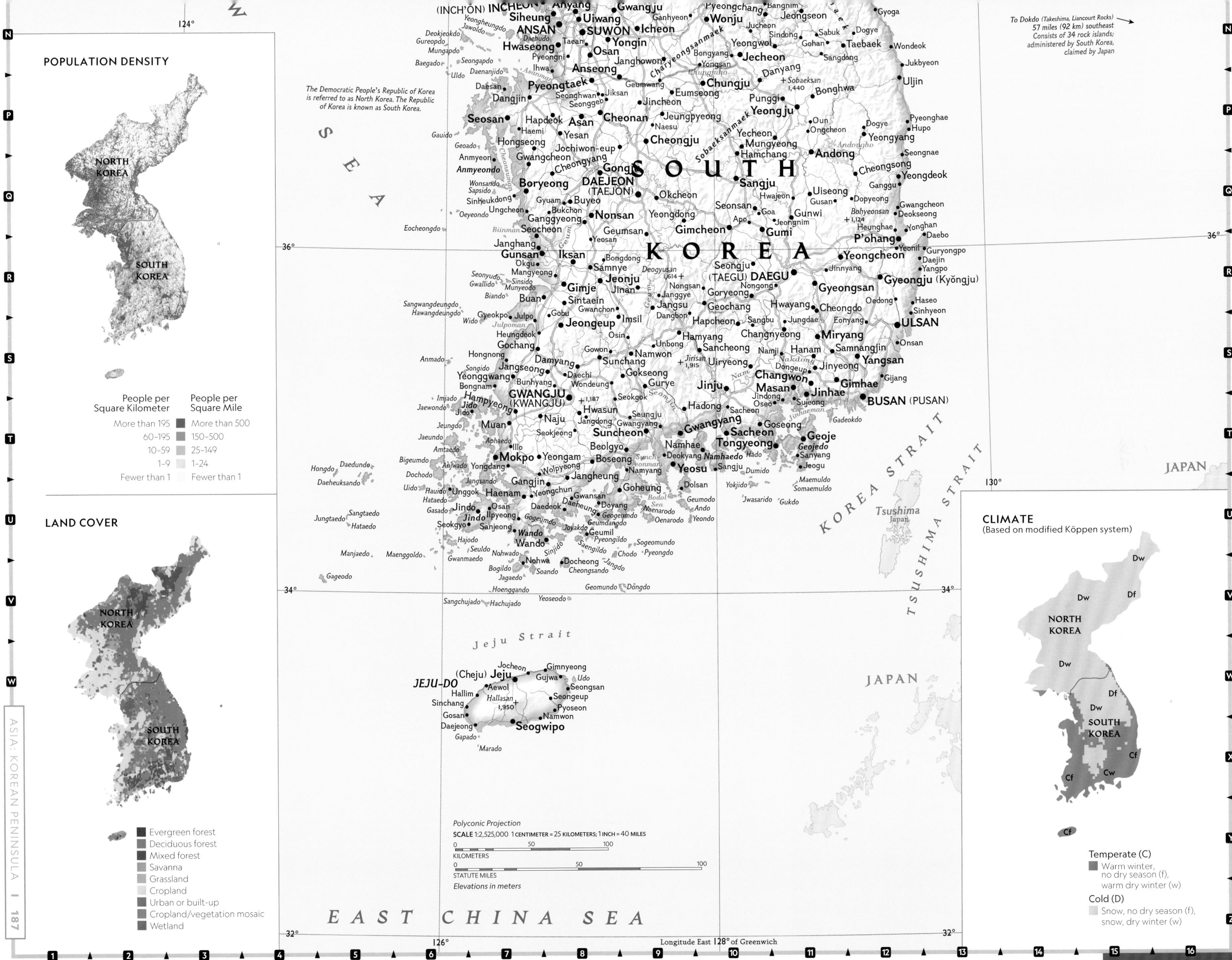

POPULATION DENSITY

NORTH KOREA

SOUTH KOREA

People per
Square Kilometer | People per
Square Mile
More than 195 | More than 500
60–195 | 150–500
10–59 | 25–149
1–9 | 1–24
Fewer than 1 | Fewer than 1

LAND COVER

NORTH KOREA

SOUTH KOREA

Evergreen forest
Deciduous forest
Mixed forest
Savanna
Grassland
Cropland
Urban or built-up
Cropland/vegetation mosaic
Wetland

CLIMATE
(Based on modified Köppen system)

NORTH KOREA

SOUTH KOREA

Dw
Df
Cf
Cw

Temperate (C)
Warm winter,
no dry season (f),
warm dry winter (w)

Cold (D)
Snow, no dry season (f),
snow, dry winter (w)

SCALE 1:2,525,000 1 CENTIMETER = 25 KILOMETERS; 1 INCH = 40 MILES

Polyconic Projection

KILOMETERS
STATUTE MILES
Elevations in meters

Longitude East 128° of Greenwich

To Dokdo (Takeshima, Liancourt Rocks)
57 miles (92 km) southeast
Consists of 34 rock islands;
administered by South Korea,
claimed by Japan

The Democratic People's Republic of Korea
is referred to as North Korea. The Republic
of Korea is known as South Korea.

JAPAN

SOUTH KOREA

S E A

EAST CHINA SEA

KOREA STRAIT

TSUSHIMA STRAIT

Jeju Strait

Tsushima
(Japan)

(INCH'ON) INCHEON
SUWON
ANSAN
SIHEUNG
Gwangju
Uiwang
Anyang
Icheon
Yongin
Osan
Anseong
Cheonan
Asan
Pyeongtaek
Gongju
DAEJEON (TAEJON)
Cheongju
Chungju
Wonju
Jecheon
Yeongju
Andong
P'ohang
Gyeongju (Kyŏngju)
Yeongdeok
Ulsan
BUSAN (PUSAN)
DAEGU (TAEGU)
Gumi
Gimcheon
Sangju
Jeonju
Iksan
Gunsan
Jeongeup
Namwon
Suncheon
GWANGJU (KWANGJU)
Mokpo
Yeosu
Changwon
Masan
Jinju
Tongyeong
Sacheon

JEJU-DO (Cheju) JEJU-DO
Jeju
Seogwipo

32°
34°
36°

124°
126°
128°
130°

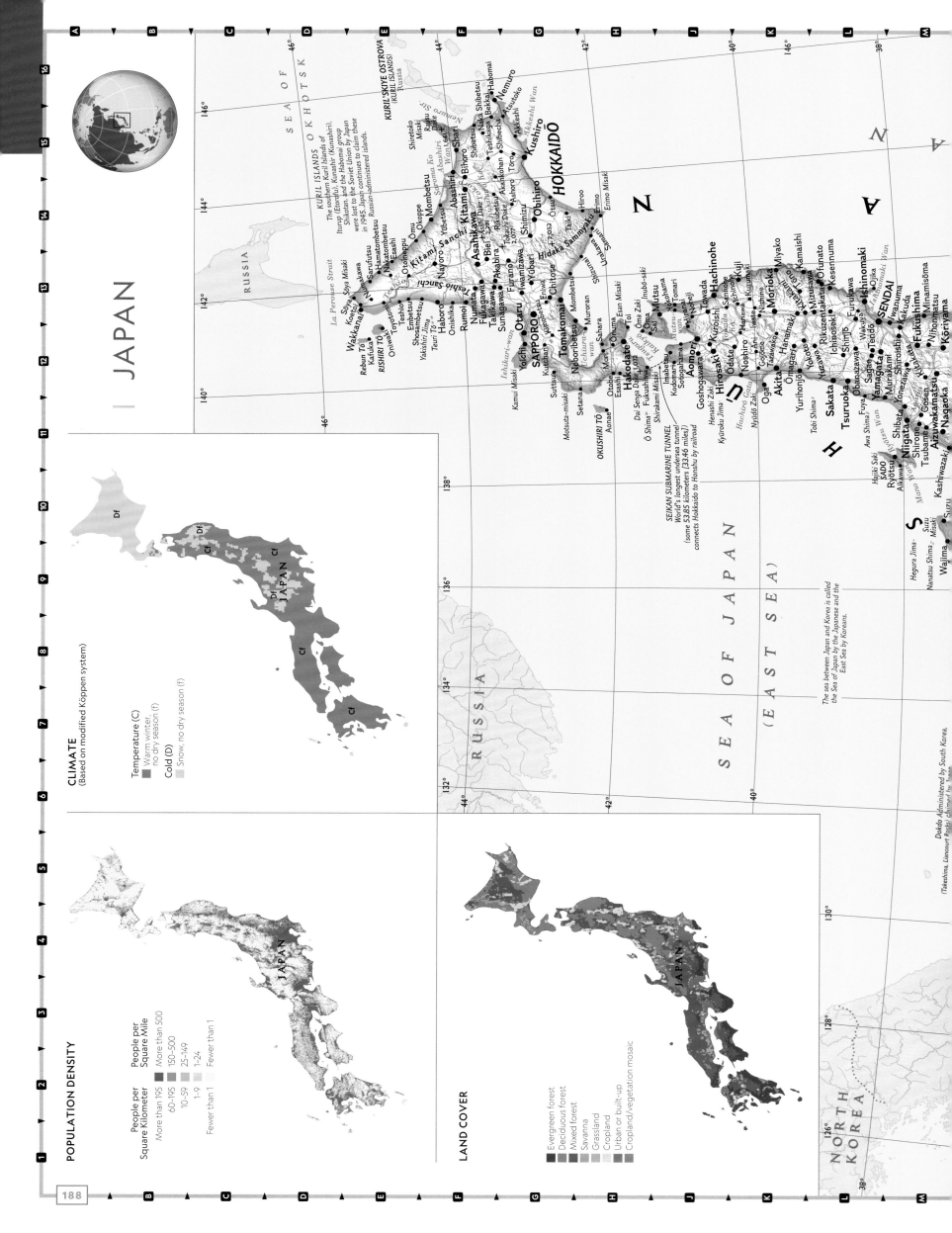

JAPAN

POPULATION DENSITY

People per Square Kilometer	People per Square Mile
More than 195	More than 500
60–195	150–500
10–59	25–149
1–9	1–24
Fewer than 1	Fewer than 1

LAND COVER

Evergreen forest
Deciduous forest
Mixed forest
Savanna
Grassland
Cropland
Urban or built-up
Cropland/vegetation mosaic

CLIMATE
(Based on modified Köppen system)

Temperature (C)
Warm winter, no dry season (f)

Cold (D)
Snow, no dry season (f)

SEIKAN SUBMARINE TUNNEL
World's longest undersea tunnel
(some 53.85 kilometers [33.46 miles])
connects Hokkaido to Honshu by railroad

The sea between Japan and Korea is called the Sea of Japan by the Japanese and the East Sea by Koreans.

KURIL ISLANDS of
The southern Kuril Islands of Iturup (Etorofu), Kunashir (Kunashiri), Shikotan, and the Habomai group were lost to the Soviet Union by Japan in 1945. Japan continues to claim these Russian-administered islands.

Dokdo Administered by South Korea.
(Takeshima, Liancourt Rocks) claimed by Japan

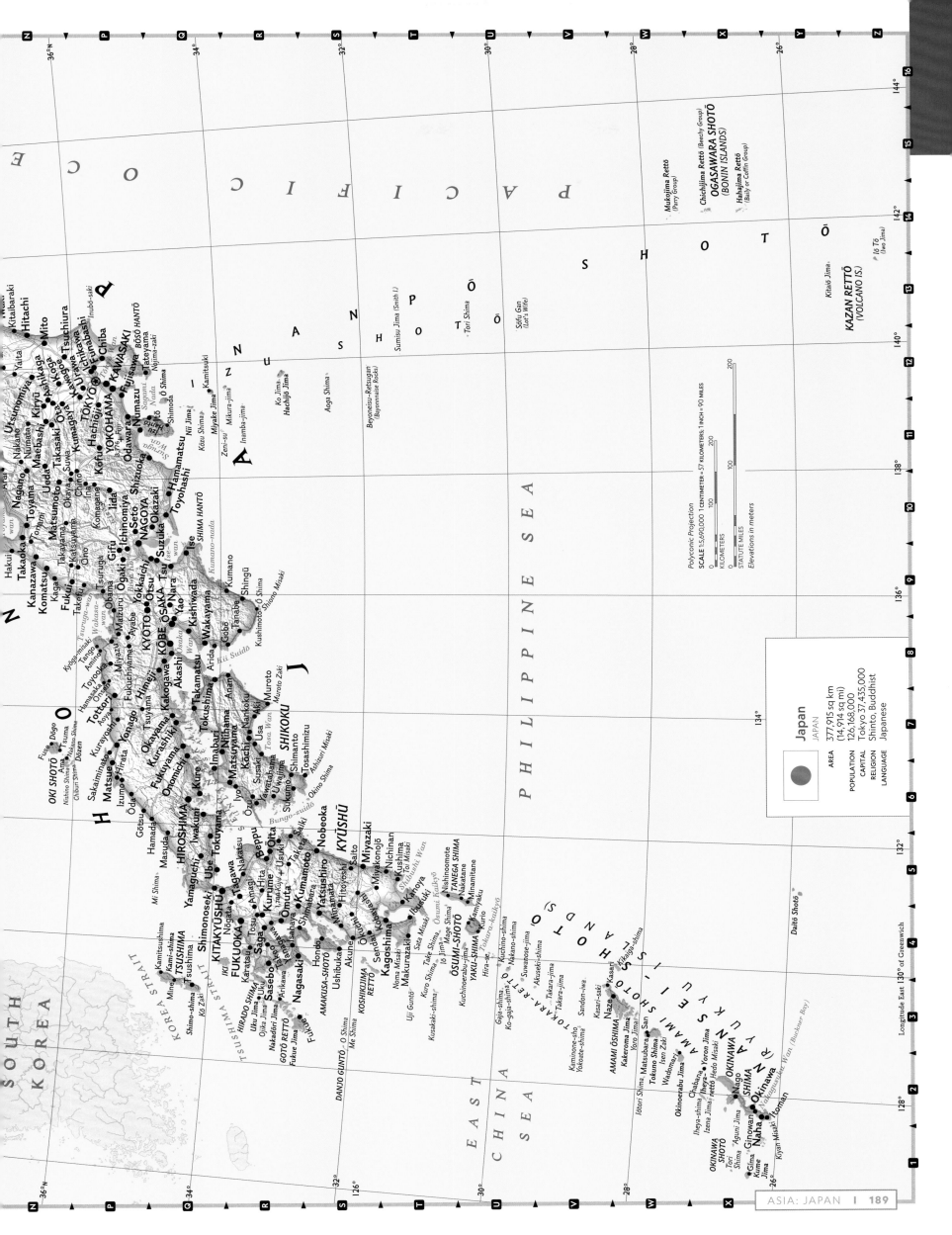

INDOCHINA

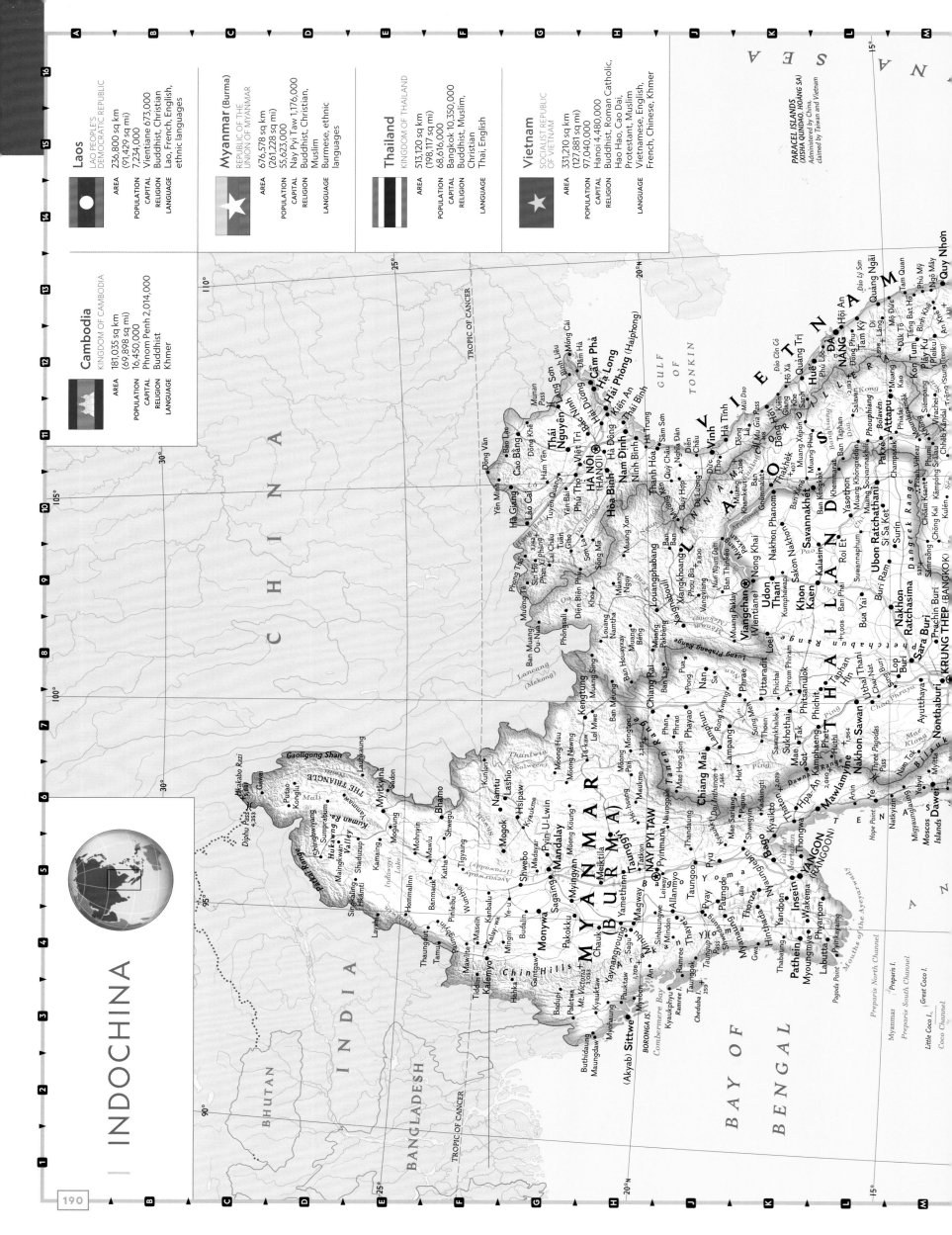

Laos
LAO PEOPLE'S DEMOCRATIC REPUBLIC
- **AREA** 236,800 sq km (91,429 sq mi)
- **POPULATION** 7,234,000
- **CAPITAL** Vientiane 673,000
- **RELIGION** Buddhist, Christian
- **LANGUAGE** Lao, French, English, ethnic languages

Myanmar (Burma)
REPUBLIC OF THE UNION OF MYANMAR
- **AREA** 676,578 sq km (261,228 sq mi)
- **POPULATION** 55,623,000
- **CAPITAL** Nay Pyi Taw 1,176,000
- **RELIGION** Buddhist, Christian, Muslim
- **LANGUAGE** Burmese, ethnic languages

Thailand
KINGDOM OF THAILAND
- **AREA** 513,120 sq km (198,117 sq mi)
- **POPULATION** 68,616,000
- **CAPITAL** Bangkok 10,350,000
- **RELIGION** Buddhist, Muslim, Christian
- **LANGUAGE** Thai, English

Vietnam
SOCIALIST REPUBLIC OF VIETNAM
- **AREA** 331,210 sq km (127,881 sq mi)
- **POPULATION** 97,040,000
- **CAPITAL** Hanoi 4,480,000
- **RELIGION** Buddhist, Roman Catholic, Cao Dai, Hoa Hao, Protestant, Muslim
- **LANGUAGE** Vietnamese, English, French, Chinese, Khmer

Cambodia
KINGDOM OF CAMBODIA
- **AREA** 181,035 sq km (69,898 sq mi)
- **POPULATION** 16,450,000
- **CAPITAL** Phnom Penh 2,014,000
- **RELIGION** Buddhist
- **LANGUAGE** Khmer

PARACEL ISLANDS (XISHA QUNDAO, HOÀNG SA) Administered by China, claimed by Taiwan and Vietnam

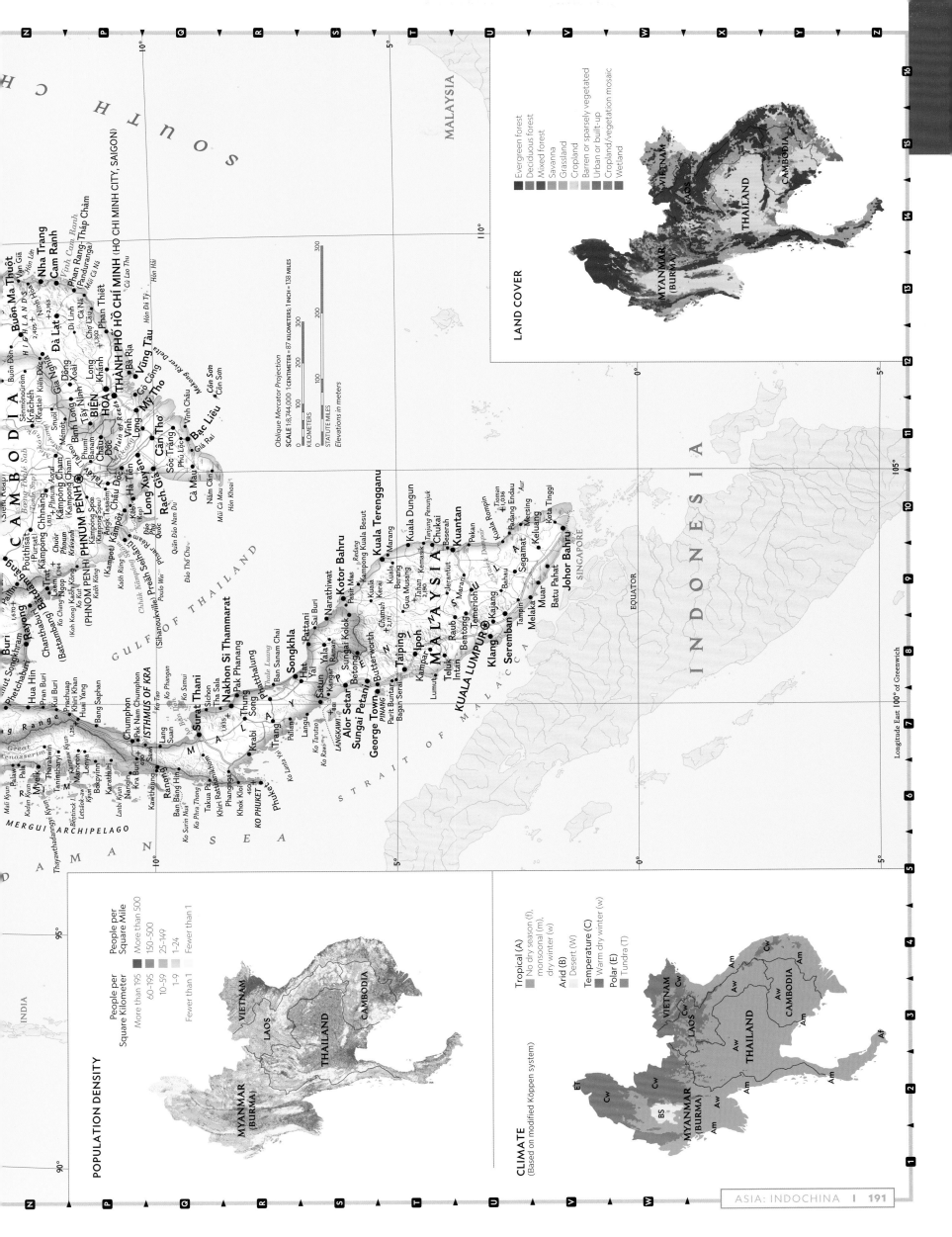

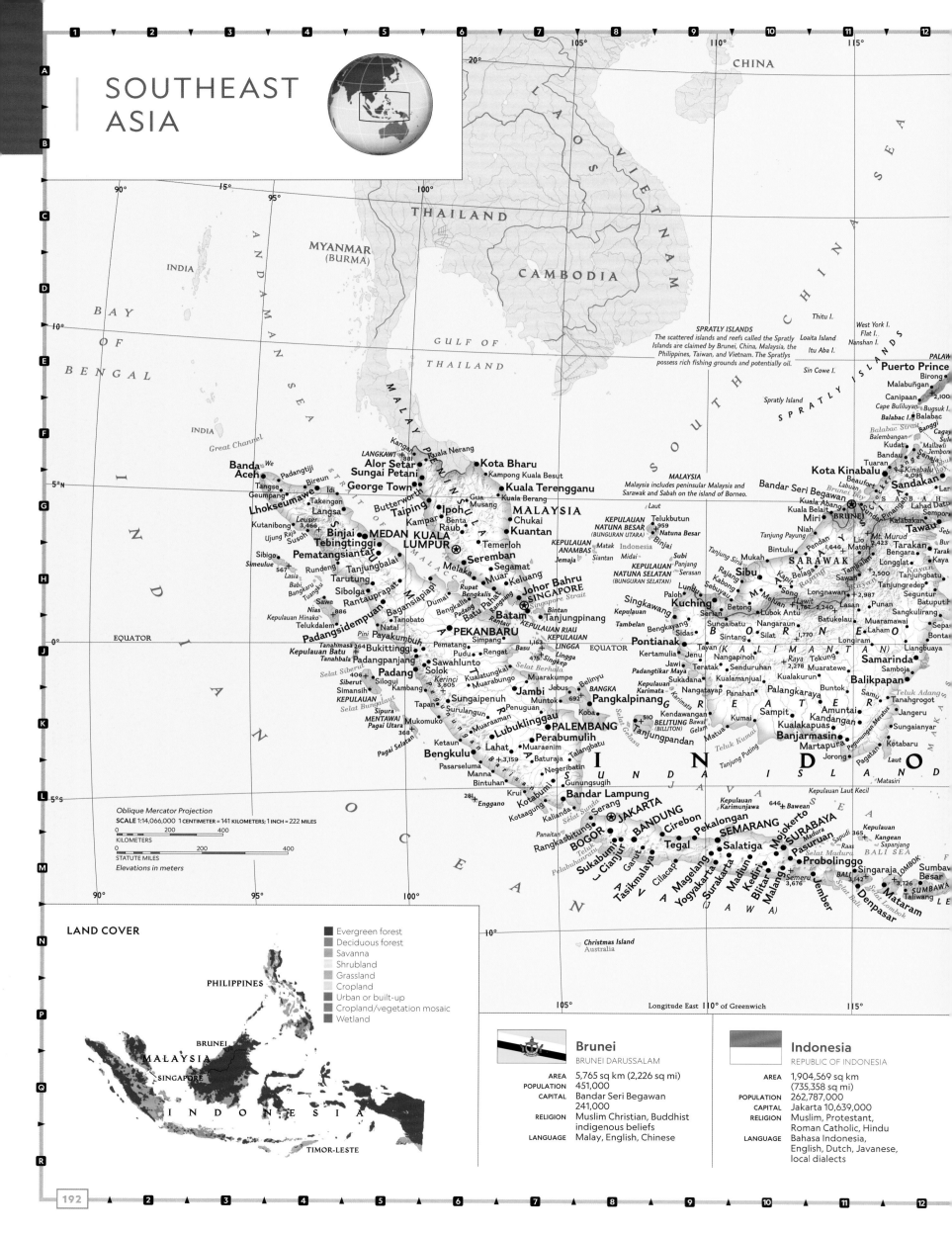

LAND COVER

- Evergreen forest
- Deciduous forest
- Savanna
- Shrubland
- Grassland
- Cropland
- Urban or built-up
- Cropland/vegetation mosaic
- Wetland

Oblique Mercator Projection
SCALE 1:14,066,000 1 CENTIMETER = 141 KILOMETERS; 1 INCH = 222 MILES
KILOMETERS
STATUTE MILES
Elevations in meters

SPRATLY ISLANDS
The scattered islands and reefs called the Spratly Islands are claimed by Brunei, China, Malaysia, the Philippines, Taiwan, and Vietnam. The Spratlys possess rich fishing grounds and potentially oil.

MALAYSIA
Malaysia includes peninsular Malaysia and Sarawak and Sabah on the island of Borneo.

Brunei
BRUNEI DARUSSALAM

AREA	5,765 sq km (2,226 sq mi)
POPULATION	451,000
CAPITAL	Bandar Seri Begawan 241,000
RELIGION	Muslim Christian, Buddhist indigenous beliefs
LANGUAGE	Malay, English, Chinese

Indonesia
REPUBLIC OF INDONESIA

AREA	1,904,569 sq km (735,358 sq mi)
POPULATION	262,787,000
CAPITAL	Jakarta 10,639,000
RELIGION	Muslim, Protestant, Roman Catholic, Hindu
LANGUAGE	Bahasa Indonesia, English, Dutch, Javanese, local dialects

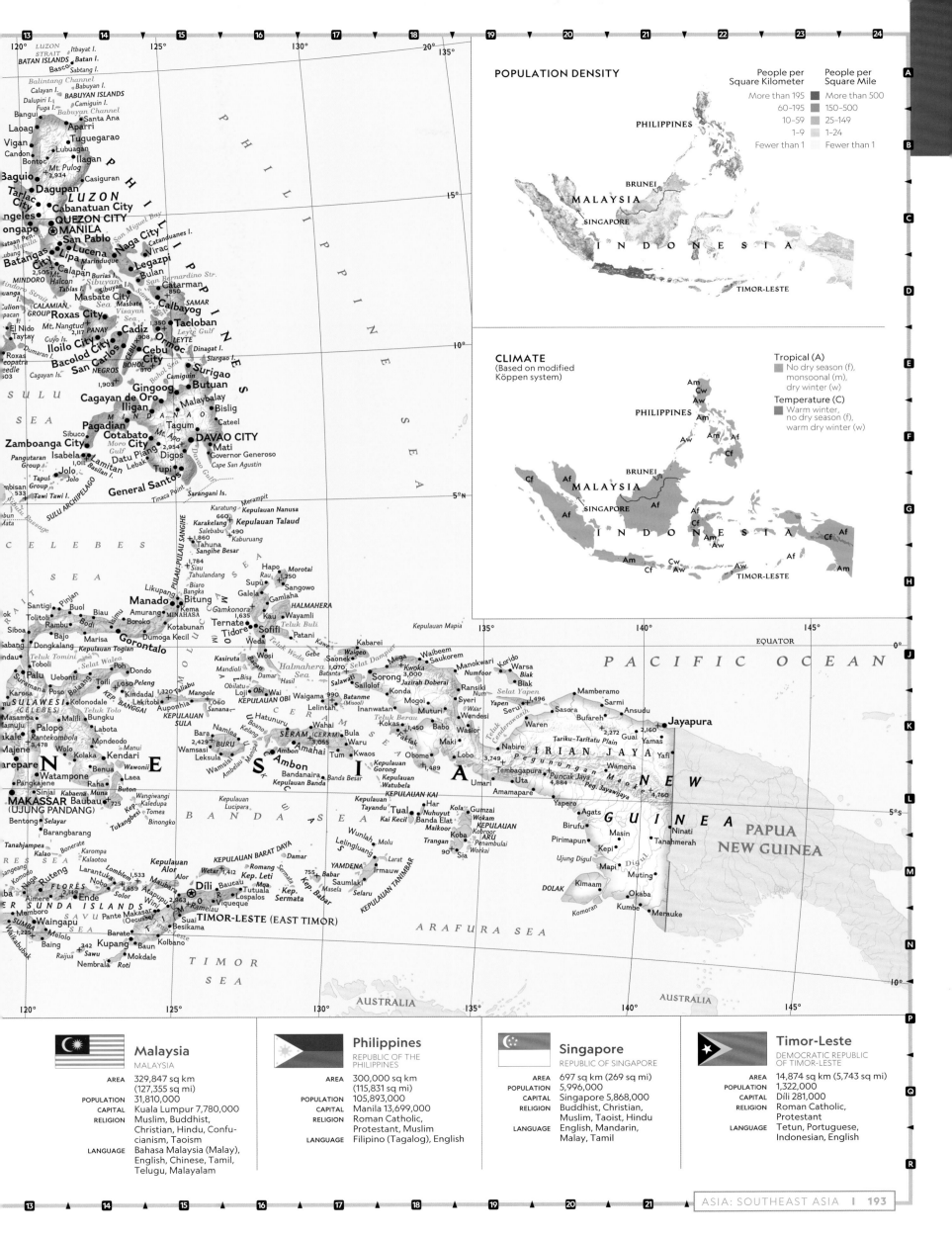

POPULATION DENSITY

	People per Square Kilometer	People per Square Mile
	More than 195	More than 500
	60–195	150–500
	10–59	25–149
	1–9	1–24
	Fewer than 1	Fewer than 1

CLIMATE
(Based on modified Köppen system)

Tropical (A) — No dry season (f), monsoonal (m), dry winter (w)

Temperature (C) — Warm winter, no dry season (f), warm dry winter (w)

Malaysia
MALAYSIA
AREA 329,847 sq km (127,355 sq mi)
POPULATION 31,810,000
CAPITAL Kuala Lumpur 7,780,000
RELIGION Muslim, Buddhist, Christian, Hindu, Confucianism, Taoism
LANGUAGE Bahasa Malaysia (Malay), English, Chinese, Tamil, Telugu, Malayalam

Philippines
REPUBLIC OF THE PHILIPPINES
AREA 300,000 sq km (115,831 sq mi)
POPULATION 105,893,000
CAPITAL Manila 13,699,000
RELIGION Roman Catholic, Protestant, Muslim
LANGUAGE Filipino (Tagalog), English

Singapore
REPUBLIC OF SINGAPORE
AREA 697 sq km (269 sq mi)
POPULATION 5,996,000
CAPITAL Singapore 5,868,000
RELIGION Buddhist, Christian, Muslim, Taoist, Hindu
LANGUAGE English, Mandarin, Malay, Tamil

Timor-Leste
DEMOCRATIC REPUBLIC OF TIMOR-LESTE
AREA 14,874 sq km (5,743 sq mi)
POPULATION 1,322,000
CAPITAL Dili 281,000
RELIGION Roman Catholic, Protestant
LANGUAGE Tetun, Portuguese, Indonesian, English

AFRICA

Female lions and cubs in the southern Serengeti of Tanzania

CONTINENT OF BEGINNINGS

RICH IN HISTORY & NATURE, AFRICA AWAITS A BRIGHT FUTURE

Fossils and tools from more than 300,000 years ago tell a remarkable tale of humankind's emergence as a species. All our *Homo sapiens* ancestors made their home in Africa before migrating to populate Asia, and later Europe, around 60,000 years ago, sometimes mixing with other hominid populations. This is the continent of beginnings.

GEOGRAPHY

ROLLING SAVANNAS, WILD CREATURES, SCORCHING DESERT

If we were to fly above the vast expanse of Africa from north to south, some 8,000 kilometers (5,000 miles), the continent would appear in three distinct bands. Hot, dry desert forms the top third of the landmass, dominated by the Sahara and the Sahel. Then, the verdant equatorial middle, with its rainforests, grasslands, and mountains. And in the bottom third, more arid land. Most of the continent falls within the tropics. Plateaus dominate Africa's physical landscape, cut by great rivers such as the Nile and the Congo; the continent has few true mountain chains. The massive volcanic peaks of Mount Kilimanjaro and Mount Kenya rise in dramatic isolation from surrounding plains.

Africa's most striking geologic feature is the East African Great Rift Valley, which runs from the Red Sea southward and forms a stunning landscape of lakes, volcanoes, and deep valleys. This is where scientists have found many of the oldest signs of human ancestry. Wildlife abounds in eastern and southern Africa, and supports ecotourism. But agriculture, hunting, deforestation, and building of roads have destroyed habitat for many of Africa's animals and thrown roadblocks into their migration paths. Hundreds of species of plants and animals live precariously close to extinction. Protected areas—nature reserves, national parks, and wilderness areas—help safeguard the land, and provide a source of income.

Sculptures of Ramses II at the temple of Abu Simbel in southern Egypt

HISTORY

WHERE HUMAN HISTORY BEGAN

From the brilliance of ancient Egypt to the misery of the slave trade, African history is human history writ large. Indigenous kingdoms such as Great Zimbabwe in southern Africa; the Mali and Songhai Empires along the Niger River; and the Kingdom of Aksum in present-day Ethiopia all had their golden periods, some lasting for many centuries. Longest lasting of all was Benin, which thrived from the 13th to the 19th centuries. From the eighth century onward, Swahili culture emerged along the coast of the Indian Ocean from the mixing of African and Arab worlds.

Colonialism and the horrific slave trade, which carried 12.5 million Africans abroad, cast a dark shadow that endures today. Portugal, England, France, Belgium, Germany, the Netherlands, Spain, and Italy held sway over parts of the continent for 500 years. Europeans carved up their African colonies in the late 19th century in a manner that cut across cultural and linguistic divisions and sowed the seeds for many of today's chronic conflicts. The mid-20th century saw a swell of independence movements: Between 1951 and 1980, 48 countries broke free from their European colonizers. Enduring challenges include lack of health care and clean water, and political and environmental instability that results in huge concentrations of refugees. Africa's 54 nations today seem to stand between hope and continued chaos.

The Zambezi River plunges over a plateau at Victoria Falls.

CULTURE

DIVERSE COMMUNITIES & TRADITIONS

The continent of Africa is home to more than 2,000 languages. Citizens of Chad belong to more than 100 different ethnic groups; the country of Togo is home to 37 different tribes that speak 39 languages. Early kingdoms in Mali, Ghana, the Swahili Coast, and other parts of Africa conducted long-distance trade, while artists of the Ife and Benin kingdoms produced masterpieces in stone, terra-cotta, and bronze. African art, especially sculpture, continues to influence world culture. And the music brought by enslaved Africans to the New World infuses the most popular sounds of today: jazz, rock, gospel, blues, reggae, samba, and hip-hop.

Traditional religion and ritual still have a powerful place in Africa. Islam is dominant in the north, where the culture has much in common with the Middle East. Christianity is strongest south of the Sahara. In areas such as Nigeria, where believers of these two great religions and diverse ethnic groups come together, tension often leads to conflict. European languages and schooling, legacies of colonialism, have had lasting effects on modern Africa. Yet far from the cities one can still find blue-turbaned Tuareg traversing the Sahara, slender Maasai on the savannas of East Africa, Bambuti (sometimes called Pygmies) in the rainforests, and San (Bushmen) living on the Kalahari Desert.

Farmers sort dried coffee cherries in Ethiopia.

ECONOMY

A CONTINENT OF WEALTH & WANT

More than half of sub-Saharan Africa's workers are employed in agriculture and over two-thirds of those workers are women. Economic life revolves around small family farms that make up 80 percent of sub-Saharan Africa's farmland. Much farmland is community- or state-owned, and many governments sell or lease agricultural land to foreign entities. Africa produces a mere 3.5 percent of the world's economic output; of the world's 28 poorest countries, 27 are in sub-Saharan Africa. The continent's economy depends on the export of cash crops such as coffee, cacao, peanuts, and palm oil.

Yet beneath Africa's crust lies oil, gold, platinum, chromium, cobalt, copper, coltan, uranium, diamonds: Raw materials the rest of the world craves to fuel cars and power ever faster electronics. Telecommunications, banking, construction, and retail are also driving growth. This has sparked a new scramble for Africa: Between 2010 and 2016, more than 320 embassies opened across the continent as nations hustled to establish or strengthen diplomatic, commercial, and military ties. The Okavango River Basin offers a compelling case study of the potential for wildlife tourism. China has signaled an interest in developing the region, and since 2015, a National Geographic team has been surveying the watershed and helping establish sustainable management for years to come.

A Tuareg teacher and student at a primary school in Mali

SNAPSHOT
Africa
FACTS & FEATURES

HIGHEST ELEVATION Kilimanjaro, Tanzania: 5,895 m (19,340 ft)

LOWEST ELEVATION Lake Assal, Djibouti: −155 m (−509 ft)

WORLD'S LONGEST RIVER Nile: 6,695 km (4,160 mi), runs through 10 countries

BIGGEST URBAN AREA Lagos, Nigeria: 14.4 million (2020)

LARGEST COUNTRY BY AREA Algeria: 2,381,741 sq km (919,595 sq mi)

SMALLEST COUNTRY BY AREA Seychelles: 455 sq km (176 sq mi)

LONGEST LAND MAMMAL MIGRATION Burchell's zebra, Botswana and Namibia: approximately 500 km (300+ mi)

OLDEST NATIONAL PARK Virunga, Democratic Republic of the Congo: est. 1925

LARGEST LAKE Lake Victoria: 69,500 sq km (26,800 sq mi)

MOST COMMON LANGUAGE Arabic

NIGERIA IS PROJECTED TO OUTPACE THE UNITED STATES AND BECOME THE THIRD MOST POPULOUS COUNTRY IN THE WORLD BY 2050. ONE IN SIX AFRICANS IS NIGERIAN.

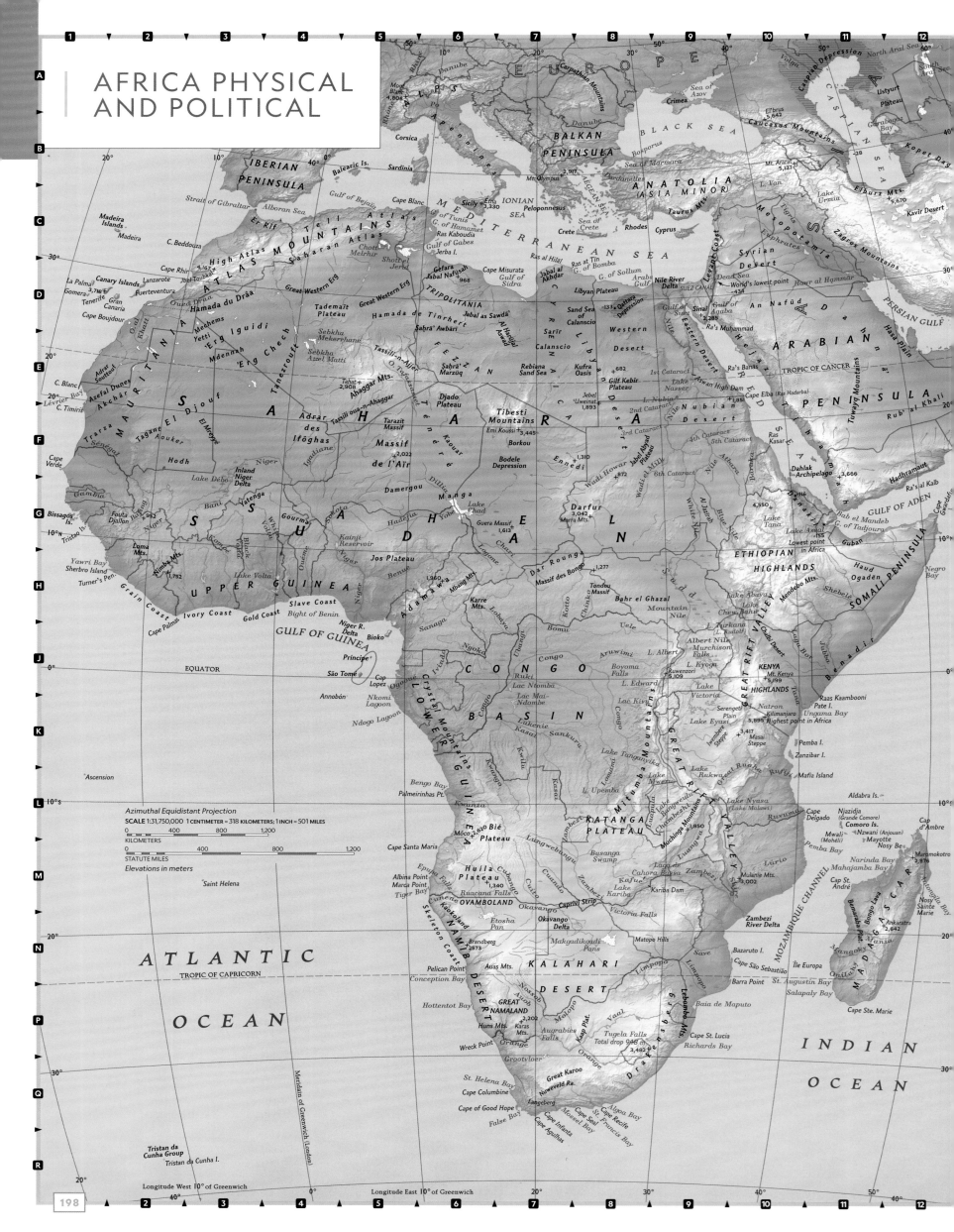

EUROPE

ATLANTIC OCEAN

INDIAN OCEAN

MEDITERRANEAN SEA

IBERIAN PENINSULA

BALKAN PENINSULA

ANATOLIA (ASIA MINOR)

ARABIAN PENINSULA

S A H A R A

MAURITANIA

S U D A N

UPPER GUINEA

GULF OF GUINEA

CONGO BASIN

LOWER GUINEA

ETHIOPIAN HIGHLANDS

KENYA HIGHLANDS

SOMALI PENINSULA

GULF OF ADEN

KATANGA PLATEAU

KALAHARI DESERT

NAMIB DESERT

GREAT NAMALAND

MADAGASCAR

MOZAMBIQUE CHANNEL

Azimuthal Equidistant Projection
SCALE 1:31,750,000 1 CENTIMETER = 318 KILOMETERS; 1 INCH = 501 MILES

KILOMETERS
0   400   800   1,200

STATUTE MILES
0   400   800   1,200

Elevations in meters

EQUATOR

TROPIC OF CANCER

TROPIC OF CAPRICORN

Longitude West 10° of Greenwich

Meridian of Greenwich (London)

Longitude East 10° of Greenwich

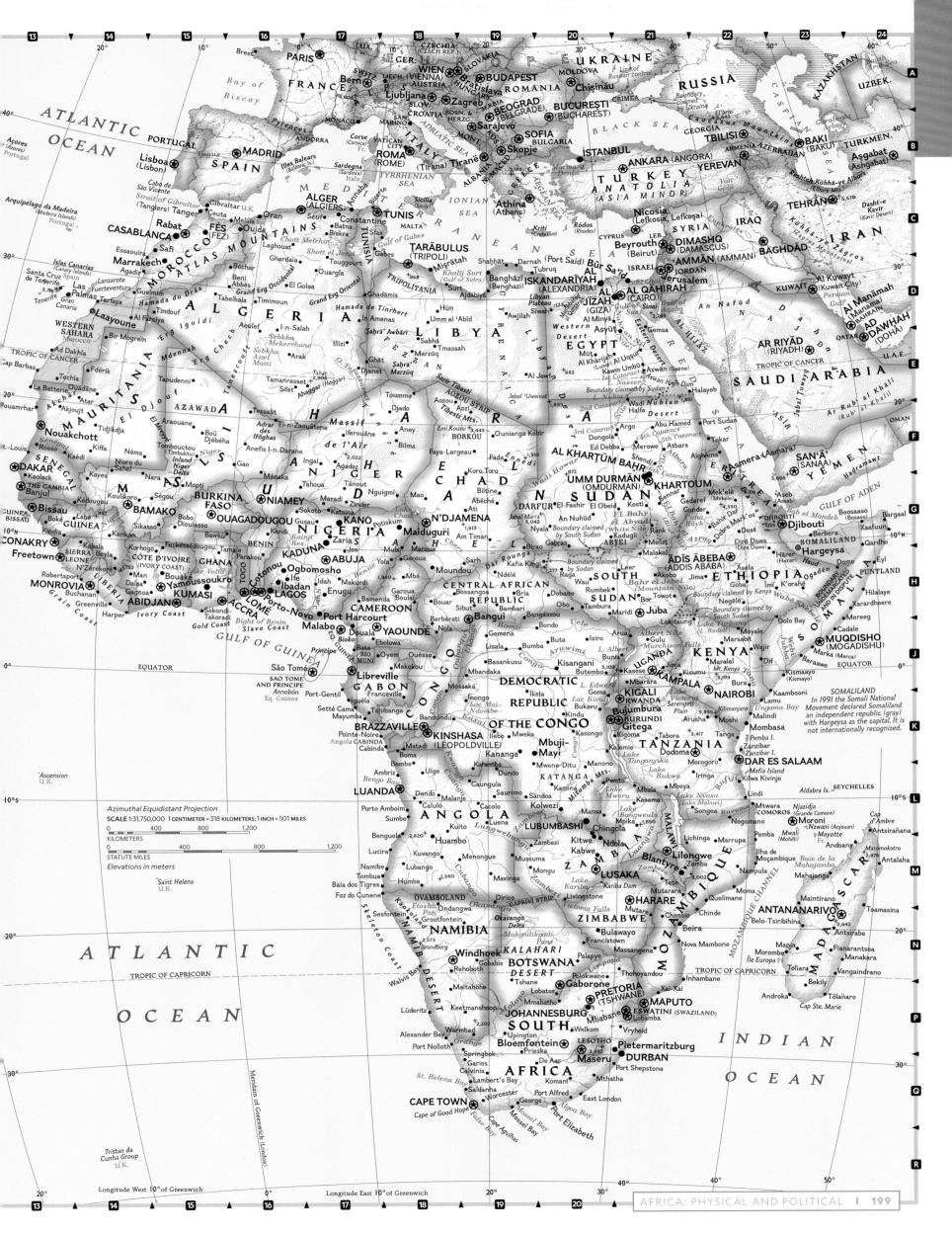

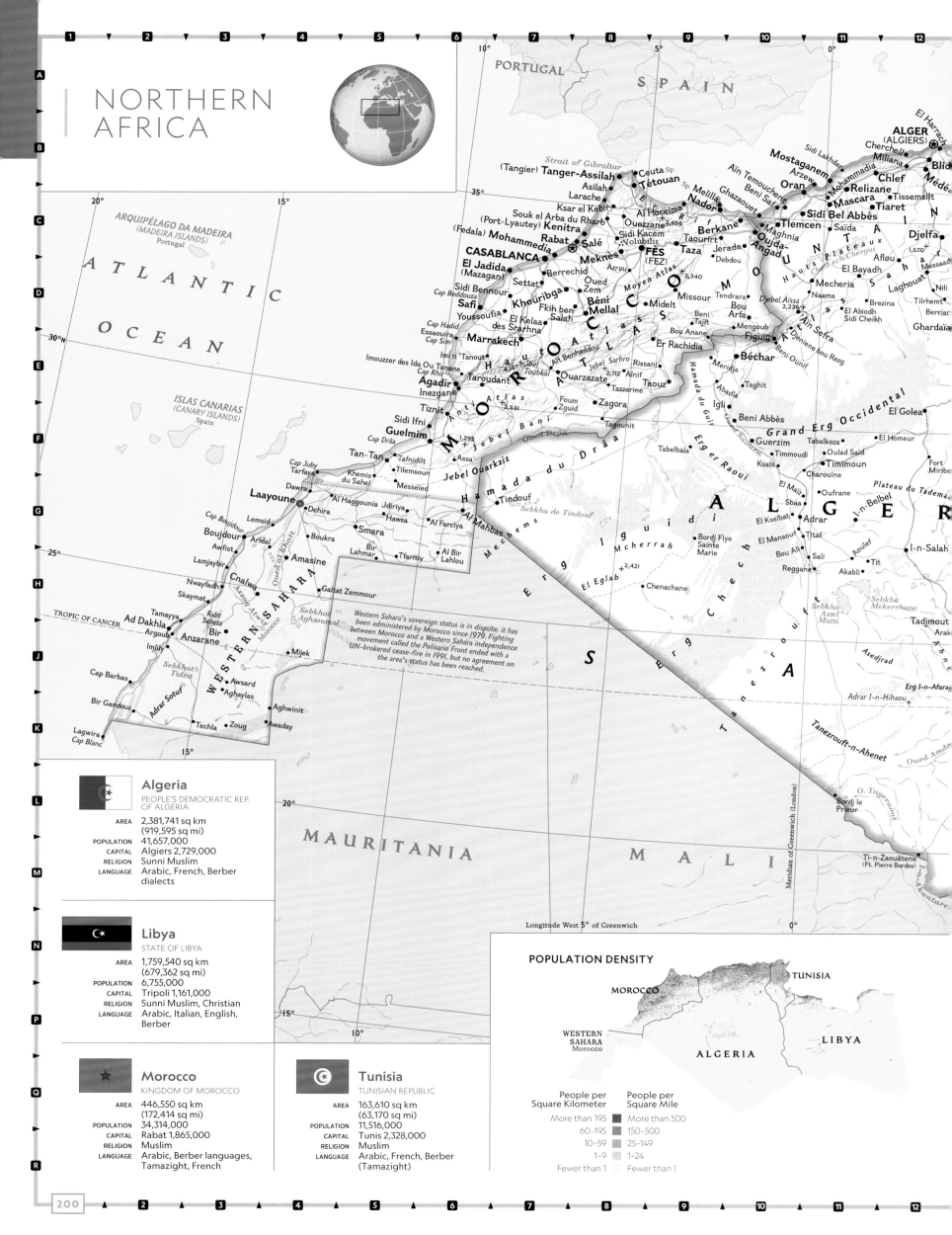

PORTUGAL

SPAIN

ATLANTIC

OCEAN

ARQUIPÉLAGO DA MADEIRA
(MADEIRA ISLANDS)
Portugal

ISLAS CANARIAS
(CANARY ISLANDS)
Spain

Strait of Gibraltar

(Tangier) Tanger-Assilah
Ceuta Sp.
Asilah
Larache
Tétouan
Sp. Melilla
Ksar el Kebir
Al Hoceima
Nador
Souk el Arba du Rharb
(Port-Lyautey) Kenitra
Ouezzane
Berkane
(Fedala) Mohammedia
Rabat Salé
Sidi Kacem
Taza
FES
(FEZ)
Oujda-
Angad
CASABLANCA
Meknes
Volubilis
Jerada
El Jadida
(Mazagan)
Berrechid
Azrou
Debdou
Settat
Oued
Zem
Béni
Mellal
Midelt
Missour
Tendrara
Khouribga
Bou
Arfa
Safi
Fkih ben
Salah
Youssoufia
El Kelaa
des Srarhna
Beni
Tajit
Bou Anane
Essaouira
Marrakech
Er Rachidia
Figuig
Imi n 'Tanout
Aït Benhaddou
Meridja
Imouzzer des Ida Ou Tanane
Jbel
Toubkal
Jebel Sarhro
Rissani
Abadla
Taroudant
Ouarzazate
Alnif
Igli
Agadir
Foum
Zguid
Tazzarine
Taouz
Inezgane
Zagora
Tiznit
Anti Atlas
Tagounit
Beni Abbès
Sidi Ifni
Jebel Bani
Oued Drâa
Guelmim
Cap Drâa
Tan-Tan
Hamada du Guir
Tafnidilt
Tilemsoun
Khemis
du Sahel
Messeied
Hamada du Drâa
Tarfaya
Tindouf
El Maïz
Dawra
Al Haggounia
Jdiriya
Al Farcia
Laayoune
Dchira
Hawza
Al Mahbas
Lemsid
Boukra
Smara
Aridal
Boujdour
Awfist
Bir
Lahmar
Al Bir
Lahlou
Lamjaybir
Tfaritiy
Nwayfadh
Cnalwa
Skaymat
Galtat Zemmour
TROPIC OF CANCER
Tamayya
Rabt
Sebeta
Ad Dakhla
Bir
Argoub Anzarane
Imlily
Mijek
Cap Barbas
Awsard
Bir Gandouz
Aghaylas
Adrar Sotuf
Aghwinit
Lagwira
Techla Zoug
Cap Blanc
Awaday

MAURITANIA

MALI

Western Sahara's sovereign status is in dispute; it has been administered by Morocco since 1979. Fighting between Morocco and a Western Sahara independence movement called the Polisario Front ended with a UN-brokered cease-fire in 1991, but no agreement on the area's status has been reached.

Longitude West 5° of Greenwich

ALGER
(ALGIERS)
El Harrach
Sidi Lakhdar
Cherchell
Mostaganem
Miliana
Oran
Arzew
Mohammadia
Relizane
Mascara
Tiaret
Tissemsilt
Sidi Bel Abbès
Tlemcen
Saïda
Djelfa
Maghnia
El Bayadh
Mecheria
Naama
Brezina
El Abiodh
Sidi Cheikh
Ghardaïa
Béchar
Beni Ounif
Grand Erg Occidental
Guerzim
El Golea
Timmoudi
El Homeur
Tabelbala
Oulad Saïd
Ksabi
Timimoun
Charouine
El Maïz
Oufrane
Bordj Flye
Sainte
Marie
Sbaa
Adrar
Reggane
In-Salah

ALGERIA

Meridian of Greenwich (London)

POPULATION DENSITY

MOROCCO
TUNISIA

WESTERN
SAHARA
Morocco

ALGERIA
LIBYA

People per Square Kilometer	People per Square Mile
More than 195	More than 500
60–195	150–500
10–59	25–149
1–9	1–24
Fewer than 1	Fewer than 1

Algeria
PEOPLE'S DEMOCRATIC REP. OF ALGERIA

AREA 2,381,741 sq km
(919,595 sq mi)
POPULATION 41,657,000
CAPITAL Algiers 2,729,000
RELIGION Sunni Muslim
LANGUAGE Arabic, French, Berber dialects

Libya
STATE OF LIBYA

AREA 1,759,540 sq km
(679,362 sq mi)
POPULATION 6,755,000
CAPITAL Tripoli 1,161,000
RELIGION Sunni Muslim, Christian
LANGUAGE Arabic, Italian, English, Berber

Morocco
KINGDOM OF MOROCCO

AREA 446,550 sq km
(172,414 sq mi)
POPULATION 34,314,000
CAPITAL Rabat 1,865,000
RELIGION Muslim
LANGUAGE Arabic, Berber languages, Tamazight, French

Tunisia
TUNISIAN REPUBLIC

AREA 163,610 sq km
(63,170 sq mi)
POPULATION 11,516,000
CAPITAL Tunis 2,328,000
RELIGION Muslim
LANGUAGE Arabic, French, Berber (Tamazight)

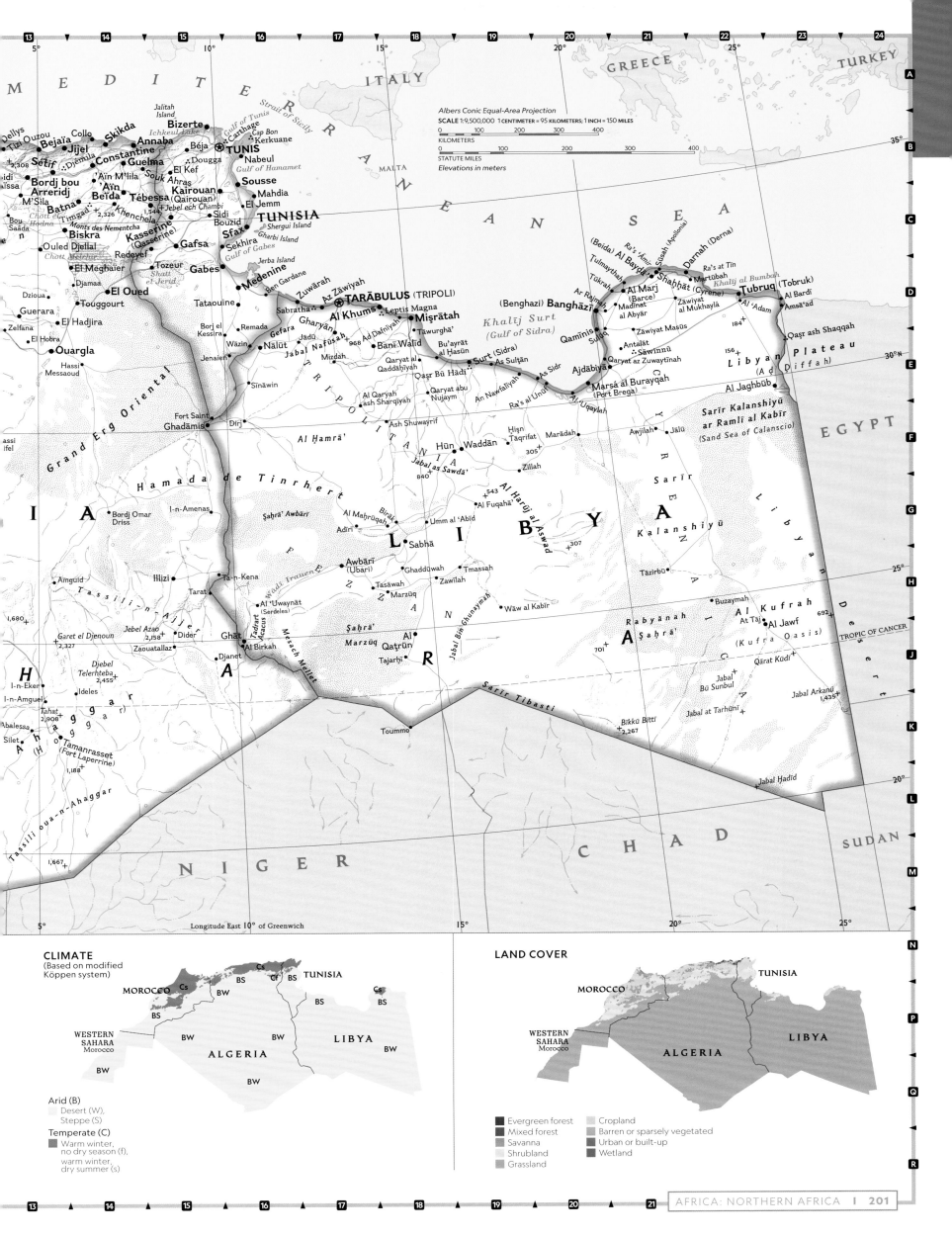

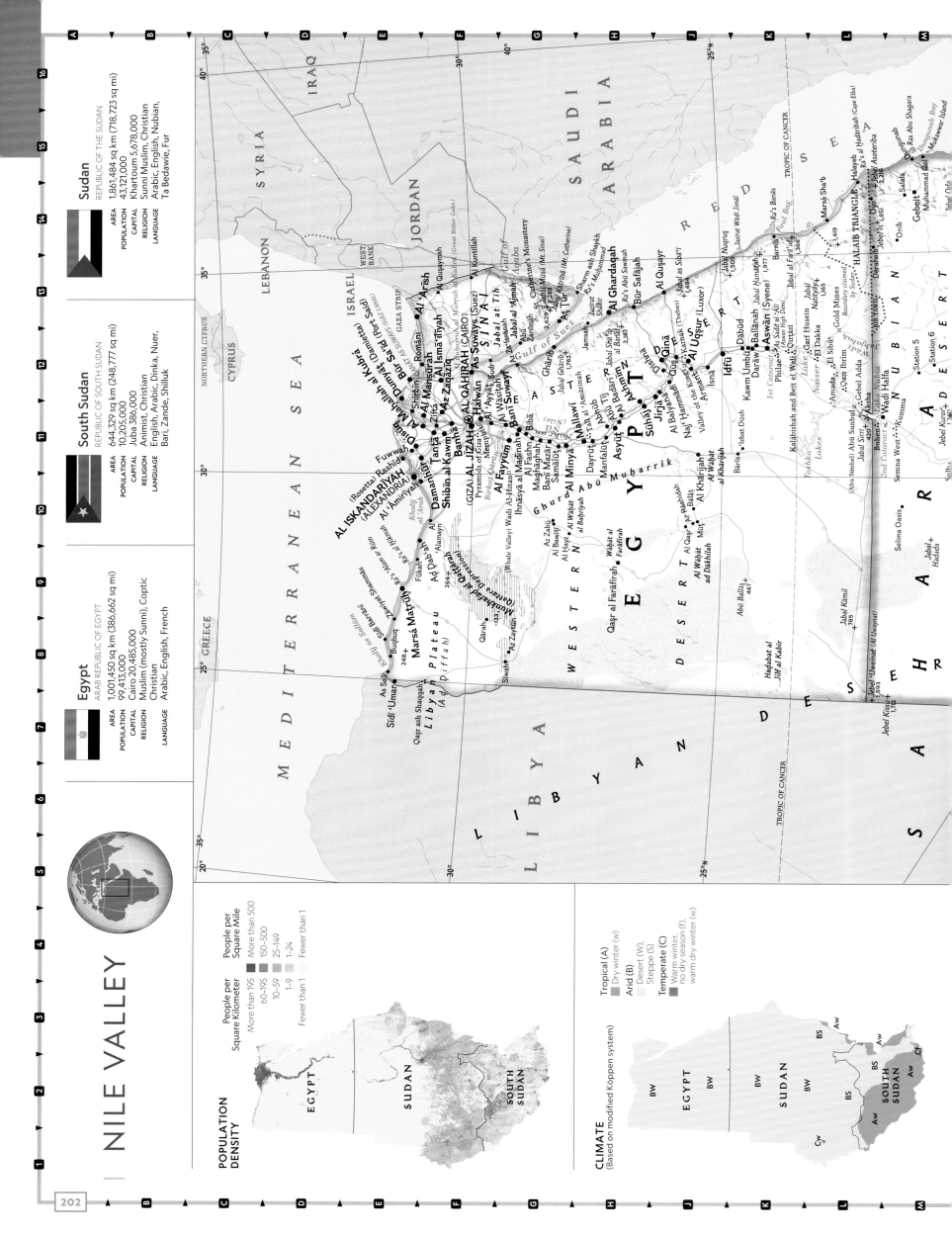

NILE VALLEY

Egypt
ARAB REPUBLIC OF EGYPT

AREA	1,001,450 sq km (386,662 sq mi)
POPULATION	99,413,000
CAPITAL	Cairo 20,485,000
RELIGION	Muslim (mostly Sunni), Coptic Christian
LANGUAGE	Arabic, English, French

South Sudan
REPUBLIC OF SOUTH SUDAN

AREA	644,329 sq km (248,777 sq mi)
POPULATION	10,205,000
CAPITAL	Juba 386,000
RELIGION	Animist, Christian
LANGUAGE	English, Arabic, Dinka, Nuer, Bari, Zande, Shilluk

Sudan
REPUBLIC OF THE SUDAN

AREA	1,861,484 sq km (718,723 sq mi)
POPULATION	43,121,000
CAPITAL	Khartoum 5,678,000
RELIGION	Sunni Muslim, Christian
LANGUAGE	Arabic, English, Nubian, Ta Bedawie, Fur

POPULATION DENSITY

People per Square Kilometer

- More than 195
- 60–195
- 10–59
- 1–9
- Fewer than 1

People per Square Mile

- More than 500
- 150–500
- 25–149
- 1–24
- Fewer than 1

CLIMATE
(Based on modified Köppen system)

Tropical (A)
- Dry winter (w)

Arid (B)
- Desert (W),
- Steppe (S)

Temperate (C)
- Warm winter, no dry season (f), warm dry winter (w)

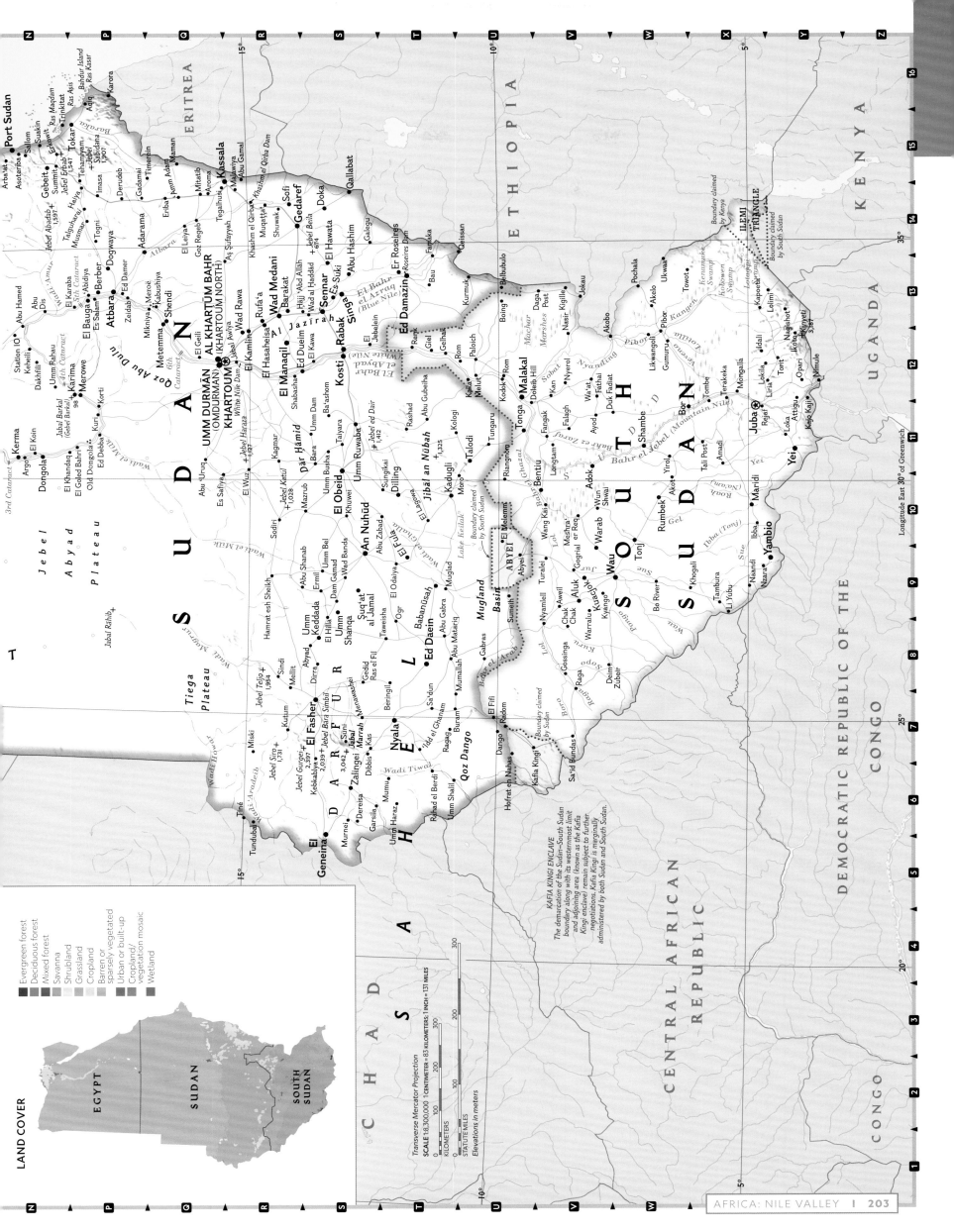

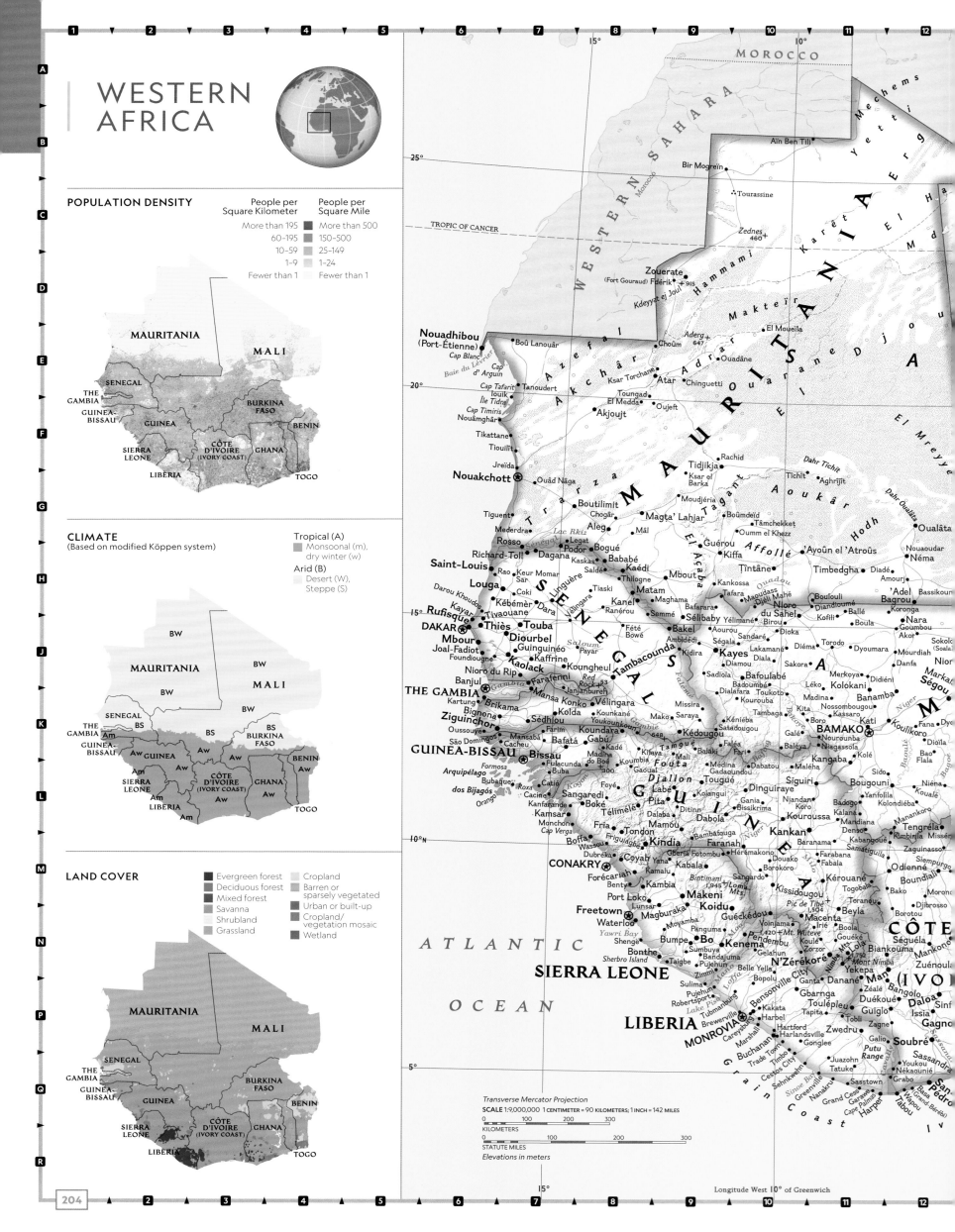

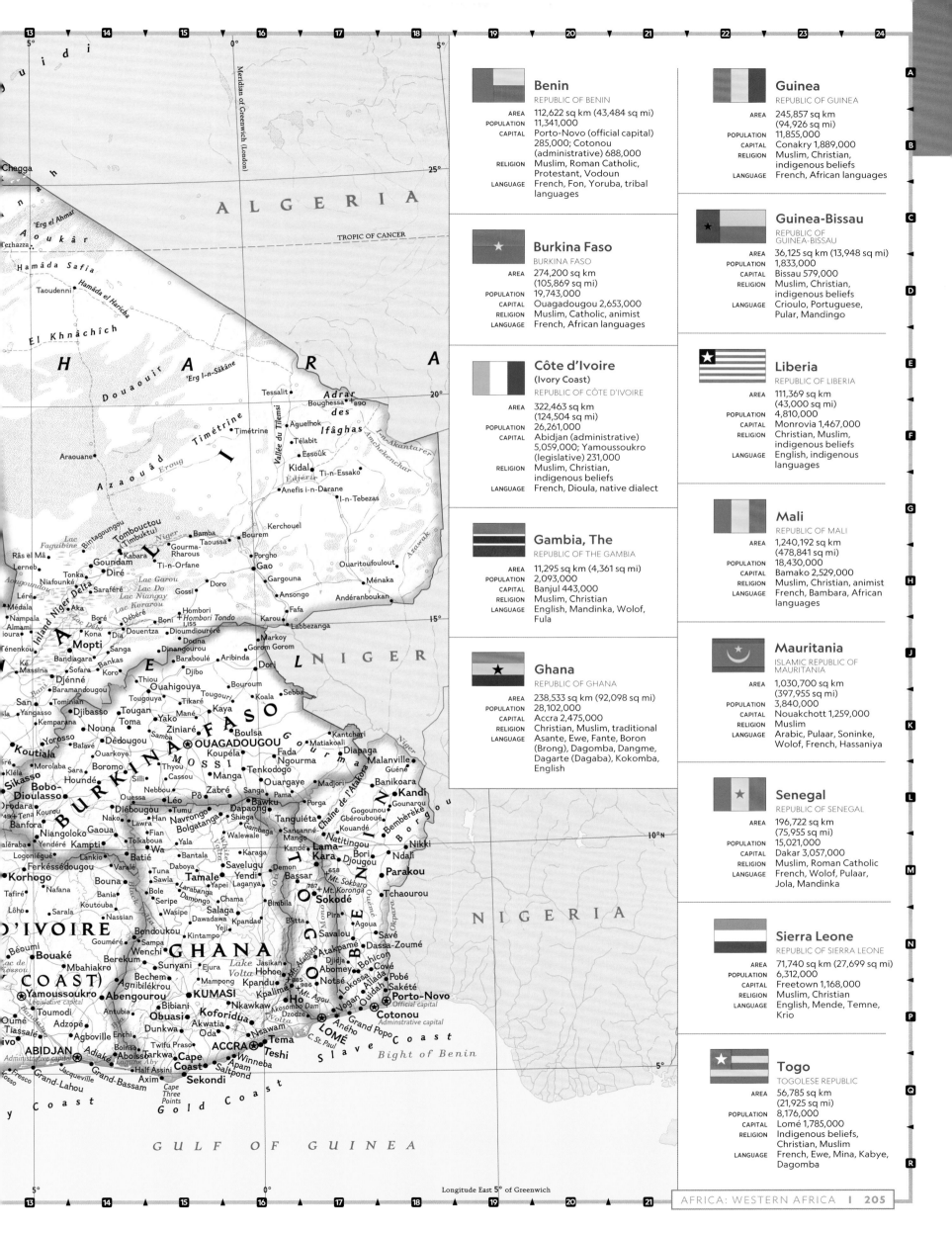

Benin
REPUBLIC OF BENIN
AREA 112,622 sq km (43,484 sq mi)
POPULATION 11,341,000
CAPITAL Porto-Novo (official capital) 285,000; Cotonou (administrative) 688,000
RELIGION Muslim, Roman Catholic, Protestant, Vodoun
LANGUAGE French, Fon, Yoruba, tribal languages

Burkina Faso
BURKINA FASO
AREA 274,200 sq km (105,869 sq mi)
POPULATION 19,743,000
CAPITAL Ouagadougou 2,653,000
RELIGION Muslim, Catholic, animist
LANGUAGE French, African languages

Côte d'Ivoire
(Ivory Coast)
REPUBLIC OF CÔTE D'IVOIRE
AREA 322,463 sq km (124,504 sq mi)
POPULATION 26,261,000
CAPITAL Abidjan (administrative) 5,059,000; Yamoussoukro (legislative) 231,000
RELIGION Muslim, Christian, indigenous beliefs
LANGUAGE French, Dioula, native dialect

Gambia, The
REPUBLIC OF THE GAMBIA
AREA 11,295 sq km (4,361 sq mi)
POPULATION 2,093,000
CAPITAL Banjul 443,000
RELIGION Muslim, Christian
LANGUAGE English, Mandinka, Wolof, Fula

Ghana
REPUBLIC OF GHANA
AREA 238,533 sq km (92,098 sq mi)
POPULATION 28,102,000
CAPITAL Accra 2,475,000
RELIGION Christian, Muslim, traditional
LANGUAGE Asante, Ewe, Fante, Boron (Brong), Dagomba, Dangme, Dagarte (Dagaba), Kokomba, English

Guinea
REPUBLIC OF GUINEA
AREA 245,857 sq km (94,926 sq mi)
POPULATION 11,855,000
CAPITAL Conakry 1,889,000
RELIGION Muslim, Christian, indigenous beliefs
LANGUAGE French, African languages

Guinea-Bissau
REPUBLIC OF GUINEA-BISSAU
AREA 36,125 sq km (13,948 sq mi)
POPULATION 1,833,000
CAPITAL Bissau 579,000
RELIGION Muslim, Christian, indigenous beliefs
LANGUAGE Crioulo, Portuguese, Pular, Mandingo

Liberia
REPUBLIC OF LIBERIA
AREA 111,369 sq km (43,000 sq mi)
POPULATION 4,810,000
CAPITAL Monrovia 1,467,000
RELIGION Christian, Muslim, indigenous beliefs
LANGUAGE English, indigenous languages

Mali
REPUBLIC OF MALI
AREA 1,240,192 sq km (478,841 sq mi)
POPULATION 18,430,000
CAPITAL Bamako 2,529,000
RELIGION Muslim, Christian, animist
LANGUAGE French, Bambara, African languages

Mauritania
ISLAMIC REPUBLIC OF MAURITANIA
AREA 1,030,700 sq km (397,955 sq mi)
POPULATION 3,840,000
CAPITAL Nouakchott 1,259,000
RELIGION Muslim
LANGUAGE Arabic, Pulaar, Soninke, Wolof, French, Hassaniya

Senegal
REPUBLIC OF SENEGAL
AREA 196,722 sq km (75,955 sq mi)
POPULATION 15,021,000
CAPITAL Dakar 3,057,000
RELIGION Muslim, Roman Catholic
LANGUAGE French, Wolof, Pulaar, Jola, Mandinka

Sierra Leone
REPUBLIC OF SIERRA LEONE
AREA 71,740 sq km (27,699 sq mi)
POPULATION 6,312,000
CAPITAL Freetown 1,168,000
RELIGION Muslim, Christian
LANGUAGE English, Mende, Temne, Krio

Togo
TOGOLESE REPUBLIC
AREA 56,785 sq km (21,925 sq mi)
POPULATION 8,176,000
CAPITAL Lomé 1,785,000
RELIGION Indigenous beliefs, Christian, Muslim
LANGUAGE French, Ewe, Mina, Kabye, Dagomba

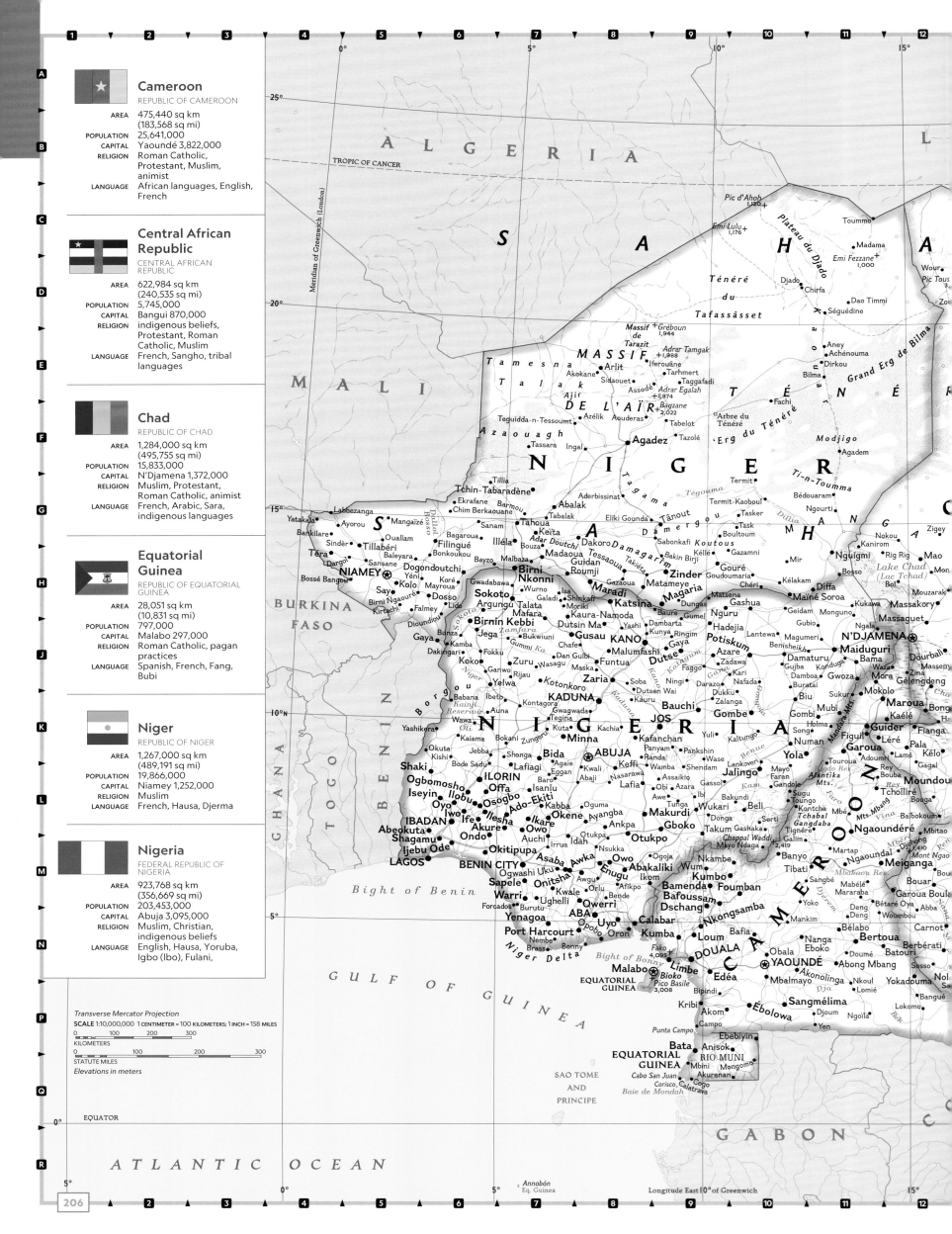

Cameroon
REPUBLIC OF CAMEROON
AREA	475,440 sq km (183,568 sq mi)
POPULATION	25,641,000
CAPITAL	Yaoundé 3,822,000
RELIGION	Roman Catholic, Protestant, Muslim, animist
LANGUAGE	African languages, English, French

Central African Republic
CENTRAL AFRICAN REPUBLIC
AREA	622,984 sq km (240,535 sq mi)
POPULATION	5,745,000
CAPITAL	Bangui 870,000
RELIGION	indigenous beliefs, Protestant, Roman Catholic, Muslim
LANGUAGE	French, Sangho, tribal languages

Chad
REPUBLIC OF CHAD
AREA	1,284,000 sq km (495,755 sq mi)
POPULATION	15,833,000
CAPITAL	N'Djamena 1,372,000
RELIGION	Muslim, Protestant, Roman Catholic, animist
LANGUAGE	French, Arabic, Sara, indigenous languages

Equatorial Guinea
REPUBLIC OF EQUATORIAL GUINEA
AREA	28,051 sq km (10,831 sq mi)
POPULATION	797,000
CAPITAL	Malabo 297,000
RELIGION	Roman Catholic, pagan practices
LANGUAGE	Spanish, French, Fang, Bubi

Niger
REPUBLIC OF NIGER
AREA	1,267,000 sq km (489,191 sq mi)
POPULATION	19,866,000
CAPITAL	Niamey 1,252,000
RELIGION	Muslim
LANGUAGE	French, Hausa, Djerma

Nigeria
FEDERAL REPUBLIC OF NIGERIA
AREA	923,768 sq km (356,669 sq mi)
POPULATION	203,453,000
CAPITAL	Abuja 3,095,000
RELIGION	Muslim, Christian, indigenous beliefs
LANGUAGE	English, Hausa, Yoruba, Igbo (Ibo), Fulani,

Transverse Mercator Projection
SCALE 1:10,000,000   1 CENTIMETER = 100 KILOMETERS; 1 INCH = 158 MILES
KILOMETERS
STATUTE MILES
Elevations in meters

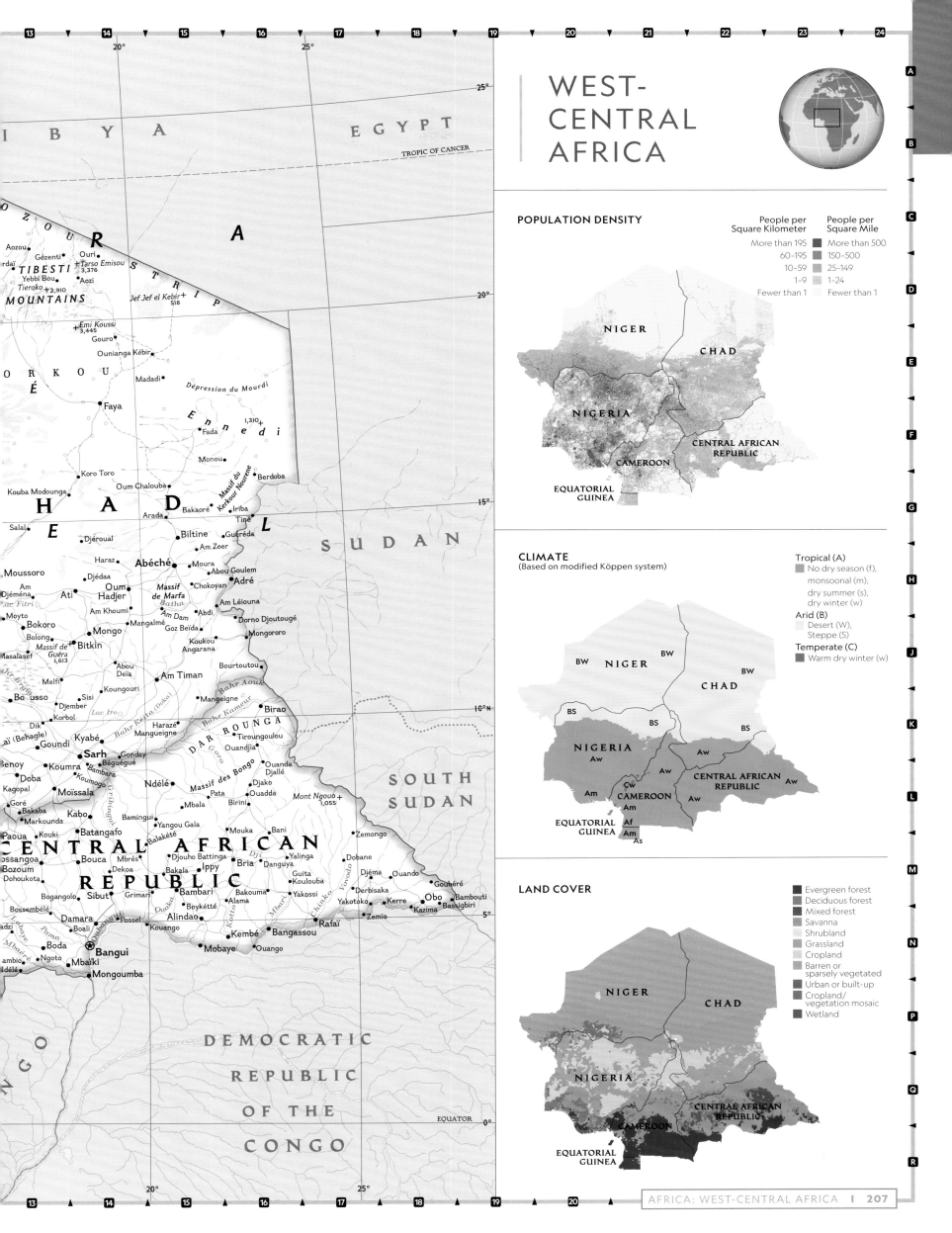

20°

25°

L I B Y A

E G Y P T

TROPIC OF CANCER

25°

20°

WEST-CENTRAL AFRICA

POPULATION DENSITY

	People per Square Kilometer	People per Square Mile
	More than 195	More than 500
	60–195	150–500
	10–59	25–149
	1–9	1–24
	Fewer than 1	Fewer than 1

NIGER

CHAD

NIGERIA

CENTRAL AFRICAN REPUBLIC

CAMEROON

EQUATORIAL GUINEA

CLIMATE
(Based on modified Köppen system)

Tropical (A)
No dry season (f), monsoonal (m), dry summer (s), dry winter (w)

Arid (B)
Desert (W), Steppe (S)

Temperate (C)
Warm dry winter (w)

BW NIGER BW

BW

CHAD

BS

BS BS

BS

NIGERIA
Aw

Aw

Aw CENTRAL AFRICAN REPUBLIC Aw

Cw
CAMEROON
Am

Aw

EQUATORIAL
GUINEA Af
Am
As

LAND COVER

- Evergreen forest
- Deciduous forest
- Mixed forest
- Savanna
- Shrubland
- Grassland
- Cropland
- Barren or sparsely vegetated
- Urban or built-up
- Cropland/ vegetation mosaic
- Wetland

NIGER

CHAD

NIGERIA

CENTRAL AFRICAN REPUBLIC

CAMEROON

EQUATORIAL GUINEA

Map labels (main map)

O Z O U

TIBESTI
MOUNTAINS

Aozou
Gézenti
Ouri
+Tarso Emisou 3,376
Yebbi Bou
Tieroko +2,910
Aozi

R A

Emi Koussi +3,445
Gouro

Ounianga Kébir

O R K O U
É

Madadi

Dépression du Mourdi

Faya

E n n e d i
1,310 +
Fada

Monou

Koro Toro

Oum Chalouba

Berdoba

Kouba Modounga
Oum Chalouba

Massif du Kerkour Nourène
Bakaoré

H A D

Arada
Iriba
Tiné

Salal

Djéroual
Biltine
Guéréda

E L

Haraz
Am Zoer

Moussoro

Abéché
Moura
Abou Goulem

S U D A N

Am
Djaména

Djédaa
Oum Hadjer

Adré
Chokoyan

Ati

Massif de Marfa
Batha
Am Léiouna

Lac Fitri

Am Khoumi
Am Dam
Abdi

Moyto

Bokoro
Mangalmé
Goz Beïda

Dorno Djoutougé

Bolong
Mongo

Koukou Angarana

Mongororo

Masalasef
Bitkin
Massif de Guéra 1,613

Melfi

Abou Deïa

Bourtoutou

Bo usso
Sisi
Djember

Am Timan

Koungouri

Bahr Aouk

Korbol
Lac Iro

Mangeigne

Birao

Dik

Bahr Kéita (Doka)

Harazé

D A R   R O U N G A

10°N

aï (Behagle)
Goundi

Kyabé
Mangueigne

Tiroungoulou

Ouandjia

Benoy
Sarh
Gondey

Bambara
Ouanda Djallé

Doba
Koumra
Béguégué

Ndélé
Djako

S O U T H
S U D A N

Kagopal
Koumogo
Moïssala

Massif des Bongo
Pata
Ouadda

Goré
Mbala
Birini

Markounda
Bamingui
Yangou Gala

Mont Ngouo + 1,055

Kabo
Balakété

Paoua
Kouki
Batangafo

Mouka
Bani

Zemongo

C E N T R A L   A F R I C A N

ossanga
Bouca
Djouho Battinga
Yalinga

Dobane

Bozoum
Dekoa
Bakala
Ippy
Bria

Danguya

Djéma

Dohoukota

Mbrés

Guita
Koulouba
Ouando

Bogangolo
Sibut
Grimari
Bambari
Bakouma
Yakossi

Derbisaka
Gouhéré

Obo
Bambouti

R E P U B L I C

Bossembélé

Alama
Boykétté

Yakotoko
Kerre
Kazima
Bassigbiri

Damara
Alindao
Kembé

Rafaï
Zemio

adi
Boali
Possel
Kouango
Bangassou

Mbaïki
Mobaye
Ouango

Boda

Bangui

Ngoto

ambio
Idélé
Mongoumba

DEMOCRATIC REPUBLIC OF THE CONGO

EQUATOR

0°

20°

25°

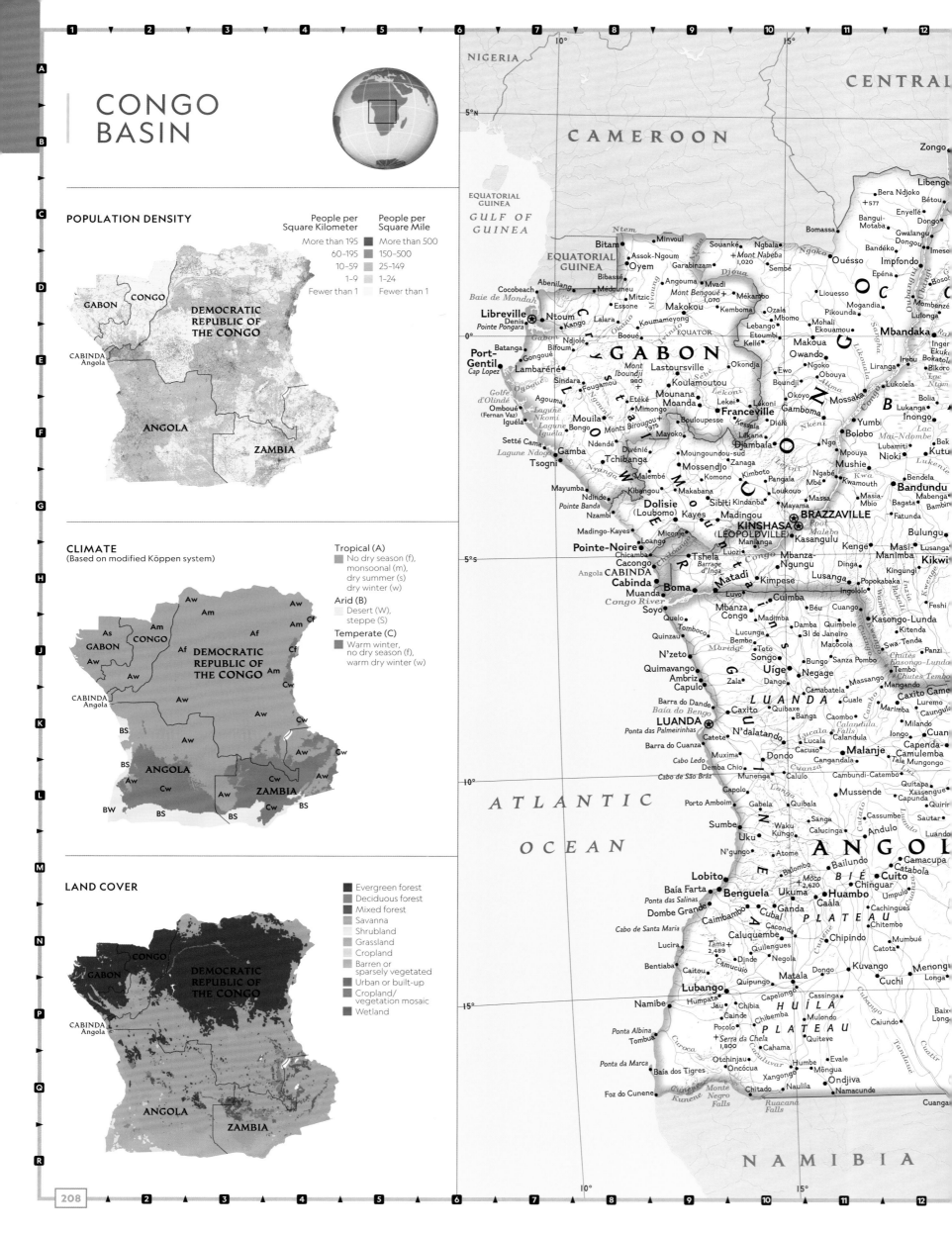

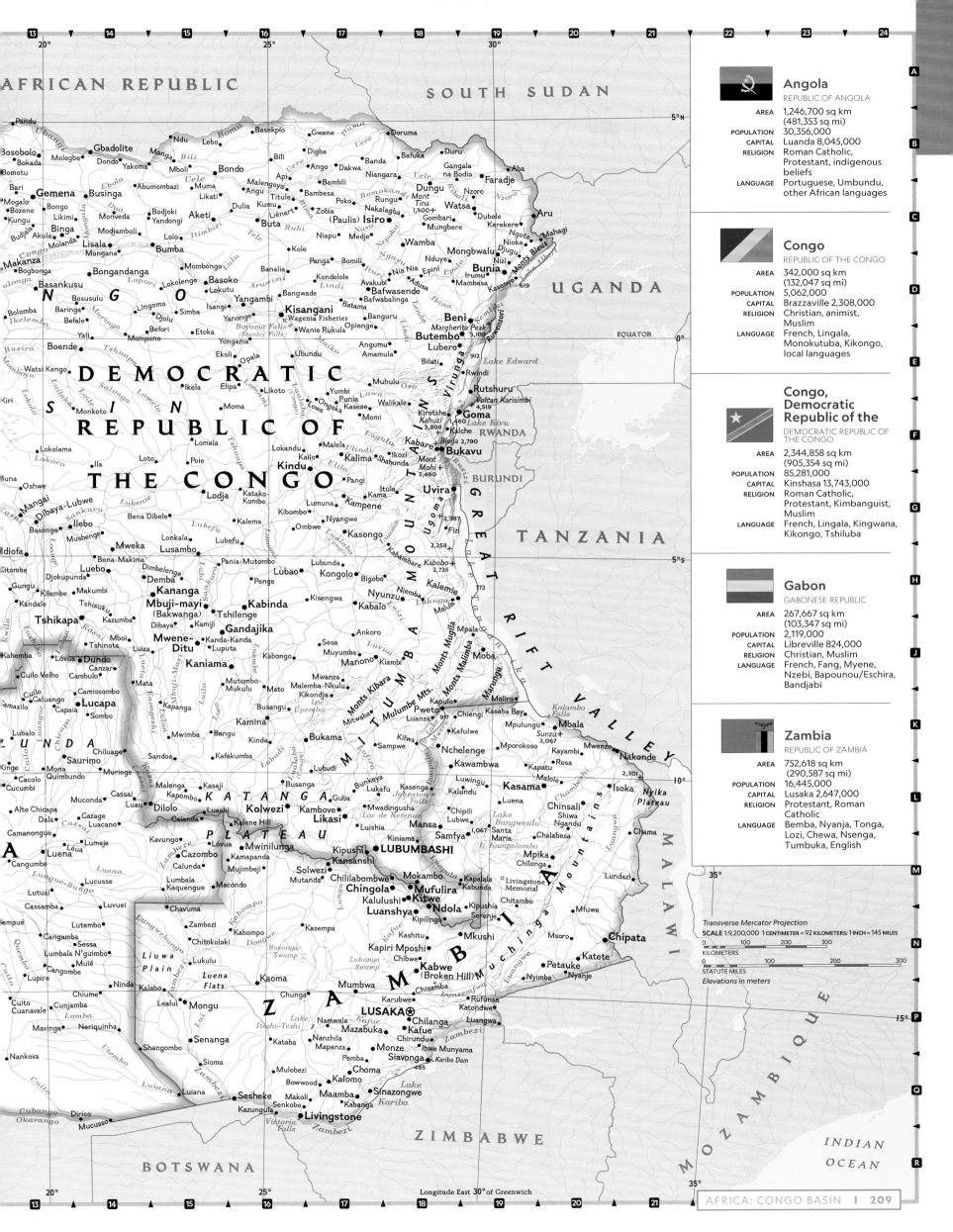

Angola
REPUBLIC OF ANGOLA

AREA	1,246,700 sq km (481,353 sq mi)
POPULATION	30,356,000
CAPITAL	Luanda 8,045,000
RELIGION	Roman Catholic, Protestant, indigenous beliefs
LANGUAGE	Portuguese, Umbundu, other African languages

Congo
REPUBLIC OF THE CONGO

AREA	342,000 sq km (132,047 sq mi)
POPULATION	5,062,000
CAPITAL	Brazzaville 2,308,000
RELIGION	Christian, animist, Muslim
LANGUAGE	French, Lingala, Monokutuba, Kikongo, local languages

Congo, Democratic Republic of the
DEMOCRATIC REPUBLIC OF THE CONGO

AREA	2,344,858 sq km (905,354 sq mi)
POPULATION	85,281,000
CAPITAL	Kinshasa 13,743,000
RELIGION	Roman Catholic, Protestant, Kimbanguist, Muslim
LANGUAGE	French, Lingala, Kingwana, Kikongo, Tshiluba

Gabon
GABONESE REPUBLIC

AREA	267,667 sq km (103,347 sq mi)
POPULATION	2,119,000
CAPITAL	Libreville 824,000
RELIGION	Christian, Muslim
LANGUAGE	French, Fang, Myene, Nzebi, Bapounou/Eschira, Bandjabi

Zambia
REPUBLIC OF ZAMBIA

AREA	752,618 sq km (290,587 sq mi)
POPULATION	16,445,000
CAPITAL	Lusaka 2,647,000
RELIGION	Protestant, Roman Catholic
LANGUAGE	Bemba, Nyanja, Tonga, Lozi, Chewa, Nsenga, Tumbuka, English

Transverse Mercator Projection
SCALE 1:9,200,000 1 CENTIMETER = 92 KILOMETERS; 1 INCH = 145 MILES
KILOMETERS
STATUTE MILES
Elevations in meters

POPULATION DENSITY

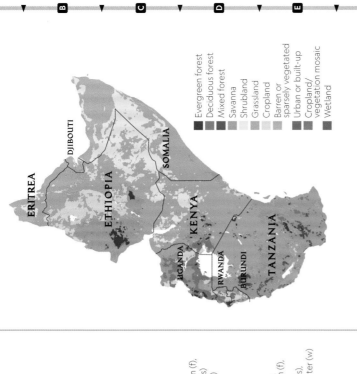

People per Square Kilometer	People per Square Mile
More than 195	More than 500
60–195	150–500
25–149	
10–59	25–149
1–9	1–24
Fewer than 1	Fewer than 1

CLIMATE
(Based on modified Köppen system)

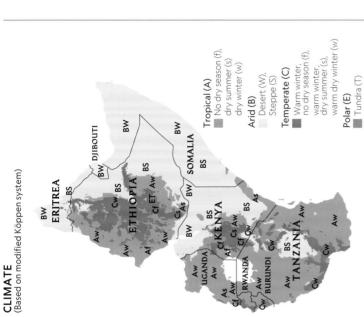

Tropical (A)
- No dry season (f), dry winter (w)

Arid (B)
- Desert (W), Steppe (S)

Temperate (C)
- Warm winter, no dry season (f), warm winter, dry summer (s), warm winter dry winter (w)

Polar (E)
- Tundra (T)

LAND COVER

- Evergreen forest
- Deciduous forest
- Mixed forest
- Savanna
- Shrubland
- Grassland
- Cropland
- Barren or sparsely vegetated
- Urban or built-up
- Cropland/vegetation mosaic
- Wetland

Burundi
REPUBLIC OF BURUNDI

AREA	27,830 sq km (10,745 sq mi)
POPULATION	11,845,000
CAPITAL	Bujumbura (commercial) 954,000; Gitega (official) 120,000
RELIGION	Roman Catholic, Protestant, Muslim
LANGUAGE	Kirundi, French, Swahili

Djibouti
REPUBLIC OF DJIBOUTI

AREA	23,200 sq km (8,958 sq mi)
POPULATION	884,000
CAPITAL	Djibouti 569,000
RELIGION	Muslim, Christian
LANGUAGE	French, Arabic, Somali, Afar

Eritrea
STATE OF ERITREA

AREA	117,600 sq km (45,406 sq mi)
POPULATION	5,971,000
CAPITAL	Asmara 929,000
RELIGION	Muslim, Coptic Christian, Roman Catholic, Protestant
LANGUAGE	Tigrigna (Tigrinya), Arabic, English, Tigre, Kunama, Afar, other Cushitic languages

Ethiopia
FEDERAL DEMOCRATIC REPUBLIC OF ETHIOPIA

AREA	1,104,300 sq km (426,372 sq mi)
POPULATION	108,386,000
CAPITAL	Addis Ababa 4,592,000
RELIGION	Ethiopian Orthodox, Muslim, Protestant, traditional beliefs
LANGUAGE	Oromo, Amharic, Somali, other indigenous languages, English, Arabic

Kenya
REPUBLIC OF KENYA

AREA	580,367 sq km (224,081 sq mi)
POPULATION	48,398,000
CAPITAL	Nairobi 4,556,000
RELIGION	Protestant, Roman Catholic, Muslim, indigenous beliefs
LANGUAGE	English, Kiswahili, indigenous languages

SOMALILAND
In 1991 the Somali National Movement declared Somaliland an independent republic with Hargeysa as the capital. It is not internationally recognized.

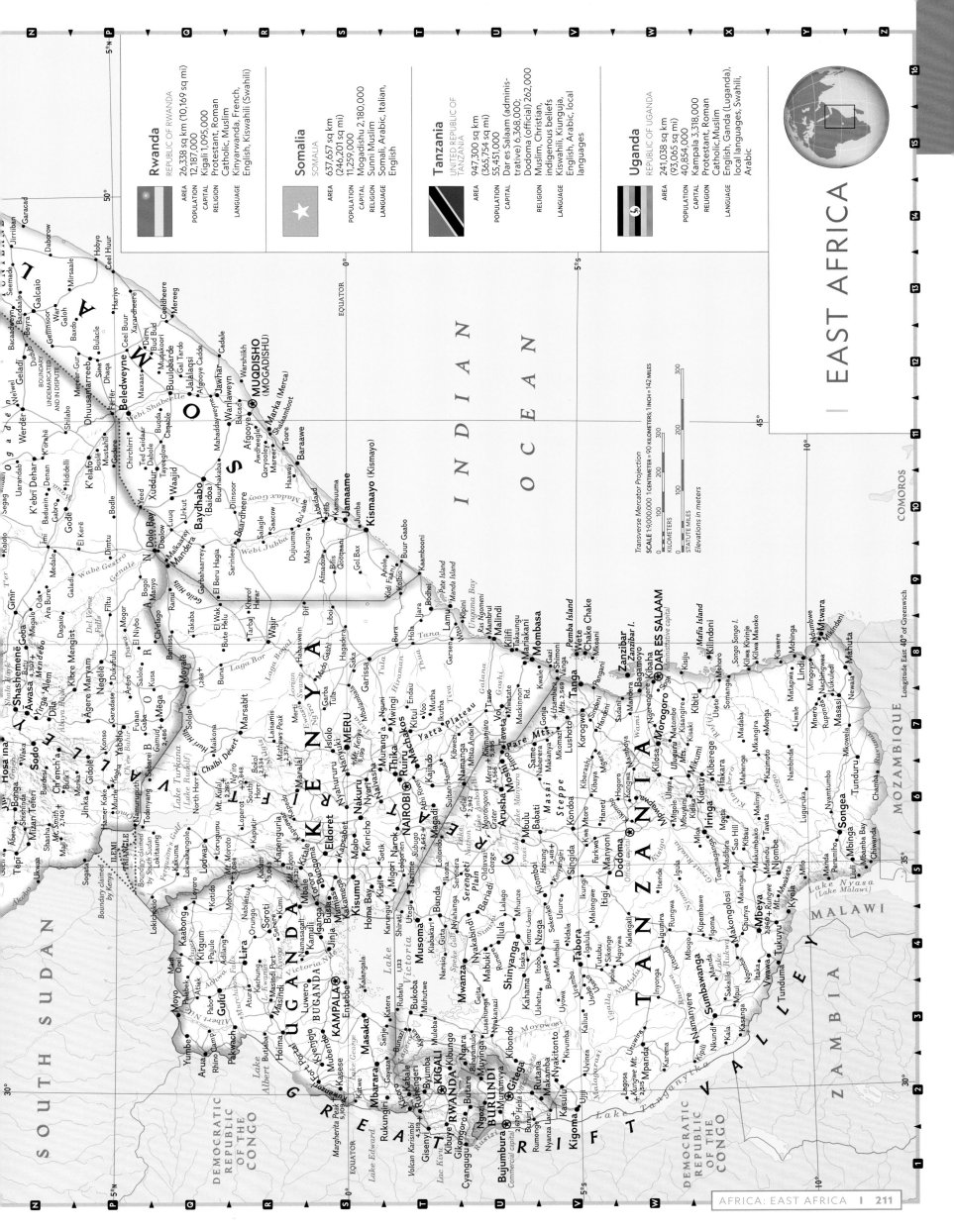

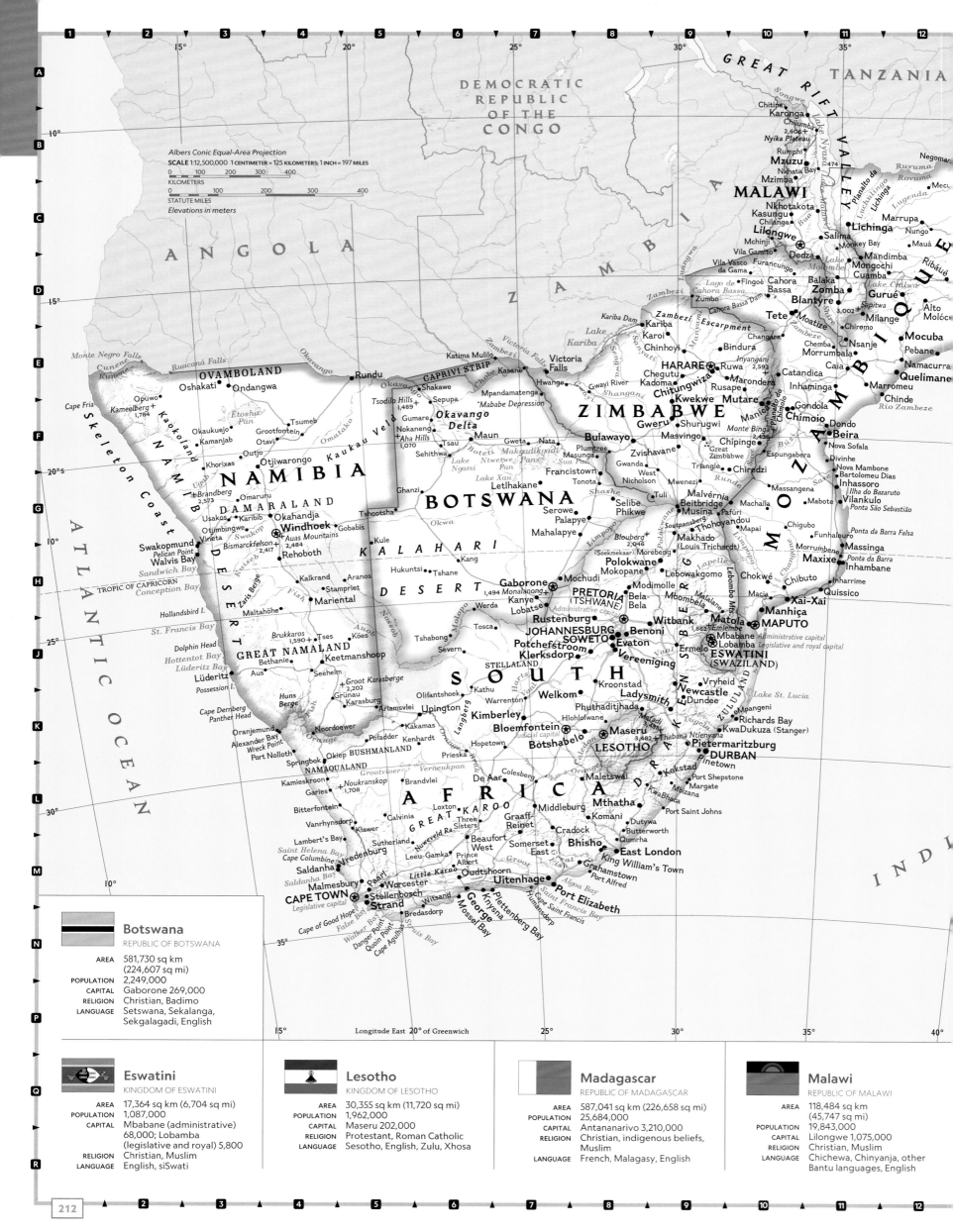

Botswana
REPUBLIC OF BOTSWANA

AREA 581,730 sq km
(224,607 sq mi)
POPULATION 2,249,000
CAPITAL Gaborone 269,000
RELIGION Christian, Badimo
LANGUAGE Setswana, Sekalanga,
Sekgalagadi, English

Eswatini
KINGDOM OF ESWATINI

AREA 17,364 sq km (6,704 sq mi)
POPULATION 1,087,000
CAPITAL Mbabane (administrative)
68,000; Lobamba
(legislative and royal) 5,800
RELIGION Christian, Muslim
LANGUAGE English, siSwati

Lesotho
KINGDOM OF LESOTHO

AREA 30,355 sq km (11,720 sq mi)
POPULATION 1,962,000
CAPITAL Maseru 202,000
RELIGION Protestant, Roman Catholic
LANGUAGE Sesotho, English, Zulu, Xhosa

Madagascar
REPUBLIC OF MADAGASCAR

AREA 587,041 sq km (226,658 sq mi)
POPULATION 25,684,000
CAPITAL Antananarivo 3,210,000
RELIGION Christian, indigenous beliefs,
Muslim
LANGUAGE French, Malagasy, English

Malawi
REPUBLIC OF MALAWI

AREA 118,484 sq km
(45,747 sq mi)
POPULATION 19,843,000
CAPITAL Lilongwe 1,075,000
RELIGION Christian, Muslim
LANGUAGE Chichewa, Chinyanja, other
Bantu languages, English

SCALE 1:12,500,000 1 CENTIMETER = 125 KILOMETERS; 1 INCH = 197 MILES
Albers Conic Equal-Area Projection
Elevations in meters

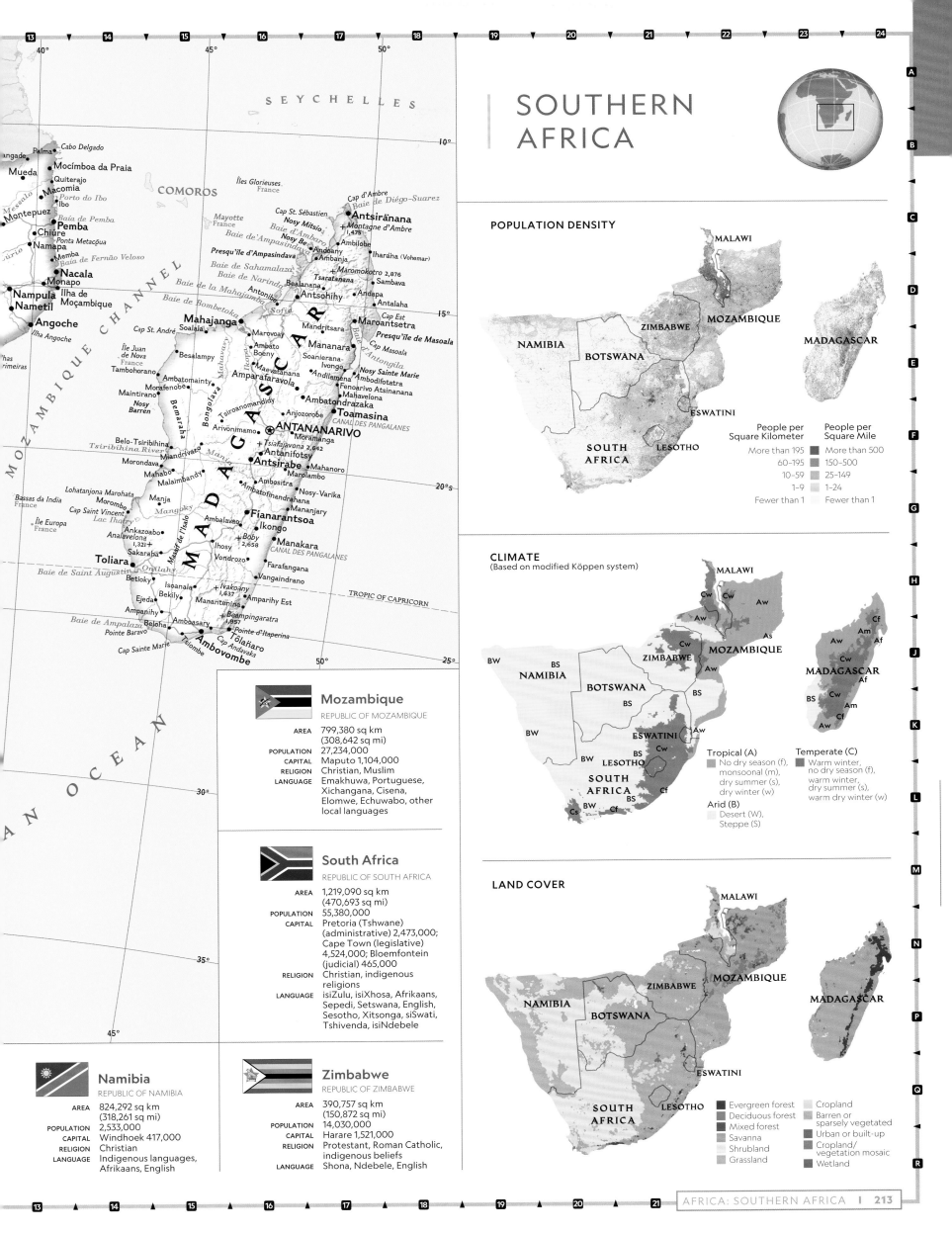

SEYCHELLES

COMOROS
Mayotte France
Îles Glorieuses France

Cabo Delgado
Palma
Mueda
Mocímboa da Praia
Quiterajo
Macomia
Porto do Ibo
Ibo
Montepuez
Messalo
Pemba
Baía de Pemba
Chiúre
Ponta Metacçua
Namapa
Memba
Nacala
Monapo
Ilha de Moçambique
Nampula
Baía de Fernão Veloso
Nametil
Angoche
Ilha Angoche

MOZAMBIQUE CHANNEL

Cap d'Ambre
Baie de Diégo-Suarez
Cap St. Sébastien
Antsiränana
Montagne d'Ambre 1,475
Ambilobe
Nosy Mitsio
Nosy Be
Andoany
Ambanja
Iharäna (Vohemar)
Baie d'Amparo
Baie d'Ampasindava
Presqu'île d'Ampasindava
Maromokotro 2,876
Sambava
Baie de Sahamalaza
Antonibe
Tsaratanana
Andapa
Antalaha
Bealanana
Antsohihy
Baie de Narinda
Cap Est
Mahajanga
Sofia
Marovoay
Mandritsara
Maroantsetra
Cap St. André
Soalala
Ambato Boeny
Presqu'île de Masoala
Cap Masoala
Besalampy
Mananara
Soanierana-Ivongo
Nosy Sainte Marie
Île Juan de Nova France
Tambohorano
Ambatomainty
Maevatanana
Andilamena
Ambodifotatra
Morafenobe
Fenoarivo Atsinanana
Maintirano
Amparafaravola
Mahavelona
Ambatondrazaka
Nosy Barren
Moramanga
Toamasina
Belo-Tsiribihina
Tsiroanomandidy
Arivonimamo
ANTANANARIVO
Tsiafajavona 2,642
CANAL DES PANGALANES
Tsiribihina River
Miandrivazo
Antanifotsy
Morondava
Mahanoro
Antsirabe
Mahabo
Malaimbandy
Ambositra
Marolambo
Ambovombe
Lohatanjona Marohata
Morombe
Ambatofinandrahana
Nosy-Varika
Analavelona 1,321
Manja
Ambalavao
Mananjary
Bassas da India France
Cap Saint Vincent
Fianarantsoa
Île Europa France
Ankazoabo
Ikongo
Boby 2,658
Analavelona
Manakara
Sakaraha
Ihosy
Toliara
Vondrozo
Farafangana
Baie de Saint Augustin
Betioky
Vangaindrano
Ivakoany 1,637
Isoanala
TROPIC OF CAPRICORN
Ejeda
Bekily
Amparihy Est
Ampanihy
Manantenina
Béampingaratra 1,957
Baie de Ampalaza
Beloha
Amboasary
Pointe d'Itaperina
Pointe Baravo
Ambovombe
Cap Andavaka
Tôlañaro
Tsiombe
Cap Sainte Marie

INDIAN OCEAN

MADAGASCAR

Mozambique
REPUBLIC OF MOZAMBIQUE

AREA	799,380 sq km (308,642 sq mi)
POPULATION	27,234,000
CAPITAL	Maputo 1,104,000
RELIGION	Christian, Muslim
LANGUAGE	Emakhuwa, Portuguese, Xichangana, Cisena, Elomwe, Echuwabo, other local languages

South Africa
REPUBLIC OF SOUTH AFRICA

AREA	1,219,090 sq km (470,693 sq mi)
POPULATION	55,380,000
CAPITAL	Pretoria (Tshwane) (administrative) 2,473,000; Cape Town (legislative) 4,524,000; Bloemfontein (judicial) 465,000
RELIGION	Christian, indigenous religions
LANGUAGE	isiZulu, isiXhosa, Afrikaans, Sepedi, Setswana, English, Sesotho, Xitsonga, siSwati, Tshivenda, isiNdebele

Namibia
REPUBLIC OF NAMIBIA

AREA	824,292 sq km (318,261 sq mi)
POPULATION	2,533,000
CAPITAL	Windhoek 417,000
RELIGION	Christian
LANGUAGE	Indigenous languages, Afrikaans, English

Zimbabwe
REPUBLIC OF ZIMBABWE

AREA	390,757 sq km (150,872 sq mi)
POPULATION	14,030,000
CAPITAL	Harare 1,521,000
RELIGION	Protestant, Roman Catholic, indigenous beliefs
LANGUAGE	Shona, Ndebele, English

SOUTHERN AFRICA

POPULATION DENSITY

MALAWI
NAMIBIA
ZIMBABWE
MOZAMBIQUE
BOTSWANA
MADAGASCAR
ESWATINI
SOUTH AFRICA
LESOTHO

People per Square Kilometer	People per Square Mile
More than 195	More than 500
60–195	150–500
10–59	25–149
1–9	1–24
Fewer than 1	Fewer than 1

CLIMATE
(Based on modified Köppen system)

MALAWI
Cw
Cw
Aw
BW
BS
NAMIBIA
Cw
ZIMBABWE
MOZAMBIQUE
As
Aw
BOTSWANA
BS
Cf
Am
Af
MADAGASCAR
Aw
BW
Cw
BS
Cw
Am
ESWATINI
Aw
Cf
BW
BS
Cw
Aw
LESOTHO
SOUTH AFRICA
BS
BW
Cs
Cf
Cf

Tropical (A)
No dry season (f), monsoonal (m), dry summer (s), dry winter (w)

Arid (B)
Desert (W), Steppe (S)

Temperate (C)
Warm winter, no dry season (f), warm winter, dry summer (s), warm dry winter (w)

LAND COVER

MALAWI
NAMIBIA
ZIMBABWE
MOZAMBIQUE
BOTSWANA
MADAGASCAR
ESWATINI
SOUTH AFRICA
LESOTHO

Evergreen forest	Cropland
Deciduous forest	Barren or sparsely vegetated
Mixed forest	Urban or built-up
Savanna	Cropland/ vegetation mosaic
Shrubland	Wetland
Grassland	

ISLANDS OF AFRICA

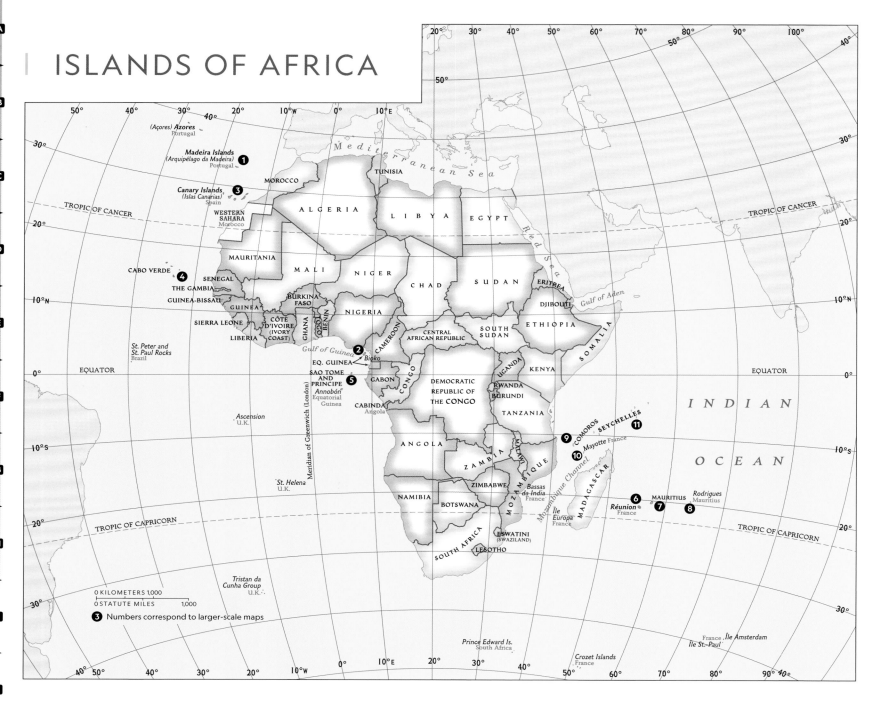

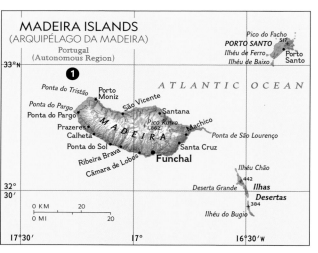

MADEIRA ISLANDS
(ARQUIPÉLAGO DA MADEIRA)

Portugal
(Autonomous Region)

Pico do Facho
PORTO SANTO
Ilhéu de Ferro
Ilhéu de Baixo
Porto Santo

ATLANTIC OCEAN

Ponta do Tristão
Porto Moniz
São Vicente
Ponta do Pargo
Ponta do Pargo
Santana
Pico Ruivo
1,862
Prazeres
Machico
Calheta
Ponta de São Lourenço
Ponta do Sol
Santa Cruz
Ribeira Brava
Funchal
Câmara de Lobos
Ilhéu Chão
442
Deserta Grande
Ilhas
Desertas
384
Ilhéu do Bugio

0 KM 20
0 MI 20

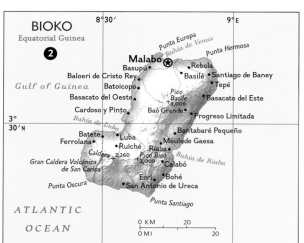

BIOKO
Equatorial Guinea

Punta Europa
Bahía de Venus
Malabo ⊛
Punta Hermosa
Basupú
Rebola
Baloeri de Cristo Rey
Basilé
Santiago de Baney
Batoicopo
Tepé
Basacato del Oeste
Pico Basilé 3,008
Basacato del Este
Cardoso y Pinto
Bao Grande
Progreso Limitada
Bahía de Lúba
Batete
Bantabaré Pequeño
Ferrolana
Luba
Moulede Gaesa
Ruiché
Caldera 2,260
Pico Biao
Bahía de Riaba
Gran Caldera Volcánica de San Carlos
2,009
Calabó
Eori
Bohé
Punta Oscura
San Antonio de Ureca
Punta Santiago

ATLANTIC OCEAN

0 KM 20
0 MI 20

CANARY ISLANDS
(ISLAS CANARIAS)

Spain
(Autonomous Community)

ATLANTIC OCEAN

Alegranza
Graciosa
Punta Fariones
Peñas del Chache 671
Haría
Caldera de Taburiente
Roque de los Muchachos 2,426
Atalaya de Femés 608
Tinajo
Teguise
LANZAROTE
Los Llanos
Santa Cruz de la Palma
Arrecife
LA PALMA
La Laguna
Playa Blanca
Playa Honda
Fuencaliente
Punta de Anaga
Solyplayas
Puerto del Carmen
TENERIFE
Puerto de la Cruz
Monte Muda 689
La Oliva
Punta de Fuencaliente
Guía de Isora
Santa Cruz de Tenerife
Antigua
Puerto del Rosario
La Orotava
Pico de Teide 3,718
Tuineje
FUERTEVENTURA
Vallehermoso
Gáldar
Arucas
GOMERA
Garajonay 1,487
Granadilla
Agaete
La Isleta
Jandía 807
Gran Tarajal
Pico de las Nieves 1,949
Los Cristianos
Las Palmas
Punta de Jandía
EL HIERRO (FERRO)
Costa del Silencio
Telde
Cape Juby
Valverde
Mogán
GRAN CANARIA
Sabinosa
Tarfaya
Malpaso 1,501
Maspalomas
MOROCCO
Punta Restinga
Morocco WESTERN SAHARA

0 KILOMETERS 75
0 STATUTE MILES 75

Cabo Verde
REPUBLIC OF CABO VERDE

AREA	4,033 sq km (1,557 sq mi)
POPULATION	568,000
CAPITAL	Praia 168,000
RELIGION	Roman Catholic, Protestant
LANGUAGE	Portuguese, Crioulo

Comoros
UNION OF THE COMOROS

AREA	2,235 sq km (863 sq mi)
POPULATION	821,000
CAPITAL	Moroni 62,000
RELIGION	Sunni Muslim
LANGUAGE	Arabic, French, Shikomoro

Mauritius
REPUBLIC OF MAURITIUS

AREA	2,040 sq km (788 sq mi)
POPULATION	1,364,000
CAPITAL	Port Louis 149,000
RELIGION	Hindu, Roman Catholic, Muslim, other Christian
LANGUAGE	Creole, Bhojpuri, French, English

All map elevations in meters

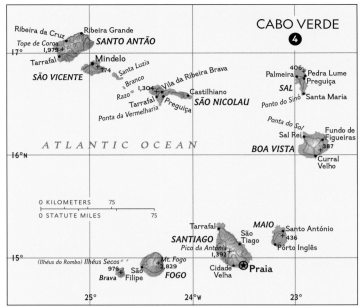

CABO VERDE
4

Ribeira da Cruz · Ribeira Grande
Tope de Coroa · *SANTO ANTÃO*
17° · 1,979
Tarrafal · Santa Luzia
Mindelo
774 · Branco
SÃO VICENTE · Razo · 1,304 · Vila da Ribeira Brava
Tarrafal · Castilhiano
Preguiça · *SÃO NICOLAU*
Ponta da Vermelharia
Ponta do Sol
Sal Rei
Fundo de Figueiras
16°N · 387
Curral Velho
BOA VISTA

Palmeira · Pedra Lume
406 · Preguiça
SAL
Santa Maria
Ponta do Sinó

ATLANTIC OCEAN

0 KILOMETERS 75
0 STATUTE MILES 75

Tarrafal · *MAIO*
SANTIAGO · São Tiago · Santo António
Pico da Antónia · 436
1,392 · Porto Inglês
15°
(Ilhéus do Rombo) Ilhéus Secos
976 · Mt. Fogo
São · 2,829
Brava · Filipe · *FOGO*
Cidade Velha · ⊛ **Praia**

25° · 24°W · 23°

SAO TOME AND PRINCIPE
5

1°N
Same scale as main map
Ilhéu Bombom · Ponta Capitão
PRÍNCIPE · Sundi · Santo António
Terreiro Velho
927 · Infante D. Henrique
Ilhéu Caroço
1° 30'N
Gulf of Guinea

0 KM 20
0 MI 20

Tinhosa Pequena
Pedras Tinhosas
Tinhosa Grande
7°30'E

Gulf of Guinea

0°30'
Ponta Cruzeiro
Rio do Ouro · *Ilhéu das Cabras*
Neves · ⊛ São Tomé
2,024 · Pico de São Tomé
Madalena · Caixão Grande
Santa Catarina · Santana
SÃO TOMÉ · Valle Formozo
Pico Cabumbé · Ribeira Afonso
1,403 · Santa Cruz
ATLANTIC · Jou · Ponta do Ló
OCEAN
Porto Alegre
0° · *Ilhéu das Rôlas* · EQUATOR

6°30' · 7°E

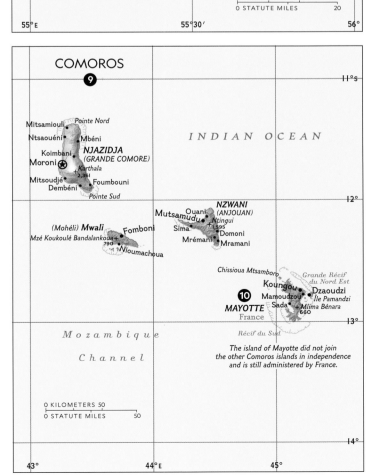

RÉUNION
France
6

Saint-Denis
Sainte-Marie
Pointe des Galets · Sainte-Suzanna
La Possession
Le Port · Saint-André
Saint-Paul
2,277 · Salazie
21°s
St.-Gilles-les-Bains · Bras-Panon
Hell-Bourg · Saint-Benoît
Trois-Bassins · 941
2,896 · Piton des Neiges · La Plaine
Cilaos · 3,071 · Sainte-Rose
1,685
RÉUNION
Saint-Leu
Les Avirons · La Plaine · Piton de la Fournaise
Entre-Deux · des Cafres · 2,631
Étang-Salé · La Rivière
Saint-Louis · Le Tampon
Saint-Pierre · Petite-Île
281 · Pointe de la Table
Saint-Joseph · Saint-Philippe

INDIAN OCEAN

21°30'

0 KILOMETERS 20
0 STATUTE MILES 20

55°E · 55°30' · 56°

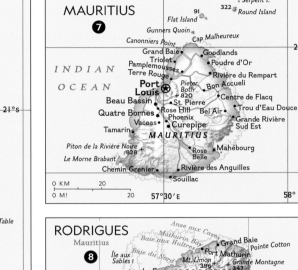

MAURITIUS
7

91 · *Serpent I.*
Flat Island · 322 · *Round Island*
Gunners Quoin
Canonniers Point · Cap Malheureux
20°s
Grand Baie · Goodlands
Triolet · Poudre d'Or
Pamplemousses
Terre Rouge · Rivière du Rempart
INDIAN · Bon Accueil
OCEAN · **Port** · Pieter · Centre de Flacq
Louis · ⊛ Both · 820
Beau Bassin · St. Pierre · Trou d'Eau Douce
Quatre Bornes · Rose Hill · Bel Air · Grande Rivière
Phoenix · Sud Est
Tamarin · Vacoas · Curepipe
MAURITIUS · Mahébourg
Piton de la Rivière Noire · Rose
Le Morne Brabant · 828 · Belle
20°30'
Chemin Grenier · Rivière des Anguilles
Souillac

0 KM 20
0 MI 20 · 57°30'E · 58°

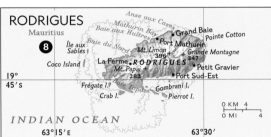

RODRIGUES
Mauritius
8

Anse aux Caves
Mathurin Bay · Grand Baie · Pointe Cotton
Baie aux Huîtres · Port Mathurin
Île aux · Baie du Nord · Mt. Limon · Grande Montagne
Sables I. · 396 · 347
Coco Island · La Ferme · *RODRIGUES* · Petit Gravier
19°45's · Mt. Papaï · 283 · Port Sud-Est
Frégate I. · Gombrani I.
Crab I. · *Pierrot I.*

INDIAN OCEAN

0 KM 4
0 MI 4

63°15'E · 63°30'

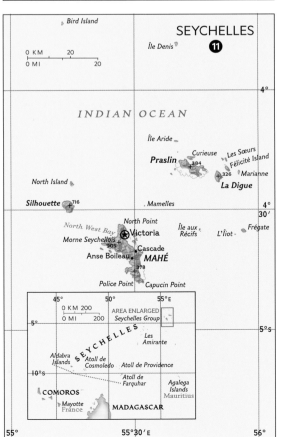

COMOROS
9

11°s
Mitsamiouli · *Pointe Nord*
Ntsaouéni · Mbéni
Koimbani · INDIAN OCEAN
NJAZIDJA
Moroni ⊛ *(GRANDE COMORE)*
Mitsoudjé · Karthala · 2,361
Dembéni · Foumbouni
Pointe Sud
12°
(Mohéli) Mwali · *NZWANI*
Mutsamudu · Ouani · *(ANJOUAN)*
Mzé Koukoulé Bandalankoua · Sima · Ntingui · Domoni
790 · Fomboni · Mrémani · 1,595
Nioumachoua · Mramani

Chissioua Mtsamboro · *Grande Récif du Nord Est*
Koungou
Mamoudzou · Dzaoudzi
Sada · *Île Pamandzi*
10 · Mlima Bénara
MAYOTTE · 660
France
13°
Mozambique
Channel

The island of Mayotte did not join
the other Comoros islands in independence
and is still administered by France.

0 KILOMETERS 50
0 STATUTE MILES 50

43° · 44°E · 45°

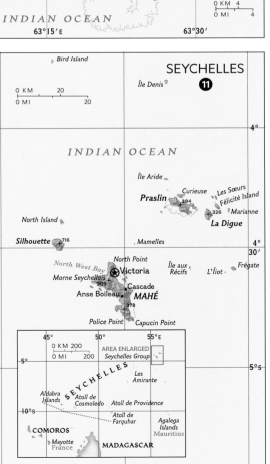

SEYCHELLES
11

Bird Island
0 KM 20
0 MI 20
Île Denis

4°
INDIAN OCEAN

Île Aride
Curieuse · *Les Sœurs*
Praslin · 384 · *Félicité Island*
326 · *Marianne*
North Island · *La Digue*
4°30'
Silhouette · 716 · *Mamelles*

North West Bay · *North Point*
⊛ **Victoria** · *Île aux* · *L'Îlot* · *Frégate*
Morne Seychellois · *Récifs*
905 · Cascade
Anse Boileau · *MAHÉ*
278
Police Point · *Capucin Point*

45° · 50° · 55°E
0 KM 200
0 MI 200
AREA ENLARGED
Seychelles Group
5°
SEYCHELLES
Les Amirante
Aldabra · *Atoll de*
Islands · *Cosmoledo* · *Atoll de Providence*
10°s · *Atoll de Farquhar* · *Agalega Islands* Mauritius
COMOROS
Mayotte · **MADAGASCAR**
France

55° · 55°30'E · 56°

Sao Tome and Principe
DEM. REP. OF SAO TOME AND PRINCIPE

AREA	964 sq km (372 sq mi)
POPULATION	204,000
CAPITAL	São Tomé 80,000
RELIGION	Roman Catholic, Protestant
LANGUAGE	Portuguese, Forro

Seychelles
REPUBLIC OF SEYCHELLES

AREA	455 sq km (176 sq mi)
POPULATION	95,000
CAPITAL	Victoria 28,000
RELIGION	Roman Catholic, Protestant, Hindu, Muslim
LANGUAGE	Seychellois Creole, English, French

St. Helena, Ascension, and Tristan da Cunha
(U.K.)
BRITISH OVERSEAS TERRITORY

AREA	308 sq km (119 sq mi)
POPULATION	8,000
CAPITAL	Jamestown 1,000
RELIGION	Protestant, Roman Catholic
LANGUAGE	English

AUSTRALIA & OCEANIA

AUSTRALIA PHYSICAL AND POLITICAL
AUSTRALIA, PAPUA NEW GUINEA, NEW ZEALAND
OCEANIA ● ISLANDS OF THE PACIFIC

The harbor and skyline of Sydney, Australia

VAST & VARIED REALM OF ISLANDS

HOME OF SEAFARING VOYAGERS & SINGULAR LIFE-FORMS

This region spreads out across the Pacific Ocean, a giant body of water that defines an immense area justly called Oceania. It includes Australasia and the island nations, atolls, and volcanoes of Melanesia, Micronesia, and Polynesia. With its distinctive animals and heritage from seafaring voyagers, Oceania's natural and human history stand apart.

GEOGRAPHY

DISTINCTIVE LANDS LARGE & SMALL

Oceania is much more than ocean. Eastern New Guinea is covered with rainforest; New Zealand has glaciers; and outlying islands are volcanic or coral atolls. The region's dominant landmass is Australia, the world's smallest, lowest, flattest, and—apart from Antarctica—driest continent. As the result of its millions of years of geographic isolation, 87 percent of the mammals in Australia, 93 percent of its reptiles, and 45 percent of its bird species are found only on that continent. This includes kangaroos, koalas, wombats, kookaburras, and emus. More than 1,500 species of fish are found in the Great Barrier Reef on Australia's northeast coast. Across the Tasman Sea are the islands of New Zealand. They have a cooler and more temperate climate, and their own array of unique flora and fauna.

With limited land and small or no human populations, the outer realms of Oceania have, for most of history, been isolated from more settled parts of the world. Palikir, capital of the Federated States of Micronesia, is a nine-hour flight from Honolulu; the 33 coral islands of Kiribati cover an expanse larger than India. These and other low-lying atolls are also vulnerable to rising seas. The Tarawa atoll, capital of Kiribati, is just a few meters above sea level, and it will likely become uninhabitable within a generation. The first animal extinction credited to climate change was the rodent *Melomys rubicola*, which lived on a single island off Australia, Bramble Cay.

Carved stone monoliths known as moai on the Polynesian Easter Island

HISTORY

CONVICTS, CANOES & A PUSH FOR CIVIL RIGHTS

The great seafaring navigators of Polynesia and Micronesia took part in the last phase of humankind's settlement of the globe, moving across the Pacific as early as 1500 B.C. Their particular genius was the development of navigational skills and canoe technology, which allowed them to travel among islands spread across thousands of kilometers of open ocean. Without compasses or sextants, they sailed the seas hundreds of years before Europeans, relying on the stars and other clues from nature. The more diverse Melanesians fished along the coasts and practiced horticulture farther inland.

Australia, at the time of Captain James Cook's 1770 visit, had been a home for at least 40,000 years to 500 Aboriginal tribes, which were largely isolated from outside influences. Using Cook's maps, Britain began colonizing the east coast in 1788 as a penal colony to relieve overcrowded English prisons. New Zealand was not populated until Polynesian Maori sailors arrived around 1300. European explorers arrived about 350 years later, and ultimately Great Britain took formal control of the land by treaty in 1840, although the Maori to this day maintain sacred lands and buildings, and Maori and English are both official languages. Australia's Aboriginal people, meanwhile, have faced more than two centuries of lost land, brutalization, and discrimination. Since the 1950s, an Aboriginal movement has pressed for full citizenship and improved education.

Lupines flank the Eglinton River in New Zealand.

CULTURE

RICH TRADITIONS, ISLAND TO ISLAND

Starting around 1200, the same time Paris's Notre-Dame was built, inhabitants of Pohnpei in Micronesia built the grand ceremonial site of Nan Madol atop an offshore reef. Indeed, indigenous communities across Oceania have nurtured distinctive languages, social structures, and observances, which persist despite widespread colonization by powers from the West. Papua New Guinea is the world's most linguistically diverse nation, with almost 850 languages spoken within its borders. New Zealand's Maori art is highly valued, and is seen in intricate architectural carvings and graceful whale bone and jade amulets.

The early Aboriginals of Australia were hunters and gatherers whose society was based on a complex network of intricate kinship relationships. A system of beliefs called the Dreaming, still vital to many Aboriginals today, found expression in song, art, and dance. Australia's populace—though still mostly of English, Irish, and Scottish ancestry—has diversified and more than doubled since the end of World War II. Today a city like Melbourne has a cosmopolitan vibe. Immigrants have arrived from Greece, Turkey, Italy, and Lebanon; substantial Asian immigration followed in the 1970s. Although the largest religious groups are Roman Catholic and Anglican, some say sport is the national religion: Australians are famous for cricket, rugby, and swimming.

In Fiji and other islands, livelihoods often revolve around fishing.

ECONOMY

HARVESTING THE LAND & THE SEA

Australia dominates all of Oceania economically as its connections to Asia and the rest of the Pacific Rim continue to grow. Mining, food processing, manufacturing, and service industries make up the bulk of the Australian economy. China and Japan are Australia's leading trade partners, and the Department of Education promotes teaching Japanese, Chinese, and Korean in Australian schools. The standard of living is high, and people have considerable leisure time—a sign for Australians of a good life.

Air transport and modern communications have shrunk distances across Oceania, enabling more exports and imports. Fiji's biggest export is bottled water, mainly sent to the United States, and its capital Suva is a major regional economic hub. New Caledonia is a leading exporter of nickel. Commercial fisheries are a leading sector in Australia and New Zealand. Harvests are small in more remote islands where subsistence fishing is part of the informal economy and livelihoods may also rely on logging and small-scale farming of such crops as banana, coconut, and sugarcane. The extraction of copper, gold, and oil brings economic security to Papua New Guinea, but that country's large-scale mining operations and palm oil production have also caused considerable environmental damage and now threaten endangered animals like orangutans.

Aboriginal musicians in Queensland, Australia

NEW ZEALAND'S PARLIAMENT HAS DECLARED THE WHANGANUI RIVER, **SACRED TO MAORI TRIBES** OF THE NORTH ISLAND, TO HAVE **"ALL THE RIGHTS, POWERS, DUTIES, AND LIABILITIES"** OF A PERSON.

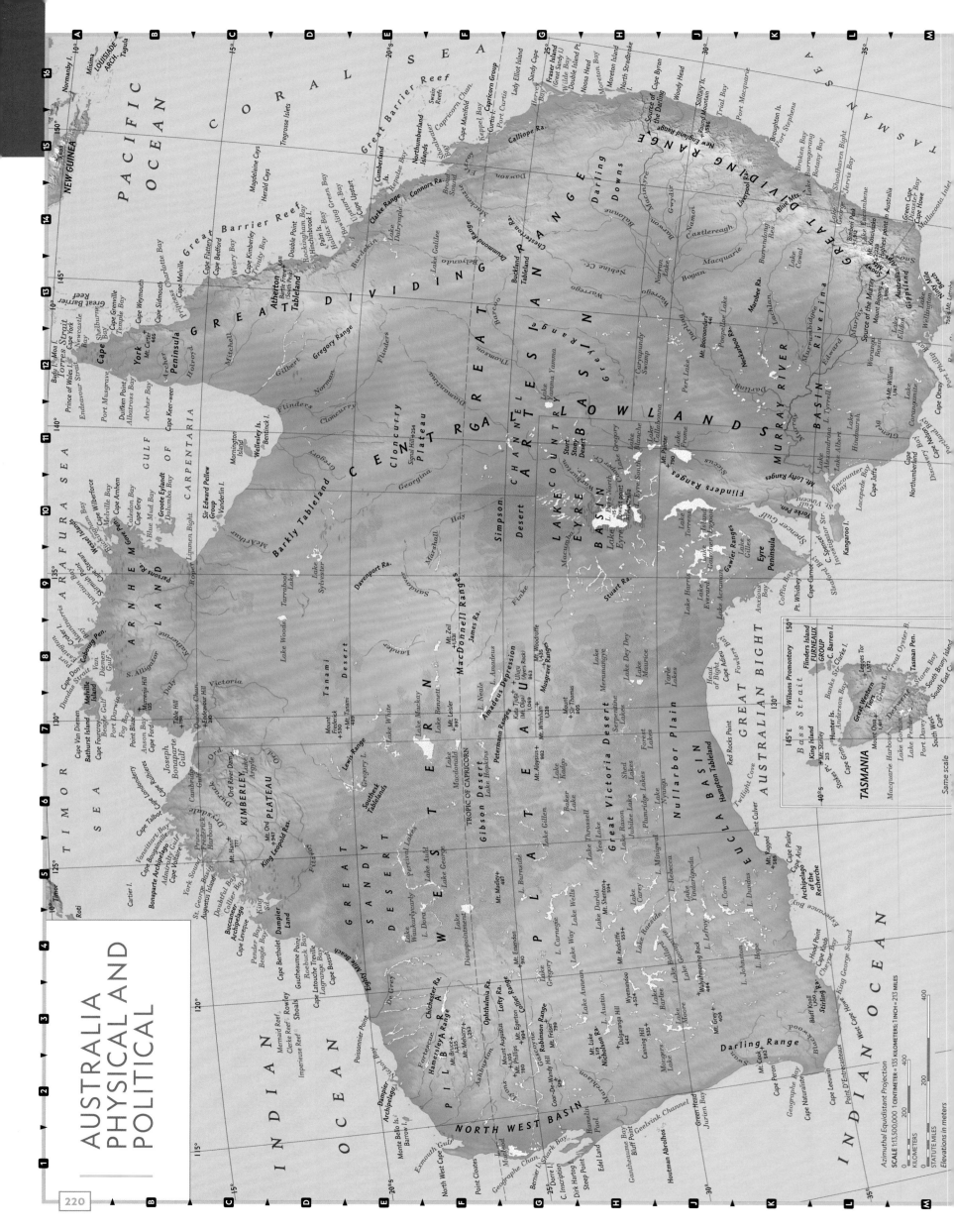

AUSTRALIA
PHYSICAL AND
POLITICAL

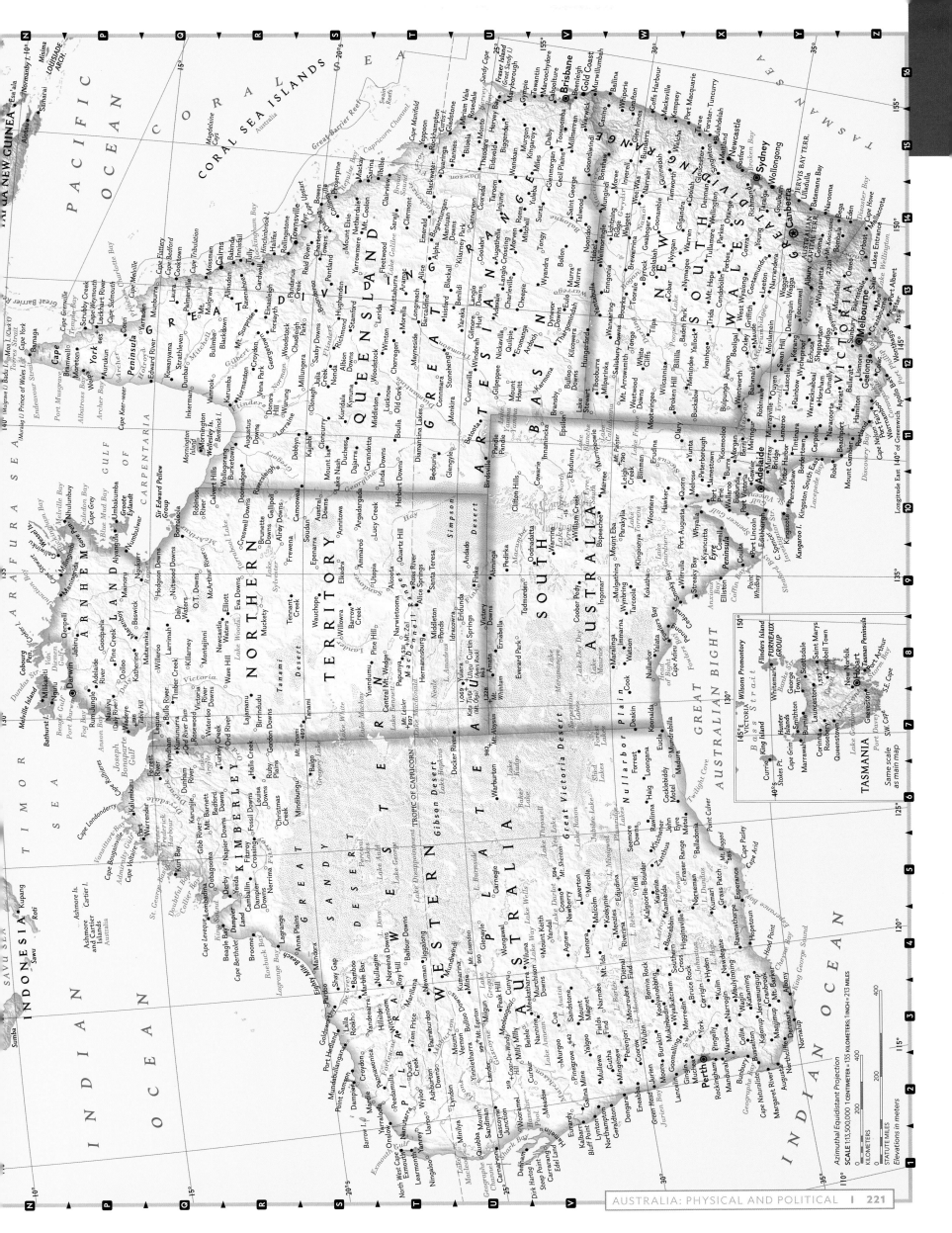

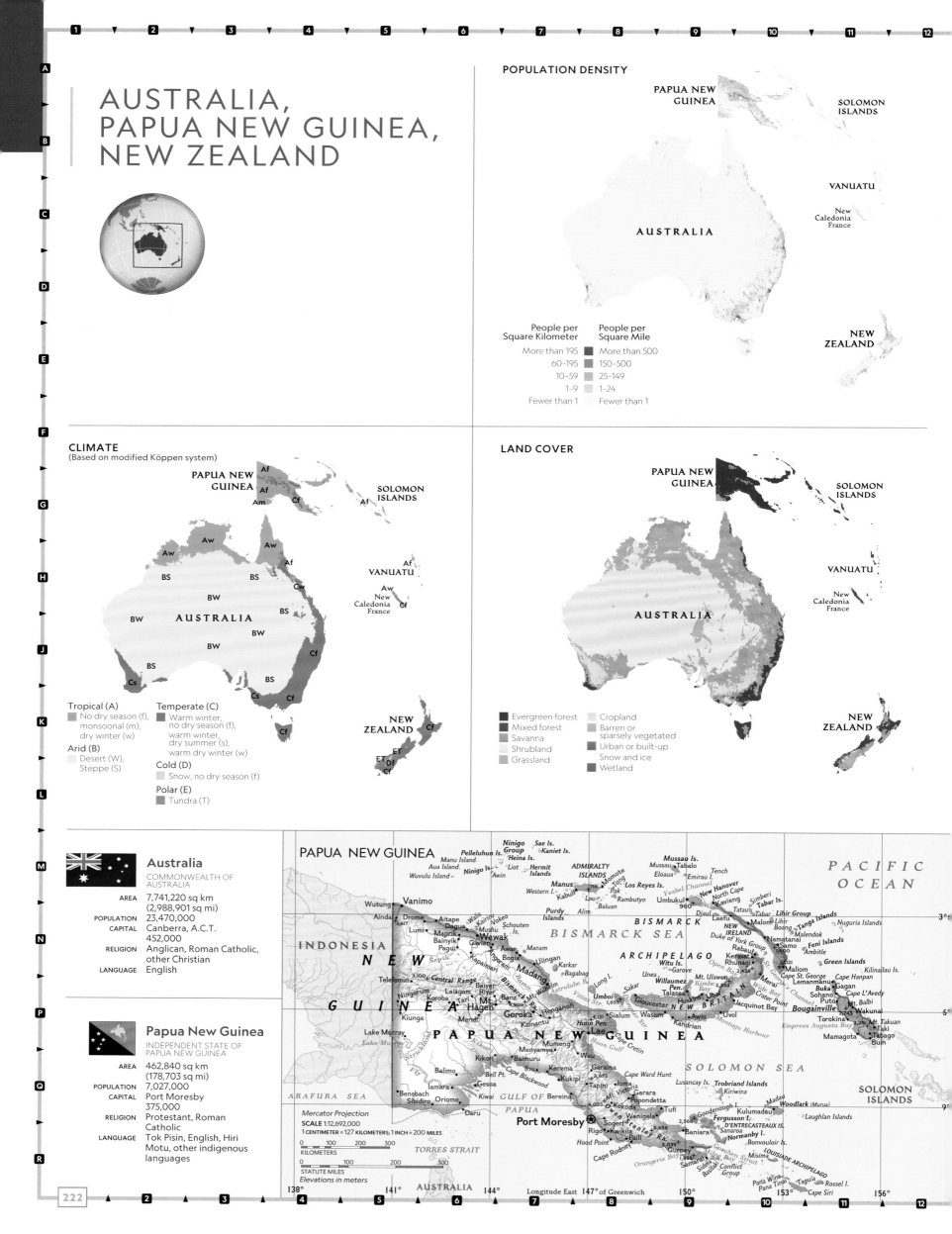

AUSTRALIA, PAPUA NEW GUINEA, NEW ZEALAND

POPULATION DENSITY

PAPUA NEW GUINEA

SOLOMON ISLANDS

AUSTRALIA

VANUATU

New Caledonia France

NEW ZEALAND

People per Square Kilometer	People per Square Mile
More than 195	More than 500
60–195	150–500
10–59	25–149
1–9	1–24
Fewer than 1	Fewer than 1

CLIMATE
(Based on modified Köppen system)

PAPUA NEW GUINEA — Af, Af, Am, Cf

SOLOMON ISLANDS — Af

Aw, Aw, Af, Cw

VANUATU — Af

Aw New Caledonia France — Cf

BS, BS, BW, BS, BW, BW, Cf, BS, BS, Cs, Cf

AUSTRALIA

Cs

NEW ZEALAND — Cf, Cf

Tropical (A)
No dry season (f), monsoonal (m), dry winter (w)

Arid (B)
Desert (W), Steppe (S)

Temperate (C)
Warm winter, no dry season (f), warm winter, dry summer (s), warm dry winter (w)

Cold (D)
Snow, no dry season (f)

Polar (E)
Tundra (T)

LAND COVER

PAPUA NEW GUINEA

SOLOMON ISLANDS

AUSTRALIA

VANUATU

New Caledonia France

NEW ZEALAND

Evergreen forest	Cropland
Mixed forest	Barren or sparsely vegetated
Savanna	Urban or built-up
Shrubland	Snow and ice
Grassland	Wetland

Australia
COMMONWEALTH OF AUSTRALIA

AREA 7,741,220 sq km (2,988,901 sq mi)
POPULATION 23,470,000
CAPITAL Canberra, A.C.T. 452,000
RELIGION Anglican, Roman Catholic, other Christian
LANGUAGE English

Papua New Guinea
INDEPENDENT STATE OF PAPUA NEW GUINEA

AREA 462,840 sq km (178,703 sq mi)
POPULATION 7,027,000
CAPITAL Port Moresby 375,000
RELIGION Protestant, Roman Catholic
LANGUAGE Tok Pisin, English, Hiri Motu, other indigenous languages

PAPUA NEW GUINEA

PACIFIC OCEAN

Ninigo Group, Sae Is., Kaniet Is., Pelleluhun Is., Mussau Is., Manu Island, Heina Is., Hermit Islands, ADMIRALTY ISLANDS, Mussau, Tabalo, Tench, Aua Island, Liot, Wuvulu Island, Ninigo Is., Awin, Western I., Kabuli, Manus, Los Reyes Is., Tong, Momote, Emirau I., Eloaua, New Hanover, North Cape, Simberi, Tabar Is., Lou, Rambutyo, Umbukul, Kavieng, Tatau, Lihir Group, Baluan, Djaul, Laefu, Boang, Tanga Islands, Nuguria Islands

Wutung, Vanimo, Aitape, Walis, Nokeo, Purdy Islands, Alim, BISMARCK SEA, NEW IRELAND, Duke of York Group, Rabaul, Malendok, Feni Islands, Ambitle, Green Islands, Kilinailau Is.

Aindai, Drome, Lumi, Dagua, Mushu, Schouten Is., Manam, ARCHIPELAGO, Witu Is., Keravat, Rhunagi, Samo, Cape St. George, Cape Hanpan

INDONESIA, Bainyik, Pagui, Gavien, Awar, Bogia, Ulingan, Karkar, Long I., Unea, Williamez Pen., Mt. Ulawun, Merai, Buka, Gagan, Cape L'Avedy

NEW, Sepik, Angoram, Ramu, Bismarck, Madang, Sakar, Umboi, Talasea, NEW BRITAIN, Sohano, Puto, Mt. Balbi, Wakunai

GUINEA, Telefomin, Central Range, Laiagam, Baiyer River, Mt. Wilhelm, Astrolabe Bay, Vitiaz Str., Gloucester, Crater Point, Bougainville, Mt. Takuan, Taki

Kiunga, Ningerum, Hagen, Goroka, Kainantu, Sialum, Wasum, Kandrian, Montagu Harbour, Empress Augusta Bay, Mamagota, Buin

Lake Murray, Mendi, Mt. Giluwe, Henganofi, Huon Pen., Lae, Huon Gulf, Cape Cretin, SOLOMON SEA

PAPUA NEW GUINEA, Kikori, Mumeng, Wau, Garaina, Lusancay Is., Trobriand Islands, Kiriwina, SOLOMON ISLANDS

Balimo, Kikori, Baimuru, Kerema, Cape Ward Hunt, Ioma, Woodlark (Murua), Laughlan Islands

Iamara, Gesoa, Bell Pt., Ihu, Tapini, Garara, Popondetta, Kulumadau, Madau

Bensbach, Sibidiro, Oriomo, Kiwi, GULF OF Bereina, Tufi, Wanigela, Goodenough I., Fergusson I., Kokoda, Sanaroa

Daru, Kwikila, Baniara, D'ENTRECASTEAUX IS., Normanby I.

ARAFURA SEA, PAPUA, Port Moresby, Rigo, Baili, Sogeri, Bonvouloir Is., Misima, Taguia

Hood Point, Cape Rodney, Gurney, Samarai, Sideia, LOUISIADE ARCHIPELAGO, Rossel I., Cape Siri

TORRES STRAIT, Orangerie Bay, Goschen Bay, Baslaki, Conflict Group, Pana Wina, Pana Tinai, Tagula

AUSTRALIA

Mercator Projection
SCALE 1:12,692,000
1 CENTIMETER = 127 KILOMETERS; 1 INCH = 200 MILES

KILOMETERS 0 100 200 300
STATUTE MILES 0 100 200 300
Elevations in meters

BISMARCK SEA

Longitude East 147° of Greenwich

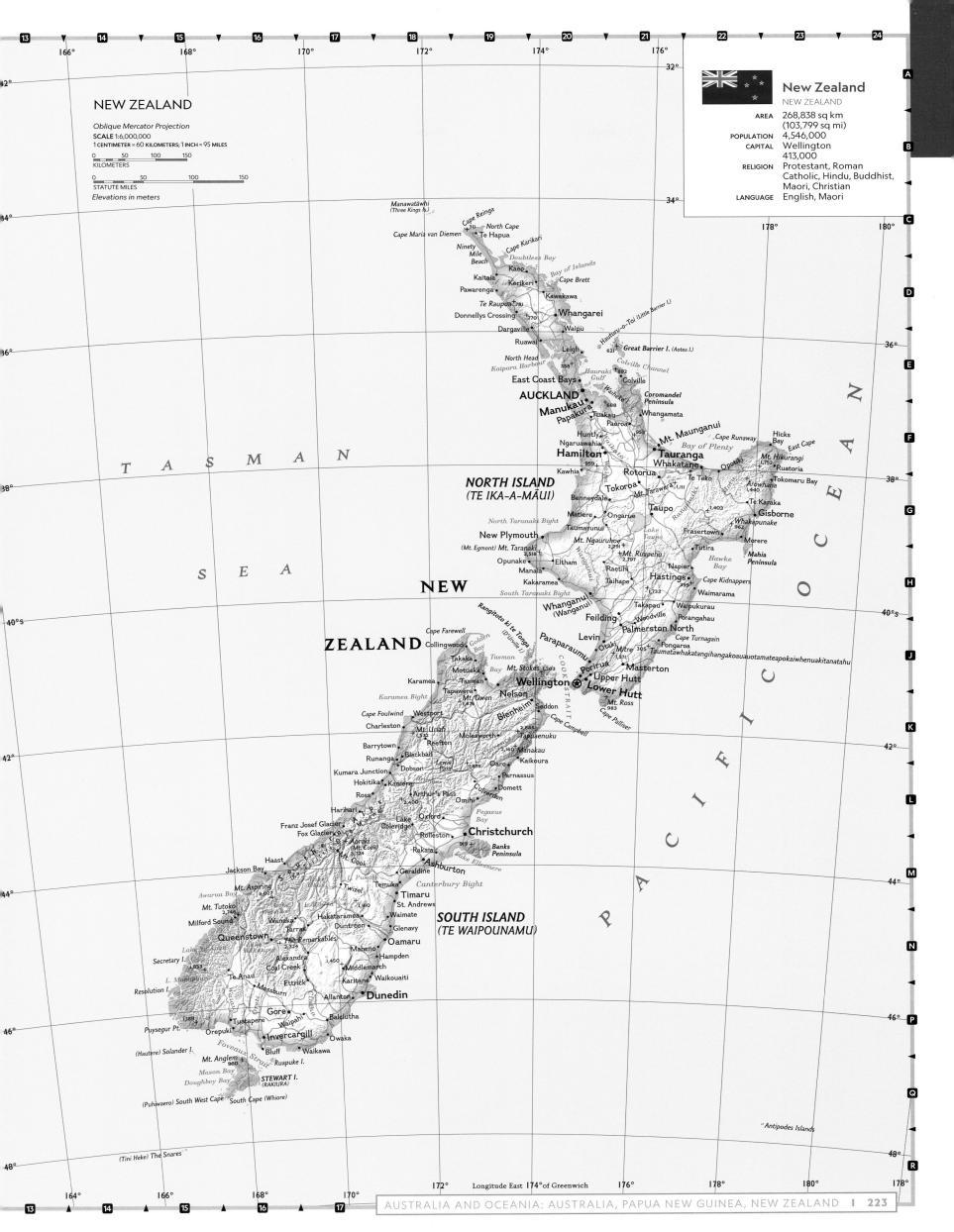

NEW ZEALAND

Oblique Mercator Projection
SCALE 1:6,000,000
1 CENTIMETER = 60 KILOMETERS; 1 INCH = 95 MILES

0   50   100   150
KILOMETERS

0   50   100   150
STATUTE MILES

Elevations in meters

New Zealand
NEW ZEALAND

AREA	268,838 sq km (103,799 sq mi)
POPULATION	4,546,000
CAPITAL	Wellington 413,000
RELIGION	Protestant, Roman Catholic, Hindu, Buddhist, Maori, Christian
LANGUAGE	English, Maori

TASMAN SEA

Manawatāwhi (Three Kings Is.)

Cape Reinga
North Cape
Cape Maria van Diemen
Te Hapua
Cape Karikari
Ninety Mile Beach
Doubtless Bay
Kaeo
Bay of Islands
Kaitaia
Kerikeri
Cape Brett
Pawarenga
Kawakawa
Te Raupua
Whangarei
Donnellys Crossing
Waipu
Dargaville
Hauturu-o-Toi (Little Barrier I.)
Ruawai
Leigh
Great Barrier I. (Aotea I.)
North Head
Kaipara Harbour
East Coast Bays
Hauraki Gulf
Colville
Waiheke I.
Colville Channel
AUCKLAND
Coromandel Peninsula
Manukau
Papakura
Tuakau
Whangamata
Paeroa
Huntly
Mt. Maunganui
Cape Runaway
Hicks Bay
Ngaruawahia
East Cape
Hamilton
Tauranga
Whakatane
Mt. Hikurangi
Kawhia
Rotorua
Opotiki
Ruatoria

NORTH ISLAND (TE IKA-A-MĀUI)

Tokoroa
Te Teko
Tokomaru Bay
Te Tarawera
Arowhana
Benneydale
Te Karaka
Matiere
Ongarue
Taupo
Gisborne
New Plymouth
Taumarunui
Mt. Ngauruhoe
Lake Taupo
Frasertown
Whakapunake
(Mt. Egmont) Mt. Taranaki
Opunake
Mt. Ruapehu
Morere
Eltham
Raetihi
Napier
Tutira
Manaia
Taihape
Hastings
Mahia Peninsula
Kakaramea
Cape Kidnappers
Hawke Bay

NEW

South Taranaki Bight
Whanganui (Wanganui)
Waimarama
Takapau
Feilding
Waipukurau
Porangahau
Levin
Palmerston North
Cape Turnagain
Pongaroa
Taumatawhakatangihangakoauauotamateapokaiwhenuakitanatahu

ZEALAND

Cape Farewell
Collingwood
Golden Bay
Rangitoto ki te Tonga (D'Urville I.)
Paraparaumu
Masterton
Takaka
Tasman Bay
Mt. Stokes
Porirua
Upper Hutt
Motueka
Wellington
Lower Hutt
Karamea
Tasman
Tapawera
Nelson
Mt. Owen
Mt. Ross
Karamea Bight
Seddon
Cape Palliser
Cape Foulwind
Westport
Blenheim
Charleston
Cape Campbell
Mt. Uriah
Molesworth
Tapuaenuku
Barrytown
Reefton
Runanga
Blackball
Manakau
Kaikoura
Kumara Junction
Dobson
Lewis Pass
Oaro
Hokitika
Kaniere
Parnassus
Ross
Arthur's Pass
Domett
Harihari
Culverden
Omihi
Pegasus Bay
Lake Coleridge
Oxford
Franz Josef Glacier
Rolleston
Fox Glacier
Aoraki (Mt. Cook)
Rakaia
Banks Peninsula
Christchurch
Haast
Cook
Ashburton
Lake Ellesmere
Jackson Bay
Geraldine
Temuka
Canterbury Bight
Mt. Aspiring
Twizel
Mt. Tutoko
Timaru
St. Andrews

SOUTH ISLAND (TE WAIPOUNAMU)

Milford Sound
Wanaka
Hakataramea
Waimate
Tarras
Duntroon
Glenavy
Queenstown
The Remarkables
Oamaru
Secretary I.
Alexandra
Maheno
Hampden
Coal Creek
Middlemarch
L. Manapouri
Te Anau
Ettrick
Karitane
Waikouaiti
Resolution I.
Messburn
Allanton
Dunedin
Gore
Balclutha
Tuatapere
Puysegur Pt.
Waipahi
Owaka
Orepuki
(Hautere) Solander I.
Invercargill
Owaka
Bluff
Waikawa
Mt. Anglem
Ruapuke I.
Mason Bay
Doughboy Bay
STEWART I. (RAKIURA)
(Puhiwaero) South West Cape   South Cape (Whiore)

TASMAN SEA

PACIFIC OCEAN

Antipodes Islands

(Tini Heke) The Snares

All map elevations in meters

Fiji
REPUBLIC OF FIJI

AREA 18,274 sq km (7,056 sq mi)
POPULATION 927,000
CAPITAL Suva 178,000
RELIGION Protestant, Hindu, Roman Catholic, Muslim
LANGUAGE English, Fijian, Hindustani

Kiribati
REPUBLIC OF KIRIBATI

AREA 811 sq km (313 sq mi)
POPULATION 109,000
CAPITAL Tarawa 64,000
RELIGION Roman Catholic, Protestant
LANGUAGE I-Kiribati, English

Marshall Islands
REPUBLIC OF THE MARSHALL ISLANDS

AREA 181 sq km (70 sq mi)
POPULATION 76,000
CAPITAL Majuro 31,000
RELIGION Protestant, Roman Catholic, Morman
LANGUAGE Marshallese, English

Micronesia
FEDERATED STATES OF MICRONESIA

AREA 702 sq km (271 sq mi)
POPULATION 104,000
CAPITAL Palikir 7,000
RELIGION Roman Catholic, Protestant
LANGUAGE English, Chuukese, Kosrean, Pohnpeian, Yapese, other indigenous languages

Nauru
REPUBLIC OF NAURU

AREA 21 sq km (8 sq mi)
POPULATION 10,000
CAPITAL Yaren (not official) 1,000
RELIGION Protestant, Roman Catholic
LANGUAGE Nauruan, English

Palau
REPUBLIC OF PALAU

AREA 459 sq km (177 sq mi)
POPULATION 22,000
CAPITAL Ngerulmud 277
RELIGION Roman Catholic, Protestant, Modekngei
LANGUAGE Palauan, English, Filipino

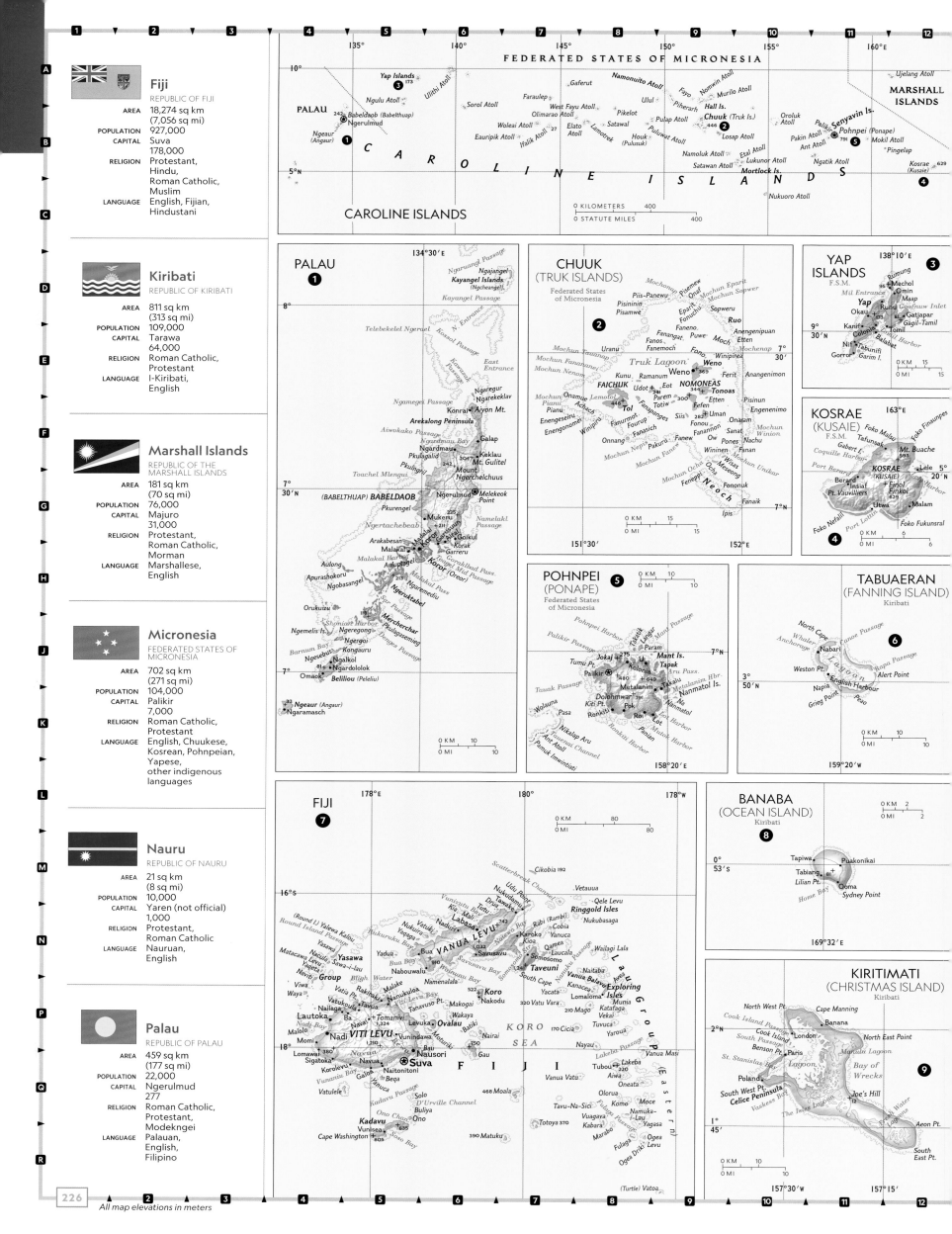

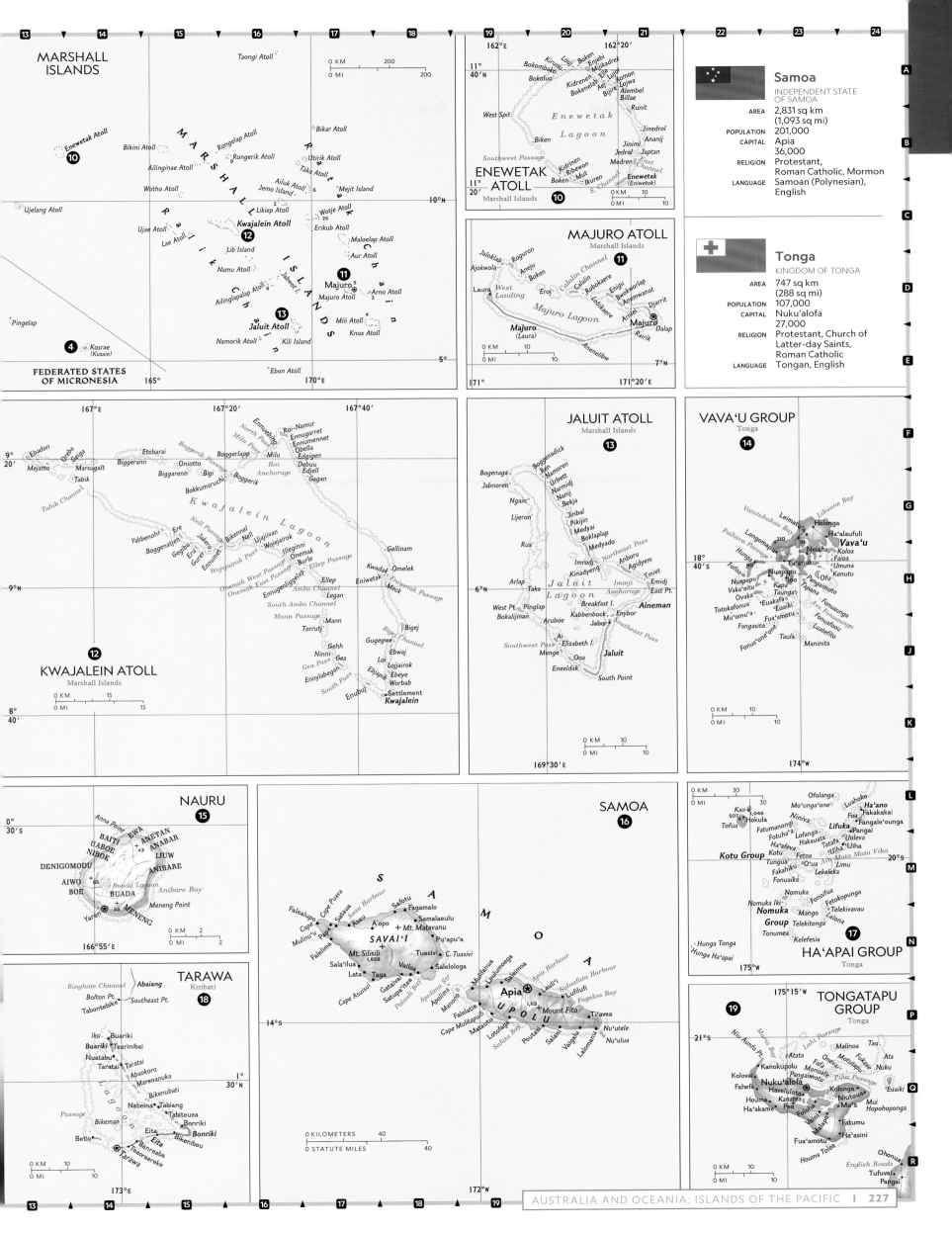

MARSHALL ISLANDS

FEDERATED STATES OF MICRONESIA

ENEWETAK ATOLL
Marshall Islands

MAJURO ATOLL
Marshall Islands

KWAJALEIN ATOLL
Marshall Islands

JALUIT ATOLL
Marshall Islands

VAVA'U GROUP
Tonga

NAURU

TARAWA
Kiribati

SAMOA

HA'APAI GROUP
Tonga

TONGATAPU GROUP
Tonga

Samoa
INDEPENDENT STATE OF SAMOA

AREA 2,831 sq km (1,093 sq mi)
POPULATION 201,000
CAPITAL Apia 36,000
RELIGION Protestant, Roman Catholic, Mormon
LANGUAGE Samoan (Polynesian), English

Tonga
KINGDOM OF TONGA

AREA 747 sq km (288 sq mi)
POPULATION 107,000
CAPITAL Nuku'alofa 27,000
RELIGION Protestant, Church of Latter-day Saints, Roman Catholic
LANGUAGE Tongan, English

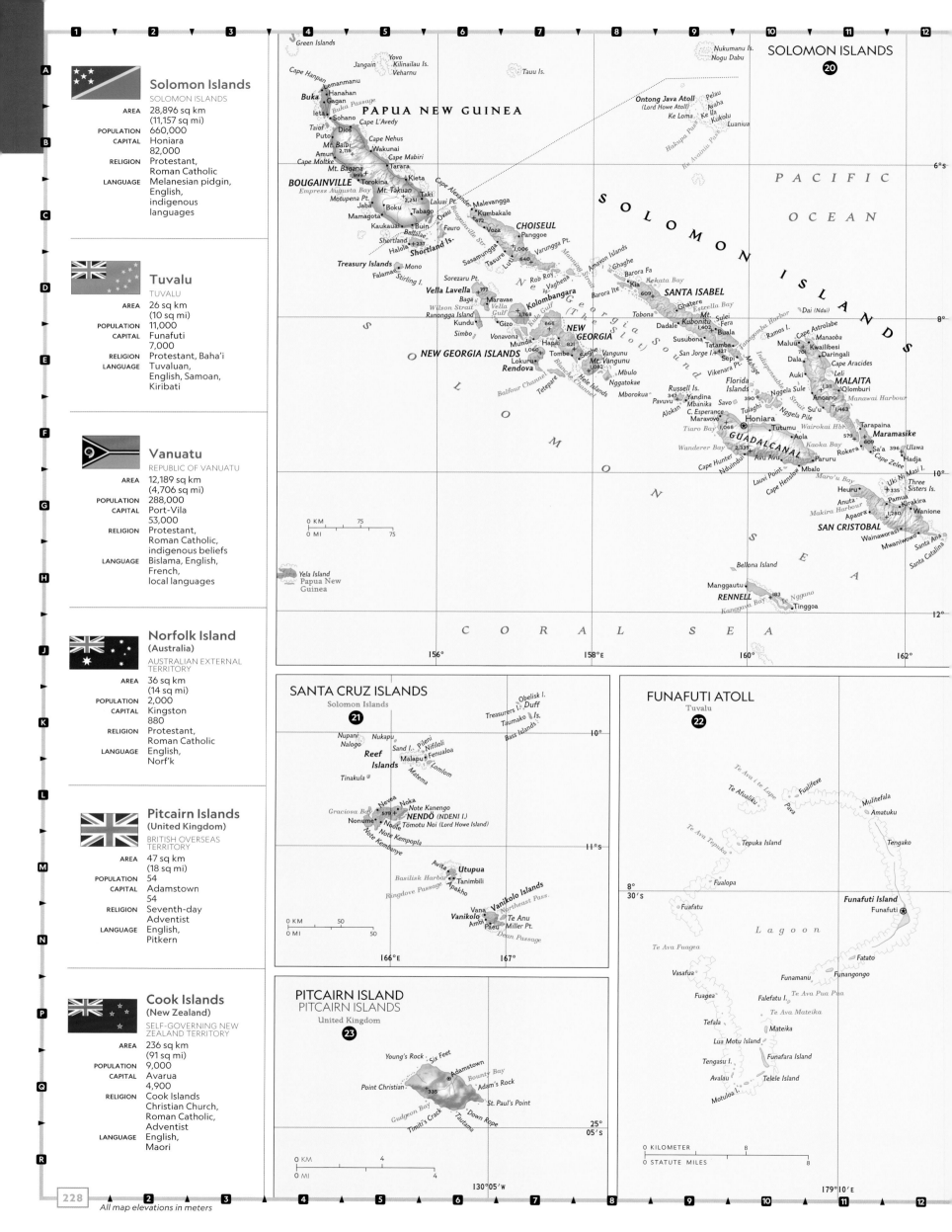

Solomon Islands
SOLOMON ISLANDS

AREA	28,896 sq km (11,157 sq mi)
POPULATION	660,000
CAPITAL	Honiara 82,000
RELIGION	Protestant, Roman Catholic
LANGUAGE	Melanesian pidgin, English, indigenous languages

Tuvalu
TUVALU

AREA	26 sq km (10 sq mi)
POPULATION	11,000
CAPITAL	Funafuti 7,000
RELIGION	Protestant, Baha'i
LANGUAGE	Tuvaluan, English, Samoan, Kiribati

Vanuatu
REPUBLIC OF VANUATU

AREA	12,189 sq km (4,706 sq mi)
POPULATION	288,000
CAPITAL	Port-Vila 53,000
RELIGION	Protestant, Roman Catholic, indigenous beliefs
LANGUAGE	Bislama, English, French, local languages

Norfolk Island
(Australia)
AUSTRALIAN EXTERNAL TERRITORY

AREA	36 sq km (14 sq mi)
POPULATION	2,000
CAPITAL	Kingston 880
RELIGION	Protestant, Roman Catholic
LANGUAGE	English, Norf'k

Pitcairn Islands
(United Kingdom)
BRITISH OVERSEAS TERRITORY

AREA	47 sq km (18 sq mi)
POPULATION	54
CAPITAL	Adamstown 54
RELIGION	Seventh-day Adventist
LANGUAGE	English, Pitkern

Cook Islands
(New Zealand)
SELF-GOVERNING NEW ZEALAND TERRITORY

AREA	236 sq km (91 sq mi)
POPULATION	9,000
CAPITAL	Avarua 4,900
RELIGION	Cook Islands Christian Church, Roman Catholic, Adventist
LANGUAGE	English, Maori

All map elevations in meters

SOLOMON ISLANDS 20

SANTA CRUZ ISLANDS 21
Solomon Islands

FUNAFUTI ATOLL 22
Tuvalu

PITCAIRN ISLAND 23
PITCAIRN ISLANDS
United Kingdom

VANUATU
24

Niue
(New Zealand)
SELF-GOVERNING NEW ZEALAND TERRITORY
AREA	260 sq km (100 sq mi)
POPULATION	2,000
CAPITAL	Alofi 1,000
RELIGION	Christian
LANGUAGE	Niuean, English

NORFOLK ISLAND
Australia
25

Point Vincent · Point Howe · Duncombe Bay
Anson Pt. · Mt. Bates · Cascade Bay
Anson Bay · 319 · Cascade
Burnt Pine · Steels Point
Rocky Point · Middlegate
Point Ross · Collins Head
Sydney Bay · Kingston · Nepean
Ball Bay
High Red Rock
278 · Philip Island

168°E
29°S

PUKAPUKA ATOLL
COOK ISLANDS
N.Z.
26

Yato · Roto
Passage · Ngake
Pukapuka
Toka San Cay
Motu Kotawa
Motu Ko

10°55′S
165°50′W

RAROTONGA
COOK ISLANDS
New Zealand
27

Teiti Point · Avatiu Harbour
Avatiu · Avarua Harbour
Nikao · Avarua
Arorangi · Matavera
Te Manga · Ngatangiia
653 · Muri · Motutapu
Koromiri
Oneroa
Tikioki · Titikaveka
Rutaki Passage · Papua Passage
Avaavaroa Passage

21°10′S
159°50′W

AITUTAKI ATOLL
COOK ISLANDS
New Zealand
28

Maungapu · Valpeka
124
Aitutaki · Akitua
Amuri · Angarei
Arutanga · Ee
Arutanga Passage · Vaipae · Mangere
Nikaupara · Papau
Te Koutu Pt. · Tautu · Tavaerua
Akaiami
Lagoon · Muritapua
Maina · Moturakau · Akaiami
Rapota · Tekopua
Tapuaetai · Motukitiu

18°55′S
159°45′W

MANIHIKI ATOLL
COOK ISLANDS
New Zealand
29

North Point
Tukao
Murihiti · Lagoon · Ngake
Tauhunu Landing · Tauhunu
Tauhunu · Atimoono
Te Puka · Hakamaru
Rangahoe · Totia · Haratini
Tikapai · Putangaroa
Motupae · Tarakite Iti
Raukotaha · Porea · Tevahavaha

10°25′S
161°W

PENRHYN
(TONGAREVA)
COOK ISLANDS
New Zealand
30

Siki Rangi Passage · Tapunui · Tokerau
Ruahara · Terae
Matunga · Takuua Passage
Te Tautua · Takuua
Taruia Passage · Tautua
Omoka · Lagoon
Moananui · Patanga
Ahu-a-miria Is.
Mangarongaro
Vaiere · Atutahi

9°S
158°W

NIUE
New Zealand
31

Hikutavake · Mutalau
Namukulu · Toi · Liha Pt.
Tuapa · 223
Makapu Pt. · Makefu · Lakepa
Alofi Bay · Makefu
Alofi · Liku
Fonuakula
Tamakautonga
Avatele Bay · Avatele · Hakupu
Tepa Pt. · Vaiea · Mata Pt.
Limufuafua Point

19°S
170°W

VANUATU area (main map)
Hiu 366 · Métoma 240 · Vétaounde 64
Tégua · Loh 155 · **Torres Islands**
Toga 240
Uréparapara 764 · Reef Islands (Rowa Islands)
Mota Lava 411
Mt. Sürétimëat 921 · Port Patteson
Vanua Lava 946 · Mota
Wasaka
BANKS ISLANDS
Santa Maria 797 · Avire · Mérig
Makéone · Méré Lava
Cape Cumberland · Marino
Olpoi 1,444 · Cape Queiros · **MAÉWO** 181
Nokuku · Lathi · Narovorovo
Malao · Kolé · Nangiré · Lolowai
ESPIRITU SANTO · Nduindui 1,496 · **Aoba**
Wusi · Tabwémasana 1,879 · Nazareth
Luganville · Aöre · Namaram
Cape Mataabé · Malo 326 · **PENTECOST**
Espiègle Bay 526 · Homa Bay · Ranwas
Norsup · Port Stanley
Lakatoro · Selwyn Strait
MALAKULA · Mégham · **AMBRYM**
Mt. Pénot 879 · Marum Volcano 1,270
Lamap · Paama · Lopévi 1,413
South West Bay · Moriu
Tomman 84 · Maskelynes Is. · Votlo · **Epi** 833 · Laïka 87
Tongoa 487
Shepherd Islands · Tongariki 521
Émaé 644 · Makura
494 Mataso I. · Étarik 155
593 Nguna · Émao 448
116 Moso · Émao
202 Lélépa · 647 · **ÉFATÉ**
Devil's Pt. · Mélé Bay · Port-Vila

Mt. Santop 886 · Potnarvin
Dillon's Bay · 802 · Cook Bay
ERROMANGO · Pilbarra Point
Aniwa 42
Lowital · Waïsisi Bay
Tanna · Waïsisi · Futuna 643
Isangel · Yasur Volcano
Mt. Tukosméra 1,084
Port Patrick
Anatom 852
Anelghowhat · Anelghowhat Bay

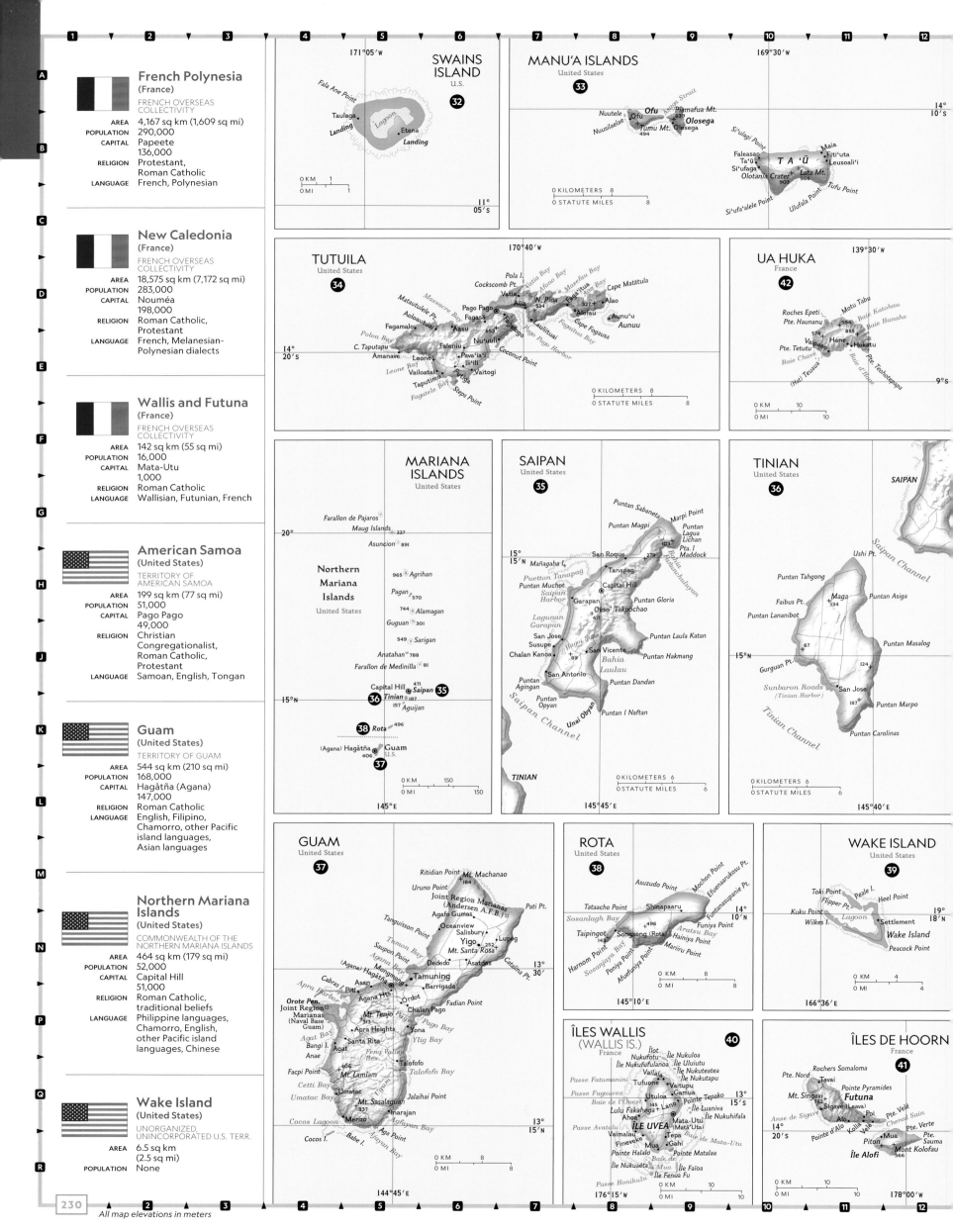

French Polynesia
(France)

FRENCH OVERSEAS COLLECTIVITY

AREA	4,167 sq km (1,609 sq mi)
POPULATION	290,000
CAPITAL	Papeete 136,000
RELIGION	Protestant, Roman Catholic
LANGUAGE	French, Polynesian

New Caledonia
(France)

FRENCH OVERSEAS COLLECTIVITY

AREA	18,575 sq km (7,172 sq mi)
POPULATION	283,000
CAPITAL	Nouméa 198,000
RELIGION	Roman Catholic, Protestant
LANGUAGE	French, Melanesian-Polynesian dialects

Wallis and Futuna
(France)

FRENCH OVERSEAS COLLECTIVITY

AREA	142 sq km (55 sq mi)
POPULATION	16,000
CAPITAL	Mata-Utu 1,000
RELIGION	Roman Catholic
LANGUAGE	Wallisian, Futunian, French

American Samoa
(United States)

TERRITORY OF AMERICAN SAMOA

AREA	199 sq km (77 sq mi)
POPULATION	51,000
CAPITAL	Pago Pago 49,000
RELIGION	Christian Congregationalist, Roman Catholic, Protestant
LANGUAGE	Samoan, English, Tongan

Guam
(United States)

TERRITORY OF GUAM

AREA	544 sq km (210 sq mi)
POPULATION	168,000
CAPITAL	Hagåtña (Agana) 147,000
RELIGION	Roman Catholic
LANGUAGE	English, Filipino, Chamorro, other Pacific island languages, Asian languages

Northern Mariana Islands
(United States)

COMMONWEALTH OF THE NORTHERN MARIANA ISLANDS

AREA	464 sq km (179 sq mi)
POPULATION	52,000
CAPITAL	Capital Hill 51,000
RELIGION	Roman Catholic, traditional beliefs
LANGUAGE	Philippine languages, Chamorro, English, other Pacific island languages, Chinese

Wake Island
(United States)

UNORGANIZED, UNINCORPORATED U.S. TERR.

AREA	6.5 sq km (2.5 sq mi)
POPULATION	None

SWAINS ISLAND
U.S. — 32

MANU'A ISLANDS
United States — 33

TUTUILA
United States — 34

UA HUKA
France — 42

MARIANA ISLANDS
United States

SAIPAN
United States — 35

TINIAN
United States — 36

GUAM
United States — 37

ROTA
United States — 38

WAKE ISLAND
United States — 39

ÎLES WALLIS (WALLIS IS.)
France — 40

ÎLES DE HOORN
France — 41

All map elevations in meters

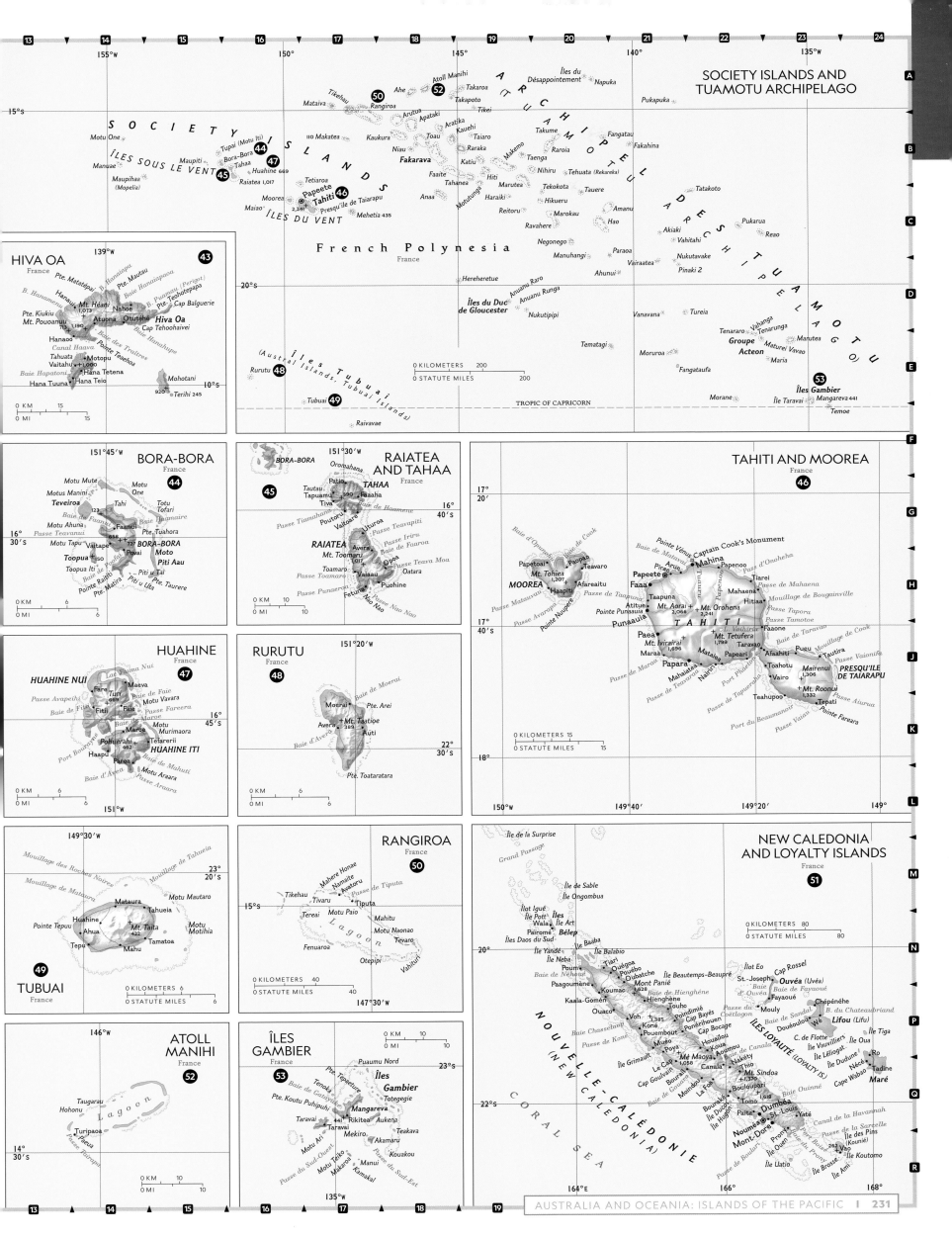

ANTARCTICA

Penguins on a small iceberg in Antarctica

A CONTINENT OF SNOW & ICE

DISCOVERING EARTH'S HISTORY LOCKED IN FROZEN LANDSCAPES

Anchoring the bottom of the world is a frozen land of austere beauty. Antarctica is the driest, highest, coldest, windiest, and least populated of Earth's seven continents. Its seas and ice, largely unvisited by humans, make up a pristine scientific laboratory to help us understand the health of our planet.

GEOGRAPHY

SEVERE CLIMATE, BUT MANY LANDSCAPES

The austral summer brings perpetual light, yet Antarctica stays cold. During the winter darkness, which can last up to six months near the South Pole, terrible storms pound the continent. The landmass itself is larger than Europe or Australia; in winter, it is ringed by a belt of sea ice that extends 18 million square kilometers (7 million square miles) around the continent, more than doubling the surface area. In the spring, melting ice coincides with the calving of huge icebergs from the Antarctic glaciers. Antarctica is a continent of valleys, lakes, islands, and even volcanoes, although many of these features are buried beneath kilometers of ice and invisible on the surface. The variety of terrain was something only dreamed of until a compilation of more than 2.5 million ice-thickness measurements revealed the startling topography below.

The Antarctic Peninsula reaches like a long arm 1,300 kilometers (800 miles) into the Southern Ocean toward the tip of South America. The surrounding seas drive circulation around the continent and contribute to the transfer of energy around the world. Around 55° south latitude, relatively warm waters from several oceans meet to create conditions that support a flourishing marine ecosystem. Seals, whales, and penguins are among the animals that have adapted extremely well to Antarctica's harsh climate.

A mountaineer skis past the wind-whipped Filchner Mountains in Queen Maud Land.

EXPLORATION

JOURNEYS TO THE ENDS OF THE EARTH

Captain James Cook, a British explorer, traveled three times into Antarctic waters between 1772 and 1775, and he was probably the first to cross the Antarctic Circle, the latitude about 66½° south of the Equator that encircles most of the continent. Though he never saw Antarctica, he believed in "a tract of land at the Pole that is the source of all the ice that is spread over this vast Southern Ocean." His observations of marine mammals in great numbers lured whalers and sealers in search of oil and skins. The first recorded sightings of the continent occurred in 1820.

Other expeditions followed. British naval officer James Clark Ross arrived in 1839 and charted the sea that now bears his name. Norwegian Roald Amundsen's 1911 expedition reached the South Pole just weeks before a separate five-man British team, led by Robert Falcon Scott. On the bitter return trip, Scott and his men all died. Ernest Shackleton's 1914–1916 British mission aimed to traverse the entire continent. Stranded when sea ice trapped and crushed their ship, his party spent more than a year on drifting ice. Then they sailed in lifeboats to the tip of the Antarctic Peninsula, where Shackleton and five others continued on another 1,300 kilometers (800 miles) in a small boat to South Georgia island. Shackleton eventually returned to Antarctica and rescued all the men left behind.

Taylor Valley is one of Antarctica's rare "dry valleys" not covered by ice and snow.

ANTARCTIC TREATY

On December 1, 1959, after a decade of secret meetings, 12 nations—Argentina, Australia, Belgium, Chile, France, Japan, New Zealand, Norway, South Africa, the Soviet Union (Russia), the United Kingdom, and the United States—signed the Antarctic Treaty, agreeing to preserve the frozen continent for peaceful scientific use and protect it from territorial sovereignty. The treaty is widely regarded as an unprecedented example of international cooperation, a major diplomatic feat during the height of Cold War rivalries; 54 nations are parties to the treaty today.

Setting forth a shared commitment to the freedom of scientific inquiry, the treaty includes a provision that scientific observations and results from Antarctica "shall be exchanged and made freely available." It also states that countries cannot make any new or expanded territorial claims. The former military and whaling bases were abandoned to be swallowed by the snow as scientists from all over the world erected new research stations and camps. Antarctica is the only continent with no indigenous population, but the ever shifting number of residents and tourists can reach more than 4,500 people in the summer. Various international protocols regulate the impact of tourism on Antarctica's fragile ecosystems—heeding one of the treaty's goals of preserving the continent as a natural reserve.

An American research base at the South Pole

Scientists extract an ice core sample.

CLIMATE

Thousands of researchers flock to the southernmost continent every summer. By studying changes in the seasonal sea-ice accumulation, for instance, they can identify and understand larger issues affecting the health of the whole planet. Among their most startling revelations was the 1985 discovery of a hole in the Earth's protective ozone layer. These findings identified a major ecological threat to the planet, and the global community swiftly responded: A global treaty known as the Montreal Protocol banned all ozone-depleting substances such as chlorofluorocarbons.

Antarctica's thick ice holds a valuable scientific record of the world, similar to the rings of a tree. Researchers drill deep into the sheets to take ice cores, whose lower layers capture what Earth's atmosphere was like hundreds of thousands of years ago. This data allows for comparison with conditions today. They can also record the ebb and flow of sea ice, which helps regulate our climate. Long-term studies showed an increase in Antarctic sea ice from 1979 to 2014, but since then scientists have seen a decrease. Antarctic glaciers, particularly in the west of the continent, have been melting at unprecedented rates. Antarctica's ice cap holds some 70 percent of the planet's freshwater. If the ice sheet were to melt completely, which is not imminent, global seas would rise by an estimated 58 meters (190 feet).

SNAPSHOT · *Antarctica* FACTS & FEATURES

HIGHEST ELEVATION Vinson Massif: 4,897 m (16,067 ft)

AVERAGE TEMPERATURE –60° Celsius (–76° Fahrenheit) in winter; –28° Celsius (–18° Fahrenheit) in summer

LOWEST TEMPERATURE RECORDED East Antarctic Plateau: –98° Celsius (–144° Fahrenheit)

MONTHS WITHOUT SUNRISE Around six months in the southernmost region

COLDEST PLACE ON EARTH Ridge A, Antarctica: annual average temperature –70° Celsius (–94° Fahrenheit)

LARGEST RESEARCH STATION McMurdo Station operated by the U.S., built in 1955

BUSIEST PENGUIN ROOKERY St. Andrew's Bay, South Georgia Is.: 400,000 king penguins

LARGEST ICEBERG RECORDED B-15, spotted in 2000: 11,000 sq km (4,250 sq mi), about the size of Connecticut

---

THE **ANTARCTIC CIRCUMPOLAR CURRENT** IS THE **LARGEST, FASTEST CURRENT** IN THE WORLD, WITH A FLOW **MORE THAN 100 TIMES** THAT OF ALL EARTH'S RIVERS COMBINED.

DYNAMIC ANTARCTICA

Antarctica constantly changes shape. The ice sheet that covers 96 percent of the continent's landmass is in flux: At summer's end, the ice measures about 3 million square kilometers (1.2 million square miles); by winter, it grows to 19 million square kilometers (7.3 million square miles). Many factors—including wind, faraway ocean currents, and water temperatures—determine the seasonal buildup and melting of Antarctica's ice, which means it can vary significantly from year to year. Permanent features such as the Vinson Massif, Antarctica's tallest mountain at 4,897 meters (16,067 feet), define the continent's landscape no matter the season. The size of Antarctica's ice sheet remained more or less stable for thousands of years despite its cyclical, seasonal changes. Over the last century, however, that balance has come undone as human-induced climate change has altered ocean dynamics. Shifting winds and warm ocean phases have eaten away at the ice more quickly than it's being replaced, causing the glaciers of West Antarctica to shrink at a startling rate and contributing to global sea-level rise. The ice shelves fringing the coast, which act like floating dams that hold back the ice, are calving huge icebergs into the ocean.

The remains of the Larsen B ice shelf, which almost entirely collapsed in 2002

A striped iceberg in Bransfield Strait on the Antarctic Peninsula

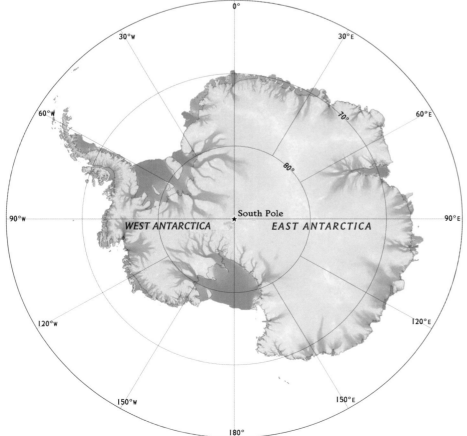

◀ ICE FLOW VELOCITY

Antarctica's massive ice sheets move slowly, but ice streams move faster—up to 10,000 meters (33,000 feet) per year. The streams function almost like frozen rivers draining ice and sediment into the sea. Areas where the underlying bedrock topography causes ice to flow faster are shown in red and dark purple. The fastest ice streams are in West Antarctica.

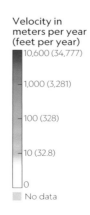

Velocity in
meters per year
(feet per year)

10,600 (34,777)

1,000 (3,281)

100 (328)

10 (32.8)

0

No data

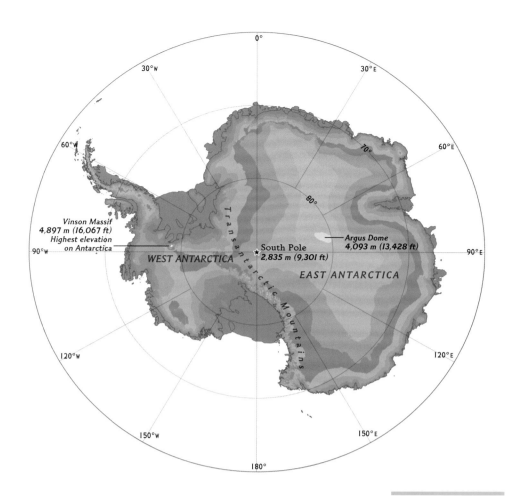

◀ SURFACE ELEVATION

Antarctica is Earth's highest continent, with an average elevation of over 1,800 meters (6,000 feet) above sea level. The peak of East Antarctica's ice sheet, Argus Dome, rivals some of the continent's highest mountain summits, but the elevation is due to the thickness of the ice rather than underlying bedrock.

Surface Elevation
in meters (feet)

4,897 (16,067)
4,000 (13,123)
3,000 (9,843)
2,000 (6,562)
1,000 (3,281)
0

ANTARCTICA'S ICE CAP IS A HEAVY ICE SHEET ALMOST 4.8 KM (3 MILES) THICK IN PLACES, AND THE WEIGHT DEPRESSES THE SOUTH POLE TO CREATE A SLIGHTLY PEAR-SHAPED EARTH.

▶ ICE SHEET THICKNESS

Snow and ice have accumulated on Antarctica for millions of years. Countless layers of summer ice and winter snow, compressed over time, create a striped appearance on glaciers. Ice is generally thinner at the continent's edges, where ice that has flowed from the interior returns to the ocean as icebergs or by melting into the sea. The red line on the map (right) connects prominent Antarctic features shown in the cross section below.

Ice Sheet Thickness
in meters (feet)

4,755 (15,600)
4,000 (13,123)
3,000 (9,843)
2,000 (6,562)
1,000 (3,281)
0

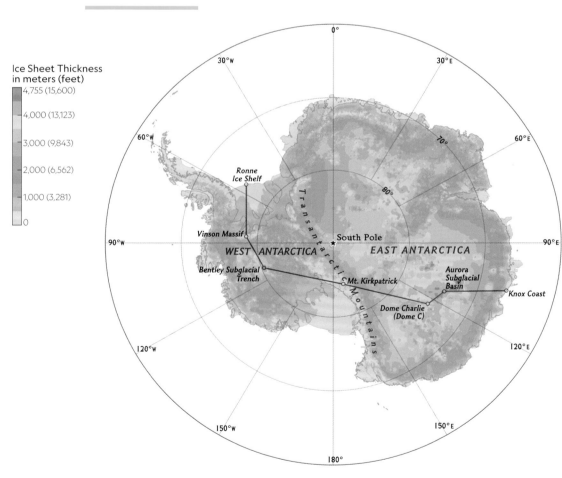

▼ CROSS SECTION

This cross section gives a sense of the volume of Antarctica's ice sheet (in blue), which is difficult to see on a flat map. It also shows how much of the continent (in brown) is below sea level, pushed down by the weight of the ice. The continent's major mountain ranges are represented as sharp spikes poking through the thick sheet of ice.

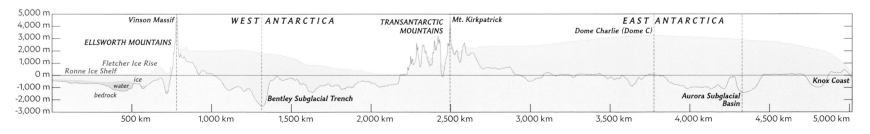

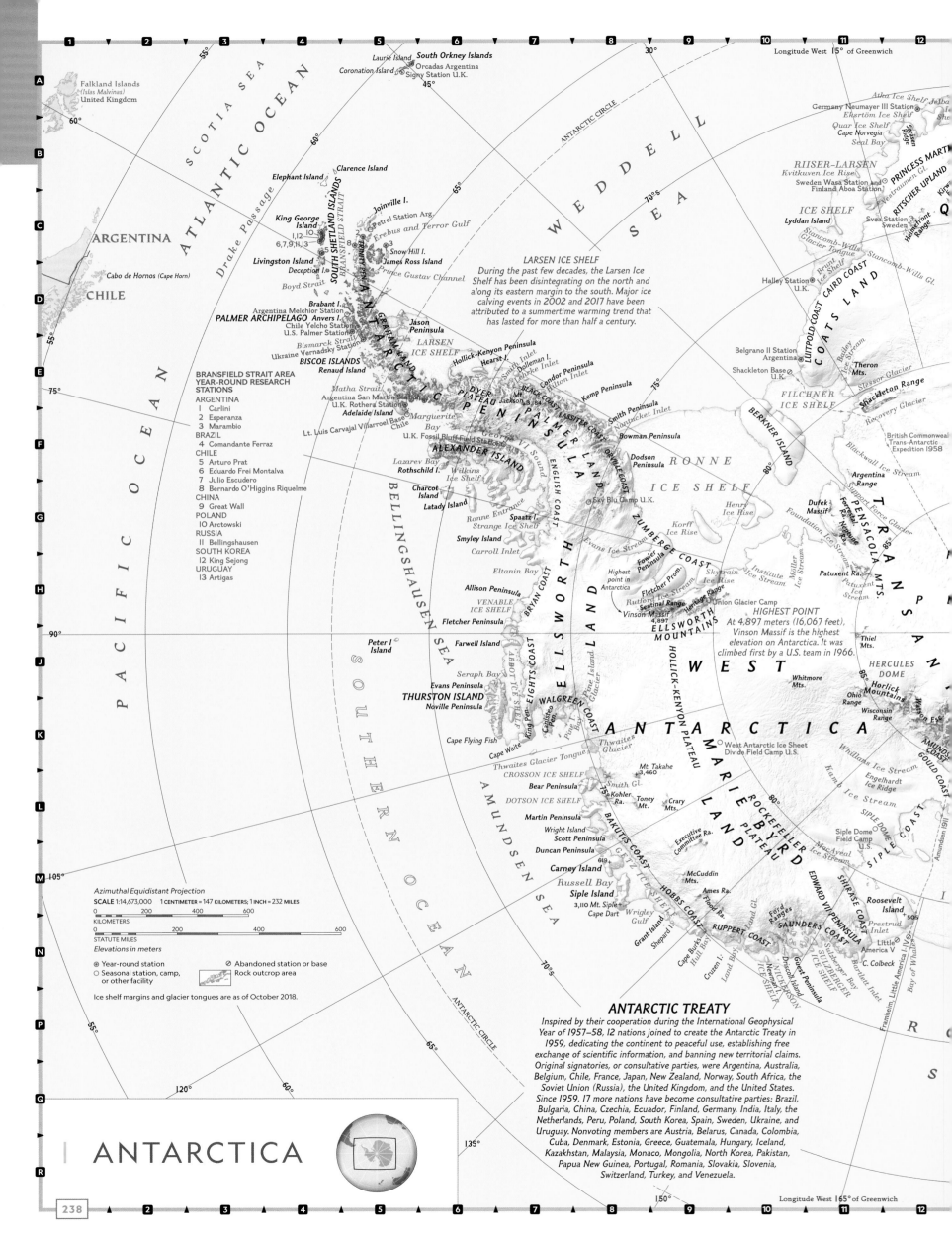

ANTARCTICA

LARSEN ICE SHELF
During the past few decades, the Larsen Ice Shelf has been disintegrating on the north and along its eastern margin to the south. Major ice calving events in 2002 and 2017 have been attributed to a summertime warming trend that has lasted for more than half a century.

HIGHEST POINT
At 4,897 meters (16,067 feet), Vinson Massif is the highest elevation on Antarctica. It was climbed first by a U.S. team in 1966.

ANTARCTIC TREATY
Inspired by their cooperation during the International Geophysical Year of 1957–58, 12 nations joined to create the Antarctic Treaty in 1959, dedicating the continent to peaceful use, establishing free exchange of scientific information, and banning new territorial claims. Original signatories, or consultative parties, were Argentina, Australia, Belgium, Chile, France, Japan, New Zealand, Norway, South Africa, the Soviet Union (Russia), the United Kingdom, and the United States. Since 1959, 17 more nations have become consultative parties: Brazil, Bulgaria, China, Czechia, Ecuador, Finland, Germany, India, Italy, the Netherlands, Peru, Poland, South Korea, Spain, Sweden, Ukraine, and Uruguay. Nonvoting members are Austria, Belarus, Canada, Colombia, Cuba, Denmark, Estonia, Greece, Guatemala, Hungary, Iceland, Kazakhstan, Malaysia, Monaco, Mongolia, North Korea, Pakistan, Papua New Guinea, Portugal, Romania, Slovakia, Slovenia, Switzerland, Turkey, and Venezuela.

BRANSFIELD STRAIT AREA YEAR-ROUND RESEARCH STATIONS
ARGENTINA
1 Carlini
2 Esperanza
3 Marambio
BRAZIL
4 Comandante Ferraz
CHILE
5 Arturo Prat
6 Eduardo Frei Montalva
7 Julio Escudero
8 Bernardo O'Higgins Riquelme
CHINA
9 Great Wall
POLAND
10 Arctowski
RUSSIA
11 Bellingshausen
SOUTH KOREA
12 King Sejong
URUGUAY
13 Artigas

Azimuthal Equidistant Projection
SCALE 1:14,673,000   1 CENTIMETER = 147 KILOMETERS; 1 INCH = 232 MILES
Elevations in meters

⊙ Year-round station
○ Seasonal station, camp, or other facility
⊘ Abandoned station or base
▨ Rock outcrop area

Ice shelf margins and glacier tongues are as of October 2018.

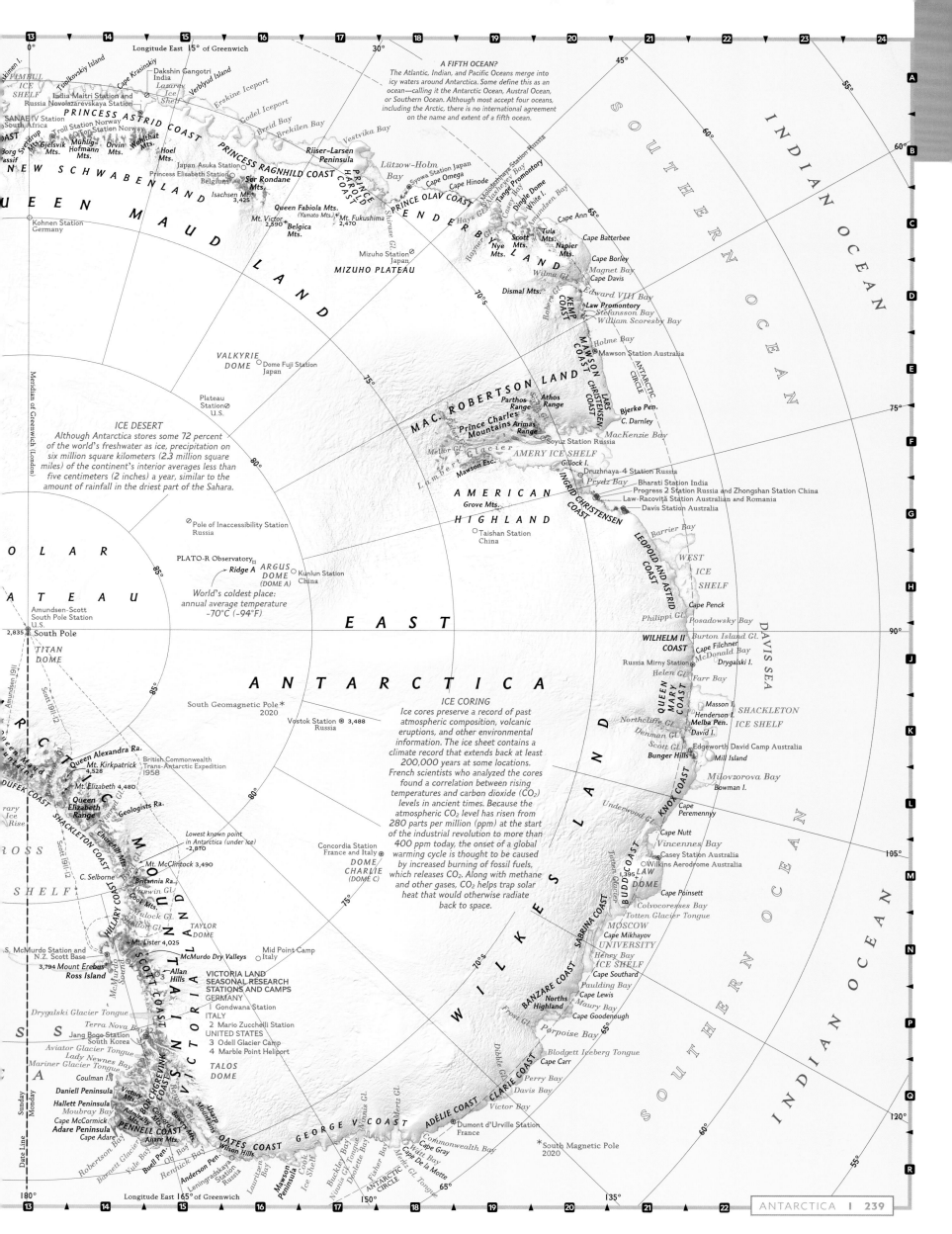

OCEANS

LIMITS OF THE OCEANS AND SEAS
PACIFIC OCEAN FLOOR ● ATLANTIC OCEAN FLOOR
INDIAN OCEAN FLOOR ● ARCTIC OCEAN FLOOR
OCEAN FLOOR AROUND ANTARCTICA

Blacktip sharks, bluefin trevallies, and twinspot snappers in a lagoon in the Pacific's Line Islands

OUR WATERY WORLD

BROAD & DEEP, THE OCEAN COVERS NEARLY THREE-QUARTERS OF OUR PLANET

Earth is a blue planet. More water than land, it is defined in nearly every way by the power and abundance of the ocean. The sea provides food and oxygen, shapes cultures and economies, and drives weather. It also holds secrets: Although the ocean teems with life, from tiny plankton to giant whales, its darkest depths are still an enigma.

| GEOGRAPHY

A UNIVERSE UNDER THE SEA

Even though Earth has one continuous body of salt water, scientists and geographers divide it into four different sections. From biggest to smallest, they are the Pacific, the Atlantic, the Indian, and the Arctic Oceans. Long mountain chains such as the East Pacific Rise and deep canyons like the Atlantic's Hudson Canyon, which stretches for 322 kilometers (200 miles), shape the dramatic topography hiding beneath the smooth surface of the sea. Ocean waters are also home to a staggering array of animals. Some show humanlike behaviors, such as whales that live in tight-knit social groups, while others seem alien, such as the bug-eyed mantis shrimp, which is capable of seeing ultraviolet light. Many make astonishing long-distance migrations, like the loggerhead sea turtles that travel from nesting beaches in Japan and Australia all the way to Mexico.

Life on Earth began in the sea. A tablespoon of ocean water may contain some millions of kinds of bacteria and a hundred million viruses that are invisible to us but that help maintain the flow of energy and nutrients that all living creatures depend on to survive. For example, the tiny bacteria *Prochlorococcus,* found in the sea and on land, produces 20 percent of oxygen in the atmosphere. Some newly discovered marine microbes have even been found to eat pollution and are used in medications.

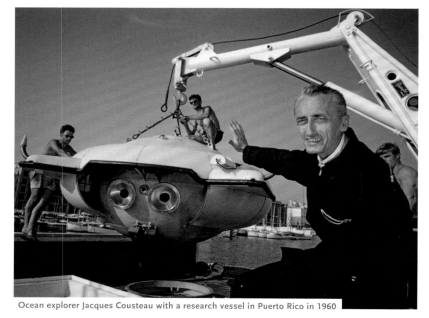

Ocean explorer Jacques Cousteau with a research vessel in Puerto Rico in 1960

| HISTORY

EXPLORING THE OCEAN

Sailors have plied the seas for centuries, guided since ancient times by their knowledge of stars, winds, and currents. Only in the last 200 years have explorers plunged beneath the surface to study the ocean depths in detail. Modern oceanography has its roots in mid-19th-century discoveries of atolls and submarine canyons. Ocean exploration made major strides as scientists used deep-diving vessels, sonar and satellite technology, and other tools to map the seafloor and to measure the diversity, abundance, and distribution of marine life. In 1943, Jacques Cousteau and Émile Gagnan developed the Aqua-Lung, allowing explorers to experience what we now call scuba diving; in 1977, Robert Ballard discovered that deep-sea hydrothermal vents support life without energy from the sun. By studying the rugged ocean floor and identifying the age of crust formed by seafloor spreading, oceanographers added key evidence to the theory of plate tectonics.

Yet scientists are only just beginning to learn about life in the deep, where the complete darkness and immense pressure make for difficult research conditions. Solving these mysteries is important. More knowledge about the ocean floor can help scientists predict earthquakes and tsunamis with greater precision. And by studying how seals hold their breath underwater, scientists have learned about how mammals respond to low-oxygen situations, informing further research into anesthesia and even space travel.

Bottlenose dolphins near the Honduran island of Roatán

CULTURE

THE SEA AS A SOURCE OF LIFE

Inside a cave in South Africa, scientists discovered the remains of a shellfish meal enjoyed by humans 164,000 years ago. Beyond sustaining our ancestors, the sea's tempests, beauty, and secrets have made it a central player in human history. The ocean connects us, serving as a bridge between continents and cultures. It has launched explorers and colonizers; ferried immigrants (and, in a great crime of maritime trade, more than 12 million enslaved Africans) to new shores; and served as a battleground in global conflicts, its dark, distant floor a graveyard for sunken warships.

We find the ocean in literature, religion, and art. There is Odysseus on his journey through the blue Aegean; a clever Maori demigod who fishes New Zealand's North Island from the sea; and Japanese artist Katsushika Hokusai's iconic woodcut print of a frothy wave. We see it in sport and thrill seeking: surfers from Australia to Hawaii and Portugal riding towering waves; divers searching for treasure and shipwrecks. The ocean is often on our plate, whether it's spicy prawns in Bangkok or spaghetti with clams in an Italian trattoria. And we find the ocean in the rhythms of daily life: Inuit hunters in Greenland skim across frozen sea ice on dogsleds, while in Southeast Asia, stateless sea nomads known as the Bajau still catch fish with spears and sleep in stilted houses that sway with the sea.

A fisherman harvesting cultivated mussels off China's Nanji Island

Hawaiian paddlers compete in a race.

ECONOMY

CURRENTS OF COMMERCE

Despite the advent of airplanes, trains, and trucks, 90 percent of global trade (by weight) still moves by sea. Merchant ships are registered to 150 countries and crewed by more than a million people. Many seafarers hail from lower-income countries, and shipbuilding and repair are among the ways the shipping industry supports port cities, many in developing economies. Offshore drilling for oil and natural gas in places like the Gulf of Mexico helps meet global energy demands, but is hazardous to marine environments and causes profound damage when explosions or spills occur.

About 97 percent of the world's fishermen and women live in developing countries, and roughly three billion people worldwide rely on seafood as their primary source of protein. But over the last several decades, large-scale commercial fisheries have placed a dangerous strain on oceans. Most populations of wild fish have been overfished, or overexploited beyond what the fish can replace through reproduction. The commercial raising and harvesting of fish and shellfish, known as aquaculture, has helped to rebuild and replenish some wild stocks of endangered species while still satisfying consumer demand. Humans have realized the oceans we'd assumed were unendingly rich are highly vulnerable—their survival depends on balancing enterprise with stewardship.

SNAPSHOT — *Oceans* **FACTS & FEATURES**

DEEPEST POINT Challenger Deep: 11 km (6.8 mi) below the Pacific Ocean surface

YOUNGEST OCEAN Indian Ocean, 65 million years old

LONGEST UNDERWATER MOUNTAIN RANGE Mid-Ocean Ridge: 60,000 km (37,000 mi)

LARGEST UNDERWATER CANYON Zhemchug Canyon, Bering Sea: 2.7 km (1.7 mi) deep

LARGEST LIVING STRUCTURE Great Barrier Reef: more than 2,000 km (1,200 mi)

BIODIVERSITY 1 million species of animals live in the ocean

NAMED SPECIES 232,811 accepted in the World Register of Marine Species, 2019

PROTECTED AREAS 7 percent of the oceans

LARGEST ANIMAL Blue whale: 25 to 32 m (82 to 105 ft) long

THE FIRST VIDEO OF A GIANT SQUID IN ITS NATURAL ENVIRONMENT WAS FILMED IN 2012 OFF THE COAST OF JAPAN AT A DEPTH OF 630 METERS (2,066 FEET).

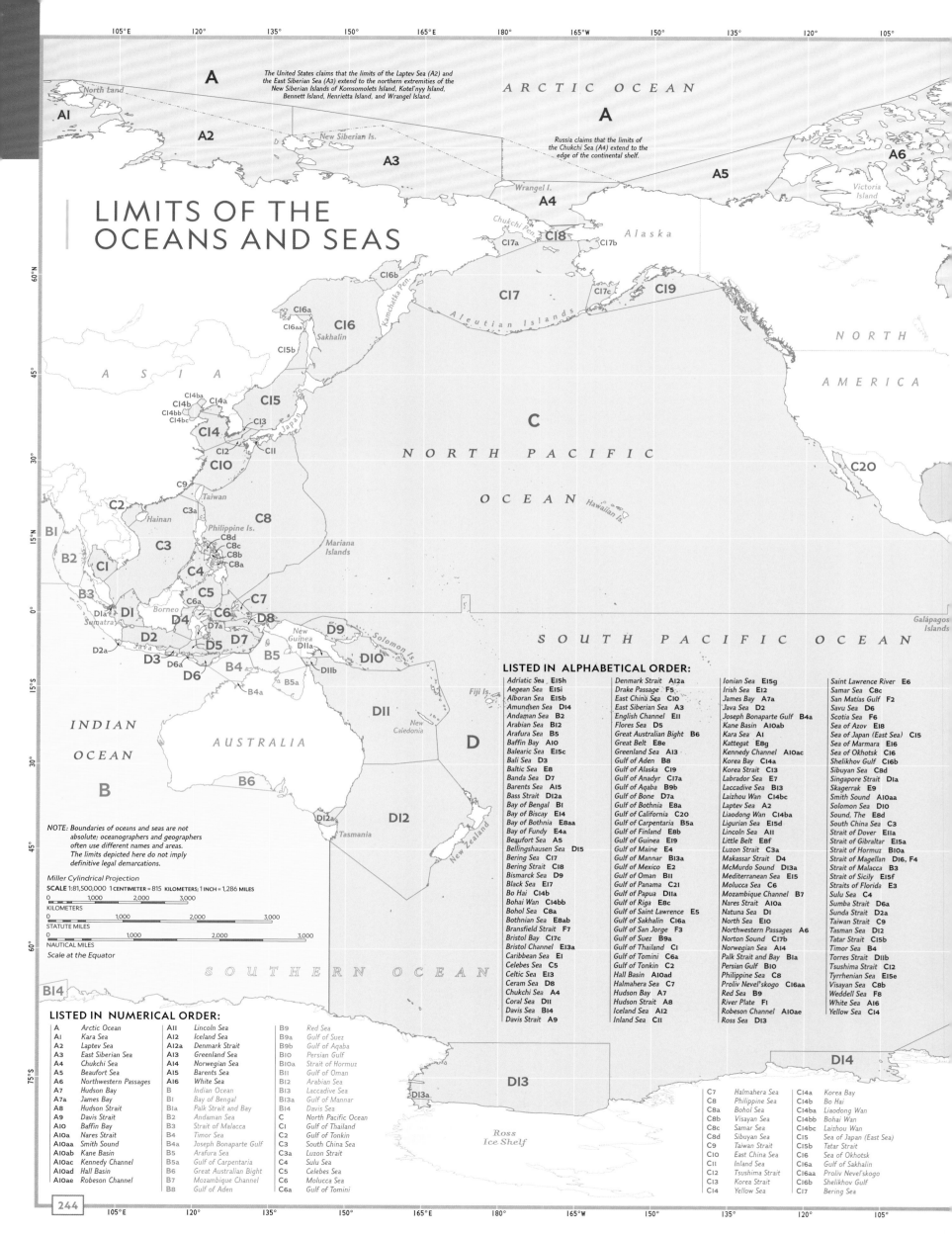

LIMITS OF THE OCEANS AND SEAS

The United States claims that the limits of the Laptev Sea (A2) and the East Siberian Sea (A3) extend to the northern extremities of the New Siberian Islands of Komsomolets Island, Kotel'nyy Island, Bennett Island, Henrietta Island, and Wrangel Island.

Russia claims that the limits of the Chukchi Sea (A4) extend to the edge of the continental shelf.

NOTE: Boundaries of oceans and seas are not absolute; oceanographers and geographers often use different names and areas. The limits depicted here do not imply definitive legal demarcations.

Miller Cylindrical Projection

SCALE 1:81,500,000 1 CENTIMETER = 815 KILOMETERS; 1 INCH = 1,286 MILES

KILOMETERS
STATUTE MILES
NAUTICAL MILES

Scale at the Equator

LISTED IN ALPHABETICAL ORDER:

Adriatic Sea E15h	Denmark Strait A12a	Ionian Sea E15g	Saint Lawrence River E6
Aegean Sea E15i	Drake Passage F5	Irish Sea E12	Samar Sea C8c
Alboran Sea E15b	East China Sea C10	James Bay A7a	San Matías Gulf F2
Amundsen Sea D14	East Siberian Sea A3	Java Sea D2	Savu Sea D6
Andaman Sea B2	English Channel E11	Joseph Bonaparte Gulf B4a	Scotia Sea F6
Arabian Sea B12	Flores Sea D5	Kane Basin A10ab	Sea of Azov E18
Arafura Sea B5	Great Australian Bight B6	Kara Sea A1	Sea of Japan (East Sea) C15
Baffin Bay A10	Great Belt E8e	Kattegat E8g	Sea of Marmara E16
Balearic Sea E15c	Greenland Sea A13	Kennedy Channel A10ac	Sea of Okhotsk C16
Bali Sea D3	Gulf of Aden B8	Korea Bay C14a	Shelikhov Gulf C16b
Baltic Sea E8	Gulf of Alaska C19	Korea Strait C13	Sibuyan Sea C8d
Banda Sea D7	Gulf of Anadyr C17a	Labrador Sea E7	Singapore Strait D1a
Barents Sea A15	Gulf of Aqaba B9b	Laccadive Sea B13	Skagerrak E9
Bass Strait D12a	Gulf of Bone D7a	Laizhou Wan C14bc	Smith Sound A10aa
Bay of Bengal B1	Gulf of Bothnia E8a	Laptev Sea A2	Solomon Sea D10
Bay of Biscay E14	Gulf of California C20	Liaodong Wan C14ba	Sound, The E8d
Bay of Bothnia E8aa	Gulf of Carpentaria B5a	Ligurian Sea E15d	South China Sea C3
Bay of Fundy E4a	Gulf of Finland E8b	Lincoln Sea A11	Strait of Dover E11a
Beaufort Sea A5	Gulf of Guinea E19	Little Belt E8f	Strait of Gibraltar E15a
Bellingshausen Sea D15	Gulf of Maine E4	Luzon Strait C3a	Strait of Hormuz B10a
Bering Sea C17	Gulf of Mannar B13a	Makassar Strait D4	Strait of Magellan D16, F4
Bering Strait C18	Gulf of Mexico E2	McMurdo Sound D13a	Strait of Malacca B3
Bismarck Sea D9	Gulf of Oman B11	Mediterranean Sea E15	Strait of Sicily E15f
Black Sea E17	Gulf of Panama C21	Molucca Sea C6	Straits of Florida E3
Bo Hai C14b	Gulf of Papua D11a	Mozambique Channel B7	Sulu Sea C4
Bohai Wan C14bb	Gulf of Riga E8c	Nares Strait A10a	Sumba Strait D6a
Bohol Sea C8a	Gulf of Saint Lawrence E5	Natuna Sea D1	Sunda Strait D2a
Bothnian Sea E8ab	Gulf of Sakhalin C16a	North Sea E10	Taiwan Strait C9
Bransfield Strait F7	Gulf of San Jorge F3	Northwestern Passages A6	Tasman Sea D12
Bristol Bay C17c	Gulf of Suez B9a	Norton Sound C17b	Tatar Strait C15b
Bristol Channel E13a	Gulf of Thailand C1	Norwegian Sea A14	Timor Sea B4
Caribbean Sea E1	Gulf of Tomini C6a	Palk Strait and Bay B1a	Torres Strait D11b
Celebes Sea C5	Gulf of Tonkin C2	Persian Gulf B10	Tsushima Strait C12
Celtic Sea E13	Hall Basin A10ad	Philippine Sea C8	Tyrrhenian Sea E15e
Ceram Sea D8	Halmahera Sea C7	Proliv Nevel'skogo C16aa	Visayan Sea C8b
Chukchi Sea A4	Hudson Bay A7	Red Sea B9	Weddell Sea F8
Coral Sea D11	Hudson Strait A8	River Plate F1	White Sea A16
Davis Sea B14	Iceland Sea A12	Robeson Channel A10ae	Yellow Sea C14
Davis Strait A9	Inland Sea C11	Ross Sea D13	

LISTED IN NUMERICAL ORDER:

A	Arctic Ocean	A11	Lincoln Sea	B9	Red Sea	C7		Halmahera Sea
A1	Kara Sea	A12	Iceland Sea	B9a	Gulf of Suez	C8		Philippine Sea
A2	Laptev Sea	A12a	Denmark Strait	B9b	Gulf of Aqaba	C8a		Bohol Sea
A3	East Siberian Sea	A13	Greenland Sea	B10	Persian Gulf	C8b		Visayan Sea
A4	Chukchi Sea	A14	Norwegian Sea	B10a	Strait of Hormuz	C8c		Samar Sea
A5	Beaufort Sea	A15	Barents Sea	B11	Gulf of Oman	C8d		Sibuyan Sea
A6	Northwestern Passages	A16	White Sea	B12	Arabian Sea	C9		Taiwan Strait
A7	Hudson Bay	B	Indian Ocean	B13	Laccadive Sea	C10		East China Sea
A7a	James Bay	B1	Bay of Bengal	B13a	Gulf of Mannar	C11		Inland Sea
A8	Hudson Strait	B1a	Palk Strait and Bay	B14	Davis Sea	C12		Tsushima Strait
A9	Davis Strait	B2	Andaman Sea	C	North Pacific Ocean	C13		Korea Strait
A10	Baffin Bay	B3	Strait of Malacca	C1	Gulf of Thailand	C14		Yellow Sea
A10a	Nares Strait	B4	Timor Sea	C2	Gulf of Tonkin	C14a		Korea Bay
A10aa	Smith Sound	B4a	Joseph Bonaparte Gulf	C3	South China Sea	C14b		Bo Hai
A10ab	Kane Basin	B5	Arafura Sea	C3a	Luzon Strait	C14ba		Liaodong Wan
A10ac	Kennedy Channel	B5a	Gulf of Carpentaria	C4	Sulu Sea	C14bb		Bohai Wan
A10ad	Hall Basin	B6	Great Australian Bight	C5	Celebes Sea	C14bc		Laizhou Wan
A10ae	Robeson Channel	B7	Mozambique Channel	C6	Molucca Sea	C15		Sea of Japan (East Sea)
		B8	Gulf of Aden	C6a	Gulf of Tomini	C15b		Tatar Strait
						C16		Sea of Okhotsk
						C16a		Gulf of Sakhalin
						C16aa		Proliv Nevel'skogo
						C16b		Shelikhov Gulf
						C17		Bering Sea

244

ARCTIC CIRCLE

ALASKA

FOXE
BASIN

BAFFIN
ISLAND

DAVIS
STRAIT

A

Southampton
Island

HUDSON STRAIT

Ungava
Peninsula

Ungava
Bay

LABRADOR
SEA

B

Bristol
Bay

GULF OF ALASKA

HUDSON
BAY

Labrador

Kodiak I.
GULF OF ALASKA
SEAMOUNT PROVINCE
Alexander
Archipelago

Patton Smts.
Gilbert Smts.
SILA FRACTURE ZONE
ALASKA PLAIN

Haida Gwaii
(Queen Charlotte Is.)

Gulf of
St. Lawrence

C

COAST AND
GEODETIC SURVEY
SEAMOUNT
PROVINCE
Sagittarius Seachannel
Aquarius Seachannel
Taurus Seachannel

TUFTS PLAIN

Moresby Seachannel
Eickelberg
Ridge
Vancouver
Island
Juan de Fuca
Ridge
Cascadia
Basin

NORTH
AMERICA

Nova Scotia
Gulf of
Maine
SCOTIAN SHELF
Georges
Bank
CONTINENTAL
SLOPE

D

SURVEYOR FRACTURE ZONE
BLANCO
F.Z.
Gorda
Ridges

NEW ENGLAND
SEAMOUNTS

E

NORTH
MENDOCINO FRACTURE ZONE
Mendocino
Ridge
PIONEER FRACTURE ZONE
Monterey
Canyon
PATTON ESCARPMENT
Cape
Hatteras
SOHM PLAIN
Bermuda Rise
Bermuda

F

PACIFIC
Mendocino Escarpment
MURRAY FRACTURE ZONE
MOONLESS MOUNTAINS
MUSICIANS SEAMOUNTS
Isla de Guadalupe
BAJA CALIFORNIA
SEAMOUNT PROVINCE
CEDROS TRENCH
Baja California
Gulf of California
FLORIDA-HATTERAS SHELF
HATTERAS
PLAIN
NORTH

OCEAN
TROPIC OF CANCER
Mississippi
River Delta
WEST FLORIDA
SHELF
Florida
Blake
Plateau
Blake
Basin
ATLANTIC

G
MOLOKAI FRACTURE ZONE
Suitcase
Seamounts
Islas Revillagigedo
GULF OF
MEXICO
MEXICO
BASIN
Great
Bahama
Bank
BAHAMA
ISLANDS
Cuba
-8,605
Atlantic Ocean's
deepest point

O'ahu
Hawai'i
Lō'ihi Seamount
KAMEHAMEHA
BASIN
Campeche Bank
Yucatan
Peninsula
GREATER ANTILLES
Jamaica
Hispaniola
PUERTO RICO TRENCH
Puerto
Rico
PUERTO RICO TRENCH
OCEAN

RIDGE
Baird
Seamounts
Snowden
Seamounts
CLARION FRACTURE ZONE
MATHEMATICIAN SEAMOUNTS
MIDDLE AMERICA TRENCH
Central America
CAYMAN TRENCH
CARIBBEAN SEA
Aves Ridge
LESSER ANTILLES
H

Kingman Reef
Palmyra Atoll
Copper Ridge
CLIPPERTON BASIN
Albatross
Plateau
Tehuantepec
Ridge
Clipperton
Seamounts
SIQUEIRES F.Z.
GUATEMALA BASIN
COCOS RIDGE
PANAMA
BASIN
Colombian Trench
Trinidad

Kiritimati (Christmas)
CLIPPERTON FRACTURE ZONE
COLÓN RIDGE
GALÁPAGOS
ISLANDS
CARNEGIE RIDGE
J

Jarvis
Island
LINE ISLANDS
Christmas Ridge
GALÁPAGOS FRACTURE ZONE
GALLEGO RISE
GRIJALVA RIDGE
EQUATOR

SOUTH
AMERICA

K
MANIHIKI
PLATEAU
Boudeuse Ridge
QUEBRADA F.Z.
GOFAR F.Z.
Manihiki Atoll
MARQUESAS
ISLANDS
Nassau
PENRHYN
BASIN
Bora-Bora
MARQUESAS FRACTURE ZONE
WILKES F.Z.
Bauer Scarp
BAUER
BASIN
GALÁPAGOS
RISE
MENDANA FRACTURE ZONE

COOK ISLANDS
SOCIETY
ISLANDS
Tahiti
Society Ridge
Tuamotu Ridge
TIKI
BASIN
Pukapuka
Pukapukae Ridge
GARRETT F.Z.
RANO RAHI
SEAMOUNTS
YUPANQUI
BASIN
MENDOZA
RISE
L

Mangaia
AUSTRAL ISLANDS
AUSTRAL SEAMOUNTS
TUAMOTU ARCHIPELAGO
Îles Maria
Îles Gambier
Henderson Island
Ducie Island
AUSTRAL FRACTURE ZONE
QUIROS FRACTURE ZONE
TROPIC OF CAPRICORN
M

AGUILA FRACTURE ZONE
Morane
Pitcairn Island
Îles Marotiri
Easter
Island
Salas y
Gómez I.
SALA Y GÓMEZ RIDGE
EASTER FRACTURE ZONE
San Félix
Island
San Ambrosio
Island

SOUTH   **PACIFIC**   **OCEAN**
RESOLUTION F.Z.
ROGGEVEEN
BASIN
ROGGEVEEN
RISE
Juan Fernández
Islands
N

SOUTHWEST
FOUNDATION SEAMOUNTS
CHILE FRACTURE ZONE
CHALLENGER FRACTURE ZONE
SELKIRK
RISE

MARLIN RISE
PACIFIC
AGASSIZ FRACTURE ZONE
VALDIVIA FRACTURE ZONE
SOUTH
ATLANTIC
OCEAN
P

San Matías Gulf

BASIN
GUAFO FRACTURE ZONE
Gulf of
San Jorge
Patagonia

Henry Trough
MENARD FRACTURE ZONE
SOUTHEAST PACIFIC
Grande
Bay
FALKLAND
ISLANDS
Q

HEEZEN FRACTURE ZONE
ELTANIN FRACTURE ZONE SYSTEM
THARP FRACTURE ZONE
BASIN
HUMBOLDT
PLAIN
West
Falkland
East
Falkland
R

UDINTSEV FRACTURE ZONE
ANTIPODES
FRACTURE ZONE
ANTARCTIC PACIFIC RIDGE
Tierra
del Fuego
Cape Horn
Burdwood Bank
Staten I.

ATLANTIC OCEAN FLOOR

A FIFTH OCEAN?

The Atlantic, Indian, and Pacific Oceans merge into icy waters around Antarctica. Some define this as an ocean—calling it the Antarctic Ocean, Austral Ocean, or Southern Ocean. While most accept four oceans, including the Arctic, there is no international agreement on the name and extent of a fifth ocean.

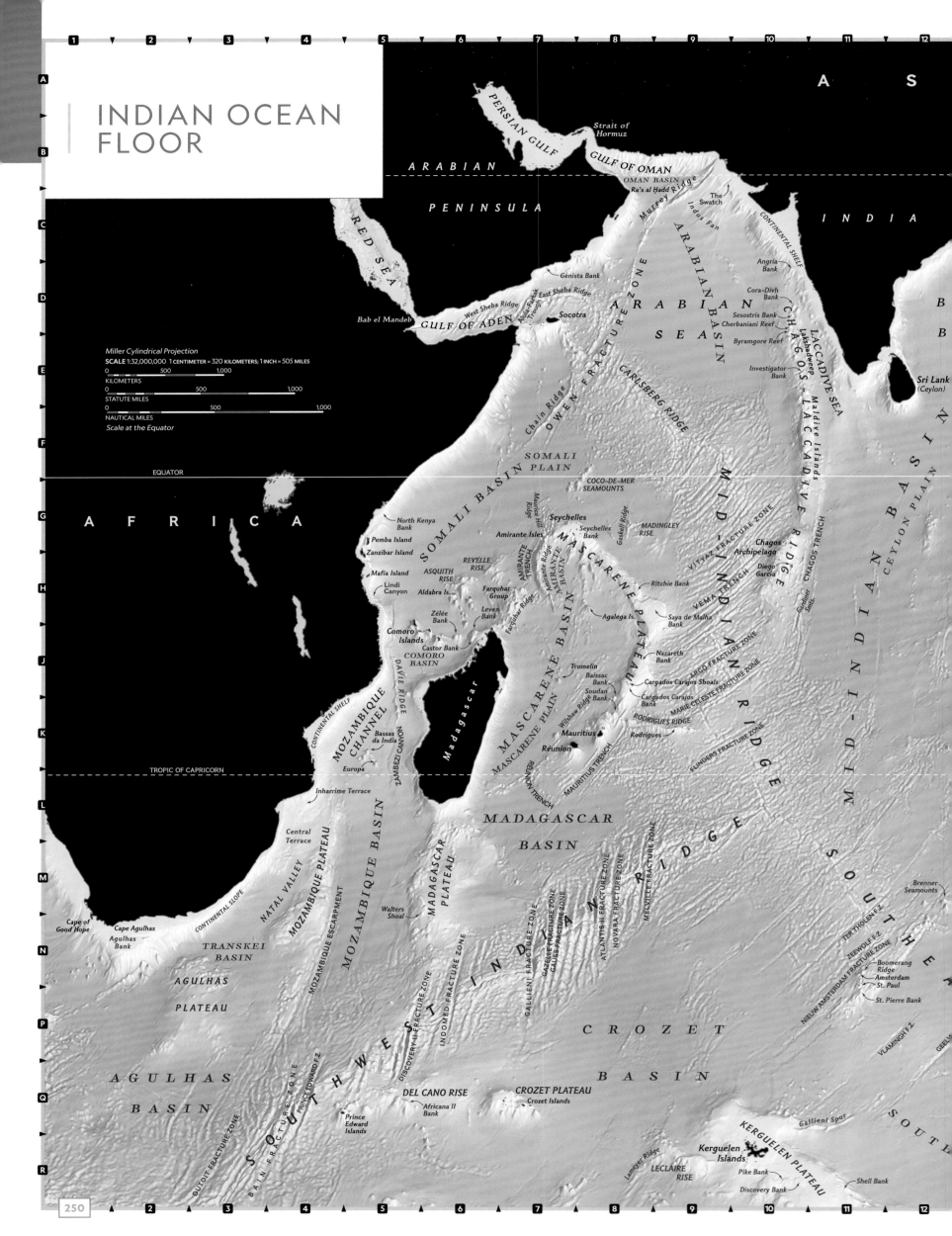

INDIAN OCEAN FLOOR

A S

A S I A

ARABIAN

PENINSULA

PERSIAN GULF

Strait of Hormuz

GULF OF OMAN

OMAN BASIN

Ra's al Hadd

Murrey Ridge

Indus Fan

CONTINENTAL SHELF

INDIA

The Swatch

RED SEA

Genista Bank

West Sheba Ridge
East Sheba Ridge
Abu-Fartak Trough

Bab el Mandeb

GULF OF ADEN

Socotra

Chain Ridge

OWEN FRACTURE ZONE

CARLSBERG RIDGE

ARABIAN

SEA

ARABIAN BASIN

Angria Bank

Cora-Divh Bank

Sesostris Bank
Cherbaniani Reef

Byramgore Reef

Investigator Bank

CHAGOS-LACCADIVE SEA

Laccadive Islands
Lakshadweep

Maldive Islands

CHAGOS-LACCADIVE

LACCADIVE SEA

LACCADIVE RIDGE

Sri Lanka (Ceylon)

B B

Miller Cylindrical Projection
SCALE 1:32,000,000 1 CENTIMETER = 320 KILOMETERS; 1 INCH = 505 MILES

0 — 500 — 1,000
KILOMETERS

0 — 500 — 1,000
STATUTE MILES

0 — 500 — 1,000
NAUTICAL MILES

Scale at the Equator

EQUATOR

SOMALI PLAIN

COCO-DE-MER SEAMOUNTS

Mauritz Ridge

MADINGLEY RISE

VITYAZ FRACTURE ZONE

MID-INDIAN RIDGE

CHAGOS-LACCADIVE RIDGE

CEYLON PLAIN

AFRICA

North Kenya Bank

Pemba Island

Zanzibar Island

Mafia Island

Lindi Canyon

SOMALI BASIN

REVELLE RISE

ASQUITH RISE

Aldabra Is.

Amirante Isles

AMIRANTE TRENCH

Amirante Ridge

Farquhar Group

Farquhar Ridge

Leven Bank

Zélée Bank

Comoro Islands

Castor Bank

COMORO BASIN

DAVIE RIDGE

Gaskell Ridge

Seychelles

Seychelles Bank

MASCARENE PLATEAU

MASCARENE BASIN

AMIRANTE BASIN

Chagos Archipelago

Diego Garcia

Ritchie Bank

Agalega Is.

Saya de Malha Bank

VEMA TRENCH

Gardiner Smts.

GARDINER CHAGOS TRENCH

ARGO FRACTURE ZONE

Nazareth Bank

Tromelin

Baissac Bank

Soudan Bank

Cargados Carajos Shoals

Cargados Carajos Bank

Wilshaw Ridge

RODRIGUES RIDGE

MARIE CELESTE FRACTURE ZONE

Rodrigues

FLINDERS FRACTURE ZONE

Mauritius

Réunion

REUNION TRENCH

MAURITIUS TRENCH

MASCARENE PLAIN

MADAGASCAR

BASIN

CONTINENTAL SHELF

MOZAMBIQUE CHANNEL

Bassas da India

Europa

TROPIC OF CAPRICORN

Inharrime Terrace

Central Terrace

NATAL VALLEY

MOZAMBIQUE PLATEAU

MOZAMBIQUE ESCARPMENT

MOZAMBIQUE BASIN

ZAMBEZI CANYON

Walters Shoal

MADAGASCAR PLATEAU

Madagascar

GALLIENI FRACTURE ZONE

GAZELLE FRACTURE ZONE

GAUSS FRACTURE ZONE

ATLANTIS II FRACTURE ZONE

NOVARA FRACTURE ZONE

MELVILLE FRACTURE ZONE

MID-INDIAN RIDGE

SOUTHEAST

Brenner Seamounts

TER THOLEN F.Z.

ZEEWOLF F.Z.

NIEUW AMSTERDAM FRACTURE ZONE

Boomerang Ridge

Amsterdam

St. Paul

St. Pierre Bank

VLAMINGH F.Z.

GEELM

SOUTH

Cape of Good Hope

Cape Agulhas
Agulhas Bank

CONTINENTAL SLOPE

TRANSKEI BASIN

AGULHAS PLATEAU

AGULHAS BASIN

DISCOVERY FRACTURE ZONE

INDOMED FRACTURE ZONE

SOUTHWEST INDIAN RIDGE

PRINCE EDWARD F.Z.

BAIN FRACTURE ZONE

DUTOIT FRACTURE ZONE

DEL CANO RISE

Prince Edward Islands

Africana II Bank

CROZET PLATEAU

Crozet Islands

CROZET

BASIN

Gallieni Spur

SOUTH

KERGUELEN PLATEAU

Kerguelen Islands

LECLAIRE RISE

Lameyer Ridge

Pike Bank

Discovery Bank

Shell Bank

I A

OF
GAL

**EAST
CHINA
SEA**

JAPAN

**NADEZHDA
BASIN**

JAPANESE
GUYOTS

IZU-OGASAWARA RISE

TROPIC OF CANCER

Swatch of
No Ground

Andaman
Islands
ALCOCK RISE

SEWELL RISE

Nicobar
Islands

**INDOCHINA
PENINSULA**

Gulf of
Tonkin

Hainan

Macclesfield
Bank

Gulf of
Thailand

Mergui
Terrace

Malay Peninsula

Strait of Malacca

ANDAMAN SEA

ANDAMAN BASIN

Taiwan Strait

Taiwan

Ryukyu Islands

RYUKYU ISLANDS

OKINAWA TROUGH

Ryukyu Ridge

RYUKYU TRENCH

Daittō Is.

Daittō Ridge

Kita
Daittō
Basin

Luzon Ridge

Luzon

SHIKOKU BASIN

Kinan Seamounts

Iwo Jima Ridge

KYUSHU-PALAU RIDGE

Two Jima Ridge

Ogasawara Is.

Ogasawara Trough

West Mariana Ridge

BONIN TROUGH

Bonin Is.

BONIN TRENCH

Ogasawara Plateau

Uda Spur

**MARCUS-WAKE
SEAMOUNTS**

PIGAFETTA BASIN

MAGELLAN SEAMOUNTS

SOUTH CHINA SEA

SOUTH CHINA BASIN

Manila Trench

**PHILIPPINE
ISLANDS**

Luzon
Plateau

**PHILIPPINE
SEA
PHILIPPINE
BASIN**

**WEST
MARIANA
BASIN**

MARIANA TROUGH

Mariana Islands

MARIANA RIDGE

MARIANA TRENCH

Guam

Challenger Deep
10,984
World's greatest
ocean depth

HMRG
Deep

PACIFIC OCEAN

**EAST
MARIANA
BASIN**

PALAWAN TROUGH

Palawan

SULU
BASIN

Mindanao

Sibutu-Basilan Ridge

Sangihe Ridge

PHILIPPINE TRENCH

Palau

PALAU TRENCH

YAP TRENCH

AYU TROUGH

WEST CAROLINE
RISE

EAURIPIK RISE

West Caroline Trough

Sorol Trough

Chuuk

CAROLINE SEAMOUNTS

CAROLINE ISLANDS

CELEBES
BASIN

**WEST
CAROLINE
BASIN**

**EAST
CAROLINE
BASIN**

MUSSAU TROUGH

**LYRA
BASIN**

SUNDA SHELF

NATUNA
SEA

Borneo

EQUATOR

I N D O N E S I A

COCOS PLAIN

Nicobar Fan

Kuenen Rise

COCOS BASIN

Mentawai Basin

Mentawai Trough

Sumatra

JAVA RIDGE

MENTAWAI RIDGE

Batu Is.

Enggano Basin

Sulawesi

**GREATER
SUNDA ISLANDS**

JAVA SEA

NORTH
BANDA
BASIN

CERAM TROUGH

**BANDA
SEA**

SOUTH BANDA
BASIN

WEBER BASIN

NISER

ARU BASIN

NEW GUINEA TRENCH

WEST MELANESIAN TRENCH

New Guinea
Basin

Manus
Basin

New Guinea

**SOLOMON
BASIN**

NINETYEAST RIDGE

OSBORN
PLATEAU

(Keeling Is.) Cocos Is.

**COCOS-KEELING
RISE**

Raitt Rise

Roo Rise

INVESTIGATOR RIDGE

JAVA SUNDA TRENCH

Vening
Melnesz
Rise

Christmas I.

VENING MELNESZ
SEAMOUNTS

CHRISTMAS RISE

Monsoon
Rise

Bartlett Seamounts

Horizon Ridge

Golden Bo'sunbird
Seamounts

-7,125
Indian Ocean's deepest point

SUNDA TROUGH

LOMBAK BASIN

JAVA RIDGE

Java

Bali

FLORES
BASIN

Lesser Sunda Islands

Savu
Basin

Timor

TIMOR TROUGH

Sahul Banks

**TIMOR
SEA**

Scott
Plateau

Sahul Shelf

Bonaparte
Basin

SAHUL SHELF

ARAFURA SEA

ARAFURA SHELF

Torres
Strait

Moresby Valley

PAPUA
PLATEAU

**CORAL SEA
BASIN**

ROO RISE

Joey
Rise

Platypus
Spur

**NORTH
AUSTRALIAN
BASIN**

Wombat
Plateau

Rowley Reefs

Gulf of
Carpentaria

CORAL SEA

QUEENSLAND
PLATEAU

Townsville Trough

GASCOYNE
PLAIN

EXMOUTH
PLATEAU

ROWLEY SHELF

Rowley Shelf

Great Barrier Reef

MARION
PLATEAU

WHARTON BASIN

ZENITH
PLATEAU

Quokka
Rise

Sonja Ridge

CUVIER
BASIN

TROPIC OF CAPRICORN

East Indiaman Ridge

Zenith Trough

Brouwer Trough

Lost Dutchman Ridge

Batavia Smt.

CUVIER
PLATEAU

Wallaby Saddle

Carnarvon Terrace

AUSTRALIA

Guilden
Draak
Smt.

HARTOG RIDGE

PERTH BASIN

BROKEN RIDGE

DIAMANTINA ESCARPMENT

DIAMANTINA

OB' TRENCH

Dordrecht Hole

Diamantina
Deep

DIAMANTINA TRENCH

NATURALISTE

NATURALISTE
PLATEAU

Naturaliste Trough

Lamar Hayes Ridge

EUCLA SHELF

**GREAT
AUSTRALIAN
BIGHT**

Eyre Terrace

Eyre
Canyon

Eucla
Canyon

CEDUNA
TERRACE

CEDUNA SLOPE

Beachport
Terrace

CONTINENTAL SLOPE

Bass Strait

CONTINENTAL SLOPE

Bass
Canyon

**TASMAN
SEA**

TASMAN PLAIN

FRACTURE ZONE

RE ZONE

**SOUTH
INDIAN
RIDGE**

**INDIAN
BASIN**

SOUTH AUSTRALIAN PLAIN

SOUTH AUSTRALIAN BASIN

CONTINENTAL SLOPE

Tasmania

**TASMAN
SEA**

South Tasman Saddle

Lowreenne Seamounts

SOUTH TASMAN RISE

East Tasman
Saddle

Tasman Escarpment

L'Atalante Valley

**EAST
TASMAN
PLATEAU**

ZEEHAEN F.Z.

HEEMSKERCK F.Z.

Australian-Antarctic

Discordance

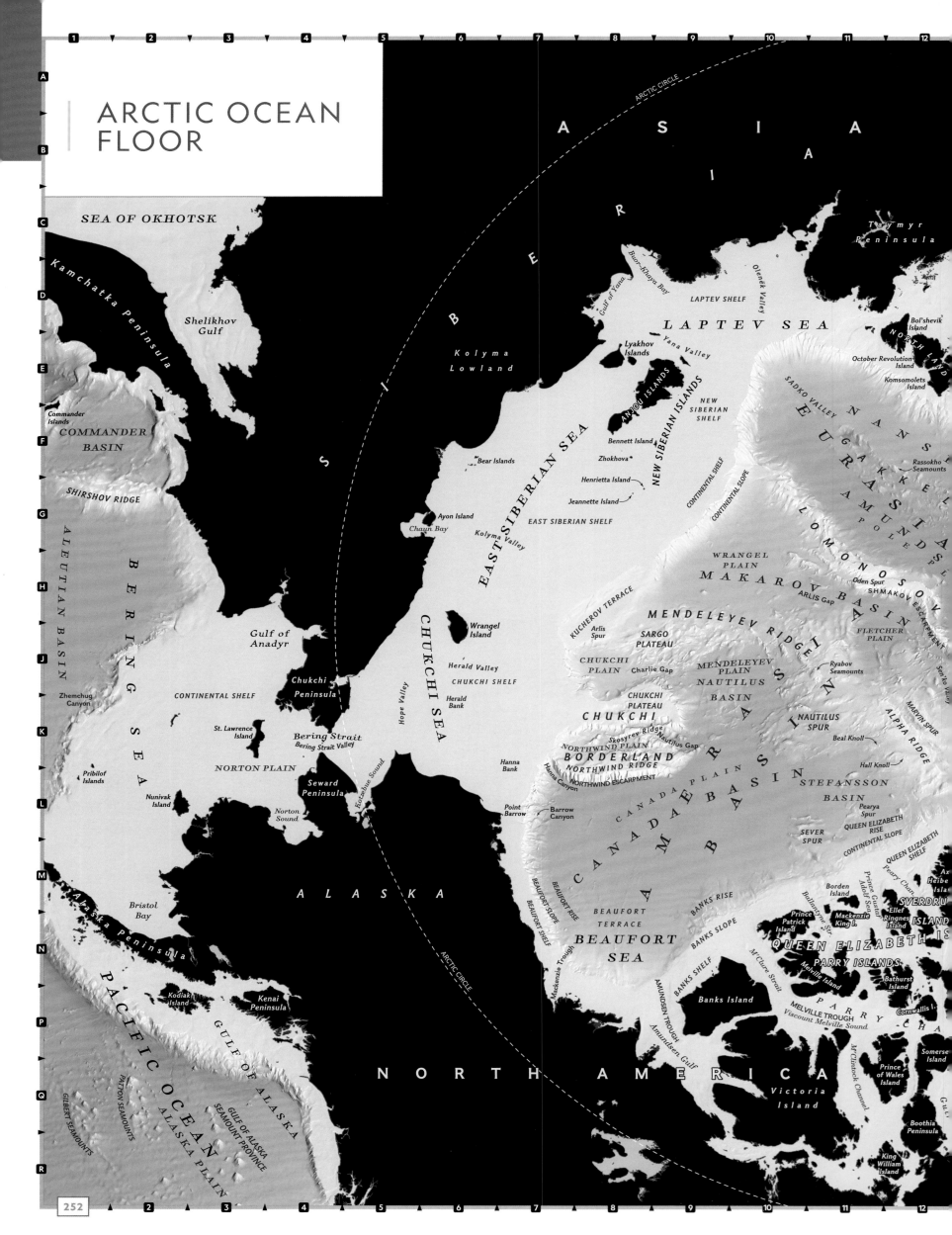

SEA OF OKHOTSK

Kamchatka Peninsula

Shelikhov Gulf

Commander Islands

COMMANDER BASIN

SHIRSHOV RIDGE

ALEUTIAN BASIN

BERING BASIN

Zhemchug Canyon

BERING SEA

CONTINENTAL SHELF

Gulf of Anadyr

Chukchi Peninsula

St. Lawrence Island

Bering Strait
Bering Strait Valley

NORTON PLAIN

Pribilof Islands

Nunivak Island

Norton Sound

Seward Peninsula

Kotzebue Sound

Bristol Bay

Kenai Peninsula

Kodiak Island

PACIFIC OCEAN

Alaska Peninsula

GULF OF ALASKA

GULF OF ALASKA SEAMOUNT PROVINCE

ALASKA PLAIN

PATTON SEAMOUNTS

GILBERT SEAMOUNTS

ALASKA

ARCTIC CIRCLE

NORTH AMERICA

A S I A

S I B E R I A

Taymyr Peninsula

Gulf of Yana

Buor-Khaya Bay

Olenëk Valley

LAPTEV SHELF

LAPTEV SEA

Kolyma Lowland

Lyakhov Islands

Yana Valley

ANJOU ISLANDS

NEW SIBERIAN ISLANDS

NEW SIBERIAN SHELF

Bennett Island

Zhokhova

Bear Islands

Henrietta Island

Jeannette Island

EAST SIBERIAN SEA

CONTINENTAL SHELF

CONTINENTAL SLOPE

EAST SIBERIAN SHELF

Ayon Island

Chaun Bay

Kolyma Valley

Wrangel Island

Herald Valley

CHUKCHI SHELF

Herald Bank

Hope Valley

CHUKCHI SEA

Hanna Bank

Point Barrow

Barrow Canyon

Mackenzie Trough

BEAUFORT SHELF

BEAUFORT SLOPE

BEAUFORT RISE

Beaufort Terrace

BEAUFORT SEA

Amundsen Trough

Amundsen Gulf

Banks Island

Victoria Island

Boothia Peninsula

King William Island

Bol'shevik Island

October Revolution Island

Komsomolets Island

NORTH LAND

SADKO VALLEY

NANSEN

EURASIA

GAKKEL

AMUNDSEN

POLE

LOMONOSOV

Rassokho Seamounts

WRANGEL PLAIN

MAKAROV BASIN

Oden Spur

ARLIS Gap

SHMAKOVE ESCARPMENT

KUCHEROV TERRACE

MENDELEYEV RIDGE

FLETCHER PLAIN

Arlis Spur

SARGO PLATEAU

MENDELEYEV PLAIN

CHUKCHI PLAIN

Charlie Gap

NAUTILUS BASIN

Ryabov Seamounts

CHUKCHI PLATEAU

MARVIN SPUR

San'ko Valley

CHUKCHI BORDERLAND

Skosyrev Ridge

Nautilus Gap

NAUTILUS SPUR

Beal Knoll

ALPHA RIDGE

NORTHWIND PLAIN

NORTHWIND RIDGE

NORTHWIND ESCARPMENT

Hanna Canyon

Hall Knoll

CANADA PLAIN

CANADA BASIN

AMERASIA BASIN

STEFANSSON BASIN

Pearya Spur

QUEEN ELIZABETH RISE

SEVER SPUR

CONTINENTAL SLOPE

QUEEN ELIZABETH SHELF

QUEEN ELIZABETH ISLANDS

BANKS RISE

BANKS SLOPE

BANKS SHELF

Borden Island

Prince Patrick Island

Mackenzie King I.

Prince Gustaf Adolf Sea

Ballantyne Str.

Ax Heibe Isla

SVERDRUP ISLAND

Ellef Ringnes Island

Parry Chan.

MELVILLE TROUGH

M'Clure Strait

Melville Island

Viscount Melville Sound

PARRY ISLANDS

Bathurst Island

Cornwallis I.

PARRY CHA

M'Clintock Channel

Prince of Wales Island

Somerset Island

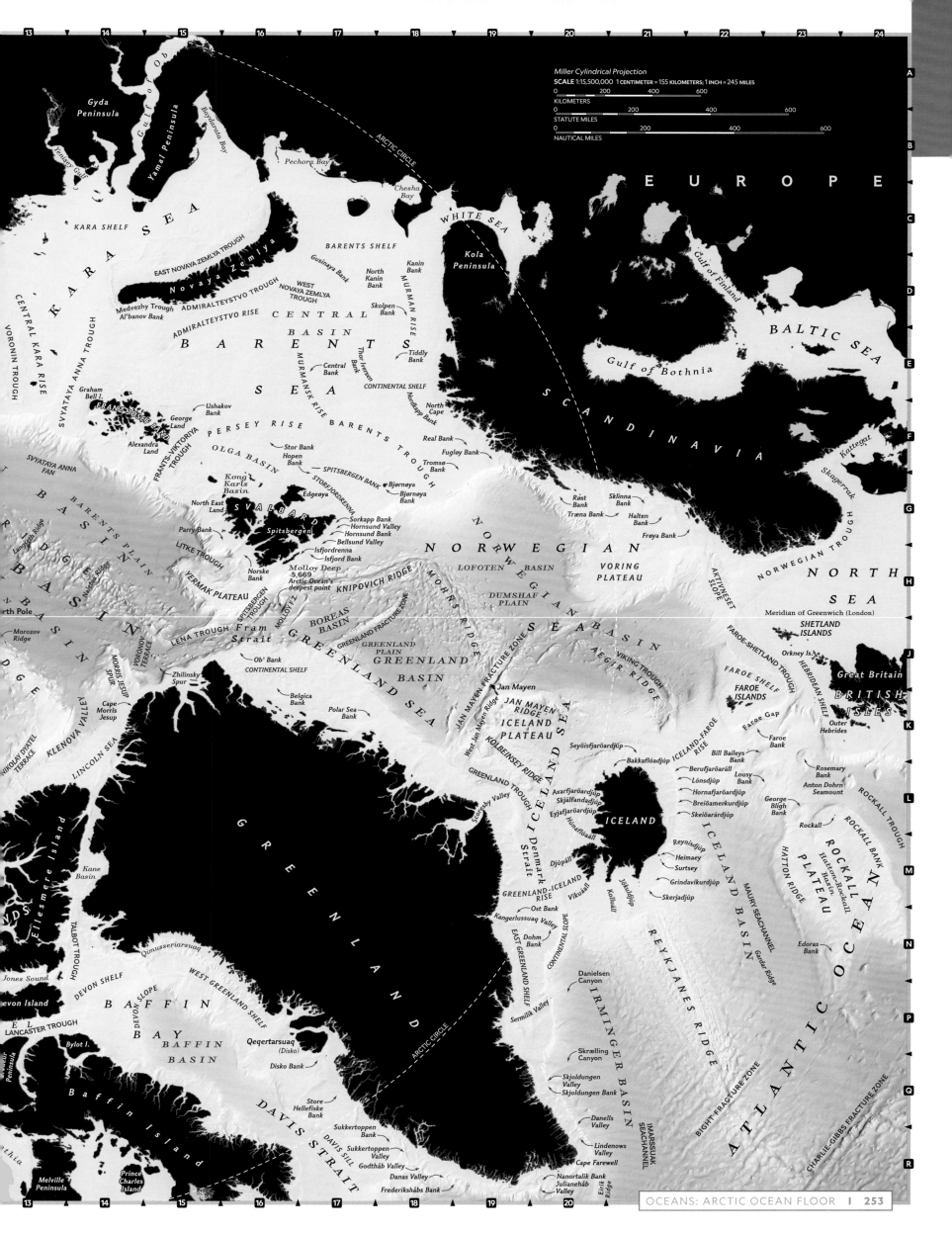

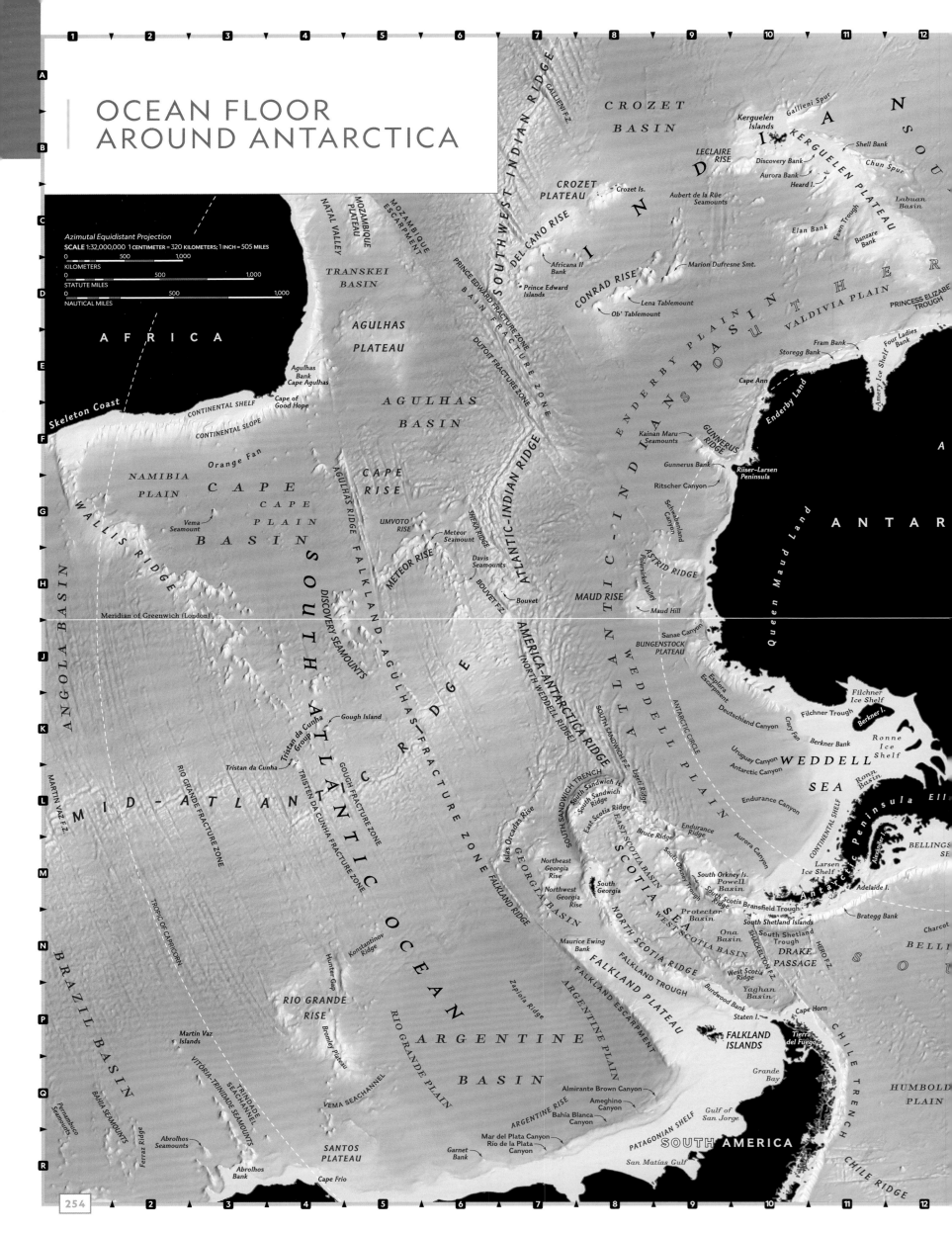

OCEAN FLOOR
AROUND ANTARCTICA

AFRICA

Skeleton Coast

CONTINENTAL SHELF

CONTINENTAL SLOPE

Cape of Good Hope

NAMIBIA PLAIN

WALLIS RIDGE

Orange Fan

Vema Seamount

CAPE BASIN

CAPE PLAIN

ANGOLA BASIN

MARTIN VAZ F.Z.

MID-ATLANTIC

BRAZIL BASIN

Pernambuco Seamounts

BAHIA SEAMOUNTS

Ferraz Ridge

Abrolhos Seamounts

Abrolhos Bank

Cape Frio

Martin Vaz Islands

VITÓRIA-TRINDADE SEAMOUNTS

TRINDADE SEACHANNEL

SANTOS PLATEAU

VEMA SEACHANNEL

RIO GRANDE RISE

Bromley Plateau

Hunter Gap

Konstantinov Ridge

TROPIC OF CAPRICORN

TRISTAN DA CUNHA FRACTURE ZONE

GOUGH FRACTURE ZONE

Tristan da Cunha

Tristan da Cunha Group

Gough Island

DISCOVERY SEAMOUNTS

SOUTH ATLANTIC RIDGE

AGULHAS RIDGE

FALKLAND-AGULHAS FRACTURE ZONE

Meridian of Greenwich (London)

TRANSKEI BASIN

NATAL VALLEY

MOZAMBIQUE PLATEAU

MOZAMBIQUE ESCARPMENT

AGULHAS PLATEAU

Agulhas Bank
Cape Agulhas

AGULHAS BASIN

CAPE RISE

UMVOTO RISE

Meteor Seamount

METEOR RISE

Davis Seamounts

SHEBA RIDGE

BOUVET F.Z.

Bouvet

ATLANTIC-INDIAN RIDGE

PRINCE EDWARD FRACTURE ZONE

DUTOIT FRACTURE ZONE

SOUTHWEST INDIAN RIDGE

GALLIENI F.Z.

CROZET BASIN

CROZET PLATEAU

DEL CAÑO RISE

Crozet Is.

Africana II Bank

Prince Edward Islands

CONRAD RISE

Lena Tablemount

Ob' Tablemount

Marion Dufresne Smt.

LECLAIRE RISE

Kerguelen Islands

Gallieni Spur

KERGUELEN PLATEAU

Discovery Bank

Shell Bank

Aurora Bank

Heard I.

Chun Spur

Elan Bank

Fawn Trough

Banzare Bank

Labuan Basin

INDIAN

SOU

VALDIVIA PLAIN

PRINCESS ELIZABE TROUGH

Fram Bank

Storegg Bank

Four Ladies Bank

Amery Ice Shelf

Cape Ann

Enderby Land

ENDERBY PLAIN

ENDERBY BASIN

ATLANTIC-INDIAN BASIN

Kainan Maru Seamounts

GUNNERUS RIDGE

Gunnerus Bank

Riiser-Larsen Peninsula

Ritscher Canyon

Schwabenland Canyon

ASTRID RIDGE

Polarsirkel Valley

MAUD RISE

Maud Hill

AMERICA-ANTARCTICA RIDGE

(NORTH WEDDELL RIDGE)

SOUTH SANDWICH TRENCH

SOUTH SANDWICH F.Z.

South Sandwich Is.

South Sandwich Ridge

East Scotia Ridge

Bruce Ridge

Islas Orcadas Rise

Northeast Georgia Rise

Northwest Georgia Rise

GEORGIA BASIN

South Georgia

Maurice Ewing Bank

NORTH SCOTIA RIDGE

Ligeia Ridge

EAST SCOTIA BASIN

Endurance Ridge

Endurance Canyon

SCOTIA SEA

WEST SCOTIA RIDGE

South Orkney Trough

South Orkney Is. Powell Basin

South Scotia Ridge

Protector Basin

Ona Basin

West Scotia Ridge

Bransfield Trough

South Shetland Islands

South Shetland Trough

SHACKLETON F.Z.

HERO F.Z.

DRAKE PASSAGE

Burdwood Bank

Yaghan Basin

Cape Horn

Staten I.

Tierra del Fuego

FALKLAND ISLANDS

FALKLAND PLATEAU

FALKLAND ESCARPMENT

FALKLAND TROUGH

Zapiola Ridge

ARGENTINE PLAIN

ARGENTINE BASIN

ARGENTINE RISE

Almirante Brown Canyon

Ameghino Canyon

Bahía Blanca Canyon

Mar del Plata Canyon
Río de la Plata Canyon

Garnet Bank

PATAGONIAN SHELF

SOUTH AMERICA

San Matías Gulf

Gulf of San Jorge

Grande Bay

Brategg Bank

Adelaide I.

Charcot

Alexander I.

Antarctic Peninsula

Ell

Larsen Ice Shelf

CONTINENTAL SHELF

BELLINGS

BELLI

HUMBOLD PLAIN

CHILE TRENCH

CHILE RIDGE

WEDDELL SEA

Ronne Ice Shelf

Ronne Basin

Berkner I.

Berkner Bank

Filchner Ice Shelf

Filchner Trough

Deutschland Canyon

Crary Fan

Uruguay Canyon

Antarctic Canyon

Aurora Canyon

ANTARCTIC CIRCLE

WEDDELL PLAIN

Explora Escarpment

Sanae Canyon

BUNGENSTOCK PLATEAU

Queen Maud Land

ANTAR

ANTARCTIC-INDIAN

THER

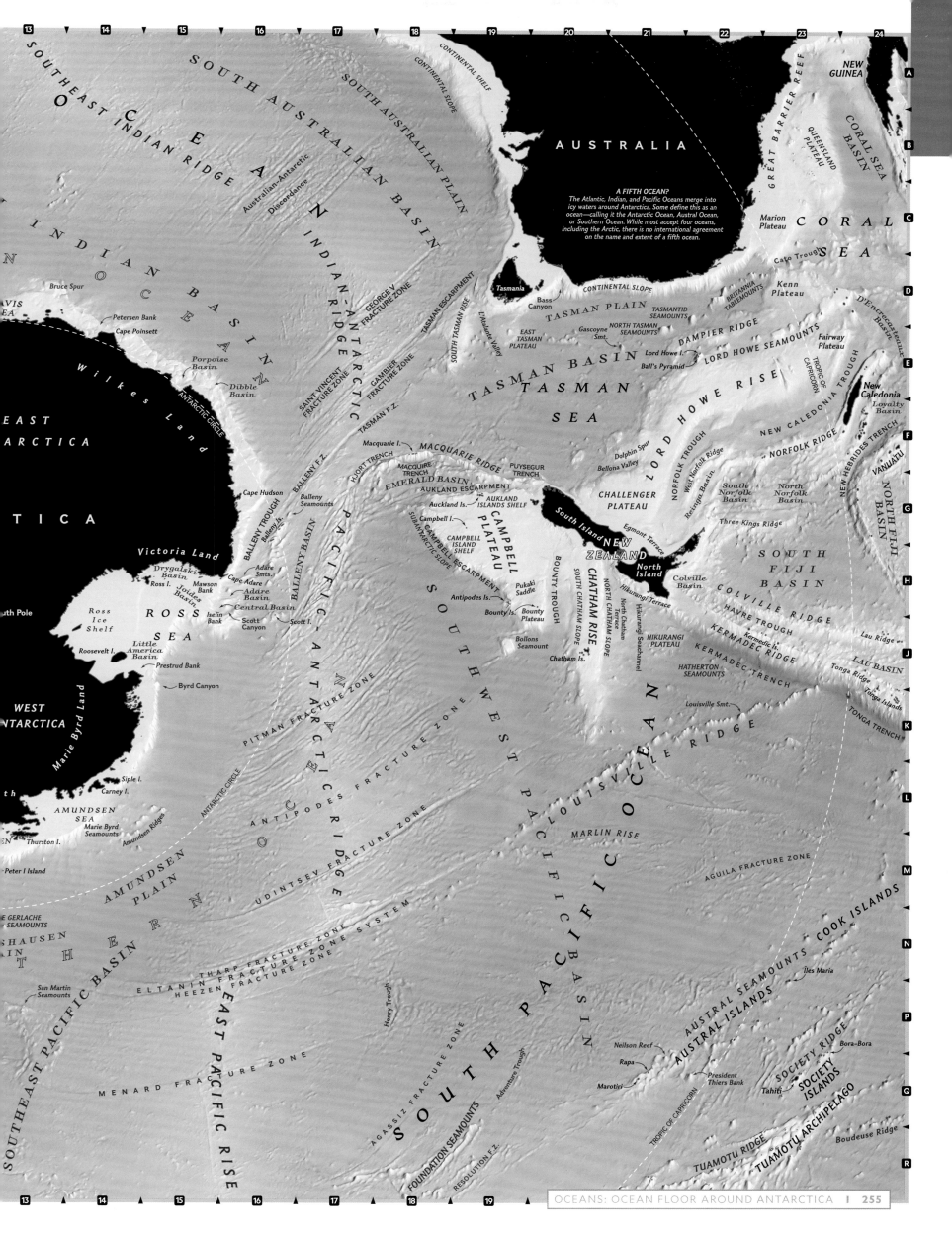

A FIFTH OCEAN?
The Atlantic, Indian, and Pacific Oceans merge into icy waters around Antarctica. Some define this as an ocean—calling it the Antarctic Ocean, Austral Ocean, or Southern Ocean. While most accept four oceans, including the Arctic, there is no international agreement on the name and extent of a fifth ocean.

SPACE

The spiral galaxy NGC 3147 is located 130 million light-years from Earth.

THE COSMIC FRONTIER

UNLOCKING THE SECRETS OF OUR SOLAR SYSTEM & BEYOND

The night sky has captivated humans since ancient times. But only in the last 60 years have we launched spacecraft and sent human explorers beyond Earth's atmosphere. From the earliest satellites to the search for life on faraway planets, our journeys into space confirm just how wondrous that night sky is.

MEASURING SPACE

DISTANCES STRETCHING BILLIONS OF YEARS

Starting with the ancient Greek astronomer Aristarchus of Samos, who first calculated the distance between Earth and the moon, scientists have been trying to measure how big the universe is—and how far the stars, planets, and galaxies are from us. We describe these distances in light-years, a unit of length equivalent to the distance that light travels in one year, which is nearly 10 trillion kilometers (6 trillion miles). We know the Milky Way galaxy is roughly 100,000 light-years across. The observable universe extends about 45 billion light-years from Earth in every direction. It takes about eight minutes for light to travel from the sun to Earth, a distance called 1 astronomical unit (AU).

Distance isn't the only way to define our view of deep space; time comes into play as well. Light from the farthest bodies in our universe takes a long time to reach us: If a galaxy is five billion light-years away, for instance, that means we are seeing it as it looked five billion years ago. As advances in technology let us see farther out into space, they also help us see back in time. In 2016, the Hubble Space Telescope calculated the distance to the farthest known galaxy yet—13.3 billion light-years, which means it existed 13.3 billion years ago, when the universe was very young.

Viewing Earth from the International Space Station

Ultraviolet light reveals the gas clouds shrouding the volatile star system Eta Carinae.

THE RACE INTO SPACE

SATELLITES, SPACE SHUTTLES & A QUEST FOR KNOWLEDGE

Humans' first forays into space began as an intense Cold War competition between the United States and the Soviet Union. The two superpowers sent satellites into space within months of each other in 1957 and 1958. The first living creature in space was a terrier mutt from Moscow named Laika. The first humans launched into space in 1961: Yuri Gagarin from the Soviet Union and Alan Shepard from the United States. On July 20, 1969, American Neil Armstrong became the first human to set foot on the moon, declaring, "That's one small step for a man, one giant leap for mankind."

The opening decades of the space age were marked by astonishing feats of aerospace engineering. The Mariner spacecraft orbited and mapped the surface of Mars. The twin Voyager spacecraft dispatched detailed pictures of the outer planets and their moons as well as taking the famous "pale blue dot" photograph of Earth from nearly 6.4 billion kilometers (4 billion miles) away. The 1980s brought the launch of NASA's space shuttle program, a 30-year effort that sent more than 300 astronauts representing 16 countries into space on 135 missions to repair satellites, conduct research, and build the International Space Station (ISS). The ISS has been continuously occupied since 2000, and it serves as a collaborative laboratory for astronauts from many different countries.

SPACE TECHNOLOGY

Rocketry, computers, advanced sensors, high-resolution cameras, human life-support systems, communications—all manner of technological advances have helped scientists study the objects in our solar system and beyond. Space missions have analyzed the atmospheres and mapped the surfaces of planets and moons and monitored mysterious solar flares. Probes have touched down on a comet, and the two trailblazing Voyager crafts have crossed into interstellar space. The James Webb Space Telescope, slated to launch in 2021, will explore the most distant objects to date by detecting infrared light, a way to glimpse back at the earliest moments of the universe.

In recent years, NASA has made the exploration of Mars a high priority, with plans for humans on Mars in the 2030s. Spacecraft have captured detailed images of the moon's surface and confirmed the presence of water ice near its poles; future plans call for surveys of the moon's south pole and the creation of permanent research outposts. Other countries—including the European Union, China, India, Japan, and Korea—and private industry have become key players in space exploration. SpaceX, for instance, is developing technology to launch cargo, crewed missions, and eventually passengers to Mars, with the long-term goal of setting up base camps for astronauts—and someday establishing a colony there.

An artist's model of the James Webb Space Telescope

The exoplanet K2-18b, which has water and conditions that could support life

EXOPLANETS

At the heart of our quest for exoplanets—planets beyond our solar system—is a search for life elsewhere in the universe. For a planet to be habitable, it needs certain key features. Once scientists know a planet exists and can be sure of its mass and radius, they assess its density and composition, atmosphere, and even wind patterns. Is it gaseous, like Jupiter and Saturn? Or rocky, like Earth and Mars? Temperature is a major factor: Scientists are looking for planets that are neither too hot nor too cold to allow the presence of liquid water. And they seek planets with atmospheres capable of supporting life. The presence of methane or oxygen could be an indicator, but scientists have also come up with chemical combinations that could support other life forms. Some exoplanets like the super-Earth 55 Cancri e—which has hints of a carbon-rich atmosphere—could have layers of graphite or diamonds, making them potential "diamond planets."

The first discovery of exoplanets came in 1992, and the field has rapidly expanded ever since: As of 2019, scientists have identified more than 4,000 confirmed exoplanets, many found with the Kepler space telescope. TESS, a Transiting Exoplanet Survey Satellite launched in 2018, is expected to find some 20,000 exoplanets in its two-year prime mission, including dozens of Earth-size planets.

SNAPSHOT | *The Solar System* FACTS & FEATURES

SHORTEST DAY Jupiter: 9.9 hours

SHORTEST YEAR Mercury: 88 days

DENSEST ATMOSPHERE Venus: at the surface, air weighs 90 times more than Earth's

HOTTEST PLANET Venus: 475° Celsius (887° Fahrenheit) surface temp

HIGHEST OBSERVED MOUNTAIN Mars's Olympus Mons, about three times as tall as Everest

MOST KNOWN SATELLITES Saturn: 82

LARGEST PLANET Jupiter: about 318 times the mass of Earth

MOST VOLCANIC ACTIVITY Jupiter's moon Io

LARGEST OBSERVED STORM Jupiter's Great Red Spot: has lasted at least 300 years

LARGEST MOON Jupiter's moon Ganymede: three-quarters the size of Mars

OUR **SUN**, THE NEAREST STAR, IS A **MILLION TIMES THE SIZE OF EARTH** IN VOLUME. IT WOULD TAKE A SPACE SHUTTLE **SEVEN MONTHS** TO FLY THERE.

MOON: NEAR SIDE

Young Earth had no moon. At some point within Earth's first 100 million years, an object roughly the size of Mars struck our planet with a great, glancing blow. Most of the rogue body and a sizable chunk of Earth were vaporized. The ensuing cloud condensed into solid particles that orbited Earth, aggregated into ever larger moonlets, and eventually combined to form our one moon. The moon is gravitationally locked with Earth, so that while it rotates, the near side always faces Earth.

Its complexity visible to the naked eye, the moon's surface includes lighter-colored highlands and darker maria, from the Latin for "seas," even though these basins hold no water. These circular depressions—in fact, basaltic lava flows more than three billion years old—make up the face of the "man in the moon."

A telescope brings into view smaller impact features, many of them scars left by objects that struck the moon long ago. The largest scars are the impact basins, ranging up to about 2,500 kilometers (1,600 miles) across. Wrinkled ridges, domed hills, and fissures distinguish the near side's landscape. The lunar mountains, called *montes* on the map, were all formed by impacts and not by tectonic plate movement, as were mountains on Earth.

Although it has rugged terrain, the south pole of the moon may prove a viable mission landing site, since the deep craters there contain water ice from comet impacts—a potential source of drinking water plus hydrogen and oxygen for fuel. The barren lunar surface is scattered with regolith—powdery rocky rubble from the constant barrage of impacts. Some believe the lunar regolith could become the building blocks of habitations on the moon.

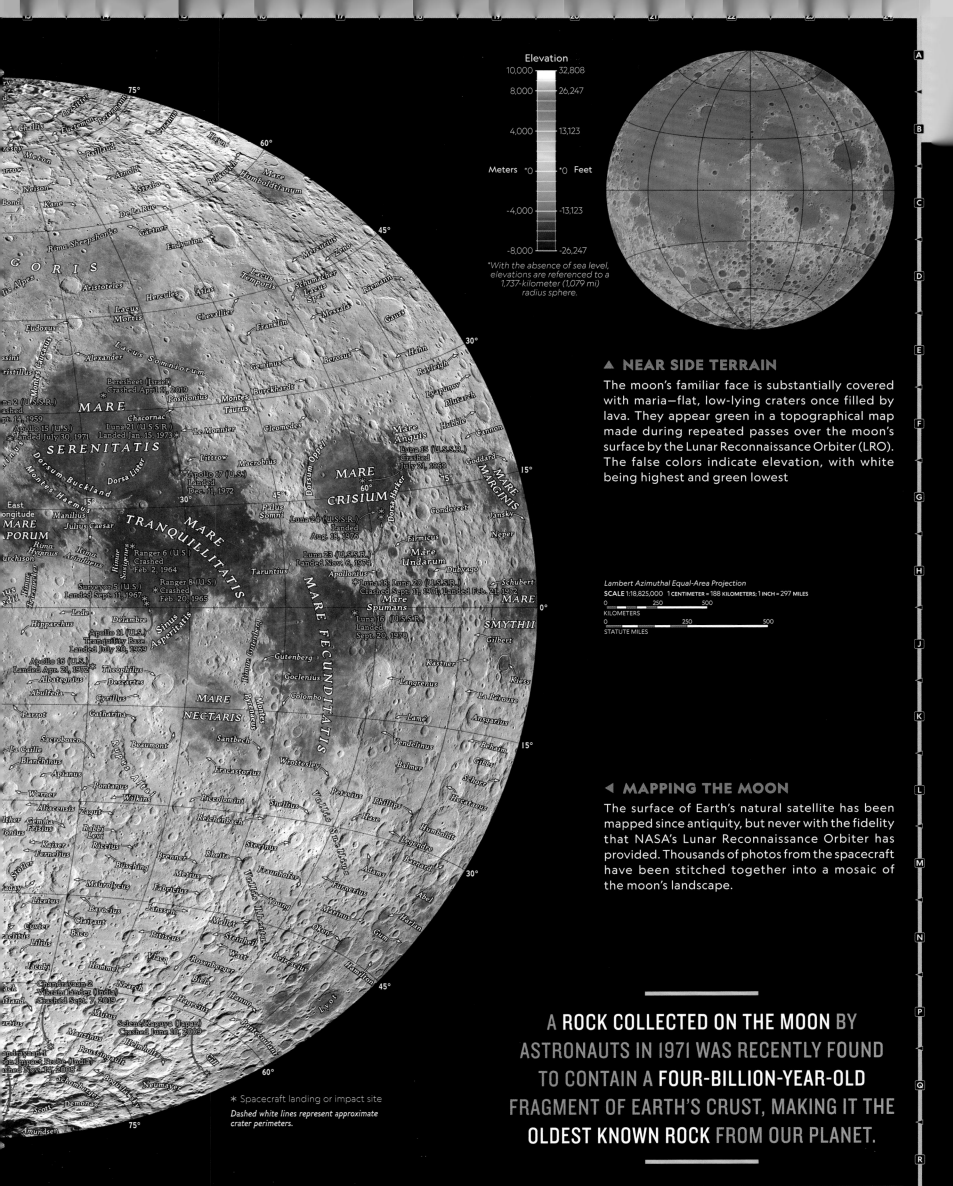

Elevation

Meters		Feet
10,000		32,808
8,000		26,247
4,000		13,123
*0		*0
-4,000		-13,123
-8,000		-26,247

*With the absence of sea level, elevations are referenced to a 1,737-kilometer (1,079 mi) radius sphere.

▲ NEAR SIDE TERRAIN

The moon's familiar face is substantially covered with maria—flat, low-lying craters once filled by lava. They appear green in a topographical map made during repeated passes over the moon's surface by the Lunar Reconnaissance Orbiter (LRO). The false colors indicate elevation, with white being highest and green lowest

Lambert Azimuthal Equal-Area Projection

SCALE 1:18,825,000   1 CENTIMETER = 188 KILOMETERS; 1 INCH = 297 MILES

```
0        250        500
KILOMETERS
0              250              500
STATUTE MILES
```

◄ MAPPING THE MOON

The surface of Earth's natural satellite has been mapped since antiquity, but never with the fidelity that NASA's Lunar Reconnaissance Orbiter has provided. Thousands of photos from the spacecraft have been stitched together into a mosaic of the moon's landscape.

A **ROCK COLLECTED ON THE MOON** BY ASTRONAUTS IN 1971 WAS RECENTLY FOUND TO CONTAIN A **FOUR-BILLION-YEAR-OLD** FRAGMENT OF EARTH'S CRUST, MAKING IT THE **OLDEST KNOWN ROCK** FROM OUR PLANET.

✳ Spacecraft landing or impact site

Dashed white lines represent approximate crater perimeters.

MOON: FAR SIDE

Many call it the "dark side of the moon," but the side we cannot see is just as sunlit as the side we always see. Still, because it always faces away from Earth and remains beyond radio reach, only since the age of space travel have we learned of its terrain, with more craters and more extreme heights and depths than the near side.

In early 2019, China's Chang'e 4 spacecraft landed on the moon's far side and began relaying back data via a satellite. Named for a Chinese goddess who lived on the moon, Chang'e is equipped with cameras and radar capable of penetrating the ground. It will explore the Von Kármán crater within the South Pole–Aitken basin, where scientists think a meteor impact during the solar system's earliest years exposed rock from deep within the moon. Research by Chang'e 4 could confirm the date of this impact and provide more insight into the history of the solar system.

▼ PHASES OF THE MOON

As the moon orbits Earth during its 29.5-day cycle, the section that is illuminated by the sun—and thus visible—changes. A full moon occurs at its orbital point farthest from the sun. A blacked-out new moon occurs when the moon is between the sun and Earth.

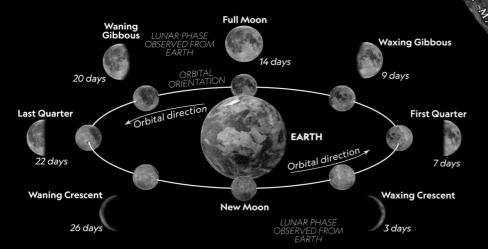

Waning Gibbous
LUNAR PHASE OBSERVED FROM EARTH
Full Moon
20 days
14 days
Waxing Gibbous
9 days
ORBITAL ORIENTATION
Last Quarter
Orbital direction
EARTH
First Quarter
22 days
Orbital direction
7 days
Waning Crescent
New Moon
Waxing Crescent
26 days
LUNAR PHASE OBSERVED FROM EARTH
3 days

Sun Direction

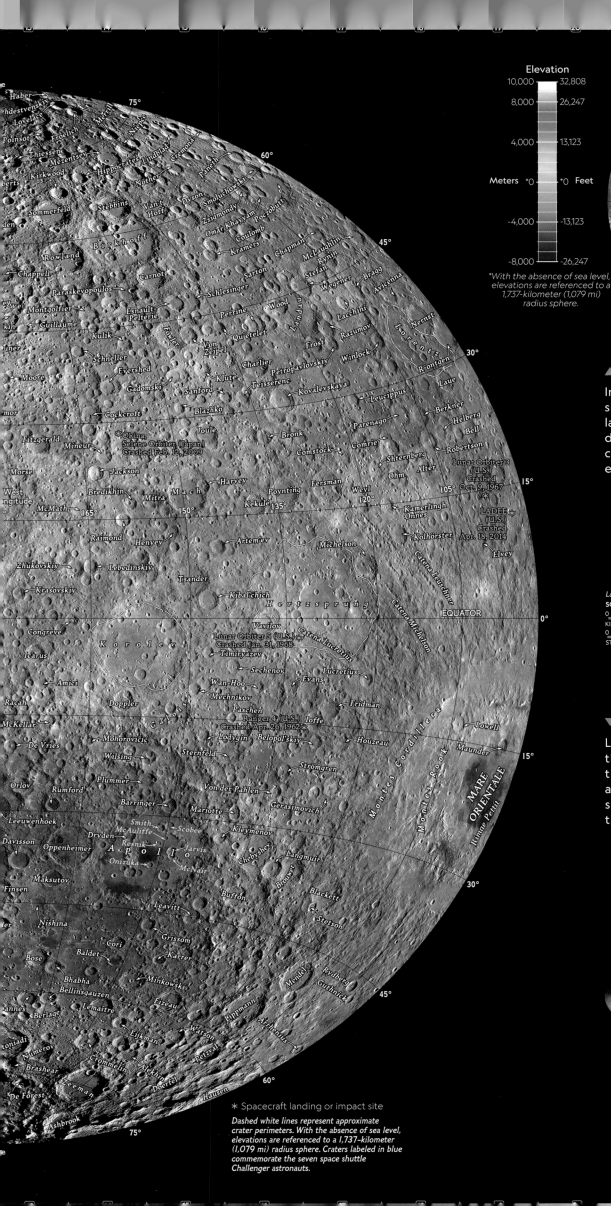

Elevation

Meters	Feet
10,000	32,808
8,000	26,247
4,000	13,123
*0	*0
-4,000	-13,123
-8,000	-26,247

With the absence of sea level, elevations are referenced to a 1,737-kilometer (1,079 mi) radius sphere.

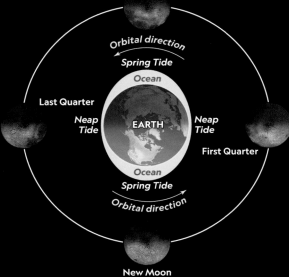

▲ FAR SIDE TERRAIN

In contrast to the near side, the crater-pocked far side—unseen until the age of space travel—lacks large maria but features much greater heights and deeper depths. The LRO's accurate elevation data could help pave the way for future landings and even moon base construction.

Lambert Azimuthal Equal-Area Projection
SCALE 1:18,825,000   1 CENTIMETER = 188 KILOMETERS; 1 INCH = 297 MILES

KILOMETERS
0 — 250 — 500

STATUTE MILES
0 — 250 — 500

▼ LUNAR INFLUENCE ON TIDES

Lunar gravity pulls Earth's oceans toward the moon; the sun's gravity also augments or dampens the tidal effect. When the sun, moon, and Earth are aligned—during a new or full moon—a more intense spring tide is produced. A neap tide occurs when the moon's pull is perpendicular to that of the sun.

Full Moon
Orbital direction
Spring Tide
Last Quarter
Neap Tide
EARTH
Ocean
Ocean
Neap Tide
First Quarter
Spring Tide
Orbital direction
New Moon
Sun Direction

✳ Spacecraft landing or impact site

Dashed white lines represent approximate crater perimeters. With the absence of sea level, elevations are referenced to a 1,737-kilometer (1,079 mi) radius sphere. Craters labeled in blue commemorate the seven space shuttle Challenger astronauts.

INNER SOLAR SYSTEM

Measured to its ultimate boundary, the sun's energy spreads to an area as large as 46 billion kilometers (30 billion miles) across, while the diameter of the star itself is just 1.4 million kilometers (865,000 miles). But the sun accounts for about 99.8 percent of the mass of the entire solar system. Everything else—planets, asteroids, meteoroids, comets, dust, and gas—are bits left over from the formation of this medium-size star about 4.5 billion years ago. We call the planets of the inner solar system the "terrestrial planets," from the scorching temperatures of Mercury to the deep winter chill of Mars. Each one has been visited and mapped in detail by spacecraft.

Mercury's surface is densely cratered. Temperature varies dramatically: It can reach 430°C (800°F) in daylight, and fall to –180°C (–290°F) at night. A nearly vertical axis means relatively little sunlight touches polar regions, where radar reveals hints of water ice.

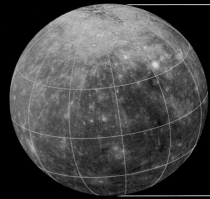

MERCURY

Average distance from the sun:	57,900,000 km
Perihelion:	46,000,000 km
Aphelion:	69,820,000 km
Revolution period:	88 days
Average orbital speed:	47.9 km/s
Average temperature:	167°C
Rotation period:	58.7 days
Equatorial diameter:	4,879 km
Mass (Earth=1):	0.055
Density:	5.43 g/cm³
Surface gravity (Earth=1):	0.38
Known satellites:	none

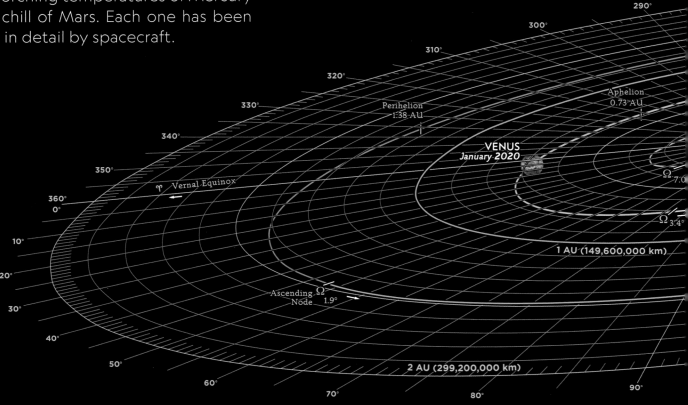

VENUS
January 2020

290°
300°
310°
320°
330°
340°
350°
360°
0°
10°
20°
30°
40°
50°
60°
70°
80°
90°

Aphelion
0.73 AU

Perihelion
1.38 AU

♈ Vernal Equinox

Ω 7.0°

Ω 3.4°

1 AU (149,600,000 km)

Ascending Ω
Node 1.9°

2 AU (299,200,000 km)

DISTANCES AMONG THE PLANETS ARE MEASURED IN **ASTRONOMICAL UNITS (AU)**. ONE AU EQUALS THE DISTANCE FROM EARTH TO THE SUN, ABOUT 150 MILLION KILOMETERS (93 MILLION MILES).

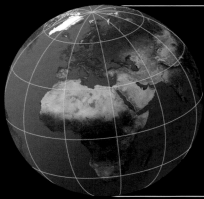

EARTH

Average distance from the sun:	149,600,000 km
Perihelion:	147,090,000 km
Aphelion:	152,100,000 km
Revolution period:	365.26 days
Average orbital speed:	29.8 km/s
Average temperature:	15°C
Rotation period:	23.9 hours
Equatorial diameter:	12,756 km
Mass:	5,973,600,000 trillion metric tons
Density:	5.52 g/cm³
Surface gravity:	9.78 m/s²
Known satellites:	1
Largest satellite:	Earth's moon

Earth is the only planet in the solar system known to support life, and it may also be the sole place where liquid water is abundant. Its heavy iron core creates a strong magnetic field, blocking the sun's constant bombardment of high-energy particles.

Venus's thick atmosphere contains mostly carbon dioxide and soaks up a great deal of the sun's energy—making it the hottest planet in the solar system. The surface has rolling plains and mountainous regions, and intense volcanism has buried many impact craters.

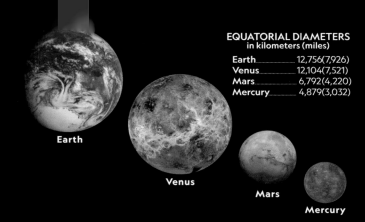

EQUATORIAL DIAMETERS
in kilometers (miles)

Earth	12,756 (7,926)
Venus	12,104 (7,521)
Mars	6,792 (4,220)
Mercury	4,879 (3,032)

Earth
Venus
Mars
Mercury

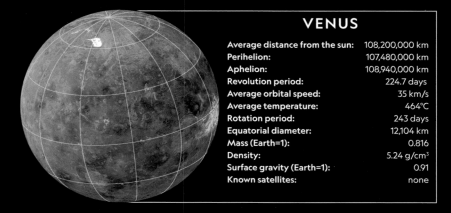

VENUS

Average distance from the sun:	108,200,000 km
Perihelion:	107,480,000 km
Aphelion:	108,940,000 km
Revolution period:	224.7 days
Average orbital speed:	35 km/s
Average temperature:	464°C
Rotation period:	243 days
Equatorial diameter:	12,104 km
Mass (Earth=1):	0.816
Density:	5.24 g/cm³
Surface gravity (Earth=1):	0.91
Known satellites:	none

▲ SIZING UP THE TERRESTRIAL PLANETS

The inner planets are shown above in proportionate size to one another. Dwarf planets are less than 3,000 kilometers (1,865 miles) in diameter—much smaller than Mercury.

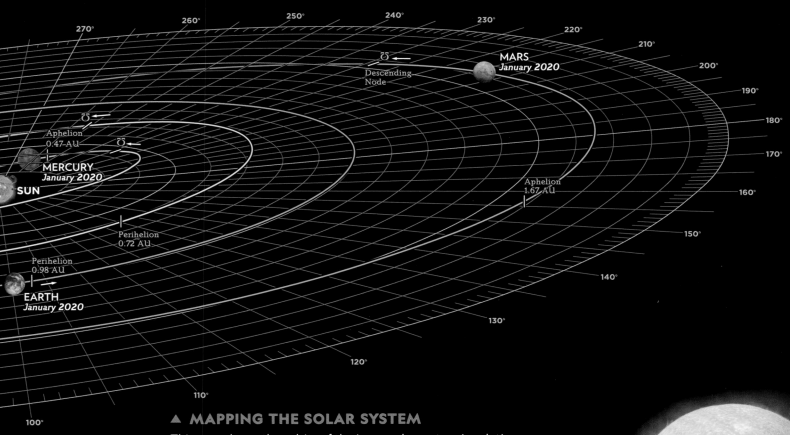

MARS
January 2020

Descending
Node

Aphelion
0.47 AU

MERCURY
January 2020

SUN

Aphelion
1.67 AU

Perihelion
0.72 AU

Perihelion
0.98 AU

EARTH
January 2020

▲ MAPPING THE SOLAR SYSTEM

This map shows the orbits of the inner solar system in relation to the plane of Earth's orbit around the sun, called the ecliptic. The blue rings show distance from the sun in astronomical units and radial lines show degrees of longitude.

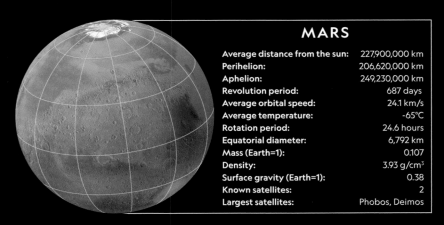

MARS

Average distance from the sun:	227,900,000 km
Perihelion:	206,620,000 km
Aphelion:	249,230,000 km
Revolution period:	687 days
Average orbital speed:	24.1 km/s
Average temperature:	-65°C
Rotation period:	24.6 hours
Equatorial diameter:	6,792 km
Mass (Earth=1):	0.107
Density:	3.93 g/cm³
Surface gravity (Earth=1):	0.38
Known satellites:	2
Largest satellites:	Phobos, Deimos

Dust storms sweep across Mars's barren surface, shrouding it in a reddish haze. Southern latitudes are rugged and heavily cratered; giant volcanoes rise above plains to the north. Mars is more like Earth than any other solar system planet is, and it could potentially harbor traces of past life.

SUN

Average surface temperature:	5,500°C
Average core temperature:	16,000,000°C
Rotation period:	24.6 days
Equatorial diameter:	1,392,000 km
Mass (Earth=1):	332,950
Density:	1.41 g/cm³
Surface gravity (Earth=1):	28.0

The sun is an average star—mid-range in temperature, energy output, and size. Its energy is the driving force of weather on every planet and moon with an atmosphere.

MARS: WESTERN HEMISPHERE

Mars is vastly colder and drier than Earth; its thin atmosphere, composed mostly of carbon dioxide, provides no protection from solar radiation. The tallest mountain— the mighty volcano Olympus Mons—is more than two and a half times the height of Earth's tallest volcanic peak, Mauna Kea. Despite these extremes, Mars and Earth are alike in one key way: liquid water. Ancient water seems to have shaped Mars's surface, which is rife with features carved by flowing water. Now scientists are focused on finding and harvesting water ice beneath the surface for human missions to the planet. NASA's Curiosity rover is monitoring unexplained fluctuations in tiny amounts of methane and oxygen in the atmosphere, a mystery tp atmospheric chemists and biologists. In coming years, several nations' space agencies plan missions to Mars to study whether it once supported life—and if it might do so again.

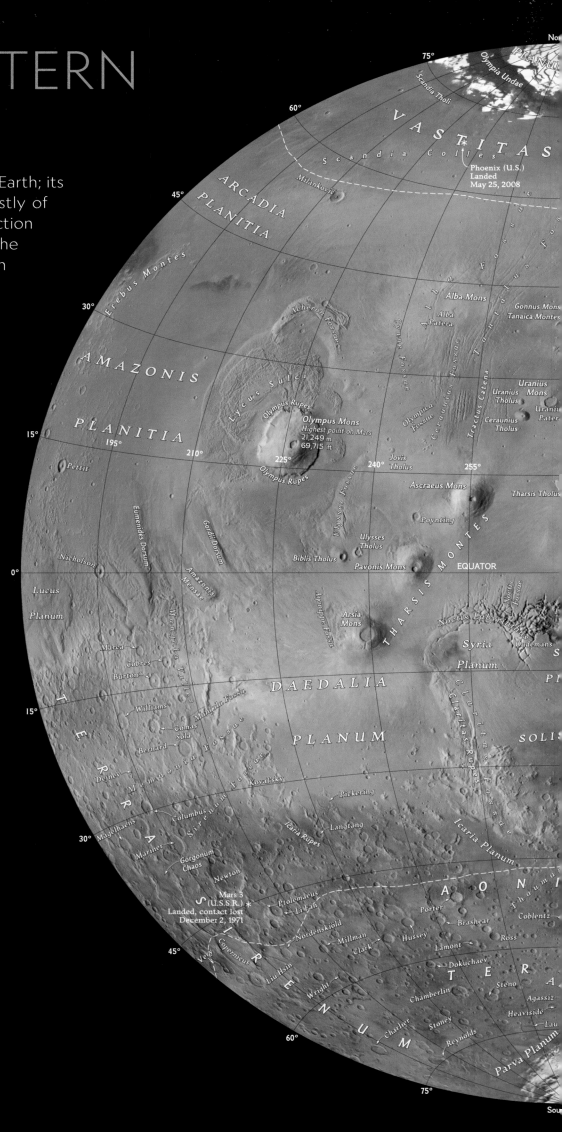

▼ PHOBOS

This irregular moon whips around Mars three times a day, orbiting just 6,000 kilometers (3,700 miles) above the surface. The small satellite is only 28.6 kilometers (16.7 miles) across on its longest axis.

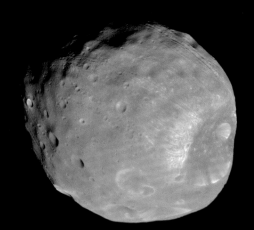

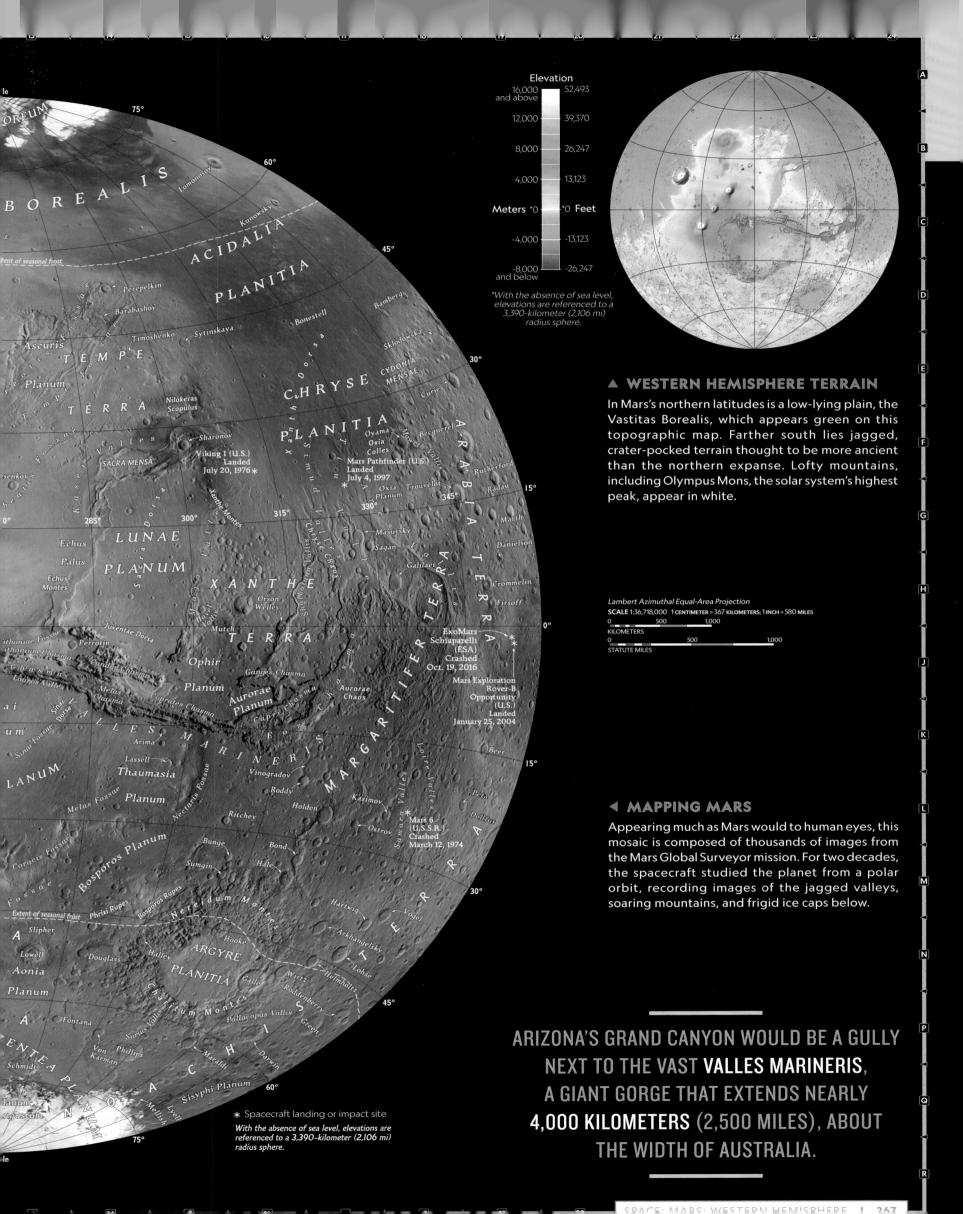

Elevation

Meters	Feet
16,000 and above	52,493
12,000	39,370
8,000	26,247
4,000	13,123
*0	*0
-4,000	-13,123
-8,000 and below	-26,247

*With the absence of sea level, elevations are referenced to a 3,390-kilometer (2,106 mi) radius sphere.

▲ WESTERN HEMISPHERE TERRAIN

In Mars's northern latitudes is a low-lying plain, the Vastitas Borealis, which appears green on this topographic map. Farther south lies jagged, crater-pocked terrain thought to be more ancient than the northern expanse. Lofty mountains, including Olympus Mons, the solar system's highest peak, appear in white.

Lambert Azimuthal Equal-Area Projection
SCALE 1:36,718,000 1 CENTIMETER = 367 KILOMETERS; 1 INCH = 580 MILES

0	500	1,000
KILOMETERS

0	500	1,000
STATUTE MILES

◄ MAPPING MARS

Appearing much as Mars would to human eyes, this mosaic is composed of thousands of images from the Mars Global Surveyor mission. For two decades, the spacecraft studied the planet from a polar orbit, recording images of the jagged valleys, soaring mountains, and frigid ice caps below.

✳ Spacecraft landing or impact site

With the absence of sea level, elevations are referenced to a 3,390-kilometer (2,106 mi) radius sphere.

ARIZONA'S GRAND CANYON WOULD BE A GULLY NEXT TO THE VAST **VALLES MARINERIS**, A GIANT GORGE THAT EXTENDS NEARLY **4,000 KILOMETERS** (2,500 MILES), ABOUT THE WIDTH OF AUSTRALIA.

MARS: EASTERN HEMISPHERE

Two notable astronomers, Eugène Antoniadi and Giovanni Schiaparelli, crafted maps of the Martian surface based on their observations in the latter half of the 19th century. They used names out of classical mythology, establishing the precedent that the International Astronomical Union (IAU) came to adopt for Mars and most of the other bodies in our solar system. Craters are named in honor of scientists, writers, and mathematicians, both ancient and modern.

The Martian landscape is both familiar and alien. All of its features, from rugged riverbeds to shifting sand dunes, are also found on Earth. Yet Mars, with its lower gravity and much thinner atmosphere, imprints its own character on these features: The volcanoes are taller, the canyons wider, and the ice caps more ephemeral than on Earth. Mars's polar caps have frozen water, like our Arctic and Antarctic, but during the winters frozen carbon dioxide also coats the poles. The search continues for conclusive evidence that life once existed on the red planet.

MARS HAS WINTERS AND SUMMERS SIMILAR TO EARTH'S. A WINTER LAYER OF FROZEN CARBON DIOXIDE—DRY ICE—APPEARS AT THE POLE TILTED AWAY FROM THE SUN AND DISAPPEARS WHEN THE SUN RETURNS.

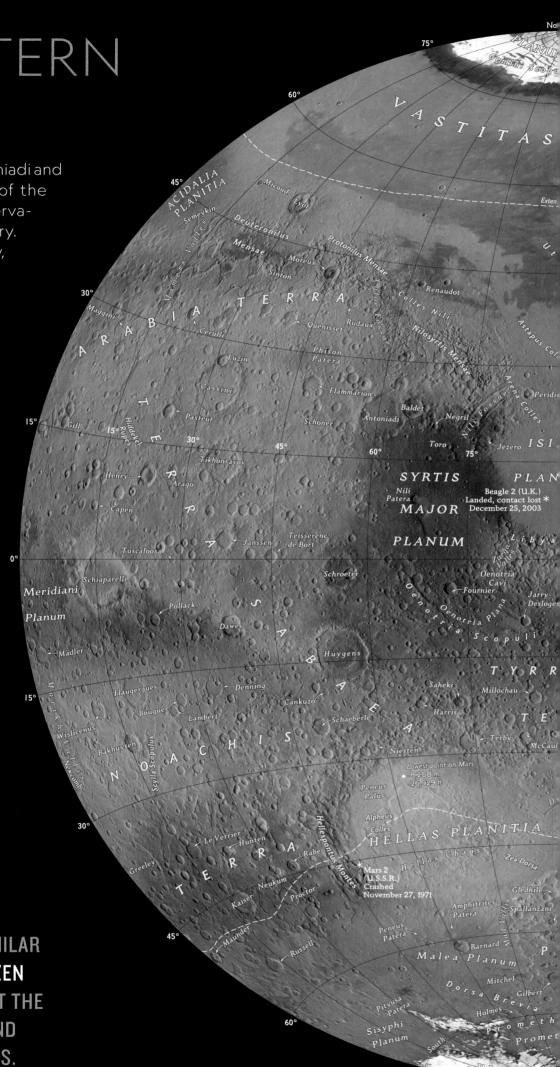

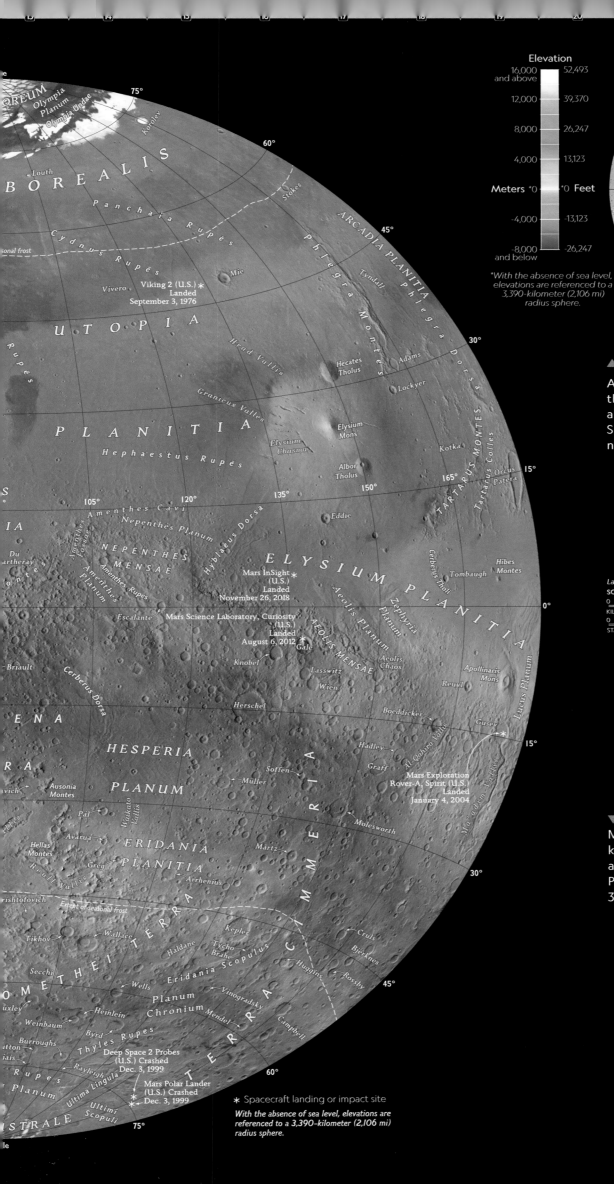

Elevation

Meters	Feet
16,000 and above	52,493
12,000	39,370
8,000	26,247
4,000	13,123
*0	*0
-4,000	-13,123
-8,000 and below	-26,247

*With the absence of sea level, elevations are referenced to a 3,390-kilometer (2,106 mi) radius sphere.

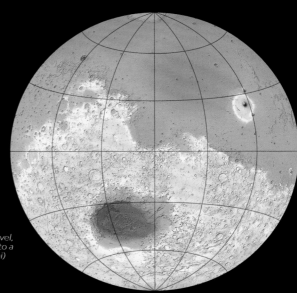

▲ EASTERN HEMISPHERE TERRAIN

Amid the cratered southern hemisphere of Mars is the giant, deep-impact basin Hellas Planitia; its abysmal depths reach –8,208 meters (–26,929 feet). Scientists theorize that the relatively smooth northern hemisphere was formed by lava flows.

Lambert Azimuthal Equal-Area Projection
SCALE 1:36,718,000 1 CENTIMETER = 367 KILOMETERS; 1 INCH = 580 MILES

0   500   1,000
KILOMETERS
0   500   1,000
STATUTE MILES

▼ DEIMOS

Mars's lumpy moon orbits the planet at 23,460 kilometers (14,577 miles) above the surface, smaller and more distant than Mars's other satellite, Phobos. It swings around the planet once every 30 hours.

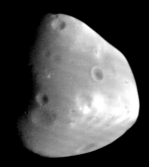

✳ Spacecraft landing or impact site

With the absence of sea level, elevations are referenced to a 3,390-kilometer (2,106 mi) radius sphere.

OUTER SOLAR SYSTEM

Our solar system has two classes of planets. Terrestrial planets, including Earth, are small with solid surfaces and mean densities that suggest an iron core surrounded by a rocky, partially molten mantle. Gas planets populate the outer solar system: very large planetary bodies consisting primarily of hydrogen and helium in gas and liquid states. All four gas planets have rings.

Rocky, icy bodies beyond Neptune's orbit make up the Kuiper belt, an icy analogue of the asteroid belt between Mars and Jupiter. Some of these objects are comparable to Pluto, no longer considered a planet. In 2006, the International Astronomical Union (IAU), an organization of professional astronomers that oversees names and classifications, redesignated Pluto, citing that while it was large enough to be roughly spherical its mass did not exert the gravity needed to clear its orbit by pulling smaller objects into itself. So now Pluto is considered a "dwarf planet."

Uranus seems to roll on its side, with its rings nearly perpendicular to those of the other gas giants. The planet's teal green surface is composed of clouds of methane gas; deeper down could be an ocean of superheated liquid water spiked with ammonia and methane.

URANUS

Average distance from the sun:	2,872,500,000 km
Perihelion:	2,741,300,000 km
Aphelion:	3,003,620,000 km
Revolution period:	83.81 years
Average orbital speed:	6.8 km/s
Average temperature:	-195°C
Rotation period:	17.2 hours
Equatorial diameter:	51,118 km
Mass (Earth=1):	14.5
Density:	1.27 g/cm³
Surface gravity (Earth=1):	0.89
Known satellites:	27
Largest satellites:	Titania, Oberon, Umbriel, Ariel

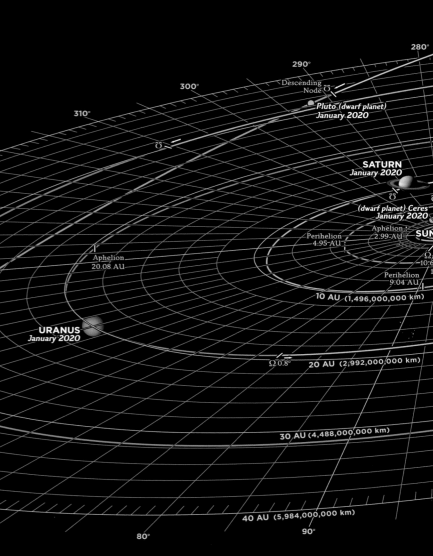

NEPTUNE

Average distance from the sun:	4,495,100,000 km
Perihelion:	4,444,450,000 km
Aphelion:	4,545,670,000 km
Revolution period:	163.84 years
Average orbital speed:	5.4 km/s
Average temperature:	-200°C
Rotation period:	16.1 hours
Equatorial diameter:	49,528 km
Mass (Earth=1):	17.1
Density:	1.64 g/cm³
Surface gravity (Earth=1):	1.12
Known satellites:	14
Largest satellite:	Triton

▲ ASTEROIDS

Asteroids are the rocky remnants of the age of planetary formation. They range from gravel-size to gigantic: Ceres, at 950 kilometers (590 miles) in diameter, is classified as a dwarf planet. The greatest concentration is in the asteroid belt between Mars and Jupiter, where Jupiter's gravity kept these objects from accreting into a planet.

Neptune is the solar system's outermost planet and the smallest of the gas giants. Its active weather systems produce unusual easterly winds, blowing in opposition to its rotation. Surrounded by a faint ring system, Neptune's depths contain ammonia, methane, water ice, and rock.

Jupiter's famous Great Red Spot, wider than Earth itself, is an intense storm that has raged for 300 years. The solar system's largest planet has 79 moons, more than any planet in the solar system. The four largest were among Galileo's first telescopic discoveries.

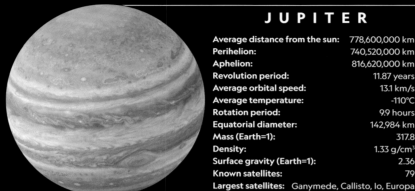

J U P I T E R

Average distance from the sun:	778,600,000 km
Perihelion:	740,520,000 km
Aphelion:	816,620,000 km
Revolution period:	11.87 years
Average orbital speed:	13.1 km/s
Average temperature:	-110°C
Rotation period:	9.9 hours
Equatorial diameter:	142,984 km
Mass (Earth=1):	317.8
Density:	1.33 g/cm³
Surface gravity (Earth=1):	2.36
Known satellites:	79
Largest satellites:	Ganymede, Callisto, Io, Europa

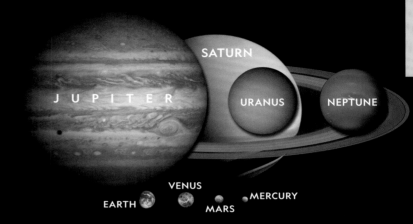

▲ PLANETARY COMPARISON

From Jupiter to Mercury, here's how the eight planets of the solar system stack up. The outer gas planets are giants when compared to the inner terrestrial planets. Each has a constellation of moons, and some—like Jupiter's Ganymede—would be large enough to qualify as planets if they orbited the sun.

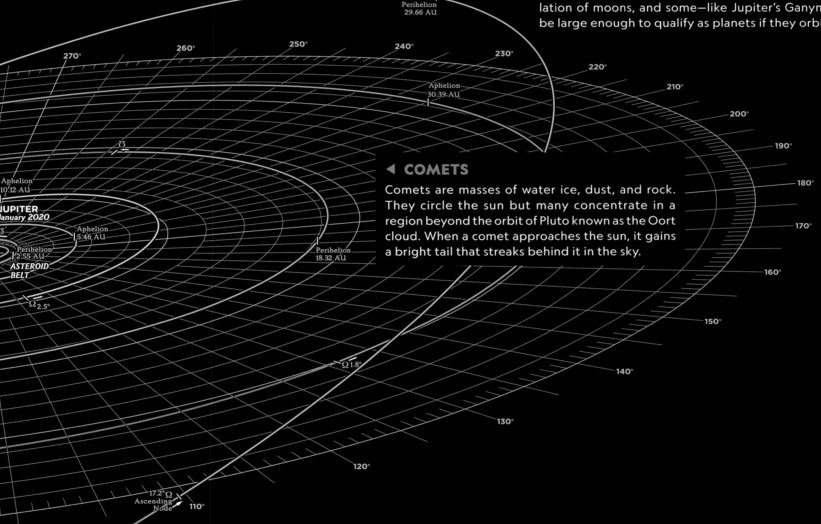

Perihelion 29.66 AU

270° 260° 250° 240° 230° 220° 210° 200° 190°

Aphelion 30.39 AU

Aphelion 10.12 AU

JUPITER January 2020

Aphelion 5.46 AU

Perihelion 2.55 AU

ASTEROID BELT

Ω 2.5°

Perihelion 18.32 AU

180°

170°

160°

150°

140°

130°

120°

110°

100°

Ω 1.8°

17.2° Ω
Ascending Node

◄ COMETS

Comets are masses of water ice, dust, and rock. They circle the sun but many concentrate in a region beyond the orbit of Pluto known as the Oort cloud. When a comet approaches the sun, it gains a bright tail that streaks behind it in the sky.

THE ONLY SPACE BODIES WITH KNOWN SURFACE LIQUID ARE EARTH (WATER) AND SATURN'S MOON TITAN (METHANE AND ETHANE.)

S A T U R N

Average distance from the sun:	1,433,500,000 km
Perihelion:	1,352,550,000 km
Aphelion:	1,514,500,000 km
Revolution period:	29.44 years
Average orbital speed:	9.7 km/s
Average temperature:	-140°C
Rotation period:	10.7 hours
Equatorial diameter:	120,536 km
Mass (Earth=1):	95.2
Density:	0.69 g/cm³
Surface gravity (Earth=1):	0.92
Known satellites:	82
Largest satellites:	Titan, Rhea, Iapetus, Dione, Tethys

Saturn's rings are one of the solar system's most majestic sights. Made of water ice particles, the rings are vast in diameter but as little as 100 meters (330 feet) thick. Saturn has a fast rotational spin of 10.7 hours; winds can reach 1,800 kilometers (1,100 miles) an hour.

COSMIC JOURNEYS

Explorations of the solar system have advanced because of human curiosity, the will to understand the universe, and sometimes as a matter of national pride. Presented here is every mission of exploration completed or under way that has a goal of studying the bodies of our solar system. These efforts are truly international, including private companies as well as governments. Recently, probes have touched down on a comet, visited distant dwarf planet Pluto, mapped the surfaces of planets and moons throughout the solar system, and delved into the mysteries of the sun.

Missions to Inner Solar System
— NASA
— U.S.S.R./Russia
— European Space Agency or other european country
— Japan
— China
— India
— Israel
— Failed mission
— Failed
— Failed
— Failed
— Failed
— Failed
— Failed

Deep Space Missions
····· Pioneer—NASA
····· Voyager—NASA
— Galileo—NASA & European Space Agency
— Cassini-Huygens— NASA & European Space Agency
— New Horizons—NASA
— Juno—NASA

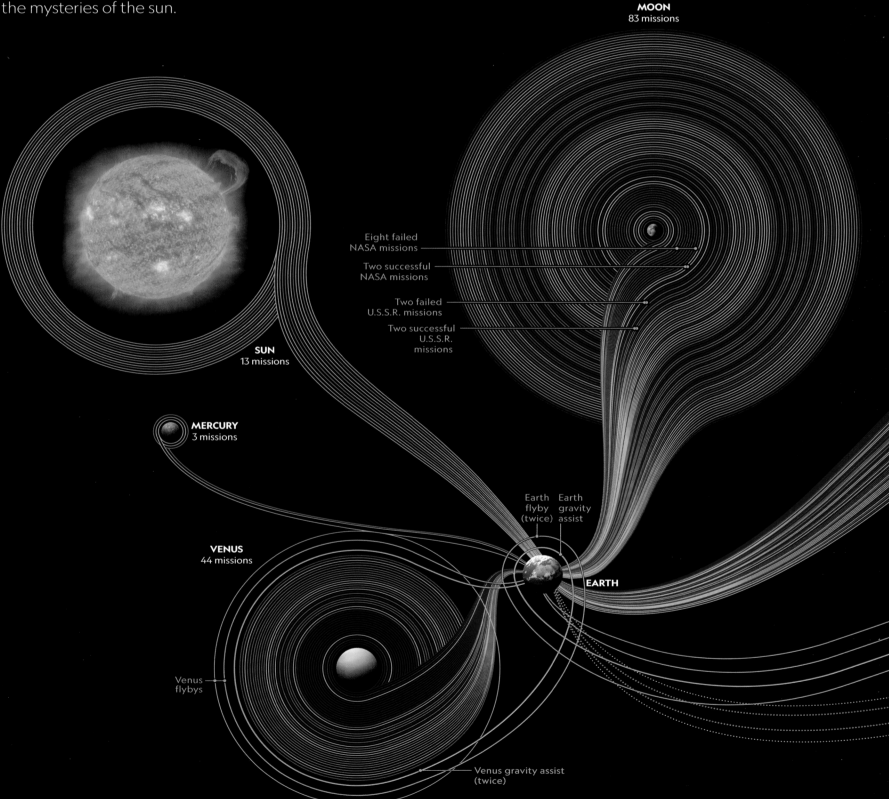

MOON
83 missions

Eight failed NASA missions

Two successful NASA missions

Two failed U.S.S.R. missions

Two successful U.S.S.R. missions

SUN
13 missions

MERCURY
3 missions

Earth flyby (twice)

Earth gravity assist

VENUS
44 missions

EARTH

Venus flybys

Venus gravity assist (twice)

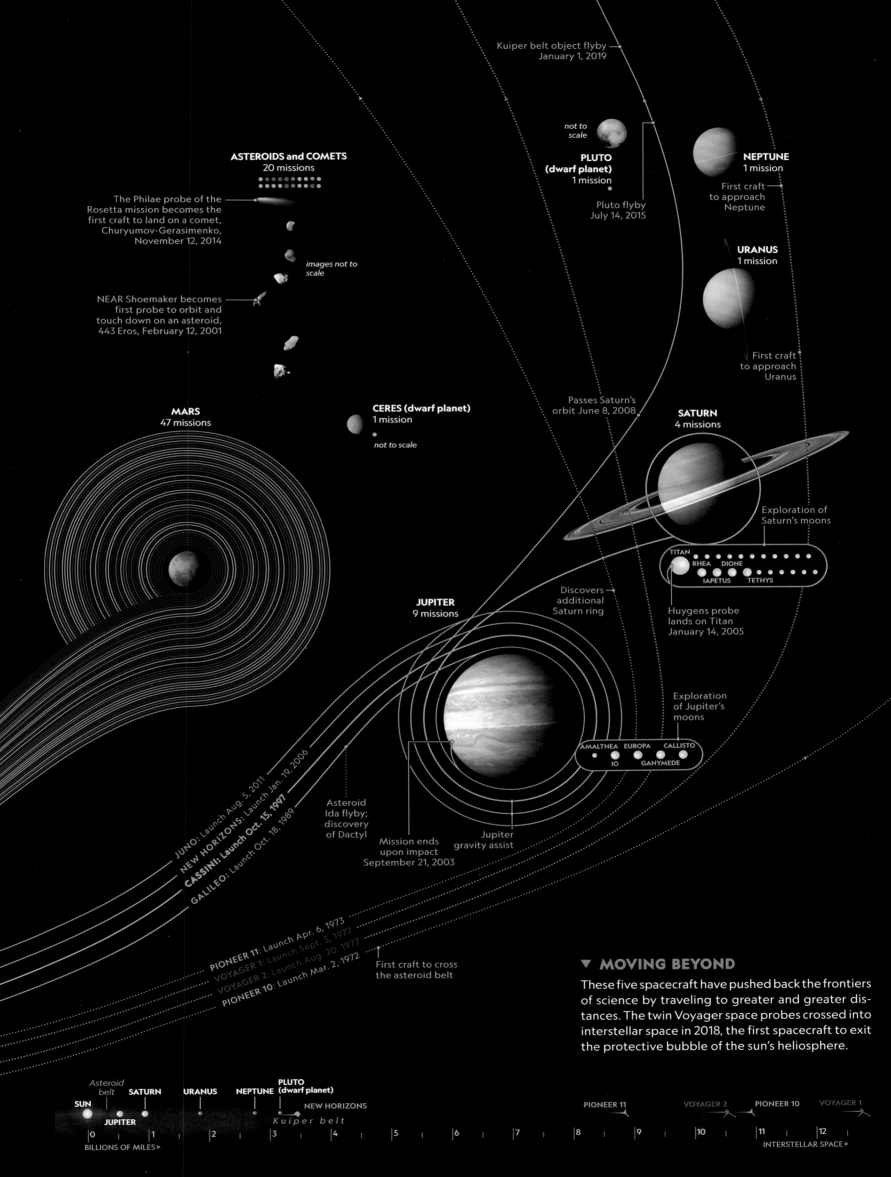

ASTEROIDS and COMETS
20 missions

The Philae probe of the Rosetta mission becomes the first craft to land on a comet, Churyumov-Gerasimenko, November 12, 2014

images not to scale

NEAR Shoemaker becomes first probe to orbit and touch down on an asteroid, 443 Eros, February 12, 2001

CERES (dwarf planet)
1 mission

not to scale

MARS
47 missions

Kuiper belt object flyby January 1, 2019

not to scale

PLUTO (dwarf planet)
1 mission

Pluto flyby July 14, 2015

NEPTUNE
1 mission

First craft to approach Neptune

URANUS
1 mission

First craft to approach Uranus

Passes Saturn's orbit June 8, 2008

SATURN
4 missions

Exploration of Saturn's moons

TITAN
RHEA   DIONE
IAPETUS   TETHYS

Huygens probe lands on Titan January 14, 2005

Discovers additional Saturn ring

JUPITER
9 missions

Exploration of Jupiter's moons

AMALTHEA   EUROPA   CALLISTO
IO   GANYMEDE

JUNO: Launch Aug. 5, 2011
NEW HORIZONS: Launch Jan. 19, 2006
CASSINI: Launch Oct. 15, 1997
GALILEO: Launch Oct. 18, 1989

Asteroid Ida flyby; discovery of Dactyl

Mission ends upon impact September 21, 2003

Jupiter gravity assist

PIONEER 11: Launch Apr. 6, 1973
VOYAGER 1: Launch Sept. 5, 1977
VOYAGER 2: Launch Aug. 20, 1977
PIONEER 10: Launch Mar. 2, 1972

First craft to cross the asteroid belt

▼ MOVING BEYOND

These five spacecraft have pushed back the frontiers of science by traveling to greater and greater distances. The twin Voyager space probes crossed into interstellar space in 2018, the first spacecraft to exit the protective bubble of the sun's heliosphere.

Asteroid belt
SUN   SATURN   URANUS   NEPTUNE   **PLUTO (dwarf planet)**
JUPITER   NEW HORIZONS
Kuiper belt

PIONEER 11   VOYAGER 2   PIONEER 10   VOYAGER 1

|0 |1 |2 |3 |4 |5 |6 |7 |8 |9 |10 |11 |12
BILLIONS OF MILES ▶   INTERSTELLAR SPACE ▶

MILKY WAY

The Milky Way is the galaxy—one of billions in the universe—to which our solar system belongs. Named for the luminous band of light visible in a dark night sky, our galaxy is made of some 200 billion stars. At its core is a bright, bar-shaped bulge of yellow and red stars containing a black hole millions of times more massive than the sun. From that center, several spiral arms sweep out to form a disk that contains younger, blue stars as well as glowing regions of star birth. Our solar system sits on the outskirts, nestled between major spiral arms.

Exploring the Milky Way galaxy requires us to reset our thinking about distances: If the sun were a bowling ball in the center of New York City, then all the planets of our solar system could be found within a dozen or so city blocks. The nearest star would require a trip to Hawaii, and to see all the other stars in our home galaxy we'd have to leave planet Earth.

PROFILE VIEW

GALACTIC HALO

GALACTIC BULGE

Globular star cluster

GALACTIC DISK

0°

30°

OUTER ARM

60°

SCUTU

SAGITTAR

Solar system orbit

P E R S E U S

90°

120°

30,000

40,000 light-years

Globular clusters

Halo

Disk

Dark matter

▲ GALAXY'S HALO

Far beyond the galactic disk, yet drawn by its gravity, lone stars and globular clusters wander the galaxy's spinning halo of gas. Regions of dark matter—a material unseen but known through its gravitational effects—extend beyond that.

OUR SUN AND THE SOLAR SYSTEM TRAVEL ALL THE WAY AROUND THE CENTER OF THE MILKY WAY—BUT IT TAKES SOME 250 MILLION YEARS TO COMPLETE THE JOURNEY.

▶ **GALACTIC PLANE**

The center line of the Milky Way's disk of stars is known as the galactic plane.

Galactic center

Galactic plane

EARTH

Ecliptic plane

SUN

300°

270°

330°

240°

FAR 3 KPC ARM

CORE

NEAR 3 KPC ARM

Direction of rotation

210°

SAGITTARIUS ARM

M–CENTAURUS ARM

10,000 light-years

YOU ARE HERE
SOLAR SYSTEM

ORION SPUR

20,000

US ARM

180°

PERSEUS ARM

ARM

Gl 687

Gl 570 A, B, C

AD Leonis

Gl 702 A, B

Wolf 359

Proxima Centauri

Eta Cassiopeiae A, B

Procyon A, B

SUN

Altair

Gl 663 A, B
Gl 664

Alpha Centauri A

Alpha Centauri B

5

Sirius A, B

10

15

Epsilon Eridani

20 light-years

▶ **THE SUN'S NEIGHBORHOOD**

There are relatively few stars near the sun, reducing risks to Earth from gravitational tugs, gamma-ray bursts, or collapsing stars called supernovae. The nearest star is Proxima Centauri, 4.24 light-years from Earth.

Gl 166 A, B, C

Tau Ceti

Epsilon Indi

Delta Pavonis

ABBREVIATIONS

MAP AND INDEX ABBREVIATIONS

A. _____ Arroio, Arroyo
A.C.T. _____ Australian Capital Territory
A.O. _____ Autonomous Oblast
A. Okr. _____ Autonomous Okrug
Adm. _____ Administrative
Af. _____ Africa
Afghan. _____ Afghanistan
Agr. _____ Agriculture
Ala. _____ Alabama
Alas. _____ Alaska
Alban. _____ Albania
Alg. _____ Algeria
Alta. _____ Alberta
Amer. _____ America-n
Amzns. _____ Amazonas
Anch. _____ Anchorage
And. & Nic. _____ Andaman and Nicobar Islands
Ant. _____ Antilles
Arch. _____ Archipelago, Archipiélago
Arg. _____ Argentina
Ariz. _____ Arizona
Ark. _____ Arkansas
Arkh. _____ Arkhangel'sk
Arm. _____ Armenia
Astrak. _____ Astrakhan'
Atl. Oc. _____ Atlantic Ocean
Aust. _____ Austria
Austral. _____ Australia
Auton. _____ Autonomous
Azerb. _____ Azerbaijan

B. _____ Baai, Baía, Baie, Bahía,Bay, Bugt-en, Buḩayrat
B. Aires _____ Buenos Aires
B.C. _____ British Columbia
B. Qazaq. _____ Batys Qazaqstan
Bashk. _____ Bashkortostan
Belg. _____ Belgium
Bol. _____ Bolivia
Bol. _____ Bol'sh-oy, -aya, -oye
Bosn. & Herzg. _____ Bosnia and Herzegovina
Br. _____ Branch
Braz. _____ Brazil
Bulg. _____ Bulgaria
Burya. _____ Buryatiya

C. _____ Cabo, Cap, Cape, Capo
C.H. _____ Court House
C.P. _____ Conservation Park
C.R. _____ Costa Rica
C.S.I. Terr. _____ Coral Sea Islands Territory
Cach. _____ Cachoeira
Calif. _____ California
Can. _____ Canada
Cap. _____ Capitán
Catam. _____ Catamarca
Cd. _____ Ciudad
Cen. Af. Rep. _____ Central African Republic
Cga. _____ Ciénaga
Chan. _____ Channel
Chap. _____ Chapada
Chech. _____ Chechnya
Chely. _____ Chelyabinsk
Chongq. _____ Chongqing Shi
Chuk. _____ Chukotskiy
Chuv. _____ Chuvashiya
Chyrv. _____ Chyrvony, -aya, -aye
Cmte. _____ Comandante
Cnel. _____ Coronel
Co.-s. _____ Cerro-s
Col. _____ Colombia
Colo. _____ Colorado
Conn. _____ Connecticut

Cord. _____ Cordillera
Corr. _____ Corrientes
Cr. _____ Creek, Crique
Croat. _____ Croatia

D. _____ Danau
D.C. _____ District of Columbia
D.F. _____ Distrito Federal
D.R.C. _____ Democratic Republic of the Congo
Del. _____ Delaware
Dem. _____ Democratic
Den. _____ Denmark
Dist. _____ District, Distrito
Dom. Rep. _____ Dominican Republic
Dr. _____ Doctor
Dz. _____ Dzong

E. _____ East-ern
E. Ríos _____ Entre Ríos
E. Santo _____ Espírito Santo
Ea. _____ Estancia
Ecua. _____ Ecuador
El Salv. _____ El Salvador
Emb. _____ Embalse
Eng. _____ England
Ens. _____ Ensenada
Entr. _____ Entrance
Eq. _____ Equatorial
Esc. _____ Escarpment
Est. _____ Estación
Est. _____ Estonia
Ét. _____ Étang
Eth. _____ Ethiopia
Eur. _____ Europe
Exp. _____ Exports
Ez. _____ Ezers

F. _____ Fiume
F.S.M. _____ Federated States of Micronesia
Falk. Is. _____ Falkland Islands
Fd. _____ Fiord, Fiordo, Fjord
Fed. _____ Federal, Federation
Fin. _____ Finland
Fk. _____ Fork
Fla. _____ Florida
Fn. _____ Fortín
Fr. _____ France, French
ft _____ feet
Ft. _____ Fort
Fy. _____ Ferry
F.Z. _____ Fracture zone

G. _____ Golfe, Golfo, Gulf
G. Altay _____ Gorno-Altay
G.R. _____ Game Reserve
Ga. _____ Georgia
Geb. _____ Gebergte, Gebirge
Gen. _____ General
Ger. _____ Germany
Gez. _____ Gezîra-t, Gezîret
Gezr. _____ Gezâir
Gl. _____ Glacier, Gletscher
Gob. _____ Gobernador
Gr. _____ Greece
Gr. _____ Gross-er
Gral. _____ General
Gt. _____ Great-er
Guang. _____ Guangdong

H.K. _____ Hong Kong
Hbr. _____ Harbor, Harbour
Hdqrs. _____ Headquarters
Heilong. _____ Heilongjiang
Hist. _____ Historic, -al
Hond. _____ Honduras

Hts. _____ Heights
Hung. _____ Hungary
Hwy. _____ Highway

I.H.S. _____ International Historic Site
I.-s. _____ Île-s, Ilha-s, Isla-s, Island-s, Isle, Isol-a, -e
Ice. _____ Iceland
Ig. _____ Igarapé
Igr. _____ Ingeniero
Ill. _____ Illinois
Ind. _____ Indiana
Ind. _____ Industry
Ind. Oc. _____ Indian Ocean
Ingush. _____ Ingushetiya
Intl. _____ International
Ire. _____ Ireland
It. _____ Italy

J. _____ Järvi, Joki
J.A.R. _____ Jewish Autonomous Region
Jab., Jeb. _____ Jabal, Jebel
Jam. _____ Jamaica
Jap. _____ Japan
Jct. _____ Jonction, Junction
Jez. _____ Jezero, Jezioro

K. _____ Kanal
Kalin. _____ Kaliningrad
Kalmy. _____ Kalmykiya
Kamchat. _____ Kamchatka
Kans. _____ Kansas
Karna. _____ Karnataka
Kaz. _____ Kazakhstan
Kemer. _____ Kemerovo
Kep. _____ Kepulauan
Kh. _____ Khor
Khabar. _____ Khabarovsk
Khak. _____ Khakasiya
Khr. _____ Khrebet
Km. _____ Kilómetro
Kól. _____ Kólpos
Kör. _____ Körfez,-i
Kos. _____ Kosovo
Kr. _____ Krasn-yy, -aya, -oye
Krasnod. _____ Krasnodar
Krasnoy. _____ Krasnoyarsk
Ky. _____ Kentucky
Kyrg. _____ Kyrgyzstan

L. _____ Lac, Lago, Lake, Límni, Loch, Lough
La. _____ Louisiana
Lab. _____ Labrador
Lag. _____ Laguna
Lakshad. _____ Lakshadweep
Latv. _____ Latvia
Ldg. _____ Landing
Leb. _____ Lebanon
Lib. _____ Libya
Liech. _____ Liechtenstein
Lith. _____ Lithuania
Lux. _____ Luxembourg

m _____ meters
M.N.M. _____ Marine National Monument
M. Gerais _____ Minas Gerais
M. Grosso _____ Mato Grosso
M. Grosso S. _____ Mato Grosso do Sul
Maced. _____ Macedonia
Madag. _____ Madagascar
Mahar. _____ Maharashtra
Mal. _____ Mal-y-y, -aya, -aye
Man. _____ Manitoba

Maran. _____ Maranhão
Maurit. _____ Mauritius
Mass. _____ Massachusetts
Md. _____ Maryland
Me. _____ Maine
Medit. Sea _____ Mediterranean Sea
Mex. _____ Mexico
Mgne. _____ Montagne
Mich. _____ Michigan
Minn. _____ Minnesota
Miss. _____ Mississippi
Mo. _____ Missouri
Mold. _____ Moldova
Mon. _____ Monument
Mont. _____ Montana
Mont. _____ Montenego
Mor. _____ Morocco
Mt.-s. _____ Mont-s, Mount-ain-s
Mte.-s. _____ Monte-s
Mti., Mtii. _____ Munţi-i
Mun. _____ Municipal
Murm. _____ Murmansk

N. _____ North-ern
NA _____ Not available
_____ Not applicable
N.B. _____ New Brunswick
N.B.P. _____ National Battlefield Park
N.B.S. _____ National Battlefield Site
N.C. _____ North Carolina
N. Dak. _____ North Dakota
N.E. _____ North East
N.H. _____ New Hampshire
N. Ire. _____ Northern Ireland
N.J. _____ New Jersey
N.M. _____ National Monument
N. Mex. _____ New Mexico
N. Mongol _____ Nei Mongol
N.M.P. _____ National Military Park
N.M.S. _____ National Marine Sanctuary
N.P. _____ National Park
N.S. _____ Nova Scotia
N.S.W. _____ New South Wales
N.T. _____ Northern Territory
N.V.M. _____ National Volcanic Monument
N.W.T. _____ Northwest Territories
N.Y. _____ New York
N.Z. _____ New Zealand
Nat. _____ National
Nat. Mem. _____ National Memorial
Nat. Mon. _____ National Monument
Nebr. _____ Nebraska
Neth. _____ Netherlands
Nev. _____ Nevada, Nevado
Nfld. & Lab. _____ Newfoundland and Labrador
Nicar. _____ Nicaragua
Nig. _____ Nigeria
Niz. Nov. _____ Nizhniy Novgorod
Nizh. _____ Nizhn-iy, -yaya, -eye
Nor. _____ Norway
Nov. _____ Nov-yy, -aya, -aye, -oye
Novg. _____ Novgorod
Novo. _____ Novosibirsk
Nr. _____ Nørre

O. _____ Ostrov, Oued
Oc. _____ Ocean
Of. _____ Oficina
Okla. _____ Oklahoma
Ont. _____ Ontario
Ør. _____ Øster
Oreg. _____ Oregon
Orenb. _____ Orenburg
Oz. _____ Ozero

P. _____ Paso, Pass, Passo
P.E.I. _____ Prince Edward Island
P.N.G. _____ Papua New Guinea
P.R. _____ Puerto Rico
Pa. _____ Pennsylvania
Pac. Oc. _____ Pacific Ocean
Pak. _____ Pakistan
Pan. _____ Panama
Pant. _____ Pantano
Para. _____ Paraguay
Pass. _____ Passage
Peg. _____ Pegunungan
Pen. _____ Peninsula, Península, Péninsule
Per. _____ Pereval
Pk. _____ Peak
Pl. _____ Planina
Plat. _____ Plateau
Pol. _____ Poland
Pol. _____ Poluostrov
Port. _____ Portugal
Pres. _____ Presidente
Prov. _____ Province, Provincial
Pt.-e. _____ Point-e
Pta. _____ Ponta, Punta, Puntan
Pto. _____ Puerto
Pul. _____ Pulau

Q. _____ Quebrada
Qnsld. _____ Queensland
Que. _____ Quebec
Qyzyl. _____ Qyzylorda

R. _____ Río, River, Rivière
R.R. _____ Railroad
R. Gr. Norte _____ Rio Grande do Norte
R. Gr. Sul _____ Rio Grande do Sul
R.I. _____ Rhode Island
R. Jan. _____ Rio de Janeiro
R. Negro _____ Río Negro
Ra.-s. _____ Range-s
Rec. _____ Recreation
Reg. _____ Region
Rep. _____ Republic
Res. _____ Reservoir, Reserve, Reservatório
Rk. _____ Rock
Rom. _____ Romania
Russ. _____ Russia

S. _____ South-ern
S.A.R. _____ Special Administrative Region
S. Af. _____ South Africa
S. Aust. _____ South Australia
S.C. _____ South Carolina
S. Dak. _____ South Dakota
S. Estero _____ Santiago del Estero
S. Ossetia _____ South Ossetia
S. Paulo _____ São Paulo
S.W. _____ Southwest
Sa.-s. _____ Serra, Sierra-s
Sal. _____ Salar, Salina
Sask. _____ Saskatchewan
Scot. _____ Scotland
Sd. _____ Sound, Sund
Sel. _____ Selat
Ser. _____ Serranía
Serb. _____ Serbia
Sev. _____ Severn-yy, -aya, -oye
Sgt. _____ Sargento
Shand. _____ Shandong
Sk. _____ Shankou
Slov. _____ Slovenia
Slovak. _____ Slovakia
Smt.-s _____ Seamount-s
Sp. _____ Spain, Spanish

Spr.-s _____ Spring-s
Sq. _____ Square
Sr. _____ Sønder
St.-e. _____ Saint-e, Sankt, Sint
St. Peter. _____ Saint Petersburg
Sta., Sto. _____ Santa, Station, Santo
Sta. Cata. _____ Santa Catarina
Sta. Cruz. _____ Santa Cruz
Stavr. _____ Stavropol'
Str.-s. _____ Straat, Strait-s
Sv. _____ Svyat-oy, -aya, -oye
Sverd. _____ Sverdlovsk
Sw. _____ Sweden
Switz. _____ Switzerland
Syr. _____ Syria

T. Fuego _____ Tierra del Fuego
Taj. _____ Tajikistan
Tas. _____ Tasmania
Tel. _____ Teluk
Tenn. _____ Tennessee
Terr. _____ Territory
Tex. _____ Texas
Tg. _____ Tanjung
Thai. _____ Thailand
Tmt.-s _____ Tablemount-s
Tocant. _____ Tocantins
Trin. _____ Trinidad
Tun. _____ Tunisia
Turk. _____ Turkey
Turkm. _____ Turkmenistan

U.A.E. _____ United Arab Emirates
U.K. _____ United Kingdom
U.N. _____ United Nations
U.S. _____ United States
Ukr. _____ Ukraine
Ulyan. _____ Ul'yanovsk
Uru. _____ Uruguay
Uzb. _____ Uzbekistan

V.I. _____ Virgin Islands
Va. _____ Virginia
Val. _____ Valley
Vdkhr. _____ Vodokhranil-ishche
Vdskh. _____ Vodoskhovy-shche
Venez. _____ Venezuela
Verkh. _____ Verkhn-iy, -yaya, -eye
Vic. _____ Victoria
Viet. _____ Vietnam
Vol. _____ Volcán, Volcano
Volg. _____ Volgograd
Voz. _____ Vozyera, -yero, -yera
Vozv. _____ Vozvyshennost'
Vr. _____ Vester
Vt. _____ Vermont
Vyal. _____ Vyaliki, -ikaya,-ikaye

W. _____ Wadi, Wâdi, Wādī, Webi
W. _____ West-ern
W. Aust. _____ Western Australia
W.H. _____ Water Hole
W. Va. _____ West Virginia
Wash. _____ Washington
Wis. _____ Wisconsin
Wyo. _____ Wyoming

Y.ar. _____ Yarımadası
Yaro. _____ Yaroslavl'
Yu. _____ Yuzhn-yy, -aya, -oye

Z.akh. _____ Zakhod-ni, -nyaya, -nye
Zal. _____ Zaliv
Zap. _____ Zapadn-yy, -aya, -oye
Zimb. _____ Zimbabwe

METRIC CONVERSIONS

QUICK REFERENCE CHART FOR METRIC TO ENGLISH CONVERSION

1 METER	1 METER = 100 CENTIMETERS
1 FOOT	1 FOOT = 12 INCHES

1 KILOMETER	1 KILOMETER = 1,000 METERS
1 MILE	1 MILE = 5,280 FEET

METERS	1	10	20	50	100	200	500	1,000	2,000	5,000	10,000
FEET	3.28084	32.8084	65.6168	164.042	328.084	656.168	1,640.42	3,280.84	6,561.68	16,404.2	32,808.4

KILOMETERS	1	10	20	50	100	200	500	1,000	2,000	5,000	10,000
MILES	0.621371	6.21371	12.42742	31.06855	62.1371	124.2742	310.6855	621.371	1,242.742	3,106.855	6,213.71

CONVERSION FROM METRIC MEASURES

SYMBOL	WHEN YOU KNOW	MULTIPLY BY	TO FIND	SYMBOL
LENGTH				
cm	centimeters	0.393701	inches	in
m	meters	3.280840	feet	ft
m	meters	1.093613	yards	yd
km	kilometers	0.621371	miles	mi
AREA				
cm^2	square centimeters	0.155000	square inches	in^2
m^2	square meters	10.76391	square feet	ft^2
m^2	square meters	1.195990	square yards	yd^2
km^2	square kilometers	0.386102	square miles	mi^2
ha	hectares	2.471054	acres	--
MASS				
g	grams	0.035274	ounces	oz
kg	kilograms	2.204623	pounds	lb
t	metric tons	1.102311	short tons	--
VOLUME				
mL	milliliters	0.061024	cubic inches	in^3
mL	milliliters	0.033814	fluid ounces	fl oz
L	liters	2.113376	pints	pt
L	liters	1.056688	quarts	qt
L	liters	0.264172	gallons	gal
m^3	cubic meters	35.31467	cubic feet	ft^3
m^3	cubic meters	1.307951	cubic yards	yd^3
TEMPERATURE				
°C	degrees Celsius (centigrade)	9/5 (or 1.8) then add 32	degrees Fahrenheit	°F

CONVERSION TO METRIC MEASURES

SYMBOL	WHEN YOU KNOW	MULTIPLY BY	TO FIND	SYMBOL
LENGTH				
in	inches	2.54	centimeters	cm
ft	feet	0.3048	meters	m
yd	yards	0.9144	meters	m
mi	miles	1.609344	kilometers	km
AREA				
in^2	square inches	6.4516	square centimeters	cm^2
ft^2	square feet	0.092903	square meters	m^2
yd^2	square yards	0.836127	square meters	m^2
mi^2	square miles	2.589988	square kilometers	km^2
--	acres	0.404686	hectares	ha
MASS				
oz	ounces	28.349523	grams	g
lb	pounds	0.453592	kilograms	kg
--	short tons	0.907185	metric tons	t
VOLUME				
in^3	cubic inches	16.387064	milliliters	mL
fl oz	fluid ounces	29.57353	milliliters	mL
pt	pints	0.473176	liters	L
qt	quarts	0.946353	liters	L
gal	gallons	3.785412	liters	L
ft^3	cubic feet	0.028317	cubic meters	m^3
yd^3	cubic yards	0.764555	cubic meters	m^3
TEMPERATURE				
°F	degrees Fahrenheit	5/9 (or 0.55556) after subtracting 32	degrees Celsius (centigrade)	°C

COMPARISONS

EARTH

PLANET FACTS

Age: Formed 4.54 billion years ago. Life appeared on its surface within a billion years.

Interior: Remains active, with a thick layer of relatively solid mantle, a liquid outer core that generates a magnetic field, and a solid iron inner core.

Mass: 5,973,600,000,000,000,000,000,000—5.9736 sextillion—metric tons (6.5848 sextillion short tons)

Total Area: 510,072,000 sq km (196,940,000 sq mi)

Surface: About 71% of the surface is covered with a saltwater ocean, the remainder consisting of continents and islands.

Land Area: 148,940,000 sq km (57,506,000 sq mi), 29.1% of total

Water Area: 361,132,000 sq km (139,434,000 sq mi), 70.9% of total

Atmosphere Composition: Dry air is 78.08% nitrogen (N_2), 20.95% oxygen (O_2), 0.93% argon (Ar), 0.038% carbon dioxide (CO_2), and 0.002% other gases. Water vapor is variable and typically about 1%.

Orbit: Earth moves around the sun once for every 366.26 times it rotates about its axis. This time period is a sidereal year, which equals 365.26 solar days.

Equatorial Diameter: 12,756 km (7,926 mi)

Polar Diameter: 12,714 km (7,900 mi)

PLANETARY EXTREMES

Hottest Place: Dalol, Danakil Depression, Ethiopia, annual average temperature 34°C (93°F)

Coldest Place: Ridge A, Antarctica, annual average temperature −70°C (−94°F)

Hottest Recorded Air Temperature: Furnace Creek Ranch (Death Valley), California, U.S., 56.7°C (134°F), October 7, 1913

Coldest Recorded Air Temperature: Antarctica, −93.2°C (−135.8°F), August 10, 2010

Wettest Place: Mawsynram, Meghalaya, India, annual average rainfall 1,187 cm (467 in)

Driest Place: Arica, Atacama Desert, Chile, rainfall barely measurable

Largest Hot Desert: Sahara, Africa, 9,000,000 sq km (3,475,000 sq mi)

Largest Ice Desert: Antarctica, 13,209,000 sq km (5,100,000 sq mi)

Largest Canyon: Grand Canyon, Colorado River, Arizona, U.S., 446 km (277 mi) long along river, 180 m (600 ft) to 29 km (18 mi) wide, about 1.6 km (1 mi) deep

Largest Coral Reef Ecosystem: Great Barrier Reef, Australia, 348,300 sq km (134,000 sq mi)

Greatest Tidal Range: Bay of Fundy, Canadian Atlantic Coast, 16 m (53 ft)

Tallest Waterfall: Angel Falls, Venezuela, 979 m (3,212 ft)

LAND

AREA OF EACH CONTINENT

	SQ KM	SQ MI	% OF LAND
Asia	44,570,000	17,208,000	30.0
Africa	30,065,000	11,608,000	20.2
North America	24,474,000	9,449,000	16.5
South America	17,819,000	6,880,000	12.0
Antarctica	13,209,000	5,100,000	8.9
Europe	9,947,000	3,841,000	6.7
Australia	7,692,000	2,970,000	5.2

LARGEST ISLANDS BY AREA

		SQ KM	SQ MI
1	Greenland	2,166,000	836,000
2	New Guinea	792,500	306,000
3	Borneo	725,500	280,100
4	Madagascar	587,000	226,600
5	Baffin Island	507,500	196,000
6	Sumatra	427,300	165,000
7	Honshu	227,400	87,800
8	Great Britain	218,100	84,200
9	Victoria Island	217,300	83,900
10	Ellesmere Island	196,200	75,800

LOWEST SURFACE POINT ON EACH CONTINENT

	METERS	FEET
Dead Sea, Asia	−434	−1,424
Lake Assal, Africa	−155	−509
Laguna del Carbón, South America	−105	−344
Death Valley, North America	−86	−282
Caspian Sea, Europe	−28	−92
Lake Eyre, Australia	−15	−49
Byrd Glacier (depression), Antarctica	−2,870	−9,416

HIGHEST POINT ON EACH CONTINENT

	METERS	FEET
Mount Everest, Asia	8,850	29,035
Cerro Aconcagua, South America	6,959	22,831
Denali (Mount McKinley), N. America	6,190	20,310
Kilimanjaro, Africa	5,895	19,340
El'brus, Europe	5,642	18,510
Vinson Massif, Antarctica	4,897	16,067
Mount Kosciuszko, Australia	2,228	7,310

MOUNTAINS AND CAVES

Tallest Mountain (above and below sea level): Mauna Kea, Hawai'i, U.S., 9,966 m (32,696 ft) above the seafloor and 4,205 m (13,796 ft) above sea level

Highest Mountain (above sea level): Mount Everest, China and Nepal border, 8,850 m (29,035 ft) above sea level

Longest Mountain Range (above sea level): Andes, South America, 7,600 km (4,700 mi)

Longest Mountain Range (above and below sea level): Mid-Ocean Ridge, 60,000 km (37,000 mi), encircles the Earth mostly along the seafloor

Largest Cave Chamber: Sarawak Chamber, Gunung Mulu National Park, Malaysia, 16 hectares and 80 meters high (40.2 acres and 260 feet)

Longest Cave System: Mammoth Cave, Kentucky, U.S., more than 644 km (400 mi) of passageways mapped

Map labels

Longitude West of Greenwich
ARCTIC
ELLESMERE ISLAND
GREENLAND
Molloy Deep −5,669 m (−18,599 ft) Arctic Ocean deepest point
Prudhoe Bay, United States
VICTORIA ISLAND
BAFFIN ISLAND
ARCTIC CIRCLE
GREENLAND SEA
Denali (Mt. McKinley), United States 6,190 m (20,310 ft) North America's highest point
Pan-American Highway World's longest road
Great Bear Lake
Great Slave Lake
Laerdal Tunnel, Norway World's longest road tunnel
GREAT BRITAIN
NORTH AMERICA
Pan-American Highway World's longest road
Lake Superior
Lake Huron
Bay of Fundy, Canada World's greatest tidal range
Sea-Me-We 3 cable World's longest submarine cable
NORTH
Furnace Creek Ranch (Death Valley), United States World's hottest recorded air temperature
Grand Canyon, United States World's largest canyon
Michigan
Mammoth Cave, United States World's longest cave system
Millau Viaduct, France World's tallest road bridge
Death Valley, United States (−282 ft) −86 m North America's lowest point
Mississippi-Missouri Drainage Basin
NORTH ATLANTIC
PACIFIC OCEAN
TROPIC OF CANCER
GULF OF MEXICO
Mauna Kea, United States World's tallest mountain (above and below sea level)
Puerto Rico Trench −8,605 m (−28,232 ft) Atlantic Ocean's deepest point
World's largest hot desert
OCEAN
CARIBBEAN SEA
Niger Drainage Basin
Angel Falls, Venezuela World's tallest waterfall
Lake Volta, Ghana World's largest reservoir by surface area
EQUATOR
Pan-American Highway World's longest road
Amazon
Amazon Drainage Basin
World's longest mountain range (above sea level)
SOUTH AMERICA
Meridian of Greenwich (London)
Arica, Chile Atacama Desert World's driest place
Paraná Drainage Basin
SOUTH
TROPIC OF CAPRICORN
Pan-American Highway World's longest road
Cerro Aconcagua, Chile (22,831 ft) 6,959 m South America's highest point
Río de la Plata
ATLANTIC
PACIFIC
30°
Laguna del Carbón, Argentina −105 m (−344 ft) South America's lowest point
OCEAN
OCEAN
Pan-American Highway World's longest road
Punta Arenas, Chile
ANTARCTIC CIRCLE
Vinson Massif (16,067 ft) 4,897 m Antarctica's highest point

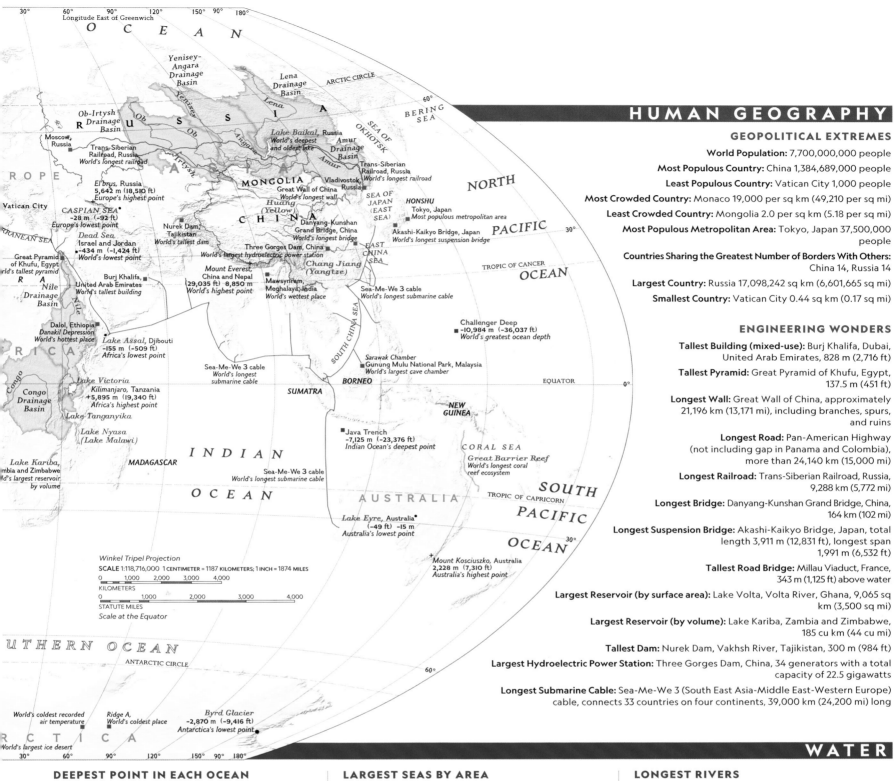

HUMAN GEOGRAPHY

GEOPOLITICAL EXTREMES

World Population: 7,700,000,000 people

Most Populous Country: China 1,384,689,000 people

Least Populous Country: Vatican City 1,000 people

Most Crowded Country: Monaco 19,000 per sq km (49,210 per sq mi)

Least Crowded Country: Mongolia 2.0 per sq km (5.18 per sq mi)

Most Populous Metropolitan Area: Tokyo, Japan 37,500,000 people

Countries Sharing the Greatest Number of Borders With Others: China 14, Russia 14

Largest Country: Russia 17,098,242 sq km (6,601,665 sq mi)

Smallest Country: Vatican City 0.44 sq km (0.17 sq mi)

ENGINEERING WONDERS

Tallest Building (mixed-use): Burj Khalifa, Dubai, United Arab Emirates, 828 m (2,716 ft)

Tallest Pyramid: Great Pyramid of Khufu, Egypt, 137.5 m (451 ft)

Longest Wall: Great Wall of China, approximately 21,196 km (13,171 mi), including branches, spurs, and ruins

Longest Road: Pan-American Highway (not including gap in Panama and Colombia), more than 24,140 km (15,000 mi)

Longest Railroad: Trans-Siberian Railroad, Russia, 9,288 km (5,772 mi)

Longest Bridge: Danyang-Kunshan Grand Bridge, China, 164 km (102 mi)

Longest Suspension Bridge: Akashi-Kaikyo Bridge, Japan, total length 3,911 m (12,831 ft), longest span 1,991 m (6,532 ft)

Tallest Road Bridge: Millau Viaduct, France, 343 m (1,125 ft) above water

Largest Reservoir (by surface area): Lake Volta, Volta River, Ghana, 9,065 sq km (3,500 sq mi)

Largest Reservoir (by volume): Lake Kariba, Zambia and Zimbabwe, 185 cu km (44 cu mi)

Tallest Dam: Nurek Dam, Vakhsh River, Tajikistan, 300 m (984 ft)

Largest Hydroelectric Power Station: Three Gorges Dam, China, 34 generators with a total capacity of 22.5 gigawatts

Longest Submarine Cable: Sea-Me-We 3 (South East Asia-Middle East-Western Europe) cable, connects 33 countries on four continents, 39,000 km (24,200 mi) long

WATER

DEEPEST POINT IN EACH OCEAN

	METERS	FEET
Challenger Deep, Pacific Ocean	-10,984	-36,037
Puerto Rico Trench, Atlantic Ocean	-8,605	-28,232
Java Trench, Indian Ocean	-7,125	-23,376
Molloy Deep, Arctic Ocean	-5,669	-18,599

AREA OF EACH OCEAN

	SQ KM	SQ MI	% OCEAN AREA
Pacific	178,800,000	69,000,000	49.5
Atlantic	91,700,000	35,400,000	25.4
Indian	76,200,000	29,400,000	21.0
Arctic	14,700,000	5,600,000	4.1

The Atlantic, Indian, and Pacific Oceans merge into icy waters around Antarctica. Some define this as an ocean— calling it the Antarctic Ocean, Austral Ocean, or Southern Ocean. While most accept four oceans, including the Arctic, there is no international agreement on the name and extent of a fifth ocean. The "Southern Ocean" extends from the Antarctic coast to 60° south latitude and includes portions of the Atlantic, Indian, and Pacific Oceans (estimated area: 20,327,000 sq km or 7,848,000 sq mi).

LARGEST SEAS BY AREA

		AREA SQ KM	AREA SQ MI	AVG. DEPTH METERS	AVG. DEPTH FEET
1	Coral Sea	4,184,000	1,615,500	2,471	8,107
2	South China Sea	3,596,000	1,388,400	1,180	3,871
3	Caribbean Sea	2,834,000	1,094,200	2,596	8,517
4	Bering Sea	2,520,000	973,000	1,832	6,010
5	Mediterranean Sea	2,469,000	953,300	1,572	5,157
6	Sea of Okhotsk	1,625,000	627,400	814	2,671
7	Gulf of Mexico	1,532,000	591,500	1,544	5,066
8	Norwegian Sea	1,425,000	550,200	1,768	5,801
9	Greenland Sea	1,158,000	447,100	1,443	4,734
10	Sea of Japan (East Sea)	1,008,000	389,200	1,647	5,404

LARGEST LAKES BY AREA *(with maximum depth)*

		SQ KM	SQ MI	METERS	FEET
1	Caspian Sea	371,000	143,200	1,025	3,363
2	Lake Superior	82,100	31,700	406	1,332
3	Lake Victoria	69,500	26,800	82	269
4	Lake Huron	59,600	23,000	229	751
5	Lake Michigan	57,800	22,300	281	922
6	Lake Tanganyika	32,600	12,600	1,470	4,823
7	Lake Baikal	31,500	12,200	1,642	5,387
8	Great Bear Lake	31,300	12,100	446	1,463
9	Lake Nyasa	28,900	11,200	695	2,280
10	Great Slave Lake	28,600	11,000	614	2,014

LONGEST RIVERS

		KILOMETERS	MILES
1	Nile, Africa	6,695	4,160
2	Amazon, South America	6,679	4,150
3	Chang Jiang (Yangtze), Asia	6,244	3,880
4	Mississippi-Missouri, N. America	5,971	3,710
5	Yenisey-Angara, Asia	5,810	3,610
6	Huang (Yellow), Asia	5,778	3,590
7	Ob-Irtysh, Asia	5,520	3,430
8	Amur, Asia	5,504	3,420
9	Lena, Asia	5,150	3,200
10	Congo, Africa	5,118	3,180

LARGEST RIVER DRAINAGE BASINS BY AREA

		SQ KM	SQ MI
1	Amazon, South America	6,145,186	2,372,670
2	Congo, Africa	3,730,881	1,440,500
3	Nile, Africa	3,254,853	1,256,706
4	Mississippi-Missouri, N. Amer.	3,202,185	1,236,370
5	Ob-Irtysh, Asia	2,972,493	1,147,686
6	Paraná, South America	2,582,704	997,188
7	Yenisey-Angara, Asia	2,554,388	986,255
8	Lena, Asia	2,306,743	890,638
9	Niger, Africa	2,261,741	873,263
10	Amur, Asia	1,929,955	745,160

FOREIGN TERMS

Aaglet _____ well
Aain _____ spring
Aauinat _____ spring
Ab _____ river, water
Ache _____ stream
Açude _____ reservoir
Ada, -sı _____ island
Adrar _____ mountain-s, plateau
Ágios _____ saint
Aguada _____ dry lake bed
Aguelt _____ water hole, well
'Ain, Aïn _____ spring, well
Aïoun-et _____ spring-s, well
Aivi _____ mountain
Akra, Akrotírio _____ cape, promontory
Alb _____ mountain, ridge
Alföld _____ plain
Alin' _____ mountain range
Alpe-n, -s _____ mountain-s
Altiplanicie _____ high plain, plateau
Alto _____ hill-s, mountain-s, ridge
Älv-en _____ river
Āmba _____ hill, mountain
Anou _____ well
Anse _____ bay, inlet
Ao _____ bay, cove, estuary
Ap _____ cape, point
Archipel, Archipiélago _____ archipelago
Arcipelago, Arkhipelag _____ archipelago
Arquipélago _____ archipelago
Arrecife-s _____ reef-s
Arroio, Arroyo _____ brook, gully, rivulet, stream
Ås _____ ridge
Ava _____ channel
Aylagy _____ gulf
'Ayn _____ spring, well

Ba _____ intermittent stream, river
Baai _____ bay, cove, lagoon
Bab _____ gate, strait
Badia _____ bay
Bælt _____ strait
Bagh _____ bay
Bahar _____ drainage basin
Bahía _____ bay
Bahr, Baḩr _____ bay, lake, river, sea, wadi
Baía, Baie _____ bay
Bajo-s _____ shoal-s
Ban _____ village
Bañado-s _____ flooded area, swamp-s
Banc, Banco-s _____ bank-s, sandbank-s, shoal-s
Band _____ dam, lake
Bandao _____ peninsula
Baño-s _____ hot spring-s, spa
Baraj-ı _____ dam, reservoir
Barra _____ bar, sandbank
Barrage, Barragem _____ dam, lake, reservoir
Barranca _____ gorge, ravine
Bazar _____ marketplace
Belentligi _____ plateau
Ben, Beinn _____ mountain
Belt _____ strait
Bereg _____ bank, coast, shore
Berg, -e _____ mountain-s
Bil _____ lake
Biq'at _____ plain, valley
Bir, Bîr, Bi'r _____ spring, well
Birket _____ lake, pool, swamp
Bjerg-e _____ mountain-s, range
Boca, Bocca _____ channel, river, mouth
Bocht _____ bay
Bodden _____ bay
Bœng _____ pond
Boğaz, -ı _____ strait
Bögeni _____ reservoir
Boka _____ gulf, mouth
Bol'sh-oy, -aya, -oye _____ big
Bolsón _____ inland basin
Boubairet _____ lagoon, lake

Bras _____ arm, branch of a stream
Braț, -ul _____ arm, branch of a stream
Bræ-er _____ glacier
Bre, -en _____ glacier, ice cap
Bredning _____ bay, broad water
Bruch _____ marsh
Bucht _____ bay
Bugt-en _____ bay
Buḩayrat, Buheirat _____ lagoon, lake, marsh
Bukhta, Bukta, Bukt-en _____ bay
Bulak, Bulaq _____ spring
Bum _____ hill, mountain
Burnu, Burun _____ cape, point
Busen _____ gulf
Buuraha _____ hill-s, mountain-s
Büyük _____ big, large

Cabeza-s _____ head-s, summit-s
Cabo _____ cape
Cachoeira _____ rapids, waterfall
Cal _____ hill, peak
Caleta _____ cove, inlet
Campo-s _____ field-s, flat country
Canal _____ canal, channel, strait
Caño _____ channel, stream
Cao Nguyên _____ plateau
Cap, Capo _____ cape
Capitán _____ captain
Càrn _____ mountain
Castillo _____ castle, fort
Catarata-s _____ cataract-s, waterfall-s
Causse _____ upland
Çay _____ brook, stream
Cay-s, Cayo-s _____ island-s, key-s, shoal-s
Cerro-s _____ hill-s, peak-s
Chaîne, Chaînons _____ mountain chain, range
Chapada-s _____ plateau, upland-s
Chedo _____ archipelago
Chenal _____ river channel
Chersónisos _____ peninsula
Chhung _____ bay
Chi _____ lake
Chiang _____ bay
Chiao _____ cape, point, rock
Ch'ih _____ lake
Chink _____ escarpment
Chott _____ intermittent salt lake, salt marsh
Chou _____ island
Chroüy _____ point
Ch'ü _____ canal
Ch'üntao _____ archipelago, islands
Chuŏr Phnum _____ mountains
Chute-s _____ cataract-s, waterfall-s
Chyrvony, -aya, -aye _____ red
Ciénaga _____ marsh
Cima _____ mountain, peak, summit
Ciudad _____ city
Co _____ lake
Col _____ pass
Collina, Colline _____ hill, mountains
Con _____ island
Cordillera _____ mountain chain
Corno _____ mountain, peak
Coronel _____ colonel
Corredeira _____ cascade, rapids
Costa _____ coast
Côte _____ coast, slope
Coxilha, Cuchilla _____ range of low hills
Crique _____ creek, stream
Csatorna _____ canal, channel
Cù Lao _____ island
Cul de Sac _____ bay, inlet

Da _____ great, greater
Daban _____ pass
Dağ, -ı, Dagh _____ mountain
Dağlar, -ı _____ mountains
Dahr _____ cliff, mesa
Dake _____ mountain, peak
Dal-en _____ valley
Dala _____ steppe

Dan _____ cape, point
Danau _____ lake
Dao _____ island
Đảo _____ island
Dar'ya _____ lake, river
Daryācheh _____ lake, marshy lake
Dasht _____ desert, plain
Dawan _____ pass
Dawḩat _____ bay, cove, inlet
Deniz, -i _____ sea
Dent-s _____ peak-s
Deo _____ pass
Deryache _____ lake
Desēt _____ hummock, island, land-tied island
Desierto _____ desert
Détroit _____ channel, strait
Dhar _____ hills, ridge, tableland
Ding _____ mountain
Distrito _____ district
Djebel _____ mountain, range
Do _____ island-s, rock-s
Doi _____ hill, mountain
Dome _____ ice dome
Dong _____ village
Dooxo _____ floodplain
Dzong _____ castle, fortress

Eiland-en _____ island-s
Eilean _____ island
Ejland _____ island-s
Elv _____ river
Embalse _____ lake, reservoir
Emi _____ mountain, rock
Enseada, Ensenada _____ bay, cove
Ér _____ rivulet, stream
Erg _____ sand dune region
Est _____ east
Estación _____ railroad station
Estany _____ lagoon, lake
Estero _____ estuary, inlet, lagoon, marsh
Estrecho _____ strait
Étang _____ lake, pond
Eylandt _____ island
Ežeras _____ lake
Ezers _____ lake

Falaise _____ cliff, escarpment
Farvand-et _____ channel, sound
Fell _____ mountain
Feng _____ mount, peak
Fiord-o _____ inlet, sound
Firn _____ snowfield
Fiume _____ river
Fjäll-et _____ mountain
Fjällen _____ mountains
Fjärd-en _____ fjord
Fjarðar, Fjörður _____ fjord
Fjeld-e _____ mountain-s, nunatak-s
Fjell-ene _____ mountain-s
Fjöll _____ mountain-s
Fjord-en _____ inlet, fjord
Fleuve _____ river
Fljót _____ large river
Flói _____ bay, marshland
Foci _____ river mouths
Főcsatorna _____ principal canal
Foko _____ point
Förde _____ fjord, gulf, inlet
Forsen _____ rapids, waterfall
Fortaleza _____ fort, fortress
Fortín _____ fortified post
Foss-en _____ waterfall
Foum _____ pass, passage
Foz _____ mouth of a river
Fuerte _____ fort, fortress
Fwafwate _____ waterfalls

Gacan-ka _____ hill, peak
Gal _____ pond, spring, water hole, well
Gang _____ harbor
Gangri _____ peak, range
Gaoyuan _____ plateau

Garaet, Gara'et _____ lake, lake bed, salt lake
Gardaneh _____ pass
Garet _____ hill, mountain
Gat _____ channel
Gata _____ bay, inlet, lake
Gattet _____ channel, strait
Gaud _____ depression, saline tract
Gave _____ mountain stream
Gebel _____ mountain-s, range
Gebergte _____ mountain range
Gebirge _____ mountains, range
Geçidi _____ mountain pass, passage
Geçit _____ mountain pass, passage
Gezâir _____ islands
Gezîra-t, Gezîret _____ island, peninsula
Ghats _____ mountain range
Ghubb-at, -et _____ bay, gulf
Giri _____ mountain
Gjiri _____ bay
Gletscher _____ glacier
Gobernador _____ governor
Gobi _____ desert
Gol _____ river, stream
Göl, -ü _____ lake
Golets _____ mountain, peak
Golf, -e, -o _____ gulf
Gor-a, -y, Gór-a, -y _____ mountain, -s
Got _____ point
Gowd _____ depression
Goz _____ sand ridge
Gran, -de _____ great, large
Gryada _____ mountains, ridge
Guan _____ pass
Guba _____ bay, gulf
Guelta _____ well
Guntō _____ archipelago
Gunung _____ mountain
Gura _____ mouth, passage
Guyot _____ table mount

Haḑabat _____ plateau
Haehyŏp _____ strait
Haff _____ lagoon
Hai _____ lake, sea
Haihsia _____ strait
Haixia _____ channel, strait
Hakau _____ reef, rock
Hakuchi _____ anchorage
Halvø, Halvøy-a _____ peninsula
Hama _____ beach
Hamada, Ḩammādah _____ rocky desert
Hamn _____ harbor, port
Hāmūn, Hamun _____ depression, lake
Hana _____ cape, point
Hantō _____ peninsula
Har _____ hill, mound, mountain
Ḩarrat _____ lava field
Hasi, Hassi _____ spring, well
Hauteur _____ elevation, height
Hav-et _____ sea
Havn, Havre _____ harbor, port
Hawr _____ lake, marsh
Hāyk' _____ lake, reservoir
He _____ canal, lake, river
Hegy, -ség _____ mountain, -s, range
Heiau _____ temple
Ho _____ lake, reservoir
Hoek _____ hook, point
Hög-en _____ high, hill
Höhe, -n _____ height, high
Høj _____ height, hill
Holm, -e, Holmene _____ island-s, islet -s
Holot _____ dunes
Hòn _____ island-s
Hor-a, -y _____ mountain, -s
Horn _____ horn, peak
Houma _____ point
Hoved _____ headland, peninsula, point
Hraun _____ lava field
Hsü _____ island
Hu _____ lake, reservoir
Huk _____ cape, point

Hüyük _____ hill, mound

Idehan _____ sand dunes
Igarapé _____ creek, stream
Île-s, Ilha-s, Illa-s, Îlot-s _____ island-s, islet-s
Îlet, Ilhéu-s _____ islet, -s
Irhil _____ mountain-s
'Irq _____ sand dune-s
Isblink _____ glacier, ice field
Is-en _____ glacier
Isebræ _____ glacier
Isfjord _____ ice fjord
Iskappe _____ ice cap
Isla-s, Islote _____ island-s, islet
Isol-a, -e _____ island, -s
Isstrøm _____ glacier, ice field
Istmo _____ isthmus
Iwa _____ island, islet, rock

Jabal, Jebel _____ mountain-s, range
Jahīl _____ lake
Järv, -i, Jaure, Javrre _____ lake
Jazā'ir, Jazīrat, Jazīreh _____ island-s
Jehīl _____ lake
Jezero, Jezioro _____ lake
Jiang _____ river, stream
Jiao _____ cape
Jibāl _____ hill, mountain, ridge
Jima _____ island-s, rock-s
Jøkel, Jökull _____ glacier, ice cap
Joki, Jokka _____ river
Jökulsá _____ river from a glacier
Jōsuji _____ lake, reservoir
Jūn _____ bay

Kaap _____ cape
Kafr _____ village
Kaikyō _____ channel, strait
Kaise _____ mountain
Kaiwan _____ bay, gulf, sea
Kanal _____ canal, channel
Kangerlua _____ fjord
Kangri _____ mountain-s, peak
Kaôh _____ island
Kap, Kapp _____ cape
Kavīr _____ salt desert
Kefar _____ village
Kënet' _____ lagoon, lake
Kep _____ cape, point
Kepulauan _____ archipelago, islands
Khalīg, Khalīj _____ bay, gulf
Khirb-at, -et _____ ancient site, ruins
Khrebet _____ mountain range
Kinh _____ canal
Klint _____ bluff, cliff
Kō _____ bay, cove, harbor
Ko _____ island, lake
Kôh _____ mountain
Koh _____ island, mountain, range
Köl-i _____ lake
Kólpos _____ gulf
Kong _____ king, mountain
Körfez, -i _____ bay, gulf
Kosa _____ spit of land
Kōtal _____ pass
Kou _____ estuary, river mouth
Kowtal-e _____ pass
Kronprince _____ crown prince
Krasn-yy, -aya, -oye _____ red
Kryazh _____ mountain range, ridge
Kuala _____ estuary, river mouth
Kuan _____ mountain pass
Kūh, Kūhhā _____ mountain-s, range
Kul', Kuli _____ lake
Kum _____ sandy desert
Kundo _____ archipelago
Kuppe _____ hill-s, mountain-s
Kust _____ coast, shore
Kyst _____ coast
Kyun _____ island

La ——— pass
Lac, Lac-ul, -us ——— lake
Lae ——— cape, point
Lago, -a ——— lagoon, lake
Lagoen, Lagune ——— lagoon
Laguna-s ——— lagoon-s, lake-s
Laht ——— bay, gulf, harbor
Laje ——— reef, rock ledge
Laut ——— sea
Lednik ——— glacier
Leida ——— channel
Lhari ——— mountain
Li ——— village
Liedao ——— archipelago, islands
Liehtao ——— archipelago, islands
Lille ——— little, small
Liman-ı ——— bay, estuary
Límni ——— lake
Ling ——— mountain-s, range
Linn ——— pool, waterfall
Lintasan ——— passage
Liqen ——— lake
Llano-s ——— plain-s
Loch, Lough ——— lake, arm of the sea
Loma-s ——— hill-s, knoll-s

Mal ——— mountain, range
Mal-yy, -aya, -oye ——— little, small
Mamarr ——— pass, path
Man ——— bay
Mar, Mare ——— large lake, sea
Marsa, Marsá ——— bay, inlet
Masabb ——— mouth of river
Massif ——— mountain-s
Mauna ——— mountain
Mēda ——— plain
Meer ——— lake, sea
Melkosopochnik ——— undulating plain
Mesa, Meseta ——— plateau, tableland
Mierzeja ——— sandspit
Minami ——— south
Mios ——— island
Misaki ——— cape, peninsula, point
Mochun ——— passage
Molsron ——— harbor
Mong ——— town, village
Mont-e, -i, -ii, -s ——— mount, -ain, -s
Montagne, -s ——— mount, -ain, -s
Montaña, -s ——— mountain, -s
More ——— sea
Morne ——— hill, peak
Morro ——— bluff, headland, hill
Motu, -s ——— islands
Mouïet ——— well
Mouillage ——— anchorage
Muang ——— town, village
Mūi ——— cape, point
Mull ——— headland, promontory
Munkhafad ——— depression
Munte ——— mountain
Munţi-i ——— mountains
Muong ——— town, village
Mynydd ——— mountain
Mys ——— cape

Nacional ——— national
Nada ——— gulf, sea
Næs, Näs ——— cape, point
Nafūd ——— area of dunes, desert
Nagor'ye ——— mountain range, plateau
Nahar, Nahr ——— river, stream
Nakhon ——— town
Namakzār ——— salt waste
Ne ——— island, reef, rock-s
Neem ——— cape, point, promontory
Nes, Ness ——— peninsula, point
Nevado-s ——— snowcapped mountain-s
Nez ——— cape, promontory
Ni ——— village
Nísi, Nísia, Nisís, Nísoi ——— island-s, islet-s
Nisídhes ——— islets
Nizhn-iy, -yaya, -eye ——— lower
Nizmennost' ——— low country
Noord ——— north
Nord-re ——— north-ern
Nørre ——— north-ern
Nos ——— cape, nose, point
Nosy ——— island, reef, rock
Nov-yy, -aya, -aye, -oye ——— new
Nudo ——— mountain
Núi ——— mountains

Numa ——— lake
Nunaa ——— area, region
Nunaat ——— area, island
Nunatak, -s, -ker ——— peak-s surrounded by ice cap
Nur ——— lake, salt lake
Nuruu ——— mountain range, ridge
Nut-en ——— peak
Nuur ——— lake

O-n, Ø-er ——— island-s
Oblast ——— administrative division, province, region
Oceanus ——— ocean
Odde-n ——— cape, point
Øer-ne ——— islands
Oglat ——— group of wells
Oguilet ——— well
Ór-os, -i ——— mountain, -s
Órmos ——— bay, port
Ort ——— place, point
Øst-er ——— east
Ostrov, -a, Ostrv-o, -a ——— island, -s
Otoci, Otok ——— islands, island
Ouadi, Oued ——— river, watercourse
Ovalığı ——— plain
Øy-a ——— island
Øyane ——— islands
Ozer-o, -a ——— lake, -s

Pää ——— mountain, point
Palus ——— marsh
Pampa-s ——— grassy plain-s
Pantà ——— lake, reservoir
Pantanal ——— marsh, swamp
Pao, P'ao ——— lake
Parbat ——— mountain
Parque ——— park
Pas, -ul ——— pass
Paso, Passo ——— pass
Passe ——— channel, pass
Pasul ——— pass
Pedra ——— rock
Pegunungan ——— mountain range
Pellg ——— bay, bight
Peña ——— cliff, rock
Pendi ——— basin
Penedo-s ——— rock-s
Péninsule ——— peninsula
Peñón ——— point, rock
Pereval ——— mountain pass
Pertuis ——— strait
Peski ——— sands, sandy region
Phnom ——— hill, mountain, range
Phou ——— mountain range
Phouphiang ——— plateau
Phu ——— mountain
Piana-o ——— plain
Pic, Pik, Piz ——— peak
Picacho ——— mountain, peak
Pico-s ——— peak-s
Pistyll ——— waterfall
Piton-s ——— peak-s
Pivdennyy ——— southern
Plaja, Playa ——— beach, inlet, shore
Planalto, Plato ——— plateau
Planina ——— mountain, plateau
Plassen ——— lake
Ploskogor'ye ——— plateau, upland
Pointe ——— point
Polder ——— reclaimed land
Poluostrov ——— peninsula
Pongo ——— water gap
Ponta, -l ——— cape, point
Ponte ——— bridge
Poolsaar ——— peninsula
Portezuelo ——— pass
Porto ——— port
Poulo ——— island-s
Praia ——— beach, seashore
Presa ——— reservoir
Presidente ——— president
Presqu'île ——— peninsula
Prins ——— prince
Prinsesse ——— princess
Prokhod ——— pass
Proliv ——— strait
Promontorio ——— promontory
Prŭsmyk ——— mountain pass
Przylądek ——— cape

Puerto ——— bay, pass, port
Pulao ——— island-s
Pulau, Pulo ——— island
Puncak ——— peak, summit, top
Punt, Punta, -n ——— point, -s
Pun ——— peak
Pu'u ——— hill, mountain
Puy ——— peak

Qā' ——— depression, marsh, mud flat
Qal'at ——— fort
Qal'eh ——— castle, fort
Qanâ ——— canal
Qārat ——— hill-s, mountain-s
Qaşr ——— castle, fort, hill
Qila ——— fort
Qiryat ——— settlement, suburb
Qolleh ——— peak
Qooriga ——— anchorage, bay
Qoz ——— dunes, sand ridge
Qu ——— canal
Quần Đảo ——— archipelago, islands
Quebrada ——— ravine, stream
Qullai ——— peak, summit
Qum-y ——— desert, sand
Qundao ——— archipelago, islands
Qurayyāt ——— hills

Raas ——— cape, point
Rabt ——— hill
Rada ——— roadstead
Rade ——— anchorage, roadstead
Rags ——— point
Ramat ——— hill, mountain
Rand ——— ridge of hills
Rann ——— swamp
Raqaba ——— wadi, watercourse
Ras, Rãs, Ra's ——— cape
Ravnina ——— plain
Récif-s ——— reef-s
Regreg ——— marsh
Represa ——— reservoir
Reservatório ——— reservoir
Restinga ——— barrier, sand area
Rettō ——— chain of islands
Ri ——— mountain range, village
Ría ——— estuary
Ribeirão ——— stream
Río, Rio ——— river
Rivière ——— river
Roca-s ——— cliff, rock-s
Roche-r, -s ——— rock-s
Rosh ——— mountain, point
Rt ——— cape, point
Rubha ——— headland
Rupes ——— scarp

Saar ——— island
Saari, Sari ——— island
Sabkha-t, Sabkhet ——— lagoon, marsh, salt lake
Sagar ——— lake, sea
Sahara, Şaḥrā' ——— desert
Sahl ——— plain
Saki ——— cape, point
Salar ——— salt flat
Salina ——— salt pan
Salin-as, -es ——— salt flat-s, salt marsh-es
Salto ——— waterfall
Sammyaku ——— mountain range
San ——— hill, mountain
San, -ta, -to ——— saint
Sandur ——— sandy area
Sankt ——— saint
Sanmaek ——— mountain range
São ——— saint
Sarīr ——— gravel desert
Sasso ——— mountain, stone
Savane ——— savanna
Scoglio ——— reef, rock
Se ——— reef, rock-s, shoal-s
Sebjet ——— salt lake, salt marsh
Sebkha ——— salt lake, salt marsh
Sebkhet ——— lagoon, salt lake
See ——— lake, sea
Selat ——— strait
Selkä ——— lake, ridge
Semenanjung ——— peninsula
Sen ——— mountain
Seno ——— bay, gulf

Sermeq ——— glacier
Sermia ——— glacier
Serra, Serranía ——— range of hills or mountains
Severn-ye, -yy, -aya, -oye ——— northern
Sgùrr ——— peak
Sha ——— island, shoal
Sha'īb ——— ravine, watercourse
Shamo ——— desert
Shan ——— island-s, mountain-s, range
Shankou ——— mountain pass
Shanmo ——— mountain range
Sharm ——— cove, creek, harbor
Shatt, Shaţţ ——— large river
Shi ——— administrative division, municipality
Shima ——— island-s, rock-s
Shō ——— island, reef, rock
Shotō ——— archipelago
Shott ——— intermittent salt lake
Shuiku ——— reservoir
Shuitao ——— channel
Shyghanaghy ——— bay, gulf
Sierra ——— mountain range
Silsilesi ——— mountain chain, ridge
Sint ——— saint
Sinus ——— bay, sea
Sjö-n ——— lake
Skarv-et ——— barren mountain
Skerry ——— rock
Slieve ——— mountain
Sø-er ——— lake-s
Sønder, Søndre ——— south-ern
Sopka ——— conical mountain, volcano
Sor ——— lake, salt lake
Sør, Sör ——— south-ern
Sory ——— salt lake, salt marsh
Spitz-e ——— peak, point, top
Sredn-iy, -yaya, -eye ——— central, middle
Stagno ——— lake, pond
Stantsiya ——— station
Stausee ——— reservoir
Stenón ——— channel, strait
Step'-i ——— steppe-s
Štít ——— summit, top
Stor-e ——— big, great
Straat ——— strait
Straum-en ——— current-s
Strelka ——— spit of land
Stretet, Stretto ——— strait
Su ——— reef, river, rock, stream
Su Anbarı ——— reservoir
Sud ——— south
Sudo ——— channel, strait
Suidō ——— channel, strait
Şummān ——— rocky desert
Sund ——— sound, strait
Sunden ——— channel, inlet, sound
Svyat-oy, -aya, -oye ——— holy, saint
Sziget ——— island

Tagh ——— mountain-s
Tai ——— coast, tide
Tall ——— hill, mound
T'an ——— lake
Tanezrouft ——— desert
Tang ——— plain, steppe
Tangi ——— peninsula, point
Tanjong, Tanjung ——— cape, point
Tao ——— island-s
Tarso ——— hill-s, mountain-s
Tassili ——— plateau, upland
Tau ——— mountain-s, range
Taūy ——— hills, mountains
Tchabal ——— mountain-s
Te Ava ——— tidal flat
Tel-l ——— hill, mound
Telok, Teluk ——— bay
Tepe, -si ——— hill, peak
Tepuí ——— mesa, mountain
Terara ——— hill, mountain, peak
Testa ——— bluff, head
Thale ——— lake
Thang ——— plain, steppe
Tien ——— lake
Tierra ——— land, region
Ting ——— hill, mountain
Tir'at ——— canal
Tó ——— lake, pool
To, Tō ——— island-s, rock-s
Tonle ——— lake

Tope ——— hill, mountain, peak
Top-pen ——— peak-s
Träsk ——— bog, lake
Tso ——— lake
Tsui ——— cape, point
Tübegi ——— peninsula
Tulu ——— hill, mountain
Tunturi-t ——— hill-s, mountain-s

Uad ——— wadi, watercourse
Udde-m ——— point
Ujong, Ujung ——— cape, point
Umi ——— bay, lagoon, lake
Ura ——— bay, inlet, lake
'Urūq ——— dune area
Uul, Uula ——— mountain, range
'Uyūn ——— springs

Vaara ——— mountain
Vaart ——— canal
Vær ——— fishing station
Vaïn ——— channel, strait
Valle, Vallée ——— valley, wadi
Vallen ——— waterfall
Valli ——— lagoon, lake
Vallis ——— valley
Vanua ——— land
Varre ——— mountain
Vatn, Vatten, Vatnet ——— lake, water
Veld ——— grassland, plain
Verkhn-iy, -yaya, -eye ——— higher, upper
Vesi ——— lake, water
Vest-er ——— west
Via ——— road
Vidda ——— plateau
Vig, Vík, Vik, -en ——— bay, cove
Vinh ——— bay
Vodokhranilishche ——— reservoir
Vodoskhovyshche ——— reservoir
Volcan, Volcán ——— volcano
Vostochn-yy, -aya, -oye ——— eastern
Vötn ——— stream
Vozvyshennost' ——— plateau, upland
Vozyera, -yero, -yera ——— lake-s
Vrchovina ——— mountains
Vrch-y ——— mountain-s
Vrh ——— hill, mountain
Vrŭkh ——— mountain
Vūng ——— bay
Vyaliki, -ikaya, -ikaye ——— big, large
Vysočina ——— highland

Wabē ——— stream
Wadi, Wâdi, Wādī ——— valley, watercourse
Wāhât, Wāḥat ——— oasis
Wald ——— forest, wood
Wan ——— bay, gulf
Water ——— harbor
Webi ——— stream
Wiek ——— cove, inlet

Xia ——— gorge, strait
Xiao ——— lesser, little

Yanchi ——— salt lake
Yang ——— ocean
Yarımadası ——— peninsula
Yazovir ——— reservoir
Yŏlto ——— island group
Yoma ——— mountain range
Yü ——— island
Yumco ——— lake
Yunhe ——— canal
Yuzhn-yy, -aya, -oye ——— southern

Zaki ——— cape, point
Zaliv ——— bay, gulf
Zan ——— mountain, ridge
Zangbo ——— river, stream
Zapadn-yy, -aya, -oye ——— western
Zatoka ——— bay, gulf
Zee ——— bay, sea
Zemlya ——— land
Zhotasy ——— mountains

TIME ZONES

DATE LINE
The date line, is an imaginary line located on or near the 180° meridian, shown on this map by a dashed black line. A person traveling west across the date line would add a day, but a person traveling east would subtract a day. The position of the date line is based on international acceptance, but it has no legal status. Island countries near the line can choose which date they will observe.

MERIDIAN OF GREENWICH
Britain's Royal Observatory at Greenwich (London) is the home of Greenwich mean time and the "prime meridian of the world" (0° longitude). In 1884, an international conference in Washington, D.C., decided on Greenwich as the location for the prime meridian.

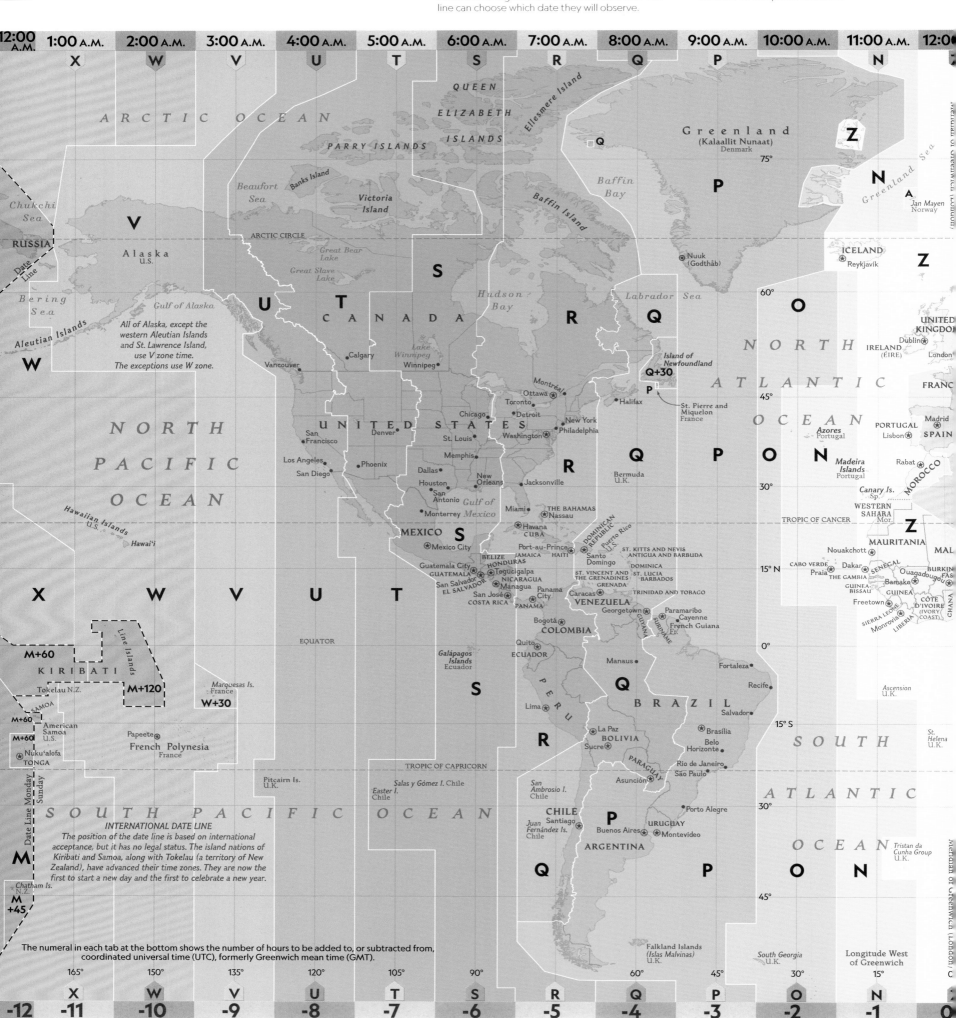

| 12:00 A.M. | 1:00 A.M. | 2:00 A.M. | 3:00 A.M. | 4:00 A.M. | 5:00 A.M. | 6:00 A.M. | 7:00 A.M. | 8:00 A.M. | 9:00 A.M. | 10:00 A.M. | 11:00 A.M. | 12:00 |

INTERNATIONAL DATE LINE
The position of the date line is based on international acceptance, but it has no legal status. The island nations of Kiribati and Samoa, along with Tokelau (a territory of New Zealand), have advanced their time zones. They are now the first to start a new day and the first to celebrate a new year.

All of Alaska, except the western Aleutian Islands and St. Lawrence Island, use V zone time. The exceptions use W zone.

The numeral in each tab at the bottom shows the number of hours to be added to, or subtracted from, coordinated universal time (UTC), formerly Greenwich mean time (GMT).

	165°	150°	135°	120°	105°	90°	60°	45°	30°	15°
	X	W	V	U	T	S	R	Q	P	O	N
-12	-11	-10	-9	-8	-7	-6	-5	-4	-3	-2	-1	0

Longitude West of Greenwich

The map outlines Earth's time zones with white lines, with each time zone covering approximately 15 degrees of longitude. Time zones are measured in reference to the Meridian of Greenwich, England (0° longitude), sometimes called the prime meridian. Time at Greenwich is known as coordinated universal time (UTC), formerly Greenwich mean time (GMT), and is the starting point in determining time worldwide. Letters on the map label

each time zone, and the corresponding numbers (with plus or minus signs) indicate the time difference from GMT/UTC. For example, the C time zone is +3. This means that when it is noon in the Greenwich Z zone, it is 3 p.m. standard time in the C zone—the time shown along the top of the map. Daylight saving time, normally one hour ahead of local standard time, is not shown on this map. Most time zones differ in one-hour increments, but

some countries choose to offset time zones by a fraction of an hour. For example, India (E time zone, +5) shows the label E+30, which means that it is 5 hours and 30 minutes ahead of GMT/UTC time. Nepal is E+45, making it 5 hours and 45 minutes ahead of GMT/UTC. Many governments choose to have their entire country in one time zone. China is the largest country with only one time zone; normally, it would be divided by five time zones.

Following Russia's annexation of Crimea, its time zone (B) was changed to Moscow time zone (C) on March 30, 2014.

APPENDIX

GEOGRAPHIC TERMS

A

abyssal plain a flat, relatively featureless region of the deep ocean floor extending from the Mid-Ocean Ridge to a continental rise or deep-sea trench

accretion the growth of a tectonic plate or landmass due to a buildup of sediment

acidification a soil-forming process where brown-earth soils transform into acid brown earth in humid temperate forest regions

alluvial fan a depositional, fan-shaped feature found where a stream or channel gradient levels out at the base of a mountain

anthrome a human-influenced biome that includes croplands, rangelands, and villages

antipode a point that lies diametrically opposite any given point on the surface of the Earth

archipelago an associated group of scattered islands in a large body of water

arid dry climate with small annual accumulated precipitation

asthenosphere the zone of viscous rock in the Earth's upper mantle, immediately below the rigid lithosphere

astronomical unit (AU) the average distance between the Earth and sun; a unit of measurement equal to 499 light-seconds

asylum seeker a refugee who has made a formal application to a government for asylum and is waiting for a decision

atmosphere the thin envelope of gases surrounding the solid Earth and comprising mostly nitrogen, oxygen, and various trace gases

atoll a circular coral reef enclosing a lagoon

B

barrier island a low-lying, sandy island parallel to a shoreline but separated from the mainland by a lagoon

basin a depression in the Earth's surface that may be dry or filled with water and sediment

bathymetry the measurement of depth within bodies of water

bay an area of a sea or other body of water bordered on three sides by a curved stretch of coastline; usually smaller than a gulf

bedrock the unweathered rock which underlies the soil and regolith or which may be exposed at the land surface

biodiversity a broad concept that refers to the variety and range of species (flora and fauna) present in an ecosystem

biofuel any liquid fuel derived from biomass; the two most common types are ethanol and biodiesel

biogeography the study of the distribution patterns of plants and animals and the processes that produce those patterns

biomass the total mass of all the organisms inhabiting a given area, or of a particular population or trophic level

biome a very large ecosystem made up of specific plant and animal communities interacting with the physical environment (climate and soil)

biosphere the realm of Earth's surface and atmosphere that includes all plant and animal life-forms

bluff a steep slope or wall of consolidated sediment adjacent to a river or its floodplain

bog soft, spongy, waterlogged ground consisting chiefly of partially decayed plant matter (peat)

boreal forest coniferous forest that extends across high northern latitudes of North America and Eurasia

breakwater a stone or concrete structure built near a shore to prevent damage to watercraft or construction

butte a tall, steep-sided, flat-topped tower of rock created by erosion and weathering

C

caldera a large, crater-like depression resulting from the explosion and collapse of a volcano

canal an artificially made channel of water used for navigation or irrigation

canopy the ceiling-like layer of branches and leaves that forms the uppermost layer of a forest

canyon a valley with very steep sides cut by erosion from flowing water; submarine canyons occur on the ocean floor

capitalism an economic system characterized by resource allocation primarily through market mechanisms; means of production are privately owned (by either individuals or corporations), and production is organized around profit maximization

carbon cycle one of the several geochemical cycles by which matter is recirculated through the lithosphere, hydrosphere, atmosphere, and biosphere

carbon neutral process a process resulting in zero net change in the balance between emission and absorption of carbon

carrying capacity the maximum number of animals and/or people a given area can support at a given time under specified levels of consumption

cartogram a map designed to present statistical information in a diagrammatic way, usually not to scale

cartography the production and study of maps and charts

chemosynthesis the process in which chemical energy is used to make organic compounds from inorganic compounds

chlorofluorocarbon a molecule of industrial origin containing chlorine, fluorine, and carbon atoms; causes severe ozone destruction

civilization a cultural concept suggesting substantial development in the form of agriculture, cities, food and labor surplus, labor specialization, social stratification, and state organization

climate the long-term behavior of the atmosphere; it includes measures of average weather conditions (e.g., temperature, humidity, precipitation, and pressure), as well as trends, cycles, and extremes

climate change the complex shifts affecting the world's climatic system caused by humans releasing heat-trapping gases into the atmosphere; the changes encompass not only rising average temperatures but also extreme weather events, rising seas, and shifting wildlife populations and habitats

collision boundary the location where two continental plates collide after the oceanic crust, which separated them, is consumed

colonialism the political, social, or economic domination of a state over another state or people

commercial fishery all of the variables involved in the activities to harvest fish for commercial profit rather than subsistence fishing for local consumption

commodity an economic good or product that can be traded, bought, or sold

composite image a product of combining two or more images

coniferous trees and shrubs with thin leaves and producing cones; also a forest or wood composed of these trees

continental divide a ridge separating watersheds that flow toward opposite sides of a continent, usually into different oceans

continental plate a thick section of the Earth's continental crust that moves over the surface of the Earth

continental shelf the submerged, offshore extension of a continent

continental slope the steeply graded seafloor connecting the edge of the continental shelf to the deep-ocean floor

convergent boundary where tectonic plates move toward each other along their common boundary, causing subduction

coral reef an offshore ridge, mainly of calcium carbonate, formed by the secretions of small marine animals

coral bleaching the whitening of a coral colony, indicative of environmental stress

core the dense, innermost layer of Earth; the outer core is liquid, while the inner core is solid

Coriolis effect the deflection of wind systems and ocean currents (as well as freely moving objects not in contact with the solid Earth) to the right in the Northern Hemisphere and to the left in the Southern Hemisphere as a consequence of the Earth's rotation

crust the rocky, relatively low-density, outermost layer of Earth

cultural diffusion the spread of cultural elements from one group to another

culture the "way of life" for a group; it is transmitted from generation to generation and involves a shared system of meanings, beliefs, values, and social relations; it also includes language, religion, clothing, music, laws, and entertainment

currency another name for money, usually applied in the context of international trade

D

dead zone oxygen-starved areas in oceans and lakes where marine life cannot be supported, often linked to runoff of excess nutrients

deciduous trees and shrubs that shed their leaves seasonally; also a forest or wood mostly composed of these trees

deforestation the complete clearance of forests by cutting and/or burning

delta a flat, low-lying, often fan-shaped region at the mouth of a river; it is composed of sediment deposited by a river entering a lake, an ocean, or another large body of water

demography the study of population statistics, changes, and trends based on various measures of fertility, mortality, and migration

denudation the overall effect of weathering, mass wasting, and erosion, which ultimately wears down and lowers the continental surface

desert a region that has little or no vegetation and averages less than 10 inches (25 cm) of precipitation a year

desertification the spread of desert conditions in arid and semiarid regions; desertification results from a combination of climatic changes and increasing human pressures in the form of overgrazing, removal of natural vegetation, and cultivation of marginal land

developed country general term for an industrialized country with a diversified and self-sustaining economy, strong infrastructure, and high standard of living, sometimes called advanced economies

developing country general term for a nonindustrialized country with a weak economy, little modern infrastructure, and low standard of living, sometimes called emerging markets

dialect a regional variation of one language, with differences in vocabulary, accent, pronunciation, and syntax

diaspora the dispersion of people from their homelands to different countries around the world

diffuse plate boundary a zone of faulting and earthquakes extending to either side of a plate boundary

digital elevation model (DEM) a digital representation of Earth's topography in which data points representing altitude are assigned coordinates and viewed spatially; sometimes called a digital terrain model (DTM)

disconformity a discontinuity in sedimentary rocks in which the rock beds remain parallel

dormant volcano an active volcano that is temporarily in repose but expected to erupt in the future

drumlin a long hummock or hill deposited and shaped under an ice sheet or very broad glacier, while the ice was still moving

E

earthquake vibrations and shock waves caused by volcanic eruptions or the sudden movement of Earth's crustal rocks along fracture zones called faults

ecosystem a group of organisms and the environment with which they interact

ecotourism the development and management of tourism such that the environment is preserved

elevation the height of a point or place above an established datum, sometimes mean sea level

El Niño a pronounced warming of the surface waters along the coast of Peru and the equatorial region of the east Pacific Ocean; it is caused by weakening (sometimes reversal) of the trade winds, with accompanying changes in ocean circulation (including cessation of upwelling in coastal waters)

emigrant a person migrating away from a country or area; an out-migrant

endangered species a species at immediate risk of extinction; the IUCN (International Union for Conservation of Nature and Natural Resources) has four categories of risk ranging from threatened to critically endangered

endemic typical of or native to a particular area, people, or environment

endogenous introduced from or originating within a given organism or system

environment the sum of the external conditions, factors, and influences that affect an organism or community

eon the largest time unit on the geologic time scale; consists of several shorter units called eras

Equator latitude 0°; an imaginary line running east and west around Earth and dividing it into two equal parts known as the Northern and Southern Hemispheres; the Equator always has 12 hours of daylight and 12 hours of darkness

equinox the time of year (usually September 22–23 and March 21–22) when the length of night and day are about equal and the sun is directly overhead at the Equator

era a major subdivision of time on the geologic time scale; consists of several shorter units called periods

erosion the general term for the removal of surface rocks and sediment by the action of water, air, ice, or gravity

escarpment a cliff or steep rock face that separates two comparatively level land surfaces

esker a long ridge of material dumped from meltwater streams running subglacially, roughly parallel to the direction of ice flow

estuary a broadened seaward end or extension of a river (usually a drowned river mouth), characterized by tidal influences and the mixing of fresh and saline water

ethnic group individuals who share distinctive origins, history, culture, and linguistic heritage

evergreen applied to a tree or shrub that has persistent leaves throughout the year

Exclusive Economic Zone (EEZ) an oceanic zone extending up to 200 nautical miles (370 km) from a shoreline, within which a coastal state claims jurisdiction over fishing, mineral exploration, and other economically important activities

exoplanet a planet orbiting a star other than the sun; also known as an extrasolar planet

F

fault a fracture or break in rock where the opposite sides are displaced relative to each other

fjord a coastal inlet that is narrow and deep and reaches far inland; it is usually formed by the sea filling in a glacially scoured valley or trough

flood basalt a huge lava flow that produces thick accumulations of basalt layers over a large area

floodplain a wide, relatively flat area adjacent to a stream or river and subject to flooding and sedimentation; it is the most preferred land area for human settlement and agriculture

food chain the feeding pattern of organisms in an ecosystem, through which energy from food passes from one level to the next in a sequence

fork the place where a river separates into branches; also may refer to one of those branches

fossil fuel fuel in the form of coal, petroleum, or natural gas derived from the remains of ancient plants and animals trapped and preserved in sedimentary rocks

free trade trade between countries that takes place free of restrictions with no tariffs and quotas

G

galaxy a collection of stars, gas, and dust bound together by gravity; there are billions of galaxies in the universe, and Earth is in the Milky Way galaxy

genocide the intentional destruction, in whole or in part, of a national, ethnic, racial, or religious group

geochemistry a branch of geology focusing on the chemical composition of earth materials

geographic information system (GIS) an integrated hardware-software system used to store, organize, analyze, manipulate, model, and display geographic information or data

geography literally means "Earth description"; as a modern academic discipline, geography is concerned with the explanation of the physical and human characteristics and patterns of Earth's surface

geomorphology the study of planetary surface features, especially the processes of landform evolution on Earth

geopolitics the study of how factors such as geography, economics, and demography affect the power and foreign policy of a state

glaciation a period of glacial advancement through the growth of continental ice sheets and/or mountain glaciers

glacier a large, natural accumulation of ice that spreads outward on the land or moves slowly down a slope or valley

globalization a social science term used to describe increased connectedness and interdependence of world cultures and economies

global positioning system (GPS) a system of artificial satellites that provides information on three-dimensional position and velocity to users at or near the Earth's surface

global warming the warming of Earth's average global temperature due to a buildup in the atmosphere of heat-trapping "greenhouse gases" (e.g., carbon dioxide and methane) released by human activities

greenhouse effect an enhanced near-surface warming that is due to certain atmospheric gases absorbing and re-radiating long-wave radiation that might otherwise have escaped to space had those gases not been present in the atmosphere

Greenwich mean time local time at Greenwich, London, situated on the prime meridian

gross domestic product (GDP) the total market value of goods and services produced by a nation's economy in a given year using global currency exchange rates

gross national income (GNI) the income derived from the capital and income belonging to nationals employed domestically or abroad

gravitational waves ripples in the fabric of space and time, usually caused by the interaction of two or more large masses

gulf a very large area of an ocean or a sea bordered by coastline on three sides

gyre a large, semicontinuous system of major ocean currents flowing around the outer margins of every major ocean basin

H

habitat the natural environment (including controlling physical factors) in which a plant or animal is usually found or prefers to exist

hemisphere half a sphere; cartographers and geographers, by convention, divide the Earth into the Northern and Southern Hemispheres at the Equator and the Eastern and Western Hemispheres at the prime meridian (longitude 0°) and 180° meridian

hot spot a localized and intensely hot region or mantle plume beneath the lithosphere; it tends to stay relatively fixed geographically as a lithospheric plate migrates over it

human geography one of the two major divisions of systematic geography; it is concerned with the spatial analysis of human population, cultures, and social, political, and economic activities

hurricane a large, rotating storm system that forms over tropical waters, with very low atmospheric pressure in the central region and winds in excess of 119 kilometers an hour (74 mph); it is called a typhoon over the western Pacific Ocean and a cyclone over the northern Indian Ocean

hydroelectric power electricity that is generated by the passage of water through a turbine, usually at a dam

hydrosphere all of the water found on, under, or over Earth's surface

hydrothermal vent a fissure on the sea floor, usually found in volcanically active areas, that emits water heated up to 400°C (750°F)

I

ice age a period of pronounced glaciation usually associated with worldwide cooling, a greater proportion of global precipitation falling as snow, and a shorter seasonal snowmelt period

iceberg a large piece of floating ice, most of which is below sea level, that has broken off a glacier

ice sheet a continuous mass of glacier ice with an area over 50,000 square kilometers (31,068 square miles)

ice shelf a sheet of ice extending over the sea from a land base, fed by snow falling on it, or from glaciers on the land

igneous the rock type formed from solidified molten rock (magma) that originates deep within Earth; the chemical composition of the magma and its cooling rate determine the final rock type

immigrant a person migrating into a particular country or area; an in-migrant

impact crater a circular depression on the surface of a planet or moon caused by the collision of another body, such as an asteroid or comet

indigenous native to or occurring naturally in a specific area or environment

infrastructure transportation and communications networks that allow goods, people, and information to flow across space

intermittent river a river that has channels that are periodically dry; amount and length of these are increasing due to climate change

internally displaced person a person who flees their home, to escape danger or persecution, but does not leave the country

international date line an imaginary line that roughly follows the 180° meridian in the Pacific Ocean; immediately west of the date line the calendar date is one day ahead of the calendar date east of the line; people crossing the date line in a westward direction lose one calendar day, while those crossing eastward gain one calendar day

irrigation the supply of water to the land by means of channels, streams, and sprinklers in order to permit the growth of crops

isthmus a relatively narrow strip of land with water on both sides and connecting two larger land areas

J

jet stream a high-speed west-to-east wind current; jet streams flow in narrow corridors within upper-air westerlies, usually at the interface of polar and tropical air

K

karst a region underlain by limestone and characterized by extensive solution features such as sinkholes, underground streams, and caves

kettle hole a depression formed when ice blocks melt; may fill with water to become a lake

L

lagoon a shallow, narrow water body located between a barrier island and the mainland, with freshwater contributions from streams and saltwater exchange through tidal inlets or breaches throughout the barrier system

landslide the usually rapid downward movement of a mass of rock, earth, or artificial fill on a slope

La Niña the pronounced cooling of equatorial waters in the eastern Pacific Ocean

latitude the distance north or south of the Equator; lines of latitude, called parallels, are evenly spaced from the Equator to the North and South Poles (from 0° to 90° N and S latitude); latitude and longitude (see below) are measured in terms of the 360 degrees of a circle and are expressed in degrees, minutes, and seconds

leeward the side away from or sheltered by the wind

lithosphere the rigid outer layer of the Earth, located above the asthenosphere and comprising the outer crust and the upper, rigid portion of the mantle

longitude the distance measured in degrees east or west of the prime meridian (0° longitude) up to 180°; lines of longitude are called meridians (compare with latitude, above)

M

macroplastics pieces of plastic debris found in the ocean (as opposed to tiny microplastic particles)

magma molten, pressurized rock in the mantle that is occasionally intruded into the lithosphere or extruded to the surface of the Earth by volcanic activity

magnetic poles the points at Earth's surface at which the geomagnetic field is vertical; the location of these points constantly changes

mantle the dense layer of Earth below the crust; the upper mantle is solid, and with the crust forms the lithosphere, the zone containing tectonic plates; the lower mantle is partially molten, making it the pliable base upon which the lithosphere "floats"

map projection the geometric system of transferring information about a round object, such as a globe, to a flat piece of paper or other surface for the purpose of producing a map with known properties and quantifiable distortion

maria volcanic plains on the moon's surface that appear to the naked eye as smooth, dark areas

meridian a north-south line of longitude used to reference distance east or west of the prime meridian (longitude 0°)

mesa a broad, flat-topped hill or mountain with marginal cliffs and/or steep slopes formed by progressive erosion of horizontally bedded sedimentary rocks

mestizo a person of mixed Spanish and indigenous ancestry in Latin America; can also generally refer to racially mixed persons

mesosphere the middle layer of Earth's atmosphere, lying above the stratosphere and below the thermosphere

metamorphic the rock type formed from preexisting rocks that have been substantially changed from their original igneous, sedimentary, or earlier metamorphic form; catalysts of this change include high heat, high pressure, hot and mineral-rich fluids, or, more commonly, some combination of these

metric ton (tonne) unit of weight equal to 1,000 kilograms, or 2,205 pounds

microbe an organism—such as bacteria, fungi, and viruses—that can be seen only with the help of a microscope

migration the movement of people or animals from one place to another

mineral an inorganic solid with a distinctive chemical composition and a specific crystal structure that affect its physical characteristics

monsoon a seasonal reversal of prevailing wind patterns, often associated with the rainy season that occurs with the onset of the southwesterly monsoon

moraine the rocks, boulders, and debris that are carried and deposited by a glacier or ice sheet

N

nation a cultural concept for a group of people bound together by a strong sense of shared values and cultural characteristics, including language, religion, and common history

national debt the debts of a country's central government, both internal (owed to residents) and external (owed to foreign lenders)

nationalism a sense of identity and belonging to a group or community associated with a particular territory

nebula a cloud of interstellar gas and dust

node a point where distinct lines or objects intersect

Normalized Difference Vegetation Index (NDVI) a measurement of plant growth density over an area of the Earth's surface, measured on a scale of 0.1 to 0.8 (low to high vegetation)

North Pole the most northerly geographic point on the Earth; the northern end of the Earth's axis of rotation; 90° N

nuclear energy energy released during nuclear fission or fusion; used at nuclear power plants to generate electricity

GEOGRAPHIC TERMS

O

oasis a fertile area with water and vegetation in a desert

ocean current the regular and persistent flow of water in the oceans, usually driven by atmospheric wind and pressure systems or by regional differences in water density (temperature, salinity)

organic relating to or derived from living things

outsourcing delegating noncore processes from within a business to an external entity such as a subcontractor

oxbow lake a crescent-shaped lake or swamp occupying a channel abandoned by a meandering river

ozone layer region of Earth's atmosphere where ozone concentration is relatively high; the ozone layer absorbs harmful ultraviolet rays from the sun

P

paleo-geographic map a map depicting the past positions of the continents, developed from historic magnetic, biological, climatological, and geologic evidence

pampas a large, relatively flat temperate grassland in South America

Pangaea the supercontinent from which today's continents are thought to have originated

peninsula a long piece of land jutting out from a larger piece of land into a body of water

perennial river a watercourse that flows throughout the year without pause

period a basic unit of time on the geologic time scale, generally 35 to 70 million years in duration; a subdivision of an era

photosynthesis process by which plants convert carbon dioxide and water to oxygen and carbohydrates

physical geography one of the two major divisions of systematic geography; the spatial analysis of the structure, process, and location of Earth's natural phenomena, such as climate, soil, plants, animals, water, and topography

pilgrimage a typically long and difficult journey to a special place, often of religious importance

plain an extensive flat-lying area characterized generally by the absence of local relief features

plate tectonics the theory that Earth's continental plates slide or shift slowly over the asthenosphere and that their interactions cause earthquakes, volcanic eruptions, movement of landmasses, and other geologic events

plateau a landform feature characterized by high elevation and gentle upland slopes

point a sharp prominence or headland on the coast that juts out into a body of water

pollution a direct or indirect process resulting from human activity; part of the environment is made potentially or actually unsafe or hazardous to the welfare of the organisms that live in it

prairie a large, relatively flat temperate grassland in North America

prime meridian the line of 0° longitude that runs through Greenwich, England, and separates the Eastern and Western Hemispheres

purchasing power parity (PPP) a method of measuring gross domestic product that compares the relative value of currencies based on what each currency will buy in its country of origin; PPP provides a good comparison between national economies and is a decent indicator of living standards

R

rain shadow the dry region on the downwind (leeward) side of a mountain range

reef a strip of rocks or sand either at or just below the surface of water

refugee a person who flees their country of origin owing to a well-founded fear of being persecuted for, for example, race, religion, or political opinion

regolith a layer of disintegrated or partly decomposed rock overlying unweathered parent materials; regolith is usually found in areas of low relief where the physical transport of debris is weak

renewable energy a source of power that can be regenerated or maintained if used at rates that do not exceed natural replenishment

Richter scale a logarithmic scale devised to represent the relative amount of energy released by an earthquake; moment magnitude has superseded the Richter scale as the preferred measurement of earthquake magnitude

rift a long, narrow trough created by plate movement at a divergent boundary

rift valley a long, structural valley formed by the lowering of a block between two parallel faults

Ring of Fire (also Rim of Fire) an arc of volcanoes and tectonic activity along the perimeter of the Pacific Ocean

runoff the movement over ground of rainwater

S

salinization the accumulation of salts in soil

sand dune a hill or ridge of sand accumulated and sorted by wind action

savanna a tropical grassland with widely spaced trees; it experiences distinct wet and dry seasons

seamount a submerged volcano rising from the ocean floor

sea stack a column of rock detached from the mainland by erosion and rising precipitously out of the sea

sedimentary the rock type formed from preexisting rocks or pieces of once living organisms; deposits accumulate on Earth's surface, generally with distinctive layering or bedding

service economy a sector of the economy based on services such as hospitality, health, financial services, and information technology

sextant a navigational instrument used to work out latitude and longitude

shrubland an open or closed stand of shrubs up to about two meters (6.6 feet) tall

smog a combination of smoke (introduced into the atmosphere by human agency) and fog (which occurs naturally)

solar energy any energy source based directly on the sun's radiation, including electricity generation

solar radiation energy emitted by the sun

solar system the sun and all the celestial bodies (planets, asteroids, comets, etc.) that revolve around it

solar wind the rapid stream of atoms and ions moving outward from the solar corona

solstice a celestial event that occurs twice a year (usually June 20–21 and December 21–22), when the sun appears directly overhead to observers at the Tropic of Cancer or the Tropic of Capricorn

sound a broad channel or passage of water connecting two larger bodies of water or separating an island from the mainland

South Pole the most southerly geographic point on the Earth; the southern end of the Earth's axis of rotation; 90° S

spreading boundary where plates move apart along their common boundary, creating a crack in the Earth's crust (typically at the Mid-Ocean Ridge), which is then filled with upwelling molten rock; also called a divergent boundary

state an area with defined and internationally acknowledged boundaries; a political unit

steppe semiarid, relatively flat plains of Europe, central Asia, and Siberia with occasional or no trees

strait a narrow passage of water that connects two larger bodies of water

stratosphere a layer of the Earth's atmosphere where temperatures rise with increasing altitude; above the troposphere

subduction the tectonic process by which the down-bent edge of one tectonic plate is forced underneath another plate

swamp a low-lying area of wetland that is usually flooded throughout the year, dominated by woody plants

T

tariff a surcharge on imports levied by a state; a form of protectionism designed to increase imports' market price and thus inhibit their consumption

tectonic plate (also lithospheric or crustal plate) a section of the Earth's rigid outer layer that moves as a distinct unit upon the plastic-like mantle materials in the asthenosphere

temperate mild or moderate climate characteristics relating to the mid-latitude, the area of Earth between the tropics and the polar regions

thermosphere an outer layer of the Earth's atmosphere, in which the temperature increases continuously with height

tide the regular rise and fall of the ocean, caused by the mutual gravitational attraction between the Earth, moon, and sun, as well as the rotation of the Earth-moon system around its center of gravity

ton unit of weight equal to 2,000 pounds (907 kg) in the U.S. or 2,240 pounds (1,016 kg) in the U.K.

tonne (see metric ton)

topography the relief features that are evident on a planetary surface

tornado a violently rotating, funnel-shaped column of air characterized by extremely low atmospheric pressures and exceptional wind speeds generated within intensethunderstorms

trade wind a wind blowing persistently from the same direction; particularly from the subtropical high-pressure centers toward the equatorial low-pressure zone

tributary a river or stream flowing into a larger river or stream

tropical warm and moist; occurring in or characteristic of the tropics

Tropic of Cancer latitude 23.5° N; the farthest northerly excursion of the sun when it is directly overhead

Tropic of Capricorn latitude 23.5° S; the farthest southerly excursion of the sun when it is directly overhead

tsunami a series of ocean waves, often very destructive along coasts, caused by the vertical displacement of the seafloor during an earthquake, submarine landslide, or volcanic eruption

tundra a zone in cold, polar regions (mostly in the Northern Hemisphere) that is transitional between the zone of polar ice and the limit of tree growth; it is usually characterized by low-lying vegetation, with extensive permafrost and waterlogged soils

U

unconformity a discontinuity in sedimentary rocks caused by erosion or nondeposition

uplift the slow, upward movement of a part of Earth's crust

upwelling the process by which water rich in nutrients rises from ocean depths toward the ocean surface; it is usually the result of diverging surface waters

urban agglomeration a group of several cities and/or towns and their suburbs

urbanization a process in which there is an increase in the percentage of people living and working in urban places compared with rural places

V

volcanism the upward movement and expulsion of molten (melted) material and gases from within the Earth's mantle onto the surface, where it cools and hardens, producing characteristic terrain

W

water cycle the continuous recirculation of water from the oceans, through the atmosphere, to the continents, through the biosphere and lithosphere, and back to the sea

watershed the drainage area of a river and its tributaries

weathering the processes or actions that cause the physical disintegration and chemical decomposition of rock and minerals

wetland an area of land covered by water or saturated by water sufficiently enough to support vegetation adapted to wet conditions; includes swamps, marshes, bogs, and fens

wilderness a natural environment that has remained essentially undisturbed by human activities and, increasingly, is protected by government or nongovernment organizations

wind energy power generated by harnessing the wind, including electricity generation

windward the side toward or unsheltered from the wind

woodland a landscape of mature trees usually spaced more widely than a forest

Z

zenith the point in the sky that is immediately overhead; also the highest point above the observer's horizon obtained by a celestial body

KEY TO FLAGS & FACTS

In this atlas, 195 independent countries, more than 50 dependencies and areas whose sovereignty is unresolved or in dispute, 13 Canadian provinces and territories, and 50 U.S. states and the District of Columbia are arranged geographically and then alphabetically by conventional short-form name—for example: Luxembourg, Oman, Uruguay. Each country's official long-form name (Grand Duchy of Luxembourg, Sultanate of Oman, Oriental Republic of Uruguay) appears beneath its short-form name. In some cases—Canada, Mongolia, Ukraine—the two names are the same.

Independent country entries feature flags as well as important statistical data. All dependencies and some areas of unresolved sovereignty provide abbreviated lists of statistical data, but they do not have flags. Following the list of countries and dependencies are separate lists for the provinces and territories of Canada and for the states of the United States.

Dependencies are nonindependent political entities associated in some way with a particular independent nation. Uninhabited islands are included in dependency lists below the sovereign countries, but only populated dependencies are profiled in this atlas. Country names in parentheses indicate the sovereign nations to which the dependencies belong.

Areas of unresolved sovereignty, sometimes informally referred to as disputed areas, are listed under the countries that claim them. These areas can be defined by international agreements or actions, but they can also involve disputed regions created by separatist activity. In many cases, the areas claim independence and are politically autonomous but lack international recognition. For example, Nagorno-Karabakh is a separatist region listed under Azerbaijan. Although Azerbaijan lost control over the area in the early 1990s, it still considers Nagorno-Karabakh to be an integral part of its sovereign territory, as do most countries and international organizations.

The statistical data for countries, provinces, states, and dependencies highlight geography and demography. These details offer a brief overview of each political entity, presenting general characteristics, and are not intended to be comprehensive. Space and source limitations dictate the amount of information included. Except where otherwise noted, all statistical data are derived from the CIA *World Factbook, cia.gov/library/publications/the-world-factbook*.

FLAGS

Sovereign flags appear above each country entry and serve as important national symbols. The colors and emblems on flags reflect the heritage and aspirations of nations. State flags appear above each U.S. state entry, and the District of Columbia. National Geographic and this atlas benefit from the research and updates provided by the Flag Institute in London, United Kingdom, and the Flags of the World website, *crwflags.com/fotw/flags/*.

AREA

Square kilometer and square mile figures report the total area of a country or territory, including all land and inland water features, as delimited by coastlines and international boundaries. Figures do not include territorial waters. Area information comes from the CIA *World Factbook* and the U.S. State Department. Figures for U.S. states and the District of Columbia come from the U.S. Census Bureau, *census.gov*; those for Canada's provinces and territories are from Statistics Canada, *statcan.gc.ca*.

POPULATION

Figures are mid-2018 estimates. Population estimates are rounded to the nearest thousand and are based on statistics from population censuses, vital statistics registration systems, or sample surveys pertaining to the recent past and on assumptions about future trends. Population numbers for U.S. states and the District of Columbia are mid-2018 estimates from the U.S. Census Bureau.

CAPITAL

The name of the capital city, or cities, for a country or dependency is followed by the city's population. The government function is shown in parentheses in cases where there are multiple capitals—for example: Amsterdam (official) 1,140,000. Most capital city populations are 2018 or 2019 estimates for urban agglomerations, which include residents of the city itself plus designated suburban areas. For U.S. capitals, figures are 2017 estimates for city proper.

LARGEST CITIES (U.S.)

The name of the largest city within each state is followed by the city's population. Most city populations are 2017 estimates from the U.S. Census Bureau.

RELIGION

The most widely practiced religions appear in rank order, though the list is not comprehensive. Words such as "traditional" or "indigenous" indicate local or regional beliefs.

LANGUAGE

The most widely spoken languages appear in rank order. A language marked with a red symbol is an official national language, but not all countries legally designate an official language.

STATE BIRD AND FLOWER (U.S.)

Official state birds and flowers represent the natural treasures of each state. Every state, as well as the District of Columbia, recognizes a bird and flower as designated by their legislature.

MAP POLICY

Nations: Issues of national sovereignty and contested borders often boil down to "de facto versus de jure" discussions. Governments and international agencies frequently make official rulings about contested regions. These de jure decisions, no matter how legitimate, are often at odds with the wishes of individuals and groups, and they often stand in stark contrast to real-world situations. The inevitable conclusion: It is simplest and best to show the world as it is—de facto—rather than as we or others wish it to be.

Place-names: Ride a barge down the Danube through the many countries it touches, and you'll hear the river called Donau, Duna, Dunaj, Dunărea, Dunav, Dunay. These are local names. This atlas uses the conventional name, Danube, on physical maps. On political maps, local names are used, with the conventional name in parentheses where space permits. Usage conventions for both foreign and domestic place-names are established by the U.S. Board on Geographic Names, a group with representatives from several federal agencies.

Page   Index Marker —   Bern, *Switz.* **152** B2

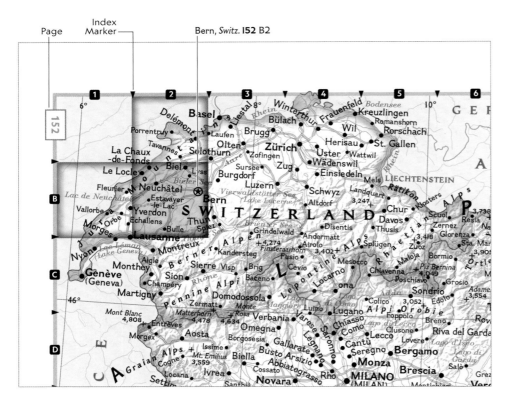

The following system is used to locate a place on a map in the *National Geographic Family Reference Atlas.* The boldface type after an entry refers to the page on which the map is found. The letter-number combination refers to the grid on which the particular place-name is located. The edge of each map is marked horizontally with numbers and vertically with letters. In between, at equally spaced intervals, are index markers (arrows). If these markers were connected with lines, each page would be divided into a grid. Take Bern, Switzerland, for example. The index entry reads "**Bern,** *Switz.* **152** B2." On page 152, Bern is located within the grid square where row B and column 2 intersect (example at left).

A place-name may appear on several maps, but the index lists only the best presentation. Usually, this means that a feature is indexed to the largest-scale map on which it appears in its entirety. (Note: Rivers are often labeled multiple times even on a single map. In such cases, the rivers are indexed to labels that are closest to their mouths.) The name of the country or continent in which a feature lies is shown in italic type and is usually abbreviated. (A full list of abbreviations appears on page 276.)

The index lists more than proper names. Some entries include a description, as in "**Elba, island,** *It.* **152** J5" and "**Amazon (Solimões), river,** *S. Amer.* **128** D5." In languages other than English, the description of a physical feature may be part of the name; e.g., the "Erg" in "**Erg Chech,** *Alg.* **200** J9" means "sand dune region." The glossary of Foreign Terms on pages 284–286 translates such terms into English.

When a feature or place can be referred to by more than one name, both may appear in the index with cross-references. For example, the entry for Cairo reads "**Cairo** *see* **Al Qāhirah,** *Egypt* **202** F11." That entry is "**Al Qāhirah (Cairo),** *Egypt* **202** F11."

1st Cataract, *Egypt* **202** K12
25 de Mayo, *Arg.* **135** N8
2nd Cataract, *Sudan* **202** L11
3l de Janeiro, *Angola* **208** J10
3rd Cataract, *Sudan* **203** N11
4th Cataract, *Sudan* **203** N12
5th Cataract, *Sudan* **203** P13
6th Cataract, *Sudan* **203** Q12
9 de Julio, *Arg.* **134** M11

A

A Cañiza, *Sp.* **146** C6
A Coruña, *Sp.* **146** A6
A Gudiña, *Sp.* **146** D7
A. B. Drachmann Bræ, glacier, *Greenland, Den.* **88** F12
A.P. Bernstorf Glacier, *Greenland, Den.* **89** R9
A.P. Olsen Land, *Greenland, Den.* **88** G13
Aachen, *Ger.* **150** F4
Aalen, *Ger.* **150** J7
Aaley, *Leb.* **172** J6
Aalsmeer, *Neth.* **148** J12
Aalst, *Belg.* **148** L11
Aappilattoq, *Greenland, Den.* **88** J5
Aare, river, *Switz.* **152** A3
Aarschot, *Belg.* **148** L12
Aasiaat (Egedesminde), *Greenland, Den.* **88** M5
Aasu, *American Samoa, U.S.* **230** D6
Abā as Saʻūd, *Saudi Arabia* **174** M9
Aba, *China* **184** K9
Aba, *Nigeria* **206** N8
Aba, *D.R.C.* **209** B19
Abaco Island, *Bahamas* **122** B9
Ābādān, *Iran* **177** K13
Ābādeh, *Iran* **177** J16
Abadla, *Alg.* **200** E9
Abaetetuba, *Braz.* **133** D13
Abaiang, *Kiribati* **227** P14
Abaji, *Nigeria* **206** L8
Abajo Peak, *Utah, U.S.* **109** P9
Abakaliki, *Nigeria* **206** M8
Abakan, *Russ.* **161** L13
Abalak, *Niger* **206** G7
Abalessa, *Alg.* **201** K13
Abancay, *Peru* **132** H5

Abaokoro, island, *Kiribati* **227** Q14
Abarkūh, *Iran* **177** J17
Abashiri Wan, *Japan* **188** F15
Abashiri, *Japan* **188** F15
Abay (Blue Nile), *Eth.* **210** L7
Abay, *Kaz.* **179** F13
Abay, river, *Eth.* **210** L6
Abaya Hāyk', *Eth.* **211** N7
Abba, *Cen. Af. Rep.* **206** N12
Abbaye, Point, *Mich., U.S.* **100** E7
Abbeville, *S.C., U.S.* **98** K7
Abbeville, *La., U.S.* **102** Q8
Abbeville, *Ala., U.S.* **103** M17
Abbeville, *Fr.* **148** M9
Abbeyfeale (Mainistir na Féile), *Ire.* **145** S3
Abbeyleix (Mainistir Laoise), *Ire.* **145** S6
Abbiategrasso, *It.* **152** E4
Abbot Ice Shelf, *Antarctica* **238** J6
Abbotsford, *B.C., Can.* **86** M8
Abbotsford, *Wis., U.S.* **100** H4
Abbottabad, *Pak.* **180** H12
'Abd al 'Azīz, Jabal, *Syr.* **172** D13
'Abd al Kūrī, island, *Yemen* **175** Q14
Abdi, *Chad* **207** H15
Abéché, *Chad* **207** H15
Abemama, island, *Kiribati* **224** G8
Abengourou, *Côte d'Ivoire* **205** P14
Abenilang, *Gabon* **208** D7
Åbenrå, *Den.* **142** N11
Abeokuta, *Nigeria* **206** L6
Ābera, *Eth.* **211** N6
Aberaeron, *Wales, U.K.* **145** T9
Aberdare, *Wales, U.K.* **145** U10
Aberdaron, *Wales, U.K.* **145** S9
Aberdaugleddau *see* **Milford Haven,** *Wales, U.K.* **145** U8
Aberdeen Lake, *Nunavut, Can.* **87** H14
Aberdeen, *Md., U.S.* **98** C14
Aberdeen, *N.C., U.S.* **98** J11
Aberdeen, *Miss., U.S.* **103** J13
Aberdeen, *S. Dak., U.S.* **106** J6
Aberdeen, *Idaho, U.S.* **108** H6
Aberdeen, *Wash., U.S.* **110** D2
Aberdeen, *Scot., U.K.* **144** J11
Aberffraw, *Wales, U.K.* **145** R9
Abergele, *Wales, U.K.* **145** R10
Abergwaun *see* **Fishguard,** *Wales, U.K.* **145** T8
Abernathy, *Tex., U.S.* **104** F12

Abert, Lake, *Oreg., U.S.* **110** K6
Abertawe *see* **Swansea,** *Wales, U.K.* **145** U9
Aberteif *see* **Cardigan,** *Wales, U.K.* **145** T9
Abertillery, *Wales, U.K.* **145** U10
Aberystwyth, *Wales, U.K.* **145** T9
Abhā, *Saudi Arabia* **174** L8
Abhar, *Iran* **177** E14
Abhē Bid Hāyk', lake, *Djibouti, Eth.* **210** K9
'Abidiya, *Sudan* **203** P13
Abidjan, *Côte d'Ivoire* **205** Q14
Abilene, *Tex., U.S.* **105** G14
Abilene, *Kans., U.S.* **107** S8
Abingdon *see* **Isla Pinta,** *Ecua.* **130** P3
Abingdon, *Va., U.S.* **98** G8
Abingdon, *Ill., U.S.* **101** Q4
Abingdon, *Eng., U.K.* **145** U12
Abiquiu, *N. Mex., U.S.* **109** R12
Abisko, *Sw.* **143** D13
Abitibi, Lake, *N. Amer.* **84** H8
Ābīy Ādī, *Eth.* **210** J8
Abīyata Hāyk', *Eth.* **211** N7
Abkhazia, republic, *Rep. of Georgia* **171** A15
Abminga, *S. Aust., Austral.* **221** U9
Abnūb, *Egypt* **202** H11
Åbo *see* **Turku,** *Fin.* **143** J15
Aboa Station, *Antarctica* **238** B12
Abohar, *India* **182** F4
Aboisso, *Côte d'Ivoire* **205** P14
Abomey, *Benin* **205** N17
Abong Mbang, *Cameroon* **206** N11
Abou Deïa, *Chad* **207** J14
Abou Goulem, *Chad* **207** H15
Abra Pampa, *Arg.* **134** D9
Abraham Lincoln Birthplace National Historic Site, *Ky., U.S.* **103** C16
Abraham's Bay, *Bahamas* **119** H16
Abreú, *Dom. Rep.* **119** L19
Abrolhos Bank, *Atl. Oc.* **249** R8
Abrolhos Seamounts, *Atl. Oc.* **249** R8
Abrolhos, Arquipélago dos, *Braz.* **133** K16
Absalom, Mount, *Antarctica* **238** E11
Absaroka Range, *Mont., Wyo., U.S.* **108** F8
Absarokee, *Mont., U.S.* **108** E9
Abşeron Yarımadası, *Azerb.* **171** D23
Abū al Abyaḍ, island, *U.A.E.* **175** G15

Abū al Khaşīb, *Iraq* **177** K13
Abū 'Alī, island, *Saudi Arabia* **174** E12
Abū Baḩr, *Saudi Arabia* **174** J11
Abū Ballāş, peak, *Egypt* **202** K9
Abū Daghmah, *Syr.* **172** D10
Abu Dhabi *see* **Abū Ẓaby,** *U.A.E.* **175** G15
Abu Dis, *Sudan* **203** N13
Abu Durba, *Egypt* **173** T2
Abu Gabra, *Sudan* **203** T8
Abu Gamal, *Sudan* **203** R15
Abu Gubeiha, *Sudan* **203** T11
Abu Hamed, *Sudan* **203** N13
Abu Hashim, *Sudan* **203** S13
Abū Kamāl, *Syr.* **172** H14
Abu Matariq, *Sudan* **203** T8
Abu Musa, island, *Iran* **177** P18
Abu Rudeis, *Egypt* **173** S2
Abū Rujmayn, Jabal, *Syr.* **172** G10
Abu Shanab, *Sudan* **203** S9
Abu Simbel *see* **Abū Sunbul,** *Egypt* **202** L11
Abū Sunbul (Abu Simbel), *Egypt* **202** L11
Abū Ṭīj, *Egypt* **202** H11
Abu 'Uruq, *Sudan* **203** Q11
Abu Zabad, *Sudan* **203** T10
Abū Ẓaby (Abu Dhabi), *U.A.E.* **175** G15
Abū Zanīmah, *Egypt* **202** G12
Abu Zenīma, *Egypt* **173** S2
Abuja, *Nigeria* **206** K8
Abumombazi, *D.R.C.* **209** C14
Abunã, *Braz.* **132** G7
Abunã, river, *S. Amer.* **132** G7
Abyad, *Sudan* **203** S8
Abyei, *Sudan* **203** U9
Abyei, region, *Sudan,* **203** U9
Ābyek, *Iran* **177** E14
Academy Glacier, *Greenland, Den.* **88** F14
Academy Gletscher, *Greenland, Den.* **88** C10
Acadia National Park, *Me., U.S.* **97** F18
Acaponeta, *Mex.* **116** G9
Acapulco, *Mex.* **116** K11
Acarai Mountains, *S. Amer.* **131** H15
Acaraú, *Braz.* **133** E16
Acarigua, *Venez.* **130** D9
Acatenango, Volcán de, *Guatemala* **117** L15

Acatlán, *Mex.* **116** J12
Accomac, *Va., U.S.* **98** E15
Accra, *Ghana* **205** P16
Accumoli, *It.* **152** J9
Achach, island, *F.S.M.* **226** F8
Acharacle, *Scot., U.K.* **144** K7
Acharnés, *Gr.* **156** J12
Achavanich, *Scot., U.K.* **144** G10
Achayvayam, *Russ.* **161** D22
Achelóos, river, *Gr.* **156** F8
Achénouma, *Niger* **206** E11
Achill Island, *Ire.* **145** P3
Achim, *Ger.* **150** C7
Achinsk, *Russ.* **161** K13
Achit Nuur, *Mongolia* **184** C6
Achna (Düzce), *Cyprus* **162** P9
Achnasheen, *Scot., U.K.* **144** H8
Achwa, river, *Uganda* **211** Q4
Acıgöl, *Turk.* **170** G5
Acıpayam, *Turk.* **170** H5
Ackerman, *Miss., U.S.* **102** K12
Ackley, *Iowa, U.S.* **107** N12
Acklins Island, *Bahamas* **119** H15
Acoma Pueblo, *N. Mex., U.S.* **109** S11
Aconcagua, Cerro, *Arg.* **134** K7
Aconcagua, Río, *Chile* **134** K6
Acquaviva, *San Marino* **163** J14
Acqui Terme, *It.* **152** F3
Acraman, Lake, *Austral.* **220** J9
Acre *see* **'Akko,** *Israel* **172** K5
Acre, river, *S. Amer.* **132** G6
Actium, battle, *Gr.* **156** G7
Açu, *Braz.* **133** F17
Ad Dabʻah, *Egypt* **202** E10
Ad Dafnīyah, *Lib.* **201** D18
Ad Dahnā', *Saudi Arabia* **174** E9
Ad Dakhla, *W. Sahara* **200** J2
Ad Dammām, *Saudi Arabia* **174** F12
Ad Dār al Ḩamrā', *Saudi Arabia* **174** E5
Ad Darb, *Saudi Arabia* **174** L7
Ad Dawādimī, *Saudi Arabia* **174** G9
Ad Dawḩah (Doha), *Qatar* **175** F13
Ad Dawr, *Iraq* **176** F10
Ad Dibdibah, *Saudi Arabia* **174** D10
Aḑ Ḑiffah *see* **Libyan Plateau,** *Lib.* **201** E23
Ad Dilam, *Saudi Arabia* **174** G11
Ad Dīwānīyah, *Iraq* **176** H11
Ad Duwayd, *Saudi Arabia* **174** C8
Ada, *Ohio, U.S.* **101** Q12
Ada, *Okla., U.S.* **105** D18

Ada, *Minn., U.S.* **106** G8
Ada, *Serb.* **154** D10
Adair, Cape, *Nunavut, Can.* **87** E17
Adak Island, *Alas., U.S.* **112** N5
Adalia *see* Antalya, *Turk.* **170** H6
Adam, *Oman* **175** H17
Adámantas, *Gr.* **157** M13
Adamawa, region, *Af.* **198** H5
Adamello, peak, *It.* **152** C6
Adam's Bridge, *Asia* **168** K7
Adam's Peak, *Sri Lanka* **183** U7
Adam's Rock, cape, *Pitcairn I.* **228** Q6
Adams, *Wis., U.S.* **100** K5
Adams, *Minn., U.S.* **106** L12
Adams, Mount, *Wash., U.S.* **110** E5
Adamstown, *Pitcairn I.* **228** Q6
Adamsville, *Tenn., U.S.* **103** G13
'Adan (Aden), *Yemen* **174** Q9
'Adan aş Şughrá, cape, *Yemen* **174** Q9
Adana, *Turk.* **170** J10
Adang, Teluk, *Indonesia* **192** K12
Adapazarı (Sakarya), *Turk.* **170** D6
Adar Doutchi, *Niger* **206** G7
Adarama, *Sudan* **203** P14
Adare Basin, *Pac. Oc.* **255** H16
Adare Peninsula, *Antarctica* **239** Q14
Adare Seamounts, *Pac. Oc.* **255** H16
Adare, Cape, *Antarctica* **239** R13
Adavale, *Qnsld., Austral.* **221** U13
Adda, river, *It.* **152** E5
Addis Ababa *see* Ādīs Ābeba, *Eth.* **210** M7
Addu Atoll, *Maldives* **183** Y3
Addy, *Wash., U.S.* **110** B8
'Adel Bagrou, *Mauritania* **204** H12
Adel, *Ga., U.S.* **99** P6
Adel, *Iowa, U.S.* **107** P11
Adel, *Oreg., U.S.* **110** L6
Adelaide Island, *Antarctica* **238** F5
Adelaide Peninsula, *Nunavut, Can.* **87** F14
Adelaide River, *N.T., Austral.* **221** P8
Adelaide, *Bahamas* **122** E2
Adelaide, *S. Aust., Austral.* **221** X10
Adélfi, island, *Gr.* **157** G13
Adélie Coast, *Antarctica* **239** G18
Aden *see* 'Adan, *Yemen* **174** Q9
Aden, Gulf of, *Ind. Oc.* **168** J3
Aderbissinat, *Niger* **206** G8
Aderg, peak, *Mauritania* **204** E9
'Ādhirīyāt, Jibāl al, *Jordan* **173** Q8
Ādī Ārk'ay, *Eth.* **210** J7
Adi Keyh, *Eritrea* **210** H8
Adi Ramets, *Eth.* **210** J7
Adiaké, *Côte d'Ivoire* **205** Q14
Adieu, Cape, *S. Aust., Austral.* **221** W8
Ādīgala, *Eth.* **210** L9
Adige, river, *It.* **152** E7
Ādīgrat, *Eth.* **210** J8
Adilabad, *India* **182** M6
Adilang, *Uganda* **211** Q4
Adin, *Calif., U.S.* **110** M5
Adinkerke, *Belg.* **148** L10
Adīrī, *Lib.* **201** G12
Adirondack Mountains, *N.Y., U.S.* **96** G10
Ādīs Ābeba (Addis Ababa), *Eth.* **210** M7
Ādīs Zemen, *Eth.* **210** K7
Adıyaman, *Turk.* **171** H13
Adjud, *Rom.* **155** D16
Adler, *Russ.* **159** U11
Admiralteystvo Rise, *Arctic Oc.* **253** E15
Admiralteystvo Trough, *Arctic Oc.* **253** D15
Admiralty Gulf, *Austral.* **220** B5
Admiralty Inlet, *Nunavut, Can.* **87** E16
Admiralty Inlet, *Wash., U.S.* **110** B4
Admiralty Island National Monument, *Alas., U.S.* **113** L22
Admiralty Islands, *P.N.G.* **222** M8
Admiralty Mountains, *Antarctica* **239** Q14
Ado-Ekiti, *Nigeria* **206** L7
Adok, *S. Sudan* **203** V11
Adoumri, *Cameroon* **206** K11
Adour, river, *Fr.* **149** X7
Adra, *Sp.* **146** M11
Adrano, *It.* **153** S11
Adrar des Ifôghas, plateau, *Alg., Mali* **205** E17
Adrar Egalah, peak, *Niger* **206** E9
Adrar I-n-Hihaou, peak, *Alg.* **200** K12

Adrar Sotuf, *W. Sahara* **200** K2
Adrar Tamgak, peak, *Niger* **206** E9
Adrar, *Alg.* **200** G11
Adrar, *Mauritania* **204** E9
Adraskan, *Afghan.* **180** J3
Adré, *Chad* **207** H16
Adria, *It.* **152** E8
Adrian, *Mich., U.S.* **101** N11
Adrian, *Tex., U.S.* **104** D11
Adrian, *Minn., U.S.* **106** L9
Adrianople *see* Edirne, *Turk.* **170** B3
Adriatic Sea, *Eur.* **140** K7
Adusa, *D.R.C.* **209** D18
Adventure Trough, *Pac. Oc.* **255** Q19
Ādwa, *Eth.* **210** J7
Adyakit, *Russ.* **161** F14
Adzopé, *Côte d'Ivoire* **205** P14
Aegean Sea, *Eur.* **140** L10
Aegina *see* Aígina, island, *Gr.* **156** K12
Aegir Ridge, *Arctic Oc.* **253** J21
Aej, island, *Marshall Is.* **227** A20
Aeon Point, *Kiribati* **226** Q12
Aewol, *S. Korea* **187** W7
Afaahiti, *Fr. Polynesia, Fr.* **231** J22
Afándou, *Gr.* **157** N20
Afar, *Eth.* **210** K9
Afareaitu, *Fr. Polynesia, Fr.* **231** H20
Affollé, *Mauritania* **204** H10
Afghanistan, *Asia* **180** J3
Afgooye Cadde, *Somalia* **211** Q12
Afgooye, *Somalia* **211** R11
'Afīf, *Saudi Arabia* **174** G8
Afikpo, *Nigeria* **206** M8
Afiq, *Israel* **172** K6
Aflou, *Alg.* **200** D12
Afmadow, *Somalia* **211** S9
Afobaka, *Suriname* **131** F17
Āfodo, *Eth.* **210** L5
Afognak Island, *Alas., U.S.* **113** L15
Afono Bay, *American Samoa, U.S.* **230** D7
Africana II Bank, *Ind. Oc.* **250** Q5
'Afrīn, *Syr.* **172** D8
Afşin, *Turk.* **170** G12
Afton, *Okla., U.S.* **105** B19
Afton, *Wyo., U.S.* **108** H8
Afuefuniya Point, *N. Mariana Is., U.S.* **230** N8
'Afula, *Israel* **172** L6
Afyonkarahisar, *Turk.* **170** F6
Áfytos, *Gr.* **156** D12
Aga Point, *Guam, U.S.* **230** Q5
Agadem, *Niger* **206** H11
Agadez, *Niger* **206** F8
Agadir, *Mor.* **200** E6
Agaete, *Sp.* **214** Q5
Agafo Gumas, *Guam, U.S.* **230** N6
Agaie, *Nigeria* **206** K7
Agana Bay, *Guam, U.S.* **230** N5
Agana Heights, *Guam, U.S.* **230** P5
Agana *see* Hagåtña, *Guam, U.S.* **230** N5
Agano, river, *Japan* **188** M11
Agar, *India* **182** J5
Āgaro, *Eth.* **210** M6
Agartala, *India* **182** J13
Agassiz Fracture Zone, *Pac. Oc.* **247** P16
Agat Bay, *Guam, U.S.* **230** P4
Agat, *Guam, U.S.* **230** P4
Agata, *Russ.* **161** G13
Agate Fossil Beds National Monument, *Nebr., U.S.* **107** N1
Agate, *Colo., U.S.* **108** M14
Agathonísi, island, *Gr.* **157** K18
Agats, *Indonesia* **193** L20
Agatti Island, *India* **183** S2
Agattu Island, *Alas., U.S.* **112** L1
Agattu Strait, *Alas., U.S.* **112** L1
Agboville, *Côte d'Ivoire* **205** P14
Ağcabǝdi, *Azerb.* **171** E21
Ağdaş, *Azerb.* **171** D21
Agde, *Fr.* **149** Y11
Agde, Cap d', *Fr.* **149** Y11
Agen, *Fr.* **149** W8
Ągere Maryam, *Eth.* **211** P7
Agfayan Bay, *Guam, U.S.* **230** Q5
Aggi, *Eth.* **210** M5
Aggtelek, *Hung.* **151** K16
Aghaylas, *W. Sahara* **200** J3
Aghrījīt, *Mauritania* **204** G11
Aghwinit, *W. Sahara* **200** K4
Agía Galíni, *Gr.* **157** Q14
Agia Napa, *Cyprus* **162** P10
Agía Pelagía, *Gr.* **156** N11

Agía Roúmeli, *Gr.* **157** Q13
Agía Triáda, ruin, *Gr.* **157** Q14
Agiá, *Gr.* **156** E11
Agiásos, *Gr.* **157** G17
Agidyem, island, *Marshall Is.* **227** H21
Aginskoye, *Russ.* **161** L17
Ágion Óros (Athos, Mount), *Gr.* **157** D13
Ágion Óros, *Gr.* **157** D13
Ágios Efstrátios, *Gr.* **157** F14
Ágios Efstrátios, island, *Gr.* **157** F14
Ágios Eirínis, Akrotírio, *Gr.* **157** E15
Ágios Geórgios *see* Ro, island, *Gr.* **157** N12
Ágios Geórgios, island, *Gr.* **157** K13
Ágios Kírykos, *Gr.* **157** K17
Ágios Nikólaos, *Gr.* **157** Q16
Agíou Órous, Kólpos, *Gr.* **157** D13
Agness, *Oreg., U.S.* **110** K2
Agnew, *W. Aust., Austral.* **221** V4
Agnibilékrou, *Côte d'Ivoire* **205** N14
Agnita, *Rom.* **155** D14
Agnone, *It.* **152** K11
Agoua, *Benin* **205** N17
Agouma, *Gabon* **208** F7
Agra, *India* **182** H6
Agrakhan Peninsula, *Eur.* **140** H15
Agrakhanskiy Poluostrov, *Russ.* **159** T15
Agrakhanskiy Zaliv, *Russ.* **159** T15
Ágreda, *Sp.* **147** D13
Ágria Gramvoúsa, island, *Gr.* **156** P12
Agrigento, *It.* **153** T10
Agrihan, island, *N. Mariana Is., U.S.* **230** H5
Agrínio, *Gr.* **156** H8
Ağrı Dağı (Mount Ararat), *Turk.* **171** E18
Agri, *It.* **171** E17
Agropoli, *It.* **153** N11
Agryz, *Russ.* **158** H16
Agua Fria National Monument, *Ariz., U.S.* **109** T7
Agua Prieta, *Mex.* **116** C8
Aguada de Pasajeros, *Cuba* **118** H8
Aguapeí, *Braz.* **132** J9
Aguas Blancas, *Chile* **134** E7
Águas Formosas, *Braz.* **133** K16
Aguascalientes, *Mex.* **116** G10
Águeda, *Port.* **146** F5
Aguelhok, *Mali* **205** F16
Aguijan, island, *N. Mariana Is., U.S.* **230** K5
Aguila Fracture Zone, *Pac. Oc.* **247** M13
Aguila, *Ariz., U.S.* **109** T6
Aguilar de Campoo, *Sp.* **146** C10
Aguilar, *Colo., U.S.* **109** P14
Aguilas, *Sp.* **147** L13
Aguja, Cabo de la, *Col.* **130** C6
Aguja, Punta, *Peru* **132** F1
Agulhas Bank, *Ind. Oc.* **250** N2
Agulhas Basin, *Ind. Oc.* **250** Q2
Agulhas Plateau, *Ind. Oc.* **250** N2
Agulhas Ridge, *Atl. Oc.* **249** U13
Aguni Jima, *Japan* **189** X1
Aha Hills, *Botswana* **212** F5
Ahaggar (Hoggar), mountains, *Alg.* **201** K13
Ahar, *Iran* **176** C12
Ahe, island, *Fr. Polynesia, Fr.* **231** A18
Ahiri, *India* **182** M7
Ahlat, *Turk.* **171** F17
Ahlatlıbel, ruins, *Turk.* **170** E8
Ahmad Wal, *Pak.* **181** N6
Ahmadabad, *India* **182** K3
Ahmadnagar, *India* **182** M4
Ahmadpur East, *Pak.* **181** N10
Ahmar Mountains, *Eth.* **210** M9
Ahnet, *Alg.* **200** J12
Ahoa, *Wallis & Futuna Is., Fr.* **230** Q8
Ahome, *Mex.* **116** E7
Ahoskie, *N.C., U.S.* **98** G13
Ahram, *Iran* **177** L15
Aḥrax Point, *Malta* **163** J21
Ahrensburg, *Ger.* **150** C8
Ahtme, *Est.* **143** J17
Ahu-a-miria Islands, *N.Z.* **229** M22
Ahua, *Fr. Polynesia, Fr.* **231** N14
Ahumoa, peak, *Hawaii, U.S.* **115** N21
Ahunui, island, *Fr. Polynesia, Fr.* **231** D21
Ahvāz (Ahwāz), *Iraq* **177** J13

Ahvenanmaa *see* Åland, island, *Fin.* **143** J14
Aḥwar, *Yemen* **174** P10
Ahwāz *see* Ahvāz, *Iraq* **177** J13
Ai, island, *Marshall Is.* **227** J20
Aïbak, *Afghan.* **180** G8
Aibonito, *P.R., U.S.* **114** P7
'Aiea, *Hawaii, U.S.* **115** H14
Aigiáli, *Gr.* **157** L16
Aígina (Aegina), island, *Gr.* **156** K12
Aígina, *Gr.* **156** K12
Aígio, *Gr.* **156** H9
Aigle, *Switz.* **152** C2
Aigoual, Mount, *Fr.* **149** W11
Aiguá, *Uru.* **134** L14
Aigurande, *Fr.* **149** T9
Aikawa, *Japan* **188** M11
Aiken, *S.C., U.S.* **98** L8
Ailinginae Atoll, *Marshall Is.* **227** B15
Ailinglapalap Atoll, *Marshall Is.* **227** D16
Ailsa Craig, *Scot., U.K.* **145** M8
Ailsa Craig, *Gibraltar, U.K.* **163** P23
Ailuk Atoll, *Marshall Is.* **227** C17
Aimere, *Indonesia* **193** M13
Aïn Beïda, *Alg.* **201** C14
Aïn Ben Tili, *Mauritania* **204** B11
'Aïn M'lila, *Alg.* **201** B14
'Aïn Sefra, *Alg.* **200** D10
Aïn Temouchent, *Alg.* **200** C10
Ainaži, *Latv.* **143** K16
Aindai, *P.N.G.* **222** N5
Aineman, island, *Marshall Is.* **227** H21
Aïnhoa, *Fr.* **149** Y5
Aínos, Óros, *Gr.* **156** J7
Aínsa, *Sp.* **147** C15
Ainsworth, *Nebr., U.S.* **107** N5
Air Force Island, *Nunavut, Can.* **87** F17
Airal, *Palau* **226** G6
Airão, *Braz.* **132** D8
Airolo, *Switz.* **152** C4
Aisey, *Fr.* **149** R12
Aisými, *Gr.* **157** B16
Aït Benhaddou, *Mor.* **200** E7
Aitape, *P.N.G.* **222** N4
Aitkin, *Minn., U.S.* **106** G11
Aitolikó, *Gr.* **156** H8
Aitutaki Atoll, *Cook Is., N.Z.* **224** K12
Aitutaki, island, *N.Z.* **229** H20
Aiud, *Rom.* **155** D13
Aiwa, island, *Fiji* **226** Q8
Aiwo, *Nauru* **227** M14
Aiwokako Passage, *Palau* **226** F6
Aix-en-Provence, *Fr.* **149** X13
Aix-les-Bains, *Fr.* **149** U14
Aix, Île d', *Fr.* **149** T6
Aixirivall, *Andorra* **162** K2
Aiyon Mountain, *Palau* **226** F6
Aizawl, *India* **182** J14
Aizenay, *Fr.* **149** S5
Aizpute, *Latv.* **143** M15
Aizuwakamatsu, *Japan* **188** M12
Aj Bogd Uul, *Mongolia* **184** E7
Ajaccio, *Fr.* **149** Y2
Ajaccio, Golfe d', *Fr.* **149** Y2
Ajajú, river, *Col.* **130** J7
Ajaria, republic, *Rep. of Georgia* **171** C16
Ajax Peak, *Idaho, Mont., U.S.* **108** E6
Ajayan Bay, *Guam, U.S.* **230** Q5
Ajdābiyā, *Lib.* **201** E20
Ajir, *Niger* **206** E7
'Ajlūn, *Jordan* **172** L6
'Ajmān, *U.A.E.* **175** F16
Ajmer, *India* **182** H4
Ajo, *Ariz., U.S.* **109** V6
Ajo, Cabo de, *Sp.* **146** B11
Ajokwola, island, *Marshall Is.* **227** D19
Ajoupa-Bouillon, *Martinique, Fr.* **123** E22
Ajrestān, *Afghan.* **180** J7
Aka, *Mali* **205** H14
Akabira, *Japan* **188** F13
Akabli, *Alg.* **200** H11
Akademil Nauk Rise, *Pac. Oc.* **246** D8
Akaiami, island, *N.Z.* **229** H21
Akaka Falls, *Hawaii, U.S.* **115** M22
Akaki, *Cyprus* **162** P9
Akamaru, island, *Fr. Polynesia, Fr.* **231** R17
Akankohan, *Japan* **188** F15
Akanthou (Akatu), *Cyprus* **162** N9
'Akāshāt, *Iraq* **176** G7

Akashi, *Japan* **189** Q8
Akaska, *S. Dak., U.S.* **106** J5
Akatu *see* Akanthou, *Cyprus* **162** N9
Akbulak, *Russ.* **160** J7
Akçaabat, *Turk.* **171** D14
Akçadağ, *Turk.* **170** G12
Akçakale, *Turk.* **171** J13
Akçakışla, *Turk.* **170** E11
Akçakoca, *Turk.* **170** C7
Akçay, *Turk.* **170** E3
Akçay, *Turk.* **170** J5
Akchâr, *Mauritania* **204** E7
Akdağ, peak, *Turk.* **170** F2
Akdağ, peak, *Turk.* **170** F2
Akdağmadeni, *Turk.* **170** E11
Akdoğan *see* Lysi, *Cyprus* **162** P9
Akeley, *Minn., U.S.* **106** G10
Akelo, *S. Sudan* **203** W13
Akera, river, *Azerb.* **171** E20
Aketi, *D.R.C.* **209** C15
Akhalts' ikhe, *Rep. of Georgia* **171** C17
Akhḍar, Jabal al, *Af.* **198** D7
Akhiok, *Alas., U.S.* **113** M15
Akhisar, *Turk.* **170** F3
Akhmīm, *Egypt* **202** H11
Akhnur, *India* **180** J13
Akhtuba, river, *Russ.* **159** Q14
Akhtubinsk, *Russ.* **159** Q14
Aki, *Japan* **189** R7
Akia, *Greenland, Den.* **89** Q5
Akiachak, *Alas., U.S.* **112** J12
Akiaki, island, *Fr. Polynesia, Fr.* **231** C21
Akimiski Island, *Nunavut, Can.* **87** M17
Akita, *Japan* **188** K12
Akitua, island, *N.Z.* **229** H20
Akjoujt, *Mauritania* **204** F8
Akkeshi Wan, *Japan* **188** G15
Akkeshi, *Japan* **188** G15
'Akko (Acre), *Israel* **172** K5
Akkuş, *Turk.* **170** D12
Aklavik, *N.W.T., Can.* **86** E9
Aklera, *India* **182** J5
'Ako'ako'a Point, *Hawaii, U.S.* **115** L21
Akobo, *S. Sudan* **203** W12
Akobo, river, *Eth.* **211** N5
Akokane, *Niger* **206** E8
Akola, *India* **182** L5
Akom, *Cameroon* **206** P9
Akonolinga, *Cameroon* **206** N10
Akor, *Mali* **204** J12
Ak'ordat, *Eritrea* **210** H7
Akosombo Dam, *Ghana* **205** P16
Akot, *S. Sudan* **203** W10
Akpatok Island, *Nunavut, Can.* **87** J19
Akqi, *China* **184** F2
Akranes, *Ice.* **142** E2
Akráta, *Gr.* **156** J10
Akron, *Ohio, U.S.* **101** P14
Akron, *Iowa, U.S.* **106** M8
Akron, *Colo., U.S.* **108** L15
Akrotiri (Western) Sovereign Base Area, *U.K.* **162** Q7
Akrotiri Bay, *Cyprus* **162** Q7
Akrotíri, *Gr.* **157** N15
Aksai Chin *see* Aksayqin, region, *China* **182** C7
Aksaray, *Turk.* **170** G9
Aksay, *Russ.* **159** Q13
Aksay, *China* **184** G7
Aksayqin (Aksai Chin), region, *China* **182** C7
Akşehir Gölü, *Turk.* **170** G7
Akşehir, *Turk.* **170** G7
Akseki, *Turk.* **170** H7
'Aksha, ruin, *Egypt* **202** L11
Aksu, *Kaz.* **178** K5
Aksu, *China* **184** E3
Aksu, river, *Turk.* **170** H6
Āksum, *Eth.* **210** J7
Aktag, peak, *China* **184** H4
Aktivneset Slope, *Atl. Oc.* **248** F12
Akugdlersssup Sermia, glacier, *Greenland, Den.* **89** Q6
Akula, *D.R.C.* **209** C13
Akulliit, *Greenland, Den.* **88** M6
Akune, *Japan* **189** S4
Akure, *Nigeria* **206** L7
Akurenan, *Eq. Guinea* **206** Q10
Akureyri, *Ice.* **142** E3
Akuseki Shima, *Japan* **189** V3
Akutan, *Alas., U.S.* **112** N10
Akwatia, *Ghana* **205** P15
Akyab *see* Sittwe, *Myanmar* **190** H3
Akyaka, *Turk.* **171** D17

Akyazı, *Turk.* **170** D6
Al 'Adam, *Lib.* **201** D22
Al Aflāj, *Saudi Arabia* **174** H10
Al Aḥmadī, *Kuwait* **174** D11
Al Ain, *U.A.E.* **175** G16
Al Akhḍar, *Saudi Arabia* **174** D5
Al 'Alamayn, *Egypt* **202** F10
Al 'Amārah, *Iraq* **176** J12
Al 'Āmirīyah, *Egypt* **202** F10
Al 'Aqabah (Aqaba), *Jordan* **173** R5
Al 'Aqīq, *Saudi Arabia* **174** J7
Al 'Arīsh, *Egypt* **202** E13
Al 'Armah, *Saudi Arabia* **174** F10
Al Arṭāwīyah, *Saudi Arabia* **174** F9
Al 'Āş (Orontes) *see* Al 'Āş, river, *Syr.* **172** F7
Al As'ad, *Saudi Arabia* **174** D5
Al Ashkharah, *Oman* **175** H18
Al 'Ayyāṭ, *Egypt* **202** F11
Al Bāb, *Syr.* **172** D9
Al Bad', *Saudi Arabia* **174** D4
Al Badārī, *Egypt* **202** H11
Al Badī', *Saudi Arabia* **174** J10
Al Bāḥah, *Saudi Arabia* **174** K7
Al Balyanā, *Egypt* **202** J12
Al Bardī, *Lib.* **201** D23
Al Başrah (Basra), *Iraq* **177** K13
Al Bawīṭī, *Egypt* **202** G10
Al Bayāḍ, *Saudi Arabia* **174** J11
Al Bayḍā' (Beida), *Lib.* **201** D21
Al Bir Lahlou, *W. Sahara* **200** H6
Al Bi'r, *Saudi Arabia* **174** D4
Al Birk, *Saudi Arabia* **174** L7
Al Birkah, *Lib.* **201** J16
Al Buḥayrah al Murrah al Kubrá (Great Bitter Lake), *Egypt* **202** F12
Al Fallūjah (Fallujah), *Iraq* **176** G10
Al Farciya, *W. Sahara* **200** G6
Al Fashn, *Egypt* **202** G11
Al Faw, *Iran* **177** K13
Al Fayyūm, *Egypt* **202** G11
Al Fuqahā', *Lib.* **201** G19
Al Furāt (Euphrates), *Iraq* **176** J11
Al Ghardaqah, *Egypt* **202** H13
Al Ghaydah, *Yemen* **175** M14
Al Ghayl, *Saudi Arabia* **174** H10
Al Ghazālah, *Saudi Arabia* **174** E7
Al Ḥadīdah, crater, *Saudi Arabia* **175** J13
Al Ḥaffah, *Syr.* **172** E7
Al Haggounia, *W. Sahara* **200** G4
Al Ḥamād, *Saudi Arabia* **174** B6
Al Ḥamīdīyah, *Syr.* **172** G7
Al Ḥamrā', *Saudi Arabia* **174** G6
Al Ḥamrā', *Lib.* **201** F17
Al Ḥamzah, *Iraq* **176** J11
Al Ḥanākīyah, *Saudi Arabia* **174** G7
Al Ḥanīyah, *Iraq* **176** L11
Al Ḥaqlānīyah, *Iraq* **176** G9
Al Ḥarīq, *Saudi Arabia* **174** H10
Al Harūj al Aswad, *Lib.* **201** G19
Al Ḥasakah, *Syr.* **172** D14
Al Ḥawrah, *Yemen* **174** P11
Al Ḥayy, *Iraq* **176** H12
Al Ḥayz, *Egypt* **202** G10
Al Ḥījānah, *Syr.* **172** K8
Al Ḥijāz (Hejaz), *Saudi Arabia* **174** E5
Al Ḥillah, *Saudi Arabia* **174** H10
Al Ḥillah, *Iraq* **176** H10
Al Hoceima, *Mor.* **200** C9
Al Ḥudaydah, *Yemen* **174** N8
Al Hufūf, *Saudi Arabia* **174** F12
Al Ikhwān (The Brothers), islands, *Yemen* **175** Q15
Al 'Irqah, *Yemen* **174** P11
Al 'Īsāwīyah, *Saudi Arabia* **174** B5
Al Iskandarīyah (Alexandria), *Egypt* **202** E10
Al Iskandarīyah, *Iraq* **176** H10
Al Ismā'īlīyah, *Egypt* **202** F12
Al Jabīn, *Yemen* **174** N8
Al Jafr, *Jordan* **173** Q7
Al Jāfūrah, *Saudi Arabia* **174** G12
Al Jaghbūb, *Lib.* **201** E23
Al Jahrā', *Kuwait* **174** D11
Al Jawādīyah, *Syr.* **172** C15
Al Jawf *see* Dawmat al Jandal, *Saudi Arabia* **174** C7
Al Jawf, *Lib.* **201** J23
Al Jazīrah, *Sudan* **203** R12
Al Jifārah, *Saudi Arabia* **174** G9
Al Jīzah (Giza), *Egypt* **202** F11
Al Jīzah, *Jordan* **172** M7
Al Jubayl, *Saudi Arabia* **174** E12
Al Jumaylīyah, *Qatar* **175** F13
Al Junaynah, *Saudi Arabia* **174** K8

Al Kahfah, *Saudi Arabia* **174** E8
Al Karak, *Jordan* **173** N6
Al Kawm, *Syr.* **172** F11
Al Khābūrah, *Oman* **175** G17
Al Khalīl (Hebron), *W. Bank* **173** N5
Al Khāliş, *Iraq* **176** G11
Al Kharfah, *Saudi Arabia* **174** H10
Al Khārijah, *Egypt* **202** J11
Al Kharj, *Saudi Arabia* **174** G11
Al Kharṭūm Baḥr (Khartoum North), *Sudan* **203** Q12
Al Khāşirah, *Saudi Arabia* **174** H9
Al Khawr, *Qatar* **175** F13
Al Khīrān, *Kuwait* **174** D11
Al Khufayfīyah, *Saudi Arabia* **174** G9
Al Khums, *Lib.* **201** D18
Al Khunn, *Saudi Arabia* **174** H12
Al Khuraybah, *Yemen* **174** N11
Al Khurmah, *Saudi Arabia* **174** J7
Al Kiswah, *Syr.* **172** J7
Al Kūfah (Kufah), *Iraq* **176** H10
Al Kufrah (Kufra Oasis), *Lib.* **201** H23
Al Kuntillah, *Egypt* **202** F13
Al Kūt, *Iraq* **176** H11
Al Kuwayt (Kuwait City), *Kuwait* **174** D11
Al Lādhiqīyah (Latakia), *Syr.* **172** F7
Al Lajā, *Syr.* **172** K7
Al Lidām, *Saudi Arabia* **174** K9
Al Līth, *Saudi Arabia* **174** K6
Al Luḥayyah, *Yemen* **174** N8
Al Luwaymī, *Saudi Arabia* **174** E8
Al Madīnah (Medina), *Saudi Arabia* **174** G6
Al Mafraq, *Jordan* **172** L7
Al Maḥallah al Kubrá, *Egypt* **202** F11
Al Mahbas, *W. Sahara* **200** G6
Al Maḥrūqah, *Lib.* **201** G18
Al Majma'ah, *Saudi Arabia* **174** F10
Al Mālikīyah, *Syr.* **172** C16
Al Manāmah (Manama), *Bahrain* **175** F13
Al Manşūrah, *Syr.* **172** E11
Al Manşūrah, *Egypt* **202** F11
Al Marj (Barce), *Lib.* **201** D21
Ma'shūqah, *Saudi Arabia* **174** K7
Al Mawşil (Mosul), *Iraq* **176** E10
Al Mayādīn, *Syr.* **172** G13
Al Mayyāh, *Saudi Arabia* **174** E8
Al Mintirib, *Oman* **175** H18
Al Minyā, *Egypt* **202** G11
Al Mismīyah, *Syr.* **172** K7
Al Mubarraz, *Saudi Arabia* **174** F12
Al Mudawwarah, *Jordan* **173** S7
Al Mukallā, *Yemen* **174** P12
Al Mukhā, *Yemen* **174** P8
Al Muwassam, *Saudi Arabia* **174** M8
Al Muwayh, *Saudi Arabia* **174** H7
Al Qābil, *Oman* **175** G16
Al Qadīmah, *Saudi Arabia* **174** H6
Al Qadmūs, *Syr.* **172** G7
Al Qāhirah (Cairo), *Egypt* **202** F11
Qā'im, *Iraq* **176** F8
Al Qa'īyah, *Saudi Arabia* **174** G8
Al Qāmishlī, *Syr.* **172** C14
Al Qārah, *Yemen* **174** P9
Al Qaryah ash Sharqīyah, *Lib.* **201** F17
Al Qaryatayn, *Syr.* **172** H9
Al Qaşīm, *Saudi Arabia* **174** F9
Al Qaşr, *Egypt* **202** J10
Al Qaṭīf, *Saudi Arabia* **174** F12
Al Qaṭrānah, *Jordan* **173** N7
Al Qaṭrūn, *Lib.* **201** J18
Al Qayşūmah, *Saudi Arabia* **174** D10
Al Qunfudhah, *Saudi Arabia* **174** K7
Al Qurayyāt, *Saudi Arabia* **174** B5
Al Qurnah, *Iraq* **176** J12
Al 'Ubaylah, *Saudi Arabia* **175** J13
Al 'Ulá, *Saudi Arabia* **174** E5
Al 'Uqaylah, *Lib.* **201** E20
Al Uqşur (Luxor), *Egypt* **202** J12
Al 'Uwaynāt (Serdeles), *Lib.* **201** H16
Al Uwaynat *see* Jebel 'Uweinat, *Sudan* **202** L7

Al 'Uwayqīlah, *Saudi Arabia* **174** C8
Al 'Uyūn, *Saudi Arabia* **174** G6
Al Wāḥat ad Dākhilah, *Egypt* **202** J9
Al Wāḥat al Baḥrīyah, *Egypt* **202** H10
Al Wāḥat al Khārijah, *Egypt* **202** J11
Al Wajh, *Saudi Arabia* **174** E4
Al Wannān, *Saudi Arabia* **174** E11
Al Wāsiţah, *Egypt* **202** G11
Al Wuday'ah, *Saudi Arabia* **174** M11
Ala, *It.* **152** D6
Alà, Monti di, *It.* **152** M4
Alabama, *U.S.* **103** K15
Alabama, river, *Ala.* **103** M14
Alabaster, *Ala.* **103** K15
Alaca Dağ, peak, *Turk.* **170** H7
Alaca, *Turk.* **170** E10
Alacahöyük, ruins, *Turk.* **170** E10
Alaçam, *Turk.* **170** C10
Alachua, *Fla.* **99** R7
Alacránes, *Arrecife, Mex.* **117** G16
Aladağ, peak, *Turk.* **170** Q12
Aladağ, peak, *Turk.* **170** H10
Alaejos, *Sp.* **146** E9
Alagir, *Russ.* **159** U13
Alagoinhas, *Braz.* **133** H17
Alagón, *Sp.* **147** D14
Alaja, *Turkm.* **178** M4
Alajuela, *C.R.* **117** N19
Alaka'i Swamp, *Hawaii, U.S.* **114** E10
Alakanuk, *Alas., U.S.* **112** H12
Alaköl, *Kaz.* **179** G17
'Alalakeiki Channel, *Hawaii, U.S.* **115** K19
Alalakh, ruins, *Turk.* **170** J11
Alama, *Cen. Af. Rep.* **207** M16
Alamagan, island, *N. Mariana Is., U.S.* **230** H5
Alameda Gardens, *Gibraltar, U.K.* **163** P23
Alameda, *Calif., U.S.* **111** S3
Alamo Heights, *Tex., U.S.* **105** L16
Alamo, *Tenn., U.S.* **102** F12
Alamo, *Nev., U.S.* **111** S12
Alamogordo, *N. Mex., U.S.* **109** U13
Alamosa, *Colo., U.S.* **109** P12
Åland (Ahvenanmaa), island, *Fin.* **143** J14
Åland Islands, *Eur.* **140** D9
Alandar, *Afghan.* **180** H5
Alandroal, *Port.* **146** J6
Alanngorsuaq (Kobberminebugt), bay, *Greenland, Den.* **89** T6
Alantika Mountains, *Cameroon, Nigeria* **206** L11
Alanya, *Turk.* **170** J7
Alao, *American Samoa, U.S.* **230** D8
Alappuzha (Alleppey), *India* **183** S5
'Alāqadārī Bālā Bulūk, *Afghan.* **180** K3
'Alāqahdārī Shāh Jōy, *Afghan.* **180** K7
'Alāqahdārī-ye Almār, *Afghan.* **180** G5
Alara, river, *Turk.* **170** J7
Alarcón, Embalse de, *Sp.* **146** G12
Alaşehir, *Turk.* **170** F4
Alasheyev Bay, *Antarctica* **239** C19
Alaska Highway, *Alas., U.S.* **113** G18
Alaska Peninsula, *Alas., U.S.* **113** M13
Alaska Plain, *Pac. Oc.* **247** C15
Alaska Range, *Alas., U.S.* **113** H16
Alaska, *U.S.* **113** G16
Alaska, Gulf of, *Alas., U.S.* **113** L18
Alassio, *It.* **152** G3
Aledo, *Ill., U.S.* **101** P3
Aleg, *Mauritania* **204** G8
Alegranza, island, *Sp.* **214** P7
Alegrete, *Braz.* **132** P10
Alejandro Selkirk Island, *S. Amer.* **128** M2
Aleksandrov Gay, *Russ.* **159** N15
Aleksandrovsk Sakhalinskiy, *Russ.* **161** J21
Alekseyevka, *Russ.* **158** L14
Alekseyevka, *Russ.* **158** L16
Alekseyevka, *Russ.* **159** N10
Aleksin, *Russ.* **158** K9
Aleksinac, *Serb.* **154** H11
Alemania, *Arg.* **134** F9
Alembel, island, *Marshall Is.* **227** A21
Alençon, *Fr.* **149** Q7
Alenquer, *Braz.* **132** D11
'Alenuihāhā Channel, *Hawaii, U.S.* **115** L20
Aleppo *see* Ḥalab, *Syr.* **172** D9

Albany, *Mo., U.S.* **107** R11
Albany, *Oreg., U.S.* **110** G3
Albany, *W. Aust., Austral.* **221** Y3
Albany, river, *N. Amer.* **84** G7
Albany, river, *Ont., Can.* **87** N17
Albardón, *Arg.* **134** J8
Albarracín, *Sp.* **147** F13
Albatross Bay, *Austral.* **220** B12
Albatross Plateau, *Pac. Oc.* **247** H19
Albemarle *see* Isla Isabela, *Ecua.* **130** Q2
Albemarle Sound, *N.C., U.S.* **98** G14
Albenga, *It.* **152** G3
Alberobello, *It.* **152** M14
Albert Lea, *Minn., U.S.* **106** L11
Albert Nile, river, *Uganda* **211** Q3
Albert Town, *Bahamas* **119** H14
Albert Town, *Jam.* **122** H8
Albert, Lake, *Austral.* **220** L11
Alberta, *Can.* **86** L10
Albertville, *Ala., U.S.* **103** H16
Albertville, *Fr.* **149** U14
Albi, *Fr.* **149** X9
Albia, *Iowa, U.S.* **107** Q12
Albin, *Wyo., U.S.* **108** K14
Albina Point, *Af.* **198** M5
Albinia, *It.* **152** J7
Albion Downs, *Qnsld., Austral.* **221** S12
Albion, *N.Y., U.S.* **96** H6
Albion, *Mich., U.S.* **100** M11
Albion, *Ill., U.S.* **101** T7
Albion, *Nebr., U.S.* **107** P7
Alboran Sea, *Sp.* **146** N10
Alborán, Isla de, *Sp.* **146** N11
Ålborg, *Den.* **142** M11
Albox, *Sp.* **146** L12
Albuquerque, *N. Mex., U.S.* **109** S12
Alburquerque, *Sp.* **146** H7
Albury, *N.S.W., Austral.* **221** Y13
Alby, *Fr.* **149** U14
Alcácer do Sal, *Port.* **146** J5
Alcalá de Guadaira, *Sp.* **146** L8
Alcalá de Henares, *Sp.* **146** F11
Alcalá la Real, *Sp.* **146** L10
Alcamo, *It.* **153** S9
Alcanar, *Sp.* **147** F15
Alcañiz, *Sp.* **147** E15
Alcántara, *Sp.* **146** G7
Alcántara, Embalse de, *Sp.* **146** G7
Alcaraz, *Sp.* **146** J12
Alcaudete, *Sp.* **146** L10
Alcázar de San Juan, *Sp.* **146** H11
Alchevs'k, *Ukr.* **159** Q10
Alcoa, *Tenn., U.S.* **103** F18
Alcoak, Cerro, *Arg.* **134** D8
Alcobaça, *Port.* **146** G5
Alcobendas, *Sp.* **146** F11
Alcock Rise, *Ind. Oc.* **251** E12
Alcoi *see* Alcoy, *Sp.* **147** J14
Alcoota, *N.T., Austral.* **221** T9
Alcorcón, *Sp.* **146** F11
Alcoutim, *Port.* **146** L6
Alcoy, *Sp.* **147** J14
Alcubierre, Sierra de, *Sp.* **147** D14
Alcúdia, *Sp.* **147** G18
Alcuéscar, *Sp.* **146** H8
Aldabra Islands, *Af.* **198** L11
Aldama, *Mex.* **116** D9
Aldan, *Russ.* **161** J18
Aldan, river, *Russ.* **161** G18
Aldanskoye Nagor'ye, *Russ.* **161** J18
Aldeadávila de la Ribera, *Sp.* **146** E7
Aldeburgh, *Eng., U.K.* **145** T16
Alderney, island, *Channel Is., U.K.* **145** X12

Aléria, *Fr.* **149** Y3
Alert Point, *Kiribati* **226** J11
Alert, *Nunavut, Can.* **87** A16
Ales, *It.* **153** P3
Alès, *Fr.* **149** W12
Alessandria, *It.* **152** E4
Ålesund, *Nor.* **142** H10
Aleutian Basin, *Pac. Oc.* **246** B11
Aleutian Islands, *Alas., U.S.* **112** N5
Aleutian Range, *Alas., U.S.* **113** M14
Aleutian Ridge, *Pac. Oc.* **246** C11
Aleutian Trench, *Pac. Oc.* **246** C10
Alevina, *Russ.* **161** F21
Alexander Archipelago, *Alas., U.S.* **113** M22
Alexander Bay, *S. Af.* **212** K4
Alexander City, *Ala., U.S.* **103** K16
Alexander Island, *Antarctica* **238** F6
Alexander, *N. Dak., U.S.* **106** F2
Alexandra, *Russ.* **160** B12
Alexandra, *N.Z.* **223** N17
Alexándreia, *Gr.* **156** C10
Alexandria Bay, *N.Y., U.S.* **96** G9
Alexandria *see* Al Iskandarīyah, *Egypt* **202** E10
Alexandria, *Va., U.S.* **98** D12
Alexandria, *Ind., U.S.* **101** Q9
Alexandria, *La., U.S.* **102** N8
Alexandria, *Minn., U.S.* **106** H9
Alexandria, *S. Dak., U.S.* **106** L7
Alexandria, *Jam.* **122** H8
Alexandria, *Rom.* **155** G15
Alexandria, Lake, *Austral.* **220** L11
Alexandroúpoli (Dedéagach), *Gr.* **157** C16
Aleysk, *Russ.* **160** L11
Alfaro, *Sp.* **147** D13
Alfatar, *Bulg.* **155** G17
Alföld, plain, *Hung.* **151** L16
Alfred Wegener Halvø, peninsula, *Greenland, Den.* **88** K6
Algarrobal, *Arg.* **134** F10
Algarrobo, *Chile* **134** G6
Algeciras, *Sp.* **146** N8
Algeciras, Bahía de *see* Gibraltar, Bay of, *Gibraltar, U.K.* **163** N21
Alger (Algiers), *Alg.* **200** B12
Algeria, *Af.* **200** G9
Algerian Basin, *Atl. Oc.* **248** J12
Algha, *Kaz.* **178** E7
Alghabas, *Kaz.* **178** E5
Alghero, *It.* **152** M3
Alghiena, *Eritrea* **210** G7
Algiers *see* Alger, *Alg.* **200** B12
Alginet, *Sp.* **147** H14
Algoa Bay, *S. Af.* **212** M7
Algoma, *Wis., U.S.* **100** J7
Algona, *Iowa, U.S.* **106** M11
Algorta, *Uru.* **134** K13
Alhambra, *Calif., U.S.* **111** X8
Alhucemas, Peñón de, *Sp.* **146** P10
Ali Sabieh, *Djibouti* **210** K10
Ali, *It.* **153** S12
Aliabad, *Iran* **177** L18
Aliákmonas, river, *Gr.* **156** D10
Alibates Flint Quarries National Monument, *Tex., U.S.* **104** C12
Alicante (Alacant), *Sp.* **147** J14
Alice Springs, *N.T., Austral.* **221** S9
Alice Town, *Bahamas* **122** B11
Alice Town, *Bahamas* **122** D5
Alice, *Tex., U.S.* **105** P16
Alice, *Qnsld., Austral.* **221** T13
Alice, Punta, *It.* **153** P14
Aliceville, *Ala., U.S.* **103** K13
Alicudi, Isola, *It.* **153** R11
Aligarh, *India* **182** G6
Alīgūdarz, *Iraq* **177** G14
'Ālika Cone, *Hawaii, U.S.* **115** P21
Alim, island, *P.N.G.* **222** N8
Alima, river, *Congo* **208** E11
Alindao, *Cen. Af. Rep.* **207** N15
Alindau, *Indonesia* **193** J13
Alingsås, *Sw.* **142** L12
Alipur Duar, *India* **182** H12
Alipur, *Pak.* **181** N10
Aliquippa, *Pa., U.S.* **96** N3
Alişar Hüyük, ruins, *Turk.* **170** E10
Alissós, *Gr.* **156** J9
Alistráti, *Gr.* **157** B13
Alivéri, *Gr.* **157** H13
All American Canal, *Calif., U.S.* **111** Z12
All Saints, *Antigua & Barbuda* **123** B21
Allada, *Benin* **205** P17
Allagash, river, *Me., U.S.* **97** B16

Allahüekber Tepe, peak, *Turk.* **171** D17
Allakaket, *Alas., U.S.* **113** E16
Allan Hills, *Antarctica* **239** N14
Allan Mountain, *Idaho, U.S.* **108** E5
Allanmyo, *Myanmar* **190** J4
Allanton, *N.Z.* **223** P17
Allaríz, *Sp.* **146** C6
Allegan, *Mich., U.S.* **100** M9
Allegheny Mountains, *U.S.* **91** G19
Allegheny Reservoir, *Pa., U.S.* **96** L5
Allegheny, river, *U.S.* **96** L5
Allen, *Arg.* **135** P8
Allen, Bog of, *Ire.* **145** R6
Allen, Lough, *Ire.* **145** P5
Allendale Town, *Eng., U.K.* **145** N11
Allendale, *S.C., U.S.* **98** M9
Allende, *Mex.* **116** D11
Allentown, *Pa., U.S.* **96** N9
Alleppey *see* Alappuzha, *India* **183** S5
Aller, river, *Ger.* **150** D8
Allerona, *It.* **152** J8
Alliance, *Ohio, U.S.* **101** P15
Alliance, *Nebr., U.S.* **107** N2
Alligator Pond, *Jam.* **122** J8
Allinge, *Den.* **143** N13
Allison Peninsula, *Antarctica* **238** H6
Alluitsup Kangerlua (Lichtenau Fjord), *Greenland, Den.* **89** T7
Alluitsup Paa (Sydprøven), *Greenland, Den.* **89** T7
Alma, *Ga., U.S.* **99** P7
Alma, *Mich., U.S.* **100** L11
Alma, *Ark., U.S.* **102** F6
Alma, *Kans., U.S.* **107** S9
Alma, *Nebr., U.S.* **107** R5
Almacelles, *Sp.* **147** D15
Almada, *Port.* **146** J4
Almaden, *Qnsld., Austral.* **221** R13
Almadén, *Sp.* **146** J9
Almami, *Mali* **205** J13
Almanor, Lake, *Calif., U.S.* **111** N5
Almansa, *Sp.* **147** J14
Almanzor, peak, *Sp.* **146** F9
Almaty, *Kaz.* **179** J15
Almazán, *Sp.* **146** E12
Almeida, *Port.* **146** F7
Almeirim, *Braz.* **132** D12
Almeirim, *Port.* **146** H5
Almelo, *Neth.* **148** H14
Almena, *Kans., U.S.* **107** R5
Almenara, *Braz.* **133** J16
Almenara, peak, *Sp.* **146** J12
Almendra, Embalse de, *Sp.* **146** E8
Almendralejo, *Sp.* **146** J8
Almere, *Neth.* **148** J12
Almería, *Sp.* **146** M12
Almería, Golfo de, *Sp.* **146** M12
Al'met'yevsk, *Russ.* **158** J16
Almina, Punta, *Sp.* **146** N9
Almira, *Wash., U.S.* **110** C7
Almirantazgo, Seno, *Chile* **135** X8
Almirante Brown Canyon, *Atl. Oc.* **249** U6
Almirante, *Pan.* **117** N20
Almo, *Idaho, U.S.* **108** J6
Almodóvar del Campo, *Sp.* **146** J10
Almodôvar, *Port.* **146** K5
Almudévar, *Sp.* **147** D14
Almuñécar, *Sp.* **146** M11
Almus Barajı, *Turk.* **170** D12
Almus, *Turk.* **170** E12
Almyrós, *Gr.* **156** F11
Alney, *Russ.* **161** F22
Alnif, *Mor.* **200** E8
Alnwick, *Eng., U.K.* **145** M12
Alo, *Wallis & Futuna Is., Fr.* **230** Q11
Alofau, *American Samoa, U.S.* **230** D8
Alofi Bay, *N.Z.* **229** Q21
Alofi, *N.Z.* **229** Q21
Alokan, island, *Solomon Is.* **228** F9
Along, *India* **182** G15
Alónnisos, *Gr.* **156** F12
Alónnisos, island, *Gr.* **156** F12
Alor Setar, *Malaysia* **191** S8
Alor, island, *Indonesia* **193** M15
Alor, Kepulauan, *Indonesia* **193** M15
Álora, *Sp.* **146** M9
Alotau, *P.N.G.* **221** N15
Aloysius, Mount, *Austral.* **220** G7
Alpena, *Mich., U.S.* **100** H12
Alpha Ridge, *Arctic Oc.* **252** K12
Alpha, *Qnsld., Austral.* **221** T13
Alpine, *Tex., U.S.* **104** K10
Alpine, *Wyo., U.S.* **108** H8
Alpine, *Ariz., U.S.* **109** T9

Alps, *Eur.* **140** J6
Alqueva, Barragem de, *Port.* **146** J6
Alroy Downs, *N.T., Austral.* **221** R10
Alsace, region, *Fr.* **149** Q15
Alsancak *see* Karavas, *Cyprus* **162** N8
Alsask, *Sask., Can.* **86** M11
Alsasua, *Sp.* **146** B12
Alston, *Eng., U.K.* **145** N11
Alt del Griu, peak, *Andorra* **162** J4
Alta Floresta, *Braz.* **132** G10
Alta Gracia, *Arg.* **134** J10
Alta, *Iowa, U.S.* **107** N9
Alta, *Nor.* **143** B14
Altagracia de Orituco, *Venez.* **130** D11
Altagracia, *Venez.* **130** C8
Altamaha, river, *Ga., U.S.* **99** P8
Altamira, *Braz.* **132** D12
Altamira, *Chile* **134** F7
Altamura, *It.* **152** M13
Altar Desert, *N. Amer.* **84** K3
Altata, *Mex.* **116** F8
Altavista, *Va., U.S.* **98** F11
Altay Mountains, *Asia* **168** F8
Altay, *China* **184** C6
Altay, *Mongolia* **184** D8
Altayskiy, *Russ.* **160** L12
Altdorf, *Switz.* **152** B4
Altea, *Sp.* **147** J15
Altenburg, *Ger.* **150** F9
Alter do Chão, *Port.* **146** H6
Altiağac, *Azerb.* **171** C22
Altiplano, region, *S. Amer.* **128** G4
Altınekin, *Turk.* **170** G8
Altıntaş, *Turk.* **170** F5
Altınyayla, *Turk.* **170** H5
Altmühl, river, *Ger.* **150** J8
Altnaharra, *Scot., U.K.* **144** G9
Alto Araguaia, *Braz.* **132** K11
Alto Chicapa, *Angola* **209** L13
Alto Garças, *Braz.* **132** J11
Alto Molócuè, *Mozambique* **212** D12
Alto Paraguai, *Braz.* **132** J10
Alto Parnaíba, *Braz.* **133** G14
Alto Purús, river, *S. Amer.* **132** G5
Alto Río Senguerr, *Arg.* **135** S7
Alto Vista, *Aruba, Neth.* **123** Q17
Alto, *Tex., U.S.* **105** H19
Alton, *Ill., U.S.* **101** S4
Altona Lagoon, *Virgin Is., U.S.* **122** Q3
Altona, *Virgin Is., U.S.* **122** N8
Altoona, *Pa., U.S.* **96** N5
Altun Shan, range, *China* **184** G5
Altun Shan, peak, *China* **184** G7
Altunhisar, *Turk.* **170** G9
Alturas, *Calif., U.S.* **110** L5
Altus, *Okla., U.S.* **105** D15
Alucra, *Turk.* **171** E13
Aluk, *S. Sudan* **203** V9
Alula-Fartak Trough, *Ind. Oc.* **250** D7
Al'Unnāb, site, *Jordan* **173** R8
Alushta, *Ukr.* **159** T8
Alva, *Okla., U.S.* **105** B15
Alvarado, *Tex., U.S.* **105** G17
Älvdalen, *Sw.* **142** J12
Alverca do Ribatejo, *Port.* **146** H4
Alvesta, *Sw.* **143** M13
Alvin, *Tex., U.S.* **105** L19
Älvkarleby, *Sw.* **143** J13
Alvorada, *Braz.* **133** H13
Alvord Desert, *Oreg., U.S.* **110** K8
Alvord Lake, *Oreg., U.S.* **110** K8
Alvord, *Tex., U.S.* **105** F16
Älvsbyn, *Sw.* **143** E14
Alwar, *India* **182** G5
Alxa Gaoyuan, *China* **184** G9
Alyangula, *N.T., Austral.* **221** P10
Alytus, *Lith.* **143** N16
Alzada, *Mont., U.S.* **108** F13
Alzamay, *Russ.* **161** K14
Alzette, river, *Lux.* **162** J10
Alzey, *Ger.* **150** H6
Alzira, *Sp.* **147** H14
Am Dam, *Chad* **207** H15
Am Djéména, *Chad* **207** H13
Am Khoumi, *Chad* **207** H14
Am Léiouna, *Chad* **207** H15
Am Timan, *Chad* **207** J15
Am Zoer, *Chad* **207** H15
Ama, *Japan* **189** N7
'Amada, ruin, *Egypt* **202** L12
Amadeus Depression, *Austral.* **220** F7
Amadeus, Lake, *Austral.* **220** F8
Amadi, *S. Sudan* **203** X11
Amadjuak Lake, *Nunavut, Can.* **87** G18
Amadora, *Port.* **146** H4
Amagi, *Japan* **189** R5

Amahai, *Indonesia* **193** K16
Amakusa Shotō, *Japan* **189** S4
Âmål, *Sw.* **142** K12
Amalfi, *It.* **152** M11
Amaliáda, *Gr.* **156** J8
Amamapare, *Indonesia* **193** L20
Amambaí, *Braz.* **132** M11
Amami Guntō, *Japan* **189** W3
Amami Ō Shima, *Japan* **189** W3
Amami, island, *Asia* **168** G13
Amamula, *D.R.C.* **209** E18
Āmanave, *American Samoa, U.S.* **230** E5
Amandola, *It.* **152** J9
Amangeldi, *Kaz.* **178** E10
Amantea, *It.* **153** Q13
Amanu, island, *Fr. Polynesia, Fr.* **231** C20
Amapá, *Braz.* **132** B12
Amarante, *Braz.* **133** F15
Amarante, *Port.* **146** E6
Amargosa Range, *Calif., Nev., U.S.* **111** U10
Amarillo, *Tex., U.S.* **104** D12
Amarkantak, *India* **182** K8
Amarwara, *India* **182** K7
Amasa, *Mich., U.S.* **100** F6
Amasine, *W. Sahara* **200** H4
Amasra, *Turk.* **170** C8
Amasya, *Turk.* **170** D11
Amata, *S. Aust., Austral.* **221** U8
Amatari, *Braz.* **132** D9
Amathous, ruin, *Cyprus* **162** Q8
Amatrice, *It.* **152** J9
Amatuku, island, *Tuvalu* **228** L11
Amatusuk Hills, *Alas., U.S.* **113** C13
Amavon Islands, *Solomon Is.* **228** D8
Amazon (Solimões), river, *S. Amer.* **128** D5
Amazon Basin, *S. Amer.* **128** E5
Amazon Fan, *Atl. Oc.* **249** N7
Amazon, *S. Amer.* **128** C8
Ambala, *India* **182** F6
Ambalangoda, *Sri Lanka* **183** U7
Ambalantota, *Sri Lanka* **183** U8
Ambalavao, *Madagascar* **213** G16
Ambanja, *Madagascar* **213** D17
Ambarchik, *Russ.* **161** C19
Ambargasta, Salinas de, *Arg.* **134** H10
Ambato Boeny, *Madagascar* **213** E16
Ambato, *Ecua.* **130** L4
Ambatofinandrahana, *Madagascar* **213** G16
Ambatomainty, *Madagascar* **213** E15
Ambatondrazaka, *Madagascar* **213** E17
Ambelau, island, *Indonesia* **193** K16
Amberg, *Ger.* **150** H9
Ambergris Caye, island, *Belize* **117** J17
Ambergris Cays, *Turks & Caicos Is., U.K.* **119** J17
Ambi, *Solomon Is.* **228** N6
Ambidédi, *Mali* **204** J9
Ambikapur, *India* **182** K9
Ambilobe, *Madagascar* **213** C17
Ambitle, island, *P.N.G.* **222** N10
Amble, *Eng., U.K.* **145** M12
Ambler, *Alas., U.S.* **113** E14
Ambo Channel, *Marshall Is.* **227** H17
Ambo, *Peru* **132** G4
Amboasary, *Madagascar* **213** J15
Ambodifotatra, *Madagascar* **213** E17
Amboise, *Fr.* **149** R8
Ambon, *Indonesia* **193** K16
Ambon, island, *Indonesia* **193** K16
Ambositra, *Madagascar* **213** G16
Ambovombe, *Madagascar* **213** J15
Amboy Crater, *Calif., U.S.* **111** W11
Amboy, *Calif., U.S.* **111** W11
Ambriz, *Angola* **208** J9
Ambrym, island, *Vanuatu* **229** G16
Amchekenchar, river, *Mali* **205** F18
Amchitka Island, *Alas., U.S.* **112** N3
Amchitka Pass, *Alas., U.S.* **112** N3
Amderma, *Russ.* **160** E11
Amdi, *India* **182** J8
Amdrup Land, *Greenland, Den.* **88** C12
Ameca, *Mex.* **116** H9
Ameghino Canyon, *Atl. Oc.* **249** U6
Ameland, island, *Neth.* **148** G13
Amelia Island, *Fla., U.S.* **99** Q9
Ameralik, fjord, *Greenland, Den.* **89** R5

America-Antarctica Ridge (North Weddell Ridge), *Atl. Oc.* **249** W10
American Falls Reservoir, *Idaho, U.S.* **108** H6
American Falls, *Idaho, U.S.* **108** H6
American Highland, *Antarctica* **239** G18
American Samoa, *Pac. Oc.* **224** J10
Americus, *Ga., U.S.* **99** N5
Amersfoort, *Neth.* **148** J13
Amery Ice Shelf, *Antarctica* **239** F19
Amery, *Wis., U.S.* **100** H2
Ames Range, *Antarctica* **238** M9
Ames, *Iowa, U.S.* **107** N11
Amfilochía, *Gr.* **156** G8
Amfípoli, *Gr.* **156** C12
Ámfissa, *Gr.* **156** H10
Amga, *Russ.* **161** H18
Amga, river, *Russ.* **161** H18
Amguema, *Russ.* **161** B20
Amguid, *Alg.* **201** H13
Amhara, *Eth.* **210** L7
Amherst, *N.S., Can.* **87** P22
Amherst, *N.Y., U.S.* **96** J5
Amherst, *Va., U.S.* **98** E11
Amherst, *Tex., U.S.* **104** E11
Amidon, *N. Dak., U.S.* **106** G2
Amiens, *Fr.* **148** N11
Amindivi Islands, *India* **183** R3
Amino, *Japan* **189** P8
Amioûn, *Leb.* **172** H6
Amir Chah, *Pak.* **181** N3
Amīr, river, *Afghan.* **180** G7
Amirante Basin, *Ind. Oc.* **250** H7
Amirante Ridge, *Ind. Oc.* **250** H7
Amirante Trench, *Ind. Oc.* **250** H7
Amistad National Recreation Area, *Tex., U.S.* **105** L13
Amistad Reservoir , *Mex., U.S.* **104** L12
Amistad, Presa de la *see* Amistad Reservoir, *Mex., U.S.* **116** D11
Amite, *La., U.S.* **102** N10
Âmli, *Nor.* **142** L11
Amlia Island, *Alas., U.S.* **112** P6
'Amm Adam, *Sudan* **203** Q14
'Ammān (Amman, Philadelphia), *Jordan* **172** M7
Amman *see* 'Ammān (Philadelphia), *Jordan* **172** M7
Ammaroo, *N.T., Austral.* **221** S9
Ammassalik Fjord, *Greenland, Den.* **89** P10
Ammassalik Ø, *Greenland, Den.* **89** P10
Ammassalik *see* Tasiilaq, *Greenland, Den.* **89** P10
Ammassivik, *Greenland, Den.* **89** T7
Ammochostos Bay, *Cyprus* **162** N10
Ammóchostos *see* Famagusta, *Cyprus* **162** P10
Amnok, river, *N. Korea* **186** G4
Āmol, *Iran* **177** E16
Amolar, *Braz.* **132** K10
Amorgós, *Gr.* **157** M16
Amorgós, island, *Gr.* **157** L16
Amoricain Shelf, *Atl. Oc.* **248** H12
Amory, *Miss., U.S.* **103** J13
Amos, *Que., Can.* **87** P18
Amourj, *Mauritania* **204** H12
Amoy *see* Xiamen, *China* **185** J14
Ampani, *India* **182** M8
Ampanihy, *Madagascar* **213** H15
Amparafaravola, *Madagascar* **213** E16
Amparihy Est, *Madagascar* **213** H16
Ampelákia, *Gr.* **156** E10
Ampezzo, *It.* **152** C9
Amplepuis, *Fr.* **149** T12
Amposta, *Sp.* **147** F15
'Amrān, *Yemen* **174** N8
Amravati, *India* **182** L6
Amreli, *India* **182** K2
Amritsar, *India* **182** E5
Amsā'ad, *Lib.* **201** D23
Amsterdam, *N.Y., U.S.* **96** J11
Amsterdam, *Neth.* **148** J12
Amstetten, *Aust.* **150** K11
Amtaedo, island, *S. Korea* **187** T6
Amu Darya (Oxus), *Turkm.* **178** N9
Amu Darya, river, *Asia* **168** F6
'Āmūdah, *Syr.* **172** C14
Amukta Island, *Alas., U.S.* **112** P7

Amun, *P.N.G.* **228** B4
Amund Ringnes Island, *Nunavut, Can.* **87** C15
Amundsen Basin, *Arctic Oc.* **252** G11
Amundsen Bay, *Antarctica* **239** C19
Amundsen Coast, *Antarctica* **238** K12
Amundsen Gulf, *N.W.T., Can.* **86** E11
Amundsen Land, *Greenland, Den.* **88** B9
Amundsen Plain, *Pac. Oc.* **255** M14
Amundsen Ridges, *Pac. Oc.* **255** M14
Amundsen Sea, *Antarctica* **238** L6
Amundsen Trough, *Arctic Oc.* **252** P9
Amundsen-Scott South Pole Station, *Antarctica* **239** J12
Amuntai, *Indonesia* **192** K11
Amur, river, *Asia* **168** E11
Amurang, *Indonesia* **193** H15
Amuri, *N.Z.* **229** H20
Amursk, *Russ.* **161** K21
Amvrakikós Kolpos, *Gr.* **156** G8
Amvrosiyivka, *Ukr.* **159** Q10
Amyderýa, *Turkm.* **178** N10
Amýntaio, *Gr.* **156** C9
An Cabhán *see* Cavan, *Ire.* **145** Q6
An Caisleán Nua *see* New Castle, *Ire.* **145** S4
An Caisleán Riabhach *see* Castlerea, *Ire.* **145** Q3
An Chathair *see* Galty Mountains, *Ire.* **145** S4
An Cheapach Mhór *see* Cappamore, *Ire.* **145** S4
An Clochán *see* Clifden, *Ire.* **145** Q3
An Craoslach *see* Creeslough, *Ire.* **145** N5
An Creagán *see* Mount Bellew, *Ire.* **145** Q4
An Daingean (Dingle), *Ire.* **145** T2
An Droichead Adhmaid *see* Woodenbridge, *Ire.* **145** S7
An Ghráinseach *see* Grange, *Ire.* **145** P4
An Gleann Garbh *see* Glengarriff, *Ire.* **145** U3
An Khê, *Vietnam* **190** M13
An Longfort *see* Longford, *Ire.* **145** Q5
An Muileann gCearr *see* Mullingar, *Ire.* **145** Q6
An Muileann Iarainn *see* Swanlinbar, *Ire.* **145** P5
An Nabk, *Syr.* **172** H8
An Nafūd, *Saudi Arabia* **174** D7
An Najaf (Najaf), *Iraq* **176** H10
An Nashshāsh, *U.A.E.* **175** H15
An Nāşirīyah (Nasiriyah), *Iraq* **176** J12
An Nawfalīyah, *Lib.* **201** E19
An Nimāş, *Saudi Arabia* **174** K7
An Nu'ayrīyah, *Saudi Arabia* **174** E11
An Nuhūd, *Sudan* **203** S9
An Nukhayb, *Iraq* **176** H9
An Ómaigh *see* Omagh, *N. Ire., U.K.* **145** N6
An Ráth *see* Charleville, *Ire.* **145** S4
An Scairbh *see* Scarriff, *Ire.* **145** R4
An Scoil *see* Skull, *Ire.* **145** U3
An Spidéal (Spiddal), *Ire.* **145** R3
An Tairbeart (Tarbert), *Scot., U.K.* **144** G6
An tAonach *see* Nenagh, *Ire.* **145** S5
An Teach Dóite *see* Maam Cross, *Ire.* **145** Q3
An Uaimh *see* Navan, *Ire.* **145** Q6
An, *Myanmar* **190** H4
Ana María, Cayos, *Cuba* **118** J10
Ana María, Golfo de, *Cuba* **118** J10
Anaa, island, *Fr. Polynesia, Fr.* **231** C18
Anabar, *Nauru* **227** M15
Anabar, river, *Russ.* **161** F15
Anabarskiy Zaliv, *Russ.* **161** E15
Anacapa Island, *Calif., U.S.* **111** X7
Anaco, *Venez.* **130** D12
Anacoco, *La., U.S.* **102** N7
Anaconda, *Mont., U.S.* **108** D6
Anacortes, *Wash., U.S.* **110** A4
Anadarko, *Okla., U.S.* **105** D16
Anadyr', *Russ.* **161** B21
Anadyr, Gulf of, *Pac. Oc.* **168** A12
Anadyr', river, *Asia* **168** B12
Anadyrskiy Zaliv, *Russ.* **161** B21
Anadyrskoye Ploskogor'ye, *Russ.* **161** C20
Anae, island, *Guam, U.S.* **230** P4

'Anaeho'omalu Bay, *Hawaii, U.S.* **115** M20
Anáfi, *Gr.* **157** N16
Anáfi, *island, Gr.* **157** N16
Anaga, Punta de, *Sp.* **214** Q5
Anaghit, *Eritrea* **210** G7
Anagni, *It.* **152** L9
Anaheim, *Calif., U.S.* **111** X8
Anahola, *Hawaii, U.S.* **114** E10
Anahuac, *Tex., U.S.* **105** L20
Anai Mudi, *peak, India* **183** S5
Anak, *N. Korea* **186** K5
Anakapalle, *India* **182** N9
Anaktuvuk Pass, *Alas., U.S.* **113** D16
Analavelona, *peak, Madagascar* **213** G15
Anamã, *Braz.* **132** E8
Anambas, Kepulauan, *Indonesia* **192** H7
Anamoose, *N. Dak., U.S.* **106** F5
Anamosa, *Iowa, U.S.* **107** N14
Anamur Burnu, *Turk.* **170** J8
Anamur, *Turk.* **170** J8
Anan, *Japan* **189** R8
Anand, *India* **182** K3
Anangenimon, *island, F.S.M.* **226** E10
Ananij, *island, Marshall Is.* **227** B21
Anantapur, *India* **183** Q6
Anantnag, *India* **182** D5
Anapa, *Russ.* **159** T9
Anápolis, *Braz.* **133** J13
Anār Darah, *Afghan.* **180** K3
Anār, *Iran* **177** J18
Anare Mountains, *Antarctica* **239** R15
Anatahan, *island, N. Mariana Is., U.S.* **230** J5
Anatolia (Asia Minor), *Asia* **168** E3
Anatom, *island, Vanuatu* **229** N18
Anatone, *Wash., U.S.* **110** E9
Añatuya, *Arg.* **134** H10
'Anbarābād, *Iran* **177** L20
Anboru, *island, Marshall Is.* **227** H21
Anbyŏn, *N. Korea* **186** J9
Ancenis, *Fr.* **149** R6
Anchor Bay, *Calif., U.S.* **111** Q2
Anchorage Whaler, *bay, Kiribati* **226** J11
Anchorage, *Alas., U.S.* **113** J16
Anchovy, *Jam.* **122** H7
Anclitas, Cayo, *Cuba* **118** K10
Ancón de Sardinas, Bahía de, *Ecua.* **130** J3
Ancona, *It.* **152** H9
Ancud, *S. Amer.* **128** N4
Ancud, *Chile* **135** Q6
Anda, *China* **185** D15
Andacollo, *Arg.* **135** N7
Andado, *N.T. Austral.* **221** U9
Andalgalá, *Arg.* **134** G9
Åndalsnes, *Nor.* **142** H10
Andalusia, *Ala., U.S.* **103** N15
Andalusia, *region, Sp.* **146** K9
Andaman and Nicobar Islands, *India* **183** R14
Andaman Basin, *Ind. Oc.* **251** E14
Andaman Islands, *India* **183** R14
Andaman Sea, *Ind. Oc.* **168** K9
Andapa, *Madagascar* **213** D17
Andelot-Blancheville, *Fr.* **149** Q13
Andenes, *Nor.* **142** C12
Andéranboukan, *Mali* **205** H18
Andermatt, *Switz.* **152** B4
Andernach, *Ger.* **150** G5
Andersen Air Force Base *see* Joint Region Marianas, *Guam, U.S.* **230** N7
Anderson Bay, *Austral.* **220** L7
Anderson Peninsula, *Antarctica* **239** R15
Anderson, *S.C., U.S.* **98** K7
Anderson, *Ind., U.S.* **101** R9
Anderson, *Mo., U.S.* **107** V11
Anderson, *Calif., U.S.* **111** N3
Anderson, *Alas., U.S.* **113** G17
Anderson, *river, N.W.T., Can.* **86** E10
Andes, *range, S. Amer.* **128** H4
Andfjorden, *bay, Nor.* **142** C12
Andicuri Bay, *Aruba, Neth.* **123** Q17
Andijon, *Uzb.* **179** L13
Andilamena, *Madagascar* **213** E17
Andırın, *Turk.* **170** H11
Andkhōy, *Afghan.* **180** F6
Ando, *S. Korea* **187** U9
Andoany, *Madagascar* **213** C17
Andoas, *Peru* **132** E3
Andomskiy, *Russ.* **158** E9

Andong, *N. Korea* **186** H9
Andong, *S. Korea* **187** Q11
Andongho, *lake, S. Korea* **187** P11
Andorra la Vella, *Andorra* **162** J3
Andorra, *Eur.* **162** E2
Andover, *Me., U.S.* **97** F15
Andøya, *island, Nor.* **142** C12
Andradina, *Braz.* **132** L12
Andreanof Islands, *Alas., U.S.* **112** N5
Andreapol', *Russ.* **158** H7
Andrée Land, *Greenland, Den.* **88** H12
Andrew Johnson National Historic Site, *Tenn., U.S.* **103** E20
Andrews, *N.C., U.S.* **98** J5
Andrews, *S.C., U.S.* **98** L11
Andrews, *Tex., U.S.* **104** H11
Andrijevica, *Montenegro* **154** J9
Andros Island, *Bahamas* **118** E11
Andros Town, *Bahamas* **118** E11
Ándros, *Gr.* **157** J14
Ándros, *island, Gr.* **157** J14
Androth Island, *India* **183** S3
Andrychów, *Pol.* **151** H15
Andryushkino, *Russ.* **161** D19
Andselv, *Nor.* **143** C13
Andújar, *Sp.* **146** K10
Andulo, *Angola* **208** M11
Anefis i-n-Darane, *Mali* **205** G16
Anegada Passage, *N. Amer.* **121** A14
Anegada, Bahía, *Arg.* **135** P11
Anegada, *island, British Virgin Is., U.K.* **121** A14
Aného, *Togo* **205** P17
Aneju, *island, Marshall Is.* **227** D19
Anelghowhat Bay, *Vanuatu* **229** N18
Anelghowhat, *Vanuatu* **229** N18
Anemwanot, *island, Marshall Is.* **227** D21
Anenelibw, *island, Marshall Is.* **227** E20
Anengenipuan, *island, F.S.M.* **226** E10
Aneta, *N. Dak., U.S.* **106** F7
Anetan, *Nauru* **227** L15
Aneto, Pico de, *Sp.* **147** C15
Aney, *Niger* **206** E11
Angamos Point, *S. Amer.* **128** J4
Angamos, Punta, *Chile* **134** E6
Angara, *river, Russ.* **161** J14
Angarei, *island, N.Z.* **229** H20
Angarsk, *Russ.* **161** L15
Angaur (Ngeaur), *island, Palau* **226** B4;K4
Änge, *Sw.* **143** H13
Ángel de la Guarda, Isla, *Mex.* **116** C6
Angel Falls, *Venez.* **130** F13
Angeles, *Philippines* **193** C13
Angelina, *river, Tex., U.S.* **105** H20
Angels Camp, *Calif., U.S.* **111** R5
Ångermanälven, *river, Sw.* **143** G13
Angers, *Fr.* **149** R7
Angerville, *Fr.* **149** Q9
Ângk Tasaôm, *Cambodia* **191** P10
Angkor, *ruins, Cambodia* **190** M10
Angla Rencuo see Ngangla Ringco, *lake, China* **184** K3
Angle Inlet, *Minn., U.S.* **106** D9
Anglesey, *island, Wales, U.K.* **145** R9
Anglet, *Fr.* **149** X5
Angleton, *Tex., U.S.* **105** M19
Anglure, *Fr.* **149** P11
Ango, *D.R.C.* **209** D17
Angoche, *Mozambique* **213** D13
Angol, *Chile* **135** N6
Angola, *Af.* **208** M10
Angola, *Ind., U.S.* **101** N10
Angola Basin, *Atl. Oc.* **249** Q12
Angola Plain, *Atl. Oc.* **249** Q12
Angoon, *Alas., U.S.* **113** L22
Angora see Ankara, *Turk.* **170** E8
Angoram, *P.N.G.* **222** N6
Angoulême, *Fr.* **149** U7
Angouma, *Gabon* **208** D9
Angra do Heroísmo, *Port.* **146** P4
Angren, *Uzb.* **178** L12
Angria Bank, *Ind. Oc.* **250** D10
Angu, *D.R.C.* **209** C16
Anguilla Cays, *Bahamas* **118** F9
Anguilla, *island, Lesser Antil.* **121** B15
Anguillita Island, *Anguilla, U.K.* **122** R10
Angujaartorfiup Nunaa, *Greenland, Den.* **89** P6
Angul, *India* **182** L10
Angumu, *D.R.C.* **209** E18
Angvik, *Nor.* **142** H10
Anholt, *island, Den.* **142** M12

Ani, *Japan* **188** K12
Aniak, *Alas., U.S.* **113** J13
Aniakchak National Monument and Preserve, *Alas., U.S.* **113** M13
Anibare Bay, *Nauru* **227** M15
Anibare, *Nauru* **227** M15
Anie, Pic d', *Fr.-Sp.* **147** B14
Animas Peak, *N. Mex., U.S.* **109** W10
Animas, *N. Mex., U.S.* **109** V10
Animas, *river, Colo., U.S.* **109** P11
Anin, *Myanmar* **190** L6
Anina, *Rom.* **154** E11
Anipemza, *Arm.* **171** D18
Anisok, *Eq. Guinea* **206** P10
Aniwa, *island, Vanuatu* **229** M18
Anjar, *India* **182** K2
Anjira, *Pak.* **181** P6
Anjou Islands, *Asia* **168** B10
Anjouan see Nzwani, *Comoros* **215** N16
Anjozorobe, *Madagascar* **213** F16
Anju, *N. Korea* **186** H6
Anjwado, *island, S. Korea* **187** T6
Ankang, *China* **184** K11
Ankara (Angora), *Turk.* **170** E8
Ankara, *river, Turk.* **170** E7
Ankaratra, *peak, Af.* **198** N12
Ankazoabo, *Madagascar* **213** G15
Ankeny, *Iowa, U.S.* **107** P11
Anklam, *Ger.* **150** B10
Ankola, *India* **183** Q4
Ankoro, *D.R.C.* **209** J17
Ankpa, *Nigeria* **206** L8
Anmado, *island, S. Korea* **187** S6
Anmyeon, *S. Korea* **187** Q7
Anmyeondo, *island, S. Korea* **187** Q7
Ann Arbor, *Mich., U.S.* **100** M12
Ann, Cape, *Mass., U.S.* **97** K15
Ann, Cape, *Antarctica* **239** C19
Anna Pink, Bahía, *Chile* **135** T5
Anna Plains, *W. Aust., Austral.* **221** R4
Anna Point, *Nauru* **227** L14
Anna, *Ill., U.S.* **101** V5
Anna, Lake, *Va., U.S.* **98** E12
Annaba, *Alg.* **201** B14
Annadale Falls, *Grenada* **123** K22
Annaka, *Japan* **189** N11
Annam Cordillera, *Asia* **168** J10
Annan, *Scot., U.K.* **145** N10
Annandale Falls, *Grenada* **123** K22
Annapolis, *Md., U.S.* **98** C14
Annapurna I, *peak, Nepal* **182** G9
Annean, Lake, *Austral.* **220** H3
Annecy, *Fr.* **149** T14
Annecy, Lac d', *Fr.* **149** U14
Annemasse, *Fr.* **149** T14
Annenskiy Most, *Russ.* **158** E9
Annigeri, *India* **183** P4
Anniston, *Ala., U.S.* **103** J16
Annitowa, *N.T., Austral.* **221** S10
Annobón, *island,, Eq. Guinea* **206**, R7
Annonay, *Fr.* **149** V12
Annopol, *Pol.* **151** F17
Annotto Bay, *Jam.* **122** H10
Anoano, *Solomon Is.* **228** E11
Anorituup Kangerlua, *fjord, Greenland, Den.* **89** T8
Anoumabo, see Grand-Lahou, *Côte d'Ivoire*
Anorituup Kangerlua, *fjord, Greenland, Den.* **89** T8
Anqing, *China* **185** K14
Ansan, *S. Korea* **187** N8
Anse Boileau, *Seychelles* **215** P20
Anse de Sigave, *bay, Wallis & Futuna Is., Fr.* **230** Q11
Anse des Pitons, *bay, St. Lucia* **123** L13
Anse d'Hainault, *Haiti* **119** M14
Anse Guyonneau, *bay, Guadeloupe, Fr.* **123** F13
Anse la Raye, *St. Lucia* **123** K13
Anse Lavoutte, *bay, St. Lucia* **123** J14
Anse Marcel, *bay, St. Martin, Fr.* **123** A15
Anse Quanery, *bay, Dominica* **123** F20
Anse-à-Foleur, *Haiti* **119** L16
Anse-à-Galets, *Haiti* **119** M16
Anse-à-Pitre, *Haiti* **119** N17
Anse-à-Veau, *Haiti* **119** M15
Anse-Bertrand, *Guadeloupe, Fr.* **123** D15
Anse-Rouge, *Haiti* **119** L16
Anseba, *river, Eritrea* **210** G7
Anselmo, *Nebr., U.S.* **107** P5
Anseong, *S. Korea* **187** P9
Anshan, *China* **185** F15

Anshun, *China* **184** M10
Ansley, *Nebr., U.S.* **107** P5
Ansó, *Sp.* **147** C14
Anson Bay, *Austral.* **220** B7
Anson Bay, *Austral.* **229** A20
Anson Point, *Austral.* **229** A20
Anson, *Me., U.S.* **97** F16
Anson, *Tex., U.S.* **105** G14
Ansongo, *Mali* **205** H16
Ansted, *W. Va., U.S.* **98** E9
Ansudu, *Indonesia* **193** K21
Ant Atoll, *island, F.S.M.* **226** B11;K7
Antakya, *Turk.* **170** J11
Antalaha, *Madagascar* **213** D17
Antalāt, *Lib.* **201** E21
Antalya (Adalia), *Turk.* **170** H6
Antalya Körfezi, *Turk.* **170** J6
Antananarivo, *Madagascar* **213** F16
Antanifotsy, *Madagascar* **213** F16
Antarctic Canyon, *Atl. Oc.* **249** Z7
Antarctic Pacific Ridge, *Pac. Oc.* **247** R14
Antarctic Peninsula, *Antarctica* **238** E5
Antarctica, *Antarctica* **236** J12
Antelope Reservoir, *Oreg., U.S.* **110** K9
Antelope, *Oreg., U.S.* **110** G5
Antequera, *Sp.* **146** L10
Anthony, *Kans., U.S.* **107** V7
Anthony, *N. Mex., U.S.* **109** V12
Anti Atlas, *Mor.* **200** F6
Anti-Lebanon Range see Al Jabal ash Sharqī, *Syr.* **172** J7
Antibes, *Fr.* **149** X15
Anticosti Island, *N. Amer.* **84** G10
Antifer, Cap d', *Fr.* **149** N7
Antigo, *Wis., U.S.* **100** H6
Antigonish, *N.S., Can.* **87** P22
Antigua and Barbuda, *N. Amer.* **121** B16
Antigua, *Sp.* **214** Q7
Antigua, *island, Lesser Antil.* **121** C16
Antikýthira, *island, Gr.* **156** P11
Antilla, *Cuba* **119** J13
Antimácheia, *Gr.* **157** L18
Antímilos, *island, Gr.* **157** M13
Antioch, *Calif., U.S.* **111** R4
Antioche, Pertuis d', *Fr.* **149** T6
Antíparos, *island, Gr.* **157** L14
Antípaxoi, *island, Gr.* **156** F6
Antipayuta, *Russ.* **160** F12
Antipodes Fracture Zone, *Pac. Oc.* **247** R13
Antipodes Islands, *N.Z.* **223** Q22
Antipodes Islands, *Pac. Oc.* **224** Q9
Antípsara, *island, Gr.* **157** H15
Antlers, *Okla., U.S.* **105** E19
Antofagasta, *Chile* **134** E6
Antofalla, Salar de, *Arg.* **134** F8
Anton Chico, *N. Mex., U.S.* **109** S13
Anton Dohrn Seamount, *Atl. Oc.* **253** L23
Anton, *Tex., U.S.* **104** F12
Antongila Bay, *Af.* **198** M12
Antónia, Pico da, *Cabo Verde* **215** D16
Antonibe, *Madagascar* **213** D16
Antonina, *Braz.* **133** L14
Antrain, *Fr.* **149** Q6
Antrim, Glens of, *N. Ire., U.K.* **145** N7
Antriol, *Bonaire, Neth.* **123** Q19
Antrodoco, *It.* **152** J9
Antropovo, *Russ.* **158** G11
Antsirabe, *Madagascar* **213** F16
Antsirañana, *Madagascar* **213** C17
Antsohihy, *Madagascar* **213** D16
Antubia, *Ghana* **205** P14
Antwerpen (Antwerp), *Belg.* **148** K12
Anuanu Raro, *island, Fr. Polynesia, Fr.* **231** D19
Anuanu Runga, *island, Fr. Polynesia, Fr.* **231** D19
Anupgarh, *India* **182** F4
Anuradhapura, *Sri Lanka* **183** T7
Anuta, *island, Solomon Is.* **228** G11
Anvers Island, *Antarctica* **238** D4
Anvik, *Alas., U.S.* **113** H13
Anvil see El Yunque, *peak, Cuba* **119** K14
Anxi, *China* **184** G7
Anxious Bay, *Austral.* **220** K9
Anyang, *China* **185** H13
Anyang, *S. Korea* **187** N8
Ánydros, *island, Gr.* **157** M16
A'nyêmaqên Shan, *China* **184** J8
Anzhero Sudzhensk, *Russ.* **160** K12
Anzhu, *Russ.* **161** D16
Anzio, *It.* **152** L8

Ao Ban Don, *bay, Thai.* **191** Q7
Aoa Bay, *American Samoa, U.S.* **230** D8
Aoba, *island, Vanuatu* **229** F15
Aoga Shima, *Japan* **189** S12
Aojing, *China* **185** L15
Aola, *Solomon Is.* **228** F10
Aoloautuai, *American Samoa, U.S.* **230** D6
Aomori, *Japan* **188** J12
Aonae, *Japan* **188** H11
Aóös, *river, Gr.* **156** D8
Aopo, *Samoa* **227** N17
Aoraki (Mount Cook), *N.Z.* **223** M17
Aóre, *island, Vanuatu* **229** F14
Aosta, *It.* **152** D2
Aotea Island see Great Barrier Island, *N.Z.* **223** E21
Aouderas, *Niger* **206** F8
Aoukâr, *Mauritania* **204** G10
Aoukâr, *Mali* **205** C13
Aouker, *island, Af.* **198** F2
Aoumou, *New Caledonia, Fr.* **231** P22
Aourou, *Mali* **204** J9
Aoya, *Japan* **189** P7
Aozi, *Chad* **207** D14
Aozou Strip, *Chad* **207** C13
Aozou, *Chad* **207** C13
Ap Iwan, Cerro, *Arg.* **135** T7
Apa, *river, S. Amer.* **132** L10
Apache, *Okla., U.S.* **105** D16
Apakho, *Solomon Is.* **228** M6
Apalachee Bay, *U.S.* **91** M18
Apalachicola Bay, *Fla., U.S.* **99** R4
Apalachicola, *Fla., U.S.* **99** R4
Apalachicola, *river, Fla., U.S.* **99** R4
Apam, *Ghana* **205** Q16
Apamea, *ruins, Syr.* **172** F7
Apaora, *Solomon Is.* **228** G11
Apaporis, *river, Col.* **130** J8
Aparri, *Philippines* **193** B14
Apataki, *island, Fr. Polynesia, Fr.* **231** B18
Apatin, *Serb.* **154** E9
Apatity, *Russ.* **160** D8
Apatzingán, *Mex.* **116** J10
Apeíranthos, *Gr.* **157** L15
Apeldoorn, *Neth.* **148** J13
Apennines, *It.* **140** K7
Apes Den, *Gibraltar, U.K.* **163** P23
Áphaea, *ruin, Gr.* **156** K12
Aphaedo, *island, S. Korea* **187** T7
Api, *D.R.C.* **209** B16
Apia Harbour, *Samoa* **227** P19
Apia, *Samoa* **227** P19
Apishapa, *river, Colo., U.S.* **109** P14
Apo, *S. Korea* **187** Q10
Apolima Strait, *Samoa* **227** P18
Apolima, *island, Samoa* **227** P18
Apollo Bay, *Vic., Austral.* **221** Z12
Apollonia see Süsah, *Lib.* **201** D21
Apollonía, *Gr.* **157** L14
Apollonia, *site, Alban.* **154** L9
Apolo, *Bol.* **132** J6
Apopka, Lake, *Fla., U.S.* **99** T8
Aporema, *Braz.* **132** C12
Apostle Islands National Lakeshore, *Wis., U.S.* **100** E4
Apostle Islands, *Wis., U.S.* **100** E4
Apóstoles, *Arg.* **134** G14
Apostolic Palace, *Vatican City* **163** P17
Apostolos Andreas Monastery, *Cyprus* **162** M12
Apostolos Andreas, Cape, *Cyprus* **162** L12
Apoteri, *Guyana* **131** G15
Appalachian Mountains, *U.S.* **91** G19
Appalachian Plateau, *U.S.* **91** G19
Appat, *Greenland, Den.* **88** L6
Appennini, *range, It.* **152** F5
Apple Valley, *Calif., U.S.* **111** W9
Applecross, *Scot., U.K.* **144** H7
Appleton City, *Mo., U.S.* **107** T11
Appleton, *Wis., U.S.* **100** J6
Appleton, *Minn., U.S.* **106** J9
Appomattox, *Va., U.S.* **98** F11
Aprilia, *It.* **152** L8
Apsheron Peninsula, *Eur.* **140** J16
Apsheronsk, *Russ.* **159** T11
'Āpua Point, *Hawaii, U.S.* **115** P22
Apuane, *Alpi, It.* **152** G5
Apucarana, *Braz.* **132** M12
Apulia, *Eur.* **140** K8

Apurashokoru, island, *Palau* **226** H5
Apure, river, *N. Amer.* **84** Q10
Apure, river, *Venez.* **130** E10
Apurímac, river, *Peru* **132** H4
Apuseni Mountains, *Rom.* **154** C12
Aqaba *see* Al 'Aqabah, *Jordan* **173** R5
Aqaba, Gulf of, *Asia* **168** F2
Aqadyr, *Kaz.* **179** F13
Aqcha *see* Āqchah, *Afghan.* **180** F6
Āqchah (Aqcha), *Afghan.* **180** F6
Aqiq, *Sudan* **203** P16
Aqköl, *Kaz.* **178** D12
Aqköl, *Kaz.* **178** J12
Aqköl, *Kaz.* **179** H14
Aqqikkol Hu, *China* **184** H4
Aqqū, *Kaz.* **179** D15
Aqqystaū, *Kaz.* **178** G4
'Aqrah, *Iraq* **176** D10
Aqsay, *Kaz.* **178** D5
Aqshataū, *Kaz.* **179** G14
Aqsū, *Kaz.* **179** D13
Aqsū, river, *Kaz.* **179** H16
Aqtaū, *Kaz.* **178** J4
Aqtöbe, *Kaz.* **178** E7
Aqtoghay, *Kaz.* **179** G16
Aquarius Seachannel, *Pac. Oc.*
 247 D14
Aquidauana, *Braz.* **132** L11
Aquin, *Haiti* **119** M15
Aquitaine Basin, *Eur.* **140** J4
Ar Rabbah, *Jordan* **173** N6
Ar Raḍwānīyah, *Iraq* **176** G10
Ar Rajmah, *Lib.* **201** D20
Ar Ramādī (Ramadi), *Iraq* **176** G10
Ar Ramthā, *Jordan* **172** L7
Ar Raqqah (Nicephorium), *Syr.*
 172 E11
Ar Rashādīyah, *Jordan* **173** P6
Ar Rāshidah, *Egypt* **202** J10
Ar Rāshidīyah, *Iraq* **176** G10
Ar Rastan (Arethusa), *Syr.* **172** G8
Ar Rawḍah, *Saudi Arabia* **174** J8
Ar Rayyān, *Qatar* **175** F13
Ar Riḥāb, *Iraq* **176** J11
Ar Rīḥānah, *Iraq* **176** F9
Ar Riyāḍ (Riyadh), *Saudi Arabia*
 174 G10
Ar Riyān, *Yemen* **174** P12
Ar Ru'ays, *Qatar* **175** F13
Ar Rub' al Khālī (Empty Quarter),
 Saudi Arabia **174** K12
Ar Rumaythah, *Iraq* **176** J11
Ar Ruṣāfah (Risāfe), ruins, *Syr.*
 172 F11
Ar Ruṣayfah, *Jordan* **172** M7
Ar Ruṭbah, *Iraq* **176** G8
Ar Ruwayshid, *Jordan* **172** L10
Ara Bure, *Eth.* **211** N9
Ara, *India* **182** H9
Arab, *Ala., U.S.* **103** H15
'Arabah, Wādī al, *Jordan* **173** Q5
Arabian Basin, *Ind. Oc.* **250** C9
Arabian Peninsula, *Asia* **168** G3
Arabian Sea, *Asia* **168** J3
Arabs Gulf, *Af.* **198** D8
Araç, *Turk.* **170** C8
Araç, river, *Turk.* **170** C9
Aracaju, *Braz.* **133** H17
Aracati, *Braz.* **133** E17
Araçatuba, *Braz.* **132** L12
Aracena, *Sp.* **146** K7
Aracena, Sierra de, *Sp.* **146** K7
Árachthos, river, *Gr.* **156** F8
Araçuaí, *Braz.* **133** K15
Arad, *Rom.* **154** D11
Arada, *Chad* **207** G15
Aradippou, *Cyprus* **162** P9
Arafura Sea, *Asia* **169** M16
Arafura Shelf, *Pac. Oc.* **246** K7
Aragarças, *Braz.* **132** J12
Arago, Cape, *Oreg., U.S.* **110** J1
Aragón, *Eur.* **140** K3
Aragua de Barcelona, *Venez.* **130** D12
Araguacema, *Braz.* **133** F13
Araguaia, river, *Braz.* **133** F13
Araguaína, *Braz.* **133** F13
Araguari, *Braz.* **133** K13
Araguari, river, *S. Amer.* **128** C8
Araguatins, *Braz.* **133** E13
Arahal, *Sp.* **146** L8
Arai, *Japan* **189** N11
Árainn (Inishmore), island, *Ire.*
 145 R3
Árainn Mhór (Aran Island), *Ire.*
 145 N4

Arak, *Alg.* **200** J12
Arāk, *Iraq* **177** G14
Arakabesan, island, *Palau* **226** G5
Arakan Yoma, *Myanmar* **190** H4
Arakht, *Afghan.* **180** E10
Aral Mangy Qaraqumy, *Kaz.* **178** G9
Aral, *Kaz.* **178** G9
Aral, *Kyrg.* **179** K14
Aralsor Köli, *Kaz.* **178** E3
Aralsor, Lake, *Eur.* **140** G14
Aralsul'fat, *Kaz.* **178** G9
Araltobe, *Kaz.* **178** E8
Aramac, *Qnsld., Austral.* **221** T13
Aramits, *Fr.* **149** Y6
Aran Arrei, *Eth.* **210** M10
Aran Island *see* Árainn Mhór, *Ire.*
 145 R3
Aran Islands, *Ire.* **145** R3
Āran, *Iran* **177** G15
Aran, island, *Scot., U.K.* **145** M8
Aranda de Duero, *Sp.* **146** D11
Aranđelovac, *Serb.* **154** G10
Arandu, *Pak.* **180** G10
Aranjuez, *Sp.* **146** G11
Aranos, *Namibia* **212** H5
Aransas Pass, *Tex., U.S.* **105** P17
Aransas, river, *Tex., U.S.* **105** N17
Araouane, *Mali* **205** F14
Arapaho National Recreation Area,
 Colo., U.S. **108** L12
Arapahoe, *Nebr., U.S.* **107** R5
Arapiraca, *Braz.* **133** G17
Arapkir, *Turk.* **171** F13
'Ar'ar, *Saudi Arabia* **174** B7
Araraquara, *Braz.* **133** L13
Ararat, *Arm.* **171** E19
Ararat, *Vic., Austral.* **221** Z12
Ararat, Mount *see* Ağrı Dağı, peak,
 Turk. **171** E18
Ararat, Mount, *Asia* **168** F4
Araria, *India* **182** H11
Araripe, Chapada do, *S. Amer.* **128** F11
Aras (Araxes), river, *Turk.* **171** E17
Aras, river, *Iran* **177** B13
Aratika, island, *Fr. Polynesia, Fr.*
 231 B18
Aratsu Bay, *N. Mariana Is., U.S.*
 230 N9
Arauá, river, *Braz.* **132** E7
Arauca, *Col.* **130** E8
Arauca, river, *S. Amer.* **130** E8
Arauco Gulf, *S. Amer.* **128** X4
Aravalli Range, *India* **182** J4
Araxá, *Braz.* **133** K13
Araxes *see* Aras, river, *Turk.* **171** E17
Arayit Dağı, peak, *Turk.* **170** F6
Arba'at, *Sudan* **203** N15
Arbatax, *It.* **153** P5
Arbazh, *Russ.* **158** G14
Arbela *see* Irbid, *Jordan* **172** L6
Arbīl, Erbil *see* Hewlêr, *Iraq* **176** E10
Arbre du Ténéré, *Niger* **206** F9
Arbroath, *Scot., U.K.* **144** K11
Arcachon, *Fr.* **149** W6
Arcachon, Bassin d', *Fr.* **149** V6
Arcadia, *Fla., U.S.* **99** V8
Arcadia, *Mich., U.S.* **100** J9
Arcadia, *Wis., U.S.* **100** J3
Arcadia, *La., U.S.* **102** L7
Arcata, *Calif., U.S.* **110** M1
Archánes, *Gr.* **157** Q15
Archangel *see* Arkhangel'sk, *Russ.*
 158 B10
Archángelos, *Gr.* **157** N20
Archbald, *Pa., U.S.* **96** M9
Archer Bay, *Austral.* **220** B12
Archer City, *Tex., U.S.* **105** F15
Archer, river, *Austral.* **220** B12
Arches National Park, *Utah, U.S.*
 109 N9
Archiac, *Fr.* **149** U7
Archialus *see* Pomorie, *Bulg.* **155** J17
Archidona, *Sp.* **146** L10
Archipiélago de Colón *see* Galápagos
 Islands, *Pac. Oc.* **225** G20
Archman, *Turkm.* **178** M6
Arcis, *Fr.* **149** P12
Arco, *Idaho, U.S.* **108** G6
Arcos de Jalón, *Sp.* **146** E12
Arcoverde, *Braz.* **133** G17
Arctic Bay, *Nunavut, Can.* **87** E16
Arctic Ocean, *Arctic Oc.* **248** A13
Arctic Red *see* Tsiigehtchic, *N.W.T., Can.* **86** D17
Arctic Village, *Alas., U.S.* **113** D17
Arctowski, station, *Antarctica* **238** C4

Ard an Rátha *see* Ardara, *Ire.* **145** N5
Arḍ aş Şawwān, *Jordan* **173** P8
Arda, river, *Bulg.* **155** K15
Ardabīl, *Iran* **177** C13
Ardahan, *Turk.* **171** D17
Ardakān, *Iran* **177** H17;K16
Ardanuç, *Turk.* **171** D16
Ardara (Ard an Rátha), *Ire.* **145** N5
Ardee (Baile Átha Fhirdhia), *Ire.*
 145 Q6
Ardencaple Seachannel, *Arctic Oc.*
 248 C11
Ardennes, Canal des, *Fr.* **149** N12
Ardennes, region, *Belg., Fr.* **148** M13
Ardestān, *Iran* **177** G16
Ardmore, *Okla., U.S.* **105** E17
Ardmore, *S. Dak., U.S.* **106** M1
Ardoch, *Qnsld., Austral.* **221** V12
Ardres, *Fr.* **148** L9
Ards Peninsula, *N. Ire., U.K.* **145** P8
Āre, *Sw.* **142** G12
Areia Branca, *Braz.* **133** E17
Arekalong Peninsula, *Palau* **226** F6
Arena, Point, *Calif., U.S.* **90** F2
Arenápolis, *Braz.* **132** H10
Arenas de San Pedro, *Sp.* **146** F9
Arenas, Punta de, *Arg.* **135** X9
Arendal, *Nor.* **142** L11
Areópoli, *Gr.* **156** M10
Arequipa, *Peru* **132** J5
Ārēro, *Eth.* **211** P7
Arès, *Fr.* **149** V6
Arethusa *see* Ar Rastan, *Syr.* **172** G8
Arévalo, *Sp.* **146** E10
Arezzo, *It.* **152** H7
Arfersiorfik, channel, *Greenland, Den.*
 89 N6
Argadargada, *N.T., Austral.* **221** S10
Argakhtakh, *Russ.* **161** D19
Arganda, *Sp.* **146** F11
Argelès-Gazost, *Fr.* **149** Y7
Argent, *Fr.* **149** R10
Argenta, *It.* **152** F7
Argentan, *Fr.* **149** P7
Argentario, Monte, *It.* **152** J7
Argentat, *Fr.* **149** V9
Argenteuil, *Fr.* **149** P10
Argentera, Capo dell', *It.* **152** M3
Argentina Range, *Antarctica* **238** F10
Argentina, *S. Amer.* **134** L10
Argentine Basin, *Atl. Oc.* **249** U7
Argentine Plain, *Atl. Oc.* **249** V6
Argentine Rise, *Atl. Oc.* **249** U6
Argentino, Lago, *Arg.* **135** V7
Argenton-Château, *Fr.* **149** S7
Argenton, *Fr.* **149** T9
Argeş, river, *Rom.* **155** F16
Arghandāb Band, *Afghan.* **180** K6
Arghandāb, river, *Afghan.* **180** K6
Arghīstān, *Afghan.* **180** L7
Argo Fracture Zone, *Ind. Oc.* **250** J9
Argo, *Sudan* **203** N11
Argolikós Kólpos, *Gr.* **156** K11
Argonne, range, *Fr.* **148** N13
Árgos Orestikó, *Gr.* **156** D8
Árgos, *Gr.* **156** K11
Argostóli, *Gr.* **156** J7
Argoub, *W. Sahara* **200** J2
Arguedas, *Sp.* **147** D13
Arguello, Point, *Calif., U.S.* **111** W5
Argun', river, *Russ.* **161** L18
Argungu, *Nigeria* **206** H6
Argus Dome (Dome A), *Antarctica*
 239 H16
Arguut, *Mongolia* **184** E9
Argyle, *Minn., U.S.* **106** E8
Argyle, Lake, *Austral.* **220** C7
Argyrus *see* Gjirokastër, *Alban.*
 154 M10
Arhus, *Den.* **142** M11
Ari Atoll, *Maldives* **183** W3
Ariamsvlei, *Namibia* **212** K5
Ariano Irpino, *It.* **152** M11
Ariari, river, *Col.* **130** H7
Aribinda, *Burkina Faso* **205** J15
Arica, *Chile* **134** B6
Arid, Cape, *Austral.* **220** K5
Arida, *Japan* **189** Q8
Aridaía, *Gr.* **156** B10
Aridal, *W. Sahara* **200** H4
Aride, Île, *Seychelles* **215** M20
Arifwala, *Pak.* **180** M12
Arīḥā (Jericho), *W. Bank* **172** M6
Arīḥā, *Syr.* **172** E8
Arikaree, river, *Colo., Nebr., U.S.*
 108 L15

Arikawa, *Japan* **189** R3
Arima, *Trin. & Tobago* **123** P22
Arimā, *Braz.* **132** E8
Arimas Range, *Antarctica* **239** F20
Arinagour, *Scot., U.K.* **144** K7
Arinos, river, *Braz.* **132** H10
Arinsal, *Andorra* **162** H2
Arinsal, river, *Andorra* **162** H2
Aripeka, *Fla., U.S.* **99** T7
Ariporo, river, *Col.* **130** F8
Aripuanã, *Braz.* **132** G9
Aripuanã, river, *Braz.* **132** F9
Ariquemes, *Braz.* **132** G8
Arîsh, Wâdi el, *Egypt* **173** Q3
Aristizábal, Cabo, *Arg.* **135** S9
Arivonimamo, *Madagascar* **213** F16
Ariyalur, *India* **183** S6
Arizaro, Salar de, *Arg.* **134** F8
Arizona City, *Ariz., U.S.* **109** V7
Arizona, *U.S.* **109** T7
Arizona, *Arg.* **134** M9
Arizoro, *S. Amer.* **128** J5
Arjeplog, *Sw.* **143** E13
Arjona, *Col.* **130** D5
Arkadak, *Russ.* **158** M12
Arkadelphia, *Ark., U.S.* **102** H7
Arkagala, *Russ.* **161** F19
Arkaig, Loch, *Scot., U.K.* **144** J8
Arkangeliskiy Fracture Zone, *Atl. Oc.*
 249 N7
Arkansas , river, *U.S.* **105** C19
Arkansas City, *Ark., U.S.* **102** J10
Arkansas City, *Kans., U.S.* **107** V8
Arkansas Post National Memorial,
 Ark., U.S. **102** H8
Arkansas Salt Fork , river,
 Kans., Okla., U.S. **105** A15
Arkansas, *U.S.* **102** G8
Arkansas, river, *U.S.* **90** H12
Arkhangel'sk (Archangel), *Russ.*
 158 B10
Arkhangel'skoye, *Russ.* **158** M11
Arkítsa, *Gr.* **156** G11
Arkoí, island, *Gr.* **157** K17
Arkona, Kap, *Ger.* **150** A10
Arkonam, *India* **183** Q7
Arktícheskiy, *Russ.* **161** C14
Arlanc, *Fr.* **149** U11
Arlap, island, *Marshall Is.* **227** H19
Arles, *Fr.* **149** X12
Arlington Heights, *Ill., U.S.* **101** N6
Arlington, *Va., U.S.* **98** C13
Arlington, *Ga., U.S.* **99** P4
Arlington, *Tex., U.S.* **105** G17
Arlington, *S. Dak., U.S.* **106** K8
Arlington, *Oreg., U.S.* **110** F6
Arlington, *Wash., U.S.* **110** B4
Arlis Gap, *Arctic Oc.* **252** H11
Arlis Spur, *Arctic Oc.* **252** J8
Arlit, *Niger* **206** F9
Arlon, *Belg.* **149** N13
Arma, *Kans., U.S.* **107** U10
Armagh, *N. Ire., U.K.* **145** P6
Armant, *Egypt* **202** J12
Armathiá, island, *Gr.* **157** Q18
Armavir, *Russ.* **159** S11
Armavir, *Arm.* **171** E18
Armenia, *Col.* **130** G5
Armenia, *Asia* **171** E19
Armenistís, *Gr.* **157** K16
Armijo, *N. Mex., U.S.* **109** S12
Armin, *N. Mex., U.S.* **109** U14
Armorican Massif, *Eur.* **140** G4
Armour, *S. Dak., U.S.* **106** M6
Armstrong, *Ont., Can.* **87** N15
Armur, *India* **182** M6
Arnaía, *Gr.* **156** D12
Arnauti, Cape, *Cyprus* **162** P5
Arnay-le-Duc, *Fr.* **149** S12
Arnedo, *Sp.* **146** C12
Árnes, *Nor.* **142** G11
Arnett, *Okla., U.S.* **105** B14
Arnhem Land, *Austral.* **220** B9
Arnhem, *Neth.* **148** J13
Arnhem, Cape, *Austral.* **220** B10
Arniel, island, *Marshall Is.* **227** D21
Árnissa, *Gr.* **156** C9
Arno Atoll, island, *Marshall Is.*
 227 D17
Arno, river, *It.* **152** G6
Arnold, *Mo., U.S.* **107** T15
Arnold, *Nebr., U.S.* **107** P5
Arnøy, island, *Nor.* **143** C13
Arnsberg, *Ger.* **150** F6
Arock, *Oreg., U.S.* **110** J9
Aroma, *Sudan* **203** Q14

Arorae, island, *Kiribati* **224** G8
Arorangi, *N.Z.* **229** E22
Arousa, Ría de, *Sp.* **146** C5
Arowhana, peak, *N.Z.* **223** G22
Arpa, river, *Turk.* **171** D17
Arpaçay, *Turk.* **171** D17
Arqalyq, *Kaz.* **178** E11
Arquipélago dos Bijagós,
 Guinea-Bissau **204** L7
Arraias, *Braz.* **133** H14
Arras, *Fr.* **148** M10
Arreau, *Fr.* **149** Y7
Arrecife, *Sp.* **200** F4
Arrecife, *Sp.* **214** P7
Arrée, Montagnes d', *Fr.* **149** Q3
Arrochar, *Scot., U.K.* **144** L9
Arromanches, *Fr.* **149** N7
Arronches, *Port.* **146** H6
Arrou, *Fr.* **149** Q8
Arrow, Lough, *Ire.* **145** P5
Arrowrock Reservoir, *Idaho, U.S.*
 108 G4
Arroyo de la Luz, *Sp.* **146** H7
Arroyo Grande, *Calif., U.S.* **111** W5
Arroyos de Mantua, *Cuba* **118** H4
Ars-en-Ré, *Fr.* **149** T5
Arsen'yev, *Russ.* **161** L21
Arshaly, *Kaz.* **179** E13
Arsiero, *It.* **152** D7
Arsikere, *India* **183** Q5
Arsuk, *Greenland, Den.* **89** T6
Árta, *Gr.* **156** F8
Artamus, island, *Indonesia* **192** K7
Artashat, *Arm.* **171** E18
Artem, *Russ.* **161** M21
Artemisa, *Cuba* **118** G6
Artemivs'k, *Ukr.* **159** Q9
Artemovsk, *Russ.* **161** L13
Artesia, *N. Mex., U.S.* **109** U14
Artesian, *S. Dak., U.S.* **106** L7
Arthur, *Nebr., U.S.* **107** P3
Arthur's Pass, *N.Z.* **223** L18
Arthur's Town, *Bahamas* **119** E13
Arthurstown (Colmán), *Ire.* **145** T6
Artigas, *Uru.* **134** J14
Artigas, station, *Antarctica* **238** C4
Art'ik, *Arm.* **171** D18
Artiste Point, *Grenada* **123** K23
Artois, *Eur.* **140** G5
Artova, *Turk.* **171** E13
Arturo Prat, station, *Antarctica*
 238 C4
Artux, *China* **184** F2
Artvin, *Turk.* **171** D16
Artyk Yuryakh, *Russ.* **161** F17
Artyk, *Turkm.* **178** N7
Aru Islands, *Asia* **168** L15
Aru Passage, *F.S.M.* **226** J9
Aru, *D.R.C.* **209** C19
Aru, Kepulauan, *Indonesia* **193** L19
Arua, *Uganda* **211** Q3
Aruanã, *Braz.* **132** J12
Aruba, island, *Lesser Antil.* **120** G8
Aruboe, island, *Marshall Is.* **227** J20
Arucas, *Sp.* **214** Q5
Arue, *Fr. Polynesia, Fr.* **231** H21
Arumã, *Braz.* **132** E8
Arumpo, *N.S.W., Austral.* **221** X12
Arusha, *Tanzania* **211** U6
Arutanga Passage, *N.Z.* **229** H20
Arutanga, *N.Z.* **229** H20
Arutua, island, *Fr. Polynesia, Fr.*
 231 B18
Aruwimi, river, *D.R.C.* **209** D16
Arvada, *Colo., U.S.* **108** L13
Arvayheer, *Mongolia* **184** D9
Arviat, *Nunavut, Can.* **87** J15
Arvidsjaur, *Sw.* **143** F13
Arvika, *Sw.* **142** K12
Arvon, Mount, *Mich., U.S.* **100** F7
Arxan, *China* **185** D14
Ary, *Russ.* **161** E16
Arys Köli, *Kaz.* **178** H10
Arys, *Kaz.* **178** K12
Arys, *Kaz.* **178** K12
Arzachena, *It.* **152** M4
Arzamas, *Russ.* **158** J12
Arzew, *Alg.* **200** B11
Arzgir, *Russ.* **159** S13
'As 'Êla, *Djibouti* **210** L9
As Sa'an, *Syr.* **172** F9
As Sabkhah, *Syr.* **172** E12
As Sadd al 'Ālī (Aswan High Dam),
 Egypt **202** K12
Aş Şafā, *Syr.* **172** K8
As Safīrah, *Syr.* **172** E9
Aş Şa'īd, *Yemen* **174** P10

Aş Şāliḥiyah, *Syr.* **172** G14
As Sall, *Egypt* **202** E8
As Salmān, *Iraq* **176** K10
As Salţ, *Jordan* **172** M6
As Samāwah (Samawah), *Iraq* **176** J11
As Sanām, *Saudi Arabia* **175** H13
Aş Şawrah, *Saudi Arabia* **174** D4
As Sidr, *Saudi Arabia* **174** G6
As Sidr, *Lib.* **201** E20
As Su'ayyirah, *Saudi Arabia* **174** E11
Aş Şufayyah, *Sudan* **203** R13
As Sukhnah, *Syr.* **172** G11
As Sulaymānīyah *see* Slêmanî, *Iraq* **176** E11
As Sulayyil, *Saudi Arabia* **174** K10
Aş Şulb, *Saudi Arabia* **174** F11
As Sulţān, *Lib.* **201** E19
Aş Şummān, *Saudi Arabia* **174** E10
Aş Şurrah, *Yemen* **174** P10
Aş Şuwār, *Syr.* **172** F14
As Suwaydā', *Syr.* **172** L8
As Suwayḩ, *Oman* **175** H18
As Suwayq, *Oman* **175** G17
As Suways (Suez), *Egypt* **202** F12
Aşa *see* Asha, *Cyprus* **162** N9
Asaba, *Nigeria* **206** M7
Asad Kêlay, *Afghan.* **180** K6
Asad, Buḩayrat al, *Syr.* **172** E10
Asadābād, *Afghan.* **180** H10
Asaga Strait, *American Samoa, U.S.* **230** B9
Aşağı Ağcakənd, *Arm.* **171** D20
Aşağıpınarbaşı, *Turk.* **170** G8
Asahi Dake, *Japan* **188** F14
Asahikawa, *Japan* **188** F13
Asaita, *Eth.* **210** N9
Āsalē, *Eth.* **210** J8
Asan, *S. Korea* **187** P8
Asan, *Guam, U.S.* **230** P5
Asanman, bay, *S. Korea* **187** P8
Asansol, *India* **182** J11
Asatdas, *Guam, U.S.* **230** N6
Asau Harbour, *Samoa* **227** N17
Asau, *Samoa* **227** N17
Āsbe Teferī, *Eth.* **210** M9
Asbury Park, *N.J., U.S.* **96** P11
Ascención, *Bol.* **132** J8
Ascension Bay, *N. Amer.* **84** N7
Ascension Fracture Zone, *Atl. Oc.* **249** Q10
Ascension, *Mex.* **116** C8
Ascension, island, *Af.* **198** L1
Aschaffenburg, *Ger.* **150** H7
Ascira, *Somalia* **210** L15
Ascó, *Sp.* **147** E15
Ascoli Piceno, *It.* **152** J10
'Aseb (Assab), *Eritrea* **210** J10
Asedjrad, *Alg.* **200** J11
Āsela, *Eth.* **210** M8
Asenovgrad, *Bulg.* **155** K14
Aşgabat (Ashgabat), *Turkm.* **178** N7
Ash Fork, *Ariz., U.S.* **109** S6
Ash Grove, *Mo., U.S.* **107** U12
Ash Shabakah, *Iraq* **176** J10
Ash Shāmīyah, *Iraq* **176** K12
Ash Sharawrah, *Saudi Arabia* **174** L11
Ash Shāriqah (Sharjah), *U.A.E.* **175** F15
Ash Sharqāţ, *Iraq* **176** E10
Ash Shawbak, *Jordan* **173** P6
Ash Shiḩr, *Yemen* **174** P12
Ash Shişar, *Oman* **175** L15
Ash Shuqayq, *Saudi Arabia* **174** L7
Ash Shurayf, *Saudi Arabia* **174** F6
Ash Shuwayrif, *Lib.* **201** F18
Asha (Aşa), *Cyprus* **162** N9
Ashāqif, Jabal al, *Jordan* **172** M9
Ashburn, *Ga., U.S.* **99** P6
Ashburton Downs, *W. Aust., Austral.* **221** T2
Ashburton, *N.Z.* **223** M18
Ashburton, river, *Austral.* **220** F2
Ashby Fort, ruin, *St. Kitts & Nevis* **123** C18
Ashchyköl, *Kaz.* **178** H11
Ashdod, *Israel* **172** M5
Ashdown, *Ark., U.S.* **102** J6
Asherton, *Tex., U.S.* **105** N14
Asheville, *N.C., U.S.* **98** H7
Ashford, *Wash., U.S.* **110** D4
Ashford, *Eng., U.K.* **145** V15
Ashgabat *see* Aşgabat, *Turkm.* **178** N7
Ashikaga, *Japan* **189** N12
Ashizuri Misaki, *Japan* **189** S6
Ashkezār, *Iran* **177** J17

Āshkhāneh, *Iran* **177** D19
Ashland, *Me., U.S.* **97** B17
Ashland, *Va., U.S.* **98** E13
Ashland, *Wis., U.S.* **100** F4
Ashland, *Ohio, U.S.* **101** P13
Ashland, *Ky., U.S.* **103** B20
Ashland, *Kans., U.S.* **107** V5
Ashland, *Nebr., U.S.* **107** Q8
Ashland, *Mont., U.S.* **108** E12
Ashland, *Oreg., U.S.* **110** K3
Ashley, *N. Dak., U.S.* **106** H6
Ashley, river, *S.C., U.S.* **98** M10
Ashmore and Cartier Islands, *Austral.* **221** P4
Ashmore Islands, *Austral.* **221** P4
Ashmyany, *Belarus* **158** K4
Ashoro, *Japan* **188** G14
Ashqelon, *Israel* **172** M4
Ashtabula, *Ohio, U.S.* **101** N15
Ashtabula, Lake, *N. Dak., U.S.* **106** G7
Ashton, *Wyo., U.S.* **108** G8
Asia Minor *see* Anatolia, *Turk.* **170** F7
Asifabad, *India* **182** M7
Asika, *India* **182** M10
Asilah, *Mor.* **200** C8
Asinara, Golfo dell', *It.* **152** M3
Asinara, Isola, *It.* **152** M3
Asipovichy, *Belarus* **158** L5
'Asīr, *Saudi Arabia* **174** K8
Aşkale, *Turk.* **171** E15
Askole, *Pak.* **180** G14
Asl, *Egypt* **173** R1
Asmar, *Afghan.* **180** H10
Asmara *see* Asmera, *Eritrea* **210** H7
Asmera (Asmara), *Eritrea* **210** H7
Āsosa, *Eth.* **210** K9
Asoteriba, *Sudan* **203** N15
Aspen, *Colo., U.S.* **108** M11
Aspermont, *Tex., U.S.* **105** F14
Aspres, *Fr.* **149** W14
Aspromonte, range, *It.* **153** R13
Asproválta, *Gr.* **156** C12
Asquith Rise, *Ind. Oc.* **250** H6
Assa, *Mor.* **200** F6
Assab *see* 'Aseb, *Eritrea* **210** J10
Assaikio, *Nigeria* **206** L9
Assaq Atwiy, river, *W. Sahara* **200** H3
Assateague Island National Seashore, *Md., U.S.* **98** D15
Asselborn, *Lux.* **162** G9
Assen, *Neth.* **148** G13
Ássiros, *Gr.* **156** C11
Assis, *Braz.* **132** L12
Assisi, *It.* **152** H8
Assodé, *Niger* **206** E9
Assok-Ngoum, *Gabon* **208** D8
Assos, ruins, *Turk.* **170** E2
Assynt, Loch, *Scot., U.K.* **144** G8
Astakída, island, *Gr.* **157** P18
Astakós, *Gr.* **156** H8
Astana *see* Nur-Sultan, *Kaz.* **178** E12
Astara, *Azerb.* **171** F22
Āstārā, *Iran* **177** C14
Asti, *It.* **152** E3
Astola Island, *Pak.* **181** S4
Astor, *Pak.* **180** G13
Astorga, *Sp.* **146** C8
Astoria, *Oreg., U.S.* **110** E2
Astra, *Arg.* **135** S9
Astrakhan', *Russ.* **159** R15
Astrid Ridge, *Atl. Oc.* **249** Y13
Astrolabe Bay, *P.N.G.* **222** P7
Ástros, *Gr.* **156** K10
Astrupa, *Russ.* **161** D14
Astudillo, *Sp.* **146** D10
Astypálaia, *Gr.* **157** M17
Astypálaia, island, *Gr.* **157** M17
Asuka Station, *Antarctica* **239** B16
Asunción, *Parag.* **132** M10
Asuncion, island, *N. Mariana Is., U.S.* **230** G5
Asunción, Punta, *Arg.* **135** N12
Asuzudo Point, *N. Mariana Is., U.S.* **230** M9
Aswān (Syene), *Egypt* **202** K12
Aswan High Dam *see* As Sadd al 'Ālī, *Egypt* **202** K12
Asyūţ, *Egypt* **202** H11
Aţ Ţafīlah, *Jordan* **173** P6
Aţ Ţā'if, *Saudi Arabia* **174** J6
Aţ Tāj, *Lib.* **201** J22
Aţ Taysīyah, *Saudi Arabia* **174** D9
At Tibnī, *Syr.* **172** F12
Aţ Ţubayq, *Saudi Arabia* **174** C5
Aţ Ţūr, *Egypt* **202** G13
At Turbah, *Yemen* **174** Q8

At-Bashy, *Kyrg.* **179** L15
Ata, island, *Tonga* **227** Q24
Ataa, *Greenland, Den.* **88** L6
Atacama Desert, *S. Amer.* **128** J5
Atacama, Desierto de, *Chile* **134** D7
Atacama, Salar de, *Chile* **134** E8
Atafu, island, *Kiribati* **224** H10
Atakpamé, *Togo* **205** N17
Atalánti, *Gr.* **156** H11
Atalaya de Femés, peak, *Sp.* **214** P7
Atammik, *Greenland, Den.* **89** Q5
Atamyrat (Kerki), *Turkm.* **178** N10
Atapupu, *Indonesia* **193** M15
Atar, *Mauritania* **204** E9
Atarfe, *Sp.* **146** L11
Atas Bogd, peak, *Mongolia* **184** E8
Atascadero, *Calif., U.S.* **111** V5
Atascosa, river, *Tex., U.S.* **105** M16
Atasū, *Kaz.* **179** F13
Atata, island, *Tonga* **227** Q23
Ataţī, *Afghan.* **180** G10
Atávyros, peak, *Gr.* **157** N19
Atbara, *Sudan* **203** P13
Atbara, river, *Sudan* **203** Q14
Atchafalaya Bay, *La., U.S.* **102** Q9
Atchafalaya, river, *La., U.S.* **102** P9
Atchison, *Kans., U.S.* **107** S10
Ateca, *Sp.* **147** E13
Áth Trasna *see* Newmarket, *Ire.* **145** T4
Ath, *Belg.* **148** L11
Athabasca, *Alta., Can.* **86** L10
Athabasca, Lake, *Alta., Sask., Can.* **86** J12
Athabasca, river, *Sask., Can.* **86** K11
Athamánio, *Gr.* **156** F9
Athboy (Baile Átha Buí), *Ire.* **145** Q6
Athena, *Oreg., U.S.* **110** E8
Athens *see* Athína, *Gr.* **156** J12
Athens, *Ga., U.S.* **98** L6
Athens, *Ohio, U.S.* **101** S14
Athens, *Ala., U.S.* **103** G15
Athens, *Tenn., U.S.* **103** F18
Athens, *Tex., U.S.* **105** H19
Atherton Tableland, *Austral.* **220** D13
Athfālah, *Saudi Arabia* **174** K7
Athi River, *Kenya* **211** T6
Athi, river, *Kenya* **211** T7
Athienou, *Cyprus* **162** P9
Athína (Athens), *Gr.* **156** J12
Athlone (Baile Átha Luain), *Ire.* **145** R5
Athol Island, *Bahamas* **122** E3
Athos Range, *Antarctica* **239** E20
Athos, Mount *see* Ágion Óros, *Gr.* **157** D13
Áthos, peak, *Gr.* **157** D13
Ati, *Chad* **207** H14
Atiak, *Uganda* **211** Q3
Atico, *Peru* **132** J4
Atikokan, *Ont., Can.* **87** P15
Atimoono, island, *N.Z.* **229** H24
Atina, *It.* **152** L10
Atitue, *Fr. Polynesia, Fr.* **231** H21
Atiu, island, *Cook Is., N.Z.* **224** K12
Atka Ice Shelf, *Antarctica* **238** A12
Atka Island, *Alas., U.S.* **112** P6
Atka, *Alas., U.S.* **112** P6
Atka, *Russ.* **161** F20
Atkarsk, *Russ.* **158** M13
Atkins, *Ark., U.S.* **102** F8
Atkinson, *Nebr., U.S.* **107** N6
Atlanta, *Ga., U.S.* **98** L5
Atlanta, *Mich., U.S.* **100** H11
Atlanta, *Tex., U.S.* **105** F20
Atlantic Beach, *N.C., U.S.* **98** J14
Atlantic Beach, *Fla., U.S.* **99** Q9
Atlantic City, *N.J., U.S.* **96** Q10
Atlantic City, *Wyo., U.S.* **108** J10
Atlantic-Indian Basin, *Atl. Oc., Ind. Oc.* **249** Y10
Atlantic-Indian Ridge, *Atl. Oc.* **249** W13
Atlantic, *N.C., U.S.* **98** J14
Atlantic, *Iowa, U.S.* **107** P10
Atlantis Fracture Zone, *Atl. Oc.* **248** K7
Atlantis II Fracture Zone, *Ind. Oc.* **250** N8
Atlas Mountains, *Alg., Mor.* **200** E8
Atlas Saharien, *Alg.* **200** D11
Atlin, *B.C., Can.* **86** G8
'Atlit, *Israel* **172** L5
Atmautluak, *Alas., U.S.* **112** J12
Atmore, *Ala., U.S.* **103** N14
Atoka, *Okla., U.S.* **105** E18

Atoll Manihi, *Fr. Polynesia, Fr.* **231** A18
Atoll Rangiroa, *Fr. Polynesia, Fr.* **231** A17
Atome, *Angola* **208** M10
Atoyac, *Mex.* **116** K11
Atrak, river, *Iran* **177** D18
Atrato, river, *Col.* **130** E5
Atrek, river, *Turkm.* **178** N5
Atri, *It.* **152** J10
Atsutoko, *Japan* **188** G15
Attalla, *Ala., U.S.* **103** H16
Attapu, *Laos* **190** M12
Attawapiskat, *Ont., Can.* **87** M17
Attent, river, *Lux.* **162** J9
Attica, *Ind., U.S.* **101** Q8
Attica, *Kans., U.S.* **107** V6
Attigu, *S. Sudan* **203** Y12
Attleboro, *Mass., U.S.* **97** L14
Attu Island, *Alas., U.S.* **112** K1
Attu, *Greenland, Den.* **89** N5
Atuel, river, *Arg.* **134** M8
Atuona, *Fr. Polynesia, Fr.* **231** D14
Atura, *Uganda* **211** R3
Atutahi, island, *N.Z.* **229** M22
Atwater, *Calif., U.S.* **111** S5
Atwood *see* Samana Cay, *Bahamas* **119** G15
Atwood, *Kans., U.S.* **107** R4
Atyraū, *Kaz.* **178** G5
Au Sable Point, *Mich., U.S.* **100** F9
Au Sable, *Mich., U.S.* **100** J12
Au Sable, river, *Mich., U.S.* **100** J11
Aua Island, *P.N.G.* **222** M6
Aūa, *American Samoa, U.S.* **230** D7
Auas Mountains, *Namibia* **212** G4
'Au'au Channel, *Hawaii, U.S.* **115** J18
Aubagne, *Fr.* **149** Y13
Aubenas, *Fr.* **149** W12
Aubert de la Rüe Seamounts, *Ind. Oc.* **254** C9
Aubiet, *Fr.* **149** X8
Aubinyà, *Andorra* **162** K2
Aubrac, Monts d', *Fr.* **149** W11
Aubry Lake, *N.W.T., Can.* **86** E10
Auburn, *N.Y., U.S.* **96** J8
Auburn, *Me., U.S.* **97** G15
Auburn, *Ill., U.S.* **101** R5
Auburn, *Ind., U.S.* **101** P10
Auburn, *Ala., U.S.* **103** L17
Auburn, *Nebr., U.S.* **107** R9
Auburn, *Wash., U.S.* **110** C4
Auburn, *Calif., U.S.* **111** Q5
Auburndale, *Fla., U.S.* **99** U8
Aubusson, *Fr.* **149** U10
Auce, *Latv.* **143** M16
Auch, *Fr.* **149** X8
Auchel, *Fr.* **148** L10
Auchi, *Nigeria* **206** M7
Auckland Islands, *N.Z.* **224** Q7
Auckland, *N.Z.* **223** E20
Audience Hall, *Vatican City* **163** R16
Audincourt, *Fr.* **149** R15
Audubon, *Iowa, U.S.* **107** P10
Augathella, *Qnsld., Austral.* **221** U13
Augsburg, *Ger.* **150** K8
August Cay, *Bahamas* **122** J4
Augusta, *Me., U.S.* **97** G16
Augusta, *Ga., U.S.* **98** L8
Augusta, *Wis., U.S.* **100** J3
Augusta, *Ark., U.S.* **102** G9
Augusta, *Ky., U.S.* **103** A18
Augusta, *Kans., U.S.* **107** U8
Augusta, *Mont., U.S.* **108** C7
Augusta, *It.* **153** T12
Augusta, *W. Aust., Austral.* **221** Y2
Augusta, Golfo di, *It.* **153** T12
Augustów, *Pol.* **151** B17
Augustus Downs, *Qnsld., Austral.* **221** R11
Augustus Island, *Austral.* **220** C5
Augustus, Mount, *Austral.* **220** G2
Aukena, island, *Fr. Polynesia, Fr.* **231** Q17
Auki, *Solomon Is.* **228** E10
Aukland Escarpment, *Pac. Oc.* **246** R10
Aukland Islands Shelf, *Pac. Oc.* **255** G19
Auld, Lake, *Austral.* **220** F5
Auletta, *It.* **152** M12
Āūlīeköl, *Kaz.* **178** D10
Aulong, island, *Palau* **226** H4
Auluptagel, island, *Palau* **226** H5
Auna, *Nigeria* **206** K6
Aunu'u, *American Samoa, U.S.* **230** D8

Aunuu, island, *American Samoa, U.S.* **230** D8
Auob , river, *Angola, S. Af.* **212** J5
Auponhia, *Indonesia* **193** K15
Aur Atoll, island, *Marshall Is.* **227** D17
Aur, island, *Malaysia* **191** U10
Auraiya, *India* **182** H7
Aurangabad, *India* **182** J9
Aurangabad, *India* **182** M4
Auray, *Fr.* **149** R4
Aurich, *Ger.* **150** C6
Aurignac, *Fr.* **149** Y8
Aurillac, *Fr.* **149** V10
Aurisina, *It.* **152** D10
Aurora Bank, *Ind. Oc.* **254** B11
Aurora Canyon, *Atl. Oc.* **249** X6
Aurora, *Ill., U.S.* **101** N6
Aurora, *Ind., U.S.* **101** S10
Aurora, *Minn., U.S.* **106** F12
Aurora, *Mo., U.S.* **107** V11
Aurora, *Nebr., U.S.* **107** Q7
Aurora, *Colo., U.S.* **108** M13
Aurukun, *Qnsld., Austral.* **221** P12
Aus, *Namibia* **212** J4
Ausa, river, *San Marino* **163** G17
Auschwitz *see* Oświęcim, *Pol.* **151** H15
Austin, *Pa., U.S.* **96** L6
Austin, *Ind., U.S.* **101** T9
Austin, *Tex., U.S.* **105** K16
Austin, *Minn., U.S.* **106** L12
Austin, *Nev., U.S.* **111** P9
Austin, Lake, *Austral.* **220** H3
Austral Downs, *N.T., Austral.* **221** S10
Austral Fracture Zone, *Pac. Oc.* **247** M17
Austral Islands (Tubuai Islands), *Fr. Polynesia, Fr.* **231** E17
Austral Seamounts, *Pac. Oc.* **247** M14
Australian Alps, *Austral.* **220** M13
Australian Capital Territory, *Austral.* **221** Y14
Australian-Antarctic Discordance, *Ind. Oc.* **251** R19
Austria, *Eur.* **150** L10
Austriahütte, site, *Aust.* **150** L10
Austvågøy, island, *Nor.* **142** D12
Autazes, *Braz.* **132** D9
Auti, *Fr. Polynesia, Fr.* **231** K17
Autlán, *Mex.* **116** H9
Autun, *Fr.* **149** S12
Auxerre, *Fr.* **149** R11
Auxonne, *Fr.* **149** S13
Auzances, *Fr.* **149** T10
Ava Fonuaunga, bay, *Tonga* **227** H23
Ava Mata Mata Vika, bay, *Tonga* **227** M23
Ava, *Mo., U.S.* **107** V13
Avaavaroa Passage, *N.Z.* **229** E23
Avaha, island, *Solomon Is.* **228** B9
Avakubi, *D.R.C.* **209** D18
Avala, peak, *Serb.* **154** F10
Avalau, island, *Tuvalu* **228** Q10
Avallon, *Fr.* **149** R11
Avalon Peninsula, *Nfld. & Lab., Can.* **87** M24
Avalon, *Calif., U.S.* **111** Y8
Avalon, Lake, *N. Mex., U.S.* **109** V14
Avanavero, *Suriname* **131** F16
Avanos, *Turk.* **170** F10
Avaqqat Kangerlua, fjord, *Greenland, Den.* **89** T8
Avarua Harbour, *N.Z.* **229** D23
Avarua, *N.Z.* **229** D23
Avatele Bay, *N.Z.* **229** Q21
Avatele, *N.Z.* **229** Q21
Avatoru, *Fr. Polynesia, Fr.* **231** M17
Avatiu Harbour, *N.Z.* **229** D23
Avatiu, *N.Z.* **229** D23
Avaz, *Iran* **177** M17
Āvāz, *Iran* **177** G21
Ávdira, ruin, *Gr.* **157** C14
Avea, island, *Fiji* **226** P8
Aveiro, *Braz.* **132** D11
Aveiro, *Port.* **146** F5
Āvej, *Iraq* **177** F14
Avellino, *It.* **152** M11
Avenal, *Calif., U.S.* **111** U6
Avera, *Fr. Polynesia, Fr.* **231** G17;K17
Aversa, *It.* **152** M11
Aves Ridge, *Atl. Oc.* **248** M5
Aves, island, *Lesser Antil.* **121** D14
Aves, Islas de, *Venez.* **130** B10
Avesta, *Sw.* **143** K13
Avezzano, *It.* **152** K9
Avgó, island, *Gr.* **156** N11

Avgonísi, island, Gr. **157** N17
Aviano, It. **152** D8
Aviator Glacier Tongue, Antarctica **239** P14
Avignon, Fr. **149** X13
Ávila, Sp. **146** F10
Avilés, Sp. **146** A8
Avire, Vanuatu **229** D15
Avissawella, Sri Lanka **183** U7
Avita, Solomon Is. **228** M6
Avlonári, Gr. **157** H13
Avoca, Iowa, U.S. **107** P9
Avon Park, Fla., U.S. **99** U9
Avon, N.C., U.S. **98** H15
Avon, river, Eng., U.K. **145** T12
Avon, river, Eng., U.K. **145** U11
Avondale, Ariz., U.S. **109** U6
Avranches, Fr. **149** P6
Avsa, Russ. **161** H13
Avu Avu, Solomon Is. **228** F10
Awa Shima, Japan **188** L11
Awaday, W. Sahara **200** K4
Awar, P.N.G. **222** N7
Āwarē, Eth. **210** M11
Awarua Bay, N.Z. **223** M16
Āwasa, Eth. **211** N7
Āwash, Eth. **210** M8
Awash, river, Eth. **210** L8
Awat, China **184** F3
Awbārī (Ubari), Lib. **201** H17
Awdheegle, Somalia **211** R11
Awe, Nigeria **206** L9
Awe, Loch, Scot., U.K. **144** K8
Aweil, S. Sudan **203** V9
Awfist, W. Sahara **200** H3
Awgu, Nigeria **206** M8
Awin, island, P.N.G. **222** M6
Awio, P.N.G. **222** P9
Awjilah, Lib. **201** F21
Awka, Nigeria **206** M7
Awsard, W. Sahara **200** J3
Ax-les-Thermes, Fr. **149** Z9
Axarfjörardjúp, valley, Arctic Oc. **253** L20
Axel Heiberg Island, Nunavut, Can. **87** B15
Axim, Ghana **205** Q15
Axinim, Braz. **132** E9
Axiós, river, Gr. **156** N13
Axtell, Nebr., U.S. **107** Q6
Ay Khanom, ruins, Afghan. **180** F9
Ay', Aruba, Neth. **123** Q17
Aya Bentih, Eth. **210** M12
Ayabe, Japan **189** P8
Ayacucho, Peru **132** H4
Ayacucho, Arg. **135** N13
Ayakkum Hu, China **184** G5
Ayaköz, Kaz. **179** F16
Ayamonte, Sp. **146** L6
Ayan, Russ. **161** H20
Ayan, Russ. **161** J15
Ayancık, Turk. **170** C10
Ayangba, Nigeria **206** L8
Ayapel, Col. **130** E6
Ayaş, Turk. **170** E8
Ayasergi see Ayios Seryios, Cyprus **162** N10
Ayat, river, Kaz. **178** D9
Ayaviri, Peru **132** J5
Aydar Kŭl, Uzb. **178** L11
Ayden, N.C., U.S. **98** H13
Aydere, Turkm. **178** M6
Aydın, Turk. **170** G3
Aydıncık, Turk. **170** J8
Āyelu Terara, peak, Eth. **210** L9
Ayerbe, Sp. **147** C14
Ayers Rock see Uluru, peak, N.T. Austral. **221** U8
Ayeyarwady (Irrawaddy), river, Myanmar **190** G5
Ayeyarwady, Mouths of the, Myanmar **190** L4
Ayina, river, Gabon **208** C9
Ayios Amvrosios (Aykuruş), Cyprus **162** N9
Ayios Seryios (Ayasergi), Cyprus **162** N10
Ayios Theodhoros (Çayırova), Cyprus **162** N10
Āykel, Eth. **210** K6
Aykhal, Russ. **161** G15
Aykino, Russ. **158** D14
Aykuruş see Ayios Amvrosios, Cyprus **162** N9
Aylesbury, Eng., U.K. **145** U13
Ayllón, Sp. **146** E11

Aylmer Lake, N.W.T. Can. **86** H12
'Ayn al 'Arab (Kobanê), Syr. **172** C10
Ayní, Taj. **178** M11
'Aynūnah, Saudi Arabia **174** D4
Ayod, S. Sudan **203** V11
Ayon Island, Asia **168** B11
Ayon, Russ. **161** C19
Ayorou, Niger **206** G4
'Ayoûn el 'Atroûs, Mauritania **204** H10
Aypolovo, Russ. **160** J11
Ayr, Scot., U.K. **145** M9
Ayr, Qnsld. Austral. **221** S14
Ayrancı, Turk. **170** H9
Ayre, Point of, Isle of Man, U.K. **145** P9
Āysha, Eth. **210** L10
Äyteke Bī, Kaz. **178** H9
Aytos, Bulg. **155** J17
Ayutla, Mex. **116** K11
Ayutthaya, Thai. **190** M8
Ayvacık, Turk. **170** E2
Ayvaj, Taj. **178** N11
Ayvalık, Turk. **170** E2
Az Zabadānī, Syr. **172** J7
Az Zabū, Egypt **202** G10
Az Za'farānah, Egypt **202** G12
Aẕ Ẕāhirīyah, W. Bank **173** N5
Az Zaqāzīq, Egypt **202** F11
Az Zarqā', Jordan **172** M7
Az Zāwiyah, Lib. **201** D17
Az Zawr, Kuwait **174** D11
Az Zaytūn, Egypt **202** G8
Az Zubayr, Iraq **177** K13
Āzāb, Afghan. **180** J4
Azahar, Costa del, Sp. **147** G16
Azamgarh, India **182** J7
Azaouâd, Mali **205** G14
Azaouagh, Niger **206** F7
Azapa, Chile **134** B6
Azara, Nigeria **206** L9
Azare, Nigeria **206** J9
Azawak, river, Af. **205** H18
A'zāz, Syr. **172** D8
Azdavay, Turk. **170** C8
Azefal Dunes, Af. **198** E1
Azefal, Mauritania, Mor. **204** E7
Azélik, Niger **206** F8
Azerbaijan, Asia **171** D21
Azincourt, Fr. **148** M9
Azle, Tex., U.S. **105** G17
Azogues, Ecua. **130** L4
Azores Plateau, Atl. Oc. **248** J9
Azores-Biscay Rise, Atl. Oc. **248** J10
Azov Upland, Eur. **140** H12
Azov, Russ. **159** R10
Azov, Russ., Ukr. **159** S9
Azrou, Mor. **200** D8
Aztec Ruins National Monument, N. Mex. U.S. **109** Q10
Aztec, N. Mex., U.S. **109** Q11
Azua, Dom. Rep. **119** M18
Azuaga, Sp. **146** J8
Azuero Peninsula, N. Amer. **84** R8
Azuero, Península de, Pan. **117** P21
Azul, Arg. **134** M12
Azurduy, Bol. **132** L8
'Azzān, Yemen **174** P11

B

Bà Rịa, Vietnam **191** Q11
Ba, Fiji **226** P5
Baabda, Leb. **172** J6
Baalbek, Leb. **172** H7
Baarbacadle, Somalia **210** M14
Baardheere, Somalia **211** R10
Baarn, Neth. **148** J12
Ba'ashom, Sudan **203** S11
Bab el Mandeb, Ind. Oc. **168** J2
Baba Burnu, Turk. **170** C7
Baba Burnu, Turk. **170** E2
Baba Dağ, peak, Turk. **170** J4
Bābā, Kōh-e, Afghan. **180** H7
Bababé, Mauritania **204** H8
Bababdağ, peak, Azerb. **171** C22
Babaeski, Turk. **170** C3
Babahoyo, Ecua. **130** L3
Babana, Nigeria **206** K6
Babanūsah, Sudan **203** T9
Babar Islands, Asia **168** M15

Babar, island, Indonesia **193** M17
Babar, Kepulauan, Indonesia **193** M17
Babati, Tanzania **211** V6
Babayevo, Russ. **158** F8
Babayurt, Russ. **159** T15
Babe Island, Guam, U.S. **230** R5
Babeldaob (Babelthaup), island, Palau **226** G5
Babelthuap see Babeldaob, island, Palau **226** G5
Babelthuap, island, Palau **226** B4
Babi, island, Indonesia **192** H4
Babinda, Qnsld. Austral. **221** R13
Babin Nos, peak, Bulg., Serb. **154** G12
Babo, Indonesia **193** K18
Bābol, Iran **177** E16
Baboquivari Peak, Ariz., U.S. **109** W6
Babruysk, Belarus **158** L5
Babu see Hezhou, China **185** N14
Babusar Pass, Pak. **182** C4
Babushkina, Russ. **158** F11
Babuyan Channel, Philippines **193** B13
Babuyan Islands, Philippines **193** A14
Babuyan, island, Philippines **193** A14
Babylon, ruin, Iraq **176** H10
Bạc Liêu, Vietnam **191** Q11
Bắc Ninh, Vietnam **190** H11
Bacaadweyn, Somalia **211** N13
Bacabal, Braz. **133** E14
Bacan, island, Indonesia **193** J16
Bacău, Rom. **155** C16
Baccarat, Fr. **149** P14
Baceno, It. **152** C3
Bachelor, Mount, Oreg., U.S. **110** H4
Bachu, China **184** F2
Back, river, Nunavut, Can. **87** G13
Bačka Palanka, Serb. **154** E9
Bačka Topola, Serb. **154** D9
Bačka, region, Serb. **154** E9
Backbone Mountain, Md., U.S. **98** C10
Bacolet, Grenada **123** L23
Bacolod City, Philippines **193** E14
Bacup, Eng. U.K. **145** Q11
Bad Axe, Mich., U.S. **100** K12
Bad Hersfeld, Ger. **150** F7
Bad Homburg, Ger. **150** G6
Bad Ischl, Aust. **150** L10
Bad Kissengen, Ger. **150** G7
Bad Mergentheim, Ger. **150** H7
Bad Reichenhall, Ger. **150** L10
Bad Säckingen, Ger. **150** L5
Bad Tölz, Ger. **150** L8
Bad, river, S. Dak., U.S. **106** L4
Badain Jaran Shamo, China **184** G9
Badajoz, Sp. **146** J7
Badalona, Sp. **147** E17
Baddeck, N.S., Can. **87** N22
Baddo, river, Pak. **181** P5
Baden Park, N.S.W. Austral. **221** X12
Baden-Baden, Ger. **150** J6
Baden, Aust. **150** K12
Baden, Eritrea **210** G7
Baderna, Croatia **154** E4
Bādgah, Afghan. **180** H6
Badger, Minn., U.S. **106** E9
Badin, Pak. **181** S8
Badlands National Park, S. Dak., U.S. **106** L2
Badlands, N. Dak., U.S. **106** G2
Badogo, Mali **204** L11
Badoumbé, Mali **204** J10
Badr Ḩunayn, Saudi Arabia **174** G6
Badu Island, Austral. **220** A12
Baduein, Eth. **211** N10
Badulla, Sri Lanka **183** U8
Badupi, Myanmar **190** G3
Baegado, island, S. Korea **187** N6
Baegamsan, peak, S. Korea **186** L9
Baengnyeongdo, island, S. Korea **186** M4
Baetic Mountains, Eur. **140** L2
Bafa Gölü, Turk. **170** G3
Bafarara, Sudan **204** H9
Bafatá, Guinea-Bissau **204** K7
Bafing, river, Af. **198** G2
Bafing, river, Guinea, Mali **204** K10
Bafoulabé, Mali **204** J10
Bafoussam, Cameroon **206** M9
Bāfq, Iran **177** J18
Bafra Burnu, Turk. **170** C11

Bafra, Turk. **170** C11
Bāft, Iran **177** L19
Bafuka, D.R.C. **209** B18
Bafwabalinga, D.R.C. **209** D17
Bafwasende, D.R.C. **209** D17
Baga, island, Solomon Is. **228** D6
Bagabag, island, P.N.G. **222** N7
Bagamoyo, Tanzania **211** W7
Bagan Serai, Malaysia **191** T8
Bagansiapiapi, Indonesia **192** H6
Bagaroua, Niger **206** G6
Bagata, D.R.C. **208** G12
Bagdad, Ariz., U.S. **109** S6
Bagdarin, Russ. **161** K16
Bagé, Braz. **132** Q11
Bages et de Sigean, Étang de, Fr. **149** Y10
Baggs, Wyo., U.S. **108** K11
Bagh a' Chaisteil (Castlebay), Scot., U.K. **144** J6
Baghdād, Iraq **176** G10
Bagheria, It. **153** S9
Bāghīn, Iran **177** K19
Baghlān, Afghan. **180** G8
Bāghrān Khūlah, Afghan. **180** J5
Bağırpaşa Dağı, peak, Turk. **171** F14
Bağışlı, Turk. **171** G18
Bagley, Minn., U.S. **106** F9
Bagnères-de-Bigorre, Fr. **149** Y7
Bagnères-de-Luchon, Fr. **149** Y8
Bago Yoma, Myanmar **190** H5
Bago, Myanmar **190** K5
Bagoé, river, Côte d'Ivoire, Mali **204** L12
Bagrām, Afghan. **180** H9
Bagrimi, Afghan. **180** H9
Baguio, Philippines **193** B13
Bagzane, peak, Niger **206** F9
Bahamas, The, N. Amer. **118** E12
Baharic-Ċagħag, Malta **163** J22
Bahau, Malaysia **191** U9
Bahawalnagar, Pak. **180** M12
Bahawalpur, Pak. **181** N11
Bahçe, Turk. **170** H11
Bahçesaray, Turk. **171** G17
Bahdur Island, Sudan **203** P16
Bäherden, Turkm. **178** M6
Bahía Banks, Ecua. **130** Q2
Bahía Blanca Canyon, Atl. Oc. **249** U6
Bahía Blanca, Arg. **135** N11
Bahía de Caráquez, Ecua. **130** K3
Bahía de Darwin, Ecua. **130** P4
Bahía Fañunchuluyan, N. Mariana Is. **230** H9
Bahia Honda Key, Fla., U.S. **99** Y9
Bahía Isabel, Ecua. **130** Q2
Bahia Laulau, N. Mariana Is. U.S. **230** J8
Bahía Laura, Arg. **135** U9
Bahia Seamounts, Atl. Oc. **249** Q9
Bahía, Islas de la, Hond. **117** K17
Bahías, Cabo dos, Arg. **135** S9
Bahir Dar, Eth. **210** K7
Bahlā, Oman **175** H17
Bahr Aouk, Cen. Af. Rep., Chad **207** J16
Bahr el 'Arab, S. Sudan, Sudan **203** U8
Bahr el Ghazal, S. Sudan **203** U10
Bahr el Jebel (Mountain Nile), S. Sudan **203** W11
Bahr Ergig, Chad **207** J13
Bahr ez Zaraf, S. Sudan **203** V11
Bahr Kameur, Cen. Af. Rep. **207** K15
Bahr Kéita (Doka), Chad **207** K14
Bahraich, India **182** G8
Bahrain, Asia **175** F13
Bāhū Kalāt, Iran **177** N23
Bai Bung, Point, Asia **168** K11
Baia de Aramă, Rom. **154** F12
Baía de Fernão Veloso, Mozambique **213** D13
Baía de Maputo, Af. **198** P9
Baía de Pemba, Mozambique **213** C13
Baía do Bengo, Angola **208** K9
Baía dos Tigres, Angola **208** Q9
Baía Farta, Angola **208** M9
Baia Mare, Rom. **155** B13
Baia Sprie, Rom. **155** B13
Baïbokoum, Chad **206** L12
Baicheng, China **184** E3
Baicheng, China **185** D15
Băicoi, Rom. **155** E15
Baidoa see Baydhabo, Somalia **211** Q10
Baie Chasseloup, New Caledonia, Fr. **231** P21
Baie Chavé, Fr. Polynesia, Fr. **230** E11

Baie d' Ouvéa, New Caledonia, Fr. **231** P22
Baie d'Ambaro, Madagascar **213** C17
Baie d'Antongila, Madagascar **213** E17
Baie d'Avea, Fr. Polynesia, Fr. **231** L14
Baie d'Avera, Fr. Polynesia, Fr. **231** K17
Baie de Ampalaza, Madagascar **213** J14
Baie de Bombetoka, Madagascar **213** D16
Baie de Canala, New Caledonia, Fr. **231** Q22
Baie de Cook, Fr. Polynesia, Fr. **231** H20
Baie de Diégo-Suarez, Madagascar **213** C17
Baie de Faanui, Fr. Polynesia, Fr. **231** G14
Baie de Faaroa, Fr. Polynesia, Fr. **231** H17
Baie de Faie, Fr. Polynesia, Fr. **231** K14
Baie de Fayaoué, New Caledonia, Fr. **231** P23
Baie de Fitii, Fr. Polynesia, Fr. **231** K14
Baie de Fort-de-France, Martinique, Fr. **123** F22
Baie de Gatayake, Fr. Polynesia, Fr. **231** Q17
Baie de Gouaro, New Caledonia, Fr. **231** Q21
Baie de Haamene, Fr. Polynesia, Fr. **231** Q17
Baie de Hienghéne, New Caledonia, Fr. **231** P21
Baie de la Mahajamba, Madagascar **213** D16
Baie de l'Ouest, Wallis & Futuna Is., Fr. **230** Q8
Baie de Mahuti, Fr. Polynesia, Fr. **231** K14
Baie de Marigot, St. Martin, Fr. **123** B14
Baie de Maroe, Fr. Polynesia, Fr. **231** K14
Baie de Mata-Utu, Wallis & Futuna Is., Fr. **230** R9
Baie de Matavai, Fr. Polynesia, Fr. **231** H21
Baie de Moerai, Fr. Polynesia, Fr. **231** K17
Baie de Mondah, Eq. Guinea, Gabon **206** Q9
Baie de Mondah, Gabon **208** D7
Baie de Mua, Wallis & Futuna Is., Fr. **230** R9
Baie de Narinda, Madagascar **213** D16
Baie de Néhoué, New Caledonia, Fr. **231** N20
Baie de Poofai, Fr. Polynesia, Fr. **231** H14
Baie de Sahamalaza, Madagascar **213** D16
Baie de Saint Augustin, Madagascar **213** H14
Baie de Sandal, New Caledonia, Fr. **231** P23
Baie de Taravao, Fr. Polynesia, Fr. **231** J23
Baie de'Ampasindava, Madagascar **213** D17
Baie des Traîtres, Fr. Polynesia, Fr. **231** E14
Baie d'Hane, Fr. Polynesia, Fr. **230** E11
Baie d'Opunohu, Fr. Polynesia, Fr. **231** H20
Baie du Chateaubriand, New Caledonia, Fr. **231** P23
Baie du Galion, Martinique, Fr. **123** E23
Baie du Lévrier, Mauritania **204** E6
Baie du Prony, New Caledonia, Fr. **231** Q23
Baie Haamaire, Fr. Polynesia, Fr. **231** G14
Baie Hanahe, Fr. Polynesia, Fr. **230** D11
Baie Hanahupe, Fr. Polynesia, Fr. **231** D14
Baie Hanaiapa, Fr. Polynesia, Fr. **231** D14
Baie Hanaiapaoa, Fr. Polynesia, Fr. **231** D14
Baie Hanamenu, Fr. Polynesia, Fr. **231** D13
Baie Hapatoni, Fr. Polynesia, Fr. **231** E13
Baie Katohau, Fr. Polynesia, Fr. **230** D11
Baie Orientale, St. Martin, Fr. **123** B15

Baie Ouinné, *New Caledonia, Fr.* **231** Q23
Baie Puamau (Perigot), *Fr. Polynesia, Fr.* **231** D15
Baie Rouge, *St. Martin, Fr.* **123** B13
Baie-Mahault, *Guadeloupe, Fr.* **123** E14
Baijiantan, *China* **184** D5
Baikal, Lake *see* Baykal, Ozero, *Russ.* **161** L16
Baikha, *Russ.* **161** H13
Baikonur Cosmodrome, *Kaz.* **178** H10
Baikonur *see* Bayqongyr, *Kaz.* **178** H9
Baikunthpur, *India* **182** K8
Baile an Chaisleáin *see* Castletownshend, *Ire.* **145** U3
Baile an Róba *see* Ballinrobe, *Ire.* **145** P3
Baile Átha Buí *see* Athboy, *Ire.* **145** Q6
Baile Átha Cliath *see* Dublin, *Ire.* **145** R7
Baile Átha Fhirdhia *see* Ardee, *Ire.* **145** Q6
Baile Átha Luain *see* Athlone, *Ire.* **145** R5
Baile Bhuirne (Ballyvourney), *Ire.* **145** T3
Baile Chruaich *see* Ballycroy, *Ire.* **145** P3
Baile Mhistéala *see* Mitchelstown, *Ire.* **145** T4
Baile na gCros *see* Castlepollard, *Ire.* **145** Q6
Bailén, *Sp.* **146** K11
Băileşti, *Rom.* **155** G13
Bailey Ice Stream, *Antarctica* **238** E11
Bailey Town, *Bahamas* **122** B11
Baileys Harbor, *Wis., U.S.* **100** H8
Bailique, *Braz.* **133** C13
Baillif, *Guadeloupe, Fr.* **123** G14
Bailundo, *Angola* **208** M11
Baily Group *see* Haha Jima Rettō, *Japan* **189** X14
Baily or Coffin Group *see* Haha Jima Rettō, *Japan* **224** C4
Baima, *China* **184** K9
Baimuru, *P.N.G.* **222** Q7
Bain Fracture Zone, *Ind. Oc.* **250** R3
Bainbridge, *Ga., U.S.* **99** Q5
Bainet, *Haiti* **119** N16
Baing, *Indonesia* **193** N13
Baingear (Bangor), *Ire.* **145** P3
Bainville, *Mont., U.S.* **108** B13
Bainyik, *P.N.G.* **222** N6
Baiona, *Sp.* **146** C5
Bā'ir, *Jordan* **173** P8
Baird Inlet, *Alas., U.S.* **112** J11
Baird Mountains, *Alas., U.S.* **113** D14
Baird Seamounts, *Pac. Oc.* **247** H13
Baird, *Tex., U.S.* **105** G15
Bairiki *see* Tarawa, *Kiribati* **227** R14
Bairin Zuoqi, *China* **185** E14
Bairnsdale, *Vic., Austral.* **221** Z13
Baissac Bank, *Ind. Oc.* **250** J8
Baiti, *Nauru* **227** M14
Baixo Longa, *Angola* **208** P12
Baixo, Ilhéu de, *Madeira, Port.* **214** L4
Baiyer River, *P.N.G.* **222** P6
Baiyin, *China* **184** H10
Baja California Seamount Province, *Pac. Oc.* **247** F17
Baja California, *Mex.* **116** D6
Baja, *Hung.* **151** N15
Bajag, *India* **182** K8
Bājgīrān, *Iran* **177** D20
Bājil, *Yemen* **174** N8
Bajmok, *Serb.* **154** D9
Bajo, *Indonesia* **193** J13
Bakaba, *Fiji* **227** L13
Bakala, *Cen. Af. Rep.* **207** M15
Bakali, river, *D.R.C.* **208** H12
Bakaoré, *Chad* **207** G15
Bakchar, *Russ.* **160** K12
Bakel, *Senegal* **204** J9
Baker City, *Oreg., U.S.* **110** G8
Baker Foreland, *Nunavut, Can.* **87** H15
Baker Island, *Pac. Oc.* **224** G9
Baker Lake, *Nunavut, Can.* **87** H14
Baker Lake, *Austral.* **220** G6
Baker, *Mont., U.S.* **108** D13
Baker, *Calif., U.S.* **111** V11
Baker, *Nev., U.S.* **111** Q13
Baker, Mount, *Wash., U.S.* **110** A4
Baker, river, *Chile* **135** U6
Bakersfield, *Calif., U.S.* **111** V7
Bakhchysaray, *Russ.* **159** T7

Bakhmach, *Ukr.* **159** N7
Bakhta, *Russ.* **161** H13
Bākhtarān *see* Kermānshāh, *Iraq* **176** G12
Baki (Baku), *Azerb.* **171** D23
Bakin Birji, *Niger* **206** H9
Bakırdağı, *Turk.* **170** G11
Bakkafjörður, *Ice.* **142** E4
Bakkaflóadjúp, valley, *Arctic Oc.* **253** K21
Bakkageröi, *Ice.* **142** E4
Bako, *Côte d'Ivoire* **204** M12
Bako, *Eth.* **210** M6
Bakony, range, *Hung.* **151** M13
Bakouma, *Cen. Af. Rep.* **207** M16
Bakoye, river, *Guinea, Mali* **204** K10
Baku *see* Baki, *Azerb.* **171** D23
Bakundi, *Nigeria* **206** L10
Bakuriani, *Rep. of Georgia* **171** C17
Bakutis Coast, *Antarctica* **238** L7
Bakwanga *see* Mbuji-mayi, *D.R.C.* **209** H15
Bal Chirāgh, *Afghan.* **180** G6
Bālā Murghāb, *Afghan.* **180** G3
Balâ, *Turk.* **170** E8
Bala, Cerros de, *S. Amer.* **128** G4
Balabac Strait, *Philippines* **192** F12
Balabac, *Malaysia* **192** F12
Balabac, island, *Philippines* **192** F12
Balabat, *F.S.M.* **226** E11
Balaclava, *Jam.* **122** H7
Balagansk, *Russ.* **161** L15
Balaghat Range, *India* **182** M5
Balaghat, *India* **182** L7
Balaguer, *Sp.* **147** D16
Balaka, *Malawi* **212** D11
Balakété, *Cen. Af. Rep.* **207** M14
Balakən, *Azerb.* **171** C20
Balakhna, *Russ.* **158** H12
Balaki, *Guinea* **204** K9
Balaklava, *Russ.* **159** T7
Balakliya, *Ukr.* **159** P9
Balakot, *Pak.* **180** H12
Balakovo, *Russ.* **158** M14
Bălan, *Rom.* **155** C15
Balancán, *Mex.* **117** J15
Balangir, *India* **182** L9
Balaözen, river, *Kaz.* **178** E3
Balashi, *Aruba, Neth.* **123** R17
Balashov, *Russ.* **158** M12
Balassagyarmat, *Hung.* **151** K15
Balâţ, *Egypt* **202** J10
Balaton, lake, *Hung.* **151** M14
Balatonszentgyörgy, *Hung.* **151** M13
Balavé, *Burkina Faso* **205** K14
Balazote, *Sp.* **146** J12
Balbina, Represa de, *Braz.* **132** D9
Balcad, *Somalia* **211** R11
Balcarce, *Arg.* **135** N13
Balchik, *Bulg.* **155** G18
Balclutha, *N.Z.* **223** P17
Bald Knob, *Ark., U.S.* **102** F9
Bald Mountain, *Nev., U.S.* **111** S11
Baldwin City, *Kans., U.S.* **107** T10
Baldwin, *Fla., U.S.* **99** Q8
Baldwinsville, *N.Y., U.S.* **96** J8
Baldwyn, *Miss., U.S.* **103** H13
Baldy Peak, *N.M., U.S.* **90** J7
Baldy, Mount, *Ariz., U.S.* **109** T9
Balearic Islands *see* Illes Balears, *Sp.* **147** G17
Balearic Sea, *Sp.* **147** G17
Baleia, Ponta da, *Braz.* **133** K16
Baleine Bay, *St. Vincent & the Grenadines* **123** J16
Balembangan, island, *Malaysia* **192** F12
Baleshare, island, *Scot., U.K.* **144** H6
Baleshwar, *India* **182** L11
Baley, *Russ.* **161** L17
Baléya, *Mali* **204** K10
Baleyara, *Niger* **206** H5
Balezino, *Russ.* **158** G16
Balfour Channel, *Solomon Is.* **228** E7
Balfour Downs, *W. Aust., Austral.* **221** T4
Balgazyn, *Russ.* **161** M13
Balgo, *W. Aust., Austral.* **221** S6
Balḩāf, *Yemen* **174** P11
Bali Sea, *Indonesia* **192** M11
Bali, *India* **182** H3
Bali, island, *Indonesia* **192** M11
Bali, Selat, *Indonesia* **192** M11
Balīkh, river, *Syr.* **172** D11
Balikpapan, *Indonesia* **192** J12
Balimo, *P.N.G.* **222** Q6

Balingen, *Ger.* **150** K6
Balintang Channel, *Philippines* **193** A13
Balık Gölü, *Turk.* **170** C11
Balıkesir, *Turk.* **170** E4
Balkan Mountains, *Eur.* **154** H12
Balkan Peninsula, *Eur.* **140** K9
Balkanabat (Nebitdag), *Turkm.* **178** M5
Balkh, *Afghan.* **180** F7
Balkhāb, *Afghan.* **180** G7
Balkhash, Lake, *Asia* **168** F7
Ball Bay, *Austral.* **229** B20
Balladonia, *W. Aust., Austral.* **221** X5
Ballalae, island, *Solomon Is.* **228** C6
Ballānah, *Egypt* **202** K12
Ballantyne Strait, *N.W.T., Can.* **87** B13
Ballarat, *Vic., Austral.* **221** Z12
Ballard, Lake, *Austral.* **220** J4
Ballari (Bellary), *India* **183** P5
Ballé, *Mali* **204** H11
Ballenas, Bahía, *Mex.* **116** D6
Ballenero Bay, *S. Amer.* **128** R5
Ballenero, Canal, *Chile* **135** Y8
Balleny Basin, *Pac. Oc.* **255** H17
Balleny Fracture Zone, *Pac. Oc.* **255** G17
Balleny Seamounts, *Pac. Oc.* **255** G16
Balleny Trough, *Pac. Oc.* **255** H16
Ballina (Béal an Átha), *Ire.* **145** P4
Ballina, *N.S.W., Austral.* **221** W16
Ballinger, *Tex., U.S.* **105** H14
Ballinrobe (Baile an Róba), *Ire.* **145** P3
Ballinskelligs Bay *see* Scealg, Bá na, *Ire.* **145** T2
Ball's Pyramid, island, *Coral Sea* **224** M6
Ballycastle, *N. Ire., U.K.* **145** M7
Ballycroy (Baile Chruaich), *Ire.* **145** P3
Ballyhaise (Beal Átha hÉis), *Ire.* **145** P6
Ballymena, *N. Ire., U.K.* **145** N7
Ballymoney, *N. Ire., U.K.* **145** P7
Ballyshannon (Béal Átha Seanaidh), *Ire.* **145** P5
Ballyvourney *see* Baile Bhuirne, *Ire.* **145** T3
Balmaceda, *Chile* **135** T7
Balmorhea, *Tex., U.S.* **104** J10
Balod, *India* **182** L8
Baloeri de Cristo Rey, *Eq. Guinea* **214** L7
Balombo, *Angola* **208** M10
Balonne, river, *Austral.* **220** H14
Balotra, *India* **182** H3
Balpyq Bī, *Kaz.* **179** H16
Balqash Köli, *Kaz.* **179** H14
Balqash, *Kaz.* **179** G14
Balrampur, *India* **182** G8
Balranald, *N.S.W., Austral.* **221** X12
Balş, *Rom.* **155** F14
Balsas, *Braz.* **133** F14
Balsas, *Braz.* **133** F14
Balsas, river, *Mex.* **116** J11
Balta, *N. Dak., U.S.* **106** E5
Balta, *Ukr.* **159** Q5
Baltasar Brum, *Uru.* **134** J13
Bălţi, *Mold.* **159** Q4
Baltic Plains, *Eur.* **140** E9
Baltic Sea, *Eur.* **140** F8
Baltimore, *Md., U.S.* **98** C13
Baltistan, region, *Pak.* **180** G13
Baltit, *Pak.* **180** F13
Baltiysk, *Russ.* **158** J1
Baluan, island, *P.N.G.* **222** M8
Baluchistan, *Iran* **177** N22
Baluchistan, region, *Iran, Pak.* **181** Q3
Balumbal, *Eth.* **210** M11
Balurghat, *India* **182** J12
Balya, *Turk.* **170** E3
Balykchy, *Kyrg.* **179** K15
Balyqshy, *Kaz.* **178** G5
Balzers, *Liech.* **162** Q2
Balho, *Djibouti* **210** K9
Bam, *Iran* **177** L20
Bama, *Nigeria* **206** J11
Bamaga, *Qnsld., Austral.* **221** N12
Bamako, *Mali* **204** K11
Bamba, *Mali* **205** G15
Bambafouga, *Guinea* **204** M9
Bambara, *Chad* **207** L14
Bambari, *Cen. Af. Rep.* **207** M15
Bamberg, *S.C., U.S.* **98** L9
Bamberg, *Ger.* **150** H8
Bambesa, *D.R.C.* **209** C16

Bambesi, *Eth.* **210** L5
Bambili, *D.R.C.* **209** C17
Bambinga, *D.R.C.* **208** G12
Bambio, *Cen. Af. Rep.* **207** N13
Bamboo, *Jam.* **122** H9
Bamboo, *W. Aust., Austral.* **221** S3
Bambouti, *Cen. Af. Rep.* **207** M18
Bamda, *China* **184** L7
Bamenda, *Cameroon* **206** M9
Bamingui, *Cen. Af. Rep.* **207** L15
Bampūr, *Iran* **177** M22
Bampūr, river, *Iran* **177** M21
Bamy, *Turkm.* **178** M6
Bāmyān, *Afghan.* **180** H8
Ban Ban, *Laos* **190** J9
Ban Bang Hin, *Thai.* **191** Q6
Ban Gnômmalat, *Laos* **190** K10
Ban Houayxay, *Laos* **190** H8
Ban Kèngkok, *Laos* **190** L11
Ban Meung, *Laos* **190** H7
Ban Muang- Ou-Nua, *Laos* **190** G8
Ban Phai, *Thai.* **190** L9
Ban Sanam Chai, *Thai.* **191** R8
Ban Taphan, *Laos* **190** L11
Ban Thinkèo, *Laos* **190** J9
Ban Xénô, *Laos* **190** K10
Banaba (Ocean Island), *Kiribati* **224** C7
Banalia, *D.R.C.* **209** D16
Banamba, *Mali* **204** K12
Banana River, *Fla., U.S.* **99** T10
Banana, *Kiribati* **226** P11
Bananal, Ilha do, *Braz.* **132** G13
Bananga, *India* **183** U15
Banango, *India* **183** U15
Bananier, *Guadeloupe, Fr.* **123** G14
Banaras *see* Varanasi, *India* **182** J9
Banas, river, *India* **182** H5
Banat, region, *Rom., Serb.* **154** D10
Banaz, *P.N.G.* **222** R9
Banbridge, *N. Ire., U.K.* **145** P7
Banbury, *Eng., U.K.* **145** T12
Banchory, *Scot., U.K.* **144** J11
Bancroft, *Iowa, U.S.* **106** M11
Banda Aceh, *Philippines* **192** G3
Banda Besar, island, *Indonesia* **193** L17
Banda Elat, *Indonesia* **193** L18
Banda Sea, *Indonesia* **193** L15
Banda, *India* **182** J6
Banda, *D.R.C.* **209** B17
Banda, Kepulauan, *Indonesia* **193** L17
Bandajuma, *Sierra Leone* **204** N9
Bandama, river, *Côte d'Ivoire* **205** P13
Bandanaira, *Indonesia* **193** L17
Bandar Murcaayo, *Somalia* **210** K14
Bandar *see* Machilipatnam, *India* **183** P8
Bandar Seri Begawan, *Brunei* **192** G11
Bandar Xarshow, *Somaliland* **210** K13
Bandar 'Abbās, *Iran* **177** M19
Bandar-e Anzalī, *Iran* **177** D14
Bandar-e Būshehr, *Iran* **177** L15
Bandar-e Chārak, *Iran* **177** N17
Bandar-e Deylam, *Iraq* **177** K14
Bandar-e Ganāveh, *Iran* **177** L15
Bandar-e Kangān, *Iran* **177** P20
Bandar-e Khamīr, *Iran* **177** N18
Bandar-e Lengeh, *Iran* **177** N18
Bandar-e Māhshahr, *Iraq* **177** K14
Bandar-e Maqām, *Iran* **177** N17
Bandar-e Mullāḩā, *Afghan.* **180** G5
Bandarban, *Bangladesh* **182** K14
Bandarbeyla, *Somalia* **210** L15
Bandau, *Malaysia* **192** F12
Bandéko, *Congo* **208** C12
Bandelier National Monument, *N. Mex., U.S.* **109** Q12
Bandera, *Tex., U.S.* **105** L15
Bandera, *Arg.* **134** H11
Banderas Bay, *N. Amer.* **84** N3
Bandiagara, *Mali* **205** J14
Bandikui, *India* **182** H5
Bandırma, *Turk.* **170** D4
Bandon, *Oreg., U.S.* **110** J1
Bandundu, *D.R.C.* **208** G12
Bandung, *Indonesia* **192** M8
Banes, *Cuba* **119** J13
Banff, *Alta., Can.* **86** M10
Banfora, *Burkina Faso* **205** L13
Bang Saphan, *Thai.* **191** P7
Banga, *Angola* **208** K10
Bangalore *see* Bengalūru, *India* **183** Q6
Bangassou, *Cen. Af. Rep.* **207** N16

Banggai, Kepulauan, *Indonesia* **193** K14
Banggi, island, *Malaysia* **192** F12
Banggong Co (Bangong Cuo), *China* **184** H2
Banghāzī (Benghazi), *Lib.* **201** D20
Banghiang, river, *Laos* **190** L11
Bangi Island, *Guam, U.S.* **230** P4
Bangka, *Indonesia* **192** K8
Bangka, island, *Indonesia* **192** K8
Bangka, island, *Indonesia* **193** H15
Bangkaru, island, *Indonesia* **192** H4
Bangkok *see* Krung Thep, *Thai.* **190** M8
Bangladesh, *Asia* **182** J12
Bangnae, *S. Korea* **186** M11
Bangnim, *S. Korea* **187** N10
Bangolo, *Côte d'Ivoire* **204** P12
Bangong Cuo *see* Banggong Co, *China* **184** H2
Bangor *see* Baingear, *Ire.* **145** P3
Bangor, *Me., U.S.* **97** F17
Bangs, *Tex., U.S.* **105** H15
Bangu, *D.R.C.* **209** K15
Bangué, *Cameroon* **206** P12
Bangui-Motaba, *Congo* **208** C12
Bangui, *Philippines* **193** B13
Bangui, *Cen. Af. Rep.* **207** N14
Banguru, *D.R.C.* **209** D17
Bangwade, *D.R.C.* **209** D16
Banhā, *Egypt* **202** F11
Banī Mazār, *Egypt* **202** G11
Banī Sharfā, *Saudi Arabia* **174** K7
Banī Suwayf, *Egypt* **202** G11
Banī Walīd, *Lib.* **201** E17
Bani, *Cen. Af. Rep.* **207** L16
Baní, *Dom. Rep.* **119** M19
Bani, river, *Mali* **205** K13
Bania, *Côte d'Ivoire* **205** M14
Baniara, *P.N.G.* **222** R9
Banihal Pass, *India* **182** D5
Banikoara, *Benin* **205** L17
Banissa, *Kenya* **211** Q8
Bāniyās, *Syr.* **172** F7
Banja Luka, *Bosn. & Herzg.* **154** F7
Banjarmasin, *Indonesia* **192** K11
Banjul, *Gambia* **204** J6
Banka, *India* **182** J11
Bankas, *Mali* **205** J14
Bankə, *Azerb.* **171** E23
Bankilare, *Niger* **206** G4
Banks Island, *B.C., Can.* **86** K7
Banks Island, *N.W.T., Can.* **86** D12
Banks Islands, *Vanuatu* **229** D15
Banks Lake, *Wash., U.S.* **110** C7
Banks Peninsula, *N.Z.* **223** M19
Banks Rise, *Arctic Oc.* **252** M9
Banks Shelf, *Arctic Oc.* **252** N9
Banks Slope, *Arctic Oc.* **252** N9
Banks Strait, *Austral.* **220** L8
Banmauk, *Myanmar* **190** F4
Bann, river, *N. Ire., U.K.* **145** N7
Bannerman Town, *Bahamas* **122** F6
Bannu, *Pak.* **180** K10
Bánovce nad Bebravou, *Slovakia* **151** J14
Banreaba, *Kiribati* **227** R14
Banská Bystrica, *Slovakia* **151** J15
Banswara, *India* **182** J4
Bantabaré Pequeño, *Eq. Guinea* **214** M7
Bantala, *Ghana* **205** M15
Bantry Bay, *Ire.* **145** U3
Bantva, *India* **182** K2
Bantval, *India* **183** R4
Banū, *Afghan.* **180** G9
Banyo, *Cameroon* **206** M10
Banz, *P.N.G.* **222** P7
Banzare Bank, *Ind. Oc.* **254** C11
Banzare Coast, *Antarctica* **239** P19
Bao Flala, *Mali* **204** K12
Baó Grande, *Eq. Guinea* **214** M7
Bảo Lạc, *Vietnam* **190** G11
Baoding, *China* **185** G13
Baoshan, *China* **184** N8
Baoshan, *China* **185** C17
Baotou, *China* **184** G11
Baoulé, river, *Mali* **204** K12
Bapatla, *India* **183** P7
Baqbaqty, *Kaz.* **179** H15
Baqty, *Kaz.* **179** G17
Ba'qūbah, *Iraq* **176** G11
Baquedano, *Chile* **134** E7
Bar Harbor, *Me., U.S.* **97** F18
Bar-le-Duc, *Fr.* **149** P13
Bar-sur-Seine, *Fr.* **149** Q12
Bar, *Montenegro* **154** K9

Bara Hoti, *India* **182** E7
Bara, *Indonesia* **193** K15
Bara, *Sudan* **203** S11
Baraawe, *Somalia* **211** R11
Barabhas (Barvas), *Scot., U.K.* **144** G7
Baraboo, *Wis., U.S.* **100** K5
Baraboulé, *Burkina Faso* **205** J15
Baracoa, *Cuba* **119** K14
Baraga, *Mich., U.S.* **100** E6
Bărăganul, plain, *Rom.* **155** F17
Barah Wazē, *Afghan.* **180** J9
Barahona, *Dom. Rep.* **119** N17
Barak, *Turk.* **170** J12
Baraka, river, *Af.* **198** F10
Barakat, *Sudan* **203** R13
Barakī, *Afghan.* **180** J8
Baramandougou, *Mali* **205** K13
Baramula, *India* **182** C4
Baran, *India* **182** J5
Baranama, *Guinea* **204** M11
Baranavichy, *Belarus* **158** L4
Baranof Island, *Alas., U.S.* **113** L22
Baranovka, *Russ.* **159** N14
Barão de Melgaço, *Braz.* **132** J10
Barat Daya, Kepulauan, *Indonesia* **193** M16
Barataria Bay, *La., U.S.* **102** R11
Barate, *Indonesia* **193** N14
Baraterre, *Bahamas* **122** E8
Barbacena, *Braz.* **133** L14
Barbacoas, *Col.* **130** J4
Barbados, *N. Amer.* **121** G18
Barbaria, Cap de, *Sp.* **147** J17
Barbastro, *Sp.* **147** D15
Barber, *Curaçao, Neth.* **123** N14
Barber's Block, *Dominica* **123** E18
Barbers Point *see* Kalealoa, *Hawaii, U.S.* **115** H14
Barberton, *Ohio, U.S.* **101** P14
Barbezieux, *Fr.* **149** U7
Barbourville, *Ky., U.S.* **103** D18
Barbuda, island, *Lesser Antil.* **121** B16
Barcaldine, *Qnsld., Austral.* **221** T13
Barce *see* Al Marj, *Lib.* **201** D21
Barcellona-Pozzo di Gotto, *It.* **153** R12
Barcelona, *Venez.* **130** C12
Barcelona, *Sp.* **147** E17
Barcelos, *Braz.* **132** D8
Barcelos, *Port.* **146** D5
Barclay Bugt, *Greenland, Den.* **88** L14
Barco, *N.C., U.S.* **98** G14
Barcoo, river, *Austral.* **220** F13
Barcs, *Hung.* **151** N13
Barda del Medio, *Arg.* **135** N8
Barda Hills, *India* **182** K1
Bardaale, *Somalia* **211** N13
Bardaï, *Chad* **207** D13
Bardas Blancas, *Arg.* **134** M7
Bardaskan, *Iran* **177** F20
Barddhaman, *India* **182** K11
Bardejov, *Slovakia* **151** H16
Bardonecchia, *It.* **152** E1
Bardsey Island, *Wales, U.K.* **145** S8
Bardsīr, *Iran* **177** K19
Bardstown, *Ky., U.S.* **103** C17
Bardwell, *Ky., U.S.* **102** D12
Bareilly, *India* **182** G7
Barentin, *Fr.* **149** N8
Barents Plain, *Arctic Oc.* **253** G14
Barents Sea, *Arctic Oc.* **253** E15
Barents Shelf, *Arctic Oc.* **253** C17
Barents Trough, *Arctic Oc.* **253** F17
Barentu, *Eritrea* **210** H7
Barfak, *Afghan.* **180** G8
Barfleur, *Fr.* **149** N6
Barfleur, Pointe de, *Fr.* **149** N6
Barg-e Matāl, *Afghan.* **180** G10
Barga, *It.* **152** G6
Bargaal, *Somalia* **210** K15
Bargarh, *India* **182** L9
Barhi, *India* **182** J10
Barhi, *India* **182** J7
Barī Kōt, *Afghan.* **180** G10
Bari, *It.* **152** L14
Bari, *D.R.C.* **209** C13
Bariadi, *Tanzania* **211** U4
Barīm (Perim), island, *Yemen* **174** Q8
Barima, river, *Guyana* **131** E14
Barinas, *Venez.* **130** D8
Baringa, *D.R.C.* **209** D14
Baripada, *India* **182** K11
Bārīs, *Egypt* **202** K11
Barisal, *Bangladesh* **182** K13
Barisan Mountains, *Asia* **168** M10

Barisan, Pegunungan, *Indonesia* **192** K6
Barito, river, *Asia* **168** M12
Barka Kana, *India* **182** K10
Barkā', *Oman* **175** G17
Barkam, *China* **184** K9
Barkhan, *Pak.* **180** M9
Barkley, Lake, *Ky., U.S.* **103** D14
Barkly Tableland, *Austral.* **220** D10
Barkol Hu, *China* **184** E7
Barkol, *China* **184** E7
Barla Dağı, peak, *Turk.* **170** G6
Bârlad, *Rom.* **155** D17
Barlee, Lake, *Austral.* **220** H4
Barletta, *It.* **152** L13
Barlinek, *Pol.* **150** D11
Barmer, *India* **182** H3
Barmou, *Niger* **206** G7
Barmouth, *Wales, U.K.* **145** S9
Bärnau, *Ger.* **150** H9
Barnaul, *Russ.* **160** L12
Barnegat Light, *N.J., U.S.* **96** Q11
Barnegat, *N.J., U.S.* **96** Q11
Barnes Ice Cap, *Nunavut, Can.* **87** F17
Barnesville, *Ga., U.S.* **98** M5
Barnesville, *Minn., U.S.* **106** G8
Barnett Glacier, *Antarctica* **239** R14
Barneville-Carteret, *Fr.* **149** N5
Barnhart, *Tex., U.S.* **105** J13
Barnīs, *Egypt* **202** K14
Barnsdall, *Okla., U.S.* **105** B18
Barnsley, *Eng., U.K.* **145** Q12
Barnstaple Bay, *Eng., U.K.* **145** V9
Barnstaple, *Eng., U.K.* **145** V9
Barnum Bay, *Palau* **226** J4
Barnwell, *S.C., U.S.* **98** L8
Baro, *Nigeria* **206** L7
Baroghil Pass, *Afghan., Pak.* **180** F12
Baron Bluff, *Virgin Is., U.S.* **122** Q2
Barora Fa, island, *Solomon Is.* **228** D8
Barora Ite, island, *Solomon Is.* **228** D8
Barpeta, *India* **182** H13
Barques, Pointe Aux, *Mich., U.S.* **100** K13
Barquisimeto, *Venez.* **130** D9
Barra do Bugres, *Braz.* **132** J10
Barra do Corda, *Braz.* **133** E14
Barra do Cuanza, *Angola* **208** K9
Barra do Dande, *Angola* **208** K9
Barra do São Manuel, *Braz.* **132** F10
Barra *see* Eilean Barraigh, island, *Scot., U.K.* **144** J6
Barra, *Braz.* **133** G15
Barra, Sound of, *Scot., U.K.* **144** J6
Barracks of Swiss Guard, *Vatican City* **163** P17
Barracuda Fracture Zone *see* Fifteen Twenty Fracture Zone, *Atl. Oc.* **248** M7
Barracuda Plain, *Atl. Oc.* **248** M5
Barrage d'Inga, dam, *D.R.C.* **208** H9
Barrancabermeja, *Col.* **130** E6
Barrancas, *Venez.* **130** D13
Barrancos, *Port.* **146** K7
Barranqueras, *Arg.* **134** G12
Barranquilla, *Col.* **130** C6
Barraouallie, *St. Vincent & the Grenadines* **123** K16
Barras, *Braz.* **133** E15
Barrax, *Sp.* **146** H12
Barre, *Vt., U.S.* **97** G13
Barreiras, *Braz.* **133** H14
Barreirinha, *Braz.* **132** D10
Barreirinhas, *Braz.* **133** D15
Barreiro, *Port., Belg.* **148** M13
Barrême, *Fr.* **149** X14
Barren, river, *Ky., U.S.* **103** D16
Barrhill, *Scot., U.K.* **145** N8
Barrie, *Ont., Can.* **87** Q18
Barrier Bay, *Antarctica* **239** G21
Barrigada, *Guam, U.S.* **230** P6
Barrington *see* Isla Santa Fe, *Ecua.* **130** Q4
Barron, *Wis., U.S.* **100** H2
Barrow Canyon, *Arctic Oc.* **252** L7
Barrow Creek, *N.T., Austral.* **221** S9
Barrow Island, *Austral.* **220** E2
Barrow-in-Furness, *Eng., U.K.* **145** P10
Barrow, *Alas., U.S.* **113** B15
Barrow, Point, *Alas., U.S.* **113** B15
Barrow, river, *Ire.* **145** S6
Barry, *Ill., U.S.* **101** R3
Barrytown, *N.Z.* **223** K18
Barshatas, *Kaz.* **179** F15
Barstow, *Calif., U.S.* **111** W9

Bartholin Land, *Greenland, Den.* **88** H13
Bartica, *Guyana* **131** F15
Bartın, *Turk.* **170** C8
Bartle Frere (South Peak), *Austral.* **220** D13
Bartlesville, *Okla., U.S.* **105** B18
Bartlett Inlet, *Antarctica* **238** N11
Bartlett Seamounts, *Ind. Oc.* **251** J16
Bartlett, *Nebr., U.S.* **107** P6
Bartolomeu Dias, *Mozambique* **212** G11
Barton, *Vt., U.S.* **97** F13
Barton, *N. Dak., U.S.* **106** E5
Bartoszyce, *Pol.* **151** B16
Bartow, *Fla., U.S.* **99** U8
Barú, Volcán, *Pan.* **117** N20
Baruun-Urt, *Mongolia* **184** D12
Baruunharaa, *Mongolia* **184** C10
Baruunsuu, *Mongolia* **184** E10
Barvas *see* Barabhas, *Scot., U.K.* **144** G7
Barwani, *India* **182** K4
Barwon, river, *Austral.* **220** J14
Barykova, *Russ.* **161** B21
Barysaw, *Belarus* **158** K5
Barysh, *Russ.* **158** K14
Basa (Grand-Bérébi), *Côte d'Ivoire* **204** Q12
Basankusu, *D.R.C.* **209** D13
Basaga, *Turkm.* **178** N10
Basanté del Este, *Eq. Guinea* **214** M8
Basanté del Oeste, *Eq. Guinea* **214** M7
Basaga, *Turkm.* **178** N10
Basatin del Este, *Eq. Guinea* **214** M8
Basco, *Philippines* **193** A14
Bascuñán, Cabo, *Chile* **134** H6
Basekpio, *D.R.C.* **209** B16
Basel, *Switz.* **152** A3
Basilaki, island, *P.N.G.* **222** R9
Basilan, island, *Philippines* **193** F14
Basile, *La., U.S.* **102** P8
Basilé, *Eq. Guinea* **214** L7
Basilé, Pico, *Eq. Guinea* **214** M7
Basiliano, *It.* **152** D9
Basilisk Harbor, *Solomon Is.* **228** M6
Basin, *Wyo., U.S.* **108** F10
Basingstoke, *Eng., U.K.* **145** V13
Basīţ, Ra's al, *Syr.* **172** E7
Başkale, *Turk.* **171** G18
Baskil, *Turk.* **171** G13
Baskunchak, *Russ.* **159** Q14
Basmat, *India* **182** M5
Basoko, *D.R.C.* **209** D15
Basongo, *D.R.C.* **209** G13
Basra *see* Al Başrah, *Iraq* **177** K13
Bass Canyon, *Pac. Oc.* **255** D20
Bass Islands, *Ohio, U.S.* **101** N13
Bass Islands, *Solomon Is.* **228** K7
Bass Strait, *Austral.* **220** K7
Bassae, ruin, *Gr.* **156** K9
Bassar, *Ghana* **205** M16
Bassas da India, *Fr.* **213** G13
Basse Terre, *Trin. & Tobago* **123** R22
Basse-Pointe, *Martinique, Fr.* **123** E22
Basse-Terre, *Guadeloupe, Fr.* **123** E14
Basse-Terre, *Guadeloupe, Fr.* **123** G14
Basseterre, *St. Kitts & Nevis* **123** B18
Bassett, *Va., U.S.* **98** G10
Bassett, *Nebr., U.S.* **107** N5
Bassigbiri, *Cen. Af. Rep.* **207** M18
Bassikounou, *Mauritania* **204** H12
Bastak, *Iran* **177** N17
Bastendorf, *Lux.* **162** H10
Basti, *India* **182** H8
Bastia, *Fr.* **149** X3
Bastogne, *Belg.* **148** M13
Bastrop, *La., U.S.* **102** K9
Bastrop, *Tex., U.S.* **105** L17
Bastuträsk, *Sw.* **143** F14
Basu, peak, *Indonesia* **192** J7
Basupú, *Eq. Guinea* **214** L7
Bat Yam, *Israel* **172** M5
Bata, *Eq. Guinea* **206** P9
Bataan Peninsula, *Philippines* **193** C13
Batabanó, Golfo de, *Cuba* **118** H6
Batagay, *Russ.* **161** F18
Batala, *India* **182** E5
Batama, *D.R.C.* **209** D17
Batamay, *Russ.* **161** G18
Batan Islands, *Philippines* **193** A14
Batan, island, *Philippines* **193** A14
Batanga, *Gabon* **208** E7
Batangafo, *Cen. Af. Rep.* **207** L14
Batangas City, *Philippines* **193** C13
Batanta, island, *Indonesia* **193** J17
Batavia Seasomunt, *Ind. Oc.* **251** L15
Batavia, *N.Y., U.S.* **96** J6

Bataysk, *Russ.* **159** R11
Bătdâmbâng (Battambang), *Thai.* **191** N9
Batemans Bay, *N.S.W. Austral.* **221** Y14
Batesburg, *S.C., U.S.* **98** L8
Batesville, *Ind., U.S.* **101** S10
Batesville, *Ark., U.S.* **102** F9
Batesville, *Miss., U.S.* **102** H11
Batesville, *Tex., U.S.* **105** M14
Batete, *Eq. Guinea* **214** M6
Bath, *N.Y., U.S.* **96** K7
Bath, *Me., U.S.* **97** G16
Bath, *Jam.* **122** J11
Bath, *St. Kitts & Nevis* **123** C18
Bath, *Eng., U.K.* **145** U11
Batha, river, *Chad* **207** H15
Bathsheba, *Barbados* **123** K19
Bathurst Inlet, *Nunavut, Can.* **87** F13
Bathurst Island, *Nunavut, Can.* **87** D14
Bathurst Island, *Austral.* **220** A7
Bathurst, *N.B., Can.* **87** N21
Bathurst, Cape, *N.W.T., Can.* **86** D10
Bathymetrists Seamounts, *Atl. Oc.* **249** N13
Batī, *Eth.* **210** K8
Batié, *Burkina Faso* **205** M14
Batiki, island, *Fiji* **226** P6
Batista, Serra da, *Braz.* **133** F16
Batman, *Turk.* **171** G15
Batna, *Alg.* **201** C14
Batoicopo, *Eq. Guinea* **214** L7
Baton Rouge, *La., U.S.* **102** P9
Batouri, *Cameroon* **206** N11
Båtsfjord, *Nor.* **143** B15
Batsí, *Gr.* **157** J14
Battambang *see* Bătdâmbâng, *Thai.* **191** N9
Batteau, *Nfld. & Lab. Can.* **87** K22
Batticaloa, *Sri Lanka* **183** T8
Battle Creek, *Mich., U.S.* **100** M10
Battle Creek, *Can., U.S.* **108** A9
Battle Harbour, *Nfld. & Lab. Can.* **87** L23
Battle Mountain, *Nev., U.S.* **111** N9
Batu Pahat, *Malaysia* **191** V9
Batu, Kepulauan, *Indonesia* **192** J5
Batu, peak, *Eth.* **211** N8
Batukelau, *Indonesia* **192** J11
Bat'umi, *Rep. of Georgia* **171** C16
Batuputih, *Indonesia* **192** J13
Baturaja, *Indonesia* **192** K7
Baturité, *Braz.* **133** E17
Batz, Île de, *Fr.* **149** P3
Bau, *Sudan* **203** T13
Bau, island, *Fiji* **226** P6
Baua, river, *Eth.* **211** N10
Baubau, *Indonesia* **193** L14
Baucau, *Timor-Leste* **193** M15
Bauchi, *Nigeria* **206** K9
Baudette, *Minn., U.S.* **106** E10
Baudó, *Col.* **128** B2
Bauer Basin, *Pac. Oc.* **247** L20
Bauer Scarp, *Pac. Oc.* **247** L19
Baugé, *Fr.* **149** R7
Bauges, range, *Fr.* **149** U14
Baun, *Indonesia* **193** N14
Baunatal, *Ger.* **150** F6
Baure, *Nigeria* **206** H9
Baures, river, *Bol.* **132** H8
Bauru, *Braz.* **133** L13
Baús, *Braz.* **132** K11
Bautino, *Kaz.* **178** H4
Bautzen, *Ger.* **150** F11
Bavaria, region, *Ger.* **150** J8
Bávaro, *Dom. Rep.* **119** M21
Bawal, island, *Indonesia* **192** K9
Bawean, island, *Indonesia* **192** L10
Bawku, *Ghana* **205** L16
Baxaya, peak, *Somalia* **210** K14
Baxdo, *Somalia* **211** P13
Baxkorgan, *China* **184** G6
Baxley, *Ga., U.S.* **99** P7
Baxoi (Baxue), *China* **184** L7
Baxter Springs, *Kans., U.S.* **107** V10
Baxter, *Minn., U.S.* **106** H10
Baxue *see* Baxoi, *China* **184** L7
Bay City, *Mich., U.S.* **100** K11
Bay City, *Tex., U.S.* **105** M19
Bay Islands, *N. Amer.* **84** P7
Bay Minette, *Ala., U.S.* **103** N14
Bay of Islands, *N.Z.* **223** D20
Bay of Plenty, *N.Z.* **223** F22
Bay of Whales, *Antarctica* **238** N11
Bay of Wrecks, *Kiribati* **226** Q11
Bay Pleasant, *Bahamas* **122** E1
Bay Saint Louis, *Miss., U.S.* **102** P12

Bay Springs, *Miss., U.S.* **102** M12
Bay'ah, *Oman* **175** F16
Bayamo, *Cuba* **118** K12
Bayan Har Shan, *China* **184** J7
Bayan Har Shankou, *China* **184** J8
Bayan Obo, *China* **184** F11
Bayan, *Mongolia* **184** D7
Bāyan, Band-e, *Afghan.* **180** H6
Bayanaūyl, *Kaz.* **179** E14
Bayandalay, *Mongolia* **184** E10
Bayanhongor, *Mongolia* **184** D9
Bayano, Lago, *Pan.* **117** N22
Bayard, *Nebr., U.S.* **107** P1
Bayard, *N. Mex., U.S.* **109** V10
Bayat, *Turk.* **170** F6
Bayboro, *N.C., U.S.* **98** H14
Bayburt, *Turk.* **171** E15
Baydarata Bay, *Arctic Oc.* **253** B15
Baydaratskaya Guba, *Russ.* **160** F11
Baydhabo (Baidoa), *Somalia* **211** Q10
Bayeux, *Fr.* **149** N6
Bayfield, *Wis., U.S.* **100** E4
Bayghanīn, *Kaz.* **178** F6
Bayḩān al Qişāb, *Yemen* **174** N10
Bayındır, *Turk.* **170** G3
Bayjī, *Iraq* **176** F10
Baykal, *Russ.* **161** L16
Baykan, *Turk.* **171** G16
Bayonnaise Rocks *see* Beyonēsu-Retsugan, *Japan* **189** S12
Bayonne, *Fr.* **149** X5
Bayou Bartholomew, river, *Ark., La., U.S.* **102** K9
Bayqongyr (Baikonur), *Kaz.* **178** H9
Bayramaly, *Turkm.* **178** N10
Bayramiç, *Turk.* **170** E2
Bayreuth, *Ger.* **150** H9
Bayt al Faqīh, *Yemen* **174** P8
Bayt Laḩm (Bethlehem), *Israel* **172** M5
Baytown, *Tex., U.S.* **105** L20
Bayzo, *Niger* **206** H6
Baza, *Sp.* **146** L12
Bāzā'ī Gunbad, *Afghan.* **180** F12
Bāzār-e Panjwā'ī, *Afghan.* **180** L6
Bāzār-e Tālah, *Afghan.* **180** G8
Bazardüzü Dağı, peak, *Azerb.* **171** C22
Bazartöbe, *Kaz.* **178** E5
Bazaruto Island, *Af.* **198** N10
Bazas, *Fr.* **149** W7
Bazber, *Eth.* **210** L5
Bazhong, *China* **184** K11
Bazkovskaya, *Russ.* **159** P11
Beach Haven, *N.J., U.S.* **96** Q11
Beach, *N. Dak., U.S.* **106** G1
Beachport, *S. Aust. Austral.* **221** Z11
Beachy Head, *Eng., U.K.* **145** V14
Beagle Bay, *Austral.* **220** C4
Beagle Bay, *W. Aust. Austral.* **221** R4
Beagle Gulf, *Austral.* **220** A7
Béal an Átha *see* Ballina, *Ire.* **145** P4
Béal an Mhuirthead (Belmullet), *Ire.* **145** P3
Beal Island, *Ire.* **145** U2
Bear Island, *Ire.* **145** U2
Bear Islands *see* Medvezh'i Ostrova, *Russ.* **161** C19
Bear Lake, *Idaho, Utah, U.S.* **108** J8
Bear Paw Mountains, *U.S.* **90** B9
Bear Peninsula, *Antarctica* **238** L8
Bear River Range, *Idaho, Utah, U.S.* **108** J7
Béar, Cap, *Fr.* **149** Z11
Bearden, *Ark., U.S.* **102** J8
Beardmore Glacier, *Antarctica* **239** L13
Beardstown, *Ill., U.S.* **101** R4
Bearnaraigh (Berneray), island, *Scot., U.K.* **144** J6
Bears Paw Mountains, *Mont., U.S.* **108** B9
Beasain, *Sp.* **146** B12
Beata, Cabo, *Dom. Rep.* **119** N17
Beata, Isla, *Dom. Rep.* **119** N17
Beatrice, *Nebr., U.S.* **107** R8
Beattock, *Scot., U.K.* **145** M10
Beatton River, *B.C., Can.* **86** J9

Beatty, *Nev. U.S.* **111** T10
Beau Bassin, *Mauritius* **215** G19
Beaucaire, *Fr.* **149** X12
Beauchêne Island, *Falk. Is. U.K.* **135** W12
Beaufort Rise, *Arctic Oc.* **252** M7
Beaufort Sea, *Arctic Oc.* **252** N8
Beaufort Shelf, *Arctic Oc.* **252** M7
Beaufort Slope, *Arctic Oc.* **252** M7
Beaufort Terrace, *Arctic Oc.* **252** N8
Beaufort West, *S. Af.* **212** M6
Beaufort, *N.C. U.S.* **98** J14
Beaufort, *S.C. U.S.* **98** M9
Beaufort, *Lux.* **162** H10
Beaufort, *Malaysia* **192** G11
Beaugency, *Fr.* **149** R9
Beaumont, *Tex. U.S.* **105** K21
Beaune, *Fr.* **149** S12
Beausoleil, *Fr.* **149** X15
Beauvais, *Fr.* **149** N9
Beauvoir, *Fr.* **149** S5
Beaver , river, *Okla. Tex. U.S.* **105** B13
Beaver Bay, *Minn. U.S.* **106** F13
Beaver City, *Nebr. U.S.* **107** R5
Beaver Creek, *Kans. U.S.* **107** R4
Beaver Creek, *Mont. U.S.* **108** B10
Beaver Dam, *Wis. U.S.* **100** L6
Beaver Dam, *Ky. U.S.* **103** C15
Beaver Falls, *Pa. U.S.* **96** N3
Beaver Island, *Mich. U.S.* **100** G9
Beaver Lake, *Ark. U.S.* **102** E6
Beaver, *Okla. U.S.* **105** B13
Beaver, *Utah, U.S.* **109** N6
Beaver, *Oreg. U.S.* **110** F2
Beaver, *Alas. U.S.* **113** E17
Beaverhead Mountains, *Mont. U.S.* **108** F6
Beaverhead, river, *Mont. U.S.* **108** F6
Beaverton, *Oreg. U.S.* **110** F3
Beawar, *India* **182** H4
Bebar, *India* **182** J1
Beccles, *Eng. U.K.* **145** S16
Bečej, *Serb.* **154** E10
Becerreá, *Sp.* **146** B7
Béchar, *Alg.* **200** E10
Becharof Lake, *Alas. U.S.* **113** L14
Bechem, *Ghana* **205** N15
Beckerich, *Lux.* **162** J9
Beckley, *W. Va. U.S.* **98** E8
Bedēsa, *Eth.* **210** M9
Bedford Downs, *W. Aust. Austral.* **221** R6
Bedford Point, *Grenada* **123** J23
Bedford, *Pa. U.S.* **96** P5
Bedford, *Va. U.S.* **98** F10
Bedford, *Ind. U.S.* **101** T9
Bedford, *Iowa, U.S.* **107** Q10
Bedford, *Eng. U.K.* **145** T13
Bedford, Cape, *Austral.* **220** C13
Bedonia, *It.* **152** F5
Bédouaram, *Niger* **206** G11
Bedourie, *Qnsld. Austral.* **221** T11
Bedous, *Fr.* **149** Y6
Beebe, *Ark. U.S.* **102** G9
Beech Grove, *Ind. U.S.* **101** R9
Beechey Group *see* Chichi Jima Rettō, *Japan* **189** X14
Beef Island, *Virgin Is. U.K.* **122** Q7
Beenleigh, *Qnsld. Austral.* **221** V16
Be'er Menuḥa, *Israel* **173** Q5
Be'er Ora, *Israel* **173** R5
Be'ér Sheva' (Beersheba), *Israel* **173** N5
Beersheba *see* Be'ér Sheva', *Israel* **173** N5
Beeville, *Tex. U.S.* **105** N17
Befale, *D.R.C.* **209** D14
Befori, *D.R.C.* **209** E14
Bega, *N.S.W. Austral.* **221** Z14
Bega, river, *Rom.* **154** D12
Begejski Kanal, *Rom.* **154** E10
Beggars Point, *Antigua & Barbuda* **123** A21
Bēgī, *Eth.* **210** L5
Bègles, *Fr.* **149** V6
Béguégué, *Cen. Af. Rep.* **207** L14
Begusarai, *India* **182** H10
Behagle *see* Laï, *Chad* **207** K13
Behbahān, *Iraq* **177** F14
Behchokǫ̀, *N.W.T. Can.* **86** H11
Behring Point, *Bahamas* **118** E11
Behshahr, *Iran* **177** E17
Bei Shan, *China* **184** F7
Bei, river, *China* **185** N13
Bei'an, *China* **185** C16
Beida *see* Al Bayḍā', *Lib.* **201** D21

Beihai, *China* **184** P12
Beilen, *Neth.* **148** H14
Beinn na Faoghla (Benbecula), island, *Scot. U.K.* **144** J6
Beipan, river, *China* **184** N10
Beipiao, *China* **185** F15
Beira, *Mozambique* **212** F11
Beirut *see* Beyrouth, *Leb.* **172** H6
Beït ed Dîne, *Leb.* **172** J6
Beitbridge, *Zimb.* **212** G9
Beiuş, *Rom.* **154** C12
Beja, *Port.* **146** K6
Béja, *Tunisia* **201** B15
Bejaïa, *Alg.* **201** B13
Bejaïa, Gulf of, *Af.* **198** C5
Béjar, *Sp.* **146** F8
Bejestān, *Iran* **177** F20
Bek, river, *Cameroon* **206** P12
Bekaa Valley, *Leb.* **172** J7
Bekaie, *D.R.C.* **208** F12
Bekdash *see* Garabogaz, *Turkm.* **178** K4
Békés, *Hung.* **151** M17
Békéscsaba, *Hung.* **151** M17
Bekily, *Madagascar* **213** H15
Bekja, island, *Marshall Is.* **227** G20
Bekkai, *Japan* **188** F15
Bekobod, *Uzb.* **178** L12
Bel Air, *Mauritius* **215** G20
Bela Vista, *Braz.* **132** L10
Bela-Bela, *S. Af.* **212** H8
Bela, *Pak.* **181** R6
Bélabo, *Cameroon* **206** N11
Belaga, *Malaysia* **192** H11
Belagavi (Belgaum), *India* **183** P4
Belalcázar, *Sp.* **146** J9
Belarus, *Eur.* **158** L5
Belaya Glina, *Russ.* **159** S11
Belaya Gora, *Russ.* **161** E18
Belaya Kalitva, *Russ.* **159** Q11
Belaya Zemlya, *Russ.* **161** C13
Belaya, *Russ.* **160** E12
Belaya, peak, *Eth.* **210** K6
Belaya, river, *Eur.* **140** D15
Belbubulo, *Sudan* **203** U13
Bełchatów, *Pol.* **151** F15
Belcher Islands, *Nunavut, Can.* **87** L17
Belcik, *Turk.* **170** E11
Belcoo, *N. Ire. U.K.* **145** P5
Belderg *see* Béal Deirg, *Ire.* **145** P3
Belding, *Mich. U.S.* **100** L10
Beledweyne, *Somalia* **211** P11
Belele, *W. Aust. Austral.* **221** U2
Belém, *Braz.* **133** D13
Belen, *N. Mex. U.S.* **109** S12
Belén, *Arg.* **134** G8
Beles, river, *Eth.* **210** L6
Belev, *Russ.* **158** L8
Belfast Lough, *N. Ire. U.K.* **145** N7
Belfast, *Me. U.S.* **97** F17
Belfast, *N. Ire. U.K.* **145** P7
Belfield, *N. Dak. U.S.* **106** G2
Bēlfodiyo, *Eth.* **210** L5
Belfort, *Fr.* **149** R14
Belgaum *see* Belagavi, *India* **183** P4
Belgica Bank, *Arctic Oc.* **253** K16
Belgica Mountains, *Antarctica* **239** C16
Belgium, *Eur.* **148** L12
Belgodère, *Fr.* **149** X2
Belgorod, *Russ.* **159** N9
Belgrade *see* Beograd, *Serb.* **154** F10
Belgrade, *Mont. U.S.* **108** E8
Belgrano II Station, *Antarctica* **238** E10
Belhaven, *N.C. U.S.* **98** H14
Beli, *Nigeria* **206** L10
Belidzhi, *Russ.* **159** U16
Beliliou (Peleliu), island, *Palau* **226** J4
Belinyu, *Indonesia* **192** K8
Belisario Domínguez, Presa, *Mex.* **117** K15
Belitung (Billiton), island, *Indonesia* **192** K8
Belize City, *Belize* **117** J17
Belize, *N. Amer.* **117** K17
Bel'kachi, *Russ.* **161** H18
Bel'kovskiy, *Russ.* **161** D16
Bell Peninsula, *Nunavut, Can.* **87** H17
Bell Point, *St. Martin, Fr.* **123** A14
Bell Point, *P.N.G.* **222** Q6
Bell Rock (Inchcape), *Scot. U.K.* **144** K11
Bell Ville, *Arg.* **134** K11
Bella Bella, *B.C. Can.* **86** K7

Bella Coola, *B.C. Can.* **86** K7
Bella Unión, *Uru.* **134** J13
Bella Vista, *Arg.* **134** G9
Bella Vista, *Arg.* **134** H12
Bellac, *Fr.* **149** T8
Bellaire, *Mich. U.S.* **100** H10
Bellaire, *Ohio, U.S.* **100** R15
Bellaire, *Tex. U.S.* **105** L19
Bellary *see* Ballari, *India* **183** P5
Bellavista, *Peru* **132** F2
Belle Fourche Reservoir, *S. Dak. U.S.* **106** K1
Belle Fourche, *S. Dak. U.S.* **106** K1
Belle Fourche, river, *S. Dak. Wyo. U.S.* **108** G12
Belle Glade, *Fla. U.S.* **99** V10
Belle Isle, *Nfld. & Lab. Can.* **87** L23
Belle Isle, Strait of, *N. Amer.* **87** L22
Belle Plaine, *Minn. U.S.* **106** K11
Belle Plaine, *Iowa, U.S.* **107** N13
Belle Yella, *Liberia* **204** N10
Belle-Anse, *Haiti* **119** M17
Belle-Île, *Fr.* **149** R4
Belledonne, Chaîne de, *Fr.* **149** U14
Bellefontaine, *Ohio, U.S.* **101** Q12
Bellefontaine, *Martinique, Fr.* **123** F22
Bellefonte, *Pa. U.S.* **96** N6
Bellegarde, *Fr.* **149** Q10
Bellegarde, *Fr.* **149** T14
Bellême, *Fr.* **149** Q8
Belleplaine, *Barbados* **123** K19
Belleville, *Ill. U.S.* **101** T4
Belleville, *Kans. U.S.* **107** R7
Belleville, *Fr.* **149** T12
Bellevue, *Ohio, U.S.* **101** P13
Bellevue, *Tex. U.S.* **105** F16
Bellevue, *Iowa, U.S.* **107** N14
Bellevue, *Nebr. U.S.* **107** Q10
Bellevue, *Wash. U.S.* **110** C4
Bellingham, *Wash. U.S.* **110** A4
Bellingshausen Plain, *Pac. Oc.* **249** X1
Bellingshausen Sea, *Antarctica* **238** H5
Bellingshausen, station, *Antarctica* **238** C4
Bellinzona, *Switz.* **152** C4
Bello, *Col.* **130** F5
Bellona Island, *Solomon Is.* **228** H10
Bellona Valley, *Pac. Oc.* **255** F21
Bellows Falls, *Vt. U.S.* **97** J13
Bellpat, *Pak.* **181** N8
Bells, *Tenn. U.S.* **102** F12
Bellsund Valley, *Arctic Oc.* **253** G17
Belluno, *It.* **152** C8
Bellville, *Tex. U.S.* **105** L18
Belmond, *Iowa, U.S.* **106** M11
Belmont, *N.C. U.S.* **98** J9
Belmont, *Miss. U.S.* **103** H13
Belmont, *Grenada* **123** L22
Belmonte, *Braz.* **133** J16
Belmonte, *Port.* **146** F7
Belmonte, *Sp.* **146** H12
Belmopan, *Belize* **117** K16
Belmullet *see* Béal an Mhuirthead, *Ire.* **145** P3
Belo Horizonte, *Braz.* **133** L14
Belo-Tsiribihina, *Madagascar* **213** F15
Belogorsk, *Russ.* **161** K19
Beloha, *Madagascar* **213** J15
Beloit, *Wis. U.S.* **100** M6
Beloit, *Kans. U.S.* **107** S7
Belomorsk, *Russ.* **158** B8
Belomorsk, *Russ.* **160** D8
Belorado, *Sp.* **146** C11
Belorechensk, *Russ.* **159** T11
Beloūsovka, *Kaz.* **179** E17
Beloye More (White Sea), *Russ.* **158** A9
Beloye More, *Russ.* **160** D8
Beloye Ozero, *Russ.* **158** F9
Beloye, Lake, *Eur.* **140** D11
Belozersk, *Russ.* **158** F9
Belpre, *Ohio, U.S.* **101** S15
Belt, *Mont. U.S.* **108** C8
Belterra, *Braz.* **132** D11
Belton Lake, *Tex. U.S.* **105** J17
Belton, *S.C. U.S.* **98** K7
Belton, *Tex. U.S.* **105** J17
Belukha, *Russ.* **160** M12
Belukha, peak, *Asia* **168** F8
Belush'ya Guba, *Russ.* **160** D10
Belvedere Courtyard, *Vatican City* **163** P14
Belvedere Marittimo, *It.* **153** P13
Belvès, *Fr.* **149** V8
Belvidere, *Ill. U.S.* **100** M6

Belvidere, *S. Dak. U.S.* **106** L4
Belvís de la Jara, *Sp.* **146** G9
Belyando, river, *Austral.* **220** F13
Belyy Island, *Asia* **168** C8
Belyy, *Russ.* **158** J7
Belyy, *Russ.* **160** E12
Belzoni, *Miss. U.S.* **102** K10
Bemaraha Plateau, *Af.* **198** M11
Bemaraha, plateau, *Madagascar* **213** F15
Bembe, *Angola* **208** J10
Bembèrèkè, *Benin* **205** M18
Bembìbre, *Sp.* **146** C8
Bemidji, *Minn. U.S.* **106** F10
Ben Gardane, *Tunisia* **201** D16
Ben Nevis, peak, *Scot. U.K.* **144** K8
Bena Dibele, *D.R.C.* **209** G15
Bena-Makima, *D.R.C.* **209** H14
Benabarre, *Sp.* **147** D15
Benadir, region, *Af.* **198** J11
Benasque, *Sp.* **147** C15
Benavente, *Sp.* **146** D9
Benavides, *Tex. U.S.* **105** P16
Benbecula *see* Beinn na Faoghla, island, *Scot. U.K.* **144** J6
Bend, *Oreg. U.S.* **110** H5
Bendals, *Antigua & Barbuda* **123** B21
Bende, *Nigeria* **206** M8
Bendela, *D.R.C.* **208** G12
Bender, *Mold.* **159** R5
Bendern, *Liech.* **162** M2
Bendigo, *Vic. Austral.* **221** Y12
Bene Beraq, *Israel* **172** M5
Benedict Fjord, *Greenland, Den.* **88** A9
Benetutti, *It.* **153** N4
Benevento, *It.* **152** M11
Bengal, Bay of, *Ind. Oc.* **168** K8
Bengalūru (Bangalore), *India* **183** Q6
Bengara, *Indonesia* **192** H12
Bengbu, *China* **185** J14
Benghazi *see* Banghāzī, *Lib.* **201** D20
Benghisa Point, *Malta* **163** L23
Bengkalis, *Indonesia* **192** J6
Bengkalis, island, *Indonesia* **192** H6
Bengkayang, *Indonesia* **192** J9
Bengkulu, *Indonesia* **192** K6
Bengo Bay, *Af.* **198** L5
Benguela, *Angola* **208** M9
Beni Abbès, *Alg.* **200** F10
Béni Mellal, *Mor.* **200** D8
Beni Ounif, *Alg.* **200** E10
Beni Saf, *Alg.* **200** C10
Beni Tajit, *Mor.* **200** D9
Beni, *D.R.C.* **209** D19
Beni, river, *Bol.* **132** H7
Benicarló, *Sp.* **147** F15
Benidorm, *Sp.* **147** J15
Benifaió, *Sp.* **147** H14
Benin, *Af.* **205** M17
Benin City, *Nigeria* **206** M7
Benisheikh, *Nigeria* **206** J11
Benítses, *Gr.* **156** F6
Benjamin Constant, *Braz.* **132** E5
Benjamin Hill, *Mex.* **116** C7
Benjamin, *Tex. U.S.* **105** F14
Benkelman, *Nebr. U.S.* **107** R3
Benlidi, *Qnsld. Austral.* **221** U13
Bennett Peak, *N. Mex. U.S.* **109** Q10
Bennett, *Russ.* **161** C17
Bennett, Lake, *Austral.* **220** F8
Bennettsville, *S.C. U.S.* **98** J10
Benneydale, *N.Z.* **223** G21
Bennington, *Vt. U.S.* **96** J12
Benoni, *S. Af.* **212** J8
Benoy, *Chad* **207** L13
Bensbach, *P.N.G.* **222** Q5
Bensheim, *Ger.* **150** H6
Benson Point, *Kiribati* **226** P10
Benson, *N.C. U.S.* **98** H12
Benson, *Minn. U.S.* **106** J9
Benson, *Ariz. U.S.* **109** V8
Bensonville City, *Liberia* **204** P10
Bent, *Iran* **177** N21
Benta, *Malaysia* **192** G6
Bentiaba, *Angola* **208** N9
Bentinck Island, *Myanmar* **191** P6
Bentinck Island, *Austral.* **220** C11
Bentiu, *S. Sudan* **203** V10
Benton Harbor, *Mich. U.S.* **101** N9
Benton Lake, *Mont. U.S.* **108** C9
Benton, *Ill. U.S.* **101** U6
Benton, *Ark. U.S.* **102** H8
Benton, *La. U.S.* **102** K6
Benton, *Ky. U.S.* **103** D13
Bentong, *Malaysia* **191** U9

Bentonville, *Ark. U.S.* **102** E6
Bent's Old Fort National Historic Site, *Colo. U.S.* **109** N15
Benty, *Guinea* **204** M8
Benua, *Indonesia* **193** L14
Benue, river, *Af.* **206** K10
Benwee Head, *Ire.* **145** P3
Beobwon, *S. Korea* **186** M8
Beograd (Belgrade), *Serb.* **154** F10
Beohari, *India* **182** J8
Beolgyo, *S. Korea* **187** T9
Béoumi, *Côte d'Ivoire* **205** N13
Beowawe, *Nev. U.S.* **111** N10
Beppu, *Japan* **189** R5
Beqa, island, *Fiji* **226** Q5
Bequia, island, *St. Vincent & the Grenadines* **123** M16
Bera Ndjoko, *Congo* **208** C11
Berane, *Montenegro* **154** J9
Berard, *F.S.M.* **226** G11
Berat, *Alban.* **154** L9
Berau, Teluk, *Indonesia* **193** K18
Berber, *Sudan* **203** P13
Berbera, *Somaliland* **210** L11
Berbérati, *Cen. Af. Rep.* **206** N12
Berbice, river, *Guyana* **131** F16
Berceto, *It.* **152** F5
Berchers Bay, *Virgin Is. U.K.* **122** Q9
Berchtesgaden, *Ger.* **150** L10
Berck, *Fr.* **148** M9
Berdoba, *Chad* **207** G16
Berdún, *Sp.* **147** C14
Berdyans'k, *Ukr.* **159** R9
Berdychiv, *Ukr.* **159** P5
Berdzor (Laçın), *Azerb.* **171** E20
Berea, *Ky. U.S.* **103** C18
Bereeda, *Somalia* **210** K15
Berehove, *Ukr.* **159** P1
Bereina, *P.N.G.* **222** Q7
Berekua, *Dominica* **123** G20
Berekum, *Ghana* **205** N14
Berens River, *Man. Can.* **87** M14
Beresford, *S. Dak. U.S.* **106** M8
Beretǎu, river, *Rom.* **154** C12
Bereznik, *Russ.* **158** C11
Berezniki, *Russ.* **160** G9
Berezovka, *Russ.* **158** A15
Berga, *Sp.* **147** D17
Bergama, *Turk.* **170** E3
Bergamo, *It.* **152** D5
Bergara, *Sp.* **146** B12
Bergen, *Nor.* **142** K9
Bergen, *Neth.* **148** H12
Bergerac, *Fr.* **149** V8
Bergues, *Fr.* **148** L10
Berhala, Selat, *Indonesia* **192** J7
Bering Glacier, *Alas. U.S.* **113** K19
Bering Island, *N. Amer.* **84** Q1
Bering Land Bridge National Preserve, *Alas. U.S.* **112** E12
Bering Sea, *Pac. Oc.* **252** H2
Bering Strait Valley, *Arctic Oc.* **252** K4
Bering Strait, *Asia, N. Amer.* **168** A12
Bering, *Russ.* **161** E23
Beringil, *Sudan* **203** T8
Beringovskiy, *Russ.* **161** B21
Berja, *Sp.* **146** M11
Berkane, *Mor.* **200** C10
Berkeley, *Calif. U.S.* **111** S3
Berkner Bank, *Atl. Oc.* **254** K11
Berkner Island, *Antarctica* **238** F9
Berlanga de Duero, *Sp.* **146** E12
Berlevåg, *Nor.* **143** A15
Berlin, *N.H. U.S.* **97** F14
Berlin, *Md. U.S.* **98** D15
Berlin, *Wis. U.S.* **100** K6
Berlin, *Ger.* **150** D10
Bermejito, river, *Arg.* **134** E11
Bermejo, river, *Arg.* **134** E10
Bermeo, *Sp.* **146** B12
Bermuda, *N. Amer.* **85** K22
Bermuda Island *see* Main Island, *Bermuda* **119** C16
Bermuda Islands, *N. Amer.* **84** K10
Bermuda Rise, *Atl. Oc.* **248** K5
Bern, *Switz.* **152** B2
Bernalda, *It.* **153** N14
Bernalillo, *N. Mex. U.S.* **109** S12
Bernardo de Irigoyen, *Arg.* **134** G15
Bernasconi, *Arg.* **135** N10
Bernau, *Ger.* **150** D10
Bernay, *Fr.* **149** P8

Bernburg, *Ger.* **150** E9
Berne, *Ind., U.S.* **101** Q10
Berner Alpen, *Switz.* **152** C2
Berneray see Bearnaraigh, island, *Scot., U.K.* **144** J6
Berneray see Eilean Bhearnaraigh, island, *Scot., U.K.* **144** H6
Bernice, *La., U.S.* **102** K8
Bernie, *Mo., U.S.* **107** V15
Bernier Bay, *Nunavut, Can.* **87** E15
Bernier Island, *Austral.* **220** G1
Bernina, Piz, *It., Switz.* **152** C5
Bernini's Colonnade, *Vatican City* **163** Q17
Bernstorff Isfjord, *Greenland, Den.* **89** R9
Berre, Étang de, *Fr.* **149** X13
Berrechid, *Mor.* **200** D7
Berri, *S. Aust., Austral.* **221** X11
Berriane, *Alg.* **201** D13
Berry Islands, *Bahamas* **118** D11
Berry, *Eur.* **140** H5
Berryville, *Ark., U.S.* **102** E7
Berthold, *N. Dak., U.S.* **106** E3
Bertholet, Cape, *Austral.* **220** D4
Berthoud Pass, *Colo., U.S.* **108** L12
Bertoua, *Cameroon* **206** N11
Bertraghboy Bay, *Ire.* **145** R3
Bertrand, *Nebr., U.S.* **107** Q5
Bertrange, *Lux.* **162** K9
Beru, island, *Kiribati* **224** G8
Berufjaröaráll, valley, *Arctic Oc.* **253** L22
Beruniy, *Uzb.* **178** L8
Beruri, *Braz.* **132** E9
Beruwala, *Sri Lanka* **183** U7
Berwick-upon-Tweed, *Eng., U.K.* **144** L11
Berwick, *La., U.S.* **102** Q9
Berwyn, *Ill., U.S.* **101** N7
Berwyn, region, *Wales, U.K.* **145** S10
Beryslav, *Ukr.* **159** R7
Besalampy, *Madagascar* **213** E15
Besançon, *Fr.* **149** R14
Beserah, *Malaysia* **191** T9
Besham Qala, *Pak.* **180** H11
Besikama, *Indonesia* **193** N15
Beşiri, *Turk.* **171** G16
Besni, *Turk.* **170** H12
Beşparmak Daği, peak, *Turk.* **170** G3
Bessarabia, region, *Mold.* **159** Q4
Besselfjord, research station, *Greenland, Den.* **88** F13
Bessemer, *Mich., U.S.* **100** F5
Bessemer, *Ala., U.S.* **103** J15
Bessines, *Fr.* **149** T9
Bestobe, *Kaz.* **179** D13
Bestuzhevo, *Russ.* **158** D12
Bestyakh, *Russ.* **161** H18
Beswick, *N.T., Austral.* **221** P8
Bet She'an (Beth Shan), *Israel* **172** L6
Betanzos, *Sp.* **146** B6
Betanzos, Ría de, *Sp.* **146** A6
Bétaré Oya, *Cameroon* **206** M11
Bete Hor, *Eth.* **210** K8
Beth Shan see Bet She'an, *Israel* **172** L6
Bethanie, *Namibia* **212** J4
Bethany Beach, *Del., U.S.* **98** D15
Bethany, *Okla., U.S.* **105** S16
Bethany, *Mo., U.S.* **107** R11
Bethel Town, *Jam.* **122** H7
Bethel, *Alas., U.S.* **112** J12
Bethesda, *Md., U.S.* **98** C13
Bethlehem see Bayt Laḥm, *Israel* **172** M5
Bethlehem, *Pa., U.S.* **96** N9
Bethlehem, *Virgin Is., U.S.* **122** Q2
Béthune, *Fr.* **148** L10
Betio, *Kiribati* **227** R14
Betioky, *Madagascar* **213** H15
Betong, *Malaysia* **191** S8
Betoota, *Qnsld., Austral.* **221** U11
Bétou, *Congo* **208** C12
Betpaqdala, *Kaz.* **178** H12
Bettembourg, *Lux.* **162** K10
Bettendorf, *Iowa, U.S.* **107** P14
Bettendorf, *Lux.* **162** H10
Bettlerjoch, peak, *Aust., Liech.* **162** Q3
Betul, *India* **182** K6
Béu, *Angola* **208** H11
Beulah, *N. Dak., U.S.* **106** F3
Bevans Town, *Bahamas* **122** J3
Beverley, *Eng., U.K.* **145** Q13
Bewcastle, *Eng., U.K.* **145** N11

Bexhill, *Eng., U.K.* **145** V15
Bexon, *St. Lucia* **123** K14
Bey Daği, peak, *Turk.* **170** E12
Bey Daği, peak, *Turk.* **170** G11
Beycesultan, ruins, *Turk.* **170** G4
Beykoz, *Turk.* **170** C5
Beyla, *Guinea* **204** N11
Beyləqan, *Azerb.* **171** E21
Bēylul, *Eritrea* **210** J10
Beyneu, *Kaz.* **178** H6
Beyonēsu-Retsugan (Bayonnaise Rocks), *Japan* **189** S12
Beypazari, *Turk.* **170** E7
Beyra, *Somalia* **211** N13
Beyrouth (Beirut), *Leb.* **172** H6
Beyşehir Gölü, *Turk.* **170** G7
Beyşehir, *Turk.* **170** G7
Beytüşşebap, *Turk.* **171** H17
Bezhetsk, *Russ.* **158** H9
Béziers, *Fr.* **149** Y11
Bəndovan Burnu, *Azerb.* **171** E23
Bərdə, *Azerb.* **171** D21
Bhadarwah, *India* **180** J14
Bhadra, *India* **182** F5
Bhadrakh, *India* **182** L11
Bhag, *Pak.* **181** N8
Bhagalpur, *India* **182** J11
Bhakkar, *Pak.* **180** L10
Bhaktapur, *Nepal* **182** G10
Bhamo, *Myanmar* **190** F6
Bhandara, *India* **182** L7
Bhanpura, *India* **182** J5
Bharati Station, *Antarctica* **239** G20
Bharatpur, *India* **182** G6
Bharatpur, *India* **182** J8
Bharuch, *India* **182** K3
Bhatapara, *India* **182** L8
Bhatarsaigh (Vatersay), island, *Scot., U.K.* **144** J6
Bhatinda, *India* **182** F5
Bhatkal, *India* **183** Q4
Bhatpara, *India* **182** K12
Bhavanipatna, *India* **182** M9
Bhavnagar, *India* **182** K3
Bhawanipatna, *India* **182** M9
Bhera, *India* **180** K12
Bhilai, *India* **182** L8
Bhilwara, *India* **182** H4
Bhima, river, *India* **182** N5
Bhimbar, *India* **180** J12
Bhind, *India* **182** H6
Bhinmal, *India* **182** J3
Bhisho, *S. Af.* **212** M8
Bholari, *Pak.* **181** S8
Bhongir, *India* **182** N6
Bhopal, *India* **182** K6
Bhreandáin, Bá (Brandon Bay), *Ire.* **145** T2
Bhubaneshwar, *India* **182** L10
Bhuj, *India* **182** J1
Bhusawal, *India* **182** L5
Bhutan, *Asia* **182** G12
Bia, river, *Côte d'Ivoire, Ghana* **205** P14
Biabou, *St. Vincent & the Grenadines* **123** L17
Biak, *Indonesia* **193** J19
Biak, island, *Indonesia* **193** J19
Biała Podlaska, *Pol.* **151** E18
Białogard, *Pol.* **150** B12
Białowieża Forest, *Eur.* **140** G9
Biały Bór, *Pol.* **151** B13
Białystok, *Pol.* **151** C17
Bianco, *It.* **153** R13
Biando, island, *S. Korea* **187** R7
Biankouma, *Côte d'Ivoire* **204** N12
Biaó, peak, *Eq. Guinea* **214** M7
Biaora, *India* **182** J5
Biaro, island, *Indonesia* **193** H15
Biarritz, *Fr.* **149** X5
Bias, *Fr.* **149** W6
Biau, *Indonesia* **193** H14
Bibā, *Egypt* **202** G11
Bibassé, *Gabon* **208** D8
Bibbiena, *It.* **152** G7
Biberach, *Ger.* **150** K7
Bibi Nani, *Pak.* **180** M7
Bibiani, *Ghana* **205** P15
Bicaz, *Rom.* **155** C15
Bichänäk Ashyrymy, *Arm.* **171** E19
Biche, *Trin. & Tobago* **123** Q23
Bichvint'a, *Rep. of Georgia* **171** A15
Bickleton, *Wash., U.S.* **110** E6
Bicknell, *Ind., U.S.* **101** T8
Bicknell, *Utah, U.S.* **109** N7
Bida, *Nigeria* **206** K7
Biddeford, *Me., U.S.* **97** H15
Bideford, *Eng., U.K.* **145** V9

Bié Plateau, *Angola* **208** M11
Bieber, *Calif., U.S.* **110** M5
Biega, peak, *D.R.C.* **209** F18
Biei, *Japan* **188** F13
Bielawa, *Pol.* **151** G13
Biele Karpaty, range, *Czechia, Slovakia* **151** J13
Bielefeld, *Ger.* **150** E6
Bieler See, *Switz.* **152** B2
Biella, *It.* **152** D3
Bielsa, *Sp.* **147** C15
Bielsk Podlaski, *Pol.* **151** D17
Bielsko-Biała, *Pol.* **151** H15
Biên Hòa, *Vietnam* **191** P12
Bienne, Lac, *Que., Can.* **87** L19
Biescas, *Sp.* **147** C14
Bifoum, *Gabon* **208** E7
Big Bay, *Mich., U.S.* **100** E7
Big Bay, *Vanuatu* **229** F14
Big Bay, *Vanuatu* **229** H16
Big Belt Mountains, *Mont., U.S.* **108** D8
Big Bend National Park, *Tex., U.S.* **104** M10
Big Black, river, *Miss., U.S.* **102** L10
Big Blue, river, *Nebr., U.S.* **107** Q8
Big Bog, swamp, *Minn., U.S.* **106** E9
Big Canyon, river, *Tex., U.S.* **104** K11
Big Creek, *Calif., U.S.* **111** T7
Big Cypress National Preserve, *Fla., U.S.* **99** W9
Big Cypress Swamp, *Fla., U.S.* **99** W9
Big Falls, *Minn., U.S.* **106** E11
Big Fork, river, *Minn., U.S.* **106** E11
Big Hatchet Peak, *N. Mex., U.S.* **109** W10
Big Hole National Battlefield, *Mont., U.S.* **108** E6
Big Hole, river, *Mont., U.S.* **108** E6
Big Island, *Nunavut, Can.* **87** H18
Big Lake, *Tex., U.S.* **104** J12
Big Mangrove, *Bahamas* **122** B11
Big Pine, *Fla., U.S.* **99** Y9
Big Pine, *Calif., U.S.* **111** T8
Big Piney, *Wyo., U.S.* **108** J9
Big Rapids, *Mich., U.S.* **100** K10
Big Sandy, *Mont., U.S.* **108** B9
Big Sandy, river, *Ariz., U.S.* **109** J9
Big Sioux, river, *S. Dak., U.S.* **106** L8
Big Snowy Mountains, *Mont., U.S.* **108** D9
Big South Fork National River and Recreation Area, *Ky., Tenn., U.S.* **103** E17
Big Spring, *Tex., U.S.* **104** H12
Big Springs, *Nebr., U.S.* **107** Q3
Big Stone City, *S. Dak., U.S.* **106** J8
Big Stone Gap, *Va., U.S.* **98** G7
Big Stone Lake, *Minn., S. Dak., U.S.* **106** J8
Big Sur, *Calif., U.S.* **111** U4
Big Thicket National Preserve, *Tex., U.S.* **105** K20
Big Timber, *Mont., U.S.* **108** E9
Big Trout Lake, *Ont., Can.* **87** M15
Big Valley Mountains, *Calif., U.S.* **110** M4
Big Wells, *Tex., U.S.* **105** N14
Big Wood Cay, *Bahamas* **118** E11
Big Wood, river, *Idaho, U.S.* **108** H6
Big, river, *N.W.T., Can.* **86** D12
Biga, *Turk.* **170** D3
Bigadiç, *Turk.* **170** E4
Bigej Channel, *Marshall Is.* **227** J18
Bigej, island, *Marshall Is.* **227** J18
Bigeumdo, island, *S. Korea* **187** T6
Bigfork, *Minn., U.S.* **106** F11
Bigfork, *Mont., U.S.* **108** B6
Biggar, *Sask., Can.* **86** M11
Biggarenn, island, *Marshall Is.* **227** F15
Biggenden, *Qnsld., Austral.* **221** U15
Biggerann, island, *Marshall Is.* **227** F14
Bighorn Canyon National Recreation Area, *Mont., Wyo., U.S.* **108** F10
Bighorn Lake, *Mont., Wyo., U.S.* **108** F10
Bighorn Mountains, *Wyo., U.S.* **108** F10
Bighorn, river, *Mont., Wyo., U.S.* **108** G10
Bight Fracture Zone, *Atl. Oc.* **248** G8
Bight of Benin, *Atl. Oc.* **205** Q18

Bight of Benin, *Af.* **206** M5
Bight of Bonny, *Af.* **206** N8
Bigi, island, *Marshall Is.* **227** G15
Bignona, *Senegal* **204** K7
Bigobo, *D.R.C.* **209** H18
Bigonville, *Lux.* **162** H9
Bihać, *Bosn. & Herzg.* **154** F6
Bihar Sharif, *India* **182** J10
Biharamulo, *Tanzania* **211** U3
Bihoro, *Japan* **188** F15
Biinman, bay, *S. Korea* **187** Q7
Bijapur see Vijayapura, *India* **182** N5
Bijapur, *India* **182** M7
Bījār, *Iran* **177** E13
Bijeljina, *Bosn. & Herzg.* **154** F9
Bijire, island, *Marshall Is.* **227** A21
Bijnor, *India* **182** F6
Bikaner, *India* **182** G4
Bikar Atoll, *Marshall Is.* **227** B17
Bikeman, island, *Kiribati* **227** Q14
Biken, island, *Marshall Is.* **227** B20
Bikenibeu, *Kiribati* **227** R15
Bikennel, island, *Marshall Is.* **227** G16
Bikenubati, island, *Kiribati* **227** Q15
Bikin, *Russ.* **161** L21
Bikini Atoll, *Marshall Is.* **227** B15
Bīkkū Bīttī, peak, *Lib.* **201** K21
Bikoro, *D.R.C.* **208** E12
Bila Tserkva, *Ukr.* **159** P5
Bilād Banī Bū 'Alī, *Oman* **175** H18
Bilaspur, *India* **182** K8
Bilati, *D.R.C.* **209** E18
Bilauktaung Range, *Thai.* **190** M7
Bilbao, *Sp.* **146** B12
Bileća, *Bosn. & Herzg.* **154** H8
Bilecik, *Turk.* **170** D5
Bīleh Savār, *Iran* **177** B13
Biləsuvar, *Azerb.* **171** E22
Bilhorod-Dnistrovs'kyy, *Ukr.* **159** R5
Bili, *D.R.C.* **209** B16
Bili, river, *D.R.C.* **209** B15
Bilibino, *Russ.* **161** C20
Bilican Dağları, peak, *Turk.* **171** F16
Bilis Qooqaani, *Somalia* **211** S9
Bill Baileys Bank, *Atl. Oc.* **248** F10
Bill of Portland, *Eng., U.K.* **145** W11
Bill Williams, river, *Ariz., U.S.* **109** T5
Bill, *Wyo., U.S.* **108** H13
Billae, island, *Marshall Is.* **227** A21
Billilla, *N.S.W. Austral.* **221** W12
Billiluna, *W. Aust., Austral.* **221** R6
Billings, *Okla., U.S.* **105** B17
Billings, *Mont., U.S.* **108** E10
Billings, *Russ.* **161** B19
Billiton see Belitung, *Indonesia* **192** K8
Billom, *Fr.* **149** U11
Billy Chinook, Lake, *Oreg., U.S.* **110** G5
Bilma, *Niger* **206** E11
Bilo Gora, mountains, *Croatia* **154** D7
Biloela, *Qnsld., Austral.* **221** U15
Bilohirs'k, *Russ.* **159** S8
Bilopillya, *Ukr.* **159** N8
Biloxi, *Miss., U.S.* **102** P12
Biltine, *Chad* **207** G15
Bilto, *Nor.* **143** C13
Bilüü, *Mongolia* **184** C6
Bilwi, *Nicar.* **117** L19
Bilzen, *Belg.* **148** L13
Bima, river, *D.R.C.* **209** C17
Bimberi Peak, *Austral.* **220** L14
Bimbila, *Ghana* **205** M16
Bimini Islands, *Bahamas* **118** D9
Bindloe see Isla Marchena, *Ecua.* **130** P3
Bindura, *Zimb.* **212** E9
Binéfar, *Sp.* **147** D15
Binəqədi, *Azerb.* **171** D23
Binford, *N. Dak., U.S.* **106** F7
Binga, *D.R.C.* **209** C13
Bingen, *Ger.* **150** G6
Binger, *Okla., U.S.* **105** D16
Bingham Channel, *Kiribati* **227** P14
Bingham, *Me., U.S.* **97** E16
Binghamton, *N.Y., U.S.* **96** L9
Bingöl Dağları, peak, *Turk.* **171** F16
Bingöl, *Turk.* **171** F14
Binh Khê, *Vietnam* **190** M13
Bình Liêu, *Vietnam* **190** G12
Binh Long, *Vietnam* **191** P11
Binjai, *Indonesia* **192** G5
Binjai, *Indonesia* **192** G8
Binongko, island, *Indonesia* **193** L15
Bintagoungou, *Mali* **205** H14
Bintan, island, *Malaysia* **192** H7

Bintimani, peak, *Sierra Leone* **204** M9
Bintuhan, *Indonesia* **192** L7
Bintulu, *Malaysia* **192** H10
Binyang, *China* **184** P11
Binzhou, *China* **185** H14
Bio Addo, *Somalia* **210** M14
Biobío, river, *Chile* **135** N6
Bioč, peak, *Montenegro* **154** H8
Bioko, island, *Eq. Guinea* **206** N9
Biokovo, peak, *Croatia* **154** H7
Bipindi, *Cameroon* **206** P9
Bi'r 'Alī, *Yemen* **174** P11
Bir Anzarane, *W. Sahara* **200** J3
Bîr el 'Abd, *Egypt* **173** N2
Bir Gandouz, *W. Sahara* **200** K2
Bîr Ḥasana, *Egypt* **173** P3
Bir Lahmar, *W. Sahara* **200** H5
Bir Mogreïn, *Mauritania* **204** B9
Bir Tawil, *Egypt, Sudan* **202** L13
Bir, island, *Marshall Is.* **182** M5
Birāk, *Lib.* **201** G18
Birao, *Cen. Af. Rep.* **207** K16
Biratnagar, *Nepal* **182** H11
Birch Mountains, *Alta., Can.* **86** K11
Birch Tree Hill, *Cayman Is., U.K.* **122** M2
Birchwood, *Alas., U.S.* **113** J16
Bird City, *Kans., U.S.* **107** R3
Bird Island see Aves, island, *Lesser Antil.* **121** D14
Bird Island, *Minn., U.S.* **106** K10
Bird Island, *Grenada* **123** J24
Bird Island, *Seychelles* **215** K19
Bird's Head Peninsula see Jazirah Doberai, *Indonesia* **193** J18
Birdsville, *Qnsld., Austral.* **221** U11
Birecik, *Turk.* **171** H13
Bireun, *Indonesia* **192** G4
Birganj, *Nepal* **182** H10
Birini, *Cen. Af. Rep.* **207** L16
Bīrjand, *Iran* **177** H21
Birkat Qārūn, lake, *Egypt* **202** G11
Birkenhead, *Eng., U.K.* **145** R10
Birkirkara, *Malta* **163** K22
Bîrlad, river, *Rom.* **155** C17
Birmingham, *Ala., U.S.* **103** J15
Birmingham, *Eng., U.K.* **145** S12
Birni Ngaouré, *Niger* **206** H5
Birni Nkonni, *Niger* **206** H7
Birnie Island, *Kiribati* **224** G10
Birnin Kebbi, *Nigeria* **206** J6
Birobidzhan, *Russ.* **161** K20
Birong, *Philippines* **192** E12
Birou, *Mali* **204** J10
Birrindudu, *N.T., Austral.* **221** R7
Birshoghyr, *Kaz.* **178** F7
Birsilpur, *India* **182** G3
Birtin, *Rom.* **154** C12
Birufu, *Indonesia* **193** L20
Biryakovo, *Russ.* **158** F11
Birżebbuġa, *Malta* **163** L23
Bisa, island, *Indonesia* **193** J16
Bisbee, *N. Dak., U.S.* **106** E6
Bisbee, *Ariz., U.S.* **109** W9
Biscarosse et de Parentis, Lac de, *Fr.* **149** W6
Biscarrosse-Plage, *Fr.* **149** W6
Biscay Plain, *Atl. Oc.* **248** H11
Biscay, Bay of, *Atl. Oc.* **140** H3
Biscayne Bay, *Fla., U.S.* **99** X10
Biscayne National Park, *Fla., U.S.* **99** X10
Bisceglie, *It.* **152** L13
Bischheim, *Fr.* **149** P15
Biscoe Islands, *Antarctica* **238** E4
Biševo, island, *Croatia* **154** H6
Bishkek, *Kyrg.* **179** K14
Bishop, *Tex., U.S.* **105** P17
Bishop, *Calif., U.S.* **111** S8
Bishopville, *S.C., U.S.* **98** K10
Bishrī, Jabal, *Syr.* **172** F12
Biskra, *Alg.* **201** C13
Bislig, *Philippines* **193** F15
Bismarck Archipelago, *P.N.G.* **222** N9
Bismarck Range, *P.N.G.* **222** P7
Bismarck Sea, *P.N.G.* **222** N8
Bismarck, *N. Dak., U.S.* **106** G4
Bismarckfelsen, peak, *Namibia* **212** G4
Bismil, *Turk.* **171** G15
Bison, *S. Dak., U.S.* **106** J3
Bissagos Islands, *Af.* **198** G1
Bissameuttack, *India* **182** M9
Bissau, *Guinea-Bissau* **204** K7
Bissen, *Lux.* **162** J10
Bissikrima, *Guinea* **204** L10

Bistcho Lake, *Alta., Can.* **86** J10
Bistineau, Lake, *La., U.S.* **102** L7
Bistreţu, *Rom.* **155** G13
Bistriţa, *Rom.* **155** C14
Bistriţa, river, *Rom.* **155** C15
Bitam, *Gabon* **208** C8
Bitkin, *Chad* **207** J14
Bitlis, *Turk.* **171** G16
Bitola, *N. Maced.* **154** L11
Bitra Island, *India* **183** R3
Bitterfontein, *S. Af.* **212** L5
Bitterroot Range, *Idaho, Mont., U.S.* **108** D5
Bitterroot, river, *Mont., U.S.* **108** D6
Biu, *Nigeria* **206** K10
Bivona, *It.* **153** S9
Biwa Ko, *Japan* **189** P9
Bixby, *Okla., U.S.* **105** C18
Bixessarri, *Andorra* **162** K2
Biya, river, *Asia* **168** E8
Biysk, *Russ.* **160** L12
Bizerte, *Tunisia* **201** B15
Bjargtangar, bay, *Atl. Oc.* **140** A3
Bjelolasica, peak, *Croatia* **154** E5
Bjelovar, *Croatia* **154** D7
Bjerge Kronprins Frederik, *Greenland, Den.* **89** N11
Bjerkø Peninsula, *Antarctica* **239** F21
Bjøllånes, *Nor.* **142** E12
Bjørnafjorden, bay, *Nor.* **142** K9
Bjørnøya Bank, *Arctic Oc.* **253** G18
Bla, *Mali* **205** K13
Black Bay Point, *Grenada* **123** K22
Black Belt, *U.S.* **91** K16
Black Canyon City, *Ariz., U.S.* **109** T7
Black Canyon of the Gunnison National Park, *Colo., U.S.* **109** N11
Black Coast, *Antarctica* **238** E6
Black Eagle, *Mont., U.S.* **108** C8
Black Forest, *Ger.* **150** K6
Black Hills, *U.S.* **90** D10
Black Irtysh, river, *Asia* **168** F8
Black Isle, *Scot., U.K.* **144** H9
Black Lake, *Mich., U.S.* **100** G11
Black Mesa, *Okla., U.S.* **104** A11
Black Mesa, *Ariz., U.S.* **109** Q8
Black Mountain, *Ky., U.S.* **103** D20
Black Mountains, *Ariz., U.S.* **109** R4
Black Mountains, *Wales, U.K.* **145** T10
Black Point *see* Kūpikipikiʻo Point, *Hawaii, U.S.* **115** H15
Black Point, *St. Vincent & the Grenadines* **123** K17
Black Range, *U.S.* **90** K8
Black River Bay, *Jam.* **122** J7
Black River Falls, *Wis., U.S.* **100** J4
Black River, *Jam.* **122** J7
Black Rock Desert, *Nev., U.S.* **110** M7
Black Rock, *Trin. & Tobago* **123** N17
Black Rock, *Ire.* **145** P2
Black Rocks, *St. Kitts & Nevis* **123** A17
Black Sea Lowland, *Eur.* **140** H11
Black Sea, *Eur.* **140** K11
Black *see* Đà, river, *Vietnam* **190** H10
Black Volta, river, *Af.* **205** M14
Black, river, *N.Y., U.S.* **96** H9
Black, river, *Wis., U.S.* **100** J4
Black, river, *Ark., Mo., U.S.* **102** F10
Black, river, *Ariz., U.S.* **109** U8
Black, river, *Jam.* **122** J7
Blackall, *Qnsld., Austral.* **221** T13
Blackball, *N.Z.* **223** K18
Blackbeard Island, *Ga., U.S.* **99** P9
Blackbeard's Tower, *Bahamas* **122** E3
Blackburn, *Eng., U.K.* **145** Q11
Blackburn, Mount, *Alas., U.S.* **113** J19
Blackdown, *Qnsld., Austral.* **221** R13
Blackduck, *Minn., U.S.* **106** F10
Blackfoot Reservoir, *Idaho, U.S.* **108** H7
Blackfoot, *Idaho, U.S.* **108** H7
Blackfoot, river, *Mont., U.S.* **108** C6
Blackhaw Seamounts, *Pac. Oc.* **247** H13
Blackmans, *Barbados* **123** K19
Blackmoor Vale, *Eng., U.K.* **145** V11
Blackpool, *Eng., U.K.* **145** Q10
Blacks Fork, river, *Utah, Wyo., U.S.* **108** K8
Blacksburg, *Va., U.S.* **98** F9
Blackshear, *Ga., U.S.* **99** P8
Blacksod Bay, *Ire.* **145** P3
Blackstone, *Va., U.S.* **98** F12
Blackville, *S.C., U.S.* **98** L9
Blackwall Ice Stream, *Antarctica* **238** F11

Blackwater, *Qnsld., Austral.* **221** T14
Blackwater, river, *Ire.* **145** T4
Blackwell, *Okla., U.S.* **105** B17
Blackwell, *Tex., U.S.* **105** H14
Blackwood, river, *Austral.* **220** L3
Blagodarnyy, *Russ.* **159** S13
Blagoevgrad, *Bulg.* **155** K13
Blagoveshchensk, *Russ.* **161** K19
Blagoyevo, *Russ.* **158** C13
Blain, *Fr.* **149** R5
Blaine, *Wash., U.S.* **110** A3
Blair, *Nebr., U.S.* **107** P9
Blaj, *Rom.* **155** D13
Blake Basin, *Atl. Oc.* **248** L3
Blake Plateau, *Atl. Oc.* **248** K3
Blake Point, *Mich., U.S.* **100** C6
Blakely, *Ga., U.S.* **99** P4
Blakeney, *Eng., U.K.* **145** R15
Blakes, *Cayman Is., U.K.* **122** N5
Blanc, Mont, *Fr., It.* **152** D2
Blanc, Mount, *Fr.* **149** U15
Blanca Bay, *S. Amer.* **128** M6
Blanca Peak, *U.S.* **90** H9
Blanca Peak, *Colo., U.S.* **109** P13
Blanca, Costa, *Sp.* **147** K15
Blanca, Portella, *Andorra, Fr.* **162** J5
Blanchard, *Okla., U.S.* **105** D16
Blanche Bay, *La., U.S.* **102** R8
Blanche Channel, *Solomon Is.* **228** E7
Blanche, Lake, *Austral.* **220** H11
Blanchisseuse, *Trin. & Tobago* **123** N22
Blanco Fracture Zone, *Pac. Oc.* **247** D17
Blanco, *Tex., U.S.* **105** L16
Blanco, Cape, *Oreg., U.S.* **110** J1
Blanco, river, *Bol.* **132** J8
Blanding, *Utah, U.S.* **109** P9
Blanes, *Sp.* **147** D18
Blangy, *Fr.* **148** M9
Blankenberge, *Belg.* **148** K10
Blanquilla, Isla, *Venez.* **130** B12
Blantyre, *Malawi* **212** D11
Blasket Island *see* Na Blascaodaí, *Ire.* **145** T2
Blaskimen Island, *Antarctica* **239** A13
Blåvands Point, *Eur.* **140** F6
Blaydon, *Eng., U.K.* **145** N12
Blaye, *Fr.* **149** V6
Blaze, Point, *Austral.* **220** B7
Blednaya, *Russ.* **160** D12
Blenheim, *N.Z.* **223** K19
Blessington Lakes, *Ire.* **145** R6
Blida, *Alg.* **200** B12
Blidinje Jezero, *Bosn. & Herzg.* **154** G7
Bligh Water, *Fiji* **226** P5
Blind River, *Ont., Can.* **87** Q17
Blinman, *S. Aust. Austral.* **221** W10
Blitar, *Indonesia* **192** M10
Blitta, *Togo* **205** N17
Block Island, *R.I., U.S.* **97** M14
Blodgett Iceberg Tongue, *Antarctica* **239** P20
Bloemfontein, *S. Af.* **212** K7
Blois, *Fr.* **149** R9
Blönduós, *Ice.* **142** E3
Bloody Bay, *Trin. & Tobago* **123** N18
Bloody Foreland, *Ire.* **145** M5
Bloomer, *Wis., U.S.* **100** H3
Bloomfield, *Ind., U.S.* **101** S8
Bloomfield, *Iowa, U.S.* **107** Q13
Bloomfield, *Mo., U.S.* **107** V15
Bloomfield, *Nebr., U.S.* **107** N7
Blooming Prairie, *Minn., U.S.* **106** L12
Bloomington, *Ill., U.S.* **101** Q5
Bloomington, *Ind., U.S.* **101** S8
Bloomington, *Tex., U.S.* **105** N18
Bloomington, *Minn., U.S.* **106** K11
Bloomsburg, *Pa., U.S.* **96** N8
Blosseville Kyst, *Greenland, Den.* **88** M13
Blossom, *Russ.* **161** B19
Blouberg, peak, *S. Af.* **212** G8
Blower Rock, *Jam.* **118** P11
Blowing Point, *Anguilla, U.K.* **122** R11
Blowing Point, *Jam.* **122** H10
Bludenz, *Aust.* **150** L7
Blue Earth, *Minn., U.S.* **106** L11
Blue Hill, *Nebr., U.S.* **107** R6
Blue Hills, *Turks & Caicos Is., U.K.* **119** H16
Blue Hole, *Jam.* **122** H11
Blue Lake, *Calif., U.S.* **110** M2
Blue Mountain Peak, *Jam.* **122** J11
Blue Mountain, *Ark., U.S.* **102** G6
Blue Mountain, *Oreg., U.S.* **110** K8

Blue Mountains, *Oreg., Wash., U.S.* **110** F8
Blue Mountains, *Jam.* **122** J10
Blue Mountains, *Austral.* **220** K14
Blue Mud Bay, *Austral.* **220** B10
Blue Nile *see* Abay, *Eth.* **210** L7
Blue Nile *see* El Bahr el Azraq, *Sudan* **203** S13
Blue Rapids, *Kans., U.S.* **107** R8
Blue Ridge, *Ga., U.S.* **98** J5
Blue Ridge, *U.S.* **98** G9
Bluebeard's Castle, *Virgin Is., U.S.* **122** M8
Bluefield, *W. Va., U.S.* **98** F8
Bluefields Bay, *Jam.* **122** H6
Bluefields, *Nicar.* **117** M19
Bluefields, *Jam.* **122** H6
Bluejoint Lake, *Oreg., U.S.* **110** K6
Bluff Knoll, *Austral.* **220** L3
Bluff Point, *Austral.* **220** H2
Bluff, *Utah, U.S.* **109** P9
Bluff, *N.Z.* **223** P16
Bluffton, *Ind., U.S.* **101** Q10
Blumenau, *Braz.* **133** N14
Blunt, *S. Dak., U.S.* **106** K5
Bly, *Oreg., U.S.* **110** K5
Blyth, *Eng., U.K.* **145** N12
Blythe, *Calif., U.S.* **111** Y12
Blytheville, *Ark., U.S.* **102** F11
Bo Hai, bay, *China* **185** G14
Bo River, *S. Sudan* **203** V9
Bo, *Sierra Leone* **204** N9
Bø, *Nor.* **142** D12
Bø, *Nor.* **142** J9
Boa Vista, *Braz.* **132** B9
Boa Vista, island, *Cabo Verde* **215** C17
Boaco, *Nicar.* **117** M18
Boali, *Cen. Af. Rep.* **207** N14
Boang, island, *P.N.G.* **222** N10
Boano, island, *Indonesia* **193** K16
Boardman, *Ohio, U.S.* **101** P15
Boaz Island, *Bermuda* **119** C14
Boaz, *Ala., U.S.* **103** H16
Bobbili, *India* **182** M9
Bobo-Dioulasso, *Burkina Faso* **205** L13
Bobovdol, *Bulg.* **154** J12
Boby, peak, *Madagascar* **213** G16
Boca Chica Key, *Fla., U.S.* **99** Y8
Boca Chica, *Dom. Rep.* **119** M19
Bôca do Acre, *Braz.* **132** F6
Bôca do Jari, *Braz.* **132** D12
Boca Grande, *Fla., U.S.* **99** W8
Boca Mavaca, *Venez.* **130** K11
Boca Raton, *Fla., U.S.* **99** W10
Boca Tabla Cave, *Curaçao, Neth.* **123** N13
Bocas del Toro, *Pan.* **117** N20
Bocay, *Nicar.* **117** L18
Bochnia, *Pol.* **151** H16
Bocholt, *Ger.* **150** E5
Bocht van Hato, bay, *Curaçao, Neth.* **123** P14
Bochum, *Ger.* **150** E5
Bock Cay, *Bahamas* **122** D7
Bocşa, *Rom.* **154** E11
Boda, *Cen. Af. Rep.* **207** N13
Bodaybo, *Russ.* **161** J16
Bodega Bay, *Calif., U.S.* **111** R2
Bodega Head, *Calif., U.S.* **111** R2
Bodele Depression, *Af.* **198** F6
Boden, *Sw.* **143** E14
Bodensee (Lake Constance), *Aust., Ger., Switz.* **150** L6
Bodhei, *Kenya* **211** T9
Bodi, *Indonesia* **193** J14
Bodī, *Mongolia* **184** E9
Bodjoki, *D.R.C.* **209** C15
Bodkin Mill, ruin, *Virgin Is., U.S.* **122** Q2
Bodle, *Eth.* **211** P10
Bodmin Moor, *Eng., U.K.* **145** W9
Bodmin, *Eng., U.K.* **145** W8
Bodø, *Nor.* **142** E12
Bodol Sea, bay, *S. Korea* **187** U9
Bodrum, *Turk.* **170** H3
Boe, *Nauru* **227** M14
Boën, *Fr.* **149** U12
Boende, *D.R.C.* **209** E14
Bœng Tônlé Sab (Tonle Sap), *Cambodia* **191** N10

Boeo, Capo, *It.* **153** S8
Boerne, *Tex., U.S.* **105** L15
Boeuf, river, *La., U.S.* **102** L9
Boffa, *Guinea* **204** M8
Bog Walk, *Jam.* **122** J9
Bogalusa, *La., U.S.* **102** N11
Bogan, river, *Austral.* **220** J13
Bogangolo, *Cen. Af. Rep.* **207** M14
Bogantungan, *Qnsld., Austral.* **221** T14
Bogbonga, *D.R.C.* **209** D13
Bogda Feng, peak, *Asia* **168** F8
Bogda Shan, *China* **184** E5
Bogenaga, island, *Marshall Is.* **227** G19
Boggenadick, island, *Marshall Is.* **227** F20
Boggenatjen, island, *Marshall Is.* **227** G15
Boggerik Passage, *Marshall Is.* **227** F15
Boggerik, island, *Marshall Is.* **227** G16
Boggerlapp, island, *Marshall Is.* **227** F16
Boghaz (Boğaz), *Cyprus* **162** N10
Bogia, *P.N.G.* **222** N7
Bogildo, island, *S. Korea* **187** V7
Bogol Manyo, *Eth.* **211** P9
Bogong, Mount, *Austral.* **220** L13
Bogor, *Indonesia* **192** M8
Bogorditsk, *Russ.* **158** L9
Bogorodsk, *Russ.* **158** J12
Bogotá, *Col.* **130** G6
Bogotol, *Russ.* **161** K13
Bogra, *Bangladesh* **182** J12
Bogué, *Mauritania* **204** H8
Bohé, *Eq. Guinea* **214** N7
Bohemia, region, *Czechia* **150** H10
Bohemian Forest, *Ger.* **150** J9
Bohicon, *Benin* **205** N17
Bohodukhiv, *Ukr.* **159** N8
Bohol Sea, *Philippines* **193** E15
Bohol, island, *Philippines* **193** E14
Bohyeonsan, peak, *S. Korea* **187** Q12
Boiaçu, *Braz.* **132** C8
Boim, *Braz.* **132** D11
Boing, *S. Sudan* **203** U13
Boinso, *Ghana* **205** P14
Bois Blanc Island, *Mich., U.S.* **100** G11
Bois, river, *S. Amer.* **128** G8
Boise City, *Okla., U.S.* **104** B11
Boise, *Idaho, U.S.* **108** G4
Bojeador, Cape, *Asia* **168** J12
Bojnūrd, *Iran* **177** D19
Bokada, *D.R.C.* **209** B13
Bokalijman, island, *Marshall Is.* **227** H19
Bokani, *Nigeria* **206** K7
Bokatola, *D.R.C.* **208** E12
Boké, *Guinea* **204** L8
Boken, island, *Marshall Is.* **227** A20; B20
Boken, island, *Marshall Is.* **227** D19
Bokenelab, island, *Marshall Is.* **227** A20
Bokkumaruchi, island, *Marshall Is.* **227** G16
Boklaplap, island, *Marshall Is.* **227** G20
Boknafjorden, bay, *Nor.* **142** K9
Bokol, *Kenya* **211** R6
Bokoluo, island, *Marshall Is.* **227** A20
Bokombako, island, *Marshall Is.* **227** A20
Bökönbaev, *Kyrg.* **179** K15
Bokoro, *Chad* **207** J13
Bokovskaya, *Russ.* **159** P11
Bokpyinn, *Myanmar* **191** P7
Boksitogorsk, *Russ.* **158** F8
Boku, *P.N.G.* **228** C5
Bokurdak, *Turkm.* **178** M7
Bol, *Chad* **206** H12
Bolaang, *Indonesia* **193** K14
Bolan Pass, *Pak.* **180** M7
Bolans, *Antigua & Barbuda* **123** B20
Bolbec, *Fr.* **149** N8
Boldumsaz, *Turkm.* **178** K8
Bole, *China* **184** D4
Bole, *Ghana* **205** M15
Bolea, *Sp.* **147** C14
Bolesławiec, *Pol.* **150** F12
Bolgatanga, *Ghana* **205** L15
Bolhrad, *Ukr.* **159** S4
Bolia, *D.R.C.* **208** F12
Bolinao, Cape, *Asia* **168** J12
Bolivar, *Tenn., U.S.* **102** G12

Boeo, Capo, *It.* **153** S8
Bolivar, *Mo., U.S.* **107** U12
Bolívar, Cerro, *Venez.* **130** E12
Bolivia, *Cuba* **118** H10
Bolivia, *S. Amer.* **132** J8
Bolkar Dağları, *Turk.* **170** H9
Bolkhov, *Russ.* **158** L8
Bollène, *Fr.* **149** W12
Bollnäs, *Sw.* **143** J13
Bollon, *Qnsld., Austral.* **221** V14
Bollons Seamount, *Pac. Oc.* **246** Q12
Bolluk Gölü, *Turk.* **170** G8
Bolnisi, *Rep. of Georgia* **171** C18
Bolobo, *D.R.C.* **208** F11
Bologna, *It.* **152** F7
Bolognesi, *Peru* **132** G4
Bologoye, *Russ.* **158** G8
Bolomba, *D.R.C.* **209** D13
Bolong, *Chad* **207** J13
Bolsena, *It.* **152** J8
Bolsena, Lago di, *It.* **152** J7
Bol'shaya Bicha, *Russ.* **160** J10
Bol'sheretsk, *Russ.* **161** G23
Bol'shevik Island, *Asia* **168** B9
Bol'shevik, *Russ.* **161** D14
Bol'shezemel'skaya Tundra, *Russ.* **160** F10
Bol'shoy Begichev, *Russ.* **161** E15
Bol'shoy Lyakhovskiy, *Russ.* **161** D17
Bol'shoy Patom, *Russ.* **161** J17
Bol'shoy Shantar, *Russ.* **161** H20
Bolsón de Mapimí, *N. Amer.* **84** M4
Bolsward, *Neth.* **148** G13
Bolt Head, *Eng., U.K.* **145** X10
Bolton Point, *Kiribati* **227** P14
Bolton, *Eng., U.K.* **145** Q11
Bolu, *Turk.* **170** D7
Boluntay, *China* **184** H6
Bolvadin, *Turk.* **170** F6
Bolzano, *It.* **152** C7
Bom Jesus da Lapa, *Braz.* **133** H15
Boma, *D.R.C.* **208** H9
Bomassa, *Congo* **208** C11
Bomba, Gulf of, *Af.* **198** D8
Bombala, *N.S.W. Austral.* **221** Z14
Bombardopolis, *Haiti* **119** L15
Bombay *see* Mumbai, *India* **182** M3
Bombom, *Sao Tome and Principe* **215** A20
Bomdila, *India* **182** G13
Bomdo, *India* **182** F15
Bomili, *D.R.C.* **209** D17
Bomotu, *D.R.C.* **209** C18
Bomokandi, river, *D.R.C.* **209** C18
Bomu, river, *D.R.C.* **209** B16
Bon Accueil, *Mauritius* **215** G20
Bona, Mount, *Alas., U.S.* **113** J19
Bonaire, island, *Lesser Antil.* **120** H10
Bonanza, *Utah, U.S.* **108** L9
Bonao, *Dom. Rep.* **119** M18
Bonaparte Archipelago, *Austral.* **220** B5
Bonaparte Basin, *Ind. Oc.* **251** J20
Bonasse, *Trin. & Tobago* **123** R21
Bonavista Bay, *N. Amer.* **84** G11
Bonavista, *Nfld. & Lab., Can.* **87** M24
Bondo, *D.R.C.* **209** B15
Bondoukou, *Côte d'Ivoire* **205** N14
Bonds Cay, *Bahamas* **118** D11
Bone, Gulf of, *Pac. Oc.* **168** M13
Bone, Teluk, *Indonesia* **224** H1
Bonefish Bay, *Bahamas* **122** E11
Bonefish Pond, *Bahamas* **122** F2
Bonerate, island, *Indonesia* **193** M13
Bonesteel, *S. Dak., U.S.* **106** M6
Bonete, Cerro, *Arg.* **134** G7
Bonga, *Eth.* **211** N6
Bongandanga, *D.R.C.* **209** D14
Bongdong, *S. Korea* **187** R8
Bonghwa, *S. Korea* **187** P11
Bongnam, *S. Korea* **187** S7
Bongo Lava, *Af.* **198** N11
Bongo, *Gabon* **208** F7
Bongo, *D.R.C.* **209** C13
Bongolava, *Madagascar* **213** F15
Bongor, *Chad* **206** K12
Bongyang, *S. Korea* **187** N10
Bonham, *Tex., U.S.* **105** F18
Boni, *Mali* **205** J15
Bonifacio, *Fr.* **149** Z2
Bonifacio, Strait of, *Fr.* **149** Z2
Bonifay, *Fla., U.S.* **99** Q3
Bonin Islands (Ogasawara Guntō), *Japan* **189** X14
Bonin Trench, *Pac. Oc.* **246** G8
Bonita Springs, *Fla., U.S.* **99** W8
Bonito, *Braz.* **132** L10

Bonkoukou, *Niger* **206** H6
Bonn, *Ger.* **150** F5
Bonne Terre, *Mo., U.S.* **107** T15
Bonners Ferry, *Idaho, U.S.* **108** A4
Bonneval, *Fr.* **149** Q9
Bonneville Dam, *Oreg., Wash., U.S.* **110** F4
Bonneville Salt Flats, *Utah, U.S.* **108** K5
Bonnie Rock, *W. Aust., Austral.* **221** W3
Bonny, *Fr.* **149** R10
Bonny, *Nigeria* **206** N8
Bonorva, *It.* **153** N3
Bonriki, *Kiribati* **227** Q15
Bonriki, island, *Kiribati* **227** R15
Bonshaw, *N.S.W., Austral.* **221** W15
Bontang, *Indonesia* **192** J12
Bonthe, *Sierra Leone* **204** N9
Bontoc, *Philippines* **193** B13
Bonvouloir Islands, *P.N.G.* **222** R10
Bony de les Neres, *Andorra* **162** J3
Booker T. Washington National Monument, *Va., U.S.* **98** F10
Booker, *Tex., U.S.* **105** B13
Boola, *Guinea* **204** N11
Booligal, *N.S.W., Austral.* **221** X12
Boomerang Ridge, *Ind. Oc.* **250** N11
Boon Point, *Antigua & Barbuda* **123** A21
Boonderoo, *Qnsld., Austral.* **221** S13
Boone, *N.C., U.S.* **98** G8
Boone, *Iowa, U.S.* **107** N11
Booneville, *Ark., U.S.* **102** G6
Booneville, *Miss., U.S.* **103** H13
Bööntsagaan Nuur, *Mongolia* **184** E8
Boonville, *N.Y., U.S.* **96** H9
Boonville, *Ind., U.S.* **101** U8
Boonville, *Mo., U.S.* **107** S12
Boorabbin, *W. Aust., Austral.* **221** W4
Boorama, *Somaliland* **210** L10
Booroondara, Mount, *Austral.* **220** J13
Boosaaso (Bosaso), *Somalia* **210** K14
Boothbay Harbor, *Me., U.S.* **97** G16
Boothia Peninsula, *Nunavut, Can.* **87** F15
Boothia, Gulf of, *Nunavut, Can.* **87** F15
Bootle Bay Village, *Bahamas* **122** J1
Booué, *Gabon* **208** E8
Bopeechee, *S. Aust., Austral.* **221** V10
Bopolu, *Liberia* **204** P10
Boppard, *Ger.* **150** G5
Bor, *Serb.* **154** G12
Bor, *Russ.* **158** H12
Bor, *Turk.* **170** G9
Bor, *S. Sudan* **203** X11
Bora-Bora, *Fr. Polynesia, Fr.* **231** B16
Bora-Bora, *Fr. Polynesia, Fr.* **231** F16;G14
Boracho Peak, *Tex., U.S.* **104** J9
Borah Peak, *Idaho, U.S.* **108** G6
Boran, *Kaz.* **179** F18
Boran, *Eth.* **211** P7
Borås, *Sw.* **142** L12
Borāzjān, *Iran* **177** L15
Borba, *Braz.* **132** E9
Borba, *Port.* **146** J6
Borborema, Planalto da, *Braz.* **133** F17
Borchgrevink Coast, *Antarctica* **239** Q15
Borçka, *Rep. of Georgia* **171** C16
Bordeaux Mountain, *Virgin Is., U.S.* **122** N11
Bordeaux, *Fr.* **149** V6
Borden Island, *N.W.T., Can.* **87** B13
Borden Peninsula, *Nunavut, Can.* **87** E16
Bordertown, *S. Aust., Austral.* **221** Y11
Borðeyri, *Ice.* **142** E2
Bordj bou Arreridj, *Alg.* **201** B13
Bordj Flye Sainte Marie, *Alg.* **200** G9
Bordj le Prieur, *Alg.* **200** L11
Bordj Omar Driss, *Alg.* **201** G14
Boré, *Mali* **205** J14
Boreas Basin, *Arctic Oc.* **253** J17
Boreray, island, *Scot., U.K.* **144** G5
Boreumdo, island, *S. Korea* **186** M7
Borg Massif, *Antarctica* **238** B12
Borgarnes, *Ice.* **142** E2
Borger, *Tex., U.S.* **104** C12
Borger, *Neth.* **148** H14
Borgholm, *Sw.* **143** M14
Borgne, Lake, *La., U.S.* **102** Q11
Borgo Maggiore, *San Marino* **163** J15
Borgo San Dalmazzo, *It.* **152** F2

Borgosesia, *It.* **152** D3
Borgou, *Benin, Nigeria* **205** L18
Borgou, *Benin, Nigeria* **206** K6
Borhoyn Tal, *Mongolia* **184** E12
Bori, *Benin* **205** N17
Borisoglebsk, *Russ.* **158** M11
Borisovo Sudskoye, *Russ.* **158** F9
Borj el Kessira, *Tunisia* **201** E16
Borjomi, *Rep. of Georgia* **171** C17
Borkou, *Chad* **207** E13
Borlänge, *Sw.* **143** J13
Bormida, *It.* **152** F3
Bormio, *It.* **152** C6
Borneo (Kalimantan), island, *Indonesia* **192** J10
Bornholm, island, *Den.* **143** N13
Bornova, *Turk.* **170** F3
Boro, *Mali* **204** K10
Boro, river, *S. Sudan* **203** V7
Borohora Shan, *China* **184** E4
Boroko, *Indonesia* **193** J14
Borokoro, *Guinea* **204** M10
Boromo, *Burkina Faso* **205** L14
Boronga Islands, *Myanmar* **190** H3
Borotou, *Côte d'Ivoire* **204** N12
Borovichi, *Russ.* **158** G8
Borovo, *Croatia* **154** E9
Borovsk, *Russ.* **158** K9
Borovskoy, *Kaz.* **178** C10
Borroloola, *N.T., Austral.* **221** Q10
Borşa, *Rom.* **155** B14
Børselv, *Nor.* **143** B14
Borsippa, ruin, *Iraq* **176** H10
Borujen, *Iran* **177** J15
Borūjerd, *Iraq* **177** G13
Boryeong, *S. Korea* **187** Q7
Boryslav, *Ukr.* **159** P2
Boryspil', *Ukr.* **159** N6
Borzya, *Russ.* **161** L17
Bosa, *It.* **153** N3
Bosanska Gradiška, *Bosn. & Herzg.* **154** E7
Bosanski Petrovac, *Bosn. & Herzg.* **154** F6
Bosanski Šamac (Šamac), *Bosn. & Herzg.* **154** E8
Bosaso *see* Boosaaso, *Somalia* **210** K14
Boscastle, *Eng., U.K.* **145** W9
Boscobel, *Wis., U.S.* **100** L4
Bose, *China* **184** N11
Boseong, *S. Korea* **187** T8
Boshan, *China* **185** H14
Bosilegrad, *Serb.* **154** J12
Bosna, river, *Bosn. & Herzg.* **154** G8
Bosnia and Herzegovina, *Eur.* **154** F7
Bōsō Hantō, *Japan* **189** P12
Bosobele, *D.R.C.* **208** D12
Bosobolo, *D.R.C.* **209** B13
Bosporus *see* İstanbul Boğazi, strait, *Turk.* **170** C5
Bossangoa, *Cen. Af. Rep.* **207** M13
Bossé Bangou, *Niger* **206** H5
Bossembélé, *Cen. Af. Rep.* **207** N13
Bossier City, *La., U.S.* **102** L6
Bosso Dallol, river, *Af.* **206** G6
Bosso, *Niger* **206** H11
Bossut, Cape, *Austral.* **220** D4
Bost *see* Lashkar Gāh, *Afghan.* **180** L5
Bostan, *Pak.* **180** M7
Bosten Hu, *China* **184** F5
Boston Bay, *Jam.* **122** J11
Boston Mountains, *Ark., U.S.* **102** F7
Boston, *Mass., U.S.* **97** K15
Boston, *Ga., U.S.* **99** Q6
Boston, *Eng., U.K.* **145** R14
Bosusulu, *D.R.C.* **209** D14
Boswell, *Okla., U.S.* **105** E18
Botabano, *Cayman Is., U.K.* **122** M2
Botany Bay, *Virgin Is., U.S.* **122** M7
Botany Bay, *Austral.* **220** L15
Boteti, river, *Botswana* **212** F6
Botev, peak, *Bulg.* **155** J14
Botevgrad, *Bulg.* **155** H13
Bothnia, Gulf of, *Fin., Sw.* **143** G14
Boticas, *Port.* **146** D6
Botoşani, *Rom.* **155** B16
Botshabelo, *S. Af.* **212** K8
Botswana, *Af.* **212** G6
Bottineau, *N. Dak., U.S.* **106** D5
Bottle Creek, *Turks & Caicos Is., U.K.* **119** H17
Botwood, *Nfld. & Lab., Can.* **87** M23
Bou Ali, *Alg.* **200** H10
Bou Anane, *Mor.* **200** E9
Bou Arfa, *Mor.* **200** D10

Boû Lanouâr, *Mauritania* **204** E7
Bou Saâda, *Alg.* **201** C13
Bouaké, *Côte d'Ivoire* **205** N13
Bouala, *Cen. Af. Rep.* **206** M12
Bouar, *Cen. Af. Rep.* **206** M12
Bouca, *Cen. Af. Rep.* **207** M14
Boudeuse Ridge, *Pac. Oc.* **247** K14
Bouga, *Chad* **206** L12
Bougainville Strait, *Solomon Is.* **228** C6
Bougainville Strait, *Vanuatu* **229** G14
Bougainville, *P.N.G.* **222** P11
Bougainville, *P.N.G.* **228** C5
Bougainville, Cape, *Austral.* **220** B6
Boughessa, *Mali* **205** F17
Bougouni, *Mali* **204** L10
Bouillante, *Guadeloupe, Fr.* **123** F14
Bouillon, *Belg.* **149** N12
Bouilly, *Fr.* **149** Q11
Boujdour, *W. Sahara* **200** G3
Boujdour, Cap, *Af.* **198** D2
Boukra, *W. Sahara* **200** G4
Boula, *Mali* **204** J11
Boulaide, *Lux.* **162** H9
Boulder City, *Nev., U.S.* **111** U12
Boulder, *Colo., U.S.* **108** L13
Boulder, *Mont., U.S.* **108** D7
Boulia, *Qnsld., Austral.* **221** T11
Boulogne-sur-Mer, *Fr.* **148** L9
Boulogne, *Fr.* **149** Y8
Boulouli, *Mauritania* **204** H11
Bouloupari, *New Caledonia, Fr.* **231** Q22
Bouloupesse, *Congo* **208** F9
Boulsa, *Burkina Faso* **205** K16
Boultoum, *Niger* **206** G9
Boûmdeïd, *Mauritania* **204** G9
Bouna, *Côte d'Ivoire* **205** M14
Boundary Peak, *Nev., U.S.* **111** S8
Boundiali, *Côte d'Ivoire* **204** M12
Boundji, *Congo* **208** E10
Bountiful, *Utah, U.S.* **108** K7
Bounty Bay, *Pitcairn I.* **228** Q6
Bounty Islands, *Pac. Oc.* **224** Q9
Bounty Plateau, *Pac. Oc.* **246** Q11
Bounty Trough, *Pac. Oc.* **246** Q11
Bourail, *New Caledonia, Fr.* **231** Q21
Bouraké, *New Caledonia, Fr.* **231** Q22
Bourbon-Lancy, *Fr.* **149** T11
Bourem, *Mali* **205** G16
Bourg-en-Bresse, *Fr.* **149** T13
Bourg-Lastic, *Fr.* **149** U10
Bourg-Madame, *Fr.* **149** Z9
Bourg-Saint-Maurice, *Fr.* **149** U15
Bourganeuf, *Fr.* **149** U9
Bourges, *Fr.* **149** S10
Bourget, Lac du, *Fr.* **149** U14
Bourgneuf, Baie de, *Fr.* **149** S5
Bourke, *N.S.W., Austral.* **221** W13
Bournemouth, *Eng., U.K.* **145** W12
Bouroum, *Burkina Faso* **205** J16
Bourscheid, *Lux.* **162** H10
Bourtoutou, *Chad* **207** J16
Bousso, *Chad* **207** K13
Boutilimit, *Mauritania* **204** G7
Bouvet Fracture Zone, *Atl. Oc.* **249** W12
Bouza, *Niger* **206** H7
Bovalino Marina, *It.* **153** R13
Bovey, *Minn., U.S.* **106** F11
Bovill, *Idaho, U.S.* **108** D4
Bovina, *Tex., U.S.* **104** E11
Bowbells, *N. Dak., U.S.* **106** D3
Bowdle, *S. Dak., U.S.* **106** J5
Bowdoin, Lake, *Mont., U.S.* **108** B11
Bowen, *Qnsld., Austral.* **221** S14
Bowers Basin, *Pac. Oc.* **246** C11
Bowers Mountains, *Antarctica* **239** Q15
Bowers Ridge, *Pac. Oc.* **246** C11
Bowie, *Md., U.S.* **98** C13
Bowie, *Tex., U.S.* **105** F16
Bowie, *Ariz., U.S.* **109** V9
Bowling Green Bay, *Austral.* **220** D14
Bowling Green, *Va., U.S.* **98** E13
Bowling Green, *Ohio, U.S.* **101** P12
Bowling Green, *Ky., U.S.* **103** D15
Bowling Green, *Mo., U.S.* **107** S14
Bowman Bay, *Nunavut, Can.* **87** G18
Bowman Island, *Antarctica* **239** L22
Bowman Peninsula, *Antarctica* **238** F8
Bowman, *N. Dak., U.S.* **106** H2
Bowwood, *Zambia* **209** Q17
Box Butte Reservoir, *Nebr., U.S.* **107** N2
Box Elder Creek, *Mont., U.S.* **108** C10

Boxelder Creek, *Mont., U.S.* **108** E13
Boyabat, *Turk.* **170** C10
Boyarka, *Russ.* **161** F14
Boyce, *La., U.S.* **102** M8
Boyd Strait, *Antarctica* **238** D4
Boyer, river, *Iowa, U.S.* **107** P9
Boykétté, *Cen. Af. Rep.* **207** M15
Boyle (Mainistir na Búille), *Ire.* **145** Q5
Boyne City, *Mich., U.S.* **100** H10
Boyne, river, *Ire.* **145** Q6
Boynton Beach, *Fla., U.S.* **99** W10
Boyoma Falls (Stanley Falls), *D.R.C.* **209** D16
Boysen Reservoir, *Wyo., U.S.* **108** H10
Boz Burun, *Turk.* **170** D4
Boz Dağ, peak, *Turk.* **170** H4
Bozashchy Tübegi, *Kaz.* **178** H4
Bozburun Dağı, peak, *Turk.* **170** H6
Bozburun, *Turk.* **170** H3
Bozcaada, island, *Turk.* **170** E2
Bozdoğan, *Turk.* **170** G4
Bozeman, *Mont., U.S.* **108** E8
Bozene, *D.R.C.* **209** C13
Bozhou, *China* **185** J13
Bozkır, *Turk.* **170** H7
Bozkurt, *Turk.* **170** C9
Bozouls, *Fr.* **149** W10
Bozoum, *Cen. Af. Rep.* **207** M13
Bozova, *Turk.* **171** H13
Bozüyük, *Turk.* **170** E5
Bra, *It.* **152** F3
Brabant Island, *Antarctica* **238** D4
Brabant, *Eur.* **140** G5
Brač, island, *Croatia* **154** H6
Bracadale, Loch, *Scot., U.K.* **144** H7
Bracciano, *It.* **152** K8
Bracciano, Lago di, *It.* **152** K8
Brackettville, *Tex., U.S.* **105** M13
Brački Kanal, *Croatia* **154** H6
Bracknell, *Eng., U.K.* **145** U13
Brad, *Rom.* **154** D12
Bradenton, *Fla., U.S.* **99** U7
Brades, *Montserrat, U.K.* **123** B23
Bradford, *Pa., U.S.* **96** L5
Bradford, *Eng., U.K.* **145** Q12
Bradley, *Ill., U.S.* **101** P7
Bradley, *S. Dak., U.S.* **106** J7
Brady, *Tex., U.S.* **105** J15
Brady, *Nebr., U.S.* **107** Q4
Bræ L. Bistrup, glacier, *Greenland, Den.* **88** F13
Braga, *Port.* **146** D5
Bragado, *Arg.* **134** L12
Bragança, *Port.* **146** D7
Braham, *Minn., U.S.* **106** H11
Brahmanbaria, *Bangladesh* **182** J13
Brahmapur, *India* **182** M10
Brahmaputra, river, *Asia* **168** J9
Brahmaur, *India* **180** K14
Brăila, *Rom.* **155** E17
Brainerd, *Minn., U.S.* **106** H10
Braintree, *Mass., U.S.* **97** K15
Braintree, *Eng., U.K.* **145** T15
Brak, river, *S. Af.* **212** K6
Bramsche, *Ger.* **150** D6
Bramwell, *Qnsld., Austral.* **221** P12
Brancaleone Marina, *It.* **153** S13
Branchville, *S.C., U.S.* **98** L9
Branco, Cape, *S. Amer.* **128** E12
Branco, island, *Cabo Verde* **215** B15
Branco, river, *N. Amer.* **84** R12
Branco, river, *Braz.* **132** C8
Brandberg, peak, *Af.* **198** N6
Brandberg, peak, *Namibia* **212** G3
Brandenburg, *Ger.* **150** D9
Brandenburg, region, *Ger.* **150** E10
Brandon Bay *see* Bhreandáin, Bá, *Ire.* **145** T2
Brandon, *Man., Can.* **87** N13
Brandon, *Fla., U.S.* **99** U8
Brandon, *Miss., U.S.* **102** L11
Brandon, *S. Dak., U.S.* **106** L8
Brandvlei, *S. Af.* **212** L5
Brandy Peak, *Oreg., U.S.* **110** K2
Branford, *Fla., U.S.* **99** R7
Bransby Point, *Montserrat, U.K.* **123** B23
Bransby, *Qnsld., Austral.* **221** V12
Bransfield Strait, *Antarctica* **238** C5
Bransfield Trough, *Atl. Oc.* **249** X5
Branson, *Mo., U.S.* **107** V12
Branson, *Colo., U.S.* **109** Q14
Brantley Lake, *N. Mex., U.S.* **109** V14
Brantley, *Ala., U.S.* **103** M16
Brantôme, *Fr.* **149** U8

Bras d'Or Lake, *N.S., Can.* **87** N22
Bras-Panon, *Réunion, Fr.* **215** G17
Brasília Legal, *Braz.* **132** E11
Brasília, *Braz.* **133** J13
Braşov, *Rom.* **155** E15
Brass, *Nigeria* **206** N7
Brasstown Bald, peak, *Ga., U.S.* **98** J6
Brategg Bank, *Pac. Oc.* **254** N11
Brateş, Lacul, *Rom.* **155** E17
Bratislava (Pressburg), *Slovakia* **151** K13
Bratsk Reservoir, *Asia* **168** E9
Bratsk, *Russ.* **161** K15
Bratskoye Vodokhranilishche, *Russ.* **161** K15
Brattleboro, *Vt., U.S.* **97** J13
Braţul Chilia, river, *Rom.* **155** E18
Braunau, *Aust.* **150** K10
Braunschweig (Brunswick), *Ger.* **150** E8
Braunton, *Eng., U.K.* **145** V9
Brava, Costa, *Sp.* **147** D19
Brava, island, *Cabo Verde* **215** E15
Brawley, *Calif., U.S.* **111** Y11
Bray (Bré), *Ire.* **145** R7
Bray Island, *Nunavut, Can.* **87** F17
Brazil Basin, *Atl. Oc.* **249** Q9
Brazil, *Ind., U.S.* **101** S8
Brazil, *S. Amer.* **132** G12
Brazilian Highlands, *S. Amer.* **128** G9
Brazoria, *Tex., U.S.* **105** M19
Brazos, river, *U.S.* **90** K12
Brazos, river, *Tex., U.S.* **105** F13
Brazos, river, *Tex., U.S.* **105** F15
Brazos, river, *Tex., U.S.* **105** K18
Brazza Seamounts, *Atl. Oc.* **249** P13
Brazzaville, *Congo* **208** G10
Brčko, *Bosn. & Herzg.* **154** F8
Bré *see* Bray, *Ire.* **145** R7
Breakers, *Cayman Is., U.K.* **122** N4
Breakfast Island, *Marshall Is.* **227** H20
Breaux Bridge, *La., U.S.* **102** P9
Breaza, *Rom.* **155** E15
Brechin, *Scot., U.K.* **144** K11
Breckenridge, *Tex., U.S.* **105** G15
Breckenridge, *Minn., U.S.* **106** H8
Breckenridge, *Colo., U.S.* **108** M12
Brecknock, Península, *Chile* **135** X7
Breda, *Neth.* **148** K12
Bredasdorp, *S. Af.* **212** N5
Bredefjord *see* Ikersuaq, *Greenland, Den.* **89** T7
Bredefjord *see* Ikersuaq, *Greenland, Den.* **89** T7
Breezewood, *Pa., U.S.* **96** P5
Bregalnica, river, *N. Maced.* **154** K12
Bregenz, *Aust.* **150** L7
Bréhat, Île de, *Fr.* **149** P4
Breid Bay, *Antarctica* **239** B16
Breiðafjörður, bay, *Ice.* **142** D2
Breil-sur-Roya, *Fr.* **149** W16
Breiöamerkurdjúp, valley, *Arctic Oc.* **253** L22
Breisach, *Fr.* **149** Q15
Brejo, *Braz.* **133** E15
Brekilen Bay, *Antarctica* **239** B16
Bremen, *Ga., U.S.* **98** L4
Bremen, *Ind., U.S.* **101** P9
Bremen, *Ger.* **150** C7
Bremerhaven, *Ger.* **150** C7
Bremerton, *Wash., U.S.* **110** C3
Bremond, *Tex., U.S.* **105** J18
Brenham, *Tex., U.S.* **105** K18
Brenner Pass, *Aust.* **150** M8
Brenner Seamounts, *Ind. Oc.* **250** M12
Brennero, *It.* **152** B7
Breno, *It.* **152** D6
Brentwood, *N.Y., U.S.* **96** N12
Brentwood, *Eng., U.K.* **145** U14
Brescia, *It.* **152** D6
Bressanone, *It.* **152** B7
Bressay, island, *Scot., U.K.* **144** C12
Bressuire, *Fr.* **149** S7
Brest, *Fr.* **149** P2
Brest, *Belarus* **158** L2
Breteuil, *Fr.* **149** N10
Breton Island, Cape, *N. Amer.* **84** H10
Breton Islands, *La., U.S.* **102** R12
Breton Sound, *La., U.S.* **102** Q12
Bretón, Cayo, *Cuba* **118** J9
Breton, Pertuis, *Fr.* **149** T5
Breu, river, *S. Amer.* **132** G5
Brevard, *N.C., U.S.* **98** J7
Breves, *Braz.* **132** D12
Brevoort Island, *Nunavut, Can.* **87** G19
Brewarrina, *N.S.W., Austral.* **221** W13
Brewer, *Me., U.S.* **97** F17

Brewers Bay, *Virgin Is., U.K.* **122** Q6
Brewers Bay, *Virgin Is., U.K.* **122** M7
Brewerville, *Liberia* **204** P10
Brewster, *Kans., U.S.* **107** S3
Brewster, *Nebr., U.S.* **107** N5
Brewster, *Wash., U.S.* **110** B6
Brewton, *Ala., U.S.* **103** N15
Brezina, *Alg.* **200** D11
Bria, *Cen. Af. Rep.* **207** M16
Briare, *Fr.* **149** R10
Briatico, *It.* **153** Q13
Bridge Point, *Bahamas* **122** D5
Bridgend (Pen-y-bont ar Ogwr),
  Wales, U.K. **145** U10
Bridgeport, *Conn., U.S.* **96** M12
Bridgeport, *W. Va., U.S.* **98** C9
Bridgeport, *Ill., U.S.* **101** T7
Bridgeport, *Ala., U.S.* **103** G16
Bridgeport, *Tex., U.S.* **105** F16
Bridgeport, *Nebr., U.S.* **107** P2
Bridgeport, *Calif., U.S.* **111** R7
Bridger Peak, *Wyo., U.S.* **108** K11
Bridgeton, *N.J., U.S.* **96** Q9
Bridgetown, *Ohio, U.S.* **101** S11
Bridgetown, *Barbados* **123** L19
Bridgewater, *N.S., Can.* **87** P22
Bridgton, *Me., U.S.* **97** G15
Bridgwater, *Eng., U.K.* **145** V11
Bridlington, *Eng., U.K.* **145** P13
Brielle, *Neth.* **148** J11
Brienne-le-Château, *Fr.* **149** Q12
Brienza, *It.* **153** N12
Brienzer See, *Switz.* **152** B3
Brig, *Switz.* **152** C3
Brigantine Cays, *Bahamas* **122** E7
Brigham City, *Utah, U.S.* **108** K7
Brighton, *Eng., U.K.* **145** V14
Brihuega, *Sp.* **146** F12
Brijuni Otoci, island, *Croatia* **154** E4
Brikama, *Gambia* **204** K6
Brindisi, *It.* **152** M15
Brinkley, *Ark., U.S.* **102** G10
Brisbane, *Qnsld., Austral.* **221** V16
Bristol Bay, *Alas., U.S.* **113** L13
Bristol Bay, *Pac. Oc.* **252** M2
Bristol Channel, *Eng., Wales, U.K.*
  145 U9
Bristol Lake, *Calif., U.S.* **111** W11
Bristol Mountains, *Calif., U.S.* **111** W11
Bristol, *Conn., U.S.* **96** L12
Bristol, *Pa., U.S.* **96** P9
Bristol, *Va., U.S.* **98** G7
Bristol, *Tenn., U.S.* **103** D20
Bristol, *S. Dak., U.S.* **106** J7
Bristol, *Eng., U.K.* **145** U11
Bristow, *Okla., U.S.* **105** C18
Britannia Range, *Antarctica* **239** M14
Britannia Tablemounts, *Pac. Oc.*
  255 D22
British Columbia, *Can.* **86** J8
British Isles, *Eur.* **140** E4
Britt, *Iowa, U.S.* **106** M11
Brittany, region, *Fr.* **149** Q5
Britton Hill, peak, *Fla., U.S.* **99** Q3
Britton, *S. Dak., U.S.* **106** H7
Brive, *Fr.* **149** V9
Briviesca, *Sp.* **146** C11
Brno, *Czechia* **151** J13
Broa, Ensenada de la, *Cuba* **118** G7
Broad Haven, *Ire.* **145** P3
Broad Sound, *Austral.* **220** F15
Broad, river, *S.C., U.S.* **98** J8
Broadstairs, *Eng., U.K.* **145** U16
Broadus, *Mont., U.S.* **108** E12
Broadview, *Sask., Can.* **86** N12
Broadwater, *Nebr., U.S.* **107** P2
Brochet, *Man., Can.* **87** K13
Brock Island, *N.W.T., Can.* **87** B13
Brocken, peak, *Ger.* **150** E8
Brockport, *N.Y., U.S.* **96** J6
Brockton, *Mass., U.S.* **97** L15
Brockway, *Pa., U.S.* **96** M5
Brockway, *Mont., U.S.* **108** C12
Brodeur Peninsula, *Nunavut, Can.*
  87 E15
Brodnica, *Pol.* **151** C15
Brody, *Ukr.* **159** N3
Brogan, *Oreg., U.S.* **110** H9
Broken Arrow, *Okla., U.S.* **105** B18
Broken Bay, *Austral.* **220** K15
Broken Bow Lake, *Okla., U.S.* **105** E20
Broken Bow, *Okla., U.S.* **105** E20
Broken Bow, *Nebr., U.S.* **107** P5
Broken Hill *see* Kabwe, *Zambia*
  209 N18
Broken Hill, *N.S.W., Austral.* **221** W11

Broken Ridge, *Ind. Oc.* **251** M14
Brokopondo, *Suriname* **131** F17
Bromley Plateau, *Atl. Oc.* **249** S8
Brønlundhus, research station,
  Greenland, Den. **88** B10
Brønlunds Grav, *Greenland, Den.*
  88 D12
Bronson, *Fla., U.S.* **99** S7
Bronte, *Tex., U.S.* **105** H14
Brookfield, *Wis., U.S.* **100** L7
Brookfield, *Mo., U.S.* **107** R12
Brookhaven, *Miss., U.S.* **102** M10
Brookings, *S. Dak., U.S.* **106** K8
Brookings, *Oreg., U.S.* **110** L1
Brooklyn Center, *Minn., U.S.* **106** J11
Brooklyn, *Iowa, U.S.* **107** P12
Brookport, *Ill., U.S.* **101** V6
Brooks Range, *Alas., U.S.* **113** D15
Brooks, *Alta., Can.* **86** M10
Brooksville, *Fla., U.S.* **99** T7
Brookville, *Pa., U.S.* **96** M4
Brookville, *Ind., U.S.* **101** S10
Broome, *W. Aust., Austral.* **221** R4
Brora, *Scot., U.K.* **144** G9
Brothers, *Oreg., U.S.* **110** H5
Broughton Islands, *Austral.* **220** K15
Brouwer Trough, *Ind. Oc.* **251** L15
Brovary, *Ukr.* **159** N6
Browerville, *Minn., U.S.* **106** H10
Brown Bay, *Virgin Is., U.S.* **122** M11
Brownfield, *Tex., U.S.* **104** F11
Browning, *Mont., U.S.* **108** B6
Brownlee Dam, *Idaho, Oreg., U.S.*
  110 G9
Browns Canyon National
  Monument, *Colo., U.S.* **109** N12
Browns Town, *Jam.* **118** M11
Browns Valley, *Minn., U.S.* **106** J8
Brownstown, *Ind., U.S.* **101** T9
Brownsville, *Tenn., U.S.* **102** F12
Brownsville, *Tex., U.S.* **105** R17
Brownsweg, *Suriname* **131** F17
Brownwood, *Tex., U.S.* **105** H15
Browse Island, *Austral.* **220** B5
Brozas, *Sp.* **146** G7
Bruce Ridge, *Atl. Oc.* **249** X8
Bruce Rock, *W. Aust., Austral.* **221** X3
Bruce Spur, *Ind. Oc.* **255** D13
Bruce, *Miss., U.S.* **102** J12
Bruce, Mount, *Austral.* **220** F3
Bruchsal, *Ger.* **150** J6
Bruciate, river, *San Marino* **163** K16
Bruck an der Mur, *Aust.* **150** L12
Brugg, *Switz.* **152** A3
Brugge, *Belg.* **148** K11
Bruini, *India* **182** F15
Brukkaros, peak, *Namibia* **212** J4
Brule, river, *Wis., U.S.* **100** G6
Brumado, *Braz.* **133** J15
Brundidge, *Ala., U.S.* **103** M16
Bruneau, *Idaho, U.S.* **108** H4
Bruneau, river, *Idaho, Nev., U.S.*
  108 H4
Brunei Bay, *Malaysia* **192** G11
Brunei, *Asia* **192** G11
Brunette Downs, *N.T., Austral.* **221** R10
Brunico, *It.* **152** B8
Brunssum, *Neth.* **148** L13
Brunswick *see* Braunschweig, *Ger.*
  150 E8
Brunswick, *Me., U.S.* **97** G16
Brunswick, *Ga., U.S.* **99** P8
Brunswick, *Ohio, U.S.* **101** P14
Brunswick, *Md., U.S.* **107** S12
Brunt Ice Shelf, *Antarctica* **238** D11
Brus Laguna, *Hond.* **117** K18
Brush, *Colo., U.S.* **108** L14
Brusque, *Braz.* **133** N13
Brüssel (Brussel) *see* Bruxelles, *Belg.*
  148 L12
Brussels *see* Bruxelles, *Belg.* **148** L12
Bruxelles (Brüssel, Brussels), *Belg.*
  148 L12
Bruzual, *Venez.* **130** E9
Bryan Coast, *Antarctica* **238** H7
Bryan, *Ohio, U.S.* **101** P11
Bryan, *Tex., U.S.* **105** K18
Bryansk, *Russ.* **158** L8
Bryant, *Ark., U.S.* **102** H8
Bryant, *S. Dak., U.S.* **106** K7
Bryce Canyon National Park,
  Utah, U.S. **109** P7
Bryson City, *N.C., U.S.* **98** J6
Bryukhovetskaya, *Russ.* **159** S10
Brzeg Dolny, *Pol.* **151** F13

Brzeg, *Pol.* **151** F13
Bua Bay, *Fiji* **226** N5
Bua Yai, *Thai.* **190** L9
Bua, *Fiji* **226** N6
Bua, river, *Malawi* **212** C10
Bu'aale, *Somalia* **211** N10
Buada Lagoon, *Nauru* **227** M14
Buada, *Nauru* **227** M14
Buala, *Solomon Is.* **228** E9
Buan, *S. Korea* **187** R8
Buariki, *Kiribati* **227** P14
Buariki, island, *Kiribati* **227** P14
Bu'ayrāt al Ḥasūn, *Lib.* **201** E18
Buba, *Guinea-Bissau* **204** L7
Bubaque, *Guinea-Bissau* **204** L7
Bucaramanga, *Col.* **130** E7
Buccament Bay, *St. Vincent & the
  Grenadines* **123** L16
Buccaneer Archipelago, *Austral.*
  220 C5
Buccaneer Point, *Bahamas* **122** B11
Buccoo Bay, *Trin. & Tobago* **123** P17
Buchan Gulf, *Nunavut, Can.* **87** E17
Buchanan, *Mich., U.S.* **101** N9
Buchanan, *Liberia* **204** P10
Bucharest *see* Bucureşti, *Rom.*
  155 F15
Bucheon, *S. Korea* **186** N8
Buchholz, *Ger.* **150** C8
Buchon, Point, *Calif., U.S.* **111** V5
Buck Island Reef National
  Monument, *Virgin Is., U.S.* **122** Q3
Buck Island, *Virgin Is., U.K.* **122** Q7
Buck Island, *Virgin Is., U.S.* **122** N9
Buck Island, *Virgin Is., U.S.* **122** Q3
Buckalow, *N.S.W., Austral.* **221** X11
Buckhannon, *W. Va., U.S.* **98** D9
Buckingham Bay, *Austral.* **220** A9
Buckland Tableland, *Austral.* **220** G13
Buckley Bay, *Antarctica* **239** R17
Buckley, *Wash., U.S.* **110** C4
Bucklin, *Kans., U.S.* **107** U5
Buckner Bay *see* Nakagusuku Wan,
  Japan **189** X2
Bucksport, *Me., U.S.* **97** F17
Bucovina, region, *Rom.* **155** A14
Bucureşti (Bucharest), *Rom.* **155** F15
Bucyrus, *Ohio, U.S.* **101** Q13
Bud Bud, *Somalia* **211** Q12
Bud, *Nor.* **142** H10
Buda, *Tex., U.S.* **105** L16
Budac, peak, *Rom.* **155** C15
Budalin, *Myanmar* **190** G4
Budapest, *Hung.* **151** L15
Búðardalur, *Ice.* **142** E2
Budaun, *India* **182** G7
Budd Coast, *Antarctica* **239** M21
Buddhas of Bamian, ruins, *Afghan.*
  180 H8
Buddusò, *It.* **152** M4
Bude Bay, *Eng., U.K.* **145** V9
Bude, *Miss., U.S.* **102** M10
Bude, *Eng., U.K.* **145** V9
Budennovsk, *Russ.* **159** S13
Búðir, *Ice.* **142** F4
Budjala, *D.R.C.* **209** C13
Budongquan, *China* **184** J7
Buell Peninsula, *Antarctica* **239** R15
Buen Tiempo, Cabo, *Arg.* **135** W8
Buena Esperanza, *Arg.* **134** L9
Buena Vista Lake Bed, *Calif., U.S.*
  111 V7
Buena Vista, *Va., U.S.* **98** E11
Buena Vista, *Ga., U.S.* **99** N5
Buena Vista, *Colo., U.S.* **109** N12
Buenaventura Bay, *S. Amer.* **128** C2
Buenaventura, *Mex.* **116** C8
Buenaventura, *Col.* **130** G5
Buendía, Embalse de, *Sp.* **146** F12
Buenos Aires, *Arg.* **134** L13
Buenos Aires, Lago, *Arg.* **135** T7
Buenos Ayres, *Trin. & Tobago* **123** R21
Bufareh, *Indonesia* **193** K20
Buff Bay, *Jam.* **122** H10
Buffalo Bill Reservoir, *Wyo., U.S.*
  108 F9
Buffalo Hump, peak, *Idaho, U.S.*
  108 E4
Buffalo National River, *Ark., U.S.*
  102 E8
Buffalo, *N.Y., U.S.* **96** J5
Buffalo, *Okla., U.S.* **105** B14
Buffalo, *Tex., U.S.* **105** J18

Buffalo, *Minn., U.S.* **106** J11
Buffalo, *S. Dak., U.S.* **106** J2
Buffalo, *Mo., U.S.* **107** U12
Buffalo, *Wyo., U.S.* **108** F11
Buffalo, river, *Ark., U.S.* **102** E8
Buford, *Ga., U.S.* **98** K5
Buftea, *Rom.* **155** F15
Buga, *Col.* **130** G5
Buganda, *Uganda* **211** S3
Buġibba, *Malta* **163** J22
Bugio, Ilhéu do, *Madeira, Port.* **214** N4
Bugsuk, island, *Philippines* **192** F12
Bugul'ma, *Russ.* **158** J16
Buguruslan, *Russ.* **158** K16
Buḩayrat ar Razāzah, lake, *Iraq*
  176 H10
Buḩayrat ath Tharthār, lake, *Iraq*
  176 G10
Buḩayrat Sadd al Qādisīyah (Lake
  Qadisiyah), *Iraq* **176** F9
Buḩayrat Sadd Şaddām, lake, *Iraq*
  176 D10
Buhen, ruin, *Sudan* **202** L11
Buhl, *Minn., U.S.* **106** F12
Buhl, *Idaho, U.S.* **108** H5
Buhuşi, *Rom.* **155** C16
Buin, *P.N.G.* **222** P11
Buin, *P.N.G.* **228** C5
Buinsk, *Russ.* **158** J14
Buir Nur, lake, *Asia* **168** F11
Buis-les-Baronnies, *Fr.* **149** W13
Bujalance, *Sp.* **146** K10
Bujanovac, *Serb.* **154** J11
Bujaraloz, *Sp.* **147** E15
Bujumbura, *Burundi* **211** U2
Buka Passage, *P.N.G.* **228** B4
Buka, island, *P.N.G.* **222** P11
Buka, island, *P.N.G.* **228** A4
Bukachacha, *Russ.* **161** L17
Bukama, *D.R.C.* **209** K17
Bukavu, *D.R.C.* **209** F18
Bukhara *see* Buxoro, *Uzb.* **178** M10
Bukhon, *S. Korea* **187** Q8
Bukene, *Tanzania* **211** V4
Bükḩān, *Iran* **176** E12
Bukanik, Maja, *Alban.* **154** L10
Bukavu, *D.R.C.* **209** K17
Bukittinggi, *Indonesia* **192** J5
Bukoba, *Tanzania* **211** T3
Bukwiuni, *Nigeria* **206** J7
Bula, *Indonesia* **193** K17
Bülach, *Switz.* **152** A4
Bulacle, *Somalia* **211** P12
Būlaevo, *Kaz.* **178** B12
Bulahdelah, *N.S.W., Austral.* **221** X15
Bulan, *Philippines* **193** D14
Bulancak, *Turk.* **171** D13
Bulandshahr, *India* **182** G6
Bulanık, *Turk.* **171** F16
Bulawayo, *Zimb.* **212** F8
Buldan, *Turk.* **170** G4
Buldir Island, *Alas., U.S.* **112** M2
Bulgan, *Mongolia* **184** C10
Bulgan, *Mongolia* **184** C9
Bulgan, *Mongolia* **184** E8
Bulgaria, *Eur.* **155** J13
Buli, Teluk, *Indonesia* **193** J16
Buliluyan, Cape, *Philippines* **192** F12
Bulimba, *Qnsld., Austral.* **221** Q12
Bull Bay, *Jam.* **122** J10
Bull Mountains, *Mont., U.S.* **108** D10
Bull Shoals Lake, *Ark., Mo., U.S.*
  102 E7
Bullaxaar, *Somaliland* **210** L11
Bulle, *Switz.* **152** B2
Bullenbaai, bay, *Curaçao, Neth.* **123** P14
Bullhead City, *Ariz., U.S.* **109** S4
Bullo River, *N.T., Austral.* **221** Q7
Bulloo Downs, *Qnsld., Austral.* **221** V12
Bulls Bay, *S.C., U.S.* **98** M11
Bulpunga, *N.S.W., Austral.* **221** X11
Bulun, *Russ.* **161** E16
Bulungu, *D.R.C.* **208** G12
Bulung'ur, *Uzb.* **178** M11
Bum La, pass, *India* **182** G13
Bumba, *D.R.C.* **209** C15
Bumbat, *Mongolia* **184** D10
Bumpe, *Sierra Leone* **204** N9
Bun Clóidí *see* Bunclody, *Ire.* **145** S6
Buna, *D.R.C.* **209** G13
Buna, *Kenya* **211** Q8
Bunazi, *Tanzania* **211** T3
Bunbury, *W. Aust., Austral.* **221** X2

Bunclody (Bun Clóidí), *Ire.* **145** S6
Bunda, *Tanzania* **211** T4
Bundarra, *N.S.W., Austral.* **221** W15
Bundi, *India* **182** H5
Bungalaut, Selat, *Indonesia* **192** K5
Bungenstock Plateau, *Atl. Oc.*
  249 Y11
Bunger Hills, *Antarctica* **239** K21
Bungku, *Indonesia* **193** K14
Bungo Strait, *Pac. Oc.* **168** G13
Bungo Suidō, *Japan* **189** R6
Bungo, *Angola* **208** J10
Bungoma, *Kenya* **211** S5
Bunguran Selatan *see* Kepulauan
  Natuna Selatan, *Indonesia* **192** H9
Bunguran Utara *see* Natuna Besar,
  Kepulauan, *Indonesia* **192** G8
Bunhyang, *S. Korea* **187** S8
Bunia, *D.R.C.* **209** D19
Bunji, *Pak.* **180** G13
Bunker, *Mo., U.S.* **107** U14
Bunkeya, *D.R.C.* **209** L17
Bunkie, *La., U.S.* **102** N8
Bunnell, *Fla., U.S.* **99** S9
Buntok, *Indonesia* **192** K11
Bünyan, *Turk.* **170** F11
Bunyu, island, *Indonesia* **192** H12
Bunza, *Nigeria* **206** J6
Buol, *Indonesia* **193** H14
Buôn Đôn, *Vietnam* **191** N12
Buôn Ma Thuột, *Vietnam* **191** N12
Buor-Khaya Bay, *Arctic Oc.* **252** D8
Buqbuq, *Egypt* **202** E8
Buqda Caqable, *Somalia* **211** Q11
Bûr Fu'ad, *Egypt* **173** N1
Bur Khaybyt, *Russ.* **161** E18
Bûr Safājah, *Egypt* **202** H13
Bûr Sa'īd (Port Said), *Egypt* **173** N1
Bûr Sa'īd (Port Said), *Egypt* **202** E12
Bur, *Russ.* **161** J15
Bura, *Kenya* **211** T8
Buraan, *Somaliland* **210** L13
Burakin, *W. Aust., Austral.* **221** W3
Buram, *Sudan* **203** T7
Burang (Pulan), *China* **184** K2
Burao *see* Burco, *Somaliland* **210** L11
Buras, *La., U.S.* **102** Q12
Buratai, *Nigeria* **206** J10
Buraydah, *Saudi Arabia* **174** F9
Burbank, *Calif., U.S.* **111** X8
Burco (Burao), *Somaliland* **210** L11
Burdalyk, *Turkm.* **178** N10
Burdekin, river, *Austral.* **220** E13
Burdesi, *Eth.* **211** N6
Burdur Gölü, *Turk.* **170** G5
Burdur, *Turk.* **170** G5
Burdwood Bank, *Atl. Oc.* **249** W5
Burē, *Eth.* **210** L6;M5
Burg, *Ger.* **150** E9
Burgas Bay, *Eur.* **140** K10
Burgas, *Bulg.* **155** J17
Burgaski Zaliv, *Bulg.* **155** J17
Burgasko Ezero, *Bulg.* **155** J17
Burgaw, *N.C., U.S.* **98** J12
Burgdorf, *Switz.* **152** B3
Burgess Hill, *Eng., U.K.* **145** V14
Burghūth, Sabkhat al, *Syr.* **172** G14
Burgos, *Sp.* **146** C11
Burgsvik, *Sw.* **143** L14
Burgundy, region, *Fr.* **149** S12
Burhaniye, *Turk.* **170** E3
Burhanpur, *India* **182** L5
Burhou, island, *Channel Is., U.K.*
  145 X11
Burias, island, *Philippines* **193** D14
Burica Point, *N. Amer.* **84** Q8
Burica, Punta, *Pan.* **117** P19
Buriram, *Thai.* **190** L9
Burkburnett, *Tex., U.S.* **105** E15
Burke, *S. Dak., U.S.* **106** M6
Burkesville, *Ky., U.S.* **103** D17
Burketown, *Qnsld., Austral.* **221** R11
Burkina Faso, *Af.* **205** L14
Burle, island, *Marshall Is.* **227** H16
Burley, *Idaho, U.S.* **108** H5
Burlingame, *Kans., U.S.* **107** T9
Burlington, *Vt., U.S.* **96** F12
Burlington, *N.C., U.S.* **98** G11
Burlington, *Wis., U.S.* **100** M6
Burlington, *Iowa, U.S.* **107** Q14
Burlington, *Kans., U.S.* **107** T9
Burlington, *Colo., U.S.* **108** M16
Burlington, *Wash., U.S.* **110** A4
Burma *see* Myanmar, *Asia* **190** H4
Burnet, *Tex., U.S.* **105** K16
Burnett Bay, *N.W.T., Can.* **86** C11

Burney, *Calif., U.S.* **110** M4
Burnham Market, *Eng., U.K.* **145** R15
Burnham-on-Sea, *Eng., U.K.* **145** V11
Burnie, *Tas., Austral.* **221** Y7
Burns Lake, *B.C., Can.* **86** K8
Burns, *Wyo., U.S.* **108** K14
Burns, *Oreg., U.S.* **110** J7
Burnside, Lake, *Austral.* **220** G5
Burnsville, *Minn., U.S.* **106** K11
Burnt Pine, *Austral.* **229** B20
Burnt, *river, Oreg., U.S.* **110** G8
Burqin, *China* **184** C5
Burra, *S. Aust. Austral.* **221** X10
Burragorang, Lake, *Austral.* **220** K14
Burray, island, *Scot., U.K.* **144** F10
Burren, region, *Ire.* **145** R4
Burrendong Reservoir, *Austral.* **220** K14
Burriana, *Sp.* **147** G15
Burrton, *Kans., U.S.* **107** U7
Burrwood, *La., U.S.* **102** R12
Bursa, *Turk.* **170** D5
Burt Lake, *Mich., U.S.* **100** H11
Burton Island Glacier, *Antarctica* **239** J22
Burton, *Mich., U.S.* **100** L12
Burton, *Nebr., U.S.* **106** M5
Buru, island, *Indonesia* **193** K16
Burūm, *Yemen* **174** P12
Burun Shibertuy, *Russ.* **161** M16
Burundi, *Af.* **211** U1
Bururi, *Burundi* **211** U2
Burutu, *Nigeria* **206** M7
Burwash Landing, *Yukon, Can.* **86** F7
Burwell, *Nebr., U.S.* **107** P6
Bury Saint Edmunds, *Eng., U.K.* **145** T15
Burylbaytal, *Kaz.* **179** H14
Burzil Pass, *Pak.* **180** H13
Busan (Pusan), *S. Korea* **187** T12
Busanga Swamp, *Zambia* **209** N16
Busanga, *D.R.C.* **209** L16
Busangu, *D.R.C.* **209** K16
Buşayrah, *Syr.* **172** F13
Bushiribana, *Aruba, Neth.* **123** Q17
Bushmanland, *S. Af.* **212** K5
Bushnell, *Ill., U.S.* **101** Q4
Bushnell, *Nebr., U.S.* **107** P1
Busia, *Kenya* **211** S5
Businga, *D.R.C.* **209** C14
Busira, river, *D.R.C.* **209** E13
Bušk Blato, lake, *Bosn. & Herzg.* **154** G7
Buslē, *Eth.* **211** P11
Buşrá ash Shām, *Syr.* **172** L7
Busselton, *W. Aust. Austral.* **221** X2
Bussière-Poitevine, *Fr. Lux.* **149** T8
Bussol', *Russ.* **161** J23
Bussoleno, *It.* **152** E2
Bustamante, Bahía, *Arg.* **135** S9
Busto Arsizio, *It.* **152** D4
Busuanga, island, *Philippines* **193** D13
Buta, *D.R.C.* **209** C16
Butare, *Rwanda* **211** U2
Butaritari, island, *Kiribati* **224** F8
Bute Helu, *Kenya* **211** Q8
Bute, Island of, *Scot., U.K.* **144** L8
Butembo, *D.R.C.* **209** E19
Buthidaung, *Myanmar* **190** H3
Butiaba, *Uganda* **211** R3
Butler Bay, *Virgin Is., U.S.* **122** Q1
Butler, *Pa., U.S.* **96** N3
Butler, *Ga., U.S.* **98** M5
Butler, *Ala., U.S.* **103** L13
Butler, *Mo., U.S.* **107** T11
Butner, *N.C., U.S.* **98** G11
Buton, island, *Indonesia* **193** L14
Butrint, *Alban.* **154** N9
Butte, *Nebr., U.S.* **106** M6
Butte, *Mont., U.S.* **108** E7
Butterworth, *Malaysia* **191** T8
Butterworth, *S. Af.* **212** L8
Button Bay, *Man., Can.* **87** K14
Button Islands, *Nfld. & Lab., Can.* **87** H20
Butuan, *Philippines* **193** E15
Buturlinovka, *Russ.* **159** N11
Butwal, *Nepal* **182** G9
Butyaalo, *Somalia* **210** K14
Buuhoodle, *Somaliland* **210** M12
Buur Gaabo, *Somalia* **211** T9
Buurhakaba, *Somalia* **211** Q11
Buxoro (Bukhara), *Uzb.* **178** M10
Buxtehude, *Ger.* **150** C7

Buxton, *N.C., U.S.* **98** H15
Buxton, *Guyana* **131** E16
Buy, *Russ.* **158** G11
Buyant, *Mongolia* **184** D12
Buyeo, *S. Korea* **187** Q8
Buynaksk, *Russ.* **159** U15
Buynī Qarah, *Afghan.* **180** F7
Buyr Nuur, *Mongolia* **185** D13
Büyük Menderes, *river, Turk.* **170** G4
Buzău, *Rom.* **155** E16
Buzău, Pasul, *Rom.* **155** E15
Buzău, river, *Rom.* **155** E16
Buzaymah, *Lib.* **201** H22
Búzi, river, *Mozambique* **212** F10
Büzmeyin, *Turkm.* **178** N7
Buzul Dağı, peak, *Turk.* **171** H18
Buzuluk, *Russ.* **158** L16
Buzuluk, *Kaz.* **178** D10
Buzzards Bay, *Mass., U.S.* **97** M15
Bwokworlap, island, *Marshall Is.* **227** D21
Byahoml', *Belarus* **158** K5
Byala Slatina, *Bulg.* **155** H13
Byala, *Bulg.* **155** H17
Byam Martin Island, *Nunavut, Can.* **87** D14
Byam Martin, Cape, *Nunavut, Can.* **87** D17
Byarezina, river, *Belarus* **158** K5
Byaroza, *Belarus* **158** L3
Byblos *see* Jbail, *Leb.* **172** H6
Bydgoszcz, *Pol.* **151** C14
Byers, *Colo., U.S.* **108** M14
Bykhaw, *Belarus* **158** L6
Bykovo, *Russ.* **159** P13
Bylas, *Ariz., U.S.* **109** U8
Bylot Island, *Nunavut, Can.* **87** E16
Bynum Reservoir, *Mont., U.S.* **108** B7
Byramgore Reef, *Ind. Oc.* **250** E10
Byrd Canyon, *Pac. Oc.* **255** J15
Byrd Glacier, *Antarctica* **239** M14
Byrock, *N.S.W., Austral.* **221** W13
Byron, Cape, *Austral.* **220** H16
Byron, Isla, *Chile* **135** U6
Bytom, *Pol.* **151** G14
Bytów, *Pol.* **151** B13
Byumba, *Rwanda* **211** T2

C

C. H. Ostenfeld Land, *Greenland, Den.* **88** G13
C. Hofman Halvø, *Greenland, Den.* **88** K12
Ca' Chiavello, river, *San Marino* **163** K16
Cà Mau, *Vietnam* **191** Q11
Cà Ná, *Vietnam* **191** P13
Caála, *Angola* **208** M11
Caazapá, *Parag.* **132** N10
Cabaiguán, *Cuba* **118** H9
Caballo Reservoir, *N. Mex., U.S.* **109** U11
Caballones, Cayo, *Cuba* **118** J9
Cabanatuan, *Philippines* **193** C13
Cabaneta, Pic de la, *Andorra, Fr.* **162** H5
Cabes Point, *Virgin Is., U.S.* **122** N9
Cabeza del Buey, *Sp.* **146** J9
Cabezas, *Bol.* **132** K8
Cabimas, *Venez.* **130** C8
Cabinda, *Angola* **208** H9
Cabinet Mountains, *Idaho, Mont., U.S.* **108** B4
Cable, *Wis., U.S.* **100** F3
Cabo, *Braz.* **133** G18
Cabo Berkeley, *Ecua.* **130** P2
Cabo Blanco, *Arg.* **135** T9
Cabo Carvoeiro, *Port.* **146** G4
Cabo de Santa Maria, *Angola* **208** N9
Cabo de São Bráz, *Angola* **208** L9
Cabo Delgado, *Mozambique* **213** B13
Cabo Douglas, *Ecua.* **130** Q2
Cabo Frio, *Braz.* **133** M15
Cabo Ledo, *Angola* **208** L9
Cabo San Juan, *Eq. Guinea* **206** Q9
Cabo San Lucas, *Mex.* **116** F7
Cabo Verde, *Af.* **214** F3
Cabonga, Réservoir, *Que., Can.* **87** P19
Cabool, *Mo., U.S.* **107** V13
Caboolture, *Qnsld. Austral.* **221** V16
Caborca, *Mex.* **116** C7

Cabot Strait, *N. Amer.* **84** H10
Cabot, *Ark., U.S.* **102** G9
Cabral, *Dom. Rep.* **119** M18
Cabras Island, *Guam, U.S.* **230** P4
Cabras, Ilhéu das, *Sao Tome and Principe* **215** C20
Cabrera Baja, Sierra de la, *Sp.* **146** C7
Cabrera, *Dom. Rep.* **119** L19
Cabrera, island, *Sp.* **147** H18
Cabril, Barragem do, *Port.* **146** G6
Cabritte Horn Point, *Virgin Is., U.S.* **122** N11
Cabrobó, *Braz.* **133** G16
Cabrón, Cabo, *Dom. Rep.* **119** L20
Čačak, *Serb.* **154** G10
Caccia, Capo, *It.* **152** M3
Cáceres, *Braz.* **132** J10
Cáceres, *Sp.* **146** H8
Cache, Peñas del, *Sp.* **214** P7
Cache, river, *Ark., U.S.* **102** E10
Cacheu, *Guinea-Bissau* **204** K7
Cachimbo, *Braz.* **132** G11
Cachingues, *Angola* **208** N11
Cachoeira do Sul, *Braz.* **132** P11
Cachoeira, *Braz.* **133** H16
Cachoeiro do Itapemirim, *Braz.* **133** L15
Cacine, *Guinea-Bissau* **204** L7
Cacolo, *Angola* **209** L13
Caconda, *Angola* **208** N10
Cacongo, *Angola* **208** H9
Cactus, *Tex., U.S.* **104** C12
Caçu, *Braz.* **132** K12
Caculé, *Braz.* **133** J15
Caculuvar, river, *Angola* **208** Q10
Cacuri, *Venez.* **130** G11
Cacuso, *Angola* **208** K11
Cadaadley, *Somaliland* **210** L11
Cadale, *Somalia* **211** Q12
Čadca, *Slovakia* **151** H14
Caddo Lake, *La., Tex., U.S.* **105** G21
Caddo, *Okla., U.S.* **105** E18
Cadí, Serra del, *Sp.* **147** C17
Cadillac, *Mich., U.S.* **100** J10
Cadillac, *Fr.* **149** W7
Cadiz Lake, *Calif., U.S.* **111** X11
Cadiz, *Ohio, U.S.* **101** Q15
Cadiz, *Ky., U.S.* **103** D14
Cadiz, *Calif., U.S.* **111** W11
Cadiz, *Philippines* **193** E14
Cádiz, *Sp.* **146** M7
Cádiz, Badia de, *Sp.* **146** M7
Cádiz, Golfo de, *Port., Sp.* **146** L7
Çadır Dağı, peak, *Turk.* **171** G17
Caen, *Fr.* **149** N7
Caerdydd *see* Cardiff, *Wales, U.K.* **145** U10
Caergybi *see* Holyhead, *Wales, U.K.* **145** R9
Caernarfon Bay, *Wales, U.K.* **145** R9
Caerphilly, *Wales, U.K.* **145** U10
Caesarea, ruins, *Israel* **172** L5
Cafayate, *Arg.* **134** F9
Cagayan de Oro, *Philippines* **193** F15
Cagayan Islands, *Philippines* **193** E13
Cagayan Sulu Island, *Malaysia* **192** F12
Cagli, *It.* **152** H8
Cagliari, *It.* **153** Q4
Cagliari, Golfo di, *It.* **153** Q4
Cagnes, *Fr.* **149** X15
Caguán, river, *Col.* **130** J6
Caha Mountains, *Ire.* **145** U3
Cahama, *Angola* **208** P10
Cahir, *Ire.* **145** S5
Cahora Bassa Dam, *Mozambique* **212** D10
Cahora Bassa, *Mozambique* **212** D10
Cahore Point, *Ire.* **145** S7
Cahors, *Fr.* **149** W9
Cahul, *Mold.* **159** S4
Caia, *Mozambique* **212** E11
Caianda, *Angola* **209** L15
Caiapônia, *Braz.* **132** J12
Caibarién, *Cuba* **118** H9
Caicara, *Venez.* **130** E11
Caicó, *Braz.* **133** F17
Caicos Islands, *Turks & Caicos Is., U.K.* **119** J17
Caicos Passage, *N. Amer.* **119** H16
Caille Island, *Grenada* **123** H24
Cailungo, *San Marino* **163** J15
Caimbambo, *Angola* **208** N10
Cainde, *Angola* **208** P10
Cainsville, *Mo., U.S.* **107** Q11

Caird Coast, *Antarctica* **238** D11
Cairngorm Mountains, *Scot., U.K.* **144** J9
Cairns, *Qnsld. Austral.* **221** R13
Cairo *see* Al Qāhirah, *Egypt* **202** F11
Cairo, *Ga., U.S.* **99** Q5
Cairo, *Ill., U.S.* **101** V5
Caiseal *see* Cashel, *Ire.* **145** S5
Caisleán an Bharraigh *see* Castlebar, *Ire.* **145** Q3
Caisleán Ghriaire *see* Castlegregory, *Ire.* **145** T2
Caitou, *Angola* **208** N9
Caiundo, *Angola* **208** P12
Caixão Grande, *Sao Tome and Principe* **215** D20
Cajamarca, *Peru* **132** F2
Cajàzeiras, *Braz.* **133** F17
Çakit, river, *Turk.* **170** H10
Çakırgöl Dağı, peak, *Turk.* **171** D14
Çakmak Dağı, peak, *Turk.* **171** E16
Çakmak, *Turk.* **170** F7
Čakovec, *Croatia* **154** C6
Çal, *Turk.* **170** G5
Calabar, *Nigeria* **206** N8
Calabash Boom, *Virgin Is., U.S.* **122** N11
Calabó, *Eq. Guinea* **206** M7
Calabozo, *Venez.* **130** D10
Calabria, *Eur.* **140** L8
Calaburras, Punta de, *Sp.* **146** M9
Calacuccia, *Fr.* **149** X2
Calafat, *Rom.* **154** G12
Calafort Ros Láir *see* Rosslare Harbour, *Ire.* **145** T7
Calahorra, *Sp.* **147** C13
Calais, *Me., U.S.* **97** E19
Calais, *Fr.* **148** L9
Calais, Pas de *see* Dover, Strait of, *Fr.* **148** L9
Calalaste, Sierra de, *Arg.* **134** F8
Calalin Channel, *Marshall Is.* **227** D20
Calalin, island, *Marshall Is.* **227** D20
Calalzo, *It.* **152** C8
Calama, *Braz.* **132** F8
Calama, *Chile* **134** D7
Calamar, *Col.* **130** D6
Calamar, *Col.* **130** H7
Calamian Group, *Philippines* **193** D13
Calamocha, *Sp.* **147** E13
Calan, *Rom.* **154** E12
Calanda, *Sp.* **147** E15
Calandula Falls, *Angola* **208** K11
Calandula, *Angola* **208** K11
Calanscio, Sand Sea of, *Af.* **198** D7
Calapan, *Philippines* **193** D13
Călăraşi, *Rom.* **155** G17
Calatayud, *Sp.* **147** E13
Calatrava, *Eq. Guinea* **206** Q9
Calayan, island, *Philippines* **193** A13
Calbayog, *Philippines* **193** D15
Calcanhar, Ponta do, *Braz.* **133** E18
Calcasieu Lake, *La., U.S.* **102** Q7
Calcasieu, river, *La., U.S.* **102** P7
Calçoene, *Braz.* **132** B12
Calcutta *see* Kolkata, *India* **182** K12
Caldas da Rainha, *Port.* **146** G4
Caldera, *Chile* **134** G6
Çaldıran, *Turk.* **171** F18
Caldwell, *Tex., U.S.* **105** K18
Caldwell, *Kans., U.S.* **107** V7
Caldwell, *Idaho, U.S.* **108** G3
Caldy Island, *Wales, U.K.* **145** U9
Caledon Bay, *Austral.* **220** B10
Caledon, river, *Lesotho, S. Af.* **212** K8
Caledonia, *Minn., U.S.* **106** L13
Caledonian Canal, *Scot., U.K.* **144** J8
Calexico, *Calif., U.S.* **111** Z11
Calf of Man (Yn Cholloo), cape, *Isle of Man, U.K.* **145** Q8
Calgary, *Alta. Can.* **86** M10
Calheta, *Port.* **146** P4
Calheta, *Madeira, Port.* **214** M2
Calhoun City, *Miss., U.S.* **102** J12
Calhoun Falls, *S.C., U.S.* **98** K7
Calhoun, *Ga., U.S.* **98** K4
Cali, *Col.* **130** H5
Calibishie, *Dominica* **123** E19
Calico Rock, *Ark., U.S.* **102** E9
Calicut *see* Kozhikode, *India* **183** R5
Caliente, *Nev., U.S.* **111** S12
California Aqueduct, *Calif., U.S.* **111** V6
California Coastal National Monument Point Arena-Stornetta Unit, *Calif., U.S.* **111** Q2
California, *Mo., U.S.* **107** T12
California, *U.S.* **111** S5

California, *Trin. & Tobago* **123** Q22
California, Golfo de, *Mex.* **116** D6
Calilegua, *Arg.* **134** E9
Calipatria, *Calif., U.S.* **111** Y11
Calistoga, *Calif., U.S.* **111** R3
Calivigny Island, *Grenada* **123** M22
Calivigny, *Grenada* **123** L22
Callabonna, Lake, *Austral.* **220** H11
Callainn *see* Callan, *Ire.* **145** S5
Callan (Callainn), *Ire.* **145** S5
Callao, *Peru* **132** H3
Callaway, *Nebr., U.S.* **107** P4
Calliaqua, *St. Vincent & the Grenadines* **123** L16
Calliope Range, *Austral.* **220** G15
Caloosahatchee, river, *Fla., U.S.* **99** V8
Calpe, *Sp.* **147** J15
Caltagirone, *It.* **153** T11
Caltanissetta, *It.* **153** T10
Çaltı Burnu, *Turk.* **170** C12
Caluango, *Angola* **209** K13
Calucinga, *Angola* **208** L11
Calulo, *Angola* **208** L10
Calumet, *Mich., U.S.* **100** E6
Calunda, *Angola* **209** M15
Caluquembe, *Angola* **208** N10
Caluso, *It.* **152** E3
Caluula, *Somalia* **210** K15
Calvert Hills, *N.T. Austral.* **221** Q10
Calvert, *Tex., U.S.* **105** J18
Calvi, *Fr.* **149** X2
Calvinia, *S. Af.* **212** L5
Calwa, *Calif., U.S.* **111** T6
Çam Burnu, *Turk.* **170** D12
Cẩm Phả, *Vietnam* **190** H12
Cam Ranh, *Vietnam* **191** N13
Camabatela, *Angola* **208** K10
Camacupa, *Angola* **208** M12
Camagüey, *Cuba* **118** J11
Camagüey, Archipiélago de, *Cuba* **118** H11
Camajuaní, *Cuba* **118** H9
Camaná, *Peru* **132** K4
Camanche Reservoir, *Calif., U.S.* **111** R5
Camanongue, *Angola* **209** L13
Camapuã, *Braz.* **132** K11
Câmara de Lobos, *Madeira. Port.* **214** M3
Camarat, Cap, *Fr.* **149** Y15
Çamardı, *Turk.* **170** G10
Camargo, *Okla., U.S.* **105** C15
Camargo, *Bol.* **132** K9
Camargue, Île de la, *Fr.* **149** X12
Camarón, Cape, *N. Amer.* **84** P8
Camarones, *Arg.* **135** S9
Camarones, Bahía, *Arg.* **135** S9
Camas Creek, *Idaho, U.S.* **108** H4
Camas Valley, *Oreg., U.S.* **110** J2
Camas, *Wash., U.S.* **110** F4
Camaxilo, *Angola* **209** K13
Camballin, *W. Aust. Austral.* **221** R5
Cambo, river, *Angola* **208** K11
Cambodia, *Asia* **191** N10
Camborne, *Eng., U.K.* **145** X8
Cambrai, *Fr.* **148** M11
Cambre, *Sp.* **146** B6
Cambria, *Calif., U.S.* **111** V5
Cambrian Mountains, *Wales, U.K.* **145** T10
Cambridge Bay, *Nunavut, Can.* **87** F13
Cambridge Bugt, bay, *Greenland, Den.* **88** J14
Cambridge Gulf, *Austral.* **220** C6
Cambridge Springs, *Pa., U.S.* **96** L3
Cambridge, *Mass., U.S.* **97** K15
Cambridge, *Md., U.S.* **98** D14
Cambridge, *Ohio, U.S.* **101** R14
Cambridge, *Minn., U.S.* **106** J11
Cambridge, *Nebr., U.S.* **107** R5
Cambridge, *Idaho, U.S.* **108** F3
Cambridge, *Jam.* **118** M11
Cambridge, *Jam.* **122** H7
Cambridge, *Eng., U.K.* **145** T14
Cambulo, *Angola* **209** J14
Cambundi-Catembo, *Angola* **208** L12
Camden Park, *St. Vincent & the Grenadines* **123** L16
Camden, *N.J., U.S.* **96** P10
Camden, *N.Y., U.S.* **96** H9
Camden, *Me., U.S.* **97** G17
Camden, *S.C., U.S.* **98** K9
Camden, *Ark., U.S.* **102** J8
Camden, *Ala., U.S.* **103** M15
Camden, *Tenn., U.S.* **103** E13
Camden, *Tex., U.S.* **105** J20
Camdenton, *Mo., U.S.* **107** U12

Çameli, *Turk.* **170** H5
Camerino, *It.* **152** H9
Cameron, *La., U.S.* **102** Q7
Cameron, *Tex., U.S.* **105** K17
Cameron, *Mo., U.S.* **107** R11
Cameron, *Ariz., U.S.* **109** R7
Cameroon, *Af.* **206** M10
Cametá, *Braz.* **133** D13
Camiguin, island, *Philippines* **193** A14
Camiguin, island, *Philippines* **193** E15
Camilla, *Ga., U.S.* **99** P5
Caminha, *Port.* **146** D5
Camiri, *Bol.* **132** L8
Camirus *see* Kámeiros, ruin, *Gr.* **157** M19
Camissombo, *Angola* **209** K14
Camlibel *see* Myrtou, *Cyprus* **162** N7
Camocim, *Braz.* **133** D16
Camooweal, *Qnsld. Austral.* **221** S10
Camopi, river, *Fr. Guiana, Fr.* **131** H18
Camorta Island, *India* **183** T15
Camp Century, *Greenland, Den.* **88** F5
Camp Crook, *S. Dak. U.S.* **106** J1
Camp Point, *Ill., U.S.* **101** R3
Camp Verde, *Ariz., U.S.* **109** S7
Campagnano di Roma, *It.* **152** K8
Campana, *Arg.* **134** L12
Campana, *It.* **153** P14
Campana, Isla, *Chile* **135** U6
Campanario, Cerro, *Arg.* **134** M7
Campania, *Eur.* **140** K7
Campbell Escarpment, *Pac. Oc.* **246** R10
Campbell Hill, peak, *Ohio, U.S.* **101** Q12
Campbell Island (Motu Ihupuku), *Pac. Oc.* **224** R7
Campbell Island Shelf, *Pac. Oc.* **255** G19
Campbell Plateau, *Pac. Oc.* **246** R10
Campbell River, *B.C. Can.* **86** L7
Campbell Town, *Tas., Austral.* **221** Y8
Campbell, *Mo., U.S.* **107** V15
Campbellsville, *Ky., U.S.* **103** C17
Campcardos, Pic de, *Fr., Sp.* **162** J6
Campeche Bank, *Atl. Oc.* **248** L2
Campeche, *Mex.* **117** H15
Campêche, *Guadeloupe, Fr.* **123** D15
Campechuela, *Cuba* **118** K11
Campidano, plain, *It.* **153** P3
Campina Grande, *Braz.* **133** F18
Campinas, *Braz.* **133** M13
Campli, *It.* **152** J10
Campo Gallo, *Arg.* **134** G10
Campo Grande, *Braz.* **132** L11
Campo Mourão, *Braz.* **132** M12
Campo, *Colo., U.S.* **109** P15
Campo, *Cameroon* **206** P9
Campoalegre, *Col.* **130** H6
Campobasso, *It.* **152** L11
Campofelice, *It.* **153** S10
Campos, *Braz.* **133** L15
Campos, region, *S. Amer.* **128** J8
Campti, *La., U.S.* **102** M7
Camrose, *Alta., Can.* **86** L10
Camsell Portage, *Sask. Can.* **86** J12
Camucuio, *Angola* **208** N9
Cân Thơ, *Vietnam* **191** Q11
Çan, *Turk.* **170** D3
Canaan, *Trin. & Tobago* **123** P17
Canacona, *India* **183** P4
Canada Basin, *Arctic Oc.* **252** M8
Cañada de Gómez, *Arg.* **134** K11
Cañada Honda, *Arg.* **134** K8
Canada Plain, *Arctic Oc.* **252** L8
Canada, *N. Amer.* **87** J14
Canadian , river, *U.S.* **105** C13
Canadian , river, *U.S.* **105** D18
Canadian Shield, *N. Amer.* **84** G7
Canadian, *Tex., U.S.* **105** C13
Canadian, river, *U.S.* **90** J11
Çanakkale Boğazi (Dardanelles), *Turk.* **170** D2
Çanakkale, *Turk.* **170** D2
Canal de la Havannah, *New Caledonia, Fr.* **231** Q23
Canal de Marchena, *Ecua.* **130** P3
Canal de Pinta, *Ecua.* **130** P3
Canal de Santa Fe, *Ecua.* **130** Q4
Canal des Pangalanes, *Madagascar* **213** F17
Canal des Pangalanes, *Madagascar* **213** H16
Canal Haava, *Fr. Polynesia, Fr.* **231** E14
Canal Isabela, *Ecua.* **130** Q3
Canala, *New Caledonia, Fr.* **231** Q22

Canalul Bega, *Serb.* **154** E10
Canandaigua, *N.Y., U.S.* **96** J7
Cananea, *Mex.* **116** C7
Canaries, *St. Lucia* **123** K13
Canarreos, Archipiélago de los, *Cuba* **118** H7
Canary Basin, *Atl. Oc.* **248** K9
Canary Islands *see* Islas Canarias, *Sp.* **214** P2
Canary Ridge, *Atl. Oc.* **248** L10
Cañas, *C.R.* **117** N18
Canastota, *N.Y., U.S.* **96** J9
Canatlán, *Mex.* **116** F9
Canaveral National Seashore, *Fla., U.S.* **99** S9
Canaveral, Cape, *Fla., U.S.* **99** T10
Canavieiras, *Braz.* **133** J16
Canberra, A.C.T., *Austral.* **221** Y14
Canby, *Minn., U.S.* **106** K9
Canby, *Calif., U.S.* **110** M5
Cancún, *Mex.* **117** H7
Çandarli Körfezi, *Turk.* **170** F3
Çandarli, *Turk.* **170** F3
Candás, *Sp.* **146** A9
Candé, *Fr.* **149** R6
Candia *see* Irákleio, *Gr.* **157** Q15
Cando, *N. Dak., U.S.* **106** E6
Candon, *Philippines* **193** B13
Cane Bay, *Virgin Is., U.S.* **122** Q2
Cane Garden Bay, *Virgin Is., U.K.* **122** Q6
Canea *see* Chaniá, *Gr.* **157** P13
Canefield, *Dominica* **123** G19
Canelles, river, *St. Lucia* **123** L14
Cañete, *Sp.* **147** G13
Caney, *Kans., U.S.* **107** V9
Canfranc, *Sp.* **147** C14
Cangamba, *Angola* **209** N13
Cangandala, *Angola* **208** L11
Cangas del Narcea, *Sp.* **146** B8
Cangas, *Sp.* **146** C5
Cangombe, *Angola* **209** N13
Canguaretama, *Braz.* **133** F18
Cangumbe, *Angola* **209** M13
Caniapiscau, Réservoir, *Que. Can.* **87** L19
Caniapiscau, river, *Que. Can.* **87** K19
Canicattì, *It.* **153** T10
Caniles, *Sp.* **146** L12
Canillo, *Andorra* **162** H3
Canillo, region, *Andorra* **162** H4
Canindé, *Braz.* **133** E16
Canipaan, *Malaysia* **192** F12
Canisteo Peninsula, *Antarctica* **238** K7
Canisteo, *N.Y., U.S.* **96** K6
Canjáyar, *Sp.* **146** L12
Çankırı, *Turk.* **170** D9
Canna, island, *Scot., U.K.* **144** J7
Cannae, ruin, *It.* **152** L13
Cannanore *see* Kannur, *India* **183** R4
Cannelton, *Ind., U.S.* **101** U8
Cannes, *Fr.* **149** X15
Canning Hill, *Austral.* **220** H3
Canning Land, *Greenland, Den.* **88** J14
Cannon Beach, *Oreg., U.S.* **110** E2
Cannon Falls, *Minn., U.S.* **106** K12
Cannonball, river, *N. Dak. U.S.* **106** H4
Cannonville, *Utah, U.S.* **109** P7
Canoas, *Braz.* **132** P12
Canoe Passage, *Kiribati* **226** J11
Canon City, *Colo., U.S.* **109** N13
Canonniers Point, *Mauritius* **215** F20
Canonsburg, *Pa., U.S.* **96** P3
Canosa di Puglia, *It.* **152** L13
Canouan, island, *St. Vincent & the Grenadines* **121** G16
Canso, Strait of, *N.S., Can.* **87** P22
Canterbury Bight, *N.Z.* **223** M18
Canterbury, *Eng., U.K.* **145** U15
Cantiles, Cayo, *Cuba* **118** H6
Canto do Buriti, *Braz.* **133** F15
Canton Lake, *Okla., U.S.* **105** B15
Canton *see* Guangzhou, *China* **185** P13
Canton, *N.Y., U.S.* **96** F9
Canton, *Pa., U.S.* **96** L7
Canton, *Ga., U.S.* **98** K5
Canton, *N.C., U.S.* **98** H7
Canton, *Ill., U.S.* **101** Q4

Canton, *Ohio, U.S.* **101** P15
Canton, *Miss., U.S.* **102** L11
Canton, *Okla., U.S.* **105** C15
Canton, *S. Dak., U.S.* **106** M8
Canton, *Mo., U.S.* **107** R13
Cantù, *It.* **152** D4
Cantwell, *Alas., U.S.* **113** H17
Canumã, *Braz.* **132** E9
Canutama, *Braz.* **132** F7
Canutillo, *Tex., U.S.* **104** H6
Cany-Barville, *Fr.* **149** N8
Canyon City, *Oreg., U.S.* **110** G7
Canyon de Chelly National Monument, *Ariz., U.S.* **109** R9
Canyon Ferry Lake, *Mont., U.S.* **108** D7
Canyon Grand, *N. Amer.* **84** K3
Canyon Lake, *Tex., U.S.* **105** L16
Canyon, *Tex., U.S.* **104** D12
Canyonlands National Park Horseshoe Canyon, *Utah, U.S.* **109** N8
Canyonlands National Park, *Utah, U.S.* **109** N9
Canyons of the Ancients National Monument, *Colo., U.S.* **109** P10
Canyonville, *Oreg., U.S.* **110** J3
Canzar, *Angola* **209** J14
Cao Bằng, *Vietnam* **190** G11
Caombo, *Angola* **208** K11
Cap Andavaka, *Madagascar* **213** J15
Cap Balguerie, *Fr. Polynesia, Fr.* **231** D15
Cap Barbas, *W. Sahara* **200** J2
Cap Bayès, *New Caledonia, Fr.* **231** P21
Cap Beddouza, *Mor.* **200** D6
Cap Blanc, *W. Sahara* **200** K1
Cap Blanc, *Mauritania, Mor.* **204** E6
Cap Bocage, *New Caledonia, Fr.* **231** P22
Cap Bon, *Tunisia* **201** B16
Cap Boujdour, *W. Sahara* **200** G3
Cap d' Arguin, *Mauritania* **204** E7
Cap d'Ambre, *Madagascar* **213** C17
Cap de Flotte, *New Caledonia, Fr.* **231** P23
Cap Dorade, *St. Martin, Fr.* **123** B14
Cap Drâa, *Mor.* **200** F5
Cap Est, *Madagascar* **213** D17
Cap Estate, *St. Lucia* **123** J14
Cap Ferré, *Martinique, Fr.* **123** G24
Cap Goulvain, *New Caledonia, Fr.* **231** Q21
Cap Hadid, *Mor.* **200** D6
Cap Juby, *Mor.* **200** F4
Cap Lopez, *Gabon* **208** E7
Cap Masoala, *Madagascar* **213** E17
Cap Rhir, *Mor.* **200** E6
Cap Rock Escarpment, *U.S.* **90** K11
Cap Rossel, *New Caledonia, Fr.* **231** N23
Cap Saint Vincent, *Madagascar* **213** G14
Cap Sainte Marie, *Madagascar* **213** J15
Cap Salomon, *Martinique, Fr.* **123** G22
Cap Sim, *Mor.* **200** E6
Cap St. André, *Madagascar* **213** E15
Cap St. Sébastien, *Madagascar* **213** C17
Cap St, Martin, *Martinique, Fr.* **123** E22
Cap Tafarit, *Mauritania* **204** E7
Cap Tehoohaivei, *Fr. Polynesia, Fr.* **231** D14
Cap Timiris, *Mauritania* **204** F7
Cap Verga, *Guinea* **204** L7
Cap-Haïtien, *Haiti* **119** L17
Capaia, *Angola* **209** K13
Capayán, *Arg.* **134** H9
Cape Adare, *Antarctica* **239** R14
Cape Agulhas, *S. Af.* **212** N5
Cape Alexander, *Solomon Is.* **228** C6
Cape Ann, *Antarctica* **239** C20
Cape Aracides, *Solomon Is.* **228** E11
Cape Astrolabe, *Solomon Is.* **228** E10
Cape Asuisui, *Samoa* **227** P17
Cape Barren Island, *Austral.* **220** L8
Cape Basin, *Atl. Oc.* **249** T13
Cape Batterbee, *Antarctica* **239** C20
Cape Blackwood, *P.N.G.* **222** Q7
Cape Borley, *Antarctica* **239** D20
Cape Breton Island, *N.S., Can.* **87** N23
Cape Brett, *N.Z.* **223** D20
Cape Burks, *Antarctica* **238** N9
Cape Campbell, *N.Z.* **223** K20
Cape Capucin, *Dominica* **123** D18
Cape Carr, *Antarctica* **239** P20

Cape Casa Cruz, *Trin. & Tobago* **123** R23
Cape Charles, *Va., U.S.* **98** F14
Cape Coast, *Ghana* **205** Q15
Cape Cod Bay, *Mass., U.S.* **97** L16
Cape Cod National Seashore, *Mass., U.S.* **97** L16
Cape Colbeck, *Antarctica* **238** N11
Cape Columbine, *S. Af.* **212** M4
Cape Coral, *Fla., U.S.* **99** W8
Cape Cornwall, *Eng., U.K.* **145** X7
Cape Cretin, *P.N.G.* **222** P8
Cape Cumberland, *Vanuatu* **229** E13
Cape Darnley, *Antarctica* **239** F21
Cape Dart, *Antarctica* **238** M8
Cape Davis, *Antarctica* **239** D20
Cape De la Motte, *Antarctica* **239** R18
Cape Dernberg, *Namibia* **212** K3
Cape Dorset, *Nunavut, Can.* **87** H17
Cape Elba *see* Ra's al Ḥadāribah, *Egypt* **202** L15
Cape Esperance, *Solomon Is.* **228** F9
Cape Farewell, *N.Z.* **223** J19
Cape Fear, river, *N.C., U.S.* **98** K12
Cape Filchner, *Antarctica* **239** J22
Cape Flying Fish, *Antarctica* **238** K7
Cape Fogausa, *American Samoa, U.S.* **230** D8
Cape Foulwind, *N.Z.* **223** K18
Cape Fria, *Namibia* **212** F1
Cape Girardeau, *Mo., U.S.* **107** U16
Cape Goodenough, *Antarctica* **239** P20
Cape Gray, *Antarctica* **239** R18
Cape Hālawa, *Hawaii, U.S.* **115** H18
Cape Hanpan, *P.N.G.* **222** N11
Cape Hanpan, *P.N.G.* **228** A4
Cape Hatteras National Seashore, *N.C., U.S.* **98** H15
Cape Henslow, *Solomon Is.* **228** F10
Cape Hinode, *Antarctica* **239** B18
Cape Hunter, *Solomon Is.* **228** F10
Cape Karikari, *N.Z.* **223** C19
Cape Kidnappers, *N.Z.* **223** H22
Cape Krasinskiy, *Antarctica* **239** A14
Cape Krusenstern National Monument, *Alas., U.S.* **113** D13
Cape Kumukahi, *Hawaii, U.S.* **115** P24
Cape L'Avedy, *P.N.G.* **222** P11
Cape L'Avedy, *P.N.G.* **228** B5
Cape Lewis, *Antarctica* **239** P20
Cape Lookout National Seashore, *N.C., U.S.* **98** J14
Cape Mabiri, *P.N.G.* **228** B5
Cape Manning, *Kiribati* **226** P11
Cape Maria van Diemen, *N.Z.* **223** C19
Cape Marquis, *St. Lucia* **123** J14
Cape Mataaébe, *Vanuatu* **229** F14
Cape Matātula, *American Samoa, U.S.* **230** D8
Cape May, *N.J., U.S.* **96** R10
Cape McCormick, *Antarctica* **239** Q14
Cape Mikhayov, *Antarctica* **239** N20
Cape Moltke, *P.N.G.* **228** B4
Cape Moule à Chique, *St. Lucia* **123** M14
Cape Mulinu'u, *Samoa* **227** N17
Cape Mulitapuili, *Samoa* **227** P19
Cape Nehus, *P.N.G.* **228** B5
Cape Norvegia, *Antarctica* **238** B12
Cape Nutt, *Antarctica* **239** L21
Cape of Good Hope, *S. Af.* **212** N5
Cape Omega, *Antarctica* **239** C18
Cape Palliser, *N.Z.* **223** K20
Cape Palmas, *Liberia* **204** Q11
Cape Penck, *Antarctica* **239** H22
Cape Peremennyy, *Antarctica* **239** L21
Cape Plain, *Atl. Oc.* **249** T13
Cape Poinsett, *Antarctica* **239** M21
Cape Puava, *Samoa* **227** N17
Cape Queiros, *Vanuatu* **229** E14
Cape Reinga, *N.Z.* **223** C19
Cape Rise, *Atl. Oc.* **249** U14
Cape Rodney, *P.N.G.* **222** R8
Cape Runaway, *N.Z.* **223** F22
Cape Saint Francis, *S. Af.* **212** M7
Cape Selborne, *Antarctica* **239** M14
Cape Shirley, *Antigua & Barbuda* **123** B21
Cape Siri, *P.N.G.* **222** R11
Cape Southard, *Antarctica* **239** N20
Cape St. George, *P.N.G.* **222** N10
Cape St. Paul, *Ghana* **205** P17
Cape Taputapu, *American Samoa, U.S.* **230** E5

Cape Three Points, *Ghana* **205** Q15
Cape Town, *S. Af.* **212** M5
Cape Tuasivi, *Samoa* **227** N18
Cape Turnagain, *N.Z.* **223** J21
Cape Verde Basin, *Atl. Oc.* **248** L8
Cape Verde Plain, *Atl. Oc.* **248** L9
Cape Verde Plateau, *Atl. Oc.* **248** M10
Cape Vincent, *N.Y., U.S.* **96** G8
Cape Wabao, *New Caledonia, Fr.* **231** Q24
Cape Waite, *Antarctica* **238** K7
Cape Ward Hunt, *P.N.G.* **222** Q8
Cape Washington, *Fiji* **226** R5
Cape York Peninsula, *Austral.* **220** B12
Cape Zelee, *Solomon Is.* **228** F11
Capella Islands, *Virgin Is., U.S.* **122** N9
Capelongo, *Angola* **208** P10
Capenda-Camulemba, *Angola* **208** K12
Capesterre-Belle-Eau, *Guadeloupe, Fr.* **123** G15
Capesterre, *Guadeloupe, Fr.* **123** G16
Capestrano, *It.* **152** K9
Capistrello, *It.* **152** K9
Capital Hill, *N. Mariana Is., U.S.* **230** H8;J5
Capitán Pablo Lagerenza, *Parag.* **132** K9
Capitan Peak, *N. Mex., U.S.* **109** T13
Capitan, *N. Mex., U.S.* **109** U13
Capitão, *Sao Tome and Principe* **215** A21
Capitol Reef National Park, *Utah, U.S.* **109** N8
Capodistria *see* Koper, *Slov.* **154** D4
Capoeira, falls, *Braz.* **132** F10
Capolo, *Angola* **208** L10
Cappadocia, region, *Turk.* **170** G10
Cappamore (An Cheapach Mhór), *Ire.* **145** S4
Cappoquin (Ceapach Choinn), *Ire.* **145** T5
Capraia, Isola di, *It.* **152** H5
Capreol, *Ont., Can.* **87** P18
Caprera, Isola, *It.* **149** X2
Capri, Isola di, *It.* **153** N10
Capricorn Channel, *Austral.* **220** F15
Capricorn Group, *Austral.* **220** F15
Caprivi Strip, *Namibia* **212** E6
Captain Cook Monument, *Hawaii, U.S.* **115** P20
Captain Cook, *Hawaii, U.S.* **115** P20
Captain Cook's Monument, *Fr. Polynesia, Fr.* **231** H22
Captieux, *Fr.* **149** W7
Captiva, *Fla., U.S.* **99** W8
Capua, *It.* **152** M10
Capucin Point, *Seychelles* **215** P20
Capulin Volcano National Monument, *N. Mex., U.S.* **109** Q14
Capulo, *Angola* **208** J9
Capunda, *Angola* **208** L12
Caquetá, river, *Col.* **130** K6
Car Nicobar, island, *India* **183** T14
Caracal, *Rom.* **155** G14
Caracaraí, *Braz.* **132** B9
Caracas, *Venez.* **130** C10
Caracasbaai, bay, *Curaçao, Neth.* **123** Q15
Caracol, *Braz.* **133** G15
Carahue, *Chile* **135** P6
Carajás, Serra dos, *Braz.* **132** E12
Caramulo, Serra do, *Port.* **146** F6
Carandotta, *Qnsld. Austral.* **221** S10
Caransebeş, *Rom.* **154** E12
Carapichaima, *Trin. & Tobago* **123** P22
Carapuse Bay, *Trin. & Tobago* **123** N18
Caratasca Lagoon, *N. Amer.* **84** P8
Caratasca, Laguna de, *Hond.* **117** K19
Carauari, *Braz.* **132** E7
Caraúbas, *Braz.* **133** F17
Caravaca de la Cruz, *Sp.* **147** K13
Caravelas, *Braz.* **133** K16
Carballiño, *Sp.* **146** C6
Carballo, *Sp.* **146** B5
Carbon Hill, *Ala., U.S.* **103** J14
Carbón, Laguna del, *Arg.* **135** V8
Carbonara, Capo, *It.* **153** Q4
Carbondale, *Pa., U.S.* **96** M9
Carbondale, *Ill., U.S.* **101** U5
Carbondale, *Colo., U.S.* **108** M11
Carboneras, *Sp.* **147** L13
Carbonia, *It.* **153** Q3
Carbonne, *Fr.* **149** Y8
Carcajou, *Alta., Can.* **86** J10

Carcans, Lac de, *Fr.* **149** V6
Carcasse, Cap-, *Haiti* **119** M14
Carcassonne, *Fr.* **149** Y10
Carche, peak, *Sp.* **147** J14
Carcross, *Yukon, Can.* **86** G7
Cardamom Mountains, *Cambodia* **191** N9
Cardamum Island *see* Kadmat Island, *India* **183** R3
Cardeña, *Sp.* **146** J10
Cárdenas, *Mex.* **116** G11
Cárdenas, *Cuba* **118** G7
Cardiff (Caerdydd), *Wales, U.K.* **145** U10
Cardigan (Aberteif), *Wales, U.K.* **145** T9
Cardigan Bay, *Wales, U.K.* **145** S9
Cardno Fracture Zone, *Atl. Oc.* **249** R9
Cardona, *Sp.* **147** D17
Cardoso y Pinto, *Eq. Guinea* **214** M7
Cardwell, *Qnsld., Austral.* **221** R13
Carei, *Rom.* **154** B12
Careiro da Várzea, *Braz.* **132** D9
Carey Øer, *Greenland, Den.* **88** E3
Carey, *Ohio, U.S.* **101** P12
Carey, Lake, *Austral.* **220** H5
Careysburg, *Liberia* **204** P10
Cargados Carajos Bank, *Ind. Oc.* **250** J8
Cargèse, *Fr.* **149** X2
Carhué, *Arg.* **134** M11
Cariati, *It.* **153** P14
Caribbean Sea, *Atl. Oc.* **248** M3
Caribou Mountains, *N. Amer.* **84** F4
Caribou, *Me., U.S.* **97** B18
Cariñena, *Sp.* **147** E13
Carinhanha, *Braz.* **133** J15
Cariparé, *Braz.* **133** H14
Caripito, *Venez.* **130** C13
Carlin, *Nev., U.S.* **111** N10
Carlingford Lough, *Ire., U.K.* **145** Q7
Carlini, station, *Antarctica* **238** C4
Carlinville, *Ill., U.S.* **101** S4
Carlisle Bay, *Jam.* **122** K9
Carlisle Bay, *Barbados* **123** L19
Carlisle, *Pa., U.S.* **96** P7
Carlisle, *Ark., U.S.* **102** G9
Carlisle, *Ky., U.S.* **103** B18
Carlisle, *Eng., U.K.* **145** N10
Carlos Manuel de Céspedes, *Cuba* **118** J10
Carlos, Isla, *Chile* **135** X7
Carlow (Ceatharlach), *Ire.* **145** S6
Carlsbad Caverns National Park, *N. Mex., U.S.* **109** V14
Carlsbad, *Tex., U.S.* **105** J13
Carlsbad, *N. Mex., U.S.* **109** V14
Carlsbad, *Calif., U.S.* **111** Y9
Carlsberg Fjord, *Greenland, Den.* **88** K14
Carlsberg Ridge, *Ind. Oc.* **250** E8
Carlshavn, bay, *Greenland, Den.* **88** H14
Carlton, *Minn., U.S.* **106** G12
Carlyle Lake, *Ill., U.S.* **101** T5
Carlyle, *Ill., U.S.* **101** T5
Carmacks, *Yukon, Can.* **86** F8
Carmagnola, *It.* **152** E3
Carmarthen (Sir Gaerfyrddin), *Wales, U.K.* **145** U10
Carmarthen Bay, *Wales, U.K.* **145** U9
Carmaux, *Fr.* **149** X10
Carmel-by-the-Sea, *Calif., U.S.* **111** T4
Carmen de Bolívar, *Col.* **130** D6
Carmen de Patagones, *Arg.* **135** Q10
Carmen, *Okla., U.S.* **105** B16
Carmen, *Isla, Mex.* **116** E7
Carmi, *Ill., U.S.* **101** U7
Carmona, *Sp.* **146** L8
Carn Domhnach *see* Carndonagh, *Ire.* **145** M6
Carnarvon Terrace, *Ind. Oc.* **251** L17
Carnarvon, *Qnsld., Austral.* **221** U14
Carndonagh (Carn Domhnach), *Ire.* **145** M6
Carnegie Ridge, *Pac. Oc.* **247** K21
Carnegie, *W. Aust., Austral.* **221** U5
Carnegie, Lake, *Austral.* **220** G4
Carney Island, *Antarctica* **238** M8
Carnic Alps, *It.* **152** C9
Carnot, *Cen. Af. Rep.* **206** N12
Carnot, Cape, *Austral.* **220** K9
Carnsore Point, *Ire.* **145** T7
Caro, *Mich., U.S.* **100** K12

Caroço, Ilhéu, *Sao Tome and Principe* **215** B21
Carol City, *Fla., U.S.* **99** X10
Carolina Beach, *N.C., U.S.* **98** K13
Carolina, *Braz.* **133** F14
Caroline Island (Millennium Island), *Kiribati* **225** H13
Caroline Islands, *F.S.M.* **226** B5
Caroline Seamounts, *Pac. Oc.* **246** J8
Caroni Swamp, *Trin. & Tobago* **123** P22
Caroni, *Trin. & Tobago* **123** P22
Caroní, river, *Venez.* **130** E13
Carora, *Venez.* **130** C9
Carp, *Nev., U.S.* **111** T12
Carpathian Mountains, *Eur.* **140** H9
Carpatho-Ukraine, region, *Ukr.* **159** P1
Carpathos *see* Kárpathos, *Gr.* **157** P18
Carpentaria, Gulf of, *Austral.* **220** B11
Carpentras, *Fr.* **149** W13
Carpi, *It.* **152** F6
Carpina, *Braz.* **133** F18
Carpio, *N. Dak., U.S.* **106** E3
Carpolac, *Vic., Austral.* **221** Y11
Carquefou, *Fr.* **149** R6
Carrabelle, *Fla., U.S.* **99** R5
Carrara, *It.* **152** G5
Carrarang, *W. Aust., Austral.* **221** V1
Carrauntoohil *see* Corrán Tuathail, peak, *Ire.* **145** T3
Carrboro, *N.C., U.S.* **98** H11
Carriacou, island, *Grenada* **121** H16
Carrick-on-Shannon (Cora Droma Rúisc), *Ire.* **145** Q5
Carrickfergus, *N. Ire., U.K.* **145** N7
Carrington, *N. Dak., U.S.* **106** F6
Carrión de los Condes, *Sp.* **146** C10
Carrizal Bajo, *Chile* **134** G6
Carrizo Creek, *N. Mex., Tex., U.S.* **109** Q15
Carrizo Plain National Monument, *Calif., U.S.* **111** V6
Carrizo Springs, *Tex., U.S.* **105** N14
Carrizozo, *N. Mex., U.S.* **109** T13
Carroll Inlet, *Antarctica* **238** G7
Carroll, *Iowa, U.S.* **107** N10
Carrollton, *Ga., U.S.* **98** L4
Carrollton, *Ill., U.S.* **101** S4
Carrollton, *Ky., U.S.* **103** A17
Carrollton, *Tex., U.S.* **105** G17
Carrollton, *Mo., U.S.* **107** S11
Carrot Bay, *Virgin Is., U.K.* **122** R5
Carrowmore Lough, *Ire.* **145** P3
Çarşamba, *Turk.* **170** D11
Çarşamba, river, *Turk.* **170** H7
Carson City, *Nev., U.S.* **111** Q6
Carson Sink, *Nev., U.S.* **111** P8
Carson, *N. Dak., U.S.* **106** H4
Carsonville, *Mich., U.S.* **100** L13
Cartagena, *Col.* **130** D5
Cartagena, *Sp.* **147** K14
Cartago, *C.R.* **117** N19
Cartago, *Col.* **130** G5
Carter, Mount, *Austral.* **220** B12
Cartersville, *Ga., U.S.* **98** K4
Carthage, *N.Y., U.S.* **96** G9
Carthage, *N.C., U.S.* **98** H11
Carthage, *Ill., U.S.* **101** Q3
Carthage, *Ark., U.S.* **102** H8
Carthage, *Miss., U.S.* **102** K12
Carthage, *Tenn., U.S.* **103** E16
Carthage, *Tex., U.S.* **105** H20
Carthage, *S. Dak., U.S.* **106** L7
Carthage, *Mo., U.S.* **107** V11
Carthage, ruin, *Tunisia* **201** B16
Cartier Island, *Austral.* **220** B5
Cartwright, *Nfld. & Lab., Can.* **87** K22
Caruaru, *Braz.* **133** G17
Carúpano, *Venez.* **130** C13
Carutapera, *Braz.* **133** D14
Caruthersville, *Mo., U.S.* **107** W16
Carvoeiro, *Braz.* **132** D8
Cary, *N.C., U.S.* **98** H11
Caryapundy Swamp, *Austral.* **220** H12
Casa Grande Ruins National Monument, *Ariz., U.S.* **109** U7
Casa Grande, *Ariz., U.S.* **109** U7
Casablanca, *Mor.* **200** D6
Casale Monferrato, *It.* **152** E3
Casamanya, Pic de, *Andorra* **162** H3
Casanare, river, *Col.* **130** F8
Casarano, *It.* **153** N15
Cascade Bay, *Austral.* **229** A20
Cascade Head, *Oreg., U.S.* **110** F2
Cascade Locks, *Oreg., U.S.* **110** F4
Cascade Range, *N. Amer.* **84** J3

Cascade Reservoir, *Idaho, U.S.* **108** F4
Cascade-Siskiyou National Monument, *Oreg., U.S.* **110** L3
Cascade, *Idaho, U.S.* **108** F4
Cascade, *Mont., U.S.* **108** C7
Cascade, *Jam.* **122** H6
Cascade, *Seychelles* **215** N20
Cascade, *Austral.* **229** A20
Cascadia Basin, *Pac. Oc.* **247** D17
Cascavel, *Braz.* **132** M11
Cascina, *It.* **152** G6
Cascorro, *Cuba* **118** J11
Casere, *It.* **152** B8
Caserta, *It.* **152** M11
Cases Velles de Formentor, *Sp.* **147** G19
Casey Bay, *Antarctica* **239** C19
Casey Station, *Antarctica* **239** M21
Casey, *Ill., U.S.* **101** S7
Cashel (Caiseal), *Ire.* **145** S5
Cashmere, *Wash., U.S.* **110** C6
Casiguran, *Philippines* **193** B14
Casilda, *Arg.* **134** K11
Casino Gardens, *Monaco* **163** D21
Casino of Monte Carlo, *Monaco* **163** D22
Casino, *N.S.W., Austral.* **221** W16
Casiquiare, river, *Venez.* **130** H11
Casma, *Peru* **132** G3
Casmalia, *Calif., U.S.* **111** W5
Casnewydd *see* Newport, *Wales, U.K.* **145** U11
Caspar, *Calif., U.S.* **111** P2
Caspe, *Sp.* **147** E15
Casper, *Wyo., U.S.* **108** H12
Caspian Depression, *Kaz., Russ.* **159** R14
Caspian Depression, *Kaz.* **178** G2
Caspian Sea, *Asia* **168** F5
Casquets, island, *Channel Is., U.K.* **145** X11
Cass City, *Mich., U.S.* **100** K12
Cass Lake, *Minn., U.S.* **106** F10
Cass, *W. Va., U.S.* **98** D10
Cassai, *Angola* **209** L14
Cassai, river, *Angola* **209** L14
Cassamba, *Angola* **209** M13
Cassel, *Fr.* **148** L10
Casselton, *N. Dak., U.S.* **106** G8
Cassiar Mountains, *B.C., Yukon, Can.* **86** H8
Cassinga, *Angola* **208** P11
Cassino, *It.* **152** L10
Cassis, *Fr.* **149** Y13
Cassou, *Burkina Faso* **205** L15
Cassumbe, *Angola* **208** L11
Cassville, *Wis., U.S.* **100** M3
Cassville, *Mo., U.S.* **107** V11
Castagneto Carducci, *It.* **152** H6
Castanhal, *Braz.* **133** D13
Castaño Nuevo, *Arg.* **134** J7
Castara, *Trin. & Tobago* **123** N18
Castaway Cay, *Bahamas* **122** C9
Casteggio, *It.* **152** E4
Castel di Sangro, *It.* **152** L10
Castelbuono, *It.* **153** S10
Castelfranco Veneto, *It.* **152** D8
Casteljaloux, *Fr.* **149** W7
Castellabate, *It.* **153** N11
Castellaccio, peak, *San Marino* **163** K15
Castellammare del Golfo, *It.* **153** S9
Castellammare, Golfo di, *It.* **153** S9
Castellaneta, *It.* **152** M14
Castellón de la Plana (Castelló de la Plana), *Sp.* **147** G15
Castelnaudary, *Fr.* **149** Y9
Castelnovo ne' Monti, *It.* **152** F6
Castelo Branco, *Port.* **146** G6
Castelo de Bode, Barragem de, *Port.* **146** G5
Castelo de Vide, *Port.* **146** H6
Castelsardo, *It.* **152** M3
Castiglione Messer Marino, *It.* **152** K11
Castilhiano, *Cabo Verde* **215** B16
Castille, region, *Sp.* **146** F10
Castillo de San Marcos National Monument, *Fla., U.S.* **99** R9
Castillo, Pampa del, *Arg.* **135** T8
Castillon, *Fr.* **149** Y8
Castillos, *Uru.* **134** L15
Castle Bruce, *Dominica* **123** F20
Castle Dale, *Utah, U.S.* **108** M8
Castle Douglas, *Scot., U.K.* **145** N9

Castle Harbour, *Bermuda* **119** B16
Castle Island, *Bahamas* **119** H14
Castle Peak, *Idaho, U.S.* **108** G5
Castle Roads, *Bermuda* **119** B16
Castle Rock, *Colo., U.S.* **108** M13
Castle Rock, *Wash., U.S.* **110** E3
Castlebar (Caisleán an Bharraigh), *Ire.* **145** Q3
Castlebay *see* Bagh a' Chaisteil, *Scot., U.K.* **144** J6
Castlebridge, *Ire.* **145** T7
Castlegregory (Caisleán Ghriaire), *Ire.* **145** T2
Castlepollard (Baile na gCros), *Ire.* **145** Q6
Castlerea (An Caisleán Riabhach), *Ire.* **145** Q3
Castlereagh, river, *Austral.* **220** J14
Castletownshend (Baile an Chaisleáin), *Ire.* **145** U3
Castlewood, *S. Dak., U.S.* **106** K8
Castor Bank, *Ind. Oc.* **250** J6
Castres, *Fr.* **149** X10
Castries, *St. Lucia* **123** J14
Castries, *Fr.* **149** X12
Castro Daire, *Port.* **146** E6
Castro Verde, *Port.* **146** K5
Castro-Urdiales, *Sp.* **146** B11
Castro, *Chile* **135** R6
Castrovillari, *It.* **153** P13
Castroville, *Tex., U.S.* **105** M15
Castuera, *Sp.* **146** J8
Cat Cays, *Bahamas* **118** D9
Cat Island, *Miss., U.S.* **102** P12
Cat Island, *Bahamas* **119** E13
Cat Lake, *Ont., Can.* **87** N15
Cat Point, *Calif., U.S.* **111** P2
Çat, *Turk.* **171** E15
Catabola, *Angola* **208** M12
Catahoula Lake, *La., U.S.* **102** M8
Çatak, *Turk.* **171** G17
Catalan Bay, *Gibraltar, U.K.* **163** P23
Catalan Bay, *Gibraltar, U.K.* **163** P23
Catalão, *Braz.* **133** K13
Catalca, *Turk.* **170** C4
Catalina Point, *Guam, U.S.* **230** N7
Catalina, *Chile* **134** F7
Catalonia, region, *Sp.* **147** D17
Catamarca, *Arg.* **134** H9
Catandica, *Mozambique* **212** E10
Catanduanes, island, *Philippines* **193** C15
Catania, Golfo di, *It.* **153** T12
Catania, *It.* **153** T12
Catania, Piana di, *It.* **153** T11
Catanzaro, *It.* **153** Q14
Cataperdís, Pic de, *Andorra, Fr.* **162** G2
Catarina, *Tex., U.S.* **105** N14
Catarman, *Philippines* **193** D15
Castanhal, *Braz.* **133** D13
Cateel, *Philippines* **193** F15
Catete, *Angola* **208** K9
Cathair na Mart *see* Westport, *Ire.* **145** Q3
Cathedral Mountain, *Tex., U.S.* **104** K10
Cathedral of Monaco, *Monaco* **163** E21
Catherine, Mount *see* Katherîna, Gebel, *Egypt* **173** T3
Cathlamet, *Wash., U.S.* **110** E3
Catingas, region, *S. Amer.* **128** E8
Catió, *Guinea-Bissau* **204** L7
Catlettsburg, *Ky., U.S.* **103** B20
Catlow Valley, *Oreg., U.S.* **110** K7
Cato Trough, *Pac. Oc.* **246** M9
Catoche, Cape, *N. Amer.* **84** N7
Catota, *Angola* **208** N12
Catonsville, *Md., U.S.* **98** C13
Catria, Monte, *It.* **152** H8
Catriló, *Arg.* **134** M10
Catrimani, *Braz.* **132** C8
Catrimani, river, *Braz.* **132** C8
Catskill Mountains, *N.Y., U.S.* **96** K10
Catskill, *N.Y., U.S.* **96** K11
Catto Cay, *Bahamas* **122** F2
Cattolica Eraclea, *It.* **153** T9
Cauca, river, *N. Amer.* **84** Q9
Cauca, river, *Col.* **130** E6
Caucasus Mountains, *Eur., Asia* **140** J14
Caucete, *Arg.* **134** J8
Caudry, *Fr.* **148** M11
Caulnes, *Fr.* **149** Q5
Caungula, *Angola* **208** K12
Cauquenes, *Chile* **134** M6

Caura, river, *Venez.* **130** F12
Caussade, *Fr.* **149** W9
Causse du Larzac, range, *Fr.* **149** X11
Cauto, river, *Cuba* **118** K12
Cavaillon, *Fr.* **149** X13
Cavalaire, *Fr.* **149** Y15
Cavalcante, *Braz.* **133** H13
Cavalese, *It.* **152** C7
Cavalier, *N. Dak., U.S.* **106** E7
Cavallo, Île, *Fr.* **149** Z3
Cavally, river, *Côte d'Ivoire, Liberia* **204** Q11
Cavan (An Cabhán), *Ire.* **145** Q6
Cave Cay, *Bahamas* **122** D7
Cave Creek, *Ariz., U.S.* **109** T7
Cave Guadirikiri, *Aruba, Neth.* **123** Q17
Cave Junction, *Oreg., U.S.* **110** K2
Cavelle Pond, *Bahamas* **122** C11
Caves, Anse aux, *Mauritius* **215** J20
Caviana, Ilha, *Braz.* **133** C13
Cavignac, *Fr.* **149** V7
Cavo, Mount, *It.* **152** L8
Cavour, *It.* **152** E2
Çavuş Burnu, *Turk.* **170** J6
Çavuşcu Gölü, *Turk.* **170** G7
Cawker City, *Kans., U.S.* **107** S6
Caxias do Sul, *Braz.* **132** P12
Caxias, *Braz.* **132** E5
Caxias, *Braz.* **133** E15
Caxito Cameia, *Angola* **208** K12
Caxito, *Angola* **208** K9
Cay Point, *Bahamas* **122** F2
Cay Sal Bank, *Bahamas* **118** F8
Çay, *Turk.* **170** F6
Cayambe, *Ecua.* **130** K4
Cayambe, peak, *Ecua.* **130** K4
Cayce, *S.C., U.S.* **98** K9
Çaycuma, *Turk.* **170** C7
Çayeli, *Turk.* **171** D15
Cayenne, *Fr. Guiana, Fr.* **131** F19
Cayeux-sur-Mer, *Fr.* **148** M9
Çayıralan, *Turk.* **170** F10
Çayırova *see* Ayios Theodhoros, *Cyprus* **162** N10
Cayman Brac, *Cayman Is., U.K.* **118** L9
Cayman Islands, *Cayman Is., U.K.* **118** L8
Cayman Trench, *Atl. Oc.* **248** M2
Caynaba, *Somaliland* **210** M12
Cayo Arenas, island, *Mex.* **117** G15
Cayon, *St. Kitts & Nevis* **123** A18
Cayos Miskitos, islands, *Nicar.* **117** L19
Cayucos, *Calif., U.S.* **111** V5
Cayuga Lake, *N.Y., U.S.* **96** K8
Cazage, *Angola* **209** L14
Cazalla de la Sierra, *Sp.* **146** K8
Cazaux et de Sanguinet, Lac de, *Fr.* **149** W6
Cazères, *Fr.* **149** Y8
Cazombo, *Angola* **209** M15
Cazorla, *Sp.* **146** K11
Ceanannas *see* Kells, *Ire.* **145** Q6
Ceann Trá (Ventry), *Ire.* **145** T2
Ceapach Choinn *see* Cappoquin, *Ire.* **145** T5
Ceara Plain, *Atl. Oc.* **249** P7
Ceara Ridge, *Atl. Oc.* **249** N7
Ceara Seamounts, *Atl. Oc.* **249** P7
Ceatharlach *see* Carlow, *Ire.* **145** S6
Cebollar, *Arg.* **134** H9
Cebu City, *Philippines* **193** E14
Cebu, island, *Philippines* **193** E14
Cecil Plains, *Qnsld., Austral.* **221** V15
Cecile Peninsula, *Kiribati* **226** Q10
Cecina, *It.* **152** H6
Cedar Bluff Reservoir, *Kans., U.S.* **107** T5
Cedar Breaks National Monument, *Utah, U.S.* **109** P6
Cedar City, *Utah, U.S.* **109** P6
Cedar Creek Reservoir, *Tex., U.S.* **105** G18
Cedar Falls, *Iowa, U.S.* **107** N12
Cedar Grove, *Antigua & Barbuda* **123** A21
Cedar Harbour, *Bahamas* **122** A8
Cedar Key, *Fla., U.S.* **99** S7
Cedar Lake, *Ind., U.S.* **101** P7
Cedar Rapids, *Iowa, U.S.* **107** N13
Cedar River, *Wis., U.S.* **100** H8
Cedar Vale, *Kans., U.S.* **107** V8
Cedar, river, *Iowa, Minn., U.S.* **106** M12
Cedartown, *Ga., U.S.* **98** K3
Cedarville, *Mich., U.S.* **100** G11
Cedarville, *Calif., U.S.* **110** L6

Cedeira, *Sp.* **146** A6
Cedillo, Embalse de, *Port., Sp.* **146** G6
Cedros Bay, *Trin. & Tobago* **123** R21
Cedros Island, *N. Amer.* **84** L3
Cedros Trench, *Pac. Oc.* **247** F18
Cedros, Isla, *Mex.* **116** C5
Ceduna Slope, *Ind. Oc.* **251** N20
Ceduna, *S. Aust., Austral.* **221** W9
Ceek, *Somaliland* **210** M11
Ceel Afweyn, *Somaliland* **210** L13
Ceel Buur, *Somalia* **211** P12
Ceel Huur, *Somalia* **211** P13
Ceelaayo, *Somaliland* **210** K13
Ceeldheere, *Somalia* **211** Q13
Ceerigaabo, *Somaliland* **210** L13
Cefalù, *It.* **153** S10
Cegléd, *Hung.* **151** L15
Cehotina, river, *Montenegro* **154** H9
Cejal, *Col.* **130** H10
Çekerek, *Turk.* **170** E10
Çekerek, river, *Turk.* **170** E10
Celano, *It.* **152** K9
Celanova, *Sp.* **146** C6
Celaya, *Mex.* **116** H11
Celebes Basin, *Pac. Oc.* **246** J5
Celebes Sea, *Indonesia* **193** H13
Celebes *see* Sulawesi, island,
  Indonesia **193** K13
Çeleken, *Turkm.* **178** L4
Celina, *Ohio, U.S.* **101** Q11
Celje, *Slov.* **154** D5
Cella, *Sp.* **147** F13
Celle, *Ger.* **150** D8
Celtic Sea, *Ire., U.K.* **145** U6
Celtic Shelf, *Atl. Oc.* **248** H11
Çemişgezek, *Turk.* **171** F13
Cenderawasih Bay, *Pac. Oc.* **168** L15
Cenderawasih, Teluk, *Indonesia*
  193 K19
Center Hill Lake, *Tenn., U.S.* **103** E16
Center Point, *Ala., U.S.* **103** J15
Center, *Tex., U.S.* **105** H20
Center, *N. Dak., U.S.* **106** G4
Centerville, *Tenn., U.S.* **103** F14
Centerville, *Tex., U.S.* **105** J18
Centerville, *S. Dak., U.S.* **106** M8
Centerville, *Iowa, U.S.* **107** Q12
Cento, *It.* **152** F7
Central African Republic, *Af.*
  207 M14
Central America, region, *N. Amer.*
  84 Q6
Central Bank, *Arctic Oc.* **253** E17
Central Basin, *Arctic Oc.* **253** D17
Central Basin, *Pac. Oc.* **255** H16
Central Brāhui Range, *Pak.* **181** P7
Central City, *Ky., U.S.* **103** D15
Central City, *Nebr., U.S.* **107** Q7
Central Greece, region, *Gr.* **156** H11
Central Highlands, *Vietnam* **190** M12
Central Kara Rise, *Arctic Oc.* **253** E13
Central Lowland, *U.S.* **91** F16
Central Lowlands, *Austral.* **220** F11
Central Makrān Range, *Pak.* **181** R3
Central Mount Wedge, *N.T. Austral.*
  221 T8
Central Pacific Basin, *Pac. Oc.*
  246 H11
Central Point, *Oreg., U.S.* **110** K3
Central Range, *Asia* **168** C13
Central Range, *P.N.G.* **222** P6
Central Russian Upland, *Eur.*
  140 F12
Central Siberian Plateau, *Asia*
  168 D9
Central Terrace, *Ind. Oc.* **250** L4
Central Valley, *Calif., U.S.* **111** N3
Central, *Alas., U.S.* **113** F18
Central, Cordillera, *Dom. Rep.* **119** L18
Centralia, *Ill., U.S.* **101** T5
Centralia, *Mo., U.S.* **107** S13
Centralia, *Wash., U.S.* **110** D3
Centre de Flacq, *Mauritius* **215** G20
Centre, *Ala., U.S.* **103** H16
Centreville, *Miss., U.S.* **102** N10
Cephalonia *see* Kefaloniá, island, *Gr.*
  156 H7
Ceprano, *It.* **152** L9
Ceram Sea, *Indonesia* **193** K17
Ceram Trough, *Pac. Oc.* **246** K6
Cerbère, *Fr.* **149** Z11
Ceres, *Braz.* **133** J13
Ceres, *Arg.* **134** H11
Cerigo *see* Kýthira, island, *Gr.*
  156 N11
Cérilly, *Fr.* **149** T10

Çerkeş, *Turk.* **170** D8
Çermik, *Turk.* **171** G14
Cernavodă, *Rom.* **155** F17
Cerralvo, Isla, *Mex.* **116** F7
Cerredo, Torre, *Sp.* **146** B9
Cerreto, *San Marino* **163** J14
Cerrillos, *N. Mex., U.S.* **109** R12
Cerrillos, *Arg.* **134** F9
Cerro Crocket, *Ecua.* **130** Q3
Cerro de Pasco, *Peru* **132** G3
Cerro San Joaquín, *Ecua.* **130** Q4
Cerro Sombrero, *Chile* **135** W8
Certaldo, *It.* **152** H6
Certers, *Andorra* **162** K3
Cervales, peak, *Sp.* **146** G9
Cervera de Pisuerga, *Sp.* **146** B10
Cervera, *Sp.* **147** D16
Cerveteri, *It.* **152** K8
Cervia, *It.* **152** F8
Cesana Torinese, *It.* **152** E1
Cesar, river, *Col.* **130** D6
Cesarò, *It.* **153** S11
Cesena, *It.* **152** G8
Cesenatico, *It.* **152** G8
Cēsis, *Latv.* **143** L17
České Budějovice, *Czechia* **150** J11
Český Krumlov, *Czechia* **150** J11
Çeşme, *Turk.* **170** F2
Cestos City, *Liberia* **204** P10
Cetinje, *Montenegro* **154** J8
Cetti Bay, *Guam, U.S.* **230** Q4
Ceuta, *Sp.* **200** B8
Cévennes, range, *Fr.* **149** W12
Cevio, *Switz.* **152** C4
Ceyhan, *Turk.* **170** H11
Ceyhan, river, *Turk.* **170** H11
Ceylanpınar, *Turk.* **171** J14
Ceylon Plain, *Ind. Oc.* **250** H12
Cabrayıl *see* Kurakhtin, *Azerb.*
  171 E21
Cəlilabad, *Azerb.* **171** E22
Chabana, *Japan* **189** X2
Chablis, *Fr.* **149** R11
Chacachacare, island, *Trin. & Tobago*
  123 P21
Chacao Channel, *S. Amer.* **128** N4
Chacao, Canal de, *Chile* **135** Q6
Chachapoyas, *Peru* **132** F3
Chacharan, *Pak.* **181** N10
Chachora, *India* **182** J5
Chachro, *Pak.* **181** S10
Chaco Austral, region, *S. Amer.*
  128 K6
Chaco Boreal, region, *S. Amer.* **128** H6
Chaco Central, region, *S. Amer.*
  128 J6
Chaco Culture National Historical
  Park, *N. Mex., U.S.* **109** R10
Chad, *Af.* **207** G14
Chadan, *Russ.* **161** L13
Chadbourn, *N.C., U.S.* **98** K11
Chaddock Bar, *Bermuda* **119** D14
Chadron, *Nebr., U.S.* **106** M2
Chadwick, *Mo., U.S.* **107** V12
Chaeryŏng, *N. Korea* **186** L6
Chaeryŏng, river, *N. Korea* **186** L6
Chafarinas, Islas, *Sp.* **146** P12
Chafe, *Nigeria* **206** J8
Chaffee, *Mo., U.S.* **107** U16
Chāgai Hills, *Pak.* **181** N4
Chagai, *Pak.* **181** N5
Chagda, *Russ.* **161** H8
Chaghcharān, *Afghan.* **180** H6
Chagoda, *Russ.* **158** F8
Chagos Archipelago, *Ind. Oc.* **168** M5
Chagos Trench, *Ind. Oc.* **250** H11
Chagos-Laccadive Ridge, *Ind. Oc.*
  250 D10
Chaguanas, *Trin. & Tobago* **123** P22
Chaguaranas, *Trin. & Tobago* **123** P21
Chagyl, *Turkm.* **178** L6
Chāh Bāgh, *Afghan.* **180** E9
Chāh Bahār, *Iran* **177** P22
Chāh-e Kirtahak, *Afghan.* **180** K3
Chahār Bāgh, *Afghan.* **180** F6
Chahār Burj, *Afghan.* **180** H3
Chahār Burjak, *Afghan.* **180** M3
Ch'aho, *N. Korea* **186** G11
Ch'ahŭng, *N. Korea* **186** G5
Chai-Nat, *Thai.* **190** L7
Chaibasa, *India* **182** K10
Ch'ail-bong, peak, *N. Korea* **186** F9
Chain Fracture Zone, *Atl. Oc.*
  249 P10
Chain of Craters, *Hawaii, U.S.* **115** P22
Chain Ridge, *Ind. Oc.* **250** F7

Chaîne de l'Atakora,
  Benin–Burkina Faso **205** L17
Chajarí, *Arg.* **134** J13
Chak Chak, *S. Sudan* **203** V8
Chakar, river, *Pak.* **181** N8
Chakaria, *Bangladesh* **182** K14
Chake Chake, *Tanzania* **211** V8
Chakhānsūr, *Afghan.* **180** L3
Chakwal, *Pak.* **180** J12
Chalabesa, *Zambia* **209** M19
Chalais, *Fr.* **149** V7
Chalan Kanoa, *N. Mariana Is., U.S.*
  230 M7
Chalan Pago, *Guam, U.S.* **230** P5
Chalbi Desert, *Kenya* **211** Q7
Ch'alch'is Terara, peak, *Eth.* **210** M6
Chaleur Bay, *N.B., Que., Can.* **87** N21
Chalhuanca, *Peru* **132** J4
Chalisgaon, *India* **182** L4
Chalkhī, *Afghan.* **180** G3
Chálki, bay, *Gr.* **157** N19
Chálki, island, *Gr.* **157** N19
Chalkída (Chalkís), *Gr.* **156** H12
Chalkís *see* Chalkída, *Gr.* **156** H12
Chalky Mount, *Barbados* **123** K19
Chalkyitsik, *Alas., U.S.* **113** E18
Challans, *Fr.* **149** S5
Challenger Deep, *Pac. Oc.* **246** H7
Challenger Fracture Zone, *Pac. Oc.*
  247 N21
Challenger Plateau, *Pac. Oc.* **246** P10
Challis, *Idaho, U.S.* **108** F5
Chalon, *Fr.* **149** S12
Châlons-en-Champagne, *Fr.* **149** P12
Chālus, *Fr.* **149** U8
Cham, *Ger.* **150** J9
Chama, *N. Mex., U.S.* **109** Q12
Chama, *Ghana* **205** M15
Chama, *Zambia* **209** L21
Chamah, peak, *Malaysia* **191** T8
Chaman, *Pak.* **180** L6
Chamba, *India* **180** K14
Chamba, *Tanzania* **211** Z6
Chambal, river, *India* **182** H5
Chamberlain Lake, *Me., U.S.* **97** C17
Chamberlain, *S. Dak., U.S.* **106** L6
Chamberlin, Mount, *Alas., U.S.*
  113 C18
Chambers Creek, *Tex., U.S.* **105** H18
Chambers, *Ariz., U.S.* **109** S9
Chambersburg, *Pa., U.S.* **96** P6
Chambéry, *Fr.* **149** U14
Chambeshi, river, *Zambia* **209** L20
Chambois, *Fr.* **149** P7
Chamburi Kalat, *Pak.* **181** R5
Chamical, *Arg.* **134** J9
Chamizal National Memorial,
  Tex., U.S. **104** H7
Chamoli, *India* **182** F7
Chamonix, *Fr.* **149** T15
Champagne, region, *Fr.* **149** P12
Champagnole, *Fr.* **149** S14
Champaign, *Ill., U.S.* **101** R6
Champasak, *Laos* **190** M11
Champéry, *Switz.* **152** C2
Champlain, Lake, *N.Y. Vt., U.S.* **96** F12
Champotón, *Mex.* **117** J15
Chamrajnagar, *India* **183** R5
Chamzinka, *Russ.* **158** K13
Chañaral, *Chile* **134** F6
Chança, river, *Port., Sp.* **146** K6
Chandalar, river, *Alas., U.S.* **113** E17
Chandeleur Islands, *La., U.S.* **103** Q13
Chandeleur Sound, *La., U.S.* **102** Q12
Chandigarh, *India* **182** E6
Chandler, *Okla., U.S.* **105** C17
Chandler, *Ariz., U.S.* **109** U7
Chandless, river, *S. Amer.* **132** G5
Chandpur, *Bangladesh* **182** K14
Chandrapur, *India* **182** M7
Chang Chenmo Range, *China, India*
  180 H16
Chang Jiang (Yangtze), river, *China*
  184 M10
Ch'angam, *N. Korea* **186** L4
Changan, *N. Korea* **186** D10
Changanacheri, *India* **183** T5
Changane, river, *Mozambique* **212** H10
Changara, *Mozambique* **212** E10
Changbai Shan, *China* **185** F16
Changchon, *S. Korea* **186** M10
Changchun, *China* **185** E16
Changde, *China* **184** L12
Ch'angdo, *N. Korea* **186** K9
Changdong, *N. Korea* **186** H9
Changdong, *S. Korea* **186** M10

Changdu *see* Qamdo, *China* **184** K8
Changgang, *N. Korea* **186** E8
Changhŭng, *N. Korea* **186** F10
Changhyŏn, *N. Korea* **186** F12
Changji, *China* **184** E5
Changjin-ho (Chosin Reservoir),
  N. Korea **186** F8
Changjin, *N. Korea* **186** F8
Changjin, river, *N. Korea* **186** E9
Changma'er *see* Changmar, *China*
  184 H2
Changmar (Changma'er), *China*
  184 H2
Changnyeong, *S. Korea* **187** S11
Changnyŏn, *N. Korea* **186** K5
Ch'angp'ung, *N. Korea* **186** L7
Ch'angp'yŏng, *N. Korea* **186** F11
Changsha, *China* **185** M13
Changshan, *China* **185** L15
Ch'angsŏng, *N. Korea* **186** F5
Changsu-ho, lake, *N. Korea* **186** L6
Changwon, *S. Korea* **187** S11
Changyŏn, *N. Korea* **186** L5
Chaniá (Canea, Cydonia), *Gr.* **157** P13
Chaníon, Kólpos, *Gr.* **157** P13
Channapatna, *India* **183** R5
Channel Country, *Austral.* **220** G11
Channel Fleeming, *Bahamas* **122** E4
Channel Islands National Park,
  Calif. U.S. **111** Y5
Channel Islands, *Calif., U.S.* **111** Y6
Channel Islands, *U.K.* **145** Y11
Channel of Massawa, *Eritrea* **210** H8
Channel Rock, *Bahamas* **118** G12
Channel-Port aux Basques,
  Nfld. & Lab., Can. **87** N23
Chantada, *Sp.* **146** C6
Chanthaburi, *Thai.* **191** N9
Chantilly, *Fr.* **149** P10
Chantrey Inlet, *Nunavut, Can.* **87** G14
Chanute, *Kans., U.S.* **107** U9
Chao Phraya, river, *Thai.* **190** L7
Chão, Ilhéu, *Madeira, Port.* **214** N4
Chapadinha, *Braz.* **133** E15
Chapaev, *Kaz.* **178** E5
Chapak Gŏżar, *Afghan.* **180** F5
Chapala, Lago de, *Mex.* **116** H10
Chapala, Lake, *N. Amer.* **84** N4
Chapcha, *Bhutan* **182** G12
Chapeco, *Braz.* **132** N12
Chapel Hill, *N.C., U.S.* **98** H11
Chapelton, *Jam.* **122** J9
Chapleau, *Ont. Can.* **87** P17
Chapman, *Ala., U.S.* **103** M15
Chapman, *Kans., U.S.* **107** S8
Chapmans, *St. Vincent & the Grenadines*
  123 K17
Chapoma, *Russ.* **158** A9
Chappal Waddi, peak,
  Cameroon, Nigeria **206** M10
Chappaquiddick Island, *Mass., U.S.*
  97 K15
Chappell, *Nebr., U.S.* **107** P2
Char, *India* **182** D6
Chara, *Russ.* **161** J17
Chara, river, *Russ.* **161** J17
Charadai, *Arg.* **134** G12
Charagua, *Bol.* **132** K8
Charambirá, Punta, *Col.* **130** G4
Charcot Fan, *Pac. Oc.* **249** X2
Charcot Island, *Antarctica* **238** G6
Charcot Land, *Greenland, Den.* **88** K12
Chard, *Eng., U.K.* **145** V11
Chardonnières, *Haiti* **119** M15
Charduar, *India* **182** G14
Chari, river, *Chad* **206** K12
Chārīkār, *Afghan.* **180** H9
Chariton, *Iowa, U.S.* **107** Q12
Chariton, river, *Iowa, Mo., U.S.*
  107 Q12
Charity, *Guyana* **131** E15
Chärjew *see* Türkmenabat, *Turkm.*
  178 M9
Charleroi, *Belg.* **148** M12
Charles City, *Iowa, U.S.* **106** M12
Charles Island, *Nunavut, Can.* **87** H18
Charles Mound, peak, *Ill., U.S.*
  100 M4
Charles Town, *W. Va., U.S.* **98** C12
Charles, Cape, *Va., U.S.* **98** F14
Charleston, *S.C., U.S.* **98** M10
Charleston, *W. Va., U.S.* **98** D8
Charleston, *Ill., U.S.* **101** S6

Charleston, *Miss., U.S.* **102** J11
Charleston, *Mo., U.S.* **107** V16
Charleston, *Oreg., U.S.* **110** J1
Charleston, *N.Z.* **223** K18
Charlestown, *Ind., U.S.* **101** T10
Charlestown, *St. Kitts & Nevis* **123** C18
Charleville (An Ráth), *Ire.* **145** S4
Charleville-Mézières, *Fr.* **149** N12
Charleville, *Qnsld., Austral.* **221** U13
Charlevoix, *Mich., U.S.* **100** H10
Charlie Gap, *Arctic Oc.* **252** J8
Charlie-Gibbs Fracture Zone, *Atl. Oc.*
  248 G8
Charlieu, *Fr.* **149** T12
Charlotte Amalie, *Virgin Is., U.S.*
  122 M8
Charlotte Harbor, *Fla., U.S.* **99** W8
Charlotte Town *see* Gouyave,
  Grenada **123** K22
Charlotte, *N.C., U.S.* **98** J9
Charlotte, *Mich., U.S.* **100** M10
Charlotte, *Tex., U.S.* **105** M15
Charlottenberg, *Sw.* **142** K12
Charlottesville, *Va., U.S.* **98** E11
Charlottetown, *P.E.I. Can.* **87** N22
Charlotteville, *Trin. & Tobago* **123** N18
Charlton Island, *Nunavut, Can.* **87** M17
Charolles, *Fr.* **149** T12
Charouine, *Alg.* **200** F11
Charsadda, *Pak.* **180** H11
Charters Towers, *Qnsld., Austral.*
  221 S13
Chartres, *Fr.* **149** Q9
Charyeongsanmaek, range, *S. Korea*
  187 P9
Chasan, *N. Korea* **186** J6
Chascomús, *Arg.* **134** M13
Chase City, *Va., U.S.* **98** F12
Chase, *Kans., U.S.* **107** T6
Chaska, *Minn., U.S.* **106** K11
Chasŏng, *N. Korea* **186** D8
Chassahowitzka, *Fla., U.S.* **99** T7
Château-Gontier, *Fr.* **149** R6
Château-la-Vallière, *Fr.* **149** R7
Château-Renault, *Fr.* **149** R8
Château-Thierry, *Fr.* **149** P11
Chateaubelair, *St. Vincent & the*
  Grenadines **123** K16
Châteaubriant, *Fr.* **149** R6
Châteaudun, *Fr.* **149** Q9
Châteaulin, *Fr.* **149** Q3
Châteauneuf-de-Randon, *Fr.*
  149 W11
Châteauneuf-en-Thymerais, *Fr.*
  149 P8
Châteauneuf-sur-Loire, *Fr.* **149** R10
Châteauroux, *Fr.* **149** S9
Châteauvillain, *Fr.* **149** Q12
Châtel, *Fr.* **149** T15
Châtelet, *Belg.* **148** M12
Châtellerault, *Fr.* **149** S8
Chatfield, *Minn., U.S.* **106** L12
Chatham Island, *Pac. Oc.* **224** P9
Chatham Islands, *Pac. Oc.* **224** P9
Chatham Rise, *Pac. Oc.* **246** P11
Chatham *see* Isla San Cristóbal, *Ecua.*
  130 Q4
Chatham Strait, *Alas., U.S.* **113** L22
Chatham, *N.B. Can.* **87** N21
Chatham, *N.Y., U.S.* **96** K11
Chatham, *Mass., U.S.* **97** L16
Chatham, *Va., U.S.* **98** F11
Chatham, *Ill., U.S.* **101** R5
Chatham, *Eng., U.K.* **145** U15
Châtillon-sur-Indre, *Fr.* **149** S8
Châtillon, *Fr.* **149** R12
Chatom, *Ala., U.S.* **103** M13
Chatra, *India* **182** J10
Chatsworth, *Ga., U.S.* **98** J4
Chattahoochee, *Fla., U.S.* **99** Q4
Chattahoochee, river, *U.S.* **91** K17
Chattanooga, *Tenn., U.S.* **103** G17
Chatyr-Tash, *Kyrg.* **179** L15
Châu Đốc, *Cambodia* **191** P11
Châu Đốc, *Vietnam* **191** P11
Chauk, *Myanmar* **190** H4
Chaukan Pass, *India* **182** G16
Chaumont, *Fr.* **149** Q13
Chaun Bay, *Arctic Oc.* **252** G6
Chaunskaya Guba, *Russ.* **161** C19
Chauny, *Fr.* **149** N11
Chaves, *Braz.* **133** C13
Chaves, *Port.* **146** D7
Chaves, Indefatigable *see* Isla Santa
  Cruz, *Ecua.* **130** Q3
Chavuma, *Zambia* **209** M15

Chaychaktū, *Afghan.* **180** G5
Chayek, *Kyrg.* **179** K14
Chayu *see* Zayü, *China* **184** L7
Ch'ayu, *N. Korea* **186** C12
Cheaha Mountain, *Ala., U.S.* **103** J16
Cheb, *Czechia* **150** H9
Cheboksarskoye Vodokhranilishche, *Russ.* **158** H13
Cheboksary, *Russ.* **158** H13
Cheboygan, *Mich., U.S.* **100** G11
Chebsara, *Russ.* **158** F10
Chechen', *Russ.* **159** T15
Chechnya, region, *Russ.* **159** T14
Cheduba Island, *Myanmar* **190** J3
Cheekha Dar, peak, *Iran, Iraq* **176** D11
Cheektowaga, *N.Y., U.S.* **96** J5
Cheepie, *Qnsld., Austral.* **221** U13
Cheeyou, *Trin. & Tobago* **123** P23
Chegga, *Mauritania* **205** B13
Chegutu, *Zimb.* **212** E9
Chehalis, *Wash., U.S.* **110** D3
Cheju Island *see* Jeju Island, *Asia* **168** G13
Cheju *see* Jeju, *S. Korea* **187** W7
Chekichler, *Turkm.* **178** N5
Ch'elago, *Eth.* **211** Q8
Chelan, *Wash., U.S.* **110** B6
Chelan, Lake, *Wash., U.S.* **110** B6
Chelforó, *Arg.* **135** P9
Chełm, *Pol.* **151** F18
Chełmno, *Pol.* **151** C14
Chelmsford, *Eng., U.K.* **145** U15
Chelsea, *Okla., U.S.* **105** B19
Cheltenham, *Eng., U.K.* **145** T12
Chelyabinsk, *Russ.* **160** J8
Chelyuskin, *Russ.* **161** D14
Chemba, *Mozambique* **212** E11
Chemchemal (Jamjamāl), *Iraq* **176** E11
Chemillé, *Fr.* **149** S6
Chemin Grenier, *Mauritius* **215** H19
Chemnitz, *Ger.* **150** G10
Chemult, *Oreg., U.S.* **110** J4
Chemung, river, *N.Y., Pa., U.S.* **96** L7
Chenab, river, *Pak.* **180** M10
Chenachane, *Alg.* **200** H8
Chenal Sain, *Wallis & Futuna Is., Fr.* **230** Q12
Chenārān, *Iran* **177** D20
Ch'ench'a, *Eth.* **211** N7
Cheney Reservoir, *Kans., U.S.* **107** U7
Cheney, *Wash., U.S.* **110** C9
Chengalpattu, *India* **183** R7
Chengchow *see* Zhengzhou, *China* **185** J13
Chengde, *China* **185** F14
Chengdu, *China* **184** L10
Chengele, *India* **182** F15
Chengkou, *China* **184** K11
Chengmai, *China* **184** Q12
Chennai (Madras), *India* **183** Q7
Chenzhou, *China* **185** N13
Cheonan, *S. Korea* **187** P8
Cheongdo, *S. Korea* **187** R11
Cheongju, *S. Korea* **187** P9
Cheongsando, island, *S. Korea* **187** V8
Cheongsong, *S. Korea* **187** Q12
Cheongyang, *S. Korea* **187** Q8
Cheonsuman, bay, *S. Korea* **187** P7
Cheorwon, *S. Korea* **186** L8
Chepén, *Peru* **132** F2
Chépénéhe, *New Caledonia, Fr.* **231** P23
Chepes, *Arg.* **134** J9
Cheptsa, river, *Russ.* **158** F16
Chequamegon Bay, *Wis., U.S.* **100** F4
Cheraw, *S.C., U.S.* **98** J10
Cherbaniani Reef, *Ind. Oc.* **250** D10
Cherbourg-Octeville, *Fr.* **149** N6
Cherchell, *Alg.* **200** B12
Cherepovets, *Russ.* **158** F9
Cherevkovo, *Russ.* **158** D12
Chéri, *Niger* **206** H10
Cherkasy, *Ukr.* **159** P6
Cherkessk, *Russ.* **159** T12
Cherlak, *Russ.* **160** K10
Chermenino, *Russ.* **158** F12
Chernaya, *Russ.* **160** E10
Cherni Vrŭkh, peak, *Bulg.* **155** J13
Chernihiv, *Ukr.* **158** M6
Chernivtsi, *Ukr.* **159** Q3
Chernobyl' *see* Chornobyl', *Ukr.* **158** M5
Chernogorsk, *Russ.* **161** L13
Chernyakhovsk, *Russ.* **158** J2
Chernyayevo, *Russ.* **161** K19
Chernyshevskiy, *Russ.* **161** H16

Cherokee Sound, *Bahamas* **122** B9
Cherokee, *Okla., U.S.* **105** B16
Cherokee, *Iowa, U.S.* **106** M9
Cherry Island *see* Anuta, *Solomon Is.* **224** H7
Cherryfield, *Me., U.S.* **97** F18
Cherskiy Range, *Asia* **168** C11
Cherskiy, *Russ.* **161** D19
Cherskogo, *Russ.* **161** F18
Chersónisos Akrotíri, *Gr.* **157** P13
Chersónisos Chalkidikís, *Gr.* **156** C11
Cherven Bryag, *Bulg.* **155** H14
Chervonohrad, *Ukr.* **159** N2
Cherykaw, *Belarus* **158** L6
Chesaning, *Mich., U.S.* **100** L11
Chesapeake Bay, *Md., Va., U.S.* **98** D14
Chesapeake, *Va., U.S.* **98** F14
Chesha Bay, *Arctic Oc.* **253** C18
Cheshskaya Guba, *Russ.* **160** E9
Cheshunt, *Eng., U.K.* **145** U14
Chester, *Pa., U.S.* **96** Q9
Chester, *S.C., U.S.* **98** J9
Chester, *W. Va., U.S.* **98** A9
Chester, *Ill., U.S.* **101** U4
Chester, *Mont., U.S.* **108** B8
Chester, *Calif., U.S.* **111** N5
Chester, *Eng., U.K.* **145** R11
Chesterfield Inlet, *Nunavut, Can.* **87** H15
Chesterfield Inlet, town, *Nunavut, Can.* **87** H15
Chesterfield, *Eng., U.K.* **145** R12
Chesterfield, Îles, *Coral Sea* **224** K6
Chesterton Range, *Austral.* **220** G14
Chestertown, *Md., U.S.* **98** C14
Chesuncook Lake, *Me., U.S.* **97** C17
Chetamale, *India* **183** S14
Chetek, *Wis., U.S.* **100** H3
Chetlat Island, *India* **183** R3
Chetumal Bay, *N. Amer.* **84** N7
Chetumal, *Mex.* **117** J17
Chetvertyy Kuril'skiy Proliv, *Russ.* **161** G23
Chevak, *Alas., U.S.* **112** H11
Chevanceaux, *Fr.* **149** V7
Chevelon Butte, peak, *Ariz., U.S.* **109** S8
Cheviot Hills, *Scot., U.K.* **145** M11
Chew Bahir, Lake, *Af.* **198** H14
Ch'ew Bahir, lake, *Eth.* **211** P6
Chewelah, *Wash., U.S.* **110** B8
Cheyenne Bottoms, *Kans., U.S.* **107** T6
Cheyenne Wells, *Colo., U.S.* **109** N16
Cheyenne, *Okla., U.S.* **105** C14
Cheyenne, *Wyo., U.S.* **108** K13
Cheyenne, river, *U.S.* **106** L3
Cheyne Bay, *Austral.* **220** L4
Chhâk Kâmpóng Saôm, bay, *Cambodia* **191** P10
Chhapra, *India* **182** H10
Chhatarpur, *India* **182** J7
Chhêb Kândal, *Cambodia* **190** M11
Chhîm, *Leb.* **172** J6
Chhindwara, *India* **182** K6
Chhukha, *Bhutan* **182** G12
Chi, river, *Thai.* **190** L9
Chiang Mai, *Thai.* **190** H9
Chiang Rai, *Thai.* **190** H7
Chiaromonte, *It.* **153** N13
Chiasso, *Switz.* **152** D4
Chiat'ura, *Rep. of Georgia* **171** B17
Chiavari, *It.* **152** F4
Chiavenna, *It.* **152** C5
Chiba, *Japan* **189** P12
Chibemba, *Angola* **208** P10
Chibia, *Angola* **208** P10
Chibougamau, *Que., Can.* **87** N19
Chiburi Shima, *Japan* **189** P7
Chibuto, *Mozambique* **212** H10
Chibwe, *Zambia* **209** N18
Chicago Heights, *Ill., U.S.* **101** P7
Chicago, *Ill., U.S.* **101** N7
Chicamba, *Angola* **208** H8
Chicapa, river, *Angola* **209** K13
Chichagof Island, *Alas., U.S.* **113** L22
Chichawatni, *Pak.* **180** M11
Chichén Itzá, ruin, *Mex.* **117** H16
Chichester Range, *Austral.* **220** F3
Chichester, *Eng., U.K.* **145** V13
Chichi Jima Rettō (Beechey Group), *Japan* **189** X14
Chickamauga Lake, *Tenn., U.S.* **103** F17
Chickamauga, *Ga., U.S.* **98** J4
Chickasaw National Recreation Area, *Okla., U.S.* **105** E17

Chickasawhay, river, *Miss., U.S.* **103** M13
Chickasha, *Okla., U.S.* **105** D16
Chicksands, *Eng., U.K.* **145** T13
Chiclayo, *Peru* **132** F2
Chico, *Calif., U.S.* **111** P4
Chico, river, *Arg.* **135** S9
Chicoma Mountain, *N. Mex., U.S.* **109** R12
Chicopee, *Mass., U.S.* **97** K13
Chief Joseph Dam, *Wash., U.S.* **110** B6
Chiefland, *Fla., U.S.* **99** S7
Chiengi, *Zambia* **209** K18
Chieri, *It.* **152** E3
Chiesanuova, *San Marino* **163** K14
Chieti, *It.* **152** K10
Chifeng, *China* **185** F14
Chignik Lagoon, *Alas., U.S.* **113** M13
Chigubo, *Mozambique* **212** G10
Chihil Abdālān, Kōh-e, *Afghan.* **180** J4
Chihuahua, *Mex.* **116** D9
Chihuahuan Desert, *Mex.* **116** C9
Chikan, *Russ.* **161** K15
Chikhachevo, *Russ.* **158** H6
Chikhli, *India* **182** L5
Chilanga, *Zambia* **209** P18
Chilanga, *Malawi* **212** C10
Chilas, *Pak.* **180** G12
Chilaw, *Sri Lanka* **183** U7
Chilca, *Peru* **132** H3
Childersburg, *Ala., U.S.* **103** K16
Childress, *Tex., U.S.* **105** E14
Chile Basin, *Pac. Oc.* **247** P22
Chile Fracture Zone, *Pac. Oc.* **247** N18
Chile Ridge, *Pac. Oc.* **254** R11
Chile Rise, *Pac. Oc.* **247** P19
Chile Trench, *Pac. Oc.* **247** Q22
Chile, *S. Amer.* **134** M6
Chilecito, *Arg.* **134** H8
Chililabombwe, *Zambia* **209** M18
Chilka Lake, *India* **182** M10
Chillán, *Chile* **134** M6
Chillicothe, *Ill., U.S.* **101** P5
Chillicothe, *Ohio, U.S.* **101** S13
Chillicothe, *Tex., U.S.* **105** E14
Chillicothe, *Mo., U.S.* **107** R11
Chiloango, river, *Angola, D.R.C.* **208** H9
Chiloé, Isla Grande de, *Chile* **135** R6
Chilonga, *Zambia* **209** M20
Chiloquin, *Oreg., U.S.* **110** K4
Chilpancingo, *Mex.* **116** K11
Chiltern Hills, *Eng., U.K.* **145** U13
Chilton, *Wis., U.S.* **100** K7
Chiluage, *Angola* **209** K14
Chilumba, *Malawi* **212** B11
Chim Berkaouane, *Niger* **206** G6
Chimacum, *Wash., U.S.* **110** B3
Chimbas, *Arg.* **134** J9
Chimborazo, peak, *Ecua.* **130** L4
Chimbote, *Peru* **132** G2
Chimboy, *Uzb.* **178** K8
Chimney Rock National Historic Site, *Nebr., U.S.* **107** P1
Chimney Rock National Monument, *Colo., U.S.* **109** O12
Chimoio, *Mozambique* **212** F10
Chin Hills, *Myanmar* **190** G4
China Point, *Calif., U.S.* **111** Z8
China, *Asia* **184** H5
Chinandega, *Nicar.* **117** M17
Chinati Peak, *Tex., U.S.* **104** L9
Chincha Alta, *Peru* **132** H3
Chinchilla de Monte Aragón, *Sp.* **147** J13
Chincoteague Bay, *Md., U.S.* **98** D15
Chincoteague, *Va., U.S.* **98** E15
Chinde, *Mozambique* **212** E11
Chindwin, river, *Myanmar* **190** E4
Chingola, *Zambia* **209** M18
Chinguar, *Angola* **208** M11
Chinguetti, *Mauritania* **204** E9
Chinhoyi, *Zimb.* **212** E9
Chiniot, *Pak.* **180** L12
Chinko, river, *Cen. Af. Rep.* **207** M17
Chinle, river, *U.S.* **109** R9
Chino Valley, *Ariz., U.S.* **109** S6
Chino, *Japan* **189** P11
Chinon, *Fr.* **149** S7
Chinook Trough, *Pac. Oc.* **246** E11
Chinook, *Mont., U.S.* **108** B9
Chinsali, *Zambia* **209** L20
Chinsong, *N. Korea* **186** D8
Chioggia, *It.* **152** E8
Chíos (Chíos), island, *Gr.* **157** H16

Chios *see* Chíos, island, *Gr.* **157** H16
Chíos, *Gr.* **157** H16
Chipata, *Zambia* **209** N20
Chipindo, *Angola* **208** N11
Chipinge, *Zimb.* **212** F10
Chipley, *Fla., U.S.* **99** Q3
Chiplun, *India* **182** N3
Chippenham, *Eng., U.K.* **145** U12
Chippewa Falls, *Wis., U.S.* **100** H3
Chippewa, Lake, *Wis., U.S.* **100** G3
Chippewa, river, *Wis., U.S.* **100** G3
Chiquián, *Peru* **132** G3
Chiquimula, *Guatemala* **117** L16
Chiquinata Bay, *S. Amer.* **128** H4
Chiquinquirá, *Col.* **130** F6
Chirchiq, *Uzb.* **178** L12
Chirchirri, *Somalia* **211** P11
Chiredzi, *Zimb.* **212** F10
Chirfa, *Niger* **206** D10
Chiricahua National Monument, *Ariz., U.S.* **109** V9
Chiricahua Peak, *Ariz., U.S.* **109** V9
Chirikof Island, *Alas., U.S.* **113** N14
Chirinda, *Russ.* **161** G14
Chiriquí, Golfo de, *Pan.* **117** P20
Chiripo, Cerro, *C.R.* **117** N19
Chirundu, *Zambia* **209** P18
Chisamba, *Zambia* **209** P18
Chisasibi, *Que., Can.* **87** M18
Chisholm, *Minn., U.S.* **106** F12
Chisht-e Sharīf, *Afghan.* **180** H4
Chishtian Mandi, *Pak.* **180** M12
Chişinău, *Mold.* **159** R4
Chişineu Criş, *Rom.* **154** C11
Chisos Mountains, *Tex., U.S.* **104** M10
Chissioua Mtsamboro, island, *Mayotte, Fr.* **215** N17
Chistopol', *Russ.* **158** J15
Chita, *Russ.* **161** L17
Chitado, *Angola* **208** Q10
Chitambo, *Zambia* **209** M19
Chitembo, *Angola* **208** N11
Chitipa, *Malawi* **212** A10
Chitokoloki, *Zambia* **209** N15
Chitose, *Japan* **188** G13
Chitradurga, *India* **183** Q5
Chitral, *Pak.* **180** G11
Chitré, *Pan.* **117** P21
Chittagong, *Bangladesh* **182** K13
Chittaurgarh, *India* **182** J4
Chittoor, *India* **183** Q6
Chitungwiza, *Zimb.* **212** E9
Chiume, *Angola* **209** P14
Chiúre, *Mozambique* **213** C13
Chivasso, *It.* **152** E3
Chivilcoy, *Arg.* **134** L12
Chiwanda, *Tanzania* **211** Z5
Chlef, *Alg.* **200** B11
Chloride, *Ariz., U.S.* **109** R5
Chợ Lầu, *Vietnam* **191** P12
Ch'o-do, island, *N. Korea* **186** K4
Chŏăm Ksant, *Cambodia* **190** M10
Chobe, river, *Angola, Botswana* **212** E6
Choc Bay, *St. Lucia* **123** J14
Chocolate Mountains, *Calif., U.S.* **111** Y12
Choctawhatchee Bay, *Fla., U.S.* **99** Q2
Choctawhatchee, river, *Fla., U.S.* **99** Q3
Chodo, island, *S. Korea* **187** U8
Chodzież, *Pol.* **151** D13
Choele Choel, *Arg.* **135** P9
Chogâr, *Mauritania* **204** G8
Ch'ogu, *N. Korea* **186** K10
Chŏgyuryŏng-sanmaek, range, *N. Korea* **186** P6
Choirokoitia, *Cyprus* **162** Q8
Choiseul Sound, *Falk. Is., U.K.* **135** W12
Choiseul, *St. Lucia* **123** L13
Choiseul, island, *Solomon Is.* **228** C7
Chŏjak, *N. Korea* **186** M5
Chojna, *Pol.* **150** D11
Chojnice, *Pol.* **151** C13
Choke Canyon Reservoir, *Tex., U.S.* **105** N16
Ch'ok'ē, mountains, *Eth.* **210** K7
Chokoloskee, *Fla., U.S.* **99** X9
Chokoyan, *Chad* **207** H15
Chokurdakh, *Russ.* **161** D18
Chokwé, *Mozambique* **212** H10
Cholet, *Fr.* **149** S6
Cholpon- Ata, *Kyrg.* **179** K15
Ch'ŏlsan, *N. Korea* **186** H4

Choluteca, *Hond.* **117** L17
Choma, *Zambia* **209** Q17
Chomutov, *Czechia* **150** G10
Chon Buri, *Thai.* **191** N8
Ch'ŏnam, *N. Korea* **186** K7
Ch'ŏnam, *N. Korea* **186** L9
Chonchi, *Chile* **135** R6
Chŏnch'ŏn, *N. Korea* **186** F7
Chone, *Ecua.* **130** K3
Chŏng Kal, *Cambodia* **190** M10
Ch'ŏngdan, *N. Korea* **186** M6
Ch'ŏngjin, *N. Korea* **186** C13
Chŏngju, *N. Korea* **186** H5
Chŏngp'yŏng, *N. Korea* **186** H9
Chongqing, *China* **184** L11
Chongsŏng, *N. Korea* **186** A13
Ch'ŏngsŏng, *N. Korea* **186** F4
Ch'ŏnma-san, peak, *N. Korea* **186** G5
Ch'ŏnma, *N. Korea* **186** G5
Ch'ŏnnae, *N. Korea* **186** J8
Chonogol, *Mongolia* **185** D13
Chonos, Archipiélago de los, *Chile* **135** S6
Ch'ŏnsu, *N. Korea* **186** D11
Chóra Sfakíon, *Gr.* **157** Q13
Chornobyl' (Chernobyl'), *Ukr.* **158** M5
Chornomors'ke, *Russ.* **159** S7
Chorregon, *Qnsld. Austral.* **221** T12
Chorrillos, *Peru* **132** H3
Chortkiv, *Ukr.* **159** P3
Ch'ŏrwŏn, *N. Korea* **186** L8
Chos Malal, *Arg.* **135** N7
Ch'osan, *N. Korea* **186** E6
Chosin Reservoir *see* Changjin-ho, *N. Korea* **186** F8
Chota Nagpur Plateau, *Asia* **168** J8
Choteau, *Mont., U.S.* **108** C7
Chott ech Chergui, lake, *Alg.* **200** C11
Chott el Hodna, lake, *Alg.* **201** C13
Chott Melrhir, lake, *Alg.* **201** C14
Choûm, *Mauritania* **204** E9
Chowan, river, *N.C., U.S.* **98** G13
Chowchilla, *Calif., U.S.* **111** T6
Choyang, *N. Korea* **186** G6
Choybalsan, *Mongolia* **184** D12
Choyr, *Mongolia* **184** D11
Christchurch, *Eng., U.K.* **145** W12
Christchurch, *N.Z.* **223** L19
Christian IV Glacier, *Greenland, Den.* **88** M12
Christiana, *Jam.* **122** H8
Christianí, island, *Gr.* **157** N15
Christiansburg, *Va., U.S.* **98** F9
Christianshåb *see* Qasigiannguit, *Greenland, Den.* **88** M6
Christiansted, *Virgin Is., U.S.* **122** Q3
Christmas Creek, *W. Aust., Austral.* **221** R6
Christmas Island *see* Kiritimati, *Kiribati* **224** G12
Christmas Lake Valley, *Oreg., U.S.* **110** J6
Christmas Ridge, *Pac. Oc.* **247** J13
Christmas Rise, *Ind. Oc.* **251** J15
Christopher, *Ill., U.S.* **101** U5
Christoval, *Tex., U.S.* **105** J13
Chrysí, island, *Gr.* **157** R16
Chrysoúpoli, *Gr.* **157** B14
Chub Cay, *Bahamas* **118** D11
Chubut, river, *Arg.* **135** R8
Chudleigh Park, *Qnsld. Austral.* **221** S13
Chudovo, *Russ.* **158** F7
Chudskoye Ozero (Peipus, Lake), *Russ.* **158** G5
Chugach Mountains, *Alas., U.S.* **113** J18
Chuginadak Island, *Alas., U.S.* **112** P8
Chugwater, *Wyo., U.S.* **108** J13
Chuhuyiv, *Ukr.* **159** P9
Chuí, *Braz.* **132** R11
Chukai, *Malaysia* **191** T9
Chukchi Borderland, *Arctic Oc.* **252** K8
Chukchi Peninsula, *Asia* **168** A12
Chukchi Plain, *Arctic Oc.* **252** J8
Chukchi Plateau, *Arctic Oc.* **252** K8
Chukchi Range, *Asia* **168** A12
Chukchi Sea, *Arctic Oc.* **252** J6
Chukchi Shelf, *Arctic Oc.* **252** J6
Chukhloma, *Russ.* **158** G11
Chukotskiy Poluostrov, *Russ.* **161** A21
Chukotskoye Nagor'ye, *Russ.* **161** B20
Chula Vista, *Calif., U.S.* **111** Z9

Chumar, India **182** D6
Chumatang, India **180** J16
Chumbicha, Arg. **134** H9
Chumikan, Russ. **161** J20
Chumphon, Thai. **191** P7
Chun Spur, Ind. Oc. **254** B11
Chuncheon, S. Korea **186** M9
Chunga, Zambia **209** P17
Chunggang, N. Korea **186** C8
Chunghwa, N. Korea **186** K6
Chungju, S. Korea **187** P10
Chungjuho, lake, S. Korea **187** P9
Chungp'yŏngjang, N. Korea **186** E10
Chŭngsan, N. Korea **186** G11
Chŭngsan, N. Korea **186** J5
Chunnel, Fr., U.K. **145** V16
Chunya, Tanzania **211** X4
Chupa, Russ. **158** A8
Chupara Point, Trin. & Tobago **123** N22
Chuprovo, Russ. **158** B12
Chuquibamba, Peru **132** J5
Chuquicamata, Chile **134** D7
Chur, Switz. **152** B5
Churachandpur, India **182** J14
Church of St. Anne, Vatican City **163** P17
Church of St. Stephen, Vatican City **163** Q15
Church Point, La., U.S. **102** P8
Churchill Falls, Nfld. & Lab., Can. **87** L21
Churchill Lake, Sask., Can. **86** L12
Churchill Lake, Me., U.S. **97** C16
Churchill Mountains, Antarctica **239** M14
Churchill, Man., Can. **87** K14
Churchill, Cape, Man., Can. **87** K15
Churchill, river, N. Amer. **87** K14
Churuguara, Venez. **130** C9
Chushul, India **182** D7
Chuska Mountains, Ariz., N. Mex., U.S. **109** R9
Chusovoy, Russ. **160** H8
Chutes Kasongo-Lunda, Angola, D.R.C. **208** J12
Chutes Tembo, Angola, D.R.C. **208** J12
Chuuk (Truk Island), F.S.M. **226** B9
Cianjur, Indonesia **192** M8
Çiçekdağı, Turk. **170** E9
Cicero, Ill., U.S. **101** N7
Ciceron, St. Lucia **123** K14
Cicia, island, Fiji **226** P8
Cidade Velha, Cabo Verde **215** E16
Cide, Turk. **170** C8
Ciechanów, Pol. **151** D16
Ciego de Ávila, Cuba **118** H10
Ciempozuelos, Sp. **146** G11
Ciénaga, Col. **130** C6
Cienfuegos, Cuba **118** H8
Cieszyn, Pol. **151** H14
Cieza, Sp. **147** K13
Çiftlik, Turk. **170** G9
Cihanbeyli, Turk. **170** F8
Ciiradhame, Somalia **210** L14
Cijara, Embalse de, Sp. **146** H9
Çikës, Maja e, Alban. **154** L10
Cikobia, island, Fiji **226** M7
Cilacap, Indonesia **192** M9
Cilaos, Réunion, Fr. **215** G16
Cilician Gates, Turk. **170** H16
Cill an Dísirt see Killadysert, Ire. **145** S3
Cill Chainnigh see Kilkenny, Ire. **145** S6
Cill Chaoi see Kilkee, Ire. **145** S3
Cill Charthaigh (Kilcar), Ire. **145** N4
Cill Choca see Kilcock, Ire. **145** R6
Cill Mhantáin see Wicklow, Ire. **145** R7
Cill Orglan see Killorglin, Ire. **145** T3
Cille Bhrighde, Scot., U.K. **144** J6
Çilov, Azerb. **171** D24
Cimarron , river, U.S. **105** B15
Cimarron, Kans., U.S. **107** U4
Cimarron, N. Mex., U.S. **109** Q13
Cimarron, river, U.S. **90** H12
Cimolais, It. **152** C8
Cimone, Monte, It. **152** G6
Cîmpina, Rom. **155** E15
Cîmpulung, Rom. **155** E14
Cinaruco, river, Venez. **130** F10
Cincinnati, Ohio, U.S. **101** S11
Cinco Balas, Cayos, Cuba **118** J9
Cîndrelu, peak, Rom. **155** E13
Çine, Turk. **170** G3

Cinéma d'Été, Monaco **163** B23
Cinn Mhara see Kinvara, Ire. **145** R4
Cinnabar Mountain, Idaho, U.S. **108** H3
Cintegabelle, Fr. **149** Y9
Cinto, Monte, Fr. **149** X2
Cioara Doiceşti, Rom. **155** F17
Čiovo, island, Croatia **154** G6
Cipangu Basin, Pac. Oc. **246** E8
Circle, Mont., U.S. **108** C12
Circle, Alas., U.S. **113** F18
Circleville, Ohio, U.S. **101** R13
Cirebon, Indonesia **192** L9
Cirencester, Eng., U.K. **145** U12
Cirò, It. **153** R13
Ciscaucasia, Eur. **140** H13
Cisco, Tex., U.S. **105** G15
Cisco, Utah, U.S. **109** N9
Cisnădie, Rom. **155** E14
Cisneros, Col. **130** F6
Cistern Point, Bahamas **118** F11
Citronelle, Ala., U.S. **103** N13
Città di Castello, It. **152** H8
Cittadella, It. **152** D7
Cittanova, It. **153** R13
City of Refuge National Historical Park see Pu'uhonua o Hōnaunau National Historical Park, Hawaii, U.S. **115** P20
City of Rocks National Reserve, Idaho, U.S. **108** J5
Ciucea, Rom. **154** C12
Ciudad Acuña, Mex. **116** D11
Ciudad Altamirano, Mex. **116** J11
Ciudad Bolívar, Venez. **130** E12
Ciudad Camargo, Mex. **116** D9
Ciudad Cortés, C.R. **117** N19
Ciudad del Carmen, Mex. **117** J15
Ciudad del Este, Parag. **132** M11
Ciudad Guayana, Venez. **130** D13
Ciudad Guerrero, Mex. **116** D8
Ciudad Guzmán, Mex. **116** H10
Ciudad Hidalgo, Mex. **116** J11
Ciudad Juárez, Mex. **116** B9
Ciudad Madero, Mex. **116** G12
Ciudad Mante, Mex. **116** G12
Ciudad Obregón, Mex. **116** D7
Ciudad Piar, Venez. **130** E13
Ciudad Real, Sp. **146** H10
Ciudad Valles, Mex. **116** G12
Ciudad Victoria, Mex. **116** G12
Ciudad-Rodrigo, Sp. **146** F7
Ciutadella de Menorca, Sp. **147** G19
Civita Castellana, It. **152** K8
Civitanova Marche, It. **152** H10
Civitavecchia, It. **152** K7
Civray, Fr. **149** T7
Çivril, Turk. **170** G5
Cizre, Turk. **171** H16
Çıldır Gölü, Turk. **171** D17
Çıldır, Turk. **171** D17
Çınar, Turk. **171** H15
Çırçır, Turk. **170** E12
Cıva Burnu, Turk. **170** C11
Clacton-on-Sea, Eng., U.K. **145** T15
Clair Engle Lake, Calif., U.S. **110** M3
Clairton, Pa., U.S. **96** P3
Clairview, Qnsld., Austral. **221** T15
Clam Lake, Wis., U.S. **100** F4
Clanton, Ala., U.S. **103** K15
Clare Island, Ire. **145** Q3
Clare, Mich., U.S. **100** K10
Claremont, N.H., U.S. **97** H13
Claremont, Jam. **122** H9
Claremore, Okla., U.S. **105** B19
Clarence Island, Antarctica **238** B4
Clarence Town, Bahamas **119** G14
Clarence, river, N.Z. **223** K19
Clarendon, Ark., U.S. **102** G10
Clarendon, Tex., U.S. **105** D13
Clarie Coast, Antarctica **239** Q19
Clarinda, Iowa, U.S. **107** Q10
Clarion Fracture Zone, Pac. Oc. **247** H15
Clarion, Pa., U.S. **96** M4
Clarion, Iowa, U.S. **106** M11
Clarión, Isla, Mex. **116** H4
Clark Basin, Atl. Oc. **249** N3
Clark Fork, Idaho, U.S. **108** B4
Clark Fork, river, Mont., U.S. **108** D6
Clark Island see Moa Island, Qnsld., Austral. **221** N12
Clark Mountain, Calif., U.S. **111** V11
Clark, S. Dak., U.S. **106** K7
Clark, Lake, Alas., U.S. **113** K15
Clarke Island, Austral. **220** L8
Clarke Range, Austral. **220** E14

Clarks Fork, river, Mont. Wyo. U.S. **108** F9
Clarks Town, Jam. **122** H8
Clarks, La., U.S. **102** L8
Clarksburg, W. Va., U.S. **98** C9
Clarksdale, Miss., U.S. **102** H10
Clarkson, Nebr., U.S. **107** P8
Clarkston, Wash., U.S. **110** D9
Clarksville, Ark., U.S. **102** F7
Clarksville, Tenn., U.S. **103** E14
Clarksville, Tex., U.S. **105** F19
Clatskanie, Oreg., U.S. **110** E2
Claude, Tex., U.S. **104** D12
Clauzetto, It. **152** C9
Clavering Ø, Greenland, Den. **88** H13
Claxton Bay, Trin. & Tobago **123** Q22
Claxton, Ga., U.S. **99** N8
Clay Center, Kans., U.S. **107** S8
Clay, Ky., U.S. **103** C14
Clayton Lake, Me., U.S. **97** B16
Clayton, N.Y., U.S. **96** G8
Clayton, Ga., U.S. **98** J6
Clayton, La., U.S. **102** M9
Clayton, Ala., U.S. **103** M17
Clayton, Okla., U.S. **105** D19
Clayton, N. Mex., U.S. **109** Q15
Cle Elum, Wash., U.S. **110** C5
Clear Island see Cléire, Ire. **145** U3
Clear Lake Reservoir, Calif., U.S. **110** L5
Clear Lake, Wis., U.S. **100** H2
Clear Lake, Iowa, U.S. **106** M11
Clear Lake, S. Dak., U.S. **106** K8
Clear Lake, Calif., U.S. **110** K3
Clear Lake, Calif., U.S. **111** Q3
Clearbrook, Minn., U.S. **106** F9
Clearfield, Pa., U.S. **96** M5
Clearfield, Utah, U.S. **108** F12
Clearmont, Wyo., U.S. **108** F12
Clearwater Mountains, Idaho, U.S. **108** D4
Clearwater, Fla., U.S. **99** U7
Clearwater, Wash., U.S. **110** C2
Clearwater, river, Alb. Sask. Can. **86** K11
Clearwater, river, Idaho, Wash., U.S. **108** D3
Cleburne, Tex., U.S. **105** G17
Cléire (Clear Island), Ire. **145** U3
Clelles, Fr. **149** V13
Clemson, S.C., U.S. **98** K7
Clendenin, W. Va., U.S. **98** D8
Cleopatra Needle, peak, Philippines **193** E13
Clermont-Ferrand, Fr. **149** U11
Clermont, Fr. **149** N10
Clermont, Qnsld., Austral. **221** T14
Clervaux, Lux. **162** G9
Clervé, river, Lux. **162** H9
Cles, It. **152** C7
Cleveland Hills, Eng., U.K. **145** P12
Cleveland, Wis., U.S. **100** K7
Cleveland, Ohio, U.S. **101** N14
Cleveland, Miss., U.S. **102** J10
Cleveland, Tenn., U.S. **103** G17
Cleveland, Okla., U.S. **105** B18
Cleveland, Tex., U.S. **105** K20
Cleveland, Mount, Mont., U.S. **108** A6
Clew Bay, Ire. **145** Q3
Clewiston, Fla., U.S. **99** V9
Clifden (An Clochán), Ire. **145** Q3
Cliff, N. Mex., U.S. **109** U10
Clifton Forge, Va., U.S. **98** E10
Clifton Hills, S. Aust., Austral. **221** U10
Clifton, Tex., U.S. **105** H17
Clifton, Kans., U.S. **107** S8
Clifton, Ariz., U.S. **109** U9
Clinch, river, Tenn. Va., U.S. **103** E19
Clingmans Dome, N.C., Tenn., U.S. **103** F19
Clint, Tex., U.S. **104** H7
Clinton, N.C., U.S. **98** J12
Clinton, S.C., U.S. **98** K8
Clinton, Ill., U.S. **101** R6
Clinton, Ind., U.S. **101** R7
Clinton, Ark., U.S. **102** F8
Clinton, Ky., U.S. **102** D12
Clinton, Miss., U.S. **102** L11
Clinton, Okla., U.S. **105** C15
Clinton, Iowa, U.S. **107** N15
Clinton, Mo., U.S. **107** T11
Clintonville, Wis., U.S. **100** J6
Clipperton Fracture Zone, Pac. Oc. **247** J15
Clipperton Seamounts, Pac. Oc. **247** H19

Clipperton, island, Pac. Oc. **225** E19
Cliza, Bol. **132** K7
Cloates, Point, Austral. **220** F1
Cloch na Rón see Roundstone, Ire. **145** Q3
Clonagh, Qnsld., Austral. **221** S11
Clonakilty Bay, Ire. **145** U4
Cloncurry Plateau, Austral. **220** E11
Cloncurry, river, Austral. **220** D11
Clonmany (Cluain Maine), Ire. **145** M6
Clonmel (Cluain Meala), Ire. **145** T5
Cloppenburg, Ger. **150** D6
Cloquet, Minn., U.S. **106** G12
Semipalatinsk Nuclear Test Site, Kaz. **179** E15
Cloud Peak, Wyo., U.S. **108** F11
Clover, S.C., U.S. **98** J9
Cloverdale, Calif., U.S. **111** Q3
Cloverport, Ky., U.S. **103** C15
Clovis, N. Mex., U.S. **109** S15
Clovis, Calif., U.S. **111** T6
Cluain Cearbán see Louisburgh, Ire. **145** Q3
Cluain Maine see Clonmany, Ire. **145** M6
Cluain Meala see Clonmel, Ire. **145** T5
Cluj-Napoca, Rom. **155** C13
Clun, Eng., U.K. **145** S10
Cluses, Fr. **149** T14
Clusone, It. **152** D5
Clutha, river, N.Z. **223** P17
Clyde Inlet, Nunavut, Can. **87** E18
Clyde River, Nunavut, Can. **87** E18
Clyde, Ohio, U.S. **101** P13
Clyde, Tex., U.S. **105** G14
Clyde, Kans., U.S. **107** S7
Clyde, Firth of, Scot., U.K. **145** M8
Cnalwa, W. Sahara **200** H3
Cnossus see Knossós, ruin, Gr. **157** V13
Coachella Canal, Calif., U.S. **111** Y11
Coachella, Calif., U.S. **111** Y11
Coahoma, Tex., U.S. **104** H12
Coakley Cay, Bahamas **122** F8
Coal Creek, N.Z. **223** N17
Coalcomán, Mex. **116** J10
Coaldale, Nev., U.S. **111** R8
Coalgate, Okla., U.S. **105** E18
Coalinga, Calif., U.S. **111** U6
Coalville, Eng., U.K. **145** S12
Coari, Braz. **132** E8
Coari, river, S. Amer. **128** E5
Coari, river, Braz. **132** E8
Coast and Geodetic Survey Seamount Province, Pac. Oc. **247** D13
Coast Mountains, N. Amer. **84** F3
Coast Range, N. Amer. **128** A5
Coast Ranges, N. Amer. **84** H3
Coastal Plain, U.S. **91** K19
Coats Island, Nunavut, Can. **87** J16
Coats Land, region, Antarctica **238** M14
Coatzacoalcos, Mex. **117** J14
Cobalt, Ont., Can. **87** P18
Cobán, Guatemala **117** K16
Cobar, N.S.W., Austral. **221** W13
Cobia, island, Fiji **226** N7
Cobija, Bol. **132** G6
Cobourg Peninsula, Austral. **220** A8
Cobre, Nev., U.S. **110** M12
Coburg Island, Nunavut, Can. **87** D16
Coburg, Ger. **150** G8
Coca, Sp. **146** E10
Cocanada see Kakinada, India **182** N8
Cochabamba, Bol. **132** K7
Cochin see Kochi, India **183** S5
Cochinoca, Arg. **134** D9
Cochinos, Bahía de, Cuba **118** H7
Cochran, Ga., U.S. **99** N6
Cochrane, Ont., Can. **87** N17
Cockburn Harbour, Turks & Caicos Is., U.K. **119** J17
Cockburn Town, Turks & Caicos Is. U.K. **119** J18
Cockburn Town, Bahamas **122** E11
Cocklebiddy Motel, W. Aust. Austral. **221** W6
Cockroach Island, Virgin Is. U.S. **122** M6
Cockscomb Point, American Samoa. U.S. **230** D7
Coco Channel, India **183** Q14

Coco-de-Mer Seamounts, Ind. Oc. **250** G8
Coco, Cayo, Cuba **118** H10
Coco, Isla del, C.R. **117** Q17
Coco, river, N. Amer. **117** L19
Cocoa Island, Mauritius **215** J19
Cocoa, Fla., U.S. **99** T9
Cocobeach, Gabon **208** D7
Coconut Point, American Samoa, U.S. **230** E7
Cocos Basin, Ind. Oc. **251** H14
Cocos Bay, Trin. & Tobago **123** Q23
Cocos Island, N. Amer. **84** R7
Cocos Island, Guam, U.S. **230** R4
Cocos Lagoon, Guam, U.S. **230** Q4
Cocos Plain, Pac. Oc. **251** G14
Cocos Ridge, Pac. Oc. **247** J21
Cocos-Keeling Rise, Ind. Oc. **251** J14
Coculus Point, Virgin Is., U.S. **122** N8
Cod Island, Nfld. & Lab., Can. **87** J21
Cod, Cape, Mass., U.S. **97** L16
Coda Cavallo, Capo, It. **152** M5
Codajás, Braz. **132** E8
Codigoro, It. **152** F8
Codlea, Rom. **155** E15
Codó, Braz. **133** E15
Codogno, It. **152** E5
Codrington, Antigua & Barbuda **121** B16
Cody, Nebr., U.S. **106** M4
Cody, Wyo., U.S. **108** F10
Coeburn, Va., U.S. **98** F7
Coen, Qnsld., Austral. **221** P12
Coeur d'Alene Lake, Idaho, U.S. **108** C3
Coeur d'Alene, Idaho, U.S. **108** C3
Coevorden, Neth. **148** H14
Coffeyville, Kans., U.S. **107** V9
Coffin Bay, Austral. **220** K9
Coffin Group see Haha Jima Rettō, Japan **189** X14
Coffs Harbour, N.S.W. Austral. **221** W15
Cofrentes, Sp. **147** H14
Coghinas, Lago del, It. **152** M4
Cognac, Fr. **149** U7
Cogne, It. **152** D2
Cogo, Eq. Guinea **206** Q9
Cogollos, Sp. **146** D11
Cohagen, Mont., U.S. **108** C11
Cohoes, N.Y., U.S. **96** J11
Coiba Island, N. Amer. **84** Q8
Coiba, Isla de, Pan. **117** P20
Coig, river, Arg. **135** W8
Coihaique, Chile **135** S7
Coimbatore, India **183** S5
Coimbra, Port. **146** F5
Coín, Sp. **146** M9
Coipasa, Salar de, Bol. **132** K6
Cojímíes, Boca de, Ecua. **130** K3
Cokato, Minn., U.S. **106** J10
Cokeville, Wyo., U.S. **108** J8
Coki, Senegal **204** H7
Colac, Vic., Austral. **221** Z12
Colatina, Braz. **133** L16
Colberger Heide, Greenland, Den. **89** R9
Colby, Kans., U.S. **107** S4
Colchester, Eng., U.K. **145** T15
Cold Bay, Alas., U.S. **112** N11
Cold Spring, Minn., U.S. **106** J10
Coldwater Creek , U.S. **104** B11
Coldwater, Mich., U.S. **101** N10
Coldwater, Kans., U.S. **107** V5
Coldwater, river, Miss., U.S. **102** H11
Colebrook, N.H., U.S. **97** F14
Coleman, Mich., U.S. **100** K11
Coleman, Tex., U.S. **105** H15
Coleraine (Cúil Rathain), N. Ire., U.K. **145** N6
Coleraine, Minn., U.S. **106** F11
Coles, Punta, Peru **132** K5
Colesberg, S. Af. **212** L7
Colfax, La., U.S. **102** M8
Colfax, Iowa, U.S. **107** P12
Colfax, Wash., U.S. **110** D8
Colfax, Calif., U.S. **111** Q5
Colhué Huapí, Lago, Arg. **135** S9
Colico, It. **152** C5
Colíder, Braz. **132** G11
Colihaut, Dominica **123** E18
Colima, Mex. **116** J9
Colinas, Braz. **133** F15
Colima, Mex. **116** J9
College Park, Ga., U.S. **98** L5
College Station, Tex., U.S. **105** K18

College, *Alas., U.S.* **113** G17
Collelongo, *It.* **152** K10
Colline, *W. Aust., Austral.* **221** X3
Collier Bay, *Austral.* **220** C5
Collier Range, *Austral.* **220** G3
Colliers Point, *Cayman Is., U.K.* **118** L7
Colliers Point, *Cayman Is., U.K.* **122** M5
Colliers, *Cayman Is., U.K.* **122** M5
Collierville, *Tenn., U.S.* **102** G12
Collingwood, *N.Z.* **223** J19
Collins Head, *Austral.* **229** B20
Collins, *Miss., U.S.* **102** M12
Collinson Peninsula, *Nunavut, Can.* **87** F14
Collinsville, *Va., U.S.* **98** G10
Collinsville, *Ill., U.S.* **101** T4
Collinsville, *Okla., U.S.* **105** B18
Collinsville, *Qnsld., Austral.* **221** S14
Collipulli, *Chile* **135** N6
Collison Point, *Virgin Is., U.K.* **122** Q8
Collo, *Alg.* **201** B14
Collonges, *Fr.* **149** T14
Collooney (Cúil Mhuine), *Ire.* **145** P4
Colmán *see* Arthurstown, *Ire.* **145** T6
Colmar, *Fr.* **149** Q15
Cologne *see* Köln, *Ger.* **150** F5
Colombey-les-Belles, *Fr.* **149** P13
Colombia, *S. Amer.* **130** G7
Colômbia, *Braz.* **133** L13
Colombian Basin, *Atl. Oc.* **248** M3
Colombian Trench, *Pac. Oc.* **247** J22
Colombo, *Sri Lanka* **183** U7
Colón Ridge, *Pac. Oc.* **247** J20
Colón, *Pan.* **117** N21
Colón, *Cuba* **118** G8
Colonarie, *St. Vincent & the Grenadines* **123** K17
Colonarie, river, *St. Vincent & the Grenadines* **123** K17
Colonel Hill, *Bahamas* **119** G14
Colonet, Cabo, *Mex.* **116** B5
Colonia Dora, *Arg.* **134** H10
Colonia, *F.S.M.* **226** E11
Colonial Beach, *Va., U.S.* **98** D13
Colonial Heights, *Va., U.S.* **98** F13
Colonial National Historical Park, *Va., U.S.* **98** F14
Colonna, Capo, *It.* **153** Q14
Colonsay, island, *Scot., U.K.* **144** L7
Colophon, ruins, *Turk.* **170** G3
Colorado City, *Tex., U.S.* **105** G13
Colorado City, *Ariz., U.S.* **109** Q6
Colorado National Monument, *Colo., U.S.* **108** M10
Colorado Plateau, *U.S.* **90** H7
Colorado River Aqueduct, *Calif., U.S.* **111** X12
Colorado Springs, *Colo., U.S.* **109** N13
Colorado, *U.S.* **108** M13
Colorado, river, *N. Amer.* **84** K3
Colorado, river, *Arg.* **135** N8
Colotepec, *Mex.* **117** K13
Colquitt, *Ga., U.S.* **99** P4
Colstrip, *Mont., U.S.* **108** E11
Columbia City, *Ind., U.S.* **101** P10
Columbia Falls, *Mont., U.S.* **108** B5
Columbia Mountains, *N. Amer.* **84** G3
Columbia Plateau, *U.S.* **90** D5
Columbia, *Pa., U.S.* **96** P8
Columbia, *Md., U.S.* **98** C13
Columbia, *N.C., U.S.* **98** H9
Columbia, *S.C., U.S.* **98** K9
Columbia, *La., U.S.* **102** L8
Columbia, *Miss., U.S.* **102** N11
Columbia, *Ky., U.S.* **103** D17
Columbia, *Tenn., U.S.* **103** F15
Columbia, *Mo., U.S.* **107** S13
Columbia, Cape, *N. Amer.* **84** B7
Columbia, Mount, *Alta., B.C., Can.* **86** L9
Columbia, river, *N. Amer.* **84** H3
Columbretes, Islas, *Sp.* **147** G16
Columbus Monument, *Bahamas* **122** F11
Columbus Point, *Trin. & Tobago* **123** P17
Columbus, *Ga., U.S.* **99** N4
Columbus, *Wis., U.S.* **100** L6
Columbus, *Ind., U.S.* **101** S9
Columbus, *Ohio, U.S.* **101** R13
Columbus, *Miss., U.S.* **103** J13
Columbus, *Tex., U.S.* **105** L18
Columbus, *N. Dak., U.S.* **106** D3
Columbus, *Kans., U.S.* **107** V10
Columbus, *Nebr., U.S.* **107** P7
Columbus, *Mont., U.S.* **108** E9

Columbus, *N. Mex., U.S.* **109** W11
Colusa, *Calif., U.S.* **111** Q4
Colville Basin, *Pac. Oc.* **255** H22
Colville Channel, *N.Z.* **223** E21
Colville Lake, *N.W.T., Can.* **86** F10
Colville Ridge, *Pac. Oc.* **246** N11
Colville, *Wash., U.S.* **110** A8
Colville, *N.Z.* **223** E21
Colville, river, *Alas., U.S.* **113** C15
Colvocoresses Bay, *Antarctica* **239** M21
Colwich, *Kans., U.S.* **107** U7
Coma Pedrosa, Pic de, *Andorra, Sp.* **162** H2
Comacchio, *It.* **152** F8
Comacchio, Valli di, *It.* **152** F8
Comallo, *Arg.* **135** Q7
Comanche, *Okla., U.S.* **105** E16
Comanche, *Tex., U.S.* **105** H15
Comandante Ferraz, station, *Antarctica* **238** C4
Comandante Fontana, *Arg.* **134** F12
Comandante Luis Piedra Buena, *Arg.* **135** V8
Comănești, *Rom.* **155** C16
Comarnic, *Rom.* **155** E15
Comaton, *Eth.* **210** M5
Comayagua, *Hond.* **117** L17
Combarbalá, *Chile* **134** J6
Combeaufontaine, *Fr.* **149** R13
Combermere Bay, *Myanmar* **190** J3
Combles, *Fr.* **148** M10
Combourg, *Fr.* **149** Q5
Comfort, *Tex., U.S.* **105** L15
Comino (Kemmuna), island, *Malta* **163** H21
Cominotto, island, *Malta* **163** H21
Commander Bay, *Pac. Oc.* **246** B10
Commander Islands *see* Komandorskiye Ostrova, *Russ.* **161** E23
Commanders Bay, *Aruba, Neth.* **123** R17
Commerce, *Ga., U.S.* **98** K6
Commerce, *Tex., U.S.* **105** F18
Commissioner's Point, *Bermuda* **119** B14
Committee Bay, *Nunavut, Can.* **87** F16
Commonwealth Bay, *Antarctica* **239** R18
Communist Peak *see* Qullai Somoniyon, peak, *Taj.* **179** M13
Como, *It.* **152** D4
Como, Lago di, *It. Switz.* **152** C4
Como, Lake, *Eur.* **140** J6
Comodoro Rivadavia, *Arg.* **135** S9
Comorin, Cape, *India* **183** T6
Comoro Basin, *Ind. Oc.* **250** J6
Comoro Islands, *Af.* **198** L11
Comoros, *Af.* **214** G8
Compiègne, *Fr.* **149** N10
Comporta, *Port.* **146** J5
Compostela, *Mex.* **116** H9
Comrat, *Mold.* **159** H4
Côn Sơn, *Vietnam* **191** Q12
Côn Sơn, island, *Vietnam* **191** Q12
Cona (Cuona), *China* **184** L6
Conakry, *Guinea* **204** M8
Concarneau, *Fr.* **149** Q3
Conceição do Maú, *Braz.* **132** B9
Conceição, *Braz.* **133** F17
Concepción del Uruguay, *Arg.* **134** K13
Concepción, *Bol.* **132** J8
Concepción, *Chile* **132** G9
Concepción, *Chile* **135** N6
Conception Bay, *Nfld. & Lab., Can.* **87** M24
Conception Bay, *Namibia* **212** H3
Conceptión del Oro, *Mex.* **116** F11
Conception Island, *Bahamas* **119** F14
Conception, Point, *Calif., U.S.* **111** W5
Conch Point, *Cayman Is., U.K.* **122** M2
Conchas Dam, *N. Mex., U.S.* **109** R14
Conchas Lake, *N. Mex., U.S.* **109** R14
Conchas, river, *N. Mex., U.S.* **109** R13
Concho, *Ariz., U.S.* **109** T9
Concho, river, *Tex., U.S.* **105** J14
Conchos, river, *Mex.* **116** D8
Conconully, *Wash., U.S.* **110** A8
Concord Mountains, *Antarctica* **239** Q15
Concord, *N.H., U.S.* **97** J14
Concord, *N.C., U.S.* **98** H9
Concord, *Calif., U.S.* **111** R4
Concord, *Grenada* **123** K22

Concordia Station, *Antarctica* **239** M18
Concordia, *Kans., U.S.* **107** S7
Concordia, *Arg.* **134** J13
Concrete, *Wash., U.S.* **110** A4
Condat, *Fr.* **149** U10
Conde, *S. Dak., U.S.* **106** J7
Condé, *Fr.* **148** M11
Condeúba, *Braz.* **133** J15
Condobolin, *N.S.W., Austral.* **221** X13
Condom, *Fr.* **149** X7
Condon, *Oreg., U.S.* **110** F6
Condor Peninsula, *Antarctica* **238** E7
Conecuh, river, *Ala., U.S.* **103** N15
Conegliano, *It.* **152** D8
Conejos, *Colo., U.S.* **109** Q12
Conero, Monte, *It.* **152** H10
Conflict Group, *P.N.G.* **222** R10
Confolens, *Fr.* **149** T8
Congaree National Park, *S.C., U.S.* **98** L9
Congo, *Af.* **208** G10
Congo, Democratic Republic of the, *Af.* **209** F14
Congo Basin, *Af.* **198** J6
Congo Cay, *Virgin Is., U.S.* **122** M10
Congo Fan, *Atl. Oc.* **249** P13
Congo *see* Lualaba, river, *D.R.C.* **209** E16;G17;K16
Congo, river, *Af.* **198** J7
Congress, *Ariz., U.S.* **109** T6
Cónico, Cerro, *Arg.* **135** R7
Conil de la Frontera, *Sp.* **146** M8
Conn, Lough, *Ire.* **145** P3
Connacht (Cúige Chonnacht), region, *Ire.* **145** Q4
Conneaut, *Ohio, U.S.* **101** N15
Connecticut, *U.S.* **96** L12
Connecticut, river, *U.S.* **97** J13
Connel, *Scot., U.K.* **144** K8
Connell, *Wash., U.S.* **110** D7
Connellsville, *Pa., U.S.* **96** P4
Connemara, region, *Ire.* **145** Q3
Connersville, *Ind., U.S.* **101** R10
Connmara, *Qnsld., Austral.* **221** T12
Connors Range, *Austral.* **220** E14
Cononaco, *Ecua.* **130** L6
Conrad Rise, *Ind. Oc.* **254** D8
Conrad, *Mont., U.S.* **108** B7
Conroe, *Tex., U.S.* **105** K19
Consdorf, *Lux.* **162** J10
Conselheiro Lafaiete, *Braz.* **133** L15
Conset Bay, *Barbados* **123** K20
Consolación del Sur, *Cuba* **118** G5
Constance, Lake *see* Bodensee, *Aust. Ger. Switz.* **150** L6
Constância dos Baetas, *Braz.* **132** E8
Constanța, *Rom.* **155** F18
Constantina, *Sp.* **146** K8
Constantine, *Alg.* **201** B14
Constitución, *Chile* **134** M6
Consuegra, *Sp.* **146** H11
Contact, *Nev., U.S.* **110** L12
Contai, *India* **182** K11
Contamana, *Peru* **132** F4
Contas, river, *S. Amer.* **128** G10
Contramaestre, *Cuba* **118** J11
Contreras, Embalse de, *Sp.* **147** G13
Contreras, Isla, *Chile* **135** W6
Contres, *Fr.* **149** R9
Contrexéville, *Fr.* **149** Q13
Contwoyto Lake, *Nunavut, Can.* **86** G12
Convention Center and Auditorium, *Monaco* **163** D22
Conversano, *It.* **152** M14
Conway Springs, *Kans., U.S.* **107** U7
Conway, *N.H., U.S.* **97** G14
Conway, *S.C., U.S.* **98** K11
Conway, *Ark., U.S.* **102** G8
Conwy, *Wales, U.K.* **145** R10
Coober Pedy, *S. Aust., Austral.* **221** V9
Cook Bay, *S. Amer.* **128** R5
Cook Bay, *Vanuatu* **229** L17
Cook Ice Shelf, *Antarctica* **239** R17
Cook Inlet, *Alas., U.S.* **113** K16
Cook Island Passage, *Kiribati* **226** P10
Cook Island, *Kiribati* **226** P10
Cook Islands, *Pac. Oc.* **224** J11
Cook Mountains, *Antarctica* **239** M14
Cook Strait, *N.Z.* **223** J20
Cook, *Minn., U.S.* **106** F12
Cook, *S. Aust., Austral.* **221** W7
Cook, Bahía, *Chile* **135** Y8
Cook, Mount, *Austral.* **220** K3
Cookes Peak, *N. Mex., U.S.* **109** V11

Cookeville, *Tenn., U.S.* **103** E17
Cooktown, *Qnsld., Austral.* **221** Q13
Coolabah, *N.S.W., Austral.* **221** W13
Coolabri, *Qnsld., Austral.* **221** U13
Coolah, *N.S.W., Austral.* **221** X14
Coolidge Dam, *Ariz., U.S.* **109** U8
Coolidge, *Tex., U.S.* **105** H18
Coolidge, *Ariz., U.S.* **109** U7
Cooma, *N.S.W., Austral.* **221** Y14
Coon Rapids, *Minn., U.S.* **106** J11
Coon Rapids, *Iowa, U.S.* **107** P10
Coonamble, *N.S.W., Austral.* **221** W14
Coondapoor *see* Kundapura, *India* **183** Q4
Cooper Creek, *Austral.* **220** H11
Cooper, *Tex., U.S.* **105** F19
Cooper's Town, *Bahamas* **122** A9
Cooperstown, *N.Y., U.S.* **96** J10
Cooperstown, *N. Dak., U.S.* **106** F7
Coor-De-Wandy Hill, *Austral.* **220** G2
Coorada, *Qnsld., Austral.* **221** U14
Coorow, *W. Aust., Austral.* **221** W2
Coos Bay, *Oreg., U.S.* **110** J2
Coosa, river, *Ala., Ga., U.S.* **103** H16
Cootamundra, *N.S.W., Austral.* **221** Y13
Copano Bay, *Tex., U.S.* **105** N17
Cope, *Colo., U.S.* **108** M15
Cope, Cabo, *Sp.* **147** L13
Copeland Island, *N. Ire., U.K.* **145** N8
Copenhagen *see* København, *Den.* **142** N12
Copertino, *It.* **153** N15
Copiapó, *Chile* **134** G7
Copiapó, Río, *Chile* **134** G6
Copinsay, island, *Scot., U.K.* **144** F11
Copparo, *It.* **152** E7
Copper Butte, *Wash., U.S.* **110** A8
Copper Center, *Alas., U.S.* **113** J18
Copper Harbor, *Mich., U.S.* **100** D7
Copper Mine Point, *Virgin Is., U.K.* **122** R8
Copper Ridge, *Pac. Oc.* **247** H14
Copper, river, *Alas., U.S.* **113** J18
Copperas Cove, *Tex., U.S.* **105** J16
Coqên (Cuoqin), *China* **184** K4
Coquille Harbor, *F.S.M.* **226** F11
Coquille, *Oreg., U.S.* **110** J2
Coquimbo, *Chile* **134** J6
Cora Droma Rúisc *see* Carrick-on-Shannon, *Ire.* **145** Q5
Cora-Divh Bank, *Ind. Oc.* **250** D10
Corabia, *Rom.* **155** G14
Coracora, *Peru* **132** J4
Corail, *Haiti* **119** M15
Coral Bay, *Virgin Is., U.S.* **122** N11
Coral Bay, *Virgin Is., U.S.* **122** M11
Coral Gables, *Fla., U.S.* **99** X10
Coral Gables, *Cayman Is., U.K.* **122** M1
Coral Harbour, *Nunavut, Can.* **87** H16
Coral Island *see* Far'ûn, *Israel* **173** R5
Coral Reef National Monument, *Virgin Is., U.S.* **122** N11
Coral Sea Basin, *Pac. Oc.* **246** L8
Coral Sea Islands, *Austral.* **221** R15
Coral Sea, *Pac. Oc.* **224** J5
Coral Springs, *Fla., U.S.* **99** W10
Corangamite, Lake, *Austral.* **220** M12
Corantijn *see* Corentyne, river, *S. Amer.* **131** G16
Corato, *It.* **152** L13
Corbeil-Essonnes, *Fr.* **149** P10
Corbières, range, *Fr.* **149** Y10
Corbin, *Ky., U.S.* **103** D18
Corby, *Eng., U.K.* **145** S13
Corcoran, *Calif., U.S.* **111** U6
Corcovado, Golfo, *Chile* **135** R6
Corcubión, *Sp.* **146** B5
Corcubión, Ría de, *Sp.* **146** B5
Corcyra *see* Kérkyra, *Gr.* **156** E6
Cordele, *Ga., U.S.* **99** N6
Cordell, *Okla., U.S.* **105** D15
Cordes, *Fr.* **149** X9
Cordillera Central, *Col.* **130** G6
Cordillera de Mérida, *Venez.* **130** E8
Cordillera Occidental, *S. Amer.* **132** H3
Cordillera Oriental, *S. Amer.* **128** H5
Córdoba, *Mex.* **117** J13
Córdoba, *Arg.* **134** J10
Córdoba, *Sp.* **146** K9
Córdoba, Sierras de, *Arg.* **134** J10
Cordouan, Île de, *Fr.* **149** U6
Cordova, *Ala., U.S.* **103** J15
Cordova, *Alas., U.S.* **113** K18

Corfu *see* Kérkyra, *Gr.* **156** E6
Corguinho, *Braz.* **132** K11
Coria del Río, *Sp.* **146** L8
Coria, *Sp.* **146** G7
Corianino, *San Marino* **163** J16
Coribe, *Braz.* **133** H14
Corigliano Calabro, *It.* **153** P13
Corinna, *Tas., Austral.* **221** Y7
Corinth Canal, *Gr.* **156** J11
Corinth *see* Kórinthos, *Gr.* **156** J11
Corinth, *N.Y., U.S.* **96** J11
Corinth, *Miss., U.S.* **103** G13
Corinth, *Grenada* **123** L23
Corisco, island, *Eq. Guinea* **206** Q9
Cork (Corcaigh), *Ire.* **145** T4
Cork Harbour, *Ire.* **145** T4
Corlay, *Fr.* **149** Q4
Corleone, *It.* **153** S9
Corleto Perticara, *It.* **153** N13
Çorlu, *Turk.* **170** C4
Cornelia, *Ga., U.S.* **98** K6
Cornell Gletscher, *Greenland, Den.* **88** H5
Cornell, *Wis., U.S.* **100** H3
Corner Brook, *Nfld. & Lab., Can.* **87** M23
Corner Seamounts, *Atl. Oc.* **248** K6
Corning, *N.Y., U.S.* **96** K7
Corning, *Ark., U.S.* **102** E10
Corning, *Iowa, U.S.* **107** Q10
Corning, *Calif., U.S.* **111** P4
Corno Grande, *It.* **152** J9
Cornwall Island, *Nunavut, Can.* **87** C15
Cornwall Point, *Bahamas* **122** B9
Cornwall, *Ont., Can.* **87** Q19
Cornwall, region, *Eng., U.K.* **145** W9
Cornwallis Island, *Nunavut, Can.* **87** D15
Cornwallis, Port, *India* **183** Q14
Coro, *Venez.* **130** C9
Coroa, Tope de, *Cabo Verde* **215** B14
Coroatá, *Braz.* **133** E15
Corocoro, *Bol.* **132** K6
Coroico, *Bol.* **132** J6
Corolla, *N.C., U.S.* **98** G15
Coromandel Coast, *India* **183** S8
Coromandel Peninsula, *N.Z.* **223** E21
Corona, *N. Mex., U.S.* **109** T13
Coronado, *S. Amer.* **84** Q7
Coronado National Memorial, *Ariz., U.S.* **109** W8
Coronado, *Calif., U.S.* **111** Z9
Coronation Gulf, *Nunavut, Can.* **86** F12
Coronation Island, *Antarctica* **238** A5
Coronel Bogado, *Parag.* **132** N10
Coronel Dorrego, *Arg.* **135** N11
Coronel Francisco Sosa, *Arg.* **135** P10
Coronel Oviedo, *Parag.* **132** M10
Coronel Pringles, *Arg.* **135** N11
Coronel Suárez, *Arg.* **135** N11
Coronel, *Chile* **135** N6
Coropuna, Nevado, *Peru* **132** J5
Corozal Point, *Trin. & Tobago* **123** P21
Corozal, *Belize* **117** J16
Corozal, *Col.* **130** D6
Corpus Christi Bay, *U.S.* **90** N12
Corpus Christi, *Tex., U.S.* **105** P17
Corral de Almaguer, *Sp.* **146** G11
Corral, *Chile* **135** P6
Corralillo, *Cuba* **118** G8
Corrán Tuathail (Carrauntoohil), peak, *Ire.* **145** T3
Corraun Peninsula, *Ire.* **145** Q3
Corredeir, falls, *Braz.* **132** F10
Corrib, Lough, *Ire.* **145** Q4
Corrientes, *Arg.* **134** G13
Corrientes, Bahía de, *Cuba* **118** H4
Corrientes, Cabo, *Mex.* **116** H9
Corrientes, Cabo, *Cuba* **118** H4
Corrientes, Cabo, *Col.* **130** F4
Corrigan, *Tex., U.S.* **105** J20
Corrigin, *W. Aust., Austral.* **221** X3
Corriverton, *Guyana* **131** F16
Corrubedo, Cabo, *Sp.* **146** C5
Corry, *Pa., U.S.* **96** L4
Corse (Corsica), island, *Fr.* **149** X2
Corse, Cap, *Fr.* **149** W3
Corse, Cap, *It.* **152** H4
Corse, Cape, *Eur.* **140** K6
Corsica *see* Corse, island, *Fr.* **149** X2
Corsica, *S. Dak., U.S.* **106** M6
Corsicana, *Tex., U.S.* **105** H18
Corte, *Fr.* **149** X2
Cortemaggiore, *It.* **152** E5
Cortez, *Colo., U.S.* **109** P10
Cortina d'Ampezzo, *It.* **152** C8

Cortland, N.Y., U.S. **96** K8
Cortona, It. **152** H8
Coruche, Port. **146** H5
Çoruh, river, Turk. **171** D15
Çorum, Turk. **170** D10
Corumbá, Braz. **132** K10
Corumbá, river, S. Amer. **128** H9
Corumbau, Ponta de, Braz. **133** K16
Corvallis, Oreg., U.S. **110** G3
Corvo, island, Port. **146** N1
Corwen, Wales, U.K. **145** R10
Coryal, Trin. & Tobago **123** P23
Corydon, Iowa, U.S. **107** Q12
Cos see Kos, island, Gr. **157** M18
Cosenza, It. **153** Q13
Coshocton, Ohio, U.S. **101** Q14
Cosmo Newberry, W. Aust., Austral.
    221 V5
Cosmopolis, Wash., U.S. **110** D2
Cosne-Cours-sur-Loire, Fr. **149** R10
Cosquín, Arg. **134** J10
Cossato, It. **152** D3
Cossé-le-Vivien, Fr. **149** Q6
Costa del Silencio, Sp. **214** R4
Costa Rica, N. Amer. **117** N18
Costeşti, Rom. **155** F14
Costilla, N. Mex., U.S. **109** Q13
Cotabato City, Philippines **193** F15
Cotagaita, Bol. **132** L7
Côte d'Ivoire (Ivory Coast),
    Af. **205** N13
Côte d'Ivoire Escarpment, Atl. Oc.
    249 N12
Côte d'Or, range, Fr. **149** S12
Coteaux, Haiti **119** M15
Cotentin, range, Fr. **149** N6
Côtes-de-Fer, Haiti **119** N15
Coton, Pointe, Mauritius **215** J21
Cotonou, Benin **205** P17
Cotopaxi, peak, Ecua. **130** K4
Cotswold Hills, Eng., U.K. **145** U11
Cottage Grove, Oreg., U.S. **110** H3
Cottbus, Ger. **150** E11
Cotter, Ark., U.S. **102** E8
Cottian Alps, Fr., It. **152** F1
Cotton Ground, St. Kitts & Nevis
    123 C18
Cotton Plant, Ark., U.S. **102** G10
Cotton Valley, La., U.S. **102** K7
Cottongarden Point, Virgin Is., U.S.
    122 Q4
Cottonwood, Idaho, U.S. **108** D3
Cottonwood, Ariz., U.S. **109** S7
Cotulla, Tex., U.S. **105** N15
Coubre, Pointe de la, Fr. **149** U6
Coudersport, Pa., U.S. **96** L6
Couëron, Fr. **149** S5
Couiza, Fr. **149** Y10
Coulee City, Wash., U.S. **110** C7
Coulee Dam, Wash., U.S. **110** B7
Coulman Island, Antarctica **239** Q14
Coulta, S. Aust., Austral. **221** X9
Council Bluffs, Iowa, U.S. **107** P9
Council Grove, Kans., U.S. **107** T8
Council, Idaho, U.S. **108** F3
Council, Russ. **158** A15
Courland, Eur. **140** F9
Couronne, Cap, Fr. **149** Y13
Courtenay, B.C., Can. **86** M7
Courtenay, Fr. **149** Q10
Courtmacsherry Bay, Ire. **145** U4
Coushatta, La., U.S. **102** L7
Coutances, Fr. **149** P6
Coutras, Fr. **149** V7
Couva, Trin. & Tobago **123** Q22
Covadonga, site, Sp. **146** B9
Covarrubias, Sp. **146** D11
Cove, Oreg., U.S. **110** F8
Cové, Benin **205** N17
Covelo, Calif., U.S. **111** P2
Coventry, Eng., U.K. **145** S12
Covered Market, Monaco **163** E20
Coves del Drac, site, Sp. **147** G19
Covilhã, Port. **146** F6
Covington, Ga., U.S. **98** L5
Covington, Va., U.S. **98** E10
Covington, Ind., U.S. **101** R7
Covington, La., U.S. **102** P11
Covington, Tenn., U.S. **102** F12
Covington, Ky., U.S. **103** A18
Covington, Okla., U.S. **105** B16
Cow and Calf Islands, Myanmar
    183 P15
Cow and Calves, islands, St. Vincent &
    the Grenadines **123** J17
Cowal, Lake, Austral. **220** K13

Cowan, Lake, Austral. **220** J5
Cowarie, S. Aust. Austral. **221** V10
Cowell, S. Aust. Austral. **221** X10
Coweta, Okla., U.S. **105** C19
Cowlitz, river, Wash. U.S. **110** D4
Cowra, N.S.W. Austral. **221** X14
Coxim, Braz. **132** K11
Coxim, river, Braz. **132** K11
Cox's Bazar, India **182** L14
Coyah, Guinea **204** M8
Coyanosa Draw, Tex., U.S. **104** J10
Cozad, Nebr., U.S. **107** Q5
Cozumel Island, N. Amer. **84** N7
Cozumel, Isla, Mex. **117** H17
Crab Island, Mauritius **215** J20
Crab Pond Point, Jam. **122** J6
Cradock, S. Af. **212** L7
Craig, Colo., U.S. **108** L11
Craig, Alas., U.S. **113** M23
Craiova, Rom. **155** F13
Cranberry Lake, N.Y., U.S. **96** G10
Cranbourne, Vic., Austral. **221** Z12
Cranbrook, B.C., Can. **86** M9
Cranbrook, W. Aust., Austral. **221** Y3
Crandon, Wis., U.S. **100** G6
Crane Beach, Barbados **123** L20
Crane Lake, Minn., U.S. **106** E12
Crane, Tex., U.S. **104** J11
Crane, Oreg., U.S. **110** J8
Craon, Fr. **149** R6
Crary Fan, Atl. Oc. **249** Z8
Crary Ice Rise, Antarctica **239** L13
Crary Mountains, Antarctica **238** L9
Crasna, Rom. **155** C17
Crasna, river, Rom. **154** B12
Crater Lake National Park, Oreg., U.S.
    110 J4
Crater Lake, Oreg. U.S. **110** J4
Crater Point, P.N.G. **222** P10
Craters of the Moon National
    Monument, Idaho, U.S. **108** H6
Crateús, Braz. **133** E16
Crato, Braz. **133** F16
Crawford, Tex., U.S. **105** J17
Crawford, Nebr., U.S. **106** M1
Crawfordsville, Ill., U.S. **101** R6
Crawfordville, Fla., U.S. **99** R5
Crazy Mountains, Mont., U.S. **108** E9
Creag Ghoraidh (Creagorry),
    Scot., U.K. **144** H6
Creagorry see Creag Ghoraidh,
    Scot., U.K. **144** H6
Crécy, Fr. **149** N11
Cree Lake, Sask., Can. **86** K12
Cree, river, Sask., Can. **86** K12
Creede, Colo., U.S. **109** P11
Creeslough (An Craoslach), Ire.
    145 N5
Creighton, Nebr., U.S. **107** N7
Creil, Fr. **149** N10
Crema, It. **152** E5
Cremona, It. **152** E5
Cres, island, Croatia **154** E4
Cresbard, S. Dak., U.S. **106** J6
Crescent City, Fla., U.S. **99** S8
Crescent City, Calif., U.S. **110** L1
Crescent Group, S. China Sea **192** B10
Crescent Junction, Utah, U.S. **109** N9
Crescent Lake, Fla., U.S. **99** S8
Crescent Lake, Oreg., U.S. **110** J4
Crescent, Okla., U.S. **105** C16
Cresco, Iowa, U.S. **106** M13
Crest, Fr. **149** V13
Creston, Iowa, U.S. **107** Q10
Crestview, Fla., U.S. **99** Q2
Creswell Downs, N.T., Austral. **221** R9
Creswell, Oreg., U.S. **110** H3
Crete see Kríti, island, Gr. **157** Q14
Crete, Nebr., U.S. **107** Q8
Crete, Sea of, Gr. **157** P15
Creus, Cap de, Sp. **147** C18
Crevillente, Sp. **147** K14
Crewe, Va., U.S. **98** F12
Crewkerne, Eng., U.K. **145** V11
Crianlarich, Scot., U.K. **144** K9
Criciúma, Braz. **132** P12
Criel-sur-Mer, Fr. **148** M9
Crimea, region, Russ. **159** S7
Crimean Mountains, Eur. **140** J12
Crinian Canal, Scot., U.K. **144** L8
Cripple Creek, Colo., U.S. **109** N13
Crisfield, Md., U.S. **98** E14
Cristalina, Braz. **133** J13
Cristóbal Colón, S. Amer. **128** A3
Crişul Alb, river, Rom. **154** D12
Crişul Negru, river, Rom. **154** C11

Crişul Repede, river, Rom. **154** C12
Crna, river, N. Maced. **154** L11
Croatia, Eur. **154** E5
Crocker Range, Malaysia **192** G11
Crockett, Tex., U.S. **105** J19
Crocus Bay, Anguilla, U.K. **122** Q11
Crocus Hill, Anguilla, U.K. **122** Q11
Crofton, Nebr., U.S. **106** M7
Croisette, Cap, Fr. **149** Y13
Croisic, Pointe du, Fr. **149** R4
Croix des Bouquets, Haiti **119** M17
Croker Island, Austral. **220** A8
Cromarty Firth, Scot., U.K. **144** H9
Crompton Point, Dominica **123** E20
Crook Point, Oreg., U.S. **110** K1
Crook, Colo., U.S. **108** K15
Crook, Eng., U.K. **145** N12
Crooked Island Passage, Bahamas
    119 G14
Crooked Island, Bahamas **119** G15
Crooked, river, Oreg., U.S. **110** H6
Crookston, Minn., U.S. **106** F8
Crosby, Minn., U.S. **106** H11
Crosby, N. Dak., U.S. **106** D2
Crosbyton, Tex., U.S. **105** F13
Cross Cays, Bahamas **122** H4
Cross City, Fla., U.S. **99** R6
Cross Keys, Jam. **122** J8
Cross Plains, Tex., U.S. **105** H15
Cross Sound, Alas., U.S. **113** K21
Crossett, Ark., U.S. **102** K9
Crossing Rocks, Bahamas **122** C9
Crosson Ice Shelf, Antarctica **238** L8
Crossville, Tenn., U.S. **103** H15
Croswell, Mich., U.S. **100** L13
Crotone, It. **153** Q14
Crow Agency, Mont., U.S. **108** E11
Crowell, Tex., U.S. **105** E14
Crowley, La., U.S. **102** P8
Crown Mountain, Virgin Is., U.S.
    122 M7
Crownpoint, N. Mex., U.S. **109** R10
Croydon, Qnsld., Austral. **221** R12
Crozet Basin, Ind. Oc. **250** P8
Crozet Plateau, Ind. Oc. **250** Q7
Cruces, Cuba **118** H8
Crump Island, Antigua & Barbuda
    123 B21
Crump Lake, Oreg., U.S. **110** K6
Cruz Alta, Braz. **132** P11
Cruz Bay, Virgin Is., U.S. **122** N10
Cruz del Eje, Arg. **134** J9
Cruz del Padre, Cayo, Cuba **118** G8
Cruz Grande, Chile **134** H6
Cruz, Cabo, Cuba **118** L11
Cruzeiro do Sul, Braz. **132** F4
Cruzeiro, Ponta, Sao Tome and Principe
    215 C20
Cruzen Island, Antarctica **238** N9
Crystal City, Tex., U.S. **105** N14
Crystal City, Mo., U.S. **107** T15
Crystal Falls, Mich., U.S. **100** G6
Crystal Lake, Ill., U.S. **100** M6
Crystal Mountains, Af. **198** J6
Crystal River, Fla., U.S. **99** S7
Crystal Springs, Miss., U.S. **102** M11
Csongrád, Hung. **151** M16
Csorna, Hung. **151** L13
Ctesiphon, ruin, Iraq **176** H11
Cù Lao Thu, island, Vietnam **191** P13
Cuale, Angola **208** K11
Cuamba, Mozambique **212** D12
Cuan Oisrí see Oysterhaven, Ire.
    145 U4
Cuangar, Angola **208** Q12
Cuango, Angola **208** H11;K12
Cuango, river, Angola **208** J11
Cuanza, river, Angola **208** L10;M12
Cuatir, river, Angola **208** Q12
Cuauhtémoc, Mex. **116** D8
Cuba, Mo., U.S. **107** T14
Cuba, N. Amer. **118** H9
Cubagua, Isla, Venez. **121** J14
Cubal, Angola **208** N10
Cubango, river, Angola **208** P11
Cubango, river, Angola **209** Q13
Çubuk, Turk. **170** E8
Cuchi, Angola **208** N11
Cuckold Point, Barbados **123** J19
Cucuí, Braz. **132** C7
Cucumbi, Angola **209** L13
Cúcuta, Col. **130** E7
Cudahy, Wis., U.S. **100** L7
Cudarebe, cape, Aruba, Neth. **123** Q17
Cuddalore, India **183** R7
Cuddapah, India **183** Q6

Cudjoe Head, Montserrat. U.K. **123** B23
Cue, W. Aust., Austral. **221** V3
Cuéllar, Sp. **146** E10
Cuenca, Ecua. **130** M4
Cuenca, Sp. **146** G12
Cuenca, Serranía de, Sp. **147** G13
Cuencamé, Mex. **116** F10
Cuernavaca, Mex. **116** J12
Cuero, Tex., U.S. **105** M17
Cuevas de Altamira, Sp. **146** B10
Cuevas del Almanzora, Sp. **147** L13
Cuevo, Bol. **132** L8
Cugir, Rom. **155** D13
Cuglieri, It. **153** N3
Cuiabá, Braz. **132** J10
Cúige Chonnacht see Connacht,
    region, Ire. **145** Q4
Cúige Laighean see Leinster,
    region, Ire. **145** R6
Cúige Mumhan see Munster,
    region, Ire. **145** T4
Cúige Uladh see Ulster, region,
    Ire., U.K. **145** N6
Cúil Mhuine see Collooney, Ire.
    145 P4
Cúil Rathain see Coleraine,
    N. Ire., U.K. **145** N6
Cuilo Velho, Angola **209** J13
Cuilo, Angola **209** K13
Cuilo, river, Angola **209** K13
Cuimba, Angola **208** H10
Cuito Cuanavale, Angola **209** P13
Cuito, Angola **208** M11
Cuito, river, Angola **209** N13;Q13
Çukurca, Turk. **171** H18
Cul de Sac, St. Maarten, Neth. **123** B14
Cul de Sac, St. Martin, Fr. **123** B15
Cul de Sac, river, St. Lucia **123** K14
Cul-de-Sac du Marin, bay,
    Martinique, Fr. **123** G23
Culberts Bay, Bahamas **122** E3
Culbertson, Nebr., U.S. **107** R4
Culbertson, Mont., U.S. **108** B13
Culebra, Sierra de la, Sp. **146** D8
Culfa, Arm. **171** F19
Culiacán, Mex. **116** F8
Culion, island, Philippines **193** D13
Cúllar-Baza, Sp. **146** L12
Cullman, Ala., U.S. **103** H15
Culpeper, Va., U.S. **98** D12
Culpepper see Isla Darwin, Ecua.
    130 N2
Culuene, river, Braz. **132** H12
Culver, Point, Austral. **220** K6
Culverden, N.Z. **223** L19
Cumalı, Turk. **170** H3
Cumana Bay, Trin. & Tobago **123** N24
Cumana Redhead, Trin. & Tobago
    123 N24
Cumaná, Venez. **130** C12
Cumanayagua, Cuba **118** H8
Cumberland Bay, St. Vincent & the
    Grenadines **123** K16
Cumberland Gap National Historical
    Park, Ky. Tenn. Va. U.S. **103** D19
Cumberland Island National
    Seashore, Ga., U.S. **99** Q9
Cumberland Islands, Austral. **220** E14
Cumberland Peninsula, Nunavut, Can.
    87 F19
Cumberland Plateau, U.S. **91** J17
Cumberland Sound, Nunavut, Can.
    87 G19
Cumberland, Md., U.S. **98** B11
Cumberland, Wis., U.S. **100** G2
Cumberland, Ky., U.S. **103** D19
Cumberland, Lake, U.S. **91** H17
Cumberland, Lake, Ky., U.S. **103** D17
Cumberland, river, U.S. **91** H17
Cumberland, river, Ky. Tenn. U.S.
    103 D16
Cumbernauld, Scot., U.K. **144** L9
Cumbrera, Cerro, Chile **135** U7
Cumbres Pass, Colo., U.S. **109** Q12
Cumbrian Mountains, Eng., U.K.
    145 P10
Cuminá, Braz. **132** D10
Cúmpas, Mex. **116** C7
Çumra, Turk. **170** H8
Cunaviche, Venez. **130** E10
Cunco, Chile **135** P6
Cunene, river, Af. **198** M6
Cuneo, It. **152** F2
Cùng Sơn, Vietnam **191** N13
Cunjamba, Angola **209** P13
Cunyu, W. Aust., Austral. **221** U4

Cuona see Cona, China **184** L6
Cuoqin see Coqên, China **184** K4
Cupica, Col. **130** F4
Cupica, Golfo de, Col. **130** F4
Curaçao, Lesser Antil. **120** H9
Curale, Eth. **211** N11
Curanilahue, Chile **135** N6
Curbur, W. Aust., Austral. **221** V2
Curcubăta, peak, Rom. **154** D12
Curecanti National Recreation
    Area, Colo., U.S. **109** N11
Curepipe, Mauritius **215** G20
Curiapo, Venez. **131** D14
Curicó, Chile **134** L6
Curieuse, island, Seychelles **215** M21
Curitiba, Braz. **133** N13
Curium, ruin, Cyprus **162** Q7
Curley Cut Cays, Bahamas **118** F11
Curoca, river, Angola **208** P9
Curral Velho, Cabo Verde **215** C17
Currawilla, Qnsld., Austral. **221** U11
Current Island, Bahamas **122** D5
Current, Bahamas **122** D5
Current, river, Ark. Mo. U.S. **107** V14
Currie, Tas., Austral. **221** Y7
Curituck Sound, N.C. U.S. **98** G15
Currituck, N.C. U.S. **98** G14
Curtea de Argeş, Rom. **155** E14
Curtici, Rom. **154** D11
Curtin Springs, N.T., Austral. **221** U8
Curtis Island, Antigua & Barbuda **220** F15
Curtis Island, Kermadec Is., N.Z.
    224 L9
Curtis, Nebr., U.S. **107** Q4
Curtis, Port, Austral. **220** F15
Curuá, river, Braz. **132** F11
Curuaés, river, Braz. **132** F11
Curuçá, Braz. **133** C13
Curuçá, river, Braz. **132** E5
Cururupu, Braz. **133** D15
Curuzú Cuatiá, Arg. **134** H13
Curvelo, Braz. **133** K14
Cusco, Peru **132** H5
Cushing, Okla., U.S. **105** C17
Cusset, Fr. **149** T11
Cusson, Pointe, Que. Can. **87** J17
Custer, S. Dak., U.S. **106** L1
Custer, Mont., U.S. **108** D11
Cut Bank, Mont., U.S. **108** B7
Cut, island, Indonesia **193** L18
Cutato, river, Angola **208** L11
Cuthbert, Ga., U.S. **99** N4
Cutler Ridge, Fla., U.S. **99** X10
Cutral Có, Arg. **135** P8
Cuttack, India **182** L10
Cuvier Basin, Ind. Oc. **251** K17
Cuvier Plateau, Ind. Oc. **251** L16
Cuxhaven, Ger. **150** B7
Cuya, Chile **134** B6
Cuyahoga Falls, Ohio, U.S. **101** P14
Cuyahoga Valley National Park,
    Ohio, U.S. **101** P14
Cuyama, river, Calif., U.S. **111** W7
Cuyo Islands, Philippines **193** E13
Cuyuni, river, S. Amer. **131** E14
Cyangugu, Rwanda **211** U1
Cyclades see Kykládes, islands, Gr.
    157 L14
Cydonia see Chaniá, Gr. **157** P13
Cymru see Wales, Wales, U.K. **145** T10
Cynthiana, Ky., U.S. **103** B18
Cyprus, Eur. **162** F6
Cyprus, island, Eur. **163** E2
Cyrenaica, Lib. **201** E21
Cyrene see Shaḩḩāt, Lib. **201** D21
Cyrus Field Bay, Nunavut, Can. **87** H19
Cythera see Kýthira, island, Gr.
    156 N11
Czaplinek, Pol. **150** C12
Czechia (Czech Republic), Eur.
    150 H10
Częstochowa, Pol. **151** F15
Człopa, Pol. **150** D12

D

Đà (Black), river, Vietnam **190** H10
Da Hinggan Ling, China **185** E14
Đà Lạt, Vietnam **191** N13
Đà Nẵng, Vietnam **190** L12
Da Qaidam, China **184** H7
Da'an, China **185** D15

Ḏaʻah, *Jordan* **173** N7
Dabat, *Eth.* **210** J7
Dabatou, *Guinea* **204** L10
Dabie Shan, *China* **185** K13
Dabola, *Guinea* **204** L9
Daborow, *Somalia* **211** N13
Daboya, *Ghana* **205** M15
Dąbrowa Górnicza, *Pol.* **151** G15
Dąbrowa Tarnowska, *Pol.* **151** G16
Dābūd, *Egypt* **202** K12
Dabus, river, *Eth.* **210** L5
Dacarbudhug, *Somaliland* **210** L11
Dachang, *China* **184** N11
Dachau, *Ger.* **150** K8
Dadale, *Solomon Is.* **228** D9
Daday, *Turk.* **170** C9
Dadda'to, *Eritrea* **210** K10
Dade City, *Fla. U.S.* **99** T8
Dadeville, *Ala. U.S.* **103** N16
Dadhar, *Pak.* **181** N7
Dadu, *Pak.* **181** Q7
Dadu, river, *China* **184** L9
Daeamsan, peak, *S. Korea* **186** L10
Daebo, *S. Korea* **187** R13
Daebudo, island, *S. Korea* **187** N7
Daecheongdo, island, *S. Korea*
    186 M4
Daechi, *S. Korea* **187** S8
Daedeok, *S. Korea* **187** U8
Daedundo, island, *S. Korea* **187** T5
Daegu (Taegu), *S. Korea* **187** R11
Daeheuksando, island, *S. Korea*
    187 U5
Daeheung, *S. Korea* **187** U8
Daejeon (Taejŏn), *S. Korea* **187** Q9
Daejeong, *S. Korea* **187** X7
Daejin, *S. Korea* **187** R12
Daenanjido, island, *S. Korea* **187** P7
Daesan, *S. Korea* **187** P7
Daeyeonpyeongdo (Yeonpyeong),
    island, *S. Korea* **186** M6
Dafní, *Gr.* **156** J9
Dafoe, *Sask. Can.* **86** M12
Daga Post, *S. Sudan* **203** V13
Dagaio, *Eth.* **211** N9
Dagana, *Senegal* **204** H7
Dagelet *see* Ulleungdo, island,
    S. Korea **186** M15
Dagö *see* Hiiumaa, island, *Est.*
    143 K15
Dagua, *P.N.G.* **222** N6
Dagupan, *Philippines* **193** C13
Dahab, *Egypt* **173** T4
Dahanu, *India* **182** L3
Dahlā, *Afghan.* **180** L6
Dahlak Archipelago, *Af.* **198** F10
Dahlak Kebir, island, *Eritrea* **210** H8
Dahlonega, *Ga. U.S.* **98** K5
Dahod, *India* **182** K4
Dahongliutan, *China* **180** F16
Ḏahr Oualâta, plateau, *Mauritania*
    204 G12
Ḏahr Tîchît, plateau, *Mauritania*
    204 F11
Dahûk *see* Dihok, *Iraq* **176** D10
Dai (Ndai), island, *Solomon Is.*
    228 D10
Dai Senga Dake, *Japan* **188** H12
Dailekh, *Nepal* **182** G8
Daimiel, *Sp.* **146** H11
Daingerfield, *Tex. U.S.* **105** F20
Daingin, Bá an *see* Dingle Bay, *Ire.*
    145 T2
Dairen *see* Dalian, *China* **185** P13
Daitō Islands, *Japan* **189** Y5
Daitō Shotō, *Japan* **224** C2
Dajabón, *Dom. Rep.* **119** L17
Dajarra, *Qnsld. Austral.* **221** S11
Dajt, *Mal. Alban.* **154** L9
Đắk Tô, *Vietnam* **190** M12
Dakar, *Senegal* **204** J6
Dakhfili, *Sudan* **203** N12
Dakingari, *Nigeria* **206** J6
Dakoank, *India* **183** U15
Dakoro, *Niger* **206** G8
Dakota City, *Nebr. U.S.* **107** N8
Ḏakovica *see* Gjakovë, *Kos.* **154** J10
Ḏakovo, *Croatia* **154** E8
Dakshin Gangotri, station, *Antarctica*
    239 A15
Ḏákura, *Nicar.* **117** L19
Dakwa, *D.R.C.* **209** B17
Dala, *Angola* **209** L13
Dala, *Solomon Is.* **228** E10
Dalaba, *Guinea* **204** L9
Dalaman, river, *Turk.* **170** H4

Dalanzadgad, *Mongolia* **184** E10
Dalap, island, *Marshall Is.* **227** E21
Dalbandin, *Pak.* **181** N5
Dalbeg, *Qnsld. Austral.* **221** S14
Dalby, *Qnsld. Austral.* **221** V15
Dale Country, *Eng. U.K.* **145** Q12
Dale, *Nor.* **142** J10
Daleville, *Ala. U.S.* **103** N16
Dalgaranga Hill, *Austral.* **220** H3
Dalhart, *Tex. U.S.* **104** C11
Dali, *Cyprus* **162** P8
Dali, *China* **184** N8
Dalian (Dairen), *China* **185** P13
Dalkhākī, *Afghan.* **180** F8
Dallas, *Tex. U.S.* **105** G17
Dallas, *Oreg. U.S.* **110** G3
Dalmatia, region, *Croatia* **154** G6
Dalmellington, *Scot. U.K.* **145** M9
Dal'negorsk, *Russ.* **161** L21
Dal'nerechensk, *Russ.* **161** L21
Daloa, *Côte d'Ivoire* **204** P12
Dalol, *Eth.* **210** J8
Dalrymple, Lake, *Austral.* **220** E14
Daltenganj, *India* **182** J9
Dalton Highway, *Alas. U.S.* **113** C17
Dalton, *Ga. U.S.* **98** J4
Dalton, *Nebr. U.S.* **107** P2
Dalumba Bay, *Austral.* **220** B10
Dalupiri, island, *Philippines* **193** A13
Dalwhinnie, *Scot. U.K.* **144** J9
Daly Bay, *Nunavut. Can.* **87** H15
Daly River *see* Nauiyu, *N.T. Austral.*
    221 P7
Daly Waters, *N.T. Austral.* **221** Q8
Daly, river, *Austral.* **220** B8
Dam Gamad, *Sudan* **203** S9
Đầm Hà, *Vietnam* **190** H12
Damagarim, *Niger* **206** H8
Daman (Damão), *India* **182** L3
Damanhûr, *Egypt* **202** F11
Damão *see* Daman, *India* **182** L3
Damar, island, *Indonesia* **193** J16
Damar, island, *Indonesia* **193** M16
Damara, *Cen. Af. Rep.* **207** N14
Damaraland, *Namibia* **212** G3
Damas Cays, *Bahamas* **118** F9
Damascus *see* Dimashq, *Syr.* **172** J7
Damaturu, *Nigeria* **206** J10
Damba, *Angola* **208** J10
Dambarta, *Nigeria* **206** J8
Damboa, *Nigeria* **206** J11
Dambulla, *Sri Lanka* **183** T7
Dame-Marie, *Haiti* **119** M14
Damergou, *Niger* **206** G9
Damergou, region, *Af.* **198** G5
Damietta *see* Dumyâţ, *Egypt* **202** E12
Damīr Qābū, *Syr.* **172** C15
Damoh, *India* **182** J7
Damongo, *Ghana* **205** M15
Dampier Downs, *W. Aust. Austral.*
    221 R5
Dampier Land, *Austral.* **220** D4
Dampier Ridge, *Pac. Oc.* **246** M9
Dampier Strait, *P.N.G.* **222** P8
Dampier, *W. Aust. Austral.* **221** S2
Dampier, Archipelago, *Austral.*
    220 E2
Dampier, Selat, *Indonesia* **193** J17
Damqawt, *Yemen* **175** M14
Damqog (Maquan), river, *China*
    184 L4
Damroh, *India* **182** F15
Damxung (Dangxiong), *China* **184** K6
Damyang, *S. Korea* **187** S8
Dan Gulbi, *Nigeria* **206** J7
Dan, river, *N.C. Va. U.S.* **98** G10
Dana Point, *Calif. U.S.* **111** Y9
Danakil, *Eth.* **210** J9
Danané, *Côte d'Ivoire* **204** N11
Danas Valley, *Atl. Oc.* **253** R18
Danbury, *Conn. U.S.* **96** M12
Danby Lake, *Calif. U.S.* **111** X12
Dandong, *China* **185** F16
Daneborg, research station,
    Greenland. Den. **88** H14
Danell Fjord, *Greenland. Den.* **89** T8
Danells Valley, *Atl. Oc.* **253** Q20
Danfa, *Mali* **204** J12
Danforth, *Me. U.S.* **97** D18
Dangbon, *S. Korea* **187** R9
Dangchang, *China* **184** J10
Dange, *Angola* **208** J10
Danger Point, *S. Af.* **212** N15
Dangila, *Eth.* **210** K6
Dangjin, *S. Korea* **187** P7
Dango, *Sudan* **203** U7

Dangrek Range, *Thai.* **190** M10
Dangriga, *Belize* **117** K17
Danguya, *Cen. Af. Rep.* **207** M16
Dangxiong *see* Damxung, *China*
    184 K6
Daniel, *Wyo. U.S.* **108** H9
Daniell Peninsula, *Antarctica* **239** Q14
Daniel's Island, *Bermuda* **119** C14
Danielsen Canyon, *Atl. Oc.* **253** N20
Danilov, *Russ.* **158** G10
Danjo Guntō, *Japan* **189** S3
Ḏank, *Oman* **175** G16
Dankhar Gömpa, *India* **182** E6
Dankov, *Russ.* **158** L10
Danmark Fjord, *Greenland. Den.* **88** C11
Danmarkshavn, research station,
    Greenland. Den. **88** F13
Dannemora, *N.Y. U.S.* **96** F11
Dansheha, *Eth.* **210** J6
Danske Øer, *Greenland. Den.* **88** E13
Dansville, *N.Y. U.S.* **96** K6
Danube River Delta, *Eur.* **140** J11
Danube *see* Donau, Duna, Dunaj,
    Dunărea, Dunav, Dunay, river, *Eur.*
    140 H7
Danube, *Rom.* **159** S5
Danube, Mouths of the, *Rom.*
    155 E19
Danville, *Va. U.S.* **98** G11
Danville, *Ill. U.S.* **101** R7
Danville, *Ky. U.S.* **103** C17
Danyang, *S. Korea* **187** P10
Danzhou, *China* **184** Q12
Đảo Côn Cỏ, island, *Vietnam* **190** K12
Đảo Lý Sơn, island, *Vietnam* **190** L13
Đảo Phú Quốc, island, *Vietnam*
    191 Q10
Đảo Thổ Chu, island, *Vietnam*
    191 Q10
Dao Timmi, *Niger* **206** D11
Dapaong, *Togo* **205** L16
Daphne, *Ala. U.S.* **103** P14
Daqing, *China* **185** D15
Dar es Salaam, *Tanzania* **211** W8
Dār Ḩamid, *Sudan* **203** R11
Dar Rounga, *Cen. Af. Rep.* **207** K15
Dar Rounga, region, *Af.* **198** H7
Dar'ā (Edrei), *Syr.* **172** L7
Dara, *Senegal* **204** H7
Dārāb, *Iran* **177** L17
Daraban, *Pak.* **180** L10
Darabani, *Rom.* **155** A16
Darah-ye Bŭm, *Afghan.* **180** G4
Darah-ye Owd, *Afghan.* **180** H7
Dārān, *Iraq* **177** H14
Darāw, *Egypt* **202** K12
Darazo, *Nigeria* **206** J9
Darbhanga, *India* **182** H10
Darby Islands, *Bahamas* **122** D7
Darby, *Mont. U.S.* **108** E5
Dardanelle, *Ark. U.S.* **102** G7
Dardanelle, Lake, *Ark. U.S.* **102** F7
Dardanelles *see* Çanakkale Boğazı,
    Turk. **170** D2
Dardere *see* Ovgos, river, *Cyprus*
    162 N7
Darende, *Turk.* **170** G12
Darfur, *Sudan* **203** S7
Dargai, *Pak.* **180** H11
Darganata, *Turkm.* **178** L9
Dargaville, *N.Z.* **223** D20
Dargeçit, *Turk.* **171** H16
Dargol, *Niger* **206** H5
Dargun, *Ger.* **150** B10
Darhan, *Mongolia* **184** C10
Darhan, *Mongolia* **184** D11
Darien, *Ga. U.S.* **99** P8
Daringali, *Solomon Is.* **228** E11
Darjiling, *India* **182** H11
Dark Head, *St. Vincent & the Grenadines*
    123 K16
Darling Downs, *Austral.* **220** H14
Darling Range, *Austral.* **220** K3
Darling, Lake, *N. Dak. U.S.* **106** E4
Darling, river, *Austral.* **220** J12
Darlington, *S.C. U.S.* **98** K10
Darlington, *Eng. U.K.* **145** P12
Darliston, *Jam.* **122** H6
Darlot, Lake, *Austral.* **220** H4
Darłowo, *Pol.* **150** B12
Darmstadt, *Ger.* **150** H6
Darnah (Derna), *Lib.* **201** D21
Darney, *Fr.* **149** Q14
Darnley, Cape, *Antarctica* **239** F20
Daroca, *Sp.* **147** E13
Daroot- Korgon, *Kyrg.* **179** M13

Darou Khoudos, *Senegal* **204** H6
Darregueira, *Arg.* **135** N10
Darreh Gaz, *Iran* **177** D20
Darrell Island, *Bermuda* **119** C15
Darrington, *Wash. U.S.* **110** B5
Darrouzett, *Tex. U.S.* **105** B14
Darsser Ort, cape, *Ger.* **150** A9
Dartmoor, plateau, *Eng. U.K.* **145** W9
Dartmouth, *N.S. Can.* **87** P22
Daru, *P.N.G.* **222** Q6
Darwājī Bālā, *Afghan.* **180** J3
Darwha, *India* **182** L6
Darwin Glacier, *Antarctica* **239** M14
Darwin, *Falk. Is. U.K.* **135** W12
Darwin, *N.T. Austral.* **221** P8
Darwin, Isla, *Galápagos Is. Ecua.*
    225 G21
Darwin, Port, *Austral.* **220** B7
Darya Khan, *Pak.* **180** L10
Daryā-ye Namak, lake, *Iran* **177** F16
Daryācheh-ye Bakhtegān, lake, *Iran*
    177 L17
Daryācheh-ye Hāmūn (Lake
    Helmand), *Iran* **177** J22
Daryācheh-ye Hāmūn Şāberī, lake,
    Iran **177** J22
Daryācheh-ye Orūmīyeh (Lake
    Urmia), *Iran* **176** C11
Daryācheh-ye Tashk, lake, *Iran*
    177 K17
Das, *Pak.* **182** C5
Dasburg, *Ger.* **150** G4
Dashkuduk, *Turkm.* **178** L4
Dasht-e Kavīr (Kavir Desert), *Iran*
    177 F18
Dasht-e Lūt (Lut Desert), *Iran*
    177 J20
Dasht, river, *Pak.* **181** R3
Daska, *Pak.* **180** K13
Daşkəsən, *Azerb.* **171** D20
Daşköpri, *Turkm.* **178** P9
Daşoguz, *Turkm.* **178** K8
Dassa-Zoumé, *Benin* **205** N17
Datça, *Turk.* **170** H3
Datia, *India* **182** H6
Datong, *China* **184** G12
Datong, *China* **184** H9
Datong, river, *China* **184** H9
Datu Piang, *Philippines* **193** F15
Datu, Cape, *Asia* **168** L11
Daud Khel, *Pak.* **180** K11
Daudnagar, *India* **182** J9
Daugård-Jensen Glacier,
    Greenland. Den. **88** K12
Daugård-Jensen Land,
    Greenland. Den. **88** D6
Daugava, river, *Latv.* **143** L17
Daugavpils, *Latv.* **143** L17
Dauli, *Somalia* **210** M15
D'Aunay Bugt, *Greenland. Den.*
    88 M14
Dauphin Island, *Miss. U.S.* **103** P13
Dauphin, *Man. Can.* **87** N13
Dauphine, *Fr.* **140** J5
Daurada, Costa, *Sp.* **147** E17
Davangere, *India* **183** Q5
Davao City, *Philippines* **193** F15
Davao Gulf, *Philippines* **193** F15
Davegoriale, *Somaliland* **210** M11
Davenport Range, *Austral.* **220** E9
Davenport, *Iowa. U.S.* **107** P14
Davenport, *Wash. U.S.* **110** C8
Daventry, *Eng. U.K.* **145** T13
Davey, Port, *Austral.* **220** M7
David Bay, *Grenada* **123** J23
David City, *Nebr. U.S.* **107** P8
David Island, *Antarctica* **239** K22
David Point, *Virgin Is. U.S.* **122** M7
David, *Pan.* **117** N20
Davidson Mountains, *Alas. U.S.*
    113 D18
Davie Ridge, *Ind. Oc.* **250** J5
Davis Bay, *Antarctica* **239** Q19
Davis Beach, *Virgin Is. U.S.* **122** Q2
Davis Dam, *Ariz. Nev. U.S.* **111** V12
Davis Mountains, *Tex. U.S.* **104** K9
Davis Sea, *Antarctica* **239** J23
Davis Seamounts, *Atl. Oc.* **254** H6
Davis Sill, *Arctic Oc.* **248** E5
Davis Station, *Antarctica* **239** G20
Davis Strait, *Arctic Oc.* **248** E5
Davis, *Okla. U.S.* **105** E17
Davis, *Calif. U.S.* **111** R4
Davis, Mount, *Pa. U.S.* **96** Q4
Davos, *Switz.* **152** B5
Davy Sund, *Greenland. Den.* **88** J14

Dawa, river, *Eth.* **211** P8
Dawadawa, *Ghana* **205** N15
Dawei, *Myanmar* **190** M6
Dawḥat Duwayhin, bay, *Saudi Arabia*
    175 G13
Dawkah, *Oman* **175** L15
Dawna Range, *Myanmar* **190** K6
Dawqah, *Saudi Arabia* **174** K7
Dawra, *W. Sahara* **200** G4
Dawros Head, *Ire.* **145** N4
Dawson Creek, *B.C. Can.* **86** K9
Dawson Springs, *Ky. U.S.* **103** D14
Dawson, *Yukon. Can.* **86** E8
Dawson, *Ga. U.S.* **99** P5
Dawson, *Tex. U.S.* **105** H18
Dawson, *Minn. U.S.* **106** K9
Dawson, *N. Dak. U.S.* **106** G5
Dawson, *Isla, Chile* **135** X8
Dawson, river, *Austral.* **220** G15
Dawukou *see* Shizuishan, *China*
    184 G10
Dawwah, *Oman* **175** J18
Dax, *Fr.* **149** X6
Daxian, *China* **184** K11
Daxue Mountains, *Asia* **168** H10
Dayr az Zawr, *Syr.* **172** F13
Dayr Sim'ān, *Syr.* **172** D8
Dayrūţ, *Egypt* **202** H11
Dayton, *Ohio. U.S.* **101** R11
Dayton, *Tenn. U.S.* **103** F17
Dayton, *Tex. U.S.* **105** L20
Dayton, *Wash. U.S.* **110** E8
Daytona Beach, *Fla. U.S.* **99** S9
Dayville, *Oreg. U.S.* **110** G7
Dazkırı, *Turk.* **170** G5
Dchira, *W. Sahara* **200** G4
De Aar, *S. Af.* **212** L7
Dê Arghīstān, river, *Afghan.* **180** L6
De Funiak Springs, *Fla. U.S.* **99** Q3
De Gerlache Seamounts, *Pac. Oc.*
    249 X1
De Grey, river, *Austral.* **220** E3
De Kalb, *Miss. U.S.* **103** K13
De Kalb, *Tex. U.S.* **105** F20
De Land, *Fla. U.S.* **99** S9
De Leon, *Tex. U.S.* **105** H16
De Long Mountains, *Alas. U.S.*
    113 D13
De Pere, *Wis. U.S.* **100** J7
De Queen, *Ark. U.S.* **102** H6
De Quincy, *La. U.S.* **102** P7
De Ridder, *La. U.S.* **102** N7
De Smet, *S. Dak. U.S.* **106** K7
De Soto, *Mo. U.S.* **107** T15
De Tour Village, *Mich. U.S.* **100** G11
De Witt, *Ark. U.S.* **102** H9
De Witt, *Iowa. U.S.* **107** N14
Dead Sea, *Israel, Jordan, W. Bank*
    173 N6
Deadhorse, *Alas. U.S.* **113** C17
Deadmans Cay, *Bahamas* **119** G14
Deadman's Cays, *Bahamas* **118** F8
Deadwood, *S. Dak. U.S.* **106** K1
Deakin, *W. Aust. Austral.* **221** W7
Deal, *Eng. U.K.* **145** U16
Deán Funes, *Arg.* **134** J10
Dean Passage, *Solomon Is.* **228** N7
Dearborn, *Mich. U.S.* **100** M12
Dease Inlet, *Alas. U.S.* **113** B15
Dease Lake, *B.C. Can.* **86** H8
Dease Strait, *Nunavut. Can.* **87** F13
Death Valley National Park Devils
    Hole, *Nev. U.S.* **111** U10
Death Valley National Park,
    Calif. Nev. U.S. **111** T9
Death Valley, *Calif. U.S.* **111** T9
Death Valley, town, *Calif. U.S.* **111** U10
Debar, *N. Maced.* **154** K10
Debark', *Eth.* **210** J7
Debdou, *Mor.* **200** C9
Débé, *Trin. & Tobago* **123** Q22
Débéré, *Mali* **205** J14
Dębica, *Pol.* **151** G17
Debin, *Russ.* **161** F20
Debir, ruins, *Israel* **173** N5
Dębno, *Pol.* **150** D11
Débo, Lake, *Af.* **198** G12
Debre Birhan, *Eth.* **210** L8
Debre Mark'os, *Eth.* **210** L7
Debre Tabor, *Eth.* **210** K7
Debre Zebīt, *Eth.* **210** K7
Debre Zeyit, *Eth.* **210** M8
Debrecen, *Hung.* **151** L17
Debuu, island, *Marshall Is.* **227** F17

Decatur, *Ga., U.S.* **98** L5
Decatur, *Ill., U.S.* **101** R6
Decatur, *Ind., U.S.* **101** Q10
Decatur, *Ala., U.S.* **103** H15
Decatur, *Tex., U.S.* **105** F17
Decatur, *Nebr., U.S.* **107** N8
Decazeville, *Fr.* **149** W10
Deccan Plateau, *Asia* **168** J7
Deccan, plateau, *India* **183** P5
Deception Island, *Antarctica* **238** D4
Dechu, *India* **182** H3
Decimomannu, *It.* **153** P4
Děčín, *Czechia* **150** G11
Decize, *Fr.* **149** S11
Decorah, *Iowa, U.S.* **106** M13
Dedéagach *see* Alexandroúpoli, *Gr.*
  157 C16
Dededo, *Guam, U.S.* **230** N6
Dedegöl Dağı, peak, *Turk.* **170** H6
Deder, *Eth.* **210** M9
Dédougou, *Burkina Faso* **205** K14
Dedovichi, *Russ.* **158** H6
Dedza, *Malawi* **212** D11
Deep Creek Lake, *Md., U.S.* **98** C10
Deep Creek Range, *Utah, U.S.*
  108 M5
Deep Creek, *Bahamas* **122** F6
Deep, river, *N. Dak., U.S.* **106** E4
Deer Isle, *Me., U.S.* **97** G17
Deer Lodge, *Mont., U.S.* **108** D7
Deer Park, *Wash., U.S.* **110** B9
Deer River, *Minn., U.S.* **106** F11
Deer Trail, *Colo., U.S.* **108** M14
Deering, *Alas., U.S.* **113** E13
Deerton, *Mich., U.S.* **100** F8
Deeth, *Nev., U.S.* **110** M11
Defiance, *Ohio, U.S.* **101** P11
Deganya, *Israel* **172** L6
Dêgê, *China* **184** K8
Degeh Bur, *Eth.* **210** M10
Deggendorf, *Ger.* **150** J10
Değirmenlik *see* Kythrea, *Cyprus*
  162 N9
Deh Bīd, *Iran* **177** K17
Deh Rāwud, *Afghan.* **180** K6
Deh-e Khāwak, *Afghan.* **180** G9
Deh-e Shū, *Afghan.* **180** M4
Dehlorān, *Iraq* **176** H12
Dehqonobod, *Uzb.* **178** N11
Dehra Dun, *India* **182** F6
Dehui, *China* **185** D16
Deichmann Fjord, *Greenland, Den.*
  88 L14
Deilgne *see* Delgany, *Ire.* **145** R7
Deim Zubeir, *S. Sudan* **203** V8
Dej, *Rom.* **155** C13
Dejen, *Eth.* **210** L7
DeKalb, *Ill., U.S.* **101** N6
Dek'emhare, *Eritrea* **210** H8
Dekle Beach, *Fla., U.S.* **99** R6
Dekoa, *Cen. Af. Rep.* **207** M14
Del Cano Rise, *Ind. Oc.* **250** Q5
Del Norte, *Colo., U.S.* **109** P12
Del Rio, *Tex., U.S.* **105** M13
Del Verme Falls, *Eth.* **211** P8
Delaford, *Trin. & Tobago* **123** N18
Delano Peak, *Utah, U.S.* **109** N6
Delano, *Calif., U.S.* **111** V7
Delaport Point, *Bahamas* **122** E2
Delarof Islands, *Alas., U.S.* **112** N4
Delavan, *Wis., U.S.* **100** M6
Delaware Bay, *Del., N.J., U.S.* **96** R9
Delaware Water Gap National
  Recreation Area, *N.J., Pa., U.S.*
  96 M10
Delaware, *U.S.* **98** C15
Delaware, *Ohio, U.S.* **101** Q13
Delaware, river, *U.S.* **96** L9
Delémont, *Switz.* **152** A2
Delfzijl, *Neth.* **148** G14
Delgany (Deilgne), *Ire.* **145** R7
Delgo, *Sudan* **203** N11
Delhi, *N.Y., U.S.* **96** K10
Delhi, *La., U.S.* **102** L9
Delhi, *India* **182** G6
Delice, river, *Turk.* **170** E9
Delicias, *Mex.* **116** D9
Delījān, *Iran* **177** G15
Deliktaş, *Turk.* **170** F12
Déline, *N.W.T., Can.* **86** F10
Delingha, *China* **184** H8
Delitzsch, *Ger.* **150** F9
Dell Rapids, *S. Dak., U.S.* **106** L8
Dellys, *Alg.* **201** B13
Delmar, *Del., Md., U.S.* **98** D15
Delmenhorst, *Ger.* **150** C7

Delos, ruin, *Gr.* **157** K15
Delphi, *Ind., U.S.* **101** Q8
Delphi, ruin, *Gr.* **156** H10
Delphos, *Ohio, U.S.* **101** Q11
Delray Beach, *Fla., U.S.* **99** W10
Delta Junction, *Alas., U.S.* **113** G18
Delta, *Utah, U.S.* **108** M6
Delta, *Colo., U.S.* **109** N10
Deltona, *Fla., U.S.* **99** S9
Demagiri, *India* **182** K14
Demanda, Sierra de la, *Sp.* **146** D11
Demba Chio, *Angola* **208** L10
Demba, *D.R.C.* **209** H14
Dembech'a, *Eth.* **210** L7
Dembéni, *Comoros* **215** M14
Dembī Dolo, *Eth.* **210** M5
Demchok, *India* **182** D7
Demerara Plain, *Atl. Oc.* **249** N6
Demerara Plateau, *Atl. Oc.* **249** N6
Deming, *N. Mex., U.S.* **109** V11
Demini, river, *Braz.* **132** C8
Demirci, *Turk.* **170** F4
Demirköprü Baraji, *Turk.* **170** F4
Demirköy, *Turk.* **170** B4
Demirtaş, *Turk.* **170** J7
Demon, *Ghana* **205** M16
Demopolis, *Ala., U.S.* **103** L14
Dempster Highway, *Yukon, Can.* **86** E9
Demre, *Turk.* **170** J5
Demyansk, *Russ.* **158** H7
Den Burg, *Neth.* **148** G12
Den Haag (The HagueO, *Neth.*
  148 J12
Den Helder, *Neth.* **148** H12
Denali (Mount McKinley), *N. Amer.*
  84 C2
Denali Highway, *Alas., U.S.* **113** H17
Denali National Park and Preserve,
  Alas., U.S. **113** G16
Denali Park, *Alas., U.S.* **113** G17
Denan, *Eth.* **211** N10
Denekamp, *Neth.* **148** H14
Deng Deng, *Cameroon* **206** N11
Denges Passage, *Palau* **226** J5
Dengkou, *China* **184** G11
Dêngqên (Dingqing), *China* **184** K7
Denham Springs, *La., U.S.* **102** P10
Denham, *W. Aust., Austral.* **221** U1
Dénia, *Sp.* **147** J15
Denigomodu, *Nauru* **227** M14
Deniliquin, *N.S.W., Austral.* **221** Y12
Denio, *Nev., U.S.* **110** L7
Denis, *Gabon* **208** D7
Denis, Île, *Seychelles* **215** L20
Denison, *Tex., U.S.* **105** F18
Denison, *Iowa, U.S.* **107** N9
Denisovskaya, *Russ.* **158** E11
Denizli, *Turk.* **170** G4
Denman Glacier, *Antarctica* **239** K21
Denman, *N.S.W., Austral.* **221** X14
Denmark Strait, *Arctic Oc.* **253** M20
Denmark, *S.C., U.S.* **98** L9
Denmark, *Eur.* **142** N11
Denmark, *W. Aust., Austral.* **221** Y3
Dennery, *St. Lucia* **123** K14
Denov, *Uzb.* **178** N11
Denpasar, *Indonesia* **192** M11
Denso, *Mali* **204** L11
Denton, *Md., U.S.* **98** C14
Denton, *Tex., U.S.* **105** F17
D'Entrecasteaux Basin, *Pac. Oc.*
  246 L3
D'Entrecasteaux Islands, *P.N.G.*
  222 Q9
D'Entrecasteaux, Point, *Austral.*
  220 L3
Denver City, *Tex., U.S.* **104** G11
Denver, *Colo., U.S.* **108** M13
Deobhog, *India* **182** M8
Deoghar, *India* **182** J11
Deogyusan, peak, *S. Korea* **187** R9
Deokjeokdo, island, *S. Korea* **187** N6
Deokseong, *S. Korea* **187** Q12
Deokyang, *S. Korea* **187** T9
Deora, *India* **182** H2
Deoria, *India* **182** H9
Deosai Mountains, *Pak.* **180** G13
Deosri, *India* **182** H13
Depoe Bay, *Oreg., U.S.* **110** G2
Deposit, *N.Y., U.S.* **96** L9
Depósito, *Braz.* **132** B9
Dépression du Mourdi, *Chad* **207** E16
Deputatskiy, *Russ.* **161** E18
Dera Bugti, *Pak.* **180** M10
Dera Ismail Khan, *Pak.* **180** L10
Dera Murad Jamali, *Pak.* **181** P8

Deraheib, *Sudan* **202** L14
Derakht-e Yaḩyā, *Afghan.* **180** L8
Derbent, *Russ.* **159** U16
Derbisaka, *Cen. Af. Rep.* **207** M17
Derby, *Kans., U.S.* **107** U8
Derby, *Eng., U.K.* **145** S12
Derby, *W. Aust., Austral.* **221** R5
Dereisa, *Sudan* **203** S6
Derg, Lough, *Ire.* **145** N5
Derg, Lough, *Ire.* **145** R4
Dergachi, *Russ.* **158** M15
Dergaon, *India* **182** H14
Derhachi, *Ukr.* **159** N9
Derik, *Turk.* **171** H15
Dermott, *Ark., U.S.* **102** J9
Derna *see* Darnah, *Lib.* **201** D21
Derravaragh, Lough, *Ire.* **145** Q6
Derri, *Somalia* **211** P12
Derry *see* Londonderry, *N. Ire., U.K.*
  145 N6
Derudeb, *Sudan* **203** P14
Derventa, *Bosn. & Herzg.* **154** E8
Derwent, river, *Austral.* **220** M7
Deryneia *see* Dherinia, *Cyprus*
  162 P10
Des Arc, *Ark., U.S.* **102** G9
Des Lacs, river, *N. Dak., U.S.* **106** D3
Des Moines, *Iowa, U.S.* **107** P11
Des Moines, *N. Mex., U.S.* **109** Q14
Des Moines, river, *Iowa, Mo., U.S.*
  107 P12
Des Plaines, *Ill., U.S.* **101** N7
Deschutes, river, *Oreg., U.S.* **110** G5
Desē, *Eth.* **210** K8
Deseado, river, *Arg.* **135** T8
Desecheo, island, *P.R., U.S.* **119** M21
Desert Center, *Calif., U.S.* **111** X12
Deserta Grande, island, *Madeira, Port.*
  214 N4
Desertas, Ilhas, *Madeira, Port.* **214** N4
Deshaies, *Guadeloupe, Fr.* **123** E13
Deshler, *Nebr., U.S.* **107** R7
Deskáti, *Gr.* **156** E9
Desloge, *Mo., U.S.* **107** T15
Desna, river, *Russ.* **158** K7
Desnudez, Punta, *Arg.* **135** N12
Desolación, Isla, *Chile* **135** X6
Despotikó, island, *Gr.* **157** L14
Dessalines, *Haiti* **119** L16
Dessau, *Ger.* **150** E9
Destruction Island, *Wash., U.S.* **110** C2
Detmold, *Ger.* **150** E7
Detroit Lake, *Oreg., U.S.* **110** G4
Detroit Lakes, *Minn., U.S.* **106** G9
Detroit, *Mich., U.S.* **100** M12
Deutschland Canyon, *Atl. Oc.* **249** Z9
Deva, *Rom.* **154** D12
Deveci Dağı, peak, *Turk.* **170** E11
Develi, *Turk.* **170** G10
Devils Lake, *N. Dak., U.S.* **106** F6
Devils Lake, town, *N. Dak., U.S.*
  106 E6
Devils Paw, mountain, *Alas., U.S.*
  113 K22
Devil's Point, *Bahamas* **119** F13
Devil's Point, *Vanuatu* **229** K16
Devils Postpile National Monument,
  Calif., U.S. **111** S7
Devils Tower National Monument,
  Wyo., U.S. **108** F13
Devils, river, *Tex., U.S.* **105** K13
Devine, *Tex., U.S.* **105** M15
Devizes, *Eng., U.K.* **145** U12
Devon Island, *Nunavut, Can.* **87** D15
Devon Shelf, *Arctic Oc.* **253** P14
Devon Slope, *Arctic Oc.* **253** X6
Devonshire Bay, *Bermuda* **119** C15
Devrek, *Turk.* **170** C7
Devrekâni, *Turk.* **170** C9
Devrukh, *India* **183** N3
Dewas, *India* **182** K5
Dewey, *Okla., U.S.* **105** A18
Dexter, *Me., U.S.* **97** E17
Dexter, *Mo., U.S.* **107** V15
Dey Dey, Lake, *Austral.* **220** H8
Deyang, *China* **184** L10
Deynau, *Turkm.* **178** M9
Dezfūl, *Iran* **177** H13
Dezhnev Bay, *N. Amer.* **84** P2
Dezhou, *China* **185** H13
Dәlidağ *see* Mets Beverratap, peak,
  Azerb. **171** E20
Dәvәçi, *Azerb.* **171** C22

Dhaka, *Bangladesh* **182** J13
Dhamār, *Yemen* **174** P9
Dhamtari, *India* **182** L8
Dhanbad, *India* **182** J10
D'Hanis, *Tex., U.S.* **105** M15
Dhanushkodi, *India* **183** T7
Dhar, *India* **182** K4
Dharan Bazar, *Nepal* **182** H11
Dharapuram, *India* **183** S6
Dharmjaygarh, *India* **182** K9
Dharmsala, *India* **182** E5
Dharwad, *India* **183** P4
Dhaulagiri I, peak, *Nepal* **182** G9
Dhavlos (Kaplıca), *Cyprus* **162** M10
Dhekelia (Eastern) Sovereign Base
  Area , *U.K.* **162** P9
Dhenkanal, *India* **182** L10
Dherinia (Deryneia), *Cyprus* **162** P10
Dhībān, *Jordan* **173** N6
Dhone, *India* **183** P6
Dhuburi, *India* **182** H12
Dhudvo, *India* **210** L14
Dhule, *India* **182** L4
Dhupgarh, peak, *India* **182** K6
Dhuudo, *Somalia* **210** L14
Dhuusamarreeb, *Somalia* **211** P12
Di Lăng, *Vietnam* **190** L13
Di Linh, *Vietnam* **191** P12
Dia, *Mali* **205** J14
Día, island, *Gr.* **157** Q15
Diable (Devils Island), Île du,
  Fr. Guiana, Fr. **131** F19
Diablo Range, *Calif., U.S.* **111** T5
Diablo, Picacho del, *N. Amer.* **84** K3
Diadé, *Mauritania* **204** H12
Diakoptó, *Gr.* **156** J10
Diala, *Mali* **204** J10
Dialafara, *Senegal* **204** K9
Diamante, *Arg.* **134** K12
Diamante, river, *Arg.* **134** L8
Diamantina Deep, *Ind. Oc.* **251** N16
Diamantina Escarpment, *Ind. Oc.*
  251 M14
Diamantina Fracture Zone, *Ind. Oc.*
  251 M14
Diamantina Lakes, *Qnsld., Austral.*
  221 T11
Diamantina Trench, *Ind. Oc.* **251** N16
Diamantina, *Braz.* **133** K15
Diamantina, Chapada, *Braz.* **133** H15
Diamantina, river, *Austral.* **220** F12
Diamantino, *Braz.* **132** H10
Diamond Craters, *Oreg., U.S.* **110** J7
Diamond Harbour, *India* **182** K11
Diamond Head, *Hawaii, U.S.* **115** H15
Diamond Mountains, *Calif., U.S.*
  111 N6
Diamond Mountains, *Nev., U.S.* **111** P11
Diamond Peak, *Oreg., U.S.* **110** J4
Diamond Rock *see* Rocher de
  Diamant, *Martinique, Fr.* **123** G22
Diamou, *Mali* **204** J10
Diandioumé, *Mali* **204** H11
Dianópolis, *Braz.* **133** G14
Diapaga, *Burkina Faso* **205** K17
Díavlos Karpáthou, *Gr.* **157** N18
Diavolo, Mount, *India* **183** R14
Dībā, *U.A.E.* **175** F16
Dibai, *India* **182** G6
Dibaya-Lubwe, *D.R.C.* **209** G13
Dibaya, *D.R.C.* **209** J15
Dibbis, *Sudan* **203** S7
Dibble Basin, *Ind. Oc.* **255** E15
Dibble Glacier, *Antarctica* **239** Q19
Diboll, *Tex., U.S.* **105** J20
Dibrugarh, *India* **182** G15
Dibsī Faraj, *Syr.* **172** E10
Dickens, *Tex., U.S.* **105** F13
Dickinson Bay, *Antigua & Barbuda*
  123 A20
Dickinson, *N. Dak., U.S.* **106** G2
Dickson City, *Pa., U.S.* **96** M9
Dickson, *Tenn., U.S.* **103** E14
Dickson, *Russ.* **161** E13
Dicle (Tigris), river, *Turk.* **171** H16
Dicle, *Turk.* **171** G14
Dicomano, *It.* **152** G7
Didao, *China* **185** D17
Dider, *Alg.* **201** J15
Dīdēsa, river, *Eth.* **210** M6
Didhav, *Eth.* **210** K9
Didiéni, *Mali* **204** J11
Dīdīmtu, *Eth.* **210** M9

Didwana, *India* **182** G4
Didymóteicho, *Gr.* **157** B17
Die, *Fr.* **149** V13
Diébougou, *Burkina Faso* **205** L14
Diego de Almagro, *Chile* **134** G7
Diego de Almagro, Isla, *Chile* **135** W6
Diego Ramírez, Islas, *Chile* **135** Y9
Diekirch, *Lux.* **162** H10
Diélé, *Congo* **208** F10
Diéma, *Mali* **204** J11
Diemal Find, *W. Aust., Austral.* **221** W4
Điện Biên Phủ, *Vietnam* **190** H9
Diễn Châu, *Vietnam* **190** J11
Dieppe Bay Town, *St. Kitts & Nevis*
  123 A17
Dieppe, *Fr.* **148** M8
Dierks, *Ark., U.S.* **102** H6
Dif, *Kenya* **211** S9
Diffa, *Niger* **206** H11
Differdange, *Lux.* **162** K9
Difnein, island, *Eritrea* **210** G8
Digba, *D.R.C.* **209** B17
Digboi, *India* **182** G15
Digby, *N.S., Can.* **87** P21
Dighton, *Kans., U.S.* **107** T4
Digne, *Fr.* **149** W14
Digoin, *Fr.* **149** T11
Digor, *Turk.* **171** D17
Digos, *Philippines* **193** F15
Digri, *Pak.* **181** S9
Digul, river, *Indonesia* **193** M21
Digura, island, *Maldives* **183** W3
Dihok (Dahūk), *Iraq* **176** D10
Diinsoor, *Somalia* **211** R10
Dijlah (Tigris), *Iraq* **176** H11
Dijon, *Fr.* **149** R13
Dik, *Chad* **207** K13
Dika, *Russ.* **161** E15
Dikhil, *Djibouti* **210** K10
Dikili, *Turk.* **170** E3
Díkti, Óros, *Gr.* **157** Q15
Dīla, *Eth.* **211** N7
Dilārām, *Afghan.* **180** K4
Dildarde Burnu, *Turk.* **170** J7
Dili, *Timor-Leste* **193** M15
Dilijan, *Arm.* **171** D19
Dillenburg, *Ger.* **150** F6
Dilley, *Tex., U.S.* **105** N15
Dillia, river, *Niger* **206** G10
Dilling, *Sudan* **203** T10
Dillingham, *Alas., U.S.* **113** L13
Dillon, *S.C., U.S.* **98** K11
Dillon, *Mont., U.S.* **108** F7
Dillon's Bay, *Vanuatu* **229** L17
Dilolo, *D.R.C.* **209** L15
Dilworth, *Minn., U.S.* **106** G8
Dimas, *Mex.* **116** F8
Dimashq (Damascus), *Syr.* **172** J7
Dimbelenge, *D.R.C.* **209** H15
Dimitrovgrad, *Bulg.* **155** K15
Dimitrovgrad, *Russ.* **158** K15
Dimitsána, *Gr.* **156** K9
Dimmitt, *Tex., U.S.* **104** E11
Dīmtu, *Eth.* **211** P9
Dinagat, island, *Philippines* **193** E15
Dinajpur, *Bangladesh* **182** H12
Dinan, *Fr.* **149** Q5
Dinangourou, *Mali* **205** J15
Dinant, *Belg.* **148** M12
Dinar, *Turk.* **170** G5
Dinara, peak, *Bosn. & Herzg., Croatia*
  154 G6
Dinara, range, *Bosn. & Herzg., Croatia*
  154 F6
Dinaric Alps, *Eur.* **154** D5
Dinas Mawddwy, *Wales, U.K.* **145** S10
Dinde, *Angola* **208** N10
Dindigul, *India* **183** S6
Dinga, *D.R.C.* **208** H11
Dingle Bay (Daingin, Bá an), *Ire.*
  145 T2
Dingle Dome, *Antarctica* **239** C19
Dingle *see* An Daingean, *Ire.* **145** T2
Dingli, *Malta* **163** L21
Dingolfing, *Ger.* **150** J9
Dingqing *see* Dêngqên, *China* **184** K7
Dinguiraye, *Guinea* **204** L10
Dinosaur National Monument,
  Colo., Utah, U.S. **108** L10
Dinosaur, *Colo., U.S.* **108** L10
Dinuba, *Calif., U.S.* **111** T7
Dioïla, *Mali* **204** K12
Dioka, *Mali* **204** J10
Diomede Islands, *Alas., U.S.* **112** E11
Dion, ruin, *Gr.* **156** D10
Dios, *P.N.G.* **228** B5

Dios, Cayos de, *Cuba* **118** H7
Dioumdiouréré, *Mali* **205** J15
Dioundiou, *Niger* **206** H6
Dioura, *Mali* **205** J13
Diourbel, *Senegal* **204** J7
Diphu Pass, *India, Myanmar* **190** C6
Dipkarpaz *see* Rizokarpaso, *Cyprus*
 162 M11
Diplo, *Pak.* **181** S9
Dir, *Pak.* **180** G11
Dirē Dawa (Dire Dawa), *Eth.* **210** L9
Dire Dawa *see* Dirē Dawa, *Eth.*
 210 L9
Diré, *Mali* **205** H14
Dírfys, Óros, *Gr.* **156** H12
Dirico, *Angola* **209** Q14
Dirj, *Lib.* **201** F16
Dirk Hartog Island, *Austral.* **220** G1
Dirkou, *Niger* **206** E11
Dirra, *Sudan* **203** S8
Dirty Devil, river, *Utah, U.S.* **109** N8
Disa, *India* **182** J3
Disappointment, Cape, *Wash., U.S.*
 110 E2
Disappointment, Lake, *Austral.*
 220 F5
Disaster Bay, *Austral.* **220** M14
Discovery Bank, *Ind. Oc.* **250** R10
Discovery Bay, *Jam.* **122** H8
Discovery Bay, *Austral.* **220** M11
Discovery II Fracture Zone, *Ind. Oc.*
 250 P5
Discovery Seamounts, *Atl. Oc.*
 249 U12
Discovery Tablemount, *Atl. Oc.*
 249 U12
Disentis, *Switz.* **152** B4
Dishnā, *Egypt* **202** J12
Disko Bank, *Atl. Oc.* **253** Q16
Disko Bugt *see* Qeqertarsuup
 Tunua, *Greenland, Den.* **88** M6
Disko *see* Qeqertarsuaq,
 Greenland, Den. **88** L5
Diskofjord, *Greenland, Den.* **88** L5
Dismal Mountains, *Antarctica*
 239 D20
Disney Cay, *Bahamas* **118** D11
Dispur, *India* **182** H13
Disteghil Sar, peak, *Pak.* **180** F13
Disūq, *Egypt* **202** E11
Ditinn, *Guinea* **204** L9
Diu, *India* **182** L2
Dīvāndarreh, *Iran* **176** E12
Divénié, *Congo* **208** F8
Divilacan Bay, *Pac. Oc.* **168** J13
Divinhe, *Mozambique* **212** F11
Divinópolis, *Braz.* **133** L14
Diviso, *Col.* **130** J4
Divo, *Côte d'Ivoire* **205** P13
Divoúnia, island, *Gr.* **157** P17
Divriği, *Turk.* **171** F13
Diwana, *Pak.* **181** R7
Dīwār, *Syr.* **172** B16
Dixie, *Idaho, U.S.* **108** E4
Dixon Entrance, *N. Amer.* **84** F3
Dixon, *Ill., U.S.* **101** N5
Dixon, *Ky., U.S.* **103** C14
Dixon, *Mo., U.S.* **107** T13
Dixon, *Mont., U.S.* **108** C5
Dixon, *Calif., U.S.* **111** R4
Dixons, *Bahamas* **122** E12
Diyadin, *Turk.* **171** E18
Diyālá, river, *Iraq* **176** F11
Diyarbakır, *Turk.* **171** G15
Dja, river, *Cameroon* **206** P11
Djado, *Niger* **206** D10
Djako, *Cen. Af. Rep.* **207** L16
Djamaa, *Alg.* **201** D14
Djambala, *Congo* **208** F10
Djanet, *Alg.* **201** J15
Djarrit, island, *Marshall Is.* **227** D21
Djaul, island, *P.N.G.* **222** N9
Djebel Aïssa, *Alg.* **200** D10
Djebel Telerhteba, *Alg.* **201** J14
Djédaa, *Chad* **207** H14
Djelfa, *Alg.* **200** C12
Djéli Mahé, *Mali* **204** H10
Djéma, *Cen. Af. Rep.* **207** M17
Djember, *Chad* **207** K13
Djémila, ruin, *Alg.* **201** B13
Djeniene bou Rezg, *Alg.* **200** D10
Djénné, *Mali* **205** J13
Djérem, river, *Cameroon* **206** M11
Djéroual, *Chad* **207** G14
Dji, river, *Cen. Af. Rep.* **207** M16
Djibasso, *Burkina Faso* **205** K14

Djibo, *Burkina Faso* **205** J15
Djibouti, *Af.* **210** K10
Djibrosso, *Côte d'Ivoire* **204** M12
Djidja, *Benin* **205** N17
Djohong, *Cameroon* **206** M12
Djokupunda, *D.R.C.* **209** H14
Djolu, *D.R.C.* **209** D15
Djoua, river, *Congo, Gabon* **208** D10
Djougou, *Benin* **205** M17
Djouho Battinga, *Cen. Af. Rep.*
 207 M15
Djoum, *Cameroon* **206** P11
Djugu, *D.R.C.* **209** D19
Djúpáll, valley, *Arctic Oc.* **253** M20
Djúpavík, *Ice.* **142** D2
Djúpivogur, *Ice.* **142** F4
Dmitriya Lapteva, *Russ.* **161** D17
Dmitriya L'govskiy, *Russ.* **158** M8
Dmitriyev L'govskiy, *Russ.* **158** J7
Dmitrov, *Russ.* **158** J9
Dnepr, river, *Russ.* **158** J7
Dnieper Lowland, *Eur.* **140** G11
Dnieper *see* Dnipro, river, *Ukr.*
 159 R7
Dnieper Upland, *Eur.* **140** H11
Dniester, river, *Mold.* **159** R5
Dnipro (Dnieper), river, *Ukr.* **159** R7
Dniprodzerzhyns'k, *Ukr.* **159** Q8
Dnipropetrovs'k, *Ukr.* **159** Q8
Dno, *Russ.* **158** G6
Dnyapro, river, *Belarus* **158** K6
Đô Lương, *Vietnam* **190** J10
Doba, *Chad* **207** L13
Dobane, *Cen. Af. Rep.* **207** M17
Dobbs, Cape, *Nunavut, Can.* **87** H16
Dobbyn, *Qnsld., Austral.* **221** S11
Doboj, *Bosn. & Herzg.* **154** F8
Dobre Miasto, *Pol.* **151** B15
Dobrich, *Bulg.* **155** G17
Dobruja, region, *Bulg., Rom.* **155** G18
Dobrush, *Belarus* **158** M6
Dobson, *N.Z.* **223** L18
Docampadó, Ensenada, *Col.* **130** G4
Doce, river, *Braz.* **133** L15
Docheong, *S. Korea* **187** V8
Dochodo, island, *S. Korea* **187** T6
Dock Junction, *Ga., U.S.* **99** P8
Docker River, *N.T., Austral.* **221** U7
Dodecanese *see* Dodekánisa, islands,
 Gr. **157** M18
Dodekánisa (Dodecanese), islands,
 Gr. **157** M18
Dodge City, *Kans., U.S.* **107** U5
Dodge, *Nebr., U.S.* **107** P8
Dodgeville, *Wis., U.S.* **100** L4
Dodman Point, *Eng., U.K.* **145** X8
Dodoma, *Tanzania* **211** W6
Dodona, ruin, *Gr.* **156** E7
Dodson Peninsula, *Antarctica* **238** F8
Dodson, *Tex., U.S.* **105** D14
Dodson, *Mont., U.S.* **108** B10
Doerun, *Ga., U.S.* **99** P5
Doetinchem, *Neth.* **148** J14
Dog Creek, *B.C., Can.* **86** L8
Dog Island, *Virgin Is., U.S.* **122** N10
Dog Rocks, islands, *Bahamas* **118** F9
Doğanşehir, *Turk.* **170** G12
Dogger Bank, *Atl. Oc.* **248** G13
Dōgo, island, *Japan* **189** N7
Dogondoutchi, *Niger* **206** H6
Doğruyol, *Turk.* **171** D17
Doğubayazıt, *Turk.* **171** E18
Dogwaya, *Sudan* **203** P13
Dogwood Point, *St. Kitts & Nevis*
 123 C19
Dogye, *S. Korea* **187** N12
Dogye, *S. Korea* **187** P12
Doha *see* Ad Dawḥah, *Qatar* **175** F13
Dohm Bank, *Atl. Oc.* **248** E9
Dohoukota, *Cen. Af. Rep.* **207** M13
Doi Inthanon, peak, *Thai.* **190** J6
Doïran, Lake *see* Dojransko Ezero,
 Gr., N. Maced. **154** L12
Doïráni, Límni, *Gr.* **156** B11
Dois Irmãos, Serra, *Braz.* **133** G16
Dojransko Ezero (Doïran, Lake),
 Gr., N. Maced. **154** L12
Doka *see* Bahr Kéita, river, *Chad*
 207 K14
Doka, *Sudan* **203** S14
Dokdo (Takeshima, Liancourt
 Rocks), *S. Korea* **188** M5
Dokkum, *Neth.* **148** G13
Dokós, island, *Gr.* **156** K12
Dolak (Yos Sudarso), island,
 Indonesia **193** M20
Doland, *S. Dak., U.S.* **106** K7

Dolbeau, *Que. Can.* **87** N19
Doldrums Fracture Zone, *Atl. Oc.*
 249 N7
Dole, *Fr.* **149** S13
Doleib Hill, *S. Sudan* **203** U11
Dolgiy Most, *Russ.* **161** K14
Dolgoshchel'ye, *Russ.* **158** A11
Doliana, *Gr.* **156** E7
Dolianova, *It.* **153** P4
Dolinsk, *Russ.* **161** K22
Dolisie (Loubomo), *Congo* **208** G9
Dolly Cays, *Bahamas* **118** F11
Dolo Bay, *Eth.* **211** Q9
Dolohmwar, peak, *F.S.M.* **226** K8
Dolomites, range, *It.* **152** C7
Dolores, *Colo., U.S.* **109** P10
Dolores, *Arg.* **134** M13
Dolores, *Uru.* **134** L13
Dolores, river, *Colo., Utah, U.S.*
 109 N10
Dolphin Head, *Namibia* **212** J3
Dolphin Spur, *Pac. Oc.* **255** F21
Dolphin, Cape, *Falk. Is., U.K.* **135** V12
Dolsan, *S. Korea* **187** U9
Domagnano, *San Marino* **163** J16
Domaniç, *Turk.* **170** E5
Domažlice, *Czechia* **150** H10
Dombás, *Nor.* **142** H11
Dombe Grande, *Angola* **208** M9
Dome A *see* Argus Dome, *Antarctica*
 239 H16
Dome Argus (Dome A), *Antarctica*
 239 H16
Dome C *see* Dome Charlie, *Antarctica*
 239 M17
Dome Charlie (Dome C), *Antarctica*
 239 M17
Dome Fuji Station, *Antarctica* **239** E16
Doméniko, *Gr.* **156** E10
Domett, *N.Z.* **223** L19
Domeyko, *Chile* **134** H6
Domeyko, Cordillera, *Chile* **134** E7
Domfront, *Fr.* **149** P6
Dominica Passage, *N. Amer.* **121** D16
Dominica, *N. Amer.* **121** E17
Dominican Republic, *N. Amer.* **119** L20
Dominion, Cape, *Nunavut, Can.* **87** G17
Domo, *Eth.* **210** M12
Domodossola, *It.* **152** C3
Domokós, *Gr.* **156** F10
Domoni, *Comoros* **215** N16
Domus de Maria, *It.* **153** Q3
Domuyo Volcano, *S. Amer.* **128** M4
Domuyo, Volcán, *Arg.* **134** M6
Don Benito, *Sp.* **146** H8
Don, Cape, *Austral.* **220** A8
Don, river, *Russ.* **159** P12
Dôn, river, *Laos* **190** L11
Donaldson, *Minn., U.S.* **106** E8
Donaldsonville, *La., U.S.* **102** P10
Donalsonville, *Ga., U.S.* **99** Q4
Donau (Danube), *Aust.* **150** K11
Donbas, region, *Ukr.* **159** Q10
Doncaster, *Eng., U.K.* **145** Q13
Dondo, *Indonesia* **193** J14
Dondo, *Angola* **208** K10
Dondo, *D.R.C.* **209** B14
Dondo, *Mozambique* **212** F11
Dondra Head, *Asia* **168** L7
Donegal Bay, *Ire.* **145** P4
Donets Basin, *Eur.* **140** H12
Donets Ridge, *Eur.* **140** H13
Donets, river, *Eur.* **140** G12
Donetsk, *Russ.* **159** Q11
Donets'k, *Ukr.* **159** Q9
Đông Hới, *Vietnam* **190** K11
Đông Khê, *Vietnam* **190** G11
Đồng Lê, *Vietnam* **190** K11
Đông Phu, *Vietnam* **190** L12
Đồng Văn, *Vietnam* **190** F10
Đông Xoài, *Vietnam* **191** P12
Donga, *Nigeria* **206** L9
Dongara, *W. Aust., Austral.* **221** W2
Dongbei (Manchuria), region, *China*
 185 C15
Dongdo, island, *S. Korea* **187** V9
Dongducheon, *S. Korea* **186** M8
Dongeup, *S. Korea* **187** S11
Dongfang, *China* **184** Q11
Donghae, *S. Korea* **186** M12
Dongjingcheng, *China* **185** D17
Dongkalang, *Indonesia* **193** J13
Dongning, *China* **185** D17
Dongo, *Angola* **208** N11
Dongo, *D.R.C.* **208** C12

Dongola, *Sudan* **203** N11
Dongou, *Congo* **208** C12
Dongsha (Pratas Island), *China*
 185 P14
Dongsheng, *China* **184** G11
Dongtai, *China* **185** J15
Dongwe, river, *Zambia* **209** N16
Dongyztaū, *Kaz.* **178** G7
Dongzhen, *China* **184** P12
Doniphan, *Mo., U.S.* **107** V14
Dønna, island, *Nor.* **142** F11
Donnelly, *Idaho, U.S.* **108** F4
Donnellys Crossing, *N.Z.* **223** D19
Donner Pass, *Calif., U.S.* **111** Q6
Donors Hill, *Qnsld., Austral.* **221** R11
Donostia-San Sebastián, *Sp.* **147** B13
Donoúsa, island, *Gr.* **157** L16
Donzenac, *Fr.* **149** V9
Doolette Bay, *Antarctica* **239** R17
Doolish *see* Douglas, *Isle of Man, U.K.*
 145 P9
Doolow, *Somalia* **211** Q9
Doomadgee, *Qnsld., Austral.* **221** R11
Doon, Loch, *Scot., U.K.* **145** M9
Door Peninsula, *Wis., U.S.* **100** J7
Dooxo Nugaaleed, *Somalia* **210** M13
Dopyeong, *S. Korea* **187** Q12
Dora, *Ala., U.S.* **103** J15
Dora, *N. Mex., U.S.* **109** T15
Dora, Lake, *Austral.* **220** F5
Dorah Pass, *Afghan., Pak.* **180** G10
Dorchester, *Eng., U.K.* **145** W11
Dorchester, Cape, *Nunavut, Can.*
 87 G17
Dordives, *Fr.* **149** Q10
Dordogne, river, *Fr.* **149** V9
Dordrecht Hole, *Ind. Oc.* **251** N15
Dordrecht, *Neth.* **148** J12
Dörgön Nuur, *Mongolia* **184** D7
Dori, *Burkina Faso* **205** J16
Dório, *Gr.* **156** K9
Dormans, *Fr.* **149** P11
Dornbirn, *Aust.* **150** L7
Dorno Djoutougé, *Chad* **207** H16
Dornoch Firth, *Scot., U.K.* **144** H9
Dornoch, *Scot., U.K.* **144** H9
Doro, *Mali* **205** H15
Dorogobuzh, *Russ.* **158** K7
Dorogorskoye, *Russ.* **158** A11
Dorohoi, *Rom.* **155** A15
Dorp Rincón, *Bonaire, Neth.* **123** Q19
Dorre Island, *Austral.* **220** G1
Dorris, *Calif., U.S.* **110** L4
Dortmund, *Ger.* **150** E6
Dörtyol, *Turk.* **170** J11
Dorūd, *Iraq* **177** G14
Dörzbach, *Ger.* **150** H7
Dos Bahías, Cape, *S. Amer.* **128** P5
Dos Hermanas, *Sp.* **146** L8
Döşemealtı, *Turk.* **170** H6
Dōshī, *Afghan.* **180** G8
Dosso, *Niger* **206** H6
Dossor, *Kaz.* **178** J4
Dostyq, *Kaz.* **179** H17
Dothan, *Ala., U.S.* **103** N17
Dotson Ice Shelf, *Antarctica* **238** L8
Douai, *Fr.* **148** M10
Douako, *Guinea* **204** M10
Douala, *Cameroon* **206** N9
Douaouir, *Mali* **205** E14
Douarnenez, *Fr.* **149** Q3
Douarnenez, Baie de, *Fr.* **149** Q2
Double Island Point, *Austral.* **220** G16
Double Mountain Fork Brazos, river,
 Tex., U.S. **105** G13
Double Point, *Austral.* **220** D13
Doubtful Bay, *Austral.* **220** C5
Doubtless Bay, *N.Z.* **223** D19
Doué, *Fr.* **149** S7
Douentza, *Mali* **205** J14
Douéoulou, *New Caledonia, Fr.* **231** P23
Dougga, ruin, *Tunisia* **201** B15
Doughboy Bay, *N.Z.* **223** Q16
Douglas (Doolish), *Isle of Man, U.K.*
 145 P9
Douglas Bay, *Dominica* **123** E18
Douglas Lake, *Tenn., U.S.* **103** E19
Douglas, *Ga., U.S.* **99** P7
Douglas, *Mich., U.S.* **100** M9
Douglas, *Wyo., U.S.* **108** H13
Douglas, *Ariz., U.S.* **109** W9
Douglass, *Kans., U.S.* **107** U8
Douglasville, *Ga., U.S.* **98** L4
Doullens, *Fr.* **148** M10
Doumé, *Cameroon* **206** N11

Douna, *Mali* **205** J15
Dourada, *S. Amer.* **128** G8
Dourados, *Braz.* **132** L11
Dourbali, *Chad* **206** J12
Douro, river, *Port., Sp.* **146** E8
Dove Bugt, *Greenland, Den.* **88** F13
Dove Creek, *Colo., U.S.* **109** P10
Dover-Foxcroft, *Me., U.S.* **97** E17
Dover, *N.H., U.S.* **97** J15
Dover, *Del., U.S.* **98** C15
Dover, *Ohio, U.S.* **101** Q15
Dover, *Eng., U.K.* **145** V16
Dover, Strait of (Pas de Calais),
 Fr., U.K. **148** L9
Dow Gonbadān, *Iran* **177** K15
Dowagiac, *Mich., U.S.* **101** N9
Dowlatābād, *Iran* **177** M19
Dowlatābād, *Afghan.* **180** F5
Dowlatyār, *Afghan.* **180** H6
Downey, *Idaho, U.S.* **108** J7
Downey, *Calif., U.S.* **111** X8
Downieville, *Calif., U.S.* **111** P5
Downpatrick Head, *Ire.* **145** P3
Downpatrick, *N. Ire., U.K.* **145** P7
Downs, *Kans., U.S.* **107** S6
Doxáto, *Gr.* **157** B13
Doyang, *S. Korea* **187** U8
Doylestown, *Pa., U.S.* **96** P9
Dōzen, island, *Japan* **189** P7
Drăgăneşti-Olt, *Rom.* **155** G14
Drăgăşani, *Rom.* **155** F14
Dragon's Mouths,
 Trin. & Tobago, Venez. **123** P21
Draguignan, *Fr.* **149** X14
Drain, *Oreg., U.S.* **110** H3
Drake Passage, *Antarctica* **238** C3
Drake Peak, *Oreg., U.S.* **110** K6
Drake, *N. Dak., U.S.* **106** F5
Drakensberg, mountains, *S. Af.*
 212 K9
Dráma, *Gr.* **157** B13
Drammen, *Nor.* **142** K11
Dranov, Ostrov, *Rom.* **155** E18
Dras, *India* **180** H14
Drasan, *Pak.* **180** F11
Drau, river, *Aust.* **150** M10
Dráva, river, *Croatia, Hung.* **154** D7
Dravograd, *Slov.* **154** C5
Drayton, *N. Dak., U.S.* **106** E8
Drépano, Akrotírio, *Gr.* **157** E13
Dresden, *Ger.* **150** F10
Dreux, *Fr.* **149** P9
Drevsjø, *Nor.* **142** H12
Drew, *Miss., U.S.* **102** J10
Drews Reservoir, *Oreg., U.S.* **110** L5
Drewsey, *Oreg., U.S.* **110** H8
Driggs, *Idaho, U.S.* **108** H8
Drin, river, *Alban.* **154** J9
Drina, river, *Bosn. & Herzg., Serb.*
 154 F9
Drinit, Gjiri i, *Alban.* **154** K9
Driscoll Island, *Antarctica* **238** N10
Driskill Mountain, *La., U.S.* **102** L7
Drøbak, *Nor.* **142** K11
Drobeta-Turnu Severin, *Rom.*
 154 F12
Drogheda (Droichead Átha), *Ire.*
 145 Q7
Drohobych, *Ukr.* **159** P2
Droichead Átha *see* Drogheda, *Ire.*
 145 Q7
Droitwich, *Eng., U.K.* **145** T11
Drome, *P.N.G.* **222** N6
Dronning Louise Land ,
 Greenland, Den. **88** F12
Dronning Louise Ø, *Greenland, Den.*
 89 T8
Drosh, *Pak.* **180** G11
Drouseia, *Cyprus* **162** P5
Drovyanoy, *Russ.* **160** E12
Drua, island, *Fiji* **226** N6
Drumheller, *Alta., Can.* **86** M10
Drummond Island, *Mich., U.S.*
 100 G12
Drummond Range, *Austral.* **220** F14
Drummond, *Mont., U.S.* **108** D6
Drumright, *Okla., U.S.* **105** C18
Druza, *Russ.* **161** F19
Druzhnaya-4 Station, *Antarctica*
 239 F20
Dry Harbour Mountains, *Jam.*
 122 H8
Dry Tortugas National Park, *Fla., U.S.*
 99 Y7
Dry Tortugas, islands, *U.S.* **91** P19
Dryanovo, *Bulg.* **155** H15

Dryden, *Ont., Can.* **87** N15
Dryden, *Tex., U.S.* **104** L11
Drygalski Basin, *Pac. Oc.* **255** H15
Drygalski Ice Tongue, *Antarctica* **239** P14
Drygalski Island, *Antarctica* **239** J22
Drysdale, river, *Austral.* **220** C6
Dschang, *Cameroon* **206** M9
Du Bois, *Pa., U.S.* **96** M5
Du Chaillu Seamounts, *Atl. Oc.* **249** P12
Du Quoin, *Ill., U.S.* **101** U5
Dua, river, *D.R.C.* **209** C14
Dŭāb-e Maykh-e Zarīn, *Afghan.* **180** G8
Duaringa, *Qnsld., Austral.* **221** T15
Duarte, Pico, *Dom. Rep.* **119** M18
Dŭbā, *Saudi Arabia* **174** E4
Dubai *see* Dubayy, *U.A.E.* **175** F15
Dubăsari, *Mold.* **159** R5
Dubawnt Lake, *Nunavut, Can.* **87** H13
Dubawnt, river, *N.W.T., Can.* **87** J13
Dubayy (Dubai), *U.A.E.* **175** F15
Dubele, *D.R.C.* **209** C19
Dubh Artach, island, *Scot., U.K.* **144** L7
Dublin (Baile Átha Cliath), *Ire.* **145** R7
Dublin, *Ga., U.S.* **98** M7
Dublin, *Tex., U.S.* **105** H16
Dubna, *Russ.* **158** H9
Dubno, *Ukr.* **159** N3
Dubois, *Idaho, U.S.* **108** G7
Dubois, *Wyo., U.S.* **108** G9
Dubovka, *Russ.* **159** P13
Dubréka, *Guinea* **204** M8
Dubrovnik (Ragusa), *Croatia* **154** J8
Dubrovytsya, *Ukr.* **158** M4
Dubuque, *Iowa, U.S.* **107** N14
Duc de Gloucester, Îles du, *Fr. Polynesia, Fr.* **225** K14
Đức Thọ, *Vietnam* **190** J11
Duchesne, *Utah, U.S.* **108** L8
Duchesne, river, *Utah, U.S.* **108** L8
Duchess, *Qnsld. Austral.* **221** S11
Ducie Island, *Pac. Oc.* **225** K17
Duck Cay, *Bahamas* **122** F8
Duck Creek, *W. Aust., Austral.* **221** T2
Duck Hill, *Miss., U.S.* **102** J11
Duck, river, *Tenn., U.S.* **103** F15
Ducktown, *Tenn., U.S.* **103** G18
Ducos, *Martinique, Fr.* **123** F23
Dudelange, *Lux.* **162** K10
Dudinka, *Russ.* **161** F13
Dudley, *Eng., U.K.* **145** S12
Dudub, *Eth.* **211** N12
Duékoué, *Côte d'Ivoire* **204** P12
Duero, river, *Sp.* **146** E9
Dufek Coast, *Antarctica* **239** L13
Dufek Massif, *Antarctica* **238** G11
Duff Islands, *Solomon Is.* **228** K7
Dufftown, *Scot., U.K.* **144** H10
Dufur, *Oreg., U.S.* **110** F5
Dugi Otok, island, *Croatia* **154** G5
Dugway, *Utah, U.S.* **108** L6
Duifken Point, *Austral.* **220** B12
Duisburg, *Ger.* **150** E5
Duitama, *Col.* **130** F7
Dujuuma, *Somalia* **211** R10
Duk Fadiat, *S. Sudan* **203** W11
Dukafulu, *Eth.* **211** P8
Duke of York Group, *P.N.G.* **222** N10
Dukku, *Nigeria* **206** K10
Dukla, *Pol.* **151** H17
Dulan, *China* **184** H8
Dulce, *N. Mex., U.S.* **109** Q11
Dulce, Golfo, *C.R.* **117** P19
Dulce, river, *Arg.* **134** H10
Dulia, *D.R.C.* **209** C16
Dulovo, *Bulg.* **155** G16
Duluth, *Minn., U.S.* **106** G12
Dūmā, *Syr.* **172** J7
Duma, river, *D.R.C.* **209** B17
Dumai, *Indonesia* **192** H6
Dumaran, island, *Philippines* **193** E13
Dumas, *Ark., U.S.* **102** J9
Dumas, *Tex., U.S.* **104** C12
Dumbarton, *Scot., U.K.* **144** L9
Dumbéa, *New Caledonia, Fr.* **231** Q22
Dumfries, *Scot., U.K.* **145** N10
Dumido, island, *S. Korea* **187** T10
Dumka, *India* **182** J11
Dumoga Kecil, *Indonesia* **193** J15
Dumont d'Urville Station, *Antarctica* **239** Q19
Dumra, *India* **182** H10

Dumshaf Plain, *Arctic Oc.* **253** H19
Dumyāţ (Damietta), *Egypt* **202** E12
Dún an Uchta *see* Eyrecourt, *Ire.* **145** R5
Dún Dealgan *see* Dundalk, *Ire.* **145** Q7
Dún Laoghaire, *Ire.* **145** R7
Dún Mánmhaí *see* Dunmanway, *Ire.* **145** U3
Dun, *Fr.* **149** N13
Duna (Danube), river, *Hung.* **151** L14
Dunaj (Danube), river, *Slovakia* **151** L14
Dunărea (Danube), river, *Rom. Serb.* **154** F11
Dunaújváros, *Hung.* **151** M15
Dunav, river, *Bulg. Rom.* **155** G16
Dunay (Danube), river, *Rom. Ukr.* **155** E18
Dunbar, *W. Va., U.S.* **98** D8
Dunbar, *Qnsld., Austral.* **221** Q12
Duncan Passage, *India* **183** S14
Duncan Peninsula, *Antarctica* **238** M8
Duncan Town, *Bahamas* **119** H13
Duncan, *Okla., U.S.* **105** E16
Duncan, *Ariz., U.S.* **109** V9
Duncans, *Jam.* **122** H8
Duncansby Head, *Scot., U.K.* **144** F10
Duncombe Bay, *Austral.* **229** A20
Dund-Urt, *Mongolia* **184** D10
Dundalk (Dún Dealgan), *Ire.* **145** Q7
Dundalk Bay, *Ire.* **145** Q7
Dundalk, *Md., U.S.* **98** C13
Dundas Peninsula, *N.W.T., Can.* **87** D13
Dundas Strait, *Austral.* **220** A8
Dundas, *Greenland, Den.* **88** F3
Dundas, Lake, *Austral.* **220** K5
Dundee, *Scot., U.K.* **144** K10
Dundee, *S. Af.* **212** K9
Dundo, *Angola* **209** J14
Dundonnell, *Scot., U.K.* **144** H8
Dundrennan, *Scot., U.K.* **145** N19
Dundrum Bay, *N. Ire., U.K.* **145** P7
Dunedin, *Fla., U.S.* **99** U7
Dunedin, *N.Z.* **223** P17
Dunes City, *Oreg., U.S.* **110** H2
Dunfermline, *Scot., U.K.* **144** L10
Dungarpur, *India* **182** J4
Dungarvan Harbour, *Ire.* **145** T5
Dungas, *Niger* **206** H9
Dungeness, cape, *Eng., U.K.* **145** V15
Dungeness, Punta, *Arg.* **135** W9
Dungiven, *N. Ire., U.K.* **145** N6
Dungu, *D.R.C.* **209** C18
Dungunab Bay, *Sudan* **202** M15
Dungunab, *Sudan* **202** M15
Dunham River, *W. Aust., Austral.* **221** Q6
Dünheger, peak, *China, Mongolia* **184** D6
Dunhua, *China* **185** E16
Dunhuang, *China* **184** G7
Dunkerque (Dunkirk), *Fr.* **148** L10
Dunkineely, *Ire.* **145** N5
Dunkirk *see* Dunkerque, *Fr.* **148** L10
Dunkirk, *N.Y., U.S.* **96** K4
Dunkirk, *Ind., U.S.* **101** Q10
Dunkwa, *Ghana* **205** P15
Dunlap, *Iowa, U.S.* **107** P9
Dunmanus Bay, *Ire.* **145** U3
Dunmanway (Dún Mánmhaí), *Ire.* **145** U3
Dunmore Town, *Bahamas* **122** D5
Dunmore, *Pa., U.S.* **96** M9
Dunn, *N.C., U.S.* **98** H12
Dunnellon, *Fla., U.S.* **99** S7
Dunning, *Nebr., U.S.* **107** P5
Dunn's River Falls, bay, *Jam.* **122** H9
Dunolly, *Vic., Austral.* **221** Y12
Dunseith, *N. Dak., U.S.* **106** D5
Dunsmuir, *Calif., U.S.* **110** M3
Duntroon, *N.Z.* **223** N17
Dunvegan, *Scot., U.K.* **144** H7
Duolun, *China* **185** F13
Dupree, *S. Dak., U.S.* **106** J3
Dupuyer, *Mont., U.S.* **108** B7
Duqm, *Oman* **175** K17
Duque de Caxias, *Braz.* **133** M15
Duque de York, Isla, *Chile* **135** V6
Duquesne, *Pa., U.S.* **96** N3
Dura Europus, ruins, *Syr.* **172** G14
Durack, river, *Austral.* **220** C6
Durance, river, *Fr.* **149** W14
Durand, *Wis., U.S.* **100** J2
Durango, *Colo., U.S.* **109** P11
Durango, *Mex.* **116** F9

Durango, *Sp.* **146** B12
Durankulak, *Bulg.* **155** G18
Durant, *Miss., U.S.* **102** K11
Durant, *Okla., U.S.* **105** E18
Duras, *Fr.* **149** W7
Durazno, *Uru.* **134** K14
Durban, *S. Af.* **212** K9
Durbuy, *Belg.* **148** M13
Đurđevac, *Croatia* **154** D7
Durduri, *Somaliland* **210** K13
Dureji, *Pak.* **181** R7
Düren, *Ger.* **150** F5
Durg, *India* **182** L8
Durham, *N.H., U.S.* **97** J15
Durham, *N.C., U.S.* **98** G11
Durham, *Eng., U.K.* **145** N12
Durlas *see* Thurles, *Ire.* **145** S5
Durlston Head, *Eng., U.K.* **145** W12
Durmitor, peak, *Eur.* **140** K8
Durness, Kyle of, *Scot., U.K.* **144** G8
Durtal, *Fr.* **149** R7
Duru Gölü, *Turk.* **170** C4
Duru, *D.R.C.* **209** B18
Đurumă, *Saudi Arabia* **174** G10
Durūz, Jabal ad, *Syr.* **172** L8
D'Urville Channel, *Fiji* **226** Q5
D'Urville Island *see* Rangitoto ki te Tonga, *N.Z.* **223** J19
Durwalē, *Eth.* **210** M10
Duşak, *Turkm.* **178** N8
Dūshākh, *Afghan.* **180** G7
Dushan, *China* **184** N11
Dushanbe, *Taj.* **178** M12
Dushanzi, *China* **184** E5
Düsseldorf, *Ger.* **150** F5
Dutch John, *Utah, U.S.* **108** K9
Dutchcap Cay, *Virgin Is., U.S.* **122** M6
Dutoit Fracture Zone, *Ind. Oc.* **250** R3
Dutse, *Nigeria* **206** J9
Dutsen Wai, *Nigeria* **206** J8
Dutsin Ma, *Nigeria* **206** J8
Dutton, Mount, *Utah, U.S.* **109** P7
Dutywa, *S. Af.* **212** L8
Duvergé, *Dom. Rep.* **119** M17
Düzce *see* Achna, *Cyprus* **162** P9
Düzce, *Turk.* **170** D7
Dvina Bay, *Arctic Oc.* **140** B11
Dvina, river, *Asia* **168** D6
Dvinskaya Guba, bay, *Russ.* **158** B10
Dvinskoy, *Russ.* **158** D12
Dwarka, *India* **182** K1
Dwight, *Ill., U.S.* **101** P6
Dworshak Reservoir, *Idaho, U.S.* **108** D4
Dyat'kovo, *Russ.* **158** L8
Dyer Bay, *N.W.T., Can.* **86** C12
Dyer Plateau, *Antarctica* **238** M8
Dyer, Cape, *Nunavut, Can.* **87** F19
Dyero, *Mali* **204** K12
Dyersburg, *Tenn., U.S.* **102** E12
Dynevor Downs, *Qnsld., Austral.* **221** V12
Dyoumara, *Mali* **204** J11
Dzaanhushuu, *Mongolia* **184** D9
Dzag, *Mongolia* **184** D9
Dzangalï, *Afghan.* **180** K7
Dzaoudzi, *Mayotte, Fr.* **215** P17
Dzavhan, river, *Mongolia* **184** D7
Dzerzhinsk, *Russ.* **158** J12
Dzhagdy Range, *Asia* **168** D11
Dzhagdy, *Russ.* **161** J19
Dzhankoy, *Russ.* **159** S7
Dzhardzhan, *Russ.* **161** F16
Dzharty-Rabat, *Taj.* **179** N14
Dzharylhach, Ostriv, *Ukr.* **159** S7
Dzhilandy, *Taj.* **179** N13
Dzhugdzhur Range, *Asia* **168** D12
Dzhugdzhur, *Russ.* **161** H20
Dzhunkun, *Russ.* **161** H16
Działdowo, *Pol.* **151** C15
Dzibilchaltún, ruin, *Mex.* **117** H16
Dzierżoniów, *Pol.* **151** G13
Dzioua, *Alg.* **201** D13
Dzodze, *Ghana* **205** P17
Dzöölön, *Mongolia* **184** B9
Dzungarian Basin *see* Junggar Pendi, *China* **184** D5
Dzungarian Gate, *Kaz.* **179** H17
Dzuunmod, *Mongolia* **184** D11

Dzvina, river, *Belarus* **158** J6
Dzyarzhynsk, *Belarus* **158** K4

E

Eads, *Colo., U.S.* **109** N15
Eagar, *Ariz., U.S.* **109** T9
Eagle Beach, *Aruba, Neth.* **123** Q17
Eagle Butte, *S. Dak., U.S.* **106** J4
Eagle Grove, *Iowa, U.S.* **107** N11
Eagle Harbor, *Mich., U.S.* **100** D7
Eagle Lake, *Me., U.S.* **97** B17
Eagle Lake, *Calif., U.S.* **111** N5
Eagle Mountain, *Minn., U.S.* **106** F14
Eagle Mountain, *Calif., U.S.* **111** X11
Eagle Nest Lake, *N. Mex., U.S.* **109** Q13
Eagle Pass, *Tex., U.S.* **105** N13
Eagle Peak, *Tex., U.S.* **104** J8
Eagle Peak, *N. Mex., U.S.* **109** T10
Eagle Peak, *Calif., U.S.* **110** M6
Eagle River, *Mich., U.S.* **100** E6
Eagle River, *Wis., U.S.* **100** G5
Eagle, *Colo., U.S.* **108** M11
Eagle, *Alas., U.S.* **113** F19
Eagles Mere, *Pa., U.S.* **96** M8
Earle, *Ark., U.S.* **102** G11
Earlimart, *Calif., U.S.* **111** U7
Earlington, *Ky., U.S.* **103** D14
Earp, *Calif., U.S.* **111** X13
Earth, *Tex., U.S.* **104** E11
Easky (Iascaigh), *Ire.* **145** P4
Easley, *S.C., U.S.* **98** J7
East Anglia, region, *Eng., U.K.* **145** T15
East Anglian Heights, *Eng., U.K.* **145** T15
East Antarctica, region, *Antarctica* **239** J18
East Aurora, *N.Y., U.S.* **96** J5
East Azores Fracture Zone, *Atl. Oc.* **248** K9
East Bay, *La., U.S.* **102** R12
East Caicos, island, *Turks & Caicos Is.* **119** H17
East Cape, *Asia* **168** A12
East Cape, *N.Z.* **223** F23
East Carbon, *Utah, U.S.* **108** M8
East Caroline Basin, *Pac. Oc.* **246** J8
East Channel, *Marshall Is.* **227** B21
East Chicago, *Ind., U.S.* **101** N7
East China Sea, *Pac. Oc.* **168** H13
East Coast Bays, *N.Z.* **223** E20
East Cote, *La., U.S.* **102** Q9
East Dubuque, *Ill., U.S.* **100** M4
East End Point, *Bahamas* **122** F6
East End Point, *Bahamas* **122** J4
East End, *Cayman Is., U.K.* **122** N5
East End, *Virgin Is., U.K.* **122** Q7
East End, *Virgin Is., U.S.* **122** N12
East Entrance, *Palau* **226** E6
East Falkland, island, *Falk. Is., U.K.* **135** W12
East Fayu, island, *F.S.M.* **226** A9
East Fork Bruneau, river, *Idaho, U.S.* **108** H4
East Fork White, river, *U.S.* **91** G17
East Frisian Islands, *Ger.* **150** B6
East Glacier Park, *Mont., U.S.* **108** B6
East Grand Forks, *Minn., U.S.* **106** F8
East Greenland Ice-core Project (EastGRIP), *Greenland, Den.* **88** G10
East Greenland Shelf, *Atl. Oc.* **248** E8
East Helena, *Mont., U.S.* **108** D7
East Indiaman Ridge, *Ind. Oc.* **251** L14
East Isaac, island, *Bahamas* **118** D10
East Jordan, *Mich., U.S.* **100** H10
East Korea Bay, *Pac. Oc.* **168** F12
East Lake, *Mich., U.S.* **100** J9
East Lansing, *Mich., U.S.* **100** M11
East Liverpool, *Ohio, U.S.* **101** Q15
East London, *S. Af.* **212** M8
East Mariana Basin, *Pac. Oc.* **246** H8
East Millinocket, *Me., U.S.* **97** D17
East Moline, *Ill., U.S.* **101** N4
East Novaya Zemlya Trough, *Arctic Oc.* **253** D15
East Orange, *N.J., U.S.* **96** N11
East Pacific Rise, *Pac. Oc.* **247** Q18
East Pen Island, *Ont., Can.* **87** L16
East Peoria, *Ill., U.S.* **101** Q5
East Point, *Ga., U.S.* **98** L5
East Point, *Virgin Is., U.S.* **122** Q4

East Point, *Marshall Is.* **227** H21
East Prairie, *Mo., U.S.* **107** V16
East Prussia, region, *Pol. Russ.* **151** B16
East Ridge, *Tenn., U.S.* **103** G17
East Saint Louis, *Ill., U.S.* **101** T4
East Scotia Basin, *Atl. Oc.* **249** W8
East Scotia Ridge, *Atl. Oc.* **254** L8
East Sea *see* Japan, Sea of, *Pac. Oc.* **168** F13
East Sheba Ridge, *Ind. Oc.* **250** D7
East Siberian Sea, *Arctic Oc.* **252** H6
East Siberian Shelf, *Arctic Oc.* **252** G7
East Takatu *see* Tacutu, river, *S. Amer.* **131** G15
East Tasman Plateau, *Pac. Oc.* **246** P8
East Tasman Saddle, *Ind. Oc.* **251** Q23
East Tawas, *Mich., U.S.* **100** J12
East Thulean Rise, *Atl. Oc.* **248** H9
East Timor *see* Timor-Leste, *Asia* **193** M15
East Wells, cape, *Bahamas* **122** A12
Eastbourne, *Eng., U.K.* **145** V14
Easter Fracture Zone, *Pac. Oc.* **247** M20
Easter Island (Isla de Pascua, Rapa Nui), *Pac. Oc.* **225** L19
Eastern Beach, *Gibraltar, U.K.* **163** N23
Eastern Desert, *Egypt* **202** G12
Eastern Ghats, range, *India* **183** R6
Eastern Point, *Bahamas* **122** E3
Eastern Sayan Mountains, *Asia* **168** E9
Eastern Group *see* Lau Group, *Fiji* **226** N8
East GRIP *see* East Greenland Ice-core Project, *Greenland, Den.* **88** G10
Eastland, *Tex., U.S.* **105** G15
Eastleigh, *Eng., U.K.* **145** V12
Eastmain, *Que., Can.* **87** M18
Eastmain, river, *Que., Can.* **87** M19
Eastman, *Ga., U.S.* **99** N6
Easton, *Pa., U.S.* **96** N9
Easton, *Md., U.S.* **98** C14
Eastpoint, *Fla., U.S.* **99** R4
Eastport, *Me., U.S.* **97** E20
Eastville, *Va., U.S.* **98** E14
Eaton Rapids, *Mich., U.S.* **100** M11
Eaton, *Ohio, U.S.* **101** R11
Eatonton, *Ga., U.S.* **98** L6
Eatonville, *Wash., U.S.* **110** D4
Eau Claire, *Wis., U.S.* **100** H3
Eau Claire, Lac a l', *N. Amer.* **84** F8
Eauripik Atoll, *F.S.M.* **226** B7
Eauripik Rise, *Pac. Oc.* **246** J7
Ebadon, island, *Marshall Is.* **227** F13
Ebbw Vale (Glyn Ebwy), *Wales, U.K.* **145** U10
Ebe Khaya, *Russ.* **161** F15
Ebebiyin, *Eq. Guinea* **206** P10
Ebenholz, *Liech.* **162** P2
Eber Gölü, *Turk.* **170** F7
Eberswalde, *Ger.* **150** D10
Ebeye, island, *Marshall Is.* **227** J18
Ebinur Hu, *China* **184** D4
Ebjapik, island, *Marshall Is.* **227** J18
Eboda *see* Horvat 'Avedat, ruins, *Israel* **173** P5
Ebola, river, *D.R.C.* **209** C14
Eboli, *It.* **152** M11
Ébolowa, *Cameroon* **206** P10
Ebon Atoll, island, *Marshall Is.* **227** E16
Ebro, Embalse del, *Sp.* **146** B11
Ebro, river, *Sp.* **147** D13
Ebwaj, island, *Marshall Is.* **227** J18
Ecbatana *see* Hamadān, *Iraq* **177** F13
Ecclesville, *Trin. & Tobago* **123** Q23
Eceabat, *Turk.* **170** D2
Echallens, *Switz.* **152** B1
Echínos, *Gr.* **157** B14
Echo Bay, *N.W.T., Can.* **86** G11
Echo, *Oreg., U.S.* **110** E7
Echternach, *Lux.* **162** J11
Echuca, *Vic., Austral.* **221** Y12
Écija, *Sp.* **146** L9
Eckenförde, *Ger.* **150** B8
Ecuador, *S. Amer.* **130** L4
Ed Daein, *Sudan* **203** T8
Ed Damazin, *Sudan* **203** T13
Ed Damer, *Sudan* **203** P13
Ed Debba, *Sudan* **203** P11
Ed Dueim, *Sudan* **203** S12
Eday, island, *Scot., U.K.* **144** E11
Eddrachillis Bay, *Scot., U.K.* **144** G8
Eddystone Rocks, *Eng., U.K.* **145** X9

Eddyville, Ky., U.S. 103 D13
Ede, Neth. 148 J13
Edéa, Cameroon 206 N9
Edel Land, Austral. 220 H1
Eden, N.C., U.S. 98 G10
Eden, Tex., U.S. 105 J14
Eden, Wyo., U.S. 108 J9
Eden, N.S.W., Austral. 221 Z14
Edenton, N.C., U.S. 98 G14
Edessa see Şanlıurfa, Turk. 171 H13
Édessa, Gr. 156 C9
Edgar, Nebr., U.S. 107 R7
Edgartown, Mass., U.S. 97 M15
Edge Island, Nor. 168 A7
Edgefield, S.C., U.S. 98 L8
Edgeley, N. Dak., U.S. 106 H6
Edgell Island, Nunavut, Can. 87 H20
Edgemont, S. Dak., U.S. 106 M1
Edgerton, Wis., U.S. 100 M5
Edgerton, Minn., U.S. 106 L9
Edgeworth David Camp, Antarctica 239 K22
Edgigen, island, Marshall Is. 227 F16
Edina, Mo., U.S. 107 R13
Edinburg, Tex., U.S. 105 R16
Edinburgh, Ind., U.S. 101 S9
Edinburgh, Scot., U.K. 144 L14
Edirne (Adrianople), Turk. 170 B3
Edison, N.J., U.S. 96 N10
Edison, Ga., U.S. 99 P4
Edisto Island, S.C., U.S. 98 M10
Edisto, river, S.C., U.S. 98 M9
Edithburgh, S. Aust., Austral. 221 Y10
Edjell, island, Marshall Is. 227 F17
Édjérir, river, Mali 205 G16
Edjudina, W. Aust., Austral. 221 W5
Edmond, Okla., U.S. 105 C17
Edmonds, Wash., U.S. 110 B4
Edmonton, Alta., Can. 86 L10
Edmore, N. Dak., U.S. 106 E6
Edmundston, N.B., Can. 87 N21
Edna, Tex., U.S. 105 M18
Edolo, It. 152 C6
Edoras Bank, Atl. Oc. 248 G9
Edrei see Dar'ā, Syr. 172 L7
Edremit Körfezi, Turk. 170 E3
Edremit, Turk. 170 E3
Edrengiyn Nuruu, Mongolia 184 E8
Eduardo Castex, Arg. 134 M10
Eduardo Frei Montalva, station, Antarctica 238 C4
Edward River, Qnsld., Austral. 221 Q12
Edward VII Peninsula, Antarctica 238 N11
Edward VIII Bay, Antarctica 239 D20
Edward, river, Austral. 220 L12
Edwards Plateau, Tex., U.S. 105 K14
Edwards, Calif., U.S. 111 W8
Edwardsville, Ill., U.S. 101 T4
Ee, island, N.Z. 229 H20
Eek, Alas., U.S. 112 K12
Eel, river, Calif., U.S. 111 N2
Éfaté, island, Vanuatu 229 J16
Effigy Mounds National Monument, Iowa, U.S. 106 M13
Effingham, Ill., U.S. 101 S6
Eflâni, Turk. 170 C8
Efuenaarukosu Point, N. Mariana Is., U.S. 230 M9
Efyra, ruin, Gr. 156 F7
Egadi, Isole, It. 153 S8
Egan Range, Nev., U.S. 111 Q12
Egedesminde see Aasiaat, Greenland, Den. 88 M5
Egegik, Alas., U.S. 113 L14
Eger, Hung. 151 K16
Egersund, Nor. 142 L10
Egerton, Mount, Austral. 220 G3
Egg Island, Bahamas 122 D4
Eggan, Nigeria 206 L7
Eggers Ø see Ittileq, Greenland, Den. 89 U8
Eğil, Turk. 171 G14
Egilsstaðir, Ice. 142 E4
Eğirdir Gölü, Turk. 170 G6
Eğirdir, Turk. 170 G6
Egiyn, river, Mongolia 184 C9
Égletons, Fr. 149 U9
Eglinton Fiord, Nunavut, Can. 87 E18
Egmont Terrace, Pac. Oc. 255 G21
Eğrigöz Dağı, peak, Turk. 170 E5
Éguzon-Chantôme, Fr. 149 T9
Egvekinot, Russ. 161 B20
Egypt, Af. 202 H8
Ehingen, Ger. 150 K7
Ehrenberg, Ariz., U.S. 109 T4

Eiao, island, Fr. Polynesia, Fr. 225 H14
Eickelberg Ridge, Pac. Oc. 247 D16
Eidsvold, Qnsld. Austral. 221 U15
Eifel, Eur. 140 G6
Eigg, island, Scot. U.K. 144 J7
Eight Degree Channel, Maldives 183 T3
Eight Mile Rock, Bahamas 122 J1
Eights Coast, Antarctica 238 J7
Eighty Mile Beach, Austral. 220 E4
Eildon, Lake, Austral. 220 M13
Eilean Barraigh (Barra), island, Scot. U.K. 144 J6
Eilean Bhearnaraigh (Berneray), island, Scot. U.K. 144 H6
Eilean Hasgeir (Haskeir), island, Scot. U.K. 144 H6
Eilean Leòdhais (Lewis), island, Scot. U.K. 144 G7
Eilean Scalpaigh (Scalpay), island, Scot. U.K. 144 H6
Einasleigh, Qnsld. Austral. 221 R13
Eindhoven, Neth. 148 K13
Einsiedeln, Switz. 152 B4
Éire see Ireland, Eur. 145 R5
Eirik Ridge, Atl. Oc. 248 G7
Eiriosgaigh (Eriskay), island, Scot. U.K. 144 J6
Eirunepé, Braz. 132 F5
Eisch, river, Lux. 162 J9
Eischen, Lux. 162 J9
Eisenach, Ger. 150 F8
Eisenerz, Aust. 150 L11
Eisenhüttenstad, Ger. 150 E11
Eisenstadt, Aust. 151 L13
Eita, Kiribati 227 R15
Eita, island, Kiribati 227 R15
Eivissa (Ibiza), Sp. 147 H17
Ejea de los Caballeros, Sp. 147 D14
Ejeda, Madagascar 213 H15
Ejin Qi, China 184 F9
Ejmiatsin, Arm. 171 E18
Ejura, Ghana 205 N15
Ekalaka, Mont., U.S. 108 E13
Ekarma, island, Russ. 161 H23
Ekerem, Turkm. 178 M5
Ekeren, Belg. 148 K12
Ekiatapskiy Khrebet, Russ. 161 B20
Ekibastuz, Kaz. 179 D14
Ekoli, D.R.C. 209 E16
Ekonda, Russ. 161 G15
Ekouamou, Congo 208 E11
Ekrafane, Niger 206 G6
Ekukola, D.R.C. 208 E12
Ekwan, river, Ont. Can. 87 M17
El Abiodh Sidi Cheikh, Alg. 200 D11
El Açaba, Mauritania 204 H9
El 'Arîsh (Rhinocolura), Egypt 173 N3
El Astillero, Sp. 146 B11
El 'Auja see Nizzana, Egypt 173 P4
El 'Auja, peak, Jordan 173 R7
El Bahr el Abyad (White Nile), Sudan 203 S12
El Bahr el Azraq (Blue Nile), Sudan 203 S13
El Banco, Col. 130 D6
El Barco de Ávila, Sp. 146 F9
El Bauga, Sudan 203 P13
El Baúl, Venez. 130 D10
El Bayadh, Alg. 200 D11
El Beru Hagia, Somalia 211 Q9
El Bureij, Gaza Strip 173 N4
El Burgo de Osma, Sp. 146 D11
El Cajon, Calif., U.S. 111 Z10
El Calafate (Lago Argentino), Arg. 135 V7
El Callao, Venez. 130 E13
El Calvario, Venez. 130 D10
El Campello, Sp. 147 J14
El Campo, Tex., U.S. 105 M18
El Carmen, Arg. 134 E9
El Ceibo, Guatemala 117 K15
El Ceibo, Mex. 117 H16
El Centro, Calif., U.S. 111 Z11
El Cerro del Aripo, Trin. & Tobago 123 P23
El Cerro, Bol. 132 K9
El Cuyo, Mex. 117 G17
El Dakka, ruin, Egypt 202 L12
El Djouf, desert, Mali, Mauritania 204 E11
El Dorado Springs, Mo., U.S. 107 U11
El Dorado, Ark., U.S. 102 K8
El Dorado, Kans., U.S. 107 U8

El Dorado, Venez. 130 E13
El Eglab, Alg. 200 H8
El Fasher, Sudan 203 S7
El Fifi, Sudan 203 U7
El Fuerte, Mex. 116 E8
El Fula, Sudan 203 T9
El Galpón, Arg. 134 F10
El Geili, Sudan 203 Q12
El Geneina, Sudan 203 S6
El Gezira, region, Af. 198 G9
El Golea, Alg. 200 F12
El Goled Bahri, Sudan 203 P11
El Grado, Sp. 147 D15
El Hadjira, Alg. 201 D13
El Hank, Af. 204 C12
El Harrach, Alg. 200 B12
El Hasaheisa, Sudan 203 R13
El Hawata, Sudan 203 S13
El Hermel, Leb. 172 H7
El Hierro (Ferro), island, Sp. 214 R2
El Hilla, Sudan 203 S9
El Hobra, Alg. 201 E13
El Homeur, Alg. 200 F11
El Jadida (Mazagan), Mor. 200 D7
El Jebelein, Sudan 203 S12
El Jemm, Tunisia 201 C16
El Kamlin, Sudan 203 R12
El Karaba, Sudan 203 P13
El Kawa, Sudan 203 S12
El Kef, Tunisia 201 B15
El Kelaa des Srarhna, Mor. 200 D7
El Khandaq, Sudan 203 P11
El Khnâchîch, Mali 205 E13
El Koin, Sudan 203 N11
El Kseïbat, Alg. 200 G10
El Kûbri, Egypt 173 Q1
El Kuntilla, Egypt 173 Q4
El Lagowa, Sudan 203 T10
El Leiya, Sudan 203 Q14
El Macao, Dom. Rep. 119 M20
El Maitén, Arg. 135 Q7
El Maïz, Alg. 200 G11
El Malpais National Monument, N. Mex., U.S. 109 S10
El Manaqil, Sudan 203 R12
El Mansour, Alg. 200 G10
El Mazâr, Egypt 173 N3
El Medda, Mauritania 204 F8
El Meghaïer, Alg. 201 D14
El Melemm, Sudan 203 U10
El Milagro, Arg. 134 J9
El Mîna, Leb. 172 G6
El Morro National Monument, N. Mex., U.S. 109 S10
El Moueïla, Mauritania 204 E10
El Mreyyé, Mauritania 204 F12
El Nido, Philippines 193 E13
El Niybo, Eth. 211 P8
El Obeid, Sudan 203 S11
El Odaiya, Sudan 203 T9
El Oued, Alg. 201 D14
El Palito, Venez. 130 C10
El Pao, Venez. 130 E12
El Paso, Ill., U.S. 101 Q6
El Paso, Tex., U.S. 104 H7
El Pedroso, Sp. 146 K8
El Portal, Calif., U.S. 111 S6
El Puerto de Santa María, Sp. 146 M8
El Qantara, Egypt 173 P1
El Quneitra see Al Qunayţirah, Syr. 172 K6
El Quseima, Egypt 173 P4
El Reno, Okla., U.S. 105 C16
El Rosario, Mex. 116 C5
El Salto, Mex. 116 F9
El Salvador, N. Amer. 117 M16
El Seibo, Dom. Rep. 119 M20
El Serrat, Andorra 162 G3
El Sibû', ruin, Egypt 202 L12
El Sombrero, Venez. 130 D10
El Suweis (Suez), Egypt 173 Q1
El Teleno, peak, Sp. 146 C8
El Thamad, Egypt 173 R4
El Tigre, Venez. 130 D12
El Tofo, Chile 134 H6
El Toro, peak, Sp. 147 G20
El Ţûr, Egypt 173 T3
El Turbio, Arg. 135 W7
El Wak, Kenya 211 Q9
El Wuz, Sudan 203 R11
El Yunque, peak, Cuba 119 K14
Elafónisos, island, Gr. 156 M11
Elafonísou, Stenón, Gr. 156 M11
Elaia, Cape, Cyprus 162 N10

Elamanchili, India 182 N9
Elan Bank, Ind. Oc. 254 C10
Elása, island, Gr. 157 Q17
Elassóna, Gr. 156 E10
Elat, Israel 173 R5
Elato Atoll, F.S.M. 226 B8
Elazığ, Turk. 171 G14
Elba, Ala., U.S. 103 M16
Elba, island, It. 152 J5
Elbasan, Alban. 154 L10
Elbe, river, Eur. 140 F7
Elbert, Mount, Colo., U.S. 108 M12
Elberta, Mich., U.S. 100 J9
Elberton, Ga., U.S. 98 K7
Elbeuf, Fr. 149 N8
Elbistan, Turk. 170 G12
Elbląg, Pol. 151 B15
Elbow Cays, Bahamas 118 F8
Elbow Lake, Minn., U.S. 106 H9
El'brus, peak, Russ. 159 U12
Elburz Mountains see Reshteh Kūhhā-ye Alborz, Iran 177 D14
Elche (Elx), Sp. 147 K14
Elchi see Wādī Mūsá, Jordan 173 Q6
Elda, Sp. 147 J14
Elderslie, Jam. 122 H7
El'dikan, Russ. 161 G19
Eldon, Iowa, U.S. 107 Q13
Eldon, Mo., U.S. 107 T12
Eldora, Iowa, U.S. 107 N12
Eldorado, Ill., U.S. 101 U6
Eldorado, Okla., U.S. 105 E14
Eldorado, Mex. 116 F8
Eldorado, Arg. 134 G15
Eldoret, Kenya 211 S5
Electra, Tex., U.S. 105 E15
Electric Mills, Miss., U.S. 103 K13
Eleftheroúpoli, Gr. 157 C13
Elektrostal', Russ. 158 J10
Eléni see Makrónisos, island, Gr. 157 K13
Elephant Butte Reservoir, N. Mex., U.S. 109 U11
Elephant Island, Antarctica 238 B4
Eleşkirt, Turk. 171 E17
Eleuthera Island, Bahamas 119 E13
Elfrida, Ariz., U.S. 109 W9
Elgin, Ill., U.S. 101 N6
Elgin, Tex., U.S. 105 K17
Elgin, N. Dak., U.S. 106 H3
Elgin, Nebr., U.S. 107 N7
Elgin, Oreg., U.S. 110 F8
Elgin, Nev., U.S. 111 S12
Elgin, Scot., U.K. 144 H10
Elgol, Scot., U.K. 144 J7
Elida, N. Mex., U.S. 109 T15
Eliki Gounda, Niger 206 G8
Elikónas (Helicon), peak, Gr. 156 H11
Elila, river, D.R.C. 209 F17
Elim, Alas., U.S. 113 F13
Elipa, D.R.C. 209 E16
Elista, Russ. 159 R13
Elizabeth City, N.C., U.S. 98 G14
Elizabeth Island, Marshall Is. 227 J20
Elizabeth, N.J., U.S. 96 N11
Elizabeth, Ill., U.S. 102 N7
Elizabeth, Cape, Me., U.S. 97 H16
Elizabeth, Cape, Asia 168 D12
Elizabethton, Tenn., U.S. 103 E20
Elizabethtown, N.C., U.S. 98 J12
Elizabethtown, Ky., U.S. 103 C16
Elizondo, Sp. 147 B13
Elk City, Okla., U.S. 105 C15
Elk Mountain, Pa., U.S. 96 L9
Elk Mountain, Wyo., U.S. 108 K12
Elk Point, Nebr., U.S. 107 N8
Elk River, Minn., U.S. 106 J11
Elk, Calif., U.S. 111 Q2
Ełk, Pol. 151 B17
Elk, river, W. Va., U.S. 98 D9
Elkader, Iowa, U.S. 106 M13
Elkedra, N.T., Austral. 221 S9
Elkhart, Ind., U.S. 101 N9
Elkhart, Kans., U.S. 107 V3
Elkhorn City, Ky., U.S. 103 C20
Elkhorn, river, Nebr., U.S. 107 N6
Elkhovo, Bulg. 155 J16
Elkin, N.C., U.S. 98 G9
Elkins, W. Va., U.S. 98 D10
Elkins, N. Mex., U.S. 109 T14
Elkland, Pa., U.S. 96 L7
Elko, Nev., U.S. 110 M11
Elkton, Md., U.S. 98 B14
Elkton, Ky., U.S. 103 D15

Ellan Vanninn see Man, Isle of, U.K. 145 P9
Elle, island, Marshall Is. 227 A20
Ellef Ringnes Island, Nunavut, Can. 87 B14
Ellen, Mount, Utah, U.S. 109 N8
Ellendale, N. Dak., U.S. 106 H6
Ellensburg, Wash., U.S. 110 D6
Ellep Passage, Marshall Is. 227 H17
Ellep, island, Marshall Is. 227 H17
Ellesmere Island, Nunavut, Can. 87 B16
Ellice Basin, Pac. Oc. 246 K11
Ellice Ridge, Pac. Oc. 246 K11
Ellijay, Ga., U.S. 98 J5
Ellington, Mo., U.S. 107 U14
Ellinwood, Kans., U.S. 107 T6
Elliot Lake, Ont. Can. 87 P17
Elliott Key, Fla., U.S. 99 X10
Elliott, N.T. Austral. 221 R9
Ellis, Kans., U.S. 107 S5
Ellis, Idaho, U.S. 108 F5
Ellisburg, N.Y., U.S. 96 H8
Elliston, S. Aust., Austral. 221 X9
Ellisville, Miss., U.S. 102 M12
Ellon, Scot., U.K. 144 H11
Ellsworth Land, Antarctica 238 H7
Ellsworth Mountains, Antarctica 238 H9
Ellsworth, Me., U.S. 97 F18
Ellsworth, Kans., U.S. 107 T7
Elm Creek, Tex., U.S. 105 M14
Elm Creek, Nebr., U.S. 107 Q5
Elma Dağı, Turk. 170 E8
Elma, Wash., U.S. 110 D3
Elmadağ, Turk. 170 E8
Elmalı, Turk. 170 H5
Elmira, N.Y., U.S. 96 L7
Elmshorn, Ger. 150 B7
Elne, Fr. 149 Z10
Eloaua, island, P.N.G. 222 M9
Elorza, Venez. 130 E9
Eloúnta, Gr. 157 Q16
Eloy, Ariz., U.S. 109 V7
Elphin, Scot., U.K. 144 G8
Elroy, Wis., U.S. 100 K4
Elsa, Tex., U.S. 105 R16
Elsberry, Mo., U.S. 107 S14
Elsinore, Utah, U.S. 109 N7
Elst, Neth. 148 J13
Eltanin Bay, Antarctica 238 H7
Eltanin Fracture Zone System, Pac. Oc. 247 R15
Eltham, N.Z. 223 H20
El'ton, Russ. 159 P14
El'ton, Russ. 159 P14
Eltopia, Wash., U.S. 110 D7
Eluru, India 182 N8
Elvas, Port. 146 H7
Elven, Fr. 149 R4
Elverum, Nor. 142 J11
Elwell, Lake, Mont., U.S. 108 B8
Elwood, Ind., U.S. 101 Q9
Elwood, Mo., U.S. 107 R10
Elwood, Nebr., U.S. 107 Q5
Elx see Elche, Sp. 147 K14
Ely, Minn., U.S. 106 F12
Ely, Nev., U.S. 111 Q12
Ely, Eng., U.K. 145 S14
Elyria, Ohio, U.S. 101 P14
Émaé, island, Vanuatu 229 J16
Émao, island, Vanuatu 229 J16
Emba, Cyprus 162 Q5
Embalse de Ricobayo, Sp. 146 D8
Embarcación, Arg. 134 E10
Embetsu, Japan 188 E13
Embi, Kaz. 178 F7
Embleton, Eng. U.K. 145 M12
Embrun, Fr. 149 W14
Emden, Ger. 150 C6
Emerald Basin, Pac. Oc. 246 R9
Emerald Island, N.W.T. Can. 87 C13
Emerald, Qnsld. Austral. 221 T14
Emery, Utah, U.S. 109 N8
Emet, Turk. 170 E5
Emi Fezzane, peak, Niger 206 D11
Emi Koussi, peak, Chad 207 E14
Emi Lulu, peak, Niger 206 C10
Emidj, island, Marshall Is. 227 H21
Emigrant Pass, Nev., U.S. 111 N10
Emilius, Mount, It. 152 D2
Emily, Minn., U.S. 106 G11
Emin, China 184 D4
Emir Dağları, peak, Turk. 170 F7
Emirau Island, P.N.G. 222 M9
Emirdağ, Turk. 170 F6
Emlembe, peak, S. Af. 212 J9

Emmastad, *Curaçao, Neth.* **123** P14
Emmaus, *Pa., U.S.* **96** N9
Emmen, *Neth.* **148** H14
Emmetsburg, *Iowa, U.S.* **106** M10
Emmett, *Idaho, U.S.* **108** G3
Emmonak, *Alas., U.S.* **112** G12
Emory Peak, *Tex., U.S.* **104** M10
eMpangeni, *S. Af.* **212** K10
Empedrado, *Arg.* **134** G13
Emperor Seamounts, *Pac. Oc.* **246** D10
Emperor Trough, *Pac. Oc.* **246** D11
Empire, *Mich., U.S.* **100** J9
Empoli, *It.* **152** G6
Emporia, *Va., U.S.* **98** F13
Emporia, *Kans., U.S.* **107** T9
Emporium, *Pa., U.S.* **96** L5
Empress Augusta Bay, *P.N.G.* **222** P11
Empress Augusta Bay, *P.N.G.* **228** C5
Empty Quarter *see* Ar Rub' al Khālī, *Saudi Arabia* **174** K12
Ems, river, *Ger.* **150** D5
'En Boqeq, *Israel* **173** N6
'En Ḥaẓeva, *Israel* **173** P5
'En Yahav, *Israel* **173** P5
Enard Bay, *Scot., U.K.* **144** G8
Encamp, *Andorra* **162** H4
Encamp, *Andorra* **162** J3
Encarnación, *Parag.* **132** N10
Enchi, *Ghana* **205** P14
Encinal, *Tex., U.S.* **105** N15
Encinitas, *Calif., U.S.* **111** Y9
Encino, *Tex., U.S.* **105** Q16
Encino, *N. Mex., U.S.* **109** S13
Encontrados, *Venez.* **130** D7
Encounter Bay, *Austral.* **220** L10
Endau, *Kenya* **211** T7
Ende, *Indonesia* **193** M14
Endeavour Hill, *Austral.* **220** C7
Endeavour Strait, *Austral.* **220** A12
Enderbury Island, *Kiribati* **224** G10
Enderby Land, *Antarctica* **239** C19
Enderby Plain, *Ind. Oc.* **254** E9
Enderlin, *N. Dak., U.S.* **106** G7
Endicott Mountains, *Alas., U.S.* **113** D16
Endicott, *N.Y., U.S.* **96** L8
Endurance Canyon, *Atl. Oc.* **249** Y7
Endurance Ridge, *Atl. Oc.* **249** X7
Eneabba, *W. Aust., Austral.* **221** W2
Eneeldak, island, *Marshall Is.* **227** J20
Enengeseiru, island, *F.S.M.* **226** F7
Enengonomei, island, *F.S.M.* **226** F8
Enewetak (Eniwetok), island, *Marshall Is.* **227** C21
Enewetak Atoll, *Marshall Is.* **227** B14
Enewetak Lagoon, *Marshall Is.* **227** B20
Enez, *Turk.* **170** C2
Enfield, *N.C., U.S.* **98** G13
Enfield, *Jam.* **122** H10
Engaño, Cabo, *Dom. Rep.* **119** M21
Engelhard, *N.C., U.S.* **98** H15
Engelhardt Ice Ridge, *Antarctica* **238** L12
Engels, *Russ.* **158** M14
Engenenima, island, *F.S.M.* **226** F10
Enggano Basin, *Ind. Oc.* **251** H16
Enggano, island, *Indonesia* **192** L6
England, *Ark., U.S.* **102** H9
England, *U.K.* **145** S13
Englewood, *Fla., U.S.* **99** V7
Englewood, *Kans., U.S.* **107** V5
Englewood, *Colo., U.S.* **108** M13
English Channel (La Manche), *Fr., U.K.* **145** X13
English Coast, *Antarctica* **238** G7
English Harbour, *Antigua & Barbuda* **123** B21
English Harbour, *Kiribati* **226** J11
English Roads, bay, *Tonga* **227** R24
Enid, *Okla., U.S.* **105** B16
Enighed, *Virgin Is., U.S.* **122** N10
Enigu, island, *Marshall Is.* **227** D21
Enipéas, river, *Gr.* **156** F10
Eniwa, *Japan* **188** G13
Eniwetak Passage, *Marshall Is.* **227** H18
Eniwetak, island, *Marshall Is.* **227** H18
Eniwetok *see* Enewetak, island, *Marshall Is.* **227** C21
Enjebi, island, *Marshall Is.* **227** A20
Enkhuizen, *Neth.* **148** H12
Ennadai, *Nunavut, Can.* **87** J13
Ennadai, *Russ.* **158** A15
Ennedi, plateau, *Chad* **207** F15

Ennell, Lough, *Ire.* **145** R5
Enngonia, *N.S.W., Austral.* **221** V13
Enning, *S. Dak., U.S.* **106** K2
Ennis (Inis), *Ire.* **145** S4
Ennis, *Tex., U.S.* **105** H18
Ennis, *Mont., U.S.* **108** E7
Enniscorthy (Inis Córthaidh), *Ire.* **145** S6
Enniskillen (Inis Ceithlann), *N. Ire., U.K.* **145** P5
Ennistimon (Inis Díomáin), *Ire.* **145** R3
Ennuebing, island, *Marshall Is.* **227** F16
Ennugarret, island, *Marshall Is.* **227** F16
Ennugenliggelap, island, *Marshall Is.* **227** H17
Ennumennet, island, *Marshall Is.* **227** F16
Ennumet, island, *Marshall Is.* **227** H15
Ennylabegan, island, *Marshall Is.* **227** J17
Enontekiö, *Fin.* **143** C14
Enriquillo, *Dom. Rep.* **119** N18
Enriquillo, Lago, *Dom. Rep.* **119** M17
Enschede, *Neth.* **148** J14
Ensenada, *Mex.* **116** B5
Enshi, *China* **184** L12
Ensisheim, *Fr.* **149** Q15
Entebbe, *Uganda* **211** S3
Enterprise, *Ala., U.S.* **103** N16
Enterprise, *Miss., U.S.* **103** L13
Enterprise, *Utah, U.S.* **109** P5
Enterprise, *Wash., U.S.* **110** F9
Entinas, Punta, *Sp.* **146** M12
Entrada, Punta, *Arg.* **135** V8
Entrance Point, *Bahamas* **122** B11
Entre Ríos, region, *S. Amer.* **128** L6
Entre-Deux, *Réunion, Fr.* **215** H16
Entrepeñas, Embalse de, *Sp.* **146** F12
Entrèves, *It.* **152** D2
Entroncamento, *Port.* **146** G5
Enubuj, island, *Marshall Is.* **227** J17
Enugu, *Nigeria* **206** M8
Enumclaw, *Wash., U.S.* **110** C4
Enurmino, *Russ.* **161** A20
Envalira, Pic d', *Andorra, Fr.* **162** J5
Envalira, Port d', *Andorra* **162** J5
Envira, *Braz.* **132** F5
Enybor, island, *Marshall Is.* **227** H21
Enyellé, *Congo* **208** C12
Eochaill *see* Youghal, *Ire.* **145** T5
Eocheongdo, island, *S. Korea* **187** Q6
Eonyang, *S. Korea* **187** S12
Eori, *Eq. Guinea* **214** N7
Eot, island, *F.S.M.* **226** E9
Epanomí, *Gr.* **156** D11
Eparit, island, *F.S.M.* **226** D9
Epéna, *Congo* **208** D12
Epenarra, *N.T., Austral.* **221** S9
Épernay, *Fr.* **149** P11
Épernon, *Fr.* **149** P9
Ephesus, ruins, *Turk.* **170** G3
Ephraim, *Utah, U.S.* **108** M7
Ephrata, *Wash., U.S.* **110** C7
Epi, island, *Vanuatu* **229** H15
Epidavros, ruin, *Gr.* **156** K11
Épinal, *Fr.* **149** Q14
Epini, *D.R.C.* **209** D18
Epirus, region, *Alban.-Gr.* **156** F7
Episkopi Bay, *Cyprus* **162** Q7
Episkopi, *Cyprus* **162** Q7
Epsilon, *Qnsld., Austral.* **221** V11
Epsom, *Eng., U.K.* **145** U14
Eptachóri, *Gr.* **156** D8
Epu Pel, *Arg.* **135** N10
Épuisay, *Fr.* **149** R8
Epulu, river, *D.R.C.* **209** D18
Epupa Falls, *Af.* **198** M5
Eqalugaarsuit, *Greenland, Den.* **89** T7
Eqlīd, *Iran* **177** J16
Equatorial Channel, *Maldives* **183** Y3
Equatorial Guinea, *Af.* **206** P8
Er Rachidia, *Mor.* **200** C9
Er Rif, range, *Af.* **198** C3
Er Roseires, *Sudan* **203** T13
Erbaa, *Turk.* **170** D11
Erçek Gölü, *Turk.* **171** F18
Erçek, *Turk.* **171** G18
Erciş, *Turk.* **171** F17
Erciyeş Daği, peak, *Turk.* **170** G10
Érd, *Hung.* **151** L15
Erdek, *Turk.* **170** D4
Erdemli, *Turk.* **170** J9

Erding, *Ger.* **150** K9
Ere, island, *Marshall Is.* **227** G15
Erebus and Terror Gulf, *Antarctica* **238** C5
Erebus, Mount, *Antarctica* **239** N14
Erech, ruin, *Iraq* **176** J11
Erechim, *Braz.* **132** N12
Ereentsav, *Mongolia* **185** C13
Ereğli, *Turk.* **170** C7
Ereğli, *Turk.* **170** H9
Erei, Monti, *It.* **153** T11
Ereikoússa, island, *Gr.* **156** E5
Erenhot, *China* **184** G12
Erenköy *see* Kokkina, *Cyprus* **162** N5
Eresós, *Gr.* **157** F16
Erétria, *Gr.* **156** H12
Ereymentaū, *Kaz.* **179** D13
Erfurt, *Ger.* **150** F8
Erg Chech, *Alg.* **200** J9
'Erg du Ténéré, *Niger* **206** F10
'Erg el Ahmar, *Alg.-Mali* **205** C13
Erg er Raoui, *Alg.* **200** F9
Erg I-n-Afarag, *Alg.* **200** J12
'Erg I-n-Sâkâne, *Mali* **205** E15
Erg Iguidi, *Alg.* **200** H7
Erg Iguidi, *Alg.-Mauritania* **204** B12
Ergani, *Turk.* **171** G14
Ergedzhey, *Russ.* **161** H16
Ergene, river, *Turk.* **170** C4
Ergun Youqi, *China* **185** B14
Ergun, river, *China* **185** B14
Eriba, *Sudan* **203** Q14
Eriboll, *Scot., U.K.* **144** F9
Eriboll, Loch, *Scot., U.K.* **144** F9
Ericht, Loch, *Scot., U.K.* **144** K9
Erick, *Okla., U.S.* **105** D14
Erie Canal, *N.Y., U.S.* **96** J9
Erie, *Pa., U.S.* **96** K3
Erie, Lake, *N. Amer.* **84** J8
Erik S. Henius Land, *Greenland, Den.* **88** B12
Erikub Atoll, *Marshall Is.* **227** C17
Erimi, *Cyprus* **162** Q7
Erimo Misaki, *Japan* **188** H14
Erimo, *Japan* **188** H14
Erin Bay, *Trin. & Tobago* **123** R21
Erin Point, *Trin. & Tobago* **123** R21
Erin, *Tenn., U.S.* **103** E14
Eriskay *see* Eiriosgaigh, island, *Scot., U.K.* **144** J6
Erisort, Loch, *Scot., U.K.* **144** G7
Eritrea, *Af.* **210** M7
Erkilet, *Turk.* **170** F10
Erkowit, *Sudan* **203** N15
Erlandsen Land, *Greenland, Den.* **88** B10
Erlangen, *Ger.* **150** H8
Erlanger, *Ky., U.S.* **103** A17
Erldunda, *N.T., Austral.* **221** U8
Erling, Lake, *Ark., U.S.* **102** K7
Ermelo, *S. Af.* **212** J9
Ermenek, *Turk.* **170** J8
Ermil, *Sudan* **203** S9
Ermióni, *Gr.* **156** K11
Ermoúpoli, *Gr.* **157** K14
Ernabella, *S. Aust., Austral.* **221** U8
Ernakulam, *India* **183** S5
Ernée, *Fr.* **149** Q6
Erode, *India* **183** R6
Eroj, island, *Marshall Is.* **227** D20
Eromanga, *Qnsld., Austral.* **221** U12
Eromango, island, *Vanuatu* **229** L17
Erris Head, *Ire.* **145** P3
Erromango, island, *Vanuatu* **229** L17
Erskine Iceport, bay, *Antarctica* **239** A15
Erskine, *Minn., U.S.* **106** F9
Ertai, *China* **184** D6
Ertis (Irtysh), river, *Kaz.* **179** D15
Ertis, *Kaz.* **179** C14
Ertix, river, *China* **184** C5
Erts, *Andorra* **162** H2
Eru, island, *Marshall Is.* **227** H15
Erudina, *S. Aust., Austral.* **221** W11
Eruh, *Turk.* **171** H16
Erwin, *N.C., U.S.* **98** H11
Erwin, *Tenn., U.S.* **103** E20
Erýmanthos, Óros, *Gr.* **156** J9
Erythrés, *Gr.* **156** H11
Erzin, *Russ.* **161** M13
Erzincan, *Turk.* **171** E14
Erzurum, *Turk.* **171** E16
Es Safiya, *Sudan* **203** R10
Es Salam, *Sudan* **203** P13
Es Suki, *Sudan* **203** S13
Esa'Ala, *P.N.G.* **221** N16

Esan Misaki, *Japan* **188** H13
Esashi, *Japan* **188** E13
Esashi, *Japan* **188** H12
Esbjerg, *Den.* **142** N11
Escalades-Engordany, *Andorra* **162** J4
Escalante, *Utah, U.S.* **109** P7
Escalante, river, *Utah, U.S.* **109** P7
Escambia, river, *Ala.-Fla., U.S.* **99** Q1
Escanaba, *Mich., U.S.* **100** G8
Escanaba, river, *Mich., U.S.* **100** F7
Escárcego, *Mex.* **117** J16
Esch, *Lux.* **162** K9
Eschen, *Liech.* **162** M2
Eschnerberg, range, *Liech.* **162** M2
Eschwege, *Ger.* **150** F7
Escocesa, Bahia, *Dom. Rep.* **119** L19
Escondido, *Calif., U.S.* **111** Y10
Escuinapa, *Mex.* **116** G9
Escuintla, *Guatemala* **117** L15
Escuintla, *Mex.* **117** L15
Esenguly, *Turkm.* **178** N5
Eṣfahān (Isfahan), *Iran* **177** H16
Esfandak, *Iran* **177** M24
Esfarāyen, *Iran* **177** D19
Esil, *Kaz.* **178** D11
Esil, river, *Kaz.* **178** C11
Eskifjörður, *Ice.* **142** F4
Eskilstuna, *Sw.* **143** K13
Eskimo Lakes, *N.W.T., Can.* **86** E10
Eskipazar, *Turk.* **170** D8
Eskişehir, *Turk.* **170** E6
Esla, river, *Sp.* **146** C9
Eslāmābād, *Iraq* **176** G12
Eslāmshahr, *Iran* **177** F15
Eşler Daği, peak, *Turk.* **170** G5
Eşme, *Turk.* **170** F4
Esmeralda, *Cuba* **118** H10
Esmeralda, Isla, *Chile* **135** U6
Esmeraldas, *Ecua.* **130** J3
Esmoriz, *Port.* **146** E5
Espagnol Point, *St. Vincent & the Grenadines* **123** J17
Espakeh, *Iran* **177** N22
Espalion, *Fr.* **149** W10
Espanola, *Mex.* **109** R12
Espanola, Isla, *Ecua.* **130** R5
Esperance Bay, *Austral.* **220** K5
Esperance, *W. Aust., Austral.* **221** X5
Esperanza, *Peru* **132** G5
Esperanza, *Arg.* **134** J11
Esperanza, station, *Antarctica* **238** C5
Espichel, Cabo, *Port.* **146** J4
Espiègle Bay, *Vanuatu* **229** G14
Espiel, *Sp.* **146** K9
Espigão Mestre, range, *S. Amer.* **128** F9
Espinhaço, Serra do, *Braz.* **133** K15
Espíritu Santo, Isla, *Mex.* **116** F7
Espiritu Santo, island, *Vanuatu* **229** F14
Espoo (Esbo), *Fin.* **143** J16
Esposende, *Port.* **146** D5
Espuña, peak, *Sp.* **147** K13
Espungabera, *Mozambique* **212** F10
Esquel, *Arg.* **135** R7
Esquina, *Arg.* **134** J12
Essaouira, *Mor.* **200** E6
Essen, *Belg.* **148** K12
Essen, *Ger.* **150** E5
Essendon, Mount, *Austral.* **220** G4
Essequibo, river, *Guyana* **131** H15
Essex Junction, *Vt., U.S.* **96** F12
Essexville, *Mich., U.S.* **100** K12
Essington, Port, *Austral.* **220** A8
Esslingen, *Ger.* **150** J7
Esso, *Russ.* **161** F22
Essone, *Gabon* **208** D8
Essoûk, *Mali* **205** F16
Estaca de Bares, Punta da, *Sp.* **146** A6
Estados, Isla de los (Staten Island), *Arg.* **135** Y10
Estahbān, *Iran* **177** L17
Estancia, *N. Mex., U.S.* **109** S12
Estância, *Braz.* **133** H17
Estany d'Engolasters, lake, *Andorra* **162** J3
Estanyó, Pic de l', *Andorra* **162** H4
Estats, Pic d', *Fr.-Sp.* **162** F1
Estavayer-le-Lac, *Switz.* **152** B2
Este, *S. Amer.* **128** L7
Este, *It.* **152** E7
Estelí, *Nicar.* **117** L18
Estella (Lizarra), *Sp.* **147** C13
Estelline, *Tex., U.S.* **105** E13
Estelline, *S. Dak., U.S.* **106** K8

Estepona, *Sp.* **146** M9
Estes Park, *Colo., U.S.* **108** L13
Estevan, *Sask., Can.* **86** N12
Estherville, *Iowa, U.S.* **106** M10
Estill, *S.C., U.S.* **98** M9
Estonia, *Eur.* **143** K16
Estrela, Serra da, *Port.* **146** F6
Estrella Bay, *Solomon Is.* **228** D9
Estrella, peak, *Sp.* **146** J11
Estremoz, *Port.* **146** H6
Eswatini (Swaziland), *Af.* **212** J9
Esztergom, *Hung.* **151** L15
et Tahta, *Leb.* **172** J6
Etadunna, *S. Aust., Austral.* **221** V10
Etal Atoll, *F.S.M.* **226** B10
Étampes, *Fr.* **149** Q9
Étang des Salines, lake, *Martinique, Fr.* **123** G23
Étang-Sale, *Réunion, Fr.* **215** H15
Étaples, *Fr.* **148** L9
Étarik, island, *Vanuatu* **229** J16
Etawah, *India* **182** H7
Etcharai, island, *Marshall Is.* **227** F14
Etéké, *Gabon* **208** F8
Etena, *Swains I., U.S.* **230** B5
Ethiopia, *Af.* **210** H7
Ethiopian College, *Vatican City* **163** Q14
Ethiopian Highlands, *Af.* **198** H9
Etive, Loch, *Scot., U.K.* **144** K8
Etna, *Wyo., U.S.* **108** H8
Etna, peak, *It.* **153** S11
Etoka, *D.R.C.* **209** E15
Etorofu *see* Iturup, island, *Russ.* **161** J23
Etosha Pan, *Namibia* **212** F3
Etoumbi, *Congo* **208** E10
Etowah, *Tenn., U.S.* **103** F18
Ettelbruck, *Lux.* **162** H10
Etten, island, *F.S.M.* **226** E10;F9
Ettlingen, *Ger.* **150** J6
Ettrick, *N.Z.* **223** N17
Eu, *Fr.* **148** M9
'Euaiki, island, *Tonga* **227** H23
'Euaiki, island, *Tonga* **227** Q24
'Euakafa, island, *Tonga* **227** H23
Euboea *see* Évvoia, island, *Gr.* **157** H13
Eucla Basin, *Austral.* **220** J6
Eucla Canyon, *Ind. Oc.* **251** N20
Eucla Shelf, *Ind. Oc.* **251** M20
Eucla, *W. Aust., Austral.* **221** W7
Euclid, *Ohio, U.S.* **101** N14
Eucumbene, Lake, *Austral.* **220** L14
Eudora, *Ark., U.S.* **102** K9
Eudunda, *S. Aust., Austral.* **221** X10
Eufaula Lake, *Okla., U.S.* **105** D19
Eufaula, *Ala., U.S.* **103** M17
Eufaula, *Okla., U.S.* **105** C19
Eugene, *Oreg., U.S.* **110** H3
Eugenia, Punta, *Mex.* **116** D5
Eugenio Bustos, *Arg.* **134** L7
Eulo, *Qnsld., Austral.* **221** V13
Eumseong, *S. Korea* **187** P9
Eunice, *La., U.S.* **102** P8
Eunice, *N. Mex., U.S.* **109** V15
Eupen, *Belg.* **148** L13
Euphrates *see* Al Furāt, *Iraq* **176** J11
Euphrates, river, *Asia* **168** F3
Eupora, *Miss., U.S.* **102** J12
Eurardy, *W. Aust., Austral.* **221** V2
Eurasia Basin, *Arctic Oc.* **252** F11
Eureka Springs, *Ark., U.S.* **102** E7
Eureka, *Nunavut, Can.* **87** B15
Eureka, *Ill., U.S.* **101** Q5
Eureka, *S. Dak., U.S.* **106** J5
Eureka, *Kans., U.S.* **107** U9
Eureka, *Mont., U.S.* **108** A5
Eureka, *Utah, U.S.* **108** L7
Eureka, *Calif., U.S.* **110** M1
Eureka, *Nev., U.S.* **111** P11
Europa Point, *Gibraltar, U.K.* **163** R23
Europa, Picos de, *Sp.* **146** B10
Europa, Punta, *Eq. Guinea* **214** L7
Eustasia Island, *Virgin Is., U.K.* **122** Q9
Eustis, *Fla., U.S.* **99** T8
Eutaw, *Ala., U.S.* **103** K14
Eutin, *Ger.* **150** B8
Eva Downs, *N.T., Austral.* **221** R9
Evale, *Angola* **208** Q11
Evans Ice Stream, *Antarctica* **238** G8
Evans Peninsula, *Antarctica* **238** J6
Evans, Mount, *Mont., U.S.* **108** E6
Evanston, *Ill., U.S.* **101** N7
Evanston, *Wyo., U.S.* **108** K8
Evansville, *Ind., U.S.* **101** U7

Evansville, *Alas., U.S.* **113** E16
Evart, *Mich., U.S.* **100** K10
Evaton, *S. Af.* **212** J8
Évdilos, *Gr.* **157** K16
Eveleth, *Minn., U.S.* **106** F12
Evensk, *Russ.* **161** E21
Everard Park, *S. Aust., Austral.* **221** U8
Everard, Lake, *Austral.* **220** J9
Everest, Mount (Sagarmāthā, Qomolangma), *China, Nep.* **184** L4
Everest, Mount, *Nepal* **182** G11
Everett, *Wash., U.S.* **110** B4
Everglades City, *Fla., U.S.* **99** X9
Everglades National Park, *Fla., U.S.* **99** X9
Everglades, The wetland, *Fla., U.S.* **99** X9
Everglades, The, *U.S.* **91** N20
Evergreen, *Ala., U.S.* **103** M15
Evesham, *Eng., U.K.* **145** T12
Évian, *Fr.* **149** T14
Évora, *Port.* **146** J6
Évreux, *Fr.* **149** P8
Évros, *river, Gr.* **157** A17
Évros, *river, Gr.* **157** C16
Evrótas, *river, Gr.* **156** L10
Évry, *Fr.* **149** P10
Evrychou, *Cyprus* **162** P7
Évvoia (Euboea), *island, Gr.* **157** H13
Ewa, *Nauru* **227** L14
Ewab *see* Kai, Kepulauan, *Indonesia* **193** L18
Ewarton, *Jam.* **122** H9
Ewaso Ng'iro, *river, Kenya* **211** S7
Ewing, *Nebr., U.S.* **107** N6
Ewo, *Congo* **208** E10
Excelsior Mountains, *Nev., U.S.* **111** R8
Excelsior Springs, *Mo., U.S.* **107** S11
Executive Committee Range, *Antarctica* **238** L9
Exeter Sound, *Nunavut, Can.* **87** F19
Exeter, *N.H., U.S.* **97** J15
Exeter, *Calif., U.S.* **111** U7
Exeter, *Eng., U.K.* **145** W10
Exmoor, *plateau, Eng., U.K.* **145** V10
Exmouth Gulf, *Austral.* **220** T1
Exmouth Plateau, *Ind. Oc.* **251** K17
Exmouth, *Eng., U.K.* **145** W10
Exmouth, *W. Aust., Austral.* **221** T1
Exochí, *Gr.* **156** B12
Exotic Garden, *Monaco* **163** E20
Explora Escarpment, *Atl. Oc.* **249** Y10
Exploring Isles, *Fiji* **226** P8
Extremadura, *Eur.* **140** K2
Exu, *Braz.* **133** F16
Exuma Sound, *Bahamas* **118** E12
Eyasi, Lake, *Af.* **198** N14
Eye Peninsula, *Scot., U.K.* **144** G7
Eye, *Eng., U.K.* **145** S15
Eyemouth, *Scot., U.K.* **144** L11
Eyjafjallajökull, *peak, Ice.* **142** F2
Eyjafjarðardjúp, *valley, Arctic Oc.* **253** L20
Eyjafjörður, *bay, Arctic Oc.* **140** A4
Eyl, *Somalia* **210** M14
Eymoutiers, *Fr.* **149** U9
Eyrarbakki, *Ice.* **142** F2
Eyre Canyon, *Ind. Oc.* **251** N19
Eyre North, Lake, *Austral.* **220** H10
Eyre Peninsula, *Austral.* **220** K9
Eyre South, Lake, *Austral.* **220** H10
Eyre Terrace, *Ind. Oc.* **251** N19
Eyre, Lake, *Austral.* **220** H10
Eyrecourt (Dún an Uchta), *Ire.* **145** R5
Eysturoy, *island, Faroe Is., Den.* **142** H6
Ezcaray, *Sp.* **146** C11
Ezequiela Ramos Mexia, Embalse, *Arg.* **135** P8
Ezhou, *China* **185** L13
Ezine, *Turk.* **170** E2
Ezion-geber (Elath), *ruins, Jordan* **173** R5

Fabala, *Guinea* **204** M11
Fabens, *Tex., U.S.* **104** H7
Faber Lake, *N.W.T., Can.* **86** G11
Fabero, *Sp.* **146** C8
Fabriano, *It.* **152** H9
Facatativá, *Col.* **130** G6
Fachi, *Niger* **206** F10
Facho, Pico do, *Madeira, Port.* **214** L5
Facpi Point, *Guam, U.S.* **230** Q4
Facture, *Fr.* **149** W6
Fada Ngourma, *Burkina Faso* **205** K16
Fada, *Chad* **207** F15
Faddeyevskiy, *Russ.* **161** D17
Fadghamī, *Syr.* **172** E14
Fadian Point, *Guam, U.S.* **230** P6
Fadiffolu Atoll, *Maldives* **183** V3
Faenza, *It.* **152** F7
Færingehavn *see* Kangerluarsoruseq, *Greenland, Den.* **89** R5
Faetano, *San Marino* **163** K17
Fafa, *Mali* **205** H16
Fafa, *island, Tonga* **227** Q23
Fafe, *Port.* **146** D6
Fafen, *river, Eth.* **210** M10
Fagaitua Bay, *American Samoa, U.S.* **230** D7
Faga'itua, *American Samoa, U.S.* **230** D7
Fagaloa Bay, *Samoa* **227** P20
Fagamalo, *Samoa* **227** N18
Fagamalo, *American Samoa, U.S.* **230** D5
Făgăraş, *Rom.* **155** D14
Fagasā, *American Samoa, U.S.* **230** D6
Fagatele Bay, *American Samoa, U.S.* **230** E6
Fagatogo, *American Samoa, U.S.* **230** D7
Fagernes, *Nor.* **142** J11
Faggo, *Nigeria* **206** J9
Fagnano, Lago, *Arg.* **135** X9
Fagurhólsmýri, *Ice.* **142** F3
Fahala, *island, Maldives* **183** X3
Fahefa, *Tonga* **227** Q22
Fahie Hill, *Virgin Is., U.K.* **122** Q6
Fahraj, *Iran* **177** L21
Faial, *island, Port.* **146** P3
Faibus Point, *N. Mariana Is., U.S.* **230** H10
Faichuk, *island, F.S.M.* **226** E8
Faie, *Fr. Polynesia, Fr.* **231** K14
Faihava Passage, *Tonga* **227** H22
Faioa, *island, Tonga* **227** H23
Fair Haven, *Vt., U.S.* **96** H12
Fair Isle, *Scot., U.K.* **144** D12
Fairbanks, *Alas., U.S.* **113** G17
Fairborn, *Ohio, U.S.* **101** R12
Fairburn, *Ga., U.S.* **98** L5
Fairbury, *Ill., U.S.* **101** Q6
Fairbury, *Nebr., U.S.* **107** R8
Fairfax, *S.C., U.S.* **98** M9
Fairfax, *Okla., U.S.* **105** B18
Fairfield, *Ill., U.S.* **101** T6
Fairfield, *Ohio, U.S.* **101** S11
Fairfield, *Ala., U.S.* **103** J15
Fairfield, *Tex., U.S.* **105** H18
Fairfield, *Iowa, U.S.* **107** Q13
Fairfield, *Idaho, U.S.* **108** G5
Fairfield, *Mont., U.S.* **108** C7
Fairfield, *Calif., U.S.* **111** R4
Fairhope, *Ala., U.S.* **103** P14
Fairmont, *W. Va., U.S.* **98** C9
Fairmont, *Minn., U.S.* **106** K12
Fairport, *N.Y., U.S.* **96** J7
Fairport, *Mich., U.S.* **100** G8
Fairview, *Tenn., U.S.* **103** F15
Fairview, *Okla., U.S.* **105** B15
Fairview, *Mont., U.S.* **108** B13
Fairview, *Utah, U.S.* **108** M7
Fairway Plateau, *Pac. Oc.* **255** E23
Fairweather, Mount, *Alas., U.S.* **113** K21
Fais, *island, F.S.M.* **224** E3
Faisalabad, *Pak.* **180** L12
Faistós (Phaestus), *ruin, Gr.* **157** Q14
Faith, *S. Dak., U.S.* **106** J3
Faizabad, *India* **182** H8
Faīzābād, *Afghan.* **180** H10
Fakahiku, *island, Tonga* **227** M23
Fakahina, *island, Fr. Polynesia, Fr.* **231** B21
Fakakakai, *Tonga* **227** L24
Fakaofo, *island, Kiribati* **224** H10
Fakarava, *island, Fr. Polynesia, Fr.* **231** B18
Fakel, *Russ.* **158** G16
Fakenham, *Eng., U.K.* **145** S15
Fakfak, *Indonesia* **193** K18

Fako, *peak, Cameroon* **206** N9
Fala Ane Point, *Swains I., U.S.* **230** A5
Falagh, *S. Sudan* **203** V11
Falakró, Óros, *Gr.* **157** B13
Falamae, *Solomon Is.* **228** D5
Falciano, *San Marino* **163** G17
Falcon Reservoir, *Mex. U.S.* **105** Q15
Falcon Reservoir, *Mex. U.S.* **116** E12
Falcon, Cape, *Oreg., U.S.* **110** E2
Falcón, Presa *see* Falcon Reservoir, *Mex. U.S.* **116** E12
Falconara Marittima, *It.* **152** H9
Falcone, Capo del, *It.* **152** M3
Falconer, *N.Y., U.S.* **96** K4
Faléa, *Mali* **204** K9
Falenberg, *Sw.* **142** M11
Falensee, *Ger.* **150** D10
Falefatu Island, *Tuvalu* **228** P10
Falelatie, *Samoa* **227** P19
Falelima, *Samoa* **227** N17
Falémé, *river, Mali, Senegal* **204** J9
Faleniu, *American Samoa, U.S.* **230** E6
Falfurrias, *Tex., U.S.* **105** Q16
Falkenberg, *Sw.* **142** M11
Falkirk, *Scot., U.K.* **144** L10
Falkland Escarpment, *Atl. Oc.* **249** V6
Falkland Islands, *S. Amer.* **135** X11
Falkland Plateau, *Atl. Oc.* **249** V6
Falkland Ridge, *Atl. Oc.* **249** V8
Falkland Sound, *Falk. Is. U.K.* **135** W12
Falkland Trough, *Atl. Oc.* **249** V6
Falkland-Agulhas Fracture Zone, *Atl. Oc.* **249** V10
Falknis, *peak, Liech. Switz.* **162** R3
Falknov *see* Sokolov, *Czechia* **150** G9
Falkonéra, *island, Gr.* **156** L12
Falköping, *Sw.* **142** L12
Fall River Lake, *Kans., U.S.* **107** U9
Fall River Mills, *Calif., U.S.* **110** M4
Fall River, *Mass., U.S.* **97** L15
Fallbrook, *Calif., U.S.* **111** Y9
Fallen Jerusalem, *island, Virgin Is., U.K.* **122** R8
Fallon, *Nev., U.S.* **111** P7
Falls City, *Nebr., U.S.* **107** R9
Fallujah *see* Al Fallūjah, *Iraq* **176** G10
Falmey, *Niger* **206** H5
Falmouth Harbour, *Antigua & Barbuda* **123** B21
Falmouth, *Me., U.S.* **97** H16
Falmouth, *Ky., U.S.* **103** A18
Falmouth, *Antigua & Barbuda* **121** C16
Falmouth, *Jam.* **122** G7
Falmouth, *Antigua & Barbuda* **123** B21
Falmouth, *Eng., U.K.* **145** X8
False Bay, *S. Af.* **212** M5
False Cape, *N. Amer.* **84** M3
False Divi Point, *India* **183** P8
False Pass, *Alas., U.S.* **112** N11
False Point, *India* **182** L11
Falset, *Sp.* **147** E16
Falso, Cabo, *Mex.* **116** F7
Falso, Cabo, *Dom. Rep.* **119** N17
Falster, *island, Den.* **142** N12
Fălticeni, *Rom.* **155** B15
Falun, *Sw.* **143** J13
Famagusta (Ammóchostos, Gazimağusa), *Cyprus* **162** P10
Famaka, *Sudan* **203** T14
Fana, *Mali* **204** K12
Fanaik, *island, F.S.M.* **226** G10
Fanan, *island, F.S.M.* **226** F10
Fanangat, *island, F.S.M.* **226** E9
Fanannon, *island, F.S.M.* **226** F10
Fanapanges, *island, F.S.M.* **226** F8
Fanasich, *island, F.S.M.* **226** F8
Fancy, *St. Vincent & the Grenadines* **123** J14
Fanemoch, *island, F.S.M.* **226** E9
Faneno, *island, F.S.M.* **226** E9
Fanew, *island, F.S.M.* **226** F9
Fangak, *S. Sudan* **203** V11
Fangale'ounga, *Tonga* **227** L24
Fangasito, *island, Tonga* **227** J23
Fangatau, *island, Fr. Polynesia, Fr.* **231** B20
Fangataufa, *island, Fr. Polynesia, Fr.* **231** E21
Fanning Island *see* Tabuaeran, *Kiribati* **224** F12
Fano, *It.* **152** G9
Fanos, *island, F.S.M.* **226** E9
Fanūdah, *Saudi Arabia* **174** F7
Fanūj, *Iran* **177** N21
Fanurmot, *island, F.S.M.* **226** F8

Farabana, *Guinea* **204** M11
Faraday Fracture Zone, *Atl. Oc.* **248** H9
Faradje, *D.R.C.* **209** B19
Farafangana, *Madagascar* **213** H16
Farafenni, *Gambia* **204** J7
Farāh, *Afghan.* **180** K3
Farāh, *river, Afghan.* **180** K3
Farallon de Medinilla, *N. Mariana Is., U.S.* **116** E12
Farallon de Pajaros, *N. Mariana Is., U.S.* **230** G5
Farallon Islands, *U.S.* **90** F2
Faranah, *Guinea* **204** M10
Faraulep Atoll, *F.S.M.* **226** A7
Fare, *Fr. Polynesia, Fr.* **231** J14
Fareham, *Eng., U.K.* **145** V13
Farewell, Cape, *N.Z.* **224** N8
Farga de Moles, *Andorra* **162** K2
Fargo, *N. Dak., U.S.* **106** G8
Farg'ona, *Uzb.* **179** L13
Fari, *Mali* **204** K10
Faribault, *Minn., U.S.* **106** K11
Faridkot, *India* **182** E4
Farim, *Guinea-Bissau* **204** K7
Farīmān, *Iran* **177** E21
Fariones, Punta, *Sp.* **214** P7
Farkhor, *Taj.* **178** H13
Farkwa, *Tanzania* **211** V5
Farmakonísi (Pharmacusa, Farmakoúsa), *island, Gr.* **157** K18
Farmersville, *Tex., U.S.* **105** F18
Farmerville, *La., U.S.* **102** K8
Farmington, *Me., U.S.* **97** F16
Farmington, *Ill., U.S.* **101** Q4
Farmington, *Mo., U.S.* **107** U15
Farmington, *Utah, U.S.* **108** K7
Farmington, *N. Mex., U.S.* **109** Q10
Farmville, *Va., U.S.* **98** F12
Farnham, *N.Y., U.S.* **96** K4
Faro, *Port.* **146** L6
Fårö, *island, Sw.* **143** L14
Faro, Punta del, *It.* **153** R12
Faro, *river, Cameroon* **206** L11
Faroe Bank, *Atl. Oc.* **248** F11
Faroe Gap, *Atl. Oc.* **248** F11
Faroe Islands (Føroyar), *Eur.* **142** H6
Faroe Shelf, *Arctic Oc. Atl. Oc.* **248** F11
Faroe-Shetland Trough, *Arctic Oc. Atl. Oc.* **248** F12
Farquhar Ridge, *Ind. Oc.* **250** J7
Farr Bay, *Antarctica* **239** J22
Farrāshband, *Iran* **177** L16
Farrell, *Pa., U.S.* **96** M3
Fārs, *Iran* **177** L17
Fársala, *Gr.* **156** F11
Fārsī, *island, Iran* **177** M14
Farsund, *Nor.* **142** L10
Fārūj, *Iran* **177** D20
Far'ūn (Coral Island), *Israel* **173** R5
Farwell Island, *Antarctica* **238** J7
Farwell, *Tex., U.S.* **104** E11
Fasā, *Iran* **177** L17
Fasano, *It.* **152** M14
Fastiv, *Ukr.* **159** N5
Fastnet Rock, *Ire.* **145** U3
Fataka (Mitre Island), *Solomon Is.* **224** J8
Fatato, *island, Tuvalu* **228** N11
Fatehabad, *India* **182** F5
Fatehgarh, *India* **182** G7
Fatehpur, *India* **182** H7
Fathai, *S. Sudan* **203** V12
Fátima, *Port.* **146** G5
Fatsa, *Turk.* **170** D12
Fatu Hiva, *island, Fr. Polynesia, Fr.* **225** H15
Fatumanongi, *island, Tonga* **227** L23
Fatumu, *Tonga* **227** Q23
Fatunda, *D.R.C.* **208** G12
Faucilles, Monts, *Fr.* **149** Q13
Faulkton, *S. Dak., U.S.* **106** J6
Fauro, *island, Solomon Is.* **228** C6
Fauske, *Nor.* **142** E12
Faversham, *Eng., U.K.* **145** U15
Favignana, Isola, *It.* **153** S8
Fawn Trough, *Ind. Oc.* **254** C11
Faxaflói, *bay, Ice.* **142** E2
Faya, *Chad* **207** F14
Fayaoué, *New Caledonia, Fr.* **231** P23
Fayd, *Saudi Arabia* **174** E8
Fayette, *Miss., U.S.* **102** M10
Fayette, *Ala., U.S.* **103** J14
Fayette, *Mo., U.S.* **107** S12
Fayetteville, *N.C., U.S.* **98** J11

Fayetteville, *W. Va., U.S.* **98** E9
Fayetteville, *Ark., U.S.* **102** E6
Fayetteville, *Tenn., U.S.* **103** G15
Fazilka, *India* **182** F4
Fdérik (Fort Gouraud), *Mauritania* **204** D9
Fear, Cape, *N.C., U.S.* **98** K12
Fécamp, *Fr.* **149** N8
Fedala *see* Mohammedia, *Mor.* **200** C7
Federal, *Arg.* **134** J13
Federally Administered Tribal Areas, *Pak.* **180** J9
Fedorovka, *Kaz.* **178** C9
Fefen, *island, F.S.M.* **226** F9
Fehmarn, *island, Ger.* **150** B9
Feijó, *Braz.* **132** F5
Feilding, *N.Z.* **223** H21
Feira de Santana, *Braz.* **133** H16
Feke, *Turk.* **170** H10
Felanitx, *Sp.* **147** H19
Feldbach, *Aust.* **150** M12
Feldberg, *peak, Ger.* **150** K6
Feldkirch, *Aust.* **150** L7
Félicité Island, *Seychelles* **215** M21
Felidu Atoll, *Maldives* **183** W3
Felipe Carrillo Puerto, *Mex.* **117** H17
Felixstowe, *Eng., U.K.* **145** T16
Fellit, *Eritrea* **210** G7
Fellowship, *Jam.* **122** J8,11
Feltre, *It.* **152** C8
Femund, lake, *Nor.* **142** H11
Fen Country, *Eng., U.K.* **145** S14
Fen, *river, China* **184** H12
Fena Valley Reservoir, *Guam, U.S.* **230** P5
Fene, *Sp.* **146** A6
Feneppi, *island, F.S.M.* **226** G9
Fengcheng, *China* **185** F16
Fengyi, *China* **184** N8
Feni Islands, *P.N.G.* **222** N10
Feno, Cap de, *Fr.* **149** Z2
Feno, Capo di, *Fr.* **149** Y2
Fenoarivo Atsinanana, *Madagascar* **213** F14
Fenualoa, *island, Solomon Is.* **228** K6
Fenuaroa, *island, Fr. Polynesia, Fr.* **231** N17
Feodosiya, *Russ.* **159** S8
Feolin Ferry, *Scot., U.K.* **144** L7
Fer, Point au, *La., U.S.* **102** R9
Fera, *island, Solomon Is.* **228** D9
Féres, *Gr.* **157** C16
Fērfēr, *Eth.* **211** P11
Fergana Valley, *Uzb.* **179** L13
Fergus Falls, *Minn., U.S.* **106** H8
Ferguson's Gulf, *Kenya* **211** Q6
Fergusson Island, *P.N.G.* **222** Q9
Feridu, *island, Maldives* **183** V3
Ferit, *island, F.S.M.* **226** E9
Ferizaj (Uroševac), *Kos.* **154** J11
Ferkéssédougou, *Côte d'Ivoire* **205** M13
Fermo, *It.* **152** H10
Fermont, *Que., Can.* **87** L20
Fermoselle, *Sp.* **146** E8
Fermoy (Mainistir Fhear Maí), *Ire.* **145** T4
Fern Gully, *Jam.* **122** H9
Fernan Vaz *see* Omboué, *Gabon* **208** F7
Fernandez Bay, *Bahamas* **122** F11
Fernandina Beach, *Fla., U.S.* **99** Q9
Fernandina, Isla, *Galápagos Is., Ecua.* **225** G21
Fernando de Noronha Ridge, *Atl. Oc.* **249** P8
Fernando de Noronha, Arquipélago de, *Braz.* **133** E19
Ferndale, *Wash., U.S.* **110** A4
Ferndale, *Calif., U.S.* **111** N1
Ferney, *Nev., U.S.* **111** P7
Ferrara, *It.* **152** F7
Ferrato, Capo, *It.* **153** Q4
Ferraz Ridge, *Atl. Oc.* **249** Q8
Ferreira do Alentejo, *Port.* **146** K5
Ferreira Gomes, *Braz.* **132** C12
Ferreñafe, *Peru* **132** F2
Ferret, Cap, *Fr.* **149** W6
Ferretta, *Fr.* **149** R15
Ferriday, *La., U.S.* **102** M9
Ferro *see* El Hierro, *island, Sp.* **214** R2
Ferro, Ilhéu de, *Madeira, Port.* **214** L4
Ferrol, *Sp.* **146** A6

F

Faaa, *Fr. Polynesia, Fr.* **231** H21
Faaaha, *Fr. Polynesia, Fr.* **231** G17
Faaite, *island, Fr. Polynesia, Fr.* **231** B18
Faanui, *Fr. Polynesia, Fr.* **231** G14
Faaone, *Fr. Polynesia, Fr.* **231** J22

Ferrolana, *Eq. Guinea* **214** M6
Ferrysburg, *Mich., U.S.* **100** L9
Fertile, *Minn., U.S.* **106** F8
Fès (Fez), *Mor.* **200** C8
Feshi, *D.R.C.* **208** H12
Fessenden, *N. Dak., U.S.* **106** F5
Festus, *Mo., U.S.* **107** T15
Fété Bowé, *Senegal* **204** J8
Fetești, *Rom.* **155** F17
Fethiye Körfezi, *Turk.* **170** J4
Fethiye, *Turk.* **170** J4
Fetlar, *island, Scot., U.K.* **144** B12
Fetoa, *island, Tonga* **227** M23
Fetokopunga, *island, Tonga* **227** N23
Fetuna, *Fr. Polynesia, Fr.* **231** H17
Feurs, *Fr.* **149** U12
Fevzipaşa, *Turk.* **170** H11
Fez *see* Fès, *Mor.* **200** C8
Fezzan, *Lib.* **201** H17
Ffryes Point, *Antigua & Barbuda* **123** B20
Fian, *Ghana* **205** L15
Fianarantsoa, *Madagascar* **213** G16
Fianga, *Chad* **206** K12
Fichē, *Eth.* **210** L7
Fidenza, *It.* **152** E5
Fields Find, *W. Aust. Austral.* **221** V3
Fier-Shegan, *Alban.* **154** L9
Fier, *Alban.* **154** L9
Fife Lake, *Mich., U.S.* **100** J10
Fife Ness, *Scot., U.K.* **144** K11
Fifteen Twenty Fracture Zone
  (Barracuda Fracture Zone), *Atl. Oc.* **248** M7
Fig Tree, *St. Kitts & Nevis* **123** C18
Figari, Capo, *It.* **152** M5
Figeac, *Fr.* **149** W9
Figueira da Foz, *Port.* **146** F5
Figueres, *Sp.* **147** C18
Figuig, *Alg.* **200** E10
Figuil, *Cameroon* **206** K11
Fiji, *Pac. Oc.* **224** J8
Fiji Plateau, *Pac. Oc.* **246** L10
Filabres, Sierra de los, *Sp.* **146** L12
Filadelfia, *Parag.* **132** L9
Filchner Ice Sheet, *Antarctica* **238** F11
Filchner Trough, *Atl. Oc.* **254** K11
Filey, *Eng., U.K.* **145** P13
Filfla, *island, Malta* **163** M22
Filiași, *Rom.* **155** F13
Filiátes, *Gr.* **156** E7
Filiatrá, *Gr.* **156** L9
Filicudi, Isola, *It.* **153** R11
Filingué, *Niger* **206** G6
Filiourí, *river, Gr.* **157** B15
Filippiáda, *Gr.* **156** F8
Fílippoi (Philippi), *ruin, Gr.* **157** B13
Filipstad, *Sw.* **142** K12
Fillmore, *Utah, U.S.* **109** N7
Fillmore, *Calif., U.S.* **111** X7
Fīltu, *Eth.* **211** P9
Filyos, *river, Turk.* **170** C8
Fimbul Ice Shelf, *Antarctica* **239** A13
Fiñana, *Sp.* **146** L12
Finarwa, *Eth.* **210** J8
Findlay, *Ohio, U.S.* **101** P12
Findochty, *Scot., U.K.* **144** H10
Fineveke, *Wallis & Futuna Is., Fr.* **230** R8
Finger Lakes, *N.Y., U.S.* **96** K7
Fíngoè, *Mozambique* **212** D10
Finike, *Turk.* **170** J5
Finisterre, Cape, *Eur.* **140** J2
Finke, *N.T. Austral.* **221** U9
Finke, *river, Austral.* **220** G9
Finland, *Minn., U.S.* **106** F13
Finland, *Eur.* **143** F15
Finland, Gulf of, *Est., Fin.* **143** J15
Finlay, *river, B.C. Can.* **86** J8
Finley Cay, *Bahamas* **122** E5
Finley, *N. Dak., U.S.* **106** F7
Finnmark Plateau, *Eur.* **140** A9
Finnsmarkvidda, *island, Nor.* **143** C14
Finol Finkol, *peak, F.S.M.* **226** G12
Finsteraarhorn, *Switz.* **152** C3
Fionn Loch, *Scot., U.K.* **144** H8
Fionnphort, *Scot., U.K.* **144** K7
Fiorentino, *San Marino* **163** K15
Fiorina, *San Marino* **163** H16
Fiorina, *river, San Marino* **163** H16
Firá, *Gr.* **157** M15
Firat, *river, Turk.* **171** G14
Fire Island National Seashore, *N.Y., U.S.* **96** N12
Firenze (Florence), *It.* **152** G7
Firenzuola, *It.* **152** G7
Firmat, *Arg.* **134** K11

Firminy, *Fr.* **149** U12
Firozabad, *India* **182** H6
First State National Monument, *Del., U.S.* **98** B14
Firth of Forth, *Atl. Oc.* **140** E4
Fish Great, *river, U.S.* **106** F5
Fish Lake, *Utah, U.S.* **109** N7
Fish, *river, Namibia* **212** H4
Fish, *river, Namibia* **212** K4
Fisher Bay, *Antarctica* **239** R18
Fisher Glacier, *Antarctica* **239** F19
Fisher Strait, *Nunavut, Can.* **87** J16
Fisher, *river, Can., U.S.* **108** B5
Fishguard (Abergwaun), *Wales, U.K.* **145** T8
Fishing Cays, *Bahamas* **122** F2
Fiskárdo, *Gr.* **156** H7
Fiskefjord *see* Niaqunngunaq, *Greenland, Den.* **89** Q5
Fiskenæsset *see* Qeqertarsuatsiaat, *Greenland, Den.* **89** R5
Fismes, *Fr.* **149** N11
Fisterra, *Sp.* **146** B5
Fisterra, Cabo, *Sp.* **146** B5
Fitchburg, *Mass., U.S.* **97** K14
Fitii, *Fr. Polynesia, Fr.* **231** K14
Fiti'uta, *American Samoa, U.S.* **230** B11
Fitzgerald, *Alta. Can.* **86** J11
Fitzgerald, *Ga., U.S.* **99** P6
Fitzroy Crossing, *W. Aust. Austral.* **221** R5
Fitzroy, *river, Austral.* **220** D5
Fitzroy, *river, Austral.* **220** F15
Fiume *see* Rijeka, *Croatia* **154** E4
Fiumicello, *river, San Marino* **163** J16
Fivaku, *island, Maldives* **183** U3
Five Island Harbour, *Antigua & Barbuda* **123** B20
Five Islands Village, *Antigua & Barbuda* **123** A20
Fizi, *D.R.C.* **209** G18
Fındıklı, *Rep. of Georgia* **171** C15
Fkih ben Salah, *Mor.* **200** D8
Flade Isblink, *Greenland, Den.* **88** B12
Flag Hill, *Virgin Is., U.S.* **122** N8
Flagler, *Colo., U.S.* **108** M15
Flagstaff Lake, *Oreg., U.S.* **110** K6
Flagstaff, *Ariz., U.S.* **109** S7
Flåm, *Nor.* **142** J10
Flambeau, *river, Wis., U.S.* **100** G4
Flamborough Head, *Eng., U.K.* **145** P14
Flaming Gorge National Recreation Area, *Utah, Wyo., U.S.* **108** K9
Flaming Gorge Reservoir, *Utah, Wyo., U.S.* **108** K9
Flamingo Cay, *Bahamas* **119** G13
Flamingo Point, *Bahamas* **119** E13
Flamingo, *Fla., U.S.* **99** Y9
Flanagan Island, *Virgin Is., U.S.* **122** N12
Flanagin Town, *Trin. & Tobago* **123** Q22
Flanders, *region, Belg., Fr.* **148** L10
Flandreau, *S. Dak., U.S.* **106** L8
Flannach Isles *see* Na h-Eileanan Flannach, *Scot., U.K.* **144** G6
Flat Island, *S. China Sea* **192** E11
Flat Island, *Mauritius* **215** F20
Flat River, *Mo., U.S.* **107** U15
Flatonia, *Tex., U.S.* **105** L17
Flattery, Cape, *Wash., U.S.* **110** B1
Flattery, Cape, *Austral.* **220** C13
Flatts Village, *Bermuda* **119** C16
Flatwoods, *Ky., U.S.* **103** A20
Flaxton, *N. Dak., U.S.* **106** D3
Fleetwood, *Eng., U.K.* **145** Q10
Fleetwood, *Qnsld., Austral.* **221** T13
Flekkefjord, *Nor.* **142** L10
Fleming Fjord, *bay, Greenland, Den.* **88** J14
Flemingsburg, *Ky., U.S.* **103** B19
Flemish Cap, *Atl. Oc.* **248** H7
Flensburg, *Ger.* **150** A7
Flers, *Fr.* **149** P7
Fletcher Peninsula, *Antarctica* **238** H6
Fletcher Plain, *Arctic Oc.* **252** J11
Fletcher Promontory, *Antarctica* **238** H9
Fleurance, *Fr.* **149** X8
Fleurier, *Switz.* **152** B1
Flight 93 National Memorial, *Pa. U.S.* **96** P4
Flin Flon, *Man., Can.* **87** L13
Flinders Fracture Zone, *Ind. Oc.* **250** K9

Flinders Island, *Austral.* **220** K8
Flinders Ranges, *Austral.* **220** J10
Flinders, *river, Austral.* **220** D11
Flint Hills, *Kans., U.S.* **107** U8
Flint Island, *Kiribati* **225** H13
Flint, *Mich., U.S.* **100** L12
Flint, *river, Ga., U.S.* **99** P5
Flipper Point, *Wake I., U.S.* **230** M11
Flisa, *Nor.* **142** J12
Flix, *Sp.* **147** E15
Flixecourt, *Fr.* **148** M9
Flocca, *river, San Marino* **163** H15
Flomaton, *Ala., U.S.* **103** N15
Flood Range, *Antarctica* **238** M9
Floodwood, *Minn., U.S.* **106** G12
Flora, *Ill., U.S.* **101** T6
Flora, *Ind., U.S.* **101** Q8
Flora, *Oreg., U.S.* **110** E9
Florala, *Ala., U.S.* **103** N15
Florange, *Fr.* **149** N14
Florence *see* Firenze, *It.* **152** G7
Florence, *S.C., U.S.* **98** K10
Florence, *Wis., U.S.* **100** G7
Florence, *Ala., U.S.* **103** G14
Florence, *Ky., U.S.* **103** A17
Florence, *Kans., U.S.* **107** T8
Florence, *Ariz., U.S.* **109** U7
Florence, *Colo., U.S.* **109** N13
Florence, *Oreg., U.S.* **110** H2
Florencia, *Col.* **130** J5
Flores Basin, *Pac. Oc.* **246** K5
Flores Sea, *Indonesia* **192** M13
Flores, *Guatemala* **117** K16
Flores, *island, Port.* **146** N1
Flores, *island, Indonesia* **193** M13
Floreşti, *Mold.* **159** Q4
Floresville, *Tex., U.S.* **105** M16
Floriano Peixoto, *Braz.* **132** G6
Floriano, *Braz.* **133** F15
Florianópolis, *Braz.* **133** N13
Florida Bay, *Fla., U.S.* **99** Y10
Florida City, *Fla., U.S.* **99** X10
Florida Islands, *Solomon Is.* **228** E10
Florida Keys, *Fla., U.S.* **99** Y9
Florida-Hatteras Shelf, *Atl. Oc.* **248** K3
Florida, *U.S.* **99** S7
Florida, *Cuba* **118** J10
Florida, *peninsula, N. Amer.* **84** L8
Florida, Straits of, *N. Amer.* **84** M8
Floridia, *It.* **153** T12
Flórina, *Gr.* **156** C8
Florissant Fossil Beds National Monument, *Colo., U.S.* **109** N13
Florissant, *Mo., U.S.* **107** S15
Florø, *Nor.* **142** J9
Flotta, *island, Scot., U.K.* **144** F10
Floyd, *N. Mex., U.S.* **109** T15
Floydada, *Tex., U.S.* **104** E12
Flumet, *Fr.* **149** U14
Fly, *river, P.N.G.* **222** Q5
Foa, *island, Tonga* **227** L24
Foča, *Bosn. & Herzg.* **154** H8
Foça, *Turk.* **170** F2
Focşani, *Rom.* **155** D16
Fofoa, *Tonga* **227** H22
Fog Bay, *Austral.* **220** B7
Foggia, *It.* **152** L12
Fogo Island, *Nfld. & Lab. Can.* **87** L23
Fogo, *island, Cabo Verde* **215** E15
Fohnsdorf, *Aust.* **150** L11
Foínikas, *Gr.* **157** K14
Fokino, *Russ.* **158** L8
Fokku, *Nigeria* **206** J6
Foko Finaunpes, *cape, F.S.M.* **226** F12
Foko Fukunsral, *cape, F.S.M.* **226** G12
Foko Malsu, *cape, F.S.M.* **226** F12
Foko Nefalil, *cape, F.S.M.* **226** G11
Folda, *bay, Nor.* **142** G11
Folégandros, *Gr.* **157** M14
Folégandros, *island, Gr.* **157** M14
Foley Island, *Nunavut, Can.* **87** F17
Foley, *Ala., U.S.* **103** P14
Foley, *Minn., U.S.* **106** J11
Foligno, *It.* **152** J8
Folkestone, *Eng., U.K.* **145** V15
Folkston, *Ga., U.S.* **99** Q8
Follett, *Tex., U.S.* **105** B14
Follonica, *It.* **152** J6
Follonica, Golfo di, *It.* **152** J6
Folly Beach, *S.C., U.S.* **98** M10
Foloha, *Tonga* **227** Q23
Folschette, *Lux.* **162** J9
Folsom, *Calif., U.S.* **111** Q5
Folteşti, *Rom.* **155** D17
Fomboni, *Comoros* **215** N15

Fomento, *Cuba* **118** H9
Fond d'Or Bay, *St. Lucia* **123** K14
Fond du Lac, *Wis., U.S.* **100** K6
Fond-du-Lac, *Sask. Can.* **86** J12
Fondi, *It.* **152** L9
Fono, *island, F.S.M.* **226** E9
Fonoifua, *island, Tonga* **227** M23
Fononuk, *island, F.S.M.* **226** G9
Fonou, *island, F.S.M.* **226** F9
Fonseca, Golfo de, *Hond.* **117** L17
Fontainebleau, *Fr.* **149** Q10
Fontana Lake, *N.C., U.S.* **98** H6
Fontaneda, *Andorra* **162** K2
Fonte Boa, *Braz.* **132** D7
Fontenelle Reservoir, *Wyo., U.S.* **108** J8
Fontenoy, *Grenada* **123** L22
Fontvieille, *Monaco* **163** F19
Fontvieille, Port of, *Monaco* **163** F21
Fonuafoou, *island, Tonga* **227** H23
Fonuaika, *island, Tonga* **227** M23
Fonuakula, *N.Z.* **229** Q21
Fonua'one'one, *island, Tonga* **227** J23
Fonuaunga, *island, Tonga* **227** H23
Fonuchu, *island, F.S.M.* **226** D9
Foppolo, *It.* **152** D5
Forbes Hill, *Bahamas* **122** F9
Forbes, *N.S.W. Austral.* **221** X14
Forbes, *N.S.W. Austral.* **221** X14
Forbesganj, *India* **182** H11
Forcados, *Nigeria* **206** M7
Forcalquier, *Fr.* **149** X14
Forchheim, *Ger.* **150** H8
Ford Ranges, *Antarctica* **238** N10
Ford, Cape, *Austral.* **220** B7
Førde, *Nor.* **142** J10
Fordyce, *Ark., U.S.* **102** J8
Forécariah, *Guinea* **204** M8
Forel, Mount, *N. Amer.* **84** C10
Foreman, *Ark., U.S.* **102** J6
Forest City, *N.C., U.S.* **98** J8
Forest City, *Iowa, U.S.* **106** M11
Forest Grove, *Oreg., U.S.* **110** F3
Forest Lakes, *Austral.* **220** H7
Forest Park, *Ga., U.S.* **98** L5
Forest Point, *Anguilla, U.K.* **122** R11
Forest, *Miss., U.S.* **102** L12
Forestville, *Que., Can.* **87** N20
Forestville, *Mich., U.S.* **100** K13
Forez, Monts du, *Fr.* **149** U11
Forfar, *Scot., U.K.* **144** K10
Forgan, *Okla., U.S.* **105** A13
Forges-les-Eaux, *Fr.* **149** N9
Forks, *Wash., U.S.* **110** B2
Forland Hochstetter, *peninsula, Greenland, Den.* **88** G13
Forlì, *It.* **152** G8
Forman, *N. Dak., U.S.* **106** H7
Formentera, *island, Sp.* **147** J17
Formiga, *Braz.* **133** L14
Formosa do Rio Prêto, *Braz.* **133** G14
Formosa, *Braz.* **133** J13
Formosa, *Arg.* **134** F17
Formosa, *island, Guinea-Bissau* **204** L7
Formosa, Serra, *Braz.* **132** G11
Forrestal Range, *Antarctica* **238** G11
Forrest City, *Ark., U.S.* **102** G10
Forrest River, *W. Aust. Austral.* **221** Q6
Forrest, *W. Aust. Austral.* **221** W7
Forsayth, *Qnsld., Austral.* **221** R12
Forssa, *Fin.* **143** H15
Forst, *Ger.* **150** E11
Forster-Tuncurry, *N.S.W. Austral.* **221** X15
Forsyth, *Ga., U.S.* **98** M5
Forsyth, *Mo., U.S.* **107** V12
Forsyth, *Mont., U.S.* **108** D11
Fort Abbas, *Pak.* **181** N12
Fort Albany, *Ont., Can.* **87** M17
Fort Antoine, *Monaco* **163** E22
Fort Atkinson, *Wis., U.S.* **100** L6
Fort Benton, *Mont., U.S.* **108** C8
Fort Bowie National Historic Site, *Ariz., U.S.* **109** V9
Fort Bragg, *Calif., U.S.* **111** P2
Fort Caroline National Memorial, *Fla., U.S.* **99** Q9
Fort Chipewyan, *Alta. Can.* **86** J11
Fort Cobb Reservoir, *Okla., U.S.* **105** D16
Fort Collins, *Colo., U.S.* **108** L13
Fort Davis National Historic Site, *Tex. U.S.* **104** K9

Fort Davis, *Tex., U.S.* **104** K9
Fort Deposit, *Ala., U.S.* **103** M15
Fort Dick, *Calif., U.S.* **110** L1
Fort Dodge, *Iowa, U.S.* **107** N11
Fort Fairfield, *Me., U.S.* **97** B18
Fort Frances, *Ont., Can.* **87** P14
Fort Fraser, *B.C. Can.* **86** K8
Fort Frederica National Monument, *Ga., U.S.* **99** P9
Fort Frederick, *Grenada* **123** L22
Fort Gaines, *Ga., U.S.* **99** P4
Fort Gibson Lake, *Okla., U.S.* **105** B19
Fort Gibson, *Okla., U.S.* **105** C19
Fort Good Hope, *N.W.T., Can.* **86** F10
Fort Gouraud *see* Fdérik, *Mauritania* **204** D9
Fort Hall, *Idaho, U.S.* **108** H7
Fort Hancock, *Tex., U.S.* **104** J7
Fort Hope, *Ont., Can.* **87** N16
Fort Jones, *Calif., U.S.* **110** L3
Fort Kent, *Me., U.S.* **97** A17
Fort Klamath, *Oreg., U.S.* **110** K4
Fort Knox, *site, Ky., U.S.* **103** C16
Fort Laperrine *see* Tamanrasset, *Alg.* **201** K13
Fort Laramie National Historic Site, *Wyo., U.S.* **108** J13
Fort Laramie, *Wyo., U.S.* **108** J13
Fort Larned National Historic Site, *Kans., U.S.* **107** T6
Fort Lauderdale, *Fla., U.S.* **99** W10
Fort Liard, *N.W.T., Can.* **86** H9
Fort Madison, *Iowa, U.S.* **107** Q14
Fort Matanzas National Monument, *Fla., U.S.* **99** R9
Fort McMurray, *Alta. Can.* **86** K11
Fort McPherson, *N.W.T., Can.* **86** E9
Fort Meade, *Fla., U.S.* **99** U8
Fort Mill, *S.C., U.S.* **98** J9
Fort Miribel, *Alg.* **200** F12
Fort Morgan, *Colo., U.S.* **108** L14
Fort Munro, *Pak.* **180** M9
Fort Myers Beach, *Fla., U.S.* **99** W8
Fort Myers, *Fla., U.S.* **99** W8
Fort Nelson, *B.C. Can.* **86** J9
Fort Ord National Monument, *Calif., U.S.* **111** T4
Fort Payne, *Ala., U.S.* **103** H16
Fort Peck Dam, *Mont., U.S.* **108** B12
Fort Peck Lake, *Mont., U.S.* **108** C11
Fort Peck, *Mont., U.S.* **108** B12
Fort Pierce, *Fla., U.S.* **99** U10
Fort Pierre Bordes *see* Ti-n-Zaouâtene, *Alg.* **200** M12
Fort Pierre, *S. Dak., U.S.* **106** K5
Fort Point, *Virgin Is., U.K.* **122** R8
Fort Portal, *Uganda* **211** S2
Fort Providence, *N.W.T., Can.* **86** H10
Fort Pulaski National Monument, *Ga., U.S.* **99** N9
Fort Raleigh National Historic Site, *N.C., U.S.* **98** G15
Fort Randall Dam, *S. Dak., U.S.* **106** M6
Fort Resolution, *N.W.T., Can.* **86** J11
Fort Rock, *Oreg., U.S.* **110** J5
Fort Royal, *Guadeloupe, Fr.* **123** E13
Fort Saint John, *B.C., Can.* **86** K9
Fort Saint, *Tunisia* **201** F15
Fort Scott National Historic Site, *Kans., U.S.* **107** U10
Fort Scott, *Kans., U.S.* **107** U10
Fort Severn, *Ont., Can.* **87** L16
Fort Shevchenko, *Kaz.* **178** H4
Fort Simpson, *N.W.T., Can.* **86** H10
Fort Smith, *N.W.T., Can.* **86** J11
Fort Smith, *Ark., U.S.* **102** F6
Fort Stanwix National Monument, *N.Y., U.S.* **96** J9
Fort Stockton, *Tex., U.S.* **104** K11
Fort Sumner, *N. Mex., U.S.* **109** S14
Fort Sumter National Monument, *S.C., U.S.* **98** M10
Fort Supply, *Okla., U.S.* **105** B14
Fort Thompson, *S. Dak., U.S.* **106** L5
Fort Union National Monument, *N. Mex., U.S.* **109** R13
Fort Union Trading Post National Historic Site, *N. Dak., U.S.* **106** E1
Fort Valley, *Ga., U.S.* **98** M5
Fort Vermilion, *Alta., Can.* **86** J10
Fort Walton Beach, *Fla., U.S.* **99** Q2
Fort Wayne, *Ind., U.S.* **101** P10
Fort Worth, *Tex., U.S.* **105** G17
Fort Yates, *N. Dak., U.S.* **106** H4
Fort Yukon, *Alas., U.S.* **113** E18

Fort-de-France, *Martinique, Fr.* **123** F22
Fort-Liberté, *Haiti* **119** L17
Fortaleza, *Braz.* **133** E17
Forteau, *Que., Can.* **87** L22
Fortescue, *river, Austral.* **220** F3
Fortezza, *It.* **152** B7
Forth, Firth of, *Scot., U.K.* **144** L10
Fortín General Díaz, *Parag.* **132** M9
Fortín Infante Rivarola, *Parag.* **132** L8
Fortín Madrejón, *Parag.* **132** L9
Fortín Mayor Alberto Gardel, *Parag.* **132** M8
Fortín Presidente Ayala, *Parag.* **132** M9
Fortín Suárez Arana, *Bol.* **132** K9
Fortress Headquarters, *Gibraltar, U.K.* **163** Q22
Fortuna Bay, *Virgin Is., U.S.* **122** M7
Fortuna, *N. Dak., U.S.* **106** D2
Fortuna, *Calif., U.S.* **111** N1
Fortune Island *see* Long Cay, *Bahamas* **119** H14
Forūr, *island, Iran* **177** N18
Fosheim Peninsula, *Nunavut, Can.* **87** B16
Fossano, *It.* **152** F2
Fossato di Vico, *It.* **152** H9
Fossil Bluff Field Station, *Antarctica* **238** F7
Fossil Butte National Monument, *Wyo., U.S.* **108** J8
Fossil Downs, *W. Aust., Austral.* **221** R6
Fossil, *Oreg., U.S.* **110** G6
Fosston, *Minn., U.S.* **106** F9
Foster Bugt, *Greenland, Den.* **88** H14
Foster, *Vic., Austral.* **221** Z13
Foster, Mount, *Can., U.S.* **113** K21
Fostoria, *Ohio, U.S.* **101** P12
Fota Terara, *peak, Eth.* **210** M7
Fotuha'a, *island, Tonga* **227** L23
Fougamou, *Gabon* **208** E8
Fougères, *Fr.* **149** Q6
Foul Bay, *Barbados* **123** L20
Foul Bay, *Egypt* **202** K14
Foula, *island, Scot., U.K.* **144** C11
Foulness Island, *Eng., U.K.* **145** U15
Foulweather, Cape, *Oreg., U.S.* **110** G2
Foum Zguid, *Mor.* **200** F7
Foumban, *Cameroon* **206** M10
Foumbouni, *Comoros* **215** M14
Foundation Ice Stream, *Antarctica* **238** H11
Foundation Seamounts, *Pac. Oc.* **247** N17
Foundiougne, *Senegal* **204** J6
Fountain City, *Wis., U.S.* **100** J3
Fountain, *Colo., U.S.* **109** N13
Fountain, *Calif., U.S.*
Four Corners, *Wyo., U.S.* **108** G14
Four Ladies Bank, *Ind. Oc.* **254** E12
Four Mountains, Islands of, *Alas., U.S.* **112** P8
Four Roads, *Trin. & Tobago* **123** P22
Fourcroy, Cape, *Austral.* **220** A7
Fourmies, *Fr.* **148** M11
Fournaise, Piton de la, *Réunion, Fr.* **215** H17
Foúrnoi, *Gr.* **157** K17
Foúrnoi, *island, Gr.* **157** K17
Fourup, *island, F.S.M.* **226** F8
Fouta Djallon, *Guinea* **204** L9
Foux, Cap-à-, *Haiti* **119** L15
Foveaux Strait, *N.Z.* **223** P16
Fowler Peninsula, *Antarctica* **238** H8
Fowler, *Ind., U.S.* **101** Q7
Fowler, *Kans., U.S.* **107** U4
Fowler, *Colo., U.S.* **109** N14
Fowlers Bay, *Austral.* **220** J8
Fowlers Bay, *S. Aust., Austral.* **221** W8
Fox Glacier, *N.Z.* **223** L17
Fox Islands, *Alas., U.S.* **112** P9
Fox Town, *Bahamas* **122** A8
Fox, *river, Ill., U.S.* **100** P6
Fox, *river, Wis., U.S.* **100** K5
Foxe Basin, *Nunavut, Can.* **87** G16
Foxe Channel, *Nunavut, Can.* **87** H17
Foxe Peninsula, *Nunavut, Can.* **87** H17
Foxpark, *Wyo., U.S.* **108** K12
Foyé, *Guinea* **204** L8
Foyle, Lough, *N. Ire., U.K.* **145** N6
Foz do Cunene, *Angola* **208** Q9
Foz do Iguaçu, *Braz.* **132** N11
Foz, *Sp.* **146** A7
Frænkel Land, *Greenland, Den.* **88** J12
Fraga, *Sp.* **147** D15
Fram Bank, *Ind. Oc.* **254** E11

Fram Peak, *Antarctica* **239** D19
Fram Strait, *Greenland, Den.* **88** C14
Franca, *Braz.* **133** L13
France, *Fr.* **149** S10
Frances Lake, *Yukon, Can.* **86** G9
Francés Viejo, Cabo, *Dom. Rep.* **119** L19
Francés, Cabo, *Cuba* **118** H4
Francés, Punta, *Cuba* **118** H5
Franceville, *Gabon* **208** F9
Francis Bay, *Virgin Is., U.S.* **122** M11
Francis Case, *S. Dak., U.S.* **106** L5
Francistown, *Botswana* **212** G8
Francs Peak, *Wyo., U.S.* **108** G9
Frank Sound, *Cayman Is., U.K.* **122** N4
Frankenberg, *Ger.* **150** F6
Frankfort, *Ind., U.S.* **100** Q8
Frankfort, *Mich., U.S.* **100** J9
Frankfort, *Ky., U.S.* **103** B17
Frankfort, *Kans., U.S.* **107** R8
Frankfurt am Main, *Ger.* **150** G6
Frankfurt an der Oder, *Ger.* **150** E11
Franklin Bay, *N.W.T., Can.* **86** D11
Franklin D. Roosevelt Lake, *Wash., U.S.* **110** B8
Franklin Lake, *Nev., U.S.* **111** N11
Franklin Mountains, *N. Amer.* **84** D4
Franklin Strait, *Nunavut, Can.* **87** E14
Franklin, *Pa., U.S.* **96** M3
Franklin, *Pa., U.S.* **96** M3
Franklin, *N.H., U.S.* **97** H14
Franklin, *N.C., U.S.* **98** J6
Franklin, *Va., U.S.* **98** F13
Franklin, *Ind., U.S.* **101** S9
Franklin, *La., U.S.* **102** Q9
Franklin, *Ky., U.S.* **103** D15
Franklin, *Tenn., U.S.* **103** F15
Franklin, *Nebr., U.S.* **107** R6
Franklinton, *La., U.S.* **102** N11
Franklinville, *N.Y., U.S.* **96** K5
Frankston, *Tex., U.S.* **105** H19
Franske Øer, *Greenland, Den.* **88** D13
Frants-Viktoriya Trough, *Arctic Oc.* **253** F15
Frantsa Iosifa, *Russ.* **160** C12
Franz Josef Glacier, *N.Z.* **223** L17
Franz Josef Land *see* Frantsa Iosifa, *Russ.* **160** C12
Franz, *Ont., Can.* **87** P17
Frasca, Capo della, *It.* **153** P3
Fraser Island (Great Sandy Island), *Austral.* **220** G16
Fraser Range, *W. Aust., Austral.* **221** W5
Fraser, Mount, *Austral.* **220** G3
Fraser, *river, N. Amer.* **84** G3
Fraserburgh, *Scot., U.K.* **144** H11
Frasertown, *N.Z.* **223** G22
Frasin, *Rom.* **155** B15
Fratta, *river, San Marino* **163** K14
Frauenfeld, *Switz.* **152** A4
Frazee, *Minn., U.S.* **106** G9
Fredensborg, *Virgin Is., U.S.* **122** Q2
Frederick E. Hyde Fjord, *Greenland, Den.* **88** A10
Frederick, *Md., U.S.* **98** C12
Frederick, *Okla., U.S.* **105** E15
Frederick, *S. Dak., U.S.* **106** H6
Frederick, Mount, *Austral.* **220** D7
Fredericksburg, *Va., U.S.* **98** D13
Fredericksburg, *Tex., U.S.* **105** K15
Fredericktown, *Mo., U.S.* **107** U15
Fredericton, *N.B., Can.* **87** P21
Frederiksdal *see* Narsaq Kujalleq, *Greenland, Den.* **89** U8
Frederikshåb Isblink, *glacier, Greenland, Den.* **89** S6
Frederikshåb *see* Paamiut, *Greenland, Den.* **89** S6
Frederikshåbs Bank, *Atl. Oc.* **253** R18
Frederikshavn, *Den.* **142** M11
Fredriksted, *Virgin Is., U.S.* **121** B13
Frederiksted, *Virgin Is., U.S.* **122** Q1
Fredonia, *N.Y., U.S.* **96** K4
Fredonia, *Kans., U.S.* **107** U9
Fredonia, *Ariz., U.S.* **109** Q6
Fredrikstad, *Nor.* **142** K11
Freedom, *Okla., U.S.* **105** B15
Freehold, *N.J., U.S.* **96** P11
Freeman, *S. Dak., U.S.* **106** M7
Freeport, *N.Y., U.S.* **96** N11
Freeport, *Me., U.S.* **97** G16
Freeport, *Ill., U.S.* **100** M5
Freeport, *Tex., U.S.* **105** M19
Freeport, *Bahamas* **118** C10
Freeport, *Bahamas* **122** J2

Freer, *Tex., U.S.* **105** P16
Freetown, *Bahamas* **122** J2
Freetown, *Jam.* **122** J9
Freetown, *Antigua & Barbuda* **123** B21
Freetown, *Sierra Leone* **204** N8
Freezeout Mountain, *Oreg., U.S.* **110** H9
Frégate Island, *Mauritius* **215** J19
Frégate, *island, Seychelles* **215** N21
Fregenal de la Sierra, *Sp.* **146** J7
Fréhel, Cap, *Fr.* **149** P5
Freiberg, *Ger.* **150** F10
Freiburg, *Ger.* **150** K5
Freidberg, *Ger.* **150** K8
Freirina, *Chile* **134** H6
Freisach, *Aust.* **150** M11
Freising, *Ger.* **150** K9
Freistadt, *Aust.* **150** K11
Freital, *Ger.* **150** F10
Fréjus, *Fr.* **149** X15
Fremont Lake, *Wyo., U.S.* **108** H9
Fremont Mountains, *Oreg., U.S.* **110** K5
Fremont, *Mich., U.S.* **100** L9
Fremont, *Ohio, U.S.* **101** P12
Fremont, *Nebr., U.S.* **107** P8
Fremont, *Calif., U.S.* **111** S4
Fremont, *river, Utah, U.S.* **109** N7
French Bay, *Bahamas* **122** F11
French Broad, *river, Tenn., U.S.* **103** F14
French Cays *see* Plana Cays, *Bahamas* **119** G15
French Guiana, *S. Amer.* **131** G18
Frenchcap Cay, *Virgin Is., U.S.* **122** P9
Frenchman Creek, *Colo.-Nebr., U.S.* **107** Q3
Frenchman Flat, *Nev., U.S.* **111** T11
Frenchman's Cay, *Virgin Is., U.K.* **122** M11
Frenchman's Cay, *Virgin Is., U.K.* **122** R5
Frenchville, *Me., U.S.* **97** A18
Fresco, *Côte d'Ivoire* **205** Q13
Fresh Water Bay, *Barbados* **123** L18
Fresh Water Lagoons, *Kiribati* **226** Q12
Fresnay, *Fr.* **149** Q7
Fresnillo, *Mex.* **116** G10
Fresno Reservoir, *Mont., U.S.* **108** B9
Fresno, *Calif., U.S.* **111** T6
Freuchen Land, *Greenland, Den.* **88** B8
Frévent, *Fr.* **148** M10
Frewena, *N.T., Austral.* **221** R9
Freyung, *Ger.* **150** J10
Fria, *Guinea* **204** L8
Friars Point, *Miss., U.S.* **102** H10
Frías, *Arg.* **134** H9
Friedrichshafen, *Ger.* **150** L7
Friend, *Nebr., U.S.* **107** Q8
Friesland, *region, Neth.* **148** G14
Frigate Bay, *St. Kitts & Nevis* **123** B18
Friguiagbé, *Guinea* **204** M8
Frio, Cabo, *S. Amer.* **128** J10
Frio, *river, Tex., U.S.* **105** M14
Friona, *Tex., U.S.* **104** D11
Frisange, *Lux.* **162** K10
Frisco City, *Ala., U.S.* **103** M14
Frisco, *Tex., U.S.* **105** F17
Frisian Islands, *Eur.* **140** F6
Frissell, Mount, *Conn., U.S.* **96** L12
Fritch, *Tex., U.S.* **104** C12
Friza, *Russ.* **161** J23
Frobisher Bay, *Nunavut, Can.* **87** H19
Frolovo, *Russ.* **159** P12
Fromberg, *Mont., U.S.* **108** E10
Frome, *Jam.* **122** H6
Frome, *Eng., U.K.* **145** V11
Frome, Lake, *Austral.* **220** J11
Frómista, *Sp.* **146** C10
Front Range, *Colo., U.S.* **108** L12
Frontenac, *Kans., U.S.* **107** U10
Frontera, *Mex.* **116** E11
Frontera, *Mex.* **117** J15
Frontignan, *Fr.* **149** X11
Frosinone, *It.* **152** L9
Frost Glacier, *Antarctica* **239** P19
Frostburg, *Md., U.S.* **98** C11
Frostproof, *Fla., U.S.* **99** U9
Frøya Bank, *Atl. Oc.* **248** G13
Frøya, *island, Nor.* **142** G10
Fruita, *Colo., U.S.* **108** M10
Fruitland, *Idaho, U.S.* **108** G3
Frý, *Gr.* **157** Q18
Fryeburg, *Me., U.S.* **97** G15

Fu, *river, China* **184** L10
Fu'an, *China* **185** M15
Fua Mulaku, *island, Maldives* **183** Y3
Fua'amotu, *Tonga* **227** R23
Fua'amotu, *island, Tonga* **227** H23
Fuafatu, *island, Tuvalu* **228** N9
Fuagea, *island, Tuvalu* **228** P9
Fualifexe, *island, Tuvalu* **228** L10
Fualopa, *island, Tuvalu* **228** M9
Fucino, Piana del, *It.* **152** K9
Fuencaliente, *Sp.* **214** Q3
Fuencaliente, Punta de, *Sp.* **214** Q3
Fuengirola, *Sp.* **146** M9
Fuente de Cantos, *Sp.* **146** J8
Fuentesaúco, *Sp.* **146** E9
Fuerte Olimpo, *Parag.* **132** L10
Fuerteventura, *island, Sp.* **214** Q7
Fuga, *island, Philippines* **193** B13
Fuglø Bank, *Atl. Oc.* **248** D14
Fuhai, *China* **184** D5
Fuidhaigh (Wiay), *island, Scot., U.K.* **144** H6
Fuikbaai, *bay, Curaçao, Neth.* **123** Q15
Fujairah, *U.A.E.* **175** F16
Fuji, *peak, Japan* **189** P11
Fujin, *China* **185** C17
Fujisawa, *Japan* **189** P12
Fukagawa, *Japan* **188** F13
Fūkah, *Egypt* **202** E9
Fukave, *island, Tonga* **227** Q24
Fukuchiyama, *Japan* **189** P8
Fukue Jima, *Japan* **189** R3
Fukue, *Japan* **189** R3
Fukui, *Japan* **189** P9
Fukuoka, *Japan* **189** R4
Fukushima, *Japan* **188** H12
Fukushima, *Japan* **188** M12
Fukushima, Mount, *Antarctica* **239** C17
Fukuyama, *Japan* **189** Q7
Fulacunda, *Guinea-Bissau* **204** L7
Fulaga Passage, *Fiji* **226** Q8
Fulaga, *island, Fiji* **226** R8
Fulda, *Minn., U.S.* **106** L9
Fulda, *Ger.* **150** G7
Fuling, *China* **184** L11
Fullarton, *Trin. & Tobago* **123** R20
Fullerton Point, *Antigua & Barbuda* **123** B20
Fullerton, *Nebr., U.S.* **107** P7
Fulton, *N.Y., U.S.* **96** H8
Fulton, *Ill., U.S.* **101** N4
Fulton, *Ky., U.S.* **102** E12
Fulton, *Miss., U.S.* **103** H13
Fulton, *Mo., U.S.* **107** S13
Fumay, *Fr.* **148** M12
Fumel, *Fr.* **149** W8
Funabashi, *Japan* **189** P12
Funafara Island, *Tuvalu* **228** Q10
Funafuti, *Tuvalu* **228** N12
Funafuti Atoll, *Tuvalu* **224** H9
Funafuti Island, *Tuvalu* **228** M12
Funamanu, *island, Tuvalu* **228** P11
Funan Gaba, *Eth.* **211** P7
Funangongo, *island, Tuvalu* **228** N11
Funchal, *Madeira, Port.* **214** M3
Fundo de Figueiras, *Cabo Verde* **215** C18
Fundy, Bay of, *N. Amer.* **84** H9
Funhalouro, *Mozambique* **212** G11
Funiya Point, *N. Mariana Is., U.S.* **230** N9
Funtanasupanie Point, *N. Mariana Is., U.S.* **230** N9
Funtua, *Nigeria* **206** J8
Furancungo, *Mozambique* **212** D10
Furano, *Japan* **188** G13
Furmanov, *Russ.* **158** H11
Furnas Reservoir, *S. Amer.* **128** H9
Furneaux Group, *Austral.* **220** L8
Furqlus, *Syr.* **172** G8
Fürstenfeldbruck, *Ger.* **150** K8
Fürstenwalde, *Ger.* **150** D11
Furth im Wald, *Ger.* **150** J10
Fürth, *Ger.* **150** H8
Furukawa, *Japan* **188** L13
Fuscaldo, *It.* **153** P13
Fuse, *Japan* **189** N7
Fushun, *China* **185** F15
Füssen, *Ger.* **150** L8
Futiga, *American Samoa, U.S.* **230** E6
Futuna, *island, Vanuatu* **229** M18
Futuna, *island, Wallis & Futuna Is. Fr.* **230** Q11
Fuwwah, *Yemen* **174** P12

Fuwwah, *Egypt* **202** E11
Fuxin, *China* **185** F15
Fuya, *Japan* **188** L12
Fuyang, *China* **185** J14
Fuyu, *China* **185** B18
Fuyuan, *China* **185** B18
Fuyun, *China* **184** D6
Fuzhou, *China* **185** M15
Füzuli *see* Karaghbyur, *Azerb.* **171** E21
Fyn, *island, Den.* **142** N11
Fyne, Loch, *Scot., U.K.* **144** L8
Fyzabad, *Trin. & Tobago* **123** R22

G

Gaal Goble, *Somalia* **210** L14
Gabbs, *Nev., U.S.* **111** Q8
Gabela, *Angola* **208** L10
Gabert Island, *F.S.M.* **226** F11
Gabes, *Tunisia* **201** D15
Gabon, *Af.* **208** E8
Gabon, *river, Gabon* **208** E7
Gaborone, *Botswana* **212** H7
Gabras, *Sudan* **203** U8
Gabriel Vera, *Bol.* **132** K7
Gabro, *Eth.* **211** N10
Gabrovo, *Bulg.* **155** H15
Gabú, *Guinea-Bissau* **204** K8
Gacanka Xaafuun Koof, *bay, Somalia* **210** L15
Gacanka Xaafuun Waq, *bay, Somalia* **210** L15
Gackle, *N. Dak., U.S.* **106** G6
Gadag, *India* **183** P5
Gadamai, *Sudan* **203** P14
Gadarwara, *India* **182** K6
Gäddede, *Sw.* **142** G12
Gadeokdo, *island, S. Korea* **187** T11
Gadifuri, *island, Maldives* **183** W3
Gadrut *see* Hadrut, *Azerb.* **171** E21
Gadsden, *Ala., U.S.* **103** H16
Gadwal, *India* **183** P6
Gadzi, *Cen. Af. Rep.* **207** N13
Gael Hamke Bugt, *bay, Greenland, Den.* **88** H14
Găești, *Rom.* **155** F15
Gaeta, *It.* **152** L10
Gaeta, Golfo di, *It.* **152** M10
Gaferut, *island, F.S.M.* **226** A7
Gaffney, *S.C., U.S.* **98** J8
Gaflei, *Liech.* **162** P2
Gafsa, *Tunisia* **201** C15
Gagal, *Chad* **206** K12
Gagan, *P.N.G.* **222** P11
Gagan, *island, Marshall Is.* **227** G17
Gagarin, *Russ.* **158** J8
Gageodo, *island, S. Korea* **187** V5
Gagil-Tamil, *peak, F.S.M.* **226** D12
Gagnoa, *Côte d'Ivoire* **205** P13
Gagra, *Rep. of Georgia* **171** A15
Gahi, *Wallis & Futuna Is. Fr.* **230** R9
Gaillac, *Fr.* **149** X9
Gaillimh *see* Galway, *Ire.* **145** R4
Gaimán, *Arg.* **135** R9
Gainesville, *Ga., U.S.* **98** K5
Gainesville, *Fla., U.S.* **99** R7
Gainesville, *Tex., U.S.* **105** F17
Gainesville, *Mo., U.S.* **107** V13
Gáios, *Gr.* **156** F6
Gairdner, Lake, *Austral.* **220** J9
Gaja Shima, *Japan* **189** U3
Gakkel Ridge, *Arctic Oc.* **252** F11
Gakona, *Alas., U.S.* **113** H18
Gakuch, *Pak.* **180** G12
Gal Tardo, *Somalia* **211** Q12
Galadi, *Nigeria* **206** H7
Galadi, *Eth.* **211** P9
Galana, *river, Kenya* **211** U8
Galap, *Palau* **226** F6
Galápagos Fracture Zone, *Pac. Oc.* **247** K15
Galápagos Rise, *Pac. Oc.* **247** J20
Galashiels, *Scot., U.K.* **145** M11
Galatás, *Gr.* **156** K12
Galați, *Rom.* **155** E17
Galatia (Mehmetcik), *Cyprus* **162** M10
Galatina, *It.* **153** N15
Galatxo, Punta del, *Sp.* **147** F16
Galax, *Va., U.S.* **98** G9
Galaxidi, *Gr.* **156** H10

Galcaio, *Somalia* **211** N13
Gáldar, *Sp.* **214** Q5
Galdhøpiggen, peak, *Nor.* **142** J10
Galé, *Mali* **204** K10
Galegu, *Sudan* **203** S14
Galela, *Indonesia* **193** H16
Galena Park, *Tex., U.S.* **105** L19
Galena, *Ill., U.S.* **100** M4
Galena, *Kans., U.S.* **107** V10
Galena, *Alas., U.S.* **113** F14
Galeota Point, *Trin. & Tobago* **123** R23
Galera Point, *Trin. & Tobago* **123** N24
Galera Point, *S. Amer.* **128** D2
Galera Point, *S. Amer.* **128** N4
Galera, Punta, *Ecua.* **130** J3
Galera, Punta, *Chile* **135** P6
Galesburg, *Ill., U.S.* **101** P4
Galesville, *Wis., U.S.* **100** K3
Galeton, *Pa., U.S.* **96** L6
Galets, Pointe des, *Réunion, Fr.* **215** F15
Gali, *Rep. of Georgia* **171** B16
Galich, *Russ.* **158** G11
Galicia, region, *Sp.* **146** C6
Galicia, region, *Pol., Ukr.* **151** G17
Galilee, Lake, *Austral.* **220** F13
Galilee, Sea of, *Israel* **172** K6
Galim, *Cameroon* **206** M10
Galina Mine, *W. Aust., Austral.* **221** V2
Galina Point, *Jam.* **122** H10
Galinoporni (Kaleburnu), *Cyprus* **162** M11
Galio, *Liberia* **204** P12
Galion, *Ohio, U.S.* **101** Q13
Galiuro Mountains, *Ariz., U.S.* **109** V8
Galiwinku, *N.T., Austral.* **221** N9
Gallarate, *It.* **152** D4
Gallatin, *Tenn., U.S.* **103** E15
Gallatin, *Mo., U.S.* **107** N11
Gallatin, river, *Mont., U.S.* **108** E8
Galle, *Sri Lanka* **183** U7
Gallego Rise, *Pac. Oc.* **247** K20
Gállego, river, *Sp.* **147** D14
Gallegos, river, *S. Amer.* **128** Q5
Galleh Dār, *Iran* **177** M16
Gallieni Fracture Zone, *Ind. Oc.* **250** N7
Gallieni Spur, *Ind. Oc.* **250** Q11
Gallinakopf, peak, *Aust., Liech.* **162** P4
Gallinas Peak, *N. Mex., U.S.* **109** T13
Gallinas, Punta, *Col.* **130** B8
Gallipoli *see* Gelibolu, *Turk.* **170** D3
Gallipoli, *It.* **153** N15
Gallipoli, *N.T., Austral.* **221** R10
Gallipolis, *Ohio, U.S.* **101** S14
Gällivare, *Sw.* **143** E13
Gallup, *N. Mex., U.S.* **109** R10
Galoa, *Fiji* **226** Q5
Galt, *Calif., U.S.* **111** R4
Galtat Zemmour, *W. Sahara* **200** H4
Galty Mountains (An Chathair), *Ire.* **145** S4
Galva, *Ill., U.S.* **101** P4
Galveston Bay, *Tex., U.S.* **105** L20
Galveston Island, *Tex., U.S.* **105** M20
Galveston, *Tex., U.S.* **105** M20
Gálvez, *Arg.* **134** K11
Galwa, *Nepal* **182** F8
Galway (Gaillimh), *Ire.* **145** R4
Galway Bay, *Ire.* **145** R3
Gamba, *Gabon* **208** F7
Gambaga, *Ghana* **205** L16
Gambēla, *Eth.* **210** M5
Gambell, *Alas., U.S.* **112** F10
Gambia Basin, *Atl. Oc.* **248** M9
Gambia Plain, *Atl. Oc.* **248** M9
Gambia, river, *Af.* **204** J7
Gambia, The, *Af.* **204** K6
Gambie, river, *Af.* **204** K8
Gambier Fracture Zone, *Pac. Oc.* **255** E18
Gamboma, *Congo* **208** F11
Gamíla *see* Týmfi, Óros, *Gr.* **156** E7
Gamış Dağı *see* Gomshasar, peak, *Azerb.* **171** D20
Gamkonora, peak, *Indonesia* **193** H16
Gamlaha, *Indonesia* **193** H16
Gamleby, *Sw.* **143** L13
Gamprin, *Liech.* **162** M2
Gamua, *Wallis & Futuna Is., Fr.* **230** Q9
Gamud, peak, *Eth.* **211** Q7
Gan Gan, *Arg.* **135** R8
Gan, island, *Maldives* **183** Y3
Gan, river, *China* **185** M14
Gana, river, *Nigeria* **206** J9
Ganado, *Tex., U.S.* **105** M18

Ganado, *Ariz., U.S.* **109** R9
Ganda, *Angola* **208** N10
Gandajika, *D.R.C.* **209** J15
Gandava, *Pak.* **181** P7
Gandhinagar, *India* **182** K3
Gandía, *Sp.* **147** H15
Gandole, *Nigeria* **206** L10
Gang, island, *Maldives* **183** X3
Ganga (Ganges), river, *India* **182** H7
Gangala na Bodio, *D.R.C.* **209** B18
Ganganagar, *India* **182** F4
Gangapur, *India* **182** H5
Gangawati, *India* **183** P5
Gangdisi Shan *see* Kangdêsê Shan, *China* **184** K3
Ganges Fan, *Ind. Oc.* **251** D13
Ganges Plain, *Asia* **168** H8
Ganges River Delta, *Asia* **168** J8
Ganges, *Fr.* **149** X11
Ganges, Mouths of the, *Bangladesh, India* **182** L12
Ganges, river, *Bangladesh, India* **182** H7
Ganggu, *S. Korea* **187** Q12
Ganggyeong, *S. Korea* **187** Q8
Ganghwa, *S. Korea* **186** M7
Gangi, *It.* **153** S10
Gangjin, *S. Korea* **187** U8
Gangneung, *S. Korea* **186** M11
Gangrenboqi Feng *see* Kangrinboqê Feng, *China* **184** J2
Gangtok, *Bhutan* **182** G12
Ganhyeon, *S. Korea* **187** N9
Gania, *Guinea* **204** M9
Gannat, *Fr.* **149** T11
Gannett Peak, *Wyo., U.S.* **108** H10
Ganseong, *S. Korea* **186** L11
Ganta, *Liberia* **204** N11
Gantgaw, *Myanmar* **190** G4
Gantheaume Bay, *Austral.* **220** H2
Ganwo, *Nigeria* **206** J6
Ganyushkino, *Kaz.* **178** G4
Ganzhou, *China* **185** M14
Gao, *Mali* **205** H16
Gaoligong Shan, *Myanmar* **190** D7
Gaoua, *Burkina Faso* **205** M14
Gaoual, *Guinea* **204** L8
Gaozhou, *China* **184** P12
Gap, *Fr.* **149** W14
Gapado, island, *S. Korea* **187** X7
Gapyeong, *S. Korea* **186** M9
Garabil Belentligi, *Turkm.* **178** P9
Garabinzam, *Congo* **208** D9
Garabogaz (Bekdash), *Turkm.* **178** K4
Garabogaz Aylagy, bay, *Turkm.* **178** K5
Garabogaz Bay, *Asia* **168** G5
Garabogazköl, *Turkm.* **178** K4
Garacad, *Somalia* **211** N14
Garadag, *Somaliland* **210** L12
Garadase, *Eth.* **211** P7
Garagum Canal, *Turkm.* **178** N9
Garagum, *Turkm.* **178** L7
Garaina, *P.N.G.* **222** Q8
Garajonay, peak, *Sp.* **214** Q4
Garanhuns, *Braz.* **133** G17
Garapan, *N. Mariana Is., U.S.* **230** H7
Garara, *P.N.G.* **222** Q8
Garavuti, *Taj.* **178** N11
Garawe, *Liberia* **204** Q11
Garba Tula, *Kenya* **211** S7
Garbahaarrey, *Somalia* **211** Q9
Garber, *Okla., U.S.* **105** B16
Garberville, *Calif., U.S.* **111** N2
Garbyang, *India* **182** F8
García de Sola, Embalse de, *Sp.* **146** H9
Garda, Lago di, *It.* **152** D6
Garda, Lake, *It.* **152** D6
Gardaneh-ye Shīnak, pass, *Iran, Iraq* **176** E11
Gardar Ridge, *Atl. Oc.* **253** N22
Garden City, *Ga., U.S.* **99** N9
Garden City, *Kans., U.S.* **107** U4
Garden Hill, *Man., Can.* **87** M14
Garden Island Bay, *La., U.S.* **102** R12
Gardēz, *Afghan.* **180** J9
Gardiner Seamounts, *Ind. Oc.* **250** H11
Gardiner, *Me., U.S.* **97** G16
Gardiner, *Mont., U.S.* **108** F8
Garessio, *It.* **152** F3
Garet el Djenoun, *Alg.* **201** J13
Garf Husein, ruin, *Egypt* **202** L12
Garfield, *Wash., U.S.* **110** D9
Gargaliánoi, *Gr.* **156** L9

Gargano, Promontorio del, *It.* **152** L12
Gargano, Testa del, *It.* **152** K13
Gargouna, *Mali* **205** H16
Garhakota, *India* **182** J7
Garibaldi, *Oreg., U.S.* **110** F2
Garies, *S. Af.* **212** L4
Garim Island, *F.S.M.* **226** E11
Garissa, *Kenya* **211** S8
Garland, *Tex., U.S.* **105** G18
Garmābāk, *Afghan.* **180** J5
Garmeh Jājarm, *Iran* **177** D19
Garmisch-Partenkirchen, *Ger.* **150** L8
Garmna (Gorumna Island), *Ire.* **145** R3
Garner, *Iowa, U.S.* **106** M11
Garnet Bank, *Atl. Oc.* **254** R6
Garnett, *Kans., U.S.* **107** T10
Garonne, river, *Fr.* **149** X8
Garoowe, *Somalia* **210** M13
Garou Boulaï, *Cen. Af. Rep.* **206** M12
Garoua, *Cameroon* **206** K11
Garove, island, *P.N.G.* **222** N9
Garreru, island, *Palau* **226** H6
Garrett Fracture Zone, *Pac. Oc.* **247** L18
Garrison, *N. Dak., U.S.* **106** F4
Garrison, *Utah, U.S.* **109** N5
Garron Point, *N. Ire., U.K.* **145** N7
Garrucha, *Sp.* **147** L13
Garry Lake, *Nunavut, Can.* **87** G14
Garrygala, *Turkm.* **178** M6
Garsen, *Kenya* **211** T8
Garsila, *Sudan* **203** S6
Gartog (Gatuo), *China* **184** L8
Garu, *India* **182** K9
Garusuun, *Palau* **226** H6
Garut, *Indonesia* **192** M8
Garve, *Scot., U.K.* **144** H9
Garwa, *India* **182** J9
Gary, *Ind., U.S.* **101** N8
Garzón, *Col.* **130** H5
Gas City, *Ind., U.S.* **101** Q10
Gas Hure Hu, *China* **184** G6
Gasa, *Bhutan* **182** G12
Gasado, island, *S. Korea* **187** U6
Gasan, island, *Marshall Is.* **227** V13
Gasconade, river, *Mo., U.S.* **107** U9
Gascony, region, *Fr.* **149** X7
Gascoyne Junction, *W. Aust., Austral.* **221** U2
Gascoyne Plain, *Ind. Oc.* **251** J17
Gascoyne Seamount, *Pac. Oc.* **255** E21
Gascoyne, river, *Austral.* **220** G2
Gåsefjord, *Greenland, Den.* **88** L13
Gåseland, *Greenland, Den.* **88** L13
Gash, river, *Eritrea* **210** H6
Gashaka, *Nigeria* **206** L10
Gashua, *Nigeria* **206** H10
Gaskell Ridge, *Ind. Oc.* **250** G8
Gaspar Grande, island, *Trin. & Tobago* **123** P21
Gasparillo, *Trin. & Tobago* **123** Q22
Gaspé Peninsula, *Que., Can.* **87** N21
Gaspé, *Que., Can.* **87** N21
Gassaway, *W. Va., U.S.* **98** D9
Gassol, *Nigeria* **206** L9
Gaston, Lake, *N.C., Va., U.S.* **98** G12
Gastonia, *N.C., U.S.* **98** J9
Gastoúni, *Gr.* **156** J8
Gastre, *Arg.* **135** Q8
Gata, Cabo de, *Sp.* **146** M12
Gata, Cape, *Cyprus* **162** R7
Gata, Sierra de, *Sp.* **146** F8
Gataivai, *Samoa* **227** N18
Gatchina, *Russ.* **158** F6
Gate City, *Va., U.S.* **98** G7
Gate, *Okla., U.S.* **105** A14
Gatehouse of Fleet, *Scot., U.K.* **145** N9
Gates of the Arctic National Park and Preserve, *Alas., U.S.* **113** D16
Gateshead, *Eng., U.K.* **145** N12
Gatesville, *Tex., U.S.* **105** J16
Gateway, *Colo., U.S.* **109** N10
Gatjapar, *F.S.M.* **226** D12
Gatlinburg, *Tenn., U.S.* **103** F19
Gatuo *see* Gartog, *China* **184** L8
Gau, island, *Fiji* **226** Q6
Gauido, island, *S. Korea* **187** P6
Gauss Fracture Zone, *Ind. Oc.* **250** N7
Gauss Halvø, peninsula, *Greenland, Den.* **88** H13
Gausta, peak, *Nor.* **142** K10
Gautheaume Point, *Austral.* **220** D4
Gavarnie, *Fr.* **149** Z7
Gavāter, *Iran* **177** P23
Gāvbandī, *Iran* **177** N16

Gavdopoúla, island, *Gr.* **157** R13
Gávdos, island, *Gr.* **157** R13
Gavien, *P.N.G.* **222** N6
Gaviota, *Calif., U.S.* **111** W6
Gavrilov Yam, *Russ.* **158** H10
Gávrio, *Gr.* **157** J14
Gawai, *Myanmar* **190** C6
Gawler Ranges, *Austral.* **220** J9
Gawler, *S. Aust., Austral.* **221** X10
Gaya, *India* **182** J10
Gaya, *Niger* **206** J6
Gaya, *Nigeria* **206** J9
Gayaza, *Uganda* **211** T2
Gayle, *Jam.* **122** H9
Gaylord, *Mich., U.S.* **100** H11
Gaylord, *Minn., U.S.* **106** K10
Gayny, *Russ.* **158** E16
Gay's Cove, *Barbados* **123** J19
Gaza City *see* Ghazzah, *Gaza Strip* **173** N4
Gaza Strip, *Asia* **173** N4
Gazak, *Iran* **177** M21
Gazak, *Afghan.* **180** G6
Gazamni, *Niger* **206** H9
Gazanjyk, *Turkm.* **178** M6
Gazaoua, *Niger* **206** H8
Gazelle Fracture Zone, *Ind. Oc.* **250** N7
Gazi, *Kenya* **211** V8
Gaziantep, *Turk.* **170** H12
Gazimağusa *see* Famagusta, *Cyprus* **162** P10
Gazipaşa, *Turk.* **170** J7
Gazojak, *Turkm.* **178** L8
Gbadolite, *D.R.C.* **209** B14
Gbarnga, *Liberia* **204** P10
Gberia Fotombu, *Sierra Leone* **204** M9
Gbérouboué, *Benin* **205** L18
Gboko, *Nigeria* **206** L9
Gdańsk, *Pol.* **151** B14
Gdańsk, Gulf of, *Pol., Russ.* **151** A14
Gdov, *Russ.* **158** G5
Gdynia, *Pol.* **151** B14
Gea Pass, *Marshall Is.* **227** J17
Gea, island, *Marshall Is.* **227** J17
Geary, *Okla., U.S.* **105** C16
Gebe, island, *Indonesia* **193** J17
Gebeit, *Sudan* **202** M14
Gebeit, *Sudan* **203** N15
Gebel Adda, ruin, *Egypt* **202** L11
Gebel Barkal *see* Jabal Barkal, *Sudan* **203** P12
Gebiz, *Turk.* **170** H6
Gebze, *Turk.* **170** D5
Gech'a, *Eth.* **211** N5
Gēdo, *Eth.* **210** M7
Gedser, *Den.* **142** N12
Geel, *Belg.* **148** K12
Geelong, *Vic., Austral.* **221** Z12
Geelvinck Fracture Zone, *Ind. Oc.* **250** P12
Geelvink Channel, *Austral.* **220** J2
Gees Gwardafuy, cape, *Somalia* **210** K15
Geesthacht, *Ger.* **150** C8
Geeveston, *Tas., Austral.* **221** Z7
Gefara, region, *Lib.* **201** E16
Géfyra, *Gr.* **156** C11
Gegibu, island, *Marshall Is.* **227** G15
Gehh, island, *Marshall Is.* **227** J17
Geidam, *Nigeria* **206** H10
Geiga, island, *Marshall Is.* **227** F14
Geilo Hills, *Kenya* **211** Q9
Geilo, *Nor.* **142** J10
Geita, *Tanzania* **211** U3
Gejiu, *China* **184** P9
Gel, river, *S. Sudan* **203** W10
Gela, *It.* **153** T11
Gela, Golfo di, *It.* **153** T10
Geladī, *Eth.* **211** N12
Gelahun, *Liberia* **204** N10
Gelai, peak, *Tanzania* **211** U6
Gelam, island, *Indonesia* **192** K9
Gelasa, Selat, *Indonesia* **192** K8
Gelendzhik, *Russ.* **158** P9
Gélengdeng, *Chad* **206** J12
Gelhak, *S. Sudan* **203** T12

Gelibolu (Gallipoli), *Turk.* **170** D3
Gellinam, island, *Marshall Is.* **227** H18
Gellinsoor, *Somalia* **211** N12
Gelsenkirchen, *Ger.* **150** E5
Gembloux, *Belg.* **148** L12
Gemena, *D.R.C.* **209** C13
Gemerek, *Turk.* **170** F11
Gemi Konaghi *see* Karavostasi, *Cyprus* **162** N7
Gemlik Körfezi, *Turk.* **170** D4
Gemlik, *Turk.* **170** D5
Gemona del Friuli, *It.* **152** C9
Genalē, river, *Eth.* **211** P9
Genç, *Turk.* **171** F15
General Acha, *Arg.* **135** N10
General Alvear, *Arg.* **134** L8
General Carrera, Lago, *Chile* **135** T7
General Conesa, *Arg.* **135** P10
General Eugenio A. Garay, *Parag.* **132** L8
General Güemes, *Arg.* **134** E9
General Guido, *Arg.* **134** M13
General Juan Madariage, *Arg.* **134** M13
General La Madrid, *Arg.* **135** N11
General Levalle, *Arg.* **134** L10
General Lorenzo Vintter, *Arg.* **135** Q10
General Paz, *Arg.* **134** G13
General Pico, *Arg.* **134** M10
General Pinedo, *Arg.* **134** G11
General Roca, *Arg.* **135** P8
General San Martín, *Arg.* **134** L13
General San Martín, *Arg.* **135** N10
General San Martín, *Arg.* **135** S7
General Santos, *Philippines* **193** F15
Genesee, river, *N.Y., U.S.* **96** K6
Geneseo, *N.Y., U.S.* **96** J6
Geneseo, *Ill., U.S.* **101** P4
Geneva *see* Genève, *Switz.* **152** C1
Geneva, *N.Y., U.S.* **96** J7
Geneva, *Ohio, U.S.* **101** N15
Geneva, *Ala., U.S.* **103** N16
Geneva, *Nebr., U.S.* **107** Q7
Geneva, *Idaho, U.S.* **108** J8
Geneva, Lake *see* Léman, Lac, *Fr.* **149** T15
Genève (Geneva), *Switz.* **152** C1
Genhe, *China* **185** B14
Genil, river, *Sp.* **146** L9
Genlis, *Fr.* **149** R13
Gennádion, *Gr.* **157** N19
Gennargentu, Monti del, *It.* **153** N4
Genoa *see* Genova, *It.* **152** F4
Genoa, *Nebr., U.S.* **107** P7
Génolhac, *Fr.* **149** W12
Genova (Genoa), *It.* **152** F4
Genova, Golfo di, *It.* **152** G4
Gent, *Belg.* **148** L11
Geoado, island, *S. Korea* **187** P7
Geochang, *S. Korea* **187** R10
Geogeumdo, island, *S. Korea* **187** U8
Geographe Bay, *Austral.* **220** K2
Geographe Channel, *Austral.* **220** G1
Geographical Society Ø, *Greenland, Den.* **88** J13
Geoje, *S. Korea* **187** T11
Geojedo, island, *S. Korea* **187** T11
Geojin, *S. Korea* **186** L11
Geologists Range, *Antarctica* **239** L14
Geomundo, island, *S. Korea* **187** V8
George Bligh Bank, *Atl. Oc.* **248** F10
George Dog Island, *Virgin Is., U.K.* **122** Q8
George Land, *Asia* **168** B8
George Rogers Clark National Historical Park, *Ind., U.S.* **101** T7
George Town, *Cayman Is., U.K.* **122** N2
George Town, *Malaysia* **191** T7
George Town, *Tas., Austral.* **221** Y7
George V Coast, *Antarctica* **239** Q17
George V Fracture Zone, *Ind. Oc.* **255** D18
George VI Sound, *Antarctica* **238** F6
George Washington Birthplace National Monument, *Va., U.S.* **98** D13
George Washington Carver National Monument, *Mo., U.S.* **107** V11
George West, *Tex., U.S.* **105** N16
George, *Russ.* **160** B12
George, *S. Af.* **212** M6
George, Lake, *N.Y., U.S.* **96** H11

George, Lake, *Fla., U.S.* **99** S8
George, Lake, *Austral.* **220** F5
George, Lake, *Austral.* **220** L14
George, river, *N. Amer.* **84** F9
George, river, *Que., Can.* **87** K20
Georges Bank, *Atl. Oc.* **248** J4
Georgetown, *Del., U.S.* **98** D15
Georgetown, *S.C., U.S.* **98** L11
Georgetown, *Ill., U.S.* **101** S20
Georgetown, *Ohio, U.S.* **101** T12
Georgetown, *Ky., U.S.* **103** B18
Georgetown, *Tex., U.S.* **105** K17
Georgetown, *Bahamas* **122** F9
Georgetown, *St. Vincent & the Grenadines* **123** K17
Georgetown, *Guyana* **131** E16
Georgetown, *Qnsld., Austral.* **221** R12
Georgia Basin, *Atl. Oc.* **249** V8
Georgia, *U.S.* **98** M6
Georgia, *Asia* **171** B17
Georgia, Strait of, *N. Amer.* **84** G3
Georgian Bay, *Ont., Can.* **87** Q18
Georgiana, *Ala., U.S.* **103** M15
Georgīevka, *Kaz.* **179** E16
Georgina, river, *Austral.* **220** E10
Georgiyevsk, *Russ.* **159** T13
Georgiyevskoye, *Russ.* **158** G12
Gera, *Ger.* **150** F9
Geráki, *Gr.* **156** L10
Geral de Goiás, Serra, *Braz.* **133** H14
Geral, *S. Amer.* **128** J8
Geraldine, *Mont., U.S.* **108** C9
Geraldine, *N.Z.* **223** M18
Geraldton, *Ont., Can.* **87** N16
Geraldton, *W. Aust., Austral.* **221** W2
Gérardmer, *Fr.* **149** G14
Gerasa *see* Jarash, *Jordan* **172** L7
Gerāsh, *Iran* **177** M17
Gerber, *Calif., U.S.* **111** N4
Gercüş, *Turk.* **171** H16
Gerede, *Turk.* **170** D7
Gereshk *see* Girishk, *Afghan.* **180** L5
Geretsried, *Ger.* **150** K8
Gering, *Nebr., U.S.* **107** N1
Gerlach, *Nev., U.S.* **111** N7
Gerlach, peak, *Eur.* **140** H9
Gerlachovsky Štit, peak, *Slovakia* **151** J16
Germania Land, *Greenland, Den.* **88** F13
Germantown, *Tenn., U.S.* **102** G11
Germany, *Eur.* **150** F8
Germering, *Ger.* **150** K8
Germfask, *Mich., U.S.* **100** F9
Germī, *Iran* **177** C13
Gerolakkos, *Cyprus* **162** N8
Geroliménas, *Gr.* **156** M10
Gerpir, *Eur.* **140** B4
Gersthofen, *Ger.* **150** K8
Gertrud Rask Land, *Greenland, Den.* **88** A9
Gerze, *Turk.* **170** C10
Gesoa, *P.N.G.* **222** Q6
Getafe, *Sp.* **146** F11
Gettysburg, *Pa., U.S.* **96** Q7
Gettysburg, *S. Dak., U.S.* **106** J5
Getz Ice Shelf, *Antarctica* **238** M8
Geum, river, *S. Korea* **187** R8
Geumdangdo, island, *S. Korea* **187** U8
Geumil, *S. Korea* **187** U8
Geumodo, island, *S. Korea* **187** U8
Geumpang, *Indonesia* **192** G4
Geumsan, *S. Korea* **187** R9
Geumwang, *S. Korea* **187** P9
Geunto, *Russ.* **161** C21
Gevaş, *Turk.* **171** G17
Gewanē, *Eth.* **210** L8
Gex, *Fr.* **149** T14
Geylegphug, *Bhutan* **182** H13
Geyve, *Turk.* **170** D6
Gézenti, *Chad* **207** D13
Gədəbəy, *Azerb.* **171** D19
Gəncə, *Azerb.* **171** D20
Ghadāmis, *Lib.* **201** F15
Ghaddūwah, *Lib.* **201** H18
Ghaghe, island, *Solomon Is.* **228** D8
Għajn Tuffieħa Bay, *Malta* **163** K21
Ghana, *Af.* **205** N15
Ghanzi, *Botswana* **212** G6
Ghar Dalum, site, *Malta* **163** L23
Gharandal, *Jordan* **173** Q5
Għarb, *Malta* **163** H19
Gharbi Island, *Tunisia* **201** C16
Ghardaïa, *Alg.* **200** D12
Ghārib, *Egypt* **202** G12

Gharo, *Pak.* **181** S7
Gharyān, *Lib.* **201** E17
Ghāt, *Lib.* **201** J16
Ghatere, *Solomon Is.* **228** D9
Għawdex *see* Gozo, island, *Malta* **163** H20
Għaxaq, *Malta* **163** L23
Ghazaouet, *Alg.* **200** C10
Ghaziabad, *India* **182** G6
Ghazipur, *India* **182** H9
Ghaznī, *Afghan.* **180** J8
Ghazzah (Gaza City), *Gaza Strip* **173** N4
Gheorghieni, *Rom.* **155** C15
Gherla, *Rom.* **155** C13
Ghilarza, *It.* **153** N3
Ghisonaccia, *Fr.* **149** Y3
Ghōriyān, *Afghan.* **180** H3
Ghōrmāch, *Afghan.* **180** G4
Ghotki, *Pak.* **181** P9
Ghow Gardān, Kōtal-e, *Afghan.* **180** H7
Ghubbat al Qamar, *Ind. Oc.* **168** J4
Ghubbat al Qamar, bay, *Yemen* **175** N14
Ghubbat Şawqirah, bay, *Oman* **175** L17
Ghurd Abū Muḥarrik, *Egypt* **202** G10
Gia Nghĩa, *Vietnam* **191** N12
Giá Rai, *Vietnam* **191** Q11
Giannitsá, *Gr.* **156** C10
Giannutri, Isola di, *It.* **152** K6
Giarre, *It.* **153** S12
Giba, *It.* **153** Q3
Gibara, *Cuba* **118** J12
Gibb River, *W. Aust., Austral.* **221** Q6
Gibbon Point, *Anguilla, U.K.* **122** Q12
Gibbon, *Nebr., U.S.* **107** Q6
Gibbonsville, *Idaho, U.S.* **108** E6
Gibb's Hill, *Bermuda* **119** C14
Gibraleón, *Sp.* **146** L7
Gibraltar Airport, *Gibraltar, U.K.* **163** N23
Gibraltar Harbour, *Gibraltar, U.K.* **163** P22
Gibraltar, *Gibraltar, U.K.* **163** F1
Gibraltar, Bay of (Bahía de Algeciras), *Gibraltar, U.K.* **163** N21
Gibraltar, Rock of, *Gibraltar, U.K.* **163** P23
Gibraltar, Strait of, *Af. Eur.* **146** N8
Gibsland, *La., U.S.* **102** L7
Gibson City, *Ill., U.S.* **101** Q6
Gibson Desert, *Austral.* **220** F6
Gīdamī, *Eth.* **210** M5
Gidar, *Pak.* **181** P6
Giddalur, *India* **183** P6
Giddings, *Tex., U.S.* **105** K17
Giel, *S. Sudan* **203** T12
Gien, *Fr.* **149** R10
Giesecke Bræer, glacier, *Greenland, Den.* **88** J6
Giessen, *Ger.* **150** F10
Gifford, *Fla., U.S.* **99** U10
Gifford, river, *Nunavut, Can.* **87** F16
Gifhorn, *Ger.* **150** D8
Gifu, *Japan* **189** P9
Gigant, *Russ.* **159** R11
Gigha Island, *Scot., U.K.* **145** M7
Giglio, Isola del, *It.* **152** K6
Gijang, *S. Korea* **187** S12
Gijón, *Sp.* **146** A9
Gikongoro, *Rwanda* **211** U2
Gila Bend Mountains, *Ariz., U.S.* **109** U6
Gila Bend, *Ariz., U.S.* **109** U6
Gila Cliff Dwellings National Monument, *N. Mex., U.S.* **109** U10
Gila, river, *Ariz., N. Mex., U.S.* **109** U5
Gilbert Islands, *Kiribati* **224** G8
Gilbert Ridge, *Pac. Oc.* **246** J10
Gilbert Seamounts, *Pac. Oc.* **247** C14
Gilbert, *Minn., U.S.* **106** F12
Gilbert, *Ariz., U.S.* **109** U7
Gilbert, river, *Austral.* **220** D12
Gilbués, *Braz.* **133** G14
Gilchrist, *Oreg., U.S.* **110** J4
Gilderen *see* Lapithos, *Cyprus* **162** N8
Gildford, *Mont., U.S.* **108** B9
Gilgandra, *N.S.W., Austral.* **221** W14
Gilgit, *Pak.* **180** G13
Gilgit, region, *Pak.* **180** G12
Gilgit, river, *Pak.* **180** G12
Gill, Lough, *Ire.* **145** P4
Gillam, *Man., Can.* **87** L14

Gillen, Lake, *Austral.* **220** G5
Gilles, Lake, *Austral.* **220** K10
Gillespie, *Ill., U.S.* **101** S4
Gillett, *Ark., U.S.* **102** H9
Gillette, *Wyo., U.S.* **108** G12
Gillock Island, *Antarctica* **239** F20
Gills Rock, *Wis., U.S.* **100** H8
Gilman, *Ill., U.S.* **101** Q7
Gilmer, *Tex., U.S.* **105** H16
Gilmore Hut, *Qnsld., Austral.* **221** U13
Gilpeppee, *Qnsld., Austral.* **221** U12
Gilroy, *Calif., U.S.* **111** T4
Gima, *Japan* **189** X1
Gīmbī, *Eth.* **210** M6
Gimcheon, *S. Korea* **187** Q10
Gimhae, *S. Korea* **187** S11
Gimhwa (Kŭmhwa), *S. Korea* **186** L9
Gimje, *S. Korea* **187** R8
Gimli, *Man., Can.* **87** N14
Gimnyeong, *S. Korea* **187** W8
Gimpo, *S. Korea* **186** M7
Gīnda‘, *Eritrea* **210** H8
Gingin, *W. Aust., Austral.* **221** X2
Gingoog, *Philippines* **193** E15
Gīnīr, *Eth.* **211** N9
Ginowan, *Japan* **189** X2
Gioia del Colle, *It.* **152** M14
Gioia Tauro, *It.* **153** R13
Gioia, Golfo di, *It.* **153** R13
Gióna, Óros, *Gr.* **156** H10
Gioúra, island, *Gr.* **157** F13
Giovinazzo, *It.* **152** L13
Gippsland, region, *Austral.* **220** M13
Gir Hills, *India* **182** L2
Girab, *India* **182** H2
Girard, *Pa., U.S.* **96** L3
Girard, *Kans., U.S.* **107** U10
Girardot, *Col.* **130** G6
Giresun, *Turk.* **171** D13
Giri, river, *D.R.C.* **208** D12
Giridih, *India* **182** J10
Girishk (Gereshk), *Afghan.* **180** L5
Girne *see* Kyrenia, *Cyprus* **162** N8
Girolata, *Fr.* **149** X2
Girona, *Sp.* **147** D18
Gironde, bay, *Fr.* **149** U6
Girvas, *Russ.* **158** D8
Gisborne, *N.Z.* **223** G22
GISP2 *see* Greenland Ice Sheet Project 2, *Greenland, Den.* **88** J10
Gissi, *It.* **152** K11
Gitega, *Burundi* **211** U2
Giulianova, *It.* **152** J10
Giurgiu, *Rom.* **155** G15
Givet, *Fr.* **148** M12
Givors, *Fr.* **149** U12
Giza *see* Al Jīzah, *Egypt* **202** F11
Gizab, *Afghan.* **180** J6
Gizhduvan, *Uzb.* **178** M10
Gizhukoe, *Kaz.* **179** E17
Gizo, *Solomon Is.* **228** D6
Giżycko, *Pol.* **151** B16
Gjakovë (Đakovica), *Kos.* **154** J10
Gjelsvik Mountains, *Antarctica* **239** B13
Gjeravica, peak, *Kos.* **154** J10
Gjilan (Gnjilane), *Kos.* **154** J11
Gjirokastër (Argyrus), *Alban.* **154** M10
Gjoa Haven, *Nunavut, Can.* **87** F14
Gjøvik, *Nor.* **142** J11
Gkreko, Cape , *Cyprus* **162** P10
Glachau, *Ger.* **150** G9
Glacier Bay National Park and Preserve, *Alas., U.S.* **113** L21
Glacier Bay, *Alas., U.S.* **113** K21
Glacier de France, *Greenland, Den.* **89** P11
Glacier National Park, *Mont., U.S.* **108** B6
Glacier Peak, *Wash., U.S.* **110** B5
Gladewater, *Tex., U.S.* **105** G20
Gladstone, *Mich., U.S.* **100** G8
Gladstone, *Qnsld., Austral.* **221** T15
Gladwin, *Mich., U.S.* **100** K11
Glåma, river, *Nor.* **142** J11
Glasco, *Kans., U.S.* **107** S7
Glasgow, *Ky., U.S.* **103** D16
Glasgow, *Mo., U.S.* **107** S12
Glasgow, *Mont., U.S.* **108** B11
Glasgow, *Jam.* **122** H6
Glasgow, *Scot., U.K.* **144** L9
Glass Buttes, *Oreg., U.S.* **110** J6
Glass Mountain, *Calif., U.S.* **110** L4
Glass Mountains, *Tex., U.S.* **104** K10
Glassboro, *N.J., U.S.* **96** Q9

Glazov, *Russ.* **158** F15
Glen Affric, *Scot., U.K.* **144** J8
Glen Arbor, *Mich., U.S.* **100** H9
Glen Canyon Dam, *Ariz., U.S.* **109** Q7
Glen Canyon National Recreation Area, *Utah, U.S.* **109** P8
Glen Innes, *N.S.W., Austral.* **221** W15
Glen Mor, *Scot., U.K.* **144** J9
Glen Rose, *Tex., U.S.* **105** H16
Glen Ullin, *N. Dak., U.S.* **106** G3
Glénan, Îles de, *Fr.* **149** R3
Glenariffe, *N. Ire., U.K.* **145** N7
Glenarm, *N. Ire., U.K.* **145** N7
Glenavy, *N.Z.* **223** N18
Glenayle, *W. Aust., Austral.* **221** U4
Glenbarr, *Scot., U.K.* **145** M7
Glencoe, *Minn., U.S.* **106** K11
Glendale, *Ariz., U.S.* **109** U7
Glendale, *Calif., U.S.* **111** X8
Glendive, *Mont., U.S.* **108** C13
Glendo Reservoir, *Wyo., U.S.* **108** H13
Glendo, *Wyo., U.S.* **108** J13
Glenelg, river, *Austral.* **220** M11
Glengarriff (An Gleann Garbh), *Ire.* **145** U3
Glengyle, *Qnsld., Austral.* **221** T11
Glenmora, *La., U.S.* **102** N8
Glenmorgan, *Qnsld., Austral.* **221** V14
Glenn Highway, *Alas., U.S.* **113** J17
Glennallen, *Alas., U.S.* **113** H18
Glenns Ferry, *Idaho, U.S.* **108** H4
Glennville, *Ga., U.S.* **99** N8
Glenrock, *Wyo., U.S.* **108** H12
Glenrothes, *Scot., U.K.* **144** L10
Glens Falls, *N.Y., U.S.* **96** H11
Glenties (Na Gleannta), *Ire.* **145** N5
Glenwood Springs, *Colo., U.S.* **108** M11
Glenwood, *Ark., U.S.* **102** H7
Glenwood, *Minn., U.S.* **106** J9
Glenwood, *Iowa, U.S.* **107** Q9
Glenwood, *N. Mex., U.S.* **109** U10
Gletscherland, *Greenland, Den.* **88** J12
Glidden, *Wis., U.S.* **100** F4
Glina, *Croatia* **154** E6
Głogów, *Pol.* **150** E12
Glorenza, *It.* **152** C6
Glossglockner, peak, *Aust.* **150** M9
Gloster, *Miss., U.S.* **102** N10
Gloucester, *Mass., U.S.* **97** K15
Gloucester, *Eng., U.K.* **145** T11
Gloucester, *N.S.W., Austral.* **221** X15
Gloucester, *P.N.G.* **222** P8
Glover Island, *Grenada* **123** M22
Gloversville, *N.Y., U.S.* **96** J10
Glubokiy, *Russ.* **159** Q11
Glubokiy, *Russ.* **159** R12
Glŭbokoe, *Kaz.* **179** E17
Glyn Ebwy *see* Ebbw Vale, *Wales, U.K.* **145** U10
Gmunden, *Aust.* **150** K10
Gniezno, *Pol.* **151** D13
Gnjilane *see* Gjilan, *Kos.* **154** J11
Gò Công, *Vietnam* **191** Q12
Goa, *India* **183** P4
Goa, *S. Korea* **187** Q10
Goalpara, *India* **182** H13
Goba, *Eth.* **211** N8
Gobabis, *Namibia* **212** G4
Gobernador Duval, *Arg.* **135** N9
Gobernador Gregores, *Chile* **135** U8
Gobi, desert, *China, Mongolia* **184** F11
Gobō, *Japan* **189** R8
Gobu, *S. Korea* **187** S8
Gochang, *S. Korea* **187** S7
Godalming, *Eng., U.K.* **145** V13
Godavari, Mouths of the, *India* **183** P8
Godavari, river, *India* **182** M5
Godē, *Eth.* **211** N10
Godel Iceport, bay, *Antarctica* **239** B16
Godere, *Eth.* **211** P11
Goderville, *Fr.* **149** N8
Godhavn *see* Qeqertarsuaq, *Greenland, Den.* **88** M5
Godhra, *India* **182** K4
Gödöllő, *Hung.* **151** L15
Godoy Cruz, *Arg.* **134** K7
Gods Lake, *Man., Can.* **87** L14
Gods, river, *Man., Can.* **87** L15
Godthåb *see* Nuuk, *Greenland, Den.* **89** Q5
Godthåb Valley, *Atl. Oc.* **253** R17

Godthåbsfjord *see* Nuup Kangerlua, *Greenland, Den.* **89** Q6
Godwin Austen *see* K2, peak, *China, Pak.* **180** G14
Goélands, Lac aux, *Que., Can.* **87** K20
Goes, *Neth.* **148** K11
Gofar Fracture Zone, *Pac. Oc.* **247** K19
Gogebic Range, *U.S.* **91** C15
Gogebic, Lake, *Mich., U.S.* **100** F5
Gogeumdo, island, *S. Korea* **187** U8
Gogland, *Russ.* **158** E5
Gognou, *Benin* **205** L18
Gogrial, *S. Sudan* **203** V9
Goha, *Eth.* **210** L5
Gohan, *S. Korea* **187** N11
Goheung, *S. Korea* **187** U9
Goianésia, *Braz.* **133** J13
Goiânia, *Braz.* **133** J13
Goikul, *Palau* **226** G6
Góis, *Port.* **146** F6
Gojeb, river, *Eth.* **211** N6
Gojōme, *Japan* **188** K12
Gojra, *Pak.* **180** L11
Gök, river, *Turk.* **170** C10
Gokak, *India* **183** P4
Gökçeada (Imbros), island, *Turk.* **170** D2
Gökdepe, *Turkm.* **178** N7
Gökova Körfezi, *Turk.* **170** H3
Gokseong, *S. Korea* **187** S9
Göksu, river, *Turk.* **170** G11
Göksu, river, *Turk.* **170** H8
Göksun, *Turk.* **170** G11
Gol Bax, *Somalia* **211** N9
Golaghat, *India* **182** H14
Golan Heights, *Israel* **172** K6
Gölbaşı, *Turk.* **170** E8
Gölbaşı, *Turk.* **170** H12
Golconda, *Nev., U.S.* **110** M9
Gölcük, *Turk.* **170** D5
Gold Beach, *Oreg., U.S.* **110** K1
Gold Coast, *Ghana* **205** Q15
Gold Coast, *Qnsld., Austral.* **221** V16
Gold Mines, *Egypt* **202** L12
Golden Bo'sunbird Seamounts, *Ind. Oc.* **251** J16
Golden Gate National Recreation Area, *Calif., U.S.* **111** R3
Golden Gate, *Calif., U.S.* **111** S3
Golden Grove, *Jam.* **122** J12
Golden Meadow, *La., U.S.* **102** R11
Golden Spike National Historic Site, *Utah, U.S.* **108** K6
Golden, *Colo., U.S.* **108** M13
Goldendale, *Wash., U.S.* **110** E5
Goldfield, *Nev., U.S.* **111** S9
Goldsboro, *N.C., U.S.* **98** H12
Goldsborough Seamounts, *Pac. Oc.* **246** H12
Goldsmith, *Tex., U.S.* **104** H11
Goldsworthy, *W. Aust., Austral.* **221** S3
Goldthwaite, *Tex., U.S.* **105** J16
Göle, *Turk.* **171** D17
Goleniów, *Pol.* **150** C11
Goleta, *Calif., U.S.* **111** W7
Golfe de Tadjoura, *Djibouti* **210** K10
Golfe d'Olindé, *Gabon* **208** E7
Golfito, *C.R.* **117** N19
Golfo Aranci, *It.* **152** M5
Goliad, *Tex., U.S.* **105** N17
Gölköy, *Turk.* **170** D12
Golmud, *China* **184** H7
Golo, river, *Fr.* **149** X3
Gölören, *Turk.* **170** G9
Golovin, *Alas., U.S.* **112** F12
Golovin, *Russ.* **158** F13
Golpāyegān, *Iraq* **177** G14
Gölpazarı, *Turk.* **170** D6
Golūbovka, *Kaz.* **179** C13
Golyam Perelik, peak, *Bulg.* **155** K14
Golyama Kamchiya, river, *Bulg.* **155** H16
Goma, *D.R.C.* **209** F19
Gombari, *D.R.C.* **209** C18
Gombe, *Nigeria* **206** K10
Gombi, *Nigeria* **206** K11
Gombrani Island, *Mauritius* **215** J20
Gomera, *D.R.C.* **200** F2
Gomera, island, *Sp.* **214** Q4
Gómez Palacio, *Mex.* **116** F10
Gomshasar (Gamış Dağı), peak, *Azerb.* **171** D20
Gonābād, *Iran* **177** F20
Gonaïves, *Haiti* **119** L16

Gonam, Russ. 161 H19
Gonam, river, Russ. 161 J18
Gonâve, Golfe de la, Haiti 119 L16
Gonâve, Île de la, Haiti 119 M15
Gonbad-e Kāvūs, Iran 177 D18
Gonda, India 182 H8
Gondar, Eth. 210 J7
Gonder, Eth. 210 K6
Gondey, Chad 207 K14
Gondia, India 182 L7
Gondola, Mozambique 212 F10
Gondomar, Port. 146 E5
Gondwana Station, Antarctica 239 P15
Gönen, Turk. 170 D3
Gonghe, China 184 H9
Gongju, S. Korea 187 Q8
Gonglee, Liberia 204 P10
Gongliu, China 184 E4
Gongola, river, Nigeria 206 K10
Gongoué, Gabon 208 E7
Gongzhuling, China 185 E16
Gonja, Tanzania 211 V7
Gónnoi, Gr. 156 E10
Gonzales, Tex., U.S. 105 L17
Gonzales, Calif., U.S. 111 U4
González Chaves, Arg. 135 N12
Good Hope Beach, Virgin Is., U.S. 122 R1
Goodenough Island, P.N.G. 222 Q9
Goodenough Land, Greenland, Den. 88 J12
Gooding, Idaho, U.S. 108 H5
Goodland, Kans., U.S. 107 S3
Goodlands, Mauritius 215 F20
Goodlettsville, Tenn., U.S. 103 E15
Goodman Bay, Bahamas 122 E2
Goodman, Wis., U.S. 100 G6
Goodman, Miss., U.S. 102 K11
Goodnews Bay, Alas., U.S. 112 K12
Goodparla, N.T., Austral. 221 P8
Goodridge, Minn., U.S. 106 E9
Goofnuw Inlet, F.S.M. 226 D12
Goomalling, W. Aust., Austral. 221 W3
Goondiwindi, Qnsld., Austral. 221 V15
Goongarrie, Lake, Austral. 220 J4
Goose Creek, S.C., U.S. 98 M10
Goose Lake, Calif., Oreg., U.S. 110 L6
Gōr Zanak, Afghan. 180 K4
Gora Ayribaba, peak, Turkm., Uzb. 178 N10
Gora Belukha, peak, Kaz. 179 E18
Gora Mus Khaya, peak, Asia 168 C11
Gorakhpur, India 182 H9
Goraklbad Passage, Palau 226 H6
Gorda Ridges, Pac. Oc. 247 E17
Gördes, Turk. 170 F4
Gordion, ruins, Turk. 170 E7
Gordo, Ala., U.S. 103 K14
Gordon Downs, W. Aust., Austral. 221 R7
Gordon, Ga., U.S. 98 M6
Gordon, Wis., U.S. 100 F3
Gordon, Nebr., U.S. 106 M3
Gordon, Lake, Austral. 220 M7
Gordon's, Bahamas 119 G14
Gordonsville, Va., U.S. 98 E12
Gore, N.Z. 223 P16
Goré, Chad 207 L13
Gorē, Eth. 210 M5
Goree, Tex., U.S. 105 F14
Görele, Turk. 171 D14
Gorey (Guaire), Ire. 145 S7
Gorgān, Iran 177 D17
Gorgona, Isla, Col. 130 H4
Gorgona, Isola di, It. 152 H5
Gorgora, Eth. 210 K7
Gorham, N.H., U.S. 97 G14
Gori Rit, Somalia 210 M13
Gori, Rep. of Georgia 171 C18
Goris, Arm. 171 E20
Gorizia, It. 152 D10
Gor'kiy Reservoir, Eur. 140 E13
Gor'kovskoye Vodokhranilishche, Russ. 158 H12
Gorlice, Pol. 151 H16
Görlitz, Ger. 150 F11
Gorman, Tex., U.S. 105 H15
Gorna Oryakhovitsa, Bulg. 155 H15
Gornja Radgona, Slov. 154 C6
Gornji Milanovac, Serb. 154 G10
Gorno Altaysk, Russ. 160 L12
Gornozavodsk, Russ. 161 K22
Gornyak, Russ. 160 L11
Goro, river, Cen. Af. Rep. 207 K15
Gorodets, Russ. 158 H12
Gorodishche, Russ. 158 L13
Goroka, P.N.G. 222 P7
Gorom Gorom, Burkina Faso 205 J16
Gorong, Kepulauan, Indonesia 193 L17
Gorontalo, Indonesia 193 J14
Gorror, F.S.M. 226 E11
Gortyn see Górtys, ruin, Gr. 157 Q14
Górtys (Gortyn), ruin, Gr. 157 Q14
Gorumna Island see Garmna, Ire. 145 R3
Góry Swietokrzyskie, range, Pol. 151 F16
Goryeong, S. Korea 187 R10
Gorzów Wielkopolski, Pol. 150 D12
Gosan, S. Korea 187 W6
Goschen Strait, P.N.G. 222 R9
Gosen, Japan 188 M11
Goseong, S. Korea 186 L9
Goseong, S. Korea 187 T10
Gosford, N.S.W., Austral. 221 X15
Goshen, Ind., U.S. 101 N9
Goshi, river, Kenya 211 U8
Goshogawara, Japan 188 J12
Goslar, Ger. 150 E8
Gosnel, Ark., U.S. 102 F11
Gospić, Croatia 154 F5
Gosselies, Belg. 148 L11
Gossi, Mali 205 H15
Gossinga, S. Sudan 203 V8
Gostivar, N. Maced. 154 K10
Gota, Eth. 210 L9
Götaland, Eur. 140 E7
Göteborg, Sw. 142 L12
Gotha, Ger. 150 F8
Gothenburg, Nebr., U.S. 107 Q5
Gotland, island, Sw. 143 L14
Goto Meer, lake, Bonaire, Neth. 123 Q18
Gotō Rettō, Japan 189 R3
Gotse Delchev, Bulg. 155 K13
Gotska Sandön, island, Sw. 143 L14
Gōtsu, Japan 189 P6
Göttingen, Ger. 150 E7
Goubéré, Cen. Af. Rep. 207 M18
Goudoumaria, Niger 206 H10
Gouéké, Guinea 204 N11
Gough Fracture Zone, Atl. Oc. 249 U10
Gouin, Réservoir, Que., Can. 87 N19
Goulburn, N.S.W., Austral. 221 Y14
Gould Coast, Antarctica 238 K12
Gould, Ark., U.S. 102 H9
Goulding Cay, Bahamas 122 E12
Gouldsboro, Me., U.S. 97 F18
Goumbou, Mali 204 J12
Gouméré, Côte d'Ivoire 205 N14
Gounarou, Benin 205 L18
Goundam, Mali 205 H14
Goundi, Chad 207 K13
Gourbeyre, Guadeloupe, Fr. 123 G14
Gouré, Niger 206 H9
Gourma-Rharous, Mali 205 H15
Gourma, region, Burkina Faso 205 K16
Gournay, Fr. 149 N9
Gourniá, ruin, Gr. 157 Q16
Gouro, Chad 207 E14
Gouverneur, N.Y., U.S. 96 G9
Gouyave (Charlotte Town), Grenada 123 K22
Gouzon, Fr. 149 T10
Gove Peninsula, Austral. 220 B10
Gove, Kans., U.S. 107 S4
Govena, Russ. 161 D22
Govenlock, Sask., Can. 86 N11
Governador Valadares, Braz. 133 K15
Government House, Monaco 163 E21
Government Palace, Vatican City 163 Q15
Governor Generoso, Philippines 193 F15
Governor's Beach, Gibraltar, U.K. 163 Q23
Governor's Harbour, Bahamas 122 E6
Governor's Residence, Gibraltar, U.K. 163 P22
Govindgarh, India 182 J8
Govorovo, Russ. 161 F16
Gowanda, N.Y., U.S. 96 K5
Gower Peninsula, Wales, U.K. 145 U9
Gowon, S. Korea 187 S8
Goya, Arg. 134 H12
Goyang, S. Korea 186 M8
Goyave, Guadeloupe, Fr. 123 F15
Goyaves, river, Guadeloupe, Fr. 123 F14
Göyçay, Azerb. 171 D21
Göygöl, Azerb. 171 D20
Göynük, Turk. 170 D6
Goz Beïda, Chad 207 J15
Goz Regeb, Sudan 203 Q14
Gozo (Għawdex), island, Malta 163 H20
Graaff-Reinet, S. Af. 212 L7
Graah Øer, Greenland, Den. 89 Q10
Graben Land, Greenland, Den. 88 K12
Grabo, Côte d'Ivoire 204 Q12
Gračac, Croatia 154 F6
Gračanica, Bosn. & Herzg. 154 F8
Graceville, Fla., U.S. 99 Q3
Graceville, Minn., U.S. 106 J8
Gracias a Dios, Cabo, Nicar. 117 K19
Graciosa Bay, Solomon Is. 228 L5
Graciosa, island, Port. 146 P4
Graciosa, island, Sp. 214 P7
Gradaús, Braz. 132 F12
Graford, Tex., U.S. 105 G16
Grafton, W. Va., U.S. 98 C10
Grafton, Wis., U.S. 100 L7
Grafton, N. Dak., U.S. 106 E7
Grafton, N.S.W., Austral. 221 W15
Grafton, Mount, Nev., U.S. 111 Q12
Graham Bell Island, Asia 168 B8
Graham Bell, Russ. 161 C13
Graham Island, B.C., Can. 86 J6
Graham Island, Nunavut, Can. 87 C15
Graham Lake, Me., U.S. 97 F18
Graham Land, region, Antarctica 238 D5
Graham, Ont., Can. 87 P15
Graham, N.C., U.S. 98 G11
Graham, Tex., U.S. 105 G15
Graham, Mount, Ariz., U.S. 109 V9
Graham's Harbour, Bahamas 122 E11
Grahamstown, S. Af. 212 M8
Graian Alps, Fr., It. 152 D2
Grain Coast, Liberia 204 Q10
Grajaú, Braz. 133 E14
Grajaú, river, S. Amer. 128 E9
Grajewo, Pol. 151 B17
Gramat, Fr. 149 V9
Grámmos, Óros, Gr. 156 D8
Grampian Mountains, Scot., U.K. 144 J9
Gran Altiplanicie Central, Arg. 135 U8
Gran Canaria, island, Sp. 214 R5
Gran Cayo Point, Trin. & Tobago 123 R23
Gran Chaco, region, S. Amer. 128 J6
Gran Laguna Salada, Arg. 135 S8
Gran Sasso d'Italia, It. 152 J10
Gran Tarajal, Sp. 214 Q7
Granada, Colo., U.S. 109 N16
Granada, Nicar. 117 M18
Granada, Sp. 146 L11
Granadilla, Sp. 214 Q4
Granbury, Tex., U.S. 105 G16
Granby, Que., Can. 87 P20
Granby, Colo., U.S. 108 L12
Grand Anse Bay, Grenada 123 L22
Grand Anse, Grenada 123 L22
Grand Bahama Island, Bahamas 122 A8
Grand Baie, Mauritius 215 F20
Grand Baie, Mauritius 215 J20
Grand Banks of Newfoundland, N. Amer. 87 N24
Grand Bay, Dominica 123 G20
Grand Bonum, peak, St. Vincent & the Grenadines 123 K16
Grand Caille Point, St. Lucia 123 K13
Grand Canal, Ire. 145 R6
Grand Canyon National Park, Ariz., U.S. 109 Q6
Grand Canyon-Parashant National Monument, Ariz., U.S. 109 Q5
Grand Canyon, Ariz., U.S. 109 R6
Grand Canyon, town, Ariz., U.S. 109 R7
Grand Case, St. Martin, Fr. 123 B14
Grand Cayman, island, Cayman Is., U.K. 118 L7
Grand Cess, Liberia 204 Q11
Grand Coulee Dam, Wash., U.S. 110 B7
Grand Coulee, Wash., U.S. 110 B8
Grand Cul-de-Sac Marin, Guadeloupe, Fr. 123 E14
Grand Erg de Bilma, Niger 206 E11
Grand Erg Occidental, Alg. 200 F11
Grand Erg Oriental, Alg. 201 F14
Grand Falls-Windsor, Nfld. & Lab., Can. 87 M23
Grand Forks, N. Dak., U.S. 106 F8
Grand Harbour, Malta 163 K23
Grand Haven, Mich., U.S. 100 L9
Grand Îlet, Guadeloupe, Fr. 123 H14
Grand Island Bay, S. Amer. 128 J9
Grand Island, N.Y., U.S. 96 J5
Grand Island, Mich., U.S. 100 F8
Grand Island, Nebr., U.S. 107 Q6
Grand Isle, La., U.S. 102 R11
Grand Junction, Colo., U.S. 108 M10
Grand Lake, Ohio, U.S. 101 Q11
Grand Lake, La., U.S. 102 Q7
Grand Ledge, Mich., U.S. 100 M10
Grand Lieu, Lac de, Fr. 149 S5
Grand Mal Bay, Grenada 123 L22
Grand Marais, Mich., U.S. 100 F9
Grand Marais, Minn., U.S. 106 F14
Grand Passage, New Caledonia, Fr. 231 M19
Grand Popo, Benin 205 P17
Grand Portage National Monument, Minn., U.S. 106 E14
Grand Portage, Minn., U.S. 106 E14
Grand Prairie, Tex., U.S. 105 G17
Grand Rapids, Man., Can. 87 M13
Grand Rapids, Mich., U.S. 100 L9
Grand Rapids, Minn., U.S. 106 G11
Grand Rivière, Martinique, Fr. 123 E22
Grand Roy, Grenada 123 K22
Grand Saline, Tex., U.S. 105 G19
Grand Staircase-Escalante National Monument, Utah, U.S. 109 P7
Grand Teton National Park, Wyo., U.S. 108 G8
Grand Teton, peak, Wyo., U.S. 108 G8
Grand Traverse Bay, Mich., U.S. 100 H10
Grand Turk, island, Turks & Caicos Is., U.K. 119 J18
Grand-Bassam, Côte d'Ivoire 205 Q14
Grand-Bérébi see Basa, Côte d'Ivoire 204 Q12
Grand-Bourg, Guadeloupe, Fr. 123 G16
Grand-Gosier, Haiti 119 M17
Grand-Lahou, Côte d'Ivoire 205 Q13
Grand, river, Mich., U.S. 100 L10
Grand, river, S. Dak., U.S. 106 J3
Grand, river, Mo., U.S. 107 R12
Grandcamp-Maisy, Fr. 149 N6
Grande Bay, Atl. Oc. 249 V4
Grande Cayemite, island, Haiti 119 M15
Grande Comore see Njazidja, Comoros 215 M14
Grande Montagne, peak, Mauritius 215 J20
Grande Prairie, Alta., Can. 86 K9
Grande Rivière du Nord, Haiti 119 L17
Grande Rivière Sud Est, Mauritius 215 G21
Grande Riviere, Trin. & Tobago 123 N23
Grande Rivière, St. Lucia 123 J14
Grande Ronde, river, Oreg., U.S. 110 F8
Grande-Anse, Guadeloupe, Fr. 123 E17
Grande-Terre, Guadeloupe, Fr. 123 E15
Grande, Bahía, Arg. 135 V8
Grande, Boca, Venez. 131 D14
Grande, Cayo, Cuba 118 J9
Grande, Cuchilla, Uru. 134 K14
Grande, river, Bol. 132 J8
Grande, river, Braz. 133 H15
Grande, river, Braz. 133 L13
Grande, river, Arg. 134 M7
Grande, Salina, Arg. 135 N8
Grandes, Salinas, Arg. 134 H9
Grandfalls, Tex., U.S. 104 J11
Grandfather Mountain, N.C., U.S. 98 G8
Grândola, Port. 146 J5
Grandview, Wash., U.S. 110 E6
Grandvilliers, Fr. 149 N9
Grange (An Ghráinseach), Ire. 145 P4
Grange Hill, Jam. 122 H6
Grange, Virgin Is., U.S. 122 Q3
Granger, Tex., U.S. 105 K17
Granger, Wyo., U.S. 108 K9
Granger, Wash., U.S. 110 E6
Grangeville, Idaho, U.S. 108 E4
Granite City, Ill., U.S. 101 T4
Granite Falls, Minn., U.S. 106 K9
Granite Mountains, Wyo., U.S. 108 H11
Granite Mountains, Calif., U.S. 111 V10
Granite Peak, Mont., U.S. 108 F9
Granite Peak, Utah, U.S. 108 L6
Granite Peak, Nev., U.S. 110 M7
Granite, Okla., U.S. 105 D15
Granite, Oreg., U.S. 110 G8
Granja, Braz. 133 E16
Granollers, Sp. 147 D17
Grant City, Mo., U.S. 107 Q10
Grant Island, Antarctica 238 N8
Grant Range, Nev., U.S. 111 R11
Grant-Kohrs Ranch National Historic Site, Mont., U.S. 108 D6
Grant, Nebr., U.S. 107 Q3
Grant, Mount, Nev., U.S. 111 R7
Grants Pass, Oreg., U.S. 110 K2
Grants, N. Mex., U.S. 109 S11
Grantsburg, Wis., U.S. 100 G2
Grantville, Ga., U.S. 98 M4
Granville Lake, Man., Can. 87 L13
Granville, N.Y., U.S. 96 H12
Granville, N. Dak., U.S. 106 E4
Granville, Fr. 149 P6
Grão Mogol, Braz. 133 J15
Grapeland, Tex., U.S. 105 J19
Grass Cay, Virgin Is., U.S. 122 M9
Grass Patch, W. Aust., Austral. 221 X5
Grass Point, Virgin Is., U.S. 122 Q4
Grass Valley, Calif., U.S. 111 Q5
Grasse, Fr. 149 X15
Grassholm Island, Wales, U.K. 145 U8
Grassrange, Mont., U.S. 108 D10
Grassy Bay, Bermuda 119 B15
Grassy Butte, N. Dak., U.S. 106 F2
Grassy Key, Fla., U.S. 99 Y9
Graus, Sp. 147 C15
Grauspitz, peak, Liech., Switz. 162 R3
Grave Peak, Idaho, U.S. 108 D5
Grave, Pointe de, Fr. 149 U6
Gravelbourg, Sask., Can. 86 N11
Gravette, Ark., U.S. 102 E6
Gravina in Puglia, It. 152 M13
Gray, Fr. 149 R13
Grayland, Wash., U.S. 110 D2
Grayling, Mich., U.S. 100 J10
Grayling, Alas., U.S. 113 H13
Grays Harbor, Wash., U.S. 110 D2
Grays Lake, Idaho, U.S. 108 H7
Grays, Eng., U.K. 145 U14
Grayson, Ky., U.S. 103 B19
Grayville, Ill., U.S. 101 U7
Graz, Aust. 150 M12
Great Artesian Basin, Austral. 220 G12
Great Australian Bight, Austral. 220 K7
Great Bacolet Bay, Grenada 123 L23
Great Bacolet Point, Grenada 123 L23
Great Bahama Bank, Atl. Oc. 248 L3
Great Barrier Island (Aotea Island), N.Z. 223 E21
Great Barrier Reef, Qnsld., Austral. 220 D14
Great Basin National Park, Nev., U.S. 111 Q12
Great Basin, U.S. 90 F5
Great Bay, N.J., U.S. 96 Q10
Great Bay, Virgin Is., U.S. 122 N9
Great Bear Lake, N.W.T., Can. 86 F11
Great Belt, Atl. Oc. 140 F7
Great Bend, Kans., U.S. 107 T5
Great Bernera, island, Scot., U.K. 144 G6
Great Bird Island, Antigua & Barbuda 123 A21
Great Bitter Lake see Al Buḩayrah al Murrah al Kubrá, Egypt 202 F12
Great Bitter Lake see Murrat el Kubra, Buheirat, Egypt 173 Q1
Great Britain, U.K. 145 P12
Great Channel, India, Indonesia 183 U15
Great Channel, Philippines 192 F3
Great Coco Island, Myanmar 190 M3
Great Courtland Bay, Trin. & Tobago 123 N17
Great Crack, Hawaii, U.S. 115 P22
Great Crater, Israel 173 P5
Great Dismal Swamp, N.C., Va., U.S. 98 G14
Great Divide Basin, Wyo., U.S. 108 J10
Great Dividing Range, Austral. 220 G14
Great Dog Island, Virgin Is., U.K. 122 Q8
Great Driffield, Eng., U.K. 145 Q13

Great Eastern Erg see Oriental, Grand Erg, Af. **198** D5
Great Exuma, Bahamas **122** E8
Great Fall, Guyana **131** F14
Great Falls, S.C. U.S. **98** K9
Great Falls, Mont. U.S. **108** C8
Great Guana Cay, Bahamas **118** C12
Great Guana Cay, Bahamas **118** F12
Great Harbour Cay, Bahamas **118** D11
Great Hungarian Plain, Eur. **140** J9
Great Inagua Island, Bahamas **119** J15
Great Indian Desert (Thar Desert), India, Pak. **181** Q9
Great Isaac, island, Bahamas **118** D10
Great Karoo, region, S. Af. **212** L6
Great Lake, Austral. **220** L7
Great Lameshur Bay, Virgin Is. U.S. **122** N11
Great Meteor Tablemount, Atl. Oc. **248** K8
Great Miami, river, U.S. **91** G17
Great Namaland, Namibia **212** J4
Great Nicobar, island, India **183** U15
Great Orme's Head, Eng. U.K. **145** R9
Great Oyster Bay, Austral. **220** L8
Great Pedro Bluff, Jam. **122** J7
Great Pedro Point, Cayman Is. U.K. **122** N3
Great Pee Dee, river, S.C. U.S. **98** L11
Great Plains, N. Amer. **84** H5
Great Pond Bay, Virgin Is. U.S. **122** Q3
Great Rift Valley, Af. **198** K10
Great River Bay, Grenada **123** K23
Great Ruaha, river, Tanzania **211** W5
Great Sacandaga Lake, N.Y. U.S. **96** J11
Great Sale Cay, Bahamas **122** H3
Great Salt Lake Desert, Utah, U.S. **108** L6
Great Salt Lake, Utah, U.S. **108** K6
Great Salt Plains Lake, Okla. U.S. **105** B16
Great Sand Dunes National Park & Preserve, Colo. U.S. **109** P13
Great Sandy Desert, U.S. **90** D4
Great Sandy Desert, Austral. **220** E5
Great Sandy Island see Fraser Island, Austral. **220** G16
Great Sitkin Island, Alas. U.S. **112** N5
Great Slave Lake, N.W.T. Can. **86** H11
Great Smoky Mountains National Park, N.C. Tenn. U.S. **103** F19
Great Smoky Mountains, U.S. **91** J18
Great Sound, Bermuda **119** C14
Great St. James Island, Virgin Is. U.S. **122** N9
Great Stirrup Cay, Bahamas **118** D11
Great Tenasserim, river, Myanmar **191** N7
Great Thatch Island, Virgin Is. U.K. **122** M11
Great Tobago Island, Virgin Is. U.K. **122** L9
Great Victoria Desert, Austral. **220** H6
Great Wall, station, Antarctica **238** C4
Great Western Erg see Occidental, Grand Erg, Af. **198** D4
Great Western Tiers, Austral. **220** L7
Great Yarmouth, Eng. U.K. **145** S16
Great Zab see Zab al Kabir, river, Iraq **176** E10
Great Zab, river, Turk. **171** H18
Great Zimbabwe, ruin, Zimb. **212** F9
Great, river, Grenada **123** K23
Greater Antilles, islands, N. Amer. **119** M13
Greater Khingan Range, Asia **168** F11
Greater Sudbury, Ont. Can. **87** P17
Greater Sunda Islands, Indonesia **192** K7
Greathead Bay, St. Vincent & the Grenadines **123** L16
Gréboun, peak, Niger **206** D9
Greci, It. **152** L12
Gredos, Sierra de, Sp. **146** F9
Greece, N.Y. U.S. **96** J6
Greece, Eur. **156** F9
Greeley, Nebr. U.S. **107** P6
Greeley, Colo. U.S. **108** L13
Greely Fjord, Nunavut, Can. **87** B15
Greem-Bell, Ostrov, Russ. **161** C13
Green Bay, U.S. **91** D16
Green Bay, Wis. U.S. **100** H7

Green Bay, Antigua & Barbuda **123** B20
Green Bay, town, Wis. U.S. **100** J7
Green Cape, Austral. **220** M14
Green Castle, Bahamas **122** F6
Green Cay, Bahamas **118** F12
Green Cay, Bahamas **122** D11
Green Cay, Virgin Is. U.K. **122** L11
Green Cay, Virgin Is. U.S. **122** Q3
Green Cove Springs, Fla. U.S. **99** R8
Green Head, Austral. **220** J2
Green Hill, Trin. & Tobago **123** P22
Green Island, Jam. **122** H6
Green Island, Antigua & Barbuda **123** B22
Green Island, Grenada **123** J24
Green Islands, P.N.G. **222** N11
Green Islands, P.N.G. **228** A4
Green Mountains, Vt. U.S. **96** H12
Green Mountains, Wyo. U.S. **108** J11
Green River, Utah, U.S. **108** M9
Green River, Wyo. U.S. **108** K9
Green Turtle Cay, Bahamas **122** A9
Green Turtle Cay, Bahamas **122** A9
Green Valley, Ariz. U.S. **109** V7
Green, river, U.S. **90** G7
Green, river, U.S. **91** H17
Green, river, Ky. U.S. **103** D15
Green, river, U.S. **108** M9
Greenbrier, river, W. Va. U.S. **98** E10
Greenbush, Mich. U.S. **100** J12
Greenbush, Minn. U.S. **106** E9
Greencastle, Pa. U.S. **96** Q6
Greencastle, Ind. U.S. **101** R8
Greeneville, Tenn. U.S. **103** E20
Greenfield, Mass. U.S. **97** K13
Greenfield, Wis. U.S. **100** L7
Greenfield, Ind. U.S. **101** R9
Greenfield, Ohio, U.S. **101** S12
Greenfield, Tenn. U.S. **103** E13
Greenfield, Iowa, U.S. **107** P10
Greenfield, Mo. U.S. **107** U11
Greenfield, Calif. U.S. **111** U4
Greenidge, Barbados **123** J18
Greenland (Kalaallit Nunaat), Greenland, Den. **88** H9
Greenland Basin, Arctic Oc. **253** J18
Greenland Environmental Observatory (Summit Station), Greenland, Den. **88** J10
Greenland Fracture Zone, Arctic Oc. **253** J17
Greenland Ice Core Project (GRIP), Greenland, Den. **88** J10
Greenland Ice Sheet Project Two (GISP2), Greenland, Den. **88** J10
Greenland Plain, Arctic Oc. **253** J18
Greenland Sea, Arctic Oc. **253** J17
Greenland Trough, Arctic Oc. **253** L19
Greenland-Iceland Rise, Arctic Oc., Atl. Oc. **253** M20
Greenland, N. Amer. **84** C9
Greenock, Scot. U.K. **144** L8
Greenport, N.Y. U.S. **97** M13
Greens Peak, Ariz. U.S. **109** T9
Greensboro, Ga. U.S. **98** L6
Greensboro, N.C. U.S. **98** G10
Greensboro, Ala. U.S. **103** K14
Greensburg, Pa. U.S. **96** P4
Greensburg, Ind. U.S. **101** S10
Greensburg, Ky. U.S. **103** D17
Greensburg, Kans. U.S. **107** U5
Greenville, Me. U.S. **97** D16
Greenville, N.C. U.S. **98** H13
Greenville, S.C. U.S. **98** J7
Greenville, Ga. U.S. **99** Q6
Greenville, Mich. U.S. **100** L10
Greenville, Ohio, U.S. **100** R11
Greenville, Ill. U.S. **101** T5
Greenville, Miss. U.S. **102** J10
Greenville, Ala. U.S. **103** M15
Greenville, Ky. U.S. **103** D15
Greenville, Tex. U.S. **105** F18
Greenville, Calif. U.S. **111** N5
Greenville, Liberia **204** Q11
Greenwood, S.C. U.S. **98** K7
Greenwood, Ind. U.S. **101** R9
Greenwood, Ark. U.S. **102** F6
Greenwood, Miss. U.S. **102** J11
Greer, S.C. U.S. **98** J7
Greers Ferry Lake, Ark. U.S. **102** F8
Gregory Lake, Austral. **220** E6
Gregory Range, Austral. **220** D12
Gregory Town, Bahamas **122** D5
Gregory, S. Dak. U.S. **106** M5
Gregory, Lake, Austral. **220** G4

Gregory, Lake, Austral. **220** H11
Gregory, river, Austral. **220** D11
Greifswald, Ger. **150** B10
Greiz, Ger. **150** G9
Grenada Bay, Grenada **123** J23
Grenada, Calif. U.S. **110** L3
Grenada, Miss. U.S. **102** J11
Grenada, N. Amer. **121** H16
Grenade, Fr. **149** X7
Grenoble, Fr. **149** V14
Grenora, N. Dak. U.S. **106** E2
Grenville Bay, Grenada **123** K23
Grenville, Grenada **121** H17
Grenville, Grenada **123** K23
Grenville, Cape, Austral. **220** A12
Grenville, Point, Wash. U.S. **110** C2
Gresham, Oreg. U.S. **110** F4
Gretna, La. U.S. **102** Q11
Grevená, Gr. **156** D8
Grevenmacher, Lux. **162** J11
Grey Islands, Nfld. & Lab. Can. **87** L23
Grey Range, Austral. **220** H12
Grey, Cape, Austral. **220** B10
Grey, Mount, Austral. **220** J3
Greybull, Wyo. U.S. **108** F10
Greybull, river, Wyo. U.S. **108** G10
Greylock, Mount, Mass. U.S. **96** K12
Grezzana, It. **152** D6
Gribanovskiy, Russ. **158** M11
Gribingui, river, Cen. Af. Rep. **207** L14
Gridley, Calif. U.S. **111** P4
Grieg Point, Kiribati **226** J11
Griffin, Ga. U.S. **98** M5
Griffith, N.S.W. Austral. **221** X13
Griggsville, Ill. U.S. **101** R3
Grijalva Ridge, Pac. Oc. **247** K21
Grim, Cape, Austral. **220** L7
Grimaldi Forum, Monaco **163** C22
Grimari, Cen. Af. Rep. **207** M15
Grimsby, Eng. U.K. **145** Q14
Grímsey, island, Ice. **142** D3
Grimshaw, Alta. Can. **86** K10
Grimstad, Nor. **142** L11
Grindávíkurdjúp, valley, Atl. Oc. **253** M21
Grindelwald, Switz. **152** B3
Grinnell Peninsula, Nunavut, Can. **87** C15
Grinnell, Iowa, U.S. **107** P12
Grintavec, peak, Slov. **154** C5
GRIP see Greenland Ice Core Project, Greenland, Den. **88** J10
Gris-Nez, Cap, Fr. **148** L9
Grise Fiord, Nunavut, Can. **87** C16
Grishkino, Russ. **160** J12
Grisslehamn, Sw. **143** J14
Gródek, Pol. **151** C18
Groenlo, Neth. **148** J14
Groesbeck, Tex. U.S. **105** J18
Groix, Île de, Fr. **149** R3
Grójec, Pol. **151** E16
Gronau, Ger. **150** D5
Groningen, Neth. **148** G14
Groom, Tex. U.S. **105** D13
Groot Bay, St. Maarten, Neth. **123** C14
Groot Karasberge, peak, Namibia **212** J5
Groot, river, S. Af. **212** M7
Groote Eylandt, Austral. **220** B10
Grootfontein, Namibia **212** F4
Grootvloer, lake, S. Af. **212** L5
Gros Islet, St. Lucia **123** J14
Gros Piton, St. Lucia **123** L13
Gros Point, Grenada **123** J22
Gros Ventre Range, Wyo. U.S. **108** H8
Gros-Morne, Haiti **119** L16
Gros-Morne, Martinique, Fr. **123** F23
Grosio, It. **152** C6
Grosse Pointe, Guadeloupe, Fr. **123** G16
Grosseto, It. **152** J6
Grossglockner, peak, Eur. **140** H7
Grossvenediger, peak, Aust. **150** L8
Grotli, Nor. **142** H10
Groton, S. Dak. U.S. **106** J7
Grottaminarda, It. **152** M11
Grotto of Lourdes, Vatican City **163** Q14
Grotto of the Observatory, Monaco **163** F19
Groupe Acteon, Fr. Polynesia, Fr. **231** E22
Grouse Creek, Utah, U.S. **108** J5
Grove City, Pa. U.S. **96** M3

Grove Hill, Ala. U.S. **103** M14
Grove Mountains, Antarctica **239** G19
Grove Place, Virgin Is. U.S. **122** Q2
Grover Beach, Calif. U.S. **111** W5
Grover, Colo. U.S. **108** K14
Groves, Tex. U.S. **105** L21
Groveton, N.H. U.S. **97** F14
Groveton, Tex. U.S. **105** J19
Groznyy, Russ. **159** T14
Grudovo, Bulg. **155** J17
Grudziądz, Pol. **151** C14
Gruinard Bay, Scot. U.K. **144** H8
Grünau, Namibia **212** K4
Grundy Center, Iowa, U.S. **107** N12
Grundy, Va. U.S. **98** F7
Gruver, Tex. U.S. **104** B12
Gryazi, Russ. **158** M10
Gryazovets, Russ. **158** G10
Gryfino, Pol. **150** C11
Grygla, Minn. U.S. **106** E9
Gua Musang, Malaysia **191** T9
Guacanayabo, Golfo de, Cuba **118** K11
Guadalajara, Mex. **116** H10
Guadalajara, Sp. **146** F11
Guadalcanal, Sp. **146** K8
Guadalcanal, Solomon Is. **228** F10
Guadalquivir, river, Sp. **146** K9
Guadalupe Island, N. Amer. **84** L2
Guadalupe Mountains National Park, Tex. U.S. **104** H8
Guadalupe Mountains, N. Mex. Tex. U.S. **104** H8
Guadalupe Peak, Tex. U.S. **104** H8
Guadalupe, Calif. U.S. **111** W5
Guadalupe, Mex. **116** C7
Guadalupe, Mex. **116** G10
Guadalupe, Isla de, Mex. **116** C4
Guadalupe, river, Tex. U.S. **105** M17
Guadalupe, Sierra de, Sp. **146** H9
Guadalupita, N. Mex. U.S. **109** R13
Guadarrama, Sierra de, Sp. **146** E10
Guadeloupe Passage, N. Amer. **121** C16
Guadeloupe, island, Lesser Antil. **121** D17
Guadiana, Bahía, Cuba **118** H4
Guadiana, river, Sp. **146** H8
Guadix, Sp. **146** L11
Guafo Fracture Zone, Pac. Oc. **247** L21
Guafo, Boca del, Chile **135** R6
Guafo, Isla, Chile **135** R5
Guai, Indonesia **193** K21
Guáimaro, Cuba **118** J11
Guainía, river, Col. **130** H9
Guaíra, Braz. **132** M11
Guaire see Gorey, Ire. **145** S7
Guajaba, Cayo, Cuba **118** H11
Guajará-Mirim, Braz. **132** G7
Guajira Peninsula, N. Amer. **84** P9
Guajira, Península de la, Col. **130** B8
Gualala, Calif. U.S. **111** Q2
Gualdicciolo, San Marino **162** J13
Gualeguay, Arg. **134** K12
Gualeguaychú, Arg. **134** K12
Gualicho, Salina, Arg. **135** P9
Guallatiri, Volcán, Chile **134** B7
Guamúchil, Mex. **116** E8
Guana Bay, St. Maarten, Neth. **123** B15
Guana Cay, Bahamas **122** F2
Guana Island, Virgin Is. U.K. **122** Q7
Guanabara Bay, S. Amer. **128** J10
Guanajay, Cuba **118** G6
Guanajuato, Mex. **116** H11
Guanambi, Braz. **133** J15
Guanapo, Trin. & Tobago **123** P23
Guanare, Venez. **130** D9
Guanare, river, Venez. **130** D9
Guandacol, Arg. **134** H8
Guane, Cuba **118** H4
Guangyuan, China **184** K10
Guangzhou (Canton), China **185** P13
Guanipa, river, Venez. **130** D13
Guano Lake, Oreg. U.S. **110** L7
Guano, Cayo, Cuba **118** H7
Guantánamo, Cuba **119** K13
Guantánamo, Bahía de, Cuba, U.S. **119** L13
Guapí, Col. **130** H4
Guapo Bay, Trin. & Tobago **123** Q21
Guaporé see Iténez, river, S. Amer. **132** H8
Guara, Sierra de, Sp. **147** C14
Guaranda, Ecua. **130** L4
Guarapari, Braz. **132** N12
Guarapuava, Braz. **132** N12
Guaratinguetá, Braz. **133** M14

Guarcino, It. **152** L9
Guarda Mor, Braz. **133** K13
Guarda, Port. **146** F7
Guardalavaca, Cuba **119** J13
Guardo, Sp. **146** C9
Guareña, Sp. **146** H8
Guarico, Punta, Cuba **119** K14
Guárico, river, Venez. **130** D10
Guasave, Mex. **116** E8
Guascama, Punta, Col. **130** H4
Guasdualito, Venez. **130** E8
Guatemala Basin, Pac. Oc. **247** J20
Guatemala City, Guatemala **117** L16
Guatemala, N. Amer. **117** K16
Guaviare, river, Col. **130** H9
Guayabal, Cuba **118** K11
Guayaguayare Bay, Trin. & Tobago **123** R23
Guayaguayare, Trin. & Tobago **123** R23
Guayaquil, Ecua. **130** L3
Guayaquil, Golfo de, Ecua. **130** M3
Guaymas, Mex. **116** D7
Guba, D.R.C. **209** L17
Guba, Eth. **210** K5
Gûbâl, Strait of, Egypt **173** U3
Guban, region, Somaliland **210** L11
Gubbio, It. **152** H8
Gubin, Pol. **150** E11
Gubio, Nigeria **206** J11
Gubkin, Russ. **159** N9
Gudalur, India **183** R5
Gudaut'a, Rep. of Georgia **171** A15
Gudermes, Russ. **159** T14
Gudgeon Bay, Pitcairn I. **228** Q5
Gudur, India **183** S7
Guebwiller, Fr. **149** Q15
Guéckédou, Guinea **204** N10
Guelma, Alg. **201** B14
Guelmim, Mor. **200** F6
Guéné, Benin **205** L18
Güeppí, Peru **132** C3
Guera Massif, peak, Af. **198** G6
Guerara, Alg. **201** D13
Guéréda, Chad **207** G15
Guéret, Fr. **149** T9
Guernsey Reservoir, Wyo. U.S. **108** J13
Guernsey, Wyo. U.S. **108** J13
Guernsey, island, Channel Is. U.K. **145** Y11
Guérou, Mauritania **204** H9
Guerzim, Alg. **200** F10
Guest Peninsula, Antarctica **238** N10
Gueydan, La. U.S. **102** P8
Gugegwe, island, Marshall Is. **227** J18
Gugu, peak, Eth. **210** M8
Guguan, island, N. Mariana Is. U.S. **230** H5
Gui, river, China **184** N12
Guía de Isora, Sp. **214** Q4
Guiana Highlands, S. Amer. **128** C6
Guiana Island, Antigua & Barbuda **123** A21
Guidan Roumji, Niger **206** H7
Guider, Cameroon **206** K11
Guienne, region, Fr. **149** W8
Guigang, China **184** P12
Guiglo, Côte d'Ivoire **204** P12
Guijuelo, Sp. **146** F8
Guilden Draak Seamount, Ind. Oc. **251** M15
Guildford, Eng. U.K. **145** V13
Guilford, Me. U.S. **97** E16
Guilin, China **184** N12
Guillaume-Delisle, Lac, Que. Can. **87** L18
Guillestre, Fr. **149** V15
Guimarães, Braz. **133** D15
Guimarães, Port. **146** D6
Guin, Ala. U.S. **103** J14
Guinchos Cay, Bahamas **118** G11
Guinea, Af. **204** L8
Guinea Basin, Atl. Oc. **249** P11
Guinea-Bissau, Af. **204** K7
Guinea Fracture Zone, Atl. Oc. **249** N9
Guinea Plain, Atl. Oc. **249** P12
Guinea Rise, Atl. Oc. **249** Q12
Guinea Seamount Chain, Atl. Oc. **249** P12
Guinea Terrace, Atl. Oc. **249** N10
Guinea, Guinea **204** L9
Güines, Cuba **118** G6
Guingamp, Fr. **149** P4
Guinguinéo, Senegal **204** J7
Güira de Melena, Cuba **118** G6

Güiria, Venez. 130 C13
Guisborough, Eng., U.K. 145 P13
Guita Koulouba, Cen. Af. Rep. 207 M16
Guitiriz, Sp. 146 B6
Guiyang, China 184 M11
Guiyang, China 185 N13
Gujar Khan, Pak. 180 J12
Gujba, Nigeria 206 J10
Gujranwala, Pak. 180 K13
Gujrat, Pak. 180 K12
Gujwa, S. Korea 187 W8
Gukdo, island, S. Korea 187 U11
Gul Bahār, Afghan. 180 H9
Gulbarga see Kalaburagi, India 182 N5
Gulbene, Latv. 143 L17
Gul'cha, Kyrg. 179 L14
Gulf Islands National Seashore, Miss., U.S. 103 P13
Gulf of Alaska Seamount Province, Pac. Oc. 247 B15
Gulf of Aqaba, Saudi Arabia 174 C4
Gulf of Aqaba, Egypt 202 G13
Gulf of Gabes, Tunisia 201 D16
Gulf of Guinea, Atl. Oc. 205 R16
Gulf of Guinea, Atl. Oc. 206 N5
Gulf of Hamamet, Tunisia 201 B16
Gulf of Mexico, Atl. Oc. 102 R7
Gulf of Papua, P.N.G. 222 Q7
Gulf of Paria, Trin. & Tobago, Venez. 123 Q20
Gulf of Sidra see Khalīj Surt, Lib. 201 D19
Gulf of Suez, Egypt 202 G12
Gulf of Tunis, Tunisia 201 B16
Gulf Shores, Ala., U.S. 103 P14
Gulfport, Miss., U.S. 102 P12
Gulin, China 184 M10
Guliston, Uzb. 178 L12
Gulja see Yining, China 184 D4
Gulkana, Alas., U.S. 113 H17
Güllük, Turk. 170 H3
Gülnar, Turk. 170 J8
Gülşehir, Turk. 170 F9
Gulu, Uganda 211 Q3
Gŭlŭbovo, Bulg. 155 J15
Gulya, Russ. 161 K18
Gulyantsi, Bulg. 155 G14
Guma see Pishan, China 184 G2
Gumare, Botswana 212 F6
Gumban, Eth. 211 N10
Gumdag, Turkm. 178 M5
Gumel, Nigeria 206 J9
Gumi, S. Korea 187 Q10
Gumla, India 182 K9
Gummi, Nigeria 206 J7
Gumuru, S. Sudan 203 W12
Gümüşhacıköy, Turk. 170 D10
Gümüşhane, Turk. 171 D14
Gümüşören, Turk. 170 G10
Gumzai, Indonesia 193 L19
Gun Bay, Cayman Is., U.K. 122 N5
Gun Cay, Bahamas 118 D9
Gun Creek, Virgin Is., U.K. 122 Q9
Gun Hill, Barbados 123 K19
Guna Terara, peak, Eth. 210 K7
Guna, India 182 J6
Gundlupet, India 183 R5
Gündoğmuş, Turk. 170 J7
Güney, Turk. 170 G4
Gungu, D.R.C. 209 H13
Gunib, Russ. 159 U15
Gunnbjørn Fjeld, peak, Greenland, Den. 88 M12
Gunnedah, N.S.W., Austral. 221 W14
Gunners Quoin, island, Mauritius 215 F20
Gunnerus Bank, Ind. Oc. 249 Y16
Gunnerus Ridge, Ind. Oc. 249 Y16
Gunnison, Utah, U.S. 108 M7
Gunnison, Colo., U.S. 109 N11
Gunsan, S. Korea 187 R8
Gunt, river, Taj. 179 N13
Guntersville Lake, Ala., U.S. 103 H16
Guntersville, Ala., U.S. 103 H16
Guntur, India 183 P7
Gunungsugih, Indonesia 192 L7
Gunwi, S. Korea 187 Q11
Gupis, Pak. 180 F12
Gura Humorului, Rom. 155 B15
Guragē, peak, Eth. 210 M7
Gurdaspur, India 180 K14
Gurdon, Ark., U.S. 102 J7
Güre, Turk. 170 F5
Gureopdo, island, S. Korea 187 N6
Gurer, island, Marshall Is. 227 H15
Gurguan Point, N. Mariana Is., U.S. 230 J10
Guri Dam, Venez. 130 E12
Guri i Topit, peak, Alban. 154 L9
Gurkovo, Bulg. 155 J15
Gurney, P.N.G. 222 R9
Gurué, Mozambique 212 D12
Gürün, Turk. 170 F12
Gurupá Island, S. Amer. 128 D8
Gurupá, Braz. 132 D12
Gurupi, Braz. 133 H13
Gurupi, Cape, S. Amer. 128 D9
Gurupi, river, Braz. 133 E14
Gurupi, Serra do, Braz. 133 E13
Gurvan Bogd Uul, Mongolia 184 E9
Gurvan Sayhan Uul, Mongolia 184 E10
Gurye, S. Korea 187 S9
Guryongpo, S. Korea 187 R13
Gus' Khrustal'nyy, Russ. 158 J11
Gusan, S. Korea 186 M11
Gusan, S. Korea 187 Q11
Gusau, Nigeria 206 J7
Gusev, Russ. 158 J2
Guşgy, Turkm. 178 P9
Gusinaya Bank, Arctic Oc. 253 D17
Gusinoozersk, Russ. 161 L15
Guspini, It. 153 P3
Güssing, Aust. 150 M12
Gustavia, St. Barthelemy, Fr. 121 B15
Gustine, Calif., U.S. 111 S5
Güstrow, Ger. 150 C9
Guta, Tanzania 211 T4
Gutâiu, peak, Rom. 155 B13
Gutcher, Scot., U.K. 144 B12
Gutenberg Castle, ruin, Liech. 162 Q1
Gütersloh, Ger. 150 E6
Gutha, W. Aust., Austral. 221 W2
Guthrie Center, Iowa, U.S. 107 P10
Guthrie, Ky., U.S. 103 D15
Guthrie, Okla., U.S. 105 C17
Guthrie, Tex., U.S. 105 F14
Guttenberg, Iowa, U.S. 106 M14
Güvem, Turk. 170 D8
Guwahati, India 182 H13
Guyana Basin, Atl. Oc. 249 N6
Guyana, S. Amer. 131 G15
Guyandotte, river, W. Va., U.S. 98 E7
Guymon, Okla., U.S. 104 B12
Guyuan, China 184 H10
Güzeloluk, Turk. 170 J9
Güzelyurt see Morphou, Cyprus 162 N7
Güzelyurt, Turk. 170 G9
G'uzor, Uzb. 178 M10
Gwa, Myanmar 190 K4
Gwabegar, N.S.W., Austral. 221 W14
Gwadabawa, Nigeria 206 H7
Gwadar, Pak. 181 S3
Gwagwada, Nigeria 206 K8
Gwalangu, D.R.C. 208 C12
Gwalior, India 182 H6
Gwallido, island, S. Korea 187 R7
Gwanchon, S. Korea 187 R8
Gwanda, Zimb. 212 F8
Gwane, D.R.C. 209 B17
Gwangcheon, S. Korea 187 Q12
Gwangcheon, S. Korea 187 Q7
Gwangju (Kwangju), S. Korea 187 T8
Gwangju, S. Korea 187 N8
Gwangmyeong, S. Korea 186 N8
Gwangyang, S. Korea 187 T9
Gwangyang, S. Korea 187 T9
Gwanmaedo, island, S. Korea 187 U6
Gwansan, S. Korea 187 U8
Gwardafuy, Cape, Asia 168 J3
Gwatar Bay, Iran 177 P23
Gwatar Bay, Iran, Pak. 181 S2
Gwayi River, Zimb. 212 E8
Gwayi, river, Zimb. 212 E8
Gweebarra Bay, Ire. 145 N4
Gweru, Zimb. 212 F9
Gweta, Botswana 212 F7
Gwinn, Mich., U.S. 100 F7
Gwinner, N. Dak., U.S. 106 H7
Gwoza, Nigeria 206 J11
Gwydir, river, Austral. 220 J14
Gyangzê (Jiangzi), China 184 L5
Gyaring Hu, China 184 J8
Gyaros, island, Gr. 157 K14
Gyda Peninsula, Asia 168 C8
Gydan see Kolymskoye Nagor'ye, range, Russ. 161 E20
Gydanskiy Poluostrov, Russ. 160 F12
Gyebangsan, peak, S. Korea 186 M11
Gyêgu see Yushu, China 184 K7
Gyeokpo, S. Korea 187 S7
Gyeonggiman, bay, S. Korea 187 N7
Gyeongju (Kyŏngju), S. Korea 187 R12
Gyeongsan, S. Korea 187 R11
Gyirong (Jilong), China 184 L3
Gyldenløve Fjord see Umiiviip Kangertiva, Greenland, Den. 89 R9
Gympie, Qnsld., Austral. 221 U16
Gyodongdo, island, S. Korea 186 M7
Gyoga, S. Korea 187 N12
Gyöngyös, Hung. 151 K15
Győr, Hung. 151 L14
Gypsum, Kans., U.S. 107 T7
Gýtheio, Gr. 156 M10
Gyuam, S. Korea 187 Q8
Gyula, Hung. 151 M17
Gyumri, Arm. 171 D18
Gyzylarbat, Turkm. 178 M6
Gyzyletrek, Turkm. 178 N5
Gyzylgaya, Turkm. 178 L6
Gyzylsuw, Turkm. 178 L4
Għira, Malta 163 K23

H

Hà Đông, Vietnam 190 H11
Hà Giang, Vietnam 190 G10
Hạ Long, Vietnam 190 H12
Hà Nội (Hanoi), Vietnam 190 H11
Hà Tiên, Cambodia 191 P10
Hà Tĩnh, Vietnam 190 J11
Hà Trung, Vietnam 190 H11
Ha'afeva, island, Tonga 227 M23
Ha'akame, Tonga 227 Q23
Ha'alaufuli, Tonga 227 G23
Ha'ano, island, Tonga 227 L24
Haapamäki, Fin. 143 G15
Haapiti, Fr. Polynesia, Fr. 231 H20
Haapsalu, Est. 143 K16
Haapu, Fr. Polynesia, Fr. 231 K14
Haarlem, Neth. 148 J12
Ha'asini, Tonga 227 R23
Haast, N.Z. 223 M16
Haaway, Somalia 211 R10
Hab, river, Pak. 181 R7
Habahe, China 184 C5
Ḩabarūt, Oman 175 M14
Habaswein, Kenya 211 R8
Habomai Islands, Russ. 161 K23
Habomai, Japan 188 F16
Haboro, Japan 188 F13
Hachijō Jima, Japan 189 R12
Hachinohe, Japan 188 J13
Hachiōji, Japan 189 P11
Hachirō Gata, Japan 188 K12
Hachita, N. Mex., U.S. 109 V10
Hachujado, island, S. Korea 187 V7
Hacıbektaş, Turk. 170 F9
Hacıqabul, Azerb. 171 D22
Haḑabat al Jilf al Kabīr, Egypt 202 K8
Haddummati Atoll, Maldives 183 X3
Hadejia, Nigeria 206 J9
Hadejia, river, Af. 198 G5
Ḩadera, Israel 172 L5
Hadhramaut, Asia 168 J3
Hadīboh, Yemen 175 Q15
Ḩadīdah, Syr. 172 D10
Ḩadīthah, Iraq 176 G9
Hadım, Turk. 170 H7
Hadja, Solomon Is. 228 F12
Hadley Bay, Nunavut, Can. 87 E13
Hado, island, S. Korea 187 T10
Hadong, S. Korea 187 T9
Ḩaḑramawt, Yemen 174 N11
Hadrian's Wall, Eng., U.K. 145 N11
Hadrut (Gadrut), Azerb. 171 E21
Hadyach, Ukr. 159 N7
Haedo, Cuchilla de, Uru. 134 K13
Haeju-man, bay, N. Korea 186 M6
Haeju, N. Korea 186 L6
Haemi, S. Korea 187 P7
Hā'ena, Hawaii, U.S. 115 N23
Haenam, S. Korea 187 U7
Haengyŏng, N. Korea 186 B13
Ḩafar al Bāţin, Saudi Arabia 174 D10
Hafik, Turk. 170 E12
Hafizabad, Pak. 180 K12
Hagadera, Kenya 211 S8
Hagåtña (Agana), Guam, U.S. 230 K5
Hagåtña (Agana), Guam, U.S. 230 N5
Hagemeister Island, Alas., U.S. 112 L12
Hagen, Ger. 150 F5
Hagerman Fossil Beds National Monument, Idaho, U.S. 108 H4
Hagerman, N. Mex., U.S. 109 U14
Hagerstown, Md., U.S. 98 B12
Häggenås, Sw. 142 G12
Hagoi Susupe, lake, N. Mariana Is., U.S. 230 J7
Hag's Head, Ire. 145 R3
Hague, Cap de la, Fr. 149 N5
Hague, The see Den Haag, Neth. 148 J12
Haguenau, Fr. 149 P15
Haha Jima Rettō (Baily or Coffin Group), Japan 189 X14
Hahoe, N. Korea 186 B14
Ha'apai Group, Tonga 224 K10
Hải Dương, Vietnam 190 H11
Hải Phòng (Haiphong), Vietnam 190 H11
Haida Gwaii (Queen Charlotte Islands), B.C., Can. 86 J6
Haifa see Ḩefa, Israel 172 K5
Haig, W. Aust., Austral. 221 W6
Haiger, Ger. 150 F6
Haikou, China 184 Q12
Haikou, China 184 N9
Haikou, China 184 Q12
Ha'iku, Hawaii, U.S. 115 J19
Ḩā'il, Saudi Arabia 174 E7
Hailar, China 185 C14
Hailey, Idaho, U.S. 108 G5
Hailun, China 185 C16
Hailuoto, island, Fin. 143 F15
Haimen, China 185 N14
Haimen, see Taizhou, China 185 L16
Hainan, China 184 G11
Hainan, island, China 184 Q12
Haines City, Fla., U.S. 99 U8
Haines Junction, Yukon, Can. 86 G7
Haines, Oreg., U.S. 110 G8
Haines, Alas., U.S. 113 K21
Hainiya Point, N. Mariana Is., U.S. 230 N9
Haiphong see Hải Phòng, Vietnam 190 H11
Haiti, N. Amer. 119 L15
Haiya, Sudan 203 P14
Haizhou Wan, China 185 J15
Hajdúböszörmény, Hung. 151 L17
Hajdúszoboszló, Hung. 151 L17
Haji Pir Pass, India, Pak. 180 J12
Hajiki Saki, Japan 188 L11
Hajinbu, S. Korea 186 M11
Ḩājj 'Abd Allāh, Sudan 203 S13
Ḩajjah, Yemen 174 N8
Ḩājjīābād, Iran 177 M19
Hajnáčka, Slovakia 151 K15
Hajnówka, Pol. 151 D18
Hajodo, island, S. Korea 187 U6
Hakalau, Hawaii, U.S. 115 M23
Hakamaru, island, N.Z. 229 J24
Hakataramea, N.Z. 223 N17
Hakauata, island, Tonga 227 M23
Hakdam, S. Korea 186 M10
Hakgok, S. Korea 186 M9
Hakha, Myanmar 190 G3
Hakkari, Turk. 171 H18
Hakodate, Japan 188 H12
Haksong, N. Korea 186 G6
Hakui, Japan 189 N9
Hakupa Pass, Solomon Is. 228 B9
Hakupu, N.Z. 229 Q22
Hala, Pak. 181 R8
Ḩalab (Aleppo), Syr. 172 D9
Ḩalabān, Saudi Arabia 174 H9
Ḩalabjah see Helebce, Iraq 176 F11
Halachó, Mex. 117 H16
Halaib Triangle, Egypt, Sudan 202 L14
Halali'i Lake, Hawaii, U.S. 114 F8
Hala'ula, Hawaii, U.S. 115 L21
Hālawa Bay, Hawaii, U.S. 115 H18
Halayeb, Egypt 202 L14
Halba Desēt, island, Eritrea 210 J10
Halba, Leb. 172 G7
Halban, Mongolia 184 C8
Halberstadt, Ger. 150 E8
Halcon, Mount, Philippines 193 D13
Haldwani, India 182 F7
Hale Center, Tex., U.S. 104 E12
Hale, Mich., U.S. 100 J12
Haleakalā Crater, Hawaii, U.S. 115 K19
Haleakalā National Park, Hawaii, U.S. 115 K20
Hale'iwa, Hawaii, U.S. 115 G14
Haleyville, Ala., U.S. 103 H14
Half Assini, Ghana 205 Q14
Half Moon Bay, Antigua & Barbuda 123 B22
Half Way Tree, Jam. 122 J10
Halfeti, Turk. 170 H12
Halfway, Oreg., U.S. 110 G9
Halifax Bay, Austral. 220 D14
Halifax, N.S., Can. 87 P22
Halifax, Va., U.S. 98 F11
Halifax, Qnsld., Austral. 221 R14
Halik Shan, China 184 E3
Halkett, Cape, Alas., U.S. 113 B16
Hálki, Gr. 156 F10
Hall Basin, Nunavut, Can. 87 A16
Hall Basin, bay, Greenland, Den. 88 C6
Hall Beach, Nunavut, Can. 87 F16
Hall Bredning, bay, Greenland, Den. 88 K13
Hall Islands, F.S.M. 226 A9
Hall Knoll, Arctic Oc. 252 K11
Hall Land, Greenland, Den. 88 C6
Hall Peninsula, Nunavut, Can. 87 H19
Hall, Aust. 150 L8
Hallandale, Fla., U.S. 99 W10
Hallasan, peak, S. Korea 187 W7
Halle, Belg. 148 L13
Halle, Ger. 150 F9
Hallein, Aust. 150 L10
Hallett Peninsula, Antarctica 239 Q14
Hallettsville, Tex., U.S. 105 L17
Halley Station, Antarctica 238 C11
Halliday, N. Dak., U.S. 106 F3
Hallim, S. Korea 187 W7
Hällnäs, Sw. 143 F14
Hallock, Minn., U.S. 106 E8
Halls Creek, W. Aust., Austral. 221 R6
Halls, Tenn., U.S. 102 F12
Hallstatt, Aust. 150 L10
Halmahera Sea, Indonesia 193 J16
Halmahera, island, Indonesia 193 H16
Halmeu, Rom. 154 A12
Halmstad, Sw. 142 M12
Halola, Solomon Is. 228 C5
Halstead, Kans., U.S. 107 U7
Halstead, Eng., U.K. 145 T15
Haltdalen, Nor. 142 H11
Halten Bank, Atl. Oc. 248 E13
Ḩalwān, Egypt 202 F11
Hàm Yên, Vietnam 190 G10
Ham, Fr. 149 N10
Ham, Chad 206 K12
Hamada de Tinrhert, Alg., Lib. 201 G15
Hamada du Drâa, Alg., Mor., W. Sahara 200 G6
Hamada du Guir, Alg. 200 E9
Hamâda el Haricha, desert, Mali 205 D14
Hamâda Safia, Mali 205 D13
Hamada, Japan 189 Q6
Hamadān (Ecbatana), Iraq 177 F13
Ḩamāh (Hamath), Syr. 172 F8
Hāmākua, Hawaii, U.S. 115 N21
Hamamatsu, Japan 189 Q10
Hamamet, Gulf of, Af. 198 C6
Hamar, Nor. 142 J11
Hamasaka, Japan 189 P8
Hamath see Ḩamāh, Syr. 172 F8
Hamatombetsu, Japan 188 E13
Hamburg, N.Y., U.S. 96 J5
Hamburg, Ark., U.S. 102 K9
Hamburg, Iowa, U.S. 107 Q9
Hamburg, Ger. 150 C8
Hamchang, S. Korea 187 Q10
Ḩamḑah, Saudi Arabia 174 L8
Ḩamdānah, Saudi Arabia 174 K6
Hamden, Conn., U.S. 96 M12
Hämeenlinna, Fin. 143 H16
Hamelin Pool, Austral. 220 H1
Hamelin, W. Aust., Austral. 221 V1
Hameln, Ger. 150 E7
Hamer Koke, Eth. 211 P6
Hamersley Range, Austral. 220 F3
Hamgyŏng-sanmaek, range, N. Korea 186 F10
Hamhŭng-man, bay, N. Korea 186 H9
Hamhŭng, N. Korea 186 G9
Hami (Kumul), China 184 F7
Hami Pendi, China 184 F6
Hamilton Bank, Atl. Oc. 248 G6
Hamilton Harbour, Bermuda 119 C15
Hamilton Inlet, Nfld. & Lab., Can. 87 K22
Hamilton, Ont., Can. 87 R18

Hamilton, *N.Y., U.S.* **96** J9
Hamilton, *Ill., U.S.* **101** Q3
Hamilton, *Ohio, U.S.* **101** S11
Hamilton, *Ala., U.S.* **103** H14
Hamilton, *Tex., U.S.* **105** H16
Hamilton, *Mo., U.S.* **107** R11
Hamilton, *Mont., U.S.* **108** D5
Hamilton, *Bermuda* **119** C15
Hamilton, *Scot., U.K.* **144** L9
Hamilton, *Vic., Austral.* **221** Z11
Hamilton, *N.Z.* **223** F20
Hamilton, Lake, *Ark., U.S.* **102** H7
Hamina, *Fin.* **143** H17
Hamirpur, *India* **182** H7
Hamju, *N. Korea* **186** H9
Hamlet, *N.C., U.S.* **98** J10
Hamlin, *Tex., U.S.* **105** G14
Hamm, *Ger.* **150** E6
Hammami, *Mauritania* **204** D9
Hammerfest, *Nor.* **143** B14
Hammon, *Okla., U.S.* **105** C15
Hammond, *Ind., U.S.* **101** N7
Hammond, *La., U.S.* **102** P10
Hammond, *Mont., U.S.* **108** E13
Hammonton, *N.J., U.S.* **96** Q10
Hampden, *N.Z.* **223** N18
Hampstead, *Dominica* **123** E19
Hampton Bays, *N.Y., U.S.* **97** N13
Hampton Butte, *Oreg., U.S.* **110** H6
Hampton Tableland, *Austral.* **220** J6
Hampton, *S.C., U.S.* **98** M9
Hampton, *Va., U.S.* **98** F14
Hampton, *Ark., U.S.* **102** J8
Hampton, *Iowa, U.S.* **106** M12
Hampyeong, *S. Korea* **187** T7
Hamrat esh Sheikh, *Sudan* **203** R9
Hamrun, *Malta* **163** K23
Hams Bay, *Virgin Is., U.S.* **122** Q1
Hams Bluff, *Virgin Is., U.S.* **122** Q1
Hāmūn Şāberī, Daryācheh-ye, *Afghan., Iran* **180** L2
Hāmūn-e Jaz Mūriān, lake, *Iran* **177** M21
Hamun-i- Mashkel, lake, *Iran* **177** L24
Hamyang, *S. Korea* **187** S9
Han, *Ghana* **205** L15
Han, river, *S. Korea* **186** M8
Hana Teio, *Fr. Polynesia* **231** E14
Hana Tetena, *Fr. Polynesia, Fr.* **231** E14
Hana Tuuna, *Fr. Polynesia* **231** E14
Hāna, *Hawaii, U.S.* **115** K20
Hanahan, *S.C., U.S.* **228** A4
Hanalei Bay, *Hawaii, U.S.* **114** E10
Hanalei, *Hawaii, U.S.* **114** E10
Hanam, *S. Korea* **186** M8
Hanam, *S. Korea* **187** S11
Hanamaki, *Japan* **188** K13
Hanamalo Point, *Hawaii, U.S.* **115** Q20
Hanamā'ulu, *Hawaii, U.S.* **114** F10
Hanang, peak, *Tanzania* **211** V5
Hanaoo, *Fr. Polynesia, Fr.* **231** E14
Hanapēpē, *Hawaii, U.S.* **114** F9
Hanau, *Ger.* **150** G6
Hanaui, *Fr. Polynesia, Fr.* **231** D14
Hanawa, *Japan* **188** K12
Hancock, *Md., U.S.* **98** B12
Hancock, *Mich., U.S.* **100** E6
Hancock, *Minn., U.S.* **106** J9
Handa Island, *Scot., U.K.* **144** G9
Handan, *China* **185** H13
Handeni, *Tanzania* **211** V7
Handsome Bay, *Virgin Is., U.K.* **122** Q8
Hane, *Fr. Polynesia, Fr.* **230** E11
Haneti, *Tanzania* **211** V6
Hanford Site, *Wash., U.S.* **110** E7
Hanford, *Calif., U.S.* **111** U6
Hangayn Nuruu, *Mongolia* **184** D8
Hangö *see* Hanko, *Fin.* **143** J15
Hangu, *Pak.* **180** J10
Hangzhou Bay, *Pac. Oc.* **168** G12
Hangzhou Wan, *China* **185** K15
Hanhöhiy Uul, *Mongolia* **184** C7
Hani, *Turk.* **171** G15
Hanimadu, island, *Maldives* **183** U3
Hankinson, *N. Dak., U.S.* **106** H8
Hanko *see* Hangö, *Fin.* **143** J15
Hanksville, *Utah, U.S.* **109** N8
Hanle, *India* **180** J16
Hann, Mount, *Austral.* **220** C5
Hanna Bank, *Arctic Oc.* **252** K7
Hanna Canyon, *Arctic Oc.* **252** L7
Hanna, *Alta., Can.* **86** M10
Hannibal, *Mo., U.S.* **107** R14
Hannover (Hanover), *Ger.* **150** D7
Hanöbukten, bay, *Sw.* **143** N13

Hanoi *see* Hà Nội, *Vietnam* **190** H11
Hanover, *Pa., U.S.* **96** Q7
Hanover, *N.H., U.S.* **97** H13
Hanover, *Kans., U.S.* **107** R8
Hanover, Isla, *Chile* **135** W6
Hans Egede Land, *Greenland, Den.* **88** B10
Hans Island, *Greenland, Den.* **88** C5
Hans Lollik Island, *Virgin Is., U.S.* **122** M8
Hans Tausen Iskappe, *Greenland, Den.* **88** B9
Hantsavichy, *Belarus* **158** L4
Hanumangarh, *India* **182** F4
Hanuy, river, *Mongolia* **184** C9
Hanyuan, *China* **184** L9
Hanzhong, *China* **184** K11
Hao, island, *Fr. Polynesia, Fr.* **231** C20
Haora, *India* **182** K12
Hā'ō'ū, *Hawaii* **115** K20
Hapai, *Solomon Is.* **228** E7
Haparanda, *Sw.* **143** E15
Hapcheon, *S. Korea* **187** S10
Hapdeok, *S. Korea* **187** P8
Hapo, *Indonesia* **193** H16
Happy Camp, *Calif., U.S.* **110** L2
Happy Valley-Goose Bay, *Nfld. & Lab., Can.* **87** L21
Happy, *Tex., U.S.* **104** D12
Haql, *Saudi Arabia* **174** C4
Har Ayrag, *Mongolia* **184** E11
Har Hu, *China* **184** H7
Har Nuur, *Mongolia* **184** C7
Har Us Nuur, *Mongolia* **184** C7
Har, *Indonesia* **193** L18
Harad, *Saudi Arabia* **174** G12
Harad, Jabal al, *Jordan* **173** R6
Haradok, *Belarus* **158** J6
Haradzishcha, *Belarus* **158** L4
Haraiki, island, *Fr. Polynesia, Fr.* **231** C19
Harald Moltke Bræ *see* Sermersuaq, glacier, *Greenland, Den.* **88** F4
Harappa, ruins, *Pak.* **180** M12
Harar *see* Härer, *Eth.* **210** M9
Harare, *Zimb.* **212** E9
Harat, island, *Eritrea* **210** H8
Haratini, island, *N.Z.* **229** J24
Haraz, *Chad* **207** H14
Harazé Mangueigne, *Chad* **207** K15
Harbel, *Liberia* **204** P10
Harbin, *China* **185** D16
Harbor Beach, *Mich., U.S.* **100** K13
Harbor Springs, *Mich., U.S.* **100** H10
Harbour Island, *Bahamas* **122** D5
Harbour Lyford Cay, *Bahamas* **122** E1
Harbours, Bay of, *Falk. Is., U.K.* **135** W12
Hard Bargain, *Bahamas* **122** B8
Harda, *India* **182** K5
Hardangerfjorden, bay, *Nor.* **142** K10
Hardenberg, *Neth.* **148** H14
Harderwijk, *Neth.* **148** J13
Hardin, *Ill., U.S.* **101** S3
Hardin, *Mont., U.S.* **108** E11
Hardman, *Oreg., U.S.* **110** F6
Hardwick, *Ga., U.S.* **98** M6
Hardy, *Ark., U.S.* **102** G6
Hardy, Península, *Chile* **135** Y9
Härer (Harar), *Eth.* **210** M9
Harewa, *Eth.* **210** L9
Hargeysa, *Somaliland* **210** L11
Harī, river, *Afghan., Iran* **180** G2
Haría, *Sp.* **214** P7
Haridwar, *India* **182** F6
Harihari, *N.Z.* **223** L17
Harij, *India* **182** J3
Hārim, *Syr.* **172** D8
Haripur, *Pak.* **180** J12
Harirud, river, *Iran* **177** E22
Harīrūd, river, *Turkm.* **178** P8
Hariyo, *Somalia* **211** P13
Harkány, *Hung.* **151** N14
Harlan, *Ky., U.S.* **103** D19
Harlan, *Iowa, U.S.* **107** P10
Harlandsville, *Liberia* **204** P10
Harlem, *Mont., U.S.* **108** B10
Harlingen, *Tex., U.S.* **105** R17
Harlingen, *Neth.* **148** G13
Harlowton, *Mont., U.S.* **108** D9
Harmancık, *Turk.* **170** E5
Harmil, island, *Eritrea* **210** G8
Harmony, *Minn., U.S.* **106** L13
Harnai, *Pak.* **180** M8
Harney Basin, *Oreg., U.S.* **110** J7
Harney Lake, *Oreg., U.S.* **110** J7

Harney Peak, *S. Dak., U.S.* **106** L1
Harnom Point, *N. Mariana Is., U.S.* **230** N8
Härnösand, *Sw.* **143** H13
Haro- Shiikh, *Somaliland* **210** M11
Haro, *Sp.* **146** C12
Haroldswick, *Scot., U.K.* **144** B12
Harpanahalli, *India* **183** P5
Harper Lake, *Calif., U.S.* **111** W9
Harper, *Kans., U.S.* **107** V7
Harper, *Liberia* **204** Q11
Harpers Ferry, *W. Va., U.S.* **98** C12
Harput, *Turk.* **171** G14
Harrah, *Yemen* **175** N13
Harrai, *India* **182** K7
Harran, *Turk.* **171** J13
Harrat al Buqūm, *Saudi Arabia* **174** J7
Harrat al Kishb, *Saudi Arabia* **174** H7
Harrat al 'Uwayrid, *Saudi Arabia* **174** E5
Harrat Khaybar, *Saudi Arabia* **174** F6
Harrat Nawāşīf, *Saudi Arabia* **174** J8
Harray, Loch of, *Scot., U.K.* **144** E10
Harricana, river, *Ont. Que. Can.* **87** N18
Harriet Tubman Underground Railroad National Monument, *Md.* **98** D15
Harriman, *Tenn., U.S.* **103** E18
Harrington Harbour, *Que. Can.* **87** M22
Harrington Sound, *Bermuda* **119** B16
Harrington, *Wash., U.S.* **110** C8
Harris, Lake, *Austral.* **220** J9
Harris, Sound of, *Scot., U.K.* **144** H6
Harrisburg, *Pa., U.S.* **96** P7
Harrisburg, *Ill., U.S.* **101** U6
Harrisburg, *Ark., U.S.* **102** F10
Harrisburg, *Nebr., U.S.* **107** P1
Harrisburg, *Oreg., U.S.* **110** G3
Harrison Bay, *Alas., U.S.* **113** B16
Harrison Point, *Barbados* **123** J18
Harrison, *Mich., U.S.* **100** K10
Harrison, *Ark., U.S.* **102** E7
Harrison, *Nebr., U.S.* **106** M1
Harrison, *Mont., U.S.* **108** C3
Harrison, Cape, *N. Amer.* **84** F10
Harrisonburg, *Va., U.S.* **98** D11
Harrisonburg, *La., U.S.* **102** M9
Harrisonville, *Mo., U.S.* **107** T10
Harrisville, *Mich., U.S.* **100** J12
Harrodsburg, *Ky., U.S.* **103** C17
Harrogate, *Eng., U.K.* **145** Q12
Harrold Pond, *Bahamas* **122** E2
Harry S. Truman Reservoir, *Mo., U.S.* **107** T11
Harşit, river, *Turk.* **171** D13
Harstad, *Nor.* **142** D12
Hart Lake, *Oreg., U.S.* **110** K6
Hart Mountain, *Oreg., U.S.* **110** K6
Hart, *Mich., U.S.* **100** K9
Hart, *Tex., U.S.* **104** E12
Hartford City, *Ind., U.S.* **101** Q10
Hartford, *Conn., U.S.* **97** L13
Hartford, *Vt., U.S.* **97** H13
Hartford, *Wis., U.S.* **100** L6
Hartford, *Mich., U.S.* **101** N9
Hartford, *Ark., U.S.* **102** G6
Hartford, *Ala., U.S.* **103** N17
Hartford, *S. Dak., U.S.* **106** L8
Hartford, *Liberia* **204** P10
Hartington, *Nebr., U.S.* **107** N8
Hartland Point, *Eng., U.K.* **145** V9
Hartley, *Tex., U.S.* **104** C11
Hartley, *Iowa, U.S.* **106** M9
Hartog Ridge, *Ind. Oc.* **251** M16
Harts, river, *S. Af.* **212** J7
Hartselle, *Ala., U.S.* **103** H15
Hartshorne, *Okla., U.S.* **105** D19
Hartsville, *S.C., U.S.* **98** K10
Hartsville, *Tenn., U.S.* **103** E16
Hartwell Lake, *Ga. S.C., U.S.* **98** K7
Hartwell, *Ga., U.S.* **98** K7
Hārūt, river, *Afghan.* **180** K2
Harūz-e Bālā, *Iran* **177** J19
Harvard, *Ill., U.S.* **100** M6
Harvard, *Nebr., U.S.* **107** Q7
Harvey, *N. Dak., U.S.* **106** F5
Harz, range, *Ger.* **150** E8
Hasa Plain, *Asia* **168** H4
Hasan Dağı, peak, *Turk.* **170** G9
Hasan Langī, *Iran* **177** M19
Hasanah, *Yemen* **175** N13
Haseo, *S. Korea* **187** R12
Hashaat, *Mongolia* **184** E10
Hashtpar, *Iran* **177** D14

Hasik, *Oman* **175** M16
Hasil, island, *Indonesia* **193** J16
Hasilpur, *Pak.* **180** M11
Haskeir *see* Eilean Hasgeir, island, *Scot., U.K.* **144** H6
Haskell, *Okla., U.S.* **105** C19
Haskell, *Tex., U.S.* **105** F14
Hassel Island, *Virgin Is., U.S.* **122** N8
Hassel Sound, *Nunavut, Can.* **87** B14
Hassi Inifel, *Alg.* **201** F13
Hassi Messaoud, *Alg.* **201** E14
Hässleholm, *Sw.* **143** M13
Hastings, *Mich., U.S.* **100** M10
Hastings, *Minn., U.S.* **106** K12
Hastings, *Nebr., U.S.* **107** Q6
Hastings, *Barbados* **123** L19
Hastings, *Eng., U.K.* **145** V15
Hastings, *N.Z.* **223** H22
Hastings, battlefied, *Eng., U.K.* **145** V15
Hasvik, *Nor.* **143** B13
Haswell, *Colo., U.S.* **109** N15
Hat *see* Teuaua, island, *Fr. Polynesia, Fr.* **230** E11
Hat Yai, *Thai.* **191** S8
Hataedo, island, *S. Korea* **187** U5
Hataedo, island, *S. Korea* **187** U6
Hatch, *N. Mex., U.S.* **109** V11
Hatchie, river, *Tenn., U.S.* **102** F12
Hateg, *Rom.* **154** E12
Hatgal, *Mongolia* **184** C9
Hatherleigh, *Eng., U.K.* **145** V9
Hatherton Seamounts, *Pac. Oc.* **255** J22
Hato Caves, *Curaçao, Neth.* **123** P14
Hato Mayor, *Dom. Rep.* **119** M20
Hatteras Plain, *Atl. Oc.* **248** K4
Hatteras, *N.C., U.S.* **98** H15
Hatteras, Cape, *N.C., U.S.* **98** H15
Hattiesburg, *Miss., U.S.* **102** N12
Hatton Ridge, *Atl. Oc.* **248** F10
Hatton-Rockall Basin, *Atl. Oc.* **253** M23
Hatton, *N. Dak., U.S.* **106** F7
Hatunuru, *Indonesia* **193** K16
Hatutu, island, *Fr. Polynesia, Fr.* **225** H15
Hatvan, *Hung.* **151** L15
Haud, *Eth. Somalia* **210** M11
Haugesund, *Nor.* **142** K9
Haugsdorf, *Aust.* **150** J12
Hauido, island, *S. Korea* **187** U6
Haukeligrend, *Nor.* **142** K10
Haukipudas, *Fin.* **143** E15
Haukivesi, lake, *Fin.* **143** G17
Haulover, *Nicar.* **117** M19
Haupt Kanal, *Liech.* **162** N2
Hauraki Gulf, *N.Z.* **223** E20
Hauser, *Oreg., U.S.* **110** H2
Haut Atlas, *Mor.* **200** E7
Haut, Isle au, *Me., U.S.* **97** G18
Hautere *see* Solander Island, island, *N.Z.* **223** P15
Hauteurs de Gâtine, *Fr.* **149** S6
Hauts Plateaux, *Alg.* **200** D11
Hauturu-o-Toi (Little Barrier Island), *N.Z.* **223** E20
Hau'ula, *Hawaii, U.S.* **115** G14
Havana *see* La Habana, *Cuba* **118** G6
Havana, *Fla., U.S.* **99** Q5
Havana, *Ill., U.S.* **101** Q5
Havant, *Eng., U.K.* **145** V13
Havasu, Lake, *Ariz., Calif., U.S.* **111** W13
Havel, river, *Ger.* **150** C10
Havelberg, *Ger.* **150** D9
Haveli, *Pak.* **180** M12
Havelock, *N.C., U.S.* **98** J14
Haveluloto, *Tonga* **227** Q23
Haven, *Kans., U.S.* **107** U7
Haverfordwest (Hwlffordd), *Wales, U.K.* **145** U8
Haverhill, *Mass., U.S.* **97** J15
Haverhill, *Eng., U.K.* **145** T14
Haveri, *India* **183** P4
Havhesten Bugt, bay, *Greenland, Den.* **89** Q9
Haviland, *Kans., U.S.* **107** U6
Havířov, *Czechia* **151** H14
Havlíčkův Brod, *Czechia* **150** H12
Havre de Grace, *Md., U.S.* **98** B14
Havre Trough, *Pac. Oc.* **255** H22
Havre-Saint-Pierre, *Que. Can.* **87** M21
Havre, *Mont., U.S.* **108** B9
Havza, *Turk.* **170** D11
Hawai'i, *U.S.* **114** H12

Hawai'i Volcanoes National Park, *Hawaii, U.S.* **115** P21
Hawaiian Islands, *Pac. Oc.* **224** C10
Hawaiian Ridge, *Pac. Oc.* **246** F11
Hawangdeungdo, island, *S. Korea* **187** R6
Hawarden, *Iowa, U.S.* **106** M8
Hawarden, *Wales, U.K.* **145** R10
Hawf, *Yemen* **175** M14
Hāwī, *Hawaii, U.S.* **115** L20
Hawick, *Scot., U.K.* **145** M11
Hawke Bay, *N.Z.* **223** H22
Hawker, *S. Aust., Austral.* **221** W10
Hawkeye Point, peak, *Iowa, U.S.* **106** M9
Hawkinsville, *Ga., U.S.* **99** N6
Hawley, *Minn., U.S.* **106** G8
Hawr al Hammār, lake, *Asia* **168** G3
Hawr ash Shuwayjah, lake, *Iraq* **176** H11
Hawthorne, *Nev., U.S.* **111** R8
Hawza, *W. Sahara* **200** G5
Haxtun, *Colo., U.S.* **108** K15
Hay River, *N.W.T. Can.* **86** H11
Hay Springs, *Nebr., U.S.* **106** M2
Hay, river, *Alta. Can.* **86** J10
Hay, river, *Austral.* **220** F10
Hayange, *Fr.* **149** N14
Hayes Center, *Nebr., U.S.* **107** Q4
Hayes Fracture Zone, *Atl. Oc.* **248** K7
Hayes Gletscher, *Greenland, Den.* **88** H5
Hayes, *S. Dak., U.S.* **106** K4
Hayes, *Jam.* **122** J9
Hayes, Mount, *Alas., U.S.* **113** G17
Hayes, river, *Man., Can.* **87** L14
Hayfork, *Calif., U.S.* **111** N2
Hayling Island, *Eng., U.K.* **145** V13
Haymā', *Oman* **175** K16
Haymana, *Turk.* **170** E8
Haynesville, *La., U.S.* **102** K7
Hayneville, *Ala., U.S.* **103** L15
Haynin, *Yemen* **174** N11
Hayrabolu, *Turk.* **170** C3
Hays Glacier, *Antarctica* **238** C19
Hays, *Kans., U.S.* **107** S5
Hays, *Mont., U.S.* **108** B10
Haysville, *Kans., U.S.* **107** U7
Haysyn, *Ukr.* **159** P5
Hayti, *Mo., U.S.* **107** W16
Hayvoron, *Ukr.* **159** Q5
Hayward, *Wis., U.S.* **100** G3
Hayward, *Calif., U.S.* **111** S4
Hazar Gölü, *Turk.* **171** G14
Hazārah Jāt, region, *Afghan.* **180** J6
Hazard, *Ky., U.S.* **103** C19
Hazārqadam, *Afghan.* **180** K7
Hazebrouck, *Fr.* **148** L10
Hazelton, *N. Dak., U.S.* **106** H5
Hazen, *Ark., U.S.* **102** G9
Hazen, *N. Dak., U.S.* **106** F4
Hazen, *Nev., U.S.* **111** P7
Hazlehurst, *Ga., U.S.* **99** N7
Hazlehurst, *Miss., U.S.* **102** M11
Hazleton, *Pa., U.S.* **96** N8
Hazro, *Turk.* **171** G15
He Devil Mountain, *Idaho, U.S.* **108** E3
Head of Barkers, cape, *Cayman Is., U.K.* **122** M2
Head of Bight, *Austral.* **220** J8
Healdsburg, *Calif., U.S.* **111** Q3
Healdton, *Okla., U.S.* **105** E17
Healy, *Kans., U.S.* **107** T4
Healy, *Alas., U.S.* **113** G17
Hearne, *Tex., U.S.* **105** J18
Hearst Island, *Antarctica* **238** E6
Hearst, *Ont., Can.* **87** N17
Heart, river, *N. Dak., U.S.* **106** G3
Heath, Pointe, *Que. Can.* **87** M22
Heavener, *Okla., U.S.* **105** D20
Hebbronville, *Tex., U.S.* **105** P15
Hebel, *Qnsld., Austral.* **221** V14
Heber City, *Utah, U.S.* **108** L7
Heber Springs, *Ark., U.S.* **102** F9
Heber, *Ariz., U.S.* **109** T8
Hebgen Lake, *Mont., U.S.* **108** F8
Hebridean Shelf, *Atl. Oc.* **248** F11
Hebrides Seamount, *Atl. Oc.* **253** L24
Hebrides, Sea of the, *Scot., U.K.* **144** J6
Hebron *see* Al Khalīl, *W. Bank* **173** N5
Hebron, *N. Dak., U.S.* **106** G3
Hebron, *N. Dak., U.S.* **107** R7
Hecate Strait, *B.C., Can.* **86** K7
Hechi, *China* **184** N11

Hecho, *Sp.* **147** C14
Hecla, *S. Dak., U.S.* **106** H7
Hecla, Cape, *Nunavut, Can.* **87** A16
Hectors River, *Jam.* **122** J12
Heddal, *Nor.* **142** K11
Hede, *Nor.* **142** H12
Hédé, *Fr.* **149** Q5
Hedley, *Tex., U.S.* **105** D13
Hedo Misaki, *Japan* **189** X2
Heel Point, *Wake I., U.S.* **230** M11
Heemskerck Fracture Zone, *Ind. Oc.* **251** R18
Heerenveen, *Neth.* **148** H13
Heezen Fracture Zone, *Pac. Oc.* **247** R16
Ḥefa (Haifa), *Israel* **172** K5
Hefei, *China* **185** K14
Heflin, *Ala., U.S.* **103** J17
Hegura Jima, *Japan* **188** M10
Heha, peak, *Burundi* **211** U2
Heide, *Ger.* **150** B7
Heidelberg, *Miss., U.S.* **102** M12
Heidelberg, *Ger.* **150** H6
Heidenheim, *Ger.* **150** J7
Heiderscheid, *Lux.* **162** H9
Heihe, *China* **185** B16
Heilbronn, *Ger.* **150** J7
Heilong Jiang, river, *China* **185** A14
Heilprin Gletscher, *Greenland, Den.* **88** E5
Heimaey, island, *Ice.* **142** F2
Heimefront Range, *Antarctica* **238** C12
Heina Islands, *P.N.G.* **222** M7
Heinerscheid, *Lux.* **162** G10
Heinola, *Fin.* **143** H16
Heisker Islands *see* Na h-Eileanan Theisgeir, *Scot., U.K.* **144** L11
Hejaz *see* Al Ḥijāz, *Saudi Arabia* **174** E5
Hejing, *China* **184** E5
Hekimhan, *Turk.* **170** F12
Hekla, peak, *Ice.* **142** F2
Hekou, *China* **185** G14
Hekou, *China* **184** P10
Helagsfjället, peak, *Sw.* **142** H12
Hele Islands, *Solomon Is.* **228** E8
Heḟebce (Ḥalabjah), *Iraq* **176** F11
Helen Glacier, *Antarctica* **239** J21
Helen Island, *Palau* **224** F2
Helena *see* Makrónisos, island, *Gr.* **157** K13
Helena-West Helena, *Ark., U.S.* **102** H10
Helena, *Ga., U.S.* **99** N7
Helena, *Okla., U.S.* **105** B16
Helena, *Mont., U.S.* **108** D7
Helengili, island, *Maldives* **183** V3
Helgoländer Bay, *Atl. Oc.* **140** F6
Helgoländer Bucht, *Ger.* **150** B6
Helheimgletscher, glacier, *Greenland, Den.* **89** P10
Helicon *see* Elikónas, peak, *Gr.* **156** H11
Hell-Bourg, *Réunion, Fr.* **215** G15
Hell, *Cayman Is. U.K.* **122** M2
Hell, *Nor.* **142** G11
Hellenic Trough, *Atl. Oc.* **248** K15
Hellevoetsluis, *Neth.* **148** J11
Hellín, *Sp.* **147** J13
Hellissandur, *Ice.* **142** E1
Hells Canyon Dam, *Idaho, Oreg., U.S.* **110** F9
Hells Canyon National Recreation Area, *Idaho, Oreg., U.S.* **110** E9
Hells Canyon, *Idaho, Oreg., U.S.* **110** F9
Helmand, river, *Afghan.* **180** K6
Helmsdale, *Scot., U.K.* **144** G10
Helper, *Utah, U.S.* **108** M8
Helsingborg, *Sw.* **142** M11
Helsinki (Helsingfors), *Fin.* **143** J16
Hemet, *Calif., U.S.* **111** W10
Hemingford, *Nebr., U.S.* **107** N2
Hempstead, *N.Y., U.S.* **96** N11
Hempstead, *Tex., U.S.* **105** L18
Henashi Zaki, *Japan* **188** J12
Hendaye, *Fr.* **149** X5
Hendek, *Turk.* **170** D6
Henderson Island, *Pac. Oc.* **225** K16
Henderson Island, *Antarctica* **239** K22
Henderson, *N.C., U.S.* **98** G12
Henderson, *Ky., U.S.* **103** C14
Henderson, *Tenn., U.S.* **103** F13
Henderson, *Tex., U.S.* **105** H20
Henderson, *Nev., U.S.* **111** U12
Hendersonville, *N.C., U.S.* **98** J7
Hendersonville, *Tenn., U.S.* **103** E15

Hendījān, *Iraq* **177** K14
Hendorābī, island, *Iran* **177** N17
Hendrik Ø, *Greenland, Den.* **88** B7
Henganofi, *P.N.G.* **222** P7
Hengduan Shan, *China* **184** L7
Hengelo, *Neth.* **148** J14
Hengshan, *China* **185** D17
Hengyang, *China* **185** M13
Heniches'k, *Ukr.* **159** S8
Henlopen, Cape, *Del., U.S.* **98** C15
Hennebont, *Fr.* **149** R4
Hennessey, *Okla., U.S.* **105** B16
Hennigsdorf, *Ger.* **150** D10
Henning, *Minn., U.S.* **106** H9
Henrietta Maria, Cape, *Ont., Can.* **87** L17
Henrietta, *Tex., U.S.* **105** F16
Henrietta, *Russ.* **161** C17
Henrik Krøyer Holme, islands, *Greenland, Den.* **88** C13
Henry Bay, *Antarctica* **239** N20
Henry Ice Rise, *Antarctica* **238** G10
Henry Kater Peninsula, *Nunavut, Can.* **87** F18
Henry Land, *Greenland, Den.* **88** L14
Henry Trough, *Pac. Oc.* **247** Q16
Henry, Cape, *Va., U.S.* **98** F14
Henryetta, *Okla., U.S.* **105** C18
Henrys Fork, river, *Idaho, U.S.* **108** G7
Heppner, *Oreg., U.S.* **110** F6
Herald Bank, *Arctic Oc.* **252** K6
Herald Cays, *Austral.* **220** C14
Herald Valley, *Arctic Oc.* **252** J6
Herald, *Russ.* **161** A19
Herāt, *Afghan.* **180** H3
Herbault, *Fr.* **149** R8
Herbert Downs, *Qnsld. Austral.* **221** T11
Herbert Ø *see* Qeqertarsuaq, *Greenland, Den.* **88** E3
Herbert, *Sask., Can.* **86** N11
Hercules Dome, *Antarctica* **238** J12
Hereford, *Tex., U.S.* **104** D11
Hereford, *Eng., U.K.* **145** T11
Hereheretue, island, *Fr. Polynesia, Fr.* **231** D19
Héﾏémakono, *Sierra Leone* **204** M9
Herford, *Ger.* **150** E6
Herington, *Kans., U.S.* **107** T8
Herisau, *Switz.* **152** A4
Hérisson, *Fr.* **149** T10
Heritage Range, *Antarctica* **238** H9
Herkimer, *N.Y., U.S.* **96** J11
Herlen, river, *Mongolia* **185** D12
Herluf Trolleland, *Greenland, Den.* **88** B11
Herlufholm Strand, *Greenland, Den.* **88** B11
Herm, island, *Channel Is. U.K.* **145** Y11
Herman, *Minn., U.S.* **106** H8
Hermann, *Mo., U.S.* **107** T14
Hermannsburg, *N.T., Austral.* **221** T8
Hermansville, *Mich., U.S.* **100** G7
Hermiston, *Oreg., U.S.* **110** E7
Hermit Islands, *P.N.G.* **222** M7
Hermleigh, *Tex., U.S.* **105** G13
Hermon, Mount, *Leb.* **172** J7
Hermosa, *S. Dak., U.S.* **106** L2
Hermosa, Punta, *Eq. Guinea* **214** L8
Hermosillo, *Mex.* **116** C7
Hernando, *Miss., U.S.* **102** G11
Hernando, *Arg.* **134** K10
Hernani, *Sp.* **147** B13
Herne, *Ger.* **150** E5
Herreid, *S. Dak., U.S.* **106** H5
Herrera de Pisuerga, *Sp.* **146** C10
Herrera del Duque, *Sp.* **146** H9
Herrera, *Arg.* **134** H10
Herrin, *Ill., U.S.* **101** U5
Herschel Island, *Yukon, Can.* **86** D9
Hershey, *Nebr., U.S.* **107** P4
Hertford, *N.C., U.S.* **98** G14
Hertford, *Eng., U.K.* **145** U14
Hervás, *Sp.* **146** F8
Hervey Bay, *Austral.* **220** G16
Hervey Bay, *Qnsld. Austral.* **221** U16
Hervey Islands, *Cook Is. N.Z.* **224** K12
Herzberg, *Ger.* **150** E10
Hesperange, *Lux.* **162** K10
Hess Rise, *Pac. Oc.* **246** E11
Hettinger, *N. Dak., U.S.* **106** H3
Heungdeok, *S. Korea* **187** S7
Heunghae, *S. Korea* **187** Q12
Heuru, *Solomon Is.* **228** G11
Hewﾏ (Arbīl, Erbil), *Iraq* **176** E10

Heze, *China* **185** J13
Hezhou, *China* **184** N12
Hezuo, *China* **184** J9
Hiawatha Gletscher, *Greenland, Den.* **88** E5
Hiawatha, *Kans., U.S.* **107** R9
Hibbing, *Minn., U.S.* **106** F11
Hicacos, Península de, *Cuba* **118** G7
Hickman, *Ky., U.S.* **102** E12
Hickman, *Ark., U.S.* **134** E10
Hickory, *N.C., U.S.* **98** H8
Hicks Bay, *N.Z.* **223** F23
Hicksville, *Ohio, U.S.* **101** P11
Hico, *Tex., U.S.* **105** H16
Hidaka Sammyaku, *Japan* **188** G14
Hidalgo del Parral, *Mex.* **116** E9
Hididelli, *Eth.* **211** N10
Hienghène, *New Caledonia, Fr.* **231** P21
Hierro (Ferro), island, *Sp.* **214** R3
Higgins Lake, *Mich., U.S.* **100** J10
Higgins, *Tex., U.S.* **105** B14
Higginsville, *Mo., U.S.* **107** S11
Higginsville, *W. Aust., Austral.* **221** W4
High Atlas, range, *Af.* **198** C3
High Cay, *Bahamas* **122** F11
High Cliff Point, *Grenada* **123** J23
High Island, *Tex., U.S.* **105** L20
High Level, *Alta., Can.* **86** J10
High Plains, *N. Amer.* **84** K5
High Point, *N.C., U.S.* **98** H10
High Point, *Bermuda* **119** D14
High Point, peak, *N.J., U.S.* **96** M10
High Red Rock, *Austral.* **229** B20
High Rock, *Bahamas* **122** J3
High Springs, *Fla., U.S.* **99** R7
High Willhays, peak, *Eng., U.K.* **145** W9
Highgate, *Jam.* **122** H10
Highland Park, *Ill., U.S.* **101** N7
Highland, *Ill., U.S.* **101** T5
Highlands, *Scot., U.K.* **144** J9
Highmore, *S. Dak., U.S.* **106** K5
Higüero, Punta, *P.R., U.S.* **119** M22
Higüey, *Dom. Rep.* **119** M20
Hiiumaa (Dagö), island, *Est.* **143** K15
Híjar, *Sp.* **147** E14
Hiko, *Nev., U.S.* **111** S11
Hikueru, island, *Fr. Polynesia, Fr.* **231** C20
Hikurangi Plateau, *Pac. Oc.* **246** P12
Hikurangi Seachannel, *Pac. Oc.* **255** J21
Hikurangi Terrace, *Pac. Oc.* **246** P11
Hikutavake, *N.Z.* **229** P21
Hilāl, Ras al, *Af.* **198** C7
Hildesheim, *Ger.* **150** E7
Hilina Pali, *Hawaii, U.S.* **115** P22
Hill City, *Minn., U.S.* **106** G11
Hill City, *S. Dak., U.S.* **106** L1
Hill City, *Kans., U.S.* **107** S5
Hill Country, *Tex., U.S.* **105** L15
Hillary Coast, *Antarctica* **239** M14
Hillman, *Mich., U.S.* **100** H11
Hillsboro, *Ill., U.S.* **101** S5
Hillsboro, *Ohio, U.S.* **101** S12
Hillsboro, *Tex., U.S.* **105** H17
Hillsboro, *N. Dak., U.S.* **106** F8
Hillsboro, *Kans., U.S.* **107** T8
Hillsboro, *N. Mex., U.S.* **109** U11
Hillsboro, *Oreg., U.S.* **110** F3
Hillsborough Bay, *Trin. & Tobago* **123** P18
Hillsborough Dam, *Trin. & Tobago* **123** N18
Hillsborough, *Grenada* **121** H17
Hillsborough, *N. Ire., U.K.* **145** P7
Hillsdale, *Mich., U.S.* **101** N11
Hillside, *W. Aust., Austral.* **221** T3
Hillsville, *Va., U.S.* **98** G9
Hillswick, *Scot., U.K.* **144** C12
Hilo Bay, *Hawaii, U.S.* **115** N23
Hilo, *Hawaii, U.S.* **115** N22
Hilo, *Hawaii, U.S.* **115** N23
Hilton Head Island, *S.C., U.S.* **99** N9
Hilton Inlet, *Antarctica* **238** E7
Hilvan, *Turk.* **171** H13
Himadítida, Límni, *Gr.* **156** C9
Himalaya, range, *Asia* **168** H7
Himatnagar, *India* **182** J3
Himeji, *Japan* **189** Q8
Himi, *N. Korea* **186** J9
Himo, *Tanzania* **211** U7
Himora, *Eth.* **210** J6
Ḥimṣ (Homs), *Syr.* **172** G8

Hinako, Kepulauan, *Indonesia* **192** J4
Hinche, *Haiti* **119** L17
Hinchinbrook Island, *Alas., U.S.* **113** K18
Hinchinbrook Island, *Austral.* **220** D13
Hinckley, *Minn., U.S.* **106** H12
Hindaun, *India* **182** H6
Hindmarsh, Lake, *Austral.* **220** L11
Hindu Kush, *Afghan., Pak.* **180** G9
Hines Creek, *Alta., Can.* **86** K10
Hines, *Oreg., U.S.* **110** J7
Hinesville, *Ga., U.S.* **99** N8
Hinganghat, *India* **182** L6
Hingol, river, *Pak.* **181** R5
Hingorja, *Pak.* **181** Q8
Hinidan, *Pak.* **181** R7
Hinksland, *Greenland, Den.* **88** K12
Hinnøya, island, *Nor.* **142** D12
Hinthada, *Myanmar* **190** K4
Hinton, *W. Va., U.S.* **98** E8
Hinton, *Okla., U.S.* **105** C16
Hippolytushoef, *Neth.* **148** H12
Hira Se, island, *Japan* **189** U4
Hirado Shima, *Japan* **189** R4
Hiraman, river, *Kenya* **211** T8
Hiranai, *Japan* **188** J12
Hirata, *Japan* **189** P6
Hirfanlı Barajı, *Turk.* **170** F9
Hiroo, *Japan* **188** H14
Hirosaki, *Japan* **188** J12
Hiroshima, *Japan* **189** Q6
Hirson, *Fr.* **148** M11
Hîrşova, *Rom.* **155** F17
Hisarönü, *Turk.* **170** C7
Ḥismā, *Saudi Arabia* **174** D4
Ḥişn Ṭāqrifat, *Lib.* **201** F19
Hispaniola, island, *N. Amer.* **119** M18
Ḥisyah, *Syr.* **172** H8
Hīt, *Iraq* **176** G9
Hita, *Japan* **189** R5
Hitachi, *Japan* **189** N12
Hitadu, island, *Maldives* **183** X3
Hiti, island, *Fr. Polynesia, Fr.* **231** B19
Hitiaa, *Fr. Polynesia, Fr.* **231** H22
Hitoyoshi, *Japan* **189** S5
Hitra, island, *Nor.* **142** G10
Hiu, island, *Vanuatu* **229** C13
Hiva Oa, island, *Fr. Polynesia, Fr.* **231** D14
Hiwassee, river, *N.C., Tenn., U.S.* **103** F17
Hizan, *Turk.* **171** G17
Hınıs, *Turk.* **171** F16
Hjälmaren, lake, *Sw.* **143** K13
Hjørring, *Den.* **142** M11
Hjort Trench, *Pac. Oc.* **255** G17
Hkakabo Razi, peak, *Myanmar* **190** C6
Hlohlolwane, *S. Af.* **212** K8
Hlohovec, *Slovakia* **151** K14
Hlukhiv, *Ukr.* **158** M7
Hlybokaye, *Belarus* **158** J5
Hněvkovice, *Czechia* **150** H11
Hnúšťa, *Slovakia* **151** J15
Ho Chi Minh City *see* Thành Phố Hồ Chí Minh (Saigon), *Vietnam* **191** P12
Hồ Xá, *Vietnam* **190** K12
Ho, *Ghana* **205** P16
Hòa Bình, *Vietnam* **190** H11
Hoàng Sa, *see* Paracel Islands, *S. China Sea* **190** K15
Hoare Bay, *Nunavut, Can.* **87** G19
Hoban, *N. Korea* **186** F9
Hobart, *Ind., U.S.* **101** P8
Hobart, *Okla., U.S.* **105** D15
Hobart, *Tas., Austral.* **221** Z8
Hobbs Coast, *Antarctica* **238** M9
Hobbs, *N. Mex., U.S.* **109** U15
Hobe Sound, *Fla., U.S.* **99** V10
Hoboksar, *China* **184** D5
Hobro, *Den.* **142** M11
Hobucken, *N.C., U.S.* **98** H14
Hobyo, *Somalia* **211** P13
Hŏch'ŏn, river, *N. Korea* **186** E10
Hochstetterbugten, *Greenland, Den.* **88** G14
Hodda, peak, *Somalia* **210** K15
Hodge, *La., U.S.* **102** L8
Hodgenville, *Ky., U.S.* **103** C16
Hodgson Downs, *N.T., Austral.* **221** Q9
Hodh, region, *Mauritania* **204** G11
Hódmezővásárhely, *Hung.* **151** M16
Hodo, *N. Korea* **186** J9
Hodonín, *Czechia* **151** J13
Hödrögö, *Mongolia* **184** C8
Hoech'ang, *N. Korea* **186** J7

Hoedic, Île, *Fr.* **149** R4
Hoel Mountains, *Antarctica* **239** B15
Hoenggando, island, *S. Korea* **187** V7
Hoengseong, *S. Korea* **187** N10
Hoeryŏng, *N. Korea* **186** B12
Hoeyang, *N. Korea* **186** K9
Hof, *Ger.* **150** G9
Höfðakaupstaður, *Ice.* **142** D3
Hoffman, *Minn., U.S.* **106** H9
Höfn, *Ice.* **142** F4
Hofrat en Nahas, *S. Sudan* **203** U7
Hofsjökull, glacier, *Ice.* **142** E3
Hog Cay, *Bahamas* **122** F10
Hog Island, *Va., U.S.* **98** E15
Hog Island, *Grenada* **123** L22
Hogansville, *Ga., U.S.* **98** M4
Hogfish Cut, *Bermuda* **119** D14
Hoggar *see* Ahaggar, mountains, *Alg.* **201** K13
Hogoro, *Tanzania* **211** W6
Hoh Xil Shan, *China* **184** H5
Hohe Tauern, range, *Aust.* **150** M9
Hohenems, *Aust.* **150** L7
Hohenwald, *Tenn., U.S.* **103** F14
Hohhot, *China* **184** F12
Hohoe, *Ghana* **205** N16
Honohu, island, *Fr. Polynesia, Fr.* **231** Q14
Hội An, *Vietnam* **190** L12
Hoima, *Uganda* **211** R3
Hoisington, *Kans., U.S.* **107** T6
Hokatu, *Fr. Polynesia, Fr.* **230** E11
Hokitika, *N.Z.* **223** L18
Hokkaido Rise *see* Zenkevich Rise, *Pac. Oc.* **246** E8
Hokkaidō, island, *Japan* **188** G14
Hokula, *Tonga* **227** L22
Hola Prystan', *Ukr.* **159** R7
Hola, *Kenya* **211** T8
Holbæk Bugt, bay, *Greenland, Den.* **88** C11
Holbrook, *Idaho, U.S.* **108** J6
Holbrook, *Ariz., U.S.* **109** S8
Hold with Hope, *Greenland, Den.* **88** H14
Holden, *Mo., U.S.* **107** T11
Holden, *Utah, U.S.* **108** M7
Holdenville, *Okla., U.S.* **105** D18
Holdrege, *Nebr., U.S.* **107** Q5
Hole in the Wall, cape, *Bahamas* **122** C9
Hōlei Pali, *Hawaii, U.S.* **115** P23
Holetown, *Barbados* **123** K18
Holguín, *Cuba* **118** J12
Holland Bay, *Jam.* **122** J12
Holland, *Mich., U.S.* **100** M9
Holland, *Eur.* **140** G5
Hollandale, *Miss., U.S.* **102** K10
Hollandsbird Island, *Namibia* **212** H3
Hollick-Kenyon Plateau, *Antarctica* **238** K9
Hollidaysburg, *Pa., U.S.* **96** N5
Hollis Reservoir, *Trin. & Tobago* **123** P23
Hollis, *Okla., U.S.* **105** D14
Hollister, *Calif., U.S.* **111** T4
Hollum, *Neth.* **148** G13
Holly Hill, *S.C., U.S.* **98** L10
Holly Ridge, *N.C., U.S.* **98** J13
Holly Springs, *Miss., U.S.* **102** G12
Hollywood, *Fla., U.S.* **99** V10
Hollywood, *Calif., U.S.* **111** X8
Holm Ø, *Greenland, Den.* **88** H5
Holma, *Nigeria* **206** K11
Holman, *N.W.T., Can.* **86** D13
Hólmavík, *Ice.* **142** D2
Holme Bay, *Antarctica* **239** E20
Holmes Rock, *Bahamas* **122** J1
Holmland, *Greenland, Den.* **88** C12
Holonga, *Tonga* **227** G23
Holopaw, *Fla., U.S.* **99** T9
Holot Ḥaluẕa, *Israel* **173** N4
Holroyd, river, *Austral.* **220** B12
Holstebro, *Den.* **142** M10
Holstein, *Iowa, U.S.* **107** N9
Holstein, *Eur.* **140** F7
Holsteinsborg *see* Sisimiut, *Greenland, Den.* **89** N5
Holston, river, *Tenn., Va., U.S.* **103** E20
Holsworthy, *Eng., U.K.* **145** V9
Holt, *Ala., U.S.* **103** K14
Holton, *Kans., U.S.* **107** S9
Holtville, *Calif., U.S.* **111** Z12
Hōlualoa, *Hawaii, U.S.* **115** N20
Holy Cross, *Alas., U.S.* **113** H13

Holy Island (Lindisfarne), *Eng., U.K.* **145** M12
Holy Island, *Wales, U.K.* **145** R9
Holyhead (Caergybi), *Wales, U.K.* **145** R9
Holyoke, *Mass., U.S.* **97** K13
Holyoke, *Colo., U.S.* **108** L16
Homa Bay, *Kenya* **211** S5
Hombori Tondo, peak, *Mali* **205** J15
Hombori, *Mali* **205** H15
Homburg, *Ger.* **150** H5
Home Bay, *Nunavut, Can.* **87** F18
Home Bay, *Kiribati* **226** M11
Home Forland, *Greenland, Den.* **88** H14
Homedale, *Idaho, U.S.* **108** G3
Homer, *N.Y., U.S.* **96** K8
Homer, *Mich., U.S.* **101** N10
Homer, *La., U.S.* **102** K7
Homer, *Alas., U.S.* **113** K16
Homerville, *Ga., U.S.* **99** Q7
Homestead National Monument of America, *Nebr., U.S.* **107** R8
Homestead, *Fla., U.S.* **99** X10
Homewood, *Ala., U.S.* **103** J15
Hominy, *Okla., U.S.* **105** B18
Hommalinn, *Myanmar* **190** E4
Homnabad, *India* **182** N5
Homo Bay, *Vanuatu* **229** G15
Homosassa Springs, *Fla., U.S.* **99** T7
Homs see Ḥimṣ, *Syr.* **172** G8
Homuz, Strait of, *Ind. Oc.* **168** H5
Homyel', *Belarus* **158** M6
Hòn Đá Tý, island, *Vietnam* **191** Q13
Hòn Hải, island, *Vietnam* **191** Q13
Hòn Khoai, island, *Vietnam* **191** R10
Hòn Lớn, island, *Vietnam* **191** N13
Honam, *N. Korea* **186** D12
Hōnaunau, *Hawaii, U.S.* **115** P20
Honavar, *India* **183** Q4
Honaz, *Turk.* **170** G5
Honda, *Col.* **130** G6
Hondo, *Tex., U.S.* **105** M15
Hondo, *N. Mex., U.S.* **109** U13
Hondo, *Japan* **189** S4
Honduras, *N. Amer.* **117** L17
Honduras, Gulf of, *N. Amer.* **117** K17
Honea Path, *S.C., U.S.* **98** K7
Hønefoss, *Nor.* **142** K11
Honesdale, *Pa., U.S.* **96** M9
Honey Grove, *Tex., U.S.* **105** F18
Honey Lake, *Calif., U.S.* **111** N6
Honeyville, *Utah, U.S.* **108** K7
Hồng (Red), river, *Vietnam* **190** G10
Hongcheon, *S. Korea* **186** N10
Hongdo, island, *S. Korea* **187** T5
Honggun, *N. Korea* **186** E10
Hongliuyuan, *China* **184** F7
Hongnong, *S. Korea* **187** S7
Hongor, *Mongolia* **184** D12
Hongseong, *S. Korea* **187** P7
Hongshui, river, *China* **184** N11
Hongwŏn-man, bay, *N. Korea* **186** G10
Hongwŏn, *N. Korea* **186** G10
Honiara, *Solomon Is.* **228** F10
Honiton, *Eng., U.K.* **145** V10
Honoka‘a, *Hawaii, U.S.* **115** M22
Honokahua, *Hawaii, U.S.* **115** J18
Honokōwai, *Hawaii, U.S.* **115** J18
Honolulu, *Hawaii, U.S.* **115** H14
Honomanū Bay, *Hawaii, U.S.* **115** J19
Honomū, *Hawaii, U.S.* **115** M23
Honshū, island, *Japan* **189** P6
Honu‘apo Bay, *Hawaii, U.S.* **115** Q21
Hood Point, *Austral.* **220** L4
Hood Point, *P.N.G.* **222** R8
Hood River, *Oreg., U.S.* **110** E4
Hood, island, see Isla Española, *Ecua.* **130** R4
Hood, Mount, *Oreg., U.S.* **110** F4
Hoodsport, *Wash., U.S.* **110** C3
Hoogeveen, *Neth.* **148** H14
Hook Head, *Ire.* **145** T6
Ho‘okena, *Hawaii, U.S.* **115** P20
Hooker, *Okla., U.S.* **105** A13
Ho‘olehua, *Hawaii, U.S.* **115** H17
Hoonah, *Alas., U.S.* **113** L22
Hooper Bay, *Alas., U.S.* **112** H11
Hooper, Cape, *Nunavut, Can.* **87** F18
Hoopeston, *Ill., U.S.* **101** Q7
Hoosier Hill, *Ind., U.S.* **101** R10
Hoover Dam, *Ariz., Nev., U.S.* **111** U12
Hoover, *Ala., U.S.* **103** J15
Hopa, *Rep. of Georgia* **171** C16

Hope Bay, *Jam.* **122** H11
Hope Mills, *N.C., U.S.* **98** J11
Hope Point, *Myanmar* **190** L6
Hope Town, *Bahamas* **122** B10
Hope Valley, *Arctic Oc.* **252** K5
Hope, *Ark., U.S.* **102** J7
Hope, *N. Dak., U.S.* **106** G7
Hope, Cape, *Nunavut, Can.* **86** F12
Hope, Lake, *Austral.* **220** K4
Hopedale, *Nfld. & Lab., Can.* **87** K21
Hopelchén, *Mex.* **117** H16
Hopen Bank, *Arctic Oc.* **253** F16
Hopetoun, *W. Aust., Austral.* **221** X4
Hopetown, *S. Af.* **212** K7
Hopewell Culture National Historical Park, *Ohio, U.S.* **101** S13
Hopewell Islands, *Nunavut, Can.* **87** K17
Hopewell, *Va., U.S.* **98** F13
Hopewell, *Jam.* **122** H6
Hopkins, Lake, *Austral.* **220** F7
Hopkins, Mount, *Ariz., U.S.* **109** W8
Hopkinsville, *Ky., U.S.* **103** D14
Hoquiam, *Wash., U.S.* **110** D2
Horasan, *Turk.* **171** E16
Hordil Sarïdag, range, *Mongolia* **184** C8
Horgos, *Serb.* **154** D10
Hörh Uul, *Mongolia* **184** F10
Horicon, *Wis., U.S.* **100** L6
Horizon Ridge, *Ind. Oc.* **251** J16
Horki, *Belarus* **158** K6
Horlick Mountains, *Antarctica* **238** J11
Horlivka, *Ukr.* **159** Q9
Ḥormak, *Iran* **177** K22
Hormoz, island, *Iran* **177** N19
Horn Island, *Miss., U.S.* **103** P13
Horn, Cape see Hornos, Cabo de, *S. Amer.* **128** R5
Hornafjörðardjúp, valley, *Atl. Oc.* **253** L22
Hornavan, lake, *Sw.* **143** E13
Hornbeck, *La., U.S.* **102** M6
Hornbrook, *Calif., U.S.* **110** L3
Hornby Hill, peak, *Nunavut, Can.* **86** F11
Hornell, *N.Y., U.S.* **96** K6
Hornos, Cabo de (Cape Horn), *Chile* **135** Y9
Hornsund Bank, *Arctic Oc.* **253** G17
Hornsund Valley, *Arctic Oc.* **253** G17
Horo Shan, *China* **184** E4
Horodnya, *Ukr.* **158** M6
Horodyshche, *Ukr.* **159** P6
Horqueta, *Parag.* **132** M10
Horse Cave, *Ky., U.S.* **103** D16
Horse Creek, *Wyo., U.S.* **108** K13
Horse Mountain, *N. Mex., U.S.* **109** T10
Horsehead Lake, *N. Dak., U.S.* **106** G5
Horseheads, *N.Y., U.S.* **96** K7
Horseshoe Beach, *Fla., U.S.* **99** S6
Horseshoe Bend, *Ark., U.S.* **102** E9
Horseshoe Seamounts, *Atl. Oc.* **248** J10
Horsham, *Eng., U.K.* **145** V14
Horsham, *Vic., Austral.* **221** Y11
Horta, *Port.* **146** P3
Horten, *Nor.* **142** K11
Horton, *Kans., U.S.* **107** R9
Horton, river, *N.W.T., Can.* **86** E11
Horvat ‘Avedat (Eboda), ruins, *Israel* **173** P5
Hosa‘ina, *Eth.* **211** N7
Hoscheid, *Lux.* **162** H10
Hosdrug, *India* **183** R4
Hoshab, *Pak.* **181** R4
Hoshangabad, *India* **182** K6
Hosingen, *Lux.* **162** H10
Hoskins, *P.N.G.* **222** P9
Hosmer, *S. Dak., U.S.* **106** J5
Hoste, Isla, *Chile* **135** Y8
Hosur, *India* **183** R6
Hot Creek Range, *Nev., U.S.* **111** R10
Hot Springs National Park, *Ark., U.S.* **102** H7
Hot Springs, *Va., U.S.* **98** E10
Hot Springs, *Ark., U.S.* **102** H7
Hot Springs, *S. Dak., U.S.* **106** L1
Hot Springs, *Mont., U.S.* **108** C5
Hot Sulphur Springs, *Colo., U.S.* **108** L12
Hot, *Thai.* **190** K6
Hotan, *China* **184** G2
Hotan, river, *China* **184** G3

Hotchkiss, *Colo., U.S.* **109** N11
Hotel de Paris, *Monaco* **163** D22
Hotel Hermitage, *Monaco* **163** D21
Hotel Métropole, *Monaco* **163** C22
Hotevilla, *Ariz., U.S.* **109** R8
Hottah Lake, *N.W.T., Can.* **86** G11
Hotte, Massif de la, *Haiti* **119** M15
Hottentot Bay, *Namibia* **212** J3
Houaïlou, *New Caledonia, Fr.* **231** P22
Houat, Île, *Fr.* **149** R4
Houffalize, *Belg.* **148** M13
Houghton, *Mich., U.S.* **100** E6
Houk (Putusuk), *F.S.M.* **224** F5
Houlton, *Me., U.S.* **97** C18
Houma Toloa, cape, *Tonga* **227** R23
Houma, *China* **184** J12
Houma, *La., U.S.* **102** Q10
Houma, *Tonga* **227** Q23
Houndé, *Burkina Faso* **205** L14
Hourtin, *Fr.* **149** V6
Houston, *Miss., U.S.* **102** J12
Houston, *Tex., U.S.* **105** L19
Houston, *Minn., U.S.* **106** L13
Houston, *Mo., U.S.* **107** U13
Houtman Abrolhos, islands, *Austral.* **220** J1
Hovd, *Mongolia* **184** C7
Hovd, *Mongolia* **184** E9
Hovd, river, *Mongolia* **184** C7
Hove, *Eng., U.K.* **145** V14
Hoven, *S. Dak., U.S.* **106** J5
Hovenweep National Monument, *Utah, U.S.* **109** P9
Hovgaard Ø, *Greenland, Den.* **88** D12
Hövsgöl Nuur, *Mongolia* **184** B9
Hövsgöl, *Mongolia* **184** E11
Howard City, *Mich., U.S.* **100** L10
Howard Lake, *N.W.T., Can.* **87** H13
Howard, *S. Dak., U.S.* **106** L7
Howard, *Kans., U.S.* **107** U9
Howe, *Idaho, U.S.* **108** G6
Howe, Cape, *Austral.* **220** M14
Howland Island, *Pac. Oc.* **224** G9
Howland, *Me., U.S.* **97** E17
Hoxie, *Ark., U.S.* **102** E10
Hoxie, *Kans., U.S.* **107** S4
Hoxud, *China* **184** E5
Hoy, island, *Scot., U.K.* **144** F10
Høyanger, *Nor.* **142** J10
Hoyerswerda, *Ger.* **150** F11
Hoyt Lakes, *Minn., U.S.* **106** F12
Hozat, *Turk.* **171** F14
Hpa-an, *Myanmar* **190** K6
Hradec Králové, *Czechia* **150** G12
Hrazdan, river, *Arm.* **171** D18
Hrebenne, *Pol.* **151** G18
Hrodna, *Belarus* **158** K3
Hron, river, *Slovakia* **151** J15
Hrubieszów, *Pol.* **151** F18
Hsi-hseng, *Myanmar* **190** H6
Hsinchu, *China* **185** N16
Hsipaw, *Myanmar* **190** G6
Hua Hin, *Thai.* **191** N7
Huachi, *China* **184** H11
Huacho, *Peru* **132** H3
Huadian, *China* **185** E16
Huahine Iti, island, *Fr. Polynesia, Fr.* **231** K14
Huahine Nui, island, *Fr. Polynesia, Fr.* **231** J14
Huahine, *Fr. Polynesia, Fr.* **231** N14
Huahine, island, *Fr. Polynesia, Fr.* **231** B16
Huai Yang, *Thai.* **191** P7
Huaibei, *China* **185** J14
Huaihua, *China* **184** M12
Huainan, *China* **185** K14
Huaiyin, *China* **185** J15
Huajuapan de León, *Mex.* **116** K12
Hualāli, peak, *Hawaii, U.S.* **115** N20
Hualapai Peak, *Ariz., U.S.* **109** S5
Hualfin, *Arg.* **134** G8
Huallaga, river, *Peru* **132** F3
Huambo, *Angola* **208** M11
Huancané, *Peru* **132** J6
Huancavelica, *Peru* **132** H4
Huancayo, *Peru* **132** H4
Huang (Yellow), river, *China* **184** H12
Huangshi, *China* **185** L13
Huangyan, *China* **185** L15
Huanren, *China* **185** F16
Huanta, *Peru* **132** H4
Huánuco, *Peru* **132** G3
Huanxian, *China* **184** H11

Huara, *Chile* **134** C7
Huaral, *Peru* **132** H3
Huaraz, *Peru* **132** G3
Huarmey, *Peru* **132** G3
Huasco, *Chile* **134** H6
Huatabampo, *Mex.* **116** E7
Huaura, *Peru* **132** H3
Huascarán, *S. Amer.* **128** F2
Hubballi, *India* **183** P4
Hubbard Lake, *Mich., U.S.* **100** H12
Hubbard, *Tex., U.S.* **105** H18
Hubbell Trading Post National Historic Site, *Ariz., U.S.* **109** R9
Hudiksvall, *Sw.* **143** H13
Hudson Bay, *N. Amer.* **84** F7
Hudson Bay, *Sask., Can.* **87** M13
Hudson Canyon, *Atl. Oc.* **248** J4
Hudson Land, *Greenland, Den.* **88** H13
Hudson Strait, *Nunavut, Can.* **248** F4
Hudson, *N.Y., U.S.* **96** K11
Hudson, *Fla., U.S.* **99** T7
Hudson, *Mich., U.S.* **101** N11
Hudson, river, *N.Y., U.S.* **96** H11
Huê, *Vietnam* **190** L12
Hu‘ehu‘e, *Hawaii, U.S.* **115** N20
Huehuetenango, *Guatemala* **117** L15
Huelma, *Sp.* **146** K11
Huelva, *Sp.* **146** L7
Huerfano, river, *Colo., U.S.* **109** P13
Huesca, *Sp.* **147** D14
Huéscar, *Sp.* **146** K12
Huevos, island, *Trin. & Tobago* **123** P21
Huff, *N. Dak., U.S.* **106** G4
Hugh Town, *Eng., U.K.* **145** X7
Hughenden, *Qnsld., Austral.* **221** S13
Hughes Bay, *Antarctica* **238** D4
Hughes, *Alas., U.S.* **113** F15
Hughes, *Ark., U.S.* **102** G11
Hugli, estuary, *India* **182** K11
Hugo, *Okla., U.S.* **105** E19
Hugoton, *Kans., U.S.* **107** V3
Hüich'ŏn, *N. Korea* **186** G7
Huíla Plateau, *Angola* **208** P11
Huinca Renancó, *Arg.* **134** L10
Hŭisaek-pong, peak, *N. Korea* **186** E8
Huîtres, Baie aux, *Mauritius* **215** J20
Huizhou, *China* **185** P13
Hukawng Valley, *Myanmar* **190** D5
Hukuntsi, *Botswana* **212** H6
Hulan, *China* **185** D16
Hulbert, *Mich., U.S.* **100** F10
Hulett, *Wyo., U.S.* **108** F13
Hulin, *China* **185** C18
Hulingol, *China* **185** D14
Hull Bay, *Virgin Is., U.S.* **122** M8
Hull Bay, *Antarctica* **238** N9
Hull, *Ill., U.S.* **101** R3
Hulst, *Neth.* **148** K11
Huludao, *China* **185** F15
Hulun Nur, *China* **185** C13
Hulyaypole, *Ukr.* **159** Q9
Huma, *China* **185** B15
Humahuaca, *Arg.* **134** E9
Humaitá, *Braz.* **132** F8
Humansdorp, *S. Af.* **212** M7
Humansville, *Mo., U.S.* **107** U11
Humbe, *Angola* **208** Q10
Humber, river, *Eng., U.K.* **145** Q13
Humble, *Tex., U.S.* **105** L19
Humboldt Bay, *Calif., U.S.* **110** M1
Humboldt Gletscher see Sermersuaq, *Greenland, Den.* **88** D5
Humboldt Plain, *Pac. Oc.* **247** R21
Humboldt, *Sask., Can.* **86** M12
Humboldt, *Tenn., U.S.* **102** F12
Humboldt, *Iowa, U.S.* **106** M11
Humboldt, *Kans., U.S.* **107** U10
Humboldt, *Nebr., U.S.* **107** R9
Humboldt, river, *Nev., U.S.* **110** M9
Hūmedān, *Iran* **177** P21
Humenné, *Slovakia* **151** J17
Humpata, *Angola* **208** P9
Humphreys Peak, *Ariz., U.S.* **109** R7
Humphrys Bay, *Antigua & Barbuda* **123** A21
Humu‘ula Saddle, *Hawaii, U.S.* **115** N22
Hūn, *Lib.* **201** F18
Húnaflóaáll, valley, *Arctic Oc.* **253** L20
Húnaflói, bay, *Ice.* **142** E2
Hunedoara, *Rom.* **154** E12
Hunga Ha‘apai, island, *Tonga* **227** N22
Hunga Tonga, island, *Tonga* **227** N22
Hunga, island, *Tonga* **227** H22

Hungary, *Eur.* **151** M15
Hŭngbong, *N. Korea* **186** G9
Hungerford, *Qnsld. Austral.* **221** V12
Hŭngnam, *N. Korea* **186** H9
Hungry Bay, *Bermuda* **119** C15
Hungry Horse Reservoir, *Mont. U.S.* **108** B6
Hungund, *India* **183** P5
Hunjiang, *China* **185** E16
Huns Berge, mountains, *Namibia* **212** K4
Huns Mountains, *Af.* **198** P6
Hunstanton, *Eng., U.K.* **145** R14
Hunsur, *India* **183** R5
Hunt, *Mongolia* **184** D9
Hunter Gap, *Atl. Oc.* **249** T9
Hunter Islands, *Austral.* **220** L7
Hunter, *N. Dak., U.S.* **106** G3
Hunter, island, *Vanuatu* **224** K8
Hunters, *Bahamas* **122** K2
Huntingburg, *Ind., U.S.* **101** T8
Huntingdon, *Pa., U.S.* **96** N6
Huntingdon, *Tenn., U.S.* **103** E13
Huntington Beach, *Calif., U.S.* **111** Y8
Huntington, *W. Va., U.S.* **98** D7
Huntington, *Ind., U.S.* **101** Q10
Huntington, *Tex., U.S.* **105** J20
Huntington, *Utah, U.S.* **108** M8
Huntington, *Oreg., U.S.* **110** G9
Huntly, *Scot., U.K.* **144** H11
Huntly, *N.Z.* **223** F20
Huntsville, *Ark., U.S.* **102** E7
Huntsville, *Ala., U.S.* **103** G15
Huntsville, *Tex., U.S.* **105** K19
Huntsville, *Mo., U.S.* **107** S12
Hunucmá, *Mex.* **117** H16
Hunza, region, *Pak.* **180** F13
Huocheng (Shuiding), *China* **184** D3
Huon Gulf, *P.N.G.* **222** P8
Huon Peninsula, *P.N.G.* **222** P8
Huon, Île, *Coral Sea* **224** J7
Hupo, *S. Korea* **187** P12
Hurdiyo, *Somalia* **210** L15
Huri Hills, *Kenya* **211** Q7
Hurley, *Wis., U.S.* **100** F4
Hurmagai, *Pak.* **181** P5
Huron Mountain, *Mich., U.S.* **100** E7
Huron, *S. Dak., U.S.* **106** K7
Huron, Lake, *N. Amer.* **84** J7
Hurtsboro, *Ala., U.S.* **103** L17
Húsavík, *Ice.* **142** E3
Huṣi, *Rom.* **155** C17
Huslia, *Alas., U.S.* **113** F14
Husum, *Ger.* **150** A7
Hutag, *Mongolia* **184** C9
Hutch Mountain, *Ariz., U.S.* **109** S7
Hutchinson Plateau, *Greenland, Den.* **88** M11
Hutchinson, *Minn., U.S.* **106** K10
Hutchinson, *Kans., U.S.* **107** U7
Huthi, *Thai.* **190** K7
Hutland, *Cayman Is., U.K.* **122** M4
Huttig, *Ark., U.S.* **102** K8
Hutuo, river, *China* **185** G13
Huvadu see Suvadiva Atoll, *Maldives* **183** X3
Huy, *Belg.* **148** L13
Hüyük, *Turk.* **170** G7
Hvammstangi, *Ice.* **142** E2
Hvannadalshnúkur, peak, *Eur.* **140** B4
Hvar (Pharus), island, *Croatia* **154** H6
Hvarski Kanal, *Croatia* **154** H6
Hveragerði, *Ice.* **142** F2
Hvíta, river, *Ice.* **142** F2
Hwaaksan, peak, *S. Korea* **186** L9
Hwacheon, *S. Korea* **186** L9
Hwadae, *S. Korea* **186** E12
Hwajeon, *S. Korea* **187** Q11
Hwajin, *N. Korea* **186** J5
Hwangjin, *N. Korea* **186** E13
Hwangju, *N. Korea* **186** K6
Hwangsuwŏn, *N. Korea* **186** F10
Hwap'yŏng, *N. Korea* **186** E8
Hwaseong, *S. Korea* **187** N8
Hwasŏng, *N. Korea* **186** E12
Hwasun, *S. Korea* **187** T8
Hwayang, *S. Korea* **187** R11
Hwlffordd see Haverfordwest, *Wales, U.K.* **145** U8
Hyangsan, *N. Korea* **186** G7
Hyannis, *Mass., U.S.* **97** L16
Hyannis, *Nebr., U.S.* **107** N3
Hyargas Nuur, *Mongolia* **184** C7
Hyde Park, *N.Y., U.S.* **96** L11

Hyden, *W. Aust., Austral.* **221** X4
Hyderabad, *Pak.* **181** R8
Hyderabad, *India* **182** N6
Hydra *see* Ýdra, *island, Gr.* **156** K12
Hydro, *Okla., U.S.* **105** C15
Hyeolli, *S. Korea* **186** M10
Hyères, *Fr.* **149** Y14
Hyères, Îles d' (Les Îles d'Or), *Fr.* **149** Y14
Hyesan, *N. Korea* **186** D10
Hyndman Peak, *Idaho, U.S.* **108** G5
Hyŏlli, *N. Korea* **186** K9
Hyrynsalmi, *Fin.* **143** F16
Hysham, *Mont., U.S.* **108** D11

I-n-Akantarer, *river, Alg., Mali* **205** F18
I-n-Amenas, *Alg.* **201** G15
I-n-Amguel, *Alg.* **201** K13
I-n-Belbel, *Alg.* **200** G11
I-n-Eker, *Alg.* **201** J13
I-n-Salah, *Alg.* **200** H12
I-n-Tebezas, *Mali* **205** G17
Iaco, *river, Braz.* **132** G6
Iaeger, *W. Va., U.S.* **98** F8
Ialomiţa, *river, Rom.* **155** F16
Iamara, *P.N.G.* **222** Q6
Iar Connaught, *region, Ire.* **145** R3
Iascaigh *see* Easky, *Ire.* **145** P4
Iaşi, *Rom.* **155** B17
Iásmos, *Gr.* **157** B15
Iauaretê, *Braz.* **132** C6
Ibadan, *Nigeria* **206** L6
Ibagué, *Col.* **130** G6
Ibapah, *Utah, U.S.* **108** L5
Ibarra, *Ecua.* **130** K4
Ibb, *Yemen* **174** P9
Ibba (Tonj), *river, S. Sudan* **203** X10
Ibba, *S. Sudan* **203** X10
Iberá, Laguna, *Arg.* **134** H13
Iberia, *Peru* **132** G6
Iberian Basin, *Atl. Oc.* **248** J10
Iberian Mountains, *Eur.* **140** K3
Iberian Peninsula, *Eur.* **140** K3
Ibeto, *Nigeria* **206** K7
Ibi, *Nigeria* **206** L9
Ibiapaba, *S. Amer.* **128** E10
Ibina, *river, D.R.C.* **209** D18
Ibipetuba, *Braz.* **133** G14
Ibiza (Iviza), *island, Sp.* **147** H17
Ibiza *see* Eivissa, *Sp.* **147** H17
Iblei, Monti, *It.* **153** T11
Ibn Hani', Ras, *Syr.* **172** E6
Ibo, *Mozambique* **213** C13
Iboperenda, *Bol.* **132** K8
Ibotirama, *Braz.* **133** H15
Ibrā', *Oman* **175** H18
'Ibrī, *Oman* **175** H16
Ibusuki, *Japan* **189** T4
Ica, *Peru* **132** J4
Ica, *river, Peru* **132** J4
Icacos Point, *Trin. & Tobago* **123** R20
Içana, *Braz.* **132** C6
Içana, *river, S. Amer.* **132** C6
Iceland Basin, *Atl. Oc.* **248** F9
Iceland Plateau, *Arctic Oc.* **253** K19
Iceland Sea, *Arctic Oc.* **253** L20
Iceland-Faroe Rise, *Arctic Oc., Atl. Oc.* **253** K22
Iceland, *Eur.* **142** F4
Ichchapuram, *India* **182** M10
Icheon, *S. Korea* **187** N9
Ichikawa, *Japan* **189** P12
Ichinohe, *Japan* **188** K13
Ichinomiya, *Japan* **189** P9
Ichinoseki, *Japan* **188** L12
Ichinskaya Sopka, *Russ.* **161** F22
Ichinskiy, *Russ.* **161** F22
Ichkeul Lake, *Tunisia* **201** B15
Ichnya, *Ukr.* **159** N7
Ich'ŏn, *N. Korea* **186** G6
Ich'ŏn, *N. Korea* **186** K8
Icó, *Braz.* **133** F17
Iconium *see* Konya, *Turk.* **170** G8
Icoraci, *Braz.* **133** D13
Icy Cape, *Alas., U.S.* **113** B13
Ida Grove, *Iowa, U.S.* **107** N9
Idabel, *Okla., U.S.* **105** E20
Idah, *Nigeria* **206** M7

Idaho City, *Idaho, U.S.* **108** G4
Idaho Falls, *Idaho, U.S.* **108** G7
Idaho, *U.S.* **108** G5
Idali, *S. Sudan* **203** X12
Idalou, *Tex., U.S.* **104** F12
Idanha-a-Nova, *Port.* **146** G7
Idanha, *Oreg., U.S.* **110** G4
'Idd el Ghanam, *Sudan* **203** T7
Idel', *Russ.* **158** B8
Ideles, *Alg.* **201** K14
Ideriyn, *river, Mongolia* **184** C8
Idfū, *Egypt* **202** J12
'Idī, *Eritrea* **210** J9
Idi, *Indonesia* **192** G4
Idil, *Turk.* **171** H16
Idiofa, *D.R.C.* **209** H13
Idlib, *Syr.* **172** E8
Idracowra, *N.T., Austral.* **221** U9
Ieper (Ypres), *Belg.* **148** L10
Ierápetra, *Gr.* **157** Q16
Ierissós, *Gr.* **156** D12
Ieta, *P.N.G.* **228** B4
Ifakara, *Tanzania* **211** X6
Ifalik Atoll, *F.S.M.* **226** B7
Ife, *Nigeria* **206** L6
Iferouâne, *Niger* **206** E8
Ifjord, *Nor.* **143** B14
Igadiane, *river, Af.* **198** F4
Igaliku, *Greenland, Den.* **89** T7
Igalula, *Tanzania* **211** V4
Iganga, *Uganda* **211** S4
Igara Paraná, *river, Col.* **130** K7
Igarka, *Russ.* **161** G13
Igatpuri, *India* **182** M3
Iğdır, *Turk.* **171** E18
Igikpak, Mount, *Alas., U.S.* **113** E15
Iglesias, *It.* **153** P3
Igli, *Alg.* **200** F10
Igloolik, *Nunavut, Can.* **87** F16
'Igma, Gebel el, *Egypt* **173** S3
Ignacio, *Colo., U.S.* **109** Q11
Ignalina, *Lith.* **143** M17
Iğneada Burnu, *Turk.* **170** B4
Iğneada, *Turk.* **170** B4
Igoma, *Tanzania* **211** X4
Igoumenítsa, *Gr.* **156** F7
Igra, *Russ.* **158** G16
Igrim, *Russ.* **160** G10
Iguaçu, Cataratas do, *Arg., Braz.* **134** F15
Iguala, *Mex.* **116** J11
Igualada, *Sp.* **147** D17
Iguatu, *Braz.* **133** F16
Iguazú Falls, *S. Amer.* **128** J7
Iguéla, *Gabon* **208** F7
Igula, *Tanzania* **211** W5
Igumira, *Tanzania* **211** W4
Igutsaat Fjord, *Greenland, Den.* **89** T8
Ih Bogd Uul, *peak, Asia* **168** F10
Iharaña (Vohemar), *Madagascar* **213** C17
Ihavandiffulu Atoll, *Maldives* **183** U3
Ihavandu, *island, Maldives* **183** U3
Ihbulag, *Mongolia* **184** E12
Iheya Rettō, *Japan* **189** X2
Iheya Shima, *Japan* **189** X2
Ihnāsyā al Madīnah, *Egypt* **202** G11
Ihosy, *Madagascar* **213** G15
Ihu, *P.N.G.* **222** Q7
Ihwa, *S. Korea* **187** N8
Ii, *Fin.* **143** E15
Iida, *Japan* **189** P10
Iisalmi, *Fin.* **143** F16
Ijara, *Kenya* **211** T8
Ijebu Ode, *Nigeria* **206** M6
Ijevan, *Arm.* **171** D19
Ijin-man, *bay, N. Korea* **186** C13
IJmuiden, *Neth.* **148** H12
IJsselmeer, *bay, Neth.* **148** H13
Ijuí, *Braz.* **132** P11
Ijuw, *Nauru* **227** M15
Ikamiut, *Greenland, Den.* **88** M6
Ikanbujmal, *Greenland, Den.* **88** M6
Ikanbujmal, *Mongolia* **184** G5
Ikare, *Nigeria* **206** L7
Ikaría, *island, Gr.* **157** K16
Ikela, *D.R.C.* **209** E15
Ikelemba, *river, D.R.C.* **209** D13
Ikeq, *channel, Greenland, Den.* **89** U8
Ikerasarsuk, *Greenland, Den.* **88** J5;M5
Ikerasaarsuk, *Greenland, Den.* **89** N5
Ikerasak, *Greenland, Den.* **88** L6
Ikerasak, *Greenland, Den.* **89** N5;Q5
Ikerasassuaq (Prins Christian Sund), *Greenland, Den.* **89** U8

Ikerssuaq, *bay, Greenland, Den.* **89** P11
Ikersuak (Bredefjord), *Greenland, Den.* **89** T7
Ikertivaq, *Greenland, Den.* **89** Q10
Ikhtiman, *Bulg.* **155** J13
Iki, *island, Japan* **189** R14
Ikizdere, *Turk.* **171** D15
Ikom, *Nigeria* **206** M8
Ikoma, *Tanzania* **211** T5
Ikongo, *Madagascar* **213** G16
Ikopa, *river, Madagascar* **213** E16
Ikoto, *S. Sudan* **203** Y12
Ikozi, *D.R.C.* **209** F18
Iksan, *S. Korea* **187** R8
Iku, *island, Kiribati* **227** P14
Ikungi, *Tanzania* **211** V5
Ikuren, *island, Marshall Is.* **227** C20
Ikutha, *Kenya* **211** T7
Ila, *D.R.C.* **209** F14
Ilagan, *Philippines* **193** B14
Īlām, *Iraq* **176** G12
Ilave, *Peru* **132** J6
Iława, *Pol.* **151** C15
Ilbilbie, *Qnsld., Austral.* **221** T15
Île Alofi, *Wallis & Futuna Is., Fr.* **230** R11
Île Ami, *New Caledonia, Fr.* **231** R24
Île Art, *New Caledonia, Fr.* **231** N20
Île Baaba, *New Caledonia, Fr.* **231** N20
Île Balabio, *New Caledonia, Fr.* **231** N20
Île Beautemps-Beaupré, *New Caledonia, Fr.* **231** N22
Île Brosse, *New Caledonia, Fr.* **231** R23
Île d'Anticosti, *Que., Can.* **87** M21
Île de la Surprise, *New Caledonia, Fr.* **231** L19
Île de Sable, *New Caledonia, Fr.* **231** M20
Île des Pins (Kounié), *New Caledonia, Fr.* **231** R24
Île Ducos, *New Caledonia, Fr.* **231** Q22
Île Dudune, *New Caledonia, Fr.* **231** P24
Île Europa, *Fr.* **213** G13
Île Faïoa, *Wallis & Futuna Is., Fr.* **230** R9
Île Fenua Fu, *Wallis & Futuna Is., Fr.* **230** R9
Île Grimault, *New Caledonia, Fr.* **231** P21
Île Hugon, *New Caledonia, Fr.* **231** Q22
Île Juan de Nova, *Fr.* **213** E14
Île Koutomo, *New Caledonia, Fr.* **231** R24
Île Lêliogat, *New Caledonia, Fr.* **231** P24
Île Luaniva, *Wallis & Futuna Is., Fr.* **230** Q9
Île Neba, *New Caledonia, Fr.* **231** N20
Île Nukuaéta, *Wallis & Futuna Is., Fr.* **230** R8
Île Nukufufulanoa, *Wallis & Futuna Is., Fr.* **230** Q9
Île Nukuhifala, *Wallis & Futuna Is., Fr.* **230** Q9
Île Nukuloa, *Wallis & Futuna Is., Fr.* **230** Q9
Île Nukutapu, *Wallis & Futuna Is., Fr.* **230** Q9
Île Nukuteatea, *Wallis & Futuna Is., Fr.* **230** Q9
Île Ongombua, *New Caledonia, Fr.* **231** M20
Île Oua, *New Caledonia, Fr.* **231** P24
Île Ouen, *New Caledonia, Fr.* **231** R23
Île Pott, *New Caledonia, Fr.* **231** N20
Île Tidra, *Mauritania* **204** F7
Île Tiga, *New Caledonia, Fr.* **231** P24
Île Tintamarre, *St. Martin, Fr.* **123** A15
Île Uatio, *New Caledonia, Fr.* **231** R23
Île Uluiutu, *Wallis & Futuna Is., Fr.* **230** Q9
Île Uvea, *Wallis & Futuna Is., Fr.* **230** Q8
Île Vauvilliers, *New Caledonia, Fr.* **231** P23
Île Yandé, *New Caledonia, Fr.* **231** N20
Île-à-la-Crosse, *Sask., Can.* **86** L12
Ile, *river, Kaz.* **179** H14
Ilebo, *D.R.C.* **209** G13
Ilemi Triangle, *S. Sudan* **203** X14
Ilemi Triangle, *Kenya* **211** P5
Îles Bélep, *New Caledonia, Fr.* **231** N20
Îles Daos du Sud, *New Caledonia, Fr.* **231** N20
Îles de la Madeleine, *Que., Can.* **87** N22
Îles de la Petite Terre, *Guadeloupe, Fr.* **123** F17
Îles du Désappointement, *Fr. Polynesia, Fr.* **231** A20

Îles du Duc de Gloucester, *Fr. Polynesia, Fr.* **231** D19
Îles du Vent, *Fr. Polynesia, Fr.* **231** C16
Îles Gambier, *Fr. Polynesia, Fr.* **231** E23
Îles Gambier, *Fr. Polynesia, Fr.* **231** Q17
Îles Loyauté (Loyalty Islands), *New Caledonia, Fr.* **231** P23
Îles Sous le Vent, *Fr. Polynesia, Fr.* **231** B15
Ilesha, *Nigeria* **206** L6
Îlet à Cabrit, *Guadeloupe, Fr.* **123** G14
Îlet à Kahouanne, *Guadeloupe, Fr.* **123** E13
Îlet Cabrits, *Martinique, Fr.* **123** G23
Îlets à Goyaves, *Guadeloupe, Fr.* **123** F13
Ilfracombe, *Eng., U.K.* **145** V9
Ilgaz Dağları, *Turk.* **170** D9
Ilgaz, *Turk.* **170** D9
Ilgın, *Turk.* **170** G7
Ilha Angoche, *Mozambique* **213** D13
Ilha de Moçambique, *Mozambique* **213** D13
Ilha do Bazaruto, *Mozambique* **212** G11
Ilhas Primeiras, *Mozambique* **213** E13
Ílhavo, *Port.* **146** F5
Ilhéus Formigas, *Port.* **146** Q6
Ilhéus, *Braz.* **133** J16
Ili, *river, China* **184** E3
Iliamna Lake, *Alas., U.S.* **113** K14
Iligan, *Philippines* **193** F15
Ili'ili, *American Samoa, U.S.* **230** E6
Ilinden, *Bulg.* **155** K13
'Ilio Point, *Hawaii, U.S.* **115** H16
Ilion, *N.Y., U.S.* **96** J10
Ilivileq *see* Nanortalik, *Greenland, Den.* **89** U7
Ilıca, *Turk.* **171** E15
Illapel, *Chile* **134** J6
Illeginni, *island, Marshall Is.* **227** H16
Illéla, *Niger* **206** G7
Illes Balears (Balearic Islands), *Sp.* **147** G17
Illichivs'k, *Ukr.* **159** R5
Illinois Peak, *U.S.* **90** B6
Illinois, *U.S.* **101** R5
Illinois, *river, Ill., U.S.* **101** Q4
Illinois, *river, Oreg., U.S.* **110** K2
Illizi, *Alg.* **201** H15
Illo, *S. Korea* **187** T7
Illorsuit, *Greenland, Den.* **88** K5
Illullip Sermia, *glacier, Greenland, Den.* **88** H5
Ilmen, *Russ.* **158** G7
Il'men', Lake, *Eur.* **140** E10
Ilmenau, *Ger.* **150** G8
Ilo, *Peru* **132** K5
Ilobu, *Nigeria* **206** L6
Iloilo City, *Philippines* **193** E14
Ilorin, *Nigeria* **206** L6
Îlot Eo, *New Caledonia, Fr.* **231** N22
Îlot Igué, *New Caledonia, Fr.* **231** M20
Îlot Nukufotu, *Wallis & Futuna Is., Fr.* **230** P9
Îlots de Bass *see* Marotiri, *Fr. Polynesia, Fr.* **225** L14
Ilovik, *island, Croatia* **154** F4
Ilpyeong, *S. Korea* **187** U7
Il'pyrskiy, *Russ.* **161** E21
Ilula, *Tanzania* **211** U4
Ilulissat (Jakobshavn), *Greenland, Den.* **88** M6
Ilwaco, *Wash., U.S.* **110** E2
Imabari, *Japan* **189** Q6
Imabetsu, *Japan* **188** J12
Imām Şāḩib, *Afghan.* **180** F8
Imandra, Lake, *Eur.* **140** B10
Imasa, *Sudan* **203** P14
Imataca, *S. Amer.* **128** B6
Imatra, *Fin.* **143** H17
Imbert, *Dom. Rep.* **119** L18
Imbros *see* Gökçeada, *island, Turk.* **170** D2
Imese, *D.R.C.* **208** C12
Imi n 'Tanout, *Mor.* **200** E6
Īmī, *Eth.* **211** N9
Imías, *Cuba* **119** K14
Imieji Anchorage, *Marshall Is.* **227** H21
İmişli, *Azerb.* **171** E22
Imjado, *island, S. Korea* **187** T6
Imjin, *river, N. Korea* **179** N10
Imlay City, *Mich., U.S.* **100** L12
Imlay, *Nev., U.S.* **111** N8
Imlily, *W. Sahara* **200** J2

Immarna, *S. Aust., Austral.* **221** W8
Immokalee, *Fla., U.S.* **99** W9
Imnaha, *Wash., U.S.* **110** F9
Imola, *It.* **152** F7
Imotski, *Croatia* **154** H7
Imouzzer des Ida Ou Tanane, *Mor.* **200** E5
Imperatriz, *Braz.* **133** E13
Imperia, *It.* **152** G3
Imperial Beach, *Calif., U.S.* **111** Z9
Imperial Dam, *Ariz., Calif., U.S.* **111** Z13
Imperial Valley, *Calif., U.S.* **111** Z11
Imperial, *Nebr., U.S.* **107** Q3
Imperial, *Calif., U.S.* **111** Z11
Impfondo, *Congo* **208** D12
Imphal, *India* **182** J14
İmralı, *island, Turk.* **170** D4
İmranlı, *Turk.* **171** E13
Imrodj, *island, Marshall Is.* **227** H20
Imsan, *N. Korea* **186** F8
Imsil, *S. Korea* **187** S9
Imtān, *Syr.* **172** L8
Ina, *Japan* **189** P10
Inamba Jima, *Japan* **189** R11
Inanwatan, *Indonesia* **193** K18
Inarajan, *Guam, U.S.* **230** Q5
Inari, *Fin.* **143** C15
Inari, lake, *Fin.* **143** C15
Inca de Oro, *Chile* **134** G7
Inca, *Sp.* **147** G18
İncekum Burnu, *Turk.* **170** J9
İncesu, *Turk.* **170** G10
Inch Island, *Ire.* **145** N6
Inchcape *see* Bell Rock, *Scot., U.K.* **144** K11
Incheon (Inch'ŏn), *S. Korea* **187** N8
Inchnadamph, *Scot., U.K.* **144** G8
Inch'ŏn *see* Incheon, *S. Korea* **187** N8
Incles, Port d', *Andorra, Fr.* **162** G5
Indalsälven, *river, Eur.* **140** C8
Indawgyi Lake, *Myanmar* **190** E5
Independence Fjord, *Greenland, Den.* **88** B10
Independence Mountains, *Nev., U.S.* **110** M10
Independence, *Iowa, U.S.* **107** N13
Independence, *Kans., U.S.* **107** V9
Independence, *Mo., U.S.* **107** S10
Independence, *Calif., U.S.* **111** T8
Inderbor, *Kaz.* **178** F5
India, *Asia* **182** J6
Indian Ocean, *Ind. Oc.* **249** U15
Indian Peak, *Utah, U.S.* **109** N5
Indian River, *Fla., U.S.* **99** U10
Indian Springs, *Nev., U.S.* **111** T11
Indian Town Point, *Antigua & Barbuda* **123** B22
Indian-Antarctic Ridge, *Ind. Oc.* **255** D17
Indiana Dunes National Lakeshore, *Ind., U.S.* **101** N8
Indiana, *Pa., U.S.* **96** N4
Indiana, *U.S.* **101** R9
Indianapolis, *Ind., U.S.* **101** R9
Indianola, *Miss., U.S.* **102** J10
Indianola, *Iowa, U.S.* **107** P11
Indianola, *Nebr., U.S.* **107** R4
Indiantown, *Fla., U.S.* **99** V10
Indigirka, *river, Asia* **168** B11
Indija, *Serb.* **154** E10
Indio, *Calif., U.S.* **111** X11
Indira Point, *India* **183** U15
Indispensable Strait, *Solomon Is.* **228** E10
Indochina Peninsula, *Asia* **168** K10
Indomed Fracture Zone, *Ind. Oc.* **250** P6
Indonesia, *Indonesia* **192** L8
Indore, *India* **182** K5
Indravati, *river, India* **182** M8
Indus Fan, *Ind. Oc.* **250** C9
Indus, Mouths of the, *Pak.* **181** S7
Indus, *river, India, Pak.* **180** L10
Inebolu, *Turk.* **170** C9
İnegöl, *Turk.* **170** E5
Ineu, *Rom.* **154** D11
Inez, *Ky., U.S.* **103** B20
Infanta, Cape, *Af.* **198** Q7
Infante D. Henrique, *Sao Tome and Principe* **215** B21
Ingal, *Niger* **206** F8
Ingende, *D.R.C.* **208** E12
Ingeniero Guillermo N. Juárez, *Arg.* **134** E11
Ingeniero Jacobacci, *Arg.* **135** Q8

Ingeniero Luiggi, *Arg.* **134** M10
Inglefield Bredning, bay,
  Greenland, Den. **88** E3
Inglefield Land, *Greenland, Den.* **88** D4
Inglesa Bay, *S. Amer.* **128** K4
Ingleside, *Tex., U.S.* **105** P17
Inglis, *Fla., U.S.* **99** S7
Ingolf Fjord, *Greenland, Den.* **88** C12
Ingololo, *D.R.C.* **208** H11
Ingolstadt, *Ger.* **150** J8
Ingomar, *S. Aust., Austral.* **221** V9
Ingraj Bazar, *India* **182** J11
Ingram, *Tex., U.S.* **105** L15
Ingrid Christensen Coast, *Antarctica*
  239 G20
Ingwiller, *Fr.* **149** P15
Inhambane, *Mozambique* **212** H11
Inhaminga, *Mozambique* **212** E11
Inharrime Terrace, *Ind. Oc.* **250** L4
Inharrime, *Mozambique* **212** H11
Inhassoro, *Mozambique* **212** G11
Inírida, *Col.* **130** G10
Inírida, river, *Col.* **130** H9
Inis Ceithlann *see* Enniskillen,
  N. Ire., U.K. **145** P5
Inis Córthaidh *see* Enniscorthy, *Ire.*
  145 S6
Inis Díomáin *see* Ennistimon, *Ire.*
  145 R3
Inis Eoáin *see* Inishannon, *Ire.*
  145 U4
Inis Gé Theas (Inishkea South),
  island, *Ire.* **145** P2
Inis Gé Thuaidh (Inishkea North),
  island, *Ire.* **145** P2
Inis Meáin (Inishmaan), island, *Ire.*
  145 R3
Inis Oírr (Inisheer), island, *Ire.*
  145 R3
Inis *see* Ennis, *Ire.* **145** S4
Inis Trá Tholl *see* Inishtrahull,
  island, *Ire.* **145** M6
Inis Trá Tholl *see* Malin Head, *Ire.*
  145 M5
Inishannon (Inis Eoáin), *Ire.* **145** U4
Inishbofin, island, *Ire.* **145** Q2
Inisheer *see* Inis Oírr, *Ire.* **145** R3
Inishkea North *see* Inis Gé Thuaidh,
  island, *Ire.* **145** P2
Inishkea South *see* Inis Gé Theas,
  island, *Ire.* **145** P2
Inishmaan *see* Inis Meáin, island, *Ire.*
  145 R3
Inishmore *see* Árainn, island, *Ire.*
  145 R3
Inishmurray, island, *Ire.* **145** P4
Inishowen Head, *Ire.* **145** M6
Inishshark, island, *Ire.* **145** Q2
Inishtrahull (Inis Trá Tholl), island,
  Ire. **145** M6
Inje, *S. Korea* **186** L10
Injune, *Qnsld., Austral.* **221** U14
Inkerman, *Qnsld., Austral.* **221** Q12
Inland Niger Delta, *Mali* **205** J13
Inland Sea *see* Seto Naikai, *Japan*
  189 Q6
Inn, river, *Aust., Ger.* **150** K10
Innaanganeq (Kap York),
  Greenland, Den. **88** F3
Innamincka, *S. Aust., Austral.* **221** V11
Inner Brass Island, *Virgin Is., U.S.*
  122 M8
Inner Hebrides, islands, *Scot., U.K.*
  144 K7
Inner Sound, *Scot., U.K.* **144** H7
Innerit, peninsula, *Greenland, Den.*
  88 K5
Inngia Isbræ, glacier, *Greenland, Den.*
  88 J3
Innisfail, *Qnsld., Austral.* **221** R13
Innoko, river, *Alas., U.S.* **113** G14
Innsbruck, *Aust.* **150** L8
Inongo, *D.R.C.* **208** F12
İnönü, *Turk.* **170** E6
Inowrocław, *Pol.* **151** D14
Insan, *N. Korea* **186** L7
Insar, *Russ.* **158** K12
Inscription, Cape, *Austral.* **220** G1
Insein, *Myanmar* **190** K5
Insiaf, *F.S.M.* **226** G11
Institute Ice Stream, *Antarctica*
  238 H10
Instow, *Eng., U.K.* **145** V9
Inta, *Russ.* **160** F10
Intendente Alvear, *Arg.* **134** L10
Interior, *S. Dak., U.S.* **106** L3

International Falls, *Minn., U.S.* **106** E11
Interview Island, *India* **183** Q14
Intiyaco, *Arg.* **134** H12
Inubō Zaki, *Japan* **189** P13
Inukjuak, *Que., Can.* **87** K17
Iñupiat Heritage Center, *Alas., U.S.*
  113 B15
Inussulik Bugt, bay, *Greenland, Den.*
  88 H5
Inuvik, *N.W.T., Can.* **86** E10
Inuya, river, *Peru* **132** G4
Inverbervie, *Scot., U.K.* **144** J11
Invercargill, *N.Z.* **223** P16
Inverell, *N.S.W., Austral.* **221** W15
Invermoriston, *Scot., U.K.* **144** J9
Inverness, *Fla., U.S.* **99** T7
Inverness, *Scot., U.K.* **144** H9
Invershin, *Scot., U.K.* **144** G9
Investigator Bank, *Ind. Oc.* **250** E10
Investigator Ridge, *Ind. Oc.* **251** H14
Investigator Strait, *Austral.* **220** L10
Inyangani, peak, *Zimb.* **212** E10
Inyo Mountains, *Calif., U.S.* **111** T8
Inyokern, *Calif., U.S.* **111** V9
Inyonga, *Tanzania* **211** W3
Inza, *Russ.* **158** K13
Inzhavino, *Russ.* **158** M12
Inzia, river, *D.R.C.* **208** H12
Iō Jima, *Japan* **189** T4
Ioánnina, *Gr.* **156** E7
Iola, *Kans., U.S.* **107** U10
Ioma, *P.N.G.* **222** Q8
Iomu (Jomu), *Tanzania* **211** U4
Iona, *Idaho, U.S.* **108** G7
Iona, island, *Scot., U.K.* **144** K7
Ione, *Oreg., U.S.* **110** F6
Ione, *Wash., U.S.* **110** A9
Ione, *Calif., U.S.* **111** R5
Ione, *Nev., U.S.* **111** Q9
Iongo, *Angola* **208** K12
Ionia, *Mich., U.S.* **100** L10
Ionian Islands, *Gr.* **140** L9
Ionian Plain, *Atl. Oc.* **248** K14
Ionian Sea, *Eur.* **140** L8
Iony, *Russ.* **161** H20
Iori, river, *Azerb., Rep. of Georgia*
  171 B19
Íos, *Gr.* **157** M15
Íos, island, *Gr.* **157** M15
Iōtori Shima, *Japan* **189** W2
Iouik, *Mauritania* **204** E7
Iowa City, *Iowa, U.S.* **107** P13
Iowa Falls, *Iowa, U.S.* **107** N12
Iowa Park, *Tex., U.S.* **105** E15
Iowa, *U.S.* **107** N12
Iowa, river, *Iowa, U.S.* **107** P12
Ipatovo, *Russ.* **159** S12
Ipiales, *Col.* **130** J4
Ipipucu, *Braz.* **132** F5
Ipixuna, river, *Braz.* **132** F4
Ipoh, *Malaysia* **191** T8
Ipole, *Tanzania* **211** V4
Iporá, *Braz.* **132** J12
Ippy, *Cen. Af. Rep.* **207** M15
İpsala, *Turk.* **170** C2
Ipswich, *S. Dak., U.S.* **106** J6
Ipswich, *Jam.* **122** H7
Ipswich, *Qnsld., Austral.* **221** V15
Ipswich, *Eng., U.K.* **145** T15
Iput', river, *Russ.* **158** L7
Iqaluit, *Nunavut, Can.* **87** H19
Iqe, *China* **184** H7
Iquique, *Chile* **134** C7
Iquitos, *Peru* **132** E4
Iraan, *Tex., U.S.* **104** K12
Iracoubo, *Fr. Guiana, Fr.* **131** F18
Irafayle, *Eritrea* **210** H8
Iraio, ruin, *Gr.* **157** K17
Irákleia, *Gr.* **157** M15
Irákleio (Candia), *Gr.* **157** Q15
Iran, *Asia* **177** H15
Īrānshahr, *Iran* **177** M22
Irapuata, *Mex.* **116** H11
Iraq, *Asia* **176** G9
Irayel', *Russ.* **158** A15
Irazú, Volcán, *C.R.* **117** N19
Irbe Strait, *Est., Latv.* **143** L15
Irbid (Arbela), *Jordan* **172** L6
Irbil, *Russ.* **160** J11
Irebu, *D.R.C.* **208** E12
Ireland (Éire), *Eur.* **145** R5
Ireland Island North, *Bermuda*
  119 B14
Ireland Island South, *Bermuda*
  119 C14
Ireland, island, *Ire., U.K.* **145** R5
Iret', *Russ.* **161** F21

Irian Jaya, *Indonesia* **193** K20
Iriba, *Chad* **207** G16
Irié, *Guinea* **204** N11
Iringa, *Tanzania* **211** X6
Iriri, river, *Braz.* **132** E13
Irish Sea, *Atl. Oc.* **140** F4
Irkeshtam, *Kyrg.* **179** M14
Irkutsk, *Russ.* **161** L15
Irmauw, *Indonesia* **193** M17
Irminger Basin, *Atl. Oc.* **248** F8
Irois Bay, *Trin. & Tobago* **123** R21
Iroise, bay, *Fr.* **149** Q2
Iron Gate Dam, *Rom., Serb.* **154** F12
Iron Gate, *Eur.* **140** J9
Iron Mountain, *Mich., U.S.* **100** G7
Iron River, *Mich., U.S.* **100** G6
Iron River, *Wis., U.S.* **100** F3
Irondequoit, *N.Y., U.S.* **96** J6
Ironton, *Ohio, U.S.* **101** T13
Ironton, *Mo., U.S.* **107** U15
Ironwood Forest National
  Monument, *Ariz., U.S.* **109** V7
Ironwood, *Mich., U.S.* **100** F4
Iroquois, *S. Dak., U.S.* **106** K7
Irpin', *Ukr.* **159** N6
Irrawaddy *see* Ayeyarwady, river,
  Myanmar **190** G5
Irrua, *Nigeria* **206** M7
Irsina, *It.* **152** M13
Irtysh *see* Ertis, river, *Kaz.* **179** D15
Irtysh, river, *Russ.* **160** H10
Irumu, *D.R.C.* **209** D19
Iruña *see* Pamplona, *Sp.* **147** C13
Irvine, *Ky., U.S.* **103** C18
Irvine, *Scot., U.K.* **145** M9
Irving, *Tex., U.S.* **105** G17
Isa Khel, *Pak.* **180** K10
Isa, *Nigeria* **206** H7
Isabel, *S. Dak., U.S.* **106** J4
Isabela de Sagua, *Cuba* **118** G9
Isabela, *Philippines* **193** F14
Isabela, Cabo, *Dom. Rep.* **119** K18
Isabela, Isla, *Galápagos Is. Ecua.*
  225 G22
Isabella, *Minn., U.S.* **106** F13
Isabella, Point, *Mich., U.S.* **100** E7
Isachsen Mountain, *Antarctica*
  239 C16
Ísafjörður, *Ice.* **142** D2
Isahaya, *Japan* **189** R4
Isaka, *Tanzania* **211** U4
Isangel, *Vanuatu* **229** M17
Isangi, *D.R.C.* **209** D16
Isanlu, *Nigeria* **206** L7
Isar, river, *Ger.* **150** K9
Íscar, *Sp.* **146** E10
Ischia, Isola d', *It.* **152** M10
Isdu, island, *Maldives* **183** X3
Ise Wan, *Japan* **189** Q9
Ise, *Japan* **189** Q9
Iselin Bank, *Pac. Oc.* **255** J15
Isen Saki, *Japan* **189** W3
Iseo, Lago d', *It.* **152** D5
Isère, Pointe, *Fr. Guiana, Fr.* **131** F18
Isernia, *It.* **152** L10
Iseyin, *Nigeria* **206** L6
Isfahan *see* Eşfahān, *Iran* **177** H16
Isfara, *Taj.* **178** L12
Isfjord Bank, *Arctic Oc.* **253** H17
Isfjordrenna, valley, *Arctic Oc.*
  253 H17
Ishikari Wan, *Japan* **188** G12
Ishim Steppe, *Kaz.* **178** B11
Ishim, *Russ.* **160** J10
Ishim, river, *Russ.* **160** J10
Ishinomaki Wan, *Japan* **188** L13
Ishinomaki, *Japan* **188** L13
Ishkāshem, *Afghan.* **180** F10
Ishkāshim, *Afghan.* **180** F10
Ishkuman, *Pak.* **180** F12
Ishpeming, *Mich., U.S.* **100** F7
Işikveren, *Turk.* **171** H17
Isili, *It.* **153** P4
Isimu, *Indonesia* **193** J14
Isiolo, *Kenya* **211** S7
Isiro (Paulis), *D.R.C.* **209** C18
Isisford, *Qnsld., Austral.* **221** T13
Işık Dağı, peak, *Turk.* **170** D8
Iskandar, *Uzb.* **178** L12
Iskappe Sukkertoppen,
  Greenland, Den. **89** P5
Iskele *see* Trikomo, *Cyprus* **162** N10
İskenderun Körfezi, *Turk.* **170** J10
İskenderun, *Turk.* **170** J11
İskilip, *Turk.* **170** D9
İskür, river, *Bulg.* **155** H14

Iskushuban, *Somalia* **210** L14
Isla Baltra, *Ecua.* **130** Q3
Isla Darwin (Culpepper), *Ecua.*
  130 N2
Isla Española (Hood), *Ecua.* **130** R4
Isla Fernandina (Narborough), *Ecua.*
  130 Q2
Isla Floreana (Santa María), *Ecua.*
  130 R3
Isla Genovesa (Tower), *Ecua.* **130** P4
Isla Isabela (Albemarle), *Ecua.*
  130 Q2
Isla Marchena (Bindloe), *Ecua.*
  130 P3
Isla Mujeres, *Mex.* **117** H17
Isla Pinta (Abingdon), *Ecua.* **130** P3
Isla Pinzón, *Ecua.* **130** Q3
Isla Rábida, *Ecua.* **130** Q3
Isla San Cristóbal (Chatham), *Ecua.*
  130 Q4
Isla Santa Cruz (Chaves,
  Indefatigable), *Ecua.* **130** Q3
Isla Santa Fe (Barrington), *Ecua.*
  130 Q4
Isla Santiago (San Salvador), *Ecua.*
  130 Q3
Isla Seymour, *Ecua.* **130** Q3
Isla Tortuga, *Ecua.* **130** R3
Isla Wolf (Wenman), *Ecua.* **130** N2
İslâhiye, *Turk.* **170** H11
Islām Qal'ah, *Afghan.* **180** H2
Islamabad, *Pak.* **180** J12
Islamkot, *Pak.* **181** S10
Islamorada, *Fla., U.S.* **99** Y10
Island Falls, *Me., U.S.* **97** C18
Island Lagoon, *Austral.* **220** J10
Island Park Reservoir, *Idaho, U.S.*
  108 F7
Island Park, *Idaho, U.S.* **108** F8
Island Pond, *Vt., U.S.* **97** F14
Islas Canarias (Canary Islands), *Sp.*
  214 P2
Islas Malvinas *see* Falkland Islands,
  S. Amer. **135** X11
Islas Orcadas Rise, *Atl. Oc.* **249** V9
Islay, island, *Scot., U.K.* **144** L7
Isle Royle National Park, *Mich., U.S.*
  100 D6
Ismâ'ilîya, *Egypt* **173** P1
Isnā, *Egypt* **202** J12
Isoanala, *Madagascar* **213** H15
Isoka, *Zambia* **209** L20
Isola di Capo Rizzuto, *It.* **153** Q14
Isolote Bay, *Trin. & Tobago* **123** R21
Isonzo, river, *It.* **152** D9
Isortoq, *Greenland, Den.* **89** Q10
Isparta, *Turk.* **170** G6
Ispas, *Turkm.* **178** M9
Ispikan, *Pak.* **181** R3
İspir, *Turk.* **171** D15
Israel, *Asia* **172** L5
Isratu, island, *Eritrea* **210** G8
Isrīyah, ruins, *Syr.* **172** F9
Issano, *Guyana* **131** F15
Issia, *Côte d'Ivoire* **204** P12
Issime, *It.* **152** D3
Issoudun, *Fr.* **149** S9
Ist, island, *Croatia* **154** F5
Istādah-ye Muqur, Āb-e, *Afghan.*
  180 K8
İstanbul Boğazi (Bosporus), *Turk.*
  170 C5
Istanbul, *Turk.* **170** C5
Istaravshan (Ura-Tyube), *Taj.*
  178 M12
Istiaía, *Gr.* **156** G11
Isto, Mount, *Alas., U.S.* **113** C18
Istokpoga, Lake, *Fla., U.S.* **99** V9
Istres, *Fr.* **149** X13
Istria, peninsula, *Croatia* **154** E4
Isua, *Greenland, Den.* **89** Q6
Isunnguata Sermia, glacier,
  Greenland, Den. **89** N6
Itabaianinha, *Braz.* **133** H17
Itabuna, *Braz.* **133** J16
Itacoatiara, *Braz.* **132** D9
Itaeté, *Braz.* **133** H16
Itaipú Dam, *Braz., Parag.* **132** M10
Itaituba, *Braz.* **132** E10
Itajaí, *Braz.* **133** N13
Itaka, *Tanzania* **211** X4
Italia, Monte, *Chile* **135** Y8
Italy, *Eur.* **152** J9
Itamarati, *Braz.* **132** F6
Itambé, Pico de, *Braz.* **133** K15
Itanagar, *India* **182** G14

Itapetinga, *Braz.* **133** J16
Itapipoca, *Braz.* **133** E16
Itapiranga, *Braz.* **132** D10
Itasca, *Tex., U.S.* **105** H17
Itasca, Lake, *Minn., U.S.* **106** G9
Itatupã, *Braz.* **132** C12
Itbayat, island, *Philippines* **193** A14
Itéa, *Gr.* **156** H10
Itende, *Tanzania* **211** W5
Iténez *see* Guaporé, river, *S. Amer.*
  132 H8
Ithaca *see* Itháki, island, *Gr.* **156** H7
Ithaca, *N.Y., U.S.* **96** K8
Ithaca, *Mich., U.S.* **100** L11
Itháki (Ithaca), island, *Gr.* **156** H7
Itháki (Vathí), *Gr.* **156** H7
Itigi, *Tanzania* **211** V5
Itilleq (Eggers Ø), *Greenland, Den.*
  89 U8
Itilleq, *Greenland, Den.* **88** E3
Itilleq, *Greenland, Den.* **89** P5;U8
Ítilo, *Gr.* **156** M10
Itimbiri, river, *D.R.C.* **209** C15
Itiquira, river, *Braz.* **132** K10
Itō, *Japan* **189** Q11
Itobo, *Tanzania* **211** V4
Itoman, *Japan* **189** Y2
Itta Bena, *Miss., U.S.* **102** J11
Ittiri, *It.* **152** M3
Ittoqqortoormiit (Scoresbysund),
  Greenland, Den. **88** H10
Itu Aba Island, *S. China Sea* **192** E11
Ituí, river, *Braz.* **132** E5
Ituiutaba, *Braz.* **133** K13
Itula, *D.R.C.* **209** G18
Ituni, *Guyana* **131** F15
Iturama, *Braz.* **132** K12
Ituri, river, *D.R.C.* **209** D17
Iturup (Etorofu), island, *Russ.*
  161 J23
Ituxi, river, *Braz.* **132** F7
Ituzaingó, *Arg.* **134** G14
Itzehoe, *Ger.* **150** B8
Iuka, *Miss., U.S.* **103** G13
Iul'tin, *Russ.* **161** B20
Ivaí, river, *Braz.* **132** M11
Ivakoany, peak, *Madagascar* **213** H15
Ivalo, *Fin.* **143** C15
Ivanhoe, *N.S.W., Austral.* **221** X12
Ivanić Grad, *Croatia* **154** D6
Ivanjica, *Serb.* **154** H10
Ivano-Frankivs'k, *Ukr.* **159** P2
Ivanovo, *Russ.* **158** H11
Ivanteyevka, *Russ.* **158** L15
Ivatsevichy, *Belarus* **158** L3
Ivaylovgrad, *Bulg.* **155** K16
Ivdel', *Russ.* **160** G9
Ivindo, river, *Af.* **198** J6
Ivindo, river, *Gabon* **208** E9
Ivisa *see* Ibiza, island, *Sp.* **147** H17
Iviza, island, *Eur.* **140** L4
Ivory Coast *see* Côte d'Ivoire,
  Af. **205** N13
Ivory Coast, *Côte d'Ivoire* **205** Q13
Ivrea, *It.* **152** D3
Ivujivik, *Que., Can.* **87** H17
Iwaki, *Japan* **189** N13
Iwakuni, *Japan* **189** Q6
Iwamizawa, *Japan* **188** G13
Iwanuma, *Japan* **188** M13
Iwate, *Japan* **188** K13
Iwembere Steppe, *Af.* **198** K9
Iwo Jima Ridge, *Pac. Oc.* **246** F7
Iwo Jima *see* Iwo To, island, *Japan*
  189 Z13
Iwo To (Iwo Jima), island, *Japan*
  189 Z13
Iwo, *Nigeria* **206** L6
Iwon-man, bay, *N. Korea* **186** G11
Iwŏn, *N. Korea* **186** G11
Ixtapa, *Mex.* **116** K10
Iyo, *Japan* **189** R6
Iz, island, *Croatia* **154** G5
Izadkhāst, *Iran* **177** J16
Izamal, *Mex.* **117** H16
'Izbat Dūsh, *Egypt* **202** K11
Izberbash, *Russ.* **159** U15
Izbica, *Pol.* **151** D14
Īzeh, *Iraq* **177** J14
Izena Shima, *Japan* **189** X2
Izhevsk, *Russ.* **158** G16
Izhma, *Russ.* **158** A15
Izhma, river, *Russ.* **158** B15
Izmayil, *Ukr.* **159** S4
Izmir (Smyrna), *Turk.* **170** F3
İzmit (Kocaeli), *Turk.* **170** D5

İznik (Nicaea), Turk. 170 D5
İznik Gölü, Turk. 170 D5
Izozog, Bañados del, S. Amer. 128 H6
Izra', Syr. 172 K7
Izu Hantō, Japan 189 Q11
Izu Islands, Asia 168 F14
Izu Shotō, Japan 189 Q12
Izu-Ogasawara Rise, Pac. Oc. 246 F7
Izu-Ogasawara Trench, Pac. Oc. 246 F8
Izumo, Japan 189 P6
Izyaslav, Ukr. 159 N4
Izyum, Ukr. 159 P9

J

J. A. D. Jensen Nunatakker, Greenland, Den. 89 R6
J. C. Christensen Land, Greenland, Den. 88 C10
J. P. Koch Fjord, Greenland, Den. 88 B8
J. P. Koch Land, Greenland, Den. 88 J5
J. Percy Priest Lake, Tenn., U.S. 103 E15
J. Strom Thurmond Reservoir, Ga., S.C., U.S. 98 L6
Jaba, P.N.G. 228 C5
Jabal ad Dukhān, Bahrain 175 F13
Jabal al 'Ajmah, Egypt 202 G13
Jabal al Farā'id, Egypt 202 K14
Jabal al Ḩijāz, Saudi Arabia 174 L8
Jabal al Lawz, Saudi Arabia 174 D4
Jabal 'Alī, U.A.E. 175 G15
Jabal an Nabī Shu'ayb, Yemen 174 N8
Jabal Arkanū, Lib. 201 K23
Jabal as Sawdā', Lib. 201 F18
Jabal as Sibā'ī, Egypt 202 J13
Jabal ash Shams, Oman 175 H17
Jabal at Tarhūnī, Lib. 201 K22
Jabal at Tīh, Egypt 202 G13
Jabal Awlya, Sudan 203 R12
Jabal Barkal (Gebel Barkal), Sudan 203 P12
Jabal Bin Ghunaymah, Lib. 201 J19
Jabal Bū Sunbul, Lib. 201 J22
Jabal Ghārib, Egypt 202 G12
Jabal Hadada, Sudan 202 M10
Jabal Ḩadīd, Lib. 201 L22
Jabal Ḩafīt, Oman, U.A.E. 175 G16
Jabal Ḩamāţah, Egypt 202 K14
Jabal Kāmil, Egypt 202 L8
Jabal Kātrīnā (Mount Catherine), Egypt 202 G13
Jabal Marrah, Sudan 203 S7
Jabal Mūsá (Mount Sinai), Egypt 202 G13
Jabal Nafūsah, Lib. 201 E17
Jabal Natityāy, Egypt 202 L13
Jabal Nuqruş, Egypt 202 J13
Jabal Raḑwá, Saudi Arabia 174 G5
Jabal Rāf, Saudi Arabia 174 D7
Jabal Rāhib, Sudan 203 P9
Jabal Sawdā', Saudi Arabia 174 L8
Jabal Shā'ib al Banāt, Egypt 202 H13
Jabal Shammar, Saudi Arabia 174 E7
Jabal Shār, Saudi Arabia 174 D4
Jabal Sinjār, Iraq 176 E9
Jabal Sirrī, Egypt 202 L11
Jabal Ţuwayq, Saudi Arabia 174 K10
Jabalpur, India 182 K7
Jabbūl, Syr. 172 E9
Jabbūl, Sabkhat a, Syr. 172 E9
Jabel Abyad Plateau, Af. 198 F8
Jabiru, N.T., Austral. 221 P8
Jablah, Syr. 172 F7
Jablaničko Jezero, Bosn. & Herzg. 154 G8
Jablonec, Czechia 150 G11
Jabnoren, island, Marshall Is. 227 G19
Jaboatão, Braz. 133 G18
Jabor, Marshall Is. 227 J20
Jabuka, island, Croatia 154 H5
Jabwot Island, Marshall Is. 227 D16
Jaca, Sp. 147 C14
Jacareacanga, Braz. 132 E10
Jaciparaná, Braz. 132 G7
Jackman, Me., U.S. 97 D15
Jackpot, Nev., U.S. 110 L12
Jacksboro, Tex., U.S. 105 F16
Jackson Bay, N.Z. 223 M16
Jackson Lake, Ga., U.S. 98 L5
Jackson Lake, Wyo., U.S. 108 G8
Jackson, Ga., U.S. 98 L5
Jackson, S.C., U.S. 98 L8
Jackson, Mich., U.S. 100 M11
Jackson, Ohio, U.S. 101 S13
Jackson, La., U.S. 102 N9
Jackson, Miss., U.S. 102 L11
Jackson, Ala., U.S. 103 M14
Jackson, Tenn., U.S. 103 F13
Jackson, Minn., U.S. 106 L10
Jackson, Mo., U.S. 107 U16
Jackson, Wyo., U.S. 108 G8
Jackson, Calif., U.S. 111 R5
Jackson, Barbados 123 K19
Jackson, Russ. 160 C12
Jacksonport, Wis., U.S. 100 H8
Jacksonville Beach, Fla., U.S. 99 R9
Jacksonville, N.C., U.S. 98 J13
Jacksonville, Fla., U.S. 99 Q8
Jacksonville, Ill., U.S. 101 R4
Jacksonville, Ark., U.S. 102 G8
Jacksonville, Ala., U.S. 103 J16
Jacksonville, Tex., U.S. 105 H19
Jacmel, Haiti 119 M16
Jacob Lake, Ariz., U.S. 109 Q7
Jacobabad, Pak. 181 P8
Jacqueville, Côte d'Ivoire 205 Q13
Jacquinot Bay, P.N.G. 222 P10
Jacumba, Calif., U.S. 111 Z11
Jadraque, Sp. 146 E12
Jādū, Lib. 201 E16
Jaén, Peru 132 F2
Jaén, Sp. 146 K11
Jaeundo, island, S. Korea 187 T6
Jaewondo, island, S. Korea 187 T6
Ja'farābād, Iran 177 B13
Jaffa, Cape, Austral. 220 L11
Jaffna, Sri Lanka 183 S7
Jagaedo, island, S. Korea 187 V7
Jagdalpur, India 182 M8
Jagdaqi, China 185 B15
Jāghir Bāzār, Syr. 172 C14
Jaghjagh, river, Syr. 172 C14
Jagodina, Serb. 154 G11
Jagraon, India 182 E5
Jagtial, India 182 M6
Jaguarão, Braz. 132 Q11
Jahanabad, India 182 J10
Jahorina, peak, Bosn. & Herzg. 154 G8
Jahrom, Iran 177 M17
Jaicós, Braz. 133 F16
Jaigarh, India 182 N3
Jaintiapur, India 182 J13
Jaipur, India 182 H5
Jaisalmer, India 182 H2
Jajpur, India 182 L10
Jakar, Bhutan 182 G13
Jakarta, Indonesia 192 L8
Jakeru, island, Marshall Is. 227 H15
Jakhau, India 182 J1
Jakobshavn Isbræ (Sermeq Kujalleq), glacier, Greenland, Den. 88 M6
Jakobshavn Isfjord, Greenland, Den. 88 M6
Jakobshavn see Ilulissat, Greenland, Den. 88 M6
Jakobstad (Pietarsaari), Fin. 143 G15
Jāl al Baţn, Iraq 176 J9
Jal, N. Mex., U.S. 109 V15
Jala Nur, China 185 C13
Jalaid Qi, China 185 D15
Jalaihai Point, Guam, U.S. 230 Q5
Jalal-Abad, Kyrg. 179 L14
Jalālābād, Afghan. 180 H10
Jalalaqsi, Somalia 211 Q12
Jalandhar, India 182 E5
Jalapa, Mex. 117 J14
Jalasjärvi, Fin. 143 G15
Jalaun, India 182 H7
Jaleswar, India 182 L11
Jalgaon, India 182 L5
Jalgaon, India 182 L5
Jalingo, Nigeria 206 L10
Jalitah Island, Tunisia 201 B15
Jalkot, Pak. 180 G12
Jaloklab, island, Marshall Is. 227 D19
Jalón, river, Sp. 147 E13
Jalor, India 182 H3
Jalpaiguri, India 182 H12
Jālq, Iran 177 M23
Jáltipan, Mex. 117 J14
Jālū, Lib. 201 F21
Jaluit Atoll, Marshall Is. 227 D16
Jaluit Lagoon, Marshall Is. 227 H20
Jaluit, island, Marshall Is. 227 J20
Jamaame, Somalia 211 S10
Jamaica Cay, Bahamas 119 G13
Jamaica, N. Amer. 118 M10
Jamaica, island, N. Amer. 84 N8
Jamalpur, Bangladesh 182 J12
Jaman Pass, Afghan., Taj. 179 N14
Jamanota, peak, Aruba, Neth. 123 R17
Jambi, Indonesia 192 K7
Jambusar, India 182 K3
James Bay, N. Amer. 87 M17
James Cistern, Bahamas 122 E6
James Point, Bahamas 122 D6
James Range, Austral. 220 F8
James Ross Island, Antarctica 238 D4
James, river, Va., U.S. 98 E11
James, river, N. Dak., U.S. 106 G6
Jameson Land, Greenland, Den. 88 K14
Jamestown National Historic Site, Va., U.S. 98 F13
Jamestown Reservoir, N. Dak., U.S. 106 G6
Jamestown, N.Y., U.S. 96 K4
Jamestown, N. Dak., U.S. 106 G6
Jamestown, S. Aust., Austral. 221 X10
Jamjamāl see Chemchemał, Iraq 176 E11
Jammu, India 182 D5
Jamnagar, India 182 K2
Jampur, Pak. 181 N10
Jämsä, Fin. 143 H15
Jamsah, Egypt 202 H13
Jamshedpur, India 182 K10
Jamuna, river, Bangladesh 182 J12
Jān Būlāq, Afghan. 180 F7
Jan Mayen Fracture Zone, Arctic Oc. 253 K19
Jan Mayen Ridge, Arctic Oc. 253 K19
Janaúba, Braz. 133 J15
Jand, Pak. 180 J11
Jandaq, Iran 177 G17
Jandia, peak, Sp. 214 Q7
Jandia, Punta de, Sp. 214 Q6
Jandiatuba, river, Braz. 132 E5
Jandola, Pak. 180 K9
Janesville, Wis., U.S. 100 M6
Jang Bogo Station, Antarctica 239 P15
Jangain, island, P.N.G. 228 A5
Jangaon, India 182 N7
Jangdo, island, S. Korea 187 V8
Jangdong, S. Korea 187 T8
Jangeru, Indonesia 192 K12
Janggye, S. Korea 187 R9
Janghang, S. Korea 187 R7
Jangheung, S. Korea 187 U8
Janghowon, S. Korea 187 N9
Jangipur, India 182 J11
Jangpyeong, S. Korea 186 M11
Jangsando, island, S. Korea 187 U7
Jangseong, S. Korea 187 S8
Jangsu, S. Korea 187 R9
Jānī Khēl, Afghan. 180 K8
Janīn, W. Bank 172 L6
Janjanbureh, Gambia 204 J7
Janów Lubelski, Pol. 151 F17
Januária, Braz. 133 J14
Jaora, India 182 J4
Japan Basin, Pac. Oc. 246 E6
Japan Rise, Pac. Oc. 246 E8
Japan Trench, Pac. Oc. 246 E8
Japan, Asia 189 R8
Japan, Sea of (East Sea), Pac. Oc. 168 F13
Japanese Guyots, Pac. Oc. 246 F8
Japtan, island, Marshall Is. 227 B21
Japurá, Braz. 132 D7
Japurá, river, N. Amer. 84 R11
Japurá, river, S. Amer. 132 D6
Jarābulus, Syr. 172 C10
Jaramillo, Arg. 135 T9
Jaranwala, Pak. 180 L12
Jarash (Gerasa), Jordan 172 L7
Jarbidge, Nev., U.S. 110 L11
Jarbidge, river, Idaho, Nev., U.S. 108 J4
Jardim, Braz. 132 L10
Jardines de la Reina, Cuba 118 K10
Jargalant Hayrahan, peak, Mongolia 184 D7
Jargalant, Mongolia 184 D8
Jargalant, Mongolia 185 D13
Jarghān, Afghan. 180 G6
Jari, river, Braz. 132 C12
Jarocin, Pol. 151 E13
Jarosław, Pol. 151 G18
Järvenpää, Fin. 143 J16
Jarvis Island, Pac. Oc. 224 G12
Järvsö, Sw. 143 H13
Jashpurnagar, India 182 K9
Jasikan, Ghana 205 N16
Jāsk, Iran 177 P20
Jasło, Pol. 151 H17
Jason Islands, Falk. Is., U.K. 135 V11
Jason Peninsula, Antarctica 238 D5
Jasonville, Ind., U.S. 101 S8
Jasper, Alta., Can. 86 L9
Jasper, Fla., U.S. 98 Q7
Jasper, Ga., U.S. 98 K5
Jasper, Ind., U.S. 101 T8
Jasper, Ark., U.S. 102 E7
Jasper, Ala., U.S. 103 J14
Jasper, Tex., U.S. 105 J21
Jastrowie, Pol. 151 C13
Jastrzębie-Zdrój, Pol. 151 H14
Jászberény, Hung. 151 L16
Jataí, Braz. 132 K12
Jati, Pak. 181 S8
Jatibonico, Cuba 118 H9
Jatobal, Braz. 133 E13
Jau, Angola 208 P10
Jaú, river, Braz. 132 D8
Jauaperi, river, Braz. 132 C9
Jaunpur, India 182 H8
Java (Jawa), island, Indonesia 192 M8
Java Ridge, Ind. Oc. 251 H16
Java Sea, Indonesia 192 L9
Java Trench (Sunda Trench), Ind. Oc. 251 G15
Javānrūd, Iraq 176 F12
Javari see Yavarí, river, S. Amer. 132 F4
Jávea (Xàbia), Sp. 147 J15
Jawa see Java, island, Indonesia 192 M8
Jawhar, Somalia 211 Q12
Jawi, Indonesia 192 J9
Jawoldo, island, S. Korea 187 N7
Jaworzno, Pol. 151 G15
Jay Em, Wyo., U.S. 108 H13
Jay, Okla., U.S. 105 B19
Jayapura, Indonesia 193 K21
Jayawijaya, Pegunungan, Indonesia 193 L20
Jayton, Tex., U.S. 105 F13
Jazā'ir az Zubayr, island, Yemen 174 N7
Jazā'ir Farasān, Saudi Arabia 174 M7
Jazā'ir Khurīyā Murīyā (Kuria Muria Islands), Oman 175 M16
Jāzān (Jīzān), Saudi Arabia 174 M8
Jazirah Doberai (Bird's Head Peninsula), Indonesia 193 J18
Jazīrat al Ḩanish al Kabīr, Yemen 174 P8
Jazīrat Antufash, Yemen 174 N7
Jazīrat Arwād, island, Syr. 172 G7
Jazīrat Jabal Zuqar, Yemen 174 P8
Jazīrat Maşīrah (Masira), Oman 175 K18
Jazīrat Maşīrah, Asia 168 J4
Jazīrat Shākir, Egypt 202 H13
Jazīrat Wādī Jimāl, Egypt 202 K14
Jazīreh-ye Khārk (Khārg Island), Iran 177 L14
Jbail (Byblos), Leb. 172 H6
Jbel Toubkal, Mor. 200 E7
Jdaïdet Ghazîr, Leb. 172 H6
Jdiriya, W. Sahara 200 G5
Jean Marie River, N.W.T., Can. 86 H10
Jean-Rabel, Haiti 119 L16
Jeanerette, La., U.S. 102 Q9
Jeanette, Russ. 161 C17
Jebba, Nigeria 206 K6
Jebel Abadab, Sudan 203 N14
Jebel Abyad Plateau, Sudan 203 P10
Jebel Asoteriba, Sudan 202 L14
Jebel Azao, Alg. 201 J15
Jebel Bani, Mor. 200 F6
Jebel Bara Simbil, Sudan 203 S7
Jebel Beila, Sudan 203 S14
Jebel ech Chambi, Tunisia 201 C15
Jebel ed Dair, Sudan 203 S11
Jebel Erbab, Sudan 203 N15
Jebel Gurgei, Sudan 203 S7
Jebel Haraza, Sudan 203 R11
Jebel Is, Sudan 202 L14
Jebel Jais, Oman, U.A.E. 175 F16
Jebel Katul, Sudan 203 R10
Jebel Kissu, Sudan 202 M7
Jebel Kuror, Sudan 202 M11
Jebel Oda, Sudan 202 M15
Jebel Ouarkziz, Alg., Mor. 200 F6
Jebel Sabidana, Sudan 203 P15
Jebel Sarhro, Mor. 200 E8
Jebel Siro, Sudan 203 R7
Jebel Teljo, Sudan 203 R8
Jebel 'Uweinat (Al Uwaynat), Sudan 202 L7
Jebri, Pak. 181 Q6
Jebus, Indonesia 192 K7
Jecheon, S. Korea 187 N10
Jeddah, Saudi Arabia 174 J6
Jedrol, island, Marshall Is. 227 B21
Jędrzejów, Pol. 151 G16
Jef Jef el Kebir, Chad 207 D15
Jefferson City, Tenn., U.S. 103 E19
Jefferson City, Mo., U.S. 107 T13
Jefferson, Ga., U.S. 98 K6
Jefferson, Wis., U.S. 100 L6
Jefferson, Tex., U.S. 105 G20
Jefferson, Iowa, U.S. 107 N10
Jefferson, Mount, Oreg., U.S. 110 G4
Jefferson, Mount, Nev., U.S. 111 Q9
Jefferson, river, Mont., U.S. 108 E7
Jeffersonville, Ind., U.S. 101 T9
Jeffrey City, Wyo., U.S. 108 J11
Jega, Nigeria 206 J6
Jégun, Fr. 149 X8
Jeju (Cheju), S. Korea 187 W7
Jeju Island (Cheju Island), Asia 168 G13
Jeju Strait, S. Korea 187 W7
Jeju-do, island, S. Korea 187 W6
Jekyll Island, Ga., U.S. 99 P9
Jelbart Ice Shelf, Antarctica 238 A12
Jeldēsa, Eth. 210 L9
Jelenia Góra, Pol. 150 F12
Jelgava (Mitau), Latv. 143 L16
Jellico, Tenn., U.S. 103 D18
Jema, river, Eth. 210 L7
Jemaja, island, Indonesia 192 H8
Jember, Indonesia 192 M11
Jembongan, island, Malaysia 192 F12
Jemez Pueblo, N. Mex., U.S. 109 R12
Jeminay, China 184 C5
Jemo Island, Marshall Is. 227 C16
Jena, Fla., U.S. 99 R6
Jena, La., U.S. 102 M8
Jena, Ger. 150 F9
Jenaien, Tunisia 201 E16
Jengish Chokusu, Tomür Feng, Pobeda Peak see Victory Peak, Kyrg. 179 K17
Jenkins, Ky., U.S. 103 C20
Jennings, La., U.S. 102 P7
Jenny Point, Dominica 123 F20
Jens Munk Island, Nunavut, Can. 87 F16
Jens Munk Ø, Greenland, Den. 89 Q9
Jensen Beach, Fla., U.S. 99 V10
Jensen, Utah, U.S. 108 L9
Jenu, Indonesia 192 J9
Jeogu, S. Korea 187 T11
Jeongeup, S. Korea 187 S8
Jeongnim, S. Korea 187 Q11
Jeongok, S. Korea 186 L8
Jeongseon, S. Korea 187 N11
Jeonju, S. Korea 187 R8
Jequié, Braz. 133 J16
Jequitinhonha, river, Braz. 133 J16
Jerada, Mor. 200 C10
Jerantut, Malaysia 191 T9
Jerba Island, Tunisia 201 D16
Jérémie, Haiti 119 M15
Jeremoabo, Braz. 133 G17
Jerer, river, Eth. 210 M10
Jerez de la Frontera, Sp. 146 M8
Jerez de los Caballeros, Sp. 146 J7
Jericho see Arīḩā, W. Bank 172 M6
Jericoacoara, S. Amer. 128 E11
Jerimoth Hill, R.I., U.S. 97 L14
Jerome, Idaho, U.S. 108 H5
Jerramungup, W. Aust., Austral. 221 X4
Jersey Bay, Virgin Is., U.S. 122 N9
Jersey City, N.J., U.S. 96 N11
Jersey Shore, Pa., U.S. 96 M7
Jersey, island, Channel Is., U.K. 145 Y12
Jerseyville, Ill., U.S. 101 S4
Jerusalem, Israel 172 M5
Jervis Bay Territory, Austral. 221 Y14
Jervis Bay, Austral. 220 L14
Jerzu, It. 153 P4
Jesenice, Slov. 154 C4
Jeseník, Czechia 151 G13
Jesi, It. 152 H9
Jessore, Bangladesh 182 K12
Jesup, Ga., U.S. 99 P8
Jesús María, Arg. 134 J10

Jetmore, *Kans., U.S.* **107** T5
Jetpur, *India* **182** K2
Jeungdo, island, *S. Korea* **187** T6
Jeungpyeong, *S. Korea* **187** P9
Jewel Cave National Monument, *S. Dak., U.S.* **106** L1
Jewell, *Iowa, U.S.* **107** N11
Jewett, *Tex., U.S.* **105** J18
Jeypore, *India* **182** M8
Jezercë, Maja, *Alban.* **154** J9
Jezioro Shiardwy, lake, *Pol.* **151** B16
Jezzîne, *Leb.* **172** J6
Jhal Jhao, *Pak.* **181** R6
Jhalawar, *India* **182** J5
Jhang, *Pak.* **180** L11
Jhansi, *India* **182** J6
Jharsuguda, *India* **182** L9
Jhatpat, *Pak.* **181** P8
Jhelum, *Pak.* **180** J12
Jhelum, river, *Pak.* **180** L11
Jhimpir, *Pak.* **181** S8
Jhudo, *Pak.* **181** S9
Jhunjhunun, *India* **182** G5
Ji-Paraná, *Braz.* **132** G8
Ji'an, *China* **185** M13
Jiaji see Qionghai, *China* **184** Q12
Jialing, river, *China* **184** K10
Jiamusi, *China* **185** C17
Jiangda see Jomda, *China* **184** K8
Jiangjunmiao, *China* **184** E6
Jiangmen, *China* **185** P13
Jiangyin, *China* **185** K15
Jiangyou, *China* **184** K10
Jiangzi see Gyangzê, *China* **184** L5
Jianli, *China* **185** L13
Jiaojiang see Taizhou, *China* **185** L16
Jiaozuo, *China* **185** J13
Jiayin, *China* **185** B17
Jiayuguan, *China* **184** G8
Jibāl an Nūbah, *Sudan* **203** T11
Jidali, *Somaliland* **210** L13
Jiddat al Ḥarāsīs, *Oman* **175** K16
Jido, *India* **182** F15
Jido, *S. Korea* **187** T7
Jido, island, *S. Korea* **187** T7
Jieshi, *China* **185** P14
Jieshou, *China* **185** J13
Jiesjavrre, lake, *Nor.* **143** C14
Jieyang, *China* **185** N14
Jiggalong, *W. Aust., Austral.* **221** T4
Jigüey, Bahía de, *Cuba* **118** H10
Jigzhi, *China* **184** K9
Jihlava, *Czechia* **150** H12
Jijel, *Alg.* **201** B13
Jijia, river, *Rom.* **155** B16
Jijiga, *Eth.* **210** M10
Jiksan, *S. Korea* **187** P8
Jilib, *Somalia* **211** S10
Jilin, *China* **185** E16
Jilong see Gyirong, *China* **184** L3
Jilong see Keelung, *China* **185** M16
Jīma, *Eth.* **211** N6
Jimaní, *Dom. Rep.* **119** M17
Jimbolia, *Rom.* **154** D10
Jiménez, *Mex.* **116** E9
Jinan, *China* **185** H14
Jinan, *S. Korea* **187** R9
Jinbal, island, *Marshall Is.* **227** G20
Jinchang, *China* **184** G9
Jincheon, *S. Korea* **187** P9
Jindo, *S. Korea* **187** U7
Jindo, island, *S. Korea* **187** U7
Jindong, *S. Korea* **187** T11
Jindřichův Hradec, *Czechia* **150** J11
Jinedrol, island, *Marshall Is.* **227** B21
Jingdezhen, *China* **185** L14
Jinggu, *China* **184** P9
Jinghe, *China* **184** D4
Jinghong, *China* **184** P9
Jingxi, *China* **184** P11
Jingyuan, *China* **184** H10
Jingzhou, *China* **184** M12
Jinhae, *S. Korea* **187** S11
Jinhaemen, bay, *S. Korea* **187** T11
Jinimi, island, *Marshall Is.* **227** B21
Jining, *China* **184** F12
Jining, *China* **185** J14
Jinja, *Uganda* **211** S4
Jinju, *S. Korea* **187** S10
Jinka, *Eth.* **211** P6
Jinnyang, *S. Korea* **187** R11
Jinsha (Yangtze), river, *China* **184** L8
Jinsha, *China* **184** M10
Jinta, *China* **184** G8
Jinyeong, *S. Korea* **187** S11
Jinzhou, *China* **185** G15

Jinzhou, *China* **185** F15
Jiparaná, river, *Braz.* **132** F8
Jipijapa, *Ecua.* **130** L3
Jirgatol, *Taj.* **179** M13
Jirisan, peak, *S. Korea* **187** S9
Jirjā, *Egypt* **202** J12
Jīroft (Sabzvārān), *Iran* **177** L20
Jirriiban, *Somalia* **211** N14
Jishou, *China* **184** M12
Jishu, *China* **185** D16
Jisr ash Shughūr, *Syr.* **172** E7
Jiu, river, *Rom.* **155** G13
Jiujiang, *China* **185** L14
Jiuquan, *China* **184** G8
Jiutai, *China* **185** E16
Jiwani, *Pak.* **181** S2
Jixi, *China* **185** D17
Jīzān see Jāzān, *Saudi Arabia* **174** M8
Jizzax, *Uzb.* **178** M11
Joaçaba, *Braz.* **132** N12
Joal-Fadiot, *Senegal* **204** J6
João Pessoa, *Braz.* **133** F18
Joaquín V. González, *Arg.* **134** F10
Jocheon, *S. Korea* **187** W7
Jochiwon-eup, *S. Korea* **187** P9
Jodar, *Sp.* **146** K11
Jodhpur, *India* **182** H3
Joensuu, *Fin.* **143** G17
Joe's Hill, *Kiribati* **226** Q11
Jōetsu, *Japan* **188** M11
Joeuf, *Fr.* **149** N14
Jogindar Nagar, *India* **180** K15
Jogipet, *India* **182** N6
Johan Petersen Bugt, *Greenland, Den.* **88** M13
Johannes V. Jensen Land, *Greenland, Den.* **88** A9
Johannesburg, *Calif., U.S.* **111** V9
Johannesburg, *S. Af.* **212** J8
John D. Rockefeller Jr. Memorial Parkway, *Wyo., U.S.* **108** G8
John Day Dam, *Oreg., Wash., U.S.* **110** E5
John Day Fossil Beds National Monument, *Oreg., U.S.* **110** G6
John Day, river, *Oreg., U.S.* **110** G6
John Eyre Motel, *W. Aust., Austral.* **221** X6
John F. Kennedy Space Center, *Fla., U.S.* **99** T10
John H. Kerr Reservoir, *N.C., Va., U.S.* **98** G12
John Martin Reservoir, *Colo., U.S.* **109** N15
John Millers, *Bahamas* **122** F6
John Murray Ø, *Greenland, Den.* **88** B8
John o'Groats, site, *Scot., U.K.* **144** F10
John Redmond Reservoir, *Kans., U.S.* **107** T9
Johns Folly Bay, *Virgin Is., U.S.* **122** N11
Johnson City, *N.Y., U.S.* **96** L8
Johnson City, *Tenn., U.S.* **103** E20
Johnson City, *Tex., U.S.* **105** K16
Johnson, *Kans., U.S.* **107** U3
Johnsonburg, *Pa., U.S.* **96** M5
Johnsons Point, *Antigua & Barbuda* **123** B20
Johnston Atoll, *Pac. Oc.* **224** D10
Johnston City, *Ill., U.S.* **101** U6
Johnston Falls, *D.R.C., Zambia* **209** L18
Johnston Seamounts, *Pac. Oc.* **246** H12
Johnston, *S.C., U.S.* **98** L7
Johnston, Lake, *Austral.* **220** K4
Johnstone, *Scot., U.K.* **144** L9
Johnstown, *N.Y., U.S.* **96** J11
Johnstown, *Pa., U.S.* **96** P4
Johor Baharu, *Malaysia* **191** V10
Joides Basin, *Pac. Oc.* **255** H15
Joigny, *Fr.* **149** Q11
Joint Region Marianas (Andersen Air Force Base), *Guam, U.S.* **230** N7
Joint Region Marianas (Naval Base Guam), *Guam, U.S.* **230** P4
Joinville Island, *Antarctica* **238** C4
Joinville, *Braz.* **133** N13
Jokaj, island, *F.S.M.* **226** J8
Jokau, *S. Sudan* **203** V13
Jøkelbugten, *Greenland, Den.* **88** E12
Jokkmokk, *Sw.* **143** E13
Jökuldjúp, valley, *Atl. Oc.* **253** M21
Joliet, *Ill., U.S.* **101** P7
Joliette, *Que., Can.* **87** P19

Jolo, *Philippines* **193** G13
Jolo, island, *Philippines* **193** G13
Jomda (Jiangda), *China* **184** K8
Jomu see Iomu, *Tanzania* **211** U4
Jones Sound, *Nunavut, Can.* **87** D15
Jones Sound, *Arctic Oc.* **253** N13
Jonesboro, *Ark., U.S.* **102** F10
Jonesboro, *La., U.S.* **102** L8
Jonesport, *Me., U.S.* **97** F19
Jonesville, *S.C., U.S.* **98** J8
Jonesville, *La., U.S.* **102** M9
Jönköping, *Sw.* **143** L13
Jonuta, *Mex.* **117** J15
Jonzac, *Fr.* **149** U7
Joplin, *Mo., U.S.* **107** V11
Joplin, *Mont., U.S.* **108** B8
Jordan Valley, *Oreg., U.S.* **110** J9
Jordan, *Minn., U.S.* **106** K11
Jordan, *Mont., U.S.* **108** C11
Jordan, *Asia* **173** P7
Jordan, river, *Jordan* **172** M6
Jorge Montt, Isla, *Chile* **135** W6
Jorhat, *India* **182** G14
Jörn, *Sw.* **143** F14
Jornada del Muerto, *N. Mex., U.S.* **109** U12
Jorong, *Indonesia* **192** K11
Jørpeland, *Nor.* **142** L10
Jos Plateau, *Af.* **198** H5
Jos, *Nigeria* **206** K9
José Batlle-y-Ordóñez, *Uru.* **134** L14
José Rodrigues, *Braz.* **132** E12
Joseph Bonaparte Gulf, *Austral.* **220** B7
Joseph, *Wash., U.S.* **110** F9
Joshimath, *India* **182** F7
Joshua Tree National Park, *Calif., U.S.* **111** X11
Josipdol, *Croatia* **154** E5
Josselin, *Fr.* **149** Q4
Jost Van Dyke, island, *Virgin Is., U.K.* **122** L10
Jotunheimen, range, *Nor.* **142** J10
Jou, *Sao Tome and Principe* **215** D19
Joulter Cays, *Bahamas* **118** D11
Jourdanton, *Tex., U.S.* **105** M16
Joutsa, *Fin.* **143** H16
Jovellanos, *Cuba* **118** G7
Joyakdo, island, *S. Korea* **187** U8
Joyeuse, *Fr.* **149** W12
Jreïda, *Mauritania* **204** F7
Juan Aldama, *Mex.* **116** F10
Juan de Fuca Ridge, *Pac. Oc.* **247** D17
Juan de Fuca, Strait of, *N. Amer.* **84** G2
Juan Fernández Islands, *S. Amer.* **128** M2
Juan José Castelli, *Arg.* **134** F12
Juan Stuven, Isla, *Chile* **135** U6
Juanjuí, *Peru* **132** F3
Juárez, *Arg.* **135** N12
Juazeiro do Norte, *Braz.* **133** F16
Juàzeiro, *Braz.* **133** G16
Juazohn, *Liberia* **204** Q11
Juba, *S. Sudan* **203** X11
Jubb al Jarrāḥ, *Syr.* **172** G9
Jubba, river, *Af.* **198** J11
Jubbah, *Saudi Arabia* **174** D7
Juberri, *Andorra* **162** K2
Jubilee Lake, *Austral.* **220** H6
Júcar, river, *Sp.* **147** H13
Júcaro, *Cuba* **118** J10
Jucheon, *S. Korea* **187** N10
Juchitán, *Mex.* **117** K14
Judaiat al Hamir, *Iraq* **176** J8
Judayyidat 'Ar'ar, *Saudi Arabia* **174** B7
Judenberg, *Aust.* **150** L11
Judith Basin, *Mont., U.S.* **108** C9
Judith Gap, *Mont., U.S.* **108** C9
Judith, river, *Mont., U.S.* **108** C9
Judsonia, *Ark., U.S.* **102** F9
Juigalpa, *Nicar.* **117** M18
Juiz de Fora, *Braz.* **133** L15
Jukbyeon, *S. Korea* **187** N12
Julesburg, *Colo., U.S.* **108** K16
Julia Creek, *Qnsld., Austral.* **221** S12
Juliaca, *Peru* **132** J6
Julian Alps, *Slov.* **152** C10
Julian, *Calif., U.S.* **111** Y10
Julianadorp, *Curaçao, Neth.* **123** P14
Julianehåb see Qaqortoq, *Greenland, Den.* **89** T7
Julianehåb Valley, *Atl. Oc.* **253** R20
Julio Escudero, station, *Antarctica* **238** C4

Julpo, *S. Korea* **187** S7
Julpoman, bay, *S. Korea* **187** S7
Jumba, *Somalia* **211** S10
Jumilla, *Sp.* **147** J13
Jumla, *Nepal* **182** F8
Jumundo, island, *S. Korea* **186** M7
Jumunjin, *S. Korea* **186** M11
Junction Bay, *Austral.* **220** A9
Junction City, *Ark., U.S.* **102** K8
Junction City, *Kans., U.S.* **107** S8
Junction City, *Oreg., U.S.* **110** H3
Junction, *Tex., U.S.* **105** K14
Junction, *Utah, U.S.* **109** N7
Jundiaí, *Braz.* **133** M13
Juneau, *Alas., U.S.* **113** K22
Jungdae, *S. Korea* **187** S11
Junggar Pendi (Dzungarian Basin), *China* **184** D5
Junglinster, *Lux.* **162** J10
Jungshahi, *Pak.* **181** S8
Jungtaedo, island, *S. Korea* **187** U5
Juniata, river, *Pa., U.S.* **96** N7
Junín de los Andes, *Arg.* **135** P7
Junín, *Peru* **132** H3
Junín, *Arg.* **134** L11
Juniper Mountain, *Oreg., U.S.* **110** J6
Junlian, *China* **184** M10
Junnar, *India* **182** M4
Juno Beach, *Fla., U.S.* **99** V10
Juntura, *Oreg., U.S.* **110** H8
Jupaguá, *Braz.* **133** H15
Jupiter, *Fla., U.S.* **99** V10
Jura Mountains, *Fr., Switz.* **152** B1
Jura, island, *Scot., U.K.* **144** L7
Jura, Sound of, *Scot., U.K.* **144** L7
Juradó, *Col.* **130** F4
Jurien Bay, *Austral.* **220** J2
Jurien, *W. Aust., Austral.* **221** W2
Jurm, *Afghan.* **180** F10
Juruá, *Braz.* **132** E7
Juruá, river, *Braz.* **132** G4
Juruena, river, *Braz.* **132** F10
Juruti, *Braz.* **132** D10
Justo Daract, *Arg.* **134** L9
Jutaí, *Braz.* **132** E6
Jutaí, river, *Braz.* **132** E6
Jutiapa, *Guatemala* **117** L16
Juticalpa, *Hond.* **117** L18
Jutland Bank, *Atl. Oc.* **248** G13
Jutland, *Eur.* **140** F6
Juventud, Isla de la, *Cuba* **118** J4
Juzennecourt, *Fr.* **149** Q12
Južna Morava, river, *Serb.* **154** H11
Juzzak, *Pak.* **181** N2
Jwasarido, island, *S. Korea* **187** U10
Jylland (Jutland), region, *Den.* **142** N11
Jyväskylä, *Fin.* **143** G16

K

K.I.V. Steenstrups Bræ, glacier, *Greenland, Den.* **89** P11
K2 (Godwin Austen, Qogir Feng), peak, *China, Pak.* **180** G14
Ka Tiritiri O Te Moana see Southern Alps, *N.Z.* **223** M17
Ka-do, island, *N. Korea* **186** H4
Ka, river, *Nigeria* **206** J7
Ka'a'awa, *Hawaii, U.S.* **115** G15
Kaabong, *Uganda* **211** Q5
Kaala-Gómén, *New Caledonia, Fr.* **231** P20
Ka'alu'alu Bay, *Hawaii, U.S.* **115** Q21
Kaamanen, *Fin.* **143** C15
Kaambooni, *Somalia* **211** T9
Kaambooni, Raas, *Af.* **198** K11
Kā'anapali, *Hawaii, U.S.* **115** J18
Kaap Plateau, *Af.* **198** P8
Kaap St. Marie, *Curaçao, Neth.* **123** P14
Kabaena, island, *Indonesia* **193** L14
Kabala, *Sierra Leone* **204** M9
Kabale, *Uganda* **211** T2
Kabalo, *D.R.C.* **209** H17
Kabambare, *D.R.C.* **209** H18
Kabanga, *Zambia* **209** N16
Kabangoué, *Côte d'Ivoire* **204** M12
Kabara, *Mali* **205** H14
Kabara, island, *Fiji* **226** Q8
Kabare, *D.R.C.* **209** F18

Kabarei, *Indonesia* **193** J17
Kabba, *Nigeria* **206** L7
Kabbenbock, island, *Marshall Is.* **227** H20
Kabena, island, *Asia* **168** M13
Kabinda, *D.R.C.* **209** H16
Kabīr, river, *Leb.* **172** G7
Kabo, *Cen. Af. Rep.* **207** L14
Kabobo, peak, *D.R.C.* **209** H18
Kabompo, *Zambia* **209** N16
Kabompo, river, *Zambia* **209** N16
Kabong, *Malaysia* **192** H10
Kabongo, *D.R.C.* **209** J16
Kaboudia, Ras, *Af.* **198** C6
Kābul (Kabul), *Afghan.* **180** H9
Kābul, river, *Afghan.* **180** H9
Kabuli, *P.N.G.* **222** M8
Kabumbé, Pico, *Sao Tome and Principe* **215** D20
Kabunda, *D.R.C.* **209** M19
Kaburuang, island, *Indonesia* **193** G16
Kabushiya, *Sudan* **203** Q13
Kabwe (Broken Hill), *Zambia* **209** N18
Kachchh, Gulf of (Gulf of Kutch), *India* **182** K1
Kachchh, Rann of (Kutch, Rann of), *India, Pak.* **181** T8
Kachia, *Nigeria* **206** K8
Kachīry, *Kaz.* **179** C14
Kachreti, *Rep. of Georgia* **171** C19
Kachung, *Uganda* **211** R4
Kaçkar Dağı, peak, *Turk.* **171** D15
Kadaingti, *Myanmar* **190** K6
Kadam, peak, *Uganda* **211** R5
Kadan Kyun, island, *Myanmar* **191** N6
Kadavu Passage, *Fiji* **226** Q5
Kadavu, island, *Fiji* **226** Q5
Kadé, *Guinea* **204** K8
Kadgo, Lake, *Austral.* **220** G6
Kading, river, *Thai.* **190** K11
Kadiri, *India* **183** Q6
Kadirli, *Turk.* **170** H11
Kadivka, *Ukr.* **159** Q10
Kadıköy, *Turk.* **170** D2
Kadınhanı, *Turk.* **170** G7
Kadmat Island (Cardamum Island), *India* **183** R3
Kadoka, *S. Dak., U.S.* **106** L3
Kadom, *Russ.* **158** K12
Kadoma, *Zimb.* **212** E9
Kadugli, *Sudan* **203** T10
Kaduna, *Nigeria* **206** K8
Kaduna, river, *Nigeria* **206** K8
Kadur, *India* **183** Q5
Kadyy, *Russ.* **158** G12
Kadzherom, *Russ.* **158** A15
Kaech'ŏn, *N. Korea* **186** H6
Kaédi, *Mauritania* **204** H8
Kaélé, *Cameroon* **206** K12
Ka'ena Point, *Hawaii, U.S.* **115** G13
Kaeo, *N.Z.* **223** D19
Kaep'ung, *N. Korea* **186** M7
Kaesŏng, *N. Korea* **186** M7
Kāf, *Saudi Arabia* **174** B5
Kafakumba, *D.R.C.* **209** K15
Kafanchan, *Nigeria* **206** K8
Kāfar Jar Ghar, range, *Afghan.* **180** K6
Kaffrine, *Senegal* **204** J7
Kafia Kingi, *S. Sudan* **203** V7
Kafue, *Zambia* **209** P18
Kafue, river, *Zambia* **209** N18;P17
Kafuka, *Japan* **188** E12
Kafulwe, *Zambia* **209** K18
Kaga, *Japan* **189** N9
Kagan, *Pak.* **180** H12
Kagera, river, *Tanzania, Uganda* **211** T3
Kağızman, *Turk.* **171** E17
Kagmar, *Sudan* **203** R11
Kagopal, *Chad* **207** L13
Kagoshima, *Japan* **189** T4
Kahakuloa, *Hawaii, U.S.* **115** J18
Kahalu'u, *Hawaii, U.S.* **115** G15
Kahalu'u, *Hawaii, U.S.* **115** N20
Kahama, *Tanzania* **211** U4
Kahan, *Pak.* **181** N8
Kahana, *Hawaii, U.S.* **115** J18
Kahemba, *D.R.C.* **209** J13
Kāhili, peak, *Hawaii, U.S.* **114** F10
Kahlotus, *Wash., U.S.* **110** D8
Kahnūj, *Iran* **177** M20
Kahoka, *Mo., U.S.* **107** Q13
Kaho'olawe, *Hawaii, U.S.* **115** K18
Kahramanmaraş, *Turk.* **170** H12
Kâhta, *Turk.* **171** H13

Kahuku Point, *Hawaii, U.S.* **115** G14
Kahuku, *Hawaii, U.S.* **115** G14
Kahuku, *Hawaii, U.S.* **115** Q21
Kahului Harbor, *Hawaii, U.S.* **115** J19
Kahului, *Hawaii, U.S.* **115** J19
Kahuzi, peak, *D.R.C.* **209** F18
Kai Kecil, island, *Indonesia* **193** L18
Kai, Kepulauan (Ewab), *Indonesia* **193** L18
Kaiama, *Nigeria* **206** K6
Kaibab Plateau, *U.S.* **90** H6
Kaibito, *Ariz., U.S.* **109** Q8
Kaieteur Falls, *Guyana* **131** F15
Kaiholena, peak, *Hawaii, U.S.* **115** Q21
Kaikoura, *N.Z.* **223** K19
Kailas Range, *Asia* **168** H7
Kaili, *China* **184** M11
Kailo, *D.R.C.* **209** F17
Kailua, *Hawaii, U.S.* **115** H15
Kailua, *Hawaii, U.S.* **115** N20
Kainan Maru Seamounts, *Ind. Oc.* **254** F9
Kainantu, *P.N.G.* **222** P7
Kainji Reservoir, *Nigeria* **206** K6
Kaipara Harbour, *N.Z.* **223** E20
Kairiru, island, *P.N.G.* **222** N6
Kairouan (Qairouan), *Tunisia* **201** C15
Kaiserslautern, *Ger.* **150** H6
Kaitaia, *N.Z.* **223** D19
Kaiwi Channel, *Hawaii, U.S.* **115** H16
Kaiyuan, *China* **185** E15
Kaiyuan, *China* **184** N9
Kaiyuh Mountains, *Alas., U.S.* **113** G14
Kajaani, *Fin.* **143** F16
Kajabbi, *Qnsld., Austral.* **221** S11
Kajakī, Band-e, *Afghan.* **180** K6
Kajang, *Malaysia* **191** U9
Kajiado, *Kenya* **211** T6
Kajo Kaji, *S. Sudan* **203** Y12
Kaka, *Turkm.* **178** N7
Kaka, *S. Sudan* **203** U12
Kākā, cape, *Hawaii, U.S.* **115** K18
Kakamas, *S. Af.* **212** K5
Kakamega, *Uganda* **211** S4
Kakana, *India* **183** T14
Kakaramea, *N.Z.* **223** H20
Kakata, *Liberia* **204** P10
Kakdwip, *India* **182** K11
Kake, *Alas., U.S.* **113** L22
Kakeroma Jima, *Japan* **189** W3
Kakhovka Reservoir, *Eur.* **140** H12
Kakhovka, *Ukr.* **159** R7
Kakhovs'ke Vodoskhovyshche, *Ukr.* **159** R7
Kakinada (Cocanada), *India* **182** N8
Kakogawa, *Japan* **189** Q8
Kakshaal Range, *China, Kyrg.* **179** L16
Kaktovik, *Alas., U.S.* **113** C18
Kakuda, *Japan* **188** M12
Kakuma, *Kenya* **211** Q5
Kala, *Tanzania* **211** X3
Kalaallit Nunaat, *Greenland, Den.* **88** H9
Kalabagh, *Pak.* **180** J10
Kalabakan, *Malaysia* **192** G12
Kalābishah and Beit el Wali, ruin, *Egypt* **202** K12
Kalabo, *Zambia* **209** P15
Kalaburagi (Gulbarga), *India* **182** N5
Kalach na Donu, *Russ.* **159** P12
Kalach, *Russ.* **159** N11
Kalae (South Cape), *Hawaii, U.S.* **115** Q21
Kalafrana, *Malta* **163** L23
Kalahari Desert, *Botswana* **212** H6
Kalāheo, *Hawaii, U.S.* **114** F10
Kal'ai Khumb, *Taj.* **179** M13
Kalalua, peak, *Hawaii, U.S.* **115** P23
Kalam, *Pak.* **180** G11
Kalama, *Wash., U.S.* **110** E3
Kalamáta, *Gr.* **156** L9
Kalamazoo, *Mich., U.S.* **100** M10
Kalamazoo, river, *Mich., U.S.* **100** M9
Kalambáka, *Gr.* **156** E9
Kalambo Falls, *Zambia* **209** K20
Kálamos, island, *Gr.* **156** H7
Kalamotí, *Gr.* **157** H16
Kalana, *Mali* **204** L11
Kalangala, *Uganda* **211** S3
Kalangali, *Tanzania* **211** W4
Kalaniopuu Basin, *Pac. Oc.* **246** G10
Kalao, island, *Indonesia* **193** M13
Kalaoa, *Hawaii, U.S.* **115** N20
Kalaotoa, island, *Indonesia* **193** M14
Kalasin, *Thai.* **190** L9

Kalat, *Pak.* **181** N7
Kalaupapa National Historical Park, *Hawaii, U.S.* **115** H17
Kalaupapa, *Hawaii, U.S.* **115** H17
Kalávryta, *Gr.* **156** J9
Kalay-wa, *Myanmar* **190** F4
Kalaýmor, *Turkm.* **178** P9
Kalbān, *Oman* **175** K18
Kalche, *D.R.C.* **209** F18
Kaldār, *Afghan.* **180** F7
Kale Burnu, *Turk.* **171** D14
Kalealoa (Barbers Point), *Hawaii, U.S.* **115** H14
Kaleburnu see Galinoporni, *Cyprus* **162** M11
Kalecik, *Turk.* **170** E9
Kaledupa, island, *Indonesia* **193** L14
Kalema, *D.R.C.* **209** G16
Kalemie, *D.R.C.* **209** H18
Kalemyo, *Myanmar* **190** F4
Kalene Hill, *Zambia* **209** L16
Kalenyy, *Kaz.* **178** E5
Kalevala, *Russ.* **158** B7
Kaleybar, *Iran* **176** C12
Kalgoorlie-Boulder, *W. Aust., Austral.* **221** W4
Kalí Límni, peak, *Gr.* **157** P18
Kaliakoúda, peak, *Gr.* **156** G9
Kaliakra, Nos, *Bulg.* **155** H18
Kalianda, *Indonesia* **192** L7
Kalihi Wai Bay, *Hawaii, U.S.* **114** E10
Kalihi Wai, *Hawaii, U.S.* **114** E10
Kalima, *D.R.C.* **209** F17
Kalimantan see Borneo, island, *Indonesia* **192** H11
Kalimpang, *India* **182** H11
Kaliningrad, *Russ.* **158** J1
Kalinino, *Russ.* **159** S10
Kalinino, *Arm.* **171** C18
Kalininsk, *Russ.* **158** M13
Kalinkavichy, *Belarus* **158** M5
Kalispell, *Mont., U.S.* **108** B5
Kalisz, *Pol.* **151** E14
Kaliua, *Tanzania* **211** V3
Kaliuwa'a see Sacred Falls, *Hawaii, U.S.* **115** G14
Kalix, *Sw.* **143** E14
Kalkan, *Turk.* **170** J5
Kalkaska, *Mich., U.S.* **100** J10
Kalkrand, *Namibia* **212** H4
Kallithéa, *Gr.* **156** J12
Kalloní, *Gr.* **157** F16
Kallonís, Kólpos, *Gr.* **157** F16
Kallsjön, lake, *Sw.* **142** G12
Kalmar, *Sw.* **143** M13
Kalnai, *India* **182** K9
Kalnik, peak, *Croatia* **154** D6
Kalo Chorio, *Cyprus* **162** Q7
Kalohi Channel, *Hawaii, U.S.* **115** J17
Kaloko-Honokōhau National Historical Park, *Hawaii, U.S.* **115** N20
Kaloli Point, *Hawaii, U.S.* **115** N23
Kalomo, *Zambia* **209** Q17
Kalpa, *India* **182** E6
Kalpeni Island, *India* **183** S3
Kalpin, *China* **184** F3
Kalpitiya, *Sri Lanka* **183** T7
Kalskag, *Alas., U.S.* **113** J13
Kalsubai, peak, *India* **182** M4
Kaltag, *Alas., U.S.* **113** G14
Kaltungo, *Nigeria* **206** K10
Kaluga, *Russ.* **158** K9
Kalulushi, *Zambia* **209** M18
Kalumburu, *W. Aust., Austral.* **221** P5
Kalundu, *Zambia* **209** L19
Kalush, *Ukr.* **159** P2
Kal'ya, *Russ.* **160** G9
Kalyan, *India* **182** M3
Kalyaxeed, *Somalia* **210** M14
Kalyazin, *Russ.* **158** H9
Kálymnos, *Gr.* **157** L18
Kálymnos, island, *Gr.* **157** L18
Kam, river, *Nigeria* **206** L10
Kama Reservoir, *Eur.* **140** C14
Kama, *D.R.C.* **209** G17
Kama, river, *Russ.* **158** E15
Kamaing, *Myanmar* **190** E5
Kamaishi, *Japan* **188** K13
Kamakal, island, *Fr. Polynesia, Fr.* **231** R17
Kamakou, peak, *Hawaii, U.S.* **115** J17
Kamalino, *Hawaii, U.S.* **114** F8
Kamalu, *Sierra Leone* **204** M9
Kaman, *Turk.* **170** F9
Kamanjab, *Namibia* **212** F3

Kamapanda, *Zambia* **209** M16
Kamar Zard, *Afghan.* **180** J3
Kamárai, *Gr.* **157** L14
Kamarān, island, *Yemen* **174** N8
Kamariótissa, *Gr.* **157** D15
Kamarod, *Pak.* **181** P4
Kambang, *Indonesia* **192** K6
Kambalda, *W. Aust., Austral.* **221** W4
Kambang, *Indonesia* **192** K6
Kambia, *Sierra Leone* **204** M8
Kambove, *D.R.C.* **209** L17
Kamchatka Peninsula, *Asia* **168** B13
Kamchatka Poluostrov, *Russ.* **161** F22
Kamchatka, Gulf of, *Pac. Oc.* **168** C13
Kamchatkiy Poluostrov, *Russ.* **161** E22
Kamchiya, river, *Bulg.* **155** H17
Kameelberg, peak, *Namibia* **212** F2
Kamehameha Basin, *Pac. Oc.* **247** G13
Kamehameha I Birthplace, *Hawaii, U.S.* **115** L20
Kámeiros (Camirus), ruin, *Gr.* **157** M19
Kamen, *Russ.* **161** G14
Kaména Voúrla, *Gr.* **156** G11
Kamenjak, Rt, *Croatia* **154** E3
Kamenka, *Russ.* **158** L12
Kamenka, *Russ.* **158** A11
Kamenka, *Russ.* **160** E9
Kamenka, *Kaz.* **178** D4
Kamensk Shakhtinskiy, *Russ.* **159** Q11
Kamensk Ural'skiy, *Russ.* **160** J9
Kamenskiy, *Russ.* **159** N13
Kamiah, *Idaho, U.S.* **108** D4
Kamieskroon, *S. Af.* **212** L4
Kamiji, *D.R.C.* **209** J15
Kamiloloa, *Hawaii, U.S.* **115** J17
Kamilonísi, island, *Gr.* **157** P17
Kamin'-Kashyrs'kyy, *Ukr.* **158** M3
Kamina, *D.R.C.* **209** K16
Kamino Shima, *Japan* **189** Q4
Kaminone Shima, *Japan* **189** V3
Kamitsuki, *Japan* **189** Q12
Kamitsushima, *Japan* **189** Q4
Kamiyaku, *Japan* **189** U4
Kamkhat Muḩaywir, peak, *Jordan* **173** N7
Kamloops, *B.C., Can.* **86** M8
Kamo, *Arm.* **171** D19
Kamōhio Bay, *Hawaii, U.S.* **115** K18
Kamoke, *Pak.* **180** K13
Kamo'oloa, *Hawaii, U.S.* **115** G14
Kampala, *Uganda* **211** S3
Kampanós, Akrotírio, *Gr.* **157** J14
Kampar, *Malaysia* **191** T8
Kampen, *Neth.* **148** H13
Kampene, *D.R.C.* **209** G17
Kamphaeng Phet, *Thai.* **190** L7
Kâmpóng Cham (Kampong Cham), *Cambodia* **191** N11
Kampong Cham see Kâmpóng Cham, *Cambodia* **191** N11
Kâmpóng Chhnăng (Kampong Chhnang), *Cambodia* **191** N10
Kampong Chhnang see Kâmpóng Chhnăng, *Cambodia* **191** N10
Kampong Kuala Besut, *Malaysia* **191** S9
Kampong Speu see Kâmpóng Spœ, *Cambodia* **191** P10
Kâmpóng Spœ (Kampong Speu), *Cambodia* **191** P10
Kampos, *Cyprus* **162** P6
Kâmpôt (Kampot), *Cambodia* **191** P10
Kampot see Kâmpôt, *Cambodia* **191** P10
Kampti, *Côte d'Ivoire* **205** M14
Kamsack, *Sask., Can.* **87** M13
Kamsar, *Guinea* **204** L7
Kamsuuma, *Somalia* **211** S10
Kamuela see Waimea, *Hawaii, U.S.* **115** M21
Kamui Misaki, *Japan* **188** G12
Kamuli, *Uganda* **211** S4
Kam'yanets'-Podil's'kyy, *Ukr.* **159** P3
Kam'yanka-Dniprovs'ka, *Ukr.* **159** R8
Kamyshin, *Russ.* **159** N13
Kamyzyak, *Russ.* **159** R15
Kan-Too Chokusu see Khan-Tengry Peak, *Kaz., Kyrg.* **179** K17
Kan, *S. Sudan* **203** V12
Kanab Creek, *Ariz., Utah, U.S.* **109** Q6
Kanab, *Utah, U.S.* **109** Q6
Kanacea, island, *Fiji* **226** P8
Kanaga Island, *Alas., U.S.* **112** N5

Kanak, river, *Turk.* **170** E10
Kanaláki, *Gr.* **156** F7
Kananga, *D.R.C.* **209** H15
Kanapou Bay, *Hawaii, U.S.* **115** K18
Kanarraville, *Utah, U.S.* **109** P6
Kanash, *Russ.* **158** J14
Kanatea, island, *Tonga* **227** Q23
Kanawha, river, *W. Va., U.S.* **98** D7
Kanazawa, *Japan* **189** N9
Kanbalu, *Myanmar* **190** F5
Kānchenjunga, peak, *Nepal* **182** G11
Kanchipuram, *India* **183** R7
Kanda-Kanda, *D.R.C.* **209** J15
Kandahār, *Afghan.* **180** L6
Kandalaksha Bay, *Eur.* **140** B10
Kandalaksha, *Russ.* **160** D8
Kandalakshskiy Zaliv, bay, *Russ.* **158** A8
Kandale, *D.R.C.* **209** H13
Kandangan, *Indonesia* **192** K11
Kandé, *Togo* **205** M17
Kandersteg, *Switz.* **152** C3
Kandhkot, *Pak.* **181** P9
Kandi, *India* **182** J11
Kandi, *Benin* **205** L18
Kandiaro, *Pak.* **181** Q8
Kandíla, *Gr.* **156** J10
Kandira, *Turk.* **170** C6
Kandrach, *Pak.* **181** R6
Kandrian, *P.N.G.* **222** P9
Kandū-ye Bālā, *Afghan.* **180** J6
Kandudu, island, *Maldives* **183** X3
Kandudu, island, *Maldives* **183** Y3
Kandy, *Sri Lanka* **183** U7
Kane Basin, *Can., Greenland, Den.* **88** D4
Kane Basin, *Arctic Oc.* **253** M14
Kane Fracture Zone, *Atl. Oc.* **248** L7
Kane Gap, *Atl. Oc.* **249** N10
Kane, *Pa., U.S.* **96** L5
Kanel, *Senegal* **204** H8
Kāne'ohe, *Hawaii, U.S.* **115** G15
Kanevskaya, *Russ.* **159** S10
Kanfarande, *Guinea* **204** L7
Kang, *Botswana* **212** H6
Kangaamiut, *Greenland, Den.* **89** P5
Kangaarsuk, *Greenland, Den.* **88** K5
Kangaatsiaq, *Greenland, Den.* **88** M5
Kangaba, *Mali* **204** L11
Kangal, *Turk.* **170** F12
Kangan, *Iran* **177** M16
Kanganpur, *Pak.* **180** L13
Kangar, *Malaysia* **191** S7
Kangaroo Island, *Austral.* **220** L10
Kangaussarssûp Sermia, glacier, *Greenland, Den.* **89** R6
Kangdêsê Shan (Gangdisi Shan), *China* **184** K3
Kangdŏk, *N. Korea* **186** C13
Kangdong, *N. Korea* **186** J7
Kangean, Kepulauan, *Indonesia* **192** L11
Kangeeak Point, *Nunavut, Can.* **87** F19
Kangen, river, *S. Sudan* **203** W13
Kangerluarsoruseq (Færingehavn), *Greenland, Den.* **89** R5
Kangerluarsuk, fjord, *Greenland, Den.* **89** P5
Kangerluluk, fjord, *Greenland, Den.* **89** T8
Kangerlussuaq Glacier, *Greenland, Den.* **88** M11
Kâmpóng Spœ (Kampong Speu), *Cambodia* **191** P10
Kangerlussuaq Valley, *Atl. Oc.* **253** N19
Kangerlussuaq, *Greenland, Den.* **89** N6
Kangerlussuaq, fjord, *Greenland, Den.* **88** K6;M12
Kangerlussuaq, fjord, *Greenland, Den.* **89** P5
Kangerlussuatsiaq (Lindenow Fjord), *Greenland, Den.* **89** T8
Kangersuatsiaq (Prøven), *Greenland, Den.* **88** J5
Kangertittivaq (Scoresby Sund), *Greenland, Den.* **88** K14
Kangertittivatsiaq, fjord, *Greenland, Den.* **89** P11
Kanggava Bay, *Solomon Is.* **228** H10
Kanggu, *N. Korea* **186** D10
Kanggye, *N. Korea* **186** E7
Kangiata Nunâta Sermia, glacier, *Greenland, Den.* **89** Q6
Kangilerngata Sermia, glacier, *Greenland, Den.* **88** L6

Kangilineq, *Greenland, Den.* **89** S6
Kangiqsualujjuaq, *Que., Can.* **87** J20
Kangiqsujuaq, *Que., Can.* **87** J18
Kangirsuk, *Que., Can.* **87** J19
Kangkar Punsum, peak, *Bhutan* **182** G13
Kangma see Kangmar, *China* **184** L5
Kangmar (Kangma), *China* **184** L5
Kangnam-sanmaek, range, *N. Korea* **186** F5
Kangnyŏng, *N. Korea* **186** M5
Kango, *Gabon* **208** E7
Kangpokpi, *India* **182** H14
Kangrinboqê Feng (Gangrenboqi Feng), *China* **184** J2
Kangsŏ, *N. Korea* **186** J5
Kangto, peak, *India* **182** G14
Kaniama, *D.R.C.* **209** J16
Kaniere, *N.Z.* **223** L18
Kaniet Islands, *P.N.G.* **222** M7
Kanif, *F.S.M.* **226** E11
Kanigiri, *India* **183** P7
Kanin Bank, *Arctic Oc.* **253** D18
Kanin Nos, *Russ.* **160** D9
Kanin Peninsula, *Eur.* **140** A12
Kanin, *Russ.* **160** E9
Kanirom, *Chad* **206** H11
Kanish, ruins, *Turk.* **170** F10
Kaniv, *Ukr.* **159** P6
Kanjarkot, site, *Pak.* **181** S9
Kanjiža, *Serb.* **154** D10
Kankakee, *Ill., U.S.* **101** P7
Kankakee, river, *Ill., Ind., U.S.* **101** P7
Kankan, *Guinea* **204** L11
Kanker, *India* **182** L8
Kankossa, *Mauritania* **204** H9
Kanmaw Kyun, island, *Myanmar* **191** P6
Kannapolis, *N.C., U.S.* **98** H9
Kannauj, *India* **182** H7
Kannur (Cannanore), *India* **183** R4
Kano, *Nigeria* **206** J8
Kano, river, *Nigeria* **206** J8
Kanokupolu, *Tonga* **227** Q22
Kanopolis, *Kans., U.S.* **107** T7
Kanorado, *Kans., U.S.* **107** S3
Kanosh, *Utah, U.S.* **109** N6
Kanoya, *Japan* **189** T5
Kanpur, *India* **182** H7
Kansanshi, *Zambia* **209** M17
Kansas City, *Kans., U.S.* **107** S10
Kansas City, *Mo., U.S.* **107** S10
Kansas, *U.S.* **107** T6
Kansas, river, *U.S.* **107** S8
Kansk, *Russ.* **161** K14
Kant, *Kyrg.* **179** K14
Kántanos, *Gr.* **156** Q12
Kantara Castle, ruin, *Cyprus* **162** N10
Kantchari, *Burkina Faso* **205** K17
Kanton, island, *Kiribati* **224** G10
Kanuku Mountains, *Guyana* **131** G15
Kanye, *Botswana* **212** H7
Kao, island, *Tonga* **227** L22
Kaôh Kông (Koh Kong), *Thai.* **191** P9
Kaôh Kŏng, island, *Cambodia* **191** P9
Kaôh Nhêk, *Cambodia* **191** N12
Kaôh Rŭng, island, *Cambodia* **191** P9
Kaohsiung, *China* **185** N16
Kaoka Bay, *Solomon Is.* **228** F10
Kaokoland, *Namibia* **212** F2
Kaolack, *Senegal* **204** J7
Kaoma, *Zambia* **209** P16
Kaongeshi, river, *D.R.C.* **209** K14
Kaouar, oasis, *Niger* **206** E11
Kap Alexander see Ullersuaq, *Greenland, Den.* **88** D3
Kap Alf Trolle, *Greenland, Den.* **88** F14
Kap Brewster see Kangikajiip Appalia, *Greenland, Den.* **88** L14
Kap Bridgman, *Greenland, Den.* **88** A10
Kap Cort Adelaer, *Greenland, Den.* **89** S9
Kap Dalton, *Greenland, Den.* **88** L14
Kap Edvard Holm, *Greenland, Den.* **89** N12
Kap Eiler Rasmussen, *Greenland, Den.* **88** B11
Kap Farvel see Nunap Isua, *Greenland, Den.* **89** U8
Kap Gustav Holm, *Greenland, Den.* **89** P11
Kap Harald Moltke, research station, *Greenland, Den.* **88** B10
Kap Louis Ussing, *Greenland, Den.* **89** N11

Kap Morris Jesup, *Greenland, Den.* **88** A9
Kap Møsting, *Greenland, Den.* **89** R9
Kap Ole Chiewitz, *Greenland, Den.* **88** A10
Kap Parry, *Greenland, Den.* **88** E3;J14
Kap Ravn, *Greenland, Den.* **88** M13
Kap Washington, *Greenland, Den.* **88** A9
Kap York *see* Innaanganeq, *Greenland, Den.* **88** F3
Kapa, island, *Tonga* **227** H23
Kapa'a, *Hawaii, U.S.* **114** E10
Kapa'au, *Hawaii, U.S.* **115** L21
Kapadvanj, *India* **182** K3
Kapaimari, *P.N.G.* **222** H23
Kapalala, *Zambia* **209** M19
Kapan, *Arm.* **171** F20
Kapanga, *D.R.C.* **209** K15
Kapatu, *Zambia* **209** L19
Kapedo, *Kenya* **211** R6
Kapela, mountains, *Croatia* **154** E5
Kapenguria, *Kenya* **211** R5
Kapfenberg, *Aust.* **150** L12
Kapingamarangi Atoll, *F.S.M.* **224** G5
Kapiri Mposhi, *Zambia* **209** N18
Kapisillit, *Greenland, Den.* **89** Q6
Kapiskau, *Ont., Can.* **87** M17
Kapit, *Malaysia* **192** H10
Kapıdağı, peak, *Turk.* **170** D4
Kaplan, *La., U.S.* **102** P8
Kaplıca *see* Dhavlos, *Cyprus* **162** M10
Kapoeta, *S. Sudan* **203** X13
Kapoho Crater, *Hawaii, U.S.* **115** P23
Kapombo, *D.R.C.* **209** L15
Kaposvár, *Hung.* **151** M14
Kappar, *Pak.* **181** R3
Kapsabet, *Kenya* **211** S5
Kapsan, *N. Korea* **186** E10
Kapsowar, *Kenya* **211** R5
Kapulena, *Hawaii, U.S.* **115** M21
Kapulo, *D.R.C.* **209** K18
Kapurthala, *India* **182** E5
Kapuskasing, *Ont., Can.* **87** N17
Kaputir, *Kenya* **211** R5
Kapydzhik, peak, *Arm.* **171** F20
Kara Burun, *Turk.* **170** J7
Kara Dağ, peak, *Turk.* **170** H8
Kara Dağ, peak, *Turk.* **171** H18
Kara Sea, *Arctic Oc.* **253** D14
Kara Shelf, *Arctic Oc.* **253** C14
Kara-Balta, *Kyrg.* **179** K14
Kara-Köl, *Kyrg.* **179** L13
Kara-Say, *Kyrg.* **179** K16
Karabiga, *Turk.* **170** D3
Karabük, *Turk.* **170** C8
Karaburun, *Turk.* **170** F2
Karaburun, cape, *Alban.* **154** M9
Karaca Dağ, peak, *Turk.* **171** H14
Karacabey, *Turk.* **170** D4
Karacadağ, *Turk.* **171** H14
Karacaköy, *Turk.* **170** C4
Karachev, *Russ.* **158** L8
Karachi, *Pak.* **181** S7
Karad, *India* **182** N4
Karaga, *Ghana* **205** M16
Karaghbyur (Füzuli), *Azerb.* **171** E21
Karaginskiy Gulf, *Pac. Oc.* **168** B13
Karaginskiy, *Russ.* **161** E22
Karahallı, *Turk.* **170** G5
Karahüyük, *Turk.* **170** G7
Karaisalı, *Turk.* **170** H10
Karaj, *Iran* **177** E15
Karakax *see* Moyu, *China* **184** G2
Karakeçi, *Turk.* **171** H14
Karakelong, island, *Indonesia* **193** G16
Karakoçan, *Turk.* **171** F14
Karakol (Przheval'sk), *Kyrg.* **179** K16
Karakoram Pass, *China, India* **180** G15
Karakoram Range, *India, Pak.* **180** F13
Karakul', *Taj.* **179** M14
Karakulino, *Russ.* **158** H16
Karam, *Russ.* **161** K15
Karaman, *Turk.* **170** H8
Karamay, *China* **184** D5
Karamea Bight, *N.Z.* **223** K18
Karamea, *N.Z.* **223** J18
Karamürsel, *Turk.* **170** D5
Karanja, *India* **182** L6
Karapınar, *Turk.* **170** H8
Karas Mountains, *Af.* **198** P7
Karasburg, *Namibia* **212** K5
Karasjok, *Nor.* **143** C14
Karasu, *Turk.* **170** C6
Karasu, river, *Turk.* **171** E15
Karasuk, *Russ.* **160** K11

Karataş, *Turk.* **170** J10
Karateki Shan, *China* **184** F2
Karatepe, *Turk.* **170** H11
Karathuri, *Myanmar* **191** P6
Karatsu, *Japan* **189** R4
Karatung, island, *Indonesia* **193** G16
Karaul, *Russ.* **161** F13
Karavas (Alsancak), *Cyprus* **162** N8
Karávia, island, *Gr.* **156** M12
Karavostasi (Gemi Konaghi), *Cyprus* **162** N7
Karavostásis, *Gr.* **157** M14
Karawanken, mountains, *Aust., Slov.* **154** C4
Karayazı, *Turk.* **171** E16
Karayün, *Turk.* **170** E12
Karbalā', *Iraq* **176** H10
Karcag, *Hung.* **151** L16
Kardámyla, *Gr.* **157** H16
Kardamýli, *Gr.* **156** L10
Kardítsa, *Gr.* **156** F9
Kardiva (Karidu), island, *Maldives* **183** V3
Kardiva Channel, *Maldives* **183** V3
Kärdla, *Est.* **143** K15
Karema, *Tanzania* **211** W2
Karesuando, *Sw.* **143** D14
Karêt, *Mauritania* **204** C11
Kārēzak, *Afghan.* **180** K2
Kargil, *India* **182** C5
Kargilik *see* Yecheng, *China* **184** F2
Kargı, *Turk.* **170** D10
Kargopol', *Russ.* **158** E10
Kari, *Nigeria* **206** H10
Kariba Dam, *Zambia, Zimb.* **212** D8
Kariba, *Zimb.* **212** D8
Karibib, *Namibia* **212** G3
Karidu *see* Kardiva, island, *Maldives* **183** V3
Karima, *Sudan* **203** P12
Karimata, island, *Indonesia* **192** K10
Karimata, Kepulauan, *Indonesia* **192** J8
Karimganj, *India* **182** J14
Karimnagar, *India* **182** M7
Karimunjawa, Kepulauan, *Indonesia* **192** L9
Karítaina, *Gr.* **156** K9
Karitane, *N.Z.* **223** N17
Karkaar, *Somalia* **210** L14
Karkar, island, *P.N.G.* **222** N7
Karkinit Gulf, *Eur.* **140** J11
Karkinits'ka Zatoka, *Ukr.* **159** S7
Karkūk (Kirkuk), *Iraq* **176** F9
Karlıova, *Turk.* **171** F15
Karlovac, *Croatia* **154** E5
Karlováşi, *Gr.* **157** J17
Karlovy Vary, *Czechia* **150** G10
Karlshamn, *Sw.* **143** M13
Karlskrona, *Sw.* **143** M13
Karlsruhe, *Ger.* **150** J6
Karlstad, *Minn., U.S.* **106** E8
Karlstad, *Sw.* **142** K12
Karmala, *India* **182** M4
Karmir Blur, ruins, *Arm.* **171** E18
Karmona, *Qnsld., Austral.* **221** V12
Karnak (Thebes), ruin, *Egypt* **202** J12
Karnali, river, *Nepal* **182** F8
Karnaphuli Reservoir, *Bangladesh* **182** K14
Karnes City, *Tex., U.S.* **105** M16
Karnobat (Polyanovgrad), *Bulg.* **155** J16
Karoi, *Zimb.* **212** E9
Karoko, *Fiji* **226** N7
Karompa, island, *Indonesia* **193** M14
Karonga, *Malawi* **212** B11
Karonie, *W. Aust., Austral.* **221** W5
Karora, *Qnsld., Austral.* **221** P16
Karosa, *Indonesia* **193** K13
Karou, *Mali* **205** J16
Karpas Peninsula, *Cyprus* **162** M11
Kárpathos (Carpathos), *Gr.* **157** P18
Kárpathos, *Gr.* **157** P18
Karpenísi, *Gr.* **156** G9
Karperó, *Gr.* **156** F8
Karpogory, *Russ.* **158** B12
Karrat Isfjord, *Greenland, Den.* **88** K5
Karre Mountains, *Af.* **198** H6
Kars, *Turk.* **171** D17
Kartal, *Turk.* **170** C5
Kartung, *Gambia* **204** K6
Karubwe, *Zambia* **209** P18
Karukh, *Afghan.* **180** H4
Karumba, *Qnsld., Austral.* **221** R11
Kārūn, river, *Iran* **177** J13

Karungu, *Kenya* **211** T5
Karunjie, *W. Aust., Austral.* **221** Q6
Karvachar (Kəlbəcər), *Azerb.* **171** E20
Karviná, *Czechia* **151** H14
Karwar, *India* **183** P4
Karwi, *India* **182** J8
Kar'yepol'ye, *Russ.* **158** A11
Karyés, *Gr.* **156** L10
Karyés, *Gr.* **157** D13
Karynzharyk, *Kaz.* **178** K5
Kárystos, *Gr.* **157** J13
Kas, *Sudan* **203** S7
Kaş, *Turk.* **170** J5
Kasaba Bay, *Zambia* **209** K19
Kasai, river, *Af.* **198** K7
Kasaji, *D.R.C.* **209** L15
Kasama, *Zambia* **209** L20
Kasane, *Namibia* **212** E7
Kasanga, *Tanzania* **211** X3
Kasangulu, *D.R.C.* **208** G10
Kasaragod, *India* **183** R4
Kasari Saki, *Japan* **189** V3
Kasari, *Japan* **189** V3
Kasba Lake, *N.W.T., Can.* **87** J13
Kasempa, *Zambia* **209** N16
Kasenga, *D.R.C.* **209** L18
Kasenye, *D.R.C.* **209** D19
Kasese, *D.R.C.* **209** F17
Kasese, *Uganda* **211** S2
Kasganj, *India* **182** G6
Kasha-Katuwe Tent Rocks National Monument, *N. Mex., U.S.* **109** R12
Kāshān, *Iran* **177** G15
Kashgar *see* Kashi, *China* **184** F2
Kashi (Kashgar), *China* **184** F2
Kashin, *Russ.* **158** H9
Kashira, *Russ.* **158** K9
Kashitu, *Zambia* **209** N18
Kashiwazaki, *Japan* **188** M11
Kashkandyow, *Afghan.* **180** F11
Kāshmar, *Iran* **177** F20
Kashmir, region, *China, India, Pak.* **180** H13
Kashmir, Vale of, *India* **180** H13
Kashmor, *Pak.* **181** P9
Kasimov, *Russ.* **158** K11
Kasiruta, island, *Indonesia* **193** J16
Kaskas, *Senegal* **204** H8
Kaskaskia, river, *Ill., U.S.* **101** S5
Kaskö, *Fin.* **143** H13
Kasongo-Lunda, *D.R.C.* **208** J11
Kasongo, *D.R.C.* **209** G17
Kásos, island, *Gr.* **157** Q18
Kaspiysk, *Russ.* **159** U15
Kaspiyskiy, *Russ.* **159** S15
Kassala, *Sudan* **203** R15
Kassándra, peninsula, *Gr.* **156** D12
Kassándras, Kólpos, *Gr.* **156** D12
Kassándreia, *Gr.* **156** E12
Kassaro, *Mali* **204** K11
Kassel, *Ger.* **150** F7
Kasserine (Qasserine), *Tunisia* **201** C15
Kassiópi, *Gr.* **156** E6
Kasson, *Minn., U.S.* **106** L12
Kastamonu, *Turk.* **170** C9
Kastaniés, *Gr.* **157** A17
Kastélli *see* Kíssamos, *Gr.* **156** P12
Kastélli, *Gr.* **157** Q15
Kastéllo, Ákra, *Gr.* **157** Q18
Kastellórizo *see* Megísti, island, *Gr.* **157** N22
Kastoriá, *Gr.* **156** C8
Kastorías, Límni, *Gr.* **156** C8
Kasulu, *Tanzania* **211** V2
Kasumkent, *Russ.* **159** V16
Kasungu, *Malawi* **212** C10
Kasur, *Pak.* **180** L13
Kata Tjuṯa (Mount Olga), *Austral.* **220** G8
Kataba, *Zambia* **209** P16
Katafaga, island, *Fiji* **226** P8
Katagum, river, *Nigeria* **206** J9
Katahdin, Mount, *Me., U.S.* **97** D17
Katako-Kombe, *D.R.C.* **209** G16
Katákolo, *Gr.* **156** K8
Katanga Plateau, *Angola, D.R.C., Zambia* **209** L16
Katanning, *W. Aust., Austral.* **221** X3
Katápola, *Gr.* **157** M16
Katchall Island, *India* **183** T15
Katera, *Uganda* **211** T3
Kateríni, *Gr.* **156** D10
Kates Needle, peak, *Alas., U.S.* **113** L23
Katete, *Zambia* **209** N20
Katha, *Myanmar* **190** F5

Katherîna, Gebel (Mount Catherine), *Egypt* **173** T3
Katherine, river, *Austral.* **220** B8
Kathiawar, peninsula, *India* **182** K2
Kathmandu, *Nepal* **182** G10
Kathu, *S. Af.* **212** J6
Kathua, *India* **180** K14
Kati, *Mali* **204** K11
Katima Mulilo, *Namibia* **212** E7
Katiu, island, *Fr. Polynesia, Fr.* **231** B19
Katla, peak, *Ice.* **142** F2
Katlanovo, *N. Maced.* **154** K11
Katmai National Park and Preserve, *Alas., U.S.* **113** L15
Katmai, Mount, *Alas., U.S.* **113** L15
Káto Achaΐa, *Gr.* **156** J8
Káto Almyrí, *Gr.* **156** J11
Kato Lakatameia, *Cyprus* **162** P8
Káto Nevrokópi, *Gr.* **156** B12
Katondwe, *Zambia* **209** P19
Katowice, *Pol.* **151** G15
Katrancık Dağı, peak, *Turk.* **170** H5
Katrine, Loch, *Scot., U.K.* **144** K9
Katsina, *Nigeria* **206** H8
Katsuyama, *Japan* **189** P9
Kattankudi, *Sri Lanka* **183** U8
Kattavía, *Gr.* **157** N19
Kattegat, strait, *Den., Sw.* **142** M12
Katun', river, *Russ.* **160** L12
Katwe, *Uganda* **211** S2
Katwijk aan Zee, *Neth.* **148** J12
Kau, *Indonesia* **193** H16
Ka'ū, *Hawaii, U.S.* **115** Q21
Kaua'i Channel, *Hawaii, U.S.* **114** F12
Kaua'i, *Hawaii, U.S.* **114** E9
Kauehi, island, *Fr. Polynesia, Fr.* **231** B18
Kaufbeuren, *Ger.* **150** K8
Kaufman, *Tex., U.S.* **105** G18
Kauhakō Crater, *Hawaii, U.S.* **115** H17
Kaukau Veld, *Botswana, Namibia* **212** F5
Kaukauna, *Wis., U.S.* **100** J6
Kaukaui, *P.N.G.* **228** C5
Kaukura, island, *Fr. Polynesia, Fr.* **231** B18
Ka'ula Gulch, *Hawaii, U.S.* **115** M22
Ka'ula, *Hawaii, U.S.* **114** G7
Kaulakahi Channel, *Hawaii, U.S.* **114** F8
Kauluoa Point, *Hawaii, U.S.* **115** P20
Kaumakani, *Hawaii, U.S.* **114** F9
Kaumalapau, *Hawaii, U.S.* **115** K17
Kaumu o Kaleihoohie, peak, *Hawaii, U.S.* **115** M21
Kaunā Point, *Hawaii, U.S.* **115** Q20
Kaunakakai Harbour, *Hawaii, U.S.* **115** J17
Kaunakakai, *Hawaii, U.S.* **115** J17
Kaunas, *Lith.* **143** M17
Kaupakulua, *Hawaii, U.S.* **115** J19
Kaupō, *Hawaii, U.S.* **115** K20
Kaura-Namoda, *Nigeria* **206** H7
Kauru, *Nigeria* **206** K8
Kautenbach, *Lux.* **162** H9
Kautokeino, *Nor.* **143** C14
Kavacha, *Russ.* **161** D22
Kavadarci, *N. Maced.* **154** K12
Kavajë, *Alban.* **154** L9
Kavak, *Turk.* **170** D11
Kavála (Neapolis), *Gr.* **157** C13
Kaválas, Kólpos, *Gr.* **157** C13
Kavalerovo, *Russ.* **161** L21
Kavali, *India* **183** P7
Kavaratti, *India* **183** S3
Kavieng, *P.N.G.* **222** M9
Kavir Desert *see* Dasht-e Kavīr, *Iran* **177** F18
Kavir Desert, *Asia* **168** G5
Kavīr-e Namak, *Iran* **177** F19
Kávo Dóro, *Gr.* **157** J14
Kávos, *Gr.* **156** F6
Kavungo, *Angola* **209** M15
Kawagoe, *Japan* **189** P12
Kawaihae Bay, *Hawaii, U.S.* **115** M20
Kawaihae, *Hawaii, U.S.* **115** M20
Kawaihoa, *Hawaii, U.S.* **114** F8
Kawaikini, peak, *Hawaii, U.S.* **114** E10
Kawakawa, *N.Z.* **223** D20
Kawambwa, *Zambia* **209** L18
Kawasak Passage, *Palau* **226** E6
Kawasaki, *Japan* **189** P12
Kawe, island, *Indonesia* **193** J17
Kawela, *Hawaii, U.S.* **115** G14
Kawela, *Hawaii, U.S.* **115** J17
Kawhia, *N.Z.* **223** F20
Kawich Peak, *Nev., U.S.* **111** R10

Kawm Umbū, *Egypt* **202** K12
Kawthoung, *Myanmar* **191** Q6
Kay, *Russ.* **158** E15
Kaya, *Indonesia* **192** H12
Kaya, *Burkina Faso* **205** K15
Kayambi, *Zambia* **209** K20
Kayangel Islands (Ngcheangel), *Palau* **226** D7
Kayangel Passage, *Palau* **226** D6
Kayanl, river, *Indonesia* **192** H11
Kayapınar, *Turk.* **171** H15
Kayar, *Senegal* **204** J5
Kaycee, *Wyo., U.S.* **108** G12
Kayenta, *Ariz., U.S.* **109** Q8
Kayes, *Mali* **204** J9
Kayes, *Congo* **208** G9
Kayl, *Lux.* **162** K9
Kaymaz, *Turk.* **170** F10
Kaynar, *Turk.* **170** F11
Kaysatskoye, *Russ.* **159** N14
Kayser Gebergte, peak, *Suriname* **131** G16
Kayseri, *Turk.* **170** F10
Kaz Dağı, peak, *Turk.* **170** E3
Kazach'ye, *Russ.* **161** E17
Kazakh Uplands, *Asia* **168** E7
Kazakh Uplands, *Kaz.* **178** E12
Kazan Rettō *see* Volcano Islands, *Japan* **189** Z13
Kazan', *Russ.* **158** J14
Kazan, river, *Nunavut, Can.* **87** H14
Kazanskaya, *Russ.* **159** P11
Kazarman, *Kyrg.* **179** L14
Kazbek, peak, *Rep. of Georgia* **171** B18
Kazhim, *Russ.* **158** E15
Kazi Ahmad, *Pak.* **181** R8
Kazima, *Cen. Af. Rep.* **207** M18
Kazimierz Dolny, *Pol.* **151** F17
Kazimoto, *Tanzania* **211** X6
Kazincbarcika, *Hung.* **151** K16
Kazumba, *D.R.C.* **209** J14
Kazungula, *Zambia* **209** Q16
Kazym Mys, *Russ.* **160** G10
Kdeyyat ej Joul, peak, *Mauritania* **204** D9
Ke Avainiu Pass, *Solomon Is.* **228** B9
Ke Ila, island, *Solomon Is.* **228** B9
Ke Loma, island, *Solomon Is.* **228** B9
Ké Massina, *Mali* **205** J13
Keadu, island, *Maldives* **183** W3
Keāhole Point, *Hawaii, U.S.* **115** N20
Kealaikahiki Channel, *Hawaii, U.S.* **115** K18
Kealakekua, *Hawaii, U.S.* **115** P20
Keālia, *Hawaii, U.S.* **114** E10
Keālia, *Hawaii, U.S.* **115** P20
Ke'anae, *Hawaii, U.S.* **115** J20
Keanapapa Point, *Hawaii, U.S.* **115** J17
Kearney, *Nebr., U.S.* **107** Q6
Keauhou, *Hawaii, U.S.* **115** N20
Keawanui Bay, *Hawaii, U.S.* **114** F8
Kêb (Kep), *Cambodia* **191** P10
Keban, *Turk.* **171** F13
Kébémèr, *Senegal* **204** H6
Kebkablya, *Sudan* **203** S7
Kebnekaise, peak, *Sw.* **143** D13
K'ebrī Dehar, *Eth.* **211** N11
Kech, river, *Pak.* **181** R4
Keçiborlu, *Turk.* **170** G6
Kecskemét, *Hung.* **151** M15
Kediri, *Indonesia* **192** M10
Kédougou, *Senegal* **204** K9
Kedvavom, *Russ.* **158** B15
Kędzierzyn-Koźle, *Pol.* **151** G14
Ke'ei, *Hawaii, U.S.* **115** P20
Keele Peak, *Yukon, Can.* **86** G9
Keele, river, *N.W.T., Can.* **86** G9
Keeler, *Calif., U.S.* **111** U8
Keelung (Jilong), *China* **185** M16
Keene, *N.H., U.S.* **97** J13
Keer-Weer, Cape, *Austral.* **220** B12
Keetmanshoop, *Namibia* **212** J4
Keewatin, *Minn., U.S.* **106** F11
Kefaloniá (Cephalonia), island, *Gr.* **156** H7
Kéfalos, *Gr.* **157** M18
Kefaluka, *Turk.* **170** H3
Keffi, *Nigeria* **206** L8
Keflavík, *Ice.* **142** E1
Keg River, *Alta., Can.* **86** J10
Kegen, *Kaz.* **179** J16
Kegha, *Eth.* **211** P6
Keheili, *Sudan* **203** N12

Kehl, *Ger.* **150** J5
Keila, *Est.* **143** J16
Keïta, *Niger* **206** G7
Keitele, lake, *Fin.* **143** G16
Keith Arm, *N.W.T., Can.* **86** F10
Keith, *Scot., U.K.* **144** H10
Keithsburg, *Ill., U.S.* **101** P3
Kejser Franz Joseph Fjord, *Greenland, Den.* **88** H13
Kekaha, *Hawaii, U.S.* **114** F9
Kékes, peak, *Hung.* **151** K16
Kekirawa, *Sri Lanka* **183** T7
Keklau, *Palau* **226** F6
Kekra, *Russ.* **161** H20
Kekri, *India* **182** H4
K'elafo, *Eth.* **211** P11
Kelai, island, *Maldives* **183** U3
Kélakam, *Niger* **206** H10
Kelang, island, *Indonesia* **193** K16
Këlcyrë, *Alban.* **154** M10
Kelefesia, island, *Tonga* **227** N23
Kelheim, *Ger.* **150** J9
Kelkit, *Turk.* **171** E14
Kelkit, river, *Turk.* **170** D12
Kellé, *Congo* **208** E10
Kéllé, *Niger* **206** H9
Keller Lake, *N.W.T., Can.* **86** G10
Kellet, Cape, *N.W.T., Can.* **86** D11
Kellits, *Jam.* **122** H9
Kells (Ceanannas), *Ire.* **145** Q6
Kelmet, *Eritrea* **210** H7
Kélo, *Chad* **206** K12
Kelowna, *B.C., Can.* **86** M9
Kelso, *Wash., U.S.* **110** E3
Kelso, *Calif., U.S.* **111** W11
Keluang, *Malaysia* **191** V9
Kem', *Russ.* **158** B8
Kem', *Russ.* **160** D8
Kema, *Indonesia* **193** H15
Kemah, *Turk.* **171** E13
Kemaliye, *Turk.* **171** F13
Kemasik, *Malaysia* **191** T9
Kembé, *Cen. Af. Rep.* **207** N16
Kembolcha, *Eth.* **210** L8
Kemboma, *Gabon* **208** D9
Kemer, *Turk.* **170** D3
Kemer, *Turk.* **170** G4
Kemer, *Turk.* **170** J5
Kemerovo, *Russ.* **160** K12
Kemi, *Fin.* **143** E15
Kemi, river, *Eur.* **140** C9
Kemijärvi, *Fin.* **143** D15
Kemijärvi, lake, *Fin.* **143** D15
Kemijoki, river, *Fin.* **143** D15
Kemlya, *Russ.* **158** K13
Kemmerer, *Wyo., U.S.* **108** J8
Kemmuna see Comino, island, *Malta* **163** H21
Kemp Coast, *Antarctica* **239** D20
Kemp Peninsula, *Antarctica* **238** F8
Kemparana, *Mali* **205** K13
Kempendyay, *Russ.* **161** H17
Kemps Bay, *Bahamas* **118** F11
Kempsey, *N.S.W., Austral.* **221** W15
Kempten, *Ger.* **150** L7
Kenai Fjords National Park, *Alas., U.S.* **113** K16
Kenai Peninsula, *Alas., U.S.* **113** K16
Kenai, *Alas., U.S.* **113** K16
Kenamuke Swamp, *S. Sudan* **203** X13
Kendal, *Eng., U.K.* **145** P11
Kendall, *N.Y., U.S.* **96** H6
Kendall, *Fla., U.S.* **99** X10
Kendall, Cape, *Nunavut, Can.* **87** H16
Kendallville, *Ind., U.S.* **101** P10
Kendari, *Indonesia* **193** K14
Kendawangan, *Indonesia* **192** K9
Kendikolu, island, *Maldives* **183** U3
Kendrapara, *India* **182** L10
Këndrevicës, Maja e, *Alban.* **154** M9
Kendrick, *Idaho, U.S.* **108** D3
Kenedy, *Tex., U.S.* **105** M16
Kenema, *Sierra Leone* **204** N9
Kenge, *D.R.C.* **208** H11
Kengtung, *Myanmar* **190** H7
Kenhardt, *S. Af.* **212** K5
Kéniéba, *Mali* **204** K9
Kenitra (Port-Lyautey), *Mor.* **200** C7
Kenli, *China* **185** H14
Kenmare (Neidín), *Ire.* **145** T3
Kenmare, *N. Dak., U.S.* **106** E3
Kenmore, *Scot., U.K.* **144** K9
Kenn Plateau, *Pac. Oc.* **246** M9
Kennebec, *S. Dak., U.S.* **106** L5
Kennebec, river, *Me., U.S.* **97** E16
Kennebunk, *Me., U.S.* **97** H15

Kennedy Channel, *Can., Greenland, Den.* **88** C5
Kennedy, Cape see Canaveral, Cape, *N. Amer.* **84** L8
Kenner, *La., U.S.* **102** Q11
Kennett, *Mo., U.S.* **107** W15
Kennewick, *Wash., U.S.* **110** E7
Keno Hill, *Yukon, Can.* **86** F8
Kenora, *Ont., Can.* **87** N14
Kenosha, *Wis., U.S.* **100** M7
Kenova, *W. Va., U.S.* **98** D7
Kent Peninsula, *Nunavut, Can.* **87** F13
Kent, *Ohio, U.S.* **101** P15
Kent, *Tex., U.S.* **104** J9
Kent, *Oreg., U.S.* **110** F5
Kent, *Wash., U.S.* **110** C4
Kentland, *Ind., U.S.* **101** Q7
Kenton, *Ohio, U.S.* **101** Q12
Kentucky Lake, *Ky., Tenn., U.S.* **103** D13
Kentucky, *U.S.* **103** C17
Kentucky, river, *U.S.* **91** G17
Kentwood, *Mich., U.S.* **100** L10
Kentwood, *La., U.S.* **102** N10
Kenutu, island, *Tonga* **227** H23
Kenya, *Af.* **211** S5
Kenya Highlands, *Af.* **198** J10
Kēōkea, *Hawaii, U.S.* **115** K19
Kēōkea, *Hawaii, U.S.* **115** P20
Keokuk, *Iowa, U.S.* **107** Q14
Keomuku, *Hawaii, U.S.* **115** J18
Keonjhargarh, *India* **182** L10
Kep see Kêb, *Cambodia* **191** P10
Kepi, *Indonesia* **193** M21
Kepino, *Russ.* **158** A10
Kępno, *Pol.* **151** F14
Keppel Bay, *Austral.* **220** F15
Kepsut, *Turk.* **170** E4
Keramotí, *Gr.* **157** C14
Keran, *Pak.* **182** C4
Kerang, *Vic., Austral.* **221** Y12
Keravat, *P.N.G.* **222** N10
Kerch Strait, *Eur.* **140** J12
Kerch, *Russ.* **159** S9
Kerchenskiy Proliv, *Ukr.* **159** S9
Kerchouel, *Mali* **205** G16
Kerekere, *D.R.C.* **209** C19
Kerema, *P.N.G.* **222** Q7
Kerempe Burnu, *Turk.* **170** B9
Keren, *Eritrea* **210** H7
Kerens, *Tex., U.S.* **105** H18
Kerguelen Plateau, *Ind. Oc.* **250** Q10
Keri, *Gr.* **156** K7
Kericho, *Kenya* **211** S5
Kerikeri, *N.Z.* **223** D20
Kerinci, peak, *Indonesia* **192** K6
Keriya see Yutian, *China* **184** G3
Kerki see Atamyrat, *Turkm.* **178** N10
Kerkíni, *Gr.* **156** B11
Kerkíni, Límni, *Gr.* **156** B11
Kerkuane, ruin, *Tunisia* **201** B16
Kérkyra (Corfu, Corcyra), *Gr.* **156** E6
Kérkyra, *Gr.* **156** E6
Kerma, *Sudan* **203** N11
Kermadec Islands, *Pac. Oc.* **224** L9
Kermadec Ridge, *Pac. Oc.* **246** N11
Kermadec Trench, *Pac. Oc.* **246** N12
Kermán, *Iran* **177** K19
Kermánsháh (Bákhtarán), *Iraq* **176** G12
Kermánsháhán, *Iran* **177** J18
Kermit, *Tex., U.S.* **104** H10
Kern, river, *Calif., U.S.* **111** V7
Kernersville, *N.C., U.S.* **98** G10
Kernville, *Calif., U.S.* **111** V8
Kéros, island, *Gr.* **157** L16
Kérouané, *Guinea* **204** M11
Kerr Mount, *Bahamas* **122** F11
Kerre, *Cen. Af. Rep.* **207** M18
Kerrobert, *Sask., Can.* **86** M11
Kerrville, *Tex., U.S.* **105** L15
Kerry Head, *Ire.* **145** S2
Kershaw, *S.C., U.S.* **98** K9
Kertamulia, *Indonesia* **192** J8
Keryneia Range, *Cyprus* **162** N9
Kerýneia see Kyrenia, *Cyprus* **162** N8
Keşan, *Turk.* **170** C3
Keşap, *Turk.* **171** D13
Kesennuma, *Japan* **188** L13
Keshod, *India* **182** L2
Keşiş Dağı, peak, *Turk.* **171** E14
Keskal, *India* **182** M8
Keskin, *Turk.* **170** E9
Kes'ma, *Russ.* **158** G9
Kessala, *Gabon* **208** F10
Kesten'ga, *Russ.* **158** A7

Keszthely, *Hung.* **151** M13
Ketanda, *Russ.* **161** G19
Ketaun, *Indonesia* **192** K6
Ketchikan, *Alas., U.S.* **113** M24
Ketchum, *Idaho, U.S.* **108** G5
Keti Bandar, *Pak.* **181** T7
Kettering, *Ohio, U.S.* **101** R11
Kettering, *Eng., U.K.* **145** S13
Kettle Falls, *Wash., U.S.* **110** A8
Kettle River Range, *Wash., U.S.* **110** B7
Keul', *Russ.* **161** K15
Keur Momar Sar, *Senegal* **204** H7
Kewanee, *Ill., U.S.* **101** P4
Kewaunee, *Wis., U.S.* **100** J7
Keweenaw Bay, *Mich., U.S.* **100** E6
Keweenaw National Historical Park, *Mich., U.S.* **100** E6
Keweenaw Peninsula, *Mich., U.S.* **100** D7
Keweenaw Point, *Mich., U.S.* **100** E7
Key Colony Beach, *Fla., U.S.* **99** Y9
Key Largo, *Fla., U.S.* **99** Y10
Key West, *Fla., U.S.* **99** Z8
Key, Lough, *Ire.* **145** P5
Keya Paha, river, *Nebr., S. Dak., U.S.* **106** M4
Keyano, *Que., Can.* **87** L19
Keyes, *Okla., U.S.* **104** B12
Keyhole Reservoir, *Wyo., U.S.* **108** F13
Keyser, *W. Va., U.S.* **98** C11
Keystone Lake, *Okla., U.S.* **91** J13
Keystone, *W. Va., U.S.* **98** F8
Keystone, *S. Dak., U.S.* **106** L2
Kežmarok, *Slovakia* **151** J16
Kәlbәcәr see Karvachar, *Azerb.* **171** E20
Khabarovsk, *Russ.* **161** K21
Khābūr, river, *Syr.* **172** D14
Khadyzhensk, *Russ.* **159** T10
Khailino, *Russ.* **161** D21
Khaïrābād, *Afghan.* **180** F7
Khairpur, *Pak.* **181** N11
Khairpur, *Pak.* **181** P8
Khakhar', *Russ.* **161** H19
Khalatse, *India* **182** C6
Khalīg el Tîna, *Egypt* **173** N1
Khalīj al 'Arab, bay, *Egypt* **202** F10
Khalīj al Bumbah, bay, *Lib.* **201** D22
Khalīj as Sallūm, bay, *Egypt* **202** E8
Khalīj Maşīrah, bay, *Oman* **175** K17
Khalīj Surt (Gulf of Sidra), *Lib.* **201** D19
Khalīj-e Chāh Bahār, bay, *Iran* **177** P22
Khalīj-e Halīleh, bay, *Iran* **177** L15
Khalīj-e Nāy Band, bay, *Iran* **177** M16
Khalūf, *Oman* **175** J17
Khambhaliya, *India* **182** K1
Khambhat, Gulf of, *India* **182** L3
Khamgaon, *India* **182** L5
Khamīs Mushayţ, *Saudi Arabia* **174** L8
Khamr, *Yemen* **174** N8
Khān Abū Shāmāt, *Syr.* **172** J8
Khān az Zābīb, *Jordan* **173** N7
Khān Shaykhūn, *Syr.* **172** F8
Khan Tängiri Shyngy, Hantengri Feng, Kan-Too Chokusu, Pik Khan-Tengri see Khan-Tengry Peak, *Kaz., Kyrg.* **179** K17
Khān Yūnis, *Gaza Strip* **173** N4
Khan-Tengry Peak (Khan Tängiri Shyngy, Hantengri Feng, Kan-Too Chokusu, Pik Khan-Tengri), *Kaz., Kyrg.* **179** K17
Khānābād, *Afghan.* **180** F9
Khanai, *Pak.* **180** M7
Khānaqāh, *Afghan.* **180** F6
Khanderi Island, *India* **182** M3
Khandwa, *India* **182** K5
Khandyga, *Russ.* **161** G18
Khanewal, *Pak.* **180** M11
Khangarh, *Pak.* **181** P11
Khaniadhana, *India* **182** J6
Khanka, *Russ.* **161** L21
Khanka, Lake, *Asia* **168** E12
Khanka, Ozero see Xingkai Hu, *China, Russ.* **185** D17
Khanozai, *Pak.* **180** M7
Khanpur, *Pak.* **181** N10
Khanty-Mansiysk, *Russ.* **160** H10
Khapalu, *Pak.* **180** G14
Kharabali, *Russ.* **159** R14
Kharagpur, *India* **182** K11
Kharan, *Pak.* **181** N6
Kharānaq, *Iran* **177** H17

Kharasavey, *Russ.* **160** E11
Khārg Island see Jazīreh-ye Khārk, *Iran* **177** L14
Khargon, *India* **182** K5
Khargūsh, *Taj.* **179** N14
Kharian, *Pak.* **180** K12
Khariar, *India* **182** L8
Kharimkotan, island, *Russ.* **161** H23
Kharkiv, *Ukr.* **159** P9
Kharlovka, *Russ.* **160** D9
Kharmanli, *Bulg.* **155** K15
Kharovsk, *Russ.* **158** F10
Khartoum North see Al Kharţūm Baḩrī, *Sudan* **203** Q12
Khartoum, *Sudan* **203** R12
Kharyyalakh, *Russ.* **161** G15
Khasavyurt, *Russ.* **159** T15
Khāsh, *Iran* **177** L22
Khāsh, *Afghan.* **180** L4
Khāsh, Dasht-e, *Afghan.* **180** K3
Khāsh, river, *Afghan.* **180** K5
Khashm el Qirba Dam, *Sudan* **203** R14
Khashm el Qirba, *Sudan* **203** R14
Khashuri, *Rep. of Georgia* **171** C17
Khasi Hills, *India* **182** H13
Khaskovo, *Bulg.* **155** K15
Khatanga Gulf, *Arctic Oc.* **168** C9
Khatanga, *Russ.* **161** F14
Khatangskiy Zaliv, *Russ.* **161** E15
Khatt, Oued al, *Af.* **198** E2
Khātūnīyah, *Syr.* **172** D15
Khatyrka, *Russ.* **161** C21
Khavda, *India* **182** J2
Khazret Sultan, peak, *Taj., Uzb.* **178** M11
Khe Sanh, *Vietnam* **190** K11
Khemis du Sahel, *Mor.* **200** G5
Khemmarat, *Thai.* **190** L11
Khenchela, *Alg.* **201** C14
Kherson, *Ukr.* **159** R7
Kherwara, *India* **182** J4
Kheta, *Russ.* **161** F14
Kheta, river, *Asia* **168** C9
Khewari, *Pak.* **181** Q8
Khimki, *Russ.* **158** J9
Khinjān, *Afghan.* **180** G8
Khipro, *Pak.* **181** R9
Khiri Ratthanikhom, *Thai.* **191** Q7
Khiwa, *Uzb.* **178** L8
Khlevnoye, *Russ.* **158** M10
Khmel'nyts'kyy, *Ukr.* **159** P4
Khoa, *Laos* **190** H9
Khoārsān, *Iran* **177** F19
Khocho, *Russ.* **161** H18
Khodzhakala, *Turkm.* **178** M6
Khogali, *S. Sudan* **203** X9
Khojak Pass, *Pak.* **180** L7
Khok Kloi, *Thai.* **191** R6
Khokhol'skiy, *Russ.* **158** M10
Khokhropar, *Pak.* **181** R10
Kholm, *Russ.* **158** H6
Kholmogory, *Russ.* **158** B10
Kholmsk, *Russ.* **161** K22
Khomeyn, *Iraq* **177** G14
Khomeynīshahr, *Iran* **177** H15
Khon Kaen, *Thai.* **190** L9
Không, *Laos* **190** M11
Khonj, *Iran* **177** M17
Khoper, river, *Russ.* **158** M12
Khor, *Iran* **177** G18
Khor, river, *Russ.* **161** K21
Khorixas, *Namibia* **212** F3
Khormūj, *Iran* **177** L15
Khorof Harar, *Kenya* **211** R9
Khorol, *Ukr.* **159** P7
Khorramābād, *Iran* **177** G13
Khorramshahr, *Iran* **177** K13
Khorugh, *Taj.* **179** N13
Khost, *Pak.* **180** M7
Khōst, *Afghan.* **180** J9
Khouribga, *Mor.* **200** D7
Khowr-e Mūsá, bay, *Iran* **177** K14
Khoyniki, *Belarus* **158** M5
Khrenovoye, *Russ.* **159** N11
Khromtaū, *Kaz.* **178** E7
Khrysokhou Bay, *Cyprus* **162** P5
Khudabad, *Pak.* **180** F13
Khuiala, *India* **182** G2
Khujand, *Taj.* **178** L12
Khūjayli, *Uzb.* **178** K7
Khulkhuta, *Russ.* **159** R14
Khulm, *Afghan.* **180** F7
Khulna, *Bangladesh* **182** K12
Khunjerab Pass, *China, Pak.* **180** F13
Khunti, *India* **182** K10
Khunzakh, *Russ.* **159** U15

Khurda, *India* **182** L10
Khurmāliq, *Afghan.* **180** K3
Khushab, *Pak.* **180** K11
Khust, *Ukr.* **159** P2
Khuwei, *Sudan* **203** S10
Khuzdar, *Pak.* **181** P6
Khvāf, *Iran* **177** F21
Khvalynsk, *Russ.* **158** L14
Khvānsār, *Iran* **177** H15
Khvorostyanka, *Russ.* **158** L15
Khvoy, *Iran* **176** C11
Khwāl, *Afghan.* **180** G6
Khwāzagak, *Afghan.* **180** H6
Khyber Pass, *Afghan., Pak.* **180** H10
Kia, *Solomon Is.* **228** D8
Kia, island, *Fiji* **226** N6
Kiabakari, *Tanzania* **211** T4
Kiakeana Point, *Hawaii, U.S.* **115** K19
Kiambi, *D.R.C.* **209** J18
Kiamichi, river, *Okla., U.S.* **105** E19
Kiana, *Alas., U.S.* **113** E13
Kiatak (Northumberland Ø), *Greenland, Den.* **88** E3
Kiáto, *Gr.* **156** J11
Kibaha, *Tanzania* **211** W8
Kibali, river, *D.R.C.* **209** C18
Kibangou, *Congo* **208** G9
Kibau, *Tanzania* **211** X5
Kibaya, *Tanzania* **211** V6
Kiberashi, *Tanzania* **211** V7
Kiberege, *Tanzania* **211** X6
Kibiti, *Tanzania* **211** X7
Kibombo, *D.R.C.* **209** G17
Kibondo, *Tanzania* **211** U2
Kibong, *N. Korea* **186** H4
Kibre Mengist, *Eth.* **211** P8
Kibungo, *Rwanda* **211** T2
Kibuye, *Rwanda* **211** T2
Kibwezi, *Kenya* **211** T7
Kičevo, *N. Maced.* **154** K10
Kichmengskiy Gorodok, *Russ.* **158** F12
Kidal, *Mali* **205** F17
Kidatu, *Tanzania* **211** X6
Kidi Faani, *Somalia* **211** T9
Kidika, *Tanzania* **211** W6
Kidira, *Senegal* **204** J9
Kidong-man, bay, *N. Korea* **186** C13
Kidrenen, island, *Marshall Is.* **227** A20
Kidrinen, island, *Marshall Is.* **227** B20
Kiel, *Wis., U.S.* **100** K7
Kiel, *Ger.* **150** B8
Kielce, *Pol.* **151** F16
Kieler Bucht, *Den., Ger.* **150** A8
Kiến An, *Vietnam* **190** H11
Kiến Đức, *Vietnam* **191** N12
Kiến Giang, *Vietnam* **190** K12
Kieta, *P.N.G.* **228** C5
Kiev see Kyiv, *Ukr.* **159** N6
Kifaya, *Guinea* **204** K8
Kiffa, *Mauritania* **204** H9
Kifisiá, *Gr.* **156** J12
Kifisós, river, *Gr.* **156** G10
Kigali, *Rwanda* **211** T2
Kigille, *S. Sudan* **203** V13
Kiǧi, *Turk.* **171** F15
Kigoma, *Tanzania* **211** V2
Kīhei, *Hawaii, U.S.* **115** K19
Kīholo Bay, *Hawaii, U.S.* **115** M20
Kii Channel, *Pac. Oc.* **168** F13
Kii Suidō, *Japan* **189** Q8
Ki'i, *Hawaii, U.S.* **114** F8
Kikaiga Shima, *Japan* **189** W4
Kikiakki, *Russ.* **160** H12
Kikinda, *Serb.* **154** D10
Kikki, *Pak.* **181** R3
Kiknur, *Russ.* **158** H13
Kikombo, *Tanzania* **211** W6
Kikondja, *D.R.C.* **209** K17
Kikori, *P.N.G.* **222** Q6
Kikori, river, *P.N.G.* **222** P6
Kikwit, *D.R.C.* **208** H12
Kil, river, *Pak.* **181** R4
Kilafors, *Sw.* **143** J13
Kilakkarai, *India* **183** T6
Kilambé, peak, *Nicar.* **117** L18
Kilar, *India* **180** J14
Kilarney Park, *Qnsld., Austral.* **221** U13
Kīlauea Caldera, *Hawaii, U.S.* **115** P22
Kīlauea, *Hawaii, U.S.* **114** E10
Kilbrannan Sound, *Scot., U.K.* **145** M8
Kilbuck Mountains, *Alas., U.S.* **113** J13
Kilcar see Cill Charthaigh, *Ire.* **145** N4
Kilchoman, *Scot., U.K.* **144** L7
Kilcock (Cill Choca), *Ire.* **145** R6
Kilcowera, *Qnsld., Austral.* **221** V12

Kildonan, *Scot., U.K.* **144** G9
Kildrummy, *Scot., U.K.* **144** J10
Kilembe, *D.R.C.* **209** H13
Kilen, *Greenland, Den.* **88** B12
Kilgore, *Tex., U.S.* **105** G20
Kilgore, *Nebr., U.S.* **106** M4
Kili Island, *Marshall Is.* **227** E16
Kilifi, *Kenya* **211** U8
Kilift, *Afghan.* **180** F6
Kilimanjaro, *Tanzania* **211** U7
Kilinailau Islands, *P.N.G.* **222** N11
Kilinailau Islands, *P.N.G.* **228** A5
Kilindoni, *Tanzania* **211** X8
Kilinochchi, *Sri Lanka* **183** T7
Kilis, *Turk.* **170** J12
Kiliya, *Ukr.* **159** S5
Kilju, *N. Korea* **186** E12
Kilkee (Cill Chaoi), *Ire.* **145** S3
Kilkenny (Cill Chainnigh), *Ire.* **145** S6
Kilkís, *Gr.* **156** B11
Killadysert (Cill an Dísirt), *Ire.* **145** S3
Killala Bay, *Ire.* **145** P4
Killarney, *N.T., Austral.* **221** Q8
Killary Harbour, *Ire.* **145** Q3
Killdeer, *N. Dak., U.S.* **106** F3
Killeen, *Tex., U.S.* **105** J16
Killiniq Island, *Nfld. & Lab., Can.* **87** J20
Killorglin (Cill Orglan), *Ire.* **145** T3
Kilmaluag, *Scot., U.K.* **144** H7
Kilmarnock, *Scot., U.K.* **144** M9
Kílnovec, peak, *Czechia* **150** G10
Kilombero, river, *Tanzania* **211** X6
Kilosa, *Tanzania* **211** W6
Kiltan Island, *India* **183** R3
Kilwa Kivinje, *Tanzania* **211** X8
Kilwa Masoko, *Tanzania* **211** X8
Kilwa, *D.R.C.* **209** K18
Kim, *Colo., U.S.* **109** P15
Kimaam, *Indonesia* **193** M20
Kimasozero, *Russ.* **158** B7
Kimball, *S. Dak., U.S.* **106** L6
Kimball, *Nebr., U.S.* **107** P1
Kimbe Bay, *P.N.G.* **222** P9
Kimberley Plateau, *Austral.* **220** C6
Kimberley, *B.C., Can.* **86** M9
Kimberley, *S. Af.* **212** K7
Kimberley, Cape, *Austral.* **220** C13
Kimberley, region, *W. Aust., Austral.* **221** R6
Kimberly, *Idaho, U.S.* **108** H5
Kimbirila, *Côte d'Ivoire* **204** L12
Kimboto, *Congo* **208** G10
Kimch'aek (Sŏngjin), *N. Korea* **186** F12
Kimhwa (Kŭmsŏng), *N. Korea* **186** L9
Kimhyŏnggwŏn, *N. Korea* **186** F10
Kimjŏngsuk (Sinp'a), *N. Korea* **186** D9
Kimmirut, *Nunavut, Can.* **87** H18
Kimobetsu, *Japan* **188** G12
Kímolos, island, *Gr.* **157** M14
Kimovsk, *Russ.* **158** K10
Kimpese, *D.R.C.* **208** H10
Kimry, *Russ.* **158** H9
Kinabalu, peak, *Malaysia* **192** G12
Kinadyeng, island, *Marshall Is.* **227** H20
Kínaros, island, *Gr.* **157** L17
Kinbrace, *Scot., U.K.* **144** G9
Kinda, *D.R.C.* **209** K16
Kindadal, *Indonesia* **193** K14
Kindanba, *Congo* **208** G10
Kinder, *La., U.S.* **102** P7
Kindia, *Guinea* **204** M8
Kindu, *D.R.C.* **209** F17
Kinel', *Russ.* **158** K15
Kineshma, *Russ.* **158** H11
King City, *Mo., U.S.* **107** R10
King City, *Calif., U.S.* **111** U5
King Cove, *Alas., U.S.* **112** N11
King George Bay, *Falk. Is., U.K.* **135** W11
King George Island, *Antarctica* **238** C4
King George Islands, *Nunavut, Can.* **87** K17
King George Sound, *Austral.* **220** L3
King George V Memorial Hospital, *Gibraltar, U.K.* **163** P23
King Island, *Alas., U.S.* **112** F11
King Island, *Austral.* **220** L2
King Khalid Military City, *Saudi Arabia* **174** E10
King Lear Peak, *Nev., U.S.* **110** M8
King Leopold Ranges, *Austral.* **220** D5
King Mountain, *Tex., U.S.* **104** J11

King Peninsula, *Antarctica* **238** K7
King Salmon, *Alas., U.S.* **113** L14
King Sejong, station, *Antarctica* **238** C4
King Sound, *Austral.* **220** C5
King William Island, *Nunavut, Can.* **87** F14
King William's Town, *S. Af.* **212** M8
Kingaroy, *Qnsld., Austral.* **221** V15
Kingfisher Lake, *Ont., Can.* **87** M15
Kingfisher, *Okla., U.S.* **105** C16
Kingisepp, *Russ.* **158** F5
Kingman Basin, *Pac. Oc.* **246** J12
Kingman Reef, *Pac. Oc.* **247** J13
Kingman, *Kans., U.S.* **107** U7
Kingman, *Ariz., U.S.* **109** S5
Kingoonya, *S. Aust., Austral.* **221** W9
Kingri, *Pak.* **180** M9
King's Bay, *Trin. & Tobago* **123** N18
Kings Canyon National Park, *Calif., U.S.* **111** T7
King's Lynn, *Eng., U.K.* **145** S14
Kings Mountain, *N.C., U.S.* **98** J8
Kings Peak, *Utah, U.S.* **108** L8
Kings Trough, *Atl. Oc.* **248** J9
Kings, river, *Calif., U.S.* **111** T7
Kingsbridge, *Eng., U.K.* **145** W10
Kingsburg, *Calif., U.S.* **111** U6
Kingscote, *S. Aust., Austral.* **221** Y10
Kingsford, *Mich., U.S.* **100** G7
Kingsland, *Ga., U.S.* **99** Q8
Kingsport, *Tenn., U.S.* **103** D20
Kingston Harbour, *Jam.* **122** J10
Kingston South East, *S. Aust., Austral.* **221** Y11
Kingston upon Hull, *Eng., U.K.* **145** Q13
Kingston, *Ont., Can.* **87** Q19
Kingston, *N.Y., U.S.* **96** L11
Kingston, *Pa., U.S.* **96** M8
Kingston, *Jam.* **122** J10
Kingston, *Austral.* **229** B20
Kingstown Bay, *St. Vincent & the Grenadines* **123** L16
Kingstown, *St. Vincent & the Grenadines* **123** L16
Kingstree, *S.C., U.S.* **98** L10
Kingsville, *Tex., U.S.* **105** P16
Kingungi, *D.R.C.* **208** H12
Kiniama, *D.R.C.* **209** M18
Kinlochewe, *Scot., U.K.* **144** H8
Kinmen (Quemoy), *China* **185** N15
Kinnear, *Wyo., U.S.* **108** H10
Kinniyai, *Sri Lanka* **183** T8
Kinshasa (Léopoldville), *D.R.C.* **208** G10
Kinsley, *Kans., U.S.* **107** U5
Kinston, *N.C., U.S.* **98** H13
Kintampo, *Ghana* **205** N15
Kintore, *Scot., U.K.* **144** J11
Kintyre, peninsula, *Scot., U.K.* **145** M8
Kinvara (Cinn Mhara), *Ire.* **145** R4
Kinyangiri, *Tanzania* **211** V5
Kinyeti, peak, *S. Sudan* **203** Y12
Kinzua, *Oreg., U.S.* **110** G6
Kioa, island, *Fiji* **226** N7
Kiokluk Mountains, *Alas., U.S.* **113** J14
Kiomboi, *Tanzania* **211** V5
Kione Maghal see Maughold Head, *Isle of Man, U.K.* **145** P9
Kiowa, *Kans., U.S.* **107** V6
Kiowa, *Colo., U.S.* **108** M14
Kīpahulu, *Hawaii, U.S.* **115** K20
Kipembawe, *Tanzania* **211** X4
Kipili, *Tanzania* **211** X2
Kipilingu, *D.R.C.* **209** N18
Kipini, *Kenya* **211** U8
Kipnuk, *Alas., U.S.* **112** K11
Kipushi, *D.R.C.* **209** M17
Kipushia, *D.R.C.* **209** M19
Kirakira, *Solomon Is.* **228** G12
Kiraz, *Turk.* **170** G4
Kirbey, *Russ.* **161** G15
Kirbymoorside, *Eng., U.K.* **145** P13
Kirbyville, *Tex., U.S.* **105** K21
Kirensk, *Russ.* **161** K15
Kiri, *D.R.C.* **209** F13
Kiribati, *Pac. Oc.* **224** G10
Kiridh, *Somaliland* **210** M12
Kirin-do, island, *S. Korea* **186** M5
Kirishi, *Russ.* **158** F7
Kiritimati (Christmas Island), *Kiribati* **224** G12
Kiriwina, island, *P.N.G.* **222** Q9
Kirk Michael (Skeylley Maayl), *Isle of Man, U.K.* **145** P9

Kirkcaldy, *Scot., U.K.* **144** L10
Kirkcolm, *Scot., U.K.* **145** N8
Kirkenes, *Nor.* **143** B15
Kirkjubæjarklaustur, *Ice.* **142** F3
Kirkland Lake, *Ont., Can.* **87** P18
Kirkland, *Tex., U.S.* **105** E14
Kirkoswald, *Scot., U.K.* **145** M8
Kirksville, *Mo., U.S.* **107** R12
Kirkuk see Karkūk, *Iraq* **176** F9
Kirkwall, *Scot., U.K.* **144** F10
Kirkwood, *Mo., U.S.* **107** T15
Kirn, *Ger.* **150** H5
Kirotshe, *D.R.C.* **209** F18
Kirov, *Russ.* **158** K8
Kirov, *Russ.* **158** F14
Kirovo Chepetsk, *Russ.* **158** F14
Kirovohrad, *Ukr.* **159** Q6
Kirovskiy, *Russ.* **159** R15
Kirs, *Russ.* **158** F15
Kirsanov, *Russ.* **158** L12
Kīrthar Range, *Pak.* **181** P7
Kirtland, *N. Mex., U.S.* **109** Q10
Kiruna, *Sw.* **143** D13
Kirunu, island, *Marshall Is.* **227** A20
Kirwan Escarpment, *Antarctica* **238** C12
Kirya, *Russ.* **158** J13
Kiryū, *Japan* **189** N11
Kisaki, *Tanzania* **211** W7
Kisangani, *D.R.C.* **209** D16
Kisangire, *Tanzania* **211** W7
Kisengwa, *D.R.C.* **209** H17
Kish, *Iran* **177** N17
Kish, island, *Iran* **177** N17
Kishanganj, *India* **182** H11
Kishangarh, *India* **182** G2
Kishangarh, *India* **182** H4
Kishi, *Nigeria* **206** K6
Kishim, *Afghan.* **180** F9
Kishiwada, *Japan* **189** Q8
Kishkeneköl, *Kaz.* **179** C13
Kisigo, river, *Tanzania* **211** W5
Kisii, *Kenya* **211** S5
Kisiju, *Tanzania* **211** W8
Kiska Island, *Alas., U.S.* **112** M2
Kiskunfélegyháza, *Hung.* **151** M15
Kiskunhalas, *Hung.* **151** M15
Kislovodsk, *Russ.* **159** T12
Kismaayo (Kismayo), *Somalia* **211** S10
Kismayo see Kismaayo, *Somalia* **211** S10
Kiso, river, *Japan* **189** P10
Kisoro, *Uganda* **211** T2
Kíssamos (Kastélli), *Gr.* **156** P12
Kissámou, Kólpos, *Gr.* **156** P12
Kissidougou, *Guinea* **204** M10
Kissimmee, *Fla., U.S.* **99** T9
Kissimmee, Lake, *Fla., U.S.* **99** U9
Kissimmee, river, *Fla., U.S.* **99** U9
Kistrand, *Nor.* **143** B14
Kisumu, *Kenya* **211** S5
Kisvárda, *Hung.* **151** K17
Kiswere, *Tanzania* **211** Y8
Kit Carson, *Colo., U.S.* **109** N15
Kita Daitō Jima, *Japan* **224** C2
Kita Iwo To, *Japan* **224** C3
Kita Iwo, island, *Japan* **189** Y13
Kita, *Mali* **204** K10
Kitaibaraki, *Japan* **189** N12
Kitakami, *Japan* **188** K13
Kitakata, *Japan* **188** M12
Kitakyūshū, *Japan* **189** Q5
Kitale, *Kenya* **211** R5
Kitami Sanchi, *Japan* **188** F14
Kitami, *Japan* **188** F14
Kit'an, *N. Korea* **186** L7
Kitchener, *Ont., Can.* **87** R18
Kitchener, *W. Aust., Austral.* **221** W5
Kitenda, *D.R.C.* **208** J12
Kitgum, *Uganda* **211** Q4
Kiti Point, *F.S.M.* **226** K8
Kiti, Cape, *Cyprus* **162** Q9
Kitimat, *B.C., Can.* **86** K7
Kitob, *Uzb.* **178** M11
Kitombe, *D.R.C.* **209** H13
Kitridge Point, *Barbados* **123** K20
Kittanning, *Pa., U.S.* **96** N4
Kittery, *Me., U.S.* **97** J15
Kittilä, *Fin.* **143** D14
Kittitas, *Wash., U.S.* **110** D6
Kittorsaq, *Greenland, Den.* **88** H5
Kitty Hawk, *N.C., U.S.* **98** G15
Kitui, *Kenya* **211** T7
Kitunda, *Tanzania* **211** W4
Kitwe, *Zambia* **209** M18

Kitzingen, *Ger.* **150** H7
Kiunga, *P.N.G.* **222** P5
Kivalina, *Alas., U.S.* **112** D12
Kivioq Fjord, *Greenland, Den.* **88** M13
Kivumba, *Tanzania* **211** V2
Kiwai, island, *P.N.G.* **222** Q6
Kiyan Misaki, *Japan* **189** Y2
Kiyāt, *Saudi Arabia* **174** L7
Kizema, *Russ.* **158** E12
Kizema, *Russ.* **160** F8
Kizlyar, *Russ.* **159** T15
Kizlyarskiy Zaliv, *Russ.* **159** S15
Kılıçkaya, *Turk.* **171** D16
Kılınç Tepesi, peak, *Turk.* **171** D13
Kırıkhan, *Turk.* **170** J11
Kırıkkale, *Turk.* **170** E9
Kırka, *Turk.* **170** F6
Kırkağaç, *Turk.* **170** E3
Kırkgeçit, *Turk.* **171** G18
Kırklar Dağı, peak, *Turk.* **171** D15
Kırklar see Tymvou, *Cyprus* **162** N9
Kırklareli, *Turk.* **170** B3
Kırşehir, *Turk.* **170** F9
Kısır Dağı, peak, *Turk.* **171** D17
Kıyıköy, *Turk.* **170** C4
Kızılay see Trakhonas, *Cyprus* **162** N8
Kızılcahamam, *Turk.* **170** D7
Kızılın, *Turk.* **171** H13
Kızılırmak, *Turk.* **170** D9
Kızılırmak, river, *Turk.* **170** F11
Kızılliman Burnu, *Turk.* **170** J8
Kızıltepe, *Turk.* **171** H15
Kjer Gletscher, glacier, *Greenland, Den.* **88** G5
Kladno, *Czechia* **150** G10
Klagenfurt, *Aust.* **150** M11
Klaipėda, *Lith.* **143** M15
Klaksvík, *Faroe Is., Den.* **142** H6
Klamath Falls, *Oreg., U.S.* **110** K4
Klamath Mountains, *Calif., U.S.* **110** M2
Klamath, *Calif., U.S.* **110** L2
Klamath, river, *Calif., Oreg., U.S.* **110** L2
Klang, *Malaysia* **191** U8
Klarälven, river, *Sw.* **142** J12
Klatovy, *Czechia* **150** H10
Klawer, *S. Af.* **212** L5
Klawock, *Alas., U.S.* **113** M23
Klein Bay, *St. Maarten, Neth.* **123** C14
Klein Bonaire, *Bonaire, Neth.* **123** Q19
Kleine Berg, peak, *Curaçao, Neth.* **123** P14
Kleitoría (Klitória), *Gr.* **156** J10
Klekovača, peak, *Bosn. & Herzg.* **154** F6
Kléla, *Mali* **205** L13
Klenova Valley, *Arctic Oc.* **253** K13
Klerksdorp, *S. Af.* **212** J8
Kleshchevo, *Russ.* **158** C10
Klesiv, *Ukr.* **158** M4
Kletskiy, *Russ.* **159** P12
Klichaw, *Belarus* **158** L5
Klickitat, *Wash., U.S.* **110** E5
Klin, *Russ.* **158** J9
Klintehamn, *Sw.* **143** L14
Klintsy, *Belarus* **158** L7
Klirou, *Cyprus* **162** P8
Klitória see Kleitoría, *Gr.* **156** J10
Kłodzko, *Pol.* **151** G12
Klondike Gold Rush National Historical Park, *Alas., U.S.* **113** K22
Klosterneuburg, *Aust.* **150** K12
Klosters, *Switz.* **152** B5
Kluczbork, *Pol.* **151** F14
Klyuchevskaya Sopka, *Russ.* **161** F22
Knaresborough, *Eng., U.K.* **145** Q12
Knezha, *Bulg.* **155** H11
Knife River, *Minn., U.S.* **106** G13
Knin, *Croatia* **154** G6
Knipovich Ridge, *Arctic Oc.* **253** H17
Knittelfeld, *Aust.* **150** L11
Knjaževac, *Serb.* **154** H12
Knob, Cape, *Austral.* **220** L4
Knockadoon Head, *Ire.* **145** T5
Knokke-Heist, *Belg.* **148** K11
Knossós (Cnossus), ruin, *Gr.* **157** Q15
Knotts Island, *N.C., U.S.* **98** G14
Knox Atoll, *Marshall Is.* **227** E18
Knox City, *Tex., U.S.* **105** F14
Knox Coast, *Antarctica* **239** L20
Knox, *Ind., U.S.* **101** P8
Knoxville, *Tenn., U.S.* **103** E18
Knoxville, *Iowa, U.S.* **107** P12
Knud Rasmussen Land, *Greenland, Den.* **88** C8
Knysna, *S. Af.* **212** M6

Ko Chang, island, *Thai.* **191** N9
Ko Gaja Shima, *Japan* **189** U3
Ko Jima, *Japan* **189** R12
Ko Kut, island, *Cambodia* **191** P9
Ko Rawi, island, *Thai.* **191** S7
Kō Saki, *Japan* **189** Q3
Ko Takara Jima, *Japan* **189** V3
Ko Tarutao, island, *Thai.* **191** S7
Ko, *Russ.* **161** K21
Koala, *Burkina Faso* **205** K16
Koam, *N. Korea* **186** J9
Koba, *Indonesia* **192** K8
Koba, *Indonesia* **193** M18
Kobanê see 'Ayn al 'Arab, *Syr.* **172** C10
Kobarid, *Slov.* **154** C4
Kobayashi, *Japan* **189** S5
Kobberminebugt see Alanngorsuaq, *Greenland, Den.* **89** T6
Kōbe, *Japan* **189** Q8
Kobo, *India* **182** G15
K'obo, *Eth.* **210** K8
Kobowen Swamp, *S. Sudan* **203** X13
Kobozha, *Russ.* **158** G8
Kobroor, island, *Indonesia* **193** L19
Kobryn, *Belarus* **158** L3
Kobuk Valley National Park, *Alas., U.S.* **113** E14
Kobuk, *Alas., U.S.* **113** E14
Kobuk, river, *Alas., U.S.* **113** E14
K'obulet'i, *Rep. of Georgia* **171** C16
Kocaeli see İzmit, *Turk.* **170** D5
Kočani, *N. Maced.* **154** K12
Koçarlı, *Turk.* **170** G3
Koçbaşı Tepe, peak, *Turk.* **171** F17
Koch Bihar, *India* **182** H12
Koch Island, *Nunavut, Can.* **87** F17
Koch Peak, *Mont., U.S.* **108** F8
Kochen'ga, *Russ.* **158** F12
Kochi (Cochin), *India* **183** S5
Kōchi, *Japan* **189** R7
Kochkor, *Kyrg.* **179** K15
Kochubey, *Russ.* **159** T15
Kodari, *Nepal* **182** G10
Kodiak Island, *Alas., U.S.* **113** M15
Kodiak, *Alas., U.S.* **113** M15
Kodinar, *India* **182** L2
Kodino, *Russ.* **158** C10
Kodinskiy, *Russ.* **161** J14
Kodok, *S. Sudan* **203** U12
Kodomari, *Japan* **188** J12
Köes, *Namibia* **212** J5
Koetoi, *Japan* **188** E13
Kofili, *Mali* **204** J11
Köflach, *Aust.* **150** M11
Kofizsdorp see Koforidua
Koforidua, *Ghana* **205** P16
Kōfu, *Japan* **189** P11
Koga, *Japan* **189** N12
Kogaluc, river, *Que., Can.* **87** K18
Kögart, *Kyrg.* **179** L14
Kogon, *Uzb.* **178** M10
Kogon, river, *Guinea* **204** L8
Koh Kong see Kaôh Kŏng, *Thai.* **191** P9
Kohala Mountains, *Hawaii, U.S.* **115** M21
Kohala, *Hawaii, U.S.* **115** M20
Kohat, *Pak.* **180** J10
Kohima, *India* **182** H15
Kohistan, region, *Pak.* **180** G11
Kohler Range, *Antarctica* **238** L8
Kohlu, *Pak.* **180** M9
Kohnen Station, *Antarctica* **239** C13
Köhne Bilgəh Burnu, *Azerb.* **171** D23
Kohtla-Järve, *Est.* **143** J17
Koidu, *Sierra Leone* **204** N10
Koilani, *Cyprus* **162** Q7
Koimbani, *Comoros* **215** M14
Koin, *N. Korea* **186** J9
Kojonup, *W. Aust., Austral.* **221** X3
Kök-Aygyr, *Kyrg.* **179** L15
Kokas, *Indonesia* **193** K18
Kokatha, *S. Aust., Austral.* **221** W9
Kokka, *Sudan* **203** N11
Kokkina (Erenköy), *Cyprus* **162** N5
Kokkola, *Fin.* **143** G15
Koko Crater, *Hawaii, U.S.* **115** H15
Koko Head, *Hawaii, U.S.* **115** H15
Koko, *Nigeria* **206** J6
Koko, *Eth.* **210** L6
Kokoda, *P.N.G.* **222** Q8
Kokomo, *Ind., U.S.* **101** Q9

Kökpekti, *Kaz.* **179** F17
Koksan, *N. Korea* **186** K7
Kökshetaū, *Kaz.* **178** C12
Kokstad, *S. Af.* **212** L9.
Köktal, *Kaz.* **179** J16
Kola Peninsula, *Eur.* **140** B10
Kola, island, *Indonesia* **193** L19
Kola, river, *Eur.* **140** A10
Kolachi, river, *Pak.* **181** P7
Kolaka, *Indonesia* **193** K14
Kolangui, *Guinea* **204** L9
Kolar, *India* **183** Q6
Kolari, *Fin.* **143** D14
Kolarovgrad *see* Shumen, *Bulg.*
155 H16
Kolayat, *India* **182** G3
Kolbano, *Indonesia* **193** N15
Kolbeinsey Ridge, *Arctic Oc.* **253** K19
Kolbio, *Kenya* **211** T9
Kolda, *Senegal* **204** K7
Kolding, *Den.* **142** N11
Kole, *D.R.C.* **209** C16
Kolé, *Mali* **204** K11
Kolé, *Vanuatu* **229** F14
Kolekole, river, *Hawaii, U.S.* **115** M22
Kolguyev Island, *Asia* **168** C7
Kolguyev, *Russ.* **160** E10
Kolhapur, *India* **182** N4
Kolī, *Eth.* **210** L6
Kolia, *Wallis & Futuna Is., Fr.* **230** Q11
Koliganek, *Alas., U.S.* **113** K14
Kolín, *Czechia* **150** H12
Kolkata (Calcutta), *India* **182** K12
Kollam (Quilon), *India* **183** T5
Kollidam, river, *India* **183** S6
Kolluáll, valley, *Atl. Oc.* **253** M21
Kolmogorovo, *Russ.* **161** J13
Köln (Cologne), *Ger.* **150** F5
Kolo, *Niger* **206** H5
Koło, *Pol.* **151** E14
Kōloa, *Hawaii, U.S.* **114** F10
Koloa, island, *Tonga* **227** H23
Kołobrzeg, *Pol.* **150** B12
Kolodnya, *Russ.* **158** K7
Kologi, *Sudan* **203** U11
Kolokani, *Mali* **204** K11
Kologriv, *Russ.* **158** F12
Kolokani, *Mali* **204** K11
Kololo, *Eth.* **211** N9
Kolombangara, island, *Solomon Is.*
228 D7
Kolomna, *Russ.* **158** K10
Kolomyya, *Ukr.* **159** P3
Kolondiéba, *Mali* **204** L12
Kolonga, *Tonga* **227** Q24
Kolonia, *F.S.M.* **226** J8
Kolonodale, *Indonesia* **193** K14
Kolovai, *Tonga* **227** Q22
Kolpashevo, *Russ.* **160** J12
Kolpino, *Russ.* **158** F6
Kol'skiy Poluostrov, *Russ.* **160** D8
Kolufuri, island, *Maldives* **183** W3
Kolumadulu Atoll, *Maldives* **183** W3
Kolwezi, *D.R.C.* **209** L16
Kolyma Lowland, *Asia* **168** B11
Kolyma Range, *Asia* **168** B12
Kolyma Valley, *Arctic Oc.* **252** G6
Kolyma, river, *Asia* **168** C12
Kolymskoye Nagor'ye (Gydan), *Russ.*
161 E20
Kom, peak, *Bulg.* **155** H13
Komagane, *Japan* **189** P10
Komandorskiye Ostrova
(Commander Islands), *Russ.*
161 E23
Komani, *S. Af.* **212** L8
Komárno, *Slovakia* **151** L14
Komatsu, *Japan* **189** N9
Kŏmdŏk-san, peak, *N. Korea* **186** E11
Kŏmdŏk, *N. Korea* **186** E11
Komo, island, *Fiji* **226** Q8
Komodo, island, *Indonesia* **193** M13
Komono, *Congo* **208** G9
Komoran, island, *Indonesia* **193** M20
Komotiní, *Gr.* **157** B15
Komovi, peak, *Montenegro* **154** J9
Kompóti, *Gr.* **156** F8
Komsomol, *Kaz.* **178** G5
Komsomolets, *Russ.* **161** C14
Komsomolets, *Kaz.* **178** C9
Komsomol'sk na Amure, *Russ.*
161 K20
Komsomol'sk-na-Ustyurte, *Uzb.*
178 J7
Komsomol'sk, *Russ.* **158** H10
Komsomol'skiy, *Russ.* **161** C19
Komusan, *N. Korea* **186** C12

Kon Tum, *Vietnam* **190** M12
Kona, *Hawaii, U.S.* **115** N20
Kona, *Mali* **205** J14
Konakovo, *Russ.* **158** H9
Konakpınar, *Turk.* **170** F12
Konawa, *Okla., U.S.* **105** D17
Konda, *Indonesia* **193** K18
Kondagaon, *India* **182** M8
Kondoa, *Tanzania* **211** V6
Kondol', *Russ.* **158** L13
Kondolole, *D.R.C.* **209** D17
Kondopoga, *Russ.* **158** D8
Kondopoga, *Russ.* **160** E7
Kondrovo, *Russ.* **158** K8
Konduga, *Nigeria* **206** J11
Kondukur, *India* **183** P7
Koné, *New Caledonia, Fr.* **231** P21
Konetsbor, *Russ.* **158** A16
Köneürgench, *Turkm.* **178** K7
Konevo, *Russ.* **158** D10
Kong Christian IX Land,
Greenland, Den. **89** N11
Kong Christian X Land,
Greenland, Den. **88** J11
Kong Frederik IX Land,
Greenland, Den. **89** N7
Kong Frederik VIII Land,
Greenland, Den. **88** F11
Kong Karls Basin, *Arctic Oc.* **253** G16
Kong Kyst Frederik VI,
Greenland, Den. **89** S8
Kong Oscar Fjord, *Greenland, Den.*
88 J13
Kong Oscar Glacier, *Greenland, Den.*
88 F5
Kong Wilhelm Land, *Greenland, Den.*
88 G13
Kong, river, *Laos* **190** L12
Kongauru, island, *Palau* **226** J4
Kongiganak, *Alas., U.S.* **112** K12
Konginskiy Khrebet, *Russ.* **161** D20
Konglu, *Myanmar* **190** D6
Kongolo, *D.R.C.* **209** H17
Kongsberg, *Nor.* **142** K11
Kongsvinger, *Nor.* **142** K12
Kongur Shan, peak, *Asia* **168** G7
Kongwa, *Tanzania* **211** W6
Konibodom, *Taj.* **178** L12
Königswinter, *Ger.* **150** F5
Konin, *Pol.* **151** E14
Konispol, *Alban.* **154** N10
Konjuh, peak, *Bosn. & Herzg.* **154** F8
Konosha, *Russ.* **158** E10
Konosha, *Russ.* **158** E7
Konotop, *Ukr.* **159** N7
Konqi, river, *China* **184** F5
Konrai, *Palau* **226** F6
Końskie, *Pol.* **151** F16
Konso, *Eth.* **211** P7
Konstantinov Ridge, *Atl. Oc.* **249** T8
Konstantinovsk, *Russ.* **159** Q11
Konstanz, *Ger.* **150** L6
Konta, *India* **182** N8
Kontagora, *Nigeria* **206** K7
Kontcha, *Cameroon* **206** L10
Konya (Iconium), *Turk.* **170** G8
Ko'oko'olau, peak, *Hawaii, U.S.*
115 N21
Kookooligit Mountains, *Alas., U.S.*
112 F10
Kookynie, *W. Aust., Austral.* **221** V4
Ko'olau Range, *Hawaii, U.S.* **115** G14
Koolyanobbing, *W. Aust., Austral.*
221 W4
Koomooloo, *S. Aust., Austral.* **221** X11
Koonalda, *S. Aust., Austral.* **221** W7
Kootenay, river, *Can., U.S.* **108** A4
Kopaonik, mountains, *Kos., Serb.*
154 H10
Kópasker, *Ice.* **142** E4
Kopbirlik, *Kaz.* **179** G15
Köpekkayası Burnu, *Turk.* **170** C8
Koper (Capodistria), *Slov.* **154** D4
Kopervik, *Nor.* **142** K9
Kopet Dag, mountains, *Asia* **168** F5
Kopište, island, *Croatia* **154** J6
Koppal, *India* **183** P5
Koprivnica, *Croatia* **154** D7
Köprü, river, *Turk.* **170** H6
Kopstal, *Lux.* **162** J10
Kop'ung, *N. Korea* **186** F6
Kopylovka, *Russ.* **160** J12
Korab, Mal., *Alban., N. Maced.* **154** K10
K'orahē, *Eth.* **211** N11
Korak, island, *Palau* **226** H6
Korangi, *Pak.* **181** S7

Korba, *India* **182** K8
Korbol, *Chad* **207** K13
Korçë, *Alban.* **154** M10
Korčula, island, *Croatia* **154** H6
Korčulanski Kanal, *Croatia* **154** H6
Korday, *Kaz.* **179** K14
Koré Mayroua, *Niger* **206** H6
Korea Bay, *China, N. Korea* **185** G15
Korea Strait, *Japan, S. Korea* **187** U11
Korea, *Asia* **168** F12
Korem, *Eth.* **210** K8
Korenovsk, *Russ.* **159** S10
Korf, *Russ.* **161** D21
Korff Ice Rise, *Antarctica* **238** G9
Korhogo, *Côte d'Ivoire* **205** M13
Kori Creek, *India* **182** J1
Korido, *Indonesia* **193** J19
Korinós, *Gr.* **156** D10
Korinthiakós Kólpos, *Gr.* **156** H10
Kórinthos (Corinth), *Gr.* **156** J11
Korissía, *Gr.* **157** K13
Kōriyama, *Japan* **188** M12
Korkuteli, *Turk.* **170** H5
Korla, *China* **184** F5
Kormacit *see* Kormakiti, *Cyprus*
162 N7
Kormakiti (Kormacit), *Cyprus* **162** N7
Kormakitis, Cape , *Cyprus* **162** N7
Körmend, *Hung.* **151** M13
Kornat, island, *Croatia* **154** G5
Koro Sea, *Fiji* **226** P7
Koro Toro, *Chad* **207** G14
Koro, *Mali* **205** J14
Koro, island, *Fiji* **226** P6
Koroba, *P.N.G.* **222** P6
Korogwe, *Tanzania* **211** V7
Korolevu, *Fiji* **226** Q5
Koromiri, island, *N.Z.* **229** E23
Koróneia, Límni, *Gr.* **156** C11
Koronga, *Mauritania* **204** H12
Koróni, *Gr.* **156** M9
Koronída, *Gr.* **157** L15
Koropí, *Gr.* **156** J12
Koror (Oreor), island, *Palau* **226** H6
Koror, *Palau* **226** H5
Korosten', *Ukr.* **159** N5
Korotkova, *Russ.* **161** K15
Korotoyak, *Russ.* **159** N10
Korsnes, *Nor.* **142** D12
Korti, *Sudan* **203** P11
Kortkeros, *Russ.* **158** D14
Kortrijk, *Belg.* **148** L11
Koryak Range, *N. Amer.* **84** P2
Koryak Range, *Asia* **168** B12
Koryakskaya Sopka, *Russ.* **161** F23
Koryakskoye Nagor'ye, *Russ.* **161** D21
Koryfí Fengári, peak, *Gr.* **157** D15
Kos (Cos), island, *Gr.* **157** M18
Kos, *Gr.* **157** L18
Kosa, *Russ.* **158** E16
Kosa, *Eth.* **210** M6
Kosan, *N. Korea* **186** K9
Kosaya Gora, *Russ.* **158** K9
Kościerzyna, *Pol.* **151** B14
Kosciusko, *Miss., U.S.* **102** K12
Kosciuszko, Mount, *Austral.* **220** L13
Koshikijima Rettō, *Japan* **189** S4
Köshim, river, *Kaz.* **178** E4
Koshk, *Afghan.* **180** H3
Košice, *Slovakia* **151** J17
Koslan, *Russ.* **158** C13
Kosmás, *Gr.* **156** K11
Kosmolets Island, *Asia* **168** B9
Koson, *Uzb.* **178** M10
Kosŏng, *N. Korea* **186** K10
Kosovo, *Eur.* **154** H10
Kosovska Mitrovica *see* Mitrovicë,
Kos. **154** H10
Kosso, *Côte d'Ivoire* **205** Q13
Kossol Passage, *Palau* **226** E6
Kosti, *Sudan* **203** S12
Kostino, *Russ.* **161** H13
Kostomuksha, *Russ.* **158** B7
Kostopil', *Ukr.* **159** N4
Kostroma, *Russ.* **158** G11
Kostromskoye Vodokhranilishche,
Russ. **158** G11
Kostyantynivka, *Ukr.* **159** Q9
Koszalin, *Pol.* **150** B12
Kot Addu, *Pak.* **180** M10
Kota Baharu, *Malaysia* **191** S9
Kota Kinabalu, *Malaysia* **192** G11

Kota Tinggi, *Malaysia* **191** V10
Kotaagung, *Indonesia* **192** L7
Kotabaru, *Indonesia* **192** K12
Kotabumi, *Indonesia* **192** L7
Kotabunan, *Indonesia* **193** J15
Kotel'nich, *Russ.* **158** G14
Kotel'nikovo, *Russ.* **159** Q12
Kotel'nyy, *Russ.* **161** D16
Köthen, *Ger.* **150** E9
Kotido, *Uganda* **211** Q5
Kotka, *Fin.* **143** H16
Kotlas, *Russ.* **158** E13
Kotlik, *Alas., U.S.* **112** G12
Kotonkoro, *Nigeria* **206** J7
Kotorska, Boka, *Montenegro* **154** J8
Kotovs'k, *Russ.* **158** M11
Kotovsk, *Ukr.* **159** Q5
Kotri, *Pak.* **181** R8
Kottayam, *India* **183** S5
Kotto, river, *Cen. Af. Rep.* **207** N16
Kotu Group, *Tonga* **227** M22
Kotu, island, *Tonga* **227** M23
Koturdepe, *Turkm.* **178** M5
Kotzebue Sound, *Alas., U.S.* **113** E13
Kotzebue, *Alas., U.S.* **113** E13
Koua, *New Caledonia, Fr.* **231** P22
Kouakou, island, *Fr. Polynesia, Fr.*
231 R18
Koualé, *Mali* **204** L12
Kouandé, *Benin* **205** L17
Kouango, *Cen. Af. Rep.* **207** N15
Kouba Modounga, *Chad* **207** G14
Koufália, *Gr.* **156** C10
Koufonísi, island, *Gr.* **157** R16
Koufonísia, island, *Gr.* **157** L15
Koukdjuak, Great Plain of the,
Nunavut, Can. **87** G18
Kouki, *Cen. Af. Rep.* **207** M13
Kouklia, *Cyprus* **162** Q6
Koukou Angarana, *Chad* **207** J15
Koúla Óros, *Gr.* **157** B14
Koulamoutou, *Gabon* **208** E9
Koulé, *Guinea* **204** N11
Koulikoro, *Mali* **204** K12
Koumac, *New Caledonia, Fr.* **231** P20
Koumameyong, *Gabon* **208** E8
Koumbia, *Guinea* **204** L8
Koumogo, *Chad* **207** L14
Koumra, *Chad* **207** L13
Koundara, *Guinea* **204** K8
Koungheul, *Senegal* **204** J7
Koungou, *Mayotte, Fr.* **215** P17
Koungouri, *Chad* **207** K14
Kounié *see* Île des Pins, island,
New Caledonia, Fr. **231** R24
Kounkané, *Senegal* **204** K8
Kountze, *Tex., U.S.* **105** K20
Kourou, *Fr. Guiana, Fr.* **131** F19
Kourouba, *Mali* **204** K10
Kouroussa, *Guinea* **204** L10
Koutiala, *Mali* **205** K13
Koutouba, *Côte d'Ivoire* **205** M14
Koutous, *Niger* **206** G9
Kouvola, *Fin.* **143** H16
Kovel', *Ukr.* **158** M3
Kovic, Baie, *Que., Can.* **87** J17
Kovin, *Serb.* **154** F11
Kovrov, *Russ.* **158** J11
Kovur, *India* **183** Q7
Kowanyama, *Qnsld., Austral.* **221** P12
Kowŏn, *N. Korea* **186** H8
Köyceğiz, *Turk.* **170** H4
Koyda, *Russ.* **158** A11
Koynas, *Russ.* **158** B13
Koytendag, *Turkm.* **178** N10
Koyuk, *Alas., U.S.* **113** F13
Koyukuk, *Alas., U.S.* **113** F14
Koyukuk, river, *Alas., U.S.* **113** E15
Koyulhisar, *Turk.* **170** E12
Kozan, *Turk.* **170** H11
Kozáni, *Gr.* **156** D9
Kozel'sk, *Russ.* **158** K8
Kozhikode (Calicut), *India* **183** R5
Kozjak, peak, *N. Maced.* **154** L11
Kozlovka, *Russ.* **159** N11
Kozlu, *Turk.* **170** C7
Kozluk, *Turk.* **171** G16
Koz'modem'yansk, *Russ.* **158** H13
Kōzu Shima, *Japan* **189** Q11
Kozyatyn, *Ukr.* **159** P5

Kra, Isthmus of, *Thai.* **191** Q7
Krabi, *Thai.* **191** R7
Krâchéh (Kratie), *Cambodia* **191** N11
Kragerø, *Nor.* **142** L11
Kragujevac, *Serb.* **154** G10
Kraków, *Pol.* **151** G15
Kralendijk, *Bonaire, Neth.* **123** Q19
Kraljevo, *Serb.* **154** G10
Král'ovský Chlmec, *Slovakia* **151** J17
Kramators'k, *Ukr.* **159** Q9
Kramfors, *Sw.* **143** H13
Kranídi, *Gr.* **156** K11
Kranj, *Slov.* **154** D4
Krasavino, *Russ.* **158** E13
Kraśnik, *Pol.* **151** F17
Krasnoarmeysk *see* Tayynsha, *Kaz.*
178 C12
Krasnoarmeyskaya, *Russ.* **159** S10
Krasnodar, *Russ.* **159** S10
Krasnogorsk, *Russ.* **161** J22
Krasnogvardeyskoye, *Russ.* **159** S11
Krasnohrad, *Ukr.* **159** P8
Krasnokamensk, *Russ.* **161** L18
Krasnoslobodsk, *Russ.* **159** P13
Krasnovishersk, *Russ.* **160** G9
Krasnoyarsk, *Russ.* **161** K13
Krasnoye, *Russ.* **158** M16
Krasnoye, *Russ.* **158** F13
Krasnozatonskiy, *Russ.* **158** D14
Krasnyy Dolginets, *Kaz.* **178** H4
Krasnyy Kholm, *Russ.* **158** G9
Krasnyy Luch, *Ukr.* **159** Q10
Krasnyy Yar, *Russ.* **159** N13
Krasnyye Baki, *Russ.* **158** H13
Kratie *see* Krâchéh, *Cambodia* **191** N11
Kraulshavn *see* Nuussuaq,
Greenland, Den. **88** H5
Krause Lagoon Channel,
Virgin Is., U.S. **122** R2
Kray Lesa, *Russ.* **161** D19
Krefeld, *Ger.* **150** F5
Kremastí, *Gr.* **157** M20
Kremen, peak, *Croatia* **154** F6
Kremenchuk, *Ukr.* **159** P7
Kremenchuts'ke Vodoskovyshche,
Ukr. **159** P7
Kremenets', *Ukr.* **159** N3
Krems, *Aust.* **150** K12
Kress, *Tex., U.S.* **104** E12
Kréstena, *Gr.* **156** K9
Kretinga, *Lith.* **143** M15
Kreuzlingen, *Switz.* **152** A4
Kribi, *Cameroon* **206** P9
Krim, peak, *Slov.* **154** D4
Krims'ke Hory, *Russ.* **159** T7
Krishna, river, *India* **183** P5
Krishnagiri, *India* **183** R6
Krishnanagar, *India* **182** K12
Kristiansand, *Nor.* **142** M10
Kristianstad, *Sw.* **143** M13
Kristiansund, *Nor.* **142** J9
Kristinestad, *Fin.* **143** H14
Kríti (Crete), island, *Gr.* **157** Q14
Krk, island, *Croatia* **154** E4
Krnov, *Czechia* **151** H13
Kroken, *Nor.* **142** F12
Krolevets', *Ukr.* **158** M7
Kromy, *Russ.* **158** M8
Kronotskaya Sopka, *Russ.* **161** F22
Kronotskiy Gulf, *Pac. Oc.* **168** C13
Kronprins Christian Land,
Greenland, Den. **88** C12
Kronshtadt, *Russ.* **158** F6
Kroonstad, *S. Af.* **212** J8
Kropotkin, *Russ.* **159** S11
Krosno Odrzańskie, *Pol.* **150** E11
Krosno, *Pol.* **151** H17
Krotoszyn, *Pol.* **151** E13
Krstača, peak, *Montenegro, Serb.*
154 H10
Krui, *Indonesia* **192** L7
Krumovgrad, *Bulg.* **155** K15
Krung Thep (Bangkok), *Thai.*
190 M8
Krupá, *Czechia* **150** G10
Krupište, *N. Maced.* **154** K12
Krupki, *Belarus* **158** K5
Krusenstern Trough, *Pac. Oc.*
246 C10
Kruševac, *Serb.* **154** G11
Krutets, *Russ.* **158** F10
Krychaw, *Belarus* **158** L6
Krylovskaya, *Russ.* **159** R11
Krymsk, *Russ.* **159** T10
Kryvichy, *Belarus* **158** K4
Kryvyy Rih, *Ukr.* **159** Q7

Ksabi, *Alg.* **200** F10
Ksar el Barka, *Mauritania* **204** G9
Ksar el Kebir, *Mor.* **200** C8
Ksar Torchane, *Mauritania* **204** E9
Kuacjok, *S. Sudan* **203** V9
Kuala Abang, *Brunei* **192** G11
Kuala Belait, *Brunei* **192** G11
Kuala Berang, *Malaysia* **191** T9
Kuala Dungun, *Malaysia* **191** T9
Kuala Kerai, *Malaysia* **191** S9
Kuala Lumpur, *Malaysia* **191** U8
Kuala Nerang, *Malaysia* **192** F6
Kuala Rompin, *Malaysia* **191** U10
Kuala Terengganu, *Malaysia* **191** T9
Kualakapuas, *Indonesia* **192** K11
Kualakurun, *Indonesia* **192** J11
Kualamanjual, *Indonesia* **192** J10
Kualapu'u, *Hawaii, U.S.* **115** H17
Kualatungkal, *Indonesia* **192** J7
Kuam-ho, lake, *N. Korea* **186** L6
Kuannersooq (Kvanefjord),
    Greenland, Den. **89** S6
Kuantan, *Malaysia* **191** U9
Kubachi, *Russ.* **159** U15
Kuban Lowland, *Eur.* **140** H13
Kuban', river, *Eur.* **140** J13
Kubeno, Lake, *Eur.* **140** D12
Kubenskoye, *Russ.* **158** F10
Kubra, Buheirat Murrat el (Great
    Bitter Lake), *Egypt* **173** Q1
Kuchaman, *India* **182** G4
Kucherov Terrace, *Arctic Oc.* **252** H8
Kuching, *Malaysia* **192** H9
Kuchino Shima, *Japan* **189** U4
Kuchinoerabu Jima, *Japan* **189** U4
Kūchnay Darwēshān, *Afghan.* **180** L5
Kuçovë, *Alban.* **154** L9
Küçüksu, *Turk.* **171** G16
Kud, river, *Pak.* **181** Q6
Kudara, *Taj.* **179** M13
Kudat, *Malaysia* **192** F12
Kudremukh, peak, *India* **183** Q4
Kudu Kuyyel', *Russ.* **161** J17
Kudu, *N. Korea* **186** G6
Kudymkar, *Russ.* **158** F16
Kuenen Rise, *Ind. Oc.* **251** G14
Kufah see Al Kūfah, *Iraq* **176** H10
Kufra Oasis see Al Kufrah, *Lib.*
    201 H23
Kufra Oasis, *Af.* **198** E8
Kufstein, *Aust.* **150** L9
Kugluktuk, *Nunavut, Can.* **86** F12
Kugul'ta, *Russ.* **159** S12
Kūh Banen, *Iran* **177** J19
Kūh-e Bazmān, *Iran* **177** M21
Kūh-e Damāvand, *Iran* **177** E16
Kūh-e Dīnār, *Iran* **177** K15
Kūh-e Fāreghān, *Iran* **177** M19
Kūh-e Garīn, *Iran* **177** G13
Kūh-e Hezār, *Iran* **177** L19
Kūh-e Nokhoch, *Iran* **177** N22
Kūh-e Sabalān, *Iran* **177** C13
Kūh-e Sīāh Khvānī, *Iran* **177** E20
Kūh-e Taftān, peak, *Iran* **177** L22
Kūh-e Zard, *Iran* **177** H14
Kūhak, *Iran* **177** M24
Kūhdasht, *Iraq* **177** G13
Kuhgrat, peak, *Liech.* **162** N3
Kūhhā-ye Bākharz, mountains, *Iran*
    177 E21
Kūhhāye Zāgros (Zagros
    Mountains), *Iran, Iraq* **176** D11
Kuhmo, *Fin.* **143** F16
Kuhn Ø, *Greenland, Den.* **88** G13
Kui Buri, *Thai.* **191** N7
Kuikui, cape, *Hawaii, U.S.* **115** K18
Kuikuina, *Nicar.* **117** L18
Kuiseb, river, *Namibia* **212** H3
Kuivaniemi, *Fin.* **143** E15
Kujang, *N. Korea* **186** G6
Kuji, *Japan* **188** K13
Kuju, peak, *Japan* **189** R5
Kūka'iau, *Hawaii, U.S.* **115** M22
Kukawa, *Nigeria* **206** H11
Kukës, *Alban.* **154** K10
Kukipi, *P.N.G.* **222** Q7
Kukolu, island, *Solomon Is.* **228** B9
Kukpowruk, river, *Alas., U.S.* **113** C13
Kūktong, *N. Korea* **186** D12
Kuku Point, *Wake I., U.S.* **230** N11
Kukui, *Hawaii, U.S.* **115** N23
Kukuihaele, *Hawaii, U.S.* **115** M21
Kukui'ula, *Hawaii, U.S.* **114** F10
Kula Gulf, *Solomon Is.* **228** D7
Kula, *Serb.* **154** E9
Kula, *Turk.* **170** F4

Kūlagīno, *Kaz.* **178** F5
Kūlani, peak, *Hawaii, U.S.* **115** P22
Kulaura, *Bangladesh* **182** J13
Kul'chi, *Russ.* **161** J20
Kuldīga, *Latv.* **143** L15
Kule, *Botswana* **212** G5
Kulebaki, *Russ.* **158** J11
Kulén, *Cambodia* **190** M10
Kulin, *W. Aust., Austral.* **221** X3
Kulkuduk, *Uzb.* **178** K9
Kulm, *N. Dak., U.S.* **106** H6
Kulmbach, *Ger.* **150** G8
Kūlob, *Taj.* **178** N12
Kuloy, river, *Russ.* **158** A11
Kulp, *Turk.* **171** G15
Kulu, *Turk.* **170** F8
Kūlua, peak, *Hawaii, U.S.* **115** P22
Kulumadau, *P.N.G.* **222** Q10
Kulunda, *Russ.* **160** L11
Kulusuk, *Greenland, Den.* **89** Q10
Kuma, river, *Russ.* **159** S14
Kumagaya, *Japan* **189** N11
Kumai, *Indonesia* **192** K10
Kumai, Teluk, *Indonesia* **192** K10
Kumamoto, *Japan* **189** R5
Kuman Range, *Myanmar* **190** E6
Kumano Nada, *Japan* **189** Q9
Kumano, *Japan* **189** R9
Kumanovo, *N. Maced.* **154** K11
Kumara Junction, *N.Z.* **223** L18
Kumarina Mine, *W. Aust., Austral.*
    221 U3
Kumarl, *W. Aust., Austral.* **221** X4
Kumasi, *Ghana* **205** P15
Kumba, *Cameroon* **206** N9
Kumbakale, *Solomon Is.* **228** C6
Kumbakonam, *India* **183** S7
Kumbe, *Indonesia* **193** M21
Kumbo, *Cameroon* **206** M10
Kŭmch'ŏn, *N. Korea* **186** K9
Kŭmch'ŏn, *N. Korea* **186** L7
Kume Jima, *Japan* **189** X1
Kŭmgang-san, peak, *N. Korea*
    186 K10
Kŭmgang, *N. Korea* **186** K10
Kŭmhwa see Gimhwa, *S. Korea* **186** L9
Kumi, *Uganda* **211** R4
Kumma, ruin, *Sudan* **202** M11
Kumphawapi, *Thai.* **190** K9
Kŭmsŏng see Kimhwa, *N. Korea*
    186 L9
Kumta, *India* **183** Q4
Kumtag Shamo, *China* **184** G6
Kumu, *D.R.C.* **209** C16
Kumukh, *Russ.* **159** U15
Kumul see Hami, *China* **184** F7
Kūmya-man see Yŏnghŭng-man,
    bay, *N. Korea* **186** J9
Kŭmya, *N. Korea* **186** H8
Kumylzhenskaya, *Russ.* **159** P12
Kumzār, *Oman* **175** F16
Kunahandu, island, *Maldives* **183** X3
Kunashir (Kunashiri), island, *Russ.*
    188 E16
Kunashiri see Kunashir, island, *Russ.*
    161 K23
Kundapura (Coondapoor), *India*
    183 Q4
Kundar, river, *Afghan., Pak.* **180** L8
Kundian, *Pak.* **180** K10
Kundu, *Solomon Is.* **228** D6
Kunduz, *Afghan.* **180** F8
Kunduz, river, *Afghan.* **180** G8
Kunene, river, *Angola* **208** Q9
Kunene, river, *Angola, Namibia* **212** E2
Kungsbacka, *Sw.* **142** M12
Kungu, *D.R.C.* **209** C13
Kungurtug, *Russ.* **161** M14
Kungwe Mountain, *Tanzania* **211** W2
Kunlon, *Myanmar* **190** F7
Kunlun Mountains, *Asia* **168** G7
Kunlun Shan, *China* **182** B6
Kunlun Shankou, *China* **184** H7
Kunlun Station, *Antarctica* **239** H17
Kunming, *China* **184** N9
Kunnamkulam, *India* **183** S5
Kunu, island, *F.S.M.* **226** E8
Kununurra, *W. Aust., Austral.* **221** Q7
Kunya, *Nigeria* **206** J8
Kuopio, *Fin.* **143** G16
Kupa, river, *Croatia* **154** E5
Kupang, *Indonesia* **193** N14
Kūpikipiki'o Point (Black Point),
    Hawaii, U.S. **115** H15
Kupino, *Russ.* **160** K11

Kupreanof Island, *Alas., U.S.* **113** M23
Kupuas, river, *Asia* **168** L12
Kup''yans'k, *Ukr.* **159** P9
Kuqa, *China* **184** E4
Kür (Kura), river, *Azerb.* **171** E22
Kür Dili, *Azerb.* **171** F23
Kür-Araz Ovalığı, *Azerb.* **171** D21
Kura see Kür, river, *Azerb.* **171** E22
Kura see Mtkvari, river,
    Rep. of Georgia **171** C18
Kurakh, *Russ.* **159** V16
Kurakhtin (Cǝbrayıl), *Azerb.* **171** E21
Kūrān Dap, *Iran* **177** N21
Kurashassayskiy, *Kaz.* **178** E7
Kurashiki, *Japan* **189** Q7
Kurayoshi, *Japan* **189** P7
Kurbağa Gölü, *Turk.* **170** G10
Kürchatov, *Kaz.* **179** E15
Kürdǝmir, *Azerb.* **171** D22
Kurdistan, *Iran, Iraq* **176** C8
Kurduna, *Pak.* **180** L7
Kurduvadi, *India* **182** N4
Kürdzhali, *Bulg.* **155** K15
Kure Beach, *N.C., U.S.* **98** K13
Kure, *Japan* **189** Q7
Küre, *Turk.* **170** C9
Kuressaare, *Est.* **143** K15
Kureyka, *Russ.* **161** G13
Kurgan, *Russ.* **160** J9
Kurganinsk, *Russ.* **159** T11
Kuri Bay, *W. Aust., Austral.* **221** Q5
Kuri, *Sudan* **203** P11
Kuria Muria Islands see Jazā'ir
    Khurīyā Murīyā, *Oman* **175** M16
Kuridala, *Qnsld., Austral.* **221** S11
Kurikka, *Fin.* **143** G15
Kuril Basin, *Pac. Oc.* **246** D8
Kuril Islands see Kuril'skiye
    Ostrova, *Russ.* **161** J23
Kuril-Kamchatka Trench, *Pac. Oc.*
    246 D8
Kuril'sk, *Russ.* **161** J23
Kuril'skiye Ostrova (Kuril Islands),
    Russ. **161** J23
Kurio, *Japan* **189** U4
Kurmuk, *Sudan* **203** U13
Kurnool, *India* **183** P6
Kuro Shima, *Japan* **189** T4
Kuroishi, *Japan* **188** J12
Kuroiso, *Japan* **189** N11
Kurshskiy Zaliv, *Lith., Russ.* **158** J1
Kursk, *Russ.* **158** M9
Kurşunlu, *Turk.* **170** D8
Kurtalan, *Turk.* **171** G16
Kurtistown, *Hawaii, U.S.* **115** N23
Kurtoğlu Burnu, *Turk.* **170** J4
Kuru, river, *S. Sudan* **203** V8
Kuruçay, *Turk.* **171** E13
Kuruktag, range, *China* **184** F5
Kurume, *Japan* **189** R4
Kurun Uraykh, *Russ.* **161** H19
Kurunegala, *Sri Lanka* **183** T7
Kuryong, river, *N. Korea* **186** G6
Kuş Gölü, *Turk.* **170** D4
Kuşadası Körfezi, *Turk.* **170** G3
Kuşadası, *Turk.* **170** G3
Kusaie see Kosrae, island, *F.S.M.*
    226 B12
Kusakaki Guntō, *Japan* **189** T3
Kushima, *Japan* **189** T5
Kushimoto, *Japan* **189** R9
Kushiro, *Japan* **188** G15
Kushk-e Kuhnah, *Afghan.* **180** H4
Kushnīya, *Israel* **172** K6
Kushtia, *Bangladesh* **182** J12
Kuskokwim Bay, *Alas., U.S.* **112** K12
Kuskokwim Mountains, *Alas., U.S.*
    113 H14
Kuskokwim, river, *Alas., U.S.* **113** H14
Kusŏng, *N. Korea* **186** G5
Kussharo Ko, *Japan* **188** F15
Kusur, *Russ.* **159** V15
Kuta, *Nigeria* **206** K7
Kütahya, *Turk.* **170** E5
K'ut'aisi, *Rep. of Georgia* **171** B17
Kutanibong, *Indonesia* **192** G4
Kutch, Gulf of see Kachchh, Gulf of,
    India **182** K1
Kutch, Rann of see Kachchh, Rann
    of, *India, Pak.* **181** T8
Kutchan, *Japan* **188** G12
Kutina, *Croatia* **154** E7
Kutno, *Pol.* **151** E15
Kutru, *India* **182** M7
Kutu, *D.R.C.* **208** F12

Kutubdia, island, *Bangladesh* **182** K13
Kutum, *Sudan* **203** R7
Kuujjuaq, *Que. Can.* **87** K19
Kuujjuarapik, *Que. Can.* **87** L18
Kuuk, *Greenland, Den.* **88** H5
Kuuli-Mayak, *Turkm.* **178** L4
Kuummiit, *Greenland, Den.* **89** P10
Kuŭp, *N. Korea* **186** J9
Kuusamo, *Fin.* **143** E16
Kuvango, *Angola* **208** N11
Kuvshinovo, *Russ.* **158** H8
Kuwait City see Al Kuwayt, *Kuwait*
    174 D11
Kuybishevskiy, *Taj.* **178** N12
Kuybyshev Reservoir, *Eur.* **140** E14
Kuybyshev, *Russ.* **160** K11
Kuybyshevskoye Vodokhranilishche,
    Russ. **158** J14
Kuytun, *China* **184** D5
Kuzey Anadolu Dağlari, *Turk.* **170** C9
Kuznetsk, *Russ.* **158** L14
Kuzomen', *Russ.* **158** A9
Kuzumaki, *Japan* **188** K13
Kvænangen, bay, *Nor.* **143** B13
Kvaløy, island, *Nor.* **143** C13
Kvanefjord see Kuannersooq,
    Greenland, Den. **89** S6
Kvarner, gulf, *Croatia* **154** F4
Kvarnerić, bay, *Croatia* **154** E4
Kvikkjokk, *Sw.* **143** E13
Kvitkuven Ice Rise, *Antarctica* **238** B11
Kwa Mtoro, *Tanzania* **211** V5
Kwa, river, *D.R.C.* **208** G11
KwaBhaca, *S. Af.* **212** L8
Kwadak, island, *Marshall Is.* **227** H18
KwaDukuza (Stanger), *S. Af.* **212** K9
Kwail (P'ungch'ŏn), *N. Korea* **186** K5
Kwailibesi, *Solomon Is.* **228** E11
Kwajalein Atoll, *Marshall Is.* **227** C16
Kwajalein Lagoon, *Marshall Is.*
    227 G16
Kwajalein, island, *Marshall Is.* **227** K18
Kwaksan, *N. Korea* **186** H5
Kwale, *Nigeria* **206** M7
Kwale, *Kenya* **211** V8
Kwali, *Nigeria* **206** L8
Kwamouth, *D.R.C.* **208** G11
Kwangje, *N. Korea* **186** J5
Kwangju see Gwangju, *S. Korea*
    187 T8
Kwangnyang-man, bay, *N. Korea*
    186 K5
Kwango, river, *Af.* **198** K6
Kwanza, river, *Af.* **198** L6
Kwaos, *Indonesia* **193** K17
Kwekwe, *Zimb.* **212** F9
Kwenge, river, *D.R.C.* **208** H12
Kwethluk, *Alas., U.S.* **112** J12
Kwidzyn, *Pol.* **151** C14
Kwigillingok, *Alas., U.S.* **112** K12
Kwikila, *P.N.G.* **222** R8
Kwilu, river, *D.R.C.* **209** J13
Kwitaro, river, *Guyana* **131** H15
Kwoka, peak, *Indonesia* **193** J18
Kyabé, *Chad* **207** K14
Kyaikto, *Myanmar* **190** K6
Kyaka, *Tanzania* **211** T3
Kyakhta, *Russ.* **161** M15
Kyancutta, *S. Aust., Austral.* **221** X9
Kyango, *S. Sudan* **203** V9
Kyaukkyi, *Myanmar* **190** J5
Kyaukme, *Myanmar* **190** G6
Kyaukphyu, *Myanmar* **190** J3
Kyauktaw, *Myanmar* **190** H3
Kybartai, *Lith.* **143** N16
Kyejŏng, *N. Korea* **186** L7
Kyela, *Tanzania* **211** Y4
Kyenjojo, *Uganda* **211** S2
Kyiv (Kiev), *Ukr.* **159** N6
Kykládes (Cyclades), islands, *Gr.*
    157 L14
Kyle, *Tex., U.S.* **105** L16
Kyllíni, *Gr.* **156** J8
Kyllíni, Óros, *Gr.* **156** J10
Kými, *Gr.* **157** H13
Kýmis, Akrotírio, *Gr.* **157** H13
Kynuna, *Qnsld., Austral.* **221** S12
Kyōga Misaki, *Japan* **189** P8
Kyŏngju see Gyeongju, *S. Korea*
    187 R12
Kyŏngsŏng, *N. Korea* **186** D12
Kyōto, *Japan* **189** Q9

Kyparissía, *Gr.* **156** L9
Kyparissiakós Kólpos, *Gr.* **156** K8
Kyrá Panagiá, island, *Gr.* **157** F13
Kyrenia (Girne, Kerýneia), *Cyprus*
    162 N8
Kyrgyz Range, *Kaz., Kyrg.* **179** K13
Kyrgyzstan, *Asia* **179** K13
Kyshtovka, *Russ.* **160** J11
Kýthira (Cerigo, Cythera), island,
    Gr. **156** N11
Kýthira, *Gr.* **156** N11
Kýthnos, *Gr.* **157** K13
Kýthnos, island, *Gr.* **157** K13
Kythrea (Değirmenlik), *Cyprus*
    162 N9
Kyubeyingde, *Russ.* **161** G16
Kyūroku Jima, *Japan* **188** J11
Kyushe, *Kaz.* **178** H6
Kyushu-Palau Ridge, *Pac. Oc.* **246** F6
Kyūshū, island, *Japan* **189** S5
Kyustendil, *Bulg.* **154** J12
Kyzyl-Adyr, *Kyrg.* **179** K13
Kyzyl-Kyya, *Kyrg.* **179** L13
Kyzyl-Suu, *Kyrg.* **179** K16
Kyzyl, *Russ.* **161** L13
Kyzylkup, *Turkm.* **178** L5

L

L'Hospitalet, *Fr.* **149** Z9
La Adela, *Arg.* **135** P10
La Almunia de Doña Godina, *Sp.*
    147 E13
La Asunción, *Venez.* **130** C12
La Banda, *Arg.* **134** G10
La Bañeza, *Sp.* **146** C8
La Barca, *Mex.* **116** H10
La Barge, *Wyo., U.S.* **108** J9
La Baule-Escoublac, *Fr.* **149** R5
La Belle, *Fla., U.S.* **99** V9
La Brea, *Trin. & Tobago* **123** Q21
La Calera, *Chile* **134** K6
La Canourgue, *Fr.* **149** W11
La Capelle, *Fr.* **148** M11
La Carolina, *Sp.* **146** J11
La Caye, *St. Lucia* **123** K14
La Ceiba, *Hond.* **117** K17
La Chambre, *Fr.* **149** U14
La Chapelle-d'Angillon, *Fr.* **149** R9
La Charité, *Fr.* **149** S10
La Châtaigneraie, *Fr.* **149** S6
La Châtre, *Fr.* **149** T9
La Chaux-de-Fonds, *Switz.* **152** B2
La Ciotat, *Fr.* **149** Y14
La Citadelle, site, *Haiti* **119** L17
La Ciudad Encantada, *Sp.* **147** G13
La Clayette, *Fr.* **149** T12
La Coche, island, *Guadeloupe, Fr.*
    123 H14
La Condamine, *Monaco* **163** E20
La Conner, *Wash., U.S.* **110** B4
La Cortinada, *Andorra* **162** H3
La Crosse, *Wis., U.S.* **100** K3
La Crosse, *Kans., U.S.* **107** T5
La Crosse, *Wash., U.S.* **110** D8
La Cruz, *Mex.* **116** F8
La Cruz, *C.R.* **117** M18
La Cygne, *Kans., U.S.* **107** T10
La Désirade, island, *Guadeloupe, Fr.*
    123 E17
La Digue, island, *Seychelles* **215** N21
La Dogana, *San Marino* **163** G16
La Dorada, *Col.* **130** F6
La Esmeralda, *Venez.* **130** G11
La Esperanza, *Cuba* **118** G5
La Fe, *Cuba* **118** H4
La Fe, *Cuba* **118** H6
La Felguera, *Sp.* **146** B9
La Fère, *Fr.* **149** N11
La Ferme, *Mauritius* **215** J20
La Ferté-Macé, *Fr.* **149** P7
La Ferté-Saint-Aubin, *Fr.* **149** R9
La Flèche, *Fr.* **149** R7
La Foa, *New Caledonia, Fr.* **231** G22
La Follette, *Tenn., U.S.* **103** E18
La Gacilly, *Fr.* **149** R5
La Gineta, *Sp.* **147** H13
La Gran Sabana, region, *Venez.*
    130 F13
La Grand' Combe, *Fr.* **149** W12
La Grande, *Oreg., U.S.* **110** F8
La Grange, *N.C., U.S.* **98** H13

La Grange, *Ky., U.S.* **103** B17
La Grange, *Tex., U.S.* **105** L18
La Grange, *Mo., U.S.* **107** R13
La Grange, *Wyo., U.S.* **108** J14
La Grave, *Fr.* **149** V14
La Grulla, *Tex., U.S.* **105** R16
La Guaira, *Venez.* **130** C10
La Guerche, *Fr.* **149** Q6
La Habana, *Cuba* **118** G6
La Harpe, *Ill., U.S.* **101** Q3
La Isleta, island, *Sp.* **214** Q6
La Jara, *Colo., U.S.* **109** P12
La Jonquera, *Sp.* **147** C18
La Junta, *Colo., U.S.* **109** P15
La Laguna, *Sp.* **214** Q5
La Leonesa, *Arg.* **134** G13
La Libertad, *Guatemala* **117** K15
La Libertad, *Ecua.* **130** L2
La Ligua, *Chile* **134** K6
La Línea de la Concepción, *Sp.* **146** N9
La Loche, *Sask., Can.* **86** K11
La Lune, *Trin. & Tobago* **123** R22
La Madrid, *Arg.* **134** G9
La Malbaie, *Que., Can.* **87** N20
La Mancha, *Eur.* **140** K3
La Manche *see* English Channel, *Fr., U.K.* **145** X13
La Marmora, Punta, *It.* **153** N4
La Marque, *Tex., U.S.* **105** L20
La Massana, *Andorra* **162** H2
La Massana, *Andorra* **162** J3
La Mesa, *N. Mex., U.S.* **109** V12
La Mesa, *Calif., U.S.* **111** Z10
La Meyze, *Fr.* **149** U8
La Montaña, region, *S. Amer.* **128** E3
La Mothe-Achard, *Fr.* **149** S5
La Oliva, *Sp.* **214** Q7
La Orotava, *Sp.* **214** Q4
La Oroya, *Peru* **132** H3
La Palma, *Pan.* **117** N22
La Palma, island, *Sp.* **214** Q5
La Paloma, *Uru.* **134** L15
La Paragua, *Venez.* **130** E12
La Paz, *Mex.* **116** F11
La Paz, *Mex.* **116** F7
La Paz, *Bol.* **132** J6
La Paz, *Arg.* **134** J12
La Paz, *Arg.* **134** K8
La Pedrera, *Col.* **130** K9
La Perouse Strait, *Russ.* **161** K22
La Perouse Strait, *Russ.* **188** D13
La Pine, *Oreg., U.S.* **110** H4
La Place, *La., U.S.* **102** P10
La Plaine des Cafres, *Réunion, Fr.* **215** H16
La Plaine, *Dominica* **123** G20
La Plaine, *Réunion, Fr.* **215** G16
La Plant, *S. Dak., U.S.* **106** J4
La Plata, *Md., U.S.* **98** D13
La Plata, *Mo., U.S.* **107** R12
La Plata, *Arg.* **134** L13
La Pobla de Segur, *Sp.* **147** C16
La Pola de Gordón, *Sp.* **146** B9
La Porte City, *Iowa, U.S.* **107** N13
La Porte, *Tex., U.S.* **105** L20
La Porte, *Ind., U.S.* **101** N10
La Possession, *Réunion, Fr.* **215** G15
La Pryor, *Tex., U.S.* **105** M14
La Puebla de Montalbán, *Sp.* **146** G9
La Purísima, *Mex.* **116** E6
La Push, *Wash., U.S.* **110** B2
La Quiaca, *Arg.* **134** D9
La Réole, *Fr.* **149** W7
La Rioja, *Arg.* **134** H8
La Rivière, *Réunion, Fr.* **215** H15
La Robla, *Sp.* **146** C9
La Roca de la Sierra, *Sp.* **146** H7
La Rocca o Guaita, *San Marino* **163** J15
La Roche-sur-Yon, *Fr.* **149** S6
La Roche, *Fr.* **149** T14
La Rochefoucauld, *Fr.* **149** U8
La Rochelle, *Fr.* **149** T6
La Roda, *Sp.* **146** H12
La Romana, *Dom. Rep.* **119** M20
La Ronge, *Sask., Can.* **86** L12
La Rubia, *Arg.* **134** J11
La Sagesse Bay, *Grenada* **123** L23
La Sal, *Utah, U.S.* **109** N9
La Salle, *Ill., U.S.* **101** P5
La Sarre, *Que., Can.* **87** P18
La Scie, *Nfld. & Lab., Can.* **87** L23
La Sènia, *Sp.* **147** F15
La Serena, *Chile* **134** J7
La Seu d'Urgell, *Sp.* **147** C16
La Seyne, *Fr.* **149** Y14

La Sila, range, *It.* **153** P13
La Sola, island, *Venez.* **121** J14
La Solana, *Sp.* **146** H11
La Sorcière, *St. Lucia* **123** K14
La Soufrière, *St. Vincent & the Grenadines* **123** J16
La Souterraine, *Fr.* **149** T9
La Spezia, *It.* **152** G5
La Suze, *Fr.* **149** R7
La Teste, *Fr.* **149** W6
La Tortuga, Isla, *Venez.* **130** C11
La Tranche-sur-Mer, *Fr.* **149** T5
La Trinité, *Martinique* **123** E23
La Tuque, *Que., Can.* **87** P19
La Unión, *Peru* **132** G3
La Unión, *Chile* **135** P6
La Unión, *Sp.* **147** K14
La Vache Bay, *Trin. & Tobago* **123** N22
La Vega, *Dom. Rep.* **119** L18
La Venta, ruin, *Mex.* **117** J14
La Verkin, *Utah, U.S.* **109** P6
La Veta, *Colo., U.S.* **109** P13
Laas Dawaco, *Somalia* **210** L14
Laascaanood, *Somaliland* **210** M13
Laasqoray, *Somalia* **210** K13
Laatzen, *Ger.* **150** D7
La'au Point, *Hawaii, U.S.* **115** J16
Laayoune, *W. Sahara* **200** G4
Labadaad, *Somalia* **211** S10
Labasa, *Fiji* **226** N6
Labbezanga, *Mali* **205** J16
Labbezanga, *Niger* **206** G4
Labé, *Guinea* **204** L9
Labe, river, *Czechia* **150** H12
Labinsk, *Russ.* **159** T11
Laborie, *St. Lucia* **123** L14
Labota, *Indonesia* **193** K14
Laboué, *Leb.* **172** H7
Labouheyre, *Fr.* **149** W6
Laboulaye, *Arg.* **134** L10
Labrador Basin, *Atl. Oc.* **248** F6
Labrador City, *Nfld. & Lab., Can.* **87** M20
Labrador Sea, *Atl. Oc.* **248** G6
Labrador Shelf, *Atl. Oc.* **248** G5
Labrador, region, *Nfld. & Lab., Can.* **87** L21
Lábrea, *Braz.* **132** F7
Labrit, *Fr.* **149** W6
Labuan Basin, *Ind. Oc.* **254** C12
Labuan, island, *Malaysia* **192** G11
Labuk Bay, *Malaysia* **192** F12
Labutta, *Myanmar* **190** L4
Lac Aougoundou, *Mali* **205** H13
Lac 'Assal (Lake Assal), *Eth.* **210** K9
Lac de Kossou, *Côte d'Ivoire* **205** N13
Lac de Retenue, *D.R.C.* **209** L17
Lac Débo, *Mali* **205** H14
Lac Do, *Mali* **205** H14
Lac du Flambeau, *Wis., U.S.* **100** G5
Lac Faguibine, *Mali* **205** H14
Lac Fauna Nui, *Fr. Polynesia, Fr.* **231** J14
Lac Fitri, *Chad* **207** H13
Lac Garou, *Mali* **205** H14
Lac Ihotry, *Madagascar* **213** G14
Lac Iro, *Chad* **207** K14
Lac Kivu, *Rwanda* **211** T1
Lac Korarou, *Mali* **205** H14
Lac La Biche, *Alta., Can.* **86** L11
Lac Mai-Ndombe, *D.R.C.* **208** F12
Lac Niangay, *Mali* **205** H14
Lac Ntomba, *D.R.C.* **208** E12
Lac Rkiz, *Mauritania* **204** G7
Lac Tchad *see* Lake Chad, *Cameroon, Chad* **206** H12
Lac Upemba, *D.R.C.* **209** K17
Lac Vaihiria, *Fr. Polynesia, Fr.* **231** J22
Lacanau-Océan, *Fr.* **149** V6
Lacanau, *Fr.* **149** V6
Lacanau, Étang de, *Fr.* **149** V6
Lacapelle-Marival, *Fr.* **149** V9
Lacaune, Monts de, *Fr.* **149** X10
Laccadive Sea, *India* **183** Q4
Lacepede Bay, *Austral.* **220** L11
Lachanás, *Gr.* **156** C11
Lachlan, river, *Austral.* **220** K13
Laçin *see* Berdzor, *Azerb.* **171** E20
Lackawanna, *N.Y., U.S.* **96** J5
Lacon, *Ill., U.S.* **101** P5
Laconia, *N.H., U.S.* **97** H14
Laconia, Gulf of, *Eur.* **140** M9
Lacoochee, *Fla., U.S.* **99** T8
Lacq, *Fr.* **149** X6
Lacre Punt, *Bonaire, Neth.* **123** R19
Ladakh Range, *India* **182** C6

Ladakh, region, *India* **182** C6
Ladik, *Turk.* **170** D11
Lādīz, *Iran* **177** L22
Ladnun, *India* **182** G4
Ladoga *see* Ladozhskoye Ozero, *Russ.* **158** F7
Ladozhskoye Ozero (Ladoga, Lake), *Russ.* **158** F7
Ladrillero, Golfo, *Chile* **135** V6
Ladron Peak, *N. Mex., U.S.* **109** T11
Ladva, *Russ.* **158** E8
Lady Ann Strait, *Nunavut, Can.* **87** D16
Lady Elliot Island, *Austral.* **220** G16
Lady Newnes Bay, *Antarctica* **239** Q14
Ladygin Yar, *Russ.* **161** F13
Ladysmith, *Wis., U.S.* **100** G3
Ladysmith, *S. Af.* **212** K9
Ladyzhenka, *Kaz.* **178** E11
Lae Atoll, *Marshall Is.* **227** C15
Lae, *P.N.G.* **222** P8
Laea, *Indonesia* **193** L14
Laefu, *P.N.G.* **222** N10
Laem Ngop, *Cambodia* **191** N9
Laeokealaikahiki, cape, *Hawaii, U.S.* **115** K18
Laerdal Tunnel, *Nor.* **142** J10
Læsø, island, *Den.* **142** M11
LaFayette, *Ga., U.S.* **98** J4
Lafayette, *Ind., U.S.* **101** Q8
Lafayette, *La., U.S.* **102** P8
Lafayette, *Ala., U.S.* **103** K17
Lafayette, *Tenn., U.S.* **103** E16
Lafia, *Nigeria* **206** L8
Lafiagi, *Nigeria* **206** L7
Laga Bogal, river, *Kenya* **211** R8
Laga Bor, river, *Kenya* **211** R8
Lagan, river, *Sw.* **143** M13
Laganá, Kólpos, *Gr.* **156** K7
Laganya, *Ghana* **205** M16
Lagdo Reservoir, *Cameroon* **206** L11
Laggan Bay, *Scot., U.K.* **145** M7
Laghouat, *Alg.* **200** D12
Lagkadás, *Gr.* **156** C11
Lago de Cahora Bassa, *Mozambique* **212** D10
Lagoa, *Port.* **146** Q5
Lagodekhi, *Rep. of Georgia* **171** C20
Lagos de Moreno, *Mex.* **116** H10
Lagos, *Port.* **146** Q3
Lagos, *Nigeria* **206** M6
Lágos, *Gr.* **157** B14
Lagosa, *Tanzania* **211** W2
Lagosta *see* Lastovo, island, *Croatia* **154** J6
Lagrange Bay, *Austral.* **220** D4
LaGrange, *Ga., U.S.* **98** M4
Lagrange, *Ind., U.S.* **101** N10
Lagrange, *W. Aust., Austral.* **221** R4
Laguépie, *Fr.* **149** W9
Lagún Jan Thiel, *Curaçao, Neth.* **123** Q15
Lagun, *Curaçao, Neth.* **123** N13
Laguna Beach, *Calif., U.S.* **111** Y9
Laguna Blanca, *Arg.* **134** F13
Laguna Dam, *Ariz., Calif., U.S.* **111** Z13
Laguna Grande, *Arg.* **135** V8
Laguna Madre, *Tex., U.S.* **105** Q17
Laguna Ojo de Liebre, *Mex.* **116** D5
Laguna Paiva, *Arg.* **134** J12
Laguna, *N. Mex., U.S.* **109** S11
Lagunan Garapan, *N. Mariana Is., U.S.* **230** H7
Lagunas, *Peru* **132** E3
Lagunas, *Chile* **134** C7
Lagune Aby, *Côte d'Ivoire, Ghana* **205** Q14
Lagune Iguéla, *Gabon* **208** F7
Lagune Ndogo, *Gabon* **208** F7
Lagune Nkomi, *Gabon* **208** F7
Lagunillas, *Bol.* **132** K8
Lagwira, *W. Sahara* **200** K1
Lahad Datu, *Malaysia* **192** G12
Lahaina, *Hawaii, U.S.* **115** J18
Laham, *Indonesia* **192** J11
Lahat, *Indonesia* **192** K7
Lahi Passage, *Tonga* **227** P23
Laḥij, *Yemen* **174** Q9
Lāhījān, *Iran* **177** D14
Lahic, *Azerb.* **171** D22
Lahn, river, *Ger.* **150** G6
Lahnstein, *Ger.* **150** G5
Lahontan Reservoir, *Nev., U.S.* **111** P7
Lahore, *Pak.* **180** L13
Lahr, *Ger.* **150** K5
Lahri, *Pak.* **181** N8
Lahti, *Fin.* **143** H16

Laï (Behagle), *Chad* **207** K13
Lai Châu, *Vietnam* **190** G9
Laiagam, *P.N.G.* **222** P6
Lā'ie, *Hawaii, U.S.* **115** G14
L'Aigle, *Fr.* **149** P8
Laïka, island, *Vanuatu* **229** H16
Laird, river, *N. Amer.* **84** E4
Lairg, *Scot., U.K.* **144** G9
Laisamis, *Kenya* **211** R7
Laiyang, *China* **185** H15
Lajamanu, *N.T., Austral.* **221** R7
Lajes das Flores, *Port.* **146** N1
Lajes, *Braz.* **132** S12
Lajes, *Port.* **146** P3
Lajitas, *Tex., U.S.* **104** M10
Lakamané, *Mali* **204** J10
Lakatoro, *Vanuatu* **229** G14
Lakaträsk, *Sw.* **143** E14
Lake Albert, *D.R.C.* **209** D19
Lake Albert, *Uganda* **211** R2
Lake Amboseli, *Kenya, Tanzania* **211** U6
Lake Andes, *S. Dak., U.S.* **106** M6
Lake Antoine, *Grenada* **123** J23
Lake Arthur, *La., U.S.* **102** P7
Lake Assal *see* Lac 'Assal, *Eth.* **210** K9
Lake Bangweulu, *Zambia* **209** L19
Lake Bridgeport, *Tex., U.S.* **105** F16
Lake Brownwood, *Tex., U.S.* **105** H15
Lake Brunner, *N.Z.* **223** L18
Lake Buchanan, *Tex., U.S.* **105** K16
Lake Butler, *Fla., U.S.* **99** R7
Lake Chad (Lac Tchad), *Cameroon, Chad* **206** H12
Lake Charles, *La., U.S.* **102** P7
Lake Chelan National Recreation Area, *Wash., U.S.* **110** B5
Lake Chilwa, *Malawi* **212** D11
Lake City, *Fla., U.S.* **98** R7
Lake City, *S.C., U.S.* **98** L10
Lake City, *Mich., U.S.* **100** J10
Lake City, *Minn., U.S.* **106** K12
Lake City, *Iowa, U.S.* **107** N10
Lake City, *Colo., U.S.* **109** P11
Lake Clark National Park and Preserve, *Alas., U.S.* **113** J15
Lake Coleridge, *N.Z.* **223** L18
Lake Corpus Christi, *Tex., U.S.* **105** N16
Lake Crystal, *Minn., U.S.* **106** L10
Lake Cunningham, *Bahamas* **122** E2
Lake District, *Eng., U.K.* **145** P10
Lake Edward, *D.R.C.* **209** E19
Lake Edward, *Uganda* **211** S2
Lake Ellesmere, *N.Z.* **223** M19
Lake Eyasi, *Tanzania* **211** U5
Lake Eyre Basin, *Austral.* **220** G10
Lake Geneva, *Wis., U.S.* **100** M6
Lake George, *Uganda* **211** S2
Lake Harbor, *Fla., U.S.* **99** V10
Lake Havasu City, *Ariz., U.S.* **109** S4
Lake Hāwea, *N.Z.* **223** M17
Lake Helmand *see* Daryācheh-ye Hāmūn, *Iran* **177** J22
Lake Itezhi-Tezhi, *Zambia* **209** P17
Lake Jackson, *Tex., U.S.* **105** M19
Lake Kampolombo, *Zambia* **209** M19
Lake Kariba, *Zambia* **209** Q18
Lake Kariba, *Zambia, Zimb.* **212** E8
Lake Keilak, *Sudan* **203** U10
Lake Kemp, *Tex., U.S.* **105** F15
Lake Killarney, *Bahamas* **122** E2
Lake Kivu, *D.R.C.* **209** F19
Lake Kwania, *Uganda* **211** R4
Lake Kyoga, *Uganda* **211** R4
Lake Lewisville, *Tex., U.S.* **105** F17
Lake Linden, *Mich., U.S.* **100** E6
Lake Livingston, *Tex., U.S.* **105** K19
Lake Louise, *Alta., Can.* **86** M9
Lake Malawi *see* Lake Nyasa, *Af.* **198** L9
Lake Manapouri, *N.Z.* **223** N15
Lake Manyara, *Tanzania* **211** U6
Lake Mead National Recreation Area, *Ariz., Nev., U.S.* **111** U12
Lake Meredith National Recreation Area, *Tex., U.S.* **104** C12
Lake Meredith, *Tex., U.S.* **104** C12
Lake Mills, *Iowa, U.S.* **106** M11
Lake Molombe, *Malawi* **212** D11
Lake Murray, *P.N.G.* **222** P5
Lake Murray, *P.N.G.* **222** P5
Lake Mweru, *D.R.C. Zambia* **209** K18
Lake Nash, *N.T., Austral.* **221** S10
Lake Nasser, *Egypt* **202** L12
Lake Natron, *Tanzania* **211** T6

Lake Ngami, *Botswana* **212** F6
Lake Nubia, *Egypt, Sudan* **202** L11
Lake Nyasa (Lake Malawi), *Af.* **198** L9
Lake O' The Cherokees, *Okla., U.S.* **105** B19
Lake O' the Pines, *Tex., U.S.* **105** G20
Lake Ohau, *N.Z.* **223** M17
Lake Onega *see* Onezhskoye Ozero, *Russ.* **158** D8
Lake Oswego, *Oreg., U.S.* **110** F3
Lake Piso, *Liberia* **204** P9
Lake Placid, *N.Y., U.S.* **96** G11
Lake Preston, *S. Dak., U.S.* **106** K7
Lake Providence, *La., U.S.* **102** K10
Lake Pukaki, *N.Z.* **223** M17
Lake Qadisiyah *see* Buḩayrat Sadd al Qādisīyah, *Iraq* **176** F9
Lake Region, *Eur.* **140** D9
Lake Roosevelt National Recreation Area, *Wash., U.S.* **110** B8
Lake Rudolf *see* Lake Turkana, *Eth., Kenya* **211** Q6
Lake Rukwa, *Tanzania* **211** X4
Lake St. Lucia, *S. Af.* **212** K10
Lake Stamford, *Tex., U.S.* **105** G14
Lake Stewart, *N.S.W. Austral.* **221** V11
Lake Tana *see* T'ana Hāyk', *Eth.* **210** K6
Lake Tanganyika, *Burundi, Tanzania* **211** W2
Lake Taupo, *N.Z.* **223** G21
Lake Tawakoni, *Tex., U.S.* **105** G18
Lake Te Anau, *N.Z.* **223** N16
Lake Texoma, *Okla., Tex., U.S.* **105** E18
Lake Travis, *Tex., U.S.* **105** K16
Lake Turkana (Lake Rudolf), *Eth., Kenya* **211** Q6
Lake Urmia *see* Daryācheh-ye Orūmīyeh, *Iran* **176** C11
Lake Victoria, *Kenya, Tanzania, Uganda* **211** T4
Lake Village, *Ark., U.S.* **102** K9
Lake Volta, *Ghana* **205** N16
Lake Vostok, station, *Antarctica* **239** K17
Lake Wakatipu, *N.Z.* **223** N16
Lake Wales, *Fla., U.S.* **99** U8
Lake Wanaka, *N.Z.* **223** M16
Lake Whitney, *Tex., U.S.* **105** H17
Lake Worth, *Fla., U.S.* **99** W10
Lake, Poopelloe, *N.S.W. Austral.* **221** W12
Lakeba Passage, *Fiji* **226** P8
Lakeba, island, *Fiji* **226** Q8
Lakefield, *Minn., U.S.* **106** L10
Lakeland, *Fla., U.S.* **99** U8
Lakeland, *Ga., U.S.* **99** Q6
Lakepa, *N.Z.* **229** P22
Lakeport, *Mich., U.S.* **100** L13
Lakeport, *Calif., U.S.* **111** Q3
Lakes Entrance, *Vic. Austral.* **221** Z13
Lakeside, *Nebr., U.S.* **107** N2
Lakeside, *Oreg., U.S.* **110** H2
Lakeview, *Mich., U.S.* **100** L10
Lakeview, *Oreg., U.S.* **110** L6
Lakeville, *Minn., U.S.* **106** K11
Lakewood, *N.J., U.S.* **96** P11
Lakewood, *Ohio, U.S.* **101** N14
Lakewood, *Colo., U.S.* **108** M13
Lakhpat, *India* **182** J1
Lakin, *Kans., U.S.* **107** U3
Lakki, *Pak.* **180** K10
Lakkí, *Gr.* **157** L18
Lakkia, *Cyprus* **162** P8
Lakonikós Kólpos, *Gr.* **156** M10
Lakota, *N. Dak., U.S.* **106** F7
Lakselv, *Nor.* **143** B14
Lakshadweep, islands, *India* **183** S2
Lala Musa, *Pak.* **180** K12
Lalago, *Tanzania* **211** U4
Lalara, *Gabon* **208** D8
Lalbenque, *Fr.* **149** W9
L'Albera, Serra de, *Sp.* **147** C18
Laléraba, *Côte d'Ivoire* **205** M13
Lalëzit, Gjiri i, *Alban.* **154** K9
Lalín, *Sp.* **146** C6
Lalitpur (Patan), *Nepal* **182** G10
Lalitpur, *India* **182** J6
Lalla Rookh, *W. Aust., Austral.* **221** S3
Lalomanu, *Samoa* **227** P20
Lalona, island, *Tonga* **227** N23
Lal'sk, *Russ.* **158** E13
Laluai Point, *P.N.G.* **228** C6

Lama-Kara, *Togo* **205** M17
Lamag, *Malaysia* **192** G12
Lamalou-les-Bains, *Fr.* **149** X11
Lamap, *Vanuatu* **229** H15
Lamar Hayes Ridge, *Ind. Oc.* **251** N16
Lamar, *Mo., U.S.* **107** U11
Lamar, *Colo., U.S.* **109** N15
Lamas, *Peru* **132** F3
Lamas, *Turk.* **170** J9
Lamballe, *Fr.* **149** Q5
Lambaréné, *Gabon* **208** E7
Lambay Island, *Ire.* **145** R7
Lambert Glacier, *Antarctica* **239** F18
Lambert Land, *Greenland, Den.* **88** D12
Lambert's Bay, *S. Af.* **212** M5
Lambertville, *Mich., U.S.* **101** N12
Lámbia, *Gr.* **156** J9
Lambton, Cape, *N.W.T., Can.* **86** D11
Lame Deer, *Mont., U.S.* **108** E11
Lamé, *Chad* **206** K12
Lamentin, *Guadeloupe, Fr.* **123** E14
Lāmerd, *Iran* **177** M17
Lameroo, *S. Aust., Austral.* **221** Y11
Lamesa, *Tex., U.S.* **104** G12
L'Ametlla, de Mar, *Sp.* **147** E16
Lameyer Ridge, *Ind. Oc.* **250** R8
Lamía, *Gr.* **156** G10
Lamitan, *Philippines* **193** F14
Lamjaybir, *W. Sahara* **200** H3
Lamoille, *Nev., U.S.* **110** M11
Lamoni, *Iowa, U.S.* **107** Q11
Lamont, *Wyo., U.S.* **108** J11
Lamotrek Atoll, *F.S.M.* **226** B8
Lamotte-Beuvron, *Fr.* **149** R9
LaMoure, *N. Dak., U.S.* **106** H7
Lampang, *Thai.* **190** J7
Lampasas, *Tex., U.S.* **105** J16
Lampertheim, *Ger.* **150** H6
Lampeter, *Wales, U.K.* **145** T9
Lamphun, *Thai.* **190** J7
Lamskoye, *Russ.* **158** L9
Lamu, *Kenya* **211** T9
Lamutskoye, *Russ.* **161** C20
Lāna'i City, *Hawaii, U.S.* **115** J17
Lāna'i, *Hawaii, U.S.* **115** J17
Lāna'ihale, peak, *Hawaii, U.S.* **115** J17
Lanbi Kyun, island, *Myanmar* **191** P6
Lancang (Mekong), river, *China* **184** P8
Lancang, *China* **184** P8
Lancaster Sound, *N. Amer.* **84** C7
Lancaster Trough, *Arctic Oc.* **253** P13
Lancaster, *Pa., U.S.* **96** P8
Lancaster, *N.H., U.S.* **97** F14
Lancaster, *S.C., U.S.* **98** J9
Lancaster, *Wis., U.S.* **100** L4
Lancaster, *Ohio, U.S.* **101** R13
Lancaster, *Tex., U.S.* **105** G17
Lancaster, *Mo., U.S.* **107** Q12
Lancaster, *Calif., U.S.* **111** W8
Lancaster, *Eng., U.K.* **145** Q11
Lance Creek, *Wyo., U.S.* **108** H13
Lancelin, *W. Aust., Austral.* **221** W2
Lanciano, *It.* **152** K11
Land Bay, *Antarctica* **238** N10
Land Glacier, *Antarctica* **238** N10
Land O'Lakes, *Wis., U.S.* **100** F5
Landau, *Ger.* **150** H6
Landay, *Afghan.* **180** M4
Landeck, *Aust.* **150** L8
Lander, *Wyo., U.S.* **108** H10
Lander, river, *Austral.* **220** E8
Landerneau, *Fr.* **149** P3
Landes, *Eur.* **140** J4
Landivisiau, *Fr.* **149** P3
Landor, *W. Aust., Austral.* **221** U2
Landquart, *Switz.* **152** B5
Landrum, *S.C., U.S.* **98** J7
Lands End, cape, *N.W.T., Can.* **86** C12
Land's End, cape, *Eng., U.K.* **145** X7
Landsberg, *Ger.* **150** K8
Landshut, *Ger.* **150** K9
Landskrona, *Sw.* **142** N12
Lanett, *Ala., U.S.* **103** K17
Lạng Sơn, *Vietnam* **190** G11
Lang Suan, *Thai.* **191** Q7
Langådia, *Gr.* **156** K9
Langanes, island, *Ice.* **142** E4
Langano Hāyk', *Eth.* **211** N7
Langar, *Taj.* **179** N14
Langar, island, *F.S.M.* **226** J8
Langberg, mountains, *S. Af.* **212** K6
Langdon, *N. Dak., U.S.* **106** E7
Langeac, *Fr.* **149** V11
Langeberg, range, *Af.* **198** Q7
Langenhangen, *Ger.* **150** D7

Langford, *S. Dak., U.S.* **106** J7
Langjökull, glacier, *Ice.* **142** E3
Langkawi, island, *Malaysia* **191** S7
Langlo Crossing, *Qnsld., Austral.*
    221 U13
Langlo Downs, *Qnsld., Austral.* **221** U13
Langlois, *Oreg., U.S.* **110** J1
Langøya, island, *Nor.* **142** D12
Langres, *Fr.* **149** R13
Langres, Plateau de, *Fr.* **149** R12
Langsa, *Indonesia* **192** G4
Langseth Ridge, *Arctic Oc.* **253** H13
Langting, *India* **182** H14
Langtry, *Tex., U.S.* **104** L12
Langu, *Thai.* **191** S7
Languedoc, region, *Fr.* **149** W11
Langzhong, *China* **184** K10
Lankio, *Côte d'Ivoire* **205** M14
Lankoveri, *Nigeria* **206** K10
Lannion, *Fr.* **149** P4
Lano, *Wallis & Futuna Is., Fr.* **230** Q9
Lanouaille, *Fr.* **149** U8
Lansdale, *Pa., U.S.* **96** P9
Lansdowne House, *Ont., Can.* **87** M16
L'Anse aux Meadows,
    Nfld. & Lab., Can. **87** L23
L'Anse, *Mich., U.S.* **100** F6
Lansing, *Mich., U.S.* **100** M11
Lansing, *Iowa, U.S.* **106** L13
Lanslebourg, *Fr.* **149** U15
Lanta Yai, Ko, *Thai.* **191** R7
Lantewa, *Nigeria* **206** J10
Lanusei, *It.* **153** P4
Lanzarote, island, *Sp.* **214** P7
Lanzhou, *China* **184** K10
Lào Cai, *Vietnam* **190** G10
Laoag, *Philippines* **193** B13
Laoha, river, *China* **185** F14
Laohekou, *China* **184** K12
Laon, *Fr.* **149** N11
Laona, *Wis., U.S.* **100** G6
Laos, *Asia* **190** J10
Lapalisse, *Fr.* **149** T11
Lapeer, *Mich., U.S.* **100** L12
Lapelle, river, *S. Af.* **212** H9
Lapinlahti, *Fin.* **143** F16
Lapithos (Gilderen), *Cyprus* **162** N8
Lapland, region, *Fin., Russ., Sw.*
    143 D14
Lappajärvi, lake, *Eur.* **140** D9
Lappeenranta, *Fin.* **143** H17
Lapseki, *Turk.* **170** D3
Laptev Sea, *Arctic Oc.* **252** D9
Laptev Shelf, *Arctic Oc.* **252** D9
Lapthal, *India* **182** F7
Lapua, *Fin.* **143** G15
L'Aquila, *It.* **152** K9
Lār, *Iran* **177** M17
Larabanga, *Ghana* **205** M15
Laracha, *Sp.* **146** B5
Larache, *Mor.* **200** C8
Lārak, island, *Iran* **177** N19
Laramie Mountains, *Wyo., U.S.*
    108 J12
Laramie Peak, *U.S.* **90** E9
Laramie, *Wyo., U.S.* **108** K13
Laramie, river, *Colo., Wyo., U.S.*
    108 J12
Larantuka, *Indonesia* **193** M14
Larat, island, *Indonesia* **193** M18
Larche, *Fr.* **149** W15
Laredo, *Tex., U.S.* **105** P14
Laredo, *Sp.* **146** B11
Largo, *Fla., U.S.* **99** U7
Largo, Cayo, *Cuba* **118** H7
Lari *see* Lhari, *China* **184** L6
Larimore, *N. Dak., U.S.* **106** F7
Lárisa, *Gr.* **156** E10
Larkana, *Pak.* **181** P8
Larnaca (Lárnaka), *Cyprus* **162** P9
Larnaka Bay, *Cyprus* **162** P9
Lárnaka *see* Larnaca, *Cyprus* **162** P9
Larne, *N. Ire., U.K.* **145** N7
Larned, *Kans., U.S.* **107** T6
Larochette, *Lux.* **162** J10
Larose, *La., U.S.* **102** Q10
Larrimah, *N.T., Austral.* **221** Q8
Lars Christensen Coast, *Antarctica*
    239 F20
Larsen Bay, *Alas., U.S.* **113** M15
Larsen Ice Shelf, *Antarctica* **238** E6
Laruri, *India* **182** H15
Larvik, *Nor.* **142** L11
Larvotto Beach, *Monaco* **163** B23
Larvotto, *Monaco* **163** B22
Larvotto, peninsula, *Monaco* **163** B23

Las Animas, *Colo., U.S.* **109** N15
Las Cabezas de San Juan, *Sp.* **146** L8
Las Caguamas, Cayo, *Cuba* **118** K10
Las Cruces, *N. Mex., U.S.* **109** V12
Las Cuevas Bay, *Trin. & Tobago*
    123 N22
Las Flores, *Arg.* **134** M12
Las Heras, *Arg.* **134** K8
Las Heras, *Arg.* **135** T8
Las Lomitas, *Arg.* **134** F12
Las Marteínas, *Cuba* **118** H4
Las Palmas, *Sp.* **200** F3
Las Palmas, *Sp.* **214** Q5
Las Plumas, *Arg.* **135** R9
Las Tablas, *Pan.* **117** P21
Las Toscas, *Arg.* **134** H12
Las Tunas, *Cuba* **118** J12
Las Varillas, *Arg.* **134** K11
Las Vegas, *N. Mex., U.S.* **109** R13
Las Vegas, *Nev., U.S.* **111** U12
Lasa *see* Lhasa, *China* **184** L5
Lasan, *Indonesia* **192** H11
Lascahobas, *Haiti* **119** M17
Lascano, *Uru.* **134** L15
Lascelles, *Vic., Austral.* **221** Y12
Lāsh-e Juwayn, *Afghan.* **180** L3
Lashio, *Myanmar* **190** G6
Lashkar Gāh (Bost), *Afghan.* **180** L5
Lashkar, *India* **182** H6
Lashkari Bazar, ruins, *Afghan.*
    180 L5
Lasia, island, *Indonesia* **192** H4
Lassen Peak, *Calif., U.S.* **111** N4
Lassen Volcanic National Park,
    Calif., U.S. **111** N5
Lassiter Coast, *Antarctica* **238** F7
Lastoursville, *Gabon* **208** E9
Lastovo (Lagosta), island, *Croatia*
    154 J6
Lastovski Kanal, *Croatia* **154** H6
Lata Mountain, *American Samoa, U.S.*
    230 B10
Lata, *Samoa* **227** N17
Latacunga, *Ecua.* **130** K4
Latady Island, *Antarctica* **238** G6
Latakia *see* Al Lādhiqīyah, *Syr.*
    172 F7
L'Atalante Valley, *Pac. Oc.* **246** Q8
Latehar, *India* **182** J9
Latheron, *Scot., U.K.* **144** G10
Lathi, island, *Vanuatu* **229** E14
Latiano, *It.* **152** M15
Latina, *It.* **152** L9
Latium, *Eur.* **140** K7
Lató, ruin, *Gr.* **157** Q15
Latouche Treville, Cape, *Austral.*
    220 D4
Latrobe, Mount, *Austral.* **220** M13
Latta, *S.C., U.S.* **98** K11
Latur, *India* **182** M5
Latvia, *Eur.* **143** L16
Lau Basin, *Pac. Oc.* **246** L12
Lau Group (Eastern), *Fiji* **226** N8
Lau Ridge, *Pac. Oc.* **246** M11
Laucala, island, *Fiji* **226** N7
Laufen, *Switz.* **152** A3
Lauge Koch Kyst, *Greenland, Den.*
    88 F5
Laughlan Islands, *P.N.G.* **222** Q10
Laughlin Peak, *N. Mex., U.S.* **109** Q14
Lauhkaung, *Myanmar* **190** E6
Lauli'i, *Samoa* **227** P20
Lauliituai, *American Samoa, U.S.*
    230 D7
Launceston, *Tas., Austral.* **221** Y8
Laupāhoehoe, *Hawaii, U.S.* **115** M22
Laura *see* Majuro, island, *Marshall Is.*
    227 E20
Laura, *Qnsld., Austral.* **221** Q13
Laura, *Marshall Is.* **227** D19
Laurel, *Del., U.S.* **98** D14
Laurel, *Miss., U.S.* **102** M12
Laurel, *Nebr., U.S.* **107** N8
Laurel, *Mont., U.S.* **108** E10
Laurens, *S.C., U.S.* **98** K8
Laurens, *Iowa, U.S.* **106** M10
Laurentian Fan, *Atl. Oc.* **248** J6
Laurentian Scarp, *N. Amer.* **84** H8
Laurentian Trough, *Atl. Oc.* **248** H5
Laurentide Scarp, *N. Amer.* **84** H9
Lauria, *It.* **153** N12
Laurie Island, *Antarctica* **238** A5
Laurinburg, *N.C., U.S.* **98** J11
Lauritzen Bay, *Antarctica* **239** R16
Laurium *see* Lávrio, *Gr.* **157** K13
Laurium, *Mich., U.S.* **100** E6

Laurot Islands *see* Laut Kecil,
    Kepulauan, *Indonesia* **192** L11
Lausanne, *Switz.* **152** C1
Laut Kecil, Kepulauan (Laurot
    Islands), *Indonesia* **192** L11
Laut, island, *Indonesia* **192** G8
Laut, island, *Indonesia* **192** K12
Lautaro, Volcán, *Chile* **135** U6
Lauterbourg, *Fr.* **149** P16
Lautoka, *Fiji* **226** P4
Lauvi Point, *Solomon Is.* **228** F10
Lava Beds National Monument,
    Calif., U.S. **110** L4
Lava Hot Springs, *Idaho, U.S.* **108** H7
Lavaca Bay, *Tex., U.S.* **105** N18
Laval, *Fr.* **149** Q6
Lāvān, island, *Iran* **177** N17
Lavardac, *Fr.* **149** W7
Lavaur, *Fr.* **149** X9
Lavelanet, *Fr.* **149** Y9
Lavello, *It.* **152** M12
Laverne, *Okla., U.S.* **105** B14
Laverton, *W. Aust., Austral.* **221** V5
Lavina, *Mont., U.S.* **108** D10
Lavit, *Fr.* **149** X8
Lavonia, *Ga., U.S.* **98** K6
Lavrentiya, *Russ.* **161** A21
Lávrio (Laurium), *Gr.* **157** K13
Law Dome, *Antarctica* **239** M21
Law Promontory, *Antarctica* **239** D20
Law-Racoviță Station, *Antarctica*
    239 G20
Lawa, river, *S. Amer.* **131** G18
Lāwa'i, *Hawaii, U.S.* **114** F10
Lawdar, *Yemen* **174** P10
Lawena, valley, *Liech.* **162** Q2
Lawit, peak, *Indonesia* **192** H10
Lawqah, *Saudi Arabia* **174** C8
Lawra, *Ghana* **205** L14
Lawrence, *Mass., U.S.* **97** J14
Lawrence, *Ind., U.S.* **101** R9
Lawrence, *Kans., U.S.* **107** S10
Lawrenceburg, *Ind., U.S.* **101** S10
Lawrenceburg, *Tenn., U.S.* **103** G14
Lawrenceville, *Ga., U.S.* **98** L5
Lawrenceville, *Va., U.S.* **98** F12
Lawrenceville, *Ill., U.S.* **101** T7
Lawton, *Okla., U.S.* **105** E16
Laylá, *Saudi Arabia* **174** H10
Layou, *St. Vincent & the Grenadines*
    123 L16
Layou, river, *Dominica* **123** F19
Layshi, *Myanmar* **190** E4
Layton, *Utah, U.S.* **108** K7
Laytonville, *Calif., U.S.* **111** P2
Lazarev Bay, *Antarctica* **238** F6
Lazarev Ice Shelf, *Antarctica* **239** A15
Lázaro Cárdenas, *Mex.* **116** J9
Lazo, *Russ.* **161** F18
Le Barcarès, *Fr.* **149** Y10
Le Barp, *Fr.* **149** W6
Le Beausset, *Fr.* **149** Y14
Le Blanc, *Fr.* **149** S8
Le Borgne, *Haiti* **119** L16
Le Bouscat, *Fr.* **149** V6
Le Cap, *New Caledonia, Fr.* **231** Q21
Le Carbet, *Martinique, Fr.* **123** F22
Le Caylar, *Fr.* **149** X11
Le Chambon-Feugerolles, *Fr.*
    149 U12
Le Creusot, *Fr.* **149** S12
Le Diamant, *Martinique, Fr.* **123** G23
Le Donjon, *Fr.* **149** T11
Le Dorat, *Fr.* **149** T8
Le Faou, *Fr.* **149** Q3
Le François, *Martinique, Fr.* **123** F23
Le Gosier, *Guadeloupe, Fr.* **123** F15
Le Grand-Lemps, *Fr.* **149** U13
Le Grau-du-Roi, *Fr.* **149** X12
Le Havre, *Fr.* **149** N7
Le Kartala, peak, *Comoros* **215** M14
Le Lamentin, *Martinique, Fr.* **123** F23
Le Lion-d'Angers, *Fr.* **149** R6
Le Locle, *Switz.* **152** B2
Le Lorrain, *Martinique, Fr.* **123** E22
Le Madonie, range, *It.* **153** S10
Le Maire, Estrecho de, *Arg.* **135** Y10
Le Mans, *Fr.* **149** Q7
Le Marin, *Martinique, Fr.* **123** G23
Le Mars, *Iowa, U.S.* **106** M9
Le Mont-Dore, *Fr.* **149** U10
Le Montet, *Fr.* **149** T10
Le Morne Brabant, cape, *Mauritius*
    215 H18
Le Morne Rouge, *Martinique, Fr.*
    123 E22

Le Moule, *Guadeloupe, Fr.* **123** E16
Le Perthus, *Fr.* **149** Z10
Le Petit-Quevilly, *Fr.* **149** N8
Le Porge, *Fr.* **149** V6
Le Port, *Réunion, Fr.* **215** G15
Le Prêcheur, *Martinique, Fr.* **123** E21
Le Puy, *Fr.* **149** V11
Le Robert, *Martinique, Fr.* **123** F23
Le Roy, *N.Y., U.S.* **96** J6
Le St. Esprit, *Martinique, Fr.* **123** F23
Le Sueur, *Minn., U.S.* **106** K10
Le Tampon, *Réunion, Fr.* **215** H16
Le Teil, *Fr.* **149** W12
Le Vauclin, *Martinique, Fr.* **123** G24
Le Verdon-sur-Mer, *Fr.* **149** U6
Leachville, *Ark., U.S.* **102** F11
Lead Mountain, *Me., U.S.* **97** E18
Lead, *S. Dak., U.S.* **106** K1
Leadore, *Idaho, U.S.* **108** F6
Leadville, *Colo., U.S.* **108** M12
Leaf, river, *Miss., U.S.* **102** M12
Leakesville, *Miss., U.S.* **103** N13
Leakey, *Tex., U.S.* **105** L14
Lealui, *Zambia* **209** P15
Leandro N. Alem, *Arg.* **134** G14
Leane, Lough, *Ire.* **145** T3
Learmonth, *W. Aust., Austral.* **221** T1
l'Eau Claire, Lac à, *Que., Can.* **87** L18
Leava *see* Sigavé, *Wallis & Futuna Is., Fr.*
    230 Q11
Leavenworth, *Kans., U.S.* **107** S10
Leavenworth, *Wash., U.S.* **110** C5
Lebak, *Philippines* **193** F15
Lebango, *Congo* **208** E10
Lebanon, *Pa., U.S.* **96** P8
Lebanon, *N.H., U.S.* **97** H13
Lebanon, *Ind., U.S.* **101** R8
Lebanon, *Ohio, U.S.* **101** S11
Lebanon, *Ky., U.S.* **103** C17
Lebanon, *Tenn., U.S.* **103** E16
Lebanon, *Kans., U.S.* **107** R6
Lebanon, *Mo., U.S.* **107** U12
Lebanon, *Oreg., U.S.* **110** G3
Lebanon, *Asia* **172** H6
Lebanon, *Leb.* **172** J6
Lebap, *Turkm.* **178** L8
Lebec, *Calif., U.S.* **111** W7
Lebedyn, *Ukr.* **159** N8
Lebo, *D.R.C.* **209** B15
Lebombo Mountains, *S. Af.* **212** H10
Lębork, *Pol.* **151** B13
Lebowakgomo, *S. Af.* **212** H9
Lebrija, *Sp.* **146** L8
Lebu, *Chile* **135** N5
Lecce, *It.* **153** N15
Lecco, *It.* **152** D5
Lecco, Lago di, *It.* **152** D5
Lechainá, *Gr.* **156** J8
Lechang, *China* **185** N13
Leclaire Rise, *Ind. Oc.* **250** R9
Lecompte, *La., U.S.* **102** N8
Ledesma, *Sp.* **146** E8
Leduck Island, *Virgin Is., U.S.* **122** N11
Ledyanaya, *Russ.* **161** C21
Lee Stocking Island, *Bahamas* **122** D8
Lee, *Nev., U.S.* **111** N11
Leech Lake, *Minn., U.S.* **106** G10
Leedey, *Okla., U.S.* **105** C15
Leeds, *Ala., U.S.* **103** J15
Leeds, *N. Dak., U.S.* **106** E6
Leeds, *Eng., U.K.* **145** Q12
Leek, *Neth.* **148** G13
Leer, *Ger.* **150** C6
Leer, *S. Sudan* **199** H20
Lees Summit, *Mo., U.S.* **107** S11
Leesburg, *Va., U.S.* **98** C12
Leesburg, *Fla., U.S.* **99** T8
Leesville, *La., U.S.* **102** N7
Leeton, *N.S.W., Austral.* **221** Y13
Leeu-Gamka, *S. Af.* **212** M6
Leeuwarden, *Neth.* **148** G13
Leeuwin, Cape, *Austral.* **220** L2
Leeward Islands, *Lesser Antil.* **121** C15
Leeward Passage, *Virgin Is., U.S.*
    122 M9
Léfini, river, *Congo* **208** F10
Lefka (Lefke), *Cyprus* **162** P7
Lefkáda (Leucas), island, *Gr.* **156** G7
Lefkáda, *Gr.* **156** G7
Lefkímmi, *Gr.* **156** F6
Lefkoniko (Geçitkale), *Cyprus* **162** N9
Lefkoşa *see* Nicosia, *Cyprus* **162** N8
Lefkosía *see* Nicosia, *Cyprus* **162** N8
Lefroy, Lake, *Austral.* **220** J4
Legan, island, *Marshall Is.* **227** H17
Legat, *Mauritania* **204** G7

Legazpi, *Philippines* **193** D14
Legges Tor, peak, *Austral.* **220** L8
Leggett, *Calif. U.S.* **111** P2
Leghorn *see* Livorno, *It.* **152** H6
Legionowo, *Pol.* **151** D16
Legnago, *It.* **152** E7
Legnano, *It.* **152** D4
Legnica, *Pol.* **150** F12
Legune, *N.T. Austral.* **221** Q7
Leh, *India* **180** H15
Lehr, *N. Dak. U.S.* **106** H6
Lehrke Inlet, *Antarctica* **238** E7
Lehrte, *Ger.* **150** D8
Lehua, *Hawaii, U.S.* **114** E8
Leiah, *Pak.* **180** L10
Leicester, *Eng., U.K.* **145** S13
Leiden, *Neth.* **148** J12
Leifear *see* Lifford, *Ire.* **145** N6
Leigh Creek, *S. Aust. Austral.* **221** W10
Leigh, *N.Z.* **223** E20
Leikanger, *Nor.* **142** J10
Leimatu'a, *Tonga* **227** G23
Leine, river, *Ger.* **150** E7
Leinster (Cúige Laighean), region, *Ire.* **145** R6
Leipsoí, *Gr.* **157** K17
Leipsoí, island, *Gr.* **157** K17
Leipzig, *Ger.* **150** F9
Leiria, *Port.* **146** G5
Leisler, Mount, *Austral.* **220** F7
Leitchfield, *Ky. U.S.* **103** C15
Leiway, *Myanmar* **190** J5
Leiyang, *China* **185** M13
Leizhou Peninsula, *Asia* **168** J11
Leizhou, *China* **184** Q12
Lékana, *Congo* **208** F10
Lekei, *Gabon* **208** F10
Lekeleka, island, *Tonga* **227** M23
Lekitobi, *Indonesia* **193** K15
Léko, *Mali* **204** J11
Lékoni, *Gabon* **208** F10
Lékoni, river, *Gabon* **208** E9
Leksula, *Indonesia* **193** K15
Lekunberri, *Sp.* **147** B13
Leland, *Miss. U.S.* **102** J10
Lel'chytsy, *Belarus* **158** M5
Lele Harbor, *F.S.M.* **226** G12
Lele, *F.S.M.* **226** F12
Leleiwi Point, *Hawaii, U.S.* **115** N23
Lélépa, island, *Vanuatu* **229** J16
Leli, island, *Solomon Is.* **228** E11
Lelingluang, *Indonesia* **193** M17
Lelintah, *Indonesia* **193** K15
Lelystad, *Neth.* **148** H13
Léman, Lac (Lake Geneva), *Fr.* **149** T15
Lemanmanu, *P.N.G.* **222** P11
Lemanmanu, *P.N.G.* **228** A4
Lemei Rock, peak, *Wash. U.S.* **110** E4
Lemesós *see* Limassol, *Cyprus* **162** Q7
Lemgo, *Ger.* **150** E7
Lemhi Range, *Idaho, U.S.* **108** F6
Lemhi, river, *Idaho, U.S.* **108** F6
Lemieux Islands, *Nunavut, Can.* **87** G19
Lemmer, *Neth.* **148** H13
Lemmon, *S. Dak. U.S.* **106** H3
Lemmon, Mount, *Ariz. U.S.* **109** V8
Lemnos *see* Límnos, island, *Gr.* **157** E14
Lemoore, *Calif. U.S.* **111** U6
Lemotol, strait, *F.S.M.* **226** F8
Lempdes, *Fr.* **149** U11
Lemsid, *W. Sahara* **200** G4
Lena River Delta, *Asia* **168** C10
Lena Tablemount, *Ind. Oc.* **254** D8
Lena Trough, *Arctic Oc.* **253** J15
Lena, river, *Russ.* **161** G17
Lena, river, *Asia* **168** C10
Lencouacq, *Fr.* **149** W7
Lenda, river, *D.R.C.* **209** D18
Lendery, *Russ.* **158** C7
Lenger, *Kaz.* **178** K12
Lenghu, *China* **184** G7
Lengua de Vaca Point, *S. Amer.* **128** L4
Lenin Peak, *Kyrg. Taj.* **179** M14
Leningrad *see* Sankt-Peterburg, *Russ.* **158** F6
Leningrad, *Taj.* **178** N12
Leningradskaya Station, *Antarctica* **239** R16
Leningradskaya, *Russ.* **159** R10
Leninogorsk *see* Ridder, *Kaz.* **179** E17
Leninogorsk, *Russ.* **158** J16
Leninsk Kuznetskiy, *Russ.* **160** K12
Leninskiy, *Russ.* **158** H13

Leninskoye, *Russ.* **158** G13
Lennox, *S. Dak. U.S.* **106** M8
Lenoir City, *Tenn. U.S.* **103** F18
Lenoir, *N.C. U.S.* **98** H8
Lenora, *Kans. U.S.* **107** R5
Lenox, *Mass. U.S.* **97** K12
Lenox, *Iowa, U.S.* **107** Q10
Lens, *Fr.* **148** M10
Lensk, *Russ.* **161** J16
Lentekhi, *Rep. of Georgia* **171** B17
Lenti, *Hung.* **151** M13
Lentini, *It.* **153** T12
Lenya, *Myanmar* **191** P7
Léo, *Burkina Faso* **205** L15
Leoben, *Aust.* **150** L11
Léogâne, *Haiti* **119** M16
Leola, *S. Dak. U.S.* **106** J6
Leominster, *Mass. U.S.* **97** K14
Leon, *Iowa, U.S.* **107** Q11
Leon, *Eur.* **140** J2
León, *Mex.* **116** H10
León, *Nicar.* **117** M17
León, *Sp.* **146** C9
Léon, *Fr.* **149** X5
León, Montes de, *Sp.* **146** C8
Leon, river, *Tex. U.S.* **105** H16
Leonard, *Tex. U.S.* **105** F18
Leonard's, *Virgin Is. U.K.* **122** Q6
Leonarisso (Ziyamet), *Cyprus* **162** M10
Leonberg, *Ger.* **150** J6
Leonding, *Aust.* **150** K11
Leone Bay, *American Samoa, U.S.* **230** E5
Leone, *American Samoa, U.S.* **230** E6
Leonídio, *Gr.* **156** L11
Leonora, *W. Aust. Austral.* **221** V4
Leopold and Astrid Coast, *Antarctica* **239** H21
Leopoldsburg, *Belg.* **148** K13
Léopoldville *see* Kinshasa, *D.R.C.* **208** G10
Leoti, *Kans. U.S.* **107** T3
Lepanto, *Ark. U.S.* **102** F11
Lepe, *Sp.* **146** L6
Lepontine Alps, *Switz.* **152** C4
Lepsi, *Kaz.* **179** G16
Leptis Magna, ruin, *Lib.* **201** D18
Lercara Friddi, *It.* **153** S10
Léré, *Mali* **205** H13
Léré, *Chad* **206** K11
Lerma, *Sp.* **146** D11
Lerma, river, *N. Amer.* **84** N4
Lerneb, *Mali* **205** H13
Léros, island, *Gr.* **157** L17
Lerik, *Azerb.* **171** F22
Lerwick, *Scot. U.K.* **144** C12
Les Abymes, *Guadeloupe, Fr.* **123** E15
Les Anses-d'Arlets, *Martinique, Fr.* **123** G22
Les Avirons, *Réunion, Fr.* **215** H15
Les Bons, *Andorra* **162** J3
Les Borges Blanques, *Sp.* **147** D16
Les Cayes, *Haiti* **119** M15
Les Coteaux, *Trin. & Tobago* **123** N17
Les Écréhou, island, *Channel Is., U.K.* **145** Y12
Les Escaldes, *Andorra* **162** J3
Les Landes, *Fr.* **149** W6
Les Sables-d'Olonne, *Fr.* **149** T5
Les Saintes, island, *Guadeloupe, Fr.* **123** H14
Les Sept Îles, *Fr.* **149** P4
Les Soeurs, islands (The Sisters), *Seychelles* **215** M21
Les Trois-Îlets, *Martinique, Fr.* **123** G22
Leş, *Rom.* **154** C11
Lesbos *see* Lésvos, island, *Gr.* **157** F16
Leshan, *China* **184** L10
Leshukonskoye, *Russ.* **158** B12
Leskovac, *Serb.* **154** H11
Leslie, *Ark. U.S.* **102** F8
Lesmont, *Fr.* **149** Q12
Lesnoy, *Russ.* **158** E15
Lesnoye, *Russ.* **158** G8
Lesosibirsk, *Russ.* **161** K13
Lesotho, *Af.* **212** K8
Lesozavodsk, *Russ.* **161** L21
L'Espérance Rock, *Kermadec Is., N.Z.* **224** M9
Lessay, *Fr.* **149** N6

Lesser Antilles, *N. Amer.* **121** B14
Lesser Caucasus, range, *Rep. of Georgia* **171** C17
Lesser Khingan Range, *Asia* **168** E11
Lesser Slave Lake, *Alta. Can.* **86** K10
Lesser Sunda Islands, *Indonesia* **192** M13
Lésvos (Lesbos, Mytilíni), island, *Gr.* **157** F16
Leszno, *Pol.* **151** E13
Letchworth, *Eng. U.K.* **145** T14
Letea, Ostrov, *Rom.* **155** E18
Lethbridge, *Alta. Can.* **86** N10
Lethem, *Guyana* **131** G15
Leti Islands, *Asia* **168** M14
Leti, Kepulauan, *Indonesia* **193** M16
Leticia, *Col.* **130** M9
Letka, *Russ.* **158** F14
Letlhakane, *Botswana* **212** G7
Letnyaya Zolotitsa, *Russ.* **158** B9
Letsâok-aw Kyun, island, *Myanmar* **191** P6
Léua, *Angola* **209** M13
Leuca, *It.* **153** P16
Leucas *see* Lefkada, island, *Gr.* **156** G7
Leucate, *Fr.* **149** Y10
Leucate, Étang de, *Fr.* **149** Y10
Leulumoega, *Samoa* **227** P19
Leuser, peak, *Indonesia* **192** G4
Leusoali'i, *American Samoa, U.S.* **230** B11
Levan, *Utah, U.S.* **108** M7
Levanger, *Nor.* **142** G11
Levant Coast, *Asia* **168** F3
Levant, Île du, *Fr.* **149** Y15
Levantine Basin, *Atl. Oc.* **248** K15
Levanto, *It.* **152** G5
Levelland, *Tex. U.S.* **104** F11
Leven Bank, *Ind. Oc.* **250** J6
Leven Pond, *Grenada* **123** J23
Leverkusen, *Ger.* **150** F5
Levet, *Fr.* **149** S10
Levice, *Slovakia* **151** K14
Levídi, *Gr.* **156** K10
Levin, *N.Z.* **223** J20
Lévis, *Que. Can.* **87** P20
Levittown, *N.Y. U.S.* **96** N12
Levittown, *Pa. U.S.* **96** P10
Lévrier Bay, *Af.* **198** E1
Levski, *Bulg.* **155** H15
Levu, Vanua, *Fiji* **224** J9
Levuka, *Fiji* **226** P6
Lewellen, *Nebr. U.S.* **107** P3
Lewes, *Del. U.S.* **98** C15
Lewes, *Eng. U.K.* **145** V14
Lewis and Clark National Historical Park, *Oreg. U.S.* **110** E2
Lewis and Clark Range, *Mont. U.S.* **108** C6
Lewis Pass, pass, *N.Z.* **223** K18
Lewis Range, *Austral.* **220** E7
Lewis *see* Eilean Leòdhais, island, *Scot. U.K.* **144** G7
Lewis Smith Lake, *Ala. U.S.* **103** H14
Lewis, *Kans. U.S.* **107** U5
Lewis, river, *Wash. U.S.* **110** E4
Lewisburg, *W. Va. U.S.* **98** E9
Lewisburg, *Tenn. U.S.* **103** F15
Lewiston, *Me. U.S.* **97** G16
Lewiston, *Idaho, U.S.* **108** D3
Lewistown, *Pa. U.S.* **96** N6
Lewistown, *Ill. U.S.* **101** Q4
Lewistown, *Mont. U.S.* **108** C9
Lewisville, *Ark. U.S.* **102** J7
Lexington Park, *Md. U.S.* **98** D14
Lexington, *N.C. U.S.* **98** H10
Lexington, *Va. U.S.* **98** E10
Lexington, *Miss. U.S.* **102** K11
Lexington, *Ky. U.S.* **103** B18
Lexington, *Tenn. U.S.* **103** F13
Lexington, *Mo. U.S.* **107** S11
Lexington, *Nebr. U.S.* **107** Q5
Lexington, *Oreg. U.S.* **110** F6
Leyte Gulf, *Philippines* **193** E15
Leyte, island, *Philippines* **193** E15
Lezha, *Russ.* **158** G10
Lezhë, *Alban.* **154** K9
Łeki, *Azerb.* **171** D21
Länkäran, *Azerb.* **171** F22
L'gov, *Russ.* **158** M8
Lhari (Lari), *China* **184** L6
Lhasa (Lasa), *China* **184** L5
Lhokseumawe, *Indonesia* **192** G4

L'Hospitalet de Llobregat, *Sp.* **147** E17
Li Yubu, *S. Sudan* **203** X9
Liancourt Rocks *see* Dokdo, *S. Korea* **188** M5
Liangbuaya, *Indonesia* **192** J12
Liannan, *China* **185** N13
Lianyungang (Xinpu), *China* **185** J14
Liao, river, *China* **185** E15
Liaodong Wan, *China* **185** F15
Liaoyuan, *China* **185** E16
Liaquatpur, *Pak.* **181** N10
Liard, river, *N.W.T. Can.* **86** H10
Liari, *Pak.* **181** R6
Lib Island, *Marshall Is.* **227** C16
Libby, *Mont. U.S.* **108** B4
Libenge, *D.R.C.* **208** C12
Liberal, *Kans. U.S.* **107** V4
Liberec, *Czechia* **150** G11
Liberia, *C.R.* **117** N18
Liberia, *Af.* **204** P10
Liberta, *Antigua & Barbuda* **123** B21
Libertador General San Martín, *Arg.* **134** E9
Liberty, *N.Y. U.S.* **96** L10
Liberty, *S.C. U.S.* **98** J7
Liberty, *Tex. U.S.* **105** L20
Liberty, *Mo. U.S.* **107** S10
Libni, Gebel, *Egypt* **173** P3
Liboi, *Kenya* **211** S9
Libourne, *Fr.* **149** V7
Libreville, *Gabon* **208** D7
Libya, *Af.* **201** G17
Libyan Desert, *Af.* **198** D7
Libyan Plateau (Aḑ Ḑiffah), *Lib.* **201** E22
Libyan Plateau, *Af.* **198** D7
Licantén, *Chile* **134** L6
Licata, *It.* **153** T10
Lice, *Turk.* **171** G15
Lichinga, *Mozambique* **212** C11
Lichtenau Fjord *see* Alluitsup Kangerlua, *Greenland, Den.* **89** T7
Lichtenfels, *Ger.* **150** G8
Licking, *Mo. U.S.* **107** U13
Licking, river, *Ky. U.S.* **103** B18
Ličko Lešće, *Croatia* **154** F5
Licosa, Punta, *It.* **153** N11
Lida, *Belarus* **158** K3
Lidgerwood, *N. Dak. U.S.* **106** H8
Lidköping, *Sw.* **142** L12
Lido di Jesolo, *It.* **152** D8
Lido di Ostia *see* Lido di Roma, *It.* **152** L8
Lido di Roma (Lido di Ostia), *It.* **152** L8
Lido, *Niger* **206** H6
Liechtenstein, *Eur.* **162** C3
Liège, *Belg.* **148** L13
Lieksa, *Fin.* **143** F17
Liénart, *D.R.C.* **209** C16
Lienz, *Aust.* **150** M9
Liepāja, *Latv.* **143** M15
Lier, *Belg.* **148** K12
Liestal, *Switz.* **152** A3
Liffey, river, *Ire.* **145** R6
Lifford (Leifear), *Ire.* **145** N6
Lifou (Lifu), island, *New Caledonia, Fr.* **231** P23
Lifu *see* Lifou, island, *New Caledonia, Fr.* **231** P23
Lifuka, island, *Tonga* **227** L24
Ligeti Ridge, *Atl. Oc.* **249** X8
Lightbourne Cay, *Bahamas* **122** J4
Lightning Ridge, *N.S.W. Austral.* **221** W14
Ligneuville, *Belg.* **148** M13
Ligny-en-Barrois, *Fr.* **149** P13
Ligurian Sea, *It.* **152** H4
Liha Point, *N.Z.* **229** P22
Lihir Group, *P.N.G.* **222** N10
Lihir, island, *P.N.G.* **222** N10
Lihu'e, *Hawaii, U.S.* **114** F10
Lijiang, *China* **184** M8
Likasi, *D.R.C.* **209** L17
Likati, *D.R.C.* **209** C15
Likiep Atoll, *Marshall Is.* **227** C16
Likimi, *D.R.C.* **209** C14
Likoto, *D.R.C.* **209** E16
Likoto, *D.R.C.* **209** E16
Likouala, river, *Congo* **208** E11
Liku, *N.Z.* **229** Q22
Likuone Bay, *Tonga* **227** G23
Likupang, *Indonesia* **193** H15
Likwangoli, *S. Sudan* **203** W12

Lilbourn, *Mo. U.S.* **107** V16
L'Île-Bouchard, *Fr.* **149** S8
L'Île-Rousse, *Fr.* **149** X2
Lilian Point, *Kiribati* **226** M11
Liliuokalani Ridge, *Pac. Oc.* **246** F12
Lille, *Fr.* **148** L10
Lillehammer, *Nor.* **142** J11
Lillesand, *Nor.* **142** L10
Lillooet, *B.C. Can.* **86** L8
Lilo Viejo, *Arg.* **134** G10
Lilongwe, *Malawi* **212** C10
Lim Fjord, *Atl. Oc.* **140** E7
Lim, river, *Serb.* **154** H9
Lima Reservoir, *Mont. U.S.* **108** F7
Lima, *Ohio, U.S.* **101** Q11
Lima, *Mont. U.S.* **108** F7
Lima, *Peru* **132** H3
Lima, river, *Port.* **146** D5
Liman Beren, lake, *Russ.* **159** R13
Liman, *Russ.* **159** R15
Liman, *Azerb.* **171** F22
Limarí, Río, *Chile* **134** J6
Limassol (Lemesós), *Cyprus* **162** Q7
Limavady, *N. Ire. U.K.* **145** N6
Limay, river, *S. Amer.* **128** N4
Limbara, Monte, *It.* **152** M4
Limbdi, *India* **182** K3
Limbe, *Cameroon* **206** N9
Limbé, *Haiti* **119** L16
Limburg, *Ger.* **150** G6
Limeira, *Braz.* **133** M13
Limenária, *Gr.* **157** C14
Limerick (Luimneach), *Ire.* **145** S4
Limestone, *Me. U.S.* **97** B18
Limfjorden, lake, *Den.* **142** M11
Limia, river, *Sp.* **146** D6
Limín *see* Thásos, *Gr.* **157** C14
Liminka, *Fin.* **143** F15
Limmen Bight, *Austral.* **220** B9
Límni, *Gr.* **156** G12
Límnos (Lemnos), island, *Gr.* **157** E14
Limoges, *Fr.* **149** U8
Limon, *Colo. U.S.* **108** M14
Limón, *C.R.* **117** N19
Limon, Mount, *Mauritius* **215** J20
Limone Piemonte, *It.* **152** F2
Limousin, *Eur.* **140** H4
Limoux, *Fr.* **149** Y10
Limpopo, river, *Af.* **198** N8
Limu, island, *Tonga* **227** M23
Limufuafua Point, *N.Z.* **229** R21
Līnah, *Saudi Arabia* **174** D9
Linapacan, island, *Philippines* **193** D13
Linares, *Mex.* **116** F11
Linares, *Chile* **134** M6
Linares, *Sp.* **146** K11
Linariá, *Gr.* **157** G14
Lincang, *China* **184** N8
Linchuan, *China* **185** M14
Lincoln City, *Oreg. U.S.* **110** F2
Lincoln Heights, *Eng. U.K.* **145** R13
Lincoln Island, island, *S. China Sea* **192** B10
Lincoln Park, *Mich. U.S.* **100** M12
Lincoln Sea, *N. Amer.* **84** B7
Lincoln, *Me. U.S.* **97** E18
Lincoln, *N.H. U.S.* **97** G14
Lincoln, *Ill. U.S.* **100** R5
Lincoln, *Mich. U.S.* **100** J12
Lincoln, *Kans. U.S.* **107** S7
Lincoln, *Nebr. U.S.* **107** Q8
Lincoln, *Calif. U.S.* **111** Q4
Lincoln, *Arg.* **134** L11
Lincoln, *Eng. U.K.* **145** R13
Lincolnton, *Ga. U.S.* **98** L7
Lincolnton, *N.C. U.S.* **98** H8
Lind, *Wash. U.S.* **110** D7
Linda Downs, *Qnsld. Austral.* **221** T10
Lindale, *Ga. U.S.* **98** K4
Lindale, *Tex. U.S.* **105** G19
Lindau, *Ger.* **150** L7
Lindbergh Bay, *Virgin Is. U.S.* **122** N8
Linden, *Ala. U.S.* **103** L14
Linden, *Tex. U.S.* **105** F20
Linden, *Guyana* **131** F15
Lindenow Fjord *see* Kangerlussuatsiaq, *Greenland, Den.* **89** T8
Lindenows Valley, *Atl. Oc.* **253** R20
Lindesnes, cape, *Nor.* **142** M10
Lindi Canyon, *Ind. Oc.* **250** H5
Lindi, *Tanzania* **211** Y8
Lindi, river, *D.R.C.* **209** D17;E18
Lindisfarne *see* Holy Island, *Eng. U.K.* **145** M12
Líndos, *Gr.* **157** N20

Lindsay, *Okla., U.S.* **105** D16
Lindsay, *Mont., U.S.* **108** C13
Lindsay, *Calif., U.S.* **111** U7
Lindsborg, *Kans., U.S.* **107** T7
Line Islands, *Pac. Oc.* **224** F12
Linfen, *China* **184** H12
Lingao, *China* **184** Q12
Lingen, *Ger.* **150** D6
Lingga, island, *Indonesia* **192** J7
Lingga, Kepulauan, *Indonesia* **192** J7
Lingle, *Wyo., U.S.* **108** J14
Lingomo, *D.R.C.* **209** D14
Linguère, *Senegal* **204** H7
Lingyuan, *China* **185** F14
Linhai, *China* **185** L15
Linhares, *Braz.* **133** L16
Linhe, *China* **184** G11
Linköping, *Sw.* **143** L13
Linn, *Tex., U.S.* **105** Q16
Linnhe, Loch, *Scot., U.K.* **144** K8
Linqing, *China* **185** H13
Linstead, *Jam.* **122** J9
Lintao, *China* **184** J10
Linton, *Ind., U.S.* **101** S8
Linton, *N. Dak., U.S.* **106** H5
Linxi, *China* **185** E14
Linxia, *China* **184** J9
Linyi, *China* **185** J14
Linz, *Aust.* **150** K11
Lio Matoh, *Malaysia* **192** H11
Lion, Golfe du, *Fr.* **149** Y12
Lionel Town, *Jam.* **122** K9
Lios na gCearrbhach *see* Lisburn,
    N. Ire., U.K. **145** P7
Lios Tuathail *see* Listowel, *Ire.*
    145 S3
Liot Point, *N.W.T., Can.* **86** D11
Liot, island, *P.N.G.* **222** M7
Liouesso, *Congo* **208** D11
Lipa, *Philippines* **193** D13
Lipari, Isola, *It.* **153** R11
Lipari, Isole (Eolie), *It.* **153** R11
Lipetsk, *Russ.* **158** M10
Lipno, *Pol.* **151** D14
Līpoa Point, *Hawaii, U.S.* **115** J18
Lipova, *Rom.* **154** D11
Lipscomb, *Tex., U.S.* **105** B14
Liptovský Mikuláš, *Slovakia* **151** J15
Lipu Pass, *China, India, Nepal* **182** F8
Lipu, *China* **184** N12
Lira, *Uganda* **211** R4
Liranga, *Congo* **208** E12
Liri, river, *It.* **152** L10
Liria, *S. Sudan* **203** Y12
Līsakovsk, *Kaz.* **178** D9
Lisala, *D.R.C.* **209** C14
Lisboa (Lisbon), *Port.* **146** H4
Lisbon Falls, *Me., U.S.* **97** G16
Lisbon *see* Lisboa, *Port.* **146** H4
Lisbon, *N. Dak., U.S.* **106** H7
Lisburn (Lios na gCearrbhach),
    N. Ire., U.K. **145** P7
Lisburne, Cape, *Alas., U.S.* **112** C12
Liscannor Bay, *Ire.* **145** R3
Lisieux, *Fr.* **149** P7
Liski, *Russ.* **159** N10
L'Isle-sur-la-Sorgue, *Fr.* **149** X13
Lismore, *Vic., Austral.* **221** Z12
Lismore, island, *Scot., U.K.* **144** K8
Lissa *see* Vis, island, *Croatia* **154** H6
Lister, Mount, *Antarctica* **239** N14
Listowel (Lios Tuathail), *Ire.* **145** S3
Lit-et-Mixe, *Fr.* **149** W6
Litang, *China* **184** L8
Litani, river, *S. Amer.* **131** H18
Lîtâni, river, *Leb.* **172** K6
Litchfield, *Ill., U.S.* **101** S5
Litchfield, *Minn., U.S.* **106** J10
Líthino, Ákra, *Gr.* **157** R14
Lithuania, *Eur.* **143** M16
Litke Trough, *Arctic Oc.* **253** H15
Litóchoro, *Gr.* **156** D10
Little Abaco, *Bahamas* **122** A8
Little America Basin, *Pac. Oc.* **255** J14
Little Andaman, island, *India* **183** S14
Little Barrier Island *see* Hauturu-o-
    Toi, *N.Z.* **223** E20
Little Bay, *Gibraltar, U.K.* **163** Q23
Little Belt Mountains, *Mont., U.S.*
    108 D9
Little Bighorn Battlefield National
    Monument, *Mont., U.S.* **108** E11
Little Bighorn, river, *Mont., Wyo., U.S.*
    108 F11
Little Bitter Lake *see* Murrat el
    Sughra, Buheirat, *Egypt* **173** Q1

Little Blue, river, *Nebr., U.S.* **107** R7
Little Cayman, island, *Cayman Is., U.K.*
    118 L8
Little Chute, *Wis., U.S.* **100** J6
Little Coco Island, *Myanmar* **190** M3
Little Colorado, river, *Ariz., U.S.*
    109 R7
Little Darby Island, *Bahamas* **122** D7
Little Dix Bay, *Virgin Is., U.K.* **122** Q8
Little Exuma, *Bahamas* **122** F9
Little Falls, *N.Y., U.S.* **96** J10
Little Falls, *Minn., U.S.* **106** H10
Little Fork, river, *Minn., U.S.* **106** E11
Little Hans Lollik Island,
    Virgin Is., U.S. **122** L8
Little Harbour, *Bahamas* **122** B9
Little Inagua Island, *Bahamas* **119** J16
Little Jost Van Dyke, island,
    Virgin Is., U.K. **122** L11
Little Karoo, mountains, *S. Af.*
    212 M6
Little London, *Jam.* **122** H6
Little Lost, river, *Idaho, U.S.* **108** F6
Little Marais, *Minn., U.S.* **106** F13
Little Missouri, river, *U.S.* **90** C10
Little Nicobar, island, *India* **183** U15
Little Powder, river, *Mont., Wyo., U.S.*
    108 F13
Little Red, river, *Ark., U.S.* **102** F8
Little Rock, *Ark., U.S.* **102** G8
Little Salt Lake, *Utah, U.S.* **109** P6
Little San Salvador, island, *Bahamas*
    119 E13
Little Sioux, river, *Iowa, U.S.* **107** N9
Little Snake, river, *Colo., Wyo., U.S.*
    108 L10
Little Sound, *Cayman Is., U.K.* **122** M3
Little Sounds (Port Royal Bay),
    Bermuda **119** C14
Little St. James Island, *Virgin Is., U.S.*
    122 N9
Little Suamico, *Wis., U.S.* **100** J7
Little Thatch Island, *Virgin Is., U.K.*
    122 M11
Little Thatch Island, *Virgin Is., U.K.*
    122 R5
Little Tobago Island, *Virgin Is., U.K.*
    122 L9
Little Tobago, *Trin. & Tobago* **123** N19
Little Wabash, river, *Ill., U.S.* **101** S6
Little, river, *Nfld. & Lab., Que., Can.*
    87 L21
Little, river, *La., U.S.* **102** M8
Little, river, *Okla., U.S.* **105** E19
Little, river, *Tex., U.S.* **105** K17
Littlefield, *Tex., U.S.* **104** E11
Littlefield, *Ariz., U.S.* **109** Q5
Littlefork, *Minn., U.S.* **106** E11
Littlehampton, *Eng., U.K.* **145** V13
Littleton, *N.H., U.S.* **97** G14
Littleton, *Colo., U.S.* **108** M13
Liuli, *Tanzania* **211** Z5
Liuwa Plain, *Zambia* **209** N15
Liuzhou, *China* **184** N12
Livadeiá, *Gr.* **156** H11
Liváki, *Gr.* **157** L14
Livádia, *Gr.* **157** M18
Līvāni, *Latv.* **143** L17
Live Oak, *Fla., U.S.* **99** R7
Livermore, *Ky., U.S.* **103** C15
Livermore, *Calif., U.S.* **111** S4
Livermore, Mount, *Tex., U.S.* **104** L12
Liverpool Bay, *N.W.T., Can.* **86** D10
Liverpool Land, *Greenland, Den.*
    88 K14
Liverpool Range, *Austral.* **220** K14
Liverpool, *N.S., Can.* **87** P22
Liverpool, *Eng., U.K.* **145** R10
Livingston Island, *Antarctica* **238** C4
Livingston, *Ala., U.S.* **103** L13
Livingston, *Tenn., U.S.* **103** E17
Livingston, *Tex., U.S.* **105** K20
Livingston, *Mont., U.S.* **108** E8
Lívingston, *Guatemala* **117** K16
Livingstone Memorial, *Zambia*
    209 M19
Livingstone, *Zambia* **209** Q16
Livny, *Russ.* **158** M9
Livonia, *Mich., U.S.* **100** M12
Livonia, *Eur.* **140** E9
Livorno (Leghorn), *It.* **152** H6
Livron, *Fr.* **149** V11
Liwale, *Tanzania* **211** Y7
Lizard Point, *Eng., U.K.* **145** X8
Lizarra *see* Estella, *Sp.* **147** C13
Ljubljana, *Slov.* **154** D5
Ljusdal, *Sw.* **143** H13

Llamara, Salar de, *Chile* **134** D7
Llandrindod *see* Llandrindod Wells,
    Wales, U.K. **145** T10
Llandrindod Wells (Llandrindod),
    Wales, U.K. **145** T10
Llanes, *Sp.* **146** B10
Llano Estacado (Staked Plain), *U.S.*
    90 K10
Llano, *Tex., U.S.* **105** K15
Llano, river, *Tex., U.S.* **105** K15
Llanos, plain, *S. Amer.* **128** B4
Llanquihue, Lago, *Chile* **135** Q6
Llavorsí, *Sp.* **147** C16
Lleida, *Sp.* **147** D16
Llerena, *Sp.* **146** J8
Lleyn Peninsula, *Wales, U.K.* **145** S9
Llica, *Bol.* **132** L6
Llívia, *Sp.* **147** C17
Llodio, *Sp.* **146** B12
Llorts, *Andorra* **162** H3
Lloydminster, *Alta., Can.* **86** L11
Llucmajor, *Sp.* **147** G18
Lluidas Vale, *Jam.* **122** J9
Llyswen, *Wales, U.K.* **145** T10
Ló, Ponta do, *Sao Tome and Principe*
    215 D20
Lô, river, *Vietnam* **190** G10
Loa, *Utah, U.S.* **109** N7
Loa, river, *Chile* **134** D7
Loaita Island, *S. China Sea* **192** E11
Loange, river, *D.R.C.* **209** H13
Loango, *Congo* **208** H8
Lobamba, *Eswatini* **212** J9
Lobatse, *Botswana* **212** H7
Lobaye, river, *Cen. Af. Rep.* **207** N13
Lobería, *Arg.* **135** N12
Lobería, *Arg.* **135** Q10
Lobikaere, island, *Marshall Is.*
    227 D20
Lobito, *Angola* **208** M9
Lobitos, *Peru* **132** E1
Lobo, *Indonesia* **193** K19
Lobos, *Arg.* **134** L12
Lobos, Cay, *Bahamas* **118** H11
Lobuya, *Russ.* **161** E19
Locana, *It.* **152** D2
Locarno, *Switz.* **152** C4
Loch Garman *see* Wexford, *Ire.*
    145 T7
Loch na Madadh (Lochmaddy),
    Scot., U.K. **144** H6
Lochcarron, *Scot., U.K.* **144** H8
Lochearnhead, *Scot., U.K.* **144** K9
Lochgilphead, *Scot., U.K.* **144** L8
Lochmaddy *see* Loch na Madadh,
    Scot., U.K. **144** H6
Lochsa, river, *Idaho, U.S.* **108** D4
Lochy, Loch, *Scot., U.K.* **144** J8
Lock Haven, *Pa., U.S.* **96** M6
Lockesburg, *Ark., U.S.* **102** H6
Lockhart River, *Qnsld., Austral.* **221** P13
Lockhart, *Tex., U.S.* **105** L17
Lockport, *N.Y., U.S.* **96** J5
Lockwood Ø, *Greenland, Den.* **88** A9
Locminé, *Fr.* **149** Q4
Locri, *It.* **153** R13
Lod (Lydda), *Israel* **172** M5
Lodeynoye Pole, *Russ.* **158** E8
Lodge Creek, *Can., U.S.* **108** A9
Lodge Grass, *Mont., U.S.* **108** E11
Lodgepole Creek, *Nebr., U.S.* **107** P1
Lodgepole, *Nebr., U.S.* **107** P2
Lodhran, *Pak.* **181** N10
Lodi, *Calif., U.S.* **111** R4
Lodi, *It.* **152** E5
Lodja, *D.R.C.* **209** G15
Lodwar, *Kenya* **211** Q5
Łódź, *Pol.* **151** E15
Loei, *Thai.* **190** K8
Lofanga, island, *Tonga* **227** M23
Loffa, river, *Liberia* **204** N10
Lofoten Basin, *Arctic Oc.* **253** H19
Lofoten, archipelago, *Nor.* **142** D12
Lofty Range, *Austral.* **220** G3
Lofty Ranges, Mount, *Austral.*
    220 K10
Logan, *W. Va., U.S.* **98** E8
Logan, *Ohio, U.S.* **101** R14
Logan, *Iowa, U.S.* **107** P9
Logan, *Kans., U.S.* **107** R5
Logan, *Utah, U.S.* **108** J7
Logan, *N. Mex., U.S.* **109** R15
Logan, Mount, *Yukon, Can.* **86** F7
Logan, Mount, *Wash., U.S.* **110** A5
Logansport, *Ind., U.S.* **101** Q9
Logansport, *La., U.S.* **102** L6

Logashkino, *Russ.* **161** D18
Logone, river, *Af.* **198** G6
Logoniégué, *Burkina Faso* **205** M13
Logroño, *Sp.* **146** C12
Loh, island, *Vanuatu* **229** C13
Lohardaga, *India* **182** K10
Loharghat, *India* **182** H13
Lohatanjona Marohata, cape,
    Madagascar **213** G14
Lohne, *Ger.* **150** D6
Lôho, *Côte d'Ivoire* **205** N13
Loi Mwe, *Myanmar* **190** H7
Loi, island, *Marshall Is.* **227** J18
Lõ'Ihi Seamount, *Pac. Oc.* **247** G14
Loile, river, *D.R.C.* **209** E14
Loire, river, *Fr.* **149** Q10
Loja, *Ecua.* **130** M4
Loja, *Sp.* **146** L10
Loji, *Indonesia* **193** K16
Lojjairok, island, *Marshall Is.* **227** J18
Lojwa, island, *Marshall Is.* **227** A21
Loka, *S. Sudan* **203** Y11
Lokandu, *D.R.C.* **209** F16
Lokichokio, *Kenya* **211** Q5
Lokila, *S. Sudan* **203** Y12
Lokitaung, *Kenya* **211** Q6
Lokolama, *D.R.C.* **209** F15
Lokolenge, *D.R.C.* **209** D15
Lokolo, river, *D.R.C.* **209** E13
Lokomo, *Cameroon* **206** P12
Lokoro, river, *D.R.C.* **209** F13
Lokossa, *Benin* **205** P17
Loks Land, *Nunavut, Can.* **87** H19
Lokuru, *Solomon Is.* **228** E7
Lokutu, *D.R.C.* **209** D15
Lokwakangole, *Kenya* **211** Q6
Lol, river, *S. Sudan* **203** V8,10
Lola, *Guinea* **204** N11
Lolgorien, *Kenya* **211** T5
Lolimi, *S. Sudan* **203** Y13
Loliondo, *Tanzania* **211** T5
Lolland, island, *Den.* **142** N12
Lolo Pass, *Idaho, Mont., U.S.* **108** D5
Lolo, *D.R.C.* **209** C15
Lolowaï, *Vanuatu* **229** F15
Lom, *Nor.* **142** J10
Lom, *Bulg.* **155** G13
Loma Mountains, *Guinea, Sierra Leone*
    204 M10
Lomaloma, *Fiji* **226** P8
Lomami, river, *D.R.C.* **209** E16;H16
Lomawai, *Fiji* **226** P4
Lombadina, *W. Aust., Austral.* **221** Q4
Lombak Basin, *Ind. Oc.* **251** H17
Lombardy, Plain of, *Eur.* **140** J6
Lombez, *Fr.* **149** X8
Lomblem Island, *Indonesia* **193** M14
Lombok, island, *Indonesia* **192** M12
Lombok, Selat, *Indonesia* **192** M11
Lomé, *Togo* **205** P17
Lomela, *D.R.C.* **209** F15
Lomela, river, *D.R.C.* **209** E14
Lometa, *Tex., U.S.* **105** J16
Lomié, *Cameroon* **206** P11
Lomlom Island, *Solomon Is.* **228** L6
Lomond, Loch, *Scot., U.K.* **144** L9
Lomonosov Ridge, *Arctic Oc.* **252** G11
Lomonosov, *Russ.* **158** F6
Lompoc, *Calif., U.S.* **111** W5
Łomża, *Pol.* **151** C17
Lonauli, *India* **182** M3
Loncoche, *Chile* **135** P6
Loncopué, *Arg.* **135** N7
London Bridge Island, *Grenada*
    123 J24
London, *Ont., Can.* **87** R18
London, *Ohio, U.S.* **101** R12
London, *Ky., U.S.* **103** D18
London, *Eng., U.K.* **145** U14
London, *Kiribati* **226** P10
Londonderry (Derry), *N. Ire., U.K.*
    145 N6
Londonderry, *St. Lucia* **123** L13
Londonderry, Cape, *Austral.* **220** B6
Londonderry, Isla, *Chile* **135** Y8
Londrina, *Braz.* **132** M12
Lone Pine, *Calif., U.S.* **111** T8
Lonerock, *Oreg., U.S.* **110** F6
Long Bay, *Jam.* **122** H5
Long Bay, *Jam.* **122** K8
Long Bay, *Virgin Is., U.K.* **122** R5
Long Bay, *Virgin Is., U.K.* **122** R5
Long Bay, *Barbados* **123** L20
Long Beach, *N.C., U.S.* **98** K12

Long Beach, *Miss., U.S.* **102** P12
Long Beach, *Wash., U.S.* **110** D2
Long Beach, *Calif., U.S.* **111** X8
Long Branch, *N.J., U.S.* **96** P11
Long Cay, *Bahamas* **119** H14
Long Creek, *Oreg., U.S.* **110** G7
Long Eaton, *Eng., U.K.* **145** S12
Long Island Sound, *U.S.* **97** M13
Long Island, *Nunavut, Can.* **87** L17
Long Island, *N.Y., U.S.* **97** N13
Long Island, *Bahamas* **119** G14
Long Island, *Antigua & Barbuda*
    123 A21
Long Island, *P.N.G.* **222** P8
Long Key, *Fla., U.S.* **99** Y10
Long Khánh, *Vietnam* **191** P12
Long Lake, *N. Dak., U.S.* **106** G5
Long Look, *Virgin Is., U.K.* **122** Q7
Long Pine, *Nebr., U.S.* **107** N5
Long Point, *Bahamas* **122** E3
Long Point, *Virgin Is., U.S.* **122** N9
Long Point, *Virgin Is., U.S.* **122** R2
Long Prairie, *Minn., U.S.* **106** H10
Long Strait, *Arctic Oc.* **168** A11
Long Xuyên, *Vietnam* **191** P11
Longa, *Angola* **208** N12
Longa, river, *Angola* **208** L10
Longboat Key, *Fla., U.S.* **99** V7
Longdenville, *Trin. & Tobago* **123** P22
Longeau, *Fr.* **149** R13
Longford (An Longfort), *Ire.* **145** Q5
Longga'er *see* Lunggar, *China*
    184 K3
Longglat, *Indonesia* **192** H12
Longiram, *Indonesia* **192** J11
Longjiang, *China* **185** C15
Longjing, *China* **185** E17
Longju, *India* **182** F14
Longlac, *Ont., Can.* **87** N16
Longmont, *Colo., U.S.* **108** L13
Longnawan, *Indonesia* **192** H11
Longomapu, *Tonga* **227** H23
Longreach, *Qnsld., Austral.* **221** T13
Longren *see* Luring, *China* **184** J3
Longs Peak, *Colo., U.S.* **108** L13
Longtam, *S. Sudan* **203** V11
Longview, *Tex., U.S.* **105** G20
Longview, *Wash., U.S.* **110** E3
Longville, *La., U.S.* **102** P7
Longwy, *Fr.* **149** N13
Longxi, *China* **184** J10
Longxian, *China* **184** J11
Longyan, *China* **185** N14
Long'yugan, *Russ.* **160** G11
Lonkala, *D.R.C.* **209** G15
Lonoke, *Ark., U.S.* **102** G9
Lonquimay, *Chile* **135** N7
Lons-le-Saunier, *Fr.* **149** S13
Lónsdjúp, valley, *Atl. Oc.* **253** L22
Loogootee, *Ind., U.S.* **101** T8
Lookout Mountain, *Tenn., U.S.*
    103 G16
Lookout Mountain, *N. Mex., U.S.*
    109 S10
Lookout Ridge, *Alas., U.S.* **113** C14
Lookout, Cape, *N.C., U.S.* **98** J14
Lookout, Cape, *Oreg., U.S.* **110** F2
Loongana, *W. Aust., Austral.* **221** W6
Loop Head, *Ire.* **145** S3
Lop Buri, *Thai.* **190** M8
Lop Nur, lake, *China* **184** F6
Lop, *China* **184** G3
Lopatino, *Russ.* **158** L13
Lopatka, *Russ.* **161** G23
Lopatka, Cape, *Asia* **168** C13
Loperot, *Kenya* **211** R6
Lopévi, island, *Vanuatu* **229** H16
Lopez Point, *Calif., U.S.* **111** U4
Lopori, river, *D.R.C.* **209** D14
Lopphavet, bay, *Nor.* **143** B13
Lopydino, *Russ.* **158** D15
Lora, Hamun-i-, *Pak.* **181** N5
Lorain, *Ohio, U.S.* **101** P14
Loraine, *Tex., U.S.* **105** G13
Loralai, *Pak.* **180** M8
Lorca, *Sp.* **147** K13
Lord Howe Atoll *see* Ontong Java
    Atoll, *Solomon Is.* **228** A9
Lord Howe Island *see* Tōmotu Noi,
    Solomon Is. **228** L5
Lord Howe Rise, *Pac. Oc.* **246** M9
Lord Howe Seamounts, *Pac. Oc.*
    246 M9
Lordsburg, *N. Mex., U.S.* **109** V10
Lorentzweiler, *Lux.* **162** J10
Lorenzo, *Tex., U.S.* **104** F12

Loreto, *Mex.* **116** E6
Loreto, *Bol.* **132** J7
Loreto, *Braz.* **133** F14
Loretto, *Tenn., U.S.* **103** G14
Lorian Swamp, *Kenya* **211** S8
Lorica, *Col.* **130** D5
Lorien Bank, *Atl. Oc.* **248** G10
Lorient, *Fr.* **149** R4
Lorimers, *Turks & Caicos Is., U.K.* **119** H17
Loris, *S.C., U.S.* **98** K11
Lormes, *Fr.* **149** R11
Lormi, *India* **182** K8
Lorn, Firth of, *Scot., U.K.* **144** K7
Lornel, Pointe de, *Fr.* **148** L9
Lörrach, *Ger.* **150** K5
Lorraine, *Qnsld., Austral.* **221** R11
Lorraine, region, *Fr.* **149** P13
Lorugumu, *Kenya* **211** Q5
Los Alamos, *N. Mex., U.S.* **109** R12
Los Andes, *Chile* **134** K7
Los Angeles Aqueduct, *Calif., U.S.* **111** V8
Los Angeles, *Calif., U.S.* **111** X8
Los Ángeles, *Chile* **135** N6
Los Banos, *Calif., U.S.* **111** T5
Los Blancos, *Arg.* **134** E11
Los Cerrillos, *Arg.* **134** K9
Los Corrales, *Sp.* **146** B10
Los Cristianos, *Sp.* **214** Q4
Los Frailes, islands, *Venez.* **121** J14
Los Gatos, *Calif., U.S.* **111** T4
Los Hermanos, island, *Venez.* **121** H13
Los Lagos, *Chile* **135** P6
Los Llanos, *Sp.* **214** Q3
Los Loros, *Chile* **134** G7
Los Lunas, *N. Mex., U.S.* **109** S12
Los Mochis, *Mex.* **116** E7
Los Osos, *Calif., U.S.* **111** V5
Los Palacios, *Cuba* **118** G5
Los Reyes Islands, *P.N.G.* **222** M8
Los Roques, Islas, *Venez.* **130** B11
Los Teques, *Venez.* **130** C10
Los Testigos, islands, *Venez.* **121** J15
Los Vilos, *Chile* **134** K6
Los Yébenes, *Sp.* **146** G10
Losap Atoll, *F.S.M.* **226** B9
Lošinj, island, *Croatia* **154** F4
Lospalos, *Timor-Leste* **193** M16
Lossiemouth, *Scot., U.K.* **144** H10
Lost Dutchmen Ridge, *Ind. Oc.* **251** L15
Lost River Range, *Idaho, U.S.* **108** G6
Lost Trail Pass, *Idaho, Mont., U.S.* **108** E6
Lostine, *Oreg., U.S.* **110** F9
Lot Harbor, *F.S.M.* **226** K9
Lot, *F.S.M.* **226** K9
Lot, river, *Fr.* **149** W11
Lota, *Chile* **135** N6
Lotagipi Swamp, *S. Sudan* **203** X13
Lotilla, river, *S. Sudan* **203** W12
Loto, *D.R.C.* **209** F15
Lotofaga, *Samoa* **227** P19
Lot's Wife *see* Sōfu Gan, island, *Japan* **189** U12
Lott, *Tex., U.S.* **105** J17
Lou, island, *P.N.G.* **222** M8
Louang Namtha, *Laos* **190** H8
Louangphabang, *Laos* **190** J9
Loubomo *see* Dolisie, *Congo* **208** G9
Loudéac, *Fr.* **149** Q4
Loudi, *China* **184** M12
Loudonville, *Ohio, U.S.* **101** Q14
Loudun, *Fr.* **149** S7
Loué, *Fr.* **149** Q7
Louga, *Senegal* **204** H7
Lougheed Island, *Nunavut, Can.* **87** C13
Louhans, *Fr.* **149** S13
Louis II Stadium, *Monaco* **163** F20
Louis Trichardt *see* Makhado, *S. Af.* **212** G9
Louis-XIV, Pointe, *Que., Can.* **87** L17
Louisa Downs, *W. Aust., Austral.* **221** R6
Louisa, *Ky., U.S.* **103** B20
Louisburg, *N.C., U.S.* **98** G12
Louisburgh (Cluain Cearbán), *Ire.* **145** Q3
Louise Boyd Land, *Greenland, Den.* **88** H12
Louisiade Archipelago, *P.N.G.* **222** R10
Louisiade Plateau, *Pac. Oc.* **246** L9
Louisiana Point, *La., U.S.* **102** Q6
Louisiana, *U.S.* **102** N8
Louisiana, *Mo., U.S.* **107** S14
Louisville Ridge, *Pac. Oc.* **246** N12

Louisville Seamount, *Pac. Oc.* **255** K22
Louisville, *Ga., U.S.* **98** M7
Louisville, *Miss., U.S.* **102** K12
Louisville, *Ky., U.S.* **103** B16
Louisville, *Nebr., U.S.* **107** Q9
Louj, island, *Marshall Is.* **227** A20
Loukhi, *Russ.* **158** A8
Loukouo, *Congo* **208** G10
Loulay, *Fr.* **149** T6
Loulé, *Port.* **146** L5
Loum, *Cameroon* **206** N9
Loup City, *Nebr., U.S.* **107** P6
Loup, river, *Nebr., U.S.* **107** P6
Lourdes-de-Blanc-Sablon, *Que., Can.* **87** L22
Lourdes, *Fr.* **149** Y7
Lousã, *Port.* **146** F6
Lousy Bank, *Atl. Oc.* **248** F10
Louth, *Eng., U.K.* **145** R14
Loutrá Aidipsoú, *Gr.* **156** G11
Loutrá Eleftherón, *Gr.* **157** C13
Loutrá Smokóvou, *Gr.* **156** F9
Loutráki, *Gr.* **156** F12
Loutráki, *Gr.* **156** F11
Louvet Point, *St. Lucia* **123** K14
Louviers, *Fr.* **149** N8
Lovango Cay, *Virgin Is., U.S.* **122** M10
Lovech, *Bulg.* **155** H14
Loveland Pass, *Colo., U.S.* **108** M12
Loveland, *Colo., U.S.* **108** L13
Lovell, *Wyo., U.S.* **108** F10
Lovelock, *Nev., U.S.* **111** N8
Lovere, *It.* **152** D5
Loving, *N. Mex., U.S.* **109** V14
Lovington, *N. Mex., U.S.* **109** U15
Lóvua, *Angola* **209** J13;M15
Lovua, river, *Angola, D.R.C.* **209** J13
Low Cay, *Bahamas* **122** F11
Low, Cape, *Nunavut, Can.* **87** H16
Lowa, *D.R.C.* **209** F17
Lowa, river, *D.R.C.* **209** E17
Lowell, *Mass., U.S.* **97** K14
Lowell, *Oreg., U.S.* **110** H3
Lowell, Lake, *Idaho, U.S.* **108** G3
Lower Bogue, *Bahamas* **122** D5
Lower Guinea, *Af.* **198** K6
Lower Hutt, *N.Z.* **223** J20
Lower Lake, *Calif., U.S.* **110** M6
Lower Lough Erne, *N. Ire., U.K.* **145** P5
Lower Matecumbe Key, *Fla., U.S.* **99** Y10
Lower Peninsula, *U.S.* **91** D17
Lower Post, *B.C., Can.* **86** H8
Lower Red Lake, *Minn., U.S.* **106** F9
Lower Valley, *Cayman Is., U.K.* **122** N3
Lowestoft, *Eng., U.K.* **145** S16
Łowicz, *Pol.* **151** E15
Lowital, *Vanuatu* **229** M17
Lowman, *Idaho, U.S.* **108** G4
Lowreenne Seamounts, *Ind. Oc.* **251** Q22
Lowville, *N.Y., U.S.* **96** H9
Loxton, *S. Af.* **212** L6
Lōy Kêlay, *Afghan.* **180** L6
Lōy Rāzay, *Afghan.* **180** L3
Loyal, Loch, *Scot., U.K.* **144** G9
Loyalty Basin, *Pac. Oc.* **246** L11
Loyalty Islands *see* Îles Loyauté, *New Caledonia, Fr.* **231** P23
Loyew, *Belarus* **158** M6
Loyola, Punta, *Arg.* **135** W8
Loyoro, *Uganda* **211** Q5
Loznica, *Serb.* **154** F9
Lozova, *Ukr.* **159** P9
Lt. Luis Carvajal Villarroel Base, station, *Antarctica* **238** F5
Lu'an, *China* **185** K14
Lua Motu Island, *Tuvalu* **228** P10
Lua, river, *D.R.C.* **208** C12
Luacano, *Angola* **209** L14
Luahoko, island, *Tonga* **227** L24
Lualaba (Congo), river, *D.R.C.* **209** E16;G17;K16
Luama, river, *D.R.C.* **209** G18
Luan, river, *China* **185** F14
Luana Point, *Jam.* **122** J6
Luanco, *Sp.* **146** A8
Luanda, *Angola* **208** K9
Luando, *Angola* **208** M12
Luando, river, *Angola* **208** L12
Luang Prabang Range, *Laos* **190** J8
Luangue, river, *Angola* **209** K13
Luangwa, *Zambia* **209** P19
Luangwa, river, *Zambia* **209** L21;N19
Luangwa, river, *Mozambique, Zambia* **212** D9

Luaniua, island, *Solomon Is.* **228** B9
Luanshya, *Zambia* **209** N18
Luanza, *D.R.C.* **209** K18
Luapula, river, *Af.* **198** L9
Luarca, *Sp.* **146** A8
Luashi, *D.R.C.* **209** L15
Luatefito, island, *Tonga* **227** J23
Luau, *Angola* **209** L14
Luba, *Eq. Guinea* **214** M7
Luba, Bahia de, *Eq. Guinea* **214** M7
Lubalo, *Angola* **209** K13
Lubamiti, *D.R.C.* **208** F12
Lubāns Ezers, lake, *Latv.* **143** L17
Lubang Island, *Philippines* **193** D13
Lubango, *Angola* **208** P9
Lubao, *D.R.C.* **209** H16
Lubartów, *Pol.* **151** E17
Lübbenau, *Ger.* **150** E11
Lubbock, *Tex., U.S.* **104** F12
Lubec, *Me., U.S.* **97** E19
Lübeck, *Ger.* **150** B8
Lubefu, *D.R.C.* **209** G16
Lubefu, river, *D.R.C.* **209** G15
Lubero, *D.R.C.* **209** E19
Lubéron, Montagne du, *Fr.* **149** X13
Lubi, river, *D.R.C.* **209** H15
Lubin, *Pol.* **150** F12
Lublin, *Pol.* **151** F17
Lubliniec, *Pol.* **151** G14
Lubny, *Ukr.* **159** P7
Lubok Antu, *Indonesia* **192** H10
Luboten, peak, *Kos.* **154** J11
Lubuagan, *Philippines* **193** B13
Lubudi, *D.R.C.* **209** L17
Lubudi, river, *D.R.C.* **209** K16
Lubukitinggau, *Indonesia* **192** K6
Lubumbashi, *D.R.C.* **209** M17
Lubunda, *D.R.C.* **209** H17
Lubwe, *Zambia* **209** L19
Lucala, *Angola* **208** K10
Lucala, river, *Angola* **208** K11
Lucapa, *Angola* **209** K14
Lucas, *Kans., U.S.* **107** S6
Lucaya, *Bahamas* **122** J2
Lucca, *It.* **152** G6
Luce Bay, *Scot., U.K.* **145** N8
Lucea, *Jam.* **122** H6
Lucedale, *Miss., U.S.* **103** N13
Lucena, *Sp.* **146** L10
Lucena, *Philippines* **193** C14
Lučenec, *Slovakia* **151** K15
Lucera, *It.* **152** L12
Lucerne, Lake *see* Vierwaldstätter See, *Switz.* **152** B3
Luchegorsk, *Russ.* **161** L21
Luchulingo, river, *Mozambique* **212** C11
Lucie, river, *Suriname* **131** G16
Lucipara, Kepulauan, *Indonesia* **193** L16
Lucira, *Angola* **208** N9
Lucknow, *India* **182** H8
Lucknow, *Qnsld., Austral.* **221** T11
Lucunga, *Angola* **208** J10
Lucusse, *Angola* **209** M14
Lucy Creek, *N.T., Austral.* **221** T10
Luda Kamchiya, river, *Bulg.* **155** H16
Ludborough, *Eng., U.K.* **145** R14
Lüdenscheid, *Ger.* **150** F6
Lüderitz Bay, *Namibia* **212** J3
Lüderitz, *Namibia* **212** J3
Ludhiana, *India* **182** E5
Ludington, *Mich., U.S.* **100** K9
Ludlow, *Vt., U.S.* **97** H13
Ludlow, *Calif., U.S.* **111** W11
Ludogorie, *Eur.* **140** K10
Ludogorsko Plato, mountains, *Bulg.* **155** G16
Ludowici, *Ga., U.S.* **99** P8
Ludus, *Rom.* **155** C13
Ludwigsburg, *Ger.* **150** J7
Ludwigshafen, *Ger.* **150** H6
Luebo, *D.R.C.* **209** H14
Luembe, river, *D.R.C.* **209** J16
Luena Flats, *Zambia* **209** N15
Luena, *Angola* **209** M13
Luena, *Zambia* **209** L19
Luena, river, *Angola* **209** M14
Lueo, *D.R.C.* **209** K15
Lueta, river, *D.R.C.* **209** J14
Lüeyang, *China* **184** K10
Lufilufi, *Samoa* **227** P20
Lufira, river, *D.R.C.* **209** K17
Lufkin, *Tex., U.S.* **105** J20
Luga, *Russ.* **158** G6
Lugano, *Switz.* **152** D4

Luganville, *Vanuatu* **229** F14
Lugenda, river, *Mozambique* **212** C12
Lugo, *Sp.* **146** B7
Lugo, *It.* **152** F7
Lugoj, *Rom.* **154** E11
Lugovoy, *Russ.* **160** H10
Lugulu, river, *D.R.C.* **209** F17
Luguruka, *Tanzania* **211** Y6
Lui, river, *Angola* **208** L12
Lui, river, *Zambia* **209** P15
Luiana, *Angola* **209** Q15
Luiana, river, *Angola* **209** Q15
Luilaka, river, *D.R.C.* **209** E13
Luilu, river, *D.R.C.* **209** K15
Luimneach *see* Limerick, *Ire.* **145** S4
Luing, island, *Scot., U.K.* **144** L8
Luino, *It.* **152** D4
Luishia, *D.R.C.* **209** L17
Luitpold Coast, *Antarctica* **238** D10
Luiza, *D.R.C.* **209** J15
Luizi, river, *D.R.C.* **209** H18
Luján, *Arg.* **134** L12
Lujor, island, *Marshall Is.* **227** A21
Lukafu, *D.R.C.* **209** L17
Lukanga Swamp, *Zambia* **209** N17
Lukanga, *D.R.C.* **208** F12
Luke, Mount, *Austral.* **220** H3
Lukenie, river, *D.R.C.* **208** F12
Lukenie, river, *D.R.C.* **209** G14
Lukeville, *Ariz., U.S.* **109** V6
Lukhī, *Afghan.* **180** L4
Lukolela, *D.R.C.* **208** E11
Lukovit, *Bulg.* **155** H14
Łuków, *Pol.* **151** E17
Lukuga, river, *D.R.C.* **209** H18
Lukuledi, *Tanzania* **211** Y7
Lukulu, *Zambia* **209** N15
Lukunor Atoll, *F.S.M.* **226** B10
Luleå, *Sw.* **143** E14
Luleälven, river, *Eur.* **140** B9
Lüleburgaz, *Turk.* **170** C3
Lules, *Arg.* **134** G9
Lüliang Shan, *China* **184** H12
Luling, *Tex., U.S.* **105** L17
Lulonga, *D.R.C.* **208** D12
Lulonga, river, *D.R.C.* **209** D13
Lulu Fakahega, peak, *Wallis & Futuna Is., Fr.* **230** Q9
Lulu, river, *D.R.C.* **209** D16
Lulua, river, *D.R.C.* **209** K15
Lumajiangdong Cuo *see* Lumaqangdong Co, *China* **184** H3
Lūmān, *Afghan.* **180** J7
Lumaqangdong Co (Lumajiangdong Cuo), *China* **184** H3
Lumbala Kaquengue, *Angola* **209** M15
Lumbala N'guimbo, *Angola* **209** N14
Lumber City, *Ga., U.S.* **99** N7
Lumberton, *N.C., U.S.* **98** J11
Lumberton, *Miss., U.S.* **102** N12
Lumbrales, *Sp.* **146** E7
Lumbres, *Fr.* **148** L9
Lumding, *India* **182** H14
Lumeje, *Angola* **209** M14
Lumi, *P.N.G.* **222** N6
Lumphăt, *Cambodia* **190** M12
Lumpkin, *Ga., U.S.* **99** N4
Lumu, *Indonesia* **193** K13
Lumuna, *D.R.C.* **209** G17
Lumut, *Malaysia* **191** T8
Luna, *N. Mex., U.S.* **109** T10
Lunavada, *India* **182** K4
Lund, *Nev., U.S.* **111** Q12
Lunda, *Angola* **209** K13
Lundazi, *Zambia* **209** M21
Lundu, *Malaysia* **192** H9
Lunga, river, *Zambia* **209** M17
Lunggar (Longga'er), *China* **184** K3
Lunglei, *India* **182** K14
Lungué-Bungo, river, *Angola* **209** M13
Lungwebungu, river, *Zambia* **209** N15
Luni, *India* **182** H3
Luni, river, *India* **182** H3
Luning, *Nev., U.S.* **111** R8
Lunino, *Russ.* **158** L13
Lunsar, *Sierra Leone* **204** M9

Lunsemfwa, river, *Zambia* **209** P19
Luntai, *China* **184** E4
Luohe, *China* **185** J13
Luorong, *China* **184** N12
Luoyang, *China* **184** J12
Luoyuan, *China* **185** M15
Luozi, *D.R.C.* **208** H10
Lupanshui, *China* **184** M10
Luperón, *Dom. Rep.* **119** L18
Lupin, *Nunavut, Can.* **86** G12
Lupire, *Angola* **209** N13
Lupog, *Guam, U.S.* **230** N6
Luputa, *D.R.C.* **209** J15
Luqa, *Malta* **163** L22
Luqiao, *China* **185** L16
Luqu, *China* **184** J9
Luray, *Va., U.S.* **98** D11
Lure, *Fr.* **149** R14
Luremo, *Angola* **208** K12
Luring (Longren), *China* **184** J3
Lúrio, river, *Mozambique* **213** C13
Lus-la-Croix-Haute, *Fr.* **149** V14
Lusahunga, *Tanzania* **211** U3
Lusaka, *Zambia* **209** P18
Lusambo, *D.R.C.* **209** H15
Lusancay Islands, *P.N.G.* **222** Q9
Lusanga, *D.R.C.* **208** H11,12
Lushiko, river, *Angola, D.R.C.* **209** J13
Lushnjë, *Alban.* **154** L9
Lushoto, *Tanzania* **211** V7
Lusignan, *Fr.* **149** T7
Lusk, *Wyo., U.S.* **108** J14
Luso, *Port.* **146** F5
Lussac-les-Châteaux, *Fr.* **149** T8
Lustenau, *Ger.* **150** L7
Lut Desert *see* Dasht-e Lūt, *Iran* **177** J20
Lut Desert, *Asia* **168** G5
Lūtak, *Iran* **177** J22
Lutcher, *La., U.S.* **102** P10
Lutembo, *Angola* **209** N14
Luti, *Solomon Is.* **228** D7
Luton, *Eng., U.K.* **145** T13
Łutselk'e, *N.W.T., Can.* **86** H12
Lutsen, *Minn., U.S.* **106** F14
Luts'k, *Ukr.* **159** N2
Lutuai, *Angola* **209** M13
Lützow-Holm Bay, *Antarctica* **239** B18
Luuq, *Somalia* **211** Q10
Luverne, *Ala., U.S.* **103** M16
Luverne, *Minn., U.S.* **106** L9
Luvo, *Angola* **208** H10
Luvua, river, *D.R.C.* **209** J17
Luvuei, *Angola* **209** M14
Luwego, river, *Tanzania* **211** Y6
Luwingu, *Zambia* **209** L19
Luxembourg, *Eur.* **162** C3
Luxembourg, *Lux.* **162** K10
Luxeuil, *Fr.* **149** R14
Luxi, *China* **184** N8
Luxor *see* Al Uqşur, *Egypt* **202** J12
Luxora, *Ark., U.S.* **102** F11
Lüyuan, *China* **184** F9
Luz, *Braz.* **133** L14
Luz, *Fr.* **149** Y7
Luza, *Russ.* **158** E13
Luzaide, *Fr.* **149** Y5
Luzern, *Switz.* **152** B3
Luzhou, *China* **184** L10
Luzon Ridge, *Pac. Oc.* **246** G5
Luzon Strait, *Philippines* **193** A13
Luzon, island, *Philippines* **193** C13
Luzy, *Fr.* **149** S11
L'viv, *Ukr.* **159** N2
Lwarah, river, *Afghan.* **180** K7
Lyady, *Russ.* **158** G5
Lyakhov Islands, *Asia* **168** B10
Lyakhovskiye, *Russ.* **161** D17
Lycksele, *Sw.* **143** F13
Lycoming, *N.Y., U.S.* **96** H8
Lydda *see* Lod, *Israel* **172** M5
Lyddan Island, *Antarctica* **238** C11
Lyell Land, *Greenland, Den.* **88** J13
Lyepyel', *Belarus* **158** K5
Lyford, *Tex., U.S.* **105** R17
Lyme Bay, *Eng., U.K.* **145** W11
Lynch, *Ky., U.S.* **103** D20
Lynch, *Nebr., U.S.* **106** M6
Lynchburg, *Va., U.S.* **98** F11
Lynden, *Wash., U.S.* **110** A4
Lyndon B. Johnson National Historical Park, *Tex., U.S.* **105** K15
Lyndon, *W. Aust., Austral.* **221** T2
Lyndonville, *Vt., U.S.* **97** F13
Lynn Haven, *Fla., U.S.* **99** R3

Lynn Lake, Man., Can. 87 L13
Lynn, Mass., U.S. 97 K15
Lynton, W. Aust., Austral. 221 V2
Lyon, Fr. 149 U12
Lyons, N.Y., U.S. 96 J7
Lyons, Ga., U.S. 99 N7
Lyons, Kans., U.S. 107 T7
Lyons, Nebr., U.S. 107 P8
Lyons, river, Austral. 220 G2
Lyra Basin, Pac. Oc. 246 J8
Lysi (Akdoğan), Cyprus 162 P9
Lyskovo, Russ. 158 J13
Lyss, Switz. 152 B2
Lysychans'k, Ukr. 159 P10
Lytton, B.C., Can. 86 M8
Lyubertsy, Russ. 158 K9
Lyubim, Russ. 158 G10
Lyubimets, Bulg. 155 K16
Lyuboml', Ukr. 158 M2
Lyubotyn, Ukr. 159 P8
Lyubytino, Russ. 158 G7
Lyucha Ongokton, Russ. 161 G15
Lyudinovo, Russ. 158 L8

M

Mā'alaea Bay, Hawaii, U.S. 115 K19
Mā'alaea, Hawaii, U.S. 115 J18
Maale (Male), Maldives 183 V3
Maam Cross (An Teach Dóite), Ire. 145 Q3
Maan Ridge, Eur. 140 B10
Ma'ān, Jordan 173 Q6
Maanselkä, region, Fin. 143 E16
Maap, F.S.M. 226 D12
Maarianhamina see Mariehamn, Fin. 143 J14
Maarmorilik, Greenland, Den. 88 K6
Ma'arrat an Nu'mān, Syr. 172 E8
Maas, river, Neth. 148 J13
Maastricht, Neth. 148 L13
Mababe Depression, Botswana 212 F6
Ma'bar, Yemen 174 N4
Mabaruma, Guyana 131 D15
Mabélé, Cameroon 206 M11
Mabenga, D.R.C. 208 G12
Mabote, Mozambique 212 G11
Mabuki, Tanzania 211 U4
Mac. Robertson Land, Antarctica 239 F19
MacAlpine Lake, Nunavut, Can. 87 G13
Macapá, Braz. 132 C12
Macará, Ecua. 130 M3
Macas, Ecua. 130 L4
Macau, Braz. 133 E17
Macau, China 185 P13
Macaúba, Braz. 132 G12
Macauley Island, Kermadec Is., N.Z. 224 L9
MacAyeal Ice Stream, Antarctica 238 M11
Macbride Head, Falk. Is., U.K. 135 W12
Macclenny, Fla., U.S. 99 R8
Macdonald, Lake, Austral. 220 F7
MacDonnell Ranges, Austral. 220 F8
Macedo de Cavaleiros, Port. 146 D7
Macedonia, region, Gr. 156 C10
Maceió, Braz. 133 G18
Maceió, Ponta do, Braz. 133 E17
Macenta, Liberia 204 N10
Macerata, It. 152 H9
Macgillycuddy's Reeks, Ire. 145 T2
Mach, Pak. 180 M7
Machagai, Arg. 134 G12
Machaíla, Mozambique 212 G10
Machakos, Kenya 211 T6
Machala, Ecua. 130 M3
Machar Marshes, S. Sudan 203 U12
Machault, Fr. 149 N12
Machecoul, Fr. 149 S5
Macheng, China 185 K13
Macherla, India 183 P7
Machias, Me., U.S. 97 F19
Machico, Madeira, Port. 214 M3
Machilipatnam (Bandar), India 183 P8
Machiques, Venez. 130 D7
Mach'ŏllyŏng-sanmaek, range, N. Korea 186 D11
Machu Picchu, ruin, Peru 132 H5
Machynlleth, Wales, U.K. 145 S10

Macia, Mozambique 212 H10
Măcin, Rom. 155 E17
Macintyre, river, Austral. 220 H14
Maçka, Turk. 171 D14
Mackay, Idaho, U.S. 108 G6
Mackay, Qnsld., Austral. 221 S15
Mackay, Lake, Austral. 220 E7
Mackenzie Bay, N.W.T., Can. 86 D9
MacKenzie Bay, Antarctica 239 F20
Mackenzie King Island, N.W.T., Can. 87 B13
Mackenzie Mountains, N.W.T., Can. 86 F9
Mackenzie Trough, Arctic Oc. 252 N7
Mackenzie, B.C., Can. 86 K9
Mackenzie, river, N.W.T., Can. 86 F10
Mackenzie, river, Austral. 220 F14
Mackinac Island, Mich., U.S. 100 G11
Mackinac, Straits of, Mich., U.S. 100 G10
Mackinaw City, Mich., U.S. 100 G10
Mackinnon Road, Kenya 211 U8
Mačkovci, Slov. 154 C6
Macksville, Kans., U.S. 107 U6
Macksville, N.S.W., Austral. 221 W15
Macleod, Lake, Austral. 220 G1
Macocola, Angola 208 J11
Macomb, Ill., U.S. 101 Q4
Macomer, It. 153 N3
Macomia, Mozambique 213 C13
Macon, Ga., U.S. 98 M6
Macon, Miss., U.S. 103 K13
Macon, Mo., U.S. 107 R13
Mâcon, Fr. 149 T12
Macondo, Angola 209 M15
Macouba, Martinique, Fr. 123 E22
Macpherson's Strait, India 183 R14
Macquarie Harbour, Austral. 220 M7
Macquarie Island, Pac. Oc. 224 R6
Macquarie Ridge, Pac. Oc. 246 R9
Macquarie, Port, Austral. 220 K15
Macquarie, river, Austral. 220 J13
MacquireTrench, Pac. Oc. 255 F18
Macroom (Maigh Chromtha), Ire. 145 T4
Macumba, river, Austral. 220 G9
Macurijes, Punta, Cuba 118 J10
Macuro, Venez. 130 C13
Macusani, Peru 132 H5
Mad, river, Calif., U.S. 111 N2
Madaba, Tanzania 211 X7
Mādabā, Jordan 172 M6
Madadi, Chad 207 E15
Madagascar Basin, Ind. Oc. 250 L7
Madagascar Plateau, Ind. Oc. 250 M6
Madagascar, Madagascar 213 F16
Madā'in Şālih, Saudi Arabia 174 E5
Madalai, Palau 226 H5
Madalena, Port. 146 P3
Madalena, Sao Tome and Principe 215 D20
Madama, Niger 206 C11
Madan, Bulg. 155 K14
Madang, P.N.G. 222 P7
Madaoua, Niger 206 H7
Madaripur, Bangladesh 182 K13
Madau, island, P.N.G. 222 Q10
Madawaska, Me., U.S. 97 A18
Madax Gooy, river, Somalia 211 R10
Madayar, Myanmar 190 G5
Maddock, N. Dak., U.S. 106 F5
Madeira Plain, Atl. Oc. 248 K9
Madeira Rise, Atl. Oc. 248 K10
Madeira, island, Madeira, Port. 214 M2
Madeira, river, S. Amer. 132 F8
Madelia, Minn., U.S. 106 L10
Madeline Island, Wis., U.S. 100 E4
Maden, Turk. 171 G14
Madera, Calif., U.S. 111 T6
Madgaon, India 183 P4
Madha, Oman 175 F16
Madhubani, India 182 H10
Madi Opei, Uganda 211 Q4
Madibira, Tanzania 211 X5
Madidi, river, Bol. 132 H6
Madill, Okla., U.S. 105 E17
Madimba, Angola 208 J10
Madina do Boé, Guinea-Bissau 204 L8
Madina, Mali 204 K11
Madīnat al Abyār, Lib. 201 D21
Madīnat ash Sha'b, Yemen 174 Q9
Madīnat ath Thawrah, Syr. 172 E11
Madīnat Zāyid, U.A.E. 175 H15
Madingley Rise, Ind. Oc. 250 G8
Madingo-Kayes, Congo 208 G8

Madingou, Congo 208 G9
Madison, Me., U.S. 97 F16
Madison, Fla., U.S. 98 Q6
Madison, Ga., U.S. 98 L6
Madison, N.C., U.S. 98 G10
Madison, W. Va., U.S. 98 E8
Madison, Wis., U.S. 100 L5
Madison, Ind., U.S. 101 T10
Madison, Minn., U.S. 106 J9
Madison, S. Dak., U.S. 106 L7
Madison, Kans., U.S. 107 T9
Madison, Nebr., U.S. 107 P7
Madisonville, Ky., U.S. 103 C14
Madisonville, Tex., U.S. 105 J19
Madiun, Indonesia 192 M10
Madjori, Burkina Faso 205 L17
Madley, Mount, Austral. 220 F5
Mado Gashi, Kenya 211 S8
Madoi, China 184 J8
Madona, Latv. 143 L17
Madras see Chennai, India 183 Q7
Madras, Oreg., U.S. 110 G5
Madre de Dios, Isla, Chile 135 V6
Madre de Dios, river, S. Amer. 132 H6
Madre, Laguna, Mex. 116 F12
Madrid Point, S. Amer. 128 H4
Madrid, Iowa, U.S. 107 P11
Madrid, Sp. 146 F11
Madridejos, Sp. 146 H11
Madriu, river, Andorra 162 J3
Madura, W. Aust., Austral. 221 W6
Madura, island, Indonesia 192 M11
Madura, Selat, Indonesia 192 M11
Madurai, India 183 S6
Madurantakam, India 183 R7
Madyan, Pak. 180 H11
Mae Hong Son, Thai. 190 J6
Mae Klong, river, Thai. 190 M7
Mae Sariang, Thai. 190 K6
Mae Sot, Thai. 190 K7
Maebashi, Japan 189 N11
Maemuldo, island, S. Korea 187 U11
Maenggoldo, island, S. Korea 187 U6
Maengjung, N. Korea 186 H6
Maengsan, N. Korea 186 H7
Maentwrog, Wales, U.K. 145 S9
Maeva, Fr. Polynesia, Fr. 231 J14
Maevatanana, Madagascar 213 E16
Maéwo, island, Vanuatu 229 F16
Mafadi, peak, S. Af. 212 K9
Mafia Island, Tanzania 211 X8
Maga, peak, N. Mariana Is., U.S. 230 H11
Magadan, Russ. 161 F20
Magadi, Kenya 211 T6
Magallanes see Strait of Magellan, S. Amer. 135 X8
Magangué, Col. 130 D6
Magaria, Niger 206 H9
Magazine Mountain, Ark., U.S. 102 G7
Magburaka, Sierra Leone 204 M9
Magdagachi, Russ. 161 K18
Magdalena Bay, N. Amer. 84 M3
Magdalena de Kino, Mex. 116 C7
Magdalena, N. Mex., U.S. 109 T11
Magdalena, Bol. 132 H8
Magdalena, Bahía, Mex. 116 F6
Magdalena, Isla, Mex. 116 E6
Magdalena, Isla, Chile 135 S6
Magdalena, Llano de la, Mex. 116 E6
Magdalena, river, N. Amer. 84 N10
Magdalena, river, S. Amer. 130 H6
Magdeburg, Ger. 150 E9
Magdelaine Cays, Austral. 220 C15
Magee, Miss., U.S. 102 M11
Magelang, Indonesia 192 M9
Magellan Rise, Pac. Oc. 246 J11
Magellan Seamounts, Pac. Oc. 246 H9
Magellan, S. Amer. 128 R5
Magens Bay, Virgin Is., U.S. 122 M8
Maggotty, Jam. 122 J7
Maghāghah, Egypt 202 G11
Maghama, Mauritania 204 H8
Maghnia, Alg. 200 C10
Magina, peak, Sp. 146 K11
Maglić, peak, Bosn. & Herzg., Montenegro 154 H8
Maglie, It. 153 N16
Magnet Bay, Antarctica 239 D20
Magnitogorsk, Russ. 160 J8
Magnolia, Ark., U.S. 102 J7
Magnolia, Miss., U.S. 102 N10

Mago, island, Fiji 226 P8
Magta' Lahjar, Mauritania 204 G8
Magumeri, Nigeria 206 J11
Magway, Myanmar 190 H4
Mahābād, Iran 176 D12
Mahabharat Range, Nepal 182 G8
Mahabo, Madagascar 213 F15
Mahad, India 182 N3
Mahaddayweyne, Somalia 211 Q12
Mahaena, Fr. Polynesia, Fr. 231 H22
Mahagi, D.R.C. 209 C20
Mahaiatea, Fr. Polynesia, Fr. 231 J22
Mahajamba Bay, Af. 198 M11
Mahajan, India 182 F4
Mahajanga, Madagascar 213 D16
Mahalapye, Botswana 212 G8
Maḩallāt, Iran 177 G15
Māhān, Iran 177 K19
Mahana Bay, Hawaii, U.S. 115 Q21
Mahanadi, river, India 182 L9
Mahanoro, Madagascar 213 F17
Mahasamund, India 182 L8
Mahaut, Dominica 123 F19
Mahavavy, river, Madagascar 213 E16
Mahavelona, Madagascar 213 E17
Mahbubnagar, India 182 N6
Mahd adh Dhahab, Saudi Arabia 174 G7
Mahdia, Guyana 131 F15
Mahdia, Tunisia 201 C16
Mahé, island, Seychelles 215 P20
Mahébourg, Mauritius 215 G20
Mahen dra Giri, peak, India 182 M9
Mahenge, Tanzania 211 X6
Mahesana, India 182 J3
Mahia Peninsula, N.Z. 223 H22
Mahige, island, Solomon Is. 228 E10
Mahilyow, Belarus 158 K6
Mahina, Fr. Polynesia, Fr. 231 H22
Mahishadal, India 182 K11
Mahitu, island, Fr. Polynesia, Fr. 231 N17
Maḩmūd-e Rāqī, Afghan. 180 H9
Mahmudiye, Turk. 170 E6
Mahnomen, Minn., U.S. 106 F9
Mahoba, India 182 J7
Mahón, Sp. 147 G20
Mahri, Pak. 181 Q7
Mahu, Fr. Polynesia, Fr. 231 N14
Mahuta, Tanzania 211 Y8
Mahuva, India 182 L2
Maia, American Samoa, U.S. 230 B11
Maiao, island, Fr. Polynesia, Fr. 231 C16
Maibong, India 182 H14
Maicao, Col. 130 C7
Maîche, Fr. 149 R14
Maicuru, river, Braz. 132 C11
Maida, It. 153 Q13
Maīdān Shahr, Afghan. 180 H8
Maidenhead, Eng., U.K. 145 U13
Maiduguri, Nigeria 206 J11
Maiduo see Mêdog, China 184 L7
Maigh Chromtha see Macroom, Ire. 145 T4
Maigualida, Sierra, Venez. 130 F11
Maihar, India 182 J7
Maiko, river, D.R.C. 209 E17
Maikona, Kenya 211 Q7
Maikoor, island, Indonesia 193 L18
Mā'ili, Hawaii, U.S. 115 G14
Mailly-le-Camp, Fr. 149 P12
Mailsi, Pak. 180 M11
Maīmanah, Afghan. 180 G5
Main Channel, Cayman Is., U.K. 122 M2
Main Island (Bermuda Island), Bermuda 119 C16
Main Ridge, Trin. & Tobago 123 N18
Main, river, Ger. 150 H8
Maine, U.S. 97 E17
Maine, Gulf of, N. Amer. 84 J9
Maingkwan, Myanmar 190 D5
Mainistir Fhear Maí see Fermoy, Ire. 145 T4
Mainistir Laoise see Abbeyleix, Ire. 145 S6
Mainistir na Búille see Boyle, Ire. 145 Q5
Mainistir na Féile see Abbeyfeale, Ire. 145 S3

Mainland, island, Scot., U.K. 144 C12
Mainland, island, Scot., U.K. 144 E10
Mainoru, N.T., Austral. 221 P9
Mainz, Ger. 150 G6
Maio, island, Cabo Verde 215 D17
Maiquetía, Venez. 130 C10
Mairenui, peak, Fr. Polynesia, Fr. 231 J23
Maisí, Cuba 119 K15
Maisí, Punta de, Cuba 119 K15
Maiskhal, island, India 182 L13
Maitland, N.S.W., Austral. 221 X15
Maitri Station, Antarctica 239 A14
Maíz, Islas del, Nicar. 117 M19
Maizhokunggar, China 184 L6
Maizuru, Japan 189 P8
Majene, Indonesia 193 K13
Majī, Eth. 211 N5
Majorca see Mallorca, island, Sp. 147 G18
Majuro (Laura), island, Marshall Is. 227 E20
Majuro Atoll, Marshall Is. 227 D17
Majuro Lagoon, Marshall Is. 227 D20
Majuro, Marshall Is. 227 D17,21
Makabana, Congo 208 G9
Mākaha, Hawaii, U.S. 115 G13
Makakilo City, Hawaii, U.S. 115 H14
Makale, Indonesia 193 K13
Makalu I, peak, Nepal 182 G11
Makamba, Burundi 211 V2
Makanya, Tanzania 211 V7
Makanza, D.R.C. 209 D13
Makapala, Hawaii, U.S. 115 L21
Makapu Point, N.Z. 229 P21
Makaroa, island, Fr. Polynesia, Fr. 231 R17
Makarov Basin, Arctic Oc. 252 H10
Makarov, Russ. 161 J22
Makarovo, Russ. 161 K15
Makar'yev, Russ. 158 G12
Makassar (Ujungpandang), Indonesia 193 L13
Makatea, island, Fr. Polynesia, Fr. 231 B17
Makawao, Hawaii, U.S. 115 J19
Make Jima, Japan 189 T5
Makefu, N.Z. 229 P21
Makemo, island, Fr. Polynesia, Fr. 231 B19
Mākena, Hawaii, U.S. 115 K19
Makeni, Sierra Leone 204 M9
Makéone, Vanuatu 229 D15
Makgadikgadi Pans, Botswana 212 F7
Makhachkala, Russ. 159 U15
Makhad, Pak. 180 J11
Makhado (Louis Trichardt), S. Af. 212 G10
Makhambet, Kaz. 178 F5
Maki, Indonesia 193 K19
Makīnsk, Kaz. 178 D12
Makira Harbour, Solomon Is. 228 G11
Makiyivka, Ukr. 159 Q10
Makkah (Mecca), Saudi Arabia 174 J6
Makkovik, Nfld. & Lab., Can. 87 K21
Mako, Senegal 204 K9
Makó, Hung. 151 M16
Makogai, island, Fiji 226 P6
Makokou, Gabon 208 D9
Mākole'ā Point, Hawaii, U.S. 115 N20
Makoli, Zambia 209 Q17
Makongolosi, Tanzania 211 X4
Makoua, Congo 208 E11
Makrá, island, Gr. 157 N16
Makrakómi, Gr. 156 G10
Makran Coast Range, Pak. 181 R3
Makrana, India 182 H4
Makrinítsa, Gr. 156 F11
Makrónisos (Eléni, Helena), island, Gr. 157 K13
Makteïr, Mauritania 204 D10
Mākū, Iran 176 B11
Makumbako, Tanzania 211 X5
Makungo, Somalia 211 S10
Makunudu Atoll (Malcolm), Maldives 183 U3
Makura, island, Vanuatu 229 J16
Makurazaki, Japan 189 T4
Makurdi, Nigeria 206 L8
Mâl, Mauritania 204 G8
Mala see Mallow, Ire. 145 T4
Mala Vyska, Ukr. 159 Q6
Malabar Coast, India 183 R4

Malabo, Eq. Guinea 206 N9
Malabo, Eq. Guinea 214 L7
Malabuñgan, Philippines 192 E12
Malacca, Malaysia 191 V9
Malacca, Strait of, Asia 191 T8
Malacky, Slovakia 151 K13
Malad City, Idaho, U.S. 108 J7
Maladzyechna, Belarus 158 K4
Málaga, Col. 130 F7
Málaga, Sp. 146 M10
Malagarasi, river, Tanzania 211 V2
Malagón, Sp. 146 H10
Malaimbandy, Madagascar 213 F15
Málainn Mhóir (Malin More), Ire. 145 P3
Malaita, island, Solomon Is. 228 E11
Malakal Harbor, Palau 226 H5
Malakal Pass, Palau 226 H5
Malakal, S. Sudan 203 U12
Malakal, Palau 226 H5
Malakanagiri, India 182 M8
Malakand, Pak. 180 H11
Malake, island, Fiji 226 P5
Malakoff, Tex., U.S. 105 H18
Malakula, island, Vanuatu 229 G14
Malalane, S. Af. 212 H9
Malam, F.S.M. 226 G12
Malang, Indonesia 192 M10
Malangali, Tanzania 211 X5
Malanje, Angola 208 K11
Malanville, Benin 205 K18
Malao, Vanuatu 229 F14
Malapo, Tonga 227 Q23
Malapu, Solomon Is. 228 K6
Malar, Pak. 181 R5
Mälaren, lake, Sw. 143 K14
Malargüe, Arg. 134 M7
Malaspina Glacier, Alas., U.S. 113 K20
Malatya, Turk. 171 G13
Malawi, Af. 212 C10
Malawi, Lake (Lake Nyasa), Af. 198 L10
Malawiya, Sudan 203 R14
Malay Peninsula, Thai. 191 R7
Malaya Vishera, Russ. 158 G7
Malayagiri, peak, India 182 L10
Malaybalay, Philippines 193 F15
Malāyer, Iraq 177 G14
Malaysia, Asia 192 G7
Malazgirt, Turk. 171 F17
Malbaza, Nigeria 206 H7
Malbork, Pol. 151 B14
Malbun, Liech. 162 Q3
Malcolm see Makunudu Atoll, Maldives 183 U3
Malcolm, W. Aust., Austral. 221 V4
Maldegem, Belg. 148 K11
Malden Island, Kiribati 224 G12
Malden, Mo., U.S. 107 V15
Maldive Islands, Maldives 183 V3
Maldives, Asia 183 V3
Maldonado, Uru. 134 L14
Maldonado, Punta, Mex. 116 K12
Male Atoll, Maldives 183 V3
Male see Maale, Maldives 183 V3
Malè, It. 152 C6
Maléas, Ákra, Gr. 156 M11
Malegaon, India 182 L4
Maléha, Guinea 204 L10
Malela, D.R.C. 209 F17
Malemba-Nkulu, D.R.C. 209 J17
Malembé, Congo 208 G8
Malendok, island, P.N.G. 222 N10
Malengoya, D.R.C. 209 C16
Malesína, Gr. 156 H11
Maletswai, S. Af. 212 L8
Malevangga, Solomon Is. 228 C6
Malgobek, Russ. 159 T13
Malhão, Serra do, Port. 146 L6
Malhargarh, India 182 J4
Malheur Lake, Oreg., U.S. 110 J7
Malheur, river, Oreg., U.S. 110 H8
Malheureux, Cape, Mauritius 215 F20
Mali Drvenik, island, Croatia 154 G6
Mali Kyun, island, Myanmar 191 N6
Mali, Af. 205 J13
Mali, Guinea 204 K9
Mali, island, Fiji 226 N6
Mali, river, Myanmar 190 D6
Mália, Gr. 157 Q15
Malili, Indonesia 193 K13
Malin Head (Inis Trá Tholl), Ire. 145 M5
Malin More see Málainn Mhóir, Ire. 145 P3
Malin, Oreg., U.S. 110 L5
Malindi, Kenya 211 U8
Malinoa, island, Tonga 227 P23
Malinyi, Tanzania 211 X6
Maliom, P.N.G. 222 N10
Malkaaray, Somalia 211 Q9
Malkapur, India 182 L5
Malkara, Turk. 170 C3
Malki, Russ. 161 F22
Malko Tŭrnovo, Bulg. 155 J17
Mallacoota Inlet, Austral. 220 M14
Mallacoota, Vic., Austral. 221 Z14
Mallawī, Egypt 202 H11
Mallawli, island, Malaysia 192 F12
Mallorca (Majorca), island, Sp. 147 G18
Mallow (Mala), Ire. 145 T4
Malmberget, Sw. 143 D13
Malmesbury, S. Af. 212 M5
Malmö, Sw. 142 N12
Malmok, cape, Bonaire, Neth. 123 Q18
Malo, island, Vanuatu 229 F14
Maloelap Atoll, Marshall Is. 227 C17
Maloja, Switz. 152 C5
Malole, Zambia 209 L20
Malolo, island, Fiji 226 P4
Malom, P.N.G. 222 N10
Malone, N.Y., U.S. 96 F11
Malonga, D.R.C. 209 L15
Malongwe, Tanzania 211 V4
Maloshuyka, Russ. 158 C9
Maloshuyka, Russ. 160 E8
Måløy, Nor. 142 J9
Malozemel'skaya Tundra, Eur. 140 A12
Malpaso, peak, Sp. 214 R3
Malpelo Island, S. Amer. 128 C1
Malportas Pond, Cayman Is., U.K. 122 M4
Mäls, Liech. 162 Q2
Malta, Mont., U.S. 108 B10
Malta, Eur. 162 F4
Malta, island, Malta 163 K22
Maltahöhe, Namibia 212 H4
Maltese Islands, Eur. 140 M7
Malton, Eng., U.K. 145 P13
Maluku see Moluccas, island, Indonesia 193 H16
Ma'lūlā, Syr. 172 J8
Malumfashi, Nigeria 206 J8
Malung, Sw. 142 J12
Maluu, Solomon Is. 228 E10
Malvan, India 183 P3
Malvern, Ark., U.S. 102 H8
Malvern, Iowa, U.S. 107 Q9
Malvern, Jam. 122 J7
Malvérnia, Mozambique 212 G10
Malyn, Ukr. 159 N5
Malyy Lyakhovskiy, Russ. 161 D17
Malyye Karmakuly, Russ. 160 D10
Mamagota, P.N.G. 222 P11
Mamagota, P.N.G. 228 C5
Mamalu Bay, Hawaii, U.S. 115 K19
Maman, Sudan 203 Q15
Mambali, Tanzania 211 V4
Mambasa, D.R.C. 209 D18
Mamberamo, Indonesia 193 K20
Mambéré, river, Cen. Af. Rep. 206 N12
Mamboya, Tanzania 211 W6
Mambrui, Kenya 211 U8
Mamelles, island, Seychelles 215 N20
Mamer, Lux. 162 K9
Mamiña, Chile 134 C7
Mammoth Cave National Park, Ky., U.S. 103 D16
Mammoth Spring, Ark., U.S. 102 E9
Mammoth, Wyo., U.S. 108 F8
Mamoiada, It. 153 N4
Mamoré, river, S. Amer. 132 G7
Mamoriá, Braz. 132 F7
Mamou, La., U.S. 102 P8
Mamou, Guinea 204 L9
Mamoudzou, Mayotte, Fr. 215 P17
Mampong, Ghana 205 N15
Mamuju, Indonesia 193 K13
Man Island, Bahamas 122 D5
Man of War Bay, Trin. & Tobago 123 N18
Man of War Settlement, Bahamas 122 B10
Man-of-War Cay, Bahamas 119 G13
Man, Côte d'Ivoire 204 N12
Man, Isle of (Ellan Vanninn), U.K. 145 P9
Mana Pass, India 182 E7
Mānā Point, Hawaii, U.S. 114 F9
Mana, Fr. Guiana, Fr. 131 F18
Mānā, Hawaii, U.S. 114 E9
Mana, river, Fr. Guiana, Fr. 131 F18
Manacapuru, Braz. 132 D9
Manacor, Sp. 147 G19
Manado, Indonesia 193 H15
Mañagaha Island, N. Mariana Is., U.S. 230 H7
Managua, Nicar. 117 M18
Managua, Lago de, Nicar. 117 M18
Manaia, N.Z. 223 H20
Manakara, Madagascar 213 G16
Manakau, peak, N.Z. 223 K19
Manākhah, Yemen 174 N8
Manam, island, P.N.G. 222 N7
Manama see Al Manāmah, Bahrain 175 F13
Manamadurai, India 183 S6
Mananara, Madagascar 213 E17
Mananjary, Madagascar 213 G16
Manankoro, Mali 204 L12
Manantenina, Madagascar 213 H16
Manaoba, island, Solomon Is. 228 E11
Manas, China 184 E5
Manassas, Va., U.S. 98 D12
Manatí, Cuba 118 J12
Manaus, Braz. 132 D9
Manavgat, Turk. 170 J6
Manawai Harbour, Solomon Is. 228 F11
Manawatāwhi (Three Kings Islands), N.Z. 223 C18
Manbij, Syr. 172 D10
Manchester, Conn., U.S. 97 L13
Manchester, N.H., U.S. 97 J14
Manchester, Ga., U.S. 98 M5
Manchester, Ky., U.S. 103 D19
Manchester, Tenn., U.S. 103 F16
Manchester, Iowa, U.S. 107 N13
Manchester, Eng., U.K. 145 R11
Manchhar Lake, Pak. 181 Q7
Manchioneal, Jam. 122 J12
Manchuria see Dongbei, region, China 185 C15
Manchurian Plain, Asia 168 F12
Manciano, It. 152 J7
Mancos, Colo., U.S. 109 P10
Mand, Pak. 181 R3
Mand, river, Iran 177 M16
Manda Island, Kenya 211 T9
Manda, Tanzania 211 X4;Y5
Mandai, Pak. 180 M8
Mandal Bay, Virgin Is., U.S. 122 M9
Mandal, Virgin Is., U.S. 122 M9
Mandal, Nor. 142 M10
Mandal, Mongolia 184 C10
Mandalay, Myanmar 190 G5
Mandalgovĭ, Mongolia 184 E10
Mandan, N. Dak., U.S. 106 G4
Mandapeta, India 182 N8
Mandara Mountains, Cameroon, Nigeria 206 K11
Mandera, Kenya 211 Q9
Mandera, Tanzania 211 W7
Mandeville, Jam. 122 J8
Mandi Burewala, Pak. 180 M12
Mandi, India 182 E6
Mandiana, Guinea 204 L11
Mandimba, Mozambique 212 D11
Mandioli, island, Indonesia 193 J16
Mandla, India 182 K7
Mandōl, Afghan. 180 G9
Mandora, W. Aust., Austral. 221 S4
Mándra, Gr. 156 J12
Mandráki, Gr. 157 M18
Mandritsa, Bulg. 155 K16
Mandritsara, Madagascar 213 E17
Mandsaur, India 182 J4
Mandu, island, Maldives 183 W3
Mandurah, W. Aust., Austral. 221 X2
Manduria, It. 153 N15
Mandvi, India 182 K1
Mané, Burkina Faso 205 K15
Manfalūţ, Egypt 202 H11
Manfredonia, It. 152 L12
Manfredonia, Golfo di, It. 152 L13
Manga, Braz. 133 J15
Manga, Burkina Faso 205 L15
Manga, Niger 206 G10
Mangabeiras, Chapada das, Braz. 133 G14
Mangai, D.R.C. 209 G13
Mangaia, island, Cook Is., N.Z. 224 K12
Mangaïzé, Niger 206 G5
Mangalia, Rom. 155 G18
Mangalmé, Chad 207 J14
Mangalore see Mangaluru, India 183 R4
Mangaluru (Mangalore), India 183 R4
Mangando, Angola 208 J11
Mangareva, island, Fr. Polynesia, Fr. 231 E23;Q17
Mangarongaro, island, N.Z. 229 M21
Mangeigne, Chad 207 K15
Mangere, island, N.Z. 229 H20
Manggautu, Solomon Is. 228 H10
Mangghyystaū, Kaz. 178 J4
Manghit, Uzb. 178 K8
Mangkalihat, Cape, Asia 168 L13
Mangnai, China 184 G6
Mango, island, Tonga 227 N23
Mangoky, river, Madagascar 213 G15
Mangole, island, Indonesia 193 K15
Mangqystaū Shyghanaghy, bay, Kaz. 178 H4
Mangrol, India 182 L2
Manguchar, Pak. 181 N7
Mangueira, Lagoa, Braz. 132 Q11
Manguinho, Ponta do, Braz. 133 G17
Mangum, Okla., U.S. 105 D14
Mangut, Russ. 161 M17
Mangyeong, S. Korea 187 R8
Manhattan, Kans., U.S. 107 S8
Manhattan, Nev., U.S. 111 R9
Manhiça, Mozambique 212 H10
Manhŭng, N. Korea 186 D8
Máni, peninsula, Gr. 156 M10
Mania, river, Madagascar 213 F15
Manianga, D.R.C. 208 H10
Manica, Mozambique 212 F10
Manicoré, Braz. 132 E9
Manicouagan, Réservoir, Que., Can. 87 M20
Manīfah, Saudi Arabia 174 E12
Manifold, Cape, Austral. 220 F15
Manihiki Atoll, Cook Is., N.Z. 224 H12
Manihiki Plateau, Pac. Oc. 247 K13
Maniitsoq (Sukkertoppen), Greenland, Den. 89 P5
Manila Bay, Philippines 193 C13
Manila, Ark., U.S. 102 F11
Manila, Utah, U.S. 108 K9
Manila, Philippines 193 C13
Manily, Russ. 161 D21
Maningrida, N.T., Austral. 221 N9
Maninita, island, Tonga 227 J23
Manipa, island, Indonesia 193 K16
Manisa, Turk. 170 F3
Manistee, Mich., U.S. 100 J9
Manistee, river, Mich., U.S. 100 J9
Manistique Lake, Mich., U.S. 100 F9
Manistique, Mich., U.S. 100 G9
Manitoba, Can. 87 L13
Manitoba, Lake, Man., Can. 87 N13
Manitou Island, Mich., U.S. 100 E7
Manitou Islands, Wis., U.S. 100 H9
Manitowoc, Wis., U.S. 100 K7
Manizales, Col. 130 G5
Manja, Madagascar 213 G15
Manjaedo, island, S. Korea 187 U5
Manjimup, W. Aust., Austral. 221 Y3
Manjra, river, India 182 N6
Mankato, Minn., U.S. 106 L11
Mankato, Kans., U.S. 107 R7
Mankim, Cameroon 206 N10
Mankono, Côte d'Ivoire 204 N12
Manley Hot Springs, Alas., U.S. 113 F16
Manlleu, Sp. 147 D17
Manly, Iowa, U.S. 106 M11
Manmad, India 182 L4
Mann Passage, Marshall Is. 227 H17
Mann, island, Marshall Is. 227 J17
Manna, Indonesia 192 L6
Mannar, Gulf of, India 183 T6
Mannheim, Ger. 150 H6
Manning Seamounts, Atl. Oc. 248 J5
Manning Strait, Solomon Is. 228 D8
Manning, S.C., U.S. 98 L10
Manning, N. Dak., U.S. 106 G3
Manning, Iowa, U.S. 107 P10
Mannington, W. Va., U.S. 98 C9
Mannu, Capo, It. 153 N3
Mano Wan, Japan 188 M11
Mano, river, Liberia, Sierra Leone 204 N9
Manoa, Bol. 132 G7
Manokotak, Alas., U.S. 113 L13
Manokwari, Indonesia 193 J19
Manono, D.R.C. 209 J17
Manono, island, Samoa 227 P19
Manoron, Myanmar 191 P7
Manp'o, N. Korea 186 E7
Manra, island, Kiribati 224 G10
Manresa, Sp. 147 D17
Mansa Konko, Gambia 204 J7
Mansa, Zambia 209 L18
Mansabá, Guinea-Bissau 204 K7
Mansehra, Pak. 180 H12
Mansel Island, Nunavut, Can. 87 J17
Manseriche, S. Amer. 128 E2
Mansfield, Pa., U.S. 96 L7
Mansfield, Ohio, U.S. 101 Q13
Mansfield, Ark., U.S. 102 G6
Mansfield, La., U.S. 102 L6
Mansfield, Mo., U.S. 107 V13
Mansfield, Eng., U.K. 145 R12
Mansfield, Vic., Austral. 221 Y13
Mansfield, Mount, Vt., U.S. 96 F12
Mansle, Fr. 149 U7
Manson, Iowa, U.S. 107 N10
Mant Islands, F.S.M. 226 J9
Mant Passage, F.S.M. 226 J9
Manta Bay, S. Amer. 128 D1
Manta, Ecua. 130 K3
Mant'ap-san, peak, N. Korea 186 E11
Manteca, Calif., U.S. 111 S5
Manteo, N.C., U.S. 98 G15
Mantes-la-Jolie, Fr. 149 P9
Manti, Utah, U.S. 108 M7
Mantiqueira, S. Amer. 128 J9
Manton, Mich., U.S. 100 J10
Mantova, It. 152 E6
Mantua, Cuba 118 H4
Mantuan Downs, Qnsld., Austral. 221 U14
Manturovo, Russ. 158 G12
Mäntyluoto, Fin. 143 H15
Manu Island, P.N.G. 222 M6
Manú, Peru 132 H5
Manuae, island, Fr. Polynesia, Fr. 231 B14
Manuhangi, island, Fr. Polynesia, Fr. 231 C20
Manu'a Islands, Amer. Samoa, U.S. 224 J10
Manui, island, Indonesia 193 K14
Manui, island, Fr. Polynesia, Fr. 231 R17
Manūjān, Iran 177 M20
Manukau, N.Z. 223 E20
Manulu Lagoon, Kiribati 226 P11
Manus, island, P.N.G. 222 M8
Many Farms, Ariz., U.S. 109 Q8
Many, La., U.S. 102 M7
Manyami, river, Mozambique, Zimb. 212 E9
Manych Gudilo, Russ. 159 R12
Manych Guidilo, Lake, Eur. 140 H13
Manyoni, Tanzania 211 V5
Manzai, Pak. 180 K9
Manzanar National Historic Site, Calif., U.S. 111 T8
Manzanares, Sp. 146 H11
Manzanilla Point, Trin. & Tobago 123 P23
Manzanillo Bay, Haiti 119 L17
Manzanillo, Mex. 116 J9
Manzanillo, Cuba 118 K11
Manzano Mountains, N. Mex., U.S. 109 S12
Manzano Peak, N. Mex., U.S. 109 S12
Manzhouli, China 185 C13
Manzil, Pak. 181 N4
Mao, Dom. Rep. 119 L18
Mao, Chad 206 H12
Maoke Mountains, Asia 168 L15
Maoke, Pegunungan, Indonesia 193 L20
Maoming, China 184 P12
Maothail see Mohill, Ire. 145 Q5
Maoudass, Mauritania 204 H10
Mapai, Mozambique 212 G10
Mapanza, Zambia 209 P17
Mapi, Indonesia 193 M21
Mapia, Kepulauan (Saint David Islands), Indonesia 193 J19
Mapleton, Iowa, U.S. 107 N9
Mapleton, Oreg., U.S. 110 H2
Mapmaker Seamounts, Pac. Oc. 246 F10
Maprik, P.N.G. 222 N6
Mapuera, river, Braz. 132 C10
Maputo, Mozambique 212 J10
Ma'qalā, Saudi Arabia 174 F11
Maqat, Kaz. 178 G5
Maqellarë, Alban. 154 K10
Maqên, China 184 J9

Maqnā, *Saudi Arabia* **174** D4
Maqshūsh, *Saudi Arabia* **174** G6
Maquan *see* Damqog, river, *China* **184** L4
Maquinchao, *Arg.* **135** Q8
Maquoketa, *Iowa, U.S.* **107** N14
Mar Chiquita, Laguna, *Arg.* **134** J10
Mar del Plata Canyon, *Atl. Oc.* **249** T6
Mar del Plata, *Arg.* **135** N13
Mar, *S. Amer.* **128** K8
Mar, *Russ.* **161** H16
Mara Rosa, *Braz.* **133** H13
Mara, *Guyana* **131** F16
Mara, *India* **182** G14
Maraa, *Fr. Polynesia, Fr.* **231** J21
Maraā, *Braz.* **132** D7
Marabá, *Braz.* **133** E13
Marabo, island, *Fiji* **226** Q8
Maracá Island, *S. Amer.* **128** C8
Maracá, Ilha de, *Braz.* **133** B13
Maracaibo Basin, *S. Amer.* **128** A3
Maracaibo, *Venez.* **130** C8
Maracaibo, Lago de, *Venez.* **130** D8
Maracaju, *S. Amer.* **128** H7
Maracaju, *Braz.* **132** L11
Maracanã, *Braz.* **133** D14
Maracas Bay, *Trin. & Tobago* **123** N22
Maracay, *Venez.* **130** C10
Marādah, *Lib.* **201** F20
Maradi, *Niger* **206** H8
Marado, island, *S. Korea* **187** X7
Marāghah, Sabkhat, *Syr.* **172** E9
Marāgheh, *Iran* **176** D12
Maragogipe, *Braz.* **133** H16
Marahuaca, Cerro, *N. Amer.* **84** Q11
Marahuaca, Cerro, *Venez.* **130** G11
Marajó, Baía de, *Braz.* **133** C13
Marajó, Ilha de, *Braz.* **133** C13
Marakei, island, *Kiribati* **224** F8
Maralal, *Kenya* **211** R6
Maralinga, *S. Aust., Austral.* **221** W8
Maramasike, island, *Solomon Is.* **228** F11
Marambio, station, *Antarctica* **238** C5
Maran, *Malaysia* **191** U9
Marana, *Ariz., U.S.* **109** V7
Maranboy, *N.T., Austral.* **221** P8
Maranchón, *Sp.* **146** E12
Marand, *Iran* **176** C12
Marang, *Malaysia* **191** T9
Maranhão, Barragem do, *Port.* **146** H6
Marano, river, *It., San Marino* **163** J17
Marañón, river, *Peru* **132** E3
Marans, *Fr.* **149** T6
Mararaba, *Cameroon* **206** M11
Marargiu, Capo, *It.* **153** N3
Maraş *see* Varosha, *Cyprus* **162** P10
Mărăşeşti, *Rom.* **155** D16
Maratea, *It.* **153** N12
Marathon, *Fla., U.S.* **99** Y9
Marathon, *Tex., U.S.* **104** K10
Marathon, battle, *Gr.* **157** J13
Marathónas, *Gr.* **157** J13
Marathus *see* 'Amrīt, ruins, *Syr.* **172** G7
Maraú, *Braz.* **133** J16
Maravae, *Solomon Is.* **228** D6
Maraval, *Trin. & Tobago* **123** P22
Marāveh Tappeh, *Iran* **177** D18
Maravillas Creek, *Tex., U.S.* **104** L10
Maravovo, *Solomon Is.* **228** F9
Marbella, *Sp.* **146** M9
Marble Bar, *W. Aust., Austral.* **221** S3
Marble Canyon, *Ariz., U.S.* **109** Q7
Marble Point Heliport, *Antarctica* **239** N14
Marburg, *Ger.* **150** F6
Marca Point, *Af.* **198** M5
Marceline, *Mo., U.S.* **107** R12
Marciana, *It.* **152** J5
Marckolsheim, *Fr.* **149** Q15
Marco, *Fla., U.S.* **99** X8
Marcola, *Oreg., U.S.* **110** H3
Marcos Juárez, *Arg.* **134** K11
Marcus Baker, Mount, *Alas., U.S.* **113** J17
Marcus Island *see* Minami Tori Shima, *Pac. Oc.* **224** C5
Marcus-Wake Seamounts, *Pac. Oc.* **246** G8
Marcy, Mount, *N.Y., U.S.* **96** G11
Mardan, *Pak.* **180** H11
Mardie, *W. Aust., Austral.* **221** S2
Mardin, *Turk.* **171** H15

Maré, island, *New Caledonia, Fr.* **231** Q24
Marea de Portillo, *Cuba* **118** L11
Marechal Taumaturgo, *Braz.* **132** G4
Maree, Loch, *Scot., U.K.* **144** H8
Mareeba, *Somalia* **211** R11
Marenanuka, island, *Kiribati* **227** Q14
Marengo, *Iowa, U.S.* **107** P13
Marennes, *Fr.* **149** C13
Marettimo, Isola, *It.* **153** S8
Mareuil, *Fr.* **149** U8
Marfa, *Tex., U.S.* **104** K9
Margaret River, *W. Aust., Austral.* **221** Y2
Margarita, Isla de, *Venez.* **130** C12
Margat Castle *see* Qal'at al Marqab, ruins, *Syr.* **172** F7
Margate, *S. Af.* **212** L9
Margeride, Monts de la, *Fr.* **149** V11
Margherita di Savoia, *It.* **152** L13
Margherita Peak, *D.R.C.* **209** D19
Margherita Peak, *Uganda* **211** S2
Marghita, *Rom.* **154** B12
Mārgō, Dasht-e, *Afghan.* **180** M3
Marguerite Bay, *Antarctica* **238** F5
Marhanets', *Ukr.* **159** Q8
Mari, ruins, *Syr.* **172** G14
Maria Bay, *Tonga* **227** Q23
Maria Islands, *St. Lucia* **123** L14
Maria, Îles, *Fr. Polynesia, Fr.* **224** K12
Maria, island, *Fr. Polynesia, Fr.* **231** E23
Marià, Pic del, *Andorra* **162** H5
Mariakani, *Kenya* **211** U8
Mariana Islands, *Asia* **168** L13
Mariana Trench, *Pac. Oc.* **246** H8
Mariana Trough, *Pac. Oc.* **246** G8
Marianna, *Fla., U.S.* **99** Q4
Marianna, *Ark., U.S.* **102** G10
Marianne, *Seychelles* **215** M21
Mariánské Lázně, *Czechia* **150** H9
Marías Islands, *N. Amer.* **84** M3
Marías, Islas, *Mex.* **116** G8
Marias, river, *Mont., U.S.* **108** B8
Ma'rib, *Yemen* **174** N9
Maribor, *Slov.* **154** C6
Maricopa, *Ariz., U.S.* **109** U7
Maricopa, *Calif., U.S.* **111** W7
Maridi, *S. Sudan* **203** X10
Marie Byrd Land, region, *Antarctica* **238** L9
Marie Byrd Seamounts, *Pac. Oc.* **255** L14
Marie Celeste Fracture Zone, *Ind. Oc.* **250** K9
Marie-Galante, island, *Guadeloupe, Fr.* **123** G16
Marié, river, *Braz.* **132** D6
Mariehamn (Maarianhamina), *Fin.* **143** J14
Mariel, *Cuba* **118** G6
Mariental, *Namibia* **212** H4
Marietta, *Ga., U.S.* **98** L5
Marietta, *Ohio, U.S.* **101** S15
Marietta, *Okla., U.S.* **105** E17
Marigot Harbour, *St. Lucia* **123** K13
Marigot, *Haiti* **119** M16
Marigot, *Dominica* **123** E20
Marigot, *Guadeloupe, Fr.* **123** F14
Marigot, *Martinique, Fr.* **123** E22
Marigot, *St. Martin, Fr.* **123** B14
Mariiru Point, *N. Mariana Is., U.S.* **230** N9
Marijampolė, *Lith.* **143** N16
Marília, *Braz.* **132** L12
Marillana, *W. Aust., Austral.* **221** T3
Marimba, *Angola* **208** K11
Marinduque, island, *Philippines* **193** D14
Marine City, *Mich., U.S.* **100** M13
Marineland, *Fla., U.S.* **99** R9
Mariner Glacier Tongue, *Antarctica* **239** Q14
Marinette, *Wis., U.S.* **100** H7
Maringá, *Braz.* **132** M12
Maringa, river, *D.R.C.* **209** D14
Marinha Grande, *Port.* **146** G5
Mario Zucchelli Station, *Antarctica* **239** P15
Marino, *It.* **152** L8
Marino, *Vanuatu* **229** E15
Marion Dufresne Seamount, *Ind. Oc.* **254** D9
Marion Plateau, *Pac. Oc.* **246** M8
Marion, *N.C., U.S.* **98** H8
Marion, *S.C., U.S.* **98** K11

Marion, *S.C., U.S.* **98** L10
Marion, *Va., U.S.* **98** G6
Marion, *Mich., U.S.* **100** K10
Marion, *Ill., U.S.* **101** U6
Marion, *Ind., U.S.* **101** Q9
Marion, *Ohio, U.S.* **101** Q12
Marion, *Ark., U.S.* **102** G11
Marion, *Ala., U.S.* **103** L14
Marion, *Ky., U.S.* **103** C13
Marion, *S. Dak., U.S.* **106** M8
Marion, *Iowa, U.S.* **107** N13
Marion, *Kans., U.S.* **107** T8
Marion, *Mo., U.S.* **107** V11
Maripa, *Venez.* **130** E11
Mariposa, *Calif., U.S.* **111** S6
Marisa, *Indonesia* **193** J14
Mariscal Estigarribia, *Parag.* **132** L9
Maritime Alps, *Fr.* **152** G2
Maritsa, river, *Bulg.* **155** K15
Mariupol', *Ukr.* **159** R9
Marīvān, *Iraq* **176** F12
Mariyets, *Russ.* **158** H15
Marjayoūn, *Leb.* **172** J6
Mark Twain Lake, *Mo., U.S.* **107** S13
Marka (Merca), *Somalia* **211** R11
Markala, *Mali* **204** J12
Markandeh, ruins, *Afghan.* **180** H7
Markaz-e Ḥukūmat-e Sulṭān-e Bakwāh, *Afghan.* **180** K4
Marked Tree, *Ark., U.S.* **102** F11
Markermeer, bay, *Neth.* **148** H12
Markham Bay, *Nunavut, Can.* **87** H18
Markit, *China* **184** F2
Markleeville, *Calif., U.S.* **111** Q6
Markópoulo, *Gr.* **156** J7
Markounda, *Cen. Af. Rep.* **207** L13
Markovo, *Russ.* **161** C21
Markoy, *Burkina Faso* **205** J16
Marks, *Miss., U.S.* **102** H11
Marks, *Russ.* **158** M14
Marksville, *La., U.S.* **102** N8
Marle, *Fr.* **149** N11
Marlette, *Mich., U.S.* **100** L12
Marlin Rise, *Pac. Oc.* **247** P14
Marlin, *Tex., U.S.* **105** J17
Marlinton, *W. Va., U.S.* **98** D10
Marlow, *Okla., U.S.* **105** D16
Marmande, *Fr.* **149** W7
Marmara Denizi, *Turk.* **170** D4
Marmara Ereğlisi (Perinthus), *Turk.* **170** C4
Marmara Gölü, *Turk.* **170** F4
Marmara, island, *Turk.* **170** D4
Marmarás, *Gr.* **156** D12
Marmári, *Gr.* **157** J13
Marmaris, *Turk.* **170** H4
Marmarth, *N. Dak., U.S.* **106** H1
Marmelos, river, *Braz.* **132** F8
Marmolada, peak, *It.* **152** C7
Marmolejo, *Sp.* **146** K10
Marne au Rhin, Canal de la, *Fr.* **149** P14
Marne, river, *Fr.* **149** P12
Maroa, *Venez.* **130** H10
Maroantsetra, *Madagascar* **213** D17
Marobee Range, *Austral.* **220** K13
Maroe, *Fr. Polynesia, Fr.* **231** K14
Marokau, island, *Fr. Polynesia, Fr.* **231** C20
Marol, *Pak.* **180** H14
Marolambo, *Madagascar* **213** F16
Maromokotro, peak, *Madagascar* **213** D17
Marondera, *Zimb.* **212** E9
Maroneía, *Gr.* **157** C15
Maroni, river, *S. Amer.* **128** B7
Maroochydore, *Qnsld., Austral.* **221** V16
Marotiri (Îlots de Bass), *Fr. Polynesia, Fr.* **225** L14
Maro'u Bay, *Solomon Is.* **228** G11
Maroua, *Cameroon* **206** K11
Marovoay, *Madagascar* **213** E16
Marpi Point, *N. Mariana Is., U.S.* **230** G9
Márpissa, *Gr.* **157** L15
Marqaköl, *Kaz.* **179** E18
Marquesas Fracture Zone, *Pac. Oc.* **247** L16
Marquesas Keys, *Fla., U.S.* **99** Z8
Marquette, *Mich., U.S.* **100** F7
Marquis Island, *Grenada* **123** K23
Marquis, *Grenada* **123** K23
Marquises (Marquesas Islands), Îles, *Fr. Polynesia, Fr.* **225** H14
Marrakech, *Mor.* **200** E7

Marrawah, *Tas., Austral.* **221** Y7
Marree, *S. Aust., Austral.* **221** V10
Marrero, *La., U.S.* **102** Q11
Marromeu, *Mozambique* **212** E11
Marrupa, *Mozambique* **212** C12
Mars Hill, *Me., U.S.* **97** C18
Mars Hill, *N.C., U.S.* **98** H7
Mars-la-Tour, *Fr.* **149** P13
Marsá al Burayqah (Port Brega), *Lib.* **201** E20
Marsá Maṭrūḥ, *Egypt* **202** E9
Marsá Sha'b, *Egypt* **202** L14
Marsa, *Malta* **163** K23
Marsabit, *Kenya* **211** R7
Marsala, *It.* **153** S8
Marsalforn, *Malta* **163** H20
Marsaxlokk Bay, *Malta* **163** L23
Marsaxlokk, *Malta* **163** L23
Marsciano, *It.* **152** J8
Marseille, *Fr.* **149** Y13
Marsfjället, peak, *Sw.* **142** F12
Marsh Harbour, *Bahamas* **122** B9
Marsh Island, *U.S.* **91** M15
Marsh Island, *La., U.S.* **102** Q8
Marshall, *Ill., U.S.* **100** S7
Marshall, *Mich., U.S.* **100** M10
Marshall, *Ark., U.S.* **102** E8
Marshall, *Tex., U.S.* **105** G20
Marshall, *Minn., U.S.* **106** K9
Marshall, *Mo., U.S.* **107** S12
Marshall, *Alas., U.S.* **112** H12
Marshall, *Liberia* **204** P10
Marshall, river, *Austral.* **220** F9
Marshall Islands, *Pac. Oc.* **224** E8
Marshall Seamounts, *Pac. Oc.* **246** H10
Marshalltown, *Iowa, U.S.* **107** N12
Marshfield, *Wis., U.S.* **100** J4
Marshfield, *Mo., U.S.* **107** U12
Marsland, *Nebr., U.S.* **107** N2
Marsugalt, island, *Marshall Is.* **227** F14
Martaban, Gulf of, *Myanmar* **190** K5
Martap, *Cameroon* **206** M11
Martapura, *Indonesia* **192** K11
Martelange, *Lux.* **162** J8
Martha's Vineyard, *Mass., U.S.* **97** M15
Martí, *Cuba* **118** J11
Martigny, *Switz.* **152** C2
Martigues, *Fr.* **149** X13
Martin Knudsen Nunatakker, *Greenland, Den.* **88** J12
Martin Peninsula, *Antarctica* **238** L8
Martin Vaz Fracture Zone, *Atl. Oc.* **249** R10
Martin, *Tenn., U.S.* **103** E13
Martin, *S. Dak., U.S.* **106** M3
Martin, *Slovakia* **151** J15
Martin, Lake, *Ala., U.S.* **103** L16
Martinez Lake, *Ariz., U.S.* **109** U4
Martinez, *Ga., U.S.* **98** L8
Martinique Passage, *N. Amer.* **121** E17
Martinique, island, *Lesser Antil.* **121** E17
Martins Ferry, *Ohio, U.S.* **101** Q15
Martinsburg, *W. Va., U.S.* **98** C12
Martinsville, *Va., U.S.* **98** G10
Martinsville, *Ind., U.S.* **101** S9
Martos, *Sp.* **146** K10
Martuba, *Lib.* **201** D22
Martuni (Xocavənd), *Azerb.* **171** E21
Martuni, *Arm.* **171** E19
Ma'rūf, *Afghan.* **180** L7
Marum Volcano, *Vanuatu* **229** G15
Marungu, *D.R.C.* **209** J19
Marutea, island, *Fr. Polynesia, Fr.* **231** B19;E23
Marv Dasht, *Iran* **177** K16
Marvão, *Port.* **146** H7
Marvin Spur, *Arctic Oc.* **252** K12
Marvine, Mount, *Utah, U.S.* **109** N7
Mārwā, *Afghan.* **180** H4
Mary Point, *Virgin Is., U.S.* **122** M10
Mary, *Turkm.* **178** N8
Maryborough, *Qnsld., Austral.* **221** U16
Maryland, *U.S.* **98** C13
Maryport, *Eng., U.K.* **145** N10
Marys, river, *Nev., U.S.* **110** M11
Marysvale, *Utah, U.S.* **109** N7
Marysville, *Ohio, U.S.* **101** Q12
Marysville, *Kans., U.S.* **107** R8
Marysville, *Wash., U.S.* **110** B4
Marysville, *Calif., U.S.* **111** Q4
Maryville, *Tenn., U.S.* **103** F18
Maryville, *Mo., U.S.* **107** R10

Marzdārān, *Iran* **177** E21
Marzo, Cabo, *Col.* **130** F4
Marzūq, *Lib.* **201** H18
Masada, ruins, *Israel* **173** N6
Masai Steppe, *Tanzania* **211** V6
Masaka, *Uganda* **211** S3
Masalasef, *Chad* **207** J13
Masallı, *Azerb.* **171** F22
Masamba, *Indonesia* **193** K13
Masan, *N. Korea* **186** K7
Masan, *S. Korea* **187** S11
Masasi, *Tanzania* **211** Y7
Masaya, *Nicar.* **117** M18
Masbate City, *Philippines* **193** D14
Masbate, island, *Philippines* **193** D14
Mascara, *Alg.* **200** C11
Mascarene Basin, *Ind. Oc.* **250** K7
Mascarene Plain, *Ind. Oc.* **250** K7
Mascarene Plateau, *Ind. Oc.* **250** G7
Masefau Bay, *American Samoa, U.S.* **230** D7
Masein, *Myanmar* **190** F4
Masela, island, *Indonesia* **193** M17
Maseru, *Lesotho* **212** K8
Masescha, *Liech.* **162** P2
Mash'abbe Sade, *Israel* **173** P5
Mashhad, *Iran* **177** E21
Mashkai, river, *Pak.* **181** Q6
Māshkel, Hāmūn-i-, *Pak.* **181** P3
Mashkel, river, *Pak.* **181** P4
Mashki Chah, *Pak.* **181** N3
Masi-Manimba, *D.R.C.* **208** H12
Masia-Mbio, *D.R.C.* **208** G11
Masin, *Indonesia* **193** M21
Masindi Port, *Uganda* **211** R3
Masindi, *Uganda* **211** R3
Masira *see* Jazīrat Maṣīrah, island, *Oman* **175** K18
Masira, Gulf of, *Ind. Oc.* **168** J4
Masisea, *Peru* **132** G4
Masjed Soleymān, *Iraq* **177** J14
Mask, Lough, *Ire.* **145** Q3
Maska, *Nigeria* **206** J8
Maskanah (Meskéné), *Syr.* **172** E10
Maskelynes Islands, *Vanuatu* **229** H15
Maslen Nos, *Bulg.* **155** J17
Maslenica, *Croatia* **154** F5
Mason Bay, *N.Z.* **223** Q16
Mason City, *Iowa, U.S.* **106** M12
Mason Hall, *Trin. & Tobago* **123** N17
Mason, *Mich., U.S.* **100** M11
Mason, *Tex., U.S.* **105** K15
Masontown, *Pa., U.S.* **96** P3
Maspalomas, *Sp.* **214** R5
Masqaṭ (Muscat), *Oman* **175** G18
Mass City, *Mich., U.S.* **100** F6
Massa Marittima, *It.* **152** H6
Massa, *Congo* **208** G10
Massachusetts Bay, *Mass., U.S.* **97** K16
Massachusetts, *U.S.* **97** K14
Massacre Bay, *American Samoa, U.S.* **230** D6
Massacre, *Dominica* **123** G19
Massafra, *It.* **152** M14
Massaguet, *Chad* **206** J12
Massakory, *Chad* **206** H12
Massangena, *Mozambique* **212** G10
Massango, *Angola* **208** J11
Massawa *see* Mits'iwa, *Eritrea* **210** H8
Massay, *Fr.* **149** S9
Massena, *N.Y., U.S.* **96** F10
Massenya, *Chad* **206** J12
Masset, *B.C., Can.* **86** J7
Massiac, *Fr.* **149** V11
Massif Central, *Fr.* **149** U10
Massif de Guéra, *Chad* **207** J14
Massif de l'Aïr, *Niger* **206** E8
Massif de l'Isalo, *Madagascar* **213** G15
Massif de Marfa, *Chad* **207** H15
Massif de Tarazit, *Niger* **206** E8
Massif des Bongo, *Cen. Af. Rep.* **207** L15
Massif du Kerkour Nourene, *Chad* **207** G16
Massillon, *Ohio, U.S.* **101** P14
Massinga, *Mozambique* **212** G11
Masson Island, *Antarctica* **239** K21
Mastābah, *Saudi Arabia* **174** J6
Maştağa, *Azerb.* **171** D23
Masterton, *N.Z.* **223** J21
Mastuj, *Pak.* **180** F11
Mastung, *Pak.* **180** M7
Mastūrah, *Saudi Arabia* **174** H6

Masty, *Belarus* **158** K3
Masuda, *Japan* **189** Q6
Masunga, *Botswana* **212** F8
Masuria, *Eur.* **140** F9
Masvingo, *Zimb.* **212** F9
Maşyāf, *Syr.* **172** F7
Mat, river, *Alban.* **154** K9
Mata Point, *N.Z.* **229** Q22
Mata-Utu (Matâ'Uta),
    Wallis & Futuna Is., Fr. **230** Q9
Mata, *D.R.C.* **209** J14
Matacawa Levu, island, *Fiji* **226** N4
Matadi, *D.R.C.* **208** H9
Matador, *Tex., U.S.* **105** E13
Matagalpa, *Nicar.* **117** L18
Matagami, *Que., Can.* **87** N18
Matagorda Bay, *Tex., U.S.* **91** N13
Matagorda Peninsula, *Tex., U.S.*
    105 M19
Matagorda, *Tex., U.S.* **105** M19
Mataiea, *Fr. Polynesia, Fr.* **231** J22
Mataiva, island, *Fr. Polynesia, Fr.*
    231 A17
Matak, island, *Indonesia* **192** H8
Matala, *Angola* **208** P10
Mátala, *Gr.* **157** R14
Matale, *Sri Lanka* **183** U7
Matam, *Senegal* **204** H8
Matameye, *Niger* **206** H8
Matamoros, *Mex.* **116** E12
Matane, *Que., Can.* **87** N21
Matanzas, *Cuba* **118** G7
Matapás see Ténaro, Akrotírio, *Gr.*
    156 M10
Matapwa, *Tanzania* **211** Y8
Matara, *Sri Lanka* **183** U7
Mataram, *Indonesia* **192** M12
Mataranka, *N.T., Austral.* **221** Q8
Mataró, *Sp.* **147** D18
Matasiri, island, *Indonesia* **192** L11
Mataso Island, *Vanuatu* **229** J16
Mataura, *Fr. Polynesia, Fr.* **231** M14
Matâ'Uta see Mata-Utu,
    Wallis & Futuna Is., Fr. **230** Q9
Matautu, *Samoa* **227** P19
Mataitulele Point,
    American Samoa, U.S. **230** D6
Matavera, *N.Z.* **229** E23
Mateguá, *Bol.* **132** H8
Matehuala, *Mex.* **116** G11
Mateika, island, *Tuvalu* **228** P10
Matelot, *Trin. & Tobago* **123** N23
Matema, island, *Solomon Is.* **228** L5
Matera, *It.* **152** M13
Matese, range, *It.* **152** L10
Mátészalka, *Hung.* **151** K17
Matfors, *Sw.* **143** H13
Matha Strait, *Antarctica* **238** E5
Mathematician Seamounts, *Pac. Oc.*
    247 H19
Mather, *Calif., U.S.* **111** S6
Mathews Peak, *Kenya* **211** R6
Mathews, *Va., U.S.* **98** E14
Mathis, *Tex., U.S.* **105** N16
Mathráki, island, *Gr.* **156** E5
Mathura, *India* **182** G6
Mathurin Bay, *Mauritius* **215** J20
Mati, *Philippines* **193** F15
Matiakoali, *Burkina Faso* **205** K17
Matiere, *N.Z.* **223** G20
Matinicus Island, *Me., U.S.* **97** G17
Matlerjoch, peak, *Aust. Liech.* **162** P4
Matli, *Pak.* **181** S8
Matlock, *Eng., U.K.* **145** R12
Mato, *D.R.C.* **209** J16
Matochkin Shar, *Russ.* **160** D11
Matola, *Mozambique* **212** J10
Matombo, *Tanzania* **211** W7
Matopo Hills, *Af.* **198** N14
Matosinhos, *Port.* **146** E5
Mátra, *Eur.* **140** H9
Maṭraḥ, *Oman* **175** G17
Matsena, *Nigeria* **206** H9
Matsu, see Mazu, *China* **185** M15
Matsubara, *Japan* **189** W3
Matsue, *Japan* **189** P7
Matsumoto, *Japan* **189** N10
Matsuyama, *Japan* **189** R6
Mattamuskeet, Lake, *N.C., U.S.*
    98 H14
Mattaponi, river, *Va., U.S.* **98** E13
Mattawamkeag, *Me., U.S.* **97** D18
Matterhorn, peak, *Nev., U.S.* **110** L11
Matterhorn, peak, *It. Switz.* **152** C3
Matthew Town, *Bahamas* **119** J14
Matthew, island, *Vanuatu* **224** K8

Matthews Peak, *U.S.* **90** H8
Matthew's Ridge, *Guyana* **131** E14
Mattili, *India* **182** M8
Mattinata, *It.* **152** L13
Mattoon, *Ill., U.S.* **101** S6
Matua, *Indonesia* **192** K9
Matua, island, *Russ.* **161** H23
Matuku, island, *Fiji* **226** R7
Matunga, island, *N.Z.* **229** L21
Matunuck, *R.I., U.S.* **97** M14
Matura Bay, *Trin. & Tobago* **123** P23
Matura, *Trin. & Tobago* **123** P23
Maturei Vavao, island, *Fr. Polynesia, Fr.*
    231 E22
Maturín, *Venez.* **130** D13
Matxitxako, Cape, *Eur.* **140** J3
Mau Ranipur, *India* **182** J7
Mauá, *Mozambique* **212** C12
Maubara, *Timor-Leste* **193** M15
Maubeuge, *Fr.* **148** M11
Maud Hill, *Atl. Oc.* **254** H8
Maud Rise, *Atl. Oc.* **249** X12
Maude, *N.S.W. Austral.* **221** X12
Maués, *Braz.* **132** D10
Maug Islands, *N. Mariana Is., U.S.*
    230 G5
Maughold Head (Kione Maghal),
    Isle of Man, U.K. **145** P9
Maui, *Hawaii, U.S.* **115** J20
Mauke, island, *Cook Is., N.Z.* **224** K12
Maukme, *Myanmar* **190** H6
Maule, river, *Chile* **134** M6
Mauléon, *Fr.* **149** S6
Maullín, *Chile* **135** Q6
Maumee, *Ohio, U.S.* **101** N12
Maumee, river, *Ohio, U.S.* **101** P11
Maumusson, Pertuis de, *Fr.* **149** U6
Maun, *Botswana* **212** F6
Mauna Kea, *Hawaii, U.S.* **115** N22
Mauna Loa Observatory, *Hawaii, U.S.*
    115 P21
Mauna Loa, *Hawaii, U.S.* **115** P21
Maunaiki, peak, *Hawaii, U.S.* **115** P22
Maunaloa, *Hawaii, U.S.* **115** H16
Maunalua Bay, *Hawaii, U.S.* **115** H15
Maungapu, peak, *N.Z.* **229** G20
Maungdaw, *Myanmar* **190** H3
Maunoir, Lac, *N.W.T., Can.* **86** F10
Maupihaa (Mopelia), island,
    Fr. Polynesia, Fr. **231** B14
Maupin, *Oreg., U.S.* **110** F5
Maupiti, island, *Fr. Polynesia, Fr.*
    231 B15
Mauren, *Liech.* **162** M2
Maurepas, Lake, *La., U.S.* **102** P10
Maures, Monts des, *Fr.* **149** Y14
Mauriac, *Fr.* **149** V10
Maurice Ewing Bank, *Atl. Oc.* **249** V7
Maurice Hill Ridge, *Ind. Oc.* **250** G7
Maurice, Lake, *Austral.* **220** H8
Mauritania, *Af.* **204** F9
Mauritania, region, *Af.* **198** F2
Mauritius, *Af.* **214** H9
Mauritius Trench, *Ind. Oc.* **250** L7
Mauron, *Fr.* **149** Q5
Maury Bay, *Antarctica* **239** P19
Maury Mountains, *Oreg., U.S.* **110** H6
Maury Seachannel, *Atl. Oc.* **248** G9
Mauston, *Wis., U.S.* **100** K4
Mauvezin, *Fr.* **149** X8
Maverick, *Ariz., U.S.* **109** T9
Mavinga, *Angola* **209** P13
Mavroli, *Cyprus* **162** P6
Mavrovouni Mine, *Cyprus* **162** P7
Mawlamyine, *Myanmar* **190** L6
Mawlite, *Myanmar* **190** H3
Mawlu, *Myanmar* **190** F5
Mawqaq, *Saudi Arabia* **174** E7
Mawshij, *Yemen* **174** P8
Mawson Coast, *Antarctica* **239** E20
Mawson Escarpment, *Antarctica*
    239 F18
Mawson Peninsula, *Antarctica*
    239 R16
Mawson Station, *Antarctica* **239** E20
Max, *N. Dak., U.S.* **106** F4
Maxaas, *Somalia* **211** P12
Maxixe, *Mozambique* **212** H11
Maxton, *N.C., U.S.* **98** J11
Maxwell Fracture Zone, *Atl. Oc.*
    248 H9
May Pen, *Jam.* **122** J9
May, Isle of, *Scot., U.K.* **144** L11
Mayabandar, *India* **183** Q14
Mayaguana Island, *Bahamas* **119** H16

Mayaguana Passage, *Bahamas*
    119 H15
Mayama, *Congo* **208** G10
Mayāmey, *Iran* **177** E18
Mayang-do, island, *N. Korea* **186** G10
Mayaro Bay, *Trin. & Tobago* **123** Q23
Maych'ew, *Eth.* **210** J8
Maycock's Bay, *Barbados* **123** J18
Maydān Ikbiz, *Syr.* **172** C8
Maydh, *Somaliland* **210** L12
Maydh, island, *Somaliland* **210** K12
Maydī, *Yemen* **174** M8
Mayenne, *Fr.* **149** Q6
Mayfa'ah, *Yemen* **174** P11
Mayfield, *Ky., U.S.* **103** D13
Mayhill, *N. Mex., U.S.* **109** U13
Maykop, *Russ.* **159** T11
Mayluu-Suu, *Kyrg.* **179** L13
Mayneside, *Qnsld., Austral.* **221** T12
Mayno Gytkino, *Russ.* **161** C21
Mayo Faran, *Nigeria* **206** L10
Mayo Ndaga, *Nigeria* **206** M10
Mayo, *Yukon, Can.* **86** F8
Mayo, *Fla., U.S.* **99** R6
Mayo, *Trin. & Tobago* **123** Q22
Mayo, Cerro, *Chile* **135** V6
Mayoko, *Congo* **208** F9
Mayorga, *Sp.* **146** D9
Mayotte, island, *Mayotte, Fr.* **215** P17
Mayreau, island, *St. Vincent & the*
    Grenadines **121** G16
Maysville, *Ky., U.S.* **103** A18
Maysville, *Mo., U.S.* **107** R11
Mayumba, *Gabon* **208** G8
Mayville, *Wis., U.S.* **100** L6
Mayville, *N. Dak., U.S.* **106** F8
Maywood, *Nebr., U.S.* **107** Q4
Mazabuka, *Zambia* **209** P18
Mazagan see El Jadida, *Mor.*
    200 D7
Mazagão, *Braz.* **132** C12
Mazamet, *Fr.* **149** X10
Mazār-e Sharīf, *Afghan.* **180** F7
Mazara del Vallo, *It.* **153** S8
Mazari Pass, *Pak.* **181** N4
Mazaruni, river, *Guyana* **131** F14
Mazatán, *Mex.* **116** D7
Mazatenango, *Guatemala* **117** L15
Mazatlán, *Mex.* **116** F8
Mazgirt, *Turk.* **171** F14
Mazirbe, *Latv.* **143** L15
Mazıdağı, *Turk.* **171** H15
Mazotos, *Cyprus* **162** Q9
Mazrub, *Sudan* **203** S10
Mazu (Matsu), *China* **185** M15
Mazyr, *Belarus* **158** M5
Mazzarino, *It.* **153** T10
Mbabane, *Eswatini* **212** J9
Mbaéré, river, *Cen. Af. Rep.* **207** N13
Mbahiakro, *Côte d'Ivoire* **205** N13
Mbaïki, *Cen. Af. Rep.* **207** N13
Mbakaou Reservoir, *Cameroon*
    206 M11
Mbala, *Cen. Af. Rep.* **207** L15
Mbala, *Zambia* **209** K20
Mbalambala, *Kenya* **211** S8
Mbale, *Uganda* **211** R5
Mbalmayo, *Cameroon* **206** N10
Mbalo, *Solomon Is.* **228** F10
Mbamba Bay, *Tanzania* **211** Z5
Mbandaka, *D.R.C.* **208** E12
Mbanika, island, *Solomon Is.* **228** F9
Mbanza Congo, *Angola* **208** H10
Mbanza-Ngungu, *D.R.C.* **208** H10
Mbarara, *Uganda* **211** S2
Mbari, river, *Cen. Af. Rep.* **207** M16
Mbé, *Cameroon* **206** L11
Mbé, *Congo* **208** G11
Mbéni, *Comoros* **215** M14
Mbéré, river, *Af.* **206** M12
Mbeya, *Tanzania* **211** X4
Mbinga, *Tanzania* **211** Z5
Mbini, *Eq. Guinea* **206** Q9
Mbitao, *Cameroon* **206** L12
Mbizana, *S. Af.* **212** L9
Mbogo, *Tanzania* **211** W4
Mboi, *D.R.C.* **209** J14
Mboli, *D.R.C.* **209** B15
Mbombela, *S. Af.* **212** H9
Mbomo, *Congo* **208** D10
Mborokua, island, *Solomon Is.* **228** E8
Mbour, *Senegal* **204** J6
Mbout, *Mauritania* **204** H9
Mbrés, *Cen. Af. Rep.* **207** M14
Mbridge, river, *Angola* **208** J9

Mbuji-mayi (Bakwanga), *D.R.C.*
    209 H15
Mbuji-Mayi, river, *D.R.C.* **209** J15
Mbulo, island, *Solomon Is.* **228** E8
Mbulu, *Tanzania* **211** U5
Mburucuyá, *Arg.* **134** G13
Mbuyuni, *Tanzania* **211** W6
McAlester, *Okla., U.S.* **105** D18
McAllen, *Tex., U.S.* **105** R16
McArthur River, *N.T., Austral.* **221** Q9
McArthur, river, *Austral.* **220** C10
McCall, *Idaho, U.S.* **108** F4
McCamey, *Tex., U.S.* **104** J11
McCammon, *Idaho, U.S.* **108** H7
McCaysville, *Ga., U.S.* **98** J5
McCleary, *Wash., U.S.* **110** C3
McClellanville, *S.C., U.S.* **98** M11
McCloud, *Calif., U.S.* **110** M4
McColl, *S.C., U.S.* **98** J10
McComb, *Miss., U.S.* **102** N10
McConaughy, Lake, *Nebr., U.S.* **107** P3
McCook, *Nebr., U.S.* **107** R4
McCormick, *S.C., U.S.* **98** L7
McCrory, *Ark., U.S.* **102** G10
McCuddin Mountains, *Antarctica*
    238 M9
McDermitt, *Nev., U.S.* **110** L9
McDonald Bay, *Antarctica* **239** J22
McDonald Peak, *Mont., U.S.* **108** C6
McDonald, *Kans., U.S.* **107** R3
McGehee, *Ark., U.S.* **102** J9
McGill, *Nev., U.S.* **111** P12
McGrath, *Alas., U.S.* **113** H15
McGregor, *Tex., U.S.* **105** J17
McGregor, *Minn., U.S.* **106** G11
McGuire, Mount, *Idaho, U.S.* **108** E5
Mcherrah, *Alg.* **200** G8
Mchinga, *Tanzania* **211** Y8
Mchinji, *Malawi* **212** C10
McIntosh, *Minn., U.S.* **106** F9
McIntosh, *S. Dak., U.S.* **106** H4
McKean Island, *Kiribati* **224** G10
McKees Rocks, *Pa., U.S.* **96** N3
McKeesport, *Pa., U.S.* **96** P3
McKenzie, *Tenn., U.S.* **103** E13
McKinley, Mount (Denali), *Alas., U.S.*
    113 H16
McKinley, Mount see Denali,
    N. Amer. **84** C2
McKinleyville, *Calif., U.S.* **110** M2
McKinney, *Tex., U.S.* **105** F18
McKinney, Lake, *Kans., U.S.* **107** U3
McKinnon, *Wyo., U.S.* **108** K9
McKittrick, *Calif., U.S.* **111** V6
McLaughlin, *S. Dak., U.S.* **106** H4
McLean, *Tex., U.S.* **105** D13
McLeans Town (Self Town), *Bahamas*
    122 J4
McLeansboro, *Ill., U.S.* **101** U6
McLeod Bay, *N.W.T., Can.* **86** H12
M'Clintock Channel, *Nunavut, Can.*
    87 E14
M'Clintock Inlet, *Nunavut, Can.*
    87 A15
McLoughlin, Mount, *Oreg., U.S.*
    110 K4
M'Clure Strait, *N.W.T., Can.* **86** C12
McMahon Line, *India* **182** G14
McMinnville, *Tenn., U.S.* **103** F16
McMinnville, *Oreg., U.S.* **110** F3
McMurdo Dry Valleys, *Antarctica*
    239 N14
McMurdo Sound, *Antarctica* **239** N14
McMurdo Station, *Antarctica* **239** N14
McNary Dam, *Oreg. Wash., U.S.* **110** E7
McPherson, *Kans., U.S.* **107** T7
McRae, *Ga., U.S.* **99** N7
McVeigh, *Ky., U.S.* **103** C20
McVille, *N. Dak., U.S.* **106** F7
Mdandu, *Tanzania* **211** X5
Mdennah, *Af.* **204** C12
Mdina, *Malta* **163** K22
Mé Maoya, peak, *New Caledonia, Fr.*
    231 P21
Me Shima, *Japan* **189** S3
Mead Point, *Anguilla, U.K.* **122** R11
Mead, Lake, *Ariz. Nev., U.S.* **111** U12
Meade Peak, *Idaho, U.S.* **108** J8
Meade, *Kans., U.S.* **107** V4
Meade, river, *Alas., U.S.* **113** C15
Meadow Lake, *Sask., Can.* **86** L11
Meadow Valley Wash, river, *Nev., U.S.*
    111 T12
Meadow, *Tex., U.S.* **104** F11
Meadow, *W. Aust. Austral.* **221** V2
Mead's Bay, *Anguilla, U.K.* **122** R10

Meadville, *Pa., U.S.* **96** L3
Meander River, *Alta. Can.* **86** J10
Meares, Cape, *Oreg., U.S.* **110** F2
Mears, *Mich., U.S.* **100** K9
Meaux, *Fr.* **149** P10
Mebane, *N.C., U.S.* **98** G11
Mecatina, river, *Nfld. & Lab., Que., Can.*
    87 L22
Mecca see Makkah, *Saudi Arabia*
    174 J6
Mecca, *Calif., U.S.* **111** Y11
Mechanicsburg, *Pa., U.S.* **96** P7
Mechelen, *Belg.* **148** L12
Mechems, region, *Af.* **198** D3
Mecheria, *Alg.* **200** D11
Mechol, *F.S.M.* **226** D12
Mecitözü, *Turk.* **170** D10
Meck, island, *Marshall Is.* **227** H18
Mecklenburger Bucht, *Ger.* **150** B9
Mecula, *Mozambique* **212** C12
Médala, *Mauritania* **205** H13
Medale, *Eth.* **211** N9
Medan, *Indonesia* **192** G5
Medanosa Point, *S. Amer.* **128** Q5
Médéa, *Alg.* **200** B12
Medellín, *Col.* **130** F5
Medenine, *Tunisia* **201** D16
Mederdra, *Mauritania* **204** G7
Medford, *Wis., U.S.* **100** H4
Medford, *Okla., U.S.* **105** A16
Medford, *Oreg., U.S.* **110** K3
Medgidia, *Rom.* **155** F18
Media Luna, Cayo, *Cuba* **118** K11
Medical Lake, *Wash., U.S.* **110** C8
Medicine Bow Mountains,
    Colo. Wyo., U.S. **108** K12
Medicine Bow Peak, *Wyo., U.S.*
    108 K12
Medicine Bow, *Wyo., U.S.* **108** J12
Medicine Hat, *Alta., Can.* **86** N11
Medicine Lake, *Mont., U.S.* **108** B13
Medicine Lodge, *Kans., U.S.* **107** V6
Medina de Pomar, *Sp.* **146** B13
Medina de Rioseco, *Sp.* **146** D9
Medina del Campo, *Sp.* **146** E9
Medina Gadaoundou, *Guinea* **204** L9
Medina see Al Madīnah, *Saudi Arabia*
    174 G6
Medina Sidonia, *Sp.* **146** M8
Medina, *N.Y., U.S.* **96** H5
Medina, *Tex., U.S.* **105** L15
Medina, *N. Dak., U.S.* **106** G6
Medina, river, *Tex., U.S.* **105** M15
Medinaceli, *Sp.* **146** E12
Medinipur, *India* **182** K11
Mediterranean Sea, *Eur.* **248** J13
Medje, *D.R.C.* **209** C17
Mednogorsk, *Russ.* **160** J7
Mednyy Island, *N. Amer.* **84** Q1
Mednyy, *Russ.* **161** E23
Médoc, region, *Fr.* **149** V6
Mêdog (Maiduo), *China* **184** L7
Medora, *N. Dak., U.S.* **106** G2
Médouneu, *Gabon* **208** D8
M'edr, *Eritrea* **210** N9
Medren, island, *Marshall Is.* **227** B21
Medvezh'i Ostrova (Bear Islands),
    Russ. **161** C19
Medvezhy Trough, *Arctic Oc.* **253** D14
Medvezh'yegorsk, *Russ.* **158** C8
Medyado, island, *Marshall Is.*
    227 H20
Medyai, island, *Marshall Is.* **227** G20
Meekatharra, *W. Aust. Austral.* **221** U3
Meeker, *Colo., U.S.* **108** L11
Meerut, *India* **182** F6
Meeteetse, *Wyo., U.S.* **108** G10
Mega, *Indonesia* **193** J18
Mêga, *Eth.* **211** Q7
Megálo Chorió, *Gr.* **157** M18
Megalo, *Eth.* **211** N9
Megalópoli, *Gr.* **156** K10
Mégara, *Gr.* **156** J12
Mégham, *Vanuatu* **229** G15
Meghri, *Arm.* **171** F20
Megiddo, *Israel* **172** L5
Megion, *Russ.* **160** H11
Megísti (Kastellórizo), island, *Gr.*
    157 N22
Mehadia, *Rom.* **154** F12
Mehamn, *Nor.* **143** A14
Mehar, *Pak.* **181** Q8
Meharry, Mount, *Austral.* **220** F3
Meherrin, river, *Va., U.S.* **98** F12
Mehetia, island, *Fr. Polynesia, Fr.*
    231 C17

Mehmetcik *see* Galatia, *Cyprus* **162** M10
Meḥola, *W. Bank* **172** L6
Mehrān, *Iraq* **176** H12
Mehrīz, *Iran* **177** J17
Mehtar Lām, *Afghan.* **180** H9
Meiganga, *Cameroon* **206** M11
Meighan Island, *Nunavut, Can.* **87** B14
Meigs, *Ga., U.S.* **99** Q5
Meihekou, *China* **185** E16
Meiktila, *Myanmar* **190** H5
Meiningen, *Ger.* **150** G8
Meissen, *Ger.* **150** F10
Meitan, *China* **184** M11
Meizhou, *China* **185** N14
Mejatto, *island, Marshall Is.* **227** F13
Mejillones del Sur, Bahía de, *Chile* **134** E5
Mejillones, *Chile* **134** E6
Mejit Island, *Marshall Is.* **227** C17
Mékambo, *Gabon* **208** D10
Mek'elē (Mekele), *Eth.* **210** J8
Mekele *see* Mek'elē, *Eth.* **210** J8
Mekhtar, *Pak.* **180** M9
Mekiro, *island, Fr. Polynesia, Fr.* **231** R17
Meknès, *Mor.* **200** C8
Mekong (Lancang, Mékôngk, Mènam), *river, Asia* **168** H9
Mekong River Delta, *Vietnam* **191** P12
Mékôngk (Mekong), *river, Cambodia* **191** N11
Mekoryuk, *Alas., U.S.* **112** J11
Melanesia, *island, Pac. Oc.* **224** G4
Melanesian Basin, *Pac. Oc.* **246** J9
Melba Peninsula, *Antarctica* **239** K22
Melbourne, *Fla., U.S.* **99** U10
Melbourne, *Vic., Austral.* **221** Z12
Melchior Station, *Antarctica* **238** D5
Mélé Bay, *Vanuatu* **229** K16
Mele, Capo, *It.* **152** G3
Melegnano, *It.* **152** E4
Melekeok Point, *Palau* **226** G6
Melenki, *Russ.* **158** J11
Melfi, *It.* **152** M12
Melfi, *Chad* **207** J13
Melfort, *Sask., Can.* **86** M12
Melgar de Fernamental, *Sp.* **146** C10
Melide, *Sp.* **146** B6
Meligalás, *Gr.* **156** L9
Melilla, *Sp.* **200** C9
Melinka, *Chile* **135** S6
Melipilla, *Chile* **134** L6
Melita *see* Mljet, *island, Croatia* **154** J7
Melitopol', *Ukr.* **159** R8
Mellansel, *Sw.* **143** G13
Melle, *Fr.* **149** T7
Melle, *Ger.* **150** D6
Mellen, *Wis., U.S.* **100** F4
Mellieħa Bay, *Malta* **163** J21
Mellieħa, *Malta* **163** J21
Mellish Rise, *Pac. Oc.* **246** L9
Mellit, *Sudan* **203** R8
Mellor Glacier, *Antarctica* **239** F18
Mělník, *Czechia* **150** G11
Melo, *Uru.* **134** K15
Mololo, *Indonesia* **193** N13
Melos *see* Mílos, *island, Gr.* **157** M13
Melrose, *Minn., U.S.* **106** J10
Melrose, *S. Aust., Austral.* **221** X10
Mels, *Switz.* **152** B5
Meltaus, *Fin.* **143** D15
Meluan, *Malaysia* **192** H10
Melun, *Fr.* **149** Q10
Melut, *S. Sudan* **203** U12
Melvich, *Scot., U.K.* **144** F9
Melville Bay, *Austral.* **220** A10
Melville Bugt *see* Qimusseriarsuaq, *Greenland, Den.* **88** C4
Melville Fracture Zone, *Ind. Oc.* **250** M8
Melville Hills, *Nunavut, N.W.T., Can.* **86** E11
Melville Island, *Nunavut, N.W.T., Can.* **87** C13
Melville Island, *Austral.* **220** A7
Melville Land, *Greenland, Den.* **88** B10
Melville Peninsula, *Nunavut, Can.* **87** G14
Melville Trough, *Arctic Oc.* **252** P10
Melville, *Sask., Can.* **86** N12
Melville, Cape, *Austral.* **220** B13
Melville, Lake, *Nfld. & Lab., Can.* **87** L22
Melvin, *Tex., U.S.* **105** J14

Melvin, Lough, *N. Ire., U.K.* **145** P5
Memba, *Mozambique* **213** D13
Memboro, *Indonesia* **193** N13
Memmingen, *Ger.* **150** K7
Mémót, *Cambodia* **191** N11
Memphis, *Tenn., U.S.* **102** G11
Memphis, *Tex., U.S.* **105** D13
Memphis, *Mo., U.S.* **107** Q13
Memphis, *Egypt* **202** F11
Mena, *Ark., U.S.* **102** G6
Mena, *Ukr.* **158** M7
Menahga, *Minn., U.S.* **106** G10
Menai Strait, *Wales, U.K.* **145** R9
Ménaka, *Mali* **205** H17
Mènam (Mekong), *river, Laos* **190** J8
Menard Fracture Zone, *Pac. Oc.* **247** Q17
Menard, *Tex., U.S.* **105** J14
Menasha, *Wis., U.S.* **100** J6
Menawashei, *Sudan* **203** S7
Mendana Fracture Zone, *Pac. Oc.* **247** L20
Mende, *Fr.* **149** W11
Mendebo Mountains, *Af.* **198** H10
Mendebo, *mountains, Eth.* **211** N8
Mendefera, *Eritrea* **210** H7
Mendeleyev Plain, *Arctic Oc.* **252** J9
Mendeleyev Ridge, *Arctic Oc.* **252** H9
Mendeleyevsk, *Russ.* **158** H16
Mendenhall, *Miss., U.S.* **102** M11
Mendi, *P.N.G.* **222** P6
Mendī, *Eth.* **210** L5
Mendip Hills, *Eng., U.K.* **145** V11
Mendocino Escarpment, *Pac. Oc.* **247** E15
Mendocino Fracture Zone, *Pac. Oc.* **247** E13
Mendocino Ridge, *Pac. Oc.* **247** E17
Mendocino, *Calif., U.S.* **111** P2
Mendocino, Cape, *U.S.* **90** E2
Mendota, *Ill., U.S.* **101** N5
Mendota, *Calif., U.S.* **111** T5
Mendoza Rise, *Pac. Oc.* **247** M20
Mendoza, *Arg.* **134** K8
Meneng Point, *Nauru* **227** M15
Meneng, *Nauru* **227** M14
Menetés, *Gr.* **157** P18
Menge, *island, Marshall Is.* **227** J20
Mengen, *Turk.* **170** D7
Mengene Daği, *peak, Turk.* **171** G18
Mengoub, *Mor.* **200** D10
Menindee, *N.S.W., Austral.* **221** X12
Menkere, *Russ.* **161** F16
Menno, *S. Dak., U.S.* **106** M7
Menominee, *Wis., U.S.* **100** H7
Menominee, *river, Mich., Wis., U.S.* **100** G7
Menomonee Falls, *Wis., U.S.* **100** L7
Menomonie, *Wis., U.S.* **100** H2
Menongue, *Angola* **208** N12
Menor, Mar, *Sp.* **147** K14
Menorca (Minorca), *island, Sp.* **147** G20
Men'shikova, *Russ.* **160** E10
Mentasta Lake, *Alas., U.S.* **113** H18
Mentawai Islands, *Asia* **168** M10
Mentawai, Kepulauan, *Indonesia* **192** K5
Mentawai Basin, *Ind. Oc.* **251** G15
Mentawai Ridge, *Ind. Oc.* **251** G15
Mentawei Trough, *Ind. Oc.* **251** G15
Menton, *Fr.* **149** X16
Mentor, *Ohio, U.S.* **101** N15
Menyamya, *P.N.G.* **222** P7
Menzies, *W. Aust., Austral.* **221** W4
Me'ona, *Israel* **172** K6
Meppel, *Neth.* **148** H13
Meppen, *Ger.* **150** D6
Mequinenza, Embalse de, *Sp.* **147** E15
Mer Rouge, *La., U.S.* **102** K9
Merai, *P.N.G.* **222** N10
Meramangye, Lake, *Austral.* **220** H8
Meramec, *river, Mo., U.S.* **107** T14
Merampit, *island, Indonesia* **193** G16
Merano, *It.* **152** C7
Meratus, Pegunungan, *Indonesia* **192** K11
Merauke, *Indonesia* **193** N21
Merca *see* Marka, *Somalia* **211** R11
Mercan Daĝlari, *peak, Turk.* **171** F14
Mercara, *India* **183** R5
Merced, *Calif., U.S.* **111** S5
Mercedes, *Tex., U.S.* **105** R16
Mercedes, *Arg.* **134** H13
Mercedes, *Arg.* **134** L12

Mercedes, *Arg.* **134** L9
Mercedes, *Uru.* **134** K13
Merceditas, *Chile* **134** H7
Mercherchar, *island, Palau* **226** J5
Mercy Bay, *N.W.T., Can.* **86** D12
Mercy, Cape, *Nunavut, Can.* **87** G19
Méré Lava, *island, Vanuatu* **229** E15
Meredith, Cape, *Falk. Is., U.K.* **135** W11
Mereeg, *Somalia* **211** Q13
Mereer-Gur, *Somalia* **211** P12
Merefa, *Ukr.* **159** P8
Mergenevo, *Kaz.* **178** E5
Mergui Archipelago, *Myanmar* **191** N6
Mergui Terrace, *Ind. Oc.* **251** E15
Meriç, *Turk.* **170** C2
Meriç, *river, Turk.* **170** C2
Mérida, *Mex.* **117** H16
Mérida, *Venez.* **130** D8
Mérida, *Sp.* **146** H8
Meriden, *Conn., U.S.* **96** M12
Meridian, *Miss., U.S.* **103** L13
Meridian, *Tex., U.S.* **105** H17
Meridian, *Idaho, U.S.* **108** G3
Meridja, *Alg.* **200** E9
Mérig, *island, Vanuatu* **229** D15
Meringur, *Vic., Austral.* **221** X11
Merir, *island, Palau* **224** F2
Meritxell, *Andorra* **162** H4
Merizo, *Guam, U.S.* **230** Q5
Merka, *Eth.* **211** P6
Merkel, *Tex., U.S.* **105** G14
Merkine, *Lith.* **143** N17
Merkoya, *Mali* **204** J11
Meroë, *ruin, Sudan* **203** Q13
Merolia, *W. Aust., Austral.* **221** V5
Merowe, *Sudan* **203** P12
Merredin, *W. Aust., Austral.* **221** W3
Merrill, *Wis., U.S.* **100** H5
Merrill, *Oreg., U.S.* **110** L4
Merrillville, *Ind., U.S.* **101** P8
Merrimack, *river, N.H., U.S.* **97** H14
Merriman, *Nebr., U.S.* **106** M3
Merritt Island, *Fla., U.S.* **99** T10
Merryville, *La., U.S.* **102** N6
Mersa Fat'ma, *Eritrea* **210** H8
Mersch, *Lux.* **162** J10
Merseburg, *Ger.* **150** F9
Mersey River, *Eng., U.K.* **145** R10
Mersin, *Turk.* **170** J9
Mersing, *Malaysia* **191** U10
Merta Road, *India* **182** H4
Mertert, *Lux.* **162** J11
Merthyr Tudful *see* Merthyr Tydfil, *Wales, U.K.* **145** U10
Merthyr Tydfil (Merthyr Tudful), *Wales, U.K.* **145** U10
Merti, *Kenya* **211** R7
Mértola, *Port.* **146** K6
Mertz Glacier Tongue, *Antarctica* **239** R18
Mertz Glacier, *Antarctica* **239** Q18
Mertzon, *Tex., U.S.* **105** J13
Meru, *Kenya* **211** S7
Méru, *Fr.* **149** N9
Meru, *peak, Tanzania* **211** U6
Merzifon, *Turk.* **170** D10
Merzig, *Ger.* **150** H4
Mesa Verde National Park, *Colo., U.S.* **109** Q10
Mesa, *Ariz., U.S.* **109** U7
Mesabi Range, *U.S.* **91** C14
Mesach Mellet, *Lib.* **201** J16
Meseong, *island, F.S.M.* **226** F9
Meseta, *Eur.* **140** K2
Meshra' er Req, *S. Sudan* **203** V10
Meskéné *see* Maskanah, *Syr.* **172** E10
Mesocco, *Switz.* **152** C4
Mesolóngi, *Gr.* **156** H8
Mesopotamia, *Iraq* **176** E8
Mesquite, *Tex., U.S.* **105** G19
Mesquite, *Nev., U.S.* **111** T13
Messaad, *Alg.* **200** D12
Messalo, *river, Mozambique* **213** C13
Messeïed, *Mor.* **200** G5
Messene, *Gr.* **156** L9
Messina, *It.* **153** R12
Messina, Stretto di, *It.* **153** R12
Messîni, *Gr.* **156** L9
Messinia, Gulf of, *Eur.* **140** M9
Messiniakós Kólpos, *Gr.* **156** L9
Mestá, *Gr.* **157** H16
Mestersvig, *research station, Greenland, Den.* **88** J13
Mestia, *Rep. of Georgia* **171** A17

Mestre, *It.* **152** D8
Mesudiye, *Turk.* **170** D12
Meta Incognita Peninsula, *Nunavut, Can.* **87** H19
Meta, *river, N. Amer.* **84** Q10
Meta, *river, S. Amer.* **130** F9
Metairie, *La., U.S.* **102** Q11
Metalanim Harbor, *F.S.M.* **226** K9
Metalanim, *F.S.M.* **226** K9
Metaline Falls, *Wash., U.S.* **110** A9
Metallifere, Colline, *It.* **152** H6
Metaline Falls, *Wash., U.S.* **110** A9
Metán, *Arg.* **134** F9
Metapontо, *It.* **153** N14
Metaxádes, *Gr.* **157** A16
Metema, *Eth.* **210** J6
Metemma, *Sudan* **203** Q13
Meteor Crater, *Ariz., U.S.* **109** S8
Meteor Rise, *Atl. Oc.* **249** U13
Meteor Seamount, *Atl. Oc.* **249** V13
Meteora, *ruin, Gr.* **156** E9
Méthana, *Gr.* **156** K12
Méthanon, *island, Gr.* **156** K12
Methóni, *Gr.* **156** L9
Metković, *Croatia* **154** H7
Metlakatla, *Alas., U.S.* **113** M24
Metolius, *Oreg., U.S.* **110** G5
Métoma, *island, Vanuatu* **229** C13
Metropolis, *Ill., U.S.* **101** V6
Mets Beverratap (Dəlidağ), *peak, Azerb.* **171** E20
Métsovo, *Gr.* **156** E8
Metter, *Ga., U.S.* **99** N8
Mettur Dam, *India* **183** R6
Metu, *Eth.* **210** M6
Metz, *Fr.* **149** P14
Meulan, *Fr.* **149** P9
Meuse, *river, Belg., Fr.* **148** M12
Mexia, *Tex., U.S.* **105** H18
Mexiana, Ilha, *Braz.* **133** C13
Mexicali, *Mex.* **116** A5
Mexican Hat, *Utah, U.S.* **109** Q9
Mexico Basin, *Atl. Oc.* **248** L1
Mexico City, *Mex.* **116** J12
Mexico, *Me., U.S.* **97** F15
Mexico, *Mo., U.S.* **107** S13
Mexico, *N. Amer.* **116** H11
Mexico, Gulf of, *N. Amer.* **84** M6
Meybod, *Iran* **177** H17
Meyrargues, *Fr.* **149** X13
Mezdra, *Bulg.* **155** H13
Mezen' Bay, *Arctic Oc.* **140** B12
Mezen', *Russ.* **158** A11
Mezen', *river, Russ.* **158** B13
Mezhdurechensk, *Russ.* **158** C13
Mezhdusharskiy, *Russ.* **160** E10
Mézidon-Canon, *Fr.* **149** P7
Mézin, *Fr.* **149** W7
Mezőtúr, *Hung.* **151** L16
Mezzolombardo, *It.* **152** C7
Mfuwe, *Zambia* **209** N20
Mgarr, *Malta* **163** H20
Mġarr, *Malta* **163** K21
Mgera, *Tanzania* **211** V7
Mgeta, *Tanzania* **211** X6
Mhunze, *Tanzania* **211** U4
Mi Shima, *Japan* **189** Q5
Miahuatlán, *Mex.* **117** K13
Miajadas, *Sp.* **146** H8
Miami Beach, *Fla., U.S.* **99** X10
Miami, *Fla., U.S.* **99** X10
Miami, *Okla., U.S.* **105** A19
Miami, *Tex., U.S.* **105** C13
Miamisburg, *Ohio, U.S.* **101** R11
Mīāndoāb, *Iran* **176** D12
Miandrivazo, *Madagascar* **213** F15
Mīāneh, *Iran* **177** D13
Miani Hor, *lake, Pak.* **181** R6
Mianrud, *Pak.* **181** P4
Mianwali, *Pak.* **180** K11
Mianyang, *China* **184** K10
Miass, *Russ.* **160** J8
Michalovce, *Slovakia* **151** J17
Miches, *Dom. Rep.* **119** M20
Michigan City, *Ind., U.S.* **101** N8
Michigan, *U.S.* **100** H10
Michigan, Lake, *N. Amer.* **84** J7
Michurinsk, *Russ.* **158** L11
Miconje, *Angola* **208** G9
Micoud, *St. Lucia* **123** L14
Micronesia, *islands, Pac. Oc.* **224** E3
Mid Point Camp, *Antarctica* **239** N16
Mid-Atlantic Ridge, *Atl. Oc.* **248** M7
Mid-Indian Basin, *Ind. Oc.* **250** L11
Mid-Indian Ridge, *Ind. Oc.* **250** O9
Mid-Pacific Mountains, *Pac. Oc.* **246** G11

Midai, *island, Indonesia* **192** H8
Midang, *N. Korea* **186** L7
Middelburg, *Neth.* **148** K11
Middle Alkali Lake, *Calif., U.S.* **110** M6
Middle America Trench, *Pac. Oc.* **247** G19
Middle Andaman, *island, India* **183** R14
Middle Bight, *Bahamas* **118** F11
Middle Caicos, *island, Turks & Caicos Is., U.K.* **119** H17
Middle Island, *St. Kitts & Nevis* **123** A17
Middle Loup, *river, Nebr., U.S.* **107** P5
Middle Passage, *Virgin Is., U.S.* **122** M9
Middle Quarters, *Jam.* **122** J7
Middle River, *Minn., U.S.* **106** E9
Middleburg, *S. Af.* **212** L7
Middlebury, *Vt., U.S.* **96** G12
Middlegate, *Austral.* **229** B20
Middlemarch, *N.Z.* **223** N17
Middleport, *Ohio, U.S.* **101** S14
Middlesboro, *Ky., U.S.* **103** D19
Middlesbrough, *Eng., U.K.* **145** P12
Middleton Ponds, *N.T., Austral.* **221** T8
Middleton, *Wis., U.S.* **100** L5
Middleton, *Qnsld., Austral.* **221** T12
Middletown, *N.Y., U.S.* **96** M10
Middletown, *Conn., U.S.* **97** M13
Middletown, *Ohio, U.S.* **101** S11
Midelt, *Mor.* **200** D8
Midi, Canal du, *Fr.* **149** Y9
Midland, *Mich., U.S.* **100** K11
Midland, *Tex., U.S.* **104** H12
Midland, *S. Dak., U.S.* **106** L4
Midlothian, *Tex., U.S.* **105** G17
Midu, *island, Maldives* **183** Y3
Midwest City, *Okla., U.S.* **105** C17
Midwest, *Wyo., U.S.* **108** H12
Midyat, *Turk.* **171** H16
Midžor, *peak, Bulg., Serb.* **154** H12
Międzyrzecz, *Pol.* **150** D12
Miélan, *Fr.* **149** X7
Mielec, *Pol.* **151** G16
Miercurea-Ciuc, *Rom.* **155** D15
Mieres, *Sp.* **146** B9
Mi'ēso, *Eth.* **210** M9
Migole, *Tanzania* **211** W6
Migori, *Kenya* **211** T5
Miguel Calmon, *Braz.* **133** H16
Migyaunglaung, *Myanmar* **190** M6
Mihalıççık, *Turk.* **170** E7
Mijas, *Sp.* **146** M9
Mijdaḥah, *Yemen* **174** P11
Mijek, *W. Sahara* **200** J4
Mijikadrek, *island, Marshall Is.* **227** A20
Mikhaylovka, *Russ.* **159** N12
Mikindani, *Tanzania* **211** Y8
Mikkeli, *Fin.* **143** H16
Mikniya, *Sudan* **203** Q13
Mikrí Préspa, Límni, *Gr.* **156** C8
Mikumi, *Tanzania* **211** W6
Mikun', *Russ.* **158** C14
Mikura Jima, *Japan* **189** R12
Mil Entrance, *F.S.M.* **226** D11
Milaca, *Minn., U.S.* **106** H11
Miladummadulu Atoll, *Maldives* **183** U3
Milagro, *Ecua.* **130** L3
Milan *see* Milano, *It.* **152** D4
Milan, *Tenn., U.S.* **103** F13
Milan, *Mo., U.S.* **107** R12
Milando, *Angola* **208** K12
Milange, *Mozambique* **212** D11
Milano (Milan), *It.* **152** D4
Milas, *Turk.* **170** H3
Milazzo, Capo di, *It.* **153** R12
Milbank, *S. Dak., U.S.* **106** J8
Milbridge, *Me., U.S.* **97** F18
Mildenhall, *Eng., U.K.* **145** S14
Mile and a Quarter, *Barbados* **123** J19
Mile Gully, *Jam.* **122** J8
Miles City, *Mont., U.S.* **108** D12
Miles, *Tex., U.S.* **105** J14
Miles, *Qnsld., Austral.* **221** V15
Mileto, *It.* **153** R13
Miletus, *ruins, Turk.* **170** H3
Milford Bay, *Trin. & Tobago* **123** P17
Milford Haven (Aberdaugleddau), *Wales, U.K.* **145** U8
Milford Lake, *Kans., U.S.* **107** S8
Milford Sound, *N.Z.* **223** N16
Milford, *Mass., U.S.* **97** K14
Milford, *Del., U.S.* **98** C15

Milford, Iowa, U.S. 106 M10
Milford, Nebr., U.S. 107 Q8
Milford, Utah, U.S. 109 N6
Milgis, river, Kenya 211 R7
Milgun, W. Aust., Austral. 221 U3
Mili Atoll, Marshall Is. 227 D17
Miliana, Alg. 200 B12
Miliés, Gr. 156 F11
Milikapiti, N.T., Austral. 221 N7
Mililani Town, Hawaii, U.S. 115 G14
Milingimbi, N.T., Austral. 221 N9
Milk River Bath, Jam. 122 K8
Milk, river, Can., U.S. 90 B8
Mil'kovo, Russ. 161 F22
Mill City, Oreg., U.S. 110 G3
Mill Island, Nunavut, Can. 87 H17
Mill Island, Antarctica 239 K21
Mill Reef, Antigua & Barbuda 123 B22
Millars Sound, Bahamas 122 F2
Millau, Fr. 149 W11
Mille Lacs Lake, Minn., U.S. 106 H11
Milledgeville, Ga., U.S. 98 M6
Millen, Ga., U.S. 98 M8
Millennium Island see Caroline Island, Kiribati 225 H13
Miller Peak, Ariz., U.S. 109 W8
Miller Point, Solomon Is. 228 N7
Miller, S. Dak., U.S. 106 K6
Millerovo, Russ. 159 P11
Millersburg, Ohio, U.S. 101 Q14
Millevaches, Plateau de, Fr. 149 U9
Milligan Cay, St. Vincent & the Grenadines 123 L17
Millington, Tenn., U.S. 102 F11
Millinocket, Me., U.S. 97 D17
Millmerran, Qnsld., Austral. 221 V15
Milltown Malbay (Sráid na Cathrach), Ire. 145 S3
Milltown, Mont., U.S. 108 D6
Millungera, Qnsld., Austral. 221 S12
Millville, N.J., U.S. 96 Q10
Milly Milly, W. Aust., Austral. 221 U2
Milly, Fr. 149 Q10
Milne Bay, P.N.G. 222 R9
Milne Land, Greenland, Den. 88 K13
Milne Seamounts, Atl. Oc. 248 J8
Milnor, N. Dak., U.S. 106 H7
Milo, Me., U.S. 97 E17
Milo, Tanzania 211 Y5
Mīlo, Eth. 210 L9
Milo, river, Guinea 204 M11
Miloli'i, Hawaii, U.S. 115 Q20
Milord Point, Virgin Is., U.S. 122 R3
Mílos (Melos), island, Gr. 157 M13
Mílos (Plaka), Gr. 157 M13
Milovzorova Bay, Antarctica 239 L22
Milparinka, N.S.W., Austral. 221 V12
Milton Keynes, Eng., U.K. 145 T13
Milton-Freewater, Oreg., U.S. 110 E8
Milton, Pa., U.S. 96 M7
Milton, Fla., U.S. 99 Q2
Miltonvale, Kans., U.S. 107 S7
Milu Pass, Marshall Is. 227 F16
Milu, island, Marshall Is. 227 F16
Milwaukee, Wis., U.S. 100 L7
Mimizan-Plage, Fr. 149 W6
Mimongo, Gabon 208 F8
Mims, Fla., U.S. 99 T9
Min, river, China 185 M15
Min, river, China 184 K10
Mina Bazar, Pak. 180 L9
Mina, Nev., U.S. 111 R8
Mīnāb, Iran 177 N19
Minahasa, Indonesia 193 H15
Minamata, Japan 189 S4
Minami To, Japan 224 C3
Minami Tori Shima (Marcus Island), Pac. Oc. 224 C5
Minamisōma, Japan 188 M13
Minamitane, Japan 189 U5
Minas, Cuba 118 J11
Minas, Uru. 134 L14
Minatitlán, Mex. 117 J14
Minbu, Myanmar 190 H4
Minco, Okla., U.S. 105 D16
Mindanao, island, Philippines 193 F15
Mindelo, Cabo Verde 215 B15
Minden, La., U.S. 102 K7
Minden, Nebr., U.S. 107 Q6
Minden, Nev., U.S. 111 Q6
Minden, Ger. 150 D7
Mindon, Myanmar 190 J4
Mindoro Strait, Philippines 193 D13
Mindoro, island, Philippines 193 D13
Mine, Japan 189 Q4
Mineloa, Tex., U.S. 105 G19
Mineral Point, Wis., U.S. 100 L4
Mineral Wells, Tex., U.S. 105 G16
Mineral'nyye Vody, Russ. 159 T13
Minersville, Utah, U.S. 109 N6
Minfeng (Niya), China 184 G3
Mingan, Que., Can. 87 M21
Mingaora, Pak. 180 H11
Mingenew, W. Aust., Austral. 221 W2
Mingəçevir Su Anbarı, Azerb. 171 C20
Mingəçevir, Azerb. 171 D21
Mingin, Myanmar 190 G4
Minglanilla, Sp. 147 H13
Mingo Cay, Virgin Is., U.S. 122 M10
Mingoyo, Tanzania 211 Y8
Mingteke Pass, Pak. 182 B5
Mingteke, China 184 G1
Mingulay Miughalaigh, island, Scot., U.K. 144 J6
Minhe, China 184 H9
Minho, river, Jam. 122 J9
Minho, river, Port., Sp. 146 D5
Minicoy Island, India 183 T3
Minidoka Internment National Monument, Idaho, U.S. 108 H5
Minidoka, Idaho, U.S. 108 H6
Minigwal, Lake, Austral. 220 J5
Minilya, W. Aust., Austral. 221 U1
Minimarg, Pak. 180 H13
Min'kovo, Russ. 158 F12
Minna, Nigeria 206 K7
Minneapolis, Minn., U.S. 106 K11
Minneapolis, Kans., U.S. 107 S7
Minnedosa, Man., Can. 87 N13
Minneola, Kans., U.S. 107 U5
Minneota, Minn., U.S. 106 K9
Minnesota, U.S. 106 H10
Minnesota, river, Minn., U.S. 106 K10
Minnewaukan, N. Dak., U.S. 106 E6
Miño, river, Sp. 146 B7
Minocqua, Wis., U.S. 100 G5
Minonk, Ill., U.S. 101 Q5
Minorca see Menorca, island, Sp. 147 G20
Minot, N. Dak., U.S. 106 E4
Minsk Mazowiecki, Pol. 151 E16
Minsk, Belarus 158 K4
Minto Inlet, N.W.T., Can. 86 E12
Minto, N. Dak., U.S. 106 E7
Minto, Lac, Que., Can. 87 K18
Minturno, It. 152 L10
Mīnū Dasht, Iran 177 D18
Minvoul, Gabon 208 C9
Minwakh, Yemen 174 M11
Mio, Mich., U.S. 100 J11
Mīr Bachah Kōṭ, Afghan. 180 H9
Mir, Belarus 158 L4
Mir, Niger 206 H10
Mira Por Vos, island, Bahamas 119 H14
Mira, Port. 146 F5
Mira, It. 152 E8
Mīrābād, Afghan. 180 M3
Mirador, Braz. 133 F15
Miraflores, Col. 130 J8
Miragoâne, Haiti 119 M16
Miram Shah, Pak. 180 K9
Miramar, Arg. 135 N13
Mirambeau, Fr. 149 U6
Mirampéllou, Kólpos, Gr. 157 Q16
Miran, China 184 G5
Miranda de Ebro, Sp. 146 C12
Miranda do Douro, Port. 146 D8
Miranda, Braz. 132 L10
Miranda, river, Braz. 132 K10
Mirandela, Port. 146 D7
Mirando City, Tex., U.S. 105 P15
Mirandola, It. 152 E6
Mirano, It. 152 D8
Mirbāṭ, Oman 175 M15
Mirebalais, Haiti 119 M17
Mirebeau, Fr. 149 S7
Mirecourt, Fr. 149 P13
Mirepoix, Fr. 149 Y9
Miri, Malaysia 192 G11
Miriam Vale, Qnsld., Austral. 221 U15
Mirim Lagoon, S. Amer. 128 L8
Mirim, Lagoa, Uru. 134 K15
Mīrjāveh, Iran 177 L22
Mirny Station, Antarctica 239 J22
Mirnyy, Russ. 161 H16
Mirpur Khas, Pak. 181 R9
Mirpur Sakro, Pak. 181 S7
Mirpur, Pak. 180 J12
Mirsaale, Somalia 211 N13
Miryang, S. Korea 187 S11
Mirzapur, India 182 J8
Misgar, Pak. 180 F13
Mishawaka, Ind., U.S. 101 N9
Misima, island, P.N.G. 222 R10
Miski, Sudan 203 R7
Miskitos, Costa de, Nicar. 117 M19
Miskolc, Hung. 151 K16
Misool, island, Indonesia 193 K17
Misquah Hills, Minn., U.S. 106 E14
Miṣrātah, Lib. 201 D18
Misséni, Mali 204 L12
Mission, Tex., U.S. 105 R16
Mission, S. Dak., U.S. 106 M4
Missira, Senegal 204 K9
Mississippi Fan, Atl. Oc. 248 L2
Mississippi River Delta, La., U.S. 103 R13
Mississippi Sound, Miss., U.S. 102 P12
Mississippi State, Miss., U.S. 103 J13
Mississippi, U.S. 102 L11
Mississippi, river, N. Amer. 84 J6
Missoula, Mont., U.S. 108 D6
Missour, Mor. 200 D9
Missouri Valley, Iowa, U.S. 107 P9
Missouri, U.S. 107 T13
Missouri, river, N. Amer. 84 H4
Mistassini, Lac, Que., Can. 87 N19
Misterbianco, It. 153 S12
Mistissini, Que., Can. 87 N19
Misty Fiords National Monument, Alas., U.S. 113 M24
Misurata, Cape, Af. 198 D6
Mitatib, Sudan 203 Q14
Mitau see Jelgava, Latv. 143 L16
Mitchell Lake, Ala., U.S. 103 K15
Mitchell, Ind., U.S. 101 T9
Mitchell, S. Dak., U.S. 106 L7
Mitchell, Nebr., U.S. 107 N1
Mitchell, Oreg., U.S. 110 G6
Mitchell, Qnsld., Austral. 221 U14
Mitchell, Mount, N.C., U.S. 98 H7
Mitchell, river, Austral. 220 C12
Mitchelstown (Baile Mhistéala), Ire. 145 T4
Mithankot, Pak. 181 N10
Mithi, Pak. 181 S9
Míthymna, Gr. 157 F16
Mitiaro, island, Cook Is., N.Z. 224 K12
Mitla Pass, Egypt 173 Q2
Mitla, Mex. 117 K13
Mito, Japan 189 N12
Mitre Island see Fataka, Solomon Is. 224 J8
Mitre, peak, N.Z. 223 J21
Mitrofanovskaya, Russ. 158 B16
Mitrovica see Mitrovicë, Kos. 154 H10
Mitrovicë (Kosovska Mitrovica, Mitrovica), Kos. 154 H10
Mitsamiouli, Comoros 215 L14
Mitsero, Cyprus 162 P8
Mits'iwa (Massawa), Eritrea 210 H8
Mitsoudjé, Comoros 215 M14
Mitú, Col. 130 J9
Mitumba Mountains, D.R.C. 209 K17
Mitwaba, D.R.C. 209 K17
Mityushikha, Russ. 160 D11
Mitzic, Gabon 208 D8
Miughalaigh, island, Scot., U.K. 144 J6
Miyake Jima, Japan 189 Q12
Miyako, Japan 188 K13
Miyakonojō, Japan 189 S5
Miyazaki, Japan 189 S5
Miyazu, Japan 189 P8
Mīzan Teferī, Eth. 211 N5
Mizdah, Lib. 201 E17
Mizen Head, Ire. 145 U3
Mizil, Rom. 155 E16
Mizo Hills, India 182 J14
Mizuho Plateau, Antarctica 239 D17
Mizuho Station, Antarctica 239 C18
Mizusawa, Japan 188 L13
Mjøsa, lake, Nor. 142 J11
Mkangira, Tanzania 211 X7
Mkoani, Tanzania 211 V8
Mkomazi, Tanzania 211 V7
Mkowela, Tanzania 211 Z7
Mkushi, Zambia 209 N19
Mladá Boleslav, Czechia 150 G11
Mladenovac, Serb. 154 F10
Mława, Pol. 151 C15
Mlima Bénara, peak, Mayotte, Fr. 215 P17
Mljet (Melita), island, Croatia 154 J7
Mljetski Kanal, Croatia 154 H7
Mloa, Tanzania 211 X5
Mnero, Tanzania 211 Y7
Mộ Đức, Vietnam 190 M13
Mo i Rana, Nor. 142 E12
Moa Island, Austral. 220 A12
Moa, Cuba 119 K14
Moa, island, Indonesia 193 M16
Moa, river, Guinea, Sierra Leone 204 N10
Moab, Utah, U.S. 109 N9
Moaco, river, Braz. 132 F6
Moala, island, Fiji 226 Q7
Moananui, island, N.Z. 229 M21
Moanda, Gabon 208 F9
Moapa, Nev., U.S. 111 T12
Moatize, Mozambique 212 D10
Moba, D.R.C. 209 J19
Mobaye, Cen. Af. Rep. 207 N15
Mobeetie, Tex., U.S. 105 C13
Moberly, Mo., U.S. 107 S13
Mobile Bay, Ala., U.S. 103 P14
Mobile Point, Ala., U.S. 103 P14
Mobile, Ala., U.S. 103 P13
Mobile, river, Ala., U.S. 103 N14
Mobridge, S. Dak., U.S. 106 J5
Moca, Dom. Rep. 119 L18
Mocajuba, Braz. 133 D13
Moce, island, Fiji 226 Q8
Moch, island, F.S.M. 226 E9
Mocha, Isla, Chile 135 N5
Mochenap, strait, F.S.M. 226 E10
Mocho Mountains, Jam. 122 J8
Mochon Point, N. Mariana Is., U.S. 230 M9
Mochonap, strait, F.S.M. 226 D9
Mochudi, Botswana 212 H7
Mochun Eparit, F.S.M. 226 D9
Mochun Fanananei, F.S.M. 226 E8
Mochun Fanew, F.S.M. 226 F9
Mochun Nenom, F.S.M. 226 E8
Mochun Nepis, F.S.M. 226 F8
Mochun Ocha, F.S.M. 226 F9
Mochun Pianu, F.S.M. 226 F7
Mochun Sopwer, F.S.M. 226 D9
Mochun Tauanap, F.S.M. 226 E8
Mochun Unikar, F.S.M. 226 F10
Mochun Winion, F.S.M. 226 F10
Mocímboa da Praia, Mozambique 213 B13
Mocksville, N.C., U.S. 98 H9
Moclips, Wash., U.S. 110 C2
Môco, peak, Angola 208 M10
Mocoa, Col. 130 J5
Moctezuma Trough, Pac. Oc. 247 H19
Moctezuma, Mex. 116 C7
Mocuba, Mozambique 212 E12
Modane, Fr. 149 V15
Model, Colo., U.S. 109 P14
Modena, It. 152 F6
Modesto, Calif., U.S. 111 S5
Modica, It. 153 U11
Modimolle, S. Af. 212 H8
Modjamboli, D.R.C. 209 C14
Modjigo, Niger 206 F11
Modot, Mongolia 184 D11
Modugno, It. 152 M14
Moe, Vic., Austral. 221 Z13
Moengo, Suriname 131 F18
Moerai, Fr. Polynesia, Fr. 231 K17
Mogadishu see Muqdisho, Somalia 211 R11
Mogadouro, Port. 146 E7
Mogalo, D.R.C. 209 C13
Mogán, Sp. 214 R5
Mogandia, Congo 208 D11
Mogaung, Myanmar 190 E6
Mogens Heinesen Fjord, Greenland, Den. 89 S8
Mogi das Cruzes, Braz. 133 M14
Mogocha, Russ. 161 K17
Mogoi, Indonesia 193 K18
Mogok, Myanmar 190 G5
Mogollon Rim, Ariz., U.S. 109 T8
Mogor, Eth. 211 P8
Mogotes, Punta, Arg. 135 N13
Mohács, Hung. 151 N15
Mohala, India 182 L7
Mohali, Congo 208 E11
Mohall, N. Dak., U.S. 106 D4
Mohammadia, Alg. 200 C11
Mohammedia (Fedala), Mor. 200 C7
Mohana, India 182 M9
Mohave, Lake, Ariz., Nev., U.S. 111 V12
Mohawk, river, N.Y., U.S. 96 J10
Mohe, China 185 A14
Mohéli see Mwali, Comoros 215 N15
Mohelnice, Czechia 151 H13
Mohenjo Daro, ruins, Pak. 181 Q8
Mohill (Maothail), Ire. 145 Q5
Mohn Glacier, Greenland, Den. 88 F4
Mohns Ridge, Arctic Oc. 253 H18
Mohnyin, Myanmar 190 E5
Mohon Peak, Ariz., U.S. 109 S5
Mohoro, Tanzania 211 X8
Mohotani, island, Fr. Polynesia, Fr. 231 E15
Mohyliv-Podil's'kyy, Ukr. 159 Q4
Moindou, New Caledonia, Fr. 231 Q22
Moineşti, Rom. 155 C16
Moirang, India 182 J14
Moirans, Fr. 149 U13
Moisie, river, Que., Can. 87 M20
Moissac, Fr. 149 W8
Moïssala, Chad 207 L13
Moita, Port. 146 J5
Mojácar, Sp. 147 L13
Mojados, Sp. 146 E10
Mojave Desert, U.S. 90 H5
Mojave National Preserve, Calif., U.S. 111 V11
Mojave, Calif., U.S. 111 W8
Mojave, river, Calif., U.S. 111 W10
Mojokerto, Indonesia 192 M10
Mokambo, D.R.C. 209 M18
Mokdale, Indonesia 193 N14
Mokil Atoll, island, F.S.M. 226 B11
Mokokchung, India 182 H15
Mokolo, Cameroon 206 K11
Mokopane, S. Af. 212 H8
Mokpo, S. Korea 187 T7
Mol, Belg. 148 K12
Mola di Bari, It. 152 M14
Molanda, D.R.C. 209 C14
Moláoi, Gr. 156 M11
Molat, island, Croatia 154 F5
Moldavia, region, Rom. 155 B16
Molde, Nor. 142 H10
Moldova Nouă, Rom. 154 F11
Moldova, Eur. 159 Q4
Moldova, river, Rom. 155 B16
Moldoveanu, peak, Rom. 155 E14
Môle Saint-Nicolas, Haiti 119 L15
Molegbe, D.R.C. 209 B14
Molesworth, N.Z. 223 K19
Molina de Aragón, Sp. 147 F13
Molina de Segura, Sp. 147 K14
Molina, Chile 134 L7
Moline, Ill., U.S. 101 P4
Moline, Kans., U.S. 107 U9
Molinière Point, Grenada 123 L22
Molino, river, San Marino 163 J16
Moliro, D.R.C. 209 K19
Moliterno, It. 153 N13
Mollendo, Peru 132 K5
Möller Ice Stream, Antarctica 238 H10
Mollerussa, Sp. 147 D16
Molloy Deep, Arctic Oc. 253 H17
Molloy Fracture Zone, Arctic Oc. 253 H16
Molo, Kenya 211 S5
Moloa'a Bay, Hawaii, U.S. 114 E10
Molodezhnaya Station, Antarctica 239 C19
Molokai Fracture Zone, Pac. Oc. 247 G14
Moloka'i, Hawaii, U.S. 115 H17
Molopo, river, Botswana, S. Af. 212 J6
Molsheim, Fr. 149 P15
Molson, Wash., U.S. 110 A7
Molu, island, Indonesia 193 M17
Molucca Sea, Indonesia 193 J15
Moluccas (Maluku), islands, Indonesia 193 K16
Molula, D.R.C. 209 H18
Moma, D.R.C. 209 F15
Moma, river, Russ. 161 E19
Mombasa, Kenya 211 U8
Mombenzélé, Congo 208 D12
Mombetsu, Japan 188 F14
Mombetsu, Japan 188 G13
Mombongo, D.R.C. 209 D15
Momboyo, river, D.R.C. 209 E13
Momence, Ill., U.S. 101 P7
Momi, D.R.C. 209 F17
Momi, Fiji 226 P4
Momote, P.N.G. 222 M8
Mompach, Lux. 162 J11
Mompono, D.R.C. 209 E14
Mompós, Col. 130 D6
Mon Repos, St. Lucia 123 K14
Møn, island, Den. 142 N12
Mona Passage, N. Amer. 119 M21

Mona Quimbundo, *Angola* **209** L13
Mona Reservoir, *Utah, U.S.* **108** M7
Mona, Isla, *P.R., U.S.* **119** N21
Monach Islands *see* Na h-Eileanan Theisgeir, *Scot., U.K.* **144** H6
Monaco Rail Road Station, *Monaco* **163** E20
Monaco-Ville, *Monaco* **163** E21
Monaco, *Eur.* **162** E3
Monaco, Port of, *Monaco* **163** D21
Monadhliath Mountains, *Scot., U.K.* **144** J9
Monaghan (Muineachán), *Ire.* **145** P6
Monahans, *Tex., U.S.* **104** J11
Monalanong, peak, *Botswana* **212** H7
Monapo, *Mozambique* **213** D13
Monarch Pass, *Colo., U.S.* **109** N12
Monari, *Afghan.* **180** K8
Moncada, *Sp.* **147** F14
Moncalieri, *It.* **152** E2
Monchegorsk, *Russ.* **160** D8
Mönchengladbach, *Ger.* **150** F4
Monchique, Serra de, *Port.* **146** L5
Monchon, *Guinea* **204** L7
Moncks Corner, *S.C., U.S.* **98** L10
Monclova, *Mex.* **116** E11
Monclova, *Mex.* **117** J16
Moncton, *N.B., Can.* **87** P21
Mondego, Cabo, *Port.* **146** F5
Mondego, Cape, *Eur.* **140** J4
Mondego, river, *Port.* **146** F6
Mondeodo, *Indonesia* **193** K14
Mondo, *Chad* **206** H12
Mondolfo, *It.* **152** G9
Mondoñedo, *Sp.* **146** B7
Mondorf, *Lux.* **162** K10
Mondovi, *Wis., U.S.* **100** J3
Mondovì, *It.* **152** F3
Mondragone, *It.* **152** M10
Mondy, *Russ.* **161** L15
Moneague, *Jam.* **122** H9
Monemvasía, *Gr.* **156** M11
Monessen, *Pa., U.S.* **96** P3
Monestier-de-Clermont, *Fr.* **149** V13
Monett, *Mo., U.S.* **107** V11
Monfalcone, *It.* **152** D9
Monforte de Lemos, *Sp.* **146** C7
Monforte, *Port.* **146** H6
Móng Cái, *Vietnam* **190** G12
Monga, *D.R.C.* **209** B15
Monga, *Tanzania* **211** X7
Mongala, river, *D.R.C.* **209** C14
Mongala, *S. Sudan* **203** X12
Mongana, *D.R.C.* **209** C14
Mongbwalu, *D.R.C.* **209** D19
Mongers Lake, *Austral.* **220** J3
Monggümp'o, *N. Korea* **186** L4
Mongmong, *Guam, U.S.* **230** N5
Mongo, *Chad* **207** J14
Mongochi, *Malawi* **212** D11
Mongo, *Cameroon* **206** Q10
Mongolia, *Asia* **184** D8
Mongolian Plateau, *Asia* **168** F10
Mongomo, *Eq. Guinea* **206** Q10
Mongororo, *Chad* **207** J16
Mongoumba, *Cen. Af. Rep.* **207** N14
Mongoy, *Russ.* **161** K17
Mongton, *Myanmar* **190** H7
Mongu, *Zambia* **209** P15
Môngua, *Angola* **208** Q11
Monguno, *Nigeria* **206** H11
Moni, *Cyprus* **162** Q8
Monistrol-sur-Loire, *Fr.* **149** V12
Monito, island, *P.R., U.S.* **119** M21
Monitor Range, *Nev., U.S.* **111** R10
Monkey Bay, *Malawi* **212** C11
Monkira, *Qnsld., Austral.* **221** U11
Monkoto, *D.R.C.* **209** F14
Monmouth, *Ill., U.S.* **101** P4
Monmouth, *Oreg., U.S.* **110** G3
Mono (Monkey Point), Punta, *Nicar.* **117** M19
Mono, *Calif., U.S.* **111** R7
Mono Lake, *Calif., U.S.* **111** R7
Mono, island, *Solomon Is.* **228** D5
Mono, river, *Togo* **205** N17
Monólithos, *Gr.* **157** N19
Monomoy Island, *Mass., U.S.* **97** L16
Monon, *Ind., U.S.* **101** Q8
Monona, *Iowa, U.S.* **106** M13
Monongahela, river, *Pa. W. Va., U.S.* **96** Q3
Monopoli, *It.* **152** M14
Monos, island, *Trin. & Tobago* **123** P21
Monou, *Chad* **207** F16
Monreal del Campo, *Sp.* **147** F13
Monroe City, *Mo., U.S.* **107** R13

Monroe Lake, *Ind., U.S.* **101** S9
Monroe, *Ga., U.S.* **98** L6
Monroe, *N.C., U.S.* **98** J9
Monroe, *Mich., U.S.* **100** N12
Monroe, *Wis., U.S.* **100** M5
Monroe, *La., U.S.* **102** L9
Monroe, *Wash., U.S.* **110** B4
Monroeville, *Ala., U.S.* **103** M14
Monrovia, *Liberia* **204** P10
Monschau, *Ger.* **150** G4
Monsoon Rise, *Ind. Oc.* **251** J15
Mönsterås, *Sw.* **143** M13
Mont Aorai, *Fr. Polynesia, Fr.* **231** H22
Mont Belvieu, *Tex., U.S.* **105** L20
Mont Bengoué, *Gabon* **208** D9
Mont Blanc, *Eur.* **140** J5
Mont Héani, *Fr. Polynesia, Fr.* **231** D14
Mont Iboundji, *Gabon* **208** E8
Mont Ivirairai, *Fr. Polynesia, Fr.* **231** J21
Mont Kolofau, *Wallis & Futuna Is., Fr.* **230** R12
Mont Mohi, *D.R.C.* **209** F18
Mont Nabeba, *Congo* **208** D10
Mont Ngaoui, *Cameroon, Cen. Af. Rep.* **206** P18
Mont Ngouo, *Cen. Af. Rep.* **206** M12
Mont Nimba, *Côte d'Ivoire, Guinea* **204** N11
Mont Orohena, *Fr. Polynesia, Fr.* **231** H22
Mont Panié, *New Caledonia, Fr.* **231** P21
Mont Pouoanuu, *Fr. Polynesia, Fr.* **231** D13
Mont Roonui, *Fr. Polynesia, Fr.* **231** J23
Mont Sindoa, *New Caledonia, Fr.* **231** Q22
Mont Sokbaro, *Benin* **205** M17
Mont Taatioe, *Fr. Polynesia, Fr.* **231** K17
Mont Taita, *Fr. Polynesia, Fr.* **231** N14
Mont Tetufera, *Fr. Polynesia, Fr.* **231** J22
Mont Tina, *D.R.C.* **209** C18
Mont Tohiea, *Fr. Polynesia, Fr.* **231** H20
Mont Toomaru, *Fr. Polynesia, Fr.* **231** H17
Mont-de-Marsan, *Fr.* **149** X6
Mont-Dore, *New Caledonia, Fr.* **231** Q23
Mont-Joli, *Que., Can.* **87** N20
Mont-Louis, *Fr.* **149** Z9
Montagne d'Ambre, *Madagascar* **213** C17
Montagne Pelée, *Martinique, Fr.* **123** E22
Montagu Harbour, *P.N.G.* **222** P9
Montague Island, *Alas., U.S.* **113** K17
Montague, *Calif., U.S.* **110** L3
Montalbán, *Sp.* **147** F14
Montalegre, *Port.* **146** D6
Montalto di Castro, *It.* **152** K7
Montana, *U.S.* **108** D9
Montana, *Bulg.* **155** H13
Montanja, *Curaçao, Neth.* **123** Q15
Montargil, Barragem de, *Port.* **146** H5
Montargis, *Fr.* **149** Q10
Montastruc-la-Conseillère, *Fr.* **149** X9
Montauban, *Fr.* **149** X9
Montauk Point, *N.Y., U.S.* **97** M14
Montbard, *Fr.* **149** R12
Montbéliard, *Fr.* **149** R14
Montblanc, *Sp.* **147** E16
Montbrison, *Fr.* **149** U12
Montceau-les-Mines, *Fr.* **149** S12
Montchanin, *Fr.* **149** S12
Montdidier, *Fr.* **149** N10
Monte Bello Islands, *Austral.* **220** E2
Monte Binga, *Mozambique, Zimb.* **212** F10
Monte Carlo Radio and Television Station, *Monaco* **163** D21
Monte Carlo, *Monaco* **163** C21
Monte Caseros, *Arg.* **134** J13
Monte Comán, *Arg.* **134** L8
Monte Cristi, *Dom. Rep.* **119** L17
Monte Dinero, *Arg.* **135** W9
Monte Giardino, *San Marino* **163** K16
Monte Lindo, river, *Parag.* **132** M9
Monte Negro Falls, *Angola* **208** Q9
Monte Negro Falls, *Angola, Namibia* **212** E2
Monte Plata, *Dom. Rep.* **119** M19
Monte Quemado, *Arg.* **134** F10
Monte San Savino, *It.* **152** H7
Monte Santu, Capo di, *It.* **153** N5

Monte Vista, *Colo., U.S.* **109** P12
Monte-Carlo Sporting Club d'Été, *Monaco* **163** B23
Montebelluna, *It.* **152** D8
Montech, *Fr.* **149** X8
Montecito, *Calif., U.S.* **111** X6
Montecristo, Isola di, *It.* **152** K5
Montefiascone, *It.* **152** J8
Montego Bay, *Jam.* **122** H7
Montego Bay, *Jam.* **122** H7
Montehermoso, *Sp.* **146** G8
Montejinni, *N.T., Austral.* **221** Q8
Montejunto, Serra de, *Port.* **146** H4
Montélimar, *Fr.* **149** W12
Montello, *Nev., U.S.* **110** M12
Montemarciano, *It.* **152** G9
Montemayor, Meseta de, *Arg.* **135** S9
Montemor-o-Novo, *Port.* **146** J5
Montemorelos, *Mex.* **116** F11
Montemuro, peak, *Port.* **146** E6
Montendre, *Fr.* **149** V7
Montenegro, *Eur.* **154** H8
Montepescali, *It.* **152** J6
Montepuez, *Mozambique* **213** C13
Montereau, *Fr.* **149** Q10
Monterey Bay, *Calif., U.S.* **111** T4
Monterey Canyon, *Pac. Oc.* **247** E17
Monterey, *Tenn., U.S.* **103** E17
Monterey, *Calif., U.S.* **111** T4
Montería, *Col.* **130** E5
Monterotondo, *It.* **152** K8
Monterrey, *Mex.* **116** F11
Montes Claros, *Braz.* **133** J15
Montesano, *Wash., U.S.* **110** D2
Montese, *It.* **152** F6
Montevallo, *Ala., U.S.* **103** K15
Montevarchi, *It.* **152** H7
Montevideo, *Minn., U.S.* **106** K9
Montevideo, *Uru.* **134** L14
Montezuma Castle National Monument, *Ariz., U.S.* **109** S7
Montezuma Creek, *Utah, U.S.* **109** P9
Montezuma, *Ga., U.S.* **99** N5
Montezuma, *Iowa, U.S.* **107** P12
Montezuma, *Kans., U.S.* **107** U4
Montfaucon, *Fr.* **149** V12
Montgenèvre, *Fr.* **149** V15
Montgomery City, *Mo., U.S.* **107** S14
Montgomery, *W. Va., U.S.* **98** C11
Montgomery, *Ala., U.S.* **103** L16
Monthey, *Switz.* **152** C2
Monti, *It.* **152** M4
Monticello, *N.Y., U.S.* **96** L10
Monticello, *Me., U.S.* **97** C18
Monticello, *Ga., U.S.* **98** L6
Monticello, *Fla., U.S.* **99** Q6
Monticello, *Ill., U.S.* **101** R6
Monticello, *Ind., U.S.* **101** Q8
Monticello, *Ark., U.S.* **102** J9
Monticello, *Miss., U.S.* **102** M13
Monticello, *Ky., U.S.* **103** D17
Monticello, *Minn., U.S.* **106** J11
Monticello, *Iowa, U.S.* **107** N14
Monticello, *Utah, U.S.* **109** P9
Montichiari, *It.* **152** E6
Montijo, *Port.* **146** J5
Montijo, *Sp.* **146** H7
Montilla, *Sp.* **146** K10
Montivilliers, *Fr.* **149** N7
Montluçon, *Fr.* **149** T10
Montmarault, *Fr.* **149** T10
Montmirail, *Fr.* **149** P11
Monto, *Qnsld., Austral.* **221** U15
Montoro, *Sp.* **146** K10
Montoursville, *Pa., U.S.* **96** M7
Montpelier, *Vt., U.S.* **97** G13
Montpelier, *Ohio, U.S.* **101** N11
Montpelier, *Idaho, U.S.* **108** J8
Montpelier, *Jam.* **122** H7
Montpellier, *Fr.* **149** X11
Montpon-Ménestérol, *Fr.* **149** V7
Montreal, *Wis., U.S.* **100** F4
Montréal, *Que., Can.* **87** P19
Montréal, *Fr.* **149** Y9
Montreuil-Bellay, *Fr.* **149** S7
Montreux, *Switz.* **152** C2
Montrevault, *Fr.* **149** R6
Montrichard, *Fr.* **149** R8
Montrose, *Colo., U.S.* **109** N11
Montrose, *S. Dak., U.S.* **106** J2
Monts Birougou, *Gabon* **208** F9
Monts Bleus, *D.R.C.* **209** D19
Monts des Nementcha, *Alg.* **201** C14
Monts Kibara, *D.R.C.* **209** M18
Monts Malimba, *D.R.C.* **209** J18
Monts Mbang, *Cameroon* **206** L11

Monts Mugila, *D.R.C.* **209** J18
Montsalvy, *Fr.* **149** V10
Montsec, Serra del, *Sp.* **147** D16
Montserrat, island, *Lesser Antil.* **121** C16
Montserrat, Monestir de, *Sp.* **147** D17
Monturull, Pic, *Andorra, Sp.* **162** K3
Monuafe, island, *Tonga* **227** Q23
Monument Valley, *Ariz., Utah, U.S.* **109** Q8
Monument, *Oreg., U.S.* **110** G7
Monumental Buttes, peak, *Idaho, U.S.* **108** C4
Monveda, *D.R.C.* **209** C14
Monywa, *Myanmar* **190** G4
Monza, *It.* **152** D4
Monze, *Zambia* **209** P17
Monzón, *Sp.* **147** D15
Moody, *Tex., U.S.* **105** J17
Mooloogool, *W. Aust., Austral.* **221** U3
Möong Hsu, *Myanmar* **190** G6
Möong Köung, *Myanmar* **190** G6
Möong Nawng, *Myanmar* **190** G6
Möong Pan, *Myanmar* **190** H6
Moonless Mountains, *Pac. Oc.* **247** F15
Moora, *W. Aust., Austral.* **221** W2
Moorcroft, *Wyo., U.S.* **108** G13
Moore Haven, *Fla., U.S.* **99** V9
Moore Town, *Jam.* **122** J11
Moore, *Okla., U.S.* **105** D17
Moore, *Tex., U.S.* **105** M15
Moore, Lake, *Austral.* **220** J3
Moorea, island, *Fr. Polynesia, Fr.* **231** C16
Moorea, island, *Fr. Polynesia, Fr.* **231** H20
Moorefield, *W. Va., U.S.* **98** C11
Mooreland, *Okla., U.S.* **105** B15
Moore's Island, *Bahamas* **122** B9
Moorhead, *Miss., U.S.* **102** J10
Moorhead, *Minn., U.S.* **106** G8
Moorish Castle, *Gibraltar, U.K.* **163** N23
Moose Jaw, *Sask., Can.* **86** N12
Moose Lake, *Minn., U.S.* **106** H12
Moosehead Lake, *Me., U.S.* **97** D16
Mooselook-meguntic Lake, *Me., U.S.* **97** F15
Moosonee, *Ont., Can.* **87** N17
Mootwingee, *N.S.W., Austral.* **221** W12
Mopelia *see* Maupihaa, island, *Fr. Polynesia, Fr.* **231** B14
Mopti, *Mali* **205** J14
Moquegua, *Peru* **132** K5
Mor Daği, peak, *Turk.* **171** H18
Mor Khun, *Pak.* **180** F13
Mora de Rubielos, *Sp.* **147** F14
Móra d'Ebre, *Sp.* **147** E16
Mora, *Minn., U.S.* **106** H11
Mora, *N. Mex., U.S.* **109** R13
Mora, *Sw.* **142** J12
Mora, *Cameroon* **206** J11
Moraĉa, river, *Montenegro* **154** J9
Moradabad, *India* **182** G6
Morafenobe, *Madagascar* **213** E15
Moraleda, Channel, *S. Amer.* **128** P4
Moraleja del Vino, *Sp.* **146** E9
Moraleja, *Sp.* **146** G7
Moramanga, *Madagascar* **213** F16
Moran, *Tex., U.S.* **105** G15
Moran, *Wyo., U.S.* **108** G8
Morane, island, *Fr. Polynesia, Fr.* **231** E22
Morant Bay, *Jam.* **122** J11
Morant Cays, *Jam.* **119** N13
Morant Point, *Jam.* **122** J12
Morar, Loch, *Scot., U.K.* **144** J7
Morava, river, *Czechia* **151** J13
Moravia, region, *Czechia* **151** H13
Morawhanna, *Guyana* **131** D15
Moray Firth, *Scot., U.K.* **144** H10
Morbi, *India* **182** K2
Morbihan, Golfe du, *Fr.* **149** R4
Mörbylånga, *Sw.* **143** M13
Morcenx, *Fr.* **149** X6
Mordovo, *Russ.* **158** M11
Mordyyakha, *Russ.* **160** F11
Moreau, river, *S. Dak., U.S.* **106** J2
Morebeng (Soekmekaar), *S. Af.* **212** H9

Morehead, *Ky., U.S.* **103** B19
Morehouse, *Mo., U.S.* **107** V16
Morelia, *Mex.* **116** H11
Morella, *Sp.* **147** F15
Morella, *Qnsld., Austral.* **221** T12
Morena, *India* **182** H6
Morena, Sierra, *Sp.* **146** K10
Moreni, *Rom.* **155** F15
Moreno Bay, *S. Amer.* **128** J4
Moreno Valley, *Calif., U.S.* **111** X9
Morere, *N.Z.* **223** G22
Moresby Island, *B.C., Can.* **86** K6
Moresby Seachannel, *Pac. Oc.* **247** D16
Moreton Bay, *Austral.* **220** H16
Moreton Island, *Austral.* **220** H16
Moreton, *Qnsld., Austral.* **221** P12
Moreuil, *Fr.* **149** N10
Morez, *Fr.* **149** T14
Morfou Bay, *Cyprus* **162** N7
Morgan City, *La., U.S.* **102** Q9
Morgan Hill, *Calif., U.S.* **111** T4
Morgan, *Utah, U.S.* **108** K7
Morgan, *S. Aust., Austral.* **221** X11
Morganfield, *Ky., U.S.* **103** C14
Morganito, *Venez.* **130** F10
Morganton, *N.C., U.S.* **98** H8
Morgantown, *W. Va., U.S.* **98** C10
Morgat, *Fr.* **149** Q2
Morges, *Switz.* **152** C1
Morgex, *It.* **152** D2
Mori, *China* **184** E6
Mori, *Japan* **188** H12
Moriah, *Trin. & Tobago* **123** N17
Moriah, Mount, *U.S.* **90** F6
Moriarty, *N. Mex., U.S.* **109** S12
Moriki, *Nigeria* **206** H7
Morin Dawa, *China* **185** C15
Morioka, *Japan* **188** K13
Moriu, *Vanuatu* **229** H16
Moriusaq, *Greenland, Den.* **88** F3
Morjärv, *Sw.* **143** E14
Mørkefjord, research station, *Greenland, Den.* **88** F13
Morkoka, river, *Russ.* **161** G16
Morlaas, *Fr.* **149** Y7
Morlaix, *Fr.* **149** P3
Mormanno, *It.* **153** P13
Mormon Lake, *Ariz., U.S.* **109** S7
Morne au Diable, *Dominica* **123** D19
Morne Constant, *Guadeloupe, Fr.* **123** G16
Morne Diablotins, *Dominica* **123** E19
Morne Macaque, *Dominica* **123** G19
Morne Raquette, *Dominica* **123** F18
Morne Seychellois, peak, *Seychelles* **215** N20
Morne Trois Pitons, *Dominica* **123** F19
Morne-à-l'Eau, *Guadeloupe, Fr.* **123** E15
Mornington Island, *Austral.* **220** C11
Mornington, *Qnsld., Austral.* **221** Q11
Mornington, Isla, *Chile* **135** V6
Moro Gulf, *Philippines* **193** F14
Moro, *Pak.* **181** Q8
Moro, *Sudan* **203** U10
Morocco, *Af.* **200** F6
Morogoro, *Tanzania* **211** W7
Morolaba, *Burkina Faso* **205** L13
Morombe, *Madagascar* **213** G14
Morón de la Frontera, *Sp.* **146** L10
Morón, *Cuba* **118** H10
Morón, *Arg.* **134** L12
Mörön, *Mongolia* **184** C9
Morona, river, *Peru* **132** E3
Morondava, *Madagascar* **213** F15
Morondo, *Côte d'Ivoire* **204** M12
Moroni, *Comoros* **215** M14
Morotai, island, *Indonesia* **193** H16
Moroto, *Uganda* **211** R5
Morozov Ridge, *Arctic Oc.* **253** J13
Morozovsk, *Russ.* **159** Q12
Morpará, *Braz.* **133** H15
Morpeth, *Eng., U.K.* **145** M12
Morphou (Güzelyurt), *Cyprus* **162** N7
Morrill, *Nebr., U.S.* **107** N1
Morrilton, *Ark., U.S.* **102** G8
Morrinhos, *Braz.* **133** K13
Morris Jesup Spur, *Arctic Oc.* **253** J14
Morris, *Ill., U.S.* **101** P6
Morris, *Okla., U.S.* **105** C18
Morris, *Minn., U.S.* **106** J9
Morrison, *Ill., U.S.* **101** N4
Morristown, *N.J., U.S.* **96** N10
Morristown, *N.Y., U.S.* **96** F9
Morristown, *Tenn., U.S.* **103** E19

Morristown, *S. Dak., U.S.* **106** H3
Morrisville, *Vt., U.S.* **97** F13
Morro Bay, *Calif., U.S.* **111** V5
Morro Point, *S. Amer.* **128** K4
Morrosquillo, Golfo de, *Col.* **130** D5
Morrumbala, *Mozambique* **212** E11
Morrumbene, *Mozambique* **212** H11
Morshansk, *Russ.* **158** L11
Morsi, *India* **182** L6
Mortagne, *Fr.* **149** Q8
Mortara, *It.* **152** E4
Morteau, *Fr.* **149** S14
Morteros, *Arg.* **134** J11
Mortes, *S. Amer.* **128** G8
Mortlock Islands, *F.S.M.* **226** B10
Morton, *Ill., U.S.* **101** Q5
Morton, *Miss., U.S.* **102** L11
Morton, *Tex., U.S.* **104** F11
Morton, *Wash., U.S.* **110** D4
Mortyq, *Kaz.* **178** E7
Moruga, *Trin. & Tobago* **123** R22
Moruroa (Mururoa), island, *Fr. Polynesia, Fr.* **231** E21
Morvan, Monts du, *Fr.* **149** S12
Morven, *Qnsld., Austral.* **221** U14
Morven, peak, *Scot., U.K.* **144** J10
Morvern, peninsula, *Scot., U.K.* **144** K7
Mosal'sk, *Russ.* **158** K8
Moscos Islands, *Myanmar* **190** M6
Moscow Canal, *Russ.* **158** J9
Moscow *see* Moskva, *Russ.* **158** J9
Moscow University Ice Shelf, *Antarctica* **239** N20
Moscow, *Idaho, U.S.* **108** D3
Mosel, river, *Ger.* **150** G5
Moselle, river, *Eur.* **140** G6
Moses Lake, *Wash., U.S.* **110** C7
Moses Lake, town, *Wash., U.S.* **110** C7
Moses Point, *Russ.* **158** G14
Moshi, *Tanzania* **211** U6
Mosinee, *Wis., U.S.* **100** H5
Mosjøen, *Nor.* **142** F12
Moskva (Moscow), *Russ.* **158** J9
Moskva, *Taj.* **178** N12
Moskva, river, *Russ.* **158** J8
Moso, island, *Vanuatu* **229** J16
Mosonmagyaróvár, *Hung.* **151** K13
Mosquero, *N. Mex., U.S.* **109** N14
Mosquitia, region, *Hond.* **117** K18
Mosquito Cays, *N. Amer.* **84** P8
Mosquito Coast, *N. Amer.* **84** Q8
Mosquito Island, *Virgin Is., U.K.* **122** Q9
Mosquito Lagoon, *Fla., U.S.* **99** T9
Mosquitos, Golfo de los, *Pan.* **117** N20
Moss Point, *Miss., U.S.* **103** P13
Moss Town, *Bahamas* **122** E8
Moss, *Nor.* **142** K11
Mossaka, *Congo* **208** E11
Mossburn, *N.Z.* **223** P16
Mossel Bay, *S. Af.* **212** M6
Mossendjo, *Congo* **208** F9
Mossi, *Burkina Faso* **205** L15
Mossman, *Qnsld., Austral.* **221** Q13
Mossoró, *Braz.* **133** E17
Most, *Czechia* **150** G10
Mosta, *Malta* **163** K22
Mostaganem, *Alg.* **200** B11
Mostar, *Bosn. & Herzg.* **154** H7
Mostardas, *Braz.* **132** Q12
Mosteiros, *Port.* **146** Q5
Móstoles, *Sp.* **146** F10
Mosul *see* Al Mawşil, *Iraq* **176** E10
Mota del Cuervo, *Sp.* **146** H12
Mota Lava, island, *Vanuatu* **229** C15
Mot'a, *Eth.* **210** K7
Mota, island, *Vanuatu* **229** D15
Motala, *Sw.* **143** L13
Motane *see* Mohotani, island, *Fr. Polynesia, Fr.* **231** H15
Motherwell, *Scot., U.K.* **144** L9
Moto Piti Aau, island, *Fr. Polynesia, Fr.* **231** H15
Motopu, *Fr. Polynesia, Fr.* **231** E14
Motril, *Sp.* **146** M11
Motru, *Rom.* **154** F12
Motru, river, *Rom.* **154** F12
Mott, *N. Dak., U.S.* **106** H3
Motta Misaki, *Japan* **188** G12
Motu Ahuna, island, *Fr. Polynesia, Fr.* **231** G14
Motu Araara, island, *Fr. Polynesia, Fr.* **231** L14

Motu Ari, island, *Fr. Polynesia, Fr.* **231** R17
Motu Ihupuku *see* Campbell Island, *Pac. Oc.* **224** R7
Motu Iti *see* Tupai, island, *Fr. Polynesia, Fr.* **231** B15
Motu Ko, island, *N.Z.* **229** E20
Motu Kotawa, island, *N.Z.* **229** E20
Motu Mautaro, island, *Fr. Polynesia, Fr.* **231** M15
Motu Motihia, island, *Fr. Polynesia, Fr.* **231** N15
Motu Murimaora, island, *Fr. Polynesia, Fr.* **231** K15
Motu Mute, island, *Fr. Polynesia, Fr.* **231** G14
Motu Naonao, island, *Fr. Polynesia, Fr.* **231** N17
Motu One, island, *Fr. Polynesia, Fr.* **231** B14;G14
Motu Paio, island, *Fr. Polynesia, Fr.* **231** M17
Motu Tabu, island, *Fr. Polynesia, Fr.* **230** D11
Motu Tapu, island, *Fr. Polynesia, Fr.* **231** G14
Motu Teiko, island, *Fr. Polynesia, Fr.* **231** R17
Motu Vavara, island, *Fr. Polynesia, Fr.* **231** K14
Motu, river, *N.Z.* **223** G22
Motueka, *N.Z.* **223** J19
Motukitiu, island, *N.Z.* **229** J21
Motul, *Mex.* **117** H16
Motuloa Island, *Tuvalu* **228** Q10
Motupae, island, *N.Z.* **229** J23
Motupae, *Peru* **132** F2
Motupena Point, *P.N.G.* **228** C5
Moturakau, island, *N.Z.* **229** H20
Moturiki, *Fiji* **226** P6
Motutapu, island, *Tonga* **227** Q24
Motutapu, island, *N.Z.* **229** E23
Mototunga, island, *Fr. Polynesia, Fr.* **231** B19
Motykleyka, *Russ.* **161** F20
Moubray Bay, *Antarctica* **239** Q14
Mouchard, *Fr.* **149** S13
Mouchoir Passage, *N. Amer.* **119** J18
Moudjéria, *Mauritania* **204** G9
Moúdros, *Gr.* **157** E15
Mouila, *Gabon* **208** F8
Mouillage de Bougainville, *Fr. Polynesia, Fr.* **231** H23
Mouillage de Cook, *Fr. Polynesia, Fr.* **231** J23
Mouillage de Mataura, *Fr. Polynesia, Fr.* **231** M14
Mouillage de Tahueia, *Fr. Polynesia, Fr.* **231** M15
Mouillage des Roches Noires, *Fr. Polynesia, Fr.* **231** M14
Mouka, *Cen. Af. Rep.* **207** L16
Moulamein, *N.S.W., Austral.* **221** Y12
Mould Bay, *N.W.T., Can.* **87** C13
Moulins-la-Marche, *Fr.* **149** P8
Moulins, *Fr.* **149** T11
Moulton, *Ala., U.S.* **103** H14
Moultrie, *S.C., U.S.* **98** L10
Moultrie, *Ga., U.S.* **99** P6
Mouly, *New Caledonia, Fr.* **231** P22
Moulyinning, *W. Aust., Austral.* **221** X3
Mounana, *Gabon* **208** F9
Mound Bayou, *Miss., U.S.* **102** J10
Mound City, *Ill., U.S.* **101** V5
Mound City, *Kans., U.S.* **107** R10
Mound City, *Mo., U.S.* **107** R10
Moundou, *Chad* **206** L12
Moundsville, *W. Va., U.S.* **98** B9
Moundville, *Ala., U.S.* **103** K14
Moúnga'one, island, *Tonga* **227** L23
Moungoundou-sud, *Congo* **208** F9
Mounlapamôk, *Laos* **190** M11
Mount Abuye Meda, *Eth.* **210** L8
Mount Afadjato, *Ghana, Togo* **205** N16
Mount Agou, *Togo* **205** P16
Mount Airy, *N.C., U.S.* **98** G9
Mount Anglem, *N.Z.* **223** P16
Mount Arrowsmith, *N.S.W., Austral.* **221** W11
Mount Aspiring, *N.Z.* **223** M16
Mount Ayr, *Iowa, U.S.* **107** Q11
Mount Bagana, *P.N.G.* **228** B5
Mount Balbi, *P.N.G.* **222** P11

Mount Balbi, *P.N.G.* **228** B5
Mount Barker, *W. Aust., Austral.* **221** Y3
Mount Barnett, *W. Aust., Austral.* **221** Q6
Mount Bates, *Austral.* **229** A20
Mount Bellew (An Creagán), *Ire.* **145** Q4
Mount Buache, *F.S.M.* **226** F12
Mount Carmel, *Pa., U.S.* **96** N8
Mount Carmel, *Ill., U.S.* **101** T6
Mount Catherine *see* Jabal Kātrīnā, *Egypt* **202** G13
Mount Cook *see* Aoraki, peak, *N.Z.* **223** M17
Mount Cook, *N.Z.* **223** M17
Mount Coolon, *Qnsld., Austral.* **221** S14
Mount Desert Island, *Me., U.S.* **97** G18
Mount Dora, *Fla., U.S.* **99** T8
Mount Dora, *N. Mex., U.S.* **109** Q15
Mount Eagle, *Virgin Is., U.S.* **122** Q2
Mount Eba, *S. Aust., Austral.* **221** W9
Mount Egmont *see* Mount Taranaki, *N.Z.* **223** H20
Mount Elgon, *Kenya, Uganda* **211** R5
Mount Elizabeth, *Antarctica* **239** L13
Mount Elsie, *Qnsld., Austral.* **221** S14
Mount Enterprise, *Tex., U.S.* **105** H20
Mount Erebus, *Antarctica* **239** N14
Mount Fito, *Samoa* **227** P20
Mount Fukushima, *Antarctica* **239** C17
Mount Gambier, *S. Aust., Austral.* **221** Z11
Mount Gimie, *St. Lucia* **123** K14
Mount Granby, *Grenada* **123** K22
Mount Gulitel, *Palau* **226** F6
Mount Hagen, *P.N.G.* **222** N9
Mount Hikurangi, *N.Z.* **223** F22
Mount Hillaby, *Barbados* **123** K19
Mount Hope, *W. Va., U.S.* **98** E9
Mount Hope, *N.Z.* **223** X13
Mount Howitt, *Qnsld., Austral.* **221** U12
Mount Ida, *W. Aust., Austral.* **221** V4
Mount Isa, *Qnsld., Austral.* **221** S11
Mount Jackson, *Antarctica* **238** E7
Mount Jewett, *Pa., U.S.* **96** L5
Mount Keith, *W. Aust., Austral.* **221** V4
Mount Kenya, *Kenya* **211** S6
Mount Kirkpatrick, *Antarctica* **239** L13
Mount Koronga, *Togo* **205** M17
Mount Kubonitu, *Solomon Is.* **228** E9
Mount Kulal, *Kenya* **211** Q6
Mount Lamlam, *Guam, U.S.* **230** Q4
Mount Liamuiga, *St. Kitts & Nevis* **123** A17
Mount Lister, *Antarctica* **239** N14
Mount Machanao, *Guam, U.S.* **230** M6
Mount Magnet, *W. Aust., Austral.* **221** V3
Mount Matavanu, *Samoa* **227** N18
Mount Maunganui, *N.Z.* **223** F21
Mount McClintock, *Antarctica* **239** M14
Mount Monier, *St. Lucia* **123** J14
Mount Moroto, *Uganda* **211** R5
Mount Morris, *Ill., U.S.* **100** N5
Mount Morris, *Mich., U.S.* **100** L12
Mount Mulgrave, *Qnsld., Austral.* **221** Q13
Mount Ngauruhoe, *N.Z.* **223** G21
Mount Ngerchelchuus, *Palau* **226** F6
Mount Obama, *Antigua & Barbuda* **123** B20
Mount Olive, *N.C., U.S.* **98** J12
Mount Olive, *Miss., U.S.* **102** M11
Mount Owen, *N.Z.* **223** K19
Mount Pénot, *Vanuatu* **229** G15
Mount Pleasant, *Mich., U.S.* **100** K10
Mount Pleasant, *Tenn., U.S.* **103** F14
Mount Pleasant, *Tex., U.S.* **105** F19
Mount Pleasant, *Iowa, U.S.* **107** Q13
Mount Pleasant, *Cayman Is., U.K.* **122** M2
Mount Rainier National Park, *Wash., U.S.* **110** D4
Mount Ross, *N.Z.* **223** K20
Mount Ruapehu, *N.Z.* **223** H21
Mount Rushmore National Memorial, *S. Dak., U.S.* **106** L2
Mount Sage, *Virgin Is., U.K.* **122** R6
Mount Saint Helens National Volcanic Monument, *Wash., U.S.* **110** E4

Mount Sandiman, *W. Aust., Austral.* **221** U2
Mount Santa Rosa, *Guam, U.S.* **230** N7
Mount Santop, *Vanuatu* **229** L17
Mount Sasalaguan, *Guam, U.S.* **230** Q5
Mount Shasta, *Calif., U.S.* **110** M3
Mount Silisili, *Samoa* **227** N18
Mount Sinai *see* Jabal Mūsá, *Egypt* **202** G13
Mount Sinai, *Grenada* **123** L23
Mount Singavi, *Wallis & Futuna Is., Fr.* **230** Q11
Mount Siple, *Antarctica* **238** M8
Mount Smith, *Eth.* **211** N6
Mount St. Catherine, *Grenada* **123** K23
Mount St. George, *Trin. & Tobago* **123** N18
Mount Sterling, *Ill., U.S.* **101** R3
Mount Sterling, *Ky., U.S.* **103** B18
Mount Stokes, *N.Z.* **223** J20
Mount Sürétiméat, *Vanuatu* **229** D14
Mount Surprise, *Qnsld., Austral.* **221** R13
Mount Takahe, *Antarctica* **238** L8
Mount Takuan, *P.N.G.* **222** P11
Mount Takuan, *P.N.G.* **228** C5
Mount Taranaki (Mount Egmont), *N.Z.* **223** H20
Mount Tarawera, *N.Z.* **223** G21
Mount Tenjo, *Guam, U.S.* **230** P5
Mount Tukosméra, *Vanuatu* **229** M17
Mount Tukoto, *N.Z.* **223** N16
Mount Ulawun, *P.N.G.* **222** N9
Mount Uriah, *N.Z.* **223** K18
Mount Vangunu, *Solomon Is.* **228** E8
Mount Vernon, *N.Y., U.S.* **96** N11
Mount Vernon, *Ill., U.S.* **101** T6
Mount Vernon, *Ind., U.S.* **101** U7
Mount Vernon, *Ohio, U.S.* **101** Q13
Mount Vernon, *Ala., U.S.* **103** N14
Mount Vernon, *Ky., U.S.* **103** C18
Mount Vernon, *Mo., U.S.* **107** V11
Mount Vernon, *Oreg., U.S.* **110** G7
Mount Vernon, *Wash., U.S.* **110** B4
Mount Vernon, *W. Aust., Austral.* **221** U3
Mount Victor, *Antarctica* **239** C16
Mount Victoria, *P.N.G.* **222** Q8
Mount Victorin, *St. Lucia* **123** L14
Mount Wilhelm, *P.N.G.* **222** P7
Mount Wuteve, *Liberia* **204** N10
Mount Wynne Bay, *St. Vincent & the Grenadines* **123** K16
Mount Zion, *Ill., U.S.* **101** R6
Mountain Brook, *Ala., U.S.* **103** J15
Mountain City, *Tenn., U.S.* **103** D21
Mountain City, *Nev., U.S.* **110** L10
Mountain Grove, *Mo., U.S.* **107** V13
Mountain Home, *Ark., U.S.* **102** E8
Mountain Home, *Idaho, U.S.* **108** H4
Mountain Lake, *Minn., U.S.* **106** L10
Mountain Nile *see* Bahr el Jebel, *S. Sudan* **203** W11
Mountain Point, *Virgin Is., U.K.* **122** Q8
Mountain View, *Ark., U.S.* **102** F9
Mountain View, *Mo., U.S.* **107** V14
Mountain View, *Wyo., U.S.* **108** K8
Mountain View, *Hawaii, U.S.* **115** N23
Mountain Village, *Alas., U.S.* **112** H12
Mountainair, *N. Mex., U.S.* **109** T12
Mountnorris Bay, *Austral.* **220** A8
Mountnorris Fjord, *Greenland, Den.* **88** H14
Mount's Bay, *Eng., U.K.* **145** X8
Moura, *Braz.* **132** D8
Moura, *Chad* **207** H15
Mourão, *Port.* **146** J6
Mourdiah, *Mali* **204** J12
Mourne Mountains, *N. Ire., U.K.* **145** P7
Mouroubra, *W. Aust., Austral.* **221** W3
Mousâ 'Alli, peak, *Djibouti, Eritrea, Eth.* **210** K10
Moussoro, *Chad* **207** H13
Moutiers, *Fr.* **149** U14
Moutsamoudou, *Comoros* **215** N16
Mouzáki, *Gr.* **156** F9
Mouzarak, *Chad* **206** H12
Moweaqua, *Ill., U.S.* **101** R5
Moÿ, *Fr.* **149** N11
Moyalē, *Kenya* **211** Q8
Moyamba, *Sierra Leone* **204** N9
Moyen Atlas, *Mor.* **200** D8

Moyo, *Uganda* **211** Q3
Moyobamba, *Peru* **132** F3
Moyowosi, river, *Tanzania* **211** V3
Moyto, *Chad* **207** J13
Moyu (Karakax), *China* **184** G2
Moyynqum, *Kaz.* **178** J12
Moyynty, *Kaz.* **179** G13
Mozambique, *Af.* **212** E11
Mozambique Basin, *Ind. Oc.* **250** N5
Mozambique Channel, *Af.* **213** E13
Mozambique Escarpment, *Ind. Oc.* **250** N4
Mozambique Plateau, *Ind. Oc.* **250** N4
Mozarlândia, *Braz.* **132** J12
Mozdok, *Russ.* **159** T13
Mozhaysk, *Russ.* **158** J8
Mozhga, *Russ.* **158** H16
Mpala, *D.R.C.* **209** J19
Mpanda, *Tanzania* **211** W3
Mpandamatenga, *Botswana* **212** E7
Mpika, *Zambia* **209** M20
Mporokoso, *Zambia* **209** K19
Mpouya, *Congo* **208** F11
Mpui, *Tanzania* **211** X3
Mpulungu, *Zambia* **209** K20
Mpwapwa, *Tanzania* **211** W6
Mqabba, *Malta* **163** L23
Mrągowo, *Pol.* **151** B16
Mrémani, *Comoros* **215** N16
Mrkojević, Zaliv, *Montenegro* **154** K9
Msida, *Malta* **163** K23
M'Sila, *Alg.* **201** C13
Msoro, *Zambia* **209** N20
Mstsislaw, *Belarus* **158** K6
Mt'a Didi Abuli, peak, *Rep. of Georgia* **171** C18
Mthatha, *S. Af.* **212** L8
Mtito Andei, *Kenya* **211** U7
Mtkvari (Kura), river, *Rep. of Georgia* **171** C18
Mtsensk, *Russ.* **158** L9
Mts'khet'a, *Rep. of Georgia* **171** C18
Mtwara, *Tanzania* **211** Y8
Mu Gia Pass, *Vietnam* **190** K11
Mu'a, *Tonga* **227** Q23
Mua, *Wallis & Futuna Is., Fr.* **230** R9,11
Muan, *S. Korea* **187** T7
Muanda, *D.R.C.* **208** H9
Muang Béng, *Laos* **190** H8
Muang Kao, *Laos* **190** M12
Muang Khamkeut, *Laos* **190** K10
Muang Khôngxédôn, *Laos* **190** L11
Muang Ngoy, *Laos* **190** H9
Muang Pakbèng, *Laos* **190** J8
Muang Paklay, *Laos* **190** K8
Muang Pakxan, *Laos* **190** J10
Muang Phin, *Laos* **190** L11
Muang Sing, *Myanmar* **190** H8
Muang Souvannakhili, *Laos* **190** L11
Muang Xianghon, *Laos* **190** J8
Muang Xépôn, *Laos* **190** K11
Muang Xon, *Laos* **190** H9
Muar, *Malaysia* **191** V9
Muaraaman, *Indonesia* **192** K6
Muarabungo, *Indonesia* **192** J6
Muaraenim, *Indonesia* **192** K7
Muarakumpe, *Indonesia* **192** J7
Muaramawai, *Indonesia* **192** J12
Muaratewe, *Indonesia* **192** J11
Mubende, *Uganda* **211** S3
Mubi, *Nigeria* **206** K11
Mubong, *N. Korea* **186** C11
Mucajaí, river, *Braz.* **132** B8
Muchachos, Roque de los, *Sp.* **214** Q3
Much'ang, *N. Korea* **186** B13
Muchea, *W. Aust., Austral.* **221** X2
Muchinga Mountains, *Zambia* **209** N19
Muchkapskiy, *Russ.* **158** M12
Muck, island, *Scot., U.K.* **144** J7
Muckety, *N.T., Austral.* **221** R9
Muckle Flugga, island, *Scot., U.K.* **144** B12
Muckle Roe, island, *Scot., U.K.* **144** C12
Muconda, *Angola* **209** L14
Mucur, *Turk.* **170** F9
Mucuripe, Ponta de, *Braz.* **133** E17
Mucusso, *Angola* **209** Q14
Mud Lake, *Idaho, U.S.* **108** G7
Muda, peak, *Sp.* **214** Q7
Mudanjiang, *China* **185** D17
Mudanya, *Turk.* **170** D4
Muḍayrib, *Oman* **175** H18
Muddy Gap, *Wyo., U.S.* **108** J11

Mudhol, *India* **183** P4
Mudurnu, *Turk.* **170** D7
Mueda, *Mozambique* **213** B13
Muéo, *New Caledonia, Fr.* **231** P21
Muertos Cays, *Bahamas* **118** F9
Mufulira, *Zambia* **209** M18
Muğanlı, *Azerb.* **171** C20
Mughalzhar Taūy, *Kaz.* **178** F8
Muğla, *Turk.* **170** H4
Muglad, *Sudan* **203** T9
Mugland Basin, *Sudan* **203** U9
Mugu, *Nepal* **182** F8
Mūḫ, Sabkhat al, *Syr.* **172** G10
Muḥammad Āghah Wuluswālī, *Afghan.* **180** H9
Muhammad Qol, *Sudan* **202** M15
Muḥayy, *Jordan* **173** P6
Mühleholz, *Liech.* **162** P2
Muhlenfels Point, *Virgin Is., U.S.* **122** N8
Mühlhausen, *Ger.* **150** F8
Mühlig-Hofmann Mountains, *Antarctica* **239** B14
Muhulu, *D.R.C.* **209** E17
Muhutwe, *Tanzania* **211** T3
Mũi Cà Mau, *Vietnam* **191** R10
Mũi Cà Ná, *Vietnam* **191** P13
Mũi Dao, *Vietnam* **190** K11
Mui Hopohoponga, cape, *Tonga* **227** Q24
Muié, *Angola* **209** N14
Muine Bheag, *Ire.* **145** S6
Muineachán see Monaghan, *Ire.* **145** P6
Muir Woods National Monument, *Calif., U.S.* **111** R3
Muisne, *Ecua.* **130** K3
Mujimbeji, *Zambia* **209** M16
Mukacheve, *Ukr.* **159** P1
Mukah, *Malaysia* **192** H10
Mukawwar Island, *Sudan* **202** M15
Mukayris, *Yemen* **174** P10
Mukeru, *Palau* **226** G6
Mukinbudin, *W. Aust., Austral.* **221** W3
Mukojima Rettō (Parry Group), *Japan* **189** W14
Mukomuko, *Indonesia* **192** K6
Mukry, *Turkm.* **178** N10
Muktinath, *Nepal* **182** G9
Mul, *India* **182** M7
Mula, river, *India* **182** M4
Muladu, island, *Maldives* **183** U3
Mulaku Atoll, *Maldives* **183** W3
Mulanie Mountains, *Af.* **198** M10
Mulberry, *Ark., U.S.* **102** F6
Mulchatna, river, *Alas., U.S.* **113** K14
Muldraugh, *Ky., U.S.* **103** C16
Mule Peak, *N. Mex., U.S.* **109** U13
Muleba, *Tanzania* **211** T3
Mulegé, *Mex.* **116** D6
Muleshoe, *Tex., U.S.* **104** E11
Mulgathing, *S. Aust., Austral.* **221** W9
Mulgrave Island see Badu Island, *Qnsld., Austral.* **221** N12
Mulhacén, peak, *Sp.* **146** L11
Mülheim, *Ger.* **150** E5
Mulhouse, *Fr.* **149** R15
Mulifanua, *Samoa* **227** P19
Mulitefala, island, *Tuvalu* **228** L11
Mull of Galloway, cape, *Scot., U.K.* **145** N8
Mull, island, *Scot., U.K.* **144** K7
Mull, Sound of, *Scot., U.K.* **144** K7
Mullan, *Idaho, U.S.* **108** C4
Mullen, *Nebr., U.S.* **107** N4
Mullens, *W. Va., U.S.* **98** E8
Mullet Peninsula, *Ire.* **145** P3
Mullett Lake, *Mich., U.S.* **100** G11
Mullewa, *W. Aust., Austral.* **221** V2
Mullingar (An Muileann gCearr), *Ire.* **145** Q6
Mullins, *S.C., U.S.* **98** K11
Mulobezi, *Zambia* **209** Q16
Mulock Glacier, *Antarctica* **239** N14
Mulondo, *Angola* **208** P10
Mulroy Bay, *Ire.* **145** M5
Multan, *Pak.* **180** M11
Mulumbe Mountains, *D.R.C.* **209** K18
Mulvane, *Kans., U.S.* **107** U8
Muma, *D.R.C.* **209** C15
Mumallah, *Sudan* **203** U8
Mumbai (Bombay), *India* **182** M3
Mumbué, *Angola* **208** N12
Mumbwa, *Zambia* **209** P17
Mumeng, *P.N.G.* **222** P7
Mumias, *Uganda* **211** S4

Mumra, *Russ.* **159** R15
Mumu, *Sudan* **203** T6
Muna, island, *Indonesia* **193** L14
Muna, river, *Russ.* **161** G16
Munaba, *India* **182** H2
Munan Pass, *China, Vietnam* **190** G11
München (Munich), *Ger.* **150** K9
Munch'ŏn, *N. Korea* **186** J9
Muncie, *Ind., U.S.* **101** R10
Muncy, *Pa., U.S.* **96** M7
Munda, *Solomon Is.* **228** E7
Mundabullangana, *W. Aust., Austral.* **221** S3
Munday, *Tex., U.S.* **105** F14
Münden, *Ger.* **150** F7
Mundiwindi, *W. Aust., Austral.* **221** T4
Mundŏk, *N. Korea* **186** H6
Mundra, *India* **182** K1
Mundrabilla, *W. Aust., Austral.* **221** W7
Munenga, *Angola* **208** L10
Munera, *Sp.* **146** H12
Mungapdo, island, *S. Korea* **187** N6
Mungbere, *D.R.C.* **209** C18
Mungeli, *India* **182** K8
Munger, *India* **182** J10
Mungindi, *N.S.W., Austral.* **221** V14
Munguba, *Braz.* **132** D12
Mungyeong, *S. Korea* **187** P10
Munia, island, *Fiji* **226** P8
Munich see München, *Ger.* **150** K9
Municipal Administrative Center, *Monaco* **163** E20
Munising, *Mich., U.S.* **100** F8
Munkhafad al Qaţţārah (Qattara Depression), *Egypt* **202** F9
Munku Sardyk, *Russ.* **161** L14
Munsa, *N. Korea* **186** H5
Munsan, *S. Korea* **186** M8
Munster (Cúige Mumhan), region, *Ire.* **145** T4
Münster, *Ger.* **150** E6
Muntok, *Indonesia* **192** K7
Munyeodo, island, *S. Korea* **187** R7
Munzur Dağları, *Turk.* **171** F14
Mu'omu'a, island, *Tonga* **227** H22
Mường Tè, *Laos* **190** G9
Mường Xén, *Vietnam* **190** J10
Muonio, *Fin.* **143** D14
Muqakoori, *Somalia* **211** Q12
Muqaţţa', *Sudan* **203** R14
Muqdisho (Mogadishu), *Somalia* **211** R11
Muqêr, *Afghan.* **180** K8
Müqtədir, *Azerb.* **171** C22
Mur-de-Bretagne, *Fr.* **149** Q4
Mur, river, *Aust.* **150** M10
Muradiye, *Turk.* **171** F18
Murakami, *Japan* **188** M12
Muralag Island see Prince of Wales Island, *Qnsld., Austral.* **221** N12
Muramvya, *Burundi* **211** U2
Murang'a, *Kenya* **211** S6
Murashi, *Russ.* **158** F14
Murashi, *Russ.* **160** G8
Murat Dağı, peak, *Turk.* **170** F5
Murat, *Fr.* **149** V10
Murat, river, *Turk.* **171** F16
Murata, *San Marino* **163** K15
Muratlı, *Turk.* **170** C3
Muravera, *It.* **153** P4
Murazzano, *It.* **152** F2
Murça, *Port.* **146** E7
Murchison Downs, *W. Aust., Austral.* **221** V3
Murchison Falls, *Af.* **198** J9
Murchison Falls, *Uganda* **211** R3
Murchison, river, *Austral.* **220** H2
Murcia, *Sp.* **147** K14
Murcuripe Point, *S. Amer.* **128** E11
Murdo, *S. Dak., U.S.* **106** L4
Mureş, river, *Rom.* **154** D11
Muret, *Fr.* **149** X9
Murfreesboro, *N.C., U.S.* **98** G13
Murfreesboro, *Ark., U.S.* **102** H7
Murgab, river, *Turkm.* **178** P9
Murgha Kibzai, *Pak.* **180** M9
Murghob, *Taj.* **179** N14
Murghob, river, *Taj.* **179** N14
Murgon, *Qnsld., Austral.* **221** U15
Murgoo, *W. Aust., Austral.* **221** V2
Murguz, peak, *Arm.* **171** D19
Muri, *N.Z.* **229** E23
Murieta, *Angola* **209** L14
Murihiti, island, *N.Z.* **229** G22
Murilo Atoll, *F.S.M.* **226** A9

Muritapua, island, *N.Z.* **229** H21
Müritz, lake, *Ger.* **150** C10
Murjek, *Sw.* **143** E14
Murle, *Eth.* **211** P6
Murman Coast, *Eur.* **140** A10
Murman Rise, *Arctic Oc.* **253** D18
Murmansk Rise, *Arctic Oc.* **253** E16
Murmansk, *Russ.* **160** D9
Murnei, *Sudan* **203** S6
Murnpeowie, *S. Aust., Austral.* **221** V11
Muro Lucano, *It.* **152** M12
Muro, Capo di, *Fr.* **149** Y2
Murom, *Russ.* **158** J11
Muroran, *Japan* **188** H12
Muros e Noia, Ría de, *Sp.* **146** B5
Muros, *Sp.* **146** B5
Muroto Zaki, *Japan* **189** R7
Muroto, *Japan* **189** R7
Murphy, *N.C., U.S.* **98** J5
Murphy, *Idaho, U.S.* **108** H3
Murphysboro, *Ill., U.S.* **101** U5
Murra Murra, *Qnsld., Austral.* **221** V13
Murrat el Sughra, Buheirat (Little Bitter Lake), *Egypt* **173** Q1
Murray Bridge, *S. Aust., Austral.* **221** Y11
Murray Fracture Zone, *Pac. Oc.* **247** F14
Murray River Basin, *Austral.* **220** K11
Murray, *S.C., U.S.* **98** K8
Murray, *Ky., U.S.* **103** E13
Murray, river, *Austral.* **220** L13
Murray's Anchorage, *Bermuda* **119** B16
Murrayville, *Vic., Austral.* **221** Y11
Murrells Inlet, *S.C., U.S.* **98** L11
Murrenja Hill, *Austral.* **220** B7
Murrey Ridge, *Ind. Oc.* **250** C8
Murro di Porco, Capo, *It.* **153** T12
Murrumbidgee, river, *Austral.* **220** K12
Murska Sobota, *Slov.* **154** C6
Mursko Središče, *Croatia* **154** C6
Murter, island, *Croatia* **154** G5
Murua see Woodlark, island, *P.N.G.* **222** Q10
Murud, *India* **182** M3
Murud, Mount, *Malaysia* **192** G11
Murukta, *Russ.* **161** G14
Mururoa see Moruroa, island, *Fr. Polynesia, Fr.* **231** E21
Murwillumbah, *N.S.W. Austral.* **221** V16
Mürzzuschlag, *Aust.* **150** L12
Muş, *Turk.* **171** F14
Musa Dağ, peak, *Turk.* **170** J11
Musa Khel Bazar, *Pak.* **180** L9
Mūsá Qal'ah, *Afghan.* **180** K5
Mûsa, Gebel (Mount Sinai), *Egypt* **173** T3
Musala, peak, *Bulg.* **155** J13
Musan, *N. Korea* **186** B12
Musaymīr, *Yemen* **174** P9
Muscat see Masqaţ, *Oman* **175** G18
Muscatine, *Iowa, U.S.* **107** P14
Muscle Shoals, *Ala., U.S.* **103** G14
Muse, *Okla., U.S.* **105** D20
Museum of Prehistoric Anthropology, *Monaco* **163** E20
Musgrave Ranges, *Austral.* **220** G8
Musgrave, Port, *Austral.* **220** A12
Mūsh Khēl, *Afghan.* **180** J8
Mushenge, *D.R.C.* **209** G14
Mushie, *D.R.C.* **208** F11
Mushu, island, *P.N.G.* **222** N6
Musicians Seamounts, *Pac. Oc.* **247** F13
Musina, *S. Af.* **212** G9
Muskegon Heights, *Mich., U.S.* **100** L9
Muskegon, *Mich., U.S.* **100** L9
Muskegon, river, *Mich., U.S.* **100** K10
Muskingum, river, *Ohio, U.S.* **101** R14
Muskogee, *Okla., U.S.* **105** C19
Muslimbagh, *Pak.* **180** M7
Muslimīyah, *Syr.* **172** D9
Musmar, *Sudan* **203** P14
Musoma, *Tanzania* **211** T4
Mussau Islands, *P.N.G.* **222** M9
Mussau, island, *P.N.G.* **222** M9
Musselburgh, *Scot., U.K.* **144** M10
Musselshell, *Mont., U.S.* **108** D10
Musselshell, river, *Mont., U.S.* **108** C11
Mussende, *Angola* **208** L11
Mustafakemalpaşa, *Turk.* **170** D4
Mustahīl, *Eth.* **211** P11
Mustang, *Nepal* **182** F9

Musters, Lago, *Arg.* **135** S8
Mustique, island, *St. Vincent & the Grenadines* **121** G17
Musu, *N. Korea* **186** C12
Musudan, *N. Korea* **186** E13
Mut, *Turk.* **170** J8
Mūţ, *Egypt* **202** J10
Mut, island, *Marshall Is.* **227** B20
Mutā Khān, *Afghan.* **180** J8
Mutá, Ponta do, *Braz.* **133** J16
Mutalau, *N.Z.* **229** P22
Mutanda, *Zambia* **209** M17
Mutare, *Zimb.* **212** F10
Mutha, *Kenya* **211** T7
Muting, *Indonesia* **193** M21
Mutki, *Turk.* **171** F16
Mutnovskaya Sopka, *Russ.* **161** F23
Mutok Harbor, *F.S.M.* **226** K8
Mutombo-Mukulu, *D.R.C.* **209** J15
Mutoray, *Russ.* **161** J14
Mutsu Wan, *Japan* **188** J13
Mutsu, *Japan* **188** J13
Muttaburra, *Qnsld., Austral.* **221** T13
Muturi, *Indonesia* **193** K18
Muxía, *Sp.* **146** B5
Muxima, *Angola* **208** K10
Muyinga, *Burundi* **211** U2
Mŭynoq, *Uzb.* **178** J7
Muyumba, *D.R.C.* **209** J17
Muzaffarabad, *Pak.* **180** H12
Muzaffargarh, *Pak.* **180** M10
Muzaffarnagar, *India* **182** F6
Muzaffarpur, *India* **182** H10
Muztag Feng, *China* **184** H5
Muztag, peak, *China* **184** H3
Mvadi, *Gabon* **208** D9
Mvoung, river, *Gabon* **208** D9
Mwadingusha, *D.R.C.* **209** L17
Mwakete, *Tanzania* **211** Y5
Mwali (Mohéli), island, *Comoros* **215** N15
Mwaniwowo, *Solomon Is.* **228** G12
Mwanza, *D.R.C.* **209** J17
Mwanza, *Tanzania* **211** U4
Mwatate, *Kenya* **211** U7
Mweka, *D.R.C.* **209** H14
Mwene-Ditu, *D.R.C.* **209** J15
Mwenezi, *Zimb.* **212** G9
Mwenzo, *Zambia* **209** K20
Mwimba, *D.R.C.* **209** K15
Mwingi, *Kenya* **211** T7
Mwinilunga, *Zambia* **209** M16
Mỹ Tho, *Vietnam* **191** P11
Myaksa, *Russ.* **158** G9
Myanaung, *Myanmar* **190** J4
Myanmar (Burma), *Asia* **190** H4
Mycenae, ruin, *Gr.* **156** K5
Myebon, *Myanmar* **190** H3
Myeik, *Myanmar* **191** N6
Myitta, *Myanmar* **190** M6
Myggbukta, *Greenland, Den.* **88** H14
Myingyan, *Myanmar* **190** G5
Myitkyinā, *Myanmar* **190** E6
Myitta, *Myanmar* **190** M6
Mykolayiv, *Ukr.* **159** R6
Mýkonos, *Gr.* **157** K15
Mýkonos, island, *Gr.* **157** K15
Mylius-Erichsen Land, *Greenland, Den.* **88** C11
Mymensingh, *Bangladesh* **182** J13
Mynbulaq, *Uzb.* **178** K9
Mynydd Preseli, peak, *Wales, U.K.* **145** T9
Myohaung, *Myanmar* **190** H3
Myohyang-san, peak, *N. Korea* **186** G7
Myohyang-sanmaek, range, *N. Korea* **186** G7
Myŏngch'ŏn, *N. Korea* **186** E12
Myoungmya, *Myanmar* **190** L4
Myrdal, *Nor.* **142** J10
Mýrdalsjökull, glacier, *Ice.* **142** F2
Myrhorod, *Ukr.* **159** P7
Mýrina, *Gr.* **157** E14
Myrtle Beach, *S.C., U.S.* **98** L11
Myrtle Creek, *Oreg., U.S.* **110** J3
Myrtle Grove, *Fla., U.S.* **99** Q1
Myrtle Point, *Oreg., U.S.* **110** J2
Mýrtos, *Gr.* **157** Q15
Myrtou (Camlibel), *Cyprus* **162** N7
Mys Shmidta, *Russ.* **161** B20
Mys Zhelaniya, *Russ.* **160** D12
Mysovaya, *Russ.* **161** D19
Mystic, *Conn., U.S.* **97** M14
Mystic, *Iowa, U.S.* **107** Q12
Mysuru, *India* **183** R5
Myszyniec, *Pol.* **151** C16

Mýtikas, *Gr.* **156** G8
Mytilene see Mytilíni, *Gr.* **157** G17
Mytilíni (Mytilene), *Gr.* **157** G17
Mytilíni see Lésvos, island, *Gr.* **157** F16
Mytishchi, *Russ.* **158** J9
Mzé Koukoulé Bandalankoua, peak, *Comoros* **215** N15
Mzimba, *Malawi* **212** B10
Mzuzu, *Malawi* **212** B11

N

Na Blascaodaí (Blasket Island), *Ire.* **145** T2
Na Gleannta see Glenties, *Ire.* **145** N5
Na h-Eileanan Flannach (Flannach Isles), *Scot., U.K.* **144** G6
Na h-Eileanan Mora (Shiant Islands), *Scot., U.K.* **144** G7
Na h-Eileanan Theisgeir (Heisker Islands), *Scot., U.K.* **144** H6
Na Sceirí see Skerries, *Ire.* **145** Q7
Na, island, *F.S.M.* **226** K9
Naafkopf, peak, *Aust., Liech. Switz.* **162** Q3
Naajat, *Greenland, Den.* **88** J5
Nā'ālehu, *Hawaii, U.S.* **115** Q21
Na'am see Roah, river, *S. Sudan* **203** X10
Naama, *Alg.* **200** D11
Naandi, *S. Sudan* **203** X9
Nabari, *Kiribati* **226** J11
Nabatîyé, *Leb.* **172** J6
Nabeina, *Kiribati* **227** Q15
Naberera, *Tanzania* **211** V6
Naberezhnyye Chelny, *Russ.* **158** H16
Nabeul, *Tunisia* **201** B16
Nabilatuk, *Uganda* **211** R5
Nabire, *Indonesia* **193** K19
Nāblus (Nablus), *W. Bank* **172** L6
Nabouwalu, *Fiji* **226** N6
Nabq Bay, site, *Egypt* **173** U4
Nacala, *Mozambique* **213** D13
Nacaome, *Hond.* **117** L17
Naches, *Wash., U.S.* **110** D5
Nachingwea, *Tanzania* **211** Y7
Náchod, *Czechia* **150** G12
Nachu, island, *F.S.M.* **226** F10
Nachvak Fiord, *Nfld. & Lab. Can.* **87** J20
Nacimiento, Lake, *Calif., U.S.* **111** V5
Nacogdoches, *Tex., U.S.* **105** H20
Nacozari de García, *Mex.* **116** C7
Nacula, island, *Fiji* **226** N4
Nadale, island, *Maldives* **183** Y3
Nadezhda Basin, *Pac. Oc.* **246** F8
Nadi Bay, *Fiji* **226** P4
Nadi, *Fiji* **226** P4
Nadiad, *India* **182** K3
Nadir, *Virgin Is. U.S.* **122** N9
Nădlac, *Rom.* **154** D10
Nador, *Mor.* **200** C9
Nadur, *Malta* **163** H20
Naduri, *Fiji* **226** N6
Nadvoitsy, *Russ.* **158** C8
Nadvoitsy, *Russ.* **160** D8
Nadym, *Russ.* **160** G11
Naenarodo, island, *S. Korea* **187** U9
Naesu, *S. Korea* **187** P9
Nafada, *Nigeria* **206** J10
Nafana, *Côte d'Ivoire* **205** M13
Náfpaktos, *Gr.* **156** H9
Náfplio, *Gr.* **156** K11
Nafūd ad Daḥy, *Saudi Arabia* **174** H10
Nag, *Pak.* **181** Q5
Naga City, *Philippines* **193** C14
Naga Hills, *India* **182** H15
Naga, *Indonesia* **193** M13
Nagano, *Japan* **189** N11
Nagaoka, *Japan* **188** M11
Nagaon, *India* **182** J10
Nagar Parkar, *Pak.* **181** S10
Nagasaki, *Japan* **189** R4
Nagaur, *India* **182** H3
Nagercoil, *India* **183** T5
Naghvarevi, *Rep. of Georgia* **171** C18
Nagir, *Pak.* **180** F13
Nagishot, *S. Sudan* **203** Y13
Nago, *Japan* **189** X2

Nagorno-Karabakh, region, Azerb. 171 E20
Nagoya, Japan 189 P10
Nagpur, India 182 L7
Nagqu (Naqu), China 184 K6
Nag's Head, St. Kitts & Nevis 123 B18
Nagwŏn, N. Korea 186 G9
Nagykanizsa, Hung. 151 M13
Nagyŏn, N. Korea 186 L5
Naha, Japan 189 X2
Nahanni Butte, N.W.T., Can. 86 H9
Nahma, Mich., U.S. 100 G8
Nahoe, Fr. Polynesia, Fr. 231 D14
Nahr an Nīl (Nile), Egypt 202 G11
Nahuei Huapí, Lago, Arg. 135 Q7
Nailaga, Fiji 226 P4
Nain, Nfld. & Lab., Can. 87 K21
Nā'īn, Iran 177 H16
Nainpur, India 182 K7
Nairai, island, Fiji 226 P6
Nairiri, Fr. Polynesia, Fr. 231 J22
Nairn, Scot., U.K. 144 H9
Nairobi, Kenya 211 T6
Naitaba, island, Fiji 226 N8
Naitonitoni, Fiji 226 Q5
Naivasha, Kenya 211 S6
Naj' Ḥammādī, Egypt 202 J12
Najaf see An Najaf, Iraq 176 H10
Najafābād, Iran 177 H15
Najd, Saudi Arabia 174 E8
Nájera, Sp. 146 C12
Najin-man, bay, N. Korea 186 B13
Najin, N. Korea 186 B13
Najrān, Saudi Arabia 174 M8
Najrān, Saudi Arabia 174 L9
Naju, S. Korea 187 T8
Naka Kharai, Pak. 181 S7
Naka Shibetsu, Japan 188 F15
Nakadōri Jima, Japan 189 R3
Nakagusuku Wan (Buckner Bay), Japan 189 X2
Nākālele Point, Hawaii, U.S. 115 J18
Nakano Shima, Japan 189 N7
Nakano Shima, Japan 189 U4
Nakano, Japan 189 N11
Nakatane, Japan 189 U5
Nakatombetsu, Japan 188 E13
Nakatsu, Japan 189 R5
Nakdong, river, S. Korea 187 S11
Nakéty, New Caledonia, Fr. 231 Q22
Nak'fa, Eritrea 210 G7
Nakhl, Egypt 173 Q3
Nakhodka, Russ. 161 M21
Nakhon Phanom, Thai. 190 K10
Nakhon Ratchasima, Thai. 190 M9
Nakhon Sawan, Thai. 190 L7
Nakhon Si Thammarat, Thai. 191 R7
Nakina, Ont., Can. 87 N16
Nakło nad Notecią, Pol. 151 C13
Naknek, Alas., U.S. 113 L14
Nako, Burkina Faso 205 L14
Nakodar, India 182 E5
Nakodu, Fiji 226 P6
Nakonde, Zambia 209 K21
Nakuru, Kenya 211 S6
Nal, Pak. 181 P6
Nalayh, Mongolia 184 D11
Nal'chik, Russ. 159 T13
Naletov Ridge, Arctic Oc. 253 H14
Nalgonda, India 182 N7
Nallıhan, Turk. 170 D7
Nalogo, island, Solomon Is. 228 K5
Nālūt, Lib. 201 E16
Năm Căn, Vietnam 191 Q11
Nam Co (Namu Cuo), China 184 K5
Nam Định, Vietnam 190 H11
Nam Ngum Dam, Laos 190 J9
Nam Tok, Thai. 190 M7
Nam, river, S. Korea 187 S10
Namacunde, Angola 208 Q11
Namacurra, Mozambique 212 E12
Namaite, island, Fr. Polynesia, Fr. 231 M17
Namaksār, lake, Afghan. 180 J2
Namanga, Kenya 211 U6
Namangan, Uzb. 179 L13
Namanyere, Tanzania 211 X3
Namapa, Mozambique 213 C13
Namaqualand, S. Af. 212 L5
Namaram, Vanuatu 229 F15
Namasagali, Uganda 211 R4
Namatanai, P.N.G. 222 N10
Nambinda, Tanzania 211 Y7
Namdae, river, N. Korea 186 E11
Namekagon, river, Wis., U.S. 100 G3
Namelakl Passage, Palau 226 G6
Namenalala, island, Fiji 226 P6
Nametil, Mozambique 213 D13
Namgia, India 182 E6
Namhae, S. Korea 187 T10
Namhaedo, island, S. Korea 187 T10
Namhŭng, N. Korea 186 F8
Namib Desert, Namibia 212 H3
Namibe, Angola 208 P9
Namibia, Af. 212 G4
Namibia Plain, Atl. Oc. 249 S13
Namiquipa, Mex. 116 D8
Namji, S. Korea 187 S11
Namlea, Indonesia 193 K16
Namoi, river, Austral. 220 J14
Nāmolokama Mountain, Hawaii, U.S. 114 E10
Namoluk Atoll, F.S.M. 226 B9
Namonuito Atoll, F.S.M. 226 A9
Namoren, island, Marshall Is. 227 G20
Namorik Atoll, Marshall Is. 227 E16
Nampa, Idaho, U.S. 108 G3
Nampala, Mali 205 J13
Namp'o, N. Korea 186 K5
Nampula, Mozambique 213 D13
Namsê Pass, Nepal 182 F8
Namsen, river, Nor. 142 F12
Namsos, Nor. 142 G11
Namtsy, Russ. 161 G18
Namtu, Myanmar 190 G6
Namu Atoll, Marshall Is. 227 D16
Namu Cuo see Nam Co, China 184 K5
Namuka- i-Lau, island, Fiji 226 Q8
Namukulu, N.Z. 229 P21
Namur, Belg. 148 L12
Namuruputh, Eth. 211 P6
Namwala, Zambia 209 P17
Namwon, S. Korea 187 S9
Namwon, S. Korea 187 W7
Namyang, N. Korea 186 F12
Namyang, S. Korea 187 T9
Nan Ling, China 185 N13
Nan, river, Thai. 190 K8
Nana, river, Cen. Af. Rep. 206 M12
Nanaimo, B.C., Can. 86 M7
Nanakru, Liberia 204 Q11
Nānākuli, Hawaii, U.S. 115 H14
Nanam, N. Korea 186 D13
Nanao, Japan 189 N10
Nanatsu Shima, Japan 188 M9
Nanchang, China 185 L14
Nanchong, China 184 L10
Nancowry Island, India 183 T15
Nancy, Fr. 149 P14
Nanda Devi, pass, India 182 F7
Nanded, India 182 M6
Nandgaon, India 182 L4
Nandurbar, India 182 L4
Nandyal, India 183 P6
Nanfeng, China 185 M14
Nanga Eboko, Cameroon 206 N10
Nanga Parbat, peak, Pak. 180 G13
Nangade, Mozambique 213 B13
Nangapinoh, Indonesia 192 J10
Nangaraun, Indonesia 192 J10
Nangatayap, Indonesia 192 K9
Nangin, Myanmar 191 P6
Nangiré, Vanuatu 229 F15
Nangis, Fr. 149 Q10
Nangnim-ho, lake, N. Korea 186 E8
Nangnim-san, peak, N. Korea 186 F8
Nangnim-sanmaek, range, N. Korea 186 E8
Nangnim, N. Korea 186 E8
Nangqên see Xangda, China 184 K7
Nangtud, Mount, Philippines 193 D14
Nanij, island, Marshall Is. 227 G20
Nanjing, China 185 K14
Nankoku, Japan 189 R7
Nankova, Angola 209 Q13
Nanmatol Islands, F.S.M. 226 K9
Nanmatol, island, F.S.M. 226 K9
Nannine, W. Aust., Austral. 221 V3
Nanning, China 184 P11
Nanortalik (Ilivileq), Greenland, Den. 89 U7
Nanortalik Bank, Atl. Oc. 253 R20
Nanpan, river, China 184 N10
Nanpo Islands, Asia 168 F14
Nanpō Shotō, Japan 189 R12
Nansei Shotō (Ryukyu Islands), Japan 189 X2
Nansen Basin, Arctic Oc. 252 F12
Nansen Fjord, Greenland, Den. 88 M12
Nansen Land, Greenland, Den. 88 B8
Nansen Sound, Nunavut, Can. 87 A14
Nanshan Island, S. China Sea 192 E11
Nansio, Tanzania 211 T4
Nant, Fr. 149 X11
Nantes, Fr. 149 S5
Nanticoke, Pa., U.S. 96 M8
Nantong, China 185 K15
Nantucket Inlet, Antarctica 238 F8
Nantucket Island, Mass., U.S. 97 M16
Nantucket Sound, Mass., U.S. 97 M16
Nantucket, Mass., U.S. 97 M16
Nantwich, Eng., U.K. 145 R11
Nānu'alele Point, Hawaii, U.S. 115 K20
Nanuku Passage, Fiji 226 N7
Nanukuloa, Fiji 226 P5
Nanumanga, island, Tuvalu 224 H8
Nanumea, island, Tuvalu 224 H8
Nanuque, Braz. 133 K16
Nanusa, Kepulauan, Indonesia 193 G16
Nanutarra, W. Aust., Austral. 221 T2
Nanyang, China 184 K12
Nanyuki, Kenya 211 S6
Nanzhila, Zambia 209 P17
Nao Nao, island, Fr. Polynesia, Fr. 231 H17
Nao, Cabo de la, Sp. 147 J15
Nao, Cape, Eur. 140 L4
Naococane, Lac, Que., Can. 87 M19
Nāomīd, Dasht-e, Afghan., Iran 180 J2
Náousa, Gr. 156 C10
Napa, Calif., U.S. 111 R3
Nāpali Coast, Hawaii, U.S. 114 E9
Napasoq, Greenland, Den. 89 Q5
Napasorsuaq Fjord, Greenland, Den. 89 S9
Naperville, Ill., U.S. 101 N6
Napia, island, Kiribati 226 J11
Napier Downs, W. Aust., Austral. 221 R5
Napier Mountains, Antarctica 239 C20
Napier, N.Z. 223 H22
Naples see Napoli, It. 152 M10
Naples, Fla., U.S. 99 W8
Naples, Tex., U.S. 105 F20
Napo see Napug, China 184 J2
Napo, river, S. Amer. 130 K5
Napoleon, Ohio, U.S. 101 P11
Napoleon, N. Dak., U.S. 106 H5
Napoleonville, La., U.S. 102 Q10
Napoli (Naples), It. 152 M10
Napoli, Golfo di, It. 152 M10
Napug (Napo), China 184 J2
Napuka, island, Fr. Polynesia, Fr. 231 A20
Nâqoûra, Leb. 172 K5
Naqu see Nagqu, China 184 K6
Nara Visa, N. Mex., U.S. 109 R15
Nara, Japan 189 Q9
Nara, Mali 204 J12
Nara, river, Pak. 181 R9
Naracoorte, S. Aust., Austral. 221 Y11
Narang, Afghan. 180 H10
Narasannapeta, India 182 M9
Narathiwat, Thai. 191 S8
Narayanganj, Bangladesh 182 J13
Narbonne, Fr. 149 Y10
Narborough see Isla Fernandina, island, Ecua. 130 Q2
Narcondam Island, India 183 Q15
Nardlunak (Skovfjorden), Greenland, Den. 89 T7
Nardò, It. 153 N15
Nares Land, Greenland, Den. 88 B8
Nares Plain, Atl. Oc. 248 L5
Nares Strait, Can., Greenland, Den. 88 D4
Narew, river, Pol. 151 C16
Narinda Bay, Af. 198 M11
Nariva Swamp, Trin. & Tobago 123 Q23
Narlı, Turk. 170 H12
Narmada, river, India 182 K4
Narmidj, island, Marshall Is. 227 G20
Narnaul, India 182 G5
Narndee, W. Aust., Austral. 221 V3
Narni, It. 152 J8
Narodnaya, Russ. 160 F10
Narodnaya, peak, Asia 168 D7
Narok, Kenya 211 T6
Narooma, N.S.W., Austral. 221 Y14
Narovorovo, Vanuatu 229 F16
Narowal, Pak. 180 K13
Narowlya, Belarus 158 M5
Narrabri, N.S.W., Austral. 221 W14
Narragansett Pier, R.I., U.S. 97 M14
Narran Lake, Austral. 220 J13
Narrandera, N.S.W., Austral. 221 Y13
Narrogin, W. Aust., Austral. 221 X3
Narrows, Va., U.S. 98 F9
Narsap Sermia, glacier, Greenland, Den. 89 Q6
Narsaq Kujalleq (Frederiksdal), Greenland, Den. 89 U8
Narsaq, Greenland, Den. 89 R5
Narsarsuaq, Greenland, Den. 89 T7
Narsinghgarh, India 182 J5
Narsipatnam, India 182 N8
Nartës, Laguna e, Alban. 154 M9
Narthákí, Óros, Gr. 156 F10
Nartkala, Russ. 159 T13
Narva, Est. 143 J17
Narvik, Nor. 142 D12
Narvskoye Vodokhranilishche, Russ. 158 F5
Narwietooma, N.T., Austral. 221 T8
Nar'yan Mar, Russ. 160 E10
Naryn Khuduk, Russ. 159 S14
Naryn Qum, Kaz. 178 F4
Naryn, Russ. 161 M14
Naryn, Kyrg. 179 K15
Naryn, river, Kyrg. 179 K13
Narynqol, Kaz. 179 J17
Näs, Sw. 142 H12
Nasarawa, Nigeria 206 L8
Nasca Ridge, Pac. Oc. 247 M22
Nasca, Peru 132 J4
Nashua, N.H., U.S. 97 J14
Nashua, Iowa, U.S. 106 M12
Nashua, Mont., U.S. 108 B12
Nashville, Ga., U.S. 99 P6
Nashville, Ill., U.S. 101 T5
Nashville, Ark., U.S. 102 H6
Nashville, Tenn., U.S. 103 E15
Nashwauk, Minn., U.S. 106 F11
Našice, Croatia 154 E8
Näsijärvi, lake, Fin. 143 H15
Nasik, India 182 L4
Nasir, S. Sudan 203 V12
Nasirabad, India 182 H4
Nasiriyah see An Nāṣirīyah, Iraq 176 J12
Naso, It. 153 R11
Nassau Bay, S. Amer. 128 R5
Nassau, Bahamas 122 E3
Nassau, island, Cook Is., N.Z. 224 H11
Nassawadox, Va., U.S. 98 E14
Nassian, Côte d'Ivoire 205 N14
Nässjö, Sw. 143 L13
Nassuttuup Nunaa, Greenland, Den. 89 N5
Nastapoka Islands, Nunavut, Can. 87 K18
Nata, Botswana 212 F7
Natal Valley, Ind. Oc. 250 M4
Natal, Braz. 133 F18
Natal, Indonesia 192 J5
Natalia, Tex., U.S. 105 M15
Natal'inskiy, Russ. 161 C22
Naṭanz, Iran 177 G16
Natara, Russ. 161 F16
Natashquan, Que., Can. 87 M22
Natashquan, river, Nfld. & Lab., Que., Can. 87 L21
Natchez, Miss., U.S. 102 M9
Natchitoches, La., U.S. 102 M7
Naternaq, Greenland, Den. 88 M6
Natewa Bay, Fiji 226 N7
Nathorst Land, Greenland, Den. 88 K13
National City, Calif., U.S. 111 Z9
National Museum of Fine Arts, Monaco 163 C21
Natitingou, Benin 205 L17
Natividad, Isla, Mex. 116 D5
Natividade, Braz. 133 H13
Natkyizin, Myanmar 190 M6
Natoma, Kans., U.S. 107 S6
Natron, Lake, Af. 198 K10
Nattavaara, Sw. 143 E14
Natuashish, Nfld. & Lab., Can. 87 K21
Natuna Besar, island, Indonesia 192 G9
Natuna Besar, Kepulauan (Bunguran Utara), Indonesia 192 G8
Natuna Islands, Asia 168 L11
Natuna Selatan, Kepulauan (Bunguran Selatan), Indonesia 192 H9
Natural Bridges National Monument, Utah, U.S. 109 P9
Naturaliste Fracture Zone, Ind. Oc. 251 M16
Naturaliste Plateau, Ind. Oc. 251 N17
Naturaliste Trough, Ind. Oc. 251 N17
Naturaliste, Cape, Austral. 220 K2
Naturita, Colo., U.S. 109 N10
Naturno, It. 152 C6
Naubinway, Mich., U.S. 100 G10
Nauiyu (Daly River), N.T., Austral. 221 P7
Naukot, Pak. 181 S9
Naulila, Angola 208 Q10
Naungpale, Myanmar 190 J6
Nauroz Kalat, Pak. 181 N6
Nauru, Pac. Oc. 224 G7
Naushahra, India 180 J13
Naushahro Firoz, Pak. 181 Q8
Nausori, Fiji 226 P5
Nauta, Peru 132 E4
Nautilus Basin, Arctic Oc. 252 J9
Nautilus Gap, Arctic Oc. 252 K9
Nautilus Spur, Arctic Oc. 252 K11
Nautla, Mex. 117 H13
Nauvoo, Ill., U.S. 101 Q3
Nava, river, D.R.C. 209 C17
Navabelitsa, Belarus 158 M6
Navadwip, India 182 K11
Navahrudak, Belarus 158 K4
Navai, Fiji 226 P5
Navajo Mountain, Utah, U.S. 109 Q8
Navajo National Monument, Ariz., U.S. 109 Q8
Navajo Reservoir, Colo., N. Mex., U.S. 109 Q11
Naval Base Guam see Joint Region Marianas, Guam, U.S. 230 P4
Navalmoral de la Mata, Sp. 146 G9
Navan (An Uaimh), Ire. 145 Q6
Navapolatsk, Belarus 158 J5
Navarin, Russ. 161 B21
Navarin, Cape, N. Amer. 84 P3
Navarin, Cape, Asia 168 A12
Navarino, Isla, Chile 135 Y9
Navarre, Eur. 140 J3
Navarrenx, Fr. 149 Y6
Navasota, Tex., U.S. 105 K18
Navasota, river, Tex., U.S. 105 J18
Navassa Island, U.S. 119 M14
Navia, Sp. 146 A7
Navidad, river, Tex., U.S. 105 M18
Naviti, island, Fiji 226 P4
Năvodari, Rom. 155 F18
Navoiy, Uzb. 178 M10
Navojoa, Mex. 116 E7
Navrongo, Ghana 205 L15
Navsari, India 182 L3
Navua, Fiji 226 Q5
Navua, river, Fiji 226 Q5
Nawá, Syr. 172 K7
Nawabshah, Pak. 181 R8
Nawada, India 182 J10
Nāwah, Afghan. 180 K8
Nawalgarh, India 182 G5
Naxçivan, region, Arm. 171 F19
Naxçıvan, Arm. 171 F19
Náxos, Gr. 157 L15
Náxos, island, Gr. 157 L15
Naxos, ruin, It. 153 S12
Naxxar, Malta 163 K22
Nāy Band, Iran 177 M16
Nay Pyi Taw, Myanmar 190 J5
Nay, Cape, Asia 168 K11
Nayak, Afghan. 180 H7
Nayau, island, Fiji 226 P8
Nayoro, Japan 188 F13
Nazaré, Braz. 133 H16
Nazaré, Port. 146 G4
Nazareth Bank, Ind. Oc. 250 J9
Nazareth see Naẓerat, Israel 172 L6
Nazareth, Vanuatu 229 F16
Nazas, Mex. 116 F9
Naze, Japan 189 W3
Naẓerat (Nazareth), Israel 172 L6
Nazik Gölü, Turk. 171 F16
Nazilli, Turk. 170 G4
Nazimovo, Russ. 161 J13
Naziya, Russ. 158 F7
Nazımiye, Turk. 171 F14
Nazran', Russ. 159 U14
Nazrēt, Eth. 210 M8
Nchelenge, Zambia 209 K18
Ndai see Dai, island, Solomon Is. 228 D10
Ndala, Tanzania 211 V4
N'dalatando, Angola 208 K10
Ndali, Benin 205 M18
Ndélé, Cen. Af. Rep. 207 L15
Ndélé, Cen. Af. Rep. 207 N13
Ndendé, Gabon 208 F8

Ndeni Island *see* Nendö, *Solomon Is.* **228** L5
Ndinde, *Gabon* **208** G8
N'Djamena, *Chad* **206** J12
Ndjolé, *Gabon* **208** E8
Ndogo Lagoon, *Af.* **198** K5
Ndola, *Zambia* **209** M18
Ndu, *D.R.C.* **209** B15
Nduindui, *Solomon Is.* **228** F10
Nduindui, *Vanuatu* **229** F15
Ndumbwe, *Tanzania* **211** Y8
Nduye, *D.R.C.* **209** D18
Néa Epídavros, *Gr.* **156** K11
Néa Michanióna, *Gr.* **156** D11
Néa Moní, *Gr.* **157** H16
Néa Moudaniá, *Gr.* **156** D11
Néa Zíchni, *Gr.* **156** B12
Neagh, Lake, *Eur.* **140** E4
Neagh, Lough, *N. Ire., U.K.* **145** N7
Neah Bay, *Wash., U.S.* **110** B2
Neale, Lake, *Austral.* **220** F7
Neápoli, *Gr.* **156** D8
Neapolis *see* Kavála, *Gr.* **157** C13
Neápolis, *Gr.* **156** M11
Near Islands, *Alas., U.S.* **112** L1
Nebbou, *Burkina Faso* **205** L15
Nebine Creek, *Austral.* **220** H13
Nebitdag *see* Balkanabat, *Turkm.* **178** M5
Neblina, *S. Amer.* **130** J11
Nebraska City, *Nebr., U.S.* **107** Q8
Nebraska, *U.S.* **107** P5
Nebrodi, Monti, *It.* **153** S11
Nécé, *New Caledonia, Fr.* **231** Q24
Necedah, *Wis., U.S.* **100** K4
Neches, river, *Tex., U.S.* **105** J20
Neckar, river, *Ger.* **150** J6
Neckarboo Range, *Austral.* **220** K12
Necker Island, *Virgin Is., U.K.* **122** P9
Necker Ridge, *Pac. Oc.* **247** G13
Necochea, *Arg.* **135** N12
Nederland, *Tex., U.S.* **105** L21
Needham's Point, *Barbados* **123** L19
Needles, *Calif., U.S.* **111** W12
NEEM Camp *see* North Greenland Eemian Ice Drilling, *Greenland, Den.* **88** F7
Neenah, *Wis., U.S.* **100** J6
Neftçala, *Azerb.* **171** E23
Nefteyugansk, *Russ.* **160** H11
Negage, *Angola* **208** J10
Negaunee, *Mich., U.S.* **100** F7
Negēlē, *Eth.* **211** P8
Negeribatin, *Indonesia* **192** L7
Negev, region, *Israel* **173** P5
Negoiul, peak, *Rom.* **155** E14
Negola, *Angola* **208** N10
Negomano, *Mozambique* **212** B12
Negombo, *Sri Lanka* **183** U7
Negonego, island, *Fr. Polynesia, Fr.* **231** C20
Negotin, *Serb.* **154** G12
Negra, Laguna, *Uru.* **134** L15
Negra, Punta, *Peru* **132** E1
Negre, Pic, *Andorra, Sp.* **162** K3
Negreira, *Sp.* **146** B5
Negrești-Oaș, *Rom.* **155** B13
Negril, *Jam.* **122** H5
Negritos, *Peru* **132** E1
Negro Bay, *Af.* **198** H12
Negro, river, *Guatemala* **117** L15
Negro, river, *S. Amer.* **130** H10
Negros, island, *Philippines* **193** E14
Negru Vodă, *Rom.* **155** G18
Nehbandān, *Iran* **177** J21
Nehe, *China* **185** C15
Neiafu, *Tonga* **227** H23
Neiba, *Dom. Rep.* **119** M17
Neiba, Bahía de, *Dom. Rep.* **119** M18
Neiden, *Nor.* **143** B15
Neidín *see* Kenmare, *Ire.* **145** T3
Neiges, Piton des, *Réunion, Fr.* **215** G16
Neihart, *Mont., U.S.* **108** D8
Neillsville, *Wis., U.S.* **100** J4
Neilson Reef, *Pac. Oc.* **255** P21
Neiva, *Col.* **130** H6
Nejo, *Eth.* **210** L5
Nekalagba, *D.R.C.* **209** C18
Nékaounié, *Côte d'Ivoire* **204** Q12
Nek'emtē, *Eth.* **210** M6
Nekoosa, *Wis., U.S.* **100** J5
Neksø, *Den.* **143** N13
Nekurandu, island, *Maldives* **183** U3
Nelemnoye, *Russ.* **161** E19
Nelidovo, *Russ.* **158** J7

Neligh, *Nebr., U.S.* **107** N7
Nel'kan, *Russ.* **161** H19
Nell Passage, *Marshall Is.* **227** G15
Nell, island, *Marshall Is.* **227** G16
Nellore, *India* **183** Q7
Nelson House, *Man., Can.* **87** L13
Nelson Island, *Alas., U.S.* **112** J11
Nelson Lagoon, *Alas., U.S.* **112** N12
Nelson Reservoir, *Mont., U.S.* **108** B11
Nelson, *B.C., Can.* **86** M9
Nelson, *Nebr., U.S.* **107** R7
Nelson, *Nev., U.S.* **111** V12
Nelson, *Eng., U.K.* **145** Q11
Nelson, *N.Z.* **223** J19
Nelson, Cape, *Austral.* **220** M11
Nelson, Estrecho, *Chile* **135** W6
Nelson, river, *Man., Can.* **87** L14
Néma, *Mauritania* **204** H12
Neman, *Russ.* **158** J2
Neman, river, *Eur.* **140** F9
Nembe, *Nigeria* **206** N7
Nembrala, *Indonesia* **193** N14
Neméa, *Gr.* **156** J10
Nemiscau, *Que., Can.* **87** M18
Nemrut Dagh, ruins, *Turk.* **171** H13
Nemrut Gölü, *Turk.* **171** G16
Nemunas, river, *Lith.* **143** M16
Nemuro Strait, *Japan* **188** F15
Nemuro, *Japan* **188** F16
Nen, river, *China* **185** B15
Nenagh (An tAonach), *Ire.* **145** S5
Nenana, *Alas., U.S.* **113** G17
Nendeln, *Liech.* **162** N2
Nendö (Ndeni Island), *Solomon Is.* **228** L5
Nenjiang, *China* **185** C15
Neoch, atoll, *F.S.M.* **226** G9
Neodesha, *Kans., U.S.* **107** U9
Neola, *Utah, U.S.* **108** L9
Néos, *Gr.* **156** D12
Neosho, *Mo., U.S.* **107** V11
Neosho, river, *U.S.* **91** H13
Nepa, *Russ.* **161** J15
Nepal, *Asia* **182** G8
Nepalganj, *India* **182** G8
Nepean, island, *Austral.* **229** B20
Nephi, *Utah, U.S.* **108** M7
Nephin Beg, peak, *Ire.* **145** P3
Nepoko, river, *D.R.C.* **209** C18
Neptune Range, *Antarctica* **238** G10
Nerchinsk, *Russ.* **161** L17
Nerchinskiy Zavod, *Russ.* **161** L18
Nerekhta, *Russ.* **158** H10
Neretva, river, *Bosn. & Herzg.* **154** H7
Neretvanski Kanal, *Croatia* **154** H7
Neria, *Greenland, Den.* **89** S6
Neriquinha, *Angola* **209** P14
Néris, *Fr.* **149** T10
Nerja, *Sp.* **146** M10
Nerpo, *Russ.* **161** J17
Nerrima, *W. Aust., Austral.* **221** R5
Nerva, *Sp.* **146** K7
Nesika Beach, *Oreg., U.S.* **110** K1
Neskaupstaður, *Ice.* **142** F4
Nespelem, *Wash., U.S.* **110** B7
Ness City, *Kans., U.S.* **107** T5
Ness, Loch, *Scot., U.K.* **144** J9
Néstos, river, *Gr.* **157** B13
Netanya, *Israel* **172** L5
Netherdale, *Qnsld., Austral.* **221** S14
Netherlands, *Eur.* **148** J13
Nettilling Lake, *Nunavut, Can.* **87** G18
Netzahualcóyotl, *Mex.* **116** J12
Neu-Ulm, *Ger.* **150** K7
Neuberg, *Ger.* **150** G5
Neubrandenburg, *Ger.* **150** C10
Neuchâtel, *Switz.* **152** B2
Neuchâtel, Lac de, *Switz.* **152** B2
Neufchâteau, *Belg.* **148** M13
Neufchâteau, *Fr.* **149** Q13
Neumagen, *Ger.* **150** G5
Neumayer III Station, *Antarctica* **238** A11
Neumünster, *Ger.* **150** B8
Neun, river, *Vietnam* **190** H10
Neunkirchen, *Ger.* **150** L12
Neunkirchen, *Ger.* **150** H5
Neuquén, *Arg.* **135** P8
Neuquén, river, *Arg.* **135** N7
Neuruppin, *Ger.* **150** D10
Neuschwanstein, site, *Ger.* **150** L8
Neuse, river, *N.C., U.S.* **98** H12
Neusiedler Lake, *Eur.* **140** H8
Neusiedler See, *Aust.* **151** L13
Neuss, *Ger.* **150** F5
Neustadt, *Ger.* **150** H6

Neustrelitz, *Ger.* **150** C10
Neuvic, *Fr.* **149** V8
Neuwied, *Ger.* **150** G5
Nevada City, *Calif., U.S.* **111** Q5
Nevada, *Iowa, U.S.* **107** N11
Nevada, *Mo., U.S.* **107** U11
Nevada, *U.S.* **111** Q10
Nevada, Sierra de, *Sp.* **146** L11
Nevea, *Solomon Is.* **228** L5
Nevel', *Russ.* **158** J6
Never, *Russ.* **161** K18
Nevers, *Fr.* **149** S11
Neves, *Sao Tome and Principe* **215** D20
Nevinnomyssk, *Russ.* **159** T12
Nevis Peak, *St. Kitts & Nevis* **123** C19
Nevis, *St. Kitts & Nevis* **123** C19
Nevis, Loch, *Scot., U.K.* **144** J8
Nevşehir, *Turk.* **170** G10
New Albany, *Ind., U.S.* **101** T9
New Albany, *Miss., U.S.* **102** H12
New Amsterdam, *Guyana* **131** F16
New Bedford, *Mass., U.S.* **97** L15
New Bern, *N.C., U.S.* **98** J13
New Bight, *Bahamas* **119** F13
New Boston, *Ohio, U.S.* **101** T13
New Boston, *Tex., U.S.* **105** F20
New Braunfels, *Tex., U.S.* **105** L16
New Britain, *Conn., U.S.* **96** L12
New Britain, *P.N.G.* **222** P9
New Brunswick, *Can.* **87** P21
New Brunswick, *N.J., U.S.* **96** N10
New Caledonia *see* Nouvelle-Calédonie, island, *New Caledonia, Fr.* **231** P20
New Caledonia Trough, *Pac. Oc.* **246** M10
New Castille, *Eur.* **140** K3
New Castle (An Caisleán Nua), *Ire.* **145** S4
New Castle, *Pa., U.S.* **96** M3
New Castle, *Ind., U.S.* **101** R10
New City, *N.Y., U.S.* **96** M11
New Delhi, *India* **182** G6
New England Range, *Austral.* **220** J15
New England Seamounts, *Atl. Oc.* **248** J5
New England, *N. Dak., U.S.* **106** G2
New Franklin, *Mo., U.S.* **107** S12
New Galloway, *Scot., U.K.* **145** N9
New Georgia Group, *Solomon Is.* **228** E7
New Georgia Sound (The Slot), *Solomon Is.* **228** D7
New Georgia, island, *Solomon Is.* **228** E7
New Glasgow, *N.S., Can.* **87** P22
New Grant, *Trin. & Tobago* **123** Q22
New Guinea Trench, *Pac. Oc.* **246** J7
New Guinea, *P.N.G.* **222** N5
New Hampshire, *U.S.* **97** H14
New Hampton, *Iowa, U.S.* **106** M12
New Hanover, island, *P.N.G.* **222** M9
New Harbor, *Me., U.S.* **97** G17
New Haven, *Conn., U.S.* **96** M12
New Haven, *Ind., U.S.* **101** P10
New Hazelton, *B.C., Can.* **86** J8
New Hebrides Trench, *Pac. Oc.* **246** L10
New Iberia, *La., U.S.* **102** Q9
New Ipswich, *N.H., U.S.* **97** J14
New Ireland, island, *P.N.G.* **222** N10
New Jersey, *U.S.* **96** P10
New Lexington, *Ohio, U.S.* **101** R14
New Lisbon, *Wis., U.S.* **100** K4
New Liskeard, *Ont., Can.* **87** P18
New London, *Conn., U.S.* **97** M13
New London, *Wis., U.S.* **100** J6
New London, *Iowa, U.S.* **107** Q13
New Madha, *Oman* **175** F14
New Madrid, *Mo., U.S.* **107** V16
New Market, *Va., U.S.* **98** D11
New Martinsville, *W. Va., U.S.* **98** C9
New Meadows, *Idaho, U.S.* **108** F4
New Mexico, *U.S.* **109** S12
New Norfolk, *Tas., Austral.* **221** Z8
New Orleans, *La., U.S.* **102** Q11
New Philadelphia, *Ohio, U.S.* **101** Q15
New Plymouth, *N.Z.* **223** G20
New Port Richey, *Fla., U.S.* **99** T7
New Prague, *Minn., U.S.* **106** K11
New Providence, island, *Bahamas* **118** E11
New Richmond, *Wis., U.S.* **100** H2

New Roads, *La., U.S.* **102** N9
New Rochelle, *N.Y., U.S.* **96** N11
New Rockford, *N. Dak., U.S.* **106** F6
New Ross (Ros Mhic Thriúin), *Ire.* **145** S6
New Salem, *N. Dak., U.S.* **106** G4
New Sandy Bay, *St. Vincent & the Grenadines* **123** J17
New Schwabenland, region, *Antarctica* **239** C14
New Siberian Islands *see* Novosibirskiye Ostrova, *Russ.* **161** D16
New Siberian Shelf, *Arctic Oc.* **252** F9
New Smyrna Beach, *Fla., U.S.* **99** S9
New South Wales, *Austral.* **221** X13
New Stuyahok, *Alas., U.S.* **113** K14
New Town, *N. Dak., U.S.* **106** E3
New Ulm, *Minn., U.S.* **106** K10
New Waverly, *Tex., U.S.* **105** K19
New York, *N.Y., U.S.* **96** N11
New York, *U.S.* **96** K8
New Zealand, *Pac. Oc.* **223** H18
Newala, *Tanzania* **211** Z8
Newark, *N.J., U.S.* **96** N11
Newark, *N.Y., U.S.* **96** J7
Newark, *Del., U.S.* **98** B14
Newark, *Ohio, U.S.* **101** R13
Newark, *Ark., U.S.* **102** F9
Newark, *Eng., U.K.* **145** R13
Newaygo, *Mich., U.S.* **100** L9
Newberg, *Oreg., U.S.* **110** F3
Newbern, *Tenn., U.S.* **102** E12
Newberry, *S.C., U.S.* **98** K8
Newberry, *Fla., U.S.* **99** R7
Newberry, *Mich., U.S.* **100** F10
Newbiggin By The Sea, *Eng., U.K.* **145** M12
Newburg, *Mo., U.S.* **107** U13
Newburgh, *N.Y., U.S.* **96** M11
Newburgh, *Scot., U.K.* **144** J11
Newbury, *Eng., U.K.* **145** U12
Newcastle Bay, *Austral.* **220** A12
Newcastle upon Tyne, *Eng., U.K.* **145** N12
Newcastle Waters, *N.T., Austral.* **221** R8
Newcastle-under-Lyme, *Eng., U.K.* **145** R11
Newcastle, *Tex., U.S.* **105** F15
Newcastle, *Wyo., U.S.* **108** G14
Newcastle, *Calif., U.S.* **111** Q5
Newcastle, *Jam.* **122** J10
Newcastle, *St. Kitts & Nevis* **123** B19
Newcastle, *S. Af.* **212** J9
Newcastle, *N.S.W., Austral.* **221** X15
Newdegate, *W. Aust., Austral.* **221** X4
Newell, *S. Dak., U.S.* **106** K2
Newenham, Cape, *Alas., U.S.* **112** L12
Newfoundland and Labrador, *Can.* **87** K21
Newfoundland Basin, *Atl. Oc.* **248** H7
Newfoundland Evaporation Basin, *Utah, U.S.* **108** K6
Newfoundland Plain, *Atl. Oc.* **248** J7
Newfoundland Ridge, *Atl. Oc.* **248** J6
Newfoundland Seamounts, *Atl. Oc.* **248** J7
Newfoundland, Island of, *Nfld. & Lab., Can.* **87** M23
Newkirk, *Okla., U.S.* **105** A17
Newkirk, *N. Mex., U.S.* **109** S14
Newlands, *Cayman Is., U.K.* **122** N3
Newman Grove, *Nebr., U.S.* **107** P7
Newman Island, *Antarctica* **238** N10
Newman, *Calif., U.S.* **111** S5
Newman, *W. Aust., Austral.* **221** T3
Newmarket (Áth Trasna), *Ire.* **145** T4
Newmarket, *Eng., U.K.* **145** T14
Newnan, *Ga., U.S.* **98** L4
Newport (Casnewydd), *Wales, U.K.* **145** U11
Newport Beach, *Calif., U.S.* **111** Y8
Newport News, *Va., U.S.* **98** F14
Newport, *Me., U.S.* **97** F17
Newport, *R.I., U.S.* **97** M14
Newport, *Vt., U.S.* **97** F13
Newport, *Ark., U.S.* **102** F10
Newport, *Ky., U.S.* **103** A18
Newport, *Tenn., U.S.* **103** E19
Newport, *Oreg., U.S.* **110** G2
Newport, *Wash., U.S.* **110** B9
Newport, *Jam.* **122** J8
Newport, *Eng., U.K.* **145** W13
Newquay, *Eng., U.K.* **145** W8

Newry, *N. Ire., U.K.* **145** P7
Newton Abbot, *Eng., U.K.* **145** W10
Newton Stewart, *Scot., U.K.* **145** N9
Newton, *Mass., U.S.* **97** K14
Newton, *Ill., U.S.* **101** S6
Newton, *Miss., U.S.* **102** L12
Newton, *Tex., U.S.* **105** J21
Newton, *Iowa, U.S.* **107** P12
Newton, *Kans., U.S.* **107** U7
Newtown Saint Boswells, *Scot., U.K.* **145** M11
Newtown, *Wales, U.K.* **145** S10
Newtownabbey, *N. Ire., U.K.* **145** N7
Newtownards, *N. Ire., U.K.* **145** P8
Newtownstewart, *N. Ire., U.K.* **145** N6
Neya, *Russ.* **158** G12
Neyrīz, *Iran* **177** L17
Neyshābūr, *Iran* **177** E20
Nez Perce National Historical Park, *Idaho, U.S.* **108** D3
Nez Perce Pass, *Idaho, Mont., U.S.* **108** E5
Nezperce, *Idaho, U.S.* **108** D4
Ngabé, *Congo* **208** G11
Ngain, island, *Marshall Is.* **227** G19
Ngajangel, island, *Palau* **226** D7
Ngake, *N.Z.* **229** D21
Ngake, island, *N.Z.* **229** H23
Ngala, *Nigeria* **206** J11
Ngalkol, *Palau* **226** J4
Ngamegei Passage, *Palau* **226** F6
Ngangla Ringco (Angla Rencuo), lake, *China* **184** K3
Nganglong Kangri, *China* **184** J2
Ngaoundal, *Cameroon* **206** M11
Ngaoundéré, *Cameroon* **206** L11
Ngara, *Tanzania* **211** U2
Ngaramasch, *Palau* **226** K4
Ngardmau Bay, *Palau* **226** F6
Ngardmau, *Palau* **226** F6
Ngardololok, *Palau* **226** J4
Ngaregur, island, *Palau* **226** E6
Ngarekeklav, island, *Palau* **226** F6
Ngaremediu, island, *Palau* **226** H5
Ngaruangl Passage, *Palau* **226** D6
Ngaruawahia, *N.Z.* **223** F20
Ngatangiia, *N.Z.* **229** E23
Ngatik Atoll, *F.S.M.* **226** B11
Ngayu, river, *D.R.C.* **209** D18
Ngbala, *Congo* **208** C10
Ngcheangel *see* Kayangel Islands, *Palau* **226** D7
Ngeaur *see* Angaur, island, *Palau* **226** B4;K4
Ngemelis Islands, *Palau* **226** J4
Ngeregong, island, *Palau* **226** J5
Ngerengere, *Tanzania* **211** W7
Ngergoi, island, *Palau* **226** J4
Ngertachebeab, bay, *Palau* **226** G5
Ngeruktabel, island, *Palau* **226** H5
Ngerulmud, *Palau* **226** B4;G6
Ngesebus, island, *Palau* **226** J4
Nggatokae, island, *Solomon Is.* **228** E8
Nggela Pile, island, *Solomon Is.* **228** F10
Nggela Sule, island, *Solomon Is.* **228** E10
Ng'iro, peak, *Kenya* **211** R6
Ngô Mây, *Vietnam* **190** M13
Ngo, *Congo* **208** F11
Ngobasangel, island, *Palau* **226** H5
Ngoïla, *Cameroon* **206** P11
Ngoko, *Congo* **208** E10
Ngoko, river, *Congo* **208** C11
Ngomba, *Tanzania* **211** X4
Ngoring Hu, *China* **184** J8
Ngorongoro Crater, *Tanzania* **211** U6
Ngote, *D.R.C.* **209** C19
Ngoto, *Cen. Af. Rep.* **207** N13
Ngounié, river, *Gabon* **208** F8
Ngourti, *Niger* **206** G11
Ngoywa, *Tanzania* **211** W4
Ngozi, *Burundi* **211** U1
NGRIP *see* North Greenland Ice Core Project, *Greenland, Den.* **88** H9
Nguigmi, *Niger* **206** H11
Nguiu, *N.T., Austral.* **221** N7
Ngukurr, *N.T., Austral.* **221** P9
Ngulu Atoll, *F.S.M.* **226** A5
Nguna, island, *Vanuatu* **229** J16
N'gungo, *Angola* **208** M10
Nguni, *Kenya* **211** T7
Nguru, *Nigeria* **206** H9
Nha Trang, *Vietnam* **191** N13
Nhamundá, *Braz.* **132** D10
Nhulunbuy, *N.T., Austral.* **221** P10

Nia Nia, *D.R.C.* **209** D18
Niafounké, *Mali* **205** H14
Niagara Falls, *N.Y. U.S.* **96** J5
Niagara, *Wis. U.S.* **100** G7
Niagara, *river, N.Y. U.S.* **96** H4
Niagassola, *Guinea* **204** K11
Niah, *Malaysia* **192** G11
Niandan Koro, *Guinea* **204** L11
Niangara, *D.R.C.* **209** B18
Niangoloko, *Burkina Faso* **205** M13
Niangua, *river, Mo. U.S.* **107** U12
Nianqingtangla Shan *see*
    Nyainqêntanglha Shan, *China*
    184 L5
Niapu, *D.R.C.* **209** C17
Niaqornaarsuk, *Greenland, Den.* **88** M5
Niaqornat, *Greenland, Den.* **88** L5
Niaqunngunaq (Fiskefjord),
    Greenland, Den. **89** Q5
Nias, *island, Indonesia* **192** H4
Niau, *island, Fr. Polynesia, Fr.* **231** B18
Nibok, *Nauru* **227** M14
Nic, *Azerb.* **171** C21
Nicaea *see* İznik, *Turk.* **170** D5
Nicaragua, *N. Amer.* **117** M18
Nicaragua, Lago de, *Nicar.* **117** M18
Nicaro, *Cuba* **119** K13
Nicastro, *It.* **153** Q13
Nice, *Fr.* **149** X15
Nicephorium *see* Ar Raqqah, *Syr.*
    172 E11
Niceville, *Fla. U.S.* **99** Q2
Nichi *see* Nyingchi, *China* **184** L7
Nichinan, *Japan* **189** T5
Nicholas Channel, *N. Amer.* **118** F8
Nicholasville, *Ky. U.S.* **103** C18
Nicholls' Town, *Bahamas* **118** E11
Nicholson Range, *Austral.* **220** H3
Nickavilla, *Qnsld. Austral.* **221** U12
Nickerson Ice Shelf, *Antarctica*
    238 N10
Nickerson, *Kans. U.S.* **107** T7
Nickol Bay, *Austral.* **220** E2
Nicobar Fan, *Ind. Oc.* **251** F14
Nicobar Islands, *India* **183** T15
Nicosia (Lefkosía, Lefkoşa), *Cyprus*
    162 N8
Nicosia, *It.* **153** S11
Nicoya, Península de, *C.R.* **117** N18
Nida, *Lith.* **143** M15
Nidzica, *Pol.* **151** C15
Niederanven, *Lux.* **162** J10
Niedere Tauern, *range, Aust.* **150** L10
Niemba, *D.R.C.* **209** H18
Niéna, *Mali* **204** L12
Nienburg, *Ger.* **150** D7
Nieuw Amsterdam Fracture Zone,
    Ind. Oc. **250** N11
Nieuw Amsterdam, *Suriname* **131** F17
Nieuw Nickerie, *Suriname* **131** F16
Nieuwpoort, *Curaçao, Neth.* **123** Q15
Nieves, Pico da las, *Sp.* **214** R5
Nif, *F.S.M.* **226** E11
Nifiloli, *island, Solomon Is.* **228** K6
Niğde, *Turk.* **170** G10
Niger Delta, *Nigeria* **206** N7
Niger Fan, *Atl. Oc.* **249** N13
Niger, *Af.* **206** F9
Niger, *river, Af.* **198** G2
Nigeria, *Af.* **206** K8
Nigríta, *Gr.* **156** C12
Nihing, *river, Pak.* **181** R3
Nihiru, *island, Fr. Polynesia, Fr.*
    231 B20
Nihonmatsu, *Japan* **188** M12
Nii Jima, *Japan* **189** Q11
Niigata, *Japan* **188** M11
Niihama, *Japan* **189** Q7
Ni'ihau, *Hawaii, U.S.* **114** F8
Nijmegen, *Neth.* **148** J13
Nīk Pey, *Iran* **177** D13
Nīk Shahr, *Iran* **177** N22
Níkaia, *Gr.* **156** F10
Nikalap Aru, *island, F.S.M.* **226** K7
Nikao, *N.Z.* **229** D23
Nikaupara, *N.Z.* **229** H20
Nikel', *Russ.* **160** C9
Nikiboko, *Bonaire, Neth.* **123** R19
Níkiti, *Gr.* **156** D12
Nikki, *Benin* **205** M18
Nikolai, *Alas. U.S.* **113** H15
Nikolay Dyatel Terrace, *Arctic Oc.*
    253 K13
Nikolayevsk na Amure, *Russ.* **161** J21
Nikolayevsk, *Russ.* **159** N13

Nikolayevskiy, *Russ.* **161** J19
Nikol'sk, *Russ.* **158** K13
Nikol'sk, *Russ.* **158** F12
Nikolski, *Alas. U.S.* **112** P8
Nikol'skoye, *Russ.* **159** Q14
Nikol'skoye, *Russ.* **161** E23
Nikopol', *Ukr.* **159** R8
Nikópolis, *ruin, Gr.* **156** G7
Niksar, *Turk.* **170** D12
Nikšić, *Montenegro* **154** J8
Nikumaroro, *island, Kiribati* **224** G10
Nil, *Russ.* **161** K18
Nilandu Atoll, *Maldives* **183** W3
Nilandu, *island, Maldives* **183** W3
Nilandu, *island, Maldives* **183** X3
Nile River Delta, *Af.* **198** D9
Nile *see* Nahr an Nīl, *Egypt* **202** G11
Nile, *river, Af.* **198** F9
Niles, *Mich. U.S.* **101** N9
Niles, *Ohio, U.S.* **101** P15
Nilgiri Hills, *India* **183** R5
Nili, *Alg.* **200** D12
Nilka, *China* **184** E4
Nima *see* Nyima, *China* **184** K4
Nimach, *India* **182** J4
Nimba Mountains, *Guinea, Liberia*
    204 N11
Nimbahera, *India* **182** J4
Nîmes, *Fr.* **149** X12
Nímos, *island, Gr.* **157** M19
Nimrod Glacier, *Antarctica* **239** L13
Nimrūd, *ruin, Iraq* **176** E10
Nimule, *S. Sudan* **203** Y12
Ninati, *Indonesia* **193** L21
Ninda, *Angola* **209** P14
Nine Degree Channel, *India* **183** T3
Ninetyeast Ridge, *Ind. Oc.* **251** L13
NinetyMile Beach, *N.Z.* **229** C19
Nineveh, *ruin, Iraq* **176** E10
Ninfas, Punta, *Arg.* **135** R10
Ningaloo, *W. Aust. Austral.* **221** T1
Ningbo, *China* **185** L16
Ningerum, *P.N.G.* **222** P5
Ningi, *Nigeria* **206** J9
Ningqiang, *China* **184** K10
Ninh Bình, *Vietnam* **190** H11
Ninh Hòa, *Vietnam* **191** N13
Ninigo Group, *P.N.G.* **222** M7
Ninigo Islands, *P.N.G.* **222** M6
Ninilchik, *Alas. U.S.* **113** K16
Niniva, *island, Tonga* **227** L23
Ninnescah, *river, Kans. U.S.* **107** U7
Ninni, *island, Marshall Is.* **227** J17
Ninnis Glacier Tongue, *Antarctica*
    238 R17
Ninnis Glacier, *Antarctica* **239** Q17
Nīnole, *Hawaii, U.S.* **115** M22
Nīnole, *Hawaii, U.S.* **115** Q21
Ninotsminda, *Rep. of Georgia* **171** C17
Ninove, *Belg.* **148** L11
Ninty Mile Beach, *Austral.* **220** M13
Nioaque, *Braz.* **132** L10
Niobrara, *river, U.S.* **90** E12
Nioghalvfjerdsfjorden, *Greenland, Den.*
    88 D12
Nioka, *D.R.C.* **209** C19
Nioki, *D.R.C.* **208** H7
Niono, *Mali* **204** J12
Nioro du Rip, *Senegal* **204** J7
Nioro du Sahel, *Mali* **204** H10
Niort, *Fr.* **149** T7
Nioumachoua, *Comoros* **215** N15
Nipawin, *Sask. Can.* **86** M12
Nipe, Bahía de, *Cuba* **119** K13
Nipigon, *Ont. Can.* **87** P16
Nipigon, Lake, *Ont. Can.* **87** N16
Nipton, *Calif. U.S.* **111** V12
Niquelândia, *Braz.* **133** J13
Niquero, *Cuba* **118** K11
Nīr, *Iran* **177** D13
Nirmal, *India* **182** M6
Niš, *Serb.* **154** H12
Nişāb, *Saudi Arabia* **174** D9
Nişāb, *Yemen* **174** P10
Nišava, *river, Serb.* **154** H12
Nishine, *Japan* **188** K13
Nishino Shima, *Japan* **189** N6
Nishinoomote, *Japan* **189** T5
Nishtūn, *Yemen* **175** N14
Nissi, *Est.* **143** K16
Nisswa, *Minn. U.S.* **106** G10
Nísyros, *island, Gr.* **157** M18
Niṭā', *Saudi Arabia* **174** E11
Niterói, *Braz.* **133** M15
Nitra, *Slovakia* **151** K14
Nitro, *W. Va. U.S.* **98** D8

Niu Aunfo Point, *Tonga* **227** Q22
Niuafo'ou, *island, Tonga* **224** J10
Niuatoputapu, *island, Tonga* **224** J10
Niue, *island, Pac. Oc.* **224** K10
Niulakita, *island, Tuvalu* **224** H9
Niutao, *island, Tuvalu* **224** H9
Niutoua, *Tonga* **227** Q24
Nixon, *Tex. U.S.* **105** M16
Nixon's Harbour, *Bahamas* **122** C11
Niya *see* Minfeng, *China* **184** G3
Niyamabad, *India* **182** M6
Nizam Sagar, *lake, India* **182** N6
Nizamabad, *India* **182** M6
Nizamghat, *India* **182** G15
Nizao, *Dom. Rep.* **119** M19
Nizhnekamsk, *Russ.* **158** H16
Nizhneudinsk, *Russ.* **161** L14
Nizhnevartovsk, *Russ.* **160** H11
Nizhneyansk, *Russ.* **161** E17
Nizhniy Baskunchak, *Russ.* **159** N15
Nizhniy Bestyakh, *Russ.* **161** G18
Nizhniy Lomov, *Russ.* **158** L12
Nizhniy Novgorod, *Russ.* **158** J12
Nizhniy Tagil, *Russ.* **160** H9
Nizhnyaya Tunguska, *river, Russ.*
    161 H13
Nizhnyaya Tura, *Russ.* **160** H9
Nizhnyaya Voch', *Russ.* **158** D16
Nizhyn, *Ukr.* **159** N6
Nizi, *D.R.C.* **209** D19
Nizip, *Turk.* **170** J12
Nizip, *river, Turk.* **170** H12
Nizwá, *Oman* **175** H17
Nizza Monferrato, *It.* **152** E3
Nizzana (El 'Auja), *Egypt* **173** P4
Njazidja (Grande Comore), *island,*
    Comoros **215** M14
Njegoš, *peak, Montenegro* **154** H8
Njombe, *Tanzania* **211** Y5
Njombe, *river, Tanzania* **211** W5
Nkambe, *Cameroon* **206** M10
Nkawkaw, *Ghana* **205** P16
Nkéni, *river, Congo* **208** F11
Nkhata Bay, *Malawi* **212** B11
Nkhotakota, *Mozambique* **212** C11
Nkomi Lagoon, *Af.* **198** K5
Nkongsamba, *Cameroon* **206** N9
Nkoul, *Cameroon* **206** P11
Nkululu, *river, Tanzania* **211** W3
Nkundi, *Tanzania* **211** X3
Nmai, *river, Myanmar* **190** D6
Noatak National Preserve, *Alas. U.S.*
    113 D14
Noatak, *Alas. U.S.* **113** D13
Nobeoka, *Japan* **189** S5
Noblesville, *Ind. U.S.* **101** R9
Nobo, *Indonesia* **193** M14
Noboribetsu, *Japan* **188** G13
Nocera Terinese, *It.* **153** Q13
Nocona, *Tex. U.S.* **105** F16
Nodales, Bahía de los, *Arg.* **135** U9
Nodaway, *river, Iowa, Mo. U.S.*
    107 R10
Nofre, Peña, *Sp.* **146** D7
Nogales, *Ariz. U.S.* **109** W8
Nogales, *Mex.* **116** B7
Nogara, *It.* **152** E7
Nōgata, *Japan* **189** R5
Nogent-le-Rotrou, *Fr.* **149** Q8
Nogent-sur-Seine, *Fr.* **149** Q11
Noginsk, *Russ.* **158** J10
Nogoyá, *Arg.* **134** K12
Nogu Dabu, *island, P.N.G.* **228** A9
Nogueira, *peak, Port.* **146** D7
Nohar, *India* **182** F4
Noheji, *Japan* **188** J13
Nohili Point, *Hawaii, U.S.* **114** E9
Nohona o Hae, *peak, Hawaii, U.S.*
    115 M21
Nohta, *India* **182** K7
Nohwa, *S. Korea* **187** V7
Nohwado, *island, S. Korea* **187** U7
Noia, *Sp.* **146** B5
Noire, Montagne, *Fr.* **149** X10
Noires, Montagnes, *Fr.* **149** Q3
Noirmoutier, Île de, *Fr.* **149** S5
Nojima Zaki, *Japan* **189** Q12
Nok Kundi, *Pak.* **181** N3
Noka, *Solomon Is.* **228** L5
Nokaneng, *Botswana* **212** F6
Nokia, *Fin.* **143** H15
Nokomis, *Ill. U.S.* **101** S5
Nokou, *Chad* **206** G12
Nola, *Cen. Af. Rep.* **206** N12
Noli, *It.* **152** F3
Nolinsk, *Russ.* **158** G14

Noma Misaki, *Japan* **189** T4
Nomans Land, *Mass. U.S.* **97** M15
Nome, *Alas. U.S.* **112** F12
Nome, *Russ.* **158** B14
Nomgon, *Mongolia* **184** F10
Nomoneas, *island, F.S.M.* **226** E9
Nomuka Group, *Tonga* **227** N23
Nomuka Iki, *island, Tonga* **227** M23
Nomuka, *island, Tonga* **227** M23
Nomwin Atoll, *island, F.S.M.* **226** A9
Nonancourt, *Fr.* **149** P8
Nonda, *Qnsld. Austral.* **221** S12
Nondalton, *Alas. U.S.* **113** K15
Nong Khai, *Thai.* **190** K9
Nongjrong, *India* **182** H14
Nongong, *S. Korea* **187** R11
Nongsan, *S. Korea* **187** R10
Nonouti, *island, Kiribati* **224** G8
Nonpareil, *Grenada* **123** J23
Nonsan, *S. Korea* **187** Q8
Nonsuch Bay, *Antigua & Barbuda*
    123 B22
Nonsuch Island, *Bermuda* **119** B16
Nonthaburi, *Thai.* **190** M8
Nonume, *Solomon Is.* **228** L5
Nonza, *Fr.* **149** X3
Noojee, *Solomon Is.* **228** L5
Noole, *Solomon Is.* **228** L5
Noonan, *N. Dak. U.S.* **106** D2
Noondoo, *Qnsld. Austral.* **221** V14
Noonan, *N. Dak. U.S.* **106** D2
Noordkaap, *Aruba, Neth.* **123** Q17
Noordoewer, *Namibia* **212** K4
Noordpunt, *Curaçao, Neth.* **123** N13
Noormarkku, *Fin.* **143** H15
Noorvik, *Alas. U.S.* **113** E13
Noosa Head, *Austral.* **220** H16
Nora Hazel Point, *Virgin Is. U.K.*
    122 R7
Nora, *island, Eritrea* **210** H8
Nora, *ruin, It.* **153** Q4
Norak, *Taj.* **178** N12
Norborne, *Mo. U.S.* **107** S11
Nord Est, Grande Récif du,
    Mayotte, Fr. **215** N17
Nord, Baie de, *Mauritius* **215** J20
Nord, Pointe, *Comoros* **215** L14
Nord, research station,
    Greenland, Den. **88** B12
Norden, *Ger.* **150** C6
Norderstedt, *Ger.* **150** C8
Nordeste, *Port.* **146** Q6
Nordfjord, *Atl. Oc.* **140** D6
Nordfjordeid, *Nor.* **142** J10
Nordhausen, *Ger.* **150** F8
Nordhorn, *Ger.* **150** D5
Nordkapp Bank, *Arctic Oc.* **253** F18
Nordkapp, *cape, Nor.* **143** B14
Nordkjosbotn, *Nor.* **143** C13
Nordli, *Nor.* **142** G12
Nordøstbugten, *bay, Greenland, Den.*
    88 K5
Nordostrundingen, *Greenland, Den.*
    88 B12
Norðoyar, *island, Faroe Is. Den.*
    142 H6
Nordre Isortoq, *fjord, Greenland, Den.*
    89 N7
Nordre Strømfjord, *Greenland, Den.*
    89 N6
Nore, *river, Ire.* **145** S6
Noreena Downs, *W. Aust. Austral.*
    221 T4
Norfolk Island, *Norfolk I. Austral.*
    224 L7
Norfolk Ridge, *Pac. Oc.* **246** M10
Norfolk Trough, *Pac. Oc.* **246** N10
Norfolk, *Va. U.S.* **98** F14
Norfolk, *Nebr. U.S.* **107** N7
Norfork Lake, *Ark. U.S.* **102** E9
Noril'sk, *Russ.* **161** F13
Normal, *Ill. U.S.* **101** Q6
Norman Island, *Virgin Is. U.K.* **122** N12
Norman Wells, *N.W.T. Can.* **86** F10
Norman, *Ark. U.S.* **102** H7
Norman, *Okla. U.S.* **105** D17
Norman, Lake, *N.C. U.S.* **98** H9
Norman, *river, Austral.* **220** D12
Normanby Island, *P.N.G.* **222** R9
Normandy, *region, Fr.* **149** P8
Norman's Pond Cay, *Bahamas* **122** D7
Normanton, *Qnsld. Austral.* **221** R12
Nornalup, *W. Aust. Austral.* **221** Y3

Ñorquincó, *Arg.* **135** Q7
Norris Lake, *Tenn. U.S.* **103** E18
Norristown, *Pa. U.S.* **96** P9
Norrköping, *Sw.* **143** L13
Norrland, *Eur.* **140** C8
Norrsundet, *Sw.* **143** J13
Norseman, *W. Aust. Austral.* **221** X5
Norsk, *Russ.* **161** K19
Norske Bank, *Arctic Oc.* **253** H16
Norske Øer, *Greenland, Den.* **88** D13
Norsup, *Vanuatu* **229** G14
Nort, *Fr.* **149** R6
Norte, Cabo, *Braz.* **133** C13
Norte, Canal do, *Braz.* **133** C13
Norte, Punta, *Arg.* **134** M13
Norte, Serra do, *Braz.* **132** G9
North Adams, *Mass. U.S.* **97** K12
North Albanian Alps, *Alban.* **154** J9
North Aleutian Slope, *Pac. Oc.*
    246 C12
North American Basin, *Atl. Oc.*
    248 K5
North Andaman, *island, India*
    183 Q14
North Aral Sea, *Kaz.* **178** H8
North Arm, *N.W.T. Can.* **86** H11
North Atlantic Ocean, *Atl. Oc.*
    248 J8
North Augusta, *S.C. U.S.* **98** L8
North Aulatsivik Island,
    Nfld. & Lab. Can. **87** J20
North Australian Basin, *Ind. Oc.*
    251 J18
North Banda Basin, *Pac. Oc.* **246** K5
North Battleford, *Sask. Can.* **86** M11
North Bay, *Ont. Can.* **87** P18
North Belcher Islands, *Nunavut, Can.*
    87 L17
North Bend, *Oreg. U.S.* **110** J2
North Bimini, *Bahamas* **122** B11
North Branch Potomac, *river,*
    Md. W. Va. U.S. **98** C10
North Branch, *Minn. U.S.* **106** J12
North Brazilian Ridge, *Atl. Oc.*
    249 P8
North Caicos, *island,*
    Turks & Caicos Is. U.K. **119** H17
North Canadian , *river, U.S.* **105** C15
North Canadian, *river, U.S.* **90** J12
North Cape, *Eur.* **140** A3
North Cape, *Eur.* **140** A9
North Cape, *P.N.G.* **222** M9
North Cape, *N.Z.* **223** C19
North Cape, *Kiribati* **226** J11
North Carolina, *U.S.* **98** H10
North Cascades National Park,
    Wash. U.S. **110** A5
North Channel, *Can. U.S.* **100** F12
North Channel, *Ire. U.K.* **145** N8
North Charleston, *S.C. U.S.* **98** M10
North Chatham Slope, *Pac. Oc.*
    255 H21
North Chatham Terrace, *Pac. Oc.*
    255 H21
North Chicago, *Ill. U.S.* **100** M7
North China Plain, *Asia* **168** G11
North Comino Channel, *Malta*
    163 H21
North Dakota, *U.S.* **106** F4
North Downs, *Eng. U.K.* **145** U14
North East Land, *Nor.* **168** A7
North East Point, *Cayman Is. U.K.*
    118 L9
North East Point, *Kiribati* **226** P11
North East, *Pac. U.S.* **96** K3
North Entrance, *Palau* **226** D6
North Fiji Basin, *Pac. Oc.* **246** L10
North Fond du Lac, *Wis. U.S.*
    100 K6
North Fork Clearwater, *river,*
    Idaho, U.S. **108** C4
North Fork Flathead, *river, Can. U.S.*
    108 A5
North Fork Payette, *river, Idaho, U.S.*
    108 G4
North Fork Red, *river, Okla. U.S.*
    105 D15
North Fork Red, *river, Tex. U.S.*
    105 C13
North Fork Salt, *river, Mo. U.S.*
    107 R13
North Fork, *Calif. U.S.* **111** T6
North Friar's Bay, *St. Kitts & Nevis*
    123 B18
North Frisian Islands, *Den. Ger.*
    150 A6

North Greenland Eemian Ice Drilling (NEEM Camp), *Greenland, Den.* **88** F7
North Greenland Ice Core Project (NGRIP), *Greenland, Den.* **88** H9
North Head, *N.Z.* **223** E20
North Horr, *Kenya* **211** Q6
North Island (Te Ika-a-Māui), *N.Z.* **223** G20
North Island, *Seychelles* **215** N19
North Kanin Bank, *Arctic Oc.* **253** D17
North Kenya Bank, *Ind. Oc.* **250** G5
North Korea, *Asia* **186** G7
North Kosovo, region, *Kos.* **154** H10
North Lakhimpur, *India* **182** G14
North Land *see* Severnaya Zemlya, *Russ.* **161** C14
North Las Vegas, *Nev., U.S.* **111** U12
North Little Rock, *Ark., U.S.* **102** G8
North Loup, river, *Nebr., U.S.* **107** N5
North Macedonia, *Eur.* **154** K10
North Magellan Rise, *Pac. Oc.* **246** H12
North Malosmadulu Atoll, *Maldives* **183** V3
North Manchester, *Ind., U.S.* **101** P9
North Mole, *Gibraltar, U.K.* **163** N22
North Myrtle Beach, *S.C., U.S.* **98** L12
North Naples, *Fla., U.S.* **99** W8
North Negril Point, *Jam.* **122** H5
North Norfolk Basin, *Pac. Oc.* **246** M10
North Olmsted, *Ohio, U.S.* **101** P14
North Pacific Ocean, *Pac. Oc.* **224** B11
North Palmetto Point, *Bahamas* **122** E6
North Pass, *Marshall Is.* **227** F16
North Perry, *Ohio, U.S.* **101** N15
North Pioa, peak, *American Samoa, U.S.* **230** D7
North Platte, *Nebr., U.S.* **107** Q4
North Platte, river, *U.S.* **90** F9
North Point, *Mich., U.S.* **100** H12
North Point, *Barbados* **123** C18
North Point, *S. Amer.* **128** M7
North Point, *Seychelles* **215** N20
North Point, *N.Z.* **229** G23
North Powder, *Oreg., U.S.* **110** F8
North Raccoon, river, *Iowa, U.S.* **107** P11
North Rock, *Bahamas* **122** A11
North Roe, *Scot., U.K.* **144** C12
North Ronaldsay, island, *Scot., U.K.* **144** E11
North Saskatchewan, river, *Alb., Sask., Can.* **86** L12
North Scotia Ridge, *Atl. Oc.* **249** W6
North Sentinel Island, *India* **183** R14
North Siberian Lowland, *Asia* **168** C9
North Side, *Cayman Is., U.K.* **122** M4
North Slope, *Alas., U.S.* **113** C16
North Solomon Trough, *Pac. Oc.* **246** K9
North Sound Estates, *Cayman Is., U.K.* **122** N3
North Sound, *Cayman Is., U.K.* **122** M2
North Sound, *Virgin Is., U.K.* **122** Q9
North Sound, *Ire.* **145** R3
North Sound, *The, Scot., U.K.* **144** E10
North Springfield, *Pa., U.S.* **96** L3
North Stradbroke Island, *Austral.* **220** H16
North Taranaki Bight, *N.Z.* **223** G20
North Tasman Seamounts, *Pac. Oc.* **255** E21
North Tonawanda, *N.Y., U.S.* **96** J5
North Truchas Peak, *N. Mex., U.S.* **109** R13
North Uist *see* Uibhist a Tuath, island, *Scot., U.K.* **144** H6
North Vanlaiphai, *India* **182** K14
North Vernon, *Ind., U.S.* **101** S10
North Walsham, *Eng., U.K.* **145** S15
North Ward, *Cayman Is., U.K.* **122** N3
North Wazīristān, region, *Pak.* **180** J9
North Weddell Ridge *see* America-Antarctica Ridge, *Atl. Oc.* **249** W10
North West Basin, *Austral.* **220** G2
North West Bay, *Seychelles* **215** N19
North West Cape, *Austral.* **220** F1
North West Point, *Cayman Is., U.K.* **118** L7
North West Point, *Cayman Is., U.K.* **122** M1

North West Point, *Kiribati* **226** P10
North West River, *Nfld. & Lab., Can.* **87** L21
North Wilkesboro, *N.C., U.S.* **98** G9
North York Moors, *Eng., U.K.* **145** P13
North, Cape, *N.S., Can.* **87** N22
Northampton, *Mass., U.S.* **97** K13
Northampton, *Eng., U.K.* **145** T13
Northampton, *W. Aust., Austral.* **221** V2
Northcliffe Glacier, *Antarctica* **238** K21
Northcliffe, *W. Aust., Austral.* **221** Y3
Northeast Cay, *Jam.* **118** P11
Northeast Georgia Rise, *Atl. Oc.* **249** V8
Northeast Greenland Ice Stream, *Greenland, Den.* **88** F11
Northeast Newfoundland Shelf, *Atl. Oc.* **248** G6
Northeast Pacific Basin, *Pac. Oc.* **247** E13
Northeast Pass, *Marshall Is.* **227** H20
Northeast Passage, *Solomon Is.* **228** N7
Northeast Point, *Bahamas* **119** G15
Northeast Point, *Jam.* **122** J11
Northeast Providence Channel, *Bahamas* **118** D12
Northeim, *Ger.* **150** E7
Northern Cyprus, *Cyprus* **162** N9
Northern Dvina, river, *Eur.* **140** C12
Northern European Plain, *Eur.* **140** G7
Northern Ireland, *N. Ire., U.K.* **145** N6
Northern Light Lake, *Can., U.S.* **106** E14
Northern Range, *Trin. & Tobago* **123** N23
Northern Sporades *see* Vóries Sporádhes, islands, *Gr.* **157** F13
Northern Territory, *Austral.* **221** R8
Northern Urals, *Eur.* **140** D13
Northfield, *Vt., U.S.* **97** G13
Northfield, *Minn., U.S.* **106** K11
Northport, *Gabon* **208** C8
Northport, *Ala., U.S.* **103** K14
Northport, *Wash., U.S.* **110** A8
Norths Highland, *Antarctica* **239** P20
Northumberland Islands, *Austral.* **220** E15
Northumberland Ø *see* Kiatak, *Greenland, Den.* **88** E3
Northumberland, Cape, *Austral.* **220** M11
Northway Junction, *Alas., U.S.* **113** H19
Northwest Angle, *Minn., U.S.* **106** D10
Northwest Atlantic Mid-Ocean Canyon, *Atl. Oc.* **248** F6
Northwest Georgia Rise, *Atl. Oc.* **249** V8
Northwest Pacific Basin, *Pac. Oc.* **246** E8
Northwest Point, *Bahamas* **122** E1
Northwest Providence Channel, *Bahamas* **118** C10
Northwest Territories, *Can.* **86** G10
Northwich, *Eng., U.K.* **145** R11
Northwind Escarpment, *Arctic Oc.* **252** L8
Northwind Plain, *Arctic Oc.* **252** K8
Northwind Ridge, *Arctic Oc.* **252** L8
Northwood, *Iowa, U.S.* **106** L12
Northwood, *N. Dak., U.S.* **106** F7
Norton Bay, *Alas., U.S.* **113** F13
Norton Plain, *Pac. Oc.* **252** L3
Norton Shores, *Mich., U.S.* **100** L9
Norton Sound, *Alas., U.S.* **112** G12
Norton, *Va., U.S.* **98** F7
Norton, *Kans., U.S.* **107** R5
Norvegia, Cape, *Antarctica* **238** B11
Norwalk, *Conn., U.S.* **96** M12
Norwalk, *Ohio, U.S.* **101** P13
Norwalk, *Calif., U.S.* **111** X8
Norway House, *Man., Can.* **87** M14
Norway, *Me., U.S.* **97** G15
Norway, *Mich., U.S.* **100** G7
Norway, *Eur.* **142** G11
Norwegian Basin, *Arctic Oc.* **248** E12
Norwegian Bay, *Nunavut, Can.* **87** C15
Norwegian Sea, *Arctic Oc.* **140** B6
Norwegian Trough, *Atl. Oc.* **248** F13
Norwich, *N.Y., U.S.* **96** K9
Norwich, *Conn., U.S.* **97** M13
Norwich, *Eng., U.K.* **145** S15
Norwood, *N.Y., U.S.* **96** F10
Norwood, *N.C., U.S.* **98** J10

Norwood, *Ohio, U.S.* **101** S11
Norwood, *Colo., U.S.* **109** N10
Noshiro, *Japan* **188** K12
Noshul', *Russ.* **158** E14
Nosivka, *Ukr.* **159** N6
Noṣratābād, *Iran* **177** K21
Noss, Isle of, *Scot., U.K.* **144** C12
Nossob , river, *Botswana, S. Af.* **212** J5
Nossombougou, *Mali* **204** K11
Nosy Barren, island, *Madagascar* **213** F15
Nosy Be, island, *Madagascar* **213** C17
Nosy Mitsio, island, *Madagascar* **213** C17
Nosy Sainte Marie, island, *Madagascar* **213** E17
Nosy-Varika, *Madagascar* **213** G16
Not'an, *N. Korea* **186** D8
Note Kanengo, cape, *Solomon Is.* **228** L5
Note Kembanye, cape, *Solomon Is.* **228** L5
Note Kempopla, cape, *Solomon Is.* **228** L5
Noteć, river, *Pol.* **151** C13
Nótios Evoïkós Kólpos, *Gr.* **156** H12
Noto Hantō, *Japan* **188** M9
Noto Peninsula, *Asia* **168** F13
Noto, *It.* **153** T12
Noto, *Japan* **188** M10
Noto, Golfo di, *It.* **153** U12
Notre Dame Bay, *Nfld. & Lab., Can.* **87** M23
Notsé, *Togo* **205** P17
Nottaway, river, *Que. Can.* **87** N18
Nottingham Island, *Nunavut, Can.* **87** H17
Nottingham, *Eng., U.K.* **145** R13
Nottoway, river, *Va., U.S.* **98** F12
Nouadhibou (Port-Étienne), *Mauritania* **204** E6
Nouakchott, *Mauritania* **204** G7
Nouâmghâr, *Mauritania* **204** F7
Nouans-les-Fontaines, *Fr.* **149** S9
Nouaoudar, *Mauritania* **204** H12
Nouart, *Fr.* **149** N12
Noukranskop, peak, *S. Af.* **212** L5
Nouméa, *New Caledonia, Fr.* **231** Q22
Nouna, *Burkina Faso* **205** K14
Nourounba, *Mali* **204** K11
Nouveau Québec, *Que. Can.* **87** L18
Nouvelle-Calédonie (New Caledonia), *New Caledonia, Fr.* **231** P20
Nova Friburgo, *Braz.* **133** M15
Nova Gorica, *Slov.* **154** D4
Nova Gradiška, *Croatia* **154** E7
Nova Iguaçu, *Braz.* **133** M14
Nova Kakhovka, *Ukr.* **159** R7
Nova Mambone, *Mozambique* **212** F11
Nova Odesa, *Ukr.* **159** R6
Nova Olinda do Norte, *Braz.* **132** E9
Nova Scotia, *Can.* **87** P22
Nova Sofala, *Mozambique* **212** F11
Nova Xavantina, *Braz.* **132** J12
Nova Zagora, *Bulg.* **155** J16
Novafeltria, *It.* **152** G8
Novalukoml', *Belarus* **158** K5
Novara Fracture Zone, *Ind. Oc.* **250** N8
Novara, *It.* **152** D4
Novaya Sibir', *Russ.* **161** C17
Novaya Vodolaga, *Ukr.* **159** P8
Novaya Zemlya, *Russ.* **160** D11
Nové Zámky, *Slovakia* **151** K14
Novelda, *Sp.* **147** J14
Noville Peninsula, *Antarctica* **238** K6
Novo Aripuanã, *Braz.* **132** E9
Novo Cruzeiro, *Braz.* **133** K15
Novo Hamburgo, *Braz.* **132** P12
Novo Mesto, *Slov.* **154** D5
Novoaleksandrovsk, *Russ.* **159** S11
Novoanninskiy, *Russ.* **159** N12
Novobod, *Taj.* **178** M12
Novocherkassk, *Russ.* **159** R11
Novodvinsk, *Russ.* **158** B10
Novohrad-Volyns'kyy, *Ukr.* **159** N4
Novoīshīmskīy, *Kaz.* **178** D11
Novokuybyshevsk, *Russ.* **158** L15
Novokuznetsk, *Russ.* **160** L12

Novolazarevskaya Station, *Antarctica* **239** A14
Novomichurinsk, *Russ.* **158** K10
Novomoskovsk, *Russ.* **158** K9
Novomoskovs'k, *Ukr.* **159** Q8
Novopokrovskaya, *Russ.* **159** S11
Novorepnoye, *Russ.* **158** M15
Novorossiysk, *Russ.* **159** T9
Novorybnoye, *Russ.* **161** F15
Novorzhev, *Russ.* **158** H6
Novoshakhtinsk, *Russ.* **159** Q11
Novosibirsk, *Russ.* **160** K12
Novosibirskiye Ostrova (New Siberian Islands), *Russ.* **161** D16
Novotitarovskaya, *Russ.* **159** S10
Novotroitsk, *Russ.* **160** J7
Novoukrayinka, *Ukr.* **159** Q6
Novouzensk, *Russ.* **159** N15
Novovolyns'k, *Ukr.* **159** N2
Novoye Ust'ye, *Russ.* **161** G20
Novozhilovskaya, *Russ.* **158** A14
Novozybkov, *Russ.* **158** M6
Novyy Buh, *Ukr.* **159** Q6
Novyy Port, *Russ.* **160** F11
Novyy Urengoy, *Russ.* **160** G12
Now Deh, *Afghan.* **180** K3
Now Zād, *Afghan.* **180** K5
Nowa Sól, *Pol.* **150** E12
Nowābād Īmj, *Afghan.* **180** E10
Nowata, *Okla., U.S.* **105** B18
Nowood, river, *Wyo., U.S.* **108** G11
Nowrangapur, *India* **182** M8
Nowshāk, peak, *Afghan., Pak.* **180** F11
Nowshera, *Pak.* **180** J11
Nowy Sącz, *Pol.* **151** H16
Nowy Targ, *Pol.* **151** H15
Noxon, *Mont., U.S.* **108** B4
Noyabr'sk, *Russ.* **160** H11
Noyon, *Fr.* **149** N10
Nozay, *Fr.* **149** R5
Nsanje, *Mozambique* **212** E11
Nsawam, *Ghana* **205** P16
Nsukka, *Nigeria* **206** M8
Ntem, *Cameroon* **208** C8
Ntingui, peak, *Comoros* **215** N16
Ntoum, *Gabon* **208** D7
Ntsaouéni, *Comoros* **215** M14
Ntwetwe Pan, *Botswana* **212** F7
Nu (Salween), river, *Asia* **168** H9
Nuapapu, *Tonga* **227** H23
Nuapapu, island, *Tonga* **227** H22
Nuatabu, *Kiribati* **227** Q14
Nubian Desert, *Sudan* **202** M12
Nubledo, *Sp.* **146** A8
Nuclear test site, *Pak.* **181** N5
Nuclear test site, *Pak.* **181** P4
Nuclear test site, *N. Korea* **186** D11
Nueces, river, *Tex., U.S.* **105** N15
Nueltin Lake, *Nunavut, Can.* **87** J14
Nuestra Señora, Bahía, *Chile* **134** F6
Nueva Gerona, *Cuba* **118** H6
Nueva Rosita, *Mex.* **116** D11
Nuevitas, *Cuba* **118** J11
Nuevo Casas Grandes, *Mex.* **116** C8
Nuevo Laredo, *Mex.* **116** F12
Nuevo Rocafuerte, *Ecua.* **130** K6
Nuevo, Bajo, *Mex.* **117** G15
Nuevo, Golfo, *Arg.* **135** R10
Nuguria Islands, *P.N.G.* **222** N11
Nui, island, *Tuvalu* **224** H9
Nukapu, island, *Solomon Is.* **228** K5
Nuku Hiva, island, *Fr. Polynesia, Fr.* **225** H15
Nuku, island, *Tonga* **227** Q24
Nuku'alofa, *Tonga* **227** Q23
Nukubasaga, island, *Fiji* **226** N8
Nukudamu, *Fiji* **226** N7
Nukufetau, island, *Tuvalu* **224** H9
Nukuira, island, *Fiji* **226** N6
Nukulaelae, island, *Tuvalu* **224** H9
Nukumanu Islands, *P.N.G.* **228** N4
Nukunonu, *Kiribati* **224** H10
Nukuoro Atoll, *F.S.M.* **226** C10
Nukus, *Uzb.* **178** K8
Nukutavake, island, *Fr. Polynesia, Fr.* **231** D21
Nukutipipi, island, *Fr. Polynesia, Fr.* **231** D19
Nulato, *Alas., U.S.* **113** F14
Nules, *Sp.* **147** G15
Nullagine, *W. Aust., Austral.* **221** T3
Nullarbor Plain, *Austral.* **221** W8
Nullarbor, *S. Aust., Austral.* **221** W8
Num, island, *Indonesia* **193** K19
Numakawa, *Japan* **188** E13
Numan, *Nigeria* **206** K10

Numata, *Japan* **188** F13
Numata, *Japan* **189** N11
Numazu, *Japan* **189** P11
Numbulwar, *N.T., Austral.* **221** P9
Numfoor, island, *Indonesia* **193** J19
Numila, *Hawaii, U.S.* **114** F10
Numto, *Russ.* **160** G11
Nun Jack Cay, *Bahamas* **122** A9
Nunap Isua (Kap Farvel), *Greenland, Den.* **89** U8
Nunap Isua, *Greenland, Den.* **88** M12
Nunarsuit, island, *Greenland, Den.* **89** T6
Nunatakavsaup Sermia, glacier, *Greenland, Den.* **88** H5
Nunavut, *Can.* **87** G14
Nuneaton, *Eng., U.K.* **145** S12
Nŭnggu, *N. Korea* **186** E9
Nungo, *Mozambique* **212** C12
Nunivak Island, *Alas., U.S.* **112** J10
Nunligran, *Russ.* **161** A21
Nuoro, *It.* **153** N4
Nupani, island, *Solomon Is.* **228** K5
Nuquí, *Col.* **130** F4
Nur Gama, *Pak.* **181** P7
Nur-Sultan (Astana), *Kaz.* **178** E12
Nūr, *Iran* **177** K15
Nura, river, *Kaz.* **178** E12
Nūrābād, *Iran* **177** K15
Nūrābād, *Iraq* **177** G13
Nuremberg *see* Nürnberg, *Ger.* **150** H8
Nurlat, *Russ.* **158** J15
Nurmes, *Fin.* **143** F16
Nürnberg (Nuremberg), *Ger.* **150** H8
Nurpur, *Pak.* **180** K11
Nurri, *It.* **153** P4
Nurse Cay, *Bahamas* **119** H13
Nusaybin, *Turk.* **171** H16
Nushagak Peninsula, *N. Amer.* **84** R4
Nushki, *Pak.* **181** N6
Nuttal, *Pak.* **181** N8
Nutwood Downs, *N.T., Austral.* **221** Q9
Nuuk (Godthåb), *Greenland, Den.* **89** Q5
Nu'ulua, island, *Samoa* **227** P20
Nuup Kangerlua (Godthåbsfjord), *Greenland, Den.* **89** Q6
Nuusilaelae, island, *American Samoa, U.S.* **230** B8
Nuussuaq (Kraulshavn), *Greenland, Den.* **88** H5
Nuussuaq, peninsula, *Greenland, Den.* **88** L6
Nu'utele, island, *Samoa* **227** P20
Nuutele, island, *American Samoa, U.S.* **230** B8
Nu'uuli, *American Samoa, U.S.* **230** E7
Nuweiba', *Egypt* **173** S4
Nuweveld Range, *S. Af.* **212** M6
Nwayfadh, *W. Sahara* **200** H3
Nyagan', *Russ.* **160** G10
Nyahanga, *Tanzania* **211** T4
Nyahururu, *Kenya* **211** S6
Nyainqêntanglha Feng, *China* **184** K5
Nyainqêntanglha Shan (Nianqingtangla Shan), *China* **184** L5
Nyakabindi, *Tanzania* **211** U4
Nyakanazi, *Tanzania* **211** U3
Nyakitonto, *Tanzania* **211** V2
Nyala, *Sudan* **203** T7
Nyamlell, *S. Sudan* **203** V8
Nyamtumbo, *Tanzania* **211** Y6
Nyanding, river, *S. Sudan* **203** V12
Nyandoma, *Russ.* **158** D10
Nyandoma, *Russ.* **160** E7
Nyanga, Lake, *Austral.* **220** J6
Nyanga, river, *Gabon* **208** F8
Nyangwe, *D.R.C.* **209** G17
Nyanje, *Zambia* **209** N19
Nyanza Lac, *Burundi* **211** V2
Nyaunglebin, *Myanmar* **190** K5
Nybergsund, *Nor.* **142** J12
Nyborg, *Den.* **142** N11
Nyborg, *Nor.* **143** B15
Nyda, *Russ.* **160** G11
Nye Mountains, *Antarctica* **239** C19
Nyeboe Land, *Greenland, Den.* **88** C7
Nyeri, *Kenya* **211** S6
Nyerol, *S. Sudan* **203** V12
Nyeying, *India* **182** F14
Nyika Plateau, *Af.* **212** B10
Nyima (Nima), *China* **184** K4
Nyimba, *Zambia* **209** N19
Nyingchi (Nichi), *China* **184** L7

Nyíregyháza, Hung. 151 K17
Nykarleby, Fin. 143 G15
Nykøbing, Den. 142 N12
Nyköping, Sw. 143 K13
Nymagee, N.S.W., Austral. 221 X13
Nymphaeum see Pínnes, Ákra, Gr. 157 D13
Nyngan, N.S.W., Austral. 221 W13
Nyoma Rap, India 180 J16
Nyoman, river, Belarus 158 K3
Nyon, Switz. 152 C1
Nyons, Fr. 149 W13
Nysa, Pol. 151 G13
Nysh, Russ. 161 J21
Nyssa, Oreg., U.S. 110 H9
Nyūdō Zaki, Japan 188 K12
Nyukhcha, Russ. 158 C12
Nyuksenitsa, Russ. 158 E12
Nyunzu, D.R.C. 209 H18
Nyurba, Russ. 161 H16
Nyuvchim, Russ. 158 D14
Nzambi, Congo 208 G8
Nzara, S. Sudan 203 X9
Nzega, Tanzania 211 V4
N'Zérékoré, Guinea 204 N11
N'zeto, Angola 208 J9
Nzoro, D.R.C. 209 C19
Nzoro, river, D.R.C. 209 C19
Nzwani (Anjouan), island, Comoros 215 N16

O

O Barco, Sp. 146 C7
O Grove, Sp. 146 C5
O Shima, Japan 189 S3
Ō Shima, Japan 188 H11
Ō Shima, Japan 189 Q12
Ō Shima, Japan 189 R9
O.T. Downs, N.T., Austral. 221 Q9
Oacoma, S. Dak., U.S. 106 L5
Oahe, Lake, N. Dak., S. Dak., U.S. 106 H4
O'ahu, Hawaii, U.S. 115 G13
Oak Bluffs, Mass., U.S. 97 M15
Oak Grove, La., U.S. 102 K9
Oak Harbor, Wash., U.S. 110 B3
Oak Hill, W. Va., U.S. 98 E8
Oak Hill, Ohio, U.S. 101 S13
Oak Lawn, Ill., U.S. 101 N7
Oak Park, Ill., U.S. 101 N7
Oak Ridge, Tenn., U.S. 103 E18
Oakdale, La., U.S. 102 N8
Oakdale, Calif., U.S. 111 S5
Oakes, N. Dak., U.S. 106 H7
Oakesdale, Wash., U.S. 110 C9
Oakhurst, Calif., U.S. 111 S6
Oakland City, Ind., U.S. 101 T7
Oakland, Md., U.S. 98 C10
Oakland, Iowa, U.S. 107 P9
Oakland, Nebr., U.S. 107 P8
Oakland, Oreg., U.S. 110 J3
Oakland, Calif., U.S. 111 S3
Oakley, Kans., U.S. 107 S4
Oakley, Idaho, U.S. 108 J5
Oakridge, Oreg., U.S. 110 H3
Oamaru, N.Z. 223 N18
Oaro, N.Z. 223 K19
Oatara, island, Fr. Polynesia, Fr. 231 H18
Oates Coast, Antarctica 239 R16
Oaxaca, Mex. 117 K13
Ob' Bank, Arctic Oc. 253 J16
Ob' Bay, Antarctica 239 R15
Ob' Tablemount, Ind. Oc. 254 D8
Ob' Trench, Ind. Oc. 251 M15
Ob, Gulf of, Arctic Oc. 168 C8
Ob', river, Russ. 160 G10
Oba, Ont., Can. 87 P17
Obala, Cameroon 206 N10
Obama, Japan 189 P9
Obanazawa, Japan 188 L12
Ōbêh, Afghan. 180 H4
Obelisk Island, Solomon Is. 228 K7
Obelisk, site, Vatican City 163 Q17
Obella, island, Marshall Is. 227 F16
Oberdrauburg, Aust. 150 M10
Oberhausen, Ger. 150 E5
Oberlin, Ohio, U.S. 101 P14
Oberlin, La., U.S. 102 P7
Oberlin, Kans., U.S. 107 R4
Oberursel, Ger. 150 G6
Obervellach, Aust. 150 M10
Oberwart, Aust. 150 L12
Obi, Nigeria 206 L9
Obi, island, Indonesia 193 K16
Obi, Kepulauan, Indonesia 193 K16
Óbidos, Braz. 132 D11
Óbidos, Port. 146 H4
Obihiro, Japan 188 G14
Obilatu, island, Indonesia 193 J16
Obil'noye, Russ. 159 Q13
Obion, Tenn., U.S. 102 E12
Oblachnaya, Russ. 161 L21
Obluch'ye, Russ. 161 L20
Obninsk, Russ. 158 K9
Obo, Cen. Af. Rep. 207 M18
Obock, Djibouti 210 K10
Obome, Indonesia 193 K18
Obong-san, peak, N. Korea 186 B13
Oborniki, Pol. 151 D13
Obouya, Congo 208 E11
Obrenovac, Serb. 154 F10
O'Brien, Oreg., U.S. 110 L2
Obrovac, Croatia 154 F5
Obruk, Turk. 170 G8
Obshchiy Syrt, Eur. 140 E15
Obskaya Guba, Russ. 160 F12
Obstrucción, Fiordo, Chile 135 W7
Obuasi, Ghana 205 P15
Ob'yachevo, Russ. 158 E14
Oca, Montes de, Sp. 146 C11
Ocala, Fla., U.S. 99 S8
Ocaña, Col. 130 E7
Ocaña, Sp. 146 G11
Ocean Cay, Bahamas 118 D10
Ocean City, N.J., U.S. 96 R10
Ocean City, Md., U.S. 98 D15
Ocean Falls, B.C., Can. 86 K7
Ocean Island see Banaba, Kiribati 224 G7
Ocean Park, Wash., U.S. 110 D2
Ocean Springs, Miss., U.S. 103 P13
Oceanographer Fracture Zone, Atl. Oc. 248 K7
Oceanographic Museum, Monaco 163 E21
Oceanside, Calif., U.S. 111 Y9
Oceanview, Guam, U.S. 230 N6
Óch, Óros, Gr. 157 J13
Ocha, island, F.S.M. 226 F9
Och'amch'ire, Rep. of Georgia 171 B16
Ochlockonee, river, Fla., Ga., U.S. 99 Q5
Ocho Rios, Jam. 122 H9
Ocilla, Ga., U.S. 99 P6
Ocmulgee National Monument, Ga., U.S. 98 M6
Ocmulgee, river, Ga., U.S. 99 N6
Ocna Mureş, Rom. 155 D13
Ocoa, Bahía de, Dom. Rep. 119 M18
Oconee, Lake, Ga., U.S. 98 L6
Oconee, river, Ga., U.S. 99 N7
Oconomowoc, Wis., U.S. 100 L6
Oconto Falls, Wis., U.S. 100 H7
Oconto, Wis., U.S. 100 H7
Ocracoke Inlet, N.C., U.S. 98 J15
Ocracoke, N.C., U.S. 98 H15
October Revolution Island, Asia 168 B9
Ocumare del Tuy, Venez. 130 C10
Oda, Ghana 205 P15
Oda, Eth. 211 N9
Ōda, Japan 189 P6
Ōdaejin, N. Korea 186 D13
Ōdate, Japan 188 K12
Odawara, Japan 189 P11
Odda, Nor. 142 K10
Odebolt, Iowa, U.S. 107 N10
Odell Glacier Camp, Antarctica 239 N15
Odem, Tex., U.S. 105 P17
Odemira, Port. 146 K5
Ödemiş, Turk. 170 G3
Oden Spur, Arctic Oc. 252 H11
Odense, Den. 142 N11
Oder, river, Ger. 150 D11
Odesa, Ukr. 159 R6
Odessa, Tex., U.S. 104 H11
Odessa, Mo., U.S. 107 S11
Odessa, Wash., U.S. 110 C7
Odienné, Côte d'Ivoire 204 M12
Odin Land, Greenland, Den. 89 R9
O'Donnell, Tex., U.S. 104 G12
Odra, river, Pol. 150 D11
Odžaci, Serb. 154 E9
Oecusse see Pante Makasar, Timor-Leste 193 N15
Oedong, S. Korea 187 R12
Oeiras, Braz. 133 F15
Oelrichs, S. Dak., U.S. 106 M2
Oelwein, Iowa, U.S. 106 M13
Oenarodo, island, S. Korea 187 U9
Oeno Island, Pac. Oc. 225 K16
Oenpelli, N.T., Austral. 221 P8
Oeta, Mount see Oíti Óros, Gr. 156 G10
Oeyeondo, island, S. Korea 187 Q6
Of, Turk. 171 D15
O'Fallon Creek, Mont., U.S. 108 D13
Offa, Nigeria 206 L6
Offenbach, Ger. 150 G6
Offenburg, Ger. 150 J6
Office of Tourism, Monaco 163 C21
Offida, It. 152 J10
Oficina Rica Aventura, Chile 134 D7
Ofidoúsa, island, Gr. 157 M16
Ofolanga, island, Tonga 227 L23
Ofu, American Samoa, U.S. 230 B8
Ofu, island, Tonga 227 H23
Ofu, island, American Samoa, U.S. 230 B8
Ōfunato, Japan 188 L13
Oga, Japan 188 K12
Ogadēn, Eth. 211 N11
Ōgaki, Japan 189 P9
Ogallala, Nebr., U.S. 107 Q3
Ogasawara Guntō see Bonin Islands, Japan 189 X14
Ogbomosho, Nigeria 206 L6
Ogden, Iowa, U.S. 107 N11
Ogden, Utah, U.S. 108 K7
Ogdensburg, N.Y., U.S. 96 F9
Ogea Driki, island, Fiji 226 R8
Ogea Levu, island, Fiji 226 R8
Ogeechee, river, Ga., U.S. 99 N8
Ogilvie Mountains, Yukon, Can. 86 E8
Oglesby, Ill., U.S. 101 P5
Oglethorpe, Ga., U.S. 99 N5
Ognev Yar, Russ. 160 J11
Ogoja, Nigeria 206 M8
Ogoki, Ont., Can. 87 N16
Ogoki, river, Ont., Can. 87 N16
Ogooué, river, Gabon 208 E7
Ogr, Sudan 203 T9
Ogulin, Croatia 154 E5
Oguma, Nigeria 206 L8
Ogwashi Uku, Nigeria 206 M7
O'Higgins, Lago, Arg., Chile 135 U7
Ohio Range, Antarctica 238 J11
Ohio, U.S. 101 Q13
Ohio, river, N. Amer. 84 K7
Oho, S. Korea 186 L11
Ohonua, Tonga 227 R24
Ohrid, N. Maced. 154 L10
Ohrid, Lake see Ohridsko Jezero, Alban., N. Maced. 154 L10
Ohridsko Jezero (Ohrid, Lake), Alban., N. Maced. 154 L10
Oía, Gr. 157 M15
Oiapoque see Oyapok, river, S. Amer. 131 G19
Oiapoque, Braz. 132 B12
Oil City, Pa., U.S. 96 M4
Oildale, Calif., U.S. 111 V7
Oilton, Okla., U.S. 105 B18
Oinoússes, island, Gr. 157 H16
Oise, river, Fr. 149 N10
Oistins, Barbados 123 L19
Ōita, Japan 189 R5
Oíti, Óros (Oeta, Mount), Gr. 156 G10
Ojika Jima, Japan 189 R3
Ojika, Japan 188 L13
Ōjin Rise, Pac. Oc. 246 E10
Ojinaga, Mex. 116 D10
Ojiya, Japan 188 M11
Ojos del Salado, Cerro, Arg. 134 G8
Oka-Don Plain, Eur. 140 F12
Oka, river, Russ. 158 K9
Okaba, Indonesia 193 M21
Okahandja, Namibia 212 G4
Okak Islands, Nfld. & Lab., Can. 87 J21
Okano, river, Gabon 208 E8
Okanogan Range, Can., U.S. 110 A6
Okanogan, Wash., U.S. 110 B7
Okanogan, river, Can., U.S. 90 A5
Okara, Pak. 180 L12
Okau, F.S.M. 226 D11
Okaukuejo, Namibia 212 F3
Okavango Delta, Botswana 212 F6
Okavango, river, Af. 198 M7
Okaya, Japan 189 P10
Okayama, Japan 189 Q7
Okazaki, Japan 189 Q10
Okcheon, S. Korea 187 Q9
Okeechobee, Fla., U.S. 99 V9
Okeechobee, Lake, Fla., U.S. 99 V9
Okeene, Okla., U.S. 105 B16
Okefenokee Swamp, Fla., Ga., U.S. 99 Q7
Okemah, Okla., U.S. 105 C18
Okene, Nigeria 206 L7
Okgu, S. Korea 187 R7
Okha, Russ. 161 H21
Okhaldhunga, Nepal 182 G10
Okhotsk Basin, Pac. Oc. 246 C8
Okhotsk, Russ. 161 G20
Okhotsk, Russ. 161 H22
Okhtyrka, Ukr. 159 N8
Oki Guntō, Japan 189 N7
Okiep, S. Af. 212 K4
Okinawa Shotō, Japan 189 X1
Okinawa Trough, Pac. Oc. 246 F6
Okinawa, Japan 188 X2
Okinawa, island, Japan 189 X2
Okino Daitō Jima, Japan 224 C2
Okino Erabu Shima, Japan 189 W2
Okino Shima, Japan 189 S6
Okinotori Shima (Parece Vela), Japan 224 D3
Okitipupa, Nigeria 206 M6
Oklahoma City, Okla., U.S. 105 C17
Oklahoma, Okla., U.S. 105 D17
Okmulgee, Okla., U.S. 105 C18
Okolona, Ky., U.S. 103 B16
Okolona, Miss., U.S. 103 J13
Okondja, Gabon 208 E9
Okoppe, Japan 188 F14
Okoyo, Congo 208 F10
Okpara, river, Benin, Nigeria 205 N18
Okso' Takpochao, peak, N. Mariana Is., U.S. 230 H8
Oksovskiy, Russ. 158 D10
Oktong, N. Korea 186 K8
Oktyabr'sk, Kaz. 178 F7
Oktyabr'skiy, Russ. 159 Q13
Oktyabr'skiy, Russ. 160 H7
Ōkuchi, Japan 189 S4
Okulovka, Russ. 158 G7
Okushiri, island, Japan 188 H11
Okuta, Nigeria 206 K6
Okwa, river, Botswana 212 G6
Okytyabr'skoye, Russ. 160 G10
Ola, Ark., U.S. 102 G7
Ólafsvík, Ice. 142 E1
Olancha Peak, Calif., U.S. 111 U8
Öland, island, Sw. 143 M14
Olary, S. Aust., Austral. 221 W11
Olathe, Kans., U.S. 107 S10
Olavarría, Arg. 134 M12
Oława, Pol. 151 F13
Olbia, It. 152 M4
Olbia, Golfo di, It. 152 M5
Old Bahama Channel, N. Amer. 118 G10
Old Bight, Bahamas 119 F13
Old Castille, Eur. 140 J3
Old Cork, Qnsld., Austral. 221 T12
Old Crow, Yukon, Can. 86 D9
Old Dongola, ruin, Sudan 203 P11
Old Forge, Pa., U.S. 96 M9
Old Fort Bay, Bahamas 122 E1
Old Harbor, Alas., U.S. 113 M15
Old Harbour, Jam. 122 J9
Old Head of Kinsale, Ire. 145 U4
Old Man Bay, Cayman Is., U.K. 122 M4
Old Man Bay, Cayman Is., U.K. 122 M4
Old Orchard Beach, Me., U.S. 97 H15
Old Road Bluff, Antigua & Barbuda 123 B20
Old Road Town, St. Kitts & Nevis 123 B17
Old Road, Antigua & Barbuda 123 B20
Old Stores, Cayman Is., U.K. 122 M2
Old Town, Me., U.S. 97 E17
Old Womans Point, Jam. 122 K8
Oldenburg, Ger. 150 C6
Oldham, Eng., U.K. 145 Q11
Olduvai Gorge, Tanzania 211 U5
Olean, N.Y., U.S. 96 K5
Olekma, river, Russ. 161 K17
Olekminsk, Russ. 161 H17
Oleksandriya, Ukr. 159 Q7
Olenegorsk, Russ. 160 D8
Olenëk Valley, Arctic Oc. 252 D10
Olenek, Russ. 161 G15
Olenek, river, Russ. 161 F16
Oleniy, Russ. 160 E12
Oléron, Île d', Fr. 149 T6
Oleśnica, Pol. 151 F13
Olga Basin, Arctic Oc. 253 F16
Olga, Mount see Kata Tjuta, Austral. 220 G8
Ölgiy, Mongolia 184 C6
Olhão da Restauração, Port. 146 L6
Oli, river, Nigeria 206 K6
Olib, island, Croatia 154 F5
Olifantshoek, S. Af. 212 K6
Olimarao Atoll, F.S.M. 226 B7
Ólimbos, Óros (Olympus, Mount), Gr. 156 D10
Olinda, Braz. 133 G18
Olite, Sp. 147 C13
Oliva, Arg. 134 K10
Olivenza, Sp. 146 J7
Olivet, Fr. 149 R9
Olivia, Minn., U.S. 106 K10
Ölkeyek, river, Kaz. 178 F9
Olla, La., U.S. 102 M8
Olmedo, Sp. 146 E10
Olmos, Peru 132 F2
Olney, Ill., U.S. 101 T7
Olney, Tex., U.S. 105 F15
'Olomburi, Solomon Is. 228 E11
Olomouc, Czechia 151 H13
Olonets, Russ. 158 E7
Olongapo, Philippines 193 C13
Oloron, Fr. 149 Y6
Olorua, island, Fiji 226 Q8
Olosega, American Samoa, U.S. 230 B9
Olosega, island, American Samoa, U.S. 230 B9
Olotania Crater, American Samoa, U.S. 230 B10
Olovyannaya, Russ. 161 L17
Olowalu, Hawaii, U.S. 115 J18
Olpoï, Vanuatu 229 E13
Olsztyn, Pol. 151 B15
Olt, river, Rom. 155 G14
Olten, Switz. 152 A3
Olteniţa, Rom. 155 G16
Olton, Tex., U.S. 104 E12
Oltu, Turk. 171 D16
Oltu, river, Turk. 171 D16
Ólvega, Sp. 147 D13
Olympia, Wash., U.S. 110 D3
Olympia, ruin, Gr. 156 K9
Olympic National Park, Wash., U.S. 110 B3
Olympos (Mount Olympus), Cyprus 162 P7
Ólympos, Gr. 157 P18
Olympus, Mount see Ólimbos, Óros, Gr. 156 D10
Olympus, Mount see Ulu Dağ, Turk. 170 E5
Olympus, Mount, Wash., U.S. 110 B2
Olyutorskiy Zaliv, Russ. 161 D22
Olyutorskiy, Russ. 161 D22
Olyutorskiy, Cape, Asia 168 B13
Ōma Zaki, Japan 188 H12
Ōma, Japan 188 H12
Ōmagari, Japan 188 K12
Omagh (An Ómaigh), N. Ire., U.K. 145 N6
Omaha Beach, Fr. 149 N6
Omaha, Nebr., U.S. 107 P9
Omaja, Cuba 118 K12
Omak Lake, Wash., U.S. 110 B7
Omak, Wash., U.S. 110 B7
Oman, Asia 175 K15
Oman Basin, Ind. Oc. 250 B8
Omaok, Palau 226 J4
Omaruru, Namibia 212 G3
Omatako, river, Namibia 212 F4
Omboué (Fernan Vaz), Gabon 208 F7
Ombu, China 184 K4
Ombwe, D.R.C. 209 G16
Omchak, Russ. 161 F20
Omdurman see Umm Durmān, Sudan 203 Q12
Omega Gardens, Cayman Is., U.K. 122 N2
Omegna, It. 152 D3
Omelek, island, Marshall Is. 227 H18
Omeo, Vic., Austral. 221 Z13
Ometepec, Mex. 116 K12
Omhājer, Eritrea 210 J6
Omihi, N.Z. 223 L19
Omin, F.S.M. 226 D12
Omo, river, Eth. 211 P6
Omodeo, Lago, It. 153 N4
Omoka, N.Z. 229 L20
Omolon, Russ. 161 D20
Omolon, river, Russ. 161 D20

Omsk, *Russ.* **160** K10
Omsukchan, *Russ.* **161** E20
Ōmu, *Japan* **188** E14
Omul, peak, *Rom.* **155** E15
Ōmuta, *Japan* **189** R4
Omutninsk, *Russ.* **158** F15
Ona Basin, *Atl. Oc.* **249** W6
Oña, *Sp.* **146** C11
Onaf, island, *F.S.M.* **226** D9
Onaga, *Kans., U.S.* **107** S9
Onalaska, *Wis., U.S.* **100** K3
Onamia, *Minn., U.S.* **106** H11
Onamue, island, *F.S.M.* **226** F8
Onancock, *Va., U.S.* **98** E14
Onaram, island, *F.S.M.* **226** F9
Onarga, *Ill., U.S.* **101** Q7
Onawa, *Iowa, U.S.* **107** N9
Onaway, *Mich., U.S.* **100** H11
Ōnay, Kōtal-e, *Afghan.* **180** H8
Oncativo, *Arg.* **134** K10
Oncócua, *Angola* **208** Q9
Ondangwa, *Namibia* **212** E3
Ondjiva, *Angola* **208** Q11
Ondo, *Nigeria* **206** L6
Öndörhaan, *Mongolia* **184** D12
One and Half Degree Channel, *Maldives* **183** X3
Oneata, island, *Fiji* **226** Q8
Onega Bay, *Arctic Oc.* **140** C11
Onega, *Russ.* **158** C9
Onega, Lake, *Eur.* **140** D11
Onega, river, *Russ.* **158** D10
Oneida Lake, *N.Y., U.S.* **96** J8
Oneida, *N.Y., U.S.* **96** J9
Oneida, *Tenn., U.S.* **103** E18
O'Neill, *Nebr., U.S.* **107** N6
Onekama, *Mich., U.S.* **100** J9
Onekotan, island, *Russ.* **161** H23
Onemak East Passage, *Marshall Is.* **227** H16
Onemak West Passage, *Marshall Is.* **227** H16
Onemak, island, *Marshall Is.* **227** H16
Oneonta, *N.Y., U.S.* **96** K10
Oneonta, *Ala., U.S.* **103** J15
Oneroa, island, *N.Z.* **229** E23
Onevai, island, *Tonga* **227** Q23
Oneşti, *Rom.* **155** D16
Onezhskaya Guba, bay, *Russ.* **158** B9
Onezhskoye Ozero (Lake Onega), *Russ.* **158** D8
Onezhskoye Ozero, *Russ.* **160** E7
Ongarue, *N.Z.* **223** G20
Ongcheon, *S. Korea* **187** P11
Ongjin-man, bay, *N. Korea* **186** M5
Ongjin, *N. Korea* **186** M5
Ongoka, *D.R.C.* **209** F17
Ongole, *India* **183** P7
Oni, *Rep. of Georgia* **171** B17
Onib, *Sudan* **202** M14
Onich, *Scot., U.K.* **144** K8
Onida, *S. Dak., U.S.* **106** K5
Onilahy, river, *Madagascar* **213** H15
Oniotto, island, *Marshall Is.* **227** F15
Onishika, *Japan* **188** F13
Onitsha, *Nigeria* **206** M7
Oniwaki, *Japan* **188** E12
Ŏnjin-sanmaek, range, *N. Korea* **186** J7
Onjŏng, *N. Korea* **186** K10
Onnang, *F.S.M.* **226** F8
Ono Channel, *Fiji* **226** Q5
Ono-i-Lau, island, *Fiji* **224** K9
Ōno, *Japan* **189** P9
Ono, island, *Fiji* **226** Q5
Onomichi, *Japan* **189** Q6
Onon, *Mongolia* **184** C11
Onon, river, *Mongolia* **184** C12
Onsan, *S. Korea* **187** S12
Onsen, *Japan* **189** P8
Onslow, *W. Aust., Austral.* **221** T2
Onsŏng, *N. Korea* **186** A13
Ont'an, *N. Korea* **186** A13
Ontario, *Can.* **87** N16
Ontario, *Oreg., U.S.* **110** H9
Ontario, *Calif., U.S.* **111** X9
Ontario, Lake, *N. Amer.* **84** J8
Ontinyent, *Sp.* **147** J14
Ontonagon, *Mich., U.S.* **100** E5
Ontong Java Atoll (Lord Howe Atoll), *Solomon Is.* **228** A9
Ontong-Java Rise, *Pac. Oc.* **246** J9
Ōnuma, *Japan* **188** H12
Ooa, island, *Marshall Is.* **227** J20
Oobagooma, *W. Aust., Austral.* **221** Q5

Oodaap Qeqertaa (Oodaaq Ø), *Greenland, Den.* **88** A10
Oodaaq Ø *see* Oodaap Qeqertaa, *Greenland, Den.* **88** A10
Oodnadatta, *S. Aust., Austral.* **221** V9
Oodweyne, *Somaliland* **210** L11
'Ō'okala, *Hawaii, U.S.* **115** M22
Oolloo, *N.T., Austral.* **221** P8
Oologah Lake, *Okla., U.S.* **105** B19
Ooma, *Kiribati* **226** M11
Oostende (Ostend), *Belg.* **148** K10
Oosterschelde, bay, *Neth.* **148** K11
Oostpunt, *Curaçao, Neth.* **123** Q16
Opala, *D.R.C.* **209** E16
Opana Point, *Hawaii, U.S.* **115** J19
Opari, *S. Sudan* **203** Y12
Oparino, *Russ.* **158** F13
Opava, *Czechia* **151** H14
Opelika, *Ala., U.S.* **103** L17
Opelousas, *La., U.S.* **102** P8
Open Bay, *P.N.G.* **222** N10
Opheim, *Mont., U.S.* **108** A12
Ophir, *Oreg., U.S.* **110** K1
Ophthalmia Range, *Austral.* **220** F3
Opienge, *D.R.C.* **209** E17
'Opihikao, *Hawaii, U.S.* **115** P23
Opinaca, river, *Que., Can.* **87** M18
Opoa, *Fr. Polynesia, Fr.* **231** H18
Opobo, *Nigeria* **206** N8
Opochka, *Russ.* **158** H5
Opole, *Pol.* **151** G14
Oporto *see* Porto, *Port.* **146** E5
Opotiki, *N.Z.* **223** F22
Opp, *Ala., U.S.* **103** N16
Oppdal, *Nor.* **142** H11
Optima Lake, *Okla., U.S.* **105** B13
Opunake, *N.Z.* **223** H19
Opuwo, *Namibia* **212** E2
Or, Les Îles d' *see* Hyères, Îles d', *Fr.* **149** Y14
Or, river, *Kaz.* **178** E7
Oracabessa, *Jam.* **122** H9
Oracle, *Ariz., U.S.* **109** V8
Oradea, *Rom.* **154** C11
Orai, *India* **182** H7
Oral, *Kaz.* **178** D5
Ŏrang, *N. Korea* **186** D12
Orange Bay, *Jam.* **122** H5
Orange Cay, *Bahamas* **118** E9
Orange City, *Iowa, U.S.* **106** M9
Orange Fan, *Atl. Oc.* **249** S13
Orange Grove, *Tex., U.S.* **105** P16
Orange Hill, *St. Vincent & the Grenadines* **123** K17
Orange Park, *Fla., U.S.* **99** R8
Orange Walk, *Belize* **117** J16
Orange, *Va., U.S.* **98** D12
Orange, *Tex., U.S.* **105** K21
Orange, *Fr.* **149** W13
Orange, *N.S.W., Austral.* **221** X14
Orange, Cabo, *Braz.* **132** B12
Orange, river, *Namibia, S. Af.* **212** K4,6
Orangeburg, *S.C., U.S.* **98** L9
Orangerie Bay, *P.N.G.* **222** R9
Orango, island, *Guinea-Bissau* **204** L7
Oranienburg, *Ger.* **150** D10
Oranjemund, *Namibia* **212** K4
Oranjestad, *St. Eustatius, Neth.* **121** B15
Oranjestad, *Aruba, Neth.* **123** Q17
Oraviţa, *Rom.* **154** F11
Orbe, *Switz.* **152** B1
Orbetello, *It.* **152** J7
Orbost, *Vic., Austral.* **221** Z13
Orcadas, station, *Antarctica* **238** A5
Orce, *Sp.* **146** K12
Orchard City, *Colo., U.S.* **109** N10
Orchila, Isla, *Venez.* **130** D3
Ord River Dam, *Austral.* **220** C7
Ord River, *W. Aust., Austral.* **221** R7
Ord, *Nebr., U.S.* **107** P6
Ord, Mount, *Austral.* **220** D5
Ord, river, *Austral.* **220** C7
Orderville, *Utah, U.S.* **109** P6
Ordes, *Sp.* **146** B6
Ordino, *Andorra* **162** H3
Ordino, *Andorra* **162** H3
Ordot, *Guam, U.S.* **230** P5
Ordu, *Turk.* **170** D12
Ordubad, *Arm.* **171** F20
Ordway, *Colo., U.S.* **109** N14
Ordzhonikidze, *Ukr.* **159** Q7
Ore Mountains, *Czechia, Ger.* **150** G9
Oreba, island, *Marshall Is.* **227** F13
Örebro, *Sw.* **143** K13

Oregon Caves National Monument, *Oreg., U.S.* **110** K2
Oregon City, *Oreg., U.S.* **110** F3
Oregon Dunes National Recreation Area, *Oreg., U.S.* **110** H2
Oregon, *Ill., U.S.* **101** N5
Oregon, *Ohio, U.S.* **101** N12
Oregon, *U.S.* **110** H5
Orekhovo-Zuyevo, *Russ.* **158** J10
Orel, *Russ.* **158** L8
Orellana, *Peru* **132** F4
Orellana, Embalse de, *Sp.* **146** H9
Orem, *Utah, U.S.* **108** L7
Orenburg, *Russ.* **160** J7
Orense *see* Ourense, *Sp.* **146** C6
Örenşehir, *Turk.* **170** F11
Oreor *see* Koror, island, *Palau* **226** H6
Orepuki, *N.Z.* **223** P16
Orestiáda, *Gr.* **157** A17
Oreti, river, *N.Z.* **223** P16
Orfanoú, Kólpos, *Gr.* **156** C12
Orford Ness, cape, *Eng., U.K.* **145** T16
Organ Pipe Cactus National Monument, *Ariz., U.S.* **109** V6
Organyà, *Sp.* **147** C16
Orgaz, *Sp.* **146** G10
Orhangazi, *Turk.* **170** D5
Orhei, *Mold.* **159** Q5
Orhon (Orkhon), river, *Mongolia* **184** C10
Orick, *Calif., U.S.* **110** M2
Oricum *see* Orikon, site, *Alban.* **154** M9
Orient, *Wash., U.S.* **110** A8
Oriente, *Arg.* **135** N12
Orihuela, *Sp.* **147** K14
Orikhiv, *Ukr.* **159** R8
Orikon (Oricum), site, *Alban.* **154** M9
Orinduik, *Guyana* **131** G14
Orinoco River Delta, *S. Amer.* **128** A6
Orinoco, river, *S. Amer.* **128** B5
Oriomo, *P.N.G.* **222** Q6
Oristano, *It.* **153** N3
Oristano, Golfo di, *It.* **153** P3
Oriximiná, *Braz.* **132** D11
Orizaba, Pico de, *Mex.* **116** J12
Orjen, peak, *Montenegro* **154** J8
Orkanger, *Nor.* **142** H11
Orkney Islands, *Scot., U.K.* **144** E10
Orland, *Calif., U.S.* **111** P3
Orlando, *Fla., U.S.* **99** T9
Orleans, *Vt., U.S.* **97** F13
Orleans, *Nebr., U.S.* **107** R5
Orléans, *Fr.* **149** R9
Orlu, *Nigeria* **206** M8
Ormara, *Pak.* **181** S5
Ormara, Ras, *Pak.* **181** S5
Orménio, *Gr.* **157** A16
Ormoc, *Philippines* **193** E15
Ormond Beach, *Fla., U.S.* **99** S9
Örnsköldsvik, *Sw.* **143** G13
Oro Grande, *Calif., U.S.* **111** W9
Orobie, Alpi, *It.* **152** C5
Orocué, *Col.* **130** G8
Orodara, *Burkina Faso* **205** L13
Orofino, *Idaho, U.S.* **108** D4
Oromahana, island, *Fr. Polynesia, Fr.* **231** F17
Oromia, *Eth.* **210** M7
Oron, *Nigeria* **206** N8
Orona, island, *Kiribati* **224** G10
Orono, *Me., U.S.* **97** E17
Oronsay, Island, *Scot., U.K.* **144** L7
Orontes, *Syr.* **172** F7
Oropesa, *Sp.* **146** G9
Oropuche Lagoon, *Trin. & Tobago* **123** Q22
Oroqen, *China* **185** B15
Orós, *Braz.* **133** F17
Orosei, *It.* **153** N5
Orosei, Golfo di, *It.* **153** N5
Orosháza, *Hung.* **151** M16
Orote Peninsula, *Guam, U.S.* **230** P4
Oroville, *Wash., U.S.* **110** A7
Oroville, *Calif., U.S.* **111** P4
Oroville, Lake, *Calif., U.S.* **111** P4
Orphan Knoll, *Atl. Oc.* **248** H7
Orr, *Minn., U.S.* **106** E12
Orsa, *Sw.* **142** J12
Orsha, *Belarus* **158** K6

Orshanka, *Russ.* **158** H14
Orsk, *Russ.* **160** J7
Orsogna, *It.* **152** K10
Orsova, *Rom.* **154** F12
Orta Nova, *It.* **152** L12
Orta, *Turk.* **170** D8
Ortaca, *Turk.* **170** H4
Ortegal, Cabo, *Sp.* **146** A6
Ortegal, Cape, *Eur.* **140** H2
Orthez, *Fr.* **149** X6
Ortigueira, *Sp.* **146** A6
Ortisei, *It.* **152** C7
Ortles, range, *It.* **152** C6
Ortonville, *Minn., U.S.* **106** J8
Orto Surt, *Russ.* **161** H17
Ortoire, river, *Trin. & Tobago* **123** Q23
Ortón, river, *Bol.* **132** G7
Ortona, *It.* **152** K11
Oruro, *Bol.* **132** K7
Orvieto, *It.* **152** J8
Orville Coast, *Antarctica* **238** F8
Orvin Mountains, *Antarctica* **238** B14
Oryakhovo, *Bulg.* **155** G13
Os Peninsula, *N. Amer.* **84** Q7
Os, river, *Andorra* **162** K2
Osa, Península de, *Pan.* **117** P19
Osage Beach, *Mo., U.S.* **107** T12
Osage City, *Kans., U.S.* **107** T9
Osage, *Iowa, U.S.* **106** M12
Osage, river, *Mo., U.S.* **107** T13
Ōsaka Wan, *Japan* **189** Q8
Ōsaka, *Japan* **189** Q8
Osakarovka, *Kaz.* **179** E13
Osakis, *Minn., U.S.* **106** H10
Osan, *S. Korea* **187** N8
Osan, *S. Korea* **187** U7
Osawatomie, *Kans., U.S.* **107** T10
Osborn Plateau, *Ind. Oc.* **251** J13
Osborne, *Kans., U.S.* **107** S6
Osceola, *Ark., U.S.* **102** F11
Osceola, *Iowa, U.S.* **107** Q11
Osceola, *Mo., U.S.* **107** U11
Osceola, *Nebr., U.S.* **107** Q7
Oschiri, *It.* **152** M4
Oscoda, *Mich., U.S.* **100** J12
Oscura Peak, *N. Mex., U.S.* **109** T12
Oscura, Punta, *Eq. Guinea* **214** N6
Ösel *see* Saaremaa, island, *Est.* **143** K15
Osen, *Nor.* **142** G11
Oseo, *S. Korea* **187** T11
Osgood Mountains, *Nev., U.S.* **110** M9
Osh, *Kyrg.* **179** L13
Oshakati, *Namibia* **212** E3
Oshawa, *Ont., Can.* **87** Q18
Oshkosh, *Wis., U.S.* **100** K6
Oshkosh, *Nebr., U.S.* **107** P2
Oshta, *Russ.* **158** E8
Oshwe, *D.R.C.* **209** G13
Osijek, *Croatia* **154** E8
Osimo, *It.* **152** H9
Osin, *S. Korea* **187** S9
Oskaloosa, *Iowa, U.S.* **107** P12
Oskarshamn, *Sw.* **143** L13
Öskemen (Ust'-Kamenogorsk), *Kaz.* **179** E17
Oskoba, *Russ.* **161** J14
Oslo, *Minn., U.S.* **106** E8
Oslo, *Nor.* **142** K11
Oslofjorden, *Atl. Oc.* **140** D7
Osmanabad, *India* **182** N5
Osmancık, *Turk.* **170** D10
Osmaniye, *Turk.* **170** H11
Osnabrück, *Ger.* **150** D6
Oso, river, *D.R.C.* **209** E18
Osogbo, *Nigeria* **206** L6
Osorno, *Chile* **135** Q6
Osorno, *Sp.* **146** C10
Osorno, Volcán, *Chile* **135** Q6
Osprey, *Fla., U.S.* **99** V7
Ossa de Montiel, *Sp.* **146** H12
Ossa, Mount, *Austral.* **220** L7
Ossabaw Island, *Ga., U.S.* **99** N9
Ossabaw Sound, *Ga., U.S.* **99** N9
Osseo, *Wis., U.S.* **100** J3
Ossetia, republic, *Rep. of Georgia* **171** B18
Ossineke, *Mich., U.S.* **100** H12
Ost Bank, *Atl. Oc.* **253** N20
Ostashkov, *Russ.* **158** H7
Ostend *see* Oostende, *Belg.* **148** K10

Østerdalen, valley, *Nor.* **142** J11
Osterholz-Scharmbeck, *Ger.* **150** C7
Östersund, *Sw.* **142** G12
Östhammar, *Sw.* **143** J14
Ostiglia, *It.* **152** E7
Ostrava, *Czechia* **151** H14
Ostróda, *Pol.* **151** C15
Ostrogozhsk, *Russ.* **159** N10
Ostrołęka, *Pol.* **151** C16
Ostrov, *Russ.* **158** H5
Ostrów Mazowiecka, *Pol.* **151** D17
Ostrów Wielkopolski, *Pol.* **151** E14
Ostrowiec Świętokrzyski, *Pol.* **151** F16
O'Sullivan Dam, *Wash., U.S.* **110** D7
Osŭm, river, *Bulg.* **155** H14
Ōsumi Kaikyō, *Japan* **189** T4
Ōsumi Shotō, *Japan* **189** U4
Osuna, *Sp.* **146** L9
Oswego, *N.Y., U.S.* **96** H8
Oswestry, *Eng., U.K.* **145** S10
Oświęcim (Auschwitz), *Pol.* **151** H15
Ōta, *Japan* **189** N11
Otaki, *Japan* **189** J20
Otar, *Kaz.* **179** J14
Otare, Cerro, *Col.* **130** H7
Otaru, *Japan* **188** G12
Otavalo, *Ecua.* **130** K4
Otavi, *Namibia* **212** F4
Otchinjau, *Angola* **208** Q10
Oţelu Roşu, *Rom.* **154** E12
Otepipi, island, *Fr. Polynesia, Fr.* **231** N17
Otgon Tenger Uul, *Mongolia* **184** D8
Othello, *Wash., U.S.* **110** D7
Othonoí, island, *Gr.* **156** E5
Óthris, Óros, *Gr.* **156** G10
Oti, river, *Ghana, Togo* **205** M16
Otish, Monts, *Que., Can.* **87** M20
Otjimbingwe, *Namibia* **212** G3
Otjiwarongo, *Namibia* **212** F3
Otobe, *Japan* **188** H12
Otoineppu, *Japan* **188** E13
Otrabanda, *Curaçao, Neth.* **123** Q14
Otradnoye, *Russ.* **161** G23
Otradnyy, *Russ.* **158** K16
Otranto, *It.* **153** N16
Otranto, Strait of, *Eur.* **140** L8
Otsego, *Mich., U.S.* **100** M9
Ōtsu, *Japan* **188** G14
Ōtsu, *Japan* **189** Q9
Ottawa Islands, *Nunavut, Can.* **87** K17
Ottawa, *Ont., Can.* **87** Q19
Ottawa, *Ill., U.S.* **101** P6
Ottawa, *Ohio, U.S.* **101** P11
Ottawa, *Kans., U.S.* **107** T10
Ottawa, river, *Ont., Que., Can.* **87** P18
Ottenby, *Sw.* **143** M14
Otter Rapids, *Ont., Can.* **87** N17
Otter Tail Lake, *Minn., U.S.* **106** H9
Ottmarsheim, *Fr.* **149** Q15
Ottumwa, *Iowa, U.S.* **107** Q12
Otukpa, *Nigeria* **206** M8
Otukpo, *Nigeria* **206** L8
Otutéhé, *Fr. Polynesia, Fr.* **231** D14
Otuzco, *Peru* **132** F2
Otway, Bahía, *Chile* **135** X7
Otway, Cape, *Austral.* **220** M12
Otway, Seno de, *Chile* **135** X8
Öztal Alps, *It.* **152** B6
Ou, river, *Laos* **190** H9
'O'ua, island, *Tonga* **227** M23
Ouachita Mountains , *Ark., Okla., U.S.* **105** D20
Ouachita Mountains, *U.S.* **91** J13
Ouachita, Lake, *Ark., U.S.* **102** G7
Ouachita, river, *U.S.* **91** K14
Ouaco, *New Caledonia, Fr.* **231** P20
Ouâd Nâga, *Mauritania* **204** G7
Ouadâne, *Mauritania* **204** E9
Ouadda, *Cen. Af. Rep.* **207** L16
Ouadou, river, *Mauritania* **204** H10
Ouagadougou, *Burkina Faso* **205** K15
Ouahigouya, *Burkina Faso* **205** K15
Ouaka, river, *Cen. Af. Rep.* **207** M15
Oualâta, *Mauritania* **204** G12
Ouallam, *Niger* **206** G5
Ouanda Djallé, *Cen. Af. Rep.* **207** K16
Ouandja, *Cen. Af. Rep.* **207** K16
Ouando, *Cen. Af. Rep.* **207** M18
Ouango, *Cen. Af. Rep.* **207** N16
Ouani, *Comoros* **215** N16
Ouarane, plateau, *Mauritania* **204** D12
Ouargaye, *Burkina Faso* **205** L16
Ouargla, *Alg.* **201** E13

Ouaritoufoulout, Mali 205 H18
Ouarkoye, Burkina Faso 205 K14
Ouarzazate, Mor. 200 E7
Oubangui, river, Af. 207 N14
Oubatche, New Caledonia, Fr. 231 N21
Oudtshoorn, S. Af. 212 M6
Oued al Khatt, W. Sahara 200 H4
Oued Amded, Alg. 200 K12
Oued Drâa, Alg., Mor. 200 F7
Oued Saoura, Alg. 200 F10
Oued Tag'eraout, Alg. 200 L12
Oued Zem, Mor. 200 D8
Ouégoa, New Caledonia, Fr. 231 N20
Ouémé, river, Benin 205 M17
Ouessa, Burkina Faso 205 L14
Ouessant, Île d', Fr. 149 P2
Ouésso, Congo 208 D11
Ouest, Pointe de l', Que., Can. 87 M21
Ouezzane, Mor. 200 C8
Oufrane, Alg. 200 G11
Ouidah, Benin 205 P17
Oujda-Angad, Alg. 200 C10
Oujeft, Mauritania 204 F9
Oulad Saïd, Alg. 200 F11
Oulainen, Fin. 143 F15
Ouled Djellal, Alg. 201 C13
Oullins, Fr. 149 U12
Oulu (Uleåborg), Fin. 143 F15
Oulu, river, Eur. 140 C9
Oulujärvi, lake, Eur. 140 C9
Oulujoki, river, Fin. 143 F15
Oulx, It. 152 E1
Oum Chalouba, Chad 207 G15
Oum Hadjer, Chad 207 H14
Oumé, Côte d'Ivoire 205 P13
Oumm el Khezz, Mauritania 204 G10
Oun, S. Korea 187 P11
Ounianga Kébir, Chad 207 E15
Our, river, Ger., Lux. 162 G10
Ouranópoli, Gr. 157 D13
Ouray, Utah, U.S. 108 L9
Ouray, Colo., U.S. 109 P11
Ourense (Orense), Sp. 146 C6
Ouri, Chad 207 D14
Ourinhos, Braz. 132 M12
Ourique, Port. 146 K5
Ouse, river, Eng., U.K. 145 P12
Oussouye, Senegal 204 K6
Oust, Fr. 149 Y8
Out Skerries, Scot., U.K. 144 C12
Outer Brass Island, Virgin Is., U.S. 122 M7
Outer Hebrides, islands, Scot., U.K. 144 H6
Outjo, Namibia 212 F3
Outokumpu, Fin. 143 G17
Ouvéa (Uvéa), island, New Caledonia, Fr. 231 N23
Ouyen, Vic., Austral. 221 Y12
Ouzouer-le-Marché, Fr. 149 R9
Ovacık, Turk. 170 J8
Ovaka, island, Tonga 227 H22
Ovalau, island, Fiji 226 P6
Ovalle, Chile 134 J6
Ovamboland, Namibia 212 E3
Ovar, Port. 146 E5
Ovau, island, Solomon Is. 228 C6
Overland Park, Kans., U.S. 107 S10
Overland Village, St. Vincent & the Grenadines 123 J17
Overstrand, Eng., U.K. 145 R15
Overton, Tex., U.S. 105 H20
Overton, Nev., U.S. 111 T12
Ovgos (Dardere), river, Cyprus 162 N7
Oviedo, Dom. Rep. 119 N18
Oviedo, Sp. 146 B8
Ovoot, Mongolia 184 E12
Ovruch, Ukr. 158 M5
Ow, island, F.S.M. 226 F9
Owaka, N.Z. 223 P17
Owando, Congo 208 E11
Owatonna, Minn., U.S. 106 L11
Owego, N.Y., U.S. 96 L8
Owel, Lough, Ire. 145 Q6
Owen Fracture Zone, Ind. Oc. 250 F7
Owen Stanley Range, P.N.G. 222 Q8
Owens Lake Bed, Calif., U.S. 111 U8
Owens, river, Calif., U.S. 111 T8
Owensboro, Ky., U.S. 103 C15
Owensville, Mo., U.S. 107 T14
Owerri, Nigeria 206 M8
Owia, St. Vincent & the Grenadines 123 J17
Owl Creek Mountains, Wyo., U.S. 108 G10
Owo, Nigeria 206 L7
Owo, Nigeria 206 M8
Owosso, Mich., U.S. 100 L11
Owyhee Mountains, Idaho, U.S. 108 H3
Owyhee, Nev., U.S. 110 L10
Owyhee, Lake, Oreg., U.S. 110 J9
Owyhee, river, Oreg., U.S. 110 J8
Oxeiá, island, Gr. 156 H8
Oxford, Pa., U.S. 96 Q8
Oxford, Ohio, U.S. 101 S11
Oxford, Miss., U.S. 102 H12
Oxford, Ala., U.S. 103 J16
Oxford, Kans., U.S. 107 V8
Oxford, Nebr., U.S. 107 R5
Oxford, Eng., U.K. 145 U13
Oxford, N.Z. 223 L18
Oxley, N.S.W., Austral. 221 X12
Oxnard, Calif., U.S. 111 X7
Oxus see Amu Darya, Turkm. 178 N9
Oyapok see Oiapoque, river, S. Amer. 131 G19
Oyem, Gabon 208 D8
Oyo, Sudan 202 L14
Oyo, Nigeria 206 L6
Oyonnax, Fr. 149 T13
Oyotung, Russ. 161 D18
Oyster Bay, Jam. 122 G7
Oyster Pond, St. Martin, Fr., St. Maarten, Neth. 123 B15
Oysterhaven (Cuan Oisrí), Ire. 145 U4
Oysterville, Wash., U.S. 110 D2
Oytal, Kaz. 179 K14
Oyyl, Kaz. 178 F6
Oyyl, river, Kaz. 178 F6
Ozala, Congo 208 D10
Özalp, Turk. 171 F18
Ozark National Scenic Riverways, Mo., U.S. 107 U14
Ozark Plateau , U.S. 105 B20
Ozark Plateau, U.S. 91 H14
Ozark, Ark., U.S. 102 F6
Ozark, Ala., U.S. 103 M17
Ozark, Mo., U.S. 107 V12
Ozarks, Lake of the, Mo., U.S. 107 T12
Ózd, Hung. 151 K16
Özen, Kaz. 178 J5
Ozera Kuyto, Russ. 158 B7
Ozernovskiy, Russ. 161 G23
Ozernoy, Mys, Russ. 161 E22
Ozernoy, Zaliv, Russ. 161 E22
Ozero Lacha, Russ. 158 E10
Ozery, Russ. 158 K10
Ozette Lake, Wash., U.S. 110 B2
Ozieri, It. 152 M4
Ozinki, Russ. 158 M15
Ozona, Tex., U.S. 105 K13
Ōzu, Japan 189 R6
Ozurget'i, Rep. of Georgia 171 C16

P

Paagoumène, New Caledonia, Fr. 231 N20
Paama, island, Vanuatu 229 H16
Paamiut (Frederikshåb), Greenland, Den. 89 S6
Paarden Bay, Aruba, Neth. 123 Q17
Paarl, S. Af. 212 M5
Pa'auhau, Hawaii, U.S. 115 M22
Pa'auilo, Hawaii, U.S. 115 M22
Paavola, Fin. 143 F15
Pabaigh (Pabbay), island, Scot., U.K. 144 H6
Pabaigh, island, Scot., U.K. 144 J6
Pabbay see Pabaigh, island, Scot., U.K. 144 H6
Pabianice, Pol. 151 E15
Pabna, Bangladesh 182 J12
Pabo, Uganda 211 Q3
Pacaraima, S. Amer. 130 G12
Pacaraimã, Serra see Pacaraima, Sierra, S. Amer. 132 B8
Pacasmayo, Peru 132 F2
Paceco, It. 153 S8
Pacheiá, island, Gr. 157 N16
Pachino, It. 153 U12
Pachuca, Mex. 116 H12
Pachyammos, Cyprus 162 N6
Pacific Beach, Wash., U.S. 110 C2
Pacific City, Oreg., U.S. 110 F2
Pacific Grove, Calif., U.S. 111 T4
Pacific-Antarctic Ridge, Pac. Oc. 255 H17
Pacy, Fr. 149 P9
Padam, India 180 J15
Padang Endau, Malaysia 191 U10
Padang, Indonesia 192 J5
Padang, island, Indonesia 192 J6
Padangpanjang, Indonesia 192 J5
Padangsidempuan, Indonesia 192 H5
Padangtiji, Indonesia 192 G4
Padangtikar Maya, island, Indonesia 192 J8
Padany, Russ. 158 C8
Paden City, W. Va., U.S. 98 C9
Paderborn, Ger. 150 E7
Padeş, peak, Rom. 154 E12
Padloping Island, Nunavut, Can. 87 F19
Padma (Ganges), river, Bangladesh 182 J12
Padmanabhapuram, India 183 T5
Padova (Padua), It. 152 E7
Padrauna, India 182 H9
Padre Island National Seashore, Tex., U.S. 105 P17
Padre Island, U.S. 91 P12
Padua see Padova, It. 152 E7
Paducah, Ky., U.S. 103 D13
Paducah, Tex., U.S. 105 E14
Paea, Fr. Polynesia, Fr. 231 J21
Paegam, N. Korea 186 D11
Paekch'ŏn, N. Korea 186 L6
Paektu-san, peak, N. Korea 186 C10
Paengma, N. Korea 186 G4
Paeroa, N.Z. 223 F21
Paestum, ruin, It. 153 N11
Paeu, Solomon Is. 228 N6
Paeua, island, Fr. Polynesia, Fr. 231 R14
Páfos see Paphos, Cyprus 162 Q5
Pafúri, S. Af. 212 G9
Pag, island, Croatia 154 F5
Pagadian, Philippines 193 F14
Pagai Selatan, island, Indonesia 192 K5
Pagai Utara, island, Indonesia 192 K5
Pagan, island, N. Mariana Is., U.S. 230 H5
Paganico, It. 152 J7
Pagasitikós Kólpos, Gr. 156 F11
Pagatan, Indonesia 192 K11
Page, N. Dak., U.S. 106 G7
Page, Ariz., U.S. 109 Q7
Pageland, S.C., U.S. 98 J10
Pager, river, Uganda 211 Q4
Paget Island, Bermuda 119 B16
Pagnag (Pana), China 184 K6
Pago Bay, Guam, U.S. 230 P6
Pago Pago Harbor, American Samoa, U.S. 230 D7
Pago Pago, American Samoa, U.S. 230 D6
Pago, river, Guam, U.S. 230 P5
Pagoda Point, Myanmar 190 L4
Pagosa Springs, Colo., U.S. 109 P11
Pagua Bay, Dominica 123 E20
Pagua Point, Dominica 123 E20
Pagui, P.N.G. 222 N6
Pāhala, Hawaii, U.S. 115 Q22
Paharpur, Pak. 180 K10
Pāhoa, Hawaii, U.S. 115 P23
Pahokee, Fla., U.S. 99 V10
Pahrump, Nev., U.S. 111 U11
Pahsimeroi, river, Idaho, U.S. 108 F6
Pahute Mesa, Nev., U.S. 111 S10
Pāi'a, Hawaii, U.S. 115 J19
Paige, Tex., U.S. 105 K17
Päijänne, lake, Fin. 143 H16
Paili, P.N.G. 222 R8
Pailĭn (Pailin), Thai. 191 N9
Pailolo Channel, Hawaii, U.S. 115 J18
Paine, Cerro, Chile 135 W7
Painesville, Ohio, U.S. 101 N15
Paint Rock, Tex., U.S. 105 J14
Painted Desert, Ariz., U.S. 109 R8
Painter, Mount, Austral. 220 J11
Paintsville, Ky., U.S. 103 C20
Païromé, New Caledonia, Fr. 231 N20
Paisley, Oreg., U.S. 110 K5
Paistunturit, peak, Fin. 143 B14
Paita, Peru 132 F1
Paita, New Caledonia, Fr. 231 Q22
Pajala, Sw. 143 D14
Pajares, Puerto de, Sp. 146 B9
Pajaros Point, Virgin Is., U.K. 122 Q10
Paju, S. Korea 186 M8
Pajule, Uganda 211 Q4
Pak Nam Chumphon, Thai. 191 P7
Pak Phanang, Thai. 191 R7
Pak, island, P.N.G. 222 M8
Pākalā Village, Hawaii, U.S. 114 F9
Pakaraima Mountains, S. Amer. 131 F14
Pakch'ŏn, N. Korea 186 C12
Pakch'ŏn, N. Korea 186 H6
Pakin Atoll, F.S.M. 226 B11
Pakistan, Asia 181 Q6
Pakleni Otoci, island, Croatia 154 H6
Pakokku, Myanmar 190 G4
Pakpattan, Pak. 180 M12
Paks, Hung. 151 M15
Pakuru, island, F.S.M. 226 F9
Pakwach, Uganda 211 R3
Pakxé, Laos 190 L11
Pal, Andorra 162 H2
Pala, Myanmar 191 N6
Pala, Chad 206 K12
Palabek, Uganda 211 Q4
Palace of Holy Office, Vatican City 163 R17
Palace of Justice, Monaco 163 E21
Palace of Monaco, Monaco 163 E21
Palace of the Tribunal, Vatican City 163 Q15
Palace Square, Monaco 163 E21
Palacios, Tex., U.S. 105 M18
Palafrugell, Sp. 147 D18
Palagruža (Pelagosa), island, Croatia 154 J6
Palaichori, Cyprus 162 P7
Palaíkastro, Gr. 157 Q17
Palaióchora, Gr. 156 Q12
Palaiokastrítsa, Gr. 156 E6
Palaiseau, Fr. 149 P10
Palakkad, India 183 S5
Palalankwe, India 183 S14
Palamás, Gr. 156 F9
Palana, Russ. 161 E21
Palanga, Lith. 143 M15
Palangkaraya, Indonesia 192 K11
Palanpur, India 182 J3
Palaoa Point, Hawaii, U.S. 115 K17
Palapye, Botswana 212 G8
Palasht, Iran 177 F15
Palatka, Fla., U.S. 99 R8
Palatka, Russ. 161 F20
Palau, Pac. Oc. 224 F2
Palau Trench, Pac. Oc. 246 J6
Palauk, Myanmar 191 N6
Palauli Bay, Samoa 227 N18
Palaw, Myanmar 191 N6
Palawan Trough, Pac. Oc. 246 J4
Palawan, island, Philippines 192 E12
Palazzolo Acreide, It. 153 T11
Paldiski, Est. 143 J16
Palel, India 182 J15
Palembang, Indonesia 192 K7
Palencia, Sp. 146 D10
Palermo, Calif., U.S. 111 P4
Palermo, It. 153 R9
Palestina, Virgin Is., U.S. 122 M11
Palestina, Chile 134 E7
Palestine, Tex., U.S. 105 H19
Paletwa, Myanmar 190 G3
P'algyŏngdae, N. Korea 186 D13
Pali, India 182 H3
Palian, Thai. 191 R7
Palikir Passage, F.S.M. 226 J8
Palikir, F.S.M. 226 B11
Palikir, F.S.M. 226 J8
Palinuro, Capo, It. 153 N12
Palioúri, Gr. 156 E12
Palioúri, Ákra, Gr. 156 E12
Palisade, Nebr., U.S. 107 R4
Paliseul, Belg. 148 M12
Palk Bay, India 183 S6
Palk Strait, India 183 S7
Palkonda, India 182 M9
Pallasovka, Russ. 159 N14
Palm Bay, Fla., U.S. 99 U10
Palm Beach, Fla., U.S. 99 V11
Palm Beach, Aruba, Neth. 123 Q17
Palm Coast, Fla., U.S. 99 S9
Palm Island, St. Vincent & the Grenadines 121 G17
Palm Islands, Austral. 220 D14
Palm Springs, Calif., U.S. 111 X10
Palma City, Cuba 118 H11
Palma de Mallorca, Sp. 147 G18
Palma del Río, Sp. 146 K9
Palma di Montechiaro, It. 153 T10
Palma Soriano, Cuba 119 K13
Palma, Mozambique 213 B13
Palma, Badia de, Sp. 147 G18
Palmanova, It. 152 D9
Palmares, Braz. 133 G18
Palmas, Braz. 132 N12
Palmas, Braz. 133 G13
Palmas, Cape, Af. 198 J2
Palmas, Golfo di, It. 153 Q3
Palmdale, Calif., U.S. 111 W8
Palmeira dos Índios, Braz. 133 G17
Palmeira, Cabo Verde 215 B17
Palmer Archipelago, Antarctica 238 D4
Palmer Land, region, Antarctica 238 F7
Palmer Station, Antarctica 238 D5
Palmer, Alas., U.S. 113 J17
Palmerston Atoll, Cook Is., N.Z. 224 J11
Palmerston North, N.Z. 223 J21
Palmerville, Qnsld., Austral. 221 Q13
Palmetto Point, Jam. 122 H10
Palmetto, Fla., U.S. 99 U7
Palmi, It. 153 R13
Palmira, Col. 130 H5
Palmiste Bay, Grenada 123 K22
Palmyra Atoll, Pac. Oc. 224 F11
Palmyra see Tadmur, Syr. 172 G10
Palmyra, N.Y., U.S. 96 J7
Palmyra, Mo., U.S. 107 R13
Palmyras Point, India 182 L11
Palo Alto Battlefield National Historic Site, Tex., U.S. 105 R17
Palo Alto, Calif., U.S. 111 S3
Palo Seco, Trin. & Tobago 123 R21
Paloh, Indonesia 192 H9
Paloich, S. Sudan 203 U12
Palomar Mountain, Calif., U.S. 111 Y10
Palopo, Indonesia 193 K13
Palos, Cabo de, Sp. 147 K14
Palos, Cape, Eur. 140 L3
Palouse, Wash., U.S. 110 D9
Palouse, river, Wash., U.S. 110 D8
Palpana, Cerro, Chile 134 D7
Paltamo, Fin. 143 F16
Palu, Turk. 171 F14
Palu, Indonesia 193 J13
Palwal, India 182 G6
Pama, Burkina Faso 205 L16
Pama, river, Cen. Af. Rep. 207 N13
Pamiers, Fr. 149 Y9
Pamir see Pomir, river, Afghan., Taj. 180 F11
Pamirs, range, Taj. 179 N14
Pamlico Sound, N.C., U.S. 98 H14
Pampa, Tex., U.S. 105 C13
Pampas, Peru 132 H4
Pampas, region, S. Amer. 128 M6
Pamplemousses, Mauritius 215 F20
Pamplona (Iruña), Sp. 147 C13
Pamplona, Col. 130 E7
Pamua, Solomon Is. 228 G12
Pamuk Imwintiati, island, F.S.M. 226 K7
Pamzal, India 182 D7
Pana see Pagnag, China 184 K6
Pana Tinai, island, P.N.G. 222 R10
Pana Wina, island, P.N.G. 222 R10
Pana, Ill., U.S. 101 S5
Panaca, Nev., U.S. 111 S12
Panacea, Fla., U.S. 99 R5
Panagyurishte, Bulg. 155 J14
Panahan, Indonesia 192 K10
Panaitan, island, Indonesia 192 M7
Panaji, India 183 P4
Panama Basin, Pac. Oc. 247 J21
Panama Canal, Pan. 117 N21
Panama City, Fla., U.S. 99 R3
Panama, Okla., U.S. 105 D20
Panama, N. Amer. 117 N21
Panamá, Pan. 117 N21
Panamá, Golfo de, Pan. 117 P21
Panama, Isthmus of, N. Amer. 84 Q8
Panamint Range, Calif., U.S. 111 U9
Panay, island, Philippines 193 E14
Pancake Range, Nev., U.S. 111 R11
Pančevo, Serb. 154 F10
Pančićev Vrh, peak, Kos., Serb. 154 H10
Pandanus Creek, Qnsld., Austral. 221 R13
Pandharkawada, India 182 M5
Pandharpur, India 182 N4
Pandhurna, India 182 L6
Pandie Pandie, S. Aust., Austral. 221 U11
Pandu, D.R.C. 209 B13

Panduranga see Phan Rang-Tháp Chàm, *Vietnam* **191** P13
Paneri, *India* **182** H13
Panevėžys, *Lith.* **143** M16
Panfilovo, *Russ.* **159** N12
Panga, *D.R.C.* **209** D17
Pangai, *Tonga* **227** L24;R24
Pangaimoto, island, *Tonga* **227** H23
Pangaimotu, island, *Tonga* **227** Q23
Pangala, *Congo* **208** G10
Pangani, *Tanzania* **211** V8
Pangéo, Óros, *Gr.* **157** C13
Panggoe, *Solomon Is.* **228** C7
Panghyŏn, *N. Korea* **186** G5
Pangi, *D.R.C.* **209** G17
Pangkajene, *Indonesia* **193** L13
Pangkalpinang, *Indonesia* **192** K8
Pangnirtung, *Nunavut, Can.* **87** G19
Pangong Tso, lake, *India* **182** D7
Pangsan, *N. Korea* **186** F4
Panguitch, *Utah, U.S.* **109** P6
Panguma, *Sierra Leone* **204** N9
Pangutaran Group, *Philippines* **193** F13
P'an'gyo, *N. Korea* **186** K8
Panhandle, *Tex., U.S.* **104** D12
Pania-Mutombo, *D.R.C.* **209** H15
Panian, island, *F.S.M.* **226** K8
Pānī'au, peak, *Hawaii, U.S.* **114** F8
Panj Poyon, *Taj.* **178** N12
Panj, river, *Afghan., Taj.* **180** E10
Panjakent, *Taj.* **178** M11
Panjang, island, *Indonesia* **192** H10
Panjgur, *Pak.* **181** Q4
Panjshayr, river, *Afghan.* **180** G9
Pankshin, *Nigeria* **206** K9
P'anmun, *N. Korea* **186** M7
Panna, *India* **182** J7
Pannawonica, *W. Aust., Austral.* **221** T2
Pano Lefkara, *Cyprus* **162** P8
Pano Panagia, *Cyprus* **162** P6
Pano Platres, *Cyprus* **162** P7
Panorama Point, peak, *Nebr., U.S.* **107** P1
Pánormos, *Gr.* **157** K14
Panovo, *Russ.* **161** J15
P'anp'yŏng, *N. Korea* **186** F6
Panruti, *India* **183** R7
Pantà de Riba-roga, *Sp.* **147** E15
Pantanal, marsh, *S. Amer.* **128** H7
Pante Makasar (Oecusse), *Timor-Leste* **193** N15
Pantelleria, *It.* **153** U7
Pantelleria, Isola di, *It.* **153** U8
Panther Head, *Namibia* **212** K3
Panvel, *India* **182** M3
Panyam, *Nigeria* **206** K9
Panzhihua, *China* **184** L9
Panzi, *D.R.C.* **208** J12
Pao, river, *Thai.* **190** K9
Paola, *Kans., U.S.* **107** T10
Paola, *Malta* **163** K23
Paoli, *Ind., U.S.* **101** T9
Paopao, *Fr. Polynesia, Fr.* **231** H20
Paoua, *Cen. Af. Rep.* **207** L13
Pāpa Bay, *Hawaii, U.S.* **115** Q20
Papa Stour, island, *Scot., U.K.* **144** C12
Papa Westray, island, *Scot., U.K.* **144** E10
Papa, *Samoa* **227** N17
Pápa, *Hung.* **151** L13
Pāpa'aloa, *Hawaii, U.S.* **115** M22
Papai, Mount, *Mauritius* **215** J20
Pāpa'ikou, *Hawaii, U.S.* **115** N23
Papakura, *N.Z.* **223** F20
Papara, *Fr. Polynesia, Fr.* **231** J21
Papau, island, *N.Z.* **229** H20
Papeari, *Fr. Polynesia, Fr.* **231** J22
Papeete, *Fr. Polynesia, Fr.* **231** C17
Papeete, *Fr. Polynesia, Fr.* **231** H21
Papenburg, *Ger.* **150** C6
Papenoo, *Fr. Polynesia, Fr.* **231** H22
Papenoo, river, *Fr. Polynesia, Fr.* **231** H22
Papetoai, *Fr. Polynesia, Fr.* **231** H20
Paphos (Páfos), *Cyprus* **162** Q5
Papíkio Óros, *Gr.* **157** B15
Pāpōhaku Beach, *Hawaii, U.S.* **115** H16
Papua New Guinea, *Pac. Oc.* **222** P6
Papua Passage, *N.Z.* **229** E23
Papun, *Myanmar* **190** K6
Papunya, *N.T., Austral.* **221** T8
Pará, river, *Braz.* **133** D13
Parabel', *Russ.* **160** J12
Paracas Peninsula, *S. Amer.* **128** G3
Paracatu, *Braz.* **133** K14

Paracel Islands (Xisha Qundao, Hoàng Sa), *S. China Sea* **190** K15
Parachinar, *Pak.* **180** J9
Parachute, *Colo., U.S.* **108** M10
Paraćin, *Serb.* **154** G11
Paracuru, *Braz.* **133** E17
Parada Point, *S. Amer.* **128** G3
Paradeep see Paradwip, *India* **182** L11
Paradera, *Aruba, Neth.* **123** Q17
Paradise Bay, *Malta* **163** J21
Paradise Island, *Bahamas* **122** E3
Paradise Point, *Bahamas* **122** A11
Paradise Valley, *Nev., U.S.* **110** L9
Paradise, *Mich., U.S.* **100** F10
Paradise, *Mont., U.S.* **108** C5
Paradise, *Calif., U.S.* **111** P4
Paradise, *Grenada* **123** K23
Paradwip (Paradeep), *India* **182** L11
Paragould, *Ark., U.S.* **102** E11
Paragua, river, *Venez.* **130** F12
Paraguá, river, *Bol.* **132** H8
Paraguaipoa, *Venez.* **130** C7
Paraguaná, Península de, *Venez.* **130** B9
Paraguay, *S. Amer.* **132** M9
Paraguay, river, *S. Amer.* **128** H7
Paraíso, *Mex.* **117** J14
Parakou, *Benin* **205** M18
Parakylia, *S. Aust., Austral.* **221** W10
Paralakhemundi, *India* **182** M9
Paralimni, *Cyprus* **162** P10
Param, island, *F.S.M.* **226** J8
Paramaribo, *Suriname* **131** F17
Paramera, Sierra de la, *Sp.* **146** F9
Paramushir, island, *Russ.* **161** G23
Paramythiá, *Gr.* **156** F7
Paran, *Israel* **173** Q5
Paraná, *Arg.* **134** J12
Paraná, *Braz.* **133** H13
Paraná, river, *S. Amer.* **128** K7
Paranaguá, *Braz.* **133** N13
Paranaguá, Baía de, *Braz.* **133** N13
Paranaíba, *Braz.* **132** K12
Paranaíba, river, *Braz.* **132** K12
Paranapanema, river, *S. Amer.* **128** J8
Paranésti, *Gr.* **157** B13
Paraoa, island, *Fr. Polynesia, Fr.* **231** C21
Paraparaumu, *N.Z.* **223** J20
Parchim, *Ger.* **150** C9
Pardes Hanna-Karkur, *Israel* **172** L5
Pardo, river, *Braz.* **132** L11
Pardo, river, *Braz.* **133** J16
Pardoo, *W. Aust., Austral.* **221** S3
Pardubice, *Czechia* **150** H12
Pare Mountains, *Tanzania* **211** U7
Parea, *Fr. Polynesia, Fr.* **231** K14
Parece Vela see Okinotori Shima, *Japan* **224** D3
Parecis, Chapada dos, *Braz.* **132** H9
Paredes de Nava, *Sp.* **146** D10
Paredón, *Mex.* **116** E11
Parem, island, *F.S.M.* **226** E9
Paren', *Russ.* **161** D21
Parepare, *Indonesia* **193** L13
Párga, *Gr.* **156** F7
Pargny, *Fr.* **149** P12
Pargo, Ponta do, *Madeira, Port.* **214** M2
Parham Peninsula, *Antigua & Barbuda* **123** A21
Parham Town, *Virgin Is., U.K.* **122** Q7
Parham, *Antigua & Barbuda* **123** B21
Paria Peninsula, *S. Amer.* **128** A6
Paria, Gulf of, *Trinidad & Tobago, Venez.* **121** K16
Paria, river, *Ariz., Utah, U.S.* **109** P7
Pariaguánviche, *Venez.* **130** D12
Parika, *Guyana* **131** E15
Parikkala, *Fin.* **143** G17
Parima, *S. Amer.* **130** G12
Pariñas, Punta, *Peru* **132** E1
Parintins, *Braz.* **132** D10
Paris Basin, *Eur.* **140** H5
Paris, *Ill., U.S.* **101** R7
Paris, *Ark., U.S.* **102** F7
Paris, *Ky., U.S.* **103** B18
Paris, *Tenn., U.S.* **103** E13
Paris, *Tex., U.S.* **105** F19
Paris, *Mo., U.S.* **107** S13
Paris, *Fr.* **149** P10
Paris, *Kiribati* **226** Q10
Parisot, *Fr.* **149** W9
Parit Buntar, *Malaysia* **191** T8
Park Falls, *Wis., U.S.* **100** G4
Park Forest, *Ill., U.S.* **101** P7

Park Range, *Colo., U.S.* **108** L12
Park Rapids, *Minn., U.S.* **106** G10
Park River, *N. Dak., U.S.* **106** E7
Park Valley, *Utah, U.S.* **108** J6
Parker Dam, *Ariz., U.S.* **109** T5
Parker Dam, *Ariz., Calif., U.S.* **111** X13
Parker, *S. Dak., U.S.* **106** M8
Parker, *Ariz., U.S.* **109** T5
Parkersburg, *W. Va., U.S.* **98** C8
Parkes, *N.S.W., Austral.* **221** X14
Parkin, *Ark., U.S.* **102** G10
Parkland, *Wash., U.S.* **110** C4
Parksley, *Va., U.S.* **98** E15
Parkston, *S. Dak., U.S.* **106** M7
Parlatuvier, *Trin. & Tobago* **123** N18
Parma, *Ohio, U.S.* **101** P14
Parma, *Idaho, U.S.* **108** G3
Parma, *It.* **152** F6
Parmakān, *Afghan.* **180** J3
Parnaguá, *Braz.* **133** G14
Parnaíba, *Braz.* **133** D16
Parnaíba, river, *Braz.* **133** E15
Parnamirim, *Braz.* **133** F16
Parnassós, peak, *Gr.* **156** H10
Parnassus, *N.Z.* **223** L19
Parnassus, peak, *Eur.* **140** L9
Párnitha, Óros, *Gr.* **156** J12
Párnonas, Óros, *Gr.* **156** L10
Pärnu, *Est.* **143** K16
Paroho, lake, *S. Korea* **186** L9
Paromay, *Russ.* **161** F17
Paropamisus Range see Silsilah-ye Safēd Kōh, *Afghan.* **180** H3
Páros, *Gr.* **157** L15
Páros, island, *Gr.* **157** L15
Parottee Point, *Jam.* **122** J7
Parowan, *Utah, U.S.* **109** P6
Parraburdoo, *W. Aust., Austral.* **221** T3
Parral, *Chile* **134** M6
Parras, *Mex.* **116** F10
Parry Bank, *Arctic Oc.* **253** G15
Parry Channel, *Nunavut, N.W.T., Can.* **87** D14
Parry Group see Muko Jima Rettō, *Japan* **189** W14
Parry Islands, *Nunavut, N.W.T., Can.* **87** C10
Parry, Cape, *N.W.T., Can.* **86** E11
Pārsābād, *Iran* **177** B13
Parshall, *N. Dak., U.S.* **106** F3
Parsons Range, *Austral.* **220** B9
Parsons, *W. Va., U.S.* **98** C10
Parsons, *Tenn., U.S.* **103** F13
Parsons, *Kans., U.S.* **107** V10
Partanna, *It.* **153** S9
Parthenay, *Fr.* **149** S7
Parthos Range, *Antarctica* **239** F19
Partizansk, *Russ.* **161** M21
Paru de Oeste (Cumina), river, *Braz.* **132** C11
Paru, river, *Braz.* **132** C11
Pārūn, *Afghan.* **180** G10
Paruru, *Solomon Is.* **228** F11
Parvatipuram, *India* **182** M9
Pas de la Casa, *Andorra* **162** J5
Pasa, island, *F.S.M.* **226** K7
Pasadena, *Tex., U.S.* **105** L19
Pasadena, *Calif., U.S.* **111** X8
Pasado, Cabo, *Ecua.* **130** K3
Pasaje, *Ecua.* **130** M3
P'asanauri, *Rep. of Georgia* **171** B18
Pasargadae, ruin, *Iran* **177** K16
Pasarseluma, *Indonesia* **192** L6
Pascagoula, *Miss., U.S.* **103** P13
Paşcani, *Rom.* **155** B16
Pasco, *Wash., U.S.* **110** E7
Pascua, Isla de see Easter Island, *Pac. Oc.* **225** L19
Pasewalk, *Ger.* **150** C11
Pasighat, *India* **182** G15
Pasinler, *Turk.* **171** E16
Pasir Mas, *Malaysia* **191** S9
Pasley, Cape, *Austral.* **220** K5
Pašman, island, *Croatia* **154** G5
Pasni, *Pak.* **181** R4
Paso de Indios, *Arg.* **135** R8
Paso de los Toros, *Uru.* **134** K14
Paso Río Mayo, *Arg.* **135** S7
Paso Robles, *Calif., U.S.* **111** V5
Pasrur, *Pak.* **180** K13
Pass Christian, *Miss., U.S.* **102** P12
Pass d'Onoheha, *Fr. Polynesia, Fr.* **231** H22
Passau, *Ger.* **150** K10
Passe Aiurua, *Fr. Polynesia, Fr.* **231** J23
Passe Araara, *Fr. Polynesia, Fr.* **231** L14

Passe Avapeihi, *Fr. Polynesia, Fr.* **231** K14
Passe Avarapa, *Fr. Polynesia, Fr.* **231** H20
Passe Avatolu, *Wallis & Futuna Is., Fr.* **230** Q8
Passe de Boulari, *New Caledonia, Fr.* **231** R22
Passe de Koné, *New Caledonia, Fr.* **231** R22
Passe de la Sarcelle, *New Caledonia, Fr.* **231** R23
Passe de Mahaena, *Fr. Polynesia, Fr.* **231** H22
Passe de Maraa, *Fr. Polynesia, Fr.* **231** J21
Passe de Taapuna, *Fr. Polynesia, Fr.* **231** H21
Passe de Tapueraha, *Fr. Polynesia, Fr.* **231** J22
Passe de Teavaraa, *Fr. Polynesia, Fr.* **231** J22
Passe de Tiputa, *Fr. Polynesia, Fr.* **231** M17
Passe du Coëtlogon, *New Caledonia, Fr.* **231** P22
Passe du Sud-Est, *Fr. Polynesia, Fr.* **231** R17
Passe du Sud-Ouest, *Fr. Polynesia, Fr.* **231** R17
Passe Fareera, *Fr. Polynesia, Fr.* **231** K14
Passe Fatumanini, *Wallis & Futuna Is., Fr.* **230** Q8
Passe Fugauvea, *Wallis & Futuna Is., Fr.* **230** Q8
Passe Honikulu, *Wallis & Futuna Is., Fr.* **230** R8
Passe Iriru, *Fr. Polynesia, Fr.* **231** G18
Passe Matauvau, *Fr. Polynesia, Fr.* **231** H20
Passe Nao Nao, *Fr. Polynesia, Fr.* **231** H17
Passe Punaeroa, *Fr. Polynesia, Fr.* **231** H17
Passe Tairapa, *Fr. Polynesia, Fr.* **231** R14
Passe Tamotoe, *Fr. Polynesia, Fr.* **231** H23
Passe Tapora, *Fr. Polynesia, Fr.* **231** H23
Passe Teava Moa, *Fr. Polynesia, Fr.* **231** H18
Passe Teavanui, *Fr. Polynesia, Fr.* **231** G14
Passe Teavapiti, *Fr. Polynesia, Fr.* **231** G18
Passe Tiamahana, *Fr. Polynesia, Fr.* **231** G17
Passe Toamaro, *Fr. Polynesia, Fr.* **231** H17
Passe Vaiau, *Fr. Polynesia, Fr.* **231** K23
Passe Vaionifa, *Fr. Polynesia, Fr.* **231** J23
Passero, Capo, *It.* **153** U12
Passo Fundo, *Braz.* **132** P11
Passos, *Braz.* **133** L14
Passu Keah, island, *S. China Sea* **192** C10
Pastavy, *Belarus* **158** J4
Pastaza, river, *S. Amer.* **128** D2
Pasto, *Col.* **130** J4
Pastora Peak, *Ariz., U.S.* **109** Q9
Pastos Bons, *Braz.* **133** F15
Pasuruan, *Indonesia* **192** M10
Pata, *Cen. Af. Rep.* **207** L15
Patagonia, *Ariz., U.S.* **109** W8
Patagonia, region, *S. Amer.* **128** Q5
Patagonian Shelf, *Atl. Oc.* **249** U5
Patan see Lalitpur, *Nepal* **182** G10
Patan, *India* **182** K7
Patan, *India* **182** K7
Patanga, island, *N.Z.* **229** L22
Patani, *Indonesia* **193** J16
Patara Shiraki, *Rep. of Georgia* **171** C20
Patchogue, *N.Y., U.S.* **96** N12
Pate Island, *Kenya* **211** T9
Paterna, *Sp.* **147** H14
Paterson, *N.J., U.S.* **96** N11
Pathankot, *India* **182** E5
Pathein, *Myanmar* **190** K4
Pathfinder Reservoir, *Wyo., U.S.* **108** J11
Pati Point, *Guam, U.S.* **230** M7
Patía, river, *Col.* **130** J4
Patiala, *India* **182** F5
Patience, Cape, *Asia* **168** D13

Patio, *Fr. Polynesia, Fr.* **231** G17
Pativilca, *Peru* **132** G3
Pātkai Range, *Myanmar* **190** D5
Pátmos, *Gr.* **157** K17
Pátmos, island, *Gr.* **157** K17
Patna, *India* **182** H10
Patnos, *Turk.* **171** F17
Patos de Minas, *Braz.* **133** K14
Patos, *Braz.* **133** F17
Patos, Lagoa dos, *Braz.* **132** Q12
Patquía, *Arg.* **134** H8
Pátra (Patras), *Gr.* **156** H9
Patraïkós Kólpos, *Gr.* **156** H9
Patras see Pátra, *Gr.* **156** H9
Patricio Lynch, Isla, *Chile* **135** U6
Patrington, *Eng., U.K.* **145** Q14
Pattani, *Thai.* **191** S8
Patten, *Me., U.S.* **97** C18
Patterson, *La., U.S.* **102** Q9
Patterson, *Calif., U.S.* **111** S5
Patteson Passage, *Vanuatu* **229** F16
Patti, *India* **182** E5
Pattoki, *Pak.* **180** L12
Patton Escarpment, *Pac. Oc.* **247** E17
Patton Seamounts, *Pac. Oc.* **247** C15
Patu, *Braz.* **133** F17
Patuakhali, *Bangladesh* **182** K13
Patuca, river, *Hond.* **117** K18
Patuxent Ice Stream, *Antarctica* **238** H11
Patuxent Range, *Antarctica* **238** H11
Patzau, *Wis., U.S.* **100** F2
Pátzcuaro, *Mex.* **116** J10
Pau, *Fr.* **149** Y7
Pauini, *Braz.* **132** F6
Pauini, river, *Braz.* **132** F6
Pauktaw, *Myanmar* **190** H3
Paulatuk, *N.W.T., Can.* **86** E10
Paulding Bay, *Antarctica* **239** P20
Paulding, *Ohio, U.S.* **101** P11
Paulhaguet, *Fr.* **149** V11
Paulhan, *Fr.* **149** X11
Paulis see Isiro, *D.R.C.* **209** C18
Paulistana, *Braz.* **133** F16
Paulo Afonso, *Braz.* **133** G17
Pauls Valley, *Okla., U.S.* **105** D17
Paungbyin, *Myanmar* **190** F4
Paungde, *Myanmar* **190** J5
Pauni, *India* **182** L7
Pa'uwela, *Hawaii, U.S.* **115** J19
Pava, island, *Tuvalu* **228** L10
Pavagada, *India* **183** Q5
Pava'ia'i, *American Samoa, U.S.* **230** E6
Pāveh, *Iraq* **176** F12
Pavia, *It.* **152** E4
Pāvilosta, *Latv.* **143** L15
Pavino, *Russ.* **158** F13
Pavlikeni, *Bulg.* **155** H15
Pavlodar, *Kaz.* **179** D14
Pavlof Volcano, *Alas., U.S.* **112** N12
Pavlovka, *Russ.* **158** L14
Pavlovo, *Russ.* **158** J12
Pavlovsk, *Russ.* **158** F6
Pavlovsk, *Russ.* **159** N11
Pavlovskaya, *Russ.* **159** S11
Pavuvu, island, *Solomon Is.* **228** F9
Paw Paw, *Mich., U.S.* **100** M9
Pawarenga, *N.Z.* **223** D19
Pawhuska, *Okla., U.S.* **105** B18
Pawnee City, *Nebr., U.S.* **107** R9
Pawnee, *Okla., U.S.* **105** B17
Pawtucket, *R.I., U.S.* **97** L14
Paxton, *Ill., U.S.* **101** R7
Paxton, *Nebr., U.S.* **107** Q3
Payakumbuh, *Indonesia* **192** J6
Payar, *Senegal* **204** J8
Payer Land, *Greenland, Den.* **88** H13
Payette Lake, *Idaho, U.S.* **108** F4
Payette, *Idaho, U.S.* **108** G3
Payette, river, *Idaho, U.S.* **108** G3
Payne, Lac, *Que., Can.* **87** J18
Paynesville, *Minn., U.S.* **106** J10
Paysandú, *Uru.* **134** K13
Payshanba, *Uzb.* **178** M10
Payson, *Ariz., U.S.* **109** T7
Pazar, *Turk.* **170** D8
Pazar, *Turk.* **170** E11
Pazar, *Turk.* **171** D15
Pazarcık, *Turk.* **170** H12
Pazin, *Croatia* **154** E4
Pazardzhik, *Bulg.* **155** J14
Pčinja, river, *N. Maced.* **154** K11
Pe Ell, *Wash., U.S.* **110** D3

Pea Ridge National Military Park, *Ark., U.S.* **102** E6
Pea, *Tonga* **227** Q23
Pea, river, *Ala., Fla., U.S.* **103** M16
Peabody Bugt, bay, *Greenland, Den.* **88** D5
Peabody, *Kans., U.S.* **107** T8
Peace River, *Alta., Can.* **86** K10
Peace, river, *Can.* **86** J11
Peach Springs, *Ariz., U.S.* **109** R5
Peacock Point, *Wake I., U.S.* **230** N12
Peak District, *Eng., U.K.* **145** R12
Peak Hill, *W. Aust., Austral.* **221** U3
Peaked Mountain, *Me., U.S.* **97** C17
Peale Island, *Wake I., U.S.* **230** M11
Peale, Mount, *Utah, U.S.* **109** N9
Peao, island, *Kiribati* **226** K11
Peard Bay, *Alas., U.S.* **113** B14
Pearl City, *Hawaii, U.S.* **115** H14
Pearl Harbor, *Hawaii, U.S.* **115** H14
Pearl River, *Asia* **168** J12
Pearl *see* Zhu, river, *China* **185** P13
Pearl, *Miss., U.S.* **102** L11
Pearl, river, *La., Miss., U.S.* **102** L11
Pearland, *Tex., U.S.* **105** L19
Pearls Rock, *Grenada* **123** K24
Pearsall, *Tex., U.S.* **105** M15
Pearson, *Ga., U.S.* **99** P7
Peary Channel, *Nunavut, Can.* **87** B14
Peary Land, *Greenland, Den.* **88** B10
Pearya Spur, *Arctic Oc.* **252** L11
Pease Bay, *Cayman Is., U.K.* **122** N3
Pease, river, *Tex., U.S.* **105** E14
Peawanuk, *Ont., Can.* **87** L16
Pebane, *Mozambique* **212** E12
Pebas, *Peru* **132** D5
Pebble Island, *Falk. Is., U.K.* **135** V12
Peć *see* Pejë, *Kos.* **154** J10
Pecan Island, *La., U.S.* **102** Q8
Pechora Basin, *Eur.* **140** A13
Pechora Bay, *Arctic Oc.* **253** B16
Pechora, *Russ.* **160** F9
Pechora, river, *Russ.* **160** F9
Pechorskoye More, *Russ.* **160** E10
Pecos, river, *N. Mex., Tex., U.S.* **104** H10
Pecos, river, *N. Mex., Tex., U.S.* **104** K12
Pecos National Historical Park, *N. Mex., U.S.* **109** R13
Pecos, *Tex., U.S.* **104** J10
Pecos, *N. Mex., U.S.* **109** R13
Pécs, *Hung.* **151** N14
Peddapalli, *India* **182** M7
Pedder, Lake, *Austral.* **220** M7
Pedernales, *Dom. Rep.* **119** N17
Pedernales, river, *Tex., U.S.* **105** K15
Pedirka, *S. Aust., Austral.* **221** U9
Pedra Azul, *Braz.* **133** J15
Pedra Lume, *Cabo Verde* **215** B17
Pedras Negras, *Braz.* **132** H8
Pedro Afonso, *Braz.* **133** G13
Pedro Betancourt, *Cuba* **118** G7
Pedro Cays, *Jam.* **118** P11
Pedro de Valdivia, *Chile* **134** D7
Pedro Juan Caballero, *Parag.* **132** L10
Pedro Luro, *Arg.* **135** P11
Pedro Osório, *Braz.* **132** Q11
Pedro, *Cayman Is., U.K.* **122** N3
Pedroso, *Port.* **146** E5
Pedroso, Sierra del, *Sp.* **146** J8
Peebles, *Ohio, U.S.* **101** S12
Peedamulla, *W. Aust., Austral.* **221** T2
Peekskill, *N.Y., U.S.* **96** M11
Peel (Purt ny h-Inshey), *Isle of Man, U.K.* **145** P9
Peel Sound, *Nunavut, Can.* **87** D14
Peel, river, *Yukon, Can.* **86** E9
Peetz, *Colo., U.S.* **108** K15
Pegasus Bay, *N.Z.* **223** L19
Pegeia, *Cyprus* **162** P5
Pego, *Sp.* **147** J15
Pehuajó, *Arg.* **134** M11
Peine, *Ger.* **150** D8
Peipus *see* Chudskoye Ozero, *Russ.* **158** G5
Peiraiás (Piraeus), *Gr.* **156** J12
Peixe, *Braz.* **133** H13
Pejë (Peć), *Kos.* **154** J10
Pekalongan, *Indonesia* **192** M9
Pekan, *Malaysia* **191** U9
Pekanbaru, *Indonesia* **192** J6
Pekelmeer, *Bonaire, Neth.* **123** R19
Pekin, *Ill., U.S.* **101** Q5
Pelabuhanratu, Teluk, *Indonesia* **192** M8

Pelado, peak, *Sp.* **147** G13
Pelagosa *see* Palagruža, island, *Croatia* **154** J6
Pelahatchie, *Miss., U.S.* **102** L11
Pelat, Mount, *Fr.* **149** W15
Pelau, island, *Solomon Is.* **228** A9
Peleaga, peak, *Rom.* **154** E12
Peleliu *see* Beliliou, island, *Palau* **226** J4
Peleng, island, *Indonesia* **193** J14
Pelham, *Ga., U.S.* **99** P5
Pelham, *Ala., U.S.* **103** K15
Pelican Island, *Virgin Is., U.K.* **122** N12
Pelican Lake, *Minn., U.S.* **106** E12
Pelican Point, *Bahamas* **122** J3
Pelican Point, *Af.* **198** N6
Pelican Point, *Namibia* **212** H3
Pelican Rapids, *Minn., U.S.* **106** G9
Pelican, *Alas., U.S.* **113** L21
Péligre, Lac de, *Haiti* **119** M17
Pelikaan Punt, *St. Maarten, Neth.* **123** C14
Pelinaío, peak, *Gr.* **157** H16
Pelister, peak, *N. Maced.* **154** L11
Pelješac, island, *Croatia* **154** H7
Pella, *Iowa, U.S.* **107** P12
Pella, ruin, *Gr.* **156** C10
Pelleluhun Islands, *P.N.G.* **222** M7
Pellston, *Mich., U.S.* **100** G10
Pelly Bay, *Nunavut, Can.* **87** F15
Pelly Crossing, *Yukon, Can.* **86** F8
Pelly, river, *Yukon, Can.* **86** G8
Peloponnese, region, *Gr.* **156** K9
Peloponnesus, *Eur.* **140** M9
Pelotas, *Braz.* **132** Q11
Pelusium, ruins, *Egypt* **173** N1
Pelvoux, Massif du, *Fr.* **149** V14
Pematang, *Indonesia* **192** J6
Pematangsiantar, *Indonesia* **192** H5
Pemba Bay, *Af.* **198** M11
Pemba Island, *Tanzania* **211** V8
Pemba, *Zambia* **209** Q17
Pemba, *Mozambique* **213** C13
Pembina, *N. Dak., U.S.* **106** D8
Pembroke (Penfro), *Wales, U.K.* **145** U8
Pembroke Fort, *Malta* **163** K22
Pembroke, *Ont., Can.* **87** Q18
Pembroke, *Ga., U.S.* **99** N8
Pembroke, *Trin. & Tobago* **123** N18
Pen-y-bont ar Ogwr *see* Bridgend, *Wales, U.K.* **145** U10
Pen, *India* **182** M3
Peña de Francia, Sierra de, *Sp.* **146** F8
Peña, Sierra de, *Sp.* **147** C14
Peñafiel, *Sp.* **146** D10
Penal, *Trin. & Tobago* **123** R22
Peñalara, peak, *Sp.* **146** F10
Peñalsordo, *Sp.* **146** J9
Penamacor, *Port.* **146** F7
Penambulai, island, *Indonesia* **193** M19
Peñaranda de Bracamonte, *Sp.* **146** E9
Peñarroya-Pueblonuevo, *Sp.* **146** J9
Peñarroya, peak, *Sp.* **147** F14
Peñas, Cabo de, *Sp.* **146** A8
Peñas, Cape, *Eur.* **140** J3
Penas, Golfo de, *Chile* **135** T6
Pend Oreille, Lake, *Idaho, U.S.* **108** B4
Pend Oreille, river, *Idaho, Wash., U.S.* **110** A9
Pendan, *Indonesia* **192** H11
Pendé, river, *Cen. Af. Rep.* **206** M12
Pendembu, *Sierra Leone* **204** N10
Pender Bay, *Austral.* **220** C4
Pender, *Nebr., U.S.* **107** N8
Pendleton, *Oreg., U.S.* **110** F7
Pendra, *India* **182** K8
Pendroy, *Mont., U.S.* **108** B7
Pendulum Øer, *Greenland, Den.* **88** G14
Penedo, *Braz.* **133** G17
Penfro *see* Pembroke, *Wales, U.K.* **145** U8
Penge, *D.R.C.* **209** H16
Penghu (Pescadores), islands, *China* **185** N15
Penha do Tapauá, *Braz.* **132** E7
Peniche, *Port.* **146** H4
Penicuik, *Scot., U.K.* **144** L10
Península del Yucatán, *N. Amer.* **117** N14
Péninsule D'Ungava, *Que., Can.* **87** J18

Peñíscola, *Sp.* **147** F15
Penmarc'h, Pointe de, *Fr.* **149** Q2
Penn Hills, *Pa., U.S.* **96** N3
Penn Yan, *N.Y., U.S.* **96** J7
Penna, Punta della, *It.* **152** K11
Penne, *It.* **152** J10
Pennell Coast, *Antarctica* **239** R14
Penner, river, *India* **183** Q6
Penneshaw, *S. Aust., Austral.* **221** Y10
Pennine Alpi, *Switz.* **152** C2
Pennington Gap, *Va., U.S.* **98** G6
Pennsylvania, *U.S.* **96** M6
Peno, *Russ.* **158** H7
Penobscot Bay, *Me., U.S.* **97** F17
Penobscot, river, *U.S.* **91** B22
Penong, *S. Aust., Austral.* **221** W8
Penonomé, *Pan.* **117** N21
Penrhyn (Tongareva), *Cook Is., N.Z.* **224** H12
Penrhyn Basin, *Pac. Oc.* **247** L14
Penrith, *Eng., U.K.* **145** N11
Pensacola Bay, *Fla., U.S.* **99** Q1
Pensacola Mountains, *Antarctica* **238** G11
Pensacola, *Fla., U.S.* **99** Q1
Pentakomo, *Cyprus* **162** Q8
Pentálofos, *Gr.* **156** D8
Pentecost, island, *Vanuatu* **229** G16
Penticton, *B.C., Can.* **86** M8
Pentland Firth, *Scot., U.K.* **144** F10
Pentland, *Qnsld., Austral.* **221** S13
Pentwater, *Mich., U.S.* **100** M9
Penuguan, *Indonesia* **192** K7
Penwell, *Tex., U.S.* **104** H11
Penza, *Russ.* **158** K9
Penzance, *Eng., U.K.* **145** X7
Penzhinskaya Guba, *Russ.* **161** E21
Peoples Creek, *Mont., U.S.* **108** B10
Peoria, *Ill., U.S.* **101** Q5
Peoria, *Ariz., U.S.* **109** U6
Pēpeʻekeo, *Hawaii, U.S.* **115** M23
Pepillo Salcedo, *Dom. Rep.* **119** L17
Pepin, Lake, *Minn., Wis., U.S.* **100** J2
Pera, *Cyprus* **162** P8
Perabumulih, *Indonesia* **192** K7
Peramiho, *Tanzania* **211** Y5
Percival Lakes, *Austral.* **220** E5
Perdido, Arroyo, *Arg.* **135** R8
Perdido, Monte, *Sp.* **147** C15
Perdido, river, *Ala., Fla., U.S.* **103** N14
Pereira, *Col.* **130** G5
Perelazovskiy, *Russ.* **159** P12
Perelyub, *Russ.* **158** M16
Peremul Par, island, *India* **183** R2
Perenjori, *W. Aust., Austral.* **221** W2
Pereslavl' Zalesskiy, *Russ.* **158** H10
Perga, ruins, *Turk.* **170** H6
Pergamino, *Arg.* **134** L12
Perham, *Minn., U.S.* **106** G9
Peri Lake, *Austral.* **220** J12
Peri, river, *Turk.* **171** F15
Péribonka, river, *Que., Can.* **87** M20
Perico, *Arg.* **134** E9
Perigoso, Canal, *Braz.* **133** C13
Perigot *see* Baie Puamau, bay, *Fr. Polynesia, Fr.* **231** D15
Périgueux, *Fr.* **149** V8
Perijá, *S. Amer.* **130** D7
Perim *see* Barīm, island, *Yemen* **174** Q8
Perinthus *see* Marmara Ereğlisi, *Turk.* **170** C4
Peristéra, island, *Gr.* **157** F13
Perito Moreno, *Arg.* **135** T7
Perkins, *Okla., U.S.* **105** C17
Perky, *Fla., U.S.* **99** Y9
Perlé, *Lux.* **162** J9
Perlevka, *Russ.* **158** M10
Perm', *Russ.* **160** H8
Pernambuco Plain, *Atl. Oc.* **249** Q9
Pernambuco Seamounts, *Atl. Oc.* **249** Q9
Pernik, *Bulg.* **154** J12
Peron, Cape, *Austral.* **220** K2
Peronit Burnu, *Rep. of Georgia* **171** C15
Péronne, *Fr.* **148** M10
Pérouges, *Fr.* **149** U13
Perpetua, Cape, *Oreg., U.S.* **110** G2
Perpignan, *Fr.* **149** Z10
Perranporth, *Eng., U.K.* **145** W8
Perry Bay, *Antarctica* **239** Q19
Perry Island, *Nunavut, Can.* **87** G13
Perry, *Fla., U.S.* **99** R6
Perry, *Ga., U.S.* **99** N6
Perry, *Okla., U.S.* **105** B17
Perry, *Iowa, U.S.* **107** P11

Perrysburg, *Ohio, U.S.* **101** N12
Perryton, *Tex., U.S.* **105** B13
Perryville, *Mo., U.S.* **107** U15
Perryville, *Alas., U.S.* **113** N13
Persepolis, ruin, *Iran* **177** K16
Perseverance Bay, *Virgin Is., U.S.* **122** M7
Persey Rise, *Arctic Oc.* **253** F16
Persian Gulf, *Ind. Oc.* **168** G4
Pertek, *Turk.* **171** F14
Perth Basin, *Ind. Oc.* **251** L15
Perth, *Scot., U.K.* **144** K10
Perth, *W. Aust., Austral.* **221** X2
Pertoúli, *Gr.* **156** F9
Peru Basin, *Pac. Oc.* **247** K21
Peru-Chile Trench, *Pac. Oc.* **247** L22
Peru, *Ill., U.S.* **101** P5
Peru, *Ind., U.S.* **101** Q9
Peru, *Nebr., U.S.* **107** Q9
Peru, *S. Amer.* **132** G4
Perugia, *It.* **152** H8
Pervari, *Turk.* **171** G17
Pervomaysk, *Russ.* **158** K12
Pervomays'k, *Ukr.* **159** Q6
Pervomayskoye, *Russ.* **158** M14
Pervoural'sk, *Russ.* **160** H8
Pervyy Kuril'skiy Proliv, *Russ.* **161** G23
Pesaro, *It.* **152** G9
Pescadores *see* Penghu, islands, *China* **185** N15
Pescara, *It.* **152** J10
Peschici, *It.* **152** K13
Peskovka, *Russ.* **158** F15
Pesmes, *Fr.* **149** R13
Pessac, *Fr.* **149** V6
Pessons, Pic dels, *Andorra* **162** J4
Pestovo, *Russ.* **158** G8
Pesu, *India* **182** H15
Petacalco Bay, *N. Amer.* **84** N4
Petacalco, Bahía, *Mex.* **116** J9
Petaḥ Tiqwa, *Israel* **172** M5
Petal, *Miss., U.S.* **102** N12
Petalión, Kólpos, *Gr.* **157** J13
Petaluma, *Calif., U.S.* **111** R3
Pétange, *Lux.* **162** K9
Petatlán, *Mex.* **116** K11
Petauke, *Zambia* **209** N20
Petén, region, *Guatemala* **117** K16
Petenwell Lake, *Wis., U.S.* **100** J5
Peter I Island, *Antarctica* **238** J5
Peter Island, *Virgin Is., U.K.* **122** R7
Peter Pond Lake, *Sask., Can.* **86** L11
Peter the Great Gulf, *Pac. Oc.* **168** F12
Peterborough, *Ont., Can.* **87** Q18
Peterborough, *N.H., U.S.* **97** J13
Peterborough, *Eng., U.K.* **145** S14
Peterborough, *S. Aust., Austral.* **221** X10
Peterhead, *Scot., U.K.* **144** H11
Petermann Bjerg, *Greenland, Den.* **88** J12
Petermann Fjord, *Greenland, Den.* **88** C6
Petermann Gletscher, *Greenland, Den.* **88** D6
Petermann Ranges, *Austral.* **220** F7
Petersburg, *Va., U.S.* **98** F14
Petersburg, *W. Va., U.S.* **98** C11
Petersburg, *Ill., U.S.* **101** R4
Petersburg, *Ind., U.S.* **101** T8
Petersburg, *Alas., U.S.* **113** L23
Petersen Bank, *Ind. Oc.* **255** D14
Petersfield, *Jam.* **122** H6
Petit Bois Island, *Miss., U.S.* **103** P13
Petit Cul-de-Sac Marin, bay, *Guadeloupe, Fr.* **123** F15
Petit Gravier, *Mauritius* **215** J21
Petit Piton, *St. Lucia* **123** L13
Petit Savane Point, *Dominica* **123** G20
Petit Trou, cape, *Trin. & Tobago* **123** P17
Petit-Bourg, *Guadeloupe, Fr.* **123** F14
Petit-Canal, *Guadeloupe, Fr.* **123** E15
Petit-Goâve, *Haiti* **119** M16
Petite Île, *Réunion, Fr.* **215** H16
Petite Rivière del' Artibonite, *Haiti* **119** L16
Petite-Terre, island, *Mayotte, Fr.* **215** P17
Petitot, river, *N. Amer.* **86** H10
Petoskey, *Mich., U.S.* **100** H10

Petra, ruins, *Jordan* **173** Q6
Petrel Station, *Antarctica* **238** C5
Petrich, *Bulg.* **155** K13
Petrified Forest National Park, *Ariz., U.S.* **109** S9
Petrila, *Rom.* **155** E13
Petrinja, *Croatia* **154** E6
Petro, *Pak.* **181** Q9
Petrolândia, *Braz.* **133** G17
Petrolia, *Tex., U.S.* **105** E16
Petrolia, *Calif., U.S.* **111** N1
Petrolina, *Braz.* **133** G16
Petropavlovsk-Kamchatskiy, *Russ.* **161** F23
Petropavlovsk, *Kaz.* **178** B11
Petrovac, *Montenegro* **154** J8
Petrovaradin, *Serb.* **154** E9
Petrovgrad *see* Zrenjanin, *Serb.* **154** E10
Petrovsk Zabaykal'skiy, *Russ.* **161** L16
Petrovsk, *Russ.* **158** L13
Petrozavodsk, *Russ.* **158** D8
Petsamo, *Eur.* **140** A10
Petukhovo, *Russ.* **160** J9
Pevek, *Russ.* **161** C19
Peyrehorade, *Fr.* **149** X6
Peza, river, *Russ.* **158** A12
Pezinok, *Slovakia* **151** K13
Pfaffenhofen, *Ger.* **150** K8
Pforzheim, *Ger.* **150** J6
Phaestus *see* Faistós, ruin, *Gr.* **157** Q14
Phagwara, *India* **182** E5
Phalodi, *India* **182** G3
Phaltan, *India* **182** L4
Phan Rang-Tháp Chàm (Panduranga), *Vietnam* **191** P13
Phan Thiêt, *Vietnam* **191** P12
Phan, *Thai.* **190** J7
Phangan, Ko, *Thai.* **191** Q7
Phangnga, *Thai.* **191** R6
Pharr, *Tex., U.S.* **105** R16
Pharus *see* Hvar, island, *Croatia* **154** H6
Phatthalung, *Thai.* **191** R7
Phayao, *Thai.* **190** J7
Phelps Lake, *N.C., U.S.* **98** H14
Phelps, *Ky., U.S.* **103** C21
Phenix City, *Ala., U.S.* **103** L17
Phetchabun Range, *Thai.* **190** L8
Phetchaburi, *Thai.* **191** N7
Phiafai, *Laos* **190** M11
Phichai, *Thai.* **190** K7
Phichit, *Thai.* **190** L8
Philadelphia *see* 'Ammān (Amman), *Jordan* **172** M7
Philadelphia, *Pa., U.S.* **96** P9
Philadelphia, *Miss., U.S.* **102** K12
Philae, ruin, *Egypt* **202** K12
Philip Island, *Austral.* **229** B20
Philip Smith Mountains, *Alas., U.S.* **113** D17
Philip, *S. Dak., U.S.* **106** L3
Philippeville, *Belg.* **148** M14
Philippi Glacier, *Antarctica* **239** H21
Philippi *see* Fílippoi, ruin, *Gr.* **157** B13
Philippi, *W. Va., U.S.* **98** C10
Philippine Basin, *Pac. Oc.* **246** G6
Philippine Islands, *Asia* **168** J13
Philippine Sea, *Pac. Oc.* **168** J14
Philippine Trench, *Pac. Oc.* **246** H5
Philippines, *Asia* **193** G13
Philippopolis *see* Plovdiv, *Bulg.* **155** J14
Philippopolis *see* Shahbā', *Syr.* **172** K8
Philipsburg, *Pa., U.S.* **96** N5
Philipsburg, *Mont., U.S.* **108** D6
Philipsburg, *St. Maarten, Neth.* **123** C15
Phillips, *Wis., U.S.* **100** G4
Phillips, Mount, *Austral.* **220** G2
Phillipsburg, *Kans., U.S.* **107** R5
Philomath, *Oreg., U.S.* **110** G3
Philpots Island, *Nunavut, Can.* **87** D16
Phippsburg, *Colo., U.S.* **108** L11
Phitsanulok, *Thai.* **190** K7
Phnom Penh *see* Phnum Pénh, *Cambodia* **191** P10
Phnum Pénh (Phnom Penh), *Cambodia* **191** P10
Phoenix Islands, *Kiribati* **224** H10
Phoenix Seamounts, *Pac. Oc.* **246** K12
Phoenix, *Ariz., U.S.* **109** U7
Phoenix, *Mauritius* **215** G20
Phong Thô, *Vietnam* **190** G9
Phôngsali, *Laos* **190** G9

Phouphiang Bolavén, *Laos* **190** L11
Phra Thong, Ko, *Thai.* **191** Q6
Phrae, *Thai.* **190** K7
Phrao, *Thai.* **190** J7
Phrom Phiram, *Thai.* **190** K7
Phsar Réam, *Cambodia* **191** P10
Phú Lộc, *Vietnam* **190** L12
Phú Lộc, *Vietnam* **191** Q11
Phù Mỹ, *Vietnam* **190** M13
Phú Thọ, *Vietnam* **190** G10
Phuket, *Thai.* **191** R6
Phuket, Ko, *Thai.* **191** R6
Phulabani, *India* **182** L9
Phulad, *India* **182** H4
Phumĭ Banam, *Cambodia* **191** P11
Phumĭ Kâmpóng Srâlau, *Cambodia* **190** M11
Phumĭ Lvéa Kraôm, *Cambodia* **191** N9
Phumĭ Thalabârĭvăt, *Cambodia* **190** M11
Phuthaditjhada, *S. Af.* **212** K8
Phyarpon, *Myanmar* **190** L5
Piacenza, *It.* **152** E5
Piamafua Mountain, *American Samoa, U.S.* **230** B9
Pianosa, Isola, *It.* **152** J5
Pianosa, Isola, *It.* **152** K12
Pianu, island, *F.S.M.* **226** F7
Piaseczno, *Pol.* **151** E16
Piatra Neamţ, *Rom.* **155** C16
Piauí, river, *S. Amer.* **128** E10
Piave, river, *It.* **152** C8
Piazza Armerina, *It.* **153** T11
Piazza Pio XII, *Vatican City* **163** Q17
Pibor, *S. Sudan* **203** W12
Pibor, river, *S. Sudan* **203** W12
Pic d'Ahoh, *Niger* **206** C10
Pic de Tibé, *Guinea* **204** M11
Pic du Paradis, *St. Martin, Fr.* **123** B14
Pic Touside, *Chad* **207** D13
Picara Point, *Virgin Is., U.S.* **122** M8
Picardy, region, *Fr.* **148** M10
Picassent, *Sp.* **147** H14
Picayune, *Miss., U.S.* **102** P11
Pichanal, *Arg.* **134** E10
Picher, *Okla., U.S.* **105** A19
Pichilemu, *Chile* **134** L6
Pickens, *Miss., U.S.* **102** K11
Pickford, *Mich., U.S.* **100** F11
Pickle Lake, *Ont., Can.* **87** N15
Pickwick Lake, *Ala., Miss., Tenn., U.S.* **103** G13
Pico Basile, *Eq. Guinea* **206** N9
Pico Fracture Zone, *Atl. Oc.* **248** J8
Pico Truncado, *Arg.* **135** T8
Pico, *It.* **152** L10
Pico, island, *Port.* **146** P4
Picos, *Braz.* **133** F16
Picos, Punta dos, *Sp.* **146** D5
Picture Gallery, *Vatican City* **163** N15
Pictured Rocks National Lakeshore, *Mich., U.S.* **100** F9
Pidarak, *Pak.* **181** R4
Piedmont, *Ala., U.S.* **103** J17
Piedmont, *Mo., U.S.* **107** U15
Piedmont, *Eur.* **140** J6
Piedmont, range, *U.S.* **91** H19
Piedra Grande, Cayo, *Cuba* **118** K10
Piedra, Monasterio de, *Sp.* **147** E13
Piedras Blancas, Point, *Calif., U.S.* **111** V4
Piedras Negras, *Mex.* **116** D11
Piedras, river, *Peru* **132** G5
Pieksämäki, *Fin.* **143** G16
Pielinen, lake, *Fin.* **143** F17
Pieniężno, *Pol.* **151** B15
Pienza, *It.* **152** H7
Piepus, Lake, *Eur.* **140** E10
Pierce, *Nebr., U.S.* **107** N7
Piéria, Óros, *Gr.* **156** D10
Pierre, *S. Dak., U.S.* **106** K5
Pierrelatte, *Fr.* **149** W12
Pierreville, *Trin. & Tobago* **123** Q23
Pierrot Island (Cat Island), *Mauritius* **215** K20
Piešťany, *Slovakia* **151** K14
Pietarsaari see Jakobstad, *Fin.* **143** G15
Pieter Both, peak, *Mauritius* **215** G20
Pietermaritzburg, *S. Af.* **212** K9
Pietra-di-Verde, *Fr.* **149** X3
Pietrosu, peak, *Rom.* **155** C14
Pietrosul, peak, *Rom.* **155** B14
Pigafetta Basin, *Pac. Oc.* **246** G8
Pigailoe (West Fayu Atoll), *F.S.M.* **226** B8

Pigeon Cay, *Bahamas* **122** B11
Pigeon Creek, *Bahamas* **122** F11
Pigeon Island, *Jam.* **122** K9
Pigeon Point, *Trin. & Tobago* **123** P17
Pigeon, *Mich., U.S.* **100** K12
Pigeon, *Guadeloupe, Fr.* **123** F13
Piggott, *Ark., U.S.* **102** E11
Pigna Courtyard, *Vatican City* **163** N16
Pigs, Bay of see Cochinos, Bahía de, *Cuba* **118** H7
Pigüé, *Arg.* **135** N11
Piha Passage, *Tonga* **227** Q24
Pihtipudas, *Fin.* **143** G16
Pihyŏn, *N. Korea* **186** G4
Piis-Panewu, island, *F.S.M.* **226** D9
Pijijiapan, *Mex.* **117** K14
Pik Pobedy see Victory Peak, *China, Kyrg.* **179** K17
Pikalevo, *Russ.* **158** F8
Pike Bank, *Ind. Oc.* **250** R10
Pikelot, island, *F.S.M.* **226** B8
Pikes Peak, *Colo., U.S.* **109** N13
Pikeville, *Ky., U.S.* **103** C20
Pikijin, island, *Marshall Is.* **227** G20
Pikiutdleq, bay, *Greenland, Den.* **89** Q9
Pikounda, *Congo* **208** D11
Piła, *Pol.* **151** C13
Pilão Arcado, *Braz.* **133** G15
Pilar, *Parag.* **132** N10
Pilar, Cabo, *Chile* **135** X6
Pilcomayo, river, *S. Amer.* **128** J6
Pileni, island, *Solomon Is.* **228** K5
Pil'gyn, *Russ.* **161** B20
Pilibhit, *India* **182** G7
Pillsbury Sound, *Virgin Is., U.S.* **122** N10
Pilón, *Cuba* **118** L11
Pilot Knob, peak, *Idaho, U.S.* **108** E4
Pilot Peak, *Nev., U.S.* **111** R8
Pilot Point, *Alas., U.S.* **113** M13
Pilot Rock, *Oreg., U.S.* **110** F7
Pilot Station, *Alas., U.S.* **112** H12
Pilottown, *La., U.S.* **102** R12
Pilsen see Plzeň, *Czechia* **150** H10
Pimentel, *Dom. Rep.* **119** L19
Pimlico Island, *Bahamas* **122** D4
Pinaki 2, island, *Fr. Polynesia, Fr.* **231** D21
Pinang, island, *Malaysia* **191** T8
Pinangah, *Malaysia* **192** G12
Pinar del Río, *Cuba* **118** G5
Pinas, *Arg.* **134** J9
Piñas, *Ecua.* **130** M3
Pinatubo, Mount, *Asia* **168** J13
Pinckneyville, *Ill., U.S.* **101** U5
Pinconning, *Mich., U.S.* **100** K11
Pindaré, river, *Braz.* **133** E14
Pinders Point, *Bahamas* **122** J1
Pindi Bhattian, *Pak.* **180** K10
Pindi Gheb, *Pak.* **180** J11
Píndos Óros, *Gr.* **156** F8
Pindus Mountains, *Alban., Gr.* **154** M10
Pine Barrens, *U.S.* **91** F21
Pine Bluff, *Ark., U.S.* **102** H9
Pine Bluffs, *Wyo., U.S.* **108** K14
Pine City, *Minn., U.S.* **106** H12
Pine Creek, *N.T., Austral.* **221** P8
Pine Hill, *N.T., Austral.* **221** T8
Pine Hills, *Fla., U.S.* **99** T9
Pine Island Bay, *Antarctica* **238** K7
Pine Island Glacier, *Antarctica* **238** J7
Pine Point, *N.W.T., Can.* **86** J11
Pine Ridge, *S. Dak., U.S.* **106** M2
Pine River, *Minn., U.S.* **106** G10
Pinedale, *Wyo., U.S.* **108** H9
Pinedale, *Calif., U.S.* **111** T6
Pinega, *Russ.* **158** B11
Pinega, river, *Russ.* **158** C12
Pinegrove, *W. Aust., Austral.* **221** V2
Pinehurst, *N.C., U.S.* **98** J11
Pinehurst, *Idaho, U.S.* **108** C4
Pineiós, river, *Gr.* **156** E10
Pineland, *Tex., U.S.* **105** J21
Pinellas Park, *Fla., U.S.* **99** U7
Pinerolo, *It.* **152** E2
Pinetop-Lakeside, *Ariz., U.S.* **109** T9
Pinetown, *S. Af.* **212** K9
Pineville, *La., U.S.* **102** N8
Pineville, *Ky., U.S.* **103** D19
Piney Buttes, *Mont., U.S.* **108** C11
Ping, river, *Thai.* **190** L7
Pingdingshan, *China* **185** J13

Pingelap Atoll, *F.S.M.* **226** B11
Pingelap Atoll, *F.S.M.* **227** D13
Pingelly, *W. Aust. Austral.* **221** X3
Pinglap, island, *Marshall Is.* **227** H19
Pingliang, *China* **184** J11
Pingtung, *China* **185** N16
Pingualuit Crater (Cratère du Nouveau-Québec), *Que., Can.* **87** J18
Pingwu, *China* **184** K10
Pingxiang, *China* **185** M13
Pingxiang, *China* **184** P11
Pinheiro, *Braz.* **133** D14
Pinhel, *Port.* **146** F7
Pini, island, *Indonesia* **192** J5
Pinjan, *Indonesia* **193** H13
Pinleibu, *Myanmar* **190** F5
Pinnacles National Park, *Calif., U.S.* **111** U5
Pinneberg, *Ger.* **150** C7
Pínnes, Ákra (Nymphaeum), *Gr.* **157** D13
Pinney's Beach, *St. Kitts & Nevis* **123** C18
Pinos-Puente, *Sp.* **146** L11
Pinos, Point, *Calif., U.S.* **111** T4
Pins, Île des (Isle of Pines), *New Caledonia, Fr.* **224** K7
Pinsk Marshes, *Belarus* **158** M4
Pinsk, *Belarus* **158** M4
Pintados, *Chile* **134** C7
Pinyug, *Russ.* **158** E13
Pioche, *Nev., U.S.* **111** R12
Piombino, *It.* **152** J6
Pioneer Fracture Zone, *Pac. Oc.* **247** E15
Pioneer Mountains, *Mont., U.S.* **108** E6
Pioner, *Russ.* **161** D14
Pionsat, *Fr.* **149** T10
Piotrków Trybunalski, *Pol.* **151** F15
Pip'agot, point, *N. Korea* **186** K5
Pipanaco, Salar de, *Arg.* **134** H9
Pipar, *India* **182** H4
Piparia, *India* **182** K6
Pipe Spring National Monument, *Ariz., U.S.* **109** Q6
Piper Peak, *Nev., U.S.* **111** S8
Pipéri, island, *Gr.* **157** F13
Pipestem Creek, river, *N. Dak., U.S.* **106** G6
Pipestone National Monument, *Minn., U.S.* **106** L8
Pipestone, *Minn., U.S.* **106** L8
Pipinas, *Arg.* **134** M13
Piplan, *Pak.* **180** K10
Piqua, *Ohio, U.S.* **101** R11
Pir Panjal Range, *India* **180** J13
Pira, *Benin* **205** N17
Piracicaba, *Braz.* **133** M13
Piracuruca, *Braz.* **133** E16
Pirae, *Fr. Polynesia, Fr.* **231** H21
Piraeus see Peiraiás, *Gr.* **156** J12
Pirallahı, *Azerb.* **171** D23
Pirané, *Arg.* **134** F12
Piranhas, *Braz.* **132** J12
Pirapora, *Braz.* **133** K14
Pirássununga, *Braz.* **133** L13
Pirimapun, *Indonesia* **193** M20
Pirin, mountains, *Bulg.* **155** K13
Piripiri, *Braz.* **133** E16
Pirmasens, *Ger.* **150** H5
Pirna, *Ger.* **150** F11
Pirogue, *Guadeloupe, Fr.* **123** G16
Pirojpur, *Bangladesh* **182** K12
Pirot, *Serb.* **154** H12
Pirsagat, river, *Azerb.* **171** D22
Pisa, *It.* **152** G6
Pisagua, *Chile* **134** C6
Pisamew, island, *F.S.M.* **226** D8
Pisaras, island, *F.S.M.* **226** A9
Piscaderabaai, bay, *Curaçao, Neth.* **123** P14
Pisciotta, *It.* **153** N12
Pisco, *Peru* **132** J3
Piscopi see Tílos, island, *Gr.* **157** M18
Písek, *Czechia* **150** H10
Pisemew, island, *F.S.M.* **226** D9
Pishan (Guma), *China* **184** G2
Pishín Jān, *Afghan.* **180** J3
Pishin, *Pak.* **180** M7
Pīshīn, *Iran* **177** N23
Pisininin, island, *F.S.M.* **226** D8
Pisinun, island, *F.S.M.* **226** F10
Pismo Beach, *Calif., U.S.* **111** V5
Piso Firme, *Bol.* **132** H8

Pissos, *Fr.* **149** W6
Pissouri, *Cyprus* **162** Q6
Pisticci, *It.* **153** N13
Pistoia, *It.* **152** G6
Pistol River, *Oreg., U.S.* **110** K1
Pisz, *Pol.* **151** C16
Pit Gorodok, *Russ.* **161** J13
Pita, *Guinea* **204** L9
Pitaga, *Nfld. & Lab., Can.* **87** L20
Pitanga, *Braz.* **132** M12
Pitav, *Afghan.* **180** F9
Pitcairn Island, *Pac. Oc.* **225** L16
Pitcairn Islands, *Pac. Oc.* **225** K16
Piteå, *Sw.* **143** F14
Piteşti, *Rom.* **155** F14
Piti u Tai, island, *Fr. Polynesia, Fr.* **231** H14
Piti u Uta, island, *Fr. Polynesia, Fr.* **231** H14
Piti, *Guam, U.S.* **230** P5
Pitlochry, *Scot., U.K.* **144** K10
Pitman Fracture Zone, *Pac. Oc.* **255** K16
Piton, peak, *Wallis & Futuna Is., Fr.* **230** R11
Pitt Island (Rangiauria), *Pac. Oc.* **224** P9
Pitti Island, *India* **183** S3
Pittsburg, *Tex., U.S.* **105** G19
Pittsburg, *Kans., U.S.* **107** U10
Pittsburgh, *Pa., U.S.* **96** N2
Pittsfield, *Mass., U.S.* **97** K12
Pittsfield, *Me., U.S.* **97** F17
Pittsfield, *Ill., U.S.* **101** R3
Pittston Farm, *Me., U.S.* **97** D16
Pittston, *Pa., U.S.* **96** M9
Piua Petrii, *Rom.* **155** F17
Piura, *Peru* **132** F2
Piva, river, *Montenegro* **154** H8
Pınarbaşı, *Turk.* **170** G11
Pkulagalid, cape, *Palau* **226** F6
Pkulagasemieg, island, *Palau* **226** J5
Pkulngril, cape, *Palau* **226** G6
Pkurengel, cape, *Palau* **226** G5
Placentia Bay, *Nfld. & Lab. Can.* **87** M24
Placentia, *Nfld. & Lab., Can.* **87** M24
Placerville, *Calif., U.S.* **111** Q5
Placetas, *Cuba* **118** H9
Placida, *Fla., U.S.* **99** V8
Plain Dealing, *La., U.S.* **102** K6
Plain of Reeds, *Vietnam* **191** P11
Plainfield, *N.J., U.S.* **96** N10
Plainfield, *Ind., U.S.* **101** R9
Plains, *Ga., U.S.* **99** N5
Plains, *Tex., U.S.* **104** F11
Plains, *Kans., U.S.* **107** V4
Plainview, *Tex., U.S.* **104** E12
Plainview, *Minn., U.S.* **106** K12
Plainview, *Nebr., U.S.* **107** N7
Plainville, *Kans., U.S.* **107** S5
Plainwell, *Mich., U.S.* **100** M10
Plaisance, *Haiti* **119** L16
Plaka see Mílos, *Gr.* **157** M13
Plakoti, Cape, *Cyprus* **162** M11
Plana Cays, *Bahamas* **119** G15
Planalto Brasileiro, *Braz.* **133** K15
Planalto da Lichinga, *Mozambique* **212** C11
Planalto do Chimoio, *Mozambique* **212** F10
Planken, *Liech.* **162** N2
Plankinton, *S. Dak., U.S.* **106** L6
Plano, *Tex., U.S.* **105** G18
Plant City, *Fla., U.S.* **99** U8
Plantation, *Fla., U.S.* **99** W10
Plaquemine, *La., U.S.* **102** P9
Plasencia, *Sp.* **146** G8
Plasstéikopf, peak, *Liech.* **162** Q3
Plata, Isla de la, *Ecua.* **130** L2
Plata, Río de la, *Arg.* **134** L13
Platamónas, *Gr.* **156** E10
Plateau du Djado, *Niger* **206** C10
Plateau du Tademaït, *Alg.* **200** G12
Plateau Fairway, *Pac. Oc.* **246** M9
Plateau Station, site, *Antarctica* **239** F15
Plato Putorana, *Russ.* **161** G14
PLATO-R Observatory, *Antarctica* **239** H16
Plato, *Col.* **130** D6
Platte, *S. Dak., U.S.* **106** M6
Platte, river, *U.S.* **90** F12

Platteville, *Wis., U.S.* **100** M4
Plattsburg, *Mo., U.S.* **107** S10
Plattsburgh, *N.Y., U.S.* **96** F11
Plattsmouth, *Nebr., U.S.* **107** Q8
Platypus Spur, *Ind. Oc.* **251** J17
Plauen, *Ger.* **150** G9
Plavsk, *Russ.* **158** L9
Playa Blanca, *Sp.* **214** Q7
Playa Dayaniguas, *Cuba* **118** H5
Playa Girón, *Cuba* **118** H8
Playa Grandi, bay, *Curaçao, Neth.* **123** N14
Playa Honda, *Sp.* **214** P7
Playas Lake, *N. Mex., U.S.* **109** W10
Plaza Huincul, *Arg.* **135** P8
Pleasant Grove, *Utah, U.S.* **108** L7
Pleasant, Lake, *Ariz., U.S.* **109** T6
Pleasanton, *Tex., U.S.* **105** M16
Pleasanton, *Kans., U.S.* **107** T10
Pleasantville, *N.J., U.S.* **96** Q10
Pleasure Ridge Park, *Ky., U.S.* **103** B16
Plei Ku (Pleiku), *Vietnam* **190** M12
Pleiku see Plei Ku, *Vietnam* **190** M12
Pléneuf, *Fr.* **149** P4
Plentywood, *Mont., U.S.* **108** A13
Plesetsk, *Russ.* **158** D10
Plestin, *Fr.* **149** P3
Plétipi, Lac, *Que., Can.* **87** M20
Plettenberg Bay, *S. Af.* **212** M6
Pleubian, *Fr.* **149** P4
Pleven, *Bulg.* **155** H14
Plevna, *Mont., U.S.* **108** D13
Pljevlja, *Montenegro* **154** H9
Ploče, *Croatia* **154** H7
Płock, *Pol.* **151** D15
Ploërmel, *Fr.* **149** Q5
Ploieşti, *Rom.* **155** F15
Plomári, *Gr.* **157** G17
Płońsk, *Pol.* **151** D15
Plouézec, *Fr.* **149** P4
Plouguerneau, *Fr.* **149** P2
Plovdiv (Philippopolis), *Bulg.* **155** J14
Plummer, *Idaho, U.S.* **108** C3
Plumridge Lakes, *Austral.* **220** H6
Plumtree, *Zimb.* **212** F8
Plymouth (former capital), ruin, *Montserrat, U.K.* **123** C23
Plymouth, *Pa., U.S.* **96** M8
Plymouth, *Vt., U.S.* **96** H12
Plymouth, *Mass., U.S.* **97** L15
Plymouth, *N.H., U.S.* **97** H14
Plymouth, *N.C., U.S.* **98** H14
Plymouth, *Ind., U.S.* **100** P9
Plymouth, *Wis., U.S.* **100** K7
Plymouth, *Trin. & Tobago* **123** N17
Plymouth, *Eng., U.K.* **145** W9
Plyusa, *Russ.* **158** G6
Plzeň (Pilsen), *Czechia* **150** H10
Pniewy, *Pol.* **150** D12
Pô, *Burkina Faso* **205** L15
Po, Foci del (Po, Mouths of the), *It.* **152** F8
Po, Mouths of the see Po, Foci del, *It.* **152** F8
Po, river, *It.* **152** E6
Pobé, *Benin* **205** N18
Pobeda, *Russ.* **161** F19
Pocahontas, *W. Va., U.S.* **98** F8
Pocahontas, *Ark., U.S.* **102** E10
Pocahontas, *Iowa, U.S.* **106** M10
Pocatello, *Idaho, U.S.* **108** H7
Pochala, *S. Sudan* **203** W13
Pocheon, *S. Korea* **186** M8
Pochep, *Russ.* **158** L7
Poch'ŏn, *N. Korea* **186** D10
Pochutla, *Mex.* **117** K13
Pocolo, *Angola* **208** P10
Pocomoke City, *Md., U.S.* **98** D15
Poconé, *Braz.* **132** J9
Pocono Mountains, *Pa., U.S.* **96** M9
Poços de Caldas, *Braz.* **133** L13
Poddor'ye, *Russ.* **158** H6
Poděbrady, *Czechia* **150** G11
Podgorica, *Montenegro* **154** J8
Podil's'ka Vysochyna, *Ukr.* **159** P4
Podkamennaya Tunguska, river, *Russ.* **161** J12
Podol'sk, *Russ.* **158** J9
Podor, *Senegal* **204** H7
Podporozh'ye, *Russ.* **158** E8
Podravska Slatina, *Croatia* **154** D7
Podsosan'ye, *Russ.* **158** C11
Pofadder, *S. Af.* **212** K5
Poggibonsi, *It.* **152** H7
Poggio Rusco, *It.* **152** E7
Pogost, *Russ.* **158** E9

Poh, *Indonesia* **193** J14
P'oha, *N. Korea* **186** E13
Pōhakueaea Point, *Hawaii, U.S.* **115** K19
P'ohang, *S. Korea* **187** R12
Pohnpei (Ponape), *F.S.M.* **226** B11
Pohnpei Harbor, *F.S.M.* **226** J8
Pohorje, *mountains, Slov.* **154** C5
Pōhue Bay, *Hawaii, U.S.* **115** Q21
Pohuerahi, *peak, Fr. Polynesia, Fr.* **231** K14
Poi, *Wallis & Futuna Is., Fr.* **230** Q11
Poie, *D.R.C.* **209** F15
Poim, *Russ.* **158** L12
Poindimié, *New Caledonia, Fr.* **231** P21
Poinsett, Cape, *Antarctica* **239** M21
Point Arena, *Calif., U.S.*
Point Blanche, *St. Maarten, Neth.* **123** C15
Point Christian, *Pitcairn I.* **228** Q5
Point Fortin, *Trin. & Tobago* **123** R21
Point Gourde, *Trin. & Tobago* **123** P21
Point Hope, *Alas., U.S.* **112** C12
Point Howe, *Austral.* **229** A20
Point Lake, *N.W.T., Can.* **86** G12
Point Lay, *Alas., U.S.* **113** C13
Point of Fort Jeudy, *Grenada* **123** L22
Point Pedro, *Sri Lanka* **183** S7
Point Pleasant, *N.J., U.S.* **96** P11
Point Pleasant, *W. Va., U.S.* **98** D7
Point Radix, *Trin. & Tobago* **123** Q24
Point Reyes National Seashore, *Calif., U.S.* **111** R2
Point Ross, *Austral.* **229** B20
Point Salines, *Grenada* **123** L21
Point Samson, *W. Aust., Austral.* **221** S2
Point Sur, *Calif., U.S.* **111** U4
Point Vauvilliers, *F.S.M.* **226** G11
Point Vincent, *Austral.* **229** A20
Point Washington, *Fla., U.S.* **99** Q3
Pointe a la Hache, *La., U.S.* **102** Q11
Pointe Allègre, *Guadeloupe, Fr.* **123** E14
Pointe Arei, *Fr. Polynesia, Fr.* **231** K17
Pointe Banda, *Gabon* **208** G8
Pointe Baravo, *Madagascar* **213** J14
Pointe d'Alo, *Wallis & Futuna Is., Fr.* **230** Q11
Pointe de la Grande Vigie, *Guadeloupe, Fr.* **123** D15
Pointe de la Rose, *Martinique, Fr.* **123** F23
Pointe d'Enfer, *Martinique, Fr.* **123** G23
Pointe des Basses, *Guadeloupe, Fr.* **123** G16
Pointe des Châteaux, *Guadeloupe, Fr.* **123** E17
Pointe d'Itaperina, *Madagascar* **213** J16
Pointe du Vieux-Fort, *Guadeloupe, Fr.* **123** G14
Pointe Falaise, *St. Martin, Fr.* **123** B14
Pointe Fareara, *Fr. Polynesia, Fr.* **231** K23
Pointe Gautier, *St. Lucia* **123** L14
Pointe Halalo, *Wallis & Futuna Is., Fr.* **230** R8
Pointe Hardy, *St. Lucia* **123** J14
Pointe Haunanu, *Fr. Polynesia, Fr.* **230** D11
Pointe Jaquot, *Dominica* **123** D19
Pointe Kiukiu, *Fr. Polynesia, Fr.* **231** D13
Pointe Koutu Puhipuhi, *Fr. Polynesia, Fr.* **231** Q17
Pointe La Ville, *St. Lucia* **123** K13
Pointe Matalaa, *Wallis & Futuna Is., Fr.* **230** R9
Pointe Matatépai, *Fr. Polynesia, Fr.* **231** D14
Pointe Matira, *Fr. Polynesia, Fr.* **231** H14
Pointe Mautau, *Fr. Polynesia, Fr.* **231** D14
Pointe Michel, *Dominica* **123** G19
Pointe Mulâtre, *Dominica* **123** G20
Pointe Nord, *St. Martin, Fr.* **123** A15
Pointe Nord, *Wallis & Futuna Is., Fr.* **230** Q11
Pointe Nuupere, *Fr. Polynesia, Fr.* **231** H20
Pointe Plum, *St. Martin, Fr.* **123** B13
Pointe Pongara, *Gabon* **208** E7
Pointe Punaauia, *Fr. Polynesia, Fr.* **231** H21
Pointe Pyramides, *Wallis & Futuna Is., Fr.* **230** Q11

Pointe Raititi, *Fr. Polynesia, Fr.* **231** H14
Pointe Ronde, *Dominica* **123** E18
Pointe Sauma, *Wallis & Futuna Is., Fr.* **230** R12
Pointe Taurere, *Fr. Polynesia, Fr.* **231** H15
Pointe Teaehoa, *Fr. Polynesia, Fr.* **231** E14
Pointe Ténos, *Martinique, Fr.* **123** E23
Pointe Teohotepapa, *Fr. Polynesia, Fr.* **230** E11
Pointe Teohotepapa, *Fr. Polynesia, Fr.* **231** D15
Pointe Tepaeture, *Fr. Polynesia, Fr.* **231** Q17
Pointe Tepako, *Wallis & Futuna Is., Fr.* **230** Q9
Pointe Tepuu, *Fr. Polynesia, Fr.* **231** N14
Pointe Tetutu, *Fr. Polynesia, Fr.* **230** E11
Pointe Toataratara, *Fr. Polynesia, Fr.* **231** L17
Pointe Tuahora, *Fr. Polynesia, Fr.* **231** G14
Pointe Velé, *Wallis & Futuna Is., Fr.* **230** Q12
Pointe Vénus, *Fr. Polynesia, Fr.* **231** H22
Pointe Verte, *Wallis & Futuna Is., Fr.* **230** Q12
Pointe-à-Gravois, *Haiti* **119** N15
Pointe-à-Peine, *Dominica* **123** F20
Pointe-à-Pierre, *Trin. & Tobago* **123** Q22
Pointe-à-Pitre, *Guadeloupe, Fr.* **123** E15
Pointe-Noire, *Guadeloupe, Fr.* **123** F13
Pointe-Noire, *Congo* **208** H8
Po'ipū, *Hawaii, U.S.* **114** F10
Poirino, *It.* **152** E3
Poissonnier Point, *Austral.* **220** E3
Poissy, *Fr.* **149** P9
Poitiers, *Fr.* **149** T7
Poitou, *Eur.* **140** H4
Poix, *Fr.* **149** N9
Pok, *F.S.M.* **226** K8
Pokhvistnevo, *Russ.* **158** K16
Pokka, *Fin.* **143** C15
Poko, *D.R.C.* **209** C17
Pola de Lena, *Sp.* **146** B8
Pola de Siero, *Sp.* **146** B9
Pola Island, *American Samoa, U.S.* **230** D7
Pola see Pula, *Croatia* **154** E4
Polacca, *Ariz., U.S.* **109** R8
Polān, *Iran* **177** P22
Poland, *Eur.* **151** E13
Poland, *Kiribati* **226** Q10
Polar Plateau, *Antarctica* **239** H13
Polar Sea Bank, *Arctic Oc.* **253** K17
Polarsirkel Valley, *Atl. Oc.* **249** Y13
Polatlı, *Turk.* **170** E7
Polatsk, *Belarus* **158** J5
Pole of Inaccessibility Station, site, *Antarctica* **239** G15
Pole Plain, *Arctic Oc.* **252** G11
Polesella, *It.* **152** E7
Poles', *Eur.* **140** G10
Policastro Bussentino, *It.* **153** N12
Policastro, Golfo di, *It.* **153** N12
Police Point, *Seychelles* **215** P20
Police, *Pol.* **150** C11
Polichnítos, *Gr.* **157** G16
Poligny, *Fr.* **149** S13
Polink Point, *Jam.* **122** K10
Polis, *Cyprus* **162** P5
Polk, *Pa., U.S.* **96** M3
Pollino, Monte, *It.* **153** P13
Pollock, *S. Dak., U.S.* **106** H5
Polnovat, *Russ.* **160** G10
Poloa Bay, *American Samoa, U.S.* **230** E5
Polohy, *Ukr.* **159** R9
Polokwane, *S. Af.* **212** H9
Polokwane, *river, S. Af.* **212** G9
Polonne, *Ukr.* **159** N4
Polson, *Mont., U.S.* **108** C5
Poltava, *Ukr.* **159** P8
Polvadera, *N. Mex., U.S.* **109** T12
Polýaigos, *island, Gr.* **157** M14
Polyarnyy, *Russ.* **160** C9
Polýgyros, *Gr.* **156** D12
Polýkastro, *Gr.* **156** B10
Polynesia, *islands, Pac. Oc.* **224** D10
Pomabamba, *Peru* **132** G3

Pombal, *Port.* **146** G5
Pomerania, region, *Ger., Pol.* **150** C11
Pomeranian Bay, *Atl. Oc.* **140** F7
Pomeroy, *Ohio, U.S.* **101** S14
Pomeroy, *Wash., U.S.* **110** D9
Pomichna, *Ukr.* **159** Q6
Pomir (Pamir), *river, Afghan., Taj.* **180** F11
Pomírla, *Rom.* **155** A15
Pomona, *Calif., U.S.* **111** X9
Pomorie (Archialus), *Bulg.* **155** J17
Pomos Point, *Cyprus* **162** N6
Pomozdino, *Russ.* **158** C15
Pompano Beach, *Fla., U.S.* **99** W10
Pompeii, ruin, *It.* **152** M11
Pompeiopolis, ruins, *Turk.* **170** J9
Pompeys Pillar National Historic Landmark, *Mont., U.S.* **108** E10
Ponape see Pohnpei, *island, F.S.M.* **226** B11
Ponca City, *Okla., U.S.* **105** B17
Ponce de Leon Bay, *Fla., U.S.* **99** X9
Ponchatoula, *La., U.S.* **102** P10
Poncin, *Fr.* **149** T13
Pond Creek, *Okla., U.S.* **105** B16
Pond Inlet, *Nunavut, Can.* **87** E17
Pondicherry see Puducherry, *India* **183** R7
Pondosa, *Calif., U.S.* **110** M4
Ponérihouen, *New Caledonia, Fr.* **231** P21
Pones, *island, F.S.M.* **226** F10
Ponferrada, *Sp.* **146** C8
Pong, *Thai.* **190** J7
Pongaroa, *N.Z.* **223** J21
Pongch'ŏn, *N. Korea* **186** L7
Ponggang, *N. Korea* **186** D12
Pongo, *river, S. Sudan* **203** W8
Pon'goma, *Russ.* **158** B8
Pongsan, *N. Korea* **186** K6
Poniya Point, *N. Mariana Is., U.S.* **230** N8
Ponnani, *India* **183** S5
Ponoy, *Russ.* **160** D9
Pons, *Fr.* **149** U6
Pont-à-Mousson, *Fr.* **149** P14
Pont-Aven, *Fr.* **149** Q3
Pont-d'Ain, *Fr.* **149** T13
Pont-de-Salars, *Fr.* **149** W10
Pont-l'Évêque, *Fr.* **149** N7
Pont-y-pwl see Pontypool, *Wales, U.K.* **145** U10
Ponta Albina, *Angola* **208** P8
Ponta da Barra Falsa, *Mozambique* **212** G11
Ponta da Barra, *Mozambique* **212** H11
Ponta da Marca, *Angola* **208** Q8
Ponta das Palmeirinhas, *Angola* **208** K9
Ponta das Salinas, *Angola* **208** M9
Ponta Delgada, *Port.* **146** Q5
Ponta do Pargo, *Madeira, Port.* **214** M2
Ponta do Pico, *Port.* **146** P3
Ponta do Sol, *Madeira, Port.* **214** M2
Ponta Grossa, *Braz.* **132** M12
Ponta Metacõua, *Mozambique* **213** C13
Ponta Porã, *Braz.* **132** L10
Ponta São Sebastião, *Mozambique* **212** G11
Pontarlier, *Fr.* **149** S14
Pontaumur, *Fr.* **149** U10
Pontchartrain, Lake, *La., U.S.* **102** P11
Pontchâteau, *Fr.* **149** R5
Ponte de Lima, *Port.* **146** D5
Ponte Vedra Beach, *Fla., U.S.* **99** R9
Ponte-Leccia, *Fr.* **149** X2
Ponteareas, *Sp.* **146** C5
Pontebba, *It.* **152** C9
Pontedecimo, *It.* **152** F4
Pontevedra, *Sp.* **146** C5
Pontevedra, Ría de, *Sp.* **146** C5
Pontiac, *Mich., U.S.* **100** M12
Pontiac, *Ill., U.S.* **101** Q6
Pontianak, *Indonesia* **192** J8
Pontifical Academy of Sciences, *Vatican City* **163** P15
Pontikonísi, *island, Gr.* **156** P12
Pontinia, *It.* **152** L9
Pontivy, *Fr.* **149** Q4
Pontoise, *Fr.* **149** P9
Pontotoc, *Miss., U.S.* **102** H12
Pontremoli, *It.* **152** F5
Ponts, *Sp.* **147** D16
Pontypool (Pont-y-pwl), *Wales, U.K.* **145** U10
Ponza, Isola di, *It.* **152** M9

Ponziane, Isole, *It.* **152** M9
Pool Malebo, *Congo, D.R.C.* **208** G11
Poole, *Eng., U.K.* **145** W12
Poolewe, *Scot., U.K.* **144** H8
Poopelloe Lake, *Austral.* **220** J12
Poopó, *Bol.* **132** K7
Poopó, Lago, *Bol.* **132** K7
Popayán, *Col.* **130** H5
Poplar Bluff, *Mo., U.S.* **107** V15
Poplar, *Mont., U.S.* **108** B13
Poplar, *river, Man., Can.* **87** M14
Poplar, *river, Can., U.S.* **108** B12
Poplarville, *Miss., U.S.* **102** N11
Popocatépetl, *peak, Mex.* **116** J12
Popokabaka, *D.R.C.* **208** H11
Popondetta, *P.N.G.* **222** Q8
Popovo, *Bulg.* **155** H16
Poprad, *Slovakia* **151** J16
Pŏptong, *N. Korea* **186** J8
Porali, *river, Pak.* **181** Q6
Porangahau, *N.Z.* **223** J21
Porbandar, *India* **182** K1
Porcuna, *Sp.* **146** K10
Porcupine Bank, *Atl. Oc.* **248** G10
Porcupine Mountains, *Mich., U.S.* **100** F5
Porcupine Plain, *Atl. Oc.* **248** H10
Porcupine Seabight, *Atl. Oc.* **248** H11
Porcupine, *river, Can., U.S.* **113** E18
Pordenone, *It.* **152** D8
Porea, *island, N.Z.* **229** J23
Poreč, *Croatia* **154** E3
Porga, *Benin* **205** L17
Porgho, *Mali* **205** H16
Porgy Bay, *Bahamas* **122** B11
Pori, *Fin.* **143** H15
Porirua, *N.Z.* **223** J20
Porlamar, *Venez.* **130** C12
Pornic, *Fr.* **149** S5
Póros, *Gr.* **156** K12
Póros, *island, Gr.* **156** K12
Porosozero, *Russ.* **158** D7
Porpoise Basin, *Ind. Oc.* **255** E15
Porpoise Bay, *Antarctica* **239** P19
Porrentruy, *Switz.* **152** A2
Porsanger, bay, *Nor.* **143** B14
Porsanger Peninsula, *Eur.* **140** A9
Porsuk, *river, Turk.* **170** E6
Port Albert, *Vic., Austral.* **221** Z13
Port Alfred, *S. Af.* **212** M8
Port Allegany, *Pa., U.S.* **96** L5
Port Allen, *La., U.S.* **102** P9
Port Angeles, *Wash., U.S.* **110** B3
Port Antonio, *Jam.* **122** H11
Port Aransas, *Tex., U.S.* **105** P17
Port Arthur, *Tex., U.S.* **105** L21
Port Arthur, *Tas., Austral.* **221** Z8
Port Augusta, *S. Aust., Austral.* **221** X10
Port Austin, *Mich., U.S.* **100** K12
Port Berard, *F.S.M.* **226** G11
Port Blair, *India* **183** R14
Port Boisé, *New Caledonia, Fr.* **231** Q23
Port Bolivar, *Tex., U.S.* **105** L20
Port Bourayne, *Fr. Polynesia, Fr.* **231** K24
Port Brega see Marsá al Burayqah, *Lib.* **201** E20
Port Burwell, *Nfld. & Lab., Can.* **87** J20
Port Castries, bay, *St. Lucia* **123** J14
Port Charlotte, *Fla., U.S.* **99** V8
Port Clinton, *Ohio, U.S.* **101** N13
Port du Beaumanoir, *Fr. Polynesia, Fr.* **231** K23
Port Eads, *La., U.S.* **102** R12
Port Elizabeth, *St. Vincent & the Grenadines* **121** G17
Port Elizabeth, *S. Af.* **212** M7
Port Ellen, *Scot., U.K.* **145** M7
Port Esquivel, *Jam.* **122** J9
Port Fairy, *Vic., Austral.* **221** Z11
Port Gibson, *Miss., U.S.* **102** M10
Port Graham, *Alas., U.S.* **113** L16
Port Harcourt, *Nigeria* **206** N8
Port Hardy, *B.C., Can.* **86** L7
Port Hawkesbury, *N.S., Can.* **87** P22
Port Hedland, *W. Aust., Austral.* **221** S3
Port Heiden, *Alas., U.S.* **113** M13
Port Henry, *N.Y., U.S.* **96** G11
Port Hope Simpson, *Nfld. & Lab., Can.* **87** L22
Port Hope, *Mich., U.S.* **100** K13
Port Howe, *Bahamas* **119** F13
Port Huron, *Mich., U.S.* **100** L13
Port Isabel, *Tex., U.S.* **105** R17
Port Jervis, *N.Y., U.S.* **96** M10
Port Kaiser, *Jam.* **122** J8

Port Kaituma, *Guyana* **131** E14
Port Láirge see Waterford, *Ire.* **145** T6
Port Laoise, *Ire.* **145** R6
Port Lavaca, *Tex., U.S.* **105** N18
Port Lincoln, *S. Aust., Austral.* **221** X9
Port Lions, *Alas., U.S.* **113** M15
Port Loko, *Sierra Leone* **204** M8
Port Lottin, *F.S.M.* **226** G11
Port Louis, *Mauritius* **215** G20
Port Macquarie, *N.S.W. Austral.* **221** X15
Port Mansfield, *Tex., U.S.* **105** Q17
Port Manvers, *Nfld. & Lab., Can.* **87** J21
Port Maria, *Jam.* **122** H10
Port Mathurin, *Mauritius* **215** J20
Port Morant, *Jam.* **122** J11
Port Moresby, *P.N.G.* **222** Q8
Port Neches, *Tex., U.S.* **105** L21
Port Nelson, *Bahamas* **119** F14
Port Nis see Port of Ness, *Scot., U.K.* **144** F7
Port Nolloth, *S. Af.* **212** K4
Port Norris, *N.J., U.S.* **96** R9
Port O'Connor, *Tex., U.S.* **105** N18
Port of Ness (Port Nis), *Scot., U.K.* **144** F7
Port of Spain, *Trin. & Tobago* **123** P22
Port Ontario, *N.Y., U.S.* **96** H8
Port Orange, *Fla., U.S.* **99** S9
Port Orchard, *Wash., U.S.* **110** C3
Port Orford, *Oreg., U.S.* **110** K1
Port Patrick, *Vanuatu* **229** N18
Port Patteson, *Vanuatu* **229** D15
Port Phaeton, *Fr. Polynesia, Fr.* **231** J22
Port Phillip Bay, *Austral.* **220** M12
Port Pirie, *S. Aust., Austral.* **221** X10
Port Royal Bay see Little Sounds, *Bermuda* **119** C14
Port Royal Sound, *S.C., U.S.* **99** N9
Port Royal, *S.C., U.S.* **99** N9
Port Royal, *Jam.* **122** J10
Port Said see Bûr Sa'îd, *Egypt* **173** N1
Port Said see Bûr Sa'îd, *Egypt* **202** E12
Port Saint Joe, *Fla., U.S.* **99** R4
Port Saint Johns, *S. Af.* **212** L9
Port Saint Lucie, *Fla., U.S.* **99** V10
Port Salerno, *Fla., U.S.* **99** V10
Port Salvador, *Falk. Is., U.K.* **135** W12
Port Saunders, *Nfld. & Lab., Can.* **87** L23
Port Shepstone, *S. Af.* **212** L9
Port South East, *Mauritius* **215** J20
Port Stanley, *Vanuatu* **229** G14
Port Stephens, *Falk. Is., U.K.* **135** W11
Port Sudan, *Sudan* **203** N15
Port Sulphur, *La., U.S.* **102** Q11
Port Talbot, *Wales, U.K.* **145** U10
Port Townsend, *Wash., U.S.* **110** B3
Port Victoria, *S. Aust., Austral.* **221** X10
Port Warrender, *W. Aust., Austral.* **221** P6
Port Washington, *Wis., U.S.* **100** L7
Port-à-Piment, *Haiti* **119** N15
Port-au-Prince see Pòtoprens, *Haiti* **119** M16
Port-Cartier, *Que., Can.* **87** M21
Port-de-Paix, *Haiti* **119** L16
Port-Étienne see Nouadhibou, *Mauritania* **204** E6
Port-Gentil, *Gabon* **208** E7
Port-la-Nouvelle, *Fr.* **149** Y11
Port-Louis, *Guadeloupe, Fr.* **123** E15
Port-Lyautey see Kenitra, *Mor.* **200** C7
Port-Menier, *Que., Can.* **87** M21
Port-Saint-Louis, *Fr.* **149** X13
Port-Salut, *Haiti* **119** N15
Port-Vila, *Vanuatu* **229** K16
Portadown, *N. Ire., U.K.* **145** P7
Portage la Prairie, *Man., Can.* **87** N13
Portage, *Wis., U.S.* **100** K5
Portage, *Ind., U.S.* **101** N8
Portage, *Mich., U.S.* **101** N10
Portageville, *Mo., U.S.* **107** V16
Portal, *N. Dak., U.S.* **106** D3
Portalegre, *Port.* **146** H6
Portales, *N. Mex., U.S.* **109** T15
Portbou, *Sp.* **147** C18
Porteira, *Braz.* **132** D10
Portel, *Braz.* **132** D12
Portel, *Port.* **146** J6
Porter Point, *St. Vincent & the Grenadines* **123** J17
Porterdale, *Ga., U.S.* **98** L5

Porterville, *Calif., U.S.* **111** U7
Porthmadog, *Wales, U.K.* **145** S9
Portimão, *Port.* **146** L5
Portiţei, Gura, *Rom.* **155** F18
Portland Bay, *Austral.* **220** M11
Portland Bight, *Jam.* **122** K9
Portland Cottage, *Jam.* **122** K9
Portland Point, *Jam.* **122** K9
Portland Rock, *island, Jam.* **118** P11
Portland, *Me., U.S.* **97** H15
Portland, *Mich., U.S.* **100** L10
Portland, *Ind., U.S.* **101** Q10
Portland, *Tenn., U.S.* **103** E15
Portland, *Tex., U.S.* **105** P17
Portland, *N. Dak., U.S.* **106** F7
Portland, *Oreg., U.S.* **110** F3
Portland, *Eng., U.K.* **145** W11
Portland, *Vic., Austral.* **221** Z11
Portloe, *Eng., U.K.* **145** X8
Portmore, *Jam.* **122** J10
Portnacroish, *Scot., U.K.* **144** K8
Portnahaven, *Scot., U.K.* **145** M7
Porto (Oporto), *Port.* **146** E5
Porto Alegre, *Braz.* **132** P12
Porto Alegre, *Sao Tome and Principe* **215** E20
Porto Amboim, *Angola* **208** L10
Porto Artur, *Braz.* **132** H11
Porto Azzurro, *It.* **152** J6
Pórto Chéli, *Gr.* **156** K11
Porto Covo, *Port.* **146** K5
Porto do Ibo, *Mozambique* **213** C13
Porto dos Gaúchos, *Braz.* **132** G10
Porto Esperidião, *Braz.* **132** J10
Porto Franco, *Braz.* **133** F14
Porto Garibaldi, *It.* **152** F8
Pórto Germenó, *Gr.* **156** J11
Porto Grande, *Braz.* **132** C12
Pôrto Inglês, *Cabo Verde* **215** D17
Porto Jofre, *Braz.* **132** K10
Pórto Koufó, *Gr.* **157** E13
Porto Moniz, *Madeira, Port.* **214** M2
Porto Murtinho, *Braz.* **132** L10
Porto Nacional, *Braz.* **133** G13
Porto Novo, *India* **183** R7
Porto Recanati, *It.* **152** H10
Porto San Giorgio, *It.* **152** H10
Porto Santo, *Madeira, Port.* **214** L5
Porto Santo, *island, Madeira, Port.* **214** L4
Porto Seguro, *Braz.* **133** K16
Porto Tolle, *It.* **152** E8
Porto Torres, *It.* **152** M3
Porto Velho, *Braz.* **132** F8
Porto-Novo, *Benin* **205** P18
Porto-Vecchio, *Fr.* **149** Y3
Porto-Vecchio, Golfe de, *Fr.* **149** Y3
Porto, *Braz.* **133** E15
Porto, *Fr.* **149** X2
Porto, Golfe de, *Fr.* **149** X2
Portobelo, *Pan.* **117** N21
Portoferraio, *It.* **152** J6
Portogruaro, *It.* **152** D9
Portola, *Calif., U.S.* **111** P5
Portoscuso, *It.* **153** Q3
Portoviejo, *Ecua.* **130** K3
Portsall, *Fr.* **149** P2
Portsmouth, *N.H., U.S.* **97** J15
Portsmouth, *N.C., U.S.* **98** J15
Portsmouth, *Va., U.S.* **98** F14
Portsmouth, *Ohio, U.S.* **101** T13
Portsmouth, *Dominica* **123** E18
Portsmouth, *Eng., U.K.* **145** V13
Portsoy, *Scot., U.K.* **144** H11
Portugal, *Eur.* **146** K6
Portuguesa, *river, Venez.* **130** D9
Porus, *Jam.* **122** J8
Porvenir, *Bol.* **132** G6
Porvenir, *Chile* **135** X8
Porvoo, *Fin.* **143** J16
Porzuna, *Sp.* **146** H10
Posadas, *Arg.* **134** G14
Posadowsky Bay, *Antarctica* **239** H22
Poschiavo, *Switz.* **152** C5
Posio, *Fin.* **143** E16
Poso, *Indonesia* **193** J13
Posof, *Rep. of Georgia* **171** C17
Possel, *Cen. Af. Rep.* **207** N14
Possession Island, *Namibia* **212** J3
Possum Kingdom Lake, *Tex., U.S.* **105** G16
Post Falls, *Idaho, U.S.* **108** B3
Post, *Tex., U.S.* **104** F12
Postville, *Nfld. & Lab., Can.* **87** K21
Postville, *Iowa, U.S.* **106** M13
Posušje, *Bosn. & Herzg.* **154** H7

Pot Mountain, *Idaho, U.S.* **108** D4
Potamoí, *Gr.* **157** B13
Potamós, *Gr.* **156** P11
Potchefstroom, *S. Af.* **212** J8
Poteau, *Okla., U.S.* **105** D20
Poteet, *Tex., U.S.* **105** M16
Potenza, *It.* **152** M12
Poth, *Tex., U.S.* **105** M16
Potholes Reservoir, *Wash., U.S.* **110** D7
P'ot'i, *Rep. of Georgia* **171** B16
Potiskum, *Nigeria* **206** J10
Potlatch, *Idaho, U.S.* **108** C3
Potnarvin, *Vanuatu* **229** L17
Potomac, *river, Md., Va., U.S.* **98** D13
Pòtoprens, *Haiti* **119** M16
Potosí, *Mo., U.S.* **107** T14
Potosí, *Bol.* **132** K7
Potrerillos, *Chile* **134** G7
Potsdam, *N.Y., U.S.* **96** F10
Potsdam, *Ger.* **150** D10
Potter, *Nebr., U.S.* **107** P1
Pottstown, *Pa., U.S.* **96** P9
Pottsville, *Pa., U.S.* **96** N8
Poudre d'Or, *Mauritius* **215** F20
Pouébo, *New Caledonia, Fr.* **231** N21
Pouembout, *New Caledonia, Fr.* **231** P21
Poughkeepsie, *N.Y., U.S.* **96** L11
Pougues, *Fr.* **149** S10
Pouilly, *Fr.* **149** R10
Poulo Wai, *island, Cambodia* **191** Q9
Poulsbo, *Wash., U.S.* **110** C3
Poulsen Fjord, *Greenland, Den.* **89** P11
Poum, *New Caledonia, Fr.* **231** N20
Poutasi, *Samoa* **227** P20
Poùthĭsăt (Pursat), *Cambodia* **191** N10
Poutoru, *Fr. Polynesia, Fr.* **231** G17
Povai, *Fr. Polynesia, Fr.* **231** G14
Povenets, *Russ.* **158** C8
Poverty Point National Monument, *La., U.S.* **102** L9
Póvoa de Varzim, *Port.* **146** D5
Povoação, *Port.* **146** Q6
Povorino, *Russ.* **159** N12
Povungnituk, Baie de, *Que., Can.* **87** J17
Powder River, *Wyo., U.S.* **108** H11
Powder, *river, Oreg., U.S.* **110** G9
Powell Basin, *Atl. Oc.* **249** X6
Powell Butte, *Oreg., U.S.* **110** H5
Powell Cay, *Bahamas* **122** A9
Powell Point, *Bahamas* **122** F6
Powell River, *B.C., Can.* **86** M8
Powell, *Wyo., U.S.* **108** F10
Powell, Lake, *Ariz., Utah, U.S.* **109** Q7
Powers Lake, *N. Dak., U.S.* **106** E3
Powers, *Oreg., U.S.* **110** J2
Poxoréo, *Braz.* **132** J11
Poya, *New Caledonia, Fr.* **231** P21
Poyang Hu, *China* **185** U14
Poygan, Lake, *Wis., U.S.* **100** J6
Poysdorf, *Aust.* **151** J13
Poza Rica, *Mex.* **116** H12
Pōzak, Jahīl-e, *Afghan.* **180** L3
Pozantı, *Turk.* **170** H10
Požarevac, *Serb.* **154** F11
Požega, *Croatia* **154** E7
Pozheg, *Russ.* **158** C15
Poznań, *Pol.* **151** D13
Pozo Alcón, *Sp.* **146** K12
Pozo Almonte, *Chile* **134** C7
Pozoblanco, *Sp.* **146** J9
Pozzallo, *It.* **153** U11
Pozzuoli, *It.* **152** M10
Pracchia, *It.* **152** G6
Prachin Buri, *Thai.* **190** M8
Prachuap Khiri Khan, *Thai.* **191** P7
Pradelles, *Fr.* **149** V11
Prado, *Braz.* **133** K16
Prague *see* Praha, *Czechia* **150** H11
Praha (Prague), *Czechia* **150** H11
Praia da Vitória, *Port.* **146** P4
Praia, *Cabo Verde* **215** E16
Prainha, *Braz.* **132** D11
Prainha, *Braz.* **132** F9
Prairie City, *Oreg., U.S.* **110** G7
Prairie Dog Creek, *Kans., U.S.* **107** S4
Prairie Dog Town Fork Red, *river, Tex., U.S.* **105** D13
Prairie du Chien, *Wis., U.S.* **100** L3
Prairie Grove, *Ark., U.S.* **102** E6
Pran Buri, *Thai.* **191** N7
Praslin, *St. Lucia* **123** K14
Praslin, *island, Seychelles* **215** M20
Prasonísion, *island, Gr.* **157** N19

Pratas Island *see* Dongsha, *China* **185** P14
Prato, *It.* **152** G7
Pratola Peligna, *It.* **152** K10
Prats-de-Mollo, *Fr.* **149** Z10
Prats, *Andorra* **162** H4
Pratt, *Kans., U.S.* **107** U6
Prattville, *Ala., U.S.* **103** L15
Prayagraj, *India* **182** M10
Prazeres, *Madeira, Port.* **214** M2
Preăh Seihânŭ (Sihanoukville), *Cambodia* **191** P10
Preah Vihear, *ruins, Cambodia* **190** M10
Preau, *Trin. & Tobago* **123** Q22
Précy-sous-Thil, *Fr.* **149** R12
Predazzo, *It.* **152** C7
Preguiça, *Cabo Verde* **215** B15
Preguiça, *Cabo Verde* **215** B17
Prehistoric Petroglyphs, *ruins, Afghan.* **180** F10
Premont, *Tex., U.S.* **105** P16
Premuda, *island, Croatia* **154** F4
Prentice, *Wis., U.S.* **100** G4
Prentiss, *Miss., U.S.* **102** M11
Prenzlau, *Ger.* **150** C10
Preobrazhenka, *Russ.* **161** J15
Preparis Island, *Myanmar* **190** M3
Preparis North Channel, *Myanmar* **190** L3
Preparis South Channel, *Myanmar* **190** M3
Přerov, *Czechia* **151** H13
Prescott, *Ark., U.S.* **102** J7
Prescott, *Ariz., U.S.* **109** S6
Prescott, *Wash., U.S.* **110** E8
Presho, *S. Dak., U.S.* **106** L5
Presidencia Roca, *Arg.* **134** F12
President Thiers Bank, *Pac. Oc.* **255** Q22
Presidente Dutra, *Braz.* **133** E15
Presidente Prudente, *Braz.* **132** L12
Presidente Roque Sáenz Peña, *Arg.* **134** G12
Presidio, *Tex., U.S.* **104** L9
Prešov, *Slovakia* **151** J16
Prespa *see* Prespansko Jezero, *Alban., N. Maced.* **154** L10
Prespa, peak, *Bulg.* **155** K14
Prespansko Jezero (Prespa, Lake), *Alban., N. Maced.* **154** L10
Presque Isle, *Me., U.S.* **97** B18
Presque Isle, *Mich., U.S.* **100** H12
Presqu'île la Caravelle, *Martinique, Fr.* **123** E23
Presqu'île d'Ampasindava, *Madagascar* **213** D17
Presqu'île de Masoala, *Madagascar* **213** E17
Presqu'île de Taiarapu, *Fr. Polynesia, Fr.* **231** J23
Presqu'île de Taiarapu, *Fr. Polynesia, Fr.* **231** C17
Pressburg *see* Bratislava, *Slovakia* **151** K13
Presteigne, *Wales, U.K.* **145** T10
Preston, *Minn., U.S.* **106** L13
Preston, *Mo., U.S.* **107** U12
Preston, *Idaho, U.S.* **108** J7
Preston, *Eng., U.K.* **145** Q11
Prestonsburg, *Ky., U.S.* **103** C20
Prestrud Bank, *Pac. Oc.* **255** S11
Prestrud Inlet, *Antarctica* **238** N11
Pretoria (Tshwane), *S. Af.* **212** H8
Préveza, *Gr.* **156** G7
Pribilof Islands, *Alas., U.S.* **112** L9
Priboj, *Serb.* **154** G9
Priboj, *Serb.* **154** J12
Příbram, *Czechia* **150** H11
Price, *Utah, U.S.* **108** M8
Price, Cape, *India* **183** Q14
Price, *river, Utah, U.S.* **108** M8
Prichard, *Ala., U.S.* **103** N13
Prickly Bay, *Grenada* **123** L22
Prickly Pear Cays, *Anguilla, U.K.* **122** Q10
Prickly Pear Island, *Virgin Is., U.K.* **122** Q9
Prickly Pear Island, *Antigua & Barbuda* **123** A21
Prickly Point, *Grenada* **123** M22
Priekule, *Latv.* **143** M15
Prieska, *S. Af.* **212** K6
Priekule, *Latv.* **143** M15
Prieta, Peña, *Sp.* **146** B10

Prievidza, *Slovakia* **151** J14
Prijedor, *Bosn. & Herzg.* **154** E6
Prijepolje, *Serb.* **154** H9
Prilep, *N. Maced.* **154** L11
Primorsko Akhtarsk, *Russ.* **159** S10
Prince Albert Peninsula, *N.W.T., Can.* **86** D12
Prince Albert Sound, *N.W.T., Can.* **86** E11
Prince Albert, *Sask., Can.* **86** M12
Prince Albert, *S. Af.* **212** M6
Prince Alfred, Cape, *N.W.T., Can.* **86** C12
Prince Charles Island, *Nunavut, Can.* **87** F17
Prince Charles Mountains, *Antarctica* **239** F18
Prince Edward Fracture Zone, *Ind. Oc.* **250** Q4
Prince Edward Island, *Can.* **87** N22
Prince Frederick Harbour, *Austral.* **220** C5
Prince George, *B.C., Can.* **86** K8
Prince Gustaf Adolf Sea, *Nunavut, Can.* **87** C14
Prince Gustav Channel, *Antarctica* **238** D5
Prince Harald Coast, *Antarctica* **239** B17
Prince of Wales Island, *Nunavut, Can.* **87** E14
Prince of Wales Island, *Alas., U.S.* **113** M23
Prince of Wales Island, *Austral.* **220** A12
Prince of Wales Strait, *N.W.T., Can.* **86** D12
Prince of Wales, Cape, *Alas., U.S.* **112** E11
Prince Olav Coast, *Antarctica* **239** C18
Prince Patrick Island, *N.W.T., Can.* **87** C13
Prince Regent Inlet, *Nunavut, Can.* **87** E15
Prince Rupert Bay, *Dominica* **123** E18
Prince Rupert, *B.C., Can.* **86** J7
Prince William Sound, *Alas., U.S.* **113** K17
Princes Town, *Trin. & Tobago* **123** Q22
Princess Anne, *Md., U.S.* **98** D15
Princess Antoinette Park, *Monaco* **163** E20
Princess Astrid Coast, *Antarctica* **239** A14
Princess Charlotte Bay, *Austral.* **220** B13
Princess Elisabeth Station, *Antarctica* **239** B16
Princess Elizabeth Trough, *Ind. Oc.* **254** D12
Princess Grace Hospital Center, *Monaco* **163** E19
Princess Martha Coast, *Antarctica* **238** B12
Princess Ragnhild Coast, *Antarctica* **239** B16
Princeton, *B.C., Can.* **86** M8
Princeton, *N.J., U.S.* **96** P10
Princeton, *W. Va., U.S.* **98** F9
Princeton, *Ill., U.S.* **101** P5
Princeton, *Ind., U.S.* **101** T7
Princeton, *Ky., U.S.* **103** D14
Princeton, *Minn., U.S.* **106** J11
Princeton, *Mo., U.S.* **107** Q11
Princeville, *Hawaii, U.S.* **114** E10
Príncipe da Beira, *Braz.* **132** H6
Príncipe, *island, Sao Tome and Principe* **215** B20
Prineville, *Oreg., U.S.* **110** G5
Pringle, *S. Dak., U.S.* **106** L1
Prins Christian Sund *see* Ikerasassuaq, *Greenland, Den.* **89** U8
Prinsesse Dagmar Ø, *Greenland, Den.* **88** B12
Prinsesse Margrethe Ø, *Greenland, Den.* **88** B12
Prinsesse Thyra Ø, *Greenland, Den.* **88** B11
Prior, Cabo, *Sp.* **146** A6
Priozersk, *Russ.* **158** E6
Pripyat' *see* Pryp'yat', *river, Ukr.* **158** M3
Priekule, *Latv.* **143** M15
Prishtinë *see* Pristina, *Kos.* **154** J11
Prislop, Pasul, *Rom.* **155** B14
Pristina (Prishtinë, Priština), *Kos.* **154** J11

Pritchett, *Colo., U.S.* **109** P15
Pritzwalk, *Ger.* **150** C9
Privas, *Fr.* **149** V12
Privol'noye, *Russ.* **161** G22
Privolzhsk, *Russ.* **158** H11
Privolzhskaya Vozvyshennost', *Russ.* **159** N13
Privolzhskiy, *Russ.* **158** M14
Prizren, *Kos.* **154** J10
Probolinggo, *Indonesia* **192** M10
Proctor, *Minn., U.S.* **106** G12
Prodromi, *Cyprus* **162** P5
Progreso Limitada, *Eq. Guinea* **214** M8
Progress 2 Station, *Antarctica* **239** G20
Progress, *Russ.* **161** K20
Prokhladnyy, *Russ.* **159** T13
Prokhod Sveti Nikola, peak, *Bulg., Serb.* **154** H12
Prokópi, *Gr.* **156** G12
Prokop'yevsk, *Russ.* **160** L12
Prokuplje, *Serb.* **154** H11
Proletarsk, *Russ.* **159** R12
Promýri, *Gr.* **156** F11
Prony, *New Caledonia, Fr.* **231** Q23
Propriá, *Braz.* **133** G17
Propriano, *Fr.* **149** Y2
Proserpine, *Qnsld., Austral.* **221** S14
Prosotsáni, *Gr.* **157** B13
Prospect Point, *Jam.* **122** J11
Prospect, *Oreg., U.S.* **110** J3
Prospect, *Cayman Is., U.K.* **122** N2
Prospect, *Barbados* **123** K18
Prosser, *Wash., U.S.* **110** E6
Prostějov, *Czechia* **151** H13
Protection, *Kans., U.S.* **107** V5
Protector Basin, *Atl. Oc.* **249** W6
Próti, *island, Gr.* **156** L8
Provadiya, *Bulg.* **155** H17
Provence, *region, Fr.* **149** X14
Providence Mountains, *Calif., U.S.* **111** W11
Providence, *R.I., U.S.* **97** L14
Providence, *Ky., U.S.* **103** C14
Providence, *Grenada* **123** L23
Providencia, Isla de, *Col.* **117** L20
Providenciales, *island, Bahamas* **119** H16
Provideniya, *Russ.* **161** A21
Provincetown, *Mass., U.S.* **97** L16
Provo, *Utah, U.S.* **108** L7
Prudhoe Bay, *Alas., U.S.* **113** C17
Prudhoe Land, *Greenland, Den.* **88** E4
Prudnik, *Pol.* **151** G13
Prune Island *see* Palm Island, *St. Vincent & the Grenadines* **121** G17
Pruszcz Gdański, *Pol.* **151** B14
Pruszków, *Pol.* **151** E16
Prut, *river, Mold., Rom.* **155** D17
Pruzhany, *Belarus* **158** L3
Prvić, *island, Croatia* **154** E5
Prydz Bay, *Antarctica* **239** G20
Pryluky, *Ukr.* **159** N7
Pryor Creek, *Mont., U.S.* **108** E10
Pryor, *Okla., U.S.* **105** B19
Pryp'yat' (Pripyat'), *river, Ukr.* **158** M3
Prypyats', *river, Eur.* **140** G10
Przemyśl, *Pol.* **151** H18
Przheval'sk *see* Karakol, *Kyrg.* **179** K16
Psachná, *Gr.* **156** H12
Psará, *Gr.* **157** H15
Psará, *island, Gr.* **157** H15
Psathoúra, *island, Gr.* **157** F13
Psérimos, *island, Gr.* **157** L18
Psiloreítis *see* Ídi, Óros, *Gr.* **157** Q14
Pskov, *Russ.* **158** G5
Pskov, Lake, *Eur.* **140** E10
Pskovskoye Ozero, *Russ.* **158** G5
Ptolemaḯda, *Gr.* **156** C9
Ptolemy Basin, *Pac. Oc.* **246** F9
Ptuj, *Slov.* **154** C6
Pua, *Thai.* **190** J8
Puakō, *Hawaii, U.S.* **115** M20
Puakonikai, *Kiribati* **226** M11
Puán, *Arg.* **135** N11
Pu'apu'a, *Samoa* **227** N18
Puaumu Nord, *island, Fr. Polynesia, Fr.* **231** Q17
Puca Urco, *Peru* **132** D5
Pucallpa, *Peru* **132** F4
Pucioasa, *Rom.* **155** E15

Puck, *Pol.* **151** A14
Pudasjärvi, *Fin.* **143** E15
Pudozh, *Russ.* **158** D9
Pudu, *Indonesia* **192** J6
Puducherry (Pondicherry), *India* **183** R7
Pudukkottai, *India* **183** S6
Puebla de Guzmán, *Sp.* **146** K6
Puebla de Sanabria, *Sp.* **146** D7
Puebla, *Mex.* **116** J12
Pueblo Bonito, *N. Mex., U.S.* **109** R11
Pueblo, *Colo., U.S.* **109** N14
Puente-Genil, *Sp.* **146** L9
Pueo Point, *Hawaii, U.S.* **114** F8
Pu'er, *China* **184** P9
Puerco, river, *Ariz., N. Mex., U.S.* **109** S9
Puerto Acosta, *Bol.* **132** J6
Puerto Aisén, *Chile* **135** S6
Puerto América, *Peru* **132** E3
Puerto Ángel, *Mex.* **117** K13
Puerto Armuelles, *Pan.* **117** P19
Puerto Ayacucho, *Venez.* **130** F10
Puerto Ayora, *Ecua.* **130** Q3
Puerto Bahía Negra, *Parag.* **132** L10
Puerto Baquerizo Moreno, *Ecua.* **130** Q4
Puerto Barrios, *Guatemala* **117** K17
Puerto Belgrano, *Arg.* **135** P11
Puerto Berrío, *Col.* **130** F6
Puerto Cabello, *Venez.* **130** C10
Puerto Carreño, *Venez.* **130** F10
Puerto Chicama, *Peru* **132** F2
Puerto Coig, *Arg.* **135** V8
Puerto Colombia, *Col.* **130** C6
Puerto Concordia, *Col.* **130** H7
Puerto Cortés, *Hond.* **117** K17
Puerto Cumarebo, *Venez.* **130** C9
Puerto de la Cruz, *Sp.* **214** Q4
Puerto de Luna, *N. Mex., U.S.* **109** S14
Puerto de Mazarrón, *Sp.* **147** L13
Puerto de Nutrias, *Venez.* **130** E9
Puerto del Carmen, *Sp.* **214** P7
Puerto del Rosario, *Sp.* **214** Q7
Puerto Deseado, *Arg.* **135** T9
Puerto Etén, *Peru* **132** F2
Puerto Heath, *Bol.* **132** H6
Puerto Jiménez, *C.R.* **117** P19
Puerto La Cruz, *Venez.* **130** C12
Puerto La Paz, *Arg.* **134** D11
Puerto Lápice, *Sp.* **146** H11
Puerto Leguízamo, *Col.* **130** K6
Puerto Libertad, *Mex.* **116** C6
Puerto Lobos, *Arg.* **135** Q9
Puerto López, *Col.* **130** G6
Puerto Madryn, *Arg.* **135** R10
Puerto Maldonado, *Peru* **132** H6
Puerto Manatí, *Cuba* **118** J12
Puerto Montt, *Chile* **135** Q6
Puerto Morelos, *Mex.* **117** H17
Puerto Natales, *Chile* **135** W7
Puerto Obaldía, *Pan.* **117** N22
Puerto Padre, *Cuba* **118** J12
Puerto Páez, *Venez.* **130** F10
Puerto Pinasco, *Parag.* **132** L10
Puerto Pirámides, *Arg.* **135** R10
Puerto Plata, *Dom. Rep.* **119** L18
Puerto Portillo, *Peru* **132** G4
Puerto Princesa, *Philippines* **192** E12
Puerto Rico, *U.S.* **114** P6
Puerto Rico, island, *N. Amer.* **119** M23
Puerto Rico Trench, *Atl. Oc.* **119** M23
Puerto San Carlos, *Chile* **135** U6
Puerto San Julián, *Arg.* **135** U9
Puerto Santa Cruz, *Arg.* **135** V8
Puerto Suárez, *Bol.* **132** K10
Puerto Tres Palmas, *Parag.* **132** L10
Puerto Vallarta, *Mex.* **116** H9
Puerto Velasco Ibarra, *Ecua.* **130** R3
Puerto Villamil, *Ecua.* **130** Q3
Puerto Wilches, *Col.* **130** E6
Puerto Williams, *Chile* **135** Y9
Puertollano, *Sp.* **146** J10
Puetton Tanapag, bay, *N. Mariana Is., U.S.* **230** H8
Pueu, *Fr. Polynesia, Fr.* **231** J23
Pueyrredó, Lago, *Arg.* **135** T7
Pugachev, *Russ.* **158** L15
Puget Sound, *U.S.* **90** B4
Pugŏ, *N. Korea* **186** C13
Puhi, *Hawaii, U.S.* **114** F10
Puhiwaero *see* South West Cape, cape, *N.Z.* **223** Q16
Puig Major, peak, *Sp.* **147** G18
Puigcerdà, *Sp.* **147** C17

Puigmal d'Err, peak, *Sp.* **147** C17
Pujehun, *Liberia* **204** P9
Pujehun, *Sierra Leone* **204** N9
Pujŏn-ho, lake, *N. Korea* **186** F9
Pujŏn, *N. Korea* **186** F9
Pujŏn, river, *N. Korea* **186** E9
Pukaki Saddle, *Pac. Oc.* **255** H19
Pukalani, *Hawaii, U.S.* **115** J19
Pukapuka Atoll, *Cook Is., N.Z.* **224** H11
Pukapuka, island, *N.Z.* **229** D20
Pukapuka, island, *Fr. Polynesia, Fr.* **231** A21
Pukapukae Ridge, *Pac. Oc.* **247** L16
Pukarua, island, *Fr. Polynesia, Fr.* **231** C22
Pukchin, *N. Korea* **186** G6
Pukch'ŏng, *N. Korea* **186** G10
Puksoozero, *Russ.* **158** D10
Pul-e 'Alam, *Afghan.* **180** J9
Pul-e Khumrī, *Afghan.* **180** G8
Pula (Pola), *Croatia* **154** E4
Pula, *It.* **153** Q4
Pula, Capo di, *It.* **153** Q4
Pulan *see* Burang, *China* **184** K2
Pulandian, *China* **185** G15
Pulap Atoll, *F.S.M.* **226** B8
Pulaski, *N.Y., U.S.* **96** H8
Pulaski, *Va., U.S.* **98** F9
Pulaski, *Tenn., U.S.* **103** G15
Puławy, *Pol.* **151** F17
Pulicat Lake, *India* **183** Q7
Puliyangudi, *India* **183** T5
Puliyankulam, *Sri Lanka* **183** T7
Pull Point, *Virgin Is., U.S.* **122** Q3
Pullman National Monument, *Ill., U.S.* **101** N7
Pullman, *Wash., U.S.* **110** D9
Pulo Anna, island, *Palau* **224** F2
Pulog, Mount, *Philippines* **193** B13
Pülümür, *Turk.* **171** F14
Pulusuk, island, *F.S.M.* **226** B8
Puluwat Atoll, *F.S.M.* **226** B8
Pumpkin Creek, *Mont., U.S.* **108** E12
Puna, *Hawaii, U.S.* **115** P23
Puná, Isla, *Ecua.* **130** M3
Punaauia, *Fr. Polynesia, Fr.* **231** H21
Punakha, *Bhutan* **182** G12
Punalu'u, *Hawaii, U.S.* **115** Q21
Punan, *Indonesia* **192** H11
Puncak Jaya, peak, *Indonesia* **193** L20
Punch, *India* **182** D4
Pune, *India* **182** M4
Punggi, *S. Korea* **187** P11
P'unggye, *N. Korea* **186** E12
P'ungsan, *N. Korea* **186** B13
P'ungsŏ-ho, lake, *N. Korea* **186** E10
P'ungsŏ, *N. Korea* **186** E10
Punia, *D.R.C.* **209** F17
Punitaqui, *Chile* **134** J6
Puno, *Peru* **132** J6
Punta Albemarle, *Ecua.* **130** P2
Punta Alta, *Arg.* **135** N11
Punta Arenas, *Chile* **135** X8
Punta Basora, *Aruba, Neth.* **123** R17
Punta Campo, *Eq. Guinea* **206** P9
Punta Cana, *Dom. Rep.* **119** M21
Punta Cristóbal, *Ecua.* **130** Q2
Punta del Este, *Uru.* **134** L14
Punta Delgada, *Arg.* **135** R10
Punta Essex, *Ecua.* **130** R2
Punta Gorda, *Fla., U.S.* **99** V8
Punta Gorda, *Belize* **117** K16
Punta Pitt, *Ecua.* **130** Q4
Punta Prieta, *Mex.* **116** C6
Punta Sur, *Ecua.* **130** R3
Punta Umbría, *Sp.* **146** L7
Puntan Agingan, *N. Mariana Is., U.S.* **230** J7
Puntan Asiga, *N. Mariana Is., U.S.* **230** H11
Puntan Carolinas, *N. Mariana Is., U.S.* **230** K11
Puntan Dandan, *N. Mariana Is., U.S.* **230** J8
Puntan Gloria, *N. Mariana Is., U.S.* **230** H8
Puntan Hakmang, *N. Mariana Is., U.S.* **230** J8
Puntan I Maddock, *N. Mariana Is., U.S.* **230** G9
Puntan I Naftan, *N. Mariana Is., U.S.* **230** K8

Puntan Lagua Lichan, *N. Mariana Is., U.S.* **230** G9
Puntan Lananibot, *N. Mariana Is., U.S.* **230** H10
Puntan Laula Katan, *N. Mariana Is., U.S.* **230** J8
Puntan Magpi, *N. Mariana Is., U.S.* **230** G8
Puntan Marpo, *N. Mariana Is., U.S.* **230** K11
Puntan Masalog, *N. Mariana Is., U.S.* **230** J11
Puntan Muchot, *N. Mariana Is., U.S.* **230** H7
Puntan Opyan, *N. Mariana Is., U.S.* **230** K7
Puntan Sabaneta, *N. Mariana Is., U.S.* **230** G9
Puntan Tahgong, *N. Mariana Is., U.S.* **230** H11
Puntarenas, *C.R.* **117** N18
Puntland, *Somalia* **211** N13
Punto Fijo, *Venez.* **130** C8
Puntudo, Cerro, *Arg.* **135** U8
Punxsutawney, *Pa., U.S.* **96** N4
Puohine, *Fr. Polynesia, Fr.* **231** H18
Pup'yŏng, *N. Korea* **186** H9
Puqi, *China* **185** L13
Puquio, *Peru* **132** J4
Puranpur, *India* **182** G7
Purari, river, *P.N.G.* **222** P7
Purcell Mountains, *Can., U.S.* **108** A4
Purcell, *Okla., U.S.* **105** D17
Purdy Islands, *P.N.G.* **222** N7
Purgatoire, river, *Colo., U.S.* **109** P14
Puri, *India* **182** M10
Purmerend, *Neth.* **148** H12
Purna, river, *India* **182** L5
Purnema, *Russ.* **158** B9
Purnia, *India* **182** H11
Pursat *see* Poŭthĭsăt, *Cambodia* **191** N10
Purt ny h-Inshey *see* Peel, *Isle of Man, U.K.* **145** P9
Purus, river, *Braz.* **132** F7
Purvis, *Miss., U.S.* **102** N12
Puryŏng, *N. Korea* **186** C13
Pusad, *India* **182** M6
Pusan *see* Busan, *S. Korea* **187** T12
Pushkin, *Russ.* **158** F6
Pushkino, *Russ.* **158** M14
Pushtah-ye Ḩasan, *Afghan.* **180** L4
Putao, *Myanmar* **190** D6
Putian, *China* **185** M15
Putilovo, *Russ.* **158** F12
Putina, *Russ.* **132** J6
Puto, *P.N.G.* **222** P11
Puto, *P.N.G.* **228** B4
Putorana Plateau, *Asia* **168** C9
Putrachoique, Cerro, *Arg.* **135** R7
Puttalam, *Sri Lanka* **183** T7
Puttelange, *Fr.* **149** P14
Puttur, *India* **183** Q7
Putumayo, river, *S. Amer.* **130** J5
Pütürge, *Turk.* **171** G13
Putusuk *see* Houk, *F.S.M.* **224** F5
Pu'uanahulu, *Hawaii, U.S.* **115** N20
Pu'uhonua o Hōnaunau National Historical Park (City of Refuge National Historical Park), *Hawaii, U.S.* **115** P20
Pu'ukoholā Heiau National Historic Site, *Hawaii, U.S.* **115** M20
Pu'umahoe, peak, *Hawaii, U.S.* **115** K19
Pu'uomoa'ula Nui, peak, *Hawaii, U.S.* **115** K18
Pu'u'ula'ula (Red Hill), *Hawaii, U.S.* **115** K19
Pu'uwai, *Hawaii, U.S.* **114** F8
Puwe, island, *F.S.M.* **226** E9
Puy de Sancy, peak, *Fr.* **149** U10
Puy-Guillaume, *Fr.* **149** U11
Puyallup, *Wash., U.S.* **110** C4
Puyang, *China* **185** J13
Puysegur Point, *N.Z.* **223** P15
Puysegur Trench, *Pac. Oc.* **255** G19
Puyun, *N. Korea* **186** C12
Pweto, *D.R.C.* **209** K18
Pya Ozero, *Russ.* **158** A7
Pya, Lake, *Eur.* **140** B10
P'yagina, *Russ.* **161** F21
Pyasinskiy Zaliv, *Russ.* **161** E13
Pyatigorsk, *Russ.* **159** T13

P'yatykhatky, *Ukr.* **159** Q7
Pyay, *Myanmar* **190** J4
Pydna, battle, *Gr.* **156** D10
Pyeongchang, *S. Korea* **187** N10
Pyeongdo, island, *S. Korea* **187** U9
Pyeonghae, *S. Korea* **187** P12
Pyeongildo, island, *S. Korea* **187** U8
Pyeongni, *S. Korea* **187** N8
Pyeongtaek, *S. Korea* **187** P8
Pyhäjoki, *Fin.* **143** F15
Pyhäselkä, lake, *Fin.* **143** G17
Pyin-U-Lwin, *Myanmar* **190** G5
Pyinkayaing, *Myanmar* **190** L4
Pyinmana, *Myanmar* **190** J5
Pýli, *Gr.* **156** F9
Pýlos, *Gr.* **156** L9
Pymatuning Reservoir, *Ohio, Pa., U.S.* **96** L3
Pyŏksŏng, *N. Korea* **186** L6
Pyŏktong, *N. Korea* **186** F5
P'yŏng-sŏng, *N. Korea* **186** J6
P'yŏnggang, *N. Korea* **186** L9
P'yŏngsan, *N. Korea* **186** F10
P'yŏngsan, *N. Korea* **186** L7
P'yŏngyang (Pyongyang), *N. Korea* **186** J6
Pyongyang *see* P'yŏngyang, *N. Korea* **186** J6
Pyoseon, *S. Korea* **187** W8
Pyote, *Tex., U.S.* **104** J10
Pyramid Lake, *Nev., U.S.* **111** P6
Pyramid Rock, island, *S. China Sea* **192** B10
Pyramids of Giza, ruin, *Egypt* **202** F11
Pyrenees, range, *Eur.* **140** J4
Pyrgí, *Gr.* **157** H16
Pýrgos, *Gr.* **156** K8
Pýrgos, *Gr.* **157** R15
Pyryatyn, *Ukr.* **159** N7
Pyshchug, *Russ.* **158** F12
Pytalovo, *Russ.* **158** H5
Pythagóreio, *Gr.* **157** K18
Pyu, *Myanmar* **190** J5

Q

Qā' Abu al Ḩusayn, *Jordan* **172** M10
Qā' al Jafr, *Jordan* **173** Q7
Qaanaaq (Thule), *Greenland, Den.* **88** E3
Qabanbay, *Kaz.* **179** G17
Qāḑub, *Yemen* **175** Q15
Qā'emshahr, *Iran* **177** E16
Qā'en, *Iran* **177** G21
Qaidam Basin, *Asia* **168** G9
Qaidam Pendi, *China* **184** H6
Qairouan *see* Kairouan, *Tunisia* **201** C15
Qala, *Malta* **163** H21
Qala, *Azerb.* **171** D23
Qal'ah-ye Fatḩ, *Afghan.* **180** M3
Qal'ah-ye Gaz, *Afghan.* **180** K5
Qal'ah-ye Now, *Afghan.* **180** H4
Qal'ah-ye Shahr, *Afghan.* **180** G6
Qal'ah, *Afghan.* **180** G7
Qalam Chashmahs, *Afghan.* **180** J5
Qalansīyah, *Yemen* **175** Q15
Qalāt (Kalat), *Afghan.* **180** K7
Qal'at al Aẓlam, ruin, *Saudi Arabia* **174** E4
Qal'at al Marqab (Margat Castle), ruins, *Syr.* **172** F7
Qal'at Bīshah, *Saudi Arabia* **174** K8
Qal'eh-ye Ganj, *Iran* **177** M20
Qalhāt, *Oman* **175** H18
Qallabat, *Sudan* **203** S14
Qamdo (Changdu), *China* **184** K8
Qamea, island, *Fiji* **226** N7
Qamīnis, *Lib.* **201** E20
Qammieh Point, *Malta* **163** J21
Qamystybas, *Kaz.* **178** H9
Qanāt as Suways (Suez Canal), *Egypt* **202** E12
Qandala, *Somalia* **210** K14
Qapshaghay Reservoir, *Kaz.* **179** J16
Qapshaghay, *Kaz.* **179** J15
Qaqortoq (Julianehåb), *Greenland, Den.* **89** T7
Qarabutaq, *Kaz.* **178** E8
Qaraçala, *Azerb.* **171** E22
Qaraghandy, *Kaz.* **179** F13
Qaraghayly, *Kaz.* **179** F14

Qarah Bāgh, *Afghan.* **180** J8
Qārah, *Egypt* **202** F8
Qaraqoynn Köli, *Kaz.* **178** H11
Qārat Kūdī, peak, *Lib.* **201** J23
Qaratal, river, *Kaz.* **179** H15
Qarataū Zhotasy, mountains, *Kaz.* **178** J11
Qarataū, *Kaz.* **178** K12
Qaratöbe, *Kaz.* **178** E6
Qaraton, *Kaz.* **178** G5
Qaraūyl, *Kaz.* **179** F16
Qarazhal, *Kaz.* **178** G12
Qarchak, *Iran* **177** F15
Qarchī Gak, *Afghan.* **180** F7
Qardho, *Somalia* **210** L14
Qarokŭl, lake, *Taj.* **179** M14
Qarqan *see* Qiemo, *China* **184** G4
Qarqan, river, *China* **184** G4
Qarqaraly, *Kaz.* **179** F14
Qarqīn, *Afghan.* **180** E6
Qarsaqbay, *Kaz.* **178** G11
Qarshi, *Uzb.* **178** M10
Qaryah-ye Mandil, *Afghan.* **180** J3
Qaryat abu Nujaym, *Lib.* **201** E18
Qaryat al Qaddāḩīyah, *Lib.* **201** E18
Qaryat al 'Ulyā, *Saudi Arabia* **174** E11
Qaryat az Zuwaytīnah, *Lib.* **201** E20
Qasigiannguit (Christianshåb), *Greenland, Den.* **88** M6
Qasinngortuup Kangertiva, bay, *Greenland, Den.* **89** S9
Qaşr al Azraq, ruin, *Jordan* **172** M8
Qaşr al Farāfirah, *Egypt* **202** H9
Qaşr al Ḩayr al Gharbī, ruins, *Syr.* **172** H9
Qaşr al Ḩayr ash Sharqī, ruins, *Syr.* **172** G11
Qaşr ash Shaqqah, ruin, *Lib.* **201** E23
Qaşr ash Shaqqah, ruin, *Egypt* **202** F8
Qaşr aṭ Ṭūbah, ruins, *Jordan* **173** N7
Qaşr Bū Hādī, ruin, *Lib.* **201** E19
Qaşr Burqu', ruins, *Jordan* **172** L10
Qaşr Ḩamām, *Saudi Arabia* **174** J10
Qaşr Ibrîm, ruin, *Egypt* **202** L12
Qaşr-e Qand, *Iran* **177** N22
Qaşr-e Shīrīn, *Iran* **176** F11
Qasserine *see* Kasserine, *Tunisia* **201** C15
Qa'ṭabah, *Yemen* **174** P9
Qaṭanā, *Syr.* **172** J7
Qatar, *Asia* **175** F13
Qattara Depression *see* Munkhafaḑ al Qaṭṭārah, *Egypt* **202** F9
Qaṭṭīnah, Buḩayrat, *Syr.* **172** G8
Qax, *Azerb.* **171** C20
Qayghy, *Kaz.* **178** E10
Qaynar, *Kaz.* **179** F15
Qayşār, *Afghan.* **180** G5
Qazaly, *Kaz.* **178** H9
Qazaq Shyghanaghy, bay, *Kaz.* **178** K4
Qazax, *Azerb.* **171** C19
Qāzī Deh, *Afghan.* **180** F11
Qazvīn, *Iran* **177** E14
Qeissan, *Sudan* **203** U14
Qele Levu, island, *Fiji* **226** M8
Qeqertaq Prins Henrik, island, *Greenland, Den.* **88** E13
Qeqertaq, *Greenland, Den.* **88** F4
Qeqertaq, *Greenland, Den.* **88** L6
Qeqertarsuaq (Disko), *Greenland, Den.* **88** L5
Qeqertarsuaq (Godhavn), *Greenland, Den.* **88** M5
Qeqertarsuaq (Herbert Ø), *Greenland, Den.* **88** E3
Qeqertarsuatsiaat (Fiskenæsset), *Greenland, Den.* **89** R5
Qeqertarsuatsiaq, island, *Greenland, Den.* **88** M5
Qeqertarsuup Tunua (Disko Bugt), bay, *Greenland, Den.* **88** M6
Qeshm, *Iran* **177** N19
Qeshm, island, *Iran* **177** N18
Qīdar, *Iran* **177** E13
Qiemo (Qarqan), *China* **184** G4
Qikiqtarjuaq, *Nunavut, Can.* **87** F19
Qikiqtarjuaq, island, *Nunavut, Can.* **87** H20
Qila Ladgasht, *Pak.* **181** P3
Qila Safed, *Pak.* **181** N2
Qila Saifullah, *Pak.* **180** M8
Qilian Shan, *China* **184** G8
Qimantag, range, *China* **184** G5

Qimusseriarsuaq (Melville Bugt), bay, *Greenland, Den.* **88** G4
Qimusseriarsuaq, bay, *Greenland, Den.* **253** N15
Qin Ling, range, *China* **184** J11
Qinā, *Egypt* **202** J12
Qin'an, *China* **184** J10
Qing Zang Gaoyuan (Tibet, Plateau of), *China* **184** J3
Qingdao, *China* **185** H15
Qinghai Hu, *China* **184** H8
Qinghe, *China* **184** D6
Qingtongxia (Xiaoba), *China* **184** H10
Qingyuan, *China* **185** N13
Qinhuangdao, *China* **185** G14
Qinzhou, *China* **184** P11
Qionghai (Jiaji), *China* **184** Q12
Qiongzhou Strait, *Pac. Oc.* **168** J11
Qiqihar, *China* **185** C15
Qīr, *Iran* **177** M16
Qira, *China* **184** G3
Qirţāsī, ruin, *Egypt* **202** K12
Qiryat Shemona, *Israel* **172** K6
Qishn, *Yemen* **175** N13
Qitai, *China* **184** E6
Qitaihe, *China* **185** D17
Qiyil, river, *Kaz.* **178** E6
Qizil Qal'ah, *Afghan.* **180** F8
Qizilqum, *Uzb.* **178** K9
Qızılağac Körfəzi, *Azerb.* **171** F22
Qobu, *Azerb.* **171** D23
Qobustan, *Azerb.* **171** D22
Qogir Feng *see* K2, *China, Pak.* **180** G14
Qom (Qum), *Iran* **177** F15
Qom, river, *Iran* **177** G15
Qomolangma *see* Everest, Mount, *China, Nep.* **184** L4
Qomsheh *see* Shahrezā, *Iran* **177** J16
Qonaqkənd, *Azerb.* **171** C22
Qo'ng'irot, *Uzb.* **178** K7
Qongyrat, *Kaz.* **179** G14
Qooqqut, *Greenland, Den.* **89** Q5
Qooriga Neegro, bay, *Somalia* **210** M14
Qoow, *Somalia* **210** K14
Qo'qon, *Uzb.* **179** L13
Qoraköl, *Uzb.* **178** M9
Qoraqalpog'iston, *Uzb.* **178** J7
Qormi, *Malta* **163** K22
Qorveh, *Iraq* **177** F13
Qoryaale, cape, *Somalia* **210** M14
Qoryooley, *Somalia* **211** R11
Qosköl, *Kaz.* **178** F11
Qosshaghyl, *Kaz.* **178** G5
Qostanay, *Kaz.* **178** C9
Qoton, *Somalia* **210** L14
Qoz Abu Dulu, *Sudan* **203** Q12
Qoz Dango, *Sudan* **203** U7
Qrendi, *Malta* **163** L22
Quakertown, *Pa., U.S.* **96** P9
Quần Đảo Nam Du, island, *Vietnam* **191** Q10
Quanah, *Tex., U.S.* **105** E14
Quảng Ngãi, *Vietnam* **190** L13
Quảng Trị, *Vietnam* **190** K12
Quanshuigou, *China* **180** G17
Quanzhou, *China* **184** N12
Quanzhou, *China* **185** N15
Quar Ice Shelf, *Antarctica* **238** B12
Quaraí, *Braz.* **132** P10
Quarantine Point, *Grenada* **123** L22
Quartier d'Orléans, *St. Martin, Fr.* **123** B15
Quartier du Colombier, *St. Martin, Fr.* **123** B14
Quartu Sant' Elena, *It.* **153** Q4
Quartz Hill, *N.T., Austral.* **221** T9
Quartzsite, *Ariz., U.S.* **109** T5
Quatre Bornes, *Mauritius* **215** G20
Quba, *Azerb.* **171** C22
Qūchān, *Iran* **177** D20
Qudaym, *Syr.* **172** F10
Quebec, *Can.* **87** M19
Québec, *Que., Can.* **87** P20
Quebracho Coto, *Arg.* **134** F10
Quebrada Fracture Zone, *Pac. Oc.* **247** K19
Queen Adelaida Archipelago, *S. Amer.* **128** R4
Queen Alexandra Range, *Antarctica* **239** L13
Queen Charlotte Islands *see* Haida Gwaii, *Can.* **86** J6
Queen Charlotte Sound, *B.C., Can.* **86** K7

Queen Elizabeth Islands, *Nunavut, N.W.T., Can.* **87** B14
Queen Elizabeth Range, *Antarctica* **239** L13
Queen Elizabeth Rise, *Arctic Oc.* **252** L11
Queen Elizabeth Shelf, *Arctic Oc.* **252** M12
Queen Fabiola Mountains (Yamato Mountains), *Antarctica* **239** C17
Queen Fabiola Mountains (Yamato Mountains), *Antarctica* **239** C17
Queen Mary Coast, *Antarctica* **239** J21
Queen Maud Gulf, *Nunavut, Can.* **87** F14
Queen Maud Land, region, *Antarctica* **239** D14
Queen Maud Mountains, *Antarctica* **239** K13
Queens Channel, *Austral.* **220** C7
Queensland Plateau, *Pac. Oc.* **246** L8
Queensland, *Austral.* **221** S12
Queenstown, *Tas., Austral.* **221** Y7
Queenstown, *N.Z.* **223** N16
Queets, *Wash., U.S.* **110** C2
Quelimane, *Mozambique* **212** E12
Quellón, *Chile* **135** R6
Quelo, *Angola* **208** J9
Quemado, *Tex., U.S.* **105** M13
Quemado, *N. Mex., U.S.* **109** T10
Quembo, river, *Angola* **209** N13
Quemoy *see* Kinmen, *China* **185** N15
Quemú Quemú, *Arg.* **134** M10
Quequén, *Arg.* **135** N12
Quercy, region, *Fr.* **149** W9
Querétaro, *Mex.* **116** H11
Quesada, *C.R.* **117** N19
Quesnel, *B.C., Can.* **86** L8
Questa, *N. Mex., U.S.* **109** Q13
Quetta, *Pak.* **180** M7
Quetzaltenango, *Guatemala* **117** L15
Quezon City, *Philippines* **193** C13
Quibala, *Angola* **208** L10
Quibaxe, *Angola* **208** K10
Quibdó, *Col.* **130** F5
Quiberon, Presqu'île de, *Fr.* **149** R4
Quijotoa, *Ariz., U.S.* **109** V7
Quilá, *Mex.* **116** F8
Quilcene, *Wash., U.S.* **110** B3
Quilengues, *Angola* **208** N10
Quillan, *Fr.* **149** Y10
Quillota, *Chile* **134** K6
Quilon *see* Kollam, *India* **183** T5
Quilpie, *Qnsld., Austral.* **221** U12
Quimavango, *Angola* **208** J9
Quimbele, *Angola* **208** J11
Quimilí, *Arg.* **134** G11
Quimper, *Fr.* **149** Q3
Quimperlé, *Fr.* **149** Q3
Quinault, *Wash., U.S.* **110** C2
Quinault, river, *Wash., U.S.* **110** C2
Quince Mil, *Peru* **132** H5
Quincy, *Mass., U.S.* **97** K15
Quincy, *Fla., U.S.* **99** Q5
Quincy, *Ill., U.S.* **101** R3
Quincy, *Calif., U.S.* **111** P5
Quines, *Arg.* **134** K9
Quinhagak, *Alas., U.S.* **112** K12
Quinn, river, *Nev., Oreg., U.S.* **110** L8
Quintanar de la Orden, *Sp.* **146** G11
Quinter, *Kans., U.S.* **107** S4
Quintin, *Fr.* **149** Q4
Quinto, *Sp.* **147** E14
Quintus Rocks, *Bahamas* **122** E4
Quinzau, *Angola* **208** J9
Quipungo, *Angola* **208** P10
Quirima, *Angola* **208** L12
Quiros Fracture Zone, *Pac. Oc.* **247** M20
Quissico, *Mozambique* **212** H11
Quitapa, *Angola* **208** L12
Quitaque, *Tex., U.S.* **105** E13
Quiterajo, *Mozambique* **213** B13
Quiteve, *Angola* **208** P10
Quitilipi, *Arg.* **134** G12
Quitman, *Ga., U.S.* **99** Q6
Quitman, *Miss., U.S.* **103** M13
Quito, *Ecua.* **130** K4
Quixadá, *Braz.* **133** E17
Qujing, *China* **184** N10
Qulan, *Kaz.* **179** K13
Qulandy, *Kaz.* **178** H8
Qulbān Banī Murrah, ruins, *Jordan* **173** Q8

Qullai Somoniyon (Communist Peak), *Taj.* **181** K18
Qullissat, *Greenland, Den.* **88** L5
Qulsary, *Kaz.* **178** G5
Qum *see* Qom, *Iran* **177** F15
Qumrha, *S. Af.* **212** M8
Quobba, *W. Aust., Austral.* **221** U1
Quogue, *N.Y., U.S.* **97** N13
Quoin Point, *S. Af.* **212** N5
Quokka Rise, *Ind. Oc.* **251** K16
Quorn, *S. Aust., Austral.* **221** W10
Qurayyāt, *Oman* **175** H18
Qūryah, *Afghan.* **180** L5
Quryq, *Kaz.* **178** J4
Qūş, *Egypt* **202** J12
Qusar, *Azerb.* **171** C22
Quşaybah, *Iraq* **176** F8
Qushui *see* Qüxü, *China* **184** L5
Qusmuryn Köli, *Kaz.* **178** D10
Qusmuryn, *Kaz.* **178** D10
Quxar (Quxia), *China* **184** L4
Quxia *see* Quxar, *China* **184** L4
Qüxü (Qushui), *China* **184** L5
Quỳ Châu, *Vietnam* **190** J10
Quỳ Hợp, *Vietnam* **190** J10
Quy Nhơn, *Vietnam* **190** M13
Quyghan, *Kaz.* **179** H14
Quyzhan, *Kaz.* **178** H5
Qvareli, *Rep. of Georgia* **171** B19
Qyzan, *Kaz.* **178** H5
Qyzylorda, *Kaz.* **178** J10

R

Raahe, *Fin.* **143** F15
Raas Binna, *Somalia* **210** K15
Raas Gabbac, *Somalia* **210** M14
Raas Khansiir, *Somaliland* **210** K12
Raas Surud, *Somaliland* **210** K13
Raas Xaafuun, *Somalia* **210** L15
Raas, island, *Indonesia* **192** M14
Raasay, island, *Scot., U.K.* **144** H7
Raasay, Sound of, *Scot., U.K.* **144** H7
Rab, island, *Croatia* **154** F5
Raba, *Indonesia* **193** M13
Rabak, *Sudan* **203** S12
Rabastens, *Fr.* **149** Y7
Rabat *see* Victoria, *Malta* **163** H20
Rabat, *Malta* **163** K22
Rabat, *Mor.* **200** C8
Rabaul, *P.N.G.* **222** N10
Rabi (Rambi), island, *Fiji* **226** N7
Rabi'ah, *Iraq* **176** D9
Rābigh, *Saudi Arabia* **174** H6
Rābor, *Iran* **177** L19
Rabt Sebeta, *W. Sahara* **200** H3
Rabyānah Şaḩrā', *Lib.* **201** J21
Racaka (Rezhaka), *China* **184** K7
Racale, *It.* **153** N15
Răcari, *Rom.* **155** F15
Racconigi, *It.* **152** F2
Raccoon Cay, *Bahamas* **119** H13
Raccoon Point, *La., U.S.* **102** R10
Race, Cape, *Nfld. & Lab., Can.* **87** M24
Raceland, *La., U.S.* **102** Q10
Rạch Giá, *Vietnam* **191** Q11
Rachaïya, *Leb.* **172** J6
Rachid, *Mauritania* **204** F9
Racine, *Wis., U.S.* **100** M7
Raco, *Mich., U.S.* **100** F10
Rada del berdi, *Sudan* **203** T6
Radcliff, *Ky., U.S.* **103** C16
Radford, *Va., U.S.* **98** F9
Radhanpur, *India* **182** J3
Radisson, *Que., Can.* **87** M18
Radnevo, *Bulg.* **155** J15
Radom, *Pol.* **151** F16
Radom, *Sudan* **203** U7
Radomsko, *Pol.* **151** F15
Radomyshl', *Ukr.* **159** N5
Radviliškis, *Lith.* **143** M16
Radzyń Podlaski, *Pol.* **151** E17
Rae Bareli, *India* **182** H8
Rae Isthmus, *Nunavut, Can.* **87** G16
Rae Lakes, *N.W.T., Can.* **86** G11
Raeford, *N.C., U.S.* **98** J11
Raeside, Lake, *Austral.* **220** H4
Raetihi, *N.Z.* **223** H20
Rafaela, *Arg.* **134** J11
Rafaḩ, *Gaza Strip* **173** N4
Rafaï, *Cen. Af. Rep.* **207** N17
Raffadali, *It.* **153** T10
Rafḩā', *Saudi Arabia* **174** C9

Rafína, *Gr.* **157** J13
Rafsanjān, *Iran* **177** K18
Raga, *S. Sudan* **203** V8
Raga, river, *S. Sudan* **203** V7
Ragachow, *Belarus* **158** L6
Ragag, *Sudan* **203** T7
Ragba (Rega), pass, *Nepal* **182** G11
Ragged Island, *Me., U.S.* **97** G17
Ragged Island, *Bahamas* **119** H13
Ragged Point, *Barbados* **123** K20
Ragged, Mount, *Austral.* **220** K5
Ragusa *see* Dubrovnik, *Croatia* **154** J8
Ragusa, *It.* **153** T11
Raha, *Indonesia* **193** L14
Rahad el Berdi, *Sudan* **203** T6
Raḩeb, Ras ir, *Malta* **163** K21
Raheita, *Eritrea* **210** J10
Rahimyar Khan, *Pak.* **181** P10
Raiatea, island, *Fr. Polynesia, Fr.* **231** B16
Raiatea, island, *Fr. Polynesia, Fr.* **231** G17
Raijua, island, *Indonesia* **193** N14
Raik Ridge, *Pac. Oc.* **246** H10
Rainbow Bridge National Monument, *Utah, U.S.* **109** P8
Rainbow City, *Ala., U.S.* **103** J16
Rainbow, *Vic., Austral.* **221** Y11
Rainier, *Oreg., U.S.* **110** E3
Rainier, Mount, *Wash., U.S.* **110** D4
Rainy Lake, *Can., U.S.* **106** E11
Rainy, river, *Can., U.S.* **106** E10
Raipur, *India* **182** L8
Rairakhol, *India* **182** L9
Rairik, island, *Marshall Is.* **227** E21
Ra'īs, *Saudi Arabia* **174** G5
Raisinghnagar, *India* **182** F4
Raitt Rise, *Ind. Oc.* **251** J14
Raivavae (Vavitu), island, *Fr. Polynesia, Fr.* **231** F17
Raiwind, *Pak.* **180** L13
Raj Nandgaon, *India* **182** L8
Raj Samund, *India* **182** J4
Rajahmundry, *India* **182** N8
Rajampet, *India* **183** Q6
Rajang, *Malaysia* **192** H9
Rajangl, river, *Malaysia* **192** H11
Rajanpur, *Pak.* **181** N10
Rajapalaiyam, *India* **183** T6
Rajasthan Canal, *India* **182** G3
Rajgarh, *India* **182** G5
Rajkot, *India* **182** K2
Rajpura, *India* **182** F5
Rakahanga Atoll, *Cook Is., N.Z.* **224** H12
Rakaia, *N.Z.* **223** M18
Rakaposhi, peak, *Pak.* **180** G13
Rakhiv, *Ukr.* **159** Q2
Rakhman Qaynary, *Kaz.* **179** E18
Rakhshan, river, *Pak.* **181** Q5
Rakhyūt, *Oman* **175** N18
Raki, river, *D.R.C.* **208** E12
Rakiraki, *Fiji* **226** P5
Rakiura *see* Stewart Island, island, *N.Z.* **223** Q16
Rakvere, *Est.* **143** J17
Raleigh, *N.C., U.S.* **98** H11
Raleigh, *Miss., U.S.* **102** L12
Ralik Chain, *Marshall Is.* **227** C15
Rām Allāh, *W. Bank* **172** M5
Ram Head, *Virgin Is., U.S.* **122** N11
Ramadi *see* Ar Ramādī, *Iraq* **176** G10
Ramah, *N. Mex., U.S.* **109** S10
Ramai (Remai), *China* **184** L5
Raman, *Thai.* **191** S8
Ramanum, island, *F.S.M.* **226** E8
Ramat Gan, *Israel* **172** M5
Rambi *see* Rabi, island, *Fiji* **226** N7
Rambu, *Indonesia* **193** J14
Rambutyo, island, *P.N.G.* **222** M8
Rameshki, *Russ.* **158** H8
Ramgarh, *India* **182** G2
Rāmhormoz, *Iraq* **177** J14
Ramla, *Israel* **172** M5
Ramlo, peak, *Eritrea* **210** J9
Ramnagar, *India* **182** F7
Râmnicu Vâlcea, *Rom.* **155** E14
Ramon, Har, *Israel* **173** Q4
Ramona, *Calif., U.S.* **111** Y10
Ramos Island, *Solomon Is.* **228** E10
Rampart, *Alas., U.S.* **113** F16
Rampur Hat, *India* **182** J11
Rampur, *India* **182** E6
Rampur, *India* **182** G7
Ramree Island, *Myanmar* **190** J3
Ramree, *Myanmar* **190** J4

Ramsay, *Mich., U.S.* **100** F5
Ramsey (Rhumsaa), *Isle of Man, U.K.* **145** P9
Ramsey Island, *Wales, U.K.* **145** U8
Ramu, *India* **182** L14
Ramu, *Kenya* **211** Q9
Ramu, river, *P.N.G.* **222** P7
Ramvik, *Sw.* **143** H13
Rancagua, *Chile* **134** L7
Ranchester, *Wyo., U.S.* **108** F11
Ranchi, *India* **182** K10
Rancho Cordova, *Calif., U.S.* **111** R4
Rancione, river, *San Marino* **163** H16
Ranco, Lago, *Chile* **135** P6
Ranco, river, *San Marino* **163** H15
Randa, *Nigeria* **206** K8
Randakoma, *Eth.* **210** H8
Randall's Cay, *Bahamas* **122** A9
Randazzo, *It.* **153** S11
Randers, *Den.* **142** M11
Randolph, *Vt., U.S.* **97** G13
Randolph, *Nebr., U.S.* **107** N7
Randolph, *Utah, U.S.* **108** K8
Ranérou, *Senegal* **204** H8
Rangahoe, island, *N.Z.* **229** H22
Rangamati, *Bangladesh* **182** K14
Rangeley, *Me., U.S.* **97** E15
Rangely, *Colo., U.S.* **108** L10
Ranger, *Tex., U.S.* **105** G15
Rangiauria *see* Pitt Island, *Pac. Oc.* **224** P9
Rangitaiki, river, *N.Z.* **223** G22
Rangitoto ki te Tonga (D'Urville Island), *N.Z.* **223** J19
Rangkasbitung, *Indonesia* **192** L8
Rangoon *see* Yangon, *Myanmar* **190** K5
Rangpur, *Bangladesh* **182** H12
Rangsang, island, *Indonesia* **192** H6
Raniganj, *India* **182** K11
Ranikhet, *India* **182** F7
Ranikot, ruins, *Pak.* **181** R8
Rankin Inlet, *Nunavut, Can.* **87** H15
Rankin, *Tex., U.S.* **104** J12
Rann of Kachchh (Rann of Kutch), lake, *India* **182** J2
Rann of Kutch *see* Rann of Kachchh, lake, *India* **182** J2
Rannes, *Qnsld., Austral.* **221** U15
Rannoch, Loch, *Scot., U.K.* **144** K9
Rano Rahi Seamounts, *Pac. Oc.* **247** L18
Ranong, *Thai.* **191** Q6
Ranongga Island, *Solomon Is.* **228** D6
Ransiki, *Indonesia* **193** J19
Rantau, island, *Indonesia* **192** J6
Rantauprapat, *Indonesia* **192** H5
Rantekombola, peak, *Indonesia* **193** K13
Rantoul, *Ill., U.S.* **101** Q7
Ranwas, *Vanuatu* **229** G16
Rao, *Senegal* **204** H6
Raoping, *China* **185** N14
Raoul Island (Sunday), *Kermadec Is., N.Z.* **224** L9
Rapa Nui *see* Easter Island, *Pac. Oc.* **225** L19
Rapa Passage, *Kiribati* **226** J11
Rapa, island, *Fr. Polynesia, Fr.* **225** L14
Rapallo, *It.* **152** F4
Rapar, *India* **182** J2
Raper, Cape, *Nunavut, Can.* **87** E18
Rapid City, *S. Dak., U.S.* **106** L2
Rapid River, *Mich., U.S.* **100** G8
Rapid, river, *Minn., U.S.* **106** E10
Rapota, island, *N.Z.* **229** H20
Rappenstein, peak, *Liech.* **162** Q3
Raquette Lake, *N.Y., U.S.* **96** H10
Raraka, island, *Fr. Polynesia, Fr.* **231** B19
Raroia, island, *Fr. Polynesia, Fr.* **231** B20
Rarotonga, island, *Cook Is., N.Z.* **224** K12
Ra's Abū Madd, *Saudi Arabia* **174** F5
Ra's Abū Qumayyiş, *Saudi Arabia* **175** G13
Ra's Abū Sawmah, *Egypt* **202** H13
Ras Abu Shagara, *Sudan* **202** M15
Ra's al 'Ayn, *Syr.* **172** C13
Ra's al Ḩadāribah (Cape Elba), *Egypt* **202** L15
Ra's al Ḩadd, *Oman* **175** H18
Ra's al Ḩikmah, *Egypt* **202** E9
Ra's al Kalb, *Yemen* **174** P12

Ra's al Khaimah, *U.A.E.* **175** F16
Ra's al Madrakah, *Oman* **175** K17
Ra's al Mish'āb, *Saudi Arabia* **174** D12
Ra's al Unūf, *Lib.* **201** E20
Ra's 'Ālam ar Rūm, *Egypt* **202** E9
Ra's 'Āmir, *Lib.* **201** D21
Ra's an Naqb, *Jordan* **173** Q6
Ra's ash Sharbatāt, *Oman* **175** L16
Ras Asis, *Sudan* **203** P16
Ra's aţ Ţarfā, *Saudi Arabia* **174** M7
Ra's at Ţīn, *Lib.* **201** D22
Ra's az Zawr, *Saudi Arabia* **174** E12
Ra's Banās, *Egypt* **202** K14
Ra's Ḍarbat 'Alī, *Oman, Yemen* **175** M14
Ras Dejen, peak, *Eth.* **210** J7
Rās el Mâ, *Mali* **205** H13
Ra's Fartak, *Yemen* **175** N14
Ra's Ḩāţibah, *Saudi Arabia* **174** H6
Ra's Jibsh, *Oman* **175** J18
Ras Kasar, *Af.* **198** F10
Rās Koh Range, *Pak.* **181** N5
Ras Maqdam, *Sudan* **203** N15
Ra's Mirbāţ, *Oman* **175** M15
Ra's Muḩammad, *Egypt* **202** H13
Ra's Musandam, *Oman* **175** F16
Ras Ngomeni, *Kenya* **211** U8
Ra's Shamrah (Ugarit), ruins, *Syr.* **172** E6
Rās Sharāţīb, *Egypt* **173** T2
Ra's Sharwayn, *Yemen* **175** N13
Ra's Shu'ab, *Yemen* **175** Q15
Rasa, Punta, *Arg.* **135** Q11
Rashad, *Sudan* **203** T11
Rashīd (Rosetta), *Egypt* **202** E11
Rasht, *Iran* **177** D14
Raška, *Serb.* **154** H10
Raso, Cabo, *Port.* **146** H4
Rason, Lake, *Austral.* **220** H5
Rasshua, island, *Russ.* **161** H23
Rasskazovo, *Russ.* **158** L11
Rassokho Seamounts, *Arctic Oc.* **252** F12
Rasua Garhi, *Nepal* **182** G10
Rat Island, *Alas., U.S.* **112** M3
Rat Islands, *Alas., U.S.* **112** M2
Ratak Chain, *Marshall Is.* **227** B17
Ratak Ridge, *Pac. Oc.* **246** H10
Ratangarh, *India* **182** G4
Rathdrum, *Idaho, U.S.* **108** B3
Rathenow, *Ger.* **150** D9
Rathlin Island, *Scot., U.K.* **145** M7
Rätikon, range, *Switz.* **152** B5
Ratingen, *Ger.* **150** F5
Ratisbon *see* Regensburg, *Ger.* **150** J9
Ratlam, *India* **182** K4
Ratnagiri, *India* **182** N3
Ratnapura, *Sri Lanka* **183** U7
Ratodero, *Pak.* **181** P8
Raton Pass, *N. Mex., U.S.* **109** Q14
Raton, *N. Mex., U.S.* **109** Q14
Rattlesnake Creek, *Oreg., U.S.* **110** K8
Rattlesnake Hills, *Wash., U.S.* **110** D6
Rattlesnake National Recreation Area, *Mont., U.S.* **108** C6
Rattray Head, *Eur.* **140** E5
Rau, island, *Indonesia* **193** H16
Raub, *Malaysia* **191** U8
Rauch, *Arg.* **134** M12
Raudal Yupurari (Devils Cataract), *Col.* **130** J8
Raufarhöfn, *Ice.* **142** D4
Raukotaha, island, *N.Z.* **229** J23
Rauma, *Fin.* **143** H15
Raurkela, *India* **182** K10
Rausu Dake, *Japan* **188** F15
Ravahere, island, *Fr. Polynesia, Fr.* **231** C20
Rāvar, *Iran* **177** J19
Ravena, *N.Y., U.S.* **96** K11
Ravenna, *Ohio, U.S.* **101** P15
Ravenna, *Nebr., U.S.* **107** Q6
Ravenna, *It.* **152** F8
Ravensburg, *Ger.* **150** K7
Ravenshoe, *Qnsld., Austral.* **221** R13
Ravensthorpe, *W. Aust., Austral.* **221** X4
Ravenswood, *W. Va., U.S.* **98** D8
Ravi, river, *Pak.* **180** L12
Ravnina, *Turkm.* **178** N9
Rawa Mazowiecka, *Pol.* **151** E15
Rāwah, *Iraq* **176** F9
Rawaki, island, *Kiribati* **224** G10
Rawalpindi, *Pak.* **180** J12
Rāwandūz *see* Ŗewanduz, *Iraq* **176** E11

Rawḍah, lake, *Syr.* **172** F14
Rawḩah, *Saudi Arabia* **174** K7
Rawicz, *Pol.* **151** E13
Rawlinna, *W. Aust., Austral.* **221** W6
Rawlins, *Wyo., U.S.* **108** J11
Rawson, *Arg.* **135** R10
Rawu, *China* **184** L7
Ray, *N. Dak., U.S.* **106** E2
Ray, Cape, *Nfld. & Lab., Can.* **87** N22
Raya, peak, *Indonesia* **192** J10
Rayachoti, *India* **183** Q6
Raychikhinsk, *Russ.* **161** L19
Raymond, *Alta., Can.* **86** N10
Raymond, *Wash., U.S.* **110** D2
Raymond, *Calif., U.S.* **111** T6
Raymondville, *Tex., U.S.* **105** R17
Rayne, *La., U.S.* **102** P8
Rayner Glacier, *Antarctica* **239** C19
Rayoke, island, *Russ.* **161** H23
Rayong, *Thai.* **191** N8
Raysūt, *Oman* **175** M15
Rayville, *La., U.S.* **102** L9
Raz, Pointe du, *Fr.* **149** Q2
Razelm, Lacul, *Rom.* **155** F18
Razgrad, *Bulg.* **155** G16
Razlog, *Bulg.* **155** K13
Razmak, *Pak.* **180** K9
Razo, island, *Cabo Verde* **215** B15
Ré, Île de, *Fr.* **149** T5
Re, river, *San Marino* **163** J14
Reading, *Pa., U.S.* **96** P8
Reading, *Jam.* **122** H7
Reading, *Eng., U.K.* **145** U13
Real Bank, *Arctic Oc.* **253** F18
Realicó, *Arg.* **134** L10
Réalmont, *Fr.* **149** X10
Reao, island, *Fr. Polynesia, Fr.* **231** C22
Rebecca, Lake, *Austral.* **220** J5
Rebiana Sand Sea, *Af.* **198** E7
Rebola, *Eq. Guinea* **214** L7
Reboly, *Russ.* **158** C7
Rebun Tō, *Japan* **188** E12
Recey, *Fr.* **149** R12
Recherche, Archipelago of the, *Austral.* **220** K5
Rechytsa, *Belarus* **158** M6
Recife, *Braz.* **133** G18
Récifs, Île aux, *Seychelles* **215** N21
Recklinghausen, *Ger.* **150** E5
Reconquista, *Arg.* **134** H12
Recovery Glacier, *Antarctica* **238** F12
Recreo, *Arg.* **134** H9
Rector, *Ark., U.S.* **102** E11
Red , river, *U.S.* **105** E15,18
Red Bank, *Tenn., U.S.* **103** G17
Red Bay, *Ala., U.S.* **103** H13
Red Bay, *Cayman Is., U.K.* **122** N2
Red Bluff Lake, *Tex., U.S.* **104** H9
Red Bluff, *Calif., U.S.* **111** N3
Red Butte, *Ariz., U.S.* **109** R7
Red Cliff, *St. Kitts & Nevis* **123** C19
Red Cloud, *Nebr., U.S.* **107** R6
Red Deer, *Alta., Can.* **86** M10
Red Devil, *Alas., U.S.* **113** J14
Red Hill *see* Pu'u'ula'ula, peak, *Hawaii, U.S.* **115** K19
Red Hills, *Kans., U.S.* **107** V5
Red Hook, *Virgin Is., U.S.* **122** N9
Red Lake Falls, *Minn., U.S.* **106** F8
Red Lake, *Ont., Can.* **87** N14
Red Lake, *Minn., U.S.* **106** F10
Red Lodge, *Mont., U.S.* **108** F9
Red Mountain, *Mont., U.S.* **108** C7
Red Oak, *Iowa, U.S.* **107** Q10
Red River of the North, river, *Can., U.S.* **90** B12
Red Rock, *Gambia, Senegal* **204** J8
Red Rock, river, *Mont., U.S.* **108** F6
Red Rocks Point, *Austral.* **220** J7
Red Sea, *Af., Asia* **268** H2
Red *see* Hông, river, *Vietnam* **190** G10
Red *see* Yuan, river, *China* **184** N9
Red Springs, *N.C., U.S.* **98** J11
Red Wing, *Minn., U.S.* **106** K12
Red, river, *Tex., U.S.* **90** K12
Redang, island, *Malaysia* **191** S9
Redange sur Attert, *Lux.* **162** J9
Redcliffe, Mount, *Austral.* **220** H4
Redding, *Calif., U.S.* **111** N3
Redeyef, *Tunisia* **201** C15
Redfield, *S. Dak., U.S.* **106** K6
Redig, *S. Dak., U.S.* **106** J2
Redlands, *Calif., U.S.* **111** X9

Redmond, *Oreg., U.S.* **110** H5
Redon, *Fr.* **149** R5
Redonda, island, *Lesser Antil.* **121** C15
Redondo, *Port.* **146** J6
Redoubt Volcano, *Alas., U.S.* **113** K15
Redruth, *Eng., U.K.* **145** W8
Redwater, river, *Mont., U.S.* **108** C12
Redwood Empire, region, *Calif., U.S.* **111** N2
Redwood Falls, *Minn., U.S.* **106** K10
Redwood National Park, *Calif., U.S.* **110** L2
Ree, Lough, *Ire.* **145** Q5
Reed City, *Mich., U.S.* **100** K10
Reeder, *N. Dak., U.S.* **106** H2
Reedley, *Calif., U.S.* **111** T7
Reedsburg, *Wis., U.S.* **100** K4
Reedsport, *Oreg., U.S.* **110** H2
Reedville, *Va., U.S.* **98** E14
Reef Bay, *Virgin Is., U.S.* **122** N11
Reef Islands *see* Rowa Islands, *Vanuatu* **229** C15
Reef Islands, *Solomon Is.* **228** K5
Reefton, *N.Z.* **223** K18
Reese, river, *Nev., U.S.* **111** N9
Refahiye, *Turk.* **171** E13
Reform, *Ala., U.S.* **103** J14
Refugio, *Tex., U.S.* **105** N17
Regência Point, *S. Amer.* **128** H10
Regensburg (Ratisbon), *Ger.* **150** J9
Reggane, *Alg.* **200** H11
Reggio di Calabria, *It.* **153** R12
Reggio nell'Emilia, *It.* **152** F6
Reghin, *Rom.* **155** C14
Regina, *Sask., Can.* **86** N12
Régina, *Fr. Guiana, Fr.* **131** G19
Registan, range, *Afghan.* **180** M5
Rehoboth Beach, *Del., U.S.* **98** C15
Rehoboth, *Namibia* **212** H4
Reid River, *Qnsld., Austral.* **221** S14
Reidsville, *Ga., U.S.* **98** N7
Reidsville, *N.C., U.S.* **98** G10
Reigate, *Eng., U.K.* **145** V14
Reims, *Fr.* **149** N11
Reina Adelaida, Archipiélago, *Chile* **135** W6
Reindeer Lake, *Sask., Man., Can.* **87** K13
Reinga Basin, *Pac. Oc.* **246** N10
Reinosa, *Sp.* **146** B10
Reisdorf, *Lux.* **162** H10
Reitoru, island, *Fr. Polynesia, Fr.* **231** C19
Rejaf, *S. Sudan* **203** X11
Rekareka *see* Tehuata, island, *Fr. Polynesia, Fr.* **231** B20
Rekata Bay, *Solomon Is.* **228** D8
Reliance, *N.W.T., Can.* **86** H12
Reliance, *Wyo., U.S.* **108** K9
Relizane, *Alg.* **200** C11
Remada, *Tunisia* **201** E16
Remai *see* Ramai, *China* **184** L5
Remanso, *Braz.* **133** G15
Remedios, *Cuba* **118** H9
Remer, *Minn., U.S.* **106** G11
Remich, *Lux.* **162** K11
Rémire, *Fr. Guiana, Fr.* **131** F19
Remiremont, *Fr.* **149** Q14
Remontnoye, *Russ.* **159** R13
Ren, island, *Marshall Is.* **227** G20
Renaud Island, *Antarctica* **238** E5
Rend Lake, *Ill., U.S.* **101** U5
Rendezvous Bay, *Virgin Is., U.S.* **122** N10
Rendova, island, *Solomon Is.* **228** E7
Rendsburg, *Ger.* **150** B7
Rengat, *Indonesia* **192** J6
Reni, *Ukr.* **159** S4
Renk, *S. Sudan* **203** T12
Renland, *Greenland, Den.* **88** K13
Rennell, island, *Solomon Is.* **228** H10
Rennes, *Fr.* **149** Q5
Rennick Bay, *Antarctica* **239** R15
Rennick Glacier, *Antarctica* **239** Q15
Reno, *Nev., U.S.* **111** P6
Renovo, *Pa., U.S.* **96** M6
Rensselaer, *Ind., U.S.* **101** P8
Renton, *Wash., U.S.* **110** C4
Renville, *Minn., U.S.* **106** K10
Repetek, *Turkm.* **178** M9
Republic, *Mich., U.S.* **100** F7
Republic, *Wash., U.S.* **110** A7
Republican, river, *U.S.* **90** G12
Repulse Bay, *Nunavut, Can.* **87** G16
Repulse Bay, *Austral.* **220** E14
Requena, *Peru* **132** E4

Requena, *Sp.* **147** H14
Requin Bay, *Grenada* **123** L23
Reşadiye, *Turk.* **170** E12
Researcher Ridge, *Atl. Oc.* **248** M6
Reserve, *La., U.S.* **102** P10
Reserve, *N. Mex., U.S.* **109** T10
Reshteh Kūhhā-ye Alborz (Elburz Mountains), *Iran* **177** D14
Resia, *It.* **152** B6
Resistencia, *Arg.* **134** G12
Reşiţa, *Rom.* **154** E11
Resolute, *Nunavut, Can.* **87** D15
Resolution Fracture Zone, *Pac. Oc.* **247** N16
Resolution Island, *N. Amer.* **84** E9
Resolution Island, *N.Z.* **223** P15
Rest, *Jam.* **122** J8
Restinga, Punta, *Sp.* **214** R3
Réthymno, *Gr.* **157** Q13
Reti, *Pak.* **181** P9
Retreat, *Jam.* **122** H9
Réunion Trench, *Ind. Oc.* **250** L7
Reus, *Sp.* **147** E16
Reutlingen, *Ger.* **150** J7
Reva, *S. Dak., U.S.* **106** J2
Reval *see* Tallinn, *Est.* **143** J16
Revelle Rise, *Ind. Oc.* **250** H6
Revelstoke, *B.C., Can.* **86** M9
Revigny, *Fr.* **149** P12
Revillagigedo Island, *Alas., U.S.* **113** M24
Revillagigedo, Islas, *Mex.* **116** H5
Revolyutsii Oktyabr'skoy, *Russ.* **161** D14
Revsbotn, *Nor.* **143** B14
Rewa, *India* **182** J8
Rewa, river, *Fiji* **226** P5
Rewanduz (Rāwandūz), *Iraq* **176** E11
Rexburg, *Idaho, U.S.* **108** G7
Rexford, *Mont., U.S.* **108** A5
Rexton, *Mich., U.S.* **100** F10
Rey Bouba, *Cameroon* **206** L11
Rey, *Iran* **177** F15
Rey, Isla del, *Pan.* **117** N21
Rey, river, *Cameroon* **206** L12
Reyes, Point, *U.S.* **90** F2
Reyes, *Bol.* **132** H6
Reyhanlı, *Turk.* **170** J11
Reykjanes Ridge, *Atl. Oc.* **253** N21
Reykjanes, *Eur.* **140** A3
Reykjavík, *Ice.* **142** E2
Reynisdjúp, valley, *Atl. Oc.* **253** M21
Reynosa, *Mex.* **116** E12
Rēzekne, *Latv.* **143** L17
Rezhaka *see* Racaka, *China* **184** K7
Rezvānshahr, *Iran* **177** D14
Rgotina, *Serb.* **154** G12
Rhaetian Alps, *Switz.* **152** C5
Rhame, *N. Dak., U.S.* **106** H2
Rhein (Rhine), river, *Switz.* **152** A3
Rheine, *Ger.* **150** D6
Rhiconich, *Scot., U.K.* **144** G8
Rhin (Rhine), *Eur.* **149** Q15
Rhine *see* Rhein, Rhin, river, *Eur.* **140** G7
Rhinelander, *Wis., U.S.* **100** G5
Rhinns of Kells, region, *Scot., U.K.* **145** N9
Rhino Camp, *Uganda* **211** Q3
Rhinocolura *see* El 'Arîsh, *Egypt* **173** N3
Rhir, Cape, *Af.* **198** D2
Rhiy di-Irīsal, cape, *Yemen* **175** Q15
Rho, *It.* **152** D4
Rhode Island, *U.S.* **97** L14
Rhodes Peak, *Idaho, U.S.* **108** D5
Rhodes *see* Ródos, island, *Gr.* **157** N20
Rhodope Mountains, *Bulg., Gr.* **155** K13
Rhône au Rhin, Canal du, *Fr.* **149** R15
Rhône, river, *Fr., Switz.* **149** W12
Rhum, Sound of, *Scot., U.K.* **144** J7
Rhumsaa *see* Ramsey, *Isle of Man, U.K.* **145** P9
Rhunagi, *P.N.G.* **222** N10
Rhydargaeau, *Wales, U.K.* **145** T9
Rhyl, *Eng., U.K.* **145** R10
Rhynie, *Scot., U.K.* **144** J10
Riaba, *Eq. Guinea* **214** M7
Riaba, Bahia de, *Eq. Guinea* **214** M7
Riachão, *Braz.* **133** F14
Rialb, river, *Andorra* **162** G3
Riang, *India* **182** G14
Riangnom, *S. Sudan* **203** U10
Riaño, *Sp.* **146** B9

Riasi, *India* **180** J13
Riau, Kepulauan, *Indonesia* **192** J7
Riaza, *Sp.* **146** E11
Rib Lake, *Wis., U.S.* **100** H4
Rib Mountain, *Wis., U.S.* **100** H5
Ribadeo, *Sp.* **146** A7
Ribáuè, *Mozambique* **212** D12
Ribeira Afonso, *Sao Tome and Principe* **215** D20
Ribeira Brava, *Madeira, Port.* **214** M2
Ribeira da Cruz, *Cabo Verde* **215** A14
Ribeira Grande, *Port.* **146** Q5
Ribeira Grande, *Cabo Verde* **215** A14
Ribeirão Preto, *Braz.* **133** L13
Ribera, *It.* **153** T9
Riberalta, *Bol.* **132** G7
Ribes de Freser, *Sp.* **147** C17
Ribewon, island, *Marshall Is.* **227** B20
Riblah, *Leb.* **172** H8
Ribniţa, *Mold.* **159** Q5
Ribnitz-Damgarten, *Ger.* **150** B9
Riccione, *It.* **152** G8
Rice Lake, *Wis., U.S.* **100** G3
Rich Hill, *Mo., U.S.* **107** R11
Rich Square, *N.C., U.S.* **98** G13
Richard-Toll, *Senegal* **204** H7
Richards Bay, *S. Af.* **212** K10
Richardson Highway, *Alas., U.S.* **113** H18
Richardson Mountains, *N.W.T., Yukon, Can.* **86** D9
Richardton, *N. Dak., U.S.* **106** G3
Richey, *Mont., U.S.* **108** C13
Richfield, *Utah, U.S.* **109** N7
Richford, *Vt., U.S.* **97** F13
Richland Center, *Wis., U.S.* **100** L4
Richland Chambers Reservoir, *Tex., U.S.* **105** H18
Richland Springs, *Tex., U.S.* **105** J15
Richland, *Ga., U.S.* **99** N5
Richland, *Mo., U.S.* **107** U13
Richland, *Oreg., U.S.* **110** G9
Richland, *Wash., U.S.* **110** E7
Richlands, *Va., U.S.* **98** F8
Richmond Peak, *St. Vincent & the Grenadines* **123** K16
Richmond, *Me., U.S.* **97** G16
Richmond, *Va., U.S.* **98** E13
Richmond, *Ind., U.S.* **101** R10
Richmond, *Ky., U.S.* **103** C18
Richmond, *Tex., U.S.* **105** L19
Richmond, *Mo., U.S.* **107** S11
Richmond, *Utah, U.S.* **108** J7
Richmond, *Calif., U.S.* **111** R3
Richmond, *Jam.* **122** H10
Richmond, *St. Vincent & the Grenadines* **123** K16
Richmond, *N.S.W., Austral.* **221** X14
Richmond, *Qnsld., Austral.* **221** S12
Richton, *Miss., U.S.* **102** N12
Richwood, *W. Va., U.S.* **98** E9
Rico, *Colo., U.S.* **109** P10
Ridā', *Yemen* **174** P9
Ridanna, *It.* **152** B7
Ridder (Lenīnogorsk), *Kaz.* **179** E17
Riddle, *Idaho, U.S.* **108** J3
Riddle, *Oreg., U.S.* **110** J2
Ridge A, *Antarctica* **239** H16
Ridgecrest, *Calif., U.S.* **111** V9
Ridgefield, *Wash., U.S.* **110** E3
Ridgeland, *S.C., U.S.* **98** M9
Ridgway, *Pa., U.S.* **96** M5
Ridgway, *Colo., U.S.* **109** N11
Riding Point, *Bahamas* **122** J3
Riding Rocks, islands, *Bahamas* **118** E10
Riesa, *Ger.* **150** F10
Rieti, *It.* **152** K9
Rieumes, *Fr.* **149** X8
Rieupeyroux, *Fr.* **149** W10
Riez, *Fr.* **149** X14
Rifle, *Colo., U.S.* **108** M11
Rifstangi, *Eur.* **140** A4
Rig Rig, *Chad* **206** H11
Rīga, *Latv.* **143** L16
Riga, Gulf of, *Est., Latv.* **143** L16
Rīgān, *Iran* **177** L21
Rigby, *Idaho, U.S.* **108** G7
Riggins, *Idaho, U.S.* **108** E4
Riglos, *Sp.* **147** C14
Rigo, *P.N.G.* **222** R8
Rigolet, *Nfld. & Lab., Can.* **87** K22
Riiser-Larsen Ice Shelf, *Antarctica* **238** C11
Riiser-Larsen Peninsula, *Antarctica* **239** B17

Rijau, *Nigeria* **206** J7
Rijeka (Fiume), *Croatia* **154** E4
Rijeka Crnojevića, *Montenegro* **154** J9
Rikaze *see* Xigazê, *China* **184** L5
Rikitea, *Fr. Polynesia, Fr.* **231** Q17
Rikubetsu, *Japan* **188** F14
Rikuzentakata, *Japan* **188** L13
Rila, mountains, *Bulg.* **155** K13
Riley, *Kans., U.S.* **107** S8
Rilly, *Fr.* **149** P11
Rimā', Jabal ar, *Jordan* **172** L8
Rimatara, island, *Fr. Polynesia, Fr.* **225** K13
Rimavská Sobota, *Slovakia* **151** K15
Rimini, *It.* **152** G8
Rîmnicu Sărat, *Rom.* **155** E16
Rimouski, *Que., Can.* **87** N20
Rincón de la Victoria, *Sp.* **146** M10
Rincón del Bonete, Lago, *Uru.* **134** K14
Rincón del Guanal, Punta, *Cuba* **118** J6
Rincon, *Ga., U.S.* **99** N9
Rinconada, *Arg.* **134** D9
Ríneia, island, *Gr.* **157** K15
Ringbu, *Nor.* **142** J11
Ringdom Gompa, *India* **180** H14
Ringdove Passage, *Solomon Is.* **228** M6
Ringgold Isles, *Fiji* **226** N8
Ringgold, *La., U.S.* **102** L7
Ringim, *Nigeria* **206** J9
Ringkøbing, *Den.* **142** N10
Ringling, *Okla., U.S.* **105** E17
Ringvassøy, island, *Nor.* **143** C13
Rink Isbræ, glacier, *Greenland, Den.* **88** K6
Rio Branco, *Braz.* **132** G6
Río Bravo del Norte *see* Rio Grande, river, *N. Amer.* **116** C9
Rio Bueno, *Jam.* **122** G8
Río Bueno, *Chile* **135** P6
Río Cauto, *Cuba* **118** K12
Rio Chama, river, *N. Mex., U.S.* **109** Q12
Río Claro, *Trin. & Tobago* **123** Q23
Río Cuarto, *Arg.* **134** K10
Rio de Janeiro, *Braz.* **133** M15
Río de la Plata Canyon, *Atl. Oc.* **249** T6
Rio Dell, *Calif., U.S.* **111** N1
Rio do Ouro, *Sao Tome and Principe* **215** C20
Río Esmeraldas, river, *Ecua.* **130** J3
Río Gallegos, *Arg.* **135** W8
Rio Grande , *Mex., U.S.* **104** K8
Rio Grande City, *Tex., U.S.* **105** R15
Río Grande del Norte National Monument, *N. Mex., U.S.* **109** Q12
Rio Grande Fracture Zone, *Atl. Oc.* **249** S10
Rio Grande Plain, *Atl. Oc.* **249** T8
Rio Grande Rise, *Atl. Oc.* **249** S8
Rio Grande Wild & Scenic River, *Tex., U.S.* **104** L11
Río Grande, *Braz.* **132** Q12
Río Grande, *Arg.* **135** X9
Rio Grande, river, *N. Amer.* **84** K4
Rio Hondo, river, *N. Mex., U.S.* **109** U13
Río Mulatos, *Bol.* **132** K7
Rio Muni, *Eq. Guinea* **206** Q9
Rio Nuevo, *Jam.* **122** H9
Rio Peñasco, river, *N. Mex., U.S.* **109** U13
Rio Puerco, river, *N. Mex., U.S.* **109** S11
Rio Rancho, *N. Mex., U.S.* **109** S12
Rio Tejo, Foz do, *Port.* **146** J4
Rio Verde de Mato Grosso, *Braz.* **132** K11
Rio Verde, *Braz.* **132** K12
Río Verde, *Mex.* **116** G11
Río Verde, *Chile* **135** W7
Rio Vouga, Foz do, *Port.* **146** F5
Rio Zambeze, river, *Mozambique* **212** F11
Río, *Gr.* **156** H9
Riobamba, *Ecua.* **130** L4
Ríohacha, *Col.* **130** C7
Riom, *Fr.* **149** U11
Rioni, river, *Rep. of Georgia* **171** B17
Rioz, *Fr.* **149** R14
Ripley, *W. Va., U.S.* **98** D8
Ripley, *Ohio, U.S.* **101** T12
Ripley, *Miss., U.S.* **102** G12

Ripley, *Tenn., U.S.* **102** F12
Ripley, *Calif., U.S.* **111** Y12
Ripoll, *Sp.* **147** C17
Ripon, *Wis., U.S.* **100** K6
Ripon, *Eng., U.K.* **145** P12
Risāfe *see* Ar Ruşāfah, ruins, *Syr.* **172** F11
Riscle, *Fr.* **149** X7
Rishikesh, *India* **182** F6
Rishiri Tō, *Japan* **188** E12
Rishon LeẒiyyon, *Israel* **172** M5
Rising Star, *Tex., U.S.* **105** H15
Rîşnov, *Rom.* **155** E15
Rison, *Ark., U.S.* **102** H8
Risør, *Nor.* **142** L11
Rissani, *Mor.* **200** E9
Ritchie Bank, *Ind. Oc.* **250** H8
Ritchie's Archipelago, *India* **183** R14
Ritidian Point, *Guam, U.S.* **230** M6
Ritscher Canyon, *Ind. Oc.* **249** Y15
Ritscher Upland, *Antarctica* **238** C12
Ritu *see* Rutog, *China* **184** H2
Ritzville, *Wash., U.S.* **110** C8
Riva del Garda, *It.* **152** D6
Riva di Tures, *It.* **152** B8
Riva, river, *San Marino* **163** K16
Rivadavia, *Chile* **134** H7
Rivas, *Nicar.* **117** M18
Rive-de-Gier, *Fr.* **149** U12
Riveira, *Sp.* **146** D10
Riviere aux Feuilles, river, *Que., Can.* **87** K18
River Plate, *Atl. Oc.* **249** T6
Rivera, *Uru.* **134** J14
Riverdale, *N. Dak., U.S.* **106** F4
Riverhead, *N.Y., U.S.* **97** N13
Riverina, *W. Aust., Austral.* **221** W4
Riverina, region, *Austral.* **220** L13
Riverside, *Wyo., U.S.* **108** K11
Riverside, *Wash., U.S.* **110** B7
Riverside, *Calif., U.S.* **111** X9
Riversleigh, *Qnsld., Austral.* **221** R11
Riverton, *Man., Can.* **87** N14
Riverton, *Wyo., U.S.* **108** H10
Rivesaltes, *Fr.* **149** Z10
Riviera, *Fr.* **140** J6
Rivière des Anguilles, *Mauritius* **215** H20
Rivière du Rempart, *Mauritius* **215** F20
Rivière Noire, Piton de la, *Mauritius* **215** G19
Rivière-du-Loup, *Que., Can.* **87** N20
Rivière-Pilote, *Martinique, Fr.* **123** G23
Rivière-Salée, *Martinique, Fr.* **123** G23
Rivne, *Ukr.* **159** N3
Rivoli, *It.* **152** E2
Riyadh *see* Ar Riyāḍ, *Saudi Arabia* **174** G10
Rize, *Turk.* **171** D15
Rizhao, *China* **185** H15
Rizokarpaso (Dipkarpaz), *Cyprus* **162** M11
Rizzuto, Capo, *It.* **153** Q14
Ro (Ágios Geórgios), island, *Gr.* **157** N22
Ro, *New Caledonia, Fr.* **231** P24
Roa, *Sp.* **146** D10
Road Bay, *Anguilla, U.K.* **122** R11
Road Town, *Virgin Is., U.K.* **122** Q6
Roah (Na'am), river, *S. Sudan* **203** X10
Roan Cliffs, *U.S.* **90** F7
Roan Plateau, *Colo., U.S.* **108** M10
Roanne, *Fr.* **149** T12
Roanoke Island, *N.C., U.S.* **98** G15
Roanoke Rapids, *N.C., U.S.* **98** G12
Roanoke, *Va., U.S.* **98** F10
Roanoke, *Ala., U.S.* **103** K17
Roanoke, river, *N.C., Va., U.S.* **98** F11
Roaringwater Bay, *Ire.* **145** U3
Rob Roy, island, *Solomon Is.* **228** D7
Robāṭ-e Khān, *Iran* **177** G18
Robbie Ridge, *Pac. Oc.* **246** L12
Robe, *S. Aust., Austral.* **221** Y11
Robersonville, *N.C., U.S.* **98** H13
Robert Glacier, *Antarctica* **239** D19
Robert Lee, *Tex., U.S.* **105** H13
Robert S. Kerr Reservoir, *Okla., U.S.* **105** C19
Robert-Bourassa, Réservoir, *Que., Can.* **87** L18
Roberts Creek Mountain, *Nev., U.S.* **111** P10
Roberts, *Idaho, U.S.* **108** G7
Robertson Bay, *Antarctica* **239** R14

Robertson Fjord *see* Siorapaluup Kangerlua, *Greenland, Den.* **88** E3
Robertsport, *Liberia* **204** P9
Robeson Channel, *Can., Greenland, Den.* **88** B6
Róbinson Crusoe, Isla, *S. Amer.* **128** L3
Robinson Range, *Austral.* **220** G3
Robinson River, *N.T., Austral.* **221** Q10
Robinson, *Ill., U.S.* **101** S7
Robinvale, *N.S.W., Austral.* **221** X12
Robokaere, island, *Marshall Is.* **227** D20
Roboré, *Bol.* **132** K9
Robson, Mount, *B.C., Can.* **86** L9
Robstown, *Tex., U.S.* **105** P17
Roby, *Tex., U.S.* **105** G13
Roca Partida, Isla, *Mex.* **116** H6
Roca Redonda, island, *Ecua.* **130** P2
Roca, Cape, *Eur.* **140** K1
Rocas Alijos, islands, *Mex.* **116** E5
Rocas, Atol das, *Braz.* **133** E18
Rocca San Casciano, *It.* **152** G7
Roccadaspide, *It.* **153** N12
Roccastrada, *It.* **152** H7
Rocha, *Uru.* **134** L15
Rochdale, *Eng., U.K.* **145** Q11
Rochedo, *Braz.* **132** K11
Rochefort, *Belg.* **148** M13
Rochefort, *Fr.* **149** U6
Rochegda, *Russ.* **158** C11
Rochelle, *Ga., U.S.* **99** N6
Rochelle, *Ill., U.S.* **101** N5
Rocher de Diamant (Diamond Rock), *Martinique, Fr.* **123** G22
Rochers Somaloma, *Wallis & Futuna Is., Fr.* **230** Q11
Roche's Bluff, *Montserrat, U.K.* **123** C24
Roches Epeti, *Fr. Polynesia, Fr.* **230** D11
Rocheservière, *Fr.* **149** S5
Rochester, *N.Y., U.S.* **96** J6
Rochester, *N.H., U.S.* **97** H15
Rochester, *Ind., U.S.* **101** P9
Rochester, *Minn., U.S.* **106** L12
Rochester, *Eng., U.K.* **145** M11
Rochester, *Eng., U.K.* **145** U15
Rock Creek Butte, *Oreg., U.S.* **110** G8
Rock Creek, *Mont., U.S.* **108** E9
Rock Falls, *Ill., U.S.* **101** N5
Rock Hill, *S.C., U.S.* **98** J9
Rock Island, *Ill., U.S.* **101** P4
Rock Lake, *N. Dak., U.S.* **106** E6
Rock Port, *Mo., U.S.* **107** Q9
Rock Rapids, *Iowa, U.S.* **106** M9
Rock River, *Wyo., U.S.* **108** J12
Rock Sound, *Bahamas* **122** F6
Rock Sound, *Bahamas* **122** F6
Rock Springs, *Mont., U.S.* **108** D12
Rock Springs, *Wyo., U.S.* **108** K9
Rock, river, *Ill. Wis., U.S.* **101** N5
Rockaway, *Oreg., U.S.* **110** F2
Rockdale, *Tex., U.S.* **105** K17
Rockefeller Plateau, *Antarctica* **238** L10
Rockford, *Ill., U.S.* **100** M5
Rockhampton, *Qnsld., Austral.* **221** T15
Rockingham Bay, *Austral.* **220** D13
Rockingham, *N.C., U.S.* **98** J10
Rockingham, *W. Aust., Austral.* **221** X2
Rockland, *Me., U.S.* **97** G17
Rockland, *Mich., U.S.* **100** F6
Rockledge, *Fla., U.S.* **99** T10
Rockly Bay, *Trin. & Tobago* **123** P17
Rockmart, *Ga., U.S.* **98** K4
Rockport, *Ind., U.S.* **101** U8
Rockport, *Tex., U.S.* **105** N17
Rockport, *Calif., U.S.* **111** P2
Rocksprings, *Tex., U.S.* **105** L14
Rockton, *Ill., U.S.* **100** M5
Rockville, *Md., U.S.* **98** C13
Rockville, *Ind., U.S.* **101** R8
Rockwell City, *Iowa, U.S.* **107** N10
Rockwood, *Me., U.S.* **97** D16
Rockwood, *Tenn., U.S.* **103** F18
Rocky Ford, *Colo., U.S.* **109** N14
Rocky Mount, *N.C., U.S.* **98** G12
Rocky Mount, *Va., U.S.* **98** F10
Rocky Mountain National Park, *Colo., U.S.* **108** L12
Rocky Mountains, *N. Amer.* **84** H4
Rocky Point, *Bahamas* **122** C9
Rocky Point, *Bahamas* **122** E11

Rocky Point, *Austral.* **229** B20
Rodanthe, *N.C., U.S.* **98** H15
Roddickton-Bide Arm, *Nfld. & Lab., Can.* **87** L23
Rodeo, *N. Mex., U.S.* **109** W10
Rodeo, *Arg.* **134** J7
Rodez, *Fr.* **149** W10
Rodney Bay, *St. Lucia* **123** J14
Rodonit, Kepi i, *Alban.* **154** K9
Ródos (Rhodes), island, *Gr.* **157** N20
Ródos, *Gr.* **157** M20
Rodrigues Ridge, *Ind. Oc.* **250** K8
Roebuck Bay, *Austral.* **220** D4
Roermond, *Neth.* **148** K13
Roes Welcome Sound, *Nunavut, Can.* **87** H16
Roeselare, *Belg.* **148** L10
Rogagua, Lago, *Bol.* **132** H7
Rogers City, *Mich., U.S.* **100** H11
Rogers Lake, *Calif., U.S.* **111** W9
Rogers, *Ark., U.S.* **102** E6
Rogers, *Tex., U.S.* **105** J17
Rogers, Mount, *Va., U.S.* **98** G8
Rogerson, *Idaho, U.S.* **108** J5
Rogersville, *Tenn., U.S.* **103** E19
Roggan, river, *Que., Can.* **87** L18
Roggeveen Basin, *Pac. Oc.* **247** N19
Roggeveen Rise, *Pac. Oc.* **247** N21
Rogliano, *Fr.* **149** W3
Rogliano, *It.* **153** Q13
Rognan, *Nor.* **142** E12
Rogue River, *Oreg., U.S.* **110** K3
Rogue, river, *Oreg., U.S.* **110** K2
Roguron, island, *Marshall Is.* **227** D19
Rohrbach, *Aust.* **150** K10
Rohri, *Pak.* **181** P8
Rohtak, *India* **182** F5
Roi Anchorage, *Marshall Is.* **227** F16
Roi Et, *Thai.* **190** L10
Roi-Namur, island, *Marshall Is.* **227** F16
Roi, *F.S.M.* **226** K8
Rojas, *Arg.* **134** L11
Rojhan, *Pak.* **181** N9
Rokera, *Solomon Is.* **228** F11
Rokiškis, *Lith.* **143** L17
Rôlas, Ilhéu das, *Sao Tome and Principe* **215** E20
Rolette, *N. Dak., U.S.* **106** E5
Rolla, *N. Dak., U.S.* **106** D5
Rolla, *Mo., U.S.* **107** U13
Rolleston, *N.Z.* **223** L18
Rolletown, *Bahamas* **122** F9
Rolleville, *Bahamas* **122** E8
Rolling Fork, *Miss., U.S.* **102** K10
Rollingstone, *Qnsld., Austral.* **221** R14
Rollins, *Mont., U.S.* **108** B5
Rolvsøya, island, *Nor.* **143** B14
Rom, *S. Sudan* **203** U12
Roma (Rome), *It.* **152** K8
Roma-Los Saenz, *Tex., U.S.* **105** R15
Roma, *Qnsld., Austral.* **221** U14
Romain, Cape, *S.C., U.S.* **98** M11
Romaine, river, *Que., Can.* **87** M21
Roman, *Rom.* **155** C16
Romanche Fracture Zone, *Atl. Oc.* **249** P9
Romanche Gap, *Atl. Oc.* **249** P10
Romang, island, *Indonesia* **193** M16
Români, *Egypt* **173** N1
Români, *Egypt* **202** F12
Romania, *Eur.* **154** D11
Romanija, peak, *Bosn. & Herzg.* **154** G8
Romano, Cape, *Fla., U.S.* **99** X8
Romano, Cayo, *Cuba* **118** H11
Romanovka, *Russ.* **161** L16
Romans, *Fr.* **149** V13
Romanshorn, *Switz.* **152** A5
Romanzof, Cape, *N. Amer.* **84** Q3
Rome *see* Roma, *It.* **152** K8
Rome, *N.Y., U.S.* **96** J9
Rome, *Ga., U.S.* **98** K4
Romilly, *Fr.* **149** Q11
Romney, *W. Va., U.S.* **98** C11
Romny, *Ukr.* **159** N6
Romodanovo, *Russ.* **158** K13
Romorantin-Lanthenay, *Fr.* **149** R9
Romsey, *Eng., U.K.* **145** V12
Rona, island, *Scot., U.K.* **144** E8
Ronan, *Mont., U.S.* **108** C6
Roncador, Serra do, *Braz.* **132** H12
Ronceverte, *W. Va., U.S.* **98** E9
Ronda, *Sp.* **146** M9
Ronde Island, *Grenada* **123** H24
Rondonópolis, *Braz.* **132** J11
Rondu, *Pak.* **180** G13

Rocky Point, *Austral.* **229** B20

Rong Kwang, *Thai.* **190** J8
Rong'an, *China* **184** N12
Rongcheng, *China* **185** H15
Ronge, Lac la, *N. Amer.* **84** G5
Rongelap Atoll, *Marshall Is.* **227** B15
Rongerik Atoll, *Marshall Is.* **227** B16
Rongjiang, *China* **184** N11
Rongxian, *China* **184** P12
Ronkiti Harbor, *F.S.M.* **226** K8
Ronkiti, *F.S.M.* **226** K8
Ronn Basin, *Atl. Oc.* **254** L11
Ronne Entrance, bay, *Antarctica* **238** G7
Ronne Ice Shelf, *Antarctica* **238** F9
Rønne, *Den.* **143** N13
Ronse, *Belg.* **148** L11
Roo Rise, *Ind. Oc.* **251** J17
Roodhouse, *Ill., U.S.* **101** S4
Roof Butte, *Ariz., U.S.* **109** Q9
Roorkee, *India* **182** F6
Roosendaal, *Neth.* **148** K12
Roosevelt Island, *Antarctica* **238** M11
Roosevelt, *Utah, U.S.* **108** L9
Roosevelt, Mount, *B.C., Can.* **86** J9
Root, river, *Minn., U.S.* **106** L13
Roper, river, *Austral.* **220** B9
Ropesville, *Tex., U.S.* **104** F12
Roquefort, *Fr.* **149** X7
Roquetas de Mar, *Sp.* **146** M12
Roraima, *S. Amer.* **131** F14
Røros, *Nor.* **142** H11
Rorschach, *Switz.* **152** A5
Rørvik, *Nor.* **142** F11
Ros Comáin *see* Roscommon, *Ire.* **145** Q5
Ros Cré *see* Roscrea, *Ire.* **145** R5
Ros Mhic Thriúin *see* New Ross, *Ire.* **145** S6
Ros Ó gCairbre *see* Ross Carbery, *Ire.* **145** U3
Rosa, *Zambia* **209** L20
Rosa, Lake, *Bahamas* **119** J15
Rosa, Monte, *It. Switz.* **152** D3
Rosal de la Frontera, *Sp.* **146** K7
Rosalia, *Wash., U.S.* **110** C9
Rosalie Bay, *Dominica* **123** F20
Rosalie, *Dominica* **123** F20
Rosario de la Frontera, *Arg.* **134** F9
Rosario de Lerma, *Arg.* **134** F9
Rosario, *Mex.* **116** G9
Rosario, *Parag.* **132** M10
Rosario, *Arg.* **134** K12
Rosário, *Braz.* **133** D15
Rosario, Cayo del, *Cuba* **118** H7
Rosarno, *It.* **153** R13
Roscoe, *Tex., U.S.* **105** G13
Roscoe, *S. Dak., U.S.* **106** J6
Roscommon (Ros Comáin), *Ire.* **145** Q5
Roscommon, *Mich., U.S.* **100** J11
Roscrea (Ros Cré), *Ire.* **145** R5
Rose Atoll, *Amer. Samoa, U.S.* **224** J11
Rose Belle, *Mauritius* **215** G20
Rose Hall, *Jam.* **122** G7
Rose Hill, *Mauritius* **215** G20
Rose Island, *Bahamas* **122** E4
Rose Peak, *Ariz., U.S.* **109** U9
Roseau, *Minn., U.S.* **106** D9
Roseau, *Dominica* **123** G19
Roseau, river, *Can., U.S.* **106** D8
Roseau, river, *Dominica* **123** G19
Roseau, river, *St. Lucia* **123** K14
Rosebery, *Tas., Austral.* **221** Y7
Rosebud Creek, *Mont., U.S.* **108** E12
Rosebud, *Tex., U.S.* **105** J17
Rosebud, *Mont., U.S.* **108** D12
Roseburg, *Oreg., U.S.* **110** J3
Rosedale, *Miss., U.S.* **102** J10
Rosedale, *Qnsld., Austral.* **221** U15
Roseires Dam, *Sudan* **203** T13
Rosemary Bank, *Atl. Oc.* **248** F11
Rosenbach, *Aust.* **150** M10
Rosenberg, *Tex., U.S.* **105** L19
Rosenheim, *Ger.* **150** K9
Rosepine, *La., U.S.* **102** N7
Roses, *Bahamas* **119** G14
Roses, *Sp.* **147** C18
Roses, Golf de, *Sp.* **147** C18
Rosetown, *Sask., Can.* **86** M11
Rosetta *see* Rashīd, *Egypt* **202** E11
Roseville, *Calif., U.S.* **111** Q4
Rosewood, *N.T., Austral.* **221** Q7
Rosholt, *S. Dak., U.S.* **106** H8
Rosht- Kala, *Taj.* **179** N13
Rosia Bay, *Gibraltar, U.K.* **163** Q22
Rosiclare, *Ill., U.S.* **101** V6

Rosignano Marittimo, *It.* **152** H6
Roşiori de Vede, *Rom.* **155** G14
Roslavl', *Russ.* **158** K7
Roslyatino, *Russ.* **158** F12
Roslyn, *Wash., U.S.* **110** C5
Rosporden, *Fr.* **149** Q3
Rosport, *Lux.* **162** J11
Ross Barnett Reservoir, *Miss., U.S.* **102** L11
Ross Carbery (Ros Ó gCairbre), *Ire.* **145** U3
Ross Dam, *Wash., U.S.* **110** A5
Ross Ice Shelf, *Antarctica* **239** M13
Ross Island, *Antarctica* **239** N13
Ross Lake National Recreation Area, *Wash., U.S.* **110** A5
Ross Lake, *Wash., U.S.* **110** A5
Ross of Mull, peninsula, *Scot., U.K.* **144** K7
Ross River, *Yukon, Can.* **86** G8
Ross River, *N.T., Austral.* **221** T9
Ross Sea, *Antarctica* **239** P13
Ross, *N.Z.* **223** L17
Rossan Point, *Ire.* **145** N4
Rossano, *It.* **153** P14
Rossel Island, *P.N.G.* **222** R11
Rosslare Harbour (Calafort Ros Láir), *Ire.* **145** T7
Rosso, *Mauritania* **204** H7
Rosso, Capo, *Fr.* **149** X2
Rossosh', *Russ.* **159** N10
Røssvatnet, lake, *Nor.* **142** F12
Rossville, *Ill., U.S.* **101** Q7
Røst Bank, *Arctic Oc.* **253** G20
Rostock, *Ger.* **150** B9
Rostov-na-Donu, *Russ.* **159** R11
Rostov, *Russ.* **158** H10
Rostrenen, *Fr.* **149** Q4
Roswell, *Ga., U.S.* **98** K5
Roswell, *N. Mex., U.S.* **109** U14
Rota see Songsong, *N. Mariana Is., U.S.* **230** N8
Rota, *Sp.* **146** M7
Rota, island, *N. Mariana Is., U.S.* **230** K5
Rotenboden, *Liech.* **162** P2
Roth, *Ger.* **150** J8
Rothera Station, *Antarctica* **238** E5
Rotherham, *Eng., U.K.* **145** R12
Rothschild Island, *Antarctica* **238** F6
Rothschild, *Wis., U.S.* **100** H5
Roti, island, *Indonesia* **193** N14
Roto, *N.Z.* **229** D21
Rotondella, *It.* **153** N13
Rotorua, *N.Z.* **223** G21
Rotterdam, *Neth.* **148** J12
Rottumeroog, island, *Neth.* **148** G14
Rottumerplaat, island, *Neth.* **148** G14
Rotuma, island, *Fiji* **224** J8
Roubaix, *Fr.* **148** L10
Rouen, *Fr.* **149** N8
Rouge, Pic de la, *Fr. Sp.* **162** G2
Rough Point, *Virgin Is., U.K.* **122** Q6
Round Island Passage, *Fiji* **226** N4
Round Island see Yalewa Kalou, *Fiji* **226** N5
Round Island, *Mauritius* **215** E21
Round Mountain, *Nev., U.S.* **111** Q9
Round Mountain, *Austral.* **220** J15
Round Rock, *Tex., U.S.* **105** K17
Round Rock, *Bahamas* **122** C11
Round Rock, *Virgin Is., U.K.* **122** R8
Roundstone (Cloch na Rón), *Ire.* **145** Q3
Roundup, *Mont., U.S.* **108** D10
Roura, *Fr. Guiana, Fr.* **131** G19
Rousay, island, *Scot., U.K.* **144** E10
Rouses Point, *N.Y., U.S.* **96** F12
Rouyn-Noranda, *Que., Can.* **87** P18
Rovaniemi, *Fin.* **143** E15
Rovdino, *Russ.* **158** D11
Rovereto, *It.* **152** D7
Rôviĕng, *Cambodia* **191** N11
Rovigo, *It.* **152** E7
Rovinj, *Croatia* **154** E4
Rovuma, river, *Mozambique, Tanzania* **212** B12
Rowa Islands (Reef Islands), *Vanuatu* **229** C15
Rowland, *N.C., U.S.* **98** K11
Rowley Island, *Nunavut, Can.* **87** F17
Rowley Reefs, *Ind. Oc.* **251** J18
Rowley Shelf, *Ind. Oc.* **251** K18
Roxa, island, *Guinea-Bissau* **204** L7
Roxas City, *Philippines* **193** D14
Roxas, *Philippines* **193** E13
Roxboro, *N.C., U.S.* **98** G11

Roxborough, *Trin. & Tobago* **123** N18
Roy Hill, *W. Aust., Austral.* **221** T3
Roy, *Mont., U.S.* **108** C10
Roy, *N. Mex., U.S.* **109** R14
Royal Island, *Bahamas* **122** D5
Royal Oak, *Mich., U.S.* **100** M12
Royal Tunbridge Wells, *Eng., U.K.* **145** V14
Royale, Isle, *Mich., U.S.* **100** D6
Royan, *Fr.* **149** U6
Royston, *Ga., U.S.* **98** K6
Roznov, *Rom.* **155** C16
Rrogozhinë, *Alban.* **154** L9
Rtishchevo, *Russ.* **158** M12
Rua, island, *Marshall Is.* **227** G20
Ruacaná Falls, *Angola, Namibia* **208** Q10
Ruahara, island, *N.Z.* **229** K22
Ruapuke Island, *N.Z.* **223** Q16
Ruatoria, *N.Z.* **223** F23
Ruawai, *N.Z.* **223** E20
Rub' al Khali, *Asia* **168** H3
Rubafu, *Tanzania* **211** T3
Rubha Robhanais, cape, *Scot., U.K.* **144** F7
Rubi, river, *D.R.C.* **209** C16
Rubicone, river, *It.* **152** G8
Rubtsovsk, *Russ.* **160** L1
Ruby Dome, peak, *U.S.* **90** F6
Ruby Mountains, *Nev., U.S.* **111** N11
Ruby Plains, *W. Aust., Austral.* **221** R6
Ruby, *Alas., U.S.* **113** G15
Ruby, river, *Mont., U.S.* **108** E7
Ruch'i, *Russ.* **158** A10
Rūdbār, *Afghan.* **180** M3
Rudder Cut Cay, *Bahamas* **122** D7
Rudnichnyy, *Russ.* **158** E15
Rudolph, *Russ.* **160** C12
Rudyard, *Mich., U.S.* **100** F11
Rue, *Fr.* **148** M9
Rufa'a, *Sudan* **203** R13
Ruffec, *Fr.* **149** T7
Rufiji, river, *Tanzania* **211** X7
Rufino, *Arg.* **134** L11
Rufisque, *Senegal* **204** J6
Rufunsa, *Zambia* **209** P19
Rugby, *N. Dak., U.S.* **106** E5
Rugeley, *Eng., U.K.* **145** S12
Rügen, island, *Ger.* **150** B10
Ruggell, *Liech.* **162** M2
Rugozero, *Russ.* **158** C7
Ruhengeri, *Rwanda* **211** T2
Ruhnu Saar, island, *Est.* **143** L16
Ruhr, river, *Ger.* **150** F6
Rui Barbosa, *Braz.* **133** H16
Rū'ī, *Afghan.* **180** G8
Rui'an, *China* **185** L15
Ruiché, *Eq. Guinea* **214** M7
Ruidosa, *Tex., U.S.* **104** L9
Ruidoso, *N. Mex., U.S.* **109** U13
Ruinas de Numancia, *Sp.* **146** D12
Ruiru, *Kenya* **211** T6
Ruivo, Pico, *Madeira, Port.* **214** M3
Ruj, peak, *Bulg. Serb.* **154** H12
Rujen, peak, *Bulg.* **154** J12
Ruki, river, *Af.* **198** J7
Rukungiri, *Uganda* **211** T2
Rukuruku Bay, *Fiji* **226** N5
Rule, *Tex., U.S.* **105** F14
Ruleville, *Miss., U.S.* **102** J10
Rulhieres, Cape, *Austral.* **220** B6
Rum Cay, *Bahamas* **119** F14
Rum Jungle, *N.T., Austral.* **221** P8
Rum Point Channel, *Cayman Is., U.K.* **122** M3
Rum Point, *Cayman Is., U.K.* **122** M3
Rum, island, *Scot., U.K.* **144** J7
Ruma, *Serb.* **154** F9
Rumbek, *D.R.C.* **203** W10
Rumelange, *Lux.* **162** K9
Rumford, *Me., U.S.* **97** F15
Rumia, *Pol.* **151** A14
Rumoi, *Japan* **188** F13
Rumonge, *Burundi* **211** U2
Rumphi, *Malawi* **212** B10
Rumung, peak, *F.S.M.* **226** D12
Runanga, *N.Z.* **223** K18
Runaway Bay, *Jam.* **122** H8
Runaway Bay, *Antigua & Barbuda* **123** K17
Runde, river, *Zimb.* **212** G10
Rundeng, *Indonesia* **192** H4
Rundu, *Namibia* **212** E5
Runere, *Tanzania* **211** U4
Runge, *Tex., U.S.* **105** M17

Rungu, *D.R.C.* **209** C18
Rungwa, *Tanzania* **211** W4
Rungwa, river, *Tanzania* **211** W3
Rungwe Mountain, *Tanzania* **211** X4
Runit, island, *Marshall Is.* **227** B21
Running Water Draw, *Tex., U.S.* **104** E11
Runu, *F.S.M.* **226** D11
Ruo, island, *F.S.M.* **226** D9
Rupat, island, *Indonesia* **192** H6
Rupert, *Idaho, U.S.* **108** H6
Rupert, river, *Que., Can.* **87** M19
Ruponda, *Tanzania* **211** Y7
Ruppert Coast, *Antarctica* **238** N10
Rupununi, river, *Guyana* **131** G15
Rurrenabaque, *Bol.* **132** J6
Rurutu, island, *Fr. Polynesia, Fr.* **231** E16
Rusanovo, *Russ.* **160** E10
Rusape, *Zimb.* **212** E10
Ruse, *Bulg.* **155** G15
Rush City, *Minn., U.S.* **106** J12
Rush Springs, *Okla., U.S.* **105** D16
Rushan, *Taj.* **179** N13
Rushden, *Eng., U.K.* **145** T13
Rushford, *Minn., U.S.* **106** L13
Rushville, *Ill., U.S.* **101** R4
Rushville, *Ind., U.S.* **101** R10
Rushville, *Nebr., U.S.* **106** M2
Rusizi, river, *D.R.C.* **209** F18
Rusizi, river, *Burundi-Rwanda* **211** U1
Rusk, *Tex., U.S.* **105** H19
Ruskin, *Fla., U.S.* **99** U7
Russas, *Braz.* **133** E17
Russell Bay, *Antarctica* **238** M8
Russell Cave National Monument, *Ala., U.S.* **103** G16
Russell Island, *Bahamas* **122** D5
Russell Islands, *Solomon Is.* **228** E9
Russell, *Kans., U.S.* **107** S6
Russellville, *Ark., U.S.* **102** F7
Russellville, *Ala., U.S.* **103** H14
Russellville, *Ky., U.S.* **103** D15
Rüsselsheim, *Ger.* **150** G6
Russi, *It.* **152** F8
Russia, *Eur. Asia* **160** H12
Russian Mission, *Alas., U.S.* **112** H12
Russian, river, *Calif., U.S.* **111** Q2
Russkaya Gavan', *Russ.* **160** D12
Russkaya Polyana, *Russ.* **160** K10
Rustāq, *Afghan.* **180** F9
Rust'avi, *Rep. of Georgia* **171** C19
Rustenburg, *S. Af.* **212** H8
Ruston, *La., U.S.* **102** L8
Rutaki Passage, *N.Z.* **229** E23
Rutana, *Burundi* **211** U2
Rute, *Sp.* **146** L10
Ruteng, *Indonesia* **193** M13
Rutford Ice Stream, *Antarctica* **238** H9
Ruth, *Nev., U.S.* **111** Q12
Rutherfordton, *N.C., U.S.* **98** J8
Ruth's Point, *Bermuda* **119** B17
Rutland Island, *India* **183** R14
Rutland, *Vt., U.S.* **96** H12
Rutog (Ritu), *China* **184** H2
Rutshuru, *D.R.C.* **209** E19
Ruvuma, river, *Mozambique, Tanzania* **211** Z7
Ruwa, *Zimb.* **212** E9
Ruwāq, Jabal ar, *Syr.* **172** J9
Ruwenzori, mountains, *D.R.C.* **209** E19
Ruwenzori, mountains, *Uganda* **211** S2
Rūzaevka, *Kaz.* **178** D11
Ruzayevka, *Russ.* **158** K13
Ružomberok, *Slovakia* **151** J15
Rwanda, *Af.* **211** T1
Rwindi, *D.R.C.* **209** E19
Ryabov Seamounts, *Arctic Oc.* **252** J11
Ryan, *Okla., U.S.* **105** E16
Ryan, Loch, *Scot., U.K.* **145** N8
Ryazan', *Russ.* **158** K10
Ryazhsk, *Russ.* **158** L10
Rybachiy, *Russ.* **158** J2
Rybinsk Reservoir, *Eur.* **140** D12
Rybinsk, *Russ.* **158** G10
Rybinskoye Vodokhranilishche, *Russ.* **158** G10
Rybnik, *Pol.* **151** G14
Rye Patch Reservoir, *Nev., U.S.* **111** N8
Rye, *Colo., U.S.* **109** P13
Ryegate, *Mont., U.S.* **108** D9

Ryl'sk, *Russ.* **158** M8
Ryōtsu Wan, *Japan* **188** M11
Ryōtsu, *Japan* **188** M11
Rypin, *Pol.* **151** C15
Rysy, peak, *Pol. Slovakia* **151** J15
Ryukyu Islands see Nansei Shotō, *Japan* **224** C2
Ryukyu Ridge, *Pac. Oc.* **246** G6
Ryukyu Trench, *Pac. Oc.* **246** G6
Ryzhovo, *Russ.* **161** E19
Rzeszów, *Pol.* **151** G17
Rzhev, *Russ.* **158** J8

S

's Hertogenbosch, *Neth.* **148** K13
Sa, *Thai.* **190** J8
Sa'a, *Solomon Is.* **228** F11
Saacow, *Somalia* **211** R10
Sa'ādatābād, *Iran* **177** M18
Saale, river, *Ger.* **150** G9
Saalfeld, *Ger.* **150** G8
Saarbrücken, *Ger.* **150** H5
Saaremaa (Ösel), island, *Est.* **143** K15
Saatta, *Eritrea* **210** G7
Saattut, *Greenland, Den.* **88** L6
Sab' Ābār, *Syr.* **172** J9
Saba Island, *Virgin Is., U.S.* **122** N7
Saba, island, *Neth.* **121** B15
Šabac, *Serb.* **154** F9
Sabadell, *Sp.* **147** D17
Sabah, *Malaysia* **192** G12
Sabana de La Mar, *Dom. Rep.* **119** L20
Sabana, Archipiélago de, *Cuba* **118** G8
Sabanalarga, *Col.* **130** C6
Sabaneta, *Dom. Rep.* **119** L18
Sabang, *Indonesia* **193** J13
Şabanözü, *Turk.* **170** D8
Sabarei, *Kenya* **211** P6
Sabaya, *Bol.* **132** K6
Sabetha, *Kans., U.S.* **107** R9
Sabhā, *Lib.* **201** G18
Şabḥā, *Jordan* **172** L7
Sabinal, *Tex., U.S.* **105** M15
Sabinal, Cayo, *Cuba* **118** J11
Sabinas Hidalgo, *Mex.* **116** E11
Sabine, river, *La. Tex., U.S.* **105** H20
Sabine Lake, *La. Tex., U.S.* **105** L21
Sabine, *Tex., U.S.* **105** L21
Sabine, river, *U.S.* **91** L14
Sabinosa, *Sp.* **214** R3
Sabkhat al Budū', *Saudi Arabia* **174** H12
Sabkhat Ḩazawzā', *Saudi Arabia* **174** B6
Sabkhat Maṭṭī, *Saudi Arabia, U.A.E.* **175** H14
Sabkhet el Bardawîl, *Egypt* **173** N2
Sable Island, *N.S., Can.* **87** P23
Sable, Cape, *N.S., Can.* **87** Q22
Sable, Cape, *Fla., U.S.* **99** Y9
Sables, Île aux (Sandy Island), *Mauritius* **215** J19
Sabonkafi, *Niger* **206** G9
Sabratha, ruin, *Lib.* **201** D17
Sabres, *Fr.* **149** W6
Sabrina Coast, *Antarctica* **239** N20
Sabtang, island, *Philippines* **193** A14
Sabugal, *Port.* **146** F7
Sabuk, *S. Korea* **187** N11
Sabunçu, *Azerb.* **171** D23
Sabura, *Pak.* **180** L7
Sabzak, Kōtal-e, *Afghan.* **180** H4
Sabzawar see Shīndand, *Afghan.* **180** J3
Sabzevār, *Iran* **177** E19
Sabzvārān see Jīroft, *Iran* **177** L20
Sac City, *Iowa, U.S.* **107** N10
Sacaca, *Bol.* **132** K7
Sacajawea Peak, *Oreg., U.S.* **110** F9
Sacedón, *Sp.* **146** F12
Săcel, *Rom.* **155** B14
Săcele, *Rom.* **155** E15
Sacheon, *S. Korea* **187** S10
Sacheon, *S. Korea* **187** T10
Sachigo, river, *Ont., Can.* **87** L15
Sachs Harbour, *N.W.T., Can.* **86** D11
Sacile, *It.* **152** D8
Sackets Harbor, *N.Y., U.S.* **96** G8

Saco, *Me., U.S.* **97** H15
Saco, *Mont., U.S.* **108** B11
Sacramento Mountains, *N. Mex., U.S.* **109** U13
Sacramento Valley, *U.S.* **90** F3
Sacramento, *Calif., U.S.* **111** R4
Sacramento, river, *Calif., U.S.* **111** N4
Sacred Falls (Kaliuwa'a), *Hawaii, U.S.* **115** G14
Sacristy and Treasury, *Vatican City* **163** Q16
Sada, *Mayotte, Fr.* **215** P17
Şa'dah, *Yemen* **174** M13
Sadani, *Tanzania* **211** W7
Saddle Hill, *St. Kitts & Nevis* **123** C19
Saddle Mountain, *Idaho, U.S.* **108** G6
Saddle Mountains, *Wash., U.S.* **110** D6
Saddle Peak, *India* **183** Q14
Şadḥ, *Oman* **175** M16
Sadiola, *Mali* **204** J9
Sadiqabad, *Pak.* **181** P9
Sadiya, *India* **182** G15
Sadko Valley, *Arctic Oc.* **252** E10
Sadlers, *St. Kitts & Nevis* **123** A17
Sado, island, *Japan* **188** M11
Sado, river, *Port.* **146** J5
Sadon, *Myanmar* **190** E6
Sadovoye, *Russ.* **159** Q13
Sa'dun, *Sudan* **203** T7
Sae Islands, *P.N.G.* **222** M7
Saebyōl, *N. Korea* **186** A13
Saelices, *Sp.* **146** G12
Saengildo, island, *S. Korea* **187** U8
Şafājah, *Saudi Arabia* **174** E6
Safata Bay, *Samoa* **227** P19
Safēd Khirs, range, *Afghan.* **180** E10
Säffle, *Sw.* **142** K12
Safford, *Ariz., U.S.* **109** U9
Safi, *Mor.* **200** D6
Şafītā, *Syr.* **172** G7
Safonovo, *Russ.* **158** K7
Safonovo, *Russ.* **158** A12
Safotu, *Samoa* **227** N18
Safranbolu, *Turk.* **170** C8
Sag Harbor, *N.Y., U.S.* **97** N13
Saga, *Japan* **189** R4
Sagae, *Japan* **188** L12
Sagaing, *Myanmar* **190** G5
Sagami Nada, *Japan* **189** Q11
Sagar, *India* **182** J6
Sagar, *India* **183** Q4
Sagarejo, *Rep. of Georgia* **171** C19
Sagarmāthā see Everest, Mount, *China, Nep.* **184** L4
Sage Creek, *Mont., U.S.* **108** A8
Saghyz, *Kaz.* **178** F6
Saghyz, river, *Kaz.* **178** F6
Saginaw Bay, *Mich., U.S.* **100** K12
Saginaw, *Mich., U.S.* **100** L11
Sagittarius Seachannel, *Pac. Oc.* **247** D14
Saglek Bank, *Atl. Oc.* **248** F5
Saglek Bay, *Nfld. & Lab., Can.* **87** J20
Sagone, *Fr.* **149** Y2
Sagone, Golfe de, *Fr.* **149** Y2
Sagra, peak, *Sp.* **146** K12
Sagres, Ponta de, *Port.* **146** L4
Sagu, *Myanmar* **190** H4
Sagua de Tánamo, *Cuba* **119** K13
Sagua la Grande, *Cuba* **118** G8
Saguache, *Colo., U.S.* **109** N12
Saguaro National Park, *Ariz., U.S.* **109** V8
Saguenay, *Que., Can.* **87** N20
Sagunto (Sagunt), *Sp.* **147** G15
Sa'gya (Sajia), *China* **184** L4
Saḩāb, *Jordan* **172** M7
Sahagún, *Sp.* **146** C9
Şaham, *Oman* **175** G16
Sahara, *Japan* **188** N11
Sahara, desert, *Af.* **198** F3
Saharan Atlas, range, *Af.* **198** C4
Saharan Fan, *Atl. Oc.* **248** L10
Saharan Seamounts, *Atl. Oc.* **248** L9
Saharanpur, *India* **182** F6
Saharsa, *India* **182** H10
Şahbuz, *Arm.* **171** F19
Sahel, region, *Af.* **198** G3
Sahiwal, *Pak.* **180** K11
Sahiwal, *Pak.* **180** M12
Sahl Rakbah, *Saudi Arabia* **174** H7
Şaḩrā' al Ḩijārah, *Iraq* **176** K10
Şaḩrā' Awbārī, *Lib.* **201** G16
Şaḩrā' Marzūq, *Lib.* **201** J17
Şahtaxtı, *Arm.* **171** F19
Sahuaripa, *Mex.* **116** D8

Sahul Banks, *Ind. Oc.* **251** J19
Sahul Shelf, *Ind. Oc.* **251** J19
Sahyŏn, *N. Korea* **186** K7
Sai Buri, *Thai.* **191** S8
Sai, *Japan* **188** H12
Sa'id Bundas, *S. Sudan* **203** V7
Saïda (Sidon), *Leb.* **172** J6
Saïda, *Alg.* **200** C11
Sa'īdābād *see* Sīrjān, *Iran* **177** L18
Saidapet, *India* **183** R7
Saidu, *Pak.* **180** H11
Saignes, *Fr.* **149** V10
Saigon *see* Thành Phố Hồ Chí Minh
   (Ho Chi Minh City), *Vietnam*
   191 P12
Saiki, *Japan* **189** R6
Sailolof, *Indonesia* **193** J17
Saimaa, lake, *Fin.* **143** H17
Saimbeyli, *Turk.* **170** G11
Saindak, *Pak.* **181** N2
Saint Abb's Head, *Scot., U.K.* **144** L11
Saint Albans, *Vt., U.S.* **96** F12
Saint Albans, *W. Va., U.S.* **98** D8
Saint Albans, *Eng., U.K.* **145** U14
Saint André, Cap, *Af.* **198** M11
Saint Andrew Bay, *Fla., U.S.* **99** R3
Saint Andrews, *Scot., U.K.* **144** K11
Saint Anthony, *Nfld. & Lab., Can.*
   87 L23
Saint Anthony, *Idaho, U.S.* **108** G7
Saint Augustin Bay, *Af.* **198** P10
Saint Augustine, *Fla., U.S.* **99** R9
Saint Austell Bay, *Eng., U.K.* **145** W8
Saint Austell, *Eng., U.K.* **145** W8
Saint Barthelemy, island, *Lesser Antil.*
   121 B15
Saint Barts *see* Saint Barthelemy,
   island, *Lesser Antil.* **121** B15
Saint Bees Head, *Eng., U.K.* **145** P10
Saint Catherines Island, *Ga., U.S.*
   99 P9
Saint Charles' Church, *Monaco*
   163 C22
Saint Charles, *Mich., U.S.* **100** L11
Saint Charles, *Minn., U.S.* **106** L13
Saint Charles, *Mo., U.S.* **107** S15
Saint Charles, *Idaho, U.S.* **108** J8
Saint Clair Shores, *Mich., U.S.*
   100 M13
Saint Clair, *Mich., U.S.* **100** L13
Saint Clair, *Mo., U.S.* **107** T14
Saint Clair, Lake, *Mich., U.S.* **100** M13
Saint Clair, river, *Can., U.S.* **100** M13
Saint Clears, *Wales, U.K.* **145** U9
Saint Cloud, *Fla., U.S.* **99** T9
Saint Cloud, *Minn., U.S.* **106** J10
Saint Combs, *Scot., U.K.* **144** H11
Saint Croix Falls, *Wis., U.S.* **100** G2
Saint Croix Island International
   Historic Site, *Can., U.S.* **97** E19
Saint Croix National Scenic
   Riverway, *Wis., U.S.* **100** G1
Saint Croix, island, *Virgin Is., U.S.*
   121 B13
Saint Croix, river, *Can., U.S.* **97** D19
Saint Croix, river, *Minn., Wis., U.S.*
   100 H1
Saint David Islands *see*
   Mapia,Kepulauan, *Indonesia* **193** J19
Saint David's Head, *Wales, U.K.* **1**
   45 T8
Saint Edward, *Nebr., U.S.* **107** P7
Saint Elias Mountains, *Can., U.S.*
   113 K20
Saint Elias, Mount, *Can., U.S.* **113** K19
Saint Eustatius, island, *Lesser Antil.*
   121 B15
Saint Finan's Bay, *Ire.* **145** T2
Saint Francis Bay, *Af.* **198** Q8
Saint Francis Bay, *S. Af.* **212** M7
Saint Francis, *Me., U.S.* **97** A17
Saint Francis, *S. Dak., U.S.* **106** M4
Saint Francis, *Kans., U.S.* **107** R3
Saint Francis, river, *Ark., Mo., U.S.*
   102 E11
Saint Gallen, *Switz.* **152** A5
Saint George Basin, *Austral.* **220** C5
Saint George Island, *Fla., U.S.* **99** S4
Saint George Island, *Alas., U.S.* **112** L9
Saint George, *S.C., U.S.* **98** L9
Saint George, *Utah, U.S.* **109** Q5
Saint George, *Qnsld. Austral.* **221** V14
Saint George, Cape, *Fla., U.S.* **99** R4
Saint George, Point, *Calif., U.S.* **110** L1
Saint George's Channel, *Ire., U.K.*
   145 T7

Saint George's Channel, *India*
   183 U15
Saint Helena Bay, *S. Af.* **212** M5
Saint Helena Fracture Zone, *Atl. Oc.*
   249 R10
Saint Helena Sound, *S.C. U.S.* **99** N10
Saint Helena, *Calif., U.S.* **111** R3
Saint Helena, island, *Af.* **198** M3
Saint Helens, *Oreg., U.S.* **110** E3
Saint Helens, *Eng., U.K.* **145** R11
Saint Helens, Mount, *Wash., U.S.*
   110 E4
Saint Helier, *Channel Is., U.K.* **145** Y12
Saint Ignace, *Mich., U.S.* **100** G10
Saint Ignatius, *Mont., U.S.* **108** C6
Saint Ives Bay, *Eng., U.K.* **145** X8
Saint James, *Minn., U.S.* **106** L10
Saint James, *Mo., U.S.* **107** T13
Saint Joe, river, *Idaho, U.S.* **108** C4
Saint John, *N.B., Can.* **87** P21
Saint John, *N. Dak., U.S.* **106** D5
Saint John, *Kans., U.S.* **107** U6
Saint John, *Wash., U.S.* **110** D9
Saint John, river, *Can., U.S.* **97** B16
Saint John's, *Nfld. & Lab., Can.*
   87 M24
Saint Johns, *Mich., U.S.* **100** L11
Saint Johns, *Ariz., U.S.* **109** T9
Saint Johns, river, *Fla., U.S.* **99** S8
Saint Johnsbury, *Vt., U.S.* **97** F13
Saint Jona Island *see* Iony, *Russ.*
   161 H20
Saint Joseph Bay, *Fla., U.S.* **99** R4
Saint Joseph, *Mich., U.S.* **101** N8
Saint Joseph, *La., U.S.* **102** M9
Saint Joseph, *Mo., U.S.* **107** R10
Saint Joseph, river, *Ind. Ohio, U.S.*
   101 P10
Saint Kilda, island, *Scot., U.K.* **144** H5
Saint Kitts and Nevis, *N. Amer.*
   121 C16
Saint Lawrence Island, *Alas., U.S.*
   112 F10
Saint Lawrence Islands National
   Park, *N.Y., U.S.* **96** F9
Saint Lawrence River, *N. Amer.*
   87 N20
Saint Lawrence, Gulf of, *Que., Can.*
   87 N22
Saint Lawrence, river, *N. Amer.*
   84 H9
Saint Louis, *Mich., U.S.* **100** L11
Saint Louis, *Mo., U.S.* **107** T15
Saint Louis, river, *Minn., U.S.* **106** G12
Saint Lucia Channel, *N. Amer.* **121** F17
Saint Lucia, *N. Amer.* **121** F17
Saint Lucia, Gildas, Pointe de, *Af.* **198** P9
Saint Magnus Bay, *Scot., U.K.* **144** C11
Saint Maries, *Idaho, U.S.* **108** C4
Saint Marks, *Fla., U.S.* **99** R5
Saint Martin's, island, *Eng., U.K.*
   145 X7
Saint Martinville, *La., U.S.* **102** P9
Saint Mary's Bay, *Nfld. & Lab., Can.*
   87 M24
Saint Mary's City, *Md., U.S.* **98** D14
Saint Mary's Loch, *Scot., U.K.*
   145 M10
Saint Marys, *Pa., U.S.* **96** M5
Saint Marys, *Ga., U.S.* **98** Q8
Saint Marys, *W. Va., U.S.* **98** C8
Saint Marys, *Ohio, U.S.* **101** Q11
Saint Marys, *Kans., U.S.* **107** S9
Saint Marys, *Tas., Austral.* **221** Y8
Saint Mary's, island, *Eng., U.K.*
   145 X7
Saint Marys, river, *Fla., Ga., U.S.*
   99 Q8
Saint Matthew Island, *Alas., U.S.*
   112 H9
Saint Matthews, *S.C., U.S.* **98** L9
Saint Michael, *Alas., U.S.* **113** G13
Saint Paul Fracture Zone, *Atl. Oc.*
   249 P9
Saint Paul Island, *Alas., U.S.* **112** L9
Saint Paul, *Minn., U.S.* **106** J12
Saint Paul, *Nebr., U.S.* **107** P6
Saint Pauls, *N.C., U.S.* **98** J11
Saint Peter Fracture Zone, *Atl. Oc.*
   249 P8
Saint Peter Port, *Channel Is., U.K.*
   145 Y11
Saint Peter, *Minn., U.S.* **106** K11
Saint Petersburg *see* Sankt-
   Peterburg, *Russ.* **158** F6
Saint Petersburg, *Fla., U.S.* **99** U7

Saint Pierre, *Mauritius* **215** G20
Saint Pierre & Miquelon, *N. Amer.*
   87 N23
Saint Pierre Bank, *Ind. Oc.* **250** P11
Saint Simons Island, *Ga., U.S.* **99** P9
Saint Stephen, *S.C., U.S.* **98** L10
Saint Thomas, *N. Dak., U.S.* **106** E7
Saint Truiden, *Belg.* **148** L12
Saint Tudwal's Islands, *Wales, U.K.*
   145 S9
Saint Vincent and the Grenadines,
   N. Amer. **121** G17
Saint Vincent Fracture Zone, *Ind. Oc.*
   255 E17
Saint Vincent Island, *Fla., U.S.* **99** R4
Saint Vincent Passage, *N. Amer.*
   121 F17
Saint Vincent, *Minn., U.S.* **106** D8
Saint Vincent, Cape, *Eur.* **140** K1
Saint Vincent, Gulf, *Austral.* **220** K10
Saint Vincent, island, *St. Vincent &* the
   Grenadines **121** G17
Saint Vith, *Belg.* **148** M13
Saint-Agrève, *Fr.* **149** V12
Saint-Aignan, *Fr.* **149** S9
Saint-Amand-Montrond, *Fr.* **149** S10
Saint-Amand, *Fr.* **148** M11
Saint-Amans-Soult, *Fr.* **149** X10
Saint-Ambroix, *Fr.* **149** W12
Saint-André, *Réunion, Fr.* **215** G16
Saint-Aubin-d'Aubigné, *Fr.* **149** Q5
Saint-Aubin-du-Cormier, *Fr.* **149** Q6
Saint-Augustin, *Que., Can.* **87** L22
Saint-Avold, *Fr.* **149** P14
Saint-Benoît, *Réunion, Fr.* **215** G17
Saint-Brieuc, *Fr.* **149** P4
Saint-Cannat, *Fr.* **149** X13
Saint-Cernin, *Fr.* **149** V10
Saint-Chély-d'Apcher, *Fr.* **149** V11
Saint-Claude, *Fr.* **149** T14
Saint-Denis, *Fr.* **149** P10
Saint-Denis, *Fr.* **149** T6
Saint-Denis, *Réunion, Fr.* **215** F16
Saint-Dié, *Fr.* **149** Q15
Saint-Dizier, *Fr.* **149** P12
Saint-Éloy-les-Mines, *Fr.* **149** T10
Saint-Étienne-du-Rouvray, *Fr.*
   149 N8
Saint-Étienne, *Fr.* **149** U12
Saint-Fargeau, *Fr.* **149** R10
Saint-Firmin, *Fr.* **149** V14
Saint-Florent, Golfe de, *It.* **152** J4
Saint-Florentin, *Fr.* **149** Q11
Saint-Flour, *Fr.* **149** V11
Saint-Germain-des-Vaux, *Fr.* **149** N5
Saint-Germain, *Fr.* **149** P9
Saint-Gildas, Pointe de, *Fr.* **149** S5
Saint-Gilles-les-Bains, *Réunion, Fr.*
   215 G15
Saint-Girons, *Fr.* **149** Y8
Saint-Guénolé, *Fr.* **149** Q2
Saint-Hillaire-du-Harcouët, *Fr.*
   149 P6
Saint-Hippolyte, *Fr.* **149** R14
Saint-Jean-d'Angély, *Fr.* **149** U6
Saint-Jean-de-Losne, *Fr.* **149** S13
Saint-Jean-de-Luz, *Fr.* **149** X5
Saint-Jean-de-Monts, *Fr.* **149** S5
Saint-Jean-Pied-de-Port, *Fr.* **149** Y6
Saint-Jean, *Réunion, Fr.* **215** J16
Saint-Joseph, *Réunion, Fr.* **215** J16
Saint-Junien, *Fr.* **149** U8
Saint-Laurent du Maroni,
   Fr. Guiana, Fr. **131** F18
Saint-Leu, *Réunion, Fr.* **215** H15
Saint-Lô, *Fr.* **149** P6
Saint-Louis du Nord, *Haiti* **119** L16
Saint-Louis, *Fr.* **149** R15
Saint-Louis, *Senegal* **204** H6
Saint-Louis, *Réunion, Fr.* **215** H15
Saint-Loup, *Fr.* **149** Q14
Saint-Lys, *Fr.* **149** X8
Saint-Maixent, *Fr.* **149** T7
Saint-Malo, *Fr.* **149** P5
Saint-Malo, Golfe de, *Fr.* **149** P5
Saint-Marc, *Haiti* **119** L16
Saint-Marc, Canal de, *Haiti* **119** M16
Saint-Marcel, Mont, *Fr. Guiana, Fr.*
   131 H19
Saint-Martin, *Fr.* **149** X11
Saint-Mathieu, *Fr.* **149** U8
Saint-Mathieu, Point, *Eur.* **140** G3
Saint-Mathieu, Pointe de, *Fr.* **149** Q2
Saint-Maurice, river, *Que., Can.*
   87 P19
Saint-Méen-le-Grand, *Fr.* **149** Q5

Saint-Nazaire, *Fr.* **149** R5
Saint-Omer, *Fr.* **148** L10
Saint-Palais, *Fr.* **149** Y6
Saint-Paul, *Réunion, Fr.* **215** G15
Saint-Paulien, *Fr.* **149** V11
Saint-Péray, *Fr.* **149** V12
Saint-Philippe, *Réunion, Fr.* **215** H17
Saint-Pierre-de-Chignac, *Fr.* **149** V8
Saint-Pierre-le-Moûtier, *Fr.* **149** S11
Saint-Pierre, *Réunion, Fr.* **215** H16
Saint-Pons, *Fr.* **149** X10
Saint-Quay-Portrieux, *Fr.* **149** P4
Saint-Quentin, *Fr.* **148** M11
Saint-QuentinPointe de, *Fr.* **148** M9
Saint-Raphaël, *Fr.* **149** X15
Saint-Savin, *Fr.* **149** T8
Saint-Sernin, *Fr.* **149** X10
Saint-Servan, *Fr.* **149** P5
Saint-Sever, *Fr.* **149** X6
Saint-Vivien, *Fr.* **149** U6
Sainte Genevieve, *Mo., U.S.* **107** T15
Sainte-Anne, *Guadeloupe, Fr.* **123** F16
Sainte-Marie, *Guadeloupe, Fr.* **123** F15
Sainte-Marie, *Martinique, Fr.* **123** E23
Sainte-Marie, *Réunion, Fr.* **215** F16
Sainte-Maxime, *Fr.* **149** X15
Sainte-Rose, *Guadeloupe, Fr.* **123** E14
Sainte-Rose, *Réunion, Fr.* **215** G17
Sainte-Suzanna, *Réunion, Fr.* **215** F16
Sainte-Suzanne, *Fr.* **149** Q7
Saintes, *Fr.* **149** U6
Saipan Channel, *N. Mariana Is., U.S.*
   230 K7
Saipan Harbor, *N. Mariana Is., U.S.*
   230 H7
Saipan, island, *N. Mariana Is., U.S.*
   230 J5
Sairang, *India* **182** J14
Saito, *Japan* **189** S5
Saivomuotka, *Sw.* **143** D14
Sajama, Nevado, *Bol.* **132** K6
Sajia *see* Sa'gya, *China* **184** L4
Sajyang Pass, *China, Myanmar* **184** M8
Saka Kalat, *Pak.* **181** Q5
Saka, *Kenya* **211** S8
Sakaiminato, *Japan* **189** P7
Sakākā, *Saudi Arabia* **174** C7
Sakakawea, Lake, *N. Dak., U.S.* **106** F3
Sakalilo, *Tanzania* **211** X3
Sakami, *Que. Can.* **87** M18
Sakami, Lac, *Que., Can.* **87** M18
Sakami, river, *Que. Can.* **87** M18
Sakar, island, *P.N.G.* **222** P8
Sakaraha, *Madagascar* **213** H15
Sakarya *see* Adapazarı, *Turk.* **170** D6
Sakarya, river, *Turk.* **170** E6
Sakata, *Japan* **188** L12
Sakçagöze, *Turk.* **170** H12
Sakchu, *N. Korea* **186** F5
Saké, *Benin* **205** P18
Sakété, *Benin* **205** P18
Sakhalin, *Russ.* **161** J21
Sakhalin, island, *Asia* **168** D13
Sākhar, *Afghan.* **180** J6
Sakishima Shotō, *Japan* **224** C1
Sakon Nakhon, *Thai.* **190** K10
Sakora, *Mali* **204** J11
Sakrand, *Pak.* **181** R8
Saky, *Russ.* **159** S7
Sal Rei, *Cabo Verde* **215** C17
Sal, Cay, *Bahamas* **118** F8
Sal, Cayo de, *Venez.* **120** H11
Sal, island, *Cabo Verde* **215** B17
Sala Consilina, *It.* **153** N12
Sala y Gómez Ridge, *Pac. Oc.*
   247 M20
Sala, *Sw.* **143** K13
Salada, Laguna, *Mex.* **116** A5
Saladas, *Arg.* **134** G13
Saladillo, *Arg.* **134** M12
Salado, river, *Arg.* **134** H11
Salado, river, *Arg.* **134** M12
Salado, river, *Arg.* **134** M9
Salaga, *Ghana* **205** N16
Salagle, *Somalia* **211** R10
Sala'ilua, *Samoa* **227** N17
Salal, *Chad* **207** G13
Salala, *Sudan* **202** M14
Şalālah, *Oman* **175** M15
Salamanca, *N.Y., U.S.* **96** K5
Salamanca, *Mex.* **116** H11
Salamanca, *Chile* **134** K6
Salamanca, *Sp.* **146** E8
Salamína, *Gr.* **156** J12
Salamis, battle, *Gr.* **156** J12
Salamis, ruin, *Cyprus* **162** N10
Salamīyah, *Syr.* **172** G8

Sālang, Kōtal-e, *Afghan.* **180** G8
Salani, *Samoa* **227** P20
Salapaly Bay, *Af.* **198** P10
Salas de los Infantes, *Sp.* **146** D11
Salas y Gómez, Isla, *Pac. Oc.* **225** L20
Salas, *Sp.* **146** B8
Salatiga, *Indonesia* **192** M9
Salavan, *Laos* **190** L11
Salavat, *Russ.* **160** J7
Salaverry, *Peru* **132** G2
Salawati, island, *Indonesia* **193** J17
Salazie, *Réunion, Fr.* **215** G16
Salbris, *Fr.* **149** R9
Salda Gölü, *Turk.* **170** H5
Saldaña, *Sp.* **146** C10
Saldanha Bay, *S. Af.* **212** M4
Saldanha, *S. Af.* **212** M5
Saldé, *Senegal* **204** H8
Sale, *Vic. Austral.* **221** Z13
Salé, *Mor.* **200** C8
Salebabu, island, *Indonesia* **193** G16
Saleimoa, *Samoa* **227** P19
Salekhard, *Russ.* **160** F11
Salelologa, *Samoa* **227** N18
Salem, *Mass., U.S.* **97** K15
Salem, *N.H., U.S.* **97** J14
Salem, *Va., U.S.* **98** F10
Salem, *W. Va., U.S.* **98** C9
Salem, *Ill., U.S.* **101** T6
Salem, *Ind., U.S.* **101** T9
Salem, *Ohio, U.S.* **101** P15
Salem, *Ark., U.S.* **102** E9
Salem, *S. Dak., U.S.* **106** L7
Salem, *Mo., U.S.* **107** U14
Salem, *Oreg., U.S.* **110** G3
Salem, *Montserrat, U.K.* **123** B23
Salem, *India* **183** R6
Salemi, *It.* **153** S9
Salerno, *It.* **152** M11
Salerno, Golfo di, *It.* **153** N11
Salford, *Eng., U.K.* **145** R11
Salgótarján, *Hung.* **151** K15
Salgueiro, *Braz.* **133** F16
Sali, *Alg.* **200** H11
Salibea, *Trin. & Tobago* **123** P23
Salibia, *Dominica* **123** E20
Salida, *Colo., U.S.* **109** N12
Salies-du-Salat, *Fr.* **149** Y8
Şalīf, *Yemen* **174** N8
Salihli, *Turk.* **170** F4
Salihorsk, *Belarus* **158** L4
Salima, *Malawi* **212** C11
Salina Bay, *Malta* **163** J22
Salina *see* Adapazarı, *Turk.* **170** D6
Salina, *Kans., U.S.* **107** T7
Salina, *Utah, U.S.* **109** N7
Salina, Isola, *It.* **153** R11
Salinas Bay, *S. Amer.* **128** F2
Salinas Peak, *N. Mex., U.S.* **109** U12
Salinas Pueblo Missions National
   Monument Gran Quivira,
   N. Mex., U.S. **109** T12
Salinas, *Calif., U.S.* **111** T4
Salinas, *Ecua.* **130** L2
Salinas, river, *Calif., U.S.* **111** U5
Saline Bay, *Trin. & Tobago* **123** P23
Saline di Volterra, *It.* **152** H6
Saline, river, *U.S.* **91** K14
Salines, Cap de ses, *Sp.* **147** H18
Salinópolis, *Braz.* **133** C14
Salins, *Fr.* **149** S14
Salisbury Downs, *N.S.W., Austral.*
   221 W12
Salisbury Island, *Nunavut, Can.* **87** H17
Salisbury Plain, *Eng., U.K.* **145** V12
Salisbury, *Md., U.S.* **98** D15
Salisbury, *N.C., U.S.* **98** H9
Salisbury, *Mo., U.S.* **107** S12
Salisbury, *Dominica* **123** F19
Salisbury, *Eng., U.K.* **145** V12
Salisbury, *Russ.* **160** C12
Salisbury, *Guam* **230** N6
Salish Mountains, *Mont. U.S.* **108** B5
Salish Sea, *Can., U.S.* **110** A3
Şalkhad, *Syr.* **172** L8
Sallent, *Sp.* **147** D17
Sallfelden, *Aust.* **150** L10
Sallisaw, *Okla., U.S.* **105** C19
Sallom, *Sudan* **203** N15
Salluit, *Que. Can.* **87** H18
Salmās, *Iran* **176** C11
Salmi, *Russ.* **158** E7
Salmon Peak, *Tex., U.S.* **105** L14
Salmon River Canyon, *Idaho, U.S.*
   108 E5

Salmon River Mountains, *Idaho, U.S.* **108** F4
Salmon, *Idaho, U.S.* **108** F6
Salmon, river, *Idaho, U.S.* **108** E5
Salo, *Fin.* **143** J15
Salo, *Cen. Af. Rep.* **206** P12
Salò, *It.* **152** D6
Salobelyak, *Russ.* **158** H14
Salole, *Eth.* **211** P8
Salome, *Ariz., U.S.* **109** T5
Salon, *Fr.* **149** X13
Salonga, river, *D.R.C.* **209** E14
Salonica *see* Thessaloníki, *Gr.* **156** C11
Salonta, *Rom.* **154** C11
Salou, Cap de, *Sp.* **147** E16
Saloum, river, *Senegal* **204** J8
Salpaus Ridge, *Eur.* **140** D9
Salpausselkä, range, *Fin.* **143** H16
Salqīn, *Syr.* **172** E8
Sal'sk, *Russ.* **159** R11
Salsomaggiore Terme, *It.* **152** F5
Salt Cay, *Turks & Caicos Is., U.K.* **119** J18
Salt Cay, *Bahamas* **122** E3
Salt Cay, *Virgin Is., U.S.* **122** M6
Salt Fork Red , river, *Okla., Tex., U.S.* **105** D13
Salt Lake City, *Utah, U.S.* **108** L7
Salt Range, *Pak.* **180** K11
Salt River Bay, *Virgin Is., U.S.* **122** Q2
Salt River Range, *Wyo., U.S.* **108** H8
Salt River, *Jam.* **122** K9
Salt, *Sp.* **147** D18
Salt, river, *Ariz., U.S.* **109** U7
Salta, *Arg.* **134** F9
Saltash, *Eng., U.K.* **145** W9
Saltburn by the Sea, *Eng., U.K.* **145** N13
Saltee Islands, *Ire.* **145** T6
Saltfjorden, bay, *Nor.* **142** E12
Salto, *Uru.* **134** J13
Salton Sea, *Calif., U.S.* **111** Y11
Saltpond, *Ghana* **205** Q15
Saltville, *Va., U.S.* **98** F8
Saluafata Harbour, *Samoa* **227** P20
Saluda, *S.C., U.S.* **98** K8
Saluda, river, *S.C., U.S.* **98** K8
Saluzzo, *It.* **152** F2
Salvación, Bahía, *Chile* **135** W6
Salvador (Bahia), *Braz.* **133** H17
Salvador, Lake, *La., U.S.* **102** Q11
Salwá, *Saudi Arabia* **175** G13
Salween (Nu, Thanlwin), river, *Asia* **168** H9
Salyan, *Azerb.* **171** E22
Salyersville, *Ky., U.S.* **103** C19
Salzburg, *Aust.* **150** L10
Salzgitter, *Ger.* **150** E8
Salzwedel, *Ger.* **150** D8
Sam Lord's Castle, *Barbados* **123** L20
Sam Rayburn Reservoir, *Tex., U.S.* **105** J20
Sâm Sơn, *Vietnam* **190** J11
Šamac *see* Bosanski Šamac, *Bosn. & Herzg.* **154** E8
Samā'il, *Oman* **175** H17
Samaipata, *Bol.* **132** K8
Samalaeulu, *Samoa* **227** N18
Samālūṭ, *Egypt* **202** G11
Samana Cay, *Bahamas* **119** G15
Samaná, Bahía de, *Dom. Rep.* **119** L19
Samaná, Cabo, *Dom. Rep.* **119** L20
Samandağ, *Turk.* **170** K11
Samani, *Japan* **188** H14
Samar Sea, *Philippines* **193** D15
Samar, island, *Philippines* **193** D15
Samara, *Russ.* **158** H15
Samara, river, *Eur.* **140** E14
Samarai, *P.N.G.* **222** R9
Samaria Gorge, *Gr.* **157** Q13
Samarinda, *Indonesia* **192** J12
Samarqand, *Uzb.* **178** M11
Sāmarrā', *Iraq* **176** G10
Samarskoe, *Kaz.* **179** E17
Samatiguila, *Côte d'Ivoire* **204** M12
Samawah *see* As Samāwah, *Iraq* **176** J11
Şamaxı, *Azerb.* **171** D22
Samba, *India* **182** D5
Samba, *Burkina Faso* **205** K15
Sambalpur, *India* **182** L9
Sambava, *Madagascar* **213** D17
Sambir, *Ukr.* **159** N2
Samboja, *Indonesia* **192** J12

Sambong, *N. Korea* **186** A13
Samborombón, Bahía, *Arg.* **134** M13
Samburg, *Russ.* **160** G12
Samch'a-do, island, *N. Korea* **186** H4
Samcheok, *S. Korea* **187** N12
Samdari, *India* **182** H3
Same, *Tanzania* **211** U7
Samer, *Fr.* **148** L9
Samfya, *Zambia* **209** L19
Samgi, *N. Korea* **186** F10
Samhae, *N. Korea* **186** C13
Samho, *N. Korea* **186** G10
Sami, *Pak.* **181** R4
Sámi, *Gr.* **156** H7
Samina, river, *Aust., Liech.* **162** N3
Saminatal, valley, *Liech.* **162** P3
Samjiyŏn, *N. Korea* **186** C10
Samnak, *N. Korea* **186** F7
Samnangjin, *S. Korea* **187** S11
Samnye, *S. Korea* **187** R8
Samo, *P.N.G.* **222** N10
Samoa, *Pac. Oc.* **224** J10
Samoa Basin, *Pac. Oc.* **246** L12
Samoa Islands, *Pac. Oc.* **224** J10
Samoded, *Russ.* **158** C10
Samokov, *Bulg.* **155** J13
Sámos, *Gr.* **157** J18
Sámos, island, *Gr.* **157** J17
Samothrace *see* Samothráki, *Gr.* **157** D15
Samothráki (Samothrace), *Gr.* **157** D15
Samothráki, *Gr.* **157** D15
Sampa, *Ghana* **205** N14
Sampacho, *Arg.* **134** K10
Sampit, *Indonesia* **192** K10
Sam'po, *N. Korea* **186** E13
Samp'o, *N. Korea* **186** E8
Sampwe, *D.R.C.* **209** K17
Sâmraông, *Cambodia* **190** M9
Samsan, *N. Korea* **186** L5
Samsŏ, *N. Korea* **186** E9
Samsŏk, *N. Korea* **186** J6
Samson, *Ala., U.S.* **103** N16
Samsu, *N. Korea* **186** D10
Samsun, *Turk.* **170** C11
Samu, *Indonesia* **192** K11
Samui, island, *Asia* **168** L10
Samui, Ko, *Thai.* **191** Q7
Samundri, *Pak.* **180** K7
Samūr, *Afghan.* **180** L3
Samur, river, *Russ.* **159** V15
Samut Prakhan, *Thai.* **190** M8
Samut Songkhram, *Thai.* **191** N7
Samyang, *N. Korea* **186** J7
San Adrián, Cabo, *Sp.* **146** B5
San Agustín, *Col.* **130** J5
San Agustin, Cape, *Philippines* **193** F15
San Ambrosio Island, *S. Amer.* **128** K2
San Andés Tuxtla, *Mex.* **117** J13
San Andreas, *Calif., U.S.* **111** R5
San Andrés del Rabanedo, *Sp.* **146** C9
San Andres Mountains, *N. Mex., U.S.* **109** U12
San Andrés, Isla de, *Col.* **117** L20
San Andros, *Bahamas* **118** E11
San Angelo, *Tex., U.S.* **105** J13
San Antonio Bay, *Tex., U.S.* **105** N18
San Antonio de los Cobres, *Arg.* **134** E9
San Antonio de Ureca, *Eq. Guinea* **214** N7
San Antonio Missions National Historical Park, *Tex., U.S.* **105** M16
San Antonio Mountain, *Tex., U.S.* **104** H8
San Antonio Oeste, *Arg.* **135** Q9
San Antonio, *Tex., U.S.* **105** L16
San Antonio, *N. Mex., U.S.* **109** T12
San Antonio, *Chile* **134** L6
San Antonio, *N. Mariana Is., U.S.* **230** J7
San Antonio, Cabo de, *Cuba* **118** H3
San Antonio, Cabo, *Arg.* **134** M13
San Antonio, Mount, *Calif., U.S.* **111** X9
San Antonio, Punta, *Mex.* **116** C5
San Antonio, river, *Tex., U.S.* **105** N17
San Augustine, *Tex., U.S.* **105** H21
San Benedetto del Tronto, *It.* **152** H14
San Benedicto, Isla, *Mex.* **116** H6
San Benito, *Tex., U.S.* **105** R17

San Bernardino Strait, *Philippines* **193** D15
San Bernardino, *Calif., U.S.* **111** X9
San Bernardo, *Chile* **134** L7
San Blas, *Mex.* **116** E8
San Blas, Archipiélago de, *Pan.* **117** N22
San Blas, Cape, *Fla., U.S.* **99** R4
San Blas, Punta, *Pan.* **117** N21
San Borja, *Bol.* **132** J7
San Candido, *It.* **152** B8
San Carlos de Bariloche, *Arg.* **135** Q7
San Carlos de Bolívar, *Arg.* **134** M11
San Carlos de Río Negro, *Venez.* **130** H10
San Carlos Reservoir, *Ariz., U.S.* **109** U8
San Carlos, *Ariz., U.S.* **109** U8
San Carlos, *Venez.* **130** D8
San Carlos, *Arg.* **134** F9
San Carlos, *Philippines* **193** E14
San Carlos, Gran Caldera Volcánica de, *Eq. Guinea* **214** M6
San Clemente, *Calif., U.S.* **111** Y9
San Clemente, *Sp.* **146** H12
San Clemente, island, *Calif., U.S.* **111** Z8
San Cristóbal de Las Casas, *Mex.* **117** K15
San Cristóbal, *Cuba* **118** G5
San Cristóbal, *Dom. Rep.* **119** M19
San Cristóbal, *Venez.* **130** E9
San Cristóbal, *Arg.* **134** J11
San Cristóbal, Isla, *Galápagos Is., Ecua.* **225** G22
San Cristobal, island, *Solomon Is.* **228** G11
San Cristoforo, *San Marino* **163** L16
San Daniele del Friuli, *It.* **152** C9
San Diego de los Baños, *Cuba* **118** G5
San Diego, *Tex., U.S.* **105** P16
San Diego, *Calif., U.S.* **111** Z9
San Diego, Cabo, *Arg.* **135** X10
San Donà di Piave, *It.* **152** D8
San Esteban de Gormaz, *Sp.* **146** D11
San Felice Circeo, *It.* **152** L9
San Felipe, *Mex.* **116** B5
San Felipe, *Col.* **130** H10
San Felipe, *Venez.* **130** C9
San Felipe, *Chile* **134** K7
San Felipe, Cayos de, *Cuba* **118** H5
San Félix Island, *S. Amer.* **128** K2
San Fernando de Apure, *Venez.* **130** E10
San Fernando de Atabapo, *Venez.* **130** G10
San Fernando, *Calif., U.S.* **111** X8
San Fernando, *Mex.* **116** F12
San Fernando, *Trin. & Tobago* **123** Q22
San Fernando, *Chile* **134** L7
San Fernando, *Sp.* **146** M8
San Francique, *Trin. & Tobago* **123** R21
San Francisco Bay, *U.S.* **90** G3
San Francisco de Paula, Cabo, *Arg.* **135** V9
San Francisco del Oro, *Mex.* **116** E9
San Francisco, *Calif., U.S.* **111** S3
San Francisco, *Arg.* **134** J11
San Francisco, *Arg.* **134** K9
San Francisco, river, *N. Mex., U.S.* **109** U10
San Gabriel Mountains National Monument, *Calif., U.S.* **111** X8
San Gabriel, *Ecua.* **130** J4
San Gabriel, river, *Tex., U.S.* **105** K17
San Gil, *Col.* **130** F7
San Giovanni in Fiore, *It.* **153** Q14
San Giovanni in Persiceto, *It.* **152** F7
San Giovanni Valdarno, *It.* **152** H7
San Giovanni, *San Marino* **163** K15
San Gorgonio Mountain, *Calif., U.S.* **111** X10
San Gregorio, *Uru.* **134** K14
San Ignacio, *Mex.* **116** D6
San Ignacio, *Bol.* **132** J9
San Ignacio, *Arg.* **134** G14
San Javier, *Arg.* **134** J12
San Javier, *Sp.* **147** K14
San Joaquin Valley, *U.S.* **90** G3
San Joaquín, *Bol.* **132** H7
San Joaquin, river, *Calif., U.S.* **111** T6
San Jon, *N. Mex., U.S.* **109** S15
San Jorge Island, *Solomon Is.* **228** E9
San Jorge, Golfo, *Arg.* **135** T9
San José de Amacuro, *Venez.* **131** D14
San José de Comondú, *Mex.* **116** E6

San José de Jáchal, *Arg.* **134** J7
San José de Las Lajas, *Cuba* **118** G6
San José de Mayo, *Uru.* **134** L13
San José de Ocuné, *Col.* **130** G8
San José del Guaviare, *Col.* **130** H7
San Jose, *N. Mex., U.S.* **109** R13
San Jose, *Calif., U.S.* **111** S4
San Jose, *N. Mariana Is., U.S.* **230** J7,11
San José, *C.R.* **117** N19
San José, *Guatemala* **117** L15
San José, Golfo, *Arg.* **135** Q10
San José, Isla, *Mex.* **116** E7
San Juan Bautista, *Calif., U.S.* **111** T4
San Juan Bautista, *Parag.* **132** N10
San Juan Capistrano, *Calif., U.S.* **111** Y9
San Juan de los Morros, *Venez.* **130** D10
San Juan del Norte, *Nicar.* **117** M19
San Juan del Sur, *Nicar.* **117** M18
San Juan Island National Historical Park, *Wash., U.S.* **110** A3
San Juan Islands National Monument, *Wash., U.S.* **110** A3
San Juan Islands, *Wash., U.S.* **110** A3
San Juan Mountains, *Colo., N. Mex., U.S.* **109** P11
San Juan y Martinez, *Cuba* **118** H5
San Juan, *Dom. Rep.* **119** M18
San Juan, *Trin. & Tobago* **123** P22
San Juan, *Peru* **132** J4
San Juan, *Arg.* **134** J8
San Juan, Cabo, *Arg.* **135** X10
San Juan, river, *Utah, U.S.* **109** P8
San Juanbaai, bay, *Curaçao, Neth.* **123** P13
San Julián, Bahía, *Arg.* **135** U9
San Julián, Gran Bajo de, *Arg.* **135** U8
San Just, Sierra de, *Sp.* **147** F14
San Justo, *Arg.* **134** J12
San Justo, *Arg.* **134** L12
San Lázaro, Cabo, *Mex.* **116** E6
San Lorenzo de El Escorial, *Sp.* **146** F10
San Lorenzo de la Parrilla, *Sp.* **146** G12
San Lorenzo, *Ecua.* **130** J3
San Lorenzo, *Parag.* **132** M10
San Lorenzo, *Arg.* **134** K11
San Lorenzo, Cabo, *Ecua.* **130** L2
San Luis de Palenque, *Col.* **130** F8
San Luis Obispo, *Calif., U.S.* **111** V5
San Luis Potosí, *Mex.* **116** G11
San Luis Reservoir, *Calif., U.S.* **111** T5
San Luis Río Colorado, *Mex.* **116** B6
San Luis, *Ariz., U.S.* **109** V4
San Luis, *Colo., U.S.* **109** P13
San Luis, *Guatemala* **117** K16
San Luis, *Cuba* **119** K13
San Luis, *Arg.* **134** K11
San Luis, Lago de, *Bol.* **132** H8
San Marcos, *Tex., U.S.* **105** L16
San Marcos, *Col.* **130** E6
San Marino, *Eur.* **163** E4
San Marino, *San Marino* **163** J15
San Marino, river, *It., San Marino* **163** J14
San Martín de los Andes, *Arg.* **135** P7
San Martin Seamounts, *Pac. Oc.* **255** P13
San Martín Station, *Antarctica* **238** E6
San Martín, *Col.* **130** G6
San Martín, *Arg.* **134** K8
San Martín, Cape, *Calif., U.S.* **111** U4
San Martín, Lago, *Chile* **135** U7
San Martín, river, *Bol.* **132** H8
San Mateo, *N. Mex., U.S.* **109** S11
San Mateo, *Calif., U.S.* **111** S3
San Matías, Golfo, *Arg.* **135** Q10
San Miguel Bay, *Philippines* **193** C14
San Miguel Creek, *Tex., U.S.* **105** M15
San Miguel de Cozumel, *Mex.* **117** H17
San Miguel de Huachi, *Bol.* **132** J7
San Miguel de Tucumán, *Arg.* **134** G9
San Miguel del Monte, *Arg.* **134** M12
San Miguel, *El Salv.* **117** L17
San Miguel, island, *Calif., U.S.* **111** X5
San Miguel, river, *Bol.* **132** J8
San Miniato, *It.* **152** G6
San Nicolás, *Arg.* **134** K12
San Nicolas, island, *Calif., U.S.* **111** Y6
San Pablo, *Philippines* **193** C14

San Pedro de las Colonias, *Mex.* **116** E10
San Pedro de Macorís, *Dom. Rep.* **119** M20
San Pedro Sula, *Hond.* **117** K17
San Pedro, *Parag.* **132** M10
San Pedro, *Arg.* **134** E9
San Pedro, *Arg.* **134** L12
San Pedro, river, *Ariz., U.S.* **109** V8
San Pedro, Sierra de, *Sp.* **146** H7
San Pietro, island, *It.* **153** N14
San Pietro, Isola di, *It.* **153** Q3
San Quintín, *Mex.* **116** B5
San Rafael del Yuma, *Dom. Rep.* **119** M20
San Rafael, *Calif., U.S.* **111** R3
San Rafael, *Bol.* **132** J9
San Rafael, *Arg.* **134** L8
San Rafael, Cabo, *Dom. Rep.* **119** L20
San Remo, *It.* **152** G2
San Roque, *Sp.* **146** N9
San Roque, *N. Mariana Is., U.S.* **230** H8
San Saba, *Tex., U.S.* **105** J15
San Saba, river, *Tex., U.S.* **105** J15
San Salvador de Jujuy, *Arg.* **134** E9
San Salvador *see* Isla Santiago, island, *Ecua.* **130** Q3
San Salvador, *El Salv.* **117** L16
San Salvador, Isla, *Galápagos Is., Ecua.* **225** G22
San Salvador, island, *Bahamas* **119** F14
San Severo, *It.* **152** L12
San Simeon, *Calif., U.S.* **111** V4
San Simon, *Ariz., U.S.* **109** V9
San Sosti, *It.* **153** P13
San Valentín, Cerro, *Chile* **135** T6
San Vicente de Alcántara, *Sp.* **146** H7
San Vicente de Barakaldo, *Sp.* **146** B12
San Vicente, *N. Mariana Is., U.S.* **230** J8
San Vincente del Caguán, *Col.* **130** H6
San Vito, Cape, *Eur.* **140** L7
San Ygnacio, *Tex., U.S.* **105** Q15
San-Pédro, *Côte d'Ivoire* **204** Q12
San, *Japan* **189** W3
San, *Mali* **205** K13
San, river, *Pol.* **151** G17
San, river, *Vietnam* **190** M12
Şan'ā' (Sanaa), *Yemen* **174** N9
Sana, river, *Bosn. & Herzg.* **154** E6
Sanaa *see* Şan'ā', *Yemen* **174** N9
Sanae Canyon, *Atl. Oc.* **249** Y12
SANAE IV, station, *Antarctica* **239** B13
Şanāfīr, island, *Saudi Arabia* **174** D3
Sanaga, river, *Af.* **198** J5
Sanak Islands, *Alas., U.S.* **112** P11
Sanam, *Niger* **206** G6
Sanām, *Saudi Arabia* **174** G9
Sanana, island, *Indonesia* **193** K15
Sanandaj, *Iraq* **176** F12
Sanaroa, island, *P.N.G.* **222** R9
Sanat, island, *F.S.M.* **226** F9
Sanāw, *Yemen* **175** L13
Sanawad, *India* **182** K5
Sancheong, *S. Korea* **187** S10
Sánchez, *Dom. Rep.* **119** L19
Sanchor, *India* **182** J3
Sanchursk, *Russ.* **158** H13
Sancti Spíritus, *Cuba* **118** H9
Sand Bluff, *Cayman Is., U.K.* **122** N5
Sand Cay, *India* **183** R2
Sand Hills, *Nebr., U.S.* **107** N3
Sand Island, *Wis., U.S.* **100** E4
Sand Island, *Solomon Is.* **228** K5
Sand Point, *Alas., U.S.* **112** N12
Sand Sea of Calanscio *see* Sarīr Kalanshiyū ar Ramlī al Kabīr, *Lib.* **201** F22
Sand Springs, *Okla., U.S.* **105** B18
Sand Springs, *Mont., U.S.* **108** C11
Sand, *Nor.* **142** K10
Sanda Island, *Scot., U.K.* **145** M8
Sandakan, *Malaysia* **192** G12
Sandaré, *Mali* **204** J10
Sanday, island, *Scot., U.K.* **144** E11
Sanders, *Ariz., U.S.* **109** S9
Sanderson, *Tex., U.S.* **104** L11
Sandersville, *Ga., U.S.* **98** M7
Sandes, *Nor.* **142** L10
Sandia Crest, *N. Mex., U.S.* **109** S12
Sandino, *Cuba* **118** H4
Sandıklı, *Turk.* **170** F6

Sandness, *Scot., U.K.* **144** C12
Sandoa, *D.R.C.* **209** K15
Sandomierz, *Pol.* **151** F17
Sandon Iwa, *Japan* **189** V4
Sandover, river, *Austral.* **220** E9
Sandovo, *Russ.* **158** G9
Sandown, *Eng., U.K.* **145** W13
Sandoy, island, *Faroe Is., Den.* **142** H6
Sandpoint, *Idaho, U.S.* **108** B4
Sandray Sanndraigh, island, *Scot., U.K.* **144** J6
Sandspit, *B.C., Can.* **86** K7
Sandstone, *Minn., U.S.* **106** H12
Sandstone, *W. Aust., Austral.* **221** V3
Sandur, *Faroe Is., Den.* **142** H6
Sandusky, *Mich., U.S.* **100** L13
Sandusky, *Ohio, U.S.* **101** P13
Sandviken, *Sw.* **143** J13
Sandwich Bay, *Nfld. & Lab., Can.* **87** K22
Sandwich Bay, *Namibia* **212** H3
Sandwich, *Ill., U.S.* **101** N6
Sandy Bay, *Jam.* **122** H6
Sandy Bay, *Gibraltar, U.K.* **163** P23
Sandy Cape, *Austral.* **220** G16
Sandy Cay, *Virgin Is., U.K.* **122** L11
Sandy Creek, *Tex., U.S.* **105** M18
Sandy Hill Bay, *Anguilla, U.K.* **122** Q12
Sandy Island, *Antigua & Barbuda* **123** A20
Sandy Island, *Grenada* **123** J24
Sandy Islet, *Austral.* **220** B4
Sandy Lake, *Ont., Can.* **87** M15
Sandy Point Town, *St. Kitts & Nevis* **123** A17
Sandy Point, *Bahamas* **122** C9
Sandy Point, *Virgin Is., U.S.* **122** R1
Sandy Point, *India* **183** S14
Sandy, *Utah, U.S.* **108** L7
Sandykgaçy, *Turkm.* **178** P9
Sanford, *Me., U.S.* **97** H15
Sanford, *Fla., U.S.* **99** T9
Sanford, *N.C., U.S.* **98** H11
Sang Bur, *Afghan.* **180** H3
Sang Sang, *Nicar.* **117** L18
Sang-e Māshah, *Afghan.* **180** J7
Sanga, *Burkina Faso* **205** L16
Sanga, *Mali* **205** J14
Sanga, *Angola* **208** L11
Sangamner, *India* **182** M4
Sangamon, river, *Ill., U.S.* **101** R5
Sangān, Kōh-e, *Afghan.* **180** J5
Sangar, *Russ.* **161** G17
Sangardo, *Guinea* **204** M10
Sangareddi, *India* **182** N6
Sangaredi, *Guinea* **204** L8
Sangbé, *Cameroon* **206** M11
Sangbu, *S. Korea* **187** S10
Sangchujado, island, *S. Korea* **187** V7
Sangdong, *S. Korea* **186** N11
Sangeang, island, *Indonesia* **193** M13
Sanger, *Tex., U.S.* **105** F17
Sanger, *Calif., U.S.* **111** T6
Sangha, river, *Congo* **208** E11
Sanghar, *Pak.* **181** R8
Sanghwagye, *S. Korea* **186** M9
Sangihe, island, *Indonesia* **193** H15
Sangihe, Kepulauan, *Indonesia* **193** H15
Sangīn, *Afghan.* **180** K5
Sangju, *S. Korea* **187** Q10
Sangju, *S. Korea* **187** T10
Sangkulirang, *Indonesia* **192** H12
Sangli, *India* **182** N4
Sangmélima, *Cameroon* **206** P10
Sangni, *N. Korea* **186** E10
Sangnong, *N. Korea* **186** F11
Sangola, *India* **182** N4
Sangolquí, *Ecua.* **130** K4
Sangowo, *Indonesia* **193** H16
Sangre de Cristo Mountains, *Colo., N. Mex., U.S.* **109** Q13
Sangre Grande, *Trin. & Tobago* **123** P23
Sangtaedo, island, *S. Korea* **187** U5
Sangt'ong, *S. Korea* **186** G9
Sangüesa, *Sp.* **147** C13
Sanguinet, *Fr.* **149** W6
Sangŭm, *N. Korea* **186** J9
Sangwangdeungdo, island, *S. Korea* **187** R6
Sanibel Island, *Fla., U.S.* **99** W8
Sanje, *Uganda* **211** T3
Sanjeong, *S. Korea* **187** U7
Sankeshwar, *India* **183** P4

Sankt George Fjord, *Greenland, Den.* **88** B7
Sankt-Peterburg (Saint Petersburg, Leningrad), *Russ.* **158** F6
Sankuru, river, *D.R.C.* **209** G14;H15
Şanlıurfa (Edessa), *Turk.* **171** H13
Sanluri, *It.* **153** P4
Sanming, *China* **185** M14
Sannat, *Malta* **163** H20
Sanndraigh, island, *Scot., U.K.* **144** J6
Sannicandro Garganico, *It.* **152** K12
Sannikova, *Russ.* **161** D17
Sannohe, *Japan* **188** J13
Sanok, *Pol.* **151** H17
Sanpoil, river, *Wash., U.S.* **110** B7
Sans Souci, *Trin. & Tobago* **123** N23
Sansane, *Niger* **206** H5
Sansanné-Mango, *Togo* **205** L16
Sansepolcro, *It.* **152** H8
Sanski Most, *Bosn. & Herzg.* **154** F6
Sansu, *N. Korea* **186** F9
Sant Ana, *Sao Tome and Principe* **215** D20
Sant' Andrea Apostolo dello Ionio, *It.* **153** R13
Sant Antoni de Portmany, *Sp.* **147** H16
Sant Carles de la Ràpita, *Sp.* **147** F16
Sant' Eufemia, Golfo di, *It.* **153** Q13
Sant Jordi, Golf de, *Sp.* **147** F16
Sant Julià de Lòria, *Andorra* **162** K2
Sant Julià de Lòria, *Andorra* **162** K2
Santa Ana, *Calif., U.S.* **111** X8
Santa Ana, *El Salv.* **117** L16
Santa Ana, *Bol.* **132** H7
Santa Ana, *Philippines* **193** B14
Santa Ana, island, *Solomon Is.* **228** G12
Santa Anna, *Tex., U.S.* **105** H15
Santa Barbara Channel, *Calif., U.S.* **111** X6
Santa Barbara, *Calif., U.S.* **111** W6
Santa Bárbara, *Mex.* **116** F9
Santa Bárbara, *Chile* **135** N6
Santa Barbara, island, *Calif., U.S.* **111** Y7
Santa Bárbara, peak, *Sp.* **146** L12
Santa Catalina, Gulf of, *Calif., U.S.* **111** Y8
Santa Catalina, island, *Calif., U.S.* **111** Y8
Santa Catalina, island, *Solomon Is.* **228** G12
Santa Catarina Island, *S. Amer.* **128** K8
Santa Catarina, *Sao Tome and Principe* **215** D19
Santa Catharina, *Curaçao, Neth.* **123** P15
Santa Clara, *Oreg., U.S.* **110** H3
Santa Clara, *Calif., U.S.* **111** S4
Santa Clara, *Cuba* **118** H9
Santa Clara, Isla, *Pac. Oc.* **225** M23
Santa Clotilde, *Peru* **132** D4
Santa Coloma, *Andorra* **162** J2
Santa Comba, *Sp.* **146** B5
Santa Cruz Cabrália, *Braz.* **133** J16
Santa Cruz da Graciosa, *Port.* **146** P4
Santa Cruz das Flores, *Port.* **146** N1
Santa Cruz de la Palma, *Sp.* **214** Q3
Santa Cruz de Mudela, *Sp.* **146** J11
Santa Cruz de Tenerife, *Sp.* **200** F3
Santa Cruz de Tenerife, *Sp.* **214** Q5
Santa Cruz del Norte, *Cuba* **118** G7
Santa Cruz del Sur, *Cuba* **118** K11
Santa Cruz do Sul, *Braz.* **132** P11
Santa Cruz Islands, *Solomon Is.* **224** H7
Santa Cruz, *Calif., U.S.* **111** T4
Santa Cruz, *Aruba, Neth.* **123** Q17
Santa Cruz, *Curaçao, Neth.* **123** N13
Santa Cruz, *Ecua.* **130** Q3
Santa Cruz, *Bol.* **132** K8
Santa Cruz, *Braz.* **133** F18
Santa Cruz, *Madeira, Port.* **214** M3
Santa Cruz, *Sao Tome and Principe* **215** D20
Santa Cruz, Isla, *Galápagos Is., Ecua.* **225** G22
Santa Cruz, island, *Calif., U.S.* **111** X6
Santa Cruz, river, *Calif., U.S.* **109** W8
Santa Cruz, river, *Arg.* **135** V8
Santa Elena Peninsula, *S. Amer.* **128** D1
Santa Elena, *Venez.* **131** G14
Santa Elena, Cabo, *C.R.* **117** N18

Santa Eugenia, *Sp.* **146** C5
Santa Eulalia del Río, *Sp.* **147** H17
Santa Fe, *N. Mex., U.S.* **109** R12
Santa Fe, *Arg.* **134** J12
Santa Fé, *Sp.* **146** L11
Santa Helena, *Braz.* **132** E10
Santa Inês, *Braz.* **133** E14
Santa Inés, Isla, *Chile* **135** X7
Santa Isabel, *Arg.* **134** M9
Santa Isabel, island, *Solomon Is.* **228** D9
Santa Lucia Range, *Calif., U.S.* **111** U4
Santa Lucía, *Cuba* **118** G5
Santa Lucía, *Arg.* **134** H12
Santa Lucía, *Arg.* **134** J8
Santa Lucía, *Uru.* **134** L14
Santa Luzia, *Port.* **146** P3
Santa Luzia, island, *Cabo Verde* **215** B15
Santa Maddalena Vallalta, *It.* **152** B8
Santa Margarita, Isla, *Mex.* **116** F6
Santa Maria Bay, *Virgin Is., U.S.* **122** M7
Santa Maria da Vitória, *Braz.* **133** H15
Santa María de Huerta, *Sp.* **146** E12
Santa Maria de Nanay, *Peru* **132** E4
Santa Maria di Leuca, Capo, *It.* **153** P16
Santa María see Isla Floreana, island, *Ecua.* **130** R3
Santa Maria, *Calif., U.S.* **111** W5
Santa Maria, *Curaçao, Neth.* **123** P14
Santa Maria, *Braz.* **132** D9
Santa Maria, *Braz.* **132** P11
Santa Maria, *Switz.* **152** C6
Santa Maria, *Zambia* **209** L19
Santa María, *Peru* **132** D4
Santa María, *Arg.* **134** G9
Santa Maria, Cabo de, *Port.* **146** L6
Santa Maria, Cape, *Bahamas* **119** F13
Santa María, Cayo, *Cuba* **118** G10
Santa María, Isla, *Galápagos Is., Ecua.* **225** G22
Santa Maria, island, *Port.* **146** R6
Santa Maria, island, *Vanuatu* **229** D14
Santa Marta de Ortigueira, Ría de, *Sp.* **146** A6
Santa Marta Grande, Cabo de, *Braz.* **133** P13
Santa Marta, *Col.* **130** C6
Santa Monica Mountains National Recreation Area, *Calif., U.S.* **111** X7
Santa Monica, *Calif., U.S.* **111** X8
Santa Olalla del Cala, *Sp.* **146** K8
Santa Paula, *Calif., U.S.* **111** X7
Santa Pola, *Sp.* **147** K14
Santa Rita do Weil, *Braz.* **132** E5
Santa Rita, *Venez.* **130** C8
Santa Rita, *Guam, U.S.* **230** P5
Santa Rosa and San Jacinto Mountains National Monument, *Calif., U.S.* **111** Y10
Santa Rosa Beach, *Fla., U.S.* **99** Q3
Santa Rosa Island, *Calif., U.S.* **111** Q2
Santa Rosa Range, *Nev., U.S.* **110** L9
Santa Rosa, *N. Mex., U.S.* **109** S14
Santa Rosa, *Calif., U.S.* **111** R3
Santa Rosa, *Hond.* **117** L16
Santa Rosa, *Curaçao, Neth.* **123** P15
Santa Rosa, *Bol.* **132** G6
Santa Rosa, *Peru* **132** G5
Santa Rosa, *Peru* **132** J5
Santa Rosa, *Arg.* **134** K9
Santa Rosa, *Arg.* **134** M10
Santa Rosa, island, *Calif., U.S.* **111** X6
Santa Rosalía, *Mex.* **116** D6
Santa Severina, *It.* **153** Q14
Santa Teresa, *N.T., Austral.* **221** T9
Santa Teresa, Embalse de, *Sp.* **146** F9
Santa Victoria, *Arg.* **134** D11
Santa Vitória do Palmar, *Braz.* **132** Q11
Santa Ynez Mountains, *Calif., U.S.* **111** W6
Santa Ynez, *Calif., U.S.* **111** W6
Santa-Manza, Golfe de, *Fr.* **149** Y3
Santana do Livramento, *Braz.* **132** P11
Santana, *Braz.* **133** H15
Santana, *Madeira, Port.* **214** M3
Santander Jiménez, *Mex.* **116** F12
Santander, *Sp.* **146** B11
Santander, Bahía de, *Sp.* **146** B11
Santanilla (Swan Is.), Islas, *Hond.* **117** J19

Sant'Antioco, Isola di, *It.* **153** Q3
Santaquin, *Utah, U.S.* **108** L7
Santarém, *Braz.* **132** D11
Santarém, *Port.* **146** H5
Santaren Channel, *Bahamas* **118** F9
Santee, river, *S.C., U.S.* **98** L11
Santhià, *It.* **152** E3
Santiago de Baney, *Eq. Guinea* **214** L8
Santiago de Compostela, *Sp.* **146** B5
Santiago de Cuba, *Cuba* **119** K13
Santiago del Estero, *Arg.* **134** G10
Santiago Ixcuintla, *Mex.* **116** G9
Santiago Papasquiaro, *Mex.* **116** F9
Santiago Peak, *Tex., U.S.* **104** L10
Santiago, *Pan.* **117** P20
Santiago, *Dom. Rep.* **119** L18
Santiago, *Chile* **134** L7
Santiago, island, *Cabo Verde* **215** D16
Santiago, Punta, *Eq. Guinea* **214** N7
Santigi, *Indonesia* **193** H13
Santo André, *Braz.* **133** M13
Santo Ângelo, *Braz.* **132** P11
Santo Antão, island, *Cabo Verde* **215** A15
Santo Antônio do Içá, *Braz.* **132** D6
Santo António, *Cabo Verde* **215** D17
Santo António, *Sao Tome and Principe* **215** A21
Santo Corazón, *Bol.* **132** K9
Santo Domingo de Silos, *Sp.* **146** D11
Santo Domingo Pueblo, *N. Mex., U.S.* **109** R12
Santo Domingo, *Dom. Rep.* **119** M19
Santo Domingo, Cay, *Bahamas* **119** J13
Santo Isabel do Rio Negro, *Braz.* **132** C7
Santo Tirso, *Port.* **146** E5
Santo Tomás, *Peru* **132** J5
Santo Tomé, *Arg.* **134** H14
Santopitar, peak, *Sp.* **146** M10
Santorini see Thíra, island, *Gr.* **157** N10
Santos Plateau, *Atl. Oc.* **249** S7
Santos, *Braz.* **133** M14
Santos, Sierra de los, *Sp.* **146** K9
Sanya, *China* **184** R12
Sanyang, *S. Korea* **187** T11
Sanyati, river, *Zimb.* **212** E8
Sanza Pombo, *Angola* **208** J11
São Bento do Norte, *Braz.* **133** E18
São Borja, *Braz.* **132** P10
São Cristóvão, *Braz.* **133** H17
São Domingos, *Guinea-Bissau* **204** K7
São Félix do Xingu, *Braz.* **132** F12
São Filipe, *Cabo Verde* **215** E15
São Francisco, Ilha de, *Braz.* **133** N13
São Francisco, river, *Braz.* **133** J14
São Gabriel da Cachoeira, *Braz.* **132** C6
São Gabriel, *Braz.* **132** P11
São Gens, peak, *Port.* **146** J6
Sao Hill, *Tanzania* **211** X5
São João da Aliança, *Braz.* **133** J13
São João da Madeira, *Port.* **146** E5
São João del Rei, *Braz.* **133** L14
São Jorge, island, *Port.* **146** P3
São José de Anauá, *Braz.* **132** C8
São José de Ribamar, *Braz.* **133** D15
São José dos Campos, *Braz.* **133** M14
São José, Baía de, *Braz.* **133** D15
São Leopoldo, *Braz.* **132** P12
São Lourenço do Sul, *Braz.* **132** Q12
São Lourenço, Ponta de, *Madeira, Port.* **214** M4
São Luís, *Braz.* **133** D15
São Luís, Ilha de, *Braz.* **133** D15
São Mamede, Pico, *Port.* **146** H6
São Marcos, Baía de, *Braz.* **133** D15
São Martinho do Porto, *Port.* **146** G4
São Mateus, *Braz.* **133** K16
São Mateus, *Port.* **146** P3
São Miguel do Araguaia, *Braz.* **132** H12
São Miguel, island, *Port.* **146** Q5
São Nicolau, island, *Cabo Verde* **215** B16
São Paulo de Olivença, *Braz.* **132** E6
São Paulo, *Braz.* **133** M13
São Raimundo Nonato, *Braz.* **133** G15
São Romão, *Braz.* **132** K6
São Roque, Cabo de, *Braz.* **133** F18
São Sebastião Island, *S. Amer.* **128** J9
São Sebastião, *Braz.* **133** N13
São Sebastião, *Port.* **146** P4
São Teotónio, *Port.* **146** K5
São Tiago, *Cabo Verde* **215** D16

São Tomé, *Sao Tome and Principe* **215** C20
São Tomé, Cabo de, *Braz.* **133** M15
São Tomé, island, *Sao Tome and Principe* **215** D19
São Tomé, Pico de, *Sao Tome and Principe* **215** D20
Sao Tome and Principe, *Af.* **214** F5
São Vicente, Cabo de, *Port.* **146** L4
São Vicente, island, *Cabo Verde* **215** B14
Saona, island, *Dom. Rep.* **119** N20
Saône, river, *Eur.* **140** H5
Saonek, *Indonesia* **193** J17
Saoner, *India* **182** L6
Saoute, Wādī, *Syr.* **172** L9
Sapanca, *Turk.* **170** D6
Sapanjang, island, *Indonesia* **192** M11
Sapele, *Nigeria* **206** M7
Sapelo Island, *Ga., U.S.* **99** P9
Sápes, *Gr.* **157** B15
Şaphane Dağı, peak, *Turk.* **170** F5
Sapiéntza, island, *Gr.* **156** M9
Sapitwa, peak, *Malawi* **212** D11
Saposoa, *Peru* **132** F3
Sapphire Mountains, *Mont., U.S.* **108** P6
Sappho, *Wash., U.S.* **110** B2
Sapporo, *Japan* **188** G13
Sapri, *It.* **153** N12
Sapsido, island, *S. Korea* **187** Q7
Sapudi, island, *Indonesia* **192** M11
Sapulpa, *Okla., U.S.* **105** C18
Saqqap Sermia, glacier, *Greenland, Den.* **89** Q6
Saqqarliit, island, *Greenland, Den.* **88** M5
Saqqez, *Iran* **176** E12
Sar Banādir, *Afghan.* **180** M5
Sar Passage, *Palau* **226** H5
Sar-e Howz, *Afghan.* **180** G5
Sar-e Pul, *Afghan.* **180** G6
Sar-i Siākiv, peak, *Iran, Iraq* **176** D11
Sara Adasy, island, *Azerb.* **171** F22
Sara, *Burkina Faso* **205** L14
Sarāb, *Iran* **177** D13
Saraburi, *Thai.* **190** M8
Saraféré, *Mali* **205** H14
Saraguro, *Ecua.* **130** M4
Sarahs, *Turkm.* **178** P8
Saraji, *Qnsld., Austral.* **221** T14
Sarakhs, *Iran* **177** D22
Sarala, *Côte d'Ivoire* **205** N13
Saranac Lake, *N.Y., U.S.* **96** G11
Saranac Lakes, *N.Y., U.S.* **96** G10
Sarandë, *Alban.* **154** N9
Sarangani Islands, *Philippines* **193** G15
Sarangpur, *India* **182** K5
Saransk, *Russ.* **158** K13
Sarapul, *Russ.* **158** H16
Sarasota Bay, *Fla., U.S.* **99** V7
Sarasota, *Fla., U.S.* **99** V7
Saratoga Springs, *N.Y., U.S.* **96** J11
Saratoga, *Wyo., U.S.* **108** K11
Saratov, *Russ.* **158** M13
Sarāvān, *Iran* **177** M23
Sarawak, *Malaysia* **192** H10
Saray, *Turk.* **170** C4
Saray, *Turk.* **171** F18
Saraya, *Senegal* **204** K9
Sarāyān, *Iran* **177** G20
Sarayköy, *Turk.* **170** G4
Sarbāz, *Iran* **177** N23
Sarbīsheh, *Iran* **177** H21
Sardar Chah, *Pak.* **181** P5
Sardegna (Sardinia), island, *It.* **153** N3
Sardinia see Sardegna, island, *It.* **153** N3
Sardino-Balearic Plain, *Atl. Oc.* **248** J13
Sardis Lake, *Miss., U.S.* **102** H11
Sardis, *Miss., U.S.* **102** H11
Sardis, ruins, *Turk.* **170** F4
Sareiserjoch, peak, *Liech.* **162** Q3
Sargent, *Nebr., U.S.* **107** P5
Sargo Plateau, *Arctic Oc.* **252** J8
Sargodha, *Pak.* **180** K11
Sarh, *Chad* **207** K14
Sarḥad, *Afghan.* **180** F12
Sārī, *Iran* **177** E16
Sariá, island, *Gr.* **157** P18

Sarigan, island, *N. Mariana Is., U.S.* **230** J5
Sarina, *Qnsld., Austral.* **221** S15
Sariñena, *Sp.* **147** D15
Sarinleey, *Somalia* **211** R10
Sarīr Calanscio, *Af.* **198** D7
Sarīr Kalanshiyū ar Ramlī al Kabīr (Sand Sea of Calanscio), *Lib.* **201** F22
Sarīr Kalanshiyū, *Lib.* **201** G21
Sarīr Tibasti, *Lib.* **201** K19
Sariwŏn, *N. Korea* **186** K6
Sarıgöl, *Turk.* **170** G4
Sarıkamış, *Turk.* **171** D17
Sarıyar Baraji, *Turk.* **170** E7
Sarıyer, *Turk.* **170** C5
Sarız, *Turk.* **170** G11
Sarjektjåkko, peak, *Sw.* **143** D13
Sark, island, *Channel Is., U.K.* **145** Y11
Sarkari Tala, *India* **182** G2
Şarkîkaraağaç, *Turk.* **170** G7
Şarkışla, *Turk.* **170** F11
Şarköy, *Turk.* **170** D3
Sarles, *N. Dak., U.S.* **106** D6
Sarman Didinte, *Eth.* **210** L10
Sarmansuyu, *Turk.* **171** G17
Sarmi, *Indonesia* **193** K20
Sarmiento, *Arg.* **135** S8
Särna, *Sw.* **142** J12
Sarnano, *It.* **152** H9
Sarny, *Ukr.* **158** M4
Sarōbī, *Afghan.* **180** H9
Saroma Ko, *Japan* **188** F14
Saronikós Kólpos, *Gr.* **156** J12
Saronno, *It.* **152** D4
Saros Körfezi, *Turk.* **170** D3
Sarowbi, *Afghan.* **180** K9
Sarpa, *Russ.* **159** Q14
Sarqan, *Kaz.* **179** H16
Sarracín, *Sp.* **146** C11
Sarrebourg, *Fr.* **149** P15
Sarreguemines, *Fr.* **149** N15
Sarria, *Sp.* **146** B7
Sarrión, *Sp.* **147** G14
Sarteano, *It.* **152** J7
Sartène, *Fr.* **149** Y2
Sarthe, river, *Fr.* **149** Q7
Sárti, *Gr.* **157** D13
Sarufutsu, *Japan* **188** E13
Saruna, *Pak.* **181** R7
Sarupsar, *India* **182** F4
Sary-Tash, *Kyrg.* **179** M14
Sarych, Mys, *Russ.* **159** T7
Saryesik- Atyraū Qumy, *Kaz.* **179** H14
Saryesik-Atyrau Desert, *Asia* **168** F7
Sarygamysh Köli, *Turkm. Uzb.* **178** K6
Saryköl, *Kaz.* **178** C10
Saryözek, *Kaz.* **179** J15
Saryqamys, *Kaz.* **178** H5
Saryqopa Köli, *Kaz.* **178** E10
Sarysay, *Kaz.* **178** F8
Saryshagan, *Kaz.* **179** H14
Sarysu, river, *Kaz.* **178** F12
Sarysū, river, *Kaz.* **178** H11
Saryzhal, *Kaz.* **179** E15
Sasamungga, *Solomon Is.* **228** C6
Sasaram, *India* **182** J9
Sasebo, *Japan* **189** R4
Saseno *see* Sazan, island, *Alban.* **154** M9
Saskatchewan, *Can.* **86** L11
Saskatchewan, river, *N. Amer.* **84** G5
Saskatoon, *Sask., Can.* **86** M12
Sason, *Turk.* **171** G16
Sasora, *Indonesia* **193** K20
Sasovo, *Russ.* **158** K11
Sassafras Mountain, *S.C., U.S.* **98** J7
Sassandra, *Côte d'Ivoire* **204** Q12
Sassandra, river, *Côte d'Ivoire* **204** P12
Sassari, *It.* **152** M3
Sassnitz, *Ger.* **150** B10
Sasstown, *Liberia* **204** Q11
Sassuolo, *It.* **152** F6
Sasu-san, peak, *N. Korea* **186** G8
Sasyqköl, *Kaz.* **179** G17
Sata-Misaki, *Japan* **189** T4
Satadougou, *Mali* **204** K9
Satala, ruins, *Turk.* **171** E14
Satanta, *Kans., U.S.* **107** U4
Satara, *India* **182** N4
Sataua, *Samoa* **227** N17
Satawal, island, *F.S.M.* **224** F4
Satawal, island, *F.S.M.* **226** B8

Satawan Atoll, *F.S.M.* **226** B9
Sätbaev, *Kaz.* **178** G11
Satbarwa, *India* **182** J9
Satilla, river, *Ga., U.S.* **99** P6
Satipo, *Peru* **132** H4
Satkania, *Bangladesh* **182** K14
Satna, *India* **182** J7
Šator, peak, *Bosn. & Herzg.* **154** G6
Sátoraljaújhely, *Hung.* **151** K17
Satpura Range, *India* **182** L4
Satu Mare, *Rom.* **154** B12
Satulung, *Rom.* **155** B13
Satun, *Thai.* **191** S7
Satupa'itea, *Samoa* **227** N18
Sauce, *Arg.* **134** J13
Sauda, *Nor.* **142** K10
Sauðárkrókur, *Ice.* **142** E3
Saudi Arabia, *Asia* **174** F8
Saugatuck, *Mich., U.S.* **100** M9
Saugerties, *N.Y., U.S.* **96** L11
Sauk Centre, *Minn., U.S.* **106** J10
Sauk City, *Wis., U.S.* **100** L5
Sauk Rapids, *Minn., U.S.* **106** J11
Saukorem, *Indonesia* **193** J18
Saulieu, *Fr.* **149** R12
Sault Sainte Marie, *Ont., Can.* **87** P17
Sault Sainte Marie, *Mich., U.S.* **100** F11
Saūmalköl, *Kaz.* **178** C11
Saumlaki, *Indonesia* **193** M17
Saumur, *Fr.* **149** S7
Saunders Coast, *Antarctica* **238** N10
Saupon Point, *Guam, U.S.* **230** N5
Saura, *Kaz.* **178** H4
Saurimo, *Angola* **209** K13
Sausalito, *Calif., U.S.* **111** S3
Sautar, *Angola* **208** L12
Sauteurs Bay, *Grenada* **123** J23
Sauteurs, *Grenada* **123** J23
Sauveterre, *Fr.* **149** X6
Sava, river, *Bosn. & Herzg. Croatia* **154** E7
Savage, *Mont., U.S.* **108** C13
Savai'i, island, *Samoa* **227** N18
Savalou, *Benin* **205** N17
Savana Island, *Virgin Is., U.S.* **122** M6
Savane des Pétrifications, *Martinique, Fr.* **123** G24
Savaneta, *Aruba, Neth.* **123** R17
Savanna Point, *Jam.* **122** H11
Savanna-la-Mar, *Jam.* **122** H6
Savanna, *Ill., U.S.* **101** N4
Savannah Bay, *Virgin Is., U.K.* **122** Q8
Savannah Sound, *Bahamas* **122** E6
Savannah, *Ga., U.S.* **99** N9
Savannah, *Tenn., U.S.* **103** G13
Savannah, *Mo., U.S.* **107** R10
Savannah, *Cayman Is., U.K.* **122** N3
Savannah, river, *U.S.* **91** K19
Savannakhét, *Laos* **190** L10
Savaştepe, *Turk.* **170** E3
Savé, *Benin* **205** N17
Save, river, *Mozambique* **212** G11
Save, river, *Zimb.* **212** F10
Sāveh, *Iran* **177** F15
Savelugu, *Ghana* **205** M16
Savenay, *Fr.* **149** R5
Saverne, *Fr.* **149** P15
Savigliano, *It.* **152** F2
Savignac-les-Églises, *Fr.* **149** V8
Savignone, *It.* **152** F4
Savinobor, *Russ.* **158** B16
Savo, island, *Solomon Is.* **228** F10
Savona, *It.* **152** F3
Savonet, *Curaçao, Neth.* **123** N13
Savonlinna, *Fin.* **143** G17
Savoonga, *Alas., U.S.* **112** F10
Savu Basin, *Pac. Oc.* **246** L5
Savu Sea, *Indonesia* **193** N14
Savukoski, *Fin.* **143** D15
Savur, *Turk.* **171** H15
Savusavu Bay, *Fiji* **226** N6
Savusavu, *Fiji* **226** N6
Sawa-i-lau, island, *Fiji* **226** N4
Sawah, *Indonesia* **192** H11
Sawahlunto, *Indonesia* **192** J6
Sawai Madhopur, *India* **182** H5
Sawankhalok, *Thai.* **190** K7
Sawatch Range, *Colo., U.S.* **108** M12
Sawe, *Indonesia* **192** H4
Sawi, *Thai.* **191** Q7
Sāwinnū, ruin, *Lib.* **201** E21
Sawla, *Ghana* **205** M15
Şawqirah, *Oman* **175** L16
Sawtooth Mountains, *Minn., U.S.* **106** F13

Sawtooth National Recreation Area, *Idaho, U.S.* **108** G5
Sawtooth Range, *Idaho, U.S.* **108** G5
Sawu, island, *Indonesia* **193** N14
Saxby Downs, *Qnsld. Austral.* **221** S12
Saxony, region, *Ger.* **150** F10
Say, *Niger* **206** H5
Saya de Malha Bank, *Ind. Oc.* **250** H9
Sayanogorsk, *Russ.* **161** L13
Sayḩūt, *Yemen* **175** N13
Saylac, *Somaliland* **210** K10
Saynshand, *Mongolia* **184** E11
Sayötesh, *Kaz.* **178** J5
Sayram Hu, *China* **184** D4
Sayre, *Pa., U.S.* **96** L8
Sayre, *Okla., U.S.* **105** D14
Say'ūn, *Yemen* **174** N12
Sayylyk, *Russ.* **161** E18
Sazaī Kalān, *Afghan.* **180** G6
Sazan (Saseno), island, *Alban.* **154** M9
Sazin, *Pak.* **180** G12
Sbaa, *Alg.* **200** G11
Scafell Pike, peak, *Eng., U.K.* **145** P10
Scalpay *see* Eilean Scalpaigh, island, *Scot., U.K.* **144** H6
Scalpay, island, *Scot., U.K.* **144** J7
Scammon Bay, *Alas., U.S.* **112** H11
Scandia, *Kans., U.S.* **107** R7
Scandinavia, *Eur.* **140** D7
Scanzano Ionico, *It.* **153** N14
Scapa Flow, *Scot., U.K.* **144** F10
Scaraben, peak, *Scot., U.K.* **144** G10
Scarba, island, *Scot., U.K.* **144** L7
Scarborough, *Trin. & Tobago* **123** P17
Scarborough, *Eng., U.K.* **145** P13
Scarinish, *Scot., U.K.* **144** K6
Scarp, island, *Scot., U.K.* **144** G6
Scarriff (An Scairbh), *Ire.* **145** R4
Scatterbreak Channel, *Fiji* **226** M7
Scealg, Bá na (Ballinskelligs Bay), *Ire.* **145** T2
Šćedro, island, *Croatia* **154** H6
Šćedro, island, *Croatia* **154** H6
Schaan, *Liech.* **162** N2
Schaanwald, *Liech.* **162** N3
Schefferville, *Que., Can.* **87** L20
Schell Creek Range, *Nev., U.S.* **111** Q12
Schellenberg, *Liech.* **162** M3
Schenectady, *N.Y., U.S.* **96** J11
Schertz, *Tex., U.S.* **105** L16
Schiermonnikoog, island, *Neth.* **148** G13
Schio, *It.* **152** D7
Schíza, island, *Gr.* **156** M9
Schleswig, *Ger.* **150** A7
Schoelcher, *Martinique, Fr.* **123** F22
Schofield, *Wis., U.S.* **100** H5
Schoinoússa, island, *Gr.* **157** L15
Schönberg, peak, *Liech.* **162** P3
Schönebeck, *Ger.* **150** E9
Schooner Cays, *Bahamas* **122** F6
Schottegat, lake, *Curaçao, Neth.* **123** P14
Schouten Islands, *P.N.G.* **222** N7
Schrems, *Aust.* **150** J11
Schroeder, *Minn., U.S.* **106** F13
Schulenburg, *Tex., U.S.* **105** L17
Schuyler, *Nebr., U.S.* **107** P8
Schwabach, *Ger.* **150** H8
Schwabenland Canyon, *Atl. Oc. Ind. Oc.* **249** Y14
Schwäbisch Hall, *Ger.* **150** J7
Schwäbische Alb, range, *Ger.* **150** K6
Schwaz, *Aust.* **150** L9
Schwechat, *Aust.* **151** K13
Schwedt, *Ger.* **150** D11
Schweinfurt, *Ger.* **150** G8
Schweitzerland, *Greenland, Den.* **89** P10
Schwerin, *Ger.* **150** C9
Schwyz, *Switz.* **152** B4
Sciacca, *It.* **153** T9
Scilla, *It.* **153** R12
Scilly, Isles of, *Eng., U.K.* **145** X7
Scioto, river, *Ohio, U.S.* **101** S13
Scipio, *Utah, U.S.* **108** M7
Scobey, *Mont., U.S.* **108** A12
Scofield Reservoir, *Utah, U.S.* **108** M8
Scooba, *Miss., U.S.* **103** K13
Scoresby Land, *Greenland, Den.* **88** K13
Scoresby Sund *see* Kangertittivaq, *Greenland, Den.* **88** K14
Scoresby Valley, *Arctic Oc.* **253** L19
Scoresbysund *see* Ittoqqortoormiit, *Greenland, Den.* **88** K14
Sebring, *Fla., U.S.* **99** U9

Scotch Bonnet, cape, *St. Kitts & Nevis* **123** B18
Scotia Sea, *Atl. Oc.* **249** W6
Scotian Shelf, *Atl. Oc.* **248** J5
Scotland Neck, *N.C., U.S.* **98** G13
Scotland, *S. Dak., U.S.* **106** M7
Scotland, *Scot., U.K.* **144** J9
Scott Base, station, *Antarctica* **239** N13
Scott Canyon, *Pac. Oc.* **255** H16
Scott City, *Kans., U.S.* **107** T4
Scott City, *Mo., U.S.* **107** U16
Scott Coast, *Antarctica* **239** N14
Scott Glacier, *Antarctica* **239** K21
Scott Islands, *B.C. Can.* **86** L7
Scott Mountains, *Antarctica* **239** C19
Scott Peninsula, *Antarctica* **238** L8
Scott Plateau, *Ind. Oc.* **251** J18
Scott Point, *Mich., U.S.* **100** G9
Scott, Mount, *Oreg., U.S.* **110** J4
Scotts Bluff National Monument, *Nebr., U.S.* **107** N1
Scottsbluff, *Nebr., U.S.* **107** N1
Scottsboro, *Ala., U.S.* **103** G16
Scottsburg, *Ind., U.S.* **101** T9
Scottsdale, *Ariz., U.S.* **109** U7
Scottsdale, *Tas., Austral.* **221** Y8
Scottsville, *Ky., U.S.* **103** D16
Scottville, *Mich., U.S.* **100** K9
Scotty's Junction, *Nev., U.S.* **111** S9
Scrabster, *Scot., U.K.* **144** F10
Scranton, *Pa., U.S.* **96** M9
Scribner, *Nebr., U.S.* **107** P8
Scrub Island, *Anguilla, U.K.* **122** Q12
Scrubby Creek, *Qnsld. Austral.* **221** P13
Sculpin Ridge, *Pac. Oc.* **247** H13
Scunthorpe, *Eng., U.K.* **145** Q13
Scuol, *Switz.* **152** B6
Scurdie Ness, cape, *Scot., U.K.* **144** K11
Scutari *see* Shkodër, *Alban.* **154** K9
Scutari, Lake *see* Skadarsko Jezero, *Alban. Montenegro* **154** J9
Scylla Glacier, *Antarctica* **239** E19
Sea Club, *Monaco* **163** B23
Sea Cow Bay, *Virgin Is., U.K.* **122** R6
Sea Grape, *Bahamas* **122** J1
Sea Island, *Ga., U.S.* **99** P9
Sea Islands, *U.S.* **91** K19
Seadrift, *Tex., U.S.* **105** N18
Seaford, *Del., U.S.* **98** D14
Seaford, *Eng., U.K.* **145** V14
Seaforth, *Jam.* **122** J11
Seaforth, Loch, *Scot., U.K.* **144** H7
Seagraves, *Tex., U.S.* **104** G11
Seaham, *Eng., U.K.* **145** N12
Seal Bay, *Antarctica* **238** B12
Seal Cay, *Bahamas* **119** H13
Seal Cays, *Turks & Caicos Is. U.K.* **119** J17
Seal Dog Island, *Virgin Is., U.K.* **122** Q8
Seal Islands, *Anguilla, U.K.* **122** Q10
Seal, river, *Man. Can.* **87** K14
Sealevel, *N.C., U.S.* **98** J14
Sealy, *Tex., U.S.* **105** L18
Searchlight, *Nev., U.S.* **111** V12
Searcy, *Ark., U.S.* **102** G9
Searles Lake, *Calif., U.S.* **111** V9
Seaside Park, *N.J., U.S.* **96** Q11
Seaside, *Oreg., U.S.* **110** E2
Seatons, *Antigua & Barbuda* **123** B21
Seattle, *Wash., U.S.* **110** C4
Sebago Lake, *Me., U.S.* **97** G15
Sebastián Vizcaíno, Bahía, *Mex.* **116** C5
Sebastian, *Fla., U.S.* **99** U10
Sebastian, Cape, *Oreg., U.S.* **110** K1
Sebastopol, *Calif., U.S.* **111** R3
Sebatik, island, *Indonesia* **192** G12
Sebba, *Burkina Faso* **205** K16
Sebdérat, *Eritrea* **210** H6
Sébé, river, *Gabon* **208** E9
Sebeka, *Minn., U.S.* **106** G10
Seben, *Turk.* **170** D7
Sebeş, *Rom.* **155** D13
Sebewaing, *Mich., U.S.* **100** K12
Sebezh, *Russ.* **158** H5
Şebinkarahisar, *Turk.* **171** E13
Sebkha Azzel Matti, *Alg.* **200** H11
Sebkha de Tindouf, *Alg.* **200** G12
Sebkha Mekerrhane, *Alg.* **200** H11
Sebkhat Aghzoumal, *W. Sahara* **200** J4
Sebkhat Tidsit, *W. Sahara* **200** J3

Sebuyau, *Malaysia* **192** H9
Sech'ŏn, *N. Korea* **186** E12
Sechura Desert, *S. Amer.* **128** E2
Sechura, Bahía de, *Peru* **132** F1
Secica, *Slovakia* **151** J13
Second Mesa, *Ariz., U.S.* **109** R8
Secondigny, *Fr.* **149** S7
Secos, Ilhéus (Ilhéus do Rombo), *Cabo Verde* **215** D15
Secretary Island, *N.Z.* **223** N15
Sécure, river, *Bol.* **132** J7
Security, *Colo., U.S.* **109** N13
Sedalia, *Mo., U.S.* **107** T12
Sedan, *Kans., U.S.* **107** V9
Sedan, *Fr.* **149** N12
Seddon, *N.Z.* **223** K20
Sedeh, *Iran* **177** G21
Sédhiou, *Senegal* **204** K7
Sedom, *Israel* **173** P6
Sedona, *Ariz., U.S.* **109** S7
Sedro Woolley, *Wash., U.S.* **110** A4
Seeheim, *Namibia* **212** J4
Seemade, *Somalia* **211** N13
Seemore Downs, *W. Aust., Austral.* **221** W6
Sées, *Fr.* **149** P7
Seevetal, *Ger.* **150** C8
Seferihisar, *Turk.* **170** F2
Seg Ozero, *Russ.* **158** C8
Seg, Lake, *Eur.* **140** C10
Segag, *Eth.* **211** N10
Ségala, *Mali* **204** J10
Segamat, *Malaysia* **191** U9
Segesta, ruin, *It.* **153** S9
Segezha, *Russ.* **158** C8
Seghe, *Solomon Is.* **228** E8
Ségou, *Mali* **204** J12
Segovia, *Sp.* **146** E10
Segré, *Fr.* **149** R6
Seguam Island, *Alas., U.S.* **112** P7
Séguédine, *Niger* **206** D11
Séguéla, *Côte d'Ivoire* **204** N12
Seguin, *Tex., U.S.* **105** L16
Seguntur, *Indonesia* **192** H12
Segura, river, *Sp.* **147** K14
Segura, Sierra de, *Sp.* **146** K12
Sehested Fjord, *Greenland, Den.* **89** R9
Sehithwa, *Botswana* **212** F6
Sehnkwehn, *Liberia* **204** Q10
Sehwan, *Pak.* **181** Q9
Seikan Submarine Tunnel, *Japan* **188** J12
Seil, island, *Scot., U.K.* **144** K8
Seiling, *Okla., U.S.* **105** B15
Seinäjoki, *Sw.* **143** G14
Seine Plain, *Atl. Oc.* **248** K11
Seine, Baie de la, *Fr.* **149** N7
Seine, river, *Fr.* **149** P9
Sekenke, *Tanzania* **211** V5
Sekhira, *Tunisia* **201** C16
Sekondi, *Ghana* **205** Q15
Sek'ot'a, *Eth.* **210** K8
Sekseūil, *Kaz.* **178** G8
Sela Dingay, *Eth.* **210** L8
Selah, *Wash., U.S.* **110** D6
Selaru, island, *Indonesia* **193** M17
Selawik Lake, *Alas., U.S.* **113** E13
Selawik, *Alas., U.S.* **113** E14
Selayar, island, *Indonesia* **193** L13
Selby, *S. Dak., U.S.* **106** J5
Selby, *Eng., U.K.* **145** Q13
Selden, *Kans., U.S.* **107** S4
Seldovia, *Alas., U.S.* **113** K16
Selendi, *Turk.* **170** F4
Selenga, river, *Russ.* **161** L15
Selenge, river, *Mongolia* **184** C10
Sélestat, *Fr.* **149** Q15
Seleucia, ruin, *Iraq* **176** G11
Self Town *see* McLeans Town, *Bahamas* **122** J4
Selfoss, *Ice.* **142** F2
Selfridge, *N. Dak., U.S.* **106** H4
Selib, *Russ.* **158** B13
Sélibaby, *Mauritania* **204** J9
Selibe Phikwe, *Botswana* **212** G8
Seliger, *Russ.* **158** H7
Seligman, *Ariz., U.S.* **109** S6
Selima Oasis, *Sudan* **202** M10
Selinus, ruin, *It.* **153** S9
Selkirk Mountains, *Can. U.S.* **108** A3
Selkirk Rise, *Pac. Oc.* **247** P21
Selkirk, *Scot., U.K.* **145** M11
Sellières, *Fr.* **149** S13
Sells, *Ariz., U.S.* **109** V7
Selma, *N.C., U.S.* **98** H12
Selma, *Ala., U.S.* **103** L15

Selma, *Calif., U.S.* **111** T6
Selmer, *Tenn., U.S.* **103** G13
Selongey, *Fr.* **149** R13
Selsey Bill, cape, *Eng., U.K.* **145** W13
Seltz, *Fr.* **149** P16
Selvas, region, *S. Amer.* **128** E5
Selway, river, *Idaho, U.S.* **108** D4
Selwyn Mountains, *Yukon, Can.* **86** F9
Selwyn Strait, *Vanuatu* **229** G16
Sema, *N. Korea* **186** L5
Semarang, *Indonesia* **192** M9
Sembé, *Congo* **208** D10
Şemdinli, *Turk.* **171** H19
Semenov, *Russ.* **158** H12
Semeru, peak, *Indonesia* **192** M10
Semey (Semipalatinsk), *Kaz.* **179** E16
Semichi Islands, *Alas., U.S.* **112** L2
Semiliki, river, *D.R.C.* **209** D19
Semiluki, *Russ.* **158** M10
Seminoe Dam, *Wyo., U.S.* **108** J11
Seminoe Reservoir, *Wyo., U.S.* **108** J11
Seminole, *Tex., U.S.* **104** G11
Seminole, *Okla., U.S.* **105** D17
Seminole, Lake, *Fla., Ga., U.S.* **99** Q4
Semipalatinsk Nuclear Test Site (Closed in 1991), *Kaz.* **179** E15
Semipalatinsk *see* Semey, *Kaz.* **179** E16
Semīrom, *Iran* **177** J15
Semisopochnoi Island, *Alas., U.S.* **112** N3
Semliki, river, *D.R.C.* **209** D19
Semmé, *Senegal* **204** H8
Semna West, ruin, *Sudan* **202** M11
Semnān, *Iran* **177** F17
Semporna, *Malaysia* **192** G12
Sên, river, *Cambodia* **190** M10
Sena Madureira, *Braz.* **132** G6
Seña, *Slovakia* **151** J17
Senaja, *Malaysia* **192** F12
Senaki, *Rep. of Georgia* **171** B16
Senanga, *Zambia* **209** P15
Senatobia, *Miss., U.S.* **102** H11
Sendai, *Japan* **188** L13
Sendai, *Japan* **189** S4
Sendhwa, *India* **182** L4
Senduruhan, *Indonesia* **192** J9
Seneca Falls, *N.Y., U.S.* **96** J7
Seneca, *S.C., U.S.* **98** K7
Seneca, *Kans., U.S.* **107** R9
Seneca, *Oreg., U.S.* **110** H7
Senegal, *Af.* **204** J8
Sénégal, river, *Mauritania, Senegal* **204** H7
Senftenberg, *Ger.* **150** F11
Sengiley, *Russ.* **158** K14
Sengwa, river, *Zimb.* **212** E8
Senhor do Bonfim, *Braz.* **133** G16
Senigallia, *It.* **152** G9
Senj, *Croatia* **154** E5
Senja, island, *Nor.* **142** C12
Senkaku-shotō (Diaoyu Qundao, Tiauyutai), islands, *Japan* **185** M17
Sen'ko Valley, *Arctic Oc.* **252** J12
Senkobo, *Zambia* **209** Q16
Senlis, *Fr.* **149** N10
Sênmônôûrôm, *Cambodia* **191** N12
Sennar, *Sudan* **203** S13
Sennecey-le-Grand, *Fr.* **149** S12
Sennen, *Eng., U.K.* **145** X7
Senneterre, *Que., Can.* **87** P18
Senorbì, *It.* **153** P4
Sens, *Fr.* **149** Q11
Sensuntepeque, *El Salv.* **117** L17
Senta, *Serb.* **154** D10
Šentilj, *Slov.* **154** C4
Sentinel Range, *Antarctica* **238** H8
Sentinel, *Okla., U.S.* **105** D15
Senyavin Islands, *F.S.M.* **226** B11
Seocheon, *S. Korea* **187** R7
Seogwipo, *S. Korea* **187** X7
Seokgok, *S. Korea* **187** T8
Seokgyo, *S. Korea* **187** U7
Seokjeong, *S. Korea* **187** T8
Seomjin, river, *S. Korea* **187** S9
Seongapdo, island, *S. Korea* **187** N6
Seongeup, *S. Korea* **187** W8
Seonggeo, *S. Korea* **187** P8
Seonghwan, *S. Korea* **187** P8
Seonginbong, peak, *S. Korea* **186** M15
Seongju, *S. Korea* **187** R10
Seongmodo, island, *S. Korea* **186** M7
Seongnae, *S. Korea* **187** Q12
Seongnam, *S. Korea* **187** N8
Seongsan, *S. Korea* **186** M10
Seongsan, *S. Korea* **187** W8
Seoni, *India* **182** K7

Seonsan, *S. Korea* **187** Q10
Seonyudo, island, *S. Korea* **187** R7
Seoraksan, peak, *S. Korea* **186** L11
Seosan, *S. Korea* **187** P7
Seoul, *S. Korea* **186** M8
Sepasu, *Indonesia* **192** J12
Sepi, *Solomon Is.* **228** E10
Sepik, river, *P.N.G.* **222** N6
Sep'o, *N. Korea* **186** K9
Sept-Îles, *Que., Can.* **87** M21
Sepúlveda, *Sp.* **146** E11
Sepupa, *Botswana* **212** F6
Sequim, *Wash., U.S.* **110** B3
Sequoia National Park, *Calif., U.S.* **111** U7
Seraing, *Belg.* **148** L13
Seram, island, *Indonesia* **193** K16
Serang, *Indonesia* **192** L8
Seraph Bay, *Antarctica* **238** J7
Serasan, island, *Indonesia* **192** H9
Sĕrba, *Eth.* **210** J8
Serbia, *Eur.* **154** G10
Serdeles *see* Al 'Uwaynāt, *Lib.* **201** H16
Serdo, *Eth.* **210** K9
Serdobsk, *Russ.* **158** L12
Serebryansk, *Kaz.* **179** E17
Sered', *Slovakia* **151** K14
Seredka, *Russ.* **158** G5
Şereflikoçhisar, *Turk.* **170** F9
Seregno, *It.* **152** D4
Seremban, *Malaysia* **191** U9
Serena, Embalse la, *Sp.* **146** H9
Serengeti Plain, *Tanzania* **211** U5
Serenje, *Zambia* **209** N19
Serere, *Uganda* **211** R4
Sergach, *Russ.* **158** J13
Sergeya Kirova, *Russ.* **161** D13
Sergino, *Russ.* **160** G10
Sergiyev Posad, *Russ.* **158** J9
Serian, *Malaysia* **192** H9
Sérifos (Sériphos), island, *Gr.* **157** L13
Sérifos, *Gr.* **157** L14
Sérignan, *Fr.* **149** Y11
Serik, *Turk.* **170** H6
Seringa, Serra da, *Braz.* **132** F12
Seriole, river, *San Marino* **163** L15
Seripe, *Ghana* **205** M15
Sériphos *see* Sérifos, island, *Gr.* **157** L13
Serkovo, *Russ.* **161** G13
Sermata, island, *Indonesia* **193** M16
Sermata, Kepulauan, *Indonesia* **193** M16
Sermeq Avannarleq, glacier, *Greenland, Den.* **88** L6
Sermeq Kujalleq *see* Jakobshavn Isbræ, glacier, *Greenland, Den.* **88** M6
Sermeq Silarleq, glacier, *Greenland, Den.* **88** L6
Sermersuaq (Harald Moltke Bræ), glacier, *Greenland, Den.* **88** B8
Sermersuaq (Humboldt Gletscher), *Greenland, Den.* **88** D5
Sermiligaaq, *Greenland, Den.* **89** P11
Sermiligaaq, fjord, *Greenland, Den.* **89** P11
Sermiligaarsuk Bræ, glacier, *Greenland, Den.* **89** S6
Sermiligaarsuk, fjord, *Greenland, Den.* **89** S6
Sermilik Bræ, glacier, *Greenland, Den.* **89** S6
Sermilik Søndre, fjord, *Greenland, Den.* **89** T8
Sermilik Valley, *Atl. Oc.* **253** P19
Sermilik, fjord, *Greenland, Den.* **89** ;S6
Sermilik, fjord, *Greenland, Den.* **89** P10
Sermilik, glacier, *Greenland, Den.* **88** L6
Sermilik, glacier, *Greenland, Den.* **89** R6
Sermilik, research station, *Greenland, Den.* **89** P10
Sernur, *Russ.* **158** H14
Seroczyn, *Pol.* **151** E17
Seroe Colorado, *Aruba, Neth.* **123** R17
Seronera, *Tanzania* **211** T5
Serov, *Russ.* **160** H9
Serowe, *Botswana* **212** G8
Serpa, *Port.* **146** K6
Serpent Island, *Mauritius* **215** E21
Serpentine Lakes, *Austral.* **220** H7

Serpent's Mouth, *Trin. & Tobago, Venez.* **123** R20
Serpukhov, *Russ.* **158** K9
Serra da Chela, *Angola* **208** P9
Serra de L'Albera, range, *Fr., Sp.* **149** Z12
Serra do Navio, *Braz.* **132** C12
Serravalle, *San Marino* **163** J16
Serre-Ponçon, Lac de, *Fr.* **149** W14
Serrera, Pic de la, *Andorra, Fr.* **162** G4
Serres, *Fr.* **149** W14
Sérres, *Gr.* **156** B12
Serrezuela, *Arg.* **134** J9
Serrières, *Fr.* **149** V12
Serrinha, *Braz.* **133** H16
Sersale, *It.* **153** Q14
Sertã, *Port.* **146** G6
Sertão, region, *S. Amer.* **128** F10
Sêrtar, *China* **184** K9
Serti, *Nigeria* **206** L10
Serui, *Indonesia* **193** K19
Sérvia, *Gr.* **156** D9
Sêrxü, *China* **184** K8
Sesa, *D.R.C.* **209** J17
Sesheke, *Zambia* **209** Q16
Sesimbra, *Port.* **146** J4
Sesklí, island, *Gr.* **157** M19
Sesostris Bank, *Ind. Oc.* **250** D10
Sessa Aurunca, *It.* **152** L10
Sessa, *Angola* **209** N14
Sestri Levante, *It.* **152** F5
Sestri Ponente, *It.* **152** F4
Sestroretsk, *Russ.* **158** F6
Sete Lagoas, *Braz.* **133** K14
Sete Quedas Falls, *S. Amer.* **128** J7
Sète, *Fr.* **149** X11
Sétif, *Alg.* **201** B13
Seto Naikai (Inland Sea), *Japan* **189** Q6
Seto, *Japan* **189** P10
Settat, *Mor.* **200** D7
Setté Cama, *Gabon* **208** F7
Settimo Torinese, *It.* **152** E3
Settlement, *Marshall Is.* **227** K18
Settlement, *Wake I.* **230** N11
Setúbal, *Port.* **146** J5
Setúbal, Baía de, *Port.* **146** J5
Seul Choix Point, *Mich., U.S.* **100** G9
Seul, Lac, *Ont., Can.* **87** N15
Seuldo, island, *S. Korea* **187** U7
Seungju, *S. Korea* **187** T9
Sevan, *Arm.* **171** D19
Sevana Lich, *Arm.* **171** D19
Sevastopol', *Russ.* **159** T7
Seven Islands Bay, *Nfld. & Lab., Can.* **87** J20
Seven Stones, island, *Eng., U.K.* **145** X7
Sever Spur, *Atl. Oc.* **252** L11
Severn, *Ont., Can.* **87** L16
Severn, *S. Af.* **212** J6
Severn, river, *N. Amer.* **84** G6
Severnaya Dvina, river, *Russ.* **158** D12
Severnaya Dvina, river, *Russ.* **160** F8
Severnaya Zemlya (North Land), *Russ.* **161** C14
Severnyy, *Russ.* **160** F11
Severo Kuril'sk, *Russ.* **161** G23
Severo Sibirskaya Nizmennost', *Russ.* **161** F14
Severobaykal'sk, *Russ.* **161** K16
Severodvinsk, *Russ.* **158** B10
Severodvinsk, *Russ.* **160** E8
Severomorsk, *Russ.* **160** D9
Sevi, *Russ.* **161** L13
Sevier Desert, *Utah, U.S.* **108** M6
Sevier Lake, *Utah, U.S.* **108** N6
Sevier, river, *Utah, U.S.* **108** M6
Sevilla (Seville), *Sp.* **146** L8
Sevilla, *Col.* **130** G5
Sevlievo, *Bulg.* **155** H15
Seward Peninsula, *Alas., U.S.* **112** E12
Seward, *Nebr., U.S.* **107** Q8
Seward, *Alas., U.S.* **113** K16
Sewell Rise, *Ind. Oc.* **251** E12
Sewell, *Chile* **134** L7
Seychelles, *Af.* **214** F8
Seychelles Bank, *Ind. Oc.* **250** G8
Seydi, *Turkm.* **178** M9
Seydişehir, *Turk.* **170** H7
Seyðisfjörður, *Ice.* **142** E4
Seyfe Gölü, *Turk.* **170** F9
Seyhan Barajı, *Turk.* **170** H10
Seyhan, river, *Turk.* **170** H10

Seyitgazi, *Turk.* **170** E6
Seylla Glacier, *Antarctica* **239** F19
Seymchan, *Russ.* **161** E20
Seymour, *Wis., U.S.* **100** J6
Seymour, *Ind., U.S.* **101** S9
Seymour, *Tex., U.S.* **105** F15
Seymour, *Iowa, U.S.* **107** Q12
Seymour, *Vic., Austral.* **221** Y12
Seyne, *Fr.* **149** W14
Seyöisfjöröardjúp, valley, *Arctic Oc.* **253** K20
Sézanne, *Fr.* **149** P11
Şəki, *Azerb.* **171** C21
Şəmkir, *Azerb.* **171** D20
Şərur, *Arm.* **171** E19
Sfântu Gheorghe, *Rom.* **155** D15
Sfântu-Gheorghe Braţul, river, *Rom.* **155** E19
Sfântu-Gheorghe, Ostrov, *Rom.* **155** E19
Sfax, *Tunisia* **201** C16
Sferracavallo, Capo, *It.* **153** P5
Shabasha, *Sudan* **203** R12
Shabla, Nos, *Bulg.* **155** G18
Shabunda, *D.R.C.* **209** F17
Shabwah, *Yemen* **174** N10
Shache (Yarkant), *China* **184** F2
Shackelton Fracture Zone, *Atl. Oc.* **249** W5
Shackleton Base, site, *Antarctica* **238** E10
Shackleton Coast, *Antarctica* **239** L13
Shackleton Ice Shelf, *Antarctica* **239** K22
Shackleton Range, *Antarctica* **238** F11
Shadehill Reservoir, *S. Dak., U.S.* **106** H3
Shaduzup, *Myanmar* **190** E5
Shafter, *Tex., U.S.* **104** L9
Shafter, *Nev., U.S.* **110** M12
Shagamu, *Nigeria* **206** M6
Shageluk, *Alas., U.S.* **113** H13
Shaghan, *Kaz.* **179** E15
Shaghan, river, *Kaz.* **178** G7
Shāh Maqşūd, Kōh-e, *Afghan.* **180** L5
Shahbā' (Philippopolis), *Syr.* **172** K8
Shahbandar, *Pak.* **181** T8
Shahdadkot, *Pak.* **181** P8
Shahdadpur, *Pak.* **181** R8
Shahdol, *India* **182** K8
Shahgarh, *India* **182** G2
Shaḩḩāt (Cyrene), *Lib.* **201** D21
Shāhīn Shahr, *Iran* **177** H15
Shahjahanpur, *India* **182** G7
Shahpur Chakar, *Pak.* **181** R8
Shahpur, *Pak.* **180** K11
Shahpur, *India* **182** N5
Shahr-e Bābak, *Iran* **177** K18
Shahr-e Kord, *Iran* **177** H15
Shahrezā (Qomsheh), *Iran* **177** J16
Shahrig, *Pak.* **180** M8
Shahrisabz, *Uzb.* **178** M11
Shāhrūd, *Iran* **177** E18
Shaighalu, *Pak.* **180** L8
Shajapur, *India* **182** K5
Shakawe, *Botswana* **212** E6
Shakhtersk, *Russ.* **161** J21
Shakhty, *Russ.* **159** Q11
Shakhun'ya, *Russ.* **158** G13
Shaki, *Nigeria* **206** L6
Shaktoolik, *Alas., U.S.* **113** G13
Shala Hāyk', *Eth.* **211** N7
Shalaamboot, *Somalia* **211** R11
Shalakusha, *Russ.* **158** D10
Shali, *Russ.* **159** U14
Shallowater, *Tex., U.S.* **104** F12
Shalqar Köli, *Kaz.* **178** E5
Shalqar, *Kaz.* **178** G8
Shal'skiy, *Russ.* **158** D9
Shaluli Shan, *China* **184** L8
Shamattawa, *Man., Can.* **87** L15
Shambe, *S. Sudan* **203** W11
Shamokin, *Pa., U.S.* **96** N8
Shamrock, *Tex., U.S.* **105** D14
Shandan, *China* **184** G9
Shandur Pass, *Pak.* **180** G11
Shangani, river, *Zimb.* **212** E8
Shangdu, *China* **184** F12
Shanghai, *China* **185** K15
Shangombo, *Zambia* **209** P14
Shangqiu, *China* **185** J13
Shangrao, *China* **185** L14
Shaniko, *Oreg., U.S.* **110** F5
Shanklin, *Eng., U.K.* **145** W13
Shannon (Sionainn), *Ire.* **145** S4
Shannon Barajı, *Turk.* **170** H10
Shannon Ø, *Greenland, Den.* **88** G14

Shannon, River, *Ire.* **145** S3
Shantar Islands, *Asia* **168** D12
Shantarskiye Ostrova, *Russ.* **161** H20
Shantou (Swatow), *China* **185** N14
Shaoguan, *China* **185** N13
Shaowu, *China* **185** M14
Shaoxing, *China* **185** L15
Shaoyang, *China* **184** M12
Shap, *Eng., U.K.* **145** P11
Shapinsay, island, *Scot., U.K.* **144** E10
Shaqrā', *Saudi Arabia* **174** F10
Shaqrā', *Yemen* **174** P10
Shar, *Kaz.* **179** E16
Sharan, *Afghan.* **180** J8
Sharbaqty, *Kaz.* **179** D15
Sharbatāt, *Oman* **175** L16
Shari, *Japan* **188** F15
Sharjah *see* Ash Shāriqah, *U.A.E.* **175** F15
Shark Bay, *Austral.* **220** G1
Sharkan, *Russ.* **158** G16
Sharlawuk, *Turkm.* **178** M5
Sharm ash Shaykh, *Egypt* **202** H13
Sharm el Sheikh, *Egypt* **173** U4
Sharon Springs, *Kans., U.S.* **107** S3
Sharon, *Pa., U.S.* **96** M3
Sharpe, *S. Dak., U.S.* **106** L5
Sharqī, Al Jabal ash (Anti-Lebanon Range), *Syr.* **172** J7
Sharuhen, ruins, *Israel* **173** N4
Shar'ya, *Russ.* **158** G13
Shasha, *Eth.* **211** N6
Shashe, river, *Botswana* **212** G8
Shashemenē, *Eth.* **211** N7
Shashi, *China* **184** L12
Shasta Lake, *Calif., U.S.* **110** M4
Shasta, Mount, *Calif., U.S.* **110** M4
Shatskiy Rise, *Pac. Oc.* **246** F9
Shaţţ al 'Arab, river, *Iraq* **177** K13
Shatt el Jerid, lake, *Tunisia* **201** D15
Shattuck, *Okla., U.S.* **105** B14
Shaunavon, *Sask., Can.* **86** N11
Shaw, *Miss., U.S.* **102** J10
Shawano, *Wis., U.S.* **100** J6
Shawinigan, *Que., Can.* **87** P19
Shawnee, *Okla., U.S.* **105** C17
Shawnee, *Kans., U.S.* **107** S10
Shay Gap, *W. Aust., Austral.* **221** S3
Shaybārā, island, *Saudi Arabia* **174** F4
Shaykh 'Uthmān, *Yemen* **174** Q9
Shayman, *Taj.* **179** N15
Shchekino, *Russ.* **158** K9
Shchigry, *Russ.* **159** N9
Shchūchīnsk, *Kaz.* **178** D12
Shebekino, *Russ.* **159** N9
Shebelē, river, *Af.* **198** H11
Shebir, *Kaz.* **178** H4
Sheboygan Falls, *Wis., U.S.* **100** K7
Sheboygan, *Wis., U.S.* **100** K7
Shed Lakes, *Austral.* **220** H6
Sheelin, Lough, *Ire.* **145** Q6
Sheep Haven, *Ire.* **145** M5
Sheep Mountain, *Colo., U.S.* **108** L11
Sheep Range, *Nev., U.S.* **111** T12
Sheerness, *Eng., U.K.* **145** U15
Shefar'am, *Israel* **172** K5
Sheffield, *Ala., U.S.* **103** G14
Sheffield, *Tex., U.S.* **104** K12
Sheffield, *Eng., U.K.* **145** R12
Sheikhupura, *Pak.* **180** L13
Sheka Ridge, *Atl. Oc.* **249** V13
Shelbina, *Mo., U.S.* **107** R13
Shelburne Bay, *Austral.* **220** A12
Shelburne, *N.S., Can.* **87** Q22
Shelby, *N.C., U.S.* **98** J3
Shelby, *Mich., U.S.* **100** K9
Shelby, *Ohio, U.S.* **100** P13
Shelby, *Miss., U.S.* **102** J10
Shelby, *Mont., U.S.* **108** B7
Shelbyville, *Ill., U.S.* **101** S6
Shelbyville, *Ind., U.S.* **101** S9
Shelbyville, *Ky., U.S.* **103** B17
Shelbyville, *Tenn., U.S.* **103** F15
Sheldon Point, *Alas., U.S.* **112** H11
Sheldon, *Iowa, U.S.* **106** M9
Shelikhov Gulf, *Pac. Oc.* **168** C12
Shelikhova, *Russ.* **161** E21
Shelikof Strait, *Alas., U.S.* **113** L15
Shell Bank, *Ind. Oc.* **250** R11
Shell Beach, *Guyana* **131** D15
Shell Lake, *Wis., U.S.* **100** G2
Shelley, *Idaho, U.S.* **108** H7
Shellman, *Ga., U.S.* **99** P5
Shelter Cove, *Calif., U.S.* **111** N1
Shelton, *Nebr., U.S.* **107** Q6
Shelton, *Wash., U.S.* **110** C3

Shemonaīkha, *Kaz.* **179** D16
Shemya Island, *Alas., U.S.* **112** L2
Shenandoah National Park, *Va., U.S.* **98** D11
Shenandoah, *Pa., U.S.* **96** N8
Shenandoah, *Va., U.S.* **98** D11
Shenandoah, *Iowa, U.S.* **107** Q9
Shenandoah, river, *Va., U.S.* **98** D11
Shendam, *Nigeria* **206** L9
Shendi, *Sudan* **203** Q13
Shenge, *Sierra Leone* **204** N8
Shenkursk, *Russ.* **158** D11
Shenmu, *China* **184** G12
Shenton, Mount, *Austral.* **220** H5
Shenzha *see* Xainza, *China* **184** K5
Shenzhen, *China* **185** P13
Sheoganj, *India* **182** J3
Sheopur, *India* **182** H5
Shepard Island, *Antarctica* **238** N9
Shepetivka, *Ukr.* **159** N4
Shepherd Islands, *Vanuatu* **229** H16
Shepherd, *Ill., U.S.* **101** R3
Shepparton, *Vic. Austral.* **221** Y13
Sheppey, Isle of, *Eng., U.K.* **145** U15
Sherard Osborn Fjord, *Greenland, Den.* **88** B7
Sherbro Island, *Sierra Leone* **204** N8
Sherbrooke, *Que. Can.* **87** P20
Sherburn, *Minn., U.S.* **106** L10
Shergui Island, *Tunisia* **201** C16
Sheridan, *Ark., U.S.* **102** H8
Sheridan, *Wyo., U.S.* **108** F11
Sheridan, Cape, *Nunavut, Can.* **87** A16
Sheringham, *Eng., U.K.* **145** R15
Sherkaly, *Russ.* **160** G10
Sherkin Island, *Ire.* **145** U3
Sherlovaya Gora, *Russ.* **161** L17
Sherman Mills, *Me., U.S.* **97** D18
Sherman Peak, *U.S.* **90** E7
Sherman, *Tex., U.S.* **105** F18
Sherobod, *Uzb.* **178** N11
Sherridon, *Man., Can.* **87** L13
Sherwood Forest, *Eng., U.K.* **145** R13
Sherwood, *N. Dak., U.S.* **106** D4
Shetland Islands (Zetland), *Scot., U.K.* **144** B12
Shetpe, *Kaz.* **178** J4
Sheyenne, *N. Dak., U.S.* **106** F6
Sheyenne, river, *N. Dak., U.S.* **106** F7
Shiant Islands *see* Na h-Eileanan Mora, islands, *Scot., U.K.* **144** G7
Shiashkotan, island, *Russ.* **161** H23
Shibām, *Yemen* **174** N11
Shibar, Kōtal-e, *Afghan.* **180** H7
Shibata, *Japan* **188** M11
Shibecha, *Japan* **188** G15
Shibetsu, *Japan* **188** F15
Shibīn al Kawm, *Egypt* **202** F11
Shibirghān, *Afghan.* **180** F6
Shibushi Wan, *Japan* **189** T5
Shidler, *Okla., U.S.* **105** A17
Shiega, *Ghana* **205** L16
Shiel Bridge, *Scot., U.K.* **144** J8
Shiel, Loch, *Scot., U.K.* **144** K8
Shīeli, *Kaz.* **178** J11
Shiḥan, *Yemen* **175** M14
Shihezi, *China* **184** E5
Shiikh, *Somaliland* **210** L11
Shijiazhuang, *China* **185** H13
Shikarpur, *Pak.* **181** P8
Shikarpur, *India* **182** H9
Shikoku Basin, *Pac. Oc.* **246** F7
Shikoku, island, *Japan* **189** R7
Shikotan, island, *Russ.* **188** F16
Shilabo, *Eth.* **211** N11
Shildon, *Eng., U.K.* **145** N12
Shīli, *Kaz.* **178** E9
Shiliguri, *India* **182** H11
Shilik, *Kaz.* **179** J16
Shiliyn Bogd Uul, *China, Mongolia* **185** E13
Shilka, *Russ.* **161** L17
Shilla, peak, *India* **182** E6
Shillong, *India* **182** H13
Shiloh National Military Park, *Tenn., U.S.* **103** G13
Shilong, *China* **185** P13
Shilovo, *Russ.* **158** K11
Shima Hantō, *Japan* **189** Q10
Shimanovsk, *Russ.* **161** K19
Shimanto, *Japan* **189** R6
Shimbiris, peak, *Somaliland* **210** L13
Shimizu, *Japan* **188** G14
Shimoda, *Japan* **189** Q11

Shimoni, *Kenya* **211** V8
Shimono Shima, *Japan* **189** Q3
Shimonoseki, *Japan* **189** Q5
Shin, Loch, *Scot., U.K.* **144** G9
Shinapaaru, *N. Mariana Is., U.S.* **230** M9
Shināş, *Oman* **175** G16
Shīndand (Sabzawar), *Afghan.* **180** J3
Shingbwiyang, *Myanmar* **190** D5
Shingphel, *Bhutan* **182** G13
Shingū, *Japan* **189** R9
Shinjō, *Japan* **188** L12
Shinkafi, *Nigeria* **206** H7
Shīnkaī, *Afghan.* **180** K7
Shīnkay Ghar, range, *Afghan.* **180** K8
Shinnston, *W. Va., U.S.* **98** C9
Shinyanga, *Tanzania* **211** U4
Shiono Misaki, *Japan* **189** R9
Ship Island, *Miss., U.S.* **102** P12
Ship Rock, peak, *N. Mex., U.S.* **109** Q10
Shiping, *China* **184** N9
Shipki Pass, *India* **182** E7
Shiprock, *N. Mex., U.S.* **109** Q10
Shīr Kūh, *Iran* **177** J17
Shirakami Misaki, *Japan* **188** H12
Shirase Coast, *Antarctica* **238** M11
Shirase Glacier, *Antarctica* **239** C17
Shirati, *Tanzania* **211** T4
Shīrāz, *Iran* **177** L16
Shirbīn, *Egypt* **202** E11
Shire, river, *Malawi, Mozambique* **212** D11
Shireet, *Mongolia* **184** D12
Shiretoko Misaki, *Japan* **188** E15
Shiriya Zaki, *Japan* **188** H13
Shiroishi, *Japan* **188** M12
Shirone, *Japan* **188** M11
Shirshov Ridge, *Pac. Oc.* **252** G1
Shīrvān, *Iran* **177** D19
Shishaldin Volcano, *Alas., U.S.* **112** N11
Shīshīnda, *Eth.* **211** N6
Shishmaref, *Alas., U.S.* **112** E12
Shishou, *China* **184** L12
Shiv, *India* **182** H2
Shivamogga, *India* **183** Q5
Shivpuri, *India* **182** H6
Shivta, ruins, *Israel* **173** P4
Shivwits Plateau, *Ariz., U.S.* **109** R5
Shiwa Ngandu, *Zambia* **209** L20
Shiyan, *China* **184** K12
Shizuishan (Dawukou), *China* **184** G10
Shizunai, *Japan* **188** H13
Shizuoka, *Japan* **189** Q11
Shkīn, *Afghan.* **180** K9
Shkodër (Scutari), *Alban.* **154** K9
Shkumbin, river, *Alban.* **154** L9
Shmakov Escarpment, *Arctic Oc.* **252** H12
Shmidta, *Russ.* **161** C14
Shoalhaven Bight, *Austral.* **220** L14
Shoals, Isles of, *N.H., U.S.* **97** J15
Shoalwater Bay, *Austral.* **220** F15
Shoghot, *Pak.* **180** G11
Sholapur, *India* **182** N5
Shonga, *Nigeria* **206** K7
Shonian Harbor, *Palau* **226** J5
Shonzhy, *Kaz.* **179** J16
Shōr Tepah, *Afghan.* **180** F7
Shōrakīān, *Afghan.* **180** K5
Sho'rchi, *Uzb.* **178** N11
Shoreham-by-Sea, *Eng., U.K.* **145** V14
Shorkot, *Pak.* **180** L11
Shortland Islands, *Solomon Is.* **228** C6
Shortland, island, *Solomon Is.* **228** C5
Shosambetsu, *Japan* **188** E13
Shoshone Falls, *U.S.* **90** E6
Shoshone Lake, *Wyo., U.S.* **108** G8
Shoshone Mountain, *Nev., U.S.* **111** T10
Shoshone Mountains, *Nev., U.S.* **111** Q9
Shoshone Range, *Nev., U.S.* **111** N9
Shoshone, *Idaho, U.S.* **108** H5
Shoshone, *Calif., U.S.* **111** U10
Shoshone, river, *Wyo., U.S.* **108** F10
Shoshoni, *Wyo., U.S.* **108** H10
Shostka, *Ukr.* **158** M7
Shotor Khūn, pass, *Afghan.* **180** H5
Shoval, *Israel* **173** N5
Show Low, *Ariz., U.S.* **109** T9
Shpola, *Ukr.* **159** P6
Shreveport, *La., U.S.* **102** L6
Shroud Cay, *Bahamas* **118** E12

Shū, *Kaz.* **179** J14
Shū, river, *Kaz.* **179** J13
Shuangcheng, *China* **185** D16
Shubarqudyq, *Kaz.* **178** F7
Shugozero, *Russ.* **158** F8
Shuiding *see* Huocheng, *China* **184** D3
Shukpa Kunzang, *India* **180** H16
Shule, *China* **184** F2
Shule, river, *China* **184** G7
Shumagin Islands, *Alas., U.S.* **112** N12
Shumanay, *Uzb.* **178** K7
Shumen (Kolarovgrad), *Bulg.* **155** H16
Shumerlya, *Russ.* **158** J13
Shumshu, island, *Russ.* **161** G23
Shun'ga, *Russ.* **158** D8
Shungay, *Kaz.* **178** E3
Shungnak, *Alas., U.S.* **113** E14
Shuozhou, *China* **184** G12
Shupenzë, *Alban.* **154** K10
Shuqualak, *Miss., U.S.* **103** K13
Shūr Gaz, *Iran* **177** L21
Shurma, *Russ.* **158** H15
Shūsf, *Iran* **177** H21
Shūsh, *Iraq* **177** H13
Shūshtar, *Iraq* **177** J13
Shuwak, *Sudan* **203** R14
Shuya, *Russ.* **158** H11
Shuya, *Russ.* **158** D8
Shuyak Island, *Alas., U.S.* **113** L15
Shwebo, *Myanmar* **190** G5
Shwedaung, *Myanmar* **190** J4
Shwegu, *Myanmar* **190** F5
Shwegyin, *Myanmar* **190** K5
Shweli, river, *Myanmar* **190** F6
Shymkent, *Kaz.* **178** K12
Shyngghyrlau, *Kaz.* **178** D6
Shyok, *India* **182** D6
Shyok, river, *Pak.* **180** G14
Sia, *Indonesia* **193** M19
Siāhān Range, *Pak.* **181** Q4
Sialkot, *Pak.* **180** K13
Sialum, *P.N.G.* **222** P8
Siantan, island, *Indonesia* **192** H8
Siapa, river, *Venez.* **130** J7
Siargao, island, *Philippines* **193** E15
Siátista, *Gr.* **156** D9
Siau, island, *Indonesia* **193** H15
Siauliai, *Lith.* **143** M16
Siavonga, *Zambia* **209** Q18
Sibari, *It.* **153** P13
Sibay, *Russ.* **160** J8
Šibenik, *Croatia* **154** G6
Siberia, region, *Asia* **168** D8
Siberut, island, *Indonesia* **192** J5
Siberut, Selat, *Indonesia* **192** J5
Sibi, *Pak.* **181** N8
Sibidiro, *P.N.G.* **222** Q6
Sibigo, *Indonesia* **192** H4
Sibiti, *Congo* **208** G9
Sibiu, *Rom.* **155** D14
Sibley, *Iowa, U.S.* **106** M9
Siboa, *Indonesia* **193** J13
Sibolga, *Indonesia* **192** H5
Sibsagar, *India* **182** G15
Sibu, *Malaysia* **192** H10
Sibuco, *Philippines* **193** F14
Sibut, *Cen. Af. Rep.* **207** M14
Sibutu Passage, *Philippines* **193** G13
Sibuyan Sea, *Philippines* **193** D14
Sibuyan, island, *Philippines* **193** D14
Sicasica, *Bol.* **132** K6
Sichon, *Thai.* **191** Q7
Sichuan Basin, *Asia* **168** H10
Sicié, Cap, *Fr.* **149** Y14
Sicilia (Sicily), island, *It.* **153** S11
Sicily *see* Sicilia, island, *It.* **153** S11
Sicily, Strait of, *It.* **153** T7
Sicuani, *Peru* **132** J5
Sidaouet, *Niger* **206** E8
Sidári, *Gr.* **156** E6
Sidas, *Indonesia* **192** J9
Siddhapur, *India* **182** J3
Siddipet, *India* **182** N6
Side, ruins, *Turk.* **170** J6
Sideia, island, *P.N.G.* **222** R9
Síderos, Ákra, *Gr.* **157** Q17
Sidi Aïssa, *Alg.* **201** C13
Sīdī Barrānī, *Egypt* **202** E8
Sidi Bel Abbès, *Alg.* **200** C11
Sidi Bennour, *Mor.* **200** D7
Sidi Bouzid, *Tunisia* **201** C15
Sidi Ifni, *Mor.* **200** F6
Sidi Kacem, *Mor.* **200** C8

Sidi Lakhdar, *Alg.* **200** B11
Sīdī 'Umar, *Egypt* **202** E8
Sidirókastro, *Gr.* **156** B12
Sidlaw Hills, *Scot., U.K.* **144** K10
Sidmouth, *Eng. U.K.* **145** W10
Sidmouth, Cape, *Austral.* **220** B13
Sidney Lanier, Lake, *Ga., U.S.* **98** K7
Sidney, *N.Y., U.S.* **96** K9
Sidney, *Ohio, U.S.* **101** Q11
Sidney, *Iowa, U.S.* **107** Q9
Sidney, *Nebr., U.S.* **107** P2
Sidney, *Mont., U.S.* **108** C13
Sido, *Mali* **204** L12
Sidon *see* Saïda, *Leb.* **172** J6
Sidorovsk, *Russ.* **160** G12
Sidra *see* Surt, *Lib.* **201** E19
Siedlce, *Pol.* **151** D17
Siegen, *Ger.* **150** F6
Siem Reap *see* Siĕmréab, *Cambodia* **191** N10
Siemiatycze, *Pol.* **151** D17
Siĕmpang, *Cambodia* **190** M11
Siempurgo, *Côte d'Ivoire* **204** M12
Siĕmréab (Siem Reap), *Cambodia* **191** N10
Siena, *It.* **152** H7
Sieradz, *Pol.* **151** E14
Sierpc, *Pol.* **151** D15
Sierra Blanca, *Tex., U.S.* **104** J8
Sierra Blanca, peak, *Tex., U.S.* **104** J8
Sierra Blanca Peak, *N. Mex., U.S.* **109** U13
Sierra Colorada, *Arg.* **135** P8
Sierra Leone, *Af.* **204** N9
Sierra Leone Basin, *Atl. Oc.* **249** N10
Sierra Leone Fracture Zone, *Atl. Oc.* **249** N8
Sierra Leone Rise, *Atl. Oc.* **249** N10
Sierra Madre del Sur, *Mex.* **116** K11
Sierra Madre Occidental, *Mex.* **116** E9
Sierra Madre Oriental, *Mex.* **116** F11
Sierra Maestra, *Cuba* **118** K12
Sierra Mojada, *Mex.* **116** E10
Sierra Nevada, *Calif., U.S.* **111** S7
Sierra Vista, *Ariz., U.S.* **109** W8
Sierre, *Switz.* **152** C2
Sífnos (Síphnos), island, *Gr.* **157** L14
Sigatoka, *Fiji* **226** Q4
Sigean, *Fr.* **149** Y10
Siġġiewi, *Malta* **163** L22
Sighetul Marmatiei, *Rom.* **155** A13
Sighişoara, *Rom.* **155** D14
Sighnaghi, *Rep. of Georgia* **171** C19
Siglufjörður, *Ice.* **142** D3
Signal Hill, *Gibraltar, U.K.* **163** P23
Signal Hill, *Austral.* **220** E11
Signal Peak, *Ariz., U.S.* **109** U5
Signy Station, *Antarctica* **238** A5
Sigourney, *Iowa, U.S.* **107** P13
Sígri, *Gr.* **157** F16
Sigüenza, *Sp.* **146** E12
Siguer, Port de, *Andorra, Fr.* **162** G3
Siguiri, *Guinea* **204** L11
Sihanoukville *see* Preăh Seihânŭ, *Cambodia* **191** P10
Siheung, *S. Korea* **187** N8
Sihora, *India* **182** K7
Siirt, *Turk.* **171** G16
Siis, island, *F.S.M.* **226** F9
Sijung, *N. Korea* **186** E7
Sikar, *India* **182** G5
Sikasso, *Mali* **205** L13
Sikeston, *Mo., U.S.* **107** V16
Sikhote Alin' Range, *Asia* **168** E13
Siki Rangi Passage, *N.Z.* **229** K20
Síkinos, *Gr.* **157** M15
Síkinos, island, *Gr.* **157** M15
Sikonge, *Tanzania* **211** V4
Sikuijivittip Apusîa, *Greenland, Den.* **89** S8
Sil, river, *Sp.* **146** C7
Sila Fracture Zone, *Pac. Oc.* **247** C14
Silat, *Indonesia* **192** J10
Silba, island, *Croatia* **154** F4
Silchar, *India* **182** J14
Side, ruins, *Turk.* **170** J6
Silesia, region, *Pol.* **151** F13
Silet, *Alg.* **201** K13
Siletitengi, Lake, *Asia* **168** E7

Silgarhi, *Nepal* **182** F8
Silhouette, island, *Seychelles* **215** N19
Silifke, *Turk.* **170** J9
Silistra, *Bulg.* **155** G17
Siljan, lake, *Sw.* **142** J12
Silkeborg, *Den.* **142** M10
Silla, *Sp.* **147** H14
Silli, *Burkina Faso* **205** L13
Siloam Springs, *Ark., U.S.* **102** E6
Silogui, *Indonesia* **192** J5
Silopi, *Turk.* **171** H17
Silsbee, *Tex., U.S.* **105** K21
Silsilah-ye Band-e Turkistān, range, *Afghan.* **180** G4
Silsilah-ye Safēd Kōh (Paropamisus Range), *Afghan.* **180** H3
Silum, *Liech.* **162** P3
Šilutė, *Lith.* **143** M15
Silvan, *Turk.* **171** G15
Silvânia, *Braz.* **133** J13
Silvassa, *India* **182** L3
Silver Bay, *Minn., U.S.* **106** F13
Silver Bell, *Ariz., U.S.* **109** V7
Silver City, *Mich., U.S.* **100** C15
Silver City, *Idaho, U.S.* **108** H3
Silver City, *N. Mex., U.S.* **109** U10
Silver City, *Nev., U.S.* **111** Q6
Silver Creek, *Oreg., U.S.* **110** H6
Silver Hill, *Montserrat, U.K.* **123** A24
Silver Lake, *Oreg., U.S.* **110** J5
Silver Peak, *Nev., U.S.* **111** S9
Silver Plain, *Atl. Oc.* **248** L4
Silver Spring, *Md., U.S.* **98** C13
Silverton, *Tex., U.S.* **104** E12
Silverton, *Colo., U.S.* **109** P11
Silverton, *Oreg., U.S.* **110** F3
Silves, *Braz.* **132** D10
Silvies, river, *Oreg., U.S.* **110** H7
Sima, *Comoros* **215** N16
Simansih, *Indonesia* **192** K5
Simav, *Turk.* **170** F4
Simav, river, *Turk.* **170** E4
Simba, *D.R.C.* **209** D15
Simberi, island, *P.N.G.* **222** M10
Simbo, island, *Solomon Is.* **228** E6
Simcoe, Lake, *Ont., Can.* **87** Q18
Simdega, *India* **182** K9
Simeria, *Rom.* **154** D12
Simeulue, island, *Indonesia* **192** H4
Simferopol', *Russ.* **159** T7
Simikot, *Nepal* **182** F8
Simiyu, river, *Tanzania* **211** U4
Simla, *Colo., U.S.* **108** M14
Simla, *India* **182** E6
Şimleu Silvaniei, *Rom.* **154** B12
Simms Point, *Bahamas* **122** E1
Simon's Point, *Bahamas* **122** E9
Simontornya, *Hung.* **151** M14
Simpang, *Indonesia* **192** J7
Simpson Bay Lagoon, *St. Martin, Fr.* **123** B14
Simpson Bay Village (Simsonbaai), *St. Maarten, Neth.* **123** B14
Simpson Bay, *St. Maarten, Neth.* **123** B14
Simpson Desert, *Austral.* **220** G10
Simpson Peninsula, *Nunavut, Can.* **87** F15
Simrishamn, *Sw.* **143** N13
Simsonbaai *see* Simpson Bay Village, *St. Maarten, Neth.* **123** B14
Simushir, island, *Russ.* **161** H23
Sin Cowe Island, *S. China Sea* **192** E11
Sin Hồ, *Vietnam* **190** Q9
Sina Dhaqa, *Somalia* **211** P12
Sinai, *Egypt* **202** F13
Sinai, peninsula, *Egypt* **173** R3
Sīnāwin, *Lib.* **201** E16
Sinazongwe, *Zambia* **209** Q17
Sinbaungwe, *Myanmar* **190** J4
Sinbukch'ŏng, *N. Korea* **186** G10
Sincelejo, *Col.* **130** D5
Sinch'ang, *N. Korea* **186** G11
Sinch'ang, *N. Korea* **186** H7
Sinchang, *S. Korea* **187** W6
Sincheon, *S. Korea* **186** M9
Sinch'ŏn, *N. Korea* **186** L5
Sinclair, *Wyo., U.S.* **108** J11
Sinclair, Lake, *Ga., U.S.* **98** M6
Sinclair's Bay, *Scot., U.K.* **144** F10
Sindara, *Gabon* **208** E8
Sindeni, *Tanzania* **211** V7
Sindèr, *Niger* **206** G5
Sindi, *Sudan* **203** R8

Sindo, *N. Korea* **186** H3
Sindong, *N. Korea* **186** E12
Sindong, *S. Korea* **187** N11
Sinegorskiy, *Russ.* **159** Q11
Sineriarsua, *Greenland, Den.* **88** F3
Sines, *Port.* **146** K5
Sines, Cabo de, *Port.* **146** K5
Sinfra, *Côte d'Ivoire* **205** P13
Sing Buri, *Thai.* **190** M8
Singa, *Sudan* **203** S13
Singapore Strait, *Asia* **192** H7
Singapore, *Asia* **191** V10
Singaraja, *Indonesia* **192** M11
Singen, *Ger.* **150** K6
Singida, *Tanzania* **211** V5
Singkaling Hkämti, *Myanmar* **190** E5
Singkawang, *Indonesia* **192** H9
Singkep, island, *Indonesia* **192** J7
Singleton, *N.S.W., Austral.* **221** X15
Sin'gye, *N. Korea* **186** K7
Sinheukdong, *S. Korea* **187** Q7
Sinhŭng, *N. Korea* **186** E9
Sinhŭng, *N. Korea* **186** G9
Sinhyeon, *S. Korea* **187** R12
Sini Vrŭkh, peak, *Bulg.* **155** K15
Siniscola, *It.* **152** M5
Sinj, *Croatia* **154** G6
Sinjai, *Indonesia* **193** L13
Sinjär, *Iraq* **176** E9
Sinjido, island, *S. Korea* **187** U8
Sinjŏn, *N. Korea* **186** E9
Sinjŏng, *N. Korea* **186** G4
Sinkiang, region, *China* **184** F4
Sinmi-do, island, *N. Korea* **186** H5
Sinnam, *S. Korea* **186** M10
Sinnamary, *Fr. Guiana, Fr.* **131** F19
Sinnar, *India* **182** M4
Sînnicolau Mare, *Rom.* **154** D10
Sinó, Ponto do, *Cabo Verde* **215** B17
Sinoe Bay, *Liberia* **204** Q11
Sinoe, Lacul, *Rom.* **155** F18
Sinop (Sinope), *Turk.* **170** C10
Sinop Burnu, *Turk.* **170** B10
Sinop, *Braz.* **132** H11
Sinope *see* Sinop, *Turk.* **170** C10
Sinp'a *see* Kimjŏngsuk, *N. Korea* **186** D9
Sinp'o, *N. Korea* **186** G10
Sinp'ung, *N. Korea* **186** D7
Sinp'yŏng, *N. Korea* **186** K8
Sinsaeng, *N. Korea* **186** M6
Sinsang, *N. Korea* **186** H9
Sinsheim, *Ger.* **150** H6
Sinsido, island, *S. Korea* **187** R7
Sinsŏngch'ŏn, *N. Korea* **186** J7
Sint Maarten, island, *Lesser Antil.* **121** B14
Sint Nicolaas, *Aruba, Neth.* **123** R17
Sintaein, *S. Korea* **187** R8
Sintang, *Indonesia* **192** J10
Sinton, *Tex., U.S.* **105** N17
Sinŭiju, *N. Korea* **186** G4
Sinujiif, *Somalia* **210** M14
Sinwŏn, *N. Korea* **186** E8
Sinwŏn, *N. Korea* **186** L6
Sinzig, *Ger.* **150** G5
Siófok, *Hung.* **151** M14
Sioma, *Zambia* **209** Q15
Sion, *Switz.* **152** C2
Sionainn *see* Shannon, *Ire.* **145** S4
Siorapaluk, *Greenland, Den.* **88** E3
Siorapaluup Kangerlua (Robertson Fjord), *Greenland, Den.* **88** E3
Sioux Center, *Iowa, U.S.* **106** M9
Sioux City, *Iowa, U.S.* **107** N8
Sioux Falls, *S. Dak., U.S.* **106** L8
Sioux Lookout, *Ont., Can.* **87** N15
Sioux Rapids, *Iowa, U.S.* **106** M10
Sipapo, Cerro, *Venez.* **130** F10
Siparia, *Trin. & Tobago* **123** R22
Síphnos *see* Sífnos, island, *Gr.* **157** L14
Siping, *China* **185** E15
Sipiwesk, *Man., Can.* **87** L14
Siple Coast, *Antarctica* **238** M11
Siple Dome Field Camp, *Antarctica* **238** L11
Siple Island, *Antarctica* **238** M7
Siple, Mount, *Antarctica* **238** M7
Sipura, island, *Indonesia* **192** K5
Siqueires Fracture Zone, *Pac. Oc.* **247** J19
Sir Edward Pellew Group, *Austral.* **220** C10
Sir Francis Drake Channel, *Virgin Is., U.K.* **122** R6

Sir Gaerfyrddin *see* Carmarthen, *Wales, U.K.* **145** U9
Sir Thomas, Mount, *Austral.* **220** G7
Sir Timothy's Hill, *St. Kitts & Nevis* **123** B18
Sira, *India* **183** Q5
Siracusa (Syracuse), *It.* **153** T12
Şiran, *Turk.* **171** E14
Sirdaryo, *Uzb.* **178** L12
Siret, river, *Rom.* **155** E17
Sirhän, Wädī as, *Jordan* **173** N9
Sīrīk, *Iran* **177** N19
Sirik, *Turk.* **171** E22
Şirvan, *Azerb.* **171** E22
Şirvan, *Turk.* **171** G16
Sisak, *Croatia* **154** E6
Sisaket, *Thai.* **190** M10
Sisi, *Chad* **207** K14
Sisian, *Arm.* **171** E20
Sisimiut (Holsteinsborg), *Greenland, Den.* **89** N5
Sisŏphôn, *Cambodia* **190** M9
Sisseton, *S. Dak., U.S.* **106** J8
Sissonne, *Fr.* **149** N11
Sīstān, *Iran* **177** J22
Sister Bay, *Wis., U.S.* **100** H8
Sisteron, *Fr.* **149** W14
Sisters, *Oreg., U.S.* **110** H4
Sistersville, *W. Va., U.S.* **98** C9
Sistine Chapel, *Vatican City* **163** P16
Sitapur, *India* **182** G7
Siteía, *Gr.* **157** Q16
Sithonías, peninsula, *Gr.* **156** D12
Sitges, *Sp.* **147** E17
Sitian, *China* **184** F7
Sítio da Abadia, *Braz.* **133** J14
Sitka National Historical Park, *Alas., U.S.* **113** L22
Sitka, *Alas., U.S.* **113** M22
Sittard, *Neth.* **148** L13
Sittwe (Akyab), *Myanmar* **190** H3
Si'ufa'alele Point, *American Samoa, U.S.* **230** C10
Si'ufaga, *American Samoa, U.S.* **230** B10
Si'ulagi Point, *American Samoa, U.S.* **230** B10
Siuri, *India* **182** J11
Siuslaw, river, *Oreg., U.S.* **110** H2
Sivac, *Serb.* **154** E9
Sivas, *Turk.* **170** E12
Siverek, *Turk.* **171** H14
Sivers'kyy Donets', river, *Ukr.* **159** P9
Sivrihisar, *Turk.* **170** E7
Sīwah, *Egypt* **202** G8
Siwalik Range, *India* **182** D5
Siwan, *India* **182** H9
Six Feet, cape, *Pitcairn I.* **228** Q6
Six Men's Bay, *Barbados* **123** J18
Six Shilling Cay, *Bahamas* **122** E4
Sixaola, *Pan.* **117** N20
Siyǝzǝn, *Azerb.* **171** C22
Sındırgı, *Turk.* **170** E4
Şırnak, *Turk.* **171** H17
Sjælland, island, *Den.* **142** N12
Sjenica, *Serb.* **154** H10
Skadarsko Jezero (Scutari, Lake), *Alban., Montenegro* **154** J9
Skærfjorden, bay, *Greenland, Den.* **88** E13
Skagafjörður, bay, *Ice.* **142** E3
Skagen, *Den.* **142** L11
Skagerrak, strait, *Den., Nor.* **142** M10
Skagit, river, *Wash., U.S.* **110** B4
Skagway, *Alas., U.S.* **113** K22
Skála, *Gr.* **156** L10
Skalistyy Golets, peak, *Russ.* **161** K17
Skåne, *Eur.* **140** F7
Skansen, *Greenland, Den.* **88** M6
Skántzoura, island, *Gr.* **157** G13
Škarda, island, *Croatia* **154** F4
Skardu, *Pak.* **180** G14
Skarżysko-Kamienna, *Pol.* **151** F16
Skawina, *Pol.* **151** H15
Skaymat, *W. Sahara* **200** H3
Skeena, river, *B.C., Can.* **86** J8
Skegness, *Eng., U.K.* **145** R14
Skeiðarárdjúp, valley, *Atl. Oc.* **253** L22

Skeleton Coast, *Namibia* **212** F2
Skellefteå, *Sw.* **143** F14
Skellefteälven, river, *Sw.* **143** F14
Skerjadjúp, valley, *Arctic Oc.* **253** M21
Skerries (Na Sceirí), *Ire.* **145** Q7
Skerryvore, island, *Scot., U.K.* **144** K6
Skeylley Maayl *see* Kirk Michael, *Isle of Man, U.K.* **145** P9
Skíathos, *Gr.* **156** F12
Skíathos, island, *Gr.* **156** F12
Skiatook, *Okla., U.S.* **105** B18
Skidmore, *Tex., U.S.* **105** N17
Skien, *Nor.* **142** M11
Skierniewice, *Pol.* **151** E15
Skikda, *Alg.* **201** B14
Skipton, *Eng., U.K.* **145** Q11
Skírmish Point, *Austral.* **220** A9
Skíros *see* Skýros, island, *Gr.* **157** G13
Skive, *Den.* **142** M11
Skjálfandadjúp, valley, *Arctic Oc.* **253** L20
Skjoldungen Bank, *Atl. Oc.* **253** Q20
Skjoldungen Valley, *Atl. Oc.* **253** Q20
Sklinna Bank, *Atl. Oc.* **248** E13
Skokholm Island, *Wales, U.K.* **145** U8
Skokie, *Ill., U.S.* **101** N7
Skolpen Bank, *Arctic Oc.* **253** D18
Skomer Island, *Wales, U.K.* **145** U8
Skomvær, island, *Nor.* **142** E12
Skópelos, *Gr.* **156** G12
Skópelos, island, *Gr.* **156** G12
Skopin, *Russ.* **158** L10
Skopje, *N. Maced.* **154** K11
Skopunarfjørður, bay, *Faroe Is., Den.* **142** H6
Skosyrev Ridge, *Arctic Oc.* **252** K8
Skoudeneshavn, *Nor.* **142** K9
Skovfjorden *see* Nardlunak, *Greenland, Den.* **89** T7
Skovfjorden, *Greenland, Den.* **89** T7
Skovorodino, *Russ.* **161** K18
Skowhegan, *Me., U.S.* **97** F16
Skrælling Canyon, *Atl. Oc.* **253** P20
Skull (An Scoil), *Ire.* **145** U3
Skunk, river, *Iowa, U.S.* **107** P13
Skvyra, *Ukr.* **159** P5
Sky Blu Camp, *Antarctica* **238** G8
Skýdra, *Gr.* **156** C10
Skye, island, *Scot., U.K.* **144** J7
Skykomish, *Wash., U.S.* **110** C5
Skýros (Skíros), island, *Gr.* **157** G13
Skýros, *Gr.* **157** G14
Skytrain Ice Rise, *Antarctica* **238** G8
Slantsy, *Russ.* **158** F5
Slater, *Mo., U.S.* **107** S12
Slatina, *Rom.* **155** F14
Slaton, *Tex., U.S.* **104** F12
Slave Coast, *Af.* **205** P17
Slavgorod, *Belarus* **158** L6
Slavonia, region, *Croatia* **154** E7
Slavonski Brod, *Croatia* **154** E8
Slavuta, *Ukr.* **159** N4
Sławno, *Pol.* **150** B12
Slayton, *Minn., U.S.* **106** L9
Sleaford Bay, *Austral.* **220** L9
Sleat, Sound of, *Scot., U.K.* **144** J7
Sleeper Islands, *Nunavut, Can.* **87** K17
Sleeping Bear Dunes National Lakeshore, *Mich., U.S.* **100** H9
Sleepy Eye, *Minn., U.S.* **106** K10
Sleetmute, *Alas., U.S.* **113** J14
Slêmanî (As Sulaymānīyah), *Iraq* **176** E11
Slessor Glacier, *Antarctica* **238** E11
Slide Mountain, *N.Y., U.S.* **96** L10
Slidell, *La., U.S.* **102** P11
Sliema, *Malta* **163** K23
Slieve Aughty Mountains, *Ire.* **145** R4
Slieve Gamph (The Ox Mountains), *Ire.* **145** P4
Sligeach *see* Sligo, *Ire.* **145** P4
Sligo (Sligeach), *Ire.* **145** P4
Sligo Bay, *Ire.* **145** P4
Slippery Rock, *Pa., U.S.* **96** M3
Slite, *Sw.* **143** L14
Sliven, *Bulg.* **155** J16
Sljeme, peak, *Croatia* **154** D6
Sloan, *Nev., U.S.* **111** U12
Slobodskoy, *Russ.* **158** F14
Slobozia, *Rom.* **155** F17
Slonim, *Belarus* **158** L3

Slough, *Eng., U.K.* **145** U13
Slovakia, *Eur.* **151** J15
Slovenia, *Eur.* **154** D4
Slov''yans'k, *Ukr.* **159** P9
Słupsk, *Pol.* **151** B13
Slutsk, *Belarus* **158** L4
Slyudyanka, *Russ.* **161** L15
Smackover, *Ark., U.S.* **102** J8
Smallan, *Pak.* **180** M8
Smallwood Reservoir, *Nfld. & Lab., Can.* **87** L21
Smalyavichy, *Belarus* **158** K5
Smara, *W. Sahara* **200** G5
Smederevo, *Serb.* **154** F10
Smethport, *Pa., U.S.* **96** L5
Smidovich, *Russ.* **160** D11
Smidovich, *Russ.* **161** K20
Smila, *Ukr.* **159** P6
Smiltene, *Latv.* **143** L17
Smith Arm, *N.W.T., Can.* **86** F10
Smith Bay, *Nunavut, Can.* **87** C16
Smith Bay, *Alas., U.S.* **113** B15
Smith Center, *Kans., U.S.* **107** R6
Smith Glacier, *Antarctica* **238** L7
Smith Inlet, *Antarctica* **238** E7
Smith Island *see* Sumisu, *Japan* **189** T12
Smith Island, *Nunavut, Can.* **87** J17
Smith Mountain Lake, *Va., U.S.* **98** F10
Smith Peninsula, *Antarctica* **238** F8
Smith River, *B.C., Can.* **86** H9
Smith River, *Calif., U.S.* **110** L1
Smith Sound, *Nunavut, Can.* **87** C16
Smith Sound, *Greenland, Den.* **88** D3
Smith, river, *Mont., U.S.* **108** C8
Smithers, *B.C., Can.* **86** K8
Smithfield, *N.C., U.S.* **98** H12
Smithfield, *Va., U.S.* **98** F14
Smithfield, *Utah, U.S.* **108** J7
Smithton, *Tas., Austral.* **221** Y7
Smithton, *Tex., U.S.* **105** L17
Smithville, *Tex., U.S.* **105** L17
Smoke Creek Desert, *Nev., U.S.* **111** N6
Smokey Hill, river, *U.S.* **90** G11
Smoky Bay, *S. Aust., Austral.* **221** W9
Smoky Hill, river, *Kans., U.S.* **107** T6
Smoky Hills, *Kans., U.S.* **107** S6
Smoky Mountains, *Idaho, U.S.* **108** G5
Smoky, river, *Alta., Can.* **86** J11
Smøla, island, *Nor.* **142** H10
Smolensk-Moscow Upland, *Eur.* **140** F11
Smolensk, *Russ.* **158** K7
Smólikas, peak, *Gr.* **156** D8
Smolyan, *Bulg.* **155** K14
Smyley Island, *Antarctica* **238** G7
Smyrna *see* İzmir, *Turk.* **170** F3
Smyrna, *Del., U.S.* **98** C14
Smyrna, *Ga., U.S.* **98** L5
Smyrna, *Tenn., U.S.* **103** E15
Snaefell (Sniaul), peak, *Isle of Man, U.K.* **145** P9
Snake Point, *Anguilla, U.K.* **122** Q12
Snake Range, *Nev., U.S.* **111** Q13
Snake River Plain, *Idaho, U.S.* **108** H6
Snake, river, *Yukon, Can.* **86** E9
Snake, river, *U.S.* **90** B5
Snare Islands (Tini Heke), *N.Z.* **223** R15
Snares Islands (Tini Heke), *N.Z.* **224** Q7
Snåsa, *Nor.* **142** G11
Sneads, *Fla., U.S.* **99** Q4
Sneek, *Neth.* **148** G13
Snezhnogorsk, *Russ.* **161** G13
Snežka, peak, *Czechia, Pol.* **150** G12
Sniaul *see* Snaefell, peak, *Isle of Man, U.K.* **145** P9
Snihurivka, *Ukr.* **159** R7
Snina, *Slovakia* **151** J17
Snøhetta, peak, *Nor.* **142** H11
Snohomish, *Wash., U.S.* **110** B4
Snoqualmie Pass, *Wash., U.S.* **110** C5
Snoqualmie, *Wash., U.S.* **110** C4
Snow Hill Island, *Antarctica* **238** C5
Snow Hill, *Md., U.S.* **98** D15
Snowden Seamounts, *Pac. Oc.* **247** H14
Snowdon, peak, *Wales, U.K.* **145** R9
Snowflake, *Ariz., U.S.* **109** T9
Snowville, *Utah, U.S.* **108** J6
Snowy Mountain, *N.Y., U.S.* **96** H10
Snowy Mountains, *Austral.* **220** L13
Snowy, river, *Austral.* **220** M13
Snug Corner, *Bahamas* **119** H15

Snuŏl, *Cambodia* **191** N11
Snyder, *Okla., U.S.* **105** D15
Snyder, *Tex., U.S.* **105** G13
Soala *see* Sokolo, *Mali* **204** J12
Soalala, *Madagascar* **213** D15
Soando, island, *S. Korea* **187** V7
Soanierana-Ivongo, *Madagascar* **213** E17
Soap Lake, *Wash., U.S.* **110** C7
Soba, *Nigeria* **206** J8
Sobaek-san, peak, *N. Korea* **186** G8
Sobaeksan, peak, *S. Korea* **187** P11
Sobaeksanmaek, range, *S. Korea* **187** Q10
Sobat, river, *S. Sudan* **203** V12
Sobinka, *Russ.* **158** J10
Sobolevo, *Russ.* **158** L16
Sobradinho, Represa de, *Braz.* **133** G15
Sobral, *Braz.* **133** E16
Sobuk, *N. Korea* **186** E8
Sóc Trăng, *Vietnam* **191** Q11
Soča, river, *Slov.* **154** C4
Sochaczew, *Pol.* **151** E15
Socheongdo, island, *S. Korea* **186** M4
Sochi, *Russ.* **159** U11
Society Islands, *Fr. Polynesia, Fr.* **231** B15
Societyridge, *Pac. Oc.* **247** L14
Socompa, *Chile* **134** E8
Socorro, *Tex., U.S.* **104** H7
Socorro, *N. Mex., U.S.* **109** T12
Socorro, *Col.* **130** F7
Socorro, Isla, *Mex.* **116** H6
Socotra *see* Suquţrá, *Yemen* **175** Q15
Soda Lake, *Calif., U.S.* **111** V11
Soda Springs, *Idaho, U.S.* **108** H7
Sodankylä, *Fin.* **143** D15
Soddy-Daisy, *Tenn., U.S.* **103** G16
Södertälje, *Sw.* **143** K14
Sodiri, *Sudan* **203** R10
Sodo, *Eth.* **211** N7
Sŏdu, *N. Korea* **186** C11
Soekmekaar *see* Morebeng, *S. Af.* **212** H9
Sofádes, *Gr.* **156** F10
Sofara, *Mali* **205** J13
Sofi, *Sudan* **203** R14
Sofia, *Bulg.* **155** J13
Sofia, river, *Madagascar* **213** D16
Sõfu Gan (Lot's Wife), island, *Japan* **189** U12
Soga, *Tanzania* **211** W7
Sogamoso, *Col.* **130** F7
Soğanlı, river, *Turk.* **170** C8
Sogata, *Eth.* **211** P5
Sogeomundo, island, *S. Korea* **187** U9
Sogeri, *P.N.G.* **222** Q8
Sogmai, *China* **184** J2
Sognefjorden, bay, *Nor.* **142** J10
Sogra, *Russ.* **158** C13
Söğüt Gölü, *Turk.* **170** H5
Söğüt, *Turk.* **170** E6
Sogzhukug (Suozhuku), *China* **184** K6
Sõhae, *N. Korea* **186** K5
Sohano, *P.N.G.* **222** P11
Sohano, *P.N.G.* **228** B4
Sohm Plain, *Atl. Oc.* **248** K5
Sŏhŭng-ho, lake, *N. Korea* **186** L6
Sŏhŭng, *N. Korea* **186** L7
Soignies, *Belg.* **148** L11
Soissons, *Fr.* **149** N11
Sojat, *India* **182** H4
Sõjosŏn-man, bay, *N. Korea* **186** J5
Sŏk-do, island, *N. Korea* **186** K5
Sokcho, *S. Korea* **186** L11
Sŏke, *Turk.* **170** G3
Sokhumi (Sukhum), *Rep. of Georgia* **171** A15
Sokodé, *Togo* **205** M17
Sokol, *Russ.* **158** F10
Sokol, *Russ.* **161** E16
Sokol, *Russ.* **161** F20
Sokółka, *Pol.* **151** C18
Sokolo (Soala), *Mali* **204** J12
Sokolov (Falknov), *Czechia* **150** G9
Sokołów Podlaski, *Pol.* **151** D17
Sokoto, *Nigeria* **206** H7
Sokoto, river, *Nigeria* **206** J6
Soksa, *N. Korea* **186** F9
Sŏksa, *N. Korea* **186** L6
Sol de Julio, *Arg.* **134** H10
Sol, Costa del, *Sp.* **146** N10
Sol, Ponta do, *Cabo Verde* **215** C17

Sola, *Cuba* **118** J11
Solander Island (Hautere), *N.Z.* **223** P15
Solano, Punta, *Col.* **130** F4
Soldeu, *Andorra* **162** H4
Soldotna, *Alas.* **113** K16
Soledad, *Calif., U.S.* **111** U4
Soledad, *Col.* **130** C6
Solenzara, *Fr.* **149** Y3
Solhan, *Turk.* **171** F15
Soligalich, *Russ.* **158** F11
Solikamsk, *Russ.* **160** G9
Solimões *see* Amazon, river, *S. Amer.* **128** D5
Solingen, *Ger.* **150** F5
Solitary Islands, *Austral.* **220** J16
Şollar, *Azerb.* **171** C22
Sollefteå, *Sw.* **143** G13
Sol'Iletsk, *Russ.* **160** J7
Sollum, Gulf of, *Af.* **198** D8
Solnechnogorsk, *Russ.* **158** J9
Solo, island, *Fiji* **226** Q5
Sologne, *Eur.* **140** H5
Solok, *Indonesia* **192** J6
Sololo, *Kenya* **211** Q7
Solomon, *Kans., U.S.* **107** S7
Solomon, river, *Kans., U.S.* **107** S7
Solomon Basin, *Pac. Oc.* **246** K8
Solomon Islands, *Pac. Oc.* **224** H6
Solomon Sea, *P.N.G.* **222** Q10
Solon Springs, *Wis., U.S.* **100** F3
Solor, island, *Indonesia* **193** M14
Solothurn, *Switz.* **152** A3
Solovetskiye Ostrova, island, *Russ.* **158** B8
Solov'yevsk, *Russ.* **161** K18
Solsona, *Sp.* **147** D16
Šolta, island, *Croatia* **154** H6
Solţānābād, *Iran* **177** E20
Sol'tsy, *Russ.* **158** G6
Soluntum, ruin, *It.* **153** S10
Solvay, *N.Y., U.S.* **96** J8
Sol'vychegodsk, *Russ.* **158** D13
Solway Firth, *Eng., Scot., U.K.* **145** N10
Solwezi, *Zambia* **209** M17
Solyanka, *Russ.* **158** M15
Solyplayas, *Sp.* **214** Q7
Soma, *Turk.* **170** E3
Somaemuldo, island, *S. Korea* **187** U11
Somain, *Fr.* **148** M11
Somali Basin, *Ind. Oc.* **250** G6
Somali Peninsula, *Af.* **198** H11
Somali Plain, *Ind. Oc.* **250** F7
Somali, *Eth.* **210** M10
Somalia, *Af.* **211** Q10
Somaliland, *Somalia* **210** L12
Somanga, *Tanzania* **211** X8
Sombo, *Angola* **209** K14
Sombor, *Croatia* **154** D9
Sombrero Channel, *India* **183** T15
Sombrero, island, *Fr.* **121** A15
Somers, *Mont., U.S.* **108** B5
Somerset East, *S. Af.* **212** M7
Somerset Island, *Nunavut, Can.* **87** E15
Somerset Island, *Bermuda* **119** C14
Somerset, *Pa., U.S.* **96** P4
Somerset, *Ky., U.S.* **103** D18
Somerset, *Colo., U.S.* **109** N11
Somerset, *Bermuda* **119** C14
Somersworth, *N.H., U.S.* **97** H15
Somerton, *Ariz., U.S.* **109** V4
Somerville, *Tenn., U.S.* **102** G12
Somerville, *Tex., U.S.* **105** K18
Someş, river, *Rom.* **155** B13
Somme, Baie de la, *Fr.* **148** M9
Somme, river, *Fr.* **148** M9
Somosomo Strait, *Fiji* **226** N7
Somosomo, *Fiji* **226** N7
Somuncurá, Meseta de, *Arg.* **135** Q9
Sơn La, *Vietnam* **190** H10
Son, river, *India* **182** J8
Sŏnbong *see* Unggi, *N. Korea* **186** B14
Sŏnch'ŏn, *N. Korea* **186** H5
Sonda, *Pak.* **181** S8
Sondar, *India* **180** J14
Sønderborg, *Den.* **142** N11
Sondershausen, *Ger.* **150** F8
Sŏndŏk, *N. Korea* **186** H9
Søndre Isortoq, fjord, *Greenland, Den.* **89** P5
Søndre Upernavik *see* Upernavik Kujalleq, *Greenland, Den.* **88** K5
Sondrio, *It.* **152** C5
Sông Câu, *Vietnam* **190** M13
Sông Mã, *Vietnam* **190** H10
Song, *Malaysia* **192** H10

Song, *Nigeria* **206** K11
Songea, *Tanzania* **211** Y6
Songeons, *Fr.* **149** N9
Sŏnggan, *N. Korea* **186** F7
Songgŏ, *N. Korea* **186** L10
Songhak, *N. Korea* **186** B12
Songhua, river, *China* **185** D16
Songhwa, *N. Korea* **186** L5
Songido, island, *S. Korea* **187** S7
Sŏngjin *see* Kimch'aek, *N. Korea* **186** F12
Sŏngjin-man, bay, *N. Korea* **186** F12
Songjin-san, peak, *N. Korea* **186** B13
Sŏngjŏn, *N. Korea* **186** K9
Songkhla, *Thai.* **191** R8
Songnim, *N. Korea* **186** K6
Songo Songo Island, *Tanzania* **211** X8
Songo, *Angola* **208** J10
Songp'a, *N. Korea* **186** F11
Songpan, *China* **184** K10
Songsong (Rota), *N. Mariana Is., U.S.* **230** N8
Songwe, river, *Malawi, Tanzania* **212** A10
Songwŏn-jŏsuji, lake, *N. Korea* **186** F6
Songwŏn, *N. Korea* **186** F6
Songyuan, *China* **185** D15
Sonja Ridge, *Ind. Oc.* **251** K16
Sonkovo, *Russ.* **158** H9
Sonmiani Bay, *Pak.* **181** S6
Sonmiani, *Pak.* **181** R7
Sonneberg, *Ger.* **150** G8
Sonoma, *Calif., U.S.* **111** R3
Sonora, *Tex., U.S.* **105** K13
Sonora, *Calif., U.S.* **111** R6
Sonoran Desert National Monument, *Ariz., U.S.* **109** U6
Sonoran Desert, *N. Amer.* **84** K3
Sonqor, *Iraq* **177** E20
Sonseca, *Sp.* **146** G10
Sonsón, *Col.* **130** F6
Sonsonate, *El Salv.* **117** L16
Sonsorol Islands, *Palau* **224** F2
Soperton, *Ga., U.S.* **99** N7
Sophia-Antipolis, *Fr.* **149** X15
Sopo, river, *S. Sudan* **203** V8
Sopot, *Pol.* **151** B14
Sopron, *Hung.* **151** L13
Sopur, *India* **180** H13
Sopweru, island, *F.S.M.* **226** D9
Sŏp'yŏng, *N. Korea* **186** K7
Sør Rondane Mountains, *Antarctica* **239** B16
Sora, *It.* **152** L10
Sorah, *Pak.* **181** Q8
Soran, *Kaz.* **179** F13
Søråsen Ridge, *Antarctica* **238** B12
Sorazocco, *San Marino* **163** J16
Sorbas, *Sp.* **146** L12
Soresina, *It.* **152** E5
Sorezaru Point, *Solomon Is.* **228** D6
Sorgono, *It.* **153** N4
Sorgues, *Fr.* **149** X13
Sorgun, *Turk.* **170** E10
Soria, *Sp.* **146** D12
Sorkapp Bank, *Arctic Oc.* **253** G17
Sorkh Kowtal Temple, ruins, *Afghan.* **180** G8
Sørkjosen, *Nor.* **143** C13
Sørli, *Nor.* **142** G12
Sorobon, *Bonaire, Neth.* **123** R19
Soroca, *Mold.* **155** Q4
Sorocaba, *Braz.* **133** M13
Sorol Atoll, *F.S.M.* **226** A6
Sorong, *Indonesia* **193** J17
Soroti, *Uganda* **211** R4
Sørøya, island, *Nor.* **143** B13
Sorsele, *Sw.* **143** F13
Sorso, *It.* **152** M3
Sortavala, *Russ.* **158** D7
Sørvágur, *Faroe Is., Den.* **142** H5
Sosanjaya Bay, *N. Mariana Is., U.S.* **230** N8
Sosanlagh Bay, *N. Mariana Is., U.S.* **230** N8
Sosna, *Russ.* **161** J15
Sosnogorsk, *Russ.* **158** B15
Sosnogorsk, *Russ.* **160** F9
Sosnovka, *Russ.* **158** H13
Sosnovka, *Russ.* **158** H15
Sosnovka, *Russ.* **158** L11
Sosnovo Ozerskoye, *Russ.* **161** L16
Sosnovo, *Russ.* **158** E6
Sosnowiec, *Pol.* **151** G15
Soso Bay, *Fiji* **226** R5
Sosso, *Cen. Af. Rep.* **206** N12

Sŏsura, *N. Korea* **186** B14
Sos'va, *Russ.* **160** G10
Sos'va, *Russ.* **160** H9
Sotik, *Kenya* **211** S5
Soto, *Curaçao, Neth.* **123** P13
Sotogahama, *Japan* **188** J12
Souanké, *Congo* **208** C10
Soubré, *Côte d'Ivoire* **204** P12
Soudan Bank, *Ind. Oc.* **250** K8
Soudan, *N.T., Austral.* **221** S10
Souflí, *Gr.* **157** B16
Soufrière Bay, *St. Lucia* **123** L13
Soufrière Hills (volcano), *Montserrat, U.K.* **123** B24
Soufrière, *Guadeloupe, Fr.* **123** G14
Soufrière, *St. Lucia* **123** K13
Souillac, *Fr.* **149** V9
Souillac, *Mauritius* **215** H20
Souk Ahras, *Alg.* **201** B14
Souk el Arba du Rharb, *Mor.* **200** C7
Soúnio, ruin, *Gr.* **157** K13
Soûr (Tyre), *Leb.* **172** K6
Sourdeval, *Fr.* **149** P6
Soure, *Braz.* **133** C13
Souris (Mouse), river, *Can., U.S.* **106** D3
Souris, river, *Can., U.S.* **90** B11
Soúrpi, *Gr.* **156** G11
Sousa, *Braz.* **133** F17
Sousel, *Port.* **146** H6
Sousse, *Tunisia* **201** C16
South Africa, *Af.* **212** J7
South Aleutian Slope, *Pac. Oc.* **246** C12
South Alligator, river, *Austral.* **220** B8
South Ambo Channel, *Marshall Is.* **227** H17
South Andaman, island, *India* **183** R14
South Aral Sea, *Kaz., Uzb.* **178** H7
South Atlantic Ocean, *Atl. Oc.* **249** S9
South Aulatsivik Island, *Nfld. & Lab., Can.* **87** K21
South Australia, *Austral.* **221** V9
South Australian Basin, *Ind. Oc.* **251** P16
South Australian Plain, *Ind. Oc.* **251** N19
South Baldy, peak, *N. Mex., U.S.* **109** T11
South Banda Basin, *Pac. Oc.* **246** K6
South Bay, *Fla., U.S.* **99** V10
South Bend, *Ind., U.S.* **101** N9
South Bend, *Wash., U.S.* **110** D2
South Bimini, *Bahamas* **122** B11
South Boston, *Va., U.S.* **98** G11
South Branch Potomac, river, *Md., W. Va., U.S.* **98** C11
South Bruny Island, *Austral.* **220** M8
South Caicos, island, *Turks & Caicos Is., U.K.* **119** J17
South Cape (Whiore), *N.Z.* **223** Q16
South Cape *see* Kalae, *Hawaii, U.S.* **115** Q21
South Cape, *Fiji* **226** N7
South Carolina, *U.S.* **98** K9
South Cay, *Jam.* **118** P11
South Channel, *Marshall Is.* **227** B21
South Charleston, *W. Va., U.S.* **98** E8
South Comino Channel, *Malta* **163** J22
South China Basin, *Pac. Oc.* **246** H4
South China Sea, *Pac. Oc.* **168** K12
South Comino Channel, *Malta* **163** J22
South Dakota, *U.S.* **106** K5
South Dome, *Greenland, Den.* **89** R8
South Downs, *Eng., U.K.* **145** V14
South East Cape, *Austral.* **220** M7
South East Mountain, *Grenada* **123** L23
South East Point, *Kiribati* **226** R12
South East Range, *St. Kitts & Nevis* **123** A17
South Fiji Basin, *Pac. Oc.* **246** M11
South Fork Owyhee, river, *Idaho, Nev., U.S.* **108** J3
South Fork Shenandoah, *Va., U.S.* **98** D11
South Fork Solomon, river, *Kans., U.S.* **107** S5
South Friar's Bay, *St. Kitts & Nevis* **123** B18
South Fulton, *Tenn., U.S.* **102** E12

South Geomagnetic Pole, *Antarctica* **239** K16
South Georgia Island, *S. Amer.* **128** R10
South Grand, river, *Mo., U.S.* **107** T11
South Haven, *Mich., U.S.* **100** M9
South Henik Lake, *Nunavut, Can.* **87** J14
South Hill, *Va., U.S.* **98** F12
South Horr, *Kenya* **211** R6
South Indian Basin, *Ind. Oc.* **250** Q12
South Island (Te Waipounamu), *N.Z.* **223** N18
South Korea, *Asia* **187** Q9
South Lake Tahoe, *Calif., U.S.* **111** Q6
South Loup, river, *Nebr., U.S.* **107** P5
South Magnetic Pole, *Antarctica* **239** R20
South Male Atoll, *Maldives* **183** W3
South Malosmadulu Atoll, *Maldives* **183** V3
South Milwaukee, *Wis., U.S.* **100** L7
South Mole, *Gibraltar, U.K.* **163** P22
South Nahanni, river, *N.W.T., Can.* **86** G9
South Negril Point, *Jam.* **122** H5
South Norfolk Basin, *Pac. Oc.* **246** N10
South Orkney Islands, *Antarctica* **238** A5
South Orkney Trough, *Atl. Oc.* **249** X7
South Ossetia, republic, *Rep. of Georgia* **171** B18
South Pacific Ocean, *Pac. Oc.* **225** M16
South Palmetto Point, *Bahamas* **122** E6
South Pass, *Marshall Is.* **227** J17
South Passage, *Kiribati* **226** P10
South Peak *see* Bartle Frere, *Austral.* **220** D13
South Pittsburg, *Tenn., U.S.* **103** G16
South Platte, river, *U.S.* **90** F11
South Point, *Barbados* **123** L19
South Point, *Marshall Is.* **227** J20
South Pole, *Antarctica* **239** J13
South Ponte Vedra Beach, *Fla., U.S.* **99** R9
South Portland, *Me., U.S.* **97** H16
South Ronaldsay, island, *Scot., U.K.* **144** F10
South San Francisco, *Calif., U.S.* **111** S3
South Sandwich Fracture Zone, *Atl. Oc.* **249** X10
South Sandwich Ridge, *Atl. Oc.* **254** L8
South Sandwich Trench, *Atl. Oc.* **249** W9
South Saskatchewan, river, *Alta., Sask., Can.* **86** N11
South Scotia Ridge, *Atl. Oc.* **249** X6
South Shetland Islands, *Antarctica* **238** C4
South Shetland Trough, *Atl. Oc., Pac. Oc.* **254** N10
South Shore, *Ky., U.S.* **103** A19
South Sioux City, *Nebr., U.S.* **107** N8
South Solomon Trough, *Pac. Oc.* **246** L9
South Sound Bluff, *Virgin Is., U.K.* **122** Q9
South Sound, *Virgin Is., U.K.* **122** Q9
South Sound, *Ire.* **145** R3
South Sudan, *Af.* **203** W8
South Taranaki Bight, *N.Z.* **223** H20
South Tasman Rise, *Ind. Oc., Pac. Oc.* **255** E19
South Tasman Saddle, *Ind. Oc.* **251** Q22
South Uist *see* Uibhist a Deas, island, *Scot., U.K.* **144** J6
South Wazīristān, region, *Pak.* **180** K9
South West Bay, *Bahamas* **122** F2
South West Bay, *Vanuatu* **229** H14
South West Cape (Puhiwaero), *N.Z.* **223** Q16
South West Cape, *Austral.* **220** M7
South West Point, *Bahamas* **122** F11
South West Point, *Cayman Is., U.K.* **122** N2
South West Point, *Kiribati* **226** Q10
South Yolla Bolly Mountains, *Calif., U.S.* **111** N2

Southampton Island, *Nunavut, Can.* **87** H16
Southampton, *N.Y., U.S.* **97** N13
Southampton, *Eng., U.K.* **145** V12
Southampton, Cape, *Nunavut, Can.* **87** J16
Southaven, *Miss., U.S.* **102** G11
Southeast Indian Ridge, *Ind. Oc.* **250** M11
Southeast Pacific Basin, *Pac. Oc.* **247** Q19
Southeast Pass, *Marshall Is.* **227** J21
Southeast Point, *Kiribati* **227** P14
Southend Reindeer, *Sask., Can.* **87** L13
Southend-on-Sea, *Eng., U.K.* **145** U15
Southend, *Scot., U.K.* **145** M8
Southern Alps (Ka Tiritiri O Te Moana), *N.Z.* **223** M17
Southern Bight, *Bahamas* **118** F11
Southern Bug, river, *Eur.* **140** H11
Southern Cross, *W. Aust., Austral.* **221** W4
Southern Indian Lake, *Man., Can.* **87** K14
Southern Ocean, **249** X9
Southern Pines, *N.C., U.S.* **98** J11
Southern Uplands, *Scot., U.K.* **145** N10
Southesk Tablelands, *Austral.* **220** E6
Southfield, *Jam.* **122** J7
Southport, *N.C., U.S.* **98** K12
Southwest Anchorage, bay, *Virgin Is., U.S.* **122** R1
Southwest Cape, *Virgin Is., U.S.* **122** R1
Southwest Cay, *Jam.* **118** P11
Southwest Indian Ridge, *Ind. Oc.* **250** R3
Southwest Pacific Basin, *Pac. Oc.* **247** N13
Southwest Pass, *Marshall Is.* **227** J20
Southwest Passage, *Marshall Is.* **227** B20
Southwest Point, *Bahamas* **122** C9
Southwest Rift, *Hawaii, U.S.* **115** P21
Southwest Rock, island, *Jam.* **118** P10
Southwold, *Eng., U.K.* **145** S16
Soutpansberg, mountains, *S. Af.* **212** G9
Souttouf, Adrar, *Af.* **198** E1
Souvigny, *Fr.* **149** T11
Sovata, *Rom.* **155** C14
Sovetsk, *Russ.* **158** G14
Sovetsk, *Russ.* **158** J2
Sovetskaya Gavan', *Russ.* **161** K21
Sovetskiy, *Russ.* **160** G10
Sovetskoye, *Russ.* **159** Q13
Soweto, *S. Af.* **212** J8
Sōya Misaki, *Japan* **188** E13
Sōya, *Japan* **188** E13
Soyang, river, *S. Korea* **186** M9
Soyeonpyeongdo, island, *S. Korea* **186** M6
Soyo, *Angola* **208** H9
Soyra, peak, *Eritrea* **210** H8
Soyuz Station, *Antarctica* **239** F20
Spa, *Belg.* **148** L13
Spaanse Baai, bay, *Curaçao, Neth.* **123** Q15
Spaanse Water, *Curaçao, Neth.* **123** Q15
Spaatz Island, *Antarctica* **238** G7
Spain, *Eur.* **146** G11
Spalding, *Nebr., U.S.* **107** P6
Spalding, *Eng., U.K.* **145** S14
Spaldings, *Jam.* **122** J8
Spanish Cay, *Bahamas* **122** A9
Spanish Fork, *Utah, U.S.* **108** L7
Spanish Lagoon, *Aruba, Neth.* **123** R17
Spanish Peak, *Oreg., U.S.* **110** G6
Spanish Town, *Jam.* **122** J10
Spanish Town, *Virgin Is., U.K.* **122** R8
Spanish Wells, *Bahamas* **122** D5
Sparkman, *Ark., U.S.* **102** J8
Sparks, *Nev., U.S.* **111** P6
Sparta *see* Spárti, *Gr.* **156** L10
Sparta, *Ga., U.S.* **98** L7
Sparta, *Ill., U.S.* **100** U5
Sparta, *Wis., U.S.* **100** K4
Sparta, *Tenn., U.S.* **103** F17
Spartanburg, *S.C., U.S.* **98** J8
Spárti (Sparta), *Gr.* **156** L10
Spartivento, Cape, *Eur.* **140** L8
Spartivento, Capo, *It.* **153** S13
Spátha, Ákra, *Gr.* **156** P12
Spearfish, *S. Dak., U.S.* **106** K1

Spearman, Tex., U.S. **105** B13
Speedwell Island, Falk. Is., U.K. **135** W12
Speightstown, Barbados **123** K18
Speke Gulf, Tanzania **211** T4
Spello, It. **152** J8
Spencer Gulf, Austral. **220** K10
Spencer, N.C., U.S. **98** H9
Spencer, W. Va., U.S. **98** D8
Spencer, Ind., U.S. **101** S8
Spencer, Iowa, U.S. **106** M10
Spencer, Nebr., U.S. **106** M6
Spencer, Idaho, U.S. **108** F7
Spencer, Cape, Austral. **220** L10
Spercheiós, river, Gr. **156** G10
Sperrin Mountains, N. Ire., U.K. **145** N6
Spétses, Gr. **156** L11
Spétses, island, Gr. **156** L11
Speyside, Trin. & Tobago **123** N19
Spezand, Pak. **180** M7
Spicer Islands, Nunavut, Can. **87** F17
Spiddal see An Spidéal, Ire. **145** R3
Spiez, Switz. **152** B3
Spíli, Gr. **157** Q14
Spilsby, Eng., U.K. **145** R14
Spīn Bōldak, Afghan. **180** L6
Spinazzola, It. **152** M13
Spirit Lake, Iowa, U.S. **106** M10
Spirit Lake, Idaho, U.S. **108** B3
Spiro, Okla., U.S. **105** C20
Spišská Nová Ves, Slovakia **151** J16
Spitsbergen Bank, Arctic Oc. **253** G17
Spitsbergen Trough, Arctic Oc. **253** H16
Spitsbergen, island, Nor. **168** A7
Spittal of Glenshee, Scot., U.K. **144** J10
Spittal, Aust. **150** M10
Split (Spalatum), Croatia **154** G6
Split Lake, Man., Can. **87** L14
Splügen, Switz. **152** C5
Spokane Valley, Wash., U.S. **110** C9
Spokane, Wash., U.S. **110** C9
Spokane, river, Idaho, Wash., U.S. **110** B8
Spoláïta, Gr. **156** G8
Spoleto, It. **152** J8
Spoloshino, Russ. **161** J16
Spoon, river, Ill., U.S. **101** P4
Spooner, Wis., U.S. **100** G3
Sporting Club, Monaco **163** D21
Spotts, Cayman Is., U.K. **122** N2
Sprague Lake, Wash., U.S. **110** C8
Sprague River, Oreg., U.S. **110** K4
Sprague, Wash., U.S. **110** C8
Sprague, river, Oreg., U.S. **110** K5
Spratly Island, S. China Sea **192** F10
Spratly Islands, S. China Sea **192** E11
Spray, Oreg., U.S. **110** G6
Spree, river, Ger. **150** E10
Spremberg, Ger. **150** F11
Spring Grove, Minn., U.S. **106** L13
Spring Hill, Tenn., U.S. **103** F15
Spring Mountains, Nev., U.S. **111** U11
Spring Valley, Ill., U.S. **101** P5
Spring Valley, Minn., U.S. **106** L12
Springbok, S. Af. **212** K4
Springdale, Ark., U.S. **102** E6
Springdale, Wash., U.S. **110** B8
Springer, N. Mex., U.S. **109** Q14
Springfield, Mass., U.S. **97** L13
Springfield, Vt., U.S. **97** H13
Springfield, Fla., U.S. **99** R3
Springfield, Ill., U.S. **101** R5
Springfield, Ohio, U.S. **101** R12
Springfield, Ky., U.S. **103** C17
Springfield, Tenn., U.S. **103** E15
Springfield, Minn., U.S. **106** K10
Springfield, S. Dak., U.S. **106** M7
Springfield, Mo., U.S. **107** V12
Springfield, Colo., U.S. **109** P15
Springfield, Oreg., U.S. **110** H3
Springhill, Ark., U.S. **102** K7
Springview, Nebr., U.S. **106** M5
Springville, N.Y., U.S. **96** K5
Spruce Knob, W. Va., U.S. **98** D10
Spruce Mountain, Nev., U.S. **111** N12
Spruce Pine, N.C., U.S. **98** H7
Spulico, Capo, It. **153** N14
Spur, Tex., U.S. **105** F13
Spurn Head, Eng., U.K. **145** Q14
Squillace, Golfo di, It. **153** Q14
Squinzano, It. **153** N15
Sráid na Cathrach see Milltown Malbay, Ire. **145** S3

Srath an Urláir see Stranorlar, Ire. **145** N5
Srebrenica, Bosn. & Herzg. **154** G9
Sredinnyy Khrebet, Russ. **161** F22
Sredne Sibirskoye Ploskogor'ye, Russ. **161** G14
Sredniy, Russ. **161** F21
Śrem, Pol. **151** E13
Sremska Mitrovica, Serb. **154** F9
Sretensk, Russ. **161** L17
Sri Jayewardenepura Kotte, Sri Lanka **183** U7
Sri Kalahasti, India **183** Q7
Sri Lanka, Asia **183** U7
Sri Lanka, island, Asia **168** L7
Sriharikota Island, India **183** Q7
Srikakulam, India **182** N9
Srinagar, India **182** D5
Srivardhan, India **182** M3
Ssanggyo, N. Korea **186** M5
St. Andrews Bay, Grenada **123** K23
St. Andrews, N.Z. **223** M18
St. Annabaai, bay, Curaçao, Neth. **123** Q14
St. Ann's Bay, Jam. **122** H9
St. Anthony's Peak, St. Kitts & Nevis **123** B18
St. Catherine Point, Bermuda **119** B16
St. Catherine's Monastery, Egypt **202** G13
St. Christoffelberg, peak, Curaçao, Neth. **123** N13
St. Christopher see St. Kitts, St. Kitts & Nevis **123** A17
St. David's Island, Bermuda **119** B16
St. David's Point, Grenada **123** L23
St. David's, Grenada **123** L23
St. Francis Bay, Namibia **212** J3
St. George Hill, Virgin Is., U.S. **122** Q1
St. George, Bermuda **119** B16
St. George's Bay, Malta **163** K23
St. George's Channel, P.N.G. **222** N10
St. George's Harbour, Grenada **123** L22
St. George's Island, Bermuda **119** B16
St. George's, Grenada **123** L22
St. Giles Islands, Trin. & Tobago **123** M19
St. James Bay, Virgin Is., U.S. **122** N9
St. James, Cayman Is., U.K. **122** N3
St. John, Virgin Is., U.S. **122** N10
St. John's Harbour, Antigua & Barbuda **123** A20
St. John's, Antigua & Barbuda **123** A20
St. John's, Montserrat, U.K. **123** B24
St. Jorisbaai, bay, Curaçao, Neth. **123** P15
St. Joseph, Dominica **123** F19
St. Joseph, Trin. & Tobago **123** Q23
St. Julian's, Malta **163** K23
St. Kitts (St. Christopher), St. Kitts & Nevis **123** A17
St. Lawrence, Tex., U.S. **104** H12
St. Lawrence, Barbados **123** L19
St. Margaret's Bay, Jam. **122** H11
St. Mark Bay, Grenada **123** J22
St. Martin's Gardens, Monaco **163** F21
St. Marys, Trin. & Tobago **123** Q22
St. Michael's Caves, Gibraltar, U.K. **163** Q23
St. Michielsbaai, bay, Curaçao, Neth. **123** P14
St. Paul's Bay, Malta **163** J21
St. Paul's Bay, Malta **163** J22
St. Paul's Islands, Malta **163** J22
St. Paul's Point, Pitcairn I. **228** Q6
St. Paul's, St. Kitts & Nevis **123** A17
St. Peter's Basilica, Vatican City **163** Q16
St. Peter's Square, Vatican City **163** Q17
St. Pierre, Trin. & Tobago **123** P21
St. Pölten, Aust. **150** K12
St. Roman, Monaco **163** A23
St. Stanislas Bay, Kiribati **226** Q10
St. Thomas Bay, Virgin Is., U.S. **122** R8
St. Thomas Harbor, Virgin Is., U.S. **122** N8
St. Thomas, Virgin Is., U.S. **122** N8
St. Veit, Aust. **150** M11
St. Wendel, Ger. **150** H5
St. Willebrordus, Curaçao, Neth. **123** P14
St, Claude, Guadeloupe, Fr. **123** G14
St, Florent, Golfe de, Fr. **149** X3

St, François, Guadeloupe, Fr. **123** E16
St, Joseph, Martinique, Fr. **123** F22
St, Joseph, New Caledonia, Fr. **231** N23
St, Louis, Guadeloupe, Fr. **123** G16
St, Louis, New Caledonia, Fr. **231** Q23
St, Pierre, Martinique, Fr. **123** E22
Sta. Martabaai, bay, Curaçao, Neth. **123** P13
Stack Skerry, Scot., U.K. **144** E9
Stade, Ger. **150** C7
Städjan, peak, Sw. **142** H12
Stadlandet, peninsula, Nor. **142** H9
Stadskanaal, Neth. **148** G14
Stadtallendorf, Ger. **150** F7
Staffa, island, Scot., U.K. **144** K7
Stafford, Kans., U.S. **107** U6
Stafford, Eng., U.K. **145** S11
Staithes, Eng., U.K. **145** P13
Staked Plain see Llano Estacado, U.S. **90** K10
Stalingrad see Volgograd, Russ. **159** P13
Stalowa Wola, Pol. **151** G17
Stambaugh, Mich., U.S. **100** G6
Stamford, Conn., U.S. **96** M12
Stamford, N.Y., U.S. **96** K10
Stamford, Tex., U.S. **105** G14
Stamford, Qnsld., Austral. **221** S12
Stampriet, Namibia **212** H4
Stamps, Ark., U.S. **102** J7
Stamsund, Nor. **142** D12
Stanberry, Mo., U.S. **107** R10
Stancomb-Wills Glacier Tongue, Antarctica **238** C11
Stancomb-Wills Glacier, Antarctica **238** D11
Standish, Mich., U.S. **100** K11
Stanford, Ky., U.S. **103** C17
Stanford, Mont., U.S. **108** C9
Stanger see KwaDukuza, S. Af. **212** K9
Stanley, Wis., U.S. **100** H4
Stanley, N. Dak., U.S. **106** E3
Stanley, Idaho, U.S. **108** F5
Stanley, Falk. Is., U.K. **135** W12
Stanley, Mount, Austral. **220** L7
Stanovaya, Russ. **161** D19
Stanovoy Khrebet, Russ. **161** J18
Stanovoy Range, Asia **168** D11
Stanovoye Nagor'ye, Russ. **161** K16
Stanton, Tex., U.S. **104** H12
Stanton, N. Dak., U.S. **106** F4
Stanton, Nebr., U.S. **107** P8
Stanwood, Wash., U.S. **110** B4
Stanyard Creek, Bahamas **118** E11
Staples, Minn., U.S. **106** H10
Stapleton, Nebr., U.S. **107** P4
Star City, Ark., U.S. **102** J9
Stara Pazova, Serb. **154** F10
Stara Zagora, Bulg. **155** J15
Staraya Russa, Russ. **158** G6
Starbuck Island, Kiribati **224** H12
Starbuck, Minn., U.S. **106** J9
Starbuck, Wash., U.S. **110** D8
Stargard Szczeciński, Pol. **150** C11
Staritsa, Russ. **158** J8
Starke, Fla., U.S. **99** R8
Starkville, Miss., U.S. **102** J12
Starobil's'k, Ukr. **159** P10
Starobin, Belarus **158** L4
Starodub, Russ. **158** M7
Starogard Gdański, Pol. **151** B14
Starokostyantyniv, Ukr. **159** N4
Staroshcherbinovskaya, Russ. **159** R10
Starozhynets', Ukr. **159** Q3
Start Point, Eng., U.K. **145** X10
Starve Gut Bay, Jam. **122** J7
Staryy Oskol, Russ. **159** N9
State College, Pa., U.S. **96** N6
Stateline, Nev., U.S. **111** Q6
Staten Island, N.Y., U.S. **96** N11
Statesboro, Ga., U.S. **99** N8
Statesville, N.C., U.S. **98** H9
Statia see Saint Eustatius, island, Lesser Antil. **121** B16
Station 10, Sudan **203** N12
Station 5, Sudan **202** M12
Station 6, Sudan **202** M12
Stauning Alper, mountains, Greenland, Den. **88** K13
Staunton, Va., U.S. **98** E11
Staunton, Ill., U.S. **101** S4
Stavanger, Nor. **142** L9

Stavropol' Plateau, Eur. **140** H14
Stavropol', Russ. **159** S12
Stavrós, Gr. **156** C12
Stavrós, Gr. **156** H7
Stavroúpoli, Gr. **157** B14
Stayton, Oreg., U.S. **110** G3
Ste. Dévôte's Church, Monaco **163** D21
Ste, Anne, Martinique, Fr. **123** G23
Ste, Luce, Martinique, Fr. **123** G23
Ste, Marguerite, Guadeloupe, Fr. **123** E15
Steamboat Springs, Colo., U.S. **108** L11
Stebbins, Alas., U.S. **112** G12
Steele Point, Virgin Is., U.K. **122** R5
Steele, N. Dak., U.S. **106** G5
Steels Point, Austral. **229** A20
Steelton, Pa., U.S. **96** P7
Steelville, Mo., U.S. **107** R14
Steenbergen, Neth. **148** K12
Steens Mountain, Oreg., U.S. **110** K7
Steensby Land, Greenland, Den. **88** E3
Steenstrup Gletscher, Greenland, Den. **88** G5
Steenwijk, Neth. **148** H13
Steep Point, Austral. **220** H1
Steese Highway, Alas., U.S. **113** F17
Stefansson Basin, Arctic Oc. **252** L11
Stefansson Bay, Antarctica **239** D20
Stefansson Island, Nunavut, Can. **87** D13
Steffen, Cerro, Arg. **135** S7
Steg, Liech. **162** P3
Steinbach, Man., Can. **87** N14
Steinfort, Lux. **162** J9
Steinhatchee, Fla., U.S. **99** R6
Steinhort, Liech. **162** P3
Steinkjer, Nor. **142** G11
Steinsel, Lux. **162** J10
Stella Maris, Bahamas **119** F13
Stellaland, S. Af. **212** J7
Stellenbosch, S. Af. **212** M5
Stendal, Ger. **150** D9
Stenón Kásou, Gr. **157** Q17
Steòrnabhagh (Stornoway), Scot., U.K. **144** G7
Stepanakert (Xankəndi), Azerb. **171** E20
Step'anavan, Arm. **171** D18
Stephen, Minn., U.S. **106** E8
Stephens, Ark., U.S. **102** J7
Stephens, Port, Austral. **220** K15
Stephenson, Wis., U.S. **100** H7
Stephenville, Nfld. & Lab., Can. **87** M23
Stephenville, Tex., U.S. **105** H16
Stepnogorsk, Kaz. **179** D13
Stepnogorsk, Kaz. —
Steps Point, American Samoa, U.S. **230** E6
Sterling City, Tex., U.S. **105** H13
Sterling Heights, Mich., U.S. **100** M12
Sterling Highway, Alas., U.S. **113** K16
Sterling, Ill., U.S. **101** N5
Sterling, N. Dak., U.S. **106** G5
Sterling, Kans., U.S. **107** T7
Sterling, Colo., U.S. **108** L15
Sterlitamak, Russ. **160** J7
Stettiner Haff (Zalew Szczeciński), Ger., Pol. **150** C10
Stettler, Alta., Can. **86** M10
Steubenville, Ohio, U.S. **101** Q16
Stevens Point, Wis., U.S. **100** J5
Stevens Village, Alas., U.S. **113** F17
Stevenson, Wash., U.S. **110** E4
Stevensville, Mont., U.S. **108** D5
Steventon, Bahamas **122** E8
Stewart Island (Rakiura), N.Z. **223** Q16
Stewart Town, Jam. **122** H8
Stewart, B.C., Can. **86** J8
Stewart, Nev., U.S. **111** Q6
Stewart, Cape, Austral. **220** A9
Stewart, Isla, Chile **135** Y8
Steyr, Aust. **150** K11
Stickney, S. Dak., U.S. **106** L6
Stigler, Okla., U.S. **105** D19
Stikine, river, B.C., Can. **86** H8
Stiles, Tex., U.S. **104** J12
Stillwater Range, Nev., U.S. **111** P8
Stillwater, Okla., U.S. **105** B17
Stillwater, Minn., U.S. **106** J12
Stilo, It. **153** R13
Stilo, Punta, It. **153** R14
Stilwell, Okla., U.S. **105** C20
Stinnett, Tex., U.S. **104** C12
Stintino, It. **152** M3
Štip, N. Maced. **154** K12

Stiring-Wendel, Fr. **149** N14
Stirling City, Calif., U.S. **111** P4
Stirling Island, Solomon Is. **228** D5
Stirling Range, Austral. **220** L3
Stirling, Scot., U.K. **144** L9
Stjørdalshalsen, Nor. **142** G11
Stobi, site, N. Maced. **154** K11
Stockdale, Tex., U.S. **105** M16
Stockerau, Aust. **150** K12
Stockholm, Sw. **143** K14
Stocking Island, Bahamas **122** E9
Stockport, Eng., U.K. **145** R11
Stockton Lake, Mo., U.S. **107** U11
Stockton-on-Tees, Eng., U.K. **145** N12
Stockton, Ill., U.S. **100** M4
Stockton, Kans., U.S. **107** S6
Stockton, Mo., U.S. **107** U11
Stockton, Calif., U.S. **111** R4
Stŏĕng Trêng (Stung Treng), Cambodia **190** M11
Stoke-on-Trent, Eng., U.K. **145** R11
Stokes Point, Austral. **220** L7
Stokes, Bahía, Chile **135** X7
Stolberg, Ger. **150** F4
Stolbovoy, Russ. **161** D17
Stolin, Belarus **158** M4
Stómio, Órmos, Gr. **156** Q12
Stonehaven, Scot., U.K. **144** J11
Stonehenge, Eng., U.K. **145** V12
Stonehenge, Qnsld., Austral. **221** T12
Stones River National Battlefield, Tenn., U.S. **103** F15
Stonewall, Miss., U.S. **103** L13
Stonewall, Okla., U.S. **105** D18
Stony Rapids, Sask., Can. **86** K12
Stony, river, Alas., U.S. **113** J14
Stor Bank, Arctic Oc. **253** F16
Stora Lulevatten, lake, Sw. **143** E13
Storavan, lake, Sw. **143** F13
Store Gletscher, Greenland, Den. **88** L6
Store Hellefiske Bank, Atl. Oc. **253** Q17
Store Koldewey, island, Greenland, Den. **88** F13
Storegg Bank, Ind. Oc. **254** E11
Støren, Nor. **142** H11
Storfjordrenna, valley, Arctic Oc. **253** G17
Storkerson Bay, N.W.T., Can. **86** C11
Storlien, Sw. **142** G11
Storm Bay, Austral. **220** M8
Storm Lake, Iowa, U.S. **107** N10
Stornoway see Steòrnabhagh, Scot., U.K. **144** G7
Storozhevsk, Russ. **158** C15
Storr's Lake, Bahamas **122** E12
Storsjön, lake, Sw. **142** H12
Storstrømmen, glacier, Greenland, Den. **88** F12
Storuman, Sw. **143** F13
Story City, Iowa, U.S. **107** N11
Story, Wyo., U.S. **108** F11
Stoughton, Wis., U.S. **100** L5
Stoyba, Russ. **161** K19
Strabane, N. Ire., U.K. **145** N6
Strait of Gibraltar, Mor. **200** B8
Strait of Hormuz, Iran, Oman **177** N19
Strait of Sicily, Tunisia **201** B16
Strakhov Fracture Zone, Atl. Oc. **249** N8
Stralsund, Ger. **150** B10
Strand, S. Af. **212** M5
Stranda, Nor. **142** H10
Strange Ice Shelf, Antarctica **238** G7
Strangers Cay, Bahamas **122** G3
Strangford Lough, N. Ire., U.K. **145** P8
Stranorlar (Srath an Urláir), Ire. **145** N5
Stranraer, Scot., U.K. **145** N8
Strasbourg, Fr. **149** P15
Strasburg, Va., U.S. **98** C11
Strasburg, Ohio, U.S. **101** Q14
Strasburg, N. Dak., U.S. **106** H5
Strassen, Lux. **162** K10
Strata Florida, Wales, U.K. **145** T10
Stratford upon Avon, Eng., U.K. **145** T12
Stratford, Tex., U.S. **104** B12
Strathleven, Qnsld., Austral. **221** Q12
Strathmore, valley, Scot., U.K. **144** K10
Stratobowl, S. Dak., U.S. **106** L2
Stratton, Me., U.S. **97** E15
Stratton, Nebr., U.S. **107** R3
Stratton, Colo., U.S. **108** M15

Straubing, Ger. 150 J9
Strawberry Mountain, Oreg., U.S. 110 G7
Strawberry Reservoir, Utah, U.S. 108 L8
Strawn, Tex., U.S. 105 G16
Streaky Bay, S. Aust., Austral. 221 X9
Streator, Ill., U.S. 101 P6
Streeter, N. Dak., U.S. 106 G6
Strelka, Russ. 161 F20
Streymoy, island, Faroe Is., Den. 142 H6
Strezhevoy, Russ. 160 H11
Stříbro, Czechia 150 H9
Strickland, river, P.N.G. 222 P6
Strindberg Land, Greenland, Den. 88 H13
Stringtown, Okla., U.S. 105 E18
Stroma, island, Scot., U.K. 144 F10
Stromboli, Isola, It. 153 Q12
Stromeferry, Scot., U.K. 144 J8
Stromness, Scot., U.K. 144 F10
Stromsburg, Nebr., U.S. 107 Q7
Strömstad, Sw. 142 L11
Strömsund, Sw. 142 G12
Strongylí (Ypsilí), island, Gr. 157 N22
Stronsay, island, Scot., U.K. 144 E11
Stroud, Okla., U.S. 105 C17
Stroud, Eng., U.K. 145 U11
Stroudsburg, Pa., U.S. 96 N9
Strovolos, Cyprus 162 N8
Struga, N. Maced. 154 L10
Struis Bay, S. Af. 212 N5
Struma, river, Bulg. 155 K13
Strumica, N. Maced. 154 K12
Struy, Scot., U.K. 144 H9
Strymónas, river, Gr. 156 C12
Stryy, Ukr. 159 P2
Stuart Lake, B.C., Can. 86 K8
Stuart Range, Austral. 220 H9
Stuart, Va., U.S. 98 G10
Stuart, Fla., U.S. 99 V10
Stuart, Nebr., U.S. 107 N6
Stubbs, St. Vincent & the Grenadines 123 L17
Stuhr, Ger. 150 C7
Stump Lake, N. Dak., U.S. 106 F7
Stumpy Point, N.C., U.S. 98 H15
Stumpy Point, Virgin Is., U.S. 122 M7
Stung Treng see Stœ̆ng Trêng, Cambodia 190 M11
Stupino, Russ. 158 K9
Sturgeon Bay, Wis., U.S. 100 H8
Sturgis, Mich., U.S. 101 N10
Sturgis, Ky., U.S. 103 C14
Sturgis, S. Dak., U.S. 106 K2
Sturt Stony Desert, Austral. 220 G11
Stuttgart, Ark., U.S. 102 H9
Stuttgart, Ger. 150 J7
Stykkishólmur, Ice. 142 E2
Stýra, Gr. 157 J13
Sua Pan, Botswana 212 F7
Suai, Timor-Leste 193 N15
Suakin, Sudan 203 N15
Suan, N. Korea 186 K7
Subantarctic Slope, Pac. Oc. 246 R10
Subaşı Dağı, peak, Turk. 171 G16
Subei, China 184 G7
Subi, island, Indonesia 192 H9
Sublette, Kans., U.S. 107 U4
Subotica, Serb. 154 D9
Subugo, peak, Kenya 211 T6
Suceava, Rom. 155 B15
Sučevići, Croatia 154 F6
Sücka, Liech. 162 P3
Sucre, Bol. 132 K7
Sud, Canal du, Haiti 119 M16
Sud, Pointe, Comoros 215 M14
Sud, Récif du, Mayotte, Fr. 215 P17
Sudak, Russ. 159 T8
Sudan, Af. 203 Q8
Sudan, Tex., U.S. 104 E11
Sudan, region, Af. 198 G5
Suday, Russ. 158 F11
Sudd, Sudan 203 V11
Suddie, Guyana 131 E15
Sudeten, range, Czechia, Pol. 150 G12
Sudr, Egypt 173 R1
Sudr, Egypt 202 F12
Suðuroy, island, Faroe Is., Den. 142 J6
Sue Wood Bay, Bermuda 119 C16
Sue, river, S. Sudan 203 W9;X9
Sueca, Sp. 147 H15
Suess Land, Greenland, Den. 88 J13
Suez Canal see Qanât as Suways, Egypt 202 E12
Suez Canal, Egypt 173 N1
Suez see As Suways, Egypt 202 F12
Suez see El Suweis, Egypt 173 Q1
Suffolk, Va., U.S. 98 F14
Sugar Land, Tex., U.S. 105 L19
Sugar Loaf, island, Grenada 123 J23
Sugarloaf Key, Fla., U.S. 99 Y9
Suğla Gölü, Turk. 170 H7
Sugu, Nigeria 206 L10
Suha, N. Korea 186 G8
Sūhāj, Egypt 202 H11
Şuḩār, Oman 175 G16
Sühbaatar, Mongolia 184 C10
Suheli Par, India 183 S3
Suhl, Ger. 150 G8
Şuhut, Turk. 170 F6
Suihua, China 185 D16
Suining, China 184 L10
Suippes, Fr. 149 P12
Suir, river, Ire. 145 T5
Suitcase Seamounts, Pac. Oc. 247 G18
Suizhou, China 185 K13
Sujawal, Pak. 181 S8
Sujeong, S. Korea 187 T11
Sukabumi, Indonesia 192 M8
Sukadana, Indonesia 192 J9
Sukagawa, Japan 188 M12
Sukch'ŏn, N. Korea 186 H6
Sŭkh, Uzb. 179 M13
Sukhinichi, Russ. 158 K8
Sukhona, river, Russ. 158 E12
Sukhothai, Thai. 190 K7
Sukhum see Sokhumi, Rep. of Georgia 171 A15
Sukkertoppen Bank, Atl. Oc. 253 R17
Sukkertoppen see Maniitsoq, Greenland, Den. 89 P5
Sukkertoppen Valley, Atl. Oc. 253 R17
Sukkur, Pak. 181 P8
Sukumo, Japan 189 R6
Sukur, Nigeria 206 K11
Sula Islands, Asia 168 L14
Sula Sgeir, island, Scot., U.K. 144 E7
Sula, Kepulauan, Indonesia 193 K15
Sula, river, Ukr. 159 N7
Sulaimān Range, Pak. 181 N9
Sulanheer, Mongolia 184 F11
Sulawesi (Celebes), island, Indonesia 193 K13
Sulb, ruin, Sudan 202 M11
Sule Skerry, island, Scot., U.K. 144 E9
Sulei, island, Solomon Is. 228 D9
Süleymanlı, Turk. 170 H11
Sulima, Sierra Leone 204 P9
Sulina, Rom. 155 E19
Sullana, Peru 132 E2
Sulligent, Ala., U.S. 103 J13
Sullivan, Ill., U.S. 101 R6
Sullivan, Ind., U.S. 101 S7
Sullivan, Mo., U.S. 107 T14
Sulmona, It. 152 K10
Sulphur Cone, Hawaii, U.S. 115 P21
Sulphur Springs Draw, Tex., U.S. 104 G11
Sulphur Springs, Tex., U.S. 105 F19
Sulphur, La., U.S. 102 P7
Sulphur, Okla., U.S. 105 E17
Sulphur, river, Tex., U.S. 105 F19
Sulţān Bāgh, Afghan. 180 J8
Sultan Dağları, Turk. 170 F6
Sultan Hamud, Kenya 211 T7
Sultanhanı, Turk. 170 G9
Sultanpur, India 182 H8
Sulu Archipelago, Philippines 193 G13
Sulu Basin, Pac. Oc. 246 H5
Sulu Sea, Philippines 193 E13
Sülüklü, Turk. 170 F7
Sülüktü, Taj. 178 M12
Suluova, Turk. 170 D11
Sulūq, Lib. 201 E20
Sulzberger Bay, Antarctica 238 N11
Sulzberger Ice Shelf, Antarctica 238 N11
Sumampa, Arg. 134 H10
Sumas, Wash., U.S. 110 A4
Sumatra, island, Asia 168 L10
Sumaúma, Braz. 132 F9
Sumayl, Iraq 176 D10
Sumba, Faroe Is., Den. 142 J6
Sumba, island, Indonesia 193 N13
Sumbawa Besar, Indonesia 192 M12
Sumbawa, island, Indonesia 192 M12
Sumbawanga, Tanzania 211 X3
Sumbe, Angola 208 L10
Sumbuya, Sierra Leone 204 N9
Sumeih, Sudan 203 U9
Sumisu (Smith Island), Japan 189 T12
Summer Lake, Oreg., U.S. 110 K5
Summerland Key, Fla., U.S. 99 Y9
Summersville, W. Va., U.S. 98 D9
Summerville, Ga., U.S. 98 K4
Summerville, S.C., U.S. 98 M10
Summit Mountain, Nev., U.S. 111 P10
Summit Peak, Colo., U.S. 109 P12
Summit Station see Greenland Environmental Observatory, Greenland, Den. 88 J10
Summit, Miss., U.S. 102 N10
Summit, Sudan 203 N15
Sumner, Iowa, U.S. 106 M13
Šumperk, Czechia 151 H13
Sumprabum, Myanmar 190 D6
Sumpter, Oreg., U.S. 110 G8
Sumqayıt, Azerb. 171 D23
Sumrall, Miss., U.S. 102 M12
Sumter, S.C., U.S. 98 K10
Sumy, Ukr. 159 N8
Sun City, Ariz., U.S. 109 U6
Sun Prairie, Wis., U.S. 100 L5
Sun, river, Mont., U.S. 108 C7
Suna, Russ. 158 G15
Sunagawa, Japan 188 G13
Sunan, N. Korea 186 J6
Sunart, Loch, Scot., U.K. 144 K8
Sunbaron Roads (Tinian Harbor), Fr. Polynesia, Fr. 230 J10
Sunburst, Mont., U.S. 108 A7
Sunbury, Pa., U.S. 96 N7
Sunbury, Vic., Austral. 221 Z12
Sunch-eonman, bay, S. Korea 187 T9
Sunchales, Arg. 134 J11
Sunchang, S. Korea 187 S8
Suncheon, S. Korea 187 T9
Sunch'ŏn, N. Korea 186 H6
Sunda Shelf, Pac. Oc. 246 J4
Sunda Strait, Ind. Oc. 168 M10
Sunda Trench see Java Trench, Ind. Oc. 251 G15
Sunda Trough, Ind. Oc. 251 H16
Sunda, Selat, Indonesia 192 L8
Sundance, Wyo., U.S. 108 F13
Sundar, Malaysia 192 G11
Sundarbans, delta, India 182 L12
Sundargarh, India 182 K9
Sunday see Raoul Island, Kermadec Is., N.Z. 224 L9
Sunderland, Eng., U.K. 145 N12
Sundi, Sao Tome and Principe 215 A20
Sündiken Dağ, peak, Turk. 170 E6
Sundown, Tex., U.S. 104 F11
Sundsvall, Sw. 143 H13
Sunflower, Mount, Kans., U.S. 107 S3
Sung Men, Thai. 190 K7
Sungai Kolok, Thai. 191 S9
Sungai Petani, Malaysia 191 S8
Sungaianyar, Indonesia 192 K12
Sungaibatu, Indonesia 192 J9
Sungaipenuh, Indonesia 192 K6
Sŭngam, N. Korea 186 D12
Sŭngho, N. Korea 186 J6
Sungikai, Sudan 203 T10
Sungurlu, Turk. 170 E9
Suni, Sudan 203 S7
Sunland Park, N. Mex., U.S. 109 W12
Sunnyside, Wash., U.S. 110 E6
Sunnyvale, Calif., U.S. 111 S4
Sunray, Tex., U.S. 104 C12
Sunset Beach, Hawaii, U.S. 115 G14
Sunset Crater Volcano National Monument, Ariz., U.S. 109 S7
Suntar Khayata, Russ. 161 G19
Suntsar, Pak. 181 R3
Sunwi-do, island, N. Korea 186 M5
Sunwu, China 185 B16
Sunyani, Ghana 205 N15
Sunzu, peak, Zambia 209 K20
Suomen Ridge, Eur. 140 C9
Suomenselkä, ridge, Fin. 143 G15
Suonenjoki, Fin. 143 G16
Suoyarvi, Russ. 158 D7
Suozhuku see Sogzhukug, China 184 K6
Supai, Ariz., U.S. 109 Q6
Superior, Wis., U.S. 100 E2
Superior, Nebr., U.S. 107 R7
Superior, Mont., U.S. 108 C5
Superior, Ariz., U.S. 109 U8
Superior, Lake, N. Amer. 84 H7
Süphan Dağı, peak, Turk. 171 F17
Support Force Glacier, Antarctica 238 G11
Supu, Indonesia 193 H16
Sup'ung-ho, lake, N. Korea 186 F5
Şuq'at al Jamal, Sudan 203 S9
Suquţrá (Socotra), island, Yemen 175 Q15
Şūr, Oman 175 H18
Sur, Point, U.S. 90 G3
Sura, river, Russ. 158 L13
Şuraabad, Azerb. 171 C23
Surab, Pak. 181 P6
Surabaya, Indonesia 192 M10
Sūrak, Iran 177 P21
Surakarta, Indonesia 192 M9
Suramana, Indonesia 193 J13
Sūrān, Iran 177 M23
Surat Thani, Thai. 191 Q7
Surat, India 182 L3
Surat, Qnsld., Austral. 221 V14
Sûre, river, Ger., Lux. 162 J11
Surf City, N.C., U.S. 98 K13
Surgères, Fr. 149 T6
Surgidero de Batabanó, Cuba 118 G6
Surgut, Russ. 160 H11
Surigao, Philippines 193 E15
Surin Nua, Ko, Thai. 191 Q6
Surin, Thai. 190 M9
Suriname, S. Amer. 131 G17
Surkh Āb, Afghan. 180 K4
Sürmene, Turk. 171 D14
Surrey, B.C., Can. 86 M8
Sursee, Switz. 152 B3
Surskoye, Russ. 158 K14
Surt (Sidra), Lib. 201 E19
Surtanahu, Pak. 181 R9
Surtsey, island, Ice. 142 F2
Suruç, Turk. 171 J13
Suruga Wan, Japan 189 Q11
Surulangun, Indonesia 192 K6
Surveyor Fracture Zone, Pac. Oc. 247 E13
Susa, ruin, Iran 177 H13
Sušac, island, Croatia 154 H6
Sūsah (Apollonia), Lib. 201 D21
Susak, island, Croatia 154 F4
Susaki, Japan 189 R7
Susamyr, Kyrg. 179 K14
Sūsangerd, Iraq 177 J13
Susanville, Calif., U.S. 111 N5
Suşehri, Turk. 171 E13
Susitna, river, Alas., U.S. 113 H17
Susoh, Indonesia 192 H4
Susquehanna, Pa., U.S. 96 L9
Susquehanna, river, U.S. 91 F20
Susques, Arg. 134 E9
Susubona, Solomon Is. 228 E9
Susuman, Russ. 161 F20
Susupe, N. Mariana Is., U.S. 230 J7
Susurluk, Turk. 170 E3
Sutak, India 182 D6
Sütçüler, Turk. 170 H6
Sutherland, Nebr., U.S. 107 P4
Sutherland, S. Af. 212 M5
Sutherlin, Oreg., U.S. 110 J3
Sutlej, river, Pak. 181 N11
Sutton-on-Sea, Eng., U.K. 145 R14
Sutton, W. Va., U.S. 98 D9
Sutton, Nebr., U.S. 107 Q7
Sutton, river, Ont., Can. 87 M16
Suttsu, Japan 188 G12
Sutwik Island, Alas., U.S. 113 M14
Su'u, Solomon Is. 228 F11
Suva, Fiji 226 Q5
Suvadiva Atoll (Huvadu), Maldives 183 X3
Suvorov, Russ. 158 K9
Suwa, Japan 189 P11
Suwa, Eritrea 210 J9
Suwałki, Pol. 151 B17
Suwannaphum, Thai. 190 L10
Suwannee, Fla., U.S. 99 S6
Suwannee, river, U.S. 91 L18
Suwanose Jima, Japan 189 U4
Suwarrow Atoll, Cook Is., N.Z. 224 J11
Suwon, S. Korea 187 N8
Suyutkino, Russ. 159 T15
Suzhou, China 185 J14
Suzhou, China 185 K15
Suzu Misaki, Japan 188 M10
Suzu, Japan 188 M10
Suzuka, Japan 189 Q9
Svalbard, islands, Nor. 168 A7
Svartenhuk Halvø see Sigguup Nunaa, peninsula, Greenland, Den. 88 K5
Svartisen, glacier, Nor. 142 E12
Svatove, Ukr. 159 P10
Svea Station, Antarctica 238 C12
Svealand, Eur. 140 D8
Sveg, Sw. 142 H12
Svendborg, Den. 142 N11
Sverdrup Channel, Nunavut, Can. 87 B14
Sverdrup Gletscher, Greenland, Den. 88 G5
Sverdrup Islands, Nunavut, Can. 87 B15
Sverdrup Mountains, Antarctica 239 B13
Sverdrup Ø, Greenland, Den. 88 B8
Svetac (Sveti Andrija), island, Croatia 154 H5
Sveti Andrija see Svetac, island, Croatia 154 H5
Svetlograd, Russ. 159 S12
Svetogorsk, Russ. 158 E6
Svilengrad, Bulg. 155 K16
Sviritsa, Russ. 158 E7
Svishtov, Bulg. 155 G15
Svobodnyy, Russ. 161 K19
Svolvær, Nor. 142 D12
Svyataya Anna Fan, Arctic Oc. 253 F14
Svyataya Anna Trough, Arctic Oc. 253 E14
Svyatoy Nos, Russ. 161 D17
Svyetlahorsk, Belarus 158 L5
Swa-Tenda, D.R.C. 208 J11
Swain Reefs, Qnsld., Austral. 221 T16
Swains Island, Pac. Oc. 224 H10
Swainsboro, Ga., U.S. 98 M7
Swakop, river, Namibia 212 G3
Swakopmund, Namibia 212 G3
Swan Hill, Vic., Austral. 221 Y12
Swan Peak, Mont., U.S. 108 C6
Swan River, Man., Can. 87 M13
Swan Valley, Idaho, U.S. 108 H8
Swan, river, Mont., U.S. 108 C6
Swan, river, Austral. 220 K2
Swanlinbar (An Muileann Iarainn), Ire. 145 P5
Swanquarter, N.C., U.S. 98 H14
Swans Island, Me., U.S. 97 G18
Swansboro, N.C., U.S. 98 J13
Swansea (Abertawe), Wales, U.K. 145 U9
Swansea Bay, Wales, U.K. 145 U9
Swanton, Vt., U.S. 96 F12
Swatch of No Ground, trough, Ind. Oc. 251 C13
Swatow see Shantou, China 185 N14
Swaziland see Eswatini, Eswatini (Swaziland) 212 J9
Sweden, Eur. 143 H13
Sweet Home, Oreg., U.S. 110 G3
Sweetgrass, Mont., U.S. 108 A7
Sweeting Cay, Bahamas 122 J4
Sweetwater Lake, N. Dak., U.S. 106 E6
Sweetwater, Tenn., U.S. 103 F18
Sweetwater, Tex., U.S. 105 G13
Sweetwater, river, Wyo., U.S. 108 J10
Swetes, Antigua & Barbuda 123 B21
Świdnica, Pol. 151 G13
Świdnik, Pol. 151 F17
Świdwin, Pol. 150 C12
Świebodzin, Pol. 150 E12
Świecie, Pol. 151 C14
Swift Current, Sask., Can. 86 N11
Swilly, Lough, Ire. 145 M5
Swindon, Eng., U.K. 145 U12
Swinford (Béal Átha na Muice), Ire. 145 P4
Świnoujście, Pol. 150 B11
Switzerland, Eur. 152 B3
Syamzha, Russ. 158 F11
Syava, Russ. 158 G13
Sychevka, Russ. 158 J8
Sydney Bay, Austral. 229 B20
Sydney Point, Kiribati 226 M11
Sydney, N.S., Can. 87 N23
Sydney, N.S.W., Austral. 221 X15
Sydprøven see Alluitsup Paa, Greenland, Den. 89 T7
Syene see Aswān, Egypt 202 K12
Syeri, Indonesia 193 K19
Syktyvkar, Russ. 158 D14
Sylacauga, Ala., U.S. 103 K16
Sylhet, Bangladesh 182 J13
Sylt, island, Ger. 150 A7
Sylva, N.C., U.S. 98 J6
Sylvania, Ga., U.S. 98 M8
Sylvania, Ohio, U.S. 101 N12
Sylvester, Ga., U.S. 99 P6

Sylvester, *Tex., U.S.* **105** G14
Sylvester, Lake, *Austral.* **220** D9
Sými, *Gr.* **157** M19
Sými, island, *Gr.* **157** M19
Synel'nykove, *Ukr.* **159** Q8
Synya, *Russ.* **160** F10
Syowa Station, *Antarctica* **239** B18
Syr Darya, river, *Asia* **168** F6
Syracuse *see* Siracusa, *It.* **153** T12
Syracuse, *N.Y., U.S.* **96** J8
Syracuse, *Kans., U.S.* **107** U3
Syracuse, *Nebr., U.S.* **107** Q8
Syria, *Asia* **172** F9
Syrian Desert, *Asia* **168** F3
Syrian Gates, *Turk.* **170** J11
Sýrna, island, *Gr.* **157** N17
Sýros, island, *Gr.* **157** K14
Sysola, river, *Russ.* **158** D14
Sysy Basa, *Russ.* **161** H18
Syumsi, *Russ.* **158** G15
Syutkya, peak, *Bulg.* **155** K14
Syzran', *Russ.* **158** L14
Szczecin, *Pol.* **150** C11
Szczecinek, *Pol.* **151** C13
Szczytno, *Pol.* **151** C16
Szeged, *Hung.* **151** M16
Székely, *Russ.* **151** K17
Székesfehérvár, *Hung.* **151** L14
Szekszárd, *Hung.* **151** M15
Szentes, *Hung.* **151** M16
Szolnok, *Hung.* **151** L16
Szombathely, *Hung.* **151** L13

T

T. Roosevelt Lake, *Ariz., U.S.* **109** T8
Tā-kaw, *Myanmar* **190** H7
Ta-n-Kena, *Alg.* **201** H15
Taakoka, island, *N.Z.* **229** E23
Taapuna, *Fr. Polynesia, Fr.* **231** H21
Tâba, *Israel* **173** R5
Tabago, *P.N.G.* **222** P11
Tabago, *P.N.G.* **228** C5
Ţābah, *Saudi Arabia* **174** E8
Tabalak, *Niger* **206** G7
Tabalo, *P.N.G.* **222** M9
Tabaquite, *Trin. & Tobago* **123** Q22
Tabar Islands, *P.N.G.* **222** N10
Tabar, island, *P.N.G.* **222** N10
Ţabas, *Iran* **177** G19
Tabelbala, *Alg.* **200** F9
Tabelkoza, *Alg.* **200** F11
Tabelot, *Niger* **206** F9
Tabernacle, *St. Kitts & Nevis* **123** A17
Tabernas, *Sp.* **146** L12
Tabiang, *Kiribati* **226** M11
Tabiang, *Kiribati* **227** Q15
Tabik Channel, *Marshall Is.* **227** G14
Tabik, island, *Marshall Is.* **227** G14
Tabiteuea, *Kiribati* **227** Q15
Tablas, island, *Philippines* **193** D14
Table Hill, *Austral.* **220** B7
Table Mountain, *S. Dak., U.S.* **106** H2
Table Rock Lake, *Mo., U.S.* **107** V11
Table, Pointe de la, *Réunion, Fr.* **215** H17
Tabontebike, *Kiribati* **227** P14
Tabor City, *N.C., U.S.* **98** K11
Tabor, *Russ.* **161** D18
Tábor, *Czechia* **150** H11
Tabora, *Tanzania* **211** V4
Tabou, *Côte d'Ivoire* **204** Q12
Tabrīz, *Iran* **176** C12
Tabuaeran (Fanning Island), *Kiribati* **224** F12
Tabūk, *Saudi Arabia* **174** D5
Tabunifi, *F.S.M.* **226** E11
Taburiente, Caldera de, *Sp.* **214** Q3
Tabwémasana, peak, *Vanuatu* **229** F14
Tacheng, *China* **184** C4
Tacloban, *Philippines* **193** D15
Tacna, *Peru* **132** K6
Tacoma, *Wash., U.S.* **110** C4
Tacuarembó, *Uru.* **134** K14
Tacutu *see* East Takutu, river, *S. Amer.* **132** B9
Tad, *Pak.* **181** R4
Tadine, *New Caledonia, Fr.* **231** Q24
Tadjmout, *Alg.* **200** J12
Tadjoura, Gulf of, *Af.* **198** G11
Tadjourah, *Djibouti* **210** K10
Tadmur (Palmyra), *Syr.* **172** G10

Tadoule Lake, *Man., Can.* **87** K14
Tadoussac, *Que., Can.* **87** N20
Tadpatri, *India* **183** Q6
Tadrart Acacus, *Alg., Lib.* **201** J16
Taean, *N. Korea* **186** K5
Taean, *S. Korea* **187** N8
Taebaek, *S. Korea* **187** N11
Taebaeksanmaek, range, *S. Korea* **186** M11
Taech'o-do, island, *N. Korea* **186** C13
T'aech'ŏn, *N. Korea* **186** G6
Taedong-man, bay, *N. Korea* **186** L5
Taedong, *N. Korea* **186** F6
Taedong, river, *N. Korea* **186** K6
Taegu *see* Daegu, *S. Korea* **187** R11
Taegwan, *N. Korea* **186** G5
Taehongdan, *N. Korea* **186** C11
Taehŭng, *N. Korea* **186** E11
Taehŭng, *N. Korea* **186** G8
Taehwa-do, island, *N. Korea* **186** H4
Taejŏn *see* Daejeon, *S. Korea* **187** Q9
Taemun, *N. Korea* **186** G4
Taenarum *see* Ténaro, Akrotírio, *Gr.* **156** M10
Taenga, island, *Fr. Polynesia, Fr.* **231** B19
Taeryŏng, river, *N. Korea* **186** G6
T'aet'an, *N. Korea* **186** L5
Taeyŏnji-bong, peak, *N. Korea* **186** C10
Tafahi, island, *Tonga* **224** J10
Tafalla, *Sp.* **147** C13
Tafara, *Mali* **204** H9
Tafelberg, peak, *Curaçao, Neth.* **123** Q15
Tafí Viejo, *Arg.* **134** G9
Tafiré, *Côte d'Ivoire* **205** M13
Tafnidilt, *Mor.* **200** F5
Tafresh, *Iraq* **177** F14
Taft, *Tex., U.S.* **105** N17
Taft, *Calif., U.S.* **111** V7
Taft, *Iran* **177** J17
Tafunsak, *F.S.M.* **226** F11
Taga, *Samoa* **227** N17
Tagama, *Niger* **206** G8
Taganrog, *Russ.* **159** R10
Taganrogskiy Zaliv, *Russ.* **159** R10
Tagant, region, *Mauritania* **204** G9
Tagawa, *Japan* **189** R5
Taggafadi, *Niger* **206** E9
Tāghaz, *Afghan.* **180** M4
Taghit, *Alg.* **200** E10
Tagliamento, river, *It.* **152** C9
Tagounit, *Mor.* **200** F8
Tagtabazar, *Turkm.* **178** P9
Taguatinga, *Braz.* **133** H14
Tague Bay, *Virgin Is., U.S.* **122** Q4
Tagula, island, *P.N.G.* **222** R10
Tagum, *Philippines* **193** F15
Tagus Basin, *Atl. Oc.* **248** J11
Tagus *see* Tajo, river, *Sp.* **147** F13
Tagus *see* Tejo, river, *Port.* **146** H5
Tahaa, island, *Fr. Polynesia, Fr.* **231** B16
Tahaa, island, *Fr. Polynesia, Fr.* **231** G17
Tahan, peak, *Malaysia* **191** T9
Tahanea, island, *Fr. Polynesia, Fr.* **231** B19
Tahat, peak, *Alg.* **201** K13
Tahe, *China* **185** A15
Ţāherī, *Iran* **177** M16
Tahi, cape, *Fr. Polynesia, Fr.* **231** G14
Tahiti, island, *Fr. Polynesia, Fr.* **231** C17
Tahiti, island, *Fr. Polynesia, Fr.* **231** J22
Tāhlāb, Dasht-i, *Pak.* **181** N3
Tahlab, river, *Iran* **177** L23
Tahlab, river, *Iran, Pak.* **181** N3
Tahlequah, *Okla., U.S.* **105** C19
Tahoe City, *Calif., U.S.* **111** Q6
Tahoe, Lake, *Calif., U.S.* **111** Q6
Tahoka, *Tex., U.S.* **104** F12
Taholah, *Wash., U.S.* **110** C2
Tahoua, *Niger* **206** G7
Tahtalı Dağ, peak, *Turk.* **170** F11
Tahtalıdağı, peak, *Turk.* **170** J6
Tahuata, island, *Fr. Polynesia, Fr.* **231** E14
Tahueia, *Fr. Polynesia, Fr.* **231** M14
Tahulandang, island, *Indonesia* **193** H15
Tahuna, *Indonesia* **193** H15
Taiaro, island, *Fr. Polynesia, Fr.* **231** B19
Taibei *see* Taipei, *China* **185** M16
Taichung, *China* **185** N16
Taigbe, *Sierra Leone* **204** N9
Taihape, *N.Z.* **223** H21
Taiki, *Japan* **188** G14

Tailai, *China* **185** D15
Tain, *Fr.* **149** V13
Tainan, *China* **185** N15
Taiof, island, *P.N.G.* **228** B4
Taipei (Taibei), *China* **185** M16
Taiping, *Malaysia* **191** T8
Taipingot, peak, *N. Mariana Is., U.S.* **230** N8
Taishan Station, *Antarctica* **239** G19
Taitao, Península de, *Chile* **135** T6
Taitung, *China* **185** N16
Taivalkoski, *Fin.* **143** E16
Taiwan Strait, *China* **185** N15
Taiwan, *China* **185** N16
Taiyara, *Sudan* **203** S11
Taiyuan, *China* **184** H12
Taizhou (Jiaojiang, Haimen), *China* **185** L16
Taizhou, *China* **185** K15
Ta'izz, *Yemen* **174** P8
Tajarhī, *Lib.* **201** J18
Tajikistan, *Asia* **178** M12
Tajima, *Japan* **188** M12
Tajo (Tagus), river, *Sp.* **147** F13
Tajrīsh, *Iran* **177** E15
Tak, *Thai.* **190** K7
Taka Atoll, *Marshall Is.* **227** B17
Taka, island, *Marshall Is.* **227** H19
Takāb, *Iran* **176** E12
Takaba, *Kenya* **211** Q8
Takahe, Mount, *Antarctica* **238** L8
Takaiu, *F.S.M.* **226** J9
Takaka, *N.Z.* **223** J19
Takamatsu, *Japan* **189** Q7
Takaoka, *Japan* **189** N10
Takapau, *N.Z.* **223** H21
Takapoto, island, *Fr. Polynesia, Fr.* **231** A18
Takara Jima, *Japan* **189** V3
Takaroa, island, *Fr. Polynesia, Fr.* **231** A19
Takasaki, *Japan* **189** N11
Takatik, island, *F.S.M.* **226** J8
Takaungu, *Kenya* **211** U8
Takayama, *Japan* **189** N10
Take Shima, *Japan* **189** T4
Takefu, *Japan* **189** P9
Takengon, *Indonesia* **192** G4
Takeo *see* Takêv, *Cambodia* **191** P10
Takeo, *Japan* **189** R4
Takeshima *see* Dokdo, *S. Korea* **188** M5
Tākestān, *Iran* **177** E14
Taketa, *Japan* **189** R5
Takêv (Takeo), *Cambodia* **191** P10
Takhiatosh, *Uzb.* **178** K8
Takhtah Pul, *Afghan.* **180** L6
Takhtakūpir, *Uzb.* **178** K8
Taki, *P.N.G.* **222** P11
Taki, *P.N.G.* **228** C5
Takiéta, *Niger* **206** H8
Takikawa, *Japan* **188** F13
Taklimakan Desert, *Asia* **168** G8
Taklimakan Shamo, *China* **184** F3
Taksimo, *Russ.* **161** K17
Takua Pa, *Thai.* **191** Q6
Takua, island, *Fr. Polynesia, Fr.* **231** B20
Takum, *India* **182** G14
Takum, *Nigeria* **206** L9
Takume, island, *Fr. Polynesia, Fr.* **231** B20
Takuu Passage, *N.Z.* **229** L22
Takuua, island, *N.Z.* **229** L22
Tal, *Pak.* **180** G11
Tala Mungongo, *Angola* **208** K12
Talagang, *Pak.* **180** K11
Talaimannar, *Sri Lanka* **183** T7
Talak, *Niger* **206** E7
Talangbatu, *Indonesia* **192** K7
Talara, *Peru* **132** E1
Talas, *Kyrg.* **179** K13
Talasea, *P.N.G.* **222** P9
Talata Mafara, *Nigeria* **206** H7
Talaud, Kepulauan, *Indonesia* **193** G16
Talavera de la Reina, *Sp.* **146** G9
Talbert, Sillon de, *Fr.* **149** P4
Talbot Inlet, *Nunavut, Can.* **87** C16
Talbot Trough, *Arctic Oc.* **253** N14
Talbot, Cape, *Austral.* **220** B6
Talbotton, *Ga., U.S.* **98** M5
Talca, *Chile* **134** M6
Talcahuano, *Chile* **134** M6
Taldyqorghan, *Kaz.* **179** H16
Taleex, *Somaliland* **210** M13
Talguharai, *Sudan* **203** P14
Talhar, *Pak.* **181** S8
Tali Post, *S. Sudan* **203** X11

Taliabu, island, *Indonesia* **193** K15
Talihina, *Okla., U.S.* **105** D19
Taliwang, *Indonesia* **192** M12
Talkeetna Mountains, *Alas., U.S.* **113** J17
Talkeetna, *Alas., U.S.* **113** H16
Tall 'Afar, *Iraq* **176** E9
Tall al Abyaḍ al 'Atiq, *Syr.* **172** D11
Tall al Aḥmar (Til Barsip), *Syr.* **172** D10
Tall al 'Amārinah, ruin, *Egypt* **202** H11
Tall Birāk, *Syr.* **172** D14
Tall Ḥalaf, ruins, *Syr.* **172** C13
Tall Kalakh, *Syr.* **172** G7
Tall Tamir, *Syr.* **172** D13
Tall Ya'rūbīyah, *Syr.* **172** C16
Talladega, *Ala., U.S.* **103** J16
Tallahassee, *Fla., U.S.* **99** Q5
Tallapoosa, *Ga., U.S.* **98** L4
Tallapoosa, river, *Ala., U.S.* **103** L16
Tallard, *Fr.* **149** W14
Tallassee, *Ala., U.S.* **103** L16
Tallinn, *Est.* **143** J16
Tallulah, *La., U.S.* **102** L10
Talmont-Saint-Hilaire, *Fr.* **149** T5
Talnakh, *Russ.* **161** F13
Talo, peak, *Eth.* **210** L7
Talodi, *Sudan* **203** U11
Talofofo Bay, *Guam, U.S.* **230** Q5
Talofofo, *Guam, U.S.* **230** Q5
Taloga, *Okla., U.S.* **105** C15
Talok, *Indonesia* **193** H13
Talos Dome, *Antarctica* **239** Q15
Taloyoak, *Nunavut, Can.* **87** F15
Talparo, *Trin. & Tobago* **123** P23
Talshand, *Mongolia* **184** E8
Talsi, *Latv.* **143** L16
Taltal, *Chile* **134** F6
Tāluqān, *Afghan.* **180** F9
Talwood, *Qnsld., Austral.* **221** V14
Tam Kỳ, *Vietnam* **190** L13
Tam Quan, *Vietnam* **190** M13
Tama, *Iowa, U.S.* **107** N12
Tama, peak, *Angola* **208** N9
Tamakautonga, *N.Z.* **229** Q21
Tamale, *Ghana* **205** M16
Tamana, island, *Kiribati* **224** G8
Tamana, peak, *Trin. & Tobago* **123** Q23
Tamanrasset (Fort Laperrine), *Alg.* **201** K13
Tamarin, *Mauritius* **215** G19
Tamatoa, *Fr. Polynesia, Fr.* **231** N14
Tamayya, *W. Sahara* **200** J2
Tamazunchale, *Mex.* **116** H12
Tambacounda, *Senegal* **204** J8
Tambaga, *Mali* **204** K10
Tambalan, *Indonesia* **192** H11
Tambelan, Kepulauan, *Indonesia* **192** H8
Tambisan, *Malaysia* **193** G13
Tambohorano, *Madagascar* **213** E15
Tambov, *Russ.* **158** L11
Tambura, *S. Sudan* **203** X9
Tamch, *Mongolia* **184** D7
Tâmchekkeţ, *Mauritania* **204** G10
Tame, *Col.* **130** F8
Tamel Aike, *Arg.* **135** U7
Tamesna, *Niger* **206** E7
Tamgué, mountains, *Guinea, Senegal* **204** K9
Tamiahua, *Mex.* **116** H12
Tammerfors *see* Tampere, *Fin.* **143** H15
Tampa Bay, *Fla., U.S.* **99** U7
Tampa, *Fla., U.S.* **99** U7
Tampere (Tammerfors), *Fin.* **143** H15
Tampico, *Mex.* **116** G12
Tampin, *Malaysia* **191** U9
Tampok, river, *Fr. Guiana, Fr.* **131** G18
Tamsagbulag, *Mongolia* **185** D13
Tamshiyacu, *Peru* **132** E4
Tamu Massif, *Pac. Oc.* **246** F9
Tamu, *Myanmar* **190** F4
Tamuning, *Guam, U.S.* **230** N5
Tamworth, *N.S.W., Austral.* **221** W15
T'an-do, island, *N. Korea* **186** H4
Tan-Tan, *Mor.* **200** F5
T'ana Häyk' (Lake Tana), *Eth.* **210** K6
Tana, *Nor.* **143** B15
Tana, river, *Nor.* **143** B15
Tana, river, *Kenya* **211** S7;T8
Tanabe, *Japan* **189** R8

Tanacross, *Alas., U.S.* **113** H18
Tanafjorden, bay, *Arctic Oc.* **140** A9
Tanaga Island, *Alas., U.S.* **112** N4
Tanagomba Harbor, *Solomon Is.* **228** E10
Tanahbala, island, *Indonesia* **192** J4
Tanahgrogot, *Indonesia* **192** K12
Tanahjampea, island, *Indonesia* **193** M13
Tanahmasa, island, *Indonesia* **192** J5
Tanahmerah, *Indonesia* **193** M21
Tanami Desert, *Austral.* **220** D8
Tanami, *N.T., Austral.* **221** S7
Tanami, Mount, *Austral.* **220** E7
Tanana, *Alas., U.S.* **113** F16
Tanana, river, *Alas., U.S.* **113** F16
Tanapag, *N. Mariana Is., U.S.* **230** H8
Tanaro, river, *It.* **152** F3
Tanaunella, *It.* **152** M5
Tanavuso Point, *Fiji* **226** P5
Tanch'ŏn, *N. Korea* **186** F11
Ţăndărei, *Rom.* **155** F17
Tandaué, river, *Angola* **208** Q12
Tandil, *S. Amer.* **128** M6
Tandil, *Arg.* **135** N12
Tandlianwala, *Pak.* **180** L12
Tando Adam, *Pak.* **181** R8
Tando Allahyar, *Pak.* **181** R8
Tando Bago, *Pak.* **181** S9
Tando Muhammad Khan, *Pak.* **181** S8
Tandur, *India* **182** M7
Tanega Shima, *Japan* **189** T5
Tanega, island, *Asia* **168** G13
Tanen Range, *Thai.* **190** J6
Tanezrouft-n-Ahenet, *Alg.* **200** K11
Tanezrouft, *Alg.* **200** J10
Tanga Islands, *P.N.G.* **222** N10
Tanga Langua, cape, *Grenada* **123** J23
Tanga, *Tanzania* **211** V8
Tangail, *Bangladesh* **182** J12
Tanganyika, Lake, *Af.* **198** K8
Tange Promontory, *Antarctica* **239** B19
Tanger-Assilah (Tangier), *Mor.* **200** B7
Tanggu, *China* **185** G14
Tanggula Range, *Asia* **168** H9
Tanggula Shan, *China* **184** J5
Tangi, *India* **182** M10
Tangier *see* Tanger-Assilah, *Mor.* **200** B7
Tango, *Japan* **189** P8
Tangra Yumco, *China* **184** K4
Tangse, *Indonesia* **192** G4
Tangshan, *China* **185** G14
Tanguiéta, *Benin* **205** L17
Tanimbar Islands, *Asia* **168** M15
Tanimbar, Kepulauan, *Indonesia* **193** M17
Tanimbili, *Solomon Is.* **228** M6
Tanjong Payong, *Malaysia* **192** G10
Tanjong Penunjok, *Malaysia* **191** T9
Tanjong Sirik, *Malaysia* **191** T16
Tanjore *see* Thanjavur, *India* **183** S6
Tanjung Puting, *Indonesia* **192** L9
Tanjungbalai, *Indonesia* **192** H5
Tanjungbatu, *Indonesia* **192** H12
Tanjungkarang-Telubketung, *Indonesia* **192** L7
Tanjungpandan, *Indonesia* **192** K8
Tanjungpinang, *Indonesia* **192** J7
Tanjungredep, *Indonesia* **192** H12
Tank, *Pak.* **180** K10
Tankovo, *Russ.* **161** J13
Tanna, island, *Vanuatu* **229** M17
Tännäs, *Sw.* **142** H12
Tannu Ola, *Russ.* **161** M13
Tannur, ruins, *Jordan* **173** P6
Tanobato, *Indonesia* **192** J5
Tanōrah, *Afghan.* **180** J3
Tanot, *India* **182** G2
Tanoudert, *Mauritania* **204** E7
Tânout, *Niger* **206** G9
Ţanţā, *Egypt* **202** F11
Tanzania, *Af.* **211** W3
Tao, Ko, *Thai.* **191** Q7
Tao, river, *China* **184** J10
Taonan, *China* **185** D15
Taongi Atoll, *Marshall Is.* **227** A16
Taos, *N. Mex., U.S.* **109** Q13
Taoudenni, *Mali* **205** D14
Taourirt, *Mor.* **200** C9
Taoussa, *Mali* **205** G16

Taouz, *Mor.* **200** E9
Tapa, *Est.* **143** J16
Tapachula, *Mex.* **117** L15
Tapajós, *river, Braz.* **132** E10
Tapak, *Indonesia* **192** K6
Tapan, *Indonesia* **192** K6
Tapana, island, *Tonga* **227** H23
Tapanahoni, river, *Suriname* **131** H17
Tapauá, river, *Braz.* **132** F7
Tapawera, *N.Z.* **223** J19
Taphan Hin, *Thai.* **190** L8
Tapi, river, *India* **182** L4
Tapiche, river, *Peru* **132** F4
Tapini, *P.N.G.* **222** Q8
Tapirapecó, *S. Amer.* **130** H12
Tapita, *Liberia* **204** P11
Tapiwa, *Kiribati* **226** M11
Taplejung, *Nepal* **182** G11
Tappahannock, *Va., U.S.* **98** E13
Taptong, *N. Korea* **186** F11
Tapuaenuku, peak, *N.Z.* **223** K19
Tapuaetai, island, *N.Z.* **229** J20
Tapuamu, *Fr. Polynesia, Fr.* **231** G17
Tapul Group, *Philippines* **193** G13
Tapunui, island, *N.Z.* **229** K21
Taputimu, *American Samoa, U.S.* **230** E6
Ţaqah, *Oman* **175** M15
Taquari, river, *Braz.* **132** K10
Tar, river, *N.C., U.S.* **98** H12
Tara, *Russ.* **160** J10
Tara, Hill of, *Ire.* **145** Q6
Tara, river, *Montenegro* **154** H9
Tarābulus (Tripoli), *Lib.* **201** D17
Taraira *see* Traíra, river, *S. Amer.* **130** K9
Tarakan, *Indonesia* **192** H12
Tarakan, island, *Indonesia* **192** H12
Tarakite Iti, island, *N.Z.* **229** J23
Taralga, *N.S.W., Austral.* **221** Y14
Tarancón, *Sp.* **146** G11
Taransay *see* Tarasaigh, island, *Scot., U.K.* **144** G6
Taranto, *It.* **153** N14
Taranto, Golfo di, *It.* **153** N14
Tarapacá, *Col.* **130** L9
Tarapaina, *Solomon Is.* **228** F11
Tarapoto, *Peru* **132** F3
Taraqu, ruins, *Afghan.* **180** M2
Tarara, *P.N.G.* **228** B5
Tarare, *Fr.* **149** U12
Tarasa Dwip, island, *India* **183** T15
Tarasaigh (Taransay), island, *Scot., U.K.* **144** G6
Tarascon, *Fr.* **149** Y9
Tarat, *Alg.* **201** H15
Tarata, *Peru* **132** K6
Taratai, *Kiribati* **227** Q14
Taratai, island, *Kiribati* **227** Q14
Tarauacá, *Braz.* **132** F5
Tarauacá, river, *Braz.* **132** F5
Taravai, *Fr. Polynesia, Fr.* **231** Q17
Taravai, island, *Fr. Polynesia, Fr.* **231** E23;Q17
Taravao, *Fr. Polynesia, Fr.* **231** J22
Taravo, river, *Fr. Polynesia, Fr.* **149** Y2
Tarawa (Bairiki), *Kiribati* **227** R14
Taraz, *Kaz.* **179** K13
Tarazit Massif, *Af.* **198** F5
Tarazona de la Mancha, *Sp.* **147** H13
Tarazona, *Sp.* **147** D13
Tarbagannakh, *Russ.* **161** G19
Tarbagatai Shan, *China, Kaz.* **184** C4
Tarbaghatay Zhotasy, *Kaz.* **179** F17
Tarbaj, *Kenya* **211** R8
Tarbat Ness, cape, *Scot., U.K.* **144** H9
Tarbert *see* An Tairbeart, *Scot., U.K.* **144** G6
Tarbert, *Scot., U.K.* **144** L8
Tarbes, *Fr.* **149** Y7
Tarboro, *N.C., U.S.* **98** H13
Tarcoola, *S. Aust., Austral.* **221** W9
Taree, *N.S.W., Austral.* **221** X15
Tarfaya, *Mor.* **200** F4
Tarfside, *Scot., U.K.* **144** J10
Targhee Pass, *Idaho, Mont., U.S.* **108** F7
Târgoviște, *Rom.* **155** F15
Târgu Jiu, *Rom.* **155** F13
Târgu-Mureș, *Rom.* **155** C14
Tarhmert, *Niger* **206** E9
Tari, *P.N.G.* **222** P6
Ţarīf, *U.A.E.* **175** G15
Tarifa, *Sp.* **146** N8
Tarifa, Punta de, *Sp.* **146** N8
Tarija, *Bol.* **132** L8
Tarikere, *India* **183** Q4

Tariku-Taritatu Plain, *Indonesia* **193** K20
Tarim Basin, *Asia* **168** G8
Tarim Pendi, *China* **184** F3
Tarim, river, *China* **184** F3
Tarīn Kōṯ, *Afghan.* **180** K6
Tarkhankut, Mys, *Russ.* **159** S6
Tarkio, *Mo., U.S.* **107** Q9
Tarko Sale, *Russ.* **160** G12
Tarkwa, *Ghana* **205** Q15
Tarma, *Peru* **132** H4
Tarn, river, *Fr.* **149** W11
Tärnaby, *Sw.* **142** F12
Tarnak, river, *Afghan.* **180** L6
Tarnobrzeg, *Pol.* **151** G17
Tarnogskiy Gorodok, *Russ.* **158** E12
Tarnów, *Pol.* **151** G16
Taroom, *Qnsld., Austral.* **221** U15
Taroudant, *Mor.* **200** E6
Tarpon Springs, *Fla., U.S.* **99** U7
Tarpum Bay, *Bahamas* **122** E6
Tarpum Bay, *Bahamas* **122** F6
Tarquinia, *It.* **152** K7
Tarrabool Lake, *Austral.* **220** D9
Tarrafal, *Cabo Verde* **215** B14
Tarrafal, *Cabo Verde* **215** B15
Tarrafal, *Cabo Verde* **215** D16
Tarragona, *Sp.* **147** E16
Tarras, *N.Z.* **223** N17
Tarsia, *It.* **153** P13
Tarso Emisou, peak, *Chad* **207** D14
Tarsus, *Turk.* **170** J10
Tarta, *Turkm.* **178** L4
Tartagal, *Arg.* **134** D10
Tartas, *Fr.* **149** X6
Tartu, *Est.* **143** K17
Ṭarṭūs (Tortosa), *Syr.* **172** G7
Tarus Seachannel, *Pac. Oc.* **247** D14
Tarutung, *Indonesia* **192** H5
Tarvisio, *It.* **152** C9
Tarxien, *Malta* **163** K23
Tas Tumus, *Russ.* **161** G17
Tasa, *N. Korea* **186** H4
Tasāwah, *Lib.* **201** H17
Tasböget, *Kaz.* **178** J10
Tasek Dampar, lake, *Malaysia* **191** U9
Taseyevo, *Russ.* **161** K13
Tāsh Guẕar, *Afghan.* **180** F7
Tash-Kömür, *Kyrg.* **179** L13
Tashanta, *Russ.* **160** M12
Tashkent *see* Toshkent, *Uzb.* **178** L12
Tasiilaq (Ammassalik), *Greenland, Den.* **89** P10
Tasikmalaya, *Indonesia* **192** M8
Tasiusaq Bugt, *Greenland, Den.* **88** J5
Tasiusaq, *Greenland, Den.* **88** J5
Tasiusaq, *Greenland, Den.* **89** T8
Task, *Niger* **206** G10
Taskan, *Russ.* **161** F20
Tasker, *Niger* **206** G10
Taskesken, *Kaz.* **179** G16
Taşköprü, *Turk.* **170** C9
Tasman Basin, *Pac. Oc.* **246** P9
Tasman Bay, *N.Z.* **223** J19
Tasman Escarpment, *Ind. Oc.* **251** Q22
Tasman Fracture Zone, *Pac. Oc.* **255** F17
Tasman Peninsula, *Austral.* **220** M8
Tasman Plain, *Pac. Oc.* **246** P8
Tasman Sea, *Pac. Oc.* **224** N6
Tasman, *N.Z.* **223** J19
Tasmania, island, *Austral.* **220** M7
Tasmantid Seamounts, *Pac. Oc.* **255** D22
Tășnad, *Rom.* **154** B12
Taşova, *Turk.* **170** D11
Tassara, *Niger* **206** F7
Tassialouc, Lac, *Que., Can.* **87** K18
Tassili oua-n-Ahaggar, *Alg.* **201** L13
Tassili-n-Ajjer, *Alg.* **201** H14
Tasure, *Solomon Is.* **228** D7
Tataacho Point, *N. Mariana Is., U.S.* **230** N8
Tatabánya, *Hung.* **151** L14
Tatafa, island, *Tonga* **227** M23
Tatakoto, island, *Fr. Polynesia, Fr.* **231** C22
Tatamba, *Solomon Is.* **228** E10
Tataouine, *Tunisia* **201** D16
Tatar Strait, *Pac. Oc.* **168** D13
Tatar Trough, *Pac. Oc.* **246** D7
Tatarbunary, *Ukr.* **159** S5

Tatarsk, *Russ.* **160** K10
Tatarskiy Proliv, *Russ.* **161** J21
Tatau, island, *P.N.G.* **222** N10
Tate, *Ga., U.S.* **98** K5
Tateyama, *Japan* **189** Q12
Tathlina Lake, *N.W.T., Can.* **86** H10
Tathlīth, *Saudi Arabia* **174** K8
Tatkon, *Myanmar* **190** H5
Tatnam, Cape, *Man., Can.* **87** L15
Tatry, range, *Pol., Slovakia* **151** J15
Tattershall, *Eng., U.K.* **145** R14
Tatuí, *Braz.* **133** M13
Tatuke, *Liberia* **204** Q11
Tatum, *N. Mex., U.S.* **109** U15
Tatvan, *Turk.* **171** G16
Ta‘ū, *American Samoa, U.S.* **230** B10
Tau, island, *Tonga* **227** P24
Ta‘ū, island, *American Samoa, U.S.* **230** B10
Tauá, *Braz.* **133** F16
Tauak Passage, *F.S.M.* **226** J8
Taubaté, *Braz.* **133** M14
Tauenai Channel, *F.S.M.* **226** K7
Tauere, island, *Fr. Polynesia, Fr.* **231** C20
Taugarau, island, *Fr. Polynesia, Fr.* **231** Q14
Tauhunu Landing, cape, *N.Z.* **229** H22
Tauhunu, *N.Z.* **229** H22
Tauhunu, island, *N.Z.* **229** H22
Taula, island, *Tonga* **227** J23
Taulaga, *Swains I., U.S.* **230** B5
Taum Sauk Mountain, *Mo., U.S.* **107** U15
Taumako, island, *Solomon Is.* **228** K7
Taumarunui, *N.Z.* **223** G20
Taumatawhakatangihangakoau-
auotamateapokaiwhenuakitan-
atahu, *N.Z.* **223** G20
Taunga, island, *Tonga* **227** H23
Taunggok, *Myanmar* **190** H4
Taunggyi, *Myanmar* **190** H5
Taungoo, *Myanmar* **190** J5
Taungup Pass, *Myanmar* **190** J4
Taunsa, *Pak.* **180** M10
Taunton, *Mass., U.S.* **97** L15
Taunton, *Eng., U.K.* **145** V10
Taupo, *N.Z.* **223** G21
Tauragė, *Lith.* **143** M16
Tauranga, *N.Z.* **223** F21
Taurus Mountains, *Asia* **168** E3
Taurus *see* Toros Dağları, *Turk.* **170** J8
Taūshyq, *Kaz.* **178** H4
Tauste, *Sp.* **147** D13
Tautama, island, *Pitcairn I.* **228** Q6
Tautau, island, *Fr. Polynesia, Fr.* **231** G17
Tautira, *Fr. Polynesia, Fr.* **231** J23
Tautu, *N.Z.* **229** H20
Tautua, island, *N.Z.* **229** L22
Tauu Islands, *P.N.G.* **228** A7
Tavaerua, island, *N.Z.* **229** H20
Tavai, *Wallis & Futuna Is., Fr.* **230** Q11
Tavan Bogd Uul, *Mongolia* **184** C6
Tavannes, *Switz.* **152** A2
Tavas, *Turk.* **170** G4
Tavda, *Russ.* **160** H9
Taverna, *It.* **153** Q14
Taverner Bay, *Nunavut, Can.* **87** G18
Tavernes de la Valldigna, *Sp.* **147** H15
Tavernes, *Fr.* **149** X14
Tavernier, *Fla., U.S.* **99** Y10
Taveta, *Kenya* **211** U7
Taveta, *Tanzania* **211** X5
Taveuni, island, *Fiji* **226** N7
Tavu-Na-Sici, island, *Fiji* **226** Q8
Tavua, *Fiji* **226** P5
Tawai, *India* **182** G16
Tawake, *Fiji* **226** N7
Tawang, *India* **182** G13
Tawas City, *Mich., U.S.* **100** J12
Tawau, *Malaysia* **192** G12
Taweisha, *Sudan* **203** T8
Tawi Tawi, island, *Philippines* **193** G13
Tāwurghā', *Lib.* **201** E18
Taxco, *Mex.* **116** J11
Taxkorgan, *China* **184** F1
Tây Ninh, *Vietnam* **191** P11
Tay, Firth of, *Scot., U.K.* **144** K11
Tay, Loch, *Scot., U.K.* **144** K9
Tayan, *Indonesia* **192** J9
Tayandu, Kepulauan, *Indonesia* **193** L18

Tāybād, *Iran* **177** F22
Tayeeglow, *Somalia* **211** Q11
Taýgetos, Óros, *Gr.* **156** L10
Taygonos, *Russ.* **161** E21
Taylakova, *Russ.* **160** J11
Taylor Dome, *Antarctica* **239** N15
Taylor Highway, *Alas., U.S.* **113** G19
Taylor, *Mich., U.S.* **100** M12
Taylor, *Tex., U.S.* **105** K17
Taylor, Mount, *N. Mex., U.S.* **109** S11
Taylors, *S.C., U.S.* **98** J7
Taylorsville, *Miss., U.S.* **102** M12
Taylorville, *Ill., U.S.* **101** S5
Taymā', *Saudi Arabia* **174** E6
Taymir, *Russ.* **161** E14
Taymyr Peninsula, *Asia* **168** C9
Taymyr, *Russ.* **161** E14
Taymyr, Lake, *Asia* **168** C9
Taypaq, *Kaz.* **178** F5
Tayshet, *Russ.* **161** K14
Taytay, *Philippines* **193** E13
Taywarah, *Afghan.* **180** J5
Tayynsha (Krasnoarmeysk), *Kaz.* **178** C12
Taz, river, *Russ.* **160** G12
Taza, *Mor.* **200** C9
Tāzah Khurmātū, *Iraq* **176** F10
Tazawako, *Japan* **188** K12
Tazolé, *Niger* **206** F9
Tazovskiy Poluostrov, *Russ.* **160** F12
Tazovskiy, *Russ.* **160** G12
Tazzarine, *Mor.* **200** E8
Tba P'arvani, *Rep. of Georgia* **171** C18
Tbilisi, *Rep. of Georgia* **171** C19
Tchabal Gangdaba, *Cameroon* **206** L11
Tchaourou, *Benin* **205** M18
Tchibanga, *Gabon* **208** F8
Tchin-Tabaradène, *Niger* **206** G7
Tcholliré, *Cameroon* **206** L11
Tczew, *Pol.* **151** B14
Te Afualiku, island, *Tuvalu* **228** L10
Te Anau, *N.Z.* **223** N16
Te Anu, island, *Solomon Is.* **228** N7
Te Ava Fuagea, *Tuvalu* **228** N9
Te Ava i te Lape, *Tuvalu* **228** L10
Te Ava Mateika, *Tuvalu* **228** P10
Te Ava Pua Pua, *Tuvalu* **228** P10
Te Ava Tepuka, *Tuvalu* **228** M9
Te Hapua, *N.Z.* **223** C19
Te Ika-a-Māui *see* North Island, island, *N.Z.* **223** G20
Te Karaka, *N.Z.* **223** G22
Te Koutu Point, *N.Z.* **229** H20
Te Manga, peak, *N.Z.* **229** E23
Te Nggano, bay, *Solomon Is.* **228** H10
Te Puka, island, *N.Z.* **229** H22
Te Raupua, peak, *N.Z.* **223** D19
Te Tautua, *N.Z.* **229** L22
Te Teko, *N.Z.* **223** F22
Te Waipounamu *see* South Island, island, *N.Z.* **223** N18
Teague, *Tex., U.S.* **105** H18
Teahupoo, *Fr. Polynesia, Fr.* **231** K23
Teaiti Point, *N.Z.* **229** D23
Teakava, island, *Fr. Polynesia, Fr.* **231** Q18
Teaoraereke, *Kiribati* **227** R14
Tearinibai, *Kiribati* **227** P14
Teavaro, *Fr. Polynesia, Fr.* **231** H20
Teberda, *Russ.* **159** U12
Tébessa, *Alg.* **201** C15
Tebingtinggi, *Indonesia* **192** H5
Techirghiol, *Rom.* **155** G18
Techla, *W. Sahara* **200** K3
Technití Límni Kremastón, *Gr.* **156** G8
Technití Límni Tavropoú, *Gr.* **156** F9
Tecka, *Arg.* **135** R7
Tecomán, *Mex.* **116** J9
Tecpan, *Mex.* **116** K11
Tecuala, *Mex.* **116** H9
Tecuci, *Rom.* **155** D17
Tecumseh, *Mich., U.S.* **101** N11
Tecumseh, *Okla., U.S.* **105** D17
Tecumseh, *Nebr., U.S.* **107** R9
Ted Ceidaar Dabole, *Somalia* **211** P11
Teec Nos Pos, *Ariz., U.S.* **109** Q9
Teel, *Mongolia* **184** D9
Tefala, island, *Tuvalu* **228** P9
Tefarerii, *Fr. Polynesia, Fr.* **231** K15
Tefé, *Braz.* **132** D7
Tefé, river, *Braz.* **132** E7
Tefenni, *Turk.* **170** H5
Tegal, *Indonesia* **192** M9

Tegalhusi, *Sudan* **203** Q14
Tegina, *Nigeria* **206** K7
Teglio, *San Marino* **163** L14
Tégouma, river, *Niger* **206** G9
Tégua, island, *Vanuatu* **229** C13
Tegucigalpa, *Hond.* **117** L17
Teguidda-n-Tessoumt, *Niger* **206** F8
Teguise, *Sp.* **214** P7
Tehachapi Mountains, *Calif., U.S.* **111** W8
Tehachapi Pass, *Calif., U.S.* **111** V8
Tehachapi, *Calif., U.S.* **111** W8
Tehamiyam, *Sudan* **203** P15
Tehek Lake, *Nunavut, Can.* **87** H14
Tehrān, *Iran* **177** E15
Tehuacán, *Mex.* **116** J12
Tehuantepec Ridge, *Pac. Oc.* **247** H20
Tehuantepec, *Mex.* **117** K13
Tehuantepec, Golfo de, *Mex.* **117** L14
Tehuantepec, Istmo de, *Mex.* **117** K14
Tehuata (Rekareka), island, *Fr. Polynesia, Fr.* **231** B20
Teide, Pico de, *Sp.* **214** Q4
Teignmouth, *Eng., U.K.* **145** W10
Tejen, *Turkm.* **178** N8
Tejo (Tagus), river, *Port.* **146** H5
Tekamah, *Nebr., U.S.* **107** P9
Tekax, *Mex.* **117** H16
Tekeli, *Kaz.* **179** H16
Tekes, *China* **184** E4
Tekezē, river, *Eth.* **210** J7
Tekirdağ, *Turk.* **170** C3
Tekirova, *Turk.* **170** J6
Tekkali, *India* **182** M9
Tekman, *Turk.* **171** E16
Teknaf, *India* **182** L14
Tekoa, *Wash., U.S.* **110** C9
Tekokota, island, *Fr. Polynesia, Fr.* **231** C20
Tekopua, island, *N.Z.* **229** H21
Tel Aviv-Yafo, *Israel* **172** M5
Tel Lakhish, ruins, *Israel* **173** N5
Télabit, *Mali* **205** F16
T'elavi, *Rep. of Georgia* **171** B19
Telde, *Sp.* **214** R5
Tele, river, *D.R.C.* **209** C16
Telebekelel Ngerael, strait, *Palau* **226** C4
Telefomin, *P.N.G.* **222** P5
Telegraph Creek, *B.C., Can.* **86** H8
Telekitonga, island, *Tonga* **227** N23
Telekivavau, island, *Tonga* **227** N23
Telele Island, *Tuvalu* **228** Q10
Telemark, *Eur.* **140** D7
Telén, *Arg.* **134** M9
Teles Pires (São Manuel), river, *Braz.* **132** F10
Telescope Peak, *Calif., U.S.* **111** U9
Telescope Point, *Grenada* **123** K23
Télimélé, *Guinea* **204** L3
Tell Atlas, range, *Af.* **198** C4
Tell City, *Ind., U.S.* **101** U8
Tell Tayinat, ruins, *Turk.* **170** J11
Teller, *Alas., U.S.* **112** F12
Tellicherry *see* Thalassery, *India* **183** R4
Telluride, *Colo., U.S.* **109** P11
Telsen, *Arg.* **135** R9
Telšiai, *Lith.* **143** M16
Teluk Intan, *Malaysia* **191** T8
Telukbutun, *Indonesia* **192** G8
Telukdalem, *Indonesia* **192** J4
Tema, *Ghana* **205** P16
Tematagi, island, *Fr. Polynesia, Fr.* **231** E21
Tembagapura, *Indonesia* **193** L20
Tembenchi, *Russ.* **161** H14
Tembo, *D.R.C.* **208** J12
Temerin, *Serb.* **154** E10
Temerloh, *Malaysia* **191** U9
Teminikov, *Russ.* **158** K12
Temirtaū, *Kaz.* **179** E13
Temoe (Timoe), island, *Fr. Polynesia, Fr.* **231** F23
Tempe, *Ariz., U.S.* **109** U7
Témpi, valley, *Gr.* **156** E10
Tempio Pausania, *It.* **152** M4
Temple Bay, *Austral.* **220** B12
Temple, *Tex., U.S.* **105** J17
Tempué, *Angola* **208** N12
Temryuk, *Russ.* **159** S9
Temryukskiy Zaliv, *Russ.* **159** S8
Temse, *Belg.* **148** K11
Temuco, *Chile* **135** P6
Temuka, *N.Z.* **223** M18

Ten Degree Channel, *India* **183** S14
Ten Sleep, *Wyo., U.S.* **108** G11
Ten Thousand Islands, *Fla., U.S.* **99** X9
Tena Kourou, peak, *Burkina Faso* **205** L13
Tena, *Ecua.* **130** K4
Tenali, *India* **183** P7
Tenao, *Monaco* **163** B22
Tenararo, island, *Fr. Polynesia, Fr.* **231** E22
Ténaro, Akrotírio (Taenarum, Matapás), *Gr.* **156** M10
Tenarunga, island, *Fr. Polynesia, Fr.* **231** E22
Tenasserim, *Myanmar* **190** L6
Tenasserim, *Myanmar* **191** N7
Tenby, *Wales, U.K.* **145** U8
Tench, island, *P.N.G.* **222** M9
Tendaho, *Eth.* **210** K9
Tende, *Fr.* **149** W16
Tendō, *Japan* **188** L12
Tendrara, *Mor.* **200** D9
Tendürek Dağı, peak, *Turk.* **171** F18
Ténenkou, *Mali* **205** J13
Ténéré du Tafassâsset, *Niger* **206** D9
Ténéré, desert, *Chad, Niger* **206** E10
Tenerife, island, *Sp.* **214** Q5
Tengako, island, *Tuvalu* **228** M12
Tengasu Island, *Tuvalu* **228** Q10
Tengchong, *China* **184** N8
Tengiz Köli, *Kaz.* **178** E12
Tengiz, Lake, *Asia* **168** E7
Tengiz, oil well, *Kaz.* **178** G5
Tengréla, *Côte d'Ivoire* **204** L12
Tengzhou, *China* **185** J14
Tenino, *Wash., U.S.* **110** D3
Tenkasi, *India* **183** T5
Tenkiller Ferry Lake, *Okla., U.S.* **105** C19
Tenkodogo, *Burkina Faso* **205** L16
Tennant Creek, *N.T., Austral.* **221** R9
Tennessee Pass, *Colo., U.S.* **108** M12
Tennessee, *U.S.* **103** F15
Tennessee, river, *U.S.* **103** E14
Tennille, *Ga., U.S.* **98** M7
Tenoka, island, *Fr. Polynesia, Fr.* **231** Q17
Tenosique, *Mex.* **117** K15
Tensaw, river, *Ala., U.S.* **103** N14
Teófilo Otoni, *Braz.* **133** K15
Teora, *It.* **152** M12
Tepa Point, *N.Z.* **229** Q21
Tepa, *Wallis & Futuna Is., Fr.* **230** R9
Tepati, *Fr. Polynesia, Fr.* **231** K23
Tepé, *Eq. Guinea* **214** L8
Tepe, peak, *Kos., Serb.* **154** J11
Tēpī, *Eth.* **211** N5
Tepic, *Mex.* **116** G9
Tepu, *Fr. Polynesia, Fr.* **231** N14
Tepuka Island, *Tuvalu* **228** M10
Ter Tholen Fracture Zone, *Ind. Oc.* **250** N11
T'er, river, *Eth.* **211** N9
Téra, *Niger* **206** H4
Terae, island, *N.Z.* **229** L22
Teraina (Washington Island), *Kiribati* **224** F12
Terakeka, *S. Sudan* **203** X12
Teramo, *It.* **152** J10
Teratak, *Indonesia* **192** J9
Tercan, *Turk.* **171** E15
Terceira, island, *Port.* **146** P4
Tereai, island, *Fr. Polynesia, Fr.* **231** N16
Terekhovka, *Belarus* **158** M6
Terekli Mekteb, *Russ.* **159** T14
Terekty, *Kaz.* **179** F18
Terenos, *Braz.* **132** L11
Teresina, *Braz.* **133** E15
Terhazza, ruin, *Mali* **205** C13
Terihi, island, *Fr. Polynesia, Fr.* **231** E15
Termas de Río Hondo, *Arg.* **134** G9
Terme, *Turk.* **170** D12
Termini Imerese, *It.* **153** S10
Termini Imerese, Golfo di, *It.* **153** S10
Términos, Laguna de, *Mex.* **117** K15
Termit-Kaoboul, *Niger* **206** G10
Termit, *Niger* **206** G10
Termita, *Russ.* **159** S14
Termiz, *Uzb.* **178** N11
Termoli, *It.* **152** K11
Ternate, *Indonesia* **193** J16
Terneuzen, *Neth.* **148** K11

Ternitz, *Aust.* **150** L12
Ternopil', *Ukr.* **159** P3
Terra Nova Bay, *Antarctica* **239** P14
Terrace Beachport, *Ind. Oc.* **251** N22
Terrace Ceduna, *Ind. Oc.* **251** N20
Terrace, *B.C., Can.* **86** K7
Terracina, *It.* **152** L9
Terralba, *It.* **153** P3
Terrassa, *Sp.* **147** D17
Terre Haute, *Ind., U.S.* **101** S7
Terre Rouge, *Mauritius* **215** F20
Terre-de-Bas, *Guadeloupe, Fr.* **123** H14
Terre-de-Bas, island, *Guadeloupe, Fr.* **123** H14
Terre-de-Haut, *Guadeloupe, Fr.* **123** H14
Terre-de-Haut, island, *Guadeloupe, Fr.* **123** H14
Terrebonne Bay, *La., U.S.* **102** R10
Terreiro Velho, *Sao Tome and Principe* **215** B21
Terrell, *Tex., U.S.* **105** G18
Terry, *Mont., U.S.* **108** D12
Tersakan Gölü, *Turk.* **170** G8
Terschelling, island, *Neth.* **148** G12
Tertenia, *It.* **153** P4
Teru, *Pak.* **180** G11
Teruel, *Sp.* **147** F14
Teseney, *Eritrea* **210** H6
Teshekpuk Lake, *Alas., U.S.* **113** B15
Teshi, *Ghana* **205** P16
Teshikaga, *Japan* **188** F15
Teshio Sanchi, *Japan* **188** F13
Teshio, *Japan* **188** E13
Teshio, river, *Japan* **188** E13
Tesiyn, river, *Mongolia* **184** C8
Teslić, *Bosn. & Herzg.* **154** F8
Teslin, *Yukon, Can.* **86** G8
Tesovo Netyl'skiy, *Russ.* **158** F6
Tessalit, *Mali* **205** E16
Tessaoua, *Niger* **206** H8
Tetas, Punta, *Chile* **134** E6
Tete, *Mozambique* **212** D10
Tetepare, island, *Solomon Is.* **228** E7
Tetere, *Russ.* **161** J14
Teteven, *Bulg.* **155** H14
Tetford, *Eng., U.K.* **145** R14
Tetiaroa, island, *Fr. Polynesia, Fr.* **231** B17
Tetlin, *Alas., U.S.* **113** H19
Teton Range, *Wyo., U.S.* **108** G8
Teton, river, *Mont., U.S.* **108** C8
Tétouan, *Mor.* **200** C8
Tetovo-Skopje, *N. Maced.* **154** K10
Tetyushi, *Russ.* **158** J14
Teuaua (Hat), island, *Fr. Polynesia, Fr.* **230** E11
Teuco, river, *Arg.* **134** F11
Teulada, *It.* **153** Q3
Teulada, Capo, *It.* **153** Q3
Teuri Tō, *Japan* **188** F13
Teutonic College, *Vatican City* **163** Q16
Tevahavaha, island, *N.Z.* **229** J23
Tevaro, island, *Fr. Polynesia, Fr.* **231** N18
Teveiroa, island, *Fr. Polynesia, Fr.* **231** G14
Tevere (Tiber), river, *It.* **152** J8
Teverya (Tiberias), *Israel* **172** K6
Tewantin, *Qnsld., Austral.* **221** V16
Texarkana, *Tex., U.S.* **105** F20
Texas City, *Tex., U.S.* **105** L20
Texas-Lousiana Shelf, *Atl. Oc.* **248** L1
Texas, *Tex., U.S.* **105** J14
Texel, island, *Neth.* **148** G12
Texhoma, *Okla., U.S.* **104** B12
Texico, *N. Mex., U.S.* **109** S15
Texline, *Tex., U.S.* **104** B11
Teykovo, *Russ.* **158** H10
Tezpur, *India* **182** H14
Tǝrtǝr, *Azerb.* **171** D21
Tfaritiy, *W. Sahara* **200** H5
Th. Sørensen Land, *Greenland, Den.* **88** K12
Tha Sala, *Thai.* **191** Q7
Thabana Ntlenyana, peak, *Lesotho* **212** K4
Thabaung, *Myanmar* **190** K4
Thaga Pass, *India* **182** E7
Thagyettaw, *Myanmar* **190** M6
Thái Bình, *Vietnam* **190** H11
Thái Hoà, *Vietnam* **190** J11
Thái Nguyên, *Vietnam* **190** G11
Thailand, *Asia* **190** L7

Thailand, Gulf of, *Pac. Oc.* **168** K10
Thakhèk, *Laos* **190** K10
Thal Desert, *Pak.* **180** L10
Thal, *Pak.* **180** J10
Thalassery (Tellicherry), *India* **183** R4
Thale Luang, lake, *Thai.* **191** R7
Thamarīt, *Oman* **175** L15
Thames, River, *Eng., U.K.* **145** U15
Thamūd, *Yemen* **174** M12
Thana, *Pak.* **181** N4
Thandaung, *Myanmar* **190** J5
Thane, *India* **182** M3
Thanet, island, *Eng., U.K.* **145** U16
Thanh Hóa, *Vietnam* **190** J11
Thành Phố Hồ Chí Minh (Ho Chi Minh City Saigon), *Vietnam* **191** P12
Thanjavur (Tanjore), *India* **183** S6
Thanlwin *see* Salween, river, *Asia* **168** H9
Thar Desert *see* Great Indian Desert, *India, Pak.* **181** Q9
Tharabwin, *Myanmar* **191** N7
Thargomindah, *Qnsld., Austral.* **221** V12
Tharp Fracture Zone, *Pac. Oc.* **247** R16
Tharros, ruin, *It.* **153** P3
Thásos (Límín), *Gr.* **157** C14
Thasos, island, *Gr.* **157** C14
Thatch Cay, *Virgin Is., U.S.* **122** M9
Thaton, *Myanmar* **190** K6
Thatta, *Pak.* **181** S8
Thau, Bassin de, *Fr.* **149** X11
Thaungdut, *Myanmar* **190** F4
Thayawthadanngyi Kyun, island, *Myanmar* **191** N6
Thayer, *Mo., U.S.* **107** V14
Thayet, *Myanmar* **190** J4
The Alley, *Jam.* **122** K9
The Barony, *Scot., U.K.* **144** E10
The Baths, *Virgin Is. U.K.* **122** R8
The Bight, *Bahamas* **122** B9
The Bluff, *Bahamas* **122** D5
The Bottom, *Saba, Neth.* **121** B15
The Brothers *see* Al Ikhwān, *Yemen* **175** Q15
The Brothers, islands, *Bahamas* **119** H13
The Cheviot, peak, *Eng., U.K.* **145** M11
The Cockpit Country, *Jam.* **122** H7
The Cow, island, *Ire.* **145** U2
The Crane, *Barbados* **123** L20
The Dalles, *Oreg., U.S.* **110** F5
The Deeps, bay, *Scot., U.K.* **144** C12
The Dogs, islands, *Virgin Is., U.K.* **122** Q8
The Galleries, *Gibraltar, U.K.* **163** N23
The Heads, *Oreg., U.S.* **110** K1
The Isles Lagoon, *Kiribati* **226** Q11
The Little Minch, bay, *Scot., U.K.* **144** H6
The Marls, islands, *Bahamas* **122** B9
The Minch, bay, *Scot., U.K.* **144** G7
The Moors, *Scot., U.K.* **145** N9
The Narrows, *Bermuda* **119** A16
The Narrows, channel, *Bahamas* **122** E3
The Narrows, channel, *Virgin Is., U.K., Virgin Is., U.S.* **122** M11
The Narrows, channel, *St. Kitts & Nevis* **123** B18
The Naze, cape, *Eng., U.K.* **145** T15
The Ox Mountains *see* Slieve Gamph, *Ire.* **145** P4
The Pas, *Man., Can.* **87** M13
The Pennines, region, *Eng., U.K.* **145** P11
The Remarkables, peak, *N.Z.* **223** N16
The Rope, cape, *Pitcairn I.* **228** Q6
The Rosses, region, *Ire.* **145** N5
The Salt Ponds, bay, *Jam.* **122** J11
The Sisters, islands, *Grenada* **123** H23
The Sisters, islands, *Trin. & Tobago* **123** N18
The Skaw, *Eur.* **140** E7
The Skerries, islands, *Wales, U.K.* **145** R9
The Slot *see* New Georgia Sound, *Solomon Is.* **228** D7
The Smalls, *Wales, U.K.* **145** U5
The Solent, bay, *Eng., U.K.* **145** W12
The Sound, *Atl. Oc.* **140** F7
The Sound, bay, *Bahamas* **122** A11
The Steppes, *Kaz.* **178** E7
The Swatch, trough, *Ind. Oc.* **250** C9

The Valley, *Anguilla, U.K.* **121** B15
The Valley, *Anguilla, U.K.* **122** Q11
The Wash, bay, *Eng., U.K.* **145** R14
The Weald, region, *Eng., U.K.* **145** V14
The Wolds, region, *Eng., U.K.* **145** Q13
Thebes *see* Karnak, ruin, *Egypt* **202** J12
Thebes *see* Thíva, *Gr.* **156** H12
Thedford, *Nebr., U.S.* **107** N4
Thelon, river, *N. Amer.* **87** H13
Theodore Roosevelt National Park Elkhorn Ranch Site, *N. Dak., U.S.* **106** F2
Theodore Roosevelt National Park North Unit, *N. Dak., U.S.* **106** F2
Theodore Roosevelt National Park South Unit, *N. Dak., U.S.* **106** G2
Theodore Roosevelt, river, *Braz.* **132** F9
Theodore, *Qnsld., Austral.* **221** U15
Theológos, *Gr.* **157** C14
Thera, ruin, *Gr.* **157** N15
Thermaïkós Kólpos, *Gr.* **156** D11
Thérmo, *Gr.* **156** H9
Thermopolis, *Wyo., U.S.* **108** G10
Thermopylae, battle, *Gr.* **156** G10
Theron Mountains, *Antarctica* **238** E11
Thessalon, *Ont., Can.* **87** Q17
Thessaloníki (Salonica), *Gr.* **156** C11
Thessaly, region, *Gr.* **156** F10
Thetford, *Eng., U.K.* **145** S15
Thiaucourt, *Fr.* **149** P13
Thibodaux, *La., U.S.* **102** Q10
Thief River Falls, *Minn., U.S.* **106** E9
Thiel Mountains, *Antarctica* **238** J11
Thielsen, Mount, *Oreg., U.S.* **110** J4
Thiene, *It.* **152** D7
Thiers, *Fr.* **149** U11
Thiès, *Senegal* **204** J6
Thika, *Kenya* **211** T6
Thilogne, *Senegal* **204** H8
Thimphu, *Bhutan* **182** G12
Thingeyri, *Ice.* **142** D2
Thio, *New Caledonia, Fr.* **231** Q22
Thiou, *Burkina Faso* **205** J14
Thíra (Santorini), island, *Gr.* **157** M15
Thirasía, island, *Gr.* **157** M15
Thirsk, *Eng., U.K.* **145** P12
Thiruvananthapuram (Trivandrum), *India* **183** T5
Thisted, *Den.* **142** M11
Thistilfjörður, bay, *Ice.* **142** E4
Thitu Island, *S. China Sea* **192** D11
Thíva (Thebes), *Gr.* **156** H12
Thjórsá, river, *Ice.* **142** F2
Thoen, *Thai.* **190** K7
Thohoyandou, *S. Af.* **212** G9
Thomas, *Okla., U.S.* **105** C15
Thomaston, *Me., U.S.* **97** G17
Thomaston, *Ga., U.S.* **98** M5
Thomasville, *Ga., U.S.* **98** Q5
Thomasville, *N.C., U.S.* **98** H10
Thomasville, *Ala., U.S.* **103** M14
Thompson Falls, *Mont., U.S.* **108** C5
Thompson Springs, *Utah, U.S.* **109** N9
Thompson Town, *Jam.* **122** J8
Thompson, *Man., Can.* **87** L14
Thompson, river, *Iowa, Mo., U.S.* **107** P11
Thomson, *Ga., U.S.* **98** L7
Thomson, river, *Austral.* **220** F12
Thongwa, *Myanmar* **190** K5
Thonon, *Fr.* **149** T14
Thonze, *Myanmar* **190** K5
Thor Iversen Bank, *Arctic Oc.* **253** E17
Thoreau, *N. Mex., U.S.* **109** S10
Thórisvatn, lake, *Ice.* **142** F2
Thorndale, *Tex., U.S.* **105** K17
Thorne Bay, *Alas., U.S.* **113** M23
Thornhill, *Scot., U.K.* **145** M10
Thornton, *Colo., U.S.* **108** L13
Thorp, *Wis., U.S.* **100** H4
Thorp, *Wash., U.S.* **110** D5
Thorpe, *Barbados* **123** K18
Thors Land, *Greenland, Den.* **89** R9
Thórshöfn, *Ice.* **142** E4
Thouars, *Fr.* **149** S7
Thrace, region, *Gr.* **157** B15
Three Forks, *Mont., U.S.* **108** E8
Three Gorges Dam, *China* **184** L12
Three Kings Islands *see* Manawatāwhi, *N.Z.* **223** C18
Three Kings Ridge, *Pac. Oc.* **246** N11
Three Lakes, *Wis., U.S.* **100** G5
Three Pagodas Pass, *Myanmar* **190** L6

Three Rivers, *Mich., U.S.* **101** N9
Three Rivers, *Tex., U.S.* **105** N16
Three Sisters Islands, *Solomon Is.* **228** G12
Three Sisters, *S. Af.* **212** L6
Three Sisters, peak, *Oreg., U.S.* **110** H4
Thrissur, *India* **183** S5
Throckmorton, *Tex., U.S.* **105** F15
Throssell, Lake, *Austral.* **220** H5
Thua, river, *Kenya* **211** T8
Thuin, *Belg.* **148** M12
Thuir, *Fr.* **149** Z10
Thule Air Base, *U.S.* **88** F3
Thule *see* Qaanaaq, *Greenland, Den.* **88** E3
Thun, *Switz.* **152** B3
Thunder Bay, *Ont., Can.* **87** P15
Thunder Bay, *Mich., U.S.* **100** H12
Thuner See, *Switz.* **152** B3
Thung Song, *Thai.* **191** R7
Thunkar, *Bhutan* **182** G13
Thuringia, *Eur.* **140** G7
Thurles (Durlas), *Ire.* **145** S5
Thurmont, *Md., U.S.* **98** B12
Thurso, *Scot., U.K.* **144** F10
Thurston Island, *Antarctica* **238** K6
Thusis, *Switz.* **152** B5
Thwaites Glacier Tongue, *Antarctica* **238** K8
Thwaites Glacier, *Antarctica* **238** K8
Thýamis, river, *Gr.* **156** E7
Thýmaina, island, *Gr.* **157** K17
Thyou, *Burkina Faso* **205** L15
Ti-n-Essako, *Mali* **205** F17
Ti-n-Orfane, *Mali* **205** H15
Ti-n-Toumma, *Niger* **206** G11
Ti-n-Zaouâtene (Fort Pierre Bordes), *Alg.* **200** M12
Tian Shan, *Asia* **168** G7
Tianjin (Tientsin), *China* **185** G14
Tianjun, *China* **184** H8
Tianshui, *China* **184** J10
Tiarei, *Fr. Polynesia, Fr.* **231** H22
Tiaret, *Alg.* **200** C11
Tiari, *New Caledonia, Fr.* **231** N20
Tiaro Bay, *Solomon Is.* **228** F9
Tiaski, *Senegal* **204** H8
Tiassalé, *Côte d'Ivoire* **205** P13
Tiauyutai *see* Senkaku-shotō, islands, *Japan* **185** M17
Ti'avea, *Samoa* **227** P20
Tibagi, river, *Braz.* **132** M12
Tibati, *Cameroon* **206** M11
Tiber Dam, *Mont., U.S.* **108** B8
Tiber *see* Tevere, river, *It.* **152** J8
Tiberias *see* Teverya, *Israel* **172** K6
Tibesti Mountains, *Chad, Lib.* **207** D13
Tibet, Plateau of *see* Qing Zang Gaoyuan, *China* **184** J3
Tibet, Plateau of, *Asia* **168** H8
Tibet, region, *China* **184** K4
Tiboku Falls, *Guyana* **131** F15
Tibooburra, *N.S.W., Austral.* **221** V12
Tibrikot, *Nepal* **182** F9
Tibugá, Golfo de, *Col.* **130** F4
Tiburón, Isla, *Mex.* **116** C6
Tîchît, *Mauritania* **204** G10
Ticonderoga, *N.Y., U.S.* **96** G11
Ticul, *Mex.* **117** H16
Tiddim, *Myanmar* **190** F4
Tiddly Bank, *Arctic Oc.* **253** E18
Tidjikja, *Mauritania* **204** F9
Tidore, *Indonesia* **193** J16
Tiega Plateau, *Sudan* **203** Q8
Tiel, *Neth.* **148** J13
Tieli, *China* **185** C16
Tieling, *China* **185** E15
Tielongtan, *China* **184** H2
Tientsin *see* Tianjin, *China* **185** G14
Tiéré, *Mali* **205** L13
Tieroko, peak, *Chad* **207** D13
Tierra Amarilla, *N. Mex., U.S.* **109** Q12
Tierra Amarilla, *Chile* **134** G7
Tierra del Fuego, *S. Amer.* **135** X9
Tietê, river, *S. Amer.* **128** J9
Tiffin, *Ohio, U.S.* **101** P12
Tifton, *Ga., U.S.* **99** P6
Tiger Bay, *Af.* **198** M5
Tignére, *Cameroon* **206** L11
Tigray, *Eth.* **210** J8
Tigre, river, *S. Amer.* **132** D3
Tigris *see* Dijlah, *Iraq* **176** H11
Tigris, river, *Asia* **168** F4
Tigssaluk, *Greenland, Den.* **89** T6

Tiguent, *Mauritania* **204** G7
Tigyaing, *Myanmar* **190** F5
Tîh, Gebel el, *Egypt* **173** R3
Tihāmah, *Asia* **168** H2
Tihāmat ash Shām, mountains, *Saudi Arabia, Yemen* **174** J6
Tijuana, *Mex.* **116** A5
Tikamgarh, *India* **182** J6
Tikapai, island, *N.Z.* **229** J22
Tikaré, *Burkina Faso* **205** K15
Tikattane, *Mauritania* **204** F7
Tikchik Lakes, *Alas., U.S.* **113** K13
Tikehau, island, *Fr. Polynesia, Fr.* **231** A17
Tikehau, island, *Fr. Polynesia, Fr.* **231** M16
Tikei, island, *Fr. Polynesia, Fr.* **231** A19
Tikhoretsk, *Russ.* **159** S11
Tikhvin, *Russ.* **158** F7
Tiki Basin, *Pac. Oc.* **247** L16
Tikopia, island, *Solomon Is.* **224** J7
Tikrīt, *Iraq* **176** F10
Tiksi, *Russ.* **161** E16
Til Barsip *see* Tall al Aḥmar, *Syr.* **172** D10
Til-Châtel, *Fr.* **149** R13
Tiladummati Atoll, *Maldives* **183** U3
Tilān, *Afghan.* **180** G5
Tilburg, *Neth.* **148** K12
Tilden, *Nebr., U.S.* **107** N7
Tilemsoun, *Mor.* **200** F5
Tillabéri, *Niger* **206** G5
Tillamook Head, *Oreg., U.S.* **110** E2
Tillamook, *Oreg., U.S.* **110** F2
Tillanchang Dwip, island, *India* **183** T15
Tillia, *Niger* **206** G6
Tilloo Cay, *Bahamas* **118** C12
Tílos (Piscopi), island, *Gr.* **157** M18
Tilpa, *N.S.W., Austral.* **221** W12
Tilrhemt, *Alg.* **200** D12
Timan Ridge, *Eur.* **140** B12
Timanskiy Kryazh, *Russ.* **160** F9
Timaru, *N.Z.* **223** M18
Timashevo, *Russ.* **158** K16
Timashevsk, *Russ.* **159** S10
Timbalier Bay, *La., U.S.* **102** R10
Timbalier Island, *La., U.S.* **102** R10
Timbaúba, *Braz.* **133** F18
Timbedgha, *Mauritania* **204** H11
Timber Creek, *N.T., Austral.* **221** Q7
Timber Lake, *S. Dak., U.S.* **106** J4
Timbío, *Col.* **130** H5
Timbo, *Liberia* **204** P10
Timbuktu *see* Tombouctou, *Mali* **205** H14
Timbun Mata, island, *Malaysia* **193** G13
Timerein, *Sudan* **203** Q15
Timétrine, *Mali* **205** F15
Timétrine, *Mali* **205** F16
Timgad, ruin, *Alg.* **201** C14
Timimoun, *Alg.* **200** F11
Timiryazevskiy, *Russ.* **160** K12
Timiš, river, *Rom.* **154** E11
Timişoara, *Rom.* **154** D11
Timiti's Crack, cape, *Pitcairn I.* **228** Q6
Timmiarmiut Fjord, *Greenland, Den.* **89** S9
Timmiarmiut, *Greenland, Den.* **89** S9
Timmins, *Ont., Can.* **87** P17
Timmonsville, *S.C., U.S.* **98** K10
Timmoudi, *Alg.* **200** F10
Timms Hill, *Wis., U.S.* **100** H4
Timoe *see* Temoe, island, *Fr. Polynesia, Fr.* **231** F23
Timor Sea, *Ind. Oc.* **168** M14
Timor Trough, *Ind. Oc.* **251** H19
Timor-Leste (East Timor), *Asia* **193** M15
Timor, island, *Asia* **193** N15
Timote, *Arg.* **134** L11
Timpanogos Cave National Monument, *Utah, U.S.* **108** L7
Timpson, *Tex., U.S.* **105** H20
Tims Ford Lake, *Tenn., U.S.* **103** G16
Timsâh, Buheirat el (Lake Timsah), *Egypt* **173** P1
Timsah, Lake *see* Buheirat el Timsâh, *Egypt* **173** P1
Timsgarry *see* Timsgearraidh, *Scot., U.K.* **144** G6
Timsgearraidh (Timsgarry), *Scot., U.K.* **144** G6
Tîn, Ras at, *Af.* **198** D8

Tinaca Point, *Philippines* **193** G15
Tinajo, *Sp.* **214** P7
Tinakula, island, *Solomon Is.* **228** L5
Tindouf, *Alg.* **200** G7
Tiné, *Sudan* **203** R6
Tiné, *Chad* **207** G16
Tineo, *Sp.* **146** B8
Tinggoa, *Solomon Is.* **228** H10
Tingo María, *Peru* **132** G3
Tinharé, Ilha de, *Braz.* **133** H16
Tinhosa Grande, *Sao Tome and Principe* **215** B20
Tinhosa Pequena, island, *Sao Tome and Principe* **215** B20
Tinhosas, *Sao Tome and Principe* **215** B20
Tini Heke *see* Snare Islands, *N.Z.* **223** R15
Tinian Harbor *see* Sunbaron Roads, *Fr. Polynesia, Fr.* **230** J10
Tinian, island, *N. Mariana Is., U.S.* **230** K5
Tinian, island, *N. Mariana Is., U.S.* **230** K7
Tinogasta, *Arg.* **134** G8
Tínos, *Gr.* **157** K15
Tínos, island, *Gr.* **157** K15
Tiñoso, Cabo, *Sp.* **147** L14
Tinsukia, *India* **182** G15
Ṭîntâne, *Mauritania* **204** H10
Tintigny, *Belg.* **149** N13
Tintina, *Arg.* **134** G11
Tintinara, *S. Aust., Austral.* **221** Y11
Tio, *Eritrea* **210** H9
Tiobraid Árann *see* Tipperary, *Ire.* **145** S5
Tioga, *Pa., U.S.* **96** L7
Tioga, *N. Dak., U.S.* **106** E2
Tioman, peak, *Malaysia* **191** U10
Tiouilît, *Mauritania* **204** F7
Tipp City, *Ohio, U.S.* **101** R11
Tipperary (Tiobraid Árann), *Ire.* **145** S5
Tipton, *Ind., U.S.* **101** Q9
Tipton, *Okla., U.S.* **105** E15
Tipton, *Iowa, U.S.* **107** P14
Tipton, *Mo., U.S.* **107** T12
Tipton, *Calif., U.S.* **111** U7
Tipton, Mount, *Ariz., U.S.* **109** R5
Tiptonville, *Tenn., U.S.* **102** E12
Tiputa, *Fr. Polynesia, Fr.* **231** M17
Tir Pul, *Afghan.* **180** H2
Tiracambu, Serra do, *Braz.* **133** E14
Tirân, island, *Saudi Arabia* **174** D3
Tīrān, Strait of, *Egypt* **173** U4
Tirana *see* Tiranë, *Alban.* **154** L9
Tiranë (Tirana), *Alban.* **154** L9
Tiraspol, *Mold.* **159** R5
Tire, *Turk.* **170** G3
Tirebolu, *Turk.* **171** D13
Tiree, island, *Scot., U.K.* **144** K6
Tîrgu Frumos, *Rom.* **155** B16
Tîrgu Lăpuş, *Rom.* **155** B13
Tîrgu Neamţ, *Rom.* **155** B15
Tîrgu Ocna, *Rom.* **155** D16
Tîrgu Secuiesc, *Rom.* **155** D15
Tirich Mir, peak, *Pak.* **180** G11
Tîrna Mare, *Rom.* **155** A13
Tîrnăveni, *Rom.* **155** C14
Tiroungoulou, *Cen. Af. Rep.* **207** K16
Tirso, river, *It.* **153** N4
Tiruchchendur, *India* **183** T6
Tiruchchirappalli, *India* **183** S6
Tirunelveli, *India* **183** T5
Tirupati, *India* **183** Q7
Tiruppattur, *India* **183** S6
Tiruppur, *India* **183** S5
Tiruvannamalai, *India* **183** R6
Tiryns, ruin, *Gr.* **156** K11
T'īs Isat Fwafwatē, waterfall, *Eth.* **210** K7
Tisa, river, *Rom.* **154** D10
Tisaiyanvilai, *India* **183** T6
Tishomingo, *Okla., U.S.* **105** E18
Tissemsilt, *Alg.* **200** C12
Tista, river, *India* **182** H12
Tisza, river, *Hung.* **151** K17
Tiszafüred, *Hung.* **151** K16
Tiszaújváros, *Hung.* **151** K16
Tit, *Alg.* **200** H11
Titaf, *Alg.* **200** H11
Titan Dome, *Antarctica* **239** J13
Titano, *San Marino* **163** J15
Titicaca, Lago, *Peru* **132** J6
Titikaveka, *N.Z.* **229** E23
Titule, *D.R.C.* **209** C16

Titusville, *Pa., U.S.* **96** L4
Titusville, *Fla., U.S.* **99** T9
Tiva, *Fr. Polynesia, Fr.* **231** G17
Tiva, river, *Kenya* **211** T7
Tivaouane, *Senegal* **204** J6
Tivaru, island, *Fr. Polynesia, Fr.* **231** M17
Tiverton, *R.I., U.S.* **97** L15
Tiverton, *Eng., U.K.* **145** V10
Tivoli, *Grenada* **123** K23
Tivoli, *It.* **152** K9
Tiyās, *Syr.* **172** G9
Tizi Ouzou, *Alg.* **201** B13
Tizimín, *Mex.* **117** H17
Tiznit, *Mor.* **200** F6
Tlaxcala, *Mex.* **116** J12
Tlemcen, *Alg.* **200** C10
Tmassah, *Lib.* **201** H19
Tmiet, island, *Marshall Is.* **227** H21
Toachel Mlengui, *Palau* **226** G5
Toadlena, *N. Mex., U.S.* **109** R10
Toagel Mid Passage, *Palau* **226** H6
Toahotu, *Fr. Polynesia, Fr.* **231** J23
Toamaro, island, *Fr. Polynesia, Fr.* **231** H17
Toamasina, *Madagascar* **213** F17
Toano, *Va., U.S.* **98** E13
Toau, island, *Fr. Polynesia, Fr.* **231** B18
Toay, *Arg.* **134** M10
Toba Kākar Range, *Pak.* **180** M7
Toba Tek Singh, *Pak.* **180** L11
Tobarra, *Sp.* **147** J13
Tobi Shima, *Japan* **188** L11
Tobi, island, *Palau* **224** F2
Tobin, Mount, *Nev., U.S.* **111** N9
Tobli, *Liberia* **204** P11
Tobol, river, *Asia* **168** E7
Toboli, *Indonesia* **193** J13
Tobol'sk, *Russ.* **160** J10
Tobolski, *Jam.* **122** H8
Tobona, island, *Solomon Is.* **228** D9
Tobruk *see* Ṭubruq, *Lib.* **201** D22
Tobyl, *Kaz.* **178** D8
Tobyl, river, *Kaz.* **178** C10;D9
Tocantínia, *Braz.* **133** G13
Tocantinópolis, *Braz.* **133** F13
Tocantins, river, *Braz.* **133** F13
Toccoa, *Ga., U.S.* **98** K6
Toco, *Trin. & Tobago* **123** N24
Toco, *Chile* **134** D7
Toconao, *Chile* **134** E8
Tocopilla, *Chile* **134** D6
Tocumwal, *N.S.W., Austral.* **221** Y13
Todenyang, *Kenya* **211** P6
Todi, *It.* **152** J8
Toding (Tuolin), *China* **184** J2
Todmorden, *S. Aust., Austral.* **221** U9
Todos os Santos Bay, *S. Amer.* **128** G11
Todos Santos, *Mex.* **116** F7
Todos Santos, Bahía, *Mex.* **116** B5
Tofino, *B.C., Can.* **86** M7
Tofte, *Minn., U.S.* **106** F13
Tofua, island, *Tonga* **227** L22
Tog Wajaale, *Somaliland* **210** L10
Toga, island, *Vanuatu* **229** C13
Togiak, *Alas., U.S.* **113** K13
Togian, Kepulauan, *Indonesia* **193** J14
Togliatti *see* Tol'yatti, *Russ.* **158** K15
Togni, *Sudan* **203** P14
Tognuf, *Eritrea* **210** G7
Togo, *Af.* **205** N17
Togobala, *Guinea* **204** M11
Togtoh, *China* **184** G12
Togyz, *Kaz.* **178** G8
Tohatchi, *N. Mex., U.S.* **109** R10
Tohma, river, *Turk.* **170** F12
Tohopekaliga, Lake, *Fla., U.S.* **99** T9
Toi Misaki, *Japan* **189** T5
Toi, *Japan* **188** H12
Toi, *N.Z.* **229** P21
Toibalawe, *India* **183** S14
Toili, *Indonesia* **193** J14
Toiyabe Range, *Nev., U.S.* **90** F5
Tojŏng-san, peak, *N. Korea* **186** C12
Tok, *Alas., U.S.* **113** G19
Toka Sand Cay, *N.Z.* **229** E19
Tokachi Dake, *Japan* **188** G14
Tōkamachi, *Japan* **189** N11
Tokar, *Sudan* **203** P15
Tokara Kaikyō, *Japan* **189** U4
Tokara Rettō, *Japan* **189** V3
Tokarevka, *Kaz.* **179** E13
Tokat, *Turk.* **170** E11
Tŏkch'ŏn, *N. Korea* **186** H7
Tokelau Basin, *Pac. Oc.* **246** K12

Tokelau, island, *Kiribati* **224** H10
Tokerau, island, *N.Z.* **229** K21
Toki Point, *Wake I., U.S.* **230** M11
Tokmak, *Ukr.* **159** R8
Tokmok, *Kyrg.* **179** K14
Toko, *Russ.* **161** J19
Tokomaru Bay, *N.Z.* **223** G23
Tokombiya, *Eritrea* **210** H7
Tokoro, river, *Japan* **188** F14
Tokoroa, *N.Z.* **223** G21
Tŏksin, *N. Korea* **186** E12
Tŏksŏng, *N. Korea* **186** G10
Toksook Bay, *Alas., U.S.* **112** J11
Toksun, *China* **184** E5
Toktogul, *Kyrg.* **179** K13
Tokung, *Indonesia* **192** J11
Tokuno Shima, *Japan* **189** W3
Tokushima, *Japan* **189** Q8
Tokuyama, *Japan* **189** Q6
Tōkyō Wan, *Japan* **189** P12
Tōkyō, *Japan* **189** P12
Tol, island, *F.S.M.* **226** F8
Tôlañaro, *Madagascar* **213** J15
Tôlanaro, cape, *Madagascar* **213** J16
Tolbo, *Mongolia* **184** C6
Tôle Bī, *Kaz.* **179** J14
Toledo Bend Reservoir, *La., Tex., U.S.* **105** H21
Toledo, *Ohio, U.S.* **101** N12
Toledo, *Iowa, U.S.* **107** N12
Toledo, *Oreg., U.S.* **110** G2
Toledo, *Wash., U.S.* **110** D3
Toledo, *Sp.* **146** G10
Toledo, Montes de, *Sp.* **146** H10
Toli, *China* **184** D4
Toliara, *Madagascar* **213** H14
Tolitoli, *Indonesia* **193** H13
Tol'ka, *Russ.* **160** H12
Tollya, *Russ.* **161** D14
Tolmezzo, *It.* **152** C9
Tolo, Teluk, *Indonesia* **193** K14
Tolob, *Scot., U.K.* **144** D12
Tolten, *Chile* **135** P6
Tolti, *Pak.* **182** C5
Tolú, *Col.* **130** D5
Toluca, *Mex.* **116** J11
Tolybay, *Kaz.* **178** D8
Tolyatti (Togliatti), *Russ.* **158** K15
Tom Price, *W. Aust., Austral.* **221** T3
Tom White, Mount, *Alas., U.S.* **113** J18
Toma, *Burkina Faso* **205** K14
Tomah, *Wis., U.S.* **100** K4
Tomahawk, *Wis., U.S.* **100** G5
Tomakomai, *Japan* **188** G13
Tomanivi, peak, *Fiji* **226** P5
Tomar, *Braz.* **132** C8
Tomar, *Port.* **146** G5
Tomari, *Japan* **188** J13
Tomaszów Lubelski, *Pol.* **151** G18
Tomaszów Mazowiecki, *Pol.* **151** E15
Tomatin, *Scot., U.K.* **144** J9
Tombador, Serra do, *Braz.* **132** H10
Tombe, *S. Sudan* **203** X12
Tombe, *Solomon Is.* **228** E7
Tombigbee, river, *Ala., Miss., U.S.* **103** J13
Tomboco, *Angola* **208** J9
Tombouctou (Timbuktu), *Mali* **205** H14
Tombstone, *Ariz., U.S.* **109** W9
Tombua, *Angola* **208** P9
Tomé, *Chile* **134** M5
Tomea, island, *Indonesia* **193** L15
Tomil Harbor, *F.S.M.* **226** E11
Tomil, *F.S.M.* **226** E12
Tomini, Gulf of, *Pac. Oc.* **168** L13
Tomini, Teluk, *Indonesia* **193** J13
Tominian, *Mali* **205** K13
Tomman, island, *Vanuatu* **229** H14
Tommot, *Russ.* **161** H18
Tomo, New Caledonia, Fr.* **231** Q22
Tomo, river, *Col.* **130** F9
Tomorit, Maja e, *Alban.* **154** M10
Tōmotu Noi (Lord Howe Island), *Solomon Is.* **228** L5
Tompa, *Russ.* **161** K16
Tompkinsville, *Ky., U.S.* **103** D16
Tomsk, *Russ.* **160** K12
Tomsyu, *Russ.* **161** E18
Tomtor Tasa, *Russ.* **161** F16
Tomtor, *Russ.* **161** F17
Tomür Feng *see* Victory Peak, *China, Kyrg.* **179** K15
Tonalá, *Mex.* **117** K14

Tonami, *Japan* **189** N10
Tonantins, *Braz.* **132** D6
Tonasket, *Wash., U.S.* **110** A7
Tonbridge, *Eng., U.K.* **145** V14
Tønder, *Den.* **142** N11
Tondon, *Guinea* **204** L8
Tondou Massif, *Af.* **198** H8
Tong, island, *P.N.G.* **222** M8
Tonga Islands, *Pac. Oc.* **224** K10
Tonga Ridge, *Pac. Oc.* **246** M12
Tonga Trench, *Pac. Oc.* **246** M12
Tonga, *S. Sudan* **203** U11
Tonga, *Pac. Oc.* **224** K10
Tongareva *see* Penrhyn, *Cook Is., N.Z.* **224** H12
Tongariki, island, *Vanuatu* **229** H16
Tongatapu Group, *Tonga* **224** K10
Tongch'ang, *N. Korea* **186** G6
T'ongch'ŏn, *N. Korea* **186** J10
Tongde, *China* **184** J9
Tongeren, *Belg.* **148** L13
Tonggo, *N. Korea* **186** G5
Tongha, *N. Korea* **186** H9
Tonghua, *China* **185** F16
Tonghŭng, *N. Korea* **186** F12
Tongjosŏn-man, bay, *N. Korea* **186** H9
Tongliao, *China* **185** E15
Tongnim-san, peak, *N. Korea* **186** G6
Tongnim, *N. Korea* **186** G4
Tongo, *N.S.W., Austral.* **221** W12
Tongoa, island, *Vanuatu* **229** H16
Tongoy, *Chile* **134** J6
Tongshi, *China* **184** Q12
Tongsin, *N. Korea* **186** G7
Tongtian (Yangtze), river, *China* **184** J7
Tongue of the Ocean, *Bahamas* **118** F11
Tongue River Reservoir, *Mont., U.S.* **108** F11
Tongue, *Scot., U.K.* **144** F9
Tongue, Kyle of, *Scot., U.K.* **144** F9
Tongue, river, *Mont., Wyo., U.S.* **108** E12
Tongy, *Qnsld., Austral.* **221** V14
Tongyeong, *S. Korea* **187** T11
Tongzi, *China* **184** M11
Tonj *see* Ibba, river, *S. Sudan* **203** X10
Tonj, *S. Sudan* **203** W10
Tonk, *India* **182** H5
Tonka, *Mali* **205** H14
Tonkawa, *Okla., U.S.* **105** B17
Tonkin, Gulf of, *Asia* **190** J12
Tonle Sap *see* Bœng Tônlé Sab, *Cambodia* **191** N10
Tonneins, *Fr.* **149** W7
Tonnerre, *Fr.* **149** R11
Tōno, *Japan* **188** K13
Tonoas, island, *F.S.M.* **226** E9
Tonopah, *Nev., U.S.* **111** R9
Tonota, *Botswana* **212** G8
Tønsberg, *Nor.* **142** K11
Tonto National Monument, *Ariz., U.S.* **109** U8
Tonumea, island, *Tonga* **227** N23
Tonya, *Turk.* **171** D14
Tooele, *Utah, U.S.* **108** L7
Toopua Iti, island, *Fr. Polynesia, Fr.* **231** H14
Toopua, island, *Fr. Polynesia, Fr.* **231** H14
Toora Khem, *Russ.* **161** L14
Tooraie, *N.S.W., Austral.* **221** W13
Toore, *Somalia* **211** R11
Toowoomba, *Qnsld., Austral.* **221** V15
Top Ozero, *Russ.* **158** A7
Top, Lake, *Eur.* **140** B10
Topaze, Baie, *Mauritius* **215** J20
Topeka, *Kans., U.S.* **107** S9
Topliţa, *Rom.* **155** C15
Topock, *Ariz., U.S.* **109** S4
Topol'čany, *Slovakia* **151** K14
Topolobampo, *Mex.* **116** E7
Toppenish, *Wash., U.S.* **110** E6
Topsail Beach, *N.C., U.S.* **98** K13
Top'yŏng, *N. Korea* **186** L5
Toquima Range, *Nev., U.S.* **111** Q9
Tor Station, *Antarctica* **239** B14
Tora Bora, site, *Afghan.* **180** H9
Tōraghūndī, *Afghan.* **180** G3
Toraigh (Tory Island), *Ire.* **145** M5
Toranou, *Côte d'Ivoire* **204** M11
Torbalı, *Turk.* **170** G3
Torbat-e Ḥeydarīyeh, *Iran* **177** F20
Torbat-e Jām, *Iran* **177** F21

Torbay, *Eng., U.K.* **145** W10
Torbert, Mount, *Alas., U.S.* **113** J16
Töre, *Sw.* **143** E14
Torez, *Ukr.* **159** Q10
Torghay, *Kaz.* **178** F9
Torghay, river, *Kaz.* **178** F9
Torgo, *Russ.* **161** J17
Torhout, *Belg.* **148** L10
Tori Shima, *Japan* **189** U12
Tori Shima, *Japan* **189** X1
Tori, *Eth.* **210** M4
Torino (Turin), *It.* **152** E2
Torit, *S. Sudan* **203** Y12
Torixoreu, *Braz.* **132** J12
Tornal'a, *Slovakia* **151** K16
Torneå see Tornio, *Fin.* **143** E15
Torneälven, river, *Eur.* **140** B9
Torneträsk, lake, *Sw.* **143** D13
Torngat Mountains, *Nfld. & Lab., Can.* **87** J20
Tornillo, *Tex., U.S.* **104** J7
Tornimparte, *It.* **152** K9
Tornio (Torneå), *Fin.* **143** E15
Torniojoki, river, *Fin. Sw.* **143** D14
Tornquist, *Arg.* **135** N11
Toro, *Sp.* **146** D9
Tōro, *Japan* **188** G15
Toro, Cerro del, *S. Amer.* **134** H7
Torodo, *Mali* **204** J11
Torokina, *P.N.G.* **222** P11
Torokina, *P.N.G.* **228** C5
Törökszentmiklós, *Hung.* **151** L16
Torom, *Russ.* **161** J20
Toronto, *Ont., Can.* **87** Q18
Toronto, *Ohio, U.S.* **101** Q16
Toronto, *Kans., U.S.* **107** U9
Toropets, *Russ.* **158** G9
Tororo, *Uganda* **211** S5
Toros Dağı, *Turk.* **170** H9
Toros Dağları (Taurus), *Turk.* **170** J8
Torquemada, *Sp.* **146** D10
Torran Rocks, *Scot., U.K.* **144** L7
Torrance, *Calif., U.S.* **111** X8
Torre Annunziata, *It.* **152** M11
Torre Astura, *It.* **152** L8
Torre de Moncorvo, *Port.* **146** E7
Torre San Giovanni, *It.* **153** N15
Torredonjimeno, *Sp.* **146** K10
Torrejón de Ardoz, *Sp.* **146** F11
Torrelavega, *Sp.* **146** B10
Torremaggiore, *It.* **152** L12
Torremolinos, *Sp.* **146** M10
Torrens, Lake, *Austral.* **220** J10
Torrent, *Sp.* **147** H14
Torreón, *Mex.* **116** F10
Torres Islands, *Vanuatu* **229** C14
Torres Strait, *Austral.* **220** A12
Torres Strait, *P.N.G.* **222** R6
Torres Vedras, *Port.* **146** H4
Torrevieja, *Sp.* **147** K14
Torrington, *Conn., U.S.* **96** L12
Torrington, *Wyo., U.S.* **108** J14
Torrutj, island, *Marshall Is.* **227** J17
Torsby, *Sw.* **142** K12
Tórshavn, *Faroe Is. Den.* **142** H6
Tortola, *Virgin Is., U.K.* **122** L12
Tortona, *It.* **152** E4
Tortosa see Ţarţūs, *Syr.* **172** G7
Tortosa, *Sp.* **147** F15
Tortosa, Cap de, *Sp.* **147** F16
Tortosa, Cape, *Eur.* **140** K4
Tortue, Île de la, *Haiti* **119** K15
Tortum, *Turk.* **171** E16
Ţorūd, *Iran* **177** F18
Torugart, *Kyrg.* **179** L15
Torul, *Turk.* **171** D14
Toruń, *Pol.* **151** D14
Tõrva, *Est.* **143** K17
Tory Island see Toraigh, *Ire.* **145** M5
Torzhok, *Russ.* **158** H14
Tōrzī, *Afghan.* **180** M6
Tosa del Braidal, peak, *Andorra* **162** J4
Tosa Wan, *Japan* **189** R7
Tosamaganga, *Tanzania* **211** X5
T'osan, *N. Korea* **186** L8
Tosashimizu, *Japan* **189** S6
Tosca, *S. Af.* **212** J7
Toscano, Arcipelago, *It.* **152** J6
Toshka Lakes, *Egypt* **202** L11
Toshkent (Tashkent), *Uzb.* **178** L12
Tosontsengel, *Mongolia* **184** C8
Tostado, *Arg.* **134** H11
Tosu, *Japan* **189** R4

Tosya, *Turk.* **170** D9
Totana, *Sp.* **147** K13
Totegegie, island, *Fr. Polynesia, Fr.* **231** Q18
Tôtes, *Fr.* **149** N8
Totia, island, *N.Z.* **229** J24
Totiw, island, *F.S.M.* **226** F9
Tot'ma, *Russ.* **158** F11
Totnes, *Eng., U.K.* **145** W10
Totness, *Suriname* **131** F17
Toto, *Angola* **208** J10
Totokafonua, island, *Tonga* **227** H22
Totora, *Bol.* **132** K7
Totoya, island, *Fiji* **226** Q7
Totten Glacier Tongue, *Antarctica* **239** N21
Totten Glacier, *Antarctica* **239** M21
Tottori, *Japan* **189** P7
Totu Tofari, island, *Fr. Polynesia, Fr.* **231** F16;G15
Touba, *Senegal* **204** J7
Toucy, *Fr.* **149** R11
Tougan, *Burkina Faso* **205** K14
Touggourt, *Alg.* **201** D14
Tougouri, *Burkina Faso* **205** K16
Tougouya, *Burkina Faso* **205** K15
Tougué, *Guinea* **204** L9
Touho, *New Caledonia, Fr.* **231** P21
Toukoto, *Mali* **204** K10
Toul, *Fr.* **149** P13
Toulépleu, *Côte d'Ivoire* **204** P11
Toulon, *Fr.* **149** Y14
Toulouse, *Fr.* **149** X9
Toummo, *Lib.* **201** E17
Toummo, *Niger* **206** C11
Toumodi, *Côte d'Ivoire* **205** P13
Toungad, *Mauritania* **204** E8
Toungo, *Nigeria* **206** L10
Touques, river, *Fr.* **149** P7
Tourassine, ruin, *Mauritania* **204** C10
Touriñán, Cabo, *Sp.* **146** B5
Tournai, *Belg.* **148** L11
Tournavista, *Peru* **132** G4
Tournay, *Fr.* **149** Y7
Touros, *Braz.* **133** F18
Touroua, *Cameroon* **206** K11
Tours, *Fr.* **149** R8
Toury, *Fr.* **149** Q9
Tovarkovskiy, *Russ.* **158** L9
Tovuz, *Azerb.* **171** D19
Towada Ko, *Japan* **188** J12
Towada, *Japan* **188** J13
Towanda, *Pa., U.S.* **96** L8
Tower see Isla Genovesa, island, *Ecua.* **130** P4
Tower, *Minn., U.S.* **106** F12
Town Hall, *Monaco* **163** E21
Town Hill, *Bermuda* **119** C16
Towner, *N. Dak., U.S.* **106** E5
Townsend, *Va., U.S.* **98** F14
Townsend, *Mont., U.S.* **108** D8
Townsville, *Qnsld., Austral.* **221** R14
Towot, *S. Sudan* **203** W13
Towson, *Md., U.S.* **98** C13
Toyah, *Tex., U.S.* **104** J10
Toyahvale, *Tex., U.S.* **104** J10
Toyama Wan, *Japan* **189** N10
Toyama, *Japan* **189** N10
Toyohashi, *Japan* **189** Q10
Toyooka, *Japan* **189** P8
Toyotomi, *Japan* **188** E13
Tozeur, *Tunisia* **201** D15
Tqvarch'eli, *Rep. of Georgia* **171** B16
Trâblous (Tripoli), *Leb.* **172** H7
Trabzon, *Turk.* **171** D14
Tracy Gletscher, *Greenland, Den.* **88** E4
Tracy, *Minn., U.S.* **106** K9
Tracy, *Calif., U.S.* **111** S4
Trade Town, *Liberia* **204** P10
Træna Bank, *Atl. Oc.* **248** E13
Traer, *Iowa, U.S.* **107** N12
Trafalgar, Cabo, *Sp.* **146** N8
Tráighlí see Tralee, *Ire.* **145** T3
Trail, *B.C., Can.* **86** M9
Traill Ø, *Greenland, Den.* **88** J13
Traíra see Taraira, river, *S. Amer.* **132** C5
Trakhonas (Kızılay), *Cyprus* **162** N8
Tralee (Tráighlí), *Ire.* **145** T3
Tralee Bay, *Ire.* **145** T3
Tranås, *Sw.* **143** L13
Trancas, *Arg.* **134** F9
Trang, *Thai.* **191** R7
Trangan, island, *Indonesia* **193** M18
Tranqueras, *Uru.* **134** J14

Trans-Amazon Highway, *Braz.* **132** F9
Trans-Canada Highway, *Can.* **86** L8
Transantarctic Mountains, *Antarctica* **239** K13
Transdniestria, region, *Mold. Ukr.* **159** R5
Transkei Basin, *Ind. Oc.* **250** N3
Transylvania, region, *Rom.* **155** C13
Transylvanian Alps, *Rom.* **154** F12
Trapani, *It.* **153** S8
Trapper Peak, *Mont., U.S.* **108** E5
Trarza, region, *Mauritania* **204** G7
Trasimeno, Lago, *It.* **152** H8
Trat, *Cambodia* **191** N9
Traun, *Aust.* **150** K11
Traunreut, *Ger.* **150** K9
Traverse City, *Mich., U.S.* **100** J9
Traverse, Lake, *Minn. S. Dak., U.S.* **106** J8
Travnik, *Bosn. & Herzg.* **154** G7
Trbovlje, *Slov.* **154** D5
Treasure Beach, *Jam.* **122** J7
Treasure Cay, *Bahamas* **122** A9
Treasurers Island, *Solomon Is.* **228** K7
Treasury Islands, *Solomon Is.* **228** D5
Třebíč, *Czechia* **150** J12
Trebinje, *Bosn. & Herzg.* **154** J8
Trebisacce, *It.* **153** P13
Trebišov, *Slovakia* **151** J17
Třeboň, *Czechia* **150** J11
Trégastel, *Fr.* **149** P3
Tregrosse Islets, *Austral.* **220** D15
Tréguier, *Fr.* **149** P4
Treignac, *Fr.* **149** U9
Treinta y Tres, *Uru.* **134** K15
Trelew, *Arg.* **135** R9
Trelleborg, *Sw.* **142** N12
Tremadoc Bay, *Wales, U.K.* **145** S9
Tremiti, Isole, *It.* **152** K12
Tremonton, *Utah, U.S.* **108** J7
Trenary, *Mich., U.S.* **100** F8
Trenčín, *Slovakia* **151** J14
Trenque Lauquen, *Arg.* **134** M11
Trento, *It.* **152** C7
Trenton, *N.J., U.S.* **96** P10
Trenton, *Mich., U.S.* **101** N12
Trenton, *Tenn., U.S.* **102** F12
Trenton, *Mo., U.S.* **107** R11
Trenton, *Nebr., U.S.* **107** R4
Trepassey, *Nfld. & Lab., Can.* **87** M24
Tres Arroyos, *Arg.* **135** N12
Tres Esquinas, *Col.* **130** J6
Três Lagoas, *Braz.* **132** L12
Três Marias Reservoir, *S. Amer.* **128** H9
Tres Montes, Golfo, *Chile* **135** T6
Tres Montes, Península, *Chile* **135** T5
Tres Picos, Cerro, *Arg.* **135** R7
Tres Puntas, Cabo, *Arg.* **135** T9
Tres Zapotes, ruin, *Mex.* **117** J13
Treshnish Isles, *Scot., U.K.* **144** K7
Treviso, *It.* **152** D8
Trevose Head, *Eng., U.K.* **145** W8
Trgovište, *Serb.* **154** J12
Trial Bay, *Austral.* **220** J15
Triangle, *Zimb.* **212** F9
Triangle, The, *Myanmar* **190** E6
Triángulos, *Arrecifes, Mex.* **117** H15
Tribulation, Cape, *Qnsld. Austral.* **221** Q13
Tribune, *Kans., U.S.* **107** T3
Tricarico, *It.* **152** M13
Tricase, *It.* **153** N16
Trichonida, Límni, *Gr.* **156** H9
Trida, *N.S.W. Austral.* **221** X13
Trie, *Fr.* **149** Y7
Trier, *Ger.* **150** H5
Triesen, *Liech.* **162** Q2
Triesenberg, *Liech.* **162** P2
Trieste, *It.* **152** D10
Triglav, peak, *Slov.* **154** C4
Trigueros, *Sp.* **146** L7
Tríkala, *Gr.* **156** E9
Tríkeri, *Gr.* **156** G11
Trikomo (Iskele), *Cyprus* **162** N10
Trincomalee, *Sri Lanka* **183** T13
Trindade Seachannel, *Atl. Oc.* **254** Q3
Třinec, *Czechia* **151** H14
Trinidad & Tobago, *N. Amer.* **121** J17
Trinidad Head, *Calif., U.S.* **110** M1
Trinidad, *Colo., U.S.* **109** P14
Trinidad, *Calif., U.S.* **110** M1
Trinidad, *Cuba* **118** H9
Trinidad, *Bol.* **132** J7
Trinidad, Golfo, *Chile* **135** V6
Trinidad, Isla, *Arg.* **135** P11

Trinidad, island, *Trinidad & Tobago* **121** K17
Trinitapoli, *It.* **152** L13
Trinity Bay, *Nfld. & Lab. Can.* **87** M24
Trinity Bay, *Austral.* **220** C13
Trinity Hills, *Trin. & Tobago* **123** R23
Trinity Islands, *Alas., U.S.* **113** N15
Trinity Peninsula, *Antarctica* **238** C5
Trinity Range, *Nev., U.S.* **111** N7
Trinity, *Tex., U.S.* **105** J19
Trinity, river, *Tex., U.S.* **105** K20
Trinity, river, *Calif., U.S.* **110** M2
Trinkat Island, *India* **183** T15
Trinkitat, *Sudan* **203** N15
Triolet, *Mauritius* **215** F20
Trion, *Ga., U.S.* **98** K4
Trionto, Capo, *It.* **153** P14
Triora, *It.* **152** G2
Trípoli, *Gr.* **156** K10
Tripoli see Tarābulus, *Lib.* **201** D17
Tripoli see Trâblous, *Leb.* **172** H7
Tripolitania, *Lib.* **201** E17
Tripp, *S. Dak., U.S.* **106** M7
Tristaina, Pic de, *Andorra, Fr.* **162** G2
Tristaina, river, *Andorra* **162** G3
Tristan da Cunha Group, islands, *Af.* **198** R2
Tristan da Cunha Island, *Af.* **198** R2
Tristão, Ponta do, *Madeira, Port.* **214** M2
Tristen da Cunha Fracture Zone, *Atl. Oc.* **249** T9
Trivandrum see Thiruvananthapuram, *India* **183** T5
Trnava, *Slovakia* **151** K13
Trobriand Islands, *P.N.G.* **222** Q9
Trofors, *Nor.* **142** F12
Trois-Bassins, *Réunion, Fr.* **215** G16
Trois-Rivières, *Que., Can.* **87** P20
Trois-Rivières, *Guadeloupe, Fr.* **123** G14
Troisdorf, *Ger.* **150** F5
Troisvierges, *Lux.* **162** G9
Troitsk, *Russ.* **160** J8
Troitsko Pechorsk, *Russ.* **158** C16
Troitskoye, *Russ.* **159** R13
Troja, *Jam.* **122** H10
Trojanów, *Pol.* **151** E17
Troll Station, *Antarctica* **239** B13
Trollhättan, *Sw.* **142** L12
Trombetas, river, *Braz.* **132** C10
Tromsø Bank, *Arctic Oc.* **253** F18
Tromsø, *Nor.* **143** C13
Trona, *Calif., U.S.* **111** V9
Tronador, Monte, *Arg.* **135** Q7
Trondheim, *Nor.* **142** G12
Trondheimsfjorden, bay, *Nor.* **142** G11
Troodos Mountains, *Cyprus* **162** P7
Troon, *Scot., U.K.* **145** M9
Trostyanets', *Ukr.* **159** N8
Trou d'Eau Douce, *Mauritius* **215** G21
Trou Gras Point, *St. Lucia* **123** K14
Troumaka, *St. Vincent & the Grenadines* **123** K16
Trout Creek, *Utah, U.S.* **108** M5
Trout Lake, *N.W.T., Can.* **86** H10
Trout Lake, *Ont., Can.* **87** N14
Trout Peak, *Wyo., U.S.* **108** F9
Trout River, *Nfld. & Lab., Can.* **87** M23
Troy Peak, *Nev., U.S.* **111** R11
Troy, *N.Y., U.S.* **96** J11
Troy, *N.C., U.S.* **98** H10
Troy, *Ohio, U.S.* **101** R11
Troy, *Ala., U.S.* **103** M16
Troy, *Kans., U.S.* **107** R10
Troy, *Mo., U.S.* **107** S14
Troy, *Mont., U.S.* **108** B4
Troy, *Oreg., U.S.* **110** E9
Troy, *Jam.* **122** H8
Troy, ruins, *Turk.* **170** D2
Troyes, *Fr.* **149** Q12
Trubchevsk, *Russ.* **158** M7
Truckee, *Calif., U.S.* **111** Q6
Truckee, river, *Calif. Nev., U.S.* **111** P6
Trujillo, *Hond.* **117** K18
Trujillo, *Venez.* **130** D8
Trujillo, *Peru* **132** G2
Truk Island see Chuuk, *F.S.M.* **226** B9
Truk Lagoon, *F.S.M.* **226** E8
Trumann, *Ark., U.S.* **102** F10
Trumbull, Mount, *Ariz., U.S.* **109** Q6
Truro, *N.S., Can.* **87** P22
Truro, *Eng., U.K.* **145** W8
Truth or Consequences, *N. Mex., U.S.* **109** U11
Trutnov, *Czechia* **150** G12

Tryavna, *Bulg.* **155** H15
Trzebnica, *Pol.* **151** F13
Tsada, *Cyprus* **162** Q6
Tsagaannuur, *Mongolia* **184** C6
Tsagan Aman, *Russ.* **159** Q14
Ts'ageri, *Rep. of Georgia* **171** B17
Tsagveri, *Rep. of Georgia* **171** C17
Tsahir, *Mongolia* **184** D8
Tsaile, *Ariz., U.S.* **109** R9
Tsaka La, peak, *India* **182** D7
Tsalka, *Rep. of Georgia* **171** C18
Tsamkanī, *Afghan.* **180** J2
Tsaratanana, peak, *Madagascar* **213** D17
Tsarevo, *Bulg.* **155** J17
Tsau, *Botswana* **212** F6
Tsavo, *Kenya* **211** U7
Tschida, Lake, *N. Dak., U.S.* **106** G3
Tsentral'nyy, *Russ.* **161** G20
Tses, *Namibia* **212** J4
Tsetsegnuur, *Mongolia* **184** D7
Tsetserleg, *Mongolia* **184** D9
Tshabong, *Botswana* **212** J6
Tshane, *Botswana* **212** H6
Tshela, *D.R.C.* **208** H9
Tshikapa, *D.R.C.* **209** J14
Tshilenge, *D.R.C.* **209** H15
Tshinota, *D.R.C.* **209** J14
Tshisuku, *D.R.C.* **209** H14
Tshootsha, *Botswana* **212** G5
Tshuapa, river, *D.R.C.* **209** E14;F16
Tshwane see Pretoria, *S. Af.* **212** H8
Tsiafajavona, peak, *Madagascar* **213** F16
Tsiigehtchic, *N.W.T., Can.* **86** E9
Tsimlyansk Reservoir, *Eur.* **140** G13
Tsimlyanskoye Vodokhranilishche, *Russ.* **159** Q12
Tsiolkovskiy Island, *Antarctica* **239** A13
Tsiombe, *Madagascar* **213** J15
Tsipanda, *Russ.* **161** H19
Tsiribihina River, *Madagascar* **213** F15
Tsiroanomandidy, *Madagascar* **213** F15
Ts'khinvali, *Rep. of Georgia* **171** B18
Tsna, river, *Russ.* **158** L11
Tsnori, *Rep. of Georgia* **171** C20
Tsodilo Hills, *Botswana* **212** F5
Tsogni, *Gabon* **208** F7
Tsu, *Japan* **189** Q9
Tsubame, *Japan* **188** M11
Tsuchiura, *Japan* **189** N12
Tsugaru Kaikyō, *Japan* **188** J12
Tsuma, *Japan* **189** N7
Tsumeb, *Namibia* **212** F4
Tsuruga Wan, *Japan* **189** P9
Tsuruga, *Japan* **189** P9
Tsuruoka, *Japan* **188** L12
Tsushima Strait, *Japan, S. Korea* **187** V13
Tsushima, *Japan* **189** Q4
Tsushima, island, *Japan* **189** Q4
Tsuyama, *Japan* **189** P7
Tsyurupyns'k, *Ukr.* **159** R7
Tuaim see Tuam, *Ire.* **145** Q4
Tuakau, *N.Z.* **223** F20
Tual, *Indonesia* **193** L18
Tuam (Tuaim), *Ire.* **145** Q4
Tuamotu Archipelago, *Fr. Polynesia, Fr.* **231** A19
Tuamotu Ridge, *Pac. Oc.* **247** L15
Tuân Giáo, *Vietnam* **190** G9
Tuangku, island, *Indonesia* **192** H4
Tu'anuku, *Tonga* **227** H23
Tuapa, *N.Z.* **229** P21
Tuapse, *Russ.* **159** T10
Tuaran, *Malaysia* **192** F12
Tuasivi, *Samoa* **227** N18
Tuatapere, *N.Z.* **223** P16
Tuath, Loch a, *Scot., U.K.* **144** G7
Tuauru, river, *Fr. Polynesia, Fr.* **231** H22
Tuba City, *Ariz., U.S.* **109** R7
Tubac, *Ariz., U.S.* **109** W8
Tubarão, *S. Amer.* **128** E11
Tubarão, *Braz.* **133** P13
Tübingen, *Ger.* **150** J6
Tubmanburg, *Liberia* **204** P10
Tubou, *Fiji* **226** Q8
Ţubruq (Tobruk), *Lib.* **201** D22
Tubuai Islands see Austral Islands, *Fr. Polynesia, Fr.* **231** E17
Tubuai, island, *Fr. Polynesia, Fr.* **231** E17
Tucacas, *Venez.* **130** C10
Tucavaca, river, *Bol.* **132** K9

Tuckerman, *Ark., U.S.* **102** F10
Tucker's Town, *Bermuda* **119** B16
Tuckerton, *N.J., U.S.* **96** Q10
Tucson, *Ariz., U.S.* **109** V8
Tucumcari, *N. Mex., U.S.* **109** S15
Tucumcari, *N. Mex., U.S.* **109** S15
Tucupita, *Venez.* **130** D13
Tucuruí, *Braz.* **133** E13
Tucuruí, Represa de, *Braz.* **133** E13
Tuddi, river, *Somalia* **210** L14
Tudela, *Sp.* **147** D13
Tuensang, *India* **182** H15
Tuffé, *Fr.* **149** Q8
Tufi, *P.N.G.* **222** Q9
Tufts Plain, *Pac. Oc.* **247** D15
Tufu Point, *American Samoa, U.S.* **230** C11
Tufuone, *Wallis & Futuna Is., Fr.* **230** Q9
Tufuvai, *Tonga* **227** R24
Tug Fork, river, *Ky. & W. Va., U.S.* **103** B20
Tugela Falls, *Af.* **198** P8
Tugela, river, *S. Af.* **212** K9
Tūghyl, *Kaz.* **179** F18
Tuguegarao, *Philippines* **193** B14
Tugur, *Russ.* **161** J20
Tui, *Sp.* **146** C5
Tuineje, *Sp.* **214** Q7
Tukangbesi Kepulauan, *Indonesia* **193** L14
Tukao, *N.Z.* **229** G23
Tukchi, *Russ.* **161** H20
Tūkrah, *Lib.* **201** D21
Tukums, *Latv.* **143** L16
Tukuyu, *Tanzania* **211** Y4
Tukzār, *Afghan.* **180** G7
Tula Mountains, *Antarctica* **239** C20
Tula, *Mex.* **116** G11
Tula, *Russ.* **158** K9
Tula, river, *Kenya* **211** S8
Tulach Mhór *see* Tullamore, *Ire.* **145** R5
Tulaghi, island, *Solomon Is.* **228** F10
Tūlak, *Afghan.* **180** J4
Tulare Lake Bed, *Calif., U.S.* **111** U6
Tulare, *Calif., U.S.* **111** U7
Tularosa, *N. Mex., U.S.* **109** U12
Tulcán, *Ecua.* **130** J4
Tulcea, *Rom.* **155** E18
Tul'chyn, *Ukr.* **159** P5
Tulelake, *Calif., U.S.* **110** L4
Tuli, *Zimb.* **212** G9
Tulia, *Tex., U.S.* **104** E12
Tulita, *N.W.T., Can.* **86** F10
Ṭūlkarm, *W. Bank* **172** L5
Tullahoma, *Tenn., U.S.* **103** F16
Tullamore (Tulach Mhór), *Ire.* **145** R5
Tullamore, *N.S.W., Austral.* **221** X13
Tulle, *Fr.* **149** V9
Tully, *Qnsld., Austral.* **221** R13
Ṭulmaythah, *Lib.* **201** D21
Tulsa, *Okla., U.S.* **105** B18
Tulu Welel, peak, *Eth.* **210** M5
Tuluá, *Col.* **130** G5
Tulun, *Russ.* **161** L14
Tum, *Indonesia* **193** K17
Tumacacori National Historical Park, *Ariz., U.S.* **109** W7
Tumaco, *Col.* **130** J4
Tumaco, Ensenada de, *Col.* **130** J4
Tumakuru, *India* **183** Q5
Tuman, river, *China* **186** A13
Tumbes, *Peru* **132** E2
Tumen, river, *N. Korea* **186** A13
Tumeremo, *Venez.* **131** E14
Tummel, Loch, *Scot., U.K.* **144** K9
Tumon Bay, *Guam, U.S.* **230** N6
Tumu Mountain, *American Samoa, U.S.* **230** B8
Tumu Point, *F.S.M.* **226** J8
Tumu, *Ghana* **205** L15
Tumucumaque, *S. Amer.* **131** H18
Tumwater, *Wash., U.S.* **110** D3
Tuna, *Ghana* **205** M15
Tunam, *N. Korea* **186** D12
Tunapuna, *Trin. & Tobago* **123** P22
Tunas de Zaza, *Cuba* **118** J9
Tunb Islands, *Iran* **177** N18
Tunceli, *Turk.* **171** F14
Tundubai, *Sudan* **203** R6
Tunduma, *Tanzania* **211** Y4
Tunduru, *Tanzania* **211** Z7
Tundzha, river, *Bulg.* **155** J15
Tunga Pass, *India* **182** F14
Tunga, *Nigeria* **206** L9

Tungabhadra, river, *India* **183** P5
Tŭngam, *N. Korea* **186** M5
Tungaru, *Sudan* **203** U11
Tungua, island, *Tonga* **227** M23
Tunguska Nizhnaya, river, *Asia* **168** D9
Tunguska Podkamennaya, *Russ.* **161** J13
Tunguskhaya, *Russ.* **161** G17
Tunica, *Miss., U.S.* **102** G11
Tunis, *Tunisia* **201** B15
Tunis, Gulf of, *Af.* **198** C6
Tunisia, *Af.* **201** C16
Tunisian Plateau, *Atl. Oc.* **248** K13
Tunja, *Col.* **130** F7
Tunnsjøen, *Nor.* **142** G11
Tununak, *Alas., U.S.* **112** J11
Tunuyán, *Arg.* **134** L7
Tuolin *see* Toding, *China* **184** J2
Tuolumne, river, *Calif., U.S.* **111** S5
Tup, *Kyrg.* **179** F18
Tupã, *Braz.* **132** L12
Tupai (Motu Iti), island, *Fr. Polynesia, Fr.* **231** B15
Tupelo National Battlefield, *Miss., U.S.* **103** H13
Tupelo, *Miss., U.S.* **103** H13
Tupi, *Philippines* **193** F15
Tupiza, *Bol.* **132** L7
Tupper Lake, *N.Y., U.S.* **96** G10
Tüpqaraghan Tübegi, *Kaz.* **178** H4
Túquerres, *Col.* **130** J4
Tura, *Russ.* **161** H14
Tura, *India* **182** H12
Turabah, *Saudi Arabia* **174** J7
Turakh, *Russ.* **161** E16
Turalei, *S. Sudan* **203** V9
Turan Lowland, *Asia* **168** F5
Ṭurāq al 'Ilab, *Syr.* **172** J10
Tur'at Maṣīrah, bay, *Oman* **175** J18
Ṭuraybīl, *Iraq* **176** H7
Ṭurayf, *Saudi Arabia* **174** B6
Turbaco, *Col.* **130** D6
Turbat, *Pak.* **181** R3
Turbo, *Col.* **130** E5
Turda, *Rom.* **155** C13
Turégano, *Sp.* **146** E10
Tureia, island, *Fr. Polynesia, Fr.* **231** D21
Turek, *Pol.* **151** E14
Tūrghān, Kōtal-e, *Afghan.* **180** E10
Türgovishte, *Bulg.* **155** H16
Turgut, *Turk.* **170** F7
Turgutlu, *Turk.* **170** F3
Turhal, *Turk.* **170** D11
Türi, *Est.* **143** K16
Turi, peak, *Fr. Polynesia, Fr.* **231** K14
Turia, river, *Sp.* **147** F14
Turiaçu, *Braz.* **133** D14
Turiaçu, Baía de, *Braz.* **133** D14
Turin *see* Torino, *It.* **152** E2
Turka, *Ukr.* **159** P2
Türkeli, island, *Turk.* **170** D3
Turkey Creek, *W. Aust., Austral.* **221** Q6
Turkey, *Tex., U.S.* **105** E13
Turkey, *Asia* **170** F6
Turkey, river, *Iowa, U.S.* **106** M13
Türkistan, *Kaz.* **178** K11
Türkmen Aylagy, bay, *Turkm.* **178** M5
Türkmen Dağı, *Turk.* **170** E6
Türkmenabat (Chärjew), *Turkm.* **178** M9
Türkmenbaşy Aylagy, bay, *Turkm.* **178** L5
Türkmenbaşy, *Turkm.* **178** L4
Turks & Caicos Islands, *N. Amer.* **119** J17
Turks Island Passage, *Turks & Caicos Is., U.K.* **119** J17
Turks Islands, *Turks & Caicos Is., U.K.* **119** J18
Turku (Åbo), *Fin.* **143** J15
Turlock, *Calif., U.S.* **111** S5
Turner Hole, bay, *Virgin Is., U.S.* **122** Q4
Turner Ø, cape, *Greenland, Den.* **88** L14
Turner, *Mont., U.S.* **108** A10
Turner's Peninsula, *Af.* **198** H1
Turnhout, *Belg.* **148** K12
Turnu Măgurele, *Rom.* **155** G14
Turnu Roşu, Pasul, *Rom.* **155** E14
Turpan Depression, *Asia* **168** G8
Turpan Pendi, *China* **184** E6
Turpan, *China* **184** E6
Turquino, Pico, *Cuba* **118** K12

Tursunzoda, *Taj.* **178** N11
Turt, *Mongolia* **184** B9
Türtkül, *Uzb.* **178** L8
Turtle Crawls, *Cayman Is., U.K.* **122** N2
Turtle Lake, *N. Dak., U.S.* **106** F4
Turtle Mountain, *N. Dak., U.S.* **106** D5
Turtle Rocks, *Bahamas* **122** C11
Turtle *see* Vatoa, island, *Fiji* **226** R9
Turugart Pass, *China, Kyrg.* **184** E2
Turyu-san, peak, *N. Korea* **186** E11
Tuscaloosa, *Ala., U.S.* **103** K14
Tuscany, *Eur.* **140** K6
Tuscarora, *Nev., U.S.* **110** M10
Tuscola, *Ill., U.S.* **101** R6
Tuscola, *Tex., U.S.* **105** H14
Tuscumbia, *Ala., U.S.* **103** G14
Tushchybas Shyghanaghy, lake, *Kaz.* **178** H8
Tuskar Rock, *Ire.* **145** T7
Tuskegee, *Ala., U.S.* **103** L16
Tussaaq, *Greenland, Den.* **88** J5
Tutak, *Turk.* **171** E17
Tutayev, *Russ.* **158** G10
Tuticorin, *India* **183** T6
Tuting, *India* **182** F15
Tutira, *N.Z.* **223** H22
Tutrakan, *Bulg.* **155** G16
Tuttle Creek Lake, *Kans., U.S.* **107** S8
Tuttlingen, *Ger.* **150** K6
Tutu, *Virgin Is., U.S.* **122** N9
Tutu, island, *Fiji* **226** N6
Tutuala, *Timor-Leste* **193** M16
Tutubu, *Tanzania* **211** V4
Tutumu, *Solomon Is.* **228** F10
Tuvalu, *Pac. Oc.* **224** H8
Tuvuca, island, *Fiji* **226** P8
Tuwayq Mountains, *Asia* **168** H3
Tuxpan, *Mex.* **116** G9
Tuxpan, *Mex.* **116** H12
Tuxtla Gutiérrez, *Mex.* **117** K14
Tuy Hòa, *Vietnam* **191** N13
Tuyên Quang, *Vietnam* **190** G10
Tūyserkān, *Iraq* **177** F13
Tuz Gölü, *Turk.* **170** F8
Tuzigoot National Monument, *Ariz., U.S.* **109** S7
Ṭūzkhūrmātū, *Iraq* **176** F10
Tuzla Gölü, *Turk.* **170** F11
Tuzla, *Bosn. & Herzg.* **154** F8
Tuzluca, *Turk.* **171** E18
Tveitsund, *Nor.* **142** L11
Tver', *Russ.* **158** H8
Tvøroyri, *Faroe Is., Den.* **142** H6
Twentynine Palms, *Calif., U.S.* **111** X11
Twilight Cove, *Austral.* **220** K6
Twin Bridges, *Mont., U.S.* **108** E7
Twin Buttes Reservoir, *Tex., U.S.* **105** J13
Twin City, *Ga., U.S.* **98** M8
Twin Falls, *Idaho, U.S.* **108** H5
Twin Peaks, *Idaho, U.S.* **108** F5
Twin Valley, *Minn., U.S.* **106** G8
Twisp, *Wash., U.S.* **110** B6
Twizel, *N.Z.* **223** M17
Two Butte Creek, *Colo., U.S.* **109** P15
Two Buttes, *Colo., U.S.* **109** P16
Two Harbors, *Minn., U.S.* **106** G13
Two Rivers, *Wis., U.S.* **100** K7
Tybee Island, *Ga., U.S.* **99** N9
Tycheró, *Gr.* **157** B16
Tychy, *Pol.* **151** G15
Tygda, *Russ.* **161** K19
Tyler, *Tex., U.S.* **105** G19
Tyler, *Minn., U.S.* **106** K9
Tylertown, *Miss., U.S.* **102** N11
Týmfi, Óros (Gamíla), *Gr.* **156** E7
Tymfristós, peak, *Gr.* **156** G9
Tymvou (Kırklar), *Cyprus* **162** N9
Tynda, *Russ.* **161** K18
Tyndall, *S. Dak., U.S.* **106** M7
Tyndaris, ruin, *It.* **153** R11
Tyndrum, *Scot., U.K.* **144** K9
Tynemouth, *Eng., U.K.* **145** N12
Tyonek, *Alas., U.S.* **113** J16
Tyre *see* Soûr, *Leb.* **172** K6
Týrnavos, *Gr.* **156** F10
Tyrnyauz, *Russ.* **159** U12
Tyrol, region, *Aust., It.* **152** C6
Tyrone, *Pa., U.S.* **96** N5
Tyrone, *Okla., U.S.* **105** A13
Tyrrell, Lake, *Austral.* **220** L12
Tyrrhenian Sea, *It.* **152** N7
Tyugyuren, *Russ.* **161** E18
Tyukalinsk, *Russ.* **160** J10
Tyumen', *Russ.* **160** J9

Tywyn, *Wales, U.K.* **145** S9
Tzia *see* Kea, island, *Gr.* **157** K13
Tzoumérka, peak, *Gr.* **156** E8

U

U, *F.S.M.* **226** J8
Ua Huka, island, *Fr. Polynesia, Fr.* **225** H15
Ua Pou, island, *Fr. Polynesia, Fr.* **225** H15
Uaboe, *Nauru* **227** M14
Uarandab, *Eth.* **211** N11
Uaroo, *W. Aust., Austral.* **221** T2
Ubá, *Braz.* **133** L15
Ubai, *P.N.G.* **222** P9
Ubangi, river, *D.R.C.* **208** D12
Ubangi, river, *D.R.C.* **209** B13
Ubari *see* Awbārī, *Lib.* **201** H17
Ubauro, *Pak.* **181** P9
Ube, *Japan* **189** Q5
Úbeda, *Sp.* **146** K11
Ubekendt Ejland, island, *Greenland, Den.* **88** K5
Uberaba, *Braz.* **133** K13
Uberlândia, *Braz.* **133** K13
Ubierna, *Sp.* **146** C11
Ubon Ratchathani, *Thai.* **190** L10
Ubundu, *D.R.C.* **209** E16
Üçajy, *Turkm.* **178** N9
Ucar, *Azerb.* **171** D21
Ucayali, river, *Peru* **132** F4
Uch, *Pak.* **181** N10
Uchiura Wan, *Japan* **188** H12
Uchiza, *Peru* **132** G3
Uchquduq, *Uzb.* **178** K9
Uda Spur, *Pac. Oc.* **246** G8
Udachnyy, *Russ.* **161** G16
Udaipur, *India* **182** J4
Udayagiri, *India* **182** M9
Uddevalla, *Sw.* **142** L12
Uderolal, *Pak.* **181** R8
Udgir, *India* **182** M5
Udhampur, *India* **182** D5
Udine, *It.* **152** D9
Udintsev Fracture Zone, *Pac. Oc.* **247** R14
Udipi, *India* **183** Q4
Udo, island, *S. Korea* **187** W8
Udon Thani, *Thai.* **190** K9
Udot, island, *F.S.M.* **226** E8
Udskaya Guba, *Russ.* **161** J20
Udu Point, *Fiji* **226** M7
Uebonti, *Indonesia* **193** J14
Ueda, *Japan* **189** N11
Uele, river, *D.R.C.* **209** B15,18
Uelen, *Russ.* **161** A21
Uelzen, *Ger.* **150** D8
Uere, river, *D.R.C.* **209** B16,17
Ufa, *Russ.* **160** H8
Ufa, river, *Eur.* **140** D15
Ufra, *Turkm.* **178** L4
Ugab, river, *Namibia* **212** G3
Ugalla, river, *Tanzania* **211** V3
Uganda, *Af.* **211** R3
Ugarit *see* Ra's Shamrah, ruins, *Syr.* **172** E6
Ughelli, *Nigeria* **206** M7
Uglich, *Russ.* **158** H9
Ugljan, island, *Croatia* **154** F5
Ugljane, *Croatia* **154** G6
Uglovoye, *Russ.* **161** K19
Ugol'nyye Kopi, *Russ.* **161** B21
Ugoma, mountains, *D.R.C.* **209** G18
Ugum, river, *Guam, U.S.* **230** Q5
Uherské Hradiště, *Czechia* **151** J13
Uhrichsville, *Ohio, U.S.* **101** Q15
Uibhist a Deas (South Uist), island, *Scot., U.K.* **144** J6
Uibhist a Tuath (North Uist), island, *Scot., U.K.* **144** H6
Uido, island, *S. Korea* **187** U6
Uig, *Scot., U.K.* **144** H7
Uíge, *Angola* **208** J10
Uijeongbu, *S. Korea* **186** M8
Ŭiju, *N. Korea* **186** G4
Uinta Mountains, *Utah, U.S.* **108** L8
Uinta, river, *Utah, U.S.* **108** L9
Uiryeong, *S. Korea* **187** S10
Uiseong, *S. Korea* **187** Q11

Uitenhage, *S. Af.* **212** M7
Uithuizen, *Neth.* **148** G14
Ujae Atoll, *Marshall Is.* **227** C15
Ujajiivan, island, *Marshall Is.* **227** G16
Ujarasussuk, *Greenland, Den.* **88** L6
Ujelang Atoll, *Marshall Is.* **226** A12
Ujelang Atoll, *Marshall Is.* **227** C13
Uji Guntō, *Japan* **189** T3
Ujiji, *Tanzania* **211** V2
Ujjain, *India* **182** K5
Ujung Raja, *Indonesia* **192** G4
Ujungpandang *see* Makassar, *Indonesia* **193** L13
Ukhiya, *India* **182** L14
Ukhrul, *India* **182** H15
Ukhta, *Russ.* **158** B15
Uki Ni Masi Island, *Solomon Is.* **228** G12
Ukiah, *Oreg., U.S.* **110** F7
Ukiah, *Calif., U.S.* **111** Q2
Ukkusissat, *Greenland, Den.* **88** K6
Ukmergė, *Lith.* **143** M17
Ukraine, *Eur.* **159** P6
Uku Shima, *Japan* **189** R3
Uku, *Japan* **189** R3
Uku, *Angola* **208** M10
Ukulahu, island, *Maldives* **183** V3
Ukuma, *Angola* **208** M10
Ukwaa, *S. Sudan* **203** W13
Ukwaa, *Eth.* **211** N5
Ula, *Turk.* **170** H4
Ulaan-Ereg, *Mongolia* **184** D11
Ulaanbaatar (Ulan Bator), *Mongolia* **184** D11
Ulaangom, *Mongolia* **184** C7
Ulaga, *Russ.* **161** F17
Ulan Bator *see* Ulaanbaatar, *Mongolia* **184** D11
Ulan Erge, *Russ.* **159** R13
Ulan Khol, *Russ.* **159** S15
Ulan Ul Hu, *China* **184** J5
Ulan-Ude, *Russ.* **161** L16
Ulan, *China* **184** H8
Ulanhot, *China* **185** D15
Ulaş, *Turk.* **170** F12
Ulawa, island, *Solomon Is.* **228** F12
Ulaya, *Tanzania* **211** W6
Ulcinj, *Montenegro* **154** K9
Uldo, island, *S. Korea* **187** P6
Uldz, *Mongolia* **184** C12
Uldz, river, *Mongolia* **184** C12
Uleåborg *see* Oulu, *Fin.* **143** F15
Uleguma, island, *Maldives* **183** U3
Ulen, *Minn., U.S.* **106** G8
Uliastay, *Mongolia* **184** D8
Ulindi, river, *D.R.C.* **209** F17
Ulingan, *P.N.G.* **222** N7
Ulithi Atoll, *F.S.M.* **226** A6
Uljin, *S. Korea* **187** P12
Uljma, *Serb.* **154** F11
Ülken Borsyq Qumy, *Kaz.* **178** G8
Ulladulla, *N.S.W., Austral.* **221** Y14
Ullapool, *Scot., U.K.* **144** H8
Ullersuaq (Kap Alexander), cape, *Greenland, Den.* **88** D3
Ulleung, *S. Korea* **187** N15
Ulleungdo (Dagelet), island, *S. Korea* **186** M15
Ulm, *Ger.* **150** K7
Ulsan, *S. Korea* **187** S12
Ulster (Cúige Uladh), region, *Ire. U.K.* **145** N6
Ulu Dağ (Mount Olympus), *Turk.* **170** E5
Ulu, *Russ.* **161** H18
UlubatGölü, *Turk.* **170** D4
Uluborlu, *Turk.* **170** G6
Ulufala Point, *American Samoa, U.S.* **230** C11
Uluguru Mountains, *Tanzania* **211** W7
Ulukışla, *Turk.* **170** H9
Ulul, island, *F.S.M.* **226** A9
Ulungur Hu, *China* **184** D5
Uluṟu (Ayers Rock), peak, *Austral.* **220** G8
Ulus, *Turk.* **170** C8
Ulva, island, *Scot., U.K.* **144** K7
Ulverston, *Eng., U.K.* **145** P10
Ul'yanovsk, *Russ.* **158** K14
Ulysses, *Kans., U.S.* **107** U3
Ulytaū, *Kaz.* **178** F11
Uman', *Ukr.* **159** P5
Uman, island, *F.S.M.* **226** F9
Umari, *Indonesia* **193** L19
Umaria, *India* **182** K7

Umarkot, *Pak.* **181** R9
Umatac Bay, *Guam, U.S.* **230** Q4
Umatac, *Guam, U.S.* **230** Q4
Umatilla, *Oreg., U.S.* **110** E7
Umatilla, *river, Oreg., U.S.* **110** F7
Umba, *Russ.* **160** D8
Umbakumba, *N.T., Austral.* **221** P10
Umboi, *island, P.N.G.* **222** P8
Umbrella Point, *Jam.* **122** G7
Umbria, *Eur.* **140** K7
Umbriatico, *It.* **153** P14
Umbukul, *P.N.G.* **222** M9
Umeå, *Sw.* **143** G14
Umeälven, *river, Sw.* **143** F13
Umfors, *Sw.* **142** F12
Umiammakku Isbræ, *glacier, Greenland, Den.* **88** K6
Umiiviip Kangertiva (Gyldenløve Fjord), *Greenland, Den.* **89** R9
Umiivik, *Greenland, Den.* **89** R9
'Umikoa, *Hawaii, U.S.* **115** M22
Umm ad Dāmī, Jabal, *Jordan* **173** S6
Umm al 'Abīd, *Lib.* **201** G18
Umm al Qaywayn, *U.A.E.* **175** F16
Umm as Samīm, *lake, Oman, Saudi Arabia* **175** J16
Umm Bel, *Sudan* **203** S9
Umm Busha, *Sudan* **203** S11
Umm Dam, *Sudan* **203** S11
Umm Durmān (Omdurman), *Sudan* **203** Q12
Umm Haraz, *Sudan* **203** T6
Umm Keddada, *Sudan* **203** S8
Umm Lajj, *Saudi Arabia* **174** F5
Umm Rahau, *Sudan* **203** N12
Umm Ruwaba, *Sudan* **203** S11
Umm Sa'īd, *Qatar* **175** G13
Umm Shalil, *Sudan* **203** T6
Umm Shanqa, *Sudan* **203** S9
Umm Urūmah, *island, Saudi Arabia* **174** F4
Umnak Island, *Alas., U.S.* **112** P9
Umpqua, *river, Oreg., U.S.* **110** J2
Umpulo, *Angola* **208** M12
Umred, *India* **182** L7
Umuna, *Tonga* **227** H23
Umvoto Rise, *Atl. Oc.* **249** V14
Una, *river, Bosn. & Herzg., Croatia* **154** E6
Unadilla, *Ga., U.S.* **99** N6
Unai Obyan, *cape, N. Mariana Is., U.S.* **230** K8
Unalakleet, *Alas., U.S.* **113** G13
Unalaska Island, *Alas., U.S.* **112** P9
Unalaska, *Alas., U.S.* **112** P10
'Unayzah, *Saudi Arabia* **174** F9
'Unayzah, Jabal, *Jordan* **172** M12
Unbong-jŏsuji, *lake, N. Korea* **186** D7
Unbong, *S. Korea* **187** S9
Ŭnch'ŏn, *N. Korea* **186** K5
Uncía, *Bol.* **132** K7
Uncompahgre Peak, *Colo., U.S.* **109** N11
Uncompahgre Plateau, *Colo., U.S.* **109** N10
Underwood Glacier, *Antarctica* **239** L21
Underwood, *N. Dak., U.S.* **106** F4
Ŭndŏk, *N. Korea* **186** B13
Unea, *island, P.N.G.* **222** N9
Unecha, *Russ.* **158** L7
Ungama Bay, *Kenya* **211** U8
Ungava Bay, *Que., Can.* **87** J19
Ungava Peninsula, *N. Amer.* **84** F8
Ungcheon, *S. Korea* **187** Q7
Unggi (Sŏnbong), *N. Korea* **186** B14
Unggok, *S. Korea* **187** U6
Unha, *N. Korea* **186** G10
Unha, *N. Korea* **186** H5
Unhŭng, *N. Korea* **186** D11
Unhŭng, *N. Korea* **186** H8
União da Vitória, *Braz.* **132** N12
Unije, *island, Croatia* **154** F4
Unimak Island, *Alas., U.S.* **112** N11
Unimak Pass, *Alas., U.S.* **112** N10
Unini, *river, Braz.* **132** D8
Union City, *Pa., U.S.* **96** L3
Union City, *Ohio, U.S.* **101** R11
Union City, *Tenn., U.S.* **102** E12
Union Glacier Camp, *Antarctica* **238** H9
Union Point, *Ga., U.S.* **98** L6
Union Springs, *Ala., U.S.* **103** L16
Union, *S.C., U.S.* **98** J8
Union, *Miss., U.S.* **102** L12
Union, *Mo., U.S.* **107** T14

Union, *Oreg., U.S.* **110** F8
Union, *Grenada* **123** J23
Uníon, Bahía, *Arg.* **135** P11
Union, *island, St. Vincent & the Grenadines* **121** G16
Uniontown, *Pa., U.S.* **96** P3
Uniontown, *Ala., U.S.* **103** L14
Unionville, *Mo., U.S.* **107** Q12
Unionville, *Nev., U.S.* **111** N8
United Arab Emirates, *Asia* **175** G14
United Estates, *Bahamas* **122** E12
United Kingdom, *Eur.* **145** P6
United States, *N. Amer.* **92** J12
United States Naval Base Guantanamo Bay, *U.S.* **119** L14
United States Range, *Nunavut, Can.* **87** A16
Unity, *Oreg., U.S.* **110** G8
University Park, *N. Mex., U.S.* **109** V12
Unnao, *India* **182** H7
Ŭnp'a, *N. Korea* **186** L6
Unp'o, *N. Korea* **186** G10
Unsan, *N. Korea* **186** G6
Unst, *island, Scot., U.K.* **144** B12
Ünye, *Turk.* **170** D12
Ŭnyul, *N. Korea* **186** K5
Unzha, *Russ.* **158** G12
Unzha, *river, Eur.* **140** D13
Uoleva, *island, Tonga* **227** M24
Upata, *Venez.* **130** E13
Upemba, Lake, *Af.* **198** L8
Upernavik Isstrøm, *glacier, Greenland, Den.* **88** J6
Upernavik Kujalleq (Søndre Upernavik), *Greenland, Den.* **88** K5
Upernavik, *Greenland, Den.* **88** J5
Upham, *N. Dak., U.S.* **106** E4
'Upolu Point, *Hawaii, U.S.* **115** L20
Upolu, *island, Samoa* **227** P19
Upper Arlington, *Ohio, U.S.* **101** R13
Upper Bogue, *Bahamas* **122** D5
Upper Darby, *Pa., U.S.* **96** P9
Upper Guinea, *region, Af.* **198** H3
Upper Hutt, *N.Z.* **223** J20
Upper Kama Upland, *Eur.* **140** C14
Upper Klamath Lake, *Oreg., U.S.* **110** K4
Upper Lake, *Calif., U.S.* **110** L6
Upper Land, *Cayman Is., U.K.* **122** M2
Upper Lough Erne, *N. Ire., U.K.* **145** P5
Upper Manzanilla, *Trin. & Tobago* **123** P23
Upper Missouri River Breaks National Monument, *Mont., U.S.* **108** C10
Upper Peninsula, *Mich., U.S.* **100** F9
Upper Red Lake, *Minn., U.S.* **106** E10
Upper Saint Clair, *Pa., U.S.* **96** P3
Upper Sandusky, *Ohio, U.S.* **101** Q12
Uppsala, *Sw.* **143** K14
Upshi, *India* **182** D6
Upstart Bay, *Austral.* **220** E14
Upstart, Cape, *Austral.* **220** E14
Upton, *Wyo., U.S.* **108** G13
'Uqayribāt, *Syr.* **172** F9
Ur, *ruin, Iraq* **176** J12
Ura-Tyube *see* Istaravshan, *Taj.* **178** M12
Urabá, Golfo de, *Col.* **130** E5
Urakawa, *Japan* **188** H14
Urambo, *Tanzania* **211** V3
Urania, *La., U.S.* **102** M8
Uranium City, *Sask., Can.* **86** J12
Uranu, *island, F.S.M.* **226** E8
Uraricoera, *Braz.* **132** B9
Uraricoera, *river, Braz.* **132** B8
Uravan, *Colo., U.S.* **109** N10
Urawa, *Japan* **189** P12
Urbana, *Ill., U.S.* **101** R7
Urbana, *Ohio, U.S.* **101** R12
Urbandale, *Iowa, U.S.* **107** P11
Urbano Noris, *Cuba* **118** K12
Urbett, *island, Marshall Is.* **227** G20
Urbino, *It.* **152** G8
Urbión, Picos de, *Sp.* **146** D12
Urdos, *Fr.* **149** Y6
Uren', *Russ.* **158** G13
Uréparapara, *island, Vanuatu* **229** C14
Urganch, *Uzb.* **178** L8
Urgūn, *Afghan.* **180** K9
Urgut, *Uzb.* **178** M11

Uri-do, *island, N. Korea* **186** H4
Uribia, *Col.* **130** C7
Uriondo, *Bol.* **132** L7
Uritskoye, *Russ.* **161** H17
Urkut, *Somalia* **211** Q10
Urla, *Turk.* **170** F2
Urlaţi, *Rom.* **155** E16
Urlings, *Antigua & Barbuda* **123** B20
Urmia *see* Orūmīyeh, *Iran* **176** D11
Uroševac *see* Ferizaj, *Kos.* **154** J11
Urt Moron, *China* **184** H6
Urtipah, *Afghan.* **180** G5
Uruana, *Braz.* **133** J13
Uruapan, *Mex.* **116** J10
Urubamba, *river, Peru* **132** H5
Urucará, *Braz.* **132** D10
Uruçuí, Serra do, *S. Amer.* **128** F10
Uruçuí, *Braz.* **133** F15
Uruguai *see* Uruguay, *river, Arg.* **134** H14
Uruguaiana, *Braz.* **132** P10
Uruguay Canyon, *Atl. Oc.* **254** K10
Uruguay, *S. Amer.* **134** K14
Uruguay, *river, S. Amer.* **128** K7
Ürümqi, *China* **184** E5
Uruno Point, *Guam, U.S.* **230** M6
Urup, *Russ.* **161** J23
Urup, *island, Asia* **168** D14
'Urūq ar Rumaylah, *Saudi Arabia* **174** J11
'Urūq Subay', *Saudi Arabia* **174** H8
Uruwira, *Tanzania* **211** W3
Uruzgān, *Afghan.* **180** K7
Uryupinsk, *Russ.* **159** N11
Ürzhar, *Kaz.* **179** G17
Urzhum, *Russ.* **158** G15
Urziceni, *Rom.* **155** F16
Usa, *Japan* **189** R7
Usa, *river, Eur.* **140** A13
Uşak, *Turk.* **170** F5
Usakos, *Namibia* **212** G3
Usambara Mountains, *Tanzania* **211** V7
Usarp Mountains, *Antarctica* **239** Q15
Usedom, *island, Ger.* **150** B10
Useko, *Tanzania* **211** V4
Useldange, *Lux.* **162** J9
Ushakov Bank, *Arctic Oc.* **253** F15
Ushakova, *Russ.* **161** C13
Ushakova, *Russ.* **161** B19
Ushakovskoye, *Russ.* **161** B19
Ūsharal, *Kaz.* **179** G17
Ushetu, *Tanzania* **211** V3
Ushi Point, *N. Mariana Is., U.S.* **230** H11
Ushibuka, *Japan* **189** S4
Üshtöbe, *Kaz.* **179** H15
Ushuaia, *Arg.* **135** X9
Ushumun, *Russ.* **161** K19
Usi, *N. Korea* **186** F6
Usinsk, *Russ.* **160** F10
Usman', *Russ.* **158** M10
Usogorsk, *Russ.* **158** C13
Usoke, *Tanzania* **211** V3
Usol'ye Sibirskoye, *Russ.* **161** L15
Usove, *Ukr.* **158** M4
Ussel, *Fr.* **149** U10
Ussing Bræer, *glacier, Greenland, Den.* **88** H5
Ussuri, *river, Russ.* **161** L21
Ussuriysk, *Russ.* **161** M21
Ussuttusoq, *fjord, Greenland, Den.* **89** Q10
Ust' Alekseyevo, *Russ.* **158** E13
Ust' Ilimpeya, *Russ.* **161** H15
Ust' Ilimsk, *Russ.* **161** K15
Ust' Ilych, *Russ.* **158** C16
Ust' Izhma, *Russ.* **158** A14
Ust' Kamo, *Russ.* **161** J14
Ust' Kulom, *Russ.* **158** D13
Ust' Kut, *Russ.* **161** K15
Ust' Labinsk, *Russ.* **159** S11
Ust' Maya, *Russ.* **161** G19
Ust' Mil', *Russ.* **161** H19
Ust' Nem, *Russ.* **158** D16
Ust' Nera, *Russ.* **161** F19
Ust' Olenek, *Russ.* **161** E16
Ust' Omchug, *Russ.* **161** F20
Ust' Ordynskiy, *Russ.* **161** L15
Ust' Pinega, *Russ.* **158** B11
Ust' Shchugor, *Russ.* **158** A16
Ust' Tsil'ma, *Russ.* **158** A14
Ust' Usa, *Russ.* **160** F10
Ust' Vym', *Russ.* **158** C14
Ust' Yansk, *Russ.* **161** E17
Ust' Yudoma, *Russ.* **161** H19
Ust'-Kamchatsk, *Russ.* **161** E22

Ust'-Kamenogorsk *see* Öskemen, *Kaz.* **179** E17
Uster, *Switz.* **152** A4
Ustí nad Labem, *Czechia* **150** G11
Ustica, *It.* **153** R9
Ustica, Isola di, *It.* **153** R9
Ustka, *Pol.* **151** A13
Ustyurt Plateau, *Kaz., Uzb.* **178** J6
Ustyuzhna, *Russ.* **158** G9
Usu, *China* **184** D4
Usui, *Japan* **189** R5
Usumacinta, *river, N. Amer.* **84** N6
Usure, *Tanzania* **211** V5
Usvyaty, *Russ.* **158** J6
Uta, *Indonesia* **193** L19
Utah Beach, *Fr.* **149** N6
Utah Lake, *Utah, U.S.* **108** L7
Utah, *U.S.* **108** M7
Ute Creek, *N. Mex., U.S.* **109** Q14
Utegi, *Tanzania* **211** T5
Utembo, *river, Angola* **209** Q14
Utete, *Tanzania* **211** X7
Uthai Thani, *Thai.* **190** L7
Uthal, *Pak.* **181** R7
Utiariti, *Braz.* **132** H10
Utica, *N.Y., U.S.* **96** J9
Utiel, *Sp.* **147** H14
Utirik Atoll, *Marshall Is.* **227** B17
Utkela, *India* **182** M9
Utnur, *India* **182** M6
Utopia, *N.T., Austral.* **221** S9
Utraula, *India* **182** G8
Utrecht, *Neth.* **148** J12
Utsunomiya, *Japan* **189** N12
Utta, *Russ.* **159** R14
Uttaradit, *Thai.* **190** K7
Utuloa, *Wallis & Futuna Is., Fr.* **230** Q8
Utupua, *island, Solomon Is.* **228** M6
Uturoa, *Fr. Polynesia, Fr.* **231** G17
Utva, *river, Kaz.* **178** E6
Utwa, *F.S.M.* **226** G11
Uummannaq Fjord, *Greenland, Den.* **88** L5
Uusikaupunki, *Fin.* **143** J15
Uva, *Russ.* **158** G16
Uvalde, *Tex., U.S.* **105** M14
Uvarovo, *Russ.* **158** M11
Uvéa *see* Ouvéa, *island, New Caledonia, Fr.* **231** N23
Uvinza, *Tanzania* **211** V2
Uvira, *D.R.C.* **209** G18
Uvol, *P.N.G.* **222** P9
Uvs Nuur, *Mongolia* **184** C7
Uwajima, *Japan* **189** R6
Uxmal, *ruin, Mex.* **117** H16
Uyar, *Russ.* **161** K13
Uyo, *Nigeria* **206** N8
Uyowa, *Tanzania* **211** V3
Uys Nuur, *lake, Asia* **168** F9
Uyuni, *Bol.* **132** L7
Uyuni, Salar de, *Bol.* **132** L6
Uzbekistan, *Asia* **178** J8
Uzbel Shankou, *Taj.* **179** M14
Uzbel Shankou, *pass, China, Taj.* **179** M14
Uzerche, *Fr.* **149** U9
Uzès, *Fr.* **149** X12
Užice, *Serb.* **154** G9
Uzlovaya, *Russ.* **158** K9
Üzümlü, *Turk.* **170** H4
Uzunköprü, *Turk.* **170** C3
Uzynköl, *Kaz.* **178** C10

#

Vaal, *river, S. Af.* **212** J7
Vaals, *Neth.* **148** L13
Vaasa, *Fin.* **143** G14
Vác, *Hung.* **151** L15
Vacaville, *Calif., U.S.* **111** R4
Vache, Île à, *Haiti* **119** N15
Vacoas, *Mauritius* **215** G20
Vadodara, *India* **182** K3
Vadsø, *Nor.* **143** B15
Vaduz Castle, *ruin, Liech.* **162** P2
Vaduz, *Liech.* **162** P2
Værøy, *island, Nor.* **142** E11
Vaga, *river, Russ.* **158** D11
Vágar, *island, Faroe Is., Den.* **142** H5
Vaghena, *island, Solomon Is.* **228** D7
Vagthus Point, *Virgin Is., U.S.* **122** R3

Váh, *river, Slovakia* **151** J14
Vahanga, *island, Fr. Polynesia, Fr.* **231** E22
Vahdat, *Taj.* **178** M12
Vahitahi, *island, Fr. Polynesia, Fr.* **231** C21
Vaiaau, *Fr. Polynesia, Fr.* **231** H17
Vaiden, *Miss., U.S.* **102** K11
Vaiea, *N.Z.* **229** Q21
Vaiere, *island, N.Z.* **229** M22
Vaigalu, *Samoa* **227** P20
Vail, *Colo., U.S.* **108** M12
Vaila, *island, Scot., U.K.* **144** C12
Vailala, *Wallis & Futuna Is., Fr.* **230** Q9
Vailoa, *Samoa* **227** N17
Vailoatai, *American Samoa, U.S.* **230** E6
Vaimalau, *Wallis & Futuna Is., Fr.* **230** R8
Vaini, *Tonga* **227** Q23
Vaipae, *N.Z.* **229** H20
Vaipaee, *Fr. Polynesia, Fr.* **230** E11
Vaipeka, *N.Z.* **229** H20
Vairaatea, *island, Fr. Polynesia, Fr.* **231** D21
Vairo, *Fr. Polynesia, Fr.* **231** J23
Vaitahu, *Fr. Polynesia, Fr.* **231** E14
Vaitape, *Fr. Polynesia, Fr.* **231** G14
Vaïtape, *Fr. Polynesia, Fr.* **231** F16
Vaitoare, *Fr. Polynesia, Fr.* **231** G17
Vaitogi, *American Samoa, U.S.* **230** E6
Vaitupu, *Wallis & Futuna Is., Fr.* **230** Q9
Vaiutukakau Bay, *Tonga* **227** G23
Vaka'eitu, *island, Tonga* **227** H22
Vakfikebir, *Turk.* **171** D14
Vakhrushev, *Russ.* **161** J22
Vakhsh, *river, Taj.* **179** M13
Vakhtan, *Russ.* **158** G13
Val-d'Or, *Que., Can.* **87** P18
Valamaz, *Russ.* **158** G15
Valcheta, *Arg.* **135** Q9
Valdagno, *It.* **152** D7
Valdahon, *Fr.* **149** S14
Valdai Hills, *Eur.* **140** E11
Valdayskaya Vozvyshennost', *Russ.* **158** H7
Valdepeñas, *Sp.* **146** J11
Valderrobres, *Sp.* **147** F15
Valdés, Península, *Arg.* **135** R10
Valdez, *Alas., U.S.* **113** J18
Valdez, *Ecua.* **130** J3
Valdivia Fracture Zone, *Pac. Oc.* **247** P20
Valdivia Plain, *Ind. Oc.* **254** D11
Valdivia, *Col.* **130** E5
Valdivia, *Chile* **135** P6
Valdosta, *Ga., U.S.* **99** Q6
Valdragone, *San Marino* **163** J15
Vale, *Oreg., U.S.* **110** H9
Valea lui Mihai, *Rom.* **154** B12
Valença, *Braz.* **133** H16
Valença, *Port.* **146** D5
Valence, *Fr.* **149** V13
Valence, *Fr.* **149** W8
Valencia Island, *Ire.* **145** T2
Valencia, *Trin. & Tobago* **123** P23
Valencia, Lake, *S. Amer.* **128** A4
Valencia, *Venez.* **130** C10
Valencia, *Sp.* **147** H14
Valencia, Golfo de, *Sp.* **147** H15
Valenciennes, *Fr.* **148** M11
Valentine, *Tex., U.S.* **104** K9
Valentine, *Nebr., U.S.* **106** M4
Valenza, *It.* **152** E4
Valera, *Venez.* **130** D8
Valga, *Est.* **143** K17
Valguarnera Caropepe, *It.* **153** T11
Valier, *Mont., U.S.* **108** B7
Valinco, Golfe de, *Fr.* **149** Y2
Valira del Nort, *river, Andorra* **162** J3
Valira d'Orient, *river, Andorra* **162** J4
Valjevo, *Serb.* **154** G9
Valka, *Latv.* **143** K16
Valkeakoski, *Fin.* **143** H15
Valkenswaard, *Neth.* **148** K12
Valkyrie Dome (Dome F, Dome Fuji), *Antarctica* **239** E16
Valladolid, *Mex.* **117** H17
Valladolid, *Sp.* **146** D10
Valle de la Pascua, *Venez.* **130** D11
Valle Fértil, Sierra del, *Arg.* **134** J8
Valle Formozo, *Sao Tome and Principe* **215** D20
Valledupar, *Col.* **130** C7

Vallée du Tîlemsi, *Mali* **205** F16
Vallegrande, *Bol.* **132** K8
Vallehermoso, *Sp.* **214** Q4
Vallejo, *Calif., U.S.* **111** R3
Vallenar, *Chile* **134** H6
Valletta, *Malta* **163** K23
Valley Center, *Kans., U.S.* **107** U7
Valley City, *N. Dak., U.S.* **106** G7
Valley Falls, *Kans., U.S.* **107** S9
Valley Falls, *Oreg., U.S.* **110** K6
Valley Mills, *Tex., U.S.* **105** H17
Valley of the Kings, ruin, *Egypt* **202** J12
Valley Station, *Ky., U.S.* **103** B16
Vallo della Lucania, *It.* **153** N12
Vallorbe, *Switz.* **152** B1
Valls, *Sp.* **147** E16
Valmeyer, *Ill., U.S.* **101** T4
Valmiera, *Latv.* **143** K17
Valmy, *Fr.* **149** P12
Valona *see* Vlorë, *Alban.* **154** M9
Valozhyn, *Belarus* **158** K4
Valparaiso, *Fla., U.S.* **99** Q2
Valparaiso, *Ind., U.S.* **101** P8
Valparaíso, *Mex.* **116** G10
Valparaíso, *Chile* **134** K6
Vals, Cape, *Asia* **168** L16
Valsad, *India* **182** L3
Valūna, valley, *Liech.* **162** Q3
Valuyevka, *Russ.* **159** R13
Valuyki, *Russ.* **159** N10
Valverde del Camino, *Sp.* **146** K7
Valverde, *Sp.* **214** R3
Van Alstyne, *Tex., U.S.* **105** F18
Van Blommestein Meer, *Suriname* **131** F17
Van Buren, *Me., U.S.* **97** A18
Van Buren, *Ark., U.S.* **102** F6
Van Diemen Gulf, *Austral.* **220** A8
Van Diemen, Cape, *Austral.* **220** A7
Van Giã, *Vietnam* **191** N13
Van Gölü, *Turk.* **171** G17
Van Horn, *Tex., U.S.* **104** J8
Van Tassell, *Wyo., U.S.* **108** H14
Van Wert, *Ohio, U.S.* **101** Q11
Van, *Turk.* **171** G17
Van, Lake, *Asia* **168** F4
Vana, *Solomon Is.* **228** N6
Vanadzor, *Arm.* **171** D18
Vanavana, island, *Fr. Polynesia, Fr.* **231** D21
Vanavara, *Russ.* **161** J15
Vanceboro, *Me., U.S.* **97** D19
Vanceboro, *N.C., U.S.* **98** H13
Vanceburg, *Ky., U.S.* **103** A19
Vancouver Island, *B.C., Can.* **86** L7
Vancouver, *B.C., Can.* **86** M8
Vancouver, *Wash., U.S.* **110** F3
Vandalia, *Ill., U.S.* **101** S5
Vandalia, *Ohio, U.S.* **101** R11
Vandalia, *Mo., U.S.* **107** S14
Vanderbilt, *Mich., U.S.* **100** H11
Vanderlin Island, *Austral.* **220** C10
Vänern, lake, *Sw.* **142** L12
Vanga, *Kenya* **211** V8
Vangaindrano, *Madagascar* **213** H16
Vangviang, *Laos* **190** J9
Vanikolo Islands, *Solomon Is.* **228** N6
Vanikolo, island, *Solomon Is.* **228** N6
Vanimo, *P.N.G.* **222** N5
Vanino, *Russ.* **161** K21
Vanj, *Taj.* **179** N13
Vankarem, *Russ.* **161** A20
Vännäs, *Sw.* **143** G14
Vannes, *Fr.* **149** R4
Vannøy, island, *Nor.* **143** C13
Vanoise, Massif de la, *Fr.* **149** U15
Vanrhynsdorp, *S. Af.* **212** L5
Vanrook, *Qnsld., Austral.* **221** Q12
Vansbro, *Sw.* **142** J12
Vansittart Bay, *Austral.* **220** B6
Vansittart Island, *Nunavut, Can.* **87** G16
Vanua Balavu, island, *Fiji* **226** P8
Vanua Lava, island, *Vanuatu* **229** D14
Vanua Levu, island, *Fiji* **226** N6
Vanua Masi, island, *Fiji* **226** Q8
Vanua Vatu, island, *Fiji* **226** Q8
Vanuatu, *Pac. Oc.* **224,** J7
Vao, *New Caledonia, Fr.* **231** R23
Varadero, *Cuba* **118** G7
Varalé, *Côte d'Ivoire* **205** M14
Varāmīn, *Iran* **177** F15
Varanasi (Banaras), *India* **182** J9
Varanger Peninsula, *Eur.* **140** A10

Varangerfjorden, bay, *Nor.* **143** B15
Varangerhalvøya, peninsula, *Nor.* **143** B15
Varano, Lago di, *It.* **152** K12
Varaždin, *Croatia* **154** D6
Varazze, *It.* **152** F4
Varberg, *Sw.* **142** M12
Vardar, river, *N. Maced.* **154** K11
Vardenis, *Arm.* **171** E19
Vardø, *Nor.* **143** B15
Varennes, *Fr.* **149** N12
Varese, *It.* **152** D4
Varilhes, *Fr.* **149** Y9
Varkaus, *Fin.* **143** G16
Varna, *Bulg.* **155** H17
Varnavino, *Russ.* **158** H12
Varnenski Zaliv, *Bulg.* **155** H17
Varnensko Ezero, *Bulg.* **155** H17
Varnville, *S.C., U.S.* **98** M9
Varosha (Maraş), *Cyprus* **162** P10
Varto, *Turk.* **171** F16
Värts Järv, lake, *Est.* **143** K17
Varungga Point, *Solomon Is.* **228** D7
Varzy, *Fr.* **149** R11
Vasafua, island, *Tuvalu* **228** N9
Vashka, river, *Russ.* **158** B12
Vasilikí, *Gr.* **156** G7
Vasiliko, *Cyprus* **162** Q8
Vasilikós, *Gr.* **156** K8
Vasilítsa, peak, *Gr.* **156** D8
Vasil'yevo, *Russ.* **158** J14
Vaskess Bay, *Kiribati* **226** Q10
Vaslui, *Rom.* **155** C17
Vassar, *Mich., U.S.* **100** L12
Västerås, *Sw.* **143** K13
Västervik, *Sw.* **143** L13
Vasto, *It.* **152** K11
Vasyl'kiv, *Ukr.* **159** N6
Vasyugan'ye, *Asia* **168** E8
Vatan, *Fr.* **149** S9
Vatersay *see* Bhatarsaigh, island, *Scot., U.K.* **144** J6
Vathí *see* Itháki, *Gr.* **156** H7
Vathí, *Gr.* **156** H12
Vatia Bay, *American Samoa, U.S.* **230** D7
Vatia Point, *Fiji* **226** P5
Vatia, *American Samoa, U.S.* **230** D7
Vatican City, *Eur.* **162** E4
Vatican Gardens, *Vatican City* **163** P15
Vatican Library, *Vatican City* **163** N16
Vatican Museums, *Vatican City* **163** N16
Vatican Radio, *Vatican City* **163** Q14
Vaticano, Capo, *It.* **153** R13
Vatnajökull, glacier, *Ice.* **142** F3
Vatneyri, *Ice.* **142** D2
Vatoa (Turtle), island, *Fiji* **226** R9
Vatra Dornei, *Rom.* **155** B15
Vättern, lake, *Sw.* **143** L13
Vatu Vara, island, *Fiji* **226** P7
Vatuki, island, *Fiji* **226** N6
Vatukoula, *Fiji* **226** P5
Vatulele, island, *Fiji* **226** Q5
Vaughan, *Nunavut, U.S.* **108** C7
Vaughn, *N. Mex., U.S.* **109** S13
Vaupés, river, *Col.* **130** J8
Vav, *India* **182** K6
Vava'u Group, *Tonga* **224** J10
Vava'u, island, *Tonga* **227** G23
Vavitu *see* Raivavae, island, *Fr. Polynesia, Fr.* **231** F17
Vavla, *Cyprus* **162** Q8
Vavuniya, *Sri Lanka* **183** T7
Vawkavysk, *Belarus* **158** L3
Växjö, *Sw.* **143** M13
Vayegi, *Russ.* **161** C21
Vaygach Island, *Asia* **168** C7
Vaygach, *Russ.* **160** E10
Vazhgort, *Russ.* **158** B12
Veblen, *S. Dak., U.S.* **106** H7
Vecchia, river, *San Marino* **163** J16
Vedea, river, *Rom.* **155** G15
Vedel Fjord, *Greenland, Den.* **88** M13
Veendam, *Neth.* **148** G14
Vega Sund, *Greenland, Den.* **88** J14
Vega, *Tex., U.S.* **104** D11
Vega, island, *Nor.* **142** F11
Vegadeo, *Sp.* **146** A7
Veghel, *Neth.* **148** K13
Vegorítida, Límni, *Gr.* **156** C9
Vegreville, *Alta., Can.* **86** L11
Veharnu, island, *P.N.G.* **228** A5
Veio, ruin, *It.* **152** K8
Vejer de la Frontera, *Sp.* **146** M8
Vekai, island, *Fiji* **226** P8

Velarde, *N. Mex., U.S.* **109** R12
Velas, *Port.* **146** P3
Velay, Monts du, *Fr.* **149** V11
Velbert, *Ger.* **150** F5
Velden, *Aust.* **150** M10
Velé, *Wallis & Futuna Is., Fr.* **230** Q11
Velebit, mountains, *Croatia* **154** F5
Velenje, *Slov.* **154** C5
Veles, *N. Maced.* **154** K11
Velestíno, *Gr.* **156** F11
Vélez de la Gomera, Peñón de , *Sp.* **146** P9
Vélez-Málaga, *Sp.* **146** M10
Vélez, *Col.* **130** F6
Velež, peak, *Bosn. & Herzg.* **154** H8
Velhas, river, *Braz.* **133** K14
Velia, ruin, *It.* **153** N12
Velika Ljubuša, peak, *Bosn. & Herzg.* **154** G7
Velika Morava, river, *Serb.* **154** F11
Velika Plana, *Serb.* **154** F11
Veliki Drvenik, island, *Croatia* **154** G6
Veliki Snežnik, peak, *Slov.* **154** D4
Veliki Troglav, peak, *Bosn. & Herzg.* **154** G6
Velikiy Novgorod, *Russ.* **158** G6
Velikiy Ustyug, *Russ.* **158** L9
Velikiye Luki, *Russ.* **158** J6
Veliko Tŭrnovo, *Bulg.* **155** H15
Velikonda Range, *India* **183** P6
Velingrad, *Bulg.* **155** K14
Vélingara, *Senegal* **204** J7
Vélingara, *Senegal* **204** K8
Velizh, *Russ.* **158** J6
Vella Gulf, *Solomon Is.* **228** D6
Vella Lavella, island, *Solomon Is.* **228** D6
Velletri, *It.* **152** L9
Vellore, *India* **183** R6
Velopoúla, island, *Gr.* **156** L12
Vel'sk, *Russ.* **158** E11
Velva, *N. Dak., U.S.* **106** E4
Vema Fracture Zone, *Atl. Oc.* **249** N7
Vema Seachannel, *Atl. Oc.* **249** T7
Vema Seamount, *Atl. Oc.* **249** T13
Vema Trench, *Ind. Oc.* **250** H9
Vena Park, *Qnsld., Austral.* **221** R12
Venable Ice Shelf, *Antarctica* **238** H6
Venado Tuerto, *Arg.* **134** L11
Venango, *Nebr., U.S.* **107** Q3
Vendôme, *Fr.* **149** R8
Vendas Novas, *Port.* **146** J5
Vendenga, *Russ.* **158** C13
Vendôme, *Grenada* **123** L22
Veneta, Laguna, *It.* **152** E8
Venetie, *Alas., U.S.* **113** E17
Venétiko, island, *Gr.* **156** M9
Veneto, *Eur.* **140** J7
Venezia (Venice), *It.* **152** E8
Venezuela, *S. Amer.* **130** E11
Venezuelan Basin, *Atl. Oc.* **248** M4
Vengurla, *India* **183** P3
Veniaminof, Mount, *Alas., U.S.* **113** N12
Venice *see* Venezia, *It.* **152** E8
Venice, *Fla., U.S.* **99** V7
Venice, *La., U.S.* **102** R12
Venice, Gulf of, *It.* **152** E9
Vening Melnesz Rise, *Ind. Oc.* **251** H15
Vening Melnesz Seamounts, *Ind. Oc.* **251** H15
Venkatapuram, *India* **182** N7
Venlo, *Neth.* **148** K13
Vénna, *Gr.* **157** B15
Vennesund, *Nor.* **142** F11
Venosa, *It.* **152** M12
Venray, *Neth.* **148** K13
Ventana, *S. Amer.* **128** M6
Ventimiglia, *It.* **152** G2
Ventnor City, *N.J., U.S.* **96** Q10
Ventoso, *San Marino* **163** H15
Ventoso, Monte, *San Marino* **163** H15
Ventotene, Isola, *It.* **152** M9
Ventoux, Mount, *Fr.* **149** W13
Ventry *see* Ceann Trá, *Ire.* **145** T2
Ventspils, *Latv.* **143** L15
Venturi, river, *Venez.* **130** G11
Ventura, *Calif., U.S.* **111** X7
Venus, Bahia de, *Eq. Guinea* **214** L7
Vera, *Arg.* **134** H12
Vera, *Sp.* **147** L13
Vera, Bahía, *Arg.* **135** R9
Veracruz, *Mex.* **117** J13
Veraval, *India* **182** L2

Verbania, *It.* **152** D4
Verblyud Island, *Antarctica* **239** A15
Vercelli, *It.* **152** E3
Verchild's Mountain, *St. Kitts & Nevis* **123** A17
Verçinin Tepesi, peak, *Turk.* **171** D15
Verdalsøra, *Nor.* **142** G11
Verde, Arroyo, *Arg.* **135** Q9
Verde, Cabo, *Af.* **198** F1
Verde, Cay, *Bahamas* **119** H14
Verde, Península, *Arg.* **135** P11
Verde, river, *Ariz., U.S.* **109** T7
Verde, river, *S. Amer.* **128** E4
Verdigre, *Nebr., U.S.* **107** N7
Verdigris, river, *U.S.* **91** H13
Verdun, *Fr.* **149** N13
Vereeniging, *S. Af.* **212** J8
Vergara, *Uru.* **134** K15
Vergato, *It.* **152** F6
Vergennes, *Vt., U.S.* **96** G12
Vergina, ruin, *Gr.* **156** D10
Vergt, *Fr.* **149** V8
Verín, *Sp.* **146** D6
Verkhneimbatsk, *Russ.* **161** H13
Verkhneye Penzhino, *Russ.* **161** D20
Verkhnyaya Amga, *Russ.* **161** H18
Verkhovazh'ye, *Russ.* **158** E11
Verkhov'ye, *Russ.* **158** L9
Verkhoyansk Range, *Asia* **168** C10
Verkhoyanskiy Khrebet, *Russ.* **161** F17
Vermelharia, Ponta da, *Cabo Verde* **215** B15
Vermelho, river, *Braz.* **132** J12
Vermenton, *Fr.* **149** R11
Vermilion Bay, *La., U.S.* **102** Q8
Vermilion Cliffs National Monument, *Ariz., U.S.* **109** Q7
Vermilion Lake, *Minn., U.S.* **106** F12
Vermilion Range, *Minn., U.S.* **106** F13
Vermillion, *S. Dak., U.S.* **106** M8
Vérmio, Óros, *Gr.* **156** C9
Vermont, *U.S.* **96** G12
Vermont, *St. Vincent & the Grenadines* **123** L16
Vernadsky Station, *Antarctica* **238** E5
Vernal, *Utah, U.S.* **108** L9
Verneuil, *Fr.* **149** P8
Vernio, *Ala., U.S.* **103** J13
Vernon, *Tex., U.S.* **105** E15
Vernon, *Fr.* **149** P9
Vernonia, *Oreg., U.S.* **110** E3
Vernou, *Guadeloupe, Fr.* **123** F14
Vero Beach, *Fla., U.S.* **99** U10
Véroia, *Gr.* **156** C10
Verona, *Miss., U.S.* **103** H13
Verona, *It.* **152** E6
Versailles, *Ky., U.S.* **103** B17
Versailles, *Mo., U.S.* **107** T12
Versailles, *Fr.* **149** P9
Vertientes, *Cuba* **118** J10
Verzy, *Fr.* **149** P12
Vescovato, *Fr.* **149** X3
Veselyy, *Russ.* **161** J17
Vesoul, *Fr.* **149** R14
Vesterålen, archipelago, *Nor.* **142** D12
Vestfjorden, bay, *Nor.* **142** E11
Vestmanna, *Faroe Is. Den.* **142** H5
Vestmannaeyjar (Westman Islands), *Ice.* **142** F2
Vestraumen Glacier, *Antarctica* **238** B12
Vestvågøy, island, *Nor.* **142** D12
Vestvika Bay, *Antarctica* **239** B17
Vesuvio, Monte, *It.* **152** M11
Vesuvius, peak, *Eur.* **140** K7
Ves'yegonsk, *Russ.* **158** G9
Veszprém, *Hung.* **151** T12
Vétaounde, island, *Vanuatu* **229** C15
Vetapalem, *India* **183** P7
Vetauaua, island, *Fiji* **226** M8
Vetralla, *It.* **152** K8
Veules-les-Roses, *Fr.* **148** M8
Veveno, river, *S. Sudan* **203** W12
Vevaru, island, *Maldives* **183** W3
Vezhen, peak, *Bulg.* **155** J14
Vezirköprü, *Turk.* **170** D10
Viacha, *Bol.* **132** J6
Viadana, *It.* **152** F5
Viana do Bolo, *Sp.* **146** C7
Viana do Castelo, *Port.* **146** D5
Vianden, *Lux.* **162** H10
Viangchan (Vientiane), *Laos* **190** K9
Viareggio, *It.* **152** G5
Vibo Valentia, *It.* **153** R13

Viborg, *Den.* **142** M11
Vibraye, *Fr.* **149** Q8
Vic-en-Bigorre, *Fr.* **149** Y7
Vic-Fézensac, *Fr.* **149** X7
Vic, *Sp.* **147** D17
Vic, Étang de, *Fr.* **149** X11
Vicente Guerrero, *Mex.* **116** B5
Vicenza, *It.* **152** D7
Vichada, river, *Col.* **130** G9
Vichuga, *Russ.* **158** H11
Vichy, *Fr.* **149** T11
Vici, *Okla., U.S.* **105** B15
Vicksburg National Military Park, *Miss., U.S.* **102** L10
Vicksburg, *Mich., U.S.* **101** N10
Vicksburg, *Miss., U.S.* **102** L10
Vicovu de Sus, *Rom.* **155** A15
Victor Bay, *Antarctica* **239** Q19
Victor Harbor, *S. Aust. Austral.* **221** Y10
Victor, Mount, *Antarctica* **239** C16
Victoria (Rabat), *Malta* **163** H20
Victoria Falls, *Zambia* **209** Q16
Victoria Falls, *Zambia, Zimb.* **212** E7
Victoria Falls, *Zimb.* **212** E7
Victoria Fjord, *Greenland, Den.* **88** B8
Victoria Island, *Nunavut, N.W.T., Can.* **87** E13
Victoria Land, region, *Antarctica* **239** P15
Victoria Nile, *Uganda* **211** S4
Victoria River Downs, *N.T., Austral.* **221** Q8
Victoria Sports Centre, *Gibraltar, U.K.* **163** N23
Victoria Strait, *Nunavut, Can.* **87** F14
Victoria, *B.C., Can.* **86** M8
Victoria, *Va., U.S.* **98** F12
Victoria, *Tex., U.S.* **105** M17
Victoria, *Kans., U.S.* **107** T6
Victoria, *Grenada* **123** J22
Victoria, *Chile* **134** C7
Victoria, *Chile* **135** N6
Victoria, *Seychelles* **215** N20
Victoria, *Austral.* **221** Z12
Victoria, Lake, *Af.* **198** J7
Victoria, Mount, *Myanmar* **190** H4
Victoria, river, *Austral.* **220** C8
Victorica, *Arg.* **134** K12
Victorica, *Arg.* **134** M9
Victorville, *Calif., U.S.* **111** W9
Victory Downs, *N.T. Austral.* **221** U8
Victory Mountains, *Antarctica* **239** Q14
Victory Peak (Jengish Chokusu, Tomür Feng, Pik Pobedy), *China, Kyrg.* **179** K17
Vicuña, *Chile* **134** J7
Vidal Plain, *Atl. Oc.* **248** M6
Vidalia, *Ga., U.S.* **99** N7
Vidalia, *La., U.S.* **102** M9
Videle, *Rom.* **155** F15
Vidigueira, *Port.* **146** J6
Vidin, *Bulg.* **154** G12
Vidisha, *India* **182** K6
Vidor, *Tex., U.S.* **105** K21
Vidzy, *Belarus* **158** J4
Viedma, *Arg.* **135** Q10
Viedma, Lago, *Arg.* **135** V7
Vielha, *Sp.* **147** C16
Vielle Case, *Dominica* **123** D19
Vienna, *W. Va., U.S.* **98** C8
Vienna, *Ga., U.S.* **99** N6
Vienna, *Ill., U.S.* **101** V6
Vienne, *Fr.* **149** U13
Vientiane *see* Viangchan, *Laos* **190** K9
Vierge Point, *St. Lucia* **123** L14
Vierwaldstätter *see* Lake of Lucerne, *Switz.* **152** B3
Vierzon, *Fr.* **149** S9
Vieste, *It.* **152** K13
Việt Trì, *Vietnam* **190** H11
Vietnam, *Asia* **190** J11
Vieux Fort, *St. Lucia* **123** L14
Vieux-Boucau, *Fr.* **149** X5
Vieux-Bourg, *Guadeloupe, Fr.* **123** E15
Vieux-Fort, *Guadeloupe, Fr.* **123** G14
Vieux-Habitants, *Guadeloupe, Fr.* **123** F14
Vigan, *Philippines* **193** B13
Vigía Chico, *Mex.* **117** H17
Vignemale, Pic de, *Fr.* **149** Y6
Vigo, *Sp.* **146** C5
Vigo, Ría de, *Sp.* **146** C5

Vihari, *Pak.* **180** M11
Viipuri *see* Vyborg, *Russ.* **158** E6
Vijayapura (Bijapur), *India* **182** N5
Vijayawada, *India* **183** P7
Vík, *Ice.* **142** F2
Vikenara Point, *Solomon Is.* **228** E10
Vikhorevka, *Russ.* **161** K14
Vikhren, peak, *Bulg.* **155** K13
Viking Bank, *Atl. Oc.* **248** F12
Viking Trough, *Arctic Oc.* **253** J21
Vikna, *Nor.* **142** F11
Víkuáall, *Atl. Oc.* **253** M20
Vila Bela da Santíssima Trindade, *Braz.* **132** J9
Vila Bittencourt, *Braz.* **132** D6
Vila da Ribeira Brava, *Cabo Verde* **215** B15
Vila de Moura, *Port.* **146** K6
Vila de Sagres, *Port.* **146** L4
Vila do Bispo, *Port.* **146** L4
Vila do Conde, *Port.* **146** E5
Vila Franca do Campo, *Port.* **146** Q5
Vila Gamito, *Mozambique* **212** C10
Vila Murtinho, *Braz.* **132** G7
Vila Nova de Gaia, *Port.* **146** E5
Vila Nova de Milfontes, *Port.* **146** K5
Vila Real, *Port.* **146** E6
Vila Vasco da Gama, *Mozambique* **212** D10
Vila Velha (Espírito Santo), *Braz.* **133** L16
Vila Velha de Rodão, *Port.* **146** G6
Vila Velha, *Braz.* **132** B12
Vila-real de los Infantes, *Sp.* **147** G14
Vilafranca del Penedès, *Sp.* **147** E17
VilagarcÍa de Arousa, *Sp.* **146** C5
Vilankulo, *Mozambique* **212** G11
Vilanova i la Geltrú, *Sp.* **147** E17
Vilar Formoso, *Port.* **146** F7
Vilhelmina, *Sw.* **143** F13
Vilhena, *Braz.* **132** H9
Viljandi, *Est.* **143** K16
Vil'kitskogo, *Russ.* **161** D14
Villa Ahumada, *Mex.* **116** C9
Villa Alberdi, *Arg.* **134** G9
Villa Ángela, *Arg.* **134** G12
Villa Atuel, *Arg.* **134** L8
Villa Cañás, *Arg.* **134** L11
Villa Dolores, *Arg.* **134** K9
Villa Grove, *Ill., U.S.* **101** R7
Villa Guillermina, *Arg.* **134** H12
Villa Krause, *Arg.* **134** K8
Villa María, *Arg.* **134** K10
Villa Martín, *Bol.* **132** L6
Villa Nueva, *Arg.* **134** K8
Villa Ocampo, *Arg.* **134** H12
Villa Regina, *Arg.* **135** P9
Villa Unión, *Mex.* **116** G8
Villa Unión, *Arg.* **134** H7
Villablino, *Sp.* **146** B8
Villacarrillo, *Sp.* **146** K11
VillacastÍn, *Sp.* **146** F10
Villach, *Aust.* **150** M10
Villacidro, *It.* **153** P3
Villafranca de los Barros, *Sp.* **146** J8
Villafranca di Verona, *It.* **152** E6
Village Sandy Ground, *Anguilla, U.K.* **122** R11
Village, *Fr. Polynesia, Fr.* **231** Q14
Villaguay, *Arg.* **134** K12
Villahermosa, *Mex.* **117** J14
Villalba, *Sp.* **146** B6
Villalonga, *Arg.* **135** P11
Villamar, *It.* **153** P4
Villamartín, *Sp.* **146** M8
Villamontes, *Bol.* **132** L8
Villanueva de Córdoba, *Sp.* **146** J10
Villanueva de los Infantes, *Sp.* **146** J11
Villanueva del Fresno, *Sp.* **146** J7
Villanueva, *Mex.* **116** G10
Villarosa, *It.* **153** S10
Villarrica, *Chile* **135** P6
Villarrica, Lago, *Chile* **135** P6
Villarrobledo, *Sp.* **146** H12
Villasalto, *It.* **153** P4
Villasimius, *It.* **153** Q4
Villatobas, *Sp.* **146** G11
Villavicencio, *Col.* **130** G7
Villaviciosa, *Sp.* **146** A9
Villazón, *Bol.* **132** L7
Ville Platte, *La., U.S.* **102** N8
Villedieu-les-Poêles, *Fr.* **149** P6
Villefranche-de-Lauragais, *Fr.* **149** Y9

Villefranche-de-Rouergue, *Fr.* **149** W9
Villefranche, *Fr.* **149** T12
Villemaur, *Fr.* **149** Q11
Villena, *Sp.* **147** J14
Villeneuve-Saint-Georges, *Fr.* **149** P10
Villeneuve-sur-Lot, *Fr.* **149** W8
Villerupt, *Fr.* **149** N13
Villeurbanne, *Fr.* **149** U13
Villingen-Schwenningen, *Ger.* **150** K6
Villisca, *Iowa, U.S.* **107** Q10
Villupuram, *India* **183** R7
Vilnius, *Lith.* **143** M17
Vilyuy, river, *Asia* **168** D10
Vilyuysk, *Russ.* **161** H17
Vilyuyskoye Vodokhranilishche, *Russ.* **161** H16
Vimoutiers, *Fr.* **149** P7
Viña del Mar, *Chile* **134** K6
Vina, *Calif., U.S.* **111** P4
Vina, river, *Cameroon* **206** L12
Vinalhaven, *Me., U.S.* **97** G17
Vinaròs, *Sp.* **147** F15
Vincennes Bay, *Antarctica* **239** M20
Vincennes, *Ind., U.S.* **101** T7
Vindelälven, river, *Sw.* **143** F13
Vindeln, *Sw.* **143** G14
Vindhya Range, *India* **182** K5
Vindrey, *Russ.* **158** K12
Vineland, *N.J., U.S.* **96** Q10
Vineta, *Namibia* **212** G3
Vịnh Cam Ranh, bay, *Vietnam* **191** N13
Vĩnh Châu, *Vietnam* **191** Q11
Vĩnh Long, *Vietnam* **191** P11
Vinh, *Vietnam* **190** J11
Vinhais, *Port.* **146** D7
Vinita, *Okla., U.S.* **105** B19
Vinkovci, *Croatia* **154** E8
Vinnytsya, *Ukr.* **159** P5
Vinson Massif, *Antarctica* **238** H9
Vinton, *La., U.S.* **102** P6
Vinton, *Iowa, U.S.* **107** N13
Vinukonda, *India* **183** P7
Vipiteno, *It.* **152** B7
Viqueque, *Timor-Leste* **193** M15
Vir, *Taj.* **179** N13
Vir, island, *Croatia* **154** F5
Virac, *Philippines* **193** C15
Virachei, *Cambodia* **190** M12
Virandozero, *Russ.* **158** C9
Viranşehir, *Turk.* **171** H14
Virawah, *Pak.* **181** S10
Virden, *Man., Can.* **87** N13
Virden, *Ill., U.S.* **101** S5
Vire, *Fr.* **149** P6
Vîrfurile, *Rom.* **154** D12
Vírgenes, Cape, *S. Amer.* **128** Q5
Virgin Gorda Peak, *Virgin Is., U.K.* **122** Q9
Virgin Gorda, island, *British Virgin Is., U.K.* **121** A14
Virgin Islands National Park, *Virgin Is., U.S.* **122** N11
Virgin Islands, U.K., U.S. **121** B13
Virgin, river, *Ariz., Nev., U.S.* **111** T13
Virginia Beach, *Va., U.S.* **98** F14
Virginia City, *Mont., U.S.* **108** E7
Virginia City, *Nev., U.S.* **111** Q6
Virginia Falls, *N.W.T., Can.* **86** H9
Virginia, *U.S.* **98** E11
Virginia, *Ill., U.S.* **101** R4
Virginia, *Minn., U.S.* **106** F12
Viroqua, *Wis., U.S.* **100** K3
Virovitica, *Croatia* **154** D7
Virrat, *Fin.* **143** H15
Virtsu, *Est.* **143** K16
Virú, *Peru* **132** G2
Virudunagar, *India* **183** S6
Virunga, mountains, *D.R.C.* **209** E18
Vis (Lissa), island, *Croatia* **154** H6
Visalia, *Calif., U.S.* **111** U7
Visayan Sea, *Philippines* **193** D14
Visby, *Sw.* **143** L14
Viscount Melville Sound, *Nunavut, N.W.T., Can.* **87** D13
Višegrad, *Bosn. & Herzg.* **154** G9
Vişeu de Sus, *Rom.* **155** B14
Viseu, *Port.* **146** F6
Vishakhapatnam, *India* **182** N9
Viški Kanal, *Croatia* **154** H6
Viso del Marqués, *Sp.* **146** J11
Viso, Monte, *It.* **152** F2
Visp, *Switz.* **152** C3
Vista Alegre, *Braz.* **132** F6

Vista, *Calif., U.S.* **111** Y9
Vistula, river, *Eur.* **140** F8
Vit, river, *Bulg.* **155** H14
Viterbo, *It.* **152** J8
Viti Levu Bay, *Fiji* **226** P5
Viti Levu, island, *Fiji* **226** P5
Vitiaz Strait, *P.N.G.* **222** P8
Vitichi, *Bol.* **132** L7
Vitigudino, *Sp.* **146** E8
Vitim, *Russ.* **161** J16
Vitim, river, *Russ.* **161** J16
Vítor, *Peru* **132** J5
Vitória da Conquista, *Braz.* **133** J16
Vitoria-Gasteiz, *Sp.* **146** C12
Vitória, *Braz.* **132** D12
Vitória, *Braz.* **133** L16
Vitré, *Fr.* **149** Q6
Vitry-le-François, *Fr.* **149** P12
Vitsyebsk, *Belarus* **158** J6
Vittangi, *Sw.* **143** D13
Vittorio Veneto, *It.* **152** D8
Vityaz Fracture Zone, *Ind. Oc.* **250** H9
Vityaz Trench, *Pac. Oc.* **246** K10
Vivario, *Fr.* **149** X2
Viveiro, *Sp.* **146** A7
Vivi, *Russ.* **161** H14
Vivi, river, *Russ.* **161** H14
Vivian, *La., U.S.* **102** K6
Viwa, island, *Fiji* **226** P4
Vizcaíno, Desierto de, *Mex.* **116** D6
Vize, *Russ.* **161** D13
Vize, *Turk.* **170** C4
Vizianagaram, *India* **182** N9
Vizinga, *Russ.* **158** D14
Vizzini, *It.* **153** T11
Vjosë, river, *Alban.* **154** M9
Vlădeasa, peak, *Rom.* **154** C12
Vladikavkaz, *Russ.* **159** U14
Vladimir, *Russ.* **158** J10
Vladīmīrovka, *Kaz.* **178** C10
Vladivostok, *Russ.* **161** M21
Vlamingh Fracture Zone, *Ind. Oc.* **250** P12
Vlasenica, *Bosn. & Herzg.* **154** G9
Vlasinje, *Bosn. & Herzg.* **154** F7
Vlieland, island, *Neth.* **148** G12
Vlorë (Valona), *Alban.* **154** M9
Vlorës, Gjiri i, *Alban.* **154** M9
Vluck Point, *Virgin Is., U.S.* **122** M7
Vöcklabruck, *Aust.* **150** K10
Voe, *Scot., U.K.* **144** C12
Vogan, *Togo* **205** P17
Vogelsberg, range, *Ger.* **150** G7
Voghera, *It.* **152** E4
Voh, *New Caledonia, Fr.* **231** P21
Vohemar *see* Iharaña, *Madagascar* **213** C17
Voi, *Kenya* **211** U7
Voinjama, *Liberia* **204** N10
Voiron, *Fr.* **149** U13
Voitsberg, *Aust.* **150** M11
Vojvodina, region, *Serb.* **154** E9
Vokeo, island, *P.N.G.* **222** N6
Volborg, *Mont., U.S.* **108** E12
Volcan Karisimbi, *D.R.C.* **209** F19
Volcan Karisimbi, *Rwanda* **211** T2
Volcán La Cumbre, *Ecua.* **130** Q2
Volcán Wolf, *Ecua.* **130** P2
Volcano Islands (Kazan Rettō), *Japan* **189** Z13
Volcano, *Hawaii, U.S.* **115** P22
Volda, *Nor.* **142** H14
Volga River Delta, *Eur.* **140** H15
Volga Upland, *Eur.* **140** F14
Volga-Baltic Waterway, *Russ.* **158** F9
Volga-Don Canal, *Russ.* **159** Q13
Volga, *S. Dak., U.S.* **106** K8
Volga, river, *Russ.* **158** L15
Volgodonsk, *Russ.* **159** Q12
Volgograd (Stalingrad), *Russ.* **159** P13
Volgograd Reservoir, *Eur.* **140** F14
Volgogradskoye Vodokhranilishche, *Russ.* **159** N13
Volímes, *Gr.* **156** J7
Volissós, *Gr.* **157** H16
Volkhov, *Russ.* **158** F7
Volkhov, river, *Eur.* **140** E10
Völklingen, *Ger.* **150** H5
Volna, *Russ.* **161** E20
Volnorez, *Russ.* **161** E13
Volnovakha, *Ukr.* **159** Q9
Volodarsk, *Russ.* **158** J12

Vologda, *Russ.* **158** F10
Vólos, *Gr.* **156** F11
Voloshka, *Russ.* **158** E10
Volquart Boon Kyst, *Greenland, Den.* **88** L14
Vol'sk, *Russ.* **158** M14
Volta White, river, *Burkina Faso, Ghana* **205** M15
Volta, river, *Ghana* **205** P16
Voltaire, Cape, *Austral.* **220** B5
Volturno, river, *It.* **152** L10
Volubilis, ruin, *Mor.* **200** C8
Vólvi, Límni, *Gr.* **156** C12
Volx, *Fr.* **149** X14
Volyn-Podolian Upland, *Eur.* **140** H10
Volzhsk, *Russ.* **158** J14
Volzhskiy, *Russ.* **159** P13
Vonavona, island, *Solomon Is.* **228** E7
Vondrozo, *Madagascar* **213** H16
Vónitsa, *Gr.* **156** G8
Voo, *Kenya* **211** T7
Vopnafjörður, *Ice.* **142** E4
Vóras Óros, *Gr.* **156** C9
Vóreios Evvoïkós Kólpos, *Gr.* **156** G11
Vóries Sporádhes (Northern Sporades), islands, *Gr.* **157** F13
Voring Plateau, *Arctic Oc.* **253** H21
Vorkuta, *Russ.* **160** F10
Vormsi, island, *Est.* **143** K16
Voronezh, *Russ.* **158** M10
Voronin Trough, *Arctic Oc.* **253** E13
Voronov Terrace, *Arctic Oc.* **253** J15
Vorontsovka, *Russ.* **161** J16
Vosges, range, *Fr.* **149** P15
Voss, *Nor.* **142** J10
Vostochnaya, *Russ.* **161** E13
Vostochnyy (Vrangel'), *Russ.* **161** M21
Vostochnyy Sayan, *Russ.* **161** L14
Vostok Island, *Kiribati* **225** H13
Vostok Station, *Antarctica* **239** K17
Votice, *Czechia* **150** H11
Votkinsk, *Russ.* **158** G16
Votlo, *Vanuatu* **229** H16
Voulx, *Fr.* **149** Q10
Voúrinos, peak, *Gr.* **156** D9
Voúxa, Ákra, *Gr.* **156** P12
Vouziers, *Fr.* **149** N12
Vovchans'k, *Ukr.* **159** N9
Vovodo, river, *Cen. Af. Rep.* **207** M17
Voyager Seamounts, *Pac. Oc.* **246** G12
Voyageurs National Park, *Minn., U.S.* **106** E12
Voyampolka, *Russ.* **161** E21
Voyvozh, *Russ.* **158** C15
Voza, *Solomon Is.* **228** C6
Vozhayel', *Russ.* **158** C14
Vozhe, *Russ.* **158** E10
Vozhega, *Russ.* **158** E10
Vozhgora, *Russ.* **158** B13
Vozhma, *Russ.* **158** F13
Voznesens'k, *Ukr.* **159** Q6
Voznesen'ye, *Russ.* **158** E8
Vozvyshennost' Lavruska, *Russ.* **161** H15
Vrachíonas, peak, *Gr.* **156** J7
Vrachnaíika, *Gr.* **156** J9
Vrang, *Taj.* **179** N13
Vrangel' *see* Vostochnyy, *Russ.* **161** M21
Vrangelya, *Russ.* **161** B19
Vranica, peak, *Bosn. & Herzg.* **154** G7
Vranje, *Serb.* **154** J11
Vranov, *Slovakia* **151** J17
Vransko Jezero, *Croatia* **154** G5
Vratsa, *Bulg.* **155** H13
Vrbas, river, *Bosn. & Herzg.* **154** F7
Vrchovina Českomoravská, range, *Czechia* **150** J12
Vredenburg, *S. Af.* **212** M5
Vršac, *Serb.* **154** E11
Vryheid, *S. Af.* **212** J9
Vuagava, island, *Fiji* **226** Q8
Vučitrn *see* Vushtrri, *Kos.* **154** J10
Vuktyl, *Russ.* **158** B16
Vulcano, Isola, *It.* **153** R11
Vulci, ruin, *It.* **152** K7
Vunaniu Bay, *Fiji* **226** Q5
Vung Tàu, *Vietnam* **191** P12
Vunidawa, *Fiji* **226** P5
Vunisea, *Fiji* **226** R5
Vunivutu Bay, *Fiji* **226** N6
Vushtrri (Vučitrn), *Kos.* **154** J10
Vwawa, *Tanzania* **211** X4

Vyartsilya, *Russ.* **158** D7
Vyatka, river, *Russ.* **158** F15
Vyatskiye Polyany, *Russ.* **158** H15
Vyazemskiy, *Russ.* **161** K21
Vyaz'ma, *Russ.* **158** K8
Vyazniki, *Russ.* **158** J11
Vyborg (Viipuri), *Russ.* **158** E6
Vychegda Lowland, *Eur.* **140** C12
Vychegda, river, *Russ.* **158** D13
Vyerkhnyadzvinsk, *Belarus* **158** J5
Vyg Ozero, *Russ.* **158** C8
Vyg, Lake, *Eur.* **140** C11
Vyksa, *Russ.* **158** J11
Vylkove, *Ukr.* **159** S5
Vym', river, *Eur.* **140** C13
Vyritsa, *Russ.* **158** F6
Vyshniy Volochek, *Russ.* **158** H8
Vyškov, *Czechia* **151** J13
Vysokaye, *Belarus* **158** L2
Vysokovsk, *Russ.* **158** J9
Vytegra, *Russ.* **158** E9

W

Wa, *Ghana* **205** M15
Waajid, *Somalia* **211** Q10
Waar, island, *Indonesia* **193** K19
Wa'at, *S. Sudan* **203** V12
Wabasca, river, *Alta., Can.* **86** K10
Wabash, *Ind., U.S.* **101** Q9
Wabash, river, *Ill., Ind., U.S.* **101** T7
Wabasha, *Minn., U.S.* **106** K13
Wabasso, *Fla., U.S.* **99** U10
Wabē Gestro, river, *Eth.* **211** P9
Wabeno, *Wis., U.S.* **100** H6
Wabi Shebele, river, *Eth.* **210** M9
Wabowden, *Man., Can.* **87** L13
Wabuk Point, *Ont., Can.* **87** L16
Waco Lake, *Tex., U.S.* **105** J17
Waco, *Tex., U.S.* **105** J17
Wad al Ḥaddād, *Sudan* **203** S13
Wad Banda, *Sudan* **203** S9
Wad Medani, *Sudan* **203** R13
Wad Rawa, *Sudan* **203** R12
Wad, *Pak.* **181** Q6
Wadamago, *Somaliland* **210** M12
Waddān, *Lib.* **201** F19
Waddenzee, bay, *Neth.* **148** G13
Waddington, Mount, *B.C., Can.* **86** L8
Wadena, *Sask., Can.* **86** M12
Wadena, *Minn., U.S.* **106** H10
Wädenswil, *Switz.* **152** B4
Wadesboro, *N.C., U.S.* **98** J10
Wadeye, *N.T., Austral.* **221** P7
Wādī adDawāsir, *Saudi Arabia* **174** K9
Wadi Al-Hitan (Whale Valley), *Egypt* **202** G10
Wadi 'Amur, *Sudan* **203** N13
Wādī ar Rimah, *Saudi Arabia* **174** F7
Wadi 'Aradeib, *Sudan* **203** R6
Wadi el Ghalla, *Sudan* **203** T9
Wadi el Milk, *Sudan* **203** P11;R9
Wadi Halfa, *Sudan* **202** L11
Wadi Howar, *Sudan* **203** Q6
Wādī Irauen, *Lib.* **201** H16
Wādī Jabjabah, *Egypt, Sudan* **202** L12
Wadi Magrur, *Sudan* **203** Q8
Wādī Mūsá (Elchi), *Jordan* **173** Q6
Wadi Tiwal, *Sudan* **203** T6
Wadley, *Ga., U.S.* **98** M7
Wadomari, *Japan* **189** W3
Wadsworth, *Ohio, U.S.* **101** P14
Wadsworth, *Nev., U.S.* **111** P7
Wadu, island, *Maldives* **183** U3
Waelder, *Tex., U.S.* **105** L17
Wafangdian, *China* **185** G15
Wagenia Fisheries, *D.R.C.* **209** D16
Wager Bay, *Nunavut, Can.* **87** G15
Wagga Wagga, *N.S.W., Austral.* **221** Y13
Wagin, *W. Aust., Austral.* **221** X3
Wagner, *S. Dak., U.S.* **106** M7
Wagoner, *Okla., U.S.* **105** C19
Wagontire Mountain, *Oreg., U.S.* **110** J6
Wągrowiec, *Pol.* **151** D13
Wah Cantonment, *Pak.* **180** J11
Wahai, *Indonesia* **193** K17
Wāḥāt al Farāfirah, *Egypt* **202** H10
Wahoo, *Nebr., U.S.* **107** P8
Wahpeton, *N. Dak., U.S.* **106** H8
Waḥx, Ras il, *Malta* **163** J21
Wai, *India* **182** N4

Wai, *Indonesia* **193** K16
Waiakoa, *Hawaii, U.S.* **115** K19
Waialua Bay, *Hawaii, U.S.* **115** G14
Waialua, *Hawaii, U.S.* **115** G14
Wai'anae Range, *Hawaii, U.S.* **115** G14
Waiau, *river, N.Z.* **223** P16
Waibeem, *Indonesia* **193** J18
Waigama, *Indonesia* **193** K17
Waigeo, *island, Indonesia* **193** J17
Waihe'e, *Hawaii, U.S.* **115** J18
Waiheke Island, *N.Z.* **223** E20
Waikabubak, *Indonesia* **193** N13
Waikahalulu Bay, *Hawaii, U.S.* **115** K18
Waikapū, *Hawaii, U.S.* **115** J18
Waikapuna Bay, *Hawaii, U.S.* **115** Q21
Waikato, *river, N.Z.* **223** F21
Waikawa, *N.Z.* **223** P16
Waikoloa, *Hawaii, U.S.* **115** M21
Waikouaiti, *N.Z.* **223** N17
Wailagi Lala, *island, Fiji* **226** N8
Wailea, *Hawaii, U.S.* **115** K19
Wailua, *Hawaii, U.S.* **114** E10
Wailua, *Hawaii, U.S.* **115** J20
Wailuku, *Hawaii, U.S.* **115** J18
Wailuku, *river, Hawaii, U.S.* **115** N22
Waimānalo Beach, *Hawaii, U.S.* **115** H15
Waimarama, *N.Z.* **223** H22
Waimate, *N.Z.* **223** N18
Waimea (Kamuela), *Hawaii, U.S.* **115** M21
Waimea Canyon, *Hawaii, U.S.* **114** E9
Waimea, *Hawaii, U.S.* **114** F9
Waimea, *Hawaii, U.S.* **115** G14
Wain, *N. Korea* **186** F6
Wainaworasi, *Solomon Is.* **228** G12
Waingapu, *Indonesia* **193** N13
Wainunu Bay, *Fiji* **226** N6
Wainwright, *Alas., U.S.* **113** B14
Wai'ōhinu, *Hawaii, U.S.* **115** Q21
Waipa, *river, N.Z.* **223** F20
Waipahi, *N.Z.* **223** P17
Waipahu, *Hawaii, U.S.* **115** H14
Waipi'o Bay, *Hawaii, U.S.* **115** M21
Waipu, *N.Z.* **223** E20
Waipukurau, *N.Z.* **223** H21
Wairokai Harbor, *Solomon Is.* **228** F11
Waïsisi Bay, *Vanuatu* **229** M17
Waïsisi, *Vanuatu* **229** M17
Waitsburg, *Wash., U.S.* **110** E8
Wajima, *Japan* **188** M9
Wajir, *Kenya* **211** R8
Waka, *Eth.* **211** N7
Wakasa Wan, *Japan* **189** P9
Wakaya, *island, Fiji* **226** P6
Wakayama, *Japan* **189** Q8
Wake Forest, *N.C., U.S.* **98** H12
Wake Island, *Wake I., U.S.* **230** N11
WaKeeney, *Kans., U.S.* **107** S5
Wakefield, *Mich., U.S.* **100** F5
Wakefield, *Kans., U.S.* **107** S8
Wakefield, *Nebr., U.S.* **107** N8
Wakefield, *Jam.* **122** H7
Wakefield, *Eng., U.K.* **145** Q12
Wakema, *Myanmar* **190** L4
Wakenaam Island, *S. Amer.* **128** B7
Wākhān, Kōh-e, *Afghan.* **180** F11
Wākhān, *region, Afghan.* **180** F11
Wakhjir Pass, *Afghan., China* **180** F13
Wakkanai, *Japan* **188** E13
Waku Kungo, *Angola* **208** M10
Wakuach, Lac, *Que., Can.* **87** K20
Wakunai, *P.N.G.* **228** B5
Wakuya, *Japan* **188** L13
Wala, *New Caledonia, Fr.* **231** N20
Walachia, *region, Rom.* **155** F14
Wałbrzych, *Pol.* **150** G12
Walcha, *N.S.W., Austral.* **221** W15
Walcott, Lake, *Idaho, U.S.* **108** H6
Wałcz, *Pol.* **150** C12
Walden, *Colo., U.S.* **108** L12
Waldkraiburg, *Ger.* **150** K9
Waldo Lake, *Oreg., U.S.* **110** H4
Waldo, *Ark., U.S.* **102** J7
Waldorf, *Md., U.S.* **98** D13
Waldport, *Oreg., U.S.* **110** G2
Waldron, *Ark., U.S.* **102** G6
Walea, Selat, *Indonesia* **193** J14
Wales Island, *Nunavut, Can.* **87** G16
Wales, *Alas., U.S.* **112** E11
Walewale, *Ghana* **205** L16
Walferdange, *Lux.* **162** J10
Walgett, *N.S.W., Austral.* **221** W14
Walgreen Coast, *Antarctica* **238** K8
Walhalla, *S.C., U.S.* **98** J6
Walhalla, *N. Dak., U.S.* **106** D7

Walikale, *D.R.C.* **209** F18
Walis, *island, P.N.G.* **222** N6
Walker Bay, *S. Af.* **212** N5
Walker Lake, *Nev., U.S.* **111** Q7
Walker, *Minn., U.S.* **106** G10
Wall, *S. Dak., U.S.* **106** L3
Walla Walla, *Wash., U.S.* **110** E8
Wallaby Saddle, *Ind. Oc.* **251** L17
Wallace, *N.C., U.S.* **98** J12
Wallace, *Nebr., U.S.* **107** Q4
Wallace, *Idaho, U.S.* **108** C4
Wallaroo, *S. Aust., Austral.* **221** X10
Wallibou, *St. Vincent & the Grenadines* **123** K16
Wallilabou, *St. Vincent & the Grenadines* **123** K16
Wallis & Futuna, *Pac. Oc.* **224** J9
Wallis Islands *see* Wallis, Îles, *Pac. Oc.* **224** J9
Wallis Ridge, *Atl. Oc.* **254** G1
Wallis, Îles (Wallis Islands), *Pac. Oc.* **224** J9
Wallops Island, *Va., U.S.* **98** E15
Wallowa Mountains, *Oreg., U.S.* **110** F9
Wallowa, *Wash., U.S.* **110** F9
Walls, *Scot., U.K.* **144** C12
Wallsend, *Eng., U.K.* **145** N12
Walney, Isle of, *Eng., U.K.* **145** P10
Walnut Canyon National Monument, *Ariz., U.S.* **109** S7
Walnut Cove, *N.C., U.S.* **98** G10
Walnut Ridge, *Ark., U.S.* **102** E10
Walong, *India* **182** F16
Walsall, *Eng., U.K.* **145** S12
Walsenburg, *Colo., U.S.* **109** P13
Walsh, *Colo., U.S.* **109** P16
Walsrode, *Ger.* **150** D7
Walter F. George Reservoir, *Ala., Ga., U.S.* **103** M17
Walterboro, *S.C., U.S.* **98** M9
Walters Shoal, *Ind. Oc.* **250** N5
Walters, *Okla., U.S.* **105** E16
Waltershausen Glacier, *Greenland, Den.* **88** H13
Walton, *N.Y., U.S.* **96** K9
Walvis Bay, *Namibia* **212** H3
Walvis Ridge, *Atl. Oc.* **249** S12
Walyahmoning Rock, *Austral.* **220** J3
Wāmā, *Afghan.* **180** H10
Wamba, *Nigeria* **206** L8
Wamba, *D.R.C.* **209** C18
Wamba, *river, D.R.C.* **208** H11
Wamego, *Kans., U.S.* **107** S9
Wamena, *Indonesia* **193** L20
Wami, *river, Tanzania* **211** W7
Wamsasi, *Indonesia* **193** K15
Wamsisi, *Indonesia* **193** K16
Wamsutter, *Wyo., U.S.* **108** K10
Wana, *Pak.* **180** K9
Wanaaring, *N.S.W., Austral.* **221** W12
Wanaka, *N.Z.* **223** N16
Wanchese, *N.C., U.S.* **98** G15
Wandel Hav, *N. Amer.* **88** B12
Wanderer Bay, *Solomon Is.* **228** F9
Wanding, *China* **184** N8
Wando, *S. Korea* **187** U8
Wando, *island, S. Korea* **187** U7
Wandoan, *Qnsld., Austral.* **221** U15
Wang Kai, *S. Sudan* **203** V10
Wanganui *see* Whanganui, *N.Z.* **223** H20
Wanganui, *river, N.Z.* **223** H20
Wangaratta, *Vic., Austral.* **221** Y13
Wangerberg, *Liech.* **162** Q2
Wangiwangi, *island, Indonesia* **193** L13
Wanie Rukula, *D.R.C.* **209** E16
Wanigela, *P.N.G.* **222** Q9
Wanione, *Solomon Is.* **228** G12
Wankaner, *India* **182** K2
Wanlaweyn, *Somalia* **211** R11
Wanlockhead, *Scot., U.K.* **145** M10
Wanning, *China* **184** Q12
Wānū, *Afghan.* **180** K6
Wapakoneta, *Ohio, U.S.* **101** Q11
Wapato, *Wash., U.S.* **110** D6
Wapello, *Iowa, U.S.* **107** P14
Wapou, *Côte d'Ivoire* **204** Q12
Wappapello, Lake, *Mo., U.S.* **107** V15
Wapsipinicon, *river, Iowa, U.S.* **107** N13
War Galoh, *Somalia* **211** N13
War, *W. Va., U.S.* **98** F8
Warab, *S. Sudan* **203** V10
Waranga Basin, *Austral.* **220** L12

Warangal, *India* **182** N7
Waras, *Afghan.* **180** J5
Waraseoni, *India* **182** L7
Warburg, *Ger.* **150** E7
Warburton Creek, *Austral.* **220** H10
Warburton, *W. Aust., Austral.* **221** U6
Ward Hunt Island, *Nunavut, Can.* **87** A15
Warden, *Wash., U.S.* **110** D7
Wardija Point, *Malta* **163** H19
Ware Shoals, *S.C., U.S.* **98** K7
Waren, *Ger.* **150** C10
Waren, *Indonesia* **193** K19
Warlubie, *Pol.* **151** C14
Warm Springs Reservoir, *Oreg., U.S.* **110** H8
Warm Springs, *Ga., U.S.* **98** M5
Warm Springs, *Va., U.S.* **98** E10
Warm Springs, *Nev., U.S.* **111** R10
Warming Land, *Greenland, Den.* **88** C7
Warminster, *Eng., U.K.* **145** V11
Warner Mountains, *Calif., Oreg., U.S.* **110** L6
Warner Robins, *Ga., U.S.* **98** M6
Warner Valley, *Oreg., U.S.* **110** K6
Waroona, *W. Aust., Austral.* **221** X3
Warora, *India* **182** L6
Warracknabeal, *Vic., Austral.* **221** Y12
Warralu, *S. Sudan* **203** V9
Warrego, *river, Austral.* **220** H13
Warren, *Pa., U.S.* **96** L4
Warren, *Mich., U.S.* **100** M12
Warren, *Ohio, U.S.* **101** P15
Warren, *Ark., U.S.* **102** J8
Warren, *Minn., U.S.* **106** E8
Warren, *N.S.W., Austral.* **221** W14
Warrender, Cape, *Nunavut, Can.* **87** D16
Warrenpoint, *N. Ire., U.K.* **145** P7
Warrensburg, *Mo., U.S.* **107** T11
Warrenton, *Ga., U.S.* **98** L7
Warrenton, *Va., U.S.* **98** D12
Warrenton, *Oreg., U.S.* **110** E2
Warrenton, *S. Af.* **212** K7
Warri, *Nigeria* **206** M7
Warrina, *S. Aust., Austral.* **221** V9
Warrington, *Fla., U.S.* **99** Q1
Warrington, *Eng., U.K.* **145** R11
Warrnambool, *Vic., Austral.* **221** Z11
Warroad, *Minn., U.S.* **106** D9
Warsa, *Indonesia* **193** J19
Warsaw, *N.C., U.S.* **98** J12
Warsaw, *Va., U.S.* **98** E13
Warsaw, *Ind., U.S.* **101** P9
Warshiikh, *Somalia* **211** R12
Warszawa (Warsaw), *Pol.* **151** D16
Warta, *river, Pol.* **151** F14
Waru, *Indonesia* **193** K17
Warwick, *R.I., U.S.* **97** L14
Warwick, *Eng., U.K.* **145** T12
Warwick, *Qnsld., Austral.* **221** V15
Wasa Station, *Antarctica* **238** B12
Wasagu, *Nigeria* **206** J7
Wasaka, *Vanuatu* **229** D15
Wasam, *Pak.* **180** F12
Wasatch Range, *Utah, U.S.* **108** K7
Wasco, *Oreg., U.S.* **110** F5
Wasco, *Calif., U.S.* **111** V7
Wase, *Nigeria* **206** K9
Waseca, *Minn., U.S.* **106** L11
Washap, *Pak.* **181** Q4
Washburn, *Me., U.S.* **97** B18
Washburn, *Wis., U.S.* **100** F14
Washburn, *N. Dak., U.S.* **106** F4
Washburn, Mount, *Wyo., U.S.* **108** F8
Washington Court House, *Ohio, U.S.* **101** R12
Washington Island *see* Teraina, *Kiribati* **224** F12
Washington Island, *Wis., U.S.* **100** H8
Washington Land, *Greenland, Den.* **88** C5
Washington, *Pa., U.S.* **96** P3
Washington, *D.C., U.S.* **98** C13
Washington, *Ga., U.S.* **98** L7
Washington, *N.C., U.S.* **98** H13
Washington, *Wis., U.S.* **100** H8
Washington, *Ill., U.S.* **101** Q5
Washington, *Ind., U.S.* **101** T8
Washington, *Iowa, U.S.* **107** P13
Washington, *Kans., U.S.* **107** R8
Washington, *Mo., U.S.* **107** T14
Washington, *Utah, U.S.* **109** Q5
Washington, *U.S.* **110** C5
Washington, Mount, *N.H., U.S.* **97** G14

Washita, *river, Okla. Tex. U.S.* **105** C14
Washita, *river, Okla. U.S.* **105** D17
Washougal, *Wash. U.S.* **110** F4
Washtucna, *Wash. U.S.* **110** D8
Washuk, *Pak.* **181** P5
Wasilla, *Alas., U.S.* **113** J16
Wasior, *Indonesia* **193** K19
Wasipe, *Ghana* **205** N15
Waskaganish, *Que. Can.* **87** N18
Wasserbillig, *Lux.* **162** J11
Wassou, *Guinea* **204** M8
Wasum, *P.N.G.* **222** P9
Waswanipi, Lac, *Que. Can.* **87** N18
Watampone, *Indonesia* **193** L13
Water Cay, *Bahamas* **122** J2
Water Cays, *Bahamas* **118** F11
Water Ground, *Cayman Is. U.K.* **122** M2
Water Island, *Virgin Is. U.S.* **122** N8
Water Valley, *Miss. U.S.* **102** H12
Waterbury, *Conn. U.S.* **96** M12
Wateree Lake, *S.C. U.S.* **98** K9
Waterford (Port Láirge), *Ire.* **145** T6
Waterford Harbour, *Ire.* **145** T6
Waterloo, *Ill. U.S.* **101** T4
Waterloo, *Iowa, U.S.* **107** N12
Waterloo, *Trin. & Tobago* **123** P22
Waterloo, *Sierra Leone* **204** N8
Waterloo, *N.T. Austral.* **221** Q7
Waterman, Isla, *Chile* **135** Y8
Watersmeet, *Mich. U.S.* **100** F6
Waterton Lakes National Park, *Alta. Can.* **108** A6
Waterton-Glacier International Peace Park, *Mont. U.S.* **108** A6
Watertown, *N.Y. U.S.* **96** G9
Watertown, *Wis. U.S.* **100** L6
Watertown, *S. Dak. U.S.* **106** K8
Waterville, *Me. U.S.* **97** F16
Waterville, *Minn. U.S.* **106** K11
Waterville, *Kans. U.S.* **107** R8
Waterville, *Wash. U.S.* **110** C6
Watford City, *N. Dak. U.S.* **106** F2
Watford, *Eng. U.K.* **145** U14
Watkins Glen, *N.Y. U.S.* **96** K7
Watling *see* San Salvador, *island, Bahamas* **119** F14
Watling's Castle, *Bahamas* **122** F11
Watonga, *Okla. U.S.* **105** C16
Watrous, *N. Mex. U.S.* **109** R13
Watsa, *D.R.C.* **209** C19
Watseka, *Ill. U.S.* **101** Q7
Watsi Kengo, *D.R.C.* **209** E13
Watson Escarpment, *Antarctica* **238** K12
Watson Lake, *Yukon, Can.* **86** H8
Watson, *S. Aust. Austral.* **221** W8
Watsonville, *Calif. U.S.* **111** T4
Watt Bay, *Antarctica* **239** R18
Watt Mountain, *Dominica* **123** G20
Watts Bar Lake, *Tenn. U.S.* **103** F17
Wattwil, *Switz.* **152** A4
Watubela, Kepulauan, *Indonesia* **193** L18
Wau, *S. Sudan* **203** W9
Wau, *P.N.G.* **222** Q8
Wau, *river, S. Sudan* **203** W8
Waubay, *S. Dak. U.S.* **106** J7
Wauchope, *N.T. Austral.* **221** S9
Wauchula, *Fla. U.S.* **99** U8
Waukarlycarly, Lake, *Austral.* **220** E4
Waukegan, *Ill. U.S.* **100** M7
Waukesha, *Wis. U.S.* **100** L6
Waukon, *Iowa, U.S.* **106** M13
Wauneta, *Nebr. U.S.* **107** Q3
Waupaca, *Wis. U.S.* **100** J6
Waupun, *Wis. U.S.* **100** K6
Waurika, *Okla. U.S.* **105** E16
Wausa, *Nebr. U.S.* **107** N7
Wausau, *Wis. U.S.* **100** H5
Wauseon, *Ohio, U.S.* **101** N11
Wautoma, *Wis. U.S.* **100** K5
Wauwatosa, *Wis. U.S.* **100** L7
Wave Hill, *N.T. Austral.* **221** R8
Waverly, *N.Y. U.S.* **96** L8
Waverly, *Va. U.S.* **98** F13
Waverly, *Ohio, U.S.* **101** S13
Waverly, *Tenn. U.S.* **103** E14
Waverly, *Iowa, U.S.* **106** M12
Waverly, *Nebr. U.S.* **107** Q8
Wāw al Kabīr, *Lib.* **201** H19
Wawa, *Ont. Can.* **87** P17
Wawa, *Nigeria* **206** K6
Waxahachie, *Tex. U.S.* **105** G17
Way, Lake, *Austral.* **220** G4

Waya, *island, Fiji* **226** P4
Wayamli, *Indonesia* **193** H16
Waycross, *Ga., U.S.* **99** P7
Wayne, *Nebr., U.S.* **107** N8
Waynesboro, *Pa., U.S.* **96** Q6
Waynesboro, *Ga., U.S.* **98** M8
Waynesboro, *Va., U.S.* **98** E11
Waynesboro, *Miss., U.S.* **103** M13
Waynesboro, *Tenn., U.S.* **103** F14
Waynesburg, *Pa., U.S.* **96** P3
Waynesville, *N.C., U.S.* **98** H6
Waynesville, *Mo., U.S.* **107** U13
Waynoka, *Okla., U.S.* **105** B15
Waza, *Cameroon* **206** J12
Wāzin, *Lib.* **201** E16
Wazirabad, *Pak.* **180** K13
Wé, *New Caledonia, Fr.* **231** P23
We, *island, Philippines* **192** F3
Weary Bay, *Austral.* **220** C13
Weatherford, *Okla., U.S.* **105** C15
Weatherford, *Tex., U.S.* **105** G16
Weaverville, *Calif., U.S.* **110** M3
Webb City, *Mo., U.S.* **107** V11
Weber Basin, *Pac. Oc.* **246** K6
Webi Jubba, *river, Somalia* **211** R10
Webi Shabeelle, *river, Somalia* **211** Q11
Webster City, *Iowa, U.S.* **107** N11
Webster Groves, *Mo., U.S.* **107** T15
Webster Springs, *W. Va., U.S.* **98** D9
Webster, *Mass., U.S.* **97** L14
Webster, *S. Dak., U.S.* **106** J7
Weda, *Indonesia* **193** J16
Weda, Teluk, *Indonesia* **193** J16
Weddell Island, *Falk. Is., U.K.* **135** W11
Weddell Plain, *Atl. Oc.* **249** Y8
Weddell Sea, *Antarctica* **238** C8
Wedemark, *Ger.* **150** D7
Wee Waa, *N.S.W., Austral.* **221** W14
Weed, *Calif., U.S.* **110** M3
Weeki Wachee, *Fla., U.S.* **99** T7
Weeping Water, *Nebr., U.S.* **107** Q9
Weert, *Neth.* **148** K13
Wegorzewo, *Pol.* **151** B16
Weiden, *Ger.* **150** H9
Weifang, *China* **185** H14
Weihai, *China* **185** G15
Weilheim, *Ger.* **150** K8
Weilmoringle, *N.S.W., Austral.* **221** V13
Weimar, *Tex., U.S.* **105** L18
Weimar, *Ger.* **150** F8
Weinan, *China* **184** J11
Weining, *China* **184** M10
Weinsberger Wald, *Aust.* **150** K11
Weipa, *Qnsld., Austral.* **221** P12
Weirton, *W. Va., U.S.* **98** B9
Weiser, *Idaho, U.S.* **108** F3
Weiser, *river, Idaho, U.S.* **108** F3
Weishan, *China* **185** J14
Weiss Lake, *Ala., Ga., U.S.* **98** K4
Weissenburg, *Ger.* **150** J8
Weissenfels, *Ger.* **150** F9
Weiswampach, *Lux.* **162** G10
Weixi, *China* **184** M8
Wejherowo, *Pol.* **151** A14
Welch, *W. Va., U.S.* **98** F8
Weldiya, *Eth.* **210** K8
Weldon, *N.C., U.S.* **98** G13
Weligama, *Sri Lanka* **183** U7
Welk'īt'ē, *Eth.* **210** M7
Welkom, *S. Af.* **212** K8
Wellesley Islands, *Austral.* **220** C11
Wellin, *Belg.* **148** L14
Wellington Bay, *Nunavut, Can.* **87** F13
Wellington, *Tex., U.S.* **105** D14
Wellington, *Kans., U.S.* **107** V7
Wellington, *Colo., U.S.* **108** L13
Wellington, *Nev., U.S.* **111** Q7
Wellington, *N.S.W., Austral.* **221** X14
Wellington, *N.Z.* **223** J20
Wellington, Isla, *Chile* **135** V6
Wellington, Lake, *Austral.* **220** M13
Wells, *Me., U.S.* **97** H15
Wells, *Tex., U.S.* **105** J20
Wells, *Minn., U.S.* **106** L11
Wells, *Nev., U.S.* **110** M12
Wells, Lake, *Austral.* **220** G5
Wellsboro, *Pa., U.S.* **96** L7
Wellston, *Ohio, U.S.* **101** S13
Wellsville, *N.Y., U.S.* **96** K6
Wellsville, *Ohio, U.S.* **101** Q15
Wellsville, *Mo., U.S.* **107** S13
Wellton, *Ariz., U.S.* **109** V5
Wels, *Aust.* **150** K11
Welsh, *La., U.S.* **102** P7
Welshpool, *Wales, U.K.* **145** S10

Welwel, *Eth.* **211** N11
Wenatchee, *Wash., U.S.* **110** C6
Wenchi, *Ghana* **205** N15
Wendeng, *China* **185** H15
Wendesi, *Indonesia* **193** K19
Wendover, *Utah, U.S.* **108** K5
Wendover, *Nev., U.S.* **110** M12
Wenman *see* Isla Wolf, *island, Ecua.* **130** N2
Weno, *F.S.M.* **226** E9
Weno, *island, F.S.M.* **226** E9
Wenshan, *China* **184** P10
Wentworth, *N.S.W., Austral.* **221** X11
Wenxian, *China* **184** K10
Wenzhou, *China* **185** L15
Werda, *Botswana* **212** H6
Werdau, *Ger.* **150** G9
Werder, *Ger.* **150** D10
Werdēr, *Eth.* **211** N11
Were Īlu, *Eth.* **210** L8
Weser, *river, Ger.* **150** E7
Weskan, *Kans., U.S.* **107** S3
Weslaco, *Tex., U.S.* **105** R16
Wesley, *Dominica* **123** E20
Wesleyville, *Pa., U.S.* **96** K3
Wessel Islands, *Austral.* **220** A10
Wessex, *region, Eng., U.K.* **145** V12
Wessington Springs, *S. Dak., U.S.* **106** L6
Wessington, *S. Dak., U.S.* **106** K6
Wesson, *Miss., U.S.* **102** M11
West Allis, *Wis., U.S.* **100** L7
West Antarctic Ice Sheet Divide Field Camp, *Antarctica* **238** K9
West Antarctica, *region, Antarctica* **238** J9
West Bank, *Asia* **172** M5
West Bay, *Fla., U.S.* **99** R3
West Bay, *La., U.S.* **102** R12
West Bay, *Cayman Is., U.K.* **122** M2
West Bay, *bay, Cayman Is., U.K.* **122** M2
West Bend, *Wis., U.S.* **100** L7
West Branch, *Mich., U.S.* **100** J11
West Branch, *Iowa, U.S.* **107** P13
West Burra, *island, Scot., U.K.* **144** D11
West Caicos, *island, Turks & Caicos Is., U.K.* **119** J16
West Cape Howe, *Austral.* **220** L3
West Caroline Basin, *Pac. Oc.* **246** J7
West Cay, *Virgin Is., U.K.* **122** M7
West Chester, *Pa., U.S.* **96** P9
West Columbia, *Tex., U.S.* **105** M19
West Des Moines, *Iowa, U.S.* **107** P11
West End Point, *Cayman Is., U.K.* **118** L8
West End Village, *Anguilla, U.K.* **122** R10
West End, *Bahamas* **122** J1
West End, *Virgin Is., U.K.* **122** R5
West European Basin, *Atl. Oc.* **248** H10
West Falkland, *island, Falk. Is., U.K.* **135** W11
West Fayu Atoll *see* Pigailoe, *F.S.M.* **226** B8
West Florida Shelf, *Atl. Oc.* **248** L2
West Fork Poplar, *river, Mont., U.S.* **108** A12
West Fork Trinity, *river, Tex., U.S.* **105** F16
West Frankfort, *Ill., U.S.* **101** U6
West Frisian Islands, *Neth.* **148** G12
West Glacier, *Mont., U.S.* **108** B6
West Grand Lake, *Me., U.S.* **97** E18
West Greenland Shelf, *Arctic Oc.* **253** N15
West Ice Shelf, *Antarctica* **239** H21
West Indies, *N. Amer.* **84** N9
West Jan Mayen Ridge, *Arctic Oc.* **253** K19
West Lafayette, *Ind., U.S.* **101** Q8
West Landing, *bay, Marshall Is.* **227** D19
West Liberty, *Ky., U.S.* **103** B19
West Liberty, *Iowa, U.S.* **107** P14
West Mariana Basin, *Pac. Oc.* **246** H7
West Melanesian Trench, *Pac. Oc.* **246** J8
West Memphis, *Ark., U.S.* **102** G10
West Monroe, *La., U.S.* **102** L8
West Nicholson, *Zimb.* **212** F9
West Norfolk Ridge, *Pac. Oc.* **246** N10
West Novaya Zemlya Trough, *Arctic Oc.* **253** D17

West Palm Beach, *Fla., U.S.* **99** V10
West Plains, *Mo., U.S.* **107** V13
West Point Lake, *Ala., Ga., U.S.* **103** K17
West Point, *N.Y., U.S.* **96** M11
West Point, *Ga., U.S.* **98** M4
West Point, *Va., U.S.* **98** E13
West Point, *Ky., U.S.* **103** B16
West Point, *Miss., U.S.* **103** J13
West Point, *Nebr., U.S.* **107** P8
West Point, *Marshall Is.* **227** H19
West Scotia Basin, *Atl. Oc.* **249** W6
West Scotia Ridge, *Atl. Oc.* **246** J7
West Sheba Ridge, *Ind. Oc.* **250** D6
West Siberian Plain, *Asia* **168** D8
West Spit, *cape, Marshall Is.* **227** B19
West Thulean Rise, *Atl. Oc.* **248** G8
West Union, *Iowa, U.S.* **106** M13
West Valley City, *Utah, U.S.* **108** L7
West Virginia, *U.S.* **98** D9
West Wyalong, *N.S.W., Austral.* **221** X13
West Yellowstone, *Mont., U.S.* **108** F8
West York Island, *S. China Sea* **192** E11
West, *Tex., U.S.* **105** H17
Westbrook, *Me., U.S.* **97** H15
Westbrook, *Tex., U.S.* **105** H13
Westby, *Wis., U.S.* **100** K4
Westcliffe, *Colo., U.S.* **109** N13
Westerhall Bay, *Grenada* **123** L23
Westerhall Point, *Grenada* **123** L22
Westerland, *Ger.* **150** A7
Westerly, *R.I., U.S.* **97** M14
Western Australia, *Austral.* **221** T4
Western Beach, *Gibraltar, U.K.* **163** N23
Western Desert, *Egypt* **202** G8
Western Dvina, *river, Eur.* **140** F10
Western Ghats, *range, India* **182** M3
Western Island, *P.N.G.* **222** M7
Western Plateau, *Austral.* **220** F6
Western Sahara, *Af.* **200** J3
Western Sayan Mountains, *Asia* **168** E9
Western, *river, Nunavut, Can.* **87** G13
Westerschelde, *bay, Neth.* **148** K11
Westerville, *Ohio, U.S.* **101** R13
Westfall, *Oreg., U.S.* **110** H8
Westfield, *Mass., U.S.* **96** L12
Westfield, *N.Y., U.S.* **96** K4
Westhope, *N. Dak., U.S.* **106** D4
Westman Islands *see* Vestmannaeyjar, *Ice.* **142** F2
Westminster, *Md., U.S.* **98** B13
Westminster, *S.C., U.S.* **98** K6
Westmorland, *Calif., U.S.* **111** Y11
Weston Point, *Kiribati* **226** J11
Weston super Mare, *Eng., U.K.* **145** U11
Weston, *W. Va., U.S.* **98** C9
Weston, *Mo., U.S.* **107** S10
Westphalia, *Eur.* **140** G6
Westport (Cathair na Mart), *Ire.* **145** Q3
Westport, *Wash., U.S.* **110** D2
Westport, *Calif., U.S.* **111** P2
Westport, *N.Z.* **223** K18
Westpunt, *Curaçao, Neth.* **123** N13
Westpunt, *point, Curaçao, Neth.* **123** N13
Westray Firth, *Scot., U.K.* **144** E10
Westray, *island, Scot., U.K.* **144** E10
Westwego, *La., U.S.* **102** Q11
Westwood Downs, *N.S.W., Austral.* **221** W11
Westwood, *Calif., U.S.* **111** N5
Wetar, *island, Indonesia* **193** M15
Wetaskiwin, *Alta., Can.* **86** L10
Wete, *Tanzania* **211** V8
Wetumka, *Okla., U.S.* **105** D18
Wetumpka, *Ala., U.S.* **103** L16
Wetzlar, *Ger.* **150** G6
Wewahitchka, *Fla., U.S.* **99** R4
Wewak, *P.N.G.* **222** N6
Wewoka, *Okla., U.S.* **105** D18
Wexford (Loch Garman), *Ire.* **145** T7
Wexford Harbour, *Ire.* **145** T7
Weyburn, *Sask., Can.* **86** N12
Weyer, *Aust.* **150** L11
Weymes Bight, *Bahamas* **122** F6
Weymouth, *Eng., U.K.* **145** W11
Weymouth, *Cape, Austral.* **220** B13
Weyprecht Fjord, *Greenland, Den.* **88** B9
Whakapunake, *peak, N.Z.* **223** G22
Whakatane, *N.Z.* **223** F22
Whale Cay, *Bahamas* **118** D11

Whale Cove, *Nunavut, Can.* **87** J15
Whale Valley *see* Wadi Al-Hitan, *Egypt* **202** G10
Whalsay, *island, Scot., U.K.* **144** C12
Whangamata, *N.Z.* **223** F21
Whanganui (Wanganui), *N.Z.* **223** H20
Whangarei, *N.Z.* **223** D20
Wharton Basin, *Ind. Oc.* **251** K14
Wharton, *Tex., U.S.* **105** M18
Wheatland Reservoir, *Wyo., U.S.* **108** J13
Wheatland, *Wyo., U.S.* **108** J13
Wheaton, *Minn., U.S.* **106** H8
Wheeler Lake, *Ala., U.S.* **103** H15
Wheeler Peak, *N. Mex., U.S.* **109** Q13
Wheeler, *Tex., U.S.* **105** C14
Wheeling, *W. Va., U.S.* **98** B9
Wheelwright, *Ky., U.S.* **103** C20
Whidbey, *Point, Austral.* **220** K9
Whiddy Island, *Ire.* **145** U3
Whillans Ice Stream, *Antarctica* **238** K11
Whinham, *Mount, Austral.* **220** G7
Whiore *see* South Cape, *cape, N.Z.* **223** Q16
Whiporie, *N.S.W., Austral.* **221** W16
Whiskeytown Shasta-Trinity National Recreation Area, *Calif., U.S.* **110** M3
Whitby, *Eng., U.K.* **145** P13
White Bird, *Idaho, U.S.* **108** E4
White Butte, *peak, N. Dak., U.S.* **106** H2
White Cliffs, *N.S.W., Austral.* **221** W12
White Cloud, *Mich., U.S.* **100** K9
White Deer, *Tex., U.S.* **105** C13
White Hall, *Ill., U.S.* **101** S4
White Hall, *Ark., U.S.* **102** H9
White Horses, *Jam.* **122** J11
White Island, *Nunavut, Can.* **87** G16
White Island, *Antarctica* **239** C19
White Lake, *La., U.S.* **102** Q8
White Mountain, *Alas., U.S.* **112** F12
White Mountain, *Russ.* **158** D14
White Mountains National Recreation Area, *Alas., U.S.* **113** F17
White Mountains, *Me., N.H., U.S.* **97** F14
White Mountains, *Calif., U.S.* **111** S8
White Nile Dam, *Sudan* **203** R12
White Nile *see* El Bahr el Abyad, *river, Sudan* **203** S12
White Nile, *river, Af.* **198** G9
White Pine, *Mich., U.S.* **100** F5
White Plains, *N.Y., U.S.* **96** M11
White River Junction, *Vt., U.S.* **97** H13
White River, *S. Dak., U.S.* **106** L4
White Salmon, *Wash., U.S.* **110** E5
White Sands National Park, *N. Mex., U.S.* **109** U12
White Sea *see* Beloye More, *Russ.* **158** A9
White Sea, *Arctic Oc.* **253** C19
White Sulphur Springs, *W. Va., U.S.* **98** E9
White Sulphur Springs, *Mont., U.S.* **108** D8
White, *S. Dak., U.S.* **106** K8
White, *Lake, Austral.* **220** E7
White, *river, Ind., U.S.* **101** T7
White, *river, Ark., U.S.* **102** E8
White, *river, Tex., U.S.* **105** F13
White, *river, S. Dak., U.S.* **106** L4
White, *river, Colo., Utah, U.S.* **108** L10
White, *river, Nev., U.S.* **111** R12
Whitefish Bay, *Mich., U.S.* **100** F10
Whitefish Point, *Mich., U.S.* **100** F10
Whitefish Point, *town, Mich., U.S.* **100** F10
Whitefish, *Mont., U.S.* **108** B5
Whitehall, *N.Y., U.S.* **96** H12
Whitehall, *Mich., U.S.* **100** L9
Whitehall, *Wis., U.S.* **100** J3
Whitehall, *Ohio, U.S.* **101** R13
Whitehall, *Mont., U.S.* **108** E7
Whitehaven, *Eng., U.K.* **145** P10
Whitehorse, *Yukon, Can.* **86** G8
Whitehouse, *Jam.* **122** J6
Whitemark, *Tas., Austral.* **221** Y8
Whiteriver, *Ariz., U.S.* **109** T9
Whitesboro, *Tex., U.S.* **105** F17
Whitesburg, *Ky., U.S.* **103** D20
Whiteville, *N.C., U.S.* **98** K12

Whiteville, *Tenn., U.S.* **102** F12
Whitewater Baldy, *peak, N. Mex., U.S.* **109** U10
Whitewater, *Wis., U.S.* **100** M6
Whitewater, *Mont., U.S.* **108** B11
Whitewright, *Tex., U.S.* **105** F18
Whithorn, *Scot., U.K.* **145** N9
Whitman Mission National Historic Site, *Wash., U.S.* **110** E8
Whitman, *Nebr., U.S.* **107** N3
Whitmire, *S.C., U.S.* **98** K8
Whitmore Mountains, *Antarctica* **238** J10
Whitney, *Mount, Calif., U.S.* **111** T8
Whittier, *Alas., U.S.* **113** J17
Wholdaia Lake, *N.W.T., Can.* **87** J13
Whyalla, *S. Aust., Austral.* **221** X10
Wiay *see* Fuidhaigh, *island, Scot., U.K.* **144** H6
Wibaux, *Mont., U.S.* **108** C13
Wichita Falls, *Tex., U.S.* **105** E16
Wichita Mountains, *Okla., U.S.* **105** D15
Wichita, *Kans., U.S.* **107** U7
Wichita, *river, Tex., U.S.* **105** F15
Wick, *Scot., U.K.* **144** G10
Wickenburg, *Ariz., U.S.* **109** T6
Wickett, *Tex., U.S.* **104** J10
Wickiup Reservoir, *Oreg., U.S.* **110** H4
Wickliffe, *Ky., U.S.* **102** D12
Wicklow (Cill Mhantáin), *Ire.* **145** R7
Wicklow Head, *Ire.* **145** R7
Wicklow Mountains, *Ire.* **145** R7
Wide Bay, *P.N.G.* **222** P10
Widnes, *Eng., U.K.* **145** R11
Wido, *island, S. Korea* **187** S7
Wieleń, *Pol.* **150** D12
Wieluń, *Pol.* **151** F14
Wien (Vienna), *Aust.* **150** K11
Wiener Neustadt, *Aust.* **150** L12
Wieprz, *river, Pol.* **151** E17
Wiesbaden, *Ger.* **150** G6
Wiggins, *Miss., U.S.* **102** N12
Wight, Isle of, *Eng., U.K.* **145** W13
Wigtown Bay, *Scot., U.K.* **145** N9
Wikieup, *Ariz., U.S.* **109** S5
Wik'ro, *Eth.* **210** J8
Wil, *Switz.* **152** A4
Wilberforce, Cape, *Austral.* **220** A10
Wilbur, *Wash., U.S.* **110** C7
Wilburton, *Okla., U.S.* **105** D19
Wilcannia, *N.S.W., Austral.* **221** W12
Wilczek, *Russ.* **160** C12
Wildcat Peak, *Nev., U.S.* **111** Q10
Wilde Bay, *Austral.* **220** G16
Wildespitze, *peak, Aust.* **150** M8
Wildrose, *N. Dak., U.S.* **106** E2
Wildwood, *N.J., U.S.* **96** R10
Wildwood, *Fla., U.S.* **99** S8
Wilhelm II Coast, *Antarctica* **239** J21
Wilhelmina Gebergte, *peak, Suriname* **131** G16
Wilhelmshaven, *Ger.* **150** C6
Wilkes Fracture Zone, *Pac. Oc.* **247** K19
Wilkes Island, *Wake I., U.S.* **230** N11
Wilkes Land, *region, Antarctica* **239** M20
Wilkes-Barre, *Pa., U.S.* **96** M8
Wilkins Aerodrome, *Antarctica* **239** M21
Wilkins Ice Shelf, *Antarctica* **238** F6
Willacoochee, *Ga., U.S.* **99** P7
Willamette, *river, Oreg., U.S.* **110** G3
Willapa Bay, *Wash., U.S.* **110** D2
Willapa Hills, *Wash., U.S.* **110** D2
Willard, *Ohio, U.S.* **101** P13
Willard, *Utah, U.S.* **108** K7
Willaumez Peninsula, *P.N.G.* **222** P9
Willcox, *Ariz., U.S.* **109** V9
Willemstad, *Curaçao, Neth.* **123** Q14
Willeroo, *N.T., Austral.* **221** Q8
William "Bill" Dannelly Reservoir, *Ala., U.S.* **103** L15
William Creek, *S. Aust., Austral.* **221** V10
William Scoresby Bay, *Antarctica* **239** D20
William, *Mount, Austral.* **220** L12
William's Cay, *Bahamas* **122** E8
Williams Island, *Bahamas* **118** E10
Williams Lake, *B.C., Can.* **86** L8
William's Town, *Bahamas* **122** F9
Williams, *Minn., U.S.* **106** E10
Williams, *Ariz., U.S.* **109** S7

Williams, *Calif., U.S.* **111** Q4
Williamsburg, *Va., U.S.* **98** F13
Williamsburg, *Ky., U.S.* **103** D18
Williamsfield, *Jam.* **122** J8
Williamson, *W. Va., U.S.* **98** D7
Williamsport, *Pa., U.S.* **96** M7
Williamston, *N.C., U.S.* **98** H13
Williamston, *S.C., U.S.* **98** K7
Williamstown, *W. Va., U.S.* **98** C8
Williamstown, *Ky., U.S.* **103** A18
Willikies, *Antigua & Barbuda* **123** B21
Willis Islets, *Austral.* **224** J5
Willis, *Tex., U.S.* **105** K19
Williston Lake, *B.C., Can.* **86** J9
Williston, *S.C., U.S.* **98** L8
Williston, *N. Dak., U.S.* **106** E2
Willits, *Calif., U.S.* **111** P2
Willmar, *Minn., U.S.* **106** J10
Willoughby Bay, *Antigua & Barbuda* **123** B21
Willow City, *N. Dak., U.S.* **106** E5
Willow Springs, *Mo., U.S.* **107** V13
Willowlake, *river, N.W.T., Can.* **86** H10
Willowra, *N.T., Austral.* **221** S8
Willows, *Calif., U.S.* **111** P3
Wills Point, *Tex., U.S.* **105** G18
Wilma Glacier, *Antarctica* **239** D20
Wilmington, *Del., U.S.* **98** B14
Wilmington, *N.C., U.S.* **98** K12
Wilmington, *Ill., U.S.* **101** P7
Wilmington, *Ohio, U.S.* **101** S12
Wilmore, *Ky., U.S.* **103** C17
Wilmot, *Ark., U.S.* **102** K9
Wilmot, *S. Dak., U.S.* **106** J8
Wilshaw Ridge, *Ind. Oc.* **250** K7
Wilson Creek, *Wash., U.S.* **110** C7
Wilson Hills, *Antarctica* **239** R16
Wilson Lake, *Ala., U.S.* **103** G14
Wilson Lake, *Kans., U.S.* **107** S6
Wilson Strait, *Solomon Is.* **228** D6
Wilson, *N.C., U.S.* **98** H12
Wilson, *Ark., U.S.* **102** F11
Wilson, *Okla., U.S.* **105** E17
Wilson, *Kans., U.S.* **107** T6
Wilson, *Mount, Colo., U.S.* **109** P10
Wilson, *Mount, Oreg., U.S.* **110** F4
Wilson's Creek National Battlefield, *Mo., U.S.* **107** V12
Wilsons Promontory, *Austral.* **220** M13
Wilton, *N. Dak., U.S.* **106** G4
Wiltz, *Lux.* **162** H9
Wiltz, *river, Lux.* **162** H9
Wiluna, *W. Aust., Austral.* **221** U4
Winamac, *Ind., U.S.* **101** P8
Winchester Bay, *Oreg., U.S.* **110** H2
Winchester, *Va., U.S.* **98** C12
Winchester, *Ind., U.S.* **101** R10
Winchester, *Ky., U.S.* **103** B18
Winchester, *Tenn., U.S.* **103** G16
Winchester, *Idaho, U.S.* **108** D3
Winchester, *Eng., U.K.* **145** V13
Wind River Range, *Wyo., U.S.* **108** H9
Wind, *river, Yukon, Can.* **86** E9
Wind, *river, Wyo., U.S.* **108** H9
Windber, *Pa., U.S.* **96** P5
Winder, *Ga., U.S.* **98** K6
Windhill Beach, *Gibraltar, U.K.* **163** Q23
Windhoek, *Namibia* **212** G4
Windom, *Minn., U.S.* **106** L10
Windorah, *Qnsld., Austral.* **221** U12
Window Rock, *Ariz., U.S.* **109** R10
Windsor Forest, *Ga., U.S.* **99** N9
Windsor Lake, *Mo., U.S.* **107** T11
Windsor, *Ont., Can.* **87** R17
Windsor, *Conn., U.S.* **97** L13
Windsor, *Vt., U.S.* **97** H13
Windsor, *N.C., U.S.* **98** G13
Windsor, *Colo., U.S.* **108** L13
Windsor, *Eng., U.K.* **145** U13
Windward Islands, *Lesser Antil.* **121** H16
Windward Passage, *N. Amer.* **84** N9
Winfield, *Ala., U.S.* **103** J14
Winfield, *Kans., U.S.* **107** V8
Wini, *Indonesia* **193** M15
Winifred, *Mont., U.S.* **108** C9
Wininen, *island, F.S.M.* **226** F9
Winipirea, *island, F.S.M.* **226** E9
Winipiru, *island, F.S.M.* **226** F8
Winisk, *river, Ont., Can.* **87** M16
Wink, *Tex., U.S.* **104** H10
Winkelman, *Ariz., U.S.* **109** U8
Winlock, *Wash., U.S.* **110** D3
Winneba, *Ghana* **205** Q16

Winnebago, *Minn., U.S.* **106** L11
Winnebago, Lake, *Wis., U.S.* **100** K6
Winnemucca Lake, *Nev., U.S.* **111** N7
Winnemucca, *Nev., U.S.* **110** M8
Winner, *S. Dak., U.S.* **106** M5
Winnfield, *La., U.S.* **102** M8
Winnibigoshish, Lake, *Minn., U.S.* **106** F10
Winnipeg, *Man., Can.* **87** N14
Winnipeg, Lake, *Man., Can.* **87** M14
Winnipegosis, *Man., Can.* **87** M13
Winnipegosis, Lake, *N. Amer.* **84** G5
Winnipesaukee, Lake, *N.H., U.S.* **97** H14
Winnsboro, *S.C., U.S.* **98** K9
Winnsboro, *La., U.S.* **102** L9
Winnsboro, *Tex., U.S.* **105** G19
Winona, *Miss., U.S.* **102** J11
Winona, *Minn., U.S.* **106** L13
Winona, *Kans., U.S.* **107** S3
Winseler, *Lux.* **162** H9
Winsen, *Ger.* **150** C8
Winslow, *Me., U.S.* **97** F16
Winslow, *Ariz., U.S.* **109** S8
Winston-Salem, *N.C., U.S.* **98** G10
Winter Garden, *Fla., U.S.* **99** T8
Winter Haven, *Fla., U.S.* **99** U8
Winter Park, *Fla., U.S.* **99** T9
Winters, *Tex., U.S.* **105** H14
Winters, *Calif., U.S.* **111** R4
Winterset, *Iowa, U.S.* **107** P11
Winterswijk, *Neth.* **148** J14
Winterthur, *Switz.* **152** A4
Winthrop, *Me., U.S.* **97** G16
Winthrop, *Minn., U.S.* **106** K10
Winthrop, *Wash., U.S.* **110** B6
Winton Heights, *Bahamas* **122** E3
Winton, *Qnsld., Austral.* **221** T12
Wippra, *Ger.* **150** E8
Wirrulla, *S. Aust., Austral.* **221** W9
Wisas, island, *F.S.M.* **226** F9
Wisbech, *Eng., U.K.* **145** S14
Wiscasset, *Me., U.S.* **97** G16
Wisconsin Dells, *Wis., U.S.* **100** K5
Wisconsin Range, *Antarctica* **238** K12
Wisconsin Rapids, *Wis., U.S.* **100** J5
Wisconsin, *U.S.* **100** J3
Wisconsin, river, *Wis., U.S.* **100** J5
Wiseman, *Alas., U.S.* **113** E16
Wishek, *N. Dak., U.S.* **106** H5
Wishram, *Wash., U.S.* **110** F5
Wismar, *Ger.* **150** B9
Wisner, *La., U.S.* **102** M9
Wisner, *Nebr., U.S.* **107** N8
Wissembourg, *Fr.* **149** P15
Witbank, *S. Af.* **212** J9
Witham, *Eng., U.K.* **145** T15
Witsand, *S. Af.* **212** M6
Witten, *S. Dak., U.S.* **106** M4
Wittenberg, *Ger.* **150** E10
Wittenberge, *Ger.* **150** D9
Wittenoom, *W. Aust., Austral.* **221** T3
Wittlich, *Ger.* **150** G5
Wittstock, *Ger.* **150** C9
Witu Islands, *P.N.G.* **222** N9
Witu, *Kenya* **211** T8
Wiwŏn, *N. Korea* **186** E7
Włocławek, *Pol.* **151** D14
Włodawa, *Pol.* **151** E18
Wöbbelin, *Ger.* **150** C9
Woburn Bay, *Grenada* **123** L22
Wohlthat Mountains, *Antarctica* **239** B14
Wojejairok Pass, *Marshall Is.* **227** H16
Wojejairok, island, *Marshall Is.* **227** H16
Wokam, island, *Indonesia* **193** L19
Wokingham, *Eng., U.K.* **145** U13
Wolauna, island, *F.S.M.* **226** K7
Wolcott, *N.Y., U.S.* **96** J7
Woleai Atoll, *F.S.M.* **226** B7
Wolf Creek , *Okla., Tex., U.S.* **105** B14
Wolf Creek Pass, *Colo., U.S.* **109** P12
Wolf Creek, *Mont., U.S.* **108** D7
Wolf Creek, *Oreg., U.S.* **110** K2
Wolf Point, *Mont., U.S.* **108** B12
Wolf, river, *Wis., U.S.* **100** H6
Wolfe City, *Tex., U.S.* **105** F18
Wolfeboro, *N.H., U.S.* **97** H14
Wolfen, *Ger.* **150** E9
Wolfenbüttel, *Ger.* **150** E8
Wolfforth, *Tex., U.S.* **104** F12
Wolfsberg, *Aust.* **150** M11
Wolfsburg, *Ger.* **150** D8
Wollaston Forland, *Greenland, Den.* **88** G14

Wollaston Lake, *Sask., Can.* **87** K13
Wollaston Peninsula, *Nunavut, N.W.T., Can.* **86** E12
Wollaston, Islas, *Chile* **135** Y9
Wollogorang, *N.T., Austral.* **221** R10
Wollongong, *N.S.W., Austral.* **221** Y14
Wolo, *Indonesia* **193** K13
Wolpyeong, *S. Korea* **187** T7
Wolstenholme Fjord, *Greenland, Den.* **88** F3
Wolstenholme Ø, *Greenland, Den.* **88** F3
Wolverhampton, *Eng., U.K.* **145** S11
Wombat Plateau, *Ind. Oc.* **251** J17
Wondeok, *S. Korea* **187** N12
Wondeung, *S. Korea* **187** S8
Wongawal, *W. Aust., Austral.* **221** U4
Wonju, *S. Korea* **187** N10
Wŏnp'yŏng, *N. Korea* **186** H9
Wŏnsan, *N. Korea* **186** F11
Wŏnsan, *N. Korea* **186** J9
Wonsando, island, *S. Korea* **187** Q7
Wonthaggi, *Vic., Austral.* **221** Z12
Wood Lake, *Nebr., U.S.* **107** N5
Wood River Lakes, *Alas., U.S.* **113** K13
Wood River, *Ill., U.S.* **101** T4
Wood River, *Nebr., U.S.* **107** Q6
Woodall Mountain, *Miss., U.S.* **103** G13
Woodbine, *Ga., U.S.* **99** Q8
Woodbridge Bay, *Dominica* **123** G19
Woodbridge, *Va., U.S.* **98** D13
Woodbridge, *Eng., U.K.* **145** T15
Woodburn, *Oreg., U.S.* **110** F3
Woodbury, *N.J., U.S.* **96** Q9
Woodenbridge (An Droichead Adhmaid), *Ire.* **145** S7
Woodford, *Grenada* **123** K22
Woodlake, *Calif., U.S.* **111** U7
Woodland Park, *Colo., U.S.* **108** M13
Woodland, *Me., U.S.* **97** E19
Woodland, *Wash., U.S.* **110** E3
Woodland, *Calif., U.S.* **111** Q4
Woodlark (Murua), island, *P.N.G.* **222** Q10
Woodroffe, Mount, *Austral.* **220** G8
Woodruff, *S.C., U.S.* **98** J8
Woodruff, *Utah, U.S.* **108** K8
Woods Hole, *Mass., U.S.* **97** M15
Woods, Lake of the, *N. Amer.* **84** H6
Woods, Lake, *Austral.* **220** D9
Woodsboro, *Tex., U.S.* **105** N17
Woodsfield, *Ohio, U.S.* **101** R15
Woodstock, *Va., U.S.* **98** D11
Woodstock, *Qnsld., Austral.* **221** R12
Woodstock, *Qnsld., Austral.* **221** T12
Woodstown, *N.J., U.S.* **96** Q9
Woodsville, *N.H., U.S.* **97** G13
Woodville, *Miss., U.S.* **102** N9
Woodville, *Tex., U.S.* **105** K20
Woodville, *N.Z.* **223** J21
Woodward, *Okla., U.S.* **105** B15
Woody Head, *Austral.* **220** J16
Woomera, *S. Aust., Austral.* **221** W10
Woonsocket, *R.I., U.S.* **97** L14
Woonsocket, *S. Dak., U.S.* **106** L7
Wooramel, *W. Aust., Austral.* **221** U1
Wooster, *Ohio, U.S.* **101** P14
Wŏram, *N. Korea* **186** M7
Worbab, island, *Marshall Is.* **227** J18
Worcester, *Mass., U.S.* **97** K14
Worcester, *Eng., U.K.* **145** T11
Worcester, *S. Af.* **212** M5
Worden, *Mont., U.S.* **108** E10
Wordie Gletscher, *Greenland, Den.* **88** H13
Workai, island, *Indonesia* **193** M19
Workington, *Eng., U.K.* **145** N10
Worland, *Wyo., U.S.* **108** G10
Wormeldange, *Lux.* **162** K11
Worms, *Ger.* **150** H6
Wortham, *Tex., U.S.* **105** H18
Worthing, *Barbados* **123** L19
Worthington, *Minn., U.S.* **106** L9
Wosi, *Indonesia* **193** J16
Wotho Atoll, *Marshall Is.* **227** C15
Wotje Atoll, *Marshall Is.* **227** C17
Woumbou, *Cameroon* **206** N11
Wour, *Chad* **206** D12
Wowoni, island, *Indonesia* **193** L14
Wragby, *Eng., U.K.* **145** R13
Wrangel Island, *Asia* **168** A11
Wrangel Plain, *Arctic Oc.* **252** H10
Wrangell-Saint Elias National Park and Preserve, *Alas., U.S.* **113** J18
Wrangell, *Alas., U.S.* **113** M23

Wrangell, Cape, *Alas., U.S.* **112** K1
Wrath, Cape, *Scot., U.K.* **144** F8
Wray, *Colo., U.S.* **108** L16
Wreck Hill, *Bermuda* **119** C14
Wreck Point, cape, *S. Af.* **212** K4
Wrecsam see Wrexham, *Wales, U.K.* **145** R10
Wrens, *Ga., U.S.* **98** M7
Wrexham (Wrecsam), *Wales, U.K.* **145** R10
Wright Brothers National Memorial, *N.C., U.S.* **98** G15
Wright Island, *Antarctica* **238** L8
Wright Patman Lake, *Tex., U.S.* **105** F20
Wright, *Wyo., U.S.* **108** G13
Wrightsville Beach, *N.C., U.S.* **98** K13
Wrightsville, *Ga., U.S.* **98** M7
Wrigley Gulf, *Antarctica* **238** M8
Wrocław, *Pol.* **151** F13
Wrzesnia, *Pol.* **151** E13
Wu, river, *China* **184** M11
Wubin, *W. Aust., Austral.* **221** W3
Wuda, *China* **184** G11
Wudāb, *Afghan.* **180** E10
Wudaoliang, *China* **184** J6
Wudu, *China* **184** K10
Wuhai, *China* **184** G11
Wuhan, *China* **185** L13
Wuhu, *China* **185** K14
Wukari, *Nigeria* **206** L9
Wular Lake, *India* **180** H13
Wulff Land, *Greenland, Den.* **88** C8
Wum, *Nigeria* **206** M9
Wun Shwai, *S. Sudan* **203** V10
Wunlah, *Indonesia* **193** M17
Wuntho, *Myanmar* **190** F5
Wupatki National Monument, *Ariz., U.S.* **109** N7
Wuppertal, *Ger.* **150** F5
Wurno, *Nigeria* **206** H7
Wurung, *Qnsld., Austral.* **221** R11
Würzburg, *Ger.* **150** H7
Wushi, *China* **184** E3
Wusi, *Vanuatu* **229** F13
Wusuli, river, *China* **185** C18
Wutung, *P.N.G.* **222** M5
Wuvulu Island, *P.N.G.* **222** M6
Wuwei, *China* **184** H9
Wuxi, *China* **185** K15
Wuyi Shan, *China* **185** M14
Wuyiling, *China* **185** C16
Wuzhong, *China* **184** H10
Wuzhou, *China* **184** N12
WWII Valor In The Pacific National Monument, *Hawaii, U.S.* **115** H14
Wyalkatchem, *W. Aust., Austral.* **221** W3
Wyandra, *Qnsld., Austral.* **221** V13
Wycheproof, *Vic., Austral.* **221** Y12
Wye, river, *Eng., Wales, U.K.* **145** T11
Wyemandoo, peak, *Austral.* **220** H3
Wyk, *Ger.* **150** A7
Wylie Lake, *N.C., S.C., U.S.* **98** J9
Wyloo, *W. Aust., Austral.* **221** T2
Wymore, *Nebr., U.S.* **107** R8
Wyndham, *W. Aust., Austral.* **221** Q6
Wyndmere, *N. Dak., U.S.* **106** H7
Wynne, *Ark., U.S.* **102** G10
Wynnewood, *Okla., U.S.* **105** D17
Wynniatt Bay, *N.W.T., Can.* **87** D13
Wyoming Range, *Wyo., U.S.* **108** H8
Wyoming, *Mich., U.S.* **100** L9
Wyoming, *U.S.* **108** H11
Wyszków, *Pol.* **151** D16
Wytheville, *Va., U.S.* **98** F9

X

X`éng, river, *Laos* **190** H9
Xaafuun, *Somalia* **210** L15
Xàbia see Jávea, *Sp.* **147** J15
Xaçmaz, *Azerb.* **171** C22
Xaghra, *Malta* **163** H20
Xai-Xai, *Mozambique* **212** H10
Xaignabouli, *Laos* **190** J8
Xainza (Shenzha), *China* **184** K5
Xalapa, *Mex.* **117** J13
Xalin, *Somaliland* **210** M13
Xangda (Nangqên), *China* **184** K7
Xangongo, *Angola* **208** Q10

Xankəndi see Stepanakert, *Azerb.* **171** E20
Xánthi, *Gr.* **157** B14
Xanthus, ruins, *Turk.* **170** J4
Xapuri, *Braz.* **132** G6
Xar Burd, *China* **184** G10
Xarardheere, *Somalia* **211** P13
Xassengue, *Angola* **208** L12
Xátiva, *Sp.* **147** H14
Xenia, *Ohio, U.S.* **101** R12
Xerxes' see Xérxis Canal, *Gr.* **157** D13
Xérxis (Xerxes') Canal, *Gr.* **157** D13
Xewkija, *Malta* **163** H20
Xi'an, *China* **184** J11
Xiamen (Amoy), *China* **185** N15
Xiangfan, *China* **184** K12
Xiangkhoang, *Laos* **190** J9
Xiangtan, *China* **185** M13
Xianyang, *China* **184** J11
Xianyou, *China* **185** N15
Xiao Hinggan Ling, *China* **185** B15
Xiaoba see Qingtongxia, *China* **184** H10
Xiaogan, *China* **185** K13
Xiapu, *China* **185** M15
Xichang, *China* **184** M9
Xifeng, *China* **184** J11
Xigazê (Rikaze), *China* **184** L5
Xiis, *Somaliland* **210** L12
Xilinhot, *China* **185** E13
Xin Barag Youqi, *China* **185** C13
Xin Barag Zouqi, *China* **185** C13
Xinge, *Angola* **209** L13
Xinghua, *China* **185** J15
Xingkai Hu (Khanka Ozero), *China, Russ.* **185** D17
Xingtai, *China* **185** H13
Xingu, river, *Braz.* **132** F12
Xinguara, *Braz.* **133** F13
Xingxingxia, *China* **184** F7
Xingyi, *China* **184** N10
Xinhe, *China* **184** E4
Xining, *China* **184** H9
Xinlong, *China* **184** L9
Xinpu see Lianyungang, *China* **185** J14
Xinqing, *China* **185** C16
Xinshao, *China* **184** M12
Xintai, *China* **185** H14
Xinxiang, *China* **185** J13
Xinyang, *China* **185** K13
Xinyu, *China* **185** M13
Xinyuan, *China* **184** E4
Xinzhou, *China* **184** G12
Xinzo de Limia, *Sp.* **146** D6
Xique Xique, *Braz.* **133** G15
Xisha Qundao, *see Paracel Islands, S. China Sea* **190** K15
Xızı, *Azerb.* **171** C22
Xocavənd see Martuni, *Azerb.* **171** E21
Xorkol, *China* **184** G6
Xuanhua, *China* **185** F13
Xuanwei, *China* **184** M10
Xudat, *Azerb.* **171** C22
Xuddur, *Somalia* **211** Q11
Xudun, *Somaliland* **210** M13
Xunke, *China* **185** B16
Xuwen, *China* **184** Q12
Xuzhou, *China* **185** J14
Xylaganí, *Gr.* **157** B15
Xylofagou, *Cyprus* **162** P10
Xylókastro, *Gr.* **156** J10

Y

Yaak, *Mont., U.S.* **108** A4
Ya'an, *China* **184** L9
Yabbenohr, island, *Marshall Is.* **227** G15
Yabēlo, *Eth.* **211** P7
Yablonovyy Khrebet, *Russ.* **161** L17
Yablonovyy Range, *Asia* **168** E10
Yabrūd, *Syr.* **172** H8
Yacata, island, *Fiji* **226** P7
Yachats, *Oreg., U.S.* **110** G2
Yacimiento Río Turbio, *Arg.* **135** W7
Yacuiba, *Bol.* **132** L8
Yacyretá Dam, *Arg., Parag.* **132** N10
Yadgir, *India* **182** N5
Yadkin, river, *N.C., U.S.* **98** G9
Yadong, *China* **184** M5
Yadua, island, *Fiji* **226** N5

Yafi, *Indonesia* **193** K21
Yagasa, island, *Fiji* **226** Q8
Yağca, *Turk.* **170** H5
Yaghan Basin, *Atl. Oc.* **249** W5
Yaguajay, *Cuba* **118** H9
Yahyalı, *Turk.* **170** G10
Yaita, *Japan* **189** N12
Yakacık, *Turk.* **170** J11
Yakapınar, *Turk.* **170** J10
Yakeshi, *China* **185** C14
Yakima, *Wash., U.S.* **110** D6
Yakima, river, *Wash., U.S.* **110** D5
Yakishiri Jima, *Japan* **188** F13
Yakmach, *Pak.* **181** N4
Yako, *Burkina Faso* **205** K15
Yakoma, *D.R.C.* **209** B14
Yakossi, *Cen. Af. Rep.* **207** M16
Yakotoko, *Cen. Af. Rep.* **207** M17
Yaksha, *Russ.* **158** C16
Yaksha, *Russ.* **160** G9
Yaku Shima, *Japan* **189** U4
Yaku, island, *Asia* **168** G13
Yakutat Bay, *Alas., U.S.* **113** K20
Yakutat, *Alas., U.S.* **113** K20
Yakutsk, *Russ.* **161** G18
Yala, *Sri Lanka* **183** U8
Yala, *Thai.* **191** S8
Yala, *Ghana* **205** M15
Yalata, *S. Aust., Austral.* **221** W8
Yale, *Mich., U.S.* **100** L13
Yale, *Okla., U.S.* **105** B17
Yalewa Kalou (Round Island), *Fiji* **226** N5
Yalgoo, *W. Aust., Austral.* **221** V3
Yali, *D.R.C.* **209** E14
Yalinga, *Cen. Af. Rep.* **207** M16
Yallahs, *Jam.* **122** J11
Yallock, *N.S.W., Austral.* **221** X12
Yalobusha, river, *Miss., U.S.* **102** J11
Yalong, river, *China* **184** L9
Yalova, *Turk.* **170** D5
Yalta, *Russ.* **159** T7
Yalvaç, *Turk.* **170** G6
Yalu, river, *China, N. Korea* **186** G4
Yamagata, *Japan* **188** L12
Yamaguchi, *Japan* **189** Q5
Yamal Peninsula, *Asia* **168** C8
Yamal, *Russ.* **160** F11
Yamas, *Indonesia* **193** K21
Yamato Mountains see Queen Fabiola Mountains, *Antarctica* **239** C17
Yambio, *S. Sudan* **203** Y9
Yambol, *Bulg.* **155** J16
Yamdena, island, *Indonesia* **193** M17
Yamethin, *Myanmar* **190** H5
Yamm, *Russ.* **158** G5
Yamma Yamma, Lake, *Austral.* **220** G11
Yammaw, *Myanmar* **190** E6
Yamoussoukro, *Côte d'Ivoire* **205** P13
Yampa, river, *Colo., U.S.* **108** L10
Yan'an, *China* **184** H11
Yana Valley, *Arctic Oc.* **252** E9
Yana, *Sierra Leone* **204** M9
Yana, Gulf of, *Arctic Oc.* **252** D8
Yana, river, *Russ.* **161** F17
Yanagawa, *Japan* **189** R4
Yanbu' al Baḥr, *Saudi Arabia* **174** G5
Yancheng, *China* **185** J15
Yandal, *W. Aust., Austral.* **221** V4
Yandeearra, *W. Aust., Austral.* **221** S3
Yandina, *Solomon Is.* **228** F9
Yandongi, *D.R.C.* **209** C15
Yandoon, *Myanmar* **190** K5
Yanfolila, *Mali* **204** L11
Yangambi, *D.R.C.* **209** D16
Yangasso, *Mali* **205** K13
Yangdŏk, *N. Korea* **186** J7
Yanggang, *N. Korea* **186** E7
Yanggu, *S. Korea* **186** L10
Yanghŭng, *N. Korea* **186** E11
Yanghwa-man, bay, *N. Korea* **186** G10
Yangī Qal'ah, *Afghan.* **180** E9
Yangī Qal'ah, *Afghan.* **180** G5
Yangjiang, *China* **185** P13
Yangju, *S. Korea* **186** M8
Yangon (Rangoon), *Myanmar* **190** K5
Yangou Gala, *Cen. Af. Rep.* **207** L15
Yangpo, *S. Korea* **187** R12
Yangsan, *S. Korea* **187** S12
Yangtze see Chang Jiang, Jinsha, river, *China* **184** M10
Yangyang, *S. Korea* **186** L11
Yanji, *China* **185** E17
Yankeetown, *Fla., U.S.* **99** S7

Yankton, *S. Dak., U.S.* **106** M7
Yano-Indigirskaya Nizmennost', *Russ.* **161** E17
Yano-Oymyakonskoye Nagor'ye, *Russ.* **161** F18
Yanonge, *D.R.C.* **209** D16
Yanqi, *China* **184** F5
Yanrey, *W. Aust., Austral.* **221** T1
Yanskiy Zaliv, *Russ.* **161** E17
Yanskiy, *Russ.* **161** F17
Yantabulla, *N.S.W. Austral.* **221** V13
Yantai, *China* **185** G15
Yantra, *river, Bulg.* **155** H15
Yanuca, *island, Fiji* **226** N7;Q5
Yao, *Japan* **189** Q9
Yaoundé, *Cameroon* **206** N10
Yap Islands, *F.S.M.* **226** A5
Yap Trench, *Pac. Oc.* **246** J7
Yap, *island, F.S.M.* **226** D11
Yapei, *Ghana* **205** M15
Yapen, *island, Indonesia* **193** K19
Yapen, Selat, *Indonesia* **193** K19
Yapero, *Indonesia* **193** L20
Yaqaga, *island, Fiji* **226** N5
Yaqeta, *island, Fiji* **226** N4
Yaqui, *river, Mex.* **116** D7
Yaquina Head, *Oreg., U.S.* **110** G2
Yara, *Cuba* **118** K12
Yaraka, *Qnsld., Austral.* **221** U12
Yaralıgöz, *peak, Turk.* **170** C9
Yaransk, *Russ.* **158** G14
Yardımcı Burnu, *Turk.* **170** J5
Yardımlı, *Azerb.* **171** F22
Yarega, *Russ.* **158** B15
Yaren, *Nauru* **227** M14
Yarensk, *Russ.* **158** D13
Yarí, *river, Col.* **130** J7
Yarīm, *Yemen* **174** P9
Yarkant *see* Shache, *China* **184** F2
Yarkant, *river, China* **184** F2
Yarle Lakes, *Austral.* **220** J8
Yarlung Zangbo (Brahmaputra), *river, China* **184** L6
Yarma, *Turk.* **170** G8
Yarmouth, *N.S. Can.* **87** Q21
Yaroslavl', *Russ.* **158** H10
Yaroua, *island, Fiji* **226** P8
Yarozero, *Russ.* **158** E10
Yarraden, *Qnsld., Austral.* **221** Q12
Yarraloola, *W. Aust., Austral.* **221** T2
Yarrowmere, *Qnsld., Austral.* **221** S13
Yartsevo, *Russ.* **158** K7
Yartsevo, *Russ.* **161** J13
Yarumal, *Col.* **130** F5
Yasawa Group, *Fiji* **226** N4
Yasawa, *island, Fiji* **226** N4
Yashi, *Nigeria* **206** J8
Yashikera, *Nigeria* **206** K6
Yashkul', *Russ.* **159** R14
Yasin, *Pak.* **180** F12
Yāsīnzaī Kêlay, *Afghan.* **180** L5
Yasothon, *Thai.* **190** L10
Yāsūj, *Iran* **177** K15
Yasun Burnu, *Turk.* **170** D12
Yasur Volcano, *Vanuatu* **229** M17
Yata, *river, Bol.* **132** G7
Yatağan, *Turk.* **170** H3
Yatakala, *Niger* **206** G4
Yaté, *New Caledonia, Fr.* **231** Q23
Yatenga, *region, Af.* **198** G3
Yates Center, *Kans., U.S.* **107** U9
Yathkyed Lake, *Nunavut, Can.* **87** J14
Yato, *N.Z.* **229** D20
Yatsushiro, *Japan* **189** S4
Yatta Plateau, *Kenya* **211** T7
Yauca, *Peru* **132** J4
Yaupi, *Ecua.* **130** L4
Yavarí *see* Javari, *river, S. Amer.* **132** F4
Yavatmal, *India* **182** L6
Yaviza, *Pan.* **117** N22
Yawatahama, *Japan* **189** R6
Yawri Bay, *Sierra Leone* **204** N8
Yaylâdağı, *Turk.* **170** K11
Yaynangyoung, *Myanmar* **190** H4
Yazd, *Iran* **177** J17
Yazılıkaya, *Turk.* **170** F6
Yazlıca Dağı, *peak, Turk.* **171** H17
Yazman, *Pak.* **181** N11
Yazoo City, *Miss., U.S.* **102** K11
Yazoo, *river, Miss., U.S.* **102** K10
Ýdra (Hydra), *island, Gr.* **156** K12
Ýdra, *Gr.* **156** K12
Ye-u, *Myanmar* **190** G5
Ye, *Myanmar* **190** L6
Yebbi Bou, *Chad* **207** D13

Yebyu, *Myanmar* **190** M6
Yecheng (Kargilik), *China* **184** F2
Yecheon, *S. Korea* **187** P11
Yecla, *Sp.* **147** J14
Yeed, *Somalia* **211** P10
Yeeda, *W. Aust., Austral.* **221** R5
Yefremov, *Russ.* **158** L9
Yeghegnadzor, *Arm.* **170** E19
Yegua Creek, *Tex., U.S.* **105** K17
Yei, *S. Sudan* **203** Y11
Yei, *river, S. Sudan* **203** X11
Yeji, *Ghana* **205** N16
Yekaterinburg, *Russ.* **160** H9
Yekateriny, *Russ.* **161** K23
Yekepa, *Liberia* **204** N11
Yela Island, *P.N.G.* **228** H4
Yelabuga, *Russ.* **158** H16
Yelan Kolenovskiy, *Russ.* **159** N11
Yelan', *Russ.* **159** N12
Yelcho Station, *Antarctica* **238** D5
Yelets, *Russ.* **158** M10
Yélimané, *Mali* **204** J10
Yelizavety, *Russ.* **161** H21
Yell Sound, *Scot., U.K.* **144** C12
Yell, *island, Scot., U.K.* **144** C12
Yellandu, *India* **182** N7
Yellow Sea, *Pac. Oc.* **168** G12
Yellow *see* Huang, *river, China* **184** H12
Yellow, *river, Ala., Fla., U.S.* **99** Q2
Yellowknife, *N.W.T. Can.* **86** H11
Yellowstone Lake, *Wyo., U.S.* **108** F8
Yellowstone National Park, *Idaho, Mont., Wyo., U.S.* **108** F8
Yellowstone, *river, Mont., Wyo., U.S.* **108** D11
Yellville, *Ark., U.S.* **102** E8
Yelm, *Wash., U.S.* **110** D3
Yel'sk, *Belarus* **158** M5
Yelwa, *Nigeria* **206** J6
Yemanzhelinsk, *Russ.* **160** J8
Yemassee, *S.C., U.S.* **98** M9
Yemen, *Asia* **174** M9
Yemetsk, *Russ.* **158** C11
Yemtsa, *Russ.* **158** C10
Yên Bái, *Vietnam* **190** G10
Yên Minh, *Vietnam* **190** F10
Yen, *Cameroon* **206** P11
Yenagoa, *Nigeria* **206** N7
Yenakiyeve, *Ukr.* **159** Q10
Yendéré, *Burkina Faso* **205** M13
Yendi, *Ghana* **205** M16
Yengisar, *China* **184** F2
Yéni, *Niger* **206** H6
Yeniçağa, *Turk.* **170** D7
Yenice, *Turk.* **170** F8
Yenice, *river, Turk.* **170** G11
Yeniceoba, *Turk.* **170** F8
Yenierenköy *see* Yialousa, *Cyprus* **162** M11
Yenişehir, *Turk.* **170** D5
Yenisey Gulf, *Arctic Oc.* **168** C8
Yenisey, *river, Asia* **168** D8
Yeniseysk, *Russ.* **161** K13
Yeniseyskiy Zaliv, *Russ.* **161** F13
Yenyuka, *Russ.* **161** J17
Yeo Lake, *Austral.* **220** H5
Yeola, *India* **182** L4
Yeoncheon, *S. Korea* **186** L8
Yeondo, *island, S. Korea* **187** U9
Yeongam, *S. Korea* **187** T7
Yeongcheon, *S. Korea* **187** R11
Yeongchun, *S. Korea* **187** U7
Yeongdeok, *S. Korea* **187** Q12
Yeongdong, *S. Korea* **187** Q9
Yeonggwang, *S. Korea* **187** S7
Yeonheungdo, *island, S. Korea* **187** N7
Yeonjongdo, *island, S. Korea* **187** N7
Yeongju, *S. Korea* **187** P11
Yeongwol, *S. Korea* **187** N11
Yeongyang, *S. Korea* **187** P12
Yeonil, *S. Korea* **187** R12
Yeonpyeong *see* Daeyeonpyeongdo, *island, S. Korea* **186** M6
Yeoryang, *S. Korea* **187** N11
Yeosan, *S. Korea* **187** R8
Yeoseodo, *island, S. Korea* **187** V8
Yeosu, *S. Korea* **187** T9
Yeovil, *Eng., U.K.* **145** V11
Yeppoon, *Qnsld., Austral.* **221** T15
Yerbogachen, *Russ.* **161** J15
Yerema, *Russ.* **161** J15
Yerevan, *Arm.* **171** E18
Yergeni Hills, *Eur.* **140** G14
Yerington, *Nev., U.S.* **111** Q7

Yerköy, *Turk.* **170** E9
Yermak Plateau, *Arctic Oc.* **253** H15
Yermo, *Calif., U.S.* **111** W10
Yeroḥam, *Israel* **173** P5
Yershov, *Russ.* **158** M15
Yesan, *S. Korea* **187** P8
Yeşilhisar, *Turk.* **170** G10
Yeşilırmak, *river, Turk.* **170** D11
Yeşilkent, *Turk.* **170** J11
Yeşilova, *Turk.* **170** H5
Yesŏng, *river, N. Korea* **186** L7
Yessentuki, *Russ.* **159** T12
Yessey, *Russ.* **161** G14
Yetti, *region, Af.* **198** E3
Yeu, Île d', *Fr.* **149** S5
Yevlax, *Azerb.* **171** D21
Yevpatoriya, *Russ.* **159** S7
Yeysk, *Russ.* **159** R10
Ygatimí, *Parag.* **132** M11
Yí, *river, Uru.* **134** K14
Yialias, *river, Cyprus* **162** P8
Yialousa (Yenierenköy), *Cyprus* **162** M11
Yibin, *China* **184** L10
Yichang, *China* **184** L12
Yichun, *China* **185** C16
Yigo, *Guam, U.S.* **230** N6
Yilan, *China* **185** C17
Yin Shan, *China* **184** F11
Yinchuan, *China* **184** H10
Yindarlgooda, Lake, *Austral.* **220** J5
Yindi, *W. Aust., Austral.* **221** W5
Yingcheng, *China* **185** K13
Yingkou, *China* **185** F15
Yingtan, *China* **185** L14
Yining (Gulja), *China* **184** D4
Yinnietharra, *W. Aust., Austral.* **221** U2
Yirga 'Alem, *Eth.* **211** N7
Yirol, *S. Sudan* **203** W11
Yiwu, *China* **184** E7
Yiyang, *China* **184** L12
Yıldız Dağları, *peak, Turk.* **170** B4
Yıldızeli, *Turk.* **170** E11
Ylig Bay, *Guam, U.S.* **230** P5
Ylivieska, *Fin.* **143** F15
Ymer Ø, *Greenland, Den.* **88** J13
Yn Cholloo *see* Calf of Man, *cape, Isle of Man, U.K.* **145** Q8
Yoakum, *Tex., U.S.* **105** M17
Yobe, *river, Af.* **198** G6
Yocona, *river, Miss., U.S.* **102** H11
Yogyakarta, *Indonesia* **192** M9
Yoichi, *Japan* **188** G12
Yokadouma, *Cameroon* **206** N12
Yokjido, *island, S. Korea* **187** U10
Yokkaichi, *Japan* **189** Q9
Yoko, *Cameroon* **206** M10
Yokoate Shima, *Japan* **189** V3
Yokohama, *Japan* **188** J13
Yokohama, *Japan* **189** P12
Yokote, *Japan* **188** K12
Yŏkp'o, *N. Korea* **186** J6
Yola, *Nigeria* **206** K10
Yolöten, *Turkm.* **178** N9
Yom, *river, Thai.* **190** K7
Yomra, *Turk.* **171** D14
Yŏmju, *N. Korea* **186** G4
Yomsŏng, *N. Korea* **186** G11
Yŏmsŏng, *N. Korea* **186** K10
Yona, *Guam, U.S.* **230** P5
Yonago, *Japan* **189** P7
Yŏnan, *N. Korea* **186** M7
Yonezawa, *Japan* **188** M12
Yongam, *N. Korea* **186** H7
Yongban, *N. Korea* **186** D12
Yongban, *N. Korea* **186** E12
Yŏngbyŏn, *N. Korea* **186** H6
Yongch'ŏn, *N. Korea* **186** G4
Yongdang, *S. Korea* **187** T7
Yongdeng, *China* **184** H10
Yongdu, *S. Korea* **186** M9
Yonggang, *N. Korea* **186** K5
Yŏnggwang, *N. Korea* **186** G9
Yongyap Pass, *India* **182** F15
Yonggye, *N. Korea* **186** B13
Yŏngha, *N. Korea* **186** E11
Yonghan, *S. Korea* **187** Q12
Yŏnghŭng-man (Kŭmya-man), *bay, N. Korea* **186** J9
Yongin, *S. Korea* **187** N8
Yongjam, *N. Korea* **186** F11
Yŏngjŏ, *N. Korea* **186** D8
Yongmunsan, *peak, S. Korea* **186** M9
Yongnim, *N. Korea* **186** F8

Yongp'o, *N. Korea* **186** H5
Yongp'o, *N. Korea* **186** J8
Yongp'ung, *N. Korea* **186** G5
Yongsan, *S. Korea* **187** N10
Yongsŏng, *N. Korea* **186** J5
Yŏngwŏn, *N. Korea* **186** H7
Yongwŏn, *N. Korea* **186** L4
Yongzhou, *China* **184** M12
Yonkers, *N.Y. U.S.* **96** N11
Yonne, *river, Fr.* **149** R11
Yŏnp'o, *N. Korea* **186** H9
Yopal, *Col.* **130** F7
Yopurga, *China* **184** F2
York Sound, *Austral.* **220** C5
York, *Pa., U.S.* **96** P7
York, *S.C., U.S.* **98** J9
York, *Ala., U.S.* **103** L13
York, *Nebr., U.S.* **107** Q7
York, *Eng., U.K.* **145** Q13
York, *W. Aust., Austral.* **221** X3
York, Cape, *Austral.* **220** A12
Yorke Peninsula, *Austral.* **220** K10
Yorkton, *Sask., Can.* **86** N12
Yorktown, *Va., U.S.* **98** F14
Yorktown, *Tex., U.S.* **105** M17
Yoro Shima, *Japan* **189** W3
Yoron Jima, *Japan* **189** X2
Yorosso, *Mali* **205** K13
Yortan Tepe, *ruins, Turk.* **170** F3
Yos Sudarso *see* Dolak, *island, Indonesia* **193** M20
Yosemite National Park, *Calif., U.S.* **111** S6
Yoshkar Ola, *Russ.* **158** H14
Yotvata, *Israel* **173** R5
You, *river, China* **184** P11
Youghal (Eochaill), *Ire.* **145** T5
Youghal Bay, *Ire.* **145** T5
Youkou, *Côte d'Ivoire* **204** Q12
Youkounkoun, *Guinea* **204** K8
Young Island, *Bahamas* **122** D7
Young Island, *St. Vincent & the Grenadines* **123** L16
Young, *N.S.W. Austral.* **221** X14
Young's Rock, *Pitcairn I.* **228** Q6
Youngstown, *Ohio, U.S.* **101** P15
Youssoufia, *Mor.* **200** D7
Youth, Isle of *see* Juventud, Isla de la, *Cuba* **118** J4
Yovan, *Taj.* **178** N12
Yovkovo, *Bulg.* **155** G17
Yovo, *island, P.N.G.* **228** A5
Yozgat, *Turk.* **170** E10
Ypáti, *Gr.* **156** G10
Ypé Jhú, *Parag.* **132** M11
Ypres *see* Ieper, *Belg.* **148** L10
Ypsilanti, *Mich., U.S.* **100** M12
Ypsilí *see* Strongylí, *island, Gr.* **157** N22
Yreka, *Calif., U.S.* **110** L3
Yrghyz, *Kaz.* **178** F8
Yrghyz, *river, Kaz.* **178** F8
Ysabel Channel, *P.N.G.* **222** M9
Ysyk-Köl, *lake, Kyrg.* **179** K15
Ysyk, Lake, *Asia* **168** F7
Ytterän, *Sw.* **142** G12
Ytyk Kyuyel', *Russ.* **161** G18
Yuan (Red), *river, China* **184** N9
Yuan, *river, China* **184** M12
Yuba City, *Calif., U.S.* **111** Q4
Yuba, *river, Calif., U.S.* **111** Q4
Yūbari, *Japan* **188** G13
Yubdo, *Eth.* **210** M5
Yūbetsu, *Japan* **188** F14
Yucatán Basin, *Atl. Oc.* **248** M2
Yucatan Channel, *N. Amer.* **84** N7
Yucca Flat, *Nev., U.S.* **111** T11
Yucca Valley, *Calif., U.S.* **111** X10
Yucca, *Ariz., U.S.* **109** S5
Yuci, *China* **184** H12
Yudino, *Russ.* **158** J14
Yuendumu, *N.T. Austral.* **221** S8
Yuexi, *China* **184** L9
Yueyang, *China* **185** L13
Yugê, *China* **184** J7
Yugorenok, *Russ.* **161** G19
Yugorskiy Poluostrov, *Russ.* **160** F11
Yukagir, *Russ.* **161** E17
Yukagirskoye Ploskogor'ye, *Russ.* **161** D19
Yukon Delta, *Alas., U.S.* **112** G12
Yukon Plateau, *N. Amer.* **84** D3
Yukon-Charley Rivers National Preserve, *Alas., U.S.* **113** F18

Yukon, *Can.* **86** F8
Yukon, *Okla., U.S.* **105** C16
Yukon, *river, N. Amer.* **84** C3
Yüksekova, *Turk.* **171** H18
Yulara, *N.T. Austral.* **221** U8
Yule Bay, *Antarctica* **239** R15
Yuleba, *Qnsld. Austral.* **221** V14
Yuli, *China* **184** F5
Yuli, *Nigeria* **206** K9
Yulin, *China* **184** P12
Yuma, *Colo., U.S.* **108** L15
Yuma, *Ariz., U.S.* **109** U4
Yumbe, *Uganda* **211** Q3
Yumbi, *D.R.C.* **208** F11
Yumbi, *D.R.C.* **209** E17
Yumen, *China* **184** G8
Yumrukaya Burnu, *Turk.* **170** B10
Yumurtalık, *Turk.* **170** J10
Yunaska Island, *Alas., U.S.* **112** P8
Yuncheng, *China* **184** J12
Yungas, *region, S. Amer.* **128** G5
Yunta, *S. Aust., Austral.* **221** X11
Yupanqui Basin, *Pac. Oc.* **247** M19
Yup'yŏng, *N. Korea* **186** C11
Yuribey, *Russ.* **160** F9
Yurihonjō, *Japan* **188** K12
Yurimaguas, *Peru* **132** F3
Yuroma, *Russ.* **158** B12
Yur'ya, *Russ.* **158** F14
Yur'yev Pol'skiy, *Russ.* **158** H10
Yur'yevets, *Russ.* **158** H12
Yushkozero, *Russ.* **158** B7
Yushu (Gyêgu), *China* **184** K7
Yushu, *China* **185** D16
Yuste, Monasterio de, *Sp.* **146** G8
Yusufeli, *Turk.* **171** D16
Yutian (Keriya), *China* **184** G3
Yuyao, *China* **185** L15
Yuzawa, *Japan* **188** L12
Yuzha, *Russ.* **158** H11
Yuzhno Sakhalinsk, *Russ.* **161** K22
Yuzhnyy, *Russ.* **159** S14
Yuzhnyy, *Russ.* **161** F21
Yuzhnyy, *Kaz.* **178** D9
Yverdon, *Switz.* **152** B1
Yvetot, *Fr.* **149** N8

Z

Z̧ahrān, *Saudi Arabia* **174** L8
Za, *river, China* **184** K7
Zaanstad, *Neth.* **148** H12
Zab al Kabir (Great Zab), *river, Iraq* **176** E10
Zabaykal'sk, *Russ.* **161** L17
Žabbar, *Malta* **163** K23
Zabīd, *Yemen* **174** P8
Zābol, *Iran* **177** J22
Zābolī, *Iran* **177** M23
Zabré, *Burkina Faso* **205** L16
Zabrze, *Pol.* **151** G14
Zacapa, *Guatemala* **117** L16
Zacatecas, *Mex.* **116** G10
Zacháro, *Gr.* **156** K9
Zackenberg, research station, *Greenland, Den.* **88** G14
Zadar (Zara), *Croatia* **154** G5
Zadawa, *Nigeria* **206** J9
Zadoi, *China* **184** K7
Zafóra, *island, Gr.* **157** N17
Zafra, *Sp.* **146** J7
Żagań, *Pol.* **150** F12
Zagne, *Côte d'Ivoire* **204** P12
Zagora, *Mor.* **200** F8
Zagorá, *Gr.* **156** F11
Zagreb, *Croatia* **154** D6
Zagros Mountains *see* Kūhhāye Zāgros, *Iran, Iraq* **176** D11
Zaguinasso, *Côte d'Ivoire* **204** M12
Zāhedān, *Iran* **177** K22
Zahirabad, *India* **182** N6
Zaḩlé, *Leb.* **172** J7
Zaječar, *Serb.* **154** G12
Zakamensk, *Russ.* **161** M15
Zakhodnyaya, *river, Belarus* **158** J6
Zākhū, *Iraq* **176** D9
Zakopane, *Pol.* **151** H15
Zákros, *Gr.* **157** Q17
Zákynthos (Zante), *island, Gr.* **156** J7
Zákynthos, *Gr.* **156** J7
Zala, *Angola* **208** J10

Zalaegerszeg, *Hung.* **151** M13
Zalanga, *Nigeria* **206** K9
Zalantun, *China* **185** C15
Zalău, *Rom.* **155** C13
Zalew Szczeciński *see* Stettiner
  Haff, *Ger., Pol.* **150** C11
Zalewo, *Pol.* **151** B15
Ẕalim, *Saudi Arabia* **174** H8
Zalingei, *Sudan* **203** S6
Zamakh, *Yemen* **174** M11
Zambeze, river, *Mozambique* **212** E11
Zambezi, river, *Af.* **198** M8
Zambezi Canyon, *Ind. Oc.* **250** K5
Zambezi Escarpment, *Zimb.* **212** D9
Zambezi River Delta, *Af.* **198** N10
Zambia, *Af.* **209** P16
Zamboanga City, *Philippines* **193** F14
Zambrów, *Pol.* **151** C17
Zamfara, river, *Nigeria* **206** J7
Zamīndāwar, region, *Afghan.* **180** K4
Zamoge, *Eth.* **210** J7
Zamora, *Ecua.* **130** M4
Zamora, *Sp.* **146** D8
Zamość, *Pol.* **151** F18
Zanaga, *Congo* **208** F9
Zanderij, *Suriname* **131** F17
Zanesville, *Ohio, U.S.* **101** R14
Zangla, *India* **180** J15
Zanjān, *Iran* **177** E13
Zante *see* Zákynthos, island, *Gr.*
  156 J7
Zanthus, *W. Aust., Austral.* **221** W5
Zanzibar Island, *Tanzania* **211** W8
Zanzibar, *Tanzania* **211** W8
Zaouatallaz, *Alg.* **201** J15
Zaoyang, *China* **185** K13
Zapadna Morava, river, *Serb.*
  154 G10
Zapadnaya Dvina, *Russ.* **158** J7
Zapadno Sibirskaya Ravnina, *Russ.*
  160 H11
Zapala, *Arg.* **135** P7
Zapata, *Tex., U.S.* **105** Q15
Zapata, Península de, *Cuba* **118** H7
Zapiga, *Chile* **134** C7
Zapiola Ridge, *Atl. Oc.* **249** U7
Zaporizhzhya, *Ukr.* **159** Q8
Zaqatala, *Azerb.* **171** C20
Zara *see* Zadar, *Croatia* **154** G5
Zara, *Turk.* **170** E12
Zarafshon Range, *Taj.* **178** M12
Zarafshon, *Uzb.* **178** L10
Zaragoza, *Sp.* **147** D14
Zarand, *Iran* **177** J19
Zaranj, *Afghan.* **180** L3
Zarautz, *Sp.* **146** B12
Zaraysk, *Russ.* **158** K10
Zaraza, *Venez.* **130** D11
Zaria, *Nigeria* **206** J8
Zaris Berge, mountains, *Namibia*
  212 H3
Zárkos, *Gr.* **156** E10
Zărneşti, *Rom.* **155** E15
Zarrin Shahr, *Iran* **177** H15
Zaruma, *Ecua.* **130** M3
Żary, *Pol.* **150** E11
Zaskar Mountains, *India* **182** D6
Zatish'ye, *Russ.* **161** D20
Zatobyl, *Kaz.* **178** C10
Zatoka, *Ukr.* **159** S5
Zavolzhsk, *Russ.* **158** H11
Zawiercie, *Pol.* **151** G15
Zawīlah, *Lib.* **201** H18
Zāwiyat al Mukhaylá, *Lib.* **201** D21
Zāwiyat Masūs, *Lib.* **201** E21
Zāwiyat Shammās, *Egypt* **202** E9
Zayaki Jangal, *Pak.* **181** P6
Zaybāk, *Afghan.* **180** F10
Zaysan Köli, *Kaz.* **179** F17
Zaysan, *Kaz.* **179** F18
Zaysan, Lake, *Asia* **168** F8
Zayü (Chayu), *China* **184** L7
Zdolbuniv, *Ukr.* **159** N3
Zealand, island, *Eur.* **140** F7
Zéalé, *Côte d'Ivoire* **204** P11
Żebbuġ, *Malta* **163** H20
Żebbuġ, *Malta* **163** K22
Zeča, island, *Croatia* **154** E4
Zednes, peak, *Mauritania* **204** C10
Zeehaen Fracture Zone, *Ind. Oc.*
  251 R17
Zeeland, *Mich., U.S.* **100** M9
Zeewolde, *Neth.* **148** J13
Zeewolf Fracture Zone, *Ind. Oc.*
  250 N11
Zeewyk Ridge, *Ind. Oc.* **251** L15

Zefat, *Israel* **172** K6
Zehak, *Iran* **177** J22
Zeidab, *Sudan* **203** P13
Zeil, Mount, *Austral.* **220** F8
Zeist, *Neth.* **148** J12
Żejtun, *Malta* **163** L23
Zélée Bank, *Ind. Oc.* **250** J6
Zelenchukskaya, *Russ.* **159** T12
Zelenodol'sk, *Russ.* **158** J14
Zelenoe, *Kaz.* **178** F5
Zelenogorsk, *Russ.* **158** E6
Zelenogradsk, *Russ.* **158** J1
Zelenokumsk, *Russ.* **159** T13
Železná Ruda, *Czechia* **150** J10
Železnik, *Serb.* **154** F10
Zelfana, *Alg.* **201** E13
Zelzate, *Belg.* **148** K11
Zemlya Georga, Ostrov, *Russ.*
  160 B12
Zemetchino, *Russ.* **158** L12
Zemio, *Cen. Af. Rep.* **207** N17
Zemongo, *Cen. Af. Rep.* **207** L17
Zencirli, site, *Turk.* **170** H11
Zeni Su, island, *Japan* **189** R11
Zenica, *Bosn. & Herzg.* **154** G8
Zenith Plateau, *Ind. Oc.* **251** K16
Zenkevich Rise (Hokkaido Rise),
  Pac. Oc. **246** E8
Zenobia, ruins, *Syr.* **172** E12
Žepče, *Bosn. & Herzg.* **154** F8
Zerah, Gowd-e, *Afghan.* **180** M3
Zermatt, *Switz.* **152** C3
Zernez, *Switz.* **152** B5
Zernograd, *Russ.* **159** R11
Zestap'oni, *Rep. of Georgia* **171** B17
Zetland *see* Shetland Islands,
  Scot., U.K. **144** B12
Zeya, *Russ.* **161** K19
Zgharta, *Leb.* **172** H7
Zgierz, *Pol.* **151** E15
Zgorzelec, *Pol.* **150** F11
Zhadong *see* Zhadün, *China* **184** K3
Zhadün (Zhadong), *China* **184** K3
Zhalpaqtal, *Kaz.* **178** E4
Zhaltyr, *Kaz.* **178** D12
Zhamog (Zhamu), *China* **184** L7
Zhamu *see* Zhamog, *China* **184** L7
Zhanatal, *Kaz.* **178** H8
Zhanga Qazan, *Kaz.* **178** E4
Zhangaly, *Kaz.* **178** G4
Zhangaözen, *Kaz.* **178** J5
Zhangaqorghan, *Kaz.* **178** J11
Zhangatas, *Kaz.* **178** J12
Zhangjiakou, *China* **185** F13
Zhangye, *China* **184** G9
Zhangzhou, *China* **185** N15
Zhanjiang, *China* **184** P12
Zhanyi, *China* **184** N10
Zhaodong, *China* **185** D16
Zhaojue, *China* **184** M9
Zhaosu, *China* **184** E3
Zhaotong, *China* **184** M10
Zhaoyuan, *China* **185** H15
Zhaqsy, *Kaz.* **178** D11
Zhari Namco (Zhari Nanmucuo),
  lake, *China* **184** K4
Zhari Nanmucuo *see* Zhari Namco,
  lake, *China* **184** K4
Zharkamys, *Kaz.* **178** F6
Zharkent, *Kaz.* **179** J16
Zharma, *Kaz.* **179** F16
Zharmysh, *Kaz.* **178** J5
Zhaslyk, *Uzb.* **178** J7
Zhayrang, *Kaz.* **178** G12
Zhayylma, *Kaz.* **178** D9
Zhayyq (Ural), river, *Kaz.* **178** F5
Zhelezīnka, *Kaz.* **179** C14
Zheleznodorozhnyy, *Russ.* **158** C14
Zheleznodorozhnyy, *Russ.* **160** F9
Zheleznogorsk, *Russ.* **158** M8
Zheleznovodsk, *Russ.* **159** T12
Zhem, river, *Kaz.* **178** G6
Zhemchug Canyon, *Pac. Oc.* **246** B12
Zhemgang, *Bhutan* **182** G13
Zhengzhou (Chengchow), *China*
  185 J13
Zhenjiang, *China* **185** K15
Zhenxiong, *China* **184** M10
Zherdevka, *Russ.* **158** M11
Zhetiqara, *Kaz.* **178** D9
Zhezqazghan, *Kaz.* **178** G11
Zhicheng, *China* **184** L12
Zhigansk, *Russ.* **161** G17
Zhigulevsk, *Russ.* **158** K15
Zhilaya Kosa, *Kaz.* **178** G5
Zhil'gur, *Russ.* **161** H18

Zhilinda, *Russ.* **161** F15
Zhilinsky Spur, *Arctic Oc.* **253** J15
Zhlobin, *Belarus* **158** L6
Zhmerynka, *Ukr.* **159** P4
Zhob, *Pak.* **180** L9
Zhob, river, *Pak.* **180** L9
Zhokhova, *Russ.* **161** C17
Zhongshan Station, *Antarctica*
  239 G20
Zhongwei, *China* **184** H10
Zhongxian, *China* **184** L11
Zhosaly, *Kaz.* **178** H10
Zhovtneve, *Ukr.* **159** R6
Zhu (Pearl), river, *China* **185** P13
Zhuji, *China* **185** L15
Zhukovka, *Russ.* **158** L7
Zhumadian, *China* **185** K13
Zhupanovo, *Russ.* **161** F23
Zhuryn, *Kaz.* **178** F7
Zhuzhou, *China* **185** M13
Zhympity, *Kaz.* **178** E5
Zhytomyr, *Ukr.* **159** N5
Ziarat, *Pak.* **180** M8
Zibo, *China* **185** H14
Zicavo, *Fr.* **149** Y2
Zielona Góra, *Pol.* **150** E12
Ziftá, *Egypt* **202** F11
Zigana Geçidi, *Turk.* **171** D14
Zigey, *Chad* **206** G12
Zigong, *China* **184** L10
Ziguinchor, *Senegal* **204** K6
Zile, *Turk.* **170** E11
Žilina, *Slovakia* **151** J14
Zillah, *Lib.* **201** F19
Zima, *Russ.* **161** L15
Zimbabwe, *Af.* **212** F9
Zimmi, *Sierra Leone* **204** N9
Zimnicea, *Rom.* **155** G15
Zimovniki, *Russ.* **159** R12
Zina, *Cameroon* **206** J12
Zinder, *Niger* **206** H9
Ziniaré, *Burkina Faso* **205** K15
Zinjibār, *Yemen* **174** Q9
Zion National Park, *Utah, U.S.* **109** P6
Zion, *Ill., U.S.* **100** M7
Zion, *St. Kitts & Nevis* **123** C19
Zirə, *Azerb.* **171** D23
Žirje, island, *Croatia* **154** G5
Zirndorf, *Ger.* **150** H8
Zittau, *Ger.* **150** F11
Živinice, *Bosn. & Herzg.* **154** F8
Ziway Hāyk', *Eth.* **210** M7
Zixing, *China* **185** M13
Ziyamet *see* Leonarisso, *Cyprus*
  162 M10
Zlarin, island, *Croatia* **154** G6
Zlatoust, *Russ.* **160** J8
Zlatoustovsk, *Russ.* **161** J20
Zlín, *Czechia* **151** J14
Złotów, *Pol.* **151** C13
Zmari Sar, peak, *Afghan.* **180** L6
Zmeica, Lacul, *Rom.* **155** F17
Zmeinogorsk, *Russ.* **160** L11
Zmiyev, *Ukr.* **159** P9
Zmiyinyy, Ostriv, *Ukr.* **159** S5
Znamenka, *Kaz.* **179** E16
Znamensk, *Russ.* **158** J2
Znam'yanka, *Ukr.* **159** Q7
Znojmo, *Czechia* **150** J12
Zobia, *D.R.C.* **209** C17
Zoetermeer, *Neth.* **148** J12
Ẕofar, *Israel* **173** Q5
Zofingen, *Switz.* **152** A4
Zoigê, *China* **184** J9
Zolochiv, *Ukr.* **159** N3
Zolochiv, *Ukr.* **159** N8
Zolotaya Gora, *Russ.* **161** K18
Zolotonosha, *Ukr.* **159** P6
Zomba, *Malawi* **212** D11
Zongo, *D.R.C.* **208** B12
Zonguldak, *Turk.* **170** C7
Zonza, *Fr.* **149** Y2
Zoo, *Monaco* **163** E20
Zor Dağ, peak, *Turk.* **171** E18
Zōr Kōṯ, *Afghan.* **180** J9
Zorritos, *Peru* **132** E2
Zorzor, *Liberia* **204** N10
Zouar, *Chad* **207** D13
Zoucheng, *China* **185** J14
Zouerate, *Mauritania* **204** D9
Zoug, *W. Sahara* **200** K3
Zrenjanin (Petrovgrad), *Serb.*
  154 E10
Zubovo, *Russ.* **158** F9
Zubtsov, *Russ.* **158** J8
Zuénoula, *Côte d'Ivoire* **204** N12

Zuera, *Sp.* **147** D14
Ẕufār, *Oman* **175** M15
Zug, *Switz.* **152** B4
Zugdidi, *Rep. of Georgia* **171** B16
Zugspitze, peak, *Aust., Ger.* **150** L8
Zula, *Eritrea* **210** H8
Zululand, *S. Af.* **212** K10
Žumberačka Gora, *Croatia, Slov.*
  154 D5
Zumberge Coast, *Antarctica* **238** G8
Zumbo, *Mozambique* **212** D9
Zungeru, *Nigeria* **206** K7
Zuni Pueblo, *N. Mex., U.S.* **109** S10
Zuni, river, *Ariz., N. Mex., U.S.* **109** S9
Zunyi, *China* **184** M11
Zuoz, *Switz.* **152** C5
Zürich, *Switz.* **152** A3
Żurrieq, *Malta* **163** L22
Zuru, *Nigeria* **206** J7
Žut, island, *Croatia* **154** G5
Zutphen, *Neth.* **148** J13
Zuwārah, *Lib.* **201** D16
Zuyevka, *Russ.* **158** F15
Zvenyhorodka, *Ukr.* **159** P6
Zvishavane, *Zimb.* **212** F9
Zvolen, *Slovakia* **151** J15
Zvornik, *Bosn. & Herzg.* **154** F9
Zwedru, *Liberia* **204** P11
Zwettl, *Aust.* **150** K12
Zwickau, *Ger.* **150** G9
Zwolle, *La., U.S.* **102** M6
Zwolle, *Neth.* **148** H13
Zygi, *Cyprus* **162** Q8
Zyryanka, *Russ.* **161** E19
Zyryanovo, *Russ.* **161** H13
Zyryanovsk, *Kaz.* **179** E17

MOON AND MARS INDEXES

LATIN EQUIVALENTS

Catena, catenae _____ chain of craters
Cavus, cavi _____ hollows, irregular steep-sided depressions, usually in arrays or clusters
Chaos, chaoses _____ distinctive area of broken terrain
Chasma, chasmata _____ deep elongated steep-sided depression
Collis, colles _____ small hills or knobs
Crater, craters _____ circular depression
Dorsum, dorsa _____ ridge
Fossa, fossae _____ long, narrow depression
Labyrinthus, labyrinthi _____ area of intersecting valleys or ridges

Lacus _____ lake; small plain; small, dark area with discrete, sharp edges
Mare, maria _____ sea; large, circular plain
Mensa, mensae _____ flat prominence with cliff-like edges
Mons, montes _____ mountain
Oceanus, oceani _____ very large, dark area
Palus, paludes _____ swamp; small plain
Patera, paterae _____ irregular crater, often with scalloped edges
Planitia, planitiae _____ low plain
Planum, plana _____ plateau or high plain
Promontorium, promontoria _____ cape; headland promontoria

Rima, rimae _____ fissure
Rupes, Rupii _____ scarp
Scopulus, scopuli _____ lobate or irregular scarp
Sinus _____ bay; small plain
Sulcus, sulci _____ subparallel furrows and ridges
Terra, terrae _____ extensive area
Tholus, tholi _____ small, dome-shaped mountain or hill
Unda, undae _____ dune
Vallis, valles _____ valley
Vastitas, vastitates _____ extensive plain

MOON INDEX

Note: Entries without a generic descriptor are craters.

Abbe, *Far Side,* **262** N12
Abel, *Near Side,* **261** M18
Abulfeda, *Near Side,* **261** K13
Abul Wáfa, *Far Side,* **262** H7
Adams, *Near Side,* **261** M17
Aestuum, Sinus, *Near Side,* **260** G12
Agricola, Montes, *Near Side,* **260** F8
Aitken, *Far Side,* **262** K12
Albategnius, *Near Side,* **261** K13
Al-Biruni, *Far Side,* **262** F6
Alden, *Far Side,* **262** L7
Alder, *Far Side,* **263** N13
Alekhin, *Far Side,* **263** Q14
Alexander, *Near Side,* **261** E14
Aliacensis, *Near Side,* **261** L13
Al-Khwarizmi, *Far Side,* **262** H7
Alpes, Montes, *Near Side,* **260** D12
Alpes, Vallis, *Near Side,* **261** D13
Alphonsus, *Near Side,* **260** K12
Altai, Rupes, *Near Side,* **261** L14
Alter, *Far Side,* **263** G18
Amici, *Far Side,* **263** J13
Amundsen, *Near Side,* **261** R13
Anaximander, *Near Side,* **260** C11
Anaximenes, *Near Side,* **260** B11
Anderson, *Far Side,* **262** G12
Anguis, Mare, *Near Side,* **261** F18
Ansgarius, *Near Side,* **261** K19
Antoniadi, *Far Side,* **263** P13
Anuchin, *Far Side,* **262** P8
Apenninus, Montes, *Near Side,* **260** G12
Apianus, *Near Side,* **261** L13
Apollo, *Far Side,* **263** M14
Apollonius, *Near Side,* **261** H17
Appleton, *Far Side,* **262** E11
Archimedes, *Near Side,* **260** F12
Ariadaeus, Rima, *Near Side,* **261** H14
Aristarchus, *Near Side,* **260** F9
Aristillus, *Near Side,* **261** F13
Aristoteles, *Near Side,* **261** D14
Arnold, *Near Side,* **261** C14
Arrhenius, *Far Side,* **263** P16
Artamonov, *Far Side,* **262** F7
Artem'ev, *Far Side,* **263** H15
Arzachel, *Near Side,* **260** K12
Ashbrook, *Far Side,* **263** Q13
Asperitatis, Sinus, *Near Side,* **261** J15
Atlas, *Near Side,* **261** D15
Australe, Mare, *Far Side,* **262** N7
Autumni, Lacus, *Near Side,* **260** K6
Avicenna, *Far Side,* **263** D18
Avogadro, *Far Side,* **262** C12

Baade, *Near Side,* **260** N8
Baade, Vallis, *Near Side,* **260** N8
Babbage, *Near Side,* **260** C10
Babcock, *Near Side,* **262** H6
Backlund, *Far Side,* **262** K7
Baco, *Near Side,* **261** N14
Baillaud, *Near Side,* **261** B14
Bailly, *Near Side,* **260** Q10
Balboa, *Near Side,* **260** F6
Baldet, *Far Side,* **263** N14
Balmer, *Near Side,* **261** L18
Barbier, *Far Side,* **262** L11
Barnard, *Near Side,* **261** M18
Barocius, *Near Side,* **261** M14
Barringer, *Far Side,* **263** L15
Barrow, *Near Side,* **261** C13
Beaumont, *Near Side,* **261** K15
Bečvář, *Far Side,* **262** J8
Behaim, *Near Side,* **261** K19
Beijerinck, *Far Side,* **262** K10
Bel'kovich, *Far Side,* **262** B9
Bel'kovich, *Near Side,* **261** B16
Bell, *Far Side,* **263** F19
Bellinsgauzen, *Far Side,* **263** P13
Belopol'skiy, *Far Side,* **263** K17
Belyaev, *Far Side,* **262** F9
Berkner, *Far Side,* **263** F18
Berlage, *Far Side,* **263** P13
Berosus, *Near Side,* **261** E17
Bettinus, *Near Side,* **260** P11
Bhabha, *Far Side,* **263** N13
Biela, *Near Side,* **261** P15
Birkeland, *Far Side,* **262** L12
Birkhoff, *Far Side,* **263** C14
Birmingham, *Near Side,* **260** C12
Blackett, *Far Side,* **263** M17
Blancanus, *Near Side,* **260** P11
Blanchinus, *Near Side,* **261** L13
Blazhko, *Far Side,* **263** F15
Boguslawsky, *Near Side,* **261** Q14
Bohr, *Near Side,* **260** G6
Bolyai, *Far Side,* **262** M9
Bonpland, *Near Side,* **260** J11
Boole, *Near Side,* **260** B10
Bose, *Far Side,* **263** N13
Boussingault, *Near Side,* **261** Q14
Bouvard, Vallis, *Near Side,* **260** N7
Boyle, *Far Side,* **262** N12
Bragg, *Far Side,* **263** D17
Brashear, *Far Side,* **263** Q13
Bredikhin, *Far Side,* **263** G14
Brenner, *Near Side,* **261** M15
Brianchon, *Near Side,* **260** B11
Bridgman, *Far Side,* **262** D10
Bronk, *Far Side,* **263** F16
Brouwer, *Far Side,* **263** M16
Buckland, Dorsum, *Near Side,* **261** G13
Buffon, *Far Side,* **263** M16

Buisson, *Far Side,* **262** J7
Bullialdus, *Near Side,* **260** K11
Bunsen, *Near Side,* **260** D7
Burckhardt, *Near Side,* **261** E17
Büsching, *Near Side,* **261** M14
Buys-Ballot, *Far Side,* **262** G12
Byrd, *Near Side,* **261** B13
Byrgius, *Near Side,* **260** L7

Cabannes, *Far Side,* **263** P13
Cajori, *Far Side,* **262** N12
Campbell, *Far Side,* **262** D11
Cannon, *Near Side,* **261** F19
Cantor, *Far Side,* **262** D8
Capuanus, *Near Side,* **260** M11
Cardanus, Rima, *Near Side,* **260** G7
Carnot, *Far Side,* **263** D15
Carpatus, Montes, *Near Side,* **260** G11
Carpenter, *Near Side,* **260** B11
Carver, *Far Side,* **262** N9
Cassegrain, *Far Side,* **262** P9
Cassini, *Near Side,* **261** E13
Catharina, *Near Side,* **261** K14
Caucasus, Montes, *Near Side,* **261** E13
Cavalerius, *Near Side,* **260** H7
Cavendish, *Near Side,* **260** L8
Chacornac, *Near Side,* **261** F15
Challis, *Near Side,* **261** B13
Chamberlin, *Far Side,* **262** P9
Champollion, *Far Side,* **262** E12
Chandler, *Far Side,* **262** E12
Chaplygin, *Far Side,* **262** J10
Chapman, *Far Side,* **263** C17
Chappell, *Far Side,* **263** D13
Charlier, *Far Side,* **263** E16
Chauvenet, *Far Side,* **262** K9
Chebyshev, *Far Side,* **263** M16
Chernyshev, *Far Side,* **262** D12
Chevallier, *Near Side,* **261** D16
Chrétien, *Far Side,* **262** N11
Clairaut, *Near Side,* **261** N13
Clavius, *Near Side,* **260** P12
Cleomedes, *Near Side,* **261** F17
Cleostratus, *Near Side,* **260** C9
Cockcroft, *Far Side,* **263** F14
Cognitum, Mare, *Near Side,* **260** J10
Colombo, *Near Side,* **261** K16
Compton, *Far Side,* **262** C9
Comrie, *Far Side,* **263** F18
Comstock, *Far Side,* **263** F17
Condorcet, *Near Side,* **261** G18
Congreve, *Far Side,* **263** J13
Copernicus, *Near Side,* **260** H11
Cordillera, Montes, **260** L7; **263** K18
Cori, *Far Side,* **263** N14

Coriolis, *Far Side,* **262** H12
Coulomb, *Far Side,* **263** C16
Cremona, **260** B10; **263** B15
Crisium, Mare, *Near Side,* **261** G17
Crocco, *Far Side,* **262** N11
Crommelin, *Far Side,* **263** Q14
Crüger, *Near Side,* **260** K7
Curie, *Far Side,* **262** L6
Curtius, *Near Side,* **261** P13
Cusanus, *Near Side,* **261** B15
Cuvier, *Near Side,* **261** N13
Cyrano, *Far Side,* **262** K11
Cyrillus, *Near Side,* **261** K14

D'Alembert, *Far Side,* **262** D11
Daedalus, *Far Side,* **262** J12
Dalton, *Near Side,* **260** G6
Danjon, *Far Side,* **262** K8
Dante, *Far Side,* **262** F12
Darwin, *Near Side,* **260** L7
Davisson, *Far Side,* **263** M13
Debye, *Far Side,* **263** D13
De Forest, *Far Side,* **263** Q13
Delambre, *Near Side,* **261** J14
De La Rue, *Near Side,* **261** C15
Dellinger, *Far Side,* **262** J9
Demonax, *Near Side,* **261** Q14
De Moraes, *Far Side,* **262** D10
Desargues, *Near Side,* **260** B10
Descartes, *Near Side,* **261** K14
De Sitter, *Near Side,* **261** B13
Deslandres, *Near Side,* **260** L12
Deutsch, *Far Side,* **262** F7
De Vries, *Far Side,* **263** K13
Doerfel, *Far Side,* **263** Q15
Donner, *Far Side,* **262** M7
Doppelmayer, *Near Side,* **260** L9
Doppler, *Far Side,* **263** K14
Dreyer, *Far Side,* **262** G6
Dryden, *Far Side,* **263** L14
Drygalski, *Near Side,* **260** R11
Dubyago, *Near Side,* **261** H18
Dunér, *Far Side,* **263** E13
Dyson, *Far Side,* **263** C15
Dziewulski, *Far Side,* **262** F6

Eddington, *Near Side,* **260** F7
Edison, *Far Side,* **262** F7
Eijkman, *Far Side,* **263** P14
Einstein, *Near Side,* **260** G6
Einthoven, *Far Side,* **262** J7
Elvey, *Far Side,* **263** G19
Emden, *Far Side,* **263** C13
Endymion, *Near Side,* **261** D15
Eötvös, *Far Side,* **262** M9
Epidemiarum, Palus, *Near Side,* **260** L10
Epigenes, *Near Side,* **260** C12
Eratosthenes, *Near Side,* **260** G11

Erro, *Far Side,* **262** H6
Esnault-Pelterie, *Far Side,* **263** D15
Espin, *Far Side,* **262** F7
Euctemon, *Near Side,* **261** B13
Eudoxus, *Near Side,* **261** E14
Euler, *Near Side,* **260** F10
Evans, *Far Side,* **263** J16
Evershed, *Far Side,* **263** E14
Excellentiae, Lacus, *Near Side,* **260** M9

Fabricius, *Near Side,* **261** M15
Fabry, *Far Side,* **262** D8
Faraday, *Near Side,* **261** M13
Fecunditatis, Mare, *Near Side,* **261** J17
Fermi, *Far Side,* **262** K8
Fernelius, *Near Side,* **261** M13
Fersman, *Far Side,* **263** G17
Finsen, *Far Side,* **263** M13
Firmicus, *Near Side,* **261** H18
Firsov, *Far Side,* **262** H7
Fitzgerald, *Far Side,* **263** F13
Fizeau, *Far Side,* **263** P15
Flammarion, *Near Side,* **260** J12
Flamsteed, *Near Side,* **260** J9
Fleming, *Far Side,* **262** G7
Florensky, *Far Side,* **262** F9
Florey, *Near Side,* **260** A12
Fourier, *Near Side,* **260** M9
Fowler, *Far Side,* **263** E15
Fracastorius, *Near Side,* **261** L15
Fra Mauro, *Near Side,* **260** J11
Franklin, *Near Side,* **261** E16
Fraunhofer, *Near Side,* **261** M16
Freundlich, *Far Side,* **262** F12
Fridman, *Far Side,* **263** K17
Frigoris, Mare, *Near Side,* **260** C12
Froelich, *Far Side,* **263** B14
Frost, *Far Side,* **263** E17
Furnerius, *Near Side,* **261** M17

Gadomski, *Far Side,* **263** E15
Gagarin, *Far Side,* **262** K10
Galois, *Far Side,* **263** K15
Galvani, *Near Side,* **260** C8
Gamow, *Far Side,* **262** C11
Ganswindt, *Far Side,* **262** Q12
Garavito, *Far Side,* **262** N11
Gärtner, *Near Side,* **261** C14
Gassendi, *Near Side,* **260** K9
Gauricus, *Near Side,* **260** M12
Gauss, *Near Side,* **261** E18
Gavrilov, *Far Side,* **262** G9
Geminus, *Near Side,* **261** E16
Gemma Frisius, *Near Side,* **261** L13
Gerard, *Near Side,* **260** D8
Gerasimovich, *Far Side,* **263** L16
Gibbs, *Near Side,* **261** L19

Gilbert, *Near Side,* **261** J19
Gill, *Near Side,* **261** Q15
Ginzel, *Far Side,* **262** G6
Goclenius, *Near Side,* **261** J16
Goddard, *Near Side,* **261** G19
Goldschmidt, *Near Side,* **260** B12
Green, *Far Side,* **262** H9
Gregory, *Far Side,* **262** H8
Grimaldi, *Near Side,* **260** J7
Grissom, *Far Side,* **263** N15
Gruemberger, *Near Side,* **260** P12
Gruithuisen, *Near Side,* **260** F10
Guericke, *Near Side,* **260** K11
Guillaume, *Far Side,* **263** E13
Gum, *Near Side,* **261** N18
Gutenberg, *Near Side,* **261** J16
Gutenberg, Rimae, *Near Side,* **261** J16
Guthnick, *Far Side,* **263** P17
Guyot, *Far Side,* **262** G7

H. G. Wells, *Far Side,* **262** D9
Haber, *Far Side,* **263** A13
Haemus, Montes, *Near Side,* **261** G13
Hagecius, *Near Side,* **261** P15
Hahn, *Near Side,* **261** E18
Hainzel, *Near Side,* **260** M10
Hale, *Far Side,* **262** Q11
Hamilton, *Near Side,* **261** N17
Hanno, *Near Side,* **261** P16
Harbinger, Montes, *Near Side,* **260** F9
Harker, Dorsa, *Near Side,* **261** G18
Harkhebi, *Far Side,* **262** D7
Harlan, *Near Side,* **261** N18
Harriot, *Far Side,* **262** E8
Hartmann, *Far Side,* **262** H9
Hartwig, *Near Side,* **260** J6
Harvey, *Far Side,* **263** G15
Hase, *Near Side,* **261** L17
Haskin, *Far Side,* **262** B11
Hausen, *Far Side,* **263** Q15
Hausen, *Near Side,* **260** Q10
Hayn, *Near Side,* **261** B15
Heaviside, *Far Side,* **262** J11
Hecataeus, *Near Side,* **261** L18
Helberg, *Far Side,* **263** F18
Helmholtz, *Near Side,* **261** Q15
Henyey, *Far Side,* **263** G15
Heraclitus, *Near Side,* **261** N13
Hercules, *Near Side,* **261** D15
Hermite, *Near Side,* **260** A12
Herodotus, *Near Side,* **260** F9
Hertz, *Far Side,* **262** G7
Hertzsprung, *Far Side,* **263** H17
Hesiodus, Rima, *Near Side,* **260** L11
Hess, *Far Side,* **262** N12
Hevelius, *Near Side,* **260** H7
Hilbert, *Far Side,* **262** L7
Hippalus, *Near Side,* **260** L10
Hipparchus, *Near Side,* **261** J13
Hippocrates, *Far Side,* **263** B14
Hirayama, *Far Side,* **262** J6
Hommel, *Near Side,* **261** N14
Hopmann, *Far Side,* **262** N11
Houzeau, *Far Side,* **263** K17
Hubble, *Near Side,* **261** F19
Humboldt, *Near Side,* **261** M18
Humboldtianum, Mare, *Near Side,* **261** C16
Humorum, Mare, *Near Side,* **260** L9
Hyginus, Rima, *Near Side,* **261** H13

Ibn Firnas, *Far Side,* **262** H8
Ibn Yunus, *Far Side,* **262** G6
Icarus, *Far Side,* **263** J13
Imbrium, Mare, *Near Side,* **260** F11
Ingenii, Mare, *Far Side,* **262** L11
Inghirami, *Near Side,* **260** N9
Inghirami, Vallis, *Near Side,* **260** N8
Insularum, Mare, *Near Side,* **260** H10
Ioffe, *Far Side,* **263** K17
Iridum, Sinus, *Near Side,* **260** D10
Isaev, *Far Side,* **262** K10

J. Herschel, *Near Side,* **260** C11
Jackson, *Far Side,* **263** G14
Jacobi, *Near Side,* **261** N13
Jansky, *Near Side,* **261** G19
Janssen, *Near Side,* **261** N15
Jarvis, *Far Side,* **263** M15
Jeans, *Far Side,* **262** P9
Jenner, *Far Side,* **262** N8
Joliot, *Far Side,* **262** E6
Joule, *Far Side,* **263** F15
Jules Verne, *Far Side,* **262** M10
Julius Caesar, *Near Side,* **261** H14

Kaiser, *Near Side,* **261** M13
Kamerlingh Onnes, *Far Side,* **263** G18
Kane, *Near Side,* **261** C14
Karpinskiy, *Far Side,* **262** B12
Karrer, *Far Side,* **263** N15
Kästner, *Near Side,* **261** J19
Keeler, *Far Side,* **262** J11
Kekulé, *Far Side,* **263** G16
Kepler, *Near Side,* **260** H9
Khvol'son, *Far Side,* **262** K7
Kibal'chich, *Far Side,* **263** H15
Kidinnu, *Far Side,* **262** E9
Kiess, *Near Side,* **261** J19
King, *Far Side,* **262** H8
Kircher, *Near Side,* **260** P11
Kirkwood, *Far Side,* **263** C13
Klaproth, *Near Side,* **260** P12
Kleymenov, *Far Side,* **263** L16
Klute, *Far Side,* **263** E15
Koch, *Far Side,* **262** M10
Kohlschütter, *Far Side,* **262** G10
Kolhörster, *Far Side,* **263** G18
Komarov, *Far Side,* **262** F10
Kondratyuk, *Far Side,* **262** K7
Konstantinov, *Far Side,* **262** G11
Korolev, *Far Side,* **263** J14
Kostinskiy, *Far Side,* **262** G7
Kovalevskaya, *Far Side,* **263** E16
Kozyrev, *Far Side,* **262** N9
Krafft, *Near Side,* **260** G7
Kramers, *Far Side,* **263** D15
Krasovskiy, *Far Side,* **263** H13
Kugler, *Far Side,* **262** P9
Kulik, *Far Side,* **263** E14
Kurchatov, *Far Side,* **262** E10
Kurchatov, Catena, *Far Side,* **262** E9

La Caille, *Near Side,* **261** L13
Lacchini, *Far Side,* **263** D17
Lade, *Near Side,* **261** J13
Lagrange, *Near Side,* **260** M7
Lamarck, *Near Side,* **260** L7
Lamb, *Far Side,* **262** N8
Lambert, *Near Side,* **260** F11
Lamé, *Near Side,* **261** K18
Lampland, *Far Side,* **262** L9
Landau, *Far Side,* **263** D16
Lane, *Far Side,* **262** J9
Langemak, *Far Side,* **262** J8
Langevin, *Far Side,* **262** E11
Langmuir, *Far Side,* **263** M16
Langrenus, *Near Side,* **261** K18
La Pérouse, *Near Side,* **261** K19
Larmor, *Far Side,* **263** F13
Laue, *Far Side,* **263** E18
Lavoisier, *Near Side,* **260** D7
Leavitt, *Far Side,* **263** M15
Lebedev, *Far Side,* **262** N8
Lebedinskiy, *Far Side,* **263** H14
Leeuwenhoek, *Far Side,* **263** L13
Legendre, *Near Side,* **261** M18
Le Gentil, *Near Side,* **260** Q11
Lehmann, *Near Side,* **260** M9
Leibnitz, *Far Side,* **262** M12
Lemaître, *Far Side,* **263** P14
Le Monnier, *Near Side,* **261** F15
Leucippus, *Far Side,* **263** F17
Leuschner, Catena, *Far Side,* **263** H18
Levi-Civita, *Far Side,* **262** L9
Ley, *Far Side,* **262** E11
Licetus, *Near Side,* **261** N13
Lilius, *Near Side,* **261** N13
Lindblad, *Far Side,* **263** B15
Lippmann, *Far Side,* **263** P16
Lipskiy, *Far Side,* **262** J12
Lister, Dorsa, *Near Side,* **261** G14
Littrow, *Near Side,* **261** G15
Lobachevskiy, *Far Side,* **262** G7
Lodygin, *Far Side,* **263** K15
Lomonosov, *Far Side,* **262** E7
Longomontanus, *Near Side,* **260** N11
Lorentz, *Far Side,* **263** E18
Love, *Far Side,* **262** J8
Lovelace, *Far Side,* **263** B13
Lowell, *Far Side,* **263** K19
Lucretius, *Far Side,* **263** J17
Lucretius, Catena, *Far Side,* **263** J17
Lundmark, *Far Side,* **262** M11
Lyapunov, *Near Side,* **261** E19
Lyman, *Far Side,* **262** P12
Lyot, *Near Side,* **261** P17

Mach, *Far Side,* **263** G15
Macrobius, *Near Side,* **261** G16
Maginus, *Near Side,* **260** N12
Mairan, *Near Side,* **260** E10
Maksutov, *Far Side,* **263** M13
Malapert, *Near Side,* **261** Q13
Mallet, *Near Side,* **261** N16
Mandel'shtam, *Far Side,* **262** H11
Manilius, *Near Side,* **261** G13
Manzinus, *Near Side,* **261** P13
Marconi, *Far Side,* **262** J10
Marginis, Mare, *Near Side,* **261** G19
Marinus, *Near Side,* **261** N17
Mariotte, *Far Side,* **263** L16
Maunder, *Far Side,* **263** K19
Maurolycus, *Near Side,* **261** M14
Maxwell, *Far Side,* **262** E7
McAuliffe, *Far Side,* **263** L15
McKellar, *Far Side,* **263** K13
McLaughlin, *Far Side,* **263** C17
McMath, *Far Side,* **263** G14
McNair, *Far Side,* **263** M15
Mechnikov, *Far Side,* **263** K15
Medii, Sinus, *Near Side,* **261** H13
Mee, *Near Side,* **260** M10
Meggers, *Far Side,* **262** F8
Meitner, *Far Side,* **262** K7
Mendel, *Far Side,* **263** N16
Mendeleev, *Far Side,* **262** H9
Mercurius, *Near Side,* **261** D16
Merrill, *Far Side,* **263** B14
Mersenius, *Near Side,* **260** L8
Meshcherskiy, *Far Side,* **262** G8
Messala, *Near Side,* **261** E17
Metius, *Near Side,* **261** M15
Meton, *Near Side,* **261** B13
Mezentsev, *Far Side,* **263** B14
Michelson, *Far Side,* **263** H17
Michelson, Catena, *Far Side,* **263** J18
Milankovič, *Far Side,* **262** B12
Miller, *Near Side,* **261** M13
Millikan, *Far Side,* **262** D9
Milne, *Far Side,* **262** M8
Mineur, *Far Side,* **263** F14
Minkowski, *Far Side,* **263** N14
Minnaert, *Far Side,* **262** P12
Mitra, *Far Side,* **263** G15
Mohorovičić, *Far Side,* **263** K14
Moiseev, *Far Side,* **262** G6
Montes Jura, *Near Side,* **260** D10
Montgolfier, *Far Side,* **263** D14
Moore, *Far Side,* **263** E13
Moretus, *Near Side,* **260** P12
Morse, *Far Side,* **263** G13
Mortis, Lacus, *Near Side,* **261** E14
Moscoviense, Mare, *Far Side,* **262** F10
Mouchez, *Near Side,* **260** B12
Murchison, *Near Side,* **261** H13
Mutus, *Near Side,* **261** P14

Nansen, *Far Side,* **262** A11
Nasmyth, *Near Side,* **260** N9
Nassau, *Far Side,* **262** L12
Nearch, *Near Side,* **261** P14
Nectaris, Mare, *Near Side,* **261** K15
Neison, *Near Side,* **261** C13
Neper, *Near Side,* **261** H19
Nernst, *Far Side,* **263** E18
Neujmin, *Far Side,* **262** L8
Neumayer, *Near Side,* **261** Q15
Newton, *Near Side,* **260** Q12
Niepce, *Far Side,* **263** B14
Nishina, *Far Side,* **263** N13
Nonius, *Near Side,* **261** M13
Nöther, *Far Side,* **263** B15
Nubium, Mare, *Near Side,* **260** K11
Numerov, *Far Side,* **263** P13
Nušl, *Far Side,* **262** F11

O'Day, *Far Side,* **262** L11
Oberth, *Far Side,* **262** C11
Obruchev, *Far Side,* **262** M11
Oenopides, *Near Side,* **260** C9
Ohm, *Far Side,* **263** G18
Oken, *Near Side,* **261** N17
Olbers, *Near Side,* **260** H6
Olcott, *Far Side,* **262** F8
Olivier, *Far Side,* **262** C10
Omar Khayyam, *Far Side,* **263** C16
Onizuka, *Far Side,* **263** M15
Oppel, Dorsum, *Near Side,* **261** G17
Oppenheimer, *Far Side,* **263** M13
Oresme, *Far Side,* **262** M12
Orientale, Mare, *Far Side,* **263** L19
Orlov, *Far Side,* **263** L13
Orontius, *Near Side,* **260** M12
Ostwald, *Far Side,* **262** H8

Paneth, *Far Side,* **263** B16
Pannekoek, *Far Side,* **262** J9
Papaleksi, *Far Side,* **262** H11
Paracelsus, *Far Side,* **262** L11
Paraskevopoulos, *Far Side,* **263** D14
Parenago, *Far Side,* **263** F18
Parkhurst, *Far Side,* **262** M7
Parrot, *Near Side,* **261** K13
Pascal, *Near Side,* **260** B11
Paschen, *Far Side,* **263** K16
Pasteur, *Far Side,* **262** K6
Pauli, *Far Side,* **262** N10
Pavlov, *Far Side,* **262** L10
Pawsey, *Far Side,* **262** D10
Peary, *Near Side,* **261** A13
Peirescius, *Near Side,* **261** N16
Pentland, *Near Side,* **261** P13
Perepelkin, *Far Side,* **262** J8
Perkin, *Far Side,* **263** D13
Perrine, *Far Side,* **263** D16
Petavius, *Near Side,* **261** L17
Petermann, *Near Side,* **261** B14
Petropavlovskiy, *Far Side,* **263** E17
Pettit, Rimae, *Far Side,* **263** M19
Petzval, *Far Side,* **263** P15
Phillips, *Near Side,* **261** L18
Philolaus, *Near Side,* **260** B12
Phocylides, *Near Side,* **260** P9
Piazzi, *Near Side,* **260** M8
Piccolomini, *Near Side,* **261** L15
Pingré, *Near Side,* **260** P9
Pirquet, *Far Side,* **262** K9
Pitatus, *Near Side,* **260** L11
Pitiscus, *Near Side,* **261** N14
Planck, *Far Side,* **262** P10
Planck, Vallis, *Far Side,* **262** P10
Plaskett, *Far Side,* **262** B12
Plato, *Near Side,* **260** D12
Plummer, *Far Side,* **263** L14
Plutarch, *Near Side,* **261** F18
Poczobutt, *Far Side,* **263** C16
Poincaré, *Far Side,* **262** N11
Poinsot, *Far Side,* **263** B13
Polzunov, *Far Side,* **262** F8
Poncelet, *Near Side,* **260** B11
Pontanus, *Near Side,* **261** L14
Pontécoulant, *Near Side,* **261** P16
Popov, *Far Side,* **262** G6
Porter, *Near Side,* **260** N12
Posidonius, *Near Side,* **261** F15
Poynting, *Far Side,* **263** G16
Prager, *Far Side,* **262** J8
Prandtl, *Far Side,* **262** P11
Procellarum, Oceanus, *Near Side,* **260** H8
Proctor, *Near Side,* **260** N12
Ptolemaeus, *Near Side,* **260** J12
Purbach, *Near Side,* **260** L12
Putredinis, Palus, *Near Side,* **261** F13
Pyrenaeus, Montes, *Near Side,* **261** K16
Pythagoras, *Near Side,* **260** C10

Quételet, *Far Side,* **263** E16

Rabbi Levi, *Near Side,* **261** M14
Racah, *Far Side,* **263** K13
Raimond, *Far Side,* **263** G14
Ramsay, *Far Side,* **262** M10
Rayleigh, *Near Side,* **261** E19
Razumov, *Far Side,* **263** E17
Réaumur, *Near Side,* **260** J12
Regiomontanus, *Near Side,* **260** L12
Reichenbach, *Near Side,* **261** L16
Reiner, *Near Side,* **260** H8
Reinhold, *Near Side,* **260** H10
Repsold, *Near Side,* **260** C9
Resnik, *Far Side,* **263** M15
Rheita, *Near Side,* **261** M16
Rheita, Vallis, *Near Side,* **261** M16
Riccioli, *Near Side,* **260** J6
Riccius, *Near Side,* **261** M14
Ricco, *Far Side,* **262** B12
Richardson, *Far Side,* **262** E7
Riemann, *Near Side,* **261** D18
Riphaeus, Montes, *Near Side,* **260** J10
Roberts, *Far Side,* **263** C13
Robertson, *Far Side,* **263** F18
Rocca, *Near Side,* **260** K7
Roche, *Far Side,* **262** M10
Röntgen, *Far Side,* **263** E18
Rook, Montes, **260** K6; **263** L18
Roris, Sinus, *Near Side,* **260** D10
Rosenberger, *Near Side,* **261** N15
Rosseland, *Far Side,* **262** M9
Rowland, *Far Side,* **263** D13
Rozhdestvenskiy, *Far Side,* **263** B13
Rumford, *Far Side,* **263** L13
Russell, *Near Side,* **260** F7
Rydberg, *Far Side,* **263** N17
Rynin, *Far Side,* **263** D17

Sacrobosco, *Near Side,* **261** K13
Saenger, *Far Side,* **262** H6
Saha, *Far Side,* **262** J6
Sanford, *Far Side,* **263** E15
Santbech, *Near Side,* **261** K16
Sarton, *Far Side,* **263** D16
Sasserides, *Near Side,* **260** M12
Saussure, *Near Side,* **260** M12
Scaliger, *Far Side,* **262** L7
Schaeberle, *Far Side,* **262** L8
Scheiner, *Near Side,* **260** P11
Schickard, *Near Side,* **260** N9
Schiller, *Near Side,* **260** N10
Schjellerup, *Far Side,* **262** C12
Schlesinger, *Far Side,* **263** D15
Schliemann, *Far Side,* **262** J10
Schlüter, *Near Side,* **260** J6
Schneller, *Far Side,* **263** E14
Schomberger, *Near Side,* **261** Q13
Schorr, *Near Side,* **261** L19
Schrödinger, *Far Side,* **262** Q11
Schrödinger, Vallis, *Far Side,* **262** Q10
Schröteri, Vallis, *Near Side,* **260** F9
Schubert, *Near Side,* **261** H19
Schumacher, *Near Side,* **261** D16
Schuster, *Far Side,* **262** H10
Schwarzschild, *Far Side,* **262** B11
Scobee, *Far Side,* **263** L15

Scoresby, *Near Side,* **261** B13
Scott, *Near Side,* **261** Q13
Seares, *Far Side,* **262** B11
Sechenov, *Far Side,* **263** J16
Segner, *Near Side,* **260** P10
Seidel, *Far Side,* **262** L10
Serenitatis, Mare, *Near Side,* **261** F14
Seyfert, *Far Side,* **262** F8
Sharonov, *Far Side,* **262** G12
Shayn, *Far Side,* **262** F12
Sheepshanks, Rima, *Near Side,* **261** D14
Shirakatsi, *Far Side,* **262** K8
Shternberg , *Far Side,* **263** F18
Siedentopf, *Far Side,* **262** F9
Sierpinski, *Far Side,* **262** L11
Sikorsky, *Far Side,* **262** Q10
Sirsalis, *Near Side,* **260** K8
Sirsalis, Rimae, *Near Side,* **260** K8
Sklodowska, *Far Side,* **262** L6
Slipher, *Far Side,* **262** D11
Smith, *Far Side,* **263** L15
Smoluchowski, *Far Side,* **263** B16
Smythii, Mare, *Near Side,* **261** H19
Snellius, *Near Side,* **261** L17
Snellius, Vallis, *Near Side,* **261** L17
Solitudinis, Lacus, *Far Side,* **262** M7
Sommerfeld, *Far Side,* **263** C13
Somni, Palus, *Near Side,* **261** G16

Somniorum, Lacus, *Near Side,* **261** E15
Sosigenes, Rimae, *Near Side,* **261** H14
South, *Near Side,* **260** C10
Spei, Lacus, *Near Side,* **261** D17
Spencer Jones, *Far Side,* **262** G11
Spumans, Mare, *Near Side,* **261** H18
St. John, *Far Side,* **262** G10
Stadius, *Near Side,* **260** H11
Stebbins, *Far Side,* **263** C14
Stefan, *Far Side,* **263** D17
Steinheil, *Near Side,* **261** N15
Sternfeld, *Far Side,* **263** K15
Stetson, *Far Side,* **263** N17
Stevinus, *Near Side,* **261** M16
Stöfler, *Near Side,* **261** M13
Störmer, *Far Side,* **262** C11
Strabo, *Near Side,* **261** C15
Stratton, *Far Side,* **262** J11
Strömgren, *Far Side,* **263** L16
Struve, *Near Side,* **260** F7
Subbotin, *Far Side,* **262** L9
Suess, Rima, *Near Side,* **260** H8
Sumner, Catena, *Far Side,* **262** D8
Sylvester, *Near Side,* **260** A12
Szilard, *Far Side,* **262** E7

Taruntius, *Near Side,* **261** H16
Taurus, Montes, *Near Side,* **261** F16

Teisserenc, *Far Side,* **263** E16
Temporis, Lacus, *Near Side,* **261** D16
Thebit, *Near Side,* **260** K12
Theophilus, *Near Side,* **261** J15
Thiessen, *Far Side,* **263** B13
Thomson, *Far Side,* **262** L11
Tikhov, *Far Side,* **262** C12
Timiryazev, *Far Side,* **263** J15
Timocharis, *Near Side,* **260** F12
Tiselius, *Far Side,* **262** H12
Titius, *Far Side,* **262** M7
Tranquillitatis, Mare, *Near Side,* **261** H15
Triesnecker, Rimae, *Near Side,* **261** H13
Trumpler, *Far Side,* **262** F12
Tsander, *Far Side,* **263** H15
Tseraskiy, *Far Side,* **262** N10
Tsiolkovskiy, *Far Side,* **262** K9
Tycho, *Near Side,* **260** M12

Ulugh Beigh, *Near Side,* **260** E7
Undarum, Mare, *Near Side,* **261** H18

Valier, *Far Side,* **262** H12
Van't Hoff, *Far Side,* **263** C15
Van de Graaff, *Far Side,* **262** L12
Van der Waals, *Far Side,* **262** N9
Van Maanen, *Far Side,* **262** E9

Vaporum, Mare, *Near Side,* **261** G13
Vasco da Gama, *Near Side,* **260** G6
Vavilov, *Far Side,* **263** J16
Vendelinus, *Near Side,* **261** K18
Vening Meinesz, *Far Side,* **262** J11
Ventris, *Far Side,* **262** L11
Veris, Lacus, *Near Side,* **260** K6
Vernadskiy, *Far Side,* **262** F9
Vertregt, *Far Side,* **262** K12
Vesalius, *Far Side,* **262** J7
Vestine, *Far Side,* **262** E7
Vetchinkin, *Far Side,* **262** H8
Vieta, *Near Side,* **260** L8
Vlacq, *Near Side,* **261** N15
Volta, *Near Side,* **260** C9
Volterra, *Far Side,* **262** C10
Von Békésy, *Far Side,* **262** D9
Von Braun, *Near Side,* **260** D8
Von der Pahlen, *Far Side,* **263** L16
Von Kármán, *Far Side,* **262** M12
Von Neumann, *Far Side,* **262** E11
Von Zeipel, *Far Side,* **263** E15

W. Bond, *Near Side,* **261** C13
Walther, *Near Side,* **261** L13
Wan-Hoo, *Far Side,* **263** J16
Wargentin, *Near Side,* **260** N9
Waterman, *Far Side,* **262** L8
Watson, *Far Side,* **263** P15
Watt, *Near Side,* **261** N15

Wegener, *Far Side,* **263** D17
Werner, *Near Side,* **261** L13
Weyl, *Far Side,* **263** G17
Wiener, *Far Side,* **262** E10
Wilhelm, *Near Side,* **260** M11
Wilkins, *Near Side,* **261** L14
Wilsing, *Far Side,* **263** K14
Wilson, *Near Side,* **260** Q11
Winlock, *Far Side,* **263** E17
Wood, *Far Side,* **263** D16
Wrottesley, *Near Side,* **261** L17
Wurzelbauer, *Near Side,* **260** M11
Wyld, *Far Side,* **262** J6

Xenophanes, *Near Side,* **260** C9

Yablochkov, *Far Side,* **262** C10
Yamamoto, *Far Side,* **262** D12
Young, *Near Side,* **261** M16

Zach, *Near Side,* **261** P13
Zagut, *Near Side,* **261** L14
Zeeman, *Far Side,* **263** Q14
Zelinskiy, *Far Side,* **262** L11
Zeno, *Near Side,* **261** D17
Zhukovskiy, *Far Side,* **263** H14
Zsigmondy, *Far Side,* **263** C16
Zucchius, *Near Side,* **260** P10
Zwicky, *Far Side,* **262** K11

MOON LANDING SITES

NEAR SIDE

Apollo 11 (U.S.)—Tranquillity Base Landed July 20, 1969 **H15**
Apollo 12 (U.S.)—Landed Nov. 19, 1969 **J10**
Apollo 14 (U.S.)—Landed Feb. 5, 1971 **J11**
Apollo 15 (U.S.)—Landed July 30, 1971 **F13**
Apollo 16 (U.S.)—Landed April 21, 1972 **J14**
Apollo 17 (U.S.)—Landed Dec. 11, 1972 **G15**
Beresheet (Israel)—Crashed April 11, 2019 **F14**
Chandrayaan-1 Moon Impact Probe(India)—
   Crashed Nov. 14, 2008 **R13**
Chandrayaan-2 Vikram lander (India)—
   Crashed Sept. 7, 2019 **P13**
Chang'e 3 (China)—Landed Dec. 14, 2013 **E11**
GRAIL A (Ebb) (U.S.)—Crashed Dec. 17, 2012 **B12**
GRAIL B (Flow) (U.S.)—Crashed Dec. 17, 2012 **B12**
Hiten (Japan)—Crashed April 10, 1993 **D11**
LCROSS, Centaur Impactor (U.S.)—
   Crashed Oct. 9, 2009 **Q12**

Luna 13 (U.S.S.R.)—Landed Dec. 24, 1966 **G8**
Luna 15 (U.S.S.R.)—Crashed July 21, 1969 **G17**
Luna 16 (U.S.S.R.)—Landed Sept. 20, 1970 **J17**
Luna 17 (U.S.S.R.)—Landed Nov. 17, 1970 **E10**
Luna 2 (U.S.S.R.)—Crashed Sept. 14, 1959 **F13**
Luna 5 (U.S.S.R.)—Crashed May 10, 1965 **H11**
Luna 7 (U.S.S.R.)—Crashed Oct. 7, 1965 **H8**
Luna 8 (U.S.S.R.)—Crashed Dec. 6, 1965 **H7**
Luna 9 (U.S.S.R.)—Crashed Feb. 3, 1966 **H7**
Luna 18 (U.S.S.R.)—Landed Sept. 11, 1971 **H17**
Luna 20 (U.S.S.R.)—Landed Feb. 21, 1972 **H17**
Luna 21 (U.S.S.R.)—Landed Jan. 15, 1973 **F15**
Luna 23 (U.S.S.R.)—Landed Nov. 6, 1974 **G18**
Luna 24 (U.S.S.R.)—Landed Aug. 18, 1976 **G18**
Lunar Prospector (U.S.)—Crashed July 31, 1999 **R12**
Ranger 6 (U.S.)—Crashed Feb. 2, 1964 **H14**
Ranger 7 (U.S.)—Crashed July 31, 1964 **J11**
Ranger 8 (U.S.)—Crashed Feb. 20, 1965 **H15**

Ranger 9 (U.S.)—Crashed Mar. 24, 1965 **K12**
Selene/Kaguya (Japan)—Crashed June 10, 2009 **Q15**
SMART-1 (ESA)—Crashed Sept. 3, 2006 **M9**
Surveyor 1 (U.S.)—Landed June 2, 1966 **J9**
Surveyor 2 (U.S.)—Crashed Sept. 22, 1966 **J12**
Surveyor 3 (U.S.)—Landed Apr. 20, 1967 **J10**
Surveyor 4 (U.S.)—Crashed July 17, 1967 **H12**
Surveyor 5 (U.S.)—Landed Sept. 11, 1967 **H14**
Surveyor 6 (U.S.)—Landed Nov. 10, 1967 **H12**
Surveyor 7 (U.S.)—Landed Jan. 10, 1968 **M12**

FAR SIDE

Chang'e 4 (China)—Landed Jan. 3, 2019 **M12**
Lunar Orbiter 1 (U.S.)—Crashed Oct. 29, 1966 **H11**
Lunar Orbiter 2 (U.S.)—Crashed Oct. 11, 1967 **H7**
Lunar Orbiter 3 (U.S.)—Crashed Oct. 9, 1967 **G19**
Lunar Orbiter 5 (U.S.)—Crashed Jan. 31, 1968 **J16**
Okina, Selene Orbiter (Japan)—Crashed Feb. 12, 2009 **F14**
Ranger 4 (U.S.)—Crashed Apr. 26, 1962 **K16**

MARS INDEX

Note: Entries without a generic descriptor are craters.

Acheron Fossae, *W. Hemis.,* **266** E9
Acidalia Planitia, *W. Hemis.,* **267** D16
Adams, *E. Hemis.,* **269** E18
Aeolis Dorsa, *E. Hemis.,* **269** J17
Aeolis Mensae, *E. Hemis.,* **269** J17
Aeolis Planum, *E. Hemis.,* **269** J17
Aeolis Serpens, *E. Hemis.,* **269** J17
Aganippe Fossa, *W. Hemis.,* **266** J9
Agassiz, *W. Hemis.,* **266** P12
Al-Qahira Vallis, *E. Hemis.,* **269** L18
Alba Fossae, *W. Hemis.,* **266** E11
Alba Mons, *W. Hemis.,* **266** E11
Alba Patera, *W. Hemis.,* **266** E11
Albor Tholus, *E. Hemis.,* **269** G17
Alpheus Colles, *E. Hemis.,* **268** M10
Amazonis Mensae, *W. Hemis.,* **266** J8
Amazonis Planitia, *W. Hemis.,* **266** F7
Amenthes Cavi, *E. Hemis.,* **269** G14
Amenthes Fossae, *E. Hemis.,* **269** H13
Amenthes Planum, *E. Hemis.,* **269** H14

Amphitrites Patera, *E. Hemis.,* **268** P11
Antoniadi, *E. Hemis.,* **268** G10
Aonia Planum, *W. Hemis.,* **267** N13
Aonia Terra, *W. Hemis.,* **266** P10
Apollinaris Mons, *E. Hemis.,* **269** J19
Arabia Terra, **267** G19; **268** E8
Arago, *E. Hemis.,* **268** H7
Arcadia Planitia, *W. Hemis.,* **266** C8
Arena Colles, *E. Hemis.,* **268** F12
Ares Vallis, *W. Hemis.,* **267** G18
Argentea Planum, *W. Hemis.,* **267** P13
Argyre Planitia, *W. Hemis.,* **267** N15
Arima, *W. Hemis.,* **267** K15
Arkhangelsky, *W. Hemis.,* **267** N17
Arrhenius, *E. Hemis.,* **269** M15
Arsia Mons, *W. Hemis.,* **266** J10
Ascraeus Mons, *W. Hemis.,* **266** G11
Ascuris Planum, *W. Hemis.,* **267** E13
Astapus Colles, *E. Hemis.,* **268** E12
Aurorae Chaos, *W. Hemis.,* **267** J17
Aurorae Planum, *W. Hemis.,* **267** K16

Ausonia Montes, *E. Hemis.,* **269** L13
Australe, Planum, *E. Hemis.,* **268** R12
Avarua, *E. Hemis.,* **269** M14

Bakhuysen, *E. Hemis.,* **268** L6
Baldet, *E. Hemis.,* **268** F10
Barabashov, *W. Hemis.,* **267** D14
Barnard, *E. Hemis.,* **268** P11
Bashkaus Valles, *W. Hemis.,* **267** M19
Becquerel, *W. Hemis.,* **267** F19
Beer, *W. Hemis.,* **267** K19
Bernard, *W. Hemis.,* **266** L8
Biblis Tholus, *W. Hemis.,* **266** H10
Bjerknes, *E. Hemis.,* **269** N17
Boeddicker, *E. Hemis.,* **269** K18
Bond, *W. Hemis.,* **267** M16
Boreum, Planum, **266** A12; **268** B12
Bosporos Planum, *W. Hemis.,* **267** M14
Bosporos Rupes, *W. Hemis.,* **267** M15
Bouguer, *E. Hemis.,* **268** L7
Brashear, *W. Hemis.,* **266** N11
Brevia, Dorsa, *E. Hemis.,* **268** P11

Briault, *E. Hemis.,* **269** J13
Burroughs, *E. Hemis.,* **269** Q13
Burton, *W. Hemis.,* **266** K7
Byrd, *E. Hemis.,* **269** P14

Campbell, *E. Hemis.,* **269** P16
Candor Chasma, *W. Hemis.,* **267** J14
Cankuzo, *E. Hemis.,* **268** K9
Capen, *E. Hemis.,* **268** H6
Capri Chasma, *W. Hemis.,* **267** K16
Caralis Chaos, *E. Hemis.,* **269** N18
Cassini, *E. Hemis.,* **268** F8
Ceraunius Fossae, *W. Hemis.,* **266** F11
Ceraunius Tholus, *W. Hemis.,* **266** F12
Cerberus Dorsa, *E. Hemis.,* **269** K14
Cerberus Tholi, *E. Hemis.,* **269** H18
Cerulli, *E. Hemis.,* **268** E8
Chamberlin, *W. Hemis.,* **266** P11
Charitum Montes, *W. Hemis.,* **267** P15
Charlier, *W. Hemis.,* **266** Q10
Chronium, Planum, *E. Hemis.,* **269** P15
Chryse Chaos, *W. Hemis.,* **267** G17

Chryse Planitia, *W. Hemis.,* **267** E17
Cimmeria, Terra, *E. Hemis.,* **269** N16
Claritas Fossae, *W. Hemis.,* **266** L11
Claritas Rupes, *W. Hemis.,* **266** L11
Clark, *W. Hemis.,* **266** N10
Coblentz, *W. Hemis.,* **266** N12
Cobres, *W. Hemis.,* **266** K7
Coloe Fossae, *E. Hemis.,* **268** E10
Columbus, *W. Hemis.,* **266** M7
Comas Sola, *W. Hemis.,* **266** L7
Copernicus, *W. Hemis.,* **266** P8
Coprates Chasma, *W. Hemis.,* **267** K15
Coracis Fossae, *W. Hemis.,* **267** M13
Crommelin, *W. Hemis.,* **267** H19
Cruls, *W. Hemis.,* **269** N17
Curie, *W. Hemis.,* **267** E18
Cyane Fossae, *W. Hemis.,* **266** E10
Cydnus Rupes, *E. Hemis.,* **269** D14
Cydonia Mensae, *W. Hemis.,* **267** E18

Daedalia Planum, *W. Hemis.,* **266** K9
Dao Vallis, *E. Hemis.,* **268** M12
Darwin, *W. Hemis.,* **267** P16

Dawes, *E. Hemis.,* **268** J8
Dejnev, *W. Hemis.,* **266** L7
Denning, *E. Hemis.,* **268** K8
Deuteronilus Mensae, *E. Hemis.,* **268** D8
Dokuchaev, *W. Hemis.,* **266** PII
Dollfus, *W. Hemis.,* **267** LI9
Douglass, *W. Hemis.,* **267** NI4
Du Martheray, *E. Hemis.,* **269** HI3

Echus Montes, *W. Hemis.,* **267** HI3
Eddie, *E. Hemis.,* **269** GI7
Eden Patera, *W. Hemis.,* **267** EI8
Elysium Chasma, *E. Hemis.,* **269** FI6
Elysium Mons, *E. Hemis.,* **269** FI7
Elysium Planitia, *E. Hemis.,* **269** HI7
Eos Chasma, *W. Hemis.,* **267** KI6
Erebus Montes, *W. Hemis.,* **266** E7
Eridania Planitia, *E. Hemis.,* **269** MI5
Eridania Scopulus, *E. Hemis.,* **269** NI6
Eumenides Dorsum, *W. Hemis.,* **266** H7

Fesenkov, *W. Hemis.,* **267** GI3
Firsoff, *W. Hemis.,* **267** HI9
Flammarion, *E. Hemis.,* **268** F9
Flaugergues, *E. Hemis.,* **268** K7
Fontana, *W. Hemis.,* **267** PI3
Fournier, *E. Hemis.,* **268** JII

Galaxias Colles, *E. Hemis.,* **269** EI6
Gale, *E. Hemis.,* **269** JI6
Galilaei, *W. Hemis.,* **267** HI8
Galle, *W. Hemis.,* **267** NI6
Ganges Chasma, *W. Hemis.,* **267** JI6
Gilbert, *E. Hemis.,* **268** PI2
Gill, *E. Hemis.,* **268** G6
Gonnus Mons, *W. Hemis.,* **266** EI2
Gordii Dorsum, *W. Hemis.,* **266** H8
Gorgonum Chaos, *W. Hemis.,* **266** N7
Graff, *E. Hemis.,* **269** LI7
Granicus Valles, *E. Hemis.,* **269** FI5
Green, *W. Hemis.,* **267** PI7
Greg, *E. Hemis.,* **269** MI4
Gusev, *E. Hemis.,* **269** KI9

Hadley, *E. Hemis.,* **269** KI8
Hadriacus Palus, *E. Hemis.,* **268** LI2
Hale, *W. Hemis.,* **267** MI6
Harris, *E. Hemis.,* **268** LII
Hartwig, *W. Hemis.,* **267** NI7
Havel Vallis, *W. Hemis.,* **267** HI5
Hecates Tholus, *E. Hemis.,* **269** EI7
Hellas Chaos, *E. Hemis.,* **268** NII
Hellas Planitia, *E. Hemis.,* **268** MII
Hellespontus Montes, *E. Hemis.,* **268** M9
Helmholtz, *W. Hemis.,* **267** NI7
Henry, *E. Hemis.,* **268** G7
Hephaestus Rupes, *E. Hemis.,* **269** FI4
Herschel, *E. Hemis.,* **269** KI6
Hesperia Planum, *E. Hemis.,* **269** LI4
Hibes Montes, *E. Hemis.,* **269** HI9

Holden, *W. Hemis.,* **267** LI7
Holmes, *E. Hemis.,* **268** QI2
Hooke, *W. Hemis.,* **267** NI5
Hrad Vallis, *E. Hemis.,* **269** EI6
Hussey, *W. Hemis.,* **266** NIO
Huxley, *E. Hemis.,* **269** PI3
Huygens, *E. Hemis.,* **268** KIO
Hyblaeus Dorsa, *E. Hemis.,* **269** HI5

Icaria Planum, *W. Hemis.,* **266** MII
Isidis Planitia, *E. Hemis.,* **268** GI2
Ismenia Patera, *E. Hemis.,* **268** D7
Ius Chasma, *W. Hemis.,* **267** JI3

Janssen, *E. Hemis.,* **268** H8
Jarry-Desloges, *E. Hemis.,* **268** JI2
Jovis Tholus, *W. Hemis.,* **266** GIO
Juventae Dorsa, *W. Hemis.,* **267** JI4

Kaiser, *E. Hemis.,* **268** N8
Kasei Valles, *W. Hemis.,* **267** FI4
Kepler, *W. Hemis.,* **269** NI6
Knobel, *E. Hemis.,* **269** JI6
Korolev, *E. Hemis.,* **269** BI4
Koval'sky, *W. Hemis.,* **266** L9
Krishtofovich, *E. Hemis.,* **269** MI3
Kunowsky, *W. Hemis.,* **267** CI6

Labeatis Fossae, *W. Hemis.,* **267** GI3
Lasswitz, *E. Hemis.,* **269** JI7
Lau, *W. Hemis.,* **266** QI2
Le Verrier, *E. Hemis.,* **268** M8
Liais, *E. Hemis.,* **269** QI3
Libya Montes, *E. Hemis.,* **268** HI2
Li Fan, *W. Hemis.,* **266** N9
Liu Hsin, *W. Hemis.,* **266** P9
Lockyer, *E. Hemis.,* **269** FI8
Lohse, *W. Hemis.,* **267** NI7
Loire Valles, *W. Hemis.,* **267** KI8
Lomonosov, *W. Hemis.,* **267** BI5
López, *E. Hemis.,* **269** KI3
Louros Valles, *W. Hemis.,* **267** JI3
Lowell, *W. Hemis.,* **267** NI3
Lucus Planum, *E. Hemis.,* **269** KI9
Lunae Planum, *W. Hemis.,* **267** GI4
Luzin, *E. Hemis.,* **268** F8
Lycus Sulci, *W. Hemis.,* **266** F8
Lyell, *W. Hemis.,* **267** QI5
Lyot, *E. Hemis.,* **268** D9

Ma'Adim Vallis, *E. Hemis.,* **269** LI9
Mädler, *E. Hemis.,* **268** K6
Mad Vallis, *E. Hemis.,* **268** PI2
Magelhaens, *W. Hemis.,* **266** M7
Maggini, *E. Hemis.,* **268** E7
Maja Valles, *W. Hemis.,* **267** HI5
Malea Planum, *E. Hemis.,* **268** PII
Mamers Valles, *E. Hemis.,* **268** D8
Mangala Fossa, *W. Hemis.,* **266** K8
Mangala Valles, *W. Hemis.,* **266** J7
Maraldi, *W. Hemis.,* **267** PI5
Marca, *W. Hemis.,* **266** K7
Mareotis Fossae, *W. Hemis.,* **267** EI3
Margaritifer Terra, *W. Hemis.,* **267** JI8
Marikh Vallis, *E. Hemis.,* **268** L6
Mariner, *W. Hemis.,* **266** M7

Marineris, Valles, *W. Hemis.,* **267** KI5
Marth, *W. Hemis.,* **267** GI9
Martz, *E. Hemis.,* **269** MI6
Masursky, *W. Hemis.,* **267** GI7
Maunder, *E. Hemis.,* **268** P8
Mawrth Vallis, *W. Hemis.,* **267** FI8
Melas Chasma, *W. Hemis.,* **267** JI4
Melas Fossae, *W. Hemis.,* **267** LI4
Mellish, *W. Hemis.,* **267** QI4
Memnonia Fossae, *W. Hemis.,* **266** L7
Meridiani Planum, **267** JI9; **268** J6
Micoud, *E. Hemis.,* **268** C9
Mie, *E. Hemis.,* **269** DI5
Milankovič, *W. Hemis.,* **266** CIO
Millman, *W. Hemis.,* **266** P9
Millochau, *E. Hemis.,* **268** KI2
Mitchel, *E. Hemis.,* **268** PI2
Molesworth, *E. Hemis.,* **269** LI7
Moreux, *E. Hemis.,* **268** E9
Müller, *E. Hemis.,* **269** LI5
Mutch, *W. Hemis.,* **267** JI5

Nectaris Fossae, *W. Hemis.,* **267** LI5
Nepenthes Mensae, *E. Hemis.,* **269** HI4
Nereidum Montes, *W. Hemis.,* **267** MI5
Newcomb, *E. Hemis.,* **268** L6
Newton, *W. Hemis.,* **266** N8
Nicholson, *W. Hemis.,* **266** H6
Niesten, *E. Hemis.,* **268** LIO
Nili, Colles, *E. Hemis.,* **268** EII
Nili Fossae, *E. Hemis.,* **268** GII
Nilokeras Scopulus, *W. Hemis.,* **267** EI5
Nilosyrtis Mensae, *E. Hemis.,* **268** FII
Noachis Terra, **267** NI7; **268** N8
Noctis Fossae, *W. Hemis.,* **266** JI2
Noctis Labyrinthus, *W. Hemis.,* **266** JII
Nordenskiöld, *W. Hemis.,* **266** P9

Oenotria Plana, *E. Hemis.,* **268** JII
Oenotria Scopulus, *E. Hemis.,* **268** JII
Okavango Valles, *E. Hemis.,* **268** E7
Olympia Undae, **266** BII; **269** BI3
Olympica Fossae, *W. Hemis.,* **266** FII
Olympus Mons, *W. Hemis.,* **266** G9
Ophir Planum, *W. Hemis.,* **267** JI5
Orcus Patera, *E. Hemis.,* **269** GI9
Orson Welles, *W. Hemis.,* **267** HI6
Oudemans, *W. Hemis.,* **266** KI2
Oxia Colles, *W. Hemis.,* **267** FI7
Oxia Planum, *W. Hemis.,* **267** GI8
Oyama, *W. Hemis.,* **267** FI7

P'Al, *E. Hemis.,* **269** LI4
Pallacopas Vallis, *W. Hemis.,* **267** PI6
Panchaia Rupes, *E. Hemis.,* **269** CI4
Parva Planum, *W. Hemis.,* **266** QI2
Pasteur, *E. Hemis.,* **268** G8
Pavonis Mons, *W. Hemis.,* **266** HIO
Perepelkin, *W. Hemis.,* **267** DI4

Peridier, *E. Hemis.,* **268** FI2
Pettit, *W. Hemis.,* **266** G6
Phillips, *W. Hemis.,* **267** PI4
Phison Patera, *E. Hemis.,* **269** F9
Phlegra Dorsa, *E. Hemis.,* **269** FI8
Phlegra Montes, *E. Hemis.,* **269** DI7
Pickering, *W. Hemis.,* **266** M9
Pityusa Patera, *E. Hemis.,* **268** QIO
Pollack, *E. Hemis.,* **268** J7
Porter, *W. Hemis.,* **266** NII
Proctor, *E. Hemis.,* **268** N9
Promethei Planum, *E. Hemis.,* **268** QI2
Promethei Rupes, *E. Hemis.,* **268** QI2
Promethei Terra, *E. Hemis.,* **269** NI4
Protonilus Mensae, *E. Hemis.,* **268** DIO
Ptolemaeus, *W. Hemis.,* **266** N9

Quenisset, *E. Hemis.,* **268** E9

Rabe, *E. Hemis.,* **268** N9
Radau, *W. Hemis.,* **267** GI9
Rayleigh, *E. Hemis.,* **269** QI3
Renaudot, *E. Hemis.,* **268** EII
Reull Vallis, *E. Hemis.,* **269** MI3
Reynolds, *W. Hemis.,* **266** QII
Robert Sharp, *E. Hemis.,* **269** JI6
Roddenberry, *W. Hemis.,* **267** PI7
Roddy, *W. Hemis.,* **267** LI6
Ross, *W. Hemis.,* **266** NI2
Rudaux, *E. Hemis.,* **268** EIO
Russell, *E. Hemis.,* **268** P9
Rutherford, *W. Hemis.,* **267** FI9

Sabaea, Terra *E. Hemis.,* **268** J8
Sacra Mensa, *W. Hemis.,* **267** FI4
Sagan, *W. Hemis.,* **267** GI7
Saheki, *E. Hemis.,* **268** KII
Samara Valles, *W. Hemis.,* **267** LI8
Savich, *E. Hemis.,* **269** LI3
Scandia Colles, *W. Hemis.,* **266** CIO
Scandia Tholi, *W. Hemis.,* **266** BII
Schaeberle, *E. Hemis.,* **268** L9
Schiaparelli, *E. Hemis.,* **268** J6
Schmidt, *W. Hemis.,* **267** QI3
Schöner, *E. Hemis.,* **268** G9
Schroeter, *E. Hemis.,* **268** J9
Secchi, *E. Hemis.,* **269** NI3
Shalbatana Vallis, *W. Hemis.,* **267** HI7
Sharonov, *W. Hemis.,* **267** FI5
Simud Valles, *W. Hemis.,* **267** GI7
Sinai Dorsa, *W. Hemis.,* **267** KI3
Sinai Fossae, *W. Hemis.,* **267** KI3
Sinai Planum, *W. Hemis.,* **266** KI2
Sirenum, Terra, *W. Hemis.,* **266** M8
Sisyphi Planum, *E. Hemis.,* **268** QIO
Sklodowska, *W. Hemis.,* **267** EI8
Slipher, *W. Hemis.,* **267** NI3
Solis Planum, *W. Hemis.,* **266** LI2
South, *E. Hemis.,* **268** QII
Steno, *W. Hemis.,* **266** PI2
Stokes, *E. Hemis.,* **269** CI6
Stoney, *W. Hemis.,* **266** QII
Surius Vallis, *W. Hemis.,* **267** PI4
Syria Planum, *W. Hemis.,* **266** KII

Syrtis Major Planum, *E. Hemis.,* **268** HII

Tanaica Montes, *W. Hemis.,* **266** EI2
Tantalus Fossae, *W. Hemis.,* **266** EI2
Tartarus Colles, *E. Hemis.,* **269** GI9
Tartarus Montes, *E. Hemis.,* **269** FI9
Teisserenc de Bort, *E. Hemis.,* **268** H9
Tempe Fossae, *W. Hemis.,* **267** EI4
Tempe Terra, *W. Hemis.,* **267** EI4
Terby, *E. Hemis.,* **268** LII
Tharsis Montes, *W. Hemis.,* **266** JIO
Tharsis Tholus, *W. Hemis.,* **266** GI2
Thaumasia Fossae, *W. Hemis.,* **266** NI2
Thaumasia Planum, *W. Hemis.,* **267** KI4
Thyles Rupes, *E. Hemis.,* **269** QI4
Tikhonravov, *E. Hemis.,* **268** G8
Tikhov, *E. Hemis.,* **269** NI3
Tithoniae Fossae, *W. Hemis.,* **267** JI3
Tithonium Chasma, *W. Hemis.,* **267** JI3
Tiu Valles, *W. Hemis.,* **267** GI7
Tombaugh, *E. Hemis.,* **269** HI8
Tractus Catena, *W. Hemis.,* **266** FII
Trouvelot, *W. Hemis.,* **267** GI8
Tyrrhena Terra, *E. Hemis.,* **268** KI2

Ultima Lingula, *E. Hemis.,* **269** QI4
Ultimi Scopuli, *E. Hemis.,* **269** QI4
Ulysses Fossae, *W. Hemis.,* **266** HIO
Ulysses Tholus, *W. Hemis.,* **266** HIO
Uranius Mons, *W. Hemis.,* **266** FI2
Uranius Patera, *W. Hemis.,* **266** FI2
Uranius Tholus, *W. Hemis.,* **266** FI2
Utopia Planitia, *E. Hemis.,* **269** EI5

Vastitas Borealis, **266** CI2; **268** CI3
Very, *W. Hemis.,* **266** P8
Vinogradov, *W. Hemis.,* **267** KI7
Vogel, *W. Hemis.,* **267** MI8

Waikato Vallis, *E. Hemis.,* **269** LI4
Wallace, *E. Hemis.,* **269** NI4
Wells, *E. Hemis.,* **269** PI4
Wien, *E. Hemis.,* **269** KI7
Williams, *W. Hemis.,* **266** K7
Wirtz, *W. Hemis.,* **267** NI6
Wislicenus, *E. Hemis.,* **268** L6
Wright, *W. Hemis.,* **266** PIO

Xanthe Montes, *W. Hemis.,* **267** GI5
Xanthe Terra, *W. Hemis.,* **267** HI6

Zarqa Valles, *E. Hemis.,* **268** JI2
Zephyria Planum, *E. Hemis.,* **269** JI8

MARS LANDING SITES

WESTERN HEMISPHERE

ExoMars Schiaparelli (ESA)—Crashed Oct. 19, 2016 **J19**
Mars 3 (U.S.S.R.)—Landed, contact lost Dec. 2, 1971 **N8**
Mars 6 (U.S.S.R.)—Crashed Mar. 12, 1974 **L18**
Mars Pathfinder (U.S.)—Landed July 4, 1997 **G17**
Opportunity (U.S.)—Landed Jan. 25, 2004 **J19**
Phoenix (U.S.)—Landed May 25, 2008 **C11**
Viking I (U.S.)—Landed July 20, 1976 **F16**

EASTERN HEMISPHERE

Beagle 2 (U.K.)—Landed, contact lost Dec. 25, 2003 **H12**
Deep Space 2 Probes (U.S.)—Crashed Dec. 3, 1999 **Q14**
Mars InSight (U.S)—Landed Nov. 26, 2018 **H16**
Mars 2 (U.S.S.R.)—Crashed Nov. 27, 1971 **N10**
Mars Polar Lander (U.S.)—Crashed Dec. 3, 1999 **Q14**
Mars Science Laboratory (Curiosity) (U.S.)—
  Landed Aug. 6, 2012 **J16**
Spirit (U.S.)—Landed Jan. 4, 2004 **K19**
Viking 2 (U.S.)—Landed Sept. 3, 1976 **D15**

CREDITS

WORLD THEMATICS

STRUCTURE OF THE EARTH
pp. 22–23

DATA SOURCES

Tectonics (main map): U.S. Geological Survey (USGS) Earthquake Hazards Program and USGS National Earthquake Information Center (NEIC). *earthquake.usgs.gov*; Smithsonian Institution, Global Volcanism Program. *volcano.si.edu*; USGS and the International Association of Volcanology and Chemistry of the Earth's Interior. *vulcan.wr.usgs.gov*

New worlds out of old: Charles Preppernau; C. R. Scotese PALEOMAP Project

ARTWORK

Looking below the surface: Chuck Carter, Eagre Games Inc.

Cutaway of the Earth: Tibor G. Tóth

THE PHYSICAL LANDSCAPE
pp. 24–25

MAP AND TEXT

Deniz Karagülle, Charlie Frye, Sean Breyer, Peter Aniello, Randy Vaughan, Dawn Wright, Environmental Systems Research Institute (ESRI)

Roger Sayre, U.S. Geological Survey (USGS)

SCULPTING EARTH'S SURFACE
pp. 26–27

ARTWORK

Types of dunes: Lawson Parker

River meanders, during glaciation, after glaciation, features of Earth's surface: Chuck Carter, Eagre Games Inc.

BIRTH STORY OF ROCKS
pp. 28–29

DATA SOURCES

Rock across the world: Global distribution of surface rock from *The National Geographic Desk Reference*. Washington, DC: The National Geographic Society, 1999; Age of oceanic crust: Müller, R. D.; Sdrolias, M., Gaina, C., and Roest, W. R., 2008, "Age spreading rates and spreading asymmetry of the world's ocean crust." Geochemistry, Geophysics, Geosystems, 9, Q04006, doi:10.1029/2007GC001743

ARTWORK

The rock cycle: Chuck Carter, Eagre Games Inc.

Reading Earth history from rocks: Chapel Design & Marketing and XNR Productions

MOSAIC OF LANDSCAPES
pp. 30–31

DATA SOURCES

Land cover classification: Boston University Department of Geography and Environment Global Land Cover Project. Source data provided by NASA's Moderate Resolution Imaging Spectroradiometer.

EARTH'S FRESHWATER
pp. 32–33

DATA SOURCES

Stopping the flow (main map): World Wildlife Fund, GWLD; Igor A. Shiklomanov, State Hydrological Institute, Russia; U.S. Geological Survey (USGS); The Nature Conservancy; University of Kassel, Watergap2, GRDC; National Snow and Ice Data Center, University of Colorado; Lehner, B., C. Reidy Liermann, C. Revenga, C. Vörösmarty, B. Fekete, P. Crouzet, P. Döll, M. Endejan, K. Frenken, J. Magome, C. Nilsson, J. C. Robertson, R. Rodel, N. Sindorf, and D. Wisser. 2011. Global Reservoir and Dam Database, Version 1 (GRanDv1): Dams, Revision 01. Palisades, NY: NASA Socioeconomic Data and Applications Center (SEDAC). *doi. org/10.7927/H4N877QK*

Water availability: Gleick, Peter. 2006. *The World's Water: The Biennial Report on Freshwater Resources*. Island Press, 2006.

OUR BLUE PLANET
pp. 34–35

DATA SOURCES

Sea surface temperature: NASA Earth Observatory

Coral reef distribution: A Global Information System for Coral Reefs. September 2019. *www.reefbase.org*

ARTWORK

Beneath the waves: Chuck Carter, Eagre Games Inc.

DEFINING CLIMATE
pp. 36–37

DATA SOURCES

Climate zones (main map): Kottek, M., J. Grieser, C. Beck, B. Rudolf, and F. Rubel, 2006: World Map of the Köppen-Geiger

climate classification updated. Meteorol. Z., 15, 259-263. doi: 10.1127/0941-2948/2006/0130

Climate extremes: Fick, S. E., and R. J. Hijmans, 2017. "Worldclim 2: New 1-km spatial resolution climate surfaces for global land areas." *International Journal of Climatology*

A WARMING PLANET
pp. 38–39

DATA SOURCES

Temperature trends (main map): Zeke Hausfather and Robert Rohde, Berkeley Earth

Melting permafrost: Permafrost Laboratory, Geophysical Institute, University of Alaska Fairbanks

Unprecedented increases: T. J. Blasing, Recent Greenhouse Gas Concentrations. United States: N. p., 2016. Web. doi:10.3334/CDIAC/atg.032; D. M. Etheridge, L. P. Steele, R. L. Langenfelds, R. J. Francey, J.-M. Barnola, and V. I. Morgan. 1998. "Historical CO2 records from the Law Dome DE08, DE08-2, and DSS ice cores." In: Trends: A compendium of data on global change. Oak Ridge, TN: U.S. Department of Energy. *cdiac.ornl.gov/trends/co2/siple.html*; National Oceanic and Atmospheric Administration (NOAA). Annual mean carbon dioxide concentrations for Mauna Loa, Hawaii. *ftp.cmdl.noaa.gov/products/trends/co2/co2_annmean_mlo.txt*; D. M. Etheridge, L. P. Steele, R. J. Francey, and R. L. Langenfelds. 2002. "Historic CH4 records from Antarctic and Greenland ice cores, Antarctic firn data, and archived air samples from Cape Grim, Tasmania." In: Trends: A compendium of data on global change. Oak Ridge, TN: U.S. Department of Energy. *cdiac.ornl.gov/trends/atm_meth/lawdome_meth.html*; National Oceanic and Atmospheric Administration (NOAA). 2016. Monthly mean CH4 concentrations for Mauna Loa, Hawaii. *ftp.cmdl.noaa.gov/data/trace_gases/ch4/flask/surface/ch4_mlo_surface-flask_1_ccgg_month.txt*

Charting sea-level change: Church, John A., and Neil J. White. 2011. "Sea-Level Rise from the Late 19th to the Early 21st Century. Surveys in Geophysics" (2011) 32:585–602. *climate.nasa.gov/vital-signs/sea-level/*

THE WORLD'S WEATHER
pp. 40–41

DATA SOURCES

Cyclone tracks: NOAA National Centers for Environmental Information; Knapp, K. R., M. C. Kruk, D. H. Levinson, H. J. Diamond, and C. J. Neumann, 2010. "The International Best Track Archive for Climate Stewardship (IBTrACS): Unifying tropical cyclone best track data." *Bulletin of the American Meteorological Society* (2010), 91:363–376. doi:10.1175/2009BAMS2755.1. *ncdc.noaa.gov*

Lightning strikes: NASA Marshall Space Flight Center Lightning Imaging Sensor (LIS), Huntsville, Alabama

ARTWORK

Weather fronts; cloud types: Chuck Carter, Eagre Games Inc.

ZONES OF LIFE ON EARTH
pp. 42–43

DATA SOURCES

Plant life on land and at sea: NASA/Goddard Space Flight Center, The SeaWiFS Project and GeoEye, Scientific Visualization Studio. NOTE: All SeaWiFS images and data presented on this web site are for research and educational use only. All commercial use of SeaWiFS data must be coordinated with GeoEye (NOTE: In January 2013, DigitalGlobe and GeoEye combined to become DigitalGlobe); Ocean Biology Processing Group (OBPG)

World's Biomass: Bar-On, Yinon M., Rob Phillips, and Ron Milo. "The biomass distribution on Earth." 2018. *Proceedings of the National Academy of Sciences of the United States of America. doi.org/10.1073/pnas.1711842115*

ARTWORK

The size of the biosphere: The COMET Program and Chapel Design & Marketing

Energy in the ocean: Chuck Carter, Eagre Games Inc.

DIVERSITY OF LIFE
pp. 44–45

DATA SOURCES

Biodiversity on land and sea (main map): Florencia Sangermano, Graduate School of Geography, Clark University; BirdLife International. *Handbook of the Birds of the World* (2017); Bird species distribution maps of the world. Version 6.0. Available online at *datazone.birdlife.org/species/requestdis*; IUCN 2018. The IUCN Red List of Threatened Species. Version 2018-2. Available online at: *www.iucnredlist.org*. Downloaded November 2018.

Species diversity: factsanddetails.com. Biodiversity, The Number of Species and Biodiversity Meetings. *factsanddetails.com/world/cat52/sub329/item1612.html*

Biodiversity hotspots: Conservation International. conservation.org, worldmap-biological-hotspot-labels. *cepf.net/our-work/biodiversity-hotspots/*

SAFE HAVENS FOR NATURE
pp. 46–47

DATA SOURCES

Low impact areas (LIA) by biome (main map): Andrew P. Jacobson, Jason Riggio, Alexander M. Tait, and Jonathan E. M. Baillie, "Global areas of low human impact ('Low Impact Areas') and fragmentation of the natural world," *Scientific Reports*, 9, Article number: 14179 (2019). *www.nature.com/articles/s41598-019-50558-6*; 2017 Ecoregions by Resolve: Eric Dinerstein, David Olson, Anup Joshi, Carly Vynne, Neil D. Burgess et al., 2017. "An Ecoregion-Based Approach to Protecting Half the Terrestrial Realm." *BioScience*, Volume 67, Issue 6, 1 June 2017, Pages 534–545. *doi.org/10.1093/biosci/bix014*

Protected areas: UNEP-WCMC and IUCN (2019), Protected Planet: The World Database on Protected Areas (WDPA)/The Global Database on Protected Areas Management Effectiveness (GD-PAME) [On-line], WDPA Dataset August 2018, Cambridge, UK: UNEP-WCMC and IUCN. *protectedplanet.net*

Fading landscapes: Andrew P. Jacobson, Jason Riggio, Alexander M. Tait, and Jonathan E. M. Baillie, "Global areas of low human impact ('Low Impact Areas') and fragmentation of the natural world," *Scientific Reports*, 9, Article number: 14179 (2019). *www.nature.com/articles/s41598-019-50558-6*

THREATS TO OUR PLANET
pp. 48–49

DATA SOURCES

Land degradation and desertification: Millennium Ecosystem Assessment. *Ecosystems and Human Well-Being, Synthesis*

Ambient air pollution: World Health Organization (WHO)

Fewer trees & forests: Hansen, M. C., P. V. Potapov, R. Moore, M. Hancher, S. A. Turubanova, A. Tyukavina, D. Thau, S. V. Stehman, S. J. Goetz, T. R. Loveland, A. Kommareddy, A. Egorov, L. Chini, C. O. Justice, and J. R. G. Townshend. 2013. "High-Resolution Global Maps of 21st-Century Forest Cover Change." *Science* 342 (15 November): 850–53. *earthenginepartners.appspot.com/science-2013-globalforest*. Accessed through Global Forest Watch. *globalforestwatch.org*; Greenpeace, University of Maryland, World Resources Institute and Transparent World. "Intact Forest Landscapes: 2000-2013." Accessed through Global Forest Watch. *globalforestwatch.org*

Global water risk: Gassert, F., M. Luck, M. Landis, P. Reig, and T. Shiao. 2014. "Aqueduct Global Maps 2.1: Constructing Decision-Relevant Global Water Risk Indicators." Working Paper. Washington, DC: World Resources Institute. Available online at: *wri.org/publication/aqueduct-global-maps-21-indicators*

HUMANS CHANGE THE PLANET
pp. 50–51

DATA SOURCES

Altered landscapes (main map): Benjamin Halpern and others, National Center for Ecological Analysis and Synthesis, University of California, Santa Barbara; Ellis, E. C., K. Klein Goldewijk, S. Siebert, D. Lightman, and N. Ramankutty. 2010

Oceans full of plastic: Lebreton, L., M. Egger, and B. Slat. A global mass budget for positively buoyant macroplastic debris in the ocean. *Nature Scientific Reports*. (2019) 9:12922

Plastic problems: Andrés Cózar Cabañas, University of Cádiz; Laurent Lebreton, Ocean Cleanup Foundation; Rachel W. Obbard, Dartmouth College; Alan J. Jamieson, Newcastle University

Human impact over time: Ellis, E. C., K. Klein Goldewijk, S. Siebert, D. Lightman, and N. Ramankutty. 2010. "Anthropogenic transformation of the biomes, 1700 to 2000." *Global Ecology and Biogeography*, 19(5):589–606

A WORLD FULL OF PEOPLE
pp. 52–53

DATA SOURCES

Population density (main map): Landscan 2018 Population Dataset created by UT-Battelle, LLC, the management and operating contractor of the Oak Ridge National Laboratory acting on behalf of the U.S. Department of Energy under Contract No. DE-AC05-00OR22725. Distributed by East View Geospatial: *geospatial.com* and East View Information Services: *eastview.com/online/landscan*

Gender of the population; as the world grows; crowded countries: United Nations Department of Economic and Social Affairs (DESA), Population Division

CITIES ON THE RISE
pp. 54–55

DATA SOURCES

Urban agglomerations (main map); urban population growth; largest cities by population: United Nations Department of Economic and Social Affairs (DESA), Population Division

POPULATION ON THE MOVE
pp. 56–57

DATA SOURCES

Migrant population (main map): United Nations Department of Economic and Social Affairs (DESA), Population Division

Displaced people due to natural disasters: Internal Displacement Monitoring Centre (IDMC)

By the numbers: Extracted from the UNHCR Population Statistics Reference Database, UNHCR

THE WORLD SPEAKS
pp. 58–59

DATA SOURCES

Language location and endangerment level (main map); major language families: Harald Hammarström, Thom Castermans, Robert Forkel, Kevin Verbeek, Michel A. Westenberg, and Bettina Speckmann. 2018. "Simultaneous Visualization of Language Endangerment and Language Description." Language Documentation & Conservation, 12, 359-392

A multilingual world: Mikael Parkvall, "Världens 100 största språk 2010" (The World's 100 Largest Languages in 2010), Nationalencyklopedin. 2010

Evolution of languages: *National Geographic Almanac of Geography.* Washington, D.C.: The National Geographic Society, 2005.

FAITH BELIEFS & PRACTICES
pp. 60–61

DATA SOURCES

Global faiths (main map); religious diversity: Central Intelligence Agency (CIA) *World Factbook*

By the numbers: Pew Research Center, April 5, 2017, "The Changing Global Religious Landscape"

QUALITY OF LIFE
pp. 62–63

DATA SOURCES

Average years of education: United Nations Development Programme (UNDP), Human Development Reports

Adult literacy: Central Intelligence Agency (CIA) *World Factbook*; Burton, James. "List of Countries By Literacy Rate." WorldAtlas, Sept. 14, 2018, *worldatlas.com/articles/the-highest-literacy-rates-in-the-world.html*

Happiness score: Helliwell, J., R. Layard, and J. Sachs (2019). *World Happiness Report 2019*, New York: Sustainable Development Solutions Network.

Press freedom: Reporters Without Borders, 2019 World Press Freedom Index. Courtesy of Reporters Sans Frontières

SEEKING WORLD WELLNESS
pp. 64–65

DATA SOURCES

Life expectancy; causes of death; communicable diseases; noncommunicable diseases: GBD compare Data Visualization, Institute for Health Metrics and Evaluation, University of Washington, 2019. Available from *vizhub.healthdata.org/gbd-compare*

The obesity pandemic; safe drinking water: World Health Organization (WHO)

THE WORLD'S HARVEST
pp. 66–67

DATA SOURCES

Land allocation (main map); crop allocation: Global Landscapes Initiative, Institute on the Environment, University of Minnesota

Percent irrigated area: Food and Agriculture Organization of the United Nations (FAO), 2008

FEEDING THE WORLD
pp. 68–69

DATA SOURCES

Sea of food: Food and Agriculture Organization of the United Nations, Total production 2017, Fishery Statistical Collection. *fao.org/fishery/statistics.* Reproduced with permission.

Meat consumption: Hannah Ritchie and Max Roser (2020) - "Meat and Dairy Production". Published online at OurWorldInData.org. Retrieved from: *ourworldindata.org/meat-production.* Licensed under a Creative Commons Attribution-4.0 International (CC BY 4.0).

Food supply; world diets: Food and Agriculture Organization of the United Nations, FAOSTAT New Food Balances. *fao.org/faostat/en/#data/FBS.* Accessed December 2019.

Food security index: Food and Agriculture Organization of the United Nations, FAOSTAT, Food Aid Shipments (WFP). *fao.org/faostat/en/#data/FA.* Accessed June 14, 2018.

BORDERS & BOUNDARIES
pp. 70–71

DATA SOURCES

Unsettled boundaries (main map): National Geographic Map Policy Committee

Maritime divisions: Flanders Marine Institute (2018). Maritime Boundaries Geodatabase: Territorial Seas (12NM), version 2. Available online at *marineregions.org/; doi.org/10.14284/313*

ARTWORK

Borders in the sea: Chuck Carter, Eagre Games Inc.

WEALTH & PROSPERITY
pp. 72–73

DATA SOURCES

GDP per capita (main map): United Nations Statistics Office, National Accounts Section. *unstats.un.org/unsd/snaama/Index*

National government debt: Central Intelligence Agency (CIA) *World Factbook*

World millionaires; nations with the most private wealth: *Credit Suisse Global Wealth Databook 2019*

Most healthy national economies: International Monetary Fund

BUYING & SELLING
pp. 74–75

DATA SOURCES

Main trading nations; merchandise exports; imports; exports: World Trade Organization Data Portal. *data.wto.org*

Major regional trade agreements: APEC, *apec.org*; ASEAN, *asean.org*; COMESA, *facebook.com/ComesaSecretariat/*; ECOWAS, *ecowas.int*; EU, *europa.eu*; MERCOSUR, *mercosur.int*; NAFTA, *ustr.gov*; SAFTA, *saarc-sec.org*

POWER SOURCES WORLDWIDE
pp. 76–77

DATA SOURCES

Energy consumption (main map); energy use trends; power without fossil fuels; renewable energy producers: U.S. Energy Information Administration. September 2019

OUR DIGITAL WORLD
pp. 78–79

DATA SOURCES

Access to the internet (main map); more phones than people: International Telecommunications Union, World Telecommunication/ICT Indicators Database. August 2019

Government restrictions to the internet: *U.S. Department of State Human Rights Report, 2018*

Undersea telecommunication cables: Telegeography Submarine Cable Map. Map is made available under the following Creative Commons License: Attribution-NonCommercial-ShareAlike 3.0 Unported (CC BY-NC-SA 3.0).

ADDITIONAL MAPS

PHYSICAL MAPS
pp. 16–17, 84, 90–91, 128, 140, 168, 198, 220

DATA SOURCES

Shaded relief and bathymetry: ETOPO1/Amante and Eakins, 2009; USGS (2004), Shuttle Radar Topography Mission (SRTM) 1 Arc Second scene Global Land Cover Facility, University of Maryland, College Park, Maryland, February 2000

REGIONAL THEMATICS
reference maps

DATA SOURCES

Population density: Landscan 2018 Population Dataset created by UT-Battelle, LLC, the management and operating contractor of the Oak Ridge National Laboratory acting on behalf of the U.S. Department of Energy under Contract No. DE-AC05-00OR22725. Distributed by East View Geospatial: *geospatial.com* and East View Information Services: *eastview.com/online/landscan*

Climate zones: Kottek, M., J. Grieser, C. Beck, B. Rudolf, and F. Rubel, 2006: World Map of the Köppen-Geiger climate classification updated. Meteorol. Z., 15, 259-263. DOI: 10.1127/0941-2948/2006/0130

Land cover classification: Boston University Department of Geography and Environment Global Land Cover Project. Source data provided by NASA's Moderate Resolution Imaging Spectroradiometer.

GREENLAND
pp. 88–89

DATA SOURCES

TOPO1/Amante and Eakins, 2009; National Snow and Ice Data Center; Danish Ministry of the Environment; Danish Meteorological Institute, Centre for Ocean and Ice; National Aeronautics and Space Administration (NASA); Moderate Resolution Imaging Spectroradiometer (MODIS) Aqua Mission. Imagery used spans October 9-24, 2018; Joughin, I., T. Moon, J. Joughin, and T. Black. 2015, 2017. MEaSUREs Annual Greenland Outlet Glacier Terminus Positions from SAR Mosaics, Version 1. Boulder, Colorado USA. NASA National Snow and Ice Data Center Distributed Active Archive Center. October 2018; Polar Geospatial Center (PGC), University of Minnesota: ArcticDEM Porter, Claire; Morin, Paul; Howat, Ian; Noh, Myoung-Jon; Bates, Brian; Peterman, Kenneth; Keesey, Scott; Schlenk, Matthew; Gardiner, Judith; Tomko, Karen; Willis, Michael; Kelleher, Cole;

Cloutier, Michael; Husby, Eric; Foga, Steven; Nakamura, Hitomi; Platson, Melisa; Wethington, Michael, Jr.; Williamson, Cathleen; Bauer, Gregory; Enos, Jeremy; Arnold, Galen; Kramer, William; Becker, Peter; Doshi, Abhijit; D'Souza, Cristelle; Cummens, Pat; Laurier, Fabien; Bojesen, Mikkel, 2018, "ArcticDEM". *doi.org/10.7910/DVN/OHHUKH*, Harvard Dataverse, V1. October 2018; Natural Resources Canada

ANTARCTICA
pp. 236–239

DATA SOURCES

Velocity: Rignot, E., J. Mouginot, and B. Scheuchl. 2017. MEaSUREs InSAR-Based Antarctica Ice Velocity Map, Version 2. Boulder, Colorado USA. NASA National Snow and Ice Data Center Distributed Active Archive Center. *dx.doi.org/10.5067/D7GK8F5J8M8R.* October 2018

Surface elevation: Byrd Polar Research Center, Ohio State University

Ice sheet thickness: Bedmap Project

Antarctica physical: British Antarctic Survey Geodata Portal; SCAR Antarctic Digital Database; Bedmap2 - Ice thickness and subglacial topographic model of Antarctica. *secure.antarctica.ac.uk/data/bedmap2/*; Scientific Committee on Antarctic Research: SCAR Composite Gazetteer of Antarctica; Polar Geospatial Center (PGC), University of Minnesota: Reference Elevation Model of Antarctica (REMA); Howat, Ian, Morin, Paul; Porter, Claire; Noh, Myong-Jong, 2018, "The Reference Elevation Model of Antarctica," Harvard Dataverse, V1. *data.pgc.umn.edu/elev/dem/setsm/REMA/*

OCEANS
pp. 244–255

MAP RELIEF

Tibor G. Tóth

SPACE

MOON
pp. 260–263

DATA SOURCES

Physical feature names: *Gazetteer of Planetary Nomenclature,* Planetary Geomatics Group of the USGS (United States Geological Survey) Astrogeology Science Center. *planetarynames.wr.usgs.gov.* Accessed March 2019.

Terrain, main map global mosaics: Lunar Reconnaissance Orbiter (LRO); NASA/JPL; University of Arizona; Johns Hopkins University Applied Physics Laboratory

Phases of the moon, lunar influence on tides: National Geographic Society; Lunar Reconnaissance Orbiter (LRO); NASA

INNER SOLAR SYSTEM
pp. 264–265

IMAGES

NASA/JPL-Caltech; Johns Hopkins University Applied Physics Laboratory; Carnegie Institution of Washington

MARS
pp. 266–269

DATA SOURCES

Physical feature names: *Gazetteer of Planetary Nomenclature,* Planetary Geomatics Group of the USGS (United States Geological Survey) Astrogeology Science Center. *planetarynames.wr.usgs.gov.* Accessed March 2019.

Global mosaics: NASA Mars Global Surveyor (MGS); National Geographic Society

Terrain: NASA Mars Global Surveyor (MGS); Mars Orbital Laser Altimeter (MOLA); Johns Hopkins University Applied Physics Laboratory

IMAGES

Phobos, Deimos: NASA/JPL-Caltech/University of Arizona

OUTER SOLAR SYSTEM
pp. 270–271

IMAGES

NASA/JPL-Caltech; Johns Hopkins University Applied Physics Laboratory; Carnegie Institution of Washington

COSMIC JOURNEYS
pp. 272–273

IMAGES

NASA Goddard Space Flight Center; NASA/JPL-Caltech; Johns Hopkins University Applied Physics Laboratory; Carnegie Institution of Washington

MILKY WAY
pp. 274–275

DATA SOURCES

Guillermo Gonzalez, Ball State University; Michael Gowanlock, Northern Arizona University; Icarus; Astrobiology; NASA/JPL; *International Journal of Astrobiology*

ARTWORK

Antoine Collignon

PHOTOGRAPHS

COVER

(Main) Charles Preppernau. Data Sources: NPP/VIIRS, NASA Goddard Space Flight Center; Blue Marble Next Generation, NASA Earth Observatory; Esri; Low Impact Areas: Andrew Jacobson, Jason Riggio, Jonathan Baillie, and Alexander M. Tait, National Geographic Society; Shuttle Radar Topography Mission (SRTM), 1 Arc Second scene Global Land Cover Facility, University of Maryland, College Park, Maryland, February 2000; General Bathymetric Chart of the Oceans (GEBCO); (LO LE), Ami Vitale/National Geographic Image Collection; (LO CTR LE), Kathleen Carney/Cavan Images; (LO CTR RT), Babak Tafreshi/National Geographic Image Collection; (LO RT), Yva Momatiuk & John Eastcott.

INTERIOR

2-3, Evgeny Vasenev/Cavan Images; 4, Ralph Lee Hopkins/National Geographic Image Collection; 5, Brent Doscher/Cavan Images; 6, Valentin Wolf/imageBROKER/Alamy Stock Photo; 7, Pasquale Vassallo; 12-13, Roman Slavik/Getty Images; 14 (LE), HagePhoto/Cavan Images; 14 (RT), bybostanci/Getty Images; 15 (LE), Beverly Joubert/National Geographic Image Collection; 15 (RT), John Stanmeyer/National Geographic Image Collection; 24 (LE), primeimages/Getty Images; 24 (RT), Hans Strand/Getty Images; 25 (LE), Martin Harvey/Getty Images; 25 (RT), Westend61/Frank Röder/Getty Images; 26 (UP), Edwin Remsberg/Getty Images; 26 (LO LE), Chen Su/Getty Images; 26 (LO RT), blickwinkel/McPHOTO/MOS/Alamy Stock Photo; 27, Sean Gallup/Getty Images; 29 (UP LE), Benjamin Van Der Spek/EyeEm/Getty Images; 29 (UP CTR), 4kodiak/Getty Images; 29 (UP RT), Bruce Dale/National Geographic Image Collection; 29 (LO), Ralph Lee Hopkins/National Geographic Image Collection; 30 (LE), Mint Images—Frans Lanting/Getty Images; 30 (RT), Raymond Haddad/Getty Images; 31 (LE), Feargus Cooney/Getty Images; 31 (RT), Guenter Guni/Getty Images; 32, rusm/Getty Images; 33, Oleksandr Rupeta/Getty Images; 34 (UP), Pete Atkinson/Getty Images; 34 (LO), WhitcombeRD/Getty Images; 37 (LE), Vitor Marigo/Cavan Images; 37 (RT), Tyler Stableford/Cavan Images; 39, David Tipling/Education Images/Getty Images; 40 (UP), Keith Ladzinski/National Geographic Image Collection; 40 (LO), Beau Van Der Graaf/EyeEm/Getty Images; 42 (UP), Ben Horton/National Geographic Image Collection; 42 (LO), Reinhard Dirscherl/Getty Images; 45 (LE), Georgette Douwma/Getty Images; 45 (RT), Sylvain Cordier/hemis.fr/Getty Images; 46, Charlie Hamilton James/National Geographic Image Collection; 47, David Aguilar Photography/Getty Images; 48 (UP), Wang Zhao/AFP via Getty Images; 48 (LO), Sierralara/Getty Images; 51, John Stanmeyer/National Geographic Image Collection; 52, Gonzalo Azumendi/Getty Images; 54, DuKai photographer/Getty Images; 55, Randy Olson/National Geographic Image Collection; 56, Jordi Ruiz Cirera/Panos Pictures; 57, John Stanmeyer/National Geographic Image Collection; 58, Michael S. Lewis/National Geographic Image Collection; 61 (UP LE), Mario Weigt/Anzenberger/Redux; 61 (UP CTR), Ami Vitale/National Geographic Image Collection; 61 (UP RT), Danielle Villasana; 61 (LO LE), Susannah Ireland/eyevine/Redux; 61 (LO CTR), Tommaso Ausili/contrasto/Redux; 61 (LO RT), Donat Sorokin/TASS via Getty Images; 62 (UP), Images By Tang Ming Tung/Getty Images; 62 (LO), Artur Widak/NurPhoto via Getty Images; 64, joSon/Getty Images; 66, Pgiam/Getty Images; 68, Brenda Tharp/Getty Images; 70, Ed Jones/AFP via Getty Images; 71, Fethi Belaid/AFP via Getty Images; 72, Brittainy Newman/The New York Times/Redux; 74, pigphoto/Getty Images; 76, Mimadeo/Getty Images; 78, John Stanmeyer/National Geographic Image Collection; 80-81, Michele Falzone/Getty Images; 82 (LE), Mike Criss/Design Pics/National Geographic Image Collection; 82 (RT), Witold Skrypczak/Getty Images; 83 (LE), Tino Soriano/National Geographic Image Collection; 83 (RT), Joe Raedle/Getty Images; 124-5, Octavio Campos Salles/Alamy Stock Photo; 126 (LE), Steve Winter/National Geographic Image Collection; 126 (RT), traumlichtfabrik/Getty Images; 127 (LE), Pablo Corral Vega/National Geographic Image Collection; 127 (RT), Aldo Pavan/Getty Images; 136-7, Suttipong Sutiratanachai/Getty Images; 138 (LE), Douglas Pearson/Getty Images; 138 (RT), Will Harding/500px/Getty Images; 139 (LE), M. Sobreira/Alamy Stock Photo; 139 (RT), Fabio Burrelli/Alamy Stock Photo; 164-5, KYON.J (www.kyonj.com); 166 (LE), Peerawat Kamklay/Getty Images; 166 (RT), Yavuz Sariyildiz/Getty Images; 167 (LE), Xinhua/eyevine/Redux; 167 (RT), Jackal Pan/Getty Images; 194-5, Chris Schmid/National Geographic Image Collection; 196 (LE), Lukas Bischoff/Alamy Stock Photo; 196 (RT), Jaroslav Havlicek/EyeEm/Getty Images; 197 (LE), Peter Langer/Design Pics/National Geographic Image Collection; 197 (RT), Ami Vitale/National Geographic Image Collection; 216-7, rudi1976/Alamy Stock Photo; 218 (LE), Johnathan Ampersand Esper/Cavan Images; 218 (RT), Stephen Alvarez/National Geographic Image Collection; 219 (LE), Rafael Ben-Ari/Alamy Stock Photo; 219 (RT), Valerii Shanin/Alamy Stock Photo; 232-3, David Merron Photography/Getty Images; 234 (LE), Alasdair Turner/Cavan Images; 234 (RT), Gordon Wiltsie/National Geographic Image Collection; 235 (LE), Galen Rowell/Getty Images; 235 (RT), Carsten Peter/National Geographic Image Collection; 236 (UP), staphy/Getty Images; 236 (LO), ad_foto/Getty Images; 240–41, Brian Skerry/National Geographic Image Collection; 242 (LE), Design Pics Inc/National Geographic Image Collection; 242 (RT), Thomas J. Abercrombie/National Geographic Image Collection; 243 (LE), Cristina Mittermeier/National Geographic Image Collection; 243 (RT), Xu Yu/Xinhua/eyevine/Redux; 256-7, NASA, ESA, S. Bianchi (Università degli Studi Roma Tre University), A. Laor (Technion-Israel Institute of Technology), and M. Chiaberge (ESA, STScI, and JHU); 258 (LE), NASA, ESA, N. Smith (University of Arizona, Tucson), and J. Morse (BoldlyGo Institute, New York); 258 (RT), NASA/Scott Kelly; 259 (LE), Northrop Grumman via NASA (https://creativecommons.org/licenses/by/2.0/legalcode); 259 (RT), NASA and ESA/Hubble.

CARTOGRAPHY: Matthew W. Chwastyk, Jerome N. Cookson, Debbie J. Gibbons, Mike McNey, Gregory Ugiansky

ADDITIONAL CARTOGRAPHY: Nat Case, INCase LLC; Maureen J. Flynn; Dianne C. Hunt; Don Larson and Michael Woodard, Mapping Specialists, Ltd.; Gus Platis; Shelley Sperry

TEXT: Libby Sander (writer), Shelley Sperry (researcher)

DESIGN: Sanaa Akkach, Debbie J. Gibbons

EDITORIAL: Theodore A. Sickley, Scott A. Zillmer (maps); Moriah Petty (text); Adrian Coakley (photos)

Since 1888, the National Geographic Society has funded more than 13,000 research, exploration, and preservation projects around the world. National Geographic Partners distributes a portion of the funds it receives from your purchase to National Geographic Society to support programs including the conservation of animals and their habitats.

National Geographic Partners
1145 17th Street NW
Washington, DC 20036-4688 USA

Get closer to National Geographic explorers and photographers, and connect with our global community. Join us today at nationalgeographic.com/join

For rights or permissions inquiries, please contact National Geographic Books Subsidiary Rights: bookrights@natgeo.com

ISBN: 978-1-4262-2144-6

The Library of Congress has cataloged the fourth edition as follows:
National Geographic Society (U.S.), author, issuing body.
  National Geographic Family reference atlas of the world / National Geographic, Washington, D.C. -- Fourth edition.
    pages cm
"Copyright © 2016"
Includes bibliographical references and indexes.
ISBN 978-1-4262-1543-8 (hardcover : alk. paper)
1. Geography--Maps. 2. Physical geography--Maps.
3. Political geography--Maps. I. Title. II. Title: Family reference atlas of the world.
G1021.N393 2016
912--dc23
                                    2015006195

Printed in Italy

20/EV/1

KEY TO
ATLAS
MAPS

RUSSIA

Alaska
112

Greenland
88

ICELAND

CANADA
86

UNITED
KINGDOM

BRITAIN AND IRELAND
144

IRELAND

BELGIUM, FRANCE, AND THE NETHERLAND

NORTH AMERICA 80-123

UNITED STATES
90-115

PORTUGAL   SP

IBÉRIAN PENINSULA
146

MOROCCO

Hawai'i
114

MEXICO

THE
BAHAMAS

GREATER ANTILLES
AND THE BAHAMAS
118

N

Western
Sahara

CUBA

DOMINICAN
REPUBLIC

LESSER ANTILLES
120

MAURITANIA

MEXICO AND
CENTRAL AMERICA
116

JAMAICA
BELIZE
HONDURAS
GUATEMALA
EL SALVADOR   NICARAGUA

HAITI

Puerto
Rico
114

ST. KITTS AND NEVIS
ANTIGUA AND BARBUDA
DOMINICA
ST. LUCIA
BARBADOS
GRENADA ST. VINCENT AND THE GRENADINES
TRINIDAD AND TOBAGO

CABO VERDE

SENEGAL
THE GAMBIA
GUINEA-BISSAU

WESTERN
AFRICA
204

BUR

GUINEA
CÔTE
D'IVOIR

COSTA RICA

PANAMA

VENEZUELA

GUYANA
SURINAME
French Guiana

SIERRA LEONE
LIBERIA

PACIFIC OCEAN FLOOR
246

NORTHERN
SOUTH AMERICA
130

COLOMBIA

ECUADOR

B R A Z I L

ATLANTIC
OCEAN
FLOOR
248

K I R I B A T I

OCEANIA
224-231

PERU

CENTRAL
SOUTH AMERICA
132

SAMOA

American
Samoa

BOLIVIA

SOUTH AMERICA 124-135

TONGA

French Polynesia

PARAGUAY

CHILE

URUGUAY

ARGENTINA

SOUTHERN
SOUTH AMERICA
134

Falkland
Islands

ROCKY
MOUNTAINS
108

NORTHERN
PLAINS
106

GREAT
LAKES
100

NORTHEAST
96

WASHINGTON

MONTANA

NORTH
DAKOTA

MINNESOTA

MAINE

OREGON

IDAHO

SOUTH
DAKOTA

WISCONSIN

M
I
C
H
I
G
A
N

VT.
N.H.
MASS.

NEW
YORK

R.I.
CONN.

WYOMING

NEBRASKA

IOWA

PA.

NEW
JERSEY

NEVADA

UTAH

COLORADO

ILLINOIS

IND.

OHIO

W. VA.

DELAWARE
MARYLAND
WASHINGTON, D.C.

CALIFORNIA

KANSAS

MISSOURI

KENTUCKY

VIRGINIA

WEST
COAST
110

ARIZONA

NEW
MEXICO

OKLAHOMA

ARK.

TENNESSEE

N.C.

S.C.

TEXAS

MISS.

ALA.

GEORGIA

SOUTH
ATLANTIC
98

LA.

F
L
O
R
I
D
A

TEXAS AND
OKLAHOMA
104

MIDDLE
SOUTH
102